Escherichia coli and *Salmonella*

CELLULAR AND MOLECULAR BIOLOGY

SECOND EDITION VOLUME 2

Escherichia coli and *Salmonella*

CELLULAR AND MOLECULAR BIOLOGY

SECOND EDITION   VOLUME 2

Editor in Chief
FREDERICK C. NEIDHARDT
University of Michigan Medical School, Ann Arbor, Michigan

Editors
ROY CURTISS III
Washington University, St. Louis, Missouri

JOHN L. INGRAHAM
University of California, Davis

EDMUND C.C. LIN
Harvard Medical School, Boston, Massachusetts

K. BROOKS LOW
Radiobiology Laboratories, Yale University School of Medicine, New Haven, Connecticut

BORIS MAGASANIK
Massachusetts Institute of Technology, Cambridge, Massachusetts

WILLIAM S. REZNIKOFF
University of Wisconsin-Madison, Madison, Wisconsin

MONICA RILEY
Marine Biological Laboratory, Woods Hole, Massachusetts

MOSELIO SCHAECHTER
San Diego State University, San Diego, California

H. EDWIN UMBARGER
Purdue University, West Lafayette, Indiana

ASM PRESS   WASHINGTON, D.C.

Copyright © 1996 ASM Press
American Society for Microbiology
1325 Massachusetts Ave., N.W.
Washington, DC 20005

Library of Congress Cataloging-in-Publication Data

Escherichia coli and Salmonella typhimurium.
Escherichia coli and Salmonella : cellular and molecular biology / editors,
Frederick C. Neidhardt . . . [et al.]. — 2nd ed.
v. <1–   > ; cm.
First ed. published under the title: Escherichia coli and Salmonella typhimurium.
Includes bibliographical references and index.
ISBN 1-55581-084-5
1. Escherichia coli. 2. Salmonella typhimurium. I. Neidhardt, Frederick C. (Frederick Carl),
1931–   . II. Title.
QR82.E6E83   1996
589.9′5—dc20   96-316
                                                                              CIP

ASM Press

Director, Patrick J. Fitzgerald
Senior Production Editor, Eleanor S. Tupper
Copyeditors and Proofreaders, Bonnie Ashbaugh, Susan Birch, Marie Smith, Yvonne Strong
Compositor, Princeton Editorial Associates, Princeton, N.J.
Printer, Courier Companies, Westford, Mass.
Interior Design, Susan Brown Schmidler
Cover and Titles, Ed Atkeson, Berg Design
Cover Image © David M. Phillips, courtesy Visuals Unlimited

Contents

Volume 1

III   Utilization of Energy for Cell Activities   1073

IV   Regulation of Gene Expression   1225

Index

Volume 2

V   Growth of Cells and Cultures   1551

VI   Genome, Genetics, and Evolution   1713

SECTION A · The Chromosome and Its Products   1715

Growth of Cells and Cultures

V

NO CHARACTERISTIC IS MORE CLOSELY ASSOCIATED WITH THE BACTERIAL cell than its remarkable ability to grow and divide rapidly and efficiently under a wide range of conditions, as well as very slowly and presumably efficiently as it must sometimes do in an infected animal host or in nutritionally lean aqueous environments. Particularly in the decades following the germinal studies of Monod on the physiology of bacterial growth, the attention of bacterial physiologists, and later of molecular biologists, was focused on the growth process: synthesis of protoplasm followed by cell division. Having learned how the building blocks required for macromolecule synthesis are synthesized and polymerized, bacteriologists gained an even deeper appreciation for the complexity and efficiency of growth.

Binary fission of a prokaryotic cell may be the simplest mode of reproduction in biology, but it presents problems of timing and topology that are still far from understood. The timing of synthetic activities in the cell cycle, the segregation of the envelope, cytosol, and nucleoid, and the formation of the septum and new poles of the daughter cells are processes that require much further exploration. Not much is known yet about how the cell knows how to define its middle. Perhaps the brightest triumph of recent years has been the role assigned to FtsZ in cell septum formation.

The concept of balanced growth—the steady state during which every extensive property of a culture increases in the same proportion and the exponential growth rate is constant—has been an organizing theme in bacterial growth studies since the time of Monod. Enteric bacteria can achieve a continuum of steady states, each reproducible for a given growth medium and temperature. In each steady state the cells assume a different structure and chemical composition according to rules that are easy to learn and describe because they can be summarized by simple monotonic relations between cell composition and growth rate at a given temperature; nevertheless, these rules have to a large extent defied precise molecular explanation.

One major unsolved problem concerns the precise variation of the translational apparatus with growth rate. By far the dominant activity of the bacterial cell is the synthesis of protein, and for a long time it has been recognized that the machinery for making protein—ribosomes, factors, activating enzymes—constitutes an increasing fraction of the cell with increasing growth rate. The mechanism has remained elusive during 3 decades of investigation, probably because (as we are beginning to accept) there is no single regulatory device, but a combination of mechanisms, both complementary and redundant, that bring about this macro adjustment of cell composition. Another set of problems concerns the timing of initiation of DNA replication and the coordination of cell division with replication. The manner in which these events are coordinately timed at different growth rates is not yet a complete story. Finally, a subject implicit in the phenomenon of alternate steady-state growth rates is the nature of growth rate-limiting reactions in the cell. After the cell completes making the myriad of adjustments that optimize its utilization of the metabolic potential of a given medium, it provides no obvious hints that suggest how the growth rate is set. Modeling the circuitry that achieves this elegance will not be easy.

The past decade has witnessed expanded interest in two areas largely ignored by growth physiologists. Stress responses are now readily recognized through global monitoring of transcription and translation and are speedily analyzed mechanistically by the standard repertoire of molecular genetics. But it is the new attention being paid to the physiology of the non-growth state that has been the most remarkable development, and this section contains a valuable and exhaustive description of stationary-phase cells. Some of the regulatory mechanisms that bring about differentiation into stationary phase are covered in a chapter on gene regulation in the previous section of the book. Surprises can be expected in this area as studies begin on genetic changes in populations in the stationary phase.

The field of growth yield and energy distribution has undergone an interesting maturation, advancing from energetic bookkeeping to defining a crisp and unanswered question. Since the accumulated evidence is now convincing that aerobic growth is a state of ATP excess, what do *E. coli* and other bacteria do with excess ATP when growing aerobically?

In spite of obstacles to production of recombinant products (such as the limited capacity for protein excretion and the absence of appropriate glycosylating and certain other protein-modifying processes), *E. coli* continues to serve important roles in biotechnology, and therefore we have included an account of this practical application of *E. coli*.

Modulation of Chemical Composition and Other Parameters of the Cell by Growth Rate

HANS BREMER AND PATRICK P. DENNIS

97

HISTORY

Schaechter et al. (121) first demonstrated that the macromolecular composition of the bacterial cell was related to its metabolic activity and that RNA-containing particles were involved in the synthesis of protein. When they examined the variations in growth and composition of *Salmonella typhimurium* cultures in different media, they realized that the cellular contents of DNA, RNA, and protein at a given temperature depended only on the growth rate and not on the nutrient supplement in the growth medium used to achieve that growth rate. They also found (i) that fast-growing bacteria are larger and contain more DNA, RNA, and protein than slow-growing bacteria, (ii) that the amounts of these macromolecules are exponential functions of growth rate, and (iii) that the exponents of these functions are different for different macromolecules. The last implies that the relative proportions of the different macromolecules change with growth rate; at a given temperature, RNA and ribosome concentration increase with increasing growth rate, DNA concentration decreases, and protein concentration remains almost constant. When the growth rate was varied by changing the temperature rather than the nutrient content of the growth medium, the DNA, RNA, and protein concentrations remained invariant.

These early studies of bacterial physiology are documented by Maaløe and Kjeldgaard (95). The statement of observations in terms of simple mathematical relationships was characteristic for the "Copenhagen approach," in which calculated constants, proportionalities, and quadratic or exponential functions suggested special control mechanisms. Many of these relationships later turned out to be more complex than originally imagined.

Nevertheless, these propositions have stimulated thought and led to more sophisticated observations; the book continues to be provocative, as many of the fundamental problems posed in it are still far from being solved.

A theoretical basis for explaining the empirical relationships of the early Copenhagen school was not available until Cooper and Helmstetter (32) derived a formula for determining the average amount of DNA per cell in an exponential culture as a function of C (the time required to replicate the chromosome), D (the time period between termination of a round of replication and the following cell division), and τ (the culture doubling time). This theory also included the important concept of overlapping rounds of chromosome replication, where a round of replication is initiated before the previous round is completed. This occurs when τ is less than C and explains how bacteria are able to grow with a doubling time shorter than the time required for chromosome replication.

Donachie (51) extended this theory by introducing the concept of the "initiation mass" (the cell mass per replication origin at the time of initiation) and derived the average mass per cell, or amount of protein per cell, as a function of C, D, τ, and an additional parameter, mass or protein per replication origin (M_O or P_O, respectively). The Cooper-Helmstetter equation predicts the amount of DNA per cell as a function of C, D, and τ; the additional parameter, P_O, links the amount of protein to the amount of DNA.

At about the same time, Schleif (122) and Maaløe (93) began to establish a theoretical relationship between the amounts of protein and RNA in the cell and the two parameters, c_p and β_r, which define ribosome function (β_r is the fraction of total ribo-

TABLE 1 Parameters related to the growth and macromolecular composition of bacterial cells

Class	No.	Parameter	Symbol	Value	Reference
I	1	Deoxyribonucleotide residues per genome	kbp/genome	4,700	4
	2	Ribonucleotide residues per rRNA precursor	nucl./prib	6,000	104
	3	Ribonucleotide residues per 70S ribosome	nucl./rib	4,566	104
	4	Amino acid residues per 70S ribosome	aa/rib	7,336	140
	5	Ribonucleotide residues per tRNA	nucl./tRNA	80	64
	6	Amino acid residues per RNA polymerase core	aa/pol	3,407	107–109
II	7	Fraction of total RNA that is stable RNA	f_s	0.98	5, 80
	8	Fraction of stable RNA that is tRNA	f_t	0.14	37, 118
	9	Fraction of active ribosomes	β_r	0.8	57
III	10	Fraction of total protein that is r-protein	α_r	0.09–0.22	Table 3
	11	Fraction of total protein that is RNA polymerase	α_p	0.009–0.01	Table 3
	12	Fraction of active RNA polymerase synthesizing rRNA and tRNA	ψ_s	0.28–0.77	Table 3
	13	Fraction of active RNA polymerase	β_p	0.15–0.32	Table 3
IV	14	Peptide chain elongation rate	c_p	12–22 aa/s	Table 3
	15	Stable RNA chain elongation rate	c_s	85 nucl./s	Table 3
	16	mRNA chain elongation rate	c_m	40–55 nucl./s	Table 3
	17	DNA chain elongation rate	c_d	500–830 nucl. bp/s	Table 3
V	18	Time to replicate the chromosome	C	40–67 min	Table 3
	19	Time between termination of replication and division	D	22–30 min	Table 3
	20	Protein per replication origin	P_O	$2.5 \times 10^8 – 4 \times 10^8$ aa	Table 2

somes actively engaged in peptide chain elongation; c_p is the rate of peptide chain elongation). Based on these relationships, Churchward et al. (27) were able to describe the global cell composition in terms of DNA, RNA, and protein content as a function of the doubling time, τ, and five additional parameters, C, D, P_O, β_r, and c_p.

(It is noteworthy that these parameters include the peptide and DNA chain elongation rates [C period], but not the RNA chain growth rate. Intuitively it would seem that all three chain elongation rates should be equally important and contribute to the DNA, RNA, and protein content. The explanation for this paradox is that the RNA chain elongation rate is implicit in the value of τ; if one asks for the composition as a function of growth rate, one assumes τ as a given parameter without asking how its value was achieved. Thus the cell composition is indeed determined by all three macromolecular chain elongation rates.)

OBSERVED CELL COMPOSITION OF *E. COLI* B/r

Cell Growth-Related Parameters

In Table 1, a number of growth-related parameters are listed that are generally useful in describing or establishing the macromolecular composition of bacterial cultures. These parameters can be divided into five classes: (i) structural parameters that are inherently constant and do not vary with the growth rate, like the number of rRNA nucleotides in a 70S ribosome; (ii) partition factors which are essentially invariant and growth rate independent, like the fraction of total RNA that is stable rRNA; (iii) other partition parameters which change as a function of the exponential growth rate and have substantial effects on cell composition, like the fraction of active RNA polymerase synthesizing rRNA and tRNA; (iv) kinetic parameters describing functional activities (the values of some of the parameters are essentially invariant, whereas others appear to approach a maximum or biological limit value, like the peptide chain elongation rate); (v) chromosome replication and cell division parameters that in general do not limit the exponential growth rate, like the C period.

Reference Units

Physiological parameters describing cell composition, like the amount or synthesis rate of a particular component, require a reference unit such as "per cell," "per cell mass," "per amount of protein," "per microgram of dry weight," etc. Except for studies dealing with the cell cycle, we recommend in most cases the use of cell mass (e.g., "per OD$_{460}$" [unit of optical density at 460 nm]) as a reference unit because its determination is simpler, faster, and more accurate than that of other units. The per-mass values may also be used to estimate the intracellular concentrations because the average cell volume per mass unit changes very little with growth conditions. In many cases the concentration is more relevant than the amounts of components per genome equivalent of DNA or per cell. Since the reference parameters are themselves subject to growth rate-dependent regulation, no single reference unit is ideal or more natural than another.

There has been a tendency in the recent literature for authors to state that rRNA or ribosomes accumulate in proportion to the square of the growth rate. This, of course, is incorrect because the unit of reference is undefined. As first pointed out by Maaløe (93), the rate of RNA accumulation per genome does increase with the square of the growth rate, μ. This reflects the fact that the amount of RNA per genome is proportional to the growth rate (at least above growth rates of 0.5 doubling per h). Since in any exponential system the synthesis rate is proportional to the amount, and the amount is already proportional to μ, it follows that the rate must be proportional to μ^2. However, since the ratio RNA/DNA reflects both chromosome replication and RNA synthesis control, this square relationship has no particular significance for ribosome control itself; the relationship no longer holds in certain replication control mutants which alter DNA content but show no change in rRNA control (27, 28).

Macromolecular Composition as a Function of Growth Rate

For physiological studies, *Escherichia coli* B/r has several advantages over other *E. coli* strains. Due to a special property of its outer membrane, this strain can be age-fractionated by the membrane elution technique (71) and used to measure cell cycle-related parameters. Helmstetter and Cooper (70) used this strain to measure the C and D periods and deduced from these measurements the relationships between chromosome replication and the cell division cycle. This strain also (i) has a lesser tendency for clumping and "snake" formation than other strains of *E. coli*, (ii) grows well in minimal media, and (iii) is free of mutations which might otherwise influence the growth rate or composition. A disadvantage is that the strain is genetically incompatible with K-12 strains because of the B and K restriction systems. Mutants deficient in B restriction or with K-12 restriction and modification are available from the authors. For these reasons, *E. coli* B/r is the preferred choice for physiological studies and composition measurements.

Table 2 lists the amounts of protein, RNA, and DNA and related physiological parameters for cultures of *E. coli* B/r growing exponentially at 37°C in different growth media at rates between 0.6 and 2.5 doublings per h. The per-mass values (top section) represent averages obtained from curves drawn as a best fit through individually measured points (28). The actual measurements fluctuate by about 15% around these curves. Most of this scatter represents a true variation from culture to culture (the contribution due to measuring errors is about ±6%, 2.5%, 5%, and 5% for protein, RNA, DNA, and cell number, respectively). Protein, RNA, and DNA were measured colorimetrically, and cell numbers were determined with an electronic particle counter as indicated in Table 2. From the per-mass values, protein and RNA per genome and protein, RNA, and DNA per cell were calculated.

The sums of the weights of protein, RNA, and DNA are proportional to the mass in OD_{460} units and correspond to 75 to 91% of the dry weight; lipids, carbohydrates, soluble metabolites, and salts represent the remaining 9 to 25% of the total dry mass. The relative proportions of the macromolecules at the different growth rates are illustrated in the bar graphs of Fig. 1. The greatest relative change is found in the RNA sector, reflecting the increasing concentration of ribosomes at higher growth rates. More ribosomes are required to support the higher rate of protein synthesis in rapidly growing cells.

The growth rate-dependent changes in the relative proportions of DNA, RNA, and protein can be described by the two ratios, RNA/protein and DNA/protein. With increasing growth rate, RNA/protein increases and DNA/protein decreases (Fig. 2a and b). The increasing RNA/protein ratio reflects the control of ribosome synthesis (see equation 18 in Table 6 below), and the decreasing DNA/protein ratio reflects the control of DNA replication (see legend to Fig. 2j). The RNA/protein ratio is proportional to the number of ribosomes per amount of protein and, therefore, is a measure for the cytoplasmic ribosome concentration. The growth rate of an exponential culture is equal to the product of ribosome concentration times the rate of ribosome function (i.e., the protein synthesis rate per average ribosome or the ribosome efficiency; 74, 122). Therefore, at a given growth rate, the protein synthesis rate per ribosome can be calculated from the RNA/protein ratio. When the growth rate increases, the rate of ribosome function approaches a maximum value, corresponding to 21 amino acids polymerized per second per active ribosome (Fig. 2c, right ordinate scale).

The number of replication origins in a culture was obtained by measuring the amount of DNA that had accumulated 50 to 80 min after treatment of a culture with rifampin. Rifampin stops initiation of replication, but allows the ongoing rounds of replication to go to completion, so that the number of completed chromosomes becomes equal to the number of functional origins present at the time of rifampin addition (128, 129). The mass per origin, M_O, was then obtained by dividing this number of completed chromosomes by the OD_{460} observed at the time of rifampin addition. The amount of protein per origin, P_O, was found from M_O by multiplication with the amount of protein per mass, P_M (Table 2).

Protein per origin, P_O, is a formal measure for the control of replication initiation; it has a meaning similar to cell mass per origin, M_O, which is proportional to the "initiation mass" defined by Donachie (51). The initiation mass is the cell mass at the time of initiation, divided by the number of replication origins at which initiation occurs, i.e., M_i/O_i (Fig. 2j), whereas M_O is the total mass in a given volume of exponential culture, divided by the number of copies of *oriC* present in that volume. Both P_O and M_O increase with growth rate and approach a constant value at growth rates above 1.5 doublings per h (Fig. 2j; $P_O = 4 \times 10^8$ amino acids per *oriC*). The exact growth rate dependence of P_O (or M_O) may depend on the strain used. A decreasing initiation mass with increasing growth rate has been reported for a K-12 strain of *E. coli* (130).

The parameter P_O links DNA replication to protein synthesis and growth. Whereas the time intervals between consecutive cell divisions vary considerably, the time intervals between consecutive initiations of rounds of replication vary very little (83). This is presumed to reflect the accumulation of a hypothetical protein that triggers initiation at a certain threshold value (17). This putative initiation protein would be made as a constant fraction of total protein synthesis, thereby linking chromosome replication to protein synthesis.

The numbers of replication termini and of forks on the chromosome were calculated from the values of the C and D periods (taken from Table 3, below). These numbers relate to the extent of chromosome branching as a result of increasing overlap in rounds of replication as the cells grow faster (Fig. 1).

Parameters Pertaining to the Macromolecular Synthesis Rates

The rates of accumulation of protein, RNA, and DNA or the rate of cell division (or of any other extensive property, X, of the system) can be calculated using the first-order rate equation $dX/dt = Xk\mu = X(\ln 2)/\tau$, where μ is in doublings per hour, τ is in minutes, and $k = (\ln 2)/60$; the rate is per minute.

For DNA, ribosomes, and protein, the rates of synthesis during periods of balanced growth are essentially equal to the rates of accumulation since their turnover is negligible (37, 38). For total RNA, however, the instantaneous synthesis rate is substantially higher than the accumulation rate because of the instability of mRNA and of spacer sequences in the primary rRNA and tRNA transcripts.

In Table 3, physiological parameters related to the macromolecular synthesis rates have been divided into three groups: parameters pertaining to (i) RNA polymerase synthesis and function, (ii) ribosome synthesis and function, and (iii) DNA synthesis and cell division. Some of these parameters were observed, and others were calculated as indicated (see Table 3 footnotes).

TABLE 2 Macromolecular composition of exponentially growing *E. coli* B/r as a function of growth rate at 37°C[a]

Parameter	Symbol	Units	τ, 100 μ, 0.6	τ, 60 μ, 1.0	τ, 40 μ, 1.5	τ, 30 μ, 2.0	τ, 24 μ, 2.5	Observed parameter(s)	Footnote
					At τ (min) and μ (doublings per h):				
Protein/mass	P_M	10^{17} aa/OD$_{460}$	6.5	5.8	5.2	5.1	5.0	P, M	b
RNA/mass	R_M	10^{16} nucl./OD$_{460}$	4.3	4.9	5.7	6.6	7.8	R, M	c
DNA/mass	G_M	10^8 genomes/OD$_{460}$	18.3	12.4	9.3	8.0	7.6	G, M	d
Cell no./mass	C_M	10^8 cells/OD$_{460}$	11.7	6.7	4.0	2.7	2.0	Cells/OD$_{460}$	e
$(P + R + G)/M$	PRD_M	μg/OD$_{460}$	149	137	129	131	136		f
Protein/genome	P_G	10^8 aa residues	3.5	4.7	5.6	6.3	6.6	P_M, G_M	
RNA/genome	R_G	10^7 nucl. residues	2.3	4.0	6.1	8.2	10.3	R_M, G_M	
Origins/genome	O_G	Dimensionless	1.25	1.32	1.44	1.58	1.73	C	g
Protein/origin	P_O	10^8 aa residues	2.8	3.6	3.9	4.0	3.8	P_G, O_G	g
Protein/cell	P_C	10^8 aa residues	5.6	8.7	13.0	18.9	25.0	P_M, C_M	
	P_C (μg)	μg/10^9 cells	100	156	234	340	450		h
RNA/cell	R_C	10^7 nucl. residues	3.7	7.3	14.3	24.4	39.0	R_M, C_M	
	R_C (μg)	μg/10^9 cells	20	39	77	132	211		h
DNA/cell	G_C	genome equiv./cell	1.6	1.8	2.3	3.0	3.8	C, D	i
	G_C (μg)	μg/10^9 cells	7.6	9.0	11.3	14.4	18.3		h
Mass/cell	M_C	OD$_{460}$ units/10^9 cells	0.85	1.49	2.5	3.7	5.0	C_M	j
	M_C (μg)	μg dry weight/10^9 cells	148	258	433	641	865	μg/OD$_{460}$	k
Sum $P + R + G$	PRD_C	μg/10^9 cells	127	204	322	486	679	P_C, R_C, G_C (in μg)	k
Origins/cell	O_C	no./cell	1.96	2.43	3.36	4.70	6.54	C, D	l
Termini/cell	T_C	no./cell	1.23	1.37	1.54	1.74	1.94	D	l
Replication forks/cell	F_C	no./cell	1.46	2.14	3.64	5.92	9.19	C, D	l

[a]Data are representative for the growth rates indicated, with an accuracy of ±10% or better. In compiling the data in this table and Table 3, we have been guided by the principle that, on average, parameters like protein, RNA, DNA, and cell number per mass, and their quotients, i.e., protein and RNA per genome, or the per-cell values, should have smooth functions of growth rate. (If two primary data were 5% off the true average, their quotient might contain a 10% error and thus make it impossible to draw a smooth line through the points.) In addition, we have checked for consistency if measurements were available from independent methods or involved theoretical relationships between different parameters. The data in our tables closely meet these criteria. For example, RNA, measured as absorption at 260 nm (A_{260}) of RNA hydrolysates, does not require a calibration standard; therefore, the RNA values are assumed to be quite accurate. Since the RNA-to-protein ratios from this table generated the same α_r curve as that determined from purified ribosomal particles (Table 3), we have confidence in both RNA and protein values. The amount of DNA per mass was measured independently with a colorimetric assay calibrated with purified *E. coli* DNA and by radioactive pyrimidine labeling of nucleic acids. The latter method gives the RNA-to-DNA nucleotide ratio which, combined with the absolute (presumably reliable) value for RNA per mass, gives DNA per mass (45). Again both methods gave essentially the same values. Thus, all three macromolecular concentrations (per mass) in this table have consistent values which are presumably accurate to better than 10% and representative for that growth rate. Representative per-cell values were more difficult to obtain, in part because the duration of the (average) D period, which affects the cell size and per-cell values, fluctuates considerably from culture to culture; this fluctuation is independent of and in addition to the variation in D from cell to cell within one culture (24). The DNA per cell values have been determined directly from DNA and cell numbers per mass and indirectly from the C and D periods. Both methods gave essentially the same DNA content of the average cell. Abbreviations: aa, amino acid; nucl., nucleotide; equiv., equivalent.

[b]Cell mass density was determined as OD$_{460}$ using a 1-cm light path (27). Protein was determined by a modification of the method of Lowry et al. (20, 92), using (weighted) bovine serum albumin as a calibration standard and assuming 5.6×10^{15} amino acid residues per μg of protein. The values shown are from Fig. 3 of Churchward et al. (29).

[c]RNA was determined as A_{260} of acid-insoluble, alkali-labile cell mass (20). One A_{260} unit at pH 2 corresponds to 5.6×10^{16} nucleotides, assuming the mole fractions of A, U, G, and C in *E. coli* stable RNA to be 0.248, 0.210, 0.324, and 0.218, respectively (102). The RNA values shown are from Churchward et al. (28).

[d]DNA per mass was calculated from DNA per cell and cells per mass: $G_M = G_C \cdot C_M$. These calculated values closely agree with direct measurements of DNA per mass (Fig. 1 of reference 29), using the colorimetric diphenylamine reaction with *E. coli* DNA as a calibration standard and assuming 1 A_{260} unit of *E. coli* DNA at pH 12 to correspond to 2.86×10^{13} kbp (for a GC content of 0.50) and the *E. coli* genome to be 4,700 kbp (4).

[e]Cell numbers were determined using a Coulter Counter with a 20-μm orifice. The values shown are from Fig. 6 of Shepherd et al. (125).

[f]The sum of the weights of protein, RNA, and DNA per mass was calculated from the sum of the weights per cell (see footnote h, this Table) and cells per mass: $PRD_M = PRD_C \cdot C_M$.

[g]The number of replication origins per genome was determined from the value of the C period (Table 3), using equation 11 in Table 5 below.

[h]The weights of protein, RNA, and DNA were calculated, assuming the average molecular weight of an amino acid residue in *E. coli* protein to be 108 (composition of *E. coli* protein from reference 133), that of an RNA nucleotide residue to be 324 (composition of *E. coli* stable RNA from reference 102), and that of a DNA base pair to be 618 (for a GC content of 0.50), respectively.

[i]The average amount of DNA per cell, in genome equivalents, was calculated from the values of C and D (Table 3), using equation 3 in Table 5 below. These calculated values agree with direct (colorimetric) measurements of DNA and cell numbers per mass (reference 28; see footnote d in this Table).

[j]The average cell mass in OD$_{460}$ units is the reciprocal of the cell number per OD$_{460}$, i.e., $M_C = 1/C_M$.

[k]The cell mass in micrograms dry weight was calculated, using the value of 173 μg per OD$_{460}$ unit of culture mass (20).

[l]The average numbers of replication origins, termini, and replication forks were determined from the values of C and D (Table 3), using equations 7, 8, and 10, respectively, of Table 5 below.

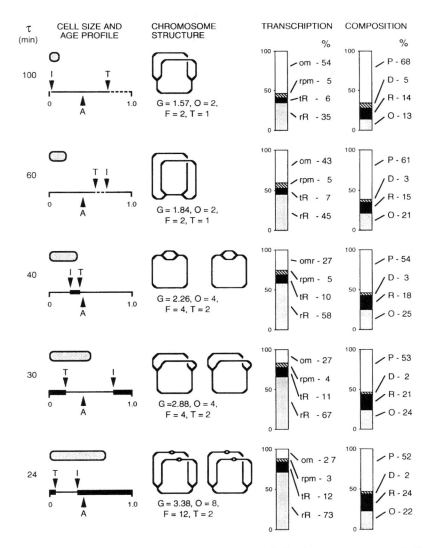

FIGURE 1 Relationships between growth rate, cell size, chromosome replication, transcription, and macromolecular composition. (Left column) Average cell size (mass per cell, Table 2) for *E. coli* B/r growing with a doubling time, τ, ranging from 100 to 24 min is depicted by the shaded ovals. An idealized cell cycle with the major cell cycle events, ranging from cell age 0.0 (a newborn daughter cell) to 1.0 (a dividing mother cell), is presented for each growth rate. The position of an average cell of age 0.41 (defined so that 50% of the cells in the population are younger and 50% are older) is indicated by A. The cell ages at initiation (I) and termination (T) of chromosome replication are also indicated. The dashed portion of the age axis indicates a period during which there is no DNA replication (no replication forks on the chromosome). The line portions represent periods where there are two forks per chromosome structure, and the heavy bar portions indicate the age periods during which there are six forks per chromosome structure. After termination, there are two chromosome structures per cell which are segregated to the daughter cells at the subsequent cell division (at age 1.0). (Center column) Structure of the replicating chromosome or chromosomes in the average cell of age 0.41. For a 24-min cell cycle (τ =

24 min), the chromosome pattern indicates that replication is being initiated and that each of these chromosome structures has multiple (six) replication forks. The amount of DNA in these structures in genome equivalents (*G*) is indicated (calculated from *C, D,* and τ in Table 2 for a cell of age 0.41). The number of origins (*O*), termini (*T*), and forks (*F*) in this average cell are also indicated. (Right columns) The synthesis rates of rRNA (rR), tRNA (tR), r-protein mRNA (rpm), and other mRNA (om), expressed as a percent of total transcription, and the macromolecular composition are illustrated in bar graph form. The stable RNA fraction of the total transcription increases with increasing growth rate, the r-protein mRNA fraction remains essentially constant, and the total mRNA fraction decreases. The proportion of the total mRNA that is r-protein mRNA clearly increases, approximately in parallel with the increase in α_r, the fraction of the total protein synthesis that is r-protein, implying that there is transcriptional control of r-protein operons (the transcription values are adapted from references 42, 63, and 89 and P. Dennis, unpublished data). Relative amounts of protein (P), DNA (D), RNA (R), and other components (O) as percent of the total cell mass are from the data in Table 2.

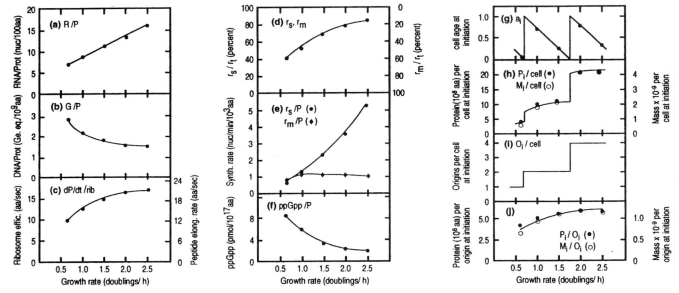

FIGURE 2 Amounts and synthesis rates of molecular components in bacteria growing exponentially at rates between 0.6 and 2.5 doublings per h. The values of the RNA-to-protein (*R/P*; panel a) and DNA-to-protein (*G/P*; panel b) ratios were calculated from lines 1, 2, and 3 in Table 2. The ribosome efficiency (i.e., the protein synthesis rate per average ribosome; panel c, left ordinate) was calculated from the number of ribosomes per cell (line 15, Table 3) and the rate of protein synthesis per cell. The latter was obtained from the amount of protein per cell (line 10, Table 2) using the first-order rate equation. The peptide chain elongation rates (panel c, right ordinate) are 1.25-fold higher than the ribosome efficiency values and account for the fact that only about 80% of the ribosomes are active at any instant. The fraction of the total RNA synthesis rate that is stable RNA or mRNA (*r_s* or *r_m*; panel d) is from line 5, Table 3. The rates of stable RNA and mRNA synthesis per amount of protein (*r_s*/P or *r_m*/P; panel e) were calculated from lines 9 and 10, Table 3, divided by the amount of protein per cell (line 10, Table 2). The ppGpp per protein value (ppGpp/*P*; panel f) is from line 11, Table 3. The cell age at which chromosome replication is initiated at *oriC* (*a_i* in fractions of a generation; panel g) is calculated from *C* and *D* (lines 23 and 24, Table 3) and equation 14 in Table 5. The protein (or mass) per cell at replication initiation (panel h) was calculated from the initiation age (*a_i*, panel g) and the cell mass immediately after cell division (age zero; i.e., *a* = 0), using equation 17 in Table 5. The latter was obtained from the average protein or mass content of cells (lines 10 or 13, respectively, Table 2), using equation 16 of Table 5. The number of replication origins at the time of replication initiation (*O_i*, panel i) was obtained from the values of *C* and *D* (Table 3), using equation 15 of Table 5. The initiation mass (panel j), given as protein (or mass) per replication origin at the time of replication initiation, was obtained as the quotient of the values for *P_i* (or *M_i*) and *O_i* shown in panels h and i.

RNA Polymerase Synthesis and Function.

RNA polymerase concentration. The instantaneous rate of transcription in the cell depends on the concentration of RNA polymerase, α_p, measured as the fraction of total protein that is RNA polymerase core enzyme (three subunits, α_2, β, and β'). The values of α_p increase with the growth rate and reflect the control of the synthesis of the β and β' subunits of RNA polymerase. Since the α subunit is in excess in *E. coli* (see Table 4), the amount of core enzyme would seem to be limited by the amount of β and β' subunit polypeptides. The synthesis of these subunits is under dual transcriptional control (i) at the level of initiation at an upstream promoter and (ii) at the level of termination-antitermination at an attenuator in front of the *rpoB* gene (6, 7, 40–42, 52). Both controls are growth rate dependent, but the mechanisms mediating these controls are poorly understood. The initiation control might involve the effector nucleotide guanosine tetraphosphate (ppGpp), whereas the read-through at the attenuator might involve an autoregulation by free or active RNA polymerase (40, 52, 96). There is also translational control of the *rpoBC* mRNA (43, 47, 48). Combining the observed α_p values with the protein per cell values of Table 2, the number of RNA polymerase molecules per cell has been calculated and was found to increase from 1,500 to 11,400 between growth rates of 0.6 and 2.5 doublings per h.

RNA polymerase activity. The total complement of RNA polymerase enzyme in the cell can be partitioned into active RNA polymerase (enzymes engaged in RNA chain elongation) and inactive RNA polymerase (DNA-bound, idling polymerase; unbound, free enzyme, ready to bind to a promoter; and truly inactive, e.g., immature enzyme). The active enzyme is determined from the rate of transcription and the RNA chain elongation rate. Due to a difference in the chain elongation rates for mRNA and rRNA (see below), the calculation contains separate components for mRNA and stable RNA. The calculation shows that only 17 to 30% of the RNA polymerase enzyme is active at any instant; this fraction, β_p, is seen to increase with growth rate (125). In ppGpp-deficient strains, up to 60% of the total RNA polymerase was found to be active; this has suggested that part of the inactive polymerase is transiently stalled at ppGpp-dependent pause sites during the synthesis of mRNA (74). Most of the remaining inactive RNA polymerase might be involved in ppGpp-independent transcriptional pausing.

Within the cell, it seems that active RNA polymerase limits the rate of transcription and that the DNA template is in excess. This was demonstrated using a mutant bacterial strain that exhibits altered DNA replication control which results in a lower

TABLE 3 Parameters pertaining to the macromolecular synthesis rates in exponentially growing *E. coli* B/r as a function of growth rate at 37°C

Parameter	Symbol	Units	At τ (min) and μ (doublings per h):					Observed parameter(s)	Footnote
			τ, 100 / μ, 0.6	τ, 60 / μ, 1.0	τ, 40 / μ, 1.5	τ, 30 / μ, 2.0	τ, 24 / μ, 2.5		
RNA polymerase protein/total protein	α_p	%	0.90	1.10	1.30	1.45	1.55	α_p	a
RNA polymerase molecules/cell	N_p	10^3 RNAP/cell	1.5	2.8	5.0	8.0	11.4	α_p, P_C	b
RNA polymerase activity	β_p	%	17	20	21	24	30	r_s, r_m, c_s, c_m, N_p	c
Active RNA polymerase per cell	N_{ap}	RNAP/cell	205	503	992	1,929	3,298		c
Stable RNA synthesized per total RNA synthesized	r_s/r_t	%	41	52	68	78	85	r_s/r_t	d
Active RNA polymerase synthesizing stable RNA	ψ_s	%	24	36	56	68	79	r_s/r_t c_s/c_m	e
rRNA chain elongation	c_s	Nucl./s	85	85	85	85	85	Indirect	f
mRNA chain elongation	c_m	Nucl./s	39	45	50	52	55	Indirect	g
Rate of stable RNA synthesis/cell	r_s	10^5 nucl./min/cell	3.0	9.9	29.0	66.4	132.5	R_C	h
Rate of mRNA synthesis/cell	r_m	10^5 nucl./min/cell	4.3	9.2	13.7	18.7	23.4	r_s, r_s/r_t	i
ppGpp concentration	ppGpp/M	pmol/OD$_{460}$	55	38	22	15	10	ppGpp/M	j
	ppGpp/P	pmol/10^{17}aa	8.5	6.6	4.2	2.9	2.0	P_M	j
r-Protein per total protein	α_r	%	9.0	11.4	14.8	17.5	21.1	P_M, R_M	k
			9	11	13.5	18.0	21.6	α_r	l
Ribosome activity	β_r	%	80	80	80	80	80	Indirect	m
Peptide chain elongation	c_p	aa residues/s	12	16	18	20	21	Indirect	n
Ribosomes/cell	N_r	10^3 ribosomes/cell	6.8	13.5	26.3	45.1	72.0	R_C, f_s, f_t	o
tRNA/cell	N_t	10^3 tRNA/cell	63	125	244	419	669	N_r, f_t	p
rrn genes/cell	N_{rrn}	Avg no./cell	12.4	15.1	20.0	26.9	35.9	C, D	q
rrn genes/genome	N_{rrn}/G	Avg no./genome	7.9	8.2	8.6	9.0	9.5	C	r
Initiation rate at *rrn* gene	i_{rrn}	Initiations/min/gene	4	10	23	39	58	N_r, N_{rrn}	s
Distance of ribosomes on mRNA	R_m/N_r	Nucl./ribosome	79	85	65	52	41	r_m, c_m, N_r	t
No. of translations per mRNA	N_{trans}	Ribosomes	27	33	49	70	93	P_C, r_m	u
DNA chain elongation	c_d	Nucl. residues/s	585	783	870	911	933	C	v
C period	C	min	67	50	45	43	42	Indirect	v
D period	D	min	30	27	25	24	23	Indirect	w

[a] The fraction of the total cell protein that is core RNA polymerase was calculated from the β and β′ content determined by sodium dodecyl sulfate-gel electrophoresis (125).

[b] The number of core RNA polymerases per cell was calculated from α_p (this table) and values of P_C (Table 2), using aa/pol (Table 1) and the relationship $N_p = P_C \cdot \alpha_p/(aa/pol)$.

[c] The fraction of active RNA polymerase was calculated from values in this table, the relationship $\beta_p = (r_s/c_s + r_m/c_m)/N_p$, and the active RNA polymerase per cell, $N_{ap} = \beta_p \cdot N_p$.

[d] The fraction of the total RNA synthesis rate that is stable RNA was determined by hybridization of pulse-labeled total RNA to an rDNA probe and correction for tRNA (119). Since the pulse-labeling time (1 min) was similar in duration to mRNA lifetimes, the r_s/r_t values shown are somewhat overestimated. This results from the underestimate of the rate of mRNA synthesis due to decay of labeled mRNA during the pulse-labeling period.

[e] The fraction of RNA polymerase synthesizing stable RNA was calculated using the relationship $\psi_s = 1/\{1 + [1/(r_s/r_t) - 1](c_s/c_m)\}$.

[f] The stable RNA (or rRNA) chain growth rate was determined from 5S rRNA or tRNA labeling after rifampin addition (99, 120, 124).

[g] The mRNA chain elongation rate was determined by analysis of pulse-labeling kinetics of RNA after size fractionation (19) and by the time lag between induction of transcription of specific mRNAs (*lacZ*, *infB*) and the appearance of specific hybridization to DNA probes from the 3′ ends of the respective genes (139).

[h] The stable RNA synthesis rate per cell was determined from the data in Tables 1 and 2 and the rate equation $r_s = R_C \cdot f_s \cdot \ln2/t$, where the factor f_s is equal to 1.2 and corrects for the 20% of the rRNA and tRNA primary transcripts that are unstable spacer or flanking sequences.

[i] The mRNA synthesis rate per cell was determined from the data in this table and the relationship $r_m = r_s \{[1/(r_s/r_t)] - 1\}$.

[j] Measurement of ppGpp was by A_{260} after separation of nucleotides by high-pressure liquid chromatography (119). ppGpp/P = (ppGpp/M)/P_M.

[k] The differential rate of r-protein synthesis was calculated from the data in Tables 1 and 2 and the relationship for α_r (equation 6, Table 5).

[l] The differential rate of r-protein synthesis was determined from measurements of the protein content of ribosomes after labeling with radioactive leucine and uridine (44, 46).

[m] The fraction of active ribosomes was measured as ribosomes in polysomes with a correction for active 70S ribosomes (57).

[n] The peptide chain elongation rate, c_p, was calculated in amino acid residues per second per ribosome from the rate of protein synthesis (*dP/dt*), the number of ribosomes (N_r), and the fraction of active ribosomes (β_r), using the relationship $dP/dt = N_r \cdot \beta_r \cdot c_p$ (see also the equivalent equation 5 in Table 5; 44, 49).

[o] The number of ribosomes per cell was determined from values in Tables 1 and 2 and the relationship $N_r = R_C \cdot f_s (1 - f_t)/(nucl./rib)$, where f_s, f_t, and nucleotides per ribosome are defined Table 1.

[p] The number of tRNAs per cell was determined from values in Tables 1 and 2 and the relationship $N_t = R_C \cdot f_s \cdot f_t/(nucl./tRNA)$, where f_t and nucleotides per tRNA are defined in Table 1.

[q] The number of rRNA genes per cell was calculated from the number of *rrn* genes per genome (this table) and the number of genomes per cell (Table 2): $N_{rrn} = (N_{rrn}/G) \cdot G_C$.

[r] The number of rRNA genes per genome was determined from the value of the C period (footnote *v*, this table) and the locations of the seven *rrn* genes (87, 89.5, 85, 72, 90.5, 57, and 5 min, respectively), using equations 11 and 12 from Table 5 below.

[s] The transcription initiation rate at each *rrn* gene was calculated from the number of ribosomes per cell and the number of *rrn* genes per cell, using the relationship $i_{rrn} = N_r \cdot (\ln2/t)/N_{rrn}$.

[t] The nucleotide distance between ribosomes on mRNA was calculated from data in this table and the relationship $R_m/N_r = r_m \cdot t_m/(\beta_r \cdot N_r)$, where R_m is the amount of mRNA and t_m is the average functional life of mRNA, assumed to be 1.0 min (5, 25).

[u] The number of translations per mRNA was calculated from data in Table 2 and this table using the relationship $N_{trans} = 3 \cdot (dP/dt)/r_m$. The factor 3 in the numerator is the coding ratio, i.e., 3 mRNA nucleotides per amino acid residue.

[v] The C period was determined from age-fractionated cultures (70), synchronized cultures (13), flow cytometric data (128), and perhaps most accurately in nonsynchronous exponential cultures by measuring the increase in the amount of DNA in the culture after treatment with rifampin or chloramphenicol (26). The values obtained by these different methods agree to within 10%. The values shown are considered to be the best average of the reported data.

[w] Like the C period, the D period was determined in age-fractionated and synchronized cultures, as well as from flow cytometric data (13, 70, 129). The D period was also determined by measuring the increase in cell number after treating exponential cultures with sodium azide; this treatment stops replication, but does not prevent the division of cells that were already in the D period at the time of the replication stop (14).

DNA concentration. In spite of the lower DNA concentration there was no change in the level of transcription, implying that the total rate of transcription is regulated at the level of active RNA polymerase in the cell and not the level of available DNA template (28).

What is the state of inactive RNA polymerase within the cell? Experiments with a minicell strain indicate that most of the (core) RNA polymerase is sequestered with the DNA (N. Shepherd, Ph.D. thesis, University of Texas, Dallas, 1979). This enzyme might either be bound nonspecifically to DNA (21), or, alternatively, it might have initiated transcription but is halted at some pausing site (81, 100), perhaps associating with termination-antitermination factors (66). It remains unclear why there is such a large excess of inactive RNA polymerase, and whether and how the partition between active and inactive enzyme is maintained.

Partitioning between stable RNA and mRNA synthesis. The active RNA polymerase enzyme can be further partitioned into the fractions engaged in the synthesis of stable RNA species (rRNA, tRNA, and their spacers; ψ_s) and of mRNA ($\psi_m = 1 - \psi_s$). This partitioning is strongly correlated with, and possibly controlled by, the nucleotide effector ppGpp (119). During periods of amino acid insufficiency, ppGpp is derived from the *relA*-dependent system and elicits the well-characterized stringent response (40, 49, 69). During exponential growth at different rates, ppGpp is derived from a *relA*-independent system (3, 56, 117) that involves a product of the *spoT* gene (73, 141).

Nomura and his coworkers have suggested that free ribosomes rather than ppGpp regulate transcription of *rrn* operons (77, 105). However, when free ribosomes were allowed to accumulate by limiting the concentration of the translation initiation factor IF2, rRNA synthesis was stimulated (30), not inhibited as predicted by the free ribosome feedback hypothesis. At the same time, ppGpp levels were reduced. Thus, limitation of IF2, like treatment with chloramphenicol, induces the equivalent of the relaxed response (39, 72, 84). Furthermore, in strains unable to synthesize ppGpp due to deletions in the genes for the two separate ppGpp synthetases, ψ_s was found to be growth rate-invariant (74). This observation supports the idea that ppGpp is involved in the partitioning of RNA polymerase into stable RNA- and mRNA-synthesizing fractions. In an in vitro system with purified RNA polymerase, ppGpp has been found to preferentially inhibit rRNA synthesis (e.g., references 136 and 137), whereas attempts to find a similar effect by free ribosomes have failed (77).

The levels of ppGpp listed in Table 3 are seen to decrease from 55 to 10 pmol per OD_{460} when the growth rate increases from 0.6 to 2.5 doublings per h. At the same time, the transcription from rRNA and tRNA genes, expressed as the fraction of the total instantaneous rate of transcription, r_s/r_t, increases from 40 to 85%. The parameters ψ_s and r_s/r_t both express the same ppGpp-correlated partitioning of RNA polymerase; the difference between these two parameters reflects the difference in the chain elongation rates for rRNA and mRNA. Since rRNA chains grow faster than mRNA chains, it follows that for equal numbers of polymerase-transcribing stable RNA and mRNA genes ($\psi_s = 0.5$), the rate of stable RNA synthesis must be somewhat greater than the rate of mRNA synthesis (i.e., $r_s/r_t > 0.5$).

Rates of stable RNA and mRNA synthesis per cell. The rate of stable RNA synthesis was calculated from the amount of RNA,

determined from its UV absorbance (Table 2). The rate of mRNA synthesis was found from its relative rate ($r_m/r_t = 1 - r_s/r_t$) and the absolute rate of stable RNA synthesis. It is seen that the stable RNA synthesis rate per cell, in particular, increases dramatically with growth rate, which accounts for the higher RNA content in rapidly growing bacteria (see Fig. 1). The same relationships are apparent when rates of stable RNA and mRNA synthesis are expressed per amount of protein (Fig. 2e).

Chain elongation rates of stable RNA and mRNA. Whereas the chain elongation rate of stable RNA is independent of the growth rate and equal to 85 nucleotides per s (99, 120, 124, 139), the chain elongation rate of mRNA appears to increase somewhat and approaches a maximum value of 55 nucleotides per s at high growth rates (19, 139). The reason for this velocity difference is not known, although it may be related to polymerase pausing or stuttering during chain elongation. The leader regions of *rrn* transcription units are known to contain strong Nus factor-dependent antitermination sites (85, 103). As a consequence, transcripts initiated at *rrn* promoters are able to transcribe through Rho protein-dependent transcription terminators and can reverse transposon-induced polarity (1, 127). Thus, for rRNA transcripts this pausing or stuttering may be minimized by the antitermination state of the RNA polymerase and account for the increased stable RNA chain elongation rate.

Ribosome Synthesis and Function.

Ribosomal components and their control. The ribosome consists of three species of RNA (16S, 23S, and 5S) and 52 species of protein. The three rRNAs are processed from a 35S primary transcript derived from seven unlinked *rrn* transcription units. The 52 different ribosomal proteins (r-proteins) are encoded by genes in about 20 different transcription units located at 14 different positions on the *E. coli* chromosome (4).

The primary regulation of ribosome synthesis is most likely the ppGpp-correlated partitioning of RNA polymerase which, at a given concentration and activity of RNA polymerase, sets the rate of transcription of rRNA and tRNA genes (36, 119; see also Fig. 1). However, even in the absence of ppGpp in bacterial strains deleted for both ppGpp synthetases (141), the stable RNA gene activity increases with increasing growth rate, apparently as a result of only an increase in the concentration of active RNA polymerase (74). There is the additional possibility that ppGpp might also influence either directly or indirectly the transcription of r-protein operons within the mRNA sector to achieve an approximate balance between the production of the rRNA and r-protein components of the ribosome (40, 49, 50). This is the transcriptional control of r-protein operons (35, 42, 63, 87, 89). It is clear from Fig. 1 that the promoter activities of rRNA operons and of r-protein operons do not respond coordinately to changes in the steady-state growth rate. Coordination is neither expected nor required if the ratio of r-protein mRNA to total mRNA is important in determining α_r.

A second mechanism in addition to the transcriptional control, used to accurately balance or fine tune the translation frequency of the 20 different mono- and polycistronic r-protein mRNAs to the availability of free rRNA, involves an autogenous translational control: specific regulatory r-proteins bind to the leader regions of their own mRNAs and inhibit further translation when they are not rapidly incorporated into assembling ribosomes (47, 55, 78, 88, 105).

During exponential growth at moderate to fast rates, turnover of ribosomal components appears to be negligible. During slow growth, some excess of newly made rRNA is degraded (63, 106); this results in a slight increase in the proportion of tRNA to rRNA during slow growth (124).

r-Protein synthesis. A measure for the synthesis of r-protein is its proportion of total protein, α_r. Values of α_r have been determined (i) from the protein content of ribosomes and (ii) from the RNA-to-protein ratio. Both methods give essentially identical values. The proportion of r-protein increases with growth rate from 9% at 0.6 to 21% at 2.5 doublings per h. It has been reported that the r-protein mRNA synthesis per total RNA synthesis rate is nearly invariant with growth rate, but, when expressed as a fraction of the rate of (total) mRNA synthesis, it increases with growth rate like α_r or rRNA synthesis (Fig. 1; 42, 63, 89). These observations support the idea that, to a first approximation, r-protein production is matched to rRNA production at the level of transcription.

Ribosomes and tRNA per cell. Given that r-protein matches rRNA and that rRNA and tRNA are synthesized in constant proportions, corresponding to nine tRNA molecules per 70S ribosome (37, 54, 124, 143), the numbers of ribosomes and of tRNA molecules per cell were calculated from the total amount of RNA. In the growth range considered, the number of ribosomes per average cell was seen to increase 10-fold from 6,700 to 71,000. This reflects both an increasing ribosome concentration (ribosomes per OD_{460} unit of cell mass) and increasing cell size (OD_{460} units per cell). For achieving rapid growth, only the ribosome concentration is relevant.

Ribosome activity. The fraction of ribosomes engaged in peptide chain elongation at any instant, β_r, has been estimated from the ribosome content of polysomes and was found to be about 80% and independent of the growth rate (57). Since the assembly and maturation of ribosomes takes about 5 min (86), it appears that, at least in fast-growing bacteria, the major portion of inactive ribosomes are particles in the final stages of ribosome assembly.

Peptide chain elongation rate. The peptide chain elongation rate has been estimated (i) from pulse-labeling kinetics of nascent polypeptides of given length (111, 142) and (ii) from the first appearance of β-galactosidase activity after induction (33, 61, 110, 111, 142). These measurements indicate that the peptide chain elongation rate increases with growth rate from about 13 to 20 amino acid residues per s between 0.6 and 2.5 doublings per h. These values are in agreement with the values calculated in Table 3 from the RNA-to-protein ratio under the assumption that 80% of the ribosomes are active at any instant.

What limits the peptide chain elongation rate during slow growth? During fast growth when the peptide chain elongation rate is maximal (20 amino acids per s; Table 3), ribosomes seem to be saturated with substrates (elongation factor Tu-GTP-aminoacyl-tRNA ternary complex), implying that the peptide chain elongation rate is limited by the rate of peptide bond formation or ribosome translocation on the mRNA. During slow growth, the (submaximal) peptide chain elongation rate is probably limited by the extent of tRNA charging (i.e., by the ratio of charged to uncharged tRNA) and to some extent by the types of codons being employed, rather than by the absolute concentration of charged tRNA. This is suggested by the observation that in a nutritional shift-up from a minimal medium to an amino acid-supplemented medium the protein synthesis rate

per average ribosome increases immediately in a stepwise fashion, before the concentrations of tRNA and EF-Tu have increased (20). During further postshift growth, the concentration of tRNA, and thus presumably also of charged tRNA, increases severalfold (e.g., about threefold in a shift from succinate minimal to glucose amino acids medium) without further increasing ribosome activity (126). It is also unlikely that ppGpp binds to and thereby inhibits elongation factor Ts function during slow growth since the peptide chain elongation rate was unaffected in a ribosome control mutant with 10-fold reduced level of ppGpp (90).

It is conceivable that in poor media, the greater proportion or abundance of mRNAs with hard-to-read codons might contribute to the reduced peptide chain elongation rate. It is well established that the utilization of codons in the genes of *E. coli* is not random (113, 123, 138). Genes expressed from strong promoters tend to contain a higher proportion of major codons, whereas genes expressed from weak promoters contain a lower proportion of major codons. With some exceptions, major codons are recognized by abundant tRNA species and minor codons are recognized by rare tRNA species (76). However the concentration of minor tRNAs is probably not limiting the rate of peptide chain elongation. Instead, each codon probably has a specific transit time for progressing through the A- and P-sites of the ribosome that depends upon the physical-chemical nature of the tRNA-mRNA (anticodon-codon) interaction. As an example, the GAA and GAG glutamic acid codons are both recognized by the abundant $tRNA_2^{Glu}$, but decoding of the prevalent GAA triplet involving strict Watson-Crick base pairing occurs threefold more rapidly than the decoding of the GAG triplet involving wobble base pairing (110, 131, 132).

Ribosomal gene dosage and gene activity. There are seven *rrn* genes on the *E. coli* chromosome; most of them map near the chromosomal origin of replication (4). The actual number of *rrn* genes per (average) cell is much greater than 7, ranging from 12 to 36, depending on the extent of chromosome branching (see Fig. 1). From this number and the rate of rRNA synthesis, the rate of initiation of rRNA chains at each *rrn* gene was calculated and was found to increase with increasing growth rate from 4/min to 61/min. These are the average values in an exponential culture. Measurements of DNA and RNA synthesis rates in age-fractionated cultures indicate that replication causes the number of *rrn* genes to fluctuate nearly twofold during the cell cycle without abrupt or concomitant changes in the rate of rRNA synthesis (36). Instead, the rate of rRNA synthesis increases by a factor of 2 in a slow and continuous manner, without perturbation, as cells progress through the division cycle; this pattern is followed regardless of when the *rrn* genes are replicated (36, 37). This implies that the rate of transcription initiation at *rrn* genes changes (i.e., decreases) nearly twofold at the time the *rrn* genes are duplicated. Accordingly, the average rate of 61 initiations per min per gene means a fluctuation from about 40 to 80 initiations per min per gene during the cell cycle for a bacterium with a 24-min doubling time (actually the fluctuation is somewhat less because rRNA genes are not all clustered at exactly the same map position). Therefore the copy number of *rrn* genes does not limit the rate of rRNA synthesis under conditions of exponential growth. In addition, if there are up to 80 initiations per min per gene in rapidly growing bacteria, the time for the formation of the open promoter complex must be less than 1 s.

These conclusions have been further corroborated by the observation that the rRNA synthesis rate is not reduced in bacteria with a mutational defect in the control of chromosome replication that leads to a 40% reduction in the concentration of all genes, including *rrn* genes (28). Furthermore, up to three *rrn* genes may be deleted from the *E. coli* chromosome without much change in the growth rate, again suggesting that the *rrn* gene dosage does not limit the rate of rRNA synthesis (31; it is to be noted, however, that contrary to the authors' interpretation, we believe that the rRNA synthesis rate per gene was not suitably measured because of the use of inappropriate reference units).

Translation frequency of mRNA. With increasing growth rate the average mRNA becomes more and more crowded with ribosomes, i.e., the average spacing of ribosomes on mRNA decreases from 120 nucleotides to 60 nucleotides in the range of growth rates considered. Here again, there is no indication that mRNA is a limiting factor for protein synthesis. Immediately after a nutritional shift-up the concentration of mRNA decreases temporarily because the increased rate of rRNA synthesis occurs partly at the expense of mRNA synthesis; at the same time, the protein synthesis rate increases (45, 102, 126). The increased spacing of ribosomes during slow growth could potentially cause some mRNA instability or premature termination of transcription (polarity), but whether this is indeed the case has not been established.

Component proteins of the transcription-translation apparatus. The protein composition of the ribosome is essentially invariant with the growth rate, and each of the 52 different r-proteins is present in one copy per 70S particle (38, 68). The only exception to this is protein L7/L12, which is present in four copies per ribosome (134). This implies that the synthesis rate of each r-protein is strictly coordinate with the synthesis of rRNA and also tRNA at growth rates above 0.5 doublings per h. At slower

growth rates there appears to be a slight excess in the synthesis rate of stable RNA (62, 63, 106); the excess rRNA is rapidly degraded, whereas the tRNA accumulates.

The synthesis rates of other components of the transcription-translation apparatus also appear to be subject to growth rate-dependent regulation (59, 65, 75, 112, 125). These components include translation initiation and elongation factors, the subunits of RNA polymerase, and the aminoacyl-tRNA synthetases (Table 4). From the available data it seems clear that the concentration of all these components increases with growth rate, but the increases might not be—and for at least some of the proteins such as RNA polymerase, for example, are not—strictly parallel with the increase in ribosome (and r-protein) concentration. In Table 4 we list the proteins which have been examined in this context and give the α_i values for each (i.e., the synthesis rate of the protein as a percent of the total protein synthesis rate) at a growth rate of 1.5 doublings per h (τ equals 40 min). In addition, the numbers of molecules of each protein per unit of mass and per ribosome are also indicated. In compiling the information in this table we have had to reinterpret or extrapolate some of the original measurements in the cited references. If the synthesis of these proteins were strictly coordinate with synthesis of ribosomes, the listed number of molecules per ribosome would remain constant and not change with changes in the steady-state growth rate.

The r-protein operons also encode the genes specifying the α, β, β', and σ subunits of RNA polymerase and the protein synthesis elongation factors Tu, Ts, and G (for review see reference 88). There is a second gene for Tu on the *E. coli* chromosome which is not in an r-protein operon; presumably two copies of this gene are required to produce the six molecules of Tu per

TABLE 4 Stoichiometric content of transcription-translation proteins in *E. coli*

Protein	Mol wt (10^3)	$\alpha_i{}^a$ ($t = 40$ min) (%)	Molecules ($\tau = 40$ min)		Reference(s)
			Per OD$_{460}$ (10^{12})	Per ribosome	
r-Protein	850	13.5	10.2	1.00	38, 44
L7/L12	12	0.81	40.8	4.00	134
EF-Tu	42	5.55	55.1	5.40	112
EF-G	84	1.66	8.2	0.80	112
EF-Ts	31	0.13	1.8	0.18	112
IF1	8	0.04	2.5	0.25	75
IF2	115	0.52	3.1	0.30	75
IF3	20	0.07	2.0	0.20	75
Leu S	100	0.12	0.5	0.05	112
Phe S-β	94	0.21	1.0	0.10	112
Lys S	58	0.11	0.8	0.08	112
Arg S	58	0.08	0.6	0.06	112
Gly S	77	0.17	0.9	0.09	112
Val S	106	0.14	0.6	0.06	112
Glu S-β	48	0.10	0.9	0.09	112
Ile S	107	0.24	1.0	0.10	112
Phe S-α	36	0.11	1.2	0.12	112
Gln S	61	0.11	0.8	0.08	112
Thr S	65	0.09	0.6	0.06	112
RNA polymerase β	150	0.52	1.4	0.14	112
RNA polymerase α	39	0.37	3.8	0.37	112
RNA polymerase, core	375	1.30	1.9	0.19	125

$^a\alpha_i$, synthesis rate of the protein as a percentage of total protein synthesis rate.

ribosome at high growth rates. The β and β′ RNA polymerase subunit genes, although cotranscribed with the L10 and L12 r-protein genes, are regulated somewhat independently by a transcription attenuator located between the upstream r-protein genes and the downstream RNA polymerase genes (see RNA polymerase synthesis and function, above).

There are about nine tRNA molecules per ribosome in exponentially growing *E. coli*, and this ratio shows little variation for growth rates above 0.5 doublings per h (Table 3). Since the peptide chain elongation rate approaches 20 amino acids per s, each tRNA is required to cycle through the ribosome on average about two times per second. Ikemura (76) has quantitated over 70% of the total tRNA population into 26 separate species, at least one for each amino acid except for proline and cysteine. For each of these 18 different amino acids there is at least one major isoacceptor which is present at a molar ratio of 0.15 to 0.60 copy per ribosome. The aminoacyl-tRNA synthetases are present at about 0.1 copy per ribosome; each synthetase molecule is therefore required to aminoacylate about 10 molecules of its cognate tRNA every second to sustain protein synthesis.

The elongation factor Tu is required for the GTPase-dependent deposition of aminoacylated tRNA into the A-site of the translating ribosome. The charging level of tRNA is about 75 to 90%. There are between two and three tRNA molecules bound to each translating ribosome (98, 116). The six copies of Tu are available for ternary complex formation with GTP and the remaining aminoacylated tRNAs. The concentration of the ternary complex required to initiate the process of amino acid addition on the translating ribosome is thus maximized.

During periods of amino acid insufficiency the synthesis of r-protein, like that of rRNA and tRNA, is subject to stringent regulation; this control is exerted at the level of transcription (40, 49, 50). Many of the genes that are cotranscribed with r-protein genes are also, as expected, stringently regulated. These include the genes encoding the elongation factors G, Tu, and Ts (8, 60, 115). In contrast, the genes encoding the β, β′, and σ subunits of RNA polymerase are not stringently regulated. The nonstringent regulation of transcription of the β and β′ genes is mediated by control at the attenuator in the L10 (*rplJL rpoBC*) operon (96). In the case of the *rpoD* (σ) gene, a new promoter signal is utilized (C. Gross, personal communication). With respect to aminoacyl-tRNA synthetases, the data on their stringent regulation are equivocal (8).

DNA Replication and Cell Division.

Chromosome replication time. The *C* period is the time interval required for the replication forks to move from the origin (*oriC*, at 84 min on the *E. coli* genetic map) to the terminus (*terC*, approximately at 36 min on the genetic map; 91). Pulse-labeling of the terminus in cells with synchronized replication has indicated that both replication forks created at every initiation event move with equal speed (65 kilobases/min for wild-type strains) clockwise and counterclockwise, respectively, with very little variation from cell to cell (9, 10). In addition, the time intervals between consecutive replications of any given section of the chromosome are very constant and equal to the mass doubling time (83, 97, 101). This suggests that both the times between consecutive initiations of rounds of replication and the replication velocities themselves are constant within a cell population.

The *C* period has been measured by Helmstetter and Cooper (70) in age-fractionated cells. Due to the considerable variability

from cell to cell in the duration of the *D* period (see below), the initiation age, termination age, and interdivision interval all vary (12). This makes the determination of *C* (and *D*) from age-fractionated or synchronous cultures somewhat inaccurate. The *C* period has also been measured in exponential cultures (i) from the relative frequencies of genes at given map locations (equation 9, Table 5 below; reference 22) and (ii) from the increase in DNA after stopping of initiation (26, 114, 144). Cooper and Helmstetter (32) estimated that the *C* period was constant (41 min) for growth rates above 1 doubling per h and increased in proportion to τ at lower growth rates. Measurements of the increase in DNA after a replication stop (26) suggest that the *C* period decreases gradually with increasing growth rate, approaching a value of 40 min in rapidly growing bacteria (Table 3).

Chromosome segregation and cell division. The *D* period is the time between termination of a round of replication and the following cell division. The cell division is believed to require the action of a protein synthesized at the time of termination of replication (79). During the *D* interval the completed chromosomes are segregated. The length of time between the completion of replication and the onset of constriction is subject to a stochastic process (13, 14). This results in a (non-Gaussian) fluctuation in the length of the *D* interval, which is the major factor contributing to cell cycle variability including (i) the variability of the initiation age (the time after cell division at which a round of chromosome replication is initiated) and (ii) the variability of the time intervals between consecutive divisions (12).

Helmstetter and Cooper (70) estimated the average *D* to be about 22 min for growth rates above 1 doubling per h and to increase as a constant fraction of the doubling time for growth rates below 1 doubling per h. The average *D* period has also been determined in exponential cultures from the increase in the cell number after a replication stop by thymine starvation or by the addition of sodium azide (14). Cells that do not terminate replication due to an experimentally induced replication stop will not divide, whereas cells that have already terminated and have made termination protein, i.e., cells in the *D* period, divide once (79). These experiments suggest that the *D* period decreases from about 30 min during slow growth to about 22 min during rapid growth.

The average *C* and *D* intervals have also been obtained by the method of flow cytometry, which measures the distribution of the amounts of DNA (labeled with a fluorescent dye) per cell in exponential cell populations (frequency of cells as a function of DNA per individual cell). In this manner, values of *C* equal to 42 min and *D* equal to 22 to 24 min were found for *E. coli* B/r A during balanced growth with a 27-min doubling time (129).

At a given doubling time, the *C* and *D* periods determine the initiation age, a_i, and the termination age, a_t (Fig. 1; see Table 5 below). Depending on the values of the three parameters, *C*, *D*, and τ, rounds of replication may be initiated at the beginning, in the middle, or near the end of the cell cycle (Fig. 1). If initiation occurs on average at the beginning of the cell cycle, it means that initiation actually occurs shortly before division in some cells of the population and shortly after in others.

In thymine-requiring bacteria, where the DNA replication velocity can be altered by changing the thymine concentration in the growth medium, the *C* period, and thus the initiation age, can be experimentally changed. This has no effect on cell growth rate or on the control of replication initiation. When the C period is extended, the time of the cell division, which occurs C plus D min after initiation, is delayed. As a consequence, cells are larger than normal, but their ribosome concentration is unaltered (28).

Replication initiation control depends on P_O, i.e., the amount of protein per origin. Changes in P_O (e.g., by mutation) do not affect the initiation age (24). This apparent paradox reflects the fact that a change in the initiation time (without a change in C and D) causes an equal change in the time of division so that the initiation age remains unaltered.

Macromolecular Composition during Growth at Different Temperatures

At 20, 25, 30, 35, and 40°C the growth rates of *E. coli* B/r in glucose minimal medium were 0.41, 0.65, 0.91, 1.18, and 1.35 doublings per h, respectively. At these temperatures the rRNA chain elongation rates were 30, 45, 59, 76, and 103 nucleotide residues per s, and the peptide chain elongation rates were 5, 8, 11, 14, and 16 amino acid residues per s, respectively (120). Most of the other physiological parameters are essentially unaltered by a change in temperature. In particular, the relative proportions of mRNA and stable RNA synthesis are the same, suggesting that the chain elongation rates of stable RNA and mRNA have equal temperature coefficients. The growth rate is expected to change as the square root of the changes in the product of the peptide and RNA chain elongation rates (equation 19, Table 6 below), which is, indeed, observed.

The C period changes with temperature in proportion to the doubling time (i.e., the ratio C/τ is constant; 24, 58), which implies identical replication fork patterns at different temperatures. The chain elongation rates for DNA, RNA, and polypeptides have thus about equal temperature coefficients, which in the absence of further regulation results in a temperature independence of the macromolecular cell composition. During the first 30 min after a temperature upshift, extensive temporary changes in the macromolecular synthesis rates have been observed (e.g., reference 120) and a new RNA polymerase sigma subunit is induced (67). These temporary perturbations constitute the heat shock response and reflect active regulation and adjustment to the postshift temperature.

MATHEMATICAL DESCRIPTION OF CELL COMPOSITION AND GROWTH

Cell Composition as a Function of the Culture Doubling Time

A number of equations have been reported that describe the macromolecular composition of an average cell in an exponential culture as a function of the culture doubling time and five additional parameters: the C and D periods, protein per origin (P_O), ribosome activity (β_r), and peptide chain elongation rate (c_p) (see History, above; 27, 32, 51, 122). These equations, reproduced in Table 5, are useful for work dealing with the cell composition and have been used for the calculation of many of the parameters in Table 3. Additional equations in Table 5 can be used to calculate the copy number of genes per cell or per genome as a function of their map location. All equations in Table 5 follow from the definitions of their constituent parameters without special, simplifying assumptions or hypotheses.

Age Distribution and the Concept of the Average Cell

The *average number of a component per cell* has to be distinguished from the number of that component *per average cell*. The first is obtained as the number per unit volume of culture, divided by the number of cells in that volume. This quotient can be any noninteger number. The latter refers to a particular *average cell*, which is defined by the fact that 50% of all cells in the population are younger and 50% are older. Because young cells are more frequent than old cells in an exponential population, the average cell has an age of 0.41, rather than 0.50. The number of any component in the average cell is always an integer; for example the number of chromosome replication origins in the average cell is equal to 2, 4, or 8, depending on the growth rate (see Fig. 1). In contrast, the average number of origins per cell (given by equation 7 in Table 5) may have any noninteger value, such as 2.43 for a growth rate of 1 doubling per h (Table 2), implying that some cells in the population have four origins, while others have only two. These average numbers of components per cell are identical to the values calculated from the formulas in Table 5.

To calculate a population average, the equation for the "ideal age distribution" (135) (meaning that all cells divide in exactly equal time intervals) has been used in the past, with integration over different age intervals. For example, the Cooper-Helmstetter equation and Donachie's equation for the average amounts of DNA and protein, respectively, per cell (equations 3 and 1 of Table 5) were originally derived under the assumption of an ideal age distribution. However, a reexamination of these equations has indicated that they are independent of any assumptions, including the assumptions of an ideal age distribution and of synchronous initiation at all origins in the cell at a given initiation age (15, 18). The formulas in Table 5 give correct values irrespective of the age distribution. In fact, the cell cycle variability has no effect on the average cell composition.

The conclusion, that the cell cycle variability has no effect on the composition and growth parameters, has been disputed by Alberghina and Mariani (2). These authors have not distinguished the doubling time, τ, defined as cell number doubling time in an exponential culture (equal to the mass doubling time), from the average interdivision interval, denoted by $\bar{\tau}$. The latter may be determined from growth curves of synchronous cultures (13) and depends on the particular subpopulation of cells for which individual division intervals are measured. Although synchronous cells have the same division age, they are generally somewhat out of phase with respect to their replication cycles. This means that a zero-age population may contain a large number of subpopulations in which the cells are in step, only in different phases, with respect to their last round of chromosome replication. Each of these subpopulations (with different phase relationships between their replication age and division age) would give a different synchronous growth curve and a different average interdivision interval despite an equal mass doubling time (12). For these reasons, τ and $\bar{\tau}$ (and, similarly, C and $\bar{C}$, and D and $\bar{D}$) must be distinguished in theory (15, 27); only with C, D, and τ are the relationships in Table 5 strictly valid. However, the extent of variability in the cell cycle is such that the differences between τ and $\bar{\tau}$, etc., amount to only a few percent (12) and, in practice, are negligible.

TABLE 5 Equations relating the cell composition in exponential cultures to basic cell cycle parameter[a]

Parameter	Symbol	Equation	Equation no.	Reference(s)
Protein/cell	P_C	$P_C = P_O \cdot 2^{(C+D)/\tau}$	1	51
RNA/cell	R_C	$R_C = K'(P_O/c_p)(1/\tau)2^{(C+D)/\tau}$, where $K' = $ (nucl./rib) $\cdot \ln2/[f_s \cdot (1-f_t) \cdot \beta_r \cdot 60]$	2	27
DNA/cell	G_C	$G_C = [\tau/(C \cdot \ln2)] \cdot [2^{(C+D)/\tau} - 2^{D/\tau}]$	3	32
Mass/cell	M_C	$M_C = k_1 \cdot P_C + k_2 \cdot R_C + k_3 \cdot G_C$, where: $\quad k_1 = 1.35 \cdot 10^{-18}$ OD$_{460}$ units per amino acid residue $\quad k_2 = 4.06 \cdot 10^{-18}$ OD$_{460}$ units per RNA nucleotide residue $\quad k_3 = 3.01 \cdot 10^{-11}$ OD$_{460}$ units per genome equivalent of DNA	4	27
Peptide chain elongation	c_p	$c_p = K'/[(R/P) \cdot \tau]$	5	44, 122
Ribosomal protein/total protein	α_r	$\alpha_r = (R/P) \cdot [($aa/ribosome$) \cdot f_s \cdot (1-f_t)/($nucl./rib$)]$	6	44, 122
Origins/cell	O_C	$O_C = 2^{(C+D)/\tau}$	7	15, 23
Termini/cell	T_C	$T_C = 2^{D/\tau}$	8	15, 23
No. of gene X/cell	X_C	$X_C = 2^{[C(1-m') + D]/\tau}$, where: $\quad m' = $ map location of gene X relative to location or replication origin $\qquad = (m + 16)/50$ for map locations (m) between 0 and 36 min $\qquad = (84 - m)/50$ for map locations between 36 and 84 min $\qquad = (m - 84)/50$ for map locations between 84 and 100 min	9	15, 23
Replication forks/cell	F_C	$F_C = 2 \cdot [2^{(C+D)/\tau} - 2^{D/\tau}]$	10	15, 23
Origins/genome	O_G	$O_G = (C/\tau) \cdot \ln2/(1-2^{-C/\tau})$	11	15, 23
No. of gene X/genome	X_G	$X_G = (O/G) \cdot 2^{-m'C/\tau}$	12	15, 23
Initiation age	a_i	$a_i = 1 + n - (C + D)/\tau$ where n is the next lower integer value of $[(C + D)/\tau]$; i.e., $n = \text{int}[(C + D)/\tau]$	13	32
Termination age	a_t	$a_t = 1 - D/\tau$	14	32
Origins per cell at initiation	O_i	$O_i = 2^n$; for a definition of n, see equation 13	15	32
Cell mass after division (a_0)	M_d	$M_d = M_C/(2 \cdot \ln2)$	16	18
Cell mass at initiation (a_i)	M_i	$M_i = M_d \cdot 2^{a_i}$	17	18

[a]See Tables 1 and 2 for definitions.

Cell Composition at a Defined Cell Age

In some instances the cell composition at certain cell ages becomes important; in particular, at the time of cell division (either shortly before or shortly after) or at the time of initiation of chromosome replication. In these cases, it is also not necessary to use the age distribution formula. Instead, the following relationships can be used. (i) The average amount of a component in the subpopulation of zero-age cells (immediately after division) is $1/(2 \cdot \ln2)$ times the average amount of that component in the population as a whole (obtained as described above). Correspondingly, the factor to obtain the average amount of the component in the cells immediately before division has twice that value, i.e., $1/(\ln2)$. (ii) The amount of protein (or cell mass) per replication origin in a cell at the time of initiation ("initiation mass" defined by Donachie [51]) is $P_O/\ln2$ (or $M_O/\ln2$), where P_O (or M_O) is the total protein (or cell mass) divided by the total number of chromosomal replication origins in a unit volume of exponential culture. (Table 2 shows only P_O, but M_O may be calculated from the data in the Table.) For the mathematical relationships dealing with the age distribution, see references 12, 15, and 16.

Parameters Determining Bacterial Growth Rate

In the preceding discussion, the cell composition was expressed as a function of the growth rate. This suggests that the composition is determined by the growth rate. In reality, the nutrients and other components in the medium, together with genetically determined structural and kinetic constants of cellular components, determine both the biochemical reaction rates (and thus the growth rate) and the cell composition. It seems likely that only a few physiological parameters limit cell growth, whereas other physiological parameters and most of the reaction rates and concentrations in a bacterium are probably not growth limiting.

The DNA concentration is not growth limiting. Maaløe and Kjeldgaard (94, 95) have discussed the idea that the amount of protein per DNA is constant and that the amount of RNA per DNA increases in direct proportion to the growth rate. These relationships seem to suggest a limitation by DNA, for example, that mRNA synthesis is limited by DNA and protein synthesis is limited by mRNA, which then results in a constant ratio of protein to DNA. Maaløe (94) argued that the apparent and exact proportionality of the ratio of RNA to DNA reflected the control of growth. Similarly, Koch (82) and Daneo-Moore and Schockman (34) have used rate constants of RNA synthesis per DNA in models for the control of RNA synthesis or growth. However, a DNA replication (initiation) mutant which has a reduced DNA concentration and therefore an increased ratio of RNA to DNA at all growth rates (because the denominator, DNA, is reduced) has an unaltered growth rate (24, 28, 29). The reduced DNA concentration is the result of an increased initiation mass (protein per *oriC*, P_O). This mutant shows that RNA synthesis and growth are not normally limited by the concentration of DNA; in the mutant, the ratio of RNA to DNA is no longer exactly proportional to the growth rate. Moreover, the amount of protein per DNA is generally not constant, even in wild-type cells (see Fig. 2b).

The ribosome concentration (measured as number of ribosomes per protein) and the protein synthesis rate per ribosome are growth limiting in any living cell whose protein turnover is negligible. These two factors determine the exponential growth rate (equation 18, Table 6). Equation 18 is identical to the one

TABLE 6 Basic parameters determining the bacterial growth rate[a]

Parameter	Symbol	Equation[b]	Equation no.
Growth rate (doublings/h)	μ	$\mu = (60/\ln2) \cdot (N_r/P) \cdot e_r$, where the ribosome efficiency, $e_r = \beta_r \cdot c_p$	18
Growth rate (doublings/h)	μ	$\mu = (60/K) \cdot [\psi_s \cdot \alpha_p \cdot \beta_p \cdot \beta_r \cdot c_s \cdot c_p]^{0.5}$, where $K = \ln2 \cdot [(\text{nucl./prib})(\text{aa/pol})/(1-f_t)]^{0.5}$	19

[a] c_s and c_p in these equations should be expressed as rates per minute to obtain the growth rate in doublings per hour. For definitions, see Table 1. (The equation 19 is from reference 11.)

[b] For a definition or explanation of symbols, see Tables 1 and 3.

used by Schleif (122) to evaluate the RNA-to-protein ratio as a function of growth rate (see also equation 5, Table 5). In his case, however, the growth rate, μ, was the independent variable and R/P was the dependent variable. By making μ the dependent parameter, we have here exchanged the roles of the two variables.

For any growth equation to be meaningful, the parameters must be constant in time. For example, in equation 18 (Table 6), both the ribosome concentration and activity must be constant. If they changed, μ would have a changing value and growth would not be exponential. A constant and given ribosome concentration results from the regulation of ribosome synthesis, which in turn involves the regulation of RNA polymerase synthesis and its partitioning into rRNA and mRNA-synthesizing enzyme (see above). These additional concepts have been taken into account in the more complex growth equation 19 (Table 6), which contains six factors: RNA polymerase concentration (α_p), RNA polymerase activity (β_p), partitioning of active RNA polymerase into stable RNA and mRNA-synthesizing enzyme (ψ_s), ribosome activity (β_r), and the chain elongation rates for stable RNA and polypeptides (c_s, c_p).

In equation 19 the parameters must again be constant in time to produce exponential growth. In the preceding discussion (see above, Observed Cell Composition of *E. coli* B/r), the constancy of physiological parameters during exponential growth was implicit in the definition of exponential growth, but if one asks for the conditions that lead to exponential growth, then this constancy cannot be taken for granted. A full explanation of why these parameters have certain, constant values under given growth conditions would solve the problem of growth control. Growth equations such as those in Table 6 identify growth-limiting parameters and predict their effect on the growth rate.

Optimal Cell Composition for Maximal Growth

Maaløe and Kjeldgaard (94, 95) have pointed out that the protein synthesis rate per average ribosome in bacteria is constant and presumably maximal under most growth conditions and, further, that this constancy is economically advantageous for the cell. Since ribosomes are more expensive than their substrates, they should always work at their maximum rate and therefore be saturated with substrates. Thus, the increased demand for protein synthesis at higher growth rates can only be achieved by increasing the ribosome concentration, since the rate of protein synthesis per ribosome is already maximal. Similarly, the increases in α_r and ψ_s with increasing growth rate can then be understood as consequences of the constant ribosome function. These arguments try to explain the changing cell composition as an expression of an optimization principle which allows the cell to achieve maximum growth at a minimum expenditure of energy.

Ehrenberg and Kurland (53) have theoretically analyzed how metabolic pathways should be designed for maximum energy efficiency. With regard to ribosomes, they concluded that an optimal utilization of energy would be achieved if both the substrate and ribosome concentration were to increase with increasing demand for protein synthesis (substrates for the ribosome are the different elongation factor Tu-aminoacyl-tRNA-GTP ternary complexes.) This can be understood as follows. At high ribosome concentrations, the substrate pools, even at saturating concentrations, would represent only a small fraction of the total mass of the protein-synthesizing system. At low ribosome concentrations the same pool would constitute a greater fraction of the total system mass, and the energy required to produce that substrate pool would no longer be negligible. The cells would then save energy by reducing the concentration of substrates, especially when this concentration is above the K_m for substrate binding, such that substantial reductions in substrate concentrations can be compensated for by only small increases in ribosome number. In support of the conclusion of Ehrenberg and Kurland, it should be noted that the ribosome function (c_p) does, indeed, increase with growth rate (Table 3). Also, the concentrations of tRNA and elongation factors Tu and Ts are not constant but increase in proportion to the ribosome concentration (Table 4). If these concentrations were above the K_m at fast growth rates, their reduction during slow growth should save energy.

It is likely that energy efficiency has played a major role in the evolution of the regulatory parameters that determine the macromolecular composition and growth rate of the cell. In addition, other principles, like rapid adaptability to a changing growth environment, have evolved at the expense of energy efficiency. In some instances an inefficiency might be a necessary byproduct of control mechanisms; for example, the "stuttering" and pausing RNA polymerase might be the result of controls that depend on the secondary structure of mRNA, like RNA chain termination, attenuator control, or control of mRNA stability.

ACKNOWLEDGMENTS

We thank Sharon Krowchuk for her role in preparing and typing this manuscript. This work was supported by grants MT6340 from the Medical Research Council of Canada to P.P.D. and R01 GM1542 from the National Institutes of Health to H.B. P.P.D. is a Fellow in the Evolutionary Biology program of the Canadian Institute for Advanced Research.

LITERATURE CITED

1. Aksoy, S., C. L. Squires, and C. Squires. 1984. Evidence for antitermination in *Escherichia coli* rRNA transcription. *J. Bacteriol.* **159**:260–264.

2. **Alberghina, L., and L. Mariani.** 1980. Analysis for a cell model for *Escherichia coli. J. Math. Biol.* **9:**389.
3. **Atherly, A.** 1979. Deletion of *relA* and *relX* has no effect on basal or carbon-downshift ppGpp synthesis, p. 53–66. *In* G. Koch and D. Richter (ed.), *Regulation of Macromolecular Synthesis by Low Molecular Weight Mediators.* Academic Press, Inc., New York.
4. **Bachmann, B.** 1990. Linkage map of *Escherichia coli* K-12, edition 8. *Microbiol. Rev.* **54:**130–197.
5. **Baracchini, E., and H. Bremer.** 1987. Determination of synthesis rate and lifetime of bacterial mRNAs. *Anal. Biochem.* **167:**245–260.
6. **Barry, G., C. Squires, and C. L. Squires.** 1980. Attenuation and processing of RNA from the *rplJL-rpoBC* transcription unit of *Escherichia coli. Proc. Natl. Acad. Sci. USA* **77:**3331–3335.
7. **Barry, G., C. L. Squires, and C. Squires.** 1979. Control features within the *rplJL-rpoBC* transcription unit of *Escherichia coli. Proc. Natl. Acad. Sci. USA* **76:**4922–4926.
8. **Blumenthal, R., P. Lemaux, F. Neidhardt, and P. Dennis.** 1976. The effects of the *relA* gene on the synthesis of amino acyl tRNA synthetase and other transcription-translation proteins in *Escherichia coli* B. *Mol. Gen. Genet.* **149:**291–296.
9. **Bouché, J. P.** 1982. Physical map of a 470 kbase-pair region flanking the terminus of DNA replication in the *Escherichia coli* K12 genome. *J. Mol. Biol.* **154:**1–20.
10. **Bouché, J. P., J. P. Gelugne, J. Louarn, J. M. Louarn, and K. Kaiser.** 1982. Relationship between the physical and genetic maps of a 470 kbase-pair region around the terminus of K12 DNA replication. *J. Mol. Biol.* **154:**21–32.
11. **Bremer, H.** 1975. Parameters affecting the rate of synthesis of ribosomes and RNA polymerase in bacteria. *J. Theor. Biol.* **53:**115–124.
12. **Bremer, H.** 1982. Variation in the generation times in *Escherichia coli* populations: its cause and implications. *J. Gen. Microbiol.* **128:**2865–2876.
13. **Bremer, H., and L. Chuang.** 1981. The cell cycle in *Escherichia coli* B/r. *J. Theor. Biol.* **88:**47–81.
14. **Bremer, H., and L. Chuang.** 1981. The cell division cycle after inhibition of chromosome replication in *Escherichia coli. J. Theor. Biol.* **93:**909–926.
15. **Bremer, H., and G. Churchward.** 1977. An examination of the Cooper-Helmstetter theory of DNA replication and its underlying assumptions. *J. Theor. Biol.* **69:**645–654.
16. **Bremer, H., and G. Churchward.** 1978. Age fractionation of bacteria by membrane elution: relation between age distribution and age profile. *J. Theor. Biol.* **74:**69–81.
17. **Bremer, H., and G. Churchward.** 1990. Control of cyclic chromosome replication in *Escherichia coli. Microbiol. Rev.* **55:**459–475.
18. **Bremer, H., G. Churchward, and R. Young.** 1979. Relation between growth and replication in bacteria. *J. Theor. Biol.* **81:**533–545.
19. **Bremer, H., and D. Yuan.** 1968. RNA chain growth rates in *Escherichia coli. J. Mol. Biol.* **38:**163–180.
20. **Brunschede, H., T. L. Dove, and H. Bremer. 1977.** Establishment of exponential growth after a nutritional shift-up in *Escherichia coli* B/r: accumulation of deoxyribonucleic acid, ribonucleic acid, and protein. *J. Bacteriol.* **129:**1020–1033.
21. **Chamberlin, M. J.** 1976. Interactions of RNA polymerase with the DNA template, p. 159–191. *In* R. Losick and M. Chamberlin (ed.), *RNA Polymerase.* Cold Spring Harbor Laboratory, Cold Spring Harbor, N.Y.
22. **Chandler, M., R. Bird, and L. Caro.** 1975. The replication time of *Escherichia coli* K12 chromosome as a function of the cell doubling time. *J. Mol. Biol.* **94:**127–132.
23. **Chandler, M. G., and R. H. Pritchard.** 1975. The effect of gene concentration and relative gene dosage on gene output in *Escherichia coli. Mol. Gen. Genet.* **138:**127–141.
24. **Choung, K.-K., E. Estiva, and H. Bremer.** 1981. Genetic and physiological characterization of a spontaneous mutant of *Escherichia coli* B/r with aberrant control of deoxyribonucleic acid replication. *J. Bacteriol.* **145:**1239–1248.
25. **Chow, J., and P. P. Dennis.** 1994. Coupling between mRNA synthesis and mRNA stability in *Escherichia coli. Mol. Microbiol.* **11:**919–932.
26. **Churchward, G., and H. Bremer.** 1977. Determination of deoxyribonucleic acid replication time in exponentially growing *Escherichia coli* B/r. *J. Bacteriol.* **130:**1206–1213.
27. **Churchward, G., H. Bremer, and R. Young.** 1982. Macromolecular composition of bacteria. *J. Theor. Biol.* **84:**651–670.
28. **Churchward, G., H. Bremer, and R. Young.** 1982. Transcription in bacteria at different DNA concentrations. *J. Bacteriol.* **150:**572–581.
29. **Churchward, G., E. Estiva, and H. Bremer.** 1981. Growth rate-dependent control of chromosome replication initiation in *Escherichia coli. J. Bacteriol.* **145:**1232–1238.
30. **Cole, J. R., C. L. Olsson, J. W. B. Hershey, M. Grunberg-Monago, and M. Nomura.** 1987. Feedback regulation of rRNA synthesis in *Escherichia coli*. Requirement for initiation factor IF2. *J. Mol. Biol.* **198:**383–392.
31. **Condon, C., S. French, C. Squires, and C. L. Squires.** 1993. Depletion of functional ribosomal RNA operons in *Escherichia coli* causes increased expression of the remaining intact copies. *EMBO J.* **12:**4305–4315.
32. **Cooper, S., and C. Helmstetter.** 1968. Chromosome replication and the division cycle of *Escherichia coli* B/r. *J. Mol. Biol.* **31:**519–540.
33. **Dalbow, D., and R. Young.** 1975. Synthesis time of b-galactosidase in *Escherichia coli* B/r as a function of growth rate. *Biochem. J.* **150:**13–20.
34. **Daneo-Moore, L., and G. D. Schockman.** 1976. The bacterial cell surface in growth and cell division, p. 653–715. *In* G. Poste and G. L. Nicholson (ed.), *The Synthesis, Assembly and Turnover of Cell Surface Components.* Elsevier/North-Holland Biomedical Press, Amsterdam.
35. **Delcuve, G., and P. P. Dennis.** 1981. An amber mutation in a ribosomal protein gene: ineffective suppression stimulates operon-specific transcription. *J. Bacteriol.* **147:**997–1001.
36. **Dennis, P. P.** 1971. Regulation of stable RNA synthesis in *Escherichia coli. Nature* (London) *New Biol.* **232:**43–47.
37. **Dennis, P. P.** 1972. Regulation of ribosomal and transfer ribonucleic acid synthesis in *Escherichia coli* B/r. *J. Biol. Chem.* **247:**2842–2845.
38. **Dennis, P. P.** 1974. In vivo stability, maturation and relative differential synthesis rates of individual ribosomal proteins in *Escherichia coli. J. Mol. Biol.* **88:**24–41.
39. **Dennis, P. P.** 1976. Effects of chloramphenicol on the transcriptional activities of ribosomal RNA and ribosomal protein genes in *Escherichia coli. J. Mol. Biol.* **108:**535–546.
40. **Dennis, P. P.** 1977. Influence of the stringent control system on the transcription of ribosomal ribonucleic acid and ribosomal protein genes in *Escherichia coli. J. Bacteriol.* **129:**580–588.
41. **Dennis, P. P.** 1977. Regulation of the synthesis and activity of a mutant RNA polymerase in *Escherichia coli. Proc. Natl. Acad. Sci. USA* **74:**5416–5420.
42. **Dennis, P. P.** 1977. Transcription patterns of adjacent segments of *Escherichia coli* containing genes coding for four 50S ribosomal proteins and the β and β′ subunits of RNA polymerase. *J. Mol. Biol.* **115:**603–625.
43. **Dennis, P. P.** 1984. Site specific deletion of regulatory sequences in a ribosomal protein-RNA polymerase operon in *E. coli:* effects on β and β′ gene expression. *J. Biol. Chem.* **259:**3203–3209.
44. **Dennis, P. P., and H. Bremer.** 1974. Differential rate of ribosomal protein synthesis in *Escherichia coli* B/r. *J. Mol. Biol.* **84:**407–422.
45. **Dennis, P. P., and H. Bremer.** 1974. Regulation of ribonucleic acid synthesis in *Escherichia coli* B/r: an analysis of a shift up. III. Stable RNA synthesis rate and ribosomal RNA chain growth rate following a shift up. *J. Mol. Biol.* **89:**233–239.
46. **Dennis, P. P., and H. Bremer.** 1974. Macromolecular composition during steady-state growth of *Escherichia coli* B/r. *J. Bacteriol.* **119:**270–281.
47. **Dennis, P. P., and N. Fiil.** 1979. Transcriptional and post-transcriptional control of RNA polymerase and ribosomal protein genes cloned on composite ColE1 plasmids in the bacterium *Escherichia coli. J. Biol. Chem.* **254:**7540–7547.
48. **Dennis, P. P., V. Nene, and R. E. Glass.** 1985. Autogenous post-transcriptional regulation of RNA polymerase β and β′ subunit synthesis in *Escherichia coli. J. Bacteriol.* **161:**803–806.
49. **Dennis, P. P., and M. Nomura.** 1974. Stringent control of ribosomal protein gene expression in *Escherichia coli. Proc. Natl. Acad. Sci. USA* **71:**3819–3823.
50. **Dennis, P. P., and M. Nomura.** 1975. Stringent control of the transcriptional activities of ribosomal protein genes in *E. coli. Nature* (London) **255:**460–465.
51. **Donachie, W.** 1968. Relationships between cell size and time of initiation of DNA replication. *Nature* (London) 219:1077–1079.
52. **Downing, W., and P. P. Dennis.** 1991. RNA polymerase activity may regulate transcription initiation and attenuation in the *rplKAJL rpoBC* operon in *E. coli. J. Biol. Chem.* **266:**1304–1311.
53. **Ehrenberg, M., and C. Kurland.** 1984. Costs of accuracy determined by a maximal growth rate constraint. *Q. Rev. Biophys.* **17:**45–82.
54. **Emilsson, V., and C. G. Kurland.** 1990. Growth rate dependence of transfer RNA abundance in *Escherichia coli. EMBO J.* **9:**4359–4366.
55. **Fallon, A. M., C. S. Jinks, M. Yamamoto, and M. Nomura.** 1979. Expression of ribosomal protein genes cloned in a hybrid plasmid in *Escherichia coli:* gene dosage effects on synthesis of ribosomal proteins and ribosomal protein messenger ribonucleic acid. *J. Bacteriol.* **138:**383–396.
56. **Fehr, S., and D. Richter.** 1981. Stringent response of *Bacillus stearothermophilus:* evidence for the existence of two distinct guanosine 3′,5′-polyphosphate synthetases. *J. Bacteriol.* **145:**68–73.
57. **Forchhammer, J., and L. Lindahl.** 1971. Growth rate of polypeptide chains as a function of the cell growth rate in a mutant of *Escherichia coli* 15. *J. Mol. Biol.* **55:**563–568.
58. **Frey, J., M. Chandler, and L. Caro.** 1981. The initiation of chromosome replication in a dnaAts46 and dnaA$^+$ strain at various temperatures. *Mol. Gen. Genet.* **182:**364–366.
59. **Furano, A.** 1975. Content of elongation factor Tu. *Proc. Natl. Acad. Sci. USA* **72:**4780–4784.
60. **Furano, A., and F. Wittel.** 1976. Syntheses of elongation factors Tu and G are under stringent control in *Escherichia coli. J. Biol. Chem.* **251:**898–901.
61. **Gausing, K.** 1972. Efficiency of protein and messenger RNA synthesis in bacteriophage T4 infected cells of *Escherichia coli. J. Mol. Biol.* **71:**529–545.
62. **Gausing, K.** 1974. Ribosomal protein in *E. coli:* rate of synthesis and pool size at different growth rates. *Mol. Gen. Genet.* **129:**61–75.
63. **Gausing, K.** 1977. Regulation of ribosome production in *Escherichia coli:* synthesis and stability of ribosomal RNA and ribosomal protein mRNA at different growth rates. *J. Mol. Biol.* **115:**335–354.
64. **Gauss, D., and M. Sprinzl.** 1983. Compilation of tRNA sequences. *Nucleic Acids Res.* **11:**r1-r53.
65. **Gordon, J.** 1970. Regulation of the *in vivo* synthesis of polypeptide chain elongation factors in *Escherichia coli. Biochemistry* **9:**912–917.
66. **Greenblatt, J.** 1984. Regulation of transcription termination in *E. coli. Can. J. Biochem.* **62:**79–88.
67. **Grossman, A. D., J. W. Erickson, and C. Gross.** 1984. The *htpR* gene product of *E. coli* is a sigma factor for heat-shock promoters. *Cell* **38:**383–390.
68. **Hardy, S.** 1975. The stoichiometry of ribosomal proteins of *Escherichia coli. Mol. Gen. Genet.* **140:**253–274.

69. Haseltine, W., R. Block, W. Gilbert, and K. Weber. 1972. MSI and MSII made on ribosomes in the idling step of protein synthesis. *Nature* (London) 238:381–384.

70. Helmstetter, C., and S. Cooper. 1968. DNA synthesis during the division cycle of rapidly growing *E. coli* B/r. *J. Mol. Biol.* 31:507–518.

71. Helmstetter, C. E., and D. J. Cummings. 1964. An improved method for the selection of bacterial cells at division. *Biochim. Biophys. Acta* 82:608–610.

72. Hernandez, V. J., and H. Bremer. 1990. Guanosine tetraphosphate (ppGpp) dependence of the growth rate control of rrnB P1 promoter activity in *Escherichia coli. J. Biol. Chem.* 265:11605–11614.

73. Hernandez, V. J., and H. Bremer. 1991. *Escherichia coli* ppGpp synthetase II activity requires spoT. *J. Biol. Chem.* 266:5991–5999.

74. Hernandez, V. J., and H. Bremer. 1993. Characterization of *Escherichia coli* devoid of ppGpp. *J. Biol. Chem.* 268:10851–10862.

75. Howe, J., and J. Hershey. 1983. Initiation factor and ribosome levels are coordinately controlled in *Escherichia coli* growing at different rates. *J. Biol. Chem.* 258:1954–1959.

76. Ikemura, T. 1981. Correlations between the abundance of *Escherichia coli* transfer RNAs and the occurrence of the respective codons in its mRNA genes. *J. Mol. Biol.* 146:1–21.

77. Jinks-Robertson, S., R. Gourse, and M. Nomura. 1983. Expression of rRNA and tRNA genes in *Escherichia coli*: evidence for feedback regulation by products of rRNA operons. *Cell* 33:865–876.

78. Johnsen, M., T. Christiansen, P. Dennis, and N. Fiil. 1982. Autogenous control: ribosomal protein L10-L12 complex binds to the leader region of its mRNA. *EMBO J.* 1:999–1004.

79. Jones, N. C., and W. D. Donachie. 1973. Chromosome replication, transcription and control of cell division in *Escherichia coli* B/r. *Nature* (London) *New Biol.* 243:100–103.

80. Kennel, D. 1968. Titration of the gene sites on DNA by DNA-RNA hybridization. II. The *Escherichia coli* chromosome. *J. Mol. Biol.* 34:85–103.

81. Kingston, R., W. Nierman, and M. Chamberlin. 1981. A direct effect of guanosine tetraphosphate on pausing of *Escherichia coli* RNA polymerase during RNA chain elongation. *J. Biol. Chem.* 256:2787–2797.

82. Koch, A. 1970. Overall control on the biosynthesis of ribosomes in growing bacteria. *J. Theor. Biol.* 28:203–231.

83. Koppes, L., and K. Nordstrom. 1986. Insertion of an R1 plasmid into the origin of replication of the *E. coli* chromosome: random timing of replication of the hybrid chromosome. *Cell* 44:117–124.

84. Lagosky, P., and F. N. Chang. 1980. Influence of amino acid starvation on guanosine 5'-diphosphate 3'-diphosphate basal level synthesis in *Escherichia coli. J. Bacteriol.* 144:499–508.

85. Li, S., C. L. Squires, and C. Squires. 1984. Antitermination of *E. coli* rRNA transcription is caused by a control region segment containing a *nut*-like sequence. *Cell* 38:851–860.

86. Lindahl, L. 1975. Intermediates and time kinetics of the *in vivo* assembly of *Escherichia coli* ribosomes. *J. Mol. Biol.* 92:15–37.

87. Lindahl, L., R. Archer, and J. Zengel. 1983. Transcription of the S10 ribosomal protein operon is regulated by an attenuator in the leader. *Cell* 33:241–248.

88. Lindahl, L., and J. Zengel. 1982. Expression of ribosomal genes in bacteria. *Adv. Genet.* 21:53–121.

89. Little, R., and H. Bremer. 1984. Transcription of ribosomal component genes and *lac* in a relA⁺/relA pair of *Escherichia coli* strains. *J. Bacteriol.* 159:863–869.

90. Little, R., J. Ryals, and H. Bremer. 1983. Physiological characterization of *Escherichia coli* rpoB mutants with abnormal control of ribosome synthesis. *J. Bacteriol.* 155:1162–1170.

91. Louarn, J., J. Patte, and J. M. Louarn. 1979. Map position of the replication terminus on the *Escherichia coli* chromosome. *Mol. Gen. Genet.* 172:7–11.

92. Lowry, O. H., N. J. Rosebrough, A. L. Farr, and R. J. Randall. 1951. Protein measurement with the Folin phenol reagent. *J. Biol. Chem.* 193:265–275.

93. Maaløe, O. 1969. An analysis of bacterial growth. *Dev. Biol.* 3(Suppl.):33–58.

94. Maaløe, O. 1979. Regulation of the protein synthesizing machinery—ribosomes, tRNA, factors and so on, p. 487–542. *In* R. Goldberger (ed.), *Biological Regulation and Development*, vol. 1. Plenum Publishing Corp., New York.

95. Maaløe, O., and N. O. Kjeldgaard. 1966. *Control of Macromolecular Synthesis*. W. A. Benjamin, New York.

96. Maher, D., and P. Dennis. 1977. In vivo transcription of *E. coli* genes coding for rRNA, ribosomal proteins and subunits of RNA polymerase: influence of the stringent control system. *Mol. Gen. Genet.* 155:203–211.

97. Meselson, M., and F. Stahl. 1958. The replication of DNA in *Escherichia coli. Proc. Natl. Acad. Sci. USA* 44:671–682.

98. Moazed, D., and H. Noller. 1989. Intermediate states in the movement of transfer RNA in the ribosome. *Nature* (London) 342:142–148.

99. Molin, S. 1976. Ribosomal RNA chain elongation rates in *Escherichia coli. Alfred Benzon Symp.* 9:333–339.

100. Müller, K., and H. Bremer. 1969. Heterogeneous initiation and termination of enzymatically synthesized ribonucleic acid. *J. Mol. Biol.* 43:89–107.

101. Newman, J., and H. Kubitschek. 1978. Variations in periodic replication of the chromosome in *Escherichia coli* B/r TT. *J. Mol. Biol.* 121:461–471.

102. Nierlich, D. P. 1972. Regulation of ribonucleic acid synthesis in growing bacterial cells. II. Control over the composition of newly made RNA. *J. Mol. Biol.* 7:765–777.

103. Nodell, J. R., and J. Greenblatt. 1993. Recognition of box A antitermination RNA by *E. coli* antitermination factor NusB and ribosomal protein S10. *Cell* 72:261–268.

104. Noller, H. F. 1984. Structure of ribosomal RNA. *Annu. Rev. Biochem.* 53:119–162.

105. Nomura, M., R. Gourse, and G. Baughman. 1984. Regulation of the synthesis of ribosomes and ribosomal components. *Annu. Rev. Biochem.* 53:75–117.

106. Norris, T., and A. Koch. 1972. Effect of growth rates on the relative rates of messenger, ribosomal and transfer RNA in *Escherichia coli. J. Mol. Biol.* 64:633–649.

107. Ovchinnikov, Y., V. Lipkin, N. Modzanov, O. Chestov, and Y. Smirnov. 1977. Primary structure of the α subunit of DNA dependent RNA polymerase from *Escherichia coli. FEBS Lett.* 76:108–111.

108. Ovchinnikov, I., G. Monastyrskaia, U. Gubanov, S. Guriev, O. Chertov, N. Modisnov, V. Grinkevich, I. Makarova, T. Marchenko, I. Polovnikova, V. Lipkin, and E. Sverdlov. 1980. Primary structure of RNA polymerase from *Escherichia coli*: nucleotide sequence of gene rpoB and amino acid sequence of β-subparticle. *Dokl. Akad. Nauk SSSR* 253:994–999.

109. Ovchinnikov, I., G. Monastyrskaia, U. Gubanov, S. Guriev, and I. Salomatina. 1981. Primary structure of RNA polymerase of *Escherichia coli*: nucleotide sequence of gene rpoC and amino acid sequence of β'-subparticle. *Dokl. Akad. Nauk SSSR* 261:763–768.

110. Pedersen, S. 1984. *Escherichia coli* ribosomes translate in vivo with variable rate. *EMBO J.* 3:2895–2898.

111. Pedersen, S. 1984. In *Escherichia coli* individual genes are translated with different rates *in vivo. Alfred Benzon Symp.* 19:101–107.

112. Pedersen, S., P. Bloch, S. Reeh, and F. Neidhardt. 1978. Patterns of protein synthesis in *E. coli*: a catalogue of the amount of 140 individual proteins at different growth rates. *Cell* 14:179–190.

113. Post, L. E., G. D. Strycharz, M. Nomura, H. Lewis, and P. P. Dennis. 1979. Nucleotide sequence of the ribosomal protein gene cluster adjacent to the gene for RNA polymerase subunit β in *Escherichia coli. Proc. Natl. Acad. Sci. USA* 76:1697–1701.

114. Pritchard, R. H., and Z. Zaritsky. 1970. Effect of thymine concentration on the replication velocity of DNA in a thymineless mutant of *E. coli. Nature* (London) 226:126–131.

115. Reeh, S., S. Pedersen, and J. Friesen. 1976. Biosynthetic regulation of individual proteins in relA⁺ and relA⁻ strains of *Escherichia coli* during amino acid starvation. *Mol. Gen. Genet.* 149:279–289.

116. Rheinberger, H.-J., H. Sternbach, and K. Nierhaus. 1981. Three tRNA bonding sites on *Escherichia coli* ribosomes. *Proc. Natl. Acad. Sci USA* 78:5310–5314.

117. Richter, D. 1979. Synthesis and degradation of the pleiotropic effector guanosine 3,5'-bis(diphosphate) in bacteria, p. 85–94. *In* G. Koch and D. Richter (ed.), *Regulation of Macromolecular Synthesis by Low Molecular Weight Mediators*. Academic Press, Inc., New York.

118. Rosset, R., J. Julian, and R. Morier. 1966. Ribonucleic acid composition of bacteria as a function of growth rate. *J. Mol. Biol.* 18:308–320.

119. Ryals, J., R. Little, and H. Bremer. 1982. Control of rRNA and tRNA synthesis in *Escherichia coli* by guanosine tetraphosphate. *J. Bacteriol.* 151:1261–1268.

120. Ryals, J., R. Little, and H. Bremer. 1982. Temperature dependence of RNA synthesis parameters in *Escherichia coli. J. Bacteriol.* 151:879–887.

121. Schaechter, E., O. Maaløe, and N. O. Kjeldgaard. 1958. Dependence on medium and temperature of cell size and chemical composition during balanced growth of *Salmonella typhimurium. J. Gen. Microbiol.* 19:592–606.

122. Schleif, R. 1967. Control of the production of ribosomal protein. *J. Mol. Biol.* 27:41–55.

123. Sharp, P. M., and W. Li. 1986. Codon usage in regulatory genes in *Escherichia coli* does not reflect selection for rare codons. *Nucleic Acids Res.* 14:7737–7749.

124. Shen, V., and H. Bremer. 1977. Rate of ribosomal ribonucleic acid chain elongation in *Escherichia coli* B/r during chloramphenicol treatment. *J. Bacteriol.* 130:1109–1116.

125. Shepherd, N. S., G. Churchward, and H. Bremer. 1980. Synthesis and activity of ribonucleic acid polymerase in *Escherichia coli. J. Bacteriol.* 141:1098–1108.

126. Shepherd, N. S., G. Churchward, and H. Bremer. 1980. Synthesis and function of ribonucleic acid polymerase and ribosomes in *Escherichia coli* B/r after a nutritional shift-up. *J. Bacteriol.* 143:1332–1344.

127. Siehnel, R. J., and E. A. Morgan. 1983. Efficient read-through of Tn9 and IS1 by RNA polymerase molecules that initiate at rRNA promoters. *J. Bacteriol.* 153:672–684.

128. Skarstad, K., E. Boye, and H. B. Steen. 1986. Timing of initiation of chromosome replication in individual *Escherichia coli* cells. *EMBO J.* 5:1711–1717.

129. Skarstad, K., H. Steen, and E. Boye. 1985. *Escherichia coli* DNA distributions measured by flow cytometry and compared with theoretical computer simulations. *J. Bacteriol.* 163:661–668.

130. Skarstad, K., H. B. Steen, T. Stokke, and E. Boye. 1994. The initiation mass of *Escherichia coli* K-12 is dependent on growth rate. *EMBO J.* 13:2097–2102.

131. Sorensen, M. A., C. G. Kurland, and S. Pedersen. 1989. Codon usage determines translation rate in *Escherichia coli. J. Mol. Biol.* 207:365–377.

132. Sorensen, M. A., and S. Pedersen. 1991. Absolute in vivo translation rates of individual codons in *Escherichia coli*. Two glutamic acid codons GAA and GAG are translated with a threefold difference in rate. *J. Mol. Biol.* 222:265–280.

133. Spahr, P. R. 1962. Amino acid composition of ribosomes from *Escherichia coli. J. Mol. Biol.* 4:395–406.

134. Subramanian, R. 1975. Copies of protein L7 and L12 and heterogeneity of the large subunit of *Escherichia coli* ribosomes. *J. Mol. Biol.* 95:1–8.

135. Sueoka, N., and Y. Yoshikawa. 1965. The chromosome of *Bacillus subtilis*. I. The theory of marker frequency analysis. *Genetics* 52:747–757.

136. **Travers, A.** 1976. Modulation of RNA polymerase specificity by ppGpp. *Mol. Gen. Genet.* **147**:225–232.

137. **van Ooyen, A., M. Gruber, and P. Jorgensen.** 1976. The mechanism of action of ppGpp on rRNA synthesis *in vitro*. *Cell* **8**:123–128.

138. **Varenne, S., J. Bue, R. Llouber, and C. Lazdunski.** 1984. Translation is a non-uniform process: effect of tRNA availability on the rate of elongation of nascent polypeptide chains. *J. Mol. Biol.* **180**:549–576.

139. **Vogel, U., and K. F. Jensen.** 1994. The RNA chain elongation rate in *Escherichia coli* depends on the growth medium. *J. Bacteriol.* **176**:2807–2813.

140. **Wittmann, H. G.** 1982. Components of the bacterial ribosome. *Annu. Rev. Biochem.* **51**:155–183.

141. **Xiao, H., M. Kalman, K. Ikehara, S. Zemel, G. Glaser, and M. Cashel.** 1991. Residual guanosine 3′,5′-bispyrophosphate synthetic activity of *relA* null mutants can be eliminated by *spoT* null mutations. *J. Biol. Chem.* **266**: 5980–5990.

142. **Young, R., and H. Bremer.** 1976. Polypeptide chain elongation rate in *Escherichia coli* B/r as a function of growth rate. *Biochem. J.* **160**: 185–194.

143. **Yuan, D., and V. Shen.** 1975. Stability of ribosomal and transfer ribonucleic acid in *Escherichia coli* B/r after treatment with ethylenedinitrilotetraacetic acid and rifampin. *J. Bacteriol.* **122**:425–432.

144. **Zaritsky, A., and R. H. Pritchard.** 1971. Replication time of the chromosome in thymineless mutants of *Escherichia coli*. *J. Mol. Biol.* **60**:65–74.

Effect of Temperature, Pressure, pH, and Osmotic Stress on Growth

JOHN L. INGRAHAM AND ALLEN G. MARR

98

INTRODUCTION

The response of *Escherichia coli* and *Salmonella typhimurium* (official designation, *Salmonella enterica* serovar Typhimurium) to their physical environment is in no respect exceptional among bacteria: these enteric organisms, classified as mesophiles with respect to temperature and as neutrophiles with respect to pH, grow over the mid range of temperatures, pH values, water activities, and pressure in which bacterial growth occurs. In this chapter we will attempt to summarize the information regarding the responses of these organisms to their physical environment, to compare these responses with those of other bacteria, and, where possible, to discuss the physical basis of the responses.

TEMPERATURE

The Arrhenius relationship (1) between the velocity (v) of chemical reactions and absolute temperature (T), $v = e^{-AE^*/RT}$ (E^* is the energy of the reaction, R is the universal gas constant, and A is a constant called the collision or the frequency factor), predicts a straight-line relationship between the logarithm of velocity and the reciprocal of absolute temperature which holds for most simple chemical reactions.

When applied to the specific growth rate, k, of bacteria as a function temperature (for this application the term μ, the temperature characteristic, is substituted for AE^*), the form of the plot is similar for almost all bacteria. Over a certain interval of temperature, commonly called the normal range, log k is linear with $1/T$. At higher and lower temperatures, the growth rate decreases progressively, approaching a vertical asymptote at both the maximum and minimum temperatures for growth.

Most bacteria, including *E. coli* and *S. typhimurium*, can grow over a range of approximately 40°C. The growth rates of most wild-type strains of *E. coli* and *S. typhimurium* respond similarly to changes in temperature; the particular response of *E. coli* B/r is shown in Fig. 1. The normal temperature range extends from 21 to 37°C, over which range μ has a value of 13,000 to 14,000 cal/mol (ca. 54,000 to 59,000 J/mol). The maximum temperature at which balanced growth can be sustained by this strain is approximately 49°C. The minimum temperature for sustained growth of *E. coli* ML30, and presumably other strains as well, lies between 7.5 and 7.8°C (88).

Measurement of the minimum temperature of growth is complicated by the pattern of growth that follows a shift in temperature to the low temperature range. After the shift, growth ceases for a period of time. Shifting an exponential-phase culture of *E. coli* ML30 growing in a minimal medium at 37°C to 10°C is followed by a 4.5-h period during which no growth occurs (69). This period increases as the shift is made to progressively lower temperatures. Thus, it is difficult to distinguish whether failure of a culture to grow after a shift to a temperature near the minimum means that the temperature is less than the minimum for growth or whether the lag in growth is extended. The minimum temperature for growth can be determined with precision by growing a culture at a low temperature somewhat above the minimum and then shifting it to progressively lower temperatures. Even this sort of experiment must be interpreted with caution, however, because incremental shifts to temperatures below the minimum are followed by prolonged transient periods of increase in mass. A shift of *E. coli* from 10 to 7.0°C is followed by a period

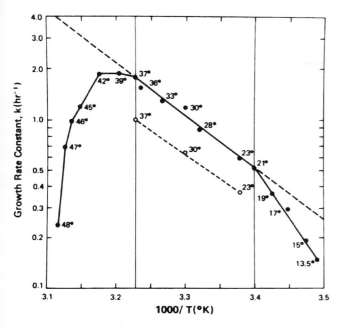

FIGURE 1 Growth rate of *E. coli* B/r as a function of temperature. The specific growth rate (k, hour^{-1}), log scale, is plotted against the inverse of absolute temperature (kelvins). Individual datum points are marked with degrees Celsius: ●, in a rich medium; ○, in a glucose-minimal medium. After Herendeen et al. (39).

of 1 day of growth during which the growth rate gradually declines.

Effect of Nutrition on the Temperature-Growth Response

In wild-type strains, the extent of the normal temperature range and the value of the temperature characteristic within it are unaffected by nutrition. The values of these parameters are the same for cultures growing in a complex medium or in a mineral salts medium with glucose; in the latter, the rate of growth is correspondingly lower at all temperatures (Fig. 1). Neither does the richness of the medium affect the minimum temperature of growth of *E. coli;* however, this growth parameter of a number of other bacteria is altered by nutrition (43). In contrast, the growth rate of several strains of *E. coli*, including K-12 strains, is markedly affected in the high temperature range (40 to 45°C) by the availability of exogenous methionine (80). Possibly all strains of *E. coli* are so affected. In the absence of exogenous methionine, growth stops at 45°C. Between 40 and 45°C, the growth rate is limited by the absence of methionine. At these temperatures the activity of the first enzyme (homoserine transsuccinylase) of the methionine biosynthetic pathway is rapidly but reversibly inhibited, presumably by the rupture of hydrophobic and hydrogen bonds (81).

Modulation of Cellular Composition by Growth Temperature

No major differences in cellular composition exist among cultures of *E. coli* grown at temperatures within the normal range. If cultures of *E. coli* growing at one of these temperatures are abruptly shifted to another, the growth rate, without a detectable transient period, assumes the value characteristic of the temperature to which the shift is made. Also in their classic studies, Schaechter et al. (84) showed that the macromolecular composition of cultures of *S. typhimurium* grown at 25°C is the same as that of cultures grown at 37°C provided that they are grown in the same medium. In contrast, a shift from the normal to the low temperature range is followed by a complex transient phase of growth: a lag followed by a period of abnormally rapid growth before the characteristic steady-state rate ensues. Similarly, a shift from a temperature in the low range to one in the normal range is followed by an extended period of growth (for about 2.3 doublings) at a low rate prior to growth at the characteristic steady-state rate. The existence of transient growth periods following shifts to or from the low temperature range suggests that cells grown within these two temperature ranges differ substantially in composition.

The pattern of proteins of *E. coli* changes significantly with growth temperatures outside the normal range (39). The changes that occur at higher temperatures are under the control of the heat shock response discussed in chapter 88. Similarly, a specific group of 13 cold shock proteins, none of which is a heat shock protein, are produced during the period of growth cessation following a shift from 37 to 10°C. Nine of these have been identified as NusA, RecA, the dihydrolipoamide acetyltransferase subunit of pyruvate dehydrogenase, polynucleotide phosphorylase, pyruvate dehydrogenase, initiation factors 2α and 2β (46, 97), the A subunit of DNA gyrase (45), and nucleoid protein, H-NS (54). A 10th polypeptide, designated CS7.4, which is undetectable at 37°C and is the first to be synthesized following a shift down, encodes a hydrophilic protein with 70 amino acids (37) and is homologous to a region of several eukaryotic DNA binding proteins (99). One of these, YB-1, is a transcriptional activator which binds specifically to the CCAAT-containing Y box of HLA class II genes (24). CS7.4 is a transcriptional activator of *hns* which also contains a CCAAT sequence in its promoter (54), as do *nusA* and *cspA* (the gene encoding CS7.4) (75). It seems likely that CS7.4 regulates the expression of cold shock proteins by acting as a transcriptional activator.

Determinants of Growth Temperature Limits

For determinants of the maximum temperature of growth, see chapter 88.

Considerable evidence suggests that the inability to synthesize protein determines the minimum temperature of growth of *E. coli* and that the sensitive step is initiation of translation. Das and Goldstein (23) showed that after shifting *E. coli* from 37 to 0°C protein synthesis slowed progressively for 4 h as 70S ribosomes accumulated. Friedman et al. (34) showed that synthesis of f-2 coat protein proceeded to completion but was not initiated at 6°C. Broeze et al. (10) showed that a shift from 37 to 5°C was followed by the accumulation of 70S ribosomes at the expense of polysomes.

PRESSURE

Increased hydrostatic pressure changes the rate and equilibrium of chemical reactions by favoring molecules with smaller molecular volumes. It slows reactions in which the molecular volume of the activated state (ΔV^*) is greater than that of the reactants and speeds reactions in which ΔV^* is smaller. Similarly, increased pressure shifts the equilibrium depending on whether the molecular volume of the products is greater or less than the volume of the reactants. In many cases molecular volume changes result from changes in solvation (92).

Since most biological reactions are far from equilibrium, the effect of pressure on rate is most evident. In a simple enzyme-catalyzed reaction obeying Michaelis-Menten kinetics, two activation volume changes occur, one associated with the formation of the enzyme-substrate complex and the other associated with the conversion of the enzyme-substrate complex to the activated state. Thus, the effect of pressure on a reaction can differ depending on the enzyme that catalyzes it. For example, Hochachka et al. (41) showed that increased hydrostatic pressure slows the fructose bisphosphatase reaction catalyzed by an enzyme from rainbow trout but speeds the same reaction catalyzed by an enzyme from the abyssal rat-tail fish. However, most biological reactions are slowed at pressures of 30 MPa or more (59). It is reasonable to assume that increased pressure favors particular pathways of protein folding, resulting in enzymes with altered activity.

Jannasch and Taylor (44) suggested that bacteria can be divided into three groups on the basis of their response to increased hydrostatic pressure: (i) those that grow at a pressure of 1 atm (0.1 MPa) and grow more slowly as hydrostatic pressure is increased (the highest pressure under with they can grown is the index of their barotolerance); (ii) those termed barophiles, which grow at 1 atm and grow more rapidly at higher pressures; and (iii) those termed obligate barophiles, which grow only under pressures greater than 1 atm. Representatives of the last two classes have been isolated and studied. One spirillum-like barophile that was isolated from a 5,800-m depth in the ocean tolerates 100 MPa but grows most rapidly over a broad range around 50 MPa and 30-fold more slowly at 1 atm. An obligate barophile isolated at a 10,000-m depth grows well at 100 MPa, grows optimally at 70 MPa, and is unable to grow at hydrostatic pressures less than 35 MPa (44). Other barophiles require pressures greater than 100 MPa for optimum growth (100).

E. coli is a moderately barotolerant organism, growing increasingly slowly as the pressure is increased above 1 atm. The maximum growth pressure for *E. coli* is about 56 MPa in complex medium and somewhat lower in minimal medium. Even pressures as low as 5 MPa cause a detectable diminution of the growth rate (59). Very high pressures are lethal to *E. coli* and other bacteria. At 37°C and 200 MPa the exponential death rate constant for *E. coli* is 0.125 min^{-1} (82).

Cellular Target of the Inhibitory Effect of High Pressure

Growth-inhibitory pressures of 69 MPa cause a virtual complete cessation of protein synthesis by *E. coli*. The pressure-sensitive steps are polysome formation and translocation; other steps of the process, including amino acid transport and transpeptidation, are pressure resistant (86, 87). Interestingly, a streptomycin-resistant strain has increased barotolerance of protein synthesis (73). The proton-translocating ATPase, flagellar formation, and function, as well as cell division and DNA and RNA syntheses, are sensitive to moderate pressures (see reference 98 for a review).

Adaptations to Elevated Hydrostatic Pressure

Welch et al. (98) studied the pattern of protein synthesis following an abrupt increase in pressure to 55 MPa on a culture of *E. coli* growing anaerobically in a minimal medium. An increase in colony-forming units ceases immediately, but after a lag of approximately 100 min, the optical density of the culture continues to increase but at a diminished rate. It is not clear whether the lag reflects a period of adaptation to elevated pressure. The total number of polypeptides synthesized at the higher pressure is greatly reduced, but 55 of them, termed pressure-induced proteins, are synthesized at an increased differential rate. Eleven pressure-induced proteins are also induced by heat shock and four are induced by cold shock; one, which undergoes the largest induction, has not been observed before. In such experiments, it is not possible to distinguish between direct effects of hydrostatic pressure and indirect effects, for example, the toxic effects of high partial pressures of CO_2.

pH

For additional information on the effects of pH, particularly effects on gene expression, see chapter 96.

In nature a wide range of proton concentrations is encountered, from pH 1 in acidic sulfur springs to pH 11 in soda lakes (53). Bacteria exist in all of these environments, although the number of species (acidophilic bacteria) that can grow in extremely acidic habitats and those (alkalophilic bacteria) that can grow in extremely alkaline ones are more restricted than those (neutrophilic bacteria) that can grow over the mid range, from about pH 5.0 to 9.0. *E. coli* and *S. typhimurium* are representative of the large neutrophilic class; they grow at a maximum rate between pH 6.0 and pH 8.0 and more slowly at half a pH unit or so beyond these limits.

The pH is an important parameter of the rate of many types of reactions, not only ionization of acids and bases but also solvolysis and oxidoreduction. As a result, enzymes and other macromolecules function optimally only over a narrow range of pH, usually close to neutrality, a range that is much more restricted than the pH range of the external environment over which growth of the bacterium is possible.

E. coli, as well as many other neutrophiles (71), has evolved remarkably effective mechanisms of homeostasis of intracellular pH. Using ^{31}P nuclear magnetic resonance, with P_i and methylphosphonate as probes, Slonczewski et al. (90) measured the internal pH (pH_i) of nongrowing *E. coli* cells as the pH of the external buffer (pH_o) was slowly decreased from 7.55 to 5.6 and then slowly increased to 8.7. pH_i changes only slightly but progressively as pH_o is raised above or below the crossover pH. These measurements have been confirmed and extended to measurements of Na^+ gradients at alkaline pH_o (72). Estimations of pH_i by measuring intracellular concentrations of permeant weak acids have given similar results for cells (7, 29, 67, 70) and vesicles (76).

The effect of pH_o on pH_i is magnified in the acid range by permeant acids and in the basic range by permeant bases. Permeant acids (or bases) are organic acids (or bases), the undissociated form of which is lipophilic and, therefore, rapidly diffuses through cell membranes. At equilibrium, the intracellular concentration of the undissociated acid (or base) should equal the extracellular concentration, which is a function of the total concentration, c_o (both dissociated and undissociated), pK, and pH_o. Permeant acids are well-known inhibitors of the growth of microorganisms. Some of them, particularly benzoic, propionic, and sorbic acids, are used as food preservatives. It has long been known that their potency as inhibitors increases as the pH_o

decreases (20), in accord with the hypothesis that the intracellular concentration of undissociated acid governs inhibition.

Permeant acids lower pH_i and have been used to manipulate this variable (47, 77). The inhibition of growth of *E. coli* by permeant acids has been found to correlate closely with pH_i (83). A. G. Marr and N. R. Eaton (unpublished results) found that the rate of growth is strictly determined by the pH_i; the specific growth rate was a sigmoid function of pH_i, with the half maximal rate at a pH_i of 7.2 and complete inhibition below a pH_i of 6.6. Conventional cultivation of *E. coli* can expose the cells to conditions of pH_o and concentrations of acetic acid, a permeant acid, sufficient to depress the pH_i significantly. At a pH_o of 6.0, 5 mM acetate reduces the specific growth rate by half.

Uncouplers of oxidative phosphorylation such as 2,4-dinitrophenol are permeant protonophores which affect the pH_i by collapsing the ΔpH. In the presence of 200 mM 2,4-dinitrophenol, pH_i is within 0.4 unit of the value of pH_o (32).

Buffering Capacity of the Cell

The buffering capacity of bacteria (B_t) has two components: the surface of the cell (B_o) and its internal contents (B_i). These are defined operationally. B_o is the buffering capacity measured by titrating suspensions of intact cells, whereas B_t is measured by titrating suspensions of cells permeabilized by detergent. B_i is a derived value, the difference between B_t and B_o. This method assumes that the pH_i is not changed during the titration of intact cells.

Careful measurements of this sort (49) comparing *E. coli* with another neutrophile, *Bacillus subtilis,* and two obligate alkalophiles, *Bacillus firmus* and *Bacillus alcalophilus,* reveal a number of significant facts. B_i is a small fraction of B_t over most of the range of pH_o. Values of B_i vary markedly among bacteria, and those for *E. coli* are relatively low at most values of pH_o. Although alkalophiles have high values of B_i in the alkaline range, *E. coli* has a low B_i over the range of pH_o from 6.0 to 8.0 in which it grows at maximal rate; the buffering capacity increases above and below this range (Fig. 2).

Since for *E. coli* the value of B_i is a small fraction of B_t and, thus, is subject to large errors, we have computed the titration curve for the cytoplasm. We assumed that the concentration of protein is equivalent to 2 M total amino acid with the amino acid composition given by Neidhardt (68), that one-third of the glutamate and aspartate residues are the respective amides, and that the average peptide contains 200 amino acids. To simplify the calculations, the weighted average of pKs was computed for three groups: (i) aspartate, glutamate, and C-terminal carboxyl, 0.135 M with pK = 4.33; (ii) histidine and N-terminal amino, 0.046 M with pK = 7.07; and (iii) arginine, lysine, tyrosine, and cysteine, 0.214 M with pK = 9.75. In most of the calculations, we used 0.05 M glutamate (pKs of 2.2, 4.2, and 9.7) to represent small molecules of the cytoplasm. Thus, the basic calculation for the j th of these *n* equilibria is given by the following:

$$pH = pK_j + \log_{10}\left(\frac{s}{a}\right)_j \quad (1)$$

in which *s* and *a* denote, respectively, the concentration of dissociated and undissociated acid. Now assume that a titrant acid (or base), *x*, is added, causing a change, ΔpH. At the new equilibrium

$$pH - \Delta pH = pK_j + \log_{10}\left(\frac{s - \Delta a}{a + \Delta a}\right)_j \quad (2a)$$

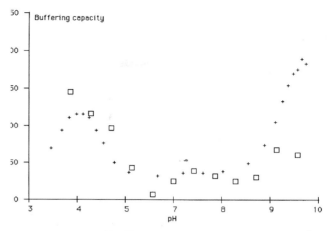

FIGURE 2 Internal buffering capacity (nanoequivalents of H^+ per pH unit per milligram of protein) of *E. coli* as a function of external pH. Squares are data of Krulwich et al. (49). Crosses are calculated from the model.

and

$$x = \sum_{j=1}^{n} \Delta a_j \quad (2b)$$

Sets of equations 2a and 2b were solved numerically by guessing a value of ΔpH and computing a corresponding estimated value of *x*; the guess was refined until the estimated value approached the actual value with an error in ΔpH of <0.001.

Buffering capacity was computed from the model with the result shown in Fig. 2. Except for higher computed values of B_i at an alkaline pH_o the computations are in reasonable agreement with the experimental results of Krulwich et al. (49). It seems likely that at an alkaline pH_o the cytoplasm is significantly titrated; i.e., pH homeostasis fails.

The model was used to simulate experimental values of pH_i for the wild type and *icd* mutants of *S. typhimurium*. The *icd* mutants accumulate about 50 mM citrate and isocitrate (52) and are more resistant to acid than is the wild type, presumably because of greater buffering capacity. Foster and Hall (33) found that at a pH_o of 3.3 the pH_i of the wild type was 4.4. According to the model, this acidification requires 103 meq of H^+ per liter of cytoplasm. With 50 mM citrate added, the model predicts a pH_i of 5.41; Foster and Hall found the pH_i of the *icd* mutant to be 5.5.

Mechanism of Maintaining pH Homeostasis

The value of pH_i is set by a number of factors: the buffering capacity of the cell, the extrusion of protons associated with respiration or hydrolysis of ATP, and the exchange of protons for Na^+ and K^+ (5). The value of pH_i is essentially independent of pH_o; thus, as the pH_o is raised, the ΔpH decreases and the membrane potential increases (5, 101). The proportion of transmembrane proton motive force expressed as a ΔpH depends upon the exchange of intracellular protons for extracellular cations (58). Since K^+ is the principal intracellular cation, it is reasonable to assume that the cytoplasmic concentration of K^+ contributes in this manner to pH homeostasis but is unlikely to be immediately regulatory since its concentration responds to external osmolarity (27).

The antiport of external protons and internal Na^+ becomes important as the value of pH_o increases above the crossover. McMorrow et al. (62) found that *E. coli* failed to grow at pH 8.5 if the concentration of Na^+ was very low (<15 mM) and that the sensitivity to high pH_o varied inversely with antiporter activity. *E. coli* has two antiporters specified by *nhaA* and *nhaB* (reviewed in reference 85). The expression of *nhaA* increases with both pH_o and the concentration of Na^+. The activity of NhaA increases from near zero as the pH_o is increased above the crossover. Thus, NhaA is a candidate regulator of pH_i. However, *nhaA nhaB* double mutants can grow at an alkaline pH_o but fail to grow if the pH_o is alkaline and the concentration of Na^+ is high (>100 mM). The means by which Na^+/H^+ antiport contributes to pH homeostasis is unclear. It cannot be merely the ionic exchange because the Na^+ gradient does not decrease monotonically with an increase in pH_o. Pan and Macnab (72), who observed this, suggest that a combination of neutral and electrogenic antiport pumps protons into the cell.

Acid Tolerance

For additional information on acid tolerance, see chapter 96.

If the pH_o is below (or above) the range compatible with normal pH homeostasis or if a permeant acid (or base) is present and the pH_o is below (or above) 7.6, the pH_i will decrease (or increase) significantly. As the pH_i decreases, the rate of growth slows and then stops. For *S. typhimurium,* a reduction in the pH_i to 5.7 is lethal (32). For *E. coli,* a pH_i below 5.0 is lethal (91). The bases of inhibition of growth and killing by low pH_i are unknown.

To be successful, an enteric pathogen must survive the low pH (pH 2 to 3) of the stomach for about 2 h (35). Although most strains of *E. coli* are considerably more resistant to the lethal effect of low pH_o than is *S. typhimurium* (38), both mount a defense against low pH_o. In principle, this defense could be an increase in the capacity of either pH homeostasis or resistance of the vital target(s) to low pH_i. In fact, both contribute to acid tolerance.

Cells of both *E. coli* and *S. typhimurium* from cultures in the stationary phase are more resistant than growing cells to acid, as they are to many other environmental stresses (reviewed in reference 89). Full resistance of stationary-phase cells requires the expression of *rpoS* (28, 37, 56). *rpoS* mutants of *S. typhimurium* acquire some acid resistance in the stationary phase at a low, but not at a neutral or high, pH_o (56). This implies that at least two mechanisms confer acid resistance to cells from cultures in the stationary phase. It is not yet known whether this resistance is due to an increased capacity for pH homeostasis or to a greater resistance of vital targets.

Growing cells are much more sensitive to acid, but acquire resistance if grown at or exposed to a low pH_o. Cells of *S. typhimurium* exposed to a pH_o of 5.8 for one generation are 100 to 1,000 times more resistant to a subsequent pH_o of 3.3 (32). This treatment was found to augment pH homeostasis, and its effect is dependent upon a functional proton-translocating ATPase (33). Resistance to a pH_o of 3.3 is also acquired by acid shock at a pH_o of 4.5; this resistance does not require augmented pH homeostasis but does require the expression of *fur* (30, 33). The augmentation of pH homeostasis by exposure to a pH_o of 5.8 apparently permits the subsequent synthesis of proteins at a pH_o

of 3.3 that are required for acid resistance and are also induced at a pH_o of 4.5.

Mutants of *S. typhimurium* that are more resistant than the wild type to direct challenge at a pH_o of 3.3 have been isolated (32). One type, referred to above, is *icd* and has a higher intracellular buffering capacity than does the wild type. Others are auxotrophic mutants which may have the resistance of cells from the stationary phase. Mutants more sensitive to acid have also been isolated (31). Some of these mutants are defective in induced pH homeostasis; others are defective in the acid shock response; others resemble in phenotype *fur* mutants; and still others are *polA,* suggesting that DNA repair is important in acid tolerance.

OSMOTIC STRESS

For additional information on osmotic stress, particularly the effect on gene expression, see chapter 77.

The effect on bacterial growth and survival of osmotic stress correlates well in most cases with a_w, which is defined by Raoult's law: $a_w = P/P_o = n_1/(n_1 + n_2)$, where P is the vapor pressure of the solution and P_o is that of the solvent, pure water; n_1 and n_2 are, respectively, the number of moles of solvent and of ideal solute. Depression of the vapor pressure is but one of the colligative properties of solutions; the others are depression of the freezing point, elevation of the boiling point, and osmotic pressure—the hydrostatic pressure (usually expressed in megapascals) which must be applied to a solution to increase the activity of solvent to equal that of pure solvent. Concentrations of actual solutes may be expressed in terms of the concentration of an ideal solute giving the same a_w and denoted as osmolal (osM). The relationship between a_w and osmolality is given by $a_w = x/(x + c)$ in which c is the concentration of solute in osM and x is the number of moles of water per liter, approximately 55.6.

Species of microorganisms differ significantly with respect to the lower limits of a_w (or upper limits of osmolality) permitting growth; the lower limit tolerated by any microorganism is about 0.62, equivalent to about 30 osM (50). If the solute is not intrinsically toxic, the lower limit is usually the same regardless of whether the solute is ionic or nonionic, although there are exceptions. *Saccharomyces rouxii* can tolerate a value of a_w of 0.845 if set by glucose but only 0.860 if set by NaCl (7). *E. coli* falls in the mid range of the tolerance to low a_w exhibited by nonhalophilic bacteria. It is more tolerant than *B. subtilis* and *Pseudomonas aeruginosa* but less so than lactobacilli and *Staphylococcus aureus.*

Osmotic Balance in the Steady State

Unlike animal cells, plant and bacterial cells (with the exception of the mollicutes) are surrounded by a cell wall that resists deformation. Bacteria maintain an intracellular osmotic pressure greater than that of the surrounding medium; osmotic equilibrium is reached by developing internal hydrostatic pressure, termed turgor pressure, such that the a_w of the cytoplasm equals that of the medium.

Maintenance of turgor pressure is essential for growth and division of the cell. Koch has proposed that the stress of turgor pressure on the bacterial cell wall is instrumental in the enlargement of the wall (48). Turgor pressure is communicated from the cytoplasmic membrane to the cell wall either directly or, in bacteria with a periplasm, by means of a gel composed of highly hydrated uncross-linked strands of peptidoglycan that fill space between the cytoplasmic membrane and the wall (40). Periplasmic proteins can diffuse within this gel.

E. coli maintains a turgor pressure of approximately 0.3 MPa over a wide range of osmolality of the external medium. It follows that the concentration of solutes in the cytoplasm must increase linearly with external osmolality in order to maintain turgor pressure. K^+ is particularly important in this respect (27). The intracellular content of K^+ increases systematically with the osmolality of the growth medium adjusted with glucose, sucrose, or NaCl. Environmental conditions other than external osmolality—temperature, pH, and external K^+—have little effect on the intracellular concentration of K^+ (27). In contrast to intracellular K^+, the concentration of intracellular Na^+ does not vary directly with the osmolality of the growth medium; the intracellular concentration of Na^+ is proportional to the external concentration of Na^+ and varies with the pH (12).

To maintain electroneutrality, intracellular K^+ must be balanced by anions. Cells of *E. coli* grown at low external osmolality (<0.2 osM) retain about two-thirds of the intracellular K^+ after hypoosmotic shock; the counterions are presumed to be macromolecular anions, particularly RNA, and thus K^+ is thought not to be osmotically active (61). Cells grown at high external osmolality accumulate substantial concentrations of glutamate (63, 95). The increase in concentration of K^+ closely parallels the concentration of glutamate (61). The accumulation of K^+ (and its counterion, glutamate) is sufficient to maintain proper turgor pressure at external osmolalities up to about 1 osM.

Most strains of *E. coli* growing at high external osmolality accumulate trehalose (93). Some strains of *E. coli* K-12 carry an amber mutation which, in the absence of suppressors, prevents trehalose accumulation (79). This and other mutations which prevent the synthesis of trehalose slow or prevent growth in minimal media at high external osmolality (36, 79). The accumulation of trehalose is accompanied by a reduction in the concentration of K^+ and glutamate (25). The replacement of ionic solutes by a nonionic solute may relieve inhibition of transcription-translation by reducing the ionic strength of the cytoplasm (74). Trehalose not only is a compatible solute permitting growth at high external osmolalities but also specifically reacts with the head groups of phospholipids on the face of membranes, thereby stabilizing the membrane (18, 19).

One cytoplasmic solute, the polyamine putrescine (1,4-diaminobutane), decreases as the external osmolality is increased. Within 5 min after the medium is increased to 0.6 osM, the intracellular concentration is reduced fivefold by excretion of putrescine into the medium (66). Putrescine is replaced by K^+. Munro et al. (66) suggested that the physiological advantage of replacing divalent putrescine with monovalent K^+ is that it allows an increase in cytoplasmic osmolality with a less than proportionate increase in intracellular ionic strength.

Thus, for *E. coli* (and presumably for *S. typhimurium*) growing in minimal medium at high external osmolality, the osmolality of the cytoplasm is increased by the accumulation of K^+, glutamate, and trehalose, and the ionic strength is reduced by the exchange of K^+ for putrescine. These adjustments permit growth up to an external osmolality of at least 2 osM (61). However, tolerance of high external osmolality can be increased by the presence of certain organic compounds, called osmotic protectants, in the medium.

Osmotic Protectants

In 1955 Christian (14) made the significant discovery that the addition of proline to the medium increased the tolerance of *Salmonella* spp. to high external osmolality. He showed that exogenous proline was essential for the growth of *Salmonella oranienburg* in a defined medium at a_w values of less than 0.97, or about 1.7 osM (15), and that proline stimulated the respiration of this and other bacteria at low a_w values (17). The proline content of cells was inversely related to the a_w value of the medium, reaching a value of 1.5 mmol/g (dry weight), or about 0.6 osM, when the a_w of the medium was 0.95, or about 3 osM (16). Thus, proline is an osmotic protectant; bacteria grow in media of higher osmolality if proline is present in the medium.

Csonka (21) isolated mutant strains of *S. typhimurium* that were resistant to L-azetidine-2-carboxylate because they overproduced L-proline. These strains grew in media of higher osmolality (0.8 M NaCl) than did their parent. In a minimal medium with 0.65 M NaCl, the specific growth rate of the parent was 0.1 h^{-1} while that of the mutant was 0.29 h^{-1}. The addition of proline to a medium containing 0.65 M NaCl doubled the growth rate of wild-type *S. typhimurium;* none of the other 19 common amino acids had this effect (21).

Other compounds known to accumulate in plants subjected to osmotic stress were tested for their ability to increase the osmotic tolerance of *E. coli*. Glycine betaine, choline, betaine aldehyde, trimethyl-γ-aminobutyrate, and proline betaine were effective (57). Subsequently β-alanine betaine, taurine betaine, γ-amino crotonic acid betaine, carnitine, choline-*O*-sulfate, 3-(*N*-morpholino)propanesulfonate, and dimethylthetin have been added to the list of osmotic protectants of *E. coli* and *S. typhimurium* (reviewed in reference 22).

Glycine betaine is the most effective osmotic protectant of *E. coli* and *S. typhimurium*. *S. typhimurium* at 30°C in a minimal medium containing 0.8 M NaCl has a specific growth rate of 0.10 h^{-1}. If the medium is supplemented with proline, the rate increases to 0.17 h^{-1}, but if supplemented with glycine betaine, the rate increases to 0.30 h^{-1} (11). Choline and betaine aldehyde are effective as a consequence of oxidation to glycine betaine. Choline is converted to betaine aldehyde by a membrane-associated, O_2-dependent enzyme, and the product, betaine aldehyde, is converted to glycine betaine by a soluble NADP-dependent enzyme. The level of both of these enzymes increases with the osmolality of the medium (57).

At high external osmolality, the concentration of osmotic protectants such as glycine betaine and endogenous compatible solutes such as trehalose is equivalent to a substantial fraction of external osmolality. Their accumulation reduces the cytoplasmic concentration of ionic solutes K^+ and glutamate (3, 13, 25). The accumulation of glycine betaine reduces the concentration of endogenous trehalose as well as ionic solutes (13, 25). The reduction in the concentration of K^+ or in ionic strength by compatible solutes may be directly responsible for an increasing growth rate through relief of inhibition of transcription-translation or other enzymatic activities (74, 94). However, Cayley et al. (13) argue that glycine betaine increases the fraction of free water.

Hyperosmotic Shock

The response to an abrupt increase in external osmolality should reveal the mechanisms underlying the osmotic homeostasis observed in the steady state. If the osmolality of a culture of *E. coli* growing in minimal medium is abruptly and substantially increased to, say, 1 osM, the cells almost instantaneously lose water by plasmolysis, increasing the turbidity of the culture; the synthesis of macromolecules is inhibited (51); and the rate of respiration decreases (42), possibly as a consequence of the reduced rate of biosynthesis.

The kinetics of change of amounts and concentrations of the principal osmolytes were measured by Dinnbier et al. (25). The amount of K^+ rises rapidly to a maximum by about 20 min. The concentration of K^+, which reflects both the amount of K^+ and the water content of the cell, reaches a maximum within a few minutes. The concentration of glutamate rises somewhat more slowly. The pH_i transiently increases, reaching a maximum of pH 8.3 after a few minutes, and then then declines rapidly to the normal value; this excursion of pH_i may reflect faster kinetics for the K^+/H^+ exchange than for the accumulation of the counterion, glutamate. Recovery from plasmolysis coincides with these events. The concentration of trehalose increases more slowly, reaching a maximum at about 1 h. The accumulation of trehalose is accompanied by a loss of part of the previously accumulated K^+ and glutamate. Finally, the addition of external proline or glycine betaine causes a loss of trehalose and further loss of K^+.

The resumption of biosynthesis lags behind the recovery from plasmolysis and coincides with the accumulation of trehalose and loss of K^+ and glutamate. This fact suggests that the high concentration of ionic (noncompatible) solutes inhibits biosynthesis.

Regulation of the Osmotic Response

The increase in the intracellular concentration of K^+ appears to be the key event in the response to hyperosmotic shock. The increased concentration results in part from plasmolysis and in part by accumulation. Accumulation is thought to result from stimulation of the K^+ transport systems, Trk and Kdp, signaled by a decrease in turgor pressure (78). However, intracellular K^+ in growing cells is dynamic: K^+ is accumulated by transport and lost by efflux. The cell membrane contains stretch-activated ion channels (102). Plasmolysis would close such channels and contribute to the accumulation of K^+.

The accumulation of glutamate depends upon the accumulation of K^+. Glutamate is not accumulated following hyperosmotic shock in a medium lacking K^+ (61). Tempest et al. (95) suggested that the accumulation of glutamate might result from an increase in the activity of glutamate dehydrogenase. They found that this enzyme has an unusually sharp dependence on pH, being maximally active at pH 8 and nearly inactive at pH 7. The alternative route for glutamate synthesis via glutamine shows a similar dependence on pH (64). The accumulation of K^+ is accompanied by a marked but transient increase in pH_i (25, 61), which would increase the rate of synthesis of glutamate by either pathway. However, McLaggan et al. (61) argue that the inhibition of biosynthesis by hyperosmotic shock is sufficient to account for the accumulation of glutamate.

Booth and Higgins (6) have proposed that the high concentration of K^+ and glutamate is the main secondary signal in adjustment to hyperosmotic shock. High ionic strength inhibits transcription from normal promoters but increases transcription from osmotically activated promoters such as *proU* (74),

possibly by release of the DNA binding protein H-NS (65). The consequence of the secondary signal is replacement of ionic solutes, K^+ and glutamate, by compatible solutes such as trehalose (by synthesis) or glycine betaine (by transport). This replacement is critical to a resumption of growth at high osmolality.

Hypoosmotic Shock

Low external osmolality per se is compatible with survival and growth of *E. coli* and *S. typhimurium;* however, a modest, abrupt decrease in external osmolality of −0.2 to −0.3 osM causes the loss of low-molecular-weight internal solutes but not macromolecules (9, 27, 96) and can be lethal (4). Amino acids and nucleotide pools (9) or previously accumulated solutes such as methylthiogalactoside or α-methylglucoside (96) are lost almost completely. If the extracellular fluid contains salts, almost all K^+ is lost (26), but if salts are absent, only a fraction is lost (61). K^+ is thought to be retained by electrostatic attraction of polyanions, principally nucleic acids (61), but can exchange with Na^+ in the extracellular fluid. Bakker (2) has found that if a culture of *E. coli* growing in 0.4 M NaCl is abruptly diluted 1:1 with water, 90% of the K^+ and glutamate and 60% of the trehalose are lost from the cells. Under this condition the loss of solutes is selective: ATP and alanine are retained.

The immediate consequence of hypoosmotic shock is a flux of water into the cell, increasing the turgor. The increased turgor would stretch the wall and the underlying membrane. Britten (8) suggested that such stretching would either cause transient "cracks" or perhaps stretch existing pores in the membrane. This behavior may reflect stretch-activated channels (102). The increased permeability is only transitory: after at most a few minutes the cells can begin to accumulate solutes (96). It seems likely that solute loss results from the opening of stretch-activated channels that may be important to normal osmoregulation. Loss of intracellular solutes by hypoosmotic shock is potentiated by low temperature (26, 96), and temperature shock under isosmotic conditions causes solute loss (55). Low temperature may decrease the fluidity of the cell membrane and thereby contribute to opening stretch-activated channels.

LITERATURE CITED

1. Arrhenius, S. 1889. Uber die Reaktionsgeschwendigkeit bei der Inversion von Rohrzucker durch Sauren. *Z. Phys. Chem.* **4**:226–248.
2. Bakker, E. P. 1992. Cell K^+ and K^+ transport systems in prokaryotes, p. 205–224. *In* E. P. Bakker (ed.), *Alkali Cation Transport Systems in Prokaryotes.* CRC Press, Boca Raton, Fla..
3. Bakker, E. P., I. R. Booth, U. Dinnbier, W. Epstein, and A. Gajewska. 1987. Evidence for multiple K^+ export systems in *Escherichia coli. J. Bacteriol.* **169**:3743–3749.
4. Bayer, M. E. 1967. Response of cell walls of *Escherichia coli* to a sudden reduction in environmental osmotic pressure. *J. Bacteriol.* **93**:1104–1112.
5. Booth, I. R. 1985. Regulation of cytoplasmic pH in bacteria. *Microbiol. Rev.* **49**:359–378.
6. Booth, I. R., and C. F. Higgins. 1990. Enteric bacteria and osmotic stress: intracellular potassium glutamate as a secondary signal of osmotic stress. *FEMS Microbiol. Rev.* **75**:239–246.
7. Booth, I. R., W. J. Mitchell, and W. A. Hamilton. 1979. Quantitative analysis of proton-linked transport systems. *Biochem. J.* **182**:687–696.
8. Britten, R. J. 1965. The concentration of small molecules with the microbial cell. *Symp. Soc. Gen. Mircobiol.* **15**:57–88.
9. Britten, R. J., and F. T. McClure. 1962. The amino acid pool in *Escherichia coli. Bacteriol. Rev.* **26**:292–353.
10. Broeze, R. J., C. J. Solomon, and D. H. Pope. 1978. Effect of low temperature on in vivo and in vitro protein synthesis in *Escherichia coli* and *Pseudomonas fluorescens. J. Bacteriol.* **134**:861–874.

11. Cairney, J., I. R. Booth, and C. F. Higgins. 1985. Osmoregulation of gene expression in *Salmonella typhimurium: proU* encodes an osmotically induced betaine transport system. *J. Bacteriol.* 164:1224–1232.

12. Castle, A. M., R. M. Macnab, and R. G. Shulman. 1986. Coupling between the sodium and proton gradients in respiring *Escherichia coli* cells measured by ^{23}Na and ^{31}P nuclear magnetic resonance. *J. Biol. Chem.* 261:7797–7806.

13. Cayley, S., B. A. Lewis, and M. T. Record. 1992. Origins of osmoprotective properties of betaine and proline in *Escherichia coli* K-12. *J. Bacteriol.* 174:1586–1595.

14. Christian, J. H. B. 1955. The influence of nutrition on the water relations of *Salmonella oranienburg. Aust. J. Biol. Sci.* 8:75–82.

15. Christian, J. H. B. 1955. The water relations of growth and respiration of *Salmonella oranienburg* at 30°C. *Aust. J. Biol. Sci.* 8:490–497.

16. Christian, J. H. B., and J. M. Hall. 1972. Water relations of *Salmonella oranienburg*: accumulation of potassium and amino acids during respiration. *J. Gen. Microbiol.* 70:497–506.

17. Christian, J. H. B., and J. A. Waltho. 1966. Water relations of *Salmonella oranienburg*: stimulation of respiration by amino acids. *J. Gen. Microbiol.* 43:345–355.

18. Crowe, J. H., L. M. Crowe, J. F. Carpenter, A. S. Rudolph, C. A. Winstrom, B. J. Spargo, and T. J. Anchordoguy. 1988. Interaction of sugars with membranes. *Biochim. Biophys. Acta Biomembr. Rev.* 947:367–384.

19. Crowe, J. H., L. M. Crowe, and D. Chapman. 1984. Preservation of membranes in anhydrobiotic organisms: the role of trehalose. *Science* 223:701–703.

20. Cruess, W. V., and P. H. Richert. 1929. Effect of hydrogen ion concentration on the toxicity of sodium benzoate to microörganisms. *J. Bacteriol.* 17:363–371.

21. Csonka, L. N. 1981. Proline over-production results in enhanced osmotolerance in *Salmonella typhimurium. Mol. Gen. Genet.* 182:82–86.

22. Csonka, L. N., and A. Hanson. 1991. Prokaryotic osmoregulation: genetics and physiology. *Annu. Rev. Microbiol.* 45:569–606.

23. Das, H. K., and A. Goldstein. 1968. Limited capacity for protein synthesis at zero degrees centigrade in *Escherichia coli. J. Mol. Biol.* 31:209–226.

24. Didier, D. K., J. Schiffenbauer, S. L. Woulfe, M. Zacheis, and B. D. Schwartz. 1988. Characterization of the cDNA encoding a protein binding to the major histocompatibility complex class II Y box. *Proc. Natl. Acad. Sci. USA* 85:7322–7326.

25. Dinnbier, U., E. Limpinsel, R. Schmid, and E. P. Bakker. 1988. Transient accumulation of potassium glutamate and its replacement by trehalose during adaptation of growing cells of *Escherichia coli* K-12 to elevated sodium chloride concentrations. *Arch. Microbiol.* 150:348–357.

26. Epstein, W., and S. G. Schultz. 1965. Cation transport in *Escherichia coli*. V. Regulation of cation content. *J. Gen. Physiol.* 49:221–234.

27. Epstein, W., and S. G. Schultz. 1966. Ion transport and osmoregulation in bacteria, p. 186–193. *In* L. B. Guze (ed.), *Microbial Protoplasts, Spheroplasts and L-Forms*. The Williams & Wilkins Co., Baltimore.

28. Fang, F. C., S. J. Libby, N. A. Buchmeier, P. C. Loewen, J. Switala, J. Harwood, and D. G. Guiney. 1993. The alternative sigma factor KatF (RpoS) regulates *Salmonella* virulence. *Proc. Natl. Acad. Sci. USA* 90:3511–3515.

29. Felle, H., J. S. Porter, and C. L. Slayman, and H. R. Kaback. 1980. Quantitative measurements of membrane potential in *Escherichia coli. Biochemistry* 19:3585–3590.

30. Foster, J. W. 1993. The acid tolerance response of *Salmonella typhimurium* involves transient synthesis of key acid shock proteins. *J. Bacteriol.* 175:1981–1987.

31. Foster, J. W., and B. Bearson. 1994. Acid-sensitive mutants of *Salmonella typhimurium* identified through a dinitrophenol lethal screening strategy. *J. Bacteriol.* 176:2596–2602.

32. Foster, J. W., and H. K. Hall. 1990. Adaptive acid tolerance response of *Salmonella typhimurium. J. Bacteriol.* 172:771–778.

33. Foster, J. W., and H. K. Hall. 1991. Inducible pH homeostasis and the acid tolerance response of *Salmonella typhimurium. J. Bacteriol.* 173:5129–5135.

34. Friedman, H., P. Lu, and A. Rich. 1971. Temperature control of initiation of protein synthesis in *Escherichia coli. J. Mol. Biol.* 61:105–121.

35. Gianella, R. A., S. A. Broitman, and M. Zamcheck. 1972. Gastric acid barrier to ingested microorganisms in man: studies in vivo and in vitro. *Gut* 13:251–256.

36. Glæver, H. M., O. B. Styrvold, I. Kaasen, and A. R. Strøm. 1988. Biochemical and genetic characterization of osmoregulatory trehalose synthesis in *Escherichia coli*. *J. Bacteriol.* 170:2841–2849.

37. Goldstein, J., N. S. Pollitt, and M. Inouye. 1990. Major cold shock protein of *Escherichia coli. Proc. Natl. Acad. Sci. USA* 87:283–287.

38. Gordon, J., and P. C. L. Small. 1993. Acid resistance in enteric bacteria. *Infect. Immun.* 61:364–367.

39. Herendeen, S. L., R. A. VanBogelen, and F. C. Neidhardt. 1979. Levels of major proteins of *Escherichia coli* during growth at different temperatures. *J. Bacteriol.* 139:185–194.

40. Hobot, J. A., E. Carlemalm, W. Villiger, and E. Kellenberger. 1984. Periplasmic gel: new concept resulting from the reinvestigation of bacterial cell envelope ultrastructure by new methods. *J. Bacteriol.* 160:143–152.

41. Hochachka, P. W., T. W. Moon, and T. Mustafa. 1972. The adaptation of enzymes to pressure in abyssal and midwater fishes, p. 175–195. *In* M. A. Sleigh and A. G. Macdonald (ed.), *The Effects of Pressure on Living Organisms*. Academic Press, London.

42. Houssin, C., N. Eynard, E. Shechter, And A. Ghazi. 1991. Effect of osmotic pressure on membrane energy-linked functions in *Escherichia coli. Biochim. Biophys. Acta* 1056:76–84.

43. Ingraham, J. L. 1962. Temperature relationships, p. 265–296. *In* I. C. Gunsalus and R. Y. Stanier (ed.), *The Bacteria*, vol. 4. Academic Press, Inc., New York.

44. Jannasch, H. W., and C. D. Taylor. 1984. Deep-sea microbiology. *Annu. Rev. Microbiol.* 38:487–514.

45. Jones, P. G., R. Krah, S. R. Tafuri, and A. P. Wolffe. 1992. DNA gyrase, CS7.4, and the cold shock response in *Escherichia coli. J. Bacteriol.* 174:5798–5802.

46. Jones, P. G., R. A. VanBogelen, and F. C. Neidhardt. 1987. Induction of proteins in response to low temperature in *Escherichia coli. J. Bacteriol.* 169:2092–2095.

47. Kihara, M., and R. M. Macnab. 1981. Cytoplasmic pH mediates pH taxis and weak-acid repellent taxis of bacteria. *J. Bacteriol.* 145:1209–1221.

48. Koch, A. L. 1983. The surface stress theory of microbial morphogenesis. *Adv. Microb. Physiol.* 24:301–366.

49. Krulwich, T. A., R. Agus, M. Schneier, and A. A. Guffanti. 1985. Buffering capacity of bacilli that grow at different pH ranges. *J. Bacteriol.* 162:768–772.

50. Kushner, D. J. 1978. Life in high salt and solute concentrations: halophilic bacteria, p. 317–368. *In* D. J. Kushner (ed.), *Microbial Life in Extreme Environments*. Academic Press, London.

51. Laimins, L. A., D. B. Rhoads, and W. Epstein. 1981. Osmotic control of *kdp* operon expression in *Escherichia coli. Proc. Natl. Acad. Sci. USA* 78:464–468.

52. Lakshmi, T. M., and R. B. Helling. 1976. Selection for citrate synthase deficiency in *icd* mutants of *Escherichia coli. J. Bacteriol.* 127:76–83.

53. Langworthy, T. A. 1978. Microbial life in extreme pH values, p. 279–315. *In* D. J. Kushner (ed.), *Microbial Life in Extreme Environments*. Academic Press, London.

54. La Teana, A., A. Brandi, M. Falconi, R. Spurio, C. L. Pon, and C. O. Gualerzi. 1991. Identification of a cold shock transcriptional enhancer of the *Escherichia coli* gene encoding nucleoid protein H-NS. *Proc. Natl. Acad. Sci. USA* 88:10907–10911.

55. Leder, I. G. 1972. Interrelated effects of cold shock and osmotic pressure on the permeability of the *Escherichia coli* membrane to permease accumulated substrates. *J. Bacteriol.* 111:211–219.

56. Lee, I. S., J. L. Slonczewski, and J. W. Foster. 1994. A low-pH-inducible, stationary-phase acid tolerance response in *Salmonella typhimurium. J. Bacteriol.* 176:1422–1426.

57. Le Rudulier, D., A. R. Strom, A. M. Dandekar, L. T. Smith, and R. C. Valentine. 1984. Molecular biology of osmoregulation. *Science* 224:1064–1068.

58. Macnab, R. M., and A. M. Castle. 1987. A variable stoichiometry model for pH homeostasis in bacteria. *Biophys. J.* 52:637–647.

59. Marquis, R. E., and P. Matsumura. 1978. Microbial life under pressure, p. 105–147. *In* D. J. Kushner (ed.), *Microbial Life in Extreme Environments*. Academic Press, London.

60. McLaggan, D., T. M. Logan, D. G. Lynn, and W. Epstein. 1990. Involvement of γ-glutamyl peptides in osmoadaptation of *Escherichia coli. J. Bacteriol.* 172:3631–3636.

61. McLaggan, D., J. Naprstek, E. T. Buurman, and W. Epstein. 1994. Interdependence of K$^+$ and glutamate accumulation during osmotic adaptation of *Escherichia coli. J. Biol. Chem.* 269:1911–1917.

62. McMorrow, I., H. A. Shuman, D. Sze, D. M. Wilson, and T. H. Wilson. 1989. Sodium/proton antiport is required for growth of *Escherichia coli* at alkaline pH. *Biochim. Biophys. Acta* 981:21–26.

63. Measures, J. C. 1975. Role of amino acids in osmoregulation of nonhalophilic bacteria. *Nature* (London) 257:398–400.

64. Meers, J. L., and D. W. Tempest. 1970. Glutamine(amide):2-oxoglutarate amino transferase oxidoreductase (NADP), an enzyme involved in the synthesis of glutamate by some bacteria. *J. Gen. Microbiol.* 64:187–194.

65. Mellies, J., R. Brems, and M. Villarejo. 1994. The *Escherichia coli proU* promoter element and its contribution to osmotically signaled transcription activation. *J. Bacteriol.* 176:3638–3645.

66. Munro, G. F., K. Hemules, J. Morgan, and W. Sauerbler. 1972. Dependence of the putrescine content of *Escherichia coli* on the osmotic strength of the medium. *J. Biol. Chem.* 247:1272–1280.

67. Navon, G., S. Ogawa, R. G. Schulman, and T. Yamane. 1977. High resolution ^{31}P nuclear magnetic resonance studies of metabolism in aerobic *Escherichia coli* cells. *Proc. Natl. Acad. Sci. USA* 74:888–891.

68. Neidhardt, F. C. 1987. Chemical composition of *Escherichia coli*, p. 3–6. *In* F. C. Neidhardt, J. L. Ingraham, K. B. Low, B. Magasanik, M. Schaechter, and H. E. Umbarger (ed.), *Escherichia coli and Salmonella typhimurium: Cellular and Molecular Biology*, vol. 1. American Society for Microbiology, Washington, D.C.

69. Ng, H., J. L. Ingraham, and A. G. Marr. 1962. Damage and derepression in *Escherichia coli* resulting from growth at low temperatures. *J. Bacteriol.* 84:331–339.

70. Padan, E., D. Zilberstein, and H. Rottenberg. 1976. The proton electrochemical gradient in *Escherichia coli* cells. *Eur. J. Biochem.* 63:533–541.

71. Padan, E., D. Zilberstein, and S. Schuldiner. 1981. pH homeostasis in bacteria. *Biochim. Biophys. Acta* 650:161–166.

72. Pan, J. W., and R. M. Macnab. 1990. Steady-state measurements of *Escherichia coli* sodium and proton potentials at alkaline pH support the hypothesis of electrogenic antiport. *J. Biol. Chem.* 265:9247–9260.

73. Pope, D. H., W. P. Smith, M. A. Orgrinc, and J. V. Landau. 1976. Protein synthesis at 680 atm: is it related to environmental origin, physiological type, or taxonomic group? *Appl. Environ. Microbiol.* 31:1001–1002.

74. Prince, W. S., and M. R. Villarejo. 1990. Osmotic control of *proU* transcription is mediated through direct action of potassium glutamate on the transcription complex. *J. Biol. Chem.* 265:17673–17679.

75. Qoronfleh, M. W., C. Debouck, and J. Keller. 1992. Identification and characterization of novel low-temperature-inducible promoters of *Escherichia coli. J. Bacteriol.* 174:7902–7909.

76. **Ramos, S., S. Schuldiner, and H. R. Kaback.** 1976. The electrochemical gradient of protons and its relationship to active transport in *Escherichia coli* vesicles. *Proc. Natl. Acad. Sci. USA* **73:**1892–1896.

77. **Repaske, D. R., and J. Adler.** 1981. Change in intracellular pH of *Escherichia coli* mediates the chemotactic response to certain attractants and repellents. *J. Bacteriol.* **145:**1196–1208.

78. **Rhodes, D. B., and W. Epstein.** 1978. Cation transport in *Escherichia coli*. IX. Regulation of K transport. *J. Gen. Physiol.* **72:**283–295.

79. **Rod, M. L., K. Y. Alam, P. R. Cunningham, and D. P. Clark.** 1988. The accumulation of trehalose by *Escherichia coli* K-12 at high osmotic pressure depends on the presence of amber suppressors. *J. Bacteriol.* **170:**3601–3610.

80. **Ron, E. Z., and B. D. Davis.** 1971. Growth rate of *Escherichia coli* at elevated temperatures: limitation by methionine. *J. Bacteriol.* **107:**391–396.

81. **Ron, E. Z., and M. Shani.** 1971. Growth rate of *Escherichia coli* at elevated temperatures: reversible inhibition of homoserine trans-succinylase. *J. Bacteriol.* **107:**397–400.

82. **Rutberg, L.** 1964. On the effects of high hydrostatic pressure on bacteria and bacteriophage. 1. Action on the reproducibility of bacteria and their ability to support growth of bacteriophage T2. *Acta Pathol. Microbiol. Scand.* **61:**81–90.

83. **Salmond, C. V., R. G. Kroll, and I. Booth.** 1984. The effect of food preservatives on pH homeostasis in *Escherichia coli. J. Gen. Microbiol.* **130:**2845–2850.

84. **Schaechter, M., O. Maaløe, and N. O. Kjeldgaard.** 1958. Dependency on medium and temperature of cell size and chemical composition during balanced growth of *Salmonella typhimurium. J. Gen. Microbiol.* **19:**592–606.

85. **Schuldiner, S., and E. Padan.** 1992. Na$^+$/H$^+$ antiporters in *Escherichia coli*, p. 25-51. *In* E. P. Bakker (ed.), *Alkali Cation Transport Systems in Prokaryotes.* CRC Press, Boca Raton, Fla.

86. **Schwarz, J. R., and J. V. Landau.** 1972. Hydrostatic pressure effects on *Escherichia coli*: site of inhibition of protein synthesis. *J. Bacteriol.* **109:**945–948.

87. **Schwarz, J. R., and J. V. Landau.** 1972. Inhibition of cell-free protein synthesis by hydrostatic pressure. *J. Bacteriol.* **112:**1222–1227.

88. **Shaw, M. K., A. G. Marr, and J. L. Ingraham.** 1971. Determination of the minimum temperature for growth of *Escherichia coli. J. Bacteriol.* **105:**683–684.

89. **Siegele, D. A., and R. Kolter.** 1992. Life after log. *J. Bacteriol.* **174:**345–348.

90. **Slonczewski, J. L., B. P. Rosen, J. R. Alger, and R. M. Macnab.** 1981. pH homeostasis in *Escherichia coli:* measurement by ^{31}P nuclear magnetic resonance of methylphosphonate and phosphate. *Proc. Natl. Acad. Sci. USA* **78:**6271–6275.

91. **Small, P., D. Blankenship, D. Welty, E. Zinser, and J. L. Slonczewski.** 1994. Acid and base resistance in *Escherichia coli* and *Shigella flexneri:* role of *rpoS* and growth pH. *J. Bacteriol.* **176:**1729–1737.

92. **Somero, G. N.** 1992. Adaptations to high hydrostatic pressure. *Annu. Rev. Physiol.* **54:**557–577.

93. **Strøm, A. R., P. Falkenberg, and B. Landfald.** 1986. Genetics of osmoregulation in *Escherichia coli:* uptake and biosynthesis of organic osmolytes. *FEMS Microbiol. Rev.* **39:**79–86.

94. **Sutherland, L., J. Cairney, M. J. Elmore, I. R. Booth, and C. F. Higgins.** 1986 Osmotic regulation of transcription: induction of the *proU* betaine transport gene is dependent on accumulation of intracellular potassium. *J. Bacteriol.* **168:**805–814.

95. **Tempest, D. W., J. L. Meers, and C. M. Brown.** 1970. Influence of environment on the content and composition of microbial free acid pools. *J. Gen. Microbiol.* **64:**171–185.

96. **Tsapis, A., and A. Kepes.** 1977. Transient breakdown of the permeability barrier of the membrane of *Escherichia coli* upon hypoosmotic shock. *Biochim. Biophys. Acta* **469:**1–12.

97. **VanBogelen, R. A., and F. C. Neidhardt.** 1990. Ribosomes as sensors of heat and cold shock in *Escherichia coli. Proc. Natl. Acad. Sci. USA* **87:**5589–5593.

98. **Welch, T. J., A. Farewell, F. C. Neidhardt, and D. Bartlett.** 1993. Stress response of *Escherichia coli* to elevated hydrostatic pressure. *J. Bacteriol.* **175:**7170–7177.

99. **Wistow, G.** 1990. Cold shock and DNA binding. *Nature* (London) **344:**823–824.

100. **Yayanos, A. A.** 1986. Evolutionary and ecological implication of the properties of deep-sea barophilic bacteria. *Proc. Natl. Acad. Sci. USA* **83:**9542–9546.

101. **Zilberstein, D., V. Agmon, S. Schuldiner, and E. Padan.** 1982. The sodium/proton antiporter is part of the pH homeostasis mechanism in *Escherichia coli. J. Biol. Chem.* **257:**3687–3691.

102. **Zoretti, M., and A. Gazi.** 1992. Stretch-activated channels in prokaryotes, p. 349–358. *In* E. P. Bakker (ed.), *Alkali Cation Transport Systems in Prokaryotes.* CRC Press, Boca Raton, Fla.

Initiation of Chromosome Replication

WALTER MESSER AND CHRISTOPH WEIGEL

99

INTRODUCTION

The genetic maps of *Escherichia coli* and *Salmonella typhimurium* (official designation, *Salmonella enterica* serovar Typhimurium) are circular. This reflects the physical structure of their single chromosomes, which are DNA rings with a circumference of 1.6 mm (44) that contain 4,720 kbp (149). A large part of the sequence of the *E. coli* chromosome is known by now.

During balanced growth, by definition, all components of the cell double in one generation. Obviously, the replication of the genome must be regulated in such a way that upon cell division, two entities can be distributed between two identical daughter cells. The replication of the chromosome is regulated at the level of initiation, as is all macromolecular synthesis of cells. This is one of the essential statements in the book by Maaloe and Kjeldgaard (185) that appeared nearly 30 years ago and has had a profound impact on research on the bacterial cell cycle: "We are therefore led to believe that the overall production of DNA, RNA, and protein is regulated by mechanisms that control the frequencies with which the synthesis of individual nucleotide and amino acid chains are initiated."

We set the stage with a brief historical introduction. A more detailed discussion of the two principal components, the replication origin *oriC* and the initiator protein DnaA, and an analysis of the biochemistry and the control of replication follow. The reader is also referred to a series of excellent recent reviews and monographs and especially the citations in the works (23, 40, 53, 61, 155, 190, 197, 281, 351). Some topics discussed by von Meyenburg and Hansen (313) in the first edition of this monograph are not discussed in the same depth here, and the reader is referred to the first edition.

HISTORICAL OVERVIEW

Replication Is a Regular Process

Besides their basic observation that DNA replicates semiconservatively, Meselson and Stahl (200) were the first to observe that DNA replication is a very regular process. Exactly one generation after the shift of a growing culture of *E. coli* from a medium containing the heavy isotope ^{15}N to a medium containing ^{14}N, all DNA is found at hybrid density. This shows that any segment of the chromosome in any cell in the population that replicates at any moment does so again exactly one generation later. It implies that successive rounds of replication are initiated at the same point on the chromosome. This basic observation was later corroborated by using pulse-labeling in combination with density shifts (164, 165) (see also reference 313).

DNA Replication Is Coupled to Growth Rate

In the same year Meselson and Stahl's article was published (1958), Schaechter et al. (264) showed that the amount of DNA per cell in *S. typhimurium* in balanced growth is dependent on the growth rate. Similar results were obtained later for different strains of *E. coli* (reviewed in reference 313). When the growth rate is increased by changing the growth medium from poor to rich (a shift-up experiment), RNA and protein synthesis adjust rapidly. The rate of DNA synthesis, however, is initially unaffected and only gradually approaches the new steady-state rate (41, 139). In the reverse situation, a shift-down, mass increase ceases, whereas DNA synthesis continues after the shift and adjusts only slowly (139). A change in growth rate thus results in a new steady-state rate of DNA synthesis only after some delay.

Mutations Affecting Initiation of Replication

Kohiyama et al. (151, 152) were the first to isolate temperature-sensitive mutants defective in DNA replication at 42°C. Depending on the kinetics of DNA synthesis following the shift to nonpermissive temperature, they were grouped into immediate-stop, i.e., elongation-type, mutants and delayed-stop mutants. Delayed-stop mutants show the phenotype expected for defects in replication initiation. Upon a temperature shift to 42°C, DNA synthesis gradually ceases and reaches a plateau; in the typical case, this plateau is 1.4 times the amount observed at the time of the shift. The first mutations of this kind, in the *dnaA* gene, are discussed extensively later. All genuine initiation mutants subsequently isolated were eventually found to map in only two genes, *dnaA* and *dnaC* (reviewed in reference 313), with one exception, the *dnaB*252 mutation (350). The product of the *dnaA* gene is the initiator protein DnaA, the DnaB protein is the replicative DNA helicase, and DnaC is required to deliver DnaB to the replication complex. Mutants in genes that affect replication initiation under special conditions, *dnaJ*, *dnaK*, and *dnaR*, are discussed later.

The Replicon

The replicon model of Jacob et al. (121) defines the minimally required elements for the regulated duplication of a DNA molecule. The replicator is the site at which replication begins and is synonymous with what we call now a replication origin. An initiator element interacts with the replicator for replication initiation and is coded for by the replicon. The DnaA protein is the initiator. The evidence for this fact and whether it acts alone or in concert with other factors are discussed below.

Mapping of *oriC*

In principle, the bacterial chromosome could consist of several replicons, i.e., could contain several origins and replicators. This is not excluded by the Meselson-Stahl experiment, which only requires that the same site(s) be used in successive initiations in the same order. However, extensive analysis has shown that the *E. coli* chromosome has one fixed replication origin, *oriC*, located at 84.3 min on the *E. coli* genetic map (see chapter 109), and that replication proceeds bidirectionally from there to the terminus, *terC* (reviewed in reference 313). Likewise, replication of the *S. typhimurium* chromosome (75) and the chromosome of *Bacillus subtilis,* first thought to be unidirectional, proceeds bidirectionally from a fixed origin (reviewed in reference 336).

The DNA segment which was first replicated after release of the replication block in a *dnaC* mutant (191) had a characteristic restriction fragment pattern which could be localized on F′ plasmids carrying the 84-min region of the chromosome (315). The diploidy of the origin region gave such F′ strains a particular phenotype (reviewed in reference 313), and it was Hiraga (105) who gave the now precisely mapped chromosomal replication origin the name *oriC*.

Minichromosomes

Dissection of the molecular events at the initiation of replication proved to be difficult or impossible, in part because of the large size of the bacterial chromosome. This problem was overcome by the cloning of *oriC* on plasmids, called minichromosomes. They were selected by their abilities to promote the autonomous replication of small DNA segments by using *oriC* as their only replication origin (204, 333). Specialized transducing phages λ*asn* were also shown to replicate autonomously as phage minichromosomes by virtue of *oriC* (212, 316, 317). Using a similar approach, the origins of members of the family *Enterobacteriaceae* other than *E. coli* were cloned and shown to function in *E. coli* (347, 349).

Minichromosomes replicate bidirectionally (199), show a similar dependence of their replication on protein and RNA synthesis, and respond to all *dna* mutants in a way similar to the way the chromosome responds (208, 317).

The nucleotide sequence of the *oriC* region was determined (42, 198, 209, 294). The construction of minichromosomes was the prerequisite for a multitude of molecular and biochemical analyses of the initiation of replication which will be discussed in the following chapters; it opened the gate to "modern" experiments.

ELEMENTS OF THE INITIATION OF REPLICATION

The interaction of DnaA protein and *oriC* results in a local unwinding of the DNA in an AT-rich region of *oriC,* the basic biochemical reaction in the initiation of replication. However, other proteins assist in this interaction, and *oriC* contains structural information for subsequent reactions, e.g., positioning of the DnaA primosome or *oriC* sequestration at the membrane. Likewise, DnaA protein has several additional functions, and understanding its regulation is central to understanding the control of initiation. Therefore, we first discuss *oriC* and DnaA protein separately and then discuss their interactions in the initiation complex.

Replication Origin, *oriC*

Chromosomal Region Containing *oriC*. The genes directly surrounding the replication origin *oriC* and their orientations are shown in Fig. 1 (reviewed in reference 313). *asnA* codes for one of the two asparagine synthases of *E. coli*. The product of *asnC* is a transcriptional activator of *asnA* and has a positive effect on posttranscriptional expression of *gidA* (153). The *mioC* gene codes for a 16-kDa protein of unknown function. The name *mioC,* for modulation of initiation at *oriC* (314), derives from the observation that *mioC* transcription influences the copy numbers of minichromosomes. Mutations in the two genes left of *oriC, gidA* and *gidB,* result in glucose inhibition of division (312), and disruption of the *gidA* gene reduces the growth rate. The *unc* or *atp* operon codes for the subunits of the membrane-bound ATP synthase. Slow growth of cells carrying F′ plasmids or minichromosomes with this region was originally interpreted to indicate a replication control region but has since been attributed to the overproduction of ATP synthase subunits.

Sequence information for the region around *oriC* has been provided by several groups (for citations, see GenBank). A 227-kbp segment centered around *oriC* is currently the largest contiguous DNA sequence available for *E. coli* (43). This region shows a remarkable symmetry. Genes tend to be transcribed divergently from *oriC* (see references in references 43 and 336), and Chi octanucleotide (5′-GCTGGTGG-3′) recombinational hot spots are oriented symmetrically with respect to *oriC* (43).

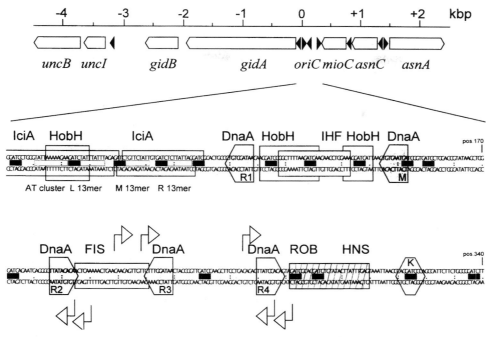

FIGURE 1 Structural organization of the *E. coli oriC* region. (Top) Genes in the vicinity of *oriC* are shown as boxes with their respective transcriptional direction; filled arrowheads mark promoters in the *oriC* region, as mentioned in the text. Genes and promoters are drawn approximately to scale (Bottom) *oriC* DNA sequence from positions 1 to 340 with structural elements and protein binding sites as mentioned in the text. Arrowheads indicate start points for DNA synthesis.

The gene for the initiator protein, *dnaA,* is close to *oriC* in most bacteria (336). However, in the case of *E. coli,* the *dnaA* gene is located 42 kbp counterclockwise of *oriC* on the genetic map (see Global Aspects of Origin Structure below).

Structure of *oriC*. The minimal DNA segment which promotes autonomous replication was found by deletion analysis of minichromosomes (204, 209) to be contained in a segment 245 bp long (232) (Fig. 1). Later, it was observed that a 12-bp AT-rich segment at the left end of *oriC* is also required for initiation (AT cluster in Fig. 1) but that it can be replaced by other AT-rich sequences (10). The minimal origin thus consists of 258 bp (coordinates 11 to 268 in Fig. 1).

Sequence analysis of the origins of replication in six other *Enterobacteriaceae* (347) and comparison to *oriC* of *E. coli* revealed that the origins contain segments that are highly conserved and are separated by segments with variable sequences but constant lengths. This finding led to the concept of binding sites separated by spacer regions in *oriC* (7). Prominent among the conserved regions are four nonpalindromic 9-bp repeats (R1 to R4 in Fig. 1), the consensus binding sites for the initiator protein DnaA, called DnaA boxes (80). DNase I footprint analysis revealed a fifth DnaA box (195), labeled M in Fig. 1. Each of the enterobacterial origins contains 9 to 14 GATC sites, 8 of them conserved in sequence and position among all these origins.

The left side of *oriC* is an AT-rich region consisting of three similar sequences, each 13 nucleotides long and each starting with GATC (35). As mentioned already, there is an additional essential AT-rich region, the AT cluster (10) (Fig. 1). A region of

helical instability, called a DNA unwinding element, was observed in the left part of *oriC* (156). Helical instability can be detected by partial unwinding in the region of the AT cluster and the left 13-mer in the absence (89) or at low concentrations (326) of Mg^{2+}. Apparently, it is the AT richness and the inherent helical instability that are important in this region. Consequently, this part can be replaced by DNA segments with similar properties but different sequences (10, 156). For the right 13-mer, however, the integrity of the 13-mer sequence is required (10, 35, 118); for the middle 13-mer, it is controversial whether the precise sequence or mere AT richness is important (10, 118).

Binding sites for the DNA-bending proteins IHF (integration host factor) (69, 244, 253b) and FIS (factor for inversion stimulation) (70, 88, 253b) are present in *oriC* (Fig. 1). Both proteins are likely to assist DnaA in the unwinding reaction.

A dimeric protein with subunits of 33 kDa called IciA binds specifically to the AT-rich 13-mer repeats (116, 119, 302). It shares amino acid homology with the LysR family of prokaryotic transcription regulators (302). Binding of IciA inhibits in vitro replication if the protein is added before DnaA protein is added. This effect is due to a block of the DnaA-dependent unwinding of the 13-mer region (118). Once this region is opened by DnaA protein, the IciA protein is without effect. IciA could thus be the long-sought-for negative regulator of initiation. However, cells with an insertion mutation in the *iciA* gene or that overproduce IciA do not show a striking phenotype, except that cells with excess IciA protein show a longer lag period when diluted into fresh medium (302). Therefore, the function of this protein is not known. This is also true for another protein binding specifically to the *oriC* region, Rob, for right origin binding. It recognizes a 26-bp site to the right of DnaA box R4 (Fig. 1) (285).

The DNA adjacent to the Rob binding site shows a bend which is considerably stronger in DNA fully methylated by the Dam methyltransferase than in hemimethylated or unmethylated DNA (Fig. 1) (138). Bending to the right of oriC is thus subject to Dam-mediated cyclic changes. Footprint experiments show that the abundant histone-like protein H-NS binds specifically to a site overlapping the Rob binding site in fully Dam methylated DNA, in agreement with its preference for curved DNA (Fig. 1) (M. Falconi, C. Weigel, W. Messer, and C. Gualerzi, unpublished data). Although this region is outside the minimal origin, H-NS binding affects oriC function.

Promoters in and around oriC. A complicated transcription pattern with an intricate regulation by DnaA protein exists in the origin region. The mioC promoter directs leftward transcription into oriC and is negatively regulated by DnaA owing to an overlapping DnaA box. Most mioC transcripts enter and traverse oriC or are terminated at specific sites within oriC. The mioC promoter requires Dam methylation for maximal activity (reviewed in reference 202). Also, part of the asnC transcripts extend into oriC or beyond, and the passage through the intergenic region between asnC and mioC of these long transcripts is regulated by DnaA (87, 266).

The gidA promoter for leftward transcription (153) has a positive effect on oriC function (10, 225). Two promoters are located in the left half within oriC: Pori-L, transcribing leftward (8, 178, 220, 271), and Pori-R1, with rightward transcription (8, 126). An additional promoter for rightward transcription, Pori-R, is outside of oriC next to DnaA box R4 (8, 178, 271). Pori-L is repressed by DnaA protein (220), but the regulation by DnaA requires the integrity of oriC DnaA boxes R2 and R4 (8). Pori-R1, on the other hand, is activated by DnaA protein. This activation is observed only with Pori-R1 embedded in an intact oriC (8). PmioC, PgidA, and Pori-L are subject to stringent control by ppGpp (225, 253). The importance of a coordinate expression of the different promoters around oriC is stressed by the recent analysis of the timing of transcription in synchronized cultures. mioC transcription is inhibited prior to initiation of chromosome replication, whereas gidA transcription is inhibited after initiation (226, 301).

oriC Mutants. Random mutagenesis of oriC gave 18 single-base changes with reduced origin function, 14 of which were in conserved oriC regions (231). However, these mutants showed nearly normal function in constructs with a functional gidA promoter, which again shows the importance of leftward transcription away from oriC (10).

A detailed functional analysis of oriC with mutations in DnaA boxes was done using oligonucleotide-directed mutagenesis. DnaA binding to the different DnaA boxes is either partially or completely abolished. All these mutants have a functional origin (108). Combinations of mutations in two or three DnaA boxes show an impaired origin function in some combinations, whereas mutation of all four DnaA boxes inactivates oriC (U. Langer and W. Messer, unpublished data).

Mutations that modify the distance between DnaA boxes have a much greater effect. Insertions or deletions of 10 bp between DnaA boxes R3 and R4 or boxes R2 and R3 give a functional origin. All changes of distances in these intervals by less or more than one helical turn inactivate oriC, suggesting that the location of the DnaA box with respect to the helix axis is important (207, 326). All 10-bp insertions in the left half of oriC between DnaA boxes R1 and R2 result in nonfunctional origins (207), demonstrating that precise distances are important in this part of oriC. Helical phasing of DnaA boxes is thus an important but not a sufficient condition for oriC function. Inactivation of oriC is also observed for a series of insertions of −6 to +16 bp, in steps of 1 bp, between the right 13-mer and DnaA box R1, because DnaA-mediated unwinding is abolished (110). Apparently, these insertions affect a site which is at the hinge for this process. On the other hand, mutants with insertions or deletions in the R3-R4 region, with functional or nonfunctional origins, show no difference from wild-type oriC organisms with respect to DnaA-mediated helical distortions in the 13-mer region (326).

In summary, base changes in individual DnaA boxes have surprisingly little effect. Any change of distances, however, inactivates oriC, except for modifications by one helical turn in the right half of oriC.

DnaA Protein

Genetics of dnaA.

The dnaA operon. The gene encoding the initiator protein DnaA has been cloned; its map location has been defined at 83.5 min on the E. coli genetic map (see chapter 109), 42 kbp counterclockwise of oriC; and the nucleotide sequence has been determined (93, 97, 230). DnaA is a basic polypeptide with a molecular weight of 52,500. dnaA is the first gene of an operon (Fig. 2) that also contains dnaN, which encodes the β subunit of DNA polymerase III holoenzyme, and recF, a gene involved in recombination (reviewed in reference 281). The next gene, gyrB, codes for the β subunit of DNA gyrase. gyrB is probably not part of the dnaA transcription unit. The arrangement and sequences of these genes as well as those of surrounding genes are highly conserved among eubacteria (223, 335, 336).

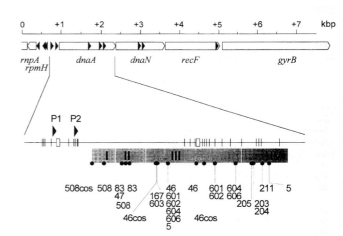

FIGURE 2 Structural organization of the E. coli dnaA region. (Top) Genes in the region flanking dnaA are indicated as boxes with their respective transcriptional directions; promoters are indicated by filled arrowheads. Genes and promoters are drawn approximately to scale. (Bottom) Positions of dnaA mutations (numbers indicate alleles; see Tables 1, 2, and 3). |, Dam methylation sites; □, DnaA box; cos, cold sensitive.

Transcription of this operon occurs from two promoters, *dnaA*1p and *dnaA*2p, located 239 and 157 bp, respectively, upstream of the *dnaA* coding sequence (97). The transcripts extend into *recF* (238, 251, 261). In addition, both *dnaN* and *recF* have their own promoters located within the preceding reading frame (5, 6, 251) (Fig. 2). A DnaA box located between the two promoters *dnaA*1p and *dnaA*2p (Fig. 2) is responsible for repression of the operon by DnaA protein (13, 14, 37, 157, 246, 321). Regulation of the *dnaA* operon is discussed in more detail below.

dnaA mutations. Temperature-sensitive mutations in the *dnaA* gene were among the first ones isolated that affect DNA replication (reviewed in reference 281). Fine mapping of *dnaA* mutations showed that mutations with similar phenotypes are clustered (92) (Tables 1, 2, and 3). Sequence analysis (98) of available temperature-sensitive *E. coli dnaA* mutations revealed that six of them carry the same mutation in the putative ATP binding site (see Fig. 3), i.e., an alanine-to-valine exchange at amino acid position 184. All six carry an additional mutation, which is different in the different alleles (Tables 1, 2, and 3; Fig. 2). Seven other temperature-sensitive alleles contain single mutations; *dnaA*83 and *dnaA*508 carry two mutations, at different positions. Four pairs of mutants carry identical mutations: *dnaA167* and *dnaA603*, *dnaA601* and *dnaA602*, *dnaA604* and *dnaA606*, and *dnaA203* and *dnaA204*.

Intragenic suppressors of the heat-sensitive *dnaA*46 and *dnaA*508 alleles were isolated. Both are cold sensitive owing to excessive initiation events at 30°C (62, 73, 133, 135). The *dnaA*46(Cs) allele was shown to contain two additional mutations (38). In *dnaA*508(Cs), *dnaA* expression is enhanced due to a mutation of the GUG start codon to an AUG (62) (Table 2). Several nonsense mutations in *dnaA* were isolated, but only one has been sequenced (136, 137, 269, 324) (Table 2).

Extragenic dnaA suppressors. Several extragenic suppressors of temperature-sensitive *dnaA* mutations have been isolated (11, 12, 19, 270, 323). Many of these suppressor mutations map in the *rpoB* gene. The observation of an allele specificity between *dnaA* and the *dnaA*-suppressing mutations (Table 1) suggests a physical interaction between the DnaA protein and RNA polymerase (11, 12, 270).

Deletion of the *topA* gene suppresses the *dnaA*46 mutation (180). Replication is still DnaA dependent, suggesting either that the *dnaA*46 product is overexpressed under these conditions or that initiation at *oriC* requires less DnaA protein in the absence of topoisomerase I.

In attempts to clone *dnaA*, several plasmids were obtained that contained genes other than *dnaA* and that were able to suppress the temperature-sensitive defect, apparently because of the high copy numbers of the clones (249, 297). One type of suppressing genes includes *groES* and *groEL*, which apparently also exhibit allele specificity in the suppression (66, 125). Suppression results in excessive chromosomal initiations, although the total amount of DnaA is not increased (134). The existence and the properties of diverse suppressors suggest a complicated pattern of interactions of DnaA with other proteins.

Among genes that suppress chromosomal temperature-sensitive *dnaA* alleles when the genes are present at high copy number are mutant *dnaA* genes themselves (314). Apparently, most temperature-sensitive *dnaA* mutants have some residual activity at 42°C, which suffices for initiation at *oriC* if the amount of functional protein is increased. This may happen by an increased expression, as in the case of *dnaA*508(Cs), increasing the number of copies of the gene or by the action of chaperones like GroE.

Mutations in a new gene, *seqA*, were selected on the basis of the interaction of SeqA with hemimethylated *oriC* DNA. *seqA* mutations suppress the defects of some temperature-sensitive *dnaA* mutations and of other mutations with compromised replication initiation like *fis* and *him*. Overexpression of SeqA suppresses the cold sensitivity of *dnaA*46(Cs) and is lethal for *fis*, *him*, or *dam* mutants. Apparently, SeqA protein affects initiation negatively, and it has been suggested that SeqA inhibits the expression or activity of DnaA (181, 311a).

Integrative suppression. The F plasmid (whole F or mini-F) integrated into the chromosome can drive chromosomal replication if replication from *oriC* is inhibited, e.g., because of a temperature-sensitive *dnaA* mutation. In such "integrative suppression," chromosome replication starts from the site of plasmid integration. Integrative suppression of *dnaA*(Ts) mutants can

TABLE 1 Characteristics of Ts alleles of *dnaA* mutants

dnaA allele(s)	Position(s) of amino acid change(s)[a]	Mutation in ATP binding site	Reversibility[b]	Cold sensitivity in merodiploid[b]	Asynchrony index[c]	Suppression by[d]:		
						rpoB902	*rpoB902*	*rpoB904*
dnaA47[e]	72		NT	NT	NT	NT	NT	NT
dnaA83[e]	72, 74		NT	NT	NT	NT	NT	NT
dnaA508	28, 80				0.3			Yes
dnaA167 and *dnaA603*[e]	157		Yes		0.4	Yes		
dnaA46	184, 252	Yes	Yes	Yes	1.3	Yes		
dnaA601 and *dnaA602*	184, 296	Yes	Yes	Yes	1.3	Yes		
dnaA604 and *dnaA606*	184, 347	Yes	Yes	Yes	1.3	Yes		
dnaA5	184, 426	Yes	Yes	Yes	1.2	Yes		
dnaA205	383				0.7	NT	NT	NT
dnaA203 and *dnaA204*	389				0.2		Yes	
dnaA211	411			NT	NT	NT	NT	NT

[a]See reference 98.
[b]See reference 92. NT, not tested.
[c]Synchrony index $A = (f_3 + f_5)/f_4$, where f_3, f_4, and f_5 are the frequencies of cells with three, four, and five chromosomes, respectively, after completion of ongoing rounds of replication. For wild-type cultures, A is ~0.1 (286).
[d]See reference 12.
[e]F. Hansen, personal communication.

TABLE 2 Intragenic suppressors of Ts alleles

dnaA allele	Amino acid positions of:	
	Ts mutation	Suppressor
dnaA46(Cs)[a]	184, 252	156, 271
dnaA508(Cs)[b]	28, 80	1 GUG to AUG

[a]See reference 38.
[b]See reference 62.

also be obtained by the integration of other plasmids, such as R1, R100, or ColV2, or of phages P1 and P2 (reviewed in reference 313). Integrative suppression is surprising, because plasmids like F themselves require DnaA for replication (94, 140). Apparently, initiation at the F origin requires less active DnaA protein. Consequently, dnaA(Null) mutants cannot be integratively suppressed by F (146), whereas they can be suppressed by the integration of R1 (25). Plasmid and phage replication will be discussed in separate chapters (see chapters 122 and 123).

A different initiation mechanism is DnaA independent, and initiations occur at origins other than oriC in induced and constitutive stable DNA replication (SDR). dnaA mutations can thus be suppressed with respect to their function at oriC by mutations in the gene for RNase H (rnhA) (9, 143, 148, 172, 304). SDR is discussed in a later chapter.

dnaA and other replication mutants. A new temperature-sensitive dna mutation called dnaR (257) is a mutation in prs, the gene for phosphoribosylpyrophosphate synthetase (258, 259). Thermosensitivity of initiation is suppressed by a mutation in the dnaA (dnaA110) or the rpoB gene. The function of DnaR in initiation is different from its synthase activity. The genetic analysis suggests that DnaR and DnaA proteins interact in vivo to initiate DNA replication under special conditions, such as elevated temperatures.

DnaK is one of the major heat shock proteins (86) and is closely related to the Hsp70 family of eukaryotic heat shock proteins (see chapter 88). The dnaK gene is in an operon with another heat shock protein, dnaJ (334), and both proteins play a central role, together with GrpE, in initiation of replication at the origin of phage λ (reviewed in reference 23). Cells with insertions or deletions in dnaK are able to grow at 30°C but are inviable at 42°C (237). Certain dnaK or dnaJ mutations are

TABLE 3 Other dnaA alleles

dnaA allele	Type
dnaA215[a]	Opal, Trp-117 to UGA
dnaA91, dnaA311, dnaA366[b]	Amber, not sequenced
dnaA17, dnaA452[c]	Amber, not sequenced
dnaA801, dnaA803, dnaA806[d]	Amber, not sequenced
dnaA101, dnaA102[a]	Neutral mutations in DnaA box, amino acids 250–252
dnaA110[e]	Suppressor of dnaR

[a]See reference 324.
[b]See reference 269.
[c]See reference 136.
[d]See reference 137.
[e]See reference 259.

specifically blocked in initiation at oriC (229, 256), and in vitro replication initiation in cell extracts from dnaK and dnaJ mutants is temperature sensitive (187). Although purified temperature-sensitive DnaA proteins DnaA46 and DnaA5 can be activated by the combined action of GrpE and DnaK (113, 114), in vivo suppression by these gene products has so far not been described. Wild-type DnaA and DnaK proteins associate in vitro (187), and DnaA-phospholipid complexes are activated by DnaK (115). Because DnaK is not required in the oriC-dependent in vitro replication system, we may speculate that the activity of DnaK in initiation is indirect. It could, for instance, be required to maintain DnaA in the active conformation at elevated temperatures.

The dnaX gene encodes both the τ and γ subunits of DNA polymerase III holoenzyme. τ is the full-length product of dnaX, whereas γ is a shorter product produced by a translational frameshift that results in premature termination (29, 71, 305). A temperature-sensitive mutation in dnaX that affects the reading frame common to τ and γ, formerly called dnaZ, can be suppressed by certain dnaA mutations (28, 320). The mechanism of this suppression is not clear at present.

Conservation of the amino acid sequence of DnaA. DnaA genes are ubiquitous among bacteria, and their amino acid sequences are highly conserved (Fig. 3). dnaA genes of Enterobacteriaceae are nearly identical, and even dnaA genes of evolutionarily very distant bacteria like Bacillus, Mycoplasma, or Cyanobacterium spp. are clearly homologous. On the basis of sequence similarities, DnaA proteins have been divided into domains with varying homologies (78, 288, 336). A short N-terminal domain with reasonable homology (domain I) is followed by a stretch that is completely dissimilar in sequence and length among the different DnaA proteins. A highly homologous domain III follows, and this domain is separated from a similarly conserved domain IV by a short stretch with lower homology (Fig. 3). Presumably, this arrangement reflects functional domains of the DnaA protein, but so far there are few data to support this possibility.

A well-conserved region at the beginning of domain III, marked "ATP binding" in Fig. 3, is a motif found in many nucleotide-binding proteins (74; see reference 281 for a review). Starting from the ATP binding motif $G-X-X-G-X-G-K-T-X_5-V$ (where X is any amino acid), additional motifs were defined that establish a relationship between DnaA, proteins of the helicase superfamily III, and transcription activators of the NtrC family (154). An especially high similarity between DnaA and NtrC enhancer binding proteins is observed in the end of domain III, with the motif NVRELEGAL (253a).

Fusion proteins consisting of DnaA protein domain IV plus either β-galactosidase or a biotin-target peptide bind oriC DNA with high specificity (Roth and Messer, unpublished data). This demonstrates that the 94 C-terminal amino acids are sufficient for specific binding to DNA. For B. subtilis DnaA protein, DNA binding was also observed in the C-terminal part (336).

Regulation of dnaA expression. Some dnaA(Ts) mutants accumulate a higher level of initiation potential than dnaA+ strains under comparable conditions (64, 100, 205, 303), which suggests that DnaA protein controls its own synthesis (100). Evidence for autoregulation of plasmid-borne dnaA genes at the transcriptional level was obtained by using transcriptional and translational fusions to monitor genes (13, 14, 37). Autoregulation of the chromosomal dnaA gene was demonstrated by using comparative S1 mapping (157). Increased levels of DnaA protein made from a DnaA-overproducing plasmid lead to transcriptional repression. A temperature shift to 42°C in *dnaA*(Ts) mutants results in derepression of transcription. Similarly, the addition of extra DnaA boxes on a plasmid increases *dnaA* expression, presumably by titrating DnaA protein away from the *dnaA* promoter (99, 157). Repression of transcription can also be shown in vitro, demonstrating that it is the binding of DnaA protein to the promoter region that is responsible for regulation (321). Evidence for autoregulation of the *B. subtilis dnaA* gene has also been presented (335), whereas the *dnaA* genes of *Pseudomonas putida* and *Synechocystis* sp. are apparently not autoregulated (120, 252a).

Autoregulation of the *E. coli dnaA* gene is exerted via a DnaA box located between the two *dnaA* promoters (93) (Fig. 2), which, however, react differentially to regulatory stimuli. The promoter *dnaA*1p is responsible for 20 to 30% of the transcripts, and the stronger downstream promoter *dnaA*2p is responsible for 70 to 80% (52, 157). At the chromosomal location, both promoters are repressed by excess DnaA protein and derepressed upon thermal inactivation of DnaA(Ts) (157). When the *dnaA* promoter region is located on plasmids, a mutation in the DnaA box between the two promoters results in an increase in *dnaA*1p and a decrease in *dnaA*2p transcription, suggesting that under these conditions, *dnaA*1p is repressed and *dnaA*2p is activated by interaction of DnaA with the DnaA box (246). In all cases, excess DnaA from a DnaA-overproducing plasmid results in a repression of both promoters (14, 37, 157, 246).

Different behaviors by the two promoters are also observed with respect to their dependence on Dam methylation. Transcription from *dnaA*2p is nearly completely blocked in the absence of Dam methylation, whereas transcription from *dnaA*1p is unaffected (39, 157). In addition to this direct effect acting differentially at the level of promoter recognition, the *dnaA* region becomes sequestered owing to hemimethylation after replication in wild-type cells (46) (see below, Dam Methylation and Replication Initiation). During this time, lasting about one-third of a generation (228), the *dnaA* promoters are not accessible to RNA polymerase. Consequently, *dnaA* transcription is inhibited for some time after the initiation of replication (46, 226, 301).

The intracellular signal nucleotide ppGpp, the mediator of the stringent response, seems to be a regulator of *dnaA* transcription, primarily for the *dnaA*2p promoter (52). The concentration of ppGpp is inversely related to growth rate (255). Consequently, *dnaA* transcription is growth rate regulated (52, 246). The *dnaA*2p promoter is completely inactive during the stationary phase of growth (246). Whether and how this growth rate regulation of transcription is reflected in a growth rate regulation of the intracellular DnaA protein concentration are controversial. Chiaramello and Zyskind (51) reported that the DnaA protein concentration is proportional to growth rate. However, in a subsequent careful analysis of the DnaA protein concentration in five different *E. coli* strains and in *S. typhimurium*, Hansen et al. (95) found the DnaA protein concentration to be constant, i.e., independent of growth rate, in all strains tested.

A mutation in the *rpoC* gene, isolated as a chromosomal copy number mutant with an increased DNA-to-mass ratio (252),

causes a more than fivefold increase in *dnaA* expression (239). Increased *dnaA* expression may also be the reason for an increased DNA/mass ratio in other *rpoB* and *rpoC* mutants (299). Such mutants show a reduced rate of RNA synthesis, and consequently, the phenotype of increased DNA/mass ratio could be mimicked by a heat shock in the presence of low concentrations of rifampin (91). This treatment also increases expression of *dnaA* (M. Wende and W. Messer, unpublished data).

The expression of *dnaA* is induced by DNA lesions that block replication (250). The effect is indirectly related to the SOS response, because the *recA*1 and *lexA*1 mutations prevent the induction of *dnaA*. However, there is no SOS (LexA) box in the promoter region of *dnaA*. Induction of *dnaA* by DNA damage may be the explanation for the observed stimulation of reinitiation at *oriC* after UV irradiation (26, 309).

The observations discussed here demonstrate that *dnaA* expression is regulated by a complicated control network. Autorepression seems to be the primary control, but in addition, *dnaA* expression is embedded in several global regulatory systems of the cell.

Biochemical Properties of the DnaA Protein. DnaA protein was initially expressed from a chromosomal DNA fragment cloned in λ, purified, and shown to bind specifically to *oriC* DNA (49). With the availability of a test for active DnaA with the *oriC*-dependent in vitro replication system (81) and the construction of suitable vectors overexpressing DnaA, the purification was successively improved (58, 80, 82, 157, 274).

Purified DnaA protein migrates a little faster in most gel systems than expected from the M_r of 52,500 which is inferred from the sequence. It has the N-terminal amino acid sequence SLSLWQQC (W. R. Jueterbock and W. Messer, unpublished data), which defines the GUG codon at nucleotide position 886 (93) as the correct start codon. DnaA protein is active as a monomer but has a strong tendency to aggregate (82).

DnaA boxes. DnaA protein binds tightly to four nonpalindromic 9-bp sequences in *oriC*, the so-called DnaA boxes, with the common sequence 5'-TTAT(C/A)CA(C/A)A, as shown by footprinting experiments (80, 108, 117, 326). A fifth box with the related sequence 5'-TCATTCACA binds DnaA in vitro with somewhat less efficiency (195). Fragments carrying the DnaA box sequence or a close match to the consensus sequence are specifically retained on nitrocellulose filters by DnaA protein (80, 82). DNase I footprinting or fragment retention was used to demonstrate DnaA binding to many such boxes in various genes: in the *mioC* and *dnaA* promoters, in the promoters of the *uvrB* and *rpoH* genes, and in M13 and the left inverted repeat of Tn*5* (reviewed in reference 281). They are also found in the origins of plasmids F, R1, pSC101, ColE1, R6K, RK2 , and phage P1 (for a review, see reference 36).

The ability of the DnaA-DnaA box complex to block transcribing RNA polymerase in vivo was used to quantify DnaA binding to DnaA boxes (268). This leads to a more relaxed definition of the DnaA box with the following consensus sequence:

$$5' - \begin{matrix} & & & A & & & & A & \\ T & T & & & A & & A & & A \\ C & C & T & T & C & C & G & C & C \\ & & C & & & & & T & \end{matrix}$$

Dissociation constants determined in vitro for DnaA boxes R1, R2, and R4 from *oriC* by using oligonucleotides and purified DnaA protein vary between K_ds of 1.0 and 50 nM. Affinity

ATP-binding region and related domains — multiple sequence alignment

Block 1 (positions to ~81/90)

```
E. coli          ...........................mSLsLWqQcLArLqdELpateFsmWIRpLQa.ELsdnTLaLyAPNrFVLDWVrdKYL   nnIngLLtsFcgadapqLrFeVgtk......   81
S. typhimurium   ...........................mSLsLWqQcLArLqdELpateFsmWIRpLQa.ELsdnTLaLyAPNrFVLDWVrdKYL   nnIngLLntFcgadapqLrFeVgtk......   81
S. marcescens    ...........................mSLsLWqQcLArLqdELpateFsmWIRpLQa.ELsdnTLaLyAPNrFVLDWVrdKYL   nnIngLLndFcgtdaplLrFeVgsk......   81
P. mirabilis     ...........................mSLsLWqhcLArLqdELpateFsmWIRpLQa.ELsdnTLaLyAPNrFVLDWVreKYI   nnInaLLvdFcgsdvpsLrFeVgnk......   81
B. aphidicola    ...........................mSLcLWqQcLdrLqsELpsteFsmWIRsLka.kLnnniLeIyAPNqFILDWkdKYL   ihFkkILqdFcgnspfIkFkVtyt......   81
P. putida        ...........................mSVelWqQcVelLrdELpaqqFntWIRpLQv.EaegdeLrVyAPNrFVLDWVneKYL   grLleLLgengsgiapaLsLlIgsrrssapr  87
B. subtilis      ..................meniLdLWnQaLAqIekkLskpsFetWmkstkahsLqgdTLtItAPNeFarDWLes..   lhL..IadtIyeltgeeLsIkFvi.......   80
S. coelicolor    .......madvpadlaavwpr.VLeQlLgegrgqgveskdehWIRrcQplaLvadTaLLavPNeFakgvLeg.rL   apI..VsetLsrecgrpIrIaItvddtagep   95
M. luteus        ............mvadqavLsswrsVVgsleddar...vsarLmgFVylaQpqgLignTLlLavPNettrEtLqgtqV   ada..LtdaLtqefreeIlLaIsidanLqpp   92
C. crescentus    mtmkggvasqdfs......aaiatacepaanVWskvcvaLkrELgdaaFgsWIapamlrEaatgdVvLvtstgiarDWIrrsaw   rrIgeLwa...ahdatgrrIdLksrlefeaa 106
R. meliloti      mrmnLatapggfrpernqsqaageKhdmrhdaLFervsArLkaqVgpdvFasWfgrLklhsVsksvVrLsvPttFLksWInnrYL  dlIttLVq...qedseiLkVeIlvr...tat 110
Synechocystis sp ............mvscenLWqQaLAiLatqLtkpaFtWIkasvLlsLgdgvatIqveNgFVLnhLq.Ksy   gpL..LMevLtdltgqeItVkLitdgl....   83
B. burgdorferi   ...........mekSknIWsliLteIkkELseeeFyvWfenLcflEsigdnIkIstPNlFhknqIekrFt   kkIkeILikngynniviVftnqppkthsnkq   90
S. citri         ...................................................................   ................................   -
M. capricolum    ...................................................................   ................................   -
```

                                                domain 1 ←→ domain 2

Block 2 (positions to ~107)

```
E. coli          .....pvtqtpqaavtsn.........vaapaqvaqtpq.............................   107
S. typhimurium   .....pvtqt1ktpvh.n.........vvapaqttttqpq..........................   107
S. marcescens    .....pitqvisqtvtas.........vssap...aapaa.........................   104
P. mirabilis     .....pvsarttesvpkt.........vthpavnstptn..........................   106
B. aphidicola    .....skekkfkknilqk.......iqn.......................................    97
P. putida        aapnapvsaavaasLaqtqahktapaaavepvavaaaepv............................   127
B. subtilis      .........pqnqdvedfmpkpqvkkavkedtsdfpq...............................   108
S. coelicolor    agpapqapqspspsrpqhryeeepeLpapgqgqgggreeyrdrdeyegygrnradqLptarpaypqeyqrpepgswprpaqqdddygwqqqrlgfperdpyaspnqepyqgqeppppyshenr 211
M. luteus        rtpssearrssLaggpsgaaapdveLppaataatsr.........ravaeeLpgfri   138
C. crescentus    agayveatpkavaaaepieivLp.........vstdaptv   138
R. meliloti      rghrptapeesvaaaaeaavvpppsrrsaaptvaiaaaav   149
Synechocystis sp .........ephsLiggess1pmettpkn..............   103
B. burgdorferi   etknpaLnetfskfdklke...................   109
S. citri         ....................................................................mnt     3
M. capricolum    ....................................................................mnl     3
```

Block 3 — ATP binding site

                                                domain 2 ←→ domain 3

```
E. coli          .............................raapstrsgWdnvpap..aept................yr   SnNvKhTFDNFV   142
S. typhimurium   .............................rvapaarsgWdnvpap..aept................yr   SnNvKhTFDNFV   142
S. marcescens    .............................rtaapsrpsWdnaaaq..pels................yr   SnVNpKhTFDNFV  139
P. mirabilis     .............................sqpvrpsWdnqpqsqLpeln................yr   SnVNpKhkFDNFV  141
B. aphidicola    .............................lnakpiWdkip..ifkkss................hr   SnINkkhsFENFI  129
P. putida        .............................lvetssrdsFdamaepaaapps.gggraeqrtvqvegaLkht   SyLNrtFTFDtFV  181
B. subtilis      .....................................................................   nmLNpKYTFDtFV  121
S. coelicolor    tsyqqdyrpqpperpsydaqrgdyeqargeyeqprgdydkprgdydqqrgdydqrgprrdLpepppgsghvhrggpvgpgpatgapgpLaaqpapatgpgept arLNpKY1FDtFV 327
M. luteus        .............................eppadvvpaanaapngngkptpappstsae.........t   SrLNdrYhFEtFV  182
C. crescentus    .............................vapsaksptq   .gLqerFTFEtFV  159
R. meliloti      .............................aaaparpvqapl.................fg   SpLdqrYgFDsFV  176
Synechocystis sp .............................a   taLNgKYTFsrFV  117
B. burgdorferi   .............................kttskeaiqniqdrikmyikkeeeeptnfkn   pfLkkrYTFENFV  153
S. citri         kelwievkeilsrdesvspeinyyyisdtnlytvsdnnclittkse...iaigvfeagLnekikniLkkLtgiqynisfeLekninkqasviskidtltennn  layyenYTFENFV  116
M. capricolum    ndilkeLklsLiankndesvyndyiktiniykkdlsnyvvvvksqfgLlaikqfrptieneikkiLkqpvnisftyeqeyqkqLektesinkdhsdiiskkn  kkVNen.TFENFV  118
```

Block 4

```
E. coli          EGkSNqLArAAARqVADNPGgA..YNPLFLY GGtGLGKT HLLHA V GNgIMar.kPNAKVVYMhSERFVqDMVkALQ..nNaIEEFKrYYRSV DaLLIDDIQFF AnKErsQEEFFHT 253
S. typhimurium   EGkSNqLArAAARqVADNPGgA..YNPLFLY GGtGLGKT HLLHA V GNgIMar.kPNAKVVYMhSERFVqDMVkALQ..nNaIEEFKrYYRSV DaLLIDDIQFF AnKErsQEEFFHT 253
S. marcescens    EGkSNqLArAAARqVADNPGgA..YNPLFLY GGtGLGKT HLLHA V GNgIMar.kaNAKVVYMhSERFVqDMVkALQ..nNaIEEFKrYYRSV DaLLIDDIQFF AnKErsQEEFFHT 250
P. mirabilis     EGkSNqLArAAARqVADNPGgA..YNPLFLY GGtGLGKT HLLHA I GNsIMer.kaNAKVVYMhSERFVqDMVkALQ..nNaIEDFKrYYRSV DaLLIDDIQFF AnKErsQEEFFHT 252
B. aphidicola    EGkSNqLArAAAsqVAkNPGns..YNPLFLY GGtGLGKT HLLHA I GNgILay.kyNvKIIYMhSERFVqDMVkALQ..nNaIEkFKlYYRSV DaLLIDDIQFF AhKErsQEEFFHT 240
P. putida        EGkSNqLArAAAWqVADNPkhg..YNPLFIY GVgGLGKT HLMHA I GNhLLkk.nPNAKVVYLhSERFVqDMVkALQ..lNaInEFKrFYRVS DaLLIDDIQFF AnKErsQEEFFHT 292
B. subtilis      iGsgNrFAhAAsLaVAEaPakA..YNPLFIY GVgGLGKT HLMHA I GHyVIdh.nPsAKVVYLsSEkFtnEFInsIr..DnkavDFrnrYRnV DVLLIDDIQFL AgKEqTQEEFFHT 232
S. coelicolor    iGASNrFAhAAAvaVAEaPakA..YNPLFIY GesGLGKT HLLHA I GHyarsl.yPgtVrVYVsSEeFtnEFInsIr..dgkgDsFrkrYReM DILLVDDIQFL AdKEsTQEEFFHT 438
M. luteus        iGSSNrFAhAAAnaVAEaPakA..YNPLFIY GesGLGKT HLLHA I GHyarrl.yPglrVrVVnSEeFtnDFInsIr..hdegasFKqvYRnV DILLIDDIQFL AdKEaTvEEFFHT 293
C. crescentus    pGpaNeFAhAvARrIAnwadg..hFNPVLFh GpyGFGKT HLLnA L aweaMrn.aPekrVVYLtaERFLstFVrAVm..drqtaaFKeelRaa DLLIIDDVhFI AgKqsTQEELFHT 270
R. meliloti      EGsSNrVAIAAARtIAEagagAvrFNPLFIs ssvGLGKT HLMqA I alaaLqs.araprVVYLtaEyFMwrFatAIr..dNdalsLKesLRnI DLLIIDDmQFL qgK.siQhEFcHl 288
Synechocystis sp wahqsh.ghAAsLaVAEsPGre..FNPLFLC GVgGLGKT HLMqA I ahyrLem.yPNAKVyVVstERFtnDLItAIr..qdnmEDFrsYYRSa DFLLIDDIQFI kgKEyTQEEFFHT 227
B. burgdorferi   iGpnNkLAynAs1sIskNPGk..YkNpcLIY GVgGLGKT HLLqs I GNkteel.hNlKILYVtaEnFLnEFFksIk..thetkkFKkkYKyL qgK.KEgiQEELFHT 243
S. citri         rGdSNheAmqAAlaVA1dlGkk..WNPLFIY GdsGLGKT HLLnA L eNkVneiyktNnrVkYLkaDeFgkiamdiLnqgheiIEaFKtsYdiy DcLLIDDIQLL AkrnkTnElFFHi 230
M. capricolum    iGsSNeqAfiAvqtVskNPGis..YNPLFIY GesGMGKT HLLkA a kNyIesnf.sdlKVsYMsgDeFarkaVdiLQkthkeIEqFKnevcqn DVLIIDDVQFL syKEkTnEiFFti 232
```

Block 5

                                                domain 3 ←→ domain 4

```
E. coli          FNALLEgNq QIILTSDR YPKEInGVEDRLKSRFgwGLtVAIEPPELETRVAILmKK.......ADEndIrLPgEVaFFIAkRLrS NVRELEGALNRVI AnAnFt...graITIDF 358
S. typhimurium   FNALLEgNq QIILTSDR YPKEInGVEDRLKSRFgwGLtVAIEPPELETRVAILmKK.......ADEndIrLPgEVaFFIAkRLrS NVRELEGALNRVI AnAnFt...graITIDF 358
S. marcescens    FNALLEgNq QIILTSDR YPKEInGVEDRLKSRFgwGLtVAIEPPELETRVAILmKK.......ADEndIrLPgEVaFFIAkRLrS NVRELEGALNRVI AnAnFt...graITIDF 355
P. mirabilis     FNALLEgNq QIILTSDR YPKEInGVEDRLKSRFeqGLtVAIEPPELETRVAILmKK.......ADEnqIqLPgEVaFFIAkRLrS NVRELEGALNRVI AnAnFt...graITIDF 357
B. aphidicola    FNALLEgNq QIILTSDR YPKEInGVEDRLKSRFgwGLtVAIDPPELETRVAILiKK.......ADEnnIvLsdEIaFFIAkRLqS NVRELEGALNRVI vnAnFt...hrsITVEF 345
P. putida        FNALLEggq QVILTSDR YPKEIeGLEERLKSRFgwGLtVAVEPPELETRVAILmKK.......ADqakVeLPhDaaFFIAqRIrS NVRELEGALkVI AhshFm...grdITIEL 397
B. subtilis      FNtLhEesK QIVIsSDR pPKEIptLEDRLrSRFeWGLitdItPPDLETRIAILrKK.......AkaegLdIPnEVMLYIAnqIdS NIRELEGALIRVV AYssLi...nkdInaDL 337
S. coelicolor    FNtLhnaNK QIVLssDR pPKqLvtLEDRLrnRFewGLitdVqPPELETRIAILrKK.......AvqeqLnaPgEVLeFIAsRIsr NIRELEGALIRVt AFAsLn...rpqVdLgL 548
M. luteus        FNtLynnNK QVVITSDl pPKqLsGFEDRLrsRFeWGLitdIpPPDLETRIAILrKK.......AEaegLvaPpEaLeYIAsRIst NIRELEGALNRVt AFAsLn...rqtVdIEL 398
C. crescentus    LtALVgeg g rVVFsaDR pPsaMteMDahLrtRLgcLEPaDrnlRLgILerKiqtLgaalgfEps..IrpEVMqFLAdRFtd sVRELEGAFNqLL fRrsFe....lsrMTLDe 381
R. meliloti      LNmLLDsaK QVVVaaDR aPwELesLDsRVrSRLqgGVaIeMEgPDYEmRLemLkrr...leaarqDDasLeIPlEILshVArnVta sgRELEGAFNqLL pqLsIEr 399
Synechocystis sp FNsLhEagK QVVVasDR aPqrIpGLqDRLiSRFsmGLiadIqvPDLETRMAILqKK.......AEydrIrLPkEVMeYIAshytS NIRELEGALiRaI AYtsLs...nvaMTVEn 332
B. burgdorferi   FNALyEDNK QLVFTcDR sPsELtnFtDRLKSRFtRGLnVdIskPnFEIRaAIVeKK.......AEEdgInVPknILnLVAqkVtt NVRDLEaAVtkLk AYidLd...nieIdIEI 369
S. citri         FNsYIEkNK QVIVITSDk YPdDLgCFEaRIiSRFsYGLsIgLDsPDFETaLkILeqkl...khqnnlgIFseEsLeFIAlnFnS dVRkLEGAIkRLL flAvMnkkpnei ITLad 340
M. capricolum    FNnFIEndK QLfFsSDk sPelLnGFDnRLitRFnmGLsIAIqkLDnkTataIIkKei......knqnikteVtnEaInFIsnyysd dVRkIkGsVsRLn fWsqqnpe.ekvITIEI 341
```

Block 6 — DNA binding region

```
E. coli          VrEaLRDLL....alqEk..LVTIDNIQKtVAEYYKIKVaDLLSKRRSRSVARPRQMAMaLaKELTNHSLPEIGDaFGGRDHTTVLHACRKIEqLReEshDIKeDFsnLIRtLss....... 467
S. typhimurium   VrEaLRDLL....alqEk..LVTIDNIQKtVAEYYKIKIaDLLSKRRSRSVARPRQMAMaLaKELTNHSLPEIGDaFGGRDHTTVLHACRKIEqLReEshDIKeDFsnLIRtLss....... 467
S. marcescens    VrEaLRDLL....alqEk..LVTIDNIQKtVAEYYKIKVaDLLSKRRSRSVARPRQMAMaLaKELTNHSLPEIGDaFGGRDHTTVLHACRKIEqLReEshDIKeDFsnLIRtLss....... 464
P. mirabilis     VrEaLRDLL....alqEk..LVTIDNIQKtVAEYYKIKVaDLLSKRRSRSVARPRQMAMaLaKELTNHSLPEIGDaFGGRDHTTVLHACRKIEqLReEshDIKeDFsnLIRtLss....... 466
B. aphidicola    VrEaLRDIL....alqEk..LVTIaNIQKtVAEYYKIKIsDLLSRRRSSVARPRQMAMaMaKELTNHSLPEIGDaFsGRDHTTVIHACRKIEkEnhDIKeDFsnLIRtLsv....... 454
P. putida        IrEsLkDLL....alQdk..LVsVDNIQrtVAEYYKIKsDLLSKRRSRSVARPRQVAMaLsKELTNHSLPEIGDmFGGRDHTTVLHACRKInelLkesDaDIreDyknLLRtLtt....... 506
B. subtilis      aaELRDII....psskpkv..ITIkeIQrvVgqqFnIKLeDFkaKkRtkSVAfPRQIAMYLsrEMTdsSLPkIGEeFGGRDHTTVIHAheKIskLlaDDeqLqqhVkeIkeqLk....... 448
S. coelicolor    teivHLkDLI....pggEdsapeITstaImgatADYFgLtVeDLcgtsRgRaLvtaRQIAMYLcrELTdISLPkIGalFGGRDHTTVMHAdrKIrnLmaErrsIynqVteLtnrIkng....... 656
M. luteus        aehvHLkDLI....tDetaheITpEIlhatgEYFnLtLeELtSkRtRtLvtaRQIAMYLrELTemSLPkIGqvLGGRDHTTVMHAdrKIreLmaErrtIynqVteLtneIkrkqrga....... 513
C. crescentus    VqaiLRphL....rsgE.kr..ITIDdIQkAtEhYgMKqaDLLSeRRnRaVARPRQaAMMLaLLTrSLPDIGrrFGGRDHTTVLHAvrRrIEaLRaEDsaLshDLetLtRkLrg....... 490
R. meliloti      VdElLghLV....nagEprr..VrIEdIQrVAkhYnVsrqELVSnRRtRvIvkPRQIAMYLsKtLTprSFPEIGrrFGGRDHTTVIHAvRKIEeLisaDtkLshEIelLkRlIne....... 507
Synechocystis sp IapvLnppV....ekvaa.....apEtIitiVAqhYqLKVeELLSnsRrReVslaRQVgMYLmrqhTd1SLPrIGEaFGGKDHTTVMysCdkItqLqqkDwEtsqtLtsLshrIniagqapes 446
B. burgdorferi   VekiIkEIIiyekettnepnnkInIENIkKiLlreLKIthkDIeghskkpeItaRhQIAMYLcrFTe1SLPDIGrrFGGKDHTtVLysinkIDrdRnnDkEInn1IteLMnIkkn....... 486
S. citri         VekaFkna.....plqnnekITpkkIkqiVADsYnItIkaMMSKsRvsnVmqaRQLAMYFcrtLldepFtrIGtEMGDTHVTMnsvkkVEahistnkEFKhlVnaIrRkIegr....... 450
M. capricolum    IsDlFRDI.....ptsklgILnVkkIkevVsEkYgIsVnaIdgKaRaSkSIvtaRhIAMYLtKEIlNHtLaqIGEeFGGRDHTTVInAeRKIEmMlkkDkqLKktVdiLknkIltk....... 451
```

depends on the DnaA box sequence and the sequence context around the box (268a).

Nucleotide binding and interaction with phospholipids. DnaA protein binds ATP and ADP with high affinity, with K_ds of 30 and 100 nM, respectively (272). The two forms bind to DNA in a similar fashion, but only ATP-DnaA is active in *oriC* replication. However, a limiting level of ATP-DnaA can be augmented by the ADP form (54). ATP function must be allosteric, because the analog ATPγS can replace ATP (272). ATP and ADP forms of DnaA are monomers in solution, but the nucleotide-free form tends to form large aggregates (55). The exchange of the bound ADP by ATP is a slow process; about 50% is exchanged in 40 min (272). DnaA is a low-activity ATPase, hydrolyzing ATP to ADP with a half-time of about 40 min. The tight binding of ATP and its slow hydrolysis, the slow exchange of ADP, and the requirement for ATP-DnaA in the initiation reaction all point to a regulatory role for these nucleotides in the initiation process.

A rapid release of bound ADP and exchange by ATP are catalyzed by phospholipids with acidic head groups (273). The state of fluidity of the phospholipid bilayer is essential for the release of nucleotides (47, 339). In the presence of *oriC* and phospholipids, DnaA protein can bind ATP and thus provide rejuvenation of an otherwise inert initiation complex. In the absence of *oriC*, phospholipids cause release of nucleotide and inactivation of DnaA (54). Aggregated inactive DnaA protein complexed with phospholipids can be activated with the help of phospholipase A_2 or DnaK protein and ATP (115, 273). One report suggests that a rapid ADP-ATP exchange and thus rejuvenation of DnaA protein can also be mediated by cyclic AMP (112). There is evidence that initiation in vivo (72) and phospholipid-mediated nucleotide exchange in vitro (339) are dependent on oleic acid. Presumably, oleic acid in the membranes provides the correct membrane fluidity for the exchange reaction.

Functions of the DnaA Protein. The primary role of DnaA protein is that of a replisome organizer. It recognizes the origin *oriC* and directs all other proteins required to form a replisome (a replication complex) to this site. This function is discussed in detail in the section on the initiation complex.

The second important role of DnaA protein in *oriC*, and in other replication origins, is its function in primosome assembly, DnaA priming. In addition, DnaA protein affects gene expression in many ways. It can act as a repressor, as discussed above for the *dnaA* gene itself, or as an activator of transcription. Within a transcription unit, binding of DnaA protein to a DnaA box may result in transcription termination. All these functions are discussed below. The proposed function of DnaA protein as a regulator of replication initiation is discussed in the section dealing with initiation control.

DnaA priming. The traditional φX174-type primosome is not responsible for replication from *oriC*, because the *oriC*-de-pendent in vitro replication system assembled from purified proteins (127) does not contain the primosomal proteins PriA, PriB, PriC, and DnaT. In addition, the closest PriA recognition sites are about 2 kb away from *oriC* (277, 293, 308). Instead, helicase and primase loading in *oriC* are mediated by DnaA protein. DnaA priming was analyzed in detail in a model system, the origin of plasmid pBR322.

Plasmid pBR322 and other ColE1-like plasmids have a DnaA box between their origin, defined by the site where the native RNA primer is processed by RNase H, and a (φX174-type) primosome assembly site (PAS). DNA polymerase I extends the processed primer, which activates the PAS on the lagging-strand template by converting it from a double-stranded to a single-stranded form (214, 260). In the absence of RNase H, extension of the RNA primer also activates PAS (234). In both cases, the assembly of the φX174-type primosome at the PAS results from the combined actions of PriA, PriB, PriC, and DnaT, which load the DnaB-DnaC (helicase) complex, followed by primase (4).

If PAS is deleted (183, 278) or φX174-type primosomal proteins are inhibited (278), then pBR322 replication becomes completely dependent on DnaA, demonstrating that DnaA protein can fulfill the combined functions of the primosomal proteins PriA (protein n′), PriB (protein n), PriC (protein n″), and DnaT (protein i). For this action, the orientation of the DnaA box and its distance to the origin are important (276). Depending on the presence or absence of RNase H, the DnaA box is in the form of a D loop or an R loop. DnaA protein binds to the double-stranded part of this loop structure, irrespective of whether it is double-stranded DNA or the DNA-RNA hybrid (235). It has been suggested that DnaA protein then displaces single-stranded binding protein (SSB) from the single-stranded part of the unwound structure and loads the DnaB helicase from a DnaB-DnaC complex in *trans* to the single strand (235).

The DnaA primosome apparently acts also when the PAS is functioning. Thermoinactivation of DnaA in *dnaA*(Ts) mutants reduces the rate of synthesis of pBR322 plasmids (1, 245), and excess DnaA protein stimulates pBR322 DNA replication in vitro (183, 278) and in vivo (50).

A stable hairpin loop with a DnaA box in the double-stranded stem acts as a DnaA-dependent complementary strand origin in M13 replication (ABC primosome), and since this structure originates from the γ origin of plasmid R6K, it may be used by this plasmid for priming (193). A DnaA box-containing hairpin loop can also be used as a lagging-strand origin for the replication of the *B. subtilis* plasmid pUB110, suggesting that DnaA priming is also active in *B. subtilis* (161).

What, then, is the function of PAS for the φX174-type primosome on the *E. coli* chromosome? If replication forks are stalled between *oriC* and a PAS about 2 kbp away from *oriC*, DNA synthesis can resume again at the PAS (277). This suggests that DnaB-helicase continues to unwind the template from these

FIGURE 3 Comparison of amino acid sequence of DnaA proteins from 15 different bacteria: *Escherichia coli* (93, 230); *Salmonella typhimurium* (288); *Serratia marcescens* (288); *Proteus mirabilis* (287); *Buchnera aphidicola* (159); *Pseudomonas putida* (76); *Bacillus subtilis* (216); *Streptomyces coelicolor* (45); *Micrococcus luteus* (77); *Caulobacter crescentus* (188); *Rhizobium meliloti* (189a); *Synechocystis* sp. (252a); *Borrelia burgdorferi* (232a); *Spiroplasma citri* (333a); *Mycoplasma capricolum* (78). Protein sequence comparison was car-ried out using the GCG Pileup program (57). Division into domains I to IV is according to Fujita et al. (77). Boxed areas indicate regions of homology of DnaA proteins with the NtrC family of transcription factors as mentioned in the text. The ATP binding site and the DNA binding regions (253a) are indicated. Uppercase letters are used if 9 amino acids are identical, and boldface uppercase letters are used if 12 amino acids are identical or show a conservative exchange; periods indicate gaps. Courtesy of Stefan Richter.

stalled forks and thereby unwinds and activates the PAS. On the basis of this observation, we may speculate that PAS and the φX174-type primosome may be part of a backup system which becomes operative if replication forks are stalled. The phenotype of *priA* mutants is compatible with this hypothesis (168, 221).

Repression by DnaA protein. Several other genes besides *dnaA* are negatively regulated at the level of transcription by DnaA protein. All of them have DnaA boxes in their promoter regions. The *mioC* gene next to *oriC* is a particularly interesting example because of the possible effects on initiation at *oriC* (reviewed in reference 281). The *rpoH* gene, encoding the heat shock σ[32] factor, is repressed by DnaA protein (322), as is *drpA*, an essential gene for global RNA and DNA synthesis (344). Indirect evidence suggests that DnaA also mediates repression of the *uvrB* gene (306). Other genes, e.g., *polA*, have DnaA boxes in their promoter regions, but repression by DnaA has so far not been demonstrated.

DnaA protein-mediated activation of gene expression. Two DnaA boxes with the same orientation are located immediately upstream of the promoter of *nrd*, the operon encoding the two subunits of ribonucleoside diphosphate reductase. Binding of DnaA protein to these sites activates the expression of the *nrd* operon (18). A similar situation is found in the promoter region of *glpD*, which encodes the aerobic glycerol-3-phosphate dehydrogenase. Overexpression of DnaA protein leads to a concomitant increase in the expression of GlpD, and DnaA directly activates the *glpD* promoter (Jueterbock and Messer, unpublished data). In this case, two DnaA boxes upstream of the promoter overlap a binding site for the cyclic AMP receptor protein. We may therefore speculate that activation is the result of structural alteration, DNA bending, or loop formation in the promoter region.

Transcription termination by DnaA protein. The *mioC* gene is transcribed from its own promoter as well as from the upstream *asnC* promoter (153). Some of the *asnC* transcripts are terminated in the intergenic region between *asnC* and *mioC* owing to the interaction of DnaA protein with the DnaA box in the *mioC* promoter region (87, 266). On the basis of this observation, several different DnaA boxes were cloned between an inducible promoter and a monitor gene, and the resulting block to transcribing RNA polymerase was used to quantify DnaA binding to DnaA boxes (268). Transcription termination is observed in only one orientation of the DnaA box (267). We may speculate that this is due to an interaction with nearby sequence elements, e.g., a DnaA box in the same orientation in the N-terminal part of the monitor gene. Such a requirement for cooperating sequences is probably the reason that only a subset of DnaA boxes leads to transcription termination upon interaction with DnaA protein. Thus, the DnaA box within the reading frame of the *E. coli dnaA* gene does not promote transcription termination (324).

DnaA-mediated regulation of the *gua* operon requires two DnaA boxes (300) and is due to transcription termination (265). Transcription termination is also the likely mechanism by which DnaA protein regulates the essential cell division genes *ftsQ* and *ftsA*. The genes contain DnaA boxes within their reading frames and are negatively controlled by DnaA (194). However, since the expression of these genes is also influenced by temperature inactivation of *dnaC*(Ts) mutants and the *dnaB252* allele, their regulation may be more complex.

DnaA protein modulates the attenuation of the *trp* operon (15). The *trp* attenuator region does not contain a DnaA box, suggesting that the attenuation results from an indirect effect.

DnaA Protein of *B. subtilis*. The DnaA protein of *B. subtilis* has been purified (79). Like the *E. coli* protein, it binds ATP with high affinity ($K_d \sim 20$ nM) and binds to DnaA boxes of the same consensus sequence (79). *E. coli* and *B. subtilis* DnaA proteins are, however, not interchangeable in vivo. Moreover, the *B. subtilis dnaA* protein, or a hybrid protein consisting of parts of the *B. subtilis* and *E. coli* proteins, is deleterious for the growth of *E. coli* (3).

Other Proteins Binding to *oriC*

Histone-like proteins. Initiation at *oriC* is stimulated in vitro by low levels of proteins HU or IHF (59, 117, 280). FIS protein may have a similar role. IHF and FIS change the DNA structure in higher-order DNA-protein complexes such as the λ integration and GIN, HIN, and CIN systems (reviewed in reference 253b). Both the IHF and the FIS proteins have one specific binding site in *oriC* (Fig. 1) and bend *oriC* upon binding (69, 70, 88, 206, 244, 253b). One dimer of IHF or FIS is enough to induce a strong bend (142, 332). HU protein binds unspecifically to DNA, but several HU dimers result in bending as well (107). Specific concentrations of HU protein stimulate the binding of IHF to its site in *oriC* (30).

Chromosomal replication is perturbed in mutants deficient in IHF or FIS. Several origins present in the same cytoplasm normally initiate simultaneously in the cell cycle (282), but mutations in *fis* or *him* result in asynchronous initiations (34). Mutants with a defect in any one of the genes for histone-like proteins are viable. Even deficiency in two histone-like proteins is possible, although growth of such cells is impaired (129). The defect in the replication of *oriC* plasmids is more severe in these mutants. Mutants deficient in both subunits of HU (*hupA*, *hupB*) are inefficiently transformed by minichromosomes, and *oriC* plasmids are unstable in such cells (227). A similar observation was made for mutants in *fis* (88). An IHF mutant (*himD*) has a more complicated phenotype. Transformation by minichromosomes is inhibited only if *polA* (69) or *hupAB* mutations (130) are also present. The functions of both IHF and FIS are required in *cis* in the *oriC* complex. Mutation of the IHF or FIS binding site inactivates *oriC* (253b).

We suggest that the greater sensitivity to defects in histone-like proteins of *oriC* on plasmids compared to *oriC* on chromosomes may be due to a stricter dependence on the status of supercoiling. Probably, the binding of HU, IHF, and FIS, the introduction of negative supercoils, and the transcription away from *oriC* all work synergistically to help DnaA protein unwind *oriC* (see below).

DNA gyrase. In agreement with such a function, gyrase was found to be an important component of the initiation machinery. Negative supercoils are a prerequisite for initiation. Additionally, gyrase is required for releasing the superhelical stress introduced by the helicase action in replication forks. Both functions are discussed in more detail below. A specific binding site for gyrase covering the *Hin*dIII site within *oriC*, which is protected from *Hin*dIII digestion by gyrase binding, was discovered (179).

Membrane attachment. The replication origin region binds specifically to membrane fractions (85, 103, 122–124, 128, 131, 132, 158, 218, 219, 328, 337, 338). The role of the membrane components is not known, except for that of a membrane fraction that binds specifically hemimethylated *oriC* DNA present directly after replication (48, 103a, 228). This is discussed in more detail later.

STAGED INITIATION AT *oriC* IN VITRO

Once the chromosomal replication origin was cloned, it became feasible to develop an in vitro replication system. This was accomplished by A. Kornberg and his group, and most of our knowledge of the biochemistry of the initiation process is derived from this *oriC*-DnaA-dependent in vitro system. Initially, in vitro replication was obtained in a crude system (81) consisting of a cell extract containing replication proteins (fraction II), a supercoiled *oriC* plasmid template, and DnaA protein.

Eventually, it was possible to obtain in vitro replication from *oriC* templates in a system assembled from purified proteins (127, 155). In addition to *oriC*, the system contains DnaA protein, RNA polymerase (RNA-P), gyrase, DNA polymerase III holoenzyme, SSB, HU, DnaB helicase, DnaC protein, and DnaG primase. The reaction was inefficient when RNA-P was present but primase was omitted, more efficient with primase alone, but of highest efficiency when both RNA-P and primase were added (224, 307). We now know that both leading- and lagging-strand primers are synthesized by primase and that RNA-P assists the initiation process by transcriptional activation (reviewed in reference 155).

Adding the different proteins sequentially and making use of environmental requirements allowed us to divide the initiation process into several successive stages, represented by different complexes (initial complex, open complex, prepriming complexes I and II, priming complex), which are followed by bidirectional replication (155) (Fig. 4).

Initial Complex

DnaA protein binds to the five DnaA boxes R1 through R4 (80) and M (195) (Fig. 1). Both ADP-DnaA and ATP-DnaA bind to linear or supercoiled DNA. However, for subsequent stages, ATP-DnaA and supercoiled *oriC* template are required. The ADP form of DnaA shows some activity but only at elevated protein levels and in the presence of a small amount of ATP-DnaA (54).

The size of the initial complex, as seen in the electron microscope, depends on the ratio of DnaA protein to *oriC*. One distinct structure is correlated with replication activity and localized to *oriC* (55). This initial complex has a compact ellipsoid structure and a size that could accommodate about 20 DnaA monomers (55, 80, 84). A 220-bp segment of *oriC* DNA is hidden within such a complex (84). The DNA seems to be wrapped around an ellipsoid with the two duplex strands crossed (55).

The phenotypes of *oriC* mutants, discussed above, suggest that the initial complex must have an ordered structure. The helical phasing and the importance of conserved distances between DnaA boxes show the importance of a correct spacing of DnaA monomers in the complex (207, 326). The unspectacular phenotypes resulting from base changes in DnaA boxes show that the actual binding efficiency of DnaA protein to individual DnaA boxes is less important (108). A model that accommodates these features as well as the minimal spacing between some DnaA boxes and the positions of IHF and FIS binding sites has been proposed (326). Its important feature is the ordered structure required for a functional initial complex.

Open Complex

The initial complex is converted to an open complex by the addition of a relatively high concentration of ATP (5 mM). In the open complex, the region of the AT-rich 13-mer repeats in the left part of *oriC* is partially unwound. This was first detected by Bramhill and Kornberg (35), who measured sensitivity to single-strand specific nuclease in this region. Gille and Messer (89) corroborated this observation in vitro and in vivo by using oxidation by $KMnO_4$ as a measure of helical distortion. They also demonstrated that this unwinding is a normal intermediate of replication initiation in vivo. Surprisingly, the two DNA strands are not equally susceptible to single-strand-specific reagents. One of them, the top strand in Fig. 1, shows a smaller region of reactivity, confined to the right 13-mer, with either nuclease or $KMnO_4$ than the other (bottom) strand shows (35, 89, 118, 326). A physical interaction between DnaA protein and the top strand (35, 340) might explain this phenomenon.

It is not known how DnaA protein binding leads to the unwinding reaction. Structural requirements are the specific sequence of the right 13-mer, the AT richness of the left and middle 13-mers, and the AT cluster left of the 13-mers (10, 35, 118). A precise spacing between the 13-mer region and DnaA box R1 is required, and even 10-bp insertions are not tolerated (110). However, variations in the distance between DnaA boxes in the right part of *oriC* (between R3 and R4) do not affect the unwinding reaction (326).

DnaA protein alone is able to unwind the 13-mer region, provided the reaction temperature is 37°C or higher. At lower temperature, assisting factors are required (280). HU protein at stimulatory levels or IHF protein allows DnaA-mediated strand opening at temperatures as low as 21°C. HU protein shows a sharp concentration optimum in the stimulation of in vitro replication (59, 280), whereas higher IHF concentrations are not inhibitory (280).

In addition to the extended unwinding in the 13-mer region, smaller helical distortions can be detected in the right part of *oriC* between DnaA boxes R2 and R4 by $KMnO_4$ oxidation or by cleavage with T4 endonuclease VII (326). These distortions have been interpreted as minibulges and may be related to the DnaA protein-mediated positioning of helicase, as discussed below.

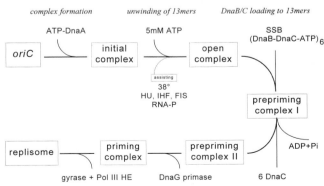

FIGURE 4 Staged in vitro replication. Pol III HE, DNA polymerase III holoenzyme.

Prepriming Complex I

DnaB is the helicase of the replicative fork (167) which is recruited by the open complex, eventually resulting in unwinding starting at *oriC* (22). It is suggestive that DnaB helicase is loaded onto the separated strands in the open complex, but rigorous experimental proof for a precise entry site is missing. Single-stranded DNA complexed with SSB is no substrate for DnaB. Therefore, we must assume that DnaA protein mediates a release of SSB followed by loading of the DnaB-DnaC complex, as has been suggested for the DnaA primosome in the pBR322 origin region (235).

DnaB loading requires a hexameric ring of DnaB complexed with six DnaC monomers, each of which binds one ATP. In the complex with DnaC, the helicase activity of DnaB is blocked (141, 162, 318, 325). A physical interaction between DnaA protein and DnaB is required at this or the following stage, as shown by interference with monoclonal antibodies and by cross-linking studies (192). Prepriming complexes can be isolated. In the electron microscope, they look somewhat bigger than open complexes (84). Besides *oriC,* they contain DnaA, DnaB, and HU (83, 84, 341). Apparently, two different prepriming complexes exist, some with DnaC (192) and some without (83, 84, 341). Release of DnaC is associated with ATP hydrolysis (318, 319) and seems to activate the helicase activity of DnaB. This reaction is similar to the DnaK-DnaJ-GrpE-mediated release of P protein from DnaB in the λ replication complex (reviewed in reference 190). Most likely, this activation of DnaB is correlated with its positioning by DnaA protein, i.e., the conversion of prepriming complex I to prepriming complex II.

Prepriming Complex II

Helicase has to exert its function at the positions where replication forks start, in the right part of *oriC,* close to DnaA boxes R2, R3, and R4 (Fig. 1) (279). Therefore, helicase must be translocated from the initial entry site to these replication start sites. Translocation could occur simply by helicase activity itself or alternatively by a direct transfer of DnaB between neighboring DNA strands within the prepriming complex from the unwound 13-mer region to minibulges in the region of the start sites (326). We assume that the translocation is coupled to the ATP-driven release of DnaC that results in activation of the helicase. DnaA-mediated priming is dependent on the orientation of the DnaA-DnaA box complex (235, 276). Therefore, we suggest positioning by a physical interaction between DnaA protein and DnaB helicase.

Priming Complex

DnaG primase interacts with DnaB helicase, resulting in the priming complex. Primase seems to play a key role in ensuring that initiation occurs in *oriC* and only in *oriC* and that two coupled replication forks that are competent for bidirectional replication are formed. An optimal concentration of primase is required, whereas low primase concentrations result in unidirectional replication from sites outside *oriC* (104).

At this stage, DnaA protein is no longer required and may be recycled to bind to a challenge *oriC* template, indicating that it may be reused (341). It has been suggested that the coordinated assembly of two coupled replication forks (104) involves the synthesis by one replication fork of a primer that then becomes the leading-strand primer for the other fork moving in the opposite direction (211). Primase acts distributively. After synthesis of the primers, it leaves the replication fork and is replaced by a new primase molecule during the next cycle of Okazaki fragment synthesis (330, 343).

The Replisome

A dimeric complex of DNA polymerase III holoenzyme assembles at each fork in an ATP-dependent mode by clamping the polymerase to the primer terminus (sliding clamp) (155, 186, 196, 222). Primers are extended, resulting in coordinate leading- and lagging-strand synthesis. DNA gyrase is required to relieve the superhelical stress.

There are conflicting reports with respect to the precise positions of the replication start sites within *oriC,* the transition sites from RNA primers to DNA. Start sites for bidirectional leading-strand synthesis were found close to DnaA boxes R2, R3, and R4 when the in vitro replication system was used (279). Mapping of chromosomal start sites in vivo, however, gave transition sites only for one leading strand, the one with the 5′→3′ orientation pointing leftward. These sites were located at several positions in the left half of *oriC,* and a few were found outside close to the left border of *oriC.* Obviously, replication in these experiments was unidirectionally leftward (106, 150). Initiation at *oriC* was aligned by temperature shifts in a *dnaC*(Ts) mutant strain. This treatment was subsequently shown to result in unidirectional synthesis leftward both in chromosomes and minichromosomes (337, 338). In contrast, without alignment, minichromosomes replicate in vivo bidirectionally from *oriC* (199). It was therefore suggested that there are different modes of replication from *oriC* (202). The dependence of direction and position of replication start sites on primase concentration is a possible explanation for unidirectional replication start sites as well as for other reports of replication starts outside of *oriC* (296). If primase is missing but gyrase is present, *oriC* plasmids are extensively unwound by the action of helicase (22). In the absence of both primase and gyrase, complexes that contain a small rather homogeneous bubble of about 400 bp can be isolated (20).

Table 4 lists the proteins involved in the chromosomal replication of *E. coli* under standard conditions, the respective genes, and the abundance of the proteins in cells.

CONTROL OF REPLICATION INITIATION IN VIVO

Initiation of replication is regulated at three different but interrelated levels: (i) initiation is precisely timed in the cell cycle; (ii) each origin initiates once and only once per cell cycle; and (iii) all origins present in the same cell initiate replication synchronously. A series of physiological experiments, to be discussed first, defined the prerequisites for initiation and culminated in the observation that bacteria initiate replication at a given cell mass per origin, called the initiation mass.

In principle, many of the actors involved in replication initiation might contribute to a regulatory element that may be considered a molecular clock (40). However, evidence is accumulating that DnaA protein is the best and only candidate molecule for the role of a molecular pacemaker. In order to exert its clock function properly, DnaA protein requires the assistance of accessory proteins, e.g., histone-like proteins, and the correct DNA topology, which in turn depends on local transcription in the *oriC* region. We shall discuss these accessory functions and then the regulatory role of DnaA protein and the different levels of control.

TABLE 4 Proteins involved in replication of the *E. coli* chromosome

Process and enzyme	Gene	Size (kDa)	No. of molecules/cell[a]
Initiation			
DnaA	*dnaA*	52	1,000
HU-α	*hupA*	9.5	
HU-β	*hupB*	9.5 } = 19	40,000
IHF-α	*himA*	11.2	
IHF-β	*himD*	10.6 } = 21.8	3,000
FIS	*fis*	2×11.2 = 22.4	30,000
RNA-P ($\alpha_2\beta\beta'\sigma$)		450	
α	*rpoA*	36.5	
β	*rpoB*	151	
β′	*rpoC*	156	
σ	*rpoD*	70	
Initiation and elongation			
Helicase	*dnaB*	6×50 = 300	20
DnaC	*dnaC*	6×29 = 174	20
Primase	*dnaG*	60	50–100
DNA gyrase	*gyrA*	2×97	
	gyrB	2×90 } = 374	
SSB	*ssb*	4×18.9 = 75.6	800
Elongation			
Polymerase III		918.4	10
$(\alpha\epsilon\theta\beta_2\tau_2)(\alpha\epsilon\theta\beta_2\gamma_2(\delta\delta'\chi\psi)_2)$			
α	*dnaE*	129.9	
τ	*dnaX*	71.1	
γ	*dnaX*	47.5	
β	*dnaN*	40.6	
δ	*holA*	34	
δ′	*holB*	32	
ε	*dnaO*	27.5	
χ	*holC*	14	
ψ	*holD*	12	
θ	*holE*	10	
Polymerase I	*polA*	103	300
Ligase	*lig*	75	300

[a]Approximate number in cells growing with one to two doublings per hour.

Physiological Aspects

When protein synthesis is blocked in a culture growing in glucose-minimal medium either by starving the culture for a required amino acid or by adding chloramphenicol, the accumulation of DNA gradually stops and reaches a plateau at about 1.4 times the amount present at the start of starvation. This increment corresponds to the amount calculated from the assumption that all rounds of replication continue to the end but no new replication rounds are initiated (184, 263). This interpretation was verified by using density shift experiments in combination with starvation (27, 165). The experiments show that protein synthesis is required for the initiation of replication but not for subsequent DNA polymerization. If protein synthesis is allowed to restart in the starved cultures, DNA synthesis lags behind protein synthesis by about one generation time (263), suggesting that one generation's worth of protein synthesis is required per initiation event.

In addition to the requirement for protein synthesis, the initiation of replication in vivo requires the action of RNA polymerase. Initiation is sensitive to the action of rifampin at a time when all required proteins are accumulated (163, 201, 348). Initiation thus depends also on untranslated de novo-synthesized RNA. A similar step was observed for replication initiation in *B. subtilis* (166).

Uncoupling of DNA and protein synthesis reveals that bacteria can accumulate the potential for replication initiation under conditions of blocked DNA synthesis with ongoing RNA and protein synthesis (reviewed in reference 313), and that the accumulated initiation potential is stable until being consumed in an initiation event. Accumulation of initiation potential can be achieved most efficiently by incubating initiation mutants at the nonpermissive temperature. Upon a shift back to the permissive temperature, all chromosomes initiate replication. Such treatment has been widely used for the characterization of initiation mutants and for aligning cells at the initiation event to achieve synchronous initiations.

DNA Topology and Transcriptional Activation

oriC Topology. Topoisomerases, particularly the counteracting activities of topoisomerase I and topoisomerase II (DNA gyrase), maintain the negative superhelicity of the bacterial chromosome. About half of the supercoils are constrained by chromatin-associated (histone-like) proteins, while the remaining unconstrained supercoils are available to facilitate transcription, replication, and site-specific recombination (241, 291, 329). It has been proposed that the *E. coli* and *S. typhimurium* chromosomes are composed of 40 to 50 distinct supercoiled domains that do not seem to vary much from each other in the degree of superhelicity, with average values of ~25 superhelical turns per kbp (213, 236, 329).

E. coli minichromosomes are maintained at a superhelical density comparable to that of the chromosome (170). As discussed above, it is difficult to maintain minichromosomes in host strains which are defective in histone-like protein HU, IHF, or FIS. These strains are viable, although with a slightly disturbed DNA replication. Since these proteins affect the local DNA topology, we may assume that *oriC* plasmids have more stringent topological requirements because they lack the possibility of balancing superhelical changes along a larger domain (170).

Deletion mutations in *topA* suppress *dnaA*(Ts) mutations (180). A possible explanation might be that *oriC* in such strains attains a topology which requires less DnaA protein for initiation. Some temperature-sensitive mutations in *gyrB*, the gene for the B subunit of DNA gyrase, exhibit an initiation phenotype (67, 68, 233). These observations stress the importance of DNA topology for the initiation event.

The precise topological requirements for initiation at *oriC* are not known, but in contrast to *oriC*, the topological requirements for DNA inversion catalyzed by HIN recombinase in *S. typhimurium* have been studied in great detail in vitro (see reference 171 and references therein). This system is likely to provide useful guidelines for future *oriC* studies. HIN binding to *hix* sites and subsequent protein-protein pairing of HIN bound to *hix* sites can be accomplished at a low superhelical ($\sigma = 0.02$) density of the template DNA. On the other hand, invertasome formation and DNA inversion can be observed only at a higher superhelical density ($\sigma = 0.06$) (171). Interestingly, like in vitro open-complex formation at *oriC,* invertasome formation for HIN-mediated DNA inversion requires HU and FIS.

Transcriptional Activation. Purified *oriC* plasmids have a superhelical density that is quite permissive for replication in vitro. In vivo, however, free superhelicity is partially constrained by bound proteins (reviewed in reference 240). Therefore, initiation of replication requires activation by transcription in vivo (163, 201, 348) and in the crude (fraction II) in vitro replication system (81). If free superhelicity is restrained by about half by protein HU, the purified in vitro replication system must also be activated by transcription by RNA-P (21, 280).

Transcriptional activation can occur by two different mechanisms. One of them is the generation of an R loop, which helps DnaA in the unwinding of the strands. The R loop may be several hundred base pairs from *oriC,* provided the region in between is not particularly difficult to melt (21, 280). These features suggest that the mechanism of activation is a propagation of base unpairing from the R loop to the 13-mer region in *oriC* (280). The second activation mechanism is due to the generation of positive supercoils in front of transcribing RNA polymerase and negative supercoils behind it (173). Transcription away from the 13-mers would thus be expected to assist in the unwinding. Apparently, it is not simple coincidence that promoters that provide transcription into *oriC* are repressed by DnaA, whereas promoters that drive transcription away from the 13-mers are activated (8, 10, 177, 220, 253, 271).

The contributions of the different promoters and transcripts are not clear. This is particularly true for the *mioC* transcription unit, which has all the features of an important regulatory element. It is repressed by DnaA protein (177, 220, 253, 271), and in minichromosomes, this repression has a positive effect on initiation at *oriC* (177, 292). This is reflected in a decrease in minichromosome copy number upon deletion of the *mioC* promoter (314) or upon replacement by a promoter that is not DnaA regulated (292, 298). Additionally, the *mioC* promoter is under stringent control, as are P*gidA* and P*ori*-L (225, 253). However, minichromosomes without the *mioC* region initiate at the same time in the cell cycle as those with *mioC* (101, 169). Different deletion mutations of the *mioC* promoter region were transferred from plasmids to the chromosome, and the cell cycle parameters of such cells were measured by flow cytometry. There was no difference between mutants and wild type, indicating that for chromosomal replication under balanced growth conditions, *mioC* transcription is dispensable (174).

In favor of a regulatory role of *mioC* and other transcripts around *oriC* is the observation of an intricate control with respect to the timing of replication. In synchronized cultures, *mioC* transcription is inhibited shortly before initiation of replication, while *gidA* transcription is stimulated. Following initiation, P*gidA* is inhibited and P*mioC* is stimulated (226, 301). These authors point to a possible correlation of the periodicity in promoter activity with the sequestration of *oriC* following replication initiation (46). DnaA-regulated transcription thus may amplify the action of DnaA protein at *oriC.*

Chromosome Replication and the Cell Cycle

C and D Times. Helmstetter and Cooper used a synchronization technique which does not disturb the physiology of the cells, called the baby machine, for measurements of cell cycle parameters in *E. coli* B/r, particularly cell division and rate of DNA replication (see chapter 102). They found that initiation occurs at a characteristic time in the cell cycle that differs with growth rate. The time between initiation and termination of replication, the C time (40 min), is constant irrespective of the growth rate, as is the time between termination and cell division, the D time (20 min), for cells growing with a generation time shorter than 60 min at 37°C. In fast-growing cells, chromosome initiation must therefore occur before the previous round of replication reaches the terminus, resulting in dichotomously branched chromosomes.

Refined analysis has since shown that the C time gradually increases with decreasing growth rate in both *E. coli* and *S. typhimurium.* More recently, flow cytometry was used to measure cell cycle parameters in exponentially growing populations (2, 284), corroborating and extending the results obtained with synchronized cultures. A more extensive description of cell cycle events is given in chapter 102.

Initiation Mass. Donachie (60) combined the classical data on DNA concentration as a function of growth rate (264) with the time of initiation observed in the Cooper and Helmstetter experiments. He concluded that *cell mass per replication origin at the time of initiation* is constant, irrespective of the growth rate of the cells. This initiation mass, or initiation volume, has since become the basis of biochemical models trying to explain the regulation of the time of initiation in the cell cycle. A recent report, however, suggests that the initiation volume may not be quite as constant as anticipated by the Donachie rule (327) (see chapter 97 for additional discussion).

Timing of Initiation

Control Circuits Based on Initiation Mass. The two basic models proposed for the control of replication in *E. coli* are reflected by two aspects of the constancy of initiation mass. Initiation may be envisioned to occur when the origin concentration (origin per mass) has decreased sufficiently or, conversely, when the mass per origin has increased sufficiently. The model for the first view, the inhibitor dilution model, was proposed by Pritchard et al. (248). It suggests the existence of an inhibitor of initiation that is synthesized at the time of initiation or shortly thereafter. Increase in cell volume results in the dilution of the inhibitor, and after its concentration is below the threshold level of 0.5, initiation can occur. Variants of this model suggest an unstable inhibitor which is synthesized constitutively (205, 247).

Although factors that affect initiation negatively have been described, e.g., IciA (119) and SeqA (181), a molecule that could be a genuine negative regulator has never been found. This makes the second type of model more attractive. Sompayrac and Maaloe (290) suggested that accumulation of an initiator is coupled to mass increase via an autoregulatory control loop. This model was later analyzed by computer simulation (189). The most likely candidate for such an initiator is DnaA protein.

It has recently been found that a mutant of *B. subtilis* that confers resistance to a protein kinase C inhibitor has an altered initiation mass (275). No connection between initiation and protein phosphorylation has been detected in *E. coli.*

DnaA Protein Regulates the Initiation Mass. Overproduction of DnaA protein from strong promoters results in a burst of initiation events, suggesting that DnaA regulates initiation positively (16, 17, 242, 243, 283, 331). The significance of these observations for the regulatory function of DnaA protein has been challenged with two arguments (see also reference 40 for a discussion): (i) in many experiments, the increased initiations did not result in an increase in the overall rate of DNA synthesis per cell; and (ii) an initiator is supposed to oscillate in order to exert its function, but such oscillation is unlikely for an autoregulated protein. However, both of these arguments have been ruled out by recent experiments. It is at least possible that the concentration of DnaA protein oscillates because the *dnaA* gene is exempt from the autoregulatory loop for about one-third of the cell cycle owing to sequestration of hemimethylated DNA (46, 301). In addition, an excessive concentration of DnaA protein results in stalling of the supernumerary newly initiated replication forks within about 50 kbp of *oriC* (17) or reduces the rate of replication fork progression. The overall rate of DNA synthesis is thereby reduced and does not reflect the number of initiated replication origins (16). In *B. subtilis*, an additional postinitiation regulatory step is operative, with replication forks stalled at discrete distances on either side of *oriC* (102). For *E. coli*, no such mechanism has been found.

At concentrations of DnaA below wild-type levels, the initiation mass is increased; i.e., cells initiate replication later in the cell cycle. Some *dnaA*(Ts) mutants show an increased initiation mass even at permissive temperature (100, 182, 303). The most convincing experiment was the change of initiation mass by the controlled expression of *dnaA* from an inducible promoter (176). We can thus safely conclude that the time of initiation in the cell cycle depends on the activity of DnaA protein in the cell.

The Initiator Titration Model. How DnaA protein might regulate initiation mass was proposed in the initiator titration model (96). It is the legal offspring of the replicon model, the autorepressor model, and later improvements of these prototype models (121, 290). At the same time, however, it represents a true second-generation model in that it uses stochastic methods for computer-based simulations of the bacterial cell cycle. In this model, all newly synthesized DnaA protein is titrated by DnaA boxes on the chromosome (initiator titration). As soon as the number of DnaA molecules exceeds the number of DnaA boxes, which also increases because of replication, initiation is supposed to occur.

Two assumptions, supported by experimental evidence, are important. (i) The final switch that triggers initiation due to suddenly available free DnaA has to be of a different kind or quality than the binding to an average DnaA box before that time point. Binding to a DnaA box with much lower affinity is an obvious possibility. The observation that the *oriC* DnaA boxes R1, R2, and R4 but not box R3 are protected by DnaA protein against methylation by dimethyl sulfate in vivo suggests that R3 might have such a pivotal role in initiation (262). (ii) Once initiation is triggered at a single *oriC*, DnaA is ejected. The sudden increase in concentration of free DnaA then leads to successive initiations at all origins in the cell in the form of a cascade (175). DnaA has to be proficient for more than one initiation, as suggested from results in in vitro initiation (341). Membrane sequestration of newly replicated *oriC* excludes them from an immediate second initiation (see below). The cascade model can, in addition, explain the synchronous initiation of minichromosomes and the chromosome.

Initiation Synchrony. Rapidly growing *E. coli* cells with multiple origins initiate the origins in each individual cell almost simulta-

neously (282). There are several conditions in which initiation synchrony is disturbed (reviewed in reference 281). In *dam*, *hobH*, and *seqA* mutants, sequestration is impaired; hence, initiation synchrony is disturbed (32, 33, 103a, 181). Temperature-sensitive *dnaA* mutants which are defective in the ATP binding site (Table 1) seem to initiate their origins at random in the cell cycle even at fully permissive temperature (282, 286). Other *dnaA* mutants are less affected in initiation synchrony (Table 1). As discussed above, mutations which affect functions that assist DnaA protein in its initiator function result in an asynchrony phenotype. This includes *fis* and *him* mutations (34) and mutations affecting the status of superhelicity in the cell (310, 311). We can conclude that optimal function of the initiation machinery is required for an uninhibited initiation cascade, which is essential for initiation synchrony.

Dam Methylation and Replication Initiation

The *E. coli* Dam methyltransferase modifies A in its recognition sequence 5′-GATC. When fully methylated DNA becomes temporarily hemimethylated upon replication, the lack of a cognate restriction enzyme in the Dam system provides the cell with a molecular monitor for newly replicated DNA. The Dam system is therefore ideally suited to ensure that all origins in an individual cell initiate once before any origin initiates a second time. In addition, Dam methylation plays a major role in cellular repair processes (mismatch repair) and is involved in the regulation of various promoters (see chapter 53 for details on Dam methylation).

The clustering of Dam sites and their positional conservation among enterobacterial replication origins suggest that Dam methylation has a role in replication control. Initially, Messer and his colleagues and Smith and colleagues found that unmethylated or fully methylated minichromosomes transform *dam* mutant strains poorly if at all (203, 289). Subsequently, it was discovered that it is hemimethylated DNA which cannot replicate in vivo and that hemimethylated DNA accumulates in *dam* mutants upon transformation with methylated plasmids (254). Whether unmethylated *oriC* plasmids are able to transform *dam* mutants is still controversial (46, 160, 203, 254, 289). Unmethylated or hemimethylated *oriC* plasmids nevertheless act as substrates for in vitro replication, albeit less efficiently than fully methylated DNA (31, 112, 160, 203, 289). *dam* mutants initiate replication at the chromosomal *oriC*, but initiation is poorly controlled. Interinitiation times are variable, as shown by density shift analysis (24) and flow cytometry (32, 175). Under conditions of controlled Dam expression in a *dam* host, the asynchrony phenotype was also revealed by flow cytometry at Dam methyltransferase concentrations that were either lower or higher than that of wild type (32). All these results emphasize that Dam methylation participates in initiation control (210).

A solution to this puzzling scenario came when Schaechter and his colleagues found that hemimethylated but not fully methylated or unmethylated *oriC* DNA specifically binds to an *E. coli* membrane fraction in vitro (228). They observed that in rapidly growing cells, *oriC* DNA remains hemimethylated after replication fork passage for about one-third of a generation before it is rapidly remethylated. Like *oriC* DNA, the promoter region of the *dnaA* gene shows a similar delay of remethylation following replication, concomitant with drastically reduced *dnaA* transcription (46). In contrast, other parts of the genome analyzed so far become remethylated within 1 to 2 min (46, 228, 295). The interpretation of these observations is that hemimethylated *oriC* becomes sequestered in the membrane and is thus inaccessible to Dam methyltransferase. The period during which a newly replicated *oriC* is refractory to a second initiation is termed the eclipse period. The autoregulatory control of *dnaA* expression is also suspended during this eclipse period. Rapid cessation of DnaA synthesis may contribute considerably to preventing secondary initiations (46, 281). Note that membrane sequestration of *oriC* not only prevents secondary initiation but also may shield the newly replicated DNA from the action of the *mutH-mutS* repair system.

A shortening of the eclipse period is the prominent phenotype of a newly discovered mutation in *E. coli*, *seqA* (181, 311a). Accordingly, *seqA* mutants exhibit an asynchrony phenotype that is most pronounced when they are grown in rich medium. Although direct evidence has not yet been obtained, Kleckner and colleagues propose for SeqA the role of a negative modulator of replication initiation in a sequestration step that precedes sequestration of newly replicated *oriC* into the membrane (181). In contrast, von Freiesleben et al. (311a) argue that a possible negative modulation of DnaA activity by SeqA protein may not depend on the methylation status of *oriC* and consequently may not be related to membrane sequestration. They found that the lethal effect of SeqA overproduction is independent of the state of methylation and that minichromosomes are more unstable in *dam* single mutants than in *dam seqA* double mutants (311a).

Sequestration of hemimethylated *oriC* DNA into the membrane was recently shown by Kohiyama and coworkers to involve HobH protein (103a). HobH was identified in membrane fractions and binds specifically to the methylated parental strand at specific sites in hemimethylated *oriC* DNA (Fig. 1). Comparable to the situation in *dam* mutants, the reduction of *oriC* membrane attachment in the *hobH* deletion mutant leads to asynchronous initiation of replication, in this case because of a missing receptor. However, the residual initiation synchrony in the *hobH* mutant precludes any key regulatory role of HobH for initiaton.

The discovery of specific binding of hemimethylated *oriC* DNA by HobH and the isolation of *seqA* as a suppressor mutation that allows efficient transformation of a *dam* host by minichromosomes provide new insight into the role of *oriC* membrane attachment (103, 181). (i) We may deduce that membrane sequestration of hemimethylated DNA is an active process likely to be regulated. (ii) A receptor mechanism which is not exhausted by the elevated copy number of minichromosomes, in addition to the chromosomal origins, can hardly contribute to any stringent chromosome or origin copy number control. (iii) Since the number of HobH-containing membrane regions does not seem to be strictly limited, there might be an additional use for these regions in the wild-type situation. They could be involved in a sequential membrane attachment of newly replicated chromosome domains, from origin to terminus, before nucleoid segregation and cell division (121). It remains to be seen whether HobH-mediated membrane sequestration is permitted or necessary for chromosome regions other than *oriC* and *dnaA*.

Following the eclipse period, *oriC* might be detached from the membrane solely by the progressive invasion of Dam methyltransferase preventing further membrane contact of then fully methylated GATC sites. The observation that Dam methyltransferase overexpression leads to a shortening of the eclipse period (203) and the observation that in *dam* mutants hemimethylated DNA is permanently sequestered and blocked in replication (46, 255) clearly argue for an active role of Dam methyltransferase prior to a new round of replication initiation. Dam-mediated initiation cycling in *E. coli* might thus allow a "sloppy" initiation control, a sort of "open the door for all origins waiting" mechanism. The finding of residual synchrony in the *hobH* and *seqA* single mutants supports this view.

Besides in *E. coli*, Dam methylation is found in *S. typhimurium* and other *Enterobacteriaceae* but not in *B. subtilis*. Intriguingly, minichromosomes can be established only at copy numbers below 1 in *B. subtilis* because of incompatibility, i.e., competition for initiation proteins (215). Apparently, a more stringent copy number control of *oriC* operates in *B. subtilis* than in *E. coli*. The recent discovery of cell cycle-regulated adenine methylation (GmANTC) in *Caulobacter crescentus* shows that DNA replication control by monitoring the methylation status of the replication origin is not a unique property of enteric bacteria (345).

INITIATION OF CHROMOSOME REPLICATION INDEPENDENT OF *oriC* AND DnaA

When normal chromosome replication is prevented by DNA damage at extreme physiological conditions or in a particular genetic background, *E. coli* cells can sustain replication by either of two bypass phenomena, termed damage-inducible SDR (iSDR) and constitutive SDR (cSDR), respectively. Neither of the two SDR pathways requires DnaA for initiation of chromosome replication. cSDR is, in addition, independent of *oriC* and adjacent sequences (for a recent review, see reference 9).

Upon induction of the SOS response in *E. coli* cells, initiation of DNA replication in the iSDR pathway depends on D-loop formation by the recombinase activity of RecA and the helicase activity of recBC at specific sequences, called *oriM*. Of the genes activated by the SOS response via the LexA regulon, solely RecA is necessary and sufficient to induce iSDR. Consequently, iSDR cannot be induced in recombinase-deficient *recA* mutants and in *recBC* mutants, while in *recD* mutants (lacking the exonuclease V activity of RecBCD enzyme), iSDR is increased. iSDR activity is, in addition, greatly increased in *ruvA, ruvB, ruvC,* and *recG* mutants, which are thought to accumulate unresolved D loops. Despite considerable effort, the initiator nuclease necessary to introduce the nick or double-stranded break at *oriM* sites for D-loop formation has not yet been found (reviewed in reference 9).

In *E. coli*, two separable *oriM* sites reside within *oriC*, while one other *oriM* is found in the *ter* region. Any distinct DNA sequence requirements for *oriM* function or additional *oriM* sites on the chromosome are presently unknown. iSDR is active in a *dnaA*::Tn*10* strain, does not require concomitant transcription and translation, and can drive DNA replication for several hours (reviewed in reference 9).

In contrast to iSDR, initiation of DNA replication in the cSDR pathway is crucially dependent on transcription and on RNase H deficiency but likewise requires recombinase activity of RecA (147). RNase H is not essential for replication initiation at *oriC* (109). Transcripts originating at various sites on the chromosome, termed *oriK*, are thought to be stabilized in the absence of RNase H, which specifically degrades RNA in RNA-DNA hybrids. While R-loop formation is sufficiently promoted by RecA in a $recA^+$ background, cSDR becomes dependent on derepression of the LexA regulon and the nick translation activity of DNA polymerase I, in particular in recombinase-deficient *recA* mutants (reviewed in reference 9). These findings emphasize the dependence of both known SDR pathways on SOS functions. The cSDR type of replication initiation may in addition resemble the *polA*-dependent replication mode of ColE1-type plasmids (144).

The *oriC* region does not contain *oriK* sequences, but at least four separate *oriK* sites have been mapped on the *E. coli* chromosome (56). Any distinct DNA sequence requirements for *oriK* function are presently unknown. *rnhA* strains can dispense with *oriC* and DnaA protein but grow poorly under these conditions (145, 148). Such mutants therefore provide an alternative to integratively suppressed strains for genetic studies.

Because iSDR and cSDR are blocked in a *priA::kan* mutant, assembly of a φX174-type primosome is thought to follow duplex opening (T. Masai and T. Kogoma, personal communication). The φX174-type primosome then loads DnaB helicase from the DnaBC complex. Priming of bidirectional replication is carried out by DnaG primase, while DNA polymerase III holoenzyme promotes chain growth (reviewed in reference 9). It is thus the where and how of duplex opening and the primosome type that discriminate iSDR and cSDR from each other and from normal replication. In cells forced to grow with cSDR, replication timing and frequency are poorly regulated (9, 148). This finding emphasizes the pivotal roles of DnaA protein and regulation of its expression for correct replication under normal growth conditions.

Although iSDR and cSDR are observed under rather extreme physiological or genetic conditions, they are not mere genetic artifacts. iSDR is part of the complex SOS response that enables cells to meet environmental challenges on viability. A cSDR-like activity was recently found in *rnh*^+ cells under certain growth conditions (108a). Kogoma and his colleagues therefore propose that iSDR and cSDR represent stress- or growth-related DNA replication activities (9). iSDR and cSDR may also be seen as the consequence of a functional and regulatory overlap of transcription, replication, and recombination. Free 3'-OH ends as created by transcription in RNA-DNA hybrids (R loops) or in double-stranded DNA (D loops) are substrates for the initial steps in either DNA replication or recombinational repair processes. To avoid or minimize errors, the competing pathways would have to allow for a switch to the other pathway during their initial steps. Concomitantly, they would provide a mutual backup system.

GLOBAL ASPECTS OF ORIGIN STRUCTURE

The arrangement of genes and the direction of their transcription have been remarkably well preserved among eubacteria even of very distant evolutionary relationships. In a comparison of sequences and transcription units in more than 20 kbp around *oriC* and *dnaA* of *B. subtilis* and *P. putida*, the major genes were conserved in sequence, position, and direction of transcription away from *dnaA* and *oriC* (223, 336). Most eubacteria seem to possess the gene sequence *dnaA-dnaN-recF-gyrB*, with *oriC* within 2 to 3 kbp of *dnaA* (77, 335). We might consider this to be the primordial origin. There are so far three exceptions to this

arrangement, in which one or more of these elements have been translocated to another region.

1. In *E. coli*, the *oriC* region is located 42 kbp from *dnaA*, and *gid* transcription is directed toward *dnaA*, thus violating the rule that transcription is away from *dnaA* (and *oriC*). It looks as if the *E. coli oriC* region has been transposed and inverted compared to the primordial origin (335, 336). The *dnaA* region of *E. coli*, however, matches perfectly the standard arrangement.

2. In *C. crescentus*, the origin is close (2 kbp) to the *dnaA* gene, but both are in a completely different environment. *oriC* is located between the *hemE* gene and a homolog of the ribosomal protein gene *rpsT*. The *dnaK* and *dnaJ* genes are downstream of *dnaA*. Transcription in this limited set of genes is away from *oriC* (188, 346). *C. crescentus* homologs of the *dnaN*, *recF*, and *gyrB* genes have also been cloned and identified. As in *E. coli* and other bacteria, they form a cluster, but this cluster is approximately 150 kbp away from *dnaA* (346).

3. The *dnaA* gene of the cyanobacterium *Synechocystis* sp. has been cloned, sequenced, and identified by virtue of its similaritiy to the *E. coli* gene (252a). It is located next to genes that are part of photosynthesis system II (*psbD*, *psbC*). No homologs of other genes normally close to *dnaA* are found. Likewise, there is no structure resembling a replication origin within 2.5 kbp. Thus, *E. coli* (and other *Enterobacteriaceae*) and *Synechocystis* sp. are so far the only bacteria in which *oriC* and *dnaA* are not next to each other.

The binding sites for DnaA protein, DnaA boxes, are also highly conserved among bacteria. In all cases analyzed, DnaA protein recognizes the same consensus sequence. The only variation is found in organisms with a high G+C content in their DNA, e.g., *Micrococcus luteus* (77), *Streptomyces lividans* (342), and *Streptomyces coelicolor* (45). DnaA boxes of these organisms frequently have G instead of A or T at the third position in the box.

The general structure of the bacterial chromosomal replication origin, *oriC,* is found in many prokaryotic origins; in plasmids like pSC101, F, R1, R6K, and RK2; and in P1, P4, λ, and lambdoid phages (reviewed in reference 36). In all these cases, iterated binding sites for an initiation protein are next to AT-rich regions. Many of these plasmids and phages contain in their origins, in addition to binding sites for their replicon-specific initiator proteins, DnaA boxes. For some of these origins, DnaA protein has a role in open-complex formation (90, 217).

Also, some eukaryotic viruses, like simian virus 40 or polyoma, show an outline in their origins similar to that of *oriC* (reviewed in reference 65). Common to all origins of this type is the formation of a higher-order complex between origin and initiator protein, the initial complex, followed by an unwinding reaction at the AT-rich region, the open complex. Because of the similarity of initiation complexes at a variety of origins and the nucleoprotein structures involved in transcription regulation or site-specific recombination, Echols proposed a common term, SNUP, for specialized nucleoprotein structure (63).

The replication origin of the bacterial chromosome, particularly *oriC* of *E. coli*, is the best-known cellular origin. However, despite intensive research over more than three decades by many groups, its complex function is not completely understood.

ACKNOWLEDGMENTS

We thank many of our colleagues for helpful suggestions and for communicating unpublished information.

LITERATURE CITED

1. **Abe, M.** 1980. Replication of ColE1 plasmid deoxyribonucleic acid in thermosensitive *dnaA* mutant of *Escherichia coli. J. Bacteriol.* **141:**1024–1030.
2. **Allman, R., T. Schjerven, and E. Boye.** 1991. Cell cycle parameters of *Escherichia coli* K-12. *J. Bacteriol.* **173:**7970–7974.
3. **Andrup, L., T. Atlung, N. Ogasawara, H. Yoshikawa, and F. G. Hansen.** 1988. Interaction of the *Bacillus subtilis* DnaA-like protein with the *Escherichia coli* DnaA protein. *J. Bacteriol.* **170:**1333–1338.
4. **Arai, K., R. L. Low, J. Kobori, J. Shlomai, and A. Kornberg.** 1981. Mechanism of DnaB action. V. Association of DnaB protein, n′ and other prepriming proteins in the primosome of DNA replication. *J. Biol. Chem.* **256:**5273–5280.
5. **Armengod, M., M. Garcia-Sogo, and E. Lambies.** 1988. Transcriptional organization of the *dnaN* and *recF* genes of *E. coli* K-12. *J. Biol. Chem.* **263:**12109–12114.
6. **Armengod, M., and E. Lambies.** 1986. Overlapping arrangement of the *recF* and *dnaN* operons of *E. coli;* positive and negative control sequences. *Gene* **43:**183–196.
7. **Asada, K., K. Sugimoto, A. Oka, M. Takanami, and Y. Hirota.** 1982. Structure of replication origin of the *Escherichia coli* K-12 chromosome: the presence of spacer sequence in the *ori* region carrying information for autonomous replication. *Nucleic Acids Res.* **10:**3745–3754.
8. **Asai, T., C.-P. Chen, T. Nagata, M. Takanami, and M. Imai.** 1992. Transcription in vivo within the replication origin of the *Escherichia coli* chromosome: a mechanism for activating initiation of replication. *Mol. Gen. Genet.* **231:**169–178.
9. **Asai, T., and T. Kogoma.** 1994. D-loops and R-loops: alternative mechanisms for the initiation of chromosome replication in *Escherichia coli. J. Bacteriol.* **176:**1807–1812.
10. **Asai, T., M. Takanami, and M. Imai.** 1990. The AT richness and *gid* transcription determine the left border of the replication origin of the *E. coli* chromosome. *EMBO J.* **9:**4065–4072.
11. **Atlung, T.** 1981. Analysis of seven *dnaA* suppressor loci in *E. coli. UCLA Symp. Mol. Cell. Biol.* **22:**297–314.
12. **Atlung, T.** 1984. Allele-specific suppression of *dnaA* (ts) mutations in *Escherichia coli. Mol. Gen. Genet.* **197:**125–178.
13. **Atlung, T., E. Clausen, and F. G. Hansen.** 1984. Autorepression of the *dnaA* gene of *Escherichia coli,* p. 199–207. *In* U. Huebscher and S. Spadari (ed.), *Proteins Involved in DNA Replication.* Plenum Press, New York.
14. **Atlung, T., E. Clausen, and F. G. Hansen.** 1985. Autoregulation of the *dnaA* gene of *Escherichia coli. Mol. Gen. Genet.* **200:**442–450.
15. **Atlung, T., and F. G. Hansen.** 1983. Effect of *dnaA* and *rpoB* mutations on attenuation in the *trp* operon of *Escherichia coli. J. Bacteriol.* **156:**985–992.
16. **Atlung, T., and F. G. Hansen.** 1993. Three distinct chromosome replication states are induced by increasing concentrations of DnaA protein in *Escherichia coli. J. Bacteriol.* **175:**6537–6545.
17. **Atlung, T., A. Loebner-Olesen, and F. G. Hansen.** 1987. Overproduction of DnaA protein stimulates initiation of chromosome and minichromosome replication in *E. coli. Mol. Gen. Genet.* **206:**51–59.
18. **Augustin, L. B., B. A. Jacobson, and J. A. Fuchs.** 1994. *Escherichia coli* Fis and DnaA proteins bind specifically to the *nrd* promoter region and affect expression of an *nrd-lac* fusion. *J. Bacteriol.* **176:**378–387.
19. **Bagdasarian, M. M., M. Izakowska, and M. Bagdasarian.** 1977. Suppression of the *dnaA* phenotype by mutations in the *rpoB* cistron of RNA polymerase in *S. typhimurium* and *E. coli. J. Bacteriol.* **130:**577–582.
20. **Baker, T. A., B. E. Funnell, and A. Kornberg.** 1987. Helicase action of dnaB protein during replication from the *Escherichia coli* chromosomal origin *in vitro. J. Biol. Chem.* **262:**6877–6885.
21. **Baker, T. A., and A. Kornberg.** 1988. Transcriptional activation of initiation of replication from the E. coli chromosomal origin: an RNA-DNA hybrid near *oriC. Cell* **55:**113–123.
22. **Baker, T. A., K. Sekimizu, B. E. Funnell, and A. Kornberg.** 1986. Extensive unwinding of the plasmid template during staged enzymatic initiation of DNA replication from the origin of the *E. coli* chromosome. *Cell* **45:**53–64.
23. **Baker, T. A., and S. H. Wickner.** 1992. Genetics and enzymology of DNA replication in *Escherichia coli. Annu. Rev. Genet.* **26:**447–477.
24. **Bakker, A., and D. W. Smith.** 1989. Methylation of GATC sites is required for precise timing between rounds of DNA replication in *Escherichia coli. J. Bacteriol.* **171:**5738–5742.
25. **Bernander, R., S. Dasgupta, and K. Nordström.** 1991. The E. coli cell cycle and the plasmid R1 replication cycle in the absence of the DnaA protein. *Cell* **64:**1145–1153.
26. **Billen, D.** 1969. Replication of the bacterial chromosome: location of new initiation sites after irradiation. *J. Bacteriol.* **97:**1169–1175.
27. **Bird, R., and K. G. Lark.** 1968. Initiation and termination of DNA replication after amino acid starvation of *E. coli* 15T$^-$. *Cold Spring Harbor Symp. Quant. Biol.* **33:**799–808.
28. **Blinkowa, A., W. G. Haldenwang, J. A. Ramsey, J. M. Henson, D. A. Mullen, and J. R. Walker.** 1983. Physiological properties of cold-sensitive suppressor mutations of a temperature-sensitive *dnaZ* mutant of *Escherichia coli. J. Bacteriol.* **153:**66–75.
29. **Blinkowa, A., and J. R. Walker.** 1990. Programmed ribosomal frameshifting generates the *Escherichia coli* DNA polymerase III gamma subunit from within the tau subunit reading frame. *Nucleic Acids Res.* **18:**1725–1729.
30. **Bonnefoy, E., and J. Rouvière-Yaniv.** 1992. HU, the major histone-like protein of *E. coli,* modulates the binding of IHF to *oriC. EMBO J.* **11:**4489–4496.
31. **Boye, E.** 1991. The hemimethylated replication origin of *Escherichia coli* can be initiated in vitro. *J. Bacteriol.* **173:**4537–4539.
32. **Boye, E., and A. Loebner-Olesen.** 1990. The role of *dam* methyltransferase in the control of DNA replication in *E. coli. Cell* **62:**981–989.
33. **Boye, E., A. Loebner-Olesen, and K. Skarstad.** 1988. Timing of chromosomal replication in *Escherichia coli. Biochim. Biophys. Acta* **951:**359–364.
34. **Boye, E., A. Lyngstadaas, A. Loebner-Olesen, K. Skarstad, and S. Wold.** 1993. Regulation of DNA replication in *Escherichia coli,* p. 15–26. *In* E. Fanning, R. Knippers, and E.-L. Winnacker (ed.), *DNA Replication and the Cell Cycle. 43. Mosbacher Kolloquium.* Springer, Berlin.
35. **Bramhill, D., and A. Kornberg.** 1988. Duplex opening by dnaA protein at novel sequences in initiation of replication at the origin of the *E. coli* chromosome. *Cell* **52:**743–755.
36. **Bramhill, D., and A. Kornberg.** 1988. A model for initiation at origins of DNA replication. *Cell* **54:**915–918.
37. **Braun, R. E., K. O'Day, and A. Wright.** 1985. Autoregulation of the DNA replication gene *dnaA* in *E. coli. Cell* **40:**159–169.
38. **Braun, R. E., K. O'Day, and A. Wright.** 1987. Cloning and characterization of *dnaA*(Cs), A mutation which leads to overinitiation of DNA replication in *Escherichia coli* K-12. *J. Bacteriol.* **169:**3898–3903.
39. **Braun, R. E., and A. Wright.** 1986. DNA methylation differentially enhances the expression of one of the two *E. coli dnaA* promoters *in vivo* and *in vitro. Mol. Gen. Genet.* **202:**246–250.
40. **Bremer, H., and G. Churchward.** 1991. Control of cyclic chromosome replication in *Escherichia coli. Microbiol. Rev.* **55:**459–475.
41. **Brunschede, H., T. L. Dove, and H. Bremer.** 1977. Establishment of exponential growth after a nutritional shift up in *Escherichia coli* B/r: accumulation of DNA, RNA, and protein. *J. Bacteriol.* **129:**1020–1033.
42. **Buhk, H. J., and W. Messer.** 1983. Replication origin region of *Escherichia coli:* nucleotide sequence and functional units. *Gene* **24:**265–279.
43. **Burland, V., G. Plunkett III, D. L. Daniels, and F. R. Blattner.** 1993. DNA sequence and analysis of 136 kilobases of the *Escherichia coli* genome: organizational symmetry around the origin of replication. *Genomics* **16:**551–561.
44. **Cairns, J.** 1963. The bacterial chromosome and its manner of replication as seen by autoradiography. *J. Mol. Biol.* **6:**208–213.
45. **Calcutt, M. J., and F. J. Schmidt.** 1992. Conserved gene arrangement in the origin region of the *Streptomyces coelicolor* chromosome. *J. Bacteriol.* **174:**3220–3226.
46. **Campbell, J. L., and N. Kleckner.** 1990. E. coli *oriC* and the *dnaA* gene promoter are sequestered from *dam* methyltransferase following the passage of the chromosomal replication fork. *Cell* **62:**967–979.
47. **Castuma, C. E., E. Crooke, and A. Kornberg.** 1993. Fluid membranes with acidic domains activate DnaA, the initiator protein of replication in *Escherichia coli. J. Biol. Chem.* **268:**24665–24668.
48. **Chakraborti, A., S. Gunji, N. Shakibai, J. Cubeddu, and L. Rothfield.** 1992. Characterization of the *Escherichia coli* membrane domain responsible for binding *oriC* DNA. *J. Bacteriol.* **174:**7202–7206.
49. **Chakraborty, T., K. Yoshinaga, H. Lother, and W. Messer.** 1982. Purification of the *E. coli dnaA* gene product. *EMBO J.* **1:**1545–1549.
50. **Chiang, C.-S., Y.-C. Xu, and H. Bremer.** 1991. Role of DnaA protein during replication of plasmid pBR322 in *Escherichia coli. Mol. Gen. Genet.* **225:**435–442.
51. **Chiaramello, A. E., and J. W. Zyskind.** 1989. Expression of *Escherichia coli dnaA* and *mioC* genes as a function of growth rate. *J. Bacteriol.* **171:**4272–4280.
52. **Chiaramello, A. E., and J. W. Zyskind.** 1990. Coupling of DNA replication to growth rate in *Escherichia coli:* a possible role for guanosine tetraphosphate. *J. Bacteriol.* **172:**2013–2019.
53. **Cooper, S.** 1991. *Bacterial Growth and Division.* Academic Press, Inc., San Diego, Calif.
54. **Crooke, E., C. E. Castuma, and A. Kornberg.** 1992. The chromosome origin of *Escherichia coli* stabilizes DnaA protein during rejuvenation by phospholipids. *J. Biol. Chem.* **267:**16779–16782.
55. **Crooke, E., R. Thresher, D. S. Hwang, J. Griffith, and A. Kornberg.** 1993. Replicatively active complexes of DnaA protein and the *Escherichia coli* chromosomal origin observed in the electron microscope. *J. Mol. Biol.* **233:**16–24.
56. **de Massy, B., O. Fayet, and T. Kogoma.** 1984. Multiple origin usage for DNA replication in *sdrA (rnh)* mutants of *Escherichia coli* K-12. *J. Mol. Biol.* **178:**227–236.
57. **Devereux, J., P. Haeberli, and O. Smithies.** 1984. A comprehensive set of sequence analysis programs for the VAX. *Nucleic Acids Res.* **12:**387–395.
58. **Diederich, L., A. Roth, and W. Messer.** 1994. A versatile plasmid vector system for the expression of genes in *Escherichia coli. BioTechniques* **16:**916–923.
59. **Dixon, N. E., and A. Kornberg.** 1984. Protein HU in the enzymatic replication of the chromosomal origin of *Escherichia coli. Proc. Natl. Acad. Sci. USA* **81:**424–428.
60. **Donachie, W. D.** 1968. Relationship between cell size and time of initiation of DNA replication. *Nature* (London) **219:**1077–1079.
61. **Drlica, K., and M. Riley.** 1990. *The Bacterial Chromosome.* American Society for Microbiology, Washington, D.C.
62. **Eberle, H., W. Van de Merwe, K. Madden, G. Kampo, L. Wright, and K. Donlon.** 1989. The nature of an intragenic suppressor of the *Escherichia coli dnaA*508 temperature-sensitive mutation. *Gene* **84:**237–245.
63. **Echols, H.** 1990. Nucleoprotein structures initiating DNA replication, transcription, and site-specific recombination. *J. Biol. Chem.* **265:**14697–14700.
64. **Evans, I. M., and H. Eberle.** 1975. Accumulation of the capacity for initiation of DNA replication in *E. coli. J. Bacteriol.* **121:**883–891.

65. Fanning, E., and R. Knippers. 1992. Structure and function of simian virus 40 large tumor antigen. *Annu. Rev. Biochem.* **61**:55–85.

66. Fayet, O., J. M. Louarn, and C. Georgopoulos. 1986. Suppression of the *E. coli dnaA*46 mutation by amplification of the *groES* and *groEL* genes. *Mol. Gen. Genet.* **202**:435–445.

67. Filutowicz, M. 1980. Requirement of DNA gyrase for the initiation of chromosome replication in *Escherichia coli* K-12. *Mol. Gen. Genet.* **177**:301–309.

68. Filutowicz, M., and P. Jonczyk. 1981. Essential role of the *gyrB* gene product in the transcriptional event coupled to *dnaA*-dependent initiation of *E. coli* chromosome replication. *Mol. Gen. Genet.* **183**:134–138.

69. Filutowicz, M., and J. Roll. 1990. The requirement of IHF protein for extrachromosomal replication of the *Escherichia coli oriC* in a mutant deficient in DNA polymerase I activity. *New Biol.* **2**:818–827.

70. Filutowicz, M., W. Ross, J. Wild, and R. L. Gourse. 1992. Involvement of Fis protein in replication of the *Escherichia coli* chromosome. *J. Bacteriol.* **174**:398–407.

71. Flower, A. M., and C. S. McHenry. 1991. Transcriptional organization of the *Escherichia coli dnaX* gene. *J. Mol. Biol.* **220**:649–658.

72. Fralick, J. A., and K. G. Lark. 1973. Evidence for the involvement of unsaturated fatty acids in initiating chromosome replication in *Escherichia coli*. *J. Mol. Biol.* **80**:459–475.

73. Frey, J., M. Chandler, and L. Caro. 1984. Overinitiation of chromosome and plasmid replication in a *dnaAcos* mutant of *Escherichia coli* K12: evidence for *dnaA*-*dnaB* interactions. *J. Mol. Biol.* **179**:171–183.

74. Fry, D. C., S. A. Kuby, and A. S. Mildvan. 1986. ATP-binding site of adenylate kinase: mechanistic implications of its homology with *ras*-encoded p21, F1-ATPase, and other nucleotide-binding proteins. *Proc. Natl. Acad. Sci. USA* **83**:907–911.

75. Fujisawa, T., and A. Eisenstark. 1973. Bi-directional chromosomal replication in *Salmonella typhimurium*. *J. Bacteriol.* **115**:168–176.

76. Fujita, M. Q., H. Yoshikawa, and N. Ogasawara. 1989. Structure of the *dnaA* region of *Pseudomonas putida*: conservation among three bacteria, *Bacillus subtilis*, *Escherichia coli* and *P. putida*. *Mol. Gen. Genet.* **215**:381–387.

77. Fujita, M. Q., H. Yoshikawa, and N. Ogasawara. 1990. Structure of the *dnaA* region of *Micrococcus luteus*: conservation and variations among eubacteria. *Gene* **93**:73–78.

78. Fujita, M. Q., H. Yoshikawa, and N. Ogasawara. 1992. Structure of the *dnaA* and *dnaA*-box region in the *Mycoplasma capricolum* chromosome: conservation and variations in the course of evolution. *Gene* **110**:17–23.

79. Fukuoka, T., S. Moriya, H. Yoshikawa, and N. Ogasawara. 1990. Purification and characterization of an initiation protein for chromosomal replication, DnaA, in *Bacillus subtilis*. *J. Biochem. (Tokyo)* **107**:732–739.

80. Fuller, R. S., B. E. Funnell, and A. Kornberg. 1984. The *dnaA* protein complex with the *E. coli* chromosomal origin (*oriC*) and other sites. *Cell* **38**:889–900.

81. Fuller, R. S., J. M. Kaguni, and A. Kornberg. 1981. Enzymatic replication of the origin of the *E. coli* chromosome. *Proc. Natl. Acad. Sci. USA* **78**:7370–7374.

82. Fuller, R. S., and A. Kornberg. 1983. Purified *dnaA* protein in initiation of replication at the *Escherichia coli* chromosomal origin of replication. *Proc. Natl. Acad. Sci. USA* **80**:5817–5821.

83. Funnell, B. E., T. A. Baker, and A. Kornberg. 1986. Complete enzymatic replication of plasmids containing the origin of the *Escherichia coli* chromosome. *J. Biol. Chem.* **261**:5616–5624.

84. Funnell, B. E., T. A. Baker, and A. Kornberg. 1987. *In vitro* assembly of a prepriming complex at the origin of the *Escherichia coli* chromosome. *J. Biol. Chem.* **262**:10327–10334.

85. Gayama, S., T. Kataoka, M. Wachi, G. Tamura, and K. Nagai. 1990. Periodic formation of the *oriC* complex of *Escherichia coli*. *EMBO J.* **9**:3761–3765.

86. Georgopoulos, C., K. Tilly, D. Drahos, and R. Hendrix. 1982. The B66.0 protein of *Escherichia coli* is the product of the *dnaK* gene. *J. Bacteriol.* **149**:1175–1177.

87. Gielow, A., C. Kücherer, R. Kölling, and W. Messer. 1988. Transcription in the region of the replication origin, *oriC*, of *Escherichia coli*. Termination of *asnC* transcripts. *Mol. Gen. Genet.* **214**:474–481.

88. Gille, H., J. B. Egan, A. Roth, and W. Messer. 1991. The FIS protein binds and bends the origin of chromosomal DNA replication, *oriC*, of *Escherichia coli*. *Nucleic Acids Res.* **19**:4167–4172.

89. Gille, H., and W. Messer. 1991. Localized unwinding and structural pertubations in the origin of replication, *oriC*, of *Escherichia coli in vitro* and *in vivo*. *EMBO J.* **10**:1579–1584.

90. Giraldo, R., and R. Diaz. 1992. Differential binding of wild-type and a mutant RepA protein to *oriR* sequence suggests a model for the initiation of plasmid R1 replication. *J. Mol. Biol.* **228**:787–802.

91. Guzman, E. C., A. Jimenez-Sanchez, E. Orr, and R. H. Pritchard. 1988. Heat stress in the presence of low RNA polymerase activity increases chromosome copy number in *Escherichia coli*. *Mol. Gen. Genet.* **212**:203–206.

92. Hansen, E. B., T. Atlung, F. G. Hansen, O. Skovgaard, and K. von Meyenburg. 1984. Fine structure genetic map and complementation analysis of mutations in the *dnaA* gene of *Escherichia coli*. *Mol. Gen. Genet.* **196**:387–396.

93. Hansen, E. B., F. G. Hansen, and K. von Meyenburg. 1982. The nucleotide sequence of the *dnaA* gene and the first part of the *dnaN* gene of *Escherichia coli* K12. *Nucleic Acids Res.* **10**:7373–7385.

94. Hansen, E. B., and M. B. Yarmolinsky. 1986. Host participation in plasmid maintenance: dependence upon *dnaA* of replicons derived from P1 and F. *Proc. Natl. Acad. Sci. USA* **83**:4423–4427.

95. Hansen, F. G., T. Atlung, R. E. Braun, A. Wright, P. Hughes, and M. Kohiyama. 1991. Initiator (DnaA) protein concentration as a function of growth rate in *Escherichia coli* and *Salmonella typhimurium*. *J. Bacteriol.* **173**:5194–5199.

96. Hansen, F. G., B. B. Christensen, and T. Atlung. 1991. The initiator titration model: computer simulation of chromosome and minichromosome control. *Res. Microbiol.* **142**:161–167.

97. Hansen, F. G., E. B. Hansen, and T. Atlung. 1982. The nucleotide sequence of the *dnaA* gene promoter and of the adjacent *rpmH* gene, coding for the ribosomal protein L34, of *Escherichia coli*. *EMBO J.* **1**:1043–1048.

98. Hansen, F. G., S. Koefoed, and T. Atlung. 1992. Cloning and nucleotide sequence determination of twelve mutant *dnaA* genes of *Escherichia coli*. *Mol. Gen. Genet.* **234**:14–21.

99. Hansen, F. G., S. Koefoed, L. Soerensen, and T. Atlung. 1987. Titration of DnaA protein by *oriC* DnaA-boxes increases *dnaA* gene expression in *Escherichia coli*. *EMBO J.* **6**:255–258.

100. Hansen, F. G., and K. V. Rasmussen. 1977. Regulation of the *dnaA* product in *E. coli*. *Mol. Gen. Genet.* **155**:219–225.

101. Helmstetter, C. E., and A. C. Leonard. 1987. Coordinate initiation of chromosome and minichromosome replication in *Escherichia coli*. *J. Bacteriol.* **169**:3489–3494.

102. Henckes, G., F. Harper, A. Levine, F. Vannier, and S. J. Seror. 1989. Overreplication of the origin region in the *dnaB37* mutant of *Bacillus subtilis*: postinitiation control of chromosomal replication. *Proc. Natl. Acad. Sci. USA* **86**:8660–8664.

103. Hendrickson, W. G., T. Kusano, H. Yamaki, R. Balakrishnan, M. King, J. Murchi, and M. Schaechter. 1982. Binding of the origin of *E. coli* to the membrane. *Cell* **30**:915–923.

103a. Herrick, J., R. Kern, S. Guha, A. Landoulsi, O. Fayet, A. Malki, and M. Kohiyama. 1994. Parental strand recognition of the DNA replication origin by the outer membrane in *Escherichia coli*. *EMBO J.* **13**:4695–4703.

104. Hiasa, H., and K. J. Marians. 1994. Primase couples leading- and lagging-strand DNA synthesis from *oriC*. *J. Biol. Chem.* **269**:6058–6063.

105. Hiraga, S. 1976. Novel F prime factors able to replicate in *E. coli* Hfr strains. *Proc. Natl. Acad. Sci. USA* **73**:198–202.

106. Hirose, S., S. Hiraga, and T. Okazaki. 1983. *Escherichia coli* origin of replication: initiation site of deoxyribonucleotide polymerisation at the replication origin of the *Escherichia coli* chromosome. *Mol. Gen. Genet.* **189**:422–431.

107. Hodges-Garcia, Y., P. J. Hagermann, and D. E. Pettijohn. 1989. DNA ring closure mediated by protein HU. *J. Biol. Chem.* **264**:14621–14623.

108. Holz, A., C. Schaefer, H. Gille, W.-R. Jueterbock, and W. Messer. 1992. Mutations in the DnaA binding sites of the replication origin of *Escherichia coli*. *Mol. Gen. Genet.* **233**:81–88.

108a. Hong, X., G. W. Cadwell, and T. Kogoma. 1995. *Escherichia coli* RecG and RecA proteins in R-loop formation. *EMBO J.* **14**:2385–2392.

109. Hong, X., and T. Kogoma. 1993. Absence of a direct role for RNase HI in initiation of DNA replication at the *oriC* site on the *Escherichia coli* chromosome. *J. Bacteriol.* **175**:6731–6734.

110. Hsu, J., D. Bramhill, and C. M. Thompson. 1994. Open complex formation by DnaA initiation protein at the *E. coli* chromosomal origin requires the 13-mers precisely spaced relative to the 9-mers. *Mol. Microbiol.* **11**:903–911.

111. Hughes, P., A. Landoulsi, and M. Kohiyama. 1988. A novel role for cAMP in the control of the activity of the *E. coli* chromosome replication initiator protein, DnaA. *Cell* **55**:343–350.

112. Hughes, P., F.-Z. Squali-Houssaini, P. Forterre, and M. Kohiyama. 1984. *In vitro* replication of a *dam* methylated and non-methylated *oriC* plasmid. *J. Mol. Biol.* **176**:155–159.

113. Hupp, T. R., and J. M. Kaguni. 1993. Activation of DnaA5 protein by GrpE and DnaK heat shock proteins in initiation of DNA replication in *Escherichia coli*. *J. Biol. Chem.* **268**:13137–13142.

114. Hupp, T. R., and J. M. Kaguni. 1993. Activation of mutant forms of DnaA protein of *Escherichia coli* by DnaK and GrpE proteins occurs prior to DNA replication. *J. Biol. Chem.* **268**:13143–13150.

115. Hwang, D. S., E. Crooke, and A. Kornberg. 1990. Aggregated dnaA protein is dissociated and activated for DNA replication by phospholipase or dnaK protein. *J. Biol. Chem.* **265**:19244–19248.

116. Hwang, D. S., and A. Kornberg. 1990. A novel protein binds a key origin sequence to block replication of an *E. coli* minichromosome. *Cell* **63**:325–331.

117. Hwang, D. S., and A. Kornberg. 1992. Opening of the replication origin of *Escherichia coli* by DnaA protein with protein HU or IHF. *J. Biol. Chem.* **267**:23083–23086.

118. Hwang, D. S., and A. Kornberg. 1992. Opposed actions of regulatory proteins, DnaA and IciA, in opening the replication origin of *Escherichia coli*. *J. Biol. Chem.* **267**:23087–23091.

119. Hwang, D. S., B. Thöny, and A. Kornberg. 1992. IciA protein, a specific inhibitor of initiation of *Escherichia coli* chromosomal replication. *J. Biol. Chem.* **267**:2209–2213.

120. Ingmer, H., and T. Atlung. 1992. Expression and regulation of a *dnaA* homologue isolated from *Pseudomonas putida*. *Mol. Gen. Genet.* **232**:431–439.

121. Jacob, F., S. Brenner, and F. Cuzin. 1963. On the regulation of DNA replication in bacteria. *Cold Spring Harbor Symp. Quant. Biol.* **28**:329–348.

122. Jacq, A., R. Kern, A. Tsugita, and M. Kohiyama. 1989. Purification and characterization of a low-molecular-weight membrane protein with affinity for the *Escherichia coli* origin of replication. *J. Bacteriol.* **171**:1409–1416.

123. Jacq, A., and M. Kohiyama. 1980. A DNA-binding protein specific for the early replicated region of the chromosome obtained from *Escherichia coli* membrane fraction. *Eur. J. Biochem.* **105:**25–31.

124. Jacq, A., M. Kohiyama, H. Lother, and W. Messer. 1983. Recognition sites for a membrane-derived DNA binding protein preparation in the *E. coli* replication origin. *Mol. Gen. Genet.* **191:**460–465.

125. Jenkins, A. J., J. B. March, I. R. Oliver, and M. Masters. 1986. A DNA fragment containing the *groE* genes can suppress mutations in the *E. coli dnaA* gene. *Mol. Gen. Genet.* **202:**446–454.

126. Junker, D. E., L. A. Rokeach, D. Ganea, A. E. Chiaramello, and J. W. Zyskind. 1986. Transcription termination within the *Escherichia coli* origin of DNA replication, *oriC. Mol. Gen. Genet.* **203:**101–109.

127. Kaguni, J. M., and A. Kornberg. 1984. Replication initiated at the origin (*oriC*) of the *E. coli* chromosome reconstituted with purified enzymes. *Cell* **38:**183–190.

128. Kambe-Honjoh, H., G. Tamura, and K. Nagai. 1992. Association of *oriC* region of *Escherichia coli* chromosome with outer membrane: effects of culture condition. *Biochem. Biophys. Res. Commun.* **187:**970–975.

129. Kano, Y., and F. Imamoto. 1990. Requirement of integration host factor (IHF) for growth of *Escherichia coli* deficient in HU protein. *Gene* **89:**133–137.

130. Kano, Y., T. Ogawa, T. Ogura, S. Hiraga, T. Okazaki, and F. Imamoto. 1991. Participation of the histone-like protein HU and of IHF in minichromosomal maintenance in *Escherichia coli. Gene* **103:**25–30.

131. Kataoka, T., S. Gayama, K. Takahashi, M. Wachi, M. Yamasaki, and K. Nagai. 1991. Only *oriC* and its flanking region are recovered from the complex formed at the time of initiation of chromosome replication in *Escherichia coli. Res. Microbiol.* **142:**155–159.

132. Kataoka, T., M. Wachi, J. Nakamura, S. Gayama, M. Yamasaki, and K. Nagai. 1993. Fully methylated *oriC* with negative superhelicity forms an *oriC*-membrane complex before initiation of chromosome replication. *Biochem. Biophys. Res. Commun.* **194:**1420–1426.

133. Katayama, T., and A. Kornberg. 1994. Hyperactive initiation of chromosomal replication *in vivo* and *in vitro* by a mutant initiator protein, DnaAcos, of *Escherichia coli. J. Biol. Chem.* **269:**12698–12703.

134. Katayama, T., and T. Nagata. 1991. Initiation of chromosomal DNA replication which is stimulated without oversupply of DnaA protein in *Escherichia coli. Mol. Gen. Genet.* **226:**491–502.

135. Kellenberger-Gujer, G., and A. Podhajska. 1978. Interactions between the plasmid lambda-dv and *Escherichia coli dnaA* mutants. *Mol. Gen. Genet.* **162:**17–22.

136. Kimura, M., T. Miki, S. Hiraga, T. Nagata, and T. Yura. 1979. Conditionally lethal amber mutations in the *dnaA* region of the *Escherichia coli* chromosome that affect chromosome replication. *J. Bacteriol.* **140:**825–834.

137. Kimura, M., T. Yura, and T. Nagata. 1980. Isolation and characterization of *Escherichia coli dnaA* amber mutants. *J. Bacteriol.* **144:**649–655.

138. Kimura, T., T. Asai, M. Imai, and M. Takanami. 1989. Methylation strongly enhances DNA bending in the replication origin region of the *Escherichia coli* chromosome. *Mol. Gen. Genet.* **219:**69–74.

139. Kjeldgaard, N. O., O. Maaloe, and M. Schaechter. 1958. The transition between different physiological states during balanced growth of *Salmonella typhimurium. J. Gen. Microbiol.* **19:**607–616.

140. Kline, B. C., T. Kogoma, J. E. Tam, and M. S. Shields. 1986. Requirement of the *Escherichia coli dnaA* gene product for plasmid F maintenance. *J. Bacteriol.* **168:**440–443.

141. Kobori, J. A., and A. Kornberg. 1982. The *Escherichia coli dnaC* gene product. III. Properties of the dnaB-dnaC protein complex. *J. Biol. Chem.* **257:**13770–13775.

142. Koch, C., and R. Kahmann. 1986. Purification and properties of the *Escherichia coli* host factor required for inversion of the G segment in bacteriophage Mu. *J. Biol. Chem.* **261:**15673–15678.

143. Kogoma, T. 1978. A novel *Escherichia coli* mutant capable of DNA replication in the absence of protein synthesis. *J. Mol. Biol.* **121:**55–69.

144. Kogoma, T. 1984. Absence of RNase H allows replication of pBR322 in *Escherichia coli* mutants lacking DNA polymerase I. *Proc. Natl. Acad. Sci. USA* **81:**7845–7849.

145. Kogoma, T. 1986. RNase H-defective mutants of *Escherichia coli. J. Bacteriol.* **166:**361–363.

146. Kogoma, T., and B. C. Kline. 1987. Integrative suppression of *dnaA*(Ts) mutations mediated by plasmid F in *E. coli* is a DnaA-dependent process. *Mol. Gen. Genet.* **210:**262–269.

147. Kogoma, T., N. L. Subia, and K. von Meyenburg. 1985. Function of ribonuclease H in initiation of DNA replication in *Escherichia coli* K-12. *Mol. Gen. Genet.* **200:**103–109.

148. Kogoma, T., and K. von Meyenburg. 1983. The origin of replication, *oriC,* and the *dnaA* protein are dispensable in stable DNA replication (*sdrA*) mutants of *Escherichia coli* K-12. *EMBO J.* **2:**463–468.

149. Kohara, Y., K. Akiyama, and K. Isono. 1987. The physical map of the whole E. coli chromosome: application of a new strategy for rapid analysis and sorting of a large genomic library. *Cell* **50:**495–508.

150. Kohara, Y., N. Tohdoh, X. W. Jiang, and T. Okazaki. 1985. The distribution and properties of RNA primed initiation sites of DNA synthesis at the replication origin of *Escherichia coli* chromosome. *Nucleic Acids Res.* **13:**6847–6866.

151. Kohiyama, M. 1968. DNA synthesis in temperature sensitive mutants in *E. coli. Cold Spring Harbor Symp. Quant. Biol.* **43:**317–324.

152. Kohiyama, M., D. Cousin, A. Ryter, and F. Jacob. 1966. Mutants thermosensibles d'*E. coli* K-12. I. Isolement et characterisation rapide. *Ann. Inst. Pasteur* **110:**465–486.

153. Kölling, R., A. Gielow, W. Seufert, C. Kücherer, and W. Messer. 1988. AsnC, a multifunctional regulator of genes located around the replication origin of *Escherichia coli, oriC. Mol. Gen. Genet.* **212:**99–104.

154. Koonin, E. V. 1993. A common set of conserved motifs in a vast variety of putative nucleic acid-dependent ATPases including MCM proteins involved in the initiation of eukaryotic DNA replication. *Nucleic Acids Res.* **21:**2541–2547.

155. Kornberg, A., and T. A. Baker. 1992. *DNA Replication.* W. H. Freeman and Co., New York.

156. Kowalski, D., and M. J. Eddy. 1989. The DNA unwinding element: a novel, *cis*-acting component that facilitates opening of the *Escherichia coli* replication origin. *EMBO J.* **8:**4335–4344.

157. Kücherer, C., H. Lother, R. Kölling, M. A. Schauzu, and W. Messer. 1986. Regulation of transcription of the chromosomal *dnaA* gene of *Escherichia coli. Mol. Gen. Genet.* **205:**115–121.

158. Kusano, T., D. Steinmetz, W. G. Hendrickson, J. Murchie, M. King, A. Benson, and M. Schaechter. 1984. Direct evidence for specific binding of the replication origin of the *Escherichia coli* chromosome to the membrane. *J. Bacteriol.* **158:**313–316.

159. Lai, C.-Y., and P. Baumann. 1992. Genetic analysis of an aphid endosymbiont DNA fragment homologous to the *rnpA-rpmH-dnaA-dnaN-gyrB* region of eubacteria. *Gene* **113:**175–181.

160. Landoulsi, A., P. Hughes, R. Kern, and M. Kohiyama. 1989. *dam* methylation and the initiation of DNA replication on *oriC* plasmids. *Mol. Gen. Genet.* **216:**217–223.

161. Langer, U., and J. C. Alonso. 1994. Genetic analysis of the DnaA-dependent priming in the initiation of lagging strand DNA synthesis of plasmid pUB110 in *Bacillus subtilis. FEMS Microbiol. Lett.* **119:**123–128.

162. Lanka, E., and H. Schuster. 1983. The DnaC protein of *Escherichia coli.* Purification, physical properties and interaction with DnaB protein. *Nucleic Acids Res.* **11:**987–997.

163. Lark, K. G. 1972. Evidence for direct involvement of RNA in the initiation of DNA replication in *E. coli* 15T. *J. Mol. Biol.* **64:**47–60.

164. Lark, K. G., and H. Renger. 1969. Initiation of DNA replication in *Escherichia coli* 15T⁻: chronological dissection of three physiological processes required for initiation. *J. Mol. Biol.* **42:**221–235.

165. Lark, K. G., T. Repko, and E. J. Hoffman. 1963. The effect of amino acid deprivation on subsequent DNA replication. *Biochim. Biophys. Acta* **76:**9–24.

166. Laurent, S. 1973. Initiation of DNA replication in a temperature sensitive mutant of *Bacillus subtilis:* evidence for a transcriptional step. *J. Bacteriol.* **116:**141–145.

167. LeBowitz, J. H., and R. McMacken. 1986. The *Escherichia coli dnaB* replication protein is a DNA helicase. *J. Biol. Chem.* **261:**4738–4748.

168. Lee, E. H., and A. Kornberg. 1991. Replication deficiencies in *priA* mutants of *Escherichia coli* lacking the primosomal replication n′ protein. *Proc. Natl. Acad. Sci. USA* **88:**3029–3032.

169. Leonard, A. C., and C. E. Helmstetter. 1986. Cell cycle-specific replication of E. *coli* minichromosomes. *Proc. Natl. Acad. Sci. USA* **83:**5101–5105.

170. Leonard, A. C., W. G. Whitford, and C. E. Helmstetter. 1985. Involvement of DNA superhelicity in minichromosome maintenance in *Escherichia coli. J. Bacteriol.* **161:**687–695.

171. Lim, H. M., and M. I. Simon. 1992. The role of negative supercoiling in Hin-mediated site-specific recombination. *J. Biol. Chem.* **267:**11176–11182.

172. Lindahl, G., and T. Lindahl. 1984. Initiation of DNA replication in *Escherichia coli:* RNase H-deficient mutants do not require the *dnaA* function. *Mol. Gen. Genet.* **196:**283–289.

173. Liu, L. F., and J. C. Wang. 1987. Supercoiling of the DNA template during transcription. *Proc. Natl. Acad. Sci. USA* **84:**7024–7027.

174. Loebner-Olesen, A., and E. Boye. 1992. Different effects of *mioC* transcription on initiation of chromosomal and minichromosomal replication in *Escherichia coli. Nucleic Acids Res.* **20:**3029–3036.

175. Loebner-Olesen, A., F. G. Hansen, K. V. Rasmussen, B. Martin, and P. L. Kuempel. 1994. The initiation cascade for chromosome replication in wild-type and Dam methyltransferase deficient *Escherichia coli* cells. *EMBO J.* **13:**1856–1862.

176. Loebner-Olesen, A., K. Skarstad, F. G. Hansen, K. von Meyenburg, and E. Boye. 1989. The DnaA protein determines the initiation mass of *Escherichia coli* K-12. *Cell* **57:**881–889.

177. Lother, H., R. Kölling, C. Kücherer, and M. A. Schauzu. 1985. Initiation of replication in *Escherichia coli* involves *dnaA* protein regulated transcription within the replication origin. *EMBO J.* **4:**555–560.

178. Lother, H., and W. Messer. 1981. Promoters in the *E. coli* replication origin. *Nature* (London) **294:**376–378.

179. Lother, H., E. Orr, and R. Lurz. 1983. DNA binding and antigenic specifications of DNA gyrase. *Nucleic Acids Res.* **12:**901–914.

180. Louarn, J., J. P. Bouche, and J. M. Louarn. 1984. Genetic inactivation of topoisomerase I suppresses a defect in initiation of chromosome replication in *E. coli. Mol. Gen. Genet.* **195:**170–174.

181. Lu, M., J. L. Campbell, E. Boye, and N. Kleckner. 1994. SeqA: a negative modulator of replication initiation in *E. coli. Cell* **77:**413–426.

182. Lycett, G. W., E. Orr, and R. H. Pritchard. 1980. Chloramphenicol releases a block in initiation of chromosome replication in a *dnaA* strain of *E. coli* K12. *Mol. Gen. Genet.* **178:**329–336.

183. Ma, D., and J. L. Campbell. 1988. The effect of dnaA protein and n′ sites on the replication of plasmid ColE1. *J. Biol. Chem.* **263:**15008–15015.

184. Maaloe, O., and P. Hanawalt. 1961. Thymine deficiency and the normal DNA replication cycle I. *J. Mol. Biol.* **3:**144–155.

185. Maaloe, O., and N. O. Kjeldgaard. 1966. *Control of Macromolecular Synthesis.* Benjamin, New York.

186. Maki, H., S. Maki, and A. Kornberg. 1988. DNA polymerase III holoenzyme of *Escherichia coli.* IV. The holoenzyme is an asymmetric dimer with twin active sites. *J. Biol. Chem.* 263:6570–6578.

187. Malki, A., P. Hughes, and M. Kohiyama. 1991. In vitro roles of *Escherichia coli* DnaJ and DnaK heat shock proteins in the replication of *oriC* plasmids. *Mol. Gen. Genet.* 225:420–426.

188. Marczynski, G. T., and L. Shapiro. 1992. Cell-cycle control of a cloned chromosomal origin of replication from *Caulobacter crescentus. J. Mol. Biol.* 226:959–977.

189. Margalit, H., R. F. Rosenberger, and N. B. Grover. 1984. Initiation of DNA replication in bacteria: analysis of an autorepressor model. *J. Theor. Biol.* 111:183–199.

189a. Margolin, W., D. Bramhill, and S. R. Long. 1995. The *dnaA* gene of *Rhizobium meliloti* lies within an unusual gene arrangement. *J. Bacteriol.* 177:2892–2900.

190. Marians, K. J. 1992. Prokaryotic DNA replication. *Annu. Rev. Biochem.* 61:673–719.

191. Marsh, R. C., and A. Worcel. 1977. A DNA fragment containing the origin of replication of the *E. coli* chromosome. *Proc. Natl. Acad. Sci. USA* 74:2720–2724.

192. Marszalek, J., and J. M. Kaguni. 1994. DnaA protein directs the binding of DnaB protein in initiation of DNA replication in *Escherichia coli. J. Biol. Chem.* 269:4883–4890.

193. Masai, H., N. Nomura, and K. Arai. 1990. The ABC-primosome. A novel priming system employing dnaA, dnaB, dnaC, and primase on a hairpin containing a dnaA box sequence. *J. Biol. Chem.* 265:15134–15144.

194. Masters, M., T. Paterson, A. G. Popplewell, T. Owen-Hughes, J. H. Pringle, and K. J. Begg. 1989. The effect of DnaA protein levels and the rate of initiation at *oriC* on transcription originating in the *ftsQ* and *ftsA* genes: in vivo experiments. *Mol. Gen. Genet.* 216:475–483.

195. Matsui, M., A. Oka, M. Takanami, S. Yasuda, and Y. Hirota. 1985. Sites of *dnaA* protein-binding in the replication origin of the *E. coli* K-12 chromosome. *J. Mol. Biol.* 184:529–533.

196. McHenry, C. S. 1986. DNA polymerase III holoenzyme of *Escherichia coli.* Components and function of a true replicative complex. *Mol. Cell. Biochem.* 66:71–85.

197. McMacken, R., L. Silver, and C. Georgopoulos. 1987. DNA replication, p. 564–612. *In* F. C. Neidhardt, J. L. Ingraham, K. B. Low, B. Magasanik, M. Schaechter, and H. E. Umbarger (ed.), *Escherichia coli and Salmonella typhimurium: Cellular and Molecular Biology.* American Society for Microbiology, Washington, D.C.

198. Meijer, M., E. Beck, F. G. Hansen, H. E. Bergmans, W. Messer, K. von Meyenburg, and H. Schaller. 1979. Nucleotide sequence of the origin of replication of the *E. coli* K-12 chromosome. *Proc. Natl. Acad. Sci. USA* 76:580–584.

199. Meijer, M., and W. Messer. 1980. Functional analysis of minichromosome replication: bidirectional and unidirectional replication from the *Escherichia coli* replication origin, *oriC. J. Bacteriol.* 143:1049–1053.

200. Meselson, M., and F. Stahl. 1958. The replication of DNA in *E. coli. Proc. Natl. Acad. Sci. USA* 44:671–682.

201. Messer, W. 1972. Initiation of DNA replication in *E. coli* B/r. Chronology of events and transcriptional control of initiation. *J. Bacteriol.* 112:7–12.

202. Messer, W. 1987. Initiation of DNA replication in *Escherichia coli. J. Bacteriol.* 169:3395–3399.

203. Messer, W., U. Bellekes, and H. Lother. 1985. Effect of *dam*-methylation on the activity of the replication origin, *oriC. EMBO J.* 4:1319–1326.

204. Messer, W., H. E. Bergmans, M. Meijer, J. E. Womack, F. G. Hansen, and K. von Meyenburg. 1978. Minichromosomes: plasmids which carry the *E. coli* replication origin. *Mol. Gen. Genet.* 162:269–275.

205. Messer, W., L. Dankwarth, R. Tippe-Schindler, J. E. Womack, and G. Zahn. 1975. Regulation of the initiation of DNA replication in *E. coli.* Isolation of iRNA and the control of iRNA synthesis. *UCLA Symp. Mol. Cell. Biol.* 3:602–617.

206. Messer, W., J. B. Egan, A. Gille, A. Holz, C. Schaefer, and B. Woelker. 1991. The complex of *oriC* DNA with the DnaA initiator protein. *Res. Microbiol.* 142:119–125.

207. Messer, W., H. Hartmann-Kühlein, U. Langer, E. Mahlow, A. Roth, S. Schaper, B. Urmoneit, and B. Woelker. 1992. The complex for replication initiation of *Escherichia coli. Chromosoma* 102:S1-S6.

208. Messer, W., B. Heimann, M. Meijer, and S. Hall. 1980. Structure of the *E. coli* replication origin. *oriC* sequences required for maintenance, establishment and bidirectional replication. *UCLA Symp. Mol. Cell. Biol.* 19:161–169.

209. Messer, W., M. Meijer, H. E. Bergmans, F. G. Hansen, K. von Meyenburg, E. Beck, and H. Schaller. 1979. Origin of replication, *oriC,* of the *Escherichia coli* K-12 chromosome: nucleotide sequence. *Cold Spring Harbor Symp. Quant. Biol.* 43:139–145.

210. Messer, W., and M. Noyer-Weidner. 1988. Timing and targeting: the biological functions of Dam methylation in *E. coli. Cell* 54:735–737.

211. Messer, W., W. Seufert, C. Schaefer, A. Gielow, H. Hartmann, and M. Wende. 1988. Functions of the DnaA protein of *Escherichia coli* in replication and transcription. *Biochim. Biophys. Acta* 951:351–358.

212. Miki, T., S. Hiraga, T. Nagata, and T. Yura. 1978. Bacteriophage lambda carrying the *E. coli* chromosomal region of the replication origin. *Proc. Natl. Acad. Sci. USA* 75:5099–5103.

213. Miller, W. G., and R. W. Simons. 1993. Chromosomal supercoiling in *Escherichia coli. Mol. Microbiol.* 10:675–684.

214. Minden, J. S., and K. J. Marians. 1985. Replication of pBR322 DNA *in vitro* with purified proteins. *J. Biol. Chem.* 260:9316–9325.

215. Moriya, S., T. Atlung, F. G. Hansen, H. Yoshikawa, and N. Ogasawara. 1992. Cloning of an autonomously replicating sequence (*ars*) from the *Bacillus subtilis* chromosome. *Mol. Microbiol.* 6:309–315.

216. Moriya, S., N. Ogasawara, and H. Yoshikawa. 1985. Structure and function of the region of the replication origin of the *Bacillus subtilis* chromosome. III. Nucleotide sequence of some 10,000 base pairs in the origin region. *Nucleic Acids Res.* 13:2251–2265.

217. Mukhopadhyay, G., K. M. Carr, J. M. Kaguni, and D. K. Chattoraj. 1993. Open-complex formation by the host initiator, DnaA, at the origin of P1 plasmid replication. *EMBO J.* 12:4547–4554.

218. Nagai, K., W. Hendrickson, R. Balakrishnan, H. Yamaki, D. Boyd, and M. Schaechter. 1980. Isolation of a replication origin complex from *Escherichia coli. Proc. Natl. Acad. Sci. USA* 77:262–266.

219. Nicolaidis, A., and I. B. Holland. 1978. Evidence for the specific association of the chromosomal origin with outer membrane fractions isolated from *Escherichia coli. J. Bacteriol.* 135:178–189.

220. Nozaki, N., T. Okazaki, and T. Ogawa. 1988. *In vitro* transcription of the origin region of replication of the *Escherichia coli* chromosome. *J. Biol. Chem.* 263:14176–14183.

221. Nurse, P., K. H. Zavitz, and K. J. Marians. 1991. Inactivation of the *Escherichia coli* PriA DNA replication protein induces the SOS response. *J. Bacteriol.* 173:6686–6693.

222. O'Donnell, M. 1987. Accessory proteins bind a primed template and mediate rapid cycling of DNA polymerase III holoenzyme from *Escherichia coli. J. Biol. Chem.* 262:16558–16565.

223. Ogasawara, N., and H. Yoshikawa. 1992. Genes and their organization in the replication origin region of the bacterial chromosome. *Mol. Microbiol.* 6:629–634.

224. Ogawa, T., T. A. Baker, A. van der Ende, and A. Kornberg. 1985. Initiation of enzymatic replication at the *Escherichia coli* chromosome: contributions of RNA polymerase and primase. *Proc. Natl. Acad. Sci. USA* 82:3562–3566.

225. Ogawa, T., and T. Okazaki. 1991. Concurrent transcription from the *gid* and *mioC* promoters activates replication of an *Escherichia coli* minichromosome. *Mol. Gen. Genet.* 230:193–200.

226. Ogawa, T., and T. Okazaki. 1994. Cell cycle-dependent transcription from the *gid* and *mioC* promoters of *Escherichia coli. J. Bacteriol.* 176:1609–1615.

227. Ogawa, T., M. Wada, Y. Kano, F. Imamoto, and T. Okazaki. 1989. DNA replication in *Escherichia coli* mutants that lack HU protein. *J. Bacteriol.* 171:5672–5679.

228. Ogden, G. B., M. J. Pratt, and M. Schaechter. 1988. The replicative origin of the *E. coli* chromosome binds to cell membranes only when hemimethylated. *Cell* 54:127–135.

229. Ohki, M., and C. L. Smith. 1989. Tracking bacterial DNA replication forks *in vivo* by pulsed field gel electrophoresis. *Nucleic Acids Res.* 17:3479–3490.

230. Ohmori, H., M. Kimura, T. Nagaha, and Y. Sakakibara. 1984. Structural analysis of the *dnaA* and *dnaN* genes of *E. coli. Gene* 28:159–170.

231. Oka, A., H. Sasaki, K. Sugimoto, and M. Takanami. 1984. Sequence organization of replication origin of the *Escherichia coli* K-12 chromosome. *J. Mol. Biol.* 176:443–458.

232. Oka, A., K. Sugimoto, M. Takanami, and Y. Hirota. 1980. Replication origin of the *Escherichia coli* K-12 chromosome: the size and structure of the minimum DNA segment carrying the information for autonomous replication. *Mol. Gen. Genet.* 178:9–20.

232a. Old, I. G., D. Margarita, and I. S. Girons. 1993. Unique genetic arrangement in the *dnaA* region of the *Borrelia burgdorferi* linear chromosome: nucleotide sequence of the *dnaA* gene. *FEMS Microbiol. Lett.* 111:109–114.

233. Orr, E., N. F. Fairweather, B. I. Holland, and R. H. Pritchard. 1979. Isolation and characterization of a strain carrying a conditional lethal mutation in the *cou* gene of *Escherichia coli* K-12. *Mol. Gen. Genet.* 177:103–112.

234. Parada, C. A., and K. J. Marians. 1989. Transcriptional activation of pBR322 DNA can lead to duplex DNA unwinding catalyzed by the *Escherichia coli* preprimosome. *J. Biol. Chem.* 264:15120–15129.

235. Parada, C. A., and K. J. Marians. 1991. Mechanism of DNA A protein-dependent pBR322 DNA replication. DNA A protein-mediated *trans*-strand loading of the DNA B protein at the origin of pBR322 DNA. *J. Biol. Chem.* 266:18895–18906.

236. Pavitt, G. D., and C. F. Higgins. 1993. Chromosomal domains of supercoiling in *Salmonella typhimurium. Mol. Microbiol.* 10:685–696.

237. Peak, K., and G. C. Walker. 1987. *Escherichia coli dnaK* null mutants are inviable at high temperature. *J. Bacteriol.* 169:283–290.

238. Pérez-Roger, I., M. García-Sogo, J. P. Navarro-Aviñó, C. López-Acedo, F. Macián, and M. E. Armengod. 1991. Positive and negative regulatory elements in the *dnaA-dnaN-recF* operon of *Escherichia coli. Biochimie* 73:329–334.

239. Petersen, S. K., and F. G. Hansen. 1991. A missense mutation in the *rpoC* gene affects chromosomal replication control in *Escherichia coli. J. Bacteriol.* 173:5200–5206.

240. Pettijohn, D. E. 1988. Histone-like proteins and bacterial chromosome structure. *J. Biol. Chem.* 263:12793–12796.

241. Pettijohn, D. E., and R. Hecht. 1973. RNA molecules bound to the folded bacterial genome stabilize DNA folds and segregate domains of supercoiling. *Cold Spring Harbor Symp. Quant. Biol.* 38:31–41.

242. Pierucci, O., C. E. Helmstetter, M. Rickert, M. Weinberger, and A. C. Leonard. 1987. Overexpression of the *dnaA* gene in *E. coli* B/r: chromosome and minichromosome replication in the presence of rifampicin. *J. Bacteriol.* 169:1871–1877.

243. Pierucci, O., M. Rickert, and C. E. Helmstetter. 1989. DnaA protein overproduction abolishes cell cycle specificity of DNA replication from oriC in Escherichia coli. J. Bacteriol. 171:3760–3766.

244. Polaczek, P. 1990. Bending of the origin of replication of E. coli by binding of IHF at a specific site. New Biol. 2:265–271.

245. Polaczek, P., and Z. Ciesla. 1984. Effect of altered efficiency of the RNA I and RNA II promoters on in vivo replication of colE1-like plasmids in Escherichia coli. Mol. Gen. Genet. 194:227–231.

246. Polaczek, P., and A. Wright. 1990. Regulation of expression of the dnaA gene in Escherichia coli: role of the two promoters and the DnaA box. New Biol. 2:574–582.

247. Pritchard, R. H. 1978. Control of DNA replication in bacteria, p. 1–26. In I. Molineux and M. Kohiyama (ed.), DNA Synthesis: Present and Future. Plenum Press, New York.

248. Pritchard, R. H., P. T. Barth, and J. Collins. 1969. Control of DNA synthesis in bacteria. Symp. Soc. Gen. Microbiol. 19:263–297.

249. Projan, S. J., and J. A. Wechsler. 1981. Isolation and analysis of multicopy extragenic suppressors of dnaA mutations. J. Bacteriol. 145:861–866.

250. Quinones, A., W.-R. Jueterbock, and W. Messer. 1991. DNA lesions that block DNA replication are responsible for the dnaA induction caused by DNA damage. Mol. Gen. Genet. 231:81–87.

251. Quinones, A., and W. Messer. 1988. Discoordinate expression in the dnaA-dnaN operon of Escherichia coli. Mol. Gen. Genet. 213:118–124.

252. Rasmussen, K. V., T. Atlung, G. Kerszman, G. E. Hansen, and F. G. Hansen. 1983. Conditional change of DNA replication control in an RNA polymerase mutant of Escherichia coli. J. Bacteriol. 154:443–451.

252a.Richter, S., and W. Messer. 1995. Genetic structure of the dnaA region of the cyanobacterium Synechocystis sp. PCC6803. J. Bacteriol. 177:4245–4251.

253. Rokeach, L. A., and J. W. Zyskind. 1986. RNA terminating within the Escherichia coli origin of replication: stringent regulation and control by DnaA protein. Cell 46:763–770.

253a.Roth, A., and W. Messer. 1995. The DNA binding domain of the initiator protein DnaA. EMBO J. 14:2106–2111.

253b.Roth, A., A. B. Urmoneit, and W. Messer. 1994. Functions of histone-like proteins in the initiation of DNA replication at oriC of Escherichia coli. Biochimie 76:917–923.

254. Russel, D. W., and N. D. Zinder. 1987. Hemimethylation prevents DNA replication in E. coli. Cell 50:1071–1079.

255. Ryals, J., R. Little, and H. Bremer. 1982. Temperature dependence of RNA synthesis parameters in Escherichia coli. J. Bacteriol. 151:879–887.

256. Sakakibara, Y. 1988. The dnaK gene of Escherichia coli functions in initiation of chromosome replication. J. Bacteriol. 170:972–979.

257. Sakakibara, Y. 1992. Novel Escherichia coli mutant, dnaR, thermosensitive in initiation of chromosome replication. J. Mol. Biol. 226:979–987.

258. Sakakibara, Y. 1992. dnaR function of the prs gene of Escherichia coli in initiation of chromosome replication. J. Mol. Biol. 226:989–996.

259. Sakakibara, Y. 1993. Cooperation of the prs and dnaA gene products for initiation of chromosome replication in Escherichia coli. J. Bacteriol. 175:5559–5565.

260. Sakakibara, Y., and J.-I. Tomizawa. 1974. Replication of colicin E1 plasmid DNA in cell extracts. II. Selective synthesis of early replicative intermediates. Proc. Natl. Acad. Sci. USA 71:1403–1407.

261. Sakakibara, Y., H. Tsukano, and T. Sako. 1981. Organization and transcription of the dnaA and dnaN genes of E. coli. Gene 13:47–55.

262. Samitt, C. E., F. G. Hansen, J. F. Miller, and M. Schaechter. 1989. In vivo studies of DnaA binding to the origin of replication of Escherichia coli. EMBO J. 8:989–993.

263. Schaechter, M. 1961. Patterns of cellular control during unbalanced growth. Cold Spring Harbor Symp. Quant. Biol. 26:53–62.

264. Schaechter, M., O. Maaloe, and N. O. Kjeldgaard. 1958. Dependency on medium and temperature of cell size and chemical composition during balanced growth of Salmonella typhimurium. J. Gen. Microbiol. 19:592–606.

265. Schaefer, C., A. Holz, and W. Messer. 1992. DnaA protein mediated transcription termination in the gua operon of Escherichia coli, p. 161–168. In P. Hughes, E. Fanning, and M. Kohiyama (ed.), DNA Replication: the Regulatory Mechanisms. Springer, Berlin.

266. Schaefer, C., and W. Messer. 1988. Termination of the E. coli asnC transcript. The DnaA protein/dnaA box complex blocks transcribing RNA polymerase. Gene 73:347–354.

267. Schaefer, C., and W. Messer. 1989. Directionality of DnaA protein/DNA interaction. Active orientation of the DnaA protein/dnaA box complex in transcription termination. EMBO J. 8:1609–1613.

268. Schaefer, C., and W. Messer. 1991. DnaA protein/DNA interaction. Modulation of the recognition sequence. Mol. Gen. Genet. 226:34–40.

268a.Schaper, S., and W. Messer. 1995. Interaction of the initiator protein DnaA of Escherichia coli with its DNA target. J. Biol. Chem. 270:17622–17626.

269. Schaus, N. A., K. O'Day, W. Peters, and A. Wright. 1981. Isolation and characterization of amber mutations in gene dnaA of E. coli K-12. J. Bacteriol. 145:904–913.

270. Schaus, N. A., K. O'Day, and A. Wright. 1981. Suppression of amber mutations in the dnaA gene of E. coli K12 by secondary mutations in rpoB. UCLA Symp. Mol. Cell. Biol. 22:315–323.

271. Schauzu, M. A., C. Kücherer, R. Kölling, W. Messer, and H. Lother. 1987. Transcripts within the replication origin, oriC, of Escherichia coli. Nucleic Acids Res. 15:2479–2497.

272. Sekimizu, K., D. Bramhill, and A. Kornberg. 1987. ATP activates dnaA protein in initiating replication of plasmids bearing the origin of the E. coli chromosome. Cell 50:259–265.

273. Sekimizu, K., and A. Kornberg. 1988. Cardiolipin activation of DnaA protein, the initiation protein in E. coli. J. Biol. Chem. 263:7131–7135.

274. Sekimizu, K., B. Y. Yung, and A. Kornberg. 1988. The dnaA protein of Escherichia coli. Abundance, improved purification, and membrane binding. J. Biol. Chem. 263:7136–7140.

275. Seror, S. J., S. Casarégola, F. Vannier, N. Zouari, M. Dahl, and E. Boye. 1994. A mutant cysteinyl-tRNA synthetase affecting timing of chromosomal replication initiation in B. subtilis and conferring resistance to a protein kinase C inhibitor. EMBO J. 13:2472–2480.

276. Seufert, W., B. Dobrinski, R. Lurz, and W. Messer. 1988. Functionality of the DnaA protein binding site in DNA replication is orientation dependent. J. Biol. Chem. 263:2719–2723.

277. Seufert, W., and W. Messer. 1986. Initiation of Escherichia coli minichromosome replication at oriC and at protein n' recognition sites. Two modes for initiating DNA synthesis in vitro. EMBO J. 5:3401–3406.

278. Seufert, W., and W. Messer. 1987. DnaA protein binding to the plasmid origin region can substitute for primosome assembly during replication of pBR322 in vitro. Cell 48:73–78.

279. Seufert, W., and W. Messer. 1987. Start sites for bidirectional in vitro replication inside the replication origin, oriC, of Escherichia coli. EMBO J. 6:2469–2472.

280. Skarstad, K., T. A. Baker, and A. Kornberg. 1990. Strand separation required for initiation of replication at the chromosomal origin of E. coli is facilitated by an RNA-DNA hybrid. EMBO J. 9:2341–2348.

281. Skarstad, K., and E. Boye. 1994. The initiator protein DnaA: evolution, properties and function. Biochim. Biophys. Acta 1217:111–130.

282. Skarstad, K., E. Boye, and H. B. Steen. 1986. Timing of initiation of chromosome replication in individual E. coli cells. EMBO J. 5:1711–1717.

283. Skarstad, K., A. Loebner-Olesen, T. Atlung, K. von Meyenburg, and E. Boye. 1989. Initiation of DNA replication in Escherichia coli after overproduction of the DnaA protein. Mol. Gen. Genet. 218:50–56.

284. Skarstad, K., H. B. Steen, and E. Boye. 1983. Cell cycle parameters of slowly growing E. coli B/r studied by flow cytometry. J. Bacteriol. 154:656–662.

285. Skarstad, K., B. Thöny, D. S. Hwang, and A. Kornberg. 1993. A novel binding protein of the origin of the Escherichia coli chromosome. J. Biol. Chem. 268:5365–5370.

286. Skarstad, K., K. von Meyenburg, F. G. Hansen, and E. Boye. 1988. Coordination of chromosome replication initiation in Escherichia coli: effects of different dnaA alleles. J. Bacteriol. 170:852–858.

287. Skovgaard, O. 1990. Nucleotide sequence of a Proteus mirabilis DNA fragment homologous to the 60K-rnpA-rpmH-dnaA-dnaN-recF-gyrB region of Escherichia coli. Gene 93:27–34.

288. Skovgaard, O., and F. G. Hansen. 1987. Comparison of dnaA nucleotide sequences of Escherichia coli, Salmonella typhimurium, and Serratia marcescens. J. Bacteriol. 169:3976–3981.

289. Smith, D. W., A. M. Garland, G. Herman, R. E. Enns, T. A. Baker, and J. W. Zyskind. 1985. Importance of state of methylation of oriC GATC sites in initiation of DNA replication in Escherichia coli. EMBO J. 4:1327–1332.

290. Sompayrac, L., and O. Maaloe. 1973. Autorepressor model for control of DNA replication. Nature New Biol. 241:133–135.

291. Steck, T. R., R. J. Franco, J.-Y. Wang, and K. Drlica. 1993. Topoisomerase mutations affect the relative abundance of many Escherichia coli proteins. Mol. Microbiol. 10:473–481.

292. Stuitje, A. R., N. de Wind, J. C. van der Spek, T. H. Pors, and M. Meijer. 1986. Dissection of promoter sequences involved in transcriptional activation of the Escherichia coli replication origin. Nucleic Acids Res. 14:2333–2344.

293. Stuitje, A. R., P. J. Weisbeek, and M. Meijer. 1984. Initiation signals for complementary strand DNA synthesis in the region of the replication origin of the Escherichia coli chromosome. Nucleic Acids Res. 12:3321–3332.

294. Sugimoto, K., A. Oka, H. Sugisaki, M. Takanami, A. Nishimura, Y. Yasuda, and Y. Hirota. 1979. Nucleotide sequence of Escherichia coli replication origin. Proc. Natl. Acad. Sci. USA 76:575–579.

295. Szyf, M., Y. Gruenbaum, S. Uriel-Shoval, and A. Razin. 1982. Studies on the biological role of DNA methylation. V. The pattern of E. coli DNA methylation. Nucleic Acids Res. 10:7247–7259.

296. Tabata, S., A. Oka, K. Sugimoto, M. Takanami, S. Yasuda, and Y. Hirota. 1983. The 245 bp oriC sequence of the E. coli chromosome directs bidirectional replication at an adjacent region. Nucleic Acids Res. 11:2617–2626.

297. Takeda, Y., and Y. Hirota. 1982. Suppressor genes of a dnaA temperature sensitive mutation in Escherichia coli. Mol. Gen. Genet. 187:67–71.

298. Tanaka, M., and S. Hiraga. 1985. Negative control of oriC plasmid replication by transcription of the oriC region. Mol. Gen. Genet. 200:21–26.

299. Tanaka, M., H. Ohmori, and S. Hiraga. 1983. A novel type of E. coli mutants with increased chromosomal copynumber. Mol. Gen. Genet. 192:51–60.

300. Tesfa-Selase, F., and W. T. Drabble. 1992. Regulation of the gua operon of Escherichia coli by the DnaA protein. Mol. Gen. Genet. 231:256–264.

301. Theisen, P., J. E. Grimwade, A. C. Leonard, J. A. Bogan, and C. E. Helmstetter. 1993. Correlation of gene transcription with the time of initiation of chromosome replication in Escherichia coli. Mol. Microbiol. 10:575–584.

302. Thöny, B., D. S. Hwang, L. Fradkin, and A. Kornberg. 1991. *iciA*, an *Escherichia coli* gene encoding a specific inhibitor of chromosomal initiation of replication *in vitro*. *Proc. Natl. Acad. Sci. USA* 88:4066–4070.

303. Tippe-Schindler, R., G. Zahn, and W. Messer. 1979. Control of the initiation of DNA replication in *E. coli*. I. Negative control of initiation. *Mol. Gen. Genet.* 168:185–195.

304. Torrey, T. A., and T. Kogoma. 1987. Genetic analysis of constitutive stable DNA replication in *rnh* mutants of *E. coli* K-12. *Mol. Gen. Genet.* 208:420–427.

305. Tsuchihashi, Z., and A. Kornberg. 1990. Translational frameshifting generates the gamma subunit of DNA polymerase III holoenzyme. *Proc. Natl. Acad. Sci. USA* 87:2516–2520.

306. van den Berg, E. A., R. H. Geerse, J. Memelink, R. A. L. Bovenberg, F. A. Magnée, and P. van de Putte. 1985. Analysis of regulatory sequences upstream of the *E. coli uvrB* gene; involvement of the DnaA protein. *Nucleic Acids Res.* 13:1829–1840.

307. van der Ende, A., T. A. Baker, T. Ogawa, and A. Kornberg. 1985. Initiation of enzymatic replication at the origin of the *E. coli* chromosome: primase as the sole priming enzyme. *Proc. Natl. Acad. Sci. USA* 82:3954–3958.

308. van der Ende, A., R. Teerstra, H. G. van der Avoort, and P. J. Weisbeek. 1983. Initiation signals for complementary DNA synthesis on single-stranded plasmid DNA. *Nucleic Acids Res.* 11:4957–4975.

309. Verma, M., K. G. Moffat, and J. B. Egan. 1989. UV irradiation inhibits initiation of DNA replication from *oriC* in *Escherichia coli*. *Mol. Gen. Genet.* 216:446–454.

310. von Freiesleben, U., and K. V. Rasmussen. 1991. DNA replication in *Escherichia coli gyrB*(Ts) mutants analysed by flow cytometry. *Res. Microbiol.* 142:223–227.

311. von Freiesleben, U., and K. V. Rasmussen. 1992. The level of supercoiling affects the regulation of DNA replication in *Escherichia coli*. *Res. Microbiol.* 143:655–663.

311a. von Freiesleben, U., K. V. Rasmussen, and M. Schaechter. 1994. SeqA limits DnaA activity in replication from *oriC* in *Escherichia coli*. *Mol. Microbiol.* 14:763–772.

312. von Meyenburg, K., and F. G. Hansen. 1980. The origin of replication, *oriC*, of the *E. coli* chromosome: genes near *oriC* and construction of *oriC* deletion mutations. *UCLA Symp. Mol. Cell. Biol.* 19:137–159.

313. von Meyenburg, K., and F. G. Hansen. 1987. Regulation of chromosome replication, p. 1555–1577. *In* F. C. Neidhardt, J. L. Ingraham, K. B. Low, B. Magasanik, M. Schaechter, and H. E. Umbarger (ed.), *Escherichia coli and Salmonella typhimurium: Cellular and Molecular Biology.* American Society for Microbiology, Washington, D.C.

314. von Meyenburg, K., F. G. Hansen, T. Atlung, L. Boe, I. G. Clausen, B. van Deurs, E. B. Hansen, B. B. Jorgensen, F. Jorgensen, L. Koppes, O. Michelsen, J. Nielsen, P. E. Pedersen, K. V. Rasmussen, E. Riise, and O. Skovgaard. 1985. Facets on the chromosomal origin of replication *oriC* of *E. coli*, p. 260–281. *In* M. Schaechter, F. C. Neidhardt, J. Ingraham, and N. O. Kjeldgaard (ed.), *The Molecular Biology of Bacterial Growth*. Jones & Bartlett, Boston.

315. von Meyenburg, K., F. G. Hansen, L. D. Nielsen, and P. Jorgensen. 1977. Origin of replication, *oriC*, of the *Escherichia coli* chromosome: mapping of genes relative to R.*EcoRI* cleavage sites in the *oriC* region. *Mol. Gen. Genet.* 158:101–109.

316. von Meyenburg, K., F. G. Hansen, L. D. Nielsen, and E. Riise. 1978. Origin of replication, *oriC*, of the *Escherichia coli* chromosome on specialized transducing phages lambda-*asn*. *Mol. Gen. Genet.* 160:287–295.

317. von Meyenburg, K., F. G. Hansen, E. Riise, H. E. Bergmans, M. Meijer, and W. Messer. 1979. Origin of replication, *oriC*, of the *E. coli* K-12 chromosome: genetic mapping and minichromosome replication. *Cold Spring Harbor Symp. Quant. Biol.* 43:121–128.

318. Wahle, E., R. S. Lasken, and A. Kornberg. 1989. The dnaB-dnaC replication protein complex of *Escherichia coli*. II. Role of the complex in mobilizing dnaB functions. *J. Biol. Chem.* 264:2469–2475.

319. Wahle, E., R. S. Lasken, and A. Kornberg. 1989. The dnaB-dnaC replication protein complex of *Escherichia coli*. I. Formation and properties. *J. Biol. Chem.* 264:2463–2468.

320. Walker, J. R., J. A. Ramsey, and W. G. Haldenwang. 1982. Interaction of the *Escherichia coli dnaA* initiation protein with the *dnaZ* polymerization protein *in vivo*. *Proc. Natl. Acad. Sci. USA* 79:3340–3344.

321. Wang, Q., and J. M. Kaguni. 1987. Transcriptional repression of the *dnaA* gene of *Escherichia coli* by dnaA protein. *Mol. Gen. Genet.* 209:518–525.

322. Wang, Q., and J. M. Kaguni. 1989. dnaA Protein regulates transcription of the *rpoH* gene of *Escherichia coli*. *J. Biol. Chem.* 264:7338–7344.

323. Wechsler, J. A., and M. Zdzienicka. 1975. Cryolethal suppressors of thermosensitive *dnaA* mutations. *UCLA Symp. Mol. Cell. Biol.* 3:624–638.

324. Wende, M., A. Quinones, L. Diederich, W.-R. Jueterbock, and W. Messer. 1991. Transcription termination in the *dnaA* gene. *Mol. Gen. Genet.* 320:486–490.

325. Wickner, S., and J. Hurwitz. 1975. Interaction of *Escherichia coli dnaB* and *dnaC(D)* gene products *in vitro*. *Proc. Natl. Acad. Sci. USA* 72:921–925.

326. Woelker, B., and W. Messer. 1993. The structure of the initiation complex at the replication origin, *oriC*, of *Escherichia coli*. *Nucleic Acids Res.* 21:5025–5033.

327. Wold, S., K. Skarstad, H. B. Steen, T. Stokke, and E. Boye. 1994. The initiation mass for DNA replication in *Escherichia coli* K-2 is dependent on growth rate. *EMBO J.* 13:2097–2102.

328. Wolf-Watz, H., and M. Masters. 1979. Deoxyribonucleic acid and outer membrane: strains diploid for the *oriC* region show elevated levels of deoxyribonucleic acid-binding protein and evidence for specific binding of the *oriC* region to outer membrane. *J. Bacteriol.* 140:50–58.

329. Worcel, A., and E. Burgi. 1972. On the structure of the folded chromosome of *Escherichia coli*. *J. Mol. Biol.* 71:127–147.

330. Wu, C. A., E. L. Zechner, J. A. Reems, C. S. McHenry, and K. J. Marians. 1992. Coordinated leading-and lagging-strand synthesis at the *Escherichia coli* replication fork. V. Primase action regulates the cycle of Okazaki fragment synthesis. *J. Biol. Chem.* 267:4074–4083.

331. Xu, Y.-C., and H. Bremer. 1988. Chromosome replication in *Escherichia coli* induced by oversupply of DnaA. *Mol. Gen. Genet.* 211:138–142.

332. Yang, C.-C., and H. Nash. 1989. The interaction of *E. coli* IHF protein with its specific binding sites. *Cell* 57:869–880.

333. Yasuda, S., and Y. Hirota. 1977. Cloning and mapping of the replication origin of *Escherichia coli*. *Proc. Natl. Acad. Sci. USA* 74:5458–5462.

333a. Ye, F., J. Renaudin, J.-M. Bové, and F. Laigret. 1994. Cloning and sequencing of the replication origin (*oriC*) of the *Spiroplasma citri* chromosome and construction of autonomously replicating artificial plasmids. *Curr. Microbiol.* 29:23–29.

334. Yochem, J., H. Uchida, M. Sunshine, H. Saito, C. Georgopoulos, and M. Feiss. 1978. Genetic analysis of two genes, *dnaJ* and *dnaK*, necessary for *Escherichia coli* and bacteriophage lambda DNA replication. *Mol. Gen. Genet.* 164:9–14.

335. Yoshikawa, H., and N. Ogasawara. 1991. Structure and function of DnaA and the DnaA-box in eubacteria: evolutionary relationships of bacterial replication origins. *Mol. Microbiol.* 5:2589–2597.

336. Yoshikawa, H., and R. G. Wake. 1993. Initiation and termination of chromosome replication, p. 507–528. *In* A. L. Sonenshein, J. A. Hoch, and R. Losick (ed.), *Bacillus subtilis and Other Gram-Positive Bacteria: Biochemistry, Physiology, and Molecular Genetics.* American Society for Microbiology, Washington, D.C.

337. Yoshimoto, M., K. Nagai, and G. Tamura. 1986. Early replicative intermediates of *E. coli* chromosome isolated from a membrane complex. *EMBO J.* 5:787–791.

338. Yoshimoto, M., K. Nagai, and G. Tamura. 1986. Asymmetric replication of an *oriC* plasmid in *E. coli*. *Mol. Gen. Genet.* 204:214–220.

339. Yung, B. Y. M., E. Crooke, and A. Kornberg. 1990. Fate of the DnaA initiator protein in replication at the origin of the *Escherichia coli* chromosome *in vitro*. *J. Biol. Chem.* 265:1282–1285.

340. Yung, B. Y.-M., and A. Kornberg. 1988. Membrane attachment activates dnaA protein, the initiation protein of chromosome replication in *Escherichia coli*. *Proc. Natl. Acad. Sci. USA* 85:7202–7205.

341. Yung, B. Y.-M., and A. Kornberg. 1989. The dnaA initiator protein bind separate domains in the replication origin of *Escherichia coli*. *J. Biol. Chem.* 264:6146–6150.

342. Zakrzewska-Czerwinska, J., and H. Schrempf. 1992. Characterization of an autonomously replication region from the *Streptomyces lividans* chromosome. *J. Bacteriol.* 174:2688–2693.

343. Zechner, E. L., C. A. Wu, and K. J. Marians. 1992. Coordinated leading-and lagging-strand synthesis at the *Escherichia coli* DNA replication fork. III. A polymerase-primase interaction governs primer size. *J. Biol. Chem.* 267:4054–4063.

344. Zhou, Z., and M. Syvanen. 1990. Identification and sequence of the *drpA* gene from *Escherichia coli*. *Mol. Gen. Genet.* 172:281–286.

345. Zweiger, G., G. Marczynski, and L. Shapiro. 1994. A *Caulobacter* DNA methyltransferase that functions only in the predivisional cell. *J. Mol. Biol.* 235:472–485.

346. Zweiger, G., and L. Shapiro. 1994. Expression of *Caulobacter dnaA* as a function of the cell cycle. *J. Bacteriol.* 176:401–408.

347. Zyskind, J. W., J. M. Cleary, W. S. A. Brusilow, N. E. Harding, and D. W. Smith. 1983. Chromosomal replication origin from the marine bacterium *Vibrio harveyi* functions in *Escherichia coli*: *oriC* consensus sequence. *Proc. Natl. Acad. Sci. USA* 80:1164–1168.

348. Zyskind, J. W., L. T. Deen, and D. W. Smith. 1977. Temporal sequence of events during the initiation process in *E. coli* deoxyribonucleic acid replication: roles of the *dnaA* and *dnaC* gene products and ribonucleic acid polymerase. *J. Bacteriol.* 129:1466–1477.

349. Zyskind, J. W., N. E. Harding, Y. Takeda, J. M. Cleary, and D. W. Smith. 1981. The DNA replication origin region of the enterobacteriaceae. *UCLA Symp. Mol. Cell. Biol.* 22:13–25.

350. Zyskind, J. W., and D. W. Smith. 1977. Novel *Escherichia coli dnaB* mutant: direct involvement of the *dnaB252* gene product in the synthesis of an origin-ribonucleic acid species during initiation of a round of deoxyribonucleic acid replication. *J. Bacteriol.* 129:1476–1486.

351. Zyskind, J. W., and D. W. Smith. 1992. DNA replication, the bacterial cell cycle, and cell growth. *Cell* 69:5–8.

Features of the Chromosomal Terminus Region

THOMAS M. HILL

100

INTRODUCTION

The convergence of two replication forks at the end of the DNA replication cycle sets in motion a series of events that are necessary for complete separation of the two newly replicated chromosomes prior to cell division. The first event is termination of DNA replication, which is defined for the purposes of this chapter as the meeting of the two replication forks and the completion of the daughter chromosomes. Termination is followed by decatenation, which removes the catenated links that result from replication termination and link the daughter chromosomes (90). In some cases, decatenation is insufficient to physically separate the daughter chromosomes, because the chromosomes are covalently joined as dimers. In these instances, site-specific or homologous recombination resolves the dimers into monomers. Following complete separation of the daughter chromosomes, the final event takes place: the active segregation of the chromosomes into daughter cells. That these events occur in the terminus region, located 180° around the circular chromosome from the origin, is not just happenstance. The terminus regions of the chromosomes of *Escherichia coli* and *Salmonella typhimurium* (official designation, *Salmonella enterica* serovar Typhimurium) contain sites which trap replication forks, ensuring that these events are always confined to the chromosomal terminus. As our knowledge has increased, it has become clear that the terminus region contains features that distinguish it from other parts of the chromosome and that facilitate the terminal events of chromosome separation.

This chapter describes our current understanding of known features of the terminus region, including the replication arrest sites, which limit the end of the DNA replication cycle to the terminus region; the *dif*-XerCD resolvase system, which promotes site-specific recombination to resolve dimeric chromosomes; and hyperrecombination in the terminus region. The issues of decatenation and chromosome segregation are dealt with in chapter 105 of this volume.

TERMINUS REGION

Masters and Broda (71) and Bird et al. (10) introduced the concept of a replication terminus by demonstrating that the two replication forks initiated from *oriC* (min 84) meet on the opposite side of the *E. coli* chromosome between the *trp* (min 28) and *his* (min 44) loci. Kuempel et al. (57–59) and Louarn et al. (66, 67) showed that the region in which the converging replication forks meet contains an impediment to replication fork progression such that replication forks traveling in clockwise or counterclockwise are inhibited between *trp* (min 28) and *manA* (min 36). The identification of a barrier to replication fork progression suggested that termination events are always confined to this section of the chromosome, called the terminus region, and that this region might contain features which facilitate the decatenation and partitioning of the daughter chromosomes.

Attempts to identify genetic elements in the terminus region that are involved in specific cellular processes initially met with limited success, despite the construction of cotransduction (11, 25) and restriction (14, 15) maps. In fact, the terminus region of the *E. coli* chromosome still remains something of an enigma. With the exception of the *dif* locus (discussed below), the entire *E. coli* terminus region can be deleted without significantly affecting cell growth, an indication that terminus region genes, including the barrier to replication forks, are not essential (35).

Relatively few genetic loci have been identified, even though the number of expressed genes per unit length in the terminus region is equivalent to that of other regions of the chromosome (72). Furthermore, several of the identified genes, such as *dicAB* (6) and *recE* (93), are apparently of bacteriophage origin and reflect the presence of several defective prophage found scattered throughout the region.

The paucity of known genes, the presence of phage remnants, and the apparent dispensability of the terminus region have given the region the reputation of a "genetic graveyard" (a phrase coined by Jean-Pierre Bouché). Nonetheless, several phenomena associated with the terminus region have now been described in molecular terms and have reaffirmed that this region of the chromosome contains unique genetic and structural elements that distinguish it from other chromosomal domains. The first of these elements to be described is the *dif* locus.

SITE-SPECIFIC RECOMBINATION AT THE *dif* LOCUS

Site-specific recombination in the terminus region plays an important role in separating covalently linked chromosomes. The *dif* (for *deletion-induced filamentation*) locus is a DNA sequence located at kb 1600 (Fig. 1) that acts as a substrate for the site-specific recombinases XerC and XerD. *dif* shares both sequence homology and functional similarity with sites found on plasmids (Table 1), such as the *cer, parB,* and *psi* sites of plasmids ColE1, CloDF13, and pSC101, respectively (21, 34, 89). The function of *dif* is to resolve dimeric chromosomes that result from an uneven number of recombination events between sister chromosomes during the replication cycle. Loss of *dif* leads to a filamentous phenotype in a subpopulation of cells, presumably because cells containing dimeric chromosomes are unable to partition the chromosomes. The *dif* locus was discovered by three laboratories, each working independently of the other, each starting from a different point, and all with results pointing to the same spot on the *E. coli* chromosome.

dif Locus

Clerget (19) found *dif* while examining the self-maintenance properties of a transposable element, Tn*2350,* which is derived from the plasmid R1. Tn*2350* can be maintained in circular form in *E. coli* but is unstable and is lost rapidly in the absence of selective pressure. The minimal region required for Tn*2350* self-maintenance has two activities: the ability to promote dimerization of plasmids in a *recA* mutant background by site-specific recombination and the ability to rescue replication-defective plasmids. The latter activity was thought to be due to a replication origin in the sequence. However, Clerget showed that this property of the self-maintenance region was due not to autonomous

replication but rather to a very efficient integration and excision of the self-maintenance sequence with a similar site in the *E. coli* chromosome. The chromosomal integration site was subsequently mapped to the terminus region, and sequence similarity between the chromosomal site and the Tn*2350* site was demonstrated. Furthermore, the self-maintenance properties of Tn*2350* were dependent on the *xerC* gene, which encodes the XerC recombinase.

Kuempel et al. (60) were examining the effects of deletions in the terminus region when they noticed that removal of the region from kb 1552 to 1621 produced cultures with a filamentous phenotype in a subpopulation of the cells. The SOS repair system was also induced in cells with this region of the chromosome deleted, and the filamentous cells displayed abnormal nucleoids. By using targeted deletions, *dif* was localized further to kb 1600. Introduction of a *dif*-containing plasmid into a *dif* mutant strain relieved the filamentous phenotype only when the plasmid was able to recombine into the chromosome, suggesting that *dif* functioned as a *cis*-acting site. Furthermore, *dif*-containing plasmids formed dimers at a high frequency, even in a *recA* mutant background, suggesting that recombination at the *dif* locus was *recA* independent.

Sherratt's group (12) found *dif* by way of studies of the site-specific recombination systems of plasmids. Many low-copy-number plasmids carry site-specific recombination systems that resolve dimeric plasmids into monomers, thereby ensuring equipartitioning of the daughter plasmids at cell division. High-copy-number plasmids, which segregate randomly, also use recombination systems to resolve multimeric plasmids, thereby increasing the number of monomeric plasmids and the probability that daughter cells will receive a plasmid at cell division. Sherratt and colleagues had previously shown that ColE1 plasmids contain a site, called *cer,* that promotes plasmid monomerization and that function of the *cer* site depends on several host genes (12, 13, 20). One of these genes, *xerC*, encodes a recombinase of the λ integrase family. When mutants of *xerC* were examined under the microscope, many of the cells showed a filamentous phenotype and aberrant nucleoid distribution similar to that observed with the Dif phenotype. The sequence homology between *cer* and *dif* (Table 1) suggested that *xerC* might also act at *dif*. This possibility was confirmed by demonstrating that *dif* promoted *xerC*-mediated recombination in plasmids and that XerC specifically bound to short oligomers containing either the *cer* or the *dif* sequence.

XerC and XerD Recombinases

The chromosomal *dif* system varies from that of plasmids in that the plasmid systems are more specific and require accessory proteins. Four genes (*argR, pepA, xerC,* and *xerD*) are required for the *cer* site

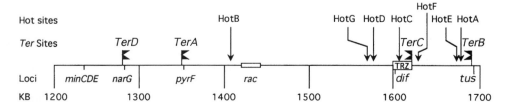

FIGURE 1 Terminus region of the *E. coli* chromosome, showing locations of the *dif* locus, where dimeric chromosomes are halted; the *Ter* sites, where replication forks are arrested by the Tus protein; and the Hot sites and TRZ, where *recA*-mediated hyperrecombination occurs.

TABLE 1 Comparison of *dif*-like recombination sites[a]

Position	Left arm		Central	Right arm	
E. coli dif	TT	GGTGCGCATAA	. . *TGTATA*	TTATGTTAAAT	CA
Plasmid R1	TT	AGTGCGCATAA	. . *TGTATA*	TTATGTTACAT	TG
Plasmid pSC101	GC	GGTGCGCGCAA	. . *GATCCA*	TTATGTTAAAC	GG
Plasmid ColE1, ColK	GC	GGTGCGTACAA	*TTAAGGGA*	TTATGGTAAAT	CC
Plasmid pMB1	GC	GGTGCGTACAA	*TTAAGGGA*	TTATGGTAAAT	CA
Plasmid pNPT16	GC	GGTGCGCGTAA	. *TGAGACG*	TTATGGTAAAT	CA
Plasmid ColN	GC	GGTGCGTACAA	. *TAAGGGA*	TTATGGTAAAT	CA
Plasmid ColDF13	GC	GGTACCGATAA	. . *GGGATG*	TTATGGTAAAT	AT
Plasmid ColA	GC	GGTACCGTTAA	. . *CGGATG*	TTATGGTAAAT	CA
Protected regions		| - - - - XerC - - - - - - -|	- - - - -XerD - - - |		

[a]*E. coli dif* sequences are compared to similar recombination sites in other replicons. The presumptive core regions flanking the central 6- to 8-bp central region are boxed. Gaps in the aligned sequences are indicated by dots. Nucleotides protected from 1,10-phenanthroline-copper footprinting by XerC and XerD are also indicated. Sequences are taken from reference 12 except for the pSC101 sequence, which is from reference 21.

in ColE1 plasmids, but only *xerC* and *xerD* are required for *dif* functioning (13). The ArgR and PepA accessory proteins presumably impart resolution selectivity to *cer* function in a manner similar to that observed for other site-specific recombination systems, such as the *res*-resolvase system of Tn*3* (88). The resolution specificity provided by the accessory proteins promotes preferential recombination between linked *cer* sites on multimeric plasmids but not between sites on separate monomeric plasmids. This resolution selectivity does not appear to exist for *dif*, thereby permitting interconversion between monomers and dimers (12).

The *xerC* gene maps to kb 4024 (min 85) and is part of a four-gene operon that includes the *dapF* (diaminopimelate epimerase) gene and two other open reading frames (ORFs). All four genes are probably transcribed from a promoter upstream of *dapF*, and all are probably expressed to the same level. The *xerC* gene encodes a protein of 298 amino acids (molecular mass, 33.8 kDa) that contains two regions with high homology to the lambda integrase family of recombinases.

The *xerD* gene was identified by Blakely et al. on the basis of its homology to the *xerC* gene (13). XerD shows 37% amino acid identity to XerC and also contains 298 amino acids, suggesting that the genes are closely related. *xerD* maps to kb 3055 (min 62) and appears to be part of the same transcriptional unit as *recJ* and *xprA*. Inactivation of *xerD* by insertional mutagenesis, as with *xerC* mutants, abolishes site-specific recombination at *dif* sites, indicating that XerD is required for *dif* functioning.

By using an oligomer containing a *dif* site, it was shown that either XerC or XerD alone is capable of binding specifically to free *dif* DNA but that each protein binds with a higher affinity to a preformed complex containing *dif* and the other protein. This suggests that binding of the two recombinases is cooperative (13). Footprinting studies of XerC-*dif* or XerD-*dif* complexes also demonstrated that each protein bound specifically to a different half of the palindromic *dif* site: XerC binds to the left half and XerD binds to the right half of a *dif* site (Table 1). However, artificial *dif* sites containing two left or two right halves of the *dif* site were inactive, indicating that binding of both XerC and XerD was absolutely required for normal *dif* functioning.

Function of *dif*

The *dif* and *xer* genes play an essential role in facilitating chromosome partitioning that is distinct from the function of topoi-

somerases. The catenated chromosomes that form at the end of the DNA replication cycle can be physically separated by topoisomerase IV as long as the sister chromosomes undergo an even number of recombination events during replication and the chromosomes exist as monomers (see chapter 105). However, if an uneven number of recombination events occurs during the replication cycle, removal of the catenates by topoisomerase IV is insufficient to physically separate the resulting dimeric chromosome (Fig. 2). The action of XerCD at the juxtaposed *dif* loci serves to monomerize the linked chromosomes. Thus, in a *dif* or *xer* mutant, the filamentous phenotype observed in a significant fraction of the population is assumed to result from cells containing linked chromosomes that cannot be partitioned.

The lack of resolution specificity at the *dif* site is somewhat curious, as there appears to be a great advantage in providing a directional bias to this process. Without resolution specificity, the action of XerCD at *dif* would be expected to dimerize monomeric chromosomes with the same efficiency as dimeric chromosomes are resolved. Thus, continued recombination at *dif* would be expected to occur regardless of whether the daughter chromosomes were monomers or dimers. What, then, provides directionality to the *dif*-XerCD system? Blakely et al. (12) proposed that the active segregation of topologically unlinked daughter chromosomes may impart directionality, for as the chromosomes pull apart, the two *dif* sites become spatially separated and can no longer function as recombination sites. As long as recombination events occur frequently and rapidly at the *dif* site, dimeric chromosomes will be converted to the monomeric form at some point during segregation and will be sufficiently separated to prevent further recombinational events.

As pointed out by Baker (3), this mechanism for *dif* resolution specificity works fine as long as *dif* is located in the last domain of the chromosome to be pulled apart. If reorganization of nucleoid structure accompanies DNA replication, then distinct domains of the nucleoid may be spatially separated quite rapidly following passage of a replication fork. Domains close to *oriC* would be expected to be segregated faster than terminal domains. Thus, placement of *dif* anywhere other than the terminal domain of the nucleoid would allow dimeric chromosomes to become trapped in that configuration, preventing chromosome segregation. This suggests that the location of the *dif* site in the terminus is not a result of happenstance but is a requirement for its function.

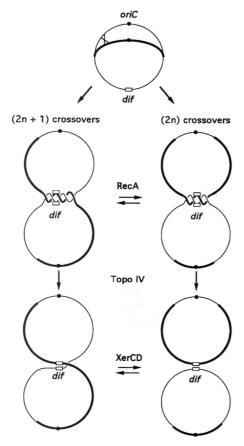

FIGURE 2 Resolution of dimeric chromosomes by the *dif* locus. Recombination between sister chromatids during replication produces monomers if an even number of crossover events occur, or dimers if an odd number of crossovers occur. Decatenation of links between newly replicated chromosomes by topoisomerases will not resolve dimeric chromosomes; for this, the XerCD resolvase must function at the *dif* site.

Recently, Tecklenburg et al. showed that the function of *dif* is in fact position dependent. When the normal *dif* locus is deleted and replaced by a new *dif* sequence located some 120 kb counterclockwise from the usual site, the new *dif* locus is unable to suppress the filamentous phenotype, even though the new site still functions as a site for XerCD-mediated recombination (91). Additional attempts to relocate the site to other points in the chromosome by using a *dif*-containing transposon failed; the only transposon insertions that relieve the *dif* phenotype are those that are located within 10 kb on either side of the original *dif* site. Thus, *dif* appears to be functional only when it is located in the terminal domain of the nucleoid, where decatenation presumably occurs. The catenates present in this region bring the *dif* sites on daughter chromosomes into juxtaposition and facilitate the monomerization of the linked genomes.

HYPERRECOMBINATION IN THE TERMINUS REGION

Recently, it was shown that the frequency of recombination in the terminus region is increased relative to that in other regions of the chromosome. It appears that this phenomenon can be attrib-

uted in part to the unique position of the terminus region with respect to chromosome replication. However, another contributing factor to hyperrecombination may be Tus-mediated arrest of DNA replication (discussed below), which produces stalled replication forks that may be highly recombinogenic. In this section, the current evidence and possible models for hyperrecombination in the terminus region are discussed.

Recombination at *Ter* Sites

Increased levels of both homologous and illegitimate recombination occur near *Ter* sites (reviewed in reference 9). Horiuchi and coworkers found eight recombinational hotspots (called Hot DNA) while searching for the replication origins utilized by *E. coli rnh* mutants during stable DNA replication. Seven of these Hot DNAs are located in the terminus region between *TerA* and *TerB* (75; Fig. 1). These sites do not function as autonomous replicons; rather, they can be isolated as plasmid-like forms because they integrate into and are excised from the chromosome at a high frequency by homologous recombination. All of the Hot DNAs possess chi activity (75). Three of the Hot DNAs are adjacent to *Ter* sites, and hyperrecombination activity at these sites depends on the presence of a functional *tus* gene (49). The other four Hot DNAs function independently of *tus*. In one case (the HotA site located near *TerB*), mutation of the chi sequence diminishes Hot activity; however, insertion of a new, properly oriented chi site between HotA and the *TerB* site restores hyperrecombination. It is postulated that arrested replication forks at *Ter* sites produce double-stranded breaks that act as an entry point for the RecBCD enzyme (49). The RecBCD enzyme then utilizes the chi site to repair the chromosome.

Replication arrest at a *Ter* site can also promote illegitimate recombination (7). In hybrid plasmids containing both pBR322 and M13 origins of replication, a *TerB* site acts as a deletion hotspot when it is oriented to arrest DNA replication initiated from either plasmid or phage origin. Approximately 80% of the deletions originating at M13 have endpoints that map within several nucleotides of the start of the *TerB* sequence, which is the precise point of arrest of leading-strand synthesis (46, 61). Furthermore, the deletion hotspot activity is dependent on the *tus* gene. Bierne et al. (7) suggest that the arrested replication fork, in conjunction with the nicked M13 origin of replication, produces structures that promote deletion formation.

Terminal Recombination Zone

Louarn et al. (68) found that the terminus has an increased frequency of RecA-mediated homologous recombination. In particular, a region near kb 1600 had a frequency of recombination 1,000-fold higher than that of other parts of the chromosome. This hyperrecombination is not sequence specific but appears to be connected to events that follow termination of DNA replication. Also, this form of recombination is not *tus* dependent and therefore does not rely on replication arrest (65, 68).

By measuring the excision of prophage from many different points around the genome, the frequency of homologous recombination at many locations on the *E. coli* chromosome was accurately determined (68). The excision frequency is uniform (10^{-5} per generation) for most of the chromosome (from min 44 clockwise to min 23), increases significantly (10- to 100-fold) at the edges of the terminus region near *TerA* and *TerB*, and reaches a maximum of 10^{-2} at min 33.8 near the *TerC* locus. The region of maximal hyperrecombination is called the terminus recombination zone (TRZ) and is located at kb 1600 to 1640 (65). Inactivation of the *recA* gene abolishes hyperrecombination, as does loss of RecBCD function. This loss of RecBCD function suggests that this region may, in part, coincide with some of the *tus*-independent Hot DNA sites (49).

Hyperrecombination in the TRZ does not appear to be associated with specific sequence elements, such as chi sites. Deletions that remove the region encompassing the TRZ (and therefore some of the *tus*-independent Hot DNA sites) do not eliminate terminal hyperrecombination; the zone of maximum recombination is simply shifted in the direction of the deletion. Furthermore, hyperrecombination activity in the TRZ is independent of the *dif* locus. Deletion of *dif* or introduction of *xerC* mutations has no effect on recombination frequency in the TRZ. These observations suggest that hyperrecombination in the TRZ is due to the position of the TRZ in the nucleoid rather than to a special recombination system (65).

Hyperrecombination in the TRZ is not linked to termination of DNA replication, even though the TRZ is proximal to *TerC,* where termination is believed to occur most frequently (discussed below). The separation of replication termination and TRZ activity was demonstrated by introducing an extra *Ter* site at kb 1400, thereby shifting the position of replication termination from kb 1600 to kb 1400. Even though the likely meeting point for replication forks was shifted by 200 kb, hyperrecombination still occurred at the same frequency at kb 1600. This indicates that hyperrecombination at TRZ is independent of replication termination and probably takes place after replication has been completed (65).

To explain the phenomenon of the TRZ, it is proposed that after the terminal stages of replication, catenation links between progeny chromosomes are "chased" to the last region of the nucleoid to be organized, which presumably includes the TRZ (65). This would happen regardless of where replication termination occurs. The driving force for chasing the catenation links to the TRZ may be the restructuring of the nucleoid into distinct domains after passage of a replication fork. The reorganization of the nucleoid would physically separate newly replicated domains of the daughter chromosomes, and if replication termination occurred outside of the terminus, the catenation links would be driven into the terminus region as the nucleoid was reorganized. Furthermore, the noninvertable zones that flank the terminus region (27, 28) may contain polarized organizing sequences for nucleoid structure, which could provide a topographical impetus to nucleoid reordering and help force the catenation links to the terminal domain of the nucleoid. As a result of these activities, the catenated progeny chromosomes would be highly susceptible to recombination in the TRZ during the final stages of nucleoid assembly and decatenation.

Role of Terminal Hyperrecombination

The physiological advantage of having elevated levels of recombination associated with the terminus region is hard to fathom, because it is difficult to imagine that the cell would benefit from persistently unstable regions of the genome. Nonetheless, there is evidence for genome instability in the terminus, as Masters and Oliver (M. Masters and I. R. Oliver, *J. Cell. Biochem. Abstr. Suppl.,* **17E**:300, 1993) showed that homology between the terminus regions of *E. coli* and other enterobacterial species is less than that at other regions of the chromosome. Also, extensive rearrangements must have occurred, because regions containing several kilobases of DNA that are present in *E. coli* appear to be missing in other enterobacteria. This suggests that insertions and deletions occur with relative frequency on an evolutionary time scale. Perhaps this is one of the reasons for the apparent dispensability of the terminus region in *E. coli;* the constant shuffling, deletion, and acquisition of genetic material by both homologous and illegitimate recombination may exclude essential genes from this region.

Perhaps hyperrecombination at the terminus is simply a price bacteria pay for having circular chromosomes. The catenated daughter chromosomes that are formed at the completion of replication serve as willing substrates for homologous recombination. Thus, the high level of terminal recombination at the completion of chromosome replication is probably not the result of design but rather is a consequence of the recombinational machinery being provided with a suitable substrate for significant periods of time.

REPLICATION ARREST IN *E. COLI* AND *S. TYPHIMURIUM*

The terminus regions of the chromosomes of *E. coli* and *S. typhimurium* contain specific sites that block the progression of DNA replication forks. These sites, called *Ter* sites (2), constitute the recognition sequence for the Tus protein, which mediates arrest of the replisome. The binding of Tus to a *Ter* site forms an asymmetric complex that impedes replication forks approaching from one direction but not the other. Thus, the *Ter* sites are oriented in the bacterial chromosome to allow completion of the DNA replication cycle, but they prevent replication forks from traveling in the terminus-to-origin direction.

It is important to emphasize that termination of DNA and replication arrest are not synonymous terms, even though the former is often used in place of the latter. Termination of DNA replication is defined as the events that occur when the two replication forks meet and complete the daughter chromosomes. Arrest of DNA replication is defined as the events that occur when a replication fork is stopped by a Tus-*Ter* complex. That termination of DNA replication and replication arrest are not equivalent has been demonstrated most clearly in strains lacking a *tus* gene. In these cells, termination of DNA apparently occurs in the absence of a functional replication arrest system, indicating that the events required for completion of the daughter chromosomes can proceed without replication forks being blocked at a *Ter* site.

Ter Sites

The *Ter* sites in the *E. coli* chromosome were identified by using strains in which DNA replication was initiated from plasmid or phage origins integrated near the terminus region (57–59, 66, 67). This approach avoided problems associated with initiation of replication from *oriC*, in which both replication forks arrive at the terminus region at approximately the same time. In the initial studies, the barrier to replication was localized to a broad area of the terminus region. Clockwise traveling replication forks were halted between *rac* (min 30) and *aroD* (min 37), and counterclockwise replication forks were arrested between *rac* and *trp* (min 28) (58, 59, 66, 67).

The impediment to DNA replication in the terminus region was separated into two distinct loci in Louarn's (24) and Kuempel's (44) laboratories. Each of the two sites was specific for the arrest of either clockwise or counterclockwise replication forks, and the two sites were well separated at the edges of the terminus region. The sites functioned in a polar fashion; that is, the *Ter* sites halted replication forks approaching from one direction but not the other. Both groups suggested that replication forks are arrested only transiently at the *Ter* sites; the arrested replication forks eventually continue past the arrest sites. The *TerA* site, which halts counterclockwise replication forks only, is located near *pyrF* at 28.5 min. A second site, which halts only clockwise replication forks, was mapped differently by the two groups; deMassy et al. (24) placed *TerB* at min 33.5, and Hill et al. (44) placed it between min 34.5 and 35.7, near *manA*. It was shown later that this apparent discrepancy reflects the presence of two distinct arrest sites for clockwise replication (*TerC* and *TerB*, respectively) on this side of the terminus region. The polarity of the *Ter* sites and their positions at the edges of the terminus region suggest that they function as a trap, allowing replication forks to enter but not exit this region.

Additional arrest sites were subsequently found in and around the terminus region (26, 39, 40, 84). Six sites are asymmetrically distributed over approximately 25% of the total chromosome and are oriented to halt DNA replication moving in the origin-to-terminus direction, with *TerB*, *TerC*, and *TerF* posi-

TABLE 2 Comparison of known *Ter* sites[a]

Site	Sequence
Chromosomal	
E. coli	
TerA	AATTAGTATGTTGTAACTAAAGT
TerB	AATAA.TA..........AAGT
TerC	ATATA.GA..........ATAT
TerD	CATTA.TA..........AATG
TerE	TTAAA.TA..........AG
TerF	CCTTC.TA.........G.CGAT
TerG	GTCAA.GA..........ACTA
S. typhimurium	
TerA	ATTAA.TA..........AAGC
Ter (amyA)	GATGA.TA..........AATG
Plasmids	
R6K*TerR1*	CTCTT.TGTGTTGTAACTAAATC
R6K*TerR2*	CTATT.AG..........CTAG
R100*TerR1*	ATTAT.AA..........CTTC
R100*TerR2*	TGTCT.AG..........AAGC
R1*TerR1*	ATTAT.AA..........CATC
R1*TerR2*	TTTTT.TG..........AATT
RepFIC*TerR1*	ATTAT.AA..........CATT
St90kb*Ter*	ATTTT.GA..........TTTG
Position	....|....|....|....|...
	5 10 15 20

[a]Known *Ter* sequences from the *E. coli* chromosome and other replicons are compared. As presented, these *Ter* sites arrest replication forks approaching from the 5′ side of the sequence and allow replication forks approaching from the 3′ side to pass unimpeded. Nucleotides that represent the core sequences identical for all *Ter* sites are indicated by dots.

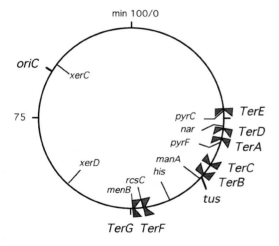

FIGURE 3 Map of the *E. coli* chromosome, showing locations of the termination sites (*TerA-TerG*) relative to locations of the origin of replication and other genetic markers. *TerA, TerD,* and *TerE* arrest counterclockwise-traveling replication forks only, whereas *TerB, TerC, TerF,* and *TerG* arrest clockwise-traveling replication forks. The location of the *tus* gene, which produces the *Ter*-binding protein Tus, is also shown.

Site	Minutes	KB
TerA	28.9	1350
TerB	36.2	1693
TerC	34.6	1619
TerD	27.5	1286
TerE	23.3	1091
TerF	48.9	2331
TerG	49.0	2345

tioned to arrest clockwise replication and *TerA, TerD,* and *TerE* positioned to arrest counterclockwise replication (Fig. 3). Recently, we identified a putative seventh *Ter* site on the basis of sequence similarity. *TerG* is located at min 49, within one of the genes for the biosynthesis of menaquinine (vitamin K$_2$) (78) and is also oriented to arrest DNA replication forks moving in the terminus-to-origin direction.

Empirical evidence for additional *Ter* sites exists. François et al. (26) identified 10 different chromosomal fragments containing homology to a oligomeric probe containing the *TerA* sequence. Most likely, these putative *Ter* sites are situated outside of the traditional terminus region, as is already the case for *TerE, TerF,* and *TerG,* and may be used rarely, if at all, during normal replication cycles.

Ter sites have also been found in *E. coli* plasmids, most notably in the R6K plasmid (30, 50, 56) and in members of the IncFII compatibility group, such as R100 and R1 (47, 50). In virtually all cases, a pair of *Ter* sites is present, and these sites are oriented to form a replication fork trap. As in the *E. coli* chromosome, these *Ter* sites also depend on the Tus protein for function.

A list of *Ter* sequences from the *E. coli* chromosome, *S. typhimurium* chromosome, and plasmids is given in Table 2. The invariant nucleotides are restricted to the G residue at position 6 and the 11-bp sequence from residues 9 through 19 (39, 47). The one exception to this rule is *TerF*, which has a G substitution at position 18. The four nucleotides upstream of the core sequence, from positions 2 through 5, are neither invariant nor protected by Tus binding but are generally A/T rich, suggesting that sequence context may influence either Tus binding or activity. A good example of the effect of context can be seen by comparing the binding affinities of Tus for the *TerB* and R6K*TerR2* sites, since both of these sites contain the consensus core sequence yet they show a 30-fold difference in the equilibrium binding constant (31; discussed below). Adjoining sequences outside the standard 23-bp *Ter* sites may also affect replication arrest activity (8).

It is worth noting that the *Ter* sites identified to date are primarily in intergenic regions, where transcriptional activity is lowest. This is probably because *Ter* sites are able to inhibit transcription (82). The orientation of a *Ter* site that blocks replication forks also halts RNA polymerase; the opposite orientation is permissive for both. Thus, a Tus-*Ter* complex located inside a transcriptional unit can interfere with RNA polymerase function if the *Ter* site is oriented to prevent transcription. By the same token, if the orientation of the *Ter* site is permissive for transcription, as occurs with the *TerF* site in the *rcsC* gene (84), RNA polymerase function would not be affected and the polymerase would displace Tus, possibly reducing the effectiveness of the *Ter* site. In fact, Tus-*Ter* complexes in intergenic regions appear to be protected from transcriptional interference by the presence of Rho-independent transcriptional terminators on the permissive side of the sites (39). It is also interesting to note that the *TerF* site, which is located within the *rcsC* gene and is oriented in the permissive direction with respect to transcription, is still functional (84). This indicates that transcriptional interference does not necessarily obviate the function of *Ter* site but may reduce its efficiency.

tus Gene

The possibility that a cell-encoded *trans*-acting factor is required for replication arrest was first raised in studies examining in vitro replication of hybrid R6K/ColE1 plasmids (4, 30). Deletion analysis of the terminus of the *E. coli* chromosome subsequently identified a region that contained a gene necessary for function of the *Ter* sites (45). Deletions that removed the *TerB* site also inactivated *TerA*, even though these two sites are separated by approximately 350 kb. Introduction of a plasmid carrying the deleted region near *TerB* restored *TerA* activity, suggesting that a *trans*-acting factor encoded by a gene in the vicinity of *TerB* is required for replication arrest. The gene is named *tus* (terminus utilization substance) or, alternatively, *tau* (53). The region encoding *tus* has two large ORFs separated by only 75 bp, with the *TerB* site situated between the two ORFs and just upstream of the second ORF. Insertional inactivation of the second ORF showed it to be the *tus* gene (42, 48), since loss of this gene was correlated in vitro with loss of a *Ter*-specific DNA-binding activity and in vivo with loss of the replication arrest function of *Ter* sites (48, 53).

Fewer than a hundred copies of Tus protein are normally present in a cell (73). This relatively low level of expression is due to several features of the *tus* gene region. First, the primary promoter for transcription of the *tus* gene, which is located immediately upstream of the gene, shows only weak homology to the consensus *E. coli* promoter sequence (82) and is probably transcribed at a low level. Second, the *TerB* site is located within the promoter region of the *tus* gene, suggesting that *tus* expression is autoregulated by binding of Tus to the *TerB* site (42, 48). Both in vivo (42, 82, 83) and in vitro (74) studies have verified this second supposition.

Upstream of *tus* and *TerB* are two genes, called *urpT* (unidentified regulatory protein/Terminus) and *uspT* (unidentified sensory protein/Terminus), that appear, on the basis of sequence information (83), to code for potential sensor-regulator proteins. *urpT* and *uspT* are not involved in replication arrest, since insertional inactivation of *uspT* has no effect on the function of the *Ter* sites (48). Both genes are transcribed in the same direction as *tus*, and the 3′ end of *uspT* is located only 75 bp from the 5′ end of *tus*. No potential transcriptional terminators are apparent in the intragenic region, and readthrough transcription of *tus* from *uspT* occurs in a *tus* mutant strain (82), albeit at a much lower level than transcription originating from the major *tus* promoter. However, in a *tus*+ strain, only a very small fraction of the transcripts originating from *uspT* traverse the *tus* gene, presumably because of the barrier posed by Tus bound to the *TerB* site. Thus, the Tus-*TerB* complex regulates *tus* expression by two methods: by occluding the major promoter and by impeding the progression of an actively transcribing RNA polymerase originating from upstream promoters.

Just downstream of and transcribed convergently with *tus* is *fumC*, the gene for a class II fumarase (33, 95). The *tus* and *fumC* genes overlap at their stop codons, and a potential Rho-independent terminator for *fumC* is positioned within *tus* (94). This suggests that transcription of *fumC* may interfere with *tus* transcription, but the effect of *fumC* on the expression of *tus* has not been examined.

The relatively long in vitro half-life of the Tus-*TerB* complex (31) suggests that expression of *tus* may be coordinated with the cell cycle, since displacement of Tus from the *TerB* site would be expected to occur only when a replication fork had passed through the *tus* gene region. However, a recent examination of *tus* expression during synchronized chromosomal replication did not reveal periodicity in the level of *tus* transcripts (C. E. Helmstetter, J. A. Bogan, P. Zhou, P. W. Theisen, H.-J. Wang, K. Welch, and J. Grimwade, *Abstracts: EMBO Workshop on the Bacterial Cell Cycle*, p. 41–43).

Tus Protein

The sequence of the *tus* gene predicts a highly basic protein of 309 amino acids (35,780 molecular mass). No significant homology to the binding motifs of other known DNA-binding proteins is found in Tus (48). Thanks to the advent of overexpression vectors and the relative ease of Tus purification (42, 46, 86), many of the biochemical and physical properties of Tus have been determined (Table 3). Two of these properties deserve comment. First, in combination with the lack of dyad symmetry in the *Ter* sequence, the fact that Tus binds as a monomer (22, 85) ensures the polar function of the protein-DNA complex, because a different face of the Tus protein will be exposed to the replication fork depending on the direction from which the replication apparatus approaches. Second, the observed isoelectric point of the native protein is 7.5 (22), which is considerably lower than the pI of 10.1 calculated from the nucleotide sequence of the *tus* gene (48). The difference between the observed and predicted pIs is presumably due to the folding of the native protein, since no posttranslational modifications of the protein that could account for this difference have been reported.

Tus has a high affinity for the *TerB* site in vitro, with an observed equilibrium binding constant of 3.4×10^{-13} M and a half-life of approximately 550 min for the Tus-*TerB* complex in optimal buffer conditions (50 mM Tris-Cl at pH 7.5, 150 mM potassium glutamate) (31). This half-life exceeds by a factor of

TABLE 3 Physical and biochemical characteristics of Tus protein

Characteristic	Value	Reference(s)
Mol wt		
SDS-PAGE[a]	36,000	22, 42, 85
S-100 gel filtration	32,000	22, 86
Ultracentrifugation	36,190	22
Calculated	35,780	42, 48
Subunit structure	Monomer	22, 85
Stoichiometry of binding	1:1	22, 85
Stokes radius	23 Å (2.3 nm)	22
Axial ratio	2	22
ε_{280}	39,000	22
Isoelectric point		
Calculated	10.1	48
Observed	7.5	22
Predicted secondary structure		22
α-Helix	40%	
β-Sheet	0%	
Random coil	45%	
Turns	15%	

[a]SDS-PAGE, sodium dodecyl sulfate-polyacrylamide gel electrophoresis.

100 that of the Lac repressor-operator complex (52), which is considered a relatively tightly binding protein. Studies of the interactions of Tus with another chromosomal site, *TerF*, and with the plasmid R6K *TerR2* site (a less efficient replication arrest site) (39, 80) demonstrated a 10- to 30-fold-lower affinity for these sites, with observed K_ds of 10^{-12} M (84) and 10^{-11} M (31), respectively. The reduced affinities of Tus for these two sites is primarily due to a faster dissociation rate of the Tus-*Ter* complexes, with half-lives of 46 and 43 min for *TerF* and R6K *TerR2*, respectively.

Footprinting studies of Tus bound to the chromosomal *TerB* site show an asymmetric distribution of protein-DNA contacts (31), which is consistent with the observed polarity of function (Fig. 4). The pattern of protected residues suggests that Tus spans the major and minor grooves of the *Ter* site, contacting primarily only one face of the double helix, and that a flexible arm of the protein extends around to the back of the helix in the major groove. Also, the region in which Tus appears to contact both strands of the double helix is on the side of the complex that impedes replication forks, whereas on the other side, which allows free passage of replication forks, only a single strand is in contact with Tus. As discussed below, this feature of Tus binding in conjunction with the stability of the protein-DNA complex may account for the observed orientation-dependent inhibition of a broad range of helicases in in vitro assays.

A protein that appears to specifically abrogate the arrest of DNA replication by Tus was reported by Natarajan et al. (73). This protein, called the anti-ter, was purified from *E. coli* cell extracts on the basis of its ability to "supershift" a Tus-*Ter* complex during gel retardation electrophoresis. When anti-ter was added to an in vitro replication system derived from crude cell extracts, the ability of Tus to halt replication of a plasmid containing a functional *Ter* site was significantly reduced. Likewise, the anti-ter protein prevented Tus-mediated inhibition of DnaB helicase activity in a helicase assay that used a short oligomeric substrate. The mode of action of the anti-ter is not known, but the protein does not appear to function by displacing Tus from

the *Ter* site, suggesting that the anti-ter may interfere with the interaction between Tus and the replication fork. The in vivo role of the anti-ter is also unknown, but possibilities for its function might include bypassing of Tus-*Ter* complexes during recombination, DNA repair, or conjugation.

Mechanism of Replication Arrest

Over the last few years, the molecular mechanism by which Tus exerts its unique ability to block the progression of replication forks has been the primary focus of studies on the Tus-*Ter* interaction. These studies have employed in vitro systems to attempt to identify the component of the replication apparatus Tus interacts with and to understand how Tus blocks DNA replication. The results from these studies support the conclusion that the target of Tus is the replicative helicase, DnaB, and that Tus prevents DnaB from unwinding the DNA strands ahead of the replication fork. In addition, Tus appears to exert its effect through protein-protein interactions with DnaB. Although the accumulated evidence clearly favors these views, conclusions about Tus function should be viewed with a modicum of skepticism, as no genetic evidence supporting the biochemical results exists.

Tus acts alone in arresting DNA replication, as shown by its ability to do so in an in vitro replication system composed entirely of purified proteins (46, 62). In such a system, Tus function depends on the orientation of the *Ter* site, faithfully mimicking its in vivo activity. These studies also demonstrated

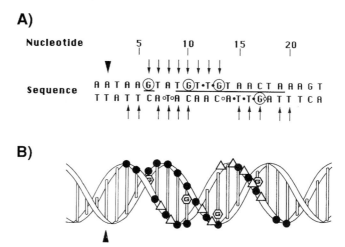

FIGURE 4 Location of protein-DNA contact sites in the Tus-*TerB* complex. (A) Circled nucleotides are guanine residues protected from dimethyl sulfate methylation. Closed circles between nucleotides indicate strong alkylation interference sites, and open circles indicate weak alkylation interference sites. Small arrows denote hydroxyl radical-protected sites. The arrowhead points to the last nucleotide of leading-strand synthesis. The heavy bar between the strands denotes the core sequence common to all known *Ter* sequences. (B) Three-dimensional representation of DNA-protein contacts in the Tus-*Ter* complex. Closed circles show the positions of hydroxyl radical-protected nucleotides, open triangles indicate positions of alkylation interference, and hexagons containing the letter G indicate dimethyl sulfoxide-protected guanine residues. (Reproduced with permission from reference 31.)

that the primary arrest sites of leading-strand replication are the first and second nucleotides of the *Ter* site (46, 61), which suggests that the DNA polymerase III holoenzyme is able to replicate very close to the Tus protein before being inhibited. However, these studies were unable to pinpoint the target of Tus; for this purpose, a simplified assay was required.

Because the replicative helicase is a likely target for Tus action, a strand displacement assay has been used to measure the effect of a Tus-*Ter* complex on helicase function. In this assay, a short, radiolabeled oligomer (30 to 60 bases) containing a *Ter* site is hybridized to a circular single-stranded DNA molecule to form a partial DNA heteroduplex. This helicase substrate is then incubated with Tus to form the Tus-*Ter* complex, and a purified DNA helicase is added to the mixture. The intrinsic unwinding activity of the helicase displaces the oligomer, which can be easily separated from the heteroduplex substrate by electrophoresis. The advantages of this assay are that (i) the direct action of Tus on the helicase can be determined because no other replication proteins are necessary and (ii) the effect of Tus on the activity of a wide variety of helicases can be tested with this simple protocol.

This assay showed that Tus acts as a polar barrier to helicases of both prokaryotic and eukaryotic origins and to helicases that unwind DNA in either the $5' \rightarrow 3'$ or the $3' \rightarrow 5'$ direction (Table 4). These observations suggest that Tus acts as an antihelicase: (i) it prevents strand separation by these helicases, and (ii) Tus function depends on the orientation of the *Ter* site, which is consistent with the observed polarity of Tus in vivo. The helicase assay also showed that a Tus-*Ter* complex inhibits strand displacement by DNA polymerases from T5 and T7 phages and the *E. coli* polymerase I large fragment (61). However, in these cases, inhibition is independent of the orientation of the *Ter* site. The endpoints of T7 DNA polymerization coincide almost exactly with the limits of the Tus footprint at the *TerB* site, suggesting that the tight binding of Tus to the *TerB* site is sufficient to halt these polymerase activities.

If Tus functions as an antihelicase, how does it inhibit helicase activity? The variety of DNA-translocating proteins blocked by Tus suggests that Tus functions as a general barrier to the progression of these proteins, a notion first advanced by Lee et al. (62). However, conflicting results regarding the activity of Tus against certain helicases, such as Rep and UvrD, confused the issue (Table 4). Also, some helicases, including the Dda helicase of T4 phage (5) and, with the appropriate helicase substrates, DnaB, are not inhibited at all by Tus (37). These observations suggest that Tus may function via specific protein-protein interactions with its target (51).

Recently, compelling evidence that supports the protein-protein interaction model has been presented. Hiasa and Marians (37) demonstrated a differential ability of Tus to inhibit DnaB depending on the type of helicase substrate used in the assay. They compared Tus function on helicase substrates with either a short or a long oligomer hybridized to the single-stranded circle. With helicase substrates containing a short (60-base) *TerB* oligomer, inhibition of DnaB was orientation dependent, which is consistent with previous reports. However, when the helicase substrate contained a *TerB* oligomer of approximately 250 bp, DnaB passed through the Tus-*TerB* complex in either orientation, indicating that the barrier posed by the Tus-*Ter* complex is insufficient to stop the progression of the helicase. To account

TABLE 4 DNA-translocating proteins inhibited by Tus-*Ter* complexes

Protein and state	Direction of unwinding	Reference(s)
Helicases inhibited		
DnaB	$5' \rightarrow 3'$	37, 51, 62
Helicase I (F plasmid)	$5' \rightarrow 3'$	41
Helicase B (mouse)	$5' \rightarrow 3'$	41
UvrD (helicase II)	$3' \rightarrow 5'$	62
Rep	$3' \rightarrow 5'$	62
PriA	$3' \rightarrow 5'$	37, 61
SV40[a] T-antigen	$3' \rightarrow 5'$	5, 41
φX174 primosome (DnaB/PriA)	$5' \rightarrow 3'$, $3' \rightarrow 5'$	37
Helicases not inhibited		
DnaB	$5' \rightarrow 3'$	37
UvrD (helicase II)	$3' \rightarrow 5'$	37, 51
Rep	$3' \rightarrow 5'$	51
Dda (T4)	$5' \rightarrow 3'$	5
Polymerases inhibited		
DNA polymerase I (Klenow)		61
T5 DNA polymerase		61
T7 DNA polymerase		61
RNA polymerase		82

[a]SV40, simian virus 40.

for this, Hiasa and Marians postulated that assays using short oligomeric substrates measure only strand displacement activity rather than true helicase activity. On longer duplex substrates, strand displacement activity gives way to true DNA unwinding, a transition that occurs somewhere between turns 2 and 10 in the duplex DNA. Thus, the characteristics of Tus binding to *TerB* account for the orientation-dependent, generalized inhibition of helicase strand displacement on short oligomeric substrates. However, the Tus-*TerB* complex cannot halt DnaB once true duplex unwinding has started, suggesting that protein-protein interactions must be an integral step in Tus-mediated arrest of DNA replication.

The model that Tus mediates replication arrest through protein-protein interactions has been strengthened further by the isolation of two classes of mutant Tus proteins. One class has a reduced affinity for the *TerB* site yet can still arrest DNA replication in vivo with a relatively high efficiency (87). One of these mutants, TusA173V, has a much shorter half-life for the Tus-*Ter* complex ($t_{1/2} = 4.5$ min versus 150 min for wild-type Tus in buffer containing 200 mM potassium glutamate), yet arrests DNA replication at approximately 75% of the efficiency of wild-type *tus*. A second class of Tus mutants show more normal binding affinities yet demonstrate a markedly reduced ability to arrest DNA replication in vivo. The most interesting of this class of mutants is TusE49K, which shows reduced arrest function, yet has a half-life that is even greater than that of the wild type ($t_{1/2} = 180$ min). This strongly suggests that binding to DNA alone cannot account for the ability of the Tus-*Ter* complex to halt replication forks.

If specific protein-protein interactions are required for Tus function, how is Tus able to inhibit such a broad array of proteins in the helicase assay? The answer probably lies in the tight binding of Tus and suggests that the inhibition of many of these helicases may be an artifact of the helicase assay. Using the stability of the Tus-*Ter* complex and the asymmetry of the pattern of contacts of Tus to the *Ter* site, Gottlieb et al. proposed a model for how Tus could function as a polar barrier to helicase unwinding (31). Although this model does not reflect the true mechanism by which Tus halts replication forks, it may explain the ability of Tus to nonspecifically inhibit unwinding activity in the helicase assay.

Possible Functions of Replication Arrest

Even though our understanding of the mechanism of Tus-mediated termination of DNA replication has advanced considerably, little progress has been made in understanding the true biological function of the Tus-*Ter* complex. Other than the simple explanation that *Ter* sites serve as one-way gates to prevent replication in the origin-to-terminus region, no other major physiological function has been ascribed to the replication arrest system. In large part, this ignorance stems from the lack of biological consequences following loss of *tus* gene function, even though several attempts have been made to determine whether *tus*⁺ cells have a selective advantage. For instance, Roecklein et al. (83) compared the growth characteristics of *tus*⁺ and *tus*::Kanʳ strains under a variety of conditions, including growth in rich and minimal media, medium shifts, anaerobic conditions, UV irradiation, and exposure to agents that inhibit DNA synthesis. None of these conditions allowed *tus*⁺ cells to significantly outgrow *tus* null mutants.

Another unknown is the frequency of replication cycles that terminate with replication forks arrested at any one of the known *Ter* sites, nor is it known how often the two replication forks simply collide without the use of a *Ter* site. In vitro replication studies using *oriC* plasmids containing a pair of *Ter* sites arranged as on the chromosome showed that virtually all replication cycles end with replication forks stalled at one or the other *Ter* site (38). However, because of the small size of the plasmid (5 kb), the frequency of *Ter* site usage may be unusually high. In the chromosome, replication cycles may not show the same bias for *Ter* site usage, as replication forks must travel over 2 Mb before arriving at the terminus.

Of the known *Ter* sites, *TerC* is probably used most often because it is almost equidistant from *oriC* (Fig. 3), and the region of the chromosome just upstream of *TerC* is replicated predominantly in the clockwise direction (65, 68). The frequent use of *TerC* may be significant, as the *dif* locus (12, 19, 60) and the TRZ (65) are both located adjacent to it. However, it is also clear that other *Ter* sites, such as *TerA* and *TerB,* are utilized in a substantial fraction of replication cycles. In exponentially growing cells, a significant population of the cells have replication forks arrested at either *TerA* and *TerB,* indicating that these sites are used frequently (79). Also, strains containing chromosomal inversions that double the distance from *oriC* to *TerB* relative to the distance from *oriC* to *TerA* use *TerA* and *TerB* at almost the same frequency. This was a surprise, because it should take twice as long to replicate the clockwise arm of the chromosome in this strain as the counterclockwise arm. This observation has not been satisfactorily explained.

Replication arrest sites located well outside of the traditional terminus region, such as *TerE* and *TerF,* are probably not used in normal replication cycles because the replication fork would have to pass through at least two other *Ter* sites located closer to the terminus region before reaching either of these sites (Fig. 3).

For example, in the studies that identified *TerF,* replication arrest at *TerF* could be demonstrated only when the region containing the *TerC* and *TerB* sites had been deleted, indicating that replication forks only rarely escape from these sites (84). The anticipated discovery of additional *Ter* sites outside the terminus region poses the question of the purpose of these nonterminus sites. Perhaps they are utilized when replication forks are initiated at origins other than *oriC*. Under these circumstances, these *Ter* sites would prevent replication in the terminus-to-origin direction but allow replication to proceed normally along the origin-to-terminus axis. Potential replication origins other than *oriC* are the *sdr* origins (1), prophage origins, or replication origins of integrated plasmids (see chapter 99 in this volume).

To date, the only examples of loss of *tus* being correlated with a phenotype are in *E. coli* strains in which completion of DNA replication is delayed or blocked by *Ter* sites. In one case, the DNA replication cycle was delayed because the chromosome was replicated predominantly in one direction (76); in the other case, inverted *Ter* sites, oriented to prevent completion of replication, were artificially introduced into the terminus region (B. Sharma and T. M. Hill, *Mol. Microbiol.,* in press). In the first instance, *oriC* was inactivated by insertion of the origin from plasmid R1, which then substituted as the chromosomal origin. In strains whose R1 origin is oriented to replicate in the counterclockwise direction (called intR1CC), the chromosome is replicated primarily in one direction; in strains with the opposite orientation of the R1 plasmid (intR1CW), replication is bidirectional, presumably because the replication fork passing through *oriC* initiates the counterclockwise fork (23). These intR1CC strains form long filaments, which subsequently pinch off small, anucleate cells (76). Presumably, the phenotype of the intR1CC strains results from the inability of the unidirectional replication fork to finish replication in coordination with cell division, since the replication fork has to pass through a series of active termination sites before the chromosome is completed. In agreement with this model, inactivation of the *tus* gene reduces filamentation (23).

In the second example, the DNA replication cycle of *E. coli* was disrupted by inserting inverted *Ter* sites into the terminus region of the chromosome to prematurely arrest DNA replication (Sharma and Hill, submitted). A pair of inverted *Ter* sites was inserted into the chromosome of a Δ*tus* strain at approximately min 31, between *TerA* and *TerC*. The resulting strain construction was called the Inv*Ter* strain. A functional *tus* gene was then introduced by transforming the Inv*Ter* strain with a plasmid carrying a copy of the *tus* gene under the control of an arabinose-inducible promoter. Induction of *tus* gene expression in the Inv*Ter* strain activates the inverted *Ter* sites and arrests replication forks after they complete all but the last 2 kb of the chromosome. In the absence of arabinose, *tus* expression is repressed and cell morphology is normal, but in its presence, the Inv*Ter* cells form filaments. This suggests that activation of the inverted *Ter* sites by induction of *tus* gene expression arrests DNA replication prior to completion of the chromosome and delays the onset of cell division.

Filamentation of Inv*Ter* cells is due to induction of the *sfiA* pathway normally associated with SOS induction and the *sfi*-independent pathway (18) as well. Thus, in replication cycles that are asymmetric, with one replication fork arriving at the terminus in advance of the other, arrest of the first fork may induce cell division inhibitors to prevent initiation of cell division until the second replication fork arrives and completes the chromosome. This function would primarily be a fine-tuning mechanism for coordinating DNA replication and cell division, and as a result, loss of *tus* gene expression would produce an uncoordinated cell cycle in only the small fraction of the population that had highly asymmetric replication cycles.

An intriguing hypothesis that links RecA activity to the function of the Tus-*Ter* system was presented by Zyskind et al. (97), who proposed that RecA may be activated by arrested replication forks at the *Ter* sites. Their hypothesis was based on their own observations, which suggested that RecA was involved with chromosome partitioning, and the observations of others, who suggested that the RecA dependence of certain integratively suppressed strains may be connected to replication arrest at the *Ter* sites (69; S. Dasgupta, R. Bernander, and K. Nordström, unpublished results). Zyskind et al. speculated that RecA is required for completion of the replication cycle when replication forks are blocked by a Tus-*Ter* complex; loss of RecA function leads to the production of anucleate cells. While this hypothesis is intriguing, examining it directly has not been possible. Also, a recent study questions the importance of RecA function following arrest of replication at a *Ter* site. Kogoma et al. (55) studied the RecA dependence of constitutive stable DNA replication and were unable to establish a link between RecA dependence and Tus activity. Thus, the relationship of RecA and replication arrest remains an interesting but uncertain proposition.

Despite our inability to identify a clear *tus* phenotype, the importance of arrest of DNA replication at sequence-specific sites is attested to not only by the high degree of similarity between the DNA sequences and the Tus proteins of *E. coli* and *S. typhimurium*, but also by the ever-increasing number of similar systems that have been identified in other organisms. The gram-positive bacterium *Bacillus subtilis* contains a functionally analogous system, in terms of both replication arrest components and their organization in the chromosome (for a review, see reference 43 or 96) In *Saccharomyces cerevisiae*, replication pausing has been associated with specific parts of the rRNA gene regions (16, 54, 63) and the centromeres (32) and is believed to depend on specific protein-DNA complexes (17, 32). Replication arrest is also associated with sequences in the rRNA genes of peas (36) and humans (64) and has been observed in bovine mitochondrial DNA at conserved sequences called termination-associated sequences, which appear to function by binding a *trans*-acting factor (70).

Replication Arrest in *S. typhimurium*

If the replication arrest system does not confer an evolutionary advantage, significant differences between the replication arrest proteins and the *Ter* sites of *S. typhimurium* and *E. coli* might have developed in the 160 to 180 million years of evolution that separate the two species (77). To examine the functional similarity of the replication arrest systems of the two bacteria, Roecklein et al. (83) introduced a plasmid containing a functional *E. coli* *TerA* site into *S. typhimurium* and showed that plasmid replication was halted at the *Ter* site. Furthermore, when a similar plasmid containing a *TerA* site in the nonfunctional orientation was introduced into *S. typhimurium*, plasmid replication was not halted, indicating that the polarity of *Ter* site function was also conserved. Hybridization with DNA probes containing the *E. coli*

TerB site or part of the *tus* gene revealed a minimum of three *Ter* sites in the *S. typhimurium* genome as well as a region that hybridized strongly to the *E. coli tus* gene. As with *E. coli*, *Ter* sequences can be found in plasmids, such as the *Salmonella* 90-kb virulence plasmid (29, 92).

The terminus regions of the chromosomes of *E. coli* and *S. typhimurium* are inverted with respect to one another, but it is reasonable to assume that the general organization of their *Ter* sites is similar. The inversion only exchanges the relative positions of the *Ter* sites but does not change their orientation with respect to the direction of DNA replication. That *Ter* site organization is conserved has been confirmed by the relationship of the *TerA* site and the *pyrF* gene in both bacteria. In *E. coli*, the *TerA* site is located 156 bp upstream of the *pyrF* start codon. In *S. typhimurium*, even though the *pyrF* gene and its upstream region show only 76 and 67% homology, respectively, to the *E. coli* sequence (83), the *TerA* site is perfectly conserved 155 bp upstream of the start of the *pyrF* gene, a testament to the importance of this site. However, not all sites are conserved, as witnessed by the divergence in the *amyA* gene (min 43), which has a putative *Ter* site in the 3′ end of the coding sequences in *S. typhimurium* but not in *E. coli* (29, 81). A possible explanation for the lack of conservation of this particular region may be that *Ter* sites located within the traditional terminus region, which would be expected to be used most often, might be more highly conserved than *Ter* sites located at other points around the chromosome.

TERMINAL DOMAIN OF THE NUCLEOID

A growing body of evidence suggests that the last domain of the nucleoid to be replicated and organized is special compared to other nucleoid domains. The uniqueness of the terminal domain is exemplified by the fact that two of its most interesting features, the TRZ and the *dif* locus, acquire function by virtue of their position in the chromosome and not by their DNA sequences. Thus, the TRZ (65) apparently exists regardless of what DNA sequences occupy the zone, and the *dif* locus (12, 19, 60) functions as a dimer resolution site only when it is located in the terminal domain of the nucleoid. In this context, the primary function of the *Ter* sites is to confine the terminal events of replication to the terminus region, thereby facilitating the final events of chromosome separation. This also explains the apparent dispensability of the replication arrest system. Loss of the *Ter* sites does not prevent the terminal replication events from occurring, but in the long run, such a loss may very well reduce the overall efficiency of cell growth and division and, consequently, the evolutionary fitness of the organism.

ACKNOWLEDGMENTS

I gratefully acknowledge all colleagues who shared in the preparation of the manuscript by providing information in advance of publication. I also extend thanks to Peter Kuempel and Rolf Bernander, who provided critical commentary on the contents of this chapter.

This work was supported by grants from the National Institutes of Health (GM43193) and the American Cancer Society (NP-828).

LITERATURE CITED

1. Asai, T., S. Sommer, A. Bailone, and T. Kogoma. 1993. Homologous recombination-dependent initiation of DNA replication from DNA damage-inducible origins in *Escherichia coli. EMBO J.* **12:**3287–3295.
2. Bachmann, B. J. 1990. Linkage map of *Escherichia coli* K-12, edition 8. *Microbiol. Rev.* **54:**130–197.
3. Baker, T. A. 1991. . . . And then there were two. *Nature* (London) **353:**794–795.

4. **Bastia, D., J. Germino, J. H. Crosa, and J. Ram.** 1981. The nucleotide sequence surrounding the replication terminus of plasmid R6K. *Proc. Natl. Acad. Sci. USA* **78:**2095–2099.
5. **Bedrosian, C. L., and D. Bastia.** 1991. *Escherichia coli* replication terminator protein impedes simian virus 40 (SV40) DNA replication fork movement and SV40 large tumor antigen helicase activity *in vitro* at a prokaryotic terminus sequence. *Proc. Natl. Acad. Sci. USA* **88:**2618–2622.
6. **Béjar, S., F. Bouché, and J.-P. Bouché.** 1988. Cell division inhibition gene *dicB* is regulated by a locus similar to lamboid bacteriophage immunity loci. *Mol. Gen. Genet.* **212:**11–19.
7. **Bierne, H., S. D. Ehrlich, and B. Michel.** 1991. The replication termination signal *terB* of the *Escherichia coli* chromosome is a deletion hot spot. *EMBO J.* **10:**2699–2705.
8. **Bierne, H., S. D. Ehrlich, and B. Michel.** 1994. Flanking sequences affect replication arrest at the *Escherichia coli* terminator *terB* in vivo. *J. Bacteriol.* **176:**4165–4167.
9. **Bierne, H., and B. Michel.** 1994. When replication forks stop. *Mol. Microbiol.* **13:**17–23.
10. **Bird, R. E., J. Louarn, J. Martuscelli, and L. Caro.** 1972. Origin and sequence of chromosome replication in *Escherichia coli. J. Mol. Biol.* **70:**549–566.
11. **Bitner, R. M., and P. L. Kuempel.** 1981. P1 transduction map spanning the replication terminus of *Escherichia coli* K12. *Mol. Gen. Genet.* **184:**208–212.
12. **Blakely, G., S. Colloms, G. May, M. Burke, and D. Sherratt.** 1991. *Escherichia coli* XerC recombinase is required for chromosomal segregation at cell division. *New Biol.* **3:**789–798.
13. **Blakely, G., G. May, R. McCulloch, L. K. Arciszewska, M. Burke, S. T. Lovett, and D. Sherratt.** 1993. Two related recombinases are required for site-specific recombination at *dif* and *cer* in *E. coli* K12. *Cell* **75:**351–361.
14. **Bouché, J. P.** 1982. Physical map of a 470 kilobase pair region flanking the terminus of DNA replication in the *Escherichia coli* K12 genome. *J. Mol. Biol.* **154:**1–20.
15. **Bouché, J. P., J. P. Gélugne, J. Louarn, J. M. Louarn, and K. Kaiser.** 1982. Relationships between the physical and genetic maps of a 470×10^3 base-pair region around the terminus of *Escherichia coli* K12 DNA replication. *J. Mol. Biol.* **154:**21–32.
16. **Brewer, B. J., and W. L. Fangman.** 1988. A replication fork barrier at the 3′ end of yeast ribosomal RNA gene. *Cell* **55:**637–643.
17. **Brewer, B. J., D. Lochshon, and W. L. Fangman.** 1992. Arrest of replication fork barrier in the rDNA of yeast occurs independently of transcription. *Cell* **71:**267–276.
18. **Burton, P., and I. B. Holland.** 1983. Two pathways of division inhibition in UV-irradiated *E. coli. Mol. Gen. Genet.* **190:**309–314.
19. **Clerget, M.** 1991. Site-specific recombination promoted by a short DNA segment of plasmid R1 and by a homologous segment in the terminus region of the *Escherichia coli* chromosome. *New Biol.* **3:**780–788.
20. **Colloms, S. D., P. Sykora, G. Szatmari, and D. J. Sherratt.** 1990. Recombination at ColE1 *cer* requires the *Escherichia coli xerC* gene product, a member of the lambda integrase family of site-specific recombinases. *J. Bacteriol.* **172:**6973–6980.
21. **Cornet, F., I. Mortier, J. Patte, and J.-M. Louarn.** 1994. Plasmid pSC101 harbors a recombination site, *psi,* which is able to resolve plasmid multimers and to substitute for the analogous chromosomal *Escherichia coli dif* site. *J. Bacteriol.* **176:**3188–3195.
22. **Coskun-Ari, F. F., A. Skokotas, G. R. Moe, and T. M. Hill.** 1994. Biophysical characteristics of Tus, the replication arrest protein of *Escherichia coli. J. Biol. Chem.* **269:**4027–4034.
23. **Dasgupta, S., R. Bernander, and K. Nordström.** 1991. In vivo effect of the *tus* gene on cell division in an *Escherichia coli* strain where chromosome replication is under control of plasmid R1. *Res. Microbiol.* **142:**177–180.
24. **deMassy, B., S. Bejar, J. Louarn, J. M. Louarn, and J. P. Bouché.** 1987. Inhibition of replication forks exiting the terminus region of the *Escherichia coli* chromosome occurs at two loci separated by 5 min. *Proc. Natl. Acad. Sci. USA* **84:**1759–1763.
25. **Fouts, K. E., and S. D. Barbour.** 1982. Insertions of transposons through the major cotransduction gap of *Escherichia coli* K-12. *J. Bacteriol.* **149:**106–113.
26. **François, V., J. Louarn, and J.-M. Louarn.** 1989. The terminus region of the *Escherichia coli* chromosome is flanked by several polar replication pause sites. *Mol. Microbiol.* **3:**995–1002.
27. **François, V., J. Louarn, J. Patte, J.-E. Rebollo, and J.-M. Louarn.** 1990. Constraints in chromosomal inversions in *Escherichia coli* are not explained by replication pausing at inverted terminator-like sequences. *Mol. Microbiol.* **4:**537–542.
28. **François, V., J. Louarn, J.-E. Rebollo, and J.-M. Louarn.** 1990. Replication termination, nondivisible zones, and structure of the *Escherichia coli* chromosome, p. 351–359. *In* K. Drlica and M. Riley (ed.), *The Bacterial Chromosome.* American Society for Microbiology, Washington, D.C.
29. **Friedrich, M. J., N. E. Kinsey, J. Vila, and R. J. Kadner.** 1993. Nucleotide sequence of a 13.9 kb segment of the 90kb virulence plasmid of *Salmonella typhimurium:* the presence of fimbrial biosynthesis genes. *Mol. Microbiol.* **8:**543–558.
30. **Germino, J., and D. Bastia.** 1981. Termination of DNA replication in vitro at a sequence-specific replication terminus. *Cell* **23:**681–687.
31. **Gottlieb, P. A., S. Wu, X. Zhang, M. Tecklenburg, P. Kuempel, and T. M. Hill.** 1992. Equilibrium, kinetic, and footprinting studies of the Tus-*ter* protein-DNA interaction. *J. Biol. Chem.* **267:**7434–7443.
32. **Greenfeder, S. A., and C. S. Newlon.** 1992. Replication forks pause at yeast centromeres. *Mol. Cell. Biol.* **12:**4056–4066.
33. **Guest, J. R., and G. C. Russell.** 1992. Complexes and complexities of the citric acid cycle in *Escherichia coli. Curr. Top. Cell. Regul.* **33:**231–247.
34. **Hakkart, M. J. J., P. J. M. van der Elzen, E. Veltkamp, and N. J. J. Nijkamp.** 1984. Maintenance of multicopy plasmid CloDF13 in *E. coli* cells: evidence for site-specific recombination at *parB. Cell* **36:**203–209.
35. **Henson, J. M., and P. L. Kuempel.** 1985. Deletion of the terminus region (340 kilobase pairs) from the chromosome of *Escherichia coli. Proc. Natl. Acad. Sci. USA* **82:**3766–3770.
36. **Hernández, P., S. S. Lamm, C. A. Bjerknes, and J. Van't Hof.** 1988. Replication termini in the rDNA of synchronized pea root cells *(Pisum sativum). EMBO J.* **7:**303–308.
37. **Hiasa, H., and K. Marians.** 1992. Differential inhibition of the DNA translocation and DNA unwinding activities of DNA helicases by the *Escherichia coli* Tus protein. *J. Biol. Chem.* **267:**11379–11385.
38. **Hiasa, H., and K. Marians.** 1994. Tus prevents overreplication of oriC plasmid DNA. *J. Biol. Chem.* **269:**26959–26968.
39. **Hidaka, M., M. Akiyama, and T. Horiuchi.** 1988. A consensus sequence of three DNA replication terminus sites on the *E. coli* chromosome is highly homologous to the *terR* sites of the R6K plasmid. *Cell* **55:**467–475.
40. **Hidaka, M., T. Kobayashi, and T. Horiuchi.** 1991. A newly identified DNA replication terminus site, *terE,* on the *Escherichia coli* chromosome. *J. Bacteriol.* **173:**391–393.
41. **Hidaka, M., T. Kobayashi, Y. Ishimi, M. Seki, T. Enomoto, M. Abdel-Monem, and T. Horiuchi.** 1992. Termination complex in *E. coli* inhibits SV40 DNA replication in vitro by impeding the action of T antigen helicase. *J. Biol. Chem.* **267:**5361–5365.
42. **Hidaka, M., T. Kobayashi, S. Takenaka, H. Takeya, and T. Horiuchi.** 1989. Purification of a DNA replication terminus (*ter*) site-binding protein in *Escherichia coli* and identification of the structural gene. *J. Biol. Chem.* **264:**21031–21037.
43. **Hill, T. M.** 1992. Arrest of bacterial DNA replication. *Annu. Rev. Microbiol.* **46:**603–633.
44. **Hill, T. M., J. M. Henson, and P. L. Kuempel.** 1987. The terminus region of the *Escherichia coli* chromosome contains two separate loci that exhibit polar inhibition of replication. *Proc. Natl. Acad. Sci. USA* **84:**1754–1758.
45. **Hill, T. M., B. J. Kopp, and P. L. Kuempel.** 1988. Termination of DNA replication in *Escherichia coli* requires a *trans*-acting factor. *J. Bacteriol.* **170:**662–668.
46. **Hill, T. M., and K. J. Marians.** 1990. *Escherichia coli* Tus protein acts to arrest the progression of DNA replication forks *in vitro. Proc. Natl. Acad. Sci. USA* **87:**2481–2485.
47. **Hill, T. M., A. J. Pelletier, M. Tecklenburg, and P. L. Kuempel.** 1988. Identification of the DNA sequence from the *E. coli* terminus region that halts replication forks. *Cell* **55:**459–466.
48. **Hill, T. M., A. J. Pelletier, M. Tecklenburg, and P. L. Kuempel.** 1989. *tus,* the trans-acting gene required for termination of DNA replication in *Escherichia coli,* encodes a DNA-binding protein. *Proc. Natl. Acad. Sci. USA* **86:**1593–1597.
49. **Horiuchi, T., Y. Fujimura, H. Nishitani, T. Kobayashi, and M. Hidaka.** 1994. The DNA replication fork blocked at the *ter* site may be an entrance for the RecBCD enzyme into duplex DNA. *J. Bacteriol.* **176:**4656–4663.
50. **Horiuchi, T., and M. Hidaka.** 1988. Core sequence of two separable terminus sites of the R6K plasmid that exhibit polar inhibition of replication is a 20 bp inverted repeat. *Cell* **54:**515–523.
51. **Khatri, G. S., T. MacAllister, P. R. Sista, and D. Bastia.** 1989. The replication terminator protein of *E. coli* is a DNA sequence-specific contra-helicase. *Cell* **59:**667–674.
52. **Khoury, A. M., H. J. Lee, M. Lillis, and P. Lu.** 1990. *lac* repressor-operator interaction: DNA length dependence. *Biochim. Biophys. Acta* **1087:**55–60.
53. **Kobayashi, T., M. Hidaka, and T. Horiuchi.** 1989. Evidence of a *ter* specific binding protein essential for the termination reaction of DNA replication in *Escherichia coli. EMBO J.* **8:**2435–2441.
54. **Kobayashi, T., M. Hidaka, M. Nishizawa, and T. Horiuchi.** 1992. Identification of a site required for DNA replication fork-blocking activity in the rRNA gene cluster in *Saccharomyces cerevisiae. Mol. Gen. Genet.* **233:**355–362.
55. **Kogoma, T., K. G. Barnard, and X. Hong.** 1994. RecA, Tus protein and constitutive stable DNA replication in *Escherichia coli rnhA* mutants. *Mol. Gen. Genet.* **244:**557–562.
56. **Kolter, R., and D. Helinski.** 1978. Activity of the replication terminus of plasmid R6K in hybrid replicons in *Escherichia coli. J. Mol. Biol.* **124:**425–441.
57. **Kuempel, P. L., and S. A. Duerr.** 1979. Chromosome replication in *Escherichia coli* is inhibited in the terminus region near the *rac* locus. *Cold Spring Harbor Symp. Quant. Biol.* **43:**563–567.
58. **Kuempel, P. L., S. A. Duerr, and P. D. Maglothin.** 1978. Chromosome replication in an *Escherichia coli dnaA* mutant integratively suppressed by prophage P2. *J. Bacteriol.* **134:**902–912.
59. **Kuempel, P. L., S. A. Duerr, and N. R. Seeley.** 1977. Terminus region of the chromosome in *Escherichia coli* inhibits replication forks. *Proc. Natl. Acad. Sci. USA* **74:**3927–3931.
60. **Kuempel, P. L., J. M. Henson, L. Dircks, M. Tecklenburg, and D. F. Lim.** 1991. *dif,* a *recA*-independent recombination site in the terminus region of the chromosome of *Escherichia coli. New Biol.* **3:**799–811.
61. **Lee, E. H., and A. Kornberg.** 1992. Features of replication fork blockage by the *Escherichia coli* replication termination-binding protein. *J. Biol. Chem.* **267:**8778–8784.
62. **Lee, E. H., A. Kornberg, M. Hidaka, T. Kobayashi, and T. Horiuchi.** 1989. *Escherichia coli* replication termination protein impedes the action of helicases. *Proc. Natl. Acad. Sci. USA* **86:**9104–9108.
63. **Linskens, M. H. K., and J. A. Huberman.** 1988. Organization of replication of ribosomal DNA in *Saccharomyces cerevisiae. Mol. Cell. Biol.* **8:**4927–4935.
64. **Little, R. D., T. H. K. Platt, and C. L. Schildkraut.** 1993. Initiation and termination of DNA replication in human rRNA genes. *Mol. Cell. Biol.* **13:**6600–6613.
65. **Louarn, J., F. Cornet, V. François, J. Patte, and J.-M. Louarn.** 1994. Hyperrecombination in the terminus of the *E. coli* chromosome: possible relation to nucleoid organization. *J. Bacteriol.* **176:**7524–7531.
66. **Louarn, J., J. Patte, and J.-M Louarn.** 1977. Evidence for a fixed termination site of chromosome replication in *Escherichia coli* K12. *J. Mol. Biol.* **115:**295–314.

67. Louarn, J., J. Patte, and J.-M. Louarn. 1979. Map position of the replication terminus on the *Escherichia coli* chromosome. *Mol. Gen. Genet.* **172:**7–11.

68. Louarn, J.-M., J. Louarn, V. François, and J. Patte. 1991. Analysis and possible role of hyperrecombination in the termination region of the *Escherichia coli* chromosome. *J. Bacteriol.* **173:**5097–5104.

69. Mao, Y.-M., Q. Sho, Q.-G. Li, and Z.-J., Sheng. 1991. *recA* dependence of replication of the *Escherichia coli* chromosome initiated by plasmid pUC13 integrated at pre-determined site. *Mol. Gen. Genet.* **225:**234–240.

70. Masden, C. S., S. C. Ghivizzani, and W. W. Hauswirth. 1993. Protein binding to a single termination-associated sequence in the mitochondrial DNA D-loop region. *Mol. Cell. Biol.* **13:**2162–2171.

71. Masters, M., and P. Broda. 1971. Evidence for the bidirectional replication of the *Escherichia coli* chromosome. *Nature* (London) *New Biol.* **232:**137–140.

72. Moir, P. D., R. Spiegelberg, I. R. Oliver, J. H. Pringle, and M. Masters. 1992. Proteins encoded by the *Escherichia coli* terminus region. *J. Bacteriol.* **174:**2102–2110.

73. Natarajan, S., S. Kaul, A. Miron, and D. Bastia. 1993. A 27kd protein of *E. coli* promotes antitermination of replication in vitro at a sequence-specific replication terminus. *Cell* **72:**113–120.

74. Natarajan, S., W. L. Kelly, and D. Bastia. 1991. Replication terminator protein of *Escherichia coli* is a transcriptional repressor of its own synthesis. *Proc. Natl. Acad. Sci. USA* **88:**3867–3871.

75. Nishitani, H., M. Hidaka, and T. Horiuchi. 1993. Specific chromosomal sites enhancing homologous recombination in *Escherichia coli* mutants defective in RNase H. *Mol. Gen. Genet.* **240:**307–14.

76. Nordström, K., R. Bernander, and S. Dasgupta. 1991. Analysis of bacterial cell cycle using strains in which chromosome replication is controlled by plasmid R1. *Res. Microbiol.* **142:**181–188.

77. Ochman, H., and A. C. Wilson. 1987. Evolutionary history of enteric bacteria, p. 1649–1654. *In* F. C. Neidhardt, J. L. Ingraham, K. B. Low, B. Magasanik, M. Schaechter, and H. E. Umbarger (ed.), *Escherichia coli and Salmonella typhimurium: Cellular and Molecular Biology.* American Society for Microbiology, Washington, D.C.

78. Palaniappan, C., V. K. Sharma, M. E. S. Hudspeth, and R. Meganathan. 1992. Menaquinone (vitamin K_2) biosynthesis: evidence that the *Escherichia coli menD* gene encodes both 2-succinyl-6-hydroxy-2,4-cyclohexadioene-1-carboxylic acid synthase and 2-ketoglutarate decarboxylase. *J. Bacteriol.* **174:**8111–8118.

79. Pelletier, A. J., T. M. Hill, and P. L. Kuempel. 1988. Location of sites that inhibit progression of replication forks in the terminus region of *Escherichia coli. J. Bacteriol.* **170:**4293–4998.

80. Pelletier, A. J., T. M. Hill, and P. L. Kuempel. 1989. Termination sites T1 and T2 from the *Escherichia coli* chromosome inhibit DNA replication in ColE1-derived plasmids. *J. Bacteriol.* **171:**1739–1741.

81. Rapa, M., I. Kawagishi, V. Müller, M. Kihara, and R. M. Macnab. 1992. *Escherichia coli* produces a cytoplasmic α-amylase, AmyA. *J. Bacteriol.* **174:**6644–6652.

82. Roecklein, B., and P. L. Kuempel. 1992. *In vivo* characterization of *tus* gene expression in *Escherichia coli. Mol. Microbiol.* **6:**1655–1661.

83. Roecklein, B., A. J. Pelletier, and P. L. Kuempel. 1991. The *tus* gene of *Escherichia coli*: autoregulation, analysis of flanking sequences, and identification of a complementary system in *Salmonella typhimurium. Res. Microbiol.* **142:**169–176.

84. Sharma, B., and T. M. Hill. 1992. *terF*, the sixth identified replication arrest site in *Escherichia coli*, is located within the *rcsC* gene. *J. Bacteriol.* **174:**7854–7858.

85. Sista, P. R., C. A. Hutchison, and D. Bastia. 1991. DNA-protein interaction at the replication termini of plasmid R6K. *Genes Dev.* **5:**74–82.

86. Sista, P. R., S. Mukherjee, P. Patel, G. S. Khatri, and D. Bastia. 1989. A host-encoded DNA-binding protein promotes termination of plasmid replication at a sequence-specific replication terminus. *Proc. Natl. Acad. Sci. USA* **86:**3026–3030.

87. Skokotas, A., M. Wrobleski, and T. M. Hill. 1994. Isolation and characterization of mutants of Tus, the replication arrest protein of *Escherichia coli. J. Biol. Chem.* **269:**20446–20455.

88. Stark, W. M., M. R. Boocock, and D. J. Sherratt. 1989. Site-specific recombination by Tn3 resolvase. *Trends Genet.* **8:**304–309.

89. Summers, D. K., and D. J. Sherratt. 1984. Multimerization of high copy number plasmids causes instability: ColE1 encodes a determinant essential for plasmid monomerization and stability. *Cell* **36:**1097–1103.

90. Sundin, O., and A. Varshavsky. 1980. Terminal stages of SV40 DNA replication proceed via multiply intertwined catenated dimers. *Cell* **21:**103–114.

91. Tecklenburg, M., A. Maumer, O. Nagappan, and P. L. Kuempel. The *dif*-resolvase locus can be replaced by a 33 basepair sequence, but function depends on location. *Proc. Natl. Acad. Sci. USA* **92:**1352–1356.

92. Tinge, S. A., and R. Curtiss. 1990. Isolation of the replication and partitioning regions of the *Salmonella typhimurium* virulence plasmid and stabilization of heterologous replicons. *J. Bacteriol.* **172:**5266–5277.

93. WIllis, D. K., L. H. Satin, and A. J. Clark. 1985. Mutation-dependent suppression of *recB21* and *recC22* by a region cloned from the Rac prophage of *Escherichia coli* K-12. *J. Bacteriol.* **162:**1166–1172.

94. Woods, S. A., J. S. Miles, R. E. Roberts, and J. R. Guest. 1986. Structural and functional relationships between fumarase and aspartase. *Biochem J.* **237:**547–557.

95. Woods, S. A., S. D. Schwartzbach, and J. R. Guest. 1988. Two chemically distinct classes of fumarase in *Escherichia coli. Biochim. Biophys. Acta* **954:**14–26.

96. Yoshikawa, H., and R. G. Wake. 1993. Initiation and termination of chromosome replication, p. 507–528. *In* A. L. Sonenshein, J. A. Hoch, and R. Losick (ed.), *Bacillus subtilis and Other Gram-Positive Bacteria: Biochemistry, Physiology, and Molecular Genetics.* American Society for Microbiology, Washington, D.C.

97. Zyskind, J. W., A. L. Svitil, W. B. Stine, M. C. Biery, and D. W. Smith. 1992. RecA protein of *Escherichia coli* and chromosome partitioning. *Mol. Microbiol.* **6:**2525–2537.

Cell Division

JOE LUTKENHAUS AND AMIT MUKHERJEE

101

INTRODUCTION

Cell division involves the partitioning of the cytoplasm into two compartments, each containing a copy of the cell's genetic information. In gram-negative bacteria such as *Escherichia coli* and *Salmonella typhimurium* (official designation, *Salmonella enterica* serovar Typhimurium) the division event, also referred to as septation or constriction, involves the circumferential invagination of the three layers of the cell envelope between the segregated chromosomes. Although many experiments indicate that septation and the processes of chromosome replication and segregation are not strictly coupled, during balanced growth they are well coordinated such that very few DNA-less cells are formed (70). In addition to this temporal regulation, mechanisms must exist to ensure that division is quantitatively and spatially regulated so that it only takes place once per cell cycle between the segregated chromosomes.

The geometrical and physical parameters pertaining to *E. coli* growth and cell division have been extensively reviewed by Donachie and Robinson (53). In a more recent review, Donachie (51) has emphasized the importance of attaining a critical cell size on the activation of the major periodic events of the cell cycle. Reviews by Nanninga (112) and Ayala et al. (6) emphasize peptidoglycan biosynthesis. The role of periseptal annuli in cell division has been reviewed by de Boer et al. (44), and the genetics of bacterial cell division have been reviewed by Bi and Lutkenhaus (22). The role of FtsZ in cell division has also been reviewed (22, 94).

CELL DIVISION GENES

Genetics of Cell Division

The identification of essential genes required for cell division involves screening conditional lethal mutants of *E. coli* as well as other

bacteria for filamentous morphology at the nonpermissive temperature. This morphology results from continued increase in cell mass in the absence of septation. Such mutants are classified as having a primary defect in DNA segregation (Par, for defect in partitioning) or septation (Fts, for filamenting temperature sensitive) (71). Par mutants are characterized by a large mass of DNA at the filament center or several unevenly distributed masses of DNA. The *par* genes include DNA replication genes and genes required to topologically separate the newly replicated chromosomes. In contrast, cells of mutants exhibiting an Fts phenotype are filamentous with regularly distributed nucleoids. Filamentation in the absence of an observable effect on DNA segregation suggests that the *fts* genes are specifically required for septum formation. Table 1 summarizes properties of the known cell division genes. For each, except *ftsW*, the essential nature of the gene and the filamentous phenotype due to its loss has been confirmed by construction of conditional null alleles (32, 38, 41, 64, 66).

Roles of the Cell Division Gene Products

FtsZ. Available evidence indicates that *ftsZ* acts early in septation before other well-studied cell division genes (93). This conclusion stems from the smooth morphology of filaments resulting from loss of *ftsZ* function (128). In contrast, filaments formed due to a loss of *ftsA* or *ftsI* function have indentations assumed to arise from abortive septation events. This order of gene action is also supported by the morphology of *fts rodA* double mutants at the nonpermissive temperature (12). *rodA* (or *pbpA*) mutants, which lack the elongation mode of peptidoglycan synthesis, have a spherical shape that amplifies the influence of septation on morphology. The *ftsZ rodA* cells are swollen without any sign of constriction,

TABLE 1   Genes required for cell division in *E. coli*

Gene	Mol wt (10^3) of gene product	No. of molecules/ cell	Cellular location	Biochemical function
ftsA	45.3	200	Inner surface of CM[a]	Suspected ATPase
ftsI(pbpB)	60	50	Periplasm/anchored to CM	Transglycosylase/transpeptidase
ftsL(mraR)	13.6	50	Periplasm/anchored to CM	Unknown
ftsQ	31.4	20	Periplasm/anchored to CM	Unknown
ftsN	35.7	ND[b]	Periplasm/anchored to CM	Unknown
ftsW	45	ND	Integral membrane protein	Unknown
ftsZ	40.3	10,000	Cytoplasm/division site	GTPase, forms filaments in vitro

[a]CM, cytoplasmic membrane.
[b]ND, not determined.

whereas combining *rodA* with an *ftsA*, *ftsI*, or *ftsQ* mutation leads to swollen cells with partial constrictions.

Immunolocalization studies show that FtsZ is localized to the division site in a pattern designated the FtsZ ring (20) (Fig. 1). During the cell cycle, FtsZ accumulates at the cytoplasmic membrane at midcell before there is visible invagination of the septum. Throughout septation, FtsZ remains at the leading edge of the septum; however, upon completion of the process, FtsZ is not retained at the new cell pole. On the basis of this dynamic behavior, the stability and abundance of FtsZ (about 10,000 to 20,000 molecules per cell), and genetics indicating FtsZ self-interaction, it was suggested that FtsZ forms a cytoskeletal element that mediates septation (20). This hypothesis, still unproven, forms the basis for ongoing experimentation regarding the structure, function, and localization of FtsZ.

So far, all evidence indicates that the *ftsZ* gene, which is highly conserved among bacteria (8, 37), is required for all bacterial cell division events. In *E. coli*, polar localized septation events, which result in minicell formation, involve the FtsZ ring (23). In the gram-positive organism *Bacillus subtilis*, *ftsZ* is essential both for vegetative division and for the asymmetric division occurring during sporulation (10). In this organism as well as several gram-positive cocci, FtsZ is localized to the leading edge of the septum (94, 145). In the filamentous organism *Strep-*

tomyces coelicolor, *ftsZ* is not essential for viability, but no septation is detected in its absence (101). This surprising result suggests that this organism can grow in the absence of septation but adds support to the hypothesis that *ftsZ* is required for septation in all eubacteria. More recently, *ftsZ* has also been cloned from a species of *Mycoplasma* and an archaebacterium (X. Wang and J. Lutkenhaus, unpublished data). Since these bacteria lack peptidoglycan, this finding has implications for the function of FtsZ (see below).

The study of FtsZ was stimulated by reports that it is a GTPase (43, 103, 117). Although FtsZ does not contain the three motifs that are present in the GTPase superfamily, it does contain a highly conserved sequence motif, GGGTGTG, that is similar to the tubulin signature motif (G/A)GGTG(S/A)G. The importance of this sequence in the ability of FtsZ to hydrolyze GTP is supported by the study of two mutant FtsZ proteins. Both FtsZ3 (GGGAGTG) and FtsZ84 (SGGTGTG) show dramatically reduced GTP binding and GTPase activities (43, 103, 117). Furthermore, the *ftsZ84* mutation alters the substrate specificity of FtsZ, allowing it to hydrolyze ATP (118). Even though FtsZ and tubulin share this signature motif, little additional sequence similarity exists. A comparison of a family of FtsZs with members of the three families of eukaryotic tubulins reveals only limited similarities (105). Additional experimentation is necessary to determine if these limited similarities comprise the GTP binding sites in these proteins.

The *E. coli* FtsZ GTPase has been characterized in three independent studies (43, 103, 117), and it was observed that the kinetics of GTP hydrolysis vary remarkably depending on the method of FtsZ purification. In one study (117), the purified FtsZ had bound GDP and the GTPase activity did not display a lag. In the other two studies (43, 103), the FtsZ had no bound GDP and the GTPase displayed a lag that was inversely proportional to the FtsZ concentration. The lag could also be suppressed by preincubation with GTP or treatments likely to affect the conformation of the protein. In another study, FtsZ isolated from *B. subtilis* did not have a lag but the GTPase activity did display a dramatic dependence on the protein concentration, with little activity below 100 μg/ml (144). These studies suggested that the GTPase activity is dependent on interaction of FtsZ molecules and raised the possibility that FtsZ used GTP to form the FtsZ ring (43, 103). This latter possibility is supported by the division defects of FtsZ3 and FtsZ84, which are deficient in interaction with GTP (18, 95). A speculative model has been

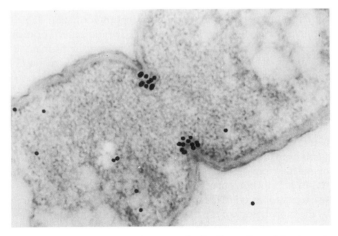

FIGURE 1   Localization of FtsZ at the leading edge of the invaginating septum by immunoelectron microscopy. Reprinted from reference 94. See also reference 20.

proposed for FtsZ localization on the basis of analogy with tubulin polymerization (94). In this model, it is suggested that the FtsZ ring is a cytoskeletal element whose formation is dependent on a GTP-GDP cycle. Furthermore, it is proposed that the assembly of FtsZ to form this structure is due to the appearance of a nucleation signal at midcell that is under cell cycle control. At present, nothing is known about the nature of this hypothetical signal.

Support for the suggestion that FtsZ functions as a cytoskeletal element comes from two in vitro studies that demonstrate that FtsZ can polymerize into filaments in the presence of guanine nucleotides (28, 105) (Fig. 2). In the first study, polymerization was observed by electron microscopy by two different methods (105). In one approach polymerized FtsZ was observed by rotary shadowing electron microscopy. The assembly of the filaments required GTP, and the diameter of the observed filaments suggested a linear polymer of FtsZ monomers: a protofilament. In a second approach, the polycation DEAE dextran, which has been used to enhance polymerization of purified tubulin (56), was found to stimulate FtsZ polymerization. Negative-stain electron microscopy revealed microtubular-like structures that might arise from the association of protofilaments. Polymerization required GTP or GDP and occurred without GTP hydrolysis. In addition, two mutant proteins were examined. FtsZ3, which is unable to support cell growth and shows a marked deficiency in interaction with GTP, was found not to polymerize. In contrast, FtsZ2, which binds GTP, displays reduced hydrolysis, and is able to support cell growth, was found to polymerize. In the second study (28), polymerization into microtubular-like structures was observed without the addition of DEAE dextran. Polymerization occurred in the presence of

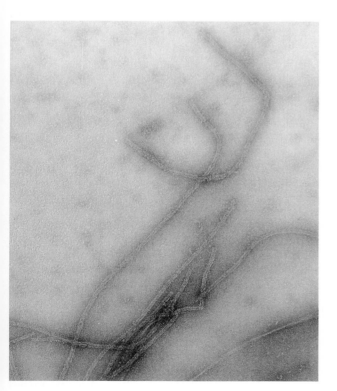

FIGURE 2   Filaments of FtsZ. Reprinted from reference 105.

high concentrations of GTP, and FtsZ could be cycled through rounds of assembly and disassembly. Also, the FtsZ84 protein, previously shown to be deficient in GTP hydrolysis, showed a reduced, albeit significant, efficiency to polymerize. Although this latter study suggested that assembly occurred quite readily into microtubular-like structures in the absence of DEAE dextran, it is now apparent that the pH had been inadvertently lowered and that assembly is enhanced at a nonphysiological, acid pH (D. Bramhill, personal communication). Thus, the protofilament may represent the physiologically relevant form of FtsZ. It should be cautioned, however, that proteins such as EF-Tu have been shown to polymerize in vitro although they are unlikely to do so in vivo (11), and filaments corresponding to FtsZ have not been observed in electron micrographs of thin sections of *E. coli*. However, if protofilaments represent the assembled form of FtsZ, they would be almost impossible to detect in vivo.

Until recently, few mutations existed in the *ftsZ* gene that might help to further elucidate its function. The classic *ftsZ84*(Ts) mutant is unable to divide and form the FtsZ ring at the nonpermissive temperature (E. Bi and J. Lutkenhaus, unpublished data), consistent with its reduced ability to polymerize in vitro (28). The *ftsZ26*(Ts) mutant is also unable to localize FtsZ at the nonpermissive temperature; however, at the permissive temperature, the cells have an altered polar morphology that correlates with an altered geometry of the FtsZ ring, not always perpendicular to the long axis of the cell (21). This result suggests that the geometry of the FtsZ ring is a property of FtsZ itself and suggests that the FtsZ ring directly determines septal growth. Thus, the function of the FtsZ ring may be analogous to that of the contractile ring in animal cells, and the invagination of the septum may be driven by motor proteins, thus far unidentified, acting on the FtsZ ring (28). In addition, the the occurrence of FtsZ in both wall-less bacteria and an archaebacterium, which lack peptidoglycan (Wang and Lutkenhaus, unpublished data), raises the possibility that FtsZ's primary function is to cause invagination of the cytoplasmic membrane, which may indirectly stimulate septal peptidoglycan synthesis.

Several studies have concluded that the level of FtsZ is rate limiting for cell division, perhaps reflecting a critical concentration required for formation of the FtsZ ring. This conclusion stems from observations on altering the levels of FtsZ. A small increase in the level of FtsZ leads to a hyperdivision activity, seen as a minicell phenotype (54), whereas lowering the level leads to a block to division (39). In several studies, a relatively small decrease (30 to 50%) appeared to be sufficient to block division (38, 58). However, in another study in which an antisense RNA was used to limit *ftsZ* expression, a 70% decrease resulted in elongated cells that were still able to achieve steady-state growth (130, 131). The increase in cell length was due to a delay in septation which correlated with a delay in nucleoid segregation. These results suggest FtsZ controls the timing of septation. It should also be noted that excess FtsZ (beyond the amount found to induce minicell formation) leads to the same phenotype as loss of FtsZ, i.e., filamentation (146). This phenomenon of the same phenotype with too little or an excess of a protein is seen with some other proteins, for example, the RepA protein of pSC101 or P1 (42, 79, 147). In these cases, it is thought that overproduction shifts the equilibrium from an active monomer to an inactive aggregate. Perhaps the overproduction of FtsZ leads to inappropriate aggregation that reduces activity.

It should be noted that there are other indications that FtsZ occupies a critical position in the cell division pathway. Most notably, FtsZ is the apparent target of the cell division inhibitors SulA, produced during the SOS response (18, 85, 92), and MinCD, a cell division inhibitor involved in placement of the division site (17, 47). Both of these inhibitors block division by preventing formation of the FtsZ ring (23). The negative effect of these inhibitors on cell division can be overcome by increasing *ftsZ* expression, either through increased gene dosage (47) or through an increase in the activity of positive activators such as *sdiA* (144). The inhibitors can also be suppressed by mutations in *ftsZ*, originally referred to as *sulB* or *sfiB* mutations (18, 92). These mutations appear to result in a form of FtsZ that is resistant to SulA.

FtsZs from a wide variety of eubacteria including gram-negative and gram-positive organisms (8, 72, 98, 152) are rich in glycine and alanine, with these residues accounting for almost 50% of the invariant residues. Also, all of the FtsZs are rich in acidic amino acid residues; the isoelectric point of the *E. coli* protein is 4.5, which makes it one of the most acidic proteins in *E. coli* (152). Comparison of the FtsZs indicates that the protein can be divided into two domains (Wang and Lutkenhaus, unpublished data): an N-terminal, GTP binding domain comprising about 80% of the protein that is highly conserved (30% invariant residues) and a C-terminal domain comprising about 20% of the protein that is highly variable except for a short segment at the extreme carboxyl end (Wang and Lutkenhaus, unpublished data). Despite the high degree of conservation, the *ftsZ* of *B. subtilis* cannot substitute for the *E. coli ftsZ;* instead, it expresses a dominant negative interference leading to filamentation (8).

FtsA. As noted above, cells lacking FtsA function form indented filaments resulting from a block after septation has initiated, suggesting that they are blocked at a step after formation of the FtsZ ring. FtsA is present at about 200 molecules per cell, and about half of the molecules are associated with the cytoplasmic membrane (115). FtsA has sequence homology to the ATP binding domain of a number of ATPases, including actin, DnaK, and sugar kinases. Similar homology is also observed between these ATPases and MreB, which is involved in cell shape determination and was previously shown to have limited homology to FtsA and DnaK (99). The homology of FtsA to this actin/DnaK/hexokinase family has been used to build a structural model (24, 122). In support of this model, FtsA has been shown to bind to an ATP affinity column and to undergo phosphorylation at a threonine that corresponds to a threonine that is phosphorylated in DnaK (122). One possibility is that FtsA acts directly on FtsZ, possibly preventing nonproductive aggregation. This suggestion stems from the observation that the ratio of FtsZ to FtsA is important for cell division to proceed (39, 50).

The location of a portion of FtsA on the cytoplasmic membrane and its importance in FtsZ function raise the possibility that FtsA acts to link the FtsZ ring and septal specific peptidoglycan biosynthesis. It has been suggested that FtsA acts at the septum, since it has been observed that abortive septa formed in the presence of FtsA3 (which is irreversibly inactivated at 42°C) are not immediately used upon return to permissive conditions (133). In contrast, abortive septa formed in the presence of other mutant FtsA proteins are readily used upon return to permissive conditions. One interpretation is that the inactivated FtsA3 protein is retained at the aborted septa, blocking newly synthesized protein from acting. On the other hand, it is possible that at the nonpermissive temperature, the FtsA3 protein acts enzymatically to produced damaged septa. FtsA may interact with penicillin-binding protein 3 (PBP3) because FtsA3 prevents the in vitro labeling of PBP3 with a radioactive penicillin derivative (132).

FtsI (PBP3). *E. coli* contains four high-molecular-weight PBPs (PBP1a, PBP1b, PBP2, and PBP3) which are thought to have transpeptidase and transglycosylase activities (55). Of these, the *ftsI* gene product, PBP3, is specifically required for septal peptidoglycan biosynthesis (123). This unique requirement for PBP3 has been demonstrated by the use of both PBP3-specific antibiotics (25) and conditional lethal mutations that map to the *ftsI* gene (125). Both ways of blocking PBP3 activity specifically block cell division and result in the formation of indented filaments with well-separated nucleoids. This result indicates that PBP3 activity is required only for septation and may be activated only during septation.

PBP3, like the other high-molecular-weight PBPs, has a single transmembrane domain that fuses a short cytoplasmic tail to the large periplasmic domain (27). It was noted that PBP3 has a potential lipoprotein modification sequence; however, biochemical and genetic studies argue that it is not a lipoprotein (68). It has also been demonstrated that PBP3 undergoes a posttranslational C-terminal cleavage that removes 11 amino acids (110), which raises the possibility that posttranslational modification is involved in regulation of the activity of PBP3. However, the functional significance of this cleavage is unclear. Constructs in which the C-terminal portion is removed function quite well, and a mutant that is unable to carry out the cleavage grows and divides normally (65, 67). Thus, a crucial question remaining is how the activity of PBP3 is topologically and temporally regulated.

FtsQ, FtsL, and FtsN. Three additional cell division genes have been identified and designated *ftsQ, ftsL (mraR),* and *ftsN* (13, 41, 64, 136). Each of the genes encodes a low-abundance membrane protein (~50 molecules per cell) that has a simple transmembrane topology. In each case, a short N-terminal cytoplasmic tail is fused through a hydrophobic transmembrane domain to a larger periplasmic domain. Even though the gene products are relatively low in abundance, the cell division process is relatively insensitive to their overproduction. Increases of 50-fold in the level of these gene products do not appear to affect the division process (13, 41, 64). This result suggests that if these gene products are directly involved in septation, they need to be activated locally.

Even though these genes have been characterized, almost nothing is known about their biochemical activities. FtsL has a potential leucine zipper motif in its periplasmic domain, raising the possibility that it could dimerize (64). In addition to a conditional null mutation, two temperature-sensitive mutations have been characterized in *ftsL* (80, 136). One mutation altering the start codon results in filamentation, similar to the null phenotype, whereas a second missense mutation results in a temperature-dependent lysis phenotype. These results indicate that lack of FtsL blocks cell division but that a defective product affects the integrity of the septation process.

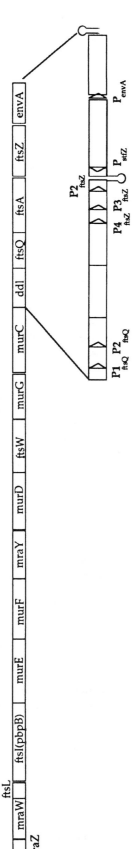

FIGURE 3 Gene organization within the 2-min region of the *E. coli* chromosome. The complete nucleotide sequence of this region has been assembled by J. A. Ayala (GenBank accession number X55034). All genes are transcribed left to right. The 3′ end of the gene cluster is a terminator just beyond the *envA* gene (indicated by the hairpin structure) (9). The 5′ boundary is not well defined. The locations of promoters that impact *ftsQAZ* expression are indicated, although evidence suggests that additional upstream promoters contribute to the expression of *ftsZ*. The putative antisense RNA for *ftsZ* would initiate at P*stfZ* and end at the terminator (indicated by the hairpin structure) located between the *ftsA* and *ftsZ* genes. This terminator only affects transcription proceeding from right to left.

FtsN was isolated as a multicopy suppressor of a temperature-sensitive *ftsA* mutation (41). Increased levels of FtsN also suppress a temperature-sensitive *ftsI* mutation, and to some extent an *ftsQ* mutation, but not a variety of other temperature-sensitive mutations, including *ftsZ84*(Ts). The mechanism of suppression is unknown. However, it is surprising that multicopy *ftsN* can suppress both a mutant protein in the cytoplasm, FtsA, and one in which the mutant domain is present in the periplasmic space, PBP3. Possibly FtsN acts in a complex with FtsA and PBP3 at the septum and the increased FtsN favors the formation of complexes destabilized by mutant proteins.

Products of Other *fts* Genes. Several other genes with the *fts* designation have been described. Among these, *ftsW* encodes a protein that is homologous to RodA, a protein that functions with PBP2 to maintain cell shape (81, 126). By analogy, it has been suggested that FtsW functions in concert with PBP3 during septum formation (78). However, morphological studies suggest that *ftsW* acts early, before PBP3 (88). A mutation in *ftsE*, located at 76 min in an operon with *ftsY* and *ftsX*, leads to filamentation (62). Since this filamentation is medium dependent, it has been questioned whether *ftsE* represents a true cell division gene (128). The FtsE protein has homology to a family of proteins involved in diverse transport processes, but its specific role is unknown (61). Initially, *ftsH* was classified as a cell division gene. However, the original mutant had two mutations, and the one responsible for filamentation was in the gene encoding PBP3 (*ftsI*) (15). It is possible that FtsH has a role in PBP3 assembly into the membrane.

Divisome, Septalsome, and Septator

The biosynthesis of the septum is surely a complex process. It must involve a subset of the cell's peptidoglycan biosynthetic machinery, mechanisms for the localized action of this machinery and for the circumferential invagination of the nascent septum. Because of the number of gene products involved, the existence of a macromolecular complex that functions to form the septum has been suggested. Several very reasonable names have been suggested for this hypothetical complex, including septalsome (73), divisome (112), and septator (139). However, the evidence to date does not allow one to distinguish between division proteins acting in sequence or as an actual complex. The evidence consists of inhibitory effects of altering the levels of one of the gene products on division, especially in mutant backgrounds (39). Examples are the inhibitory effect of increasing *ftsQ*, *ftsL*, or *ftsI* in various mutant backgrounds (*ftsZ*, *ftsA*, and *ftsI*) but not in the wild type and the already noted inhibitory effects of varying the levels of FtsZ and FtsA and how they can counteract each other (39, 86).

Gene Organization within the Large 2-Min Cluster

Many of the mutations that have been isolated that affect cell division map at 2 min within the large *mra* cluster. This region has been completely sequenced, and genetic and biochemical studies have delineated the functions of many of these genes (Fig. 3). Within this cluster, most of the genes function either in the biosynthesis of the precursor for peptidoglycan (UDP-*N*-acetylmuramyl pentapeptide) or in its transport across the membrane or have the *fts* designation, indicating that their primary role is in cell division. An exception is the *envA* gene located at the 3′ end of this cluster. *envA* has long been considered to have a role in cell separation (53); however, it has recently been shown to encode a gene required for an early step in lipopolysaccharide biosynthesis (116). In addition, upstream of *ftsL* there are two open reading frames of unknown function (6). The tight juxtaposition of genes in this cluster, all oriented in the same direction, indicates that these genes may be cotranscribed. This possibility is consistent with the location of the only transcriptional terminator in this region, just downstream of the *envA* gene (9). Thus, transcription beginning anywhere within the cluster could be expected to continue to this terminator. Despite this simple organization, it is clear that many promoters are scattered throughout the region, providing possibilities for differential expression. It is likely that regulation of this region, only beginning to be addressed, is quite complex.

Regulation of the *ftsQAZ* Cluster. Increasing the gene dosage of the *ftsQAZ* cluster leads to an increase in the level of these gene products, resulting in a hyperdivision activity expressed as a minicell phenotype (146). This observation, along with other observations that raising the level of *ftsA* or *ftsZ* alone is deleterious, suggests that the level of these proteins normally has to be maintained within a narrow range (39).

On the basis of complementation, the use of promoter detection vehicles, and primer extension experiments, the promoters in the *ftsQAZ* region have been identified (Fig. 3) (5, 119, 120, 127, 153). Two promoters, P1$_{ftsQ}$ and P2$_{ftsQ}$, are located upstream of *ftsQ* within the *ddl* gene and transcribe *ftsQ*, *ftsA*, *ftsZ*, and *envA* (5). Although these genes are coordinately expressed from these promoters, the levels of the gene products vary as a result of different efficiencies of translation (104) and additional promoters located downstream. The two *ftsQ* promoters are differentially regulated. Expression from P2$_{ftsQ}$ occurs throughout growth, and expression from P1$_{ftsQ}$, a so-called gearbox promoter, increases as the growth rate declines (5). Gearbox promoters are defined as those exhibiting a level of expression that is inversely proportional to the growth rate of the culture (5). Expression from P1$_{ftsQ}$ is dependent on *rpoS*, and increased expression from this promoter may be responsible for increasing *ftsQAZ* expression as the growth rate declines. This increased rate of expression could in part be responsible for an increased rate of division that results in smaller cells as cultures enter stationary phase (91). Expression from P2$_{ftsQ}$ is controlled by *sdiA*, a member of the *luxR* family of transcriptional regulators (57, 144). Cloning *sdiA* on a multicopy plasmid leads to increased expression from P2$_{ftsQ}$, which results in a minicell phenotype and suppression of the block to septation due to overexpression of *minCD* (144). The physiological role of *sdiA* is unknown, however, as inactivation of *sdiA* does not affect the pattern of division under standard laboratory conditions. Nonetheless, it seems likely that these promoters allow the cell to coordinately adjust the level of *ftsQAZ* expression in response to changes in growth rate and other physiological conditions and thereby adjust cell size. For example, it is likely that the *rpoS*-dependent reduction in cell length that occurs during entry into stationary phase is due in part to *rpoS* acting directly or indirectly at P1$_{ftsQ}$ (91). It is also possible that the presumed increased expression of *ftsZ* due to increased amounts of *rcsB* or *rcsF* occurs through either P1$_{ftsQ}$ or P2$_{ftsQ}$ (59, 60). This needs to be further clarified.

In addition to the upstream promoters that transcribe *ftsZ* coordinately with *ftsQ* and *ftsA*, *ftsZ* is also expressed from promoters located within the *ftsA* gene (96, 119, 127). Originally four transcriptional start points were identified within *ftsA* by primer extension analysis (5). However, one of these ends, corresponding to P1$_{ftsZ}$, is generated by RNA processing (C. A. M. Kaymevang and J.-P. Bouche, personal communication). Of the remaining three promoters, P2$_{ftsZ}$ accounts for most of *ftsZ* expression arising from promoters within the *ftsA* gene (58). Recent studies from Vicente's laboratory (58) using a PCR transcription titration assay to quantitate promoter activity at the resident chromosomal location demonstrate that the P2$_{ftsZ}$ promoter undergoes periodic activation during the cell cycle coincident with the initiation of DNA replication. In contrast, Zhou and Helmstetter (154), using a quantitative S1 hybridization assay, found that all promoters for *ftsZ*, including those upstream of *ftsQ*, exhibited cell cycle periodicity. They concluded that the periodicity resulted from an inhibiton of transcription as replication forks pass the 2-min region. This aspect of *ftsZ* regulation needs to be further clarified because it may be critical for the observed coordination between DNA replication and cell division (130). One attractive candidate for temporal regulation of *ftsZ*, *dnaA*, does not appear to play such a role (58). There is some evidence that ppGpp may activate *ftsZ* expression and by a process of elimination operate at P2$_{ftsZ}$; it does not operate at the four upstream promoters (142).

The regulation of expression of *ftsZ* may be even more complex, since it is affected by natural antisense RNAs. The first antisense RNA for *ftsZ*, designated *dicF*, is part of a defective prophage present in some, but not all, *E. coli* strains as well as other bacteria (26). No physiological condition that leads to expression of this antisense RNA is known, although it has been used experimentally to manipulate the level of FtsZ (130, 131). More recently, indirect evidence for a second putative antisense RNA has been presented. This antisense RNA, designated *stfZ*, is produced by antisense transcription of the 5′ end of *ftsZ*. The antisense RNA initiates at a promoter that must lie just inside the 5′ end of the *ftsZ* gene and ends at a terminator, which is active only in the antisense direction, located between *ftsZ* and *ftsA* (49). This RNA inhibits division when the gene for *stfZ* (but not the entire *ftsZ* gene) is present at high copy number and at high temperature. This same requirement for high temperature for effective inhibition of division is also seen with *dicF* (130). Whether *stfZ* plays a physiological role is unknown.

Expression of *ftsI* (*pbpB*) and *ftsL*. The *ftsI* gene is located near the 5′ end of the 2-min cluster just downstream of *ftsL* (*mraR*) (111). Both of these genes are weakly expressed, resulting in approximately 50 molecules per cell (64, 124). Two additional open reading frames of unknown function lie just upstream of *ftsL*, and their proximity to *ftsL* makes it likely that they are cotranscribed with *ftsL* and *ftsI*. In fact, available evidence suggests that expression of *ftsI* is dependent on sequences far upstream of *ftsL* (66). The level of PBP3, as monitored by β-lactam binding, is constant during the cell cycle, suggesting that PBP3 is activated during septation (149). It is known that the level of PBP3 is affected by the *mre* locus, which is involved in shape determination (143). A mutation in the *mreB* gene leads to an increase in the level of PBP3, and spherical morphology and deletion of the entire *mre* locus lead to even further increases in PBP3. The regulatory mechanism for this effect is unknown. It is also unknown if the spherical morphology is due to the increase in PBP3 or to other effects of inactivation of the *mre* locus.

INFLUENCE OF DNA REPLICATION AND SEGREGATION ON DIVISION

SulA-Mediated Inhibition of Cell Division

In *E. coli*, blocking DNA replication results in inhibition of cell division by at least two mechanisms (31). The first and best understood mechanism involves induction of SulA (SfiA), a component of the SOS response (75, 76). A rapid increase in SulA occurs following DNA damage, leading to a rapid block to cell division. This block is transient because as DNA damage is repaired and the SOS response is again repressed, SulA disappears through rapid degradation by the *lon* protease (103). Genetic and biochemical evidence indicates that FtsZ is the target of SulA (18, 85, 92), and immunoelectron microscopy indicates that SulA inhibits formation of the FtsZ ring (23), thus blocking cell division at an early step. Once SulA is removed by proteolysis, the FtsZ ring is able to form and cells divide, even without additional FtsZ synthesis (23, 97). Thus, SulA is an effective reversible inhibitor, in part because it is an unstable protein that is readily degraded by the *lon* protease and in part because SulA does not irreversibly damage the division apparatus. An attractive possibility is that SulA blocks FtsZ polymerization.

Since the SulA-mediated inhibition of cell division is lethal when enhanced in a *lon* background, suppressors can be readily selected. Most suppressor mutations that do not inactivate SulA map to the *ftsZ* gene (63). A number of these mutations have been characterized and found to be scattered throughout the *ftsZ* gene (18). One common property of the corresponding mutant proteins is that they display a reduced GTPase activity (40). However, other mutant proteins with reduced GTPase activity (43, 117), such as FtsZ84, are not resistant to SulA (18), suggesting that selection for SulA^r yields a special class of mutations. One suggestion is that such mutations alter the FtsZ protein, yielding a form of the protein which is resistant to interference by SulA.

Non-SulA-Mediated Inhibition, the Par Phenotype

Even in the absence of SulA, however, cell division is at least transiently blocked whenever DNA replication or segregation is affected (31). Almost any perturbation of DNA metabolism leads to some degree of filamentation and in some cases minicell formation. When DNA replication or segregation is completely blocked, filamentation occurs, but after a delay some degree of division resumes, away from the DNA mass, producing DNA-less cells (107). Thus, the DNA replication-segregation cycle and the cell division cycle are not obligatorily coupled (16, 114). However, interference with the DNA replication-segregation cycle prevents division from occurring at midcell and causes at least some filamentation before division resumes in DNA-free regions. One well-studied example is thymine starvation, which blocks DNA replication and leads to filamentation and, after a lag, to the production of DNA-less cells (84). In the presence of SulA, cell division is more efficiently blocked and fewer DNA-less cells are produced. Thus, conditional lethal mutations in any of the DNA replication genes will probably lead to filamentation and the production of DNA-less cells at the nonpermissive temperature, especially if SulA is not induced (107). Although the block to division could be a direct consequence of blocking replication, it may be due to the absence of DNA termination and segregation. In addition, inactivation of genes required for topo-logical resolution of the replicated chromosomes results in a Par phenotype. Such genes include those encoding for gyrase (77, 87), topoisomerase IV (1), and the *dif* system (89; chapters 100 and 105 in this volume).

Other Genes Affecting Segregation and Division

The thinking about segregation of chromosomes has been dominated by the replicon model, which invokes attachment of the chromosomes to the membrane and segregation resulting from zonal growth between the attachment points (82). However, several studies suggest that *E. coli* has a system for the rapid displacement of replicated chromosomes (52, 69). It should be noted that this conclusion is still controversial, as the methodology for measuring nucleoid positioning has been questioned (138). This topic is further discussed in chapter 105.

A candidate for a protein involved in an active segregation is MukB, which has characteristics of a force-generating protein (113). The *mukB* gene was identified following an ingenious screen designed to look for mutations producing anucleate cells (70). The phenotype of a *mukB* null mutant at low temperature includes the production of normal-size anucleate cells (5%) and cells that have twice the normal content of DNA. In addition, some cells are observed in which septation has apparently cleaved the chromosome, a rare phenomenon for *E. coli*. This phenotype at low temperature is distinct from that of *par* mutants that have defects in untangling the newly replicated chromosomes described above, since division appears to occur at the normal time and place. However, at higher temperatures, *mukB* is essential and the phenotype resembles Par, with filamentous cells containing extended nucleoids (113). This phenotype cannot be traced to lack of topoisomerase activity or SOS induction, indicating it may be directly due to a lack of *mukB*.

Deletion of *dnaK* also results in a Par phenotype (29, 30). At 30°C, the *dnaK* gene is not absolutely essential; however, an obvious Par phenotype is apparent. Since the primary role of DnaK is in the protein folding pathway, it is likely that a protein involved in partition requires DnaK. Interestingly, the filamentous morphology of DnaK deletion mutants can be suppressed by increasing the level of FtsZ, suggesting that the segregation defect is also suppressed (29).

There are several observations that appear to link *ftsZ* to nucleoid segregation. Reducing the level of expression of *ftsZ* fivefold by expressing the antisense RNA, *dicF*, results in an increase in cell volume and retarded nucleoid separation (130). This was a surprising result, since it had previously been shown that cell division was quite sensitive to smaller changes in FtsZ levels and that nucleoid segregation appeared quite normal as cells filament in the complete absence of FtsZ function (38). To explain the effect of reduced FtsZ on nucleoid separation, Tetart et al. (130) suggested that FtsZ inhibits nucleoid segregation until a threshold level that is able to carry out division is reached. Upon reaching this threshold level, the FtsZ may take on a new form that is able to initiate division but can no longer inhibit nucleoid segregation. In light of results on FtsZ localization (20) and in vitro assembly of FtsZ (28, 105), this new form could be the FtsZ ring. Some caution must be taken in interpreting these results, however, as *dicF* could have effects in addition to reducing *ftsZ* expression. Also, segregation is normal in the *ftsZ84*(Ts) mutant at the nonpermissive temperature (29, 71) even though it must be close to the threshold for division, since increasing the level of the mutant protein less than twofold can rescue division (144). On the other hand, it is has been observed that the Par phenotypes seen in *dnaK* and *min* mutants are reversed by over-expressing FtsZ (19, 29) and that the *ftsZ84*(Ts) mutation can be suppressed by mutations in *gyrB* (121).

Four genes have been identified on the basis of resistance to camphor, which is known to cause increased ploidy in eukaryotic cells (134). Mutations in these genes, designated *mbrA* to *mbrD*, confer resistance to camphor, temperature sensitivity, and sensitivity to growth on rich media. An examination of the morphology of the *mbrA4*, *mbrC17*, and *mbrD19* mutants after a shift from minimal to rich medium revealed that the cells filamented and contained ribbons of DNA, indicating retarded segregation. The results suggest that these mutants have difficulty in adjusting the rate of chromosome segregation and cell division to the new growth rate; they cannot keep up with the rate of cell mass increase. Interestingly, segregation and cell division are restored as cells enter stationary phase, where the decrease in growth rate may allow DNA segregation and cell division to catch up. On the basis of similar morphology of a mutation in *murI* and map location, it has been suggested that *mbrC* is identical to *murI* (7), a gene required for the biosynthesis of D-glutamic acid, a specific component of peptidoglycan (54).

LOCALIZATION AND FORMATION OF THE DIVISION SITE

In cells of *E. coli* or *S. typhimurium* in balanced growth, cell division occurs at the center of the long axis of the cell; however, the mechanism for positioning the division event at midcell is unknown. Models for positioning the division event encompass two extreme possibilities and suffer from a lack of corroborating biochemical or genetic data. In one model, potential division sites preexist on the cell envelope and division at midcell occurs by selection of the topologically correct site (33, 44). In a second model, no sites exist on the cell envelope and positioning of the division event is determined by replication and segregation of the DNA (150). In any event, the *min* locus affects placement of the division site, and continued analysis of this locus may lead to the underlying mechanism.

The *min* Locus

The placement of the division site is dramatically affected by mutations at the *min* locus, which consists of three genes, *minC*, *minD*, and *minE* (46). The classic *min* mutation of Adler et al. (2) occurs in *minD* (109) and results in a minicell phenotype, namely, division events at the poles to produce anucleate minicells. Deletion of the entire *min* locus results in the same phenotype, stressing that *min* is not required for the process of septation but rather is required for its efficient localization at midcell (46). An early hypothesis for *min* function was that it inhibited cell division from occurring at the cell poles, the old division sites, thus restricting division to midcell (129). This hypothesis received support from experiments demonstrating that a division inhibitor was encoded by the *min* locus (46). By analyzing the effects of expressing various combinations of the *min* genes, it was revealed that *minC* and *minD* encode a bipartite division inhibitor. Subsequent work has shown that MinC is the portion of the inhibitor that is proximal to the division apparatus (47, 48) and that MinD is a membrane-bound ATPase that activates MinC (45). It appears that this ATPase activity is required for MinD function.

Importantly, the activity and topological specificity of MinCD are somehow determined by the ratio of MinCD to a third component, MinE, a small polypeptide consisting of 88 amino acids (46). When MinE is in excess, the inhibitory effect of MinCD is canceled and a minicell phenotype is observed, division occurring at midcell or the cell poles (46). When MinCD is in excess, its inhibitory activity is dominant and cell division is blocked at all sites. However, when the ratio of MinE to MinCD is normal, the typical division pattern is observed, with septation only occurring at midcell. How this combination of MinCDE achieves this topological specificity is unknown. The target of the MinCD inhibitor appears to be FtsZ, since increasing the level of FtsZ and certain mutations in *ftsZ* can suppress the effect of MinCD (17, 47). Also, MinCD, like SulA, can inhibit the formation of the FtsZ ring (23). Interestingly, MinC can also combine with DicB, encoded by a defective prophage, to become an inhibitor without topological specificity (47, 90). Since this inhibitor is nonresponsive to MinE, this finding, along with other evidence, implies that the topological specificity of the MinCDE complex is imparted by MinE.

An additional unexpected feature of the *min* phenotype is the appearance of a fraction of cells in the population with abnormally segregated nucleoids (3, 83, 106) and branched cells, with the frequency of branching dependent on the genetic background and the medium (4). The observed abnormal DNA segregation raises the issue of whether the primary function of *min* is associated with chromosomal segregation, the effect on septal positioning being secondary. Although the suspected interaction of *min* with FtsZ favors a more direct effect on septation, the possibility that FtsZ can also affect segregation (130) emphasizes that more needs to be learned about the activities of all proteins involved.

Periseptal Annuli

In the periseptal annuli model, the placement of the division event is determined by preexisting annuli, circumferential zones of adhesion between the cytoplasmic membrane and the murein-outer membrane layers (44). Periseptal annuli flanking a completed septum were first detected by electron microscopy of a mutant of *S. typhimurium* that forms chains of cells (100). The existence of periseptal annuli raised the question of whether their formation preceded the initiation of septation or was the result of the septation process. If their formation preceded septation, they were likely to be involved in determining the division site. By determining the positions of plasmolysis bays, used as markers for the boundaries of annuli, it was observed that newborn cells growing in rich media contain incomplete annuli (indicated by the presence of partial plasmolysis bays) at midcell (33). As cells of increasing length are analyzed, the annuli at midcell are complete and additional new annuli appear to arise at midcell and migrate to the one-quarter and three-quarters positions. Following cell division, these latter annuli appear to be retained and become the future periseptal annuli. If the interpretation of these observations is correct, it suggests that the annuli are morphological precursors of the division site.

The relevance of plasmolysis bays in the biogenesis of the division site has been questioned in another study in which it was observed that plasmolysis bays occur primarily at new cell poles but internal bays occur fairly randomly along the cell's long axis (108). Thus, the positioning of plasmolysis bays may be artifactual and determined mainly by hydrodynamic effects. In addition, the random position of plasmolysis bays in *ftsZ* mutant filaments did not correlate with the rather discrete division of the filaments upon a return to the permissive temperature.

More recently, Cook and Rothfield have reexamined the positioning of plasmolysis bays in *ftsZ* and *ftsA* mutant filaments (35, 36). Significantly, it was observed that plasmolysis bays occurred at regular intervals along the cell length in *ftsZ* mutant filaments, in contrast to earlier results reporting random positioning (34). The earlier results with the *ftsZ* mutant were obfuscated by using conditions that did not completely block division at the nonpermissive conditions. The earlier result was interpreted as *ftsZ* having a primary role in positioning periseptal annuli (44); however, these recent results rule out this role for *ftsZ*. The discrete positioning of plasmolysis bays in *ftsZ* mutant filaments, which are smooth cylinders and therefore not subjected to the hydrodynamic effects present in indented filaments, suggests that the bays arise from attachment of the cytoplasmic membrane to the cell wall.

Nucleoid Occlusion

In a second model, the nucleoid occlusion model proposed by Woldringh et al. (150), the location of the division event is determined as a result of DNA replication and segregation. This model arose from observations on the division pattern in DNA replication mutants (107). When DNA replication or segregation is blocked, even in the absence of the SOS response, septation is inhibited in the area of the nucleoids although septation, after some delay, occurs in DNA-free regions of the cell. These division events producing anucleate cells do not occur at a fixed distance from the poles but rather occur randomly in the anucleate regions. As cells resume DNA replication, septation takes place close to the nucleoid, suggesting that the actively replicating nucleoid stimulates septation nearby. In this model, it is suggested that the nucleoid exerts a negative effect on septation but that the replicating nucleoid exerts a positive effect. When nucleoids segregate, the positive effect can locally overcome the negative effect, leading to septation. Furthermore, it is suggested that these positive and negative effects directly influence the topography of peptidoglycan synthesis, leading to septation. Although this model suggests similarities between prokaryotes and eukaryotes, with chromosome segregation playing a critical role in determining the site of cell division, it is a difficult model to test.

PEPTIDOGLYCAN SYNTHESIS AND SEPTATION

The process of septation involves the invagination of the cytoplasmic membrane, with the accompanying ingrowth of the peptidoglycan layer and outer membrane. The invagination of the outer membrane may be passive and not required for the division of the cytoplasmic contents into two distinct compartments. In contrast, there is considerable evidence that ingrowth of the peptidoglycan layer is critical for the septation event, and it is not clear if the division of the cytoplasm into two compartments can occur without it. Thus, a crucial event in septation may be the activation of septal peptidoglycan synthesis.

Topology of Peptidoglycan Synthesis

The topology of peptidoglycan biosynthesis during the cell cycle has been examined by in situ autoradiography to monitor [*meso*-^{3}H]diaminopimelic acid incorporation into peptidoglycan (148). In nondividing cells, there is diffuse incorporation along the entire length of the cell, whereas in dividing cells, incorporation occurs mainly at midcell at the leading edge of the constriction. This observation suggesting localized synthetic activity raises questions as to how the biosynthetic activity switches from diffuse to septal biosynthesis and whether there is a difference in composition between the lateral cell wall and the polar caps. Despite intense efforts using current separation techniques, no muropeptide that is specific to the septum has been identified (112), even though septation specifically requires PBP3.

PBP3 is required for septation but not for initiation of this event (112). Filaments that are formed because of lack of PBP3 activity contain blunt constrictions that appear to arise from septation events that initiate but are then aborted because of the lack of PBP3 activity. Examination of the topology of diaminopimelic acid incorporation in such filaments revealed that diaminopimelic acid is incorporated as such blunt constrictions are formed, but once they are formed, there is no further localized incorporation. This result suggests that a penicillin-insensitive step may occur early in septation before PBP3 is involved. This step has been designated penicillin-insensitive peptidoglycan biosynthesis (112).

Substrates for PBP3

One approach to the regulation of PBP3 activity has been to try to identify interacting proteins genetically by looking for extragenic suppressors of a temperature-sensitive mutation in *ftsI*, the gene encoding PBP3. One such attempt revealed the unexpected finding that an increase in the level of PBP5 (a DD-carboxypeptidase) compensated for the reduced PBP3 activity at the nonpermissive temperature (14). The suppression required the temperature-sensitive protein, suggesting that mutant PBP3 was not being bypassed but that its activity was somehow being increased. Further study revealed that other manipulations that could be expected to increase the level of the tripeptide acceptor for peptidoglycan synthesis also suppressed a temperature-sensitive allele of *ftsI* (14). This finding led to the hypothesis that PBP3 prefers the tripeptide acceptor as a substrate and suggests that changes in the ratio of the tri- to pentapeptide acceptor may be important in switching cells from an elongation mode of peptidoglycan growth to a septal mode of growth. This result also suggests that the initial goal, identification of interacting proteins, will be difficult to achieve by this approach.

Role of PBP2

For wild-type *E. coli* to be viable and maintain its rod shape (chapter 68), it requires the activity of a second high-molecular-weight PBP, PBP2, plus RodA, the product of the adjacent *rodA* gene (123, 125). RodA, an integral membrane protein, appears required for the in vitro enzymatic activity of PBP2, and so the two are thought to function in concert (81). Some mutants of *E. coli* are able to grow and divide in the absence of PBP2 and exhibit a spherical morphology (141). This bypass of the PBP2 requirement can be accomplished by either increasing the level of ppGpp or increasing the level of FtsZ (142). One possible explanation for these two results is that spherical cells, which have a larger volume, require an increased level of FtsZ and that ppGpp is able to stimulate *ftsZ* expression (140, 142). If a portion of *ftsZ* expression is dependent on ppGpp, it may explain the filamentous phenotype of a mutant deleted for *spoT* and *relA* which totally lacks ppGpp (151) (see chapter 92).

Cell Separation

As the leading edge of the septum invaginates, the newly synthesized peptidoglycan is cleaved to form the new poles of the two daughter cells. The enzyme(s) responsible for this topologically restricted cleavage of the peptidoglycan is not known. As stated earlier, the *envA* gene, long a candidate for this role since a mutation in this gene led to growth of the cells in chains, is involved in an early step in lipopolysaccharide biosynthesis and must affect the hydrolytic cleavage indirectly (116). *E. coli* contains at least 10 hydrolytic activities, including two soluble lytic transglycosylases, a membrane-bound lytic transglycosylase, endopeptidases, and *N*-acetylmuramyl-L-alanine amidases (74). Which of these enzymes might be responsible for cleaving the septal peptidoglycan is not known. Deletion of the gene encoding the major soluble lytic transglycosylase or inactivation of *amiB* encoding one of the *N*-acetylmuramyl-L-alanine amidases does not have a significant effect on cell morphology (135, 137). The lack of a conditional lethal mutant with the expected chain morphology raises the possibility that this activity, which must be well regulated to prevent lysis, is redundant, with more than one enzyme able to carry it out.

CONCLUDING REMARKS

Although progress has been made over the past few years in our understanding of bacterial cell division, this aspect of bacterial physiology remains obscure compared with other cellular processes in *E. coli* and *S. typhimurium*. The continued application of techniques based on genetics along with the more recent introduction of biochemical analyses of the gene products that are essential or those that dramatically affect the process should reveal details about the overall process. Thus, the progress made recently in the biochemical analysis of the *ftsZ*, *ftsA*, and *min* gene products is a foundation on which future research can be built. For example, the continued physiological characterization of *ftsZ* mutants along with the biochemical analyses of an array of FtsZ mutant proteins should help to clarify the role of FtsZ in cell division and its interactions with other proteins. As revealed with FtsZ, the localization of the protein can have important implications in its function. Although other cell division proteins are less abundant and therefore undetectable by immunolocalization, the development of more sensitive methods would extend this important facet of the biology of cell division. Thus, there is hope that the study of cell division, which has long remained refractory to molecular and biochemical analyses, will begin to yield details that will be both fascinating and novel.

ACKNOWLEDGMENTS

We thank colleagues for sharing unpublished work and sending unpublished manuscripts.

LITERATURE CITED

1. Adams, D. E., E. M. Shekhtman, E. L. Zechiedrich, M. B. Schmid, and N. R. Cozzarelli. 1992. The role of topoisomerase IV in partitioning bacterial replicons and the structure of catenated intermediates in DNA replication. *Cell* **71:**277–288.
2. Adler, H. I., W. D. Fischer, A. Cohen, and A. A. Hardigree. 1967. Miniature *Escherichia coli* cells deficient in DNA. *Proc. Natl. Acad. Sci. USA* **57:**321–326.
3. Akerland, T., R. Bernander, and K. Nordstrom. 1992. Cell division in *E. coli minB* mutants. *Mol. Microbiol.* **6:**2073–2083.
4. Akerlund, T., K. Nordstrom, and R. Bernander. 1994. Branched *Escherichia coli* cells. *Mol. Microbiol.* **10:**849–858.
5. Aldea, M., T. Garrido, J. Pla, and M. Vicente. 1990. Division genes in *Escherichia coli* are expressed coordinately to cell septum requirements by gearbox promoters. *EMBO J.* **9:**3787–3794.
6. Ayala, J. A., T. Garrido, M. A. de Pedro, and M. Vicente. 1994. Molecular biology of bacterial septation, p. 73–101. *In* J.-M. Ghuysen and R. Hakenbeck (ed.), *Bacterial Cell Wall*. Elsevier Science Publishing, New York.
7. Baliko, G., and P. Venetianer. 1993. An *Escherichia coli* gene in search of a function: phenotypic effects of the gene recently identified as *murI*. *J. Bacteriol.* **175:**6571–6577.
8. Beall, B., M. Lowe, and J. Lutkenhaus. 1988. Cloning and characterization of *Bacillus subtilis* homologs of *Escherichia coli* cell division genes *ftsZ* and *ftsA*. *J. Bacteriol.* **170:**4855–4864.
9. Beall, B., and J. Lutkenhaus. 1987. Sequence analysis, transcriptional organization, and insertional mutagenesis of the *envA* gene of *Escherichia coli*. *J. Bacteriol.* **169:**5408–5415.
10. Beall, B., and J. Lutkenhaus. 1991. FtsZ in *Bacillus subtilis* is required for vegetative septation and for asymmetric septation during sporulation. *Genes Dev.* **5:**447–455.
11. Beck, B. D. 1979. Polymerization of the bacterial elongation factor for protein synthesis, EF-Tuf. *Eur. J. Biochem.* **97:**495–502.
12. Begg, K. J., and W. D. Donachie. 1985. Cell shape and division in *Escherichia coli*: experiments with shape and division mutants. *J. Bacteriol.* **163:**615–622.
13. Begg, K. J., G. F. Hatfull, and W. D. Donachie. 1980. Identification of new genes in a cell envelope-cell division gene cluster of *Escherichia coli*: cell division gene *ftsQ*. *J. Bacteriol.* **144:**435–437.
14. Begg, K. J., A. Takasuga, D. H. Edwards, S. J. Dewar, B. G. Spratt, and W. D. Donachie. 1990. The balance between different peptidoglycan precursors determines whether *Escherichia coli* cells will elongate or divide. *J. Bacteriol.* **172:**6697–6703.
15. Begg, K. J., T. Tomoyasu, W. D. Donachie, M. Khattar, H. Niki, K. Yamanaka, S. Hiraga, and S. Ogura. 1993. *Escherichia coli* mutant Y16 is a double mutant carrying thermosensitive *ftsH* and *ftsI* mutations. *J. Bacteriol.* **174:**2416–2417.
16. Bernander, R., and K. Nordstrom. 1990. Chromosome replication does not trigger cell division in *E. coli*. *Cell* **60:**365–374.
17. Bi, E., and J. Lutkenhaus. 1990. Interaction between the *min* locus and *ftsZ*. *J. Bacteriol.* **172:**5610–5616.
18. Bi, E., and J. Lutkenhaus. 1990. Analysis of *ftsZ* mutations that confer resistance to the cell division inhibitor SulA (SfiA). *J. Bacteriol.* **172:**5602–5609.
19. Bi, E., and J. Lutkenhaus. 1990. FtsZ regulates frequency of cell division in *Escherichia coli*. *J. Bacteriol.* **172:**2765–2768.
20. Bi, E., and J. Lutkenhaus. 1991. FtsZ ring structure associated with division in *Escherichia coli*. *Nature* (London) **354:**161–164.
21. Bi, E., and J. Lutkenhaus. 1992. Isolation and characterization of *ftsZ* alleles that affect septal morphology. *J. Bacteriol.* **174:**5414–5423.
22. Bi, E., and J. Lutkenhaus. 1992. Genetics of bacterial cell division, p. 123–152. *In* S. Mohan, C. Dow, and J. A. Cole (ed.), *Prokaryotic Structure and Function: a New Perspective*. Cambridge University Press, Cambridge.
23. Bi, E., and J. Lutkenhaus. 1993. Cell division inhibitors SulA and MinCD prevent formation of the FtsZ ring. *J. Bacteriol.* **175:**1118–1125.
24. Bork, P., C. Sander, and A. Valencia. 1994. An ATPase domain common to prokaryotic cell cycle proteins, sugar kinases, actin and hsp70 heat shock proteins. *Proc. Natl. Acad. Sci. USA* **89:**7290–7294.
25. Botta, G. A., and J. T. Park. 1981. Evidence for involvement of penicillin-binding protein 3 in murein synthesis during septation but not during cell elongation. *J. Bacteriol.* **145:**333–340.
26. Bouche, F., and J.-P. Bouche. 1989. Genetic evidence that DicF, a second division inhibitor, encoded by the *Escherichia coli dicB* operon, is probably RNA. *Mol. Microbiol.* **3:**991–994.
27. Bowler, L. D., and B. G. Spratt. 1989. Membrane topology of penicillin-binding protein 3 of *Escherichia coli*. *Mol. Microbiol.* **3:**1277–1286.
28. Bramhill, D., and C. M. Thompson. 1994. GTP-dependent polymerization of *Escherichia coli* FtsZ protein to form tubules. *Proc. Natl. Acad. Sci. USA* **91:**5813–5817.
29. Bukau, B., and G. C. Walker. 1989. Cellular defects caused by deletion of the *Escherichia coli dnaK* gene indicate roles for heat shock protein in normal metabolism. *J. Bacteriol.* **171:**2337–2346.
30. Bukau, B., and G. C. Walker. 1989. Δ*dnaK52* mutants of *Escherichia coli* have defects in chromosome segregation and plasmid maintenance at normal growth temperatures. *J. Bacteriol.* **171:**6030–6038.
31. Burton, P., and I. B. Holland. 1983. Two pathways of division inhibition in UV-irradiated *E. coli*. *Mol. Gen. Genet.* **189:**128–132.
32. Carson, M., J. Barondess, and J. Beckwith. 1991. The FtsQ protein of *Escherichia coli*: membrane topology, abundance, and cell division phenotype due to overproduction and insertion mutations. *J. Bacteriol.* **173:**2187–2195.
33. Cook, W. R., F. Kepes, D. Joseleau-Petit, T. J. MacAlister, and L. I. Rothfield. 1987. A proposed mechanism for the generation and localization of new division sites during the division cycle of *Escherichia coli*. *Proc. Natl. Acad. Sci. USA* **84:**7144–7148.
34. Cook, W. R., and L. I. Rothfield. 1991. Biogenesis of cell division sites in *ftsA* and *ftsZ* filaments. *Res. Microbiol.* **142:**321–324.
35. Cook, W. R., and L. I. Rothfield. 1994. Early stages in the development of the *E. coli* division site. *Mol. Microbiol.* **14:**485–495.
36. Cook, W. R., and L. I. Rothfield. 1994. Development of the cell division site in FtsA⁻ filaments. *Mol. Microbiol.* **14:**497–503.

37. **Corton, J. C., Jr., J. E. Ward, and J. Lutkenhaus.** 1987. Analysis of cell division *ftsZ* (*sulB*) from gram-negative and gram-positive bacteria. *J. Bacteriol.* **169:**1–7.
38. **Dai, K., and J. Lutkenhaus.** 1991. *ftsZ* is an essential cell division gene in *Escherichia coli. J. Bacteriol.* **173:**3500–3506.
39. **Dai, K., and J. Lutkenhaus.** 1992. The proper ratio of FtsZ to FtsA is required for cell division to occur in *Escherichia coli. J. Bacteriol.* **174:**6145–6151.
40. **Dai, K., A. Mukherjee, and J. Lutkenhaus.** 1994. Mutations in *ftsZ* that confer resistance to SulA affect the interaction of FtsZ with GTP. *J. Bacteriol.* **176:**130–136.
41. **Dai, K., Y. Xu, and J. Lutkenhaus.** 1993. Cloning and characterization of *ftsN,* an essential cell division gene in *Escherichia coli,* isolated as a multicopy suppressor of *ftsA12*(Ts). *J. Bacteriol.* **175:**3790–3797.
42. **Dasgupta, S., G. Mukhopadhyay, P. P. Papp, M. S. Lewis, and D. K. Chattoraj.** 1993. Activation of DNA binding by the monomeric form of the P1 replication initiator RepA by heat shock proteins DnaJ and DnaK. *J. Mol. Biol.* **232:**23–34.
43. **de Boer, P., R. Crossley, and L. Rothfield.** 1992. The essential bacterial cell-division protein FtsZ is a GTPase. *Nature* (London) **359:**254–256.
44. **de Boer, P. A. J., W. R. Cook, and L. I. Rothfield.** 1990. Bacterial cell division. *Annu. Rev. Genet.* **24:**249–274.
45. **de Boer, P. A. J., R. E. Crossley, R. E. Hand, and L. I. Rothfield.** 1991. The MinD protein is a membrane ATPase required for the correct placement of the *Escherichia coli* division site. *EMBO J.* **10:**4371–4380.
46. **de Boer, P. A. J., R. E. Crossley, and L. I. Rothfield.** 1989. A division inhibitor and a topological specificity factor coded for by the minicell locus determine the proper placement of the division site in *Escherichia coli. Cell* **56:**641–649.
47. **de Boer, P. A. J., R. E. Crossley, and L. I. Rothfield.** 1990. Central role of the *Escherichia coli minC* gene product in two different division-inhibition systems. *Proc. Natl. Acad. Sci. USA* **87:**1129–1133.
48. **de Boer, P. A. J., R. E. Crossley, and L. I. Rothfield.** 1992. Roles of MinC and MinD in the site-specific septation block mediated by the MinCDE system of *Escherichia coli. J. Bacteriol.* **174:**63–70.
49. **Dewar, S., and W. D. Donachie.** 1993. Antisense transcription of the *ftsZ-ftsA* gene junction inhibits cell division in *Escherichia coli. J. Bacteriol.* **175:**7097–7101.
50. **Dewar, S. J., K. J. Begg, and W. D. Donachie.** 1992. Inhibition of cell division initiation by an imbalance in the ratio of FtsA to FtsZ. *J. Bacteriol.* **174:**6314–6316.
51. **Donachie, W. D.** 1993. The cell cycle of *Escherichia coli. Annu. Rev. Microbiol.* **47:**199–230.
52. **Donachie, W. D., and K. J. Begg.** 1989. Chromosome partition in *Escherichia coli* requires postreplication protein synthesis. *J. Bacteriol.* **171:**5405–5409.
53. **Donachie, W. D., and A. C. Robinson.** 1987. Cell division: parameter values and the process, p. 1578–1593. *In* F. C. Neidhardt, J. L. Ingraham, K. B. Low, B. Magasanik, M. Schaechter, and H. E. Umbarger (ed.), *Escherichia coli and Salmonella typhimurium: Cellular and Molecular Biology.* American Society for Microbiology, Washington, D.C.
54. **Doublet, P., J. van Heijenoort, J.-B. Bohin, and D. Mengin-Lecreulx.** 1992. Identification of the *Escherichia coli murI* gene, which is required for the biosynthesis of D-glutamic acid, a specific component of bacterial peptidoglycan. *J. Bacteriol.* **174:**5772–5779.
55. **Englebert, S., A. E. Kharroubi, G. Piras, B. Joris, J. Coyette, M. Nguyen-Disteche, and J.-M. Ghuysen.** 1993. Modular design of the bi (multi?)-functional penicillin-binding proteins, p. 319–334. *In* M. A. de Pedro, J.-V. Holtje, and W. Loffelhardt (ed.), *Bacterial Growth and Lysis: Metabolism and Structure of the Bacterial Sacculus.* Plenum Press, New York.
56. **Erickson, H. P., and W. A. Voter.** 1975. Polycation-induced assembly of purified tubulin. *Proc. Natl. Acad. Sci. USA* **73:**2813–2817.
57. **Fuqua, W. C., S. C. Winans, and E. P. Greenberg.** 1994. Quorum sensing in bacteria: the LuxR-LuxI family of cell density-responsive transcriptional regulators. *J. Bacteriol.* **176:**269–277.
58. **Garrido, T., M. Sanchez, P. Palacios, and M. Vicente.** 1993. Transcription of *ftsZ* oscillates during the cell cycle of *Escherichia coli. EMBO J.* **12:**3957–3965.
59. **Gervais, F. G., and G. R. Drapeau.** 1992. Identification, cloning, and characterization of *rcsF,* a new regulator gene for exopolysaccharide synthesis that suppresses the division mutation *ftsZ84* in *Escherichia coli* K-12. *J. Bacteriol.* **174:**8016–8022.
60. **Gervais, F. G., P. Phoenix, and G. R. Drapeau.** 1992. The *rcsB* gene, a positive regulator of colanic acid biosynthesis in *Escherichia coli,* is also an activator of *ftsZ* expression. *J. Bacteriol.* **174:**3964–3971.
61. **Gibbs, T. W., D. R. Gill, and G. P. C. Salmond.** 1992. Localized mutagenesis of the *ftsYEX* operon: conditionally lethal missense substitutions om the FtsE cell division protein of *Escherichia coli* are similar to those found in the cystic fibrosis transmembrane conductance regulator protein (CFTR) of human patients. *Mol. Gen. Genet.* **234:**121–128.
62. **Gill, D. R., G. F. Hatfull, and G. P. C. Salmond.** 1986. A new cell division operon in *Escherichia coli. Mol. Gen. Genet.* **205:**134–145.
63. **Gottesman, S., E. Halpern, and P. Trisler.** 1981. Role of *sulA* and *sulB* in filamentation by *lon* mutants of *Escherichia coli* K-12. *J. Bacteriol.* **148:**265–273.
64. **Guzman, L.-M., J. J. Barondess, and J. Beckwith.** 1992. FtsL, an essential cytoplasmic membrane protein involved in cell division in *Escherichia coli. J. Bacteriol.* **174:**7716–7728.
65. **Hara, H., Y. Nishimura, J.-I. Kato, H. Suzuki, H. Nagasawa, A. Suzuki, and Y. Hirota.** 1989. Genetic analysis of processing involving C-terminal cleavage in penicillin-binding protein of *Escherichia coli. J. Bacteriol.* **171:**5882–5889.
66. **Hara, H., and J. T. Park.** 1993. A far upstream region is required for expression of the *ftsI* gene coding for penicillin-binding protein 3 of *Escherichia coli,* p. 303–308. *In* M. A. de Pedro, J.-V. Holtje, and W. Loffelhardt (ed.), *Bacterial Growth and Lysis: Metabolism and Structure of the Bacterial Sacculus.* Plenum Press, New York.
67. **Hara, H., Y. Yamamoto, A. Higashitani, H. Suzuki, and Y. Nishimura.** 1991. Cloning, mapping, and characterization of the *Escherichia coli prc* gene involved in C-terminal processing of penicillin-binding protein 3. *J. Bacteriol.* **173:**4799–4813.
68. **Hayashi, S., H. Hara, H. Suzuki, and Y. Hirota.** 1988. Lipid modification of the *Escherichia coli* penicillin-binding protein 3. *J. Bacteriol.* **170:**5392–5395.
69. **Hiraga, S., T. Ogura, H. Niki, C. Ichinose, and H. Mori.** 1990. Positioning of replicated chromosomes in *Escherichia coli. J. Bacteriol.* **172:**31–39.
70. **Hiraga, S., T. Ogura, H. Niki, C. Ichinose, H. Mori, B. Ezaki, and A. Jaffe.** 1989. Chromosome partitioning in *Escherichia coli:* novel mutants producing anucleate cells. *J. Bacteriol.* **171:**1496–1505.
71. **Hirota, Y., A. Ryter, and F. Jacob.** 1968. Thermosensitive mutants of *Escherichia coli* affected in the process of DNA synthesis and cellular division. *Cold Spring Harbor Symp. Quant. Biol.* **33:**677–694.
72. **Holden, P. R., J. F. Y. Brookfield, and P. Jones.** 1993. Cloning and characterization of an *ftsZ* homologue from a bacterial symbiont of *Drosophila melanogaster. Mol. Gen. Genet.* **240:**213–220.
73. **Holland, I. B.** 1987. Genetic analysis of the *E. coli* division clock. *Cell* **48:**361–362.
74. **Holtje, J.-V., and E. I. Tuomanen.** 1991. The murein hydrolases of *Escherichia coli:* properties, functions and impact on the course of infections in vivo. *J. Gen. Microbiol.* **137:**441–454.
75. **Huisman, O., and R. D'Ari.** 1981. An inducible DNA replication-cell division coupling mechanism of *Escherichia coli. Nature* (London) **290:**797–799.
76. **Huisman, O., R. D'Ari, and S. Gottesman.** 1984. Cell-division control in *Escherichia coli:* specific induction of the SOS function SfiA protein is sufficient to block septation. *Proc. Natl. Acad. Sci. USA* **81:**4490–4494.
77. **Hussain, K., E. J. Elliot, and G. P. C. Salmond.** 1987. The ParD^{-} mutant of *Escherichia coli* also carries a *gyrA*$_{am}$ mutation. The complete sequence of *gyrA. Mol. Microbiol.* **1:**259–273.
78. **Ikeda, M., T. Sato, M. Wachi, H. K. Jung, F. Ishino, Y. Kobayashi, and M. Matsuhashi.** 1989. Structural similarity among *Escherichia coli* FtsW and RodA proteins and *Bacillus subtilis* SpoVE protein, which function in cell division, cell elongation, and spore formation, respectively. *J. Bacteriol.* **171:**6375–6378.
79. **Ingmer, H., and S. N. Cohen.** 1993. Excess intracellular concentration of the pSC101 RepA protein interferes with both plasmid DNA replication and partitioning. *J. Bacteriol.* **175:**7834–7841.
80. **Ishino, F., H. K. Jung, M. Ikeda, M. Doi, M. Wachi, and M. Matsuhashi.** 1989. New mutations *fts-36, lts-33,* and *ftsW* clustered in the *mra* region of the *Escherichia coli* chromosome induce thermosensitive cell growth and division. *J. Bacteriol.* **171:**5523–5530.
81. **Ishino, F., W. Park, S. Tomioka, S. Tamaki, and I. Takase.** 1986. Peptidoglycan synthetic activities in membranes in *Escherichia coli* caused by overproduction of penicillin-binding protein 2 and RodA protein. *J. Biol. Chem.* **261:**7024–7031.
82. **Jacob, F., S. Brenner, and F. Cuzin.** 1963. On the regulation of DNA replication in bacteria. *Cold Spring Harbor Symp. Quant. Biol.* **28:**329–348.
83. **Jaffe, A., R. D'Ari, and S. Hiraga.** 1988. Minicell-forming mutants of *Escherichia coli:* production of minicells and anucleate rods. *J. Bacteriol.* **170:**3094–3101.
84. **Jaffe, A., R. D'Ari, and V. Norris.** 1986. SOS-independent coupling between DNA replication and cell division in *Escherichia coli. J. Bacteriol.* **165:**66–71.
85. **Jones, C. A., and I. B. Holland.** 1985. The role of the SulB (FtsZ) protein in division inhibition during the SOS response in *E. coli:* FtsZ stabilizes the inhibitor SulA in maxicells. *Proc. Natl. Acad. Sci. USA* **82:**6045–6049.
86. **Jung, H. K., F. Ishino, and M. Matsuhashi.** 1989. Inhibition of growth of *ftsQ, ftsA,* and *ftsZ* mutant cells of *Escherichia coli* by amplification of a chromosomal region encompassing closely aligned cell division and cell growth genes. *J. Bacteriol.* **171:**6379–6382.
87. **Kato, J., Y. Nishimura, and H. Suzuki.** 1989. *Escherichia coli parA* is an allele of the *gyrB* gene. *Mol. Gen. Genet.* **217:**178–181.
88. **Khattar, M., K. J. Begg, and W. D. Donachie.** 1994. Identification of FtsW and characterization of a new *ftsW* division mutant of *Escherichia coli. J. Bacteriol.* **176:**7140–7147.
89. **Kuempel, P., J. M. Henson, L. Dircks, M. Tecklenburg, and D. F. Lim.** 1991. *dif,* a *recA*-independent recombination site in the terminus region of the chromosome of *Escherichia coli. New Biol.* **3:**799–811.
90. **Labie, C., F. Bouche, and J.-P. Bouche.** 1990. Minicell-forming mutants of *Escherichia coli:* suppression of both DicB- and MinD-dependent division inhibition by inactivation of the *minC* gene product. *J. Bacteriol.* **172:**5852–5855.
91. **Lange, R., and R. Hengge-Aronis.** 1991. Growth phase-regulated expression of *bolA* and morphology of stationary-phase *Escherichia coli* cells are controlled by the novel sigma factor σ^{S}. *J. Bacteriol.* **173:**4474–4481.
92. **Lutkenhaus, J.** 1983. Coupling of DNA replication and cell division: *sulB* is an allele of *ftsZ. J. Bacteriol.* **154:**1339–1346.
93. **Lutkenhaus, J.** 1990. Regulation of cell division in *E. coli. Trends Genet.* **6:**22–25.
94. **Lutkenhaus, J.** 1993. FtsZ ring in bacterial cytokinesis. *Mol. Microbiol.* **9:**404–409.
95. **Lutkenhaus, J., H. Wolf-Watz, and W. D. Donachie.** 1980. Organization of genes in the *ftsA-envA* region of the *Escherichia coli* genetic map and identification of a new *fts* locus (*ftsZ*). *J. Bacteriol.* **142:**615–620.
96. **Lutkenhaus, J., and H. C. Wu.** 1980. Determination of transcriptional units and gene products from the *ftsA* region of *Escherichia coli. J. Bacteriol.* **143:**1281–1288.

97. Maguin, E., J. Lutkenhaus, and R. D'Ari. 1986. Reversibility of SOS-associated division inhibition in *Escherichia coli*. *J. Bacteriol.* 166:733–738.

98. Margolin, W., J. C. Corbo, and S. R. Long. 1991. Cloning and characterization of a *Rhizobium meliloti* homolog of the *Escherichia coli* cell division gene *ftsZ*. *J. Bacteriol.* 173:5822–5830.

99. Matsuhashi, M., M. Wachi, and F. Ishimo. 1990. Machinery for cell growth and division: penicillin-binding proteins and other proteins. *Res. Microbiol.* 141:89–103.

100. McAlister, T. J., B. MacDonald, and L. I. Rothfield. 1983. The periseptal annulus: an organelle associated with cell division in gram negative bacteria. *Proc. Natl. Acad. Sci. USA* 80:1372–1376.

101. McCormick, J. R., E. P. Su, A. Driks, and R. Losick. 1994. Growth and viability of *Streptomyces coelicolor* mutant for the cell division gene *ftsZ*. *Mol. Microbiol.* 14:243–254.

102. Mitzusawa, S., and S. Gottesman. 1983. Protein degradation in *Escherichia coli*: the *lon* gene controls the stability of SulA protein. *Proc. Natl. Acad. Sci. USA* 80:358–362.

103. Mukherjee, A., K. Dai, and J. Lutkenhaus. 1993. E. coli cell division protein FtsZ is a guanine nucleotide binding protein. *Proc. Natl. Acad. Sci. USA* 90:1053–1057.

104. Mukherjee, A., and W. D. Donachie. 1990. Differential translation of cell division proteins. *J. Bacteriol.* 172:6106–6111.

105. Mukherjee, A., and J. Lutkenhaus. 1994. Guanine nucleotide-dependent assembly of FtsZ into filaments. *J. Bacteriol.* 176:2754–2758.

106. Mulder, E., M. El'Bouhali, E. Pas, and C. L. Woldringh. 1990. The *Escherichia coli min* mutation resembles *gyrB* in defective nucleoid segregation and decreased negative supercoiling of plasmids. *Mol. Gen. Genet.* 221:87–93.

107. Mulder, E., and C. L. Woldringh. 1989. Actively replicating nucleoids influence the positioning of division sites in DNA-less cell forming filaments of *Escherichia coli*. *J. Bacteriol.* 171:4304–4314.

108. Mulder, E., and C. L. Woldringh. 1993. Plasmolysis bays in *Escherichia coli*: are they related to development and positioning of division sites? *J. Bacteriol.* 175:2241–2247.

109. Mulder, E., C. L. Woldringh, F. Tetart, and J.-P. Bouche. 1992. New *minC* mutations suggest different interactions of the same region of division inhibitor MinC with proteins specific for *minD* and *dicB* coinhibition pathways. *J. Bacteriol.* 174:35–39.

110. Nagasawa, H., Y. Sakagami, A. Suzuki, H. Suzuki, H. Hara, and Y. Hirota. 1989. Determination of the cleavage site involved in C-terminal processing of penicillin-binding protein 3 of *Escherichia coli*. *J. Bacteriol.* 171:5890–5893.

111. Nakamura, M., I. N. Maruyama, M. Soma, J.-I. Kato, H. Suzuki, and Y. Hirota. 1983. On the process of cellular division in *Escherichia coli*: nucleotide sequence of the gene for penicillin-binding protein 3. *Mol. Gen. Genet.* 191:1–9.

112. Nanninga, N. 1991. Cell division and peptidoglycan assembly in *Escherichia coli*. *Mol. Microbiol.* 5:791–795.

113. Niki, H., A. Jaffe, R. Imamura, T. Ogura, and S. Hiraga. 1991. The new gene *mukB* codes for a 177 kd protein with coiled-coil domains involved in chromosome partitioning of *E. coli*. *EMBO J.* 10:183–193.

114. Nordstrom, K., R. Bernander, and S. Dasgupta. 1991. The *Escherichia coli* cell cycle: one cycle or multiple independent processes that are co-ordinated? *Mol. Microbiol.* 5:769–774.

115. Pla, J., A. Dopazo, and M. Vicente. 1990. The native form of FtsA, a septal protein of *Escherichia coli*, is located in the cytoplasmic membrane. *J. Bacteriol.* 172:5097–5102.

116. Raetz, C. R. H. 1993. Bacterial endotoxins: extraordinary lipids that activate eucaryotic signal transduction. *J. Bacteriol.* 175:5745–5753.

117. RayChaudhuri, D., and J. T. Park. 1992. *Escherichia coli* cell-division gene *ftsZ* encodes a novel GTP-binding protein. *Nature* (London) 359:251–254.

118. RayChaudhuri, D., and J. T. Park. 1994. A point mutation converts *Escherichia coli* FtsZ septation GTPase to an ATPase. *J. Biol. Chem.* 269:22941–22944.

119. Robinson, A. C., D. J. Kenan, G. F. Hatfull, N. F. Sullivan, R. Spiegelberg, and W. D. Donachie. 1984. DNA sequence and transcriptional organization of essential cell division genes *ftsQ* and *ftsA* of *Escherichia coli*: evidence for overlapping transcriptional units. *J. Bacteriol.* 160:546–555.

120. Robinson, A. C., D. J. Kenan, J. Sweeney, and W. D. Donachie. 1986. Further evidence for overlapping transcriptional units in an *Escherichia coli* cell envelope-cell division gene cluster: DNA sequence and transcriptional organization of the *ddl ftsQ* region. *J. Bacteriol.* 167:809–817.

121. Ruberti, I., F. Crescenzi, L. Paolozzi, and P. Ghelardini. 1991. A class of *gyrB* mutants, substantially unaffected in DNA topology, suppresses the *Escherichia coli* K12 *ftsZ84* mutation. *Mol. Microbiol.* 5:1065–1072.

122. Sanchez, M., A. Valencia, M.-J. Ferrandiz, C. Sander, and M. Vicente. 1994. Corelation between the structure and biochemical activities of FtsA, an essential cell division protein of the actin family. *EMBO J.* 13:4919–4925.

123. Spratt, B. G. 1975. Distinct penicillin-binding proteins involved in the division, elongation and shape of *Escherichia coli* K-12. *Proc. Natl. Acad. Sci. USA* 72:2999–3003.

124. Spratt, B. G. 1977. Properties of penicillin-binding proteins of *Escherichia coli* K12. *Eur. J. Biochem.* 72:341–352.

125. Spratt, B. G. 1977. Temperature-sensitive cell division mutants of *Escherichia coli* with thermolablie penicillin binding proteins. *J. Bacteriol.* 131:293–305.

126. Stoker, N. G., J. M. Pratt, and B. Spratt. 1983. Identification of the *rodA* gene product of *Escherichia coli*. *J. Bacteriol.* 155:854–859.

127. Sullivan, N. F., and W. D. Donachie. 1984. Overlapping functional units in a cell division gene cluster in *Escherichia coli*. *J. Bacteriol.* 158:1198–1201.

128. Taschner, P. E., P. G. Huls, E. Pas, and C. L. Woldringh. 1988. Division behavior and shape changes in isogenic *ftsZ*, *ftsQ*, *ftsA*, *pbpB*, and *ftsE* cell division mutants of *Escherichia coli* during temperature shift experiments. *J. Bacteriol.* 170:1533–1540.

129. Teather, R. M., J. F. Collins, and W. D. Donachie. 1974. Quantal behavior of a diffusible factor which initiates septum formation at potential division sites in *Escherichia coli*. *J. Bacteriol.* 118:407–413.

130. Tetart, F., R. Albigot, A. Conter, E. Mulder, and J. P. Bouche. 1992. Involvement of FtsZ in coupling of nucleoid separation with septation. *Mol. Microbiol.* 6:621–627.

131. Tetart, F., and J. P. Bouche. 1992. Regulation of the expression of the cell-cycle gene *ftsZ* by DicF antisense RNA. Division does not require a fixed number of FtsZ molecules. *Mol. Microbiol.* 6:615–620.

132. Tormo, A., J. A. Ayala, M. A. de Pedro, and M. Vicente. 1986. Interaction of FtsA and PBP3 proteins in the *Escherichia coli* septum. *J. Bacteriol.* 166:985–992.

133. Tormo, A., and M. Vicente. 1984. The *ftsA* gene product participates in formation of the *Escherichia coli* septum structure. *J. Bacteriol.* 157:779–784.

134. Trun, N. J., and S. Gottesman. 1990. On the bacterial cell cycle: *Escherichia coli* mutants with altered ploidy. *Genes Dev.* 4:2036–2047.

135. Tsui, H.-C. T., G. Zhao, G. Geng, H.-C. E. Leung, and M. E. Winkler. 1994. The *mutL* repair gene of *Escherichia coli* K-12 forms a superoperon with a gene encoding a new cell-wall amidase. *Mol. Microbiol.* 11:189–202.

136. Ueki, M., M. Wachi, H. K. Jung, F. Ishino, and M. Matsuhashi. 1992. *Escherichia coli mraR* gene involved in cell growth and division. *J. Bacteriol.* 174:7841–7843.

137. Ursinus, A., and J.-V. Holtje. 1994. Purification and properties of a membrane-bound lytic transglycosylase from *Escherichia coli*. *J. Bacteriol.* 176:338–343.

138. Van Helvoort, J. M. L. M., and C. L. Woldringh. 1994. Nucleoid partitioning in *Escherichia coli* during steady state growth and upon recovery from chloramphenicol treatment. *Mol. Microbiol.* 13:577–583.

139. Vicente, M., S. R. Kushner, T. Garrido, and M. Aldea. 1991. The role of the 'gearbox' in the transcription of essential genes. *Mol. Microbiol.* 5:2085–2091.

140. Vinella, D., and R. D'Ari. 1994. Thermoinducible filamentation in *Escherichia coli* due to altered RNA polymerase β subunit is suppressed by high levels of ppGpp. *J. Bacteriol.* 176:966–972.

141. Vinella, D., R. D'Ari, A. Jaffe, and P. Bouloc. 1992. Penicillin-binding protein 2 is dispensable in *Escherichia coli* when ppGpp synthesis is induced. *EMBO J.* 11:1493–1501.

142. Vinella, D., D. Joseleau-Petit, D. Thevenet, P. Bouloc, and R. D'Ari. 1993. Penicillin-binding protein 2 inactivation in *Escherichia coli* results in cell division inhibition, which is relieved by FtsZ overexpression. *J. Bacteriol.* 175:6704–6710.

143. Wachi, M., and M. Matsuhashi. 1989. Negative control of cell division by *mreB*, a gene that functions in determining the rod shape of *Escherichia coli* cells. *J. Bacteriol.* 171:3123–3127.

144. Wang, X., P. A. de Boer, and L. I. Rothfield. 1991. A factor that positively regulates cell division by activating transcription of the major cluster of essential cell division genes of *Escherichia coli*. *EMBO J.* 10:3363–3372.

145. Wang, X., and J. Lutkenhaus. 1993. FtsZ protein of *Bacillus subtilis* is localized at the division site and has GTPase activity that is dependent upon FtsZ concentration. *Mol. Microbiol.* 9:435–442.

146. Ward, J. E., Jr., and J. Lutkenhaus. 1985. Overproduction of FtsZ induces minicell formation in *Escherichia coli*. *Cell* 42:941–949.

147. Wickner, S., J. Hoskins, and K. McKenney. 1991. Monomerization of RepA dimers by heat shock proteins activates binding to DNA replication origin. *Proc. Natl. Acad. Sci. USA* 88:7903–7907.

148. Wientjes, F. B., and N. Nanninga. 1989. Rate and topography of peptidoglycan synthesis during cell division in *Escherichia coli*: concept of a leading edge. *J. Bacteriol.* 171:3412–3419.

149. Wientjes, F. B., A. J. M. Olijhoek, U. Schwarz, and N. Nanninga. 1983. Labeling pattern of major penicillin-binding proteins of *Escherichia coli* during the division cycle. *J. Bacteriol.* 153:1287–1293.

150. Woldringh, C. L., E. Mulder, P. G. Huls, and N. Vischer. 1991. Toporegulation of bacterial division according to the nucleoid occlusion model. *Res. Microbiol.* 142:309–320.

151. Xiao, H., M. Kalman, K. Ikehara, S. Zemel, G. Glaser, and M. Cashel. 1991. Residual guanosine 3′,5′-bispyrophosphate synthetic activity of *relA* null mutants can be eliminated by spoT null mutations. *J. Biol. Chem.* 266:5980–5990.

152. Yi, Q.-M., and J. Lutkenhaus. 1985. The nucleotide sequence of the essential cell-division gene *ftsZ* of *Escherichia coli*. *Gene* 36:241–247.

153. Yi, Q.-M., S. Rockenbach, J. E. Ward, Jr., and J. Lutkenhaus. 1985. Structure and expression of the cell division genes *ftsQ*, *ftsA* and *ftsZ*. *J. Mol. Biol.* 184:399–412.

154. Zhou, P., and C. E. Helmstetter. 1994. Relationship between *ftsZ* gene expression and chromosome replication in *Escherichia coli*. *J. Bacteriol.* 176:6100–6106.

Timing of Synthetic Activities in the Cell Cycle

CHARLES E. HELMSTETTER

102

INTRODUCTION

Cell division in *Escherichia coli* is an end and not a beginning. The biosynthetic steps leading to division begin well before the previous division at all growth rates. This chapter presents information on the timing and duration of events that the cell undergoes as it duplicates its constituents and eventually divides by binary fission to form two essentially identical daughter cells. During the course of this duplication process, there are four clearly discernible actions: initiation and termination of chromosome replication, and initiation and termination of cell division. The timing of synthetic processes will be related to these acts whenever possible.

THE DIVISION CYCLE

Slowly Growing Cells

The first portion of this chapter will describe the means to determine the relationship between chromosome replication and cell division for cells growing at any rate. It is essential to understand this relationship, and the manner in which it varies with growth rate, to fully understand the basis for the timing of the events to be described later. This relationship is determined by applying the "$I + C + D$ rule" (66) to a hypothetical cell containing a single nonreplicating chromosome, where I is the time required to achieve the capacity for initiation of chromosome replication (i.e., the interinitiation time), C is the time for a round of chromosome replication, and D is the time between completion of chromosome replication and the subsequent cell division. According to this rule, cells start their duplication by performing the biosyntheses needed for initiation of chromosome replication, and when that is completed in I minutes, chromosome replication begins and the cells complete division $C + D$ minutes later. An example of the construction of chromosome replication and division patterns for a case in which I equals 70 min, C equals

40 min, and D equals 20 min is shown in Fig. 1. The numerical values for I, C, and D were chosen arbitrarily to simplify the analysis, but these are realistic values for *E. coli*, as will be shown in subsequent sections. The construction begins at the far left of the figure with a hypothetical cell containing a single, nonreplicating chromosome. The sawtoothed line represents the time, I, required for this cell to progress through the steps necessary for initiation of chromosome replication. This is then followed by chromosome replication (solid line), and the cell divides D minutes after completion of chromosome replication (dotted line). Each division is indicated by a vertical line. The second ($I + C + D$) sequence is shown below the first sequence, with the cell beginning to prepare for the second initiation at the moment the first initiation takes place. The same sequence is then repeated a third time. Note that the sequences do not interact. The $I + C + D$ rule simply states that once a cell initiates a round of chromosome replication, a new $I + C + D$ sequence is inaugurated. Since the construction begins with a single hypothetical cell at the far left of the diagram, two cells are formed at the first division, four at the second division, and so on. Therefore, if the number of cells were plotted versus time, this construction results in an idealized, synchronously dividing population.

Designations for additional periods of the duplication process are shown above the second $I + C + D$ sequence. B is the time between cell division and the initiation of chromosome replication. U is the time between initiation of chromosome replication and the initial appearance of visible cellular constriction during the process of division. T is the time between initiation of visible constriction of the cell envelope and cell separation. The significance of these intervals and their durations will be discussed in subsequent sections of this chapter.

From the construction in Fig. 1 the pattern of chromosome replication during the division cycle can be determined. The simplest manner to determine the pattern will be described, in

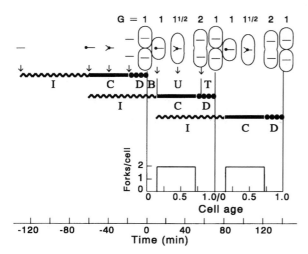

FIGURE 1 Construction of chromosome replication and cell division patterns for *E. coli* with $I = 70$ min, $C = 40$ min, and $D = 20$ min. The construction is based on the $I + C + D$ rule, beginning at the left (−130 min) with a hypothetical cell containing a single chromosome. Preparation for chromosome replication requires I minutes (sawtoothed line), chromosome replication occupies C minutes (solid line), and the cell divides D minutes later (dotted line). At the time chromosome replication begins (−60 min), preparation for the next $I + C + D$ sequence commences below the first. The long vertical lines indicate divisions. The chromosome configurations are shown across the top as linear structures. A filled circle at the left end of a chromosome indicates initiation of replication. The filled circle, corresponding to a replication fork, then proceeds along the chromosome. The configurations between 0 and 70 min, i.e., between the first two divisions, indicate the chromosome replication patterns during the division cycle of cells growing under these conditions. Only one cell is followed through the division cycle. The number of cells formed at successive division would be two, four, etc. The amount of DNA per cell is shown at the top in terms of the number of chromosome equivalents of DNA (*G*), where 1 *G* is the amount of DNA in one nonreplicating chromosome. The rate of chromosome replication during the division cycle is shown in the lower portion of the figure in terms of the number of replication forks per cell. Since replication is actually bidirectional with two forks proceeding in opposite directions, the values are twice the number of replication forks shown in the linear chromosome configurations.

order to facilitate similar constructions for other growth rates not presented in this chapter. The chromosome is a circular structure which is replicated bidirectionally from a unique origin, *oriC*, located at 84 min on the genetic map (chapter 99). However, it is very difficult to draw these circular replicating structures, particularly at rapid growth rates. Since the goal of this presentation is to explain the means to easily determine cell cycle relationships at any growth rate, chromosomes will be shown as linear structures which replicate from left to right. In effect, one half of the actual chromosome will be considered. The chromosome replication and cell division patterns are shown along the top of the figure. These are obtained by following the construction from left to right and drawing the replicating chromosomes as indicated. At the extreme left (−130 min),

the hypothetical cell contains a single chromosome. At the end of I (−60 min), replication of the chromosome is initiated as indicated by the filled circle at the left end of the chromosome. The replicating chromosome is then shown at 1/2 of the C period (−40 min). Upon termination of replication, two chromosomes are present and the cell divides D minutes later such that each daughter cell receives a single chromosome. Only one of the two daughter cells formed at each division is followed during each successive division cycle. All of the constructions to the left of the first division (at time 0) are totally hypothetical. After this first division, a true (albeit ideal) synchronous culture is developed as indicated by the cell outlines surrounding the chromosomes. Ten minutes after division, the cells have completed preparation for the next initiation event in the second duplication sequence, a new round of replication is initiated, and the cell divides again $C + D$ (or $U + T$) minutes later. The relationship between chromosome replication and the division cycle is obtained by observing the chromosome replication pattern between the first and second (or second and third) divisions. In the example shown, chromosome replication initiates 10 min after division, the chromosome replicates during the next 40 min, and then there is a period devoid of DNA replication during the last 20 min of the division cycle. The amount of chromosomal DNA per cell, expressed as the number of chromosome equivalents of DNA (*G*), can be easily determined from the stick figure chromosomes as shown at the top.

DNA replication during the cycle can also be estimated from this construction in terms of replication forks per cell, as shown in the lower portion of the frame. The plot also indicates the rate of chromosome replication, if it is assumed that the rate of DNA chain elongation is constant during C (3). Although there may be some variation in chain elongation rate, particularly in slower growing cells (90) and in the terminus region (chapter 100), the $I + C + D$ construction only defines the time for a round of replication and does not require specification of the rate of fork movement.

The important point to note from Fig. 1 is that the relationship between chromosome replication and the cell division cycle consists of a repeating series of linear sequences that overlap. The duplication of the bacterial cell is not cyclic because the duplication sequences do not repeat one after the other. On the other hand, the preparation for initiation of chromosome replication is continuous and cyclic since duplication can also be described as the cyclic achievement of the capacity to initiate chromosome replication, followed each time by cell division $C + D$ minutes later. Thus, the synthetic events required for division begin before the previous division, and the interdivision time (τ) is determined by the interinitiation time (I).

Rapidly Growing Cells

The relationship between chromosome replication and cell division in cells growing at any rate can be determined by similar applications of the $I + C + D$ rule. Figure 2 shows a second example, in which I equals 30 min, C equals 40 min, and D equals 20 min. The construction is performed exactly as shown in Fig. 1, beginning with a hypothetical cell containing a single chromosome. Again, the chromosome replication patterns and the rate of chromosome replication during the division cycle are shown at the top and bottom of the figure, respectively. Since I is less than C, a new round of replication begins before the previous

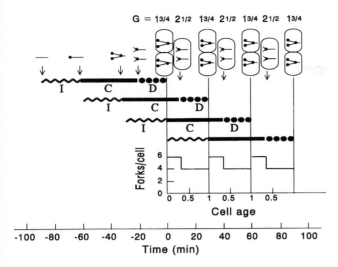

FIGURE 2 Construction of chromosome replication and cell division patterns for $I = 30$ min, $C = 40$ min, and $D = 20$ min (see legend to Fig. 1 for details).

round has terminated, resulting in the appearance of multiple levels of replication forks per chromosome. This situation obtains because each time cells have completed preparing for initiation, a new round of replication begins, irrespective of the status of the cell with respect to ongoing chromosome replication or division. There is a minimum interval allowable between replication forks (the eclipse period), but this will not be considered until later in this chapter.

There is one final important issue that distinguishes fast-growing from slow-growing cells. When τ is less than $C + D$, there is more than one chromosomal origin at the time of initiation of replication. The model described to this point assumes that at the time of initiation, replication starts at all origins essentially simultaneously. Based on the distribution of initiation in the cell cycle (30, 32, 58, 61) and on studies with flow cytometry (reviewed in reference 143), this is a fair assumption for most cells. However, it is not always the case. Some mutations, such as *dam* (4, 10, 99), *dnaA* (98, 143), and *seqA* (101), result in asynchronous initiation.

I AND *B* PERIODS

The remainder of this chapter surveys measurements of the timing of the division cycle periods described schematically in Fig. 1 and 2, and the kinetics of biosynthetic processes during these periods. Table 1 shows a list of measured values for the various cell cycle periods for *E. coli* strains B/r and K-12. The B/r strain was subdivided into the A, K, and F substrains when it was observed that stocks maintained in different collections possessed distinctly different cell division properties (65, 161). Since inclusion of every reported value for each period is beyond the scope of a useful table, the listed values show the results of a majority of representative measurements performed in different laboratories. Values for *I* are not given because it has been infrequently measured and, as indicated in Fig. 1 and 2, during balanced growth the interdivision time (τ) is equal to, and determined by, the interinitiation time. It should be noted that unlike the ideal cultures represented in Fig. 1 and 2, individual cells in

a growing population possess variances in the lengths of the periods of the division cycle (see chapter 103). It is important to consider the extents and distributions of these variances when evaluating the significance of measured parameter values. The numerical values for the periods will differ depending upon whether they represent averages over an entire batch culture, averages of individual cells in a culture, or averages from synchronous cultures (11, 61). These differences, which are likely to be 10% or less, are not specifically indicated in the data since the accuracy with which cell cycle period measurements can be performed does not warrant such distinctions.

Temporal Expression of Genes Involved in Chromosome Replication

The biosynthetic processes which take place during *I*, that is, the requirements for initiation of replication, are the subject of chapter 99 and will not be discussed in detail here except as regards the timing of synthesis of some components. RNA and protein synthesis are required throughout most of the *I* period under most growth conditions (14, 60, 66, 96, 97, 106), as expected for synthesis of the gene products that are involved in the formation of the complex of components which initiate DNA polymerization at *oriC*. The question is: are these required components synthesized continuously or periodically in the division cycle? Presumably, initiation timing could depend on the temporal regulation of formation of components of the initiation complex.

There have been several reports suggesting that certain genes in *E. coli* are expressed periodically with respect to either chromosome replication or the cell division cycle. Especially notable among these are genes that encode products involved in an aspect of chromosome replication (19, 121, 147, 148, 151). One of these is the *dnaA* gene. As described in chapter 99, the product of the *dnaA* gene acts early in initiation by binding to four sites (DnaA boxes) in *oriC* to participate in the opening of the origin to start replication. In fact, the timing of initiation of replication during steady-state growth may be determined by the binding of a fixed quantity of active DnaA protein to *oriC* (98). Thus, the timing of formation of the protein could conceivably participate in initiation timing. Expression of the *dnaA* gene has been found to fluctuate in the cell cycle (19, 121, 151). This fluctuation is due to an inhibition of transcription for an interval immediately after the gene replicates, which can last up to one-third of the interdivision time.

The inhibition of *dnaA* transcription after the gene replicates is due to the presence of a GATC sequence within the *dnaA* promoter. This is the recognition sequence for *dam* methyltransferase which methylates the N^6 position of adenine. It has been found that newly replicated, hemimethylated DNA binds to the cell membrane. Ogden et al. (122) were first to show this binding with *oriC* DNA. Originally it was thought that *oriC* bound to outer membrane, but it now appears that it binds to a unique membrane fraction (22), mediated through a protein, SeqA, which binds hemimethylated DNA (101). Campbell and Kleckner (19) subsequently reported that the time course of the period of hemimethylation varied for different genes, from about 1.5 to 12 min at 37°C. During this interval the genes are presumed to be sequestered in the membrane and unavailable for transcription.

TABLE 1 Values for cell cycle periods in *E. coli* growing at 37°C

Strain[a]	τ[c]	B	C	D	C+D	T	Method[b]	Reference
B/r A	22.5					12	b	154
	25–60		42	22	64		c	65
	25–60					10–12	a,b	161
	27		43	23	66		d	145
	28		38	24	62		a	26
	29		38				a	24
	43		45	25			c	12
	46		46	31	79		a	26
	47		48				a	24
	51		52				a	24
	60		43	27	70		d	145
	62					11	b	154
	70	0	45	25	70		c	65
	73	11			62		d	144
	75	1	50	25	74		c	65
	80	2	52	26	79		c	65
	90		53	15			h	93
	95	24			71		d	144
	100		67	31			a	25
	109	5	82	22		16	e	90
	113	34			79		d	144
	117			39		23	a,b	161
	119		71	11			h	93
	120			33		18	a,b	144
	120	0	80	40			c	65
	120		64				a	24
	125					18	b	154
	135	4	83	48		15	e	90
	166			34		30	a,b	161
	220			53		36	a,b	161
	236	118			118		d	144
	293	147			147		d	144
	330	138	144	48			d	145
	419		178	9			h	93
	685		231	23			h	93
	1,020	816	102	102	204		d	144
B/r K	25–60		42	14			c	65
	25–60					10–12	a,b	161
	27					10	b	154
	38			23			a	91
	60		43	15			h	93
	70			26			a	91
	88					17	b	154
	90		54			8	e	88
	92			26			a	91
	95	30	49	16	65		c	65
	96		53	16			h	93
	100	22	58	20	80		c	65
	100	39	55	6		10	e	90
	100					17	b	154
	120		66	14			h	93
	140	40	75	23	100		c	65
	150	40	75	20	110		c	65
	150			21		13	a,b	161
	150			25			a	91
	154		71	20			h	93
	156		72	17			h	93
	171		84	13			h	93
	176		87	13			h	93
	180		75			14	e	88
	182			25			a	91
	200			20		17	a,b	161

TABLE 1 *Continued*

Strain[a]	Time (min) τ[c]	B	C	D	C+D	T	Method[b]	Reference
	210	86	106	17		16	e	90
	212		106	10			h	93
	231	92			139		d	144
	315			33			a	91
	328		140	13			h	93
	380		170	6			h	93
	400	220	60	120	180		d	144
	623		223	22			h	93
	745		324	9			h	93
	960	576	144	240	384		d	144
B/r F	25–60		42	16			c	65
	70	11	43	16			c	65
	82					15	b,c	87
	85	19	47	19			c	65
	120	40	60	20			c	65
	165	61				26	b,c	87, 154
	240	83	132	25		27	e	90
K-12								
CF34	19					7	b	154
MC1000	21		41	27			d	2
W1485	22–220		39–47				g	23
CR34	27–219		39–45				g	95
CM735	29		44	40			d	2
AB1157	28		55	23			d	2
OV-2	30		37	16		13	a	132
Wildtype	40		40	20	60		c	63
KN126	48 (30°C)			34	31		a	—[d]
SP45	50 (30°C)			25	35		a	—[d]
CR34	52		53				a	166
M182	54		41	20	61		c	63
OV-2	54		43	42		12	a	132
CR34	55		43	18	61		c	63
AB1157	55		52	24	76		c	63
MC1000	68		46	21			d	2
3000	72		48	24		7–9	f	54
AB1157	79			24			a	91
3000	103		62	41		10	f	54
CM735	111		64	43			d	2
AB1157	113		77	40			d	2
AB1157	125			19			a	91
CR34	130					14	b	154
PAT84	170 (30°C)		106	100		29	e	89
AB1157	272			28			a	91

[a]In some experiments thymine-requiring strains were used, in which case the values shown are for experiments employing cultures grown in high thymine concentrations or in thymine plus deoxynucleoside.

[b]The methods employed for analyses of the cell cycle parameters are the following: a, residual division and DNA replication (ΔG) during exposure to chloramphenicol or thymine starvation, and/or G-to-cell mass ratio; b, percentage of constricted cells, c, baby-machine analysis; d, flow cytometry; e, autoradiography and electron microscopy; f, sucrose gradient synchrony; g, gene frequency; h, radioactive decay of [^{125}I]iododeoxyuridine in DNA.

[c]The growth of the cultures was measured in various ways, including the doubling time (T_d) of exponential-phase cultures, the average interdivision time in synchronous cultures, the mean generation time during membrane-bound growth, and the volume doubling times for chemostat cultures. Since no distinction has been made between these growth conditions or methods for determining growth rates, the Greek letter τ, usually reserved for generation time, is used synonymously, for convenience, to designate the doubling times, generation times, or interdivision times of the culture.

[d]D. Buchnik and A. Zaritsky, *EMBO Workshop*, p. III-26, 1984.

Transcription of the *gidA* gene, located immediately to the left of the origin of chromosome replication, displays an oscillatory pattern identical to the *dnaA* gene (121, 151). The *mioC* gene, adjacent to *oriC* but on the right side, also exhibits a strong periodicity in expression, but with considerably different kinetics. *mioC* transcription is shut off prior to initiation of chromosome replication and apparently remains off for several minutes after initiation of replication (121, 151). The shutoff of the *mioC* gene is probably due to DnaA protein binding upstream of its promoter, and this repression of transcription may be necessary for initiation of replication. Transcription of the genes encoding ribonucleotide reductase has also been shown to fluctuate in the cycle, with a maximum in expression coincident with initiation of replication (147, 148). Finally, there are a few promoters within *oriC* that might also be involved in initiation (chapter 99). These transcripts could show periodicities in the cycle, but information on their expression has yet to be reported.

It is conceivable that these transcriptional periodicities play roles in regulating initiation of replication such that genes required for replication are maximally expressed just prior to replication, and then shut down after initiation as part of the mechanism to prevent premature reinitiation of replication. Conversely, genes whose transcripts might prevent initiation, such as *mioC*, might need to be shut down for replication to start at *oriC*. As regards formation of the DnaA protein itself, it is likely that there would be a decrease in its rate of synthesis during inhibition of *dnaA* transcription. However, this could be a small change in rate, and in fact, no dramatic change in DnaA formation was detected in synchronized cells (139). Furthermore, no dramatic cell cycle-dependent variations in synthesis of individual proteins in general have been detected (40, 105). Thus, the general conclusion is that most proteins required for initiation of replication are produced continuously with only minor changes in their rates of synthesis.

Interinitiation Intervals

One of the earliest findings with regard to I was the apparent relationship between chromosome replication and cell mass in *Salmonella typhimurium* and *E. coli* (56, 106, 141). It was suggested that the mean cell size per chromosomal origin at initiation of replication is constant (39) and that initiation takes place when this critical size is reached (39, 59, 62, 131). Measurements of cell size, usually cell mass as assayed by absorbance at 450 or 460 nm, indicated that the average size per chromosomal origin at initiation was essentially the same in cells of a given strain growing with generation times between about 20 and 60 min at 37°C, but that it may differ in more slowly growing cells (26, 60, 129) or after a period of protein synthesis inhibition (51). Recently, variations in cell mass at initiation have been reported based on measurements by flow cytometry (160a), but again there has been little change in cell size per *oriC* detected at initiation for cells growing with doubling times in the 20- to 60-min range. Furthermore, there is a positive correlation between cell size and initiation in a given population, and initiation takes place within a distribution of cell sizes which is decidedly narrower than the distribution of cell ages at initiation (12, 84, 87, 88, 90, 117). This fixed relationship between cell size and initiation, *when it exists,* is likely a reflection of the specific macromolecular synthesis that fixes the length of I and could be indicative of a constant active DnaA protein concentration at initiation.

The information in Table 1 deals with cell cycle parameters in normal cells in which chromosome replication initiates from *oriC*. Chromosome replication can also take place, and cells can divide, when sites other than *oriC* are employed as origins. In one example, the requirement for *oriC* can be circumvented by mutations in *rnh* which eliminate RNase activity and lead to initiation at alternative sites designated *oriK* (86, 157). As a second example, chromosome replication can also be driven from integrated plasmid replication origins when *oriC* is nonfunctional (6, 7, 100, 120, 153, 158). *C* and *D* appear longer in some plasmid F-controlled replication (153) and in *rnh* (156) mutants. When plasmid R1 was integrated in *oriC*, and replaced it as an origin, division seemed fairly normal, as did DNA content, consistent with little change in cell cycle parameters (7). Finally, *C* is also longer when DnaA concentration is increased above normal levels in cells replicating from *oriC*, due in part to slow replication close to the origin (3, 98).

The minimum I is the time between two successive initiations when all positive-acting macromolecules which could normally be rate-limiting for initiation are present in excess, and all *trans*-acting negative elements are disengaged. Theoretically, this minimum interval might establish the minimum I, and possibly the minimum τ, in a given strain. Studies on this subject generally involve analyses of mutants which are temperature sensitive for initiation at *oriC*, especially *dnaA* and *dnaC* mutants, which produce gene products that are nonfunctional at nonpermissive temperature. When such mutants are shifted from permissive to nonpermissive temperature, initiation of chromosome replication ceases but potential for initiation accumulates in the cells (20, 46, 47, 57, 64, 125, 152). When employing mutant alleles in which the activity of the gene product is thermoreversible, this potential can be expressed by shifting to permissive temperature, resulting in multiple rounds of reinitiation of chromosome replication. The interval between the first and second initiations is approximately 25 min at 30°C and 30 min at 25°C (46, 47, 64). Similarly, when potential is accumulated during thymine starvation of a *thy* mutant, the interval between subsequent initiations is about 12 min at 37°C (165).

This is the minimal interval between successive initiations, also referred to as the "eclipse" period. Part of this interval is accounted for by the time required for the newly synthesized, hemimethylated *oriC* to become fully methylated since it is buried in the membrane and inert for initiation while in that state. As expected, overexpression of *dam* methyltransferase decreased the interval (109). However, the period of hemimethylated *oriC* only accounts for about half of the eclipse. The explanation for the remainder is unknown, although it has been suggested that it is the time needed to shut down *mioC* transcription (151). In any case, this eclipse period, when *oriC* cannot function, probably prevents premature reinitiation of replication when DnaA activity may be very high in the vicinity of *oriC* (10, 94, 110). It should be noted that *oriC* reportedly also binds membrane before initiation and that this association is involved positively in the initiation event (50, 80, 81).

At the opposite extreme, when the interdivision time is long, i.e., greater than $C + D$, a gap in DNA synthesis exists between cell division and initiation of chromosome replication, which is designated *B*. As seen from Table 1, the *B* period can occupy a very large fraction of the division cycle in very slowly growing cells. The values given are those in the referenced reports, but B can also be obtained for the other growth rates listed by subtracting $(C + D)$ from τ. All three substrains of B/r possess B periods during slow growth, although the length of the period during moderately slow growth ($T < 150$ min) is often less in strain B/r A than in substrains B/r K and B/r F. The considerable variation in the measured values for B in substrain B/r A is probably a real effect of different growth conditions, i.e., glucose-limited chemostat versus exponential or synchronous growth. The B period, however, does not have a unique physiological significance since it is simply a variable portion of the I period designated by $I - (C + D)$. The obvious similarity between the B period in the cell cycle of slowly growing bacteria and the G1 phase found in most eukaryotic cells has led to a number of theoretical and experimental comparisons of slow-growing bacteria and animal cells (29). As with the B period in bacteria, the entire G1 phase in eukaryotes is not an essential part of the mitotic cycle, and a portion of it, at least, simply represents the growth period required between successive initiations of chromosomal DNA replication (the S phase).

C AND D PERIODS

Durations of C and D

The time for a round of chromosome replication (the C period) depends upon the growth medium and the growth temperature. Table 1 shows a summary of a number of measured values of C in cultures growing at a variety of rates at 37°C. In *E. coli* B/r A, B/r F, and B/r K, there is little variation in the length of the C period in cells growing with doubling times between about 20 and 60 min at 37°C, with an average value being approximately 42 min. It can also be seen that C increases in cells growing under steady-state conditions in batch cultures with longer generation times. This may not be the case under non-steady-state conditions such as in cells grown in chemostats. As a means to compare the change in C as a function of growth rate for *E. coli* B/r A, B/r K, and B/r F, the values of C for the three substrains reported in the references listed in Table 1 are plotted in Fig. 3. Although the data are from experiments in which C was measured by a variety of techniques in batch-grown cultures, the consensus shape of the curve showing C as a function of growth rate is clearly evident. There is little change in the C period for both strains as the growth rate decreases from 2.5 to 1.0 doublings per h, and then there is a gradual and continuous increase in the length of C. The data in Fig. 3 indicate that there is little detectable difference between the lengths of the C periods in the three strains.

There is more variability in C for the K-12 strains listed in the table. This is because most entries are for different K-12 strains, or derivatives of strains, as indicated. Each of these could be very different as regards genetic and physiological make-up. This is in contrast to the listings for B/r, in which the same substrain was used for every study. However, if an individual K-12 strain were considered, it is likely that the constancy pattern of C with growth rate would be similar to that for the B/r substrains (see, e.g., AB1157). The actual values, however, could be different, especially in derivatives that have been subjected to numerous exposures to mutagens (chapter 133). Although there are a few measurements indicating a constancy of C during slower growth, based on gene frequency determinations, the accuracy of such methods for slow-growth studies has been questioned

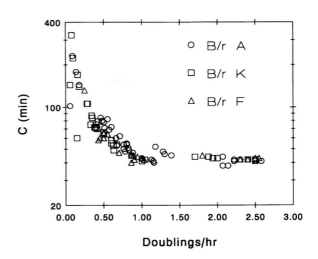

FIGURE 3 Duration of the C period in *E. coli* B/r A, B/r K, and B/r F as a function of growth rate at 37°C. The values for C were reported in the references listed in Table 1.

(24), and the bulk of the evidence indicates that C increases gradually as the doubling time is increased above 60 min at 37°C.

The table also shows that the D period does not change significantly in a given strain growing with doubling times between about 20 and 60 min at 37°C, similar to the constancy of the C period. The duration of the D period is partly determined by the time required for cell invagination, and the apparent constancy could be due to a compensatory relationship between cell diameter and the rate of invagination at different growth rates. In strain B/r A the D period is 22 to 25 min long over this range of growth rates, whereas in other strains it can vary between about 15 and 40 min. During slower growth, the D period increases, but again to a greater extent in some strains than others. Some of this variation can be accounted for by differences in techniques used to perform the measurements. For instance, division in the presence of inhibitors of DNA, RNA, or protein synthesis has sometimes been used. The possibility exists that in some strains and at some growth rates, protein synthesis might be required during the early stages of D for division to ensue (12, 161). Nevertheless, the majority of the division process in D can be completed in the absence of protein synthesis.

Equations Relating Cell Cycle Parameters to C and D

From values of C, D, and τ, it is possible to calculate reasonable values for various cell cycle parameters in a growing culture, and a number of equations have been derived for this purpose.

The average chromosomal DNA content per cell in an exponentially growing culture is given by: $\overline{G} = (\tau/C \ln 2)[2^{(C + D)/\tau} - 2^{D/\tau}]$, where $\overline{G}$ is the average number of chromosome equivalents of DNA per cell, a chromosome equivalent being the mass of DNA corresponding to a single nonreplicating chromosome.

The average cell age, in fractions of the division cycle, at replication of a specific gene on the chromosome is given by: $a_x = (n + 1)\tau - [(1 - x)C + D]$, where n is the smallest integer so that $(n + 1)\tau \geq [(1 - x)C + D]$ and x is the fraction of the C period at which the gene replicates.

Consequently, average cell age at initiation of replication is: $a_i = (n + 1)\tau - (C + D)$, and average cell age at termination of replication is: $a_t = (n + 1)\tau - D$.

The mean number of copies of a gene per cell in an exponential-phase culture is: $X = 2^{[(1 - x)C + D]/\tau}$.

Consequently, the mean number of origins per cell is: $X_O = 2^{(C + D)/\tau}$, and the mean number of termini per cell is: $X_T = 2^{D/\tau}$.

Highly accurate calculations depend upon information on the variability in the measured values (25), but quite reasonable calculations can be made by simply employing the average values given in Table 1 or Fig. 3. The duration of the D period varies considerably among individual cells in a culture, and cell cycle variability in general, at least when τ is less than $(C + D)$, is a consequence of this distribution in the length of D (11–13). In slow-growing cultures, $\tau > (C + D)$, it has been shown that the B period is also broadly distributed (87). Thus, the D and B periods of the cell cycle may be the most variable phases, although they are not independent and have compensatory variations (118).

U AND T PERIODS

The next issue to consider is the division process and its relationship to chromosome replication. It is often convenient to describe division timing, and the relationship between chromosome replication and initiation of division, with respect to the timing of visible cell invagination. However, the steps in the division process appear to actually start long before the visible binary fission of the cell begins. Since U is the time between initiation of chromosome replication and initiation of visible cell constriction, and T is the time between initiation of visible constriction and cell separation, then $C + D = U + T$ by definition. T is reasonably constant at 10 to 12 min over the same range of growth rates in which D is also constant (Table 1), possibly for the same reason. T increases in length with increasing generation time, but to different extents in different strains. It has generally been observed that the duration of the T period is less than, or at most equal to, the duration of the D period. As shown in the table, however, there are a few instances in which T appears to be somewhat longer than D. The significance of the relationship between T and D will be considered later when the coupling between chromosome replication and cell division is discussed.

Kinetics of Cell Growth

The kinetics of cell growth, particularly in relation to U and T, require consideration of changes in cell dimensions during growth (i.e., volume, surface area, length, and diameter) and the rate of total macromolecular syntheses (i.e., total cell mass, cytoplasm, membrane, and peptidoglycan). Many early analyses of growth focused on measurements of the kinetics of cell elongation during the division cycle, since the diameter of the rod-shaped E. coli cell changes little if at all during growth (31, 155), but more recent work has centered on the kinetics of growth of total cell mass. Findings and proposals on this topic can generally be fit into two categories (see references 30 and 116). The first possibility is a constant rate of increase, with a doubling in rate at a specific time in the division cycle (a bilinear pattern). Indeed, models for cell cycle control have often been based on the idea that sites of cellular growth might double at a specific time in the division cycle (33, 42, 45, 53, 92, 126, 130, 136, 167, 168). The second possibility is an exponential increase during the cell cycle.

There now appears to be a consensus that the cytoplasm and/or total mass of the cell increases exponentially during the division cycle (30, 31, 83, 116). This may also be the case with cell volume during the division cycle, and in fact, there has been a recent suggestion that volume growth responds directly to cytoplasmic growth and thus would be indistinguishable from exponential (48). Length growth could well have a more complex pattern due to the interruption of side wall growth by pole formation during invagination.

In principle, the kinetics of synthesis of individual components of the cell should be easier to determine than the kinetics of growth of the entire structure, since rates of synthesis can be measured directly, thereby permitting accurate information on synthetic patterns. Measurements of the rate of protein synthesis during the cycle, based on uptake of a radioactive amino acid, have clearly shown that it increases exponentially (30, 48). Similar approaches have been used to investigate cell cycle timing of the synthesis of the major components of the cell envelope: peptidoglycan and membrane. The peptidoglycan sacculus of E. coli, consisting of a single network of long glycan chains and short peptide chains covalently linked together, completely encloses the entire cell between the cytoplasmic membrane and the outer membrane (chapter 6). Synthesis of this covalently closed structure during the division cycle is particularly intriguing because it involves the breakage and reformation of the covalent bonds between the polysaccharide units. Based on measurements of the incorporation of radioactive peptidoglycan precursors (usually diaminopimelic acid), the rate of synthesis during side wall formation increases continuously, probably exponentially, in direct proportion to protein synthesis (48, 69, 111, 124, 160). Although there had been some early evidence for zonal growth of peptidoglycan in the cell surface (138, 142), it now is evident that it takes place diffusely at multiple sites over the cell (15–17, 36, 85, 115, 156, 160, 162). When invagination begins, there is increased synthesis at this position and decreased side wall synthesis (31, 48, 160, 162), with the main incorporation at the leading edge (160).

It is difficult to precisely define the pattern of membrane synthesis since there has been some disparity in results, probably due to the use of synchronously dividing cultures for many of the experiments and to the well-known disturbances of growth properties which can take place upon synchronization. Furthermore, it is often difficult to distinguish between a stimulation in synthesis at a specific time in the cycle (e.g., a doubling in rate of synthesis) versus a repression in synthesis at an earlier time. The general finding is that membrane phospholipid synthesis is higher toward the end of the cycle (21, 27, 48, 55, 77, 78, 123, 127, 129). However, no consistent relationship has been found between rates of membrane synthesis and initiation of I, U, or T periods, although there could be an increase at initiation of invagination. It may be that membrane synthesis and peptidoglycan synthesis go hand-in-hand, both in response to cytoplasmic growth (31, 48). Similar considerations apply to the synthesis of proteins in the envelope of E. coli. A number of envelope proteins, particularly from the outer membrane, have been reported to increase in rate of synthesis (or activity) at discrete times in the division cycle (9, 27, 55, 129). Again, however, no consistent relationships have been established with either initiation of chromosome replication or the initial stages of cell division.

Initiation of Cell Division

Several years ago, Rothfield and coworkers (28, 107, 137) proposed the existence of zones of envelope differentiation which could be involved in division timing and placement. The zones, called periseptal annuli, were reported to consist of rings of adhesion of outer membrane, peptidoglycan, and inner membrane which flank the developing constriction site during the process of division. These adhesion zones separate the division site from the rest of the cell. It was further proposed that the annuli form well before division actually starts, such that a newborn cell already contains the annuli centrally located in the cell to be used in the next division. New annuli appear to form at the site of existing annuli and move toward the cell quarters as they develop. Thus, a cell late in the cycle has one set bracketing the ongoing constriction, and two new annuli at the centers of the nascent daughter cells. This continuous formation of the zones is consistent with the finding that membrane fractions enriched for membrane-peptidoglycan adhesion zones are formed continuously in the cell cycle (79). If the model proves correct, then initiation of annuli synthesis could represent the first stage of initiation of division, long before the beginning of the observable T period. Furthermore, initiation of chromosome replication and initiation of annuli synthesis could be associated, and this association could represent the biosynthetic definition of $U + T$.

The preceding idea was based on the appearance and organization of spaces or "bays" between the cytoplasmic membrane and the cell wall during hyperosmotic shock. Whereas Rothfield and coworkers consistently found an organized pattern to these plasmolysis bays (28, 28a, 107, 137), in agreement with their model, Mulder and Woldringh (114) reported random localization of the bays, which is not consistent with the model. Woldringh (160b) has proposed an alternative model based solely on the physical properties of the cell envelope to explain the positioning of plasmolysis bays. Additional work will be needed to resolve these differences and validate the involvement of these structures in an early stage of constriction-site localization.

Another identifiable event in the bacterial cell cycle is the initiation of cell invagination during the division process. Shortly after completion of a round of chromosome replication, invagination begins in midcell, leading to cell fission about 20 min later. Among the numerous gene products required for the division process (41, 44, 108), the product of the *ftsZ* gene is essential for division, acts early in the initiation of division (5, 34, 72, 102), and is found localized in a ring around the cell at the constriction site (8). It remains at the leading edge of the constriction and then dissociates when division is completed. Initiation of cell division could be set by the periodic formation of the FtsZ ring, and it is suggested that the level of FtsZ dictates the frequency of cell division (103, 104).

If FtsZ ring formation is critical for division, then it is of considerable interest to inquire as to the mechanism that might control the timing of its formation. It is conceivable that temporal expression of the gene plays a role in this periodicity of FtsZ action. The *ftsZ* gene resides within a contiguous group of cell division genes and is transcribed from several promoters located in upstream genes (1, 37, 134, 135, 159, 164). A number of laboratories have examined cell cycle expression of the gene. Recently, Garrido et al. (49) and Zhou and Helmstetter (169) examined total *ftsZ* expression in the cell cycle and found a two-

to threefold fluctuation in transcript levels. Although maximal expression occurred near the time of initiation of replication, the periodicity is most likely due to inhibition of transcription at the time of gene replication, similar to the *dnaA* and *gidA* genes, and not to a coupling of transcription to initiation of chromosome replication. Earlier studies on *ftsZ* transcription employing transcriptional fusions to *lacZ* reached conflicting conclusions. It was reported that transcription was restricted to the time of cell division (38) or that it occurred throughout the cycle with a doubling in rate at the time of initiation of replication (133), but neither study measured total, intrinsic *ftsZ* transcription. In any case, the periodicity in transcript levels would not be expected to cause a major change in FtsZ protein in the cycle. It appears most likely that ring formation is due to timed self-assembly of the structure from preexisting molecules in the cell (104).

COUPLING BETWEEN REPLICATION AND DIVISION

The last issue to consider is the control of the timing of cell division. Chromosome replication and cell division must be coupled, because chromosomeless cells are rarely detected, but the question is: how are they coupled and what actually times cell division? One important observation in this regard concerns the behavior of cells with respect to cell division when DNA replication is inhibited. When replication is inhibited in the absence of induction of the SOS response, which blocks division via inhibition of FtsZ (chapter 89), cell division continues for D minutes and then it stops, at least temporarily. From these and related experiments (e.g., references 18, 52, and 62), it has been concluded that termination of chromosome replication is normally required for division. However, division resumes after its initial blockage during inhibition of chromosome replication in certain temperature-sensitive DNA replication mutants (67, 68, 73, 146, 149) and in mutants defective in SOS functions (70, 71, 74, 75). Studies on the kinetics of division during inhibition of DNA replication showed that cells with completed chromosomes divided, and then there was a delay in cell division while the cells continued to elongate. After this delay, chromosomeless cells of roughly normal size began forming from the DNA-free ends of the elongated cells. This continued division in the absence of chromosome replication leads to the inescapable conclusion that cell division can proceed in the absence of concurrent chromosome replication. It is thus likely that the normal coupling between replication and division is mediated by a negative control such that division in midcell is prevented until the chromosome has completed replication and at least begun to segregate from the cell center (41, 43, 66). This could be a consequence of a veto-type inhibitory effect which prevents division by a negative influence of the chromosomal mass on invagination. Indeed, it has been shown that peptidoglycan synthesis is lower in the nucleoid-containing portion of filaments (113). The release of the veto would coincide with the appearance of a DNA-free zone in the cell center as the chromosomes vacate the site of the developing invagination. In some slowly growing cells, invagination can begin before nucleoid separation, suggesting that the veto may be exerted subsequent to initiation of constriction (43, 66, 150). Furthermore, cell pairs with one member containing about two chromosomal DNA copies and the other containing a small amount of DNA are observed in some mutants (e.g., reference 119), suggesting a guillotine-like behavior of constriction when it is completed near the periphery of a nucleoid.

Based on the preceding, and on similar information that has accumulated over the past two decades, the processes leading to division and chromosome replication probably follow separate but interactive pathways, as suggested by Jones and Donachie (76; reviewed in references 41 and 140), such that the steps leading to cell division progress in parallel with chromosome replication. In a recent study on this topic, division was examined in cells in which a portion of the left end of *oriC* was replaced with a temperature-controllable R1 plasmid (7, 120). When timing of initiation was varied, relative to cell mass, the cells divided at normal size, lending strong support to the concept of parallel pathways. The pathways could begin simultaneously (e.g., at the end of *I*) or one might be an offshoot of the other, and they are coupled due to the veto effect of the unsegregated chromosomes. But the question remains: how is the division site localized in the cell? As described in chapter 101, the *minB* operon encodes topological specificity factors that locate division to a site in midcell—but what determines the placement of this site? One view is that it is determined by the envelope/division pathway exclusively, i.e., the pathway localizes the upcoming division site at midcell, and the only function of chromosome replication/segregation is to prevent its full activity until the chromosomes have segregated. The periseptal annuli could serve this site localization function, for instance. A second view is that a positive division signal directly couples envelope synthesis and division to DNA replication and segregation. This "nucleoid occlusion" model (112, 163) states that in addition to the inhibitory effect of the DNA mass on division, there is a positive effect of completion of chromosome replication that stimulates division between the chromosomes. In support of this idea, the localization of division during formation of chromosomeless cells in filaments appears to be influenced by the positions of the chromosomes. When they were replicating, division was nearby, whereas when they were not, division was more random in the DNA-free regions (112, 113). Thus, the chromosomeless cells formed in filaments are not necessarily of normal "newborn" size, as had been reported earlier (35, 68, 75, 149), but their existence clearly shows that completion of chromosome replication does not "trigger" division. The chromosomes may simply fine-tune division site placement.

If the dual-pathway concept is correct, then the timing of *U* and *T* relative to *C* and *D*, and evidence for positive correlations between cell sizes at initiation of *C* and *T* (see references 82 and 88), might depend on the extent of overlap of the pathways in a given strain or growth state. In some instances invagination could be inaugurated before the end of the *C* period, but completion of division would await termination of replication and segregation of the chromosome. On the other hand, if the *U* period were longer than the *C* period, the timing and variations in initiation of the *T* period would be functions solely of properties of the cell division pathway. This hypothesis could explain a number of interactions between replication and division which have been summarized in this chapter. For instance, a lengthy interval of protein synthesis is required for division to take place (about 40 min at 37°C; 128), and this interval could define *U*. Thus, a requirement for protein synthesis at the beginning of the *D* period for some strains to divide is anticipated when *U* is longer than *C*. This would not be the case when *C* is longer than, or equal to, *U*. Conversely, the finding that certain strains divide at a smaller-than-normal size after completion of chromosome replication in the absence of protein synthesis

(51) would be anticipated if *C* were longer than *U*. In any case, the majority of steps in the division process are probably timed by their own pathway, rather than the chromosome pathway, with the exception that segregation of completed chromosomes is normally a prerequisite for division to be completed.

LITERATURE CITED

1. Aldea, M., T. Garrido, J. Pla, and M. Vicente. 1990. Division genes in *Escherichia coli* are expressed coordinately to cell septum requirements by gearbox promoters. *EMBO J.* **9:**3787–3794.
2. Allman, R., T. Schjerven, and E. Boye. 1991. Cell cycle parameters of *Escherichia coli* K-12. *J. Bacteriol.* **173:**7970–7974.
3. Atlung, T., and F. G. Hansen. 1993. Three distinct chromosome replication states are induced by increasing concentrations of DnaA protein in *Escherichia coli.* *J. Bacteriol.* **175:**6537–6545.
4. Bakker, A., and D. W. Smith. 1989. Methylation of GATC sites is required for precise timing between rounds of DNA replication in *Escherichia coli.* *J. Bacteriol.* **171:**5738–5742.
5. Begg, K. J., and W. D. Donachie. 1985. Cell shape and division in *Escherichia coli:* experiments with shape and division mutants. *J. Bacteriol.* **163:**615–622.
6. Bernander, R., A. Merryweather, and K. Nordström. 1989. Overinitiation of replication of the *Escherichia coli* chromosome from an integrated runaway-replication derivative of plasmid R1. *J. Bacteriol.* **171:**674–683.
7. Bernander, R., and K. Nordström. 1990. Chromosome replication does not trigger cell division in *E. coli.* *Cell* **60:**365–374.
8. Bi, E., and J. Lutkenhaus. 1991. FtsZ ring structure associated with division in *Escherichia coli.* *Nature* (London) **354:**161–164.
9. Boyd, A., and I. B. Holland. 1979. Regulation of synthesis of surface protein in the cell cycle of *E. coli* B/r. *Cell* **18:**287–296.
10. Boye, E., and A. Løbner-Olesen. 1990. The role of dam methyltransferase in the control of DNA replication in *E. coli.* *Cell* **62:**981–989.
11. Bremer, H. 1982. Variation of generation times in *Escherichia coli* populations: its cause and implications. *J. Gen. Microbiol.* **128:**2865–2876.
12. Bremer, H., and L. Chuang. 1981. The cell cycle in *Escherichia coli* B/r. *J. Theor. Biol.* **88:**47–81.
13. Bremer, H., and L. Chuang. 1981. Cell division after inhibition of chromosome replication in *Escherichia coli.* *J. Theor. Biol.* **93:**909–926.
14. Bremer, H., and G. Churchward. 1977. Deoxyribonucleic acid synthesis after inhibition of initiation of rounds of replication in *Escherichia coli* B/r. *J. Bacteriol.* **130:**692–697.
15. Burman, L. G., and J. T. Park. 1984. Molecular model for the elongation of the murein sacculus of *Escherichia coli.* *Proc. Natl. Acad. Sci. USA* **81:**1844–1848.
16. Burman, L. G., J. Raichler, and J. T. Park. 1983. Evidence for diffuse growth of the cylindrical portion of the *Escherichia coli* murein sacculus. *J. Bacteriol.* **155:**983–988.
17. Burman, L. G., J. Raichler, and J. T. Park. 1983. Evidence for multisite growth of *Escherichia coli* murein involving concomitant endopeptidase and transpeptidase activities. *J. Bacteriol.* **156:**386–392.
18. Burton, P., and I. B. Holland. 1983. Two pathways of division inhibition in UV-irradiated *E. coli.* *Mol. Gen. Genet.* **190:**128–132.
19. Campbell, J. L., and N. Kleckner. 1990. *E. coli oriC* and the *dnaA* gene promoter are sequestered from *dam* methyltransferase following passage of the chromosomal replication fork. *Cell* **62:**967–979.
20. Carl, P. L. 1970. *Escherichia coli* mutants with temperature-sensitive synthesis of DNA. *Mol. Gen. Genet.* **109:**107–122.
21. Carty, C. E., and L. O. Ingram. 1981. Lipid synthesis during the *Escherichia coli* cell cycle. *J. Bacteriol.* **145:**472–478.
22. Chakraborti, A., S. Gunji, N. Shakibai, J. Cubeddu, and L. Rothfield. 1992. Characterization of the *Escherichia coli* membrane domain responsible for binding *oriC* DNA. *J. Bacteriol.* **174:**7202–7206.
23. Chandler, M., R. E. Bird, and L. Caro. 1975. The replication time of the *Escherichia coli* K12 chromosome as a function of cell doubling time. *J. Mol. Biol.* **94:**127–132.
24. Churchward, G., and H. Bremer. 1977. Determination of deoxyribonucleic acid replication time in exponentially growing *Escherichia coli* B/r. *J. Bacteriol.* **130:**1206–1213.
25. Churchward, G., H. Bremer, and R. Young. 1982. Macromolecular composition of bacteria. *J. Theor. Biol.* **94:**651–670.
26. Churchward, G., E. Estiva, and H. Bremer. 1981. Growth rate-dependent control of chromosome replication initiation in *Escherichia coli.* *J. Bacteriol.* **145:**1232–1238.
27. Churchward, G., and I. B. Holland. 1976. Envelope synthesis during the division cycle in *Escherichia coli* B/r. *J. Mol. Biol.* **105:**245–261.
28. Cook, W. R., F. Joseleau-Petit, T. J. MacAlister, and L. I. Rothfield. 1987. Proposed mechanism for generation and localization of new division sites during the division cycle of *Escherichia coli.* *Proc. Natl. Acad. Sci. USA* **84:**7144–7148.
28a. Cook, W. R., and L. I. Rothfield. 1984. Early stages of the *Escherichia coli* cell division site. *Mol. Microbiol.* **14:**485–495.
29. Cooper, S. 1979. A unifying model for the G1 period in prokaryotes and eukaryotes. *Nature* (London) **280:**17–19.
30. Cooper, S. 1991. Bacterial growth and division. Academic Press, San Diego.

31. Cooper, S. 1991. Synthesis of the cell surface during the division cycle of rod-shaped, gram-negative bacteria. *Microbiol. Rev.* **55:**649–674.

32. Cooper, S., and C. E. Helmstetter. 1968. Chromosome replication and the division cycle of *Escherichia coli* B/r. *J. Mol. Biol.* **31:**519.

33. Cullum, J., and M. Vicente. 1977. Cell growth and length distribution in *Escherichia coli. J. Bacteriol.* **134:**330–337.

34. Dai, K., and J. Lutkenhaus. 1991. *ftsZ* is an essential cell division gene in *Escherichia coli. J. Bacteriol.* **173:**3500–3506.

35. D'Ari, R., E. Magnin, P. Bouloc, A. Jaffé, A. Robin, J.-C. Liebart, and D. Joseleau-Petit. 1990. Aspects of cell regulation. *Res. Microbiol.* **141:**9–16.

36. de Jonge, B. L. M., F. B. Wientjes, I. Jurida, F. Driehuis, J. T. M. Wouters, and N. Nanninga. 1989. Peptidoglycan synthesis during the cell cycle of *Escherichia coli*: composition and mode of insertion. *J. Bacteriol.* **171:**5783–5794.

37. dePedro, M. A., J. E. Llamas, and J. L. Canovas. 1975. A timing control of cell division in *Escherichia coli. J. Gen. Microbiol.* **91:**307–314.

38. Dewar, S. J., V. Kagan-Zur, K. J. Begg, and W. D. Donachie. 1989. Transcriptional regulation of cell division genes in *Escherichia coli. Mol. Microbiol.* **3:**1371–1377.

39. Donachie, W. D. 1968. Relationship between cell size and time of initiation of DNA replication. *Nature* (London) **219:**1077–1079.

40. Donachie, W. D. 1979. The cell cycle of *Escherichia coli*, p.11–35. *In* J. H. Parish (ed.), *Developmental Biology of Prokaryotes.* University of California Press, Berkeley.

41. Donachie, W. D. 1993. The cell cycle of *Escherichia coli. Annu. Rev. Microbiol.* **47:**199–230.

42. Donachie, W. D., and K. J. Begg. 1970. Growth of the bacterial cell. *Nature* (London) **227:**1220–1225.

43. Donachie, W. D., and K. J. Begg. 1989. Cell length, nucleoid separation, and cell division of rod-shaped and spherical cells of *Escherichia coli. J. Bacteriol.* **171:**4633–4639.

44. Donachie, W. D., K. J. Begg, and N. F. Sullivan. 1984. Morphogenes of *Escherichia coli*, p. 27–62. *In* R. Losick and L. Shapiro (ed.), *Microbial Development.* Cold Spring Harbor Laboratory, Cold Spring Harbor, N.Y.

45. Donachie, W. D., K. J. Begg, and M. Vicente. 1976. Cell length, cell growth and cell division. *Nature* (London) **264:**328–333.

46. Eberle, H., N. Forrest, J. Hrynyszyn, and J. Van Knapp. 1982. Regulation of DNA synthesis and capacity for initiation of DNA temperature sensitive mutants of *Escherichia coli.* I. Reinitiation and chain elongation. *Mol. Gen. Genet.* **186:**57–65.

47. Evans, I. M., and H. Eberle. 1965. Accumulation of the capacity for initiation of deoxyribonucleic acid replication in *Escherichia coli. J. Bacteriol.* **121:**883–891.

48. Gally, D., K. Bray, and S. Cooper. 1993. Synthesis of peptidoglycan and membrane during the division cycle of rod-shaped, gram negative bacteria. *J. Bacteriol.* **175:**3121–3130.

49. Garrido, T., M. Sanchez, P. Palacios, M. Aldea, and M. Vicente. 1993. Transcription of *ftsZ* oscillates during the cell cycle of *Escherichia coli. EMBO J.* **12:**3957–3965.

50. Gayama, S., T. Kataoka, M. Wachi, G. Tamura, and K. Nagai. 1990. Periodic formation of the *oriC* complex of *Escherichia coli. EMBO J.* **9:**3761–3765.

51. Grossman, N., and E. Z. Ron. 1980. Initiation of deoxyribonucleic acid replication in *Escherichia coli* B: uncoupling from mass/deoxyribonucleic acid ratio. *J. Bacteriol.* **143:**100–104.

52. Grossman, N., E. Rosner, and E. Z. Ron. 1989. Termination of DNA replication is required for cell division in *Escherichia coli. J. Bacteriol.* **171:**74–79.

53. Grover, N. B., C. L. Woldringh, A. Zaritsky, and R. F. Rosenberger. 1977. Elongation of rod-shaped bacteria. *J. Theor. Biol.* **67:**181–193.

54. Gudas, L. J., and A. Pardee. 1974. Deoxyribonucleic acid synthesis during the division cycle of *Escherichia coli*: a comparison of strains B/r, K-12, 15, and 15T⁻ under conditions of slow growth. *J. Bacteriol.* **117:**1216–1223.

55. Hakenbeck, R., and W. Messer. 1977. Oscillations in the synthesis of cell wall components in synchronized cultures of *Escherichia coli. J. Bacteriol.* **129:**1234–1238.

56. Hanawalt, P. C., O. Maaløe, D. J. Cummings, and M. Schaechter. 1961. The normal DNA replication cycle. II. *J. Mol. Biol.* **3:**156–165.

57. Hansen, F. G., and K. V. Rasmussen. 1977. Regulation of the dnaA product in *Escherichia coli. Mol. Gen. Genet.* **155:**219–225.

58. Helmstetter, C. E. 1967. Rate of DNA synthesis during the division cycle of *Escherichia coli* B/r. *J. Mol. Biol.* **24:**416.

59. Helmstetter, C. E. 1969. Regulation of chromosome replication and cell division in *Escherichia coli*, p. 15–35. *In* G. M. Padilla, G. L. Whitson, and I. L. Cameron (ed.), *The Cell Cycle, Gene-Enzyme Interactions.* Academic Press, Inc., New York.

60. Helmstetter, C. E. 1974. Initiation of chromosome replication in *Escherichia coli*. I. Requirements for RNA and protein synthesis at different growth rates. *J. Mol. Biol.* **83:**1–19.

61. Helmstetter, C. E., and S. Cooper. 1968. DNA synthesis during the division cycle of rapidly growing *Escherichia coli* B/r. *J. Mol. Biol.* **31:**507–518.

62. Helmstetter, C. E., S. Cooper, O. Pierucci, and E. O. Revelas. 1968. On the bacterial life sequence. *Cold Spring Harbor Symp. Quant. Biol.* **33:**809.

63. Helmstetter, C. E., C. Eenhuis, P. Theisen, J. Grimwade, and A. C. Leonard. 1992. Improved bacterial baby machine: application to *Escherichia coli* K-12. *J. Bacteriol.* **174:**3445–3449.

64. Helmstetter, C. E., and C. A. Krajewski. 1982. Initiation of chromosome replication in *dnaA* and *dnaC* mutants of *Escherichia coli* B/rF. *J. Bacteriol.* **149:**685–693.

65. Helmstetter, C. E., and O. Pierucci. 1976. DNA synthesis during the division cycle of three substrains of *E. coli* B/r. *J. Mol. Biol.* **102:**477–486.

66. Helmstetter, C. E., O. Pierucci, M. Weinberger, M. Holmes, and M. S. Tang. 1979. Control of cell division in *Escherichia coli*, p. 517–579. *In* L. N. Ornston and J. R. Sokatch (ed.), *The Bacteria,* vol. VII. Academic Press, Inc., New York.

67. Hirota, Y., F. Jacob, A. Ryter, G. Buttin, and T. Nakai. 1968. On the process of cellular division in *E. coli*. 1. Asymmetrical cell division and production of DNA-less bacteria. *J. Mol. Biol.* **35:**175–192.

68. Hirota, Y., and M. Ricard. 1972. Production of DNA-less bacteria, p. 29–50. *In* S. Bonotto, R. Kirchman, R. Goutier, and J. R. Maisin (ed.), *Biology and Radiobiology of Anucleate Systems,* vol. 1. Academic Press, Inc., New York.

69. Hoffman, B., W. Messer, and U. Schwarz. 1972. Regulation of polar cap formation in the life cycle of *Escherichia coli. J. Supramol. Struct.* **1:**29–37.

70. Howe, W. E., and D. W. Mount. 1975. Production of cells without deoxyribonucleic acid during thymidine starvation of *lexA*⁻ cultures of *Escherichia coli* K-12. *J. Bacteriol.* **124:**1113–1121.

71. Howe, W. E., and D. W. Mount. 1978. Analysis of cell division in single clones of *Escherichia coli* K-12 *lexA* mutant. *J. Bacteriol.* **133:**1278–1281.

72. Huisman, O., R. D'Ari, and S. Gottesman. 1984. Cell-division control in *Escherichia coli*: specific induction of the SOS function *sfiA* protein is sufficient to block septation. *Proc. Natl. Acad. Sci. USA* **81:**4490–4494.

73. Inouye, M. 1969. Unlinking of cell division from deoxyribonucleic acid replication in a temperature-sensitive deoxyribonucleic acid synthesis mutant of *Escherichia coli. J. Bacteriol.* **99:**842–850.

74. Inouye, M. 1971. Pleiotropic effect of *recA* gene of *Escherichia coli*: uncoupling of cell division from deoxyribonucleic acid replication. *J. Bacteriol.* **106:**539–542.

75. Jaffé, A., R. D'Ari, and V. Norris. 1986. SOS-independent coupling between DNA replication and cell division in *Escherichia coli. J. Bacteriol.* **165:**66–71.

76. Jones, N. C., and W. D. Donachie. 1973. Chromosome replication, transcription and control of cell division in *Escherichia coli. Nature* (London) *New Biol.* **243:**100–103.

77. Joseleau-Petit, D., F. Kepes, and A. Kepes. 1984. Cyclic changes of the rate of phospholipid synthesis during synchronous growth of *Escherichia coli. Eur. J. Biochem.* **139:**605–611.

78. Joseleau-Petit, D., F. Képès, L. Peutat, R. D'Ari, and A. Képès. 1987. DNA replication, initiation, doubling of rate of phospholipid synthesis, and cell division in *Escherichia coli. J. Bacteriol.* **169:**3701–3706.

79. Joseleau-Petit, D., F. Képès, L. Peutat, R. D'Ari, and L. I. Rothfield. 1990. Biosynthesis of a membrane adhesion zone fraction throughout the cell cycle of *Escherichia coli. J. Bacteriol.* **172:**6573–6575.

80. Kambe-Honjoh, H., G. Tamura, and K. Nagai. 1992. Association of *oriC* region of *Escherichia coli* chromosome with outer membrane: effects of culture condition. *Biochem. Biophys. Res. Commun.* **187:**970–975.

81. Kataoka, T., M. Wachi, J. Nakamura, S. Gayama, M. Yamasaki, and K. Nagai. 1993. Fully methylated *oriC* with negative superhelicity forms an *oriC*-membrane complex before initiation of chromosome replication. *Biochem. Biophys. Res. Commun.* **194:**1420–1426.

82. Koch, A. L. 1977. Does initiation of chromosome replication regulate cell division? *Adv. Microb. Physiol.* **16:**49–98.

83. Koch, A. L. 1993. Biomass growth rate during the prokaryotic cell cycle. *Crit. Rev. Microbiol.* **19:**17–42.

84. Koch, A. L., and M. L. Higgins. 1982. Cell cycle dynamics inferred from the static properties of cells in balanced growth. *J. Gen. Microbiol.* **128:**2877–2892.

85. Koch, A. L., R. W. H. Verwer, and N. Nanninga. 1982. Incorporation of diaminopimelic acid into the old poles of *Escherichia coli. J. Gen. Microbiol.* **128:**2893–2898.

86. Kogoma, T. 1978. A novel *Escherichia coli* mutant capable of DNA replication in the absence of protein synthesis. *J. Mol. Biol.* **121:**55–69.

87. Koppes, L. J. H., M. Meyer, H. B. Oonk, M. A. de Jong, and N. Nanninga. 1980. Correlation between size and age at different events in the cell division cycle of *Escherichia coli. J. Bacteriol.* **143:**1241–1252.

88. Koppes, L. J. H., and N. Nanninga. 1980. Positive correlation between size at initiation of chromosome replication in *Escherichia coli* and size at initiation of cell constriction. *J. Bacteriol.* **143:**89–99.

89. Koppes, L. J. H., N. Overbeeke, and N. Nanninga. 1978. DNA replication pattern and cell wall growth in *Escherichia coli* PAT84. *J. Bacteriol.* **133:**1053–1061.

90. Koppes, L. J. H., C. L. Woldringh, and N. Nanninga. 1978. Size variations and correlation of different cell cycle events in slow-growing *Escherichia coli. J. Bacteriol.* **134:**423–433.

91. Kubitschek, H. E. 1974. Estimation of the D period from residual division after exposure of exponential phase bacteria to chloramphenicol. *Mol. Gen. Genet.* **135:**123–130.

92. Kubitschek, H. E. 1981. Bilinear cell growth of *Escherichia coli. J. Bacteriol.* **148:**730–733.

93. Kubitschek, H. E., and C. N. Newman. 1978. Chromosome replication during the division cycle in slowly growing, steady-state cultures of three *Escherichia coli* B/r strains. *J. Bacteriol.* **136:**179–190.

94. Landouisi, A., A. Malki, R. Kern, M. Kohiyama, and P. Hughes. 1990. The *E. coli* cell surface specifically prevents the initiation of DNA replication at *oriC* on hemimethylated DNA templates. *Cell* **63:**1053–1060.

95. Lane, H. E. D., and D. T. Denhardt. 1975. The *rep* mutation. IV. Slower movement of replication forks in *Escherichia coli rep* strains. *J. Mol. Biol.* **97:**99–112.

96. Lark, K. G. 1973. Initiation and termination of bacterial deoxyribonucleic acid replication in low concentrations of chloramphenicol. *J. Bacteriol.* **113:**1066–1069.

97. Lark, K. G., and H. Renger. 1969. Initiation of DNA replication in *Escherichia coli* 15T⁻: chronological dissection of three physiological processes required for initiation. *J. Mol. Biol.* 42:221–235.

98. Løbner-Olesen, A., K. Skarstad, F. G. Hansen, K. von Meyenburg, and E. Boye. 1989. The DnaA protein determines the initiation mass of *Escherichia coli* K-12. *Cell* 57:881–889.

99. Louarn, J., V. Francois, and J.-M. Louarn. 1990. Chromosome replication pattern in *dam* mutants of *Escherichia coli. Mol. Gen Genet.* 221:291–294.

100. Louarn, J., J. Patte, and J. M. Louarn. 1982. Suppression of *Escherichia coli dnaA46* mutations by integration of plasmid R100.1 derivatives: constraints imposed by the replication terminus. *J. Bacteriol.* 151:657–667.

101. Lu, M., J. L. Campbell, E. Boye, and N. Kleckner. 1994. SeqA: a negative modulator of replication initiation in *E. coli. Cell* 77:413–426.

102. Lutkenhaus, J. F. 1983. Coupling of DNA replication and cell division: *sulB* is an allele of *ftsZ. J. Bacteriol.* 154:1339–1346.

103. Lutkenhaus, J. 1990. Regulation of cell division in *E. coli. Trends Genet.* 6:22–25.

104. Lutkenhaus, J. 1993. FtsZ ring in bacterial cytokinesis. *Mol. Microbiol.* 9:403–409.

105. Lutkenhaus, J. F., B. A. Moore, M. Masters, and W. D. Donachie. 1979. Individual proteins are synthesized continuously throughout the *Escherichia coli* cell cycle. *J. Bacteriol.* 138:352–360.

106. Maaløe, O., and N. O. Kjeldgaard. 1966. *Control of Macromolecular Synthesis.* W. A. Benjamin, New York.

107. MacAlister, T. J., B. MacDonald, and L. I. Rothfield. 1983. The periseptal annulus: an organelle associated with cell division. *Proc. Natl. Acad. Sci. USA* 80:1372–1376.

108. Mendelson, N. H. 1982. Bacterial growth and division: genes, structures, forces, and clocks. *Microbiol. Rev.* 46:341–375.

109. Messer, W., V. Bellekes, and H. Lother. 1985. Effect of dam methylation on the activity of the *E. coli* replication origin, oriC. *EMBO J.* 4:1327–1332.

110. Messer, W., and M. Noyer-Weidner. 1988. Timing and targeting: the biological functions of *dam* methylation in *E. coli. Cell* 54:735–737.

111. Mirelman, D., Y. Yashouv-Gan, Y. Nuchamovitz, S. Rozenhak, and E. Ron. 1978. Murein biosynthesis during a synchronous cell cycle of *Escherichia coli* B. *J. Bacteriol.* 134:458–461.

112. Mulder, E., and C. L. Woldringh. 1989. Actively replicating nucleoids influence positioning of division sites in *Escherichia coli* filaments forming cells lacking DNA. *J. Bacteriol.* 171:4303–4314.

113. Mulder, E., and C. L. Woldringh. 1991. Autoradiographic analysis of diaminopimelic acid incorporation in filamentous cells of *Escherichia coli*: repression of peptidoglycan synthesis around the nucleoid. *J. Bacteriol.* 173:4751–4756.

114. Mulder, E., and C. L. Woldringh. 1993. Plasmolysis bays in *Escherichia coli*: are they related to development and positioning of division sites? *J. Bacteriol.* 175:2241–2247.

115. Nanninga, N. 1991. Cell division and peptidoglycan assembly in *Escherichia coli. Mol. Microbiol.* 5:791–795.

116. Nanninga, N., F. B. Wientjes, E. Mulder, and C. L. Woldringh. 1992. Envelope growth in *Escherichia coli*: spatial and temporal organization, p. 185–221. *In* S. Mohan, C. Dow, and J. A. Cole (ed.), *Prokaryotic Structure and Function: a New Perspective.* Cambridge University Press, London.

117. Nanninga, N., C. L. Woldringh, and L. J. H. Koppes. 1982. Growth and division of *Escherichia coli*, p. 225–270. *In* C. Nicolini (ed.), *Cell Growth.* Plenum Press, London.

118. Newman, C. N., and H. E. Kubitschek. 1978. Variation in periodic replication of the chromosome in *Escherichia coli* B/rTT. *J. Mol. Biol.* 121:461–471.

119. Niki, H., A. Jaffé, R. Imamura, T. Ogura, and S. Hiraga. 1991. The new gene *mukB* codes for a 177 kd protein with coiled-coil domains involved in chromosome partitioning of *E. coli. EMBO J.* 10:183–193.

120. Nordström, K., R. Bernander, and S. Dasgupta. 1991. The *Escherichia coli* cell cycle: one cycle or multiple independent processes that are co-ordinated? *Mol. Microbiol.* 5:769–774.

121. Ogawa, T., and T. Okazaki. 1994. Cell cycle-dependent transcription from the *gid* and *mioC* promoters of *Escherichia coli. J. Bacteriol.* 176:1609–1615.

122. Ogden, G. B., M. J. Pratt, and M. Schaechter. 1988. The replicative origin of the *E. coli* chromosome binds to cell membranes only when hemimethylated. *Cell* 54:127–135.

123. Ohki, M. 1972. Correlation between metabolism of phosphatidylglycerol and membrane synthesis in *Escherichia coli. J. Mol. Biol.* 68:249–264.

124. Olijhoek, A. J. M., S. Klencke, E. Pas, N. Nanninga, and U. Schwarz. 1982. Volume growth, murein synthesis, and murein cross-linkage during the division cycle of *Escherichia coli* PA 3092. *J. Bacteriol.* 152:1248–1254.

125. Orr, E., P. A. Meacock, and R. H. Pritchard. 1978. Genetic and physiological properties of an *Escherichia coli* strain carrying the *dnaA* mutation T46, p. 85–99. *In* I. Molineux and M. Kohiyama (ed.), *DNA Synthesis, Present and Future.* Plenum Publishing Corp., New York.

126. Pierucci, O. 1978. Dimensions of *Escherichia coli* at various growth rates: model for envelope growth. *J. Bacteriol.* 135:559–574.

127. Pierucci, O. 1979. Phospholipid synthesis during the cell division cycle of *Escherichia coli. J. Bacteriol.* 138:453–460.

128. Pierucci, O., and C. E. Helmstetter. 1969. Chromosome replication, protein synthesis and cell division in *E. coli* B/r. *Fed. Proc.* 28:1755.

129. Pierucci, O., M. Melzer, C. Querini, M. Rickert, and C. Krajewski. 1981. Comparison among patterns of macromolecular synthesis in *Escherichia coli* B/r at growth rates of less and more than one doubling per hour at 37°C. *J. Bacteriol.* 148:684–696.

130. Previc, E. P. 1970. Biochemical determination of bacterial morphology and geometry of cell division. *J. Theor. Biol.* 27:471–497.

131. Pritchard, R. H., P. T. Barth, and J. Collins. 1969. Control of DNA synthesis in bacteria. *Symp. Soc. Gen. Microbiol.* 19:263–297.

132. Puyet, A., and J. Canovas. 1989. Changes of *Escherichia coli* cell cycle parameters during fast growth and throughout growth with limiting amounts of thymine. *Arch. Microbiol.* 152:578–583.

133. Robin, A., D. Joseleau-Petit, and R. D'Ari. 1990. Transcription of the *ftsZ* gene and cell division in *Escherichia coli. J. Bacteriol.* 172:1392–1399.

134. Robinson, A. C., D. J. Kenan, G. F. Hatfull, N. F. Sullivan, R. Spiegelberg, and W. D. Donachie. 1984. DNA sequence and transcriptional organization of essential cell division genes *ftsQ* and *ftsA* of *Escherichia coli*: evidence for overlapping transcriptional units. *J. Bacteriol.* 160:546–555.

135. Robinson, A. C., D. J. Kenan, J. Sweeney, and W. D. Donachie. 1986. Further evidence for overlapping transcriptional units in an *Escherichia coli* cell envelope-cell division gene cluster: DNA sequence and transcriptional organization of the *ddl ftsQ* region. *J. Bacteriol.* 167:809–817.

136. Rosenberger, R. F., N. B. Grover, A. Zaritsky, and C. L. Woldringh. 1978. Surface growth in rod-shaped bacteria. *J. Theor. Biol.* 73:711–721.

137. Rothfield, L. I., P. A. J. de Boer, and W. R. Cook. 1990. Localization of septation sites. *Res. Microbiol.* 141:57–63.

138. Ryter, A., Y. Hirota, and U. Schwarz. 1973. Process of cellular division in *Escherichia coli* murein. *J. Mol. Biol.* 78:185–195.

139. Sakakibara, Y., and S. Yuasa. 1982. Continuous synthesis of the *dnaA* gene product of *Escherichia coli* in the cell cycle. *Mol. Gen. Genet.* 186:87–94.

140. Sargent, M. G. 1978. Surface extension and the cell cycle in prokaryotes. *Adv. Microb. Physiol.* 18:105–176.

141. Schaechter, M., O. Maaloe, and N. O. Kjeldgaard. 1958. Dependency on medium and temperature of cell size and chemical composition during balanced growth of *Salmonella typhimurium. J. Gen. Microbiol.* 19:592–606.

142. Schwarz, U., A. Ryter, A. Rambach, R. Hellio, and Y. Hirota. 1975. Process of cellular division in *Escherichia coli*: differentiation of growth zones in the sacculus. *J. Mol. Biol.* 98:749–759.

143. Skarstad, K., and E. Boye. 1994. The initiator protein DnaA: evolution, properties and function. *Biochim. Biophys. Acta* 1217:111–130.

144. Skarstad, K., H. B. Steen, and E. Boye. 1983. Cell cycle parameters of slowly growing *Escherichia coli* studied by flow cytometry. *J. Bacteriol.* 154:656–662.

145. Skarstad, K., H. B. Steen, and E. Boye. 1985. *Escherichia coli* DNA distributions measured by flow cytometry and compared with theoretical computer simulations. *J. Bacteriol.* 163:661–668.

146. Spratt, B. G., and R. J. Rowbury. 1971. Physiological and genetical studies on a mutant of *Salmonella typhimurium* which is temperature-sensitive for DNA synthesis. *Mol. Gen. Genet.* 114:35–49.

147. Sun, L., and J. A. Fuchs. 1992. *Escherichia coli* ribonucleotide reductase expression is cell cycle regulated. *Mol. Biol. Cell* 3:1095–1105.

148. Sun, L., B. A. Jacobson, B. S. Dien, F. Srienc, and J. A. Fuchs. 1994. Cell cycle regulation of the *Escherichia coli nrd* operon: requirement for a *cis*-acting upstream AT-rich sequence. *J. Bacteriol.* 176:2415–2426.

149. Tang, M.-S., and C. E. Helmstetter. 1980. Coordination between chromosome replication and cell division in *Escherichia coli. J. Bacteriol.* 141:1148–1156.

150. Tetart, F., R. Albigot, A. Conter, E. Mulder, and J.-P. Bouche. 1992. Involvement of FtsZ in coupling of nucleoid separation with septation. *Mol. Microbiol.* 6:621–627.

151. Theisen, P. W., J. E. Grimwade, A. C. Leonard, J. A. Bogan, and C. E. Helmstetter. 1993. Correlation of gene transcription with the time of initiation of chromosome replication in *Escherichia coli. Mol. Microbiol.* 10:575–584.

152. Tippe-Schindler, R., G. Zahn, and W. Messer. 1979. Control of the initiation of DNA replication in *Escherichia coli*. I. Negative control of initiation. *Mol. Gen. Genet.* 168:185–195.

153. Tresguerres, E. F., C. Nieto, I. Casquero, and J. L. Canovas. 1986. Host cell variations resulting from F plasmid-controlled replication of the *Escherichia coli* chromosome. *J. Bacteriol.* 165:424–427.

154. Trueba, F. J., O. M. Neijssel, and C. L. Woldringh. 1982. Generality of the growth kinetics of the average individual cell in different bacterial populations. *J. Bacteriol.* 150:1048–1055.

155. Trueba, F. J., and C. L. Woldringh. 1980. Changes in cell diameter during the division cycle of *Escherichia coli. J. Bacteriol.* 142:869–878.

156. Verwer, R. W. H., and N. Nanninga. 1980. Pattern of *meso*-DL-2,6-diaminopimelic acid incorporation during the division cycle of *Escherichia coli. J. Bacteriol.* 144:327–336.

157. von Meyenburg, K., E. Boye, K. Skarstad, L. Koppes, and T. Kogoma. 1987. Mode of initiation of constitutive stable DNA replication in RNase H-defective mutants of *Escherichia coli* K-12. *J. Bacteriol.* 169:2650–2658.

158. von Meyenburg, K., and F. G. Hansen. 1980. The origin of replication, *oriC*, of the *Escherichia coli* chromosome: genes near *oriC* and construction of *oriC* deletion mutations. *ICN-UCLA Symp. Mol. Cell. Biol.* 19:137–159.

159. Wang, X., P. A. J. de Boer, and L. I. Rothfield. 1991. A factor that positively regulates cell division by activating transcription of the major cluster of essential cell division genes of *Escherichia coli. EMBO J.* 10:3363–3372.

160. Wientjes, F. B., and N. Nanninga. 1989. Rate and topography of peptidoglycan synthesis during cell division in *Escherichia coli*: concept of a leading edge. *J. Bacteriol.* 171:3412–3419.

160a. Wold, S., K. Skarstad, H. B. Steen, T. Stokke, and E. Boye. 1994. The initiation mass for DNA replication in *Escherichia coli* K-12 is dependent on growth rate. *EMBO J.* **13:**2097–2102.

160b. Woldringh, C. L. 1994. Significance of plasmolysis spaces as markers for periseptal annuli and adhesion sites. *Mol. Microbiol.* **14:**597–607.

161. Woldringh, C. L., M. A. de Jong, W. van den Berg, and L. Koppes. 1977. Morphological analysis of the division cycle of two *Escherichia coli* substrains during slow growth. *J. Bacteriol.* **131:**270–279.

162. Woldringh, C. L., P. Huls, E. Pas, G. J. Brakenhoff, and N. Nanninga. 1987. Topography of peptidoglycan synthesis during elongation and polar cap formation in a cell division mutant of *Escherichia coli* MC4100. *J. Gen. Microbiol.* **133:**575–586.

163. Woldringh, C. L., E. Mulder, P. G. Huls, and N. Vischer. 1991. Toporegulation of bacterial division according to the nucleoid occlusion model. *Res. Microbiol.* **142:**309–320.

164. Yi, Q.-M., S. Rockenbach, J. E. Ward, Jr., and J. Lutkenhaus. 1985. Structure and expression of the cell division genes *ftsQ*, *ftsA*, and *ftsZ*. *J. Mol. Biol.* **184:**399–412.

165. Zaritsky, A. 1975. Rate stimulation of deoxyribonucleic acid synthesis after inhibition. *J. Bacteriol.* **122:**841–846.

166. Zaritsky, A., and R. H. Pritchard. 1971. Replication time of the chromosome in thymineless mutants of *Escherichia coli*. *J. Mol. Biol.* **60:**65–74.

167. Zaritsky, A., and R. H. Pritchard. 1973. Changes in cell size and shape associated with changes in the replication time of the chromosome of *Escherichia coli*. *J. Bacteriol.* **114:**824–837.

168. Zaritsky, A., R. F. Rosenberger, J. Naaman, C. L. Woldringh, and N. B. Grover. 1982. Growth and form in bacteria. *Commun. Mol. Cell. Biophys.* **1:**237–260.

169. Zhou, P., and C. E. Helmstetter. 1994. Relationship between *ftsZ* gene expression and chromosome replication in *Escherichia coli*. *J. Bacteriol.* **176:**6100–6106.

Similarities and Differences of Individual Bacteria within a Clone

ARTHUR L. KOCH

103

INTRODUCTION

Seemingly, two sister bacteria are biological entities as closely identical as can be conceived. In this chapter, their differences are scrutinized. The examination will extend to the differences between subcultures and subclones of *Escherichia coli*. Important concepts have also come from the study of other eubacteria. These differences presumably have a biological basis and can be subdivided under the major divisions listed in Table 1 of strategies to cope with environmental challenges. A few of the variations are due to cell systematic changes during the cell cycle and to idiosyncratic mutational events; the rest concern features that have evolved to allow the bacteria to cope with unpredictable variations in their environment. Importantly, these include mechanisms to generate nongenetic differences between sister cells to ensure that one or the other survives potential catastrophes and preserves the strain. Key among them is an ability to enter into a Dauer modification state, referred to by Dow et al. (15) as the shutdown state.

The global life strategy of a bacterium is characterized by its expression of a diversity which allows it to invade those niches and habitats obligate for its lifestyle. Its strategy can achieve the variability necessary in both the short and long terms to respond to a range of rare catastrophes, engage in genetic exchange to a controlled degree with other kinds of organisms, balance mutagenic forces with repair systems so that accumulation of mutations is consistent with both remaining the same and changing innovatively, and possibly enable invasion of habitats that are not necessary for the cell's cycle of growth and continued propagation (such as opportunistic infections of a compromised host).

First a catalog of the types of variability found in nature is given. Then some theoretical ideas for the evolutionary processes in prokaryotes are presented. Finally, the cell cycle is analyzed to understand some of the variability in comparing two cells taken at random from an asynchronous culture.

CLONAL HETEROGENEITY

Presence of Duplicated Regions of the Chromosome

Although most of the bacterial chromosome is single-copy DNA, there are some duplicate genes even in the normal wild-type genome. The most notable cases are the occurrence of seven copies of the DNA coding for rRNA and for other components

TABLE 1 Strategies to cope with environmental challenges

Mass responses of growing cultures
  Induced enzyme formation
  Heat shock: σ^{32} and σ^{e}
  Ethanol resistance
  Oxidative stress resistance
  SOS response to ionizing and UV radiations
  Excessive amounts of a needed substance
Mass responses of cultures approaching the stationary phase
  Development of motility
  Development of transformability
  Production of antibiotics
  Production of a variety of secondary metabolites
  Formation of forms resistant to heat
  Formation of forms resistant to drying
  Formation of forms resistant to starvation
Apparently inappropriate responses of rare individuals that provide
    resistance
  To:
    Ionizing radiation
    UV radiation
    The antibody response of a mammalian host
  By:
    Inappropriate sporulation
    Inappropriate formation of resistant forms
    Inappropriate partial reactivation of cryptic pathways
    Inversion of a genetic region

of the protein synthesis machinery in the standardized *E. coli*. These homologous regions permit unequal crossing over between sister chromosomes, leading to an equal number of deficiencies and duplications of the regions between ribosomal genes (1). Both of these recombinant cells are eventually lost, except under very selective conditions. For example, the genotypes with duplications can survive and experience selection, causing further recombination and multiplication of duplicated genes (35). Usually a growing population has large duplications (two or more copies) in about 1% of the cells (J. R. Roth, personal communication). These cells can have properties quite different from those of cells with the nonduplicated genome. Note that the cells with an abnormal composition usually grow more slowly and can be an additional source of the slowly growing component of the population that generates a skewed age-at-division distribution.

Inversion Switches

Phase variations in *Salmonella typhimurium* (official designation, *Salmonella enterica* serovar Typhimurium) and phage Mu are well-studied cases of special genetic regions that are selectively inverted to yield other gene products and different mutant phenotypes. A similar system exists in bacteriophage P1. These systems are related to each other and to transposon Tn*A* (66). In these cases, a promoter initiates RNA transcription in an inverted or not normally oriented genetic structure, thus making different products. The value of such systems for long-term survival of the organism is in providing a mechanism to reversibly switch to an alternative phenotype, not simply convert by an all-or-none change to a different state irreversibly. Depending on the human host of the bacterium or the bacterial host for the phage, different states are positively selected. However, the alternate state appears with a high probability (10^{-3} to 10^{-5} per generation). Such systems need not have the same rate of inversion in the two directions, because the gene catalyzing the inversion is under the control of different promoters in the two states. Clearly, it would

not benefit an organism to have inappropriate switching at inappropriate rates, as would happen if two such systems were to function within the cells of a bacterial culture (cross talk?). However, the number of such switching systems is probably quite limited because of the close genetic relationship among the systems that have been studied and because the protein catalyzing inversion in some cases can act on other systems. Again, this circumstance could explain a positive skewness of the age-at-division distribution if some proportion of the cells in the population were in a form that grew more slowly.

Radiation-Resistant Fraction of Populations

Early in the history of radiation biology, Luria and Laterjet (59) discovered that although the survival curves of microorganisms grown under different treatment conditions might have quite differently shaped shoulders on the standard semilogarithmic plots, in all cases the survivorship at high doses was unexpectedly high. At moderate doses, the curves exhibited a first-order region, but the straight line did not extend to very low survival values. Instead, at high doses the line curved, and survivorship was essentially independent of dose. Thus, a resistant fraction of cells remained at values between 10^{-2} and 10^{-4} of the initial number of viable cells. This response has been seen with UV light, X rays, and even tritium suicide (61). In all these cases, it appears that a fraction of the population is in a resistant and temporarily nongrowing state. This state of affairs makes good evolutionary and ecological sense: a certain proportion of the bacteria that are formed are somehow destined to be quiescent. In this state, they do not grow and are resistant to many environmental stresses. Eventually they do grow and produce normal cultures and colonies. It is as if the culture of gram-negative organisms allots a small fraction of its productivity during balanced growth to form the equivalent of the spore stage of gram-positive organisms. This process has been little studied, no doubt because it is so rare. But in many ways it is a strategy superior to the complete conversion of a culture upon starvation into endospores: it is an a priori calculated diversification to form products that are a hedge against radiation damage as well as other difficulties that might arise. This phenomenon is yet a third possibility to explain the positive skewness of the age-at-division distribution.

Phage Yield and Cell Size Variation

The first reference (of which I am aware) to the importance of biological variability was a report by Delbrück (13) in 1945. He assessed the variability of the burst size of viruses from one-step growth curves and found it to be much larger than the variability of cell lengths in the population. This finding implies that either the virus replication is highly variable or the bacteria are highly variable in characteristics other than length.

Nutritional Generation of Cellular Heterogeneity

The kinetics of lactose metabolism was quite confusing until it was recognized that one needed to distinguish between enzyme formation averaged over the entire culture and that within individual cells. Benzer (5) realized this and set out to measure the variability between bacteria within a culture that had been shifted to lactose as a carbon and energy source. He stopped enzyme induction by phage infection. Under this circumstance, only those bacteria that had on hand the enzymes necessary for lactose metabolism could support the growth of the virus and later lyse, liberating infectious particles and β-galactosidase. With this system as an assay, Benzer was able to show that after a culture of bacteria has been exposed to lactose for a few minutes (when the level of β-galactosidase is still low), the enzyme levels of individual cells are very different from each other. Only a few cells had become sufficiently induced to be able to use lactose as a carbon and energy source and thereby support normal phage reproduction. The bulk of the original population, which remained uninduced, supported a limited growth of phage, and that only after a delay. Benzer showed that after a shift to lactose, the heterogeneity of the population first increased and then decreased. When a poorly used substrate, methyl-β-galactopyranoside, was substituted for lactose, the response was uniform among the cells throughout the induction process.

How were the bacteria that were more easily induced different from the rest? With the knowledge accumulated since 1953, many of the factors involved can now be specified (3, 4, 73). It is most probable that in this small, easily induced portion of the population, an uninduced (random) transcription event of the *lac* operon had recently occurred. Although such untriggered mRNA syntheses are rare under noninducing conditions, each such event results in a number of translation products. Consequently, the background level of β-galactosidase and permease in these few cells is high, and thus the enzyme level in the population is quite heterogeneous. Lactose is accumulated much more rapidly in the few cells with higher than average levels of permease, and there the level of β-galactosidase is sufficiently high that enough transglycosylation events take place to form the actual inducers, allolactose (6-*o*-β-D-galactopyranosyl-D-glucose) (21) and glycerol-β-D-galactopyranoside (A. W. Hsie, personal communication). This starts the avalanche of events leading to increased transport, transglycosylation, induction, and production of galactose and glucose as carbon and energy sources. The process is autocatalytic in that it creates the cellular growth response, leading to an increase of the induced group of cells and further heterogeneity in the culture. However, after the lactose-metabolizing cells become the dominant part of the population, cell-to-cell variability decreases.

On this basis, the major source of variability comes from the chance event of the release of the bound repressor from the operator DNA of the *lac* operon. These kinetics have been well studied, both theoretically (30, 60, 73) and experimentally (2). The fact that there are few molecules of tetrameric repressor per cell is sufficient to explain the situation that creates the heterogeneity in β-galactosidase activity in partially induced populations.

Quasi-Genetic Variation of Growth Rate

In an early but well-designed experiment, Hughes (18) incubated a single cell for 3 h to obtain a clone. He then spread a portion of the clone on nutrient agar and after 3 h selected both a small and a large microcolony. From these second and third clones, he similarly selected a fourth and fifth, again on the basis of colony size. The diameter distributions obtained from 100 microcolonies derived from a further subcloning of each of these five colonies showed large variability within each subclone, while their mean sizes were significantly different, demonstrating short-term "inheritance" of colony size. It might not be surprising that one can select slowly growing variants. The fact that it is as easy to select faster-growing variants is possibly more surprising. However, the key point is that this inheritance is not permanent. But the degree of variation present in the cells of a microcolony that has had only 3 h at 37°C under nominally anaerobic conditions to grow from a single cell is quite surprising. In this time, one cell could have grown to at most 500 cells, and many of these were shown to have growth properties markedly, but temporarily, different from those of the progenitor cell. Axelrod et al. have used essentially the same technique to show that mouse cell lines exhibit this same phenomenon (2).

All-or-None Formation of Induced Enzymes

With the development of the gratuitous inducer thiomethyl-β-D-galactopyranoside (TMG) and the technique of chemostat culture, a kinetic study of conversion of cells to the induced state could be made. Novick and Weiner (60) designed a medium containing a level of TMG that failed to induce uninduced cells but maintained the state of induction of previously induced cells that had permease and could accumulate the TMG. Consequently, these authors were able to measure the rate at which cells first become induced. Note that their technique gives quite different information than does a simple estimate of the general level of *lac* operon products. They showed that the transformation to the induced state was a first-order process in the bacterial concentration but a high-order process (the 8.4 power of TMG concentration) for the inducer concentration. This means that while the pseudo-first-order rate constant for recruitment to enzyme production rises rapidly as the external TMG concentration is changed, from the cell's point of view the process is a stochastic one like the decay of a radioactive atom; i.e., it depends on the rare chance of the binding of the inducer to the repressor, changing the latter's conformation, and leading to an opening of the operator site. Once that happens, induction becomes autocatalytic until that cell and its descendants are fully induced.

Quiescent Cells in Low-Dilution-Rate Chemostats

Our test for this phenomenon (50) depended on two different assays for β-galactosidase. One was the usual assay in which the bacteria are lysed with sodium dodecyl sulfate; the other was the hydrolysis of *o*-nitro-β-D-galactopyranoside by dense suspensions of permease-negative intact bacterial cells.

If the induction response is heterogeneous, then the population will consist of a few cells that contain most of the β-galactosidase and a majority of cells that contain very little. Although all of the enzyme molecules will function at a high rate in the lysis assay because the *o*-nitro-β-D-galactopyranoside is available at nearly saturating concentrations, they will not function in the whole-cell assay because the reaction rate of the enzyme within the cell will be limited by the passive diffusion of substrate through the membrane of these (permease-negative) organisms. On the other hand, if after a short time from the inducer addition (allowing for the penetration of the inducer, the start of transcription, and the completion of the first translation products) the enzyme molecules are uniformly distributed among all cells, then the hydrolysis rate of the whole cells will be maximal because the permeation of *o*-nitro-β-D-galactopyranoside through a combined total surface area of all the cells is much greater than it would be through the surfaces of a smaller subset of these cells.

With this test, it was shown that all cells in chemostat populations growing with doubling times of up to 13 h had a uniform and rapid response. On the other hand, cultures growing more slowly were heterogeneous. The kinetic results were interpreted as follows. In low-dilution-rate cultures, each cell varies in its instantaneous rate of protein synthesis. A particular cell may be temporarily quiescent (possibly accumulating reserves) but will at some time self-activate and synthesize proteins at the normal rate. In fact, it was found that the same step time was characteristic of both cells from a faster-growing chemostat culture and cells growing in unlimited medium, otherwise of the same composition. For a 24-h chemostat culture, only about one-third of the cells are inactive at any instance of time, but all cells are active over a 3-h period. This type of heterogeneity could not be due to inadequacy of mixing in the chemostat culture. It is probably controlled by minor fluctuations in the energy resources available to individual organisms and the time since the last activation of protein synthesis (also see the discussions in references 22, 23, and 31).

Individuality of Cellular Chemotaxis

It is now well understood that bacteria collectively and individually swim in an ecologically favorable direction by the following process. They proceed in a more or less constant direction for a time (69); however, sooner or later they change direction, more or less at random. The frequency of the "tumbles" depends on the chemoreception process and is usually responsive to whether the conditions are satisfactory for the cell's needs. The operant philosophy of the motility system apparently is, if things are satisfactory, don't rock the boat (i.e., don't change course).

The tumbling response is controlled by the level of methylation of key proteins and is an all-or-none response which is caused by reversing the direction of rotation of the cell's flagella (69). Spudich and Koshland (68) observed that upon a shift to favorable conditions, i.e., the presence of 0.5 mM serine as an attractant, individual cells of a tumbling mutant strain exhibited a lag period before a tumbling event. Eventually the cells engaged in a few tumbling events and then tumbled continuously. The key point of the Spudich and Koshland study was that individual cells had significantly different lag times and that the lag time did not vary systematically with cell size; the lag appeared to be a characteristic of an individual cell. The same individuality was shown with wild-type cells, although the assay is more difficult. In repeated trials, particular cells maintained a quite constant time until tumbling.

Mechanistically the cells respond to the stimulus by activating the methylating (CheR) and demethylating (CheB) enzymes that function until the protein substrate occupies enough of the critical sites to control a sufficient number of flagella. If the rate constant of the process is k and if there are r critical sites per cell that must be altered out of a total of n, the fraction of cells that have not yet tumbled is described by the binomial distribution, with $p = e^{-kt}$ and $q = 1 - e^{-kt}$. This is basically the model that Rahn (64) used to describe bacterial division kinetics. The binomial expression would describe the case if k, r, and n were exactly the same for every cell in the population. In fact, the number of functional flagella does vary, and thus r and n may vary from cell to cell. It could also be the case that the number of enzymes affecting the state of methylation is small and varies stochastically from cell to cell.

Variability of the Growth Rate in Broth

With a computer-linked turbidimetric system constructed to monitor bacterial growth accurately and continuously, Wang and Koch (74) observed a phenomenon which probably could not be detected in any other way. The experimental measuring system was capable of measuring the growth rate to better than 1% in a 3-min period. The conclusion of their paper is important because without knowledge of this phenomenon, false conclusions would be made about what might appear to reflect bacterial individuality. It was found that the growth rate of wild-type bacteria on ordinary Difco nutrient broth exhibited a large and systematic, but temporary, slowing when a particular bacterial density was reached. It was shown that the cultures had depleted something in the growth medium and therefore became "shifted down." The cells, however, quickly readjusted physiologically so that the growth rate increased to nearly the original value. The implication is that if samples are removed from a culture which is apparently in balanced growth and growing in rich medium, they could have radically different instantaneous properties if taken at slightly different time points.

FORMULATION OF THE NEED FOR VARIABILITY OF MANY DIFFERENT KINDS

The First Cell Had No Versatility and No Backup Mechanisms

The earliest living organism must have had a repertoire of catalytic abilities that was limited but sufficient to enable it to survive and multiply (38, 45, 54). The tool chest of the earliest cells capable of adaptive evolution had to include three processes: first, a mechanism to duplicate information-bearing substances, including, possibly, clays, mineral surfaces, inorganic and organic membranes, proteinaceous polymers, RNA, and DNA; second, a mechanism to transduce energy into a form that it could exploit from an exergonic chemical reaction using reactants present in the environment; and third, a mechanism to foster cell division. This earliest cell may have depended on the environment for other needs (38).

Thus, many required items had to be present in the environment at the right level for the cells to survive. Such a primitive cell, however, could engage in Darwinian evolution, thus making it possible for mutants advanced in these three properties to supplant the prior population. Qualitatively new processes could also have arisen, and thus the cells became more versatile and were less rigidly dependent on the environment. These improvements would have led to an expansion of world biomass and a wider range of habitat.

Elaboration, Refinement, and Development of Survival Mechanisms

The vast bulk of the thousands of steps involved in the cell processes common to all cells, no matter to which kingdom or domain they belong, apparently evolved before stable diversity developed (Fig. 1). The logical explanation of this unity is that evolution was essentially monophyletic throughout this period, and although there were many offshoots at any one time, all but one evolutionary branch died out. Thus, the world biosphere was essentially a monoculture up to the time of the last universal ancestor. The initial split into three kingdoms (or domains) must have occurred because the founder organisms used quite different, noncompeting strategies (45). This chapter focuses on only certain enteric organisms, but they have hidden in their genomes processes inherited from the earlier monophyletic time. Significant here are the systems to deal with challenges, the last item in the center section of Fig. 1.

Although the generation of stable diversity occurred late compared with the development of a highly developed cell physiology, many mechanisms to cope with environmental fluctuations developed much earlier. The heat shock system, for example, is ubiquitous, and the genes are homologous across all phyla. Conversely, resistance to environmental fluctuations in water activity developed after stable diversity was well established. For this adaptation, some organisms perfected sporulation, some developed mechanisms to generate or sequester osmoprotectants, etc. The sporadic and unique occurrence of these protection mechanisms suggests that the dependence of life on water is so essential that ways to overcome its lack were very difficult to devise and developed late in specialist organisms. Similarly, ways to survive and grow at a range of temperatures, salinities, and pH values had only feeble starts before diversity became well established, but then specialists in narrower, more extreme, environments did arise. It has been argued that high-temperature forms, particularly of the domain *Archaea*, came first (75, 76), but one would have thought that such an origin would have left its mark on many proteins from organisms now growing at more moderate temperatures.

Enteric organisms, whose biomass is thought to be largely restricted to the colon and feces of animals, have developed an astounding ability to adapt and to cope with the necessary fluctuations during the host cycle (40). I will use the term "host cycle" to include the circular process by which coliforms move from one gastrointestinal tract to the external environment and then establish themselves in another colon. The term "host cycle," of course, includes the "culture cycle," i.e., the process set in motion when a stationary culture is diluted into fresh rich medium and grows again into a stationary culture. That term, in turn, includes the "cell cycle," i.e., the process set in motion when a cell divides and each newborn cell grows and divides

again into two newborn cells. In addition, we need something to include longer-term changes, like response to an ice age or to a time when certain antibiotics are temporarily abundant in the ecosystems. Thus, layers of regulatory systems must be coded for.

Consequently, to think about the variability of a modern prokaryotic species, including its inheritance from the past and the strategies that it uses to survive today, we must consider the time to express that variability (44). Some cellular controls must function within a small fraction of a second, while others depend on reactivating cellular processes evolved many millions of years ago that are needed only occasionally. To consider them, I will first discuss some thought experiments that, if carried out, would eliminate such mechanisms, and I will then examine cases of living organisms that have lost certain of these mechanisms.

TWO THOUGHT EXPERIMENTS

The Glucose-Limited "Perennial" Chemostat

Imagine that a single cell of a modern bacterial species with no plasmids or cryptic viruses (i.e., truly gnotobiotic) was inoculated into a constant, continuously mixed environment and cultured for a very long time (say, billions of years). Specifically, imagine a "perennial" chemostat culture (39, 41, 47, 48) with one starting organism growing continuously limited by one nutrient—say, *E. coli* with a low level of glucose in a minimal salts medium. (Imagine that unspecified means are then used to prevent adherence of the organism to the walls of the vessel.) In this perennial austerity, the culture would remain as a monoculture because there is only a single nonsubstitutable resource available. Therefore, in this truly extreme case, the competitive exclusion principle of Gause will ensure that any favorable mutation will displace the parental kind. Thus, the system would remain as a monoculture, although there would be population turnovers and temporarily there would be transients with mixed

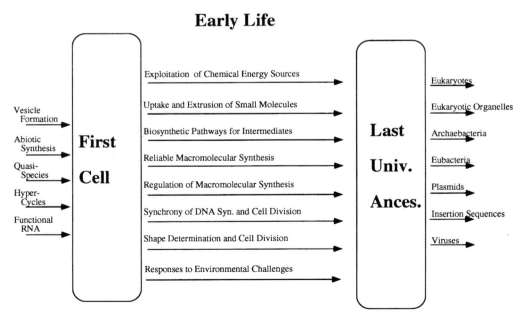

Early Life

FIGURE 1   Evolution of the early cell.

genotypes present. Under such conditions, diversity could not develop. Moreover, there would be no development of symbiosis or antibiosis. Plasmidlike entities might develop as permanent parts of the cell, but development of the ability to transmit host genes to other cells would not serve a useful purpose. This is because transmission of genes between cells would not be selected for because of the occurrence of population turnovers which would regenerate a homogeneous population and by periodic selection (see reference 32) eliminate diversity, and thus the movement of genes from cell to cell would not increase fitness.

Thus, long-term selection would produce cells that have achieved an effective and efficient glucose transport system so that in the final state, neither more units of phosphotransferase in the membrane nor a higher specific activity (or a qualitatively different mechanism) would improve their ability to scavenge glucose from the environment. The steady-state concentration of free glucose in the medium would be extremely small. I chose this example because the extant coliforms have almost reached this state. Actually, modern E. coli is not far from the limiting state in which every glucose molecule that diffuses to it is consumed (27, 34, 56). Even so, the most rate-limiting step in the uptake and metabolism of glucose is the diffusion through the porins in the outer membrane during the steady state of uptake (48, 56), and thus the outer membrane would be discarded during our continuous culture. As a second example, modern enteric organisms change size depending on the growth rate, which in turn depends on the nutrients in the culture medium; the cells are smaller in poor medium. To increase its surface-to-volume ratio, under extended chemostat growth, the cells should evolve to be as small as possible under constraints fixed by other cellular processes. Also, the cells should become as elongated and narrow as mechanistically possible to increase further their surface-to-volume ratio. Furthermore, long-term evolution in the chemostat will favor a mechanism whereby the sister cells will not adhere to each other but rapidly separate after cell division. If a cell separates itself from others, it will maximize its rate of glucose uptake. Motility and chemotaxis in principle can increase the rate at which glucose impinges on the cell, but mathematical analysis showed (39) that if the cells remain well separated, motility of the isolated cell would not help the bacterium scavenge a low-molecular-weight substrate in its environment because diffusion of the nutrient is rapid compared with the speed of chemotaxis. The modern organism has adapted to the fluctuations in its environment at considerable cost compared with the fixed needs in the hypothetical chemostat.

The Perpetual Turbidostat

Now imagine that the same experiment had been carried out in a perfect device that continuously measured the turbidity (or biomass in some other way) of a well-stirred culture, added growth medium of constant composition in response to cell growth, and removed an equivalent volume of culture simultaneously to maintain the cell density. The same starting organism during the long evolutionary time would learn how to grow faster and faster in that given medium. In a rich medium under favorable conditions, it would grow very rapidly. In the language of the ecologist, the evolved strain would be a pure r-strategist; i.e., all of its systems would become adapted and dedicated to growing as fast as possible. Basically, this strategy entails austerity in order

to survive with smaller amounts of protein per new genome and to increase efficiency in order to streamline the process of protein synthesis so that it is faster per ribosome (32, 39, 42, 47). Consequently, the "doodads" that are inherently present in the ribosome to protect it against antibiotic agents that are no longer in the environment of the organism could be eliminated with significant simplification. With respect to the function of protein synthesis itself, there is a balance point between simplifying or retaining the changes built in during the organism's history even though they could protect against a rechallenge that in the chemostat would never happen.

The Evolutionary Product of Long-Term Continuous Culture

Genes That Would Be Deleted. Now let us focus on the cellular changes that would develop independent of the details of how the constant continuous culture was carried out. Our hypothetical organism would lose many chromosomal regions present in modern organism; these are listed in Table 2. Certain systems would be refined to a sophisticated degree; these are listed in Table 3. Changes that could possibly arise in the regulation of protein synthesis are grouped together in Table 4.

The genome would become much smaller, largely because many open reading frames can be omitted (Table 1). Furthermore, in an unchanging environment, the organism does not need to possess alternative metabolic pathways and could eliminate currently utilizable inducible catabolic enzymes and systems, such as the *lac* operon, and cryptic, inactivated genes, such as the evolved β-galactosidase gene. It would need no genes to protect itself from temporary nutritional starvation or overfeeding, from lowered or elevated temperature, or from dehydration. Host cycle genes now needed for survival during the necessary transit from one gastrointestinal tract to the next could also be eliminated. Selection for smaller genomes would allow shorter genome replication times. This may appear as a minor issue

---

TABLE 2 Deletions that would occur after long-term continuous culture

"Junk" DNA
Fragments of old genes
Fragments of old lysogenic viruses, insertion sequences
Repeated DNA
Stimulon DNA
Recombination (but not repair)
Heat shock (in part)
SOS (in part)
Starvation protection
Oxidation (in part, for some culture conditions)
Yet unknown stimulons
Inducible pathways not needed during continuous culture
For carbon sources
For nitrogen sources
For inorganic ions
For facultative ability to use oxygen and/or nitrate
For surviving outside the gastrointestinal tract
For metabolic conversions
Adherence mechanisms
Cryptic pathways
Evolved β-galactosidase
Genetic regions needed for "last-ditch" genetic changes
Genetic regions needed for genetic transformation
Pathways to switch to a different metabolism
Pathways protecting against antibiotics, both known and unknown

TABLE 3 Pathways and processes that would be further
refined by long-term continuous culture

All proteins would become streamlined.
  Made smaller
  Become more efficient catalysts
Translation would become faster.
The dynamic range of the control of translation would become smaller.
The genome would be maintained without change, after most
    streamlining has taken place.
  By accurate repair
  By prevention of mututational events
The cell cycle would become more accurately controlled.
  Age and mass at division more precisely fixed
  Partition of cytoplasm at division more even

because the amount of a cell's resources devoted to chromosomal synthesis is small. However, any saving in time would be
positively selected. Those mechanisms maintained and refined
as listed in Table 3 would not need much more coding capacity.
The simplification of the control of translation, itemized in
Table 4, would also lead to a decrease in the size of the genome.
Conservatively, we can estimate that the genome could be less
than 1/10 the size of that in the modern "worldly" and "cau-
tious" organisms. It would have a genome considerably smaller
than that of the most streamlined mycoplasma found thus far
(47).

**Processes That Would Be Refined during Long-Term Continu-
ous Culture.** The cardinal aspect of the organisms generated by
these selection processes is that they would be essentially perfectly adapted to their environment. The bacteria would be
stultified; they would carry out only those processes constantly
needed but would do so very efficiently. They therefore would
need to have amplified and refined conservative mechanisms
preventing genetic change. Because of the way our hypothetical
continuous culture system was set up, the usual genetic mechanisms for transfer of genes from organism to organism were
eliminated from the start by commencing with a true gnotobiotic system containing only the host chromosome and no plasmids or activatable lysogenic viruses. Once evolution had
reached the point that physical and chemical conditions in the
environment had become rate limiting in all essential aspects,
then further genetic improvements could no longer be selected
and any mutations would create only deleterious genotypes.
Consequently, the cells would evolve so that mutational events
would be prevented as far as possible and the effects of unavoid-

TABLE 4 Possible simplifications of the regulation of
protein synthesis

Have several dedicated RNA polymerases with different frequencies of
    initiation
Use only repressor-type control
Use only transcriptional activators
Use only attenuation
Use only variants of the Shine-Dalgarno sequences
Use only variations in the region between a Shine-Dalgarno sequence
    and the initiating AUG
Reduce turnover of mRNA
Discard the use of the catabolite repression mechanism
Delete the −10 and −35 control regions

able mutations would be eliminated, again as much as physically
and chemically possible. The development of mechanisms for
more accurate repair and prevention of mutations would be a
very slow process because (i) improvement in the repair systems
is not directly selected and (ii) as the accuracy of repair improves,
the mutation rate would become smaller and change in the
repair system, as well any other system, would be slow. Changes
in the DNA would continue to appear as a result of ionizing or
nonionizing electromagnetic radiation unless the long-term
cultures were carried out deep in the earth in the dark. There
would also be biochemical mistakes. Radioactive decay of naturally occurring potassium would cause unavoidable mutations.
Mutations so produced could be minimized only by repair of the
genetic damage or outgrowth of the defective cells.

Streamlining the Control of Protein Synthesis. Protein synthesis is subject to a very large number of regulations and
controls that make the kinds of proteins being formed attuned
to the external and internal environments. The continuous-
culture organism would need no adjustable regulation and a
much simpler physiology would suffice. Table 3 lists some possibilities. It may be that an encyclopedic knowledge of the entire
text of this volume would allow a better judgment of which
regulatory mechanisms would survive the winnowing of these
varieties of control mechanisms.

TRUE OLIGOTROPHS

Description

Mycoplasmas and VBNC (viable but not culturable) organisms may
be almost the equivalent of the organisms that would arise after
long-term continuous culture. A particularly important study of
marine ultramicroorganisms (oligobacteria) has been carried out by
Button's group (9). These organisms appear to die when grown in
rich media containing even as little as 5 mg of amino acids per liter.
Because biomasses are very small, special circumstances are needed
to show that they are alive. This was done by diluting a sample into
sterilized seawater and observing that after a long time at a low
temperature, the number of particles possessing double-stranded
nucleic acid (measured by a flow cytometer) increased to the steady-
state level of the original sample. With this technique, pure strains
could be isolated. One of these, oligobacterium RB1, had 1,120 kb
of DNA per genome, compared with the 4,700 kb of DNA in the *E.
coli* genome. It would appear that the ecosystem that is the highly
oligotrophic oceans is a very constant environment with a heterotrophic biota that has adapted by producing a very small genome size
and has retained almost no ability to cope with a surfeit of amino
acids. Clearly this is an extreme case of "substrate-activated death"
(62). Mycoplasmas, particularly phytoplasmas, can have a genome
size of only 450 kb, or 10% of that of *E. coli*. They live intracellularly
in plant tissues and must be able to survive both in an insect vector
and in plant tissue. Without this need and if plant growth were
continuous and phytoplasma dispersal were effective, we can imagine that the genome would be smaller still.

Effects of the SOS System and Other Mechanisms
Protecting the Genome

One of the number of genetic regulatory systems that modify the
rules of the timing of cell division is the SOS system. The *sfiA*
(*sulA*) gene of the SOS system (19) inhibits cell division when
DNA has been damaged or when the chromosome cannot repli-

cate. In addition, *sfiC,* a non-SOS gene, and genes such as the *ccd* gene of the low-copy-number F plasmid (20) have similar effects. All of these genes are normally repressed, but a basal level of activity could function upon occasion either inappropriately or even appropriately when some accidental damage to the DNA had occurred during normal balanced growth. The unusual activation of these genetic functions would cause a few cells in a growing population to "filament," i.e., to continue elongation without making constrictions or dividing. If the stimulus does not continue, such temporarily inhibited cells eventually recover, become repaired, and belatedly divide, giving rise to a few cells that take an extraordinarily long time to divide.

The frequency of these special cells in a bacterial population can be estimated from the frequency of nonnucleated cells. Normally, nonnucleated cells with approximately normal birth length constitute about 0.1% of the population in a culture in balanced growth. Recent experiments (19) have established that when the *sfiA* system does not work, 0.5% nonnucleated cells are found, and when both the *sfiA* and *sfiC* systems are inoperative, 0.7% nonnucleated cells are found. These results imply that in wild-type bacteria, 0.5 to 0.6% of scheduled cell divisions do not actually take place until some later time because of unusual SOS action, thereby distorting the distributions of age at division and size at division at their high ends. Consequently, a sufficient explanation of the skewness of the age-at-division distribution is the occasional operation of the SOS system acting in a temporary fashion even when not triggered by sufficient DNA damage to cause mass blockage of cell division in the population.

SYSTEMATIC CHANGES WITH GROWTH

I will now turn to the question of how cells differ from each other even though they originate from the same recently cloned culture.

The Canonical Case

When a cell divides, the two daughter cells are different from the parental cell from which they arose. Let us first consider the canonical model, in which every cell behaves like every other. This model has been used a great deal in physiological studies of various aspects of the microbial cell cycle. Although the model, predicated on precise division, is a poor representation of actual cultures, it would probably apply quite well to the hypothetical products of the continuous cultures considered above because evolution would select for even more precision in the evenness of division of cells. This idealized model has been used for two purposes: (i) calculating the average content of cell constituents per cell (e.g., the average DNA content per cell) and (ii) calculating the average time spent in a cell cycle phase from the fraction of the cells in the population showing a certain morphological or autoradiographic character. To cut through the mathematical complications (33, 43, 53, 48), in this ideal case a single cell taken at random from a growing population will have a mean age of 0.44 (with a standard deviation of ± 0.29) of the age at division (or the doubling time, symbolized by T_2). Many cellular functions, such as the rate of protein synthesis, increase continuously during the cell cycle. Other functions, such as those having to do with chromosomal replication or cell division, change discontinuously. Consequently, if mass growth is continuous and increases in a monotonic fashion, the distribution of cell masses or sizes will have a mean size of 0.693 (with a standard deviation of ±0.14) of the size at division (critical size) (43, 53). The DNA content depends on the age in the cell cycle when chromosome replication starts, at age *B,* and when it ends, at age *B+C,* where *B* is the interval before initiation of chromosome replication and *C* is its duration. The age when DNA synthesis starts can also be expressed as $T_2 - D - C$, where *D* is the interval between chromosome completion and division. In sum, even with clocklike precision there is considerable variability, in age, size, and DNA content, from cell to cell in a balanced growing population.

The canonical age and cell size distributions are applicable to a population of cells that grow according to some deterministic law (28) (i.e., every cell of a given size and age class does what every other cell of that class does according to same explicit rule) and in which every cell divides at a critical age into two identical daughter cells. The distribution of ages was first derived by Euler in the 1700s and independently rediscovered many times since (see reference 30). Its formula is given by

$$\phi(a)\,da = 2\mu e^{-\mu a}\,da: \qquad 0 \le a \le \ln 2/\mu = T_2$$

where μ is the specific growth rate, *a* is the age since birth, and $\phi(a)\,da$ is the frequency of cells whose ages are between *a* and *a + da*. The canonical age distribution is depicted in Fig. 2; its main characteristic is that the frequency decreases exponentially twofold from birth to division. This twofold factor results because one cell divides into exactly two daughters. The logic is simple: in cells that are just about to divide, whatever their number, there will be twice as many at a time infinitesimally later when they have divided into two newborn cells.

The distribution of cell sizes or masses in many ways is more useful than the distribution of ages (43, 49, 53). For the usual case of exponential growth, where the rate of biomass formation increases continuously in proportion to the biomass, a cell would increase in size twice as fast when it was about to divide as when it was just born. Consequently, within a given fixed mass range, *dm,* fewer cells will be found in a cell size class approaching division than in the class of newly separated cells. Therefore, there would be an additional twofold decrease because the rate of passage through a given size increment for those large and rapidly growing cells about to divide, relative to the rate of passage through the same size increment for newborn cells, is twice as great. The formula for this canonical distribu-

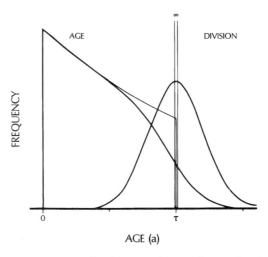

FIGURE 2 Extant age distribution and age-at-division distribution.

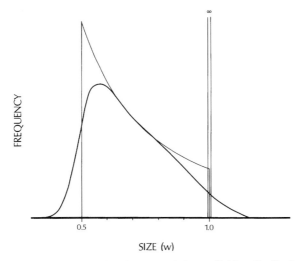

FIGURE 3 Extant size distribution and size-at-division distribution.

tion where the frequency drops fourfold between birth and division is given by $\Theta(m)\,dm = c/m^2\,dm$ as long as $c/2 \leq m \leq c$, where m is the cellular biomass and c is the critical size at cell division and all division events produce two identical daughters of mass $c/2$. This inverse-square distribution is shown in Fig. 3.

Variability in the Age, Size, and Evenness at Division

The age-at-division distribution is measured by watching cells arise, grow, and divide under the microscope or examining time-lapse films. If one could examine a population of cells during balanced growth at an instant of time and follow each cell forward and backward in time, one could construct both an extant age distribution and the momentary age-at-division distribution equivalent to those shown in Fig. 2. The extant size distribution is directly observable, and the size-at-division distribution is approximated by the distribution of cells showing constrictions. Figure 4 shows an example taken from the very many available in the publications from Nanninga's laboratory in Amsterdam (58). The evenness of the division process culled from the many studies of the Amsterdam group has been brought together in the thesis of Trueba (70) (see also reference 17).

As far back as 1932 (24), it was clear that the age at division varied a good deal from one cell to another in a population of

bacteria in good exponential growth. Typically the coefficient of variation (CV) of the age at division of random cells is about 20% (6, 29, 53, 65, 71). This high variability and the detailed shape of the distribution, particularly the positive skewness of the distribution, were much studied in the early literature and have been used in numerous speculative discussions. Some of the variability may be genetic, but most is something else. A little may be due to death and dying, but physiologic Daurer modifications or a change to the shutdown cell state is more important. In these early studies, however, the variation was presumed to be entirely stochastic. It was assumed that the broad skewing of available distributions reflected a small number of random discrete events that were needed to complete the cell cycle. Thus, it was suggested that for *E. coli*, there were 25 independent gene duplication events that had to be completed for a cell to be capable of division (64) or that there were 20 sequential processes that had to be completed (25, 26). Although from the modern perspective such interpretations seem naive, these papers were important in the development of the theory of stochastic processes and may be important for understanding the consequences of some of the phenomena discussed below.

At a later time, when cellular biochemical processes were better understood, a deterministic model for cell growth (53) was proposed as a simple extension of the canonical model discussed above. The extension assumed that the size at division of cells varied by a small amount as a result of inherent "biological variation" and that this generated a larger variation of age at division. Thus, the CV of the age at division is larger than the CV of the size at division because two cell division events (birth and division) determine the age at division, and only one event, division, determines the size at division.

Subsequently, the deterministic model has been modified in several minor ways and has been often renamed (11, 12, 36, 72). It has served as a useful way to study the cell cycle without the problems attendant to synchronization procedures (51, 52). A recent version of this model accounts for the age and size distributions as well as for correlations between the ages at division of mother and daughter cells and of sister cells (46). Other related properties of the population have also been calculated, but the key issue is that the variability of the size at division, no matter what its source, is the direct determinant of the broadness of the age-at-division distribution. However, positive skewness, which was the other frequently observed salient feature of the age-at-division distribution observed in most, but not all, of the relevant studies, is not so easily explained. Analytical and computer studies have shown that no matter whether the size-at-division distribution is approximately Gaussian (and, therefore, presumably due to many sources of variation) or due to a single stochastic event (negative exponential distribution), the resultant age-at-division distributions and the extant mass and extant age distributions change very little (43, 46). Control of the instant of cell division may trivially be dependent on fluctuations in the shear forces due to local stirring of cultures in aerated suspension or on the fluctuations in the mechanical stresses that develop in cells which grow when affixed to an agar surface. Variation in measured cell size at division may be trivially attributable to artifacts due to the cell's orientation when viewed in the microscope and not to factors important or real with respect to the cell (63). These various possibilities affect the shape of the size-at-division distribution directly. However, because of the central limit theorem of statistics, the other two relevant distributions, the size at

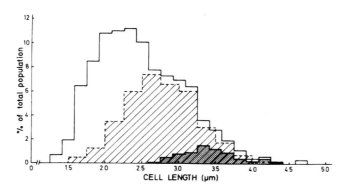

FIGURE 4 Experimentally determined size distribution.

birth of cells and the distribution of ages in the entire population, are less affected.

The distribution of cell sizes at the instant of cell division is called the momentary distribution, and its CV is here designated q. The actual distribution could be measured directly only if a number of individual growing cells were followed through the cycle and their size, volume, or mass was noted at the exact instant of division. Practically, q is estimated from the distribution of cells in a population of cells showing evidence of impending division (46, 51, 52).

Possible Causes of the Positive Skewness of the Age-at-Division Distribution

In analyses of the cell cycle from the statistical viewpoint by Rahn (64) and Kendall (25, 26), the focus was on the observation that the age-at-division distribution was positively skewed in the samples of data available to them. With the development of the transition probability theory by Smith and colleagues (6, 67), another representation of the age-at-division distribution, called the alpha plot, became popular. An alpha plot is a semilogarithmic plot of the fraction of cells that have yet to divide. A linear tail is a precise prediction of the transition probability model in its original form that assumes that there is only one stochastic process involved. Many sets of data that did have a linear tail were found. Of course, any age-at-division distribution that is moderately positively skewed translates into an alpha plot that has a linear tail. Therefore, many instances of a positively skewed age-at-division distribution have been presented in the literature.

Since positive skewness is not predicted by the deterministic theory and its variants, it is important to establish the basis of the more commonly observed skewness. One suggestion (36) is that a small fraction of the cellular population is abnormal, i.e., pathological or defective because of a temporary error in the control of cell processes, and does not divide according to the same rules as does the bulk of the population. These could also be cells in an altered state that in the older literature were called Daurer modifications. Because of their rareness, such unusual events would be hard to study directly, but there are many phenomena observed in special ways that would cause the age-at-division distribution to have positive skewness.

Correlation of the Ages of Relatives

The age at division of one sister cell can differ from that of the other under presumably identical conditions. Experimentally, the correlation coefficient of the sister:sister age at division, CC_{ss}, is much less than unity; it is typically +0.5 to +0.8. Similarly, the correlation of the age at division of a mother, CC_{md}, with either of her daughters ranges from –0.5 to a slightly positive value. More distantly related cells in a clone are even more divergent in age at division and their correlation coefficients are closer to 0. For the theoretical case where the rate of formation of cell biomass is directly proportion to amount of cell biomass, these correlation coefficients are $CC_{ss} = +0.5$ and $CC_{md} = –0.5$. Further studies of how the correlation coefficients depend on evenness of the division process can be found in reference 46.

Distribution of the Oldest Cell Parts

Certain parts of the cell are subject to little or no turnover or regeneration. For example, a DNA strand created during semi-conservative replication serves as a template in subsequent generations but remains inviolate in that act. Although the action of gyrase opens the chain, it also closes it without material change of the residues involved. On the other hand, repair systems (except photoreactivation) replace the material in portions of the chain. Excision repair replaces parts of the chain with new material, while recombination and postreplication repair exchange material from other strands. Nevertheless, the greatest portion of the DNA remains intact.

The other candidate for immortality is the cell wall. For *Streptococcus* spp., it is unquestionable that wall growth takes place in zones; after a pole is formed, it remains unchanged and does not turn over constituent molecules (10). (At least, no autolytic activity occurs unless the cell is stressed [e.g., temporary inhibition with chloramphenicol followed by removal of the chloramphenicol].) Although it was thought that *E. coli* represented a different case (55), it is now clear that the outer two layers of the pole wall are metabolically inert (57).

The distribution of poles or DNA strands of various ages is readily derived for the canonical case. If the formed strands were completely stable and if the culture had undergone balanced asynchronous growth for a long time, then the number of cells, N_T, that arose by division and formed a new pole at a prior time, T, would be $N_T = Ne^{-\mu T}$, where N is the number of cells per unit volume at the time the culture is sampled. Since a frequency distribution must sum to unity, the frequency of cells in which the oldest pole or DNA strand has an age of T is $\phi(T) = \mu c e^{-\mu T}$. Notice that μ replaces the 2μ of the canonical extant age distribution.

A different formula is necessary for cases in which cell senescence is slow and cells with very old poles or old DNA strands die. There are other cases in which the material in the cells is replaced gradually, but eventually the original cell part is no longer identifiable by morphological or radiochemical criteria. If this occurs after T_c doublings, then $\mu(T)$ has a zero value outside the range $0 \leq T \leq T_c$ and $\mu(T) = \mu e^{-\mu T}/(1 - e^{-\mu T_c})$.

All of these distributions are positively skewed. The presence of poles or DNA strands of disparate ages could be important for certain cellular properties but not for others. These phenomena are clearly evident in the cellular process of fission yeasts and bacteria with specialized life cycles, such as *Caulobacter crescentus* (7, 16) and *Rhodomicrobium vannielii* (15).

Donachie (14) had argued quite persuasively that wall growth in *E. coli* is zonal and parallels the behavior of the fission yeast *Schizosaccharomyces pombe*. But my coworkers and I (37, 55), Burman et al. (8), and, most critically, Woldringh et al. (77) have demonstrated that wall growth is dispersive under the usual growth conditions. However, as Crawford Dow has pointed out (personal communication), there is no convincing evidence of dispersive growth during wall elongation of *E. coli* growing more slowly than a doubling time of 1 h. He points out that under conditions of very slow growth, *E. coli* may grow in a zonal manner and we not know it.

MECHANISMS FOR LONG-TERM ADAPTATION WITHOUT GENETIC LOSS

The last universal ancestor evidently had a very elaborate and refined cell physiology, and its descendants have improved on it. Bacteria, because of their short cell cycles, have therefore created for themselves an especially difficult problem that no amount of refinement of the cellular machinery to cope with a single envi-

TABLE 5 Optimum strategy for a organism that
reproduces by binary division in an uncertain
world

Strategies allowing successive multiple choices
  Reversible mutations
  Inversions of gene regions
  Storage of genetic information in plasmids
  Duplication and inactivation without losing the structural information
Evolved survival pathways to outwait a challenge
  Sporulation
  Reactivate silenced pathways
  Catastrophe kit components
  Protection from starvation killing
Obtaining genes from other organisms via:
  Plasmids
  Viruses
  DNA transformation

ronment will solve. Conditions change, and during the many
generations of growth that occur under a new set of constraints,
mutation and selection will lead to adaptation; however, seemingly most of the potential evolutionary advances would delete
the improvements suitable to the previous set of conditions and,
upon reestablishment of these cells, would be at a dead end. In
responding to the need for change, and the much longer term
needed to enable a return to the original state, the organisms
must have evolved, in a still longer time frame, techniques that
would allow such mutational reversibility. A number of the possibilities are presented in Table 5 and outline the next phases of
the study of the basic strategy that bacteria use for long-term
survival.

LITERATURE CITED

1. Anderson, R. P., and J. R. Roth. 1977. Tandem genetic duplications in phage and bacteria. *Annu. Rev. Microbiol.* 31:473–505.
2. Axelrod, D. E., Y. Gusev, and T. Kuczek. 1993. Persistence of cell cycle times over many generations as determined by heritability of colony sizes of *ras* oncogene-transformed and non-transformed cells. *Cell Prolif.* 26:235–249.
3. Barkley, M. D., and S. Bourgeois. 1979. Repressor recognition of operator and effectors, p. 177–120. *In* J. H. Miller and W. S. Reznikoff (ed.), *The Operon*. Cold Spring Harbor Laboratory, Cold Spring Harbor, N.Y.
4. Beckwith, J. R., and D. Zipser (ed.). 1970. *The Lactose Operon*. Cold Spring Harbor Laboratory, Cold Spring Harbor, N.Y.
5. Benzer, S. 1953. Induced synthesis of enzymes in bacteria analyzed at the cellular level. *Biochim. Biophys. Acta* 11:383–395.
6. Brooks, R. F., D. C. Bennett, and J. A. Smith. 1980. Mammalian cell cycles need two random transitions. *Cell* 19:493–504.
7. Brun, Y. V., G. Marczynski, and L. Shapiro. 1994. The expression of asymmetry during caulobacter cell differentiation. *Annu. Rev. Biochem.* 63:419–450.
8. Burman, L. G., J. Raichler, and J. T. Park. 1983. Evidence for diffuse growth of the cylindrical portion of the *Escherichia coli* murein sacculus. *J. Bacteriol.* 155:983–988.
9. Button, D. K., F. Schut, P. Quang, R. Martin, and B. R. Robertson. 1993. Viability and isolation of marine bacteria by dilution culture: theory, procedures, and initial results. *Appl. Environ. Microbiol.* 59:881–891.
10. Cole, R. M., and J. J. Hahn. 1962. Cell wall replication in *Streptococcus pyrogenes*. *Science* 135:722–724.
11. Cooper, S. 1984. Application of the continuum model to the clock model of the cell division cycle, p. 209–218. *In* L. N. Edmunds, Jr. (ed.), *Cell Cycle Clocks*. Marcel Dekker, Inc., New York.
12. Cooper, S. 1984. The continuum model as a unified description of the division cycle of eukaryotes and prokaryotes, p. 7–18. *In* P. Nurse and E. Steiblova (ed.), *The Microbial Cell Cycle*. CRC Press, Inc., Boca Raton, Fla.
13. Delbrück, M. 1945. The burst size distribution in the growth of bacterial viruses (bacteriophages). *J. Bacteriol.* 50:131–135.
14. Donachie, W. D. 1968. Relationship between cell size and time of initiation of DNA replication. *Nature* (London) 219:1077–1079.
15. Dow, C. S., R. Whittenbury, and N. G. Carr. 1983. The 'shut down' cell or 'growth precursor' cell—an adaptation for survival in a potentially hostile environment. *Symp. Soc. Gen. Microbiol.* 34:187–247.
16. Ely, B., and L. Shapiro. 1984. Regulation of cell differentiation in *Caulobacter crescentus*, p. 1–26. *In* R. Losick and L. Shapiro (ed.), *Microbial Development*. Cold Spring Harbor Laboratory, Cold Spring Harbor, N.Y.
17. Grover, N. B., C. L. Woldringh, and L. J. H. Koppes. 1987. Elongation and surface extension of individual cells of *Escherichia coli* B/r: comparison of theoretical and experimental size distributions. *J. Theor. Biol.* 129:337–348.
18. Hughes, W. H. 1955. The inheritance of differences in growth rate in *Escherichia coli*. *J. Gen. Microbiol.* 12:265–268.
19. Jaffé, A., and R. D'Ari. 1985. Regulation of chromosome segregation in *Escherichia coli. Ann. Inst. Pasteur Microbiol.* 136A:159–164.
20. Jaffé, A., T. Ogura, and S. Hiraga. 1985. Effects of the *ccd* function of the F plasmid on bacterial growth. *J. Bacteriol.* 163:841–849.
21. Jobe, A., and S. Bourgeols. 1972. *lac* repressor-operator interaction. VI. The natural inducer of the *lac* operon. *J. Mol. Biol.* 69:397–408.
22. Kaprelyants, A. S., J. C. Gottshal, and D. B. Kell. 1993. Dormancy in non-sporulating bacteria. *FEMS Microbiol. Rev.* 104:271–286.
23. Kaprelyants, A. S., and D. B. Kell. 1993. Dormancy in stationary-phase cultures of *Micrococcus luteus*: flow cytometric analysis of starvation and resuscitation. *Appl. Environ. Microbiol.* 59:3187–3196.
24. Kelly, C. D., and O. Rahn. 1932. The growth rate of individual bacterial cells. *J. Bacteriol.* 23:147–153.
25. Kendall, D. G. 1948. On the role of variable generation time in the development of a stochastic birth process. *Biometrika* 35:316–330.
26. Kendall, D. G. 1952. On the choice of mathematical models to represent normal bacterial growth. *J. R. Stat. Soc. B* 14:41–44.
27. Koch, A. L. 1960. Encounter efficiency of coliphage-bacterium interaction. *Biochim. Biophys. Acta* 39:311–318.
28. Koch, A. L. 1966. On evidence supporting a deterministic process of bacterial growth. *J. Gen. Microbiol.* 43:1–5.
29. Koch, A. L. 1966. Distribution of cell size in growing cultures of bacteria and the applicability of the Collins-Richmond principle. *J. Gen. Microbiol.* 45:409–417.
30. Koch, A. L. 1967. The tight binding nature of the repressor-operon interaction. *J. Theor. Biol.* 16:166–186.
31. Koch, A. L. 1971. The adaptive responses of *Escherichia coli* to a feast and famine existence. *Adv. Microb. Physiol.* 6:147–217.
32. Koch, A. L. 1974. The pertinence of the periodic selection phenomenon to pro-caryote evolution. *Genetics* 77:127–142.
33. Koch, A. L. 1976. How bacteria face depression, recession, and derepression. *Perspect. Biol. Med.* 20:44–63.
34. Koch, A. L. 1979. Microbial growth in low concentrations of nutrients, p. 261–279. *In* M. Shilo (ed.), *Strategies in Microbial Life in Extreme Environments, Dahlem Konferenzen-1978*. Verlag Chemie, Berlin.
35. Koch, A. L. 1980. Selection and recombination in populations containing tandem multiple genes. *J. Mol. Evol.* 14:273–285.
36. Koch, A. L. 1980. Does variability of the cell cycle result from one or many chance events? *Nature* (London) 286:80–82.
37. Koch, A. L. 1983. The surface stress theory of microbial morphogenesis. *Adv. Microb. Physiol.* 24:301–366.
38. Koch, A. L. 1985. Primeval cells: possible energy-generating and cell-division mechanisms. *J. Mol. Evol.* 21:270–277.
39. Koch, A. L. 1985. The macroeconomics of bacterial growth. *Symp. Soc. Gen. Microbiol.* 1985:1–42.
40. Koch, A. L. 1987. Why *Escherichia coli* should be renamed *Escherichia ilei*, p. 300–305. *In* A. Torriani-Gorini, F. Rothman, S. Silver, A. Wright, and E. Yagil (ed.), *Phosphate Metabolism and Cellular Regulation in Microorganisms*. American Society for Microbiology, Washington, D.C.
41. Koch, A. L. 1987. Evolution from the point of view of *Escherichia coli*, p. 85–103. *In* P. Calow (ed.), *Evolutionary Physiological Ecology*. Cambridge University Press, Cambridge.
42. Koch, A. L. 1988. Why can't a cell grow infinitely fast? *Can. J. Microbiol.* 34:421–426.
43. Koch, A. L. 1993. Biomass growth rate during the cell cycle. *Crit. Rev. Microbiol.* 19:17–42.
44. Koch, A. L. 1993. Microbial genetic responses to extreme challenges. *J. Theor. Biol.* 160:1–21.
45. Koch, A. L. 1994. Development and diversification of the last universal ancestor. *J. Theor. Biol.* 168:269–280.
46. Koch, A. L. 1995. The default growth strategy of bacteria. *Proceedings of the Third International Conference on Mathematical Population Biology*.
47. Koch, A. L. 1995. *Bacterial Evolution, Growth and Form*. Chapman & Hall, New York.
48. Koch, A. L. 1995. The Monod model and its alternatives. *In* J. A. Robinson and G. A. Milliken (ed.), *Mathematical Models in Microbial Ecology*. Chapman & Hall, New York.
49. Koch, A. L., and G. Blumberg. 1976. Distribution of bacteria in the velocity gradient centrifuge. *Biophys.* J. 16:389–405.
50. Koch, A. L., and R. Coffman. 1970. Diffusion, permeation, or enzyme limitation: a probe for the kinetics of enzyme induction. *Biotechnol. Bioeng.* 12:651–677.
51. Koch, A. L., and M. L. Higgins. 1982. Cell cycle dynamics inferred from the static properties of cells in balanced growth. *J. Gen. Microbiol.* 128:2877–2892.
52. Koch, A. L., and M. L. Higgins. 1991. Simulation of the streptococcal population dynamics, p. 577–591. *In* O. Arino, D. E. Axelrod, and M. Kimmel (ed.), *Mathematical Population Dynamics, Second International Conference*. Marcel Dekker, New York.
53. Koch, A. L., and M. Schaechter. 1962. A model for statistics of the cell division process. *J. Gen. Microbiol.* 29:435–454.
54. Koch, A. L., and T. M. Schmidt. 1991. The first cellular bioenergetic process: primitive generation of a protonmotive force. *J. Mol. Evol.* 33:297–304.
55. Koch, A. L., R. W. H. Verwer, and N. Nanninga. 1982. Incorporation of diaminopimelic acid into the old poles of *Escherichia coli. J. Bacteriol.* 128:2893–2898.
56. Koch, A. L., and C. H. Wang. 1982. How close to the theoretical diffusion limit do bacterial uptake systems function? *Arch. Microbiol.* 131:36–42.
57. Koch, A. L., and C. L. Woldringh. 1994. The metabolic inertness of the pole wall of a Gram-negative rod. *J. Theor. Biol.* 171:415–425.

58. **Koppes, L. J. H., C. L. Woldringh, and N. Nanninga.** 1978. Size variations and correlation of different cell cycle events in slow-growing *Escherichia coli. J. Bacteriol.* **134:**423–433.

59. **Luria, S. E., and R. Laterjet.** 1947. Ultraviolet irradiation of bacteriophage during intracellular growth. *J. Bacteriol.* **53:**149–163.

60. **Novick, A., and M. Weiner.** 1957. Enzyme induction as an all-or-none phenomenon. *Proc. Natl. Acad. Sci. USA* **43:**553–556.

61. **Pachler, P. F., A. L. Koch, and M. Schaechter.** 1965. Continuity of DNA synthesis in *Escherichia coli. J. Mol. Biol.* **11:**650–653.

62. **Postgate, J. R.** 1975. Death in macrobes and microbes. *Symp. Soc. Gen. Microbiol.* **26:**1–18.

63. **Powell, E. O.** 1958. An outline of the pattern of bacterial generation times. *J. Gen. Microbiol.* **18:**382–417.

64. **Rahn, O.** 1931–1932. A chemical explanation of the variability of the growth rate. *J. Gen. Physiol.* **15:**257–277.

65. **Schaechter, M., J. P. Williamson, J. R. Hood, Jr., and A. L. Koch.** 1962. Growth, cell, and nuclear divisions in some bacteria. *J. Gen. Microbiol.* **29:**421–434.

66. **Silverman, M., and M. Simon.** 1983. Phase variation and related systems, p. 537–557. *In* J. A. Shapiro (ed.), *Mobile Genetic Elements.* Academic Press, Inc., Orlando, Fla.

67. **Smith, J. A., and L. Martin.** 1973. Do cells cycle? *Proc. Natl. Acad. Sci. USA* **70:**1263–1267.

68. **Spudich, J. L., and D. E. Koshland, Jr.** 1976. Non-genetic individuality: chance and the single cell. *Nature* (London) **262:**467–471.

69. **Stock, J., and D. E. Koshland, Jr.** 1984. Sensory adaptation mechanisms in swarm development, p. 117–131. *In* R. Losick and L. Shapiro (ed.), *Microbial Development.* Cold Spring Harbor Laboratory, Cold Spring Harbor, N.Y.

70. **Trueba, F. J.** 1981. A morphometric analysis of *Escherichia coli* and other rod-shaped bacteria. Ph.D. dissertation. University of Amsterdam, Amsterdam, The Netherlands.

71. **Tyson, J. J.** 1985. The coordination of cell growth and division—intentional or incidental? *BioEssay* **2:**72–77.

72. **Tyson, J. J., and K. B. Hannsgen.** 1985. The distribution of cell size and generation time in a model of the cell cycle incorporating size control and random transitions. *J. Theor. Biol.* **113:**29–62.

73. **Von Hippel, P. H.** 1979. On the molecular bases of the specificity of interaction of transcriptional proteins with genome DNA, p. 279–347. *In* R. F. Goldberger (ed.), *Biological Regulation and Development,* vol. I. *Gene Expression.* Plenum Publishing Corp., New York.

74. **Wang, C. H., and A. L. Koch.** 1978. Constancy of growth on simple and complex media. *J. Bacteriol.* **136:**969–975.

75. **Wheelis, M. L., O. Kandler, and C. R. Woese.** 1992. On the nature of global classification. *Proc. Natl. Acad. Sci. USA* **89:**2930–2934.

76. **Woese, C. R., O. Kandler, and L. Wheelis.** 1990. Towards a natural system of organisms: proposed for the domains Archaea, Bacteria, and Eucaryae. *Proc. Natl. Acad. Sci. USA* **87:**4576–4579.

77. **Woldringh, C. L., P. Huls, E. Pas, G. H. Brakenhoff, and N. Nanninga.** 1987. Topography of peptidoglycan synthesis during elongation and polar cap formation in a cell division mutant of *Escherichia coli* MC43100. *J. Gen. Microbiol.* **133:**5754–5586.

Segregation of Cell Structures

STEPHEN COOPER

104

ANALYSIS OF SEGREGATION IN BACTERIA

A newborn cell on its trajectory toward division must double all of its components. How much of the new material (or old material) present in a dividing cell goes to one daughter cell and how much goes to the other? Between the two extremes of completely dispersive segregation (old and new material go equally to both daughter cells) and a completely conservative segregation (all new material goes to one cell and all old material goes to the other), other possibilities arise. If the cell were simply an amorphous blob of cytoplasm that divided randomly at division, a simple equipartition of all cell components at division would be expected. But the cell is not that simple, and each of the three major cell components—the DNA, the cell surface, and the cell cytoplasm—has a different segregation pattern.

If each molecule of the cell had a different segregation pattern, or if there were a large number of patterns for many different molecules, the description and understanding of the segregation rules would pose an insuperable problem. Fortunately, one can simplify the problem by apportioning the material of the cell into three categories. These three categories—the cytoplasm, the genome, and the cell surface—have been used before to describe and simplify the growth pattern of the cell (22). A particular molecule is considered a member of a category or group by its location and not necessarily by its biochemical properties.

The history of segregation analysis began when Van Tubergen and Setlow (82) studied the behavior of a number of macro-molecules as cells grew and divided. Cells were labeled with radioactive precursors of cell components, and the labeled cells were "chased" or grown for a number of generations in unlabeled medium. At selected times, the cells were fixed and autoradiographed, and the number of grains per cell was determined. If there were an equipartition of material, daughter cells would be expected to receive equal shares of the original labeled material, and a Poisson distribution of grains would be expected. A non-Poissonian distribution would suggest a segregation process other than equipartition. By determining the number of generations prior to achieving a non-Poissonian distribution, the number of independent segregating subunits in the original cell could be estimated. Van Tubergen and Setlow (82) concluded from the observation of a non-Poissonian distribution of grains in thymidine-labeled cells that only a limited number of chromosomes or DNA strands were present in the bacterial cell, and these units were conserved after synthesis. Later studies confirmed and extended this work (15, 59). These DNA units segregated as stable units and did not subdivide with further growth; this result is consistent with the semiconservative replication of the bacterial DNA. In contrast, when protein or cell wall was specifically labeled, a dispersive or nonconservative segregation pattern was observed.

Forro and Wertheimer (34, 35) expanded on these results by autoradiographing microcolonies formed by pulse-labeled or long-term-labeled cells. After long-term labeling, microcolonies had two heavily labeled cells and two lightly labeled cells. These results were consistent with the cells having one or two replicating chromosomes, equivalent to cells growing with a 40-min interdivision time (22). Because the cells were not arranged in any order when the microcolonies were formed, the spatial pattern of DNA segregation could not be determined. This point has been the object of considerable effort and will be analyzed below.

These basic observations will now be elaborated to present further evidence regarding the details of segregation of different cell components. Each analysis will be presented in the following form. First a simple description of our current understanding of the basic segregation pattern will be presented, omitting various details for the sake of clarity. This will be followed by a discussion of the experimental basis for the proposed segregation pattern. Finally, an analysis of alternative views will be presented.

SEGREGATION OF CYTOPLASM

Cytoplasm is generally believed to be dispersed randomly and nonconservatively to each of the daughter cells (Fig. 1). The evidence for this pattern of segregation comes from two sources. First, Van Tubergen and Setlow (82) demonstrated that amino acid- and uridine-labeled material segregated randomly and dispersively. However, even diaminopimelic acid (DAP)-labeled material gave a dispersive segregation pattern, even though cell wall does not have a completely dispersive segregation pattern (see below). More persuasive evidence for dispersive, random segregation of cytoplasm is that cells labeled and placed on a membrane elution apparatus ("baby machine") divide their material equally, such that the radioactivity per cell is halved at each generation (19). No departure from this pattern is seen over a period of at least 10 generations. This is strong evidence that protein segregates in a random manner (22; unpublished results).

Random cytoplasmic segregation is also supported by the absence of other proposals as to how any nonrandom patterns could arise. There is no morphological evidence for compartments or structures that could produce a nonrandom or conservative segregation pattern.

SEGREGATION OF THE BACTERIAL CELL SURFACE

Arrangement of Peptidoglycan on the *Escherichia coli* Surface

The peptidoglycan of the bacterial cell is a meshlike structure made of glycan chains cross-linked by chains of amino acids. As reviewed elsewhere (22 [specifically chapter 6], 23; this volume, chapter 6), the glycan chains appear to go circumferentially around the outside of the rod-shaped bacterial cell, much like

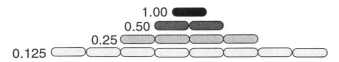

FIGURE 1 Random segregation of cytoplasmic material at division. At each division, the material is partitioned equally to the daughter cells, leading to a halving of the concentration of older material at each generation. The numbers indicate the expected amount of material, from the original cell, present in each of the cells at each generation.

hoops encircling a barrel. Electron microscopic analysis of partially digested cell walls (85) and the results of controlled sonication studies (83) suggested that the strands of the cylindrical side wall are indeed arranged in a hooplike pattern. When cell wall fragments were observed, the strands went preferentially in one direction, perpendicular to the long axis of the cell. Growth in cell length, according to this structure, occurs by the insertion of new hoops between the existing hoops. Other arrangements such as helical glycan strands are difficult to visualize. If the glycan strands were parallel to the long axis, then strand insertion would lead to cells growing in width rather than length, and a helical arrangement would lead to changing angles of helix as new strands were inserted within the helices. The arrangement of strands at the pole is unknown at this time.

Although there may not be a perfect alignment, as has been argued by Koch (56), a hoop arrangement is the best way to allow lengthwise extension of the growing cell. New strands are inserted between preexisting glycan chains, and the cell grows in a lengthwise direction between divisions. If the glycan chains were placed in the axial direction—i.e., in the lengthwise direction of the cell—then insertion of new strands between preexisting glycan chains would lead to an increase in cell circumference. It is clear from even the earliest observations of living cells in the light microscope (75) that rod-shaped cells grow lengthwise, which is strong a priori support of the hooplike strand arrangement model.

Biochemical analysis indicates that glycan strands are relatively short and cannot extend around the entire circumference of the cell (41). Therefore, the peptidoglycan layer is made up of short overlapping strands that collectively go completely around the cell. The absence of long-range order in this arrangement does not change the fundamental conclusion that the insertion of new glycan strands between resident glycan strands leads to the growth of the cell in the axial or lengthwise direction.

Cell wall segregation brings up many questions. Where does the old wall go at division, where is the new wall synthesized and inserted, and what is the inheritance pattern of peptidoglycan over a number of division cycles? These questions suggest that the segregation of preexisting material is determined by the topological insertion pattern of new material. Segregation is the mirror image of the location and insertion mode of new wall material. An example of the relationship of insertion to segregation is that a thoroughly conservative mode of surface segregation would lead to an insertion pattern whereby only half of the growing cell would have surface label following a pulse-label.

Idealized Pattern of Surface Synthesis

Cell growth leads to both conservative and dispersive modes of segregation because the poles and the side wall behave differently in their segregation patterns. This can be shown with cells fully labeled in the peptidoglycan by a specific label such as DAP. The poles act as conserved segregation units, with both labeled poles going to separate daughter cells. The side wall segregates in a dispersive manner: as new strands are introduced between old strands, the labeled side wall material is diluted by new material. Thus, after one generation, the side wall is uniformly labeled at half the specific activity of the side wall in the initial cell. (An important exception to uniform labeling and insertion in the side wall is noted below.) This idealized pattern of cell surface segre-

FIGURE 2   Segregation of peptidoglycan. For a cell with the surface totally labeled (black lines), there is a random segregation of the side wall material (indicated by the decreasing shading at each generation). The old pole material is conserved (indicated by solid black line). The new side walls are completely unlabeled (indicated by the white material). Thus, three distinct areas of the cell are postulated: the older poles where the density of label does not change, new poles which are unlabeled, and side walls where the density of label halves at each generation.

gation (Fig. 2) is supported by a large amount of experimental evidence.

The segregation pattern is a reflection of the synthetic pattern of the cell wall. The surface of *E. coli* and other rod-shaped bacteria is considered to be a cylinder capped by two hemispheres. These polar hemispheres are stable, with no growth occurring in them once they have formed (57, 58). Side wall synthesis, on the other hand, is spread throughout the side wall and does not appear to occur in any zonal pattern. If we know that new material is inserted, at least in the side wall area, between any preexisting pair of hoops, then by definition we must have a dispersive pattern of side wall segregation. The most compelling evidence indicates that the overall pattern is uniform incorporation over the side wall (89, 90).

From this model, a number of questions arise. For example, is there a sharp distinction between the side wall and the pole, or is this merely an anthropomorphic construct that does not actually exist for the cell? Given that there are poles, we can ask whether the poles are completely stable or whether there is some turnover or metabolism of polar peptidoglycan. With regard to the side wall, is insertion completely random over the length of the cell, or are there some preferred sites of strand insertion or some connections between strands that are resistant to the intrusion of a new one?

Location of Insertion of Peptidoglycan on the Cell Surface

The earliest experiments on cell wall growth used fluorescent antibodies to label the cell surface. These studies indicated that in *E. coli*, the side wall grows diffusely with no conserved areas (4, 17, 66). Additional autoradiographic evidence showed that new cell wall material could be inserted over the entire surface of the cell (11, 12). Similar results were obtained for the matrix protein attached to the peptidoglycan (5).

There have been numerous reports that peptidoglycan is not stable and is actually released to the medium (24, 37, 39). This is a complication for segregation studies that is difficult to evaluate at this time, since the amount of release reported varies between 7 and 50% per generation.

The clearest demonstration of this random pattern of insertion is the autoradiographic evidence showing that there are no apparent zones of preferred synthesis prior to invagination (90). Biochemical support came from studies of the peptidoglycan acceptor-donor radioactivity ratio (ADRR), a technique that

measures the pattern of strand insertion into the cell wall. These studies also indicated that there were no conserved areas of the cell wall and that new material could be inserted between any two strands (10, 26). (See discussion of ADRR below.)

The question of whether cell wall growth is symmetrical—i.e., whether it is equal in both of the prospective two new daughter cells or occurs primarily or entirely in only one—is an important problem. The symmetrical nature of cell growth was conclusively demonstrated (at least for peptidoglycan) by the sophisticated statistical analysis of Verwer and Nanninga (84). They analyzed the distribution of radioactive DAP on each of the two cell halves of dividing cells. They determined whether differences in the number of grains in one half of the cell were due solely to statistical variation or whether there was a significant difference that would fit an asymmetrical pattern of peptidoglycan synthesis. Their results showed that synthesis of cell wall in the two cell halves was the same, indicating that there is no conservation of material, with one daughter cell receiving a majority of the new peptidoglycan material.

Additional evidence for random insertion over the side wall comes from membrane elution experiments demonstrating that there is no sudden drop in the elution of peptidoglycan label from the membrane as would be expected if there were a conservative mode of synthesis (M. Ma and S. Cooper, unpublished results). Since elution of label can continue for up to 10 generations, it can be concluded that insertion, and hence peptidoglycan segregation, is random with no large conserved subunits. (To be rigorous about this analysis, it should be stated that any regular and precise pattern of zonal growth and conservation of zones is excluded by this experiment; zonal growth that is completely random and not associated with any particular cellular location during the division cycle may not be excluded by this experiment.)

At some time in the life of the cell, new poles are formed in the center of the cell; these poles are made up of completely new material. Thus, the newborn cell is composed of three zones of cell surface: an older pole made in some previous generation, a new pole just made, and a cylindrical surface in between, made up of a mixture of old and new strands.

Evidence for Zonal Growth

Although a nonzonal pattern of synthesis is indicated by the data cited above, the first two decades of the study of bacterial cell wall growth were dominated by the idea of growth zones, i.e., localized areas of cell surface growth. There are three sources for this idea. One is the early recognition that in *Streptococcus* spp., the cell wall is a rigid structure that grows with one zone of wall growth; old cell wall is not metabolized, and new material is not inserted within it. This zonal pattern of growth was readily extended to gram-negative bacteria. The second source of the idea of zonal growth is the analogy to DNA synthesis. DNA replication is regulated by the insertion of new replication points at the origins of DNA. DNA segregation is conservative, as befits our understanding of the mode of DNA replication. This model was applied to cell wall synthesis, suggesting that there may be growth zones and that at some time or times during the cycle, new growth zones are inserted. Such a zonal mode of growth would lead to a conservative or semiconservative mode of segregation. The third element that supported the idea of zonal growth was the proposal of the replicon model to explain DNA segregation. In the absence of a visible mitotic apparatus, it was proposed that the regular segregation of DNA at division could be explained by the binding of DNA strands to the cell surface, with cell wall growth taking place between the bound DNA strands (50, 51). The wall growth between the surface-bound DNA strands could lead to separation and sequestration of DNA in the two new daughter cells. It was believed that zonal growth, particularly in the center of the cell, was an important requirement for DNA segregation.

The experimental support for zonal growth came mainly from autoradiographic experiments. When cells were labeled for a short time with DAP and analyzed for the location of grains by autoradiography using the electron microscope, it was found that for cells of all sizes there was a preferential location of zones of incorporation in a relatively narrow band at the center of the cell (70). This result implied a preferential zone of growth in the center of the cell. New zones would appear in the new daughter cells at some later time. Immunofluorescence analysis also supported the insertion of discrete zones of synthesis in the side wall of rod-shaped cells (16). When pulse-chase experiments were performed, the evidence for zones was ambiguous (77), and it was concluded that there was a randomization of the material in the initial zone. At the same time that autoradiography suggested a zonal growth mechanism, kinetic measurements of cell wall synthesis during the division cycle indicated that in the middle of the cycle there was a sudden doubling in the rate of peptidoglycan synthesis (48, 67). This result was supportive of the idea of zonal growth. In retrospect, the early indications of a central zone of growth were probably due to elevated incorporation at the new septum.

Donachie and Begg proposed the unit cell model as a specific type of zonal growth (31). This model proposes that the cells grow only from one pole, producing one daughter cell with completely new and one with old peptidoglycan in the side wall. Support for the unit cell model came from microscope observations of cells growing in only one direction. One difficulty with these experiments is that one cannot eliminate a preferential attachment of one cell pole to the substrate and free movement of the other pole, which could produce the appearance of growth in one direction. However, this model got additional support from the finding (6, 7) that phage attachment sites are inserted asymmetrically on the cell surface. Verwer and Nanninga (84) disproved the unit cell model, at least for peptidoglycan, by analyzing the distribution of radioactive DAP on each of the two halves of dividing cells. The unit cell model is also incompatible with the pattern of elution of DAP from cells bound to the membrane in a membrane elution experiment (18).

Is the Pole Metabolized?

A sophisticated analysis of earlier DAP incorporation experiments suggested that there could be a small amount of incorporation of label into old poles (57). A subsequent reanalysis of the data from high-resolution, electron microscopic, autoradiographic studies (using a computer program that corrected for effects of normalization of cell lengths) indicated that the poles, once formed, were extremely stable (58).

An interesting result with regard to the question of pole stability is the report that the major outer membrane protein, OmpA, is concentrated at the poles (8). Cephalexin prevented the insertion of this protein at the poles. Removal of the antibi-

otic allowed the randomly dispersed protein to migrate to the pole and possibly to the newly forming septum. This result suggests some metabolism of the pole after its formation.

Growth of the pole in a narrow band, termed the "leading edge" model, has been proposed on the basis of autoradiographs indicating that at all stages of pole growth, the width of labeled material incorporated into the pole is invariant (87).

Peptidoglycan Segregation at the Strand Level

This discussion has dealt primarily with the segregation and insertion of peptidoglycan at the level of visible areas of growth. At a deeper level, we may consider whether strand insertion occurs between every strand or whether there are some connections between strands that cannot be broken. The ADRR method has been introduced to answer this question (10, 21, 26). The ADRR is a measure of how much labeled DAP is in the donor penta- or tetrapeptide, compared to the acceptor position, during a short period of labeling. A zero ADRR indicates that all of the radioactivity is in the donor position and that new strand insertion into peptidoglycan is occurring by single-strand insertion. Some ADRR measurements are consistent with a two-strand insertion mechanism (10), while others indicate a single-strand insertion mechanism (26). There has been disagreement regarding the meaning of the ADRR values (21, 24, 30, 33). Anyone interested in using ADRR analysis should become familiar with these papers.

Höltje (49) has proposed a model in which three strands are inserted at the same time that one strand is removed from the peptidoglycan. His model explains the maintenance of the specific shape of the bacterial cell. By having the newly inserted strands follow along a previously existing strand, cell circumference is kept constant. A processive degradation of a resident strand, coupled with the simultaneous insertion of new strands, coordinates growth while maintaining cell shape. Höltje's proposal also accounts for the turnover of peptidoglycan. No biochemical results are supportive of this model (chapter 6, this volume). Furthermore, this model is inconsistent with ADRR measurements indicating single-strand insertion (26).

It has been observed that the degree of cross-linking increases after insertion (9). This is a surprising result, since it is hard to imagine how new cross-links can be formed once a strand is inserted. The removal of the final D-alanine from the peptidoglycan pentapeptide by a carboxypeptidase precludes cross-linking, as the energy for transpeptidation is no longer available. Furthermore, as the peptidoglycan appears to be under stress due to turgor pressure (55), it is probable (but as yet unproven) that cross-links cannot be introduced between strands that are physically too far apart. One explanation for the observed increase in cross-linking is that there may be a replacement of low-cross-linked strands by more highly cross-linked strands. With a uniform stress over the surface, the less densely cross-linked strands will be stressed more at each cross-link. This is because the stress, which is constant over the surface of the cell, is spread among fewer cross-links in a low-density cross-linking region than in a more densely cross-linked region. Thus, the low-density cross-linked regions will be preferentially replaced by new strands of peptidoglycan material. Since, by definition, the less densely cross-linked strands have a cross-linking value below the average of the inserting strands, there would be an increase in cross-linking of the resident peptidoglycan as it ages (22). A numerical example can illustrate this process. Consider a peptidoglycan with an average of 25% cross-linking made up of equal amounts of 20, 25, and 30% regions. A newly inserted strand has an average 25% cross-linking. If this new strand preferentially replaces a 20% region, the average degree of cross-linking would increase. This "evolutionary" mechanism allows the cell to continue to strengthen its peptidoglycan as the peptidoglycan drifts toward higher cross-linking values.

Zonal versus Dispersive Growth of the Side Wall Peptidoglycan: Conclusion

The conclusion of this review is that growth of the side wall peptidoglycan of gram-negative, rod-shaped bacteria is dispersive. Zonal growth has been reported, but my conclusion is that the side wall is subdivisible down to the molecular level, with no evidence for major zones of growth or conservation of material.

Segregation of Membrane Lipids

The measured pattern of membrane synthesis during the division cycle, using glycerol and palmitic acid as membrane labels, is similar to the pattern of peptidoglycan synthesis during the division cycle (36). A simplified view of membrane synthesis is that it is made in response to the increase of surface as determined by the growth of the peptidoglycan layer. One idea that arises from this observation is that the segregation of membrane will be similar to the segregation of peptidoglycan. But because peptidoglycan is a more rigid structure, and membrane may be more fluid, there could be differences in segregation patterns between membrane and peptidoglycan.

Early support for the dispersive segregation of membranes comes from density shift experiments (81, 88). In experiments formally analogous to the Meselson-Stahl experiments with DNA, it was observed that the membrane grows in a nonconservative manner. These experiments were sensitive enough to find a small amount of conservation of membrane during growth. In more measurements over many more generations, Green and Schaechter (40) labeled cells with glycerol and used autoradiography to demonstrate that membrane segregation was dispersive for eight generations. Thereafter, the presence of conserved units of membrane (e.g., floating islands of original membrane from the labeled cell) was indicated by analysis of the label over the individual cells. Calculations indicated that there are approximately 256 independently segregating subunits in the cells. (In this experiment, Arthur Koch in a footnote extended the Poisson distribution to cells of different original size and with the age distribution of a growing culture. This equation includes the important ideas that cells of all ages are not represented equally in a culture and that the cells in the culture are not all of the same size.)

Segregation of Membrane Proteins

Studies of the segregation of membrane proteins have produced findings that support different and apparently irreconcilable models. Some experiments have suggested a zonal growth model, while others have proposed a dispersive and nonconservative mode of protein insertion.

A zonal pattern of growth has been suggested for both the inner (29) and outer (2, 3, 51, 53, 54, 81) membranes. Kepes and Autissier looked for the unequal segregation of β-galactoside

permease as a membrane marker among the progeny after cessation of enzyme synthesis (2, 3, 53, 54). Studies on cytochromes supported a nondispersive segregation pattern (78). In contrast, dispersive segregation was found in studies of the membrane-bound anaerobic nitrate reductase in *E. coli* (13). Additional experiments on β-galactoside permease (1) and phage receptors (60) support a conservative insertion segregation mode.

Segregation of membrane components has also been studied by pulse labeling and autoradiography. Finding that the labeled material is inserted diffusely over the surface is supportive evidence for random segregation. Electron microscopic studies of the insertion of outer membrane proteins indicated that there was no clear zonal growth (45, 78, 86). The protein studied, the LamB protein, was inserted diffusely over the entire surface. Analysis of the potential for diffusion of proteins within the outer membrane indicates that this does not account for the observed homogeneous distribution of pulse-inserted outer membrane proteins (86). This work supports the random segregation model of surface segregation.

On the other hand, earlier work on other outer membrane proteins (OmpF and OmpA) using autoradiography supported the zonal insertion model (6, 8). In addition, other studies of membrane proteins using a fluorescent antibody supported zonal insertion (29). Studies of outer membrane receptor proteins for phage lambda (71) or T6 (6, 7) supported a zonal model of surface growth as well. At this time it is difficult to reconcile these experiments with others suggesting that there is a symmetrical synthesis of the bacterial cell surface (85).

There is no reliable information on the segregation of flagella and fimbriae. It is generally assumed, without any strong evidence either way, that the fimbriae and flagella are made at random points on the cell surface and are segregated along with their nearby cell surface peptidoglycan or membrane components.

Summary View of Membrane Segregation

How can this apparently contradictory and confusing set of results be summarized to give a unified picture of membrane growth during the division cycle? My personal conclusion is that there is no significant zonal growth in the side wall of gram-negative rod-shaped bacteria for the membrane or peptidoglycan.

As a final thought, the cell may not benefit from a zonal mode of growth. A diffuse model of surface growth avoids the accumulation of patches of material that are old and which may have various errors of aging. With the continuous insertion of newer material next to older material, the cell surface can be constantly strengthened and errors can be repaired in a manner analogous to proofreading for informational macromolecules. Whether errors in a zone could be repaired by noninsertional means has yet to be determined.

SEGREGATION OF DNA

During unhindered, exponential, balanced bacterial growth, essentially every cell in a bacterial culture is viable. This implies that every cell in a culture contains a genome, since termination of replication occurs sometime prior to division, and therefore at least two complete genomes will be present in the dividing *E. coli* (chapter 102, this volume). The problem of nucleoid segregation is to define how the two genomes are apportioned so that each daughter cell always gets one genome. Nucleoid separation and partition of completed nucleoids to two daughter cells have been discussed in detail elsewhere (46, 61, 63, 72, 74; chapter 105, this volume).

I will now turn to a discussion of the inheritance and segregation of the individual strands of DNA to daughter cells. It is not always appreciated that every cell can receive a genome with perfect fidelity, while at the same time individual strands may segregate nonrandomly. As will be seen, this analysis of strand segregation gives an unexpected insight into potential mechanisms for the movement of DNA into newly formed daughter cells.

Experimental Analysis of Strand Segregation

Lin et al. (62) analyzed strand segregation by using the chain-forming methylcellulose (Methocel) technique. This method allows the growth of *E. coli* in chains so that the cells retain their respective order in the chain. A random segregation pattern was observed (62). We now understand that this observation was due to the fact that the cells being analyzed in these experiments contained up to eight or more labeled strands, and this complexity precluded any observation of nonrandom segregation. Subsequently, in a technical tour de force, Pierucci and Zuchowski (69) demonstrated that segregation of strands is nonrandom. These authors studied cells with known chromosome configurations (i.e., cells with either two or four strands) and determined the locations of the labeled strands in chains of cells formed in Methocel. They compared their data with a number of models and proposed that in each cell, one strand segregates randomly and one segregates nonrandomly. Nonrandom segregation was also found in membrane elution experiments (68).

A simplified approach utilizing presegregation of labeled DNA prior to chain formation confirmed the nonrandom pattern but led to a different conclusion. Tritiated thymidine-labeled cells were allowed to grow in unlabeled medium for a few generations prior to chain formation; these cells had only one labeled strand, and consequently only one cell in a chain was labeled (28). It was observed that in a chain of four cells, the outermost cell was preferentially labeled. Nonrandom segregation was greatest at slow growth rates, with a more random pattern appearing at faster growth rates. The nonrandom pattern fit a rule stating that strands go preferentially to the same pole that they went to in the previous generation. Because the probability that a strand will continue in the same direction it went previously is greater than 0.5, the model was called the strand-inertia model. There was no permanent association of any DNA strand with either pole (27). Thus, nonrandom strand segregation is probabilistic rather than deterministic in nature (Fig. 3).

A more accurate way to measure the nonrandom segregation of DNA has been introduced with a long-term, automated-ratio method using the baby machine (Cooper and Ma, unpublished results). Cells are labeled with tritiated thymidine and a ^{14}C-amino acid, placed on a membrane, and eluted for up to 10 generations. The amino acid label per cell is halved each generation. After the third generation, the thymidine label per cell is reduced by less than half each generation. What is happening is that a constant fraction of the extant thymidine is eluted each generation, but this amount is less than half of the amount of bound label. Thus, the label in the bound cells is not decreasing by half each generation. This pattern of radioactivity release is due to the nonrandom segregation pattern. Because of the dif-

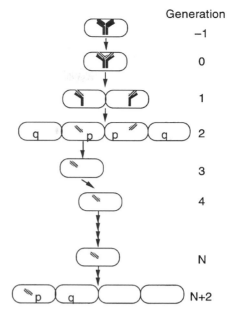

FIGURE 3 Nonrandom DNA strand segregation. Consider an exponentially growing culture with the DNA labeled at generation 0. Before labeling, there is one replicating but unlabeled chromosome. At the instant of labeling (generation 0), a cell has two labeled strands (as the cell has one replicating chromosome), as indicated by the striped lines. The corresponding unlabeled strands are indicated by the solid lines. The cells will grow for a number of generations (N) and then be allowed to form chains in Methocel. At N+2 generations (a four-cell chain), one can determine the proportion of chains labeled in the inner and outer positions. Here only the labeled strands are shown, as these are the strands observed by autoradiography. More cell chains are labeled in the outer position (p) than in the inner position (q). Although this result varies slightly with growth rate, the basic mechanism and underlying quantitative pattern are constant at all growth rates. Equivalent results are seen in cells forming chains immediately after pulse-labeling. The analysis is somewhat different, as there are two labeled cells in a four-cell chain. More inner cells than outer cells are labeled, since the unlabeled complementary DNA strands (not seen in autoradiography) are acting in the same manner as the strands observed in the N+2 generation. Because of equipartition, the labeled strands, which segregate away from the unlabeled strands, now go to the inner positions of the four-cell chains. This is because the older, unlabeled strands go preferentially to the outer positions of four-cell chains as shown in the N+2 generation.

ferences in segregation of the protein (random) and the DNA (nonrandom), the ratio of tritiated thymidine to ^{14}C label increases. The slope of this ratio, plotted on a semilogarithmic graph, is a direct and sensitive measure of the degree of nonrandom segregation. Segregation values from 55/45 up to 60/40 have been obtained by this method (Cooper and Ma, unpublished results). Strand segregation is nonrandom and probabilistic in nature and, as we shall see below, serves as an important indicator that the cell surface is an agent of strand segregation.

The question can be raised as to whether the apparently slight deviation from 50/50 is important, since an all-or-none segregation pattern may have been expected. As will be seen below, it is just this slight deviation from randomness, obtained through rather precise measurements, that makes this finding important. This value fits with a model that relates nonrandom DNA segregation to the asymmetry of the cell in terms of cell wall insertion and segregation. It is to this model that we now turn.

Helmstetter-Leonard Model for DNA Strand Segregation

An elegant mechanistic explanation for nonrandom DNA segregation has been proposed by Helmstetter and Leonard (42, 43). When cells were pulse-labeled and allowed to form chains immediately with no presegregation of the labeled DNA, the center cells of four-cell chains were preferentially labeled. This is exactly what is expected from the presegregation finding that the labeled strands preferentially go to the outer cells of a four-cell chain. Helmstetter and Leonard noted that segregation of these new strands preferentially to the center cells in the four-cell chains mimicked the synthesis and segregation of cell wall material. If one distinguishes between old and new cell wall material, new and old material is randomly placed over the side wall, new material appears preferentially in the newly forming poles, and the outer poles of the cell contain all of the older cell wall material (Fig. 2). At division, the side wall has been intercalated with new cell wall material such that the amounts of old and new material in the side wall are the same in the two daughter cells. In the center of the dividing cell, however, the new poles of the daughter cells have been synthesized from new material. At the outer ends of the dividing cell, the poles are made up solely of old material, with no new material inserted in the poles during the previous generation of growth. Thus, in every *E. coli* cell one pole is brand new, having been formed at the previous division; the other pole is older. Helmstetter and Leonard proposed that this asymmetry is the basis for the asymmetric segregation of DNA. Their model states that a newly synthesized DNA strand tends to bind preferentially to, and to segregate with, new cell wall. Segregation will then be nonrandom. The degree of nonrandomness in this model is determined by the fraction of the total surface that is pole material. The greater the fraction of pole material, the more nonrandom the segregation pattern. Thus, a short, thick cell will give a more nonrandom pattern than a similarly sized cell that is thin and long. An alternative formulation is that older strands preferentially bind to older cell wall material. The ultimate result is the same, but the biochemical tests might be different. Helmstetter et al. have presented other formulations whereby the nonrandom segregation pattern is discussed in terms of excluded volumes and excluded areas (43, 44). The binding of the new DNA strand to the new cell wall material occurs only once, and for one generation. Afterwards, the DNA strand acts in reaction to the segregation pattern of the complementary, newer daughter strand.

This model is consistent with seemingly contradictory experimental results. The degree of nonrandom segregation decreased with increasing growth rate when measured by the presegregation Methocel method (28). In contrast, an invariant degree of nonrandom segregation was observed with the membrane elution method (68). The explanation of this apparent paradox has been presented in detail elsewhere (22). Briefly, in the cell population as a whole, segregation appears more random with increasing growth rate because the proportion of symmetrical segregation to asymmetrical segregation increases. In the Methocel method, there is an increase in the *proportion* of cells with random segregation at higher growth rates. This is because poles present in these cells were both made during the period just before strand attachment and thus, from the strand's point of view, are equivalent. With the membrane elution method, cells are continuously lost from the membrane, and at one particular division there is a clear indication of nonrandom segregation without the interference of these symmetrical cells.

The important point is to relate DNA strand segregation to the cell surface. In the slowly growing cell, the daughter cell DNA "chooses" between an old pole, present during DNA synthesis, and a new pole, made after DNA synthesis ended. In the more rapidly growing culture, the four cells produced during the time of DNA replication and nucleoid partition are divided between two kinds of cells: (i) cells with an old (original) pole and a new pole and (ii) cells with two "new" poles in the sense that they were made during the period of replication and segregation. Neither of the poles in these cells was made before the initiation that is related to the segregation event. Thus, some of the DNA would segregate asymmetrically (the cell has both a very old pole and a new pole), and some would segregate symmetrically (there are two new poles).

The explanatory power of the Helmstetter-Leonard model should not be overlooked. It takes two contradictory results, accepts both of them, and explains them with a unified model that fits the data (22). As the Helmstetter-Leonard model is based on the relative shape of cells, the constancy of segregation values (obtained by the membrane elution method) implies that cell shape at different growth rates is invariant. A constant cell shape has been invoked to explain a large variety of other data (22). If the segregation model of Helmstetter and Leonard is proven to be correct, then the constant segregation data (68) support a constant cell shape (22).

Segregation is more nonrandom at lower temperatures (25). One implication of this result, in view of the Helmstetter-Leonard model, is that at lower temperatures the cells should be squatter than at higher temperatures. This change in shape would increase the area of the cell devoted to pole, thus making DNA segregation more nonrandom. When the length-to-width ratio was determined in cells growing at different temperatures, there was indeed a change in shape; cells decreased their length more than their width (80). This result supports the shape determination of nonrandom segregation (44).

Minichromosome Segregation

The bacterial chromosome exhibits nonrandom strand segregation as well as equipartition of chromosomes. Equipartition means that even though strand segregation is nonrandom with regard to direction, each cell always has one chromosome. In contrast, it has been proposed that the minichromosome (an autonomously replicating circle of DNA controlled by the normal bacterial origin) exhibits nonrandom segregation but does not exhibit equipartition (42). This leads to the appearance of minichromosomeless cells and explains how, in the face of cell cycle-specific replication of the minichromosome, there is a drift toward higher minichromosome content in cultures under continuous antibiotic selection (22, 42, 52, 76).

A different model explaining the nonrandom segregation of DNA has been considered (14). It was proposed that the nonrandom segregation of the DNA, while not perfectly deterministic, could be described as "nonrandom, with a certain degree of randomness." A critical analysis of this model has been presented (22).

Relationship of DNA Strand Segregation to Cell Envelope Segregation

The first experiments on DNA segregation defined the number of units to be considered during a segregation experiment. At the time the early experiments were performed, it was just becoming known that there were but a few units of segregating genetic material in the cell; i.e., the cell did not contain a large number of randomly assorting chromosomes. This result is subtly related to the proposal by Jacob et al. (50) of the replicon model. From observations that there was an equipartition of DNA to daughter cells, they postulated an attachment of DNA to the cell surface; they proposed that the cell surface was analogous to the mitotic apparatus of higher cells (72, 74; M. Schaechter and U. von Freiesleben, personal communication). One of the predictions of the replicon model is that the DNA may be permanently attached to one of the poles of the bacterial cell. If the attachment of DNA to the surface were permanent, and if the attachment site was a fixed position in the cell relative to the poles of the cell, then a particular strand would always segregate toward one pole at division. This permanent attachment has not been observed. There does appear, however, to be some relationship of cell surface growth to the segregation of DNA. There is very likely a "bacterial equivalent of mitosis" (74; Schaechter and von Freiesleben, personal communication), but it does not appear to have the same mechanical relationship to DNA separation as the complex mitotic apparatus of eukaryotic cells. Because side wall growth of the rod-shaped E. coli is dispersive, the DNA attachment to both side wall and polar material leads to more probabilistic predictions.

Do Nucleoids Jump at Termination?

It has been proposed that the positioning of newly formed nucleoids, from the center of the cell at 0.5 to new positions at 0.25 and 0.75, occurs with a sudden movement or "jump" that cannot be accounted for by cell growth (32, 47). Løbner-Olesen and Kuempel (63) have proposed an alternative view of the apparently rapid movement of nucleoids away from the center of the cell whereby origins of chromosomes are separated continuously after initiation. It is the final formation of the separated nucleoid that makes them appear to jump to new locations. The original experiments of Schaechter et al. (73) produced data indicating that nucleoid formation was a continuous process (i.e., new daughter nucleoids are being formed prior to termination) and that the nucleoids observed were connected until termination occurred (25).

CELL POLARITY

A recent report has suggested that even E. coli, which is believed to be the quintessentially symmetrical cell, may be functionally asymmetric with regard to its cell surface (65). The observation leading to this proposal was the finding, by immunoelectron microscopy, of chemoreceptor complexes clustered predominantly at the older cell poles. Thus, the cell is asymmetrical with respect to the location of particular molecules.

This finding of polarity in E. coli has led to the proposal that polarity is a general and important phenomenon (64). In contrast, one may view the observed polarity as merely the inevitable consequence of the fact that all rod-shaped cells have poles of different ages. More generally, all bacteria have one old and one new pole. It may be the age of the pole that determines its chemical constitution; i.e., there may be no inherent function, meaning, or purpose to the observed polarity. When the younger pole eventually matures and becomes an older pole, it will gain the same chemical characteristics that were found in the original older pole. In this view, even the complex life cycle of Caulobacter crescentus is a simple elaboration of the basic cell cycle pattern of E. coli (20, 22). The linear sequence of events—from initiation of DNA replication, to replication, to partition and segregation, and finally to pole formation—is not complete at the separation of cells but continues as the pole matures after cell division. This maturation is not as evident in E. coli as in C. crescentus, with its sequence from a bald pole, to a pole with a flagellum, to the formation of a stalk. This maturation has been given the term "E period" since it logically follows the B, C, and D periods of DNA synthesis and segregation (20, 22). The point of the E-period proposal is that the poles forming in the center of the cell are younger than the poles at the ends of the cell. Such an unavoidable age difference may lead to the observation of polar differentiation. This holds true for E. coli as well as for C. crescentus.

METHODOLOGICAL CONSIDERATIONS

It is generally believed that topological models require optical (e.g., light or electron microscopic) observations as a source of experimental support. This is not the case, as can be seen by the successful study of the segregation of each of the cell components using membrane elution methods rather than optical methods. It is possible to infer segregation patterns from membrane elution studies by reconstructing the segregation pattern from the radioactivity eluted at each generation.

PROBLEMS REMAINING FOR SEGREGATION OF CELL STRUCTURES

There are a number of results in the literature that do not conform to the idealized models presented for the segregation of each of the parts of the cell. These experiments give results that disagree with the general model (such as those proposing zonal growth or unit cell growth). How should we handle these early data? Should we continue to spend a great deal of time with these older results when newer, improved methodologies have led to different results? In particular, how should older results be viewed when newer results are explained in satisfying molecular terms?

I propose that the data to date, after choices are made between different and perhaps opposing results, fit the general models presented here and summarized in Fig. 1 to 3. Segregation of the cytoplasm is dispersive, segregation of the cell wall is understood in terms of conserved poles and dispersive, nonzonal side wall segregation, and strand segregation of DNA is nonrandom and follows, in a formal way, the pattern of wall synthesis and segregation.

Our understanding of cytoplasm and cell surface synthesis enables us to understand the segregation pattern of these cell components. The mechanisms determining the nonrandom

DNA segregation pattern, however, remain a mystery. The primary remaining problem with regard to understanding the apportionment of material to daughter cells at division relates to DNA segregation.

ACKNOWLEDGMENTS

Comments and suggestions regarding this review were given by Arthur Koch, William Donachie, Moselio Schaechter, Conrad Woldringh, and Jochen Höltje. The thoughtful, creative, and collaborative editing by Alexandra Cooper has made this article a joy to write. I thank her for this and look forward to many more collaborative efforts.

LITERATURE CITED

1. Autissier, F., A. Jaffe, and A. Kepes. 1971. Segregation of galactoside permease, a membrane marker during growth and cell division in *Escherichia coli*. *Mol. Gen. Genet.* 112:275–288.
2. Autissier, F., and A. Kepes. 1971. Segregation of membrane markers during cell division in *Escherichia coli*. II. Segregation of lac-permease and mel-permease studied with a penicillin technique. *Biochim. Biophys. Acta* 249:611–615.
3. Autissier, F., and A. Kepes. 1972. Ségrégation de marqueurs membranaires au cours de la croissance et de la division d'*Escherichia coli*. 3. Utilisation de marqueurs variés: perméases, phosphotransferases, oxydoreductases membranaires. *Biochimie* 54:93–101.
4. Beachey, E. H., and R. M. Cole. 1966. Cell wall replication in *Escherichia coli*, studied by immunofluorescence and immunoelectron microscopy. *J. Bacteriol.* 92:1245–1251.
5. Begg, K. J. 1978. Cell surface growth in *Escherichia coli*: distribution of matrix protein. *J. Bacteriol.* 135:307–310.
6. Begg, K. J., and W. D. Donachie. 1973. Topography of outer membrane growth in *E. coli*. *Nature* (London) *New Biol.* 245:38–39.
7. Begg, K. J., and W. D. Donachie. 1977. Growth of the *Escherichia coli* cell surface. *J. Bacteriol.* 129:1524–1536.
8. Begg, K. J., and W. D. Donachie. 1984. Concentration of a major outer membrane protein at the cell poles in *Escherichia coli*. *J. Gen. Microbiol.* 130:2339–2346.
9. Burman, L. G., and J. T. Park. 1983. Changes in the composition of *Escherichia coli* murein as it ages during exponential growth. *J. Bacteriol.* 155:447–453.
10. Burman, L. G., and J. T. Park. 1984. Molecular model for the elongation of the murein sacculus of *Escherichia coli*. *Proc. Natl. Acad. Sci. USA* 81:1844–1848.
11. Burman, L. G., J. Raichler, and J. T. Park. 1983. Evidence for diffuse growth of the cylindrical portion of the *Escherichia coli* murein sacculus. *J. Bacteriol.* 155:983–988.
12. Burman, L. G., J. Raichler, and J. T. Park. 1983. Evidence for multisite growth of *Escherichia coli* murein involving concomitant endopeptidase and transpeptidase activities. *J. Bacteriol.* 156:386–392.
13. Cadenas, E., and P. B. Garland. 1979. Synthesis of cytoplasmic membrane during growth and division of *Escherichia coli*. Dispersive behaviour of respiratory nitrate reductase. *Biochem. J.* 184:45–50.
14. Canovas, J. L., E. F. Tresguerres, A. E. Yousif, J. F. Lopez-Saez, and M. H. Navarette. 1984. DNA segregation in *Escherichia coli* cells with 5-bromodeoxyuridine-substituted nucleoids. *J. Bacteriol.* 158:128–133.
15. Caro, L. G. 1970. Chromosome replication in *Escherichia coli*. 3. Segregation of chromosomal strands in multi-forked replication. *J. Mol. Biol.* 48:329–338.
16. Chung, K. L., R. Z. Hawirko, and P. K. Isaac. 1964. Cell wall replication. II. Cell wall growth and cross wall formation of *Escherichia coli* and *Streptococcus faecalis*. *Can. J. Microbiol.* 10:473–482.
17. Cole, R. M. 1964. Cell wall replication in *Salmonella typhosa*. *Science* 143:820–822.
18. Cooper, S. 1988. Leucine uptake and protein synthesis are exponential during the division cycle of *Escherichia coli* B/r. *J. Bacteriol.* 170:436–438.
19. Cooper, S. 1988. Rate and topography of cell wall synthesis during the division cycle of *Salmonella typhimurium*. *J. Bacteriol.* 170:422–430.
20. Cooper, S. 1990. An alternative view of the *Caulobacter crescentus* division cycle pattern with application to cell differentiation and cell-cycle-specific synthesis. *Proc. R. Soc. London Ser. B* 242:197–200.
21. Cooper, S. 1990. Relationship between the acceptor/donor radioactivity ratio and cross-linking in bacterial peptidoglycan: application to surface synthesis during the division cycle. *J. Bacteriol.* 172:5506–5510.
22. Cooper, S. 1991. *Bacterial Growth and Division: Biochemistry and Regulation of the Division Cycle of Prokaryotes and Eukaryotes*. Academic Press, Inc., San Diego, Calif.
23. Cooper, S. 1991. Synthesis of the cell surface during the division cycle of rod-shaped, gram-negative bacteria. *Microbiol. Rev.* 55:649–674.
24. Cooper, S., D. Gally, Y. Suneoka, M. Penwell, K. Caldwell, and K. Bray. 1993. Peptidoglycan synthesis in *Salmonella typhimurium*, p. 161–168. *In* M. A. De Pedro, J.-V. Höltje, and W. Loffelhardt (ed.), *Bacterial Growth and Lysis*. FEMS Symposium 65. Plenum Press, New York.
25. Cooper, S., and C. E. Helmstetter. 1968. Chromosome replication and the division cycle of *Escherichia coli* B/r. *J. Mol. Biol.* 31:519–540.
26. Cooper, S., M.-L. Hsieh, and B. Guenther. 1988. Mode of peptidoglycan synthesis in *Salmonella typhimurium*: single-strand insertion. *J. Bacteriol.* 170:3509–3512.
27. Cooper, S., M. Schwimmer, and S. Scanlon. 1978. Probabilistic behavior of DNA segregation in *Escherichia coli*. *J. Bacteriol.* 124:60–65.
28. Cooper, S., and M. Weinberger. 1977. Medium-dependent variation of deoxyribonucleic acid segregation in *Escherichia coli*. *J. Bacteriol.* 130:118–127.
29. Davison, M. T., and P. B. Garland. 1983. Immunochemical demonstration of zonal growth of the cell envelope of *Escherichia coli*. *Eur. J. Biochem.* 130:589–597.
30. De Jonge, B. L., F. B. Wientjes, I. Jurida, F. Driehuis, J. T. M. Wouters, and N. Nanninga. 1989. Peptidoglycan synthesis during the cell cycle of *Escherichia coli*: composition and mode of insertion. *J. Bacteriol.* 171:5783–5794.
31. Donachie, W. D., and K. J. Begg. 1970. Growth of the bacterial cell. *Nature* (London) 227:1220–1224.
32. Donachie, W. D., and K. J. Begg. 1989. Chromosome partition in *Escherichia coli* requires postreplication protein synthesis. *J. Bacteriol.* 171:5404–5409.
33. Driehuis, F., B. L. M. De Jonge, and N. Nanninga. 1992. Cross-linkage and cross-linking of peptidoglycan in *Escherichia coli*: definition, determination, and implications. *J. Bacteriol.* 174:2028–2031.
34. Forro, F., Jr. 1965. Autoradiographic studies of bacterial chromosome replication in amino-acid deficient *Escherichia coli* 15T⁻. *Biophys. J.* 5:629–649.
35. Forro, F., Jr., and S. A. Wertheimer. 1960. The organization and replication of deoxyribonucleic acid in thymine deficient strains of *Escherichia coli*. *Biochim. Biophys. Acta* 40:9–21.
36. Gally, D., K. Bray, and S. Cooper. 1993. Synthesis of peptidoglycan and membrane during the division cycle of rod-shaped, gram-negative bacteria. *J. Bacteriol.* 175:3121–3130.
37. Goodell, E. W. 1985. Recycling of murein in *Escherichia coli*. *J. Bacteriol.* 163:305–310.
38. Goodell, E. W., and U. Schwarz. 1983. Cleavage and resynthesis of peptide cross-bridges in *Escherichia coli* murein. *J. Bacteriol.* 156:136–140.
39. Goodell, E. W., and U. Schwarz. 1985. Release of cell wall peptides into culture medium by exponentially growing *Escherichia coli*. *J. Bacteriol.* 162:391–397.
40. Green, E. W., and M. Schaechter. 1972. The mode of segregation of the bacterial cell membrane. *Proc. Natl. Acad. Sci. USA* 69:2312–2316.
41. Harz, H., K. Burgdorf, and J.-V. Höltje. 1990. Isolation and separation of the glycan strands from murein of *Escherichia coli* by reversed-phase high-performance liquid chromatography. *Anal. Biochem.* 190:120–128.
42. Helmstetter, C. E., and A. C. Leonard. 1987. Mechanism for chromosome and minichromosome segregation in *Escherichia coli*. *J. Mol. Biol.* 197:195–204.
43. Helmstetter, C. E., and A. C. Leonard. 1990. Involvement of cell shape in the replication and segregation of chromosomes in *Escherichia coli*. *Res. Microbiol.* 141:30–39.
44. Helmstetter, C. E., A. C. Leonard, and J. E. Grimwade. 1992. Relationships between chromosome segregation, cell shape, and temperature in *Escherichia coli*. *J. Theor. Biol.* 159:261–266.
45. Hiemstra, H., N. Nanninga, C. L. Woldringh, M. Inouye, and B. Witholt. 1987. Distribution of newly synthesized lipoprotein over the outer membrane and the peptidoglycan sacculus of an *Escherichia coli* lac-lpp strain. *J. Bacteriol.* 169:5434–5444.
46. Hiraga, S. 1992. Chromosome and plasmid partition in *Escherichia coli*. *Annu. Rev. Biochem.* 61:283–306.
47. Hiraga, S., T. Ogura, H. Niki, C. Ichinose, and H. Mori. 1990 Positioning of replicated chromosomes in *Escherichia coli*. *J. Bacteriol.* 172:31–39.
48. Hoffman, B., W. Messer, and U. Schwarz. 1972. Regulation of polar cap formation in the life cycle of *Escherichia coli*. *J. Supramol. Struct.* 1:29–37.
49. Höltje, J.-V. 1993. "Three for one"—a simple growth mechanism that guarantees a precise copy of the thin, rod-shaped murein sacculus of *Escherichia coli*, p. 419–426. *In* M. A. De Pedro, J.-V. Höltje, and W. Loffelhardt (ed.), *Bacterial Growth and Lysis*. FEMS Symposium 65. Plenum Press, New York.
50. Jacob, F., S. Brenner, and F. Cuzin. 1963. On the regulation of DNA replication in bacteria. *Cold Spring Harbor Symp. Quant. Biol.* 28:329–347.
51. Jacob, F., A. Ryter, and F. Cuzin. 1966. On the association between DNA and membrane in bacteria. *Proc. R. Soc. London Ser. B* 164:267–278.
52. Jensen, M. R., A. Løbner-Olesen, and K. V. Rasmussen. 1990. *Escherichia coli* minichromosomes: random segregation and absence of copy number control. *J. Mol. Biol.* 215:257–265.
53. Kepes, A., and F. Autissier. 1972. Topology of membrane growth in bacteria. *Biochim. Biophys. Acta* 265:443–469.
54. Kepes, A., and F. Autissier. 1974. Membrane growth and cell division in *Escherichia coli*, p. 383–406. *In* A. R. Kolber and M. Kohiyama (ed.), *Mechanism and Regulation of DNA Replication*. Plenum Press, New York.
55. Koch, A. L. 1983. The surface stress theory of microbial morphogenesis. *Adv. Microb. Physiol.* 24:301–366.
56. Koch, A. L. 1988. The sacculus, a non-woven, carded, stress-bearing fabric, p. 43–59. *In* P. Actor, L. Daneo-Moore, M. Higgins, M. R. J. Salton, and G. D. Shockman (ed.), *Antibiotic Inhibition of Bacterial Cell Surface Assembly and Function*. American Society for Microbiology, Washington, D.C.
57. Koch, A. L., R. W. H. Verwer, and N. Nanninga. 1982. Incorporation of diaminopimelic acid into the old poles of *Escherichia coli*. *J. Gen. Microbiol.* 128:2893–2898.
58. Koch, A. L., and C. L. Woldringh. 1995. The inertness of the poles of a Gram-negative rod. *J. Theor. Biol.* 171:415–425.
59. Lark, K. G., and R. E. Bird. 1965. Segregation of the conserved units of DNA in *Escherichia coli*. *Proc. Natl. Acad. Sci. USA* 54:1444–1450.
60. Leal, J., and H. Marcovich. 1971. Segregation of phage receptors T6 during cell division in *Escherichia coli* K12. *Ann. Inst. Pasteur* 120:467–474.
61. Leonard, A. C., and C. Helmstetter. 1989. Replication and segregation control of *Escherichia coli* chromosomes, p. 65–94. *In* K. W. Adolph (ed.), *Chromosomes: Eukaryotic, Prokaryotic, and Viral*, vol. 2. CRC Press, Boca Raton, Fla.
62. Lin, E. C. C., Y. Hirota, and F. Jacob. 1971. On the process of cellular division in *Escherichia coli*. VI. Use of a methocel autoradiographic method for the study of cellular division in *Escherichia coli*. *J. Bacteriol.* 108:375–385.
63. Løbner-Olesen, A., and P. L. Kuempel. 1992. Chromosome partitioning in *Escherichia coli*. *J. Bacteriol.* 174:7883–7889.

64. Maddock, J. R., M. R. K. Alley, and L. Shapiro. 1993. Polarized cells, polar actions. *J. Bacteriol.* **175:**7125–7129.

65. Maddock, J. R., and L. Shapiro. 1993. Polar location of the chemoreceptor complex in the *Escherichia coli* cell. *Science* **259:**1717–1723.

66. May, J. W. 1963. The distribution of cell-wall label during growth and division of *Salmonella typhimurium. Exp. Cell Res.* **31:**217–220.

67. Olijhoek, A. J. M., S. Klencke, E. Pas, N. Nanninga, and U. Schwarz. 1982. Volume growth, murein synthesis, and murein cross-linkage during the division cycle of *Escherichia coli* PA3092. *J. Bacteriol.* **152:**1248–1254.

68. Pierucci, O., and C. E. Helmstetter. 1976. Chromosome segregation in *Escherichia coli* B/r at various growth rates. *J. Bacteriol.* **128:**708–716.

69. Pierucci, O., and C. Zuchowski. 1973. Non-random segregation of DNA strands in *Escherichia coli* B/r. *J. Mol. Biol.* **80:**477–503.

70. Ryter, A., Y. Hirota, and U. Schwarz. 1973. Process of cellular division in *Escherichia coli:* growth pattern of *E. coli* murein. *J. Mol. Biol.* **78:**185–195.

71. Ryter, A., H. Shuman, and M. Schwartz. 1975. Integration of the receptor for bacteriophage lambda in the outer membrane of *Escherichia coli:* coupling with cell division. *J. Bacteriol.* **122:**295–301.

72. Schaechter, M. 1990. The bacterial equivalent of mitosis, p. 313–322. *In* K. Drlica and M. Riley (ed.), *The Bacterial Chromosome.* American Society for Microbiology, Washington, D.C.

73. Schaechter, M., O. Maaløe, and N. O. Kjeldgaard. 1958. Dependency on medium and temperature of cell size and chemical composition during balanced growth of *Salmonella typhimurium. J. Gen. Microbiol.* **19:**592–606.

74. Schaechter, M., and U. von Freiesleben. 1993. The equivalent of mitosis in bacteria, p. 61–73. *In* J. S. Heslop-Harrison and R. B. Flavell (ed.), *John Innes Symposium on the Chromosome.* BIOS Scientific Publishers Ltd., Oxford, UK.

75. Schaechter, M., J. P. Williamson, J. R. Hood, Jr., and A. L. Koch. 1962. Growth, cell and nuclear divisions in some bacteria. *J. Gen. Microbiol.* **29:**421–434.

76. Schurr, T., and N. B. Grover. 1990. Analysis of a model for minichromosome segregation in *Escherichia coli. J. Theor. Biol.* **146:**395–406.

77. Schwarz, U., A. Ryter, A. Rambach, R. Hellio, and Y. Hirota. 1975. Process of cellular division in *Escherichia coli:* differentiation of growth zones in the sacculus. *J. Mol. Biol.* **98:**749–759.

78. Scott, R. I., and R. K. Poole. 1987. Evidence for unequal segregation of cytochromes at cell division in *Escherichia coli. FEMS Microbiol. Lett.* **43:**1–4.

79. Smit, J., and H. Nikaido. 1978. Outer membrane of gram-negative bacteria. XVIII. Electron microscopic studies on porin insertion sites and growth of cell surface of *Salmonella typhimurium. J. Bacteriol.* **136:**687–702.

80. Trueba, F. J., E. A. Van Spronsen, J. Traas, and C. L. Woldringh. 1982. Effects of temperature on the size and shape of *Escherichia coli* cells. *Arch. Microbiol.* **131:**235–240.

81. Tsukagoshi, N., P. Fielding, and C. F. Fox. 1971. Membrane assembly in *Escherichia coli.* I. Segregation of preformed and newly formed membrane into daughter cells. *Biochem. Biophys. Res. Commun.* **44:**497–502.

82. Van Tubergen, R. P., and R. B. Setlow. 1961. Quantitative radioautographic studies on exponentially growing cultures of *Escherichia coli.* The distribution of parental DNA, RNA, protein and cell wall among progeny cells. *Biophys. J.* **1:**589–625.

83. Verwer, R. W. H., E. H. Beachey, W. Keck, A. M. Stoub, and J. E. Poldermans. 1980. Oriented fragmentation of *Escherichia coli* sacculi by sonication. *J. Bacteriol.* **141:**327–332.

84. Verwer, R. W. H., and N. Nanninga. 1980. Pattern of *meso*-DL-2,6-diaminopimelic acid incorporation during the division cycle of *Escherichia coli. J. Bacteriol.* **144:**327–336.

85. Verwer, R. W. H., N. Nanninga, W. Keck, and U. Schwarz. 1978. Arrangement of glycan chains in the sacculus of *Escherichia coli. J. Bacteriol.* **136:**723–729.

86. Vos-Scheperkeuter, G. H., E. Pas, G. J. Brakenhoff, N. Nanninga, and B. Witholt. 1984. Topography of the insertion of LamB protein into the outer membrane of *Escherichia coli* wild-type and *lac-lamB* cells. *J. Bacteriol.* **159:**440–447.

87. Wientjes, F. B., and N. Nanninga. 1989. Rate and topography of peptidoglycan synthesis during cell division in *Escherichia coli:* concept of a leading edge. *J. Bacteriol.* **171:**3412–3419.

88. Wilson, G., and C. F. Fox. 1971. Membrane assembly in *Escherichia coli.* II. Segregation of preformed and newly formed membrane into cells and minicells. *Biochem. Biophys. Res. Commun.* **44:**503–509.

89. Woldringh, C. L., P. Huls, N. Nanninga, E. Pas, P. E. M. Taschner, and F. B. Wientjes. 1988. Autoradiographic analysis of peptidoglycan synthesis in shape and division mutants of *Escherichia coli* MC1400, p. 66–78. *In* P. Actor, L. Daneo-Moore, M. L. Higgins, M. R. J. Salton, and G. D. Shockman (ed.), *Antibiotic Inhibition of Bacterial Cell Surface Assembly and Function.* American Society for Microbiology, Washington, D.C.

90. Woldringh, C. L., P. Huls, E. Pas, G. J. Brakenhoff, and N. Nanninga. 1987. Topography of peptidoglycan synthesis during elongation and polar cap formation in a cell division mutant of *Escherichia coli* MC4100. *J. Gen. Microbiol.* **133:**575–586.

Nucleoid Segregation

MOLLY B. SCHMID AND ULRIK VON FREIESLEBEN

105

INTRODUCTION

Bacterial nucleoids segregate with great fidelity into the daughter cells that arise during cell division. Less than 1 in 30,000 wild-type *Escherichia coli* cells lacks DNA, as seen under a microscope (46). Faithful nucleoid segregation, therefore, depends on a precise mechanism that ensures the equipartition of nucleoids among daughter cells as well as the coordination between DNA replication and cell division (for details, see chapters 99 to 102). The process involves a number of distinct steps, namely, termination of DNA replication, decatenation of the daughter chromosomes, and nucleoid separation, the spatial repositioning of daughter nucleoids at both sides of the cell's midline. The first two of these steps are better understood than the last one, but, in general, the study of this fundamental cell cycle event has proven to be quite difficult and our state of knowledge has proven to be limited. The genes required for this process have not been fully identified, and the role of several known proteins, while plausible, has not been definitely established. There are several recent reviews on this topic (28, 29, 31, 43, 44, 65, 90, 95).

One of the reasons why this fundamental subject has proven to be so elusive is that mutants with lowered segregation fidelity are not necessarily lethal. Mutants that lack highly faithful segregation remain viable in artificial media, although presumably they are at a disadvantage under natural conditions. This suggests that, in bacteria, multiple systems contribute to the overall fidelity of nucleoid segregation. This is in stark contrast to the situation in eukaryotes, in which a clearly "dedicated system," the mitotic apparatus, ensures faithful distribution of daughter chromosomes at cell division. Whether bacteria possess more than one functionally dedicated system that work in combination to assure highly faithful nucleoid segregation or accomplish this using multifunctional proteins with roles in other cell cycle functions remains uncertain. There is precedent for functionally redundant mechanisms in the faithful segregation of large, low-copy plasmids, such as F, R, and P1. These plasmids have both a primary segregation system (*sopA, sopB,* and *sopC* in F) and a mechanism to prevent host cell division (*ccd*) until proper partition of the plasmids is accomplished (71, 74).

In this chapter we review the process of bacterial nucleoid segregation as it is known in *E. coli* and *Salmonella typhimurium* (official designation, *Salmonella enterica* serovar Typhimurium). We will not explore nucleoid segregation in other bacterial species such as *Bacillus subtilis* or *Caulobacter crescentus,* where developmental programs can lead to dimorphic nucleoids and nonidentical daughter cells, even though such systems provide good insight into the segregation process (69, 118); nor do we review the faithful segregation mechanisms of the low-copy-number plasmids of *E. coli* and *S. typhimurium*. Proteins and DNA sites necessary for segregation of the F factors, R factors, and P1 phage have been extensively studied and reviewed (44) and may indeed share components with the nucleoid segregation apparatus. Especially interesting (and regrettably not included here) are the *ccd* genes which inactivate gyrase action until the F factor is appropriately partitioned to the daughter nucleoids (71, 74).

Unlike in the eukaryotes, segregation of daughter DNA in bacteria is achieved without involving an obvious cellular ultrastructure such as a mitotic spindle. The first attempt to address how the process is carried out in bacteria was included in the classic postulation of the replicon model by Jacob et al. (55). They suggested that the rigid bacterial cell envelope serves as the functional analog of the eukaryotic spindle and that the attachment of the daughter nucleoids to this structure suffices to separate them. They postulated that new cell envelope material is laid down between the nucleoids (and segregates in a "semiconservative" fashion), which, upon closer examination, turns out not to be the case (see chapter 104). This is a "passive" segregation model because it does not involve any dedicated proteins that can generate a mechanical force.

A variety of methods have been employed in the search for answers to the following fundamental questions about nucleoid segregation in *E. coli* and *S. typhimurium*. Which genes are required? What are the functions of the relevant gene products? How do these gene products interact? What are the steps of the nucleoid segregation process? How is coordination with DNA replication and cell division achieved? The analysis of nucleoid segregation has required techniques of cell biology—microscopy, cell sorting, and specific genetic selections and screens—in addition to the general tools of genetics, biochemistry, and molecular biology.

MICROSCOPIC OBSERVATIONS

Light Microscopy

Nucleoids of living cells can be seen under a phase-contrast microscope, which allows observation and recording of the gross morphological changes that occur during growth (72). For visualization of the nucleoid, the refractive index of the medium must be altered to nearly match that of the cytoplasm, or a fluorescent DNA-binding dye, such as 4′,6-diamidino-2-phenylindole (DAPI) or Hoechst no. 33342 (16), can be used. Fluorescent dyes are generally taken up after fixation (46) or permeabilization (97), but they can also be incorporated into living cells (116). Advances in fluorescent probe technology and digitized microscopy have allowed some greater resolution of intracellular compartments in bacteria (63, 118).

Time lapse photomicrographs of living *E. coli* cells (72) show that nucleoids change shape and separate without undergoing the cycles of condensation and decondensation characteristic of mitotic chromosomes in eukaryotes. Eventually, the sister nucleoids separate completely and move to more polar positions in the cell. Repositioning of the sister nucleoids takes 6 to 8 min of the 30-min generation time of fast-growing cells and occurs before the onset of septum formation. The timing of the microscopically visible events in nucleoid separation does not vary greatly from cell to cell and has a coefficient of variation of about 20% (96). A reason why the timing of nucleoid segregation may be quite precise is that all cells terminate nucleoid replication at nearly the same time, at least in synchronized cultures (13).

The actual temporal order of nucleoid separation, termination of replication, and formation of the septum depends on the growth rate of the cells. Under conditions of slow growth, cell constriction (septum initiation) often occurs before visible nucleoid separation, while under conditions of rapid growth, most cells reverse the order, with nucleoid segregation preceding cell constriction (113). Termination of replication cannot be visualized but in slowly growing cells is presumed to occur before the time of visible nucleoid separation and in rapidly growing cells is presumed to occur concurrent with nucleoid separation (113).

Electron Microscopy

The development of electron microscopic methods to study cellular structures and nucleoid shape in bacterial cells is reviewed in chapter 4. Fixation procedures can alter the cellular and nucleoid morphology (see chapter 4). A recent advance in fixation methodology relies on ultralow temperature rather than chemical cross-linking. The cryofixation freeze-substitution technique has resulted in images believed to closely resemble the unfixed cells (51). However, this point is still being discussed (6).

In thin sections of cryofixed bacteria, nucleoids do not appear as the compact body seen with conventional fixation but as a multilobate structure with projections extended over a considerable distance (see chapter 12). There is little in this morphology to directly suggest aspects of the segregation process. A few observations carried out with an electron microscope may, however, be relevant to nucleoid segregation. Although there is no cytologically observable spindle in wild-type *E. coli* or *S. typhimurium* cells, mutants that overproduce certain proteins (e.g., CafA [see below]) result in intracellular filaments (85). In addition, other species have filamentous structures that may prove to have functional homology with the eukaryotic spindle (8). The mesosome, a structure previously thought to be associated with the nucleoid, is considered to be a fixation artifact (51; see chapter 4).

Introduction of Macromolecules into Cells

Small molecules and small proteins can be introduced into viable cells (99). It has been possible to show, for example, that the histone-like protein HU, when added externally, is distributed throughout the ribosome-free nucleoid space, whereas non-DNA binding proteins localize in the ribosome-filled space (99). While this technique has not been widely exploited thus far, it may provide an important tool for the complementation of nucleoid segregation mutants by purified proteins.

CELL SORTING TECHNIQUES

Flow Cytometry

The distribution of DNA content in members of a population can be studied by flow cytometry, using a fluorescent DNA-specific dye, such as mithramycin (14). Single cells are counted and measured on the basis of both DNA content (fluorescence intensity) and cell mass (light scattering). This important new technique has proven to be valuable for assessing the heterogeneity of DNA content in populations of exponentially growing wild-type cultures (101, 102), as well as in the characterization of mutant strains (64, 110, 112). This technique has not yet been used to determine whether segregation of the nucleoids of multinucleated, fast-growing cells takes place with the same degree of synchrony as the initiation of their replication (101).

"Baby Machine"

A uniform population of newborn cells can be obtained from an exponentially growing culture by using membrane elution techniques (see chapter 102 for details). Such a baby machine relies on the adsorption of cells to a nitrocellulose membrane and the release of only one of the daughter cells formed upon division (40). Newborn cells are detached from the membrane and eluted into the medium flowing through the membrane. This method has been used in conjunction with a microscopic technique to determine the cellular polarity of the segregating nucleoids (see chapter 104). The results generally indicate that there is a preference toward nonrandom segregation toward one of the cell poles.

GENETIC SCREENS FOR SEGREGATION MUTANTS

Several methods have been used to identify *E. coli* and *S. typhimurium* mutants defective in nucleoid segregation. Two different cytological phenotypes have been used to determine deviations from faithful nucleoid segregation. One consists of cells with a centrally located nucleoid containing multiple genome equivalents, the so-called partitioning or *par* mutants (Fig. 1). The other phenotype leads to the production of DNA-less, anucleate cells. In addition, a genetic selection based on the resistance of polyploid cells of many species to camphor vapors (84) has been used to identify mutations in nonessential genes that cause increased DNA content.

"Partitioning" Mutants

Mutants with nucleoid partitioning defects (*par*) have been found within the collections of conditional lethal mutants of *E. coli* (48, 49, 60, 91) and *S. typhimurium* (97), using microscopic screening of cells placed under nonpermissive conditions. These mutants are relatively rare members of the filamentous temperature-sensitive (*fts*) class that continue DNA replication but leave the newly replicated daughter nucleoids centrally located in filamentous cells (50). In some cases the mutant populations contain

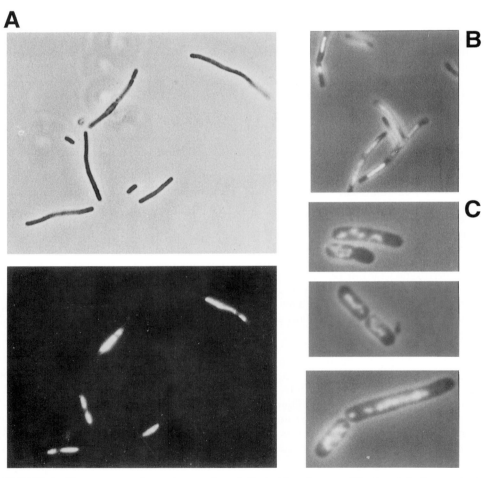

FIGURE 1 Fluorescence photomicrographs of *S. typhimurium* partitioning-defective mutants stained with DAPI. (A) Heat-sensitive strain SE5171 carrying the *parC171* mutation, grown under nonpermissive conditions. Two photographs allow comparison of the phase-contrast image and the DAPI fluorescence image. (B and C) Photomicrographs taken while viewing both the fluorescence and phase-contrast images provides increased photographic sensitivity. (B) Heat-sensitive strain SE7784 carrying the *parC281* mutation, grown under nonpermissive conditions. (C) Certain mutations, such as those causing partial inhibition of *ftsZ*, can cause increased cell size (107), which provides additional resolution. The figure shows strain SE5758, which carries *parE206* and at least one additional suppressor mutation.

cells that are highly varied both in cell dimensions and in nucleoid morphology (97, 120).

The genes identified thus far by screening conditional lethal mutant collections do not encode proteins involved in the generation of mechanical force. However, the set of genes likely to be involved in nucleoid segregation is not yet complete. The initial search through the E. coli collection of temperature-sensitive mutants (49) did not identify the parC and parE mutants, which were found within the same collection 20 years after the initial screening (59, 60). Additional essential genes with a role in nucleoid segregation may still remain unidentified. Microscopic screening through a conditional lethal mutant collection of S. typhimurium has identified several additional par mutants whose mutations are other than the known parA to parF genes (67, 97; J. Fukayama, D. Sekula, and M. B. Schmid, unpublished data).

DNA-Less Cell Producers

Some conditional lethal mutants produce anucleate, DNA-less cells, indicating at a minimum a loss of coordination between nucleoid segregation and cell division (50, 103). The proportion of anucleate cells in a mutant population may be difficult to determine because, even in the absence of a mechanism that ensures equipartition, there may be physical or mechanical constraints to the localization of two nucleoids in one daughter cell.

One class of mutants (in the min genes) leads to the production of "minicells," small, spherical DNA-less cells (2). It would appear that, since nucleoid segregation proceeds otherwise normally in such mutants, these are not directly relevant to nucleoid segregation. As discussed below, this has turned out not to be the case.

Another class of mutants produces a high percentage of DNA-less normal-sized cells. An ingenious method to specifically identify nonlethal mutants that produce anucleate cells has been developed by Hiraga and coworkers (46). The starting strain does not express the lacZ gene because the high level of lac repressor represses the plasmid-encoded lacZ gene. In DNA-less cells the loss of the chromosomally encoded lac repressor allows expression of the plasmid lacZ gene. Thus, mutants that frequently segregate DNA-less cells have a blue or mixed colony color on 5-bromo-4-chloro-3-indolyl-β-D-galactopyranoside plates. As a control for the screening procedure, Hiraga showed that min mutants make blue colonies with this method. These colonies arose because ~50% of the cell number in a minB colony are DNA-less, although they account for only ~5% of the cell mass in the colony, since they are of smaller than average size. On the basis of this result, it was estimated that the screen could detect mutants that segregate as few as 5% of normal-sized DNA-less cells, which is the actual frequency observed with the mutants. This genetic method identifies nonlethal mutants that produce anucleate cells.

Polyploidy

Trun and Gottesman (108, 109) and Trun et al. (110) used resistance to camphor vapors to identify mutants that have lost the ability to maintain only one chromosome per cell. The rationale for this genetic selection remains elusive; however, polyploid mutants in several species have been identified by this procedure (84). Several E. coli genes that can mutate to give an Mbr (mothball resistant) phenotype have been identified. By several criteria, many of these mutant strains show increased ploidy. Whether these mbr mutations cause a primary defect in the nucleoid segregation process or in the regulation of replication initiation remains uncertain.

GENES INVOLVED IN SEGREGATION

Many genes with potential roles in nucleoid segregation are known through genetic analysis of mutants. However, since some of these mutants produce only a small percentage of DNA-less cells, it is uncertain whether these genes play a primary role in nucleoid segregation or merely relax the normally tight coupling between nucleoid segregation and cell division.

The first phase of nucleoid segregation, the resolution through decatenation and monomer formation, depends on genes with a known biochemical function, such as topoisomerases and resolvases. The remaining genes cannot be so definitively classified. Necessary to the understanding of nucleoid segregation will be the biochemical characterization of these gene products. The genes identified in this section remain contenders for playing functional roles in nucleoid segregation.

DNA Replication Genes

Several conditional lethal DNA replication mutants, dnaA (50, 80), dnaB (54), dnaG (39), dnaK (16), dnaX (80), and dnaZ (117), accumulate normal-sized DNA-less cells under nonpermissive growth conditions. DNA-less cells are thought to arise in these mutants because they undergo cell division without nucleoid segregation. The DNA-less cells are not the result of degradation of the DNA (52). This uncoupling between cell division and nucleoid segregation takes place despite the existence of several mechanisms that normally inhibit septation when DNA replication stops (26). The existence of these multiple systems for inhibiting septation makes the appearance of DNA-less cells in certain conditional dna(Ts) mutants perplexing. Most likely, DNA-less cells result from the release of cell division inhibition. However, the case of dnaG suggests that, in certain genes, only specific alleles will have nucleoid segregation defects. Two mutant alleles of dnaG lie within a small region in the C-terminal domain of DnaG and have a partitioning-defective phenotype (39).

Cell Division Genes

Mutations in essential cell division proteins have been isolated as temperature-sensitive (fts) mutants that filament at the nonpermissive temperature. Most of these mutants have well-spaced nucleoids and no apparent nucleoid segregation defect. However, mutants in two distinct fts loci, ftsB and ftsC, display slightly disturbed nucleoid segregation (88), producing a small percentage of DNA-less cells. Further work showed that these two fts mutants have replication-related defects. The ftsB mutant is an allele of nrdB, which encodes the B_1 subunit of ribonucleotide reductase which converts ribonucleotides to deoxyribonucleotides (106). Mutants in the second subunit of ribonucleotide reductase, encoded by nrdA, were originally isolated as dnaF mutants. Thus, the nucleoid segregation defect in ftsB mutants may be another case of a replication block causing abnormal septum formation. An ftsC mutant also segregates normal-sized DNA-less cells and has been shown to replicate DNA at a decreased rate.

Tetart et al. (107) proposed a model whereby nucleoid segregation occurs as the result of a decrease in the levels of a protein directly involved in septum formation, FtsZ (see below). The amount of DNA/nucleoid appears to be increased in strains with decreased FtsZ expression. It is proposed that FtsZ may inhibit nucleoid segregation when free in the cytoplasm, before it acts to initiate septum formation. This conflicts with previous results suggesting no role for FtsZ in nucleoid segregation on the basis of results using a temperature-sensitive allele (23). However, FtsZ is a complex protein, and a specific missense allele may have a phenotype dissimilar to the lack of the protein altogether.

Mutations in the *min* genes (see chapter 101) cause the formation of DNA-less minicells. In addition, populations of these mutant strains include filamentous cells, 10 to 20% of which have abnormal nucleoid distributions, indicating loss of nucleoid segregation fidelity (4, 56). Certain *minB* mutant alleles produce DNA-less cells that are heterogeneous in size (56), whereas other *min* mutants lead to the production of normal-sized cells lacking any DNA (57). The MinD protein has homology with the plasmid-encoded SopA protein (27, 76), which is required for faithful partition of mini-F plasmids. In addition, loss of Min protein function causes loss of negative superhelicity, while overexpression of Min proteins causes increased negative superhelicity (79). This correlation between the level of Min proteins and DNA supercoiling may be related to the SopA-dependent changes in mini-F plasmid superhelicity that has been observed (10).

The *divA*, *divB*, and *divC* mutations of *E. coli* cause production of DNA-less cells but only in *fts* and *dna* mutants (49). These mutations by themselves cause no discernible phenotype. Mutants with analogous phenotypes were found in *S. typhimurium* (98, 103).

Muk

The *muk* mutants have been identified through Hiraga's screen that identifies colonies with a high percentage of anucleate cells. The authors called these mutants *muk*, from "mukaku" or "anucleate" in Japanese. Five *muk* genes have been described and partially characterized. These are *mukA* (*tolC*), *mukB*, *mukC*, *mukD* (43–46), and *mukF* (S. Hiraga, personal communication).

The MukB protein has been most extensively characterized. This 177-kDa protein has two globular domains connected by an α-helical coiled-coil region. The N-terminal globular region has an ATP-binding consensus motif and weak homology with rat dynamin. The C-terminal globular region has three putative zinc finger motifs that are hypothesized to bind DNA (81). MukB protein shows ATP/GTP binding activity in the presence of Zn^{2+} and adsorbs to a calf thymus DNA-cellulose column (44). Electron microscopy of purified MukB protein has shown that it has a structure similar to that of the heavy chains of eukaryotic myosin and kinesin (44). It has been suggested that MukB is a motor that drives the nucleoids along a cytoskeletal system of filaments spanning the length of the cell (44).

MukB null mutants form normal-sized cells, 5% of which are anucleate and, for reasons not known, are viable at a low temperature only. DNA replication and timing of initiation are normal in these mutants. F plasmids partition normally into both the nucleate and anucleate cells, indicating that *mukB* affects chromosomal partitioning only (35). Multicopy suppressors of the high-temperature lethality phenotype of the *mukB106* mutation have been identified (*msmA*, *msmB*, and *msmC*) (119). Two of these also suppress anucleate cell formation (*msmB* and

msmC). None of them could suppress both phenotypes in a *mukB* null mutant.

Mbr

Four unlinked *mbr* loci have been identified by the selection for polyploidy (108–110). The *mbrA* to *mbrD* mutants have an increased DNA/protein ratio and altered cytology, including both DNA-less cells and heterogeneous nucleoid distribution. These characteristics suggest that the wild-type *mbr* genes play a role in nucleoid segregation. Many of the *mbr* mutants showed conditional growth at a high temperature or on enriched medium, which allowed further characterization of the mutations in these strains. The *mbrD* mutation may be allelic with *rpoB* on the basis of genetic mapping. The *mbrA* mutant is not viable when coupled with a mutation in RNase H (*rnh*), suggesting that initiation from the alternate replication origins, *oriK* (which occurs in *rnh* mutants), causes inviability.

RecA

A direct role for *recA* in segregation has been suggested by Zyskind et al. (120), who found a high number of normal-sized anucleate cells in *recA* and *recA recD* double mutants. These anucleate cells were assumed to arise from a direct segregation defect, contrasting with conclusions of previous experiments (18–21). However, Skarstad and Boye (100) have reported that the high number of anucleate cells in a *recA recD* double mutant may still be explained by selective degradation of individual chromosomes because *recD* mutants exhibit substantial exonuclease activity (89). Supporting the original interpretation of Zyskind et al. is the finding that a mutant in four genes (*recA recD recJ xonA*) lacking demonstrable exonuclease activity still produces anucleate cells (W. Wackernagel and J. Zyskind, personal communication).

HU

The major "histone-like" proteins in *E. coli* and *S. typhimurium* are HUa and HUb. Together with H-NS and other DNA binding proteins, these proteins bind DNA, affecting the ability of DNA to function as a substrate for transcription, replication, and recombination. A strain carrying mutations in both *hupA* and *hupB*, encoding the two HU subunit proteins, produces a high percentage of DNA-less cells and aberrant nucleoid structures (53). As suggested in a recent review by Løbner-Olesen and Kuempel (65), this points to the importance of nucleoid structure in proper segregation of daughter chromosomes, a point that deserves further attention.

The following three genes encode proteins that may form filamentous structures involved in nucleoid segregation.

FtsZ

FtsZ is essential for septation and cell division (for details, see chapter 101). FtsZ is uniformly distributed in the cytoplasm of nondividing cells but becomes concentrated in a ring-like structure at the site of septum formation just prior to cell division (9). FtsZ protein shows several similarities to eukaryotic tubulin: it binds GTP and GDP and hydrolyzes GTP to GDP (25, 77, 87). FtsZ contains a tubulin-like signature motif and assembles into filaments and tubules (15, 78). At the moment, FtsZ is the best candidate for a prokaryotic homolog of a cytoskeletal protein.

CafA

Overexpression of a gene, *cafA*, located at 71 min on the *E. coli* chromosome map causes formation of cytoplasmic axial filament (85). The amino acid sequence of the CafA protein has similarity to members of the eukaryotic myosin and kinesin family, suggesting a cytoskeletal nature. However, mutants defective in *cafA* do not show any phenotype.

FtsA

The appearance of axial filamentous structures has been reported when a mutant FtsA protein is overproduced (37). A structural homology between FtsA and actin has been suggested (12, 92). Mutants in FtsA do not show defects in segregation, although they have a clear defect in appropriate septum formation.

CELLULAR STRUCTURES INVOLVED IN SEGREGATION

Cell Membrane

The classical model for chromosome segregation of Jacob et al. (55) proposes that the daughter nucleoids attach to the cell membrane on both sides of the incipient septum. Subsequent growth of the membrane and elongation of the cell between these attachment sites would ensure proper segregation. Lamentably, the membrane is not synthesized in such a semiconservative mode. Rather, it has been shown that the membrane of *E. coli* is assembled by insertion of new material at many sites (115). However, this does not exclude a role of the membrane in chromosome segregation.

There is considerable evidence that the chromosome is attached to the membrane at the replicative origin, but probably for a portion of the cell cycle only. It has been shown that the replicative origin, *oriC*, is attached to the membrane in vivo prior to initiation (58). In vitro, *oriC* DNA binds specifically to an outer membrane fraction (83). This specific affinity is limited to newly made origin DNA and due to its unique state of methylation. For a substantial period after replication, newly made origin DNA is hemimethylated, that is, the new DNA strand has not had a chance to become methylated by the major methyltransferase of *E. coli*, the Dam enzyme. This enzyme methylates the adenine of the sequence GATC, which is found in unusual abundance in the origin of *E. coli* and other enteric bacteria (typically, 11 GATCs in 245 bp [121]). Binding of hemimethylated *oriC* DNA takes place mainly with a specific fraction that is separable on sucrose density gradients (22). Methylation of the origin has been shown to be significantly delayed when compared with the average time required to methylate GATCs elsewhere on the chromosome (17, 83). This delay has been ascribed to sequestration of *oriC* in the membrane. While this sequestration appears to be important for the proper timing of replication by preventing premature initiations, it could also serve as an initial step in segregation (for more details on the role of methylation in the timing of replication initiation, see chapter 99). Sequestration of *oriC* lasts about 10 min, which could be sufficient time to give the daughter chromosomes their sense of direction and ensure that each becomes destined to occupy one of the cell halves.

Attachment to the membrane is not limited to the replicative origin. Early studies have shown that there are several sites, perhaps situated randomly along the DNA, that are attached to the membrane. The number of these sites has been estimated to be between 20 and 80 per nucleoid (for a review, see reference 62). It has been proposed that this kind of membrane attachment may be longer lasting than that of the origin and that it may play a role in segregation by ensuring that each incipient daughter nucleoid remains in its own cell half until septum formation is completed (93).

Several problems arise, however, when trying to imply a role of *oriC* membrane attachment in segregation. *E. coli dam* mutants, which do not contain hemimethylated DNA and hence do not sequester *oriC*, do not show segregation defects such as anucleate cell formation (112). Also, if *oriC* acts as a centromere-like site, the presence of high numbers of minichromosomes (plasmids using *oriC* as their only origin of replication) would be expected to interfere with segregation of the chromosomal origin. This is not the case, and moreover the minichromosomes themselves segregate in a random fashion.

Do centromere-like loci exist on the *E. coli* chromosome? We do not know. Every attempt so far to identify such sites has failed. One approach has been to try to clone sequences from the chromosome that would stabilize otherwise unstable plasmid replicons such as partitionless mini-F (D. Lane, personal communication). Centromere-like sites are known to exist for unit-copy plasmids such as P, F, and R1 (5, 24, 44). The *cis*-acting DNA sites *parS* in P1, *sopC* in F, and *parC* in R1 are required for accurate plasmid partitioning. These sites are also capable of stabilizing minichromosomes if the plasmid-encoded partition proteins are provided in *trans*.

Because termination of replication is a key event in segregation, the terminus region is a good candidate for containing a centromere-like site. Strains with deletions that remove up to 330 kb from the terminus region exhibit the phenotype associated with the absence of the *dif* locus, but other partitioning errors are not apparent (41). Thus, aside from the *dif* site's role in chromosome resolution (see below), no locus in the terminus region has been shown to be important for nucleoid segregation.

Periseptal Annuli

The periseptal annuli are future sites of septum formation that can be observed microscopically. These membrane regions extend around the cell and are found at the cell midpoint and at the cell poles. The periseptal annuli are believed to create a separate compartment of the periplasm at these regions (36). The periseptal annuli are observed in electron micrographs, as well as by phase-contrast microscopy, when plasmolyzed cells are used. Observing plasmolysis bays, the periseptal annuli were hypothesized to move to the 1/4 and 3/4 positions within the cell as the cell elongated. Thus, these structures move by an unknown mechanism to similar positions within the cell as the nucleoids do. Further evidence comes from *ftsZ* mutants. In *ftsZ* mutant strains, the periseptal annuli are randomly distributed (26), and in cells underexpressing FtsZ, the nucleoids fail to separate consistently (107). However, the specific location of plasmolysis bays, hence the existence of periseptal annuli, has been called into question (114).

Cell Poles

The cell poles are former sites of cell division; one of the poles is the site of the most recently completed cell division. As such, there may be remnants of a mechanoforce-generating push mechanism from a previous cell division or active participants in a mechanoforce-generating pull mechanism. There are unique periplasmic and cell membrane proteins in the cell poles of *E. coli* (33, 69). These proteins include the maltose-binding protein and chemoreceptor proteins (70) and the Era protein (3). Era protein is an essential GTP-binding protein that is localized at the cell poles and the cell midpoint and halfway between the poles and midpoint (1/4 and 3/4 positions [38]). The cellular localization makes it interesting; however, mutants with a nucleoid segregation defect have not been described. It is possible that finding proteins at the poles is due to previous localization at the septum from which the cell poles are derived and may thus be involved in septation rather than nucleoid segregation.

STEPS IN NUCLEOID SEGREGATION

After termination of replication, the two daughter chromosome face two problems in order to segregate: they must resolve dimeric chromosomes produced by homologous recombination between daughter chromosomes and must unlink the catenanes that result from replication of the DNA duplex.

Decatenation

To achieve physical separation after replication of the right-handed DNA duplex, the two circular daughter chromosomes must undergo complete topological unlinking. Four topoisomerases capable of aiding this topological unlinking are known in *E. coli*, allowing for the participation of several mechanisms. Much of the unlinking probably occurs during the elongation phase of replication, presumably accomplished by DNA gyrase. Any remaining interlocks must be resolved at or after the termination phase of replication. Intertwined completed duplex molecules require a type 2 DNA topoisomerase for resolution. *E. coli* and *S. typhimurium* have two type 2 topoisomerases, DNA gyrase (encoded by *gyrA* and *gyrB*) and topoisomerase IV (encoded by *parC* and *parE*). The unlinking of gapped molecules, on the other hand, can additionally be carried out by type 1 topoisomerases, either topoisomerase I (encoded by *topA*) or topoisomerase III (encoded by *topB*). The type 1 topoisomerase, topoisomerase III, is required to achieve resolution of replicated daughter plasmids in vitro (30, 42, 75).

Mutant alleles of the genes encoding the type 2 topoisomerase DNA gyrase, *gyrA* (*parD*) and *gyrB* (*parA*), were identified among the original *par* mutations (48, 49, 82, 91). The role for DNA gyrase as the enzyme responsible for decatenation was further strengthened by finding that isolated nucleoids from a *gyrB* mutant strain grown under nonpermissive conditions were doublets and that these doublet nucleoids could be resolved into singlets in vitro by the addition of purified DNA gyrase (105).

Recent data have cast doubt on the conclusion that DNA gyrase decatenates the newly replicated chromosomes. A new type 2 topoisomerase enzyme was discovered in a search through conditional lethal mutants (59, 60, 67, 97, 104). The enzyme, topoisomerase IV, is encoded by the linked genes *parC* and *parE*. The enzyme has been purified and has a strong decatenase activity (86). In vivo experiments demonstrated that small plasmids remain catenated after a shift to nonpermissive growth conditions in *parC* and *parE* mutants but not in *gyrA* and *gyrB* mutant strains (1). The catenated plasmids were shown to have right-handed interlinks, a critical feature resulting from the separation of daughter DNA strands during DNA

replication. These experiments demonstrate that topoisomerase IV, not DNA gyrase, is the major replication decatenase in bacterial cells. However, they do not explain why gyrase mutants also have a cytological nucleoid segregation defect. Potentially, small plasmids are an imperfect model for nucleoid replication and segregation. The domain structure of the nucleoid might impose additional topological constraints not found in small plasmids. The mutants suggest that both gyrase and topoisomerase IV are required for complete resolution of daughter nucleoids. Two different unlinking reactions may be accomplished by gyrase and topoisomerase IV; both may be necessary for complete nucleoid decatenation. Alternatively, gyrase may play an active role in the segregation process after decatenation or may affect the expression of topoisomerase IV, causing the segregation defect as a secondary phenotype.

Resolution of Dimeric Chromosomes

A reciprocal recombination between two circular molecules results in a single dimeric molecule; if this process were neither prevented nor resolved, nucleoid segregation could not proceed. Since homologous recombination between sister chromosomes can create these dimers, a mechanism is necessary to ensure ultimate resolution to two monomeric circular molecules. This subject is described in more detail in chapter 100. The site-specific recombination accomplished by the XerCD proteins, the ArgR and PepA proteins, and the *dif* site is required for proper cell division and faithful segregation in strains that are capable of homologous recombination (11, 61, 73). Cultures of recombination-proficient strains that lack the *dif* site include filamentous cells, 4% DNA-less cells, and 10% of cells with abnormal nucleoid morphology (61). The interpretation is that only the abnormal cells have failed to properly resolve the replicated chromosome to monomers.

In addition to the site-specific recombination event, a region of high homologous recombination neighbors the terminus, potentially aiding in the separation of sister chromosomes (66).

Nucleoid Separation

Two independent experiments have shown that nucleoids do not separate if protein synthesis is inhibited (32, 47). These experiments are variously interpreted to show that the synthesis of a specific new protein is necessary for nucleoid separation or that protein synthesis per se is required. This hypothesis has been further modified to suggest a protein synthesis-mediated link between DNA and the membrane. Membrane proteins in the process of simultaneous transcription-translation-protein localization could provide a DNA-membrane bridge, serving a role in proper nucleoid separation (68).

There is still controversy about whether the daughter nucleoids separate abruptly (32, 47) or gradually (111). Microscopic observations have suggested that nucleoids move in a gradual manner during cell elongation (111). The distance between the edge of the nucleoid and the cell pole remained constant, supporting gradual movement of the nucleoids rather than abrupt movement. These experiments are consistent with a passive separation mechanism, in which attachment of the DNA to a fixed cellular location accomplishes nucleoid separation as the cell elongates. However, nucleoids that were prevented from segregating by inhibition of protein synthesis, upon release of this inhibition, segregate more rapidly than the cells divide (32,

47). These experiments support the notion of an active movement of nucleoids, which requires a machinery for accomplishing nucleoid separation. Begg and Donachie (7) suggested as an alternative that this rapid movement could be accomplished by releasing the nucleoids from a possible membrane attachment and subsequent repulsion between the daughter nucleoids. The small size of cells and fixation-dependent changes in nucleoid shape make these very difficult experiments. In addition, it should be kept in mind that inhibition of protein synthesis results in rearrangement of the nucleoid shape, leading even to the fusion of individual nucleoids within a cell (34, 94).

CONCLUSION

The best-understood steps in the process of nucleoid segregation have proven to be the early ones that are required for decatenation and dimer resolution. The subsequent steps of faithful nucleoid partitioning have proven to be more difficult to study, to the point that it has been difficult to postulate even moderately detailed models. A number of promising findings have lacked subsequent substantiation. For example, there is neither genetic nor physiological evidence to validate the early postulation that attachment of the nucleoid to the cell membrane plays a direct role in the segregation process. This simply reflects our lack of knowledge and does not by itself mean that the cell envelopes may not be involved.

Evidence for the involvement of structural proteins is beginning to accumulate and to suggest an active process for bacterial nucleoid segregation. The strongest evidence arises from the production of normal-sized anucleate cells in mutants defective in MukB, a protein with similarities to eukaryotic proteins involved in the movement of organelles, and from the biochemical and cytological characterization of the FtsZ protein. The identification of additional proteins involved in nucleoid segregation is necessary for further understanding of nucleoid segregation and its interrelationship with events of the bacterial cell cycle. The characterization of these proteins may come from genetic methods as well as biochemical studies. The further biochemical characterization of the identified proteins should provide additional information that will allow the formulation of specific molecular models of nucleoid segregation. A great deal of hard work is still required to provide a molecular understanding of the process by which bacteria segregate their genomes with such high fidelity.

LITERATURE CITED

1. Adams, D. E., E. M. Shekhtman, E. L. Zechiedruch, M. B. Schmid, and N. R. Cozzarelli. 1992. The role of topoisomerase IV in partitioning bacterial replicons and the structure of catenated intermediates in DNA replication. *Cell* **71**:277–288.
2. Adler, H. I., W. D. Fisher, A. Cohen, and A. A. Hardigree. 1967. Miniature *Escherichia coli* cells deficient in DNA. *Proc. Natl. Acad. Sci. USA* **57**:321–326.
3. Ahnn, J., P. E. March, H. E. Takiff, and M. Inouye. 1986. A GTP-binding protein of *Escherichia coli* has homology to yeast RAS proteins. *Proc. Natl. Acad. Sci. USA* **83**:8849–8853.
4. Akerlund, T., R. Bernander, and K. Nordstrom. 1992. Cell division in *Escherichia coli minB* mutants. *Mol. Microbiol.* **6**:2073–2083.
5. Austin, S. 1988. Plasmid partitioning. *Plasmid* **20**:1–9.
6. Bayer, M. E. 1991. Zones of membrane adhesion in the cryofixed envelope of *Escherichia coli*. *J. Struct. Biol.* **107**:268–280.
7. Begg, K. J., and W. D. Donachie. 1991. Experiments on chromosome separation in *Escherichia coli*. *New Biol.* **3**:475–486.
8. Bermudes, D., G. Hinkle, and L. Margulis. 1994. Do prokaryotes contain microtubules? *Microbiol. Rev.* **58**:387–400.
9. Bi, E. F., and J. Lutkenhaus. 1991. FtsZ ring structure associated with division in *Escherichia coli*. *Nature* (London) **354**:161–164.
10. Biek, D. P., and J. Shi. 1994. A single 43-bp *sopC* repeat of plasmid mini-F is sufficient to allow assembly of a functional nucleoprotein partition complex. *Proc. Natl. Acad. Sci. USA* **91**:8027–8031.
11. Blakely, G., S. Colloms, G. May, M. Burke, and D. Sherratt. 1991. *Escherichia coli* XerC recombinase is required for chromosomal segregation at cell division. *New Biol.* **3**:789–798.
12. Bork, A., C. Sander, and A. Valencia. 1992. An ATPase domain common to prokaryotic cell cycle proteins, sugar kinases, actin and hsp70 heat shock proteins. *Proc. Natl. Acad. Sci. USA* **89**:7290–7294.
13. Bouche, J. P. 1982. Physical map of a 470×10^3 base-pair region flanking the terminus of DNA replication in the *Escherichia coli* K-12 genome. *J. Mol. Biol.* **154**:1–20.
14. Boye, E., H. B. Steen, and K. Skarstad. 1983. Flow cytometry of bacteria: a promising tool in experimental and clinical microbiology. *J. Gen. Microbiol.* **129**:973–980.
15. Bramhill, D., and C. M. Thompson. 1994. GTP-dependent polymerization of *Escherichia coli* FtsZ protein to form tubules. *Proc. Natl. Acad. Sci. USA* **91**:5813–5817.
16. Bukau, B., and G. C. Walker. 1989. ΔdnaK52 mutants of *Escherichia coli* have defects in chromosome segregation and plasmid maintenance at normal growth temperatures. *J. Bacteriol.* **171**:6030–6038.
17. Campbell, J. L., and N. Kleckner. 1990. *E. coli oriC* and the *dnaA* gene promoter are sequestered from *dam* methyltransferase following the passage of the chromosomal replication fork. *Cell* **62**:967–979.
18. Capaldo, F. N., and S. D. Barbour. 1973. Isolation of the nonviable cells produced during normal growth of recombination-deficient strains of *Escherichia coli* K-12. *J. Bacteriol.* **115**:928–936.
19. Capaldo, F. N., and S. D. Barbour. 1975. DNA content, synthesis and integrity in dividing and non-dividing cells of rec– strains of *Escherichia coli* K12. *J. Mol. Biol.* **91**:53–66.
20. Capaldo, F. N., and S. D. Barbour. 1975. The role of the *rec* genes in the viability of *Escherichia coli* K12. *Basic Life Sci.* **5**:405–418.
21. Capaldo, F. N., G. Ramsey, and S. D. Barbour. 1974. Analysis of the growth of recombination-deficient strains of *Escherichia coli* K-12. *J. Bacteriol.* **118**:242–249.
22. Chakraborti, A., S. Gunji, N. Shakibai, J. Cubeddu, and L. Rothfield. 1992. Characterization of the *Escherichia coli* membrane domain responsible for binding *oriC* DNA. *J. Bacteriol.* **174**:7202–7206.
23. Dai, K., and J. Lutkenhaus. 1991. *ftsZ* is an essential cell division gene in *Escherichia coli*. *J. Bacteriol.* **173**:3500–3506.
24. Dam, M., and K. Gerdes. 1994. Partitioning of plasmid R1: ten direct repeats flanking the *parA* promoter constitute a centromere-like partition site *parC*, that expresses incompatibility. *J. Mol. Biol.* **236**:1289–1298.
25. de Boer, P., R. Crossley, and L. Rothfield. 1992. The essential bacterial cell-division protein FtsZ is a GTPase. *Nature* (London) **359**:254–256.
26. de Boer, P. A., W. R. Cook, and L. I. Rothfield. 1990. Bacterial cell division. *Annu. Rev. Genet.* **24**:249–274.
27. de Boer, P. A., R. E. Crossley, A. R. Hand, and L. I. Rothfield. 1991. The MinD protein is a membrane ATPase required for the correct placement of the *Escherichia coli* division site. *EMBO J.* **10**:4371–4380.
28. de Boer, P. A. J. 1993. Chromosome segregation and cytokinesis in bacteria. *Curr. Opin. Cell Biol.* **5**:232–237.
29. de Boer, P. A. J., W. R. Cook, and L. I. Rothfield. 1990. Bacterial cell division. *Annu. Rev. Genet.* **24**:249–274.
30. DiGate, R. J., and K. J. Marians. 1989. Molecular cloning and DNA sequence analysis of *Escherichia coli topB*, the gene encoding topoisomerase III. *J. Biol. Chem.* **264**:17924–17930.
31. Donachie, W. D. 1993. The cell cycle of *Escherichia coli*. *Annu. Rev. Microbiol.* **47**:199–230.
32. Donachie, W. D., and K. J. Begg. 1989. Chromosome partition in *Escherichia coli* requires postreplication protein synthesis. *J. Bacteriol.* **171**:5405–5409.
33. Dvorak, H. F., B. K. Wetzel, and L. A. Heppel. 1970. Biochemical and cytochemical evidence for the polar concentration of periplasmic enzymes in a "minicell" strain of *Escherichia coli*. *J. Bacteriol.* **104**:543–548.
34. Dworsky, P., and M. Schaechter. 1973. Effect of rifampin on the structure and membrane attachment of the nucleoid of *Escherichia coli*. *J. Bacteriol.* **116**:1364–1374.
35. Ezaki, B., T. Ogura, H. Niki, and S. Hiraga. 1991. Partitioning of a mini-F plasmid into anucleate cells of the *mukB* null mutant. *J. Bacteriol.* **173**:6643–6646.
36. Foley, M., J. M. Brass, J. Birmingham, W. R. Cook, P. B. Garland, C. F. Higgins, and L. I. Rothfield. 1989. Compartmentalization of the periplasm at cell division sites in *Escherichia coli* as shown by fluorescence photobleaching experiments. *Mol. Microbiol.* **3**:1329–1336.
37. Gayda, R. C., M. C. Henk, and D. Leong. 1992. C-shaped cells caused by expression of an *ftsA* mutation in *Escherichia coli*. *J. Bacteriol.* **174**:5362–5370.
38. Gollop, N., and P. E. March. 1991. Localization of the membrane binding sites of Era in *Escherichia coli*. *Res. Microbiol.* **142**:301–307.
39. Grompe, M., J. Versalovic, T. Koeuth, and J. R. Lupski. 1991. Mutations in the *Escherichia coli dnaG* gene suggest coupling between DNA replication and chromosome partitioning. *J. Bacteriol.* **173**:1268–1278.
40. Helmstetter, C. E., C. Eenhuis, P. Theisen, J. Grimwade, and A. C. Leonard. 1992. Improved bacterial baby machine: application to *Escherichia coli* K-12. *J. Bacteriol.* **174**:3445–3449.

41. Henson, J. M., and P. L. Kuempel. 1985. Deletion of the terminus region (340 kilobase pairs of DNA) from the chromosome of *Escherichia coli. Proc. Natl. Acad. Sci. USA* 82:3766–3770.

42. Hiasa, H., and K. J. Marians. 1994. Topoisomerase III, but not topoisomerase I, can support nascent chain elongation during theta-type DNA replication. *J. Biol. Chem.* 269:32655–32659.

43. Hiraga, S. 1990. Partitioning of nucleoids. *Res. Microbiol.* 141:50–56.

44. Hiraga, S. 1992. Chromosome and plasmid partition in *Escherichia coli. Annu. Rev. Biochem.* 61:283–306.

45. Hiraga, S., H. Niki, R. Imamura, T. Ogura, K. Yamanaka, J. Feng, B. Ezaki, and A. Jaffe. 1991. Mutants defective in chromosome partitioning in *E. coli. Res. Microbiol.* 142:189–194.

46. Hiraga, S., H. Niki, T. Ogura, C. Ichinose, H. Mori, B. Ezaki, and A. Jaffe. 1989. Chromosome partitioning in *Escherichia coli*: novel mutants producing anucleate cells. *J. Bacteriol.* 171:1496–1505.

47. Hiraga, S., T. Ogura, H. Niki, C. Ichinose, and H. Mori. 1990. Positioning of replicated chromosomes in *Escherichia coli. J. Bacteriol.* 172:31–39.

48. Hirota, Y., and F. Jacob. 1966. Production of bacteria without DNA. *C.R. Acad. Sci. Ser. D* 263:1619–1621.

49. Hirota, Y., F. Jacob, A. Ryter, G. Buttin, and T. Nakai. 1968. On the process of cellular division in *Escherichia coli*. I. Asymmetrical cell division and production of deoxyribonucleic acid-less bacteria. *J. Mol. Biol.* 35:175–192.

50. Hirota, Y., A. Ryter, and F. Jacob. 1968. Thermosensitive mutants of *E. coli* affected in the processes of DNA synthesis and cellular division. *Cold Spring Harbor Symp. Quant. Biol.* 33:677–693.

51. Hobot, J. A., M.-A. Bjornsti, and E. Kellenberger. 1987. Use of on-section immunolabeling and cryosubstitution for studies of bacterial DNA distribution. *J. Bacteriol.* 169:2055–2062.

52. Howe, W. E., and D. W. Mount. 1975. Production of cells without deoxyribonucleic acid during thymidine starvation of *lexA⁻* cultures of *Escherichia coli* K-12. *J. Bacteriol.* 124:1113–1121.

53. Huisman, O., M. Faelen, D. Girard, A. Jaffe, A. Toussaint, and J. Rouviere-Yaniv. 1989. Multiple defects in *Escherichia coli* mutants lacking HU protein. *J. Bacteriol.* 171:3704–3712.

54. Inouye, M. 1969. Unlinking of cell division from deoxyribonucleic acid replication in a temperature-sensitive deoxyribonucleic acid synthesis mutant of *Escherichia coli. J. Bacteriol.* 99:842–850.

55. Jacob, F., S. Brenner, and F. Cuzin. 1963. On the regulation of DNA replication in bacteria. *Cold Spring Harbor Symp. Quant. Biol.* 28:329–348.

56. Jaffe, A., R. D'Ari, and S. Hiraga. 1988. Minicell-forming mutants of *Escherichia coli*: production of minicells and anucleate rods. *J. Bacteriol.* 170:3094–3101.

57. Jaffe, A., R. D'Ari, and V. Norris. 1986. SOS-independent coupling between DNA replication and cell division in *Escherichia coli. J. Bacteriol.* 165:66–71.

58. Kataoka, T., S. Gayama, K. Takahashi, M. Wachi, M. Yamasaki, and K. Nagai. 1991. Only *oriC* and its flanking region are recovered from the complex formed at the time of initiaion of chromosome replication in *Escherichia coli. Res. Microbiol.* 142:155–159.

59. Kato, J., Y. Nishimura, R. Imamura, H. Niki, S. Hiraga, and H. Suzuki. 1990. New topoisomerase essential for chromosome segregation in *E. coli. Cell* 63:393–404.

60. Kato, J., Y. Nishimura, M. Yamada, H. Suzuki, and Y. Hirota. 1988. Gene organization in the region containing a new gene involved in chromosome partition in *Escherichia coli. J. Bacteriol.* 170:3967–3977.

61. Kuempel, P. L., J. M. Henson, L. Dircks, M. Tecklenburg, and D. F. Lim. 1991. *dif*, a *recA*-independent recombination site in the terminus region of the chromosome of *Escherichia coli. New Biol.* 3:799–811.

62. Leibowitz, P. J., and M. Schaechter. 1975. The attachment of the bacterial chromosome to the cell membrane. *Int. Rev. Cytol.* 41:1–28.

63. Lewis, P. J., C. E. Nwoguh, M. R. Barer, C. R. Harwood, and J. Errington. 1994. Use of digitized video microscopy with a fluorogenic enzyme substrate to demonstrate cell- and compartment-specific gene expression in *Salmonella enteritidis* and *Bacillus subtilis. Mol. Microbiol.* 13:655–662.

64. Løbner-Olesen, A., F. G. Hansen, K. V. Rasmussen, B. Martin, and P. L. Kuempel. 1994. The initiation cascade for chromosome replication in wild-type and Dam methyltransferase deficient *Escherichia coli* cells. *EMBO J.* 13:1856–1862.

65. Løbner-Olesen, A. and P. L. Kuempel. 1992. Chromosome partitioning in *Escherichia coli. J. Bacteriol.* 174:7883–7889.

66. Louarn, J. M., J. Louarn, V. François, and J. Patte. 1991. Analysis and possible role of hyperrecombination in the termination region of the *Escherichia coli* chromosome. *J. Bacteriol.* 173:5097–5104.

67. Luttinger, A. L., A. L. Springer, and M. B. Schmid. 1991. A cluster of genes that affects nucleoid segregation in *Salmonella typhimurium. New Biol.* 3:687–697.

68. Lynch, A. S., and J. C. Wang. 1993. Anchoring of DNA to the bacterial cytoplasmic membrane through cotranscriptional synthesis of polypeptides encoding membrane proteins or proteins for export: a mechanism of plasmid hypernegative supercoiling in mutants deficient in DNA topoisomerase I. *J. Bacteriol.* 175:1645–1655.

69. Maddock, J. R., M. R. K. Alley, and L. Shapiro. 1993. Polarized cells, polar actions. *J. Bacteriol.* 175:7125–7129.

70. Maddock, J. R., and L. Shapiro. 1993. Polar location of the chemoreceptor complex in the *Escherichia coli* cell. *Science* 259:1717–1723.

71. Maki, S., S. Takiguchi, T. Miki, and T. Horiuchi. 1992. Modulation of DNA supercoiling activity of *Escherichia coli* DNA gyrase by F plasmid proteins. Antago-

72. Mason, D. J., and D. M. Powelson. 1956. Nuclear division as observed in live bacteria by a new technique. *J. Bacteriol.* 71:474–479.

73. McCulloch, R., L. W. Coggins, S. D. Colloms, and D. J. Sherratt. 1994. Xer-mediated site-specific recombination at *cer* generates Holliday junctions in vivo. *EMBO J.* 13:1844–1855.

74. Miki, T., J. A. Park, K. Nagao, N. Murayama, and T. Horiuchi. 1992. Control of segregation of chromosomal DNA by sex factor F in *Escherichia coli*. Mutants of DNA gyrase subunit A suppress *letD* (*ccdB*) product growth inhibition. *J. Mol. Biol.* 225:39–52.

75. Minden, J. S., and K. J. Marians. 1986. *Escherichia coli* topoisomerase I can segregate replicating pBR322 daughter DNA molecules in vitro. *J. Biol. Chem.* 261:11906–11917.

76. Motallebi-Veshareh, M., D. A. Rouch, and C. M. Thomas. 1990. A family of ATPases involved in active partitioning of diverse bacterial plasmids. *Mol. Microbiol.* 4:1455–1463.

77. Mukherjee, A., K. Dai, and J. Lutkenhaus. 1993. *Escherichia coli* cell division protein FtsZ is a guanine nucleotide binding protein. *Proc. Natl. Acad. Sci. USA* 90:1053–1057.

78. Mukherjee, A., and J. Lutkenhaus. 1994. Guanine nucleotide-dependent assembly of FtsZ into filaments. *J. Bacteriol.* 176:2754–2758.

79. Mulder, E., M. ElBouhali, E. Pas, and C. L. Woldringh. 1990. The *Escherichia coli minB* mutation resembles *gyrB* in defective nucleoid segregation and decreased negative supercoiling of plasmids. *Mol. Gen. Genet.* 221:87–93.

80. Mulder, E., and C. L. Woldringh. 1989. Actively replicating nucleoids influence positioning of division sites in *Escherichia coli* filaments forming cells lacking DNA. *J. Bacteriol.* 171:4303–4314.

81. Niki, H., R. Imamura, M. Kitaoka, K. Yamanaka, T. Ogura, and S. Hiraga. 1992. *E. coli* MukB protein involved in chromosome partition forms a homodimer with a rod-and-hinge structure having DNA binding and ATP/GTP binding activities. *EMBO J.* 11:5101–5109.

82. Norris, V., T. Alliotte, A. Jaffe, and R. D'Ari. 1986. DNA replication termination in *Escherichia coli parB* (a *dnaG* allele), *parA*, and *gyrB* mutants affected in DNA distribution. *J. Bacteriol.* 168:494–504.

83. Ogden, G. B., M. J. Pratt, and M. Schaechter. 1988. The replicative origin of the *E. coli* chromosome binds to cell membranes only when hemimethylated. *Cell* 54:127–35.

84. Ogg, J. E., and M. R. Zelle. 1957. Isolation and characterization of a large cell possibly polyploid strain of *Escherichia coli. J. Bacteriol.* 74:477–484.

85. Okada, Y., M. Wachi, A. Hirata, K. Suzuki, K. Nagai, and M. Matsuhashi. 1994. Cytoplasmic axial filaments in *Escherichia coli* cells: possible function in the mechanisms of chromosome segregation and cell division. *J. Bacteriol.* 176:917–922.

86. Peng, H., and K. J. Marians. 1993. Decatenation activity of topoisomerase IV during *oriC* and pBR322 DNA replication *in vitro. Proc. Natl. Acad. Sci. USA* 90:8571–8575.

87. RayChaudhuri, D., and J. T. Park. 1992. *Escherichia coli* cell-division gene *ftsZ* encodes a novel GTP-binding protein. *Nature* (London) 359:251–254.

88. Ricard, M., and Y. Hirota. 1973. Process of cellular division in *Escherichia coli*: physiological study on thermosensitive mutants defective in cell division. *J. Bacteriol.* 116:314–322.

89. Rinken, R., B. Thoms, and W. Wackernagel. 1992. Evidence that *recBC*-dependent degradation of duplex DNA in *Escherichia coli recD* mutants involves DNA unwinding. *J. Bacteriol.* 174:5424–5429.

90. Rothfield, L. I. 1994. Bacterial chromosome segregation. *Cell* 77:963–966.

91. Ryter, A., Y. Hirota, and F. Jacob. 1968. DNA-membrane complex and nuclear segregation in bacteria. *Cold Spring Harbor Symp. Quant. Biol.* 33:669–676.

92. Sanchez, M., A. Valencia, M.-J. Ferrandiz, C. Sander, and M. Vicente. 1994. Correlation between the structure and biochemical activities of *ftsA*, an essential cell division protein of the actin family. *EMBO J.* 13:4919–4925.

93. Schaechter, M. 1990. The bacterial equivalent of mitosis, p. 313–322. *In* K. Drlica and M. Riley (ed.), *The Bacterial Chromosome*. American Society for Microbiology, Washington, D.C.

94. Schaechter, M., and V. O. Lang. 1961. Direct observation of fusion of bacterial nuclei. *J. Bacteriol.* 81:667–668.

95. Schaechter, M., and U. von Freiesleben. 1993. The equivalent of mitosis in bacteria, p. 61–73. *In* J. Heslop-Harrison and R. B. Flavell (ed.), *The Chromosome*. BIOS Scientific Publishers, Ltd., Oxford.

96. Schaechter, M., J. P. Williamson, J. R. Hood, and A. L. Koch. 1962. Growth, cell and nuclear divisions in some bacteria. *J. Gen. Microbiol.* 29:421–434.

97. Schmid, M. B. 1990. A locus affecting nucleoid segregation in *Salmonella typhimurium. J. Bacteriol.* 172:5416–5424.

98. Shannon, K. P., B. G. Spratt, and R. J. Rowbury. 1972. Cell division and the production of cells lacking nuclear bodies in a mutant of *Salmonella typhimurium. Mol. Gen. Genet.* 118:185–197.

99. Shellman, V. L., and D. E. Pettijohn. 1991. Introduction of proteins into living bacterial cells: distribution of labeled HU protein in *Escherichia coli. J. Bacteriol.* 173:3047–3059.

100. Skarstad, K., and E. Boye. 1993. Degradation of individual chromosomes in *recA* mutants of *Escherichia coli. J. Bacteriol.* 175:5505–5509.

101. Skarstad, K., E. Boye, and H. B. Steen. 1986. Timing of initiation of chromosome replication in individual *Escherichia coli* cells. *EMBO J.* 5:1711–1717.

nistic actions of LetA (CcdA) and LetD (CcdB) proteins. *J. Biol. Chem.* 267:12244–12251.

102. Skarstad, K., H. B. Steen, and E. Boye. 1985. *Escherichia coli* DNA distributions measured by flow cytometry and compared with theoretical computer simulations. *J. Bacteriol.* **163**:661–668.

103. Spratt, B. G., and R. J. Rowbury. 1971. Physiological and genetical studies on a mutant of *Salmonella typhimurium* which is temperature-sensitive for DNA synthesis. *Mol. Gen. Genet.* **114**:35–49.

104. Springer, A. L., and M. B. Schmid. 1993. Molecular characterization of the *Salmonella typhimurium parE* gene. *Nucleic Acids Res.* **21**:1805–1809.

105. Steck, T. R., and K. Drlica. 1984. Bacterial chromosome segregation: evidence for DNA gyrase involvement in decatenation. *Cell* **36**:1081–1088.

106. Taschner, P. E., J. G. Verest, and C. L. Woldringh. 1987. Genetic and morphological characterization of *ftsB* and *nrdB* mutants of *Escherichia coli. J. Bacteriol.* **169**:19–25.

107. Tetart, F., R. Albigot, A. Conter, E. Mulder, and J. P. Bouche. 1992. Involvement of FtsZ in coupling of nucleoid separation with septation. *Mol. Microbiol.* **6**:621–627.

108. Trun, N. J., and S. Gottesman. 1990. On the bacterial cell cycle: *Escherichia coli* mutants with altered ploidy. *Genes Dev.* **4**:2036–2047.

109. Trun, N. J., and S. Gottesman. 1991. Characterization of *Escherichia coli* mutants with altered ploidy. *Res. Microbiol.* **142**:195–200.

110. Trun, N. J., S. Gottesman, and A. Lobner-Olesen. 1991. Analysis of *Escherichia coli* mutants with altered DNA content. *Cold Spring Harbor Symp. Quant. Biol.* **56**:353–358.

111. van Helvoort, J. M. L. M., and C. L. Woldringh. 1994. Nucleoid partitioning in *Escherichia coli* during steady-state growth and upon recovery from chloramphenicol treatment. *Mol. Microbiol.* **13**:577–583.

112. Vinella, D., A. Jaffe, R. D'Ari, M. Kohiyama, and P. Hughes. 1992. Chromosome partitioning in *Escherichia coli* in the absence of *dam*-directed methylation. *J. Bacteriol.* **174**:2388–2390.

113. Woldringh, C. L. 1976. Morphological analysis of nuclear separation and cell division during the life cycle of *Escherichia coli. J. Bacteriol.* **125**:248–257.

114. Woldringh, C. L. 1994. Significance of plasmolysis spaces as markers for periseptal annuli and adhesion sites. *Mol. Microbiol.* **14**:597–607.

115. Woldringh, C. L., P. Hus, E. Pas, G. J. Brakenhoff, and N. Nanninga. 1987. Topography of peptidoglycan synthesis during elongation and polar cap formation in a cell division mutant of *Escherichia coli* MC4100. *J. Gen. Microbiol.* **133**:575–586.

116. Woldringh, C. L., and N. Nanninga. 1985. Structure of the nucleoid and cytoplasm in the intact cell, p. 161–197. *In* N. Nanninga (ed.), *Molecular Cytology of Escherichia coli.* Academic Press, London.

117. Woldringh, C. L., J. A. C. Valkendurg, E. Pas, P. E. M. Taschner, P. Huls, and F. B. Wientjes. 1985. Physiological and geometrical conditions for cell division in *Escherichia coli. Ann. Inst. Pasteur Microbiol.* **1**:131–138.

118. Wu, L. J., and J. Errington. 1994. *Bacillus subtilis* SpoIIIE protein required for DNA segregation during asymmetric cell division. *Science* **264**:572–575.

119. Yamanaka, K., T. Mitani, T. Ogura, H. Niki, and S. Hiraga. 1994. Cloning, sequencing, and characterization of multicopy suppressors of a *mukB* mutation in *Escherichia coli. Mol. Microbiol.* **13**:301–312.

120. Zyskind, J., A. L. Svitil, W. B. Stine, M. C. Biery, and D. W. Smith. 1992. RecA protein of *Escherichia coli* and chromosome partitioning. *Mol. Microbiol.* **6**:2525–2537.

121. Zyskind, J. W., J. M. Cleary, W. S. A. Brusilow, N. E. Harding, and D. W. Smith. 1983. Chromosomal replication origin from the marine bacterium *Vibrio harveyi* functions in *Escherichia coli: oriC* consensus sequence. *Proc. Natl. Acad. Sci. USA* **80**:1164–1168.

Morphological and Physiological Changes during Stationary Phase

GJALT W. HUISMAN, DEBORAH A. SIEGELE,
MARÍA M. ZAMBRANO, AND ROBERTO KOLTER

106

INTRODUCTION

Upon depletion of essential nutrients from the medium, the growth rate of a bacterial culture slows down and evenually reaches zero. At this point the culture has entered stationary phase. The term "stationary phase" should be understood as an operational term indicating that the culture is displaying no net increase in cell number. However, the physiology of cells in a stationary-phase culture is heterogeneous and depends completely on the conditions that resulted in the cessation of growth. Starvation for an essential nutrient, perhaps most often the carbon source, has been used experimentally as a tool to study the physiology of nongrowing cells. Changes resulting from carbon source starvation, however, should not be assumed to hold for other forms of entry into stationary phase. When the term stationary phase is used here, the reader should keep in mind that the observation described may only apply for the particular conditions used in each experiment and that generalizations may not be valid.

The transition from rapid growth with a generation time of less than 20 min to stationary phase is accompanied by adaptation of the culture to the new conditions. During growth, energy-demanding processes convert nutrients primarily to nucleic acids, proteins, and other cell constituents to enable cell growth and cell division. Rapidly growing cells rely on inducible responses to express defense mechanisms against particular stress conditions. In stationary phase, energy generation is usually limited (79), making inducible responses less immediate and less effective. It thus makes sense for organisms such as *Escherichia coli* and *Salmonella typhimurium* (*Salmonella enterica* serovar Typhimurium) to develop increased resistances against many potential environmental assaults at the onset of starvation. Such a program is advantageous to the cell because it can cope with environmental stresses instantaneously without requiring new protein synthesis. In this way, precious energy equivalents are preserved and chances for survival are not impaired.

A possible consequence of the development of a highly resistant state is that resumption of growth may take more time since the protective measures have to be dismantled. Within a population, cells that have not fully expressed their increased resistances might respond more quickly to conditions supporting growth and could have an advantage in resuming growth. The identification of mutants that take over stationary-phase cultures supports this possibility (178). One potential result is that a culture that is repeatedly switched from starvation to growth conditions loses its clonal identity and evolves into a mixture of cell genotypes. Some conditions appear to select mutants that can take over the population (178), while other conditions select for balanced mixtures of cells that complement each other (135). The observations that starved cultures can quickly lose their clonality should undoubtedly warn workers in this field to exercise extreme care when storing bacterial strains in agar stabs and plates.

This chapter describes the changes that *E. coli* and *S. typhimurium* undergo when starved. We summarize the changes (schematically represented in Fig. 1) that occur in the different subcellular compartments and in the genome and describe genes that have been found to be essential to survive in stationary phase. Several additional reviews describing bacteria in stationary phase and gene expression in slowly growing cells have appeared recently (65, 72, 81, 101, 102, 145).

PHENOTYPIC CHANGES DURING STATIONARY PHASE

The Envelope: Inner Membrane

The inner membrane of gram-negative cells forms the most important barrier between the cytoplasm and the extracellular medium. This membrane is also a key component of the transport mechanisms that allow the selective flow of molecules into and out of the cell and is the main facilitator in maintaining the proton motive force. The membrane protects the cell from extracellular chemical and physical assaults and is the major scaffold on which the cell shape-determining peptidoglycan molecule is synthesized. It is therefore not surprising that this cellular component undergoes major changes when cells enter stationary phase.

The fatty acid composition of the membrane changes dramatically in response to starvation. The membranes of both *E. coli* and *S. typhimurium* contain monounsaturated fatty acids, predominantly palmitoleic acid ($C_{16:1}$) and oleic acid ($C_{18:1}$) (chapter 37, this book). The relative abundance of these unsaturated fatty acids declines more than 10-fold during starvation, from almost 50% to less than 5% (42). Until recently, no polyunsaturated fatty acids had been observed in *E. coli* or *S. typhimurium*; they had only been analyzed during exponential phase or early stationary phase. A recent study of fatty acid content after 72 h of carbon starvation indicated the presence of linoleic acid (18:2), a polyunsaturated fatty acid (126); however, this report could not be confirmed (26). The role of unsaturated fatty acids in the membrane during stationary phase remains unknown, although an unsaturated fatty acid auxotrophic mutant was shown to develop extreme sensitivity to low osmolarity exclusively when starved (140).

The reduction of monounsaturated fatty acids observed at the onset of starvation is accompanied by a corresponding increase in the levels of cyclopropyl fatty acid derivatives (157). This increase of cyclopropyl derivatives from 10% during growth to nearly 50% of total fatty acid is accounted for by C_{17} and C_{19} cyclopropane fatty acids (174). The conversion of unsaturated fatty acids to their cyclopropyl derivatives is catalyzed by cyclopropyl fatty acid synthase, a membrane-bound enzyme encoded by the *cfa* gene (157, 168). Synthesis of cyclopropyl fatty acids appears to be oxygen sensitive because increased aeration leads to a decreased cyclopropyl fatty acid formation (116). Mutants lacking *cfa* have been constructed and subjected to many different extreme environments and treatments, such as prolonged incubation in stationary phase or exposure to drying, detergents, heavy metals, low pH, or high salt concentrations (157). Such mutants showed no clear phenotype except for being less resistant to repeated cycles of freezing and thawing (63).

The phospholipid head group composition has also been found to change in response to starvation. The levels of phosphatidylglycerol and phosphatidylserine decrease in stationary phase, whereas cardiolipin levels increase (1, 104). A gene important for cardiolipin synthesis (*cls*) has been cloned from *E. coli* and analyzed (113). As is the case for mutants lacking *cfa*, no growth phenotypes have been found for mutants lacking *cls*, however, such strains still had low levels of cardiolipin, probably synthesized by phosphatidylserine synthase (encoded by *pss*). Attempts to introduce a *cls* mutation into a *pss*-null strain have been unsuccesful, suggesting that cardiolipin is essential for suvival (113).

While no firm conclusions can be drawn regarding the physiological consequences of these growth phase-dependent changes in membrane composition, fluorescence polarization studies suggest that membranes from stationary-phase cells are in a highly ordered state that may render them more damage resistant (149, 150). It has been suggested that this highly ordered state may result not only from changes in lipid composition of the membrane, but also from starvation-induced increases in the levels of polyamines (151) and allysine crosslinks among membrane proteins (106).

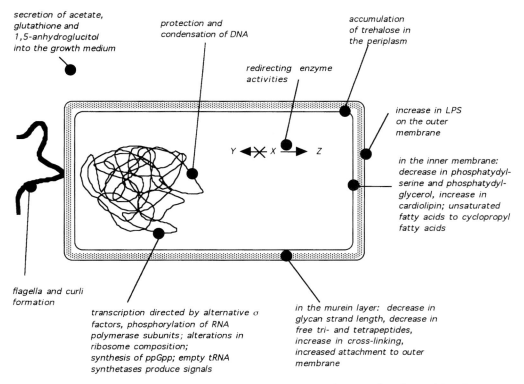

FIGURE 1 Summary of phenotypic changes in stationary-phase *E. coli* and *S. typhimurium*.

The Envelope: Outer Membrane

The composition of the outer membrane also changes at the onset of stationary phase; the amount of exposed lipopolysaccharide on the outer surface of the outer membrane increases (75), thus increasing the charge at the surface of the cell and thereby possibly increasing its resistance to membrane-damaging agents. Additionally, more compounds containing sulfhydryl substituents are directed to the outer membrane and eventually excreted (93, 94, 120, 147). Whether this phenomenon serves as a protective mechanism remains to be elucidated.

There is an overall decrease in the amount of proteins in the outer membrane (6); however, synthesis of some outer membrane-associated proteins is induced during stationary phase. Important new outer membrane proteins that are synthesized include flagellin (10), the curli subunit CsgA (118), and the Pap-related pili encoded by *htrE* and *ecpD* (127). Flagellin synthesis peaks at the onset of stationary phase, but then declines rapidly (10). The synthesis of attachment organelles and increased motility can be seen both as important strategies to search for alternative nutrient sources and to escape conditions that do not support growth. Chemotaxis is by far much more energy-consuming than surface attachment. The cells' diminished proton motive force decreases their ability to generate ATP (79), which may be the reason why flagellar synthesis is only transiently turned on at the onset of stationary phase in batch cultures (10).

In contrast, surface attachment organelles such as curli pili can provide the cell, at a lower cost in energy, with a physically secure location in which to survive long periods of nutrient scarcity. Attachment to host cells, in conjunction with toxin production or cell entry, can also be a way to increase the local nutrient concentration. However, conditions within the host cannot be directly compared to those in batch cultures. It is possible that to successfully contact and invade eukaryotic cells, organisms such as *S. typhimurium* alter their surface composition by alternating the expression of genes, some of which are expressed during growth in culture (82, 98) while others are induced during stationary phase (46, 64, 114).

At the periplasmic side of the outer membrane, the amount of lipoprotein bound to the peptidoglycan layer increases during the transition from logarithmic growth to stationary phase (170). Recently, a gene encoding a new lipoprotein (*nlp*) was described (74, 87) which, interestingly, is organized in an operon with *rpoS*, the gene encoding the stationary phase-specific sigma factor (156). The increased connection between wall and outer membrane may reflect the cell's need to protect its outer structure under harsh environments. In doing so it may sacrifice some flexibility that could be useful during rapid growth, but gain in the overall stability of the envelope.

The Envelope: Peptidoglycan and Periplasm

The peptidoglycan of *E. coli* and *S. typhimurium* consists of a single molecule in which an oligosaccharide is cross-linked by specific oligopeptides (chapter 6). During growth the peptidoglycan wall consists of two to three layers (representing 0.75% of the cellular weight) which are increased to four to five layers during stationary phase (90, 105). In addition to the increase in thickness, composition also changes during stationary phase. The length of the glycan strands decreases and the amounts and type of peptide constituents change. Free tri- and tetrapeptides

decrease with a corresponding increase in peptide-peptide and peptide-lipoprotein cross-linking (122, 170).

The changes in peptidoglycan structure are not accompanied by a net increase in wall components, but are due to a turnover process in which the penicillin-binding proteins (PBPs) play an important role (17). During starvation the translocation of cell wall precursors across the membrane as isoprenoid lipid-linked intermediates slows down (50). The activities of some peptidoglycan biosynthetic enzymes and the levels of PBPs change (105, 163). The specific activities of the D-glutamic acid and D-alanyl-D-alanine-adding enzymes vary little with growth phase, whereas the specific activities of UDP-N-acetylglucosamine transferase and UDP-N-acetylglucosamine-enolpyruvate reductase and the diaminopimelic acid-adding enzymes decrease 20 to 50% in late stationary phase (105). PBP 6 is thought to be a D-alanine carboxypeptidase I that functions in the stabilization of peptidoglycan specifically during stationary phase (20, 163).

The expression of proteins involved in peptidoglycan alterations at the onset of stationary phase is regulated by different factors, including cyclic AMP (cAMP)-CAP, ppGpp, and the stationary phase-specific sigma factor σ^S (4, 18, 28, 40, 85). For example, expression of PBP 2 is regulated by ppGpp (165), but its link to the septation apparatus appears to be mediated by a factor whose expression is cAMP-CAP dependent (28). Expression of both PBP 3 and PBP 6 is σ^S regulated. When cells enter stationary phase, levels of PBP 3 decrease, whereas PBP 6 levels increase 2- to 10-fold (40). The increase in PBP 6 is particularly relevant since it has been proposed to play a role in stabilization of stationary-phase peptidoglycan analogous to the sporulation-specific PBP 5a from *Bacillus subtilis* (163).

Other stationary phase-dependent changes in the periplasm include the accumulation of membrane-derived oligosaccharides (83, 138) and trehalose (164), both thought to act as osmoprotectants (15). Expression of the genes encoding enzymes involved in trehalose metabolism is σ^S dependent (67). Trehalose is synthesized by the *otsAB* gene products and degraded by trehalase encoded by *treA* (19, 58, 78). Other periplasmic proteins that are expressed as part of the σ^S regulon are acidic phosphatase, AppA (31, 160), and the *osmY* gene product, whose function remains unknown (175, 176).

Extracellular Environment

When *E. coli* enters stationary phase, a number of metabolites are excreted into the medium. Acetate is the major metabolite that is generally found at the end of exponential phase when cells are grown on glucose and which is rapidly used at the onset of stationary phase (43). Other metabolites that have been found to accumulate when cells cease growing are reduced glutathione and other thiol compounds (93, 94, 120). Recently, the transient accumulation of 1,5-anhydroglucitol early in stationary phase was observed (144). The metabolic pathways for biosynthesis and degradation of this molecule remain unknown, but 1,5-anhydroglucitol has been found to be produced by many different organisms, including humans, suggesting it may be a ubiquitous signaling molecule (144). Another interesting finding is that glucose-grown *cya* or *crp* mutants excrete L-glutamate into the medium upon starvation (91), demonstrating the intricacies of changes in metabolic pathways at the onset of stationary phase.

Several lines of evidence suggest that an external metabolite may trigger the expression of a starvation response by activation of the σ^S regulon. Different reports have suggested that weak acids such as acetate and benzoate induce expression of *rpoS* when growing cells are resuspended in medium lacking a carbon source (109, 141). However, a signaling role for acetate or benzoate in cells entering stationary phase by nutrient depletion has never been shown. In contrast, homoserine lactone has recently been shown to act as an inducer of the σ^S regulon (73). This may be analogous to the inducer role of extracellular acylated homoserine lactone derivatives in a number of cell density-dependent phenomena such as bioluminescence in *Vibrio fischeri* and *Vibrio harveyi*, Ti plasmid transfer in *Agrobacterium tumefaciens*, elastase synthesis in *Pseudomonas aeruginosa*, and antibiotic and exoenzyme biosynthesis in *Erwinia carotovora* (for a recent review, see reference 51). Unmodified homoserine lactone in *E. coli* may be an intracellular switch to signal its nutritional status, or a modified form may be excreted to signal starvation to the population. However, no excreted form of homoserine lactone has been characterized from *E. coli* supernatants yet.

The Cytoplasm: DNA/Nucleoid

During growth in rich medium, *E. coli* has on average more than four chromosomes per cell (111). During the final cell divisions, new rounds of replication are no longer initiated, and at the onset of stationary phase the cells are typically found to have a single chromosome (100). The synthesis of several DNA-binding proteins, including H-NS (or H1a), IHF, and Dps, which regulate gene expression and are involved in compacting or protecting DNA against physical or chemical attack, has been shown to be growth phase regulated (7, 38, 39).

H-NS is a DNA-binding protein that was discovered as a regulator for the expression of several genes (35, 68). Mutants lacking H-NS have extremely pleiotropic phenotypes (69). H-NS expression is autoregulated during logarithmic growth, but this repression ceases when cultures enter stationary phase and σ^s-independent transcription of *hns* is induced 10-fold (38, 45, 161). H-NS compacts DNA both in vivo and in vitro (162). At the *lacUV5* promoter, H-NS stimulates transcription initiation directly, decreasing the rate of open complex formation (152).

There appears to be a connection between H-NS DNA binding and the differential promoter recognition by RNA polymerase with σ^{70} and σ^S. At subsaturating concentrations, H-NS has specificity for curved DNA (119, 158, 171), and some σ^S-dependent promoters have intrinsic curvature (44). There are at least two examples of σ^S-dependent promoters, P_{csgA} (12, 117) and P_{mcc} (108), whose σ^S dependence can be relieved by null mutations in *hns*. This suggests that H-NS may keep these promoters in a conformation that prevents σ^{70}-RNA polymerase from recognizing them. This may explain why, when using linear templates in in vitro transcription assays, promoter discrimination by $E\sigma^{70}$ and $E\sigma^S$ has not been clearly observed (156). The *proU* promoter may be another example of the interplay between H-NS and σ^S since H-NS regulates the level of P_{proU} supercoiling (68) and *proU* expression is controlled by σ^S in stationary phase (99).

Integration host factor (IHF) was originally identified because of its role in λ integrative recombination, and was later shown to be a pleiotropic regulator of gene expression (49). In contrast to H-NS, IHF binds to specific sites in the *E. coli* genome and induces bending of the DNA (60). In contrast to

this low number of specific IHF-binding sites, Ditto et al. (39) have reported that the IHF copy number during growth is around 10,000 to 15,000, which increases six times when cells enter stationary phase. It may be that IHF, like H-NS, occupies specific sites at low concentrations but loses specificity at higher concentrations. Several reports have indicated that IHF can compensate for the absence of another highly expressed DNA-binding protein, HU (39). HU is involved in multiple cellular processes (136), and its expression with respect to growth phase regulation has not been reported.

Dps was discovered as a protein synthesized to high levels after *E. coli* had been in stationary phase for over 3 days (7). Dps appears to bind DNA nonspecifically (possibly as a dodecamer) and may prevent oxidative damage caused by hydrogen peroxide (155). However, in addition to this protective role, a mutation in *dps* has a pleiotropic effect on the expression of several proteins as judged by two-dimensional gel electrophoresis. *dps* is expressed in stationary phase as part of the σ^S regulon, but its expression also depends on IHF (8). Generally *dps* is not expressed during growth but is induced after hydrogen peroxide treatment in an OxyR-dependent manner (8).

The Cytoplasm: Protein Synthesis

At the transition into stationary phase, changes occur in the transcriptional and translational processes that have a direct impact on the physiology of the cell. Besides the appearance of the new sigma factor, σ^S, the core of RNA polymerase is modified, apparently by phosphorylation (121). Reduction in the number of ribosomes may be a consequence of their being scavenged for nutrients, and starvation also results in temporarily empty acceptor sites at the ribosomes, which initiates the *relA*-dependent synthesis of ppGpp (chapter 92). In addition, carbon starvation causes a *relA*-independent increase in ppGpp, probably due to the action of the *spoT* gene product. Increased ppGpp levels induce the stringent response and affect many cellular processes including increasing σ^S levels (55). Perhaps to preserve ribosomes, *E. coli* synthesizes a ribosome modulating factor, encoded by *rmf*, that is involved in the dimerization of ribosomes (167). Absence of the 55-amino-acid Rmf protein reduces the viability of the strain in stationary phase. Expression of *rmf* increases 15-fold upon entry into stationary phase but is *rpoS* independent (172).

As the rate of protein synthesis decreases during starvation, key signal molecules may be synthesized as a natural consequence of imbalances in the pools of free amino acids and the transient accumulation of intermediates in amino acid biosynthesis (73). Several tRNA synthetases have been shown to bind such intermediates and generate cyclic compounds (76). While these noncognate reactions have been studied from the perspective of proofreading or editing during the charging reaction, it is possible that they serve a physiological role in generating starvation-induced signals. For example, methionine-tRNA synthetase binds homocysteine, adenylates it, and releases AMP and homocysteine thiolactone (76) (Fig. 2). Similarly, isoleucine-, valine-, and lysine-tRNA synthetases bind homoserine (an intermediate in methionine, threonine, and isoleucine biosynthesis [chapters 27, 32, and 33]) and generate homoserine lactone in vitro (76) (Fig. 2). This latter observation is particularly significant because acylated homoserine lactone derivatives are ubiquitous high-cell-density signaling molecules in bacteria

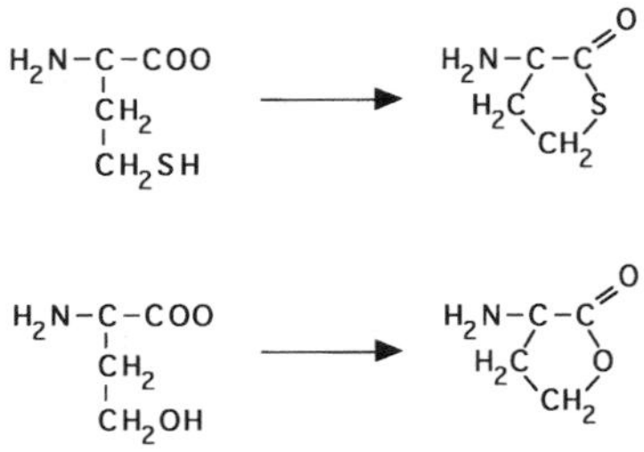

homocysteine thiolactone

homoserine lactone

FIGURE 2 Synthesis of homocysteine thiolactone and homoserine lactone is catalyzed by tRNA synthetases.

(51), and homoserine lactone itself has been shown to be involved in induction of *rpoS* expression (73).

In addition to amino acid levels, tRNA concentrations also fluctuate with growth phase. *E. coli* has five different tRNA genes that specify leucine codons. The *leuX* tRNA recognizes one of the minor *leu* codons, and its expression increases with decreasing growth rate (137). Strains that harbor an amber suppressor allele of *leuX* (*supP*) have a pleiotropic phenotype in stationary phase. For example, induction of microcin B17 production is greatly decreased (E. I. Vivas and R. Kolter, unpublished data), and such strains grow well within the mouse intestine but cannot be maintained there (112). Interestingly, such *supP* mutant strains can colonize the intestine. As long as they are increasing in numbers within the animal they behave as wild-type strains. But once bacterial counts no longer increase, the *supP* strain is lost. These effects are specific for *supP*; amber suppressor alleles of other tRNAs had no effect in this system. It may be that the frequency and position of rare codons is a general regulatory mechanism for achieving growth phase-dependent control of gene expression (23).

The Cytoplasm: Metabolic Circuits

At the onset of stationary phase the bacterial cell redirects its physiology in order to survive. The proton motive force is the main engine in growing cells for energy generation. At the onset of stationary phase, however, the proton motive force declines slightly and energy generation proceeds at a slower rate (79). Redirecting metabolism to scavenge any potential nutrient from the medium or within the cell may increase survival.

Cells growing aerobically express primarily one cytochrome oxidase, encoded by the *cyo* locus (14). Continuation of the efficient utilization of oxygen as terminal electron acceptor during stationary phase is governed by the expression of two alternative cytochrome oxidases, encoded by *cyd* (61) and *cyx* or *appAB* (32). Transcription of *cyd* depends on *arcA/arcB* (56), and transcription of *cyx* depends on *rpoS* (32). The significance of having three different terminal oxidation systems is underscored by the finding that a *cyo cyd* double mutant is less sensitive to oxygen than a mutant that lacks all three oxidases (32). Apparently, evolutionary pressure has resulted in *E. coli* maintaining three independently regulated systems for survival under aerobic conditions. While the subtle physiological consequences of expressing different cytochrome oxidases during starvation remain unclear, it is interesting that a mutant defective in the proper assembly of cytochrome *d* is unable to reinitiate growth under aerobic conditions at 37°C (146).

As discussed above, *E. coli* excretes metabolites during growth that are taken up again and used at the onset of stationary phase. When *E. coli* grows on glucose it excretes large amounts of acetate (up to 85 mM) (96). This apparent waste of nutrients is a result of an inefficient tricarboxylic acid (TCA) cycle in which the α-ketoglutarate dehydrogenase complex is inactive (9). Consequently, the TCA cycle is not a cyclic process under those conditions and instead proceeds as two pathways: one from oxaloacetate to succinate and the other from oxaloacetate and acetyl coenzyme A (acetyl CoA) to α-ketoglutarate (9). Apparently, *E. coli* does not need all the energy that can be obtained from glucose and saves some of it in the growth medium in the form of acetate until later stages where carbon becomes limiting. Upon entrance into stationary phase, *E. coli* has to conserve energy and carbon sources, and those reactions that waste carbon building blocks are redirected (89). For instance, the complete TCA cycle runs, and to prevent the complete loss of carbon in the form of CO_2 during the conversion of isocitrate via α-ketoglutarate to succinyl CoA the glyoxylate shunt is activated (chapter 21). In this pathway isocitrate is converted to succinate and glyoxylate rather than being converted to α-ketoglutarate by isocitrate dehydrogenase. Glyoxylate then reenters the TCA cycle by condensing with acetyl CoA to form malate. This anapleurotic route saves carbon and thus is expected to be important during stationary phase. The fact that addition of acetate to stationary-phase cultures leads to loss of isocitrate dehydrogenase activity supports this idea (53, 54). Regulation of isocitrate dehydrogenase activity by phosphorylation/dephosphorylation is performed by the *aceK* gene product (88).

Studies on expression of *katE*, the gene encoding catalase HPII, suggested that TCA cycle intermediates may play a pivotal role in sensing the growth stage of *E. coli* (95). This gene is under the control of σ^s, and its expression increases 10-fold at the onset of stationary phase. However, a similar increase is found when *E. coli* is grown in minimal medium on TCA cycle intermediates. Whether this increase reflects a relation between growth rate, which is lower on these carbon sources, and σ^s expression, or is a consequence of the presence of TCA cycle intermediates, remains to be elucidated.

Recently there have been numerous reports of genes whose products are involved in metabolic pathways in stationary-phase cells. In connection to efficient metabolism of acetate, the identification of the gene that encodes pyruvate-formate lyase (*pfl*) (128) and the gene encoding pyruvate oxidase (*pox*) (22) as being expressed preferentially in stationary phase underscores the importance of central metabolic pathways in stationary-phase cells.

S-Adenosylmethionine (SAM) is perhaps the most important methyl donor in the cell (chapter 33), and its synthesis is of vital importance for the cell. The enzymes involved in SAM biosynthesis represent an example of redundant metabolic functions. Two adjacent genes, *metK* and *metX*, encode SAM synthases (139). MetK is preferentially active during growth in rich medium, and MetX is expressed during growth in poor medium and during stationary phase. Although mutants in any one of the two genes are viable, a double mutant cannot be made without the appearance of revertants.

A well-known phenotype of stationary-phase *E. coli* is its ability to accumulate glycogen as a carbon reserve (30). Glycogen is synthesized from glucose in a three-step pathway involving glucose kinase, glucose-6-phosphate ADP transferase, and glycogen synthase. Branching enzymes can introduce 1,6-glycosidic bonds that result in a highly branched polyglucose molecule. Glycogen synthesis is regulated at several different levels by (i) allosteric interactions of metabolites with glucose-6-phosphate UDP transferase (123, 124); (ii) genetic regulation of the *glgCAB* operon by cAMP, CRP, and ppGpp (133); (iii) regulation of expression of the putative priming protein GlgS by *rpoS* (66); and (iv) the product of the *csr* gene (132). Expression of the first gene in the gluconeogenic pathway, *pckA*, is also growth phase regulated and regulated by catabolite repression (59).

A second important storage polymer in stationary-phase *E. coli* is polyphosphate. Polyphosphate kinase catalyzes the elongation reaction in which the γ-phosphate group from ATP is added to an existing polyphosphate molecule (2). A polyphosphatase is involved its degradation (3). Strains deficient in polyphosphate accumulation are impaired in survival during prolonged periods of starvation (27). Polyphosphate is proposed to play an essential role in the uptake of DNA by *E. coli* in the transformation process as part of a complex with poly(3-hydroxybutyrate) (71, 130). Recently, it has been shown that the polyphosphate/poly(3-hydroxybutyrate) complex functions as a voltage-gated Ca^{2+} channel reminiscent of the Ca^{2+} channels found in eukaryotes (R. N. Reusch and L. Bramble, personal communication). The σ^S-dependent appearance of this complex in stationary phase also suggests a role for this structure under nongrowing conditions (G. W. Huisman and R. Kolter, unpublished data).

Upon exhaustion of a carbon source, *E. coli* expresses several systems dedicated to the uptake of sugars such as galactose, ribose, and maltodextrins (5, 34). The induction of such operons, which is due to the accumulation of endogenous inducer and cAMP, improves the sugar-scavenging ability of a nearly starved cell (33). Chemostat cultures operating at low steady-state levels of glucose have been shown to accumulate galactose to levels as high as 0.2 mM, which acts as endogenous inducer of the *mgl* system. The facts that LamB is involved in the uptake of several sugars (34) and that Mgl and LamB both constitute high-affinity glucose uptake systems illustrate how *E. coli* initially responds to diminishing availability of a carbon source. Possibly, the transition from exponential growth to stationary phase is a "hunger" phase in which bacteria turn on high-affinity transport systems for many nutrient sources (33).

When *E. coli* is unable to synthesize amino acids efficiently de novo due to starvation, it degrades endogenous proteins and peptides (129). Degradation of proteins during carbon starvation is accomplished in part by the Clp protease (29). Clp consists of two subunits; expression of the ClpP subunit is growth phase independent, whereas ClpA is expressed when the growth rate slows down (80). This catabolic pathway to convert existing proteins to amino acids and then to new proteins requires energy for which the NADH dehydrogenase I, encoded by the *nuo* operon, is important (11). Mutations in *nuo* also result in a competitive disadvantage in mixed cultures during stationary phase (177) and result in poor growth on some amino acids (125). NADH dehydrogenase I activity confers on the cell a greater efficiency in converting the energy potential from reduced NADH into proton motive force (52). The finding that *nuo* mutants have severe phenotypes during stationary phase,

but virtually no phenotype during rapid growth, indicates that efficient energy generation is much more crucial for starved cells than for rapidly growing cells (177).

Because the degradation of proteins to synthesize new proteins is a burden on the energy capacity of the cell, this system needs to be well controlled. The finding that the *rpoH* regulon, which includes DnaK, is involved in stationary-phase survival suggests that recognition of misfolded and unfolded proteins is important to preserve cell integrity (77). Additionally, DnaK is essential in carbon starvation survival and plays a role in the expression of other starvation-specific proteins (153). During exponential growth, DnaK functions together with DnaJ and GrpE as a molecular chaperone (57). A stationary phase-specific DnaJ homolog (CbpA) has recently been discovered (173). Expression of CbpA is σ^S dependent and may be important in protein turnover in stationary-phase cells. Additionally, stationary-phase *E. coli* express an enzyme that recognizes and repairs proteins in which isoaspartyl residues have been introduced (92). The corresponding gene, *pcm*, has been cloned and characterized and, interestingly, was found to be located just upstream of *rpoS*, possibly in the same operon (74).

To ensure that *E. coli* has adequate reserve material to resume fatty acid biosynthesis as quickly as possible, it preserves the required cofactor. Acyl carrier protein, which is bound to all fatty acids during their biosynthesis, is protected from degradation during stationary phase by the formation of mixed disulfides with glutathione (131). Similarly, the fatty acid catabolic cofactor CoA is joined to glutathione (93).

The Cytoplasm: Regulatory Changes

The regulatory processes on which all of the phenotypic changes described above rely can be divided into two categories, *rpoS* dependent and *rpoS* independent. The *rpoS* regulon is the subject of the chapter by Hengge-Aronis (chapter 93). The second category involves other pathways through which genes are specifically expressed in stationary phase. A number of stationary phase-induced genes have been found to be expressed in a σ^S-independent manner; several are regulated by cAMP-CRP (142). Examples of these are the *glgCAB* operon and *cstA* (66, 143). The increase in cytochrome *d* oxidase expression as cells approach stationary phase is dependent on *arcA/arcB*, which appear to sense the rate of flow of electrons through the electron transport chain (56). Some starvation-induced proteins may respond to increases in ppGpp levels in an *rpoS*-independent fashion. Microcin B17 is an example of a gene whose induction in stationary phase is not dependent on σ^S, cAMP/CRP, or ppGpp (18, 25; D. Siegele, unpublished data).

Several other regulatory genes that respond to starvation or regulate starvation-inducible genes have been identified. Among these are *uspA*, which is induced under every stress condition thus far tested (115); *mprA*, which in high copy represses growth phase induction of the *mcb* and *mcc* operons (36, 37); and *csr*, which encodes a carbon storage regulator (132). Mutations in *csr* affect not only the biosynthesis of glycogen, but also metabolism of certain carbon sources and cell adhesive properties (132). Analogous to *mprA*, the *rspAB* operon was also identified as a locus that when overexpressed represses the expression of a stationary-phase gene, namely, σ^S (73). The similarity of the encoded proteins with catabolic enzymes suggested the involvement of small molecules in starvation signaling. Such approaches are expected to give further insights into the signals that sense the state of individual cells or bacterial cultures in general.

A number of procedures have recently been used to identify genes that are induced in stationary phase or proteins that are specifically expressed in this growth phase. Analysis of new proteins synthesized at the onset of stationary phase has resulted in a general idea of how many proteins are specifically expressed (62, 86), and through reverse genetics, these proteins have been identified and characterized (7). Random *lacZ* fusions have been generated in the *E. coli* chromosome, and those that were induced in stationary phase were selected (*cstA, osmY,* and others) (62, 169). A different procedure was recently reported by Chuang et al. and consisted of probing the ordered Kohara phage library with RNA isolated from stationary-phase or stress-challenged *E. coli* (24).

A characteristic of stationary phase-inducible genes is that often multiple factors are involved in regulating their expression and conditions other than starvation induce their expression. For example, *aidB* (whose function remains unknown) can be induced by alkylating agents in an *ada*-dependent manner, but it is also expressed independently of *ada* in stationary phase (166). Using *lacZ* fusions, Lange et al. (84) observed that Lrp, IHF, and cAMP-CRP all affected the σ^S-dependent expression of *osmY*, a gene that can also be induced by high osmolarity, independent of all these effectors. The *dps* gene, encoding a DNA-binding protein, is transcribed in stationary phase by RNA polymerase with σ^S in conjunction with IHF and can be expressed upon oxidative challenge with hydrogen peroxide during growth in an OxyR-dependent manner, presumably by RNA polymerase with σ^{70} (8). The cross-regulation of gene expression in nongrowing *E. coli* is further demonstrated by the fact that *mprA* is a high-copy repressor of both σ^S-dependent (*mcc* and *proU*) and σ^S-independent (*mcb*) genes (36, 37).

GENOTYPIC CHANGES IN STATIONARY-PHASE *E. COLI* CULTURES

Dynamics of Stationary-Phase Cultures

Prolonged incubation of batch cultures of *E. coli* results in the apparent death (as measured by the loss of plating ability) of some of the cells. Surprisingly, under some conditions the initial death rate slows down after a few days and a fraction of the cells remain able to form colonies for months and even years (154, 159). Studies in which surviving cells of old cultures were mixed with fresh stationary-phase cultures led to the discovery of the GASP (growth advantage in stationary phase) phenotype (178). This phenotype results from mutations that allow some cells to grow when the majority of the population is dying and represents an unexpected strategy of microbial survival during prolonged starvation. Thus, starved *E. coli* cultures are not necessarily static but can undergo population changes as GASP mutants take over the population. Mutations in *rpoS* that result in reduced σ^S activity can confer on cells the GASP phenotype. However, mutations in other genes can also result in the same phenotype.

Not only can mutations in different loci confer the GASP phenotype, but successive rounds of selection can take place during prolonged incubation of a single culture (178). Population changes had been known to occur during continuous growth of *E. coli* cultures (41). This phenomenon, termed periodic selection, is the result of the successive appearance of mutants with a growth advantage over the parent strain during

continuous growth in chemostats (13, 103). However, the population takeovers observed in stationary-phase cultures due to GASP mutations occur much more rapidly than the population shifts reported for continually growing cultures.

Bacteria have often been used as experimental systems to study evolution. The large population numbers and short generation times make them well suited for such studies. Much of the work has focused on the adaptation of bacteria to nutrient-limited conditions. Some of the genetic changes that result in a selective advantage under these conditions include gene duplications (148) and transposition events (107, 110). Continuous subculturing of cells selects for changes that increase cell fitness (16, 70). In most experiments, the fraction of mutants increases slowly and they are able to take over only after prolonged growth, involving hundreds to thousands of generations. This contrasts sharply with the short incubation periods and numbers of generations needed for mutants with a GASP phenotype to take over stationary-phase batch cultures (178). Given that bacteria spend most of their existence under conditions of very low nutrient availability, the discovery of the GASP phenotype should have important implications for studies on microbial ecology and evolution.

The dynamic state of stationary-phase cultures was discovered by mixing differentially marked cell populations. But during prolonged incubation of a "pure" culture, GASP mutants with a competitive advantage replace the original population under the strong selective pressure imposed by starvation. Figure 3 depicts what can be assumed to be occurring in stationary-phase cultures that are seeded with a single population type. An advantageous mutation may occur during growth of the population which allows mutants to utilize more efficiently nutrients released by dying cells. As a result, the original population is rapidly replaced by the fitter strain. In stationary phase the number of viable counts in a culture is the sum of the growing mutants and the survivors of the original population. The kinetics of growth and the resulting population takeover will vary depending on the incubation conditions of the culture, but they have been observed in a variety of media and using many different strains, including fresh clinical isolates and many different bacterial species (A. Tormo, unpublished data; M. M. Zambrano, Ph.D. Thesis, Harvard University, Cambridge, Mass., 1993; E. Zinser and R. Kolter, unpublished data). Variability from culture to culture can also be expected depending on the nature of the particular mutation involved. At a particular point during incubation, a culture is a mixed population of cells determined by the amount of population takeover that has occurred. Additional incubation can bring about successive rounds of selection in which cells with a growth advantage over the parental strain can result in new population changeovers.

Postselection Mutations

One of the most exciting, as well as controversial, subjects dealing with stationary-phase *E. coli* has been the study of the origin of mutants under nonlethal selection conditions. The fluctuation analyses of Luria and Delbrück provided much of the impetus for the dominant hypothesis that mutations arise spontaneously and at random during nonselective growth prior to selective plating (97). However, the number and time of appearance of mutants under nonlethal selections did not fit well with the

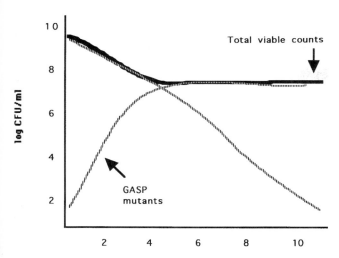

FIGURE 3 Population changes in stationary-phase cultures.

predictions that all mutations arise prior to selection, and several observations of the unexpected appearance of "late-arising" mutations were reported over the years. In 1988, a paper by Cairns et al. brought the question of the origin of mutations in bacteria back to the limelight (21). The observations presented went directly against dogma, suggesting that mutations could also arise among populations of starved cells and, most importantly, that the process appeared to be "adaptive" in that the mutations that arose under those conditions were those that conferred a growth advantage on the cell. Since the publication of that paper, numerous reports have appeared arguing for and against the adaptive nature of these mutations, and many theories have been offered to explain the phenomenon. It is not the intention here to provide a critical review of this literature; this has been done extensively elsewhere (47). Our attempt here is simply to summarize the current state of the field.

The emphasis has recently switched from attempts to develop theories that explain the phenomenon to characterization of the types of mutations arising postselection and the gene products involved in the process. The results from these efforts make it clear that, in contrast to mutations arising during growth, postselection mutagenesis involves the major recombination pathway (RecABC) and often results in simple deletions in homopolymeric runs (48, 134). This indicates that there are underlying mechanistic differences in the origin of mutations in growing versus nongrowing cells.

What about the "adaptive" nature of these mutations? The jury is still out on that question. Suffice it to end this chapter with an editorial note. Most of the work done in this field thus far has made the implicit assumption that the physiological state of a starved bacterium is unchanged whether or not the selective conditions are present. For example, the overall metabolic state of a Lac⁻ cell is assumed to be the same during starvation in the presence or absence of lactose. As more studies of the physiology of nongrowing bacteria are carried out in the future, it may become evident that even though the presence of the selective condition (e.g., lactose) does not allow growth, it changes the physiology of the cell. Such changes may help explain why postselection mutations appear adaptive.

LITERATURE CITED

1. **Ahlers, J., and T. Gunther.** 1975. Phospholipids and ATPase activity of wild-type and ATPase deficient and uncoupled mutants of *E. coli*. *Z. Naturforsch.* 30:832–834.
2. **Ahn, K., and A. Kornberg.** 1990. Polyphosphate kinase from *Escherichia coli*. Purification and demonstration of a phosphoenzyme intermediate. *J. Biol. Chem.* 265:11734–11739.
3. **Akiyama, M., E. Crooke, and A. Kornberg.** 1993. An exopolyphosphatase of *Escherichia coli*: the enzyme and its *ppx* gene in a polyphosphate operon. *J. Biol. Chem.* 268:633–639.
4. **Aldea, M., T. Garrido, C. Hernandez-Chico, M. Vicente, and S. R. Kushner.** 1989. Induction of a growth-phase-dependent promoter triggers transcription of *bolA*, an *Escherichia coli* morphogene. *EMBO J.* 8:3923–3931.
5. **Alexander, D. M., K. Damerau, and A. C. St. John.** 1993. Carbohydrate uptake genes in *Escherichia coli* are induced by carbon starvation. *Curr. Microbiol.* 27:335–340.
6. **Allen, R. J., and G. K. Scott.** 1979. Biosynthesis and turnover of outer-membrane proteins in *Escherichia coli* ML308–225. *Biochem. J.* 182:407–412.
7. **Almirón, M., A. Link, D. Furlong, and R. Kolter.** 1992. A novel DNA binding protein with regulatory and protective roles in starved *E. coli*. *Genes Dev.* 6:2646–2654.
8. **Altuvia, S., M. Almirón, G. Huisman, R. Kolter, and G. Storz.** 1994. The *dps* promoter is activated by OxyR during growth and by IHF and σ^S in stationary phase. *Mol. Microbiol.* 13:265–272.
9. **Amarasingham, C. R., and B. D. Davis.** 1965. Regulation of α-ketoglutarate dehydrogenase formation in *Escherichia coli*. *J. Biol. Chem.* 240:3664–3668.
10. **Amsler, C. D., M. Cho, and P. Matsumura.** 1993. Multiple factors underlying the maximum motility of *Escherichia coli* as cultures enter post-exponential growth. *J. Bacteriol.* 175:6238–6244.
11. **Archer, C. D., X. Wang, and T. Elliott.** 1993. Mutants defective in the energy-conserving NADH dehydrogenase of *Salmonella typhimurium* identified by a decrease in energy-dependent proteolysis after carbon starvation. *Proc. Natl. Acad. Sci. USA* 90:9877–9881.
12. **Arnqvist, A., A. Olsen, and S. Normark.** 1994. σ^S-dependent growth phase induction of the *csgAB* promoter in *Escherichia coli* can be achieved *in vivo* by σ^70 in the absence of the nucleoid-associated protein H-NS. *Mol. Microbiol.* 13:1021–1032.
13. **Atwood, K. C., L. K. Schneider, and F. J. Ryan.** 1951. Periodic selection in *Escherichia coli*. *Proc. Natl. Acad. Sci. USA* 37:146–155.
14. **Au, D. C.-T., R. M. Lorence, and R. B. Gennis.** 1985. Isolation and characterization of an *Escherichia coli* mutant lacking the cytochrome *o* terminal oxidase. *J. Bacteriol.* 161:123–127.
15. **Back, J. F., D. Oakenfull, and M. B. Smith.** 1979. Increased thermal stability of proteins in the presence of sugars and polyols. *Biochemistry* 18:5191–5196.
16. **Bennett, A. F., K. M. Dao, and R. E. Lenski.** 1990. Rapid evolution in response to high temperature-selection. *Nature* (London) 346:79–81.
17. **Blasco, B., A. G. Pisabarro, and M. A. de Pedro.** 1988. Peptidoglycan biosynthesis in stationary-phase cells of *Escherichia coli*. *J. Bacteriol.* 170:5224–5228.
18. **Bohannon, D. E., N. Connell, J. Keener, A. Tormo, M. Espinosa-Urgel, M. M. Zambrano, and R. Kolter.** 1991. Stationary-phase-inducible "gearbox" promoters: differential effects of *katF* mutations and role of σ^70. *J. Bacteriol.* 173:4482–4492.
19. **Boos, W., U. Ehmann, E. Bremer, A. Middendorf, and P. Postma.** 1987. Trehalase of *Escherichia coli*. Mapping and cloning of its structural gene and identification of the enzyme as a periplasmic protein induced under high osmolarity growth conditions. *J. Biol. Chem.* 262:13212–13218.
20. **Buchanan, C. E., and M. O. Sowell.** 1982. Synthesis of penicillin-binding protein 6 by stationary-phase *Escherichia coli*. *J. Bacteriol.* 151:491–494.
21. **Cairns, J., J. Overbaugh, and S. Miller.** 1988. The origin of mutants. *Nature* (London) 335:142–145.
22. **Chang, Y.-Y., A.-Y. Wang, and J. E. Cronan, Jr.** 1994. Expression of *Escherichia coli* pyruvate oxidase (PoxB) depends on the sigma factor encoded by the *rpoS(katF)* gene. *Mol. Microbiol.* 11:1019–1028.
23. **Chen, G. F., and M. Inouye.** 1990. Suppression of the negative effect of minor arginine codons on gene expression; preferential usage of minor codons within the first 25 codons of the *Escherichia coli* genes. *Nucleic Acids Res.* 18:1465–1473.
24. **Chuang, S. E., D. L. Daniels, and F. R. Blattner.** 1993. Global regulation of gene expression in *Escherichia coli*. *J. Bacteriol.* 175:2026–2036.
25. **Connell, N., Z. Han, F. Moreno, and R. Kolter.** 1987. An *E. coli* promoter induced by the cessation of growth. *Mol. Microbiol.* 1:195–201.
26. **Cronan, J. E., Jr., and C. O. Rock.** 1994. The presence of linoleic acid in *Escherichia coli* cannot be confirmed. *J. Bacteriol.* 176:3069–3071.
27. **Crooke, E., M. Akiyama, N. N. Rao, and A. Kornberg.** 1994. Genetically altered levels of inorganic polyphosphate in *Escherichia coli*. *J. Biol. Chem.* 269:6290–6295.
28. **D'Ari, R., A. Jaffe, P. Bouloc, and A. Robin.** 1988. Cyclic AMP and cell division in *Escherichia coli*. *J. Bacteriol.* 170:65–70.
29. **Damerau, K., and A. C. St. John.** 1993. Role of Clp protease subunits in degradation of carbon starvation proteins in *Escherichia coli*. *J. Bacteriol.* 175:53–63.
30. **Damotte, M., J. Cattaneo, N. Sigal, and J. Puig.** 1968. Mutants of *Escherichia coli* K12 altered in their ability to store glycogen. *Biochem. Biophys. Res. Commun.* 32:916–920.

31. Dassa, E., M. Cahu, B. Desjoyaux-Cherel, and P. L. Boquet. 1982. The acid phosphatase with optimum pH of 2.5 of *Escherichia coli*. Physiological and biochemical study. *J. Biol. Chem.* 257:6669–6676.

32. Dassa, J., H. Fsihi, C. Marck, M. Dion, M. Kieffer-Bontemps, and P. L. Boquet. 1991. A new oxygen-regulated operon in *Escherichia coli* comprises the genes for a putative third cytochrome oxidase and for pH 2.5 acid phosphatase (*appA*). *Mol. Gen. Genet.* 229:341–352.

33. Death, A., and T. Ferenci. 1994. Between feast and famine: endogenous inducer synthesis in the adaptation of *Escherichia coli* to growth with limiting carbohydrates. *J. Bacteriol.* 176:5101–5107.

34. Death, A., L. Notley, and T. Ferenci. 1993. Derepression of LamB protein facilitates outer membrane permeation of carbohydrates into *Escherichia coli* under conditions of nutrient stress. *J. Bacteriol.* 175:1475–1483.

35. Defez, R., and M. D. Felice. 1981. Cryptic operon for β-glucoside metabolism in *Escherichia coli* K12: genetic evidence for a regulatory protein. *Genetics* 97:11–25.

36. del Castillo, I., J. M. Gómez, and F. Moreno. 1990. *mprA*, an *Escherichia coli* gene that reduces growth-phase-dependent synthesis of microcins B17 and C7 and blocks osmoinduction of *proU* when cloned on a high-copy-number plasmid. *J. Bacteriol.* 172:437–445.

37. del Castillo, I., J. E. González-Pastor, J. L. S. Millán, and F. Moreno. 1991. Nucleotide sequence of the *Escherichia coli* regulatory gene *mprA* and construction and characterization of *mprA*-deficient mutants. *J. Bacteriol.* 173:3924–3929.

38. Dersch, P., K. Schmidt, and E. Bremer. 1993. Synthesis of the *Escherichia coli* K-12 nucleoid-associated DNA-binding protein H-NS is subjected to growth-phase control and autoregulation. *Mol. Microbiol.* 8:875–889.

39. Ditto, M. D., D. Roberts, and R. A. Weisberg. 1994. Growth phase variation of integration host factor level in *Escherichia coli*. *J. Bacteriol.* 176:3738–3748.

40. Dougherty, T. J., and M. J. Pucci. 1994. Penicillin-binding proteins are regulated by *rpoS* during transitions in growth states of *Escherichia coli*. *Antimicrob. Agents Chemother.* 38:205–210.

41. Dykhuizen, D. E., and D. L. Hartl. 1983. Selection in chemostats. *Microbiol. Rev.* 47:150–168.

42. El-Khani, M. A., and R. J. Stretton. 1981. Effect of growth medium on the lipid composition of log and stationary phase cultures of *Salmonella typhimurium*. *Microbios* 31:161–169.

43. El-Mansi, E. M., and W. H. Holms. 1989. Control of carbon flux to acetate excretion during growth of *Escherichia coli* in batch and continuous cultures. *J. Gen. Microbiol.* 135:2875–2883.

44. Espinosa-Urgel, M., and A. Tormo. 1993. Sigma s-dependent promoters in *Escherichia coli* are located in DNA regions with intrinsic curvature. *Nucleic Acids Res.* 21:3667–3670.

45. Falconi, M., N. P. Higgins, R. Spurio, C. L. Pon, and C. O. Gualerzi. 1993. Expression of the gene encoding the major bacterial nucleoid protein h-ns is subject to transcriptional auto-repression. *Mol. Microbiol.* 10:273–282.

46. Fang, F. C., S. J. Libby, N. A. Buchmeier, P. C. Loewen, J. Switala, J. Harwood, and D. G. Guiney. 1992. The alternative σ factor katF (rpoS) regulates *Salmonella* virulence. *Proc. Natl. Acad. Sci. USA* 89:11978–11982.

47. Foster, P. L. 1993. Adaptive mutation: the uses of adversity. *Annu. Rev. Microbiol.* 47:467–504.

48. Foster, P. L., and J. M. Trimarchi. 1994. Adaptive reversion of a frameshift mutation in *Escherichia coli* by simple base deletions at homopolymeric runs. *Science* 265:407–409.

49. Friedman, D. I. 1988. Integration host factor: a protein for all reasons. *Cell* 55:545–554.

50. Fujisaki, S., T. Nishino, and H. Katsuki. 1986. Biosynthesis of isoprenoids in intact cells of *Escherichia coli*. *J. Biochem.* 99:1137–1146.

51. Fuqua, W. C., S. C. Winans, and E. P. Greenberg. 1994. Quorum sensing in bacteria: the LuxR-LI family of cell density-responsive transcriptional regulators. *J. Bacteriol.* 176:269–275.

52. Garcia-Horsman, J. A., B. Barquera, J. Rumbley, J. Ma, and R. B. Gennis. 1994. The superfamily of heme-copper respiratory oxidases. *J. Bacteriol.* 176:5587–5600.

53. Garnak, M., and H. C. Reeves. 1979. Phosphorylation of isocitrate dehydrogenase of *Escherichia coli*. *Science* 203:1111–1112.

54. Garnak, M., and H. C. Reeves. 1979. Purification and properties of phosphorylated isocitrate dehydrogenase of *Escherichia coli*. *J. Biol. Chem.* 254:7915–7920.

55. Gentry, D. R., V. J. Hernández, L. H. Nguyen, D. B. Jensen, and M. Cashel. 1993. Synthesis of the stationary-phase sigma factor σ^s is positively regulated by ppGpp. *J. Bacteriol.* 175:7982–7989.

56. Georgiou, C. D., T. J. Dueweke, and R. B. Gennis. 1988. Regulation of expression of the cytochrome *d* terminal oxidase in *Escherichia coli* is transcriptional. *J. Bacteriol.* 170:961–966.

57. Georgopoulos, C. 1992. The emergence of the chaperone machines. *Trends Biochem. Sci.* 17:295–299.

58. Giaever, H. M., O. B. Styrvold, I. Kaasen, and A. R. Strom. 1988. Biochemical and genetic characterization of osmoregulatory trehalose synthesis in *Escherichia coli*. *J. Bacteriol.* 170:2841–2849.

59. Goldie, H. 1984. Regulation of transcription of the *Escherichia coli* phosphoenolpyruvate carboxykinase locus: studies with *pck-lacZ* operon fusions. *J. Bacteriol.* 159:832–836.

60. Goodrich, J. A., M. L. Schwartz, and W. R. McClure. 1990. Searching for and predicting the activity of sites for DNA binding proteins: compilation and analysis

61. of the binding sites for *Escherichia coli* integration host factor (IHF). *Nucleic Acids Res.* 18:4993–5000.

61. Green, G. N., H. Fang, R.-J. Lin, G. Newton, M. Mather, C. D. Georgiou, and R. B. Gennis. 1988. The nucleotide sequence of the *cyd* locus encoding the two subunits of the cytochrome d terminal oxidase complex of *Escherichia coli*. *J. Biol. Chem.* 263:13138–13143.

62. Groat, R. G., J. E. Schultz, E. Zychlinsky, A. Bockman, and A. Matin. 1986. Starvation proteins in *Escherichia coli*: kinetics of synthesis and role in starvation survival. *J. Bacteriol.* 168:486–493.

63. Grogan, D. W., and J. E. Cronan, Jr. 1986. Characterization of *Escherichia coli* mutants completely defective in synthesis of cyclopropane fatty acids. *J. Bacteriol.* 166:872–877.

64. Gulig, P. A., H. Danbara, D. G. Guiney, A. J. Lax, F. Norel, and M. Rhen. 1993. Molecular analysis of *spv* virulence genes of the *Salmonella* virulence plasmids. *Mol. Microbiol.* 7:825–830.

65. Hengge-Aronis, R. 1993. Survival of hunger and stress: the role of *rpoS* in early stationary phase gene regulation in *E. coli*. *Cell* 72:165–168.

66. Hengge-Aronis, R., and D. Fischer. 1992. Identification and molecular analysis of *glgS*, a novel growth-phase-regulated and *rpoS*-dependent gene involved in glycogen synthesis in *Escherichia coli*. *Mol. Microbiol.* 6:1877–1886.

67. Hengge-Aronis, R., W. Klein, R. Lange, M. Rimmele, and W. Boos. 1991. Trehalose synthesis genes are controlled by the putative sigma factor encoded by *rpoS* and are involved in stationary-phase thermotolerance in *Escherichia coli*. *J. Bacteriol.* 173:7918–7924.

68. Higgins, C. F., C. J. Dorman, D. A. Stirling, L. Waddell, I. R. Booth, G. May, and E. Bremer. 1988. A physiological role for DNA supercoiling in the osmotic regulation of gene expression in *S. typhimurium* and *E. coli*. *Cell* 52:569–584.

69. Higgins, C. F., J. C. Hinton, C. S. Hulton, T. Owen-Hughes, G. D. Pavitt, and A. Seirafi. 1990. Protein H1: a role for chromatin structure in the regulation of bacterial gene expression and virulence? *Mol. Microbiol.* 4:2007–2012.

70. Hill, C. W., and J. A. Gray. 1988. Effects of chromosomal inversion on cell fitness in *Escherichia coli* K-12. *Genetics* 119:771–778.

71. Huang, R., and R. N. Reusch. 1995. Genetic competence in *Escherichia coli* requires poly-β-hydroxybutyrate/calcium polyphosphate membrane complexes and certain divalent cations. *J. Bacteriol.* 176:486–490.

72. Huisman, G., and R. Kolter. 1993. Regulation of gene expression at the onset of stationary phase in *Escherichia coli*, p. 21–40. *In* P. J. Piggot, C. P. Moran, and P. Youngman (ed.), *Regulation of Bacterial Differentiation*. American Society for Microbiology, Washington, D.C.

73. Huisman, G. W., and R. Kolter. 1994. Sensing starvation: a homoserine lactone-dependent signaling pathway in *Escherichia coli*. *Science* 265:537–539.

74. Ichikawa, J. K., C. Li, J. Fu, and S. Clarke. 1994. A gene at 59 minutes on the *Escherichia coli* chromosome encodes a lipoprotein with unusual amino acid repeat sequences. *J. Bacteriol.* 176:1630–1638.

75. Ivanov, A. I., and V. M. Fomchenkov. 1989. Relation between the damaging effect of surface-active compounds on *Escherichia coli* cells and the phase of culture growth. *Mikrobiologiia* 58:969–975.

76. Jakubowski, H., and E. Goldman. 1992. Editing of errors in selection of amino acids for protein synthesis. *Microbiol. Rev.* 56:412–429.

77. Jenkins, D. E., E. A. Auger, and A. Matin. 1991. Role of RpoH, a heat shock regulator protein, in *Escherichia coli* carbon starvation protein synthesis and survival. *J. Bacteriol.* 173:1992–1996.

78. Kaasen, I., P. Falkenberg, O. B. Styrvold, and A. R. Strom. 1992. Molecular cloning and physical mapping of the *otsBA* genes, which encode the osmoregulatory trehalose pathway of *Escherichia coli*: evidence that transcription is activated by KatF (AppR). *J. Bacteriol.* 174:889–898.

79. Kashket, E. R. 1981. Effects of aerobiosis and nitrogen source on the proton motive force in growing *Escherichia coli* and *Klebsiella pneumoniae* cells. *J. Bacteriol.* 146:377–384.

80. Katayama, Y., A. Kasahara, H. Kuraishi, and F. Amano. 1990. Regulation of activity of an ATP-dependent protease, Clp, by the amount of a subunit, ClpA, in the growth of *Escherichia coli* cells. *J. Biochem.* 108:37–41.

81. Kolter, R., D. A. Siegele, and A. Tormo. 1993. The stationary phase of the bacterial life cycle. *Annu. Rev. Microbiol.* 47:855–874.

82. Kusters, J. G., G. A. Mulders-Kremers, C. E. van Doornik, and B. A. van der Zeyst. 1993. Effects of multiplicity of infection, bacterial protein synthesis, and growth phase on adhesion to and invasion of human cell lines by *Salmonella typhimurium*. *Infect. Immun.* 61:5013–5020.

83. Lacroix, J. M., I. Loubens, M. Tempete, B. Menichi, and J. P. Bohin. 1991. The *mdoA* locus of *Escherichia coli* consists of an operon under osmotic control. *Mol. Microbiol.* 5:1745–1753.

84. Lange, R., M. Barth, and R. Hengge-Aronis. 1993. Complex transcriptional control of the sigma s-dependent stationary-phase-induced and osmotically regulated *osmY* (*csi-5*) gene suggests novel roles for Lrp, cyclic AMP (cAMP) receptor protein-cAMP complex, and integration host factor in the stationary-phase response of *Escherichia coli*. *J. Bacteriol.* 175:7910–7917.

85. Lange, R., and R. Hengge-Aronis. 1991. Growth phase-regulated expression of *bolA* and morphology of stationary-phase *Escherichia coli* cells are controlled by the novel sigma factor σ^s. *J. Bacteriol.* 173:4474–4481.

86. Lange, R., and R. Hengge-Aronis. 1991. Identification of a central regulator of stationary-phase gene expression in *Escherichia coli*. *Mol. Microbiol.* 5:49–59.

87. Lange, R., and R. Hengge-Aronis. 1994. The *nlpD* gene is located in an operon with *rpoS* on the *Escherichia coli* chromosome and encodes a novel lipoprotein with a potential function in cell wall formation. *Mol. Microbiol.* 13:733–743.

88. LaPorte, D. C., and T. Chung. 1985. A single gene codes for the kinase and phosphatase which regulate isocitrate dehydrogenase. *J. Biol. Chem.* 260:15291–15297.

89. LaPorte, D. C., P. E. Thorsness, and D. E. Koshland, Jr. 1985. Compensatory phosphorylation of isocitrate dehydrogenase. A mechanism for adaptation to the intracellular environment. *J. Biol. Chem.* 260:10563–10568.

90. Leduc, M., C. Frehel, E. Siegel, and J. Van Heijenoort. 1989. Multilayered distribution of peptidoglycan in the periplasmic space of *Escherichia coli. J. Gen. Microbiol.* 135:1243–1254.

91. Leung, K. L., and H. Yamazaki. 1980. Extracellular accumulation of L-glutamate in adenylyl cyclase deficient or cyclic AMP receptor protein deficient mutants of *Escherichia coli. Can. J. Microbiol.* 26:718–721.

92. Li, C., and S. Clarke. 1992. A protein methyltransferase specific for altered aspartyl residues is important in *Escherichia coli* stationary-phase survival and heat-shock resistance. *Proc. Natl. Acad. Sci. USA* 89:9885–9889.

93. Loewen, P. C. 1977. Identification of a coenzyme A–glutathione disulfide (DSI), a modified coenzyme A disulfide (DSII), and a NADPH-dependent coenzyme A–glutathione disulfide reductase in *E. coli. Can. J. Biochem.* 55:1019–1027.

94. Loewen, P. C. 1979. Levels of glutathione in *Escherichia coli. Can. J. Biochem.* 57:107–111.

95. Loewen, P. C., J. Switala, and B. L. Triggs-Raine. 1985. Catalases HPI and HPII in *Escherichia coli* are induced independently. *Arch. Biochem. Biophys.* 243:144–149.

96. Luli, G. W., and W. R. Strohl. 1990. Comparison of growth, acetate production, and acetate inhibition of *Escherichia coli* strains in batch and fed-batch fermentations. *Appl. Environ. Microbiol.* 56:1004–1011.

97. Luria, S. E., and M. Delbrück. 1943. Mutations of bacteria from virus sensitivity to virus resistance. *Genetics* 28:491–511.

98. MacBeth, K. J., and C. A. Lee. 1993. Prolonged inhibition of bacterial protein synthesis abolishes *Salmonella* invasion. *Infect. Immun.* 61:1544–1546.

99. Manna, D., and J. Gowrishankar. 1994. Evidence for involvement of proteins HU and RpoS in transcription of the osmoresponsive *proU* operon in *Escherichia coli. J. Bacteriol.* 176:5378–5384.

100. Mason, C. A., and T. Egli. 1993. Dynamics of microbial growth in the decelerating and stationary phase of batch cultures, p. 81–102. *In* S. Kjelleberg (ed.), *Starvation in Bacteria.* Plenum Press, New York.

101. Matin, A. 1991. The molecular basis of carbon-starvation-induced general resistance in *Escherichia coli. Mol. Microbiol.* 5:3–10.

102. Matin, A., E. A. Auger, P. H. Blum, and J. E. Schultz. 1989. Genetic basis of starvation survival in nondifferentiating bacteria. *Annu. Rev. Microbiol.* 43:293–316.

103. McDonald, D. J. 1955. Segregation of the selective advantage obtained through orthoselection in *Escherichia coli. Genetics* 40:937–950.

104. McGarrity, J. T., and J. B. Armstrong. 1975. The effect of salt on phospholipid fatty acid composition in *Escherichia coli* K-12. *Biochim. Biophys. Acta* 398:258–264.

105. Mengin-Lecreulx, D., and J. van Heijenoort. 1985. Effect of growth conditions on peptidoglycan content and cytoplasmic steps of its biosynthesis in *Escherichia coli. J. Bacteriol.* 163:208–212.

106. Mirelman, D., and R. C. Siegel. 1979. Oxidative deamination of ε-aminolysine residues and formation of Schiff base cross-linkages in cell envelopes of *Escherichia coli. J. Biol. Chem.* 254:571–574.

107. Modi, R. I., L. H. Castilla, S. Puskas-Rozsa, R. B. Helling, and J. Adams. 1992. Genetic changes accompanying increased fitness in evolving populations of *Escherichia coli. Genetics* 130:241–249.

108. Moreno, F., J. L. S. Millan, I. del Castillo, J. M. Gomez, M. C. Rodriguez-Sainz, J. E. Gonzalez-Pastor, and L. Diaz-Guerra. 1992. *Escherichia coli* genes regulating the production of microcins MccB17 and MccC7, p. 3–13. *In* R. James, C. Lazdunski, and F. Pattus (ed.), *Bacteriocins, Microcins and Lantibiotics.* Springer Verlag, Berlin.

109. Mulvey, M. R., J. Switala, A. Borys, and P. C. Loewen. 1990. Regulation of transcription of *katE* and *katF* in *Escherichia coli. J. Bacteriol.* 172:6713–6720.

110. Naas, T., M. Blot, W. M. Fitch, and W. Arber. 1994. Insertion sequence-related genetic variation in resting *Escherichia coli* K-12. *Genetics* 136:721–730.

111. Neidhardt, F. C., J. L. Ingraham, and M. Schaechter. 1990. *Physiology of the Bacterial Cell: a Molecular Approach.* Sinauer, Sunderland, Mass.

112. Newman, J. V., R. Kolter, D. C. Laux, and P. S. Cohen. 1994. The role of *leuX* in *Escherichia coli* colonization of the streptomycin-treated mouse large intestine. *Microb. Pathog.* 17:301–311.

113. Nishijima, S., Y. Asami, N. Uetake, S. Yamagoe, A. Ohta, and I. Shibuya. 1988. Disruption of the *Escherichia coli cls* gene responsible for cardiolipin synthesis. *J. Bacteriol.* 170:775–780.

114. Norel, F., V. Robbe-Saule, M. Y. Popoff, and C. Coynault. 1992. The putative sigma factor KatF (RpoS) is required for the transcription of the *Salmonella typhimurium* virulence gene *spvB* in *Escherichia coli. FEMS Microbiol. Lett.* 78:271–276.

115. Nyström, T., and F. C. Neidhardt. 1992. Cloning, mapping and nucleotide sequencing of a gene encoding a universal stress protein in *Escherichia coli. Mol. Microbiol.* 6:3187–3198.

116. Ohlrogge, J. B., F. D. Gunstone, I. A. Ismail, and W. E. Lands. 1976. Positional specificity of cyclopropane ring formation from cis-octadecanoic acid isomers in *Escherichia coli. Biochim. Biophys. Acta* 431:257–267.

117. Olsén, A., A. Arnqvist, M. Hammar, S. Sukupolvi, and S. Normark. 1993. The RpoS sigma factor relieves H-NS-mediated transcriptional repression of *csgA*, the subunit gene of fibronectin-binding curli in *Escherichia coli. Mol. Microbiol.* 7:523–536.

118. Olsén, A., A. Jonsson, and S. Normark. 1989. Fibronectin binding mediated by a novel class of surface organelles on *Escherichia coli. Nature* (London) 338:652–655.

119. Owen-Hughes, T. A., G. D. Pavitt, D. S. Santos, J. M. Sidebotham, C. S. Hulton, J. C. Hinton, and C. F. Higgins. 1992. The chromatin-associated protein H-NS interacts with curved DNA to influence DNA topology and gene expression. *Cell* 71:255–265.

120. Owens, R. A., and P. E. Hartman. 1986. Export of glutathione by some widely used *Salmonella typhimurium* and *Escherichia coli* strains. *J. Bacteriol.* 168:109–114.

121. Ozaki, M., A. Wada, N. Fujita, and A. Ishihama. 1991. Growth phase-dependent modification of RNA polymerase in *Escherichia coli. Mol. Gen. Genet.* 230:17–23.

122. Pisabarro, A. G., M. A. de Pedro, and D. Vazquez. 1985. Structural modifications in the peptidoglycan of *Escherichia coli* associated with changes in the state of growth of the culture. *J. Bacteriol.* 161:238–242.

123. Preiss, J. 1984. Bacterial glycogen synthesis and its regulation. *Annu. Rev. Microbiol.* 38:419–458.

124. Preiss, J., and T. Romeo. 1989. Physiology, biochemistry and genetics of bacterial glycogen synthesis. *Adv. Microb. Physiol.* 30:183–233.

125. Pruss, B. M., J. M. Nelms, C. Park, and A. J. Wolfe. 1994. Mutations in NADH:ubiquinone oxidoreductase of *Escherichia coli* affect growth on mixed amino acids. *J. Bacteriol.* 176:2143–2150.

126. Rabinowitch, H. D., D. Sklan, D. H. Chace, R. D. Stevens, and I. Fridovich. 1993. *Escherichia coli* produces linoleic acid during late stationary phase. *J. Bacteriol.* 175:5324–5328.

127. Raina, S., D. Missiakis, L. Baird, S. Kumar, and C. Georgopoulos. 1993. Identification and transcriptional analysis of the *Escherichia coli htrE* operon which is homologous to *pap* and related pilin operons. *J. Bacteriol.* 175:5009–5021.

128. Rasmussen, L. J., P. L. Moller, and T. Atlung. 1991. Carbon metabolism regulates expression of the *pfl* (pyruvate formate-lyase) gene in *Escherichia coli. J. Bacteriol.* 173:6390–6397.

129. Reeve, C. A., A. T. Bockman, and A. Matin. 1984. Role of protein degradation in the survival of carbon-starved *Escherichia coli* and *Salmonella typhimurium. J. Bacteriol.* 157:758–763.

130. Reusch, R. N., and H. L. Sadoff. 1988. Putative structure and functions of a poly-beta-hydroxybutyrate/calcium polyphosphate channel in bacterial plasma membranes. *Proc. Natl. Acad. Sci. USA* 85:4176–4180.

131. Rock, C. O., J. E. Cronan, Jr., and I. M. Armitage. 1981. Molecular properties of acyl carrier protein derivatives. *J. Biol. Chem.* 256:2669–2674.

132. Romeo, T., M. Gong, M. Y. Liu, and A. M. Brun-Zinkernagel. 1993. Identification and molecular characterization of *csrA*, a pleiotropic gene from *Escherichia coli* that affects glycogen biosynthesis, gluconeogenesis, cell size, and surface properties. *J. Bacteriol.* 175:4744–4755.

133. Romeo, T., and J. Preiss. 1989. Genetic regulation of glycogen biosynthesis in *Escherichia coli*: in vitro effects of cyclic AMP and guanosine 5′-diphosphate 3′-diphosphate and analysis of in vivo transcripts. *J. Bacteriol.* 171:2773–2782.

134. Rosenberg, S. M., S. Longerich, P. Gee, and R. S. Harris. 1994. Adaptive mutation by deletions in small mononucleotide repeats. *Science* 265:405–407.

135. Rosenzweig, R. F., R. R. Sharp, D. S. Treves, and J. Adams. 1994. Microbial evolution in a simple unstructured environment: genetic differentiation in *Escherichia coli. Genetics* 137:903–917.

136. Rouviere-Yaniv, J., and F. Gros. 1975. Characterization of a novel, low-molecular-weight DNA-binding protein from *Escherichia coli. Proc. Natl. Acad. Sci. USA* 72:3428–3432.

137. Rowley, K. B., R. M. Elford, I. Roberts, and W. M. Holmes. 1993. In vivo regulatory responses of four *Escherichia coli* operons which encode leucyl-tRNAs. *J. Bacteriol.* 175:1309–1315.

138. Rumley, M. K., H. Therisod, A. C. Weissborn, and E. P. Kennedy. 1992. Mechanisms of regulation of the biosynthesis of membrane-derived oligosaccharides in *Escherichia coli. J. Biol. Chem.* 267:11806–11810.

139. Satischandran, C., J. C. Taylor, and G. D. Markham. 1993. Isozymes of S-adenosylmethionine synthetase are encoded by tandemly duplicated genes in *Escherichia coli. Mol. Microbiol.* 9:835–846.

140. Scaravaglio, O. R., R. N. Farias, and E. M. Massa. 1993. Starved cells of the fatty acid auxotroph *Escherichia coli* AK7 develop abnormal sensitivity to media with low osmolarity. *Appl. Environ. Microbiol.* 59:2760–2762.

141. Schellhorn, H. E., and V. L. Stones. 1992. Regulation of *katF* and *katE* in *Escherichia coli* K-12 by weak acids. *J. Bacteriol.* 174:4769–4776.

142. Schultz, J. E., G. I. Latter, and A. Matin. 1988. Differential regulation by cyclic AMP of starvation protein synthesis in *Escherichia coli. J. Bacteriol.* 170:3903–3909.

143. Schultz, J. E., and A. Matin. 1991. Molecular and functional characterization of a carbon starvation gene of *Escherichia coli. J. Mol. Biol.* 218:129–140.

144. Shiga, Y., H. Mizuno, and H. Akanuma. 1993. Conditional synthesis and utilization of 1,5-anhydroglucitol in *Escherichia coli. J. Bacteriol.* 175:7138–7141.

145. Siegele, D. A., and R. Kolter. 1992. Life after log. *J. Bacteriol.* 174:345–348.

146. Siegele, D. A., and R. Kolter. 1993. Isolation and characterization of an *Escherichia coli* mutant defective in resuming growth after starvation. *Genes Dev.* 7:2629–2640.

147. Smirnova, G. V., and O. N. Oktiabr'skii. 1990. Change in the level of SH-compounds in *Escherichia coli* and *Bacillus subtilis* cultures during transition to the stationary phase. *Mikrobiologiia* 59:387–393.

148. Sonti, R. V., and J. R. Roth. 1989. Role of gene duplications in the adaptation of *Salmonella typhimurium* to growth on limiting carbon sources. *Genetics* 123:19–28.

149. Souzu, H. 1982. *Escherichia coli* B membrane stability related to cell growth phase. Measurement of temperature dependent physical state change of the membrane over a wide range. *Biochim. Biophys. Acta* 691:161–170.

150. Souzu, H. 1986. Fluorescence polarization studies on *Escherichia coli* membrane stability and its relation to the resistance of the cell to freeze-thawing. I. Membrane stability in cells of differing growth phase. *Biochim. Biophys. Acta* 861:353–360.

151. Souzu, H. 1986. Fluorescence polarization studies on *Escherichia coli* membrane stability and its relation to the resistance of the cell to freeze-thawing. II. Stabilization of the membranes by polyamines. *Biochim. Biophys. Acta* 861:361–367.

152. Spassky, A., S. Rimsky, H. Garreau, and H. Buc. 1984. H1a, an *E. coli* DNA-binding protein which accumulates in stationary phase, strongly compacts DNA in vitro. *Nucleic Acids Res.* 12:5321–5340.

153. Spence, J., A. Cegielska, and C. Georgopoulos. 1990. Role of *Escherichia coli* heat shock proteins DnaK and HtpG (C62.5) in response to nutritional deprivation. *J. Bacteriol.* 172:7157–7166.

154. Steinhaus, E. A., and J. M. Birkeland. 1939. Studies on the life and death of bacteria. I. The senescent phase in aging cultures and the probable mechanisms involved. *J. Bacteriol.* 38:249–261.

155. Storz, G., and S. Altuvia. 1994. OxyR regulon. *Methods Enzymol.* 234:217–223.

156. Tanaka, K., Y. Takayanagi, N. Fujita, A. Ishihama, and H. Takahashi. 1993. Heterogeneity of the principal σ factor in *Escherichia coli:* the *rpoS* gene product, σ^{38}, is a second principal σ factor of RNA polymerase in stationary-phase *Escherichia coli. Proc. Natl. Acad. Sci. USA* 90:3511–3515.

157. Taylor, F., and J. E. Cronan, Jr. 1976. Selection and properties of *Escherichia coli* mutants defective in the synthesis of cyclopropane fatty acids. *J. Bacteriol.* 125:518–523.

158. Tippner, D., H. Afflerbach, C. Bradaczek, and R. Wagner. 1994. Evidence for a regulatory function of the histone-like *Escherichia coli* protein H-NS in ribosomal RNA synthesis. *Mol. Microbiol.* 11:589–604.

159. Tormo, A., M. Almirón, and R. Kolter. 1990. surA, an *Escherichia coli* gene essential for survival in stationary phase. *J. Bacteriol.* 172:4339–4347.

160. Touati, E., E. Dassa, and P. L. Boquet. 1986. Pleiotropic mutations in *appR* reduce pH 2.5 acid phosphatase expression and restore succinate utilisation in CRP-deficient strains of *Escherichia coli. Mol. Gen. Genet.* 202:257–264.

161. Ueguchi, C., M. Kakeda, and T. Mizuno. 1993. Autoregulatory expression of the *Escherichia coli hns* gene encoding a nucleoid protein: H-NS functions as a repressor of its own transcription. *Mol. Gen. Genet.* 236:171–178.

162. Ueguchi, C., and T. Mizuno. 1993. The *Escherichia coli* nucleoid protein H-NS functions directly as a transcriptional repressor. *EMBO J.* 12:1039–1046.

163. van der Linden, M. P., L. de Haan, M. A. Hoyer, and W. Keck. 1992. Possible role of *Escherichia coli* penicillin-binding protein 6 in stabilization of stationary-phase peptidoglycan. *J. Bacteriol.* 174:7572–7578.

164. van Laere, A. 1989. Trehalose, reserve and/or stress metabolite? *FEMS Microbiol. Rev.* 63:201–210.

165. Vinella, D., R. D'Ari, and A. Jaff. 1992. Penicillin binding protein 2 is dispensable in *Escherichia coli* when ppGpp synthesis is induced. *EMBO J.* 11:1493–1501.

166. Volkert, M. R., L. I. Hajec, and D. C. Nguyen. 1989. Induction of the alkylation-inducible *aidB* gene of *Escherichia coli* by anaerobiosis. *J. Bacteriol.* 171:1196–1198.

167. Wada, A., Y. Yamazaki, N. Fujita, and A. Ishihama. 1990. Structure and probable genetic location of a "ribosome modulation factor" associated with 100S ribosomes in stationary-phase *Escherichia coli* cells. *Proc. Natl. Acad. Sci. USA* 87:2657–2661.

168. Wang, A.-Y., D. W. Grogan, and J. E. Cronan, Jr. 1992. Cyclopropane fatty acid synthase of *Escherichia coli:* deduced amino acid sequence, purification, and studies of the enzyme active site. *Biochemistry* 31:11020–11028.

169. Weichart, D., R. Lange, N. Henneberg, and R. Hengge-Aronis. 1993. Identification and characterization of stationary phase-inducible genes in *Escherichia coli. Mol. Microbiol.* 10:407–420.

170. Wensink, J., N. Gilden, and B. Witholt. 1982. Attachment of lipoprotein to the murein of *Escherichia coli. Eur. J. Biochem.* 122:587–590.

171. Yamada, H., S. Muramatsu, and T. Mizuno. 1990. An *Escherichia coli* protein that preferentially binds to sharply curved DNA. *J. Biochem.* 108:420–425.

172. Yamagishi, M., H. Matsushima, A. Wada, M. Sakagami, N. Fujita, and A. Ishihama. 1993. Regulation of the *Escherichia coli rmf* gene encoding the ribosome modulation factor: growth phase- and growth rate-dependent control. *EMBO J.* 12:625–630.

173. Yamashino, T., M. Kakeda, C. Ueguchi, and T. Mizuno. 1994. An analogue of the DnaJ molecular chaperone whose expression is controlled by σ^{s} during the stationary phase and phosphate starvation in *Escherichia coli. Mol. Microbiol.* 13:475–483.

174. Yatvin, M. B., J. J. Gipp, D. R. Klessig, and W. H. Dennis. 1986. Hyperthermic sensitivity and growth stage in *Escherichia coli. Radiat. Res.* 106:78–88.

175. Yim, H. H., R. L. Brems, and M. Villarejo. 1994. Molecular characterization of the promoter of *osmY*, an *rpoS*-dependent gene. *J. Bacteriol.* 176:100–107.

176. Yim, H. H., and M. Villarejo. 1992. *osmY*, a new hyperosmotically inducible gene, encodes a periplasmic protein in *Escherichia coli. J. Bacteriol.* 174:3637–3644.

177. Zambrano, M. M., and R. Kolter. 1993. *Escherichia coli* mutants lacking NADH dehydrogenase I have a competitive disadvantage in stationary phase. *J. Bacteriol.* 175:5642–5647.

178. Zambrano, M. M., D. A. Siegele, M. Almirón, A. Tormo, and R. Kolter. 1993. Microbial competition: *Escherichia coli* mutants that take over stationary phase cultures. *Science* 259:1757–1760.

Growth Yield and Energy Distribution

OENSE M. NEIJSSEL, M. JOOST TEIXEIRA DE MATTOS, AND
DAVID W. TEMPEST

107

INTRODUCTION

Although quantitative studies of microbial growth and of growth energetics were in progress some 60 years ago or more (52, 59), it is widely acknowledged that present-day theories regarding the relationships between substrate concentration, growth rate, and yield value stem from the classical studies of Monod (33). In these studies, quantitative measurements of the growth of *Escherichia coli, Salmonella typhimurium,* and *Bacillus subtilis* in batch cultures revealed that the equivalent dry weight of organisms formed per gram of carbon substrate metabolized (the yield value) was remarkably constant. Thus, when growing in a simple salts medium on a variety of related carbon substrates (hexoses, pentoses, polyalcohols, and disaccharides), *E. coli* expressed yield values that ranged between 0.21 and 0.28 g (equivalent dry weight) of cells formed per g of substrate consumed. Correspondingly cultures of *S. typhimurium* expressed slightly (though consistently) lower values. The constancy of these data indicated the presence of mechanisms that precisely partition the flow of carbon substrate between catabolic (energy-generating) and anabolic (energy-consuming) processes such as to allow growth to proceed with a fixed overall efficiency. In this connection, however, Monod (33) realized that not all of the energy generated by catabolism necessarily would be consumed by anabolic processes and postulated that a small amount might be needed for cell maintenance ("ration d'entretien"). But, from his observations, he was forced to conclude that, with actively growing cultures, the latter requirement was too small to be detected by the methods then available.

The relationship between substrate concentration, growth rate, and respiration rate was subsequently studied by Schulze and Lipe (45), who used continuous-culture techniques that allowed growth rate to be varied over a wide range. From these studies on an unnamed strain of *E. coli,* they concluded that the maintenance rate of glucose consumption was indeed small but

sufficient to affect markedly the yield value of a glucose-limited chemostat culture at low dilution rates. When corrected for maintenance, the (maximum) growth yield was found to be 0.53 g of cells per g of glucose. This value was more than double that reported by Monod (33), but is in accord with many subsequent measurements made with different strains of *E. coli* growing in glucose-limited chemostat culture (Table 1).

The results obtained by Schulze and Lipe (45) confirmed and extended those of Herbert (18) and Marr et al. (30), who had shown that the yield value with respect to the carbon substrate consumed was not a constant but varied with the growth rate (Table 1). A similar variation of oxygen yield was found, not surprisingly, and was also ascribed to a requirement of metabolic energy for cell maintenance (41). Thus, it became clear that meaningful comparisons of yield values could only be made by taking maintenance into account. Unfortunately, the precise nature of these maintenance processes and their minimum energetic requirements were then (as now) largely unresolved. Hence, it is still not clear whether maintenance energy requirements are quantitatively independent of growth rate, as originally assumed by Pirt (41) and others (17, 18, 30), or whether they vary with growth rate, as is now thought probable (38, 42, 59, 62). Thus it is not easy to assess the significance of small differences in the yield value expressed by related organisms growing under comparable conditions, but large consistent differences ought to be interpretable in physiological terms.

It is the purpose of this chapter to concentrate on those conditions that markedly affect the growth yield with respect to carbon substrate and oxygen consumption (Y_{sub} and Y_{O_2}, respectively) or ATP generation (Y_{ATP}) and to assess their physiological implications. It is useful to consider first the overall relationship between catabolism and anabolism which the yield value embodies. The uptake of a carbon substrate and its subsequent metabolism generate intermediary metabolites and reductant.

TABLE 1 Glucose consumption rates ($q_{glucose}$) and yield values ($Y_{glucose}$) obtained with glucose-limited cultures of several strains of *E. coli*[a]

Dilution rate	C(PC-1000)		B/r		NCTC 9001		Unnamed	
	$q_{glucose}$	$Y_{glucose}$	$q_{glucose}$	$Y_{glucose}$	$q_{glucose}$	$Y_{glucose}$	$q_{glucose}$	$Y_{glucose}$
0.1	1.56	64.1	1.50	66.7	1.47	70.9	1.25	80.0
0.3	4.08	73.5	3.93	76.4	3.95	76.0	3.56	84.3
0.5	6.60	75.8	6.36	78.7	6.45	77.5	5.92	84.5
0.7	9.12	76.8	8.86	79.0	8.93	78.4	8.27	84.6
Apparent $Y_{glucose}^{max}$ [c]	79.4		81.5		79.8		85.4	
Apparent $m_{glucose}$ [d]	0.30		0.26		0.17		0.07	

The *E. coli* strain[b] header spans all four strain groups.

[a] To facilitate direct comparison, the reported values at each dilution rate were derived from linear plots of $q_{glucose}$ versus dilution rate.
[b] $q_{glucose}$ in nanomoles of glucose per per hour per gram (dry weight) of cells; $Y_{glucose}$ in grams (dry weight) of cells formed per mole of glucose consumed. Sources of data: C(PC-1000), unpublished results; B/r, M. P. M. Leegwater, personal communication; NCTC 9001, reference 35; unnamed strain, reference 45.
[c] Calculated maximum growth yield (i.e., corrected for maintenance).
[d] Maintenance rate of glucose consumption; the extrapolated rate of glucose consumption at zero growth rate.

Aerobically, part of the reductant so formed is oxidized by the respiratory chain to generate ATP; this ATP, along with the remaining reductant, is used to convert intermediary metabolites into monomers and polymers. A shortfall of reductant or ATP can be avoided by oxidation of some of the intermediary metabolites to CO_2. However, because the concentrations of the main carriers of reductant (the pyridine nucleotides) and of energy (the adenine nucleotides) are relatively low (15), surplus reductant or energy must be rapidly disposed of. Thus, with organisms growing aerobically a potential redox imbalance can be circumvented by a higher respiration rate, but if respiration is coupled to ATP generation, surplus energy is generated that cannot be stored as adenine nucleotides. This is not a hypothetical problem, for there are at least two circumstances in which catabolism is extensively dissociated from anabolism: first, washed cell suspensions often oxidize carbon substrate at a high rate without growing; second, carbon substrate-sufficient cultures frequently catabolize substrate far more rapidly than corresponding carbon-limited cultures (4, 34, 37, 38, 53, 54). In both of these cases, excess energy is generated that must be dissipated at a high rate (presumably as heat) by growth-unassociated processes. Therefore, microbial cells must have a capacity either to uncouple respiration from ADP phosphorylation or to turn over the ATP pool at a high rate in the absence of biosynthesis. This assumption raises questions regarding the regulation of energy fluxes in microorganisms and the nature of energy coupling between catabolic and anabolic processes in actively growing cells. To answer these questions, we consider the following: (i) the mechanisms of energy generation and how they might be caused to vary, (ii) the multifarious nature of energy-consuming processes, and (iii) the possible nature of energy-spilling reactions.

ENERGY-GENERATING SYSTEMS

Oxidative Phosphorylation

The efficiency of respiratory energy conservation (moles of ATP equivalents generated per 0.5 mol of oxygen reduced) depends upon the composition of the respiratory chain and the number of proton-translocating segments. Extensive studies of the aero-

bic respiratory chain components of several *E. coli* strains (24, 26) suggest that they lack a *c*-type cytochrome and hence contain only two proton-translocating segments. A schematic representation of the aerobic respiratory chain of *E. coli* is shown in Fig. 1. (For more detailed information, see chapter 17). Briefly, NADH can be oxidized via either of two dehydrogenases, NADH dehydrogenase 1 (NDH-1) and NADH dehydrogenase 2 (NDH-2). NDH-1 catalyzes the generation of a proton motive force during NADH oxidation, whereas NDH-2 does not. From the quinone pool there are two branches that transport electrons to oxygen. Under fully aerobic conditions a *bo*-type oxidase is operational, and this enzyme has an H^+/e ratio of 2. At low oxygen tensions, the *bd*-type oxidase, which has an H^+/e ratio of 1, operates.

Apart from the NADH-linked dehydrogenases and those linked to flavins, *E. coli* and *S. typhimurium* have been found to synthesize a glucose dehydrogenase apo-enzyme that is active in the presence of 2,7,9-tricarboxy-1H-pyrrolo-(2,3f)-quinoline quinone (PQQ) (3, 20, 21). The active center of this enzyme faces the periplasmic space, and the activity of glucose dehydrogenase of *E. coli* or *S. typhimurium* can be reconstituted when PQQ is added to the medium (3, 20, 21). Thus, in the presence of PQQ, and not in its absence, whole cells or cell homogenates of these organisms oxidize glucose directly to gluconate. Strains of *Klebsiella pneumoniae*, in contrast, are able to synthesize PQQ and have a functionally competent glucose dehydrogenase (3,

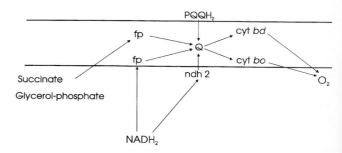

FIGURE 1 Organization of the aerobic respiratory chain in *E. coli*.

39). It has now been established unequivocally that $PQQH_2$ reacts with the quinone pool, because a *cyo* as well as a *cyd* mutant of *E. coli* oxidize glucose to gluconate in the presence of PQQ (R. de Jonge, M. J. Teixeira de Mattos, and O. M. Neijssel, unpublished data).

It is interesting to know how the electron transport fluxes are partitioned over these different branches in the growing wild-type organism. This problem was investigated by Calhoun et al. (5), who determined the specific rates of oxygen consumption of wild-type and mutant strains of *E. coli* in which one or two genes encoding components of the respiratory chain (*ndh*, *cyo*, or *cyd*) were deleted. The organisms were grown in a fully aerobic, glucose-limited chemostat culture at different dilution rates. According to the classical theory, the growth rate (μ), specific rate of oxygen consumption (q_{O_2}), and growth yield on oxygen (Y_{O_2}) are interrelated according to: $\mu = Y_{O_2} \cdot q_{O_2}$, and when maintenance requirement is taken into account, $\mu = Y_{O_2}^{max} \cdot (q_{O_2} - q_m)$, where q_m is the specific rate of oxygen consumption at zero growth rate.

The results obtained by Calhoun et al. (5) (Fig. 2; Table 2) lead to interesting conclusions. First, the specific rates of oxygen consumption by the *ndh cyo* strain and the *ndh* strain differ significantly. Assuming that the synthesis of cytochrome *bd* by the *ndh* mutant, like that of the wild type, is repressed under fully aerobic conditions, the results show that the cytochrome *bo* branch conserves more energy than the cytochrome *bd* branch. But the *bd* branch must contribute to energy conservation, because the growth yield on oxygen of the *ndh cyo* strain is higher than would be expected if no energy were conserved. In this latter case, the specific rates of oxygen consumption of the *ndh cyo* strain should have been twofold higher than those of the *ndh* strain.

Another surprising result was the observation that NDH-2 catalyzes part of the electron transport in the wild-type organism. This could be deduced from the fact that an *ndh* mutant had lower oxygen uptake rates (= higher growth yields) than the wild type when both organisms were grown at similar growth rates. The possible physiological significance of this result will be discussed later, but this result strongly suggests that maximiza-

TABLE 2 Parameters of the lines of regression representing the best linear fits for dependence of the specific rate of oxygen consumption on the dilution rate of different strains of *E. coli*[a]

Strain	Relevant genotype	Slope	Intercept
Wild type		20.10 ± 0.91	3.20 ± 0.49
GO103	*cyd*	23.68 ± 0.69	1.71 ± 0.40
GO104	*cyo*	24.65 ± 1.36	3.35 ± 0.99
MWC215	*ndh*	20.11 ± 0.84	1.87 ± 0.40
MWC217	*ndh cyo*	27.54 ± 1.29	2.91 ± 0.59

[a]Strains were grown in aerobic, glucose-limited chemostat culture. The slope is the incremental increase in O_2 required per gram of biomass as the growth rate is increased (millimoles of O_2 per gram of biomass [$Y_{O_2}^{max}$]). The intercept is the extrapolated specific oxygen consumption rate at zero growth rate (q_m). Data from Calhoun et al. (5).

tion of growth yield has not been a selective factor during the evolution of *E. coli*.

The effect of the presence of PQQ in the medium on growth yields of *E. coli* has been studied by Hommes et al. (22). When *E. coli* was grown in a glucose-limited chemostat culture, no effect of PQQ could be observed. This is not surprising because the PQQ-linked glucose dehydrogenase has a rather low affinity for glucose ($K_m = 0.9$ to 5 mM [1, 21]), and the glucose concentration in glucose-limited cultures is in the micromolar region (45). When the organism was grown under carbon-excess conditions (glucose concentration, >5 mM), addition of PQQ to the growth medium invariably caused the production of gluconate, with a concomitant stoichiometric increase in the specific rates of oxygen and glucose consumption (thus, lower growth yields on oxygen and glucose, respectively), but no further effects could be noted. This result shows again that when heterotrophic organisms are grown under carbon-excess conditions their rate of consumption of the carbon and energy source is not strictly proportional to growth rate.

Substrate-Level Phosphorylation

When growing anaerobically on glucose, in the absence of an added electron acceptor, *E. coli* effects a mixed-acid fermentation in which the principal products are lactate, ethanol, acetate, and formate (or H_2 plus CO_2 at acid pH values). Thus, energy is conserved (as ATP) principally at the levels of 3-phosphoglycerate kinase, pyruvate kinase, and acetate kinase (Fig. 3). Significant amounts of succinate derived from endogenously generated fumarate also are often found (Table 3), and, because fumarate reduction is coupled to respiratory chain-linked oxidation of NADH or formate, the formation of succinate is accompanied by the generation of a proton motive force. It follows, therefore, that whereas the formation of acetate is accompanied by a net production of 2 mol of ATP per mol, the formation of lactate, ethanol, and succinate is accompanied by just 1 mol of ATP per mol.

It would be advantageous (in terms of net ATP gain) to ferment glucose solely to acetate and formate. Redox considerations, however, require a concomitant formation of a product more reduced than lactate because acetate production is accom-

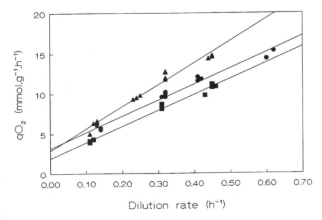

FIGURE 2 Specific rates of oxygen consumption (q_{O_2}) of *E. coli* strains growing in aerobic, glucose-limited chemostat culture: GR70N (wild type [●]), MWC215 (*ndh* [■]), and MWC217 (*ndh cyo* [▲]). Data of Calhoun et al. (5).

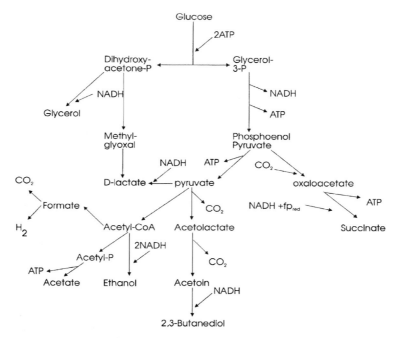

FIGURE 3 Pathways of product formation in *E. coli* grown anaerobically on glucose. Data are those of Gottschalk (12).

panied by a net formation of NADH. Production of ethanol fulfills this requirement in *E. coli* and *K. pneumoniae*. Hence, in fermenting glucose, a maximal ATP gain would be achieved in producing equal amounts of ethanol and acetate and would amount to 3 mol of ATP per mol of glucose fermented. It follows, therefore, that because anaerobic cultures generally are considered to be limited in their growth by the availability of energy (ATP), one would expect glucose to be fermented entirely to ethanol and acetate (plus formate, of course). Surprisingly, this was not found with batch cultures of *E. coli* (Table 3) or *Aerobacter cloacae* (19) or with glucose-sufficient chemostat cultures of *K. pneumoniae* (54). This condition was apparent only with glucose-limited chemostat cultures of *E. coli*, though small amounts of succinate and lactate were still formed (47). Taken at face value,

these observations suggest that organisms growing on glucose anaerobically in batch culture (as in glucose-sufficient chemostat culture) are not energy limited. On the other hand, a high rate of cell synthesis, such as is manifest in batch cultures, demands a high rate of ATP generation, and the presence of branched fermentation capacity in an organism may actually serve to facilitate this high rate of ATP generation, albeit with a concomitant decrease in the efficiency of catabolic energy conservation (in terms of net moles of ATP formed per mole of glucose fermented).

ENERGY-CONSUMING REACTIONS

We now review the major energy-consuming processes that lead directly to the synthesis of biomass or extracellular products.

Cell Synthesis

The principal known energy-consuming reactions occurring in the growing microbial cell are those associated with solute transport, monomer formation, and most particularly, polymer synthesis, which all lead to a net increase in biomass. A detailed analysis of the theoretical ATP requirements for these processes, for organisms growing in a simple salts medium on a variety of carbon substrates, was provided by Stouthamer (48, 49) (Table 4).

On the basis of a rather atypical macromolecular composition of *E. coli* (i.e., a high polysaccharide content and a relatively low protein content) it was concluded that ATP needed for cell synthesis varied from 34.8 mmol/g (dry weight) of cells for growth on glucose to 99.5 mmol/g (dry weight) of cells for growth on acetate. Translated into yield coefficients, these values gave theoretical Y_{ATP}^{max} values ranging from 28.8 to 10 g (dry weight) of cells per mol of ATP. These values are, respectively, 207% and 43% higher than the experimentally derived values

TABLE 3 Mixed-acid fermentation of glucose as effected by *E. coli*[a]

Product or recovery	mmol of product/100 mmol of glucose fermented at pH:	
	6.2	7.8
Acetic acid	36.5	38.7
Acetoin	0.06	0.19
2,3-Butanediol	0.3	0.26
Ethanol	49.8	50.5
Formic acid	2.43	86.0
Glycerol	1.42	0.32
Lactic acid	79.5	70.0
Succinic acid	10.7	14.8
Carbon dioxide	75.0	0.26
Carbon recovery (%)	92.0	90.0

[a]Data of Blackwood et al. (2).

TABLE 4 Theoretical ATP requirement of *E. coli* growing in a simple salts medium with different carbon and energy sources[a]

Product or parameter	ATP requirement (mmol/g of cells) of cells growing on[b]:			
	Glucose	Malate	Lactate	Acetate
Protein	20.5	28.5	33.9	42.7
Polysaccharide	2.05	5.1	7.1	9.2
RNA	5.86	7.0	8.5	10.1
Lipid	0.15	2.5	2.7	5.0
DNA	1.05	1.3	1.6	1.9
Transport Y_{ATP}^{max}	5.21	20.0	20.0	30.6
Theoretical	28.8	15.4	13.4	10.0
Experimentally assessed $Y_{ATP}^{max c}$	13.9		9.5	7.1

[a]Data of Stouthamer (49).
[b]The macromolecular composition of the cells was taken to be 52.4% protein, 16.6% polysaccharide, 15.7% RNA, 9.4% lipid, and 3.2% DNA.
[c]Reference 49.

for Y_{ATP}^{max} for *E. coli* growing aerobically on glucose and acetate, assuming two sites of respiratory chain energy conservation (5, 11). Hence either the energy requirements for transport and for monomer and polymer synthesis had been grossly underestimated or, as seems more probable, a substantial amount of energy is needed for growth-associated processes other than those specified above.

It is now known that certain biosynthetic processes are energetically more expensive than thought previously. For instance, the ATP consumption by molecular chaperones during the folding of proteins had not been taken into account (16). Martin et al. (31) determined the ATP requirement of GroEL during the folding of denatured bovine liver rhodanese in vitro. They arrived at the figure of 130 ± 20 molecules of ATP hydrolyzed per molecule of rhodanese folded at GroEL in the presence of GroES. This would account for 5 to 10% of the total ATP requirement for rhodanese synthesis. If this result were the same for all proteins synthesized in the *E. coli* cell, and this is highly unlikely, the total costs of protein synthesis as calculated by Stouthamer have to be increased by only 5 to 10%.

A similar discrepancy between theoretical and probable Y_{ATP}^{max} values previously had been noted by Gunsalus and Shuster (13). They postulated dissipation of energy by ATPase mechanisms. Other authors (14, 64) also addressed this question and postulated that a major part of this discrepancy is explained by thermodynamic effects (see below).

We are aware of no definitive experiments that identify unequivocally how this apparent excess energy is consumed. There are, however, other processes that consume energy and lead to an apparent lowering of the yield values.

Extracellular Polymer Formation

Whereas most of the polymers synthesized by microbial cells are contained within the plasma membrane and envelope structure, some may be secreted into the medium. This is particularly the case with polysaccharides, which under some conditions and with particular species may account for a substantial part of the carbon substrate consumed. For example, ammonia-limited chemostat cultures of *K. pneumoniae* growing on glucose at a rate of $0.17 h^{-1}$ excreted more than one-third of the substrate consumed into the medium as polysaccharide (37). With similarly limited cultures of *E. coli* (PC-1000), also growing at a low rate ($0.18 h^{-1}$), less extracellular polysaccharide was excreted, but polysaccharide still accounted for 14% of the total glucose consumed (Table 5). Though polyglucose is energetically less expensive to synthesize than protein, RNA, and DNA (i.e., 12.4 mmol of ATP per g as compared with, respectively, 39.1, 37.3, and 33.0 mmol of ATP per g [49]), substantial production of this compound would consume a significant portion of the ATP generated by catabolism.

Proteins are not commonly secreted by *E. coli* and *K. pneumoniae*. Nevertheless, it is conceivable that enzymes and binding proteins normally located in the periplasmic space may, under some conditions, diffuse into the extracellular fluids. Thus, protein was found to accumulate in the extracellular fluids of chemostat cultures of *K. pneumoniae* that were sulfate, magnesium, or potassium limited when growing on glucose at a low rate (57). The extracellular protein from the sulfate-limited culture was shown not to be a product of lysis because it was composed of an excess of acidic amino acids over basic amino acids and contained virtually no methionine or cysteine.

Excretion of Low-Molecular-Weight Metabolites

Batch cultures of *E. coli*, or carbon substrate-limited chemostat cultures growing aerobically in a defined simple salts medium, generally metabolize substrate solely to cells and CO_2. However, carbon substrate-sufficient chemostat cultures often metabolize substrate inefficiently, accumulating substantial amounts of products in the medium (Table 5). Acetate almost invariably is found (37), along with various amounts of pyruvate, 2-oxoglutarate, and D-lactate. Hence, the yield value with respect to the carbon substrate is markedly lowered. In general, formation of these so-called overflow metabolites does not consume energy. On the contrary, these metabolites often (perhaps invariably) are associated with an increased energy flux, as manifested by an increased rate of respiration.

When cells are growing anaerobically by fermentation, accumulation of metabolites in the extracellular fluids is a sine qua non. However, when growing at a fixed rate, glucose-sufficient chemostat cultures of *E. coli* and *K. pneumoniae* consumed glucose at enhanced rates as compared to glucose-limited cultures

TABLE 5 Comparison of glucose consumed and products formed by variously limited aerobic chemostat cultures of *E. coli* PC-1000[a]

Limiting and substrate and dilution rate (h⁻¹)	Amt of glucose used (mmol of carbon)	Amt of product formed (mmol of carbon)[b]:						Carbon recovery (%)	$Y_{glucose}$ (g/mol)
		CO_2	Acetic acid	Pyruvic acid	D-Lactic acid	2-Oxyglutaric acid	Extracellular polysaccharide		
Glucose									
0.18	40.5	16.1						89	74.1
0.30	40.1	15.7						89	74.8
Ammonia									
0.18	84.6	29.2	9.8	4.6		6.3	11.9	97	35.5
0.34	54.9	17.1	5.3			5.2	9.1	103	54.6
Potassium									
0.15	162	62.9	20.2	23.1	4.9		3.8	83	18.6
0.30	107	35.1	16.5	2.9	0.8		20.0	89	28.1

[a]To facilitate comparison, all values for glucose used and products formed have been adjusted to a cell production rate of 20 mmol of carbon per h (37).
[b]The amount of cells formed was 20 mmol of carbon under all conditions shown.

(54; J. L. Snoep, Ph.D. thesis, University of Amsterdam, Amsterdam, The Netherlands, 1992). Under conditions of glucose excess, a branched fermentation pattern was manifest, with a concomitant lowering of the overall efficiency of ATP generated (i.e., net moles of ATP formed per mole of glucose fermented). Nevertheless, this decrease in the efficiency factor was smaller than the increase in glucose consumption rate, and again, just as with aerobic cultures, the apparent Y_{ATP} values were low.

It remains a matter for speculation whether, in these anaerobic cultures, the formation of some products is energetically favorable, neutral, or unfavorable. For example, the formation of D-lactate might be accompanied by a net ATP gain (as would be the case if it arose from the reduction of pyruvate or if lactate excretion generated a proton motive force [32]), or it might cause a severe energy drain if it arose via dihydroxyacetone phosphate and methylglyoxal (8, 9).

ENERGY-DISSIPATING PROCESSES

In view of the above considerations, there must be endergonic processes that do not lead to a net increase of biomass or the excretion of a metabolic product into the culture fluid. We call these energy-dissipating processes.

Maintenance Functions

As mentioned, Monod (33) conceived that a small amount of energy must be dissipated in the maintenance of cell integrity and viability. But although this concept now has been repeatedly validated experimentally, it still is not clear what precise physiological functions constitute maintenance. This issue is clouded and confused by the widely disparate maintenance values cited in the literature for different strains of the same species growing in putatively identical media (Table 1), as well as for a single strain growing on a range of different carbon substrates (17). In this latter study, exceptionally high values of the apparent oxygen consumption rate for maintenance (m_O) were reported for the growth of *E. coli* B on glycerol, succinate, glutamate, and acetate (respectively, 10.2, 13.1, 17.1, and 25.9 mmol of O per h per g [dry weight] of cells) as compared with values of 0.9 and 1.8 mmol of O per h per g (dry weight) of cells for growth on glucose and galactose, respectively. Moreover, at the other extreme, microcalorimetric studies of glucose-limited chemostat cultures of

E. coli K-12 (25) indicated a maintenance rate of heat evolution of only 0.02 kcal/h per g (dry weight) of cells (1 cal = 4.184 J), which would equate with an oxygen consumption rate of no more than 0.36 mmol of O per h per g (dry weight) of cells. As yet, there is no rational explanation for these widely differing maintenance rates. If the quantitatively major components of maintenance were turnover of macromolecules, maintenance of ion gradients, and motility, there is no obvious reason why the maintenance requirement of an acetate-limited culture should be almost 30-fold higher than that of a glucose-limited one.

High apparent maintenance rates of oxygen or carbon substrate consumption or both are commonplace with chemostat cultures in which growth is limited by the availability of an anabolic substrate (e.g., source of nitrogen, sulfur, or phosphorus) or an essential cation (K⁺ or Mg²⁺). Here, the question becomes how the uptake of carbon substrate can be modulated to meet the low biosynthetic and bioenergetic demands of cell synthesis. Because washed suspensions of bacteria, which clearly cannot grow, often oxidize substrates such as glucose at a high rate, one may conclude that such regulation, if present, is by no means stringent. It seems reasonable that the high rates of oxygen and/or carbon substrate consumption by slowly growing carbon substrate-sufficient chemostat cultures do not necessarily result from some enhanced specific maintenance energy requirement, but more likely from partial uncoupling of catabolism from anabolism.

To analyze yield values in physiological terms one needs to know more about mechanisms that might dissociate catabolism from anabolism and how they might be modulated. We address these questions below.

Regulatory Processes

Regulatory processes are an important class of energy-consuming reactions that do not contribute directly to net biomass synthesis. One could consider energy consumption by these reactions as part of the maintenance energy requirement, but in view of their special nature we will deal with these regulatory processes separately.

Energy Flux Regulation. Several researchers have proposed that biological energy converters have evolved toward maximal energy output (biomass formation) and consequently function at reduced efficiency (14, 28, 50). This hypothesis is based on the application of nonequilibrium thermodynamics to energy conversion in biological systems. Westerhoff and van Dam (63) have analyzed this problem in detail, and they, like their predecessors, concluded that maximal efficiency of energy generation is possible only when the rate of energy generation is zero. Maximal output power is attained at an efficiency of free energy conservation of 50%. This could explain the large difference between calculated and experimentally derived growth yields (see above). The interesting question therefore arises of whether and how the efficiency of energy conversion is regulated.

There are two mechanisms by which a microbe could decrease its efficiency of energy conservation: first, ATPase reactions, and second, mechanisms that lower the efficiency of energy generation by the respiratory chain or substrate-level phosphorylation. Several lines of evidence indicate that at high proton motive force values, energy dissipation occurs by leakage of protons through the membrane into the cytoplasm or by a slip in the proton pumps (for review, see reference 64). The uncoupled NADH dehydrogenase (NDH-2) could also play a role: electron transport catalyzed by this enzyme would lead to a 50% lower efficiency of energy generation when cytochrome bo is the terminal oxidase and 2/3 lower when cytochrome bd is the terminal oxidase. Experimental evidence for this physiological role of NDH-2 is not available. The ndh strain of $E. coli$ used by Calhoun et al. (5) grew in a glucose-limited chemostat culture without apparent difficulty, but its response to sudden changes in the concentration of the growth-limiting nutrient and its behavior under energy-excess conditions were not tested.

Under fermentative conditions, NADH oxidation and ADP phosphorylation can be uncoupled by the reactions of the methylglyoxal bypass (Fig. 3) (53). Methylglyoxal synthase, the first enzyme of the pathway, is strongly inhibited by moderate concentrations of phosphate (<1 mM); it has been suggested that this enzyme recycles phosphate when the intracellular phosphate concentration is low (9). In slowly growing glucose-limited anaerobic cultures of $K. pneumoniae$, the products of glucose fermentation were acetate and ethanol (plus formate, CO_2, and H_2). But when suddenly exposed to a cell-saturating concentration of glucose, 50% of the extra glucose was converted to D-lactate. Because in the short term (20 min) there was a large increase in the glucose consumption rate but no increase in the growth rate, and because the key enzymes of the methylglyoxal bypass were present in substantial amounts, it was concluded that the formation of D-lactate occurred exclusively via methylglyoxal. The other products formed, after the glucose

pulse, were succinate and 2,3-butanediol, and by drawing out the fermentation scheme (Fig. 4) it became clear that this extra glucose was indeed being fermented with no concomitant net formation of ATP. In other words, fermenting this extra glucose through the methylglyoxal bypass generates no ATP. Because the enzymes that effect a conversion of dihydroxyacetone phosphate to D-lactate, via methylglyoxal, are formed constitutively in $E. coli$ (8, 9), there is little doubt that this organism also can totally uncouple glycolysis from ADP phosphorylation. The fact that $E. coli$ as well as $K. pneumoniae$ can synthesize D-lactate from either pyruvate (via a nicotinamide adenine dinucleotide [NAD]-linked lactate dehydrogenase [51]) or dihydroxyacetone phosphate (via methylglyoxal synthase and glyoxylase) renders difficult an accurate assessment of the energy balance in those anaerobic cultures in which this metabolite is a major product. For example, a homolactic fermentation of glucose (to D-lactate) could yield, maximally, 2 mol of ATP per mol of glucose (if only the NAD-linked lactate dehydrogenase was active) or, minimally, zero ATP if 50% of the lactate arose via methylglyoxal. Anaerobically, no more than 50% of the glucose carbon can be converted to D-lactate via the methylglyoxal bypass, since the conversion of glucose to the two triose phosphates consumes 2 mol of ATP and this must be recovered by glycolysis. Aerobically, however, all the glucose carbon could be converted to D-lactate via methylglyoxal, and the D-lactate could then be oxidized to pyruvate to generate ATP for phosphorylating glucose and fructose 6-phosphate. Indeed, since the oxidation of D-lactate to pyruvate by the respiratory chain-linked dehydrogenase invokes just one site of energy conservation, glucose could be aerobically catabolized solely to pyruvate without any net gain of biologically usable energy.

Relevant in the context mentioned above is the observation that glucose-limited aerobic chemostat cultures of $K. pneumoniae$ and $E. coli$ can immediately accelerate the rate of glucose consumption when suddenly relieved of the growth limitation (36). There was a concomitant marked increase in the respiration rate but no corresponding increase in the growth rate, indicating that excess energy was being dissipated. Significantly, the extra glucose consumed was not oxidized completely; substantial amounts of pyruvate and acetate were excreted into the medium (M. P. M. Leegwater, Ph.D. thesis, University of Amsterdam, Amsterdam, The Netherlands, 1983). Clearly, if the pyruvate arose from D-lactate one could partially account for the energetic uncoupling; however, the formation of acetate from pyruvate would yield NADH, and therefore a question arises as to whether this could be oxidized without generating a proton

3 Glucose + 6 ATP ⟶

    3 Glyceraldehyde 3-phosphate + 3 Dihydroxyacetone phosphate + 6 ADP

3 Glyceraldehyde 3-phosphate + 3 P_i + 3 ADP ⟶

    1 Succinate + 1 2,3-Butanediol + 6 ATP + 1 CO_2 + H_2O

3 Dihydroxyacetone phosphate ⟶ 3 D-lactate + 3 P_i

Sum: 3 Glucose ⟶ 3 D-lactate + 1 Succinate + 1 2,3-Butanediol + 1 CO_2 + 1 H_2O

FIGURE 4 Pattern of fermentation of glucose by $K. pneumoniae$ implicating no net synthesis of ATP.

motive force. We know of no such mechanism or postulated mechanism, but both *K. pneumoniae* and *E. coli* have constitutively two enzymes (an NAD-linked lactate dehydrogenase that generates D-lactate and a D-lactate dehydrogenase which is flavoprotein) that together can oxidize NADH without invoking site 1 of the respiratory chain. The NAD-linked lactate dehydrogenase is homotropic with respect to pyruvate; hence, an accumulation of pyruvate within the cell would promote this bypass reaction.

Metabolic Flux Regulation. Several biochemical pathways serve more than one function. For example, the respiratory chain generates energy and reoxidizes cofactors. Similarly, glycolysis provides reducing equivalents, energy via substrate level phosphorylation, carbon skeletons for anabolism, and pyruvate as fuel for the tricarboxylic acid cycle. One can imagine growth conditions in which these different functions are not optimally compatible. The coupling of these functions could be loosened if there were a mechanism to regulate the flux through glycolysis such that one process (say, anabolism) is provided optimally with the essential ingredients while another (say, energy generation) is diminished. Futile cycles might optimize the regulation of metabolic fluxes (27, 40). When two compounds are interconverted by two reactions in which one consumes more energy than the other generates, there is futile cycling which dissipates energy without changing metabolite concentrations.

One possible futile cycle in *E. coli* is the interconversion of fructose 6-phosphate and fructose 1,6-bisphosphate, effected by phosphofructokinase and fructose-1,6-bisphosphatase, respectively (6, 10). However, little if any futile cycling could be observed in batch cultures growing on either glucose or gluconeogenic substrates. But the growth conditions used were not those which one would expect to favor extensive futile cycling. Scrutton and Utter (46) found that phosphofructokinase is inhibited by ATP and activated by AMP, whereas fructose-1,6-bisphosphatase is inhibited by AMP and activated by ATP. Hence, rapid ATP generation by respiration when ATP utilization by biosynthetic reactions is constrained might well activate this futile cycle.

E. coli synthesizes both glutamine synthetase and glutaminase, which also could cause futile cycling (43). Because glutamine synthetase is highly active in ammonia-limited cells (29) and because significant amounts of glutaminase also are present (56), futile cycling must occur unless the activity of glutaminase is constrained in some way. Activities of both enzymes are markedly influenced by the energy charge (43); however, the patterns of regulation suggest that when the energy charge is high the glutaminase activity would be low (i.e., about 20% of the maximal activity), though significant. A possible explanation for the presence of glutaminase in ammonia-limited cells is that, by releasing ammonia, it allows the key enzyme of ammonia assimilation (glutamine synthetase) to continue working at a significant rate in the virtual absence of exogenous ammonia. This, one might argue, permits the uptake system to remain highly active during transient changes in the supply of growth-limiting nutrient. This hypothesis could be tested by determining whether the intracellular pool level of ammonia is maintained at a significant value after an interruption in the supply of nutrients to an ammonia-limited culture. To the best of our knowledge, this experiment has not been performed.

If some futile cycles are involved in regulating the rate of uptake of growth-limiting nutrients, one might expect similar reactions to be associated with common nutrients such as sulfate, phosphate, or potassium. Indeed, it has been shown that *E. coli* has two enzymes [adenosine-5'-phosphosulfate kinase and 3'(2'),5'-diphosphonucleoside 3'-phosphohydrolase] that are potentially capable of acting as a futile cycle (61). Similarly, phosphatases are synthesized to high levels by phosphate-limited cells (60).

The specific rates of oxygen consumption of potassium-limited cultures of *K. pneumoniae* or *E. coli* are extremely high (4, 58). The same effect can be observed in glucose-limited cultures with a low [K^+] input, and the respiration rate varies and appears to correlate closely with the transmembrane K^+ gradient (23). It was proposed that much respiratory energy was dissipated as a consequence of an induced K^+ leakage current (58). Subsequently, Mulder et al. (34) postulated that futile cycling of K^+ ions in K-limited *E. coli* cells was caused by the uptake of this ion via the Kdp system (the high-affinity K^+ uptake system, which is derepressed under these growth conditions) and leakage via the low-affinity Trk system.

Buurman et al. (4) investigated this cycle further and found that the very high rates of oxygen consumption (17.1 mmol of O_2 per g [dry weight] of cells per h, at $D = 0.3$ h^{-1}) observed in potassium-limited chemostat cultures of *E. coli* (with ammonium chloride as the nitrogen source) are caused by the Kdp system, which also pumps NH_4^+ ions into the cell. This can be explained by the similar size and charge of the K^+ and NH_4^+ ions (ionic radii: 1.33 Å [0.133 nm] and 1.43 Å, respectively). Intracellularly, NH_4^+ dissociates into NH_3 and a proton. Since the cell membrane of *E. coli* is permeable to NH_3, this molecule will diffuse out if the extracellular [NH_3] is lower. This is the case when the culture pH value is lower than the cytoplasmic pH, or when the Kdp activity is high. In addition, the maintenance of cytoplasmic pH homeostasis requires extrusion of the proton formed by the dissociation of NH_4^+, an energy-consuming process. When alanine was used as the nitrogen source (at concentrations that did not cause the production of ammonia due to deamination), the specific rate of oxygen consumption of a potassium-limited *E. coli* culture was similar to those of phosphate or sulfate-limited cultures (10 to 12 mmol of O_2 per g [dry weight] of cells per h at $D = 0.3$ h^{-1}). Similarly, a mutant strain, in which the Kdp was not active, showed the same lower oxygen uptake rates, indicating that futile cycling of NH_4^+ ions no longer occurred.

Signal Transduction. Changes in the growth environment can be detected by two-component regulatory systems (see chapter 80). These systems consist of a sensor protein, embedded in the membrane, which can phosphorylate an activator protein that subsequently binds at the appropriate place in the genome. Thus detecting a signal from the growth environment costs energy, although no data on the amount of energy consumed by these systems are known. Possibly, processes such as DNA editing, protein folding, and signal transduction are very expensive terms energetically and are responsible for a large part of the difference between the energetic costs of polymer synthesis and the synthesis of a living cell.

GENERAL CONCLUSIONS

Microbial growth is the product of a large number of interconnected enzyme-catalyzed reactions; the fact that, in any particular environment, cell synthesis proceeds with a more or less constant efficiency indicates that a substantial measure of control must be exercised over the fluxes of intermediary metabolites and precursor substances involved in polymer synthesis. Moreover, in chemoheterotrophic organisms such as *E. coli* and *S. typhimurium*, processes of regulation are further complicated by the fact that the energy needed for biosynthesis necessarily must be derived from the breakdown of carbon substrate that simultaneously is being assimilated into cell substance. Hence, one might expect mechanisms operating in these cells to precisely apportion the flow of intermediary metabolites between catabolic and anabolic reactions. Thus, control systems might act at specific branch points between catabolic and anabolic pathways which would be "tuned" to the overall energy status of the cell. That some such controls indeed are present within the microbial cell is abundantly obvious from the widely reported involvement of adenine nucleotides as control elements in intermediary metabolism (7). Indeed, the mode of action of these regulatory processes (e.g., allosteric effectors) suggests a stringent coupling between ATP synthesis and growth which manifests itself in a precise yield value. However, such a concept is untenable because of observations extending back over many years (13, 44, 55); energy (ATP) generation can occur at a high rate when cell synthesis is severely constrained. Clearly, there is no obligatory coupling between catabolism and anabolism, and herein lies the source of much difficulty in attempting to interpret yield data in energetic terms. It is obvious, therefore, that further progress in evaluating the energetics of microbial growth (particularly aerobic growth) hinges critically on the acquisition of a better understanding both of those energy-spilling processes extant within the cell and of their associated regulatory mechanisms. In this chapter we have attempted to make a start along these lines.

LITERATURE CITED

1. Ameyama, M., M. Nonobe, E. Shinagawa, K. Matsushita, K. Takimoto, and O. Adachi. 1986. Purification and characterization of the quinoprotein D-glucose dehydrogenase apoenzyme from *Escherichia coli*. *Agric. Biol. Chem.* 50:49–57.
2. Blackwood, A. C., A. C. Neish, and G. A. Ledingham. 1956. Dissimilation of glucose at controlled pH values by pigmented and non-pigmented strains of *Escherichia coli*. *J. Bacteriol.* 72:497–499.
3. Bouvet, O. M. M., P. Lenormand, and P. A. D. Grimont. 1989. Taxonomic diversity of the D-glucose oxidation pathway in the *Enterobacteriaceae*. *Int. J. Syst. Bacteriol.* 39:61–67.
4. Buurman, E. T., M. J. Teixeira de Mattos, and O. M. Neijssel. 1991. Futile cycling of ammonium ions via the high affinity uptake system (Kdp) of *Escherichia coli*. *Arch. Microbiol.* 155:391–395.
5. Calhoun, M. W., K. L. Oden, R. B. Gennis, M. J. Teixeira de Mattos, and O. M. Neijssel. 1993. Energetic efficiency of *Escherichia coli*: effects of mutations in components of the aerobic respiratory chain. *J. Bacteriol.* 175:3020–3025.
6. Chambost, J. P., and D. G. Fraenkel. 1980. The use of 6-^{14}C-labelled glucose to assess futile cycling in *Escherichia coli*. *J. Biol. Chem.* 255:2867–2869.
7. Chapman, A. G., and D. E. Atkinson. 1977. Adenine nucleotide concentrations and turnover rates. Their correlation with biological activity in bacteria and yeast. *Adv. Microb. Physiol.* 15:253–306.
8. Cooper, R. A. 1984. Metabolism of methylglyoxal in microorganisms. *Annu. Rev. Microbiol.* 38:49–68.
9. Cooper, R. A., and A. Anderson. 1970. The formation and catabolism of methylglyoxal during glycolysis in *Escherichia coli*. *FEBS Lett.* 11:273–276.
10. Daldal, F., and D. G. Fraenkel. 1983. Assessment of a futile cycle involving reconversion of fructose 6-phosphate to fructose 1,6-bisphosphate during gluconeogenic growth of *Escherichia coli*. *J. Bacteriol.* 153:390–394.
11. Farmer, I. S., and C. W. Jones. 1976. The energetics of *Escherichia coli* during aerobic growth in continuous culture. *Eur. J. Biochem.* 67:115–122.
12. Gottschalk, G. 1979. *Bacterial Metabolism*. Springer-Verlag, New York.
13. Gunsalus, I. C., and C. W. Shuster. 1961. Energy yielding metabolism in bacteria, p. 1–58. *In* I. C. Gunsalus and R. Y. Stanier (ed.), *The Bacteria*, vol. 2. Academic Press, Inc., New York.
14. Harder, W., J. P. van Dijken, and J. A. Roels. 1981. Utilization of energy in methylotrophs, p. 258–269. *In* H. Dalton (ed.), *Microbial Growth on C₁ Compounds*. Heyden & Son Ltd., London.
15. Harrison, D. E. F. 1976. The regulation of respiration rate in growing bacteria. *Adv. Microb. Physiol.* 14:423–313.
16. Hartl, F.-U., R. Hlodan, and T. Langer. 1994. Molecular chaperones in protein folding: the art of avoiding sticky situations. *Trends Biochem. Sci.* 19:20–25.
17. Hempfling, W. P., and S. E. Mainzer. 1975. Effects of varying the carbon source limiting growth on yield and maintenance characteristics of *Escherichia coli* in continuous culture. *J. Bacteriol.* 123:1076–1087.
18. Herbert, D. 1958. Some principles of continuous culture, p. 381–396. *In Recent Progress in Microbiology*. Proceedings of the VII International Congress of Microbiology, Stockholm.
19. Hernandez, E., and M. J. Johnson. 1967. Anaerobic growth yields of *Aerobacter cloacae* and *Escherichia coli*. *J. Bacteriol.* 94:991–995.
20. Hommes, R. W. J., W. A. M. Loenen, O. M. Neijssel, and P. W. Postma. 1986. Galactose metabolism in *gal* mutants of *Salmonella typhimurium* and *Escherichia coli*. *FEMS Microbiol. Lett.* 36:187–190.
21. Hommes, R. W. J., P. W. Postma, O. M. Neijssel, D. W. Tempest, P. Dokter, and J. A. Duine. 1984. Evidence of a quinoprotein glucose dehydrogenase apoenzyme in several strains of *Escherichia coli*. *FEMS Microbiol. Lett.* 24:329–333.
22. Hommes, R. W. J., J. A. Simons, J. L. Snoep, P. W. Postma, D. W. Tempest, and O. M. Neijssel. 1991. Quantitative aspects of glucose metabolism by *Escherichia coli* B/r, grown in the presence of pyrroloquinoline quinone. *Antonie van Leeuwenhoek* 60:373–382.
23. Hueting, S., T. de Lange, and D. W. Tempest. 1979. Energy requirement for the maintenance of the transmembrane potassium gradient in *Klebsiella aerogenes* NCTC 418: a continuous culture study. *Arch. Microbiol.* 123:183–188.
24. Ingledew, W. J., and R. K. Poole. 1984. The respiratory chains of *Escherichia coli*. *Microbiol. Rev.* 48:222–271.
25. Ishikawa, Y., and M. Shoda. 1983. Calorimetric analysis of *Escherichia coli* in continuous culture. *Biotechnol. Bioeng.* 25:1817–1827.
26. Jones, C. W. 1977. Aerobic respiratory systems in bacteria, p. 23–59. *In* B. A. Haddock and W. A. Hamilton (ed.), *Microbial Energetics: 27th Symposium of the Society for General Microbiology*. Cambridge University Press, Cambridge.
27. Katz, J., and R. Rognstad. 1978. Futile cycling in glucose metabolism. *Trends Biochem. Sci.* 3:171–174.
28. Kedem, O., and S. R. Caplan. 1965. Degree of coupling and its relation to efficiency of energy conversion. *Trans. Faraday Soc.* 61:1897–1911.
29. Magasanik, B., M. J. Prival, and J. E. Brenchley. 1973. Glutamine synthetase, regulator of the synthesis of glutamate-forming enzymes, p. 65–70. *In* S. Prusiner and E. R. Stadtman (ed.), *The Enzymes of Glutamine Metabolism*. Academic Press, Inc., New York.
30. Marr, A. G., E. H. Nilson, and D. J. Clark. 1963. The maintenance requirement of *Escherichia coli*. *Ann. N.Y. Acad. Sci.* 102:536–548.
31. Martin, J., T. Langer, R. Boteva, A. Schramel, A. L. Horwich, and F.-U. Hartl. 1991. Chaperonin-mediated protein folding at the surface of groEL through a 'molten globule'-like intermediate. *Nature* (London) 352:36–42.
32. Michels, P. A. M., J. P. J. Michels, J. Boonstra, and W. N. Konings. 1979. Generation of electrochemical proton gradient in bacteria by the excretion of metabolic end-products. *FEMS Microbiol. Lett.* 5:357–364.
33. Monod, J. 1942. Recherches sur la croissance des cultures bactériennes. Hermann, Editeurs des Sciences et des Arts, Paris.
34. Mulder, M. M., M. J. Teixeira de Mattos, P. W. Postma, and K. van Dam. 1986. Energetic consequences of multiple potassium uptake systems in *Escherichia coli*. *Biochim. Biophys. Acta* 851:223–228.
35. Neijssel, O. M., G. P. M. A. Hardy, J. C. Lansbergen, D. W. Tempest, and R. W. O'Brien. 1980. Influence of growth environment on the phosphoenolpyruvate:glucose phosphotransferase activities of *Escherichia coli* and *Klebsiella aerogenes*: a comparative study. *Arch. Microbiol.* 125:175–179.
36. Neijssel, O. M., S. Hueting, and D. W. Tempest. 1977. Glucose transport capacity is not the rate-limiting step in the growth of some wild-type strains of *Escherichia coli* and *Klebsiella aerogenes* in chemostat culture. *FEMS Microbiol. Lett.* 2:1–3.
37. Neijssel, O. M., and D. W. Tempest. 1975. The regulation of carbohydrate metabolism in *Klebsiella aerogenes* NCTC 418 organisms growing in chemostat culture. *Arch. Microbiol.* 106:251–258.
38. Neijssel, O. M., and D. W. Tempest. 1976. Bioenergetic aspects of aerobic growth of *Klebsiella aerogenes* NCTC 418 in carbon limited and carbon sufficient chemostat culture. *Arch. Microbiol.* 107:215–221.
39. Neijssel, O. M., D. W. Tempest, P. W. Postma, J. A. Duine, and J. Frank. 1983. Glucose metabolism by K$^+$ limited *Klebsiella aerogenes*: evidence for the involvement of a quinoprotein glucose dehydrogenase. *FEMS Microbiol. Lett.* 20:35–39.
40. Newsholme, E. A. 1978. Substrate cycles: their metabolic, energetic and thermic consequences in man. *Biochem. Soc. Symp.* 43:183–205.
41. Pirt, S. J. 1965. The maintenance energy of bacteria in growing cultures. *Proc. R. Soc. London Ser. B* 163:224–231.
42. Pirt, S. J. 1982. Maintenance energy: a general model for energy-limited and energy-sufficient growth. *Arch. Microbiol.* 113:300–302.
43. Prusiner, S. 1973. Glutaminases of *Escherichia coli*: properties, regulation and evolution, p. 293–316. *In* S. Prusiner and E. R. Stadtman (ed.), *The Enzymes of Glutamine Metabolism*. Academic Press, Inc., New York.
44. Rosenberger, R. F., and S. R. Elsden. 1960. The yields of *Streptococcus faecalis* grown in continuous culture. *J. Gen. Microbiol.* 22:726–739.
45. Schulze, K. L., and R. S. Lipe. 1964. Relationship between substrate concentration, growth rate and respiration rate of *Escherichia coli* in continuous culture. *Arch. Mikrobiol.* 48:1–20.
46. Scrutton, M. C., and M. F. Utter. 1968. The regulation of glycolysis and glyconeogenesis in animal tissues. *Annu. Rev. Biochem.* 37:249–302.
47. Snoep, J. L., M. R. de Graef, A. H. Westphal, A. de Kok, M. J. Teixeira de Mattos, and O. M. Neijssel. 1993. Differences in sensitivity to NADH of purified pyruvate dehydrogenase complexes of *Enterococcus faecalis*, *Lactococcus lactis*, *Azotobacter vinelandii*, and *Escherichia coli*: implications for their activity in vivo. *FEMS Microbiol. Lett.* 114:279–284.

48. **Stouthamer, A. H.** 1977. Energetic aspects of the growth of microorganisms, p. 285–315. *In* B. A. Haddock and W. A. Hamilton (ed.), *Microbial Energetics: 27th Symposium of the Society for General Microbiology.* Cambridge University Press, Cambridge.

49. **Stouthamer, A. H.** 1979. The search for correlation between theoretical and experimental growth yields. *Int. Rev. Biochem.* **21**:1–47.

50. **Stucki, J. W.** 1980. The optimal efficiency and the economic degrees of coupling of oxidative phosphorylation. *Eur. J. Biochem.* **109**:269–283.

51. **Tarmy, E. M., and N. O. Kaplan.** 1968. Chemical characterization of D-lactate dehydrogenase from *Escherichia coli* B. *J. Biol. Chem.* **243**:2579–2586.

52. **Teissier, G.** 1936. Les lois quantitatives de la croissance. *Ann. Physiol. Veg.* (Paris) **12**:527–586.

53. **Teixeira de Mattos, M. J., H. Streekstra, and D. W. Tempest.** 1984. Metabolic uncoupling of substrate-level phosphorylation in anaerobic glucose-limited chemostat cultures of *Klebsiella aerogenes* NCTC 418. *Arch. Microbiol.* **139**:260–264.

54. **Teixeira de Mattos, M. J., and D. W. Tempest.** 1983. Metabolic and energetic aspects of the growth of *Klebsiella aerogenes* NCTC 418 on glucose in anaerobic chemostat culture. *Arch. Microbiol.* **134**:80–85.

55. **Tempest, D. W.** 1978. The biochemical significance of microbial growth yields: a reassessment. *Trends Biochem. Sci.* **3**:180–184.

56. **Tempest, D. W., J. L. Meers, and C. M. Brown.** 1970. Synthesis of glutamate in *Aerobacter aerogenes* by a hitherto unknown route. *Biochem. J.* **117**:405–407.

57. **Tempest, D. W., and O. M. Neijssel.** 1978. Eco-physiological aspects of microbial growth in aerobic nutrient-limited environments. *Adv. Microb. Ecol.* **3**:105–153.

58. **Tempest, D. W., and O. M. Neijssel.** 1984. The status of Y_{ATP} and maintenance energy as biologically interpretable phenomena. *Annu. Rev. Microbiol.* **38**:459–486.

59. **Terroine, E., and R. Wurmser.** 1922. L'énergie de croissance. I. Le développement de l'*Aspergillus niger. Bull. Soc. Chim. Biol.* **4**:519.

60. **Torriani, A.** 1960. Influence of inorganic phosphate in the formation of phosphatases by *Escherichia coli. Biochim. Biophys. Acta* **38**:460–469.

61. **Tsang, M. L. S., and J. Schiff.** 1976. Sulfate reducing pathways in *Escherichia coli* involving bound intermediates. *J. Bacteriol.* **125**:923–933.

62. **van Schie, B. J., K. J. Hellingwerf, J. P. van Dijken, M. G. L. Elferink, J. M. van Dijl, J. G. Kuenen, and W. N. Konings.** 1985. Energy transduction by electron transfer via a pyrrolo-quinoline quinone-dependent glucose dehydrogenase in *Escherichia coli*, *Pseudomonas aeruginosa*, and *Acinetobacter calcoaceticus* (var. *lwoffi). J. Bacteriol.* **163**:493–499.

63. **Westerhoff, H. V., and K. van Dam.** 1987. *Thermodynamics and Control of Biological Free-Energy Transduction.* Elsevier, Amsterdam.

Escherichia coli Recombinant DNA Technology

JAMES R. SWARTZ

108

INTRODUCTION

In 1973, the first gene was cloned; in 1977, the first recombinant DNA protein was produced. These events have given birth to a new industry driven by exciting applications and fueled by a tremendous body of knowledge accumulated from decades of research. *Escherichia coli*, one of the most intensively studied organisms, has played a dominant role. It was the demonstration organism for most of the significant advances, and it continues to play a major role in commercial applications. This chapter provides a brief history as well as a current summary of this technology.

By design, this chapter is somewhat unusual for this compendium. It is intended to cover the commercial applications of science rather than science itself. However, the application of scientific knowledge to useful purposes often generates new knowledge; some of it quite basic. This is certainly the case for recombinant DNA (rDNA) biotechnology, and this chapter provides many examples.

Because of the focus of this chapter on applied science, many patents, in addition to journal articles and book chapters, are referenced. The patent references identify intellectual property and expand the breadth of coverage. However, this chapter is not and cannot be a complete account of all applications and all useful science and technology related to *E. coli* rDNA technology. I have attempted to describe and provide examples of the most prominent applications and the most basic and useful science and technology. Much of the chapter deals with the production of rDNA proteins as products, mostly pharmaceuticals. This application is, to date, the most commercially important one, and it is hoped that this focus will adequately convey the science and technology which have allowed *E. coli* to deliver so many benefits through rDNA technology.

History of Basic Techniques

It is difficult to say when the history of rDNA technology really began—perhaps in 1944 with the discovery by Avery and colleagues that DNA was the "transforming principle," perhaps with the elucidation of the DNA structure in 1953 by Watson and Crick and the deciphering of the genetic code in the mid-1960s, perhaps with the description of restriction endonucleases by Arber in 1962. However, the potential power of rDNA biotechnology did not become apparent until the first cloning of a gene in 1973 in the laboratories of Boyer and Cohen (40, 51, 153; S. N. Cohen and H. W. Boyer, U.S. patent 4,468,464, August, 1984). Those initial results have formed the basis of a multibillion dollar industry with the potential to affect all of our lives.

Other key events are listed in Table 1. The first recombinant protein to be produced in the laboratory was somatostatin (103). It was produced in the cytoplasm of *E. coli* as part of a larger fusion protein. The mature polypeptide was then released by cyanogen bromide to produce a molecule recognizable by antisomatostatin antibodies and with the ability to inhibit the release of growth hormone from rat pituitary cells. Although this was a clear demonstration of the technology's potential, it was not a commercial success. Insulin, the next protein to be expressed, was. The A and B chains of human insulin were separately produced in fusion proteins (84) and were released by cyanogen bromide cleavage. The two chains were then assembled into active human insulin by an in vitro folding reaction. This was a major accomplishment in molecular biology and in protein biochemistry.

TABLE 1 Key events in the history of *E. coli* rDNA technology

Year	Event
1944	DNA is found to be the transforming principle
1953	DNA structure is determined
1962	Restriction endonucleases are described
1966	Genetic code is deciphered
1973	First cloned gene is produced
1977	First recombinant protein is produced
1982	First rDNA protein pharmaceutical is approved
1985	Phage display technology is described

The first mature protein to be produced was human growth hormone (83). Although the recombinant protein contained an extra methionine residue at its N terminus, it proved to be fully active after extraction, solubilization, and a gentle in vitro folding reaction. These accomplishments set the stage for a rapid series of successes in expressing heterologous proteins in the cytoplasm of *E. coli.*

However, rDNA proteins produced cytoplasmically often accumulate as insoluble inclusion bodies (refractile particles). They also often have an extra methionine at the N terminus. To avoid these limitations, researchers began to investigate the secretion of heterologous proteins into the periplasmic space. The first attempts also used fusion proteins (188, 203; T. J. Silhavy, H. A. Shuman, J. Beckwith, and M. Schwartz, U.S. patent 4,336,336, June, 1982; W. Gilbert, S. A. Broome, L. J. Villa-Komaroff, and A. A. Efstratiadis, European patent 6694, January, 1980), but subsequent work demonstrated that mature proteins could be transported to and recovered from the periplasmic space (88, 98; W. Gilbert and K. Talmadge, European patent 38,182, October, 1981; G. L. Gray and H. L. Heyneker, European patent 0,127,305, May, 1984). Although there are many examples of heterologous proteins which fold properly in the periplasmic space (41, 99, 161, 212), secreted proteins often are deposited as periplasmic inclusion bodies (30, 218). Thus, the history of *E. coli* rDNA technology has continually been marked by the need to solubilize and fold aggregated proteins. As will be described later, a significant degree of success has been achieved.

In 1985, Smith reported the first example of phage display technology (195). This technique uses bacteriophage to display a family of closely related polypeptides by fusing the structural genes for the desired polypeptides to the gene for a bacteriophage coat protein. The affinity of the desired protein for its ligand, receptor, substrate, or antibody can be used to separate the phage containing the desired gene from the population. This technique has been further developed (18, 140, 144) into a powerful screening technique to identify rare proteins or peptides in large libraries and to simultaneously isolate the corresponding gene.

The history of *E. coli* rDNA technology would not be complete without mentioning a companion and often competitive technology, the expression of heterologous proteins in mammalian cells. Two of the protein pharmaceuticals targeted in the early 1980s, erythropoietin and tissue plasminogen activator, are relatively complex, glycosylated proteins. Attempts to produce active erythropoietin and tissue plasminogen activator with *E. coli* were not successful, and it became apparent that the glycosylation chains conferred desirable biochemical properties

(63, 85). Since then, significant investment has been made in mammalian cell technology for the production of heterologous proteins. Mammalian cell culture, primarily using immortalized Chinese hamster ovary cells, has become a successful technology. Although these developments have limited the use of *E. coli* in making high-value protein pharmaceuticals, *E. coli* continues to be an important producer of recombinant proteins. (In fact, it is interesting that a tissue plasminogen activator analog which may have more desirable qualities is now being produced by *E. coli* and is in the latter stages of development [135].)

Biotechnology Applications

In the Laboratory. *E. coli*-based rDNA technology has provided and continues to provide the foundation for the entire rDNA industry, especially with respect to basic laboratory techniques. Whether applied to basic research or to the development of commercial products, most laboratory techniques depend on *E. coli* technology (176). DNA libraries are generally constructed in *E. coli.* Most DNA manipulations are done with *E. coli* plasmids. DNA sequencing generally uses *E. coli* to produce DNA of the proper size and composition for efficient sequencing. Protein production is generally first tested in *E. coli,* and the organism provides an excellent expression system for protein design and screening (48), especially with phage display technology (18, 136). Finally, *E. coli* extracts can be used effectively for the in vitro synthesis of both natural and "unnatural" (i.e., with mutated amino acid sequences or with modified amino acids) proteins (65).

Pathway Engineering. In the late 1970s and early 1980s, as the power of rDNA technology became apparent, many applications were envisioned. Because of their high unit value, pharmaceutical proteins became the primary focus. However, several early companies, including Amgen, Cetus, and Engenics, began projects in which the rDNA protein was not the product. Instead, the rDNA proteins were intended to either amplify or alter existing biosynthetic pathways to overexpress existing or new metabolites. This type of endeavor became known as pathway engineering or, more generally, metabolic engineering (9). More recently, these approaches have been applied to enhancing the ability of *E. coli* to overexpress heterologous proteins (7, 47, 57, 162) and to designing microorganisms to biodegrade pollutants (217).

Two of the first examples of rDNA pathway engineering in *E. coli* addressed the production of tryptophan (2, 3) and indigo (69). In the first case, endogenous, rate-limiting enzymes were modified and overexpressed to increase the synthetic rate of an existing biosynthetic pathway. In the second example, overexpression of an enzyme from a different organism created a new pathway which produces indigo, a useful product which otherwise is not formed.

Another interesting example of pathway engineering is the production of ethanol by *E. coli* (5, 32, 102). Here the basic glycolytic pathway is modified for ethanol production by overexpressing an efficient alcohol dehydrogenase and a pyruvate carboxylase with a high affinity for pyruvate. Together, these modifications allow the efficient conversion of cellulose hydrolysates into ethanol.

In general, pathway engineering has been applied to the production of small biomolecules which have relatively low unit value. Because of correspondingly large production volumes, large capital investments are required. Often the competing production technology is highly evolved. The new technology must deliver high specific productivities and high product yields from inexpensive, readily available substrates, and the process must be stable and easily controlled. Also, the market must be reasonably stable. To date, rDNA technology has made only modest inroads into this very competitive business. However, early problems such as plasmic instability (101) have motivated the development of new techniques (191, 229) and new approaches (204). Also, there is now an increased understanding that the supporting metabolic processes, as well as the target pathway (9), may need to be augmented. This realization, in fact, is motivating the search for a better understanding of metabolic control mechanisms (128). As this knowledge base grows, it is likely that there will begin to be successful examples of production processes for small biomolecules which are based on *E. coli* rDNA pathway engineering.

Recombinant Protein Production. This is the most direct and, to date, most successful application of rDNA technology in *E. coli*. As can be seen in Table 2, *E. coli* rDNA technology has fostered a significant new industry. The field is dominated by human pharmaceuticals, a situation that is unlikely to change in the near future. Four of the first five products have become "blockbuster" products. Many more *E. coli*-produced, protein pharmaceuticals are now in development.

As important as the revenues are to companies and investors, a more dramatic value is realized by those who enjoy the benefits brought by these products. Diabetics now no longer need to fear producing antibodies to animal insulins. Growth hormone-deficient children can avoid the pains of dwarfism without the fear of contracting Kreutzfeld-Jacob syndrome. Children with chronic granulomatous disease can hope for a normal life thanks to the benefits of gamma interferon therapy. Thousands who must endure chemotherapy or radiation therapy for cancer can expect more rapid recoveries with fewer infections because of granulocyte colony-stimulating factor. These products are bringing hope and life itself to millions.

It is interesting to examine the biochemistry of these products in more detail. Although insulin is dimeric, in general the early products were small proteins with relatively simple tertiary structures. In most cases, production depended on in vitro folding. In many cases, it still does. However, *E. coli* rDNA technology is no longer limited to simple, small proteins.

In many cases, in vivo folding is possible. Perhaps the most dramatic example is the cytoplasmic production of properly folded, active human hemoglobin (96). Coexpression of the α- and β-globin polypeptides results in an assembled, soluble tetrameric protein of 50 kDa containing the required heme group. More frequently, it is necessary to transport the protein out of the cytoplasm in order to obtain in vivo folding. This is the case, for example, for Fab antibody fragments which fold properly and form several essential disulfide bonds when secreted into the periplasmic space (24, 165). There are now many examples of successful periplasmic folding of secreted rDNA proteins (99, 104, 117, 220). In addition, improved technology has been developed for in vitro protein folding, when required (50, 174).

These accumulated technologies have greatly expanded the range of proteins that can be successfully produced in *E. coli*. The technology can be applied to the production of human and animal vaccines, animal pharmaceuticals, enzymes, bioadhesives, and biomaterials (82). In the early days of *E. coli* rDNA protein production, it was not uncommon for protein production to cost thousands of dollars per gram of active protein. Now, a product such as bovine growth hormone is being produced by *E. coli* technology for

TABLE 2 Sales of rDNA proteins in 1993

Product	Year of U.S. approval	1993 sales (world estimates, million $)
Produced by *E. coli*		
Insulin	1982	560
Human growth hormone	1985	983
Alpha interferon	1986	1,576
Gamma interferon	1990	5
Rennin (chymosin)	1990	Proprietary
Granulocyte colony-stimulating factor	1991	719
Interleukin-2	1992	36
Bovine growth hormone	1993[a]	
Beta interferon	1993	≈434[b]
Total		4,313
Produced by mammalian cells		
Orthoclone OKT3	1986	100
Tissue plasminogen activator	1987	236
Erythropoietin	1989	1,806
Factor VIII	1992	41
DNase	1993	88[c]
Total		2,183
Produced by yeast cells		
Insulin	1987	≈600
Hepatitis B vaccine	1986	725
Total		1,325

[a]Sales began in February 1994.
[b]Includes 1993 sales in Japan plus expected first-year sales in the United States (approved in October 1993).
[c]Includes first-year sales (approved in December 1993).

a price-sensitive market. As judged from selling prices, it appears that production costs may be less than $10/g. These low costs require extensive process development and the economies of large-scale production, but we now know that they are feasible. Thus, *E. coli* rDNA protein production technology has progressed to an amazing extent since it produced the first rDNA protein. Several rapidly developing technologies are in competition. These include mammalian cell, insect cell, and animal and plant production of recombinant proteins. It is likely that each will be important, and it will be interesting to see what role evolves for each over time. In many ways, the *E. coli* technology is the most developed. Although it is certainly not the answer for all applications, it appears likely that *E. coli* technology will continue to be of major importance for the production of recombinant proteins.

Why *E. coli*?

The best answer to this question lies in the preceding discussion and in Table 2. Nonetheless, it is useful to reflect on both the strengths and limitations of the technology.

Since *E. coli* has been the dominant microbiological research target for biochemists, geneticists, and microbial physiologists for decades, it was logical that it would be the first target for molecular biologists. Because the first rDNA work as well as the first applications utilized *E. coli*, approaches based on this organism established a considerable head start in accumulating relevant knowledge and demonstrated utility. However, the real strength of *E. coli* rDNA technology can be found in its versatility and its potential for enabling the production of proteins and efficient metabolic pathways at low cost.

Versatility. *E. coli* is clearly the organism with which we can most effectively combine an extensive base of genetic and physiologic knowledge, a well-developed ability to quickly alter the organism, and the ability to quickly assess the consequences of these alterations.

As this compendium shows, a tremendous knowledge base for *E. coli* has accumulated after decades of intensive study. With this knowledge has come the ability to quickly and precisely modify the chromosome (91, 168) and to implement plasmid-based modifications (76, 176). Because of the organism's fast growth rate and ease of culture, we can quickly assess the consequences of any changes. Alteration of the chromosome has allowed the implementation of such improvements as a reduction in protease activity (13), the avoidance of amino acid analog incorporation (27), alteration of promoter control (S. Bass and J. R. Swartz, U.S. patent 5,304,472, April, 1994), stabilization of plasmid copy number (191), alteration of basic metabolic carbon flow (47), and the formation of intracellular disulfide bonds (58), to name just a few examples.

Modification of the expression vector is rapid and simple, and the desired number of gene copies can be quickly and reliably achieved (12, 182). This capability allows the rapid characterization of a variety of changes affecting transcription, translation, protein export, and even protein folding (221).

With the combination of the strong knowledge base, ease of manipulation, and rapid assessment of effects, *E. coli* technology provides an unparalleled versatility in optimizing rDNA protein production or enhanced metabolic function.

Potential for Low-Cost Production. To achieve low-cost production, several factors must come together. We need high-volume production (low capital costs), reasonable raw material costs (inexpensive substrates and high product yields), inexpensive isolation and purification procedures, and reliable process performance.

E. coli has favorable intrinsic characteristics relevant to each of these requirements. It has a rapid growth rate (at least an order of magnitude greater than rates of mammalian cells) and the corresponding ability to rapidly produce heterologous proteins as well as to rapidly metabolize substrates for small molecule production. *E. coli* can be grown to very high cell densities, often exceeding 50 g (dry weight)/liter (226, 230). rDNA proteins can be accumulated at levels up to 50% of total cell protein (112, 184), and poly-β-hydroxybutyrate has been accumulated at up to 80% of the cell dry weight (193). *E. coli* has evolved to survive a wide variety of environmental conditions and does not require expensive medium components. The simplicity of the medium and the high metabolic rates also encourage the use of computer control for reliable process performance. These capabilities allow potential productivities and product yields which are significantly higher than those achieved with other expression systems.

The expense involved in extracting, folding (if necessary), and purifying the rDNA proteins is often cited as a disadvantage for *E. coli*. Indeed, that can be true. However, as new methods are being developed for protein isolation and folding (94, 174, 199), this factor is becoming less important. Indeed, the rDNA proteins which have required low-cost production—insulin, human hemoglobin, bovine growth hormone, chymosin, and insulin-like growth factor I (IGF-I)—have all been manufactured by using *E. coli* technology. This is true despite the fact that the four administered parenterally must be highly purified to avoid possible adverse immunologic reactions and that four of the five require in vitro folding.

Even in the very competitive area of small-biomolecule production, recent advances in metabolic pathway engineering (162) are very promising. Again, the rapid metabolic rates and the tremendous knowledge base coupled with the ability to quickly improve the organism appear to offer significant opportunities for low-cost production.

Limitations. The most obvious current limitation for *E. coli* is its inability to glycosylate proteins. The inability to properly fold many rDNA proteins in vivo can also limit its usefulness for some applications. Although the endotoxic lipopolysaccharide of *E. coli* is sometimes mentioned as a disadvantage, in practice it has not been a serious problem. Finally, for many bioconversions and metabolic engineering applications, *E. coli* may be too sensitive to the toxicity of reagents or products as well as to hostile environments. Obviously, *E. coli* will not be the answer for all applications.

rDNA PROTEIN PRODUCTION: THE BIOLOGICAL TECHNOLOGY

Overview

The production of rDNA pharmaceutical proteins, the most developed of the *E. coli* rDNA technologies, will be used here as a vehicle to summarize the general technology. In the time since the first rDNA protein was expressed, a tremendous literature has developed to describe relevant science and technology. Tasks that were once difficult, unreliable, and tedious can now be done easily and quickly with commerically available kits and reagents. Although a significant portion of relevant knowledge is protected as trade secrets, an even larger portion has been made public through patents or through open publication by academic and industrial scientists.

Expression Vectors

Although it is possible to produce recombinant proteins from genes integrated into the *E. coli* chromosome, the vast majority of applications utilize plasmid-based expression. The plasmid must contain three key elements to be useful: (i) an origin of replication to allow autonomous plasmid replication, (ii) an element providing selective pressure to allow cell transformation and plasmid retention, and (iii) a convenient locus or loci for promoter and gene insertion. A variety of plasmids are now available commercially that allow convenient insertion and expression of foreign DNA.

Origin of Replication. The most commonly used origin of replication is that derived from the plasmid ColE1. It is used in pBR322 and its derivatives and in the pUC plasmids (166). Each plasmid is maintained within a characteristic range of number of copies per cell according to the interaction between two plasmid-encoded RNA molecules, RNA I and RNA II (123). RNA I binds to RNA II, inhibiting the maturation of RNA II into a form which is required as a template for plasmid replication. Stronger RNA I-RNA II binding therefore results in lower copy number. This interaction is enhanced by a plasmid-expressed protein called Rop or Rom and is reduced by RNase E cleavage of RNA I (130). For example, pBR322 is normally present at 15 to 20 copies per cell, but when the *rop* gene is deleted as in the pUC plasmids, the copy number increases to 500 to 700 (176). The RNA I-RNA II interaction is also responsible for plasmid incompatibility.

Because the plasmid replication reactions are now relatively well understood, plasmid copy number can be easily modified (154, 182). However, more is not necessarily better. High plasmid copy number can negatively affect host cell metabolism (10). The transcription rate for the desired mRNA is a function of both gene dosage and promoter function. Use of a strong, controllable promoter is normally better for achieving high transcription rates than the use of high-copy-number plasmids.

Alternatively, a so-called runaway plasmid can be used (125, 170; B. E. Uhlen, K. Nordstrom, S. Molin, and P. Gustaffson, U.S. patent 4,487,835, December, 1984). The plasmid is maintained at low copy number during cell growth and is induced to increase copy number at the time of product induction. Although impressive results have been obtained with this approach (71), such complicated regulation is usually not necessary to obtain high product yields.

Selection for Plasmid Introduction and Retention. The plasmid must be designed to provide a growth advantage to the cell under certain conditions in order to select plasmid-bearing cells. Such positive selection may also be necessary to counteract any disadvantage conferred by plasmid presence or by plasmid-driven protein expression. Although the presence of plasmids similar to pBR322 (15 to 20 copies per cell) does not appear to significantly reduce the cell growth rate, high-copy-number plasmids and plasmids expressing significant levels of recombinant proteins do place a significant burden on cell metabolim (23, 26, 44,

160). Therefore, it is advantageous to maintain a selective pressure for plasmid retention and also to use a controllable promoter to minimize the disadvantage of plasmid presence during cell growth.

The standard method for plasmid selection is to use an ample, pBR322 encodes for resistance both to β-lactam antibiotics such as ampicillin and to tetracycline (29). Resistance to other antibiotics, such as chloramphenicol, can also be used (28). While this approach has been successful for years, it should be used carefully. For example, since the mechanism for resistance to β-lactam antibiotics is their degradation, depletion of the antibiotic may remove selective pressure during longer fermentation processes (110, 164). Also, β-lactam antibiotics are not allowed in the production of human pharmaceuticals. Although tetracycline is allowed and is not degraded, the tetracycline resistance mechanism may affect cellular metabolism, with a possible reduction in product yields. The tetracycline resistance conferred by pBR322 and similar plasmids is effected by a plasmid-encoded membrane protein which expels tetracycline from the cytoplasm (142). This resistance gene is classified as a class C gene and thus differs from the more thoroughly studied class B resistance gene in the Tn*10* transposon (146). The Tn*10*-associated resistance is conferred by a proton antiporter which expels a Mg^{2+}-tetracycline complex (223, 224). Unfortunately, we do not yet know whether the class C tetracycline resistance of pBR322 uses the same mechanism. It is known, however, that the pBR322-associated tetracycline resistance can render the cell more sensitive to aminoglycoside antibiotics (146) and better able to transport potassium (62). It is likely that tetracycline resistance has other effects as well, some of which may affect recombinant protein expression and secretion.

Because antibiotic resistance is not always a satisfactory selection, other methods have been devised. These include insertion of mini-F DNA from F factor (228, 229), insertion of the partition locus from plasmid pSC101 (143, 192), use of the *parB* locus of plasmid R1 (79), use of a plasmid-borne repressor of a chromosomally inserted lethal gene (173; C. L. Hershberger and P. R. Rosteck, Jr., U.S. patent 4,650,761, March, 1987), and use of a plasmid-borne *valS* gene in a temperature-sensitive *valS* host (191), among others. Each of these methods offers certain advantages, but plasmid stability requires more than plasmid retention (68, 101). If plasmid presence confers a sufficient disadvantage, the plasmid may become genetically modified. In these cases, it is more effective to lessen the negative impact of plasmid presence (for example, by using tightly controlled promoters on lower-copy-number plasmids) rather than to introduce stronger selection for plasmid retention. Nonetheless, there is still room for developing improved means of plasmid selection and retention. An effective method should allow inexpensive plasmid retention for large-scale production and should not adversely affect metabolic processes related to protein production.

Protein Expression Elements

If the host cell and the plasmid can be classified as the rDNA infrastructure, then the protein expression elements—the promoter, ribosome binding site (RBS), secretion signal sequence, structural gene, and transcription terminator—embody the core of rDNA technology. Although there is still significant room for improvement, a strong body of technology has been assembled for optimizing these components. The field has been reviewed by Glick and Pasternak (82), Shatzman (185), Balbas and Bolivar (12), and more recently Georgiou (76).

Elements Related to Transcription. The promoter is one of the most important elements for recombinant protein production. Ideally, an effective promoter should be regulated for minimal expression during growth and for rapid transcription after induction. The induction signal should be easily applied and inexpensive and should not adversely affect other metabolic processes. Occasionally, it also is useful to modulate promoter induction in order to optimize expression rates for improved protein folding and secretion (30, 120). The promoters normally used are the *lacUV5*, *trp*, *tac*, *lpp-lac*, lambda p_R, and *phoA* promoters. The T7 promoter offers an interesting alternative; its induction is effected by expressing a new RNA polymerase rather than by directly inducing or derepressing the promoter.

The *lacUV5*, *tac*, and *lpp-lac* promoters vary in the efficiency of the RNA polymerase binding site but share the same operator region. All are inducible by galactosides, most commonly by isopropyl-β-D-thiogalactopyranoside (IPTG) or by lactose. IPTG is a strong, gratuitous inducer, but it is relatively expensive for commercial application and may adversely affect cellular metabolism (122). Lactose is less expensive but requires β-galactosidase activity for conversion to allolactose, the true inducer (17). It is important to realize that overexpression of the lac repressor (*lacI*q) is necessary to avoid premature expression from multicopy plasmids and that induction of the lactose permease may adversely affect the proton motive force across the cytoplasmic membrane (1, 64). Despite these limitations, the *lac*-based promoters are often used very successfully. They are strong, are easily induced, and offer the rare capability of being effectively modulated at various inducer concentrations (30).

The *trp* promoter is also commonly used but is somewhat more difficult to control (225). Induction normally requires tryptophan depletion but is often premature, and control may require tryptophan addition, overexpression of the *trp* repressor, or both. The regulation may also be complicated by induction of tryptophanase activity, which then rapidly degrades the inhibitor. Addition of indole-3-acrylic acid provides strong induction. Although sometimes problematic, the *trp* promoter often allows effective accumulation of rDNA proteins which are not toxic to cell metabolism (184).

The lambda p_R promoter is a strong promoter which is controlled by a heat-sensitive repressor (134). The p_L promoter is similar. These promoters are normally induced by raising the culture temperature to 42°C to denature the repressor. While this is effective in inducing strong protein expression, the temperature increase may adversely affect protein folding and will also induce heat shock proteins which may degrade rDNA proteins. The *phoA* promoter is a strong promoter which is easily induced by phosphate limitation (214). Although phosphate limitation may also induce protease expression (198) and cause depletion of ribosome pools (55), a specific mutation in the *phoS* protein which senses periplasmic phosphate concentrations can allow promoter induction without complete phosphate starvation (Bass and Swartz, U.S. patent 5,304,472).

A number of other promoters have also been developed for expression of rDNA proteins. Some combine different features of several systems. For example, Studier et al. (201) have placed the T7 RNA polyerase under control of the *lac* promoter with the T7 promoter placed in front of the structural gene for the desired protein. IPTG induction then initiates rapid and preferential transcription of the desired gene.

The amount of effort expended on the development of controllable promoters attests to their importance. However, there is still room for improvement, especially for large-scale production of proteins inhibitory to growth as well as for systems in which modulated synthesis is beneficial. The induction mechanism should be inexpensive, easily and rapidly applied, and without effect on other cellular processes.

The other element related to transcription is the transcription terminator. The terminator stops the RNA polymerase from extending the message beyond the desired gene. Several have been studied. They often depend on RNA secondary structure and/or helper proteins (45, 141). Although not necessary, they may be beneficial, especially if continued transcription of downstream genes negatively affects other plasmid properties such as copy number or drug resistance (12).

Factors Affecting Translation. Translation of the mRNA by the ribosome is a complex and important step for rDNA protein production. Efficient ribosomal binding to the message is critical. Codon selection for the structural gene may affect translation efficiency, especially for the section encoding the N terminus, and may also affect translational accuracy. Finally, an effective translational stop signal is required.

Ringquist et al. offer recommendations for efficient ribosomal binding (171). A strong RBS (UAAGGAGG) and an optimal length of approximately nine A-rich nucleotides between the RBS and the AUG initiation codon are beneficial. It is also important to minimize possible mRNA secondary structures near the RBS (59, 177). Surprisingly, the nucleotide sequence which encodes the N terminus of the protein can have a profound effect on the efficiency of translation initiation, apparently well beyond the effect predicted by mRNA secondary structure. Devlin et al. increased expression of human granulocyte colony-stimulating factor from undetectable levels to 17% of soluble protein by altering all possible G and C residues at the 5′ end of the gene to A and T (60). Reducing the G-C content was also beneficial for a bovine growth hormone fusion protein (98). Possibly the most dramatic example was presented by Seow et al. (184), who increased human tumor necrosis factor beta accumulation from undetectable levels to 34% of cellular protein by altering several N-terminal codons and using codons preferred for *E. coli* genes.

In general, codon usage by itself has, at most, a modest effect on rDNA protein production. Although Sorensen et al. showed that rare codons may decrease the rate of translation by up to sixfold (197), the results of Ernst and Kawashima suggest that this has little effect on protein production (70). In contrast, codon usage can affect translational fidelity. Seetharam et al. showed that the use of the rare AGA codon for arginine resulted in a significant substitution of lysine for arginine in rapidly translated IGF-I (181). This mistranslation was abolished by using CGT, a preferred codon for *E. coli*. Finally, translation termination seems to be favored by using the stop codon UAA (12), the codon most commonly used for highly expressed *E. coli* genes.

The specialized ribosome approach of Hui and deBoer also deserves mention. They designed a new RBS as well as a mutated ribosome to recognize it (100). When the mutated ribosome is expressed, it efficiently recognizes only the mutated RBS and thereby provides selective tanslation of the desired protein. This approach has recently been reviewed by Leipold and Dhurjati (127).

Secretion Signals. The export of rDNA proteins into the periplasm or outside of the cell offers several potential advantages: a high probability of having the correct N terminus, an improved probability of obtaining proper folding of disulfide-containing proteins upon export to the higher redox potential of the periplasm, protection from the proteases that reside in the cytoplasm, and the possibility of a more selective protein isolation. In almost all cases, a secretion signal sequence is required at the N terminus of the nascent polypeptide to allow export of the nascent polypeptide across the cytoplasmic membrane. Izard and Kendall have recently reviewed the field (105).

The choice of signal sequence can sometimes be critical for rDNA protein export. Voss et al. tested eight different signal sequences for the export of human alpha 2c interferon (212). Only the *STII* signal sequence derived from the heat-stable enterotoxin of *E. coli* allowed detectable export. There are three important portions of the signal sequence: the net positive charge near the N terminus, the central hydrophobic region, and the C-terminal region recognized by leader peptidase (105). The signal sequences can vary significantly in length, but most have about 24 amino acids. Interestingly, the primary sequence of the protein to be translocated may also affect export. A particularly interesting experiment showed that alteration of the N-terminal region of alkaline phosphatase resulted in impaired export (175). Export could be restored, however, by increasing the hydrophobicity of the central region of the signal sequence. The authors suggest that increased signal sequence hydrophobicity may offer a general advantage for the export of foreign proteins in *E. coli*.

Host Characteristics

The *E. coli* host organism provides many essential functions for the production of rDNA proteins. Its importance for process development and optimization rivals that of the expression vector. Nevertheless, the production host has received less attention. For efficient production of rDNA proteins, the host must perform beyond its evolved capabilities. It must supply the protein synthetic machinery as well as the building blocks and energy required to rapidly express a new, foreign protein. It also must be able to store significant quantities of that foreign protein. In addition, we often want the cell to efficiently export and fold foreign proteins through a complicated process specifically evolved for native proteins. We also want the cell to suspend its highly evolved mechanism for dealing with unwanted proteins, namely, proteolysis. Clearly, we are asking a lot, and it is a testament to the versatility of the organism that it has been able to deliver significant successes without major modifications. However, over the years, improvements and increased understanding have been gained.

Throughout the development of *E. coli* rDNA technology, the most commonly used strains have been *E. coli* K-12 derivatives. K-12 strains have been used in laboratory research since 1947, when Tatum and Lederberg discovered genetic recombination by using a K-12 strain. Thereafter, K-12 strains were favored by geneticists, and K-12 derivatives were used for the first recombinant DNA experiments (40, 51, 153).

However, *E. coli* B has been favored by *E. coli* physiologists (156) because of its faster growth rate in minimal medium and the collection of *E. coli* B bacteriophages which can be used as tools for genetic manipulation. As the rDNA technology developed, the molecular biologists continued to use K-12 strains as hosts. The experience gained with K-12 caused *E. coli* B and other strains to fade into the background. This trend was strengthened by the National Institutes of Health guidelines for work with recombinant organisms. The safety of K-12 strains was more actively investigated, and these strains were then given preferential treatment by the guidelines. Large-scale work requires approval by local institutional biosafety committees, which may be reluctant to approve other strains without the same level of safety information that exists for K-12.

There are occasional reports that other strains may offer advantages (201). For example, in addition to exhibiting improved growth on minimal medium, *E. coli* B does not suffer from valine toxicity (210) and is naturally deficient in the *lon* and OmpT proteases (163, 201). However, defined mutations can be introduced into the K-12 strains to avoid both valine toxicity and expression of proteases. K-12 strains are likely to continue to be dominant.

What Host Attributes Do We Want? It will be advantageous for the host to have (i) the ability to provide a surplus of protein synthetic apparatus (building blocks and energy), particularly during the early stationary phase, (ii) precise control over the recombinant promoter function, (iii) a deficiency of harmful protease(s), (iv) resistance to bacteriophage, (v) an enhanced ability to make large quantities of recombinant protein with an accurate primary sequence, (vi) an adequate supply of helper proteins for efficient protein secretion and folding, (vii) improved in vivo protein folding or inclusion body formation, (viii) improved product release, either during or after the fermentation, and (ix) the ability to provide cell extracts which are amenable to efficient folding and purification of the product. The host can be modified to enhance each of these attributes, either by chromosomal changes or by plasmid-based overexpression of chromosomal genes. A complete survey is beyond the scope of this chapter. Instead, examples will be given for improving basic metabolism, promoter control, and product stability.

Improved Basic Metabolism. The effective production of high levels of recombinant proteins places significant demands on the production organism. Efficient production is ideally achieved by growing the cell culture quickly to a high cell density, e.g., 30 to 50 g (dry weight)/liter or higher, and then inducing product formation (226). In almost all fermentors, the high specific metabolic capacity of *E. coli* causes the maximum oxygen assimilation rate to exceed the fermentor's oxygen delivery capacity at a cell density of approximately 10 to 20 g/liter. To avoid oxygen depletion thereafter, it is common to limit the carbon and energy source, normally glucose. As the cell mass increases to higher levels, this limitation results in lower specific growth rates. The culture therefore enters what may be termed an early stationary phase, as illustrated in Fig. 1. The demands on the organism are increased when the promoter is induced, initiating the metabolic demands of rDNA protein expression. If the rDNA promoter is induced by limiting nutrients such as phosphate or tryptophan or by increasing the temperature, the cell's metabolic condition

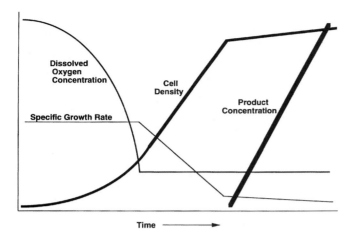

FIGURE 1   Idealized growth and production kinetics for high-cell-density cultures.

is further changed. Finally, the rDNA protein may be toxic. Although careful process optimization and control are important in limiting the metabolic demands, process performance can also be improved by using an organism better able to support high-level rDNA protein production.

One complication of high-density growth and expression is the accumulation of acetate to toxic concentrations when glucose is the carbon and energy source. Continuous culture experiments showed that acetate is formed under aerobic conditions even when glucose is the limiting nutrient (66, 147, 169). Evidently, the rate of pyruvate formation exceeds either the amphibolic (66) or respiration (6) capacity of the organism when the growth rate exceeds 60% of its maximum. Thus, acetate accumulation is difficult to avoid in a batch fermentation. Acetate can begin to inhibit rDNA protein production at approximately 40 mM and growth at 100 mM (106, 133). Even if the accumulated acetate does not reach these levels, it is rapidly assimilated when glucose sufficiently limits growth (Fig. 1). Since acetate is taken up in its protonated form, the resulting pH rise may disrupt metabolism.

A number of researchers have tested host modifications to limit acetate formation by either slowing the rate of pyruvate formation or channeling excess pyruvate into less harmful compounds. Chou et al. (47) introduced a mutation in the *ptsG* gene, which encodes enzyme II of the glucose phosphotransferase system. The mutant produced less acetate and more recombinant protein in batch culture. Interestingly, although its growth rate was reduced in a defined medium, the growth rate was not altered in complex media.

Acetate is formed from acetyl coenzyme A by the consecutive action of phosphotransacetylase and acetate kinase (Fig. 2). Work at Cetus Corp. demonstrated that mutational inactivation of phosphotransacetylase decreased acetate formation and increased the formation of human interleukin-2 in a high-cell-density fermentation (19). However, Diaz-Ricci et al. showed that inactivation of the acetate-forming enzymes caused lactate and pyruvate to accumulate (61). Expression of recombinant pyruvate decarboxylase and alcohol dehydrogenase reduced the pyruvate accumulation but resulted in significant ethanol formation. In later work, Bailey and coworkers showed that divert-

ing glucose 6-phosphate to glycogen formation decreased acetate formation without significant pyruvate accumulation (57). In yet another interesting approach, Aristidou et al. overexpressed an acetolactate synthase gene to convert excess pyruvate into acetoin, a compound 50-fold less toxic than acetate (7). Less acetate accumulated, and growth in shake flasks was significantly improved. The growth advantage in a well-aerated fermentation, however, was less dramatic.

Although the measures described above were all successful in limiting the formation of acetate, they may not always be necessary. Often, high-cell-density fermentations become limited for glucose before acetate reaches inhibitory concentrations. Under glucose limitation, the culture may catabolize the accumulated acetate before induction of product synthesis. If the pH increase can be controlled or is not harmful, the cost of transient acetate accumulation may be minimal. Thus, alteration of the glycolytic pathway may not be necessary for efficient rDNA product formation.

Another class of modifications addresses the effectiveness of aerobic respiration in a large industrial fermentor. Uniform delivery of oxygen throughout a large bioreactor is difficult because of the high respiration rates of *E. coli,* a nonuniform mixing environment, and the low solubility of oxygen. In at least one case (J. R. Swartz, U.S. patent 5,342,763, August, 1995), spatial heterogeneity in the dissolved oxygen concentration caused the organism to rapidly switch from cytochome *o* oxidase to cytochrome *d* oxidase. (Cytochrome *o* oxidase has a lower affinity for oxygen but is more efficient in transferring protons than cytochrome *d* oxidase [35].) The switch caused the dissolved oxygen concentration in the vessel to drop rapidly to zero. Under certain circumstances, the dissolved oxygen level recovered, causing a switch back to the cytochrome *o* oxidase and setting the stage for another precipitous drop in dissolved oxygen. This instability in environmental and metabolic control can disrupt the progress of large-scale cultures but can be prevented by using a host with a disabling mutation in either cytochrome oxidase.

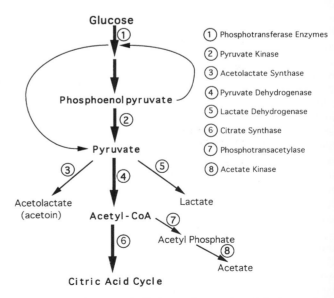

FIGURE 2   Glucose assimilation pathways. Acetyl-CoA, acetyl coenzyme A.

In another modification to the respiration pathway, Khosla and Bailey cloned the gene for a hemoglobin protein expressed by an aerobic, pond-dwelling organism, *Vitreoscilla* sp., into *E. coli* (115). Expression of the protein improved both growth and protein production under oxygen-limited conditions (116). Even in large bioreactors with adequate bulk dissolved oxygen concentrations, the cloned hemoglobin may protect the organism when it encounters regions that are oxygen depleted.

Improved Promoter Control. The host organism can play a major role in contributing both to promoter control and to minimizing the effect of promoter induction. Probably the best known example is the use of a strain which overexpresses the *lac* repressor to control promoters employing the *lac* operator (30, 33). Similarly, the *trp* repressor can be overexpressed to avoid premature expression from the *trp* promoter (215). A more unusual approach was used by Bass and Swartz for control of the alkaline phosphatase promoter (Bass and Swartz, U.S. patent 5,304,472). The protein which senses extracellular phosphate concentration, PstS (PhoS), was mutated to decrease its affinity for phosphate. Thus, the promoter is induced at growth-permitting phosphate concentrations, and continuous phosphate feeding can be used without repressing the promoter. The result is higher rDNA product yields.

Increasing Product Stability. Often, a serious challenge to rDNA product accumulation is degradation by *E. coli* proteases. One possible solution is to express a fusion protein that is more resistant to proteolysis (103). Another is to promote the incorporation of the target protein into insoluble inclusion bodies (46). Expression of protease inhibitors has also been evaluated (189). However, an increasingly important approach is to modify the host organism to reduce its proteolytic activity (145).

There are now several excellent reviews of *E. coli* proteases and their effects on rDNA proteins (14, 67, 87, 138). For intracellular accumulation of rDNA proteins, mutations in the *lon* (la) and *clp* proteases and in *rpoH* (*htpR*) have been most useful. The *lon* protease is the major ATP-stimulated intracellular protease (138, 139). The *clp* protease complex, also stimulated by ATP, contributes to the proteolysis of foreign proteins (113). Both *lon* and *clp* activities can be eliminated by introducing inactivating mutations, but removal of *lon* activity can cause cell mucoidy and UV sensitivity. Secondary mutations are often incorporated to avoid these phenotypes, as described by Gottesman (87).

The *rpoH* (*htpR*) gene encodes the σ^{32} factor, which stimulates production of heat shock proteins (89, 157). Since a subset of these proteins are proteases (including the *lon* protease), mutations in *htpR* can reduce intracellular proteolysis, even in strains deficient in the *lon* protease (34). Since the *htpR* gene product is necessary for growth above 22°C (231), a variety of approaches have been used to decrease its activity during rDNA protein production without adversely affecting cell growth and product synthesis (87, 145).

rDNA proteins transported into the periplasm are threatened by a different set of proteases. These include HtrA (DegP), OmpT, protease III, and Prc (Tsp). Significant improvements in product accumulation have resulted from mutations which eliminate these proteases, either singly or in combination (13, 145, 200; G. Georgiou and F. Baneyx, U.S. patent 5,264,365, November, 1993; J. Beckwith and K. L. Strauch, International Patent Application WO 8805821, August, 1988). However, an *htrA* deletion causes temperature sensitivity. This can be suppressed (11), but increasing the number of protease lesions can lead to depressed growth rates, leaky outer membranes, and other undesirable phenotypes (145). The reasons for these limitations are not well understood.

Other Modifications. Many other modifications have been or could be used: mutations in the *fhuA* (*tonA*) gene confer resistance to infection by T-odd bacteriophage (92), nonessential proteins which tend to copurify with the product protein can be eliminated, pathways which make nonnatural amino acids such as norleucine can be eliminated (27), and chaperonin or other helper proteins can be overexpressed (126).

Phage Display Technology

A recent subset of *E. coli* rDNA technology allows the display of polypeptides on the surface of bacteriophage. Typically, a library of closely related genes for the activity of interest is fused to the gene encoding the surface protein of a filamentous bacteriophage virion. The virions which display polypeptides having the greatest affinity for an immobilized ligand can then be isolated. The isolated virions will also contain the desired DNA. When first introduced in 1985 by Smith (195), the technique was limited to smaller polypeptides. The affinity selection was also limited because more than a single fusion protein was displayed on each phage. These limitations were both addressed by Bass et al. (18), who used a "phagemid" vector containing two origins of replication, one for *E. coli* and the other for M13, a filamentous phage. Into this vector was inserted a gene encoding the fusion of the gene III coat protein and human growth hormone. Infection by a helper phage expressing the normal gene III protein and careful control of fusion protein expression allowed the production of phagemid particles which displayed approximately four normal coat proteins and at most only one molecule of the fusion protein per phage. Thus, on each recombinant phage virion, only one molecule of growth hormone was displayed, and this was found to be properly folded. The system was then used to isolate growth hormone mutants with increased affinity for immobilized growth hormone receptor.

The phage display technique has been used to screen for better protease substrates (136), enzyme inhibitors (172), and other receptor ligands (144). However, its most exciting and highly evolved application is the selection of antibody fragments with desired selectivity and affinity. This technique was first described by McCafferty et al. in 1990 (140) and has been developed rapidly into a powerful technology for screening DNA libraries isolated from both immunized and nonimmunized animals (216). A recent publication describes the development of a human antibody Fab fragment by using a rodent monoclonal antibody as a template and selecting first for a hybrid antibody with the desired properties and then for a fully human antibody (107).

rDNA PROTEIN PRODUCTION: PROCESS TECHNOLOGY

Overview

There is nearly as much flexibility in developing the production process as in developing the engineered organism. In some cases, the protein itself can be modified, either for easier production or to have more desirable properties. The organism, process, and product options depend on many factors, such as the economic value of the product, the importance of having the natural molecule, the size and complexity of the protein, the propensity of the protein to be folded in vivo versus in vitro, the sensitivity of the protein to posttranslational modifications, the required homogeneity of the product, and other factors.

The following sections describe major process technologies as well as technology related to specific process steps.

General Process Options

The ideal production system would secrete and fold the mature protein so that it could be isolated from either the periplasm or the surrounding medium. However, the present technology accomplishes this for a limited number of proteins. Many proteins are readily exported to the periplasmic space but do not fold properly (218, 221). Others are difficult to secrete and may be more efficiently accumulated in the cytoplasm. The properties of a protein that limit its secretion and folding in *E. coli* are as yet poorly understood. Although size and disulfide bond complexity are important, results both with model systems (30, 175) and with potential products (220) suggest that other factors are important as well. These same results and a recent review (221) suggest that organism and process modifications can successfully overcome some limitations to protein secretion and folding. In other cases, it may be more beneficial to alter the protein's primary amino acid sequence or to cotranslate it as a fusion protein.

Fusion Proteins. When protein stability, secretion, or in vivo protein folding is problematic, it often has been productive to turn to the first rDNA production technology, the use of fusion proteins. Both somatostatin (103) and insulin (84) peptides were found to be excellent substrates for intracellular proteases, but both could be accumulated when attached to a larger protein, β-galactosidase. Although it was necessary to release the desired polypeptide by proteolytic cleavage, the product could be obtained with acceptable yields.

Fusion protein technology has progressed significantly since those early experiments. The fusion partner can confer many benefits in addition to resistance to degradation, most notably easier product isolation and purification either by antibody recognition (80, 97, 152) or by other interactions (8, 194). As described in reviews by Ford et al. (73) and by Nilsson et al. (159), fusion proteins have also been used to improve protein solubility and folding, to facilitate secretion into the periplasm and beyond, to anchor the protein either on the surface of a bacteriophage or the surface of the *E. coli* cell (74, 75; G. Georgiou and J. A. Francisco, International Patent Application WO 93/10214, May, 1993), and to produce proteins with new or improved therapeutic properties. Many patents have been granted on such innovations.

The principal limitation to the use of fusion proteins is the specific proteolytic cleavage required to liberate the product. Initially, cyanogen bromide was used to cleave the fusion protein at methionine residues (84, 103). However, if the desired product contains internal methionine residues, this approach is difficult at best. Other workers have used a variety of chemicals, proteases, and peptidases (36). One of the more interesting approaches uses a genetically engineered version of subtilisin (37,

38) to increase specificity by requiring the substrate to provide one of the amino acid residues of the enzyme's catalytic triad.

Cytoplasmic Accumulation. The first rDNA products were deposited in the cytoplasmic space. This continues to be a viable way to produce recombinant proteins. Although product can accumulate up to 50% of the cell's protein, it is usually not soluble or properly folded (178). Because of its relatively low oxidation potential, the cytoplasm does not promote the proper folding of most proteins with disulfide bonds. Also, proteins accumulated in the cytoplasm often have an extra N-terminal methionine and may be aggressively proteolyzed.

However, there are now many examples of successful production using cytoplasmic accumulation. If the expressed protein rapidly aggregates to form inclusion bodies, proteolysis is reduced (46). Also, product deposited in inclusion bodies is easily separated from other cell components by differential centrifugation (72, 180) as an initial purification step. Alternatively, if the product is stable to proteolysis or if proteolysis can be controlled by the use of a protease-deficient host, soluble product can accumulate. Schein and Noteborn (179) and a variety of other investigators have shown that lower temperature often favors product solubility. Kopetzki et al. (120) also showed that slower expression rates resulted in less cytoplasmic aggregation, a result in agreement with the model expressed by Mitraki and King (149) that describes inclusion bodies as aggregates of protein folding intermediates. Cytoplasmic inclusion bodies can also be formed from properly folded protein (206), but this apparently is relatively rare.

The removal of the N-terminal methionine often is incomplete but sometimes can be enhanced by overexpression of the *E. coli* methionine amino peptidase (21). The efficency of the methionine removal depends on the side chain length of the second amino acid (95), with shorter chains allowing more efficient methionine removal. However, even with a short adjacent residue, rapid aggregation may inhibit methionine removal.

Periplasmic Accumulation or Export. Early attempts to translocate rDNA proteins out of the cytoplasm and into the periplasm and beyond established feasibility (187, 188, 202, 203), but examples of commercial application have come slowly. Nutropin brand human growth hormone, approved in 1993, may have been the first. However, the secretion into the periplasmic space offers several potential advantages (41, 88, 99, 117, 202). Because the signal sequence is precisely removed, the proper N terminus is usually obtained. Deposition of the protein in the more highly oxidized periplasmic environment offers a greater chance for proper protein folding. The product, if soluble, may be more easily extracted, and its accumulation is less likely to affect cell metabolism. Finally, the periplasmic environment is more easily manipulated to further increase the degree of protein folding and solubility. Although the periplasm contains many proteases, hosts deficient in several of them are viable and are less likely to degrade secreted proteins (145).

The use of mammalian signal sequences to drive protein translocation can be effective (88, 203), but native *E. coli* signal sequences may be more efficient (41, 99). Commonly used signal sequences are the *phoA, lamB, ompA,* and *STII* leader sequences. For some proteins, the choice of signal sequence may be important, as suggested by Wong et al. (218; E. Wong and M. L. Bittner, U.S. patent 5,084,384, January, 1992) and by a recent report by Voss et al. (212). Pugsley has recently reviewed the secretion pathway in *E. coli* (167), and Izard and Kendall have reviewed the role of signal sequences (105). Although some proteins are difficult to secrete from *E. coli*, work by Rusch and Kendall suggests that increasing the hydrophobicity of the signal sequence may help (175).

There are now several examples of disulfide-containing proteins that fold properly when secreted into the periplasm. They include human growth hormone (88, 99), human alpha interferon (150, 212), epidermal growth factor (104), antibody fragments (25, 190), and bovine growth hormone (117). However, transport into the periplasm does not necessarily lead to proper folding. Aggregation of rDNA proteins occurs in the periplasm as well (30, 42, 218). The use of slower expression rates and the addition of nonmetabolizable sugars may discourage aggregation (30). But when product aggregation cannot be avoided, the aggregated protein may still be isolated and folded into its active conformation (42). The aggregated product often can be isolated by differential centrifugation. However, periplasmic aggregates may be more difficult to isolate in this manner because of their irregular shape (31) and possible interaction with other periplasmic components (211). In these cases, in situ solubilization and aqueous two-phase extraction may be better (94).

Various approaches have been used to release secreted polypeptide into the medium (118, 121). Although this potentially offers easier product isolation, some proteins may not be stable when exposed to the highly active gas-liquid interfaces of an intensely aerated microbial culture. A few proteins can leak into the medium without host or protein modifications. These include single-chain antibodies (25, 165), IGF-I (218), and *E. coli* β-lactamase (77). However, this happens infrequently and often destabilizes the outer membrane. The outer membrane can be intentionally destabilized by cell mutations or specific effectors to release secreted proteins during production. However, significant benefits from such efforts have not been reported, probably because they tend to negatively impact cell metabolism.

The most effective examples of product release have employed fusion proteins. When IGF-I is fused to a portion of staphylococcal protein A, the fusion protein is efficiently delivered to the surrounding medium (152). Specialized secretion systems such as the one for pullulanase (121) can also be used. Finally, L-form cells of *Proteus mirabilis* release secreted chymosin directly into the medium (118). Although these technologies may offer advantages for special cases, they have not come into general use.

Protein Folding In Vivo. Proper folding of the expressed polypeptide is a critical part of rDNA protein production. Although a few proteins fold properly when they accumulate in the cytoplasm, and host cell mutations may allow disulfide bond formation (58), most naturally secreted proteins do not fold efficiently in the cytoplasm (178). Thus, most of the effort on in vivo protein folding has focused on the periplasm. Not only does the periplasm have the higher oxidation potential required for disulfide bond formation, but it also is more easily manipulated without compromising other metabolic functions. However, as pointed out by a recent review (221), the work in this area is relatively recent and is not highly evolved.

Because the mammalian proteins that we wish to secrete and fold are naturally folded in the endoplasmic reticulum (ER) of eukaryotic cells, the dominant theme has been to model the periplasmic space after the lumen of the ER. Fortunately, *E. coli* already provides several important properties, as shown in Table 3. A periplasmic prolyl isomerase has been found in *E. coli* which is similar to the one found in the ER (131). To date, however, overexpression of this enzyme has not enhanced folding of rDNA proteins (119, 221). At least two proteins with disulfide isomerase activity, DsbA (16, 111) and DsbC (148), are naturally present in the periplasm. The principal role of DsbA and DsbC appears to be the donation of disulfide bonds to nascent proteins (15, 186), and mutants deficient in DsbA do not efficiently fold either endogenous or recombinant proteins in the periplasm (119, 161). DsbA-mediated catalysis of disulfide rearrangement at neutral pH has been demonstrated (109), but DsbA is a weak disulfide isomerase compared with the mammalian enzyme. Nonetheless, Wunderlich and Glockshuber obtained improved folding of a secreted protease inhibitor when DsbA was overexpressed and reduced glutathione was added to the medium (222). DsbA overexpression did not improve folding for secreted antibody fragments (119), but it did improve the folding of secreted T-cell receptor fragments when combined with induction of the heat shock response at low temperature (220), suggesting a possible interaction with periplasmic chaperonin proteins, perhaps similar to the effect of the chaperonins in the eukaryotic ER. Although these results are encouraging, they underscore the complexity of the in vivo folding of proteins with disulfide bonds.

Protein Folding In Vitro. As mentioned earlier, in vitro protein folding has been an important contributor to the success of *E. coli* rDNA technology. Even though a large patent and nonpatent literature now exists, the power of this technology often is not fully appreciated. A recent review by Rudolph (174) concludes that "therefore, *in vitro* folding of any recombinant protein deposited in inclusion bodies will likely be successful." Occasionally, the inclusion body can be converted into properly folded protein in a single step (42, 199; L. A. Bentle, J. W. Mitchell, and

TABLE 3 Protein export components

Component	Eukaryotes[a]	*E. coli*
Prolyl isomerase	Yes	Yes
Disulfide isomerase	Yes (PDI)	Yes (DsbA, DsbC)
Signal recognition particle	Yes	Yes (unknown function)
Receiving compartment chaperonins	Yes (BIP, calnexin)	? (PapD)
Redox control	Yes	Unknown
Glycosylation	Yes	No

[a] PDI, protein disulfide isomerase; BIP, immunoglobulin heavy-chain binding protein, also known as GRP78.

S. B. Storrs, U.S. patent 4,652,630, March, 1981; C. Y. Chang and J. R. Swartz, U.S. patent 5,288,931, February, 1994). However, the process is usually more complicated.

Fortunately, the new rDNA industry was able to call upon a strong existing body of knowledge about protein folding (52) which propelled the early successes in rDNA protein production. Typically, the nascent rDNA protein is isolated in the form of inclusion bodies. It is then solubilized by using chaotropic agents such as urea, guanidine hydrochloride, and detergents in the presence of a reducing agent such as dithiothreitol or β-mercaptoethanol (56; A. J. S. Jones, K. C. Olson, and S. J. Shire, U.S. patent 4,512,922, April, 1985; K. C. Olson, U.S. patent 4,518,526, May, 1985; S. E. Builder and J. R. Ogez, U.S. patent 4,620,948, November, 1986). Alternatively, the sulfhydryl groups can be treated with sodium sulfite in a process called sulfitolysis to break intermolecular disulfide bonds and facilitate solubilization (39; R. B. Wetzel, U.S. patent 4,599,197, July, 1986; J. L. Bobbitt and J. V. Manetta, European Patent Office [EPO] 0 361 830, April, 1990).

The solution is then diluted or dialyzed to produce a suitable folding environment. The objective is to obtain proper protein folding with high yields at high protein concentrations in order to use smaller equipment (less capital cost) and to conserve reagents. However, at higher concentrations, denatured and partially folded polypeptides are more likely to aggregate. Two approaches have been taken to control aggregation. The first optimizes the folding environment to discourage aggregation. Detergents or agents such as polyethylene glycol (50) can be used to shield exposed hydrophobic regions which contribute to aggregation. Optimization of the ionic environment can also be useful (93). The second approach limits the concentration of partially folded species by slowly adding the denatured polypeptide at a rate which matches the rate of protein folding (R. Rudolph and S. Fischer, U.S. patent 4,933,434, June, 1990).

It is often important to control the redox potential of the solution to encourage disulfide bond formation and rearrangement (213). This can be effected by using redox buffers such as oxidized and reduced glutathione and conducting the folding reaction at high pH (174). However, these agents are expensive, and good folding results can often be obtained by using oxygen as the oxidant in the presence of small concentrations of Cu^{2+} (132).

Available space limits a more comprehensive coverage of this highly evolved technology. The reader is directed to the review by Rudolph (174) and to two recent books (49, 78). The use of the technology for the large-scale production of two cost-sensitive products, bovine growth hormone and human insulin, demonstrates the practicality of large-scale, low-cost in vitro protein folding.

Specific Process Steps

After the product form has been chosen, the plasmid has been designed, and the host organism has been determined, there is a significant body of technology that addresses the performance of the production process. General guidelines for small-scale fermentations have been offered elsewhere (129). Thus, this section will focus on technologies developed for large-scale, commercial applications.

Transformation and Organism Storage. The transformation of *E. coli* with plasmids has now become routine and is well described in the molecular cloning manual by Sambrook et al. (176). Although the $CaCl_2$-based protocol is usually adequate, electroporation may be more effective in some cases. It may be important to transform a host which is restriction negative (*hsdR*) but modification positive. *E. coli* MM294 is a popular cloning strain for this reason. Additionally, it is often important to use a host background and a selection medium that inhibit expression of the cloned gene. For example, *E. coli* JM101 is often used as a cloning strain for expression systems using the *lac* operator region. JM101 contains a *lacI*q element to overexpress the *lac* repressor and avoid unintentional product expression from a multicopy plasmid.

A variety of methods can be used for organism storage. Lyophilization is best for permanent storage but is less convenient than freezing in liquid nitrogen or at −70°C (4). Although freezing in liquid nitrogen will preserve most bacteria for more than 30 years (158), freezing at −70°C is adequate for several years. The culture should be made approximately 15% in glycerol or 5 to 10% in dimethyl sulfoxide concentration before freezing. No special precautions are necessary to thaw for recovery.

Inoculum Preparation. Little has been written about this part of the process, but it can have an important influence on the final outcome. Luria-Bertani broth (LB medium; 10 g of tryptone, 5 g of yeast extract, and 10 g of NaCl per liter) containing an antibiotic for plasmid selection is often used to prepare the initial inoculum. This medium is convenient to make and usually allows rapid recovery of frozen cultures. It supports growth to 2 to 4 units of optical density at 550 nm in baffled shake flasks. Inoculum culture volumes that are 0.5 to 5% of the next fermentation volume are normally used. In practice, a broad variety of inoculum media can be used, and it may be advantageous to customize the medium to better support growth, avoid pH changes, avoid product expression, and encourage a rapid adjustment to the next stage. pH drift and oxygen depletion deserve special attention. The degree of pH change during the flask culture can be influenced by the type of carbon and energy source used (for example, glucose versus amino acids) and by the addition of pH buffers. Oxygen transfer for shake flask cultures can also limit growth and cause unwanted metabolic changes. It is best to use baffled flasks with closures that allow oxygen transfer (209).

For large-scale fermentations, several inoculum stages may be necessary. The culture entering the final production vessel should consistently be at the same density and in the same metabolic state for process reproducibility. It is usually good practice to avoid severe metabolic challenges when the culture is transferred to the next stage of culture. For example, the inoculum should be grown at the same temperature as the next stage. If the inoculum is grown in a rich medium such as LB, it may be beneficial to add a small amount of yeast extract and protein hydrolysate to the next stage as well, even if growth on a defined medium is desired during product expression.

The Fermentation Process. Environmental and metabolic influences can have major effects on both the yield and the quality of the product. These include obvious influences such as temperature, pH, and dissolved oxygen tension as well as others which are less obvious, such as the preinduction growth rate (53, 54) and the amount of yeast extract in the medium (43). Obviously, high product concentrations (grams per liter) and volumetric productivities (grams per liter per hour) lower fermentation costs by minimizing capital, raw material, and labor costs. However, because of the relative efficiency of *E. coli* rDNA expression, product isolation and purification costs can exceed fermentation costs by severalfold. Thus, to obtain an optimal overall process, the fermentation should be optimized for efficient and inexpensive downstream processing.

Proteins are easily modified during production, and altered forms of the product can be difficult to separate. Minimizing mistranslation and reactions such as protein oxidation, deamidation, and proteolysis can contribute significantly to lowering production costs, especially for human pharmaceuticals. Such optimization requires rapid and accurate assays. High-pressure liquid chromatography often provides a powerful tool to monitor both product concentration and quality in complex samples.

Much of the work done on large-scale process development has been done by industry and has not been published. However, academic laboratories have examined relevant issues, and several industrial publications and patents contribute to the literature. *E. coli* rDNA fermentation processes are usually designed with separate growth and expression phases. By avoiding product expression during the initial growth period, product toxicities are avoided, rDNA protein expression does not compete for metabolic resources, and the duration of product exposure to modification reactions can be minimized.

As discussed earlier, the chief challenges during the initial growth period are avoiding acetate accumulation and oxygen depletion. Yee and Blanch (226) have summarized methods used to obtain high-cell density *E. coli* cultures. Although oxygen depletion can be avoided by sparging with pure oxygen (81) or by slowing metabolic rates with reduced temperature (20), the preferred method is to limit the culture for its carbon and energy source, usually glucose (226). As the culture progresses to high density, glucose limitation has the additional advantage of limiting acetate formation. The use of other carbon and energy sources such as sucrose (81) or glycerol (P. M. Keith and W. Cain, U.S. patent 5,104,796, April, 1992) may help to limit acetate formation, but substrate limitation or pure oxygen sparging is still needed to avoid dissolved oxygen depletion.

A variety of media, both defined and complex, have been used for high-cell density rDNA protein production (226, 230). Fieschko and Ritch (71) described their methods for designing a defined medium to support high cell density. Other published recipes (129, 227) are similar, with the major difference depending on the mode of pH control. If pH is not controlled by NH$_4$OH additions, more NH$_4^+$ is added to the initial medium. Growth is relatively insensitive to NH$_4^+$ concentrations below 170 mM (205). Ionic strength (106, 183) as well as the concentrations of specific ions (43) may be important, but the effects appear to vary with the production system. Dividing the required nutrients between the initial medium and medium feeds can be used to more precisely control the fermentation environment (71).

Complex nitrogen sources such as hydrolyzed proteins and yeast extract are most frequently used in complex media. Tsai et al. reported a 10-fold increase in intracellular human IGF-I accumulation by increasing yeast extract and tryptone additions (208). Since they induced product expression by increasing temperature to 42°C, and since Buell et al. saw a similar increase in IGF-I accumulation with *lon* and *htpR* mutations (34), it is reasonable to hypothesize that the complex nitrogen additions reduced product proteolysis. Chang et al. observed that the addition of yeast extract stimulated rDNA protein secretion (43).

The use of complex nitrogen components requires special precautions. For high-cell-density cultures, it is unlikely that sufficient amino acids will be added to supply all requirements. Since the rate of amino acid utilization can be quite high, amino acid depletion is correspondingly abrupt, particularly if complex nitrogen feeds are not used (90). These depletions cause periods of amino acid starvation which, at best, delay the culture and, more probably, cause metabolic changes such as protease induction (138).

Complex nitrogen feeds can play a role in improving product quality, for example, in avoiding norleucine substitution for methionine. Leucine can be fed to repress the pathway producing norleucine, methionine can be fed to saturate the methionyl-tRNA synthase, or leucine can be fed to a leucine auxotroph which is unable to make norleucine (27, 207; D. P. Brunner, G. C. Harbour, R. J. Kirschner, J. F. Pinner, and R. L. Garlick, WO 89/07651, August, 1989). Leucine feeding may have a separate effect as well (151), especially in light of the leucine-induced regulon in *E. coli* (137). Clearly, complex nitrogen sources can have significant effects on rDNA protein production. Unfortunately, with the exceptions of decreasing proteolysis and norleucine incorporation, the mechanisms for these effects are largely unknown.

The environmental conditions during product induction can have a significant effect on product formation. Kopetzki et al. (120) observed that the formation of active α-glucosidase was significantly improved at temperatures and pHs which were not optimal for growth. Interestingly, Curless et al. (53, 54) have found that the preinduction growth rate affects subsequent production. Unfortunately, optimal conditions for induction and product accumulation are not consistent and vary with the product, its disposition (secreted or cytoplasmic, soluble or inclusion body), and the expression system. Reports such as those of Wood and Peretti (219) and Bailey (10) have begun to describe the metabolic dependencies and influences of rDNA product expression. However, there is still much to learn. Particularly important are factors influencing proteolysis, protein secretion, and protein folding.

Isolation and Purification of the Product. Optimal isolation and purification methods depend on the initial disposition of the product and the biochemical characteristics of both the product and the contaminants. In general, inclusion bodies are isolated by differential centrifugation after cell disruption. Soluble proteins may require cell disruption or, if secreted to the periplasmic space, may be extracted by osmotic shock. Chemical permeabilization of the outer membrane has also been used (155) to recover soluble periplasmic proteins, and Hart et al. (94) have used in situ solubilization and extraction for periplasmic inclusion bodies. Freeze-thaw cycles have also been reported to extract recombinant proteins (108).

Purification technology for recombinant proteins has received much attention, and representative technologies are described in a recent book (124). Purification processes generally begin with relatively low-resolution steps which reduce the liquid volume and remove the bulk of the contaminants. Such steps are precipitation, liquid-liquid extraction, and the use of general adsorption columns. Higher-resolution steps are then used to remove contaminants which derive from the host organism, the process steps, or from the product itself. These steps typically take advantage of a variety of specific properties of the product such as molecular size, hydrophobicity, specific affinities, and electrical charge and its dependence on pH. Although many authors suggest that removal of endotoxic lipopolysaccharides might be a limitation for *E. coli*-derived pharmaceuticals, this has not been the case. The steps required for removal of contaminating proteins usually are more than adequate for removal of host-derived lipopolysaccharides. However, lipopolysaccharide contamination from other organisms which contaminate purification columns or purification raw materials can be a problem, and endotoxin levels require careful attention independent of the producing organism.

The success of this technology is indicated by the characteristics of its products. Host proteins are often removed to the part-per-million levels. In fact, new analytical technology was required to allow measurements of this sensitivity. Modified product molecules such as oxidized or deamidated products are more difficult to remove but are also less likely to be problematic. These are generally removed to less than 1% of the desired product form. In fact, because *E. coli*-produced rDNA protein products are simpler and more easily analyzed than their glycosylated counterparts, they are usually significantly more homogeneous than proteins produced by mammalian cells.

Regulatory Considerations

Regulatory requirements play a large role in the use of rDNA organisms and especially in the production of human pharmaceuticals. Although the National Institutes of Health guidelines for the use of recombinant organisms have been relaxed for *E. coli* K-12, large-scale production generally requires that release of the live organism be minimized. The bioreactors are operated with containment filters for the exhaust gases, and the organism must be killed before containment is broken. This may be effected before, after, or coincident with product removal. Cell inactivation methods include treatment with high or low pH, addition of toxic agents, and heat inactivation. Since most of the treatments that kill bacteria also denature proteins, this step must be developed and tested carefully to avoid modifying the product or complicating its recovery.

Protein pharmaceuticals must be manufactured according to current Good Manufacturing Practice (cGMP) as defined by the Food and Drug Administration. In general, cGMP requires testing and control of all raw materials, validation of all equipment and procedures to ensure proper operation (including cleaning and sterilization), documentation of all procedures, repeated training of personnel, careful control of all steps in the process, and careful testing of process intermediates and final products. Compliance with cGMP requires an enormous effort. However, although precautions are sometimes taken to extremes, this type of system is important in protecting the quality of rDNA pharmaceuticals.

The input of Food and Drug Administration officials often helps determine acceptable process and product characteristics. The officials may specify direct requirements or may suggest safety or efficacy tests. It is usually necessary to demonstrate that the organism is genetically stable, that an adequate assay for contaminating cell proteins exists, that the production process is reproducible, and that the analytical procedures are adequate to determine the purity and the correct biochemical characteristics of the product. However, it is not yet accepted that analytical procedures can fully characterize relevant properties of an rDNA protein pharmaceutical, even one produced in *E. coli*. Thus, the production process must be carefully set and controlled. It is often necessary to repeat animal and human testing before process improvements can be implemented.

As the product moves through the three stages of testing in human subjects, the requirements become more stringent. For example, a change in the manufacturing process after the product has entered phase III trials will probably require additional clinical trials to demonstrate that the product is still equally safe and efficacious. Thus, it is important to define an acceptable and reproducible process as early as possible. Since it is always important to develop new products quickly, this often forces the process developer to fix upon a process which is reproducible but which may not be fully optimized.

FUTURE DIRECTIONS

New and Improved Applications

It is always difficult to see into the future, but it can be useful to estimate what could take place. *E. coli* biotechnology has clearly established a formidable record of accomplishment, and many of the same applications will continue. In addition, recent progress in modeling and understanding metabolic pathways would suggest that *E. coli* rDNA may be used more for bioconversions and for the production of small biomolecules, particularly those with chiral centers. If methods that allow *E. coli* to function in more hostile environments could be found, application of the technology might be broader.

Even for the present applications, there is significant opportunity for innovation, for example, better plasmid selection without affecting cell metabolism, improved promoters which allow modulated expression and inexpensive induction without adversely affecting host cell metabolism, and more specific and less expensive methods for fusion protein cleavage.

It is also possible that more complex proteins will be made with *E. coli*. Even those that require posttranslational modifications such as glycosylation could be made if technology were developed for accurate in vitro modification. For example, synthetic polymers could be designed and attached at specific sites to produce biochemical characteristics the same as (or better than) those of the natural molecule. Modification methods for increasing molecular size by attaching polyethylene glycol polymers already exist (86, 114, 196). If such polymers could routinely be attached at specific sites, complex, modified proteins could be designed and produced as well-defined, homogeneous pharmaceuticals. Alternatively, in vitro glycosylation using immobilized glycosyltransferases might be feasible if the substrates and catalysts could be economically produced. However, so far there has not been sufficient economic motivation to implement these developments.

Better Knowledge and Control of Cell Physiology

With knowledge comes power. Although we have considerable knowledge about *E. coli* physiology, we have little knowledge of cellular responses to rDNA protein production or of many factors which affect recombinant protein expression, secretion, and folding. Why does N-terminal codon selection have such a strong effect on translation? How do protease deletions affect cell metabolism, and can the effects be suppressed? What are the properties of the N terminus of a protein that limit its secretion? What are the biochemical properties of the periplasmic space that influence protein folding? What determines the cell's response to rDNA protein production? Which proteases and chaperonins are induced and why? Why do complex nitrogen sources often provide substantial benefits? It is exciting to look forward to the time when the answers to these and many more questions will be known.

ACKNOWLEDGMENTS

I thank the many people who contributed to this chapter, especially my editor, John Ingraham, for overall guidance, encouragement, and countless corrections. Roger Hart, Jeff Cleland, Brad Snedecor, and Norm Lin helped with pertinent references, and John Joly and John Jost provided many helpful suggestions.

LITERATURE CITED

1. Ahmed, S., and I. R. Booth. 1983. The effect of β-galactosides on the protonmotive force and growth of *Escherichia coli. J. Gen. Microbiol.* **129:**2521–2529.
2. Aiba, S., T. Imanka, and H. Tsunekawa. 1980. Enhancement of tryptophan production by *Escherichia coli* as an application of genetic engineering. *Biotechnol. Lett.* **2:**525–530.
3. Aiba, S., H. Tsunekawa, and T. Imanaka. 1982. New approach to tryptophan production by *Escherichia coli*: genetic manipulation of composite plasmids in vitro. *Appl. Environ. Microbiol.* **43:**289–297.
4. Alexander, M., P. M. Daggett, R. Gherna, S. Jong, and F. Simione, Jr. 1980. Laboratory manual on preservation freezing and freeze-drying as applied to algae, bacteria, fungi, and protozoa, p. 1–57. *In* H. Hatt (ed.), *American Type Culture Collection Methods.* American Type Culture Collection, Rockville, Md.
5. Altherthum, F., and L. O. Ingram. 1989. Efficient ethanol production from glucose, lactose, and xylose by recombinant *Escherichia coli. Appl. Environ. Microbiol.* **55:**1943–1948.
6. Anderson, A., and K. von Meyenburg. 1980. Are growth rates of *Escherichia coli* in batch cultures limited by respiration? *J. Bacteriol.* **144:**114–123.
7. Aristidou, A. A., K. Y. San, and G. N. Bennett. 1994. Modification of central metabolic pathway in *Escherichia coli* to reduce acetate accumulation by heterologous expression of the *Bacillus subtilis* acetolactate synthase gene. *Biotechnol. Bioeng.* **44:**944–951.
8. Arnold, F. H. 1991. Metal-affinity separations: a new dimension in protein processing. *Bio/Technology* **9:**151–156.
9. Bailey, J. E. 1991. Toward a science of metabolic engineering. *Science* **252:**1668–1675.
10. Bailey, J. E. 1993. Host-vector interactions in *Escherichia coli. Adv. Biochem. Eng. Biotechnol.* **48:**29–52.
11. Baird, L., B. Lipinska, S. Raina, and C. Georgopoulos. 1991. Identification of the *Escherichia coli sohB* gene, a multicopy suppressor of the HtrA (DegP) null phenotype. *J. Bacteriol.* **173:**5763–5770.
12. Balbas, P., and F. Bolivar. 1990. Design and construction of expression plasmid vectors in *Escherichia coli. Methods Enzymol.* **185:**14–37.
13. Baneyx, F., and G. Georgiou. 1991. Construction and characterization of *Escherichia coli* strains deficient in multiple secreted proteases: protease III degrades high-molecular-weight substrates in vivo. *J. Bacteriol.* **173:**2696–2703.
14. Baneyx, F., and G. Georgiou. 1992. Expression of proteolytically sensitive polypeptides in *Escherichia coli*, p. 69–108. *In* T. J. Ahern and M. C. Manning (ed.), *Stability of Protein Pharmaceuticals*, part A. *Chemical and Physical Pathways of Protein Degradation.* Plenum Press, New York.
15. Bardwell, J. C. A., J. O. Lee, G Jander, N. Martin, D. Belin, and J. Beckwith. 1993. A pathway for disulfide bond formation *in vivo. Proc. Natl. Acad. Sci. USA* **90:**1038–1042.
16. Bardwell, J. C. A., K. McGovern, and J. Beckwith. 1991. Identification of a protein required for disulfide bond formation *in vivo. Cell* **67:**581–589.
17. Barkley, M. D., and S. Bourgeois. 1978. Repressor recognition of operator and effectors, p. 177–220. *In* J. H. Miller and W. S. Reznikoff (ed.), *The Operon.* Cold Spring Harbor Laboratory, Cold Spring Harbor, N.Y.
18. Bass, S., R. Green, and J. A. Wells. 1990. Hormone phage: an enrichment method for variant proteins with altered binding properties. *Proteins Struct. Funct. Genet.* **8:**309–314.
19. Bauer, K. A., A. Ben-Bassat, M. Dawson, V. T. de la Puente, and J. O. Neway. 1990. Improved expression of human interleukin-2 in high-cell-density fermentation cultures of *Escherichia coli* K-12 by a phosphotransacetylase mutant. *Appl. Environ. Microbiol.* **56:**1296–1302.
20. Bauer, S., and M. D. White. 1976. Pilot scale exponential growth of *Escherichia coli* W to high cell concentration with temperature variation. *Biotechnol. Bioeng.* **18:**839–846.
21. Ben-Bassat, A., K. Bauer, S. Y. Chang, K. Myambo, A. Boosman, and S. Chang. 1987. Processing of the initiation methionine from proteins: properties of the *Escherichia coli* methionine aminopeptidase and its gene structure. *J. Bacteriol.* **169:**751–757.
22. Bentley, W. E., N. Mirjalili, D. C. Andersen, R. H. Davis, and D. Kompala. 1990. Plasmid-encoded protein: the principal factor in the metabolic burden associated with recombinant bacteria. *Biotechnol. Bioeng.* **35:**668–681.
23. Betenbaugh, M. J., C. Beaty, and P. Dhurjati. 1989. Effects of plasmid amplification and recombinant gene expression on the growth kinetics of recombinant *E. coli. Biotechnol. Bioeng.* **33:**1425–1436.
24. Better, M., C. P. Chang, R. R. Robinson, and A. H. Horwitz. 1988. *Escherichia coli* secretion of an active chimeric antibody fragment. *Science* **240:**1041–1043.
25. Better, M., and A. H. Horwitz. 1993. *In vivo* expression of correctly folded antibody fragments from microorganisms, p. 203–217. *In* J. L. Cleland (ed.), *Protein Folding: In Vivo and In Vitro.* American Chemical Society, Washington, D.C.
26. Birnbaum, S., and J. E. Bailey. 1991. Plasmid presence changes the relative levels of many host cell proteins and ribosome components in recombinant *Escherichia coli. Biotechnol. Bioeng.* **37:**736–745.
27. Bogosian, G., B. N. Violand, E. J. Dorward-King, W. E. Workman, P. E. Jung, and J. F. Kane. 1989. Biosynthesis and incorporation into protein of norleucine by *Escherichia coli. J. Biol. Chem.* **264:**531–539.
28. Bolivar, F. 1978. Construction and characterization of new cloning vehicles. III. Derivatives of plasmid pBR322 carrying EcoR1 sites for selection of EcoR1 generated recombinant DNA molecules. *Gene* **4:**121–136.
29. Bolivar, F., R. L. Rodriguez, P. J. Greene, M. O. Betlach, H. L. Heyneker, H. W. Boyer, I. H. Crosa, and S. Falkow. 1977. Construction and characterization of new cloning vehicles. II. A multiple cloning system. *Gene* **2:**95–113.
30. Bowden, G. A., and G. Georgiou. 1990. Folding and aggregation of β-lactamase in the periplasmic space of *Escherichia coli. J. Biol. Chem.* **265:**16760–16766.
31. Bowden, G. A., A. M. Paredes, and G. Georgiou. 1991. Structure and morphology of protein inclusion bodies in *Escherichia coli. Bio/Technology* **9:**725–730.
32. Brau, B., and H. Sahm. 1986. Cloning and expression of the structural gene for pyruvate decarboxylase of *Zymomonas mobilis* in *Escherichia coli. Arch. Microbiol.* **144:**296–301.
33. Brent, R., and M. Ptashne. 1981. Mechanism of action of the *lexA* gene product. *Proc. Natl. Acad. Sci. USA* **78:**4204–4208.
34. Buell, G. M. F. Schulz, G. Selzer, A. Chollet, N. Rao Movva, D. Semon, S. Escanez, and E. Kawashima. 1985. Optimizing the expression in *E. coli* of a synthetic gene encoding somatomedin-C (IGF-I). *Nucleic Acids Res.* **13:**1923–1938.
35. Calhoun, M. W., K. L. Oden, R. B. Gennis, M. J. Teixeira de Mattos, and O. M. Neijssel. 1993. Energetic efficiency of *Escherichia coli*: effects of mutations in components of the aerobic respiratory chain. *J. Bacteriol.* **175:**3020–3025.
36. Carter, P. 1990. Site-specific proteolysis of fusion proteins, p. 181–193. *In* M. R. Ladisch, R. C. Willson, C. D. C. Painton, and S. E. Builder (ed.), *Protein Purification: from Molecular Mechanisms to Large-Scale Processes.* American Chemical Society, Washington, D.C.
37. Carter, P., L. Abrahmsen, and J. A. Wells. 1991. Probing the mechanism and improving the rate of substrate-assisted catalysis in subtilisin BPN'. *Biochemistry* **30:**6142–6148.
38. Carter, P., B. Nilsson, J. P. Burnier, D. Burdick, and J. A. Wells. 1989. Engineering subtilisin BPN' for site-specific proteolysis. *Proteins Struct. Funct. Genet.* **6:**240–248.
39. Chan, W. W. C. 1968. A method for the complete S sulfonation of cysteine residues in proteins. *Biochemistry* **7:**4247–4253.
40. Chang, A. C. Y., and S. N. Cohen. 1974. Genome construction between bacterial species in vitro: replication and expression of *Staphylococcus* plasmid genes in *Escherichia coli. Proc. Natl. Acad. Sci. USA* **71:**1030–1034.
41. Chang, C. N., M. Rey, B. Bochner, H. Heyneker, and G. Gray. 1987. High-level secretion of human growth hormone by *Escherichia coli. Gene* **55:**189–196.
42. Chang, J. Y., and J. R. Swartz. 1993. Single-step solubilization and folding of IGF-1 aggregates from *Escherichia coli*, p. 178–188. *In* J. L. Cleland (ed.), *Protein Folding: In Vivo and In Vitro.* American Chemical Society, Washington, D.C.
43. Chang, J. Y. H., R. C. Pai, W. F. Bennett, R. G. Keck, and B. R. Bochner. 1986. Culture medium effects on periplasmic secretion of human growth hormone by *Escherichia coli*, p. 324–329. *In* L. Leive (ed.), *Microbiology—1986.* American Society for Microbiology, Washington, D.C.
44. Cheah, U. E., W. A. Weigand, and B. C. Stark. 1987. Effects of recombinant plasmid size on cellular processes in *Escherichia coli. Plasmid* **18:**127–134.
45. Cheng, S. W. C., E. C. Lynch, K. R. Leason, D. L. Court, B. A. Shapiro, and D. I Freidman. 1991. Functional importance of sequence in the stem-loop of a transcription terminator. *Science* **254:**1205–1207.
46. Cheng, Y. S. E., D. Y. Kwoh, T. J. Kwoh, B. C. Soltvedt, and D. Zipser. 1981. Stabilization of a degradable protein by its overexpression in *Escherichia coli. Gene* **14:**121–130.
47. Chou, C. H., G. N. Bennett, and K. Y. San. 1994. Effect of modified glucose uptake using genetic engineering techniques on high-level recombinant protein production in *Escherichia coli* dense cultures. *Biotechnol. Bioeng.* **44:**952–960.
48. Clackson, T., and J. A. Wells. 1995. A hot spot of binding energy in a hormone receptor interface. *Science* **267:**383–386.
49. Cleland, J. L. (ed.). 1993. *Protein Folding: In Vivo and In Vitro.* American Chemical Society, Washington, D.C.
50. Cleland, J. L., S. E. Builder, J. R. Swartz, M. Winkler, J. Y. Chang, and D. I. C. Wang. 1992. Polyethylene glycol enhanced protein refolding. *Bio/Technology* **9:**1013–1019.

51. Cohen, S. N., A. C. Y. Chang, H. W. Boyer, and R. B. Helling. 1973. Construction of biologically functional bacterial plasmids *in vitro*. *Proc. Natl. Acad. Sci. USA* **70:**3240–3244.

52. Creighton, T. E. 1978. Experimental studies of protein folding and unfolding. *Prog. Biophys. Mol. Biol.* **33:**231–297.

53. Curless, C., J. Pope, and L. Tsai. 1990. Effect of preinduction specific growth rate on recombinant alpha consensus interferon synthesis in *Escherichia coli*. *Biotechnol. Prog.* **6:**149–152.

54. Curless, C. E., J. Pope, L. Loredo, and L. B. Tsai. 1994. Effect of preinduction specific growth rate on secretion of granulocyte macrophage colony stimulating factor by *Escherichia coli*. *Biotechnol. Prog.* **10:**467–471.

55. Davis, B. D., S. M. Luger, and P. C. Tai. 1986. Role of ribosome degradation in the death of starved *Escherichia coli* cells. *J. Bacteriol.* **166:**439–445.

56. De Bernardez-Clark, E., and G. Georgiou. 1991. Inclusion bodies and recovery of proteins from the aggregated state, p. 1–20. *In* G. Georgiou and E. De Bernardez-Clark (ed.), *Protein Refolding*. American Chemical Society, Washington D.C.

57. Dedhia, N. N., T. Hottinger, and J. E. Bailey. 1994. Overproduction of glycogen in *Escherichia coli* blocked in the acetate pathway improves cell growth. *Biotechnol. Bioeng.* **44:**132–139.

58. Derman, A. I., W. A. Prinz, D. Belin, and J. Beckwith. 1993. Mutations that allow disulfide bond formation in the cytoplasm of *Escherichia coli*. *Science* **262:**1744–1747.

59. De Smit, M. H., and J. Van Duin. 1990. Secondary structure of the ribosome binding site determines translational efficiency: a quantitative analysis. *Proc. Natl. Acad. Sci. USA* **87:**7668–7672.

60. Devlin, P., R. Drummond, P. Toy, D. Mark, K. Watt, and J. Devlin. 1988. Alteration of amino-terminal codons of human granulocyte-colony-stimulating factor increases expression levels and allows efficient processing by methionine aminopeptidase in *E. coli. Gene* **65:**13–22.

61. Diaz-Ricci, J. C., L. Regan, and J. E. Bailey. 1991. Effect of alteration of the acetic acid synthesis pathway on the fermentation pattern of *Escherichia coli*. *Biotechnol. Bioeng.* **38:**1318–1324.

62. Dosch, D. C., F. F. Salvacion, and W. Epstein. 1984. Tetracycline resistance element of pBR322 mediates potassium transport. *J. Bacteriol.* **160:**1188–1190.

63. Dubé, S., J. W. Fisher, and J. S. Powell. 1988. Glycosylation at specific sites of erythropoietin is essential for biosynthesis, secretion, and biological function. *J. Biol. Chem.* **263:**17516–17521.

64. Dykhuizen, D., and D. Hartl. 1978. Transport by the lactose permease of *Escherichia coli* as the basis of lactose killing. *J. Bacteriol.* **135:**876–882.

65. Ellman, J., D. Mendel, S. Anthony-Cahill, C. J. Noren, and P. G. Schultz. 1991. Biosynthetic methods for introducing unnatural amino acids site-specifically into proteins. *Methods Enzymol.* **202:**301–336.

66. El-Mansi, E. M. T., and W. H. Holms. 1989. Control of carbon flux to acetate excretion during growth of *Escherichia coli* in batch and continuous cultures. *J. Gen. Microbiol.* **135:**2875–2883.

67. Enfors, S. O. 1992. Control of *in vivo* proteolysis in the production of recombinant proteins. *Trends Biotechnol.* **10:**310–315.

68. Ensley, B. D. 1986. Stability of recombinant plasmids in industrial organisms. *Crit. Rev. Biotechnol.* **4:**263–277.

69. Ensley, B. D., B. J. Ratzkin, T. D. Osslund, M. J. Simon, L. P. Wackett, and D. T. Gibson. 1983. Expression of naphthalene oxidation genes in *Escherichia coli* results in the biosynthesis of indigo. *Science* **222:**167–169.

70. Ernst, J. F., and E. Kawashima. 1988. Variations in codon usage are not correlated with heterologous gene expression in *Saccharomyces cerevisiae* and *Escherichia coli*. *J. Biotechnol.* **7:**1–10.

71. Fieschko, J., and T. Ritch. 1986. Production of human alpha consensus interferon in recombinant *Escherichia coli*. *Chem. Eng. Commun.* **45:**229–240.

72. Fish, N. M., and M. Hoare. 1988. Recovery of protein inclusion bodies. *Biochem. Soc. Trans.* **16:**102–104.

73. Ford, C. F., I. Suominen, and C. E. Glatz. 1991. Fusion tails for the recovery and purification of recombinant proteins. *Protein Expr. Purif.* **2:**95–107.

74. Francisco, J. A., C. F. Earhart, and G. Georgiou. 1992. Transport and anchoring of β-lactamase to the external surface of *Escherichia coli*. *Proc. Natl. Acad. Sci. USA* **89:**2713–2717.

75. Fuchs, P., F. Breitling, S. Dübel, T. Seehaus, and M. Little. 1991. Targeting recombinant antibodies to the surface of *Escherichia coli*: fusion to a peptidoglycan associated lipoprotein. *Bio/Technology* **9:**1369–1372.

76. Georgiou, G. 1995. Expression of proteins in bacteria, p. 101–127. *In* J. L. Cleland and C. S. Craik (ed.), *Principles and Practice of Protein Engineering*. John Wiley & Sons, New York.

77. Georgiou, G., J. J. Chalmers, and M. L. Shuler. 1985. Continuous immobilized recombinant protein production from *E. coli* capable of selective protein excretion: a feasibility study. *Biotechnol. Prog.* **1:**75–79.

78. Georgiou, G. and E. De Bernardez-Clark (ed.). 1991. *Protein Refolding*. American Chemical Society, Washington, D.C.

79. Gerdes, K. 1988. The *parB* (*hok/sok*) locus of plasmid R1: a general purpose plasmid stabilization system. *Bio/Technology* **6:**1402–1405.

80. Germino, J., and D. Bastia. 1884. Rapid purification of a cloned gene product by genetic fusion and site-specific proteolysis. *Proc. Natl. Acad. Sci. USA* **81:**4692–4696.

81. Gleister, I. E., and S. Bauer. 1981. Growth of *E. coli* W to high cell concentration by oxygen level linked control of carbon source concentration. *Biotechnol. Bioeng.* **23:**1015–1021.

82. Glick, B. R., and J. J. Pasternak. 1994. *Molecular Biotechnology: Principles and Applications of Recombinant DNA*. American Society for Microbiology, Washington, D.C.

83. Goeddel, D. V., H. L. Heyneker, T. Hozumi, R. Arentzen, K. Itakura, D. G. Yansura, M. J. Ross, G. Miozzari, R. Crea, and P. H. Seeburg. 1979. Direct expression in *Escherichia coli* of a DNA sequence coding for human growth hormone. *Nature* (London) **281:**544–548.

84. Goeddel, D. V., D. G. Kleid, F. Bolivar, H. L. Heyneker, D. G. Yansura, R. Crea, T. Hirose, A. Kraszewski, K. Itakura, and A. D. Riggs. 1979. Expression in *Escherichia coli* of chemically synthesized genes for human insulin. *Proc. Natl. Acad. Sci. USA* **76:**106–110.

85. Goochee, C. F., M. J. Gramer, D. C. Anderson, J. B. Bahr, and J. R. Rasmussen. 1992. The oligosaccharides of glycoproteins: factors affecting their synthesis and their influence on glycoprotein properties, p. 199–240. *In* P. Todd, S. K. Sikdar, and M. Bier (ed.), *Frontiers in Bioprocessing II*. American Chemical Society, Washington, D.C.

86. Goodson, R. J., and N. V. Katre. 1990. Site-directed pegylation of recombinant interleukin-2 at its glycosylation site. *Bio/Technology* **8:**343–346.

87. Gottesman, S. 1990. Minimizing proteolysis in *Escherichia coli*: genetic solutions. *Methods Enzymol.* **185:**119–129.

88. Gray, G. L., J. S. Baldridge, K. S. McKeown, H. L. Heyneker, and C. N. Chang. 1985. Periplasmic production of correctly processed human growth hormone in *Escherichia coli*: natural and bacterial signal sequences are interchangeable. *Gene* **39:**247–254.

89. Grossman, A. D., J. W. Erickson, and C. A. Gross. 1984. The *htpR* gene product of *E. coli* is a sigma factor for heat shock promoters. *Cell* **38:**383–390.

90. Gschaedler, A., and J. Boudrant. 1994. Amino acid utilization during batch and continuous cultures of *Escherichia coli* on a semi-synthetic medium. *J. Biotechnol.* **37:**235–251.

91. Hamilton, C. M., M. Aldea, B. K. Washburn, P. Babitzke, and S. R. Kushner. 1989. New method for generating deletions and gene replacements in *Escherichia coli*. *J. Bacteriol.* **171:**4617–4622.

92. Hantke, K., and V. Braun. 1978. Functional interaction of the *tonA/tonB* receptor system in *Escherichia coli*. *J. Bacteriol.* **135:**190–197.

93. Hart, R. A., D. M. Giltinan, P. M. Lester, D. H. Reifsnyder, J. R. Ogez, and S. E. Builder. 1994. Effect of environment on insulin-like growth factor I refolding selectivity. *Biotechnol. Appl. Biochem.* **20:**217–232.

94. Hart, R. A., P. M. Lester, D. H. Reifsnyder, J. R. Ogez, and S. E. Builder. 1994. Large scale, in situ isolation of periplasmic IGF-I from *E. coli*. *Bio/Technology* **12:**1113–1117.

95. Hirel, P. H., J. M. Schmitter, P. Dessen, G. Fayat, and S. Blanquet. 1989. Extent of N-terminal methionine excision from *Escherichia coli* proteins is governed by the side-chain length of the penultimate amino acid. *Proc. Natl. Acad. Sci. USA* **86:**8247–8251.

96. Hoffman, S. J., D. L. Looker, J. M. Roehrich, P. E. Cozart, S. L. Durfee, J. L. Tedesco, and G. L. Stetler. 1990. Expression of fully functional tetrameric human hemoglobin in *Escherichia coli*. *Proc. Natl. Acad. Sci. USA* **87:**8521–8525.

97. Hopp, T. P., K. S. Pricket, V. L. Price, R. T. Libby, C. J. March, D. P. Cerretti, D. L. Urdal, and P. J. Conlon. 1988. A short polypeptide marker sequence useful for recombinant protein identification and purification. *Bio/Technology* **6:**1204–1210.

98. Hsiung, H. M., and W. C. MacKellar. 1987. Expression of bovine growth hormone derivative in *Escherichia coli* and the use of the derivatives to produce natural sequence growth hormone by cathepsin C cleavage. *Methods Enzymol.* **153:**390–401.

99. Hsiung, H. M., N. G. Mayne, and G. W. Becker. 1986. High-level expression, efficient secretion and folding of human growth hormone in *Escherichia coli*. *Bio/Technology* **4:**991–995.

100. Hui, A., and H. A. deBoer. 1987. Specialized ribosome system: preferential translation of a single mRNA species by a subpopulation of mutated ribosomes in *Escherichia coli*. *Proc. Natl. Acad. Sci. USA* **84:**4762–4766.

101. Imanaka, T., H. Tsunekawa, and S. Aiba. 1980. Phenotypic stability of *trp* operon recombinant plasmids in *Escherichia coli*. *J. Gen. Microbiol.* **118:**253–261.

102. Ingram, L. O., and T. Conway. 1988. Expression of different levels of ethanologenic enzymes from *Zymomonas mobilis* in recombinant strains of *Escherichia coli*. *Appl. Environ. Microbiol.* **54:**397–404.

103. Itakura, K., H. Hirose, R. Crea, A. D. Riggs, H. L. Heyneker, F. Bolivar, and H. W. Boyer. 1977. Expression in *Escherichia coli* of a chemically synthesized gene for the hormone somatostatin. *Science* **198:**1056–1063.

104. Ito, A., T. Katoh, H. Gomi, F. Kishimoto, H. Agui, S. Ogino, K. Yoda, M. Yamasaki, and G. Tamura. 1986. Expression and secretion of human epidermal growth factor in *Escherichia coli*. *Agric. Biol. Chem.* **50:**1381–1388.

105. Izard, J. W., and D. A. Kendall. 1994. Signal peptides: exquisitely designed transport promoters. *Mol. Microbiol.* **13:**765–773.

106. Jensen, E. B., and S. Carlsen. 1990. Production of recombinant human growth hormone in *Escherichia coli*: expression of different precursors and physiological effects of glucose, acetate, and salts. *Biotechnol. Bioeng.* **36:**1–11.

107. Jespers, J. S., A. Roberts, S. M. Mahler, G. Winter, and H. R. Hoogenboom. 1994. Guiding the selection of human antibodies from phage display repertoires to a single epitope of an antigen. *Bio/Technology* **12:**899–903.

108. Johnson, B. H., and M. H. Hecht. 1994. Recombinant proteins can be isolated from *E. coli* cells by repeated cycles of freezing and thawing. *Bio/Technology* **12:**1357–1360.

109. Joly, J. C., and J. R. Swartz. 1994. Protein folding activities of *Escherichia coli* protein disulfide isomerase. *Biochemistry* 33:4231–4236.

110. Jung, G., P. Denefle, J. Becquart, and J. F. Mayaux. 1988. High-cell density fermentation studies of recombinant *Escherichia coli* strains expressing human interleukin-1β. *Ann. Inst. Pasteur Microbiol.* 139:129–146.

111. Kamitani, S., Y. Akiyama, and K. Ito. 1992. Identification and characterization of an *Escherichia coli* gene required for the formation of correctly folded alkaline phosphatase, a periplasmic enzyme. *EMBO J.* 11:57–62.

112. Kapralek, F., P. Jecmen, J. Sedlacek, M. Fabry, and S. Zadrazil. 1991. Fermentation conditions for high-level expression of the tac-promoter-controlled calf prochymosin cDNA in *Escherichia coli* HB101. *Biotechnol. Bioeng.* 37:71–79.

113. Katayama, Y., S. Gottesman, J. Pumphrey, S. Rudikoff, W. P. Clark, and M. R. Maurizi. 1988. The two-component, ATP-dependent Clp protease of *Escherichia coli. J. Biol. Chem.* 263:15226–15236.

114. Katre, N. V., M. J. Knauf, and W. J. Laird. 1987. Chemical modification of recombinant interleukin-2 by polyethylene glycol increases its potency in the murine Meth A sarcoma model. *Proc. Natl. Acad. Sci. USA* 84:1487–1491.

115. Khosla, C., and J. E. Bailey. 1988. Heterologous expression of a bacterial hemoglobin improves the growth properties of recombinant *Escherichia coli. Nature* (London) 331:633–635.

116. Khosla, C., J. E. Curtis, J. DeModena, U. Rinas, and J. E. Bailey. 1990. Expression of intracellular hemoglobin improves protein synthesis in oxygen-limited *Escherichia coli. Bio/Technology* 8:849–853.

117. Klein, B. K., S. R. Hill, C. S. Devine, E. Rowold, C. E. Smith, S. Galosy, and P. O. Olins. 1991. Secretion of active bovine somatotropin in *Escherichia coli. Bio/Technology* 9:869–872.

118. Klessen, C., K. H. Schmidt, J. Gumbert, H. H. Grosse, and H. Malke. 1989. Complete secretion of activable bovine prochyosin by genetically engineered L forms of *Proteus mirabilis. Appl. Environ. Microbiol.* 55:1009–1015.

119. Knappik, A. C. Krebber, and A. Plückthun. 1993. The effect of folding catalysts on the in vivo folding process of different antibody fragments expressed in *Escherichia coli. Bio/Technology* 11:77–83.

120. Kopetzki, E., G. Schumacher, and P. Buckel. 1989. Control of formation of active soluble or inactive insoluble baker's yeast alpha-glucosidase PI in *Escherichia coli* by induction and growth conditions. *Mol. Gen. Genet.* 216:149–155.

121. Kornacker, M. G., and A. P. Pugsley. 1990. The normally periplasmic enzyme β-lactamase is specifically and efficiently translocated through the *Escherichia coli* outer membrane when it is fused to the cell-surface enzyme pullulanase. *Mol. Microbiol.* 4:1101–1109.

122. Kosinski, M. J., U. Rinas, and J. E. Bailey. 1992. Isopropyl-β-D-thiogalactopyranoside influences the metabolism of *Escherichia coli. Appl. Microbiol. Biotechnol.* 36:782–784.

123. Lacatena, R. M., and G. Cesareni. 1981. Base pairing of RNA I with its complementary sequence in the primer precursor inhibits ColE1 replication. *Nature* (London) 294:623–626.

124. Ladisch, M. R., R. C. Willson, C. D. C. Painton, and S. E. Builder. 1990. *Protein Purification: from Molecular Mechanisms to Large-Scale Processes.* American Chemical Society, Washington, D.C.

125. Larson, J., K. Gerdes, J. Light, and S. Molin. 1984. Low-copy-number plasmid-cloning vectors amplifiable by derepression of an inserted foreign promoter. *Gene* 28:45–54.

126. Lee, S. C., and P. O. Olins. 1992. Effect of overproduction of heat shock chaperones GroESL and DnaK on human procollagenase production in *Escherichia coli. J. Biol. Chem.* 267:2849–2852.

127. Leipold, R. J., and P. Dhurjati. 1993. Specialized ribosomes in *Escherichia coli. Biotechnol. Prog.* 9:443–449.

128. Liao, J. C., and J. Delgado. 1993. Advances in metabolic control analysis. *Biotechnol. Prog.* 9:221–233.

129. Lin, N. S., and J. R. Swartz. 1992. Production of heterologous proteins from recombinant DNA *Escherichia coli* in bench fermentors. *Methods* (Orlando) 4:159–168.

130. Lin-Chao, S., and S. N. Cohen. 1991. The rate of processing and degradation of antisense RNAI regulates the replication of ColE1-type plasmids *in vivo. Cell* 65:1233–1242.

131. Liu, J., and C. T. Walsh. 1990. Peptidyl-prolyl cis-trans-isomerase from *Escherichia coli*: a periplsmic homolog of cyclophilin that is not inhibited by cyclosporin A. *Proc. Natl. Acad. Sci. USA* 87:4028–4032.

132. Lu, H. S., C. L. Clogston, L. A. Merewether, L. O. Narhi, and T. C. Boone. 1993. Role of disulfide bonds in folding recombinant human granulocyte colony stimulating factor produced in *Escherichia coli*, p. 189–202. In J. L. Cleland (ed.), *Protein Folding: In Vivo and In Vitro.* American Chemical Society, Washington, D.C.

133. Luli, G. W., and W. R. Strohl. 1990. Comparison of growth, acetate production, and acetate inhibition of *Escherichia coli* strains in batch and fed-batch fermentations. *Appl. Environ. Microbiol.* 56:1004–1011.

134. Maniatis, T., M. Ptashne, K. Backman, D. Kleid, S. Flashman, A. Jeffrey, and R. Maurer. 1975. Recognition sequences of repressor and polymerase in the operators of bacteriophage lambda. *Cell* 5:109–113.

135. Martin, U., S. Fischer, U. Kohnert, R. Opitz, R. Rudolph, G. Sponer, A. Stern, and K. Strein. 1991. Thrombolysis with an *Escherichia coli*-produced recombinant plasminogen activator (BM 01.022) in the rabbit model of jugular vein thrombosis. *Thromb. Haemostasis* 65:560–564.

136. Matthews, D. J., and J. A. Wells. 1993. Substrate phage: selection of protease substrates by monovalent phage display. *Science* 260:1113–1117.

137. Matthews, G. 1994. The leucine-responsive regulatory protein, a global regulator of metabolism in *Escherichia coli. Microbiol. Rev.* 58:401–425.

138. Maurizi, M. R. 1992. Proteases and protein degradation in *Escherichia coli. Experientia* 48:178–201.

139. Maurizi, M. R., P. Trisler, and S. Gottesman. 1985. Insertional mutagenesis of the *lon* gene in *Escherichia coli: lon* is dispensable. *J. Bacteriol.* 164:1124–1135.

140. McCafferty, J., A. D. Griffiths, G. Winter, and D. J. Chiswell. 1990. Phage antibodies: filamentous phage displaying antibody variable domains. *Nature* (London) 348:552–554.

141. McDowell, J. C., J. W. Roberts, D. J. Jin, and C. Gross. 1994. Determination of intrinsic transcription termination efficiency by RNA polymerase elongation rate. *Science* 266:822–825.

142. McMurry, I., R. E. Petrucci, Jr., and S. B. Levy. 1980. Active efflux of tetracycline encoded by four genetically different tetracycline resistance determinants in *Escherichia coli. Proc. Natl. Acad. Sci. USA* 77:3974–3977.

143. Meacock, P. A., and S. N. Cohen. 1980. Partitioning of bacterial plasmids during cell division: a cis-acting locus that accomplishes stable plasmid inheritance. *Cell* 20:529–542.

144. Medynski, D. 1994. Phage display: all dressed up and ready to role. *Bio/Technology* 12:1134–1136.

145. Meerman, H. J., and G. Georgiou. 1994. Construction and characterization of a set of *E. coli* strains deficient in all known loci affecting the proteolytic stability of secreted recombinant proteins. *Bio/Technology* 12:1107–1110.

146. Merlin, T. L., G. E. Davis, W. L. Anderson, R. K. Moyzis, and J. K. Griffith. 1989. Aminoglycoside uptake increased by *tet* gene expression. *Antimicrob. Agents Chemother.* 33:1549–1552.

147. Meyer, H. P., C. Leist, and A. Fiechter. 1984. Acetate formation in continuous culture of *Escherichia coli* K12 D1 on defined and complex media. *J. Biotechnol.* 1:355–358.

148. Missiakas, D., C. Georgopoulos, and S. Raina. 1994. The *Escherichia coli dsbC* (*xprA*) gene encodes a periplasmic protein involved in disulfide bond formation. *EMBO J.* 13:2013–2020.

149. Mitraki, A., and J. King. 1989. Protein folding intermediates and inclusion body formation. *Bio/Technology* 7:690–697.

150. Miyake, T., T. Oka, T. Mishizawa, F. Misoka, T. Fuwa, K. Yoda, M. Yamasaki, and G. Tamura. 1985. Secretion of human interferon-alpha induced by using secretion vectors containing a promoter and signal sequence of alkaline phosphatase gene of *Escherichia coli. J. Biochem.* 97:1429–1436.

151. Mizutani, S. H. Mori, S. Shimizu, K. Sakaguchi, and T. Kobayashi. 1986. Effect of amino acid supplement on cell yield and gene product in *Escherichia coli* harboring plasmid. *Biotechnol. Bioeng.* 28:204–209.

152. Moks, T., L. Abrahmsen, E. Holmgren, M. Bilich, A. Olsson, M. Uhlen, G. Pohl, C. Sterky, H. Hultberg, S. Josephson, A. Holmgren, H. Jörnvall, and B. Nilsson. 1987. Expression of human insulin-like growth factor I in bacteria: use of optimized gene fusion vectors to facilitate protein purification. *Biochemistry* 26:5239–5244.

153. Morrow, J. F., S. N. Cohen, A. C. Y. Chang, H. W. Boyer, H. M. Goodman, and R. B. Helling. 1974. Replication and transcription of eukaryotic DNA in *Escherichia coli. Proc. Natl. Acad. Sci. USA* 71:1743–1747.

154. Moser, D. R., and J. L. Campbell. 1983. Characterization and complementation of pMB1 copy number mutant: effect of RNA I gene dosage on plasmid copy number and incompatibility. *J. Bacteriol.* 154:809–818.

155. Naglak, T. J., and H. Y. Wang. 1990. Recovery of a foreign protein from the periplasm of *Escherichia coli* by chemical permeabilization. *Enzyme Microb. Technol.* 12:603–611.

156. Neidhardt, F. C. 1987. Chemical composition of the cell, p. 3–6. In F. C. Neidhardt, J. L. Ingraham, K. B. Low, B. Magasanik, M. Schaechter, and H. E. Umbarger (ed.), *Escherichia coli and Salmonella typhimurium: Cellular and Molecular Biology.* American Society for Microbiology, Washington, D.C.

157. Neidhardt, F. C., R. A. VanBogelen, and E. T. Lau. 1983. Molecular cloning and expression of a gene that controls the high-temperature regulon of *Escherichia coli. J. Bacteriol.* 153:597–603.

158. Nierman, W. C., and T. Feldblyum. 1985. Cryopreservation of cultures that contain plasmids. *Dev. Ind. Microbiol.* 26:423–434.

159. Nilsson, B., G. Forsberg, T. Moks, M. Hartmannis, and M. Uhlen. 1992. Fusion proteins in biotechnology and structural biology. *Curr. Opin. Struct. Biol.* 2:569–575.

160. Noack, D., M. Roth, R. Geuther, G. Müller, K. Undisz, C. Hoffmeie, and S. Gaspar. 1981. Maintenance and genetic stability of vector plasmids pBR322 and pBR325 in *Escherichia coli* K12 strains grown in a chemostat. *Mol. Gen. Genet.* 184:121–124.

161. Ostermeier, M., and G. Georgiou. 1994. The folding of bovine pancreatic trypsin inhibitor in the *Escherichia coli* periplasm. *J. Biol. Chem.* 269:21072–21077.

162. Patnaik, R., and J. C. Liao. 1994. Engineering of *Escherichia coli* central metabolism for aromatic metabolite production with near theoretical yield. *Appl. Environ. Microbiol.* 60:3903–3908.

163. Phillips, T. A., R. A. VanBogelen, and F. C. Neidhardt. 1984. *lon* gene product of *Escherichia coli* is a heat-shock protein. *J. Bacteriol.* 159:283–287.

164. Pierce, J., and S. Gutteridge. 1985. Large-scale preparation of ribulosebisphosphate carboxylase from a recombinant system in *Escherichia coli* characterized by extreme plasmid instability. *Appl. Environ. Microbiol.* 49:1094–1100.

165. Plückthun, A., and A. Skerra. 1989. Expression of functional antibody Fv and Fab fragments in Escherichia coli. *Methods Enzymol.* 178:497–515.

166. Polisky, B. 1988. ColE1 replication control circuitry: sense from antisense. *Cell* 55:929–932.

167. Pugsley, A. P. 1993. The complete general secretory pathway in gram-negative bacteria. *Microbiol. Rev.* 57:50–108.

168. Reid, J. L., and A. Collmer. 1987. An *nptI-scaB-sacR* cartridge for constructing directed, unmarked mutations in gram-negative bacteria by marker exchange-eviction mutagenesis. *Gene* 57:239–246.

169. Reiling, H. E., H. Laurila, and A. Fiechter. 1985. Mass culture of *Escherichia coli*: medium developments for low and high census cultivation of *Escherichia coli* B/r in minimal and complex media. *J. Biotechnol.* 2:191–206.

170. Remaut, E., H. Tsao, and W. Fiers. 1983. Improved plasmid vectors with a thermoinducible expression and temperature-regulated runaway replication. *Gene* 22:103–113.

171. Ringquist, S., S. Shinedling, D. Barrick, L. Green, J. Binkley, G. D. Stormo, and L. Gold. 1992. Translation initiation in *Escherichia coli*: sequences within the ribosome-binding site. *Mol. Microbiol.* 6:1219–1229.

172. Roberts, B. L., W. Markland, A. C. Ley, R. B. Kent, D. W. White, S. K. Guterman, and R. C. Ladner. 1992. Directed evolution of a protein: selection of potent neutrophil inhibitors displayed on M13 fusion phage. *Proc. Natl. Acad. Sci. USA* 89:2419–2433.

173. Rosteck, P. R., Jr., and C. L. Hershberger. 1983. Selective retention of recombinant plasmids coding for human insulin. *Gene* 25:29–38.

174. Rudolph, R. 1996. Successful protein folding on an industrial scale, p. 283–298. *In* J. L. Cleland and C. S. Craik (ed.), *Principles and Practice of Protein Engineering.* John Wiley & Sons, New York.

175. Rusch, S. L., and D. A. Kendall. 1994. Transport of an export-defective protein by a highly hydrophobic signal peptide. *J. Biol. Chem.* 269:1243–1248.

176. Sambrook, J., E. F. Fritsch, and T. Maniatis. 1989. *Molecular Cloning: a Laboratory Manual,* 2nd ed. Cold Spring Harbor Laboratory Press, Cold Spring Harbor, N.Y.

177. Schauder, B., and J. E. G. McCarthy. 1989. The role of bases upstream of the Shine-Dalgarno region and in the coding sequence in the control of gene expression in *Escherichia coli*: translation and stability of mRNA's *in vivo. Gene* 78:59–72.

178. Schein, C. H. 1989. Production of soluble recombinant proteins in bacteria. *Bio/Technology* 7:1141–1149.

179. Schein, C. H., and M. H. M. Noteborn. 1988. Formation of soluble recombinant proteins in *Escherichia coli* is favored by lower growth temperature. *Bio/Technology* 6:291–294.

180. Schoner, R. G., L. F. Ellis, and B. E. Schoner. 1985. Isolation and purification of protein granules from *Escherichia coli* cells overproducing bovine growth hormone. *Bio/Technology* 3:151–154.

181. Seetharam, R., R. A. Heeren, E. Y. Wong, S. R. Braford, B. K. Klein, S. Aykent, C. E. Kotts, K. J. Mathis, B. F. Bishop, M. J. Jennings, C. E. Smith, and N. R. Siegel. 1988. Mistranslation in IGF-1 during over-expression of the protein in *Escherichia coli* using a synthetic gene containing low frequency codons. *Biochem. Biophys. Res. Commun.* 155:518–523.

182. Seo, J. H., and J. E. Bailey. 1985. Effects of recombinant plasmid content on growth properties and cloned gene product formation in *Escherichia coli. Biotechnol. Bioeng.* 27:1668–1674.

183. Seo, J. H., A. I. Löffler, and J. E. Bailey. 1988. A parametric study of cloned fusion protein expression in *Escherichia coli. Biotechnol. Bioeng.* 32:725–730.

184. Seow, H. F., C. R. Goh, L. Krishnan, and A. G. Porter. 1989. Bacterial expression, facile purification and properties of recombinant human lymphotoxin (tumor necrosis factor beta). *Bio/Technology* 7:363–368.

185. Shatzman, A. R. 1990. Gene expression using gram-negative bacteria. *Curr. Opin. Biotechnol.* 1:5–11.

186. Shevchik, V. E., G. Condemine, and J. Robert-Baudouy. 1994. Characterization of DsbC, a periplasmic protein of *Erwinia chrysanthemi* and *Escherichia coli* with disulfide isomerase activity. *EMBO J.* 13:2007–2012.

187. Silhavy, T. J., M. J. Casadaban, H. A. Shuman, and J. R. Beckwith. 1976. Conversion of β-galactosidase to a membrane-bound state by gene fusion. *Proc. Natl. Acad. Sci. USA* 73:3423–3427.

188. Silhavy, T. J., H. A. Shuman, J. Beckwith, and M. Schwartz. 1977. Use of gene fusions to study outer membrane protein localization in *Escherichia coli. Proc. Natl. Acad. Sci. USA* 74:5411–5415.

189. Simon, L. D., B. Randolph, N. Irwin, and G. Binkowski. 1983. Stabilization of proteins by a bacteriophage T4 gene cloned in *Escherichia coli. Proc. Natl. Acad. Sci. USA* 80:2059–2062.

190. Skerra, A., and A. Plückthun. 1988. Assembly of a functional immunoglobulin Fv fragment in *Escherichia coli. Science* 240:1038–1041.

191. Skogman, G., and J. Nilsson. 1984. Temperature-dependent retention of a tryptophan-operon-bearing plasmid in *Escherichia coli. Gene* 31:117–122.

192. Skogman, G., J. Nilsson, and P. Gustafsson. 1983. The use of a partition locus to increase stability of tryptophan-operon-bearing plasmids in *Escherichia coli. Gene* 23:105–115.

193. Slater, S. C., W. H. Voige, and D. E. Dennis. 1988. Cloning and expression in *Escherichia coli* of the *Alcaligenes eutrophus* H16 poly-β-hydroxybutyrate biosynthetic pathway. *J. Bacteriol.* 170:4431–4436.

194. Smith, D. B., and K. S. Johnson. 1988 One-step purification of polypeptides expressed in *Escherichia coli* as fusions with glutathione S-transferase. *Gene* 67:31–40.

195. Smith, G. P. 1985. Filamentous fusion phage: novel expression vectors that display cloned antigens on the virion surface. *Science* 228:1315–1317.

196. Smith, R. A. G., J. M. Dewdney, R. Fears, and G. Poste. 1993. Chemical derivatization of therapeutic proteins. *Trends Biotechnol.* 111:397–403.

197. Sorensen, M. A., C. G. Kurland, and S. Pedersen. 1989. Codon usage determines translation rate in *Escherichia coli. J. Mol. Biol.* 207:365–377.

198. St. John, A. C., and A. L. Goldberg. 1980. Effects of starvation for potassium and other inorganic ions on protein degradation and ribonucleic acid synthesis in *Escherichia coli. J. Bacteriol.* 143:1223–1233.

199. Storrs, S. B., and T. M. Przybycien. 1991. Commercial-scale refolding of recombinant methionyl bovine somatotropin, p. 197–205. *In* G. Georgiou and E. De Bernardez-Clark (ed.), *Protein Refolding.* American Chemical Society, Washington, D.C.

200. Strauch, K. L., and J. Beckwith. 1988. An *Escherichia coli* mutation preventing degradation of abnormal periplasmic proteins. *Proc. Natl. Acad. Sci. USA* 85:1576–1580.

201. Studier, F. W., A. H. Rosenberg, J. J. Dunn, and J. W. Dubendorff. 1990. Use of T7 polymerase to direct expression of cloned genes. *Methods Enzymol.* 185:60–89.

202. Talmadge, K., J. Kaufman, and W. Gilbert. 1980. Bacteria mature preproinsulin to proinsulin. *Proc. Natl. Acad. Sci. USA* 77:3988–3992.

203. Talmadge, K., S. Stahl, and W. Gilbert. 1980. Eukaryotic signal sequence transports insulin antigen in *Escherichia coli. Proc. Natl. Acad. Sci. USA* 77:3369–3373.

204. Terasawa, M., M. Fukushima, Y. Kurusu, and H. Yukawa. 1990. L-Tryptophan production by the application of high expressed tryptophanase in *Escherichia coli. Proc. Biochem. Int.* 25:172–175.

205. Thompson, B. G., M. Kole, and D. F. Gerson. 1985. Control of ammonium concentration in *Escherichia coli. Biotechnol. Bioeng.* 27:818–824.

206. Tokatlidis, K., P. Dhurjati, J. MIllet, P. Beguin, and J. P. Aubert. 1991. High activity of inclusion bodies formed in *Escherichia coli* overproducing *Clostridium thermocellum* endoglucanase D. *FEBS Lett.* 282:205–208.

207. Tsai, B. B. H. S. Lu, W. C. Kenney, C. C. Curless, M. L. Klein, P. H. Lai, D. M. Fenton, B. W. Altrock, and M. B. Mann. 1988. Control of misincorporation of de novo synthesized norleucine into recombinant interleukin-2 in *E. coli. Biochem. Biophys. Res. Commun.* 156:733–739.

208. Tsai, L. B., M. Mann, F. Morris, C. Rotgers, and D. Fenton. 1987. The effect of organic nitrogen and glucose on the production of recombinant human insulin-like growth factor in high cell density *Escherichia coli* fermentations. *J. Ind. Microbiol.* 2:181–187.

209. Tunac, J. B. 1989. High-aeration capacity shake-flask system. *J. Ferment. Bioeng.* 68:157–159.

210. Umbarger, H. E. 1987. Biosynthesis of the branched-chain amino acids, p. 352–367. *In* F. C. Neidhardt, J. L. Ingraham, K. B. Low, B. Magasanik, M. Schaechter, and H. E. Umbarger (ed.), *Escherichia coli and Salmonella typhimurium: Cellular and Molecular Biology.* American Society for Microbiology, Washington, D.C.

211. Valax, P., and G. Georgiou. 1993. Molecular characterization of β-lactamase inclusion bodies produced in *Escherichia coli*. 1. Composition. *Biotechnol. Prog.* 9:539–547.

212. Voss, T., E. Falkner, H. Ahorn, E. Krystek, I. Maurer-Fogy, G. Bodo, and R. Hauptmann. 1994. Periplasmic expression of human interferon-alpha2c in *Escherichia coli* results in a correctly folded molecule. *Biochem. J.* 298:719–725.

213. Walker, K. W., and H. F. Gilbert. 1994. Effect of redox environment on the *in vitro* and *in vivo* folding of RTEM-1β-lactamase and *Escherichia coli* alkaline phosphatase. *J. Biol. Chem.* 269:28487–28493.

214. Wanner, B. L. 1993. Gene regulation by phosphate in enteric bacteria. *J. Cell. Biochem.* 51:47–54.

215. Warne, S. R., C. M. Thomas, M. E. Nugent, and W. C. A. Tacon. 1986. Use of a modified *Escherichia coli* trpR gene to obtain tight regulation of high-copy-number expression vectors. *Gene* 46:103–112.

216. Winter, G., A. D. Griffiths, R. E. Hawkins, and H. R. Hoogenboom. 1994. Making antibodies by phage display technology. *Annu. Rev. Immunol.* 12:433–455.

217. Winter, R. B., K. M. Yen, and B. D. Ensley. 1989. Efficient degradation of trichloroethylene by a recombinant *Escherichia coli. Bio/Technology* 7:282–285.

218. Wong, E. Y., R. Seetharam, C. E. Kotts, R. A. Heeren, B. K. Klein, S. R. Braford, K. J. Mathis, B. F. Bishop, N. R. Siegel, C. E. Smith, and W. C. Tacon. 1988. Expression of secreted insulin-like growth factor-1 in *Escherichia coli. Gene* 68:193–203.

219. Wood, T. K., and S. W. Peretti. 1991. Effect of chemically-induced, cloned-gene expression on protein synthesis in *E. coli. Biotechnol. Bioeng.* 38:397–412.

220. Wülfing, C., and A. Plückthun. 1994. Correctly folded T-cell receptor fragments in the periplasm of *Escherichia coli*: influence of folding catalysts. *J. Mol. Biol.* 242:655–669.

221. Wülfing, C., and A. Plückthun. 1994. Protein folding in the periplasm of *Escherichia coli. Mol. Microbiol.* 12:685–692.

222. Wunderlich, M., and R. Glockshuber. 1993. In vivo control of redox potential during protein folding catalyzed by bacterial protein disulfide-isomerase (DsbA). *J. Biol. Chem.* 268:24547–24550.

223. Yamaguchi, A., N. Ono, T. Akasaka, T. Noumi, and T. Sawai. 1990. Metal-tetracycline/H+ antiporter of *Escherichia coli* encoded by a transposon, Tn10. *J. Biol. Chem.* 265:15525–15530.

224. Yamaguchi, A., T. Udagawa, and T. Sawai. 1990. Transport of divalent cations with tetracycline as mediated by the transposon Tn10-encoded tetracycline resistance protein. *J. Biol. Chem.* 265:4809–4813.

225. Yansura, D. G., and D. J. Henner. 1990. Use of *Escherichia coli* trp promoter for direct expression of proteins. *Methods Enzymol.* 185:54–60.

226. **Yee, L., and H. W. Blanch.** 1992. Recombinant protein expression in high cell density fed-batch cultures of *Escherichia coli*. *Bio/Technology* **10:**1550–1556.

227. **Yee, L., and H. W. Blanch.** 1993. Recombinant trypsin production in high cell density fed-batch cultures in *Escherichia coli*. *Biotechnol. Bioeng.* **41:**781–790.

228. **Yukawa, H., Y. Kurusu, M. Shimazu, M. Terasawa, A. Ohta, and I. Shibuya.** 1985. Stabilization by the mini-F fragment of a pBR322 derivative bearing the tryptophan operon in *Escherichia coli*. *Agric. Biol. Chem.* **49:**3619–3622.

229. **Yukawa, H., Y. Kurusu, M. Shimazu, H. Yamagata, and M. Terasawa.** 1988. Stabilization of an *E. coli* plasmid by a mini-F fragment of DNA. *J. Ind. Microbiol.* **2:**323–328.

230. **Zabriskie, D. W., and E. J. Arcuri.** 1986. Factors influencing productivity of fermentations employing recombinant microorganisms. *Enzyme Microb. Technol.* **8:**706–717.

231. **Zhou, Y. N., N. Kusulawa, J. W. Erickson, C. A. Gross, and T. Yura.** 1988. Isolation and characterization of *Escherichia coli* mutants that lack the heat shock sigma factor (σ^{32}). *J. Bacteriol.* **170:**3640–3649.

Genome, Genetics, and Evolution

VI

THE GENETICS OF ENTERIC BACTERIA DOMINATES THE CURRENT EDITION of this work. The assignment of 40 chapters to this subject is appropriate because genetics has been a major factor in maintaining the position of *E. coli* and its close relatives as favorite subjects for biological inquiry. The importance of genetics derives only in part from the power of classical genetic analysis in metabolic, structural, and physiological studies; it is also due in some measure to the particularly immediate and participatory role of the genome in the life of prokaryotic cells. Not only are all unicellular organisms the equivalent of germ line cells in multicellular plants and animals, but also, uniquely, in bacterial cells the genetic material is in intimate contact with the cytosol and is attached to the surrounding envelope. Signals reporting the nature of the exterior world and the physiological status of the cell interior are received directly by the chromosome, and rapid adjustments are made in the flow of information to the ribosomes. The entire program of gene expression for growth, non-growth, pathogenesis, or whatever, is continuously renewed and updated on a minute-by-minute basis thanks to the rapid turnover of mRNA.

A fascinating and complex picture is emerging of the cell's ability to assess damage to the genome and deal with a multitude of types of deleterious lesions in DNA or its precursors. Processes ranging from in situ repair to complex damage-stimulated recombination events involving replication are all being actively investigated to relate currently identified genes or gene functions to their in vivo roles. Other types of recombination, including site-specific and transpositional, further broaden the cell's capacity to regulate gene expression and the movement of genetic elements.

The monumental task of coordinating the physical and linkage maps of *E. coli* is merely one of the efforts spurred on by the genome sequencing project. Efforts to analyze the presence of repeated sequences, insertion sequence elements, and cryptic prophages, to identify all of the protein products of the genome and to learn functional and phylogenetic information about them, and to begin to discern the pathways of evolution of the enterobacteria, their genes, enzymes, and metabolic pathways—all these efforts have been accelerated by the organized nucleotide sequencing of the *E. coli* genome. Molecular sequence analysis is bringing tools of informatics to bear on extracting the inherent information content of the genome. Comparative sequence information now supplements information on protein polymorphism in illuminating both the clonal structure of bacterial populations and the relative contributions of recombination, mutation, and genetic drift to the genetic variation of these populations. The increasing interest in integrative molecular biology, including the explicit goal of integrating molecular information of regulon control into predictive models of the cell, will rely heavily on knowledge of the total genome. For these reasons there is probably no organism for which total genomic sequencing makes more sense in terms of immediate payoff.

Linkage Map of *Escherichia coli* K-12, Edition 9

MARY K. B. BERLYN, K. BROOKS LOW, AND KENNETH E. RUDD

109

INTRODUCTION

We present two different representations of the genetic map of *Escherichia coli* K-12, the linkage map of known genes and other functional sites (Fig. 1; p. 1717) and the physical map showing restriction sites, physically mapped genes, and open reading frames (Fig. 2; p. 1723). In addition, an appendix shows the locations of Tn insertions found in the Gross laboratory collection of mapping strains (3949) and other widely distributed strains, recalibrating the insertion sites based on the coordinates in the Fig. 1 and 2 maps. These locations are placed on an abridged circular map in Fig. 3 (Appendix, p. 1812; Fig. 3, p. 1815). Thus three different levels of resolution of the *E. coli* K-12 map can be examined.

The linkage map in classical form, presented as Fig. 1, includes genes located primarily by restriction, sequence, and cotransduction data reported in the literature and databases. It combines known genes identified and mapped within long, contiguous regions sequenced as part of the *E. coli* sequencing project (372, 548, 548a, 903, 3359, 3995); genes placed on the physical map of *E. coli* (2208, 2209, 3097) by restriction and sequence comparisons as part of the EcoSeq project (Fig. 2); genes placed on previous editions of the *E. coli* linkage maps (173, 175); and additional genes reported in recent literature but not yet placed on any of these previous maps.

The linkage map of Fig. 1 includes 1,958 genes and about 40 other chromosomal markers, such as phage attachment sites, defective-phage elements, replication origins and termini, and other features traditionally included on the published linkage map. These loci are listed, and briefly described, with literature citations in Table 1 (p. 1754). Figure 1 and Table 1 do not include open reading frames (ORFs) of unknown function or of putative function inferred by sequence homology with known genes in other organisms. These can be found on the physical map (Fig. 2).

The Fig. 1 map places the genes that can also be found in Fig. 2 on the right-hand side of the line. On the left-hand side are genes not physically mapped, and therefore not connected to a specific point on the axis, and some physically or otherwise precisely mapped genes not yet included in this version of the physical map. The latter are connected by a line to the axis. As in previous editions of the *E. coli* linkage map, an asterisk indicates that the gene is not precisely located with respect to near neighbors and parentheses indicate that the location is even more uncertain and that the gene is located only within that general region. Also shown on the left-hand side in boldface followed by a colon are operon names that are distinct from any gene name within the operon.

Updates of map information are available in electronic form from several sites. These include the World Wide Web server for the *E. coli* Genetic Stock Center (CGSC) at URL http://cgsc.biology.yale.edu, its gopher server (under Connecticut, or at cgsc.biology.yale.edu), and the Apt forms interface to the Sybase server (see chapter 134 of this book); the National Center for Biotechnology Information ftp site for EcoSeq and EcoMap, ncbi.nlm.nih.gov/repository/Eco/EcoMap7; and the ftp site for the sequencing project at the University of Wisconsin, Ecoliftp.genetics.wisc.edu/pub.

MAP UNITS

Since the 1976 recalibration of the linkage map in terms of minutes required for time of entry of markers in interrupted conjugation experiments, the standard representation of the map has used the basic units of minutes and length of 100 min (176). This has been a convenient and accepted coordinate system for the map, and although the current map units are based on restriction and sequence data rather than time of entry, we retain the term "minute" for one-hundredth of the length of the chromosome. Both the CGSC database and EcoMap use as "left endpoints" the counterclockwise boundary of the coding region, and genes in Fig. 1 are placed approximately at these coordinates, with the higher-resolution map of Fig. 2 providing more exact placement.

NOMENCLATURE

The standard genetic nomenclature for *E. coli* is that of Demerec et al. (1966 [reference 994]) as subsequently amended through use and as described in the Instructions to Authors for the *Journal of Bacteriology* (see also *Trends in Genetics, Genetics Nomenclature Guide*, 1995 [4075a]). This map, like those preceding it, follows those nomenclatural conventions. Accordingly, we have adhered to a three-letter lowercase mnemonic symbol, with an uppercase letter added when there are two or more genes in that mnemonic category. If authors have added an uppercase letter for a gene in a single-instance category, we have used that published four-letter symbol. For attachment sites, noncoding features of the chromosome, etc., the same standard has not been used, and we have continued to use the variable-length symbols historically applied to these sites. We have continued the convention proposed for sites of termination of replication and repetitive sequences, using italicized symbols with the first letter uppercase, e.g., *TerB, RhsA* (175).

Many gene names have been changed by investigators since the 1990 map was published. When those changes were part of a systematic revision of nomenclature (often aimed at clarifying usage and resolving conflicts) for a group of related genes and were in compliance with the current *E. coli* gene nomenclature system, or were changed for compelling mnemonic reasons or for resolution of redundancy or conflict, also in conformance with the standard system, we have adopted those changes. We have not adopted and *we wish to discourage changes of valid preexisting names proposed by authors simply because they believe that theirs is a more apt or accurate mnemonic.* For example, a previously published name based on the pathway or phenotype *is valid and should not be replaced,* simply as a matter of course, by an alternate mnemonic based on the name of the enzyme encoded by the gene, once that functional information has been determined.

In a few cases, we have been compelled to use a new name, despite apparent validity of the original name, simply because the new name has been widely adopted in the literature. In a number of cases, a new gene has been assigned a symbol which has already been used or is simultaneously proposed for another gene, with the two mnemonics having entirely different meanings. These names have had to be resolved, usually by changing the newer assignment. In a few cases, a uniquely named gene has been shown later to belong to a category for which a symbol already exists, and that symbol has been used instead of the earlier assignment.

Some synonymy is unavoidable, since a gene under study may be named and described in print before its identity to a known gene is discovered. However, a common practice in the recent literature seems to allow publication of an author's initial name for a gene even if its identity to a known gene has been discovered before publication, and that creates unnecessary synonymy. Alternate gene symbols are listed in column 4 of Table 1 (p. 1754) and Table 2 (p. 1806) provides an alphabetized list of such symbols with the cross-reference to the symbol used in Table 1.

There is a standing tradition of coordinating gene symbols between the CGSC and the *Salmonella* Genetic Stock Center to avoid assignment of the same symbol to different genes in the two organisms or assignment of different symbols to homologous genes, insofar as this coordination is feasible. We have not, however, changed names of *E. coli* genes in order to extend this tradition to other bacteria or other organisms. The desirability and feasibility of uniform nomenclature conventions for all bacteria or other microbial groupings are currently only topics of discussion and conjecture, and changes to enhance similarities in an ad hoc, piecemeal fashion seem counterproductive at present. Readers are reminded that symbol changes create discontinuity with previous literature concerning genes, with even more serious ramifications for allele designations, since unique allele numbers are assigned on the basis of the three-letter mnemonic, and changes in symbol may necessitate renumbering of alleles as well.

PHYSICAL MAP CONSTRUCTION

Figure 2 (p. 1723) depicts an integrated physical map of the *E. coli* chromosome, EcoMap7. A set of sequenced *E. coli* genes (through August 1994) are aligned to a physical map derived from the original high-resolution restriction map produced by Kohara et al. (2209) using MapSearch, a restriction map alignment program (3634, 3635). The restriction sites derived from the DNA sequences are then used to replace the original gel-derived data in the aligned region, creating a mosaic restriction map containing both gel-derived and DNA sequence-derived restriction sites (3632). A previous version, Eco-Map5, has been published (3630). EcoMap6 was distributed on computer diskettes and described briefly (3631). EcoMap6 was also incorporated into a new restriction map viewing and editing program, GeneScape (433). EcoMap7 is used as the basis for gene map coordinates in this edition 9 of the genetic map, using the left endpoint (start codon for clockwise protein genes; stop codon for counterclockwise protein genes) as the map coordinate for those genes located on EcoMap7.

ORFs for which expression or function has not been demonstrated are not included in Table 1 and Fig. 1, but are depicted in Fig. 2. A number of new putative genes of *E. coli* have been systematically identified in the unannotated regions of GenBank entries (412, 413) and these, along with all other unnamed ORFs on this physical map, have been given provisional designations beginning with the letter "y" (3631; *Trends in Genetics, Genetics Nomenclature Guide* [4075a], p. 5–8). Therefore, for this system, new genes should not be assigned gene names that start with "y." Authors can obtain "y" designations prior to publication by contacting one of us (K.E.R.). A total of 2,642 named loci are shown on EcoMap7.

In addition to the sequenced genes and ORFs, Fig. 2 depicts the restriction map of the genome, the predicted positions of the Kohara miniset clones (2208), the regions of sequenced DNA used to construct EcoMap7, and the position and direction of transcription for the sequenced genes. Approximately 60% of the restriction sites in EcoMap7 are derived from DNA sequence, with the remainder coming from the original Kohara et al. map. It is now standard procedure to assign genes to these clones using hybridization filters that contain the entire miniset (3108).

Space limitations have precluded a description of the GenBank entries used to create EcoMap7 and this information will be published elsewhere (K. E. Rudd et al., manuscript in preparation) and can be downloaded via anonymous ftp from the ncbi.nlm.nih.gov/repository/Eco/EcoMap7 directory.

ACKNOWLEDGMENTS

This work was supported by National Science Foundation grants DIR9315421 and DIR9010005 to M.B. and by U.S. Public Health Service grant CA39238 to K.B.L. We thank Stanley Letovsky and Peter Kalamarides for devising and revising software and scripts that allowed direct retrieval and formatting of all the various Table 1 data and the drawing of the Fig. 1 map directly from the CGSC database; we thank Webb Miller, Gerard Bouffard, and Craig Werner for the development of the software used to maintain and display EcoMap7; and we give special thanks to Elise Low for providing highly skillful layout work for Fig. 1, proofreading cross-checks on the tables, and encouraging our sanity during the course of this work. Thanks to James Bryan

(Continued on page 1816)

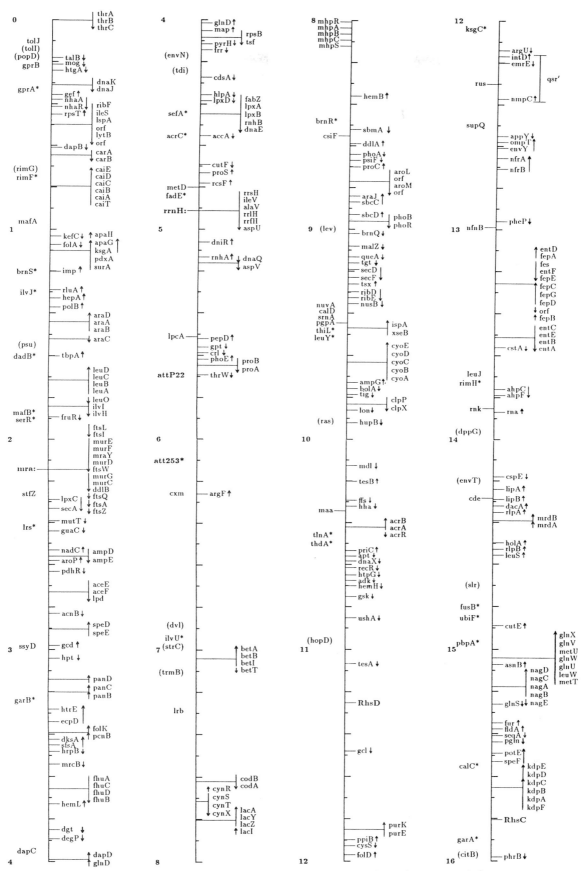

FIGURE 1 Linear drawing of circular linkage map of *E. coli* K-12. Symbols defined in Table 1. Arrows show direction of transcription.

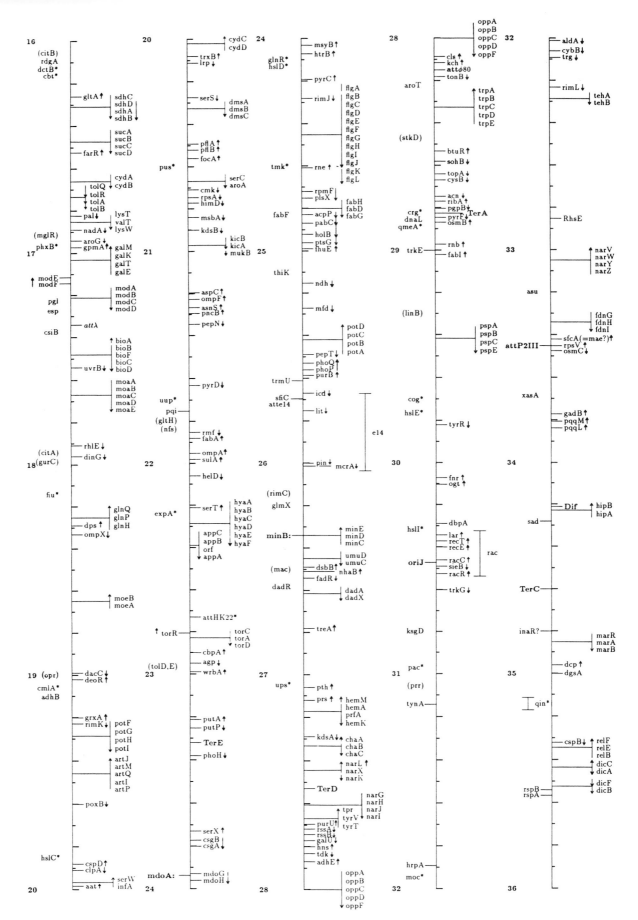

FIGURE 1 *Continued*

1718

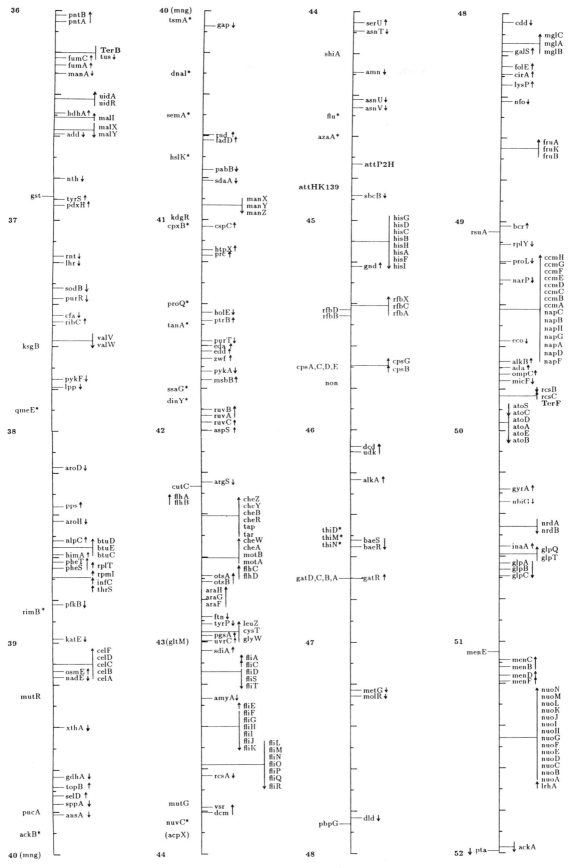

FIGURE 1 *Continued*

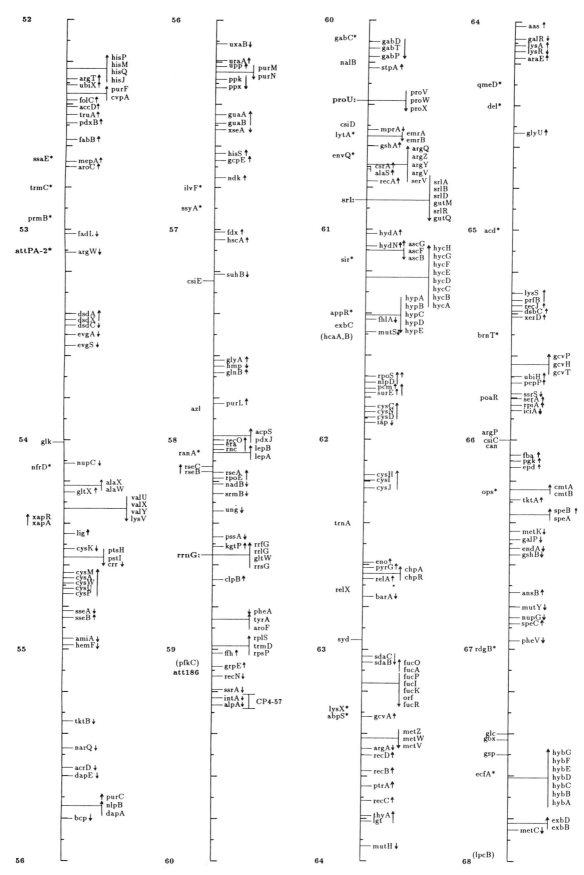

FIGURE 1 *Continued*

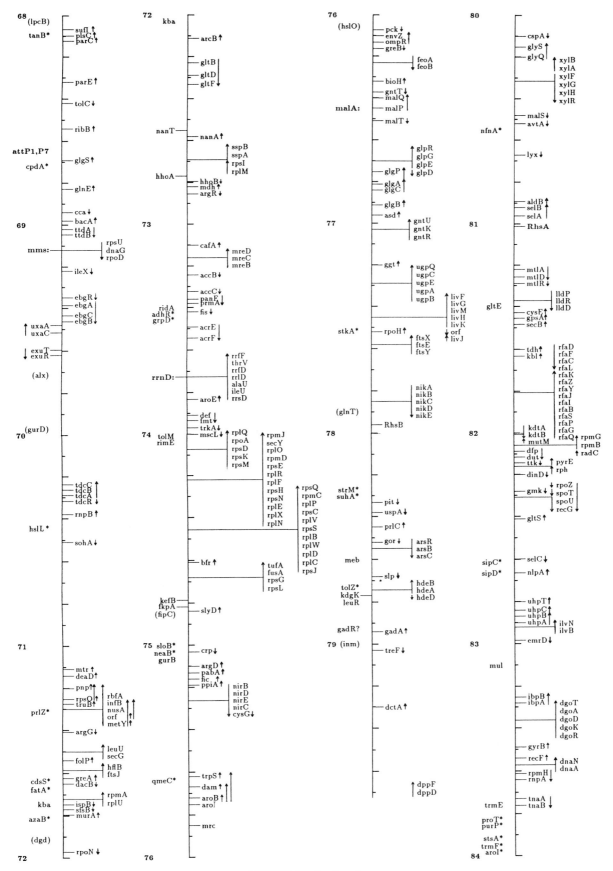

FIGURE 1 *Continued*

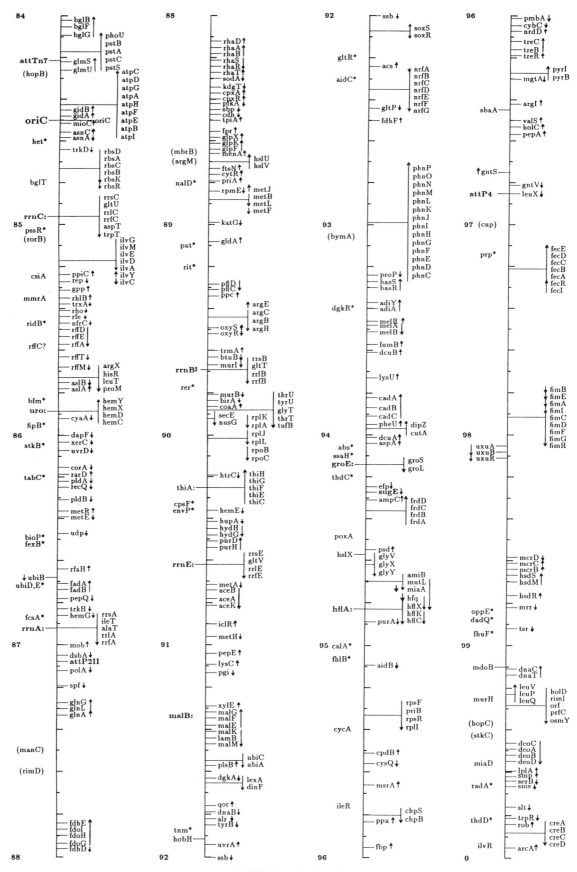

FIGURE 1 Continued

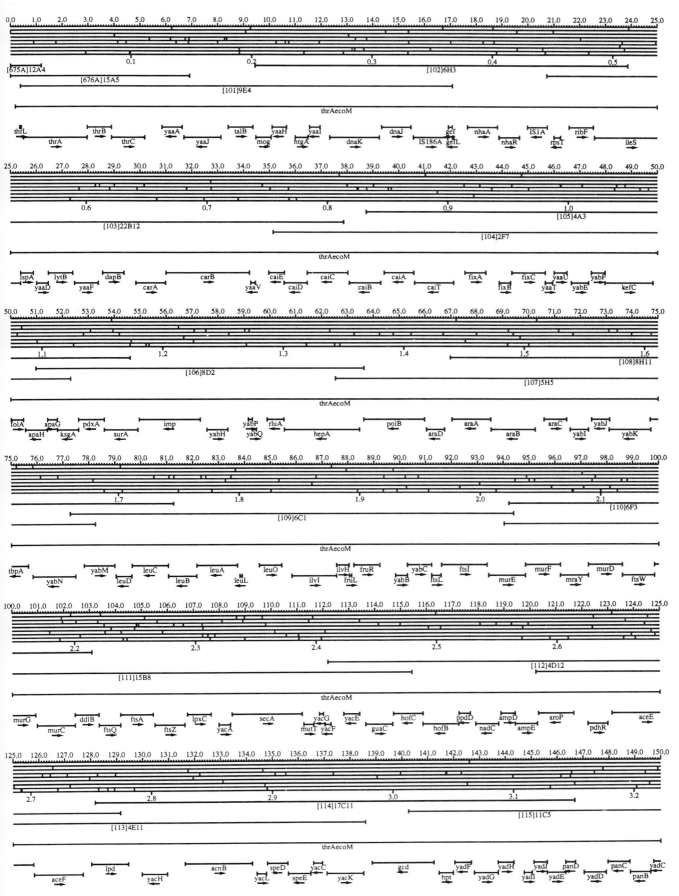

FIGURE 2 See p. 1754.

1723

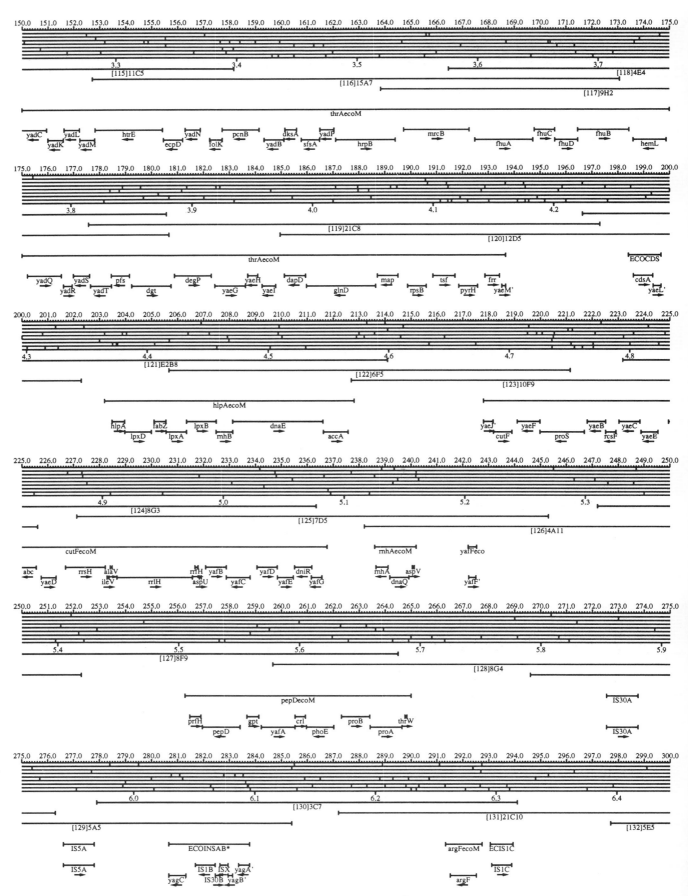

FIGURE 2 Continued

1724

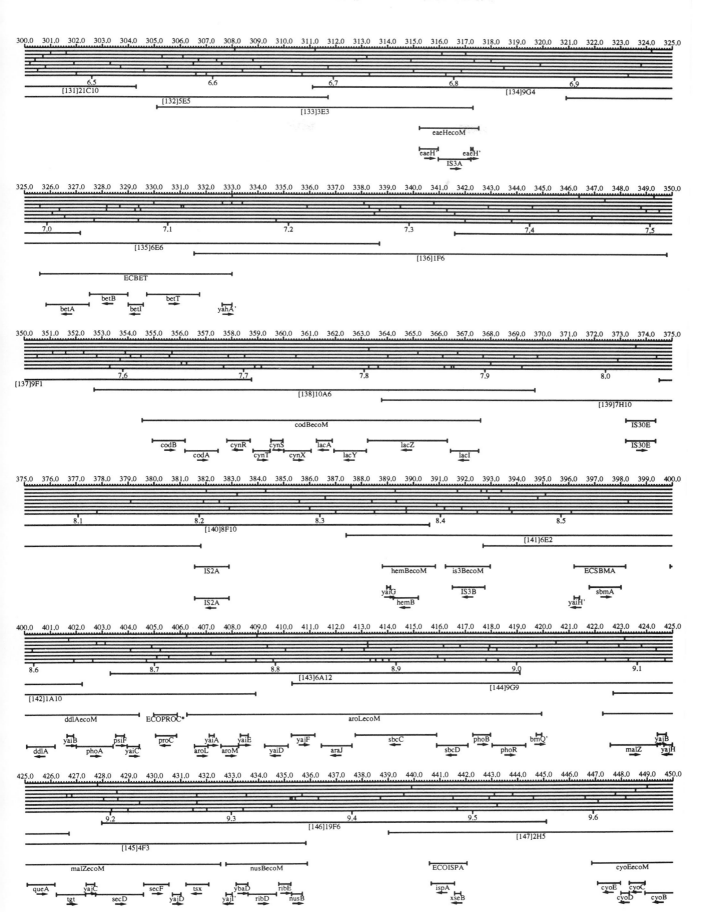

FIGURE 2 *Continued*

1725

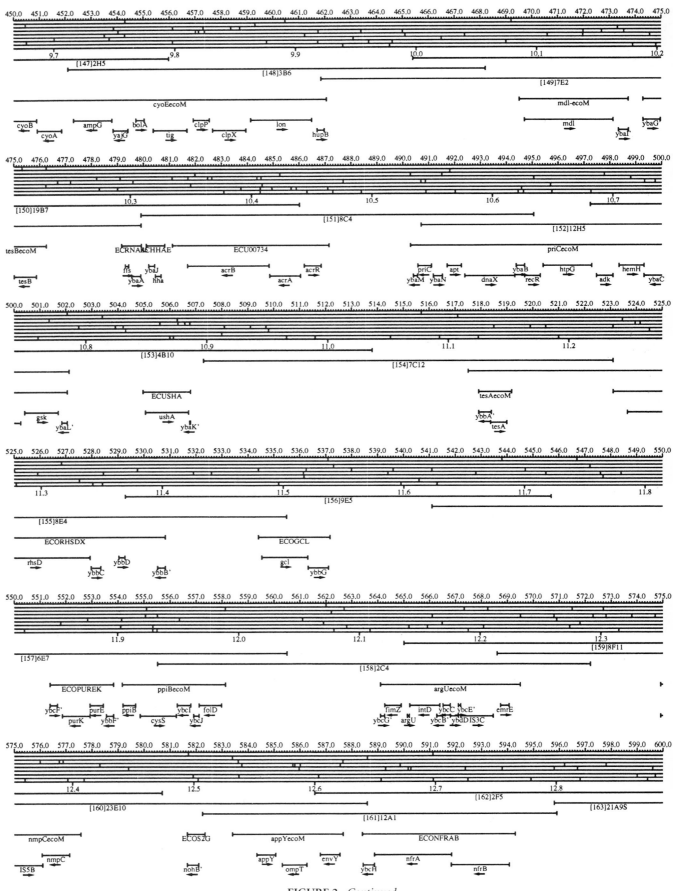

FIGURE 2 *Continued*

1726

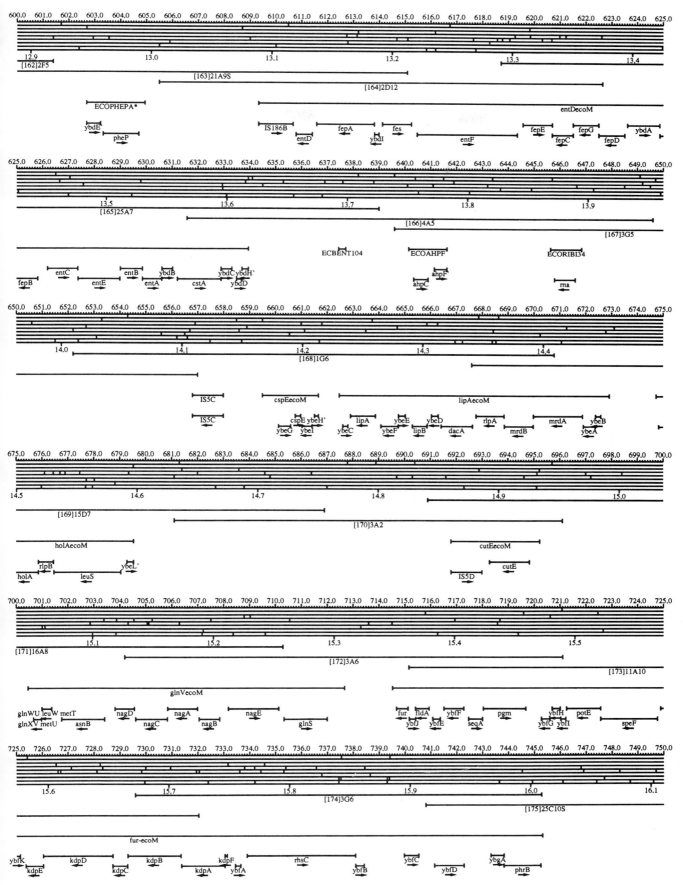

FIGURE 2 *Continued*

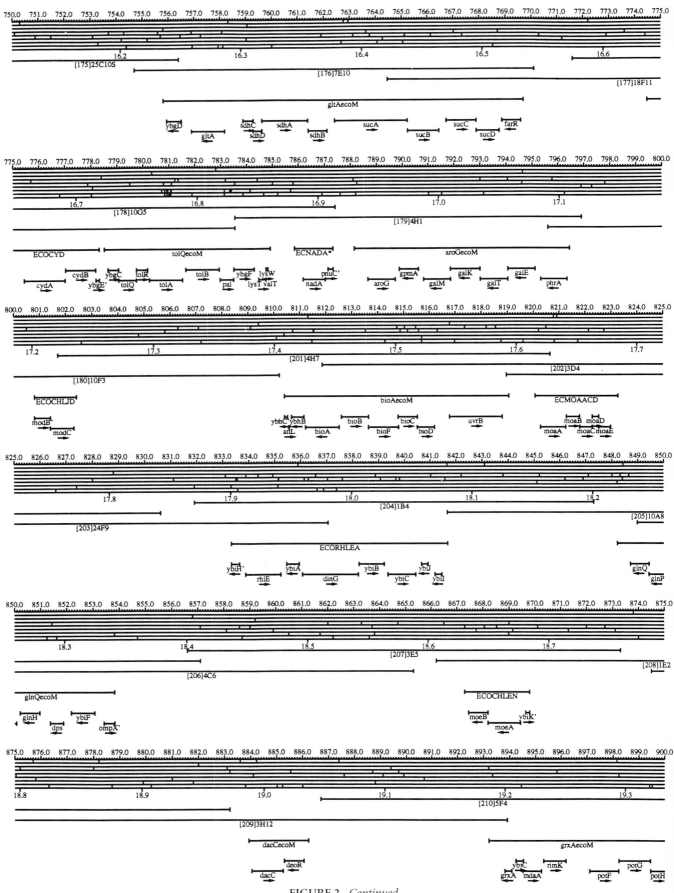

FIGURE 2 *Continued*

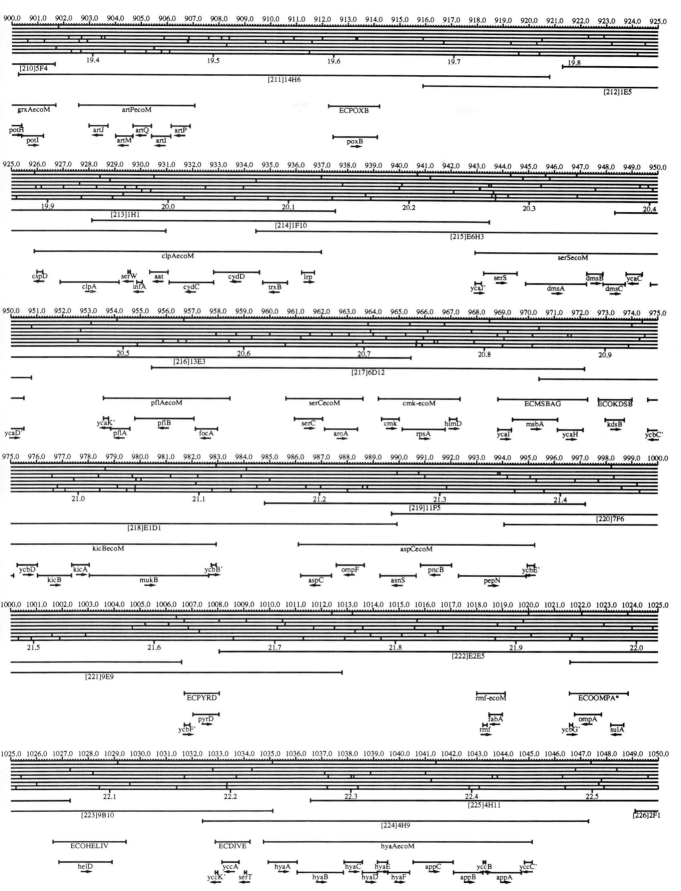

FIGURE 2 *Continued*

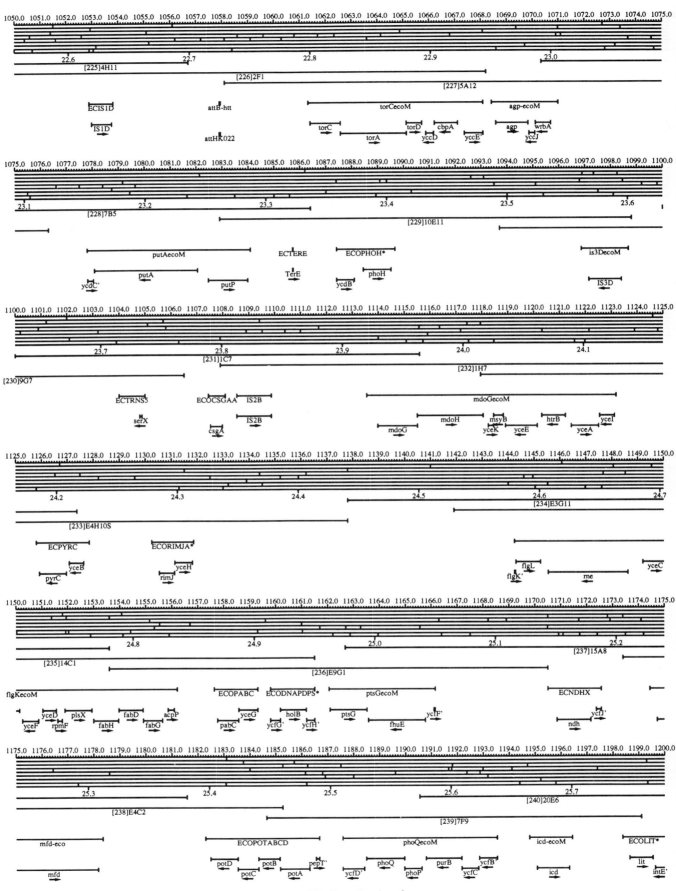

FIGURE 2 Continued

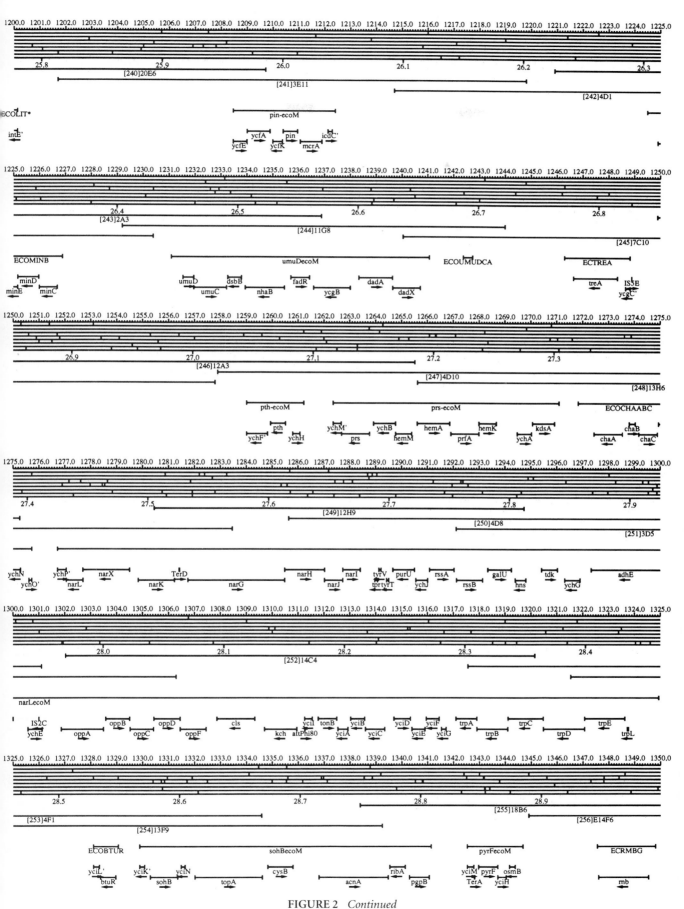

FIGURE 2 *Continued*

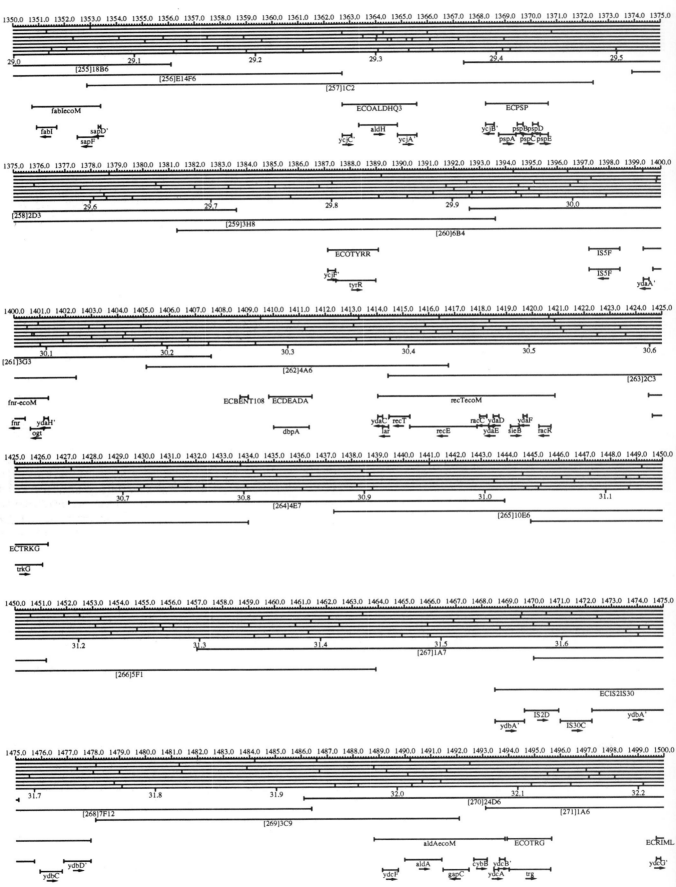

FIGURE 2 *Continued*

1732

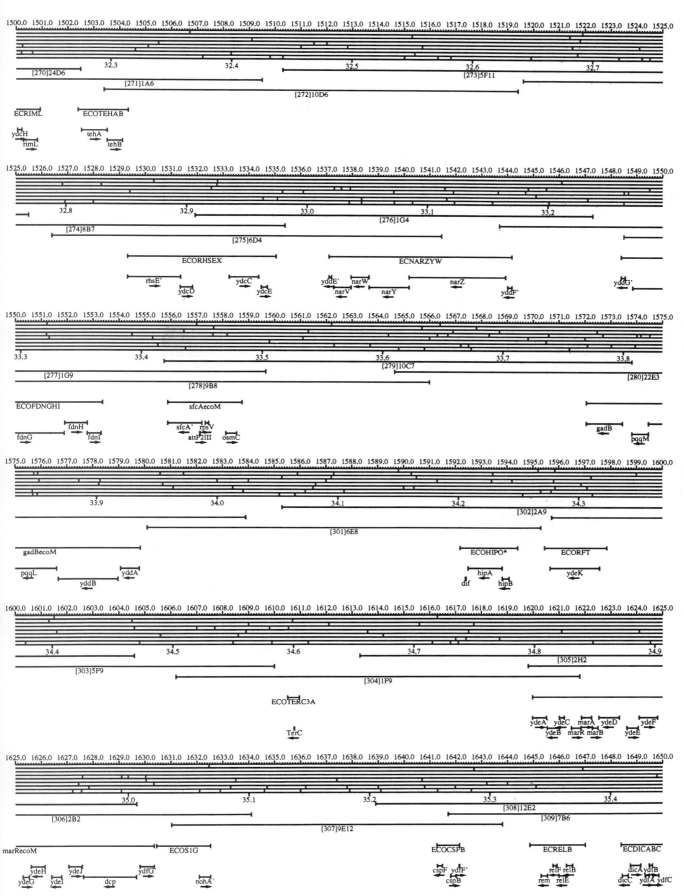

FIGURE 2 Continued

1733

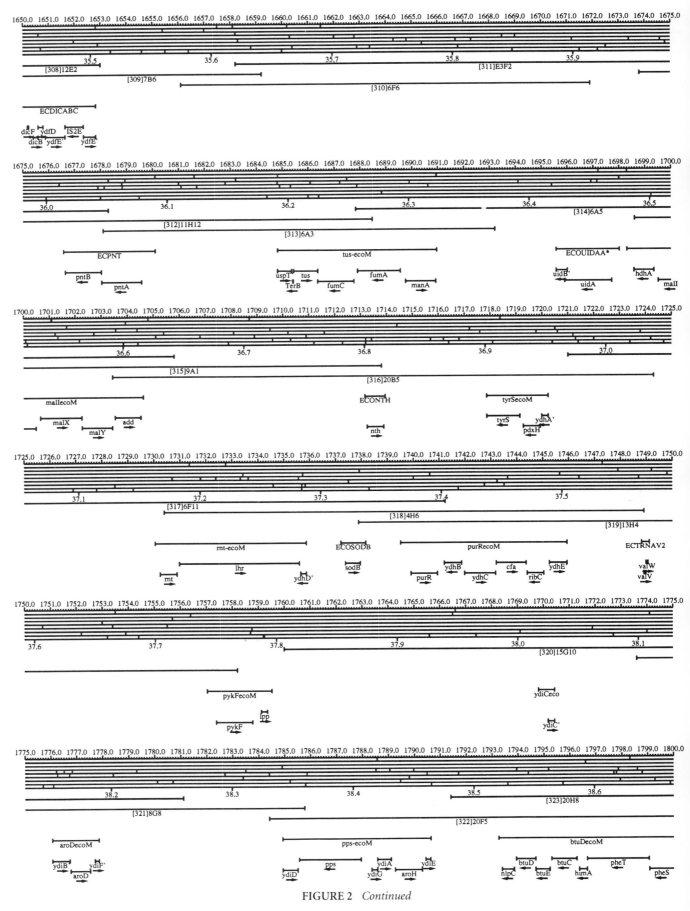

FIGURE 2 *Continued*

1734

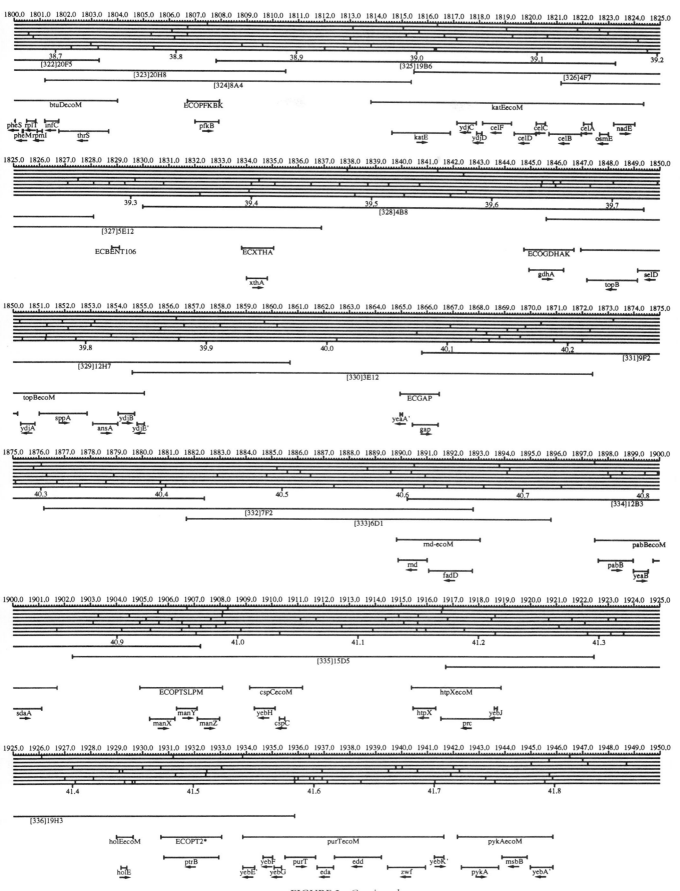

FIGURE 2 *Continued*

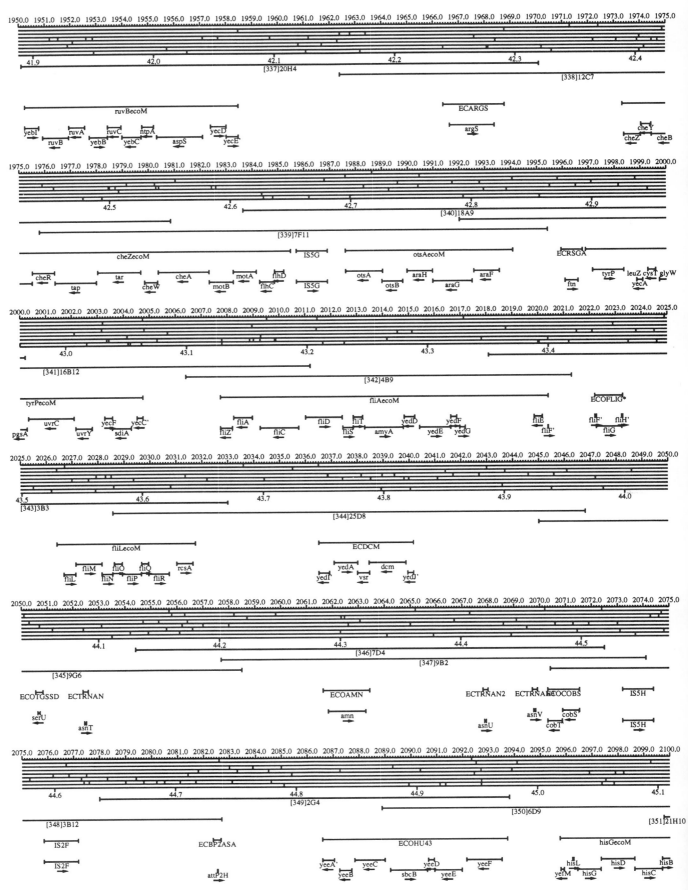

FIGURE 2 *Continued*

1736

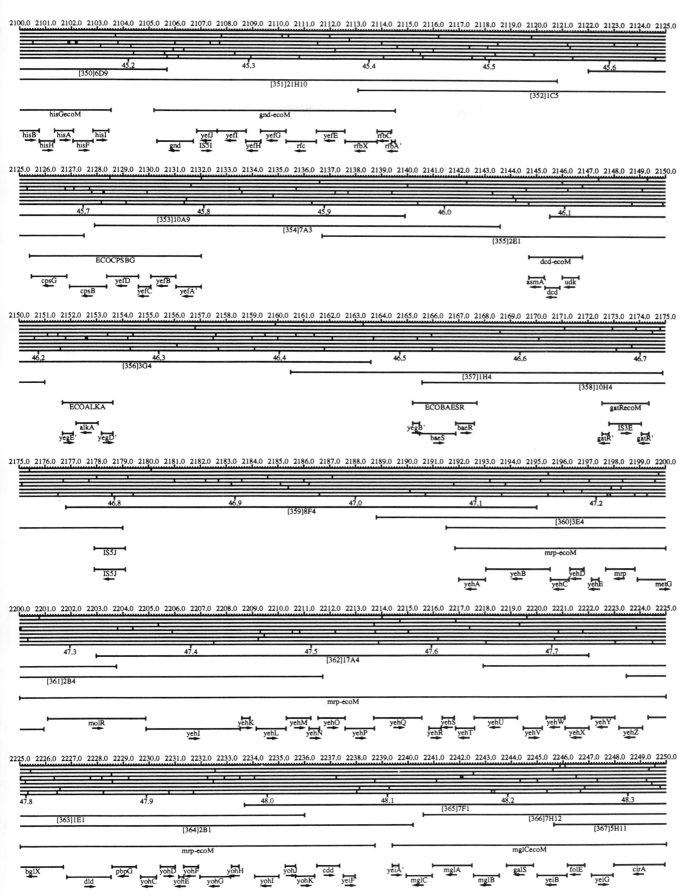

FIGURE 2 *Continued*

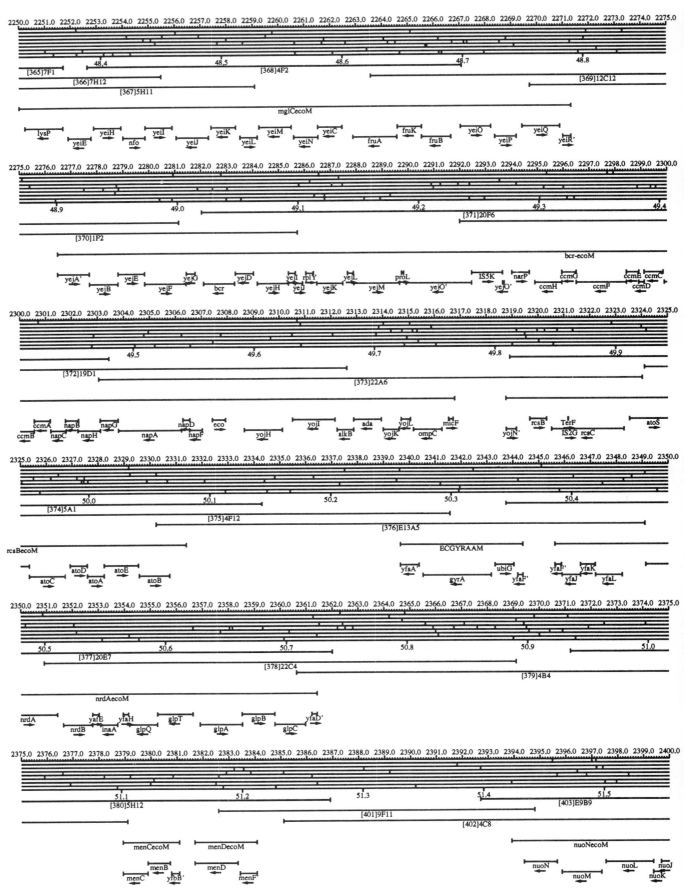

FIGURE 2 *Continued*

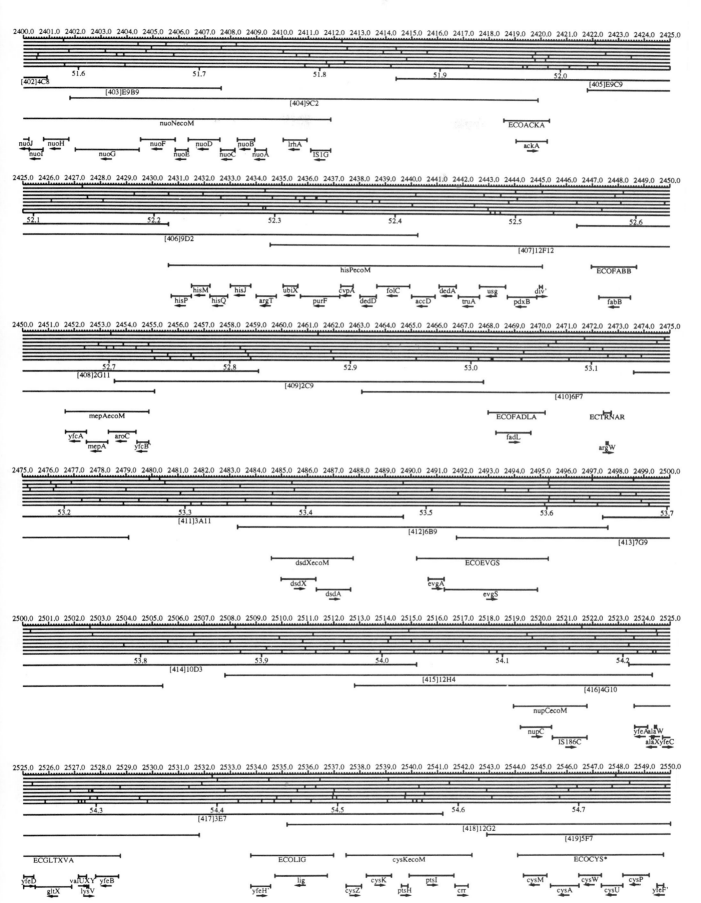

FIGURE 2 *Continued*

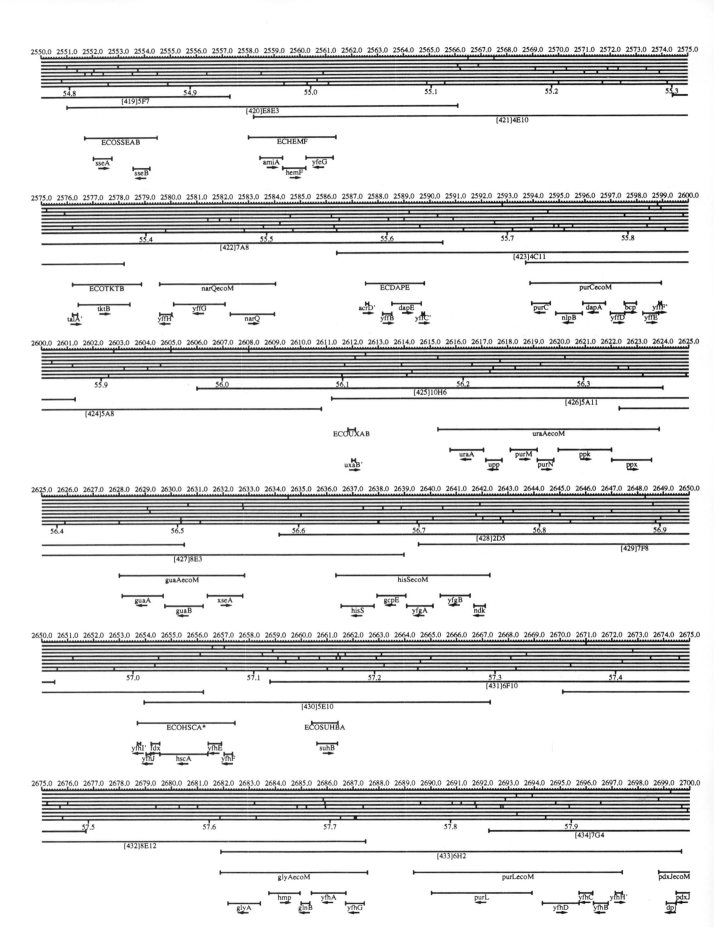

FIGURE 2 *Continued*

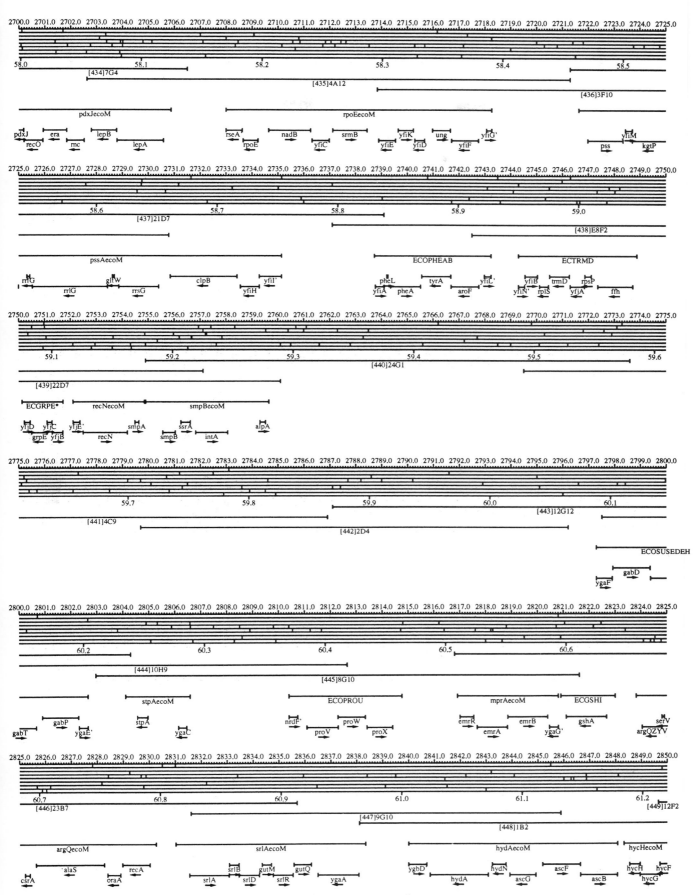

FIGURE 2 *Continued*

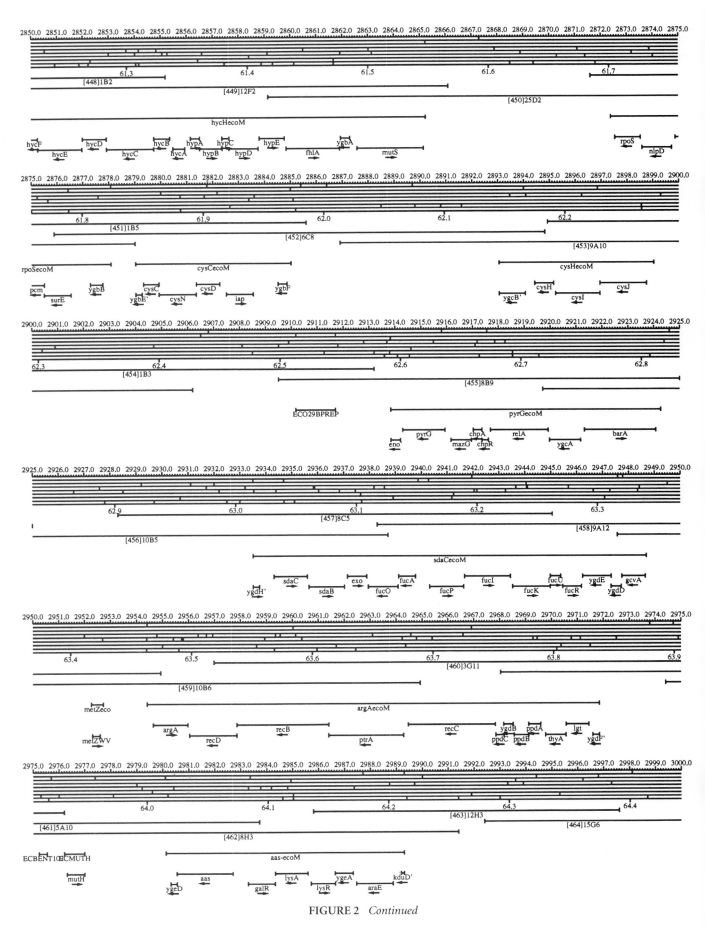

FIGURE 2 *Continued*

1742

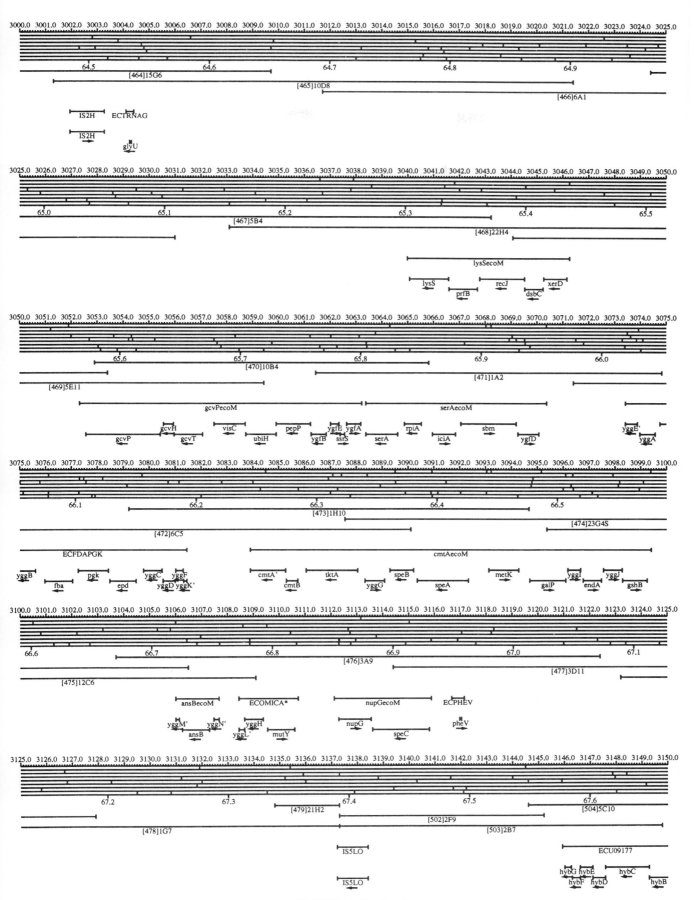

FIGURE 2 *Continued*

1743

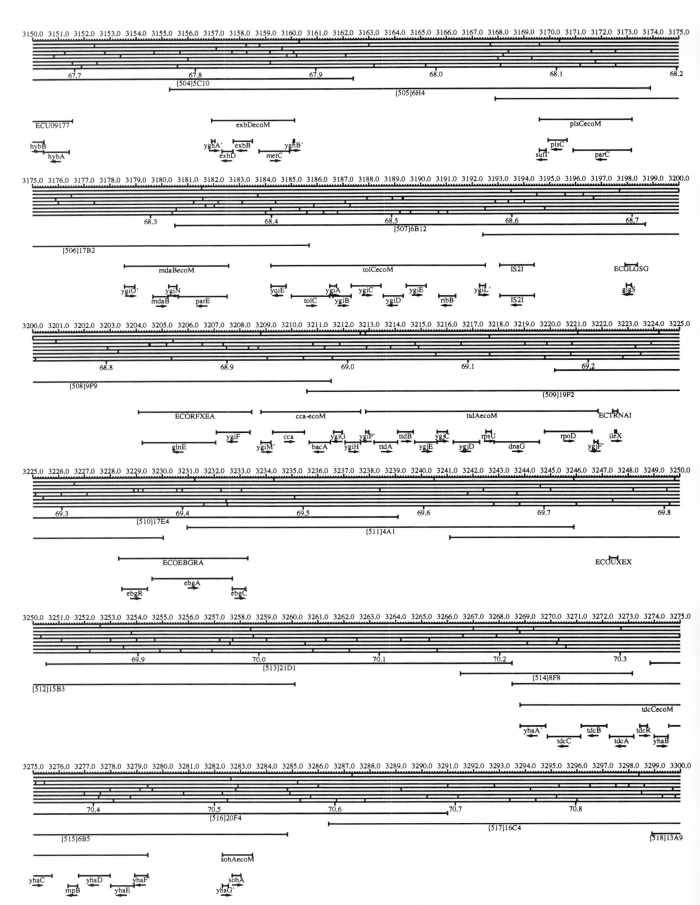

FIGURE 2 *Continued*

1744

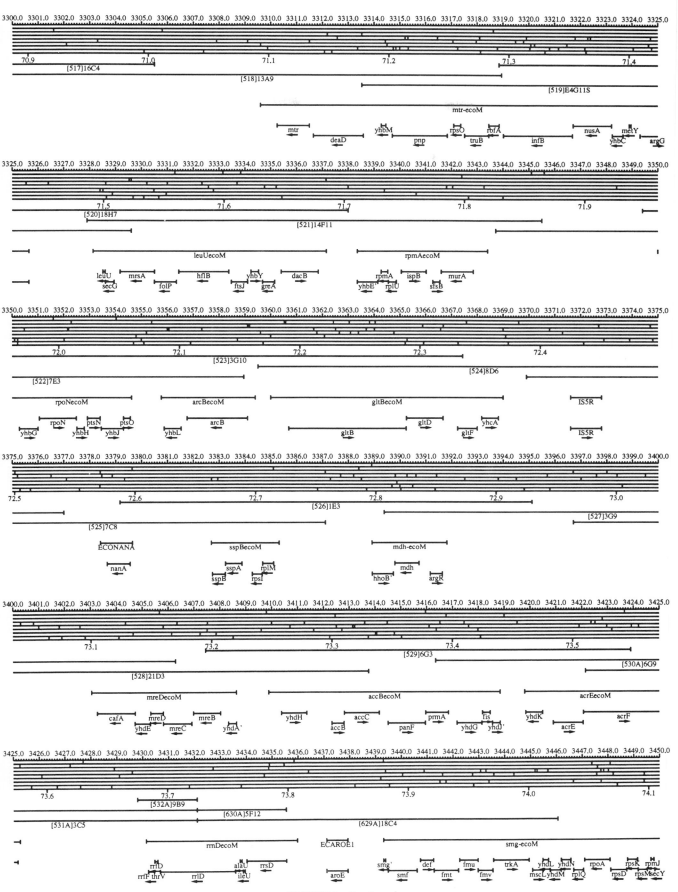

FIGURE 2 *Continued*

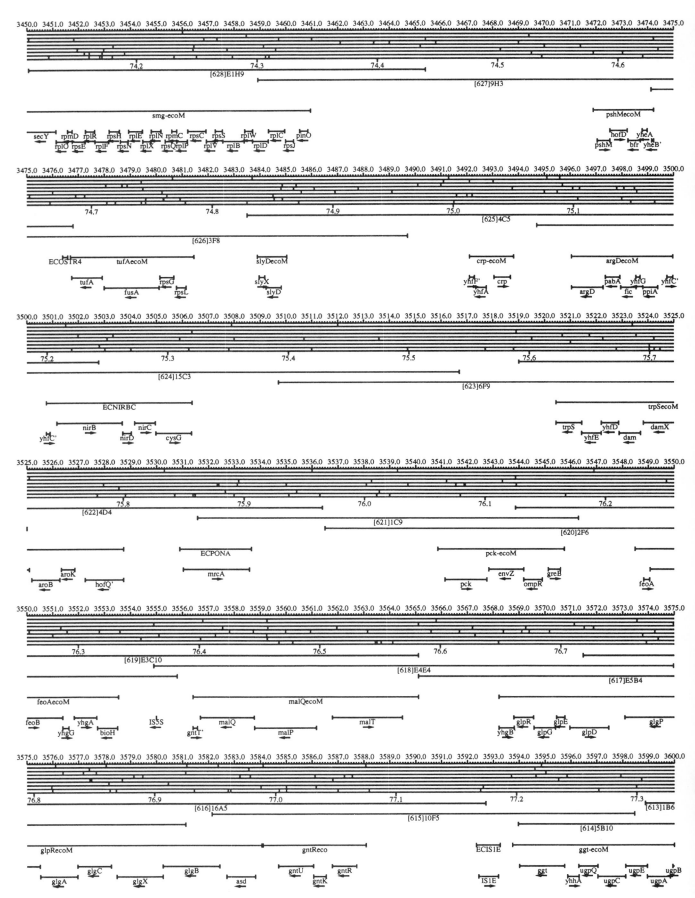

FIGURE 2 *Continued*

1746

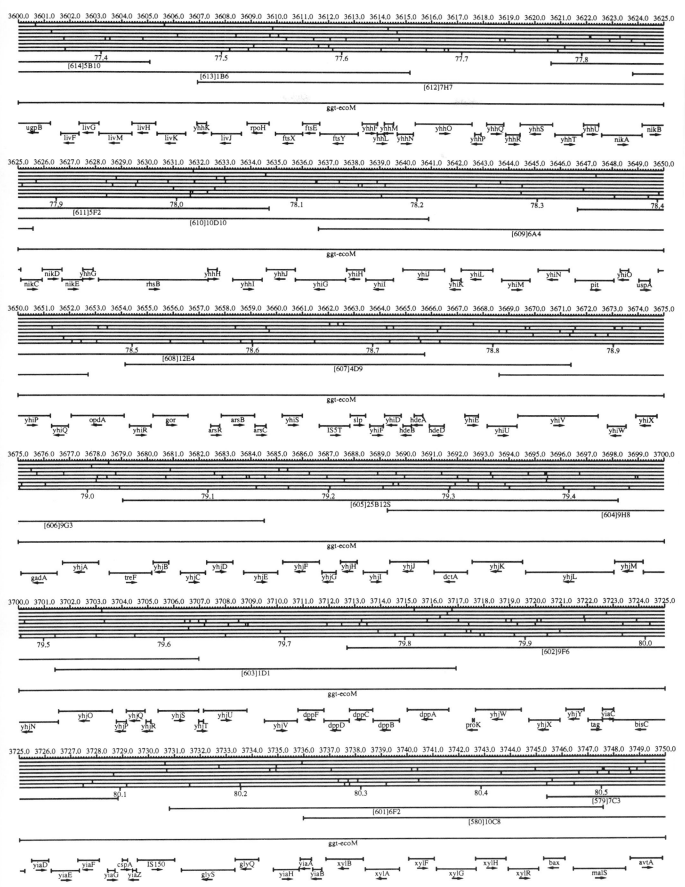

FIGURE 2 *Continued*

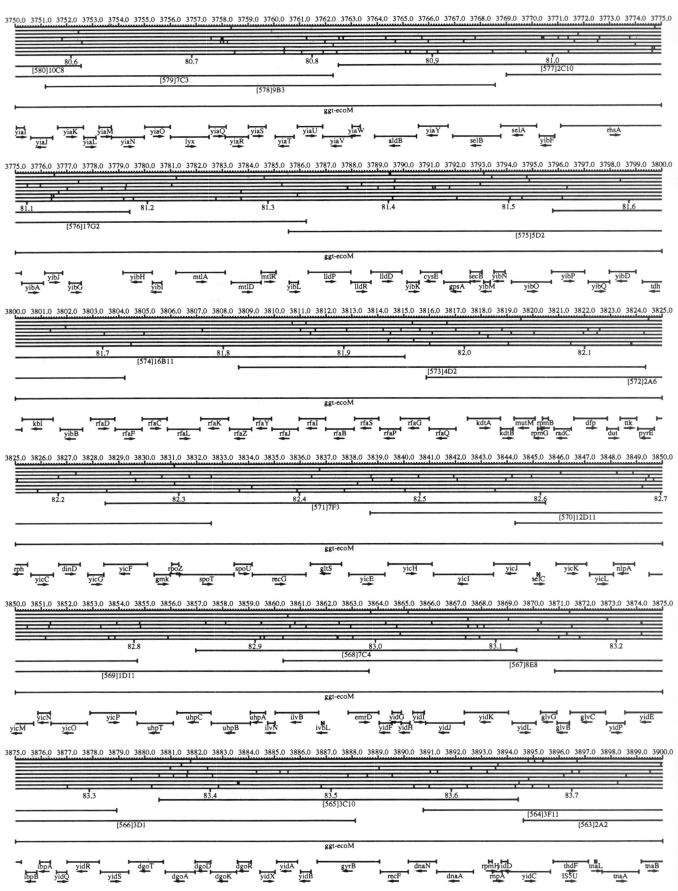

FIGURE 2 *Continued*

1748

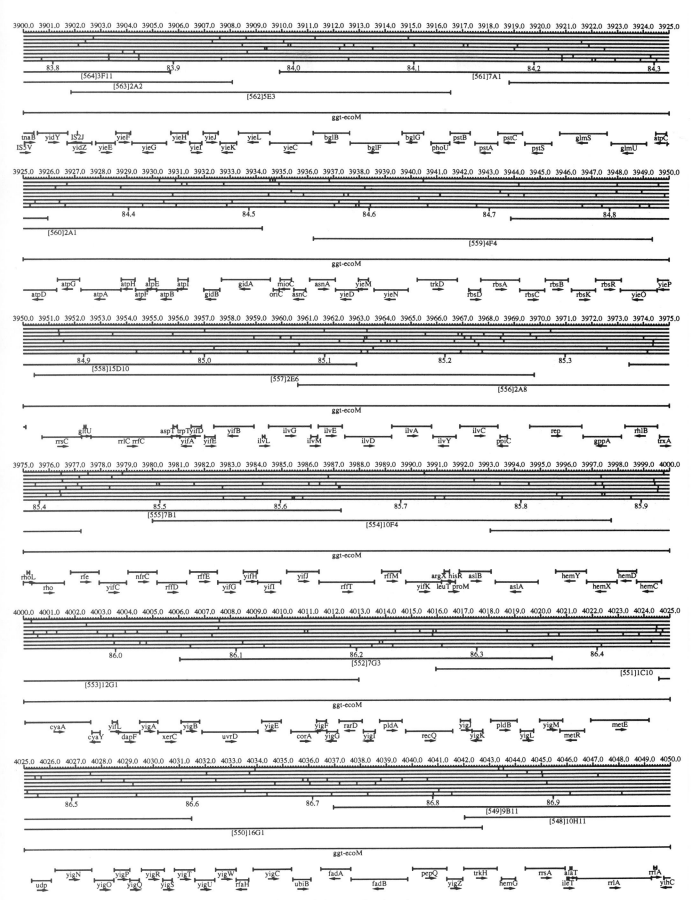

FIGURE 2 Continued

1749

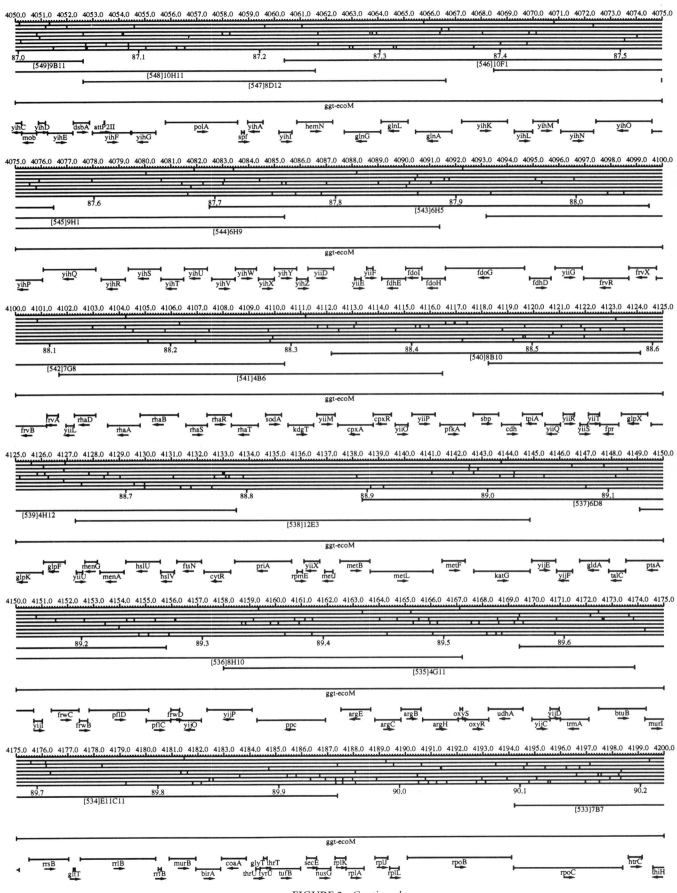

FIGURE 2 *Continued*

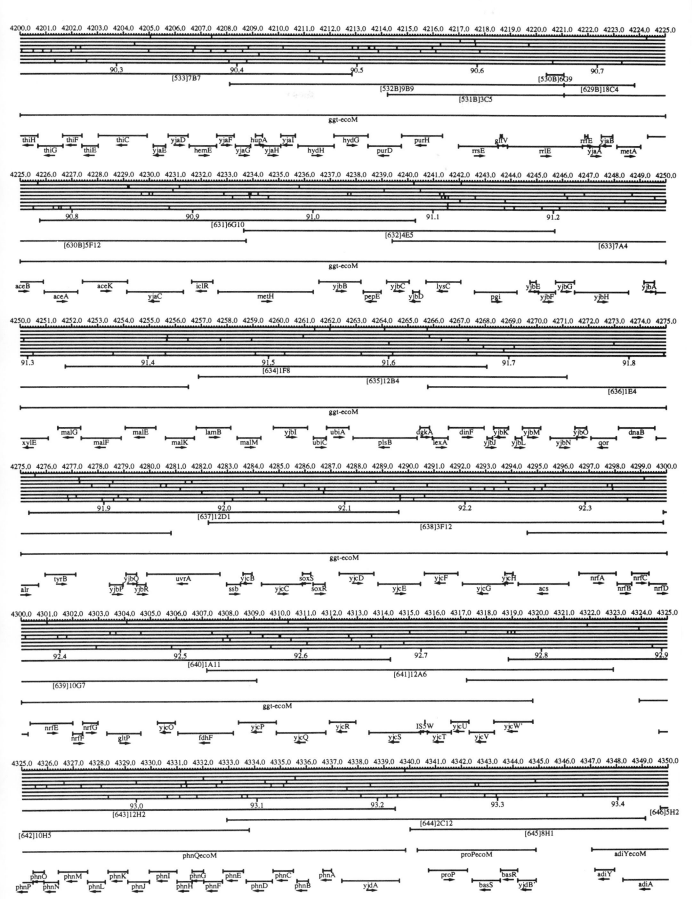

FIGURE 2 *Continued*

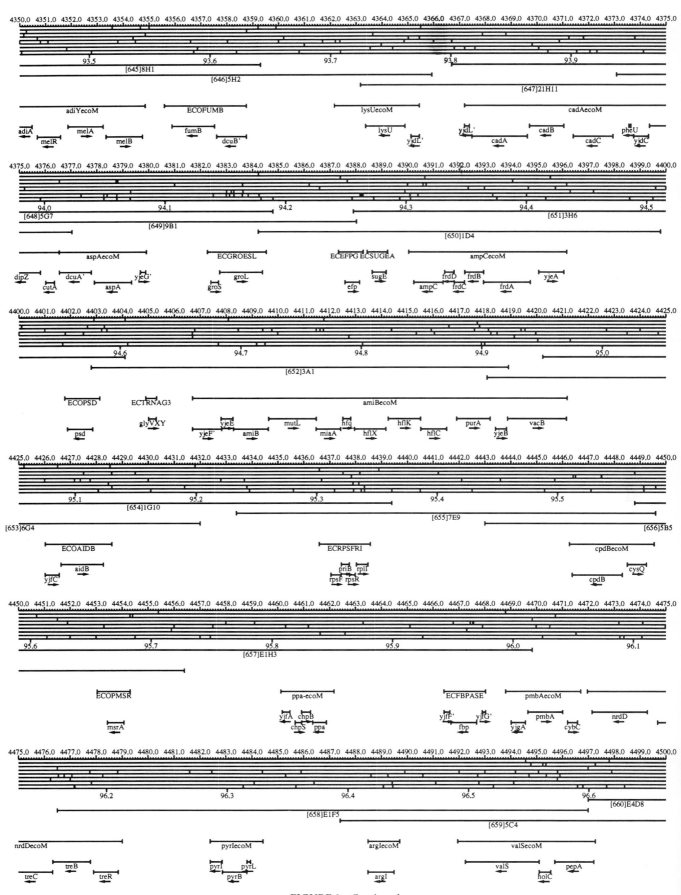

FIGURE 2 *Continued*

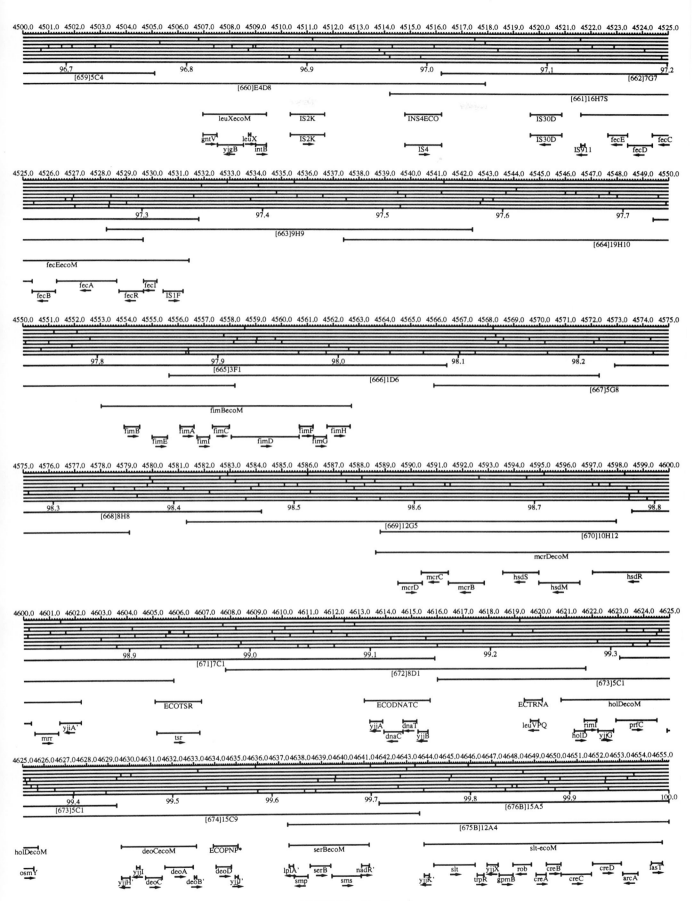

FIGURE 2 *Continued*

1753

FIGURE 2 EcoMap7, an integrated physical map of the *E. coli* chromosome. This map depicts sites for eight restriction enzymes (top line to bottom line: *Bam*HI, *Hin*dIII, *Eco*RI, *Eco*RV, *Bgl*I, *Kpn*I, *Pst*I, *Pvu*II). Above the restriction map are position coordinates in kilobases; immediately below the map are minute coordinates. The first set of spanning lines below the map (with numbers) represent the endpoints of the Kohara miniset clones. The second set of spanning lines (with locus names) represent the EcoSeq7 DNA sequence entries. The third set of spanning lines (with gene names) depict the positions and orientations of the sequences genes and ORFs. Some EcoSeq7 locus names are the same as the GenBank locus names when there are no overlapping entries. GenBank locus names that are followed by an asterisk were modified slightly before being included in EcoSeq7. This figure was created using the PrintMap Postscript drawing program which implements the Plasmid Description Language (PDL) developed by Craig Werner (3635).

TABLE 1 *E. coli* genes and replication- or phage-related sites[a]

Symbol	Mnemonic[b] for symbol	Position (min)[c]	Synonyms; enzyme or phenotype affected[d]	References[e]
aas	Acyl-ACP synthase	64.0	Membrane phospholipid turnover and incorporation of fatty acids and lipophospholipids; 2-acyl-glycerophosphoethanolamine acyltransferase acyl-ACP synthetase; bifunctional inner membrane protein	1802, 1945, CG-29780
aat	Amino acyl-tRNA-protein transferase	20.0	Aminoacyl-tRNA-protein-transferase; EC 2.3.2.6	863, 3923, CG-1054
abpS	Arginine-binding protein	63.3	Low-affinity transport system for arginine and ornithine; periplasmic binding protein	634, CG-18562
abs	Antibiotic sensitivity	94.1	Sensitivity and permeability to antibiotics and dyes	736a, CG-18559
accA	Acetyl-CoA carboxylase	4.6	Acetyl-CoA carboxylase α-carboxyltransferase subunit; EC 6.4.1.2	2443, 2444, CG-29829
accB	Acetyl-CoA carboxylase	73.3	*fabE;* acetyl-CoA carboxylase, biotin carboxyl carrier protein; EC 6.4.1.2	59, 2230, 2442, 2444, 2614, 2962, 4154, CG-796
accC	Acetyl-CoA carboxylase	73.3	*fabG;* acetyl-CoA carboxylase, biotin carboxylase subunit; EC 6.4.1.2	2230, 2442, 2444, CG-29834
accD	Acetyl-CoA carboxylase	52.4	*dedB, usg;* acetyl-CoA carboxylase, β-carboxyltransferase subunit; EC 6.4.1.2	382, 2444, 2445, 2976, 3118, CG-28570
acd	Acetaldehyde-CoA dehydrogenase	65.0	Acetaldehyde-CoA dehydrogenase; EC 1.2.1.10	740, CG-1053
aceA	Acetate	90.8	*icl;* utilization of acetate; isocitrate lyase; EC 4.1.3.1	372, 2197, 2648, 2708, 3544, 4416a, CG-1052
aceB	Acetate	90.7	*mas;* utilization of acetate; malate synthase A; EC 4.1.3.2	372, 566, 567, 817, 2648, 4416a, CG-1051
aceE	Acetate	2.6	*aceE1;* acetate requirement; pyruvate dehydrogenase (decarboxylase component) E1p; EC 1.2.4.1	1513, 1514, 1517, 2849, 4016, 4069, CG-1050
aceF	Acetate	2.7	*aceE2;* acetate requirement; pyruvate dehydrogenase (dihydrolipoyltransacetylase component, E2p); EC 1.6.4.3, EC 2.3.1.12	56, 1557, 2849, 3651, 4016, 4068, CG-26530
aceK	Acetate	90.8	Isocitrate dehydrogenase kinase/phosphatase	372, 729, 816, 1309, 1871, 2197, 2332–2334, 2708, CG-1770
ackA	Acetate kinase	52.0	Fluoroacetate sensitivity; acetate kinase; EC 2.7.2.1	514, 1508, 2421, 2709, CG-1048
ackB	Acetate kinase	39.9	Acetate kinase activity; EC 2.7.2.1	3279, CG-1047
acnA	Aconitase	28.7	Aconitase A; EC 4.2.1.2	1332, 1333, 1498a, 3428, 3429, CG-28218
acnB	Aconitase	2.8	Aconitase B; EC 4.2.1.2	O, 1498a, CG-36955
acp	ACP	44.0	Holo-acyl carrier protein (ACP) synthase; EC 2.7.8.7	3363, CG-1046
acpP	ACP	24.8	ACP	1946, 3491, 4408, 4409, CG-31871
acpS	ACP	58.0	*dpj;* holo-ACP synthase; EC 2.7.8.7	2313, 2315a, 4195, CG-32953
acrA	Acridine	10.4	*Mb, lir, mbl, mtcA, sipB;* sensitivity to acriflavine, phenethyl alcohol, and SDS	782, 1687, 2593, 3449, CG-1045
acrB	Acridine	10.4	*acrE* (beware: name changed from *acrE* to *acrB,* and *envC* renamed *acrE; acrB* also previously used for a class of *gyrB* alleles); drug susceptibility; permeability of outer membrane; probably multidrug efflux pump	2593, CG-35806
acrC	Acridine	4.6	Sensitivity to acriflavine; transmembrane protein	3001, CG-1044
acrD	Acridine	55.6	Sensitivity to acriflavine	440, CG-35697
acrE	Acridine	73.5	*envC* (*acrE* also used earlier for what is now called *acrB*); mutants show anomalous cell formation, chain formation, multidrug resistance; role in septum formation	2183, 2184, CG-813
acrF	Acridine	73.5	*envD;* osmotically remedial envelope defect; multidrug resistance; lipoprotein with signal peptide	2183, 2184, CG-33608

TABLE 1 *Continued*

Symbol	Mnemonic[b] for symbol	Position (min)[c]	Synonyms; enzyme or phenotype affected[d]	References[e]
acrR	Acridine	10.4	Regulatory protein for *acrA* and *acrB*	2593, CG-35809
acs	Acetyl-CoA synthetase	92.2	Acetyl-CoA synthetase 2; EC 6.2.1.1	372, 2283, 4665, CG-34317
ada	Adaptive (response)	49.7	O^6-Methylguanine-DNA methyltransferase, inducible; DNA repair system protecting against methylating and alkylating agents; transcription factor	15, 997, 1161, 1971, 2108, 2229, 2402, 2599, 2666, 2877a, 2990, 3668, 3688, 3839–3841, 3900, 4186, 4238, CG-1043
add	Adenosine deaminase	36.6	Adenosine deaminase; EC 3.5.4.4	648, 737, 1988, CG-1042
adhB	Alcohol dehydrogenase	19.1	Alcohol dehydrogenase	K, 738, CG-36932
adhE	Alcohol/acetaldehyde dehydrogenase	27.9	*ana, adhC;* multifunctional; alcohol dehydrogenase (EC 1.1.1.), acetaldehyde dehydrogenase, CoA-linked, and regulatory deactivase for pyruvate-formate lyase	864, 1418, 1539, 2152, 2406, 2546, 3279, CG-1041
adhR	Alcohol/acetaldehyde dehydrogenase	73.4	Regulatory gene for *adhE* and *acd*	698, 741, CG-18556
adiA	Arginine decarboxylase, induced	93.4	*adi;* arginine decarboxylase, inducible by acid; homology with three other decarboxylases, CadA, SpeC, and SpeF	1619, 3658, 3903, 3904, 3906, 4086, 4538, CG-34495
adiY	Arginine decarboxylase, induced	93.4	Regulates *adiA*	U02384, CG-35597
adk	Adenylate kinase	10.7	*dnaW, plsA;* adenylate kinase; EC 2.7.4.3; pleiotropic effects on glycerol-3-P acyltransferase activity	215, 522, 1154, 1686, 2883, CG-1040
agp	Acid glucose-1-phosphatase	23.0	Periplasmic; homology with *appA*	924, 3409–3412, CG-31830
ahpC	Alkyl hydroperoxide	13.8	Member superoxide stress regulon (*oxyR* regulation); alkyl hydroperoxide reductase small subunit	3971, 4102, 4361, CG-31190
ahpF	Alkyl hydroperoxide	13.8	Member superoxide stress regulon (*oxyR* regulation); alkyl hydroperoxide reductase large subunit	3971, 4102, CG-31194
aidB	Alkylating agent induction	95.1	Induced by alkylating agents	2322, 4449, 4451, CG-18553
aidC	Alkylating agent induction	92.3	Induced by alkylating agents	4451, CG-18550
alaS	Alanine	60.7	*act, ala-act lovB;* alanyl-tRNA synthetase (ligase) 1B; EC 6.1.1.7	1689, 2080, 2861–2863, 3444–3446, 3590, 4435, CG-1039
alaT	Alanine	86.9	*talA;* alanine tRNA 1B; in *rrnA* operon	903, 2225, CG-1038
alaU	Alanine	73.8	*talD;* alanine tRNA 1B; in *rrnD* operon	2225, CG-1037
alaV	Alanine	4.9	Alanine tRNA 1B in *rrnH* operon	1120, CG-1036
alaW	Alanine	54.2	*alaW*α; alanine tRNA 2 (tandemly duplicated gene; see *alaW*)	518–520, 2225, CG-32851
alaX	Alanine	54.2	*alaW*β; alanine tRNA 2 (tandemly duplicated gene; see *alaX*)	518, CG-18547
aldA	Aldehyde dehydrogenase	32.0	*ald;* aldehyde dehydrogenase, NAD linked	687, 1712, 3453, CG-17767
aldB	Aldehyde dehydrogenase	80.9	Aldehyde dehydrogenase	4690, 4691, CG-35668
alkA	Alkylation	46.2	*aidA;* 3-methyl-adenine DNA glycosylase II, inducible	745, 1161, 2699, 2991–2993, 3900, 4238, 4451, CG-1035
alkB	Alkylation	49.7	*aidD;* DNA repair system specific for alkylated DNA	665, 997, 2108, 2109, 2229, 4448, 4451, CG-18544
alpA	Activation of Lon protease	59.3	Element of cryptic prophage CP4-57; positive transcriptional regulator of *intA* (*slpA*), required for activation of Alp protease, which suppresses *lon* phenotypes	2173, 4309, CG-33086
alr	Alanine racemase	91.8	Alanine racemase; EC 5.1.1.1	372, 2288, 2465, 3022, CG-1034
alx	Alkaline-induced expression	(69.7)	pH-regulated locus; induced in alkaline media	332, CG-36513
amiA	Amidase	5.5	*N*-Acetylmuramyl-L-alanine amidase activity	4205, 4281, 4316, CG-18541
amiB	Amidase	94.7	Hypersensitivity to osmotic shock; cell wall amidase; EC 3.5.1.28?	4343, 4344, CG-34432
amn	AMP nucleosidase	44.3	AMP nucleosidase; EC 3.2.2.4	2411, 2412, CG-17764
ampC	Ampicillin	94.3	*ampA;* β-lactamase; penicillin resistance	337, 771, 1096, 1097, 1499, 1500, 1963–1965, 3120, CG-1033
ampD	Ampicillin	2.6	Repressor of AmpC	1762, 2479, CG-30478
ampE	Ampicillin	2.6	Membrane protein; ampicillin resistance	1762, 2479, CG-30481
ampG	Ampicillin	9.7	Membrane protein; ampicillin resistance	2480, CG-31027
amyA	Amylase	43.3	α-Amylase, cytoplasmic	2123, 3463, 3464, CG-30745
ansA	Asparaginase	39.8	L-Asparaginase I, cytoplasmic; EC 3.5.1.1	983, 1980, 4030, CG-1030
ansB	Asparaginase	66.7	L-Asparaginase II; EC 3.5.1.1	402, 1001, 1473, 1608, 1975, 2629, 2630, 3307, CG-30045
apaG	Ade-P-Ade	1.1	Expressed as part of complex *ksgA* operon	358, 3557, 4787, CG-30310
apaH	Ade-P-Ade	1.1	Stress response; member complex operon *surA-apaH;* diadenosine tetraphosphatase	358, 1174, 1530, 1997, 2415, 2771, 3557, 4787, CG-17761
appA	Acid (poly)phosphatase	22.4	Acid phosphatase, pH 2.5, exopolyphosphatase; EC 3.1.3.2, EC 3.6.1.11	407, 920, 923, 924, 1475, 4300, CG-17758
appB	Acid (poly)phosphatase	22.4	*cyxB;* putative additional cytochrome oxidase subunit	152, 923, CG-31813
appC	Acid (poly)phosphatase	22.4	*cyxA;* putative additional cytochrome oxidase subunit	152, 923, CG-31810

(Table continues)

TABLE 1 *Continued*

Symbol	Mnemonic[b] for symbol	Position (min)[c]	Synonyms; enzyme or phenotype affected[d]	References[e]
appR	Acid (poly)phosphatase	61.4	Regulation of *appA*	4301, CG-18538
appY	Acid(poly)phosphatase	12.6	Putative DNA-binding protein, possible third global regulator of energy metabolism genes	152, 153, 1752, 2144, CG-31146
apt	Adenine phosphoribosyltransferase	10.6	Adenine phosphoribosyltransferase; EC 2.4.2.7	554, 1228, 1699, 1700, CG-1029
araA	Arabinose	1.4	L-Arabinose isomerase; EC 5.3.1.4	2373, 2374, 2386, 4787, CG-1028
araB	Arabinose	1.5	L-Ribulokinase; EC 2.7.1.16	1086, 1390, 2120, 2373, 2374, 2386, 2387, 2881, 3152, 3973, 4787, CG-1027
araC	Arabinose	1.5	Regulatory gene; transcriptional activator and repressor for *ara* genes	611, 1086, 2217, 2374, 2387, 2517, 2880, 3896, 4490, 4787, CG-1026
araD	Arabinose	1.4	L-Ribulosephosphate 4-epimerase; EC 5.1.3.4	400, 670, 671, 1938, 2373, 2374, 2386, 2865, 4787, CG-1025
araE	Arabinose	64.2	Low-affinity L-arabinose transport system; L-arabinose proton symport	2216, 2217, 2607, 2626, 2627, 4099, CG-1024
araF	Arabinose	42.8	L-Arabinose-binding protein	732, 1671, 1672, 1747, 1772, 1773, 2216, 2217, 2244, 3062, 3454, 3835, 3836, CG-1023
araG	Arabinose	42.8	High-affinity L-arabinose transport system	1671, 1672, 1772, 1773, 2216, 2217, 2244, 3836, CG-1022
araH	Arabinose	42.8	High-affinity L-arabinose transport system; membrane protein	1671, 1672, 1772, 1773, 2749, 3836, CG-18535
araJ	Arabinose	8.8	Arabinose-inducible, function unknown (not a transport gene)	1672, 2560, 3509, CG-29861
arcA	Aerobic pathways control	100	*dye, fexA, msp, seg, sfrA*; global regulatory gene for genes in aerobic pathways	563, 564, 823, 1073, 1289, 1618, 1928, 1929, 2471, 3756, 3942, 4215, CG-831
arcB	Aerobic pathways control	72.1	Histidine kinase that phosphorylates and activates ArcA	1899, 1926a, 1929, 1931, 2471, 3756, CG-29063
argA	Arginine	63.5	*Arg1, Arg2, argB*; growth on acetylornithine; amino acid acetyltransferase; *N*-acetylglutamate synthase; EC 2.3.1.1	507, 654, 838, 1090, 3052, 3736, CG-1021
argB	Arginine	89.5	Acetylglutamate kinase; EC 2.7.2.8	372, 838, 839, 2909, 3273, CG-1020
argC	Arginine	89.4	*Arg2, argH*; *N*-acetyl-α-glutamyl-P reductase; EC 1.2.1.38	285, 372, 838, 839, 2785, 2909, 3273, 3328, CG-1019
argD	Arginine	75.1	*Arg1, argG*; acetylornithine β-aminotransferase; EC 2.6.1.11	290, 654, 838, 1652, 2140, 3546, CG-1018
argE	Arginine	89.4	*Arg4, argA*; acetylornithine deacetylase; EC 3.5.1.16	285, 372, 450, 839, 2785, 2909, 3328, CG-1017
argF	Arginine	6.3	*Arg5, argD*; ornithine transcarbamylase; EC 2.1.3.3; duplicate gene; see *argL*	654, 839, 1172, 1552, 1807, 1981, 2391, 2906, 2907, 3329, 4362, 4405, 4407, 4763, CG-1016
argG	Arginine	71.4	*Arg6, argE*; argininosuccinate synthetase; EC 6.3.4.5	654, 4406, CG-1015
argH	Arginine	89.5	Argininosuccinate lyase; EC 4.3.2.1	372, 839, 2909, 3273, CG-1014
argL	Arginine	96.4	Ornithine transcarbamylase; EC 2.1.3.3; duplicate gene; see *argF*	275, 2285, 3329, 3598, 4405, 4407, CG-1013
argM	Arginine	88.7	Acetylornithine transaminase; cryptic, may be duplicate of *argD*	3546, CG-18532
argP	Arginine	66.0	Canavanine sensitivity; transport of arginine, ornithine, and lysine	1570, 2595, 3324, 3601, CG-1012
argQ	Arginine	60.7	*argVδ*; arginine tRNA 2; tandemly quadruplicated gene *argQVYZ*	CG-35604
argR	Arginine	72.8	*xerA, Rarg; cer*-mediated site-specific recombination; *arg* regulon repressor, autogenously regulated	242, 290, 654, 1095, 1226, 2409, 2468, 3297, 4088, 4089, 4265, CG-1011
argS	Arginine	42.3	*lov*; arginyl-tRNA synthetase; EC 6.1.1.19	1140, 1736, 3047, 4435, CG-1010
argT	Arginine	52.3	Lysine-, arginine-, ornithine-binding protein	3118, CG-18529
argU	Arginine	12.1	*dnaY, pin*; arginine tRNA 4	484, 667, 1248, 1324, 2225, 2482, 2963, 3757, 4006, CG-17755
argV	Arginine	60.7	ArgVα; arginine tRNA; tandemly quadruplicated gene *argQVYZ*	1258, CG-11650
argW	Arginine	53.1	Arginine tRNA 5	2225, CG-17752
argX	Arginine	85.7	Arginine tRNA 3	903, 1803, 2225, CG-17749
argY	Arginine	60.7	*argVβ*; arginine tRNA 2; tandemly quadruplicated gene *argQVYZ*	CG-35610
argZ	Arginine	60.7	*argVα*; arginine tRNA 2; tandemly quadruplicated gene *argQVYZ*	CG-35607
aroA	Aromatic	20.7	3-Enolpyruvylshikimate-5-P synthetase; EC 2.5.1.19	1083, 1841, 1842, CG-1008
aroB	Aromatic	75.7	3-Dehydroquinate synthase; EC 4.6.1.3	1999, 2851, CG-1007

TABLE 1 *Continued*

Symbol	Mnemonic[b] for symbol	Position (min)[c]	Synonyms; enzyme or phenotype affected[d]	References[e]
aroC	Aromatic	52.7	Chorismate synthase; EC 4.6.1.4	650, 4581, CG-1006
aroD	Aromatic	38.2	3-Dehydroquinate dehydratase; EC 4.2.1.10	408, 660, 981, 1082, 2172, CG-1005
aroE	Aromatic	73.8	Dehydroshikimate reductase; EC 1.1.1.25	108, 2775, CG-1004
aroF	Aromatic	58.9	DAHP synthetase (tyrosine repressible); EC 4.1.2.15	760, 1336, 1698, 1813, 3924, 4067, CG-1003
aroG	Aromatic	16.9	Member TyrR regulon, single TyrR box, not Tyr mediated; DAHP synthetase (phenylalanine repressible); EC 4.1.2.15	933, 934, 1659, 1813, 1876, 2715, 4067, CG-1002
aroH	Aromatic	38.4	DAHP synthetase (tryptophan repressible); EC 4.1.2.15	408, 933, 1082, 1814, 2936, 3493, 4067, 4833, CG-1001
aroI	Aromatic	84.0	Function unknown	1388a, CG-1000
aroK	Aromatic	75.8	Shikimate kinase I	1477, 2520, 2669, 4578, CG-30224
aroL	Aromatic	8.7	Member TyrR regulon; shikimate kinase II; EC 2.7.1.71	975, 976, 2351, 2852, 4578, CG-999
aroM	Aromatic	8.8	Unknown function, regulated by *aroR*; member of TyrR regulon	975, 976, CG-18523
aroP	Aromatic	2.6	Member TyrR regulon; general aromatic amino acid transport	508, 730, 1514, 1517, 1611, 1761, 2329, 2330, 3565, CG-998
aroT	Aromatic	28.2	*aroR, trpR;* indole acrylic acid sensitivity; transport of aromatic amino acids, alanine and glycine	4263, CG-997
arsB	Arsenate resistance	78.6	*arsF;* resistance to arsenate, arsenite, and antimonite	604, 1028, 3995, CG-35996
arsC	Arsenate resistance	78.6	*arsG;* resistance to arsenate, arsenite, and antimonite	604, 1028, 3995, CG-35999
arsR	Arsenate resistance	78.6	*arsE;* resistance to arsenate, arsenite, and antimonite	604, 1028, 3995, CG-35993
artI	Arginine transport	19.5	Arginine periplasmic transport system	4620, CG-31674
artJ	Arginine transport	19.4	Arginine periplasmic transport system	4620, CG-31664
artM	Arginine transport	19.4	Arginine periplasmic transport system	4620, CG-31667
artP	Arginine transport	19.5	Arginine periplasmic transport system	4620, CG-31677
artQ	Arginine transport	19.4	Arginine periplasmic transport system	4620, CG-31670
ascB	Arbutin, salicin, cellobiose	61.2	*sac;* member of cryptic *asc* operon, paralogous to cryptic *bglFB* genes; 6-P-β-glucosidase	1561, 1562, 3268, CG-33236
ascF	Arbutin, salicin, cellobiose	61.1	*sac;* member of cryptic *asc* operon activated when repressor AscG is mutated	1561, 1562, 3268, CG-33233
ascG	Arbutin, salicin, cellobiose	61.1	Repressor of cryptic *asc* operon; when inactivated, *asc* operon expressed; paralogous to *galR*	1561, 1562, 3268, CG-33230
asd	Aspartate semialdehyde dehydrogenase	77.0	*dap, hom;* aspartate semialdehyde dehydrogenase; EC 1.2.1.11	1634, 1635, 3183, 3821, CG-996
aslA	Arylsulfatase-like	85.8	*gppB;* arylsulfatase homolog that suppresses mutation *gpp*	903, CG-33957
aslB	Arylsulfatase-like	85.8	Putative regulator of *aslA*	903, CG-33960
asnA	Asparagine	84.6	Asparagine synthetase A; EC 6.3.1.1	536, 548, 1734, 1828, 3007, CG-995
asnB	Asparagine	15.1	Asparagine synthetase B; EC 6.3.1.1	1828, 3348, 3834, CG-994
asnC	Asparagine	84.5	Regulatory gene for *asnA, asnC,* and *gidA*	536, 548, 702, 963, 2212, 2213, 3007, CG-18520
asnS	Asparagine	21.3	*lcs, tss;* asparaginyl-tRNA synthetase; EC 6.1.1.22	105, 106, 112, 1888, 2608, CG-993
asnT	Asparagine	44.1	Asparagine tRNA	1248, 2225, CG-992
asnU	Asparagine	44.4	Asparagine tRNA	2225, CG-17746
asnV	Asparagine	44.5	Asparagine tRNA	2225, CG-17743
aspA	Aspartate	94.0	Aspartate ammonia-lyase (aspartase); EC 4.3.1.1	1515, 4182, 4651, CG-991
aspC	Aspartate	21.2	Aspartate aminotransferase; EC 2.6.1.1	292, 1244, 1891, 2231, 2290, 2642, 4744, CG-990
aspS	Aspartate	42.0	*tls;* aspartyl-tRNA synthetase	1141, 1316, 3881, 4183, CG-32508
aspT	Aspartate	85.0	*tasC;* triplicated gene, aspartate tRNA 1	903, 2225, 4777, CG-989
aspU	Aspartate	5.0	In *rrnH* operon, triplicated gene, aspartate tRNA 1	1120, 2225, CG-988
aspV	Aspartate	5.2	Triplicated gene aspartate tRNA 1	1778, 2225, CG-987
asu	Asparagine utilization	33.2	Utilization of asparagine as sole nitrogen source (Note also that *asu* was used for "antisuppressor" phenotype, for a series of alleles found to be mutations at known loci, and use of this symbol for new "antisuppressor" genes is sometimes reintroduced.) See *trmU* for *asuE*	698, CG-18517
atoA	Acetoacetate	50.0	Short-chain fatty acid degradation; acetyl-CoA:acetoacetyl-CoA transferase, β subunit; EC 2.8.3.-	Y, 2238, CG-986
atoB	Acetoacetate	50.0	Acetyl-CoA acetyltransferase; EC 2.3.1.9	Y, 1974, CG-985
atoC	Acetoacetate	50.0	Az (antizyme); positive regulatory gene; response regulator, see *atoS*	Y, 595a, 1973, 1974, 3283, CG-984
atoD	Acetoacetate	50.0	Acetyl-CoA:acetoacetyl-CoA transferase	Y, 1974, CG-18514
atoE	Acetoacetate	50.0	*ato* operon gene of unknown function	Y, CG-37064

(Table continues)

TABLE 1 *Continued*

Symbol	Mnemonic[b] for symbol	Position (min)[c]	Synonyms; enzyme or phenotype affected[d]	References[e]
atoS	Acetoacetate	50.0	Sensor kinase in two-component system with atoC response regulator	Y, CG-37064
atpA	ATP	84.4	papA, uncA; membrane-bound ATP synthase, F_1 sector, α subunit; EC 3.6.1.34	548, 1349, 1667, 2064, 2112, 2199, 3078, 4051, 4484, CG-33
atpB	ATP	84.4	papD, uncB; membrane-bound ATP synthase, F_0 sector, subunit a; EC 3.6.1.34	548, 577, 1348, 2066, 2428, 3078, 3079, 3764, 4432, 4455, 4484, 4826, CG-32
atpC	ATP	84.3	papG, uncC; membrane-bound ATP synthase, F_1 sector, ε subunit; EC 3.6.1.34	548, 2063, 2738, 3280, 3727, 4484, 4689, CG-31
atpD	ATP	84.3	papB, uncD; membrane-bound ATP synthase, F_1 sector, β subunit; EC 3.6.1.34	548, 1937, 2063, 2064, 3727, 4484, 4619, CG-30
atpE	ATP	84.4	papH, uncE; membrane-bound ATP synthase, F_0 sector, subunit c: DCCD$^-$; EC 3.6.1.34	548, 1207, 1348, 1667, 1962, 2067, 2738–2740, 3078, 3121, 3764, 4467, 4484, CG-29
atpF	ATP	84.4	papF, uncF; membrane-bound ATP synthase, F_0 sector, subunit b: EC 3.6.1.34	548, 1348, 1667, 2066, 2596, 2746, 3078, 3379, 4484, CG-28
atpG	ATP	84.3	papC, uncG; membrane-bound ATP synthase, F_1 sector, γ subunit; EC 3.6.1.34	548, 1667, 2064, 2199, 3727, 4484, CG-27
atpH	ATP	84.4	papE, uncH; membrane-bound ATP synthase, F_1 sector, δ subunit; EC 3.6.1.34	104, 548, 1348, 1667, 2596, 3078, 4484, CG-26
atpI	ATP	84.4	uncI; membrane-bound ATP synthase subunit, F_1F_0-type proton-ATPase; EC 3.6.1.34	525, 548, 1348, 2004, 2065, 2738, 2739, 2903, 3079, 3281, 3378, 3796–3798, 4455, 4484, 4826, CG-18511
att186	Attachment	59.1	Integration site for phage 186	CG-972
att253	Attachment	6.1	Integration site for phage 253	3372, CG-18508
atte14	Attachment	25.7	Attachment site for element e14	1471, CG-983
attHK139	Attachment	44.8	Attachment site for phage HK139	1012, CG-981
attHK022	Attachment	22.7	atthtt; attachment site for phage HK022	1011, 2218, 2977, 3945, 4692, CG-982
attλ	Attachment	17.3	att92, att434; integration site for phages λ, 82, and 434	1806, CG-980
attP1, P7	Attachment	68.7	loxB; integration site for phages P1 and P7	697, CG-979
attP22	Attachment	5.7	ata; integration site for phage P22	1769, 2481, CG-975
attP2H	Attachment	44.7	Integration site H for phage P2	CG-978
attP2II	Attachment	87.1	Integration site II for phage P2	
attP2III	Attachment	33.5	Integration site III for phage P2	CG-32179
attP4	Attachment	96.9	Integration site for phage P4	581, 3327a, CG-976
attPA-2	Attachment	53.1	Integration site for phage PA-2	CG-974
attφ80	Attachment	28.2	Integration site for phage φ80	3667, CG-973
attTn7	Attachment	84.2	Site of high-frequency integration of transposon Tn7	1482a, 2761
avtA	Alanine-valine transaminase	80.5	Alanine-α-ketoisovalerate transaminase, transaminase C	292, 1171, 2505, 3995, 4505, 4576, CG-971
azaA	Azaserine	44.6	Resistance or sensitivity to azaserine	4600, CG-970
azaB	Azaserine	71.7	Resistance or sensitivity to azaserine	4600, CG-969
azl	Azaleucine	57.9	Regulation of ilv and leu genes; azaleucine resistance	CG-967
bacA	Bacitracin resistance	69.0	Lipid kinase activity	576, 857, 4762, CG-29739
baeR	Bacterial adaptive (envZ regulator)	46.6	Putative sensor/regulator family member; suppresses envZ and phoR/creC mutations	2979, CG-30764
baeS	Bacterial adaptive (envZ regulator)	46.5	Regulatory protein	2979, CG-30761
barA	Bacterial adaptive response	62.8	Signal transducer; member of EnvZ-OmpR family of regulatory factors; has sensory kinase and response regulator domains	1899, 2980, CG-33320
basR	Bacterial adaptive sensor	93.3	Two-component regulatory system, homologous with OmpR-EnvZ family	2979, CG-28168
basS	Bacterial adaptive sensor	93.3	Regulatory gene; see basR	2979, CG-28171
bcp	Bacterioferritin comigratory protein	55.8	Comigrates with bacterioferritin on gels	101, CG-33035
bcr	Bicyclomycin resistance	49.0	bicA, bicR, sur, suxA; bicyclomycin resistance, affects sulfathiazole-sulfonamide resistance; transmembrane protein	284, 3073, CG-32582
betA	Betaine	7.0	Choline dehydrogenase	94, 448, 2315, 4118, CG-17740
betB	Betaine	7.0	Betaine aldehyde dehydrogenase; EC 1.2.1.8	94, 448, 2315, CG-17737
betI	Betaine	7.1	Regulatory gene? possible repressor in choline regulation of bet genes	2315, CG-30692
betT	Betaine	7.1	High-affinity transport for choline, driven by proton motive force	94, 2315, CG-18505
bfm	BF23 multiplication	85.8	Phage BF23 multiplication	3920a, CG-966

TABLE 1 *Continued*

Symbol	Mnemonic[b] for symbol	Position (min)[c]	Synonyms; enzyme or phenotype affected[d]	References[e]
bfr	Bacterioferritin	74.6	Bacterioferritin	100, 103, CG-32528
bglB	β-Glucoside	84.0	blgA; growth on arbutin or salicin; phospho-β-glucosidase B, cryptic	548, 2618, 3416, 3530, 3799, 3801, CG-964
bglF	β-Glucoside	84.1	bglC, blgB; BglG kinase and phosphatase, cryptic	455, 548, 2618, 3799, 3801, CG-18502
bglG	β-Glucoside	84.1	bglC, bglS; positive regulatory gene; sequence-specific RNA-binding protein; suppresses termination, regulated by phosphorylation	79, 166, 548, 1790, 2618, 2619, 3530, 3531, 3799–3801, CG-963
bglT	β-Glucoside	84.8	bglE; regulatory gene for P-β-glucosidase A synthesis	3759, CG-961
bioA	Biotin	17.4	Biotin biosynthesis; diaminopelargonic acid synthetase	14, 220, 223, 2252, 2472, 3218, 3921, 4171, CG-959
bioB	Biotin	17.5	Biotin biosynthesis; dethiobiotin→biotin	14, 220, 223, 2252, 2472, 3218, 4171, CG-958
bioC	Biotin	17.5	Biotin biosynthesis; step prior to pimeloyl-CoA	14, 3218, 4171, CG-957
bioD	Biotin	17.5	Biotin biosynthesis; dethiobiotin synthetase	14, 3218, 4171, CG-956
bioF	Biotin	17.5	Biotin biosynthesis; 7-keto-8-aminopelargonic acid synthetase	14, 3218, 4171, CG-955
bioH	Biotin	76.3	bioB; biotin biosynthesis	220, 3147, 3821, CG-954
bioP	Biotin	86.5	bir, birB; biotin transport	592, 1265, CG-953
birA	Biotin retention	89.8	bioR, dhbB; biotin-[acetyl-CoA carboxylase] holoenzyme synthetase, both biosynthetic role and repressor	220–222, 372, 542, 592, 1798, 2647, 3248, 3249, CG-952
bisC	Biotin sulfoxide	80.0	Biotin sulfoxide reductase; structural gene	982, 3327, 3995, CG-951
bolA	Bolus	9.8	Morphogene; affects cell division and growth; possible role in switching between cell elongation and septation; produces stable spherical cells when overexpressed	48–50, 1674, 2325, CG-31032
brnQ	Branched chain	9.0	hrbA; valyl, glycylvaline sensitivity; transport system 1 for Ile, Leu, and Val	1501, 1502, 2637, CG-950
brnR	Branched chain	8.5	Valyl, glycylvaline sensitivity; component of transport system 1 and 2 for Ile, Leu, and Val	1501, CG-949
brnS	Branched chain	1.2	Valyl, glycylvaline sensitivity; transport system for Ile, Leu, and Val	1501, CG-948
brnT	Branched chain	65.5	Low-affinity transport system for Ile	1848a, CG-947
btuB	B12 uptake	89.6	bfe, cer; receptor for vitamin B$_{12}$, E colicins, and phage BF23	157, 372, 562, 999, 1057, 1225, 1504, 1505, 1534, 1660, 1816, 2030, 2248, 2909, 3435, CG-946
btuC	B12 uptake	38.6	Vitamin B$_{12}$ transport	962, 1007, 1281, 2028, 3550, CG-945
btuD	B12 uptake	38.5	Vitamin B$_{12}$ transport, membrane associated	962, 1007, 1281, 2028, CG-18499
btuE	B12 uptake	38.6	Member of btu operon with no role in B$_{12}$ transport	962, 1281, 2028, 3550, CG-18496
btuR	B12 uptake	28.5	Regulatory gene for btuB	2577, 2581, CG-18493
bymA	Bypass maltose	(93.0)	Bypass of maltose permease at malB	1744, CG-944
cadA	Cadaverine	93.8	Lysine decarboxylase; EC 4.1.1.18	2799, 2800, 3658, 3906, 4178, 4528, CG-943
cadB	Cadaverine	93.9	Transport of lysine and cadaverine; probably arginine/ornithine antiporter	2799, 2800, 4528, CG-34228
cadC	Cadaverine	93.9	Regulatory gene	2799, 4528, CG-34231
cafA	Cytoplasmic axial filaments	73.1	Affects cell division and growth; overproduction of CafA gives bundles of cytoplasmic filaments, chained cells, and minicells	3180, 4465, CG-31358
cai	Carnitine inducible	0.8	Cluster; this cluster also includes three carnitine-inducible genes located clockwise from the genes above and transcribed in the opposite direction with homology to known proteins but unknown function, referred to as fixABC in reference cited	1108, CG-38208, CG-37071
caiA	Carnitine inducible	0.8	Carnitine metabolism, an oxidoreductase	1108, 4787, CG-36825
caiB	Carnitine inducible	0.8	Carnitine dehydratase	1108, 1109, 4787, CG-36800
caiC	Carnitine inducible	0.8	Crotonobetaine/carnitine-CoA ligase	1108, 4787, CG-36828
caiD	Carnitine inducible	0.8	Putative enoyl hydratase/isomerase with carnitine racemase activity	1108, 4787, CG-36831
caiE	Carnitine inducible	0.8	Stimulates carnitine racemase activity of CaiD and the dehydrogenase, CaiB	1108, 4787, CG-36834
caiT	Carnitine inducible	0.8	Putative carnitine/betaine transport	1108, 4787, CG-36822
calA	Calcium	95.0	Transport of calcium	477, CG-941
calC	Calcium	15.6	Transport of calcium	477, CG-940
calD	Calcium	9.4	Transport of calcium	477, CG-939
can	Canavanine	66.0	Canavanine resistance	CG-938
carA	Carbamoylphosphate	0.6	arg+ura, cap, pyrA; carbamoylphosphate synthase (glutamine-hydrolyzing) small subunit; EC 6.3.5.5	438, 840, 1391, 3140, 3330, 4787, CG-936

(Table continues)

TABLE 1 *Continued*

Symbol	Mnemonic[b] for symbol	Position (min)[c]	Synonyms; enzyme or phenotype affected[d]	References[e]
carB	Carbamoylphosphate	0.7	*arg+ura, cap, pyrA*; carbamoylphosphate synthase (ammonia), large subunit; EC 6.3.4.16	438, 840, 1391, 3140, 3330, 4787, CG-935
cbpA	Curved-DNA-binding protein	22.9	Sequence similarity to *dnaJ*; DNA-binding protein, recognized curved DNA sequence	4355, CG-31822
cbt	Colicin B (and D) tolerance	16.2	Chloroacetate and colicin B and D sensitivity; dicarboxylate-binding protein production	3441, CG-934
cca	CCA tRNA terminus	68.9	tRNA nucleotidyltransferase; EC 2.7.7.25	857, 4428, CG-933
ccmA	Cytochrome *c* maturation	49.4	Cytochrome *c*-related ABC transporter, ATPase subunit	CC, FF, 712, 713, CG-36574
ccmB	Cytochrome *c* maturation	49.4	Cytochrome *c*-related ABC transporter, ATPase subunit	CC, FF, 712, 713, CG-36577
ccmC	Cytochrome *c* maturation	49.4	Cytochrome *c*-related ABC transporter, heme binding subunit	CC, FF, 712, 713, CG-36581
ccmD	Cytochrome *c* maturation	49.4	Cytochrome *c*-related	CC, FF, 712, 713, CG-36584
ccmE	Cytochrome *c* maturation	49.4	Cytochrome *c*-related	CC, FF, 712, 713, CG-36587
ccmF	Cytochrome *c* maturation	49.3	Cytochrome *c*-related	CC, FF, 712, 713, CG-36590
ccmG	Cytochrome *c* maturation	49.3	Cytochrome *c*-related thioredoxin homolog	CC, FF, 712, 713, CG-36594
ccmH	Cytochrome *c* maturation	49.3	Cytochrome *c*-related	CC, FF, 712, 713, CG-36597
cdd	Deoxycytidine deaminase	48.0	Deoxycytidine deaminase; EC 3.5.4.5	406, 1273, 1578, 1688, 2017, 2018, 2839, 3145, 4178, 4379, 4728, CG-932
cde	Constitutive *dam* expression	14.3	Growth rate control of *dam* gene expression; near or within *lipB*	3487a
cdh	CDP diglyceride hydrolase	88.5	CDP diglyceride hydrolase	43, 541, 1665, 1856, 3359, CG-931
cdsA	CDP diglyceride synthase	4.3	CDP diglyceride synthase	1320, 1858, CG-930
cdsS	CDP diglyceride synthase	71.6	Stability of CdsA activity	1321, CG-18490
celA	Cellobiose	39.1	Transport of cellobiose, arbutin, and salicin	3269, 3253, CG-34873
celB	Cellobiose	39.1	Phosphotransferase system enzyme IIcel, PEP dependent	2260, 2261, 3269, 3523, CG-18487
celC	Cellobiose	39.1	Transport of cellobiose, arbutin, and salicin; phosphotransferase system enzyme IIIcel, PEP dependent	1560, 2260, 2261, 3269, 3523, CG-18484
celD	Cellobiose	39.1	Negative regulatory protein for *cel*	2260, 2261, 3269, 3523, CG-18481
celF	Cellobiose	39.1	Phospho-β-glucosidase B	2260, 2261, 3269, 3523, CG-17734
cfa	Cyclopropane fatty acid	37.5	*cdfA*; cyclopropane fatty acid synthase	1489–1491, 4497, CG-10810
cfcA	Control frequency of cell division	79.9	Controls cell division frequency per round of DNA replication	3096, CG-36615
chaA	Ca^{2+}/H^{+} antiporter	27.3	Ca^{2+}/H^{+} antiporter	1934, 3175, 3239, CG-30293
chaB	Ca^{2+}/H^{+} antiporter	27.4	Accessory and regulatory protein for *chaA*	Q, CG-37193
chaC	Ca^{2+}/H^{+} antiporter	27.4	Accessory and regulatory protein for *chaA*	Q, CG-37196
cheA	Chemotaxis	42.5	Chemotactic signal transduction; flagellar regulon member; autophosphorylating histidine kinase	437, 1359, 2206, 2973, 3202, 3701, 3968, 3985, 4224, 4639, CG-928
cheB	Chemotaxis	42.4	Chemotactic signal transduction; flagellar regulon member; protein methylesterase	437, 447, 900, 1633, 1702, 2269, 2973, 3968, 4077, CG-927
cheR	Chemotaxis	42.4	*cheX*; chemotactic signal transduction; flagellar regulon member; protein methyltransferase (in chemotactic response)	447, 1633, 2868, 2973, 3968, CG-926
cheW	Chemotaxis	42.5	Chemotactic signal transduction; flagellar regulon member	1359, 2206, 2973, 3968, CG-925
cheY	Chemotaxis	42.4	Chemotactic signal transduction; flagellar regulon member	437, 447, 749, 1633, 1702, 2707, 2973, 3714, 3968, CG-924
cheZ	Chemotaxis	42.4	Chemotactic signal transduction; flagellar regulon member	447, 1633, 2973, 3968, CG-923
chpA	Chromosomal homolog of *pem*	62.7	*chpAK, mazF*; cell division, growth inhibitor; toxic protein	2695, CG-33287
chpB	Chromosomal homolog of *pem*	95.8	*chpBK*; cell division, growth inhibitor	2304, 2695, 2696, 4558, CG-33290
chpR	Chromosomal homolog of *pem*	62.7	*chpAl, mazE*; suppresses growth inhibitor ChpA; antagonizes toxic effect of ChpA	2695, 2822, CG-33283
chpS	Chromosomal homolog of *pem*	95.8	*chpBl*; suppresses growth inhibitor ChpB	2695, 2696, 4558, CG-33293
cirA	Colicin I resistance/receptor	48.3	*feuA*; colicin I receptor production	403, 1480, 1481, 2839, 3037, 3435, 4062, 4655, CG-916
citA	Citrate	(18.0)	Citrate transport system, cryptic gene	1558, CG-18469
citB	Citrate	(16.0)	Citrate transport system, cryptic gene	1558, CG-18466
clpA	Caseinolytic protease	19.9	Clp ATP-dependent protease, ATP-binding subunit	894, 1432, 1433, 2112, 3870, CG-31293
clpB	Caseinolytic protease	58.7	ClpB protease, ATP dependent, chaperone; EC 1.17.4.-, EC 3.4.21.-	1292, 1433, 2177, 3263, 3368, 3894, 4040, 4642, CG-32875

TABLE 1 *Continued*

Symbol	Mnemonic[b] for symbol	Position (min)[c]	Synonyms; enzyme or phenotype affected[d]	References[e]
clpP	Caseinolytic protease	9.8	F21.5, LopP; ClpP ATP-dependent protease proteolytic subunit	134, 894, 1431, 1433, 2264, 2719–2721, 3870, 4167, 4635, 4761, CG-31280
clpX		9.8	LopC; ClpX protease, which actives ClpP	1431, 4167, 4635, 4761, CG-31287
cls	Cardiolipin synthase	28.1	*nov*; cardiolipin synthase	1642, 1729, 1730, 1845, 1932, 2850, 3095, 3170, 3477, CG-915
cmk	CMP kinase	20.7	*mssA*; CMP kinase	1274, 3292, 3802, 4720, CG-31736
cmlA	Chloramphenicol	19.1	Resistance or sensitivity to chloramphenicol	173, 3510, 3922, CG-914
coaA	CoA	89.9	*panK, rts*; pantothenate kinase	372, 1220, 4002, 4003, 4388, CG-17731
codA	Cytosine deaminase	7.7	Cytosine deaminase; EC 3.5.4.1	21, 904, 4391, CG-913
codB	Cytosine deaminase	7.6	Cytosine transport	21, 904, CG-912
cog	Control of *ompG*	30.3	Controls production of porin-like protein OmpG	2872a, CG-37297
corA	Cobalt resistance	86.2	Mg^{2+} transport, system I	51, 903, 2654, 3148, 3149, 3260, 3359, 3986, 4534, CG-911
CP4-57	Cryptic phage	59.2	Cryptic prophage element; see *intA* and *alpA*,	
cpdA	Cyclic nucleotide phosphodiesterase	68.8	Affects requirement for cAMP during growth on maltose; 3',5'-cAMP phosphodiesterase; EC 3.1.4.17	EE
cpdB	Cyclic nucleotide phosphodiesterase	95.5	2',3'-cyclic nucleotide 2'-phosphodiesterase; EC 3.1.4.16	244, 1967, 2497, 2498, CG-909
cpsA	Capsular polysaccharide (CPS) synthesis	45.8	Colanic acid (CPS) biosynthesis	4314, CG-18463
cpsB	CPS synthesis	45.7	Colanic acid (CPS) biosynthesis; mannose 1-P guanyltransferase	113a, 4314, CG-18460
cpsC	CPS synthesis	45.8	Colanic acid (CPS) biosynthesis	4314, CG-18457
cpsD	CPS synthesis	45.8	Colanic acid (CPS) biosynthesis	4314, CG-18454
cpsE	CPS synthesis	45.8	Colanic acid (CPS) biosynthesis	4314, CG-18451
cpsF	CPS synthesis	90.3	Colanic acid (CPS) biosynthesis	4314, CG-18448
cpsG	CPS synthesis	45.8	Colanic acid (CPS) biosynthesis, phosphomannomutase isozyme	113a, 1968a, CG-37429
cpxA	Conjugative plasmid expression	88.3	*ecfB, eup, ssd*; phage Q resistance; membrane sensor in two-component signal transduction	43, 44, 1967, 2751, 2752, 2921, 3067, 3344, 3359, 3475, 3940, 3941, 3943, 4262, 4536, CG-908
cpxB	Conjugative plasmid expression	41.0	Phage Q resistance; membrane protein	2751, 2752, 3940, CG-907
cpxR	Conjugative plasmid expression	88.4	Cognate regulator for membrane sensor CpxA in two-component signal transduction	1038, 3359, 4536, CG-34166
creA	Catabolite regulation	99.9	Transcribed with *cre* operon; function unknown	74, 3959, CG-34809
creB	Catabolite regulation	99.9	*phoM-orf2*; structurally homologous to and upstream of *creC* (*phoM*)	74, 75, 4509, CG-34803
creC	Catabolite regulation	99.9	*phoM*; sensor in *Pho* regulon	74, 75, 1074, 2565, 2638, 2639, 4284, 4508, 4509, 4511, 4514, 4518, CG-395
creD	Catabolite regulation	99.9	*cet* (colicin E2 tolerance)	74, 1073, 1074, 1245, 1422, 1525, CG-929
crg	Cold-resistant growth	28.8	Cold-resistant growth	B, 2126a, CG-18445
crl	Curli	5.6	Affects expression of cryptic *csgA* genes; regulatory protein for curli	127, 3196, 3226, 3431, CG-30625
crp	cAMP receptor protein	75.0	*cap, csm*; cAMP receptor protein	15, 26, 136, 235, 245, 270, 479, 819, 820, 890, 1364, 1370, 1518, 1581, 2211, 2871, 3161, 3181, 3447, 4020, 4380, CG-906
crr		54.6	*gsr, iex, tgs*; phosphocarrier protein for glucose of PTS; III^{Glc}	488, 490, 518, 519, 958–960, 1255, 1560, 2768, 3050, 3272, 3665, CG-905
csgA	Curlin σ^S (stationary phase)-dependent growth	23.8	Cryptic gene for surface fibers; induced in stationary phase, σ^S-dependent transcription; curlin subunit	126, 127, 1674, 2324, 3194–3196, CG-30620
csgB	Curlin σ^S (stationary phase)-dependent growth	23.8	Minor curlin?	126, CG-36735
csiA	Carbon starvation induced	85.3	Stationary-phase inducible protein	P, 4544, CG-36892
csiB	Carbon starvation induced	17.4	Stationary-phase inducible protein	P, 1674, 2324, 4544, CG-36895
csiC	Carbon starvation induced	66	Stationary-phase inducible protein	P, 4544, CG-36898
csiD	Carbon starvation induced	60.5	Stationary-phase inducible protein	P, 4544, CG-36904
csiE	Carbon starvation induced	57.3	Stationary-phase inducible protein	P, 4544, CG-36901
csiF	Carbon starvation induced	8.6	Stationary-phase inducible protein	P, 4544, CG-36998
cspA	Cold shock protein	80.1	Similarity with Y-box DNA-binding proteins of eukaryotes; transcription factor; cold shock protein CS7.4; cold shock enhancement of *hns*	1413, 2009, 2298, 2390, 3995, 4204, CG-29540

(Table continues)

TABLE 1 *Continued*

Symbol	Mnemonic[b] for symbol	Position (min)[c]	Synonyms; enzyme or phenotype affected[d]	References[e]
cspB	Cold shock protein	35.3	Cold shock-induced protein, with similarity to CspA	2390, CG-32231
cspC	Cold shock protein	41.0	*msmB;* multicopy suppresses *mukB* mutants	2008, 2390, CG-35339
cspD	Cold shock protein	19.9	Homology with CspA but not cold shock induced	1043, 1432, 2390, CG-31688
cspE	Cold shock protein	14.2	*msmC;* multicopy suppresses *mukB* mutants	2008, 4719, CG-31528
csrA	Carbon storage regulator	60.7	Regulatory gene inhibiting glycogen biosynthesis	2506, 3589, 3590, CG-34504
cstA	Carbon starvation	13.6	Starvation-induced stress response protein	377, 1486, 2500, 2700, 3816, 4114, CG-31179
cup	Carbohydrate uptake	97.0	Uptake of carbohydrates	2621, CG-18442
cutA	Cu transport tolerance/utilization	94.0	*cycY;* copper sensitivity; possible role in cytochrome *c* maturation	251, 1235, 1896, 3693, CG-34216
cutC	Cu transport tolerance/utilization	42.3	Copper sensitivity	1529, CG-36974
cutE	Cu transport tolerance/utilization	14.9	*lnt;* copper sensitivity; apolipoprotein *N*-acetyltransferase	1176, 1528, 3582, CG-31471
cutF	Cu transport tolerance/utilization	4.7	*nlpE;* copper sensitivity	1529, 3987a, CG-35748
cvpA	Colicin V production	52.4	*dedE;* in *purF* operon; affects colicin V production	1176, 2631, 3118, CG-32727
cxm	Carbon-xylose metabolism	6.3	*cxr;* affects cAMP control of D-xylose utilization; methyl glyoxal synthesis	9, CG-903
cyaA	Cyclase, adenylate	85.9	Adenylate cyclase; EC 4.6.1.1	24, 27, 28, 51, 204, 478, 903, 987, 2238, 2281, 3447, 3624–3626, CG-902
cybB	Cytochrome *b*	32.1	Cytochrome b_{561}	2956, 2957, 3002, CG-17728
cybC	Cytochrome *b*	96.1	Cytochrome b_{561}	1917, 2368, 3088, 4317, CG-34583
cycA	Cycloserine	95.4	*dagA;* D-alanine, D-serine, glycine permease; loss confers cycloserine resistance	818, 3558, 4520, CG-900
cydA	Cytochrome *d*	16.7	Cytochrome *d* terminal oxidase, polypeptide subunit I	585, 823, 1081, 1289, 1372, 1454–1457, 3070, CG-10369
cydB	Cytochrome *d*	16.7	Cytochrome *d* terminal oxidase, polypeptide subunit II	585, 823, 1372, 1454, 1456–1458, 3070, CG-9469
cydC	Cytochrome *d*	20.0	*mdrA, mdrH, surB;* cytochrome *d* terminal oxidase, possibly heme d component	863, 1373, 3923, 3933, CG-17725
cydD	Cytochrome *d*	20.0	ABC membrane transporter	247, 988, 990, 3369–3371, 3933, CG-31720
cynR	Cyanase; cyanate metabolism	7.7	Positive regulator	93, 2317, 4143, CG-31255
cynS	Cyanase; cyanate metabolism	7.7	Cyanate aminohydrolase; EC 3.5.5.3	93, 705, 1521, 2317, 4141, 4143, 4144, CG-15267
cynT	Cyanase; cyanate metabolism	7.7	Carbonic anhydrase	93, 1522, 2254, 4142, CG-31258
cynX	Cyanase; cyanate metabolism	7.8	Member of *cyn* operon; apparently hydrophobic protein of unknown function	93, 4142, CG-31261
cyoA	Cytochrome *o* oxidase	9.7	Cytochrome *o* ubiquinol oxidase subunit II	155, 156, 585, 694–696, 823, 2480, 2790, 2864, 2867, 3003, CG-18439
cyoB	Cytochrome *o* oxidase	9.6	Cytochrome *o* ubiquinol oxidase subunit I	155, 585, 694–696, 823, 2864, 2867, 3003, CG-30997
cyoC	Cytochrome *o* oxidase	9.6	Cytochrome *o* ubiquinol oxidase subunit III	155, 585, 694–696, 823, 2864, 2867, CG-31005
cyoD	Cytochrome *o* oxidase	9.6	Cytochrome *o* ubiquinol oxidase subunit IV	155, 585, 694–696, 823, 2864, CG-31008
cyoE	Cytochrome *o* oxidase	9.6	Cytochrome *o* oxidase subunit; protoheme IX farnesyltransferase	155, 585, 694–696, 823, 2864, 3672, CG-31014
cysA	Cysteine	54.7	Sulfate permease; chromate resistance	518, 519, 1390, 3665, 3953, 3954, CG-898
cysB	Cysteine	28.7	Positive regulator of *cys* regulon	1955, 3215, 3428, 4237, CG-897
cysC	Cysteine	61.9	*cys* regulon member; P-adenylylsulfate kinase; EC 2.7.1.25	1829, 2429, 2430, 3745, CG-896
cysD	Cysteine	61.9	*cys* regulon member; ATP:sulfate adenylyltransferase; EC 2.7.7.4	1829, 2429, 2430, 2645, CG-895
cysE	Cysteine	81.4	Serine acetyltransferase; EC 2.3.1.30	1000, 3995, 4235, CG-894
cysG	Cysteine	75.3	Converts uroporphyrinogen III to siroheme, a prosthetic group for nitrite and sulfite reductases, required for cysteine synthesis and nitrite reduction; uroporphyrinogen III methyltransferase	270, 765, 1949, 2597, 2598, 3286, 3287, 4013, 4522, 4668, CG-893
cysH	Cysteine	62.2	*cys* regulon member; adenylylsulfate reductase; EC 1.8.99.2	1829, 2265, 2266, 2432, CG-892
cysI	Cysteine	62.2	*cysQ; cys* regulon member; sulfite reductase, α subunit; EC 1.8.1.2	1829, 2266, 2432, 4236, 4668, CG-891
cysJ	Cysteine	62.2	*cysP; cys* regulon member; sulfite reductase, β (flavoprotein) subunit; EC 1.8.1.2	1829, 2432, 3214, 4236, 4668, CG-890
cysK	Cysteine	54.5	*cysZ;* cysteine synthase A; EC 4.2.99.8	329, 414, 488, 518, 519, 567, 958, 1208, 1688, 2423, 3272, 3665, 3954, CG-889
cysM	Cysteine	54.7	O-Acetylserine sulfhydrolase B; EC 4.2.99.8	518, 519, 3665, 3953, 3954, CG-17722

TABLE 1 *Continued*

Symbol	Mnemonic[b] for symbol	Position (min)[c]	Synonyms; enzyme or phenotype affected[d]	References[e]
cysN	Cysteine	61.9	ATP sulfurylase (ATP:sulfate adenylyltransferase)	2429, 2430, CG-18436
cysP	Cysteine	54.7	Periplasmic sulfate-binding protein	1800, 3953, CG-27367
cysQ	Cysteine	95.6	*amt*; mutants require sulfite or cysteine during aerobic growth	1168, 1966, 1967, 2497, 3058, 3313, CG-34409
cysS	Cysteine	11.9	Cysteinyl-tRNA synthetase; EC 6.1.1.16	163, 388, 1142, 1787, CG-888
cysT	Cysteine	43.0	Cysteine tRNA; EC 6.1.1.16	1248, 2225, CG-17719
cysU	Cysteine	54.7	*cysT*; sulfate transport system permease protein	3953, CG-37093
cysW	Cysteine	54.7	Membrane-bound sulfate transport protein	3953, CG-27371
cytR		88.8	Regulatory gene for *deo* operon, *udp*, *cdd*, CytR; autogenously regulated	211, 212, 1381, 2378, 2951, 3291, 3359, 3996, 3997, 4381, CG-887
dacA	D-Alanine carboxypeptidase	14.3	*pfv*; D-alanine carboxypeptidase IA; PBP; EC 3.4.12.11	494, 496, 2644, 2701, 2703, 3101, 4025, 4094, 4187, 4397, CG-886
dacB	D-Alanine carboxypeptidase	71.7	D-Alanine carboxypeptidase IB; PBP; EC 3.4.12.11	2239, 2702, 2929, 3885, 3948, 4194, CG-885
dacC	D-Alanine carboxypeptidase	19.0	PBP 6; EC 3.4.17.8	262, 496, 3318, CG-34706
dadA	D-Amino acid dehydrogenase	26.6	*dadR*; D-amino acid dehydrogenase subunit	2273, 2522, 4590–4593, CG-884
dadB	D-Amino acid dehydrogenase	1.6	*alnA*; D-amino acid dehydrogenase subunit	1265, CG-883
dadQ	D-Amino acid dehydrogenase	98.8	*alnR*; regulator of *dad* regulon	1265, CG-882
dadR	D-Amino acid dehydrogenase	26.6	*alnR*; regulatory gene	2522, 4593, CG-18433
dadX	D-Amino acid dehydrogenase	26.6	*msuA?*; alanine racemase; EC 5.1.1.1	2522, 4590, CG-17716
dam	DNA adenine methylase	75.7	DNA adenine methylase	133, 449, 493, 1544, 1564, 1999, 2302, 2323, 2520, 2760, 3349, 4436, 4674, CG-881
dapA	Diaminopimelate	55.8	Lysine biosynthesis; dihydrodipicolinate synthase; EC 4.2.1.52	439, 3540, 4266, CG-880
dapB	Diaminopimelate	0.6	Lysine biosynthesis; dihydrodipicolinate reductase; EC 1.3.1.26	438, 441, 442, 2600, 4787, CG-879
dapC	Diaminopimelate	4.0	Lysine biosynthesis; tetrahydropicolinate succinylase	540, CG-878
dapD	Diaminopimelate	4.0	Lysine biosynthesis; succinyl-diaminopimelate aminotransferase	277, 977, 978, 1296, 3539, 4402, CG-877
dapE	Diaminopimelate	55.6	*dapB*; lysine biosynthesis; N-succinyl-diaminopimelate deacylase	440, 3267, 3540, 4663, CG-876
dapF	Diaminopimelate	86.0	Lysine biosynthesis; diaminopimelate epimerase	787, 903, 3537, 3538, CG-17713
dbpA	DEAD-box protein	30.3	Putative member of DEAD box family of RNA helicases/ATPases; specifically activated by 23S rRNA	1300, 1864, 2045, 3791, CG-32058
dcd	dCTP deaminase	46.1	*paxA*; dCTP deaminase; EC 3.5.4.13	3056, 4504, CG-875
dcm	DNA cytosine methylation	43.8	*mec*; DNA cytosine methylase	309, 1280, 1583, 3998, CG-874
dcp	Dipeptidyl carboxypeptidase	35.0	Dipeptidyl carboxypeptidase II; EC 3.4.15.1	254, 1005, 1681, CG-873
dctA	Dicarboxylic acid transport	79.2	Transport system for dicarboxylic acids, C4 amino acids	3766, 3995, CG-872
dctB	Dicarboxylic acid transport	16.1	Transport system for dicarboxylic acids, C4 amino acids	3766, CG-871
dcuA	Dicarboxylate uptake	94.0	*genA*; membrane protein, C4-dicarboxylate transporter, anaerobic	272, 3956, 4182, 4651, CG-34476
dcuB	Dicarboxylate uptake	93.6	*genF*; membrane protein, C4-dicarboxylate transporter, anaerobic	272, 3956, CG-34479
ddlA	D-Alanine ligase	8.6	D-Alanine:D-alanine ligase, ADP forming	39, 4800, CG-30966
ddlB	D-Alanine ligase	2.2	*ddl*; D-alanine:D-alanine ligase	39, 1870, 2587, 2588, 3569, 4787, CG-870
deaD	DEAD box protein	71.1	*mssB*; dosage-dependent suppressor of temperature-sensitive *rpsB* mutations; putative RNA helicase; DEAD box motif	2045, 3296, 3791, 4291, CG-33472
def	Deformylase	73.9	*fms*; peptide deformylase, N-formylmethionylaminoacyl-tRNA deformylase; EC 3.4.11.-, EC 3.5.1.27	1520, 2731, 2781–2783, CG-33619
degP	Degradative protease	3.9	*htrA*; stress-response, sigma-E-dependent DegP serine protease, EC 3.4.99	632, 1296, 1731, 2488, 2490, 2491, 3455, 3855, 4111, 4682, CG-30554
del	Deletion	64.4	Affects frequency of IS1-mediated deletions	CG-869
deoA	Deoxyribose	99.5	*tpp-75*; thymidine phosphorylase; EC 2.4.2.4	AA, 20, 899, 1217, 3997, 4375–4377, 4380, 4493, CG-868
deoB	Deoxyribose	99.5	*drm, thyR, tlr*; deoxyribouratase (phosphopentomutase); EC 2.7.5.6	20, 1217, 2529, 3577, 4374, 4375, 4377, CG-867
deoC	Deoxyribose	99.5	*dra, thyR, tlr*; deoxyribose-P aldolase; EC 4.1.2.4	47, 898, 899, 1217, 2529, 3577, 3997, 4373, 4374, 4377, 4380, CG-866
deoD	Deoxyribose	99.5	*pup*; purine-nucleoside phosphorylase; EC 2.4.2.1	20, 1217, 1688, 1701, 2337, 3577, 4377, CG-865
deoR	Deoxyribose	19.0	*nucR, tsc, nupG*; regulatory gene for *deo* operon	2926, 2951, 3922, 4378, CG-864

(Table continues)

TABLE 1 *Continued*

Symbol	Mnemonic[b] for symbol	Position (min)[c]	Synonyms; enzyme or phenotype affected[d]	References[e]
dfp	DNA synthesis flavoprotein	82.1	*dnaS, dut;* flavoprotein affecting DNA and pantothenate metabolism	548, 2574, 4022, 4023, CG-18430
dgd	D-Galactose dehydrogenase	(71.7)	D-Galactose dehydrogenase production	4677, CG-863
dgkA	Diglyceride kinase	91.6	Diglyceride kinase	372, 2459, 2460, 2537, CG-862
dgkR	Diglyceride kinase	93.4	Level of diglyceride kinase	3461, CG-861
dgoA	D-Galactonate	83.4	2-Oxo-3-deoxygalactonate 6-P aldolase; EC 4.1.2.21	169, CG-36891
dgoD	D-Galactonate	83.4	Galactonate dehydratase; EC 4.2.1.6	169, CG-859
dgoK	D-Galactonate	83.4	2-Oxo-3-deoxygalactonate kinase; EC 2.7.1.58	169, CG-858
dgoR	D-Galactonate	83.4	Regulatory	169, 808, CG-857
dgoT	D-Galactonate	83.3	Galactonate transport	169, CG-856
dgsA	D-Glucosamine	35.0	Affects enzyme IIA/B of PTS	2922, 3579, CG-855
dgt	dGTP	3.9	*optA;* dGTP triphosphohydrolase; EC 3.1.5.1	977, 1296, 3455, 3456, 3863, 4682, CG-30546
dicA	Division control	35.4	Regulatory gene for *dicB*	264, 265, CG-18427
dicB	Division control	35.5	Inhibition of cell division	213, 264, 265, 589, CG-18424
dicC	Division control	35.4	Regulatory of *dicB*	264, 265, CG-18421
dicF	Division control	35.5	Small RNA that inhibits cell division of Kim prophage; DicF antisense RNA	430, 1177, 1178, 4245, CG-32240
dif	Deletion-induced filamentation	34.2	Recombination site in terminus	354, 356, 754, 812, 2272, 2410, 2548, 4233
dinD	Damage inducible	82.2	*pcsA;* repair enzyme	548, 2575, 3167, 3401, 3403, CG-33582
dinF	Damage inducible	91.7	Repair enzyme, induced by UV and mitomycin; subject to *recA* and *lex* regulation	372, 2149, 2267, 2840, CG-854
dinG	Damage inducible	18.0	Repair enzyme; SOS-related damage-inducible gene	2235, 2426, 2427, 3165, CG-31247
dinY	Damage inducible	41.8	Repair enzyme; damage-inducible in *lexA* (Def) background	3308, CG-36880
dipZ	Disulfide isomerase	94.0	*cycZ, dsbD, cutA2;* may be involved in cytochrome maturation; see *ccm* genes	251, 850, 3694, CG-34213
dksA	*dnaK* suppressor	3.4	*msmA;* multicopy suppresses *mukB* mutant, suppresses temperature-sensitive growth and filamentation of *dnaK* mutant	1296, 2070, 2128, CG-30521
dld	D-Lactate dehydrogenase	47.8	*ldh;* D-lactate dehydrogenase, FAD enzyme; EC 1.1.1.28	594, 2018, 3640, 3891, 4326, CG-852
dmsA	DMSO reductase	20.3	Anaerobic DMSO reductase subunit A	329, 330, 2790, 3615, 4310, CG-17710
dmsB	DMSO reductase	20.4	Anaerobic DMSO reductase (probable Fe-S binding) subunit	329, 591, 3615, CG-31724
dmsC	DMSO reductase	20.4	Anaerobic DMSO reductase membrane-bound subunit C	329, 3615, 4552, CG-31727
dnaA	DNA	83.6	DNA biosynthesis initiation binding protein	192, 548, 1259, 1374, 1587–1590, 1592, 1758, 1832, 1843, 2170, 2196, 2521, 2692, 2816, 2841, 2955, 3299, 3451, 3452, 3684, 3687, 3960, 4506, 4566, 4768, 4782, 4784, 4785, 4838, CG-851
dnaB	DNA	91.8	*groP, grpA;* DNA biosynthesis, chain elongation	372, 641, 1376, 1526, 2460, 2527, 3022, 3242, 3833, CG-850
dnaC	DNA	99.1	*dnaD;* DNA biosynthesis, initiation, and chain elongation	H, 2234, 2685, 2686, 3023, 3620, CG-849
dnaE	DNA	4.5	*polC, sdgC* (suppressor of *dnaG* mutation); DNA polymerase III, α subunit	H, 277, 401, 527, 1203, 1257, 2444, 2897, 3144, 3895, 4280, 4565, 4621, CG-373
dnaG	DNA	69.1	*dnaP, parB, sdgA;* primer synthesis for initiation of leading- and lagging-strand synthesis; primase	67, 557, 1494, 2111, 2584–2586, 2958, 3009, 3023, 3122, 3620, 3970, 4050, 4232, 4428, 4636, CG-847
dnaI	DNA	40.3	DNA biosynthesis	CG-846
dnaJ	DNA	0.3	*groP, grpC;* stress-related DNA biosynthesis, responsive to heat shock; chaperone, with *dnaK*	218, 1315, 1376, 1449, 1898, 1919, 2452, 2453, 3160–3162, 3678, 3902, 4112, 4476, 4586, 4587, 4589, 4787, CG-845
dnaK	DNA	0.3	*gro, groP, groPAB, groPC, groPF, grpC, grpF, seg;* stress-related DNA biosynthesis, responsive to heat shock proteins, autoregulated; Hsp70-type chaperone, with *dnaJ*	214, 410, 539, 1164, 1315, 1449, 1898, 1919, 2069, 2336, 2451–2454, 2741, 2742, 2885, 3160, 3162, 3241, 3681, 3902, 4112, 4180, 4476, 4586, 4587, 4589, 4787, CG-844
dnaL	DNA	28.9	DNA biosynthesis	3864a, CG-843

TABLE 1 *Continued*

Symbol	Mnemonic[b] for symbol	Position (min)[c]	Synonyms; enzyme or phenotype affected[d]	References[e]
dnaN	DNA	83.6	*lexA* regulon; DNA polymerase III holoenzyme, β subunit	10, 121, 122, 357, 545, 548, 1588, 2170, 3144, 3166, 3497, 3498, 3683, 3684, 3687, 4179, 4782, CG-842
dnaQ	DNA	5.1	*mutD; lexA* regulon; DNA polymerase III ε subunit	833, 834, 1023, 1094, 1661, 1777, 2632, 2683, 3116, 3144, 3771, 3931, 4185, CG-840
dnaT	DNA	99.1	DNA replication primosomal protein I	2685, 2686, 3023, CG-839
dnaX	DNA	10.6	*dnaZ*; τ and γ subunits of DNA polymerase III holoenzyme; DNA elongation factor III	374, 1228–1230, 1623, 1811, 2205, 2388, 2389, 2634, 2945, 3144, 3620, 4339–4341, 4758, CG-838
dniR	Dissimilatory nitrite reductase	5.1	Regulator for hexaheme nitrite reductase (cytochrome c_{552}) expression	2037, CG-30876
dppA	Dipeptide permease	79.8	*tpp?*; uptake of dipeptides	5, 951, 1415, 3199, 3285, 3995, CG-35111
dppB	Dipeptide permease	79.8	*tpp?*; uptake of dipeptides	5, 3285, 3984, 3995, 4425, CG-33771
dppC	Dipeptide permease	79.8	Uptake of dipeptides	5, 3295, 3984, 3995, 4425, CG-33768
dppD	Dipeptide permease	79.7	Uptake of dipeptides	3995, 4425, CG-33765
dppF	Dipeptide permease	79.7	Uptake of dipeptides	5, 3995, CG-33752
dppG	Dipeptide permease	14.0	*dpp;* first gene described as specific dipeptide permease; published as *dpp*	3285, CG-835
dps	DNA-binding protein, starvation	18.3	*pexB;* starvation-induced (stationary phase), membrane-associated DNA-binding protein (regulatory and protective effects, e.g., starvation-induced H_2O_2 resistance to H_2O_2 and oxidative damage)	66, 1243, 2534, 3114, CG-31650
dsbA	Disulfide bond	87.1	*iarA, ppfA;* disulfide oxidoreductase, periplasmic; EC 5.3.4.1	37, 216, 217, 268, 883, 2051, 2681, 3359, 3436, 3437, 4681, 4718, CG-34063
dsbB	Disulfide bond	26.5	*iarB;* DTT sensitivity phenotype	60, 216, 1960, 2876, 3334, CG-31933
dsbC	Disulfide bond	65.4	*xprA;* periplasmic protein affecting disulfide bond in periplasmic space; a disulfide oxidase	2554, 2877, 3899, CG-33355
dsdA	D-Serine deaminase	53.4[f]	D-Serine deaminase	W, 40, 411, 610, 883, 2656, 2753, 2754, 3250, CG-834
dsdC	D-Serine deaminase	53.4[f]	D-Serine regulatory gene	W, 40, 411, 610, 2753, 2877, 3250, CG-833
dsdX	D-Serine deaminase	53.4[f]	Unknown function	W, CG-35717
dut	dUTPase	82.1	*dnaS, sof;* deoxyuridine triphosphatase; EC 3.6.1.23	548, 2573, 2574, 4023, 4226, CG-832
dvl		(6.9)	Sensitivity to SDS and toluidine blue plus light	4482, CG-18418
e14		25.8	Defective prophage element; includes loci *sfiC, lit, pin, mcrA*	491, 492, 2077, 2617, 3342, 3343, 3480, 4394, CG-18409
ebgA	Evolved β-galactosidase	69.4	Growth on lactose; cryptic gene for phospho-β-D-galactosidase, α subunit	595, 1252, 1559, 4096, 4097, CG-830
ebgB	Evolved β-galactosidase	69.5	Cryptic permease; possibly homology with *lacY*	1559, 4097, CG-18415
ebgC	Evolved β-galactosidase	69.4	Cryptic gene for phospho-β-D-galactosidase, β subunit	1559, CG-18412
ebgR	Evolved β-galactosidase	69.4	Regulatory gene for *ebg* operon	1559, 4097, CG-829
ecfA	Energy-coupling factor	67.6	May be *metC*; has pleiotropic effects on coupling of active transport with metabolic energy	CG-828
eco	Ecotin	49.6	Ecotin, a serine protease inhibitor	1152, 2379, 2758, 2759, CG-32630
ecpD	*E. coli papD* homolog	3.3	Possible pilin chaperone	1296, 3473, CG-30513
eda	Entner-Doudoroff aldolase	41.6	*kdgA, kga;* growth on gluconate; 2-keto-3-deoxygluconate 6-P aldolase 2-keto-4-hydroxyglutarate aldolase; EC 4.1.2.14	612, 805, 1101, 1258, 3282, 4439, 4440, CG-826
edd	Entner-Doudoroff aldolase	41.6	Phosphogluconate dehydratase; EC 4.2.1.12	612, 805, 1101, 1256, 1258, CG-825
efp	Elongation factor P	94.3	Elongation factor P, prokaryotic	111, CG-34470
emrA	E-multidrug resistance	60.5	Sensitivity to uncouplers; multidrug resistance pump, homologous to facilitators of proton motive force-dependent translocase; homology to HlyD protein family	2424, 2532, 2533, CG-33259
emrB	E-multidrug resistance	60.6	Membrane protein involved with multidrug resistance	2532, 2533, CG-33262
emrD	E-multidrug resistance	83	Adaptation to low energy shock; putative multidrug resistance pump	548, 3030, CG-36938
emrE	E-multidrug resistance	12.2	*mvrC, envB, mon, rodY;* multidrug resistance; effects on envelope, cell shape, and methylviologen sensitivity	2911, 3443, 4752, CG-36935
endA	Endonuclease	66.5	DNA-specific endonuclease I	1972, 4657, CG-824

(Table continues)

TABLE 1 *Continued*

Symbol	Mnemonic[b] for symbol	Position (min)[c]	Synonyms; enzyme or phenotype affected[d]	References[e]
eno	Enolase	62.6	Enolase; EC 4.2.1.11	148, 4567, CG-823
entA	Enterochelin	13.5	2,3-Dihydro-2,3-dihydroxybenzoate dehydrogenase	1222, 2310, 2311, 2500, 2986, 2987, CG-822
entB	Enterochelin	13.5	2,3-Dihydro-2,3-dihydroxybenzoate synthetase	1222, 2310, 2311, 2500, 2986, 2987, 3235, CG-821
entC	Enterochelin	13.5	Isochorismate synthetase	480, 1222, 1510, 2310, 2311, 2502, 2986, 2987, 3235, 3236, 3728, CG-820
entD	Enterochelin	13.1	Enterochelin synthetase, component D	124, 762, 1223, 1497, 2310, 2311, CG-819
entE	Enterochelin	13.5	Enterochelin synthetase, component E	1222, 2310, 2311, 2500, 2986, 2987, 3728, 4043, CG-818
entF	Enterochelin	13.2	Enterochelin synthetase, component F	1222, 2310, 2311, 3311, 3312, 3643, CG-817
envN	Envelope	(4.2)	Affects envelope; defects osmotically remedied	1100, CG-811
envP	Envelope	90.4	Affects envelope; defects osmotically remedied	1100, CG-810
envQ	Envelope	60.7	Affects envelope; defects osmotically remedied	1100, CG-809
envT	Envelope	(14.2)	Affects envelope; defects osmotically remedied	CG-808
envY	Envelope	12.6	Envelope protein; thermoregulation of porin biosynthesis	2578, 2579, CG-18406
envZ	Envelope	76.1	*ompB, perA, tpo;* regulatory gene for production of outer membrane proteins; inner membrane osmosensor protein	633, 790, 791, 855, 1337, 1338, 1446, 1565, 1566, 2462, 2576, 2891–2894, 3562, 3641, 4276, 4346, 4496, 4516, 4531, 4683, 4741, CG-807
epd	Erythrose-4-P dehydrogenase	66.1	*gapB;* erythrose-4-P dehydrogenase	54, 1049, 4816, CG-32089
era	*E. coli ras*-like	58.0	*sdgE* (suppressor of *dnaG* mutation); essential gene; GTP-binding protein	H, 22, 683, 1414, 2407, 2658, 4004, 4372, CG-29010
esp	Efficiency site for phage	17.3	Site for efficient packaging of phage T1	CG-805
evgA	*E. coli* homolog of virulence gene	53.5	Environmentally responsive two-component regulatory system; in multicopy, they induce *ompC* expression	4367, CG-32763
evgS	*E. coli* homolog of virulence gene	53.5	Environmentally responsive two-component regulatory system; in multicopy, they induce *ompC* expression	4367, CG-32766
exbB	Export	67.8	Uptake of enterochelin; colicin sensitivity or resistance; cytoplasmic membrane protein involved in transport across outer membrane; homology to TolQ	462, 467, 1110, 1111, 1215, 1536, 2057, 2082, 2162, 3441, 3958, CG-804
exbC	Export	61.2	Expression of export proteins; determines colicin sensitivity	1536, 3441, CG-803
exbD	Export	67.8	Expression of export proteins; cytoplasmic membrane protein involved in transport across outer membrane; homology to TolR	467, 1110, 1215, 1679, 2055, 2162, CG-33384
expA	Export	22.2	Expression of export proteins	919, CG-802
exuR	Hexuronate	69.6	Negative regulator for *exu* regulon (*exuT, uxaAC, uxuB*)	1820, 1822, 2697, 3377, 3554, 3559, CG-801
exuT	Hexuronate	69.6	Transport of hexuronates; *exu* regulon member	1820, 2697, 2698, 3377, CG-800
fabA	Fatty acid biosynthesis	21.9	β-Hydroxydecanoylthioester dehydrase; EC 4.2.1.60	845, 846, 1685, CG-799
fabB	Fatty acid biosynthesis	52.6	*fabC;* β-ketoacyl-ACP synthase I; EC 2.3.1.41	139, 1341, 1342, 2122, 3934, CG-798
fabD	Fatty acid biosynthesis	24.8	Malonyl-CoA-ACP transacylase; EC 2.3.1.39	1341, 2615, 3491, 4429, CG-797
fabF	Fatty acid biosynthesis	24.9	*cvc, fabJ, vtr;* β-ketoacyl-ACP synthase II; EC 2.3.1.41	785, 1341, 1342, 2613, 3935, CG-795
fabG	Fatty acid biosynthesis	24.8	Reuse of former symbol for *accC*; 3-ketoacyl-ACP reductase; EC 1.1.1.100	2230, 3158, 3491, CG-31865
fabH	Fatty acid biosynthesis	24.8	ACP synthase III	3158, 4332, CG-31860
fabI	Fatty acid biosynthesis	29.0	*envM;* enoyl-ACP reductase, NADH dependent; EC 1.3.1.9	295, 296, CG-812
fabZ	Fatty acid biosynthesis	4.4	May be *sefA*?; sequence similarity to *fabA*, probably hydroxymyristoyl-ACP dehydrase	778, 1243, 2899a, CG-30597
fadA	Fatty acid degradation	86.7	*oldA;* thiolase I; EC 2.3.1.16	736, 903, 1030, 2995, 4026, 4732, 4733, 4735, 4736, CG-794
fadB	Fatty acid degradation	86.7	*oldB;* multifunctional polypeptide; 3-hydroxyacyl-CoA dehydrogenase:3-hydroxyacyl-CoA epimerase, Δ³-cis-Δ²-*trans* enoyl-CoA isomerase; enoyl-CoA hydratase; fatty acid oxidation complex, α subunit; EC 1.1.1.35, EC 4.2.1.17, EC 5.1.2.3, EC 5.3.3.8	903, 1030, 2995, 4007, 4026, 4732, 4734–4736, 4740, CG-793
fadD	Fatty acid degradation	40.6	Acyl-CoA synthetase; EC 6.2.1.3	350, 1299, CG-792
fadE	Fatty acid degradation	4.8	Electron transport flavoprotein of β-oxidation	736, CG-791
fadL	Fatty acid degradation	53.0	*ttr;* sensitivity to phage T2; fatty acid transport protein, outer membrane	348, 349, 351, 1604, 2916, 3670, CG-790

TABLE 1 *Continued*

Symbol	Mnemonic[b] for symbol	Position (min)[c]	Synonyms; enzyme or phenotype affected[d]	References[e]
fadR	Fatty acid degradation	26.5	*dec, ole, thdB;* negative regulator of *fad* regulon, positive regulator of *fabA*	3, 736, 1029, 1031, 1032, 1685, 1693, 2648, 3946, 3947, 4590, CG-789
farR	Fatty acid responsive regulatory	16.5	Fatty acid/fatty acyl CoA-responsive DNA-binding protein; may modulate citric acid cycle	530, 3449a, CG-36837
fatA	Fatty acid	71.7	Utilization of *trans* unsaturated fatty acids	1006, CG-18403
fba	Fructose bisphosphate aldolase	66.1	*ald, fda;* fructose bisphosphate aldolase	53, 54, 299, 3950, CG-786
fbp	Fructose bisphosphate phosphatase	96.0	*fdp;* fructose bisphosphatase; EC 3.1.3.11	1576, 2199, 2662, 3844, CG-784
fcsA		86.9	Affects cell division and septation	2271a, CG-787
fdhD	Formate dehydrogenase	88.0	Affects formate dehydrogenase-N, homologous to *fdnC* and *fdnB* of *Salmonella* spp.	2650, 3359, 3783, CG-33972
fdhE	Formate dehydrogenase	87.8	Affects formate dehydrogenase-N	2650, 3359, 3783, 3784, CG-33978
fdhF	Formate dehydrogenase	92.5	Formate dehydrogenase-N selenopolypeptide subunit; EC 1.2.2.1; part of formate hydrogen-lyase complex	164, 165, 336, 372, 668, 3289, 4670, 4671, 4831, 4832, CG-18400
fdnG	Formate dehydrogenase-N	33.3	Formate oxidation coupled to nitrate reduction; major anaerobic respiratory path; formate dehydrogenase-N major subunit	287–289, 3458, 3548, 4078, 4081, CG-32160
fdnH	Formate dehydrogenase-N	33.3	Formate dehydrogenase-N iron-sulfur subunit; nitrate inducible	287–289, 3458, 3548, 4078, 4081, CG-32164
fdnI	Formate dehydrogenase-N	33.4	Formate dehydrogenase-N cytochrome subunit	287–289, 3458, 3548, 4078, 4081, CG-32168
fdoG	Formate dehydrogenase-O	87.9	Formate dehydrogenase-O major subunit	3359, CG-33985
fdoH	Formate dehydrogenase-O	87.9	Formate dehydrogenase-O iron-sulfur subunit	3359, CG-33988
fdoI	Formate dehydrogenase-O	87.9	Formate dehydrogenase-O cytochrome b_{556} subunit	3359, CG-33991
fdx	Ferredoxin	57.0	Ferredoxin	2129, 2198, 4172, 4173, CG-32984
fecA	Iron (citrate dependent)	97.2	Citrate-dependent iron transport, outer membrane receptor	1595, 1840, 3422, 4403, 4420, 4477, 4829, CG-783
fecB	Iron (citrate dependent)	97.2	Citrate-dependent iron transport, periplasmic protein	1840, 3422, 4052, 4420, 4829, CG-782
fecC	Iron (citrate dependent)	97.2	Citrate-dependent iron transport	4052, 4403, 4420, CG-35620
fecD	Iron (citrate dependent)	97.2	Citrate-dependent iron transport, membrane protein	3422, 4052, 4420, CG-18394
fecE	Iron (citrate dependent)	97.2	Citrate-dependent iron transport	4052, 4403, 4420, CG-35617
fecI	Iron (citrate dependent)	97.3	Regulatory gene mediating induction by iron	2536, 4403, 4420, CG-35630
fecR	Iron (citrate dependent)	97.3	Regulatory gene mediating induction by iron	4403, 4420, CG-35627
feoA	Ferrous iron transport	76.2	Ferrous iron transport	2053, CG-28964
feoB	Ferrous iron	76.2	Membrane protein, ferrous iron transport	2053, CG-28967
fepA	Ferrienterobactin permease	13.1	*cbr, cbt;* outer membrane component of ferribactin transport system	123, 1223, 1595, 1830, 2580, 3236, 3311, 3326, 3602, CG-18388
fepB	Ferrienterobactin permease	13.4	Periplasmic component of ferrienterobactin transport system	465, 480, 690, 1118, 1510, 3236, 3325, 3326, 3602, CG-18385
fepC	Ferrienterobactin permease	13.3	Cytoplasmic membrane component ferrienterobactin transport	690, 3236, 3325, 3892, CG-4987
fepD	Ferrienterobactin permease	13.4	Ferrienterobactin permease, membrane-bound subunit	690, 1830, 3236, 3892, CG-18382
fepE	Ferrienterobactin permease	13.3	Ferric enterobactin (enterochelin) uptake	3236, 3643, CG-18379
fepG	Ferrienterobactin permease	13.4	Ferrienterobactin permease, membrane-bound subunit	690, 3892, CG-31162
fes	Iron	13.2	Enterochelin esterase	479, 1223, 1830, 2310, 2311, 3311, 3312, CG-780
fexB	F-exclusion	86.5	Affects ArcA phenotype	2408, CG-778
ffh	54-kDa homolog	59.0	Homolog of mammalian signal recognition particle 54-kDa protein; functions in protein export	238, 298, 510, 568, 2568, 2658, 3316, 3376, 3532, 3593, CG-33006
ffs	4.5S	10.3	4.5S RNA; component of RNP homologous to mammalian signal recognition particle	238, 513, 1804, 1894, 2227, 3290, 3532, CG-18373
fhlA	Formate hydrogen-lyase	61.4	Transcription factor, formate hydrogen-lyase system, global formate regulator, activator for *fdhF, hyc,* and *hyp* operons	1770, 1771, 1930, 2718, 3613, 3721, 3722, 3777, 3779, CG-18370
fhlB	Formate hydrogen-lyase	95	Regulatory gene, transcriptionally activated by FhlA	2718, CG-36851
fhuA	Ferric hydroxamate uptake	3.6	*T1, T5rec, tonA;* outer membrane protein receptor for ferrichrome, colicin M, and phages T1, T5, and φ80	465, 466, 468, 547, 826, 827, 1184, 1296, 1595, 2029, 2160–2162, 2247, 2563, 2564, CG-777
fhuB	Ferric hydroxamate uptake	3.7	Hydroxamate-dependent iron uptake, cytoplasmic membrane component	465, 547, 1184, 1296, 2029, 2245, 2246, 3430, 3817, CG-776
fhuC	Ferric hydroxamate uptake	3.7	Hydroxamate-dependent iron uptake, cytoplasmic membrane component	465, 547, 825, 1184, 1296, 3817, CG-11187
fhuD	Ferric hydroxamate uptake	3.7	Hydroxamate-dependent iron uptake, cytoplasmic membrane component	465, 547, 825, 1184, 1296, 2247, CG-11184

(Table continues)

TABLE 1 *Continued*

Symbol	Mnemonic[b] for symbol	Position (min)[c]	Synonyms; enzyme or phenotype affected[d]	References[e]
fhuE	Ferric hydroxamate uptake	25.0	Iron uptake; outer membrane receptor for ferric-rhodotorulic acid	1594, 1595, 3749, 3750, CG-18367
fhuF	Ferric hydroxamate uptake	99.0	Ferric hydroxamate transport	1596, CG-18364
fic	Filamentation-cAMP	75.1	Cell division; filamentation in presence of cAMP	1624, 2127, 2219, 4304, 4368, CG-18361
fimA	Fimbriae	97.9	*fimD, pilA;* fimbrin (pilin) type 1, major structural subunit	6, 7, 375, 1115, 1269, 1270, 2186, 2191, 3206, 3819, 4012, CG-18358
fimB	Fimbriae	97.8	*pil;* recombinase?; regulatory gene for *fimA*	375, 1053, 1115, 2187, 2191, 2743, 3819, CG-18355
fimC	Fimbriae	97.9	*pil, pilB;* biosynthesis of fimbriae; periplasmic chaperone for type 1 fimbriae	R, 375, 2002, 2188, 2191, CG-18352
fimD	Fimbriae	97.9	*pil, pilC;* biosynthesis (export and assembly) of type 1 fimbrial outer membrane protein	375, 2190, 2191, CG-18349
fimE	Fimbriae	97.9	*pilH;* recombinase?; regulatory gene for expression of *fimA*	375, 1053, 1115, 1312, 2187, 2743, 3819, CG-18346
fimF	Fimbriae	98.0	*pilD;* fimbrial morphology; fimbrin type 1 minor component	375, 2189, 2190, 3652, CG-18343
fimG	Fimbriae	98.0	*pilD;* pilus length; perhaps inhibits polymerization; fimbrin type 1 minor component	375, 2189, 3652, CG-18340
fimH	Fimbriae	98.0	*pilE;* major fimbrial subunit, membrane-specific adhesin	375, 1610, 2189, 2263, CG-18337
fimI	Fimbriae	97.9	Fimbria-related	2186, CG-35633
fipB	F1 phage	86.0	Morphogenesis of phage F1	2542, CG-18334
fipC	F1 phage	(74.9)	Morphogenesis of phage F1	2542, CG-18331
fis	Factor for inversion stimulation	73.4	DNA-binding protein that bends DNA, causing site-specific DNA inversion plus direct interaction with RNA polymerase, stimulating rRNA operons and ribosome synthesis	158, 196, 303, 420, 1205, 1399, 1424, 1995, 2203, 2249, 3063, 3089, 3092, 3216, 3609, 4130, 4422, 4554, 4690, 4781, 4791, CG-18328
fiu	Ferric iron uptake	18.1	Outer membrane protein, ferric iron uptake	1594, CG-18325
fkpA	FK506-binding protein	74.8	Protein that binds immunosuppressive drug FK506; *mip*-like protein	1779, CG-36042
fldA	Flavodoxin	15.4	Flavodoxin	3211, CG-31496
flgA	Flagella	24.4	*flaU;* flagellar synthesis; flagellar-regulon member	2222, 2604a, CG-756
flgB	Flagella	24.4	*flbA;* flagellar synthesis; flagellar-regulon member	2604a, CG-750
flgC	Flagella	24.4	*flaW;* flagellar synthesis; flagellar-regulon member; basal-body protein	2604a, CG-754
flgD	Flagella	24.4	*flaV;* flagellar synthesis; flagellar-regulon member; basal-body rod modification	2604a, CG-755
flgE	Flagella	24.5	*flaK;* flagellar-regulon member; flagellar-hook subunit protein	2604, 2604a, CG-766
flgF	Flagella	24.5	*flaX;* flagellar synthesis; flagellar-regulon member; basal-body rod protein	2604a, CG-753
flgG	Flagella	24.5	*flaL;* flagellar synthesis; flagellar-regulon member; basal-body rod protein	2604a, CG-765
flgH	Flagella	24.5	*flaY;* flagellar synthesis; basal-body L-ring protein	2003, 2604, 2604a, CG-752
flgI	Flagella	24.5	*flaM;* flagellar regulon member; basal-body P-ring protein	2003, 2604a, CG-764
flgJ	Flagella	24.6	*flaZ;* flagellar synthesis; flagellar-regulon member	2604, 2604a, CG-751
flgK	Flagella	24.6	*flaS;* flagellar synthesis; flagellar-regulon member; hook-associated protein 1	2604, 2604a, CG-758
flgL	Flagella	24.6	*flaT;* flagellar synthesis; flagellar-regulon member; hook-associated protein 1	620, 2604a, CG-757
flhA	Flagella	42.4	*flaH;* flagellar synthesis; flagellar-regulon member	2604a, CG-768
flhB	Flagella	42.4	*flaG;* flagellar synthesis; flagellar-regulon member	2604a, CG-769
flhC	Flagella	42.6	*flaI;* flagellar synthesis; regulatory gene	233, 2507, 2604a, CG-767
flhD	Flagella	42.6	*flbB;* regulatory gene, flagellum-specific σ factor, transcriptional activator of Fla class II operons	233, 2507, 2604a, CG-749
fliA	Flagella	43.1	*flaD, rpoF;* regulation of late gene expression; RNA polymerase sigma transcription factor for Fla class 3a and 3b operons (flagellar regulon)	I, 1900, 2033, 2222, 2535, 2604a, CG-771
fliC	Flagella	43.2	*flaF, hag;* flagellar-regulon member; flagellin, structural gene	1580, 2296, 4168, 2604a, CG-649
fliD	Flagella	43.2	*flbC;* flagellar-regulon member; member of axial family of structural proteins; hook-associated protein 2	1580, 2123, 2604a, 4168, CG-748
fliE	Flagella	43.4	*flaN;* flagellar synthesis; basal-body component	2604a, 2941, 3464, CG-763
fliF	Flagella	43.4	*flaBI;* flagellar basal-body M-ring protein	234, 2221, 2223, 2604a, CG-17707

TABLE 1 *Continued*

Symbol	Mnemonic[b] for symbol	Position (min)[c]	Synonyms; enzyme or phenotype affected[d]	References[e]
fliG	Flagella	43.4	*flaBII;* motor switching and energizing	234, 2221, 2223, 2604a, 3588, CG-18322
fliH	Flagella	43.5	*flaBIII;* flagellar biosynthesis	234, 2221, 2223, 2604a, CG-18319
fliI	Flagella	43.5	*flaC;* flagellar biosynthesis	234, 2604a, CG-772
fliJ	Flagella	43.5	*flaO;* flagellar biosynthesis	234, 2604a, CG-762
fliK	Flagella	43.5	*flaE;* hook filament junction; controls hook length	234, 2604a, CG-770
fliL	Flagella	43.5	*flaAI;* flagellar biosynthesis	234, 2286, 2604a, 2641, CG-18316
fliM	Flagella	43.5	*cheC, flaA;* flagellar synthesis; motor switching and energizing	750, 2286, 2604a, 2641, CG-774
fliN	Flagella	43.6	*motD;* flagellar switch protein; motor switching and energizing	2604a, 2641, CG-18313
fliO	Flagella	43.6	*fibD;* flagellar synthesis; flagellar-regulon member	2604a, 2640, 2641, CG-17510
fliP	Flagella	43.6	*flaR;* flagellar synthesis; flagellar-regulon member	2604a, 2640, 2641, CG-759
fliQ	Flagella	43.6	*flaQ* flagellar synthesis; flagellar-regulon member	2604a, 2640, 2641, CG-760
fliR	Flagella	43.6	*flaP;* flagellar synthesis; flagellar-regulon member	2604a, 2640, 2641, CG-761
fliS	Flagella	43.2	Flagellar synthesis; flagellar-regulon member	2123, 2604a, CG-30736
fliT	Flagella	43.2	Flagellar synthesis; flagellar-regulon member	2123, 2604a, CG-30739
flu	Flagella	44.5	Affects surface properties, piliation, and colonial morphology; unstable gene	2604a, CG-746
fmt	fMet-tRNA formyltransferase	73.9	Methionyl-tRNA formyltransferase; EC 2.1.2.9	1519, 1520, 2775, 2783, CG-33616
fnr	Fumarate-nitrate-reductase	30.1	*frdB, nirA, nirR;* regulatory gene for nitrite and nitrate reductases, hydrogenase, and fumarate reductase	102, 269, 270, 823, 1138, 1289, 1459, 1618, 1678, 2471, 2790, 3756, 3764, 3888, 3889, 4020, 4021, 4303, CG-745
focA	Formate channel	20.6	Membrane protein	3754, 4148, CG-31732
folA	Folate	1.1	*tmrA;* trimethoprim resistance; dihydrofolate reductase; EC 1.5.1.3	97, 171, 278, 358, 1224, 1935, 3596, 3893, 3975–3977, 4098, 4787, CG-744
folC	Folate	52.4	Dihydrofolate:folylpolyglutamate synthetase; EC 6.3.2.12	382, 2151, 2169, 3118, 3448, CG-17704
folD	Folate	12.0	*ads;* methenyltetrahydrofolate dehydrogenase/cyclohydrolase; EC 1.5.1.5, EC 3.5.4.9	867, 1026, CG-31127
folE	Folate	48.3	GTP cyclohydrolase I	2121, 3789, 3806, CG-32718
folK	Folate	3.4	Dihydro-hydroxymethylpterin pyrophosphokinase	1296, 2504, 4197, 4198, CG-29572
folP	Folate	71.5	Dihydropteroate synthase; EC 2.5.1.15	4197, CG-29566
fpr	Ferredoxin reductase	88.6	*mvrA;* methylviologen (paraquat) resistance; member of anaerobic ribonucleotide reductase system, *soxR* regulon; ferredoxin NADP+ reductase, anaerobic	319, 2910, 3359, 4325, CG-18127
frdA	Fumarate reductase	94.4	Fumarate reductase, anaerobic, flavoprotein subunit; EC 1.3.99.1	373, 766, 770–773, 1097, 1499, 2005, 2528, 3813, CG-742
frdB	Fumarate reductase	94.4	Fumarate reductase, anaerobic, iron-sulfur protein subunit; EC 1.3.99.1	770, 771, 773, 1499, 2005, 2528, 4569, CG-741
frdC	Fumarate reductase	94.3	Fumarate reductase membrane anchor polypeptide; EC 1.3.99.1	771, 1499, 2005, 2401, CG-740
frdD	Fumarate reductase	94.3	Fumarate reductase membrane anchor polypeptide; EC 1.3.99.1	771, 1499, 2005, 2401, CG-739
frr	Factor for ribosome recycling (release)	4.1	Essential gene; dissociates ribosomes from mRNA after termination of translation; ribosome recycling factor	1296, 1852, 1853, 1961, 3912, 4722, CG-30590
fruA	Fructose	48.6	*ptsF;* fructosephosphotransferase enzyme II	1147, 1357, 3424, CG-350
fruB	Fructose	48.7	*fruF;* fructosephosphotransferase enzyme III	1357, 2241, 3204, 3522, CG-17515
fruK	Fructose	48.7	*fpk, fruF;* fructose-1-P kinase; EC 2.7.1.3	403, 1357, 2839, 3204, CG-743
fruR	Fructose	1.9	*fruC, shl;* regulatory gene; possibly repressor of *fru* operon	1357, 1957, 2242, 2366, 4787, CG-18304
ftn	Ferritin	42.9	*gen-165, rsgA;* ferritin	1812, 1943, CG-32521
ftsA	Filamentation temperature sensitivity	2.2	*divA;* cell division, septation	49, 260, 410, 806, 880, 954, 1002, 1008, 1009, 1352, 2000, 2587, 2588, 2692, 3567, 3568, 4132, 4500, 4754, 4755, 4787, CG-738
ftsE	Filamentation temperature sensitivity	77.6	Cell division; ATP-binding protein	1395, 3691, 3995, CG-736
ftsI	Filamentation temperature sensitivity	2.0	*pbpB, sep;* essential cell division gene; septum peptidoglycan synthetase; PBP 3	260, 444, 1322, 1788, 2978, 3006, 4218, 4356, 4787, CG-423
ftsJ	Filamentation temperature sensitivity	71.6	Affects cell division and growth; heat inducible	38, 207, 1690, 3155, 4289, 4290, CG-33511

(Table continues)

TABLE 1   *Continued*

Symbol	Mnemonic[b] for symbol	Position (min)[c]	Synonyms; enzyme or phenotype affected[d]	References[e]
ftsL	Filamentation temperature sensitivity	2.0	Essential gene; affects cell division and growth, growth of wall at septum; cytoplasmic membrane protein	227, 1545, 1906, 2022, 3006, 4356, CG-30402
ftsN	Filamentation temperature sensitivity	88.7	*msgA;* essential gene; cell division and growth in multicopy suppresses *ftsA12* mutation	882, 3359, CG-34153
ftsQ	Filamentation temperature sensitivity	2.2	Cell division; growth of wall at septum	49, 260, 261, 1002, 1050, 2692, 3025, 3568, 3569, 4101, 4755, 4787, CG-734
ftsW	Filamentation temperature sensitivity	2.1	Cytoplasmic membrane protein required for expression of PBP 2; homology to *rodA*	1868, 1870, 2153, 4787, CG-30409
ftsX	Filamentation temperature sensitivity	77.6	*ftsS;* cell division	1395, 3691, 3995, CG-18298
ftsY	Filamentation temperature sensitivity	77.6	Cell division	238, 1395, 1396, 2569, 2658, 3995, CG-18295
ftsZ	Filamentation temperature sensitivity	2.3	*sfiB, sulB;* cell division and growth; appears to be required for initiation of cell septation; GTP-binding protein	49, 246, 260, 313, 315–317, 806, 879–881, 945, 1002, 1008, 1339, 1383, 1540, 2000, 3025, 3338, 3495, 3628, 4132, 4244, 4754, 4755, 4787, CG-143
fucA	Fucose	63.1	*fucC;* L-fuculose-1-P aldolase	637, 686, 688, 804, 2559, 3963, 4825, CG-17701
fucI	Fucose	63.2	L-Fucose isomerase	686, 688, 2559, 4825, CG-10878
fucK	Fucose	63.2	L-Fuculose kinase; EC 2.7.1.51	410, 686, 688, 2559, 4825, CG-10881
fucO	Fucose	63.1	L-1,2-Propanediol oxidoreductase	686, 688, 804, 2559, 3868, 3963, 4825, CG-17698
fucP	Fucose	63.2	*prd;* fucose permease	637, 686, 688, 2559, 3963, 4825, CG-10875
fucR	Fucose	63.3	Positive regulatory protein for *fuc* regulon	688, 2559, 3962, 4825, CG-10884
fumA	Fumarate	36.3	Fumerase A, aerobic	1512, 1513, 2847, 2848, 4652, CG-18292
fumB	Fumarate	93.6	Regulatory gene?; fumarase B, anaerobic	272, 1512, 1513, CG-18289
fumC	Fumarate	36.2	Member of *soxRS* regulon; fumarase C	1512, 2487, 2848, 4651, 4652, CG-15273
fur	Ferric uptake regulation	15.4	Ferric iron uptake; negative regulatory gene	184, 185, 836, 955, 1595, 1618, 1678, 3760, 4215, CG-18286
fusA	Fusidic acid	74.7	*far;* fusidic acid resistance; protein chain elongation factor G	A, 41, 1786, 1990, 3221, 3387, 4447, 4760, 4805, CG-732
fusB	Fusidic acid	14.8	Fusidic acid resistance; pleiotropic effects on RNA synthesis and ribosomes, ribosomal protein S6	A, CG-731
gabC	GABA	60	Utilization of GABA as nitrogen source; regulatory for *gab*?	1060, 2821a, CG-730
gabD	GABA	60.1	Succinate-semialdehyde dehydrogenase, NADP dependent; EC 1.2.1.16	237, 3077, CG-729
gabP	GABA	60.2	Membrane protein GABA permease	3077, CG-728
gabT	GABA	60.1	Aminobutyrate aminotransferase; EC 2.6.1.19	236, 237, 3818, CG-727
gadA	Glutamate decarboxylase	78.9	*gadS*?; glutamate decarboxylase; EC 4.1.1.15	2655, 3974, 3995, CG-32191
gadB	Glutamate decarboxylase	33.8	Glutamate decarboxylase; EC 4.1.1.15	3974, 4766, CG-32194
gadR	Glutamate decarboxylase	79	Regulatory gene	2583, 2664, CG-726
galE	Galactose	17.1	UDP-galactose 4-epimerase; hexose-1-P uridylyltransferase; EC 2.7.7.12	25, 241, 558, 559, 972, 2349, 2400, CG-724
galK	Galactose	17.0	Galactokinase; EC 2.7.1.6	972, 3664, 3780, CG-723
galM	Galactose	17.0	Aldose-1-epimerase (mutarotase)	434, 2687, CG-35171
galP	Galactose	66.5	D-Galactose/H$^+$ symporter	3564, CG-722
galR	Galactose	64.1	Repressor of *galETK* operon; regulates low-affinity transport; Gal repressor	890, 1801, 2403, 2607, 3664, 4107, 4457, 4545, 4547, CG-721
galS	Galactoside	48.2	*mglD*?; utilization of methyl galactoside, represses *mgl*, induced *lac;* negative regulator of high-affinity transport; Gal isorepressor	2943, 4546, 4548, CG-499
galT	Galactose	17.0	*galB;* galactose-1-P uridylyltransferase; EC 2.7.7.10	815, 972, 2400, 2403, CG-720
galU	Galactose	27.8	Regulatory locus for low-affinity transport system; glucose-1-P uridylyltransferase; EC 2.7.7.9	1146, 1266, 1507, 2406, 3869, 4354, 4560, CG-719
gap	Glyceraldehyde phosphate dehydrogenase	40.1	*gad;* glyceraldehyde 3-P dehydrogenase; EC 1.2.1.12	458, 1049, 2353, 3049, CG-718
garA	Glucarate	15.9	D-Glucarate utilization	3561, CG-717
garB	Glucarate	3.2	D-Glucarate utilization	3561, CG-716
gatA	Galactitol	46.7	Galactitol-specific enzyme IIA of PTS	T, 406, 993, 2018, 2484, 2839, 3519, CG-715
gatB	Galactitol	46.7	Galactitol-specific enzyme IIB of PTS	T, CG-32559
gatC	Galactitol	46.7	Galactitol-specific enzyme IIC of PTS	T, 406, 2018, 2484, 2839, CG-714

TABLE 1 *Continued*

Symbol	Mnemonic[b] for symbol	Position (min)[c]	Synonyms; enzyme or phenotype affected[d]	References[e]
gatD	Galactitol	46.7	Galactitol-1-P dehydrogenase	T, 406, 2018, 2484, 2839, CG-713
gatR	Galactitol	46.7	Regulatory gene for *gat*	T, 2228, CG-32562
gcd	Glucose dehydrogenase	3.0	Inner membrane protein, glucose dehydrogenase (pyrroloquinoline-quinone); EC 1.1.99.17	755, 1296, 4700, 4705, CG-30502
gcl	Glyoxylate carboligase	11.5	Glyoxylate carboligase; EC 4.1.1.47	646, CG-31107
gcpE	Gene coding for protein (E)	56.7	Apparently essential gene for unknown function	190, 1268, CG-32909
gcvA	Glycine cleavage	63.3	Regulatory *trans*-acting gene effecting glycine-induced activation of glycine cleavage operon; may be activator in presence of glycine, repressor in presence of inosine	4646, 4617, CG-28676
gcvH	Glycine cleavage	65.6	Glycine cleavage, carrier of aminomethyl group	3182, 4055, 4056, 4058, CG-28664
gcvP	Glycine cleavage	65.6	Glycine dehydrogenase (decarboxylating); EC 1.4.4.2	3182, CG-28661
gcvT	Glycine cleavage	65.6	Aminomethyltransferase, tetrahydrofolate dependent; EC 2.1.2.10	3182, 4055, 4057, CG-28667
gdhA	Glutamate dehydrogenase	39.6	Glutamate dehydrogenase	248, 831, 1662, 2766, 3713, 4389, 4390, CG-712
gef	Gene expression fatal	0.4	Toxic membrane polypeptide, present as multimeric complex	3397–3400, 4293, 4787, CG-30252
ggt	γ-Glutamyltranspeptidase	77.2	γ-Glutamyltranspeptidase; EC 2.3.2.2	1617, 3995, 4155–4157, CG-18280
gidA	Glucose-inhibited division	84.5	Glucose effects on cell division (chromosome replication?)	144, 536, 548, 2199, 3150, 4127, 4484, CG-18277
gidB	Glucose-inhibited division	84.5	Glucose effects on cell division (chromosome replication?)	548, 2199, 3079, 3150, 4484, CG-18274
glc	Glycolate	67.4	Malate synthase G; EC 4.1.3.2	2901, CG-711
gldA	Glycerol dehydrogenase	89.1	Glycerol dehydrogenase; NAD$^+$ dependent	372, 4324, CG-34147
glgA	Glycogen	76.8	Glycogen synthase; EC 2.4.1.21	181, 2034, 2279, 2345, 3183, CG-710
glgB	Glycogen	76.9	1,4-α-Glucan branching enzyme; EC 2.4.1.18	182, 2345, 3591, CG-709
glgC	Glycogen	76.8	Glucose-1-P adenylyltransferase; EC 2.7.7.27	181, 1385, 2278, 2280, 2345, 2824, 2825, 3183, 3591, 3592, CG-708
glgP	Glycogen	76.8	α-Glucan phosphorylase; EC 2.4.1.1	716, 3591, CG-33653
glgS	Glycogen	68.7	Sigma-S (stationary phase)-controlled, cAMP-controlled regulator of glycogen metabolism	1673, CG-29727
glk	Glucokinase	54.0	Glucokinase; EC 2.7.1.2	518, 866a, CG-707
glmS	Glucosamine	84.2	Glucosamine-6-P synthase; EC 2.6.1.16	548, 3353, 4484, CG-706
glmU	Glucosamine	84.3	Bifunctional enzyme for consecutive steps: glucosamine-1-P acetyltransferase and *N*-acetylglucosamine-1-P uridyltransferase	548, 2805, 2806, 4484, CG-29106
glmX	Glucosamine	26.2	Affects suppression of *glmS* mutations by *nagB*	4446, CG-36538
glnA	Glutamine	87.3	Glutamine synthetase; EC 6.3.1.2	179, 788, 829–831, 1191, 1537, 2604, 2605, 2870, 2970, 3246, 3359, 3521, 3571, 3616, 4360, CG-705
glnB	Glutamine	57.7	Regulatory protein P$_{II}$ for glutamine synthetase	150, 535, 1636, 1757, 2501, 2612, 4001, 4402, 4419, CG-32933
glnD	Glutamine	4.0	Uridylyltransferase, acting on P$_{II}$ (GlnB)	1296, 4401, 4402, CG-704
glnE	Glutamine	68.8	Glutamine synthetase adenylyltransferase; EC 2.7.7.42; activates glutamine synthetase (GlnA)	4402, CG-33468
glnG	Glutamine	87.3	*gln*, *ntrC*; nitrogen regulator I	179, 2604–2606, 2870, 3246, 3247, 3359, CG-702
glnH	Glutamine	18.3	Periplasmic glutamine-binding protein	3113, 3114, CG-18271
glnL	Gllutamine	87.3	*glnR*, *ntrB*; bifunctional protein kinase/phosphatase nitrogen regulator II	150, 151, 179, 685, 2605, 2606, 2755, 2870, 3091, 3114, 3246, 3359, 3521, 3571, 4360, CG-701
glnP	Glutamine	18.3	Glutamine transport; methionine sulfoximine resistance; L-glutamate periplasmic binding protein	2693, 3114, CG-700
glnQ	Glutamine	18.2	Glutamine high-affinity transport system	3114, CG-18268
glnR	Glutamine	24.1	Glutamine transport	699, CG-699
glnS	Glutamine	15.3	Glutaminyl-tRNA synthetase; EC 6.1.1.18	286, 1625, 1741, 3348, 3349, 3581, 4357, 4723, CG-698
glnT	Glutamine	77.9	Levels of glutamine tRNA 1 and glutamine synthetase	2909a, CG-697
glnU	Glutamine	15.1	*su*$_B$, *supB* (ochre [UAA] suppression); *glnU*α; tandemly duplicated genes *glnUW*; glutamine tRNA 1	1248, 1887, 2225, 2999, 3000, 3348, CG-696
glnV	Glutamine	15.1	*Su*$_{II}$, *Su*2, *supE* (amber [UAG] suppressor); *glnV*α; duplicate genes *glnVX*, glutamine tRNA 2	1248, 1886, 1887, 2225, 2999, 3000, 3348, CG-695
glnW	Glutamine	15.1	*su*$_B$, *supB* (ochre [UAA] suppression); *glnU*β; duplicate genes *glnUW*; glutamine tRNA 1	1248, 1887, 2225, 2999, 3000, 3348, CG-37440

(Table continues)

TABLE 1 *Continued*

Symbol	Mnemonic[b] for symbol	Position (min)[c]	Synonyms; enzyme or phenotype affected[d]	References[e]
glnX	Glutamine	15.1	Su$_{II}$, Su2, supE (amber [UAG] suppressor), glnVβ; duplicate genes glnUX; glutamine tRNA 2	1248, 1886, 1887, 2225, 2999, 3000, 3348, CG-37443
glpA	Glycerol P	50.6	Glycerol-3-P dehydrogenase (anaerobic), large subunit; EC 1.1.99.5	769, 1107, 2341, 3815, CG-694
glpB	Glycerol P	50.7	sn-Glycerol-3-P dehydrogenase (anaerobic), membrane anchor subunit; EC 1.1.99.5	769, 1107, CG-17695
glpC	Glycerol P	50.7	sn-Glycerol-3-P dehydrogenase (anaerobic), small subunit; EC 1.1.99.5	769, 1107, GC-17692
glpD	Glycerol P	76.7	glyD; sn-glycerol-3-P dehydrogenase (aerobic); EC 1.1.99.5	159, 717, 2403, 3821, 3827, 3828, 4750, CG-693
glpE	Glycerol P	76.7	Member of glp regulon	717, 718, 3828, CG-18265
glpF	Glycerol P	88.6	Glycerol facilitator	2961, 3257, 3359, 4162, 4561, CG-692
glpG	Glycerol P	76.7	Member of glp regulon	718, 3828, CG-18262
glpK	Glycerol P	88.6	Glycerol kinase; EC 2.7.1.30	43, 297, 410, 802, 1834, 2961, 3310, 3359, 3671, 3821, 4209, 4428, 4561, CG-691
glpQ	Glycerol P	50.6	Glycerol-3-P diesterase, periplasmic	159, 1112, 2339, 2341, 4283, CG-690
glpR	Glycerol P	76.7	Repressor of glp operon	718, 3821, 3825, 3828, CG-688
glpT	Glycerol P	50.6	sn-Glycerol-3-P permease	1107, 1112, 1538, 1577, 2339, 2341, 2393, 2846, 3815, 4421, 4611, 4706, CG-689
glpX	Glycerol P	88.6	May be involved in glycerol metabolism but not needed for growth on glycerol	2910, 3359, 4325, CG-30884
gltA	Glutamate	16.3	glut, icdB; citrate synthase; EC 4.1.3.7	312, 376, 1509, 1826, 2312, 3262, 4015, 4595, 4643, CG-687
gltB	Glutamate	72.2	aspB; glutamate synthase, large subunit; EC 1.4.1.13	628, 831, 1327, 1428, 2555, 3192, CG-686
gltD	Glutamate	72.3	aspB; glutamate synthase, small subunit	628, 1327, 1428, 3192, CG-17689
gltE	Glutamate	81.0	Affects glutamyl-tRNA synthetase activity	2965a, CG-685
gltF	Glutamate	72.3	Regulatory gene?	628, 629, CG-18259
gltH	Glutamate	21.8	Growth on glutamate	2663, 2664, 3766, CG-684
gltM	Glutamate	43.0	Affects glutamyl-tRNA synthetase activity	2965a, CG-683
gltP	Glutamate	92.4	Proton-glutamate-aspartate transport protein	372, 2572, 4279, 4487, CG-34330
gltR	Glutamate	92.2	Growth on glutamate	2664, CG-682
gltS	Glutamate	82.4	Glutamate permease	548, 979, 980, 1057, 1156, 2044, 2046, 2567, 2663, 2664, 2866, 3766, 4476, CG-681
gltT	Glutamate	89.7	tgtB; glutamate tRNA 2	501, 503, 2225, 3339, 3462, CG-680
gltU	Glutamate	84.9	tgtC; glutamate tRNA 2	2225, CG-679
gltV	Glutamate	90.6	tgtE; glutamate tRNA 2	2225, GC-678
gltW	Glutamate	58.6	Glutamate tRNA 2	1120, 2225, CG-677
gltX	Glutamate	54.3	Glutamyl-tRNA synthetase; EC 6.1.1.17	476, 518–520, 2256, 3720, 3979, CG-676
glyA	Glycine	57.6	Serine hydroxymethyltransferase; EC 2.1.2.1	3340, 3341, 4053, 4369, 4419, CG-675
glyQ	Glycine	80.2	glySa; glycyl-tRNA synthetase, α subunit; EC 6.1.1.14	2145, 3995, 4541, CG-33775
glyS	Glycine	80.2	act, gly, glySb; glycyl-tRNA synthetase, β subunit; EC 6.1.1.14	481, 2145, 3995, 4541, CG-674
glyT	Glycine	89.9	sumA (UGA suppression); glycine tRNA 2	83, 1815, 2225, 2383, 3611, CG-673
glyU	Glycine	64.5	suA36, sufD, sumA, sumB, supT (UGA suppression); glycine tRNA 1	2225, CG-672
glyV	Glycine	94.6	ins, mutA (UGA suppression); tandemly triplicate glyVXY; glycine tRNA 3	BB[1], 1248, 2225, CG-35636
glyW	Glycine	43.0	ins, mutC; glycine tRNA 3	BB[1], 2225, 2832, 4348, 4349, CG-670
glyX	Glycine	94.6	Tandemly triplicate glyVXY; glycine tRNA 3	1248, 2225, CG-35639
glyY	Glycine	94.6	Tandemly triplicate glyVXY; glycine tRNA 3	1248, 2225, CG-35642
gmk	GMP kinase	82.3	Guanylate kinase; EC 2.7.4.8	548, 1365, CG-33871
gnd	Gluconate dehydrogenase	45.2	Gluconate-6-P dehydrogenase; EC 1.1.1.43	3033, 3034, 3056, 4146, 4637, CG-669
gntK	Gluconate	77.0	Gluconokinase, thermoresistant	L, 957, CG-35465
gntR	Gluconate	77.1	Regulatory gene for edd; transport and phosphorylation of gluconate	L, 172, 1169, 3821, 4835, CG-667
gntS	Gluconate	96.8	Secondary gluconate transport system	172, 548a, CG-666
gntT	Gluconate	76.4	gntM, usgA; high-affinity transport of gluconate	L, 1916, 3375, CG-668
gntU	Gluconate	77.0	Low-affinity gluconate transport protein	L, 957, CG-35461
gntV	Gluconate	96.8	Gluconate kinase, thermosensitive	L, 1916, CG-17686
gor	Glutathione oxidoreductase	78.5	Glutathione oxidoreductase; EC 1.6.4.2	939, 1123, 1148, 1474, 3995, CG-665
gox	Glycolate oxidase	67.4	Glycolate oxidase	2901, CG-36912
gpmA	Phosphoglyceromutase	17.0	Phosphoglycerate mutase	434, 933, CG-35790

TABLE 1 *Continued*

Symbol	Mnemonic[b] for symbol	Position (min)[c]	Synonyms; enzyme or phenotype affected[d]	References[e]
gpp	Guanosine pentaphosphatase	85.3	Guanosine pentaphosphatase activity; exopolyphosphatase	903, 2045, 2133, 3525, CG-664
gprA	Growth of phage	0.3	Replication of certain lambdoid phage	3700, CG-15896
gprB	Growth of phage	0.2	Replication of certain lambdoid phage	3700, CG-15890
gpsA	Glycerol P	81.5	*sn*-Glycerol-3-P dehydrogenase [NAD(P)$^+$]; EC 1.1.1.94	3995, CG-663
gpt	Guanine-xanthine phosphotransferase	5.6	*glyD, gpp, gxu;* guanine-xanthine phosphoribosyltransferase; EC 2.4.2.22	51, 516, 903, 1429, 1953, 2944, 3128, 3419, 3535, 3536, CG-662
greA	Growth regulation	71.6	Transcription elongation factor	71, 418, 419, 1145, 3127, 4007, 4008, 4059, CG-33506
greB	Growth regulation	76.2	Transcription elongation factor	71, 419, 3127, CG-35448
groE	Growth of phage	94.1	*hdh, mop, tabB;* locus including *groL* and *groS;* morphogenesis of phage, chaperone	
groL	Growth of phage	94.2	*groEL, mopB;* morphogenesis of phage; chaperone for assembly of enzyme complexes; large subunit of GroE chaperone	551, 639, 1046, 1183, 1375, 1377, 1449, 1668, 2294, 2336, 2365, 2842, 3045, 3898, 4180, 4270, 4400, 4575, 4803, CG-492
groS	Growth of phage	94.1	*groES, mopA;* morphogenesis of phage; chaperone for assembly of enzyme complexes; small subunit of GroE chaperone	203, 1046, 1377, 1449, 1668, 1688, 2294, 2309, 2336, 2842, 3045, 4180, 4270, 4400, 4803, CG-493
grpD	Growth after prophage induction?	73.5	Survives induction of mutant λ phage	3678, CG-661
grpE	Growth after prophage induction?	59.1	Nucleotide exchange factor, heat inducible	1315, 1898, 2453, 2489, 3678, 3902, 4112, 4587, 4662, CG-660
grxA	Glutaredoxin	19.2	Glutaredoxin	1763, 1764, 2258, 3644, 3650, 3989, 3990, CG-17683
gshA	Glutathione	60.6	γ-Glutamyl-cysteine synthetase; EC 6.3.2.2	1334, 1464, 1568, 4524, CG-659
gshB	Glutathione	66.6	Glutathione synthetase; EC 6.3.2.3	941, 1531, CG-33370
gsk	Guanosine kinase	10.8	Guanosine kinase	1606, 1797, 2883, CG-658
gsp	Glutathionylspermidine	67.5	Glutathionylspermidine synthetase/amidase	395a, CG-36859
gst	Glutathione *S*-transferase	36.9		3094, CG-37217
guaA	Guanine	56.5	GMP synthetase; EC 6.3.4.1	4242, 4254, 4255, 4269, 4384, 4795, CG-657
guaB	Guanine	56.5	IMP dehydrogenase; EC 1.2.1.14	2137, 4242, 4254, 4255, 4268, 4384, CG-656
guaC	Guanine	2.4	GMP reductase; EC 1.6.6.8	99, 1296, 2898, 3565, CG-655
gurB	Glucuronide	75.1	Utilization of glucuronides	3124, CG-654
gurC	Glucuronide	18.0	Utilization of glucuronides	3124, CG-653
gurD	Glucuronide	70.0	Utilization of glucuronides	3124, CG-652
gutM	Glucitol	60.9	Part of *srl* (sorbitol = glucitol) operon	4704, CG-33198
gutQ	Glucitol	60.9	Part of *srl* (sorbitol = glucitol) operon; putative ATP-binding protein	731, 4707, CG-33206
gyrA	Gyrase	50.3	*hisW, nalA, parD;* nalidixic acid resistance; member of cold shock response regulon; DNA gyrase, subunit A	1027, 1200, 1264, 1569, 1584, 1780, 1839, 2009, 2259, 2810–2812, 2839, 2843, 2879, 3008, 3203, 3503, 3504, 3633, 3674, 4110, 4161, 4533, 4764, CG-651
gyrB	Gyrase	83.5	*Cou, acrB, himB, hisU, nalC, nalD, pcbA, parA;* novobiocin, coumermycin resistance; DNA gyrase, subunit B	10, 11, 426, 548, 803, 986, 1027, 1200, 1264, 1279, 1361, 1569, 1765, 2810–2812, 2879, 3008, 3208, 3503, 3628, 3633, 3674, 4202, 4292, 4438, 4453, 4711, CG-650
hcaA	Hydrocinnamic acid	(61.5)	3-Phenylpropionate dioxygenase	548b, CG-35007
hcaB	Hydrocinnamic acid	(61.5)	3-Phenylproprionate-2′,3′-dihydrodiol dehydrogenase	548b, CG-35010
hdeA	H-NS determined expression	78.7	Periplasmic, unknown function, σ^S-dependent promoter	126, 3995, 4765, 4766, CG-33734
hdeB	H-NS determined expression	78.7	Periplasmic, unknown function, σ^S-dependent promoter	126, 3995, 4765, 4766, CG-33731
hdeD	H-NS determined expression	78.8	Periplasmic, unknown function, σ^S-dependent promoter	3995, 4765, 4766, CG-33737
hdhA	Hydroxysteroid dehydrogenase	36.5	*hsdH;* 7-α-hydroxysteroid dehydrogenase; EC 1.1.1.159	4770, 4771, CG-32248
helD	Helicase	22.1	*srjB;* helicase IV	2527, 2796, 2797, 4644, CG-31769
hemA	Hemin	27.2	Neomycin sensitivity; hemin biosynthesis; glutamyl tRNA dehydrogenase	684, 1070, 1872, 1956, 2435, 3742, 4426, 4427, CG-648
hemB	Hemin	8.4	*ncf;* 5-aminolevulinate dehydratase; EC 4.2.1.24	16, 1093, 2425, 2436, 2437, 2744, 2745, 3146, 4017, CG-647

(Table continues)

TABLE 1 *Continued*

Symbol	Mnemonic[b] for symbol	Position (min)[c]	Synonyms; enzyme or phenotype affected[d]	References[e]
hemC	Hemin	85.9	*popE;* neomycin sensitivity; porphobilinogen deaminase; EC 4.3.1.8	52, 903, 1551, 2011, 2013, 2318, 2550, 2744, 2853, 2854, 3736, 3741, 4258, CG-646
hemD	Hemin	85.9	Neomycin sensitivity; uroporphyrinogen III cosynthase	52, 903, 2011, 2012, 3736, 3741, CG-645
hemE	Hemin	90.4	Uroporphyrinogen decarboxylase; EC 4.1.1.37	372, 1882, 3099, 3738, CG-644
hemF	Hemin	55.0	*popB, sec;* coproporphyrinogen III oxidase; EC 1.3.3.3	2744, 4316, CG-643
hemG	Hemin	86.9	Neomycin sensitivity; protoporphyrinogen oxidase activity	903, 2996, 3100, 3737, 3740, CG-642
hemH	Hemin	10.7	*popA, visA;* photosensitivity, probably involving protoporphyrin IX photochemical reactions; ferrochelatase; EC 4.99.1.1	1288, 2883, 2884, 2997, 2998, CG-641
hemK	Hemin	27.2	Heme biosynthesis	3024
hemL	Hemin	3.7	*gsa, popC;* glutamate-1-semialdehyde aminotransferase; EC 5.4.3.8	1296, 1482, 1873, 1874, CG-372
hemM	Hemin	27.2	May affect aminolevulinate synthesis	684, 1872, 1956, 3383, CG-31940
hemX	Hemin	85.9	Uroporphyrinogen III methylase	52, 903, 3739, CG-33965
hemY	Hemin	85.9	Member of *uro* operon,	
hepA	Helicase-like protein	1.3	Downstream of *polB*, not under *lexA* control; sequence similarity to helicases	409, 1938, 2426, 4787, CG-29491
het	Heterogeneous gene size	84.6	*cop;* possibly structural gene for DNA-binding protein	3153, 4638, CG-640
hflB	High-frequency lysogenization	71.6	*ftsH, mrsC;* controls cell growth, septum formation, lambda development; essential inner membrane protein	1690, 3155, 4290, CG-735
hflC	High-frequency lysogenization	94.9	*hflA;* HflA protein complex cleaves lambda *c*II; protease	206, 691, 3106, CG-17520
hflK	High-frequency lysogenization	94.8	*hflA;* HflA protein complex cleaves lambda *c*II; protease	206, 691, 3106, CG-639
hflX	High-frequency lysogenization	94.8	*hflA;* HflX GTPase, putative	206, 691, 2039, 3106, CG-34423
hfq	Host factor for Qβ	94.8	HF-I, host factor for phage Qβ; RNA-binding protein	609, 2039, 2040, 3106, 4342, CG-34450
hha	High hemolysin activity	10.3	Down-regulates gene expression; stimulates transposition events; histone-like protein	1406, 2844, 3081, CG-31084
hhoA	Htr homolog	72.8	A serine protease	E, 4443, CG-36675
hhoB	Htr homolog	72.8	*htrH;* a serine protease	E, 443, CG-36678
himA	Host infection with mutant lambda	38.6	*hid;* site-specific recombination; sequence-specific DNA-binding transcriptional activator; IHF, α subunit	397, 429, 653, 792, 1279, 1303, 1392, 1393, 1479, 2072, 2287, 2375, 2769, 2770, 2794, 2795, 2856–2859, 3031, 3244, 3245, 3298, 3484, 4678, CG-638
himD	Host infection with mutant lambda	20.8	*hip;* sequence-specific DNA-binding transcriptional activator; IHF, β subunit	397, 429, 1219, 1303, 1392, 1393, 1479, 1571, 2072, 2158, 2287, 2375, 2795, 2857, 2858, 3031, 3244, 3245, 3298, 3484, CG-637
hipA	High persistence	34.2	Probable cell division role	343, 344, 2794, 2934, 2935, 3770, CG-18244
hipB	High persistence	34.2	Probable cell division role	343, 344, 2794, CG-32207
hisA	Histidine	45.1	N-(5′-Phospho-L-ribosylformimino)-5-amino-1-(5′-phosphoribosyl)-4-imidazolecarboxamide isomerase; EC 5.3.1.16	606, CG-636
hisB	Histidine	45.1	Bifunctional enzyme imidazoleglycerolphosphate dehydratase, histidinol-P phosphatase; EC 3.1.3.15, EC 4.2.1.19	606, 704, 1484, CG-635
hisC	Histidine	45.1	Histidinol-P aminotransferase; EC 2.6.1.9	606, 1483, 1484, CG-634
hisD	Histidine	45.1	Histidinol dehydrogenase; EC 1.1.1.23	524, 606, 703, 2019, CG-633
hisF	Histidine	45.2	Cyclase	606, CG-631
hisG	Histidine	45.0	ATP phosphoribosyltransferase; EC 2.4.2.17	57, 524, 606, 1287, 2839, 4423, CG-630
hisH	Histidine	45.1	Amidotransferase	606, CG-629
hisI	Histidine	45.2	Formerly *hisE* and *hisI;* bifunctional enzyme phosphoribosyl-AMP cyclohydrolase, phosphoribosyl-ATP pyrophosphatase; EC 3.5.4.19, EC 3.6.1.31	606, 703, CG-628
hisJ	Histidine	52.3	Histidine-binding protein of high-affinity histidine transport system	D, 2421, CG-627
hisM	Histidine	52.2	Histidine transport; membrane protein M	2255, CG-25879

TABLE 1 *Continued*

Symbol	Mnemonic[b] for symbol	Position (min)[c]	Synonyms; enzyme or phenotype affected[d]	References[e]
hisP	Histidine	52.2	Histidine permease; inner membrane receptor protein P	2255, 2421, CG-626
hisQ	Histidine	52.3	Histidine transport	D, CG-32721
hisR	Histidine	85.7	*hisT;* histidine tRNA	426, 903, 1200, 1803, 2225, CG-625
hisS	Histidine	56.6	histidyl-tRNA synthetase; EC 6.1.1.21	1113, 1114, 1268, CG-624
hlpA	Histone-like protein	4.4	*ompH, firA, skp;* histone-like protein HLP-I; in some cases, confused with adjacent *lpxD* gene	1, 2, 778, 1017, 1738, 1739, 1749, 2343, 4259, CG-18241
hmp	Hemoprotein	57.7	*fsrB;* hemoglobin-like flavoprotein	102, 3340, 4419, CG-32917
hns	Histone-like protein (HN-S)	27.8	*bglY, cur, drc, drdX, drs, fimG, mysA, osmZ, pilG, topX, virR;* HN-S (H1a); involved in chromosome organization in the nucleoid; diverse mutant phenotypes affecting transcription, transposition, inversion, and cryptic-gene expression; DNA-binding protein, histone-like	228, 300–302, 864, 1019, 1052, 1170, 1420, 1714, 1715, 1827, 2039, 2298, 2308, 2398, 2399, 2406, 2727, 3366, 3790, 3906, 4012, 4019, 4035, 4354, 4698, 4699, CG-960
hobH	Hemimethylated *oriC* binding	91.9	Outer membrane protein that binds hemimethylated *oriC*	1695, CG-36725
holA	Holoenzyme *pol*	14.5	DNA polymerase III, δ subunit	614, 1042, 3144, 3201, 4187, CG-31431
holB	Holoenzyme *pol*	24.9	DNA polymerase III, δ subunit	615, 1042, 3144, 3201, CG-31447
holC	Holoenzyme *pol*	96.6	DNA polymerase, χ subunit	617, 3144, 4685, 4686, CG-31456
holD	Holoenzyme *pol*	99.3	DNA polymerase, ψ subunit	616, 3144, 4685, 4686, 4767, CG-31460
holE	Holoenzyme *pol*	41.4	DNA polymerase III, θ subunit	613, 3144, 3967, 4116, CG-31439
hop	Host plasmid maintenance	—	*hop* is also in use for ORFs "homologous to Pil genes in *Pseudomonas*"; name "*hof*" has been substituted for these in Fig. 2	
hopB	Host plasmid maintenance	(84.2)	Required for maintenance of stable mini-F plasmid	3083, CG-37343
hopC	Host plasmic maintenance	(99.4)	Required for maintenance of stable mini-F plasmid	3083, CG-37346
hopD	Host plasmid maintenance	(11)		4474, CG-37349
hpt	HGPRT	3.0	Hypoxanthine-guanine phosphoribosyltransferase; EC 2.4.2.8	516, 1296, 1988, CG-622
hrpA	Helicase-related protein	31.9	RNA-helicase-like; similarity to eukaryotic DEAH family proteins	2915, CG-36987
hrpB	Helicase-related protein	3.5	RNA-helicase-like; similarity to eukaryotic DEAH family proteins and AraC	1296, 2915, CG-36926
hscA	Heat shock cognate	57.0	Member of Hsp70 family	2129, 3838, CG-32977
hsdM	Host specificity	98.7	*hs, hsm, hsp, rrm;* host DNA modification; DNA methylase M	2523, 3673, CG-621
hsdR	Host specificity	98.8	*hs, hsp, hsr, rm;* host DNA restriction; endonuclease R	2523, 3673, 4481, CG-620
hsdS	Host specificity	98.7	*hss;* specificity determinant for *hsdM* and *hsdR*	1436, 2523, 2969, 3673, CG-619
hslC	Heat shock loci	19.9	Expressed as member of heat shock regulon	725, CG-36762
hslD	Heat shock loci	24.1	Expressed as member of heat shock regulon	725, CG-36765
hslE	Heat shock loci	29.4	Expressed as member of heat shock regulon	725, CG-37190
hslI	Heat shock loci	30.3	Expressed as member of heat shock regulon	725, CG-37446
hslK	Heat shock loci	40.7	Expressed as member of heat shock regulon	725, CG-36945
hslL	Heat shock loci	70.7	Expressed as member of heat shock regulon	725, CG-37449
hslO	Heat shock loci	76.0	Heat shock protein G21.0	725, CG-36769
hslU	Heat shock loci	88.7	*htpI, clpY;* protease?; heat shock protein D48.5	725, 726, 1307, 3359, CG-34157
hslV	Heat shock loci	88.7	*htpO;* expressed as member of heat shock regulon	725, 726, 1307, 3959, CG-34160
hslX	Heat shock loci	94.8	Expressed as member of heat shock regulon	725, CG-36941
htgA	High-temperature growth	0.2	*htpY;* heat-inducible protein HtpY	965, 1958, 2875, 4787, CG-30233
htpG	Heat shock protein	10.6	Heat shock protein C62.5; chaperone	215, 2974, 4575, CG-17680
htpX	Heat shock protein	41.2	Expressed as member of heat shock regulon	2243, CG-32320
htrB	High-temperature requirement	24.1	Actually not under heat shock regulation; membrane protein affecting cell division and growth and high-temperature survival	2084, 2085, 2088, CG-31878
htrC	High-temperature requirement	90.2	Essential for growth at high temperature, under heat shock (sigma-32) regulation	372, 2088, 3470, CG-34265
htrE	High-temperature requirement	3.3	Sequence homology with pilin porin protein PapC	1296, 3473, CG-30510
hupA	HU-protein	90.4	HU-2, HU-α, histone-like protein	372, 397, 429, 2072–2074, 2210, 2307, 2793, 3920, CG-34243
hupB	HU-protein	9.9	*depA, hopD;* HU-1, HU-β, histone-like protein	397, 429, 1423, 2072, 2075, 2076, 2210, 2307, 2652, 2793, 3154, 4100, CG-18235
hyaA	Hydrogenase 1	22.2	Hydrogenase 1 small subunit [NiFe]	2808, 2809, CG-31781
hyaB	Hydrogenase 1	22.3	Hydrogenase 1 large subunit [NiFe]	2808, 2809, CG-31784
hyaC	Hydrogenase 1	22.3	Possible membrane-spanning protein of *hya* operon	2808, 2809, CG-31788
hyaD	Hydrogenase 1	22.3	Processing of HyaA and HyaB proteins	2808, 2809, CG-31791

(Table continues)

TABLE 1 *Continued*

Symbol	Mnemonic[b] for symbol	Position (min)[c]	Synonyms; enzyme or phenotype affected[d]	References[e]
hyaE	Hydrogenase 1	22.3	Processing of HyaA and HyaB proteins	2808, 2809, CG-31794
hyaF	Hydrogenase 1	22.3	Nickel incorporation in hydrogenase 1 proteins	2808, 2809, CG-31797
hybA	Hydrogenase 2	67.7	*hydL, hydC;* probably hydrogenase 2 [Ni Fe], small subunit	2381, 2807, 4091, CG-33407
hybB	Hydrogenase 2	67.7	Electron transport chain, membrane protein; probably hydrogenase 2 cytochrome *b*-type component	2807, CG-33414
hybC	Hydrogenase 2	67.6	Hydrogenase 2 [Ni Fe] large subunit, probably	2807, CG-33418
hybD	Hydrogenase 2	67.6	Probable processing element for hydrogenase 2	2807, CG-33421
hybE	Hydrogenase 2	67.6	Member of *hyb* operon, function undefined	2807, CG-33424
hybF	Hydrogenase 2	67.6	Regulatory gene; may moderate hydrogenase 2 levels	2807, CG-33427
hybG	Hydrogenase 2	67.6	May affect maturation of hydrogenase 2 large subunit	2807, CG-33403
hycA	Hydrogenase 3	61.3	Regulatory gene for *hyc* and *hyp;* counteracts transcription activation by FhlA	387, 1770, 2589, 3751, CG-33143
hycB	Hydrogenase 3	61.3	Formate-hydrogenlyase system; formate regulon member; small subunit hydrogenase 3?	387, 3613, 3751, CG-33169
hycC	Hydrogenase 3	61.4	Formate-hydrogenlyase system; formate regulon member; small subunit hydrogenase 3?	387, 3751, CG-33166
hycD	Hydrogenase 3	61.3	Hydrogenase 3 subunit	387, 3751, CG-33162
hycE	Hydrogenase 3	61.2	Hydrogenase 3 subunit, large, precursor	387, 3612, 3751, CG-33159
hycF	Hydrogenase 3	61.2	Hydrogenase 3 subunit	387, 3751, CG-33154
hycG	Hydrogenase 3	61.2	Hydrogenase 3 subunit	387, 3751, CG-33151
hycH	Hydrogenase 3	61.2	Required for conversion from precursor HycE to large subunit of hydrogenase large subunit	387, 2590, 3751, CG-33140
hydA	Hydrogenase	61.0	Hydrogenase activity	2381, 3277, 3721, 4716, CG-18232
hydG	Hydrogenase	90.5	Two-component regulation of hydrogenase 3 activity (with HydH)	372, 3306, 4093, CG-34252
hydH	Hydrogenase	90.5	Two-component regulation of hydrogenase 3 activity (sensor kinase of HydG)	372, 4093, CG-34249
hydN	Hydrogenase	61.1	Iron-sulfur protein required for hydrogenase activity	D14422, CG-33521
hypA	Hydrogenase	61.4	Formate-hydrogenlyase system; guanine-nucleotide-binding protein, functions as Ni donor for hydrogenase 3 large subunit	1770, 1950, 2589, 2590, 3613, 3751, CG-33104
hypB	Hydrogenase	61.4	*hydE;* formate-hydrogenlyase system; formate regulon member	659, 1770, 1950, 2590, 2628, 3613, CG-33110
hypC	Hydrogenase	61.4	Formate-hydrogenlyase system; formate regulon member	1770, 1950, 2590, 3613, CG-33113
hypD	Hydrogenase	61.4	*hydF;* formate-hydrogenlyase system; formate regulon member	1770, 1950, 2590, 3613, CG-33116
hypE	Hydrogenase	61.4	*hydB;* formate-hydrogenlyase system; formate regulon member	1770, 1950, 2381, 2590, 3613, 3721, 3723, 4530, CG-33119
iap	Isozymic alkaline phosphatase	61.9	Aminopeptidase, presumably, that generates alkaline phosphatase isozyme	1908, 2430, 3018, CG-616
ibpA	Inclusion body protein	83.3	*hslT, htpN;* chaperone, heat-inducible protein of HSP20 family	63, 548, 726, CG-33847
ibpB	Inclusion body protein	83.3	*hslS, htpE;* chaperone, heat-inducible protein of HSP20 family	63, 548, 726, CG-33843
icd	Isocitrate dehydrogenase	25.7	*icdE;* isocitrate dehydrogenase, NADP$^+$-specific; EC 1.1.1.42	593, 1664, 1693, 1718, 1833, 1835, 4264, CG-615
iciA	Inhibitor chromosome initiation	65.9	Replication initiation inhibitor, binds to 13-mers at *oriC*	1795, 1844, 4260, CG-28646
iclR		90.9	Repressor of *aceBA* operon; isocitrate lyase; EC 4.1.3.1	372, 1309, 2648, 3043, 3283, 4145, CG-614
ileR	Isoleucine	95.8	*avr, flrA;* Ile repressor; regulates *thr* and *ilv* operons	1992, 1993, 4558, CG-15914
ileS	Isoleucine	0.5	Isoleucyl-tRNA synthetase; EC 1.1.1.5	GG, 1884, 2050, 2860, 3513, 4169, 4277, 4540, 4697, 4710, 4779, 4787, CG-613
ileT	Isoleucine	86.9	Triplicated *ileTUV* isoleucine tRNA 1	903, 2225, CG-612
ileU	Isoleucine	73.8	Triplicated *ileTUV* isoleucine tRNA 1	2225, CG-611
ileV	Isoleucine	4.9	Triplicated *ileTUV* isoleucine tRNA 1	1120, 2225, CG-610
ileX	Isoleucine	69.2	Isoleucine tRNA 2	2964, CG-17677
ilvA	Isoleucine-valine (requirement)	85.2	*ile;* threonine deaminase; EC 4.2.1.16	51, 835, 903, 1068, 1116, 1453, 1607, 1809, 2358, 3695, 4458, 4562, CG-609
ilvB	Isoleucine-valine (requirement)	82.9	Acetolactate synthase I, valine sensitive; EC 4.1.3.18	548, 645, 935, 1275, 1276, 1415, 1622, 2031, 2297, 3068, 3069, 3887, 4153, 4556, 4564, CG-608
ilvC	Isoleucine-valine (requirement)	85.2	*ilvA;* ketol-acid reductoisomerase; EC 1.1.1.86	51, 243, 325, 903, 4562, CG-607
ilvD	Isoleucine-valine (requirement)	85.1	Dihydroxyacid dehydratase; EC 4.2.1.9	51, 835, 903, 1068, 1453, 1607, 1809, 2289, 2358, CG-606

TABLE 1 *Continued*

Symbol	Mnemonic[b] for symbol	Position (min)[c]	Synonyms; enzyme or phenotype affected[d]	References[e]
ilvE	Isoleucine-valine (requirement)	85.1	*ilvC, ilvJ;* branched-chain amino acid aminotransferase; EC 2.6.1.42	137, 291, 292, 583, 835, 903, 1068, 1453, 1809, 1890, 2289, 2357, 2358, 2541, 4124, 4371, 4563, CG-605
ilvF	Isoleucine-valine (requirement)	56.8	Production of valine-resistant acetolactate synthase activity, cryptic	935, CG-604
ilvG	Isoleucine-valine (requirement)	85.1	Acetolactate synthase II, valine insensitive, cryptic; EC 4.1.3.18	12, 51, 291, 810, 835, 903, 1068, 1279, 1351, 1453, 1809, 2289, 2354–2358, 2558, 3029, 3210, 3244, 3245, 3298, 4123, 4124, 4371, CG-603
ilvH	Isoleucine-valine (requirement)	1.9	*brnP;* acetolactate synthase II, valine sensitive; EC 4.1.3.18	935, 1415, 1957, 2297, 2366, 3345, 4038, 4039, 4153, 4556, 4787, CG-602
ilvI	Isoleucine-valine (requirement)	1.8	Acetolactate synthase II, valine sensitive; EC 4.1.3.18	935, 1415, 1620, 2297, 3345, 4038, 4039, 4153, 4556, 4787, CG-601
ilvJ	Isoleucine-valine (requirement)	1.3	Acetolactate synthase IV, valine resistant; EC 4.1.3.18	935, 1948, 3570, CG-600
ilvM	Isoleucine-valine (requirement)	85.1	Acetohydroxy acid synthase II, valine-insensitive small subunit; EC 4.1.3.18	835, 903, 1809, 2289, 2358, 2540, 2558, 4563, CG-18214
ilvN	Isoleucine-valine (requirement)	82.9	Acetohydroxy acid synthase I, valine-sensitive small subunit; EC 4.1.3.18	548, 645, 1275, 1276, 1282, 4556, 4564, CG-15441
ilvR	Isoleucine-valine (requirement)	99.9	Positive regulatory gene of *thr* and *ilv* operons	1993, CG-15917
ilvU	Isoleucine-valine (requirement)	6.9	Regulation of *ileS* and modification of isoleucine tRNA 2 and valine tRNA 2	1182, CG-599
ilvY	Isoleucine-valine (requirement)	85.2	Positive regulatory locus for *ilvC*	51, 325, 903, 3695, 4562, CG-598
imp	Increased membrane permeability	1.3	*ostA;* permeability of outer membrane to large maltodextrins; antibiotic and detergent sensitivity, stress induced	F, 113, 1296, 3699, CG-36754
inaA	Inducible by acid	50.6	Protein induced by acid and independent of SoxRS regulation	607, 4583, CG-32692
inaR	Inducible by acid	34.8	Regulates *inaA;* same as *marA* or *soxZ*?	4583, CG-32696
infA	Initiation factor	20.0	Protein chain initiation factor IF1	862, 863, 3367, 3718, 3923, 4583, CG-17674
infB	Initiation factor	71.3	*ssyG;* protein chain initiation factor IF2	1450, 1667, 1902, 2299, 3011, 3136, 3354, 3357, 3659, 3661, 3719, 3907–3909, CG-597
infC	Initiation factor	38.7	*fit?, srjA;* protein chain initiation factor IF3	461, 560, 561, 917, 943, 1121, 1556, 2730, 3358, 3415, 3660, 4033, 4570, 4676, CG-596
inm	Insensitive to nitrosoguanidine mutagenesis	79.0	Susceptibility to mutagenesis by nitrosoguanidine	3639, CG-595
intA	Integrase	59.2	*slpA;* from defective prophage CP4-57	2173, 4309, CG-33082
intD	Integrase	12.2	*int, int(qsr′);* integrase locus within defective prophage *qsr′*	A, 2963
ispA	Isoprenoid synthesis	9.5	Farnesyl diphosphate synthase; EC 2.5.1.1	1293, 1294, CG-30978
ispB	Isoprenoid synthesis	71.8	*cel;* octaprenyl diphosphate synthase	140, 719, 1979, CG-35712
katE	Catalase	39.0	Catalase hydroperoxidase III	2524, 2526, 4456, CG-593
katG	Catalase	89.0	Bifunctional; catalase-hydrogen peroxidase, HPII	372, 2526, 3359, 4311, 4312, CG-14983
kba	Ketose-bis-P aldolase	72.0	Ketose-bis-P aldolase	1, 2402a, CG-592
kbl	Ketobutyrate ligase	81.6	2-Amino-3-ketobutyrate CoA ligase (glycine acetyltransferase); EC 2.3.1.29	128, 2938, 3490, 3995, CG-18205
kch	K⁺ channel	28.1	Putative potassium channel protein	2850, CG-35309
kdgK	Ketodeoxygluconate	78.8	Ketodeoxygluconokinase; EC 2.7.1.45	3406, CG-591
kdgR	Ketodeoxygluconate	41.0	Regulator of *kdgK, kdgT,* and *eda*	3406, CG-590
kdgT	Ketodeoxygluconate	88.3	Ketodeoxygluconate transport system, structural gene	2651, 3359, 3406, CG-589
kdpA	Potassium dependence	15.7	*kac;* no growth on low K⁺; K⁺-translocating ATPase Kdp subunit	68, 855, 1703, 3016, 4129, CG-588
kdpB	Potassium dependence	15.7	*kac;* high-affinity potassium transport system	68, 855, 1703, 3016, 3442, 4129, CG-587
kdpC	Potassium dependence	15.7	*kac;* high-affinity potassium transport system	68, 885, 1703, 3016, 4129, CG-586
kdpD	Potassium dependence	15.6	*kac;* sensor kinase for K⁺-*kdp* system	3014–3016, 3364, 4442, 4483, CG-585
kdpE	Potassium dependence	15.6	Transcriptional effector of *kdp* operon	145, 3014–3016, 3364, 4442, 4483, CG-31571
kdpF	Potassium dependence	15.8	Inner membrane protein in K⁺ transport	68, 1703, CG-35751
kdsA	KDO synthesis	27.3	3-Deoxy-D-manno-octulosonic acid 8-P synthetase	4633, 4634, CG-17671
kdsB	KDO synthesis	20.9	CMP-3-deoxy-D-manno-octulosonate cytidylyltransferase	395, 1411, 1412, CG-18202
kdtA	KDO transfer (to lipid A)	82.0	CMP-deoxy-D-manno-octulosonate-lipid A transferase; EC 2.4.99.-	752, 753, 3793, 3995, CG-33819

(Table continues)

TABLE 1 *Continued*

Symbol	Mnemonic[b] for symbol	Position (min)[c]	Synonyms; enzyme or phenotype affected[d]	References[e]
kdtB	KDO transfer (to lipid A)	82.0	CMP-deoxy-D-manno-octulosonate-lipid A transferase; EC 2.4.99.-	752, 3594, 3945, CG-33900
kefB	K⁺ efflux	74.9	*trkB*; NEM-activatable K⁺/H⁺ antiporter	193, 1137a, CG-83
kefC	K⁺ efflux	1.0	*trkC*; NEM-activatable K⁺/H⁺ antiporter	193, 1058, 1137a, 2954, 3976, 4787, CG-82
kgtP	Ketoglutarate	58.5	*witA*; α-ketoglutarate permease	45, 3856–3861, CG-32867
kicA	Kill cell	21.0	Killing protein	1192, 3086, CG-31751
kicB	Kill cell	21.0	Suppressor of killing protein	1192, CG-31748
ksgA	Kasugamycin	1.1	S-Adenosylmethionine-6-N′,N′-adenosyl (rRNA) dimethyltransferase	95–97, 358, 3557, 4392, 4787, CG-583
ksgB	Kasugamycin	37.6	High-level resistance to kasugamycin	1020, 1249, 1490, 4010, CG-582
ksgC	Kasugamycin	12.0	Resistance to kasugamycin; affects ribosomal protein S2	4769, CG-581
ksgD	Kasugamycin	30.8	Resistance to kasugamycin	1249, CG-580
lacA	Lactose	7.8	Thiogalactoside acetyltransferase; EC 2.3.1.18	1251, 1648, CG-579
lacI	Lactose	7.9	Repressor protein of *lac* operon	305, 588, 1173, 1319, 1421, 1783, 2331, 2392, 2404, 2538, 2539, 2673, 2942, 3123, 3346, 3671, 3987, CG-578
lacY	Lactose	7.8	Galactoside permease (M protein)	320, 529, 1105, 1808, 2026, 2171, 2942, 3198, 3669, 3987, 4496, 4724, CG-577
lacZ	Lactose	7.8	β-D-Galactosidase; EC 3.2.1.23	587, 636, 1252, 1253, 1356, 1552, 1694, 1951, 2047, 2331, 2682, 2716, 2780, 2817, 2959, 2960, 4479, CG-576
lamB	Lambda	91.4	*malB*; in *malB* cluster; phage lambda receptor protein; maltose high-affinity uptake system	372, 469, 511, 649, 751, 789, 1129, 1647, 1653, 1743, 2562, 3822, 4248, CG-575
lar	Ral-inverse restriction alleviation	30.4	Element of defective prophage *rac*; Ral-like activity; restriction alleviation and modification enhancement	2170a, 4292a, CG-37053
lepA	Leader peptidase	58.1	GTP-binding protein	1016, 2293, 2658, 2660, 2661, CG-18199
lepB	Leader peptidase	58.1	Leader peptidase, signal peptidase I	328, 346, 347, 885, 2661, 4140, 4640, CG-573
leuA	Leucine	1.8	α-Isopropylmalate synthase; EC 4.1.3.12	938, 1362, 4571, 4787, CG-572
leuB	Leucine	1.7	β-Isopropylmalate dehydrogenase; EC 1.1.1.85	938, 2175, 4787, CG-571
leuC	Leucine	1.7	α-Isopropylmalate isomerase subunit	938, 2175, 4787, CG-570
leuD	Leucine	1.7	α-Isopropylmalate isomerase subunit	938, 4787, CG-569
leuJ	Leucine	13.7	*flr*; regulation of *leu* and *ilv* operons	1992, CG-18196
leuO	Leucine	1.8	Regulates CadC?	1621, 1675, 3905, 4787, CG-35918
leuP	Leucine	99.2	*leuV*β; tandemly triplicate *leuPQV* leucine tRNA; also duplicate with *leuT*	
leuQ	Leucine	99.2	*leuV*γ; tandemly triplicate *leuVPQ* leucine tRNA; also duplicate with *leuT*	
leuR	Leucine	78.8	Regulates level of leucyl-tRNA synthetase	4245a, CG-567
leuS	Leucine	14.5	Leucyl-tRNA synthetase; EC 6.1.1.4	1613, 4025, 4603, CG-566
leuT	Leucine	85.7	Leucine tRNA 1, duplicate locus to tandemly triplicated *leuVQP*	903, 1803, 2225, CG-565
leuU	Leucine	71.5	Leucine tRNA 2	2225, 4480, CG-564
leuV	Leucine	99.2	*leuV*α; tandemly triplicate *leuVPQ* leucine tRNA 1; also duplicate with *leuT*	1079, 1248, 2225, CG-563
leuW	Leucine	15.1	*feeB*; leucine tRNA 3	674, 2225, 2999, 3000, 3348, CG-562
leuX	Leucine	96.9	*Su-6*, *supP* (amber [UAG] suppression); leucine tRNA 5	125, 2225, 3117, 3327a, 4261, 4773, CG-561
leuY	Leucine	9.5	Regulates level of leucyl-tRNA synthetase	2336a, CG-560
leuZ	Leucine	43.0	Leucine tRNA 4	2225, CG-17668
lev	Levallorphan	9.0	Resistance to levallorphan	893, CG-559
lexA	Lambda excision	91.6	*exrA*, *recA*, *spr*, *umuA*; global regulator for SOS (*lexA*) operon; represses ca. 20 SOS genes	372, 457, 474, 475, 1775, 2494, 2495, 2671, 2840, 4249, 4410, CG-558
lgt	(Pro)lipoprotein glyceryl transferase	63.8	*umpA*; essential membrane protein phosphatidylglycerol:prolipoprotein diacylglyceryl transferase	267, 1317, 3724, 4599, CG-33350
lhr	Long helicase-related protein	37.2	Probable ATP-dependent helicase	1810, CG-35720
lig	Ligase	54.5	*dnaL*, *pdeC*; DNA ligase	519, 1360, 1908, 1909, CG-557
linB	Lincomycin	(29.3)	High-level lincomycin resistance	114, CG-556
lipA	Lipoate	14.2	Protein of lipoate biosynthesis	645, 1626, 1627, 3505, 4415, CG-31534
lipB	Lipoate	14.3	Lipoyl-protein ligase	2924, 3505, 4025, 4415, CG-555

TABLE 1 *Continued*

Symbol	Mnemonic[b] for symbol	Position (min)[c]	Synonyms; enzyme or phenotype affected[d]	References[e]
lit	Late induced, T4	25.8	Locus in defective prophage e14; late-gene expression of phage T4	807, 2077, 2078, CG-554
livF	Leucine, isoleucine, and valine (transport)	77.4	Membrane protein	13, 3995, CG-33713
livG	Leucine, isoleucine, and valine (transport)	77.4	hrbB, hrbC, hrbD; high-affinity branched-chain amino acid transport system; membrane component	13, 1611, 3041, 3230, 3995, 4725, CG-553
livH	Leucine, isoleucine, and valine (transport)	77.4	hrbB, hrbC, hrbD; valine resistance; high-affinity branched-chain amino acid transport system; membrane component	13, 90, 3040, 3230, 3995, 4725, CG-552
livJ	Leucine, isoleucine, and valine (transport	77.5	hrbB, hrbC, hrbD; valine resistance; periplasmic binding protein, high-affinity branched-chain amino acid transport system	13, 90, 110, 2319, 3223, 3230, 3995, 4272, 4604, 4725, CG-551
livK	Leucine, isoleucine, and valine (transport)	77.5	hrbB, hrbC, hrbD; valine resistance; leucine-specific periplasmic binding protein, high-affinity branched-chain amino acid transport	13, 90, 110, 2319, 3040, 3041, 3230, 3231, 3995, 4272, 4603, 4725, CG-550
livM	Leucine, isoleucine, and valine (transport)	77.4	High-affinity branched-chain amino acid transport	13, 3040, 3041, 3995, CG-18190
lldD	L-Lactose dehydrogenase	81.4	lct, lctD; member of Arc regulon L-lactate dehydrogenase; FMN dependent; EC 1.1.1.27	U, 1039, 1926, 2475, 3891, 3995, CG-574
lldP	L-Lactose dehydrogenase	81.3	lctP; L-lactate permease	U, 1039, 1926, 2475, 3891, 3995, CG-35161
lldR	L-Lactose dehydrogenase	81.4	lctR; regulatory gene	U, 1039, 2475, 3891, 3995, CG-35164
lon	Long form	9.9	LON protease; DNA-binding, ATP-dependent protease LA	17, 77, 435, 706, 728, 1037, 1216, 1353, 2675, 2722, 2902, 3807, 4251, 4575, 4801, 4802, CG-547
lpcA	LPS core	5.5	Phage T3, P1, T4, and T7 resistance; deficiency in conjugation	1684, 4391, CG-546
lpcB	LPS core	68.0	mrc, pon; phage T4, T7 resistance; novobiocin sensitivity	4200, CG-545
lpd	Lipoamide dehydrogenase	2.8	dhl; lipoamide dehydrogenase (NADH); EC 1.8.1.4	1514, 1517, 2329, 2330, 4016, 4063, 4070, CG-544
lplA	Lipoate-protein ligase	99.6	slr; selenolipoic acid resistant; lipoate-protein ligase A	2923, 2924, 3060, CG-33052
lpp	Lipoprotein	37.8	mlpA; murein lipoprotein structural gene	463, 598, 1363, 1490, 1597, 1895, 2762, 3004, 3005, 4139, 4836, CG-543
lpxA	Lipid A expression?	4.4	UDP-N-acetylglucosamine acetyltransferase	778, 851, 852, 1311, 4602, CG-17665
lpxB	Lipid A expression?	4.4	pgsB; lipid A disaccharide synthase	778, 851, 852, 4280, CG-404
lpxC	Lipid A expression?	2.3	envA; essential gene; cell envelope and cell separation	246, 2587, 2588, 2987, 3793, 4133, 4754, 4787, CG-815
lpxD	Lipid A expression?	4.4	fir, firA, skp, ssc, omsA; UDP-3-O-(R-3-hydroxymyristoyl)-glucosamine N-acyltransferase; in some cases, confused with adjacent hlpA gene	1, 2, 778, 1017, 1018, 1657, 1738, 1739, 2143, 3627, 3793, 4459, CG-775
lrb	L-Ribose	7.3	L-Ribose utilization; L-ribose reductase, NADPH-linked activity	4313, CG-36791
lrhA	Lys-R homolog	51.8	genR; regulatory protein, similar to LysR family of regulators	396a, CG-32685
lrp	Leucine-responsive regulatory protein	20.1	ihb, livR, lss, lstR, oppl, rblA, mbf; methylation blocking factor; utilization of D-leucine; high-affinity branched-chain amino acid transport system; global regulatory DNA-binding protein, positive and negative regulator, inhibits DNA methylation at specific GATC sites in bound promoters	AA, 91, 452, 453, 1313, 1355, 1585, 2273, 2473, 3066, 3978, 4347, 4605, CG-549
lrs		2.4	Level of leucine tRNA	2330, 4365, CG-542
lspA	(Pro)lipoprotein signal peptidase	0.6	Prolipoprotein signal peptidase; EC 3.4.99.-	442, 1884, 2050, 2860, 2953, 3513, 4277, 4697, 4709, 4710, 4779, 4787, CG-11326
lysA	Lysine	64.1	Diaminopimelate decarboxylase; EC 4.1.1.20	689, 954, 4107, 4109, CG-540
lysC	Lysine	91.1	apk; aspartokinase II; EC 1.1.1.3	372, 445, 625, 626, CG-539
lysP	Lysine	48.3	cadR; lysine permease	3042, 3373, 4062, 4178, CG-942
lysR	Lysine	64.1	Transcriptional regulator; defines LysR family of regulators	1675, 4108, 4109, CG-18187
lysS	Lysine	65.3	herC, asuD; lysyl-tRNA synthetase, constitutive	655, 1125, 1316, 1343, 1737, 1924, 2124–2126, 2417, 4131a, CG-17662
lysT	Lysine	16.9	Suβ (ochre [UAA] suppression), lysTα, supG, supL; one of triplicated (lysT,V,W) loci for lysine tRNA	518, 1248, 2225, 3418, 4774, CG-537

(Table continues)

TABLE 1 *Continued*

Symbol	Mnemonic[b] for symbol	Position (min)[c]	Synonyms; enzyme or phenotype affected[d]	References[e]
lysU	Lysine	93.7	Lysyl-tRNA synthetase, inducible	655, 742, 1355, 1924, 1925, 2124, 2416, 2417, 2473, 4412, CG-16693
lysV	Lysine	54.3	*supN* (ochre [UAA] suppression); one of triplicated (*lysT,V,W*) loci for lysine tRNA	518, 519, 2225, 4358, CG-132
lysW	Lysine	16.8	*lysT*β; one of triplicated (*lysT,V,W*) loci for lysine tRNA; see *lysT* for suppressor phenotypes	518, 1248, 2225, 3418, 4774, CG-28352
lysX	Lysine	63.3	Lysine excretion	1974a, CG-536
lytA	Lytic	60.5	Tolerance to beta-lactams; autolysis	1605, 3913, CG-18184
lytB	Lytic	0.6	Penicillin tolerance and stringent response effects	442, 1533, 4787, CG-34696
lyx	Lyxose	80.7	Novel pathway for utilization of lyxose via xylulose through rhamnose and pentose phosphate pathways; xylulose kinase	3995, CG-36908
maa	Maltose acetylase	10.4	*mac;* maltose transacetylase	G, 456, 4009, CG-36808
mac	Macrolide	(26.5)	Macrolide resistance, erythromycin resistance; see also *maa*	4009, CG-535
mae	Malic enzyme	33.4	*sfcA*??; malic enzyme, NAD linked; EC 1.1.1.38	3712a, CG-36710
mafA	Maintenance of F	0.8	Maintenance of F-like plasmids	3153, 32119, 4472–4475, CG-534
mafB	Maintenance of F	1.9	Maintenance of F-like plasmids	4472, CG-533
malA	Maltose	76.5	*malQPT* cluster	CG-23561
malB	Maltose	91.3	*malG-malM* cluster	CG-18924
malE	Maltose	91.4	*malB;* substrate recognition for transport and chemotaxis; maltose-binding protein, periplasmic	255–258, 372, 784, 877, 1087, 1193, 1260, 2763, 3169, 3519, CG-532
malF	Maltose	91.3	*malB;* maltose transport complex, inner membrane-spanning subunit	372, 930, 1286, 2756, 2757, 2763, 3671, CG-531
malG	Maltose	91.3	*malB;* maltose/maltodextrin transport complex, inner membrane-spanning subunit	372, 918, 921, 922, 930, 1193, 1263, 3671, CG-530
malI	Maltose	36.5	Regulatory gene with homology to repressors LacI, GalR, CytR	1106, 3515, 3516, CG-18181
malK	Maltose	91.4	*malB;* maltose transport complex, ATP-binding subunit	255, 257, 258, 372, 789, 877, 930, 964, 1193, 1400, 1656, 3169, 3671, 4248, 4491, CG-529
malM	Maltose	91.5	*malB, molA;* periplasmic protein	372, 751, 1401, CG-18178
malP	Maltose	76.5	*malA, blu;* maltodextrin phosphorylase; EC 2.4.1.1	967, 1742, 1744, 1745, 3251, 3252, 3438, 3468, 3578, 3821, CG-528
malQ	Maltose	76.4	*malA;* amylomaltase; EC 2.4.1.25	1742, 1744, 1745, 3438, 3821, CG-527
malS	Maltose	80.5	*malA;* α-amylase, periplasmic	1272, 3795, 3995, CG-17659
malT	Maltose	76.5	*malA;* lambda sensitivity; positive regulatory gene for *mal* regulon	775, 908, 967–969, 1193, 1742, 1745, 3468, 3469, 3541, CG-526
malX	Maltose	36.5	Regulated by *malI;* PTS enzyme II homolog	3515, 3516, CG-32254
malY	Maltose	36.6	Affects induction of maltose system	3515, CG-32257
malZ	Maltose	9.1	Maltodextrin glucosidase	2560, 3529, 4213, CG-29868
manA	Mannose	36.3	*pmi;* mannose-6-P isomerase; EC 5.3.1.8	363, 1988, 2403, 2847, 2848, 3579, CG-525
manC	Mannose	(87.5)	*mni;* D-mannose isomerase	4072, CG-524
manX	Mannose	40.9	*gptB, mpt, ptsL, ptsM, ptsX;* mannose PTS, protein II-A (III)	1149, 1151, 3255, 4601, CG-17656
manY	Mannose	41.0	*pel, ptsM, ptsP;* penetration of phage lambda	1149, 1151, 3255, 4601, CG-18175
manZ	Mannose	41.0	*gptB, mpt, ptsM, ptsX;* mannosephosphotransferase enzyme IIB	1151, 3255, 4601, CG-346
map	Methionine aminopeptidase	4.1	Essential gene; methionine aminopeptidase; EC 3.4.11.18	273, 642, 1296, 1732, CG-30569
marA	Multiple antibiotic resistance	34.8	*cpxB?, soxQ;* resistance to tetracycline and other antibiotics; transcriptional activator of multiple antibiotic resistance system	120, 763, 764, 1314, 1369, 1463, 1547, 3605, CG-18172
marB	Multiple antibiotic resistance	34.9	Regulatory gene for *mar*	120, 763, 1314, 3605, CG-32217
marR	Multiple antibiotic resistance	34.8	*cfxB, soxQ;* repressor of *mar* operon	120, 763, 2751, 2752, 3605, 3940, CG-30935
mbrB	Mothball resistant	88.7	Resistance to camphor vapors; coupling of cell division and DNA replication, link between growth rate and partitioning of chromosomes	4320–4322, CG-36929
mcrA	Modified-cytosine restriction	26.0	*rglA;* locus in defective prophage e14; restriction of DNA at 5-methylcytosine residues	1723, 3480, 3481, 3488, CG-10961
mcrB	Modified-cytosine restriction	98.6	*rglB;* component of McrBC restriction system; restriction of DNA at 5-methylcytosine residues	1025, 2270, 3478–3481, 3606–3608, 4822, CG-4978
mcrC	Modified-cytosine restriction	98.6	Modifies specificity of McrB restriction	1025, 2270, 3478, 3479, 3607, 4821, CG-34613
mcrD	Modified-cytosine restriction	98.6	Inhibits McrE restriction	3479, CG-34616
mdh	Malate dehydrogenase	72.8	Malate dehydrogenase; EC 1.1.1.37	1194, 1688, 2733, 4152, 4443, CG-523

TABLE 1 *Continued*

Symbol	Mnemonic[b] for symbol	Position (min)[c]	Synonyms; enzyme or phenotype affected[d]	References[e]
mdoA	Membrane-derived oligosaccharide	23.9	*mdoGH* cluster	
mdoB	Membrane-derived oligosaccharide	99.1	Phosphoglycerol transferase I activity	1197, 1947, CG-18169
mdoG	Membrane-derived oligosaccharide	23.9	*mdoA;* periplasmic membrane-derived oligosaccharide synthesis	386, 1198, 2301, 2549, CG-31849
mdoH	Membrane-derived oligosaccharide	24.0	*mdoA;* membrane glycosyltransferase	386, 1198, 2301, 2549, CG-31852
meb	*malE* bypass	78.6	Suppressor of *malE secB*-defective transport of *mal*-binding protein,	
melA	Melibiose	93.5	*mel-7;* α-galactosidase; EC 3.2.1.22	1390, 1582, 2463, 3911, 4538, CG-522
melB	Melibiose	93.5	*mel-4;* thiomethylgalactoside permease II	427, 1390, 1582, 3405, 4613, 4748, CG-521
melR	Melibiose	93.5	Regulatory gene	631, 4537, 4538, CG-18166
menA	Menaquinone	88.7	Vitamin K_2 biosynthesis; methylmenaquinone formation	V, 3359, CG-520
menB	Menaquinone	51.1	Vitamin K_2 biosynthesis; 1,4-dihydroxy-2-naphthoate synthase	1508, 1516, 3877, 3890, CG-519
menC	Menaquinone	51.1	Vitamin K_2 biosynthesis; *o*-succinylbenzoate synthase II	1508, 1516, 3876, 3890, CG-518
menD	Menaquinone	51.2	Vitamin K_2 biosynthesis; *o*-succinylbenzoate synthase I; EC 4.1.3.–	1508, 1516, 3249, 3374, 3890, CG-517
menE	Menaquinone	51.1	Vitamin K_2 biosynthesis; *o*-succinylbenzoate-CoA synthase	V, 3890, CG-17653
menF	Menaquinone	51.2	Vitamin K_2 biosynthesis; a menaquinone pathway-specific isochorismate synthase	V, 3374, CG-35955
mepA	Murein peptidase	52.7	Murein DD-endopeptidase	1867, 2135, CG-17650
metA	Methionine	90.7	Homoserine transsuccinylase; EC 2.3.1.46	334, 335, 372, 1078, 2830, 2831, CG-516
metB	Methionine	88.9	*met₁, met-1;* cystathionine γ-synthase; EC 4.2.99.9	428, 1077, 1469, 1470, 2174, 2268, 2461, 2679, 3359, 3676, 4793, CG-515
metC	Methionine	67.9	Cystathionine β-lyase; EC 4.4.1.8	266, CG-514
metD	Methionine	4.8	Methionine methoxamine sensitivity	2027, CG-513
metE	Methionine	86.4	Tetrahydropteroyltriglutamate methyltransferase; EC 2.1.1.14	51, 571, 903, 1417, 2725, 3184, 4559, CG-512
metF	Methionine	89.0	5,10-Methylenetetrahydrofolate reductase; EC 1.1.1.68	1077, 1469, 2174, 2268, 3186, 3359, 3676, 3677, 4054, 4308, 4793, CG-511
metG	Methionine	47.2	Ethionine sensitivity; methionyl-tRNA synthetase	116, 905–907, 1246, 1247, 1791, 2653, 2784, 4804, CG-510
metH	Methionine	90.9	B_{12}-dependent homocysteine-N^5-methyltetrahydrofolate transmethylase	201, 202, 372, 571, 1072, 3184–3186, 4559, CG-509
metJ	Methionine	88.9	Methionine sulfoximine plus α-methylmethionine sensitivity; transcriptional repressor	658, 1077, 1469, 2027, 2174, 2256, 2268, 2461, 3359, 3676, 3972, 4559, 4793, CG-508
metK	Methionine	66.4	Ethionine sensitivity; methionine adenosyltransferase; EC 2.5.1.6	451, 658, 1468, 1553, 2256, 2672, 2905, 3746, 3747, CG-507
metL	Methionine	88.9	*metM;* aspartokinase II-homoserine dehydrogenase II	929, 1077, 1469, 1470, 2174, 2268, 3359, 3676, 4792, 4793, CG-506
metR	Methionine	86.4	Positive regulatory gene for *metE* and *metH* and autogenous regulation	903, 2725, 2726, 3186, 4364, 4559, CG-18163
metT	Methionine	15.1	*metTα;* duplicate gene (*metTU*) methionine tRNA_m	1248, 1887, 2225, 2999, 3000, 3348, CG-505
metU	Methionine	15.1	*metTβ;* duplicate gene (*metTU*) methionine tRNA_m	1248, 1887, 2225, 2999, 3000, 3348, CG-28349
metV	Methionine	63.4	*metZβ;* triplicate repeat *metZWV;* initiator methionine tRNA_fl	2147, 2148, 2225, 2981, CG-35659
metW	Methionine	63.4	Triplicate repeat *metZWV;* initiator methionine tRNA_fl	2147, 2148, 2225, 2981, CG-35656
metY	Methionine	71.4	Methionine tRNA_f2	1450, 1902, 1903, 2146, 2225, 2981, CG-504
metZ	Methionine	63.4	*metZα;* triplicate repeated *metZWV;* initiator methionine tRNA_fl	2147, 2148, 2225, 2981, CG-503
mfd	Mutation frequency decline	25.2	Transcription repair coupling factor	X, 3193, 3847–3852, CG-35179
mglA	Methyl-galactoside	48.1	PMG; *mglP;* methyl-galactoside transport and galactose taxis; cytoplasmic membrane protein	403, 406, 732, 1133, 1600, 1748, 2839, 3056, 3205, 3392, 3617, 3810, CG-502

(Table continues)

TABLE 1 *Continued*

Symbol	Mnemonic[b] for symbol	Position (min)[c]	Synonyms; enzyme or phenotype affected[d]	References[e]
mglB	Methyl-galactoside	48.2	PMG; galactose-binding protein; receptor for galactose taxis	732, 1133, 1600, 1748, 2625, 2839, 3012, 3056, 3205, 3392, 3617, 3810, 3820, 3835, 4460–4462, CG-501
mglC	Methyl-galactoside	48.1	PMG; *mglP;* methyl-galactoside transport and galactose taxis	732, 1133, 1600, 1748, 2839, 3056, 3205, 3392, 3617, CG-500
mglR	Methyl-galactoside	17.0	R-MG; regulatory gene	1319a, CG-498
mgtA	Magnesium transport	96.3	*corB;* cobalt resistance; magnesium transport	548a, 3260, 4534, CG-497
mhpA	*meta*-Hydroxyphenylpropionic acid utilizing	8.0	3-Phenylpropionic acid degradation pathway; 3-(3-hydroxyphenyl)propionate 2-hydroxylase	548b, 549, CG-29024
mhpB	*meta*-Hydroxyphenylpropionic acid utilizing	8.1	3-(2,3-Dihydroxyphenyl)propionate dioxygenase	548b, 549, CG-29027
mhpC	*meta*-Hydroxyphenylpropionic acid utilizing	8.1	3-Phenylpropionic acid degradation pathway	548b, 549, CG-29030
mhpR	*meta*-Hydroxyphenylpropionic acid utilizing	8.0	Regulatory gene for 3-phenylpropionic acid degradation pathway	548b, 549, CG-29021
mhpS	*meta*-Hydroxyphenylpropionic acid utilizing	8.1	Regulatory gene for 3-phenylpropionic acid degradation pathway	548b, 549, CG-29033
miaA	Me-isopentyl-adenine	94.8	*trpX;* 2-methylthio-N^6-isopentyladenosine tRNA hypermodification	544, 573, 800, 801, 1310, 2039, 3211, 4344, CG-18160
miaD	Me-isopentyl-adenine	99.5	Suppresses leaky *miaA* mutation	801, CG-35254
micF	mRNA-interfering cRNA	49.8	*stc;* member of *soxRS* regulon; regulatory antisense RNA affecting *ompF*	85, 801, 837, 1894, 2873, 2890, 4191, CG-18157
minB	Minicell	26.3	Formation of minicells containing no DNA; position of division septum	
minC	Minicell	26.3	Inhibition of FtsZ ring at potential division site	314, 948, 949, 3150, CG-31329
minD	Minicell	26.3	Affects cell division and growth; a membrane ATPase that activates MinC	314, 925, 946, 948, 2300, CG-31326
minE	Minicell	26.3	Reverses inhibition by MinC of FtsZ ring	925, 948, CG-31317
mioC	Minimal origin	84.5	Initiation of chromosome replication; transcription of 16-kDa protein proceeds through *oriC*	536, 548, 2518, 2519, 2779, 3150, 4127, CG-18154
mmrA	Minimal medium recovery	85.4	May be *rhlB* (RNA helicase motif); recovery in rich medium following UV; postreplication repair; ATP-dependent RNA helicase or RNA-dependent ATPase, DEAD-box family	903, 2045, 3872
mms	Macromolecular synthesis	69.1	Complex operon, macromolecular synthesis	CG-18151
mng	Manganese	40.0	Manganese resistance	3939, CG-494
moaA	Molybdenum	17.6	*bisA, chlA;* MPT synthesis; chlorate resistance protein A	14, 982, 1206, 1403, 1994, 3520, 3866, CG-922
moaB	Molybdenum	17.7	MPT synthesis; chlorate resistance protein B	3556, 3866, CG-31222
moaC	Molybdenum	17.7	MPT synthesis; chlorate resistance protein C	3556, 3866, CG-31225
moaD	Molybdenum	17.7	*chlM;* MPT synthesis; chlorate resistance	1994, 3556, 3866, 4228, CG-18475
moaE	Molybdenum	17.7	MTP synthesis; chlorate resistance	3556, 3866, CG-31228
mob	Molybdenum	87.0	*chlB, narB;* MPT guanine dinucleotide synthesis; chlorate resistance	14, 173, 627, 982, 1206, 1403, 2316, 3254, 3359, 3520, 3615, 3866, CG-921
moc	Modification of CCA	32	Modification of CCA trinucleotide at 3′ end of tRNA	3142, CG-36923
modA	Molybdenum	17.2	Molybdate uptake; chlorate resistance	1989, 3501, 3866, CG-18478
modB	Molybdenum	17.2	*chlJ;* molybdate uptake; chlorate resistance	1989, 3501, 3866, CG-18478
modC	Molybdenum	17.2	*chlD, narD;* molybdate uptake; chlorate resistance	14, 173, 982, 1206, 1403, 1989, 3501, 3866, CG-920
modD	Molybdenum	17.2	Molybdate uptake; chlorate resistance	1989, 3501, 3866, CG-18478
modE	Molybdenum	17.1	Molybdate uptake	Z, CG-37366
modF	Molybdenum	17.2	Molybdate uptake	Z, CG-37369
moeA	Molybdenum	18.7	*bisB, chlE;* MPT synthesis; chlorate resistance	14, 173, 982, 1206, 1403, 1994, 2316, 3111, 3278, 3520, 3866, 4082, CG-919
moeB	Molybdenum	18.6	*chlN;* MPT synthesis; chlorate resistance	1994, 3111, 3866, 4228, CG-18472
mog	Molybdenum	0.2	*bisD, chlG;* unknown function; chlorate resistance	14, 173, 982, 1206, 1403, 1958, 3866, 4082, 4787, CG-917
molR	Molybdate	47.3	Unknown function possibly related to molybdate	Z, 2382, CG-32567
motA	Motility	42.6	*flaJ;* flagellar-regulon member; flagellar rotation	352, 966, 3968, 4045, CG-491
motB	Motility	42.6	*flaJ;* flagellar-regulon member; flagellar rotation	353, 727, 3968, 4045, CG-490
mprA	Microcin peptide regulation	60.5	*emrR;* controls level of microcin synthesis; negative regulation of EmrAB	984, 985, 2532, 2533, CG-33255
mraY	Murein cluster a	2.1	UDP-*N*-acetylmuramoyl-pentapeptide:undecaprenyl-PO$_4$ phosphatase; EC 2.7.8.13	1869, 1871a, CG-30466
mrcA	Murein cluster c	75.9	*ponA;* PBP 1A	495, 1907, 2134, 4194, CG-484

TABLE 1 *Continued*

Symbol	Mnemonic[b] for symbol	Position (min)[c]	Synonyms; enzyme or phenotype affected[d]	References[e]
mrcB	Murein cluster c	3.5	*pbpF, ponB*; peptidoglycan synthetase; PBP 1Bs	495, 601, 1296, 2117, 2134, 2994, 3585, 4158, 4194, 4203, 4827, CG-483
mrdA	Murein cluster d	14.4	*pbpA*; PBP; cell shape	146, 147, 260, 526, 2713, 4025, 4094, 4095, 4199, 4435, 4437, CG-18148
mrdB	Murein cluster d	14.4	*rodA*; cell shape; mecillinam sensitivity	146, 147, 260, 2015, 2713, 4094, 4095, 4187, 4435, CG-18145
mreB	Murein cluster e; mecillinam resistance	73.2	*envB, mon, rodY*; mecillinam resistance; cell shape and antibiotic sensitivity; affects division versus elongation; rod shape-determining protein	410, 1034, 1942, 2536a, 4463, 4464, 4466, 4573a, CG-31343
mreC	Murein cluster e; mecillinam resistance	73.2	Cell division and growth; mecillinam resistance; rod shape-determining protein	1034, 4463, CG-31348
mreD	Murein cluster e; mecillinam resistance	73.2	Mecillinam resistance; rod shape-determining protein	4463, 4465, CG-31351
mrr	Methyl-purine restriction	98.8	Restriction of methylated adenine	1654, 4481, CG-18139
msbA	Multicopy suppressor of *htrB*	20.8	ABC transporter homology; involved in biogenesis of outer membrane	2087, CG-31743
msbB	Multicopy suppressor of *htrB*	41.8	Role in outer membrane structure or function	1131, 2086, CG-32359
mscL	Mechanosensitive channel	74.0		3785, 4131, CG-35417
msrA	Methionine sulfoxide reductase	95.7	Methionine sulfoxide reductase	3466, 3467, CG-34400
msyB	Multicopy suppressor of *secY*	24.0		4354, CG-34718
mtlA	Mannitol	81.2	*mtlC* (promoter/operator); mannitol-specific enzyme II of PTS	940, 1982, 2369–2371, 3275, 3393, 3995, 4000, 4130, 4708, CG-481
mtlD	Mannitol	81.3	Mannitol-1-P dehydrogenase; EC 1.1.1.17	940, 1199, 1982, 2268, 3126, 3995, CG-479
mtlR	Mannitol	81.3	Mannitol repressor	1199, 1982, 3995, CG-36774
mtr	Methyltryptophan	71.1	Member of *tyrR* regulon; high-affinity Trp permease	1640, 1641, 1727, 3731, CG-478
mukB	Mukaku (anucleate)	21.0	Required for chromosome partitioning; DNA binding; kinesin-like motor protein?	621, 1165, 1725, 1728, 3084, 3086, CG-31516
mul	Mutability lambda	83.1	Mutability in UV-radiated lambda phage	4468, CG-477
murA	Murein	71.8	*mrbA, murZ*; phosphomycin resistance; peptidoglycan synthesis; UDP-*N*-acetylglucosamine enoylpyruvyltransferase; EC 2.5.1.7	505, 2677, 3433, CG-33518
murB	Murein	89.8	*mrb*; UDP-*N*-acetylglucosaminyl-3-enolpyruvate reductase; EC 1.1.1.158	283, 372, 1798, 3433, CG-34141
murC	Murein	2.2	L-Alanine-adding enzyme	1870, 2587, 2588, 4787, CG-476
murD	Murein	2.1	UDP-*N*-acetylmuramoyl-L-alanine:D-glutamate ligase; EC 6.3.2.9	1868, 1869, 2801, 2804, 3420, 4787, CG-30450
murE	Murein	2.0	*meso*-Diaminopimelate-adding enzyme	2835, 2836, 4210, 4787, CG-475
murF	Murein	2.0	*mra*; D-alanyl:D-alanine-adding enzyme	3271, 4787, CG-474
murG	Murein	2.1	UDP-NAc-glucosamine:NAc-muramyl-(pentapeptide) pyrophosphoryl-undecaprenol NAc-glucosamine transferase	543, 1870, 2802, 2803, 3690, 4787, CG-473
murH	Murein	99.5	Terminal stage in peptidoglycan synthesis, incorporating disaccharide peptide units into wall	878, 3107, CG-18136
murI	Murein	89.7	*mbrC, dga, glr*; synthesis of D-glutamate, essential peptidoglycan component; EC 5.1.1.3	372, 501, 1055–1057, 4775, CG-29401
mutG	Mutator	43.8	C-to-T mutation; distinct from *vsr*	3638, CG-28933
mutH	Mutator	63.9	*mutR, prv*; methyl-directed mismatch repair	833, 1447, 1448, 1661, 3496, 3931, CG-471
mutL	Mutator	94.7	Methyl-directed mismatch repair	276, 801, 833, 1661, 3496, 3931, 4343, 4344, CG-470
mutM	Mutator	82.0	*fpg*; repair system; GC-to-TA transversions; formamidopyrimidine-DNA glycosylase	391, 392, 548, 570, 753, 779, 833, 1189, 1452, 1661, 3931, 3995, CG-18133
mutR	Mutator	39.3	Decreased deletion formation between short repeats	4585, CG-36749
mutS	Mutator	61.5	*ant ?, fdv* (=formate dehydrogenase 2 activity?); methyl-directed mismatch repair	833, 1661, 3496, 3721, 3778, 3931, 4675, 4751, G-469
mutT	Mutator	2.4	AT-to-GC transversions	32, 310, 311, 833, 1254, 1296, 1661, 2633, 3792, 3931, 4787, CG-468
mutY	Mutator	66.8	*micA*; GC-to-TA transversions; adenine glycosylase; G-A repair	833, 1661, 2384, 3071, 3460, 3931, 4327, 4328, CG-18130
nadA	NAD	16.9	Quinolinate synthetase, A protein	1218, 2715, CG-467
nadB	NAD	58.2	Quinolinate synthetase, B protein	1218, 3845, CG-466
nadC	NAD	2.5	Quinolinate phosphoribosyltransferase	1296, 1514, 1517, 1762, 1818, 2329, 2330, 2479, 3565, CG-465

(Table continues)

TABLE 1 *Continued*

Symbol	Mnemonic[b] for symbol	Position (min)[c]	Synonyms; enzyme or phenotype affected[d]	References[e]
nadE	NAD	39.2	efg, ntrL; NAD synthetase, ammonia dependent	64, 2166, 4609, 4610, CG-28576
nagA	N-Acetylglucosamine	15.2	N-Acetylglucosamine-6-P deacetylase; EC 3.5.1.25	2403, 3300, 3348, 3350, 4444, 4582, CG-464
nagB	N-Acetylglucosamine	15.2	glmD; glucosamine-6-P deaminase; EC 5.3.1.10	3348, 3350, 3353, 3581, 4444, 4446, 4582, CG-463
nagC	N-Acetylglucosamine	15.1	nagR; nag operon transcriptional regulator	3300, 3350–3352, 4444, CG-31479
nagD	N-Acetylglucosamine	15.1	Function unknown, but expressed as part of nag operon	3300, 3350, 4444, CG-36240
nagE	N-Acetylglucosamine	15.2	ptsN; N-acetylglucosamine-specific enzyme II of PTS	2010, 2403, 3301, 3348, 3581, 4445, 4582, CG-462
nalB	Nalidixic acid	60.2	Sensitivity to nalidixic acid	1584, CG-460
nalD	Nalidixic acid	88.8	Penetration of nalidixic acid and glycerol through outer membrane	1799, CG-18124
nanA	N-Acetylneuraminate	72.6	N-Acetylneuraminate lyase (aldolase) (EC 4.1.3.3)	3174, 4433, CG-17647
nanT		72.6	Sialic acid transport	4433, CG-18121
napA	Nitrate reductase, periplasmic	49.5	Nitrate reductase homolog	CC, 712, 713, CG-36550
napB	Nitrate reductase, periplasmic	49.4	Cytochrome c homolog	CC, 712, 713, CG-36560
napC	Nitrate reductase, periplasmic	49.4	Cytochrome c homolog	CC, 712, 713, CG-36566
napD	Nitrate reductase, periplasmic	49.5		CC, 712, 713, CG-36547
napF	Nitrate reductase, periplasmic	49.6	Ferredoxin homolog	CC, 712, 713, CG-36544
napG	Nitrate reductase, periplasmic	49.5	Ferredoxin homolog	CC, 712, 713, CG-36553
napH	Nitrate reductase, periplasmic	49.5	Ferredoxin homolog	CC, 712, 713, CG-36556
narG	Nitrate reductase, nitrate regulation	27.5	chlC, narC; nitrate reductase α subunit; EC 1.7.99.4	368, 370, 1041, 1098, 1550, 2406, 2440, 2441, 2765, 3115, 3595, 3992, 3993, CG-459
narH	Nitrate reductase, nitrate regulation	27.6	chlC; nitrate reductase β subunit; EC 1.7.99.4	368, 370, 1098, 2406, 2441, 3992, 3993, CG-18118
narI	Nitrate reductase, nitrate regulation	27.7	chlI; cytochrome b_{NR}, structural gene, γ subunit	370, 399, 707, 1550, 2406, 2441, 3209, 3595, 3992, 3993, 4082, CG-458
narJ	Nitrate reductase, nitrate regulation	27.7	Nitrate reductase δ subunit	368, 370, 371, 1076, 2406, 3992, CG-18115
narK	Nitrate reductase, nitrate regulation	27.5	Nitrate/nitrite antiporter (probably)	995, 2406, 3112, 3115, 4083, CG-18112
narL	Nitrate reductase, nitrate regulation	27.4	frdR, narR; regulatory protein	1041, 1102, 1523, 1927, 2042, 2043, 2406, 2441, 2471, 3112, 3243, 4083, 4084, CG-18109
narP	Nitrate reductase, nitrate regulation	49.3	Regulatory protein	3458, 4079, CG-32598
narQ	Nitrate reductase, nitrate regulation	55.5	Nitrate sensor-transmitter protein, anaerobic respiratory path; functionally redundant with narX	700, 701, 3457, 3814, CG-36333
narV	Nitrate reductase, nitrate regulation	33.0	Cryptic nitrate reductase II, γ subunit	369–371, CG-32136
narW	Nitrate reductase, nitrate regulation	33.0	Cryptic nitrate reductase II, δ subunit	369–371, CG-32139
narX	Nitrate reductase, nitrate regulation	27.5	frdR, narR; nitrate sensor-transmitter protein; functionally redundant with narQ	2042, 2043, 2406, 2471, 3112, 3115, 4083, 4084, CG-18106
narY	Nitrate reductase, nitrate regulation	33.1	Cryptic nitrate reductase II, β subunit	369–371, CG-32142
narZ	Nitrate reductase, nitrate regulation	33.1	Cryptic nitrate reductase II, α subunit	369–371, 398, CG-18103
ndh	NADH dehydrogenase	25.2	Respiratory NADH dehydrogenase	585, 4776, CG-457
ndk	Nucleoside diphosphate kinase	56.8	Nucleoside diphosphate kinase; EC 2.7.4.6	1572, 1574, 3494, 4813, CG-32926
neaB	Neamine	75.1	Neamine sensitivity	596, CG-456
nfnA	Nitrofurantoin	80.6	Nitrofurantoin sensitivity	3744, CG-18100
nfnB	Nitrofurantoin	13	nfsB; resistance to nitrofurantoin; a nitroreductase	470, 2734, 2829, 3744, CG-18097
nfo	Endonuclease four	48.4	Members of soxRS regulon; endonuclease IV	865, 2419, 3725, CG-14161
nfrA	N4 (phage) resistant	12.7	Outer membrane protein, proposed as structural receptor for N4 adsorption	2156, 2157, CG-31154
nfrB	N4 (phage) resistant	12.7	Phage N4 susceptibility; membrane protein	2156, 2157, CG-31157
nfrC	N4 (phage) resistant	85.5	Phage N4 susceptibility	903, 2154, CG-33915
nfrD	N4 (phage) resistant	54.2	Phage N4 susceptibility	2154, CG-36113
nfsA	Nitrofurazone sensitivity	21.7	Nitrofuran reductase I activity B	470, 2734, CG-454
nhaA	Na$^+$/H$^+$ antiporter	0.4	ant; antiporter; stress response adaptation to high salinity and pH; Na$^+$/H$^+$ antiporter	1379, 1407, 1881, 2089, 2090, 3176, 3238, 3239, 3465, 4181, 4247, 4787, CG-15893
nhaB	Na$^+$/H$^+$ antiporter	26.5	Regulator of intracellular pH; Na$^+$/H$^+$ antiporter	2131, 3176, 3238, 3239, 3333, 3334, 4247, CG-30269
nhaR	Na$^+$/H$^+$ antiporter	0.4	Positive regulator of Na$^+$-dependent transcription of nhaA; DNA-binding protein for nhaA promoter; member of LysR family of regulatory proteins	608, 1675, 2602, 3239, 3465, 4787, CG-30281

TABLE 1   *Continued*

Symbol	Mnemonic[b] for symbol	Position (min)[c]	Synonyms; enzyme or phenotype affected[d]	References[e]
nikA	Nickel	77.8	*hydC, hydD;* affects formate hydrogen-lyase activity; hydrogenase and hydrogenase-related fumarase	1305, 3038, 3995, 4092, 4669, 4672, 4673, CG-18226
nikB	Nickel	77.9	*hydC, hydD;* affects formate hydrogen-lyase activity; hydrogenase and hydrogenase-related fumarase	3038, 3995, 4673, CG-33180
nikC	Nickel	77.9	*hydC, hydD;* affects formate hydrogen-lyase activity; hydrogenase and hydrogenase-related fumarase	3038, 3995, 4673, CG-33183
nikD	Nickel	77.9	*hydC, hydD;* affects formate hydrogen-lyase activity; hydrogenase and hydrogenase-related fumarase	3038, 3995, 4673, CG-33186
nikE	Nickel	77.9	*hydD, lipP?;* formate hydrogen-lyase activity	3038, 3995, 4673, CG-18223
nirB	Nitrite reductase	75.2	Nitrite reductase [NAD(P)H] subunit; EC 1.6.6.4	269, 270, 1949, 1968, 2597, 2598, 3287, CG-451
nirC	Nitrite reductase	75.3	Membrane protein affecting nitrite reductase [NAD(P)H] activity	270, 1602, 3287, CG-452
nirD	Nitrite reductase	75.3	Nitrite reductase [NAD(P)H] subunit; EC 1.6.6.4	270, 3287, CG-35425
nirE	Nitrite reductase	75.3	Transcribed ORF; translation not observed	1602, 3287, CG-35428
nlpA	New lipoprotein	82.7	Lipoprotein in outer membrane vesicles	548, 1987, 4713, 4778, CG-35704
nlpB	New lipoprotein	55.7	Lipoprotein in outer membrane vesicles, nonessential	439, 4266, CG-33030
nlpC	New lipoprotein	38.5	Lipoprotein	HH, CG-35802
nlpD	New lipoprotein	61.7	May function in cell wall formation	1851, 2327, CG-33225
nmpC	New membrane protein	12.4	*phmA;* locus of *qsr'*-defective prophage; porin, outer membrane; not expressed in wild-type K-12 because of IS5 insertion	366, 367, 783, 1717
non	Nonmucoid	45.8	Affects capsule formation	3460a, CG-446
nrdA	Nucleotide reductase	50.5	*dnaF;* ribonucleoside diphosphate reductase subunit B1; EC 1.17.4.1	4, 607, 618, 1291, 3090, 3347, 3957, 4135, 4350, 4706, CG-443
nrdB	Nucleotide reductase	50.5	*ftsB;* ribonucleoside diphosphate reductase subunit B2; EC 1.17.4.1	607, 618, 756, 1117, 1291, 2257, 3119, 3347, 3692, 4217, 4706, CG-444
nrdD	Nucleotide reductase	96.1	Ribonucleotide reductase, anaerobic	4138, CG-34589
nrfA	Nitrite reduction, formate dependent	92.3	Formate-dependent nitrite reduction; tetraheme cytochrome c_{552}	372, 912, 1838, 1896, 3243, CG-34336
nrfB	Nitrite reduction, formate dependent	92.3	Formate-dependent nitrite reduction; pentaheme cytochrome *c*	372, 913, 1838, CG-34364
nrfC	Nitrite reduction, formate dependent	92.3	Formate-dependent nitrite reduction; nonheme iron-sulfur protein, probably transmembrane	372, 913, 1838, CG-34352
nrfD	Nitrite reduction, formate dependent	92.4	Formate-dependent nitrite reduction; transmembrane protein similar to NADH-quinone oxide-reductase	372, 913, 1838, CG-34355
nrfE	Nitrite reduction, formate dependent	92.4	Formate-dependent nitrite reduction; membrane protein	372, 913, 1838, CG-34358
nrfF	Nitrite reduction, formate dependent	92.4	Formate-dependent nitrite reduction; periplasmic protein with putative role in synthesis of NrfA and NrfB cytochromes	372, 913, 1838, CG-34361
nrfG	Nitrite reduction, formate dependent	92.4	Required for Nrf pathway; function unknown	372, 1838, 3243, CG-34345
nth	Endonuclease three	36.8	Endonuclease specific for apurinic and/or apyrimidinic sites; endonuclease III	139, 187, 866, 4557, CG-13070
nuoA	NADH:ubiquinone oxidoreductase	51.8	NADH dehydrogenase I subunit; EC 1.6.99.-	396a, 4549, CG-32679
nuoB	NADH:ubiquinone oxidoreductase	51.7	NADH dehydrogenase I subunit; EC 1.6.99.-	396a, 4549, CG-32676
nuoC	NADH:ubiquinone oxidoreductase	51.7	NADH dehydrogenase I subunit; EC 1.6.99.-	396a, 4549, CG-32673
nuoD	NADH:ubiquinone oxidoreductase	51.7	NADH dehydrogenase I subunit; EC 1.6.99.-	396a, 4549, CG-32670
nuoE	NADH:ubiquinone oxidoreductase	51.7	NADH dehydrogenase I subunit; EC 1.6.99.-	396a, 4549, CG-32667
nuoF	NADH:ubiquinone oxidoreductase	51.7	NADH dehydrogenase I subunit; EC 1.6.99.-	396a, 3432, 4549, 4796, CG-32664
nuoG	NADH:ubiquinone oxidoreductase	51.6	NADH dehydrogenase I subunit; EC 1.6.99.-	396a, 3432, 4549, CG-32661
nuoH	NADH:ubiquinone oxidoreductase	51.6	NADH dehydrogenase I subunit; EC 1.6.99.-	396a, 3432, 4549, CG-32658
nuoI	NADH:ubiquinone oxidoreductase	51.6	NADH dehydrogenase I subunit; EC 1.6.99.-	396a, 3432, 4549, CG-32655
nuoJ	NADH:ubiquinone oxidoreductase	51.6	NADH dehydrogenase I subunit; EC 1.6.99.-	396a, 3432, 4549, CG-32652
nuoK	NADH:ubiquinone oxidoreductase	51.5	NADH dehydrogenase I subunit; EC 1.6.99.-	396a, 3432, 4549, CG-32649

(Table continues)

TABLE 1 *Continued*

Symbol	Mnemonic[b] for symbol	Position (min)[c]	Synonyms; enzyme or phenotype affected[d]	References[e]
nuoL	NADH:ubiquinone oxidoreductase	51.5	NADH dehydrogenase I subunit; EC 1.6.99.-	396a, 3432, 4549, CG-32646
nuoM	NADH:ubiquinone oxidoreductase	51.4	NADH dehydrogenase I subunit; EC 1.6.99.-	396a, 4549, 4796, CG-32639
nuoN	NADH:ubiquinone oxidoreductase	51.4	NADH dehydrogenase I subunit; EC 1.6.99.-	396a, 584, 4549, CG-29354
nupC	Nucleoside permease	54.1	*cru;* transport of nucleosides except guanosine	518, 565, 841, 2220, 2949, 2950, 4814, CG-443
nupG	Nucleoside permease	66.9	Transport of nucleosides	2948–2950, 3071, 4573, CG-442
nusA	N (lambda protein) utilization substance	71.4	Survives lambda prophage induction; transcription termination and antitermination L factor	219, 914, 1397, 1398, 1450, 1466, 1467, 1474, 1878, 1902, 1903, 1922, 1985, 2291, 3011, 3354, 3357, 3679, 4519, 4815, CG-441
nusB	N (lambda protein) utilization substance	9.4	*groNB, ssaD, ssyB;* survives lambda prophage induction; transcription termination factor	219, 828, 1278, 1328, 1378, 1878, 1901, 2688, 3189, 3883, 4165, 4223, 4519, CG-440
nusG	N (lambda protein) utilization substance	89.9	Stabilizes lambda-N-NusA-RNAP antitermination complex	372, 553, 914, 1067, 2433, 2434, 2486, 3044, 4134, CG-31308
nuvA	Near UV	9.4	Near-UV radiation sensitivity; uridine thiolation factor A activity	578, 1179, 2492, 3483, CG-439
nuvC	Near UV	43.9	Uridine thiolation factor C activity	3653, CG-438
ogt	O-Methylguanine transferase	30.1	O^6-Methylguanine-DNA methyltransferase, constitutive	8, 1609, 2667, 3394–3396, 3500, CG-31984
ompA	Outer membrane protein	22.0	*con, tolG, tut;* outer membrane protein 3a (II*;G;d), structural gene	250, 263, 681, 768, 774, 926, 1126, 1127, 1371, 1462, 1493, 1646, 2917, 2918, 2930–2932, 4126, CG-437
ompC	Outer membrane protein	49.7	*meoA, par;* outer membrane protein 1b (1b:c)	1969, 2611, 2872, 2888, 2889, 2892, 3082, 3109, 3748, 4424, 4741, CG-436
ompF	Outer membrane protein	21.2	*cmlB, coa, cry, tolF;* outer membrane protein 1a (Ia;b;F); porin	245, 679, 884, 1888, 1889, 1969, 2642, 2872, 2892, 2971, 2972, 3082, 3109, 3437, 3484, 3510, 3604, 3748, 4229, 4345, 4424, CG-435
ompR	Outer membrane protein	76.1	*cry, envZ, ompB;* activator protein for osmoregulation of porins OmpC and OmpF	487, 790, 837, 855, 1338, 1446, 1565, 1566, 2058, 2462, 2611, 2891–2894, 3027, 3826, 4230, 4346, 4531, 4683, 4741, CG-434
ompT	Outer membrane protein	12.6	Outer membrane protein 3b, protease VII; cleaves T7 RNA polymerases, Ada, SecY	891, 1422, 1487, 1488, 1586, 1669, 1752, 3641, 4128, CG-4984
ompX	Outer membrane protein	18.3	Outer membrane protein, with role in inducing RNAP-σ^E production	2773, CG-35913
opdA	Oligopeptidase	78.5	*prlC;* oligopeptidase A	797, 798, 1128, 3305, 3995, 4323, CG-18031
oppA	Oligopeptide permease	28.0	Oligopeptide transport; periplasmic binding protein	AA, 98, 1746, 2096, 2101, 2152, 2405, 3259, CG-18094
oppB	Oligopeptide permease	28.0	Oligopeptide transport	AA, 98, 1746, 2101, 2405, CG-18091
oppC	Oligopeptide permease	28.0	Oligopeptide transport	AA, 98, 1746, CG-18088
oppD	Oligopeptide permease	28.0	Oligopeptide transport	AA, 98, 1746, CG-18085
oppE	Oligopeptide permease	98.9	Oligopeptide transport	98, CG-18082
oppF	Oligopeptide permease	28.1	Oligopeptide transport; ATP hydrolysis	AA, CG-35799
opr	*rpo* backwards?	19.0	Rate of degradation of aberrant RNA polymerase subunit proteins	3864b, CG-18079
ops	Overproduction of (exo)polysaccharide	66.3	Level of exopolysaccharide production	4830, CG-18076
oriC	Origin of replication	84.5	*poh;* origin of DNA replication	138, 141, 144, 192, 536, 849, 1350, 1589, 1725, 2025, 2110, 2199, 2420, 2547, 2692, 2705, 2816, 3177–3179, 3758, 3864, 3999, 4117, 4175, 4484, 4837, 4838, CG-431
oriJ	Origin of replication	30.5	Locus in defective prophage *rac*	1013, 1015, 2035, CG-430
osmB	Osmotically inducible	28.9	OsmB lipoprotein	1381, 2023, 2024, CG-31965
osmC	Osmotically inducible	33.5		1542, CG-32184
osmE	Osmotically inducible	39.2	*anr*	1543, CG-36628
osmY	Osmotically inducible	99.4	*csi-5;* periplasmic protein, σ^S dependent (stationary phase induced)	2324, 4756, 4757, CG-34640
otsA	Osmoregulatory trehalose synthesis	42.7	Trehalose P synthase; EC 2.4.1.15	1387, 2749, CG-18073
otsB	Osmoregulatory trehalose synthesis	42.7	Trehalose P phosphatase; EC 3.1.3.12	1387, 2749, CG-18070

TABLE 1 *Continued*

Symbol	Mnemonic[b] for symbol	Position (min)[c]	Synonyms; enzyme or phenotype affected[d]	References[e]
oxyR	Oxygen	89.5	*mor, momR;* bifunctional regulatory protein sensor for oxidative stress	393, 723, 2274, 2275, 4103, 4104, 4211, 4212, 4216, 4521, CG-28841
oxyS	Oxygen	89.5	Activator for genes that detoxify oxidative damage; a small RNA	DD, 4104, CG-28844
pabA	*para*-Aminobenzoate	75.1	Sulfonamide resistance; *p*-aminobenzoate biosynthesis	516, 2079, 2127, 3619, 4304, CG-429
pabB	*para*-Aminobenzoate	40.8	Sulfonamide resistance; *p*-aminobenzoate biosynthesis	516, 1416, 4749, CG-428
pabC	*para*-Aminobenzoate	24.9	*p*-Aminobenzoate biosynthesis; aminodeoxychorismate lyase	1460, 1461, CG-31889
pac	Phenylacetate	31.0	Phenylacetate degradation	809, CG-18067
pal	Peptidoglycan-associated lipoprotein	16.8	*excC;* essential lipoprotein of unknown function associated with peptidoglycan	678, 1234, 2359a, 2360, 2414, CG-35129
panB	Pantothenate	3.2	Ketopantoate hydroxymethyltransferase; EC 4.1.2.12	847, 2001, CG-427
panC	Pantothenate	3.2	Pantothenate synthetase; EC 6.3.2.1	847, CG-426
panD	Pantothenate	3.1	Aspartate 1-decarboxylase; EC 4.1.1.11	847, CG-425
panF	Pantothenate	73.4	Pantothenate permease (symporter)	1944, 3524, 4387, CG-10818
parC	Partition	68.1	Cell partitioning; topoisomerase IV subunit A	777, 1725, 2115, 2116, 2118, 3295, 4076, 4428, CG-33440
parE	Partition	68.3	Cell partitioning; topoisomerase IV subunit B	1725, 2115, 2118, 3295, 4428, CG-33451
pat	Putrescine aminotransferase	89.1	Putrescine aminotransferase	3865, CG-18064
pbpG	PBP	47.9	Membrane protein; PBP 7	1670, CG-36349
pck	PEP carboxykinase	76.1	PEP carboxykinase (ATP); EC 4.1.1.49	980, 1408–1410, 2774, CG-422
pcm	Protein carboxyl methyltransferase	61.8	Repair of isoaspartyl residues in damaged proteins; L-isoaspartyl protein carboxyl methyltransferase; EC 2.1.1.77	779, 1290, CG-33221
pcnB	Plasmid copy number	3.4	Controls plasmid copy number; poly(A)polymerase I; cell division and chromosome replication	597, 1296, 1638a, 2504, 2543, 2657, 2690, 2691, 4689a, CG-13584
pdhR	Pyruvate dehydrogenase	2.6	Pyruvate dehydrogenase repressor	1628, 3450, 4069, CG-30488
pdxA	Pyridoxine	1.1	Isoniazid resistance; pyridoxine biosynthesis	96, 97, 998, 3557, 4787, CG-420
pdxB	Pyridoxine	52.5	Isoniazid resistance; pyridoxine biosynthesis; erythronate-4-P dehydrogenase?	130–132, 998, 3808, CG-419
pdxH	Pyridoxine	36.9	Isoniazid resistance; pyridoxine biosynthesis; pyridoxine-P oxidase	998, 1490, 2314, CG-417
pdxJ	Pyridoxine	58.0	Isoniazid resistance; pyridoxine biosynthesis	2313, 4195, CG-416
pepA	Peptides	96.6	*xerB, carP;* amino-exopeptidase A	651, 652, 1226, 2747, 3600, 4087, 4088, CG-30215
pepD	Peptides	5.5	Peptidase D, a dipeptidase	1680, 1682–1684, 2182, 2855, 3285, CG-415
pepE	Peptides	91.0	α-Aspartyl dipeptidase, peptidase E; EC 3.4.11.-	372, 796, CG-34261
pepN	Peptides	21.3	Aminopeptidase N	197–199, 1196, 1233, 1750, 2344, 2735–2737, 3285, CG-414
pepP	Peptides	65.7	Proline aminopeptidase II	2997, 4772, CG-28682
pepQ	Peptides	86.8	Proline dipeptidase; EC 3.4.13.9	903, 2996, CG-34071
pepT	Peptides	25.5	Putative peptidase T	1304, 2531, CG-31914
pfkA	Phosphofructokinase	88.4	6-Phosphofructokinase I; EC 2.7.1.11	886, 1665, 1666, 3359, 3578, CG-413
pfkB	Phosphofructokinase	38.8	Suppresses *pfkA;* 6-phosphofructokinase, Pfk-2	792, 886–888, CG-412
pflA	Pyruvate formate-lyase	20.5	*act;* pyruvate formate lyase I activase	3573, 3752, CG-35839
pflB	Pyruvate formate-lyase	20.5	Induced by anaerobiosis, multiple promoters coordinately induced; pyruvate formate lyase I	3288, 3573, 3752–3754, 3756, 4478, CG-410
pflC	Pyruvate formate-lyase	89.3	Pyruvate formate lyase II activase	372, CG-35847
pflD	Pyruvate formate-lyase	89.2	Pyruvate formate lyase II	372, CG-35843
pgi	Phosphoglucose isomerase	91.1	Glucosephosphate isomerase; EC 5.3.1.9	372, 1277, 1285, 3578, 3982, CG-409
pgk	Phosphoglycerate kinase	66.1	Phosphoglycerate kinase; EC 2.7.2.3	54, 3048, CG-408
pgl	Phosphogluconolactonase	17.2	*blu;* 6-phosphogluconolactonase; EC 3.1.1.31	3578, CG-407
pgm	Phosphoglucomutase	15.4	*blu;* phosphoglucomutase; EC 5.4.4.2	2556, 2557, 3578, CG-406
pgpA	Phosphatidylglycerophosphate phosphatase	9.5	Nonessential phosphatidylglycerophosphate phosphatase, membrane bound	1302, 1855, 1857, CG-17644
pgpB	Phosphatidylglycerolphosphate phosphatase	28.8	Nonessential phosphatidylglycerophosphate phosphatase, membrane bound	1302, 1854, 1857, CG-17641
pgsA	Phosphotidylglycerophosphate synthase	43.0	Phosphatidylglycerophosphate synthetase; EC 2.7.8.5	1063, 1419, 3172, 4348, 4366, CG-405
pheA	Phenylalanine	58.8	Fluorophenylalanine resistance; chorismate mutase P, prephenate dehydratase; EC 5.4.99.5, EC 4.2.1.51	1346, 1347, 1442, 1443, 1813, 3128, CG-403
pheP	Phenylalanine	13.0	Phenylalanine-specific permease	3319, 3320, 4577, CG-402

(Table continues)

TABLE 1 *Continued*

Symbol	Mnemonic[b] for symbol	Position (min)[c]	Synonyms; enzyme or phenotype affected[d]	References[e]
pheS	Phenylalanine	38.7	*phe-act;* phenylalanyl-tRNA synthetase α subunit; EC 6.1.1.20	1121, 1181, 2080, 2105–2107, 2769, 2770, 3355, 3356, 3358, 4032–4034, 4676, CG-400
pheT	Phenylalanine	38.6	Phenylalanyl-tRNA synthetase β subunit; EC 6.1.1.20	1121, 1181, 2769, 2770, 3355, 3356, 3358, 4032, 4033, 4676, CG-399
pheU	Phenylalanine	94.0	*pheR, pheW;* phenylalanine tRNA	574, 1310, 1345, 1441, 1443, 2225, 3337, 3820, CG-18061
pheV	Phenylalanine	67.0	Phenylalanine tRNA	572, 575, 2225, 3337, 4615, CG-18058
phnC	Phosphonate	93.1	Phosphonate utilization, cryptic in K-12; member of *pho* regulon; phosphonate transporter subunit I	666, 2635, 2819, 2820, 4512, 4515, CG-34553
phnD	Phosphonate	93.1	*psiD;* phosphonate utilization (cryptic); member of *pho* regulon; phosphonate transporter subunit, periplasmic	666, 2635, 2819–2821, 4469, 4512, 4515, CG-17638
phnE	Phosphonate	93.1	Phosphonate utilization (cryptic); member of *pho* regulon; phosphonate transporter subunit, integral membrane component	666, 2635, 2819, 2820, 4515, CG-34550
phnF	Phosphonate	93.1	Phosphonate utilization (cryptic); putative regulatory gene, member *pho* regulon; cryptic in K-12	666, 2635, 2820, 4515, CG-34547
phnG	Phosphonate	93.1	Phosphonate utilization (cryptic); membrane-bound carbon-phosphorus lyase complex subunit	666, 2635, 2820, 4515, CG-34544
phnH	Phosphonate	93.0	Phosphonate utilization (cryptic); membrane-bound carbon-phosphorus lyase complex subunit	666, 2635, 2819, 2820, 4515, CG-34541
phnI	Phosphonate	93.0	Phosphonate utilization (cryptic); membrane-bound carbon-phosphorus lyase complex subunit	666, 2635, 2819, 2820, 4515, CG-34538
phnJ	Phosphonate	93.0	Phosphonate utilization (cryptic); membrane-bound carbon-phosphorus lyase complex subunit	666, 2635, 2819, 2820, 4515, CG-34535
phnK	Phosphonate	93.0	Phosphonate utilization (cryptic); membrane-bound carbon-phosphorus lyase complex subunit	666, 2635, 2819, 2820, 4515, CG-34532
phnL	Phosphonate	93.0	Phosphonate utilization (cryptic); membrane-bound carbon-phosphorus lyase complex subunit	666, 2635, 2820, 4515, CG-34529
phnM	Phosphonate	92.9	Phosphonate utilization (cryptic); membrane-bound carbon-phosphorus lyase complex subunit	666, 2635, 2820, 4515, CG-34526
phnN	Phosphonate	92.9	Phosphonate utilization (cryptic); membrane-bound carbon-phosphorus lyase complex subunit	666, 2635, 2820, 4515, CG-34523
phnO	Phosphonate	92.9	Phosphonate utilization (cryptic); probably regulatory gene for carbon-phosphorus lyase complex, member of *pho* regulon	666, 2635, 2819, 2820, 4515, CG-34520
phnP	Phosphonate	92.9	Phosphonate utilization (cryptic); membrane-bound carbon-phosphorus lyase complex subunit	666, 2635, 2819, 2820, 4515, CG-34514
phoA	Phosphate	8.6	Alkaline phosphatase; EC 3.1.3.1	294, 390, 454, 640, 1075, 1552, 1704, 1892, 1893, 2159, 2817, 3732, 3733, 3927, 4514, 4517, CG-398
phoB	Phosphate	9.0	*phoRc, phoT;* positive response regulator for *pho* regulon; response regulator of two-component system	390, 1552, 2636, 2639, 3919, 4282, 4287, 4510, 4513, 4514, 4517, CG-397
phoE	Phosphate	5.6	*ompE;* outer membrane porin, pore protein E	950, 1003, 1630, 1969, 3082, 3128, 3226, 3487, 4114, 4115, 4285, 4287, 4288, CG-396
phoH	Phosphate	23.4	*psiH;* member of *pho* regulon; P starvation-induced	2167, 2818, 4510, CG-31841
phoP	Phosphate	25.6	Sensor in two-component regulatory system with *phoQ*	1492, 2094, CG-31919
phoQ	Phosphate	25.5	Two-component regulatory system with *phoP*	2094, CG-31922
phoR	Phosphate	9.0	*R1pho, nmpB, phoR1;* positive and negative regulatory gene for *pho* regulon; sensor of two-component system	2637, 2639, 3919, 4282, 4286, 4510, 4513, 4514, 4611, 4701, CG-394
phoU	Phosphate	84.1	*phoT;* P uptake, high-affinity P-specific transport system, regulatory gene	73, 76, 548, 3017, 4060, 4149, 4150, CG-18055
phrB	Photoreactivation	16.0	Deoxyribodipyrimidine photolyase; EC 4.1.99.3	1836, 2449, 2545, 3711, 3712, 4714, CG-391
phxB	Phage φX	17.0	Adsorption of phage φX154	2952, CG-389
pin	Prophage-derived inversion	26.0	Locus in defective prophage e14; calcium-binding protein required for initiation of chromosome replication	1134, 2295, 3342, 3343, 4394, CG-18049
pit	P$_i$ transport	78.3	Low-affinity P$_i$ transport	1123, 1124, 2385, 3995, 4024, CG-385
pldA	Phospholipase, detergent resistant	86.2	Detergent-resistant phospholipase A activity	51, 903, 952, 953, 1603, 1759, 1760, 1897, 2201, 4452, CG-384
pldB	Phospholipase, detergent resistant	86.3	Lysophospholipase L2	51, 903, 1760, 2081, 2201, 2202, CG-5001

TABLE 1 *Continued*

Symbol	Mnemonic[b] for symbol	Position (min)[c]	Synonyms; enzyme or phenotype affected[d]	References[e]
plsB	Phospholipid synthesis	91.6	Glycerol P acyltransferase activity	372, 2459, 2460, 3072, 3765, CG-382
plsC	Phospholipid synthesis	68.1	Affects partitioning; called *parF* in *Salmonella* sp.; 1-acyl-*sn*-glycerol-3-P acyltransferase; EC 2.3.1.51	776, 777, CG-33443
plsX	Phospholipid synthesis	24.7	Glycerol P auxotrophy in *plsB* background	2340, 3158, CG-18046
pmbA	Peptide MccB17	96.0	Antibiotic peptide MccB17 maturation	3575, CG-34577
pncA	Pyridine nucleotide cycle	39.8	*nam*; nicotinamide deamidase; EC 3.5.1.19	3258, CG-381
pncB	Pyridine nucleotide cycle	21.3	Nicotinate phosphoribosyltransferase; EC 2.4.2.11	4679, CG-380
pnp	Polynucleotide phosphorylase	71.2	Polynucleotide phosphorylase; EC 2.7.7.8	1160, 1405, 1755, 2763, 2764, 3380–3382, 3511, 3512, 3560, 4188, 4190, CG-379
pntA	Pyridine nucleotide transhydrogenase	36.1	Pyridine nucleotide transhydrogenase α subunit; EC 1.6.1.1	18, 19, 743, 744, CG-18043
pntB	Pyridine nucleotide transhydrogenase	36.0	Pyridine nucleotide transhydrogenase β subunit; EC 1.6.1.1	18, 19, 743, 744, CG-18040
poaR	Proline oxidase	65.8	Regulation of proline oxidase production	A, CG-377
polA	Polymerase	87.1	*resA*; DNA repair synthesis; a 3′-to-5′ polymerase and a 5′-to-3′ and 3′-to-5′ exonuclease; DNA polymerase I; EC 2.7.7.7	515, 2021, 2138, 2139, 2970, 3359, 3642, 4621, 4742, CG-375
polB	Polymerase	1.4	*dinA*; DNA polymerase II; EC 2.7.7.7	400, 401, 669–671, 1153, 1938, 1939, 2253, 2426, 3916, 4565, 4787, CG-374
pocD	Porphyrin	(0.2)	5-Aminolevulinate dehydratase; EC 4.2.1.24	2744, 3406a, CG-371
potA	Putrescine-ornithine transporter	25.5	ATP-binding membrane protein; putrescine/spermidine-ornithine transporter	1304, 2097, 2531, CG-31899
potB	Putrescine-ornithine transporter	25.4	Membrane protein, channel-forming protein for spermidine uptake	1304, 2097, 2531, CG-31902
potC	Putrescine-ornithine transporter	25.4	Membrane protein, channel-forming protein for spermidine uptake	1304, 2097, 2531, CG-31905
potD	Putrescine-ornithine transporter	25.4	Spermidine-binding membrane protein	1304, 2097, 2531, CG-31908
potE	Putrescine-ornithine transporter	15.5	Putrescine-lysine antiporter	2098–2100, CG-31552
potF	Putrescine-ornithine transporter	19.3	Apparent periplasmic putrescine-specific binding protein	3335, CG-31694
potG	Putrescine-ornithine transporter	19.3	Apparent nucleotide-binding subunit of putrescine-ornithine transporter	3335, CG-31697
potH	Putrescine-ornithine transporter	19.3	Transmembrane-spanning subunit of putrescine-ornithine transporter	3335, CG-31700
potI	Putrescine-ornithine transporter	19.3	Apparent transmembrane-spanning subunit of putrescine-ornithine transporter	3335, CG-31703
poxA	Pyruvate oxidase	94.2	Regulator of *poxB*	643, CG-370
poxB	Pyruvate oxidase	19.6	Pyruvate oxidase; EC 1.2.2.2	644, 645, 647, 1444, 1445, 3502, CG-369
ppa	Pyrophosphatase	95.8	Inorganic pyrophosphatase; EC 3.6.1.1	672, 2068, 2303–2306, CG-34394
ppc	PEP carboxylase	89.3	*asp, glu*; PEP carboxylase; EC 4.1.1.31	372, 1295, 2785, 2909, 3657, 4239, 4240, CG-368
ppiA	Peptidylprolyl isomerase	75.2	*rot*, a rotamase; peptidylprolyl-*cis-trans*-isomerase A	795, 1186, 1624, 2127, 2499, 2503, 3120a, 3811, 4304, CG-31121
ppiB	Peptidylprolyl isomerase	11.9	A rotamase, peptidylprolyl-*cis-trans*-isomerase B	163, 795, 1624, 3811, CG-31117
ppiC	Peptidylprolyl isomerase	85.3	*parvA*; peptidylprolyl-*cis-trans*-isomerase C	903, 3636, CG-35829
ppk	Polyphosphate kinase	56.3	Polyphosphate (linear P_i linked by high-energy bonds) kinase	30, 31, 848, 2240, 3980, CG-32894
pps	PEP synthase	38.4	*ppsA*; PEP synthase	1358, 3080, CG-367
ppx	Exopolyphosphatase	56.3	Exopolyphosphatase	31, 2240, 3525, CG-32899
pqi	Paraquat inducible	21.8	Induced by paraquat, regulated by SoxRS	2207, CG-36161
pqqL	Pyrroloquinoline quinone	33.8	Redox cofactor, pyrroloquinoline quinone synthesis, cryptic in K-12	342, 4351, CG-35854
pqqM	Pyrroloquinoline quinone	33.8	Pyrroloquinoline quinone, cryptic in K-12	342, 4351, CG-35857
prc	PBP protease, C terminal	41.2	*tsp*; carboxy-terminal protease for PBP 3	1598, 1599, 2978, 3853, 3937, 3938, CG-32334
prfA	Protein release factor	27.2	*asuA?, sueB, uar, ups?*; peptide chain release factor 1	624, 842, 2372, 3654, 3655, 4131a, 4426, CG-14922
prfB	Protein release factor	65.3	*supK*; peptide chain release factor 2	842, 1044, 1045, 2125, 2372, 3580, CG-17635
prfC	Protein release factor	99.3	*tos*; peptide chain release factor 3	1476, 2845, CG-34633
priA	Primosome	88.8	DNA helicase; primosome factor Y, a.k.a. protein n′	385, 1655, 2378, 2527, 3131, 3133, 3220, 3359, 4799, CG-27611
priB	Primosome	95.3	Primosomal protein n	62, 3804, 4798, CG-29153
priC	Primosome	10.5	Primosomal protein n″	4798, CG-29161

(Table continues)

TABLE 1 *Continued*

Symbol	Mnemonic[b] for symbol	Position (min)[c]	Synonyms; enzyme or phenotype affected[d]	References[e]
prlZ	Protein localization	71.3	Suppresses export defects in signal sequence mutations	4542, CG-36421
prmA	Posttranslational ribosomal protein modification	73.4	Methyltransferase for 50S subunit L11	196, 1944, 4417, CG-366
prmB	Posttranslational ribosomal protein modification	53.0	Methylation of 50S subunit L3	789a, CG-365
proA	Proline	5.7	*pro-1;* γ-glutamyl P reductase; EC 1.2.1.41	290, 1003, 1182, 1552, 1630, 1631, 1683, 1733, 2622, 3051, 4391, CG-364
proB	Proline	5.6	*pro-2;* γ-glutamyl kinase; EC 2.7.2.11	290, 854, 1003, 1182, 1552, 1630, 1631, 1733, 2622, 3051, 4391, CG-363
proC	Proline	8.7	*pro-2, pro-3;* pyrroline-5-carboxylate reductase; EC 1.5.1.2	390, 976, 1004, 1552, 2745, CG-362
proK	Proline	79.9	*proV;* proline tRNA 1	1248, 2225, 2271, 3995, CG-17632
proL	Proline	49.2	*proW;* proline tRNA 2	2225, CG-17629
proM	Proline	85.7	*proU;* proline tRNA 3	903, 2225, CG-17626
proP	Proline	93.2	Low-affinity transport system; proline permease, minor	861, 1437, 1438, 2729, 2789, 2813, 4049, CG-361
proQ	Proline	41.4	Sensitivity to toxic proline analogs and control of proline porter II activity; positive regulator of proline porter II	2863a, CG-36840
proS	Proline	4.7	*drp;* prolyl-tRNA synthetase; EC 1.1.1.15	389, 1139, 2364, 4824, CG-360
proT	Proline	83.8	Proline transport carrier protein, putative	2928, CG-359
proU	Proline	60.4	*osrA; proVWX* operon; high-affinity transport system for glycine,	
proV	Proline	60.4	*proU;* high-affinity transport system for glycine; glycine betaine-binding protein	855, 928, 1071, 1167, 1248, 1437, 1439, 1440, 1498, 2479, 2561, 2728, 2729, 3227, 3261, 3423, 3485, 4090, 4151, 4503, CG-18022
proW	Proline	60.4	*proU;* high-affinity transport system for glycine, betaine, and proline	1167, 1437, 1439, 1440, 1498, 2729, 4151, CG-18019
proX	Proline	60.4	*proU;* high-affinity transport system for glycine, betaine, and proline	230, 1167, 1437, 1439, 1440, 1498, 2729, 4151, CG-35653
prp	Propionate	97.0	Growth on propionate	2130, 4027, CG-358
prr	Pyrroline	31.0	γ-Aminobutyraldehyde (pyrroline) dehydrogenase activity	3865, CG-18016
prs	PRPP synthetase	27.1	*dnaR;* PRPP synthetase; EC 2.7.6.1	443, 1792–1794, 1796, 3682, CG-357
psd	Phosphatidylserine decarboxylase	94.6	Phosphatidylserine decarboxylase	1065, 2438, 2439, 3901, CG-356
psiF	P starvation induced	8.7	*pho* regulon member, requires PhoR sensor, PhoB response regulator Pst transport system	640, 2818, 457, CG-18013
pspA	Phage shock protein	29.4	Negative regulatory gene for stress-induced σ^{54}-dependent phage-shock-protein operon	485, 486, 4553, CG-32009
pspB	Phage shock protein	29.4	Regulatory gene, cooperatively with PspC activates expression of operon	486, 4553, CG-32012
pspC	Phage shock protein	29.4	Positive regulatory gene, cooperatively with PspB	486, 4553, CG-32015
pspE	Phage shock protein	29.4	Expressed in response to stress as part of *psp* operon, but also transcribed independently under normal conditions	485, 486, 4553, CG-32018
pssA	Phosphatidylserine synthase	58.5	Phosphatidylserine synthase; EC 2.7.8.8	739, 973, 1064, 3171, CG-355
pssR	Phosphatidylserine synthase	85	Regulatory gene	4011, CG-18010
pstA	P-specific transport	84.2	*R2pho, phoR2b, phoT;* high-affinity P-specific transport	73, 76, 548, 3487, 4060, 4150, 4532, CG-18007
pstB	P-specific transport	84.1	*phoT;* high-affinity P-specific transport, cytoplasmic ATP-binding protein	73, 76, 548, 3487, 4060, 4150, 4532, CG-18004
pstC	P-specific transport	84.2	*phoW;* high-affinity P-specific transport; cytoplasmic membrane component	73, 76, 548, 3487, 4060, 4150, 4532, CG-18001
pstS	P-specific transport	84.2	*R2pho, nmpA, phoR2a, phoS;* high-affinity P-specific transport; periplasmic P-binding protein	73, 76, 548, 1936, 2422, 2566, 2616, 2914, 4060, 4149, 4150, CG-17998
psu	Pleiotropic suppressor	(1.6)	Temporary designation for pleiotropic suppressor gene; oxolinic acid resistance	1135, CG-37018
pta	Phosphotransacetylase	52	Phosphotransacetylase; EC 2.3.1.8	514, 1508, 1527, 2041, 2421, 2709, 2710, CG-353
pth	Peptidyl tRNA hydrolase	27.1	*rap;* required for phage lambda growth; peptidyl-tRNA hydrolase	1308, 1326, 1503, 2966, CG-352
ptrA	Protease three	63.6	Protease III	206, 252, 253, 693, 746, 1089, 1090, 1212, 1213, CG-351
ptrB	Protease "three"	41.5	Protease II	2059, 2060, CG-32339
ptsG	Phosphotransferase system (PTS)	25.0	*CR, car, cat, gpt, umg;* glucosephosphotransferase enzyme II	436, 537, 1150, 2786, 3637, 3966, CG-349

TABLE 1 *Continued*

Symbol	Mnemonic[b] for symbol	Position (min)[c]	Synonyms; enzyme or phenotype affected[d]	References[e]
ptsH	PTS	54.6	*ctr, hpr;* phosphohistidinoprotein-hexose phosphotransferase; EC 2.7.1.69	488–490, 519, 567, 958–961, 1147, 1255, 3272, 3407, 3637, 3665, 3805, 4550, CG-348
ptsI	PTS	54.6	*ctr;* PEP-protein PTS enzyme I	78, 488–490, 519, 567, 958–961, 1137, 1146, 1255, 2403, 3272, 3637, 3665, 4421, CG-347
purA	Purine	94.9	*ade;* member *pur* regulon; adenylosuccinate synthetase; EC 6.3.4.4	279, 1040, 1639, 2689, 4641, CG-345
purB	Purine	25.6	*ade;* member *pur* regulon; adenylosuccinate lyase; EC 4.3.2.2	1638, 1663, 2689, CG-344
purC	Purine	55.7	*ade_g;* phosphoribosylaminoimidazole-succinocarboxamide synthetase; EC 6.3.2.6	439, 2689, 2826, 3267, 4266, CG-343
purD	Purine	90.5	*adth_a;* phosphoribosylglycinamide synthetase; EC 6.3.4.13	23, 372, 692, 1221, 2689, 3306, CG-342
purE	Purine	11.9	*Pur2, ade3, ade_f;* phosphoribosylaminoimidazole carboxylase, catalytic subunit; EC 4.1.1.21	1552, 2049, 2689, 2826, 4267, 4526, CG-341
purF	Purine	52.3	*ade_ub, purC;* amidophosphoribosyltransferase; EC 2.4.2.14	2631, 2689, 3118, 3697, 3981, 4336, 4337, 4794, CG-340
purH	Purine	90.5	Phosphoribosylaminoimidazolecarboxamide formyltransferase; EC 2.1.2.3	23, 372, 1221, 2689, 3306, CG-338
purK	Purine	11.9	*purE2;* phosphoribosylaminoimidazole carboxylase, CO_2-fixing subunit; EC 4.1.1.21	2049, 2826, 4267, 4526, CG-17995
purL	Purine	57.8	*purI;* phosphoribosylformylglycinamide synthetase; EC 6.3.5.3	1789, 2689, 3698, 3769, CG-336
purM	Purine	56.2	*adth_ub, purG;* phosphoribosylaminoimidazole synthetase; EC 6.3.3.1	1789, 2689, 3979, 3980, CG-335
purN	Purine	56.3	*ade_c;* 5'-phosphoribosyglycinamide (GAR) transformylase; EC 2.1.2.2; see *purT*, duplicate enzyme function	65, 1883, 3980, CG-17623
purP	Purine	83.8	High-affinity adenine transport	555, CG-17992
purR	Purine	37.4	Regulatory gene for *pur* regulon; purine repressor	714, 715, 921, 1637, 1639, 2163, 2798, 3587, CG-17989
purT	Purine	41.6	GAR transformylase, non-folate-requiring (GAR-transferase 2); see *purN*, duplicate enzyme function	2676, 3137, CG-32348
purU	Purine	27.7	*tgs;* balances pools of FH4 and C1-FH4; formyltetrahydrofolate hydrolase	422, 423, 2984, 2985, CG-35231
pus	Reverse of *sup*	20.6	Reverses amber suppression and accentuates *relB* mutation	1022, CG-334
putA	Proline utilization	23.2	*poaA;* bifunctional; proline dehydroge nase; EC 1.5.99.8	506, 2647, 2899, 3013, 4049, 4315, 4638, 4647, 4684, CG-333
putP	Proline utilization	23.3	Major proline/Na^+, Li^+ symport protein	1579, 2899, 3012, 3013, 3168, 4049, 4315, 4645–4647, 4726, 4727, CG-332
pykA	Pyruvate kinase	41.7	Pyruvate kinase A (II), EC 2.7.1.40	1340, 2086, 2363, 4382, 4383, CG-32363
pykF	Pyruvate kinase	37.8	Fructose-stimulated pyruvate kinase (I); EC 2.7.1.40	1340, 3159, 4018, 4383, CG-17620
pyrB	Pyrimidine	96.3	Aspartate transcarbamylase, catalytic subunit; EC 2.1.3.2	1036, 1766, 2199, 2233, 2418, 3039, 3284, 3586, 3598, 4352, CG-330
pyrC	Pyrimidine	24.2	Dihydroorotase; EC 3.5.2.3	180, 504, 1978, 4005, 4614, CG-329
pyrD	Pyrimidine	21.6	Dihydroorotate oxidase; EC 1.3.3.1	1978, 2338, CG-328
pyrE	Pyrimidine	82.1	Orotate phosphoribosyltransferase; EC 2.4.2.10	86, 87, 548, 1976, 3401–3404, 4226, CG-327
pyrF	Pyrimidine	28.9	Orotidine-5'-P decarboxylase; EC 4.1.1.23	381, 1047, 1978, 4353, CG-326
pyrG	Pyrimidine	62.6	CTP synthetase; EC 6.3.4.2	148, 3323, 4567, 4568, CG-325
pyrH	Pyrimidine	4.1	*smbA;* UMP kinase	1296, 3969a, 4720–4722, 4762, CG-324
pyrI	Pyrimidine	96.3	Aspartate transcarbamylase, regulatory subunit; EC 2.1.3.2	1036, 1187, 1766, 2418, 3039, 3284, 3323, 3598, 4535, 4594, CG-323
qin		35.1	*kim;* cryptic lambdoid phage	432, 1155, 2250, CG-17529
qmeA	Q? membrane effect	28.9	*gfs;* glycine resistance, unspecified membrane defect	4588, CG-322
qmeC	Q? membrane effect	75.6	Glycine resistance; penicillin sensitivity; unspecified membrane defect	4588, CG-321
qmeD	Q? membrane effect	64.3	Glycine resistance; penicillin sensitivity; unspecified membrane defect	4588, CG-320
qmeE	Q? membrane effect	37.9	Glycine resistance	4588, CG-319
qor	Quinone oxidoreductase	91.8	Quinone oxidoreductase, NAD(P)H dependent	372, CG-36654
qsr'	QSR-gene replacement, defective prophage	12.2	DLP12; defective, Q-independent lambdoid prophage; includes *intD* and *nmpD*	2034a, CG-19200

(Table continues)

TABLE 1 *Continued*

Symbol	Mnemonic[b] for symbol	Position (min)[c]	Synonyms; enzyme or phenotype affected[d]	References[e]
queA	Queuosine/queuine	9.1	Synthesis of queuine in tRNA; S-adenosylmethionine:tRNA ribosyltransferase-isomerase	2560, 3529, 3965, CG-29889
rac	Recombination activation-defective prophage	30.3	*sbcA*; defective prophage; see *lar, recE, recT, racC, sieB*	331, 482, 1013, 1014, 1159, 2035, 2036, 4606, 4607, CG-318
racC	Recombination activation-defective prophage	30.5	Element of defective prophage *rac*	724, CG-32076
racR	Recombination activation	30.5	Element of defective prophage *rac*	J, 724, CG-37433
radA	Radiation	99.7	*sms*?; sensitivity to gamma and UV radiation	1032a, CG-15905
radC	Radiation	82.1	Sensitivity to radiation	548, 1188–1190, CG-13913
ranA	RNA metabolism (?)	57.9	Affects RNA metabolism?	A, CG-317
rarD	Recombination and repair	86.2	Chloramphenicol resistance	903, CG-34089
ras	Radiation sensitive	10.0	Sensitivity to UV and X rays	4485, 4486, CG-316
rbfA	Ribosome-binding factor	71.3	Overexpression suppresses cold-sensitive 16S rRNA; ribosome-binding factor	895, 3719, CG-36919
rbsA	Ribose	84.7	*rbsP, rbsT*; ribose transport, ATP-binding protein	271, 534, 548, 1865, 2544, CG-12082
rbsB	Ribose	84.8	*prlB, rbsP*; D-ribose periplasmic binding protein	271, 548, 1128, 1485, 1767, 1865, 2544, 2933, CG-12092
rbsC	Ribose	84.7	*rbsP, rbsT*; D-ribose high-affinity transport system	271, 548, 1865, 2544, CG-12089
rbsD	Ribose	84.7	*rbsP*; D-ribose high-affinity transport system	271, 548, 3781, CG-314
rbsK	Ribose	84.8	Ribokinase; EC 2.7.1.15	548, 1767, 1865, 2544, CG-315
rbsR	Ribose	84.8	Regulatory gene	548, 2544, 2723, 2724, CG-12086
rcsA	Regulation capsule (colanic acid) synthesis	43.6	Positive regulator for capsule synthesis and colanic acid biosynthesis	1434, 4106, 4295, CG-17980
rcsB	Regulation capsule (colanic acid) synthesis	49.8	Positive regulator for capsule synthesis and colanic acid biosynthesis	483, 1383, 1434, 4105, 4106, 4295, CG-17977
rcsC	Regulation capsule (colanic acid) synthesis	49.9	Negative regulator of colanic acid synthesis, component of capsule; controls sliminess; contains TerE; probable histidine kinase	483, 1434, 4105, 4295, CG-17974
rcsF	Regulation capsule (colanic acid) synthesis	4.8	Regulates exopolysaccharide synthesis, colanic acid biosynthesis; overexpression gives mucoid phenotype and increases capsule synthesis	1382, CG-29845
rdgA	RecA-dependent growth	16.1	Dependence of growth on *recA* gene product	1283, CG-17971
rdgB	RecA-dependent growth	67.0	Growth and viability dependent on RecA	759, CG-17968
recA	Recombination	60.8	*lexB, umuB, zab, srf*; general recombination and DNA repair; DNA pairing and strand exchange; role in cleavage of LexA repressor and SOS mutagenesis	15, 475, 733, 1088, 1774, 2232, 2335, 2347, 2348, 2493, 2495, 2609, 2610, 2939, 3434, 3459, 3706, 4250, 4450, 4503, 4608, 4820, CG-312
recB	Recombination	63.5	*rorA*; recombination and repair; RecBCD enzyme subunit (exonuclease V)	380, 661, 733, 1090, 1158, 1212, 1708, 2527, 2694, 3549, 3736, CG-311
recC	Recombination	63.7	Recombination and repair; RecBCD enzyme subunit (exonuclease V)	661, 733, 1090, 1214, 1222, 1708, 2527, 2694, 3549, 3736, CG-310
recD	Recombination	63.5	*hopE*; recombination and repair; RecBCD enzyme subunit (exonuclease V)	80, 321, 661, 733, 1090, 1211, 1214, 1222, 1708, 2512, 2527, 2694, 3087, 3549, 3736, CG-310
recE	Recombination	30.4	Recombination and repair; locus in defective prophage *rac*, activated by *sbcA* mutation; degrades one strand, 5'-3' in duplex DNA (exonuclease VIII)	340, 724, 733–735, 1159, 1250, 2035, 2036, 2620, 4607, CG-309
recF	Recombination	83.5	*uvrF*; recombination and repair	10, 122, 142, 357, 548, 733, 1478, 2609, 2610, 3497, 3498, 3633, 3716, 3717, 4715, CG-308
recG	Recombination	82.4	*spoV*?; branch migration of Holliday junctions; junction-specific DNA helicase; see also *ruvABC*; recombination and repair	548, 2046, 2511, 2514–2516, 2982, 4579, CG-307
recJ	Recombination	65.4	Single-stranded-DNA-specific exonuclease, 5'-3'	733, 2512, 2551–2554, CG-17965
recN	Recombination	59.1	*radB*; *lexA* regulon; recombination and repair	2512, 2513, 3322, 3614, 3729, CG-10872
recO	Recombination	58.0	Recombination and DNA repair; DNA-binding protein	2214, 2313, 2571, 2925, 4196, CG-17962
recQ	Recombination	86.2	Recombination and DNA repair; presynaptic stage of recombination; *lexA* regulon; RecQ helicase	51, 903, 1897, 2527, 2797, 3020, 3021, 4363, CG-17959
recR	Recombination	10.6	Recombination and DNA repair	2623, 2624, 4753, CG-31049
recT	Recombination	30.4	Locus in defective prophage *rac*; activated by *sbcA* mutation; DNA-annealing protein	735, 1567, 2215, 2292, CG-32070
relA	Relaxed	62.7	Required for synthesis of ppGpp during stringent response to amino acid starvation; ATP:GTP 3'-pyrophosphotransferase; EC 2.7.6.5	208, 210, 901, 1201, 1306, 1615, 1691, 1919, 2007, 2420, 2822, 2823, 3483, 3812, 4160, 4234, 4411, 4687, 4789, 4790, CG-306

TABLE 1 *Continued*

Symbol	Mnemonic[b] for symbol	Position (min)[c]	Synonyms; enzyme or phenotype affected[d]	References[e]
relB	Relaxed	35.4	Stringent/relaxed response; regulation of RNA synthesis	249, 1020, CG-305
relE	Relaxed	35.4	Stringent/relaxed response; function unknown	249, 1380, CG-17956
relF	Relaxed	35.4	Stringent/relaxed response; function unknown	249, 1380, 3397, 3398, CG-17953
relX	Relaxed	62.7	Control of synthesis of ppGpp	3256a, CG-304
rep	Replicase	85.3	*mbrA, dasC, mmrA;* Rep helicase, a single-stranded DNA-dependent ATPase	51, 243, 318, 580, 903, 1068, 1394, 2527, 3994, 4243, CG-303
rer	Resistance to radiation	89.8	Resistance to UV and gamma radiation	4041, CG-302
rfaB	Rough	81.9	UDP-galactose:(glucosyl)LPS-1,6-galactosyltransferase	160, 844, 3264, 3413, 3414, 3594, 3793, 3794, 3995, CG-17617
rfaC	Rough	81.7	LPS core biosynthesis; proximal hexose; heptosyltransferase	263, 673, 2193, 3793, 3794, 3995, CG-300
rfaD	Rough	81.7	*htrM;* heat inducible; LPS biosynthesis; permits growth at high temperatures; DL-glycero-D-mannoheptose epimerase	780–782, 2088, 2193, 3293, 3471, 3594, 3793, 3794, 3995, CG-299
rfaF	Rough	81.7	ADP-heptose:LPS heptosyltransferase I	673, 3265, 3293, 3793, 3794, 3952, 3995, CG-28434
rfaG	Rough	82.0	LPS core biosynthesis; glucosyltransferase I	160, 752, 844, 3264, 3265, 3414, 3594, 3793, 3794, 3995, CG-15583
rfaH	Rough	86.6	*sfrB;* regulates LPS core biosynthesis; positive regulatory gene, transcriptional activator for LPS core and pIII	186, 844, 903, 3514, 3793, CG-164
rfaI	Rough	81.9	UDP-D-galactose; (glucosyl)LPS-1,3-D-galactosyltransferase	844, 3264, 3413, 3594, 3793, 3794, 3995, CG-17614
rfaJ	Rough	81.8	UDP-D-glucose:(galactosyl)LPS glucosyltransferase	844, 3264, 3413, 3414, 3594, 3793, 3794, 3995, CG-17611
rfaK	Rough	81.8	Modification of LPS core prior to attachment of O antigen	2192, 2193, 3793, 3794, 3995, CG-33791
rfaL	Rough	81.8	LPS core biosynthesis; O-antigen ligase	2192, 2193, 3793, 3794, 3995, CG-28438
rfaP	Rough	81.9	LPS core biosynthesis; phosphorylation of core heptose	263, 3264, 3265, 3414, 3594, 3793, 3794, 3995, CG-298
rfaQ	Rough	82.0	Heptose region of LPS core	752, 2192, 3264, 3265, 3331, 3413, 3414, 3793, 3794, 3995, CG-33801
rfaS	Rough	81.9	LPS core, not affecting attachment of O antigen	160, 2192, 2194, 3413, 3793, 3794, 3995, CG-28949
rfaY	Rough	81.8	LPS core biosynthesis	2193, 3793, 3794, 3995, CG-33828
rfaZ	Rough	81.8	LPS core biosynthesis	2192, 2193, 3793, 3794, 3995, CG-33824
rfbA	Rough	45.4	*som;* TDP-glucose pyrophosphorylase	2195, 4073, 4747, CG-297
rfbB	Rough	45.4	*som;* TDP-glucose oxidoreductase (α-TDP-glucose-4,6-dehydratase)	2195, 4073, 4747, CG-296
rfbC	Rough	45.4	TDP-4-deoxyrhamnose-3,5-epimerase	2195, 4073, 4747, CG-38129
rfbD	Rough	45.4	*som;* TDP-rhamnose synthetase (α-TDP-4 dehydrorhamnose reductase)	2195, 4073, 4747, CG-295
rfbX	Rough	45.4	Hydrophobic protein involved in O-antigen assembly	2195, 4747, CG-37353
rfe	Rough	85.4	Tunicamycin sensitivity; synthesis of common antigen and O antigen; UDP-GlcNAc:undecaprenylphosphate GlcNAc-1-P transferase	903, 2776, 2777, 3173, CG-294
rffA	Rough	85.6	Synthesis of enterobacterial common antigen	229, 2777, CG-33935
rffC	Rough	85.6	Synthesis of enterobacterial common antigen and chain elongation	229, 2777, CG-33938
rffD	Rough	85.5	Synthesis of enterobacterial common antigen; UDP-ManNAcA dehydrogenase	903, 2154, 2776–2778, 3173, CG-293
rffE	Rough	85.5	Synthesis of enterobacterial common antigen; UDP-GlcNAc-2-epimerase	903, 2776–2778, CG-33925
rffM	Rough	85.7	Synthesis of enterobacterial common antigen; UDP-ManNAcA transferase	903, 2776–2778, CG-33948
rffT	Rough	85.6	Synthesis of enterobacterial common antigen; Fuc4NAc transferase	903, 2776, 2777, CG-33930
rhaA	Rhamnose	88.2	L-Rhamnose isomerase; EC 5.3.1.14	2908, 3359, 4275, CG-292
rhaB	Rhamnose	88.2	Rhamnulokinase; EC 2.7.1.5	2908, 3359, 4275, CG-291
rhaD	Rhamnose	88.1	Rhamnulosephosphate aldolase D; EC 4.1.2.19	2908, 3359, 4275, CG-289
rhaR	Rhamnose	88.2	*rhaC;* positive regulatory gene	1325, 3359, 4220, 4275, CG-290
rhaS	Rhamnose	88.2	*rhaC;* positive regulatory gene	3359, 4275, CG-17950
rhaT	Rhamnose	88.3	Rhamnose permease	195, 1325, 3102, 3359, 4219, 4220, CG-34185

(Table continues)

TABLE 1 *Continued*

Symbol	Mnemonic[b] for symbol	Position (min)[c]	Synonyms; enzyme or phenotype affected[d]	References[e]
rhlB	RNA helicase like	85.3	Protein with ATP-dependent RNA helicase-like motif; may be *mmrA*	903, 2045, CG-36971
rhlE	RNA helicase like	17.9	DEAD-box protein family; ATP-dependent RNA helicase-like protein	2045, 3165, CG-33907
rho	Termination factor Rho	85.4	*nitA, nusD, psuA, rnsC, sun, tsu;* polarity suppressor; transcription termination factor Rho	51, 58, 243, 473, 512, 582, 903, 915, 1885, 1983, 1985, 2467, 2527, 2706, 3173, 3332, 3555, 3846, 4101, 4243, 4371, CG-288
RhsA	Recombination hot spot	81.0	Repetitive sequence responsible for duplications within chromosome	1195, 2474, 3662, 3995, 4441, 4819, CG-17947
RhsB	Recombination hot spot	78.0	Repetitive sequence responsible for duplications within chromosome	1195, 2474, 3662, 3995, 4819, CG-17944
RhsC	Recombination hot spot	15.8	Repetitive sequence responsible for duplications within chromosome	1195, 3662, 4819, CG-17941
RhsD	Recombination hot spot	11.3	Repetitive sequence responsible for duplications within chromosome	1195, 3662, 3663, 4819, CG-17938
RhsE	Recombination hot spot	32.9	Repetitive sequence responsible for duplications within chromosome	3663, 4819, CG-33720
ribA	Riboflavin	28.8	Riboflavin biosynthesis; GTP-cyclohydrolase II; EC 3.5.4.20	200, 3428, 3542, CG-287
ribB	Riboflavin	68.5	*htrP, luxH*-like; riboflavin biosynthesis; 3,4-dihydroxy-2-butanone 4-P synthase	200, 3472, 3543, 4738, CG-286
ribC	Riboflavin	37.5	Riboflavin synthase, α chain	200, CG-11923
ribD	Riboflavin	9.3	*ribG;* riboflavin biosynthesis; a deaminase	4223, CG-35685
ribE	Riboflavin	9.3	*ribH;* riboflavin synthase, β chain	4223, CG-35688
ribF	Riboflavin	0.5	Flavokinase and FAD synthetase	C, CG-35833
ridA	Rifampin dependence	73.4	Rifampin resistance and dependence	876, CG-285
ridB	Rifampin dependence	85.5	Transcription and translation; rifampin sensitivity	874, CG-17935
rimB	Ribosomal modification	38.9	50S ribosomal subunit maturation	1490, CG-284
rimC	Ribosomal modification	26.0	50S ribosomal subunit maturation	527a, CG-283
rimD	Ribosomal modification	87.6	50S ribosomal subunit maturation	527a, CG-282
rimE	Ribosomal modification	74.0	Ribosomal protein modification	2775, CG-281
rimF	Ribosomal modification	0.8	*res;* ribosomal modification	1340a, CG-280
rimG	Ribosomal modification	0.7	*ramB;* modification of 30S ribosomal subunit protein S4	4828a, CG-279
rimH	Ribosomal modification	13.7	*stsB;* ribosomal modification	1996, CG-278
rimI	Ribosomal modification	99.3	Modification of 30S ribosomal subunit protein S18; acetylation of N-terminal alanine	616, 1912, 4767, CG-277
rimJ	Ribosomal modification	24.3	*tcp;* modification of 30S ribosomal subunit protein S5; acetylation of N-terminal alanine	1959, 4584, 4767, CG-276
rimK	Ribosomal modification	19.2	*nek;* modifies ribosomal protein S6 by adding glutamic acid residues	2071, CG-31682
rimL	Ribosomal modification	32.2	Modification of 30S ribosomal subunit protein L7; acetylation of N-terminal serine	1914, 4207, CG-275
rit	Ribosomal thermolability	89.2	Affects thermolability of 50S ribosomal subunit	3199a, CG-274
rlpA	(Rare?) lipoprotein	14.3	Minor lipoprotein	2713, 4187, CG-17932
rlpB	(Rare?) lipoprotein	14.5	Minor lipoprotein	4187, CG-17929
rluA	rRNA, large, uridine modification	1.3	Dual-specificity pseudouridine synthase for 23S rRNA and tRNA[Phe]	4659, CG-36966
rmf	Ribosome modulation factor	21.9	Associated with 100S dimers of 70S ribosomes found in stationary cells	846, 4471, 4712, CG-31760
rna	RNase A	13.9	*rnsA;* cleaves phosphodiester bond between two nucleotides; RNase I	1676, 2767, CG-273
rnb	RNase B	29.0	mRNA degradation; RNase II	1048, 4828, CG-272
rnc	RNase C	58.0	Cleaves double-stranded RNA; RNase III	22, 683, 2048, 2659, 2838, 3032, 4042, 4527, CG-271
rnd	RNase D	40.6	Processes tRNA precursors; RNase D	1299, 4797, 4810–4812, CG-270
rne	RNase E	24.6	*ams;* enzyme complex for RNA processing, mRNA turnover, and maturation of 5S RNA; RNase E	170, 619–621, 639, 663, 664, 747, 748, 811, 1104, 2603, 2750, 2788, 2838, 2874, 2937, 3200, 3281, 3492, 4042, 4214, CG-269
rnhA	RNase H	5.1	*cer, dasF, herA, rnh, sdrA, sin;* degrades RNA of DNA-RNA hybrids; participates in DNA replication; RNase HI	117, 599, 600, 603, 834, 1777, 1905, 2061, 2062, 2113, 2632, 2989, 3116, 3151, 4296, CG-268
rnhB	RNase H	4.5	Degrades RNA of DNA-RNA hybrids; RNase HII; EC 3.1.26.4	1918, 4280, CG-30604
rnk	Regulator nucleoside-diphosphate kinase	13.9	Regulator of nucleoside diphosphate kinase; suppresses *Pseudomonas aeruginosa algR2*	3782
rnpA	RNase P	83.6	Component of tRNA 4,55S RNA-processing RNase, RNase P, protein component	548, 1590, 1591, 2176, 2448, 2838, 3237, 3290, 4042, CG-267

TABLE 1 *Continued*

Symbol	Mnemonic[b] for symbol	Position (min)[c]	Synonyms; enzyme or phenotype affected[d]	References[e]
rnpB	RNase P	70.4	RNase P, RNA component	115, 911, 1506, 1546, 1894, 2132, 2176, 2226, 2227, 2448, 2838, 2927, 3237, 3290, 3302, 3507, 3508, 3685, 4529, CG-266
rnt	RNase T	37.2	RNase that degrades tRNA; RNase T; EC 3.1.13	623, 1810, 3240, CG-32271
rob	Right *oriC* binding	99.8	*oriC*-binding protein; binds to right border of *oriC*	3959, CG-34661
rorB	Roentgen resistance	85.0	Sensitivity to ionizing radiation, mitomycin C	970, 971, CG-36611
rph	RNase PH	82.2	RNase PH	87, 548, 1976, 1977, 2141, 2142, 3213, 3402, 3403, CG-33892
rpiA	Ribose P isomerase	65.8	Ribose P isomerase (constitutive); EC 5.3.1.6	1795, CG-264
rplA	Ribosomal protein, large	90.0	50S ribosomal subunit protein L1	82, 372, 459, 517, 1066, 1823, 2485, 2496, 2594, 3388, 3482, 4074, 4257, CG-263
rplB	Ribosomal protein, large	74.3	50S ribosomal subunit protein L2	4834, CG-262
rplC	Ribosomal protein, large	74.3	50S ribosomal subunit protein L3	2965, 3188, 4834, CG-261
rplD	Ribosomal protein, large	74.3	*eryA;* erythromycin sensitivity; 50S ribosomal subunit protein L4	709, 1267, 2477, 4622, 4808, 4834, CG-260
rplE	Ribosomal protein, large	74.2	50S ribosomal subunit protein L5	635, 677, 3187, CG-259
rplF	Ribosomal protein, large	74.2	Gentamicin sensitivity; 50S ribosomal subunit protein L6	533, 635, 676, CG-258
rplI	Ribosomal protein, large	95.3	50S ribosomal subunit protein L9	3804, CG-257
rplJ	Ribosomal protein, large	90.0	50S ribosomal subunit protein L10	82, 231, 232, 372, 517, 757, 758, 1062, 1066, 1202, 1756, 1823, 2485, 2496, 2594, 3065, 3309, 3388, 3482, 4074, 4231, CG-256
rplK	Ribosomal protein, large	90.0	*relC;* kasugamycin sensitivity; 50S ribosomal subunit protein L11	82, 372, 517, 720, 869, 896, 1033, 1066, 1067, 1823, 2485, 2496, 3065, 3388, 3482, 4074, 4231, CG-255
rplL	Ribosomal protein, large	90.0	50S ribosomal subunit protein L7/L12	82, 231, 232, 372, 571, 1066, 1756, 1823, 2485, 2496, 2594, 3065, 3309, 3388, 3482, 4074, 4241, CG-254
rplM	Ribosomal protein, large	72.7	50S ribosomal subunit protein L13	872, 1915, 2792, CG-253
rplN	Ribosomal protein, large	74.2	50S ribosomal subunit protein L14	635, 2476, 2717, 2912, 3385, CG-252
rplO	Ribosomal protein, large	74.1	50S ribosomal subunit protein L15	635, 1402, 1921, CG-251
rplP	Ribosomal protein, large	74.2	50S ribosomal subunit protein L16	499, 4834, CG-250
rplQ	Ribosomal protein, large	74.0	50S ribosomal subunit protein L17	259, 635, 2476, 2775, 3384, CG-249
rplR	Ribosomal protein, large	74.2	50S ribosomal subunit protein L18	502, 635, CG-248
rplS	Ribosomal protein, large	59.0	50S ribosomal subunit protein L19	498, 568, 569, CG-247
rplT	Ribosomal protein, large	38.7	*pdzA;* 50S ribosomal subunit protein L20	1181, 4032, 4632, CG-17606
rplU	Ribosomal protein, large	71.7	50S ribosomal subunit protein L21	1650, 1979, CG-246
rplV	Ribosomal protein, large	74.3	*eryB;* erythromycin sensitivity; 50S ribosomal subunit protein L22	709, 4622, 4627, 4834, CG-245
rplW	Ribosomal protein, large	74.3	50S ribosomal subunit protein L23	4626, 4834, CG-244
rplX	Ribosomal protein, large	74.2	50S ribosomal subunit protein L24	635, 873, 3187, 4623, CG-243
rplY	Ribosomal protein, large	49.1	50S ribosomal subunit protein L25	339, 1061, 4359, CG-242
rpmA	Ribosomal protein, large	71.7	50S ribosomal subunit protein L27	680, 1979, CG-241
rpmB	Ribosomal protein, large	82.1	50S ribosomal subunit protein L28	548, 1913, 2384, 4023, 4629, CG-240
rpmC	Ribosomal protein, large	74.2	50S ribosomal subunit protein L29	338, 4834, CG-239
rpmD	Ribosomal protein, large	74.1	50S ribosomal subunit protein L30	635, 3551, CG-238
rpmE	Ribosomal protein, large	88.8	50S ribosomal subunit protein L31	497, 871, 3359, CG-237
rpmF	Ribosomal protein, large	24.7	50S ribosomal subunit protein L32	1959, 4208, 4628, CG-17605
rpmG	Ribosomal protein, large	82.1	50S ribosomal subunit protein L33	548, 1913, 2384, 4023, 4631, CG-236
rpmH	Ribosomal protein, large	83.6	*rimA, ssaF;* 50S ribosomal subunit protein L34	548, 675, 1590, 1591, 3189, CG-235
rpmI	Ribosomal protein, large	38.7	50S ribosomal subunit protein A (L35)	2095, 3660, 4470, CG-17602
rpmJ	Ribosomal protein, large	74.1	50S ribosomal subunit protein X (L36)	635, 3384, 4470, CG-17599
rpoA	RNA polymerase	74.1	*phs, sez;* phage P2 *vir1* resistance; RNA polymerase, α subunit; EC 2.7.7.6	259, 635, 1390, 1532, 1629, 1860, 1861, 1863, 2530, 2775, 3224, 3233, 3234, 3384, 3386, 3621, 3622, 4147, 4256, CG-234
rpoB	RNA polymerase	90.0	*ftsR, groN, nitB, rif, ron, stl, stv, tabD, sdgB, mbrD;* streptovaricin, rifampin, streptolydigin sensitivity; RNA polymerase, β subunit; EC 2.7.7.6	82, 149, 161, 210, 231, 232, 372, 517, 992, 1066, 1166, 1376, 1386, 1756, 1823, 1984–1986, 2038, 2111, 2320, 2380, 2485, 2496, 2594, 2680, 3065, 3225, 3233, 3234, 3388, 3482, 3572, 3776, 4074, 4075, 4234, 4370, 4434, 4786, CG-233

(Table continues)

TABLE 1 *Continued*

Symbol	Mnemonic[b] for symbol	Position (min)[c]	Synonyms; enzyme or phenotype affected[d]	References[e]
rpoC	RNA polymerase	90.1	tabD; RNA polymerase, β subunit; EC 2.7.7.6	82, 231, 232, 372, 517, 1066, 1756, 1985, 2038, 2485, 2496, 2594, 3065, 3222, 3225, 3233, 3234, 3482, 4037, 4074, 4075, CG-232
rpoD	RNA polymerase	69.2	alt; RNA polymerase, σ^{70} subunit (transcription of genes for proteins induced at high temperature); EC 2.7.7.6	546, 556, 557, 1035, 1330, 1397, 1398, 1494, 2451, 2457, 2535, 2584, 2586, 3009, 3010, 3233, 3234, 3932, 3944, 4166, 4232, 4428, CG-231
rpoE	RNA polymerase	58.2	RNA polymerase σ^E subunit (transcription in response to heat shock and oxidative stress)	1731, 2536, 2772, 3474, 3618, 4780, CG-36181
rpoH	RNA polymerase	77.5	fam, hin, htpR; RNA polymerase σ^{32} subunit; EC 2.7.7.6	539, 579, 1144, 1315, 1495, 1496, 2069, 2102, 2321, 2975, 3046, 3327, 3995, 4180, 4272, 4273, 4338, 4410, 4788, 4823, CG-618
rpoN	RNA polymerase	72.0	glnF, ntrA; RNA polymerase σ^{60} subunit; transcription of genes controlled by N-source availability	628, 1735, 1831, 1877, 3743, CG-17926
rpoS	RNA polymerase	61.7	abrD, dpeB, katF, nur; RNA polymerase σ^S subunit, σ^S (stationary phase)	1900, 1933, 2326, 2525, 2652, 2946, 2947, 3680, 3696, 3768, 4206, 4301, 4302, 4431, 4449, CG-18208
rpoZ	RNA polymerase	82.3	spoS; in spoT operon; RNA polymerase, ω subunit	548, 1366–1368, 1859, 2514, 3735, CG-28445
rpsA	Ribosomal protein, small	20.7	ssyF; 30S ribosomal subunit protein S1	721, 1083, 2179, 2180, 3292, 3802, 3803, CG-230
rpsB	Ribosomal protein, small	4.1	30S ribosomal subunit protein S2	81, 277, 394, 1296, 2342, 4291, 4361, 4624, CG-229
rpsC	Ribosomal protein, small	74.2	30S ribosomal subunit protein S3	460, 4834, CG-228
rpsD	Ribosomal protein, small	74.1	ramA, sud$_2$; streptomycin sensitivity, dependence; 30S ribosomal subunit protein S4	259, 635, 1616, 3384, 3386, 3517, 3772, 4253, CG-227
rpsE	Ribosomal protein, small	74.2	eps, spc, spcA; spectinomycin sensitivity; 30S ribosomal subunit protein S5	635, 931, 1616, 2886, 4625, CG-226
rpsF	Ribosomal protein, small	95.3	sdgH (suppressor of dnaG mutation); 30S ribosomal subunit protein S6	H, 1740, 3804, CG-225
rpsG	Ribosomal protein, small	74.8	K12; 30S ribosomal subunit protein S7	1990, 3387, 3518, CG-224
rpsH	Ribosomal protein, small	74.2	30S ribosomal subunit protein S8	61, 635, 3187, 4656, CG-223
rpsI	Ribosomal protein, small	72.7	30S ribosomal subunit protein S9	682, 875, 1915, CG-17596
rpsJ	Ribosomal protein, small	74.3	nusE; 30S ribosomal subunit protein S10	828, 991, 2476, 2477, 2688, 3188, 4694, 4808, 4834, CG-222
rpsK	Ribosomal protein, small	74.1	30S ribosomal subunit protein S11	259, 635, 2054, 3384, CG-221
rpsL	Ribosomal protein, small	74.8	strA; 30S ribosomal subunit protein S12	1301, 1817, 3385, 3387, 4271, CG-220
rpsM	Ribosomal protein, small	74.1	30S ribosomal subunit protein S13	259, 635, 1180, 2476, 2478, 2878, 3384, CG-219
rpsN	Ribosomal protein, small	74.2	30S ribosomal subunit protein S14	635, CG-218
rpsO	Ribosomal protein, small	71.3	secC; 30S ribosomal subunit protein S15	1160, 2913, 3354, 3380, 3382, 3511, 3512, 3719, 4189, 4190, CG-217
rpsP	Ribosomal protein, small	59.0	30S ribosomal subunit protein S16	568, 569, 570, 4414, CG-216
rpsQ	Ribosomal protein, small	74.2	neaA; neamine sensitivity; 30S ribosomal subunit protein S17	596, 3385, 4695, 4834, CG-215
rpsR	Ribosomal protein, small	95.3	Kasugamycin sensitivity; 30S ribosomal subunit protein S18	869, 3804, 4693, CG-214
rpsS	Ribosomal protein, small	74.3	30S ribosomal subunit protein S19	4696, 4834, CG-213
rpsT	Ribosomal protein, small	0.5	sup$_{S20}$; 30S ribosomal subunit protein S20	2050, 2601, 2602, 4630, 4710, 4787, CG-212
rpsU	Ribosomal protein, small	69.1	30S ribosomal subunit protein S21	557, 870, 1143, 2584, 2586, 4413, 4428, CG-211
rpsV	Ribosomal protein, small	33.5	30S ribosomal protein S22	2620, CG-32198
rrfA	rRNA, 5S	87.0	5S rRNA of rrnA operon	417, 1120, CG-210
rrfB	rRNA, 5S	89.8	5S rRNA of rrnB operon	501, 503, CG-209
rrfC	rRNA, 5S	85.0	5S rRNA of rrnC operon	4777, CG-208
rrfD	rRNA, 5S	73.7	5S rRNA of rrnD operon	1080, CG-207
rrfE	rRNA, 5S	90.7	5S rRNA of rrnE operon	417, 1120, CG-206
rrfF	rRNA, 5S	73.7	rrfDβ, rrvD; 5S rRNA of rrnD operon	1080, CG-33643
rrfG	rRNA, 5S	58.5	5S rRNA of rrnG operon	1120, CG-205
rrfH	rRNA, 5S	5.0	5S rRNA of rrnH operon	1120, CG-204
rrlA	rRNA large, 23S	86.9	23S rRNA of rrnA operon	209, 417, 903, 1120, 4007, CG-203
rrlB	rRNA large, 23S	89.7	23S rRNA of rrnB operon	500, 501, 503, 853, 1615, 1691, 2016, 2397, 3063, 3064, 3163, 3421, 3609, 3656, 4246, 4789, CG-202
rrlC	rRNA large, 23S	84.9	23S rRNA of rrnC operon	810, 2014, 3884, CG-201

TABLE 1 *Continued*

Symbol	Mnemonic[b] for symbol	Position (min)[c]	Synonyms; enzyme or phenotype affected[d]	References[e]
rrlD	rRNA large, 23S	73.7	23S rRNA of *rrnD* operon	1080, CG-200
rrlE	rRNA large, 23S	90.6	23S rRNA of *rrnE* operon	417, 1120, 3306, CG-199
rrlG	rRNA large, 23S	58.6	23S rRNA of *rrnG* operon	45, 46, 1120, 3857, 3884, CG-198
rrlH	rRNA large, 23S	4.9	23S rRNA of *rrnH* operon	1120, CG-197
rrnA	rRNA	86.9	rRNA operon	CG-23593
rrnB	rRNA	89.7	*cqsE, rrnB1;* rRNA operon	CG-195
rrnC	rRNA	84.9	*cqsB;* rRNA operon	CG-19503
rrnD	rRNA	73.7	*cqsD;* rRNA operon	CG-193
rrnE	rRNA	90.6	*rrnD1;* rRNA operon	CG-192
rrnG	rRNA	58.5	rRNA operon	CG-10891
rrnH	rRNA	4.9	rRNA operon	CG-190
rrsA	rRNA small, 16S	86.9	16S rRNA of *rrnA* operon	903, 944, CG-189
rrsB	rRNA small, 16S	89.7	16S rRNA of *rrnB* operon	293, 416, 420, 501, 503, 856, 1384, 1615, 3164, 3207, 3253, 4790, 4791, CG-188
rrsC	rRNA small, 16S	84.9	16S rRNA of *rrnC* operon	810, 2014, 3884, CG-187
rrsD	rRNA small, 16S	73.8	16S rRNA of *rrnD* operon	2328, CG-186
rrsE	rRNA small, 16S	90.6	16S rRNA of *rrnE* operon	944, 2458, CG-185
rrsG	rRNA small, 16S	58.6	16S rRNA of *rrnG* operon	1120, 3894, CG-184
rrsH	rRNA small, 16S	4.9	16S rRNA of *rrnH* operon	1120, CG-183
rseA	Regulation of σ^E	58.2	*mclA;* membrane protein regulator of σ^E activity	N, CG-37241
rseB	Regulation of σ^E	58.2	Bonds *rseA*	N, CG-37244
rseC	Regulation of σ^E	58.2	Putative negative control of σ^E activity	N, CG-37247
rspA	Repression of σ^S-regulated proteins	35.5	Regulates expression of σ^S-dependent proteins	1824, CG-36959
rspB	Repression of σ^S-regulated proteins	35.5	Member of *rsp* operon, function unknown	1824, CG-36962
rssA	Regulator of σ^S	27.7	Regulates expression of σ^S-dependent proteins	P, CG-37468
rssB	Regulator of σ^S	27.7	Two-component response regulator, posttranscriptional control	P, CG-37471
rsuA	rRNA, small, pseudouridine	49.1	16S RNA pseudouridine 516 synthase	4658, CG-36034
rus	*ruv* suppressor	12.3	Resolves Holliday structures	2649, 3879, CG-32471
ruvA	Resistance UV	41.9	*lexA* regulon; Holliday junction recognition	154, 280, 281, 1940, 2515, 3878, 3918, 3926, 4183, 4225, 4329, 4330, CG-17923
ruvB	Resistance UV	41.9	*lexA* regulon; branch migration of Holliday structures	280, 281, 2515, 3878, 3918, 4225, 4329, 4330, CG-17920
ruvC	Resistance UV	42.0	Not SOS regulated; endonuclease that resolves Holliday structure; RuvC endonuclease	143, 282, 799, 1084, 1085, 1941, 2509, 3878, 3880, 3882, 4183, 4225, 4572, CG-32474
sad	Succinate semialdehyde dehydrogenase	34.3	Succinate-semialdehyde dehydrogenase, NAD dependent; EC 1.2.1.24	2665, 3961, CG-180
sbaA	Serine and branched amino acids	96.5	Regulation of serine and branched-chain amino acid metabolism	897, CG-179
sbcB	Suppression of *recBC*	44.9	*xonA;* suppresses *recB recC* mutations; exonuclease I	3056, 3314, 3315, 3417, 3715, 4146, CG-178
sbcC	Suppression of *recBC*	8.9	Cosuppressor with *sbcB* of *recB recC* mutations	638, 2362, 2508, 2510, 3026, 3509, CG-17917
sbcD	Suppression of *recBC*	8.9	Cosuppressor with *sbcB* of *recBC* mutations	1389, 2362, 3026, CG-30972
sbmA	Sensitivity to B17 microcin	8.5	Sensitivity to microcin B17; methyl malonyl-CoA mutase (*mcm*)	2350, 4762, CG-17914
sbp	Sulfate-binding protein	88.5	Periplasmic sulfate-binding protein	1665, 3359, CG-17911
sdaA	Serine deaminase	40.8	L-Serine deaminase	3868, 4119, 4121, CG-32312
sdaB	Serine deaminase	63.1	L-Serine deaminase L-SD2	3867, 3868, 4120, CG-33324
sdaC	Serine deaminase	63.0	Regulator of L-SD2; putative serine transporter	3867, 4120, CG-33329
sdhA	Succinate dehydrogenase	16.3	Succinate dehydrogenase flavoprotein subunit; EC 1.3.99.1	585, 909, 4595, 4643, CG-17908
sdhB	Succinate dehydrogenase	16.4	Succinate dehydrogenase iron-sulfur protein; EC 1.3.99.1	585, 909, 4595, 4643, CG-17905
sdhC	Succinate dehydrogenase	16.3	*cybA;* membrane anchor subunit of succinate dehydrogenase, cytochrome b_{556}; EC 1.3.99.1	585, 2957, 4643, CG-17902
sdhD	Succinate dehydrogenase	16.3	Succinate dehydrogenase subunit, hydrophobic; EC 1.3.99.1	585, 4595, 4643, CG-17899
sdiA	Suppress division inhibitors	43.0	Suppresses inhibitory effect of MinC/MinD division inhibitor; positive regulator enhancing transcription of one of *ftsQAZ* promoters	3875, 4507, CG-30919
secA	Secretory	2.3	*azi, pea, prlD;* translocation ATPase for protein export	29, 205, 246, 324, 521, 1204, 1242, 2236, 2464, 3190, 3191, 3792, 4787, CG-176

(Table continues)

TABLE 1 *Continued*

Symbol	Mnemonic[b] for symbol	Position (min)[c]	Synonyms; enzyme or phenotype affected[d]	References[e]
secB	Secretory	81.5	Protein export; chaperone SecB	69, 246, 324, 1260, 2165, 2276, 2277, 2365, 3408, 3486, 3792, 3995, 4589, CG-17896
secD	Secretory	9.2	Membrane component of protein export complex	324, 1328, 1329, 2560, 2712, 4125, CG-17893
secE	Secretory	89.9	*prlG;* inner membrane protein involved in protein secretion (with SecY)	29, 168, 323, 324, 372, 521, 1059, 1067, 1227, 1998, 2711, 2968, 3104, 3361, 3439, 3545, 3762, 3763, 4125, 4222, CG-34108
secF	Secretory	9.2	Membrane protein with protein secretion function	324, 1329, 2560, 2712, 3360, 3362, 3666, 4125, CG-29893
secG	Secretory	71.5	p12 cytoplasmic membrane protein involved with protein export	889, 3103, 3105, CG-33484
secY	Secretory	74.1	*prlA;* multispanning membrane protein, translocator of proteins (with SecE)	33–36, 167, 322–324, 521, 635, 1128, 1227, 1260, 1920, 1921, 1923, 3197, 3212, 3439, 3686, 3910, 3914, 3925, 4163, CG-18037
sefA	Septum formation	4.4	May be *fabZ*?; affects septum formation	3120a, CG-175
selA	Selenium	81.0	*fdhA* (formate dehydrogenase activity); conversion of seryl-tRNA$_{UCA}$ into selenocysteyl-tRNA$_{UCA}$; selenocysteine synthase	378, 1238, 2395, 3995, CG-785
selB	Selenium	80.9	*fdhA;* selenium metabolism; novel translation factor specific for selenocysteyl-tRNA$_{UCA}$	225, 378, 1237, 1239, 2395, 3995, 4046, CG-17890
selC	Selenium	82.6	*fdhC;* selenocysteine tRNA	225, 226, 378, 1237, 1987, 2225, 2395, 2396, 4046, CG-17887
selD	Selenium	39.7	Donor of reduced selenium in selenocysteic tRNA and in protein; selenophosphate synthase	378, 1103, 1237, 2164, 2394, 3755, 4047, CG-32292
semA	Sensitivity to microcin	40.5	Sensitivity to microcin E492	3440, CG-17884
seqA	Sequestration	15.4	DNA biosynthesis; negative modulator of initiation of replication	191, 2556, 4454, CG-31506
serA	Serine	65.8	Phosphoglycerate dehydrogenase; EC 1.1.1.95	3143, 4274, 4297, CG-173
serB	Serine	99.6	Phosphoserine phosphatase; EC 3.1.3.3	3059, 3577, CG-172
serC	Serine	20.6	*pdxC, pdxF;* phosphoserine aminotransferase; EC 2.6.1.52	998, 1083, 1335, 3059, 4399, CG-171
serR	Serine	1.9	Regulates level of seryl-tRNA synthetase	4245a, CG-170
serS	Serine	20.3	Serine hydroxamate sensitivity; seryl-tRNA synthetase; EC 6.1.1.11	329, 1614, 2363, 4297, CG-169
serT	Serine	22.2	*divE;* serine tRNA 1	2225, 4201, CG-168
serU	Serine	44.1	*su₁, Su-1, ftsM, supD, supH* (amber [UAG] suppression); serine tRNA 2	2225, 2367, 4061, 4261, CG-167
serV	Serine	60.7	*supD* (ochre [UAG] suppression); serine tRNA 3	2225, CG-166
serW	Serine	20.0	Duplicate gene; serine tRNA 5	2225, 3718, CG-17881
serX	Serine	23.7	*serW;* duplicate gene; serine tRNA 5	2225, CG-17878
sfcA	*sbc* fusion c	33.4	May be *mae,* malic enzyme, NAD linked; EC 1.1.1.38	2119, 2620, CG-32176
sfiC	Septum formation inhibition	25.7	Locus in defective prophage e14; cell division inhibition	868, 1825, 2617, CG-17875
sfsA	Sugar fermentation stimulation	3.5	Overexpression increases amylomaltase 10-fold; putative regulator of maltose metabolism	1296, 2070, 2128, CG-30524
sfsB	Sugar fermentation stimulation	71.8	*nlp, sfs1;* regulation of sugar fermentation enzymes?	162, 719, CG-35693
shiA	Shikimate	44.2	Shikimate and dehydroshikimate permease	1162, CG-163
sieB		30.5	Locus in defective prophage *rac*	1177, CG-3208
sipC	Suppressor of increased permeability	82.6	Reverses vancomycin sensitivity caused by plasmid TraT; outer membrane mutant suppressor	3449, CG-36866
sipD	Suppressor of increased permeability	82.6	Reverses vancomycin sensitivity caused by plasmid TraT; outer membrane mutant suppressor	3449, CG-36869
sir	SOS-independent repair	61.0	SOS-independent repair of mitomycin-induced damage	2282, CG-36505
sloB	Slow growth	75.0	Slow growth rate; tolerance to amidinopenicillin and nalidixic acid	B, 2536a, 4573a, CG-162
slp	Stationary-phase lipoprotein	78.7	C starvation and stationary phase inducible; outer membrane lipoprotein	55, 3854, 3995, CG-33726
slr	Selenolipoic acid resistant	(14.7)	Suppresses lipoate requirement of *lipA* strains; unknown function in synthesis of lipoic acid	3506, CG-36705
slt	Soluble lytic transglycosylase	99.8	Lytic murein transglycosylase, major autolysin	304, 1130, 1524, 3951, CG-34819
slyD	Sensitivity to lysis	74.8	FK-506-BP-like lysis protein for φX174; probable rotamase, peptidyl prolyl *cis-trans* isomerase	3599, 4680, CG-35441
smp	Serine B-contraposed membrane protein	99.6	Flanks *serB* gene; promoter divergently overlaps *serB* promoter some homology with EF-Ts and riboprotein L4	3060, 3061, CG-34649

TABLE 1 *Continued*

Symbol	Mnemonic[b] for symbol	Position (min)[c]	Synonyms; enzyme or phenotype affected[d]	References[e]
sms	Sensitivity to methylmethane sulfonate	99.7	Appears to be cotranscribed with *serB* but not related to serine metabolism; sequence similarities with ATP-dependent proteases	3057, CG-34653
sodA	SOD	88.3	Member of *soxRS* regulon; Mn-superoxide dismutase	605, 794, 1185, 1325, 1618, 2988, 3359, 3426, 3427, 3983, 4064, 4193, 4215, 4220, 4298, 4299, CG-17593
sodB	SOD	37.3	Fe-superoxide dismutase	605, 1185, 1490, 3054, 3773, 3983, 4048, 4065, CG-15256
sohA	Suppressor of *htr*	70.5	*prlF;* suppressor of *htrA;* putative protease	188, 2155, 3988, CG-35411
sohB	Suppressor of *htr*	28.6	Suppressor of *htr;* homology with inner membrane protease IV, which digests cleaved signal peptides	189, CG-31951
soxR	Superoxide	92.1	Regulatory protein of *soxRS* regulon; induces nine-protein superoxide regulon when intracellullar superoxide levels increase; dual regulatory system with *soxS*	72, 372, 996, 1465, 1713, 2487, 3129, 3130, 4331, 4666, 4667, CG-27798
soxS	Superoxide	92.1	*soxRS* regulatory system; homology with C-terminal regions of AraC positive regulatory family; dual regulatory system with *soxR*	72, 372, 996, 2450, 2487, 3129, 3130, 4666, 4667, CG-27801
speA	Spermidine	66.4	Biosynthetic arginine decarboxylase; EC 4.1.1.19	451, 2905, 3746, CG-161
speB	Spermidine	66.4	Agmatinase; EC 3.5.3.11	451, 2905, 3746, 4170, CG-160
speC	Spermidine	66.9	Ornithine decarboxylase; EC 4.1.1.17	451, 3746, CG-159
speD	Spermidine	2.9	*S*-Adenosylmethionine decarboxylase; EC 4.1.1.50	107, 1296, 2869, 4176, 4177, 4688, CG-158
speE	Spermidine	2.9	Spermidine synthase	1296, 2869, 4176, 4177, 4688, CG-17590
speF	Spermidine	15.5	Ornithine decarboxylase, inducible	2099, 2100, CG-31559
spf	Spot 42	87.2	"Spot 42" RNA	1894, 2020, 2227, 3365, 3534, CG-157
spoT	Spot ("magic spot")	82.3	Guanosine 5′-diphosphate, 3′-diphosphate pyrophosphatase; [p]ppGpp synthetase II activity	374, 3735, 4411, 4687, CG-156
spoU	Spot ("magic spot")	82.4	rRNA methylase, putative	548, 2237, 3735, 4687, CG-33877
sppA	Signal peptide peptidase	39.8	Protease IV; signal peptide peptidase	1850, 4159, CG-13763
srlA	Sorbitol	60.8	*gutA;* sorbitol (glucitol) utilization; D-glucitol-specific enzyme II of PTS	415, 3519, 4608, 4702, 4703, CG-155
srlB	Sorbitol	60.9	*gutB;* D-glucitol (sorbitol)-specific enzyme III of PTS	1560, 4702, 4703, CG-11886
srlD	Sorbitol	60.9	*gutD;* sorbitol (glucitol)-6-P dehydrogenase; EC 1.1.1.140	3126, 4608, 4702, 4703, CG-153
srlR	Sorbitol	60.9	*gutR;* regulatory gene for *srl*	4608, 4704, CG-152
srmB	Suppressor ribosomal mutant	58.3	*rbaB, rhlA;* ATP-dependent RNA helicase	2045, 3093, 3791, CG-32859
srnA	Stable RNA	9.4	Degradation of stable RNA	3167a, CG-151
ssaE	Suppression of *secA*	52.7	Suppresses *secA* mutations	3189, CG-17869
ssaG	Suppression of *secA*	41.8	Suppresses *secA* mutations	3189, CG-17866
ssaH	Suppression of *secA*	94.1	Suppresses *secA* mutations	3189, CG-17863
ssb	Single-strand binding	92.0	*exrB, lexC; lex* regulon member; single-strand DNA-binding protein	306, 372, 538, 622, 656, 1404, 2309, 2347, 2674, 2827, 2828, 3707, 3708, 3788, 4543, 4596, CG-150
sseA	Sensitivity to serine	54.8	Enhances serine sensitivity (homoserine dehydrogenase inhibition) on lactate; rhodanese-like protein (thiosulfate sulfur transferase?)	1573, CG-33023
sseB	Sensitivity to serine	54.9	Enhances serine sensitivity (homoserine dehydrogenase inhibition) on lactate, weaker enhancement than *sseA*	1573, CG-33026
sspA	Stringent starvation protein	72.7	*pog;* stress response protein	1297, 1298, 3782, 2862, 4597, 4598, CG-17860
sspB	Stringent starvation protein	72.7	Stress response protein	4597, 4598, CG-33594
ssrA	Small stable RNA	59.2	*sipB;* 10S RNA (nonribosomal)	662, 2227, 2837, 3156, 3157, 3527, CG-35783
ssrS	Small stable RNA	65.8	Stable 6S RNA	1805, 2227, 2369, 4042, CG-17857
ssyA	Suppression of *secY*	56.9	Suppressor of *secY* mutation	CG-17854
ssyD	Suppression of *secY*	3.0	Suppressor of *secY* mutation	3909, CG-17848
stfZ	*fts* in reverse	2.3	Antisense RNA that blocks *ftsZ* mRNA translation and inhibits cell division	1010, CG-36124
stkA	Suppressor of transposase killing	77.5	Suppresses cell aberrations and death caused by overexpression of Tn5 transposase	4555, CG-36083
stkB	Suppressor of transposase killing	86.8	Suppresses cell aberrations and death caused by overexpression of Tn5 transposase	4555, CG-36093
stkC	Suppressor of transposase killing	(99.4)	Suppresses cell aberrations and death caused by overexpression of Tn5 transposase	4555, CG-36087

(Table continues)

TABLE 1 *Continued*

Symbol	Mnemonic[b] for symbol	Position (min)[c]	Synonyms; enzyme or phenotype affected[d]	References[e]
stkD	Suppressor of transposase killing	(28.6)	Suppresses cell aberrations and death caused by overexpression of Tn5 transposase	4555, CG-36090
stpA	Suppressor of *td* phenotype	60.2	*hnsB;* hns-like protein; suppresses T4 *td* mutant	3903, 4809, CG-34492
strC	Streptomycin	7.0	*strB;* low-level streptomycin resistance; modifies ribosome structure	3563, CG-149
strM	Streptomycin	78.3	Control of ribosomal ambiguity	3712b, CG-148
stsA		83.9	Altered RNase activity	2404a, CG-147
sucA	Succinate	16.4	*lys, met;* α-ketoglutarate dehydrogenase (decarboxylase component)	532, 909, 910, 4015, 4016, 4595, 4643, CG-146
sucB	Succinate	16.4	*lys, met;* dihydrolipoamide succinyltransferase component of α-ketoglutarate dehydrogenase; EC 2.3.1.61	532, 3566, 4014–4016, CG-145
sucC	Succinate	16.5	Succinyl-CoA synthetase β subunit	531, 532, 2582, 4014, 4016, CG-17845
sucD	Succinate	16.5	Succinyl-CoA synthetase α subunit	531, 532, 2582, CG-17842
sufI	Suppressor of *ftsI*	68.1	*sui;* repressor of *ftsI*	777, 2116, CG-33437
sugE	Suppressor of *gro*	94.3	Suppresses a *groL* mutation and mimics effects of *gro* overexpression	1472, CG-34460
suhA	Suppressor of heat shock	78.3	Induction of heat shock genes	4273, CG-17839
suhB	Suppressor of heat shock	57.2	Inositol monophosphatase; EC 3.1.3.25	631, 2704, 3327, 3908, 4743, CG-32968
sulA	Suppressor of *lon*	22.0	*sfiA;* inhibits cell division and *ftsZ* ring formation	250, 767, 774, 1271, 1345, 1991, 2895, CG-144
sup	Suppressor designations		Class of mutations; renamed as individual tRNA loci; see synonym list (Table 2)	
supQ	Suppressor	12.5	Suppressor	3652a, CG-129
surA	Survival	1.2	Affects stationary-phase survival	S, 3557, 3636, 4294, 4787, CG-30326
surE	Survival	61.8	Affects stationary-phase survival	2431, CG-33215
syd	*secY*(d-1) suppression	63	Interacts with SecY	3915, CG-36980
tabC	T4 (abortion?)	86.2	Mutants fail to support growth of T4	4184, CG-125
tag	3-Methyladenine glycosylase	80.0	3-Methyladenine DNA glycosylase, constitutive	745, 1161, 2091, 3689, 3839, 3995, 4066, CG-124
talB	Transaldolase	0.18	Transaldolase B	4787, CG-36818
tanA	Trehalose-anaerobic	41.5	Anaerobic growth on trehalose	
tanB	Trehalose-anaerobic	68	Anaerobic growth on trehalose	
tap	Taxis protein	42.5	Flagellar-regulon member; chemotactic membrane receptor gene, receptor for aspartate; methyl-accepting chemotaxis protein IV, peptide receptor	447, 2262, 2868, 3968, 3969, 4499, CG-123
tar	Taxis to aspartate and repellents	42.5	*cheM;* flagellar-regulon member; chemotactic signal transducer protein; methyl-accepting chemotaxis protein II, aspartate chemoreceptor	4499, 4741, CG-122
tbpA	Thiamin-binding protein	1.6	Thiamin-binding protein	4787, CG-34691
tdcA	Threonine dehydratase, catabolic	70.3	LysR family of regulatory proteins; transcriptional activator of *tdc* operon	1318, 1426, 1555, 3829–3832, 4678, CG-29702
tdcB	Threonine dehydratase, catabolic	70.3	*tdc;* threonine dehydratase; EC 4.2.1.16	927, 1425, 1426, 2168, 3829, 3831, 3832, 4678, CG-17587
tdcC	Threonine dehydratase, catabolic	70.2	Anaerobically inducible L-threonine L-serine permease, membrane associated	1427, 3829, 3831, 3832, 4678, CG-29709
tdcR	Threonine dehydratase, catabolic	70.3	Positive regulatory protein for threonine dehydratase, TdcB	3829–3832, CG-29692
tdh	Threonine dehydrogenase	81.6	Threonine dehydrogenase	129, 3489, 3490, 3995, CG-17584
tdi	Transduction inhibition	(4.2)	Affects transduction, transformation, and rates of mutation	4044, CG-121
tdk	Thymidine kinase	27.8	Thymidine kinase	345, 379, 1726, 1862, 2406, CG-120
tehA	Tellurite resistance	32.3	Resistance to tellurite	4227, 4492, CG-32106
tehB	Tellurite resistance	32.3	Resistance to tellurite	4227, 4492, CG-32109
TerA	Terminus	28.8	Terminus of DNA replication; replication fork inhibition	956, 1261, 1709, 1719–1721, 1776, 3294, CG-17836
TerB	Terminus	36.2	*psrB;* terminus of DNA replication; replication fork inhibition	326, 327, 956, 1261, 1435, 1709, 1719–1721, 3294, CG-17833
TerC	Terminus	34.6	*psrA;* terminus of DNA replication; replication fork inhibition	274, 431, 1261, 1709, 1721, CG-17830
TerD	Terminus	27.5	Terminus of DNA replication; replication fork inhibition	1261, 1709, CG-17827
TerE	Terminus	23.3	Terminus of DNA replication; replication fork inhibition	1710, CG-29805
TerF	Terminus	49.9	A sixth replication arrest site, located within *rcsC*	3871, CG-29289
tesA	Thioesterase	11.1	*apeA;* acyl-CoA thioesterase I; also protease I	710, 711, 1849, CG-31102
tesB	Thioesterase	10.2	Thioesterase II	2983, 3028, CG-17581
tgt	tRNA-guanine transglycosylase	9.2	tRNA-guanine transglycosylase	1323, 1329, 3110, 3528, 3529, CG-118
thdA	Thiophene degradation	10.5	Degradation of furans and thiophenes; may be *tlnA*?	3, CG-17824

TABLE 1 *Continued*

Symbol	Mnemonic[b] for symbol	Position (min)[c]	Synonyms; enzyme or phenotype affected[d]	References[e]
thdC	Thiophene degradation	94.2	Degradation of furans and thiophenes	3, CG-17821
thdD	Thiophene degradation	99.8	Degradation of furans and thiophenes	3, CG-15908
thiA	Thiamin cluster	90.2	Hydroxyethylthiazole synthesis	
thiC	Thiamin (and thiazole)	90.3	Thiamin pyrimidine moiety biosynthesis; hydroxymethylpyrimidine synthesis	372, 4416, CG-117
thiD	Thiamin (and thiazole)	46.6	Phosphomethylpyrimidine kinase	1879, 3019, CG-114
thiE	Thiamin (and thiazole)	90.3	*thiA;* thiamin-thiazole moiety synthesis	372, 4416, CG-34301
thiF	Thiamin (and thiazole)	90.3	*thiA;* thiamin-thiazole moiety synthesis	372, 4416, CG-34298
thiG	Thiamin (and thiazole)	90.2	*thiA;* thiamin-thiazole moiety synthesis	372, 4416, CG-34295
thiH	Thiamin (and thiazole)	90.2	*thiA;* thiamin-thiazole moiety synthesis	372, 4416, CG-34292
thiK	Thiamin (and thiazole)	25.1	Thiamin kinase	1880, 3019, CG-113
thiL	Thiamin (and thiazole)	9.5	Thiamin monophosphate kinase	1880, 3019, CG-112
thiM	Thiamin (and thiazole)	46.6	Hydroxyethylthiazole kinase; EC 2.7.1.50	2887, 3019, CG-34309
thiN	Thiamin (and thiazole)	46.6	Hydroxymethylpyrimidine kinase	3019
thrA	Threonine	0.0	*HS; thrD;* aspartokinase I-homoserine dehydrogenase I; EC 1.1.1.3, EC 2.7.2.4	821, 902, 1525, 2114, 2592, 3928, 4787, 4792, CG-111
thrB	Threonine	0.1	Homoserine kinase; EC 2.7.1.39	821, 822, 1525, 3675, 4787, CG-110
thrC	Threonine	0.1	Threonine synthase; EC 4.2.99.2	821, 822, 1525, 3274, 4787, CG-109
thrS	Threonine	38.7	Autogenously regulated; threonyl-tRNA synthetase; EC 6.1.1.3	1284, 2647, 2653, 2730, 3035, 3242, 3356, 3358, 3415, 4031, 4033, 4476, 4570, 4676, CG-108
thrT	Threonine	89.9	Threonine tRNA 3	83, 1815, 2225, 2383, 3611, 4395, CG-107
thrU	Threonine	89.9	Threonine tRNA 4	83, 1815, 2225, 2383, 3611, 4422, CG-106
thrV	Threonine	73.7	Threonine tRNA 1 in *rrnD*	793, 1080, 1120, 2225, CG-105
thrW	Threonine	5.7	Threonine tRNA 2	793, 892, 2225, CG-10363
thyA	Thymine	63.8	Aminopterin, trimethoprim resistance; thymidylate synthetase; EC 2.1.1.45	267, 630, 1090, 1214, 1697, 1707, 2833, 3629, 3736, CG-104
tig	Trigger factor	9.8	Trigger factor; chaperone	48, 1540, CG-35681
tktA	Transketolase	66.3	Transketolase; EC 2.2.1.1	4028, 4029, 4170, 4817, CG-103
tktB	Transketolase	55.3	Transketolase; EC 2.2.1.1	1866, 4817, CG-29992
tlnA	Thiolutin	10.5	*tlnI;* resistance or sensitivity to thiolutin; may be *thdA?*	1651, 3955, CG-102
tmk	Thymidine kinase	24.6	Deoxythymidine kinase/thymidylate kinase	333, 942, CG-17578
tnaA	Tryptophanase	83.7	*ind, tnaR;* tryptophanase; EC 4.1.99.1	548, 974, 2032, 2841, 3317, 3730, 4080, 4085, 4278, 4525, 4783, CG-101
tnaB	Tryptophanase	83.8	*trpP;* low-affinity Trp permease	548, 974, 1099, 3229, 3730, 4085, CG-69
tnm	Tn migration	91.9	Transposition of Tn*9* and other transposons, development of phage Mu	1875, 2516, 4580, CG-100
tolA	Tolerance	16.8	*cim, excC, lky, tol-2;* bacteriocin tolerant; tolerance to group A colicins and single-stranded filamentous DNA phage; required for outer membrane integrity; membrane-spanning protein	464, 467, 602, 932, 1234, 1245, 2413, 2414, 4136, 4137, 4252, 4539, CG-99
tolB	Tolerance	16.8	*lky, lkyA* (leakage of periplasmic proteins), *tol-3;* azaleucine resistant; bacteriocin tolerant; tolerance to colicins E2, E, A, and K	92, 464, 602, 932, 1234, 1245, 2359–2361, 2414, 4136, 4137, CG-98
tolC	Tolerance	68.4	*colE1-i, mtcB, refl, weeA, toc* (topoisomerase compensation), *mukA;* bacteriocin tolerant; specific tolerance to colicin E1; affects segregation of daughter chromosomes; outer membrane porin	464, 602, 932, 1245, 1548, 1549, 1728, 1728a, 2919, 2920, 3085, 3217, 3476, 4428, 4495, 4737, 4738, CG-97
tolD	Tolerance	23.0	Bacteriocin tolerant; sensitivity to colicins E2 and E3, ampicillin	464, 550, 932, 1245, CG-96
tolE	Tolerance	23.0	Bacteriocin tolerant; sensitivity to colicins E2 and E3, ampicillin	464, 602, 932, 1245, CG-95
tolI	Tolerance	0.1	Bacteriocin tolerant; sensitivity to colicins Ia and Ib	464, 602, 932, CG-94
tolJ	Tolerance	0.1	Bacteriocin tolerant; sensitivity to L, A, S4, E, and K	464, 602, 932, 1245, CG-93
tolM	Tolerance	74.0	*cmt;* mutant phenotype: high-level tolerance to colicin M	464, 602, 932, 1245, 1575, 2775, 3761, CG-92
tolQ	Tolerance	16.7	*fii, tolP?;* tolerance to group A colicins and single-stranded DNA filamentous phage; required for integrity of cell envelope; inner membrane protein	462, 467, 2056, 2940, 3958, 4136, 4137, 4252, 4430, 4539, CG-17815
tolR	Tolerance	16.8	Tolerance to group A colicins and single-stranded DNA filamentous phage; required for integrity of cell envelope; inner membrane protein	467, 1679, 2056, 4137, 4252, 4539, CG-17812

(Table continues)

TABLE 1 *Continued*

Symbol	Mnemonic[b] for symbol	Position (min)[c]	Synonyms; enzyme or phenotype affected[d]	References[e]
tolZ	Tolerance	78.8	Tolerance to colicins E2, E3, D, Ia and Ib; generates chemical proton gradient	2714, CG-17809
tonB	T1	28.2	*T1rec, exbA;* sensitivity to phages T1, φ80, and colicins; uptake of chelated iron and cyanocobalamin; energy transducer	109, 239, 464, 467, 1051, 1215, 1430, 1451, 2028, 2082, 2083, 2162, 3389–3391, 3597, 3936, 4305, 4306, CG-90
topA	Topoisomerase	28.6	*supX;* DNA topoisomerase I, omega protein I	1027, 1069, 1904, 2405, 2470, 2591, 2668, 2896, 3203, 3215, 4071, 4318, 4319, 4333–4335, 4501, 4502, CG-89
topB	Topoisomerase	39.7	Topoisomerase III	1024, 3809, CG-29988
torA	Trimethylamine oxide reductase	22.8	Molybdo-protein trimethylamine *N*-oxide reductase	2787, 3276, 4355, CG-17575
torC	Trimethylamine oxide reductase	22.8	*c*-Type cytochrome	2787, CG-29977
torD	Trimethylamine oxide reductase	22.9	Cotranscribed with *tor*	2787, CG-29980
torR	Trimethylamine oxide reductase	22.8	Regulatory gene	3945, CG-36408
tpiA	Triose-F isomerase	88.5	Triosephosphate isomerase	1665, 3321, 3359, CG-88
tpr	Protamine-like protein	27.7	Protamine-like protein	70, 3610, CG-87
treA	Trehalose	26.8	*tre;* trehalase, periplasmic	404, 1541, 2775, 3526, CG-17572
treB	Trehalose	96.2	IITre, trehalose-specific PTS enzyme II	405, 4113, CG-34601
treC	Trehalose	96.1	Trehalose metabolism (stress protectant; osmoprotectant); trehalose-6-P hydrolase	405, 3547, 4113, 4138, CG-34597
treF	Trehalose	79.0	Cytoplasmic trehalase	3995, CG-35861
treR	Trehalose	96.2	Repressor of *treA,B,C*	G, 2185
trg	Taxis to ribose and galactose	32.1	Flagellar-regulon member; methyl-accepting chemotaxis protein III, ribose receptor	340, 341, 396, 1601, 1632, 2775, 2900, CG-85
trkA	Transport of K$^+$	74.0	Major constitutive potassium transport system; inner cytoplasmic membrane protein; potassium transport membrane protein subunit	424, 1136, 1137a, 1575, 2775, 3785, CG-84
trkD	Transport of K$^+$	84.6	*kup;* major constitutive potassium transport system	372, 425, 548, 1137a, 3781, CG-81
trkE	Transport of K$^+$	29.0	Major constitutive potassium transport system	1137a, CG-80
trkG	Transport of K$^+$	30.6	Major constitutive potassium transport system; membrane protein, with perhaps no activity in anaerobically grown bacteria	1054, 1136, 3786, 3787, 4307, CG-29963
trkH	Transport of K$^+$	86.8	Major constitutive potassium transport system; potassium transport membrane protein subunit; involved in binding of TrkA to membrane; mutants require elevated potassium for growth	903, 2996, 3787, CG-34066
trmA	tRNA methyltransferase	89.6	tRNA (uracil-5)-methyltransferase; EC 2.1.1.35	A, 372, 1534, 1535, 2483, 3134, 3135, 3304, CG-79
trmB	tRNA methyltransferase	(7)	tRNA(guanine-7)-methyltransferase; EC 2.1.1.33	2669a, CG-78
trmC	tRNA methyltransferase	52.8	5-Methylaminoethyl-2-thiouridine in tRNA	342a, 1554, 2669a, CG-77
trmD	tRNA methyltransferase	59.0	tRNA(guanine-7)-methyltransferase; EC 2.1.1.33	342a, 568, 569, 1554, 1751, CG-76
trmE	tRNA methyltransferase	83.7	*thdF* (multicopy effects on thiophene oxidation?); tRNA base modification; 5-methylaminoethyl-2-thiouridine biosynthesis	42, 548, 1122, CG-17806
trmF	tRNA methyltransferase	84.0	5-Methylaminoethyl-2-thiouridine biosynthesis	1122, CG-17803
trmU	tRNA methyltransferase	25.6	*asuE;* tRNA base modification, 2-thiouridine synthesis	BB[2], 3580a, 4131b, CG-37613
trnA	tRNA	62.4	Level of several tRNAs	699, CG-75
trpA	Tryptophan	28.3	Tryptophan synthase subunit A; EC 4.2.1.20	1132, 1430, 1511, 1724, 2469, 2715, 3076, 3580, 4164, 4660, 4661, 4746, CG-74
trpB	Tryptophan	28.3	Tryptophan synthase subunit B; EC 4.2.1.20	824, 843, 1132, 1231, 1430, 1716, 1724, 2715, 4746, 4818, CG-73
trpC	Tryptophan	28.3	Bifunctional enzyme *N*-(5-phosphoribosyl)anthranilate isomerase, indole-3-glycerol P synthetase; EC 4.1.1.48	722, 1132, 1430, 1724, 1782, 1785, 2715, 3774, 4612, 4746, CG-72
trpD	Tryptophan	28.4	Bifunctional enzyme anthranilate synthase component II, phosphoribosyl anthranilate transferase, glutamine amidotransferase; EC 4.1.3.27, EC 2.4.2.18	1132, 1430, 1724, 1781, 1782, 1784, 2446, 2715, 3074, 4746, 4795, CG-71
trpE	Tryptophan	28.4	*anth, tryD, tryp-4;* anthranilate synthase component I; EC 4.1.3.27	1132, 1430, 1658, 1724, 2447, 2715, 3074, 3075, 3232, 4746, CG-70
trpR	Tryptophan	99.8	5-Methyltryptophan resistance; regulator of *trp* operon and *aroH;* autogenously regulated; tryptophan repressor protein	194, 384, 1524, 1525, 2570, 2732, 2936, 3577, 3951, 4745, CG-68
trpS	Tryptophan	75.6	Tryptophanyl-tRNA synthetase; EC 6.1.1.2	383, 1563, 1564, 3584, 4618, CG-67
trpT	Tryptophan	85.0	*su7* (UGA suppression), *su8, supU* (amber [UAG] suppression), *supV* (ochre [UAA] suppression); tryptophan tRNA	903, 2225, 4777, CG-66

TABLE 1 *Continued*

Symbol	Mnemonic[b] for symbol	Position (min)[c]	Synonyms; enzyme or phenotype affected[d]	References[e]
truA	tRNA uridine (modified to pseudouridine)	52.5	*asuC, hisT, leuK;* pseudouridine synthase, anticodon stem and loop specific	130–132, 523, 2052, 2684, 3118, 3266, 3837, 4131a, 4370, CG-623
truB	tRNA uridine (modified to pseudouridine)	71.3	Pseudouridine synthase, tRNA Psi-55 specific	3132, CG-35528
trxA	Thioredoxin	85.4	*fipA, tsnC;* thioredoxin	51, 903, 1091, 1753, 1754, 1832, 2251, 2466, 2467, 2670, 2706, 3645–3647, 3650, 3991, 4488, CG-65
trxB	Thioredoxin	20.1	Thioredoxin reductase	989, 990, 1568, 3648, 3649, 3674, 4361, CG-17569
tsf	EF-Ts	4.1	EF-Ts; elongation factor for transcription, stable	81, 277, 394, 977, 1296, 2342, 4722, CG-64
tsmA	Thymine suppression modifier	40.0	Affects efficiency of the suppression of nonsense and frameshift mutations by thymine-requiring strains	1696, CG-36603
tsr	Taxis to serine and repellents	98.9	*cheD;* serine chemoreceptor; methyl-accepting chemotactic membrane receptor protein I	445, 446, 586, 1645, 2136, 320, 3533, 4498, CG-63
tsx	T6	9.3	*T6rec, nupA;* T6 resistance, colicin K resistance; nucleoside channel	471, 472, 1593, 2560, 3435, CG-62
ttdA	Tartrate dehydratase	69.0	L-Tartrate dehydratase subunit; EC 4.2.1.32	3053, 3499, CG-33459
ttdB	Tartrate dehydratase	69.0	L-Tartrate dehydratase subunit; EC 4.2.1.32	3053, 3499, CG-33462
tufA	Tu factor, EF-Tu	74.7	Duplicate genes *tufA,B;* EF-Tu subunit; elongation factor, unstable	83, 84, 372, 1846–2848, 1952, 2006, 2346, 2882, 4393, 4398, 4551, 4760, 4806, 4807, CG-61
tufB	Tu factor, EF-Tu	89.9	Duplicate genes *tufA,B;* EF-Tu subunit; elongation factor, unstable	82, 83, 372, 1815, 1846–1848, 1952, 2006, 2181, 2346, 2383, 2485, 2882, 4192, 4231, 4393, 4395, 4398, 4422, 4551, 4760, CG-60
tus	Terminus utilization substance	36.2	*tau;* inhibition of replication at *Ter;* DNA-binding protein for *Ter* sites; mutants are defective for *Ter* binding activity and do not block progress of replication fork at *Ter* sites; DNA sequence-specific contrahelicase	916, 1262, 1435, 1705, 1711, 1720, 1722, 2152a, 2200, 2376, 2377, 3036, 3294, 3576, 3576a, CG-17800
tynA	Tyramine	31.2	*maoA;* tyramine oxidase; EC 1.4.3.4	14, 173, 982, 1206, 1403, 1994, 2316, 2967, 3520, 3556, 3866, CG-59
tyrA	Tyrosine	58.9	Member of TyrR regulon; chorismate mutase T, prephenate dehydrogenase; EC 5.4.99.5, EC 1.3.1.12	1813, 2715, CG-58
tyrB	Tyrosine	91.9	Member of TyrR regulon; tyrosine aminotransferase; EC 2.6.1.5	292, 372, 1244, 1954, 2288, 4731, CG-57
tyrP	Tyrosine	42.9	Member of TyrR regulon; tyrosine-specific transport system	2103, 2104, 3730, 4142, 4577, 4653, 4654, CG-56
tyrR	Tyrosine	29.8	TyrR regulon repressor; regulates *aroF, aroG, tyrA,* and aromatic amino acid transport systems; autoregulatory repression of *tyrR* not mediated by tyrosine; TyrR repressor	118, 509, 590, 708, 761, 813, 814, 858–860, 2351, 2715, 2936, 3336, 4489, 4591, 4729, 4730, CG-55
tyrS	Tyrosine	36.9	Tyrosyl-tRNA synthetase; EC 6.1.1.1	224, 1021, 2314, 2855, 3897, CG-54
tyrT	Tyrosine	27.7	*su*III, *Su-3, Su-4, supC* (ochre [UAA] suppression), *supF* (amber [UAG] suppression); *tyrT*α; tandemly duplicated *tyrTV;* tyrosine tRNA 1	420, 421, 2225, 2406, 3610, 3611, CG-53
tyrU	Tyrosine	89.9	*sup15B, supM* (ochre [UAA] suppression), *supZ* (amber [UAG] suppression); tyrosine tRNA 2	83, 1815, 2225, 2383, 3611, CG-52
tyrV	Tyrosine	27.7	*tyrT, tyrT*β; tandemly duplicated *tyrTV;* tyrosine tRNA 1; see *tyrT* for suppressor phenotypes	3611, CG-51
ubiA	Ubiquinone	91.6	4-Hydroxybenzoate polyprenyltransferase	372, 2465, 2494, 2791, 3072, 3098, 3929, 4664, CG-50
ubiB	Ubiquinone	86.6	*fadI, fre, fsrC;* 2-octaprenylphenol→2-octaprenyl-6-methoxyphenol; flavin reductase; EC 1.6.99?	903, 1236, 1344, 4036, CG-49
ubiC	Ubiquinone	91.5	Chorismate lyase	372, 2465, 3072, 3098, 3929, 4664, CG-48
ubiD	Ubiquinone	86.6	3-Octoprenyl-4-hydroxybenzoate→ 2-octaprenylphenol	834a, CG-47
ubiE	Ubiquinone	86.6	2-Octoprenyl-6-methoxy-1,4-benzoquinone→ 2-octoprenyl-3-methyl-6-methoxy-1,4-benzoquinone	4776a, CG-46
ubiF	Ubiquinone	14.9	2-Octoprenyl-3-methyl-6-methoxy-1,4-benzoquinone→ 2-octoprenyl-3-methyl-5-hydroxy-6-methoxy-1,4-benzoquinone	786, CG-45
ubiG	Ubiquinone	50.3	2-Octoprenyl-3-methyl-5-hydroxy-6-methoxy-1,4-benzoquinone→ ubiquinone 8	1388, 1839, 3347, 4665, CG-44

(Table continues)

TABLE 1   *Continued*

Symbol	Mnemonic[b] for symbol	Position (min)[c]	Synonyms; enzyme or phenotype affected[d]	References[e]
ubiH	Ubiquinone	65.7	*visB*; 2-octoprenyl-6-methoxyphenol→ 6-methoxy-1,4-benzoquinone-1,4-benzoquinone	2997, CG-43
ubiX	Ubiquinone	52.3	*dedF*; probable polyprenyl *p*-hydroxybenzoate carboxylase	3118, CG-17797
udk	Uridine kinase	46.1	Uridine/cytidine kinase; EC 2.7.1.48	2018, 3055, 3056, CG-42
udp	Uridine phosphorylase	86.5	Uridine phosphorylase; EC 2.4.2.3	51, 903, 1688, 3323, 3245, 4494, CG-41
ugpA	Uptake of glycerol P	77.3	*psiB, psiC*; glycerol P transport system, integral membrane protein	1163, 2093, 2846, 3228, 3425, 3823, 3824, 3826, 3995, 4122, CG-40
ugpB	Uptake of glycerol P	77.3	*psiB, psiC*; periplasmic binding protein of *sn*-glycerol-3-P transport system	2846, 3228, 3823, 3824, 3826, 3995, 4122, CG-39
ugpC	Uptake of glycerol P	77.3	*sn*-Glycerol-3-P transport system; permease; member of ABC family	1656, 3228, 3823, 3824, 3995, 4122, CG-17794
ugpE	Uptake of glycerol P	77.3	*sn*-Glycerol-3-P transport system; membrane protein	3288, 3823, 3824, 3995, 4122, CG-17791
ugpQ	Uptake of glycerol P	77.3	Glycerophosphoryl diester phosphodiesterase, cytosolic; EC 3.1.4.-	528, 2092, 3288, 3995, 4122, 4283, CG-33697
uhpA	Utilization and uptake of hexose P	82.9	Response regulator (two-component regulatory system) required for *uhpT* transcription	548, 1282, 1911, 2031, 3887, 4574, CG-15437
uhpB	Utilization and uptake of hexose P	82.9	Membrane protein controlling UhpA activity, sensor kinase	548, 1282, 1911, 4574, CG-15448
uhpC	Utilization and uptake of hexose P	82.8	Membrane protein controlling UhpA activity in concert with UhpB	548, 1282, 1910, 1911, 2031, 3887, 4574, CG-15449
uhpT	Utilization and uptake of hexose P	82.8	Fosfomycin sensitivity; sugar P transport system; transport protein for hexose phosphates	548, 1157, 1282, 1538, 1911, 2031, 2646, 2814, 2846, 3886, 3887, 4421, 4574, CG-37
uidA	Hexuronides	36.4	*gurA, gusA*; β-D-glucuronidase; EC 3.2.1.31	363, 364, 1175, 1970, 3775, CG-36
uidR	Hexuronides	36.5	*gusR*; regulatory gene	361, 363–365, CG-35
umuC	UV mutator	26.5	*uvm*; SOS mutagenesis; UV induction of mutations, error-prone repair; forms complex with UmuD and UmuD′	491, 947, 1119, 2178, 2204, 2678, 3303, 3842, 3843, 3917, 4648, 4650, CG-34
umuD	UV mutator	26.5	*uvm*; SOS mutagenesis; inducible mutagenesis; error-prone repair; processed to UmuD ; forms complex with UmuC	947, 1119, 2178, 2204, 2678, 3303, 3842, 3843, 3917, 4648–4650, CG-17788
ung	Uracil-nucleic acid glycosylase	58.3	Uracil-DNA-glycosylase	4418, CG-25
upp	Uracil phosphoribosyltransferase	56.2	*uraP*; uracil phosphoribosyltransferase; EC 2.4.2.9	89, 1768, CG-24
ups	(Up suppression?)	27.1	Efficiency of nonsense suppressors; see *prfA*	832, 3654, CG-23
uraA	Uracil	56.2	Uracil-concentration dependence in *pyr* mutant strains; Ura ABC transporter	88, CG-33043
ushA	UDP-sugar hydrolase	10.9	UDP-glucose hydrolase (5 -nucleotidase)	552, 832, CG-22
uspA	Universal stress protein	78.4	Global regulatory gene for stress response; transcriptionally activated, using σ^{70} during growth arrest, changing pattern of proteins produced	1688, 3138, 3139, 3995, CG-28127
uup		21.8	Precise excision of insertion elements	1768, CG-17785
uvrA	UV resistance	91.9	*dar*; *lex* regulon member; repair of UV damage to DNA; excision nuclease subunit A	177, 372, 457, 1837, 2150, 2527, 3704, 3705, 3707, CG-21
uvrB	UV resistance	17.5	*dar-1, dar-6*; DNA repair; excision nuclease subunit B; DNA-dependent ATPase I and helicase II	119, 178, 1232, 2527, 3256, 3702, 3704, 3709, 4171, 4396, CG-20
uvrC	UV resistance	43.0	DNA repair; multicopy causes mucoidy; excision nuclease subunit C	1240, 1241, 2904, 3125, 3703, 3704, 3710, 3873–3875, 4348, 4404, 4759, CG-19
uvrD	UV resistance	86.1	*dar-2, dda, mutU, pdeB, rad, recL, uvr-502, uvrE, srjC*; UV sensitivity and increased rates of spontaneous mutagenesis; DNA-dependent ATPase I-DNA helicase II	51, 135, 787, 903, 1092, 1209, 1210, 1706, 2284, 2527, 2654, 1797, 3148, 3149, 3930, 4221, 4253, 4717, CG-18
uxaA	Utilization of hexuronate galacturonate	69.5	*exu* regulon member; galacturonate utilization; altronate hydrolase; EC 4.2.1.7	870, 1819, 3377, 3554, CG-17
uxaB	Utilization of hexuronate galacturonate	56.1	*exu* regulon member; galacturonate utilization; altronate oxidoreductase; EC 1.1.1.58	359, 360, 1821, 2698, CG-16
uxaC	Utilization of hexuronate galacturonate	69.5	Uronate isomerase; EC 5.3.1.12	870, 1819, 2697, 2698, 3377, 3554, CG-15
uxuA	Utilization of hexuronate galacturonate	98.1	Mannonate hydrolase; EC 4.2.1.8	362, 1269, 3553, 3559, CG-14
uxuB	Utilization of hexuronate galacturonate	98.1	Mannonate oxidoreductase; EC 1.1.1.57	362, 1270, 3553, 3559, CG-13
uxuR	Utilization of hexuronate galacturonate	98.1	Regulatory gene for *uxuBA* operon	3552, 3553, 3559, CG-12

TABLE 1 *Continued*

Symbol	Mnemonic[b] for symbol	Position (min)[c]	Synonyms; enzyme or phenotype affected[d]	References[e]
valS	Valine	96.5	*val-act;* valyl-tRNA synthetase; EC 6.1.1.9	183, 1612, 1643, 1644, 3964, CG-11
valT	Valine	16.9	Duplicate gene with triplicated *valUXY* locus; valine tRNA 1	2225 4774, CG-10
valU	Valine	54.3	*valUα;* tandemly triplicate *valUXY* and also duplicate of *valT;* valine tRNA 1	518–520, 2225, 3141, CG-17782
valV	Valine	37.6	*val* tRNA 2B	2225, CG-17563
valW	Valine	37.6	*val* tRNA 2A	2225, CG-17560
valX	Valine	54.3	*valUβ;* tandemly triplicate *valUXY* and also duplicate of *valT;* valine tRNA 1	518, CG-28696
valY	Valine	54.3	*valUγ;* tandemly triplicate *valUXY* and also duplicate of *valT;* valine tRNA 1	518, CG-28699
vsr	Very short (patch) repair	43.8	DNA repair system; endonuclease for certain mismatches	307, 308, 1583, 1677, 2455, 2456, 2815, 3998, CG-28927
wrbA	W(Trp) repressor binding	23.0	Regulatory gene; affects association of *trp* repressor and operator	4739, CG-31836
xapA	Xanthosine P	54.4	*pndA;* xanthosine phosphorylase	518, 565, CG-9
xapR	Xanthosine P	54.4	*pndR;* regulatory gene for *xapA*	518, 565, CG-8
xasA	Extreme acid sensitivity	33.7	Allows survival in extreme acid conditions	BB, CG-36518
xerC	*cer*-specific recombination	86.0	Recombinase, site specific; acting at *cer* and *dif* sites	354–356, 787, 903, 2748, CG-30184
xerD	*cer*-specific recombination	65.4	*xprB;* recombinase, site specific; acting at *cer* and *dif* sites	355, 356, 2554, CG-30201
xseA	Exonuclease VII	56.5	Exonuclease VII, large subunit	657, 4385, CG-6
xseB	Exonuclease VII	9.5	Exonuclease VII, small subunit	M, 1293, 4385, 4386, CG-17557
xthA	Exonuclease III	39.4	Exonuclease III	3583, 3726, CG-7
xylA	Xylose	80.3	D-Xylose isomerase; EC 5.3.1.5	240, 481, 1279, 2352, 2643, 3603, 3767, 3995, CG-5
xylB	Xylose	80.3	Xylulokinase; EC 2.7.1.17	481, 2352, 2643, 3603, 3995, CG-4
xylE	Xylose	91.3	Xylose-proton symport	372, 936, 937, 11263, 2626, CG-17776
xylF	Xylose	80.3	*xylT;* xylose-binding protein, transport system	937, 3603, 3995, CG-17773
xylG	Xylose	80.4	Putative xylose transport ATP-binding protein	3995, CG-33780
xylH	Xylose	80.4	Putative xylose transport membrane protein	3995, CG-33783
xylR	Xylose	80.4	Regulatory gene	2643, 3603, 3995, CG-3
zwf	Zwischenferment	41.7	Member of *soxRS* regulon; glucose 6-P dehydrogenase; EC 1.1.1.49	612, 805, 3623, CG-2

[a]We have retained the table format and some of the phenotypic descriptions from previous editions of the map (e.g., Bachmann [175]). Abbreviatons: ABC, ATP-binding cassette; ACP, acyl carrier protein; cAMP, cyclic AMP; CoA, coenzyme A; CPS, capsular polysaccharide; DCCD, *N,N* -dicyclohexylcarbodiimide; DEAD box, Asp-Glu-Ala-Asp box; DMSO, dimethyl sulfoxide; DTT, dithiothreitol; FAD, flavin adenine dinucleotide; fMet, formylmethionyl; FMN, flavin mononucleotide; GABA, γ-aminobutyric acid; GAR, glycinamide ribonucleotide; IHF, integration host factor; KDO, 2-keto-3-deoxyoctulosonic acid; LPS, lipopolysaccharide; MPT, molybdopterin; NAc, *N*-acetyl; NEM, *N*-ethylmaleimide; ORF, open reading frame; P, phosphate; PBP, penicillin-binding protein; PEP, phosphoenolpyruvate; PRPP, 5-phosphoribosyl-1-pyrophosphate; PTS, phosphotransferase system; RNP, ribonucleoprotein; SDS, sodium dodecyl sulfate; SOD, superoxide dismutase.

[b]The descriptive phrase from which the symbol was derived.

[c]Map location in minutes (see Fig. 1 and 2).

[d]Synonyms for the symbols that have been used in the literature; the gene product, if known; the Enzyme Nomenclature Commission (EC) number where appropriate; and/or the phenotypic trait affected.

[e]The prefix CG indicates the unique identifier of this gene or other site in the CGSC database. Although the database is easily queried by name or synonym, with synonyms linked to the primary name, this unchanging identifier is useful in unusual circumstances for tracking gene name changes that may have resulted in homonyms among the alternative names or any very complicated history of name changes. Additional references: A indicates that additional references can be found in Table 1 of Bachmann's map (reference 174); B indicates that additional references can be found in reference 175; C refers to chapter 40 of this book (Bacher et al.). Personal communications: D, G. Ames; E, S. Bass; F, S. Benson; G, W. Boos; H, R. Britton and J. Lupski; I, M. Chamberlin; J, A. J. Clark; K, D. Clark; L, T. Conway; M, D. Diaz; N, C. Gross; O, J. R. Guest; P, R. Hengge-Aronis; Q, D. Ivey; R, P. Klemm; S, R. Kolter; T, J. Lengeler; U, E. C. C. Lin; V, R. Meganathan; W, M. Norregaard-Madsen, E. McFall, and P. Valentin-Hansen; X, A. Sancar; Y, M. Shanley; Z, K. Shanmugam; AA, S. Short; BB, J. Slonczewski; BB[1], G. Slupska, C. Scheimer, R. Lloyd, and J. H. Miller; BB[2], D. Söll; CC, V. Stewart; DD, G. Storz; EE, W.-J. Tang; FF, L. Thony-Meyer and H. Hennecke; GG, T. Webster; HH, H. Wu.

[f]The order of the *dsd* loci shown on the physical map (Fig. 2) differs from the order reported in Fig. 1. This is thought to reflect strain differences for this apparently unstable region (M. Norregaard-Madsen, E. McFall, and P. Valentin-Hansen, personal communication).

TABLE 2 Alternate gene designations

Synonym	Table 1 symbol
abrD	rpoS
aceE1	aceE
aceE2	aceF
acrB	gyrB
acrE	acrB
act	alaS, glyS, pflA
ade3	purE
Ad₄	purA
ade_c	purN
ade_f	purE
ade_g	purC
ade_h	purB
ade_k	purA
ade_ub	purF
adhC	adhE, regulatory region
adi	adiA
ads	folD
adth_a	purD
adth_ub	purM
aeg-46.5	ccm-nap operon
aidA	alkA
aidD	alkB
ala-act	alaS
alaWα	alaW
alaWβ	alaX
ald	fba
alnA	dadB
alnR	dadQ,R
alt	rpoD
ampA	ampC
ams	rne
amt	cysQ
ana	adhE
anr	osmE
ant	nhaA
ant?	mutR
anth	trpE
apeA	tesA
apk	lysC
arg	car
Arg1	argA,D
Arg2	argA,C
Arg4	argE
Arg5	argF
Arg6	argG
argA	argE
argB	argA
argD	argF
argE	argG
argG	argD
argH	argC
arg+ura	carA,B
argVα	argV
argVβ	argY
argVδ	argQ
argVγ	argZ
aroR	aroT
arsE	arsR
arsF	arsB
arsG	arsC
asp	ppc
aspB	gltB,D
asuA?	prfA
asuC	truA
asuD	lysS
asuE?	trmU
ata	attP22
atthtt	attHK022
att92	attλ
att434	attλ

TABLE 2 *Continued*

Synonym	Table 1 symbol
avr	ileR
Az	atoC
azi	secA
bfe	btuB
bglC	bglF,G
bglE	bglT
bglS	bglG
bglY	hns
bicA	bcr
bicR	bcr
bioB	bioH
bioR	birA
bir	bioP
birB	bioP
bisA	moaA
bisB	moeA
bisD	mog
blgA	bglB
blgB	bglF
blu	malP, pgl, pgm
brnP	ilvH
cadR	lysP
cap	carA,B, crp
capR	lon
car	ptsG
carP	pepA
cat	ptsG
cbr	fepA
cbt	fepA
cdfA	cfa
cel	ispB
cer	btuB, rnhA
cet	creD
cfxB	marR
cheC	fliM
cheD	tsr
cheM	tar
cheX	cheR
chlA	moaA
chlB	mob
chlC	narG,H
chlD	modC
chlE	moeA
chlG	mog
chlI	narI
chlJ	modB
chlM	moaD
chlN	moeB
chpAI	chpR
chpAK	chpA
chpBI	chpS
chpBK	chpB
cim	tolA
clpY	hslU
cmlB	ompF
cmt	tolM
coa	ompF
colE1-i	tolC
con	ompA
cop	het
corB	mgtA
Cou	gyrB
cpxB?	marA
cqsB	rrnC
cqsD	rrnD
cqsE	rrnB
CR	ptsG
cru	nupC
cry	ompF,R
csi-5	osmY

TABLE 2 *Continued*

Synonym	Table 1 symbol
csm	crp
ctr	ptsH,I
cur	hns
cutA2	dipZ
cvc	fabF
cxr	cxm
cybA	sdhC
cycY	cutA
cycZ	dipZ
cysP	cysJ
cysQ	cysI
cysZ	cysK
cyxA	appC
cyxB	appB
dadR	dadA
dagA	cycA
dap	asd
dapB	dapE
dar	uvrA
dar-1	uvrB
dar-2	uvrD
dar-6	uvrB
dasC	rep
dasF	rnhA
dda	uvrD
ddl	ddlB
dec	fadR
dedB	accD
dedE	cvpA
dedF	ubiX
deg	lon
dga	murI
dhbB	birA
dhl	lpd
dinA	polB
dir	lon
divA	ftsA
divE	serT
DLP12	qsr'
dnaD	dnaC
dnaF	nrdA
dnaL	lig
dnaP	dnaG
dnaR	prs
dnaS	dfp, dut
dnaW	adk
dnaY	argU
dnaZ	dnaX
dpeA	hupB
dpeB	rpoS
dpp	dppG
dra	deoC
drc	hns
drdX	hns
drm	deoB
drp	proS
drs	hns
dsbD	dipZ
dut	dfp
dye	arcA
ecfB	cpxA
EF-Ts	tsf
efg	nadE
emrR	mprA
envA	lpxC
envB	mreB, emrE
envC	acrE
envD	acrF
envM	fabI
envZ	ompR

TABLE 2 *Continued*

Synonym	Table 1 symbol
eps	rpsE
eryA	rplD
eryB	rplV
eup	cpxA, ecfB?
exbA	tonB
excC	pal, tolA
exrA	lexA
exrB	ssb
F21.5	clpP
fabC	fabB
fabE	accB
fabG	accC
fabJ	fabF
fadI	ubiB
fam	rpoH
far	fusA
fda	fba
fdhA	selA,B
fdhC	selC
fdp	fbp
fdv	mutS
feeB	leuW
feuA	cirA
fexA	arcA
fii	tolQ
fimD	fimA
fimG	hns
fipA	trxA
fir	lpxD
firA	hlpA, lpxD
fit?	infC
flaA	fliM
flaAI	fliL
flaBI	fliF
flaBII	fliG
flaBIII	fliH
flaC	fliI
flaD	fliA
flaE	fliK
flaF	fliC
flaG	flhB
flaH	flhA
flaI	flhC
flaJ	motA,B
flaK	flgE
flaL	flgG
flaM	flgI
flaN	fliE
flaO	fliJ
flaP	fliR
flaQ	fliQ
flaR	fliP
flaS	flgK
flaT	flgL
flaU	flgA
flaV	flgD
flaW	flgC
flaX	flgF
flaY	flgH
flaZ	flgJ
flbA	flgB
flbB	flhD
flbC	fliD
flbD	fliO
flr	leuJ
flrA	ileR
fms	def
fpg	mutM
fpk	fruK
frdB	fnr

(Table continues)

TABLE 2   *Continued*

Synonym	Table 1 symbol
frdR	*narL,X*
fre	*ubiB*
fruC	*fruR*
fruF	*fruB,K*
fsrB	*hmp*
fsrC	*ubiB*
ftsB	*nrdB*
ftsH	*hflB*
ftsM	*serU*
ftsR	*rpoB*
ftsS	*ftsX*
fucC	*fucA*
gad	*gap*
gadS?	*gadA*
galB	*galT*
gapB	*epd*
gen-165	*ftn*
genA	*dcuA*
genF	*dcuB*
genR	*lrhA*
glmD	*nagB*
glnF	*rpoN*
glnR	*glnL*
glnT	*glnG*
glnUα	*glnU*
glnUβ	*glnW*
glnVα	*glnV*
glnVβ	*glnX*
glr	*murI*
glu	*ppc*
glut	*gltA*
gly	*glyS*
glyD	*glpD, gpt*
glySα	*glyQ*
glySβ	*glyS*
glyVα	*glyV*
glyVβ	*glyX*
glyVγ	*glyY*
gntM	*gntT*
gpp	*gpt*
gppB	*aslA*
gpt	*ptsG*
gptB	*manX,Z*
gro	*dnaK*
groEL	*groL*
groES	*groS*
groN	*rpoB*
groNB	*nusB*
groP	*dnaB,J,K*
groPAB	*dnaK*
groPC	*dnaK*
groPF	*dnaK*
grpA	*dnaB*
grpC	*dnaJ, dnaK*
grpF	*dnaK*
gsa	*hemL*
gsr	*crr*
gts	*qmeA*
gurA	*uidA*
gusA	*uidA*
gusR	*uidR*
gut	*srl*
gutA	*srlA*
gutB	*srlB*
gutD	*srlD*
gutR	*srlR*
gxu	*gpt*
hag	*fliC*
hdh	*groE*
herA	*rnhA*
herC	*lysS*

TABLE 2   *Continued*

Synonym	Table 1 symbol
hflA	*hflC,K,X*
hid	*himA*
himB	*gyrB*
hin	*rpoH*
hip	*himD*
hisE	*hisI*
hisT	*hisR, truA*
hisU	*gyrB*
hisW	*gyrA*
hnsB	*stpA*
hom	*asd*
hopD	*hupB*
hopE	*recD*
hpr	*ptsH*
hrbA	*brnQ*
hrbB	*livG,H,J,K*
hrbC	*livG,H,J,K*
hrbD	*livG,H,J,K*
Hs	*thrA*
hs	*hsdM,R*
hsdH	*hdhA*
hslS	*ibpB*
hslT	*ibpA*
hsm	*hsdM*
hsp	*hsd, hsdM,R*
hsr	*hsdR*
hss	*hsdS*
htpE	*ibpA*
htpI	*hslU*
htpN	*ibpA*
htpO	*hslV*
htpR	*rpoH*
htpY	*htgA*
htrA	*degP*
htrH	*hhoB*
htrP	*ribB*
hyd	*nik*
hydB	*hypE*
hydC	*hybA, nik, nikA,B,C,D*
hydD	*nikA,B,C,D,E*
hydE	*hypB*
hydF	*hypD*
hydL	*hybA*
iarA	*dsbA*
iarB	*dsbB*
icdB	*gltA*
icdE	*icd*
icl	*aceA*
iex	*crr*
ihb	*lrp*
ile	*ilvA*
ilvA	*ilvC*
ilvC	*ilvE*
ilvJ	*ilvE*
ind	*tnaA*
ins	*glyV,W*
int(qsr')	*intD*
ior	*recB*
irk	*hns*
K12	*rpsG*
kac	*kdpA,B,C,D*
katF	*rpoS*
kdgA	*eda*
kga	*eda*
kim	*qin*
kup	*trkD*
lcs	*asnS*
lct	*lldD*
lctD	*lldD*
lctP	*lldP*
lctR	*lldR*

TABLE 2 *Continued*

Synonym	Table 1 symbol
ldh	dld
leuK	truA
leuVα	leuV
leuVβ	leuP
leuVγ	leuQ
lexB	recA
lexC	ssb
lipP?	nikE
lir	acrA
livR	lrp
lky	tolA,B
lkyA	tolB
lnt	cutE
LopC	clpX
LopP	clpP
lov	argS
lovB	alaS
loxB	attP1,P7
lss	lrp
lstR	lrp
luxH-like	ribB
lys	sucA,B
lysTα	lysT
lysTβ	lysW
mae?	sfcA?
malA	malQ,P,T
malB	malK,E,F,G,M, lamB
maoA	tynA
mas	aceB
mazE	chpR
mazF	chpA
Mb	acrA
mbf	lrp
mbl	acrA
mbrA	rep
mbrC	murI
mbrD	rpoB
mclA	rseA
mdoA	mdoG,H
mdrA	cydC
mdrH	cydC
mec	dcm
mel-4	melB
mel-7	melA
meoA	ompC
met	sucA,B
met-1	metB
metM	metL
metTα	metT
metTβ	metU
metZα	metZ
metZβ	metV
mglD?	galS
mglP	mglA,C
micA	mutY
mlpA	lpp
mmrA	rep
mni	manC
molA	malM
mon	mreB, emrE
mop	groE
mopA	groS
mopB	groL
motD	fliN
mpt	manX,Z
mra	murC,F
mrbA	murA
mrc	lpcB
mrsC	hflB
msgA	ftsN

TABLE 2 *Continued*

Synonym	Table 1 symbol
msmA	dksA
msmB	cspC
msmC	cspE
msp	arcA
mssA	cmk
mssB	deaD
msuA?	dadX
msyA	hns
mtcA	acrA
mtcB	tolC
mtlC	mtlAp,o
muc	lon
mukA	tolC
murZ	murA
mutA	glyV
mutC	glyW
mutD	dnaQ
mutR	mutH
mutU	uvrD
mvrA	fpr
mvrC	emrE
nagR	nagC
nalA	gyrA
nalC	gyrB
nalD	gyrB
nam	pncA
narB	mob
narC	narG
narD	modC
narR	narL,X
ncf	hemB
neaA	rpsQ
nek	rimK
nfsB	nfnB
nirA	fnr
nirR	fnr
nitA	rho
nitB	rpoB
nlp	sfsB
nlpE	cutF
nmpA	pst, pstS
nmpB	phoR
nov	cls
ntrA	rpoN
ntrB	glnL
ntrC	glnG
ntrL	nadE
nucR	deoR
nupA	tsx
nupG	deoR
nur	rpoS
nusD	rho
nusE	rpsJ
old	fad
oldA	fadA
oldB	fadB
ole	fadR
ompB	envZ, ompR
ompE	phoE
ompH	hlpA
omsA	lpxD
oppI	lrp
optA	dgt
osmZ	hns
osrA	proU
ostA	imp
panK	coaA
papA	atpA
papB	atpD
papC	atpG

(Table continues)

TABLE 2 *Continued*

Synonym	Table 1 symbol
papD	*atpB*
papE	*atpH*
papF	*atpF*
papG	*atpC*
papH	*atpE*
par	*ompC*
parA	*gyrB*
parB	*dnaG*
parD	*gyrA*
parvA	*ppiC*
paxA	*dcd*
pbpA	*mrdA*
pbpB	*ftsI*
pbpF	*mrcB*
pcbA	*gyrB*
pcsA	*dinD*
pdeB	*uvrD*
pdeC	*lig*
pdxC	*serC*
pdxF	*serC*
pdzA	*rplT*
pea	*secA*
pel	*manY*
perA	*envZ*
pexB	*dps*
pfv	*dacA*
pgsB	*lpxB*
phe-act	*pheS*
pheR	*pheU*
pheW	*pheU*
phmA	*nmpC*
phoM	*creC*
phoM-orf2	*creB*
phoR1	*phoR*
phoR2a	*pstS*
phoR2b	*pstA*
phoRc	*phoB*
phoS	*pstS*
phoT	*phoB,U, pstA,B*
phoW	*pstC*
phs	*rpoA*
pil	*fim, fimB,C,D*
pilA	*fimA*
pilB	*fimC*
pilC	*fimD*
pilD	*fimF,G*
pilE	*fimH*
pilG	*hns*
pilH	*fimE*
pin	*argU*
plsA	*adk*
PMG	*mgl*
pmi	*manA*
pndA	*xapA*
pndR	*xapR*
poaA	*putA*
pog	*sspA*
poh	*oriC*
polC	*dnaE*
pon	*lpcB*
ponA	*mrcA*
ponB	*mrcB*
popA	*hemH*
popB	*hemF*
popC	*hemL*
popE	*hemC*
ppfA	*dsbA*
ppsA	*pps*
prd	*fucP*
prlA	*secY*
prlB	*rbsB*

TABLE 2 *Continued*

Synonym	Table 1 symbol
prlC	*opdA*
prlD	*secA*
prlF	*sohA*
prlG	*secE*
Pro2	*proC*
pro2	*proB*
pro3	*proC*
proL	*proA*
proU	*proM; proVWX operon*
proV	*proK*
proW	*proL*
prv	*mutH*
psiB	*ugpA,B*
psiC	*ugpA,B*
psiD	*phn, phnD*
psiH	*phoH*
psrA	*TerC*
psrB	*TerB*
psuA	*rho*
ptsF	*fruA*
ptsL	*manX*
ptsM	*man, manX,Y,Z*
ptsN	*nagE*
ptsP	*manY*
ptsX	*manX,Z*
pup	*deoD*
Pur2	*purE*
purC	*purF*
purE2	*purK*
purG	*purM*
purI	*purL*
pyrA	*car, carA,B*
R1pho	*phoR*
R2pho	*pstA,S*
Rarg	*argR*
rad	*uvrD*
radB	*recN*
ramA	*rpsD*
ramB	*rimG*
rap	*pth*
rbaB	*srmB*
rblA	*lrp*
rbsP	*rbsA,B,C,D*
rbsT	*rbsA,C*
recL	*lexA*
recL	*uvrD*
refI	*tolC*
refII	*creD*
relC	*rplK*
res	*rimF*
resA	*polA*
rglA	*mcrA*
rglB	*mcrB*
rhaC	*rhaR,S*
rhlA	*srmB*
ribG	*ribD*
ribH	*ribE*
rif	*rpoB*
rimA	*rpmH*
rm	*hsdM,R*
rnh	*rnhA*
rnsA	*rna*
R-MG	*mglR*
rnsC	*rho*
rodA	*mrdB*
rodY	*mreB*
ron	*rpoB*
rorA	*recB*
rot	*ppiA*
rpoF	*fliA*
rrfDβ	*rrfF*

TABLE 2 *Continued*

Synonym	Table 1 symbol
rrnB1	*rrnB*
rrnD1	*rrnE*
rrvD	*rrfF*
rsgA	*ftn*
rts	*coaA*
sac	*asc, ascB,F*
sbcA	*rac*
sbl	*srl*
sdg	*rpoB*
sdgA	*dnaG*
sdgB	*rpoB*
sdrA	*rnhA*
sec	*hemF*
secC	*rpsO*
sefA?	*fabZ?*
seg	*arcA, dnaK*
sep	*ftsI*
serW	*serX*
sez	*rpoA*
sfcA??	*maeA, sfiA, sulA*
sfiB	*ftsZ*
sfrA	*arcA*
sfrB	*rfaH*
sfs1	*sfsB*
shl	*fruR*
sin	*rnhA*
sipB	*acrA, ssrA*
skp	*hlpA, lpxD*
slpA	*intA*
slr	*lplA*
smbA	*pyrH*
sof	*dut*
som	*rfb*
soxQ	*marA,R*
spc	*rpsE*
spcA	*rpsE*
spoS	*rpoZ*
spoV?	*recG*
spr	*lexA*
srf	*recA*
srjA	*infC*
srjB	*helD*
srjC	*uvrD*
ssaD	*nusB*
ssaF	*rpmH*
ssc	*lpxD*
ssd	*cpxA*
ssyB	*nusB*
ssyF	*rpsA*
ssyG	*infB*
stc	*micF*
stl	*rpoB*
strA	*rpsL*
strB	*strC*
stsB	*rimH*
stv	*rpoB*
Su-1	*serU*
su₁	*serU*
Su2	*glnV*
Su-3	*tyrT*
Su-4	*tyrT*
Su-6	*leuX*
su7	*trpT*
su8	*trpT*
suA36	*glyU*
suB	*glnU*
Suβ	*lysT*
sud₂	*rpsD*
sueB	*prfA*
sufD	*glyU*

TABLE 2 *Continued*

Synonym	Table 1 symbol
sui	*sufI*
Su_II	*glnV*
su_III	*tyrT*
sulB	*ftsZ*
sumA	*glyT,U*
sumB	*glyU*
sun	*rho*
sup	*topA*
sup15B	*tyrU*
supB	*glnU*
supC	*tyrT*
supD	*serU,V*
supE	*glnV*
supF	*tyrT*
supG	*lysT*
supH	*serU*
supK	*prfB*
supL	*lysT*
supM	*tyrU*
supN	*lysV,U*
supP	*leuX*
sup_s20	*rpsT*
supT	*glyU*
supU	*trpT*
supV	*trpT*
supX	*topA*
sur	*bcr*
surB	*cydC*
suxA	*bcr*
T1	*fhuA*
T1rec	*tonB*
T5rec	*fhuA*
T6rec	*tsx*
tabB	*groE*
tabD	*rpoB,C*
talC	*alaT*
talD	*alaU*
tasC	*aspT*
tau	*tus*
tcp	*rimJ*
tdc	*tdcB*
tgs	*crr, purU*
tgtB	*gltT*
tgtC	*gltU*
tgtE	*gltV*
thdB	*fadR*
thdF	*trmE*
thiA	*thiE,F,G,H*
thiB	*thi*
thyR	*deoB,C*
tlnI	*tlnA*
tlr	*deoB,C*
tls	*aspS*
tmrA	*folA*
tnaR	*tnaA*
toc	*tolC*
tol-2	*tolA*
tol-3	*tolB*
tolF	*ompF*
tolG	*ompA*
tolP?	*tolQ*
tonA	*fhuA*
topX	*hns*
tos	*prfC*
tpo	*envZ*
tpp?	*dppA,B,C,D,E,F?*
tpp-75	*deoA*
tre	*treA*
trkB	*kefB*
trkC	*kefC*

(Table continues)

TABLE 2 *Continued*

Synonym	Table 1 symbol
trpP	*tnaB*
trpR	*aroT*
trpX	*miaA*
tryD	*trpE*
tryp-4	*trpE*
tsc	*deoR*
tsnC	*trxA*
tsp	*prc*
tss	*asnS*
tsu	*rho*
ttr	*fadL*
tut	*ompA*
tyrT	*tyrV*
tyrTα	*tyrT*
tyrTβ	*tyrV*
uar	*prfA*
umg	*ptsG*
umpA	*lgt*
umuA	*lexA*
umuB	*recA*
uncA	*atpA*
uncB	*atpB*
uncC	*atpC*
uncD	*atpD*
uncE	*atpE*
uncF	*atpF*
uncG	*atpG*
uncH	*atpH*

TABLE 2 *Continued*

Synonym	Table 1 symbol
uncI	*atpI*
ups?	*prfA*
ura	*car*
uraP	*upp*
usg	*accD*
usgA	*gntT*
uvm	*umuC,D*
uvr502	*uvrD*
uvrE	*uvrD*
uvrF	*recF*
val-act	*valS*
valUα	*valU*
valUβ	*valX*
valUγ	*valY*
virR	*hns*
visA	*hemH*
visB	*ubiH*
vtr	*fabF*
weeA	*tolC*
witA	*kgtP*
xerA	*argR*
xerB	*pepA*
xonA	*sbcB*
xprA	*dsbC*
xprB	*xerD*
xylT	*xylF*
zab	*recA*

APPENDIX
Circular Map and Recalibrated Map Positions of Transposon Insertion Sites
MITCHELL SINGER AND K. BROOKS LOW

Many transposon insertions for the *E. coli* K-12 genome have been used for mapping new genes. In particular, the set described by Singer et al. (3949) has been used and cited a number of times and should continue to be very useful. Accordingly, the transposon insertion sites for this set of strains have been compared with the new coordinates reported in the accompanying chapter, and the locations have been recalculated by using cotransduction frequencies for nearby genes as was done before (3949). Both the old and the new values are given in Table 3, and the locations of the insertions are shown outside the circular map drawn in Fig. 3. The strains

carrying these insertions are available from the CGSC (see chapter 134 of this volume). Other transposon insertions in known genes available from the CGSC are indicated in Fig. 3 by loci drawn inside the circle. The CGSC also maintains strains with other transposon insertions in intergenic regions (not shown in Fig. 3).

ACKNOWLEDGMENT

We thank E. Low for the graphics, editing, and word processing.

TABLE 3 Corrected transposon insertion genotypes and map locations

Strain	New genotype	Old genotype[a]	Map location (min) Old[a]	New
CAG18442	*thr-34*::Tn*10*	*thr-34*::Tn*10*	0.00	0.0
CAG18425	*thr-3091*::Tn*10kan*	*thr-3091*::Tn*10kan*		
CAG12093	*car-96*::Tn*10*[b]	*car-96*::Tn*10*	0.75	0.7
CAG18620	*car-3092*::Tn*10kan*[b]	*car-3092*::Tn*10kan*		
CAG12095	*ab-3051*::Tn*10*	*zac-3051*::Tn*10*	2.00	1.9
CAG12131	*zab-3093*::Tn*10kan*	*zac-3093*::Tn*10kan*		
CAG12025	*zad-220*::Tn*10*	*zad-220*::Tn*10*	3.50	3.2
CAG12105	*zad-3094*::Tn*10kan*	*zad-3094*::Tn*10kan*		
CAG18436	*zae-502*::Tn*10*	*zae-502*::Tn*10*	4.75	4.4
CAG18580	*zae-3095*::Tn*10kan*	*zae-3095*::Tn*10kan*	4.75	4.6
CAG18447	*proAB81*::Tn*10*	*proAB81*::Tn*10*	6.25	5.7

TABLE 3 *Continued*

Strain	New genotype	Old genotype[a]	Map location (min)	
CAG18515	*proAB3096*::Tn*10kan*	*proAB3096*::Tn*10kan*		
CAG18633	*zag-3198*::Tn*10kan*	*zag-3198*::Tn*10kan*	6.75	6.6
CAG12080	*zah-281*::Tn*10*	*zah-281*::Tn*10*	7.75	7.5
CAG18439	***lacI42*::Tn*10* *lacZU118***	***lacI42*::Tn*10***	**8.00**	**7.9**
CAG18420	***lacI3098*::Tn*10kan* *lacZU118***	***lacI3098*::Tn*10kan***		
CAG18091	*zaj-305*::Tn*10 proC*	*zaj-3053*::Tn*10*	9.00	9.0
CAG12148	***tsx-247*::Tn*10***	***tsx-247*::Tn*10***	**9.50**	**9.3**
CAG18413	***tsx-3100*::Tn*10kan***	***tsx-3100*::Tn*10kan***		
CAG12017	*zba-3054*::Tn*10*	*zba-3054*::Tn*10*	10.50	10.1
CAG12107	*zba-3101*::Tn*10kan*	*zba-3101*::Tn*10kan*		
CAG12154	*zba-3055*::Tn*10*	*zbb-3055*::Tn*10*	11.50	10.9
CAG12171	***purE79*::Tn*10***	***purE79*::Tn*10***	**12.25**	**11.9**
CAG18566	***purE3200*::Tn*10kan***	***purE3200*::Tn*10kan***		
CAG12021	*zbd-3105*::Tn*10*	*zbd-3105*::Tn*10*	13.25	13.1
CAG12116	*abd-3200*::Tn*10kan*	*zbd-3200*::Tn*10kan*		
CAG12149	*zbe-601*::Tn*10*	*zbd-601*::Tn*10*	14.50	14.3
CAG12077	*zbe-280*::Tn*10*	*zbe-280*::Tn*10*	15.00	14.7
CAG12123	*zbe-3105*::Tn*10kan*	*zbe-3105*::Tn*10kan*		
CAG18433	*zbf-3057*::Tn*10*	*zbf-3057*::Tn*10*	16.25	15.9
CAG18514	*zbf-3106*::Tn*10kan*	*zbf-3106*::Tn*10kan*		
CAG12147	***nadA57*::Tn*10***	***nadA57*::Tn*10***	**16.75**	**16.9**
CAG18341	***nadA3052*::Tn*10kan***	***nadA3052*::Tn*10kan***		
CAG18493	*zbi-29*::Tn*10*	*zbh-29*::Tn*10*	17.75	18.0
CAG18531	*zbi-3108*::Tn*10kan*	*zbh-3108*::Tn*10kan*		
CAG12034	*zbi-3058*::Tn*10*	*zbi-3058*::Tn*10*	18.75	18.7
CAG18478	*zbj-1230*::Tn*10*	*zbj-1230*::Tn*10*	20.00	19.8
CAG18528	*zbj-3110*::Tn*10kan*	*zbj-3110*::Tn*10kan*		
CAG12094	*zcb-3059*::Tn*10*	*zcb-3059*::Tn*10*	21.00	21.1
CAG18466	*zcc-282*::Tn*10*	*zcc-282*::Tn*10*	22.25	22.6
CAG18703	***putP5*::Tn*5***	***putP5*::Tn*5***	**22.75**	**23.3**
CAG12078	*zce-726*::Tn*10*	*zce-726*::Tn*10*	24.25	24.6
CAG12124	*zce-3113*::Tn*10kan*	*zce-3113*::Tn*10kan*		
CAG18463	*zcf-117*::Tn*10*	*zcf-117*::Tn*10*	25.25	25.5
CAG18516	*zcf-3114*::Tn*10kan*	*zcf-3114*::Tn*10kan*		
CAG18497	***fadR13*::Tn*10***	***fadR13*::Tn*10***	**25.75**	**26.6**
CAG18544	***fadR3115*::Tn*10kan***	***fadR3115*::Tn*10kan***		
CAG12016	*zch-3060*::Tn*10*	*zcg-3060*::Tn*10*	26.75	27.4
CAG12106	*zch-3166*::Tn*10kan*	*zcg-3116*::Tn*10kan*		
CAG12169	*zci-506*::Tn*10*	*zch-506*::Tn*10*	27.25	28.0
CAG18551	*zci-3117*::Tn*10kan*	*zch-3117*::Tn*10kan*		
CAG18458	***trpB83*::Tn*10***			**28.3**
CAG18579	***trpB3193*::Tn*10kan***			
CAG12028	*zcj-223*::Tn*10*	*zci-233*::Tn*10*	28.50	29.0
CAG12111	*zcj-3118*::Tn*10kan*	*zci-3118*::Tn*10kan*		
CAG12081	*zda-3061*::Tn*10*	*zcj-3061*::Tn*10*	29.50	30.4
CAG12026	***trg-2*::Tn*10***	***trg-2*::Tn*10***	**31.00**	**32.1**
CAG12108	***trg-3120*::Tn*10kan***	***trg-3120*::Tn*10kan***		
CAG18461	*zdd-235*::Tn*10*	*zdc-235*::Tn*10*	32.00	33.3
CAG18576	*zdd-3117*::Tn*10kan*	*zdc-3117*::Tn*10kan*		
CAG12027	*zde-230*::Tn*9*	*zdd-230*::Tn*9*	32.75	34.0
CAG18637	*zdg-3062*::Tn*5*	*zdf-3062*::Tn*5*	34.50	36.0
CAG18462	*zdh-603*::Tn*10*	*zdg-603*::Tn*10*	35.75	37.1
CAG18567	*zdh-3121*::Tn*10kan*	*zdg-3121*::Tn*10kan*		
CAG12151	*zdi-925*::Tn*10*	*zdh-925*::Tn*10*	37.50	38.6
CAG18568	*zdi-3122*::Tn*10kan*	*zdh-3122*::Tn*10kan*		
CAG18464	*zdj-276*::Tn*10*	*zdi-276*::Tn*10*	38.25	39.4
CAG18465	*zea-225*::Tn*10*	*zdj-225*::Tn*10*	39.50	40.3
CAG18578	*zea-3124*::Tn*10kan*	*zdj-3124*::Tn*10kan*		
CAG12074	*zea-3068*::Tn*10*	*zea-3068*::Tn*10*	40.25	40.9
CAG12122	*zea-3125*::Tn*10kan*	*zea-3125*::Tn*10kan*		
CAG18486	***eda-51*::Tn*10***	***eda-51*::Tn*10***	**40.75**	**41.6**
CAG18561	***eda-3126*::Tn*10kan***	***eda-3126*::Tn*10kan***		
CAG12156	***uvrC279*::Tn*10***	***uvrC279*::Tn*10***	**42.25**	**43.0**
CAG18451	*zed-3069*::Tn*10*	*zed-3069*::Tn*10*	43.00	43.9
CAG18563	*zed-3128*::Tn*10kan*	*zed-3128*::Tn*10kan*		
CAG12099	*zef-3129*::Tn*10*	*zee-3129*::Tn*10*	44.25	45.4
CAG12176	*zef-3189*::Tn*10kan*	*zee-3189*::Tn*10kan*		
CAG12179	***mgl-500*::Tn*10***	***mgl-500*::Tn*10***	**45.75**	**48.2**
CAG12098	*zei-722*::Tn*10*	*zeg-722*::Tn*10*	46.5	48.3

(Table continues)

TABLE 3 *Continued*

Strain	New genotype	Old genotype[a]	Map location (min)	
CAG12100	zei-3130::Tn10kan	zeg-3130::Tn10kan		
CAG12177	zej-298::Tn10	zeh-298::Tn10	47.75	49.6
CAG18577	zej-3142::Tn10kan	zeh-3142::Tn10kan		
CAG12178	zfa-723::Tn10	zei-723::Tn10	48.50	50.4
CAG12183	zfa-3143::Tn10kan	zei-3143::Tn10kan		
CAG18484	zfb-223::Tn10	zej-223::Tn10	49.50	51.5
CAG18483	**fadL771::Tn10**	**fadL771::Tn10**	**50.50**	**53.0**
CAG18467	zfd::Tn10	zfb-1::Tn10	51.00	53.4
CAG18522	zfd-3135::Tn10kan	zfb-3135::Tn10kan		
CAG18468	**nupC510::Tn10**	**nupC510::Tn10**	**51.75**	**54.1**
CAG18565	**nupC3146::Tn10kan**	**nupC3146::Tn10kan**		
CAG18632	zff-3071::Tn10kan	zfc-3071::Tn10kan	52.75	55.3
CAG18470	**purC80::Tn10**	**purC80::Tn10**	**53.25**	**55.7**
CAG18524	**purC3137::Tn10kan**	**purC::Tn10kan**		
CAG188631	zfg-3138::Tn10kan	zfe-3138::Tn10kan	53.50	56.1
CAG18469	**gua-26::Tn10**	**gua-26::Tn10**	**54.00**	**56.5**
CAG18481	zfh-208::Tn10	zff-208::Tn10	54.75	57.5
CAG18570	zfh-3139::Tn10kan	zff-3139::Tn10kan		
CAG18480	**nadB51::Tn10**	**nadB51::Tn10**	**55.75**	**58.2**
CAG18412	**nadB3140::Tn10kan**	**nadB3140::Tn10kan**		
CAG12158	**pheA18::Tn10**	**pheA18::Tn10**	**56.75**	**58.9**
CAG18608	**pheA3141::Tn10kan**	**pheA3141::Tn10kan**		
CAG18642	zfj-3131::Tn10	zfh-3131::Tn10	57.50	59.4
CAG18562	zga-3143::Tn10kan	zfi-3143::Tn10kan	58.25	60.5
CAG12173	**cysC95::Tn10**	**cysC95::Tn10**	**59.25**	**61.9**
CAG12182	**cysC3152::Tn10kan**	**cysC3152::Tn10kan**		
CAG12079	**fuc-3072::Tn10**	**fuc-3072::Tn10**	**60.25**	**63.2**
CAG12115	**fuc-3154::Tn10kan**	**fuc-3154-::Tn10kan**		
CAG12135	**recD::Tn10**			**63.5**
CAG18709	zge-3074::Tn10	zgc-3074::Tn10	62.00	64.7
CAG12168	zgf-210::Tn10	zgd-210::Tn10	63.50	66.0
CAG18604	zgf-3156::Tn10kan	zgd-3156::Tn10kan		
CAG18452	**nupG511::Tn10**	**nupG511::Tn10**	**64.25**	**66.9**
CAG18559	**nupG3157::Tn10kan**	**nupG3157::Tn10kan**		
CAG18475	**metC162::Tn10**	**metC162::Tn10**	**65.00**	**67.9**
CAG18527	**metC3158::Tn10kan**	**metC3158::Tn10kan**		
CGA12184	**tolC210::Tn10**	**tolC210::Tn10**	**66.25**	**68.4**
CAG12152	zgj-3075::Tn10	zgh-3075::Tn10	67.00	69.5
CAG18574	zgj-3159::Tn10kan	zgh-3159::Tn10kan		
CAG12072	zha-203::Tn10	zgj-203::Tn10	68.75	70.9
CAG12127	zha-3198::Tn10kan	zgj-3198::Tn10kan		
CAG12153	zhc-6::Tn10	zha-6::Tn10	70.00	72.1
CAG18605	zhc-3168::Tn10kan	zha-3168::Tn10kan		
CAG12071	zhd-3082::Tn10	zhb-3082::Tn10	71.75	73.3
CAG12159	zhd-9::Tn10	zhc-9::Tn10	72.00	73.8
CAG18606	zhd-3::Tn10kan	zhc-3170::Tn10kan		
CAG12075	zhe-3083::Tn10	zhd-3083::Tn10	72.75	74.7
CAG12133	zhe-3171::Tn10kan	zhd-3171::Tn10kan		
CAG18452	zhf-3085::Tn10kan	zhe-3085::Tn10kan	74.50	75.8
CAG22119	zhg-50::Tn10	zhf-50::Tn10	75.75	76.9
CAG18638	zhh-21::Tn10	zhg-3086::Tn10	76.50	77.8
CAG18639	zhi-3087::Tn10kan	zhi-3087::Tn10kan	77.75	78.9
CAG18640	zhj-3076::Tn10	zhj-3076::Tn10	78.50	79.7
CAG12175	zia-3077::Tn10kan	zia-3077::Tn10kan	80.00	81.0
CAG12163	zib-207::Tn10	zib-207::Tn10	80.75	81.8
CAG18569	zib-3160::Tn10kan	zib3160-::Tn10kan		
CAG18492	zic-4901::Tn10	zic-4901::Tn10	81.75	82.7
CAG18499	zid-501::Tn10	zid-501::Tn10	83.00	83.9
CAG18558	zid-3162::Tn10kan	zid-3162::Tn10kan		
CAG18501	zie-296::Tn10	zie-296::Tn10	83.75	84.5
CAG18592	zie-3163::Tn10kan	zie-3163::Tn10kan		
CAG18431	**ilv-500::Tn10**	**ilv-500::Tn10**	**84.50**	**85.2**
CAG18599	**ilv-3164::Tn10kan**	**ilv-3164::Tn10kan**		
CAG18491	**metE3079::Tn10**	**metE3079::Tn10**	**85.50**	**86.4**
CAG18496	**fadAB101::Tn10**	**fadAB101::Tn10**	**86.25**	**86.7**
CAG18557	**fadAB3165::Tn10kan**	**fadAB3165::Tn10kan**		
CAG18495	zih-35::Tn10	zih-35::Tn10	87.00	87.4
CAG18601	zih-3166::Tn10kan	zih-3166::Tn10kan		
CAG18636	zih-3088::Tn10kan	zii-3088::Tn10kan	87.50	87.9
CAG18477	zij-501::Tn10	zij-501::Tn10	88.50	89.1

TABLE 3 *Continued*

Strain	New genotype	Old genotype[a]	Map location (min)	
CAG12185	*argE86*::Tn*10*	*argE86*::Tn*10*	89.50	89.4
CAG18500	*thi-39*::Tn*10*	*thi-39*::Tn*10*	90.25	90.3
CAG18618	*thi-3178*::Tn*10kan*	*thi-3178*::Tn*10kan*		
CAG18498	*zjb-504*::Tn*10*	*zjb-504*::Tn*10*	90.75	91.0
CAG18615	*zjb-3179*::Tn*10kan*	*zjb-3179*::Tn*10kan*		
CAG12164	*malF3089*::Tn*10*	*malF3089*::Tn*10*	91.50	91.4
CAG18609	*malF3180*::Tn*10kan*	*malF3180*::Tn*10kan*		
CAG18630	*zjc-3181*::Tn*10kan*	*zjc-3181*::Tn*10kan*	92.50	92.5
CAG18488	*zjd-2231*::Tn*10*	*zjd-2231*::Tn*10*	93.75	93.7
CAG18571	*zjd-3182*::Tn*10kan*	*zjd-3182*::Tn*10kan*		
CAG18427	*zje-2241*::Tn*10*	*zje-2241*::Tn*10*	94.50	94.1
CAG12073	*cycA30*::Tn*10*	*cycA30*::Tn*10*	95.75	95.5
CAG12114	*cycA3185*::Tn*10kan*	*cycA3185*::Tn*10kan*		
CAG12019	*zjg-920*::Tn*10*	*zjh-920*::Tn*10*	96.75	96.2
CAG12110	*zjg-3186*::Tn*10kan*	*zjh-3186*::Tn*10kan*		
CAG18429	*zjh-6*::Tn*10*	*zji-6*::Tn*10*	98.25	97.7
CAG18610	*zjh-3187*::Tn*10kan*	*zji-3187*::Tn*10kan*		
CAG18430	*zji-202*::Tn*10*	*zjj-202*::Tn*10*	99.50	98.6
CAG18619	*zji-3188*::Tn*10kan*	*zjj-3188*::Tn*10kan*		

[a]Referring to what was published by Singer et al. (3949). Paired strains have insertions at identical sites.
[b]*car* alleles are probably point mutations very tightly linked to the transposon insertions.

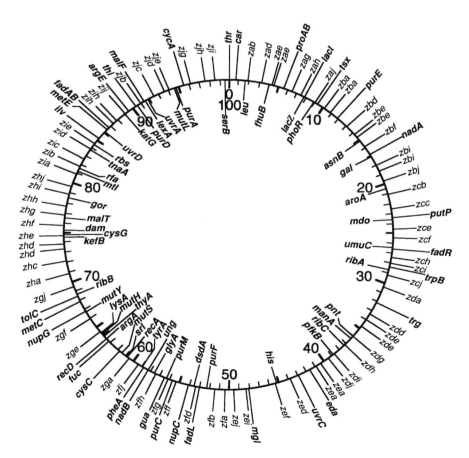

FIGURE 3 Circular map of *E. coli* K-12 showing positions of transposon insertions. Those indicated outside the circle are from Singer et al. (3949), now available from the CGSC. Those indicated inside the circle, as well as others not shown, are additional examples available from the CGSC.

for help with the required Macintosh reformatting of Table 2 and Linda Mattice for proofreading work on tables and figures. We thank Amos Bairoch and Bobby Baum for information and helpful comments regarding Table 1. This chapter continues the tradition of *E. coli* K-12 linkage map editions so capably constructed by Barbara J. Bachmann over the past 20 years and her predecessors for *E. coli* maps dating back to 1958. Our debt to these previous maps is obvious. And of course we are indebted to scientists too numerous to mention for helpful discussions, preprints, and personal communications of map-related information.

LITERATURE CITED

1. Aasland, R., J. Coleman, A. L. Holck, C. L. Smith, C. R. Raetz, and K. Kleppe. 1988. Identity of the 17-kilodalton protein, a DNA-binding protein from *Escherichia coli*, and the *firA* gene product. *J. Bacteriol.* **170:**5916–5918.

2. Aasland, R., and C. Smith. 1988. An efficient manual method for aligning DNA restriction map data on very large genomic restriction maps. *Nucleic Acids Res.* **16:**10392.

3. Abdulrashid, N., and D. P. Clark. 1987. Isolation and genetic analysis of mutations allowing the degradation of furans and thiophenes by *Escherichia coli*. *J. Bacteriol.* **169:**1267–1271.

4. Aberg, A., S. Hahne, M. Karlsson, A. Larsson, M. Ormo, A. Ahgren, and B. M. Sjoberg. 1989. Evidence for two different classes of redox-active cysteines in ribonucleotide reductase of *Escherichia coli*. *J. Biol. Chem.* **264:**12249–12252.

5. Abouhamad, W. N., M. Manson, M. M. Gibson, and C. F. Higgins. 1991. Peptide transport and chemotaxis in *Escherichia coli* and *Salmonella typhimurium*: characterization of the dipeptide permease (Dpp) and the dipeptide-binding protein. *Mol. Microbiol.* **5:**1035–1047.

6. Abraham, J. M., C. S. Freitag, J. R. Clements, and B. I. Eisenstein. 1985. An invertible element of DNA controls phase variation of type 1 fimbriae of *Escherichia coli*. *Proc. Natl. Acad. Sci. USA* **82:**5724–5727.

7. Abraham, S. N., M. Land, S. Ponniah, R. Endres, D. L. Hasty, and J. P. Babu. 1992. Glycerol-induced unraveling of the tight helical conformation of *Escherichia coli* type 1 fimbriae. *J. Bacteriol.* **174:**5145–5148.

8. Abril, N., C. Hera, E. Alejandre, J. A. Rafferty, G. P. Margison, and C. Pueyo. 1994. Effect of *ogt* expression on mutation induction by methyl-, ethyl- and propylmethanesulphonate in *Escherichia coli* K12 strains. *Mol. Gen. Genet.* **242:**744–748.

9. Ackerman, R. S., N. R. Cozzarelli, and W. Epstein. 1974. Accumulation of toxic concentrations of methylglyoxal by wild-type *Escherichia coli* K-12. *J. Bacteriol.* **119:**357–362.

10. Adachi, T., K. Mizuuchi, R. Menzel, and M. Gellert. 1984. DNA sequence and transcription of the region upstream of the *E. coli gyrB* gene. *Nucleic Acids Res.* **12:**6389–6395.

11. Adachi, T., M. Mizuuchi, E. A. Robinson, E. Appella, M. H. O'Dea, M. Gellert, and K. Mizuuchi. 1987. DNA sequence of the *E. coli gyrB* gene: application of a new sequencing strategy. *Nucleic Acids Res.* **15:**771–784.

12. Adams, C. W., M. Rosenberg, and G. W. Hatfield. 1985. Analysis of in vivo RNA transcription products of the *ilvGEDA* attenuator region of *Escherichia coli* K12. *J. Biol. Chem.* **260:**8538–8544.

13. Adams, M. D., L. M. Wagner, T. J. Graddis, R. Landick, T. K. Antonucci, A. L. Gibson, and D. L. Oxender. 1990. Nucleotide sequence and genetic characterization reveal six essential genes for the LIV-I and LS transport systems of *Escherichia coli*. *J. Biol. Chem.* **265:**11436–11443.

14. Adhya, S., P. Cleary, and A. Campbell. 1968. A deletion analysis of prophage lambda and adjacent genetic regions. *Proc. Natl. Acad. Sci. USA* **61:**956–962.

15. Adhya, S., and S. Garges. 1990. Positive control. *J. Biol. Chem.* **265:**10797–10800.

16. Adler, H., R. Mural, and B. Suttle. 1992. Oxygen sensitivity of an *Escherichia coli* mutant. *J. Bacteriol.* **174:**2072–2077.

17. Adler, H. I., W. D. Fisher, A. A. Hardigree, and G. E. Stapleton. 1965. Repair of radiation-induced damage to the cell division mechanism of *Escherichia coli*. *J. Bacteriol.* **91:**737–742.

18. Ahmad, S., N. A. Glavas, and P. D. Bragg. 1992. A mutation at Gly314 of the beta subunit of the *Escherichia coli* pyridine nucleotide transhydrogenase abolishes activity and affects the NADP(H)-induced conformational change. *Eur. J. Biochem.* **207:**733–739.

19. Ahmad, S., N. A. Glavas, and P. D. Bragg. 1992. Subunit interactions involved in the assembly of pyridine nucleotide transhydrogenase in the membranes of *Escherichia coli*. *J. Biol. Chem.* **267:**7007–7012.

20. Ahmad, S. I., and R. H. Pritchard. 1969. A map of four genes specifying enzymes involved in catabolism of nucleosides and deoxynucleosides in *Escherichia coli*. *Mol. Gen. Genet.* **104:**351–359.

21. Ahmad, S. I., and R. H. Pritchard. 1972. Location of gene specifying cytosine deaminase in *Escherichia coli*. *Mol. Gen. Genet.* **118:**323–325.

22. Ahnn, J., P. E. March, H. E. Takiff, and M. Inouye. 1986. A GTP-binding protein of *Escherichia coli* has homology to yeast RAS proteins. *Proc. Natl. Acad. Sci. USA* **83:**8849–8853.

23. Aiba, A., and K. Mizobuchi. 1989. Nucleotide sequence analysis of genes *purH* and *purD* involved in the de novo purine nucleotide biosynthesis of *Escherichia coli*. *J. Biol. Chem.* **264:**21239–21246.

24. Aiba, H. 1985. Transcription of the *Escherichia coli* adenylate cyclase gene is negatively regulated by cAMP-cAMP receptor protein. *J. Biol. Chem.* **260:**3063–3070.

25. Aiba, H., S. Adhya, and B. de Crombrugghe. 1981. Evidence for two functional *gal* promoters in intact *Escherichia coli* cells. *J. Biol. Chem.* **256:**11905–11910.

26. Aiba, H., S. Fujimoto, and N. Ozaki. 1982. Molecular cloning and nucleotide sequencing of the gene for *E. coli* cAMP receptor protein. *Nucleic Acids Res.* **10:**1345–1361.

27. Aiba, H., M. Kawamukai, and A. Ishihama. 1983. Cloning and promoter analysis of the *Escherichia coli* adenylate cyclase gene. *Nucleic Acids Res.* **11:**3451–3465.

28. Aiba, H., K. Mori, M. Tanaka, T. Ooi, A. Roy, and A. Danchin. 1984. The complete nucleotide sequence of the adenylate cyclase gene of *Escherichia coli*. *Nucleic Acids Res.* **12:**9427–9440.

29. Akimaru, J., S. Matsuyama, H. Tokuda, and S. Mizushima. 1991. Reconstitution of a protein translocation system containing purified SecY, SecE, and SecA from *Escherichia coli*. *Proc. Natl. Acad. Sci. USA* **88:**6545–6549.

30. Akiyama, M., E. Crooke, and A. Kornberg. 1992. The polyphosphate kinase gene of *Escherichia coli*. Isolation and sequence of the *ppk* gene and membrane location of the protein. *J. Biol. Chem.* **267:**22556–22561.

31. Akiyama, M., E. Crooke, and A. Kornberg. 1993. An exopolyphosphatase of *Escherichia coli*. The enzyme and its *ppx* gene in a polyphosphate operon. *J. Biol. Chem.* **268:**633–639.

32. Akiyama, M., T. Horiuchi, and M. Sekiguchi. 1987. Molecular cloning and nucleotide sequence of the *mutT* mutator of *Escherichia coli* that causes A:T to C:G transversion. *Mol. Gen. Genet.* **206:**9–16.

33. Akiyama, Y., T. Inada, Y. Nakamura, and K. Ito. 1990. SecY, a multispanning integral membrane protein, contains a potential leader peptidase cleavage site. *J. Bacteriol.* **172:**2888–2893.

34. Akiyama, Y., and K. Ito. 1985. The SecY membrane component of the bacterial protein export machinery: analysis by new electrophoretic methods for integral membrane proteins. *EMBO J.* **4:**3351–3356.

35. Akiyama, Y., and K. Ito. 1987. Topology analysis of the SecY protein, an integral membrane protein involved in protein export in *Escherichia coli*. *EMBO J.* **6:**3465–3470.

36. Akiyama, Y., and K. Ito. 1990. SecY protein, a membrane-embedded secretion factor of *E. coli*, is cleaved by the *ompT* protease in vitro. *Biochem. Biophys. Res. Commun.* **167:**711–715.

37. Akiyama, Y., S. Kamitani, N. Kusukawa, and K. Ito. 1992. In vitro catalysis of oxidative folding of disulfide-bonded proteins by the *Escherichia coli dsbA* (*ppfA*) gene product. *J. Biol. Chem.* **267:**22440–22445.

38. Akiyama, Y., Y. Shirai, and K. Ito. 1994. Involvement of FtsH in protein assembly into and through the membrane. II. Dominant mutations affecting FtsH functions. *J. Biol. Chem.* **269:**5225–5229.

39. al-Bar, O. A., C. D. O'Connor, I. G. Giles, and M. Akhtar. 1992. D-Alanine:D-alanine ligase of *Escherichia coli*. Expression, purification and inhibitory studies on the cloned enzyme. *Biochem. J.* **282:**747–752.

40. Alaeddinoglu, N. G., and H. P. Charles. 1979. Transfer of a gene for sucrose utilization into *Escherichia coli* K12, and consequent failure of expression of genes for D-serine utilization. *J. Gen. Microbiol.* **110:**47–59.

41. Alakhov, Y. B., N. V. Dovgas, L. P. Motuz, L. M. Vinokurov, and Y. A. Ovchinnikov. 1981. The primary structure of the elongation factor G from *Escherichia coli*: amino acid sequence of the C-terminal domain. *FEBS Lett.* **126:**183–186.

42. Alam, K. Y., and D. P. Clark. 1991. Molecular cloning and sequence of the *thdF* gene, which is involved in thiophene and furan oxidation by *Escherichia coli*. *J. Bacteriol.* **173:**6018–6024.

43. Albin, R., and P. M. Silverman. 1984. Physical and genetic structure of the *glpK-cpxA* interval of the *Escherichia coli* K-12 chromosome. *Mol. Gen. Genet.* **197:**261–271.

44. Albin, R., R. Weber, and P. M. Silverman. 1986. The Cpx proteins of *Escherichia coli* K12. Immunologic detection of the chromosomal *cpxA* gene product. *J. Biol. Chem.* **261:**4698–4705.

45. Albrechtsen, B., B. M. Ross, C. Squires, and C. L. Squires. 1991. Transcriptional termination sequence at the end of the *Escherichia coli* ribosomal RNA G operon: complex terminators and antitermination. *Nucleic Acids Res.* **19:**1845–1852.

46. Albrechtsen, B., C. L. Squires, S. Li, and C. Squires. 1990. Antitermination of characterized transcriptional terminators by the *Escherichia coli rrnG* leader region. *J. Mol. Biol.* **213:**123–134.

47. Albrechtsen, H., and S. I. Ahmad. 1980. Regulation of the synthesis of nucleoside catabolic enzymes in *Escherichia coli*: further analysis of a *deo* Oc mutant strain. *Mol. Gen. Genet.* **179:**457–460.

48. Aldea, M., T. Garrido, C. Hernandez-Chico, M. Vicente, and S. R. Kushner. 1989. Induction of a growth-phase-dependent promoter triggers transcription of *bolA*, an *Escherichia coli* morphogene. *EMBO J.* **8:**3923–3931.

49. Aldea, M., T. Garrido, J. Pla, and M. Vicente. 1990. Division genes in *Escherichia coli* are expressed coordinately to cell septum requirements by gearbox promoters. *EMBO J.* **9:**3787–3794.

50. Aldea, M., C. Hernandez-Chico, A. G. de la Campa, S. R. Kushner, and M. Vicente. 1988. Identification, cloning, and expression of *bolA*, an *ftsZ*-dependent morphogene of *Escherichia coli*. *J. Bacteriol.* **170:**5169–5176.

51. Aldea, M., V. F. Maples, and S. R. Kushner. 1988. Generation of a detailed physical and genetic map of the *ilv-metE-udp* region of the *Escherichia coli* chromosome. *J. Mol. Biol.* **200:**427–438.

52. **Alefounder, P. R., C. Abell, and A. R. Battersby.** 1988. The sequence of *hemC*, *hemD* and two additional *E. coli* genes. *Nucleic Acids Res.* **16:**9871.

53. **Alefounder, P. R., S. A. Baldwin, R. N. Perham, and N. J. Short.** 1989. Cloning, sequence analysis and over-expression of the gene for the class II fructose 1,6-bisphosphate aldolase of *Escherichia coli. Biochem. J.* **257:**529–534.

54. **Alefounder, P. R., and R. N. Perham.** 1989. Identification, molecular cloning and sequence analysis of a gene cluster encoding the class II fructose 1,6-bisphosphate aldolase, 3-phosphoglycerate kinase and a putative second glyceraldehyde 3-phosphate dehydrogenase of *Escherichia coli. Mol. Microbiol.* **3:**723–732.

55. **Alexander, D. M., and A. C. St John.** 1994. Characterization of the carbon starvation-inducible and stationary phase-inducible gene *slp* encoding an outer membrane lipoprotein in *Escherichia coli. Mol. Microbiol.* **11:**1059–1071.

56. **Ali, S. T., and J. R. Guest.** 1990. Isolation and characterization of lipoylated and unlipoylated domains of the E2p subunit of the pyruvate dehydrogenase complex of *Escherichia coli. Biochem. J.* **271:**139–145.

57. **Alifano, P., M. S. Carlomagno, and C. B. Bruni.** 1992. Location of the *hisGDCBHAFI* operon on the physical map of *Escherichia coli. J. Bacteriol.* **174:**3830–3831.

58. **Alifano, P., F. Rivellini, D. Limauro, C. B. Bruni, and M. S. Carlomagno.** 1991. A consensus motif common to all Rho-dependent prokaryotic transcription terminators. *Cell* **64:**553–563.

59. **Alix, J. H.** 1989. A rapid procedure for cloning genes from lambda libraries by complementation of *E. coli* defective mutants: application to the *fabE* region of the *E. coli* chromosome. *DNA* **8:**779–789.

60. **Alksne, L. E., D. Keeney, and B. A. Rasmussen.** 1995. A mutation in either *dsbA* or *dsbB*, a gene encoding a component of a periplasmic disulfide bond-catalyzing system, is required for high-level expression of the *Bacteroides fragilis* metallo-β-lactamase, CcrA, in *Escherichia coli. J. Bacteriol.* **177:**462–464.

61. **Allen, G., and B. Wittmann-Liebold.** 1978. The amino acid sequence of the ribosomal protein S8 of *Escherichia coli. Hoppe-Seyler's Z. Physiol. Chem.* **359:**1509–1525.

62. **Allen, G. C., Jr., and A. Kornberg.** 1991. The *priB* gene encoding the primosomal replication n protein of *Escherichia coli. J. Biol. Chem.* **266:**11610–11613.

63. **Allen, S. P., J. O. Polazzi, J. K. Gierse, and A. M. Easton.** 1992. Two novel heat shock genes encoding proteins produced in response to heterologous protein expression in *Escherichia coli. J. Bacteriol.* **174:**6938–6947.

64. **Allibert, P., J. C. Willison, and P. M. Vignais.** 1987. Complementation of nitrogen-regulatory (*ntr*-like) mutations in *Rhodobacter capsulatus* by an *Escherichia coli* gene: cloning and sequencing of the gene and characterization of the gene product. *J. Bacteriol.* **169:**260–271.

65. **Almassy, R. J., C. A. Janson, C. C. Kan, and Z. Hostomska.** 1992. Structures of apo and complexed *Escherichia coli* glycinamide ribonucleotide transformylase. *Proc. Natl. Acad. Sci. USA* **89:**6114–6118.

66. **Almiron, M., A. J. Link, D. Furlong, and R. Kolter.** 1992. A novel DNA-binding protein with regulatory and protective roles in starved *Escherichia coli. Genes Dev.* **6:**2646–2654.

67. **Almond, N., V. Yajnik, P. Svec, and G. N. Godson.** 1989. An *Escherichia coli* cis-acting antiterminator sequence: the *dnaG* nut site. *Mol. Gen. Genet.* **216:**195–203.

68. **Altendorf, K., A. Siebers, and W. Epstein.** 1992. The KDP ATPase of *Escherichia coli. Ann. N.Y. Acad. Sci.* **671:**228–243.

69. **Altman, E., C. A. Kumamoto, and S. D. Emr.** 1991. Heat-shock proteins can substitute for SecB function during protein export in *Escherichia coli. EMBO J.* **10:**239–245.

70. **Altman, S., P. Model, G. H. Dixon, and N. A. Wosnick.** 1981. An *E. coli* gene coding for a protamine-like protein. *Cell* **26:**299–304.

71. **Altmann, C. R., D. E. Solow-Cordero, and M. J. Chamberlin.** 1994. RNA cleavage and chain elongation by *Escherichia coli* DNA-dependent RNA polymerase in a binary enzyme. RNA complex. *Proc. Natl. Acad. Sci. USA* **91:**3784–3788.

72. **Amabile-Cuevas, C. F., and B. Demple.** 1991. Molecular characterization of the *soxRS* genes of *Escherichia coli*: two genes control a superoxide stress regulon. *Nucleic Acids Res.* **19:**4479–4484.

73. **Amemura, M., K. Makino, H. Shinagawa, A. Kobayashi, and A. Nakata.** 1985. Nucleotide sequence of the genes involved in phosphate transport and regulation of the phosphate regulon in *Escherichia coli. J. Mol. Biol.* **184:**241–250.

74. **Amemura, M., K. Makino, H. Shinagawa, and A. Nakata.** 1986. Nucleotide sequence of the *phoM* region of *Escherichia coli*: four open reading frames may constitute an operon. *J. Bacteriol.* **168:**294–302.

75. **Amemura, M., K. Makino, H. Shinagawa, and A. Nakata.** 1990. Cross talk to the phosphate regulon of *Escherichia coli* by PhoM protein: PhoM is a histidine protein kinase and catalyzes phosphorylation of PhoB and PhoM open reading frame 2. *J. Bacteriol.* **172:**6300–6307.

76. **Amemura, M., H. Shinagawa, K. Makino, N. Otsuji, and A. Nakata.** 1982. Cloning of and complementation tests with alkaline phosphatase regulatory genes (*phoS* and *phoT*) of *Escherichia coli. J. Bacteriol.* **152:**692–701.

77. **Amerik, A. I., L. G. Chistiakov, N. I. Ostroumova, A. I. Gurevich, and V. K. Antonov.** 1988. Klonirovanie, ekspressiia i struktura funktsional'no aktivnogo ukorochennogo gena lon *Escherichia coli. Bioorg. Khim.* **14:**408–411.

78. **Ammer, J., M. Brennenstuhl, P. Schindler, J. V. Holtje, and H. Zahner.** 1979. Phosphorylation of streptozotocin during uptake via the phosphoenolpyruvate:sugar phosphotransferase system in *Escherichia coli. Antimicrob. Agents Chemother.* **16:**801–807.

79. **Amster-Choder, O., and A. Wright.** 1990. Regulation of activity of a transcriptional anti-terminator in *E. coli* by phosphorylation in vivo. *Science* **249:**540–542.

80. **Amundsen, S. K., A. F. Taylor, A. M. Chaudhury, and G. R. Smith.** 1986. *recD*: the gene for an essential third subunit of exonuclease V. *Proc. Natl. Acad. Sci. USA* **83:**5558–5562.

81. **An, G., D. S. Bendiak, L. A. Mamelak, and J. D. Friesen.** 1981. Organization and nucleotide sequence of a new ribosomal operon in *Escherichia coli* containing the genes for ribosomal protein S2 and elongation factor Ts. *Nucleic Acids Res.* **9:**4163–4172.

82. **An, G., and J. D. Friesen.** 1980. Characterization of promoter-cloning plasmids: analysis of operon structure in the *rif* region of *Escherichia coli* and isolation of an enhanced internal promoter mutant. *J. Bacteriol.* **144:**904–916.

83. **An, G., and J. D. Friesen.** 1980. The nucleotide sequence of *tufB* and four nearby tRNA structural genes of *Escherichia coli. Gene* **12:**33–39.

84. **An, G., J. S. Lee, and J. D. Friesen.** 1982. Evidence for an internal promoter preceding *tufA* in the *str* operon of *Escherichia coli. J. Bacteriol.* **149:**548–553.

85. **Andersen, J., N. Delihas, K. Ikenaka, P. J. Green, O. Pines, O. Ilercil, and M. Inouye.** 1987. The isolation and characterization of RNA coded by the *micF* gene in *Escherichia coli. Nucleic Acids Res.* **15:**2089–2101.

86. **Andersen, J. T., K. F. Jensen, and P. Poulsen.** 1991. Role of transcription pausing in the control of the *pyrE* attenuator in *Escherichia coli. Mol. Microbiol.* **5:**327–333.

87. **Andersen, J. T., P. Poulsen, and K. F. Jensen.** 1992. Attenuation in the *rph-pyrE* operon of *Escherichia coli* and processing of the dicistronic mRNA. *Eur. J. Biochem.* **206:**381–390.

88. **Andersen, P. S., D. Frees, R. Fast, and B. Mygind.** 1995. Uracil uptake in *Escherichia coli* K-12: isolation of *uraA* mutants and cloning of the gene. *J. Bacteriol.* **177:**2008–2013.

89. **Andersen, P. S., J. M. Smith, and B. Mygind.** 1992. Characterization of the *upp* gene encoding uracil phosphoribosyltransferase of *Escherichia coli* K12. *Eur. J. Biochem.* **204:**51–56.

90. **Anderson, J. J., and D. L. Oxender.** 1977. *Escherichia coli* transport mutants lacking binding protein and other components of the branched-chain amino acid transport systems. *J. Bacteriol.* **130:**384–392.

91. **Anderson, J. J., S. C. Quay, and D. L. Oxender.** 1976. Mapping of two loci affecting the regulation of branched-chain amino acid transport in *Escherichia coli* K-12. *J. Bacteriol.* **126:**80–90.

92. **Anderson, J. J., J. M. Wilson, and D. L. Oxender.** 1979. Defective transport and other phenotypes of a periplasmic "leaky" mutant of *Escherichia coli* K-12. *J. Bacteriol.* **140:**351–358.

93. **Anderson, P. M., Y. C. Sung, and J. A. Fuchs.** 1990. The cyanase operon and cyanate metabolism. *FEMS Microbiol. Rev.* **7:**247–252.

94. **Andresen, P. A., I. Kaasen, O. B. Styrvold, G. Boulnois, and A. R. Strom.** 1988. Molecular cloning, physical mapping and expression of the *bet* genes governing the osmoregulatory choline-glycine betaine pathway of *Escherichia coli. J. Gen. Microbiol.* **134:**1737–1746.

95. **Andresson, O. S., and J. E. Davies.** 1980. Some properties of the ribosomal RNA methyltransferase encoded by *ksgA* and the polarity of *ksgA* transcription. *Mol. Gen. Genet.* **179:**217–222.

96. **Andresson, O. S., and J. E. Davies.** 1980. Genetic organization and restriction enzyme cleavage map of the *ksgA-pdxA* region of the *Escherichia coli* chromosome. *Mol. Gen. Genet.* **179:**211–216.

97. **Andresson, O. S., and J. E. Davies.** 1980. Isolation and characterization of lambda transducing phages for the *E. coli* genes *ksgA* and *pdxA. Mol. Gen. Genet.* **179:**201–209.

98. **Andrews, J. C., and S. A. Short.** 1985. Genetic analysis of *Escherichia coli* oligopeptide transport mutants. *J. Bacteriol.* **161:**484–492.

99. **Andrews, S. C., and J. R. Guest.** 1988. Nucleotide sequence of the gene encoding the GMP reductase of *Escherichia coli* K12. *Biochem. J.* **255:**35–43.

100. **Andrews, S. C., P. M. Harrison, and J. R. Guest.** 1989. Cloning, sequencing, and mapping of the bacterioferritin gene (*bfr*) of *Escherichia coli* K-12. *J. Bacteriol.* **171:**3940–3947.

101. **Andrews, S. C., P. M. Harrison, and J. R. Guest.** 1991. A molecular analysis of the 53.3 minute region of the *Escherichia coli* linkage map. *J. Gen. Microbiol.* **137:**361–367.

102. **Andrews, S. C., D. Shipley, J. N. Keen, J. B. Findlay, P. M. Harrison, and J. R. Guest.** 1992. The haemoglobin-like protein (HMP) of *Escherichia coli* has ferrisiderophore reductase activity and its C-terminal domain shares homology with ferredoxin NADP+ reductases. *FEBS Lett.* **302:**247–252.

103. **Andrews, S. C., J. M. Smith, J. R. Guest, and P. M. Harrison.** 1989. Amino acid sequence of the bacterioferritin (cytochrome b1) of *Escherichia coli* K12. *Biochem. Biophys. Res. Commun.* **158:**489–496.

104. **Angov, E., T. C. Ng, and W. S. Brusilow.** 1991. Effect of the delta subunit on assembly and proton permeability of the F0 proton channel of *Escherichia coli* F1F0 ATPase. *J. Bacteriol.* **173:**407–411.

105. **Anselme, J., and M. Hartlein.** 1989. Asparaginyl-tRNA synthetase from *Escherichia coli* has significant sequence homologies with yeast aspartyl-tRNA synthetase. *Gene* **84:**481–485.

106. **Anselme, J., and M. Hartlein.** 1991. Tyr-426 of the *Escherichia coli* asparaginyl-tRNA synthetase, an amino acid in a C-terminal conserved motif, is involved in ATP binding. *FEBS Lett.* **280:**163–166.

107. **Anton, D. L., and R. Kutny.** 1987. *Escherichia coli* S-adenosylmethionine decarboxylase. Subunit structure, reductive amination, and NH₂-terminal sequences. *J. Biol. Chem.* **262**:2817–2822.

108. **Anton, I. A., and J. R. Coggins.** 1988. Sequencing and overexpression of the *Escherichia coli aroE* gene encoding shikimate dehydrogenase. *Biochem. J.* **249**:319–326.

109. **Anton, M., and K. J. Heller.** 1991. Functional analysis of a C-terminally altered TonB protein of *Escherichia coli. Gene* **105**:23–29.

110. **Antonucci, T. K., R. Landick, and D. L. Oxender.** 1985. The leucine binding proteins of *Escherichia coli* as models for studying the relationships between protein structure and function. *J. Cell. Biochem.* **29**:209–216.

111. **Aoki, H., S. L. Adams, D. G. Chung, M. Yaguchi, S. E. Chuang, and M. C. Ganoza.** 1991. Cloning, sequencing and overexpression of the gene for prokaryotic factor EF-P involved in peptide bond synthesis. *Nucleic Acids Res.* **19**:6215–6220.

112. **Aoki, H., P. J. Yaworsky, S. D. Patel, D. Margolin-Brzezinski, K. S. Park, and M. C. Ganoza.** 1992. The asparaginyl-tRNA synthetase gene encodes one of the complementing factors for thermosensitive translation in the *Escherichia coli* mutant strain, N4316. *Eur. J. Biochem.* **209**:511–521.

113. **Aono, R., T. Negishi, and H. Nakajima.** 1994. Cloning of organic solvent tolerance gene *ostA* that determines *n*-hexane tolerance level in *Escherichia coli. Appl. Environ. Microbiol.* **60**:4624–4626.

113a.**Aoyama, K., A. M. Haase, and P. R. Reeves.** 1994. Evidence for effect of random genetic drift on G+C content after lateral transfer of fucose pathway genes to *Escherichia coli* K-12. *Mol. Biol. Evol.* **11**:829–838.

114. **Apirion, D.** 1967. Three genes that affect *Escherichia coli* ribosomes. *J. Mol. Biol.* **30**:255–275.

115. **Apirion, D., and N. Watson.** 1980. A second gene which affects the RNA processing enzyme ribonuclease P of *Escherichia coli. FEBS Lett.* **110**:161–163.

116. **Archibold, E. R., and L. S. Williams.** 1973. Regulation of methionyl-transfer ribonucleic acid synthetase formation in *Escherichia coli* and *Salmonella typhimurium. J. Bacteriol.* **114**:1007–1013.

117. **Arendes, J., P. L. Carl, and A. Sugino.** 1982. A mutation in the *rnh*-locus of *Escherichia coli* affects the structural gene for RNase H. Examination of the mutant and wild type protein. *J. Biol. Chem.* **257**:4719–4722.

118. **Argaet, V. P., T. J. Wilson, and B. E. Davidson.** 1994. Purification of the *Escherichia coli* regulatory protein TyrR and analysis of its interactions with ATP, tyrosine, phenylalanine, and tryptophan. *J. Biol. Chem.* **269**:5171–5178.

119. **Arikan, E., M. S. Kulkarni, D. C. Thomas, and A. Sancar.** 1986. Sequences of the *E. coli uvrB* gene and protein. *Nucleic Acids Res.* **14**:2637–2650.

120. **Ariza, R. R., S. P. Cohen, N. Bachhawat, S. B. Levy, and B. Demple.** 1994. Repressor mutations in the *marRAB* operon that activate oxidative stress genes and multiple antibiotic resistance in *Escherichia coli. J. Bacteriol.* **176**:143–148.

121. **Armengod, M. E., M. Garcia-Sogo, I. Perez-Roger, F. Macian, and J. P. Navarro-Avino.** 1991. Tandem transcription termination sites in the *dnaN* gene of *Escherichia coli. J. Biol. Chem.* **266**:19725–19730.

122. **Armengod, M. E., and E. Lambies.** 1986. Overlapping arrangement of the *recF* and *dnaN* operons of *Escherichia coli;* positive and negative control sequences. *Gene* **43**:183–196.

123. **Armstrong, S. K., C. L. Francis, and M. A. McIntosh.** 1990. Molecular analysis of the *Escherichia coli* ferric enterobactin receptor FepA. *J. Biol. Chem.* **265**:14536–14543.

124. **Armstrong, S. K., G. S. Pettis, L. J. Forrester, and M. A. McIntosh.** 1989. The *Escherichia coli* enterobactin biosynthesis gene, *entD:* nucleotide sequence and membrane localization of its protein product. *Mol. Microbiol.* **3**:757–766.

125. **Arnardottir, A., S. Thorbjarnardottir, and G. Eggertsson.** 1980. Mapping of the *supP* (Su6⁺) amber suppressor gene in *Escherichia coli. J. Bacteriol.* **141**:977–978.

126. **Arnqvist, A., A. Olsen, and S. Normark.** 1994. Sigma S-dependent growth-phase induction of the *csgBA* promoter in *Escherichia coli* can be achieved in vivo by sigma 70 in the absence of the nucleoid-associated protein H-NS. *Mol. Microbiol.* **13**:1021–1032.

127. **Arnqvist, A., A. Olsen, J. Pfeifer, D. G. Russell, and S. Normark.** 1992. The Crl protein activates cryptic genes for curli formation and fibronectin binding in *Escherichia coli* HB101. *Mol. Microbiol.* **6**:2443–2452.

128. **Aronson, B. D., P. D. Ravnikar, and R. L. Somerville.** 1988. Nucleotide sequence of the 2-amino-3-ketobutyrate coenzyme A ligase (*kbl*) gene of *E. coli. Nucleic Acids Res.* **16**:3586.

129. **Aronson, B. D., R. L. Somerville, B. R. Epperly, and E. E. Dekker.** 1989. The primary structure of *Escherichia coli* L-threonine dehydrogenase. *J. Biol. Chem.* **264**:5226–5232.

130. **Arps, P. J., C. C. Marvel, B. C. Rubin, D. A. Tolan, E. E. Penhoet, and M. E. Winkler.** 1985. Structural features of the *hisT* operon of *Escherichia coli* K-12. *Nucleic Acids Res.* **13**:5297–5315.

131. **Arps, P. J., and M. E. Winkler.** 1987. An unusual genetic link between vitamin B6 biosynthesis and tRNA pseudouridine modification in *Escherichia coli* K-12. *J. Bacteriol.* **169**:1071–1079.

132. **Arps, P. J., and M. E. Winkler.** 1987. Structural analysis of the *Escherichia coli* K-12 *hisT* operon by using a kanamycin resistance cassette. *J. Bacteriol.* **169**:1061–1070.

133. **Arraj, J. A., and M. G. Marinus.** 1983. Phenotypic reversal in *dam* mutants of *Escherichia coli* K-12 by a recombinant plasmid containing the *dam⁺* gene. *J. Bacteriol.* **153**:562–565.

134. **Arribas, J., and J. G. Castano.** 1993. A comparative study of the chymotrypsin-like activity of the rat liver multicatalytic proteinase and the ClpP from *Escherichia coli. J. Biol. Chem.* **268**:21165–21171.

135. **Arthur, H. M., and P. B. Eastlake.** 1983. Transcriptional control of the *uvrD* gene of *Escherichia coli. Gene* **25**:309–316.

136. **Artman, M., and S. Werthamer.** 1974. Use of streptomycin and cyclic adenosine 5′-monophosphate in the isolation of mutants deficient in CAP protein. *J. Bacteriol.* **120**:542–544.

137. **Asada, K., S. Nakatani, and M. Takanami.** 1985. Cloning of the contiguous 165-kilobase-pair region around the terminus of *Escherichia coli* K-12 DNA replication. *J. Bacteriol.* **163**:398–400.

138. **Asada, K., K. Sugimoto, A. Oka, M. Takanami, and Y. Hirota.** 1982. Structure of replication origin of the *Escherichia coli* K-12 chromosome: the presence of spacer sequences in the *ori* region carrying information for autonomous replication. *Nucleic Acids Res.* **10**:3745–3754.

139. **Asahara, H., P. M. Wistort, J. F. Bank, R. H. Bakerian, and R. P. Cunningham.** 1989. Purification and characterization of *Escherichia coli* endonuclease III from the cloned *nth* gene. *Biochemistry* **28**:4444–4449.

140. **Asai, K., S. Fujisaki, Y. Nishimura, T. Nishino, K. Okada, T. Nakagawa, M. Kawamukai, and H. Matsuda.** 1994. The identification of *Escherichia coli ispB* (*cel*) gene encoding the octaprenyl diphosphate synthase. *Biochem. Biophys. Res. Commun.* **202**:340–345.

141. **Asai, T., C. P. Chen, T. Nagata, M. Takanami, and M. Imai.** 1992. Transcription in vivo within the replication origin of the *Escherichia coli* chromosome: a mechanism for activating initiation of replication. *Mol. Gen. Genet.* **231**:169–178.

142. **Asai, T., and T. Kogoma.** 1994. The RecF pathway of homologous recombination can mediate the initiation of DNA damage-inducible replication of the *Escherichia coli* chromosome. *J. Bacteriol.* **176**:7113–7114.

143. **Asai, T., S. Sommer, A. Bailone, and T. Kogoma.** 1993. Homologous recombination-dependent initiation of DNA replication from DNA damage-inducible origins in *Escherichia coli. EMBO J.* **12**:3287–3295.

144. **Asai, T., M. Takanami, and M. Imai.** 1990. The AT richness and *gid* transcription determine the left border of the replication origin of the *E. coli* chromosome. *EMBO J.* **9**:4065–4072.

145. **Asha, H., and J. Gowrishankar.** 1993. Regulation of *kdp* operon expression in *Escherichia coli:* evidence against turgor as signal for transcriptional control. *J. Bacteriol.* **175**:4528–4537.

146. **Asoh, S., H. Matsuzawa, F. Ishino, J. L. Strominger, M. Matsuhashi, and T. Ohta.** 1986. Nucleotide sequence of the *pbpA* gene and characteristics of the deduced amino acid sequence of penicillin-binding protein 2 of *Escherichia coli* K12. *Eur. J. Biochem.* **160**:231–238.

147. **Asoh, S., H. Matsuzawa, M. Matsuhashi, and T. Ohta.** 1983. Molecular cloning and characterization of the genes (*pbpA* and *rodA*) responsible for the rod shape of *Escherichia coli* K-12: analysis of gene expression with transposon Tn5 mutagenesis and protein synthesis directed by constructed plasmids. *J. Bacteriol.* **154**:10–16.

148. **Atherly, A. G.** 1979. *Escherichia coli* mutant containing a large deletion from *relA* to *argA. J. Bacteriol.* **138**:530–534.

149. **Atkinson, B. L., and M. E. Gottesman.** 1992. The *Escherichia coli rpoB60* mutation blocks antitermination by coliphage HK022 Q-function. *J. Mol. Biol.* **227**:29–37.

150. **Atkinson, M. R., and A. J. Ninfa.** 1992. Characterization of *Escherichia coli glnL* mutations affecting nitrogen regulation. *J. Bacteriol.* **174**:4538–4548.

151. **Atkinson, M. R., and A. J. Ninfa.** 1993. Mutational analysis of the bacterial signal-transducing protein kinase/phosphatase nitrogen regulator II (NRII or NtrB). *J. Bacteriol.* **175**:7016–7023.

152. **Atlung, T., and L. Brondsted.** 1994. Role of the transcriptional activator AppY in regulation of the *cyx appA* operon of *Escherichia coli* by anaerobiosis, phosphate starvation, and growth phase. *J. Bacteriol.* **176**:5414–5422.

153. **Atlung, T., A. Nielsen, and F. G. Hansen.** 1989. Isolation, characterization, and nucleotide sequence of *appY,* a regulatory gene for growth-phase-dependent gene expression in *Escherichia coli. J. Bacteriol.* **171**:1683–1691.

154. **Attfield, P. V., F. E. Benson, and R. G. Lloyd.** 1985. Analysis of the *ruv* locus of *Escherichia coli* K-12 and identification of the gene product. *J. Bacteriol.* **164**:276–281.

155. **Au, D. C., and R. B. Gennis.** 1987. Cloning of the *cyo* locus encoding the cytochrome *o* terminal oxidase complex of *Escherichia coli. J. Bacteriol.* **169**:3237–3242.

156. **Au, D. C., R. M. Lorence, and R. B. Gennis.** 1985. Isolation and characterization of an *Escherichia coli* mutant lacking the cytochrome *o* terminal oxidase. *J. Bacteriol.* **161**:123–127.

157. **Aufrere, R., M. Tempete, and J. P. Bohin.** 1986. Regulation of expression of the gene for vitamin B12 receptor cloned on a multicopy plasmid in *Escherichia coli. Mol. Gen. Genet.* **205**:358–365.

158. **Augustin, L. B., B. A. Jacobson, and J. A. Fuchs.** 1994. *Escherichia coli* Fis and DnaA proteins bind specifically to the *nrd* promoter region and affect expression of an *nrd-lac* fusion. *J. Bacteriol.* **176**:378–387.

159. **Austin, D., and T. J. Larson.** 1991. Nucleotide sequence of the *glpD* gene encoding aerobic *sn*-glycerol 3-phosphate dehydrogenase of *Escherichia coli* K-12. *J. Bacteriol.* **173**:101–107.

160. **Austin, E. A., J. F. Graves, L. A. Hite, C. T. Parker, and C. A. Schnaitman.** 1990. Genetic analysis of lipopolysaccharide core biosynthesis by *Escherichia coli* K-12: insertion mutagenesis of the *rfa* locus. *J. Bacteriol.* **172**:5312–5325.

161. **Austin, S., and J. Scaife.** 1970. A new method for selecting RNA polymerase mutants. *J. Mol. Biol.* **49:**263–267.

162. **Autexier, C., and M. S. DuBow.** 1992. The *Escherichia coli* Mu/D108 phage *ner* homologue gene (*nlp*) is transcribed and evolutionarily conserved among the Enterobacteriaceae. *Gene* **114:**13–18.

163. **Avalos, J., L. M. Corrochano, and S. Brenner.** 1991. Cysteinyl-tRNA synthetase is a direct descendant of the first aminoacyl-tRNA synthetase. *FEBS Lett.* **286:**176–180.

164. **Axley, M. J., A. Bock, and T. C. Stadtman.** 1991. Catalytic properties of an *Escherichia coli* formate dehydrogenase mutant in which sulfur replaces selenium. *Proc. Natl. Acad. Sci. USA* **88:**8450–8454.

165. **Axley, M. J., D. A. Grahame, and T. C. Stadtman.** 1990. *Escherichia coli* formate-hydrogen lyase. Purification and properties of the selenium-dependent formate dehydrogenase component. *J. Biol. Chem.* **265:**18213–18218.

166. **Aymerich, S., and M. Steinmetz.** 1992. Specificity determinants and structural features in the RNA target of the bacterial antiterminator proteins of the BglG/SacY family. *Proc. Natl. Acad. Sci. USA* **89:**10410–10414.

167. **Baba, T., A. Jacq, E. Brickman, J. Beckwith, T. Taura, C. Ueguchi, Y. Akiyama, and K. Ito.** 1990. Characterization of cold-sensitive *secY* mutants of *Escherichia coli.* *J. Bacteriol.* **172:**7005–7010.

168. **Baba, T., T. Taura, T. Shimoike, Y. Akiyama, T. Yoshihisa, and K. Ito.** 1994. A cytoplasmic domain is important for the formation of a SecY-SecE translocator complex. *Proc. Natl. Acad. Sci. USA* **91:**4539–4543.

169. **Babbitt, P. C., G. T. Mrachko, M. S. Hasson, G. W. Huisman, R. Kolter, D. Ringe, G. A. Petsko, G. L. Kenyon, and J. A. Gerlt.** 1995. A functionally diverse enzyme superfamily that abstracts the alpha protons of carboxylic acids. *Science* **267:**1159–1161.

170. **Babitzke, P., and S. R. Kushner.** 1991. The Ams (altered mRNA stability) protein and ribonuclease E are encoded by the same structural gene of *Escherichia coli.* *Proc. Natl. Acad. Sci. USA* **88:**1–5.

171. **Baccanari, D. P., D. Stone, and L. Kuyper.** 1981. Effect of a single amino acid substitution on *Escherichia coli* dihydrofolate reductase catalysis and ligand binding. *J. Biol. Chem.* **256:**1738–1747.

172. **Bachi, B., and H. L. Kornberg.** 1975. Genes involved in the uptake and catabolism of gluconate by *Escherichia coli.* *J. Gen. Microbiol.* **90:**321–335.

173. **Bachmann, B. J.** 1983. Linkage map of *Escherichia coli* K-12, edition 7. *Microbiol. Rev.* **47:**180–230.

174. **Bachmann, B. J.** 1987. Linkage map of *Escherichia coli* K-12, edition 7, p. 807–876. *In* F. C. Neidhardt, J. L. Ingraham, K. B. Low, B. Magasanik, M. Schaechter, and H. E. Umbarger (ed.), *Escherichia coli and Salmonella typhimurium: Cellular and Molecular Biology.* American Society for Microbiology, Washington, D.C.

175. **Bachmann, B. J.** 1990. Linkage map of *Escherichia coli* K-12, edition 8. *Microbiol. Rev.* **54:**130–197.

176. **Bachmann, B. J., K. B. Low, and A. L. Taylor.** 1976. Recalibrated linkage map of *Escherichia coli* K-12. *Bacteriol. Rev.* **40:**116–167.

177. **Backendorf, C., R. Olsthoorn, and P. van de Putte.** 1989. Superhelical stress restrained in plasmid DNA during repair synthesis initiated by the UvrA, B and C proteins in vitro. *Nucleic Acids Res.* **17:**10337–10351.

178. **Backendorf, C., H. Spaink, A. P. Barbeiro, and P. van de Putte.** 1986. Structure of the uvrB gene of *Escherichia coli.* Homology with other DNA repair enzymes and characterization of the *uvrB5* mutation. *Nucleic Acids Res.* **14:**2877–2890.

179. **Backman, K., Y. M. Chen, and B. Magasanik.** 1981. Physical and genetic characterization of the *glnA–glnG* region of the *Escherichia coli* chromosome. *Proc. Natl. Acad. Sci. USA* **78:**3743–3747.

180. **Backstrom, D., R. M. Sjoberg, and L. G. Lundberg.** 1986. Nucleotide sequence of the structural gene for dihydroorotase of *Escherichia coli* K12. *Eur. J. Biochem.* **160:**77–82.

181. **Baecker, P. A., C. E. Furlong, and J. Preiss.** 1983. Biosynthesis of bacterial glycogen. Primary structure of *Escherichia coli* ADP-glucose synthetase as deduced from the nucleotide sequence of the *glgC* gene. *J. Biol. Chem.* **258:**5084–5088.

182. **Baecker, P. A., E. Greenberg, and J. Preiss.** 1986. Biosynthesis of bacterial glycogen. Primary structure of *Escherichia coli* 1,4-α-D-glucan:1,4-α-D-glucan 6-α-D-(1,4-α-D-glucano)-transferase as deduced from the nucleotide sequence of the glgB gene. *J. Biol. Chem.* **261:**8738–8743.

183. **Baer, M., K. B. Low, and D. Soll.** 1979. Regulation of the biosynthesis of aminoacyl-transfer ribonucleic acid synthetases and of transfer ribonucleic acid in *Escherichia coli.* V. Mutants with increased levels of valyl-transfer ribonucleic acid synthetase. *J. Bacteriol.* **139:**165–175.

184. **Bagg, A., and J. B. Neilands.** 1985. Mapping of a mutation affecting regulation of iron uptake systems in *Escherichia coli* K-12. *J. Bacteriol.* **161:**450–453.

185. **Bagg, A., and J. B. Neilands.** 1987. Ferric uptake regulation protein acts as a repressor, employing iron(II) as a cofactor to bind the operator of an iron transport operon in *Escherichia coli.* *Biochemistry* **26:**5471–5477.

186. **Bailey, M. J., V. Koronakis, T. Schmoll, and C. Hughes.** 1992. *Escherichia coli* HlyT protein, a transcriptional activator of haemolysin synthesis and secretion, is encoded by the *rfaH* (*sfrB*) locus required for expression of sex factor and lipopolysaccharide genes. *Mol. Microbiol.* **6:**1003–1012.

187. **Bailly, V., and W. G. Verly.** 1987. *Escherichia coli* endonuclease III is not an endonuclease but a beta-elimination catalyst. *Biochem. J.* **242:**565–572.

188. **Baird, L., and C. Georgopoulos.** 1990. Identification, cloning, and characterization of the *Escherichia coli sohA* gene, a suppressor of the *htrA* (*degP*) null phenotype. *J. Bacteriol.* **172:**1587–1594.

189. **Baird, L., B. Lipinska, S. Raina, and C. Georgopoulos.** 1991. Identification of the *Escherichia coli sohB* gene, a multicopy suppressor of the HtrA (DegP) null phenotype. *J. Bacteriol.* **173:**5763–5770.

190. **Baker, J., D. B. Franklin, and J. Parker.** 1992. Sequence and characterization of the gcpE gene of *Escherichia coli.* *FEMS Microbiol. Lett.* **73:**175–180.

191. **Baker, T. A.** 1994. Replication initiation. A new controller in *Escherichia coli.* *Curr. Biol.* **4:**945–946.

192. **Baker, T. A., and A. Kornberg.** 1988. Transcriptional activation of initiation of replication from the *E. coli* chromosomal origin: an RNA-DNA hybrid near oriC. *Cell* **55:**113–123.

193. **Bakker, E. P., I. R. Booth, U. Dinnbier, W. Epstein, and A. Gajewska.** 1987. Evidence for multiple K$^+$ export systems in *Escherichia coli.* *J. Bacteriol.* **169:**3743–3749.

194. **Balbinder, E., R. Callahan, P. P. McCann, J. C. Cordaro, A. R. Weber, A. M. Smith, and F. Angelosanto.** 1970. Regulatory mutants of the tryptophan operon of *Salmonella typhimurium.* *Genetics* **66:**31–53.

195. **Baldoma, L., J. Badia, G. Sweet, and J. Aguilar.** 1990. Cloning, mapping and gene product identification of *rhaT* from *Escherichia coli* K12. *FEMS Microbiol. Lett.* **60:**103–107.

196. **Ball, C. A., R. Osuna, K. C. Ferguson, and R. C. Johnson.** 1992. Dramatic changes in Fis levels upon nutrient upshift in *Escherichia coli.* *J. Bacteriol.* **174:**8043–8056.

197. **Bally, M., M. Foglino, M. Bruschi, M. Murgier, and A. Lazdunski.** 1986. Nucleotide sequence of the promoter and amino-terminal encoding region of the *Escherichia coli pepN* gene. *Eur. J. Biochem.* **155:**565–569.

198. **Bally, M., M. Murgier, and A. Lazdunski.** 1984. Cloning and orientation of the gene encoding aminopeptidase N in *Escherichia coli.* *Mol. Gen. Genet.* **195:**507–510.

199. **Bally, M., M. Murgier, J. Tommassen, and A. Lazdunski.** 1984. Physical mapping of the gene for aminopeptidase N in *Escherichia coli* K12. *Mol. Gen. Genet.* **193:**190–191.

200. **Bandrin, S. V., P. M. Rabinovich, and A. I. Stepanov.** 1983. Three linkage groups of the genes of riboflavin biosynthesis in *Escherichia coli.* *Genetika* **19:**1419–1425.

201. **Banerjee, R. V., N. L. Johnston, J. K. Sobeski, P. Datta, and R. G. Matthews.** 1989. Cloning and sequence analysis of the *Escherichia coli metH* gene encoding cobalamin-dependent methionine synthase and isolation of a tryptic fragment containing the cobalamin-binding domain. *J. Biol. Chem.* **264:**13888–13895.

202. **Banerjee, R. V., and R. G. Matthews.** 1990. Cobalamin-dependent methionine synthase. *FASEB J.* **4:**1450–1459.

203. **Baneyx, F., and A. A. Gatenby.** 1992. A mutation in GroEL interferes with protein folding by reducing the rate of discharge of sequestered polypeptides. *J. Biol. Chem.* **267:**11637–11644.

204. **Bankaitis, V. A., and P. J. Bassford, Jr.** 1982. Regulation of adenylate cyclase synthesis in *Escherichia coli:* studies with *cya-lac* operon and protein fusion strains. *J. Bacteriol.* **151:**1346–1357.

205. **Bankaitis, V. A., and P. J. Bassford, Jr.** 1985. Proper interaction between at least two components is required for efficient export of proteins to the *Escherichia coli* cell envelope. *J. Bacteriol.* **161:**169–178.

206. **Banuett, F., and I. Herskowitz.** 1987. Identification of polypeptides encoded by an *Escherichia coli* locus (*hflA*) that governs the lysis-lysogeny decision of bacteriophage lambda. *J. Bacteriol.* **169:**4076–4085.

207. **Banuett, F., M. A. Hoyt, L. McFarlane, H. Echols, and I. Herskowitz.** 1986. *hflB*, a new *Escherichia coli* locus regulating lysogeny and the level of bacteriophage lambda cII protein. *J. Mol. Biol.* **187:**213–224.

208. **Baracchini, E., and H. Bremer.** 1988. Stringent and growth control of rRNA synthesis in *Escherichia coli* are both mediated by ppGpp. *J. Biol. Chem.* **263:**2597–2602.

209. **Baracchini, E., and H. Bremer.** 1991. Control of rRNA synthesis in *Escherichia coli* at increased *rrn* gene dosage. Role of guanosine tetraphosphate and ribosome feedback. *J. Biol. Chem.* **266:**11753–11760.

210. **Baracchini, E., R. Glass, and H. Bremer.** 1988. Studies in vivo on *Escherichia coli* RNA polymerase mutants altered in the stringent response. *Mol. Gen. Genet.* **213:**379–387.

211. **Barbier, C. S., and S. A. Short.** 1985. Studies on *deo* operon regulation in *Escherichia coli:* cloning and expression of the *cytR* structural gene. *Gene* **36:**37–44.

212. **Barbier, C. S., and S. A. Short.** 1992. Amino acid substitutions in the CytR repressor which alter its capacity to regulate gene expression. *J. Bacteriol.* **174:**2881–2890.

213. **Barcellini-Couget, S., P. Galucci, and J. P. Bouche.** 1988. *Escherichia coli dicB* operon includes a truncated IS2 element. *Nucleic Acids Res.* **16:**10388.

214. **Bardwell, J. C., and E. A. Craig.** 1984. Major heat shock gene of *Drosophila* and the *Escherichia coli* heat-inducible *dnaK* gene are homologous. *Proc. Natl. Acad. Sci. USA* **81:**848–852.

215. **Bardwell, J. C., and E. A. Craig.** 1987. Eukaryotic Mr 83,000 heat shock protein has a homologue in *Escherichia coli.* *Proc. Natl. Acad. Sci. USA* **84:**5177–5181.

216. **Bardwell, J. C., J. O. Lee, G. Jander, N. Martin, D. Belin, and J. Beckwith.** 1993. A pathway for disulfide bond formation in vivo. *Proc. Natl. Acad. Sci. USA* **90:**1038–1042.

217. **Bardwell, J. C., K. McGovern, and J. Beckwith.** 1991. Identification of a protein required for disulfide bond formation in vivo. *Cell* **67:**581–589.

218. **Bardwell, J. C., K. Tilly, E. Craig, J. King, M. Zylicz, and C. Georgopoulos.** 1986. The nucleotide sequence of the *Escherichia coli* K12 dnaJ$^+$ gene. A gene that encodes a heat shock protein. *J. Biol. Chem.* **261:**1782–1785.

219. Barik, S., and A. Das. 1990. An analysis of the role of host factors in transcription antitermination in vitro by the Q protein of coliphage lambda. *Mol. Gen. Genet.* 222:152–156.

220. Barker, D. F., and A. M. Campbell. 1980. Use of *bio-lac* fusion strains to study regulation of biotin biosynthesis in *Escherichia coli. J. Bacteriol.* 143:789–800.

221. Barker, D. F., and A. M. Campbell. 1981. The *birA* gene of *Escherichia coli* encodes a biotin holoenzyme synthetase. *J. Mol. Biol.* 146:451–467.

222. Barker, D. F., and A. M. Campbell. 1981. Genetic and biochemical characterization of the *birA* gene and its product: evidence for a direct role of biotin holoenzyme synthetase in repression of the biotin operon in *Escherichia coli. J. Mol. Biol.* 146:469–492.

223. Barker, D. F., J. Kuhn, and A. M. Campbell. 1981. Sequence and properties of operator mutations in the *bio* operon of *Escherichia coli. Gene* 13:89–102.

224. Barker, D. G., C. J. Bruton, and G. Winter. 1982. The tyrosyl-tRNA synthetase from *Escherichia coli*. Complete nucleotide sequence of the structural gene. *FEBS Lett.* 150:419–423.

225. Baron, C., and A. Bock. 1991. The length of the aminoacyl-acceptor stem of the selenocysteine-specific tRNA (Sec) of *Escherichia coli* is the determinant for binding to elongation factors SELB or Tu. *J. Biol. Chem.* 266:20375–20379.

226. Baron, C., J. Heider, and A. Bock. 1990. Mutagenesis of *selC*, the gene for the selenocysteine-inserting tRNA species in *E. coli*: effects on in vivo function. *Nucleic Acids Res.* 18:6761–6766.

227. Barondess, J. J., M. Carson, L. M. Guzman Verduzco, and J. Beckwith. 1991. Alkaline phosphatase fusions in the study of cell division genes. *Res. Microbiol.* 142:295–299.

228. Barr, G. C., N. N. Bhriain, and C. J. Dorman. 1992. Identification of two new genetically active regions associated with the *osmZ* locus of *Escherichia coli*: role in regulation of *proU* expression and mutagenic effect of *cya*, the structural gene for adenylate cyclase. *J. Bacteriol.* 174:998–1006.

229. Barr, K., and P. D. Rick. 1993. Physical map location of the *rffC* and *rffA* genes of *Escherichia coli. J. Bacteriol.* 175:5738–5739.

230. Barron, A., J. U. Jung, and M. Villarejo. 1987. Purification and characterization of a glycine betaine binding protein from *Escherichia coli. J. Biol. Chem.* 262:11841–11846.

231. Barry, G., C. Squires, and C. L. Squires. 1980. Attenuation and processing of RNA from the *rplJL–rpoBC* transcription unit of *Escherichia coli. Proc. Natl. Acad. Sci. USA* 77:3331–3335.

232. Barry, G., C. L. Squires, and C. Squires. 1979. Control features within the *rplJL-rpoBC* transcription unit of *Escherichia coli. Proc. Natl. Acad. Sci. USA* 76:4922–4926.

233. Bartlett, D. H., B. B. Frantz, and P. Matsumura. 1988. Flagellar transcriptional activators FlbB and FlaI: gene sequences and 5′ consensus sequences of operons under FlbB and FlaI control. *J. Bacteriol.* 170:1575–1581.

234. Bartlett, D. H., and P. Matsumura. 1984. Identification of *Escherichia coli* region III flagellar gene products and description of two new flagellar genes. *J. Bacteriol.* 160:577–585.

235. Barton, J. W., and T. Melton. 1987. Generation of deletions in the 3′-flanking sequences of the *Escherichia coli crp* gene that induce cyclic AMP suppressor functions. *J. Bacteriol.* 169:654–659.

236. Bartsch, K., R. Dichmann, P. Schmitt, E. Uhlmann, and A. Schulz. 1990. Stereospecific production of the herbicide phosphinothricin (glufosinate) by transamination: cloning, characterization, and overexpression of the gene encoding a phosphinothricin-specific transaminase from *Escherichia coli. Appl. Environ. Microbiol.* 56:7–12.

237. Bartsch, K., A. von Johnn-Marteville, and A. Schulz. 1990. Molecular analysis of two genes of the *Escherichia coli gab* cluster: nucleotide sequence of the glutamate:succinic semialdehyde transaminase gene (*gabT*) and characterization of the succinic semialdehyde dehydrogenase gene (*gabD*). *J. Bacteriol.* 172:7035–7042.

238. Bassford, P., J. Beckwith, K. Ito, C. Kumamoto, S. Mizushima, D. Oliver, L. Randall, T. Silhavy, P. C. Tai, and B. Wickner. 1991. The primary pathway of protein export in *E. coli. Cell* 65:367–368.

239. Bassford, P. J., Jr., C. Bradbeer, R. J. Kadner, and C. A. Schnaitman. 1976. Transport of vitamin B12 in *tonB* mutants of *Escherichia coli. J. Bacteriol.* 128:242–247.

240. Batt, C. A., A. C. Jamieson, and M. A. Vandeyar. 1990. Identification of essential histidine residues in the active site of *Escherichia coli* xylose (glucose) isomerase. *Proc. Natl. Acad. Sci. USA* 87:618–622.

241. Bauer, A. J., I. Rayment, P. A. Frey, and H. M. Holden. 1992. The molecular structure of UDP-galactose 4-epimerase from *Escherichia coli* determined at 2.5 A resolution. *Proteins* 12:372–381.

242. Baumberg, S. 1970. Acetylhistidine as substrate for acetylornithinase: a new system for the selection of arginine regulation mutants in *Escherichia coli. Mol. Gen. Genet.* 106:162–173.

243. Baylor, N. W., A. L. Williams, and N. Cofie. 1983. Molecular characterization of *ilvC* specialized transducing phages of *Escherichia coli* K-12. *Mol. Gen. Genet.* 191:347–352.

244. Beacham, I. R., and S. Garrett. 1980. Isolation of *Escherichia coli* mutants (cpdB) deficient in periplasmic 2′:3′-cyclic phosphodiesterase and genetic mapping of the *cpdB* locus. *J. Gen. Microbiol.* 119:31–34.

245. Beacham, I. R., R. Kahana, L. Levy, and E. Yagil. 1973. Mutants of *Escherichia coli* K-12 "cryptic," or deficient in 5′-nucleotidase (uridine diphosphate-sugar hy-

drolase) and 3′-nucleotidase (cyclic phosphodiesterase) activity. *J. Bacteriol.* 116:957–964.

246. Beall, B., and J. Lutkenhaus. 1987. Sequence analysis, transcriptional organization, and insertional mutagenesis of the *envA* gene of *Escherichia coli. J. Bacteriol.* 169:5408–5415.

247. Bebbington, K. J., and H. D. Williams. 1993. Investigation of the role of the *cydD* gene product in production of a functional cytochrome *d* oxidase in *Escherichia coli. FEMS Microbiol. Lett.* 112:19–24.

248. Becerril, B., F. Valle, E. Merino, L. Riba, and F. Bolivar. 1985. Repetitive extragenic palindromic (REP) sequences in the *Escherichia coli gdhA* gene. *Gene* 37:53–62.

249. Bech, F. W., S. T. Jorgensen, B. Diderichsen, and O. H. Karlstrom. 1985. Sequence of the *relB* transcription unit from *Escherichia coli* and identification of the *relB* gene. *EMBO J.* 4:1059–1066.

250. Beck, E., and E. Bremer. 1980. Nucleotide sequence of the gene *ompA* coding the outer membrane protein II of *Escherichia coli* K-12. *Nucleic Acids Res.* 8:3011–3027.

251. Beck, R., H. Crooke, M. Jarsch, J. Cole, and H. Burtscher. 1994. Mutation in *dipZ* leads to reduced production of active human placental alkaline phosphatase in *Escherichia coli. FEMS Microbiol. Lett.* 124:209–214.

252. Becker, A. B., and R. A. Roth. 1992. An unusual active site identified in a family of zinc metalloendopeptidases. *Proc. Natl. Acad. Sci. USA* 89:3835–3839.

253. Becker, A. B., and R. A. Roth. 1993. Identification of glutamate-169 as the third zinc-binding residue in proteinase III, a member of the family of insulin-degrading enzymes. *Biochem. J.* 292:137–142.

254. Becker, S., and R. Plapp. 1992. Location of the *dcp* gene on the physical map of *Escherichia coli. J. Bacteriol.* 174:1698–1699.

255. Bedouelle, H. 1983. Mutations in the promoter regions of the *malEFG* and *malK-lamB* operons of *Escherichia coli* K12. *J. Mol. Biol.* 170:861–882.

256. Bedouelle, H., P. J. Bassford, Jr., A. V. Fowler, I. Zabin, J. Beckwith, and M. Hofnung. 1980. Mutations which alter the function of the signal sequence of the maltose binding protein of *Escherichia coli. Nature* (London) 285:78–81.

257. Bedouelle, H., and M. Hofnung. 1982. A DNA sequence containing the control regions of the *malEFG* and *malK-lamB* operons in *Escherichia coli* K12. *Mol. Gen. Genet.* 185:82–87.

258. Bedouelle, H., U. Schmeissner, M. Hofnung, and M. Rosenberg. 1982. Promoters of the *malEFG* and *malK-lamB* operons in *Escherichia coli* K12. *J. Mol. Biol.* 161:519–531.

259. Bedwell, D., G. Davis, M. Gosink, L. Post, M. Nomura, H. Kestler, J. M. Zengel, and L. Lindahl. 1985. Nucleotide sequence of the alpha ribosomal protein operon of *Escherichia coli. Nucleic Acids Res.* 13:3891–3903.

260. Begg, K. J., and W. D. Donachie. 1985. Cell shape and division in *Escherichia coli*: experiments with shape and division mutants. *J. Bacteriol.* 163:615–622.

261. Begg, K. J., G. F. Hatfull, and W. D. Donachie. 1980. Identification of new genes in a cell envelope-cell division gene cluster of *Escherichia coli*: cell division gene *ftsQ. J. Bacteriol.* 144:435–437.

262. Begg, K. J., A. Takasuga, D. H. Edwards, S. J. Dewar, B. G. Spratt, H. Adachi, T. Ohta, H. Matsuzawa, and W. D. Donachie. 1990. The balance between different peptidoglycan precursors determines whether *Escherichia coli* cells will elongate or divide. *J. Bacteriol.* 172:6697–6703.

263. Beher, M. G., and C. A. Schnaitman. 1981. Regulation of the OmpA outer membrane protein of *Escherichia coli. J. Bacteriol.* 147:972–985.

264. Bejar, S., and J. P. Bouche. 1985. A new dispensable genetic locus of the terminus region involved in control of cell division in *Escherichia coli. Mol. Gen. Genet.* 201:146–150.

265. Bejar, S., K. Cam, and J. P. Bouche. 1986. Control of cell division in *Escherichia coli*. DNA sequence of *dicA* and of a second gene complementing mutation *dicA1, dicC. Nucleic Acids Res.* 14:6821–6833.

266. Belfaiza, J., C. Parsot, A. Martel, C. B. de la Tour, D. Margarita, G. N. Cohen, and I. Saint-Girons. 1986. Evolution in biosynthetic pathways: two enzymes catalyzing consecutive steps in methionine biosynthesis originate from a common ancestor and possess a similar regulatory region. *Proc. Natl. Acad. Sci. USA* 83:867–871.

267. Belfort, M., G. Maley, J. Pedersen-Lane, and F. Maley. 1983. Primary structure of the *Escherichia coli thyA* gene and its thymidylate synthase product. *Proc. Natl. Acad. Sci. USA* 80:4914–4918.

268. Belin, P., E. Quemeneur, and P. L. Boquet. 1994. A pleiotropic acid phosphatase-deficient mutant of *Escherichia coli* shows premature termination in the *dsbA* gene. Use of *dsbA::phoA* fusions to localize a structurally important domain in DsbA. *Mol. Gen. Genet.* 242:23–32.

269. Bell, A. I., J. A. Cole, and S. J. Busby. 1990. Molecular genetic analysis of an FNR-dependent anaerobically inducible *Escherichia coli* promoter. *Mol. Microbiol.* 4:1753–1763.

270. Bell, A. I., K. L. Gaston, J. A. Cole, and S. J. Busby. 1989. Cloning of binding sequences for the *Escherichia coli* transcription activators, FNR and CRP: location of bases involved in discrimination between FNR and CRP. *Nucleic Acids Res.* 17:3865–3874.

271. Bell, A. W., S. D. Buckel, J. M. Groarke, J. N. Hope, D. H. Kingsley, and M. A. Hermodson. 1986. The nucleotide sequences of the *rbsD*, *rbsA*, and *rbsC* genes of *Escherichia coli* K12. *J. Biol. Chem.* 261:7652–7658.

272. Bell, P. J., S. C. Andrews, M. N. Sivak, and J. R. Guest. 1989. Nucleotide sequence of the FNR-regulated fumarase gene (*fumB*) of *Escherichia coli* K-12. *J. Bacteriol.* 171:3494–3503.

273. Ben-Bassat, A., K. Bauer, S. Y. Chang, K. Myambo, A. Boosman, and S. Chang. 1987. Processing of the initiation methionine from proteins: properties of the *Escherichia coli* methionine aminopeptidase and its gene structure. *J. Bacteriol.* 169:751–757.

274. Ben-Neria, T., and E. Z. Ron. 1991. A DNA replication gene maps near *terC* in *Escherichia coli* K12. *Mol. Gen. Genet.* 226:315–317.

275. Bencini, D. A., J. E. Houghton, T. A. Hoover, K. F. Foltermann, J. R. Wild, and G. A. O'Donovan. 1983. The DNA sequence of *argI* from *Escherichia coli* K12. *Nucleic Acids Res.* 11:8509–8518.

276. Bende, S. M., and R. H. Grafstrom. 1991. The DNA binding properties of the MutL protein isolated from *Escherichia coli*. *Nucleic Acids Res.* 19:1549–1555.

277. Bendiak, D. S., and J. D. Friesen. 1981. Organization of genes in the four minute region of the *Escherichia coli* chromosome: evidence that *rpsB* and *tsf* are co-transcribed. *Mol. Gen. Genet.* 181:356–362.

278. Bennett, C. D., J. A. Rodkey, J. M. Sondey, and R. Hirschmann. 1978. Dihydrofolate reductase: the amino acid sequence of the enzyme from a methotrexate-resistant mutant of *Escherichia coli*. *Biochemistry* 17:1328–1337.

279. Benson, C. E., S. H. Love, and C. N. Remy. 1970. Inhibition of de novo purine biosynthesis and interconversion by 6-methylpurine in *Escherichia coli*. *J. Bacteriol.* 101:872–880.

280. Benson, F., S. Collier, and R. G. Lloyd. 1991. Evidence of abortive recombination in *ruv* mutants of *Escherichia coli* K12. *Mol. Gen. Genet.* 225:266–272.

281. Benson, F. E., G. T. Illing, G. J. Sharples, and R. G. Lloyd. 1988. Nucleotide sequencing of the *ruv* region of *Escherichia coli* K-12 reveals a LexA regulated operon encoding two genes. *Nucleic Acids Res.* 16:1541–1549.

282. Benson, F. E., and S. C. West. 1994. Substrate specificity of the *Escherichia coli* RuvC protein. Resolution of three- and four-stranded recombination intermediates. *J. Biol. Chem.* 269:5195–5201.

283. Benson, T. E., J. L. Marquardt, A. C. Marquardt, F. A. Etzkorn, and C. T. Walsh. 1993. Overexpression, purification, and mechanistic study of UDP-N-acetyl-enolpyruvylglucosamine reductase. *Biochemistry* 32:2024–2030.

284. Bentley, J., L. S. Hyatt, K. Ainley, J. H. Parish, R. B. Herbert, and G. R. White. 1993. Cloning and sequence analysis of an *Escherichia coli* gene conferring bicyclomycin resistance. *Gene* 127:117–120.

285. Beny, G., A. Boyen, D. Charlier, W. Lissens, A. Feller, and N. Glansdorff. 1982. Promoter mapping and selection of operator mutants by using insertion of bacteriophage Mu in the *argECBH* divergent operon of *Escherichia coli* K-12. *J. Bacteriol.* 151:62–67.

286. Beresten, S., M. Jahn, and D. Soll. 1992. Aminoacyl-tRNA synthetase-induced cleavage of tRNA. *Nucleic Acids Res.* 20:1523–1530.

287. Berg, B. L., C. Baron, and V. Stewart. 1991. Nitrate-inducible formate dehydrogenase in *Escherichia coli* K-12. II. Evidence that a mRNA stem-loop structure is essential for decoding opal (UGA) as selenocysteine. *J. Biol. Chem.* 266:22386–22391.

288. Berg, B. L., J. Li, J. Heider, and V. Stewart. 1991. Nitrate-inducible formate dehydrogenase in *Escherichia coli* K-12. I. Nucleotide sequence of the *fdnGHI* operon and evidence that opal (UGA) encodes selenocysteine. *J. Biol. Chem.* 266:22380–22385.

289. Berg, B. L., and V. Stewart. 1990. Structural genes for nitrate-inducible formate dehydrogenase in *Escherichia coli* K-12. *Genetics* 125:691–702.

290. Berg, C. M., and J. J. Rossi. 1974. Proline excretion and indirect suppression in *Escherichia coli* and *Salmonella typhimurium*. *J. Bacteriol.* 118:928–934.

291. Berg, C. M., K. J. Shaw, and D. E. Berg. 1980. The *ilvG* gene is expressed in *Escherichia coli* K-12. *Gene* 12:165–170.

292. Berg, C. M., M. D. Wang, N. B. Vartak, and L. Liu. 1988. Acquisition of new metabolic capabilities: multicopy suppression by cloned transaminase genes in *Escherichia coli*. *Gene* 65:195–202.

293. Berg, K. L., C. L. Squires, and C. Squires. 1987. In vivo translation of a region within the *rrnB* 16S rRNA gene of *Escherichia coli*. *J. Bacteriol.* 169:1691–1701.

294. Berg, P. E. 1981. Cloning and characterization of the *Escherichia coli* gene coding for alkaline phosphatase. *J. Bacteriol.* 146:660–667.

295. Bergler, H., G. Hogenauer, and F. Turnowsky. 1992. Sequences of the *envM* gene and of two mutated alleles in *Escherichia coli*. *J. Gen. Microbiol.* 138:2093–2100.

296. Bergler, H., P. Wallner, A. Ebeling, B. Leitinger, S. Fuchsbichler, H. Aschauer, G. Kollenz, G. Hogenauer, and F. Turnowsky. 1994. Protein EnvM is the NADH-dependent enoyl-ACP reductase (FabI) of *Escherichia coli*. *J. Biol. Chem.* 269:5493–5496.

297. Berman-Kurtz, M., E. C. Lin, and D. P. Richey. 1971. Promoter-like mutant with increased expression of the glycerol kinase operon of *Escherichia coli*. *J. Bacteriol.* 106:724–731.

298. Bernstein, H. D., D. Zopf, D. M. Freymann, and P. Walter. 1993. Functional substitution of the signal recognition particle 54-kDa subunit by its *Escherichia coli* homolog. *Proc. Natl. Acad. Sci. USA* 90:5229–5233.

299. Berry, A., and K. E. Marshall. 1993. Identification of zinc-binding ligands in the class II fructose-1,6-bisphosphate aldolase of *Escherichia coli*. *FEBS Lett.* 318:11–16.

300. Bertin, P., P. Lejeune, C. Colson, and A. Danchin. 1992. Mutations in *bglY*, the structural gene for the DNA-binding protein H1 of *Escherichia coli*, increase the expression of the kanamycin resistance gene carried by plasmid pGR71. *Mol. Gen. Genet.* 233:184–192.

301. Bertin, P., P. Lejeune, C. Laurent-Winter, and A. Danchin. 1990. Mutations in *bglY*, the structural gene for the DNA-binding protein H1, affect expression of several *Escherichia coli* genes. *Biochimie* 72:889–891.

302. Bertin, P., E. Terao, E. H. Lee, P. Lejeune, C. Colson, A. Danchin, and E. Collatz. 1994. The H-NS protein is involved in the biogenesis of flagella in *Escherichia coli*. *J. Bacteriol.* 176:5537–5540.

303. Betermier, M., V. Lefrere, C. Koch, R. Alazard, and M. Chandler. 1989. The *Escherichia coli* protein, Fis: specific binding to the ends of phage Mu DNA and modulation of phage growth. *Mol. Microbiol.* 3:459–468.

304. Betzner, A. S., and W. Keck. 1989. Molecular cloning, overexpression and mapping of the *slt* gene encoding the soluble lytic transglycosylase of *Escherichia coli*. *Mol. Gen. Genet.* 219:489–491.

305. Beyreuther, K., K. Adler, E. Fanning, C. Murray, A. Klemm, and N. Geisler. 1975. Amino-acid sequence of *lac* repressor from *Escherichia coli*. Isolation, sequence analysis and sequence assembly of tryptic peptides and cyanogen-bromide fragments. *Eur. J. Biochem.* 59:491–509.

306. Beyreuther, K., V. Berthold-Schmidt, and K. Geider. 1982. Biological activity and a partial amino-acid sequence of *Escherichia coli* DNA-binding protein I isolated from overproducing cells. *Eur. J. Biochem.* 123:415–420.

307. Bhagwat, A. S., and M. McClelland. 1992. DNA mismatch correction by very short patch repair may have altered the abundance of oligonucleotides in the *E. coli* genome. *Nucleic Acids Res.* 20:1663–1668.

308. Bhagwat, A. S., A. Sohail, and M. Lieb. 1988. A new gene involved in mismatch correction in *Escherichia coli*. *Gene* 74:155–156.

309. Bhagwat, A. S., A. Sohail, and R. J. Roberts. 1986. Cloning and characterization of the *dcm* locus of *Escherichia coli* K-12. *J. Bacteriol.* 166:751–755.

310. Bhatnagar, S. K., and M. J. Bessman. 1988. Studies on the mutator gene, *mutT* of *Escherichia coli*. Molecular cloning of the gene, purification of the gene product, and identification of a novel nucleoside triphosphatase. *J. Biol. Chem.* 263:8953–8957.

311. Bhatnagar, S. K., L. C. Bullions, and M. J. Bessman. 1991. Characterization of the *mutT* nucleoside triphosphatase of *Escherichia coli*. *J. Biol. Chem.* 266:9050–9054.

312. Bhayana, V., and H. W. Duckworth. 1984. Amino acid sequence of *Escherichia coli* citrate synthase. *Biochemistry* 23:2900–2905.

313. Bi, E., K. Dai, S. Subbarao, B. Beall, and J. Lutkenhaus. 1991. FtsZ and cell division. *Res. Microbiol.* 142:249–252.

314. Bi, E., and J. Lutkenhaus. 1990. Interaction between the *min* locus and *ftsZ*. *J. Bacteriol.* 172:5610–5616.

315. Bi, E., and J. Lutkenhaus. 1990. FtsZ regulates frequency of cell division in *Escherichia coli*. *J. Bacteriol.* 172:2765–2768.

316. Bi, E., and J. Lutkenhaus. 1992. Isolation and characterization of *ftsZ* alleles that affect septal morphology. *J. Bacteriol.* 174:5414–5423.

317. Bi, E. F., and J. Lutkenhaus. 1991. FtsZ ring structure associated with division in *Escherichia coli*. *Nature* (London) 354:161–164.

318. Bialkowska-Hobrzanska, H., and D. T. Denhardt. 1984. The *rep* mutation. VII. Cloning and analysis of the functional *rep* gene of *Escherichia coli* K-12. *Gene* 28:93–102.

319. Bianchi, V., P. Reichard, R. Eliasson, E. Pontis, M. Krook, H. Jornvall, and E. Haggard-Ljungquist. 1993. *Escherichia coli* ferredoxin NADP$^+$ reductase: activation of *E. coli* anaerobic ribonucleotide reduction, cloning of the gene (*fpr*), and overexpression of the protein. *J. Bacteriol.* 175:1590–1595.

320. Bibi, E., and H. R. Kaback. 1992. Functional complementation of internal deletion mutants in the lactose permease of *Escherichia coli*. *Proc. Natl. Acad. Sci. USA* 89:1524–1528.

321. Biek, D. P., and S. N. Cohen. 1986. Identification and characterization of *recD*, a gene affecting plasmid maintenance and recombination in *Escherichia coli*. *J. Bacteriol.* 167:594–603. (Erratum, 168:1051.)

322. Bieker, K. L., G. J. Phillips, and T. J. Silhavy. 1990. The *sec* and *prl* genes of *Escherichia coli*. *J. Bioenerg. Biomembr.* 22:291–310.

323. Bieker, K. L., and T. J. Silhavy. 1990. PrlA (SecY) and PrlG (SecE) interact directly and function sequentially during protein translocation in *E. coli*. *Cell* 61:833–842.

324. Bieker, K. L., and T. J. Silhavy. 1990. The genetics of protein secretion in *E. coli*. *Trends Genet.* 6:329–334.

325. Biel, A. J., and H. E. Umbarger. 1981. Mutations in the *ilvY* gene of *Escherichia coli* K-12 that cause constitutive expression of *ilvC*. *J. Bacteriol.* 146:718–724.

326. Bierne, H., S. D. Ehrlich, and B. Michel. 1991. The replication termination signal *terB* of the *Escherichia coli* chromosome is a deletion hot spot. *EMBO J.* 10:2699–2705.

327. Bierne, H., S. D. Ehrlich, and B. Michel. 1994. Flanking sequences affect replication arrest at the *Escherichia coli* terminator TerB in vivo. *J. Bacteriol.* 176:4165–4167.

328. Bilgin, N., J. I. Lee, H. Y. Zhu, R. Dalbey, and G. von Heijne. 1990. Mapping of catalytically important domains in *Escherichia coli* leader peptidase. *EMBO J.* 9:2717–2722.

329. Bilous, P. T., S. T. Cole, W. F. Anderson, and J. H. Weiner. 1988. Nucleotide sequence of the *dmsABC* operon encoding the anaerobic dimethylsulphoxide reductase of *Escherichia coli*. *Mol. Microbiol.* 2:785–795.

330. Bilous, P. T., and J. H. Weiner. 1988. Molecular cloning and expression of the *Escherichia coli* dimethyl sulfoxide reductase operon. *J. Bacteriol.* 170:1511–1518.

331. Binding, R., G. Romansky, R. Bitner, and P. Kuempel. 1981. Isolation and properties of Tn*10* insertions in the *rac* locus of *Escherichia coli*. *Mol. Gen. Genet.* 183:333–340.

332. Bingham, R. J., K. S. Hall, and J. L. Slonczewski. 1990. Alkaline induction of a novel gene locus, *alx*, in *Escherichia coli*. *J. Bacteriol.* 172:2184–2186.

333. Binkley, J. P., and P. L. Kuempel. 1986. Genetic mapping in *Escherichia coli* of *tmk*, the locus for dTMP kinase. *J. Bacteriol.* **168**:1457–1458.

334. Biran, D., N. Brot, H. Weissbach, and E. Z. Ron. 1995. Heat shock-dependent transcriptional activation of the *metA* gene of *Escherichia coli*. *J. Bacteriol.* **177**:1374–1379.

335. Biran, D., S. Michaeli, G. Segal, and E. Z. Ron. 1992. Location of the *metA* gene on the physical map of *Escherichia coli*. *J. Bacteriol.* **174**:5753–5754.

336. Birkmann, A., F. Zinoni, G. Sawers, and A. Bock. 1987. Factors affecting transcriptional regulation of the formate-hydrogen-lyase pathway of *Escherichia coli*. *Arch. Microbiol.* **148**:44–51.

337. Bishop, R. E., and J. H. Weiner. 1992. Coordinate regulation of murein peptidase activity and AmpC beta-lactamase synthesis in *Escherichia coli*. *FEBS Lett.* **304**:103–108.

338. Bitar, K. G. 1975. The primary structure of the ribosomal protein L29 from *Escherichia coli*. *Biochim. Biophys. Acta* **386**:99–106.

339. Bitar, K. G., and B. Wittmann-Liebold. 1975. The primary structure of the 5s rRNA binding protein L25 of *Escherichia coli* ribosomes. *Hoppe-Seyler's Z. Physiol. Chem.* **356**:1343–1352.

340. Bitner, R. M., and P. L. Kuempel. 1981. P1 transduction map spanning the replication terminus of *Escherichia coli* K12. *Mol. Gen. Genet.* **184**:208–212.

341. Bitner, R. M., and P. L. Kuempel. 1982. P1 transduction mapping of the *trg* locus in *rac*+ and *rac* strains of *Escherichia coli* K-12. *J. Bacteriol.* **149**:529–533.

342. Biville, F., E. Turlin, and F. Gasser. 1991. Mutants of *Escherichia coli* producing pyrroloquinoline quinone. *J. Gen. Microbiol.* **137**:1775–1782.

342a.Bjork, G. R., and K. Kjellin-Straby. 1978. *Escherichia coli* mutants with defects in the biosynthesis of 5-methylamino-methyl-2-thiouridine or 1-methylguanosine in their tRNA. *J. Bacteriol.* **133**:508–517.

343. Black, D. S., B. Irwin, and H. S. Moyed. 1994. Autoregulation of *hip*, an operon that affects lethality due to inhibition of peptidoglycan or DNA synthesis. *J. Bacteriol.* **176**:4081–4091.

344. Black, D. S., A. J. Kelly, M. J. Mardis, and H. S. Moyed. 1991. Structure and organization of *hip*, an operon that affects lethality due to inhibition of peptidoglycan or DNA synthesis. *J. Bacteriol.* **173**:5732–5739.

345. Black, M. E., and D. E. Hruby. 1991. Nucleotide sequence of the *Escherichia coli* thymidine kinase gene provides evidence for conservation of functional domains and quaternary structure. *Mol. Microbiol.* **5**:373–379.

346. Black, M. T. 1993. Evidence that the catalytic activity of prokaryote leader peptidase depends upon the operation of a serine-lysine catalytic dyad. *J. Bacteriol.* **175**:4957–4961.

347. Black, M. T., J. G. Munn, and A. E. Allsop. 1992. On the catalytic mechanism of prokaryotic leader peptidase 1. *Biochem. J.* **282**:539–543.

348. Black, P. N. 1988. The *fadL* gene product of *Escherichia coli* is an outer membrane protein required for uptake of long-chain fatty acids and involved in sensitivity to bacteriophage T2. *J. Bacteriol.* **170**:2850–2854.

349. Black, P. N. 1991. Primary sequence of the *Escherichia coli fadL* gene encoding an outer membrane protein required for long-chain fatty acid transport. *J. Bacteriol.* **173**:435–442.

350. Black, P. N., C. C. DiRusso, A. K. Metzger, and T. L. Heimert. 1992. Cloning, sequencing, and expression of the *fadD* gene of *Escherichia coli* encoding acyl coenzyme A synthetase. *J. Biol. Chem.* **267**:25513–25520.

351. Black, P. N., S. F. Kianian, C. C. DiRusso, and W. D. Nunn. 1985. Long-chain fatty acid transport in *Escherichia coli*. Cloning, mapping, and expression of the *fadL* gene. *J. Biol. Chem.* **260**:1780–1789.

352. Blair, D. F., and H. C. Berg. 1991. Mutations in the MotA protein of *Escherichia coli* reveal domains critical for proton conduction. *J. Mol. Biol.* **221**:1433–1442.

353. Blair, D. F., D. Y. Kim, and H. C. Berg. 1991. Mutant MotB proteins in *Escherichia coli*. *J. Bacteriol.* **173**:4049–4055.

354. Blakely, G., S. Colloms, G. May, M. Burke, and D. Sherratt. 1991. *Escherichia coli* XerC recombinase is required for chromosomal segregation at cell division. *New Biol.* **3**:789–798.

355. Blakely, G., G. May, R. McCulloch, L. K. Arciszewska, M. Burke, S. T. Lovett, and D. J. Sherratt. 1993. Two related recombinases are required for site-specific recombination at *dif* and *cer* in *E. coli* K12. *Cell* **75**:351–361.

356. Blakely, G. W., and D. J. Sherratt. 1994. Interactions of the site-specific recombinases XerC and XerD with the recombination site *dif*. *Nucleic Acids Res.* **22**:5613–5620.

357. Blanar, M. A., S. J. Sandler, M. E. Armengod, L. W. Ream, and A. J. Clark. 1984. Molecular analysis of the *recF* gene of *Escherichia coli*. *Proc. Natl. Acad. Sci. USA* **81**:4622–4626.

358. Blanchin-Roland, S., S. Blanquet, J. M. Schmitter, and G. Fayat. 1986. The gene for *Escherichia coli* diadenosine tetraphosphatase is located immediately clockwise to *folA* and forms an operon with *ksgA*. *Mol. Gen. Genet.* **205**:515–522.

359. Blanco, C., and M. Mata-Gilsinger. 1986. A DNA sequence containing the control sites for the *uxaB* gene of *Escherichia coli*. *J. Gen. Microbiol.* **132**:697–705.

360. Blanco, C., M. Mata-Gilsinger, and P. Ritzenthaler. 1983. Construction of hybrid plasmids containing the *Escherichia coli uxaB* gene: analysis of its regulation and direction of transcription. *J. Bacteriol.* **153**:747–755.

361. Blanco, C., M. Mata-Gilsinger, and P. Ritzenthaler. 1985. The use of gene fusions to study the expression of *uidR*, a negative regulatory gene of *Escherichia coli* K-12. *Gene* **36**:159–167.

362. Blanco, C., P. Ritzenthaler, and A. Kolb. 1986. The regulatory region of the *uxuAB* operon in *Escherichia coli* K12. *Mol. Gen. Genet.* **202**:112–119.

363. Blanco, C., P. Ritzenthaler, and M. Mata-Gilsinger. 1982. Cloning and endonuclease restriction analysis of *uidA* and *uidR* genes in *Escherichia coli* K-12: determination of transcription direction for the *uidA* gene. *J. Bacteriol.* **149**:587–594.

364. Blanco, C., P. Ritzenthaler, and M. Mata-Gilsinger. 1985. Nucleotide sequence of a regulatory region of the *uidA* gene in *Escherichia coli* K12. *Mol. Gen. Genet.* **199**:101–105.

365. Blanco, C., P. Ritzenthaler, and M. Mata-Gilsinger. 1986. Negative dominant mutations of the *uidR* gene in *Escherichia coli*: genetic proof for a cooperative regulation of *uidA* expression. *Genetics* **112**:173–182.

366. Blasband, A. J., W. R. Marcotte, Jr., and C. A. Schnaitman. 1986. Structure of the *lc* and *nmpC* outer membrane porin protein genes of lambdoid bacteriophage. *J. Biol. Chem.* **261**:12723–12732.

367. Blasband, A. J., and C. A. Schnaitman. 1987. Regulation in *Escherichia coli* of the porin protein gene encoded by lambdoid bacteriophages. *J. Bacteriol.* **169**:2171–2176.

368. Blasco, F., C. Iobbi, G. Giordano, M. Chippaux, and V. Bonnefoy. 1989. Nitrate reductase of *Escherichia coli*: completion of the nucleotide sequence of the *nar* operon and reassessment of the role of the alpha and beta subunits in iron binding and electron transfer. *Mol. Gen. Genet.* **218**:249–256.

369. Blasco, F., C. Iobbi, J. Ratouchniak, V. Bonnefoy, and M. Chippaux. 1990. Nitrate reductases of *Escherichia coli*: sequence of the second nitrate reductase and comparison with that encoded by the *narGHJI* operon. *Mol. Gen. Genet.* **222**:104–111.

370. Blasco, F., F. Nunzi, J. Pommier, R. Brasseur, M. Chippaux, and G. Giordano. 1992. Formation of active heterologous nitrate reductases between nitrate reductases A and Z of *Escherichia coli*. *Mol. Microbiol.* **6**:209–219.

371. Blasco, F., J. Pommier, V. Augier, M. Chippaux, and G. Giordano. 1992. Involvement of the *narJ* or *narW* gene product in the formation of active nitrate reductase in *Escherichia coli*. *Mol. Microbiol.* **6**:221–230.

372. Blattner, F. R., V. Burland, G. Plunkett, H. J. Sofia, and D. L. Daniels. 1993. Analysis of the *Escherichia coli* genome. IV. DNA sequence of the region from 89.2 to 92.8 minutes. *Nucleic Acids Res.* **21**:5408–5417.

373. Blaut, M., K. Whittaker, A. Valdovinos, B. A. Ackrell, R. P. Gunsalus, and G. Cecchini. 1989. Fumarate reductase mutants of *Escherichia coli* that lack covalently bound flavin. *J. Biol. Chem.* **264**:13599–13604.

374. Blinkowa, A. L., and J. R. Walker. 1990. Programmed ribosomal frameshifting generates the *Escherichia coli* DNA polymerase III gamma subunit from within the tau subunit reading frame. *Nucleic Acids Res.* **18**:1725–1729.

375. Blomfield, I. C., M. S. McClain, and B. I. Eisenstein. 1991. Type 1 fimbriae mutants of *Escherichia coli* K12: characterization of recognized afimbriate strains and construction of new *fim* deletion mutants. *Mol. Microbiol.* **5**:1439–1445.

376. Bloxham, D. P., C. J. Herbert, I. G. Giles, and S. S. Ner. 1983. The use of bacteriophage M13 carrying defined fragments of the *Escherichia coli gltA* gene to determine the location and structure of the citrate synthase promoter region. *Mol. Gen. Genet.* **191**:499–506.

377. Blum, P. H., S. B. Jovanovich, M. P. McCann, J. E. Schultz, S. A. Lesley, R. R. Burgess, and A. Matin. 1990. Cloning and in vivo and in vitro regulation of cyclic AMP-dependent carbon starvation genes from *Escherichia coli*. *J. Bacteriol.* **172**:3813–3820.

378. Bock, A., K. Forchhammer, J. Heider, and C. Baron. 1991. Selenoprotein synthesis: an expansion of the genetic code. *Trends Biochem. Sci.* **16**:463–467.

379. Bockamp, E. O., R. Blasco, and E. Vinuela. 1991. *Escherichia coli* thymidine kinase: nucleotide sequence of the gene and relationships to other thymidine kinases. *Gene* **101**:9–14.

380. Boehmer, P. E., and P. T. Emmerson. 1992. The RecB subunit of the *Escherichia coli* RecBCD enzyme couples ATP hydrolysis to DNA unwinding. *J. Biol. Chem.* **267**:4981–4987.

381. Boeke, J. D., F. LaCroute, and G. R. Fink. 1984. A positive selection for mutants lacking orotidine-5′-phosphate decarboxylase activity in yeast: 5-fluoro-orotic acid resistance. *Mol. Gen. Genet.* **197**:345–346.

382. Bognar, A. L., C. Osborne, and B. Shane. 1987. Primary structure of the *Escherichia coli folC* gene and its folylpolyglutamate synthetase-dihydrofolate synthetase product and regulation of expression by an upstream gene. *J. Biol. Chem.* **262**:12337–12343.

383. Bogosian, G., P. V. Haydock, and R. L. Somerville. 1983. Indolmycin-mediated inhibition and stimulation of transcription at the *trp* promoter of *Escherichia coli*. *J. Bacteriol.* **153**:1120–1123.

384. Bogosian, G., R. L. Somerville, K. Nishi, Y. Kano, and F. Imamoto. 1984. Transcription of the *trpR* gene of *Escherichia coli*: an autogeneously regulated system studied by direct measurements of mRNA levels in vivo. *Mol. Gen. Genet.* **193**:244–250.

385. Boheler, K. R., L. Carrier, C. Chassagne, D. de la Bastie, J. J. Mercadier, and K. Schwartz. 1991. Regulation of myosin heavy chain and actin isogenes expression during cardiac growth. *Mol. Cell Biochem.* **104**:101–107.

386. Bohin, J. P., and E. P. Kennedy. 1984. Mapping of a locus (*mdoA*) that affects the biosynthesis of membrane-derived oligosaccharides in *Escherichia coli*. *J. Bacteriol.* **157**:956–957.

387. Bohm, R., M. Sauter, and A. Bock. 1990. Nucleotide sequence and expression of an operon in *Escherichia coli* coding for formate hydrogenlyase components. *Mol. Microbiol.* **4**:231–243.

388. Bohman, K., and L. A. Isaksson. 1979. Temperature-sensitive mutants in cysteinyl-tRNA ligase of *E. coli* K-12. *Mol. Gen. Genet.* **176**:53–55.

389. Bohman, K., and L. A. Isaksson. 1980. A temperature-sensitive mutant in prolinyl-tRNA ligase of *Escherichia coli* K-12. *Mol. Gen. Genet.* **177**:603–605.

390. Boidol, W., M. Simonis, M. Topert, and G. Siewert. 1982. Recombinant plasmids with genes for the biosynthesis of alkaline phosphatase of *Escherichia coli*. *Mol. Gen. Genet.* **185**:510–512.

391. Boiteux, S., and O. Huisman. 1989. Isolation of a formamidopyrimidine-DNA glycosylase (*fpg*) mutant of *Escherichia coli* K12. *Mol. Gen. Genet.* **215**:300–305.

392. Boiteux, S., T. R. O'Connor, and J. Laval. 1987. Formamidopyrimidine-DNA glycosylase of *Escherichia coli*: cloning and sequencing of the *fpg* structural gene and overproduction of the protein. *EMBO J.* **6**:3177–3183.

393. Bolker, M., and R. Kahmann. 1989. The *Escherichia coli* regulatory protein OxyR discriminates between methylated and unmethylated states of the phage Mu *mom* promoter. *EMBO J.* **8**:2403–2410.

394. Bollen, A., R. Lathe, A. Herzog, D. Denicourt, J. P. Lecocq, L. Desmarez, and R. Lavalle. 1979. A conditionally lethal mutation of *Escherichia coli* affecting the gene coding for ribosomal protein S2 (*rpsB*). *J. Mol. Biol.* **132**:219–233.

395. Bolling, T. J., and W. Mandecki. 1990. An *Escherichia coli* expression vector for high-level production of heterologous proteins in fusion with CMP-KDO synthetase. *BioTechniques* **8**:488–492.

395a. Bollinger, J., J. M. Kwon, G. W. Huisman, R. Kolter, and C. T. Walsh. 1995. Glutathionylspermidine metabolism in *Escherichia coli*: purification, cloning, and overproduction and characterization of a bifunctional glutathionylspermidine synthetase/amidase. *J. Biol. Chem.* **270**:14031–14041.

396. Bollinger, J., C. Park, S. Harayama, and G. L. Hazelbauer. 1984. Structure of the Trg protein: homologies with and differences from other sensory transducers of *Escherichia coli*. *Proc. Natl. Acad. Sci. USA* **81**:3287–3291.

396a. Bongaerts, J., S. Zoske, U. Weidner, and G. Unden. 1995. Transcriptional regulation of proton translocating NADH dehydrogenase genes (*nuoA-N*) of *Escherichia coli* by electron acceptors, electron donors and gene regulators. *Mol. Microbiol.* **16**:521–534.

397. Bonnefoy, E., and J. Rouviere-Yaniv. 1991. HU and IHF, two homologous histone-like proteins of *Escherichia coli*, form different protein-DNA complexes with short DNA fragments. *EMBO J.* **10**:687–696.

398. Bonnefoy, V., J. F. Burini, G. Giordano, M. C. Pascal, and M. Chippaux. 1987. Presence in the 'silent' terminus region of the *Escherichia coli* K12 chromosome of cryptic gene(s) encoding a new nitrate reductase. *Mol. Microbiol.* **1**:143–150.

399. Bonnefoy-Orth, V., M. Lepelletier, M. C. Pascal, and M. Chippaux. 1981. Nitrate reductase and cytochrome b nitrate reductase structural genes as parts of the nitrate reductase operon. *Mol. Gen. Genet.* **181**:535–540.

400. Bonner, C. A., S. Hays, K. McEntee, and M. F. Goodman. 1990. DNA polymerase II is encoded by the DNA damage-inducible *dinA* gene of *Escherichia coli*. *Proc. Natl. Acad. Sci. USA* **87**:7663–7667.

401. Bonner, C. A., P. T. Stukenberg, M. Rajagopalan, R. Eritja, M. O'Donnell, K. McEntee, H. Echols, and M. F. Goodman. 1992. Processive DNA synthesis by DNA polymerase II mediated by DNA polymerase III accessory proteins. *J. Biol. Chem.* **267**:11431–11438.

402. Bonthron, D. T. 1990. L-Asparaginase II of *Escherichia coli* K-12: cloning, mapping and sequencing of the *ansB* gene. *Gene* **91**:101–105.

403. Boos, W., C. Bantlow, D. Benner, and E. Roller. 1983. *cir*, a gene conferring resistance to colicin I maps between *mgl* and *fpk* on the *Escherichia coli* chromosome. *Mol. Gen. Genet.* **191**:401–406.

404. Boos, W., U. Ehmann, E. Bremer, A. Middendorf, and P. Postma. 1987. Trehalase of *Escherichia coli*. Mapping and cloning of its structural gene and identification of the enzyme as a periplasmic protein induced under high osmolarity growth conditions. *J. Biol. Chem.* **262**:13212–13218.

405. Boos, W., U. Ehmann, H. Forkl, W. Klein, M. Rimmele, and P. Postma. 1990. Trehalose transport and metabolism in *Escherichia coli*. *J. Bacteriol.* **172**:3450–3461.

406. Boos, W., I. Steinacher, and D. Engelhardt-Altendorf. 1981. Mapping of *mglB*, the structural gene of the galactose-binding protein of *Escherichia coli*. *Mol. Gen. Genet.* **184**:508–518.

407. Boquet, P. L., C. Manoil, and J. Beckwith. 1987. Use of TnphoA to detect genes for exported proteins in *Escherichia coli*: identification of the plasmid-encoded gene for a periplasmic acid phosphatase. *J. Bacteriol.* **169**:1663–1669.

408. Borg-Olivier, S. A., D. Tarlinton, and K. D. Brown. 1987. Defective regulation of the phenylalanine biosynthetic operon in mutants of the phenylalanyl-tRNA synthetase operon. *J. Bacteriol.* **169**:1949–1953.

409. Bork, P., and E. V. Koonin. 1993. An expanding family of helicases within the 'DEAD/H' superfamily. *Nucleic Acids Res.* **21**:751–752.

410. Bork, P., C. Sander, and A. Valencia. 1992. An ATPase domain common to prokaryotic cell cycle proteins, sugar kinases, actin, and hsp70 heat shock proteins. *Proc. Natl. Acad. Sci. USA* **89**:7290–7294.

411. Bornstein-Forst, S. M., E. McFall, and S. Palchaudhuri. 1987. In vivo D-serine deaminase transcription start sites in wild-type *Escherichia coli* and in *dsdA* promoter mutants. *J. Bacteriol.* **169**:1056–1060.

412. Borodovsky, M., E. V. Koonin, and K. E. Rudd. 1994. New genes in old sequence: a strategy for finding genes in the bacterial genome. *Trends Biochem. Sci.* **19**:309–313.

413. Borodovsky, M., K. E. Rudd, and E. V. Koonin. 1994. Intrinsic and extrinsic approaches for detecting genes in a bacterial genome. *Nucleic Acids Res.* **22**:4756–4767.

414. Boronat, A., P. Britton, M. C. Jones-Mortimer, H. L. Kornberg, L. G. Lee, D. Murfitt, and F. Parra. 1984. Location on the *Escherichia coli* genome of a gene specifying O-acetylserine (thiol)-lyase. *J. Gen. Microbiol.* **130**:673–685.

415. Boronat, A., M. C. Jones-Mortimer, and H. L. Kornberg. 1982. A specialized transducing phage, lambda psrlA, for the sorbitol phosphotransferase of *Escherichia coli* K12. *J. Gen. Microbiol.* **128**:605–611.

416. Boros, I., E. Csordas-Toth, A. Kiss, I. Kiss, I. Torok, A. Udvardy, K. Udvardy, and P. Venetianer. 1983. Identification of two new promoters probably involved in the transcription of a ribosomal RNA gene of *Escherichia coli*. *Biochim. Biophys. Acta* **739**:173–180.

417. Boros, I., A. Kiss, and P. Venetianer. 1979. Physical map of the seven ribosomal RNA genes of *Escherichia coli*. *Nucleic Acids Res.* **6**:1817–1830.

418. Borukhov, S., A. Polyakov, V. Nikiforov, and A. Goldfarb. 1992. GreA protein: a transcription elongation factor from *Escherichia coli*. *Proc. Natl. Acad. Sci. USA* **89**:8899–8902.

419. Borukhov, S., V. Sagitov, and A. Goldfarb. 1993. Transcript cleavage factors from *E. coli*. *Cell* **72**:459–466.

420. Bosch, L., L. Nilsson, E. Vijgenboom, and H. Verbeek. 1990. FIS-dependent transactivation of tRNA and rRNA operons of *Escherichia coli*. *Biochim. Biophys. Acta* **1050**:293–301.

421. Bosl, M., and H. Kersten. 1991. A novel RNA product of the *tyrT* operon of *Escherichia coli*. *Nucleic Acids Res.* **19**:5863–5870.

422. Bosl, M., and H. Kersten. 1994. Organization and functions of genes in the upstream region of *tyrT* of *Escherichia coli*: phenotypes of mutants with partial deletion of a new gene (*tgs*). *J. Bacteriol.* **176**:221–231.

423. Bosl, M. R. 1993. Genetic map of the *tyrT* region of *Escherichia coli* from 27.1 to 27.7 minutes based exclusively on sequence data. *J. Bacteriol.* **175**:7751–7753.

424. Bossemeyer, D., A. Borchard, D. C. Dosch, G. C. Helmer, W. Epstein, I. R. Booth, and E. P. Bakker. 1989. K+-transport protein TrkA of *Escherichia coli* is a peripheral membrane protein that requires other *trk* gene products for attachment to the cytoplasmic membrane. *J. Biol. Chem.* **264**:16403–16410.

425. Bossemeyer, D., A. Schlosser, and E. P. Bakker. 1989. Specific cesium transport via the *Escherichia coli* Kup (TrkD) K+ uptake system. *J. Bacteriol.* **171**:2219–2221.

426. Bossi, L. 1983. The *hisR* locus of *Salmonella*: nucleotide sequence and expression. *Mol. Gen. Genet.* **192**:163–170.

427. Botfield, M. C., K. Naguchi, T. Tsuchiya, and T. H. Wilson. 1992. Membrane topology of the melibiose carrier of *Escherichia coli*. *J. Biol. Chem.* **267**:1818–1822.

428. Botfield, M. C., D. M. Wilson, and T. H. Wilson. 1990. The melibiose carrier of *Escherichia coli*. *Res. Microbiol.* **141**:328–331.

429. Boubrik, F., E. Bonnefoy, and J. Rouviere-Yaniv. 1991. HU and IHF: similarities and differences. In *Escherichia coli*, the lack of HU is not compensated for by IHF. *Res. Microbiol.* **142**:239–247.

430. Bouche, F., and J. P. Bouche. 1989. Genetic evidence that DicF, a second division inhibitor encoded by the *Escherichia coli dicB* operon, is probably RNA. *Mol. Microbiol.* **3**:991–994.

431. Bouche, J. P. 1982. Physical map of a 470×10^3 base-pair region flanking the terminus of DNA replication in the *Escherichia coli* K12 genome. *J. Mol. Biol.* **154**:1–20.

432. Bouche, J. P., J. P. Gelugne, J. Louarn, J. M. Louarn, and K. Kaiser. 1982. Relationships between the physical and genetic maps of a 470×10^3 base-pair region around the terminus of *Escherichia coli* K12 DNA replication. *J. Mol. Biol.* **154**:21–32.

433. Bouffard, G., J. Ostell, and K. E. Rudd. 1992. GeneScape: a relational database of *Escherichia coli* genomic map data for Macintosh computers. *Comput. Appl. Biosci.* **8**:563–567.

434. Bouffard, G. G., K. E. Rudd, and S. L. Adhya. 1994. Dependence of lactose metabolism upon mutarotase encoded in the *gal* operon in *Escherichia coli*. *J. Mol. Biol.* **244**:269–278.

435. Bouloc, P., A. Jaffe, and R. D'Ari. 1989. The *Escherichia coli lov* gene product connects peptidoglycan synthesis, ribosomes and growth rate. *EMBO J.* **8**:317–323. (Erratum, **8**:1290.)

436. Bouma, C. L., N. D. Meadow, E. W. Stover, and S. Roseman. 1987. II-BGlc, a glucose receptor of the bacterial phosphotransferase system: molecular cloning of *ptsG* and purification of the receptor from an overproducing strain of *Escherichia coli*. *Proc. Natl. Acad. Sci. USA* **84**:930–934.

437. Bourret, R. B., J. Davagnino, and M. I. Simon. 1993. The carboxy-terminal portion of the CheA kinase mediates regulation of autophosphorylation by transducer and CheW. *J. Bacteriol.* **175**:2097–2101.

438. Bouvier, J., J. C. Patte, and P. Stragier. 1984. Multiple regulatory signals in the control region of the *Escherichia coli carAB* operon. *Proc. Natl. Acad. Sci. USA* **81**:4139–4143.

439. Bouvier, J., A. P. Pugsley, and P. Stragier. 1991. A gene for a new lipoprotein in the *dapA-purC* interval of the *Escherichia coli* chromosome. *J. Bacteriol.* **173**:5523–5531.

440. Bouvier, J., C. Richaud, W. Higgins, O. Bogler, and P. Stragier. 1992. Cloning, characterization, and expression of the *dapE* gene of *Escherichia coli*. *J. Bacteriol.* **174**:5265–5271.

441. Bouvier, J., C. Richaud, F. Richaud, J. C. Patte, and P. Stragier. 1984. Nucleotide sequence and expression of the *Escherichia coli dapB* gene. *J. Biol. Chem.* **259**:14829–14834.

442. Bouvier, J., and P. Stragier. 1991. Nucleotide sequence of the *lsp-dapB* interval in *Escherichia coli*. *Nucleic Acids Res.* **19**:180.

443. Bower, S. G., K. W. Harlow, R. L. Switzer, and B. Hove-Jensen. 1989. Characterization of the *Escherichia coli prsA1*-encoded mutant phosphoribosylpyrophosphate synthetase identifies a divalent cation-nucleotide binding site. *J. Biol. Chem.* **264**:10287–10291.

444. Bowler, L. D., and B. G. Spratt. 1989. Membrane topology of penicillin-binding protein 3 of *Escherichia coli*. *Mol. Microbiol.* **3**:1277–1286.

445. Boy, E., F. Borne, and J. C. Patte. 1979. Isolation and identification of mutants constitutive for aspartokinase III synthesis in *Escherichia coli* K 12. *Biochimie* **61**:1151–1160.

446. Boyd, A., K. Kendall, and M. I. Simon. 1983. Structure of the serine chemoreceptor in *Escherichia coli*. *Nature* (London) **301**:623–626.

447. Boyd, A., A. Krikos, and M. Simon. 1981. Sensory transducers of *E. coli* are encoded by homologous genes. *Cell* **26**:333–343.

448. Boyd, L. A., L. Adam, L. E. Pelcher, A. McHughen, R. Hirji, and G. Selvaraj. 1991. Characterization of an *Escherichia coli* gene encoding betaine aldehyde dehydrogenase (BADH): structural similarity to mammalian ALDHs and a plant BADH. *Gene* **103**:45–52.

449. Boye, E., M. G. Marinus, and A. Lobner-Olesen. 1992. Quantitation of Dam methyltransferase in *Escherichia coli*. *J. Bacteriol.* **174**:1682–1685.

450. Boyen, A., D. Charlier, J. Charlier, V. Sakanyan, I. Mett, and N. Glansdorff. 1992. Acetylornithine deacetylase, succinyldiaminopimelate desuccinylase and carboxypeptidase G2 are evolutionarily related. *Gene* **116**:1–6.

451. Boyle, S. M., G. D. Markham, E. W. Hafner, J. M. Wright, H. Tabor, and C. W. Tabor. 1984. Expression of the cloned genes encoding the putrescine biosynthetic enzymes and methionine adenosyltransferase of *Escherichia coli* (*speA, speB, speC* and *metK*). *Gene* **30**:129–136.

452. Braaten, B. A., L. B. Blyn, B. S. Skinner, and D. A. Low. 1991. Evidence for a methylation-blocking factor (*mbf*) locus involved in *pap* pilus expression and phase variation in *Escherichia coli*. *J. Bacteriol.* **173**:1789–1800.

453. Braaten, B. A., J. V. Platko, M. W. van der Woude, B. H. Simons, F. K. De Graaf, J. M. Calvo, and D. A. Low. 1992. Leucine-responsive regulatory protein controls the expression of both the *pap* and *fan* pili operons in *Escherichia coli*. *Proc. Natl. Acad. Sci. USA* **89**:4250–4254.

454. Bradshaw, R. A., F. Cancedda, L. H. Ericsson, P. A. Neumann, S. P. Piccoli, M. J. Schlesinger, K. Shriefer, and K. A. Walsh. 1981. Amino acid sequence of *Escherichia coli* alkaline phosphatase. *Proc. Natl. Acad. Sci. USA* **78**:3473–3477.

455. Bramley, H. F., and H. L. Kornberg. 1987. Nucleotide sequence of *bglC*, the gene specifying enzyme II *bgl* of the PEP:sugar phosphotransferase system in *Escherichia coli* K12, and overexpression of the gene product. *J. Gen. Microbiol.* **133**:563–573.

456. Brand, B., and W. Boos. 1991. Maltose transacetylase of *Escherichia coli*. Mapping and cloning of its structural gene, *mac*, and characterization of the enzyme as a dimer of identical polypeptides with a molecular weight of 20,000. *J. Biol. Chem.* **266**:14113–14118.

457. Brandsma, J. A., J. Stoorvogel, C. A. van Sluis, and P. van de Putte. 1982. Effect of *lexA* and *ssb* genes, present on a *uvrA* recombinant plasmid, on the UV survival of *Escherichia coli* K-12. *Gene* **18**:77–85.

458. Branlant, G., and C. Branlant. 1985. Nucleotide sequence of the *Escherichia coli gap* gene. Different evolutionary behavior of the NAD⁺-binding domain and of the catalytic domain of D-glyceraldehyde-3-phosphate dehydrogenase. *Eur. J. Biochem.* **150**:61–66.

459. Brauer, D., and I. Ochsner. 1978. The primary structure of protein L1 from the large ribosomal subunit of *Escherichia coli*. *FEBS Lett.* **96**:317–321.

460. Brauer, D., and R. Roming. 1979. The primary structure of protein S3 from the small ribosomal subunit of *Escherichia coli*. *FEBS Lett.* **106**:352–357.

461. Brauer, D., and B. Wittmann-Liebold. 1977. The primary structure of the initiation factor IF-3 from *Escherichia coli*. *FEBS Lett.* **79**:269–275.

462. Braun, V. 1989. The structurally related *exbB* and *tolQ* genes are interchangeable in conferring *tonB*-dependent colicin, bacteriophage, and albomycin sensitivity. *J. Bacteriol.* **171**:6387–6390.

463. Braun, V., and V. Bosch. 1972. Sequence of the murein-lipoprotein and the attachment site of the lipid. *Eur. J. Biochem.* **28**:51–69.

464. Braun, V., J. Frenz, K. Hantke, and K. Schaller. 1980. Penetration of colicin M into cells of *Escherichia coli*. *J. Bacteriol.* **142**:162–168.

465. Braun, V., R. Gross, W. Koster, and L. Zimmermann. 1983. Plasmid and chromosomal mutants in the iron(III)-aerobactin transport system of *Escherichia coli*. Use of streptonigrin for selection. *Mol. Gen. Genet.* **192**:131–139.

466. Braun, V., K. Hantke, and W. Stauder. 1977. Identification of the *sid* outer membrane receptor protein in *Salmonella typhimurium* SL1027. *Mol. Gen. Genet.* **155**:227–229.

467. Braun, V., and C. Herrmann. 1993. Evolutionary relationship of uptake systems for biopolymers in *Escherichia coli*: cross-complementation between the TonB-ExbB-ExbD and the TolA-TolQ-TolR proteins. *Mol. Microbiol.* **8**:261–268.

468. Braun, V., H. Killmann, and R. Benz. 1994. Energy-coupled transport through the outer membrane of *Escherichia coli* small deletions in the gating loop convert the FhuA transport protein into a diffusion channel. *FEBS Lett.* **346**:59–64.

469. Braun-Breton, C., and M. Hofnung. 1981. In vivo and in vitro functional alterations of the bacteriophage lambda receptor in *lamB* missense mutants of *Escherichia coli* K-12. *J. Bacteriol.* **148**:845–852.

470. Breeze, A. S., and E. E. Obaseiki-Ebor. 1983. Mutations to nitrofurantoin and nitrofurazone resistance in *Escherichia coli* K12. *J. Gen. Microbiol.* **129**:99–103.

471. Bremer, E., P. Gerlach, and A. Middendorf. 1988. Double negative and positive control of *tsx* expression in *Escherichia coli*. *J. Bacteriol.* **170**:108–116.

472. Bremer, E., A. Middendorf, J. Martinussen, and P. Valentin-Hansen. 1990. Analysis of the *tsx* gene, which encodes a nucleoside-specific channel-forming protein (Tsx) in the outer membrane of *Escherichia coli*. *Gene* **96**:59–65.

473. Brennan, C. A., and T. Platt. 1991. Mutations in an RNP1 consensus sequence of Rho protein reduce RNA binding affinity but facilitate helicase turnover. *J. Biol. Chem.* **266**:17296–17305.

474. Brent, R., and M. Ptashne. 1980. The *lexA* gene product represses its own promoter. *Proc. Natl. Acad. Sci. USA* **77**:1932–1936.

475. Brent, R., and M. Ptashne. 1981. Mechanism of action of the *lexA* gene product. *Proc. Natl. Acad. Sci. USA* **78**:4204–4208.

476. Breton, R., H. Sanfacon, I. Papayannopoulos, K. Biemann, and J. Lapointe. 1986. Glutamyl-tRNA synthetase of *Escherichia coli*. Isolation and primary structure of the *gltX* gene and homology with other aminoacyl-tRNA synthetases. *J. Biol. Chem.* **261**:10610–10617.

477. Brey, R. N., and B. P. Rosen. 1979. Properties of *Escherichia coli* mutants altered in calcium/proton antiport activity. *J. Bacteriol.* **139**:824–834.

478. Brickman, E., L. Soll, and J. Beckwith. 1973. Genetic characterization of mutations which affect catabolite-sensitive operons in *Escherichia coli*, including deletions of the gene for adenyl cyclase. *J. Bacteriol.* **116**:582–587.

479. Brickman, T. J., and M. A. McIntosh. 1992. Overexpression and purification of ferric enterobactin esterase from *Escherichia coli*. Demonstration of enzymatic hydrolysis of enterobactin and its iron complex. *J. Biol. Chem.* **267**:12350–12355.

480. Brickman, T. J., B. A. Ozenberger, and M. A. McIntosh. 1990. Regulation of divergent transcription from the iron-responsive *fepB-entC* promoter-operator regions in *Escherichia coli*. *J. Mol. Biol.* **212**:669–682.

481. Briggs, K. A., W. E. Lancashire, and B. S. Hartley. 1984. Molecular cloning, DNA structure and expression of the *Escherichia coli* D-xylose isomerase. *EMBO J.* **3**:611–616.

482. Brikun, I., K. Suziedelis, and D. E. Berg. 1994. DNA sequence divergence among derivatives of *Escherichia coli* K-12 detected by arbitrary primer PCR (random amplified polymorphic DNA) fingerprinting. *J. Bacteriol.* **176**:1673–1682.

483. Brill, J. A., C. Quinlan-Walshe, and S. Gottesman. 1988. Fine-structure mapping and identification of two regulators of capsule synthesis in *Escherichia coli* K-12. *J. Bacteriol.* **170**:2599–2611.

484. Brinkmann, U., R. E. Mattes, and P. Buckel. 1989. High-level expression of recombinant genes in *Escherichia coli* is dependent on the availability of the *dnaY* gene product. *Gene* **85**:109–114.

485. Brissette, J. L., M. Russel, L. Weiner, and P. Model. 1990. Phage shock protein, a stress protein of *Escherichia coli*. *Proc. Natl. Acad. Sci. USA* **87**:862–866.

486. Brissette, J. L., L. Weiner, T. L. Ripmaster, and P. Model. 1991. Characterization and sequence of the *Escherichia coli* stress-induced *psp* operon. *J. Mol. Biol.* **220**:35–48.

487. Brissette, R. E., K. Tsung, and M. Inouye. 1992. Mutations in a central highly conserved non-DNA-binding region of OmpR, an *Escherichia coli* transcriptional activator, influence its DNA-binding ability. *J. Bacteriol.* **174**:4907–4912.

488. Britton, P., A. Boronat, D. A. Hartley, M. C. Jones-Mortimer, H. L. Kornberg, and F. Parra. 1983. Phosphotransferase-mediated regulation of carbohydrate utilization in *Escherichia coli* K12: location of the *gsr* (*tgs*) and *iex* (*crr*) genes by specialized transduction. *J. Gen. Microbiol.* **129**:349–356.

489. Britton, P., L. G. Lee, D. Murfitt, A. Boronat, M. C. Jones-Mortimer, and H. L. Kornberg. 1984. Location and direction of transcription of the *ptsH* and *ptsI* genes on the *Escherichia coli* K12 genome. *J. Gen. Microbiol.* **130**:861–868.

490. Britton, P., D. Murfitt, F. Parra, M. C. Jones-Mortimer, and H. L. Kornberg. 1982. Phosphotransferase-mediated regulation of carbohydrate utilisation in *Escherichia coli* K12: identification of the products of genes on the specialised transducing phages lambda *iex* (*crr*) and lambda *gsr* (*tgs*). *EMBO J.* **1**:907–911.

491. Brody, H., A. Greener, and C. W. Hill. 1985. Excision and reintegration of the *Escherichia coli* K-12 chromosomal element e14. *J. Bacteriol.* **161**:1112–1117.

492. Brody, H., and C. W. Hill. 1988. Attachment site of the genetic element e14. *J. Bacteriol.* **170**:2040–2044.

493. Brooks, J. E., R. M. Blumenthal, and T. R. Gingeras. 1983. The isolation and characterization of the *Escherichia coli* DNA adenine methylase (*dam*) gene. *Nucleic Acids Res.* **11**:837–851.

494. Broome-Smith, J., and B. G. Spratt. 1984. An amino acid substitution that blocks the deacylation step in the enzyme mechanism of penicillin-binding protein 5 of *Escherichia coli*. *FEBS Lett.* **165**:185–189.

495. Broome-Smith, J. K., A. Edelman, S. Yousif, and B. G. Spratt. 1985. The nucleotide sequences of the *ponA* and *ponB* genes encoding penicillin-binding protein 1A and 1B of *Escherichia coli* K12. *Eur. J. Biochem.* **147**:437–446.

496. Broome-Smith, J. K., I. Ioannidis, A. Edelman, and B. G. Spratt. 1988. Nucleotide sequences of the penicillin-binding protein 5 and 6 genes of *Escherichia coli*. *Nucleic Acids Res.* **16**:1617.

497. Brosius, J. 1978. Primary structure of *Escherichia coli* ribosomal protein L31. *Biochemistry* **17**:501–508.

498. Brosius, J., and U. Arfsten. 1978. Primary structure of protein L19 from the large subunit of *Escherichia coli* ribosomes. *Biochemistry* **17**:508–516.

499. Brosius, J., and R. Chen. 1976. The primary structure of protein L16 located at the peptidyltransferase center of *Escherichia coli* ribosomes. *FEBS Lett.* **68**:105–109.

500. Brosius, J., T. J. Dull, and H. F. Noller. 1980. Complete nucleotide sequence of a 23S ribosomal RNA gene from *Escherichia coli. Proc. Natl. Acad. Sci. USA* 77:201–204.

501. Brosius, J., T. J. Dull, D. D. Sleeter, and H. F. Noller. 1981. Gene organization and primary structure of a ribosomal RNA operon from *Escherichia coli. J. Mol. Biol.* 148:107–127.

502. Brosius, J., E. Schiltz, and R. Chen. 1975. The primary structure of the 5S RNA binding protein L18 from *Escherichia coli* ribosomes. *FEBS Lett.* 56:359–361.

503. Brosius, J., A. Ullrich, M. A. Raker, A. Gray, T. J. Dull, R. R. Gutell, and H. F. Noller. 1981. Construction and fine mapping of recombinant plasmids containing the *rrnB* ribosomal RNA operon of *E. coli. Plasmid* 6:112–118.

504. Brown, D. C., and K. D. Collins. 1991. Dihydroorotase from *Escherichia coli*. Substitution of Co(II) for the active site Zn(II). *J. Biol. Chem.* 266:1597–1604.

505. Brown, E. D., E. I. Vivas, C. T. Walsh, and R. Kolter. 1995. MurA (MurZ), the enzyme that catalyzes the first committed step in peptidoglycan biosynthesis, is essential in *Escherichia coli. J. Bacteriol.* 177:4194–4197.

506. Brown, E. D., and J. M. Wood. 1992. Redesigned purification yields a fully functional PutA protein dimer from *Escherichia coli. J. Biol. Chem.* 267:13086–13092.

507. Brown, K., P. W. Finch, I. D. Hickson, and P. T. Emmerson. 1987. Complete nucleotide sequence of the *Escherichia coli argA* gene. *Nucleic Acids Res.* 15:10586.

508. Brown, K. D. 1970. Formation of aromatic amino acid pools in *Escherichia coli* K-12. *J. Bacteriol.* 104:177–188.

509. Brown, K. D., and R. L. Somerville. 1971. Repression of aromatic amino acid biosynthesis in *Escherichia coli* K-12. *J. Bacteriol.* 108:386–399.

510. Brown, S. 1991. 4.5S RNA: does form predict function? *New Biol.* 3:430–438.

511. Brown, S. 1992. Engineered iron oxide-adhesion mutants of the *Escherichia coli* phage lambda receptor. *Proc. Natl. Acad. Sci. USA* 89:8651–8655.

512. Brown, S., B. Albrechtsen, S. Pedersen, and P. Klemm. 1982. Localization and regulation of the structural gene for transcription-termination factor rho of *Escherichia coli. J. Mol. Biol.* 162:283–298.

513. Brown, S., and M. J. Fournier. 1984. The 4.5 S RNA gene of *Escherichia coli* is essential for cell growth. *J. Mol. Biol.* 178:533–550.

514. Brown, T. D., M. C. Jones-Mortimer, and H. L. Kornberg. 1977. The enzymic interconversion of acetate and acetyl-coenzyme A in *Escherichia coli. J. Gen. Microbiol.* 102:327–336.

515. Brown, W. E., K. H. Stump, and W. S. Kelley. 1982. *Escherichia coli* DNA polymerase I. Sequence characterization and secondary structure prediction. *J. Biol. Chem.* 257:1965–1972.

516. Bruce, I., J. Hardy, and K. A. Stacey. 1984. Potentiation by purines of the growth-inhibitory effects of sulphonamides on *Escherichia coli* K12 and the location of the gene which mediates this effect. *J. Gen. Microbiol.* 130:2489–2495.

517. Bruckner, R., and H. Matzura. 1981. In vivo synthesis of a polycistronic messenger RNA for the ribosomal proteins L11, L1, L10 and L7/12 in *Escherichia coli. Mol. Gen. Genet.* 183:277–282.

518. Brun, Y. V., R. Breton, P. Lanouette, and J. Lapointe. 1990. Precise mapping and comparison of two evolutionarily related regions of the *Escherichia coli* K-12 chromosome. Evolution of *valU* and *lysT* from an ancestral tRNA operon. *J. Mol. Biol.* 214:825–843.

519. Brun, Y. V., and J. Lapointe. 1990. Locations of genes in the 52-minute region on the physical map of *Escherichia coli* K-12. *J. Bacteriol.* 172:4746–4747.

520. Brun, Y. V., H. Sanfacon, R. Breton, and J. Lapointe. 1990. Closely spaced and divergent promoters for an aminoacyl-tRNA synthetase gene and a tRNA operon in *Escherichia coli*. Transcriptional and post-transcriptional regulation of *gltX*, *valU* and *alaW. J. Mol. Biol.* 214:845–864.

521. Brundage, L., C. J. Fimmel, S. Mizushima, and W. Wickner. 1992. SecY, SecE, and band 1 form the membrane-embedded domain of *Escherichia coli* preprotein translocase. *J. Biol. Chem.* 267:4166–4170.

522. Brune, M., R. Schumann, and F. Wittinghofer. 1985. Cloning and sequencing of the adenylate kinase gene (*adk*) of *Escherichia coli. Nucleic Acids Res.* 13:7139–7151.

523. Bruni, C. B., V. Colantuoni, L. Sbordone, R. Cortese, and F. Blasi. 1977. Biochemical and regulatory properties of *Escherichia coli* K-12 *hisT* mutants. *J. Bacteriol.* 130:4–10.

524. Bruni, C. B., A. M. Musti, R. Frunzio, and F. Blasi. 1980. Structural and physiological studies of the *Escherichia coli* histidine operon inserted into plasmid vectors. *J. Bacteriol.* 142:32–42.

525. Brusilow, W. S., A. C. Porter, and R. D. Simoni. 1983. Cloning and expression of *uncI*, the first gene of the *unc* operon of *Escherichia coli. J. Bacteriol.* 155:1265–1270.

526. Bryan, S. K., and R. E. Moses. 1984. Map location of the *pcbA* mutation and physiology of the mutant. *J. Bacteriol.* 158:216–221.

527. Bryan, S. K., and R. E. Moses. 1992. Interallelic complementation of *dnaE*(Ts) mutations. *J. Bacteriol.* 174:4850–4852.

527a. Bryant, R. E., and P. S. Sypherd. 1974. Genetic analysis of cold-sensitive ribosome maturation mutants of *Escherichia coli. J. Bacteriol.* 117:1082–1092.

528. Brzoska, P., and W. Boos. 1988. Characteristics of a *ugp*-encoded and *phoB*-dependent glycerophosphoryl diester phosphodiesterase which is physically dependent on the *ugp* transport system of *Escherichia coli. J. Bacteriol.* 170:4125–4135.

529. Buchel, D. E., B. Gronenborn, and B. Muller-Hill. 1980. Sequence of the lactose permease gene. *Nature* (London) 283:541–545.

530. Buck, D., and J. R. Guest. 1989. Overexpression and site-directed mutagenesis of the succinyl-CoA synthetase of *Escherichia coli* and nucleotide sequence of a gene (g30) that is adjacent to the *suc* operon. *Biochem. J.* 260:737–747.

531. Buck, D., M. E. Spencer, and J. R. Guest. 1985. Primary structure of the succinyl-CoA synthetase of *Escherichia coli. Biochemistry* 24:6245–6252.

532. Buck, D., M. E. Spencer, and J. R. Guest. 1986. Cloning and expression of the succinyl-CoA synthetase genes of *Escherichia coli* K12. *J. Gen. Microbiol.* 132:1753–1762.

533. Buckel, P., A. Buchberger, A. Bock, and H. G. Wittmann. 1977. Alteration of ribosomal protein L6 in mutants of *Escherichia coli* resistant to gentamicin. *Mol. Gen. Genet.* 158:47–54.

534. Buckel, S. D., A. W. Bell, J. K. Rao, and M. A. Hermodson. 1986. An analysis of the structure of the product of the *rbsA* gene of *Escherichia coli* K12. *J. Biol. Chem.* 261:7659–7662.

535. Bueno, R., G. Pahel, and B. Magasanik. 1985. Role of *glnB* and *glnD* gene products in regulation of the *glnALG* operon of *Escherichia coli. J. Bacteriol.* 164:816–822.

536. Buhk, H. J., and W. Messer. 1983. The replication origin region of *Escherichia coli*: nucleotide sequence and functional units. *Gene* 24:265–279.

537. Buhr, A., G. A. Daniels, and B. Erni. 1992. The glucose transporter of *Escherichia coli*. Mutants with impaired translocation activity that retain phosphorylation activity. *J. Biol. Chem.* 267:3847–3851.

538. Bujalowski, W., and T. M. Lohman. 1991. Monomers of the *Escherichia coli* SSB-1 mutant protein bind single-stranded DNA. *J. Mol. Biol.* 217:63–74.

539. Bukau, B., and G. C. Walker. 1990. Mutations altering heat shock specific subunit of RNA polymerase suppress major cellular defects of *E. coli* mutants lacking the DnaK chaperone. *EMBO J.* 9:4027–4036.

540. Bukhari, A. I., and A. L. Taylor. 1971. Genetic analysis of diaminopimelic acid- and lysine-requiring mutants of *Escherichia coli. J. Bacteriol.* 105:844–854.

541. Bulawa, C. E., and C. R. Raetz. 1984. Isolation and characterization of *Escherichia coli* strains defective in CDP-diglyceride hydrolase. *J. Biol. Chem.* 259:11257–11264.

542. Buoncristiani, M. R., P. K. Howard, and A. J. Otsuka. 1986. DNA-binding and enzymatic domains of the bifunctional biotin operon repressor (BirA) of *Escherichia coli. Gene* 44:255–261.

543. Bupp, K., and J. van Heijenoort. 1993. The final step of peptidoglycan subunit assembly in *Escherichia coli* occurs in the cytoplasm. *J. Bacteriol.* 175:1841–1843.

544. Burdett, V. 1993. tRNA modification activity is necessary for Tet(M)-mediated tetracycline resistance. *J. Bacteriol.* 175:7209–7215.

545. Burgers, P. M., A. Kornberg, and Y. Sakakibara. 1981. The *dnaN* gene codes for the beta subunit of DNA polymerase III holoenzyme of *Escherichia coli. Proc. Natl. Acad. Sci. USA* 78:5391–5395.

546. Burgess, R. R., C. A. Gross, W. Walter, and P. A. Lowe. 1979. Altered chemical properties in three mutants of *E. coli* RNA polymerase sigma subunit. *Mol. Gen. Genet.* 175:251–257.

547. Burkhardt, R., and V. Braun. 1987. Nucleotide sequence of the *fhuC* and *fhuD* genes involved in iron(III) hydroxamate transport: domains in FhuC homologous to ATP-binding proteins. *Mol. Gen. Genet.* 209:49–55.

548. Burland, V., G. Plunkett, D. L. Daniels, and F. R. Blattner. 1993. DNA sequence and analysis of 136 kilobases of the *Escherichia coli* genome: organizational symmetry around the origin of replication. *Genomics* 16:551–561.

548a. Burland, V. D., G. Plunkett, H. J. Sofia, D. L. Daniels, and F. R. Blattner. 1995. Analysis of the *Escherichia coli* genome VI: DNA sequence of the region from 92.8 through 100 minutes. *Nucleic Acids Res.* 23:2105–2119.

548b. Burlingame, R. P. 1983. The degradation of 3-phenylpropionic acid by *Escherichia coli* K-12. Ph.D. thesis, University of Minnesota, St. Paul.

549. Burlingame, R. P., L. Wyman, and P. J. Chapman. 1986. Isolation and characterization of *Escherichia coli* mutants defective for phenylpropionate degradation. *J. Bacteriol.* 168:55–64.

550. Burman, L. G., and K. Nordstrom. 1971. Colicin tolerance induced by ampicillin or mutation to ampicillin resistance in a strain of *Escherichia coli* K-12. *J. Bacteriol.* 106:1–13.

551. Burnett, B. P., A. L. Horwich, and K. B. Low. 1994. A carboxy-terminal deletion impairs the assembly of GroEL and confers a pleiotropic phenotype in *Escherichia coli* K-12. *J. Bacteriol.* 176:6980–6985.

552. Burns, D. M., and I. R. Beacham. 1986. Nucleotide sequence and transcriptional analysis of the *E. coli ushA* gene, encoding periplasmic UDP-sugar hydrolase (5'-nucleotidase): regulation of the *ushA* gene, and the signal sequence of its encoded protein product. *Nucleic Acids Res.* 14:4325–4342.

553. Burova, E., S. C. Hung, V. Sagitov, B. L. Stitt, and M. E. Gottesman. 1995. *Escherichia coli* NusG protein stimulates transcription elongation rates in vivo and in vitro. *J. Bacteriol.* 177:1388–1392.

554. Burton, K. 1977. Transport of adenine, hypoxanthine and uracil into *Escherichia coli. Biochem. J.* 168:195–204.

555. Burton, K. 1983. Transport of nucleic acid bases into *Escherichia coli. J. Gen. Microbiol.* 129:3505–3513.

556. Burton, Z., R. R. Burgess, J. Lin, D. Moore, S. Holder, and C. A. Gross. 1981. The nucleotide sequence of the cloned *rpoD* gene for the RNA polymerase sigma subunit from *E. coli* K12. *Nucleic Acids Res.* 9:2889–2903.

557. Burton, Z. F., C. A. Gross, K. K. Watanabe, and R. R. Burgess. 1983. The operon that encodes the sigma subunit of RNA polymerase also encodes ribosomal protein S21 and DNA primase in *E. coli* K12. *Cell* 32:335–349.

558. Busby, S., H. Aiba, and B. de Crombrugghe. 1982. Mutations in the *Escherichia coli* operon that define two promoters and the binding site of the cyclic AMP receptor protein. *J. Mol. Biol.* 154:211–227.

559. Busby, S., and M. Dreyfus. 1983. Segment-specific mutagenesis of the regulatory region in the *Escherichia coli* galactose operon: isolation of mutations reducing the initiation of transcription and translation. *Gene* 21:121–131.

560. Butler, J. S., M. Springer, J. Dondon, M. Graffe, and M. Grunberg-Manago. 1986. *Escherichia coli* protein synthesis initiation factor IF3 controls its own gene expression at the translational level in vivo. *J. Mol. Biol.* 192:767–780.

561. Butler, J. S., M. Springer, and M. Grunberg-Manago. 1987. AUU-to-AUG mutation in the initiator codon of the translation initiation factor IF3 abolishes translational autocontrol of its own gene (*infC*) in vivo. *Proc. Natl. Acad. Sci. USA* 84:4022–4025.

562. Buxton, R. S. 1971. Genetic analysis of *Escherichia coli* K12 mutants resistant to bacteriophage BF23 and the E-group colicins. *Mol. Gen. Genet.* 113:154–156.

563. Buxton, R. S., and L. S. Drury. 1983. Cloning and insertional inactivation of the dye (*sfrA*) gene, mutation of which affects sex factor F expression and dye sensitivity of *Escherichia coli* K-12. *J. Bacteriol.* 154:1309–1314.

564. Buxton, R. S., and L. S. Drury. 1984. Identification of the *dye* gene product, mutational loss of which alters envelope protein composition and also affects sex factor F expression in *Escherichia coli* K-12. *Mol. Gen. Genet.* 194:241–247.

565. Buxton, R. S., K. Hammer-Jespersen, and P. Valentin-Hansen. 1980. A second purine nucleoside phosphorylase in *Escherichia coli* K-12. I. Xanthosine phosphorylase regulatory mutants isolated as secondary-site revertants of a *deoD* mutant. *Mol. Gen. Genet.* 179:331–340.

566. Byrne, C., H. W. Stokes, and K. A. Ward. 1988. Nucleotide sequence of the *aceB* gene encoding malate synthase A in *Escherichia coli*. *Nucleic Acids Res.* 16:9342.

567. Byrne, C. R., R. S. Monroe, K. A. Ward, and N. M. Kredich. 1988. DNA sequences of the *cysK* regions of *Salmonella typhimurium* and *Escherichia coli* and linkage of the *cysK* regions to *ptsH*. *J. Bacteriol.* 170:3150–3157.

568. Bystrom, A. S., K. J. Hjalmarsson, P. M. Wikstrom, and G. R. Bjork. 1983. The nucleotide sequence of an *Escherichia coli* operon containing genes for the tRNA(m1G)methyltransferase, the ribosomal proteins S16 and L19 and a 21-K polypeptide. *EMBO J.* 2:899–905.

569. Bystrom, A. S., A. von Gabain, and G. R. Bjork. 1989. Differentially expressed *trmD* ribosomal protein operon of *Escherichia coli* is transcribed as a single polycistronic mRNA species. *J. Mol. Biol.* 208:575–586.

570. Cabrera, M., Y. Nghiem, and J. H. Miller. 1988. mutM, a second mutator locus in *Escherichia coli* that generates G·C→T·A transversions. *J. Bacteriol.* 170:5405–5407.

571. Cai, X. Y., M. E. Maxon, B. Redfield, R. Glass, N. Brot, and H. Weissbach. 1989. Methionine synthesis in *Escherichia coli*: effect of the MetR protein on *metE* and *metH* expression. *Proc. Natl. Acad. Sci. USA* 86:4407–4411.

572. Caillet, J. 1990. Genetic mapping of *pheV*, an *Escherichia coli* gene for tRNA(Phe). *Mol. Gen. Genet.* 220:317–319.

573. Caillet, J., and L. Droogmans. 1988. Molecular cloning of the *Escherichia coli* *miaA* gene involved in the formation of delta 2-isopentenyl adenosine in tRNA. *J. Bacteriol.* 170:4147–4152.

574. Caillet, J., and D. Pages. 1991. Precise physical mapping of the *Escherichia coli* *pheU* transcription unit. *FEBS Lett.* 292:45–47.

575. Caillet, J., J. A. Plumbridge, and M. Springer. 1985. Evidence that *pheV*, a gene for tRNAPhe of *E. coli*, is transcribed from tandem promoters. *Nucleic Acids Res.* 13:3699–3710.

576. Cain, B. D., P. J. Norton, W. Eubanks, H. S. Nick, and C. M. Allen. 1993. Amplification of the *bacA* gene confers bacitracin resistance to *Escherichia coli*. *J. Bacteriol.* 175:3784–3789.

577. Cain, B. D., and R. D. Simoni. 1989. Proton translocation by the F1F0 ATPase of *Escherichia coli*. Mutagenic analysis of the a subunit. *J. Biol. Chem.* 264:3292–3300.

578. Caldeira de Araujo, A., and A. Favre. 1986. Near ultraviolet DNA damage induces the SOS responses in *Escherichia coli*. *EMBO J.* 5:175–179.

579. Calendar, R., J. W. Erickson, C. Halling, and A. Nolte. 1988. Deletion and insertion mutations in the *rpoH* gene of *Escherichia coli* that produce functional sigma 32. *J. Bacteriol.* 170:3479–3484.

580. Calendar, R., B. Lindqvist, G. Sironi, and A. J. Clark. 1970. Characterization of REP-mutants and their interaction with P2 phage. *Virology* 40:72–83.

581. Calendar, R., E. Ljungquist, G. Deho, D. C. Usher, R. Goldstein, P. Youderian, G. Sironi, and E. W. Six. 1981. Lysogenization by satellite phage P4. *Virology* 113:20–38.

582. Calhoun, D. H., L. Traub, J. W. Wallen, J. E. Gray, and S. K. Guterman. 1984. Location of the *rho* gene and characterization of lambda *ilv-gal* derivatives of lambda *ilv-rho* bacteriophage. *Mol. Gen. Genet.* 193:205–209.

583. Calhoun, D. H., J. W. Wallen, L. Traub, J. E. Gray, and H. F. Kung. 1985. Internal promoter in the *ilvGEDA* transcription unit of *Escherichia coli* K-12. *J. Bacteriol.* 161:128–132.

584. Calhoun, M. W., and R. B. Gennis. 1993. Demonstration of separate genetic loci encoding distinct membrane-bound respiratory NADH dehydrogenases in *Escherichia coli*. *J. Bacteriol.* 175:3013–3019.

585. Calhoun, M. W., G. Newton, and R. B. Gennis. 1991. Physical map locations of genes encoding components of the aerobic respiratory chain of *Escherichia coli*. *J. Bacteriol.* 173:1569–1570.

586. Callahan, A. M., B. L. Frazier, and J. S. Parkinson. 1987. Chemotaxis in *Escherichia coli*: construction and properties of lambda *tsr* transducing phage. *J. Bacteriol.* 169:1246–1253.

587. Calos, M. P., and J. H. Miller. 1980. Molecular consequences of deletion formation mediated by the transposon Tn9. *Nature* (London) 285:38–41.

588. Calos, M. P., and J. H. Miller. 1980. DNA sequence alteration resulting from a mutation impairing promoter function in the *lac* repressor gene. *Mol. Gen. Genet.* 178:225–227.

589. Cam, K., S. Bejar, D. Gil, and J. P. Bouche. 1988. Identification and sequence of gene *dicB*: translation of the division inhibitor from an in-phase internal start. *Nucleic Acids Res.* 16:6327–6338.

590. Camakaris, H., and J. Pittard. 1982. Autoregulation of the *tyrR* gene. *J. Bacteriol.* 150:70–75.

591. Cammack, R., and J. H. Weiner. 1990. Electron paramagnetic resonance spectroscopic characterization of dimethyl sulfoxide reductase of *Escherichia coli*. *Biochemistry* 29:8410–8416.

592. Campbell, A., R. Chang, D. Barker, and G. Ketner. 1980. Biotin regulatory (*bir*) mutations of *Escherichia coli*. *J. Bacteriol.* 142:1025–1028.

593. Campbell, A., S. J. Schneider, and B. Song. 1992. Lambdoid phages as elements of bacterial genomes (integrase/phage21/*Escherichia coli* K-12/*icd* gene). *Genetica* 86:259–267.

594. Campbell, H. D., B. L. Rogers, and I. G. Young. 1984. Nucleotide sequence of the respiratory D-lactate dehydrogenase gene of *Escherichia coli*. *Eur. J. Biochem.* 144:367–373.

595. Campbell, J. H., J. A. Lengyel, and J. Langridge. 1973. Evolution of a second gene for beta-galactosidase in *Escherichia coli*. *Proc. Natl. Acad. Sci. USA* 70:1841–1845.

595a.Canellakis, E. S., A. A. Paterakis, S. C. Huang, C. A. Panagiotidis, and D. A. Kyriakidis. 1993. Identification, cloning and nucleotide sequencing of the ornithine decarboxylase antizyme gene of *Escherichia coli*. *Proc. Natl. Acad. Sci. USA* 90:7129–7133.

596. Cannon, M., T. Cabezon, and A. Bollen. 1974. Mapping of neamine resistance: identification to two genetic loci, *nea A* and *nea B*. *Mol. Gen. Genet.* 130:321–326.

597. Cao, G. J., and N. Sarkar. 1992. Identification of the gene for an *Escherichia coli* poly(A) polymerase. *Proc. Natl. Acad. Sci. USA* 89:10380–10384.

598. Cao, G. J., and N. Sarkar. 1992. Poly(A) RNA in *Escherichia coli*: nucleotide sequence at the junction of the *lpp* transcript and the polyadenylate moiety. *Proc. Natl. Acad. Sci. USA* 89:7546–7550.

599. Cao, Y., and T. Kogoma. 1993. Requirement for the polymerization and 5′–3′ exonuclease activities of DNA polymerase I in initiation of DNA replication at *oriK* sites in the absence of RecA in *Escherichia coli rnhA* mutants. *J. Bacteriol.* 175:7254–7259.

600. Cao, Y., R. R. Rowland, and T. Kogoma. 1993. DNA polymerase I and the bypassing of RecA dependence of constitutive stable DNA replication in *Escherichia coli rnhA* mutants. *J. Bacteriol.* 175:7247–7253.

601. Caparros, M., J. L. Torrecuadrada, and M. A. de Pedro. 1991. Effect of D-amino acids on *Escherichia coli* strains with impaired penicillin-binding proteins. *Res. Microbiol.* 142:345–350.

602. Cardelli, J., and J. Konisky. 1974. Isolation and characterization of an *Escherichia coli* mutant tolerant to colicins Ia and Ib. *J. Bacteriol.* 119:379–385.

603. Carl, P. L., L. Bloom, and R. J. Crouch. 1980. Isolation and mapping of a mutation in *Escherichia coli* with altered levels of ribonuclease H. *J. Bacteriol.* 144:28–35.

604. Carlin, A., W. Shi, S. Dey, and B. P. Rosen. 1995. The *ars* operon of *Escherichia coli* confers arsenical and antimonial resistance. *J. Bacteriol.* 177:981–986.

605. Carlioz, A., and D. Touati. 1986. Isolation of superoxide dismutase mutants in *Escherichia coli*: is superoxide dismutase necessary for aerobic life? *EMBO J.* 5:623–630.

606. Carlomagno, M. S., L. Chiariotti, P. Alifano, A. G. Nappo, and C. B. Bruni. 1988. Structure and function of the *Salmonella typhimurium* and *Escherichia coli* K-12 histidine operons. *J. Mol. Biol.* 203:586–606.

607. Carlson, J., J. A. Fuchs, and J. Messing. 1984. Primary structure of the *Escherichia coli* ribonucleoside diphosphate reductase operon. *Proc. Natl. Acad. Sci. USA* 81:4294–4297.

608. Carmel, O., N. Dover, O. Rahav-Manor, P. Dibrov, D. Kirsch, R. Karpel, S. Schuldiner, and E. Padan. 1994. A single amino acid substitution (Glu134-Ala) in NhaR1 increases the inducibility by Na⁺ of the product of *nhaA*, a Na⁺/H⁺ antiporter gene in *Escherichia coli*. *EMBO J.* 13:1981–1989.

609. Carmichael, G. G., K. Weber, A. Niveleau, and A. J. Wahba. 1975. The host factor required for RNA phage Qβ RNA replication in vitro. Intracellular location, quantitation, and purification by polyadenylate-cellulose chromatography. *J. Biol. Chem.* 250:3607–3612.

610. Carothers, A. M., E. McFall, and S. Palchaudhuri. 1980. Physical mapping of the *Escherichia coli* D-serine deaminase region: contiguity of the *dsd* structural and regulatory genes. *J. Bacteriol.* 142:174–184.

611. Carra, J. H., and R. F. Schleif. 1993. Variation of half-site organization and DNA looping by AraC protein. *EMBO J.* 12:35–44.

612. Carter, A. T., B. M. Pearson, J. R. Dickinson, and W. E. Lancashire. 1993. Sequence of the *Escherichia coli* K-12 *edd* and *eda* genes of the Entner-Doudoroff pathway. *Gene* 130:155–156.

613. Carter, J. R., M. A. Franden, R. Aebersold, D. R. Kim, and C. S. McHenry. 1993. Isolation, sequencing and overexpression of the gene encoding the theta subunit of DNA polymerase III holoenzyme. *Nucleic Acids Res.* 21:3281–3286.

614. Carter, J. R., M. A. Franden, R. Aebersold, and C. S. McHenry. 1992. Molecular cloning, sequencing, and overexpression of the structural gene encoding the delta subunit of *Escherichia coli* DNA polymerase III holoenzyme. *J. Bacteriol.* 174:7013–7025.

615. Carter, J. R., M. A. Franden, R. Aebersold, and C. S. McHenry. 1993. Identification, isolation, and characterization of the structural gene encoding the delta subunit of *Escherichia coli* DNA polymerase III holoenzyme. *J. Bacteriol.* 175:3812–3822.

616. Carter, J. R., M. A. Franden, R. Aebersold, and C. S. McHenry. 1993. Identification, isolation, and overexpression of the gene encoding the *psi* subunit of DNA polymerase III holoenzyme. *J. Bacteriol.* 175:5604–5610.

617. Carter, J. R., M. A. Franden, J. A. Lippincott, and C. S. McHenry. 1993. Identification, molecular cloning and characterization of the gene encoding the *chi* subunit of DNA polymerase III holoenzyme of *Escherichia coli.* *Mol. Gen. Genet.* 241:399–408.

618. Casado, C., M. Llagostera, and J. Barbe. 1991. Expression of *nrdA* and *nrdB* genes of *Escherichia coli* is decreased under anaerobiosis. *FEMS Microbiol. Lett.* 67:153–157.

619. Casaregola, S., M. Chen, N. Bouquin, V. Norris, A. Jacq, M. Goldberg, S. Margarson, M. Tempete, S. Mckenna, H. Sweetman, et al. 1991. Analysis of a myosin-like protein and the role of calcium in the *E. coli* cell cycle. *Res. Microbiol.* 142:201–207.

620. Casaregola, S., A. Jacq, D. Laoudj, G. McGurk, S. Margarson, M. Tempete, V. Norris, and I. B. Holland. 1992. Cloning and analysis of the entire *Escherichia coli* ams gene. *ams* is identical to *hmp1* and encodes a 114 kDa protein that migrates as a 180 kDa protein. *J. Mol. Biol.* 228:30–40.

621. Casaregola, S., V. Norris, M. Goldberg, and I. B. Holland. 1990. Identification of a 180 kD protein in *Escherichia coli* related to a yeast heavy-chain myosin. *Mol. Microbiol.* 4:505–511.

622. Casas-Finet, J. R., M. I. Khamis, A. H. Maki, and J. W. Chase. 1987. Tryptophan 54 and phenylalanine 60 are involved synergistically in the binding of *E. coli* SSB protein to single-stranded polynucleotides. *FEBS Lett.* 220:347–352.

623. Case, L. M., X. N. Chen, and M. P. Deutscher. 1989. Localization of the *Escherichia coli* rnt gene encoding RNase T by using a combination of physical and genetic mapping. *J. Bacteriol.* 171:5736–5737.

624. Caskey, C. T., W. C. Forrester, W. Tate, and C. D. Ward. 1984. Cloning of the *Escherichia coli* release factor 2 gene. *J. Bacteriol.* 158:365–368.

625. Cassan, M., C. Parsot, G. N. Cohen, and J. C. Patte. 1986. Nucleotide sequence of *lysC* gene encoding the lysine-sensitive aspartokinase III of *Escherichia coli* K12. Evolutionary pathway leading to three isofunctional enzymes. *J. Biol. Chem.* 261:1052–1057.

626. Cassan, M., J. Ronceray, and J. C. Patte. 1983. Nucleotide sequence of the promoter region of the *E. coli* lysC gene. *Nucleic Acids Res.* 11:6157–6166.

627. Casse, F. 1970. Mapping of the gene *chlB* controlling membrane bound nitrate reductase and formic hydrogen-lyase activities in *Escherichia coli* K 12. *Biochem. Biophys. Res. Commun.* 39:429–436.

628. Castano, I., F. Bastarrachea, and A. A. Covarrubias. 1988. *gltBDF* operon of *Escherichia coli.* *J. Bacteriol.* 170:821–827.

629. Castano, I., N. Flores, F. Valle, A. A. Covarrubias, and F. Bolivar. 1992. *gltF,* a member of the *gltBDF* operon of *Escherichia coli,* is involved in nitrogen-regulated gene expression. *Mol. Microbiol.* 6:2733–2741.

630. Caster, J. H. 1967. Selection of thymine-requiring strains from *Escherichia coli* on solid medium. *J. Bacteriol.* 94:1804.

631. Caswell, R., J. Williams, A. Lyddiatt, and S. Busby. 1992. Overexpression, purification and characterization of the *Escherichia coli* MelR transcription activator protein. *Biochem. J.* 287:493–499.

632. Cavard, D., C. Lazdunski, and S. P. Howard. 1989. The acylated precursor form of the colicin A lysis protein is a natural substrate of the DegP protease. *J. Bacteriol.* 171:6316–6322.

633. Cavard, D., J. M. Pages, and C. J. Lazdunski. 1982. A protease as a possible sensor of environmental conditions in *E. coli* outer membrane. *Mol. Gen. Genet.* 188:508–512.

634. Celis, R. T. 1982. Mapping of two loci affecting the synthesis and structure of a periplasmic protein involved in arginine and ornithine transport in *Escherichia coli* K-12. *J. Bacteriol.* 151:1314–1319.

635. Cerretti, D. P., D. Dean, G. R. Davis, D. M. Bedwell, and M. Nomura. 1983. The *spc* ribosomal protein operon of *Escherichia coli:* sequence and cotranscription of the ribosomal protein genes and a protein export gene. *Nucleic Acids Res.* 11:2599–2616.

636. Chakerian, A. E., and K. S. Matthews. 1992. Effect of *lac* repressor oligomerization on regulatory outcome. *Mol. Microbiol.* 6:963–968.

637. Chakrabarti, T., Y. M. Chen, and E. C. Lin. 1984. Clustering of genes for L-fucose dissimilation by *Escherichia coli.* *J. Bacteriol.* 157:984–986.

638. Chalker, A. F., D. R. Leach, and R. G. Lloyd. 1988. *Escherichia coli* sbcC mutants permit stable propagation of DNA replicons containing a long palindrome. *Gene* 71:201–205.

639. Chanda, P. K., M. Ono, M. Kuwano, and H. Kung. 1985. Cloning, sequence analysis, and expression of alteration of the mRNA stability gene (*ams⁺*) of *Escherichia coli.* *J. Bacteriol.* 161:446–449.

640. Chang, C. N., W. J. Kuang, and E. Y. Chen. 1986. Nucleotide sequence of the alkaline phosphatase gene of *Escherichia coli.* *Gene* 44:121–125.

641. Chang, S. F., D. Ng, L. Baird, and C. Georgopoulos. 1991. Analysis of an *Escherichia coli* dnaB temperature-sensitive insertion mutation and its cold-sensitive extragenic suppressor. *J. Biol. Chem.* 266:3654–3660.

642. Chang, S. Y., E. C. McGary, and S. Chang. 1989. Methionine aminopeptidase gene of *Escherichia coli* is essential for cell growth. *J. Bacteriol.* 171:4071–4072.

643. Chang, Y. Y., and J. E. Cronan, Jr. 1982. Mapping nonselectable genes of *Escherichia coli* by using transposon Tn*10:* location of a gene affecting pyruvate oxidase. *J. Bacteriol.* 151:1279–1289.

644. Chang, Y. Y., and J. E. Cronan, Jr. 1983. Genetic and biochemical analyses of *Escherichia coli* strains having a mutation in the structural gene (*poxB*) for pyruvate oxidase. *J. Bacteriol.* 154:756–762.

645. Chang, Y. Y., J. E. Cronan, Jr., S. J. Li, K. Reed, T. Vanden Boom, and A. Y. Wang. 1991. Locations of the *lip,* *poxB,* and *ilvBN* genes on the physical map of *Escherichia coli.* *J. Bacteriol.* 173:5258–5259.

646. Chang, Y. Y., A. Y. Wang, and J. E. Cronan, Jr. 1993. Molecular cloning, DNA sequencing, and biochemical analyses of *Escherichia coli* glyoxylate carboligase. An enzyme of the acetohydroxy acid synthase-pyruvate oxidase family. *J. Biol. Chem.* 268:3911–3919.

647. Chang, Y. Y., A. Y. Wang, and J. E. Cronan, Jr. 1994. Expression of *Escherichia coli* pyruvate oxidase (PoxB) depends on the sigma factor encoded by the *rpoS* (*katF*) gene. *Mol. Microbiol.* 11:1019–1028.

648. Chang, Z. Y., P. Nygaard, A. C. Chinault, and R. E. Kellems. 1991. Deduced amino acid sequence of *Escherichia coli* adenosine deaminase reveals evolutionarily conserved amino acid residues: implications for catalytic function. *Biochemistry* 30:2273–2280.

649. Charbit, A., J. Ronco, V. Michel, C. Werts, and M. Hofnung. 1991. Permissive sites and topology of an outer membrane protein with a reporter epitope. *J. Bacteriol.* 173:262–275. (Erratum, 173:8014.)

650. Charles, I. G., H. K. Lamb, D. Pickard, G. Dougan, and A. R. Hawkins. 1990. Isolation, characterization and nucleotide sequences of the *aroC* genes encoding chorismate synthase from *Salmonella typhi* and *Escherichia coli.* *J. Gen. Microbiol.* 136:353–358.

651. Charlier, D., G. Hassanzadeh, A. Kholti, D. Gigot, and N. Glansdorff. 1995. *carP,* involved in pyrimidine regulation of the *Escherichia coli* carbamoylphosphate synthetase operon, encodes a sequence-specific DNA-binding protein identical to XerB and PepA, also required for resolution of ColE1 multimers. *J. Mol. Biol.* 250:392–406.

652. Charlier, D., N. Huysveld, M. Roovers, and N. Glansdorff. 1994. On the role of the *Escherichia coli* integration host factor (IHF) in repression at a distance of the pyrimidine specific promoter P1 of the *carAB* operon. *Biochimie* 76:1041–1051.

653. Charlier, D., M. Roovers, D. Gigot, N. Huysveld, A. Pierard, and N. Glansdorff. 1993. Integration host factor (IHF) modulates the expression of the pyrimidine-specific promoter of the *carAB* operons of *Escherichia coli* K12 and *Salmonella typhimurium* LT2. *Mol. Gen. Genet.* 237:273–286.

654. Charlier, D., M. Roovers, F. Van Vliet, A. Boyen, R. Cunin, Y. Nakamura, N. Glansdorff, and A. Pierard. 1992. Arginine regulon of *Escherichia coli* K-12. A study of repressor-operator interactions and of in vitro binding affinities versus in vivo repression. *J. Mol. Biol.* 226:367–386.

655. Charlier, J., and R. Sanchez. 1987. Lysyl-tRNA synthetase from *Escherichia coli* K12. Chromatographic heterogeneity and the *lysU*-gene product. *Biochem. J.* 248:43–51.

656. Chase, J. W., J. J. L'Italien, J. B. Murphy, E. K. Spicer, and K. R. Williams. 1984. Characterization of the *Escherichia coli* SSB-113 mutant single-stranded DNA-binding protein. Cloning of the gene, DNA and protein sequence analysis, high pressure liquid chromatography peptide mapping, and DNA-binding studies. *J. Biol. Chem.* 259:805–814.

657. Chase, J. W., B. A. Rabin, J. B. Murphy, K. L. Stone, and K. R. Williams. 1986. *Escherichia coli* exonuclease VII. Cloning and sequencing of the gene encoding the large subunit (*xseA*). *J. Biol. Chem.* 261:14929–14935.

658. Chattopadhyay, M. K., A. K. Ghosh, and S. Sengupta. 1991. Control of methionine biosynthesis in *Escherichia coli* K12: a closer study with analogue-resistant mutants. *J. Gen. Microbiol.* 137:685–691.

659. Chaudhuri, A., and A. I. Krasna. 1987. Isolation of genes required for hydrogenase synthesis in *Escherichia coli.* *J. Gen. Microbiol.* 133:3289–3294.

660. Chaudhuri, S., K. Duncan, L. D. Graham, and J. R. Coggins. 1991. Identification of the active-site lysine residues of two biosynthetic 3- dehydroquinases. *Biochem. J.* 275:1–6.

661. Chaudhury, A. M., and G. R. Smith. 1985. Role of *Escherichia coli* RecBC enzyme in SOS induction. *Mol. Gen. Genet.* 201:525–528.

662. Chauhan, A. K., and D. Apirion. 1989. The gene for a small stable RNA (10Sa RNA) of *Escherichia coli.* *Mol. Microbiol.* 3:1481–1485.

663. Chauhan, A. K., and D. Apirion. 1991. The *rne* gene is the structural gene for the processing endoribonuclease RNase E of *Escherichia coli.* *Mol. Gen. Genet.* 228:49–54.

664. Chauhan, A. K., A. Miczak, L. Taraseviciene, and D. Apirion. 1991. Sequencing and expression of the *rne* gene of *Escherichia coli.* *Nucleic Acids Res.* 19:125–129.

665. Chen, B. J., P. Carroll, and L. Samson. 1994. The *Escherichia coli* AlkB protein protects human cells against alkylation-induced toxicity. *J. Bacteriol.* 176:6255–6261.

666. Chen, C. M., Q. Z. Ye, Z. M. Zhu, B. L. Wanner, and C. T. Walsh. 1990. Molecular biology of carbon-phosphorus bond cleavage. Cloning and sequencing of the *phn* (*psiD*) genes involved in alkylphosphonate uptake and C-P lyase activity in *Escherichia coli* B. *J. Biol. Chem.* 265:4461–4471.

667. **Chen, G. F.** 1990. Suppression of the negative effect of minor arginine codons on gene expression; preferential usage of minor codons within the first 25 codons of the *Escherichia coli* genes. *Nucleic Acids Res.* 18:1465–1473.

668. **Chen, G. T., M. J. Axley, J. Hacia, and M. Inouye.** 1992. Overproduction of a selenocysteine-containing polypeptide in *Escherichia coli*: the *fdhF* gene product. *Mol. Microbiol.* 6:781–785.

669. **Chen, H., S. K. Bryan, and R. E. Moses.** 1989. Cloning the *polB* gene of *Escherichia coli* and identification of its product. *J. Biol. Chem.* 264:20591–20595.

670. **Chen, H., C. B. Lawrence, S. K. Bryan, and R. E. Moses.** 1990. Aphidicolin inhibits DNA polymerase II of *Escherichia coli*, an alpha-like DNA polymerase. *Nucleic Acids Res.* 18:7185–7186.

671. **Chen, H., Y. Sun, T. Stark, W. Beattie, and R. E. Moses.** 1990. Nucleotide sequence and deletion analysis of the *polB* gene of *Escherichia coli*. *DNA Cell Biol.* 9:631–635.

672. **Chen, J., A. Brevet, M. Fromant, F. Leveque, J. M. Schmitter, S. Blanquet, and P. Plateau.** 1990. Pyrophosphatase is essential for growth of *Escherichia coli*. *J. Bacteriol.* 172:5686–5689.

673. **Chen, L., and W. G. Coleman, Jr.** 1993. Cloning and characterization of the *Escherichia coli* K-12 *rfa-2* (*rfaC*) gene, a gene required for lipopolysaccharide inner core synthesis. *J. Bacteriol.* 175:2534–2540.

674. **Chen, M. X., N. Bouquin, V. Norris, S. Casaregola, S. J. Seror, and I. B. Holland.** 1991. A single base change in the acceptor stem of tRNA(3Leu) confers resistance upon *Escherichia coli* to the calmodulin inhibitor, 48/80. *EMBO J.* 10:3113–3122.

675. **Chen, R.** 1976. The sequence determination of a protein in a micro scale: the sequence analysis of ribosomal protein L34 of *Escherichia coli*. *Hoppe-Seyler's Z. Physiol. Chem.* 357:873–886.

676. **Chen, R., U. Arfsten, and U. Chen-Schmeisser.** 1977. The primary structure of protein L6 from the aminoacyl-tRNA binding site of the *Escherichia coli* ribosome. *Hoppe-Seyler's Z. Physiol. Chem.* 358:531–535.

677. **Chen, R., and G. Ehrke.** 1976. The primary structure of the 5 S RNA binding protein L5 of *Escherichia coli* ribosomes. *FEBS Lett.* 69:240–245.

678. **Chen, R., and U. Henning.** 1987. Nucleotide sequence of the gene for the peptidoglycan-associated lipoprotein of *Escherichia coli* K12. *Eur. J. Biochem.* 163:73–77.

679. **Chen, R., C. Kramer, W. Schmidmayr, U. Chen-Schmeisser, and U. Henning.** 1982. Primary structure of major outer-membrane protein I (*ompF* protein, porin) of *Escherichia coli* B/r. *Biochem. J.* 203:33–43.

680. **Chen, R., L. Mende, and U. Arfsten.** 1975. The primary structure of protein L27 from the peptidyl-tRNA binding site of *Escherichia coli* ribosomes. *FEBS Lett.* 59:96–99.

681. **Chen, R., W. Schmidmayr, C. Kramer, U. Chen-Schmeisser, and U. Henning.** 1980. Primary structure of major outer membrane protein II (*ompA* protein) of *Escherichia coli* K-12. *Proc. Natl. Acad. Sci. USA* 77:4592–4596.

682. **Chen, R., and B. Wittmann-Liebold.** 1975. The primary structure of protein S9 from the 30S subunit of *Escherichia coli* ribosomes. *FEBS Lett.* 52:139–140.

683. **Chen, S. M., H. E. Takiff, A. M. Barber, G. C. Dubois, J. C. Bardwell, and D. L. Court.** 1990. Expression and characterization of RNase III and Era proteins. Products of the *rnc* operon of *Escherichia coli*. *J. Biol. Chem.* 265:2888–2895.

684. **Chen, W., C. S. Russell, Y. Murooka, and S. D. Cosloy.** 1994. 5-Aminolevulinic acid synthesis in *Escherichia coli* requires expression of *hemA*. *J. Bacteriol.* 1761:2743–2746.

685. **Chen, Y. M., K. Backman, and B. Magasanik.** 1982. Characterization of a gene, *glnL*, the product of which is involved in the regulation of nitrogen utilization in *Escherichia coli*. *J. Bacteriol.* 150:214–220.

686. **Chen, Y. M., Z. Lu, and E. C. Lin.** 1989. Constitutive activation of the *fucAO* operon and silencing of the divergently transcribed *fucPIK* operon by an IS5 element in *Escherichia coli* mutants selected for growth on L-1,2-propanediol. *J. Bacteriol.* 171:6097–6105.

687. **Chen, Y. M., Y. Zhu, and E. C. Lin.** 1987. NAD-linked aldehyde dehydrogenase for aerobic utilization of L-fucose and L-rhamnose by *Escherichia coli*. *J. Bacteriol.* 169:3289–3294.

688. **Chen, Y. M., Y. Zhu, and E. C. Lin.** 1987. The organization of the *fuc* regulon specifying L-fucose dissimilation in *Escherichia coli* K12 as determined by gene cloning. *Mol. Gen. Genet.* 210:331–337.

689. **Chenais, J., C. Richaud, J. Ronceray, H. Cherest, Y. Surdin-Kerjan, and J. C. Patte.** 1981. Construction of hybrid plasmids containing the *lysA* gene of *Escherichia coli*: studies of expression in *Escherichia coli* and *Saccharomyces cerevisiae*. *Mol. Gen. Genet.* 182:456–461.

690. **Chenault, S. S., and C. F. Earhart.** 1991. Organization of genes encoding membrane proteins of the *Escherichia coli* ferrienterobactin permease. *Mol. Microbiol.* 5:1405–1413.

691. **Cheng, H. H., P. J. Muhlrad, M. A. Hoyt, and H. Echols.** 1988. Cleavage of the cII protein of phage lambda by purified HflA protease: control of the switch between lysis and lysogeny. *Proc. Natl. Acad. Sci. USA* 85:7882–7886.

692. **Cheng, Y. S., Y. Shen, J. Rudolph, M. Stern, J. Stubbe, K. A. Flannigan, and J. M. Smith.** 1990. Glycinamide ribonucleotide synthetase from *Escherichia coli*: cloning, overproduction, sequencing, isolation, and characterization. *Biochemistry* 29:218–227. (Erratum, 29:5220.)

693. **Cheng, Y. S., D. Zipser, C. Y. Cheng, and S. J. Rolseth.** 1979. Isolation and characterization of mutations in the structural gene for protease III (*ptr*). *J. Bacteriol.* 140:125–130.

694. **Chepuri, V., and R. B. Gennis.** 1990. The use of gene fusions to determine the topology of all of the subunits of the cytochrome *o* terminal oxidase complex of *Escherichia coli*. *J. Biol. Chem.* 265:12978–12986.

695. **Chepuri, V., L. Lemieux, D. C. Au, and R. B. Gennis.** 1990. The sequence of the *cyo* operon indicates substantial structural similarities between the cytochrome *o* ubiquinol oxidase of *Escherichia coli* and the aa3-type family of cytochrome *c* oxidases. *J. Biol. Chem.* 265:11185–11192.

696. **Chepuri, V., L. Lemieux, J. Hill, J. O. Alben, and R. B. Gennis.** 1990. Recent studies of the cytochrome *o* terminal oxidase complex of *Escherichia coli*. *Biochim. Biophys. Acta* 1018:124–127.

697. **Chesney, R. H., and E. Adler.** 1982. Chromosomal location of *att* P7, the *recA*-independent p7 integration site used in the suppression of *Escherichia coli dnaA* mutations. *J. Bacteriol.* 150:1400–1404.

698. **Chesney, R. H., P. Sollitti, and D. R. Vickery.** 1985. Identification of a new locus in the *Escherichia coli* cotransduction gap that represents a new genetic component of the L-asparagine utilization system. *J. Gen. Microbiol.* 131:2079–2085.

699. **Cheung, A., S. Morgan, K. B. Low, and D. Soll.** 1979. Regulation of the biosynthesis of aminoacyl-transfer ribonucleic acid synthetases and of transfer ribonucleic acid in *Escherichia coli*. VI. Mutants with increased levels of glutaminyl-transfer ribonucleic acid synthetase and of glutamine transfer ribonucleic acid. *J. Bacteriol.* 139:176–184.

700. **Chiang, R. C., R. Cavicchioli, and R. P. Gunsalus.** 1992. Identification and characterization of *narQ*, a second nitrate sensor for nitrate-dependent gene regulation in *Escherichia coli*. *Mol. Microbiol.* 6:1913–1923.

701. **Chiang, R. C., R. Cavicchioli, and R. P. Gunsalus.** 1992. Physical map location of the *narQ* gene of *Escherichia coli*. *J. Bacteriol.* 174:7882.

702. **Chiaramello, A. E., and J. W. Zyskind.** 1989. Expression of *Escherichia coli dnaA* and *mioC* genes as a function of growth rate. *J. Bacteriol.* 171:4272–4280.

703. **Chiariotti, L., P. Alifano, M. S. Carlomagno, and C. B. Bruni.** 1986. Nucleotide sequence of the *Escherichia coli hisD* gene and of the *Escherichia coli* and *Salmonella typhimurium hisIE* region. *Mol. Gen. Genet.* 203:382–388.

704. **Chiariotti, L., A. G. Nappo, M. S. Carlomagno, and C. B. Bruni.** 1986. Gene structure in the histidine operon of *Escherichia coli*. Identification and nucleotide sequence of the *hisB* gene. *Mol. Gen. Genet.* 202:42–47.

705. **Chin, C. C., P. M. Anderson, and F. Wold.** 1983. The amino acid sequence of *Escherichia coli* cyanase. *J. Biol. Chem.* 258:276–282.

706. **Chin, D. T., S. A. Goff, T. Webster, T. Smith, and A. L. Goldberg.** 1988. Sequence of the *lon* gene in *Escherichia coli*. A heat-shock gene which encodes the ATP-dependent protease La. *J. Biol. Chem.* 263:11718–11728.

707. **Chippaux, M., V. Bonnefoy-Orth, J. Ratouchniak, and M. C. Pascal.** 1981. Operon fusions in the nitrate reductase operon and study of the control gene *nir R* in *Escherichia coli*. *Mol. Gen. Genet.* 182:477–479.

708. **Chippaux, M., D. Giudici, A. Abou-Jaoude, F. Casse, and M. C. Pascal.** 1978. Laboratoire de Chimie Bacterienne C.N.R.S., Marseille, France. *Mol. Gen. Genet.* 160:225–229.

709. **Chittum, H. S., and W. S. Champney.** 1994. Ribosomal protein gene sequence changes in erythromycin-resistant mutants of *Escherichia coli*. *J. Bacteriol.* 176:6192–6198.

710. **Cho, H., and J. E. Cronan, Jr.** 1993. *Escherichia coli* thioesterase I, molecular cloning and sequencing of the structural gene and identification as a periplasmic enzyme. *J. Biol. Chem.* 268:9238–9245.

711. **Cho, H., and J. E. Cronan, Jr.** 1994. "Protease I" of *Escherichia coli* functions as a thioesterase in vivo. *J. Bacteriol.* 176:1793–1795.

712. **Choe, M., and W. S. Reznikoff.** 1991. Anaerobically expressed *Escherichia coli* genes identified by operon fusion techniques. *J. Bacteriol.* 173:6139–6146.

713. **Choe, M., and W. S. Reznikoff.** 1993. Identification of the regulatory sequence of anaerobically expressed locus *aeg-46.5*. *J. Bacteriol.* 175:1165–1172.

714. **Choi, K. Y., and H. Zalkin.** 1992. Structural characterization and corepressor binding of the *Escherichia coli* purine repressor. *J. Bacteriol.* 174:6207–6214.

715. **Choi, K. Y., and H. Zalkin.** 1994. Role of the purine repressor hinge sequence in repressor function. *J. Bacteriol.* 176:1767–1772.

716. **Choi, Y. L., M. Kawamukai, R. Utsumi, H. Sakai, and T. Komano.** 1989. Molecular cloning and sequencing of the glycogen phosphorylase gene from *Escherichia coli*. *FEBS Lett.* 243:193–198.

717. **Choi, Y. L., S. Kawase, M. Kawamukai, H. Sakai, and T. Komano.** 1991. Regulation of *glpD* and *glpE* gene expression by a cyclic AMP-cAMP receptor protein (cAMP-CRP) complex in *Escherichia coli*. *Biochim. Biophys. Acta* 1088:31–35.

718. **Choi, Y. L., S. Kawase, T. Nishida, H. Sakai, T. Komano, M. Kawamukai, R. Utsumi, Y. Kohara, and K. Akiyama.** 1988. Nucleotide sequence of the *glpR* gene encoding the repressor for the glycerol-3-phosphate regulon of *Escherichia coli* K12. *Nucleic Acids Res.* 16:7732.

719. **Choi, Y. L., T. Nishida, M. Kawamukai, R. Utsumi, H. Sakai, and T. Komano.** 1989. Cloning and sequencing of an *Escherichia coli* gene, *nlp*, highly homologous to the *ner* genes of bacteriophages Mu and D108. *J. Bacteriol.* 171:5222–5225.

720. **Choli, T.** 1989. Structural properties of ribosomal protein L11 from *Escherichia coli*. *Biochem. Int.* 19:1323–1338.

721. **Christiansen, L., and S. Pedersen.** 1981. Cloning, restriction endonuclease mapping and post-transcriptional regulation of *rpsA*, the structural gene for ribosomal protein S1. *Mol. Gen. Genet.* 181:548–551.

722. **Christie, G. E., and T. Platt.** 1980. Gene structure in the tryptophan operon of *Escherichia coli*. Nucleotide sequence of *trpC* and the flanking intercistronic regions. *J. Mol. Biol.* 142:519–530.

723. Christman, M. F., G. Storz, and B. N. Ames. 1989. OxyR, a positive regulator of hydrogen peroxide-inducible genes in *Escherichia coli* and *Salmonella typhimurium*, is homologous to a family of bacterial regulatory proteins. *Proc. Natl. Acad. Sci. USA* 86:3484–3488.

724. Chu, C. C., A. Templin, and A. J. Clark. 1989. Suppression of a frameshift mutation in the *recE* gene of *Escherichia coli* K-12 occurs by gene fusion. *J. Bacteriol.* 171:2101–2109.

725. Chuang, S. E., and F. R. Blattner. 1993. Characterization of twenty-six new heat shock genes of *Escherichia coli. J. Bacteriol.* 175:5242–5252.

726. Chuang, S. E., V. Burland, G. Plunkett, D. L. Daniels, and F. R. Blattner. 1993. Sequence analysis of four new heat-shock genes constituting the *hslTS/ibpAB* and *hslVU* operons in *Escherichia coli. Gene* 134:1–6.

727. Chun, S. Y., and J. S. Parkinson. 1988. Bacterial motility: membrane topology of the *Escherichia coli* MotB protein. *Science* 239:276–278.

728. Chung, C. H., and A. L. Goldberg. 1981. The product of the *lon* (*capR*) gene in *Escherichia coli* is the ATP-dependent protease, protease La. *Proc. Natl. Acad. Sci. USA* 78:4931–4935.

729. Chung, T., D. J. Klumpp, and D. C. LaPorte. 1988. Glyoxylate bypass operon of *Escherichia coli*: cloning and determination of the functional map. *J. Bacteriol.* 170:386–392.

730. Chye, M. L., and J. Pittard. 1987. Transcription control of the *aroP* gene in *Escherichia coli* K-12: analysis of operator mutants. *J. Bacteriol.* 169:386–393.

731. Cieslewicz, M. J., S. M. Steenbergen, and E. R. Vimr. 1993. Cloning, sequencing, expression, and complementation analysis of the *Escherichia coli* K1 *kps* region 1 gene, *kpsE*, and identification of an upstream open reading frame encoding a protein with homology to GutQ. *J. Bacteriol.* 175:8018–8023.

732. Clark, A. F., and R. W. Hogg. 1981. High-affinity arabinose transport mutants of *Escherichia coli*: isolation and gene location. *J. Bacteriol.* 147:920–924.

733. Clark, A. J. 1991. *rec* genes and homologous recombination proteins in *Escherichia coli. Biochimie* 73:523–532.

734. Clark, A. J., L. Satin, and C. C. Chu. 1994. Transcription of the *Escherichia coli recE* gene from a promoter in Tn5 and IS50. *J. Bacteriol.* 176:7024–7031.

735. Clark, A. J., V. Sharma, S. Brenowitz, C. C. Chu, S. Sandler, L. Satin, A. Templin, I. Berger, and A. Cohen. 1993. Genetic and molecular analyses of the C-terminal region of the *recE* gene from the Rac prophage of *Escherichia coli* K-12 reveal the *recT* gene. *J. Bacteriol.* 175:7673–7682.

736. Clark, D. 1981. Regulation of fatty acid degradation in *Escherichia coli*: analysis by operon fusion. *J. Bacteriol.* 148:521–526.

736a. Clark, D. 1984. Novel antibiotic hypersensitive mutants of *Escherichia coli*. Genetic mapping and chemical characterization. *FEMS Microbiol. Lett.* 21:189–195.

737. Clark, D., and J. E. Cronan, Jr. 1980. *Escherichia coli* mutants with altered control of alcohol dehydrogenase and nitrate reductase. *J. Bacteriol.* 141:177–183.

738. Clark, D. P. 1989. The fermentation pathways of *Escherichia coli. FEMS Microbiol. Rev.* 63:223–224.

739. Clark, D. P., and J. P. Beard. 1979. Altered phospholipid composition in mutants of *Escherichia coli* sensitive or resistant to organic solvents. *J. Gen. Microbiol.* 113:267–274.

740. Clark, D. P., and J. E. Cronan, Jr. 1980. Acetaldehyde coenzyme A dehydrogenase of *Escherichia coli. J. Bacteriol.* 144:179–184.

741. Clark, D. P., and M. L. Rod. 1987. Regulatory mutations that allow the growth of *Escherichia coli* on butanol as carbon source. *J. Mol. Evol.* 25:151–158.

742. Clark, R. L., and F. C. Neidhardt. 1990. Roles of the two lysyl-tRNA synthetases of *Escherichia coli*: analysis of nucleotide sequences and mutant behavior. *J. Bacteriol.* 172:3237–3243.

743. Clarke, D. M., and P. D. Bragg. 1985. Cloning and expression of the transhydrogenase gene of *Escherichia coli. J. Bacteriol.* 162:367–373.

744. Clarke, D. M., T. W. Loo, S. Gillam, and P. D. Bragg. 1986. Nucleotide sequence of the *pntA* and *pntB* genes encoding the pyridine nucleotide transhydrogenase of *Escherichia coli. Eur. J. Biochem.* 158:647–653.

745. Clarke, N. D., M. Kvaal, and E. Seeberg. 1984. Cloning of *Escherichia coli* genes encoding 3-methyladenine DNA glycosylases I and II. *Mol. Gen. Genet.* 197:368–372.

746. Claverie-Martin, F., M. R. Diaz-Torres, and S. R. Kushner. 1987. Analysis of the regulatory region of the protease III (*ptr*) gene of *Escherichia coli* K-12. *Gene* 54:185–195.

747. Claverie-Martin, F., M. R. Diaz-Torres, S. D. Yancey, and S. R. Kushner. 1989. Cloning of the altered mRNA stability (*ams*) gene of *Escherichia coli* K-12. *J. Bacteriol.* 171:5479–5486.

748. Claverie-Martin, F., M. R. Diaz-Torres, S. D. Yancey, and S. R. Kushner. 1991. Analysis of the altered mRNA stability (*ams*) gene from *Escherichia coli*. Nucleotide sequence, transcriptional analysis, and homology of its product to MRP3, a mitochondrial ribosomal protein from *Neurospora crassa. J. Biol. Chem.* 266:2843–2851.

749. Clegg, D. O., and D. E. Koshland, Jr. 1984. The role of a signaling protein in bacterial sensing: behavioral effects of increased gene expression. *Proc. Natl. Acad. Sci. USA* 81:5056–5060.

750. Clegg, D. O., and D. E. Koshland, Jr. 1985. Identification of a bacterial sensing protein and effects of its elevated expression. *J. Bacteriol.* 162:398–405.

751. Clement, J. M., and M. Hofnung. 1981. Gene sequence of the lambda receptor, an outer membrane protein of *E. coli* K12. *Cell* 27:507–514.

752. Clementz, T. 1992. The gene coding for 3-deoxy-manno-octulosonic acid transferase and the *rfaQ* gene are transcribed from divergently arranged promoters in *Escherichia coli. J. Bacteriol.* 174:7750–7756.

753. Clementz, T., and C. R. Raetz. 1991. A gene coding for 3-deoxy-D-manno-octulosonic-acid transferase in *Escherichia coli*. Identification, mapping, cloning, and sequencing. *J. Biol. Chem.* 266:9687–9696.

754. Clerget, M. 1991. Site-specific recombination promoted by a short DNA segment of plasmid R1 and by a homologous segment in the terminus region of the *Escherichia coli* chromosome. *New Biol.* 3:780–788.

755. Cleton-Jansen, A. M., N. Goosen, O. Fayet, and P. van de Putte. 1990. Cloning, mapping, and sequencing of the gene encoding *Escherichia coli* quinoprotein glucose dehydrogenase. *J. Bacteriol.* 172:6308–6315.

756. Climent, I., B. M. Sjoberg, and C. Y. Huang. 1992. Site-directed mutagenesis and deletion of the carboxyl terminus of *Escherichia coli* ribonucleotide reductase protein R2. Effects on catalytic activity and subunit interaction. *Biochemistry* 31:4801–4807.

757. Climie, S. C., and J. D. Friesen. 1987. Feedback regulation of the *rplJL-rpoBC* ribosomal protein operon of *Escherichia coli* requires a region of mRNA secondary structure. *J. Mol. Biol.* 198:371–381.

758. Climie, S. C., and J. D. Friesen. 1988. In vivo and in vitro structural analysis of the *rplJ* mRNA leader of *Escherichia coli*. Protection by bound L10-L7/L12. *J. Biol. Chem.* 263:15166–15175.

759. Clyman, J., and R. P. Cunningham. 1987. *Escherichia coli* K-12 mutants in which viability is dependent on *recA* function. *J. Bacteriol.* 169:4203–4210.

760. Cobbett, C., B. Dickson, and L. Farmer. 1989. The role of a static bend in the DNA of the *aroF* regulatory region of *Escherichia coli. Gene* 75:185–191.

761. Cobbett, C. S., and J. Pittard. 1980. Formation of a lambda (Tn10) *tyrR*+ specialized transducing bacteriophage from *Escherichia coli* K-12. *J. Bacteriol.* 144:877–883.

762. Coderre, P. E., and C. F. Earhart. 1989. The *entD* gene of the *Escherichia coli* K12 enterobactin gene cluster. *J. Gen. Microbiol.* 135:3043–3055.

763. Cohen, S. P., H. Hachler, and S. B. Levy. 1993. Genetic and functional analysis of the multiple antibiotic resistance (*mar*) locus in *Escherichia coli. J. Bacteriol.* 175:1484–1492.

764. Cohen, S. P., L. M. McMurry, and S. B. Levy. 1988. *marA* locus causes decreased expression of OmpF porin in multiple-antibiotic-resistant (Mar) mutants of *Escherichia coli. J. Bacteriol.* 170:5416–5422.

765. Cole, J. A., B. M. Newman, and P. White. 1980. Biochemical and genetic characterization of *nirB* mutants of *Escherichia coli* K 12 pleiotropically defective in nitrite and sulphite reduction. *J. Gen. Microbiol.* 120:475–483.

766. Cole, S. T. 1982. Nucleotide sequence coding for the flavoprotein subunit of the fumarate reductase of *Escherichia coli. Eur. J. Biochem.* 122:479–484.

767. Cole, S. T. 1983. Characterisation of the promoter for the LexA regulated *sulA* gene of *Escherichia coli. Mol. Gen. Genet.* 189:400–404.

768. Cole, S. T., E. Bremer, I. Hindennach, and U. Henning. 1982. Characterisation of the promoters for the *ompA* gene which encodes a major outer membrane protein of *Escherichia coli. Mol. Gen. Genet.* 188:472–479.

769. Cole, S. T., K. Eiglmeier, S. Ahmed, N. Honore, L. Elmes, W. F. Anderson, and J. H. Weiner. 1988. Nucleotide sequence and gene-polypeptide relationships of the *glpABC* operon encoding the anaerobic sn-glycerol-3-phosphate dehydrogenase of *Escherichia coli* K-12. *J. Bacteriol.* 170:2448–2456.

770. Cole, S. T., T. Grundstrom, B. Jaurin, J. J. Robinson, and J. H. Weiner. 1982. Location and nucleotide sequence of *frdB*, the gene coding for the iron-sulphur protein subunit of the fumarate reductase of *Escherichia coli. Eur. J. Biochem.* 126:211–216.

771. Cole, S. T., and J. R. Guest. 1979. Production of a soluble form of fumarate reductase by multiple gene duplication in *Escherichia coli* K12. *Eur. J. Biochem.* 102:65–71.

772. Cole, S. T., and J. R. Guest. 1980. Genetic and physical characterization of lambda transducing phages (lambda *frdA*) containing the fumarate reductase gene of *Escherichia coli* K12. *Mol. Gen. Genet.* 178:409–418.

773. Cole, S. T., and J. R. Guest. 1980. Amplification of fumarate reductase synthesis with lambda *frdA* transducing phages and orientation of *frdA* gene expression. *Mol. Gen. Genet.* 179:377–385.

774. Cole, S. T., and N. Honore. 1989. Transcription of the *sulA-ompA* region of *Escherichia coli* during the SOS response and the role of an antisense RNA molecule. *Mol. Microbiol.* 3:715–722.

775. Cole, S. T., and O. Raibaud. 1986. The nucleotide sequence of the *malT* gene encoding the positive regulator of the *Escherichia coli* maltose regulon. *Gene* 42:201–208.

776. Coleman, J. 1990. Characterization of *Escherichia coli* cells deficient in 1-acyl-sn-glycerol-3-phosphate acyltransferase activity. *J. Biol. Chem.* 265:17215–17221.

777. Coleman, J. 1992. Characterization of the *Escherichia coli* gene for 1-acyl-sn-glycerol-3-phosphate acyltransferase (*plsC*). *Mol. Gen. Genet.* 232:295–303.

778. Coleman, J., and C. R. Raetz. 1988. First committed step of lipid A biosynthesis in *Escherichia coli*: sequence of the *lpxA* gene. *J. Bacteriol.* 170:1268–1274.

779. Coleman, S. H., and D. G. Wild. 1991. Location of the *fpg* gene on the *Escherichia coli* chromosome. *Nucleic Acids Res.* 19:3999.

780. Coleman, W. G., Jr. 1983. The *rfaD* gene codes for ADP-L-glycero-D-mannoheptose-6-epimerase. An enzyme required for lipopolysaccharide core biosynthesis. *J. Biol. Chem.* 258:1985–1990.

781. Coleman, W. G., Jr., and K. S. Deshpande. 1985. New *cysE-pyrE*-linked *rfa* mutation in *Escherichia coli* K-12 that results in a heptoseless lipopolysaccharide. *J. Bacteriol.* 161:1209–1214.

782. Coleman, W. G., Jr., and L. Leive. 1979. Two mutations which affect the barrier function of the *Escherichia coli* K-12 outer membrane. *J. Bacteriol.* 139:899–910.

783. **Coll, J. L., M. Heyde, and R. Portalier.** 1994. Expression of the *nmpC* gene of *Escherichia coli* K-12 is modulated by external pH. Identification of *cis*-acting regulatory sequences involved in this regulation. *Mol. Microbiol.* **12:**83–93.

784. **Collier, D. N., and P. J. Bassford, Jr.** 1989. Mutations that improve export of maltose-binding protein in SecB⁻ cells of *Escherichia coli. J. Bacteriol.* **171:**4640–4647.

785. **Collins, M. L., D. E. Mallon, and R. A. Niederman.** 1980. Assessment of *Rhodopseudomonas sphaeroides* chromatophore membrane asymmetry through bilateral antiserum adsorption studies. *J. Bacteriol.* **143:**221–230.

786. **Collis, C. M., and G. W. Grigg.** 1989. An *Escherichia coli* mutant resistant to phleomycin, bleomycin, and heat inactivation is defective in ubiquinone synthesis. *J. Bacteriol.* **171:**4792–4798.

787. **Colloms, S. D., P. Sykora, G. Szatmari, and D. J. Sherratt.** 1990. Recombination at ColE1 *cer* requires the *Escherichia coli xerC* gene product, a member of the lambda integrase family of site-specific recombinases. *J. Bacteriol.* **172:**6973–6980.

788. **Colombo, G., and J. J. Villafranca.** 1986. Amino acid sequence of *Escherichia coli* glutamine synthetase deduced from the DNA nucleotide sequence. *J. Biol. Chem.* **261:**10587–10591.

789. **Colonna, B., and M. Hofnung.** 1981. *rho* mutations restore *lamB* expression in *E. coli* K12 strains with an inactive *malB* region. *Mol. Gen. Genet.* **184:**479–483.

789a.**Colson, C., S. W. Glover, N. Symonds, and K. A. Stacey.** 1965. The location of the genes for host-controlled modification and restriction in *Escherichia coli* L-12. *Genetics* **52:**1043–1050.

790. **Comeau, D. E., K. Ikenaka, K. L. Tsung, and M. Inouye.** 1985. Primary characterization of the protein products of the *Escherichia coli ompB* locus: structure and regulation of synthesis of the OmpR and EnvZ proteins. *J. Bacteriol.* **164:**578–584.

791. **Comeau, D. E., and M. Inouye.** 1988. A novel method for the cloning of chromosomal mutations in a single step: isolation of two mutant alleles of *envZ*, an osmoregulatory gene from *Escherichia coli. Mol. Gen. Genet.* **213:**166–169.

792. **Comer, M. M.** 1981. Gene organization around the phenylalanyl-transfer ribonucleic acid synthetase locus in *Escherichia coli. J. Bacteriol.* **146:**269–274.

793. **Comer, M. M.** 1982. Threonine tRNAs and their genes in *Escherichia coli. Mol. Gen. Genet.* **187:**132–137.

794. **Compan, I., and D. Touati.** 1993. Interaction of six global transcription regulators in expression of manganese superoxide dismutase in *Escherichia coli* K-12. *J. Bacteriol.* **175:**1687–1696.

795. **Compton, L. A., J. M. Davis, J. R. Macdonald, and H. P. Bachinger.** 1992. Structural and functional characterization of *Escherichia coli* peptidyl-prolyl *cis-trans* isomerases. *Eur. J. Biochem.* **206:**927–934.

796. **Conlin, C. A., T. M. Knox, and C. G. Miller.** 1994. Cloning and physical map position of an alpha-aspartyl dipeptidase gene, *pepE*, from *Escherichia coli. J. Bacteriol.* **176:**1552–1553.

797. **Conlin, C. A., and C. G. Miller.** 1993. Location of the *prlC* (*opdA*) gene on the physical map of *Escherichia coli. J. Bacteriol.* **175:**5731–5732.

798. **Conlin, C. A., N. J. Trun, T. J. Silhavy, and C. G. Miller.** 1992. *Escherichia coli prlC* encodes an endopeptidase and is homologous to the *Salmonella typhimurium opdA* gene. *J. Bacteriol.* **174:**5881–5887.

799. **Connolly, B., C. A. Parsons, F. E. Benson, H. J. Dunderdale, G. J. Sharples, R. G. Lloyd, and S. C. West.** 1991. Resolution of Holliday junctions in vitro requires the *Escherichia coli ruvC* gene product. *Proc. Natl. Acad. Sci. USA* **88:**6063–6067.

800. **Connolly, D. M., and M. E. Winkler.** 1989. Genetic and physiological relationships among the *miaA* gene, 2-methylthio-N⁶-(delta-2-isopentenyl)-adenosine tRNA modification, and spontaneous mutagenesis in *Escherichia coli* K-12. *J. Bacteriol.* **171:**3233–3246.

801. **Connolly, D. M., and M. E. Winkler.** 1991. Structure of *Escherichia coli* K-12 *miaA* and characterization of the mutator phenotype caused by *miaA* insertion mutations. *J. Bacteriol.* **173:**1711–1721.

802. **Conrad, C. A., G. W. Stearns, W. E. Prater, J. A. Rheiner, and J. R. Johnson.** 1984. Characterization of a *glpK* transducing phage. *Mol. Gen. Genet.* **193:**376–378.

803. **Contreras, A., and A. Maxwell.** 1992. *gyrB* mutations which confer coumarin resistance also affect DNA supercoiling and ATP hydrolysis by *Escherichia coli* DNA gyrase. *Mol. Microbiol.* **6:**1617–1624.

804. **Conway, T., and L. O. Ingram.** 1989. Similarity of *Escherichia coli* propanediol oxidoreductase (*fucO* product) and an unusual alcohol dehydrogenase from *Zymomonas mobilis* and *Saccharomyces cerevisiae. J. Bacteriol.* **171:**3754–3759.

805. **Conway, T., K. C. Yi, S. E. Egan, R. E. Wolf, Jr., and D. L. Rowley.** 1991. Locations of the *zwf, edd,* and *eda* genes on the *Escherichia coli* physical map. *J. Bacteriol.* **173:**5247–5248.

806. **Cook, W. R., and L. I. Rothfield.** 1991. Biogenesis of cell division sites in *ftsA* and *ftsZ* filaments. *Res. Microbiol.* **142:**321–324.

807. **Cooley, W., K. Sirotkin, R. Green, and L. Synder.** 1979. A new gene of *Escherichia coli* K-12 whose product participates in T4 bacteriophage late gene expression: interaction of *lit* with the T4-induced polynucleotide 5′-kinase 3′-phosphatase. *J. Bacteriol.* **140:**83–91.

808. **Cooper, R. A.** 1978. The utilisation of D-galactonate and D-2-oxo-3-deoxygalactonate by *Escherichia coli* K-12. Biochemical and genetical studies. *Arch. Microbiol.* **118:**199–206.

809. **Cooper, R. A., D. C. Jones, and S. Parrott.** 1985. Isolation and mapping of *Escherichia coli* K12 mutants defective in phenylacetate degradation. *J. Gen. Microbiol.* **131:**2753–2757.

810. **Coppola, G., F. Huang, J. Riley, J. L. Cox, P. Hantzopoulos, L. B. Zhou, and D. H. Calhoun.** 1991. Sequence and transcriptional activity of the *Escherichia coli* K-12 chromosome region between *rrnC* and *ilvGMEDA. Gene* **97:**21–27.

811. **Cormack, R. S., J. L. Genereaux, and G. A. Mackie.** 1993. RNase E activity is conferred by a single polypeptide: overexpression, purification, and properties of the *ams/rne/hmp1* gene product. *Proc. Natl. Acad. Sci. USA* **90:**9006–9010.

812. **Cornet, F., I. Mortier, J. Patte, and J. M. Louarn.** 1994. Plasmid pSC101 harbors a recombination site, *psi*, which is able to resolve plasmid multimers and to substitute for the analogous chromosomal *Escherichia coli* site *dif. J. Bacteriol.* **176:**3188–3195.

813. **Cornish, E. C., V. P. Argyropoulos, J. Pittard, and B. E. Davidson.** 1986. Structure of the *Escherichia coli* K12 regulatory gene *tyrR*. Nucleotide sequence and sites of initiation of transcription and translation. *J. Biol. Chem.* **261:**403–410.

814. **Cornish, E. C., B. E. Davidson, and J. Pittard.** 1982. Cloning and characterization of *Escherichia coli* K-12 regulator gene *tyrR. J. Bacteriol.* **152:**1276–1279.

815. **Cornwell, T. L., S. L. Adhya, W. S. Reznikoff, and P. A. Frey.** 1987. The nucleotide sequence of the *galT* gene of *Escherichia coli. Nucleic Acids Res.* **15:**8116.

816. **Cortay, J. C., F. Bleicher, C. Rieul, H. C. Reeves, and A. J. Cozzone.** 1988. Nucleotide sequence and expression of the *aceK* gene coding for isocitrate dehydrogenase kinase/phosphatase in *Escherichia coli. J. Bacteriol.* **170:**89–97.

817. **Cortay, J. C., D. Negre, A. Galinier, B. Duclos, G. Perriere, and A. J. Cozzone.** 1991. Regulation of the acetate operon in *Escherichia coli*: purification and functional characterization of the IclR repressor. *EMBO J.* **10:**675–679.

818. **Cosloy, S. D.** 1973. D-Serine transport system in *Escherichia coli* K-12. *J. Bacteriol.* **114:**679–684.

819. **Cossart, P., and B. Gicquel-Sanzey.** 1982. Cloning and sequence of the *crp* gene of *Escherichia coli* K 12. *Nucleic Acids Res.* **10:**1363–1378.

820. **Cossart, P., E. A. Groisman, M. C. Serre, M. J. Casadaban, and B. Gicquel-Sanzey.** 1986. *crp* genes of *Shigella flexneri*, *Salmonella typhimurium*, and *Escherichia coli. J. Bacteriol.* **167:**639–646.

821. **Cossart, P., M. Katinka, and M. Yaniv.** 1981. Nucleotide sequence of the *thrB* gene of *E. coli*, and its two adjacent regions; the *thrAB* and *thrBC* junctions. *Nucleic Acids Res.* **9:**339–347.

822. **Cossart, P., M. Katinka, M. Yaniv, I. Saint Girons, and G. N. Cohen.** 1979. Construction and expression of a hybrid plasmid containing the *Escherichia coli thrA* and *thrB* genes. *Mol. Gen. Genet.* **175:**39–44.

823. **Cotter, P. A., and R. P. Gunsalus.** 1992. Contribution of the *fnr* and *arcA* gene products in coordinate regulation of cytochrome *o* and *d* oxidase (*cyoABCDE* and *cydAB*) genes in *Escherichia coli. FEMS Microbiol. Lett.* **70:**31–36.

824. **Cotton, R. G., and I. P. Crawford.** 1972. Tryptophan synthetase B 2 subunit. Application of genetic analysis to the study of primary structure. *J. Biol. Chem.* **247:**1883–1891.

825. **Coulton, J. W., P. Mason, and D. D. Allatt.** 1987. *fhuC* and *fhuD* genes for iron (III)-ferrichrome transport into *Escherichia coli* K-12. *J. Bacteriol.* **169:**3844–3849.

826. **Coulton, J. W., P. Mason, D. R. Cameron, G. Carmel, R. Jean, and H. N. Rode.** 1986. Protein fusions of beta-galactosidase to the ferrichrome-iron receptor of *Escherichia coli* K-12. *J. Bacteriol.* **165:**181–192.

827. **Coulton, J. W., P. Mason, and M. S. DuBow.** 1983. Molecular cloning of the ferrichrome-iron receptor of *Escherichia coli* K-12. *J. Bacteriol.* **156:**1315–1321.

828. **Court, D. L., T. A. Patterson, T. Baker, N. Costantino, X. Mao, and D. I. Friedman.** 1995. Structural and functional analyses of the transcription-translation proteins NusB and NusE. *J. Bacteriol.* **177:**2589–2591.

829. **Covarrubias, A. A., and F. Bastarrachea.** 1983. Nucleotide sequence of the *glnA* control region of *Escherichia coli. Mol. Gen. Genet.* **190:**171–175.

830. **Covarrubias, A. A., M. Rocha, F. Bolivar, and F. Bastarrachea.** 1980. Cloning and physical mapping of the *glnA* gene of *Escherichia coli* K-12. *Gene* **11:**239–251.

831. **Covarrubias, A. A., R. Sanchez-Pescador, A. Osorio, F. Bolivar, and F. Bastarrachea.** 1980. ColE1 hybrid plasmids containing *Escherichia coli* genes involved in the biosynthesis of glutamate and glutamine. *Plasmid* **3:**150–164.

832. **Cowman, A., and I. R. Beacham.** 1980. Molecular cloning of the gene (*ush*) from *Escherichia coli* specifying periplasmic UDP-sugar hydrolase (5′-nucleotidase). *Gene* **12:**281–286.

833. **Cox, E. C., and T. C. Gibson.** 1974. Selection for high mutation rates in chemostats. *Genetics* **77:**169–184.

834. **Cox, E. C., and D. L. Horner.** 1986. DNA sequence and coding properties of *mutD* (*dnaQ*) a dominant *Escherichia coli* mutator gene. *J. Mol. Biol.* **190:**113–117.

834a.**Cox, G. B., I. G. Young, L. M. McCann, and F. Gibson.** 1969. Biosynthesis of ubiquinone in *Escherichia coli* K-12; location of genes affecting the metabolism of 3-octaprenyl-4-hydroxybenzoic acid and 2-octaprenyl-phenol. *J. Bacteriol.* **99:**450–458.

835. **Cox, J. L., B. J. Cox, V. Fidanza, and D. H. Calhoun.** 1987. The complete nucleotide sequence of the *ilvGMEDA* cluster of *Escherichia coli* K-12. *Gene* **56:**185–198.

836. **Coy, M., and J. B. Neilands.** 1991. Structural dynamics and functional domains of the *fur* protein. *Biochemistry* **30:**8201–8210.

837. **Coyer, J., J. Andersen, S. A. Forst, M. Inouye, and N. Delihas.** 1990. *micF* RNA in *ompB* mutants of *Escherichia coli*: different pathways regulate *micF* RNA levels in response to osmolarity and temperature change. *J. Bacteriol.* **172:**4143–4150.

838. **Crabeel, M., D. Charlier, R. Cunin, A. Boyen, N. Glansdorff, and A. Pierard.** 1975. Accumulation of arginine precursors in *Escherichia coli*: effects on growth,

enzyme repression, and application to the forward selection of arginine auxotrophs. *J. Bacteriol.* 123:898–904.

839. Crabeel, M., D. Charlier, R. Cunin, and N. Glansdorff. 1979. Cloning and endonuclease restriction analysis of *argF* and of the control region of the *argECBH* bipolar operon in *Escherichia coli*. *Gene* 5:207–231.

840. Crabeel, M., D. Charlier, G. Weyens, A. Feller, A. Pierard, and N. Glansdorff. 1980. Use of gene cloning to determine polarity of an operon: genes *carAB* of *Escherichia coli*. *J. Bacteriol.* 143:921–925.

841. Craig, J. E., Y. Zhang, and M. P. Gallagher. 1994. Cloning of the *nupC* gene of *Escherichia coli* encoding a nucleoside transport system, and identification of an adjacent insertion element, IS *186*. *Mol. Microbiol.* 11:1159–1168.

842. Craigen, W. J., R. G. Cook, W. P. Tate, and C. T. Caskey. 1985. Bacterial peptide chain release factors: conserved primary structure and possible frameshift regulation of release factor 2. *Proc. Natl. Acad. Sci. USA* 82:3616–3620.

843. Crawford, I. P., B. P. Nichols, and C. Yanofsky. 1980. Nucleotide sequence of the *trpB* gene in *Escherichia coli* and *Salmonella typhimurium*. *J. Mol. Biol.* 142:489–502.

844. Creeger, E. S., T. Schulte, and L. I. Rothfield. 1984. Regulation of membrane glycosyltransferases by the *sfrB* and *rfaH* genes of *Escherichia coli* and *Salmonella typhimurium*. *J. Biol. Chem.* 259:3064–3069.

845. Cronan, J. E., Jr., and E. P. Gelmann. 1973. An estimate of the minimum amount of unsaturated fatty acid required for growth of *Escherichia coli*. *J. Biol. Chem.* 248:1188–1195.

846. Cronan, J. E., Jr., W. B. Li, R. Coleman, M. Narasimhan, D. de Mendoza, and J. M. Schwab. 1988. Derived amino acid sequence and identification of active site residues of *Escherichia coli* beta-hydroxydecanoyl thioester dehydrase. *J. Biol. Chem.* 263:4641–4646.

847. Cronan, J. E., Jr., K. J. Littel, and S. Jackowski. 1982. Genetic and biochemical analyses of pantothenate biosynthesis in *Escherichia coli* and *Salmonella typhimurium*. *J. Bacteriol.* 149:916–922.

848. Crooke, E., M. Akiyama, N. N. Rao, and A. Kornberg. 1994. Genetically altered levels of inorganic polyphosphate in *Escherichia coli*. *J. Biol. Chem.* 269:6290–6295.

849. Crooke, E., D. S. Hwang, K. Skarstad, B. Thony, and A. Kornberg. 1991. *E. coli* minichromosome replication: regulation of initiation at *oriC*. *Res. Microbiol.* 142:127–130.

850. Crooke, H., and J. Cole. 1995. The biogenesis of c-type cytochromes in *Escherichia coli* requires a membrane-bound protein, DipZ, with a protein disulfide isomerase-like domain. *Mol. Microbiol.* 15:1139–1150.

851. Crowell, D. N., M. S. Anderson, and C. R. Raetz. 1986. Molecular cloning of the genes for lipid A disaccharide synthase and UDP-*N*-acetylglucosamine acyltransferase of *Escherichia coli*. *J. Bacteriol.* 168:152–159.

852. Crowell, D. N., W. S. Reznikoff, and C. R. Raetz. 1987. Nucleotide sequence of the *Escherichia coli* gene for lipid A disaccharide synthase. *J. Bacteriol.* 169:5727–5734.

853. Csiszar, K., T. Lukacsovich, and P. Venetianer. 1990. Regulatory elements downstream of the promoter of an rRNA gene of *E. coli*. *Biochim. Biophys. Acta* 1050:312–316.

854. Csonka, L. N. 1981. Proline over-production results in enhanced osmotolerance in *Salmonella typhimurium*. *Mol. Gen. Genet.* 182:82–86.

855. Csonka, L. N., and A. D. Hanson. 1991. Prokaryotic osmoregulation: genetics and physiology. *Annu. Rev. Microbiol.* 45:569–606.

856. Csordas-Toth, E., I. Boros, and P. Venetianer. 1979. Structure of the promoter region for the *rrnB* gene in *Escherichia coli*. *Nucleic Acids Res.* 7:2189–2197.

857. Cudny, H., J. R. Lupski, G. N. Godson, and M. P. Deutscher. 1986. Cloning, sequencing, and species relatedness of the *Escherichia coli cca* gene encoding the enzyme tRNA nucleotidyltransferase. *J. Biol. Chem.* 261:6444–6449.

858. Cui, J., and R. L. Somerville. 1992. Physical map location and transcriptional orientation of the *tyrR* gene of *Escherichia coli* K-12. *J. Bacteriol.* 174:3832–3833.

859. Cui, J., and R. L. Somerville. 1993. A mutational analysis of the structural basis for transcriptional activation and monomer-monomer interaction in the TyrR system of *Escherichia coli* K-12. *J. Bacteriol.* 175:1777–1784.

860. Cui, J., and R. L. Somerville. 1993. The TyrR protein of *Escherichia coli*, analysis by limited proteolysis of domain structure and ligand-mediated conformational changes. *J. Biol. Chem.* 268:5040–5047.

861. Culham, D. E., B. Lasby, A. G. Marangoni, J. L. Milner, B. A. Steer, R. W. van Nues, and J. M. Wood. 1993. Isolation and sequencing of *Escherichia coli* gene *proP* reveals unusual structural features of the osmoregulatory proline/betaine transporter, ProP. *J. Mol. Biol.* 229:268–276.

862. Cummings, H. S., and J. W. Hershey. 1994. Translation initiation factor IF1 is essential for cell viability in *Escherichia coli*. *J. Bacteriol.* 176:198–205.

863. Cummings, H. S., J. F. Sands, P. C. Foreman, J. Fraser, and J. W. Hershey. 1991. Structure and expression of the *infA* operon encoding translational initiation factor IF1. Transcriptional control by growth rate. *J. Biol. Chem.* 266:16491–16498.

864. Cunningham, P. R., and D. P. Clark. 1986. The use of suicide substrates to select mutants of *Escherichia coli* lacking enzymes of alcohol fermentation. *Mol. Gen. Genet.* 205:487–493.

865. Cunningham, R. P., S. M. Saporito, S. G. Spitzer, and B. Weiss. 1986. Endonuclease IV (*nfo*) mutant of *Escherichia coli*. *J. Bacteriol.* 168:1120–1127.

866. Cunningham, R. P., and B. Weiss. 1985. Endonuclease III (*nth*) mutants of *Escherichia coli*. *Proc. Natl. Acad. Sci. USA* 82:474–478.

866a. Curtis, S. J., and W. Epstein. 1975. Phosphorylation of D-glucose in *Escherichia coli* mutants defective in glucosephosphotransferase, mannosephosphotransferase, and glucokinase. *J. Bacteriol.* 122:1189–1199.

867. D'Ari, L., and J. C. Rabinowitz. 1991. Purification, characterization, cloning, and amino acid sequence of the bifunctional enzyme 5,10-methylenetetrahydrofolate dehydrogenase/5,10-methenyltetrahydrofolate cyclohydrolase from *Escherichia coli*. *J. Biol. Chem.* 266:23953–23958.

868. D'Ari, R., and O. Huisman. 1983. Novel mechanism of cell division inhibition associated with the SOS response in *Escherichia coli*. *J. Bacteriol.* 156:243–250.

869. Dabbs, E. R. 1978. Kasugamycin-dependent mutants of *Escherichia coli*. *J. Bacteriol.* 136:994–1001.

870. Dabbs, E. R. 1980. The gene for ribosomal protein S21, *rpsU*, maps close to *dnaG* at 66.5 min on the *Escherichia coli* chromosomal linkage map. *J. Bacteriol.* 144:603–607.

871. Dabbs, E. R. 1981. The gene for ribosomal protein L31, *rpmE*, is located at 88.5 minutes on the *Escherichia coli* chromosomal linkage map. *J. Bacteriol.* 148:379–382.

872. Dabbs, E. R. 1982. The gene for ribosomal protein L13, *rplM*, is located near *argR*, at about 70 minutes on the *Escherichia coli* chromosomal linkage map. *J. Bacteriol.* 149:779–782.

873. Dabbs, E. R. 1982. A spontaneous mutant of *Escherichia coli* with protein L24 lacking from the ribosome. *Mol. Gen. Genet.* 187:453–458.

874. Dabbs, E. R. 1982. Three additional loci of rifampicin dependence in *Escherichia coli*. *Mol. Gen. Genet.* 187:519–522.

875. Dabbs, E. R. 1983. *Escherichia coli* kasugamycin dependence arising from mutation at the *rpsI* locus. *J. Bacteriol.* 153:709–715.

876. Dabbs, E. R., and K. Looman. 1981. An antibiotic dependent conditional lethal mutant with a lesion affecting transcription and translation. *Mol. Gen. Genet.* 184:224–229.

877. Dahl, M. K., E. Francoz, W. Saurin, W. Boos, M. D. Manson, and M. Hofnung. 1989. Comparison of sequences from the *malB* regions of *Salmonella typhimurium* and *Enterobacter aerogenes* with *Escherichia coli* K12: a potential new regulatory site in the interoperonic region. *Mol. Gen. Genet.* 218:199–207.

878. Dai, D., and E. E. Ishiguro. 1988. *murH*, a new genetic locus in *Escherichia coli* involved in cell wall peptidoglycan biosynthesis. *J. Bacteriol.* 170:2197–2201.

879. Dai, K., and J. Lutkenhaus. 1991. *ftsZ* is an essential cell division gene in *Escherichia coli*. *J. Bacteriol.* 173:3500–3506.

880. Dai, K., and J. Lutkenhaus. 1992. The proper ratio of FtsZ to FtsA is required for cell division to occur in *Escherichia coli*. *J. Bacteriol.* 174:6145–6151.

881. Dai, K., A. Mukherjee, Y. Xu, and J. Lutkenhaus. 1994. Mutations in *ftsZ* that confer resistance to SulA affect the interaction of FtsZ with GTP. *J. Bacteriol.* 176:130–136.

882. Dai, K., Y. Xu, and J. Lutkenhaus. 1993. Cloning and characterization of *ftsN*, an essential cell division gene in *Escherichia coli* isolated as a multicopy suppressor of *ftsA12*(Ts). *J. Bacteriol.* 175:3790–3797.

883. Dailey, F. E., and H. C. Berg. 1993. Mutants in disulfide bond formation that disrupt flagellar assembly in *Escherichia coli*. *Proc. Natl. Acad. Sci. USA* 90:1043–1047.

884. Dairi, T., K. Inokuchi, T. Mizuno, and S. Mizushima. 1985. Positive control of transcription initiation in *Escherichia coli*. A base substitution at the Pribnow box renders *ompF* expression independent of a positive regulator. *J. Mol. Biol.* 184:1–6.

885. Dalbey, R. E. 1991. Leader peptidase. *Mol. Microbiol.* 5:2855–2860.

886. Daldal, F. 1983. Molecular cloning of the gene for phosphofructokinase-2 of *Escherichia coli* and the nature of a mutation, *pfkB1*, causing a high level of the enzyme. *J. Mol. Biol.* 168:285–305.

887. Daldal, F. 1984. Nucleotide sequence of gene *pfkB* encoding the minor phosphofructokinase of *Escherichia coli* K-12. *Gene* 28:337–342.

888. Daldal, F., and D. G. Fraenkel. 1981. Tn*10* insertions in the *pfkB* region of *Escherichia coli*. *J. Bacteriol.* 147:935–943.

889. Dallas, W. S., I. K. Dev, and P. H. Ray. 1993. The dihydropteroate synthase gene, *folP*, is near the leucine tRNA gene, *leuU*, on the *Escherichia coli* chromosome. *J. Bacteriol.* 175:7743–7744.

890. Dalma-Weiszhausz, D. D., and M. Brenowitz. 1992. Interactions between DNA-bound transcriptional regulators of the *Escherichia coli gal* operon. *Biochemistry* 31:6980–6989.

891. Dalphin, M. E., B. A. Fane, M. O. Skidmore, and M. Hayashi. 1992. Proteolysis of bacteriophage φX174 prohead accessory protein gpB by *Escherichia coli* OmpT protease is not essential for phage maturation in vivo. *J. Bacteriol.* 174:2404–2406.

892. Dalrymple, B., and J. S. Mattick. 1986. Genes encoding threonine tRNAs with the anticodon CGU from *Escherichia coli* and *Pseudomonas aeruginosa*. *Biochem. Int.* 13:547–553.

893. Dame, J. B., and B. M. Shapiro. 1976. Use of polymyxin B, levallorphan, and tetracaine to isolate novel envelope mutants of *Escherichia coli*. *J. Bacteriol.* 127:961–972.

894. Damerau, K., and A. C. St. John. 1993. Role of Clp protease subunits in degradation of carbon starvation proteins in *Escherichia coli*. *J. Bacteriol.* 175:53–63.

895. Dammel, C. S., and H. F. Noller. 1995. Suppression of a cold-sensitive mutation in 16S rRNA by overexpression of a novel ribosome-binding factor, RbfA. *Genes Dev.* 9:626–637.

896. **Danchin, A.** 1977. A new technique for selection of sensitive and auxotrophic mutants of *E. coli*: isolation of a strain sensitive to an excess of one-carbon metabolites. *Mol. Gen. Genet.* **150**:293–299.

897. **Danchin, A., and L. Dondon.** 1980. Serine sensitivity of *Escherichia coli* K 12: partial characterization of a serine resistant mutant that is extremely sensitive to 2- ketobutyrate. *Mol. Gen. Genet.* **178**:155–164.

898. **Dandanell, G., and K. Hammer.** 1985. Two operator sites separated by 599 base pairs are required for *deoR* repression of the *deo* operon of *Escherichia coli*. *EMBO J.* **4**:3333–3338.

899. **Dandanell, G., and K. Hammer.** 1991. *deoP1* promoter and operator mutants in *Escherichia coli*: isolation and characterization. *Mol. Microbiol.* **5**:2371–2376.

900. **Dang, C. V., M. Niwano, J. Ryu, and B. L. Taylor.** 1986. Inversion of aerotactic response in *Escherichia coli* deficient in *cheB* protein methylesterase. *J. Bacteriol.* **166**:275–280.

901. **Daniel, J., and A. Danchin.** 1979. Involvement of cyclic AMP and its receptor protein in the sensitivity of *Escherichia coli* K 12 toward serine: excretion of 2-ketobutyrate, a precursor of isoleucine. *Mol. Gen. Genet.* **176**:343–350.

902. **Daniel, J., and I. Saint-Girons.** 1982. Attenuation in the threonine operon: effects of amino acids present in the presumed leader peptide in addition to threonine and isoleucine. *Mol. Gen. Genet.* **188**:225–227.

903. **Daniels, D. L., G. Plunkett, V. Burland, and F. R. Blattner.** 1992. Analysis of the *Escherichia coli* genome: DNA sequence of the region from 84.5 to 86.5 minutes. *Science* **257**:771–778.

904. **Danielsen, S., M. Kilstrup, K. Barilla, B. Jochimsen, and J. Neuhard.** 1992. Characterization of the *Escherichia coli codBA* operon encoding cytosine permease and cytosine deaminase. *Mol. Microbiol.* **6**:1335–1344.

905. **Dardel, F., G. Fayat, and S. Blanquet.** 1984. Molecular cloning and primary structure of the *Escherichia coli* methionyl-tRNA synthetase gene. *J. Bacteriol.* **160**:1115–1122.

906. **Dardel, F., M. Panvert, S. Blanquet, and G. Fayat.** 1991. Locations of the *metG* and *mrp* genes on the physical map of *Escherichia coli*. *J. Bacteriol.* **173**:3273.

907. **Dardel, F., M. Panvert, and G. Fayat.** 1990. Transcription and regulation of expression of the *Escherichia coli* methionyl-tRNA synthetase gene. *Mol. Gen. Genet.* **223**:121–133.

908. **Dardonville, B., and O. Raibaud.** 1990. Characterization of *malT* mutants that constitutively activate the maltose regulon of *Escherichia coli*. *J. Bacteriol.* **172**:1846–1852.

909. **Darlison, M. G., and J. R. Guest.** 1984. Nucleotide sequence encoding the iron-sulphur protein subunit of the succinate dehydrogenase of *Escherichia coli*. *Biochem. J.* **223**:507–517.

910. **Darlison, M. G., M. E. Spencer, and J. R. Guest.** 1984. Nucleotide sequence of the *sucA* gene encoding the 2-oxoglutarate dehydrogenase of *Escherichia coli* K12. *Eur. J. Biochem.* **141**:351–359.

911. **Darr, S. C., K. Zito, D. Smith, and N. R. Pace.** 1992. Contributions of phylogenetically variable structural elements to the function of the ribozyme ribonuclease P. *Biochemistry* **31**:328–333.

912. **Darwin, A., H. Hussain, L. Griffiths, J. Grove, Y. Sambongi, S. Busby, and J. Cole.** 1993. Regulation and sequence of the structural gene for cytochrome c552 from *Escherichia coli*: not a hexahaem but a 50 kDa tetrahaem nitrite reductase. *Mol. Microbiol.* **9**:1255–1265.

913. **Darwin, A., P. Tormay, L. Page, L. Griffiths, and J. Cole.** 1993. Identification of the formate dehydrogenases and genetic determinants of formate-dependent nitrite reduction by *Escherichia coli* K12. *J. Gen. Microbiol.* **139**:1829–1840.

914. **Das, A.** 1992. How the phage lambda N gene product suppresses transcription termination: communication of RNA polymerase with regulatory proteins mediated by signals in nascent RNA. *J. Bacteriol.* **174**:6711–6716.

915. **Das, A., D. Court, and S. Adhya.** 1976. Isolation and characterization of conditional lethal mutants of *Escherichia coli* defective in transcription termination factor rho. *Proc. Natl. Acad. Sci. USA* **73**:1959–1963.

916. **Dasgupta, S., R. Bernander, and K. Nordstrom.** 1991. In vivo effect of the *tus* mutation on cell division in an *Escherichia coli* strain where chromosome replication is under the control of plasmid R1. *Res. Microbiol.* **142**:177–180.

917. **Dass, S. B., and R. Jayaraman.** 1985. Intragenic suppression of the temperature-sensitivity caused by a mutation in a gene controlling transcription (*fit*) in *Escherichia coli*. *Mol. Gen. Genet.* **198**:299–303.

918. **Dassa, E.** 1993. Sequence-function relationships in MalG, an inner membrane protein from the maltose transport system in *Escherichia coli*. *Mol. Microbiol.* **7**:39–47.

919. **Dassa, E., and P. L. Boquet.** 1981. ExpA: a conditional mutation affecting the expression of a group of exported proteins in *Escherichia coli* K-12. *Mol. Gen. Genet.* **181**:192–200.

920. **Dassa, E., and P. L. Boquet.** 1985. Identification of the gene *appA* for the acid phosphatase (pH optimum 2.5) of *Escherichia coli*. *Mol. Gen. Genet.* **200**:68–73.

921. **Dassa, E., and M. Hofnung.** 1985. Sequence of gene *malG* in *E. coli* K12: homologies between integral membrane components from binding protein-dependent transport systems. *EMBO J.* **4**:2287–2293.

922. **Dassa, E., and S. Muir.** 1993. Membrane topology of MalG, an inner membrane protein from the maltose transport system of *Escherichia coli*. *Mol. Microbiol.* **7**:29–38.

923. **Dassa, J., H. Fsihi, C. Marck, M. Dion, M. Kieffer-Bontemps, and P. L. Boquet.** 1991. A new oxygen-regulated operon in *Escherichia coli* comprises the genes for a putative third cytochrome oxidase and for pH 2.5 acid phosphatase (*appA*). *Mol. Gen. Genet.* **229**:341–352.

924. **Dassa, J., C. Marck, and P. L. Boquet.** 1990. The complete nucleotide sequence of the *Escherichia coli* gene *appA* reveals significant homology between pH 2.5 acid phosphatase and glucose-1-phosphatase. *J. Bacteriol.* **172**:5497–5500.

925. **Dassain, M., and J. P. Bouche.** 1994. The *min* locus, which confers topological specificity to cell division, is not involved in its coupling with nucleoid separation. *J. Bacteriol.* **176**:6143–6145.

926. **Datta, D. B., B. Arden, and U. Henning.** 1977. Major proteins of the *Escherichia coli* outer cell envelope membrane as bacteriophage receptors. *J. Bacteriol.* **131**:821–829.

927. **Datta, P., T. J. Goss, J. R. Omnaas, and R. V. Patil.** 1987. Covalent structure of biodegradative threonine dehydratase of *Escherichia coli*: homology with other dehydratases. *Proc. Natl. Acad. Sci. USA* **84**:393–397.

928. **Dattananda, C. S., and J. Gowrishankar.** 1989. Osmoregulation in *Escherichia coli*: complementation analysis and gene-protein relationships in the *proU* locus. *J. Bacteriol.* **171**:1915–1922.

929. **Dautry-Varsat, A., L. Sibilli-Weill, and G. N. Cohen.** 1977. Subunit structure of the methionine-repressible aspartokinase II-homoserine dehydrogenase II from *Escherichia coli* K12. *Eur. J. Biochem.* **76**:1–6.

930. **Davidson, A. L., H. A. Shuman, and H. Nikaido.** 1992. Mechanism of maltose transport in *Escherichia coli*: transmembrane signaling by periplasmic binding proteins. *Proc. Natl. Acad. Sci. USA* **89**:2360–2364.

931. **Davies, J., P. Anderson, and B. D. Davis.** 1965. Inhibition of protein synthesis by spectinomycin. *Science* **149**:1096–1098.

932. **Davies, J. K., and P. Reeves.** 1975. Genetics of resistance to colicins in *Escherichia coli* K-12: cross-resistance among colicins of group A. *J. Bacteriol.* **123**:102–117.

933. **Davies, W. D., and B. E. Davidson.** 1982. The nucleotide sequence of *aroG*, the gene for 3-deoxy-D-arabinoheptulosonate-7-phosphate synthetase (*phe*) in *Escherichia coli* K12. *Nucleic Acids Res.* **10**:4045–4048.

934. **Davies, W. D., J. Pittard, and B. E. Davidson.** 1985. Cloning of *aroG*, the gene coding for phospho-2-keto-3-deoxy-heptonate aldolase (*phe*), in *Escherichia coli* K-12, and subcloning of the *aroG* promoter and operator in a promoter-detecting plasmid. *Gene* **33**:323–331.

935. **Davis, E. J., J. M. Blatt, E. K. Henderson, J. J. Whittaker, and J. H. Jackson.** 1977. Valine-sensitive acetohydroxy acid synthases in *Escherichia coli* K-12: unique regulation modulated by multiple genetic sites. *Mol. Gen. Genet.* **156**:239–249.

936. **Davis, E. O., and P. J. Henderson.** 1987. The cloning and DNA sequence of the gene *xylE* for xylose-proton symport in *Escherichia coli* K12. *J. Biol. Chem.* **262**:13928–13932.

937. **Davis, E. O., M. C. Jones-Mortimer, and P. J. Henderson.** 1984. Location of a structural gene for xylose-H$^+$ symport at 91 min on the linkage map of *Escherichia coli* K12. *J. Biol. Chem.* **259**:1520–1525.

938. **Davis, M. G., and J. M. Calvo.** 1977. Isolation and characterization of λ *pleu* bacteriophages. *J. Bacteriol.* **129**:1078–1090.

939. **Davis, N. K., S. Greer, M. C. Jones-Mortimer, and R. N. Perham.** 1982. Isolation and mapping of glutathione reductase-negative mutants of *Escherichia coli* K12. *J. Gen. Microbiol.* **128**:1631–1634.

940. **Davis, T., M. Yamada, M. Elgort, and M. H. Saier, Jr.** 1988. Nucleotide sequence of the mannitol (*mtl*) operon in *Escherichia coli*. *Mol. Microbiol.* **2**:405–412.

941. **Daws, T., C. J. Lim, and J. A. Fuchs.** 1989. In vitro construction of *gshB::kan* in *Escherichia coli* and use of *gshB::kan* in mapping the *gshB* locus. *J. Bacteriol.* **171**:5218–5221.

942. **Daws, T. D., and J. A. Fuchs.** 1984. Isolation and characterization of an *Escherichia coli* mutant deficient in dTMP kinase activity. *J. Bacteriol.* **157**:440–444.

943. **De Bellis, D., D. Liveris, D. Goss, S. Ringquist, and I. Schwartz.** 1992. Structure-function analysis of *Escherichia coli* translation initiation factor IF3: tyrosine 107 and lysine 110 are required for ribosome binding. *Biochemistry* **31**:11984–11990.

944. **de Boer, H. A., S. F. Gilbert, and M. Nomura.** 1979. DNA sequences of promoter regions for rRNA operons *rrnE* and *rrnA* in *E. coli*. *Cell* **17**:201–209.

945. **de Boer, P., R. Crossley, and L. Rothfield.** 1992. The essential bacterial cell-division protein FtsZ is a GTPase. *Nature* (London) **359**:254–256.

946. **de Boer, P. A., R. E. Crossley, A. R. Hand, and L. I. Rothfield.** 1991. The MinD protein is a membrane ATPase required for the correct placement of the *Escherichia coli* division site. *EMBO J.* **10**:4371–4380.

947. **de Boer, P. A., R. E. Crossley, and L. I. Rothfield.** 1988. Isolation and properties of *minB*, a complex genetic locus involved in correct placement of the division site in *Escherichia coli*. *J. Bacteriol.* **170**:2106–2112.

948. **de Boer, P. A., R. E. Crossley, and L. I. Rothfield.** 1989. A division inhibitor and a topological specificity factor coded for by the minicell locus determine proper placement of the division septum in *E. coli*. *Cell* **56**:641–649.

949. **de Boer, P. A., R. E. Crossley, and L. I. Rothfield.** 1990. Central role for the *Escherichia coli minC* gene product in two different cell division-inhibition systems. *Proc. Natl. Acad. Sci. USA* **87**:1129–1133.

950. **de Cock, H., W. Overeem, and J. Tommassen.** 1992. Biogenesis of outer membrane protein PhoE of *Escherichia coli*. Evidence for multiple SecB-binding sites in the mature portion of the PhoE protein. *J. Mol. Biol.* **224**:369–379.

951. **De Felice, M., J. Guardiola, A. Lamberti, and M. Iaccarino.** 1973. *Escherichia coli* K-12 mutants altered in the transport systems for oligo- and dipeptides. *J. Bacteriol.* **116**:751–756.

952. **de Geus, P., I. Van Die, H. Bergmans, J. Tommassen, and G. de Haas.** 1983. Molecular cloning of *pldA*, the structural gene for outer membrane phospholipase of *E. coli* K12. *Mol. Gen. Genet.* **190**:150–155.

953. de Geus, P., H. M. Verheij, N. H. Riegman, W. P. Hoekstra, and G. H. de Haas. 1984. The pro- and mature forms of the *E. coli* K-12 outer membrane phospholipase A are identical. *EMBO J.* **3**:1799–1802.

954. de Jonge, B. L. 1990. Isogenic variants of *Escherichia coli* with altered morphology have peptidoglycan with identical muropeptide composition. *J. Bacteriol.* **172**:4682–4684.

955. De Lorenzo, V., M. Herrero, F. Giovannini, and J. B. Neilands. 1988. Fur (ferric uptake regulation) protein and CAP (catabolite-activator protein) modulate transcription of *fur* gene in *Escherichia coli*. *Eur. J. Biochem.* **173**:537–546.

956. de Massy, B., S. Bejar, J. Louarn, J. M. Louarn, and J. P. Bouche. 1987. Inhibition of replication forks exiting the terminus region of the *Escherichia coli* chromosome occurs at two loci separated by 5 min. *Proc. Natl. Acad. Sci. USA* **84**:1759–1763.

957. de Rekarte, U. D., M. Cortes, A. Porco, G. Nino, and T. Isturiz. 1994. Mutations affecting gluconate catabolism in *Escherichia coli*. Genetic mapping of loci for the low affinity transport and the thermoresistant gluconokinase. *J. Basic Microbiol.* **34**:363–370.

958. De Reuse, H., and A. Danchin. 1988. The *ptsH*, *ptsI*, and *crr* genes of the *Escherichia coli* phosphoenolpyruvate-dependent phosphotransferase system: a complex operon with several modes of transcription. *J. Bacteriol.* **170**:3827–3837.

959. De Reuse, H., E. Huttner, and A. Danchin. 1984. Analysis of the *ptsH-ptsI-crr* region in *Escherichia coli* K-12: evidence for the existence of a single transcriptional unit. *Gene* **32**:31–40.

960. De Reuse, H., A. Kolb, and A. Danchin. 1992. Positive regulation of the expression of the *Escherichia coli pts* operon. Identification of the regulatory regions. *J. Mol. Biol.* **226**:623–635.

961. De Reuse, H., A. Roy, and A. Danchin. 1985. Analysis of the *ptsH-ptsI-crr* region in *Escherichia coli* K-12: nucleotide sequence of the *ptsH* gene. *Gene* **35**:199–207.

962. de Veaux, L. C., D. S. Clevenson, C. Bradbeer, and R. J. Kadner. 1986. Identification of the *btuCED* polypeptides and evidence for their role in vitamin B12 transport in *Escherichia coli*. *J. Bacteriol.* **167**:920–927.

963. de Wind, N., M. de Jong, M. Meijer, and A. R. Stuitje. 1985. Site-directed mutagenesis of the *Escherichia coli* chromosome near *oriC*: identification and characterization of *asnC*, a regulatory element in *E. coli* asparagine metabolism. *Nucleic Acids Res.* **13**:8797–8811.

964. Dean, D. A., A. L. Davidson, and H. Nikaido. 1990. The role of ATP as the energy source for maltose transport in *Escherichia coli*. *Res. Microbiol.* **141**:348–352.

965. Dean, D. O., and R. James. 1991. Identification of a gene, closely linked to *dnaK*, which is required for high-temperature growth of *Escherichia coli*. *J. Gen. Microbiol.* **137**:1271–1277.

966. Dean, G. E., R. M. Macnab, J. Stader, P. Matsumura, and C. Burks. 1984. Gene sequence and predicted amino acid sequence of the *motA* protein, a membrane-associated protein required for flagellar rotation in *Escherichia coli*. *J. Bacteriol.* **159**:991–999.

967. Debarbouille, M., P. Cossart, and O. Raibaud. 1982. A DNA sequence containing the control sites for gene *malT* and for the *malPQ* operon. *Mol. Gen. Genet.* **185**:88–92.

968. Debarbouille, M., and M. Schwartz. 1979. The use of gene fusions to study the expression of *malT* the positive regulator gene of the maltose regulon. *J. Mol. Biol.* **132**:521–534.

969. Debarbouille, M., and M. Schwartz. 1980. Mutants which make more *malT* product, the activator of the maltose regulon in *Escherichia coli*. *Mol. Gen. Genet.* **178**:589–595.

970. Debenham, P. G., and M. B. Webb. 1988. The isolation and preliminary characterisation of a novel *Escherichia coli* mutant *rorB* with enhanced sensitivity to ionising radiation. *Mol. Gen. Genet.* **215**:161–164.

971. Debenham, P. G., M. B. Webb, and J. Law. 1988. The cloning of the *rorB* gene of *Escherichia coli*. *Mol. Gen. Genet.* **215**:156–160.

972. Debouck, C., A. Riccio, D. Schumperli, K. McKenney, J. Jeffers, C. Hughes, M. Rosenberg, M. Heusterspreute, F. Brunel, and J. Davison. 1985. Structure of the galactokinase gene of *Escherichia coli*, the last (?) gene of the *gal* operon. *Nucleic Acids Res.* **13**:1841–1853.

973. DeChavigny, A., P. N. Heacock, and W. Dowhan. 1991. Sequence and inactivation of the *pss* gene of *Escherichia coli*. Phosphatidylethanolamine may not be essential for cell viability. *J. Biol. Chem.* **266**:5323–5332.

974. Deeley, M. C., and C. Yanofsky. 1981. Nucleotide sequence of the structural gene for tryptophanase of *Escherichia coli* K-12. *J. Bacteriol.* **147**:787–796.

975. DeFeyter, R. C., B. E. Davidson, and J. Pittard. 1986. Nucleotide sequence of the transcription unit containing the *aroL* and *aroM* genes from *Escherichia coli* K-12. *J. Bacteriol.* **165**:233–239.

976. DeFeyter, R. C., and J. Pittard. 1986. Genetic and molecular analysis of *aroL*, the gene for shikimate kinase II in *Escherichia coli* K-12. *J. Bacteriol.* **165**:226–232.

977. Degryse, E. 1991. Polymorphism in the *dgt-dapD-tsf* region of *Escherichia coli* K-12 strains. *Gene* **102**:141–142.

978. Degryse, E. 1991. Development of stable, genetically well-defined conditionally viable *Escherichia coli* strains. *Mol. Gen. Genet.* **227**:49–51.

979. Deguchi, Y., I. Yamato, and Y. Anraku. 1989. Molecular cloning of *gltS* and *gltP*, which encode glutamate carriers of *Escherichia coli* B. *J. Bacteriol.* **171**:1314–1319.

980. Deguchi, Y., I. Yamato, and Y. Anraku. 1990. Nucleotide sequence of *gltS*, the Na+/glutamate symport carrier gene of *Escherichia coli* B. *J. Biol. Chem.* **265**:21704–21708.

981. Deka, R. K., C. Kleanthous, and J. R. Coggins. 1992. Identification of the essential histidine residue at the active site of *Escherichia coli* dehydroquinase. *J. Biol. Chem.* **267**:22237–22242.

982. del Campillo-Campbell, A., and A. Campbell. 1982. Molybdenum cofactor requirement for biotin sulfoxide reduction in *Escherichia coli*. *J. Bacteriol.* **149**:469–478.

983. Del Casale, T., P. Sollitti, and R. H. Chesney. 1983. Cytoplasmic L-asparaginase: isolation of a defective strain and mapping of *ansA*. *J. Bacteriol.* **154**:513–515.

984. del Castillo, I., J. M. Gomez, and F. Moreno. 1990. *mprA*, an *Escherichia coli* gene that reduces growth-phase-dependent synthesis of microcins B17 and C7 and blocks osmoinduction of *proU* when cloned on a high-copy-number plasmid. *J. Bacteriol.* **172**:437–445.

985. del Castillo, I., J. E. Gonzalez-Pastor, J. L. San Millan, and F. Moreno. 1991. Nucleotide sequence of the *Escherichia coli* regulatory gene *mprA* and construction and characterization of *mprA*-deficient mutants. *J. Bacteriol.* **173**:3924–3929.

986. del Castillo, I., J. L. Vizan, M. C. Rodriguez-Sainz, and F. Moreno. 1991. An unusual mechanism for resistance to the antibiotic coumermycin A1. *Proc. Natl. Acad. Sci. USA* **88**:8860–8864.

987. Delaney, J. M. 1990. A *cya* deletion mutant of *Escherichia coli* develops thermotolerance but does not exhibit a heat-shock response. *Genet. Res.* **55**:1–6.

988. Delaney, J. M., D. Ang, and C. Georgopoulos. 1992. Isolation and characterization of the *Escherichia coli htrD* gene, whose product is required for growth at high temperatures. *J. Bacteriol.* **174**:1240–1247.

989. Delaney, J. M., and C. Georgopoulos. 1992. Physical map locations of the *trxB*, *htrD*, *cydC*, and *cydD* genes of *Escherichia coli*. *J. Bacteriol.* **174**:3824–3825.

990. Delaney, J. M., D. Wall, and C. Georgopoulos. 1993. Molecular characterization of the *Escherichia coli htrD* gene: cloning, sequence, regulation, and involvement with cytochrome *d* oxidase. *J. Bacteriol.* **175**:166–175.

991. Delcuve, G., and P. P. Dennis. 1981. An amber mutation in a ribosomal protein gene: ineffective suppression stimulates operon-specific transcription. *J. Bacteriol.* **147**:997–1001.

992. Delcuve, G., W. Downing, H. Lewis, and P. P. Dennis. 1980. Nucleotide sequence of the proximal portion of the RNA polymerase beta subunit gene of *Escherichia coli*. *Gene* **11**:367–373.

993. Delidakis, C. E., M. C. Jones-Mortimer, and H. L. Kornberg. 1982. A mutant inducible for galactitol utilization in *Escherichia coli* K12. *J. Gen. Microbiol.* **128**:601–604.

994. Demerec, M., E. A. Adelberg, A. J. Clark, and P. E. Hatman. 1966. A proposal for a uniform nomenclature in bacterial genetics. *Genetics* **54**:61–76.

995. DeMoss, J. A., and P. Y. Hsu. 1991. NarK enhances nitrate uptake and nitrite excretion in *Escherichia coli*. *J. Bacteriol.* **173**:3303–3310.

996. Demple, B., and C. F. Amabile-Cuevas. 1991. Redox redux: the control of oxidative stress responses. *Cell* **67**:837–839.

997. Demple, B., B. Sedgwick, P. Robins, N. Totty, M. D. Waterfield, and T. Lindahl. 1985. Active site and complete sequence of the suicidal methyltransferase that counters alkylation mutagenesis. *Proc. Natl. Acad. Sci. USA* **82**:2688–2692.

998. Dempsey, W. B., and L. J. Arcement. 1971. Identification of the forms of vitamin B6 present in the culture media of "vitamin B6 control" mutants. *J. Bacteriol.* **107**:580–582.

999. Deneer, H. G., and G. B. Spiegelman. 1987. *Bacillus subtilis* rRNA promoters are growth rate regulated in *Escherichia coli*. *J. Bacteriol.* **169**:995–1002.

1000. Denk, D., and A. Bock. 1987. L-Cysteine biosynthesis in *Escherichia coli*: nucleotide sequence and expression of the serine acetyltransferase (*cysE*) gene from the wild-type and a cysteine-excreting mutant. *J. Gen. Microbiol.* **133**:515–525.

1001. Derst, C., J. Henseling, and K. H. Rohm. 1992. Probing the role of threonine and serine residues of *E. coli* asparaginase II by site-specific mutagenesis. *Protein Eng.* **5**:785–789.

1002. Descoteaux, A., and G. R. Drapeau. 1987. Regulation of cell division in *Escherichia coli* K-12: probable interactions among proteins FtsQ, FtsA, and FtsZ. *J. Bacteriol.* **169**:1938–1942.

1003. Deutch, A. H., K. E. Rushlow, and C. J. Smith. 1984. Analysis of the *Escherichia coli proBA* locus by DNA and protein sequencing. *Nucleic Acids Res.* **12**:6337–6355.

1004. Deutch, A. H., C. J. Smith, K. E. Rushlow, and P. J. Kretschmer. 1982. *Escherichia coli* delta 1-pyrroline-5-carboxylate reductase: gene sequence, protein overproduction and purification. *Nucleic Acids Res.* **10**:7701–7714.

1005. Deutch, C. E., and R. L. Soffer. 1978. *Escherichia coli* mutants defective in dipeptidyl carboxypeptidase. *Proc. Natl. Acad. Sci. USA* **75**:5998–6001.

1006. DeVeaux, L. C., J. E. Cronan, Jr., and T. L. Smith. 1989. Genetic and biochemical characterization of a mutation (*fatA*) that allows *trans*-unsaturated fatty acids to replace the essential *cis*-unsaturated fatty acids of *Escherichia coli*. *J. Bacteriol.* **171**:1562–1568.

1007. DeVeaux, L. C., and R. J. Kadner. 1985. Transport of vitamin B12 in *Escherichia coli*: cloning of the *btuCD* region. *J. Bacteriol.* **162**:888–896.

1008. Dewar, S. J., K. J. Begg, and W. D. Donachie. 1992. Inhibition of cell division initiation by an imbalance in the ratio of FtsA to FtsZ. *J. Bacteriol.* **174**:6314–6316.

1009. Dewar, S. J., and W. D. Donachie. 1990. Regulation of expression of the *ftsA* cell division gene by sequences in upstream genes. *J. Bacteriol.* **172**:6611–6614.

1010. Dewar, S. J., and W. D. Donachie. 1993. Antisense transcription of the *ftsZ-ftsA* gene junction inhibits cell division in *Escherichia coli*. *J. Bacteriol.* **175**:7097–7101.

1011. **Dhillon, T. S.** 1981. Temperate coliphage HK022: virions, DNA, one-step growth, attachment site, and the prophage genetic map. *J. Gen. Virol.* **55:**487–492.

1012. **Dhillon, T. S., A. P. Poon, Y. W. Hui, and E. K. Dhillon.** 1982. Lambdoid coliphage HK139 integrates between *his* and *supD*. *J. Virol.* **44:**716–719.

1013. **Diaz, R., P. Barnsley, and R. H. Pritchard.** 1979. Location and characterisation of a new replication origin in the *E. coli* K12 chromosome. *Mol. Gen. Genet.* **175:**151–157.

1014. **Diaz, R., and K. Kaiser.** 1981. Rac⁻ *E. coli* K12 strains carry a preferential attachment site for lambda *rev. Mol. Gen. Genet.* **183:**484–489.

1015. **Diaz, R., and R. H. Pritchard.** 1978. Cloning of replication origins from the *E. coli* K12 chromosome. *Nature* (London) **275:**561–564.

1016. **Dibb, N. J., and P. B. Wolfe.** 1986. *lep* operon proximal gene is not required for growth or secretion by *Escherichia coli. J. Bacteriol.* **166:**83–87.

1017. **Dicker, I. B., and S. Seetharam.** 1991. Cloning and nucleotide sequence of the *firA* gene and the *firA200*(Ts) allele from *Escherichia coli. J. Bacteriol.* **173:**333–344.

1018. **Dicker, I. B., and S. Seetharam.** 1992. What is known about the structure and function of the *Escherichia coli* protein FirA? *Mol. Microbiol.* **6:**817–823.

1019. **Diderichsen, B.** 1980. *cur-1*, a mutation affecting a phenotype of *sup⁺* strains of *Escherichia coli. Mol. Gen. Genet.* **180:**425–428.

1020. **Diderichsen, B.** 1981. Improved mapping of *ksgB* and integration of transposons near *relB* and *terC* in *Escherichia coli. J. Bacteriol.* **146:**409–411.

1021. **Diderichsen, B., and G. De Hauwer.** 1980. Improved mapping of the *tyrS* locus in *Escherichia coli. Mol. Gen. Genet.* **178:**647–650.

1022. **Diderichsen, B., and L. Desmarez.** 1980. Variations in phenotype of *relB* mutants of *Escherichia coli* and the effect of *pus* and *sup* mutations. *Mol. Gen. Genet.* **180:**429–437.

1023. **DiFrancesco, R., S. K. Bhatnagar, A. Brown, and M. J. Bessman.** 1984. The interaction of DNA polymerase III and the product of the *Escherichia coli* mutator gene, *mutD. J. Biol. Chem.* **259:**5567–5573.

1024. **DiGate, R. J., and K. J. Marians.** 1989. Molecular cloning and DNA sequence analysis of *Escherichia coli* topB, the gene encoding topoisomerase III. *J. Biol. Chem.* **264:**17924–17930.

1025. **Dila, D., E. Sutherland, L. Moran, B. Slatko, and E. A. Raleigh.** 1990. Genetic and sequence organization of the *mcrBC* locus of *Escherichia coli* K-12. *J. Bacteriol.* **172:**4888–4900.

1026. **Dimri, G. P., G. F. Ames, L. D'Ari, and J. C. Rabinowitz.** 1991. Physical map location of the *Escherichia coli* gene encoding the bifunctional enzyme 5,10-methylene-tetrahydrofolate dehydrogenase/5,10-methenyl-tetrahydrofolate cyclohydrolase. *J. Bacteriol.* **173:**5251.

1027. **DiNardo, S., K. A. Voelkel, R. Sternglanz, A. E. Reynolds, and A. Wright.** 1982. *Escherichia coli* DNA topoisomerase I mutants have compensatory mutations in DNA gyrase genes. *Cell* **31:**43–51.

1028. **Diorio, C., J. Cai, J. Marmor, R. Shinder, and M. S. DuBow.** 1995. An *Escherichia coli* chromosomal *ars* operon homolog is functional in arsenic detoxification and is conserved in gram-negative bacteria. *J. Bacteriol.* **177:**2050–2056.

1029. **DiRusso, C. C.** 1988. Nucleotide sequence of the *fadR* gene, a multifunctional regulator of fatty acid metabolism in *Escherichia coli. Nucleic Acids Res.* **16:**7995–8009.

1030. **DiRusso, C. C.** 1990. Primary sequence of the *Escherichia coli fadBA* operon, encoding the fatty acid-oxidizing multienzyme complex, indicates a high degree of homology to eucaryotic enzymes. *J. Bacteriol.* **172:**6459–6468.

1031. **DiRusso, C. C., T. L. Heimert, and A. K. Metzger.** 1992. Characterization of FadR, a global transcriptional regulator of fatty acid metabolism in *Escherichia coli.* Interaction with the *fadB* promoter is prevented by long chain fatty acyl coenzyme A. *J. Biol. Chem.* **267:**8685–8691.

1032. **DiRusso, C. C., and W. D. Nunn.** 1985. Cloning and characterization of a gene (*fadR*) involved in regulation of fatty acid metabolism in *Escherichia coli. J. Bacteriol.* **161:**583–588.

1032a. **Diver, W. P., N. J. Sargentini, and K. C. Smith.** 1982. A mutation (*radA100*) in *Escherichia coli* that selectively sensitizes cells grown in rich medium to X- or UV-radiation, or methyl methanesulphonate. *Int. J. Radiat. Biol.* **42:**339–346.

1033. **Dognin, M. J., and B. Wittmann-Liebold.** 1980. Purification and primary structure determination of the N-terminal blocked protein, L11, from *Escherichia coli* ribosomes. *Eur. J. Biochem.* **112:**131–151.

1034. **Doi, M., M. Wachi, F. Ishino, S. Tomioka, M. Ito, Y. Sakagami, A. Suzuki, and M. Matsuhashi.** 1988. Determinations of the DNA sequence of the *mreB* gene and of the gene products of the *mre* region that function in formation of the rod shape of *Escherichia coli* cells. *J. Bacteriol.* **170:**4619–4624.

1035. **Dombroski, A. J., W. A. Walter, M. T. Record, Jr., D. A. Siegele, and C. A. Gross.** 1992. Polypeptides containing highly conserved regions of transcription initiation factor sigma 70 exhibit specificity of binding to promoter DNA. *Cell* **70:**501–512.

1036. **Donahue, J. P., and C. L. Turnbough, Jr.** 1990. Characterization of transcriptional initiation from promoters P1 and P2 of the *pyrBI* operon of *Escherichia coli* K12. *J. Biol. Chem.* **265:**19091–19099.

1037. **Donch, J., and J. Greenberg.** 1968. Genetic analysis of *lon* mutants of strain K-12 of *Escherichia coli. Mol. Gen. Genet.* **103:**105–115.

1038. **Dong, J., S. Iuchi, H. S. Kwan, Z. Lu, and E. C. Lin.** 1993. The deduced amino-acid sequence of the cloned *cpxR* gene suggests the protein is the cognate regulator for the membrane sensor, CpxA, in a two-component signal transduction system of *Escherichia coli. Gene* **136:**227–230.

1039. **Dong, J. M., J. S. Taylor, D. J. Latour, S. Iuchi, and E. C. Lin.** 1993. Three overlapping *lct* genes involved in L-lactate utilization by *Escherichia coli. J. Bacteriol.* **175:**6671–6678.

1040. **Dong, Q., and H. J. Fromm.** 1990. Chemical modification of adenylosuccinate synthetase from *Escherichia coli* by pyridoxal 5′-phosphate. Identification of an active site lysyl residue. *J. Biol. Chem.* **265:**6235–6240.

1041. **Dong, X. R., S. F. Li, and J. A. DeMoss.** 1992. Upstream sequence elements required for NarL-mediated activation of transcription from the *narGHJI* promoter of *Escherichia coli. J. Biol. Chem.* **267:**14122–14128.

1042. **Dong, Z., R. Onrust, M. Skangalis, and M. O'Donnell.** 1993. DNA polymerase III accessory proteins. I. *holA* and *holB* encoding delta and delta′. *J. Biol. Chem.* **268:**11758–11765.

1043. **Doniger, J., D. Landsman, M. A. Gonda, and G. Wistow.** 1992. The product of *unr*, the highly conserved gene upstream of N-*ras*, contains multiple repeats similar to the cold-shock domain (CSD), a putative DNA-binding motif. *New Biol.* **4:**389–395.

1044. **Donly, B. C., C. D. Edgar, F. M. Adamski, and W. P. Tate.** 1990. Frameshift autoregulation in the gene for *Escherichia coli* release factor 2: partly functional mutants result in frameshift enhancement. *Nucleic Acids Res.* **18:**6517–6522.

1045. **Donly, C., J. Williams, C. Richardson, and W. Tate.** 1990. Frameshifting at the internal stop codon within the mRNA for bacterial release factor-2 on eukaryotic ribosomes. *Biochim. Biophys. Acta* **1050:**283–287.

1046. **Donnelly, C. E., and G. C. Walker.** 1992. Coexpression of UmuD′ with UmuC suppresses the UV mutagenesis deficiency of *groE* mutants. *J. Bacteriol.* **174:**3133–3139.

1047. **Donovan, W. P., and S. R. Kushner.** 1983. Cloning and physical analysis of the *pyrF* gene (coding for orotidine-5′- phosphate decarboxylase) from *Escherichia coli* K-12. *Gene* **25:**39–48.

1048. **Donovan, W. P., and S. R. Kushner.** 1983. Amplification of ribonuclease II (*rnb*) activity in *Escherichia coli* K- 12. *Nucleic Acids Res.* **11:**265–275.

1049. **Doolittle, R. F., D. F. Feng, K. L. Anderson, and M. R. Alberro.** 1990. A naturally occurring horizontal gene transfer from a eukaryote to a prokaryote. *J. Mol. Evol.* **31:**383–388.

1050. **Dopazo, A., P. Palacios, M. Sanchez, J. Pla, and M. Vicente.** 1992. An amino-proximal domain required for the localization of FtsQ in the cytoplasmic membrane, and for its biological function in *Escherichia coli. Mol. Microbiol.* **6:**715–722.

1051. **Dorman, C. J., G. C. Barr, N. Ni Bhriain, and C. F. Higgins.** 1988. DNA supercoiling and the anaerobic and growth phase regulation of *tonB* gene expression. *J. Bacteriol.* **170:**2816–2826.

1052. **Dorman, C. J., N. Ni Bhriain, and C. F. Higgins.** 1990. DNA supercoiling and environmental regulation of virulence gene expression in *Shigella flexneri. Nature* (London) **344:**789–792.

1053. **Dorman, C. J., and C. F. Higgins.** 1987. Fimbrial phase variation in *Escherichia coli*: dependence on integration host factor and homologies with other site-specific recombinases. *J. Bacteriol.* **169:**3840–3843.

1054. **Dosch, D. C., G. L. Helmer, S. H. Sutton, F. F. Salvacion, and W. Epstein.** 1991. Genetic analysis of potassium transport loci in *Escherichia coli*: evidence for three constitutive systems mediating uptake potassium. *J. Bacteriol.* **173:**687–696.

1055. **Doublet, P., J. van Heijenoort, J. P. Bohin, and D. Mengin-Lecreulx.** 1993. The *murI* gene of *Escherichia coli* is an essential gene that encodes a glutamate racemase activity. *J. Bacteriol.* **175:**2970–2979.

1056. **Doublet, P., J. van Heijenoort, and D. Mengin-Lecreulx.** 1992. Identification of the *Escherichia coli murI* gene, which is required for the biosynthesis of D-glutamic acid, a specific component of bacterial peptidoglycan. *J. Bacteriol.* **174:**5772–5779.

1057. **Dougherty, T. J., J. A. Thanassi, and M. J. Pucci.** 1993. The *Escherichia coli* mutant requiring D-glutamic acid is the result of mutations in two distinct genetic loci. *J. Bacteriol.* **175:**111–116.

1058. **Douglas, R. M., J. A. Roberts, A. W. Munro, G. Y. Ritchie, A. J. Lamb, and I. R. Booth.** 1991. The distribution of homologues of the *Escherichia coli* KefC K(⁺)-efflux system in other bacterial species. *J. Gen. Microbiol.* **137:**1999–2005.

1059. **Douville, K., M. Leonard, L. Brundage, K. Nishiyama, H. Tokuda, S. Mizushima, and W. Wickner.** 1994. Band 1 subunit of *Escherichia coli* preprotein translocase and integral membrane export factor P12 are the same protein. *J. Biol. Chem.* **269:**18705–18707.

1060. **Dover, S., and Y. S. Halpern.** 1974. Genetic analysis of the gamma-aminobutyrate utilization pathway in *Escherichia coli* K-12. *J. Bacteriol.* **117:**494–501.

1061. **Dovgas, N. V., L. F. Markova, T. A. Mednikova, L. M. Vinokurov, Y. B. Alakhov, and Y. A. Ovchinnikov.** 1975. The primary structure of the 5 S RNA binding protein L25 from *Escherichia coli* ribosomes. *FEBS Lett.* **53:**351–354.

1062. **Dovgas, N. V., L. M. Vinokurov, I. S. Velmoga, Y. B. Alakhov, and Y. A. Ovchinnikov.** 1976. The primary structure of protein L10 from *Escherichia coli* ribosomes. *FEBS Lett.* **67:**58–61.

1063. **Dowhan, W.** 1992. Phosphatidylglycerophosphate synthase from *Escherichia coli. Methods Enzymol.* **209:**313–321.

1064. **Dowhan, W.** 1992. Phosphatidylserine synthase from *Escherichia coli. Methods Enzymol.* **209:**287–298.

1065. **Dowhan, W., and Q. X. Li.** 1992. Phosphatidylserine decarboxylase from *Escherichia coli. Methods Enzymol.* **209:**348–359.

1066. **Downing, W. L., and P. P. Dennis.** 1987. Transcription products from the *rplKAJL-rpoBC* gene cluster. *J. Mol. Biol.* **194:**609–620.

1067. **Downing, W. L., S. L. Sullivan, M. E. Gottesman, and P. P. Dennis.** 1990. Sequence and transcriptional pattern of the essential *Escherichia coli secE-nusG* operon. *J. Bacteriol.* **172:**1621–1627.

1068. **Driver, R. P., and R. P. Lawther.** 1985. Physical analysis of deletion mutations in the *ilvGEDA* operon of *Escherichia coli* K-12. *J. Bacteriol.* **162:**598–606.

1069. **Drlica, K.** 1992. Control of bacterial DNA supercoiling. *Mol. Microbiol.* **6:**425–433.

1070. **Drolet, M., L. Peloquin, Y. Echelard, L. Cousineau, and A. Sasarman.** 1989. Isolation and nucleotide sequence of the *hemA* gene of *Escherichia coli* K12. *Mol. Gen. Genet.* **216:**347–352.

1071. **Druger-Liotta, J., V. J. Prange, D. G. Overdier, and L. N. Csonka.** 1987. Selection of mutations that alter the osmotic control of transcription of the *Salmonella typhimurium proU* operon. *J. Bacteriol.* **169:**2449–2459.

1072. **Drummond, J. T., R. R. Loo, and R. G. Matthews.** 1993. Electrospray mass spectrometric analysis of the domains of a large enzyme: observation of the occupied cobalamin-binding domain and redefinition of the carboxyl terminus of methionine synthase. *Biochemistry* **32:**9282–9289.

1073. **Drury, L. S., and R. S. Buxton.** 1985. DNA sequence analysis of the *dye* gene of *Escherichia coli* reveals amino acid homology between the Dye and OmpR proteins. *J. Biol. Chem.* **260:**4236–4242.

1074. **Drury, L. S., and R. S. Buxton.** 1988. Identification and sequencing of the *Escherichia coli cet* gene which codes for an inner membrane protein, mutation of which causes tolerance to colicin E2. *Mol. Microbiol.* **2:**109–119.

1075. **DuBose, R. F., D. E. Dykhuizen, and D. L. Hartl.** 1988. Genetic exchange among natural isolates of bacteria: recombination within the *phoA* gene of *Escherichia coli.* *Proc. Natl. Acad. Sci. USA* **85:**7036–7040.

1076. **Dubourdieu, M., and J. A. DeMoss.** 1992. The *narJ* gene product is required for biogenesis of respiratory nitrate reductase in *Escherichia coli.* *J. Bacteriol.* **174:**867–872.

1077. **Duchange, N., M. M. Zakin, P. Ferrara, I. Saint-Girons, I. Park, S. V. Tran, M. C. Py, and G. N. Cohen.** 1983. Structure of the *metJBLF* cluster in *Escherichia coli* K12. Sequence of the *metB* structural gene and of the 5′- and 3′-flanking regions of the *metBL* operon. *J. Biol. Chem.* **258:**14868–14871.

1078. **Duclos, B., J. C. Cortay, F. Bleicher, E. Z. Ron, C. Richaud, I. Saint Girons, and A. J. Cozzone.** 1989. Nucleotide sequence of the *metA* gene encoding homoserine trans-succinylase in *Escherichia coli.* *Nucleic Acids Res.* **17:**2856.

1079. **Duester, G., R. K. Campen, and W. M. Holmes.** 1981. Nucleotide sequence of an *Escherichia coli* tRNA (Leu 1) operon and identification of the transcription promoter signal. *Nucleic Acids Res.* **9:**2121–2139.

1080. **Duester, G. L., and W. M. Holmes.** 1980. The distal end of the ribosomal RNA operon *rrnD* of *Escherichia coli* contains a tRNA1thr gene, two 5s rRNA genes and a transcription terminator. *Nucleic Acids Res.* **8:**3793–3807.

1081. **Dueweke, T. J., and R. B. Gennis.** 1990. Epitopes of monoclonal antibodies which inhibit ubiquinol oxidase activity of *Escherichia coli* cytochrome *d* complex localize functional domain. *J. Biol. Chem.* **265:**4273–4277.

1082. **Duncan, K., S. Chaudhuri, M. S. Campbell, and J. R. Coggins.** 1986. The overexpression and complete amino acid sequence of *Escherichia coli* 3-dehydroquinase. *Biochem. J.* **238:**475–483.

1083. **Duncan, K., and J. R. Coggins.** 1986. The *serC-aroA* operon of *Escherichia coli.* A mixed function operon encoding enzymes from two different amino acid biosynthetic pathways. *Biochem. J.* **234:**49–57.

1084. **Dunderdale, H. J., F. E. Benson, C. A. Parsons, G. J. Sharples, R. G. Lloyd, and S. C. West.** 1991. Formation and resolution of recombination intermediates by *E. coli* RecA and RuvC proteins. *Nature* (London) **354:**506–510.

1085. **Dunderdale, H. J., G. J. Sharples, R. G. Lloyd, and S. C. West.** 1994. Cloning, overexpression, purification, and characterization of the *Escherichia coli* RuvC Holliday junction resolvase. *J. Biol. Chem.* **269:**5187–5194.

1086. **Dunn, T. M., and R. Schleif.** 1984. Deletion analysis of the *Escherichia coli* ara PC and PBAD promoters. *J. Mol. Biol.* **180:**201–204.

1087. **Duplay, P., H. Bedouelle, A. Fowler, I. Zabin, W. Saurin, and M. Hofnung.** 1984. Sequences of the *malE* gene and of its product, the maltose-binding protein of *Escherichia coli* K12. *J. Biol. Chem.* **259:**10606–10613.

1088. **Dutreix, M., B. Burnett, A. Bailone, C. M. Radding, and R. Devoret.** 1992. A partially deficient mutant, *recA1730*, that fails to form normal nucleoprotein filaments. *Mol. Gen. Genet.* **232:**489–497.

1089. **Dykstra, C. C., and S. R. Kushner.** 1985. Physical characterization of the cloned protease III gene from *Escherichia coli* K-12. *J. Bacteriol.* **163:**1055–1059.

1090. **Dykstra, C. C., D. Prasher, and S. R. Kushner.** 1984. Physical and biochemical analysis of the cloned *recB* and *recC* genes of *Escherichia coli* K-12. *J. Bacteriol.* **157:**21–27.

1091. **Dyson, H. J., G. P. Gippert, D. A. Case, A. Holmgren, and P. E. Wright.** 1990. Three-dimensional solution structure of the reduced form of *Escherichia coli* thioredoxin determined by nuclear magnetic resonance spectroscopy. *Biochemistry* **29:**4129–4136.

1092. **Easton, A. M., and S. R. Kushner.** 1983. Transcription of the *uvrD* gene of *Escherichia coli* is controlled by the *lexA* repressor and by attenuation. *Nucleic Acids Res.* **11:**8625–8640.

1093. **Echelard, Y., J. Dymetryszyn, M. Drolet, and A. Sasarman.** 1988. Nucleotide sequence of the *hemB* gene of *Escherichia coli* K12. *Mol. Gen. Genet.* **214:**503–508.

1094. **Echols, H., C. Lu, and P. M. Burgers.** 1983. Mutator strains of *Escherichia coli, mutD* and *dnaQ,* with defective exonucleolytic editing by DNA polymerase III holoenzyme. *Proc. Natl. Acad. Sci. USA* **80:**2189–2192.

1095. **Eckhardt, T.** 1980. Isolation of plasmids carrying the arginine repressor gene *argR* of *Escherichia coli* K12. *Mol. Gen. Genet.* **178:**447–452.

1096. **Edlund, T., T. Grundstrom, and S. Normark.** 1979. Isolation and characterization of DNA repetitions carrying the chromosomal beta-lactamase gene of *Escherichia coli* K-12. *Mol. Gen. Genet.* **173:**115–125.

1097. **Edlund, T., and S. Normark.** 1981. Recombination between short DNA homologies causes tandem duplication. *Nature* (London) **292:**269–271.

1098. **Edwards, E. S., S. S. Rondeau, and J. A. DeMoss.** 1983. *chlC* (*nar*) operon of *Escherichia coli* includes structural genes for alpha and beta subunits of nitrate reductase. *J. Bacteriol.* **153:**1513–1520.

1099. **Edwards, R. M., and M. D. Yudkin.** 1982. Location of the gene for the low-affinity tryptophan-specific permease of *Escherichia coli.* *Biochem. J.* **204:**617–619.

1100. **Egan, A. F., and R. R. Russell.** 1973. Conditional mutations affecting the cell envelope of *Escherichia coli* K-12. *Genet. Res.* **21:**139–152.

1101. **Egan, S. E., R. Fliege, S. Tong, A. Shibata, R. E. Wolf, Jr., and T. Conway.** 1992. Molecular characterization of the Entner-Doudoroff pathway in *Escherichia coli:* sequence analysis and localization of promoters for the *edd-eda* operon. *J. Bacteriol.* **174:**4638–4646.

1102. **Egan, S. M., and V. Stewart.** 1991. Mutational analysis of nitrate regulatory gene *narL* in *Escherichia coli* K-12. *J. Bacteriol.* **173:**4424–4432.

1103. **Ehrenreich, A., K. Forchhammer, P. Tormay, B. Veprek, and A. Bock.** 1992. Selenoprotein synthesis in *E. coli.* Purification and characterisation of the enzyme catalysing selenium activation. *Eur. J. Biochem.* **206:**767–773.

1104. **Ehretsmann, C. P., A. J. Carpousis, and H. M. Krisch.** 1992. Specificity of *Escherichia coli* endoribonuclease RNase E: in vivo and in vitro analysis of mutants in a bacteriophage T4 mRNA processing site. *Genes Dev.* **6:**149–159.

1105. **Ehring, R., K. Beyreuther, J. K. Wright, and P. Overath.** 1980. In vitro and in vivo products of *E. coli* lactose permease gene are identical. *Nature* (London) **283:**537–540.

1106. **Ehrmann, M., and W. Boos.** 1987. Identification of endogenous inducers of the *mal* regulon in *Escherichia coli.* *J. Bacteriol.* **169:**3539–3545.

1107. **Ehrmann, M., W. Boos, E. Ormseth, H. Schweizer, and T. J. Larson.** 1987. Divergent transcription of the *sn*-glycerol-3-phosphate active transport (*glpT*) and anaerobic *sn*-glycerol-3-phosphate dehydrogenase (*glpA glpC glpB*) genes of *Escherichia coli* K-12. *J. Bacteriol.* **169:**526–532.

1108. **Eichler, K., F. Bourgis, A. Buchet, H. P. Kleber, and M. A. Mandrand-Berthelot.** 1994. Molecular characterization of the *cai* operon necessary for carnitine metabolism in *Escherichia coli. Mol. Microbiol.* **13:**775–786.

1109. **Eichler, K., W. H. Schunck, H. P. Kleber, and M. A. Mandrand-Berthelot.** 1994. Cloning, nucleotide sequence, and expression of the *Escherichia coli* gene encoding carnitine dehydratase. *J. Bacteriol.* **176:**2970–2975.

1110. **Eick-Helmerich, K., and V. Braun.** 1989. Import of biopolymers into *Escherichia coli:* nucleotide sequences of the *exbB* and *exbD* genes are homologous to those of the *tolQ* and *tolR* genes, respectively. *J. Bacteriol.* **171:**5117–5126.

1111. **Eick-Helmerich, K., K. Hantke, and V. Braun.** 1987. Cloning and expression of the *exbB* gene of *Escherichia coli* K-12. *Mol. Gen. Genet.* **206:**246–251.

1112. **Eiglmeier, K., W. Boos, and S. T. Cole.** 1987. Nucleotide sequence and transcriptional startpoint of the *glpT* gene of *Escherichia coli:* extensive sequence homology of the glycerol-3- phosphate transport protein with components of the hexose-6-phosphate transport system. *Mol. Microbiol.* **1:**251–258.

1113. **Eisenbeis, S. J., and J. Parker.** 1981. Strains of *Escherichia coli* carrying the structural gene for histidyl-tRNA synthetase on a high copy-number plasmid. *Mol. Gen. Genet.* **183:**115–122.

1114. **Eisenbeis, S. J., and J. Parker.** 1982. The nucleotide sequence of the promoter region of *hisS,* the structural gene for histidyl-tRNA synthetase. *Gene* **18:**107–114.

1115. **Eisenstein, B. I., D. S. Sweet, V. Vaughn, and D. I. Friedman.** 1987. Integration host factor is required for the DNA inversion that controls phase variation in *Escherichia coli. Proc. Natl. Acad. Sci. USA* **84:**6506–6510.

1116. **Eisenstein, E.** 1991. Cloning, expression, purification, and characterization of biosynthetic threonine deaminase from *Escherichia coli. J. Biol. Chem.* **266:**5801–5807.

1117. **Elgren, T. E., J. B. Lynch, C. Juarez-Garcia, E. Munck, B. M. Sjoberg, and L. Que, Jr.** 1991. Electron transfer associated with oxygen activation in the B2 protein of ribonucleotide reductase from *Escherichia coli. J. Biol. Chem.* **266:**19265–19268.

1118. **Elkins, M. F., and C. F. Earhart.** 1989. Nucleotide sequence and regulation of the *Escherichia coli* gene for ferrienterobactin transport protein FepB. *J. Bacteriol.* **171:**5443–5451.

1119. **Elledge, S. J., and G. C. Walker.** 1983. Proteins required for ultraviolet light and chemical mutagenesis. Identification of the products of the *umuC* locus of *Escherichia coli. J. Mol. Biol.* **164:**175–192.

1120. **Ellwood, M., and M. Nomura.** 1982. Chromosomal locations of the genes for rRNA in *Escherichia coli* K-12. *J. Bacteriol.* **149:**458–468.

1121. **Elseviers, D., P. Gallagher, A. Hoffman, B. Weinberg, and I. Schwartz.** 1982. Molecular cloning and regulation of expression of the genes for initiation factor 3 and two aminoacyl-tRNA synthetases. *J. Bacteriol.* **152:**357–362.

1122. **Elseviers, D., L. A. Petrullo, and P. J. Gallagher.** 1984. Novel *E. coli* mutants deficient in biosynthesis of 5-methylaminomethyl-2-thiouridine. *Nucleic Acids Res.* **12:**3521–3534.

1123. **Elvin, C. M., N. E. Dixon, and H. Rosenberg.** 1986. Molecular cloning of the phosphate (inorganic) transport (*pit*) gene of *Escherichia coli* K12. Identification of the *pit*+ gene product and physical mapping of the *pit-gor* region of the chromosome. *Mol. Gen. Genet.* **204:**477–484.

1124. Elvin, C. M., C. M. Hardy, and H. Rosenberg. 1985. Pi exchange mediated by the GlpT-dependent *sn*-glycerol-3-phosphate transport system in *Escherichia coli*. J. Bacteriol. 161:1054–1058.

1125. Emmerich, R. V., and I. N. Hirshfield. 1987. Mapping of the constitutive lysyl-tRNA synthetase gene of *Escherichia coli* K-12. J. Bacteriol. 169:5311–5313.

1126. Emory, S. A., and J. G. Belasco. 1990. The *ompA* 5′ untranslated RNA segment functions in *Escherichia coli* as a growth-rate-regulated mRNA stabilizer whose activity is unrelated to translational efficiency. J. Bacteriol. 172:4472–4481.

1127. Emory, S. A., P. Bouvet, and J. G. Belasco. 1992. A 5′-terminal stem-loop structure can stabilize mRNA in *Escherichia coli*. Genes Dev. 6:135–148.

1128. Emr, S. D., S. Hanley-Way, and T. J. Silhavy. 1981. Suppressor mutations that restore export of a protein with a defective signal sequence. Cell 23:79–88.

1129. Emr, S. D., J. Hedgpeth, J. M. Clement, T. J. Silhavy, and M. Hofnung. 1980. Sequence analysis of mutations that prevent export of lambda receptor, an *Escherichia coli* outer membrane protein. Nature (London) 285:82–85.

1130. Engel, H., B. Kazemier, and W. Keck. 1991. Murein-metabolizing enzymes from *Escherichia coli*: sequence analysis and controlled overexpression of the *slt* gene, which encodes the soluble lytic transglycosylase. J. Bacteriol. 173:6773–6782.

1131. Engel, H., A. J. Smink, L. van Wijngaarden, and W. Keck. 1992. Murein-metabolizing enzymes from *Escherichia coli*: existence of a second lytic transglycosylase. J. Bacteriol. 174:6394–6403.

1132. Enger-Valk, B. E., H. L. Heyneker, R. A. Oosterbaan, and P. H. Pouwels. 1980. Construction of new cloning vehicles with genes of the tryptophan operon of *Escherichia coli* as genetic markers. Gene 9:69–85.

1133. Englesberg, E., J. Irr, J. Power, and N. Lee. 1965. Positive control of enzyme synthesis by gene C in the L-arabinose system. J. Bacteriol. 90:946–957.

1134. Enomoto, M., K. Oosawa, and H. Momota. 1983. Mapping of the *pin* locus coding for a site-specific recombinase that causes flagellar-phase variation in *Escherichia coli* K-12. J. Bacteriol. 156:663–668.

1135. Ephrati-Elizur, E. 1993. A mutation in a new gene of *Escherichia coli*, *psu*, requires secondary mutations for survival: *psu* mutants express a pleiotropic suppressor phenotype. J. Bacteriol. 175:207–213.

1136. Epstein, W., E. Buurman, D. McLaggan, and J. Naprstek. 1993. Multiple mechanisms, roles and controls of K⁺ transport in *Escherichia coli*. Biochem. Soc. Trans. 21:1006–1010.

1137. Epstein, W., S. Jewett, and C. F. Fox. 1970. Isolation and mapping of phosphotransferase mutants in *Escherichia coli*. J. Bacteriol. 104:793–797.

1137a. Epstein, W., and B. S. Kim. 1971. Potassium transport loci in *Escherichia coli* K-12. J. Bacteriol. 108:639–644.

1138. Eraso, J. M., and G. M. Weinstock. 1992. Anaerobic control of colicin E1 production. J. Bacteriol. 174:5101–5109.

1139. Eriani, G., M. Delarue, O. Poch, J. Gangloff, and D. Moras. 1990. Partition of tRNA synthetases into two classes based on mutually exclusive sets of sequence motifs. Nature (London) 347:203–206.

1140. Eriani, G., G. Dirheimer, and J. Gangloff. 1989. Isolation and characterization of the gene coding for *Escherichia coli* arginyl-tRNA synthetase. Nucleic Acids Res. 17:5725–5736.

1141. Eriani, G., G. Dirheimer, and J. Gangloff. 1990. Aspartyl-tRNA synthetase from *Escherichia coli*: cloning and characterisation of the gene, homologies of its translated amino acid sequence with asparaginyl- and lysyl-tRNA synthetases. Nucleic Acids Res. 18:7109–7118.

1142. Eriani, G., G. Dirheimer, and J. Gangloff. 1991. Cysteinyl-tRNA synthetase: determination of the last *E. coli* aminoacyl- tRNA synthetase primary structure. Nucleic Acids Res. 19:265–269.

1143. Erickson, B. D., Z. F. Burton, K. K. Watanabe, and R. R. Burgess. 1985. Nucleotide sequence of the *rpsU-dnaG-rpoD* operon from *Salmonella typhimurium* and a comparison of this sequence with the homologous operon of *Escherichia coli*. Gene 40:67–78.

1144. Erickson, J. W., V. Vaughn, W. A. Walter, F. C. Neidhardt, and C. A. Gross. 1987. Regulation of the promoters and transcripts of *rpoH*, the *Escherichia coli* heat shock regulatory gene. Genes Dev. 1:419–432.

1145. Erie, D. A., O. Hajiseyedjavadi, M. C. Young, and P. H. von Hippel. 1993. Multiple RNA polymerase conformations and GreA: control of the fidelity of transcription. Science 262:867–873.

1146. Eriksson-Grennberg, K. R., K. Nordstrom, and P. Englund. 1971. Resistance of *Escherichia coli* to penicillins. IX. Genetics and physiology of class II ampicillin-resistant mutants that are galactose negative or sensitive to bacteriophage C21, or both. J. Bacteriol. 108:1210–1223.

1147. Erlagaeva, R. S., T. N. Bol'shakova, N. A. Kyzylova, and V. N. Gershanovich. 1987. Analysis of mutations affecting the expression of catabolite-sensitive operons in *Escherichia coli* K12 mutants defective in the HPr-component of the carbohydrate transport system. Mol. Gen. Mikrobiol. Virusol. 21:43–47. (In Russian.)

1148. Ermler, U., and G. E. Schulz. 1991. The three-dimensional structure of glutathione reductase from *Escherichia coli* at 3.0 A resolution. Proteins 9:174–179.

1149. Erni, B., and B. Zanolari. 1985. The mannose-permease of the bacterial phosphotransferase system. Gene cloning and purification of the enzyme IIMan/III-Man complex of *Escherichia coli*. J. Biol. Chem. 260:15495–15503.

1150. Erni, B., and B. Zanolari. 1986. Glucose-permease of the bacterial phosphotransferase system. Gene cloning, overproduction, and amino acid sequence of enzyme IIGlc. J. Biol. Chem. 261:16398–16403.

1151. Erni, B., B. Zanolari, and H. P. Kocher. 1987. The mannose permease of *Escherichia coli* consists of three different proteins. Amino acid sequence and

1152. Erpel, T., P. Hwang, C. S. Craik, R. J. Fletterick, and M. E. McGrath. 1992. Physical map location of the new *Escherichia coli* gene *eco*, encoding the serine protease inhibitor ecotin. J. Bacteriol. 174:1704.

1153. Escarceller, M., J. Hicks, G. Gudmundsson, G. Trump, D. Touati, S. Lovett, P. L. Foster, K. McEntee, and M. F. Goodman. 1994. Involvement of *Escherichia coli* DNA polymerase II in response to oxidative damage and adaptive mutation. J. Bacteriol. 176:6221–6228.

1154. Esmon, B. E., C. R. Kensil, C. H. Cheng, and M. Glaser. 1980. Genetic analysis of *Escherichia coli* mutants defective in adenylate kinase and *sn*-glycerol 3-phosphate acyltransferase. J. Bacteriol. 141:405–408.

1155. Espion, D., K. Kaiser, and C. Dambly-Chaudiere. 1983. A third defective lambdoid prophage of *Escherichia coli* K12 defined by the lambda derivative, lambda qin111. J. Mol. Biol. 170:611–633.

1156. Essenberg, R. C. 1984. Use of homocysteic acid for selecting mutants at the *gltS* locus of *Escherichia coli* K12. J. Gen. Microbiol. 130:1311–1314.

1157. Essenberg, R. C., and H. L. Kornberg. 1977. Location of the gene specifying hexose phosphate transport (*uhp*) on the chromosome of *Escherichia coli*. J. Gen. Microbiol. 99:157–169.

1158. Estevenon, A. M., B. Martin, and N. Sicard. 1985. Characterization of a mutation conferring radiation sensitivity, *ior*, located close to the gene coding for deoxycytidine deaminase in *Escherichia coli*. Mol. Gen. Genet. 200:132–137.

1159. Evans, R., N. R. Seeley, and P. L. Kuempel. 1979. Loss of *rac* locus DNA in merozygotes of *Escherichia coli* K12. Mol. Gen. Genet. 175:245–250.

1160. Evans, S., and P. P. Dennis. 1985. Promoter activity and transcript mapping in the regulatory region for genes encoding ribosomal protein S15 and polynucleotide phosphorylase of *Escherichia coli*. Gene 40:15–22.

1161. Evensen, G., and E. Seeberg. 1982. Adaptation to alkylation resistance involves the induction of a DNA glycosylase. Nature (London) 296:773–775.

1162. Ewart, C. D., D. A. Jude, J. L. Thain, and W. W. Nichols. 1995. Frequency and mechanism of resistance to antibacterial action of ZM 240401, (6S)-6-fluoroshikimic acid. Antimicrob. Agents Chemother. 39:87–93.

1163. Ezaki, B., H. Mori, T. Ogura, and S. Hiraga. 1990. Possible involvement of the *ugpA* gene product in the stable maintenance of mini-F plasmid in *Escherichia coli*. Mol. Gen. Genet. 223:361–368.

1164. Ezaki, B., T. Ogura, H. Mori, H. Niki, and S. Hiraga. 1989. Involvement of DnaK protein in mini-F plasmid replication: temperature-sensitive *seg* mutations are located in the *dnaK* gene. Mol. Gen. Genet. 218:183–189.

1165. Ezaki, B., T. Ogura, H. Niki, and S. Hiraga. 1991. Partitioning of a mini-F plasmid into anucleate cells of the *mukB* null mutant. J. Bacteriol. 173:6643–6646.

1166. Ezekiel, D. H., and J. E. Hutchins. 1968. Mutations affecting RNA polymerase associated with rifampicin resistance in *Escherichia coli*. Nature (London) 220:276–277.

1167. Faatz, E., A. Middendorf, and E. Bremer. 1988. Cloned structural genes for the osmotically regulated binding-protein-dependent glycine betaine transport system (ProU) of *Escherichia coli* K- 12. Mol. Microbiol. 2:265–279.

1168. Fabiny, J. M., A. Jayakumar, A. C. Chinault, and E. M. Barnes, Jr. 1991. Ammonium transport in *Escherichia coli*: localization and nucleotide sequence of the *amtA* gene. J. Gen. Microbiol. 137:983–989.

1169. Faik, P., and H. L. Kornberg. 1973. Isolation and properties of E. coli mutants affected in gluconate uptake. FEBS Lett. 32:260–264.

1170. Falconi, M., M. T. Gualtieri, A. La Teana, M. A. Losso, and C. L. Pon. 1988. Proteins from the prokaryotic nucleoid: primary and quaternary structure of the 15-kD *Escherichia coli* DNA binding protein H-NS. Mol. Microbiol. 2:323–329.

1171. Falkinham, J. O. 1979. Identification of a mutation affecting an alanine-alpha-ketoisovalerate transaminase activity in *Escherichia coli* K-12. Mol. Gen. Genet. 176:147–149.

1172. Falmagne, P., D. Portetelle, and V. Stalon. 1985. Immunological and structural relatedness of catabolic ornithine carbamoyltransferases and the anabolic enzymes of enterobacteria. J. Bacteriol. 161:714–719.

1173. Farabaugh, P. J. 1978. Sequence of the *lacI* gene. Nature (London) 274:765–769.

1174. Farr, S. B., D. N. Arnosti, M. J. Chamberlin, and B. N. Ames. 1989. An *apaH* mutation causes ApppppA to accumulate and affects motility and catabolite repression in *Escherichia coli*. Proc. Natl. Acad. Sci. USA 86:5010–5014.

1175. Farrell, L. B., and R. N. Beachy. 1990. Manipulation of beta-glucuronidase for use as a reporter in vacuolar targeting studies. Plant Mol. Biol. 15:821–825.

1176. Fath, M. J., H. K. Mahanty, and R. Kolter. 1989. Characterization of a *purF* operon mutation which affects colicin V production. J. Bacteriol. 171:3158–3161.

1177. Faubladier, M., and J. P. Bouche. 1994. Division inhibition gene *dicF* of *Escherichia coli* reveals a widespread group of prophage sequences in bacterial genomes. J. Bacteriol. 176:1150–1156.

1178. Faubladier, M., K. Cam, and J. P. Bouche. 1990. *Escherichia coli* cell division inhibitor DicF-RNA of the *dicB* operon. Evidence for its generation in vivo by transcription termination and by RNase III and RNase E-dependent processing. J. Mol. Biol. 212:461–471.

1179. Favre, A., V. Chams, and A. Caldeira de Araujo. 1986. Photosensitized UVA light induction of the SOS response in *Escherichia coli*. Biochimie 68:857–864.

1180. Faxen, M., A. Walles-Granberg, and L. A. Isaksson. 1994. Antisuppression by a mutation in rpsM(S13) giving a shortened ribosomal protein S13. Biochim. Biophys. Acta 1218:27–34.

1181. Fayat, G., J. F. Mayaux, C. Sacerdot, M. Fromant, M. Springer, M. Grunberg-Manago, and S. Blanquet. 1983. *Escherichia coli* phenylalanyl-tRNA synthetase operon region. Evidence for an attenuation mechanism. Identification of the gene for the ribosomal protein L20. *J. Mol. Biol.* 171:239–261.

1182. Fayerman, J. T., M. C. Vann, L. S. Williams, and H. E. Umbarger. 1979. *ilvU*, a locus in *Escherichia coli* affecting the derepression of isoleucyl-tRNA synthetase and the RPC-5 chromatographic profiles of tRNAIle and tRNAVal. *J. Biol. Chem.* 254:9429–9440.

1183. Fayet, O., T. Ziegelhoffer, and C. Georgopoulos. 1989. The *groES* and *groEL* heat shock gene products of *Escherichia coli* are essential for bacterial growth at all temperatures. *J. Bacteriol.* 171:1379–1385.

1184. Fecker, L., and V. Braun. 1983. Cloning and expression of the *fhu* genes involved in iron(III)-hydroxamate uptake by *Escherichia coli*. *J. Bacteriol.* 156:1301–1314.

1185. Fee, J. A. 1991. Regulation of *sod* genes in *Escherichia coli*: relevance to superoxide dismutase function. *Mol. Microbiol.* 5:2599–2610.

1186. Fejzo, J., F. A. Etzkorn, R. T. Clubb, Y. Shi, C. T. Walsh, and G. Wagner. 1994. The mutant *Escherichia coli* F112W cyclophilin binds cyclosporin A in nearly identical conformation as human cyclophilin. *Biochemistry* 33:5711–5720.

1187. Feller, A., A. Pierard, N. Glansdorff, D. Charlier, and M. Crabeel. 1981. Mutation of gene encoding regulatory polypeptide of aspartate carbamoyltransferase. *Nature* (London) 292:370–373.

1188. Felzenszwalb, I., S. Boiteux, and J. Laval. 1992. Identification of the *radC102* mutation. Order of the genes in the 81.5-82.0 min region of the *Escherichia coli* chromosome. *Nucleic Acids Res.* 20:366.

1189. Felzenszwalb, I., S. Boiteux, and J. Laval. 1992. Molecular cloning and DNA sequencing of the *radC* gene of *Escherichia coli* K-12. *Mutat. Res.* 273:263–269.

1190. Felzenszwalb, I., N. J. Sargentini, and K. C. Smith. 1984. Characterization of a new radiation-sensitive mutant, *Escherichia coli* K-12 *radC102*. *Radiat. Res.* 97:615–625.

1191. Feng, J., M. R. Atkinson, W. McCleary, J. B. Stock, B. L. Wanner, and A. J. Ninfa. 1992. Role of phosphorylated metabolic intermediates in the regulation of glutamine synthetase synthesis in *Escherichia coli*. *J. Bacteriol.* 174:6061–6070.

1192. Feng, J., K. Yamanaka, H. Niki, T. Ogura, and S. Hiraga. 1994. New killing system controlled by two genes located immediately upstream of the *mukB* gene in *Escherichia coli*. *Mol. Gen. Genet.* 243:136–147.

1193. Ferenci, T. 1980. Methyl-α-maltoside and 5-thiomaltose: analogs transported by the *Escherichia coli* maltose transport system. *J. Bacteriol.* 144:7–11.

1194. Fernley, R. T., S. R. Lentz, and R. A. Bradshaw. 1981. Malate dehydrogenase: isolation from *E. coli* and comparison with the eukaryotic mitochondrial and cytoplasmic forms. *Biosci. Rep.* 1:497–507.

1195. Feulner, G., J. A. Gray, J. A. Kirschman, A. F. Lehner, A. B. Sadosky, D. A. Vlazny, J. Zhang, S. Zhao, and C. W. Hill. 1990. Structure of the *rhsA* locus from *Escherichia coli* K-12 and comparison of *rhsA* with other members of the *rhs* multigene family. *J. Bacteriol.* 172:446–456.

1196. Feutrier, J., M. Lepelletier, M. C. Pascal, and M. Chippaux. 1982. Tn10 insertions directed in the *pyr D-ser C* region and improved mapping of *pep N* in *Escherichia coli* K12. *Mol. Gen. Genet.* 185:518–519.

1197. Fiedler, W., and H. Rotering. 1985. Characterization of an *Escherichia coli* mdoB mutant strain unable to transfer *sn*-1-phosphoglycerol to membrane-derived oligosaccharides. *J. Biol. Chem.* 260:4799–4806.

1198. Fiedler, W., and H. Rotering. 1988. Properties of *Escherichia coli* mutants lacking membrane-derived oligosaccharides. *J. Biol. Chem.* 263:14684–14689.

1199. Figge, R. M., T. M. Ramseier, and M. H. Saier, Jr. 1994. The mannitol repressor (MtlR) of *Escherichia coli*. *J. Bacteriol.* 176:840–847.

1200. Figueroa, N., N. Wills, and L. Bossi. 1991. Common sequence determinants of the response of a prokaryotic promoter to DNA bending and supercoiling. *EMBO J.* 10:941–949.

1201. Fiil, N., and J. D. Friesen. 1968. Isolation of "relaxed" mutants of *Escherichia coli*. *J. Bacteriol.* 95:729–731.

1202. Fiil, N. P., J. D. Friesen, W. L. Downing, and P. P. Dennis. 1980. Post-transcriptional regulatory mutants in a ribosomal protein-RNA polymerase operon of *E. coli*. *Cell* 19:837–844.

1203. Fijalkowska, I. J., and R. M. Schaaper. 1993. Antimutator mutations in the alpha subunit of *Escherichia coli* DNA polymerase III: identification of the responsible mutations and alignment with other DNA polymerases. *Genetics* 134:1039–1044.

1204. Fikes, J. D., and P. J. Bassford, Jr. 1989. Novel *secA* alleles improve export of maltose-binding protein synthesized with a defective signal peptide. *J. Bacteriol.* 171:402–409.

1205. Filutowicz, M., W. Ross, J. Wild, and R. L. Gourse. 1992. Involvement of Fis protein in replication of the *Escherichia coli* chromosome. *J. Bacteriol.* 174:398–407.

1206. Fimmel, A. L., and B. A. Haddock. 1979. Use of *chlC-lac* fusions to determine regulation of gene *chlC* in *Escherichia coli* K-12. *J. Bacteriol.* 138:726–730.

1207. Fimmel, A. L., D. A. Jans, L. Langman, L. B. James, G. R. Ash, J. A. Downie, A. E. Senior, F. Gibson, and G. B. Cox. 1983. The F₁F₀-ATPase of *Escherichia coli*. Substitution of proline by leucine at position 64 in the c-subunit causes loss of oxidative phosphorylation. *Biochem. J.* 213:451–458.

1208. Fimmel, A. L., and R. E. Loughlin. 1977. Isolation and characterization of *cysK* mutants of *Escherichia coli* K12. *J. Gen. Microbiol.* 103:37–43.

1209. Finch, P., and P. T. Emmerson. 1983. Nucleotide sequence of the regulatory region of the *uvrD* gene of *Escherichia coli*. *Gene* 25:317–323.

1210. Finch, P. W., and P. T. Emmerson. 1984. The nucleotide sequence of the *uvrD* gene of *E. coli*. *Nucleic Acids Res.* 12:5789–5799.

1211. Finch, P. W., A. Storey, K. Brown, I. D. Hickson, and P. T. Emmerson. 1986. Complete nucleotide sequence of *recD*, the structural gene for the alpha subunit of exonuclease V of *Escherichia coli*. *Nucleic Acids Res.* 14:8583–8594.

1212. Finch, P. W., A. Storey, K. E. Chapman, K. Brown, I. D. Hickson, and P. T. Emmerson. 1986. Complete nucleotide sequence of the *Escherichia coli* recB gene. *Nucleic Acids Res.* 14:8573–8582.

1213. Finch, P. W., R. E. Wilson, K. Brown, I. D. Hickson, and P. T. Emmerson. 1986. Complete nucleotide sequence of the *Escherichia coli* ptr gene encoding protease III. *Nucleic Acids Res.* 14:7695–7703.

1214. Finch, P. W., R. E. Wilson, K. Brown, I. D. Hickson, A. E. Tomkinson, and P. T. Emmerson. 1986. Complete nucleotide sequence of the *Escherichia coli* recC gene and of the *thyA-recC* intergenic region. *Nucleic Acids Res.* 14:4437–4451.

1215. Fischer, E., K. Gunter, and V. Braun. 1989. Involvement of ExbB and TonB in transport across the outer membrane of *Escherichia coli*: phenotypic complementation of *exb* mutants by overexpressed *tonB* and physical stabilization of TonB by ExbB. *J. Bacteriol.* 171:5127–5134.

1216. Fischer, H., and R. Glockshuber. 1993. ATP hydrolysis is not stoichiometrically linked with proteolysis in the ATP-dependent protease La from *Escherichia coli*. *J. Biol. Chem.* 268:22502–22507.

1217. Fischer, M., and S. A. Short. 1982. The cloning of the *Escherichia coli* K-12 deoxyribonucleoside operon. *Gene* 17:291–298.

1218. Flachmann, R., N. Kunz, J. Seifert, M. Gutlich, F. J. Wientjes, A. Laufer, and H. G. Gassen. 1988. Molecular biology of pyridine nucleotide biosynthesis in *Escherichia coli*. Cloning and characterization of quinolinate synthesis genes *nadA* and *nadB*. *Eur. J. Biochem.* 175:221–228.

1219. Flamm, E. L., and R. A. Weisberg. 1985. Primary structure of the *hip* gene of *Escherichia coli* and of its product, the beta subunit of integration host factor. *J. Mol. Biol.* 183:117–128.

1220. Flamm, J. A., J. D. Friesen, and A. J. Otsuka. 1988. The nucleotide sequence of the *Escherichia coli rts* gene. *Gene* 74:555–558.

1221. Flannigan, K. A., S. H. Hennigan, H. H. Vogelbacker, J. S. Gots, and J. M. Smith. 1990. Purine biosynthesis in *Escherichia coli* K12: structure and DNA sequence studies of the *purHD* locus. *Mol. Microbiol.* 4:381–392.

1222. Fleming, T. P., M. S. Nahlik, and M. A. McIntosh. 1983. Regulation of enterobactin iron transport in *Escherichia coli*: characterization of *ent*::Mu d(Apr *lac*) operon fusions. *J. Bacteriol.* 156:1171–1177.

1223. Fleming, T. P., M. S. Nahlik, J. B. Neilands, and M. A. McIntosh. 1985. Physical and genetic characterization of cloned enterobactin genomic sequences from *Escherichia coli* K-12. *Gene* 34:47–54.

1224. Flensburg, J., and O. Skold. 1987. Massive overproduction of dihydrofolate reductase in bacteria as a response to the use of trimethoprim. *Eur. J. Biochem.* 162:473–476.

1225. Fletcher, A. J., A. P. Pugsley, and R. E. Glass. 1986. Investigation of the regulation of the *Escherichia coli* btuB gene using operon fusions. *J. Gen. Microbiol.* 132:2643–2646.

1226. Flinn, H., M. Burke, C. J. Stirling, and D. J. Sherratt. 1989. Use of gene replacement to construct *Escherichia coli* strains carrying mutations in two genes required for stability of multicopy plasmids. *J. Bacteriol.* 171:2241–2243.

1227. Flower, A. M., R. C. Doebele, and T. J. Silhavy. 1994. PrlA and PrlG suppressors reduce the requirement for signal sequence recognition. *J. Bacteriol.* 176:5607–5614.

1228. Flower, A. M., and C. S. McHenry. 1986. The adjacent *dnaZ* and *dnaX* genes of *Escherichia coli* are contained within one continuous open reading frame. *Nucleic Acids Res.* 14:8091–8101.

1229. Flower, A. M., and C. S. McHenry. 1990. The gamma subunit of DNA polymerase III holoenzyme of *Escherichia coli* is produced by ribosomal frameshifting. *Proc. Natl. Acad. Sci. USA* 87:3713–3717.

1230. Flower, A. M., and C. S. McHenry. 1991. Transcriptional organization of the *Escherichia coli dnaX* gene. *J. Mol. Biol.* 220:649–658.

1231. Fluri, R., L. E. Jackson, W. E. Lee, and I. P. Crawford. 1971. Tryptophan synthetase 2 subunit. Primary structure of the pyridoxyl peptide from the *Escherichia coli* enzyme. *J. Biol. Chem.* 246:6620–6624.

1232. Fogliano, M., and P. F. Schendel. 1981. Evidence for the inducibility of the *uvrB* operon. *Nature* (London) 289:196–198.

1233. Foglino, M., S. Gharbi, and A. Lazdunski. 1986. Nucleotide sequence of the *pepN* gene encoding aminopeptidase N of *Escherichia coli*. *Gene* 49:303–309.

1234. Fognini-Lefebvre, N., J. C. Lazzaroni, and R. Portalier. 1987. *tolA, tolB* and *excC*, three cistrons involved in the control of pleiotropic release of periplasmic proteins by *Escherichia coli* K12. *Mol. Gen. Genet.* 209:391–395.

1235. Fong, S. -T., J. Camakaris, and B. T. O. Lee. 1995. Molecular genetics of a chromosomal locus involved in copper tolerance in *Escherichia coli* K-12. *Mol. Microbiol.* 15:1127–1137.

1236. Fontecave, M., J. Coves, and J. L. Pierre. 1994. Ferric reductases or flavin reductases? *Biometals* 7:3–8.

1237. Forchhammer, K., W. Leinfelder, and A. Bock. 1989. Identification of a novel translation factor necessary for the incorporation of selenocysteine into protein. *Nature* (London) 342:453–456.

1238. Forchhammer, K., W. Leinfelder, K. Boesmiller, B. Veprek, and A. Bock. 1991. Selenocysteine synthase from *Escherichia coli*. Nucleotide sequence of the gene (*selA*) and purification of the protein. *J. Biol. Chem.* 266:6318–6323.

1239. **Forchhammer, K., K. P. Rucknagel, and A. Bock.** 1990. Purification and biochemical characterization of SELB, a translation factor involved in selenoprotein synthesis. *J. Biol. Chem.* **265:**9346–9350.

1240. **Forster, J. W., and P. Strike.** 1985. Organisation and control of the *Escherichia coli uvrC* gene. *Gene* **35:**71–82.

1241. **Forster, J. W., and P. Strike.** 1988. Analysis of the regulatory elements of the *Escherichia coli uvrC* gene by construction of operon fusions. *Mol. Gen. Genet.* **211:**531–537.

1242. **Fortin, Y., P. Phoenix, and G. R. Drapeau.** 1990. Mutations conferring resistance to azide in *Escherichia coli* occur primarily in the *secA* gene. *J. Bacteriol.* **172:**6607–6610.

1243. **Foster, S. J.** 1993. Purification and characterization of an 'actomyosin' complex from *Escherichia coli* W3110. *FEMS Microbiol. Lett.* **110:**295–298.

1244. **Fotheringham, I. G., S. A. Dacey, P. P. Taylor, T. J. Smith, M. G. Hunter, M. E. Finlay, S. B. Primrose, D. M. Parker, and R. M. Edwards.** 1986. The cloning and sequence analysis of the *aspC* and *tyrB* genes from *Escherichia coli* K12. Comparison of the primary structures of the aspartate aminotransferase and aromatic aminotransferase of *E. coli* with those of the pig aspartate aminotransferase isoenzymes. *Biochem. J.* **234:**593–604.

1245. **Foulds, J., and C. Barrett.** 1973. Characterization of *Escherichia coli* mutants tolerant to bacteriocin JF246: two new classes of tolerant mutants. *J. Bacteriol.* **116:**885–892.

1246. **Fourmy, D., Y. Mechulam, S. Brunie, S. Blanquet, and G. Fayat.** 1991. Identification of residues involved in the binding of methionine by *Escherichia coli* methionyl-tRNA synthetase. *FEBS Lett.* **292:**259–263.

1247. **Fourmy, D., T. Meinnel, Y. Mechulam, and S. Blanquet.** 1993. Mapping of the zinc binding domain of *Escherichia coli* methionyl-tRNA synthetase. *J. Mol. Biol.* **231:**1068–1077.

1248. **Fournier, M. J., and H. Ozeki.** 1985. Structure and organization of the transfer ribonucleic acid genes of *Escherichia coli* K-12. *Microbiol. Rev.* **49:**379–397.

1249. **Fouts, K. E., and S. D. Barbour.** 1981. Transductional mapping of *ksgB* and a new Tn*5*-induced kasugamycin resistance gene, *ksgD*, in *Escherichia coli* K-12. *J. Bacteriol.* **145:**914–919.

1250. **Fouts, K. E., T. Wasie-Gilbert, D. K. Willis, A. J. Clark, and S. D. Barbour.** 1983. Genetic analysis of transposon-induced mutations of the Rac prophage in *Escherichia coli* K-12 which affect expression and function of *recE*. *J. Bacteriol.* **156:**718–726.

1251. **Fowler, A. V., M. A. Hediger, R. E. Musso, and I. Zabin.** 1985. The amino acid sequence of thiogalactoside transacetylase of *Escherichia coli*. *Biochimie* **67:**101–108.

1252. **Fowler, A. V., and P. J. Smith.** 1983. The active site regions of *lacZ* and *ebg* beta-galactosidases are homologous. *J. Biol. Chem.* **258:**10204–10207.

1253. **Fowler, A. V., and I. Zabin.** 1978. Amino acid sequence of beta-galactosidase. XI. Peptide ordering procedures and the complete sequence. *J. Biol. Chem.* **253:**5521–5525.

1254. **Fowler, R. G., J. A. Erickson, and R. J. Isbell.** 1994. Activity of the *Escherichia coli mutT* mutator allele in an anaerobic environment. *J. Bacteriol.* **176:**7727–7729.

1255. **Fox, D. K., K. A. Presper, S. Adhya, S. Roseman, and S. Garges.** 1992. Evidence for two promoters upstream of the *pts* operon: regulation by the cAMP receptor protein regulatory complex. *Proc. Natl. Acad. Sci. USA* **89:**7056–7059.

1256. **Fradkin, J. E., and D. G. Fraenkel.** 1971. 2-Keto-3-deoxygluconate 6-phosphate aldolase mutants of *Escherichia coli*. *J. Bacteriol.* **108:**1277–1283.

1257. **Fradkin, L. G., and A. Kornberg.** 1992. Prereplicative complexes of components of DNA polymerase III holoenzyme of *Escherichia coli*. *J. Biol. Chem.* **267:**10318–10322.

1258. **Fraenkel, D. G., and S. Banerjee.** 1972. Deletion mapping of *zwf*, the gene for a constitutive enzyme, glucose 6-phosphate dehydrogenase in *Escherichia coli*. *Genetics* **71:**481–489.

1259. **Fralick, J. A.** 1991. Studies on the alteration of chromosome copy number and cell division potential in a *dnaA* mutant of *Escherichia coli*. *Mol. Gen. Genet.* **229:**175–180.

1260. **Francetić, O., M. P. Hanson, and C. A. Kumamoto.** 1993. *prlA* suppression of defective export of maltose-binding protein in *secB* mutants of *Escherichia coli*. *J. Bacteriol.* **175:**4036–4044.

1261. **Francois, V., J. Louarn, and J. M. Louarn.** 1989. The terminus of the *Escherichia coli* chromosome is flanked by several polar replication pause sites. *Mol. Microbiol.* **3:**995–1002.

1262. **Francois, V., J. Louarn, J. Patte, J. E. Rebollo, and J. M. Louarn.** 1990. Constraints in chromosomal inversions in *Escherichia coli* are not explained by replication pausing at inverted terminator-like sequences. *Mol. Microbiol.* **4:**537–542.

1263. **Francoz, E., and E. Dassa.** 1988. 3′ end of the *malEFG* operon in *E. coli:* localization of the transcription termination site. *Nucleic Acids Res.* **16:**4097–4109.

1264. **Frank-Kamenetskii, M.** 1989. Gene transcription. Waves of DNA supercoiling. *Nature* (London) **337:**206.

1265. **Franklin, F. C., W. A. Venables, and H. J. Wijsman.** 1981. Genetic studies of D-alanine-dehydrogenase-less mutants of *Escherichia coli* K12. *Genet. Res.* **38:**197–208.

1266. **Franklin, N. C.** 1969. Mutation in *gal U* gene of *E. coli* blocks phage P1 infection. *Virology* **38:**189–191.

1267. **Freedman, L. P., J. M. Zengel, R. H. Archer, and L. Lindahl.** 1987. Autogenous control of the S10 ribosomal protein operon of *Escherichia coli:* genetic dissection of transcriptional and posttranscriptional regulation. *Proc. Natl. Acad. Sci. USA* **84:**6516–6520.

1268. **Freedman, R., B. Gibson, D. Donovan, K. Biemann, S. Eisenbeis, J. Parker, and P. Schimmel.** 1985. Primary structure of histidine-tRNA synthetase and characterization of *hisS* transcripts. *J. Biol. Chem.* **260:**10063–10068.

1269. **Freitag, C. S., J. M. Abraham, J. R. Clements, and B. I. Eisenstein.** 1985. Genetic analysis of the phase variation control of expression of type 1 fimbriae in *Escherichia coli*. *J. Bacteriol.* **162:**668–675.

1270. **Freitag, C. S., and B. I. Eisenstein.** 1983. Genetic mapping and transcriptional orientation of the *fimD* gene. *J. Bacteriol.* **156:**1052–1058.

1271. **Freudl, R., G. Braun, N. Honore, and S. T. Cole.** 1987. Evolution of the enterobacterial *sulA* gene: a component of the SOS system encoding an inhibitor of cell division. *Gene* **52:**31–40.

1272. **Freundlieb, S., and W. Boos.** 1986. Alpha-amylase of *Escherichia coli*, mapping and cloning of the structural gene, *malS*, and identification of its product as a periplasmic protein. *J. Biol. Chem.* **261:**2946–2953.

1273. **Frick, L., C. Yang, V. E. Marquez, and R. Wolfenden.** 1989. Binding of pyrimidin-2-one ribonucleoside by cytidine deaminase as the transition-state analogue 3,4-dihydrouridine and the contribution of the 4-hydroxyl group to its binding affinity. *Biochemistry* **28:**9423–9430.

1274. **Fricke, J., J. Neuhard, R. A. Kelln, and S. Pedersen.** 1995. The *cmk* gene encoding cytidine monophosphate kinase is located in the *rpsA* operon and is required for normal replication rate in *Escherichia coli*. *J. Bacteriol.* **177:**517–523.

1275. **Friden, P., J. Donegan, J. Mullen, P. Tsui, M. Freundlich, L. Eoyang, R. Weber, and P. M. Silverman.** 1985. The *ilvB* locus of *Escherichia coli* K-12 is an operon encoding both subunits of acetohydroxyacid synthase I. *Nucleic Acids Res.* **13:**3979–3993.

1276. **Friden, P., T. Newman, and M. Freundlich.** 1982. Nucleotide sequence of the *ilvB* promoter-regulatory region: a biosynthetic operon controlled by attenuation and cyclic AMP. *Proc. Natl. Acad. Sci. USA* **79:**6156–6160.

1277. **Friedberg, I.** 1972. Localization of phosphoglucose isomerase in *Escherichia coli* and its relation to the induction of the hexose phosphate transport system. *J. Bacteriol.* **112:**1201–1205.

1278. **Friedman, D. I., M. Baumann, and L. S. Baron.** 1976. Cooperative effects of bacterial mutations affecting lambda N gene expression. I. Isolation and characterization of a *nusB* mutant. *Virology* **73:**119–127.

1279. **Friedman, D. I., E. J. Olson, D. Carver, and M. Gellert.** 1984. Synergistic effect of *himA* and *gyrB* mutations: evidence that *him* functions control expression of *ilv* and *xyl* genes. *J. Bacteriol.* **157:**484–489.

1280. **Friedman, S.** 1982. Bactericidal effect of 5-azacytidine on *Escherichia coli* carrying *Eco*RII restriction-modification enzymes. *J. Bacteriol.* **151:**262–268.

1281. **Friedrich, M. J., L. C. de Veaux, and R. J. Kadner.** 1986. Nucleotide sequence of the *btuCED* genes involved in vitamin B12 transport in *Escherichia coli* and homology with components of periplasmic-binding-protein-dependent transport systems. *J. Bacteriol.* **167:**928–934.

1282. **Friedrich, M. J., and R. J. Kadner.** 1987. Nucleotide sequence of the *uhp* region of *Escherichia coli*. *J. Bacteriol.* **169:**3556–3563.

1283. **Froehlich, B., and W. Epstein.** 1981. *Escherichia coli* mutants in which transcription is dependent on *recA* function. *J. Bacteriol.* **147:**1117–1120.

1284. **Frohler, J., A. Rechenmacher, J. Thomale, G. Nass, and A. Bock.** 1980. Genetic analysis of mutations causing borrelidin resistance by overproduction of threonyl-transfer ribonucleic acid synthetase. *J. Bacteriol.* **143:**1135–1141.

1285. **Froman, B. E., R. C. Tait, and L. D. Gottlieb.** 1989. Isolation and characterization of the phosphoglucose isomerase gene from *Escherichia coli*. *Mol. Gen. Genet.* **217:**126–131.

1286. **Froshauer, S., and J. Beckwith.** 1984. The nucleotide sequence of the gene for *malF* protein, an inner membrane component of the maltose transport system of *Escherichia coli*. Repeated DNA sequences are found in the *malE-malF* intercistronic region. *J. Biol. Chem.* **259:**10896–10903.

1287. **Frunzio, R., C. B. Bruni, and F. Blasi.** 1981. In vivo and in vitro detection of the leader RNA of the histidine operon of *Escherichia coli* K-12. *Proc. Natl. Acad. Sci. USA* **78:**2767–2771.

1288. **Frustaci, J. M., and M. R. O'Brian.** 1993. The *Escherichia coli visA* gene encodes ferrochelatase, the final enzyme of the heme biosynthetic pathway. *J. Bacteriol.* **175:**2154–2156.

1289. **Fu, H. A., S. Iuchi, and E. C. Lin.** 1991. The requirement of ArcA and Fnr for peak expression of the *cyd* operon in *Escherichia coli* under microaerobic conditions. *Mol. Gen. Genet.* **226:**209–213.

1290. **Fu, J. C., L. Ding, and S. Clarke.** 1991. Purification, gene cloning, and sequence analysis of an L-isoaspartyl protein carboxyl methyltransferase from *Escherichia coli*. *J. Biol. Chem.* **266:**14562–14572.

1291. **Fuchs, J. A., and H. O. Karlstrom.** 1976. Mapping of *nrdA* and *nrdB* in *Escherichia coli* K-12. *J. Bacteriol.* **128:**810–814.

1292. **Fuge, E. K., and S. B. Farr.** 1993. AppppA-binding protein E89 is the *Escherichia coli* heat shock protein ClpB. *J. Bacteriol.* **175:**2321–2326.

1293. **Fujisaki, S., H. Hara, Y. Nishimura, K. Horiuchi, and T. Nishino.** 1990. Cloning and nucleotide sequence of the *ispA* gene responsible for farnesyl diphosphate synthase activity in *Escherichia coli*. *J. Biochem.* **108:**995–1000.

1294. **Fujisaki, S., T. Nishino, H. Katsuki, H. Hara, Y. Nishimura, and Y. Hirota.** 1989. Isolation and characterization of an *Escherichia coli* mutant having temperature-sensitive farnesyl diphosphate synthase. *J. Bacteriol.* **171:**5654–5658.

1295. **Fujita, N., T. Miwa, S. Ishijima, K. Izui, and H. Katsuki.** 1984. The primary structure of phosphoenolpyruvate carboxylase of *Escherichia coli*. Nucleotide sequence of the *ppc* gene and deduced amino acid sequence. *J. Biochem.* **95:**909–916.

1296. Fujita, N., H. Mori, T. Yura, and A. Ishihama. 1994. Systematic sequencing of the *Escherichia coli* genome: analysis of the 2.4–4.1 min (110,917–193,643 bp) region. *Nucleic Acids Res.* 22:1637–1639.

1297. Fukuda, R., A. Nishimura, and H. Serizawa. 1988. Genetic mapping of the *Escherichia coli* gene for the stringent starvation protein and its dispensability for normal cell growth. *Mol. Gen. Genet.* 211:515–519.

1298. Fukuda, R., R. Yano, T. Fukui, T. Hase, A. Ishihama, and H. Matsubara. 1985. Cloning of the *Escherichia coli* gene for the stringent starvation protein. *Mol. Gen. Genet.* 201:151–157.

1299. Fulda, M., E. Heinz, and F. P. Wolter. 1994. The *fadD* gene of *Escherichia coli* K12 is located close to *rnd* at 39.6 min of the chromosomal map and is a new member of the AMP-binding protein family. *Mol. Gen. Genet.* 242:241–249.

1300. Fuller-Pace, F. V., S. M. Nicol, A. D. Reid, and D. P. Lane. 1993. DbpA: a DEAD box protein specifically activated by 23s rRNA. *EMBO J.* 12:3619–3626.

1301. Funatsu, G., M. Yaguchi, and B. Wittmann-Liebold. 1977. Primary stucture of protein S12 from the small *Escherichia coli* ribosomal subunit. *FEBS Lett.* 73:12–17.

1302. Funk, C. R., L. Zimniak, and W. Dowhan. 1992. The *pgpA* and *pgpB* genes of *Escherichia coli* are not essential: evidence for a third phosphatidylglycerophosphate phosphatase. *J. Bacteriol.* 174:205–213.

1303. Funnell, B. E. 1988. Participation of *Escherichia coli* integration host factor in the P1 plasmid partition system. *Proc. Natl. Acad. Sci. USA* 85:6657–6661.

1304. Furuchi, T., K. Kashiwagi, H. Kobayashi, and K. Igarashi. 1991. Characteristics of the gene for a spermidine and putrescine transport system that maps at 15 min on the *Escherichia coli* chromosome. *J. Biol. Chem.* 266:20928–20933.

1305. Furuya, N., and T. Komano. 1995. Specific binding of the NikA protein to one arm of 17-base-pair inverted repeat sequences within the *oriT* region of plasmid R64. *J. Bacteriol.* 177:46–51.

1306. Gaal, T., and R. L. Gourse. 1990. Guanosine 3′-diphosphate 5′-diphosphate is not required for growth rate-dependent control of rRNA synthesis in *Escherichia coli. Proc. Natl. Acad. Sci. USA* 87:5533–5537.

1307. Gage, D. J., and F. C. Neidhardt. 1993. Adaptation of *Escherichia coli* to the uncoupler of oxidative phosphorylation 2,4-dinitrophenol. *J. Bacteriol.* 175:7105–7108.

1308. Galindo, J. M., G. Guarneros, and F. M. De La Vega. 1994. Open reading frames flanking the peptidyl-tRNA hydrolase-encoding gene of *Escherichia coli. Gene* 151:153–156.

1309. Galinier, A., F. Bleicher, D. Negre, G. Perriere, B. Duclos, A. J. Cozzone, and J. C. Cortay. 1991. Primary structure of the intergenic region between *aceK* and *iclR* in the *Escherichia coli* chromosome. *Gene* 97:149–150.

1310. Gallagher, P. J., I. Schwartz, and D. Elseviers. 1984. Genetic mapping of *pheU*, an *Escherichia coli* gene for phenylalanine tRNA. *J. Bacteriol.* 158:762–763.

1311. Galloway, S. M., and C. R. Raetz. 1990. A mutant of *Escherichia coli* defective in the first step of endotoxin biosynthesis. *J. Biol. Chem.* 265:6394–6402.

1312. Gally, D. L., J. A. Bogan, B. I. Eisenstein, and I. C. Blomfield. 1993. Environmental regulation of the *fim* switch controlling type 1 fimbrial phase variation in *Escherichia coli* K-12: effects of temperature and media. *J. Bacteriol.* 175:6186–6193.

1313. Gally, D. L., T. J. Rucker, and I. C. Blomfield. 1994. The leucine-responsive regulatory protein binds to the fim switch to control phase variation of type 1 fimbrial expression in *Escherichia coli* K-12. *J. Bacteriol.* 176:5665–5672.

1314. Gambino, L., S. J. Gracheck, and P. F. Miller. 1993. Overexpression of the MarA positive regulator is sufficient to confer multiple antibiotic resistance in *Escherichia coli. J. Bacteriol.* 175:2888–2894.

1315. Gamer, J., H. Bujard, and B. Bukau. 1992. Physical interaction between heat shock proteins DnaK, DnaJ, and GrpE and the bacterial heat shock transcription factor sigma 32. *Cell* 69:833–842.

1316. Gampel, A., and A. Tzagoloff. 1989. Homology of aspartyl- and lysyl-tRNA synthetases. *Proc. Natl. Acad. Sci. USA* 86:6023–6027.

1317. Gan, K., K. Sankaran, M. G. Williams, M. Aldea, K. E. Rudd, S. R. Kushner, and H. C. Wu. 1995. The *umpA* gene of *Escherichia coli* encodes phosphatidylglycerol:prolipoprotein diacylglyceryl transferase (*lgt*) and regulates thymidylate synthase levels through translational coupling. *J. Bacteriol.* 177:1879–1882.

1318. Ganduri, Y. L., S. R. Sadda, M. W. Datta, R. K. Jambukeswaran, and P. Datta. 1993. TdcA, a transcriptional activator of the *tdcABC* operon of *Escherichia coli*, is a member of the LysR family of proteins. *Mol. Gen. Genet.* 240:395–402.

1319. Ganem, D., J. H. Miller, J. G. Files, T. Platt, and K. Weber. 1973. Reinitiation of a *lac* repressor fragment at a condon other than AUG. *Proc. Natl. Acad. Sci. USA* 70:3165–3169.

1319a. Ganesan, A. K., and B. Rotman. 1966. Transport systems for galactose and galactosides in *Escherichia coli*. I. Genetic determination and regulation of the methylgalactoside permease. *J. Mol. Biol.* 16:42–50.

1320. Ganong, B. R., J. M. Leonard, and C. R. Raetz. 1980. Phosphatidic acid accumulation in the membranes of *Escherichia coli* mutants defective in CDP-diglyceride synthetase. *J. Biol. Chem.* 255:1623–1629.

1321. Ganong, B. R., and C. R. Raetz. 1983. pH-sensitive CDP-diglyceride synthetase mutants of *Escherichia coli*: phenotypic suppression by mutations at a second site. *J. Bacteriol.* 153:731–738.

1322. Garcia del Portillo, F., M. A. de Pedro, and J. A. Ayala. 1991. Identification of a new mutation in *Escherichia coli* that suppresses a *pbpB* (Ts) phenotype in the presence of penicillin-binding protein 1B. *FEMS Microbiol. Lett.* 68:7–13.

1323. Garcia, G. A., K. A. Koch, and S. Chong. 1993. tRNA-guanine transglycosylase from *Escherichia coli*. Overexpression, purification and quaternary structure. *J. Mol. Biol.* 231:489–497.

1324. Garcia, G. M., P. K. Mar, D. A. Mullin, J. R. Walker, and N. E. Prather. 1986. The *E. coli dnaY* gene encodes an arginine transfer RNA. *Cell* 45:453–459.

1325. Garcia-Martin, C., L. Baldoma, J. Badia, and J. Aguilar. 1992. Nucleotide sequence of the *rhaR-sodA* interval specifying *rhaT* in *Escherichia coli. J. Gen. Microbiol.* 138:1109–1116.

1326. Garcia-Villegas, M. R., F. M. De La Vega, J. M. Galindo, M. Segura, R. H. Buckingham, and G. Guarneros. 1991. Peptidyl-tRNA hydrolase is involved in lambda inhibition of host protein synthesis. *EMBO J.* 10:3549–3555.

1327. Garciarrubio, A., E. Lozoya, A. Covarrubias, and F. Bolivar. 1983. Structural organization of the genes that encode two glutamate synthase subunits of *Escherichia coli. Gene* 26:165–170.

1328. Gardel, C., S. Benson, J. Hunt, S. Michaelis, and J. Beckwith. 1987. secD, a new gene involved in protein export in *Escherichia coli. J. Bacteriol.* 169:1286–1290.

1329. Gardel, C., K. Johnson, A. Jacq, and J. Beckwith. 1990. The *secD* locus of *E. coli* codes for two membrane proteins required for protein export. *EMBO J.* 9:3209–3216.

1330. Gardella, T., H. Moyle, and M. M. Susskind. 1989. A mutant *Escherichia coli* sigma 70 subunit of RNA polymerase with altered promoter specificity. *J. Mol. Biol.* 206:579–590.

1331. Gardina, P., C. Conway, M. Kossman, and M. Manson. 1992. Aspartate and maltose-binding protein interact with adjacent sites in the Tar chemotactic signal transducer of *Escherichia coli. J. Bacteriol.* 174:1528–1536.

1332. Gardner, P. R., and I. Fridovich. 1991. Superoxide sensitivity of the *Escherichia coli* aconitase. *J. Biol. Chem.* 266:19328–19333.

1333. Gardner, P. R., and I. Fridovich. 1992. Inactivation-reactivation of aconitase in *Escherichia coli*. A sensitive measure of superoxide radical. *J. Biol. Chem.* 267:8757–8763.

1334. Gardner, P. R., and I. Fridovich. 1993. Effect of glutathione on aconitase in *Escherichia coli. Arch. Biochem. Biophys.* 301:98–102.

1335. Garnant, M. K., and G. V. Stauffer. 1984. Construction and analysis of plasmids containing the *Escherichia coli serB* gene. *Mol. Gen. Genet.* 193:72–75.

1336. Garner, C. C., and K. M. Herrmann. 1985. Operator mutations of the *Escherichia coli aroF* gene. *J. Biol. Chem.* 260:3820–3825.

1337. Garrett, S., R. K. Taylor, and T. J. Silhavy. 1983. Isolation and characterization of chain-terminating nonsense mutations in a porin regulator gene, *envZ. J. Bacteriol.* 156:62–69.

1338. Garrett, S., R. K. Taylor, T. J. Silhavy, and M. L. Berman. 1985. Isolation and characterization of ΔompB strains of *Escherichia coli* by a general method based on gene fusions. *J. Bacteriol.* 162:840–844.

1339. Garrido, T., M. Sanchez, P. Palacios, M. Aldea, and M. Vicente. 1993. Transcription of *ftsZ* oscillates during the cell cycle of *Escherichia coli. EMBO J.* 12:3957–3965.

1340. Garrido-Pertierra, A., and R. A. Cooper. 1983. Evidence for two distinct pyruvate kinase genes in *Escherichia coli* K-12. *FEBS Lett.* 162:420–422.

1340a. Garvin, R. T., and L. Gorini. 1975. A new gene for ribosomal restriction in *Escherichia coli. Mol. Gen. Genet.* 137:73–78.

1341. Garwin, J. L., A. L. Klages, and J. E. Cronan, Jr. 1980. Structural, enzymatic, and genetic studies of beta-ketoacyl-acyl carrier protein synthases I and II of *Escherichia coli. J. Biol. Chem.* 255:11949–11956.

1342. Garwin, J. L., A. L. Klages, and J. E. Cronan, Jr. 1980. Beta-ketoacyl-acyl carrier protein synthase II of *Escherichia coli*. Evidence for function in the thermal regulation of fatty acid synthesis. *J. Biol. Chem.* 255:3263–3265.

1343. Gatti, D. L., and A. Tzagoloff. 1991. Structure and evolution of a group of related aminoacyl-tRNA synthetases. *J. Mol. Biol.* 218:557–568.

1344. Gaudu, P., D. Touati, V. Niviere, and M. Fontecave. 1994. The NAD(P)H:flavin oxidoreductase from *Escherichia coli* as a source of superoxide radicals. *J. Biol. Chem.* 269:8182–8188.

1345. Gavini, N., and B. E. Davidson. 1990. The *pheR* gene of *Escherichia coli* encodes tRNA(Phe), not a repressor protein. *J. Biol. Chem.* 265:21527–21531.

1346. Gavini, N., and B. E. Davidson. 1990. *pheAo* mutants of *Escherichia coli* have a defective *pheA* attenuator. *J. Biol. Chem.* 265:21532–21535.

1347. Gavini, N., and L. Pulakat. 1991. Role of ribosome release in the basal level of expression of the *Escherichia coli* gene *pheA. J. Gen. Microbiol.* 137:679–684.

1348. Gay, N. J., and J. E. Walker. 1981. The *atp* operon: nucleotide sequence of the promoter and the genes for the membrane proteins, and the delta subunit of *Escherichia coli* ATP-synthase. *Nucleic Acids Res.* 9:3919–3926.

1349. Gay, N. J., and J. E. Walker. 1981. The *atp* operon: nucleotide sequence of the region encoding the alpha-subunit of *Escherichia coli* ATP-synthase. *Nucleic Acids Res.* 9:2187–2194.

1350. Gayama, S., T. Kataoka, M. Wachi, G. Tamura, and K. Nagai. 1990. Periodic formation of the *oriC* complex of *Escherichia coli. EMBO J.* 9:3761–3765.

1351. Gayda, D. J., T. D. Leathers, J. D. Noti, F. J. Smith, J. M. Smith, C. S. Subrahmanyam, and H. E. Umbarger. 1980. Location of the multivalent control site for the *ilvEDA* operon of *Escherichia coli. J. Bacteriol.* 142:556–567.

1352. Gayda, R. C., M. C. Henk, and D. Leong. 1992. C-shaped cells caused by expression of an *ftsA* mutation in *Escherichia coli. J. Bacteriol.* 174:5362–5370.

1353. Gayda, R. C., P. E. Stephens, R. Hewick, J. M. Schoemaker, W. J. Dreyer, and A. Markovitz. 1985. Regulatory region of the heat shock-inducible *capR* (*lon*) gene: DNA and protein sequences. *J. Bacteriol.* 162:271–275.

1354. **Gayda, R. C., L. T. Yamamoto, and A. Markovitz.** 1976. Second-site mutations in *capR* (*lon*) strains of *Escherichia coli* K-12 that prevent radiation sensitivity and allow bacteriophage lambda to lysogenize. *J. Bacteriol.* **127:**1208–1216.

1355. **Gazeau, M., F. Delort, P. Dessen, S. Blanquet, and P. Plateau.** 1992. *Escherichia coli* leucine-responsive regulatory protein (Lrp) controls lysyl-tRNA synthetase expression. *FEBS Lett.* **300:**254–258.

1356. **Gebler, J. C., R. Aebersold, and S. G. Withers.** 1992. Glu-537, not Glu-461, is the nucleophile in the active site of (*lac Z*) beta-galactosidase from *Escherichia coli. J. Biol. Chem.* **267:**11126–11130.

1357. **Geerse, R. H., C. R. Ruig, A. R. Schuitema, and P. W. Postma.** 1986. Relationship between pseudo-HPr and the PEP:fructose phosphotransferase system in *Salmonella typhimurium* and *Escherichia coli. Mol. Gen. Genet.* **203:**435–444.

1358. **Geerse, R. H., J. van der Pluijm, and P. W. Postma.** 1989. The repressor of the PEP:fructose phosphotransferase system is required for the transcription of the *pps* gene of *Escherichia coli. Mol. Gen. Genet.* **218:**348–352.

1359. **Gegner, J. A., D. R. Graham, A. F. Roth, and F. W. Dahlquist.** 1992. Assembly of an MCP receptor, CheW, and kinase CheA complex in the bacterial chemotaxis signal transduction pathway. *Cell* **70:**975–982.

1360. **Gellert, M., and M. L. Bullock.** 1970. DNA ligase mutants of *Escherichia coli. Proc. Natl. Acad. Sci. USA* **67:**1580–1587.

1361. **Gellert, M., M. H. O'Dea, T. Itoh, and J. Tomizawa.** 1976. Novobiocin and coumermycin inhibit DNA supercoiling catalyzed by DNA gyrase. *Proc. Natl. Acad. Sci. USA* **73:**4474–4478.

1362. **Gemmill, R. M., J. W. Jones, G. W. Haughn, and J. M. Calvo.** 1983. Transcription initiation sites of the leucine operons of *Salmonella typhimurium* and *Escherichia coli. J. Mol. Biol.* **170:**39–59.

1363. **Gennity, J. M., H. Kim, and M. Inouye.** 1992. Structural determinants in addition to the amino-terminal sorting sequence influence membrane localization of *Escherichia coli* lipoproteins. *J. Bacteriol.* **174:**2095–2101.

1364. **Gent, M. E., S. Gartner, A. M. Gronenborn, R. Sandulache, and G. M. Clore.** 1987. Site-directed mutants of the cAMP receptor protein—DNA binding of five mutant proteins. *Protein Eng.* **1:**201–203.

1365. **Gentry, D., C. Bengra, K. Ikehara, and M. Cashel.** 1993. Guanylate kinase of *Escherichia coli* K-12. *J. Biol. Chem.* **268:**14316–14321.

1366. **Gentry, D., H. Xiao, R. Burgess, and M. Cashel.** 1991. The omega subunit of *Escherichia coli* K-12 RNA polymerase is not required for stringent RNA control in vivo. *J. Bacteriol.* **173:**3901–3903.

1367. **Gentry, D. R., and R. R. Burgess.** 1986. The cloning and sequence of the gene encoding the omega subunit of *Escherichia coli* RNA polymerase. *Gene* **48:**33–40.

1368. **Gentry, D. R., and R. R. Burgess.** 1989. *rpoZ*, encoding the omega subunit of *Escherichia coli* RNA polymerase, is in the same operon as *spoT. J. Bacteriol.* **171:**1271–1277.

1369. **George, A. M., and S. B. Levy.** 1983. Gene in the major cotransduction gap of the *Escherichia coli* K-12 linkage map required for the expression of chromosomal resistance to tetracycline and other antibiotics. *J. Bacteriol.* **155:**541–548.

1370. **George, S. E., and T. Melton.** 1986. Cloning and molecular characterization of *csm* mutations allowing expression of catabolite-repressible operons in the absence of exogenous cyclic AMP. *J. Bacteriol.* **166:**533–540.

1371. **Georgellis, D., S. Arvidson, and A. von Gabain.** 1992. Decay of *ompA* mRNA and processing of 9S RNA are immediately affected by shifts in growth rate, but in opposite manners. *J. Bacteriol.* **174:**5382–5390.

1372. **Georgiou, C. D., T. J. Dueweke, and R. B. Gennis.** 1988. Beta-galactosidase gene fusions as probes for the cytoplasmic regions of subunits I and II of the membrane-bound cytochrome *d* terminal oxidase from *Escherichia coli. J. Biol. Chem.* **263:**13130–13137.

1373. **Georgiou, C. D., H. Fang, and R. B. Gennis.** 1987. Identification of the *cydC* locus required for expression of the functional form of the cytochrome *d* terminal oxidase complex in *Escherichia coli. J. Bacteriol.* **169:**2107–2112.

1374. **Georgopoulos, C.** 1989. The *E. coli dnaA* initiation protein: a protein for all seasons. *Trends Genet.* **5:**319–321.

1375. **Georgopoulos, C., and D. Ang.** 1990. The *Escherichia coli groE* chaperonins. *Semin. Cell Biol.* **1:**19–25.

1376. **Georgopoulos, C. P.** 1971. Bacterial mutants in which the gene N function of bacteriophage lambda is blocked have an altered RNA polymerase. *Proc. Natl. Acad. Sci. USA* **68:**2977–2981.

1377. **Georgopoulos, C. P., R. W. Hendrix, A. D. Kaiser, and W. B. Wood.** 1972. Role of the host cell in bacteriophage morphogenesis: effects of a bacterial mutation on T4 head assembly. *Nature New. Biol.* **239:**38–41.

1378. **Georgopoulos, C. P., J. Swindle, F. Keppel, M. Ballivet, R. Bisig, and H. Eisen.** 1980. Studies on the *E. coli groNB* (*nusB*) gene which affects bacteriophage lambda *N* gene function. *Mol. Gen. Genet.* **179:**55–61.

1379. **Gerchman, Y., Y. Olami, A. Rimon, D. Taglicht, S. Schuldiner, and E. Padan.** 1993. Histidine-226 is part of the pH sensor of NhaA, a Na^+/H^+ antiporter in *Escherichia coli. Proc. Natl. Acad. Sci. USA* **90:**1212–1216.

1380. **Gerdes, K., F. W. Bech, S. T. Jorgensen, A. Lobner-Olesen, P. B. Rasmussen, T. Atlung, L. Boe, O. Karlstrom, S. Molin, and K. von Meyenburg.** 1986. Mechanism of postsegregational killing by the *hok* gene product of the *parB* system of plasmid R1 and its homology with the *relF* gene product of the *E. coli relB* operon. *EMBO J.* **5:**2023–2029.

1381. **Gerlach, P., P. Valentin-Hansen, and E. Bremer.** 1990. Transcriptional regulation of the *cytR* repressor gene of *Escherichia coli:* autoregulation and positive control by the cAMP/CAP complex. *Mol. Microbiol.* **4:**479–488.

1382. **Gervais, F. G., and G. R. Drapeau.** 1992. Identification, cloning, and characterization of *rcsF*, a new regulator gene for exopolysaccharide synthesis that suppresses the division mutation *ftsZ84* in *Escherichia coli* K-12. *J. Bacteriol.* **174:**8016–8022.

1383. **Gervais, F. G., P. Phoenix, and G. R. Drapeau.** 1992. The *rcsB* gene, a positive regulator of colanic acid biosynthesis in *Escherichia coli,* is also an activator of *ftsZ* expression. *J. Bacteriol.* **174:**3964–3971.

1384. **Ghosh, B., E. Grzadzielska, P. Bhattacharya, E. Peralta, J. DeVito, and A. Das.** 1991. Specificity of antitermination mechanisms. Suppression of the terminator cluster T1-T2 of *Escherichia coli* ribosomal RNA operon, *rrnB,* by phage lambda antiterminators. *J. Mol. Biol.* **222:**59–66.

1385. **Ghosh, P., C. Meyer, E. Remy, D. Peterson, and J. Preiss.** 1992. Cloning, expression, and nucleotide sequence of *glgC* gene from an allosteric mutant of *Escherichia coli* B. *Arch. Biochem. Biophys.* **296:**122–128.

1386. **Ghysen, A., and M. Pironio.** 1972. Relationship between the *N* function of bacteriophage lambda and host RNA polymerase. *J. Mol. Biol.* **65:**259–272.

1387. **Giaever, H. M., O. B. Styrvold, I. Kaasen, and A. R. Strom.** 1988. Biochemical and genetic characterization of osmoregulatory trehalose synthesis in *Escherichia coli. J. Bacteriol.* **170:**2841–2849.

1388. **Gibert, I., M. Llagostera, and J. Barbe.** 1988. Regulation of *ubiG* gene expression in *Escherichia coli. J. Bacteriol.* **170:**1346–1349.

1388a. **Gibson, F., and J. Pittard.** 1968. Pathways of biosynthesis of aromatic amino acids and vitamins and their control in microorganisms. *Bacteriol. Rev.* **32:**465–492.

1389. **Gibson, F. P., D. R. Leach, and R. G. Lloyd.** 1992. Identification of *sbcD* mutations as cosuppressors of *recBC* that allow propagation of DNA palindromes in *Escherichia coli* K-12. *J. Bacteriol.* **174:**1222–1228.

1390. **Giffard, P. M., and I. R. Booth.** 1988. The *rpoA341* allele of *Escherichia coli* specifically impairs the transcription of a group of positively-regulated operons. *Mol. Gen. Genet.* **214:**148–152.

1391. **Gigot, D., M. Crabeel, A. Feller, D. Charlier, W. Lissens, N. Glansdorff, and A. Pierard.** 1980. Patterns of polarity in the *Escherichia coli car AB* gene cluster. *J. Bacteriol.* **143:**914–920.

1392. **Giladi, H., M. Gottesman, and A. B. Oppenheim.** 1990. Integration host factor stimulates the phage lambda p_L promoter. *J. Mol. Biol.* **213:**109–121.

1393. **Giladi, H., S. Koby, M. E. Gottesman, and A. B. Oppenheim.** 1992. Supercoiling, integration host factor, and a dual promoter system, participate in the control of the bacteriophage lambda p_L promoter. *J. Mol. Biol.* **224:**937–948.

1394. **Gilchrist, C. A., and D. T. Denhardt.** 1987. *Escherichia coli rep* gene: sequence of the gene, the encoded helicase, and its homology with *uvrD. Nucleic Acids Res.* **15:**465–475.

1395. **Gill, D. R., G. F. Hatfull, and G. P. Salmond.** 1986. A new cell division operon in *Escherichia coli. Mol. Gen. Genet.* **205:**134–145.

1396. **Gill, D. R., and G. P. Salmond.** 1990. The identification of the *Escherichia coli ftsY* gene product: an unusual protein. *Mol. Microbiol.* **4:**575–583.

1397. **Gill, S. C., S. E. Weitzel, and P. H. von Hippel.** 1991. *Escherichia coli* sigma 70 and NusA proteins. I. Binding interactions with core RNA polymerase in solution and within the transcription complex. *J. Mol. Biol.* **220:**307–324.

1398. **Gill, S. C., T. D. Yager, and P. H. von Hippel.** 1991. *Escherichia coli* sigma 70 and NusA proteins. II. Physical properties and self-association states. *J. Mol. Biol.* **220:**325–333.

1399. **Gille, H., J. B. Egan, A. Roth, and W. Messer.** 1991. The FIS protein binds and bends the origin of chromosomal DNA replication, *oriC,* of *Escherichia coli. Nucleic Acids Res.* **19:**4167–4172.

1400. **Gilson, E., H. Nikaido, and M. Hofnung.** 1982. Sequence of the *malK* gene in *E. coli* K12. *Nucleic Acids Res.* **10:**7449–7458.

1401. **Gilson, E., J. P. Rousset, A. Charbit, D. Perrin, and M. Hofnung.** 1986. *malM,* a new gene of the maltose regulon in *Escherichia coli* K12. I. *malM* is the last gene of the *malK-lamB* operon and encodes a periplasmic protein. *J. Mol. Biol.* **191:**303–311.

1402. **Giorginis, S., and R. Chen.** 1977. The primary structure of protein L15 located at the peptidyltransferase center of *Escherichia coli* ribosomes. *FEBS Lett.* **84:**347–350.

1403. **Glaser, J. H., and J. A. DeMoss.** 1972. Comparison of nitrate reductase mutants of *Escherichia coli* selected by alternative procedures. *Mol. Gen. Genet.* **116:**1–10.

1404. **Glassberg, J., R. R. Meyer, and A. Kornberg.** 1979. Mutant single-strand binding protein of *Escherichia coli:* genetic and physiological characterization. *J. Bacteriol.* **140:**14–19.

1405. **Goddard, J. M., D. Caput, S. R. Williams, and D. W. Martin, Jr.** 1983. Cloning of human purine-nucleoside phosphorylase cDNA sequences by complementation in *Escherichia coli. Proc. Natl. Acad. Sci. USA* **80:**4281–4285.

1406. **Godessart, N., F. J. Munoa, M. Regue, and A. Juarez.** 1988. Chromosomal mutations that increase the production of a plasmid-encoded haemolysin in *Escherichia coli. J. Gen. Microbiol.* **134:**2779–2787.

1407. **Goldberg, E. B., T. Arbel, J. Chen, R. Karpel, G. A. Mackie, S. Schuldiner, and E. Padan.** 1987. Characterization of a Na^+/H^+ antiporter gene of *Escherichia coli. Proc. Natl. Acad. Sci. USA* **84:**2615–2619.

1408. **Goldie, A. H., and B. D. Sanwal.** 1980. Genetic and physiological characterization of *Escherichia coli* mutants deficient in phosphoenolpyruvate carboxykinase activity. *J. Bacteriol.* **141:**1115–1121.

1409. **Goldie, A. H., and B. D. Sanwal.** 1981. Temperature-sensitive mutation affecting synthesis of phosphoenolpyruvate carboxykinase in *Escherichia coli. J. Bacteriol.* **148:**720–723.

1410. **Goldie, H., and V. Medina.** 1990. Physical and genetic analysis of the phosphoenolpyruvate carboxykinase (*pckA*) locus from *Escherichia coli* K12. *Mol. Gen. Genet.* **220**:191–196.

1411. **Goldman, R. C., T. J. Bolling, W. E. Kohlbrenner, Y. Kim, and J. L. Fox.** 1986. Primary structure of CTP:CMP-3-deoxy-D-manno-octulosonate cytidylyltransferase (CMP-KDO synthetase) from *Escherichia coli*. *J. Biol. Chem.* **261**:15831–15835.

1412. **Goldman, R. C., and W. E. Kohlbrenner.** 1985. Molecular cloning of the structural gene coding for CTP:CMP-3-deoxy- manno-octulosonate cytidylyltransferase from *Escherichia coli* K-12. *J. Bacteriol.* **163**:256–261.

1413. **Goldstein, J., N. S. Pollitt, and M. Inouye.** 1990. Major cold shock protein of *Escherichia coli*. *Proc. Natl. Acad. Sci. USA* **87**:283–287.

1414. **Gollop, N., and P. E. March.** 1991. Localization of the membrane binding sites of Era in *Escherichia coli*. *Res. Microbiol.* **142**:301–307.

1415. **Gollop, N., H. Tavori, and Z. Barak.** 1982. Acetohydroxy acid synthase is a target for leucine containing peptide toxicity in *Escherichia coli*. *J. Bacteriol.* **149**:387–390.

1416. **Goncharoff, P., and B. P. Nichols.** 1984. Nucleotide sequence of *Escherichia coli pabB* indicates a common evolutionary origin of *p*-aminobenzoate synthetase and anthranilate synthetase. *J. Bacteriol.* **159**:57–62.

1417. **Gonzalez, J. C., R. V. Banerjee, S. Huang, J. S. Sumner, and R. G. Matthews.** 1992. Comparison of cobalamin-independent and cobalamin-dependent methionine synthases from *Escherichia coli*: two solutions to the same chemical problem. *Biochemistry* **31**:6045–6056.

1418. **Goodlove, P. E., P. R. Cunningham, J. Parker, and D. P. Clark.** 1989. Cloning and sequence analysis of the fermentative alcohol-dehydrogenase-encoding gene of *Escherichia coli*. *Gene* **85**:209–214.

1419. **Gopalakrishnan, A. S., Y. C. Chen, M. Temkin, and W. Dowhan.** 1986. Structure and expression of the gene locus encoding the phosphatidylglycerophosphate synthase of *Escherichia coli*. *J. Biol. Chem.* **261**:1329–1338.

1420. **Goransson, M., B. Sonden, P. Nilsson, B. Dagberg, K. Forsman, K. Emanuelsson, and B. E. Uhlin.** 1990. Transcriptional silencing and thermoregulation of gene expression in *Escherichia coli*. *Nature* (London) **344**:682–685.

1421. **Gordon, A. J., P. A. Burns, D. F. Fix, F. Yatagai, F. L. Allen, M. J. Horsfall, J. A. Halliday, J. Gray, C. Bernelot-Moens, and B. W. Glickman.** 1988. Missense mutation in the *lacI* gene of *Escherichia coli*. Inferences on the structure of the repressor protein. *J. Mol. Biol.* **200**:239–251.

1422. **Gordon, G., R. C. Gayda, and A. Markovitz.** 1984. Sequence of the regulatory region of *omp T*, the gene specifying major outer membrane protein a (3b) of *Escherichia coli* K-12: implications for regulation and processing. *Mol. Gen. Genet.* **193**:414–421.

1423. **Goshima, N., K. Kohno, F. Imamoto, and Y. Kano.** 1990. HU-1 mutants of *Escherichia coli* deficient in DNA binding. *Gene* **96**:141–145.

1424. **Gosink, K. K., W. Ross, S. Leirmo, R. Osuna, S. E. Finkel, R. C. Johnson, and R. L. Gourse.** 1993. DNA binding and bending are necessary but not sufficient for Fis-dependent activation of *rrnBp*$_1$. *J. Bacteriol.* **175**:1580–1589.

1425. **Goss, T. J., and P. Datta.** 1984. *Escherichia coli* K-12 mutation that inactivates biodegradative threonine dehydratase by transposon Tn5 insertion. *J. Bacteriol.* **158**:826–831.

1426. **Goss, T. J., and P. Datta.** 1985. Molecular cloning and expression of the biodegradative threonine dehydratase gene (*tdc*) of *Escherichia coli* K12. *Mol. Gen. Genet.* **201**:308–314.

1427. **Goss, T. J., H. P. Schweizer, and P. Datta.** 1988. Molecular characterization of the *tdc* operon of *Escherichia coli* K-12. *J. Bacteriol.* **170**:5352–5359.

1428. **Gosset, G., E. Merino, F. Recillas, G. Oliver, B. Becerril, and F. Bolivar.** 1989. Amino acid sequence analysis of the glutamate synthase enzyme from *Escherichia coli* K-12. *Protein Seq. Data Anal.* **2**:9–16.

1429. **Gots, J. S., C. E. Benson, and S. R. Shumas.** 1972. Genetic separation of hypoxanthine and guanine-xanthine phosphoribosyltransferase activities by deletion mutations in *Salmonella typhimurium*. *J. Bacteriol.* **112**:910–916.

1430. **Gottesman, S., and J. R. Beckwith.** 1969. Directed transposition of the arabinose operon: a technique for the isolation of specialized transducing bacteriophages for any *Escherichia coli* gene. *J. Mol. Biol.* **44**:117–127.

1431. **Gottesman, S., W. P. Clark, V. de Crecy-Lagard, and M. R. Maurizi.** 1993. ClpX, an alternative subunit for the ATP-dependent Clp protease of *Escherichia coli*. Sequence and in vivo activities. *J. Biol. Chem.* **268**:22618–22626.

1432. **Gottesman, S., W. P. Clark, and M. R. Maurizi.** 1990. The ATP-dependent Clp protease of *Escherichia coli*. Sequence of *clpA* and identification of a Clp-specific substrate. *J. Biol. Chem.* **265**:7886–7893.

1433. **Gottesman, S., C. Squires, E. Pichersky, M. Carrington, M. Hobbs, J. S. Mattick, B. Dalrymple, H. Kuramitsu, T. Shiroza, T. Foster, et al.** 1990. Conservation of the regulatory subunit for the Clp ATP-dependent protease in prokaryotes and eukaryotes. *Proc. Natl. Acad. Sci. USA* **87**:3513–3517.

1434. **Gottesman, S., P. Trisler, and A. Torres-Cabassa.** 1985. Regulation of capsular polysaccharide synthesis in *Escherichia coli* K-12: characterization of three regulatory genes. *J. Bacteriol.* **162**:1111–1119.

1435. **Gottlieb, P. A., S. Wu, X. Zhang, M. Tecklenburg, P. Kuempel, and T. M. Hill.** 1992. Equilibrium, kinetic, and footprinting studies of the Tus-Ter protein- DNA interaction. *J. Biol. Chem.* **267**:7434–7443.

1436. **Gough, J. A., and N. E. Murray.** 1983. Sequence diversity among related genes for recognition of specific targets in DNA molecules. *J. Mol. Biol.* **166**:1–19.

1437. **Gowrishankar, J.** 1985. Identification of osmoresponsive genes in *Escherichia coli*: evidence for participation of potassium and proline transport systems in osmoregulation. *J. Bacteriol.* **164**:434–445.

1438. **Gowrishankar, J.** 1986. *proP*-mediated proline transport also plays a role in *Escherichia coli* osmoregulation. *J. Bacteriol.* **166**:331–333.

1439. **Gowrishankar, J.** 1989. Nucleotide sequence of the osmoregulatory *proU* operon of *Escherichia coli*. *J. Bacteriol.* **171**:1923–1931. (Erratum, **172**:1165, 1990.)

1440. **Gowrishankar, J., P. Jayashree, and K. Rajkumari.** 1986. Molecular cloning of an osmoregulatory locus in *Escherichia coli*: increased *proU* gene dosage results in enhanced osmotolerance. *J. Bacteriol.* **168**:1197–1204.

1441. **Gowrishankar, J., and J. Pittard.** 1982. Molecular cloning of *pheR* in *Escherichia coli* K-12. *J. Bacteriol.* **152**:1–6.

1442. **Gowrishankar, J., and J. Pittard.** 1982. Construction from Mu d1 (*lac* Apr) lysogens of lambda bacteriophage bearing promoter-*lac* fusions: isolation of lambda p*pheA-lac*. *J. Bacteriol.* **150**:1122–1129.

1443. **Gowrishankar, J., and J. Pittard.** 1982. Regulation of phenylalanine biosynthesis in *Escherichia coli* K-12: control of transcription of the *pheA* operon. *J. Bacteriol.* **150**:1130–1137.

1444. **Grabau, C., and J. E. Cronan, Jr.** 1984. Molecular cloning of the gene (*poxB*) encoding the pyruvate oxidase of *Escherichia coli*, a lipid-activated enzyme. *J. Bacteriol.* **160**:1088–1092.

1445. **Grabau, C., and J. E. Cronan, Jr.** 1986. Nucleotide sequence and deduced amino acid sequence of *Escherichia coli* pyruvate oxidase, a lipid-activated flavoprotein. *Nucleic Acids Res.* **14**:5449–5460.

1446. **Graeme-Cook, K. A., G. May, E. Bremer, and C. F. Higgins.** 1989. Osmotic regulation of porin expression: a role for DNA supercoiling. *Mol. Microbiol.* **3**:1287–1294.

1447. **Grafstrom, R. H., and R. H. Hoess.** 1983. Cloning of *mutH* and identification of the gene product. *Gene* **22**:245–253.

1448. **Grafstrom, R. H., and R. H. Hoess.** 1987. Nucleotide sequence of the *Escherichia coli mutH* gene. *Nucleic Acids Res.* **15**:3073–3084.

1449. **Gragerov, A., E. Nudler, N. Komissarova, G. A. Gaitanaris, M. E. Gottesman, and V. Nikiforov.** 1992. Cooperation of GroEL/GroES and DnaK/DnaJ heat shock proteins in preventing protein misfolding in *Escherichia coli*. *Proc. Natl. Acad. Sci. USA* **89**:10341–10344.

1450. **Granston, A. E., D. L. Thompson, and D. I. Friedman.** 1990. Identification of a second promoter for the *metY-nusA-infB* operon of *Escherichia coli*. *J. Bacteriol.* **172**:2336–2342.

1451. **Gratia, J. P.** 1966. Studies on defective lysogeny due to chromosomal deletion in *Escherichia coli*. I. Single lysogens. *Biken J.* **9**:77–87.

1452. **Graves, R. J., I. Felzenszwalb, J. Laval, and T. R. O'Connor.** 1992. Excision of 5′-terminal deoxyribose phosphate from damaged DNA is catalyzed by the Fpg protein of *Escherichia coli*. *J. Biol. Chem.* **267**:14429–14435.

1453. **Gray, J. E., D. C. Bennett, H. E. Umbarger, and D. H. Calhoun.** 1982. Physical and genetic localization of *ilv* regulatory sites in lambda *ilv* bacteriophages. *J. Bacteriol.* **149**:1071–1081.

1454. **Green, G. N., H. Fang, R. J. Lin, G. Newton, M. Mather, C. D. Georgiou, and R. B. Gennis.** 1988. The nucleotide sequence of the *cyd* locus encoding the two subunits of the cytochrome *d* terminal oxidase complex of *Escherichia coli*. *J. Biol. Chem.* **263**:13138–13143.

1455. **Green, G. N., and R. B. Gennis.** 1983. Isolation and characterization of an *Escherichia coli* mutant lacking cytochrome *d* terminal oxidase. *J. Bacteriol.* **154**:1269–1275.

1456. **Green, G. N., J. E. Kranz, and R. B. Gennis.** 1984. Cloning the *cyd* gene locus coding for the cytochrome *d* complex of *Escherichia coli*. *Gene* **32**:99–106.

1457. **Green, G. N., R. G. Kranz, R. M. Lorence, and R. B. Gennis.** 1984. Identification of subunit I as the cytochrome b_{558} component of the cytochrome *d* terminal oxidase complex of *Escherichia coli*. *J. Biol. Chem.* **259**:7994–7997.

1458. **Green, G. N., R. M. Lorence, and R. B. Gennis.** 1986. Specific overproduction and purification of the cytochrome b_{558} component of the cytochrome *d* complex from *Escherichia coli*. *Biochemistry* **25**:2309–2314.

1459. **Green, J., M. Trageser, S. Six, G. Unden, and J. R. Guest.** 1991. Characterization of the FNR protein of *Escherichia coli*, an iron-binding transcriptional regulator. *Proc. R. Soc. Lond. [Biol.]* **244**:137–144.

1460. **Green, J. M., W. K. Merkel, and B. P. Nichols.** 1992. Characterization and sequence of *Escherichia coli pabC*, the gene encoding aminodeoxychorismate lyase, a pyridoxal phosphate-containing enzyme. *J. Bacteriol.* **174**:5317–5323.

1461. **Green, J. M., and B. P. Nichols.** 1991. *p*-Aminobenzoate biosynthesis in *Escherichia coli*. Purification of aminodeoxychorismate lyase and cloning of *pabC*. *J. Biol. Chem.* **266**:12971–12975.

1462. **Green, P. J., and M. Inouye.** 1984. Roles of the 5′ leader region of the *ompA* mRNA. *J. Mol. Biol.* **176**:431–442.

1463. **Greenberg, J. T., J. H. Chou, P. A. Monach, and B. Demple.** 1991. Activation of oxidative stress genes by mutations at the *soxQ/cfxB/marA* locus of *Escherichia coli*. *J. Bacteriol.* **173**:4433–4439.

1464. **Greenberg, J. T., and B. Demple.** 1986. Glutathione in *Escherichia coli* is dispensable for resistance to H_2O_2 and gamma radiation. *J. Bacteriol.* **168**:1026–1029.

1465. **Greenberg, J. T., P. Monach, J. H. Chou, P. D. Josephy, and B. Demple.** 1990. Positive control of a global antioxidant defense regulon activated by superoxide-generating agents in *Escherichia coli*. *Proc. Natl. Acad. Sci. USA* **87**:6181–6185.

1466. **Greenblatt, J., J. Li, S. Adhya, D. I. Friedman, L. S. Baron, B. Redfield, H. F. Kung, and H. Weissbach.** 1980. L factor that is required for beta-galactosidase synthesis is the *nusA* gene product involved in transcription termination. *Proc. Natl. Acad. Sci. USA* 77:1991–1994.

1467. Greenblatt, J., M. McLimont, and S. Hanly. 1981. Termination of transcription by *nusA* gene protein of *Escherichia coli*. *Nature* (London) 292:215–220.

1468. Greene, R. C., J. S. Hunter, and E. H. Coch. 1973. Properties of *metK* mutants of *Escherichia coli* K-12. *J. Bacteriol.* 115:57–67.

1469. Greene, R. C., J. H. Krueger, and J. R. Johnson. 1982. Localization of the *metJBLF* gene cluster of *Escherichia coli* in lambda *met* transducing phage. *Mol. Gen. Genet.* 187:401–404.

1470. Greene, R. C., and A. A. Smith. 1984. Insertion mutagenesis of the *metJBLF* gene cluster of *Escherichia coli* K-12: evidence for an *metBL* operon. *J. Bacteriol.* 159:767–769.

1471. Greener, A., and C. W. Hill. 1980. Identification of a novel genetic element in *Escherichia coli* K-12. *J. Bacteriol.* 144:312–321.

1472. Greener, T., D. Govezensky, and A. Zamir. 1993. A novel multicopy suppressor of a *groEL* mutation includes two nested open reading frames transcribed from different promoters. *EMBO J.* 12:889–896.

1473. Greenquist, A. C., and J. C. Wriston, Jr. 1972. Chemical evidence for identical subunits in L-asparaginase from *Escherichia coli* B. *Arch. Biochem. Biophys.* 152:280–286.

1474. Greer, S., and R. N. Perham. 1986. Glutathione reductase from *Escherichia coli*: cloning and sequence analysis of the gene and relationship to other flavoprotein disulfide oxidoreductases. *Biochemistry* 25:2736–2742.

1475. Greiner, R., U. Konietzny, and K. D. Jany. 1993. Purification and characterization of two phytases from *Escherichia coli*. *Arch. Biochem. Biophys.* 303:107–113.

1476. Grentzmann, G., D. Brechemier-Baey, V. Heurgue, L. Mora, and R. H. Buckingham. 1994. Localization and characterization of the gene encoding release factor RF3 in *Escherichia coli*. *Proc. Natl. Acad. Sci. USA* 91:5848–5852.

1477. Griffin, H. G., and M. J Gasson. 1995. The gene (*aroK*) encoding shikimate kinase I from *Escherichia coli*. *DNA Seq.* 5:195–197.

1478. Griffin, T. J., and R. D. Kolodner. 1990. Purification and preliminary characterization of the *Escherichia coli* K-12 recF protein. *J. Bacteriol.* 172:6291–6299.

1479. Griffo, G., A. B. Oppenheim, and M. E. Gottesman. 1989. Repression of the lambda *pcin* promoter by integrative host factor. *J. Mol. Biol.* 209:55–64.

1480. Griggs, D. W., K. Kafka, C. D. Nau, and J. Konisky. 1990. Activation of expression of the *Escherichia coli* cir gene by an iron-independent regulatory mechanism involving cyclic AMP-cyclic AMP receptor protein complex. *J. Bacteriol.* 172:3529–3533.

1481. Griggs, D. W., B. B. Tharp, and J. Konisky. 1987. Cloning and promoter identification of the iron-regulated *cir* gene of *Escherichia coli*. *J. Bacteriol.* 169:5343–5352.

1482. Grimm, B., A. Bull, and V. Breu. 1991. Structural genes of glutamate 1-semialdehyde aminotransferase for porphyrin synthesis in a cyanobacterium and *Escherichia coli*. *Mol. Gen. Genet.* 225:1–10.

1482a.Gringauz, E., K. A. Orle, C. S. Waddell, and N. L. Craig. 1988. Recognition of *Escherichia coli* attTn7 by transposon Tn7: lack of specific sequence requirements at the point of Tn7 insertion. *J. Bacteriol.* 170:2832–2840.

1483. Grisolia, V., M. S. Carlomagno, A. G. Nappo, and C. B. Bruni. 1985. Cloning, structure, and expression of the *Escherichia coli* K-12 hisC gene. *J. Bacteriol.* 164:1317–1323.

1484. Grisolia, V., A. Riccio, and C. B. Bruni. 1983. Structure and function of the internal promoter (*hisBp*) of the *Escherichia coli* K-12 histidine operon. *J. Bacteriol.* 155:1288–1296.

1485. Groarke, J. M., W. C. Mahoney, J. N. Hope, C. E. Furlong, F. T. Robb, H. Zalkin, and M. A. Hermodson. 1983. The amino acid sequence of D-ribose-binding protein from *Escherichia coli* K12. *J. Biol. Chem.* 258:12952–12956.

1486. Groat, R. G., J. E. Schultz, E. Zychlinsky, A. Bockman, and A. Matin. 1986. Starvation proteins in *Escherichia coli*: kinetics of synthesis and role in starvation survival. *J. Bacteriol.* 168:486–493. (Erratum, 169:3866, 1987.)

1487. Grodberg, J., and J. J. Dunn. 1988. *ompT* encodes the *Escherichia coli* outer membrane protease that cleaves T7 RNA polymerase during purification. *J. Bacteriol.* 170:1245–1253.

1488. Grodberg, J., M. D. Lundrigan, D. L. Toledo, W. F. Mangel, and J. J. Dunn. 1988. Complete nucleotide sequence and deduced amino acid sequence of the *ompT* gene of *Escherichia coli* K-12. *Nucleic Acids Res.* 16:1209.

1489. Grogan, D. W., and J. E. Cronan, Jr. 1983. Use of lambda phasmids for deletion mapping of non-selectable markers cloned in plasmids. *Gene* 22:75–83.

1490. Grogan, D. W., and J. E. Cronan, Jr. 1984. Genetic characterization of the *Escherichia coli* cyclopropane fatty acid (*cfa*) locus and neighboring loci. *Mol. Gen. Genet.* 196:367–372.

1491. Grogan, D. W., and J. E. Cronan, Jr. 1984. Cloning and manipulation of the *Escherichia coli* cyclopropane fatty acid synthase gene: physiological aspects of enzyme overproduction. *J. Bacteriol.* 158:286–295.

1492. Groisman, E. A, F. Heffron, and F. Solomon. 1992. Molecular genetic analysis of the *Escherichia coli* phoP locus. *J. Bacteriol.* 174:486–491.

1493. Gromiha, M. M., and P. K. Ponnuswamy. 1993. Prediction of transmembrane beta-strands from hydrophobic characteristics of proteins. *Int. J. Pept. Protein Res.* 42:420–431.

1494. Gross, C. A., F. R. Blattner, W. E. Taylor, P. A. Lowe, and R. R. Burgess. 1979. Isolation and characterization of transducing phage coding for sigma subunit of *Escherichia coli* RNA polymerase. *Proc. Natl. Acad. Sci. USA* 76:5789–5793.

1495. Grossman, A. D., J. W. Erickson, and C. A. Gross. 1984. The *htpR* gene product of *E. coli* is a sigma factor for heat-shock promoters. *Cell* 38:383–390.

1496. Grossman, A. D., Y. N. Zhou, C. Gross, J. Heilig, G. E. Christie, and R. Calendar. 1985. Mutations in the *rpoH* (*htpR*) gene of *Escherichia coli* K-12 phenotypically suppress a temperature-sensitive mutant defective in the sigma 70 subunit of RNA polymerase. *J. Bacteriol.* 161:939–943.

1497. Grossman, T. H., M. Tuckman, S. Ellestad, and M. S. Osburne. 1993. Isolation and characterization of *Bacillus subtilis* genes involved in siderophore biosynthesis: relationship between *B. subtilis* sfpo and *Escherichia coli* entD genes. *J. Bacteriol.* 175:6203–6211.

1498. Grothe, S., R. L. Krogsrud, D. J. McClellan, J. L. Milner, and J. M. Wood. 1986. Proline transport and osmotic stress response in *Escherichia coli* K-12. *J. Bacteriol.* 166:253–259. (Erratum, 167:1098.)

1498a.Gruer, M. J., and J. R. Guest. 1994. Two genetically-distinct and differentially-regulated aconitases (AcnA and AcnB) in *Escherichia coli*. *Microbiology* 140:2531–2541.

1499. Grundstrom, T., and B. Jaurin. 1982. Overlap between *ampC* and *frd* operons on the *Escherichia coli* chromosome. *Proc. Natl. Acad. Sci. USA* 79:1111–1115.

1500. Grundstrom, T., B. Jaurin, T. Edlund, and S. Normark. 1980. Physical mapping and expression of hybrid plasmids carrying chromosomal beta-lactamase genes of *Escherichia coli* K-12. *J. Bacteriol.* 143:1127–1134.

1501. Guardiola, J., M. De Felice, T. Klopotowski, and M. Iaccarino. 1974. Mutations affecting the different transport systems for isoleucine, leucine, and valine in *Escherichia coli* K-12. *J. Bacteriol.* 117:393–405.

1502. Guardiola, J., and M. Iaccarino. 1971. *Escherichia coli* K-12 mutants altered in the transport of branched-chain amino acids. *J. Bacteriol.* 108:1034–1044.

1503. Guarneros, G., G. Machado, P. Guzman, and E. Garay. 1987. Genetic and physical location of the *Escherichia coli* rap locus, which is essential for growth of bacteriophage lambda. *J. Bacteriol.* 169:5188–5192.

1504. Gudmundsdottir, A., P. E. Bell, M. D. Lundrigan, C. Bradbeer, and R. J. Kadner. 1989. Point mutations in a conserved region (TonB box) of *Escherichia coli* outer membrane protein BtuB affect vitamin B_{12} transport. *J. Bacteriol.* 171:6526–6533.

1505. Gudmundsdottir, A., C. Bradbeer, and R. J. Kadner. 1988. Altered binding and transport of vitamin B_{12} resulting from insertion mutations in the *Escherichia coli* btuB gene. *J. Biol. Chem.* 263:14224–14230.

1506. Guerrier-Takada, C., and S. Altman. 1992. Reconstitution of enzymatic activity from fragments of M1 RNA. *Proc. Natl. Acad. Sci. USA* 89:1266–1270.

1507. Guest, J. R. 1969. Biochemical and genetic studies with nitrate reductase C-gene mutants of *Escherichia coli*. *Mol. Gen. Genet.* 105:285–297.

1508. Guest, J. R. 1979. Anaerobic growth of *Escherichia coli* K12 with fumarate as terminal electron acceptor. Genetic studies with menaquinone- and fluoroacetate-resistant mutants. *J. Gen. Microbiol.* 115:259–271.

1509. Guest, J. R. 1981. Hybrid plasmids containing the citrate synthase gene (*gltA*) of *Escherichia coli* K12. *J. Gen. Microbiol.* 124:17–23.

1510. Guest, J. R., S. J. Angier, and G. C. Russell. 1989. Structure, expression, and protein engineering of the pyruvate dehydrogenase complex of *Escherichia coli*. *Ann. N.Y. Acad. Sci.* 573:76–99.

1511. Guest, J. R., G. R. Drapeau, B. C. Carlton, and C. Yanofsky. 1967. The amino acid sequence of the A protein (alpha subunit) of the tryptophan synthetase of *Escherichia coli*. *J. Biol. Chem.* 242:5442–5446.

1512. Guest, J. R., J. S. Miles, R. E. Roberts, and S. A. Woods. 1985. The fumarase genes of *Escherichia coli*: location of the *fumB* gene and discovery of a new gene (*fumC*). *J. Gen. Microbiol.* 131:2971–2984.

1513. Guest, J. R., and R. E. Roberts. 1983. Cloning, mapping, and expression of the fumarase gene of *Escherichia coli* K-12. *J. Bacteriol.* 153:588–596.

1514. Guest, J. R., R. E. Roberts, and P. E. Stephens. 1983. Hybrid plasmids containing the pyruvate dehydrogenase complex genes and gene-DNA relationships in the 2 to 3 minute region of the *Escherichia coli* chromosome. *J. Gen. Microbiol.* 129:671–680.

1515. Guest, J. R., R. E. Roberts, and R. J. Wilde. 1984. Cloning of the aspartase gene (*aspA*) of *Escherichia coli*. *J. Gen. Microbiol.* 130:1271–1278.

1516. Guest, J. R., and D. J. Shaw. 1981. Molecular cloning of menaquinone biosynthetic genes of *Escherichia coli* K12. *Mol. Gen. Genet.* 181:379–383.

1517. Guest, J. R., and P. E. Stephens. 1980. Molecular cloning of the pyruvate dehydrogenase complex genes of *Escherichia coli*. *J. Gen. Microbiol.* 121:277–292.

1518. Guidi-Rontani, C., A. Danchin, and A. Ullmann. 1981. Isolation and characterization of an *Escherichia coli* mutant affected in the regulation of adenylate cyclase. *J. Bacteriol.* 148:753–761.

1519. Guillon, J. M., Y. Mechulam, S. Blanquet, and G. Fayat. 1993. Importance of formylability and anticodon stem sequence to give a tRNA^{Met} an initiator identity in *Escherichia coli*. *J. Bacteriol.* 175:4507–4514.

1520. Guillon, J. M., Y. Mechulam, J. M. Schmitter, S. Blanquet, and G. Fayat. 1992. Disruption of the gene for Met-tRNA^{fMet} formyltransferase severely impairs growth of *Escherichia coli*. *J. Bacteriol.* 174:4294–4301.

1521. Guilloton, M., and F. Karst. 1987. Isolation and characterization of *Escherichia coli* mutants lacking inducible cyanase. *J. Gen. Microbiol.* 133:645–653.

1522. Guilloton, M. B., J. J. Korte, A. F. Lamblin, J. A. Fuchs, and P. M. Anderson. 1992. Carbonic anhydrase in *Escherichia coli*. A product of the *cyn* operon. *J. Biol. Chem.* 267:3731–3734.

1523. Gunsalus, R. P., L. V. Kalman, and R. R. Stewart. 1989. Nucleotide sequence of the *narL* gene that is involved in global regulation of nitrate controlled respiratory genes of *Escherichia coli*. *Nucleic Acids Res.* 17:1965–1975.

1524. **Gunsalus, R. P., and C. Yanofsky.** 1980. Nucleotide sequence and expression of *Escherichia coli trpR*, the structural gene for the *trp* aporepressor. *Proc. Natl. Acad. Sci. USA* 77:7117–7121.

1525. **Gunsalus, R. P., G. Zurawski, and C. Yanofsky.** 1979. Structural and functional analysis of cloned deoxyribonucleic acid containing the *trpR-thr* region of the *Escherichia coli* chromosome. *J. Bacteriol.* 140:106–113.

1526. **Gunther, E., M. Bagdasarian, and H. Schuster.** 1984. Cloning of the *dnaB* gene of *Escherichia coli*: the *dnaB* gene of *groPB534* and *groPB612* and the replication of phage lambda. *Mol. Gen. Genet.* 193:225–230.

1527. **Gupta, S., and D. P. Clark.** 1989. *Escherichia coli* derivatives lacking both alcohol dehydrogenase and phosphotransacetylase grow anaerobically by lactate fermentation. *J. Bacteriol.* 171:3650–3655.

1528. **Gupta, S. D., K. Gan, M. B. Schmid, and H. C. Wu.** 1993. Characterization of a temperature-sensitive mutant of *Salmonella typhimurium* defective in apolipoprotein N-acyltransferase. *J. Biol. Chem.* 268:16551–16556.

1529. **Gupta, S. D., B. T. O. Lee, J. Camakaria, and H. C. Wu.** 1995. Identification of *cutC* and *cutF* (*nlpE*) genes involved in copper tolerance in *Escherichia coli*. *J. Bacteriol.* 177:4207–4215.

1530. **Guranowski, A., H. Jakubowski, and E. Holler.** 1983. Catabolism of diadenosine 5′,5‴-P_1,P_4-tetraphosphate in procaryotes. Purification and properties of diadenosine 5′,5‴-P_1,P_4-tetraphosphate (symmetrical) pyrophosphohydrolase from *Escherichia coli* K12. *J. Biol. Chem.* 258:14784–14789.

1531. **Gushima, H., S. Yasuda, E. Soeda, M. Yokota, M. Kondo, and A. Kimura.** 1984. Complete nucleotide sequence of the *E. coli* glutathione synthetase *gsh-II*. *Nucleic Acids Res.* 12:9299–9307.

1532. **Gussin, G. N., C. Olson, K. Igarashi, and A. Ishihama.** 1992. Activation defects caused by mutations in *Escherichia coli rpoA* are promoter specific. *J. Bacteriol.* 174:5156–5160.

1533. **Gustafson, C. E., S. Kaul, and E. E. Ishiguro.** 1993. Identification of the *Escherichia coli lytB* gene, which is involved in penicillin tolerance and control of the stringent response. *J. Bacteriol.* 175:1203–1205.

1534. **Gustafsson, C., P. H. Lindstrom, T. G. Hagervall, K. B. Esberg, and G. R. Bjork.** 1991. The *trmA* promoter has regulatory features and sequence elements in common with the rRNA P1 promoter family of *Escherichia coli*. *J. Bacteriol.* 173:1757–1764.

1535. **Gustafsson, C., and S. R. Warne.** 1992. Physical map of the *oxyR-trmA* region (minute 89.3) of the *Escherichia coli* chromosome. *J. Bacteriol.* 174:7878–7879.

1536. **Guterman, S. K., and L. Dann.** 1973. Excretion of enterochelin by *exbA* and *exbB* mutants of *Escherichia coli*. *J. Bacteriol.* 114:1225–1230.

1537. **Guterman, S. K., G. Roberts, and B. Tyler.** 1982. Polarity in the *glnA* operon: suppression of the Reg⁻ phenotype by *rho* mutations. *J. Bacteriol.* 150:1314–1321.

1538. **Guth, A., R. Engel, and B. E. Tropp.** 1980. Uptake of glycerol 3-phosphate and some of its analogs by the hexose phosphate transport system of *Escherichia coli*. *J. Bacteriol.* 143:538–539.

1539. **Gutheil, W. G., B. Holmquist, and B. L. Vallee.** 1992. Purification, characterization, and partial sequence of the glutathione-dependent formaldehyde dehydrogenase from *Escherichia coli*: a class III alcohol dehydrogenase. *Biochemistry* 31:475–481.

1540. **Guthrie, B., and W. Wickner.** 1990. Trigger factor depletion or overproduction causes defective cell division but does not block protein export. *J. Bacteriol.* 172:5555–5562.

1541. **Gutierrez, C., M. Ardourel, E. Bremer, A. Middendorf, W. Boos, and U. Ehmann.** 1989. Analysis and DNA sequence of the osmoregulated *treA* gene encoding the periplasmic trehalase of *Escherichia coli* K12. *Mol. Gen. Genet.* 217:347–354.

1542. **Gutierrez, C., and J. C. Devedjian.** 1991. Osmotic induction of gene *osmC* expression in *Escherichia coli* K12. *J. Mol. Biol.* 220:959–973.

1543. **Gutierrez, C., S. Gordia, and S. Bonassie.** 1995. Characterization of the osmotically inducible gene *osmE* of *Escherichia coli* K12. *Mol. Microbiol.* 16:553–563.

1544. **Guyot, J. B., J. Grassi, U. Hahn, and W. Guschlbauer.** 1993. The role of the preserved sequences of Dam methylase. *Nucleic Acids Res.* 21:3183–3190.

1545. **Guzman, L. M., J. J. Barondess, and J. Beckwith.** 1992. FtsL, an essential cytoplasmic membrane protein involved in cell division in *Escherichia coli*. *J. Bacteriol.* 174:7716–7728.

1546. **Haas, E. S., D. P. Morse, J. W. Brown, F. J. Schmidt, and N. R. Pace.** 1991. Long-range structure in ribonuclease P RNA. *Science* 254:853–856.

1547. **Hachler, H., S. P. Cohen, and S. B. Levy.** 1991. *marA*, a regulated locus which controls expression of chromosomal multiple antibiotic resistance in *Escherichia coli*. *J. Bacteriol.* 173:5532–5538.

1548. **Hackett, J., R. Misra, and P. Reeves.** 1983. The TolC protein of *Escherichia coli* K12 is synthesised in a precursor form. *FEBS Lett.* 156:307–310.

1549. **Hackett, J., and P. Reeves.** 1983. Primary structure of the *tolC* gene that codes for an outer membrane protein of *Escherichia coli* K12. *Nucleic Acids Res.* 11:6487–6495.

1550. **Hackett, N. R., and P. D. Bragg.** 1983. Membrane cytochromes of *Escherichia coli chl* mutants. *J. Bacteriol.* 154:719–727.

1551. **Hadener, A., P. R. Alefounder, G. J. Hart, C. Abell, and A. R. Battersby.** 1990. Investigation of putative active-site lysine residues in hydroxymethylbilane synthase. Preparation and characterization of mutants in which (a) Lys-55, (b) Lys-59 and (c) both Lys-55 and Lys-59 have been replaced by glutamine. *Biochem. J.* 271:487–491.

1552. **Hadley, R. G., M. Hu, M. Timmons, K. Yun, and R. C. Deonier.** 1983. A partial restriction map of the *proA-purE* region of the *Escherichia coli* K12 chromosome. *Gene* 22:281–287.

1553. **Hafner, E. W., C. W. Tabor, and H. Tabor.** 1977. Isolation of a *metK* mutant with a temperature-sensitive S-adenosylmethionine synthetase. *J. Bacteriol.* 132:832–840.

1554. **Hagervall, T. G., and G. R. Bjork.** 1984. Genetic mapping and cloning of the gene (*trmC*) responsible for the synthesis of tRNA (mnm⁵s²U)methyltransferase in *Escherichia coli* K12. *Mol. Gen. Genet.* 196:201–207.

1555. **Hagewood, B. T., Y. L. Ganduri, and P. Datta.** 1994. Functional analysis of the *tdcABC* promoter of *Escherichia coli*: roles of TdcA and TdcR. *J. Bacteriol.* 176:6214–6220.

1556. **Haggerty, T. J., and S. T. Lovett.** 1993. Suppression of *recJ* mutations of *Escherichia coli* by mutations in translation initiation factor IF3. *J. Bacteriol.* 175:6118–6125.

1557. **Hale, G., and R. N. Perham.** 1980. Amino acid sequence around lipoic acid residues in the pyruvate dehydrogenase multienzyme complex of *Escherichia coli*. *Biochem. J.* 187:905–908.

1558. **Hall, B. G.** 1982. Chromosomal mutation for citrate utilization by *Escherichia coli* K-12. *J. Bacteriol.* 151:269–273.

1559. **Hall, B. G., P. W. Betts, and J. C. Wootton.** 1989. DNA sequence analysis of artificially evolved *ebg* enzyme and *ebg* repressor genes. *Genetics* 123:635–648. (Erratum, 124:791, 1990.)

1560. **Hall, B. G., and P. M. Sharp.** 1992. Molecular population genetics of *Escherichia coli*: DNA sequence diversity at the *celC*, *crr*, and *gutB* loci of natural isolates. *Mol. Biol. Evol.* 9:654–665.

1561. **Hall, B. G., and L. Xu.** 1992. Nucleotide sequence, function, activation and evolution of the cryptic *asc* operon of *Escherichia coli* K12. *Mol. Biol. Evol.* 9:688–706.

1562. **Hall, B. G., L. Xu, and H. Ochman.** 1991. Physical map location of the *asc* (formerly *sac*) operon of *Escherichia coli* K-12. *J. Bacteriol.* 173:5250.

1563. **Hall, C. V., M. VanCleemput, K. H. Muench, and C. Yanofsky.** 1982. The nucleotide sequence of the structural gene for *Escherichia coli* tryptophanyl-tRNA synthetase. *J. Biol. Chem.* 257:6132–6136.

1564. **Hall, C. V., and C. Yanofsky.** 1981. Cloning and characterization of the gene for *Escherichia coli* tryptophanyl-transfer ribonucleic acid synthetase. *J. Bacteriol.* 148:941–949.

1565. **Hall, M. N., and T. J. Silhavy.** 1981. The *ompB* locus and the regulation of the major outer membrane porin proteins of *Escherichia coli* K12. *J. Mol. Biol.* 146:23–43.

1566. **Hall, M. N., and T. J. Silhavy.** 1981. Genetic analysis of the *ompB* locus in *Escherichia coli* K-12. *J. Mol. Biol.* 151:1–15.

1567. **Hall, S. D., M. F. Kane, and R. D. Kolodner.** 1993. Identification and characterization of the *Escherichia coli* RecT protein, a protein encoded by the *recE* region that promotes renaturation of homologous single-stranded DNA. *J. Bacteriol.* 175:277–287. (Erratum, 175:1211.)

1568. **Haller, B. L., and J. A. Fuchs.** 1984. Mapping of *trxB*, a mutation responsible for reduced thioredoxin reductase activity. *J. Bacteriol.* 159:1060–1062.

1569. **Hallett, P., A. J. Grimshaw, D. B. Wigley, and A. Maxwell.** 1990. Cloning of the DNA gyrase genes under *tac* promoter control: overproduction of the gyrase A and B proteins. *Gene* 93:139–142.

1570. **Halsall, D. M.** 1975. Overproduction of lysine by mutant strains of *Escherichia coli* with defective lysine transport systems. *Biochem. Genet.* 13:109–124.

1571. **Haluzi, H., D. Goitein, S. Koby, I. Mendelson, D. Teff, G. Mengeritsky, H. Giladi, and A. B. Oppenheim.** 1991. Genes coding for integration host factor are conserved in gram-negative bacteria. *J. Bacteriol.* 173:6297–6299.

1572. **Hama, H., N. Almaula, C. G. Lerner, S. Inouye, and M. Inouye.** 1991. Nucleoside diphosphate kinase from *Escherichia coli*; its overproduction and sequence comparison with eukaryotic enzymes. *Gene* 105:31–36.

1573. **Hama, H., T. Kayahara, W. Ogawa, M. Tsuda, and T. Tsuchiya.** 1994. Enhancement of serine-sensitivity by a gene encoding rhodanese-like protein in *Escherichia coli*. *J. Biochem.* 115:1135–1140.

1574. **Hama, H., C. Lerner, S. Inouye, and M. Inouye.** 1991. Location of the gene (*ndk*) for nucleoside diphosphate kinase on the physical map of the *Escherichia coli* chromosome. *J. Bacteriol.* 173:3276.

1575. **Hamann, A., D. Bossemeyer, and E. P. Bakker.** 1987. Physical mapping of the K⁺ transport *trkA* gene of *Escherichia coli* and overproduction of the TrkA protein. *J. Bacteriol.* 169:3138–3145.

1576. **Hamilton, W. D., D. A. Harrison, and T. A. Dyer.** 1988. Sequence of the *Escherichia coli* fructose-1,6-bisphosphatase gene. *Nucleic Acids Res.* 16:8707.

1577. **Hammelburger, J. W., and G. A. Orr.** 1983. Interaction of *sn*-glycerol 3-phosphorothioate with *Escherichia coli*: effect on cell growth and metabolism. *J. Bacteriol.* 156:789–799.

1578. **Hammer-Jespersen, K., and A. Munch-Petersen.** 1973. Mutants of *Escherichia coli* unable to metabolize cytidine: isolation and characterization. *Mol. Gen. Genet.* 126:177–186.

1579. **Hanada, K., T. Yoshida, I. Yamato, and Y. Anraku.** 1992. Sodium ion and proline binding sites in the Na⁺/proline symport carrier of *Escherichia coli*. *Biochim. Biophys. Acta* 1105:61–66.

1580. **Hanafusa, T., A. Sakai, A. Tominaga, and M. Enomoto.** 1989. Isolation and characterization of *Escherichia coli hag* operator mutants whose *hag48* expression has become repressible by a *Salmonella* H1 repressor. *Mol. Gen. Genet.* 216:44–50.

1581. Hanamura, A., and H. Aiba. 1991. Molecular mechanism of negative autoregulation of *Escherichia coli crp* gene. *Nucleic Acids Res.* 19:4413–4419.

1582. Hanatani, M., H. Yazyu, S. Shiota-Niiya, Y. Moriyama, H. Kanazawa, M. Futai, and T. Tsuchiya. 1984. Physical and genetic characterization of the melibiose operon and identification of the gene products in *Escherichia coli. J. Biol. Chem.* 259:1807–1812.

1583. Hanck, T., N. Gerwin, and H. J. Fritz. 1989. Nucleotide sequence of the *dcm* locus of *Escherichia coli* K12. *Nucleic Acids Res.* 17:5844.

1584. Hane, M. W., and T. H. Wood. 1969. *Escherichia coli* K-12 mutants resistant to nalidixic acid: genetic mapping and dominance studies. *J. Bacteriol.* 99:238–241.

1585. Haney, S. A., J. V. Platko, D. L. Oxender, and J. M. Calvo. 1992. Lrp, a leucine-responsive protein, regulates branched-chain amino acid transport genes in *Escherichia coli. J. Bacteriol.* 174:108–115.

1586. Hanke, C., J. Hess, G. Schumacher, and W. Goebel. 1992. Processing by OmpT of fusion proteins carrying the HlyA transport signal during secretion by the *Escherichia coli* hemolysin transport system. *Mol. Gen. Genet.* 233:42–48.

1587. Hansen, E. B., T. Atlung, F. G. Hansen, O. Skovgaard, and K. von Meyenburg. 1984. Fine structure genetic map and complementation analysis of mutations in the *dnaA* gene of *Escherichia coli. Mol. Gen. Genet.* 196:387–396.

1588. Hansen, E. B., F. G. Hansen, and K. von Meyenburg. 1982. The nucleotide sequence of the *dnaA* gene and the first part of the *dnaN* gene of *Escherichia coli* K-12. *Nucleic Acids Res.* 10:7373–7385.

1589. Hansen, F. G., B. B. Christensen, and T. Atlung. 1991. The initiator titration model: computer simulation of chromosome and minichromosome control. *Res. Microbiol.* 142:161–167.

1590. Hansen, F. G., E. B. Hansen, and T. Atlung. 1982. The nucleotide sequence of the *dnaA* gene promoter and of the adjacent *rpmH* gene, coding for the ribosomal protein L34, of *Escherichia coli. EMBO J.* 1:1043–1048.

1591. Hansen, F. G., E. B. Hansen, and T. Atlung. 1985. Physical mapping and nucleotide sequence of the *rnpA* gene that encodes the protein component of ribonuclease P in *Escherichia coli. Gene* 38:85–93.

1592. Hansen, F. G., S. Koefoed, and T. Atlung. 1992. Cloning and nucleotide sequence determination of twelve mutant *dnaA* genes of *Escherichia coli. Mol. Gen. Genet.* 234:14–21.

1593. Hantke, K. 1976. Phage T6—colicin K receptor and nucleoside transport in *Escherichia coli. FEBS Lett.* 70:109–112.

1594. Hantke, K. 1983. Identification of an iron uptake system specific for coprogen and rhodotorulic acid in *Escherichia coli* K12. *Mol. Gen. Genet.* 191:301–306.

1595. Hantke, K. 1984. Cloning of the repressor protein gene of iron-regulated systems in *Escherichia coli* K12. *Mol. Gen. Genet.* 197:337–341.

1596. Hantke, K. 1987. Selection procedure for deregulated iron transport mutants (*fur*) in *Escherichia coli* K 12: *fur* not only affects iron metabolism. *Mol. Gen. Genet.* 210:135–139.

1597. Hantke, K., and V. Braun. 1973. Covalent binding of lipid to protein. Diglyceride and amide-linked fatty acid at the N-terminal end of the murein-lipoprotein of the *Escherichia coli* outer membrane. *Eur. J. Biochem.* 34:284–296.

1598. Hara, H., Y. Nishimura, J. Kato, H. Suzuki, H. Nagasawa, A. Suzuki, and Y. Hirota. 1989. Genetic analyses of processing involving C-terminal cleavage in penicillin-binding protein 3 of *Escherichia coli. J. Bacteriol.* 171:5882–5889.

1599. Hara, H., Y. Yamamoto, A. Higashitani, H. Suzuki, and Y. Nishimura. 1991. Cloning, mapping, and characterization of the *Escherichia coli prc* gene, which is involved in C-terminal processing of penicillin-binding protein 3. *J. Bacteriol.* 173:4799–4813.

1600. Harayama, S., J. Bollinger, T. Iino, and G. L. Hazelbauer. 1983. Characterization of the *mgl* operon of *Escherichia coli* by transposon mutagenesis and molecular cloning. *J. Bacteriol.* 153:408–415.

1601. Harayama, S., P. Engstrom, H. Wolf-Watz, T. Iino, and G. L. Hazelbauer. 1982. Cloning of *trg*, a gene for a sensory transducer in *Escherichia coli. J. Bacteriol.* 152:372–383.

1602. Harborne, N. R., L. Griffiths, S. J. Busby, and J. A. Cole. 1992. Transcriptional control, translation and function of the products of the five open reading frames of the *Escherichia coli nir* operon. *Mol. Microbiol.* 6:2805–2813.

1603. Hardaway, K. L., and C. S. Buller. 1979. Effect of ethylenediaminetetraacetate on phospholipids and outer membrane function in *Escherichia coli. J. Bacteriol.* 137:62–68.

1604. Hardesty, C., C. Ferran, and J. M. DiRienzo. 1991. Plasmid-mediated sucrose metabolism in *Escherichia coli*: characterization of *scrY*, the structural gene for a phosphoenolpyruvate-dependent sucrose phosphotransferase system outer membrane porin. *J. Bacteriol.* 173:449–456.

1605. Harkness, R. E., and E. E. Ishiguro. 1983. Temperature-sensitive autolysis-defective mutants of *Escherichia coli. J. Bacteriol.* 155:15–21.

1606. Harlow, K. W., P. Nygaard, and B. Hove-Jensen. 1995. Cloning and characterization of the *gsk* gene encoding guanosine kinase of *Escherichia coli. J. Bacteriol.* 177:2236–2240.

1607. Harms, E., E. Higgins, J. W. Chen, and H. E. Umbarger. 1988. Translational coupling between the *ilvD* and *ilvA* genes of *Escherichia coli. J. Bacteriol.* 170:4798–4807.

1608. Harms, E., A. Wehner, H. P. Aung, and K. H. Rohm. 1991. A catalytic role for threonine-12 of E. coli asparaginase II as established by site-directed mutagenesis. *FEBS Lett.* 285:55–58.

1609. Harris, L. C., P. M. Potter, and G. P. Margison. 1992. Site directed mutagenesis of two cysteine residues in the *E. coli ogt* O^6-alkylguanine DNA alkyltransferase protein. *Biochem. Biophys. Res. Commun.* 187:425–431.

1610. Harris, S. L., D. A. Elliott, M. C. Blake, L. M. Must, M. Messenger, and P. E. Orndorff. 1990. Isolation and characterization of mutants with lesions affecting pellicle formation and erythrocyte agglutination by type 1 piliated *Escherichia coli. J. Bacteriol.* 172:6411–6418.

1611. Harrison, L. I., H. N. Christensen, M. E. Handlogten, D. L. Oxender, and S. C. Quay. 1975. Transport of L-4-azaleucine in *Escherichia coli. J. Bacteriol.* 122:957–965.

1612. Hartlein, M., R. Frank, and D. Madern. 1987. Nucleotide sequence of *Escherichia coli* valyl-tRNA synthetase gene *valS. Nucleic Acids Res.* 15:9081–9082.

1613. Hartlein, M., and D. Madern. 1987. Molecular cloning and nucleotide sequence of the gene for *Escherichia coli* leucyl-tRNA synthetase. *Nucleic Acids Res.* 15:10199–10210.

1614. Hartlein, M., D. Madern, and R. Leberman. 1987. Cloning and characterization of the gene for *Escherichia coli* seryl-tRNA synthetase. *Nucleic Acids Res.* 15:1005–1017.

1615. Harvey, S., C. W. Hill, C. Squires, and C. L. Squires. 1988. Loss of the spacer loop sequence from the *rrnB* operon in the *Escherichia coli* K-12 subline that bears the *relA1* mutation. *J. Bacteriol.* 170:1235–1238.

1616. Hasenbank, R., C. Guthrie, G. Stoffler, H. G. Wittmann, L. Rosen, and D. Apirion. 1973. Electrophoretic and immunological studies on ribosomal proteins of 100 *Escherichia coli* revertants from streptomycin dependence. *Mol. Gen. Genet.* 127:1–18.

1617. Hashimoto, W., H. Suzuki, S. Nohara, and H. Kumagai. 1992. *Escherichia coli* gamma-glutamyltranspeptidase mutants deficient in processing to subunits. *Biochem. Biophys. Res. Commun.* 189:173–178.

1618. Hassan, H. M., and H. C. Sun. 1992. Regulatory roles of Fnr, Fur, and Arc in expression of manganese-containing superoxide dismutase in *Escherichia coli. Proc. Natl. Acad. Sci. USA* 89:3217–3221.

1619. Hassani, M., D. H. Pincus, G. N. Bennett, and I. N. Hirshfield. 1992. Temperature-dependent induction of an acid-inducible stimulon of *Escherichia coli* in broth. *Appl. Environ. Microbiol.* 58:2704–2707.

1620. Haughn, G. W., C. H. Squires, M. DeFelice, C. T. Largo, and J. M. Calvo. 1985. Unusual organization of the *ilvIH* promoter of *Escherichia coli. J. Bacteriol.* 163:186–198.

1621. Haughn, G. W., S. R. Wessler, R. M. Gemmill, and J. M. Calvo. 1986. High A+T content conserved in DNA sequences upstream of *leuABCD* in *Escherichia coli* and *Salmonella typhimurium. J. Bacteriol.* 166:1113–1117.

1622. Hauser, C. A., and G. W. Hatfield. 1983. Nucleotide sequence of the *ilvB* multivalent attenuator region of *Escherichia coli* K12. *Nucleic Acids Res.* 11:127–139.

1623. Hawker, J. R., Jr., and C. S. McHenry. 1987. Monoclonal antibodies specific for the tau subunit of the DNA polymerase III holoenzyme of *Escherichia coli*. Use to demonstrate that tau is the product of the *dnaZX* gene and that both it and gamma, the *dnaZ* gene product, are integral components of the same enzyme assembly. *J. Biol. Chem.* 262:12722–12727.

1624. Hayano, T., N. Takahashi, S. Kato, N. Maki, and M. Suzuki. 1991. Two distinct forms of peptidylprolyl-*cis-trans*-isomerase are expressed separately in periplasmic and cytoplasmic compartments of *Escherichia coli* cells. *Biochemistry* 30:3041–3048.

1625. Hayase, Y., M. Jahn, M. J. Rogers, L. A. Sylvers, M. Koizumi, H. Inoue, E. Ohtsuka, and D. Soll. 1992. Recognition of bases in *Escherichia coli* tRNA^{Gln} by glutaminyl-tRNA synthetase: a complete identity set. *EMBO J.* 11:4159–4165.

1626. Hayden, M. A., I. Huang, D. E. Bussiere, and G. W. Ashley. 1992. The biosynthesis of lipoic acid. Cloning of *lip*, a lipoate biosynthetic locus of *Escherichia coli. J. Biol. Chem.* 267:9512–9515.

1627. Hayden, M. A., I. Y. Huang, G. Iliopoulos, M. Orozco, and G. W. Ashley. 1993. Biosynthesis of lipoic acid: characterization of the lipoic acid auxotrophs *Escherichia coli* W1485-*lip2* and JRG33-*lip9. Biochemistry* 32:3778–3782.

1628. Haydon, D. J., M. A. Quail, and J. R. Guest. 1993. A mutation causing constitutive synthesis of the pyruvate dehydrogenase complex in *Escherichia coli* is located within the *pdhR* gene. *FEBS Lett.* 336:43–47.

1629. Hayward, R. S., K. Igarashi, and A. Ishihama. 1991. Functional specialization within the alpha-subunit of *Escherichia coli* RNA polymerase. *J. Mol. Biol.* 221:23–29.

1630. Hayzer, D. J. 1983. Sub-cloning of the wild-type *proAB* region of the *Escherichia coli* genome. *J. Gen. Microbiol.* 129:3215–3225.

1631. Hayzer, D. J., and T. Leisinger. 1980. The gene-enzyme relationships of proline biosynthesis in *Escherichia coli. J. Gen. Microbiol.* 118:287–293.

1632. Hazelbauer, G. L., P. Engstrom, and S. Harayama. 1981. Methyl-accepting chemotaxis protein III and transducer gene *trg. J. Bacteriol.* 145:43–49.

1633. Hazelbauer, G. L., C. Park, and D. M. Nowlin. 1989. Adaptational "crosstalk" and the crucial role of methylation in chemotactic migration by *Escherichia coli. Proc. Natl. Acad. Sci. USA* 86:1448–1452.

1634. Haziza, C., M. Cassan, and J. C. Patte. 1982. Identification of the promoter of the *asd* gene of *Escherichia coli* using in vitro fusion with the *lac* operon. *Biochimie* 64:227–230.

1635. Haziza, C., P. Stragier, and J. C. Patte. 1982. Nucleotide sequence of the *asd* gene of *Escherichia coli*: absence of a typical attenuation signal. *EMBO J.* 1:379–384.

1636. He, B., K. Y. Choi, and H. Zalkin. 1993. Regulation of *Escherichia coli glnB, prsA,* and *speA* by the purine repressor. *J. Bacteriol.* 175:3598–3606.

1637. He, B., A. Shiau, K. Y. Choi, H. Zalkin, and J. M. Smith. 1990. Genes of the *Escherichia coli pur* regulon are negatively controlled by a repressor-operator interaction. *J. Bacteriol.* 172:4555–4562.

1638. **He, B., J. M. Smith, and H. Zalkin.** 1992. *Escherichia coli purB* gene: cloning, nucleotide sequence, and regulation by *purR*. *J. Bacteriol.* **174:**130–136.

1639. **He, B., and H. Zalkin.** 1994. Regulation of *Escherichia coli purA* by purine repressor, one component of a dual control mechanism. *J. Bacteriol.* **176:**1009–1013.

1639a.**He, L., F. Soderbom, E. G. Wagner, U. Binnie, N. Binns, and M. Masters.** 1993. PcnB is required for the rapid degradation of RNAI, the antisense RNA that controls the copy number of ColE1-related plasmids. *Mol. Microbiol.* **9:**1131–1142.

1640. **Heatwole, V. M., and R. L. Somerville.** 1991. The tryptophan-specific permease gene, *mtr*, is differentially regulated by the tryptophan and tyrosine repressors in *Escherichia coli* K-12. *J. Bacteriol.* **173:**3601–3604.

1641. **Heatwole, V. M., and R. L. Somerville.** 1991. Cloning, nucleotide sequence, and characterization of *mtr*, the structural gene for a tryptophan-specific permease of *Escherichia coli* K-12. *J. Bacteriol.* **173:**108–115.

1642. **Heber, S., and B. E. Tropp.** 1991. Genetic regulation of cardiolipin synthase in *Escherichia coli*. *Biochim. Biophys. Acta* **1129:**1–12.

1643. **Heck, J. D., and G. W. Hatfield.** 1988. Valyl-tRNA synthetase gene of *Escherichia coli* K12. Primary structure and homology within a family of aminoacyl-tRNA synthetases. *J. Biol. Chem.* **263:**868–877.

1644. **Heck, J. D., and G. W. Hatfield.** 1988. Valyl-tRNA synthetase gene of *Escherichia coli* K12. Molecular genetic characterization. *J. Biol. Chem.* **263:**857–867.

1645. **Hedblom, M. L., and J. Adler.** 1980. Genetic and biochemical properties of *Escherichia coli* mutants with defects in serine chemotaxis. *J. Bacteriol.* **144:**1048–1060.

1646. **Hedges, R. W., and K. P. Shannon.** 1984. Resistance to apramycin in *Escherichia coli* isolated from animals: detection of a novel aminoglycoside-modifying enzyme. *J. Gen. Microbiol.* **130:**473–482.

1647. **Hedgpeth, J., J. M. Clement, C. Marchal, D. Perrin, and M. Hofnung.** 1980. DNA sequence encoding the NH_2-terminal peptide involved in transport of lambda receptor, an *Escherichia coli* secretory protein. *Proc. Natl. Acad. Sci. USA* **77:**2621–2625.

1648. **Hediger, M. A., D. F. Johnson, D. P. Nierlich, and I. Zabin.** 1985. DNA sequence of the lactose operon: the *lacA* gene and the transcriptional termination region. *Proc. Natl. Acad. Sci. USA* **82:**6414–6418.

1649. **Heiland, I., D. Brauer, and B. Wittmann-Liebold.** 1976. Primary structure of protein L10 from the large subunit of *Escherichia coli* ribosomes. *Hoppe-Seyler's Z. Physiol. Chem.* **357:**1751–1770.

1650. **Heiland, I., and B. Wittmann-Liebold.** 1979. Amino acid sequence of the ribosomal protein L21 of *Escherichia coli*. *Biochemistry* **18:**4605–4612.

1651. **Heiman, C., and C. G. Miller.** 1978. *Salmonella typhimurium* mutants lacking protease II. *J. Bacteriol.* **135:**588–594.

1652. **Heimberg, H., A. Boyen, M. Crabeel, and N. Glansdorff.** 1990. *Escherichia coli* and *Saccharomyces cerevisiae* acetylornithine aminotransferase: evolutionary relationship with ornithine aminotransferase. *Gene* **90:**69–78.

1653. **Heine, H. G., J. Kyngdon, and T. Ferenci.** 1987. Sequence determinants in the *lamB* gene of *Escherichia coli* influencing the binding and pore selectivity of maltoporin. *Gene* **53:**287–292.

1654. **Heitman, J., and P. Model.** 1987. Site-specific methylases induce the SOS DNA repair response in *Escherichia coli*. *J. Bacteriol.* **169:**3243–3250.

1655. **Heitman, J., and P. Model.** 1991. SOS induction as an in vivo assay of enzyme-DNA interactions. *Gene* **103:**1–9.

1656. **Hekstra, D., and J. Tommassen.** 1993. Functional exchangeability of the ABC proteins of the periplasmic binding protein-dependent transport systems Ugp and Mal of *Escherichia coli*. *J. Bacteriol.* **175:**6546–6552.

1657. **Helander, I. M., B. Lindner, U. Seydel, and M. Vaara.** 1993. Defective biosynthesis of the lipid A component of temperature-sensitive *firA* (*omsA*) mutant of *Escherichia coli*. *Eur. J. Biochem.* **212:**363–369.

1658. **Held, W. A., and O. H. Smith.** 1970. Mechanism of 3-methylanthranilic acid derepression of the tryptophan operon in *Escherichia coli*. *J. Bacteriol.* **101:**209–217.

1659. **Held, W. A., and O. H. Smith.** 1970. Regulation of the *Escherichia coli* tryptophan operon by early reactions in the aromatic pathway. *J. Bacteriol.* **101:**202–208.

1660. **Heller, K., and R. J. Kadner.** 1985. Nucleotide sequence of the gene for the vitamin B12 receptor protein in the outer membrane of *Escherichia coli*. *J. Bacteriol.* **161:**904–908.

1661. **Helling, R. B.** 1968. Selection of a mutant of *Escherichia coli* which has high mutation rates. *J. Bacteriol.* **96:**975–980.

1662. **Helling, R. B.** 1990. The glutamate dehydrogenase structural gene of *Escherichia coli*. *Mol. Gen. Genet.* **223:**508–512.

1663. **Helling, R. B., and B. S. Adams.** 1970. Nalidixic acid-resistant auxotrophs of *Escherichia coli*. *J. Bacteriol.* **104:**1027–1029.

1664. **Helling, R. B., and J. S. Kukora.** 1971. Nalidixic acd-resistant mutants of *Escherichia coli* deficient in isocitrate dehydrogenase. *J. Bacteriol.* **105:**1224–1226.

1665. **Hellinga, H. W., and P. R. Evans.** 1985. Nucleotide sequence and high-level expression of the major *Escherichia coli* phosphofructokinase. *Eur. J. Biochem.* **149:**363–373.

1666. **Hellinga, H. W., and P. R. Evans.** 1987. Mutations in the active site of *Escherichia coli* phosphofructokinase. *Nature* (London) **327:**437–439.

1667. **Hellmuth, K., G. Rex, B. Surin, R. Zinck, and J. E. McCarthy.** 1991. Translational coupling varying in efficiency between different pairs of genes in the central region of the atp operon of *Escherichia coli*. *Mol. Microbiol.* **5:**813–824.

1668. **Hemmingsen, S. M., C. Woolford, S. M. van der Vies, K. Tilly, D. T. Dennis, C. P. Georgopoulos, R. W. Hendrix, and R. J. Ellis.** 1988. Homologous plant and bacterial proteins chaperone oligomeric protein assembly. *Nature* (London) **333:**330–334.

1669. **Henderson, T. A., P. M. Dombrosky, and K. D. Young.** 1994. Artifactual processing of penicillin-binding proteins 7 and 1b by the OmpT protease of *Escherichia coli*. *J. Bacteriol.* **176:**256–259.

1670. **Henderson, T. A., A. M. Templin, and K. D. Young.** 1995. Identification and cloning of the gene encoding penicillin-binding protein 7 of *Escherichia coli*. *J. Bacteriol.* **177:**2074–2079.

1671. **Hendrickson, W., and K. E. Rudd.** 1992. Physical map location of the *araFGH* operon of *Escherichia coli*. *J. Bacteriol.* **174:**3836–3837.

1672. **Hendrickson, W., C. Stoner, and R. Schleif.** 1990. Characterization of the *Escherichia coli araFGH* and *araJ* promoters. *J. Mol. Biol.* **215:**497–510.

1673. **Hengge-Aronis, R., and D. Fischer.** 1992. Identification and molecular analysis of *glgS*, a novel growth-phase-regulated and *rpoS*-dependent gene involved in glycogen synthesis in *Escherichia coli*. *Mol. Microbiol.* **6:**1877–1886.

1674. **Hengge-Aronis, R., R. Lange, N. Henneberg, and D. Fischer.** 1993. Osmotic regulation of *rpoS*-dependent genes in *Escherichia coli*. *J. Bacteriol.* **175:**259–265.

1675. **Henikoff, S., G. W. Haughn, J. M. Calvo, and J. C. Wallace.** 1988. A large family of bacterial activator proteins. *Proc. Natl. Acad. Sci. USA* **85:**6602–6606.

1676. **Henikoff, S., and J. G. Henikoff.** 1994. Protein family classification based on searching a database of blocks. *Genomics* **19:**97–107.

1677. **Hennecke, F., H. Kolmar, K. Brundl, and H. J. Fritz.** 1991. The *vsr* gene product of *E. coli* K-12 is a strand- and sequence-specific DNA mismatch endonuclease. *Nature* (London) **353:**776–778.

1678. **Hennecke, H.** 1990. Regulation of bacterial gene expression by metal-protein complexes. *Mol. Microbiol.* **4:**1621–1628.

1679. **Hennessey, E. S., and J. K. Broome-Smith.** 1994. Two related bacterial membrane proteins, ExbD and ToIR, have opposite transmembrane charge dipolarity. *Mol. Microbiol.* **11:**417. (Letter.)

1680. **Henrich, B., H. Backes, J. R. Klein, and R. Plapp.** 1992. The promoter region of the *Escherichia coli pepD* gene: deletion analysis and control by phosphate concentration. *Mol. Gen. Genet.* **232:**117–125.

1681. **Henrich, B., S. Becker, U. Schroeder, and R. Plapp.** 1993. *dcp* gene of *Escherichia coli*: cloning, sequencing, transcript mapping, and characterization of the gene product. *J. Bacteriol.* **175:**7290–7300.

1682. **Henrich, B., U. Monnerjahn, and R. Plapp.** 1990. Peptidase D gene (*pepD*) of *Escherichia coli* K-12: nucleotide sequence, transcript mapping, and comparison with other peptidase genes. *J. Bacteriol.* **172:**4641–4651.

1683. **Henrich, B., and R. Plapp.** 1991. Locations of the genes from *pepD* through *proA* on the physical map of the *Escherichia coli* chromosome. *J. Bacteriol.* **173:**7407–7408.

1684. **Henrich, B., U. Schroeder, R. W. Frank, and R. Plapp.** 1989. Accurate mapping of the *Escherichia coli* pepD gene by sequence analysis of its 5′ flanking region. *Mol. Gen. Genet.* **215:**369–373.

1685. **Henry, M. F., and J. E. Cronan, Jr.** 1991. *Escherichia coli* transcription factor that both activates fatty acid synthesis and represses fatty acid degradation. *J. Mol. Biol.* **222:**843–849.

1686. **Henson, J. M., A. Blinkowa, and J. R. Walker.** 1982. The *Escherichia coli dnaW* mutation is an allele of the *adk* gene. *Mol. Gen. Genet.* **186:**488–492.

1687. **Henson, J. M., and J. R. Walker.** 1982. Genetic analysis of *acrA* and *lir* mutations of *Escherichia coli*. *J. Bacteriol.* **152:**1301–1302.

1688. **Henzel, W. J., T. M. Billeci, J. T. Stults, S. C. Wong, C. Grimley, and C. Watanabe.** 1993. Identifying proteins from two-dimensional gels by molecular mass searching of peptide fragments in protein sequence databases. *Proc. Natl. Acad. Sci. USA* **90:**5011–5015.

1689. **Herlihy, W. C., N. J. Royal, K. Biemann, S. D. Putney, and P. R. Schimmel.** 1980. Mass spectra of partial protein hydrolysates as a multiple phase check for long polypeptides deduced from DNA sequences: NH_2-terminal segment of alanine tRNA synthetase. *Proc. Natl. Acad. Sci. USA* **77:**6531–6535.

1690. **Herman, C., T. Ogura, T. Tomoyasu, S. Hiraga, Y. Akiyama, K. Ito, R. Thomas, R. D'Ari, and P. Bouloc.** 1993. Cell growth and lambda phage development controlled by the same essential *Escherichia coli* gene, *ftsH/hflB*. *Proc. Natl. Acad. Sci. USA* **90:**10861–10865.

1691. **Hernandez, V. J., and H. Bremer.** 1990. Guanosine tetraphosphate (ppGpp) dependence of the growth rate control of *rrnB* P_1 promoter activity in *Escherichia coli*. *J. Biol. Chem.* **265:**11605–11614.

1692. **Hernandez, V. J., and H. Bremer.** 1991. *Escherichia coli* ppGpp synthetase II activity requires *spoT*. *J. Biol. Chem.* **266:**5991–5999.

1693. **Herold, C., and E. A. Birge.** 1990. Location of *icdA* and *fadR* on the physical map of *Escherichia coli*. *J. Bacteriol.* **172:**6618.

1694. **Herrchen, M., and G. Legler.** 1984. Identification of an essential carboxylate group at the active site of *lacZ* beta-galactosidase from *Escherichia coli*. *Eur. J. Biochem.* **138:**527–531.

1695. **Herrick, J., R. Kern, S. Guha, A. Landoulsi, O. Fayet, A. Malki, and M. Kohiyama.** 1994. Parental strand recognition of the DNA replication origin by the outer membrane in *Escherichia coli*. *EMBO J.* **13:**4695–4703.

1696. **Herrington, M. B., J. Basso, M. Faraci, and C. Autexier.** 1991. Modification of the suppressor phenotype of thymine requiring strains of *Escherichia coli*. *Genet. Res.* **58:**185–192.

1697. **Herrington, M. B., A. Kohli, and P. H. Lapchak.** 1984. Suppression by thymidine-requiring mutants of *Escherichia coli* K-12. *J. Bacteriol.* **157:**126–129.

1698. Herrmann, K. M., J. Shultz, and M. A. Hermodson. 1980. Sequence homology between the tyrosine-sensitive 3-deoxy-D-arabino-heptulosonate 7-phosphate synthase from *Escherichia coli* and hemerythrin from Sipunculida. *J. Biol. Chem.* 255:7079–7081.

1699. Hershey, H. V., R. Gutstein, and M. W. Taylor. 1982. Cloning and restriction map of the *E. coli apt* gene. *Gene* 19:89–92.

1700. Hershey, H. V., and M. W. Taylor. 1986. Nucleotide sequence and deduced amino acid sequence of *Escherichia coli* adenine phosphoribosyltransferase and comparison with other analogous enzymes. *Gene* 43:287–293.

1701. Hershfield, M. S., S. Chaffee, L. Koro-Johnson, A. Mary, A. A. Smith, and S. A. Short. 1991. Use of site-directed mutagenesis to enhance the epitope-shielding effect of covalent modification of proteins with polyethylene glycol. *Proc. Natl. Acad. Sci. USA* 88:7185–7189.

1702. Hess, J. F., K. Oosawa, N. Kaplan, and M. I. Simon. 1988. Phosphorylation of three proteins in the signaling pathway of bacterial chemotaxis. *Cell* 53:79–87.

1703. Hesse, J. E., L. Wieczorek, K. Altendorf, A. S. Reicin, E. Dorus, and W. Epstein. 1984. Sequence homology between two membrane transport ATPases, the Kdp-ATPase of *Escherichia coli* and the Ca^{2+}-ATPase of sarcoplasmic reticulum. *Proc. Natl. Acad. Sci. USA* 81:4746–4750.

1704. Heyde, M., and R. Portalier. 1982. New pleiotropic alkaline phosphatase-negative mutants of *Escherichia coli* K-12. *J. Bacteriol.* 151:529–533.

1705. Hiasa, H., and K. J. Marians. 1992. Differential inhibition of the DNA translocation and DNA unwinding activities of DNA helicases by the *Escherichia coli* Tus protein. *J. Biol. Chem.* 267:11379–11385.

1706. Hickson, I. D., H. M. Arthur, D. Bramhill, and P. T. Emmerson. 1983. The *E. coli uvrD* gene product is DNA helicase II. *Mol. Gen. Genet.* 190:265–270.

1707. Hickson, I. D., K. E. Atkinson, and P. T. Emmerson. 1982. Molecular cloning and amplification of the gene for thymidylate synthetase of *E. coli. Gene* 18:257–260.

1708. Hickson, I. D., and P. T. Emmerson. 1981. Identification of the *Escherichia coli recB* and *recC* gene products. *Nature* (London) 294:578–580.

1709. Hidaka, M., M. Akiyama, and T. Horiuchi. 1988. A consensus sequence of three DNA replication terminus sites on the *E. coli* chromosome is highly homologous to the *terR* sites of the R6K plasmid. *Cell* 55:467–475.

1710. Hidaka, M., T. Kobayashi, and T. Horiuchi. 1991. A newly identified DNA replication terminus site, TerE, on the *Escherichia coli* chromosome. *J. Bacteriol.* 173:391–393.

1711. Hidaka, M., T. Kobayashi, S. Takenaka, H. Takeya, and T. Horiuchi. 1989. Purification of a DNA replication terminus (*ter*) site-binding protein in *Escherichia coli* and identification of the structural gene. *J. Biol. Chem.* 264:21031–21037.

1712. Hidalgo, E., Y. M. Chen, E. C. Lin, and J. Aguilar. 1991. Molecular cloning and DNA sequencing of the *Escherichia coli* K-12 *ald* gene encoding aldehyde dehydrogenase. *J. Bacteriol.* 173:6118–6123.

1713. Hidalgo, E., and B. Demple. 1994. An iron-sulfur center essential for transcriptional activation by the redox-sensing SoxR protein. *EMBO J.* 13:138–146.

1714. Higgins, C. F., C. J. Dorman, D. A. Stirling, L. Waddell, I. R. Booth, G. May, and E. Bremer. 1988. A physiological role for DNA supercoiling in the osmotic regulation of gene expression in *S. typhimurium* and *E. coli. Cell* 52:569–584.

1715. Higgins, C. F., J. C. Hinton, C. S. Hulton, T. Owen-Hughes, G. D. Pavitt, and A. Seirafi. 1990. Protein H1: a role for chromatin structure in the regulation of bacterial gene expression and virulence? *Mol. Microbiol.* 4:2007–2012.

1716. Higgins, W., E. W. Miles, and T. Fairwell. 1980. Location of three active site residues in the NH_2-terminal sequence of the beta 2 subunit tryptophan synthase from *Escherichia coli. J. Biol. Chem.* 255:512–517.

1717. Highton, P. J., Y. Chang, W. R. Marcotte, Jr., and C. A. Schnaitman. 1985. Evidence that the outer membrane protein gene *nmpC* of *Escherichia coli* K-12 lies within the defective *qsr'* prophage. *J. Bacteriol.* 162:256–262.

1718. Hill, C. W., J. A. Gray, and H. Brody. 1989. Use of the isocitrate dehydrogenase structural gene for attachment of e14 in *Escherichia coli* K-12. *J. Bacteriol.* 171:4083–4084.

1719. Hill, T. M., J. M. Henson, and P. L. Kuempel. 1987. The terminus region of the *Escherichia coli* chromosome contains two separate loci that exhibit polar inhibition of replication. *Proc. Natl. Acad. Sci. USA* 84:1754–1758.

1720. Hill, T. M., B. J. Kopp, and P. L. Kuempel. 1988. Termination of DNA replication in *Escherichia coli* requires a *trans*-acting factor. *J. Bacteriol.* 170:662–668.

1721. Hill, T. M., A. J. Pelletier, M. L. Tecklenburg, and P. L. Kuempel. 1988. Identification of the DNA sequence from the *E. coli* terminus region that halts replication forks. *Cell* 55:459–466.

1722. Hill, T. M., M. L. Tecklenburg, A. J. Pelletier, and P. L. Kuempel. 1989. *tus*, the trans-acting gene required for termination of DNA replication in *Escherichia coli*, encodes a DNA-binding protein. *Proc. Natl. Acad. Sci. USA* 86:1593–1597.

1723. Hiom, K., and S. G. Sedgwick. 1991. Cloning and structural characterization of the *mcrA* locus of *Escherichia coli. J. Bacteriol.* 173:7368–7373.

1724. Hiraga, S. 1969. Operator mutants of the tryptophan operon in *Escherichia coli. J. Mol. Biol.* 39:159–179.

1725. Hiraga, S. 1992. Chromosome and plasmid partition in *Escherichia coli. Annu. Rev. Biochem.* 61:283–306.

1726. Hiraga, S., K. Igarashi, and T. Yura. 1967. A deoxythymidine kinase-deficient mutant of *Escherichia coli*. I. Isolation and some properties. *Biochim. Biophys. Acta* 145:41–51.

1727. Hiraga, S., K. Ito, T. Matsuyama, H. Ozaki, and T. Yura. 1968. 5-Methyltryptophan-resistant mutations linked with the arginine G marker in *Escherichia coli. J. Bacteriol.* 96:1880–1881.

1728. Hiraga, S., H. Niki, R. Imamura, T. Ogura, K. Yamanaka, J. Feng, B. Ezaki, and A. Jaffe. 1991. Mutants defective in chromosome partitioning in *E. coli. Res. Microbiol.* 142:189–194.

1728a.Hiraga, S., H. Niki, T. Ogura, C. Ichinose, H. Mori, B. Ezak, and A. Jaffe. 1989. Chromosome partitioning in *Escherichia coli*: novel mutants producing anucleate cells. *J. Bacteriol.* 171:1496–505.

1729. Hiraoka, S., H. Matsuzaki, and I. Shibuya. 1993. Active increase in cardiolipin synthesis in the stationary growth phase and its physiological significance in *Escherichia coli. FEBS Lett.* 336:221–224.

1730. Hiraoka, S., K. Nukui, N. Uetake, A. Ohta, and I. Shibuya. 1991. Amplification and substantial purification of cardiolipin synthase of *Escherichia coli. J. Biochem.* 110:443–449.

1731. Hiratsu, K., M. Amemura, H. Nashimoto, H. Shinagawa, and K. Makino. 1995. The *rpoE* gene of *Escherichia coli*, which encodes sigma E, is essential for bacterial growth at high temperature. *J. Bacteriol.* 177:2918–2922.

1732. Hirel, P. H., M. J. Schmitter, P. Dessen, G. Fayat, and S. Blanquet. 1989. Extent of N-terminal methionine excision from *Escherichia coli* proteins is governed by the side-chain length of the penultimate amino acid. *Proc. Natl. Acad. Sci. USA* 86:8247–8251.

1733. Hirota, Y., M. Inuzuka, and M. Tomoeda. 1966. Elective selection of proline-requiring mutants. *J. Bacteriol.* 91:2392.

1734. Hirota, Y., S. Yasuda, M. Yamada, A. Nishimura, K. Sugimoto, H. Sugisaki, A. Oka, and M. Takanami. 1979. Structural and functional properties of the *Escherichia coli* origin of DNA replication. *Cold Spring Harbor Symp. Quant. Biol.* 43:129–138.

1735. Hirschman, J., P. K. Wong, K. Sei, J. Keener, and S. Kustu. 1985. Products of nitrogen regulatory genes *ntrA* and *ntrC* of enteric bacteria activate *glnA* transcription in vitro: evidence that the *ntrA* product is a sigma factor. *Proc. Natl. Acad. Sci. USA* 82:7525–7529.

1736. Hirshfield, I. N., P. C. Horn, D. A. Hopwood, W. K. Maas, and R. DeDeken. 1968. Studies on the mechanism of repression of arginine biosynthesis in *Escherichia coli*. 3. Repression of enzymes of arginine biosynthesis in arginyl-tRNA synthetase mutants. *J. Mol. Biol.* 35:83–93.

1737. Hirshfield, I. N., and P. C. Zamecnik. 1972. Thiosine-resistant mutants of *Escherichia coli* K-12 with growth-medium-dependent lysl-tRNA synthetase activity. I. Isolation and physiological characterization. *Biochim. Biophys. Acta* 259:330–343.

1738. Hirvas, L., J. Coleman, P. Koski, and M. Vaara. 1990. Bacterial 'histone-like protein I' (HLP-I) is an outer membrane constituent? *FEBS Lett.* 262:123–126.

1739. Hirvas, L., P. Koski, and M. Vaara. 1991. The *ompH* gene of *Yersinia enterocolitica*: cloning, sequencing, expression, and comparison with known enterobacterial ompH sequences. *J. Bacteriol.* 173:1223–1229.

1740. Hitz, H., D. Schafer, and B. Wittmann-Liebold. 1977. Determination of the complete amino-acid sequence of protein S6 from the wild-type and a mutant of *Escherichia coli. Eur. J. Biochem.* 75:497–512.

1741. Hoben, P., N. Royal, A. Cheung, F. Yamao, K. Biemann, and D. Soll. 1982. *Escherichia coli* glutaminyl-tRNA synthetase. II. Characterization of the *glnS* gene product. *J. Biol. Chem.* 257:11644–11650.

1742. Hofnung, M., A. Jezierska, and C. Braun-Breton. 1976. *lamB* mutations in *E. coli* K12: growth of lambda host range mutants and effect of nonsense suppressors. *Mol. Gen. Genet.* 145:207–213.

1743. Hofnung, M., E. Lepouce, and C. Braun-Breton. 1981. General method for fine mapping of the *Escherichia coli* K-12 *lamB* gene: localization of missense mutations affecting bacteriophage lambda adsorption. *J. Bacteriol.* 148:853–860.

1744. Hofnung, M., and M. Schwartz. 1971. Mutations allowing growth on maltose of *Escherichia coli* K 12 strains with a deleted *malT* gene. *Mol. Gen. Genet.* 112:117–132.

1745. Hofnung, M., M. Schwartz, and D. Hatfield. 1971. Complementation studies in the maltose-A region of the *Escherichia coli* K12 genetic map. *J. Mol. Biol.* 61:681–694.

1746. Hogarth, B. G., and C. F. Higgins. 1983. Genetic organization of the oligopeptide permease (*opp*) locus of *Salmonella typhimurium* and *Escherichia coli. J. Bacteriol.* 153:1548–1551.

1747. Hogg, R. W., and M. A. Hermodson. 1977. Amino acid sequence of the L-arabinose-binding protein from *Escherichia coli* B/r. *J. Biol. Chem.* 252:5135–5141.

1748. Hogg, R. W., C. Voelker, and I. Von Carlowitz. 1991. Nucleotide sequence and analysis of the *mgl* operon of *Escherichia coli* K12. *Mol. Gen. Genet.* 229:453–459.

1749. Holck, A., and K. Kleppe. 1988. Cloning and sequencing of the gene for the DNA-binding 17K protein of *Escherichia coli. Gene* 67:117–124.

1750. Holmes, G. E., and N. R. Holmes. 1986. Accumulation of DNA damages in aging *Paramecium tetraurelia. Mol. Gen. Genet.* 204:108–114.

1751. Holmes, W. M., C. Andraos-Selim, I. Roberts, and S. Z. Wahab. 1992. Structural requirements for tRNA methylation. Action of *Escherichia coli* tRNA(guanosine-1) methyltransferase on $tRNA_1^{Leu}$ structural variants. *J. Biol. Chem.* 267:13440–13445.

1752. Holmes, Z., and U. Henning. 1994. Location of the *ompT* gene relative to that of *appY* on the physical map of the *Escherichia coli* chromosome. *Mol. Gen. Genet.* 242:363–364.

1753. Holmgren, A. 1968. Thioredoxin. 6. The amino acid sequence of the protein from *Escherichia coli* B. *Eur. J. Biochem.* 6:475–484.

1754. Holmgren, A., B. O. Soderberg, H. Eklund, and C. I. Branden. 1975. Three-dimensional structure of *Escherichia coli* thioredoxin-S2 to 2.8 A resolution. *Proc. Natl. Acad. Sci. USA* 72:2305–2309.

1755. Holowachuk, E. W., and J. D. Friesen. 1982. Isolation of a recombinant lambda phage carrying *nusA* and surrounding region of the *Escherichia coli* K-12 chromosome. *Mol. Gen. Genet.* **187**:248–253.

1756. Holowachuk, E. W., J. D. Friesen, and N. P. Fiil. 1980. Bacteriophage lambda vehicle for the direct cloning of *Escherichia coli* promoter DNA sequences: feedback regulation of the *rplJL-rpoBC* operon. *Proc. Natl. Acad. Sci. USA* **77**:2124–2128.

1757. Holtel, A., and M. Merrick. 1988. Identification of the *Klebsiella pneumoniae glnB* gene: nucleotide sequence of wild-type and mutant alleles. *Mol. Gen. Genet.* **215**:134–138.

1758. Holz, A., C. Schaefer, H. Gille, W. R. Jueterbock, and W. Messer. 1992. Mutations in the DnaA binding sites of the replication origin of *Escherichia coli*. *Mol. Gen. Genet.* **233**:81–88.

1759. Homma, H., T. Kobayashi, N. Chiba, K. Karasawa, H. Mizushima, I. Kudo, K. Inoue, H. Ikeda, M. Sekiguchi, and S. Nojima. 1984. The DNA sequence encoding *pldA* gene, the structural gene for detergent-resistant phospholipase A of *E. coli*. *J. Biochem.* **96**:1655–1664.

1760. Homma, H., T. Kobayashi, Y. Ito, I. Kudo, K. Inoue, H. Ikeda, M. Sekiguchi, and S. Nojima. 1983. Identification and cloning of the gene coding for lysophospholipase L2 of *E. coli* K-12. *J. Biochem.* **94**:2079–2081.

1761. Honore, N., and S. T. Cole. 1990. Nucleotide sequence of the *aroP* gene encoding the general aromatic amino acid transport protein of *Escherichia coli* K-12: homology with yeast transport proteins. *Nucleic Acids Res.* **18**:653. (Erratum, **18**:1332.)

1762. Honore, N., M. H. Nicolas, and S. T. Cole. 1989. Regulation of enterobacterial cephalosporinase production: the role of a membrane-bound sensory transducer. *Mol. Microbiol.* **3**:1121–1130.

1763. Hoog, J. O., H. Jornvall, A. Holmgren, M. Carlquist, and M. Persson. 1983. The primary structure of *Escherichia coli* glutaredoxin. Distant homology with thioredoxins in a superfamily of small proteins with a redox-active cystine disulfide/cysteine dithiol. *Eur. J. Biochem.* **136**:223–232.

1764. Hoog, J. O., H. von Bahr-Lindstrom, H. Jornvall, and A. Holmgren. 1986. Cloning and expression of the glutaredoxin (*grx*) gene of *Escherichia coli*. *Gene* **43**:13–21.

1765. Hooper, D. C., J. S. Wolfson, C. Tung, K. S. Souza, and M. N. Swartz. 1989. Effects of inhibition of the B subunit of DNA gyrase on conjugation in *Escherichia coli*. *J. Bacteriol.* **171**:2235–2237.

1766. Hoover, T. A., W. D. Roof, K. F. Foltermann, G. A. O'Donovan, D. A. Bencini, and J. R. Wild. 1983. Nucleotide sequence of the structural gene (*pyrB*) that encodes the catalytic polypeptide of aspartate transcarbamoylase of *Escherichia coli*. *Proc. Natl. Acad. Sci. USA* **80**:2462–2466.

1767. Hope, J. N., A. W. Bell, M. A. Hermodson, and J. M. Groarke. 1986. Ribokinase from *Escherichia coli* K12. Nucleotide sequence and overexpression of the *rbsK* gene and purification of ribokinase. *J. Biol. Chem.* **261**:7663–7668.

1768. Hopkins, J. D., M. Clements, and M. Syvanen. 1983. New class of mutations in *Escherichia coli* (*uup*) that affect precise excision of insertion elements and bacteriophage Mu growth. *J. Bacteriol.* **153**:384–389.

1769. Hoppe, I., and J. Roth. 1974. Specialized transducing phages derived from salmonella phage P22. *Genetics* **76**:633–654.

1770. Hopper, S., M. Babst, V. Schlensog, H. M. Fischer, H. Hennecke, and A. Bock. 1994. Regulated expression in vitro of genes coding for formate hydrogenlyase components of *Escherichia coli*. *J. Biol. Chem.* **269**:19597–19604.

1771. Hopper, S., and A. Bock. 1995. Effector-mediated stimulation of ATPase activity by the sigma 54-dependent transcriptional activator FHLA from *Escherichia coli*. *J. Bacteriol.* **177**:2798–2803.

1772. Horazdovsky, B. F., and R. W. Hogg. 1987. High-affinity L-arabinose transport operon. Gene product expression and mRNAs. *J. Mol. Biol.* **197**:27–35.

1773. Horazdovsky, B. F., and R. W. Hogg. 1989. Genetic reconstitution of the high-affinity L-arabinose transport system. *J. Bacteriol.* **171**:3053–3059.

1774. Horii, T., T. Ogawa, and H. Ogawa. 1980. Organization of the *recA* gene of *Escherichia coli*. *Proc. Natl. Acad. Sci. USA* **77**:313–317.

1775. Horii, T., T. Ogawa, and H. Ogawa. 1981. Nucleotide sequence of the *lexA* gene of *E. coli*. *Cell* **23**:689–697.

1776. Horiuchi, T., and Y. Fujimura. 1995. Recombinational rescue of the stalled DNA replication fork: a model based on analysis of an *Escherichia coli* strain with a chromosome region difficult to replicate. *J. Bacteriol.* **177**:783–791.

1777. Horiuchi, T., H. Maki, M. Maruyama, and M. Sekiguchi. 1981. Identification of the *dnaQ* gene product and location of the structural gene for RNase H of *Escherichia coli* by cloning of the genes. *Proc. Natl. Acad. Sci. USA* **78**:3770–3774.

1778. Horiuchi, T., T. Nagasawa, K. Takano, and M. Sekiguchi. 1987. A newly discovered tRNA^{fAsp} gene (*aspV*) of *Escherichia coli* K12. *Mol. Gen. Genet.* **206**:356–361.

1779. Horne, S. M., and K. D. Young. 1995. *Escherichia coli* and other species of the Enterobacteriaceae encode a protein similar to the family of Mip-like FK506 binding proteins. *Arch. Microbiol.* **63**:357–365.

1780. Horowitz, D. S., and J. C. Wang. 1987. Mapping the active site tyrosine of *Escherichia coli* DNA gyrase. *J. Biol. Chem.* **262**:5339–5344.

1781. Horowitz, H., G. E. Christie, and T. Platt. 1982. Nucleotide sequence of the *trpD* gene, encoding anthranilate synthetase component II of *Escherichia coli*. *J. Mol. Biol.* **156**:245–256.

1782. Horowitz, H., and T. Platt. 1982. Identification of *trp-p2*, an internal promoter in the tryptophan operon of *Escherichia coli*. *J. Mol. Biol.* **156**:257–267.

1783. Horowitz, H., and T. Platt. 1982. A termination site for LacI transcription is between the CAP site and the *lac* promoter. *J. Biol. Chem.* **257**:11740–11746.

1784. Horowitz, H., and T. Platt. 1983. Initiation in vivo at the internal *trp p2* promoter of *Escherichia coli*. *J. Biol. Chem.* **258**:7890–7893.

1785. Horowitz, H., J. Van Arsdell, and T. Platt. 1983. Nucleotide sequence of the *trpD* and *trpC* genes of *Salmonella typhimurium*. *J. Mol. Biol.* **169**:775–797.

1786. Hou, Y., Y. P. Lin, J. D. Sharer, and P. E. March. 1994. In vivo selection of conditional-lethal mutations in the gene encoding elongation factor G of *Escherichia coli*. *J. Bacteriol.* **176**:123–129.

1787. Hou, Y. M., K. Shiba, C. Mottes, and P. Schimmel. 1991. Sequence determination and modeling of structural motifs for the smallest monomeric aminoacyl-tRNA synthetase. *Proc. Natl. Acad. Sci. USA* **88**:976–980.

1788. Houba-Herin, N., H. Hara, M. Inouye, and Y. Hirota. 1985. Binding of penicillin to thiol-penicillin-binding protein 3 of *Escherichia coli*: identification of its active site. *Mol. Gen. Genet.* **201**:499–504.

1789. Houlberg, U., B. Hove-Jensen, B. Jochimsen, and P. Nygaard. 1983. Identification of the enzymatic reactions encoded by the *purG* and *purI* genes of *Escherichia coli*. *J. Bacteriol.* **154**:1485–1488.

1790. Houman, F., M. R. Diaz-Torres, and A. Wright. 1990. Transcriptional antitermination in the *bgl* operon of *E. coli* is modulated by a specific RNA binding protein. *Cell* **62**:1153–1163.

1791. Hountondji, C., J. M. Schmitter, C. Beauvallet, and S. Blanquet. 1990. Mapping of the active site of *Escherichia coli* methionyl-tRNA synthetase: identification of amino acid residues labeled by periodate-oxidized tRNA^{fMet} molecules having modified lengths at the 3'-acceptor end. *Biochemistry* **29**:8190–8198.

1792. Hove-Jensen, B. 1983. Chromosomal location of the gene encoding phosphoribosylpyrophosphate synthetase in *Escherichia coli*. *J. Bacteriol.* **154**:177–184.

1793. Hove-Jensen, B. 1985. Cloning and characterization of the *prs* gene encoding phosphoribosylpyrophosphate synthetase of *Escherichia coli*. *Mol. Gen. Genet.* **201**:269–276.

1794. Hove-Jensen, B., K. W. Harlow, C. J. King, and R. L. Switzer. 1986. Phosphoribosylpyrophosphate synthetase of *Escherichia coli*. Properties of the purified enzyme and primary structure of the *prs* gene. *J. Biol. Chem.* **261**:6765–6771.

1795. Hove-Jensen, B., and M. Maigaard. 1993. *Escherichia coli rpiA* gene encoding ribose phosphate isomerase A. *J. Bacteriol.* **175**:5628–5635.

1796. Hove-Jensen, B., and P. Nygaard. 1982. Phosphoribosylpyrophosphate synthetase of *Escherichia coli*: identification of a mutant enzyme. *Eur. J. Biochem.* **126**:327–332.

1797. Hove-Jensen, B., and P. Nygaard. 1989. Role of guanosine kinase in the utilization of guanosine for nucleotide synthesis in *Escherichia coli*. *J. Gen. Microbiol.* **135**:1263–1273.

1798. Howard, P. K., J. Shaw, and A. J. Otsuka. 1985. Nucleotide sequence of the *birA* gene encoding the biotin operon repressor and biotin holoenzyme synthetase functions of *Escherichia coli*. *Gene* **35**:321–331.

1799. Hrebenda, J., H. Heleszko, K. Brzostek, and J. Bielecki. 1985. Mutation affecting resistance of *Escherichia coli* K12 to nalidixic acid. *J. Gen. Microbiol.* **131**:2285–2292.

1800. Hryniewicz, M., A. Sirko, A. Palucha, A. Bock, and D. Hulanicka. 1990. Sulfate and thiosulfate transport in *Escherichia coli* K-12: identification of a gene encoding a novel protein involved in thiosulfate binding. *J. Bacteriol.* **172**:3358–3366.

1801. Hsieh, M., P. Hensley, M. Brenowitz, and J. S. Fetrow. 1994. A molecular model of the inducer binding domain of the galactose repressor of *Escherichia coli*. *J. Biol. Chem.* **269**:13825–13835.

1802. Hsu, L., S. Jackowski, and C. O. Rock. 1991. Isolation and characterization of *Escherichia coli* K-12 mutants lacking both 2-acyl-glycerophosphoethanolamine acyltransferase and acyl-acyl carrier protein synthetase activity. *J. Biol. Chem.* **266**:13783–13788.

1803. Hsu, L. M., H. J. Klee, J. Zagorski, and M. J. Fournier. 1984. Structure of an *Escherichia coli* tRNA operon containing linked genes for arginine, histidine, leucine, and proline tRNAs. *J. Bacteriol.* **158**:934–942.

1804. Hsu, L. M., J. Zagorski, and M. J. Fournier. 1984. Cloning and sequence analysis of the *Escherichia coli* 4.5 S RNA gene. *J. Mol. Biol.* **178**:509–531.

1805. Hsu, L. M., J. Zagorski, Z. Wang, and M. J. Fournier. 1985. *Escherichia coli* 6S RNA gene is part of a dual-function transcription unit. *J. Bacteriol.* **161**:1162–1170.

1806. Hsu, P. L., W. Ross, and A. Landy. 1980. The lambda phage *att* site: functional limits and interaction with Int protein. *Nature* (London) **285**:85–91.

1807. Hu, M., and R. C. Deonier. 1981. Mapping of IS*1* elements flanking the *argF* gene region on the *Escherichia coli* K-12 chromosome. *Mol. Gen. Genet.* **181**:222–229.

1808. Huang, A. M., J. I. Lee, S. C. King, and T. H. Wilson. 1992. Amino acid substitution in the lactose carrier protein with the use of amber suppressors. *J. Bacteriol.* **174**:5436–5441.

1809. Huang, F., G. Coppola, and D. H. Calhoun. 1992. Multiple transcripts encoded by the *ilvGMEDA* gene cluster of *Escherichia coli* K-12. *J. Bacteriol.* **174**:4871–4877.

1810. Huang, S., and M. P. Deutscher. 1992. Sequence and transcriptional analysis of the *Escherichia coli rnt* gene encoding RNase T. *J. Biol. Chem.* **267**:25609–25613.

1811. Hubscher, U., and A. Kornberg. 1979. The delta subunit of *Escherichia coli* DNA polymerase III holoenzyme is the *dnaX* gene product. *Proc. Natl. Acad. Sci. USA* **76**:6284–6288.

1812. Hudson, A. J., S. C., Andrews, C. Hawkins, J. M. Williams, M. Izuhara, F. C. Meldrum, S. Mann, P. M. Harrison, and J. R. Guest. 1993. Overproduction, purification and characterization of the *Escherichia coli* ferritin. *Eur. J. Biochem.* **218**:985–995.

1813. **Hudson, G. S., and B. E. Davidson.** 1984. Nucleotide sequence and transcription of the phenylalanine and tyrosine operons of *Escherichia coli* K12. *J. Mol. Biol.* **180**:1023–1051.

1814. **Hudson, G. S., P. Rellos, and B. E. Davidson.** 1991. Two promoters control the *aroH* gene of *Escherichia coli*. *Gene* **102**:87–91.

1815. **Hudson, L., J. Rossi, and A. Landy.** 1981. Dual function transcripts specifying tRNA and mRNA. *Nature* (London) **294**:422–427.

1816. **Hufton, S. E., R. J. Ward, N. A. Bunce, J. T. Armstrong, A. J. Fletcher, and R. E. Glass.** 1995. Structure-function analysis of the vitamin B$_{12}$ receptor of *Escherichia coli* by means of informational suppression. *Mol. Microbiol.* **15**:381–393.

1817. **Hughes, D., and R. H. Buckingham.** 1991. The nucleotide sequence of *rpsL* and its flanking regions in *Salmonella typhimurium*. *Gene* **104**:123–124.

1818. **Hughes, K. T., A. Dessen, J. P. Gray, and C. Grubmeyer.** 1993. The *Salmonella typhimurium nadC* gene: sequence determination by use of Mud-P22 and purification of quinolinate phosphoribosyltransferase. *J. Bacteriol.* **175**:479–486.

1819. **Hugouvieux-Cotte-Pattat, N., and J. Robert-Baudouy.** 1981. Isolation of fusions between the *lac* genes and several genes of the *exu* regulon: analysis of their regulation, determination of the transcription direction of the *uxaC-uxaA* operon, in *Escherichia coli* K-12. *Mol. Gen. Genet.* **182**:279–287.

1820. **Hugouvieux-Cotte-Pattat, N., and J. Robert-Baudouy.** 1982. Determination of the transcription direction of the *exuT* gene in *Escherichia coli* K-12: divergent transcription of the *exuT-uxaCA* operons. *J. Bacteriol.* **151**:480–484.

1821. **Hugouvieux-Cotte-Pattat, N., and J. Robert-Baudouy.** 1983. Determination of the transcription direction of the *uxaB* gene, in *Escherichia coli* K12. *Mol. Gen. Genet.* **189**:334–336.

1822. **Hugovieux-Cotte-Pattat, N., and J. Robert-Baudouy.** 1982. Regulation and transcription direction of *exuR*, a self-regulated repressor in *Escherichia coli* K-12. *J. Mol. Biol.* **156**:221–228.

1823. **Hui, I., K. Maltman, R. Little, S. Hastrup, M. Johnsen, N. Fiil, and P. Dennis.** 1982. Insertions of transposon Tn5 into ribosomal protein PNA polymerase operons. *J. Bacteriol.* **152**:1022–1032.

1824. **Huisman, G. W., and R. Kolter.** 1994. Sensing starvation: a homoserine lactone–dependent signaling pathway in *Escherichia coli*. *Science* **265**:537–539.

1825. **Huisman, O., R. D'Ari, and S. Gottesman.** 1984. Cell-division control in *Escherichia coli*: specific induction of the SOS function SfiA protein is sufficient to block septation. *Proc. Natl. Acad. Sci. USA* **81**:4490–4494.

1826. **Hull, E. P., M. E. Spencer, D. Wood, and J. R. Guest.** 1983. Nucleotide sequence of the promoter region of the citrate synthase gene (*gltA*) of *Escherichia coli*. *FEBS Lett.* **156**:366–370.

1827. **Hulton, C. S., A. Seirafi, J. C. Hinton, J. M. Sidebotham, L. Waddell, G. D. Pavitt, T. Owen-Hughes, A. Spassky, H. Buc, and C. F. Higgins.** 1990. Histone-like protein H1 (H-NS), DNA supercoiling, and gene expression in bacteria. *Cell* **63**:631–642.

1828. **Humbert, R., and R. D. Simoni.** 1980. Genetic and biomedical studies demonstrating a second gene coding for asparagine synthetase in *Escherichia coli*. *J. Bacteriol.* **142**:212–220.

1829. **Hunt, C. L., V. Colless, M. T. Smith, D. O. Molasky, M. S. Malo, and R. E. Loughlin.** 1987. Lambda transducing phage and clones carrying genes of the *cysJIHDC* gene cluster of *Escherichia coli* K12. *J. Gen. Microbiol.* **133**:2707–2717.

1830. **Hunt, M. D., G. S. Pettis, and M. A. McIntosh.** 1994. Promoter and operator determinants for fur-mediated iron regulation in the bidirectional *fepA-fes* control region of the *Escherichia coli* enterobactin gene system. *J. Bacteriol.* **176**:3944–3955.

1831. **Hunt, T. P., and B. Magasanik.** 1985. Transcription of *glnA* by purified *Escherichia coli* components: core RNA polymerase and the products of *glnF*, *glnG*, and *glnL*. *Proc. Natl. Acad. Sci. USA* **82**:8453–8457.

1832. **Hupp, T. R., and J. M. Kaguni.** 1988. Suppression of the *Escherichia coli* dnaA46 mutation by a mutation in *trxA*, the gene for thioredoxin. *Mol. Gen. Genet.* **213**:471–478.

1833. **Hurley, J. H., A. M. Dean, J. L. Sohl, D. E. Koshland, Jr., and R. M. Stroud.** 1990. Regulation of an enzyme by phosphorylation at the active site. *Science* **249**:1012–1016.

1834. **Hurley, J. H., H. R. Faber, D. Worthylake, N. D. Meadow, S. Roseman, D. W. Pettigrew, and S. J. Remington.** 1993. Structure of the regulatory complex of *Escherichia coli* IIIGlc with glycerol kinase. *Science* **259**:673–677.

1835. **Hurley, J. H., P. E. Thorsness, V. Ramalingam, N. H. Helmers, D. E. Koshland, Jr., and R. M. Stroud.** 1989. Structure of a bacterial enzyme regulated by phosphorylation, isocitrate dehydrogenase. *Proc. Natl. Acad. Sci. USA* **86**:8635–8639.

1836. **Husain, I., and A. Sancar.** 1987. Photoreactivation in *phr* mutants of *Escherichia coli* K-12. *J. Bacteriol.* **169**:2367–2372.

1837. **Husain, I., B. Van Houten, D. C. Thomas, and A. Sancar.** 1986. Sequences of *Escherichia coli* uvrA gene and protein reveal two potential ATP binding sites. *J. Biol. Chem.* **261**:4895–4901.

1838. **Hussain, H., J. Grove, L. Griffiths, S. Busby, and J. Cole.** 1994. A seven-gene operon essential for formate-dependent nitrite reduction to ammonia by enteric bacteria. *Mol. Microbiol.* **12**:153–163.

1839. **Hussain, K., E. J. Elliott, and G. P. Salmond.** 1987. The *parD*⁻ mutant of *Escherichia coli* also carries a *gyrAam* mutation. The complete sequence of *gyrA*. *Mol. Microbiol.* **1**:259–273.

1840. **Hussein, S., K. Hantke, and V. Braun.** 1981. Citrate-dependent iron transport system in *Escherichia coli* K-12. *Eur. J. Biochem.* **117**:431–437.

1841. **Huynh, Q. K.** 1991. 5-Enolpyruvylshikimate-3-phosphate synthase from *Escherichia coli*—the substrate analogue bromopyruvate inactivates the enzyme by modifying Cys-408 and Lys-411. *Arch. Biochem. Biophys.* **284**:407–412.

1842. **Huynh, Q. K.** 1993. Photo-oxidation of 5-enolpyruvoylshikimate-3-phosphate synthase from *Escherichia coli*: evidence for a reactive imidazole group (His385) at the herbicide glyphosate-binding site. *Biochem. J.* **290**:525–530.

1843. **Hwang, D. S., E. Crooke, and A. Kornberg.** 1990. Aggregated *dnaA* protein is dissociated and activated for DNA replication by phospholipase or *dnaK* protein. *J. Biol. Chem.* **265**:19244–19248.

1844. **Hwang, D. S., B. Thony, and A. Kornberg.** 1992. IciA protein, a specific inhibitor of initiation of *Escherichia coli* chromosomal replication. *J. Biol. Chem.* **267**:2209–2213.

1845. **Hwang, Y. W., R. Engel, and B. E. Tropp.** 1984. Correlation of 3,4-dihydroxybutyl 1-phosphonate resistance with a defect in cardiolipin synthesis in *Escherichia coli*. *J. Bacteriol.* **157**:846–856.

1846. **Hwang, Y. W., P. G. McCabe, M. A. Innis, and D. L. Miller.** 1989. Site-directed mutagenesis of the GDP binding domain of bacterial elongation factor Tu. *Arch. Biochem. Biophys.* **274**:394–403.

1847. **Hwang, Y. W., and D. L. Miller.** 1987. A mutation that alters the nucleotide specificity of elongation factor Tu, a GTP regulatory protein. *J. Biol. Chem.* **262**:13081–13085.

1848. **Hwang, Y. W., A. Sanchez, and D. L. Miller.** 1989. Mutagenesis of bacterial elongation factor Tu at lysine 136. A conserved amino acid in GTP regulatory proteins. *J. Biol. Chem.* **264**:8304–8309.

1848a.**Iaccarino, M., J. Guardiola, and M. DeFelice.** 1978. On the permeability of biological membranes. *J. Membr. Sci.* **3**:287–302.

1849. **Ichihara, S., Y. Matsubara, C. Kato, K. Akasaka, and S. Mizushima.** 1993. Molecular cloning, sequencing, and mapping of the gene encoding protease I and characterization of proteinase and proteinase-defective *Escherichia coli* mutants. *J. Bacteriol.* **175**:1032–1037.

1850. **Ichihara, S., T. Suzuki, M. Suzuki, and S. Mizushima.** 1986. Molecular cloning and sequencing of the *sppA* gene and characterization of the encoded protease IV, a signal peptide peptidase, of *Escherichia coli*. *J. Biol. Chem.* **261**:9405–9411.

1851. **Ichikawa, J. K., C. Li, J. Fu, and S. Clarke.** 1994. A gene at 59 minutes on the *Escherichia coli* chromosome encodes a lipoprotein with unusual amino acid repeat sequences. *J. Bacteriol.* **176**:1630–1638.

1852. **Ichikawa, S., and A. Kaji.** 1989. Molecular cloning and expression of ribosome releasing factor. *J. Biol. Chem.* **264**:20054–20059.

1853. **Ichikawa, S., M. Ryoji, Z. Siegfried, and A. Kaji.** 1989. Localization of the ribosome-releasing factor gene in the *Escherichia coli* chromosome. *J. Bacteriol.* **171**:3689–³695.

1854. **Icho, T.** 1↑ ↓8. Membrane-bound phosphatases in *Escherichia coli*: sequence of the *pgpB* gene and dual subcellular localization of the *pgpB* product. *J. Bacteriol.* **170**:5117–5124.

1855. **Icho, T.** 1988. Membrane-bound phosphatases in *Escherichia coli*: sequence of the *pgpA* gene. *J. Bacteriol.* **170**:5110–5116.

1856. **Icho, T., C. E. Bulawa, and C. R. Raetz.** 1985. Molecular cloning and sequencing of the gene for CDP-diglyceride hydrolase of *Escherichia coli*. *J. Biol. Chem.* **260**:12092–12098.

1857. **Icho, T., and C. R. Raetz.** 1983. Multiple genes for membrane-bound phosphatases in *Escherichia coli* and their action on phospholipid precursors. *J. Bacteriol.* **153**:722–730.

1858. **Icho, T., C. P. Sparrow, and C. R. Raetz.** 1985. Molecular cloning and sequencing of the gene for CDP-diglyceride synthetase of *Escherichia coli*. *J. Biol. Chem.* **260**:12078–12083.

1859. **Igarashi, K., N. Fujita, and A. Ishihama.** 1989. Promoter selectivity of *Escherichia coli* RNA polymerase: omega factor is responsible for the ppGpp sensitivity. *Nucleic Acids Res.* **17**:8755–8765.

1860. **Igarashi, K., N. Fujita, and A. Ishihama.** 1990. Sequence analysis of two temperature-sensitive mutations in the alpha subunit gene (*rpoA*) of *Escherichia coli* RNA polymerase. *Nucleic Acids Res.* **18**:5945–5948.

1861. **Igarashi, K., A. Hanamura, K. Makino, H. Aiba, T. Mizuno, A. Nakata, and A. Ishihama.** 1991. Functional map of the alpha subunit of *Escherichia coli* RNA polymerase: two modes of transcription activation by positive factors. *Proc. Natl. Acad. Sci. USA* **88**:8958–8962.

1862. **Igarashi, K., S. Hiraga, and T. Yura.** 1967. A deoxythymidine kinase deficient mutant of *Escherichia coli*. II. Mapping and transduction studies with phage phi 80. *Genetics* **57**:643–654.

1863. **Igarashi, K., and A. Ishihama.** 1991. Bipartite functional map of the E. coli RNA polymerase alpha subunit: involvement of the C-terminal region in transcription activation by cAMP-CRP. *Cell* **65**:1015–1022.

1864. **Iggo, R., S. Picksley, J. Southgate, J. McPheat, and D. P. Lane.** 1990. Identification of a putative RNA helicase in *E. coli*. *Nucleic Acids Res.* **18**:5413–5417.

1865. **Iida, A., S. Harayama, T. Iino, and G. L. Hazelbauer.** 1984. Molecular cloning and characterization of genes required for ribose transport and utilization in *Escherichia coli* K-12. *J. Bacteriol.* **158**:674–682.

1866. **Iida, A., S. Teshiba, and K. Mizobuchi.** 1993. Identification and characterization of the *tktB* gene encoding a second transketolase in *Escherichia coli* K-12. *J. Bacteriol.* **175**:5375–5383.

1867. **Iida, K., Y. Hirota, and U. Schwarz.** 1983. Mutants of *Escherichia coli* defective in penicillin-insensitive murein DD-endopeptidase. *Mol. Gen. Genet.* **189**:215–221.

1868. Ikeda, M., T. Sato, M. Wachi, H. K. Jung, F. Ishino, Y. Kobayashi, and M. Matsuhashi. 1989. Structural similarity among *Escherichia coli* FtsW and RodA proteins and *Bacillus subtilis* SpoVE protein, which function in cell division, cell elongation, and spore formation, respectively. *J. Bacteriol.* 171:6375–6378.

1869. Ikeda, M., M. Wachi, F. Ishino, and M. Matsuhashi. 1990. Nucleotide sequence involving *murD* and an open reading frame ORF-Y spacing *murF* and *ftsW* in *Escherichia coli*. *Nucleic Acids Res.* 18:1058.

1870. Ikeda, M., M. Wachi, H. K. Jung, F. Ishino, and M. Matsuhashi. 1990. Nucleotide sequence involving *murG* and *murC* in the *mra* gene cluster region of *Escherichia coli*. *Nucleic Acids Res.* 18:4014.

1871. Ikeda, T., and D. C. LaPorte. 1991. Isocitrate dehydrogenase kinase/phosphatase: *aceK* alleles that express kinase but not phosphatase activity. *J. Bacteriol.* 173:1801–1806.

1871a. Ikeda, M., M. Wachi, H. K. Jung, F. Ishino, and M. Matsuhashi. 1991. The *Escherichia coli mraY* gene encoding UDP-*N*-acetylmuramoyl-pentapeptide:undecaprenyl-phosphate phospho-*N*-acetylmuramoyl-pentapeptide transferase. *J. Bacteriol.* 173:1021–1026.

1872. Ikemi, M., K. Murakami, M. Hashimoto, and Y. Murooka. 1992. Cloning and characterization of genes involved in the biosynthesis of delta-aminolevulinic acid in *Escherichia coli*. *Gene* 121:127–132.

1873. Ilag, L. L., and D. Jahn. 1992. Activity and spectroscopic properties of the *Escherichia coli* glutamate 1-semialdehyde aminotransferase and the putative active site mutant K265R. *Biochemistry* 31:7143–7151.

1874. Ilag, L. L., D. Jahn, G. Eggertsson, and D. Soll. 1991. The *Escherichia coli hemL* gene encodes glutamate 1-semialdehyde aminotransferase. *J. Bacteriol.* 173:3408–3413.

1875. Ilyina, T. S., E. V. Nechaeva, Y. M. Romanova, and G. B. Smirnov. 1981. Isolation and mapping of *Escherichia coli* K12 mutants defective in Tn9 transposition. *Mol. Gen. Genet.* 181:384–389.

1875a. Ilyina, T. S., Y. M. Romanova, and G. B. Smirnov. 1981. The effect of *tnm* mutations of *Escherichia coli* K12 on transposition of various movable genetic elements. *Mol. Gen. Genet.* 183:376–379.

1876. Im, S. W., and J. Pittard. 1971. Phenylalanine biosynthesis in *Escherichia coli* K-12: mutants derepressed for chorismate mutase P-prephenate dehydratase. *J. Bacteriol.* 106:784–790.

1877. Imaishi, H., M. Gomada, S. Inouye, and A. Nakazawa. 1993. Physical map location of the *rpoN* gene of *Escherichia coli*. *J. Bacteriol.* 175:1550–1551. (Erratum, 175:3244.)

1878. Imamoto, F., and Y. Nakamura. 1986. *Escherichia coli* proteins involved in regulation of transcription termination: function, structure, and expression of the *nusA* and *nusB* genes. *Adv. Biophys.* 21:175–192.

1879. Imamura, N., and H. Nakayama. 1981. *thiD* locus of *Escherichia coli*. *Experientia* 37:1265–1266.

1880. Imamura, N., and H. Nakayama. 1982. *thiK* and *thiL* loci of *Escherichia coli*. *J. Bacteriol.* 151:708–717.

1881. Inaba, K., T. Kuroda, T. Shimamoto, T. Kayahara, M. Tsuda, and T. Tsuchiya. 1994. Lithium toxicity and Na$^+$(Li$^+$)/H$^+$ antiporter in *Escherichia coli*. *Biol. Pharm. Bull.* 17:395–398.

1882. Ineichen, G., and A. J. Biel. 1993. Location of the *hemE* gene on the physical map of *Escherichia coli*. *J. Bacteriol.* 175:7749–7750.

1883. Inglese, J., J. M. Smith, and S. J. Benkovic. 1990. Active-site mapping and site-specific mutagenesis of glycinamide ribonucleotide transformylase from *Escherichia coli*. *Biochemistry* 29:6678–6687.

1884. Innis, M. A., M. Tokunaga, M. E. Williams, J. M. Loranger, S. Y. Chang, S. Chang, and H. C. Wu. 1984. Nucleotide sequence of the *Escherichia coli* prolipoprotein signal peptidase (*lsp*) gene. *Proc. Natl. Acad. Sci. USA* 81:3708–3712.

1885. Inoko, H., K. Shigesada, and M. Imai. 1977. Isolation and characterization of conditional-lethal *rho* mutants of *Escherichia coli*. *Proc. Natl. Acad. Sci. USA* 74:1162–1166.

1886. Inokuchi, H., M. Kodaira, F. Yamao, and H. Ozeki. 1979. Identification of transfer RNA suppressors in *Escherichia coli*. II. Duplicate genes for tRNA$_2^{Gln}$. *J. Mol. Biol.* 132:663–677.

1887. Inokuchi, K., H., F. Yamao, H. Sakano, and H. Ozeki. 1979. Identification of transfer RNA suppressors in *Escherichia coli*. I. Amber suppressor su^{+2}, an anticodon mutant of tRNA$_2^{Gln}$. *J. Mol. Biol.* 132:649–662.

1888. Inokuchi, K., H. Furukawa, K. Nakamura, and S. Mizushima. 1984. Characterization by deletion mutagenesis in vitro of the promoter region of *ompF*, a positively regulated gene of *Escherichia coli*. *J. Mol. Biol.* 178:653–668.

1889. Inokuchi, K., N. Mutoh, S. Matsuyama, and S. Mizushima. 1982. Primary structure of the *ompF* gene that codes for a major outer membrane protein of *Escherichia coli* K-12. *Nucleic Acids Res.* 10:6957–6968.

1890. Inoue, K., S. Kuramitsu, K. Aki, Y. Watanabe, T. Takagi, M. Nishigai, A. Ikai, and H. Kagamiyama. 1988. Branched-chain amino acid aminotransferase of *Escherichia coli*: overproduction and properties. *J. Biochem.* 104:777–784.

1891. Inoue, K., S. Kuramitsu, A. Okamoto, K. Hirotsu, T. Higuchi, and H. Kagamiyama. 1991. Site-directed mutagenesis of *Escherichia coli* aspartate aminotransferase: role of Tyr70 in the catalytic processes. *Biochemistry* 30:7796–7801.

1892. Inouye, H., W. Barnes, and J. Beckwith. 1982. Signal sequence of alkaline phosphatase of *Escherichia coli*. *J. Bacteriol.* 149:434–439.

1893. Inouye, H., S. Michaelis, A. Wright, and J. Beckwith. 1981. Cloning and restriction mapping of the alkaline phosphatase structural gene (*phoA*) of *Escherichia coli* and generation of deletion mutants in vitro. *J. Bacteriol.* 146:668–675.

1894. Inouye, M., and N. Delihas. 1988. Small RNAs in the prokaryotes: a growing list of diverse roles. *Cell* 53:5–7.

1895. Inouye, S., S. Wang, J. Sekizawa, S. Halegoua, and M. Inouye. 1977. Amino acid sequence for the peptide extension on the prolipoprotein of the *Escherichia coli* outer membrane. *Proc. Natl. Acad. Sci. USA* 74:1004–1008.

1896. Iobbi-Nivol, C., H. Crooke, L. Griffiths, J. Grove, H. Hussain, J. Pommier, V. Mejean, and J. A. Cole. 1994. A reassessment of the range of *c*-type cytochromes synthesized by *Escherichia coli* K-12. *FEMS Microbiol. Lett.* 119:89–94.

1897. Irino, N., K. Nakayama, and H. Nakayama. 1986. The *recQ* gene of *Escherichia coli* K12: primary structure and evidence for SOS regulation. *Mol. Gen. Genet.* 205:298–304.

1898. Ishiai, M., C. Wada, Y. Kawasaki, and T. Yura. 1992. Mini-F plasmid mutants able to replicate in *Escherichia coli* deficient in the DnaJ heat shock protein. *J. Bacteriol.* 174:5597–5603.

1899. Ishige, K., S. Nagasawa, S. Tokishita, and T. Mizuno. 1994. A novel device of bacterial signal transducers. *EMBO J.* 1354:5195–5202.

1900. Ishihama, A. 1993. Protein-protein communication within the transcription apparatus. *J. Bacteriol.* 175:2483–2489.

1901. Ishii, S., E. Hatada, T. Maekawa, and F. Imamoto. 1984. Molecular cloning and nucleotide sequencing of the *nusB* gene of *E. coli*. *Nucleic Acids Res.* 12:4987–4995.

1902. Ishii, S., M. Ihara, T. Maekawa, Y. Nakamura, H. Uchida, and F. Imamoto. 1984. The nucleotide sequence of the cloned *nusA* gene and its flanking region of *Escherichia coli*. *Nucleic Acids Res.* 12:3333–3342.

1903. Ishii, S., K. Kuroki, and F. Imamoto. 1984. tRNA$_{f2}^{Met}$ gene in the leader region of the *nusA* operon in *Escherichia coli*. *Proc. Natl. Acad. Sci. USA* 81:409–413.

1904. Ishii, S., T. Murakami, and K. Shishido. 1992. A pSC101-par sequence-mediated study on the intracellular state of supercoiling of the pBR322 genome in *Escherichia coli* DNA topoisomerase I deletion mutant. *FEMS Microbiol. Lett.* 72:115–120.

1905. Ishikawa, K., S. Kimura, S. Kanaya, K. Morikawa, and H. Nakamura. 1993. Structural study of mutants of *Escherichia coli* ribonuclease HI with enhanced thermostability. *Protein Eng.* 6:85–91.

1906. Ishino, F., H. K. Jung, M. Ikeda, M. Doi, M. Wachi, and M. Matsuhashi. 1989. New mutations *fts-36*, *lts-33*, and *ftsW* clustered in the *mra* region of the *Escherichia coli* chromosome induce thermosensitive cell growth and division. *J. Bacteriol.* 171:5523–5530.

1907. Ishino, F., K. Mitsui, S. Tamaki, and M. Matsuhashi. 1980. Dual enzyme activities of cell wall peptidoglycan synthesis, peptidoglycan transglycosylase and penicillin-sensitive transpeptidase, in purified preparations of *Escherichia coli* penicillin-binding protein 1A. *Biochem. Biophys. Res. Commun.* 97:287–293.

1908. Ishino, Y., H. Shinagawa, K. Makino, M. Amemura, and A. Nakata. 1987. Nucleotide sequence of the *iap* gene, responsible for alkaline phosphatase isozyme conversion in *Escherichia coli*, and identification of the gene product. *J. Bacteriol.* 169:5429–5433.

1909. Ishino, Y., H. Shinagawa, K. Makino, S. Tsunasawa, F. Sakiyama, and A. Nakata. 1986. Nucleotide sequence of the *lig* gene and primary structure of DNA ligase of *Escherichia coli*. *Mol. Gen. Genet.* 204:1–7.

1910. Island, M. D., and R. J. Kadner. 1993. Interplay between the membrane-associated UhpB and UhpC regulatory proteins. *J. Bacteriol.* 175:5028–5034.

1911. Island, M. D., B. Y. Wei, and R. J. Kadner. 1992. Structure and function of the *uhp* genes for the sugar phosphate transport system in *Escherichia coli* and *Salmonella typhimurium*. *J. Bacteriol.* 174:2754–2762.

1912. Isono, K., and S. Isono. 1980. Ribosomal protein modification in *Escherichia coli*. II. Studies of a mutant lacking the N-terminal acetylation of protein S18. *Mol. Gen. Genet.* 177:645–651.

1913. Isono, K., J. Schnier, and M. Kitakawa. 1980. Genetic fine structure of the *pyrE* region containing the genes for ribosomal proteins L28 and L33 in *Escherichia coli*. *Mol. Gen. Genet.* 179:311–317.

1914. Isono, S., and K. Isono. 1981. Ribosomal protein modification in *Escherichia coli*. III. Studies of mutants lacking an acetylase activity specific for protein L12. *Mol. Gen. Genet.* 183:473–477.

1915. Isono, S., S. Thamm, M. Kitakawa, and K. Isono. 1985. Cloning and nucleotide sequencing of the genes for ribosomal proteins S9 (*rpsI*) and L13 (*rplM*) of *Escherichia coli*. *Mol. Gen. Genet.* 198:279–282.

1916. Isturiz, T., E. Palmero, and J. Vitelli-Flores. 1986. Mutations affecting gluconate catabolism in *Escherichia coli*. Genetic mapping of the locus for the thermosensitive gluconokinase. *J. Gen. Microbiol.* 132:3209–3219.

1917. Itagaki, E., and L. P. Hager. 1968. The amino acid sequence of cytochrome b562 of *Escherichia coli*. *Biochem. Biophys. Res. Commun.* 32:1013–1019.

1918. Itaya, M. 1990. Isolation and characterization of a second RNase H (RNase HII) of *Escherichia coli* K-12 encoded by the *rnhB* gene. *Proc. Natl. Acad. Sci. USA* 87:8587–8591.

1919. Itikawa, H., H. Fujita, and M. Wada. 1986. High temperature induction of a stringent response in the *dnaK*(Ts) and *dnaJ*(Ts) mutants of *Escherichia coli*. *J. Biochem.* 99:1719–1724.

1920. Ito, K. 1990. Structure, function, and biogenesis of SecY, an integral membrane protein involved in protein export. *J. Bioenerg. Biomembr.* 22:353–367.

1921. Ito, K., D. P. Cerretti, H. Nashimoto, and M. Nomura. 1984. Characterization of an amber mutation in the structural gene for ribosomal protein L15, which impairs the expression of the protein export gene, *secY*, in *Escherichia coli*. *EMBO J.* 3:2319–2324.

1922. Ito, K., K. Egawa, and Y. Nakamura. 1991. Genetic interaction between the beta′ subunit of RNA polymerase and the arginine-rich domain of *Escherichia coli* nusA protein. *J. Bacteriol.* 173:1492–1501.

1923. Ito, K., Y. Hirota, and Y. Akiyama. 1989. Temperature-sensitive *sec* mutants of *Escherichia coli*: inhibition of protein export at the permissive temperature. *J. Bacteriol.* 171:1742–1743.

1924. Ito, K., K. Kawakami, and Y. Nakamura. 1993. Multiple control of *Escherichia coli* lysyl-tRNA synthetase expression involves a transcriptional repressor and a translational enhancer element. *Proc. Natl. Acad. Sci. USA* 90:302–306.

1925. Ito, K., T. Oshima, T. Mizuno, and Y. Nakamura. 1994. Regulation of lysyl-tRNA synthetase expression by histone-like protein H-NS of *Escherichia coli*. *J. Bacteriol.* 176:7383–7386.

1926. Iuchi, S., A. Aristarkhov, J. M. Dong, J. S. Taylor, and E. C. Lin. 1994. Effects of nitrate respiration on expression of the Arc-controlled operons encoding succinate dehydrogenase and flavin-linked L-lactate dehydrogenase. *J. Bacteriol.* 176:1695–1701.

1926a. Iuchi, S., D. C. Cameron, and E. C. Lin. 1989. A second global regulator gene (*arcB*) mediating repression of enzymes in aerobic pathways of *Escherichia coli*. *J. Bacteriol.* 171:868–873.

1927. Iuchi, S., and E. C. Lin. 1987. The *narL* gene product activates the nitrate reductase operon and represses the fumarate reductase and trimethylamine N-oxide reductase operons in *Escherichia coli*. *Proc. Natl. Acad. Sci. USA* 84:3901–3905.

1928. Iuchi, S., and E. C. Lin. 1988. *arcA* (*dye*), a global regulatory gene in *Escherichia coli* mediating repression of enzymes in aerobic pathways. *Proc. Natl. Acad. Sci. USA* 85:1888–1892.

1929. Iuchi, S., and E. C. Lin. 1992. Purification and phosphorylation of the Arc regulatory components of *Escherichia coli*. *J. Bacteriol.* 174:5617–5623.

1930. Iuchi, S., and E. C. Lin. 1993. Adaptation of *Escherichia coli* to redox environments by gene expression. *Mol. Microbiol.* 9:9–15.

1931. Iuchi, S., Z. Matsuda, T. Fujiwara, and E. C. Lin. 1990. The *arcB* gene of *Escherichia coli* encodes a sensor-regulator protein for anaerobic repression of the *arc* modulon. *Mol. Microbiol.* 4:715–727.

1932. Ivanisevic, R., M. Milic, D. Ajdic, J. Rakonjac, and D. J. Savic. 1995. Nucleotide sequence, mutational analysis, transcriptional start site, and product analysis of *nov*, the gene which affects *Escherichia coli* K-12 resistance to the gyrase inhibitor novobiocin. *J. Bacteriol.* 177:1766–1771.

1933. Ivanova, A., M. Renshaw, R. V. Guntaka, and A. Eisenstark. 1992. DNA base sequence variability in *katF* (putative sigma factor) gene of *Escherichia coli*. *Nucleic Acids Res.* 20:5479–5480.

1934. Ivey, D. M., A. A. Guffanti, J. Zemsky, E. Pinner, R. Karpel, E. Padan, S. Schuldiner, and T. A. Krulwich. 1993. Cloning and characterization of a putative Ca^{2+}/H$^+$ antiporter gene from *Escherichia coli* upon functional complementation of Na$^+$/H$^+$ antiporter-deficient strains by the overexpressed gene. *J. Biol. Chem.* 268:11296–11303.

1935. Iwakura, M., Y. Shimura, and K. Tsuda. 1982. Cloning of dihydrofolate reductase gene of *Escherichia coli* K12. *J. Biochem.* 91:1205–1212.

1936. Iwakura, M., Y. Shimura, and K. Tsuda. 1982. Isolation of DNA fragment containing *phoS* gene of *Escherichia coli* K-12. *J. Biochem.* 92:615–622.

1937. Iwamoto, A., H. Omote, H. Hanada, N. Tomioka, A. Itai, M. Maeda, and M. Futai. 1991. Mutations in Ser174 and the glycine-rich sequence (Gly149, Gly150, and Thr156) in the beta subunit of *Escherichia coli* H$^+$-ATPase. *J. Biol. Chem.* 266:16350–16355.

1938. Iwasaki, H., Y. Ishino, H. Toh, A. Nakata, and H. Shinagawa. 1991. *Escherichia coli* DNA polymerase II is homologous to alpha-like DNA polymerases. *Mol. Gen. Genet.* 226:24–33.

1939. Iwasaki, H., A. Nakata, G. C. Walker, and H. Shinagawa. 1990. The *Escherichia coli* polB gene, which encodes DNA polymerase II, is regulated by the SOS system. *J. Bacteriol.* 172:6268–6273.

1940. Iwasaki, H., T. Shiba, A. Nakata, and H. Shinagawa. 1989. Involvement in DNA repair of the *ruvA* gene of *Escherichia coli*. *Mol. Gen. Genet.* 219:328–331.

1941. Iwasaki, H., M. Takahagi, T. Shiba, A. Nakata, and H. Shinagawa. 1991. *Escherichia coli* RuvC protein is an endonuclease that resolves the Holliday structure. *EMBO J.* 10:4381–4389.

1942. Iwaya, M., C. W. Jones, J. Khorana, and J. L. Strominger. 1978. Mapping of the mecillinam-resistant, round morphological mutants of *Escherichia coli*. *J. Bacteriol.* 133:196–202.

1943. Izuhara, M., K. Takamune, and R. Takata. 1991. Cloning and sequencing of an *Escherichia coli* K12 gene which encodes a polypeptide having similarity to the human ferritin H subunit. *Mol. Gen. Genet.* 225:510–513.

1944. Jackowski, S., and J. H. Alix. 1990. Cloning, sequence, and expression of the pantothenate permease (*panF*) gene of *Escherichia coli*. *J. Bacteriol.* 172:3842–3848.

1945. Jackowski, S., P. D. Jackson, and C. O. Rock. 1994. Sequence and function of the *aas* gene in *Escherichia coli*. *J. Biol. Chem.* 269:2921–2928.

1946. Jackowski, S., and C. O. Rock. 1987. Altered molecular form of acyl carrier protein associated with beta-ketoacyl-acyl carrier protein synthase II (*fabF*) mutants. *J. Bacteriol.* 169:1469–1473.

1947. Jackson, B. J., J. P. Bohin, and E. P. Kennedy. 1984. Biosynthesis of membrane-derived oligosaccharides: characterization of *mdoB* mutants defective in phosphoglycerol transferase I activity. *J. Bacteriol.* 160:976–981.

1948. Jackson, J. H., E. J. Davis, A. C. Madu, and S. E. Braxter. 1981. Three-factor reciprocal cross mapping of a gene that causes expression of feedback-resistant

1949. Jackson, R. H., A. Cornish-Bowden, and J. A. Cole. 1981. Prosthetic groups of the NADH-dependent nitrite reductase from *Escherichia coli* K12. *Biochem. J.* 193:861–867.

1950. Jacobi, A., R. Rossmann, and A. Bock. 1992. The *hyp* operon gene products are required for the maturation of catalytically active hydrogenase isoenzymes in *Escherichia coli*. *Arch. Microbiol.* 158:444–451.

1951. Jacques, N., J. Guillerez, and M. Dreyfus. 1992. Culture conditions differentially affect the translation of individual *Escherichia coli* mRNAs. *J. Mol. Biol.* 226:597–608.

1952. Jacquet, E., and A. Parmeggiani. 1989. Substitution of Val20 by Gly in elongation factor Tu. Effects on the interaction with elongation factors Ts, aminoacyl-tRNA and ribosomes. *Eur. J. Biochem.* 185:341–346.

1953. Jagadeeswaran, P., C. R. Ashman, S. Roberts, and J. Langenberg. 1984. Nucleotide sequence and analysis of deletion mutants of the *Escherichia coli* gpt gene in plasmid pSV2 gpt. *Gene* 31:309–313.

1954. Jager, J., T. Solmajer, and J. N. Jansonius. 1992. Computational approach towards the three-dimensional structure of *E. coli* tyrosine aminotransferase. *FEBS Lett.* 306:234–238.

1955. Jagura-Burdzy, G., and D. Hulanicka. 1981. Use of gene fusions to study expression of *cysB*, the regulatory gene of the cysteine regulon. *J. Bacteriol.* 147:744–751.

1956. Jahn, D., U. Michelsen, and D. Soll. 1991. Two glutamyl-tRNA reductase activities in *Escherichia coli*. *J. Biol. Chem.* 266:2542–2548.

1957. Jahreis, K., P. W. Postma, and J. W. Lengeler. 1991. Nucleotide sequence of the *ilvH-fruR* gene region of *Escherichia coli* K12 and *Salmonella typhimurium* LT2. *Mol. Gen. Genet.* 226:332–336.

1958. James, R., D. O. Dean, and J. Debbage. 1993. Five open reading frames upstream of the *dnaK* gene of *E. coli*. *DNA Seq.* 3:327–332.

1959. Janda, I., M. Kitakawa, and K. Isono. 1985. Gene *rpmF* for ribosomal protein L32 and gene *rimJ* for a ribosomal protein acetylating enzyme are located near *pyrC* (23.4 min) in *Escherichia coli*. *Mol. Gen. Genet.* 201:433–436.

1960. Jander, G., N. L. Martin, and J. Beckwith. 1994. Two cysteines in each periplasmic domain of the membrane protein DsbB are required for its function in protein disulfide bond formation. *EMBO J.* 1354:5121–5127.

1961. Janosi, L., I. Shimizu, and A. Kaji. 1994. Ribosome recycling factor (ribosome releasing factor) is essential for bacterial growth. *Proc. Natl. Acad. Sci. USA* 91:4249–4253.

1962. Jans, D. A., A. L. Fimmel, L. Langman, L. B. James, J. A. Downie, A. E. Senior, G. R. Ash, F. Gibson, and G. B. Cox. 1983. Mutations in the *uncE* gene affecting assembly of the c-subunit of the adenosine triphosphatase of *Escherichia coli*. *Biochem. J.* 211:717–726.

1963. Jaurin, B., and T. Grundstrom. 1981. *ampC* cephalosporinase of *Escherichia coli* K-12 has a different evolutionary origin from that of beta-lactamases of the penicillinase type. *Proc. Natl. Acad. Sci. USA* 78:4897–4901.

1964. Jaurin, B., T. Grundstrom, T. Edlund, and S. Normark. 1981. The *E. coli* beta-lactamase attenuator mediates growth rate-dependent regulation. *Nature* (London) 290:221–225.

1965. Jaurin, B., T. Grundstrom, and S. Normark. 1982. Sequence elements determining *ampC* promoter strength in *E. coli*. *EMBO J.* 1:875–881.

1966. Jayakumar, A., S. J. Hwang, J. M. Fabiny, A. C. Chinault, and E. M. Barnes, Jr. 1989. Isolation of an ammonium or methylammonium ion transport mutant of *Escherichia coli* and complementation by the cloned gene. *J. Bacteriol.* 171:996–1001.

1967. Jayakumar, A., K. E. Rudd, J. M. Fabiny, and E. M. Barnes, Jr. 1991. Localization of the *Escherichia coli* amtA gene to 95.8 minutes. *J. Bacteriol.* 173:1572–1573.

1968. Jayaraman, P. S., T. C. Peakman, S. J. Busby, R. V. Quincey, and J. A. Cole. 1987. Location and sequence of the promoter of the gene for the NADH-dependent nitrite reductase of *Escherichia coli* and its regulation by oxygen, the Fnr protein and nitrite. *J. Mol. Biol.* 196:781–788.

1968a. Jayaratne, P., D. Bronner, P. R. MacLachlan, C. Dodgson, N. Kido, and C. Whitfield. 1994. Cloning and analysis of duplicated *rfbM* and *rfbK* genes involved in the formation of GDP-mannose in *Escherichia coli* O9_K30 and participation of *rfb* genes in the synthesis of the group I K30 capsular polysaccharide. *J. Bacteriol.* 176:3126–3139.

1969. Jeanteur, D., J. H. Lakey, and F. Pattus. 1991. The bacterial porin superfamily: sequence alignment and structure prediction. *Mol. Microbiol.* 5:2153–2164.

1970. Jefferson, R. A., S. M. Burgess, and D. Hirsh. 1986. β-Glucuronidase from *Escherichia coli* as a gene-fusion marker. *Proc. Natl. Acad. Sci. USA* 83:8447–8451.

1971. Jeggo, P. 1979. Isolation and characterization of *Escherichia coli* K-12 mutants unable to induce the adaptive response to simple alkylating agents. *J. Bacteriol.* 139:783–791.

1972. Jekel, M., and W. Wackernagel. 1994. Location of the *endA* gene coding for endonuclease I on the physical map of the *Escherichia coli* K-12 chromosome. *J. Bacteriol.* 176:1550–1551.

1973. Jenkins, L. S., and W. D. Nunn. 1987. Regulation of the *ato* operon by the *atoC* gene in *Escherichia coli*. *J. Bacteriol.* 169:2096–2102.

1974. Jenkins, L. S., and W. D. Nunn. 1987. Genetic and molecular characterization of the genes involved in short-chain fatty acid degradation in *Escherichia coli*: the *ato* system. *J. Bacteriol.* 169:42–52.

1974a. Jenkins, S. J., C. A. Sparkes, and M. C. Jones-Mortimer. 1974. A gene involved in lysine excretion in *Escherichia coli* K-12. *Heredity* 32:409–412.

1975. Jennings, M. P., and I. R. Beacham. 1990. Analysis of the *Escherichia coli* gene encoding L-asparaginase II, *ansB*, and its regulation by cyclic AMP receptor and FNR proteins. *J. Bacteriol.* **172:**1491–1498.

1976. Jensen, K. F. 1993. The *Escherichia coli* K-12 "wild types" W3110 and MG1655 have an *rph* frameshift mutation that leads to pyrimidine starvation due to low *pyrE* expression levels. *J. Bacteriol.* **175:**3401–3407.

1977. Jensen, K. F., J. T. Andersen, and P. Poulsen. 1992. Overexpression and rapid purification of the *orfE/rph* gene product, RNase PH of *Escherichia coli. J. Biol. Chem.* **267:**17147–17152.

1978. Jensen, K. F., J. N. Larsen, L. Schack, and A. Sivertsen. 1984. Studies on the structure and expression of *Escherichia coli pyrC, pyrD,* and *pyrF* using the cloned genes. *Eur. J. Biochem.* **140:**343–352.

1979. Jeong, J. H., M. Kitakawa, S. Isono, and K. Isono. 1993. Cloning and nucleotide sequencing of the genes, *rplU* and *rpmA,* for ribosomal proteins L21 and L27 of *Escherichia coli.* DNA Seq. **4:**59–67.

1980. Jerlstrom, P. G., D. A. Bezjak, M. P. Jennings, and I. R. Beacham. 1989. Structure and expression in *Escherichia coli* K-12 of the L-asparaginase I-encoding *ansA* gene and its flanking regions. *Gene* **78:**37–46.

1981. Jessop, A. P., and C. Clugston. 1985. Amplification of the ArgF region in strain HfrP4X of *E. coli* K-12. *Mol. Gen. Genet.* **201:**347–350.

1982. Jiang, W., L. F. Wu, J. Tomich, M. H. Saier, Jr., and W. G. Niehaus. 1990. Corrected sequence of the mannitol (*mtl*) operon in *Escherichia coli. Mol. Microbiol.* **4:**2003–2006.

1983. Jin, D. J., R. R. Burgess, J. P. Richardson, and C. A. Gross. 1992. Termination efficiency at *rho*-dependent terminators depends on kinetic coupling between RNA polymerase and *rho. Proc. Natl. Acad. Sci. USA* **89:**1453–1457.

1984. Jin, D. J., and C. A. Gross. 1988. Mapping and sequencing of mutations in the *Escherichia coli rpoB* gene that lead to rifampicin resistance. *J. Mol. Biol.* **202:**45–58.

1985. Jin, D. J., and C. A. Gross. 1989. Three *rpoBC* mutations that suppress the termination defects of *rho* mutants also affect the functions of *nusA* mutants. *Mol. Gen. Genet.* **216:**269–275.

1986. Jin, D. J., and C. A. Gross. 1991. RpoB8, a rifampicin-resistant termination-proficient RNA polymerase, has an increased K_m for purine nucleotides during transcription elongation. *J. Biol. Chem.* **266:**14478–14485.

1987. Jin, T., K. E. Rudd, and M. Inouye. 1992. The *nlpA* lipoprotein gene is located near the *selC* tRNA gene on the *E. coli* chromosome. *J. Bacteriol.* **174:**3822–3823.

1988. Jochimsen, B., P. Nygaard, and T. Vestergaard. 1975. Location on the chromosome of *Escherichia coli* of genes governing purine metabolism. Adenosine deaminase (*add*), guanosine kinase (*gsk*) and hypoxanthine phosphoribosyltransferase (*hpt*). *Mol. Gen. Genet.* **143:**85–91.

1989. Johann, S., and S. M. Hinton. 1987. Cloning and nucleotide sequence of the *chlD* locus. *J. Bacteriol.* **169:**1911–1916.

1990. Johanson, U., and D. Hughes. 1992. Comparison of the complete sequence of the *str* operon in *Salmonella typhimurium* and *Escherichia coli. Gene* **120:**93–98.

1991. Johnson, B. F. 1977. Fine structure mapping and properties of mutations suppressing the *lon* mutation in *Escherichia coli* K-12 and B strains. *Genet. Res.* **30:**273–286.

1992. Johnson, D. I., and R. L. Somerville. 1983. Evidence that repression mechanisms can exert control over the *thr, leu,* and *ilv* operons of *Escherichia coli* K-12. *J. Bacteriol.* **155:**49–55.

1993. Johnson, D. I., and R. L. Somerville. 1984. New regulatory genes involved in the control of transcription initiation at the *thr* and *ilv* promoters of *Escherichia coli* K-12. *Mol. Gen. Genet.* **195:**70–76.

1994. Johnson, M. E., and K. V. Rajagopalan. 1987. Involvement of *chlA, E, M,* and *N* loci in *Escherichia coli* molybdopterin biosynthesis. *J. Bacteriol.* **189:**117–125.

1995. Johnson, R. C., C. A. Ball, D. Pfeffer, and M. I. Simon. 1988. Isolation of the gene encoding the Hin recombinational enhancer binding protein. *Proc. Natl. Acad. Sci. USA* **85:**3484–3488.

1996. Johnson, S. C., N. Watson, and D. Apirion. 1976. A lethal mutation which affects the maturation of ribosomes. *Mol. Gen. Genet.* **147:**29–37.

1997. Johnstone, D. B., and S. B. Farr. 1991. AppppA binds to several proteins in *Escherichia coli,* including the heat shock and oxidative stress proteins DnaK, GroEL, E89, C45 and C40. *EMBO J.* **10:**3897–3904.

1998. Joly, J. C., M. R. Leonard, and W. T. Wickner. 1994. Subunit dynamics in *Escherichia coli* preprotein translocase. *Proc. Natl. Acad. Sci. USA* **91:**4703–4707.

1999. Jonczyk, P., R. Hines, and D. W. Smith. 1989. The *Escherichia coli dam* gene is expressed as a distal gene of a new operon. *Mol. Gen. Genet.* **217:**85–96.

2000. Jones, C. A., and I. B. Holland. 1984. Inactivation of essential division genes, *ftsA, ftsZ,* suppresses mutations at *sfiB,* a locus mediating division inhibition during the SOS response in *E. coli. EMBO J.* **3:**1181–1186.

2001. Jones, C. E., J. M. Brook, D. Buck, C. Abell, and A. G. Smith. 1993. Cloning and sequencing of the *Escherichia coli panB* gene, which encodes ketopantoate hydroxymethyltransferase, and overexpression of the enzyme. *J. Bacteriol.* **175:**2125–2130.

2002. Jones, C. H., J. S. Pinkner, A. V. Nicholes, L. N. Slonim, S. N. Abraham, and S. J. Hultgren. 1993. FimC is a periplasmic PapD-like chaperone that directs assembly of type 1 pili in bacteria. *Proc. Natl. Acad. Sci.* **90:**8397–8401.

2003. Jones, C. J., M. Homma, and R. M. Macnab. 1987. Identification of proteins of the outer (L and P) rings of the flagellar basal body of *Escherichia coli. J. Bacteriol.* **169:**1489–1492.

2004. Jones, H. M., C. M. Brajkovich, and R. P. Gunsalus. 1983. In vivo 5′ terminus and length of the mRNA for the proton-translocating ATPase (*unc*) operon of *Escherichia coli. J. Bacteriol.* **155:**1279–1287.

2005. Jones, H. M., and R. P. Gunsalus. 1985. Transcription of the *Escherichia coli* fumarate reductase genes (*frdABCD*) and their coordinate regulation by oxygen, nitrate, and fumarate. *J. Bacteriol.* **164:**1100–1109.

2006. Jones, M. D., T. E. Petersen, K. M. Nielsen, S. Magnusson, L. Sottrup-Jensen, K. Gausing, and B. F. Clark. 1980. The complete amino-acid sequence of elongation factor Tu from *Escherichia coli. Eur. J. Biochem.* **108:**507–526.

2007. Jones, P. G., J. Cashel, G. Glaser, and F. C. Neidhardt. 1992. Function of a relaxed-like state following temperature downshifts in *Escherichia coli. J. Bacteriol.* **174:**3903–3914.

2008. Jones, P. G., and M. Inouye. 1994. The cold-shock response—a hot topic. *Mol. Microbiol.* **11:**811–818.

2009. Jones, P. G., R. Krah, S. R. Tafuri, and A. P. Wolffe. 1992. DNA gyrase, CS7.4, and the cold shock response in *Escherichia coli. J. Bacteriol.* **174:**5798–5802.

2010. Jones-Mortimer, M. C., and H. L. Kornberg. 1980. Amino-sugar transport systems of *Escherichia coli* K12. *J. Gen. Microbiol.* **117:**369–376.

2011. Jordan, P. M., B. I. Mgbeje, A. F. Alwan, and S. D. Thomas. 1987. Nucleotide sequence of *hemD,* the second gene in the *hem* operon of *Escherichia coli* K-12. *Nucleic Acids Res.* **15:**10583.

2012. Jordan, P. M., B. I. Mgbeje, S. D. Thomas, and A. F. Alwan. 1988. Nucleotide sequence for the *hemD* gene of *Escherichia coli* encoding uroporphyrinogen III synthase and initial evidence for a *hem* operon. *Biochem. J.* **249:**613–616.

2013. Jordan, P. M., and S. C. Woodcock. 1991. Mutagenesis of arginine residues in the catalytic cleft of *Escherichia coli* porphobilinogen deaminase that affects dipyrromethane cofactor assembly and tetrapyrrole chain initiation and elongation. *Biochem. J.* **280:**445–449.

2014. Jorgensen, P., J. Collins, N. Fiil, and K. von Meyenbourg. 1978. A ribosomal RNA gene, *rrnC,* of *Escherichia coli,* mapped by specialized transducing lambda dilv and lambda drbs phages. *Mol. Gen. Genet.* **163:**223–228.

2015. Joris, B., G. Dive, A. Henriques, P. J. Piggot, and J. M. Ghuysen. 1990. The life-cycle proteins RodA of *Escherichia coli* and SpoVE of *Bacillus subtilis* have very similar primary structures. *Mol. Microbiol.* **4:**513–517.

2016. Josaitis, C. A., T. Gaal, W. Ross, and R. L. Gourse. 1990. Sequences upstream of the −35 hexamer of *rrnB* P1 affect promoter strength and upstream activation. *Biochim. Biophys. Acta* **1050:**307–311.

2017. Josephsen, J., and K. Hammer-Jespersen. 1981. Fusion of the *lac* genes to the promotor for the cytidine deaminase gene of *Escherichia coli* K-12. *Mol. Gen. Genet.* **182:**154–158.

2018. Josephsen, J., K. Hammer-Jespersen, and T. D. Hansen. 1983. Mapping of the gene for cytidine deaminase (*cdd*) in *Escherichia coli* K-12. *J. Bacteriol.* **154:**72–75.

2019. Jovanovic, G., T. Kostic, and D. J. Savic. 1990. Nucleotide and amino acid polymorphism in the gene for L-histidinol dehydrogenase of *Escherichia coli* K12. *Nucleic Acids Res.* **18:**3634.

2020. Joyce, C. M., and N. D. Grindley. 1982. Identification of two genes immediately downstream from the *polA* gene of *Escherichia coli. J. Bacteriol.* **152:**1211–1219.

2021. Joyce, C. M., W. S. Kelley, and N. D. Grindley. 1982. Nucleotide sequence of the *Escherichia coli polA* gene and primary structure of DNA polymerase I. *J. Biol. Chem.* **257:**1958–1964.

2022. Jung, H. K., F. Ishino, and M. Matsuhashi. 1989. Inhibition of growth of *ftsQ, ftsA,* and *ftsZ* mutant cells of *Escherichia coli* by amplification of a chromosomal region encompassing closely aligned cell division and cell growth genes. *J. Bacteriol.* **171:**6379–6382.

2023. Jung, J. U., C. Gutierrez, F. Martin, M. Ardourel, and M. Villarejo. 1990. Transcription of *osmB,* a gene encoding an *Escherichia coli* lipoprotein, is regulated by dual signals. Osmotic stress and stationary phase. *J. Biol. Chem.* **265:**10574–10581.

2024. Jung, J. U., C. Gutierrez, and M. R. Villarejo. 1989. Sequence of an osmotically inducible lipoprotein gene. *J. Bacteriol.* **171:**511–520.

2025. Junker, D. E., Jr., L. A. Rokeach, D. Ganea, A. Chiaramello, and J. W. Zyskind. 1986. Transcription termination within the *Escherichia coli* origin of DNA replication, *oriC. Mol. Gen. Genet.* **203:**101–109.

2026. Kaback, H. R. 1990. The *lac* permease of *Escherichia coli:* a prototypic energy-transducing membrane protein. *Biochim. Biophys. Acta* **1018:**160–162.

2027. Kadner, R. J. 1977. Transport and utilization of D-methionine and other methionine sources in *Escherichia coli. J. Bacteriol.* **129:**207–216.

2028. Kadner, R. J. 1990. Vitamin B_{12} transport in *Escherichia coli:* energy coupling between membranes. *Mol. Microbiol.* **4:**2027–2033.

2029. Kadner, R. J., K. Heller, J. W. Coulton, and V. Braun. 1980. Genetic control of hydroxamate-mediated iron uptake in *Escherichia coli. J. Bacteriol.* **143:**256–264.

2030. Kadner, R. J., and G. L. Liggins. 1973. Transport of vitamin B_{12} in *Escherichia coli:* genetic studies. *J. Bacteriol.* **115:**514–521.

2031. Kadner, R. J., and D. M. Shattuck-Eidens. 1983. Genetic control of the hexose phosphate transport system of *Escherichia coli:* mapping of deletion and insertion mutations in the *uhp* region. *J. Bacteriol.* **155:**1052–1061.

2032. Kagamiyama, H., H. Matsubara, and E. E. Snell. 1972. The chemical structure of tryptophanase from *Escherichia coli.* 3. Isolation and amino acid sequence of the tryptic peptides. *J. Biol. Chem.* **247:**1576–1586.

2033. Kagawa, H., N. Ono, M. Enomoto, and Y. Komeda. 1984. Bacteriophage chi sensitivity and motility of *Escherichia coli* K-12 and *Salmonella typhimurium* Fla⁻ mutants possessing the hook structure. *J. Bacteriol.* **157:**649–654.

2034. **Kaidow, A., T. Kataoka, M. Wachi, A. Takada, M. Yamasaki, and K. Nagai.** 1992. The 55-kilodalton protein in an *oriC* complex fraction is glycogen synthase. *J. Bacteriol.* 174:5454–5456.

2034a. **Kaiser, K.** 1980. The origin of Q-independent derivatives of phage lambda. *Mol. Gen. Genet.* 179:547–554.

2035. **Kaiser, K., and N. E. Murray.** 1979. Physical characterisation of the "Rac prophage" in *E. coli* K12. *Mol. Gen. Genet.* 175:159–174.

2036. **Kaiser, K., and N. E. Murray.** 1980. On the nature of *sbcA* mutations in *E. coli* K 12. *Mol. Gen. Genet.* 179:555–563.

2037. **Kajie, S., R. Ideta, I. Yamato, and Y. Anraku.** 1991. Molecular cloning and DNA sequence of *dniR*, a gene affecting anaerobic expression of the *Escherichia coli* hexaheme nitrite reductase. *FEMS Microbiol. Lett.* 67:205–211.

2038. **Kajitani, M., R. Fukuda, and A. Ishihama.** 1980. Autogenous and post-transcriptional regulation of *Escherichia coli* RNA polymerase synthesis in vitro. *Mol. Gen. Genet.* 179:489–496.

2039. **Kajitani, M., and A. Ishihama.** 1991. Identification and sequence determination of the host factor gene for bacteriophage Q beta. *Nucleic Acids Res.* 19:1063–1066.

2040. **Kajitani, M., A. Kato, A. Wada, Y. Inokuchi, and A. Ishihama.** 1994. Regulation of the *Escherichia coli hfq* gene encoding the host factor for phage Q beta. *J. Bacteriol.* 176:531–534.

2041. **Kakuda, H., K. Hosono, K. Shiroishi, and S. Ichihara.** 1994. Identification and characterization of the *ackA* (acetate kinase A)-*pta* (phosphotransacetylase) operon and complementation analysis of acetate utilization by an *ackA-pta* deletion mutant of *Escherichia coli. J. Biochem.* 116:916–922.

2042. **Kalman, L. V., and R. P. Gunsalus.** 1988. The *frdR* gene of *Escherichia coli* globally regulates several operons involved in anaerobic growth in response to nitrate. *J. Bacteriol.* 170:623–629.

2043. **Kalman, L. V., and R. P. Gunsalus.** 1989. Identification of a second gene involved in global regulation of fumarate reductase and other nitrate-controlled genes for anaerobic respiration in *Escherichia coli. J. Bacteriol.* 171:3810–3816.

2044. **Kalman, M., D. R. Gentry, and M. Cashel.** 1991. Characterization of the *Escherichia coli* K12 *gltS* glutamate permease gene. *Mol. Gen. Genet.* 225:379–386.

2045. **Kalman, M., H. Murphy, and M. Cashel.** 1991. *rhlB*, a new *Escherichia coli* K-12 gene with an RNA helicase-like protein sequence motif, one of at least five such possible genes in a prokaryote. *New Biol.* 3:886–895.

2046. **Kalman, M., H. Murphy, and M. Cashel.** 1992. The nucleotide sequence of *recG*, the distal *spo* operon gene in *Escherichia coli* K-12. *Gene* 110:95–99.

2047. **Kalnins, A., K. Otto, U. Ruther, and B. Muller-Hill.** 1983. Sequence of the *lacZ* gene of *Escherichia coli. EMBO J.* 2:593–597.

2048. **Kameyama, L., L. Fernandez, D. L. Court, and G. Guarneros.** 1991. RNase III activation of bacteriophage lambda N synthesis. *Mol. Microbiol.* 5:2953–2963.

2049. **Kamholz, J., J. Keyhani, and J. S. Gots.** 1986. Molecular cloning and characterization of the *purE* operon of *Escherichia coli. Gene* 44:55–62.

2050. **Kamio, Y., C. K. Lin, M. Regue, and H. C. Wu.** 1985. Characterization of the *ileS-lsp* operon in *Escherichia coli*. Identification of an open reading frame upstream of the *ileS* gene and potential promoter(s) for the *ileS-lsp* operon. *J. Biol. Chem.* 260:5616–5620.

2051. **Kamitani, S., Y. Akiyama, and K. Ito.** 1992. Identification and characterization of an *Escherichia coli* gene required for the formation of correctly folded alkaline phosphatase, a periplasmic enzyme. *EMBO J.* 11:57–62.

2052. **Kammen, H. O., C. C. Marvel, L. Hardy, and E. E. Penhoet.** 1988. Purification, structure, and properties of *Escherichia coli* tRNA pseudouridine synthase I. *J. Biol. Chem.* 263:2255–2263.

2053. **Kammler, M., C. Schon, and K. Hantke.** 1993. Characterization of the ferrous iron uptake system of *Escherichia coli. J. Bacteriol.* 175:6212–6219.

2054. **Kamp, R., and B. Wittmann-Liebold.** 1980. Primary structure of protein S11 from *Escherichia coli* ribosomes. *FEBS Lett.* 121:117–122.

2055. **Kampfenkel, K., and V. Braun.** 1992. Membrane topology of the *Escherichia coli* ExbD protein. *J. Bacteriol.* 174:5485–5487.

2056. **Kampfenkel, K., and V. Braun.** 1993. Membrane topologies of the TolQ and TolR proteins of *Escherichia coli*: inactivation of TolQ by a missense mutation in the proposed first transmembrane segment. *J. Bacteriol.* 175:4485–4491.

2057. **Kampfenkel, K., and V. Braun.** 1993. Topology of the ExbB protein in the cytoplasmic membrane of *Escherichia coli. J. Biol. Chem.* 268:6050–6057.

2058. **Kanamaru, K., and T. Mizuno.** 1992. Signal transduction and osmoregulation in *Escherichia coli*: a novel mutant of the positive regulator, OmpR, that functions in a phosphorylation-independent manner. *J. Biochem.* 111:425–430.

2059. **Kanatani, A., T. Masuda, T. Shimoda, F. Misoka, X. S. Lin, T. Yoshimoto, and D. Tsuru.** 1991. Protease II from *Escherichia coli*: sequencing and expression of the enzyme gene and characterization of the expressed enzyme. *J. Biochem.* 110:315–320.

2060. **Kanatani, A., T. Yoshimoto, H. Nagai, K. Ito, and D. Tsuru.** 1992. Location of the protease II gene (*ptrB*) on the physical map of the *Escherichia coli* chromosome. *J. Bacteriol.* 174:7881.

2061. **Kanaya, S., and R. J. Crouch.** 1983. DNA sequence of the gene coding for *Escherichia coli* ribonuclease H. *J. Biol. Chem.* 258:1276–1281.

2062. **Kanaya, S., S. Kimura, C. Katsuda, and M. Ikehara.** 1990. Role of cysteine residues in ribonuclease H from *Escherichia coli*. Site-directed mutagenesis and chemical modification. *Biochem. J.* 271:59–66.

2063. **Kanazawa, H., T. Kayano, T. Kiyasu, and M. Futai.** 1982. Nucleotide sequence of the genes for beta and epsilon subunits of proton-translocating ATPase from *Escherichia coli. Biochem. Biophys. Res. Commun.* 105:1257–1264.

2064. **Kanazawa, H., T. Kayano, K. Mabuchi, and M. Futai.** 1981. Nucleotide sequence of the genes coding for alpha, beta and gamma subunits of the proton-translocating ATPase of *Escherichia coli. Biochem. Biophys. Res. Commun.* 103:604–612.

2065. **Kanazawa, H., K. Mabuchi, and M. Futai.** 1982. Nucleotide sequence of the promoter region of the gene cluster for proton-translocating ATPase from *Escherichia coli* and identification of the active promotor. *Biochem. Biophys. Res. Commun.* 107:568–575.

2066. **Kanazawa, H., K. Mabuchi, T. Kayano, T. Noumi, T. Sekiya, and M. Futai.** 1981. Nucleotide sequence of the genes for F0 components of the proton-translocating ATPase from *Escherichia coli*: prediction of the primary structure of F0 subunits. *Biochem. Biophys. Res. Commun.* 103:613–620.

2067. **Kanazawa, H., K. Mabuchi, T. Kayano, F. Tamura, and M. Futai.** 1981. Nucleotide sequence of genes coding for dicyclohexylcarbodiimide-binding protein and the alpha subunit of proton-translocating ATPase of *Escherichia coli. Biochem. Biophys. Res. Commun.* 100:219–225.

2068. **Kaneko, S., T. Ichiba, N. Hirano, and A. Hachimori.** 1993. Modification of tryptophan 149 of inorganic pyrophosphatase from *Escherichia coli. Int. J. Biochem.* 25:233–238.

2069. **Kanemori, M., H. Mori, and T. Yura.** 1994. Induction of heat shock proteins by abnormal proteins results from stabilization and not increased synthesis of sigma 32 in *Escherichia coli. J. Bacteriol.* 176:5648–5653.

2070. **Kang, P. J., and E. A. Craig.** 1990. Identification and characterization of a new *Escherichia coli* gene that is a dosage-dependent suppressor of a *dnaK* deletion mutation. *J. Bacteriol.* 172:2055–2064.

2071. **Kang, W. K., T. Icho, S. Isono, M. Kitakawa, and K. Isono.** 1989. Characterization of the gene *rimK* responsible for the addition of glutamic acid residues to the C-terminus of ribosomal protein S6 in *Escherichia coli* K12. *Mol. Gen. Genet.* 217:281–288.

2072. **Kano, Y., T. Ogawa, T. Ogura, S. Hiraga, T. Okazaki, and F. Imamoto.** 1991. Participation of the histone-like protein HU and of IHF in minichromosomal maintenance in *Escherichia coli. Gene* 103:25–30.

2073. **Kano, Y., K. Osato, M. Wada, and F. Imamoto.** 1987. Cloning and sequencing of the HU-2 gene of *Escherichia coli. Mol. Gen. Genet.* 209:408–410.

2074. **Kano, Y., M. Wada, and F. Imamoto.** 1988. Genetic characterization of the gene *hupA* encoding the HU-2 protein of *Escherichia coli. Gene* 69:331–335.

2075. **Kano, Y., M. Wada, T. Nagase, and F. Imamoto.** 1986. Genetic characterization of the gene *hupB* encoding the HU-1 protein of *Escherichia coli. Gene* 45:37–44.

2076. **Kano, Y., S. Yoshino, M. Wada, K. Yokoyama, M. Nobuhara, and F. Imamoto.** 1985. Molecular cloning and nucleotide sequence of the HU-1 gene of *Escherichia coli. Mol. Gen. Genet.* 201:360–362.

2077. **Kao, C., E. Gumbs, and L. Snyder.** 1987. Cloning and characterization of the *Escherichia coli lit* gene, which blocks bacteriophage T4 late gene expression. *J. Bacteriol.* 169:1232–1238.

2078. **Kao, C., and L. Snyder.** 1988. The *lit* gene product which blocks bacteriophage T4 late gene expression is a membrane protein encoded by a cryptic DNA element, e14. *J. Bacteriol.* 170:2056–2062.

2079. **Kaplan, J. B., and B. P. Nichols.** 1983. Nucleotide sequence of *Escherichia coli pabA* and its evolutionary relationship to *trp(G)D. J. Mol. Biol.* 168:451–468.

2080. **Kaplan, S., and D., Anderson.** 1968. Selection of temperature-sensitive activating enzyme mutants in *Escherichia coli. J. Bacteriol.* 95:991–997.

2081. **Karasawa, K., I. Kudo, T. Kobayashi, H. Homma, N. Chiba, H. Mizushima, K. Inoue, and S. Nojima.** 1991. Lysophospholipase L1 from *Escherichia coli* K-12 overproducer. *J. Biochem.* 109:288–293.

2082. **Karlsson, M., K. Hannavy, and C. F. Higgins.** 1993. ExbB acts as a chaperone-like protein to stabilize TonB in the cytoplasm. *Mol. Microbiol.* 8:389–396.

2083. **Karlsson, M., K. Hannavy, and C. F. Higgins.** 1993. A sequence-specific function for the N-terminal signal-like sequence of the TonB protein. *Mol. Microbiol.* 8:379–388.

2084. **Karow, M., O. Fayet, A. Cegielska, T. Ziegelhoffer, and C. Georgopoulos.** 1991. Isolation and characterization of the *Escherichia coli htrB* gene, whose product is essential for bacterial viability above 33°C in rich media. *J. Bacteriol.* 173:741–750.

2085. **Karow, M., and C. Georgopoulos.** 1991. Sequencing, mutational analysis, and transcriptional regulation of the *Escherichia coli htrB* gene. *Mol. Microbiol.* 5:2285–2292.

2086. **Karow, M., and C. Georgopoulos.** 1992. Isolation and characterization of the *Escherichia coli msbB* gene, a multicopy suppressor of null mutations in the high-temperature requirement gene *htrB. J. Bacteriol.* 174:702–710.

2087. **Karow, M., and C. Georgopoulos.** 1993. The essential *Escherichia coli msbA* gene, a multicopy suppressor of null mutations in the *htrB* gene, is related to the universally conserved family of ATP-dependent translocators. *Mol. Microbiol.* 7:69–79.

2088. **Karow, M., S. Raina, C. Georgopoulos, and O. Fayet.** 1991. Complex phenotypes of null mutations in the *htr* genes, whose products are essential for *Escherichia coli* growth at elevated temperatures. *Res. Microbiol.* 142:289–294.

2089. **Karpel, R., T. Alon, G. Glaser, S. Schuldiner, and E. Padan.** 1991. Expression of a sodium proton antiporter (NhaA) in *Escherichia coli* is induced by Na$^+$ and Li$^+$ ions. *J. Biol. Chem.* 266:21753–21759.

2090. **Karpel, R., Y. Olami, D. Taglicht, S. Schuldiner, and E. Padan.** 1988. Sequencing of the gene *ant* which affects the Na$^+$/H$^+$ antiporter activity in *Escherichia coli. J. Biol. Chem.* 263:10408–10414.

2091. Karran, P., T. Lindahl, I. Ofsteng, G. B. Evensen, and E. Seeberg. 1980. *Escherichia coli* mutants deficient in 3-methyladenine-DNA glycosylase. *J. Mol. Biol.* 140:101–127.

2092. Kasahara, M., K. Makino, M. Amemura, and A. Nakata. 1989. Nucleotide sequence of the *ugpQ* gene encoding glycerophosphoryl diester phosphodiesterase of *Escherichia coli* K-12. *Nucleic Acids Res.* 17:2854.

2093. Kasahara, M., K. Makino, M. Amemura, A. Nakata, and H. Shinagawa. 1991. Dual regulation of the *ugp* operon by phosphate and carbon starvation at two interspaced promoters. *J. Bacteriol.* 173:549–558.

2094. Kasahara, M., A. Nakata, and H. Shinagawa. 1992. Molecular analysis of the *Escherichia coli phoP-phoQ* operon. *J. Bacteriol.* 174:492–498.

2095. Kashiwagi, K., and K. Igarashi. 1987. Nonspecific inhibition of *Escherichia coli* ornithine decarboxylase by various ribosomal proteins: detection of a new ribosomal protein possessing strong antizyme activity. *Biochim. Biophys. Acta* 911:180–190.

2096. Kashiwagi, K., A. Miyaji, S. Ikeda, T. Tobe, C. Sasakawa, and K. Igarashi. 1992. Increase of sensitivity to aminoglycoside antibiotics by polyamine-induced protein (oligopeptide-binding protein) in *Escherichia coli*. *J. Bacteriol.* 174:4331–4337.

2097. Kashiwagi, K., S. Miyamoto, E. Nukui, H. Kobayashi, and K. Igarashi. 1993. Functions of *potA* and *potD* proteins in spermidine-preferential uptake system in *Escherichia coli*. *J. Biol. Chem.* 268:19358–19363.

2098. Kashiwagi, K., S. Miyamoto, F. Suzuki, H. Kobayashi, and K. Igarashi. 1992. Excretion of putrescine by the putrescine-ornithine antiporter encoded by the *potE* gene of *Escherichia coli*. *Proc. Natl. Acad. Sci. USA* 89:4529–4533.

2099. Kashiwagi, K., T. Suzuki, F. Suzuki, T. Furuchi, H. Kobayashi, and K. Igarashi. 1991. Coexistence of the genes for putrescine transport protein and ornithine decarboxylase at 16 min on *Escherichia coli* chromosome. *J. Biol. Chem.* 266:20922–20927.

2100. Kashiwagi, K., R. Watanabe, and K. Igarashi. 1994. Involvement of ribonuclease III in the enhancement of expression of the *speF-potE* operon encoding inducible ornithine decarboxylase and polyamine transport protein. *Biochem. Biophys. Res. Commun.* 200:591–597.

2101. Kashiwagi, K., Y. Yamaguchi, Y. Sakai, H. Kobayashi, and K. Igarashi. 1990. Identification of the polyamine-induced protein as a periplasmic oligopeptide binding protein. *J. Biol. Chem.* 265:8387–8391.

2102. Kashlev, M. V., A. I. Gragerov, and V. G. Nikiforov. 1989. Heat shock response in *Escherichia coli* promotes assembly of plasmid encoded RNA polymerase betasubunit into RNA polymerase. *Mol. Gen. Genet.* 216:469–474.

2103. Kasian, P. A., B. E. Davidson, and J. Pittard. 1986. Molecular analysis of the promoter operator region of the *Escherichia coli* K-12 *tyrP* gene. *J. Bacteriol.* 167:556–561.

2104. Kasian, P. A., and J. Pittard. 1984. Construction of a *tyrP-lac* operon fusion strain and its use in the isolation and analysis of mutants derepressed for *tyrP* expression. *J. Bacteriol.* 160:175–183.

2105. Kast, P., and H. Hennecke. 1991. Amino acid substrate specificity of *Escherichia coli* phenylalanyl-tRNA synthetase altered by distinct mutations. *J. Mol. Biol.* 222:99–124.

2106. Kast, P., B. Keller, and H. Hennecke. 1992. Identification of the *pheS5* mutation, which causes thermosensitivity of *Escherichia coli* mutant NP37. *J. Bacteriol.* 174:1686–1689.

2107. Kast, P., C. Wehrli, and H. Hennecke. 1991. Impaired affinity for phenylalanine in *Escherichia coli* phenylalanyl- tRNA synthetase mutant caused by Gly-to-Asp exchange in motif 2 of class II tRNA synthetases. *FEBS Lett.* 293:160–163.

2108. Kataoka, H., and M. Sekiguchi. 1985. Molecular cloning and characterization of the *alkB* gene of *Escherichia coli*. *Mol. Gen. Genet.* 198:263–269.

2109. Kataoka, H., Y. Yamamoto, and M. Sekiguchi. 1983. A new gene (*alkB*) of *Escherichia coli* that controls sensitivity to methyl methane sulfonate. *J. Bacteriol.* 153:1301–1307.

2110. Kataoka, T., S. Gayama, K. Takahashi, M. Wachi, M. Yamasaki, and K. Nagai. 1991. Only *oriC* and its flanking region are recovered from the complex formed at the time of initiation of chromosome replication in *Escherichia coli*. *Res. Microbiol.* 142:155–159.

2111. Katayama, T., Y. Murakami, C. Wada, H. Ohmori, T. Yura, and T. Nagata. 1989. Genetic suppression of a *dnaG* mutation in *Escherichia coli*. *J. Bacteriol.* 171:1485–1491.

2112. Katayama, Y., S. Gottesman, J. Pumphrey, S. Rudikoff, W. P. Clark, and M. R. Maurizi. 1988. The two-component, ATP-dependent Clp protease of *Escherichia coli*. Purification, cloning, and mutational analysis of the ATP-binding component. *J. Biol. Chem.* 263:15226–15236.

2113. Katayanagi, K., M. Miyagawa, M. Matsushima, M. Ishikawa, S. Kanaya, M. Ikehara, T. Matsuzaki, and K. Morikawa. 1990. Three-dimensional structure of ribonuclease H from *E. coli*. *Nature* (London) 347:306–309.

2114. Katinka, M., P. Cossart, L. Sibilli, I. Saint-Girons, M. A. Chalvignac, G. Le Bras, G. N. Cohen, and M. Yaniv. 1980. Nucleotide sequence of the *thrA* gene of *Escherichia coli*. *Proc. Natl. Acad. Sci. USA* 77:5730–5733.

2115. Kato, J., Y. Nishimura, R. Imamura, H. Niki, S. Hiraga, and H. Suzuki. 1990. New topoisomerase essential for chromosome segregation in *E. coli*. *Cell* 63:393–404.

2116. Kato, J., Y. Nishimura, M. Yamada, H. Suzuki, and Y. Hirota. 1988. Gene organization in the region containing a new gene involved in chromosome partition in *Escherichia coli*. *J. Bacteriol.* 170:3967–3977.

2117. Kato, J., H. Suzuki, and Y. Hirota. 1984. Overlapping of the coding regions for alpha and gamma components of penicillin-binding protein 1 b in *Escherichia coli*. *Mol. Gen. Genet.* 196:449–457.

2118. Kato, J., H. Suzuki, and H. Ikeda. 1992. Purification and characterization of DNA topoisomerase IV in *Escherichia coli*. *J. Biol. Chem.* 267:25676–25684.

2119. Katsuki, H., K. Takeo, K. Kameda, and S. Tanaka. 1967. Existence of two malic enzymes in *Escherichia coli*. *Biochem. Biophys. Res. Commun.* 27:331–336.

2120. Katz, L. 1970. Selection of *araB* and *araC* mutants of *Escherichia coli* B-r by resistance to ribitol. *J. Bacteriol.* 102:593–595.

2121. Katzenmeier, G., C. Schmid, J. Kellermann, F. Lottspeich, and A. Bacher. 1991. Biosynthesis of tetrahydrofolate. Sequence of GTP cyclohydrolase I from *Escherichia coli*. *Biol. Chem. Hoppe-Seyler* 372:991–997.

2122. Kauppinen, S., M. Siggaard-Andersen, and P. von Wettstein-Knowles. 1988. β-Ketoacyl-ACP synthase I of *Escherichia coli*: nucleotide sequence of the *fabB* gene and identification of the cerulenin binding residue. *Carlsberg Res. Commun.* 53:357–370.

2123. Kawagishi, I., V. Muller, A. W. Williams, V. M. Irikura, and R. M. Macnab. 1992. Subdivision of flagellar region III of the *Escherichia coli* and *Salmonella typhimurium* chromosomes and identification of two additional flagellar genes. *J. Gen. Microbiol.* 138:1051–1065.

2124. Kawakami, K., K. Ito, and Y. Nakamura. 1992. Differential regulation of two genes encoding lysyl-tRNA synthetases in *Escherichia coli*: *lysU*-constitutive mutations compensate for a *lysS* null mutation. *Mol. Microbiol.* 6:1739–1745.

2125. Kawakami, K., Y. H. Jonsson, G. R. Bjork, H. Ikeda, and Y. Nakamura. 1988. Chromosomal location and structure of the operon encoding peptide-chain-release factor 2 of *Escherichia coli*. *Proc. Natl. Acad. Sci. USA* 85:5620–5624.

2126. Kawakami, K., S. Naito, N. Inoue, Y. Nakamura, H. Ikeda, and H. Uchida. 1989. Isolation and characterization of *herC*, a mutation of *Escherichia coli* affecting maintenance of ColE1. *Mol. Gen. Genet.* 219:333–340.

2126a.Kawamoto, S., S. Tokuyama, K. Aoyama, S. Yashima, and Y. Eguchi. 1984. Genetic mapping of cold resistance gene of *Escherichia coli*. *Agric. Biol. Chem.* 48:5620–5624.

2127. Kawamukai, M., H. Matsuda, W. Fujii, R. Utsumi, and T. Komano. 1989. Nucleotide sequences of *fic* and *fic-1* genes involved in cell filamentation induced by cyclic AMP in *Escherichia coli*. *J. Bacteriol.* 171:4525–4532.

2128. Kawamukai, M., R. Utsumi, K. Takeda, A. Higashi, H. Matsuda, Y. L. Choi, and T. Komano. 1991. Nucleotide sequence and characterization of the *sfs-1* gene: *sfs-1* is involved in CRP*-dependent *mal* gene expression in *Escherichia coli*. *J. Bacteriol.* 173:2644–2648.

2129. Kawula, T. H., and M. J. Lelivelt. 1994. Mutations in a gene encoding a new Hsp70 suppress rapid DNA inversion and *bgl* activation, but not *proU* derepression, in *hns-1* mutant *Escherichia coli*. *J. Bacteriol.* 176:610–619.

2130. Kay, W. W. 1972. Genetic control of the metabolism of propionate by *Escherichia coli* K12. *Biochim. Biophys. Acta* 264:508–521.

2131. Kayahara, T., P. Thelen, W. Ogawa, K. Inaba, M. Tsuda, E. B. Goldberg, and T. Tsuchiya. 1992. Properties of recombinant cells capable of growing on serine without NhaB Na$^+$/H$^+$ antiporter in *Escherichia coli*. *J. Bacteriol.* 174:7482–7485.

2132. Kazakov, S., and S. Altman. 1991. Site-specific cleavage by metal ion cofactors and inhibitors of M1 RNA, the catalytic subunit of RNase P from *Escherichia coli*. *Proc. Natl. Acad. Sci. USA* 88:9193–9197.

2133. Keasling, J. D., L. Bertsch, and A. Kornberg. 1993. Guanosine pentaphosphate phosphohydrolase of *Escherichia coli* is a long-chain exopolyphosphatase. *Proc. Natl. Acad. Sci. USA* 90:7029–7033.

2134. Keck, W., B. Glauner, U. Schwarz, J. K. Broome-Smith, and B. G. Spratt. 1985. Sequences of the active-site peptides of three of the high-M_r penicillin-binding proteins of *Escherichia coli* K-12. *Proc. Natl. Acad. Sci. USA* 82:1999–2003.

2135. Keck, W., A. M. van Leeuwen, M. Huber, and E. W. Goodell. 1990. Cloning and characterization of *mepA*, the structural gene of the penicillin-insensitive murein endopeptidase from *Escherichia coli*. *Mol. Microbiol.* 4:209–219.

2136. Kehry, M. R., and F. W. Dahlquist. 1982. The methyl-accepting chemotaxis proteins of *Escherichia coli*. Identification of the multiple methylation sites on methyl-accepting chemotaxis protein I. *J. Biol. Chem.* 257:10378–10386.

2137. Keller, J. A., and L. D. Simon. 1987. Isolation and analysis of *Escherichia coli* mutants that allow increased replication of bacteriophage lambda. *J. Bacteriol.* 169:1585–1592.

2138. Kelley, W. S. 1980. Mapping of the *polA* locus of *Escherichia coli* K12: genetic fine structure of the cistron. *Genetics* 95:15–38.

2139. Kelley, W. S., and H. J. Whitfield. 1971. Purification of an altered DNA polymerase from an *E. coli* strain with a *pol* mutation. *Nature* (London) 230:33–36.

2140. Kelln, R. A., and G. A. O'Donovan. 1976. Isolation and partial characterization of an *argR* mutant of *Salmonella typhimurium*. *J. Bacteriol.* 128:528–535.

2141. Kelly, K. O., and M. P. Deutscher. 1992. Characterization of *Escherichia coli* RNase PH. *J. Biol. Chem.* 267:17153–17158.

2142. Kelly, K. O., N. B. Reuven, Z. Li, and M. P. Deutscher. 1992. RNase PH is essential for tRNA processing and viability in RNase-deficient *Escherichia coli* cells. *J. Biol. Chem.* 267:16015–16018.

2143. Kelly, T. M., S. A. Stachula, C. R. Raetz, and M. S., Anderson. 1993. The *firA* gene of *Escherichia coli* encodes UDP-3-O-(R-3- hydroxymyristoyl)-glucosamine N-acyltransferase. The third step of endotoxin biosynthesis. *J. Biol. Chem.* 268:19866–19874.

2144. Kemp, E. H., N. P. Minton, and N. H. Mann. 1987. Complete nucleotide sequence and deduced amino acid sequence of the M5 polypeptide gene of *Escherichia coli*. *Nucleic Acids Res.* 15:3924.

2145. Keng, T., T. A. Webster, R. T. Sauer, and P. Schimmel. 1982. Gene for *Escherichia coli* glycyl-tRNA synthetase has tandem subunit coding regions in the same reading frame. *J. Biol. Chem.* 257:12503–12508.

2146. Kenri, T., F. Imamoto, and Y. Kano. 1992. Construction and characterization of an *Escherichia coli* mutant deficient in the *metY* gene encoding tRNA$_{f2}^{Met}$: either tRNA$_{f1}^{Met}$ or tRNA$_{f2}^{Met}$ is required for cell growth. *Gene* 114:109–114.

2147. Kenri, T., F. Imamoto, and Y. Kano. 1994. Three tandemly repeated structural genes encoding tRNA$_{f1}^{Met}$ in the *metZ* operon of *Escherichia coli* K-12. *Gene* 138:261–262.

2148. Kenri, T., K. Kohno, N. Goshima, F. Imamoto, and Y. Kano. 1991. Construction and characterization of an *Escherichia coli* mutant with a deletion of the *metZ* gene encoding tRNA$_{f1}^{Met}$. *Gene* 103:31–36.

2149. Kenyon, C. J., and G. C. Walker. 1980. DNA-damaging agents stimulate gene expression at specific loci in *Escherichia coli*. *Proc. Natl. Acad. Sci. USA* 77:2819–2823.

2150. Kenyon, C. J., and G. C. Walker. 1981. Expression of the *E. coli uvrA* gene is inducible. *Nature* (London) 289:808–810.

2151. Keshavjee, K., C. Pyne, and A. L. Bognar. 1991. Characterization of a mutation affecting the function of *Escherichia coli* folylpolyglutamate synthetase-dihydrofolate synthetase and further mutations produced in vitro at the same locus. *J. Biol. Chem.* 266:19925–19929.

2152. Kessler, D., I. Leibrecht, and J. Knappe. 1991. Pyruvate-formate-lyase-deactivase and acetyl-CoA reductase activities of *Escherichia coli* reside on a polymeric protein particle encoded by *adhE*. *FEBS Lett.* 281:59–63.

2152a.Khatri, G. S., T. MacAllister, P. R. Sista, and D. Bastia. 1989. The replication terminator protein of *Escherichia coli* is a DNA sequence-specific contra-helicase. *Cell* 59:667–674.

2153. Khattar, M. M., K. J. Begg, and W. D. Donachie. 1994. Identification of FtsW and characterization of a new *ftsW* division mutant of *Escherichia coli*. *J. Bacteriol.* 176:7140–7147.

2154. Kiino, D. R., R. Licudine, K. Wilt, D. H. Yang, and L. B. Rothman-Denes. 1993. A cytoplasmic protein, NfrC, is required for bacteriophage N4 adsorption. *J. Bacteriol.* 175:7074–7080.

2155. Kiino, D. R., G. J. Phillips, and T. J. Silhavy. 1990. Increased expression of the bifunctional protein PrlF suppresses overproduction lethality associated with exported β-galactosidase hybrid proteins in *Escherichia coli*. *J. Bacteriol.* 172:185–192.

2156. Kiino, D. R., and L. B. Rothman-Denes. 1989. Genetic analysis of bacteriophage N4 adsorption. *J. Bacteriol.* 171:4595–4602.

2157. Kiino, D. R., M. S. Singer, and L. B. Rothman-Denes. 1993. Two overlapping genes encoding membrane proteins required for bacteriophage N4 adsorption. *J. Bacteriol.* 175:7081–7085.

2158. Kikuchi, A., E. Flamm, and R. A. Weisberg. 1985. An *Escherichia coli* mutant unable to support site-specific recombination of bacteriophage lambda. *J. Mol. Biol.* 183:129–140.

2159. Kikuchi, Y., K. Yoda, M. Yamasaki, and G. Tamura. 1981. The nucleotide sequence of the promoter and the amino-terminal region of alkaline phosphatase structural gene (*phoA*) of *Escherichia coli*. *Nucleic Acids Res.* 9:5671–5678.

2160. Killmann, H., R. Benz, and V. Braun. 1993. Conversion of the FhuA transport protein into a diffusion channel through the outer membrane of *Escherichia coli*. *EMBO J.* 12:3007–3016.

2161. Killmann, H., and V. Braun. 1992. An aspartate deletion mutation defines a binding site of the multifunctional FhuA outer membrane receptor of *Escherichia coli* K-12. *J. Bacteriol.* 174:3479–3486.

2162. Killmann, H., and V. Braun. 1994. Energy-dependent receptor activities of *Escherichia coli* K-12: mutated TonB proteins alter FhuA receptor activities to phages T5, T1, phi 80 and to colicin M. *FEMS Microbiol. Lett.* 119:71–76.

2163. Kilstrup, M., L. M. Meng, J. Neuhard, and P. Nygaard. 1989. Genetic evidence for a repressor of synthesis of cytosine deaminase and purine biosynthesis enzymes in *Escherichia coli*. *J. Bacteriol.* 171:2124–2127.

2164. Kim, I. Y., Z. Veres, and T. C. Stadtman. 1992. *Escherichia coli* mutant SELD enzymes. The cysteine 17 residue is essential for selenophosphate formation from ATP and selenide. *J. Biol. Chem.* 267:19650–19654.

2165. Kim, J., Y. Lee, C. Kim, and C. Park. 1992. Involvement of SecB, a chaperone, in the export of ribose-binding protein. *J. Bacteriol.* 174:5219–5227.

2166. Kim, S., and D. L. Wulff. 1990. Location of an *ntr*-like gene on the physical map of *Escherichia coli*. *J. Bacteriol.* 172:6619.

2167. Kim, S. K., T. Makino, M. Amemura, H. Shinagawa, and A. Nakata. 1993. Molecular analysis of the *phoH* gene, belonging to the phosphate regulon in *Escherichia coli*. *J. Bacteriol.* 175:1316–1324.

2168. Kim, S. S., and P. Datta. 1982. Chemical characterization of biodegradative threonine dehydratases from two enteric bacteria. *Biochim. Biophys. Acta* 706:27–35.

2169. Kimlova, L. J., C. Pyne, K. Keshavjee, J. Huy, G. Beebakhee, and A. L. Bognar. 1991. Mutagenesis of the *folC* gene encoding folylpolyglutamate synthetase-dihydrofolate synthetase in *Escherichia coli*. *Arch. Biochem. Biophys.* 284:9–16.

2170. Kimura, M., T. Miki, S. Hiraga, T. Nagata, and T. Yura. 1979. Conditionally lethal amber mutations in the *dnaA* region of the *Escherichia coli* chromosome that affect chromosome replication. *J. Bacteriol.* 140:825–834.

2170a.King, G., and N. E. Murray. 19XX. Restriction alleviation and modification enhancement by the Rac prophage of *Escherichia coli* K-12. *Mol. Microbiol.* 16:769–777.

2171. King, S. C., C. L. Hansen, and T. H. Wilson. 1991. The interaction between aspartic acid 237 and lysine 358 in the lactose carrier of *Escherichia coli*. *Biochim. Biophys. Acta* 1062:177–186.

2172. Kinghorn, J. R., M. Schweizer, N. H. Giles, and S. R. Kushner. 1981. The cloning and analysis of the *aroD* gene of *E. coli* K-12. *Gene* 14:73–80.

2173. Kirby, J. E., J. E. Trempy, and S. Gottesman. 1994. Excision of a P4-like cryptic prophage leads to Alp protease expression in *Escherichia coli*. *J. Bacteriol.* 176:2068–2081.

2174. Kirby, T. W., B. R. Hindenach, and R. C. Greene. 1986. Regulation of in vivo transcription of the *Escherichia coli* K-12 *metJBLF* gene cluster. *J. Bacteriol.* 165:671–677.

2175. Kirino, H., M. Aoki, M. Aoshima, Y. Hayashi, M. Ohba, A. Yamagishi, T. Wakagi, and T. Oshima. 1994. Hydrophobic interaction at the subunit interface contributes to the thermostability of 3-isopropylmalate dehydrogenase from an extreme thermophile, *Thermus thermophilus*. *Eur. J. Biochem.* 220:275–281.

2176. Kirsebom, L. A., and S. Altman. 1989. Reaction in vitro of some mutants of RNase P with wild-type and temperature-sensitive substrates. *J. Mol. Biol.* 207:837–840.

2177. Kitagawa, M., C. Wada, S. Yoshioka, and T. Yura. 1991. Expression of ClpB, an analog of the ATP-dependent protease regulatory subunit in *Escherichia coli*, is controlled by a heat shock sigma factor (sigma 32). *J. Bacteriol.* 173:4247–4253.

2178. Kitagawa, Y., E. Akaboshi, H. Shinagawa, T. Horii, H. Ogawa, and T. Kato. 1985. Structural analysis of the *umu* operon required for inducible mutagenesis in *Escherichia coli*. *Proc. Natl. Acad. Sci. USA* 82:4336–4340.

2179. Kitakawa, M., L. Blumenthal, and K. Isono. 1980. Isolation and characterization of specialized transducing lambda phages carrying ribosomal protein genes of *Escherichia coli*. *Mol. Gen. Genet.* 180:343–349.

2180. Kitakawa, M., and K. Isono. 1982. An amber mutation in the gene *rpsA* for ribosomal protein S1 in *Escherichia coli*. *Mol. Gen. Genet.* 185:445–447.

2181. Kjeldgaard, M., and J. Nyborg. 1992. Refined structure of elongation factor EF-Tu from *Escherichia coli*. *J. Mol. Biol.* 223:721–742.

2182. Klein, J., B. Henrich, and R. Plapp. 1986. Cloning and expression of the *pepD* gene of *Escherichia coli*. *J. Gen. Microbiol.* 132:2337–2343.

2183. Klein, J. R., B. Henrich, and R. Plapp. 1991. Molecular analysis and nucleotide sequence of the *envCD* operon of *Escherichia coli*. *Mol. Gen. Genet.* 230:230–240.

2184. Klein, J. R., and R. Plapp. 1992. Locations of the *envCD* genes on the physical map of the *Escherichia coli* chromosome. *J. Bacteriol.* 174:3828–3829.

2185. Klein, W., R. Horlacher, and W. Boos. 1995. Molecular analysis of *treB* encoding the *Escherichia coli* enzyme II specific for trehalose. *J. Bacteriol.* 177:4043–4052.

2186. Klemm, P. 1984. The *fimA* gene encoding the type-1 fimbrial subunit of *Escherichia coli*. Nucleotide sequence and primary structure of the protein. *Eur. J. Biochem.* 143:395–399.

2187. Klemm, P. 1986. Two regulatory *fim* genes, *fimB* and *fimE*, control the phase variation of type 1 fimbriae in *Escherichia coli*. *EMBO J.* 5:1389–1393.

2188. Klemm, P. 1992. FimC, a chaperone-like periplasmic protein of *Escherichia coli* involved in biogenesis of type 1 fimbriae. *Res. Microbiol.* 143:831–838.

2189. Klemm, P., and G. Christiansen. 1987. Three *fim* genes required for the regulation and mediation of adhesion of *Escherichia coli* type 1 fimbriae. *Mol. Gen. Genet.* 208:439–445.

2190. Klemm, P., and G. Christiansen. 1990. The *fimD* gene required for cell surface localization of *Escherichia coli* type 1 fimbriae. *Mol. Gen. Genet.* 220:334–338.

2191. Klemm, P., B. J. Jorgensen, I. Van Die, H. de Ree, and H. Bergmans. 1985. The *fim* genes responsible for synthesis of type 1 fimbriae in *Escherichia coli*, cloning and genetic organization. *Mol. Gen. Genet.* 199:410–414.

2192. Klena, J. D., R. S. Ashford, and C. A. Schnaitman. 1992. Role of *Escherichia coli* K-12 *rfa* genes and the *rfp* gene of *Shigella dysenteriae* 1 in generation of lipopolysaccharide core heterogeneity and attachment of O antigen. *J. Bacteriol.* 174:7297–7307.

2193. Klena, J. D., E. Pradel, and C. A. Schnaitman. 1992. Comparison of lipopolysaccharide biosynthesis genes *rfaK*, *rfaL*, *rfaY*, and *rfaZ* of *Escherichia coli* K-12 and *Salmonella typhimurium*. *J. Bacteriol.* 174:4746–4752.

2194. Klena, J. D., E. Pradel, and C. A. Schnaitman. 1993. The *rfaS* gene, which is involved in production of a rough form of lipopolysaccharide core in *Escherichia coli* K-12, is not present in the *rfa* cluster of *Salmonella typhimurium* LT2. *J. Bacteriol.* 175:1524–1527.

2195. Klena, J. D., and C. A. Schnaitman. 1993. Function of the *rfb* gene cluster and the *rfe* gene in the synthesis of O antigen by *Shigella dysenteriae* 1. *Mol. Microbiol.* 9:393–402.

2196. Kline, B. C., T. Kogoma, J. E. Tam, and M. S. Shields. 1986. Requirement of the *Escherichia coli dnaA* gene product for plasmid F maintenance. *J. Bacteriol.* 168:440–443.

2197. Klumpp, D. J., D. W. Plank, L. J. Bowdin, C. S. Stueland, T. Chung, and D. C. LaPorte. 1988. Nucleotide sequence of *aceK*, the gene encoding isocitrate dehydrogenase kinase/phosphatase. *J. Bacteriol.* 170:2763–2769.

2198. Knoell, H. E., and J. Knappe. 1974. *Escherichia coli* ferredoxin, an iron-sulfur protein of the adrenodoxin type. *Eur. J. Biochem.* 50:245–252.

2199. Knott, V., D. J. Rees, Z. Cheng, and G. G. Brownlee. 1988. Randomly picked cosmid clones overlap the *pyrB* and *oriC* gap in the physical map of the *E. coli* chromosome. *Nucleic Acids Res.* 16:2601–2612.

2200. Kobayashi, T., M. Hidaka, and T. Horiuchi. 1989. Evidence of a *ter* specific binding protein essential for the termination reaction of DNA replication in *Escherichia coli*. *EMBO J.* 8:2435–2441.

2201. **Kobayashi, T., I. Kudo, H. Homma, K. Karasawa, K. Inoue, H. Ikeda, and S. Nojima.** 1985. Gene organization of *pldA* and *pldB*, the structural genes for detergent-resistant phospholipase A and lysophospholipase L2 of *Escherichia coli*. *J. Biochem.* **98:**1007–1016.

2202. **Kobayashi, T., I. Kudo, K. Karasawa, H. Mizushima, K. Inoue, and S. Nojima.** 1985. Nucleotide sequence of the *pldB* gene and characteristics of deduced amino acid sequence of lysophospholipase L2 in *Escherichia coli*. *J. Biochem.* **98:**1017–1025.

2203. **Koch, C., J. Vandekerckhove, and R. Kahmann.** 1988. *Escherichia coli* host factor for site-specific DNA inversion: cloning and characterization of the *fis* gene. *Proc. Natl. Acad. Sci. USA* **85:**4237–4241.

2204. **Koch, W. H., D. G. Ennis, A. S. Levine, and R. Woodgate.** 1992. *Escherichia coli umuDC* mutants: DNA sequence alterations and UmuD cleavage. *Mol. Gen. Genet.* **233:**443–448.

2205. **Kodaira, M., S. B. Biswas, and A. Kornberg.** 1983. The *dnaX* gene encodes the DNA polymerase III holoenzyme tau subunit, precursor of the gamma subunit, the *dnaZ* gene product. *Mol. Gen. Genet.* **192:**80–86.

2206. **Kofoid, E. C., and J. S. Parkinson.** 1991. Tandem translation starts in the *cheA* locus of *Escherichia coli*. *J. Bacteriol.* **173:**2116–2119.

2207. **Koh, Y. S., and J. H. Roe.** 1995. Isolation of a novel paraquat-inducible (*pqi*) gene regulated by the *soxRS* locus in *Escherichia coli*. *J. Bacteriol.* **177:**2673–2678.

2208. **Kohara, Y.** 1990. Correlation between the physical and genetic maps of the *Escherichia coli* K-12 chromosome, p. 29–42. *In* K. Drlica and M. Riley (ed.), *The Bacterial Chromosome*. American Society for Microbiology, Washington, D.C.

2209. **Kohara, Y., K. Akiyama, and K. Isono.** 1987. The physical map of the whole *E. coli* chromosome: application of a new strategy for rapid analysis and sorting of a large genomic library. *Cell* **50:**495–508.

2210. **Kohno, K., M. Wada, Y. Kano, and F. Imamoto.** 1990. Promoters and autogenous control of the *Escherichia coli hupA* and *hupB* genes. *J. Mol. Biol.* **213:**27–36.

2211. **Kolb, A., S. Busby, H. Buc, S. Garges, and S. Adhya.** 1993. Transcriptional regulation by cAMP and its receptor protein. *Annu. Rev. Biochem.* **62:**749–795.

2212. **Kolling, R., A. Gielow, W. Seufert, C. Kucherer, and W. Messer.** 1988. AsnC, a multifunctional regulator of genes located around the replication origin of *Escherichia coli, oriC*. *Mol. Gen. Genet.* **212:**99–104.

2213. **Kolling, R., and H. Lother.** 1985. AsnC: an autogenously regulated activator of asparagine synthetase A transcription in *Escherichia coli*. *J. Bacteriol.* **164:**310–315.

2214. **Kolodner, R., R. A. Fishel, and M. Howard.** 1985. Genetic recombination of bacterial plasmid DNA: effect of RecF pathway mutations on plasmid recombination in *Escherichia coli*. *J. Bacteriol.* **163:**1060–1066.

2215. **Kolodner, R., S. D. Hall, and C. Luisi-DeLuca.** 1994. Homologous pairing proteins encoded by the *Escherichia coli recE* and *recT* genes. *Mol. Microbiol.* **11:**23–30.

2216. **Kolodrubetz, D., and R. Schleif.** 1981. L-Arabinose transport systems in *Escherichia coli* K-12. *J. Bacteriol.* **148:**472–479.

2217. **Kolodrubetz, D., and R. Schleif.** 1981. Regulation of the L-arabinose transport operons in *Escherichia coli*. *J. Mol. Biol.* **151:**215–227.

2218. **Kolot, M., and E. Yagil.** 1994. Position and direction of strand exchange in bacteriophage HK022 integration. *Mol. Gen. Genet.* **245:**623–627.

2219. **Komano, T., R. Utsumi, and M. Kawamukai.** 1991. Functional analysis of the *fic* gene involved in regulation of cell division. *Res. Microbiol.* **142:**269–277.

2220. **Komatsu, Y., and K. Tanaka.** 1972. A showdomycin-resistant mutant of *Escherichia coli* K-12 with altered nucleoside transport character. *Biochim. Biophys. Acta* **288:**390–403.

2221. **Komeda, Y.** 1982. Fusions of flagellar operons to lactose genes on a mu *lac* bacteriophage. *J. Bacteriol.* **150:**16–26.

2222. **Komeda, Y.** 1986. Transcriptional control of flagellar genes in *Escherichia coli* K-12. *J. Bacteriol.* **168:**1315–1318.

2223. **Komeda, Y., K. Kutsukake, and T. Iino.** 1980. Definition of additional flagellar genes in *Escherichia coli* K12. *Genetics* **94:**277–290.

2224. **Komeda, Y., M. Silverman, and M. Simon.** 1977. Genetic analysis of *Escherichia coli* K-12 region I flagellar mutants. *J. Bacteriol.* **131:**801–808.

2225. **Komine, Y., T. Adachi, H. Inokuchi, and H. Ozeki.** 1990. Genomic organization and physical mapping of the transfer RNA genes in *Escherichia coli* K12. *J. Mol. Biol.* **212:**579–598.

2226. **Komine, Y., and H. Inokuchi.** 1991. Precise mapping of the *rnpB* gene encoding the RNA component of RNase P in *Escherichia coli* K-12. *J. Bacteriol.* **173:**1813–1816.

2227. **Komine, Y., and H. Inokuchi.** 1991. Physical map locations of the genes that encode small stable RNAs in *Escherichia coli*. *J. Bacteriol.* **173:**5252.

2228. **Komoda, Y., M. Enomoto, and A. Tominaga.** 1991. Large inversion in *Escherichia coli* K-12 1485IN between inversely oriented IS*3* elements near *lac* and *cdd*. *Genetics* **129:**639–645.

2229. **Kondo, H., Y. Nakabeppu, H. Kataoka, S. Kuhara, S. Kawabata, and M. Sekiguchi.** 1986. Structure and expression of the *alkB* gene of *Escherichia coli* related to the repair of alkylated DNA. *J. Biol. Chem.* **261:**15772–15777.

2230. **Kondo, H., K. Shiratsuchi, T. Yoshimoto, T. Masuda, A. Kitazono, D. Tsuru, M. Anai, M. Sekiguchi, and T. Tanabe.** 1991. Acetyl-CoA carboxylase from *Escherichia coli*: gene organization and nucleotide sequence of the biotin carboxylase subunit. *Proc. Natl. Acad. Sci. USA* **88:**9730–9733.

2231. **Kondo, K., S. Wakabayashi, T. Yagi, and H. Kagamiyama.** 1984. The complete amino acid sequence of aspartate aminotransferase from *Escherichia coli*: se-

2232. quence comparison with pig isoenzymes. *Biochem. Biophys. Res. Commun.* **122:**62–67.

2232. **Konforti, B. B., and R. W. Davis.** 1992. ATP hydrolysis and the displaced strand are two factors that determine the polarity of RecA-promoted DNA strand exchange. *J. Mol. Biol.* **227:**38–53.

2233. **Konigsberg, W. H., and L. Henderson.** 1983. Amino acid sequence of the catalytic subunit of aspartate transcarbamoylase from *Escherichia coli*. *Proc. Natl. Acad. Sci. USA* **80:**2467–2471.

2234. **Koonin, E. V.** 1992. DnaC protein contains a modified ATP-binding motif and belongs to a novel family of ATPases including also DnaA. *Nucleic Acids Res.* **20:**1997.

2235. **Koonin, E. V.** 1993. *Escherichia coli dinG* gene encodes a putative DNA helicase related to a group of eukaryotic helicases including Rad3 protein. *Nucleic Acids Res.* **21:**1497.

2236. **Koonin, E. V., and A. E. Gorbalenya.** 1992. Autogenous translation regulation by *Escherichia coli* ATPase SecA may be mediated by an intrinsic RNA helicase activity of this protein. *FEBS Lett.* **298:**6–8.

2237. **Koonin, E. V., and K. E. Rudd.** 1993. SpoU protein of *Escherichia coli* belongs to a new family of putative rRNA methylases. *Nucleic Acids Res.* **21:**5519.

2238. **Koop, A. H., M. Hartley, and S. Bourgeois.** 1984. Analysis of the *cya* locus of *Escherichia coli*. *Gene* **28:**133–146.

2239. **Korat, B., H. Mottl, and W. Keck.** 1991. Penicillin-binding protein 4 of *Escherichia coli*: molecular cloning of the *dacB* gene, controlled overexpression, and alterations in murein composition. *Mol. Microbiol.* **5:**675–684.

2240. **Kornberg, A.** 1995. Inorganic polyphosphate: toward making a forgotten polymer unforgettable. *J. Bacteriol.* **177:**491–496.

2241. **Kornberg, H.** 1986. The roles of HPr and FPr in the utilization of fructose by *Escherichia coli*. *FEBS Lett.* **194:**12–15.

2242. **Kornberg, H. L., and C. M. Elvin.** 1987. Location and function of *fruC*, a gene involved in the regulation of fructose utilization by *Escherichia coli*. *J. Gen. Microbiol.* **133:**341–346.

2243. **Kornitzer, D., D. Teff, S. Altuvia, and A. B. Oppenheim.** 1991. Isolation, characterization, and sequence of an *Escherichia coli* heat shock gene, *htpX*. *J. Bacteriol.* **173:**2944–2953.

2244. **Kosiba, B. E., and R. Schleif.** 1982. Arabinose-inducible promoter from *Escherichia coli*. Its cloning from chromosomal DNA, identification as the *araFG* promoter and sequence. *J. Mol. Biol.* **156:**53–66.

2245. **Koster, W., and B. Bohm.** 1992. Point mutations in two conserved glycine residues within the integral membrane protein FhuB affect iron(III) hydroxamate transport. *Mol. Gen. Genet.* **232:**399–407.

2246. **Koster, W., and V. Braun.** 1986. Iron hydroxamate transport of *Escherichia coli*: nucleotide sequence of the *fhuB* gene and identification of the protein. *Mol. Gen. Genet.* **204:**435–442.

2247. **Koster, W., and V. Braun.** 1990. Iron (III) hydroxamate transport into *Escherichia coli*. Substrate binding to the periplasmic FhuD protein. *J. Biol. Chem.* **265:**21407–21410.

2248. **Koster, W., A. Gudmundsdottir, M. D. Lundrigan, A. Seiffert, and R. J. Kadner.** 1991. Deletions or duplications in the BtuB protein affect its level in the outer membrane of *Escherichia coli*. *J. Bacteriol.* **173:**5639–5647.

2249. **Kostrewa, D., J. Granzin, C. Koch, H. W. Choe, S. Raghunathan, W. Wolf, J. Labahn, R. Kahmann, and W. Saenger.** 1991. Three-dimensional structure of the *E. coli* DNA-binding protein FIS. *Nature* (London) **349:**178–180.

2250. **Kotani, H., A. Kawamura, A. Takahashi, M. Nakatsuji, N. Hiraoka, K. Nakajima, and M. Takanami.** 1992. Site-specific dissection of *E. coli* chromosome by lambda terminase. *Nucleic Acids Res.* **20:**3357–3360.

2251. **Kotani, H., and K. Nakajima.** 1992. Cloning and sequence of thioredoxin gene of *Salmonella typhimurium* LT2. *Nucleic Acids Res.* **20:**1424.

2252. **Kotval, J., A. Campbell, G. Konopa, and W. Szybalski.** 1982. Leftward transcription in the *Escherichia coli bio* operon does not require products of the rightward transcript. *Gene* **17:**219–222.

2253. **Kow, Y. W., G. Faundez, S. Hays, C. A. Bonner, M. F. Goodman, and S. S. Wallace.** 1993. Absence of a role for DNA polymerase II in SOS-induced translesion bypass of φX174. *J. Bacteriol.* **175:**561–564.

2254. **Kozliak, E. I., M. B. Guilloton, M. Gerami-Nejad, J. A. Fuchs, and P. M., Anderson.** 1994. Expression of proteins encoded by the *Escherichia coli cyn* operon: carbon dioxide-enhanced degradation of carbonic anhydrase. *J. Bacteriol.* **176:**5711–5717.

2255. **Kraft, R., and L. A. Leinwand.** 1987. Sequence of the complete P protein gene and part of the M protein gene from the histidine transport operon of *Escherichia coli* compared to that of *Salmonella typhimurium*. *Nucleic Acids Res.* **15:**8568.

2256. **Kraus, J., D. Soll, and K. B. Low.** 1979. Glutamyl-gamma-methyl ester acts as a methionine analogue in *Escherichia coli*: analogue resistant mutants map at the *metJ* and *metK* loci. *Genet. Res.* **33:**49–55.

2257. **Kren, B., and J. A. Fuchs.** 1987. Characterization of the *ftsB* gene as an allele of the *nrdB* gene in *Escherichia coli*. *J. Bacteriol.* **169:**14–18.

2258. **Kren, B., D. Parsell, and J. A. Fuchs.** 1988. Isolation and characterization of an *Escherichia coli* K-12 mutant deficient in glutaredoxin. *J. Bacteriol.* **170:**308–315.

2259. **Kreuzer, K. N., and N. R. Cozzarelli.** 1979. *Escherichia coli* mutants thermosensitive for deoxyribonucleic acid gyrase subunit A: effects on deoxyribonucleic acid replication, transcription, and bacteriophage growth. *J. Bacteriol.* **140:**424–435.

2260. **Kricker, M., and B. G. Hall.** 1984. Directed evolution of cellobiose utilization in *Escherichia coli* K12. *Mol. Biol. Evol.* **1:**171–182.

2261. Kricker, M., and B. G. Hall. 1987. Biochemical genetics of the cryptic gene system for cellobiose utilization in *Escherichia coli* K12. *Genetics* 115:419–429.

2262. Krikos, A., N. Mutoh, A. Boyd, and M. I. Simon. 1983. Sensory transducers of *E. coli* are composed of discrete structural and functional domains. *Cell* 33:615–622.

2263. Krogfelt, K. A., H. Bergmans, and P. Klemm. 1990. Direct evidence that the FimH protein is the mannose-specific adhesin of *Escherichia coli* type 1 fimbriae. *Infect. Immun.* 58:1995–1998.

2264. Kroh, H. E., and L. D. Simon. 1990. The ClpP component of Clp protease is the sigma 32-dependent heat shock protein F21.5. *J. Bacteriol.* 172:6026–6034.

2265. Krone, F. A., G. Westphal, H. E. Meyer, and J. D. Schwenn. 1990. PAPS-reductase of *Escherichia coli*. Correlating the N-terminal amino acid sequence with the DNA of gene *cysH*. *FEBS Lett.* 260:6–9.

2266. Krone, F. A., G. Westphal, and J. D. Schwenn. 1991. Characterisation of the gene *cysH* and of its product phospho-adenylylsulphate reductase from *Escherichia coli*. *Mol. Gen. Genet.* 225:314–319.

2267. Krueger, J. H., S. J. Elledge, and G. C. Walker. 1983. Isolation and characterization of Tn5 insertion mutations in the *lexA* gene of *Escherichia coli*. *J. Bacteriol.* 153:1368–1378.

2268. Krueger, J. H., J. R. Johnson, R. C. Greene, and M. Dresser. 1981. Structural studies of lambda transducing bacteriophage carrying bacterial deoxyribonucleic acid from the *metBJLF* region of the *Escherichia coli* chromosome. *J. Bacteriol.* 147:612–621.

2269. Krueger, J. K., J. Stock, and C. E. Schutt. 1992. Evidence that the methylesterase of bacterial chemotaxis may be a serine hydrolase. *Biochim. Biophys. Acta* 1119:322–326.

2270. Kruger, T., C. Grund, C. Wild, and M. Noyer-Weidner. 1992. Characterization of the *mcrBC* region of *Escherichia coli* K-12 wild-type and mutant strains. *Gene* 114:1–12.

2271. Kuchino, Y., F. Mori, and S. Nishimura. 1985. Structure and transcription of the tRNA$_1^{Pro}$ gene from *Escherichia coli*. *Nucleic Acids Res.* 13:3213–3220.

2271a. Kudo, T., K. Nagai, and G. Tamura. 1977. Characteristics of a cold-sensitive cell division mutant of *Escherichia coli* K-12. *Agric. Biol. Chem.* 41:97–107.

2272. Kuempel, P. L., J. M. Henson, L. Dircks, M. Tecklenburg, and D. F. Lim. 1991. *dif*, a *recA*-independent recombination site in the terminus region of the chromosome of *Escherichia coli*. *New Biol.* 3:799–811.

2273. Kuhn, J., and R. L. Somerville. 1971. Mutant strains of *Escherichia coli* K12 that use D-amino acids. *Proc. Natl. Acad. Sci. USA* 68:2484–2487.

2274. Kullik, I., J. Stevens, M. B. Toledano, and G. Storz. 1995. Mutational analysis of the redox-sensitive transcriptional regulator OxyR: regions important for DNA binding and multimerization. *J. Bacteriol.* 177:1285–1291.

2275. Kullik, I., M. B. Toledano, L. A. Tartaglia, and G. Storz. 1995. Mutational analysis of the redox-sensitive transcriptional regulator OxyR: regions important for oxidation and transcriptional activation. *J. Bacteriol.* 177:1275–1284.

2276. Kumamoto, C. A. 1991. Molecular chaperones and protein translocation across the *Escherichia coli* inner membrane. *Mol. Microbiol.* 5:19–22.

2277. Kumamoto, C. A., and A. K. Nault. 1989. Characterization of the *Escherichia coli* protein-export gene *secB*. *Gene* 75:167–175.

2278. Kumar, A., P. Ghosh, Y. M. Lee, M. A. Hill, and J. Preiss. 1989. Biosynthesis of bacterial glycogen. Determination of the amino acid changes that alter the regulatory properties of a mutant *Escherichia coli* ADP-glucose synthetase. *J. Biol. Chem.* 264:10464–10471.

2279. Kumar, A., C. E. Larsen, and J. Preiss. 1986. Biosynthesis of bacterial glycogen. Primary structure of *Escherichia coli* ADP-glucose:alpha-1,4-glucan, 4-glucosyltransferase as deduced from the nucleotide sequence of the *glgA* gene. *J. Biol. Chem.* 261:16256–16259.

2280. Kumar, A., T. Tanaka, Y. M. Lee, and J. Preiss. 1988. Biosynthesis of bacterial glycogen. Use of site-directed mutagenesis to probe the role of tyrosine 114 in the catalytic mechanism of ADP-glucose synthetase from *Escherichia coli*. *J. Biol. Chem.* 263:14634–14639.

2281. Kumar, S. 1976. Properties of adenyl cyclase and cyclic adenosine 3′,5′-monophosphate receptor protein-deficient mutants of *Escherichia coli*. *J. Bacteriol.* 125:545–555.

2282. Kumaresan, K. R., and R. Jayaraman. 1990. The *sir* locus of *Escherichia coli*: a gene involved in SOS-independent repair of mitomycin C-induced DNA damage. *Mutat. Res.* 235:85–92.

2283. Kumari, S., R. Tishel, M. Eisenbach, and A. J. Wolfe. 1995. Cloning, characterization, and functional expression of *acs*, the gene which encodes acetyl coenzyme A synthetase in *Escherichia coli*. *J. Bacteriol.* 177:2878–2886.

2284. Kumura, K., and M. Sekiguchi. 1984. Identification of the *uvrD* gene product of *Escherichia coli* as DNA helicase II and its induction by DNA-damaging agents. *J. Biol. Chem.* 259:1560–1565.

2285. Kuo, L. C., A. W. Miller, S. Lee, and C. Kozuma. 1988. Site-directed mutagenesis of *Escherichia coli* ornithine transcarbamoylase: role of arginine-57 in substrate binding and catalysis. *Biochemistry* 27:8823–8832. (Erratum, 28:4522, 1989.)

2286. Kuo, S. C., and D. E. Koshland, Jr. 1986. Sequence of the *flaA* (*cheC*) locus of *Escherichia coli* and discovery of a new gene. *J. Bacteriol.* 166:1007–1012.

2287. Kur, J., N. Hasan, and W. Szybalski. 1992. Integration host factor (IHF) binds to many sites in the A+T-rich b2 region of phage lambda DNA. *Gene* 111:1–9.

2288. Kuramitsu, S., K. Inoue, T. Ogawa, H. Ogawa, and H. Kagamiyama. 1985. Aromatic amino acid aminotransferase of *Escherichia coli*: nucleotide sequence of the *tyrB* gene. *Biochem. Biophys. Res. Commun.* 133:134–139.

2289. Kuramitsu, S., T. Ogawa, H. Ogawa, and H. Kagamiyama. 1985. Branched-chain amino acid aminotransferase of *Escherichia coli*: nucleotide sequence of the *ilvE* gene and the deduced amino acid sequence. *J. Biochem.* 97:993–999.

2290. Kuramitsu, S., S. Okuno, T. Ogawa, H. Ogawa, and H. Kagamiyama. 1985. Aspartate aminotransferase of *Escherichia coli*: nucleotide sequence of the *aspC* gene. *J. Biochem.* 97:1259–1262.

2291. Kurihara, T., and Y. Nakamura. 1983. Cloning of the *nusA* gene of *Escherichia coli*. *Mol. Gen. Genet.* 190:189–195.

2292. Kusano, K., N. K. Takahashi, H. Yoshikura, and I. Kobayashi. 1994. Involvement of RecE exonuclease and RecT annealing protein in DNA double-strand break repair by homologous recombination. *Gene* 138:17–25.

2293. Kushiro, M., M. Shimizu, and K. Tomita. 1987. Molecular cloning and sequence determination of the *tuf* gene coding for the elongation factor Tu of *Thermus thermophilus* HB8. *Eur. J. Biochem.* 170:93–98.

2294. Kusukawa, N., and T. Yura. 1988. Heat shock protein GroE of *Escherichia coli*: key protective roles against thermal stress. *Genes Dev.* 2:874–882.

2295. Kutsukake, K., T. Nakao, and T. Iino. 1985. A gene for DNA invertase and an invertible DNA in *Escherichia coli* K-12. *Gene* 34:343–350.

2296. Kuwajima, G., J. Asaka, T. Fujiwara, K. Node, and E. Kondo. 1986. Nucleotide sequence of the *hag* gene encoding flagellin of *Escherichia coli*. *J. Bacteriol.* 168:1479–1483.

2297. La Cara, F., and M. De Felice. 1979. Chromatographic detection of the acetohydroxy acid synthase isoenzymes of *Escherichia coli* K-12. *Biochem. Biophys. Res. Commun.* 91:319–324.

2298. La Teana, A., A. Brandi, M. Falconi, R. Spurio, C. L. Pon, and C. O. Gualerzi. 1991. Identification of a cold shock transcriptional enhancer of the *Escherichia coli* gene encoding nucleoid protein H-NS. *Proc. Natl. Acad. Sci. USA* 88:10907–10911.

2299. Laalami, S., C. Sacerdot, G. Vachon, K. Mortensen, H. U. Sperling-Petersen, Y. Cenatiempo, and M. Grunberg-Manago. 1991. Structural and functional domains of *E. coli* initiation factor IF2. *Biochimie* 73:1557–1566.

2300. Labie, C., F. Bouche, and J. P. Bouche. 1990. Minicell-forming mutants of *Escherichia coli*: suppression of both DicB- and MinD-dependent division inhibition by inactivation of the *minC* gene product. *J. Bacteriol.* 172:5852–5855.

2301. Lacroix, J. M., I. Loubens, M. Tempete, B. Menichi, and J. P. Bohin. 1991. The *mdoA* locus of *Escherichia coli* consists of an operon under osmotic control. *Mol. Microbiol.* 5:1745–1753.

2302. Laengle-Rouault, F., G. Maenhaut-Michel, and M. Radman. 1986. GATC sequence and mismatch repair in *Escherichia coli*. *EMBO J.* 5:2009–2013.

2303. Lahti, R., M. Perala, P. Heikinheimo, T. Pitkaranta, E. Kukko-Kalske, and J. Heinonen. 1991. Characterization of the 5′ flanking region of the *Escherichia coli ppa* gene encoding inorganic pyrophosphatase: mutations in the ribosome-binding site decrease the level of *ppa* mRNA. *J. Gen. Microbiol.* 137:2517–2523.

2304. Lahti, R., T. Pitkaranta, E. Valve, I. Ilta, E. Kukko-Kalske, and J. Heinonen. 1988. Cloning and characterization of the gene encoding inorganic pyrophosphatase of *Escherichia coli* K-12. *J. Bacteriol.* 170:5901–5907.

2305. Lahti, R., K. Pohjanoksa, T. Pitkaranta, P. Heikinheimo, T. Salminen, P. Meyer, and J. Heinonen. 1990. A site-directed mutagenesis study on *Escherichia coli* inorganic pyrophosphatase. Glutamic acid-98 and lysine-104 are important for structural integrity, whereas aspartic acids-97 and -102 are essential for catalytic activity. *Biochemistry* 29:5761–5766.

2306. Lahti, R., T. Salminen, S. Latonen, P. Heikinheimo, K. Pohjanoksa, and J. Heinonen. 1991. Genetic engineering of *Escherichia coli* inorganic pyrophosphatase. Tyr55 and Tyr141 are important for the structural integrity. *Eur. J. Biochem.* 198:293–297.

2307. Laine, B., D. Kmiecik, P. Sautiere, G. Biserte, and M. Cohen-Solal. 1980. Complete amino-acid sequences of DNA-binding proteins HU-1 and HU-2 from *Escherichia coli*. *Eur. J. Biochem.* 103:447–461.

2308. Laine, B., P. Sautiere, A. Spassky, and S. Rimsky. 1984. A DNA-binding protein from *E. coli* isolation, characterization and its relationship with proteins H1 and B1. *Biochem. Biophys. Res. Commun.* 119:1147–1153.

2309. Laine, P. S., and R. R. Meyer. 1992. Interaction of the heat shock protein GroEL of *Escherichia coli* with single-stranded DNA-binding protein: suppression of *ssb*-113 by *groEL*46. *J. Bacteriol.* 174:3204–3211.

2310. Laird, A. J., D. W. Ribbons, G. C. Woodrow, and I. G. Young. 1980. Bacteriophage Mu-mediated gene transposition and in vitro cloning of the enterochelin gene cluster of *Escherichia coli*. *Gene* 11:347–357.

2311. Laird, A. J., and I. G. Young. 1980. Tn5 mutagenesis of the enterochelin gene cluster of *Escherichia coli*. *Gene* 11:359–366.

2312. Lakshmi, T. M., and R. B. Helling. 1976. Selection for citrate synthase deficiency in *icd* mutants of *Escherichia coli*. *J. Bacteriol.* 127:76–83.

2313. Lam, H. M., E. Tancula, W. B. Dempsey, and M. E. Winkler. 1992. Suppression of insertions in the complex *pdxJ* operon of *Escherichia coli* K-12 by *lon* and other mutations. *J. Bacteriol.* 174:1554–1567.

2314. Lam, H. M., and M. E. Winkler. 1992. Characterization of the complex *pdxH-tyrS* operon of *Escherichia coli* K-12 and pleiotropic phenotypes caused by *pdxH* insertion mutations. *J. Bacteriol.* 174:6033–6045.

2315. Lamark, T., I. Kaasen, M. W. Eshoo, P. Falkenberg, J. McDougall, and A. R. Strom. 1991. DNA sequence and analysis of the *bet* genes encoding the osmoregulatory choline-glycine betaine pathway of *Escherichia coli*. *Mol. Microbiol.* 5:1049–1064.

2315a.Lambalot, R. H., and C. T. Walsh. 1995. Cloning, overproduction, and characterization of the *Escherichia coli* holo-acyl carrier protein synthase. *J. Biol. Chem.* 270:24658–24661.

2316. Lambdren, P. R., and J. R. Guest. 1976. A novel method for isolating chlorate-resistant mutants of *Escherichia coli* K12 by anaerobic selection on a lactate plus fumarate medium. *J. Gen. Microbiol.* 93:173–176.

2317. Lamblin, A. F., and J. A. Fuchs. 1993. Expression and purification of the *cynR* regulatory gene product: CynR is a DNA-binding protein. *J. Bacteriol.* 175:7990–7999.

2318. Lander, M., A. R. Pitt, P. R. Alefounder, D. Bardy, C. Abell, and A. R. Battersby. 1991. Studies on the mechanism of hydroxymethylbilane synthase concerning the role of arginine residues in substrate binding. *Biochem. J.* 275:447–452.

2319. Landick, R., and D. L. Oxender. 1985. The complete nucleotide sequences of the *Escherichia coli* LIV-BP and LS-BP genes. Implications for the mechanism of high-affinity branched-chain amino acid transport. *J. Biol. Chem.* 260:8257–8261.

2320. Landick, R., J. Stewart, and D. N. Lee. 1990. Amino acid changes in conserved regions of the beta-subunit of *Escherichia coli* RNA polymerase alter transcription pausing and termination. *Genes Dev.* 4:1623–1636.

2321. Landick, R., V. Vaughn, E. T. Lau, R. A. VanBogelen, J. W. Erickson, and F. C. Neidhardt. 1984. Nucleotide sequence of the heat shock regulatory gene of *E. coli* suggests its protein product may be a transcription factor. *Cell* 38:175–182.

2322. Landini, P., L. I. Hajec, and M. R. Volkert. 1994. Structure and transcriptional regulation of the *Escherichia coli* adaptive response gene *aidB*. *J. Bacteriol.* 176:6583–6589.

2323. Landoulsi, A., P. Hughes, R. Kern, and M. Kohiyama. 1989. *dam* methylation and the initiation of DNA replication on *oriC* plasmids. *Mol. Gen. Genet.* 216:217–223.

2324. Lange, R., M. Barth, and R. Hengge-Aronis. 1993. Complex transcriptional control of the sigma s-dependent stationary-phase-induced and osmotically regulated *osmY* (*csi-5*) gene suggests novel roles for Lrp, cyclic AMP (cAMP) receptor protein-cAMP complex, and integration host factor in the stationary-phase response of *Escherichia coli*. *J. Bacteriol.* 175:7910–7917.

2325. Lange, R., and R. Hengge-Aronis. 1991. Growth phase-regulated expression of *bolA* and morphology of stationary-phase *Escherichia coli* cells are controlled by the novel sigma factor sigma S. *J. Bacteriol.* 173:4474–4481.

2326. Lange, R., and R. Hengge-Aronis. 1991. Identification of a central regulator of stationary-phase gene expression in *Escherichia coli*. *Mol. Microbiol.* 5:49–59.

2327. Lange, R., and R. Hengge-Aronis. 1994. The *nlpD* gene is located in an operon with *rpoS* on the *Escherichia coli* chromosome and encodes a novel lipoprotein with a potential function in cell wall formation. *Mol. Microbiol.* 13:733–743.

2328. Langert, W., M. Meuthen, and K. Mueller. 1991. Functional characteristics of the *rrnD* promoters of *Escherichia coli*. *J. Biol. Chem.* 266:21608–21615.

2329. Langley, D., and J. R. Guest. 1977. Biochemical genetics of the alpha-keto acid dehydrogenase complexes of *Escherichia coli* K12: isolation and biochemical properties of deletion mutants. *J. Gen. Microbiol.* 99:263–276.

2330. Langley, D., and J. R. Guest. 1978. Biochemical genetics of the alpha-keto acid dehydrogenase complexes of *Escherichia coli* K12: genetic characterization and regulatory properties of deletion mutants. *J. Gen. Microbiol.* 106:103–117.

2331. Langridge, J. 1969. Mutations conferring quantitative and qualitative increases in beta-galactosidase activity in *Escherichia coli*. *Mol. Gen. Genet.* 105:74–83.

2332. LaPorte, D. C., and T. Chung. 1985. A single gene codes for the kinase and phosphatase which regulate isocitrate dehydrogenase. *J. Biol. Chem.* 260:15291–15297.

2333. LaPorte, D. C., C. S. Stueland, and T. P. Ikeda. 1989. Isocitrate dehydrogenase kinase/phosphatase. *Biochimie* 71:1051–1057.

2334. LaPorte, D. C., P. E. Thorsness, and D. E. Koshland, Jr. 1985. Compensatory phosphorylation of isocitrate dehydrogenase. A mechanism for adaptation to the intracellular environment. *J. Biol. Chem.* 260:10563–10568.

2335. Larminat, F., C. Cazaux, M. Germanier, and M. Defais. 1992. New mutations in and around the L2 disordered loop of the RecA protein modulate recombination and/or coprotease activity. *J. Bacteriol.* 174:6264–6269.

2336. LaRossa, R. A., and T. K. Van Dyk. 1991. Physiological roles of the DnaK and GroE stress proteins: catalysts of protein folding or macromolecular sponges? *Mol. Microbiol.* 5:529–534.

2336a.LaRossa, R., G. Vögeli, K. B. Low, and D. Söll. 1977. Regulation of biosynthesis of aminoacyl-tRNA synthetases and of tRNA in *Escherichia coli*. II. Isolation of regulatory mutants affecting leucyl-tRNA synthetase levels. *J. Mol. Biol.* 117:1033.

2337. Larsen, J. E., B. Albrechtsen, and P. Valentin-Hansen. 1987. Analysis of the terminator region after the *deoCABD* operon of *Escherichia coli* K-12 using a new class of single copy number operon-fusion vectors. *Nucleic Acids Res.* 15:5125–5140.

2338. Larsen, J. N., and K. F. Jensen. 1985. Nucleotide sequence of the *pyrD* gene of *Escherichia coli* and characterization of the flavoprotein dihydroorotate dehydrogenase. *Eur. J. Biochem.* 151:59–65.

2339. Larson, T. J., J. S. Cantwell, and A. T. van Loo-Bhattacharya. 1992. Interaction at a distance between multiple operators controls the adjacent, divergently transcribed *glpTQ-glpACB* operons of *Escherichia coli* K-12. *J. Biol. Chem.* 267:6114–6121.

2340. Larson, T. J., D. N. Ludtke, and R. M. Bell. 1984. sn-Glycerol-3-phosphate auxotrophy of *plsB* strains of *Escherichia coli*: evidence that a second mutation, *plsX*, is required. *J. Bacteriol.* 160:711–717.

2341. Larson, T. J., G. Schumacher, and W. Boos. 1982. Identification of the *glpT*-encoded sn-glycerol-3-phosphate permease of *Escherichia coli*, an oligomeric integral membrane protein. *J. Bacteriol.* 152:1008–1021.

2342. Lathe, R., A. Bollen, and R. Herzog. 1981. Revised location of the *Escherichia coli* gene coding for ribosomal protein S2. *Mol. Gen. Genet.* 182:178–179.

2343. Lathe, R., H. Buc, J. P. Lecocq, and E. K. Bautz. 1980. Prokaryotic histone-like protein interacting with RNA polymerase. *Proc. Natl. Acad. Sci. USA* 77:3548–3552.

2344. Latil, M., M. Murgier, A. Lazdunski, and C. Lazdunski. 1976. Isolation and genetic mapping of *Escherichia coli* aminopeptidase mutants. *Mol. Gen. Genet.* 148:43–47.

2345. Latil-Damotte, M., and C. Lares. 1977. Relative order of *glg* mutations affecting glycogen biosynthesis in *Escherichia coli* K12. *Mol. Gen. Genet.* 150:325–328.

2346. Laursen, R. A., J. J. L'Italien, S. Nagarkatti, and D. L. Miller. 1981. The amino acid sequence of elongation factor Tu of *Escherichia coli*. The complete sequence. *J. Biol. Chem.* 256:8102–8109.

2347. Lavery, P. E., and S. C. Kowalczykowski. 1992. A postsynaptic role for single-stranded DNA-binding protein in *recA* protein-promoted DNA strand exchange. *J. Biol. Chem.* 267:9315–9320.

2348. Lavery, P. E., and S. C. Kowalczykowski. 1992. Enhancement of *recA* protein-promoted DNA strand exchange activity by volume-occupying agents. *J. Biol. Chem.* 267:9307–9314.

2349. Lavigne, M., M. Herbert, A. Kolb, and H. Buc. 1992. Upstream curved sequences influence the initiation of transcription at the *Escherichia coli* galactose operon. *J. Mol. Biol.* 224:293–306.

2350. Lavina, M., A. P. Pugsley, and F. Moreno. 1986. Identification, mapping, cloning and characterization of a gene (*sbmA*) required for microcin B17 action on *Escherichia coli* K12. *J. Gen. Microbiol.* 132:1685–1693.

2351. Lawley, B., and A. J. Pittard. 1994. Regulation of *aroL* expression by TyrR protein and Trp repressor in *Escherichia coli* K-12. *J. Bacteriol.* 176:6921–6930.

2352. Lawlis, V. B., M. S. Dennis, E. Y. Chen, D. H. Smith, and D. J. Henner. 1984. Cloning and sequencing of the xylose isomerase and xylulose kinase genes of *Escherichia coli*. *Appl. Environ. Microbiol.* 47:15–21.

2353. Lawrence, J. G., H. Ochman, and D. L. Hartl. 1991. Molecular and evolutionary relationships among enteric bacteria. *J. Gen. Microbiol.* 137:1911–1921.

2354. Lawther, R. P., D. H. Calhoun, C. W. Adams, C. A. Hauser, J. Gray, and G. W. Hatfield. 1981. Molecular basis of valine resistance in *Escherichia coli* K-12. *Proc. Natl. Acad. Sci. USA* 78:922–925.

2355. Lawther, R. P., D. H. Calhoun, J. Gray, C. W. Adams, C. A. Hauser, and G. W. Hatfield. 1982. DNA sequence fine-structure analysis of *ilvG* (IlvG⁺) mutations of *Escherichia coli* K-12. *J. Bacteriol.* 149:294–298.

2356. Lawther, R. P., and G. W. Hatfield. 1980. Multivalent translational control of transcription termination at attenuator of *ilvGEDA* operon of *Escherichia coli* K-12. *Proc. Natl. Acad. Sci. USA* 77:1862–1866.

2357. Lawther, R. P., B. Nichols, G. Zurawski, and G. W. Hatfield. 1979. The nucleotide sequence preceding and including the beginning of the *ilvE* gene of the *ilvGEDA* operon of *Escherichia coli* K-12. *Nucleic Acids Res.* 7:2289–2301.

2358. Lawther, R. P., R. C. Wek, J. M. Lopes, R. Pereira, B. E. Taillon, and G. W. Hatfield. 1987. The complete nucleotide sequence of the *ilvGMEDA* operon of *Escherichia coli* K-12. *Nucleic Acids Res.* 15:2137–2155. (Errata, 15:9108 and 16:3602, 1988.)

2359. Lazzaroni, J. C., N. Fognini-Lefebvre, and R. C. Portalier. 1986. Cloning of the *lkyB* (*tolB*) gene of *Escherichia coli* K12 and characterization of its product. *Mol. Gen. Genet.* 204:285–288.

2359a.Lazzaroni, J. C., N. Fognini-Lefebvre, and R. Portalier. 1989. Cloning of the *excC* and *excD* genes involved in the release of periplasmic proteins by *Escherichia coli* K12. *Mol. Gen. Genet.* 218:460–464.

2360. Lazzaroni, J. C., and R. Portalier. 1992. The *excC* gene of *Escherichia coli* K-12 required for cell envelope integrity encodes the peptidoglycan-associated lipoprotein (PAL). *Mol. Microbiol.* 6:735–742.

2361. Lazzaroni, J. C., and R. C. Portalier. 1981. Genetic and biochemical characterization of periplasmic-leaky mutants of *Escherichia coli* K-12. *J. Bacteriol.* 145:1351–1358.

2362. Leach, D. R., R. G. Lloyd, and A. F. Coulson. 1992. The SbcCD protein of *Escherichia coli* is related to two putative nucleases in the UvrA superfamily of nucleotide-binding proteins. *Genetica* 87:95–100.

2363. Leberman, R., M. Hartlein, and S. Cusack. 1991. *Escherichia coli* seryl-tRNA synthetase: the structure of a class 2 aminoacyl-tRNA synthetase. *Biochim. Biophys. Acta* 1089:287–298.

2364. Lech, K. F., C. H. Lee, R. R. Isberg, and M. Syvanen. 1985. New gene in *Escherichia coli* K-12 (*drpA*): does its product play a role in RNA synthesis? *J. Bacteriol.* 162:117–123.

2365. Lecker, S., R. Lill, T. Ziegelhoffer, C. Georgopoulos, P. J. Bassford, Jr., C. A. Kumamoto, and W. Wickner. 1989. Three pure chaperone proteins of *Escherichia coli*—SecB, trigger factor and GroEL—form soluble complexes with precursor proteins in vitro. *EMBO J.* 8:2703–2709.

2366. Leclerc, G., G. Noel, and G. R. Drapeau. 1990. Molecular cloning, nucleotide sequence, and expression of *shl*, a new gene in the 2-minute region of the genetic map of *Escherichia coli*. *J. Bacteriol.* 172:4696–4700.

2367. Leclerc, G., C. Sirard, and G. R. Drapeau. 1989. The *Escherichia coli* cell division mutation *ftsM1* is in *serU*. *J. Bacteriol.* 171:2090–2095.

2368. Lederer, F., A. Glatigny, P. H. Bethge, H. D. Bellamy, and F. S. Matthew. 1981. Improvement of the 2.5 A resolution model of cytochrome b_{562} by redetermining the primary structure and using molecular graphics. *J. Mol. Biol.* 148:427–448.

2369. Lee, C. A., M. J. Fournier, and J. Beckwith. 1985. *Escherichia coli* 6S RNA is not essential for growth or protein secretion. *J. Bacteriol.* 161:1156–1161.

2370. **Lee, C. A., G. R. Jacobson, and M. H. Saier, Jr.** 1981. Plasmid-directed synthesis of enzymes required for D-mannitol transport and utilization in *Escherichia coli*. *Proc. Natl. Acad. Sci. USA* **78:**7336–7340.

2371. **Lee, C. A., and M. H. Saier, Jr.** 1983. Mannitol-specific enzyme II of the bacterial phosphotransferase system. III. The nucleotide sequence of the permease gene. *J. Biol. Chem.* **258:**10761–10767.

2372. **Lee, C. C., Y. Kohara, K. Akiyama, C. L. Smith, W. J. Craigen, and C. T. Caskey.** 1988. Rapid and precise mapping of the *Escherichia coli* release factor genes by two physical approaches. *J. Bacteriol.* **170:**4537–4541.

2373. **Lee, D. H., L. Huo, and R. Schleif.** 1992. Repression of the *araBAD* promoter from *araO1*. *J. Mol. Biol.* **224:**335–341.

2374. **Lee, D. H., and R. F. Schleif.** 1989. In vivo DNA loops in *araCBAD*: size limits and helical repeat. *Proc. Natl. Acad. Sci. USA* **86:**476–480.

2375. **Lee, E. C., L. M. Hales, R. I. Gumport, and J. F. Gardner.** 1992. The isolation and characterization of mutants of the integration host factor (IHF) of *Escherichia coli* with altered, expanded DNA-binding specificities. *EMBO J.* **11:**305–313.

2376. **Lee, E. H., and A. Kornberg.** 1992. Features of replication fork blockage by the *Escherichia coli* terminus- binding protein. *J. Biol. Chem.* **267:**8778–8784.

2377. **Lee, E. H., A. Kornberg, M. Hidaka, T. Kobayashi, and T. Horiuchi.** 1989. *Escherichia coli* replication termination protein impedes the action of helicases. *Proc. Natl. Acad. Sci. USA* **86:**9104–9108.

2378. **Lee, E. H., H. Masai, G. C. Allen, Jr., and A. Kornberg.** 1990. The *priA* gene encoding the primosomal replicative n′ protein of *Escherichia coli*. *Proc. Natl. Acad. Sci. USA* **87:**4620–4624.

2379. **Lee, H. R., J. H. Seo, O. M. Kim, C. S. Lee, S. W. Suh, Y. M. Hong, K. Tanaka, A. Ichihara, D. B. Ha, and C. H. Chung.** 1991. Molecular cloning of the ecotin gene in *Escherichia coli*. *FEBS Lett.* **287:**53–56.

2380. **Lee, J., M. Kashlev, S. Borukhov, and A. Goldfarb.** 1991. A beta subunit mutation disrupting the catalytic function of *Escherichia coli* RNA polymerase. *Proc. Natl. Acad. Sci. USA* **88:**6018–6022.

2381. **Lee, J. H., P. Patel, P. Sankar, and K. T. Shanmugam.** 1985. Isolation and characterization of mutant strains of *Escherichia coli* altered in H2 metabolism. *J. Bacteriol.* **162:**344–352.

2382. **Lee, J. H., J. C. Wendt, and K. T. Shanmugam.** 1990. Identification of a new gene, *molR*, essential for utilization of molybdate by *Escherichia coli*. *J. Bacteriol.* **172:**2079–2087.

2383. **Lee, J. S., G. An, J. D. Friesen, and N. P. Fiil.** 1981. Location of the *tufB* promoter of *E. coli*: cotranscription of *tufB* with four transfer RNA genes. *Cell* **25:**251–258.

2384. **Lee, J. S., G. An, J. D. Friesen, and K. Isono.** 1981. Cloning and the nucleotide sequence of the genes for *Escherichia coli* ribosomal proteins L28 (*rpmB*) and L33 (*rpmG*). *Mol. Gen. Genet.* **184:**218–223.

2385. **Lee, K. L., and F. T. Kenney.** 1970. Induction of alanine transaminase by adrenal steroids in cultured hepatoma cells. *Biochem. Biophys. Res. Commun.* **40:**469–475.

2386. **Lee, N., W. Gielow, R. Martin, E. Hamilton, and A. Fowler.** 1986. The organization of the *araBAD* operon of *Escherichia coli*. *Gene* **47:**231–244.

2387. **Lee, N. L., W. O. Gielow, and R. G. Wallace.** 1981. Mechanism of *araC* autoregulation and the domains of two overlapping promoters, Pc and PBAD, in the L-arabinose regulatory region of *Escherichia coli*. *Proc. Natl. Acad. Sci. USA* **78:**752–756.

2388. **Lee, S. H., P. Kanda, R. C. Kennedy, and J. R. Walker.** 1987. Relation of the *Escherichia coli dnaX* gene to its two products—the tau and gamma subunits of DNA polymerase III holoenzyme. *Nucleic Acids Res.* **15:**7663–7675.

2389. **Lee, S. H., and J. R. Walker.** 1987. *Escherichia coli* DnaX product, the tau subunit of DNA polymerase III, is a multifunctional protein with single-stranded DNA-dependent ATPase activity. *Proc. Natl. Acad. Sci. USA* **84:**2713–2717.

2390. **Lee, S. J., A. Xie, W. Jiang, J. P. Etchegaray, P. G. Jones, and M. Inouye.** 1994. Family of the major cold-shock protein, CspA (CS7.4), of *Escherichia coli*, whose members show a high sequence similarity with the eukaryotic Y-box binding proteins. *Mol. Microbiol.* **11:**833–839.

2391. **Legrain, C., P. Halleux, V. Stalon, and N. Glansdorff.** 1972. The dual genetic control of ornithine carbamolytransferase in *Escherichia coli*. A case of bacterial hybrid enzymes. *Eur. J. Biochem.* **27:**93–102.

2392. **Lehming, N., J. Sartorius, B. Kisters-Woike, B. von Wilcken-Bergmann, and B. Muller-Hill.** 1990. Mutant *lac* repressors with new specificities hint at rules for protein-DNA recognition. *EMBO J.* **9:**615–621. (Erratum, **9:**1674.)

2393. **Leifer, Z., R. Engel, and B. E. Tropp.** 1977. Transport of 3,4-dihydroxybutyl-1-phosphonate, an analogue of *sn*-glycerol 3-phosphate. *J. Bacteriol.* **130:**968–971.

2394. **Leinfelder, W., K. Forchhammer, B. Veprek, E. Zehelein, and A. Bock.** 1990. In vitro synthesis of selenocysteinyl-tRNA(UCA) from seryl-tRNA(UCA): involvement and characterization of the *selD* gene product. *Proc. Natl. Acad. Sci. USA* **87:**543–547.

2395. **Leinfelder, W., K. Forchhammer, F. Zinoni, G. Sawers, M. A. Mandrand-Berthelot, and A. Bock.** 1988. *Escherichia coli* genes whose products are involved in selenium metabolism. *J. Bacteriol.* **170:**540–546.

2396. **Leinfelder, W., E. Zehelein, M. A. Mandrand-Berthelot, and A. Bock.** 1988. Gene for a novel tRNA species that accepts L-serine and cotranslationally inserts selenocysteine. *Nature* (London) **331:**723–725.

2397. **Leirmo, S., and R. L. Gourse.** 1991. Factor-independent activation of *Escherichia coli* rRNA transcription. I. Kinetic analysis of the roles of the upstream activator region and supercoiling on transcription of the *rrnB* P1 promoter in vitro. *J. Mol. Biol.* **220:**555–568.

2398. **Lejeune, P., P. Bertin, C. Walon, K. Willemot, C. Colson, and A. Danchin.** 1989. A locus involved in kanamycin, chloramphenicol and L-serine resistance is located in the *bglY-galU* region of the *Escherichia coli* K12 chromosome. *Mol. Gen. Genet.* **218:**361–363.

2399. **Lejeune, P., and A. Danchin.** 1990. Mutations in the *bglY* gene increase the frequency of spontaneous deletions in *Escherichia coli* K-12. *Proc. Natl. Acad. Sci. USA* **87:**360–363.

2400. **Lemaire, H. G., and B. Muller-Hill.** 1986. Nucleotide sequences of the *galE* gene and the *galT* gene of *E. coli*. *Nucleic Acids Res.* **14:**7705–7711.

2401. **Lemire, B. D., J. J. Robinson, and J. H. Weiner.** 1982. Identification of membrane anchor polypeptides of *Escherichia coli* fumarate reductase. *J. Bacteriol.* **152:**1126–1131.

2402. **Lemotte, P. K., and G. C. Walker.** 1985. Induction and autoregulation of *ada*, a positively acting element regulating the response of *Escherichia coli* K-12 to methylating agents. *J. Bacteriol.* **161:**888–895.

2402a.**Lengeler, J.** 1977. Analysis of mutations affecting the dissimilation of galactitol (dulcitol) in *Escherichia coli* K-12. *Mol. Gen. Genet.* **152:**83–98.

2403. **Lengeler, J.** 1980. Characterisation of mutants of *Escherichia coli* K12, selected by resistance to streptozotocin. *Mol. Gen. Genet.* **179:**49–54.

2404. **Lengeler, J., and H. Steinberger.** 1978. Analysis of the regulatory mechanisms controlling the synthesis of the hexitol transport systems in *Escherichia coli* K12. *Mol. Gen. Genet.* **164:**163–169.

2404a.**Lennette, E. T., and D. Apirion.** 1971. Genetic analysis of an *Escherichia coli* syndrome. *J. Bacteriol.* **108:**1322–1328.

2405. **Lenny, A. B., and P. Margolin.** 1980. Locations of the *opp* and *supX* genes of *Salmonella typhimurium* and *Escherichia coli*. *J. Bacteriol.* **143:**747–752.

2406. **Leonardo, M. R., and D. P. Clark.** 1991. Locations of genes in the *nar-adhE* region of the *Escherichia coli* K-12 chromosome. *J. Bacteriol.* **173:**1574–1575.

2407. **Lerner, C. G., and M. Inouye.** 1991. Pleiotropic changes resulting from depletion of Era, an essential GTP-binding protein in *Escherichia coli*. *Mol. Microbiol.* **5:**951–957.

2408. **Lerner, T. J., and N. D. Zinder.** 1982. Another gene affecting sexual expression of *Escherichia coli*. *J. Bacteriol.* **150:**156–160.

2409. **Lesinger, T., D. Haas, and M. P. Hegarty.** 1972. Indospicine as an arginine antagonist in *Escherichia coli* and *Pseudomonas aeruginosa*. *Biochim. Biophys. Acta* **262:**214–219.

2410. **Leslie, N. R., and D. J. Sherratt.** 1995. Site-specific recombination in the replication terminus region of *Escherichia coli*: functional replacement of *dif*. *EMBO J.* **14:**1561–1570.

2411. **Leung, H. B., K. L. Kvalnes-Krick, S. L. Meyer, J. K. deRiel, and V. L. Schramm.** 1989. Structure and regulation of the AMP nucleosidase gene (*amn*) from *Escherichia coli*. *Biochemistry* **28:**8726–8733.

2412. **Leung, H. B., and V. L. Schramm.** 1984. The structural gene for AMP nucleosidase. Mapping, cloning, and overproduction of the enzyme. *J. Biol. Chem.* **259:**6972–6978.

2413. **Levengood, S. K., W. F. Beyer, Jr., and R. E. Webster.** 1991. TolA: a membrane protein involved in colicin uptake contains an extended helical region. *Proc. Natl. Acad. Sci. USA* **88:**5939–5943.

2414. **Levengood, S. K., and R. E. Webster.** 1989. Nucleotide sequences of the *tolA* and *tolB* genes and localization of their products, components of a multistep translocation system in *Escherichia coli*. *J. Bacteriol.* **171:**6600–6609.

2415. **Leveque, F., S. Blanchin-Roland, G. Fayat, P. Plateau, and S. Blanquet.** 1990. Design and characterization of *Escherichia coli* mutants devoid of Ap4N-hydrolase activity. *J. Mol. Biol.* **212:**319–329.

2416. **Leveque, F., M. Gazeau, M. Fromant, S. Blanquet, and P. Plateau.** 1991. Control of *Escherichia coli* lysyl-tRNA synthetase expression by anaerobiosis. *J. Bacteriol.* **173:**7903–7910.

2417. **Leveque, F., P. Plateau, P. Dessen, and S. Blanquet.** 1990. Homology of *lysS* and *lysU*, the two *Escherichia coli* genes encoding distinct lysyl-tRNA synthetase species. *Nucleic Acids Res.* **18:**305–312.

2418. **Levin, H. L., and H. K. Schachman.** 1985. Regulation of aspartate transcarbamoylase synthesis in *Escherichia coli*: analysis of deletion mutations in the promoter region of the *pyrBI* operon. *Proc. Natl. Acad. Sci. USA* **82:**4643–4647.

2419. **Levin, J. D., R. Shapiro, and B. Demple.** 1991. Metalloenzymes in DNA repair. *Escherichia coli* endonuclease IV and *Saccharomyces cerevisiae* Apn1. *J. Biol. Chem.* **266:**22893–22898.

2420. **Levine, A., F. Vannier, M. Dehbi, G. Henckes, and S. J. Seror.** 1991. The stringent response blocks DNA replication outside the *ori* region in *Bacillus subtilis* and at the origin in *Escherichia coli*. *J. Mol. Biol.* **219:**605–613.

2421. **LeVine, S. M., F. Ardeshir, and G. F. Ames.** 1980. Isolation and characterization of acetate kinase and phosphotransacetylase mutants of *Escherichia coli* and *Salmonella typhimurium*. *J. Bacteriol.* **143:**1081–1085.

2422. **Levitz, R., A. Klar, N. Sar, and E. Yagil.** 1984. A new locus in the phosphate specific transport (PST) region of *Escherichia coli*. *Mol. Gen. Genet.* **197:**98–103.

2423. **Levy, S., and A. Danchin.** 1988. Phylogeny of metabolic pathways: O-acetylserine sulphydrylase A is homologous to the tryptophan synthase beta subunit. *Mol. Microbiol.* **2:**777–783.

2424. **Lewis, K.** 1994. Multidrug resistance pumps in bacteria: variations on a theme. *Trends Biochem. Sci.* **19:**119–123.

2425. **Lewis, L. A., K. B. Li, A. Gousse, F. Pereira, N. Pacheco, S. Pierre, P. Kodaman, and S. Lawson.** 1991. Genetic and molecular analysis of spontaneous respiratory deficient (res⁻) mutants of *Escherichia coli* K-12. *Microbiol. Immunol.* **35:**289–301.

2426. **Lewis, L. K., M. E. Jenkins, and D. W. Mount.** 1992. Isolation of DNA damage-inducible promoters in *Escherichia coli*: regulation of *polB* (*dinA*), *dinG*, and *dinH* by LexA repressor. *J. Bacteriol.* **174:**3377–3385.

2427. **Lewis, L. K., and D. W. Mount.** 1992. Interaction of LexA repressor with the asymmetric *dinG* operator and complete nucleotide sequence of the gene. *J. Bacteriol.* **174:**5110–5116.

2428. **Lewis, M. J., J. A. Chang, and R. D. Simoni.** 1990. A topological analysis of subunit alpha from *Escherichia coli* F_1F_0-ATP synthase predicts eight transmembrane segments. *J. Biol. Chem.* **265:**10541–10550.

2429. **Leyh, T. S., J. C. Taylor, and G. D. Markham.** 1988. The sulfate activation locus of *Escherichia coli* K12: cloning, genetic, and enzymatic characterization. *J. Biol. Chem.* **263:**2409–2416.

2430. **Leyh, T. S., T. F. Vogt, and Y. Suo.** 1992. The DNA sequence of the sulfate activation locus from *Escherichia coli* K-12. *J. Biol. Chem.* **267:**10405–10410.

2431. **Li, C., J. K. Ichikawa, J. J. Ravetto, H. C. Kuo, J. C. Fu, and S. Clarke.** 1994. A new gene involved in stationary-phase survival located at 59 minutes on the *Escherichia coli* chromosome. *J. Bacteriol.* **176:**6015–6022.

2432. **Li, C., H. D. Peck, Jr., and A. E. Przybyla.** 1987. Cloning of the 3′-phosphoadenylyl sulfate reductase and sulfite reductase genes from *Escherichia coli* K-12. *Gene* **53:**227–234.

2433. **Li, J., R. Horwitz, S. McCracken, and J. Greenblatt.** 1992. NusG, a new *Escherichia coli* elongation factor involved in transcriptional antitermination by the N protein of phage lambda. *J. Biol. Chem.* **267:**6012–6019.

2434. **Li, J., S. W. Mason, and J. Greenblatt.** 1993. Elongation factor NusG interacts with termination factor rho to regulate termination and antitermination of transcription. *Genes Dev.* **7:**161–172.

2435. **Li, J. M., C. S. Russell, and S. D. Cosloy.** 1989. Cloning and structure of the *hemA* gene of *Escherichia coli* K-12. *Gene* **82:**209–217.

2436. **Li, J. M., C. S. Russell, and S. D. Cosloy.** 1989. The structure of the *Escherichia coli hemB* gene. *Gene* **75:**177–184.

2437. **Li, J. M., H. Umanoff, R. Proenca, C. S. Russell, and S. D. Cosloy.** 1988. Cloning of the *Escherichia coli* K-12 *hemB* gene. *J. Bacteriol.* **170:**1021–1025.

2438. **Li, Q. X., and W. Dowhan.** 1988. Structural characterization of *Escherichia coli* phosphatidylserine decarboxylase. *J. Biol. Chem.* **263:**11516–11522.

2439. **Li, Q. X., and W. Dowhan.** 1990. Studies on the mechanism of formation of the pyruvate prosthetic group of phosphatidylserine decarboxylase from *Escherichia coli*. *J. Biol. Chem.* **265:**4111–4115.

2440. **Li, S., T. Rabi, and J. A. DeMoss.** 1985. Delineation of two distinct regulatory domains in the 5′ region of the *nar* operon of *Escherichia coli*. *J. Bacteriol.* **164:**25–32.

2441. **Li, S. F., and J. A. DeMoss.** 1987. Promoter region of the *nar* operon of *Escherichia coli*: nucleotide sequence and transcription initiation signals. *J. Bacteriol.* **169:**4614–4620.

2442. **Li, S. J., and J. E. Cronan, Jr.** 1992. The gene encoding the biotin carboxylase subunit of *Escherichia coli* acetyl-CoA carboxylase. *J. Biol. Chem.* **267:**855–863.

2443. **Li, S. J., and J. E. Cronan, Jr.** 1992. The genes encoding the two carboxyltransferase subunits of *Escherichia coli* acetyl-CoA carboxylase. *J. Biol. Chem.* **267:**16841–16847.

2444. **Li, S. J., and J. E. Cronan, Jr.** 1993. Growth rate regulation of *Escherichia coli* acetyl coenzyme A carboxylase, which catalyzes the first committed step of lipid biosynthesis. *J. Bacteriol.* **175:**332–340.

2445. **Li, S. J., C. O. Rock, and J. E. Cronan, Jr.** 1992. The *dedB* (*usg*) open reading frame of *Escherichia coli* encodes a subunit of acetyl-coenzyme A carboxylase. *J. Bacteriol.* **174:**5755–5757.

2446. **Li, S. L., J. Hanlon, and C. Yanofsky.** 1974. Structural homology of the glutamine amidotransferase subunits of the anthranilate synthetases of *Escherichia coli*, *Salmonella typhimurium* and *Serratia marcescens*. *Nature* (London) **248:**48–50.

2447. **Li, S. L., J. Hanlon, and C. Yanofsky.** 1974. Separation of anthranilate synthetase components I and II of *Escherichia coli*, *Salmonella typhimurium*, and *Serratia marcescens* and determination of their amino-terminal sequences by automatic Edman degradation. *Biochemistry* **13:**1736–1744.

2448. **Li, Y., C. Guerrier-Takada, and S. Altman.** 1992. Targeted cleavage of mRNA in vitro by RNase P from *Escherichia coli*. *Proc. Natl. Acad. Sci. USA* **89:**3185–3189.

2449. **Li, Y. F., and A. Sancar.** 1990. Active site of *Escherichia coli* DNA photolyase: mutations at Trp277 alter the selectivity of the enzyme without affecting the quantum yield of photorepair. *Biochemistry* **29:**5698–5706.

2450. **Li, Z., and B. Demple.** 1994. SoxS, an activator of superoxide stress genes in *Escherichia coli*. Purification and interaction with DNA. *J. Biol. Chem.* **269:**18371–18377.

2451. **Liberek, K., T. P. Galitski, M. Zylicz, and C. Georgopoulos.** 1992. The DnaK chaperone modulates the heat shock response of *Escherichia coli* by binding to the sigma 32 transcription factor. *Proc. Natl. Acad. Sci. USA* **89:**3516–3520.

2452. **Liberek, K., C. Georgopoulos, and M. Zylicz.** 1988. Role of the *Escherichia coli* DnaK and DnaJ heat shock proteins in the initiation of bacteriophage lambda DNA replication. *Proc. Natl. Acad. Sci. USA* **85:**6632–6636.

2453. **Liberek, K., J. Marszalek, D. Ang, C. Georgopoulos, and M. Zylicz.** 1991. *Escherichia coli* DnaJ and GrpE heat shock proteins jointly stimulate ATPase activity of DnaK. *Proc. Natl. Acad. Sci. USA* **88:**2874–2878.

2454. **Liberek, K., D. Skowyra, M. Zylicz, C. Johnson, and C. Georgopoulos.** 1991. The *Escherichia coli* DnaK chaperone, the 70-kDa heat shock protein eukaryotic equivalent, changes conformation upon ATP hydrolysis, thus triggering its dissociation from a bound target protein. *J. Biol. Chem.* **266:**14491–14496.

2455. **Lieb, M.** 1991. Spontaneous mutation at a 5-methylcytosine hotspot is prevented by very short patch (VSP) mismatch repair. *Genetics* **128:**23–27.

2456. **Lieb, M., and S. Rehmat.** 1995. Very short patch repair of T:G mismatches in vivo: importance of context and accessory proteins. *J. Bacteriol.* **177:**660–666.

2457. **Liebke, H., C. Gross, W. Walter, and R. Burgess.** 1980. A new mutation, *rpoD800*, affecting the sigma subunit of *E. coli* RNA polymerase is allelic to two other sigma mutants. *Mol. Gen. Genet.* **177:**277–282.

2458. **Liebke, H., and G. Hatfull.** 1985. The sequence of the distal end of the *E. coli* ribosomal RNA *rrnE* operon indicates conserved features are shared by *rrn* operons. *Nucleic Acids Res.* **13:**5515–5525.

2459. **Lightner, V. A., R. M. Bell, and P. Modrich.** 1983. The DNA sequences encoding *plsB* and *dgk* loci of *Escherichia coli*. *J. Biol. Chem.* **258:**10856–10861.

2460. **Lightner, V. A., T. J. Larson, P. Tailleur, G. D. Kantor, C. R. Raetz, R. M. Bell, and P. Modrich.** 1980. Membrane phospholipid synthesis in *Escherichia coli*. Cloning of a structural gene (*plsB*) of the *sn*-glycerol-3-phosphate acyl/transferase. *J. Biol. Chem.* **255:**9413–9420.

2461. **Liljestrand-Golden, C. A., and J. R. Johnson.** 1984. Physical organization of the *metJB* component of the *Escherichia coli* K-12 *metJBLF* gene cluster. *J. Bacteriol.* **157:**413–419.

2462. **Liljestrom, P., I. Laamanen, and E. T. Palva.** 1988. Structure and expression of the *ompB* operon, the regulatory locus for the outer membrane porin regulon in *Salmonella typhimurium* LT-2. *J. Mol. Biol.* **201:**663–673.

2463. **Liljestrom, P. L., and P. Liljestrom.** 1987. Nucleotide sequence of the *melA* gene, coding for alpha-galactosidase in *Escherichia coli* K-12. *Nucleic Acids Res.* **15:**2213–2220.

2464. **Lill, R., K. Cunningham, L. A. Brundage, K. Ito, D. Oliver, and W. Wickner.** 1989. SecA protein hydrolyzes ATP and is an essential component of the protein translocation ATPase of *Escherichia coli*. *EMBO J.* **8:**961–966.

2465. **Lilley, P. E., N. P. Stamford, S. G. Vasudevan, and N. E. Dixon.** 1993. The 92-min region of the *Escherichia coli* chromosome: location and cloning of the *ubiA* and *alr* genes. *Gene* **129:**9–16.

2466. **Lim, C. J., D. Geraghty, and J. A. Fuchs.** 1985. Cloning and nucleotide sequence of the *trxA* gene of *Escherichia coli* K-12. *J. Bacteriol.* **163:**311–316.

2467. **Lim, C. J., B. Haller, and J. A. Fuchs.** 1985. Thioredoxin is the bacterial protein encoded by *fip* that is required for filamentous bacteriophage f1 assembly. *J. Bacteriol.* **161:**799–802.

2468. **Lim, D. B., J. D. Oppenheim, T. Eckhardt, and W. K. Maas.** 1987. Nucleotide sequence of the *argR* gene of *Escherichia coli* K-12 and isolation of its product, the arginine repressor. *Proc. Natl. Acad. Sci. USA* **84:**6697–6701.

2469. **Lim, W. K., S. K. Sarkar, and J. K. Hardman.** 1991. Enzymatic properties of mutant *Escherichia coli* tryptophan synthase alpha-subunits. *J. Biol. Chem.* **266:**20205–20212.

2470. **Lima, C. D., J. C. Wang, and A. Mondragon.** 1994. Three-dimensional structure of the 67K N-terminal fragment of *E. coli* DNA topoisomerase I. *Nature* (London) **367:**138–146.

2471. **Lin, E. C., and S. Iuchi.** 1991. Regulation of gene expression in fermentative and respiratory systems in *Escherichia coli* and related bacteria. *Annu. Rev. Genet.* **25:**361–387.

2472. **Lin, K. C., A. Campbell, and D. Shiuan.** 1991. Binding characteristics of *Escherichia coli* biotin repressor-operator complex. *Biochim. Biophys. Acta* **1090:**317–325.

2473. **Lin, R., B. Ernsting, I. N. Hirshfield, R. G. Matthews, F. C. Neidhardt, R. L. Clark, and E. B. Newman.** 1992. The *lrp* gene product regulates expression of *lysU* in *Escherichia coli* K-12. *J. Bacteriol.* **174:**2779–2784.

2474. **Lin, R. J., M. Capage, and C. W. Hill.** 1984. A repetitive DNA sequence, *rhs*, responsible for duplications within the *Escherichia coli* K-12 chromosome. *J. Mol. Biol.* **177:**1–18.

2475. **Lin, R. J., and C. W. Hill.** 1983. Mapping the *xyl*, *mtl*, and *lct* loci in *Escherichia coli* K-12. *J. Bacteriol.* **156:**914–916.

2476. **Lindahl, L., F. Sor, R. H. Archer, M. Nomura, and J. M. Zengel.** 1990. Transcriptional organization of the S10, *spc* and alpha operons of *Escherichia coli*. *Biochim. Biophys. Acta* **1050:**337–342.

2477. **Lindahl, L., and J. M. Zengel.** 1979. Operon-specific regulation of ribosomal protein synthesis in *Escherichia coli*. *Proc. Natl. Acad. Sci. USA* **76:**6542–6546.

2478. **Lindemann, H., and B. Wittmann-Liebold.** 1977. Primary structure of protein S13 from the small subunit of *Escherichia coli* ribosomes. *Hoppe-Seyler's Z. Physiol. Chem.* **358:**843–863.

2479. **Lindquist, S., M. Galleni, F. Lindberg, and S. Normark.** 1989. Signalling proteins in enterobacterial AmpC beta-lactamase regulation. *Mol. Microbiol.* **3:**1091–1102.

2480. **Lindquist, S., K. Weston-Hafer, H. Schmidt, C. Pul, G. Korfmann, J. Erickson, C. Sanders, H. H. Martin, and S. Normark.** 1993. AmpG, a signal transducer in chromosomal beta-lactamase induction. *Mol. Microbiol.* **9:**703–715.

2481. **Lindsey, D. F., C. Martinez, and J. R. Walker.** 1992. Physical map location of the *Escherichia coli* attachment site for the P22 prophage (*attP22*). *J. Bacteriol.* **174:**3834–3835.

2482. **Lindsey, D. F., D. A. Mullin, and J. R. Walker.** 1989. Characterization of the cryptic lambdoid prophage DLP12 of *Escherichia coli* and overlap of the DLP12 integrase gene with the tRNA gene *argU*. *J. Bacteriol.* **171:**6197–6205.

2483. **Lindstrom, P. H., D. Stuber, and G. R. Bjork.** 1985. Genetic organization and transcription from the gene (*trmA*) responsible for synthesis of tRNA (uracil-5)-methyltransferase by *Escherichia coli*. *J. Bacteriol.* **164:**1117–1123.

2484. Link, C. D., and A. M. Reiner. 1983. Genotypic exclusion: a novel relationship between the ribitol-arabitol and galactitol genes of *E. coli*. *Mol. Gen. Genet.* 189:337–339.

2485. Linn, T., M. Goman, and J. Scaife. 1979. Lambda transducing bacteriophage carrying deletions of the *argCBH-rpoBC* region of the *Escherichia coli* chromosome. *J. Bacteriol.* 140:479–489.

2486. Linn, T., and J. Greenblatt. 1992. The NusA and NusG proteins of *Escherichia coli* increase the in vitro readthrough frequency of a transcriptional attenuator preceding the gene for the beta subunit of RNA polymerase. *J. Biol. Chem.* 267:1449–1454.

2487. Liochev, S. I., and I. Fridovich. 1992. Fumarase C, the stable fumarase of *Escherichia coli*, is controlled by the *soxRS* regulon. *Proc. Natl. Acad. Sci. USA* 89:5892–5896.

2488. Lipinska, B., O. Fayet, L. Baird, and C. Georgopoulos. 1989. Identification, characterization, and mapping of the *Escherichia coli htrA* gene, whose product is essential for bacterial growth only at elevated temperatures. *J. Bacteriol.* 171:1574–1584.

2489. Lipinska, B., J. King, D. Ang, and C. Georgopoulos. 1988. Sequence analysis and transcriptional regulation of the *Escherichia coli grpE* gene, encoding a heat shock protein. *Nucleic Acids Res.* 16:7545–7562.

2490. Lipinska, B., S. Sharma, and C. Georgopoulos. 1988. Sequence analysis and regulation of the *htrA* gene of *Escherichia coli*: a sigma 32-independent mechanism of heat-inducible transcription. *Nucleic Acids Res.* 16:10053–10067.

2491. Lipinska, B., M. Zylicz, and C. Georgopoulos. 1990. The HtrA (DegP) protein, essential for *Escherichia coli* survival at high temperatures, is an endopeptidase. *J. Bacteriol.* 172:1791–1797.

2492. Lipsett, M. N. 1978. Enzymes producing 4-thiouridine in *Escherichia coli* tRNA: approximate chromosomal locations of the genes and enzyme activities in a 4-thiouridine-deficient mutant. *J. Bacteriol.* 135:993–997.

2493. Little, J. W. 1979. Construction and characterization of a plasmid coding for a fragment of the *Escherichia coli recA* protein. *Mol. Gen. Genet.* 177:13–22.

2494. Little, J. W. 1980. Isolation of recombinant plasmids and phage carrying the *lexA* gene of *Escherichia coli* K-12. *Gene* 10:237–247.

2495. Little, J. W., D. W. Mount, and C. R. Yanisch-Perron. 1981. Purified *lexA* protein is a repressor of the *recA* and *lexA* genes. *Proc. Natl. Acad. Sci. USA* 78:4199–4203.

2496. Little, R., N. P. Fiil, and P. P. Dennis. 1981. Transcriptional and post-transcriptional control of ribosomal protein and ribonucleic acid polymerase genes. *J. Bacteriol.* 147:25–35.

2497. Liu, J., and I. R. Beacham. 1990. Transcription and regulation of the *cpdB* gene in *Escherichia coli* K12 and *Salmonella typhimurium* LT2: evidence for modulation of constitutive promoters by cyclic AMP-CRP complex. *Mol. Gen. Genet.* 222:161–165.

2498. Liu, J., D. M. Burns, and I. R. Beacham. 1986. Isolation and sequence analysis of the gene (*cpdB*) encoding periplasmic 2′,3′-cyclic phosphodiesterase. *J. Bacteriol.* 165:1002–1010.

2499. Liu, J., C. M. Chen, and C. T. Walsh. 1991. Human and *Escherichia coli* cyclophilins: sensitivity to inhibition by the immunosuppressant cyclosporin A correlates with a specific tryptophan residue. *Biochemistry* 30:2306–2310.

2500. Liu, J., K. Duncan, and C. T. Walsh. 1989. Nucleotide sequence of a cluster of *Escherichia coli* enterobactin biosynthesis genes: identification of *entA* and purification of its product 2,3-dihydro-2,3-dihydroxybenzoate dehydrogenase. *J. Bacteriol.* 171:791–798.

2501. Liu, J., and B. Magasanik. 1993. The *glnB* region of the *Escherichia coli* chromosome. *J. Bacteriol.* 175:7441–7449.

2502. Liu, J., N. Quinn, G. A. Berchtold, and C. T. Walsh. 1990. Overexpression, purification, and characterization of isochorismate synthase (EntC), the first enzyme involved in the biosynthesis of enterobactin from chorismate. *Biochemistry* 29:1417–1425.

2503. Liu, J., and C. T. Walsh. 1990. Peptidyl-prolyl cis-trans-isomerase from *Escherichia coli*: a periplasmic homolog of cyclophilin that is not inhibited by cyclosporin A. *Proc. Natl. Acad. Sci. USA* 87:4028–4032.

2504. Liu, J. D., and J. S. Parkinson. 1989. Genetics and sequence analysis of the *pcnB* locus, an *Escherichia coli* gene involved in plasmid copy number control. *J. Bacteriol.* 171:1254–1261.

2505. Liu, L., W. Whalen, A. Das, and C. M. Berg. 1987. Rapid sequencing of cloned DNA using a transposon for bidirectional priming: sequence of the *Escherichia coli* K-12 *avtA* gene. *Nucleic Acids Res.* 15:9461–9469.

2506. Liu, M. Y., H. Yang, and T. Romeo. 1995. The product of the pleiotropic *Escherichia coli* gene *csrA* modulates glycogen biosynthesis via effects on mRNA stability. *J. Bacteriol.* 177:2663–2672.

2507. Liu, X., and P. Matsumura. 1994. The FlhD/FlhC complex, a transcriptional activator of the *Escherichia coli* flagellar class II operons. *J. Bacteriol.* 176:7345–7351.

2508. Lloyd, R. G. 1991. Linkage distortion following conjugational transfer of *sbcC*+ to *recBC sbcBC* strains of *Escherichia coli*. *J. Bacteriol.* 173:5694–5698.

2509. Lloyd, R. G. 1991. Conjugational recombination in resolvase-deficient *ruvC* mutants of *Escherichia coli* K-12 depends on *recG*. *J. Bacteriol.* 173:5414–5418.

2510. Lloyd, R. G., and C. Buckman. 1985. Identification and genetic analysis of *sbcC* mutations in commonly used *recBC sbcB* strains of *Escherichia coli* K-12. *J. Bacteriol.* 164:836–844.

2511. Lloyd, R. G., and C. Buckman. 1991. Genetic analysis of the *recG* locus of *Escherichia coli* K-12 and of its role in recombination and DNA repair. *J. Bacteriol.* 173:1004–1011.

2512. Lloyd, R. G., and C. Buckman. 1991. Overlapping functions of *recD*, *recJ*, and *recN* provide evidence of three epistatic groups of genes in *Escherichia coli* recombination and repair. *Biochimie* 73:313–320.

2513. Lloyd, R. G., S. M. Picksley, and C. Prescott. 1983. Inducible expression of a gene specific to the RecF pathway for recombination in *Escherichia coli* K 12. *Mol. Gen. Genet.* 190:162–167.

2514. Lloyd, R. G., and G. J. Sharples. 1991. Molecular organization and nucleotide sequence of the *recG* locus of *Escherichia coli* K-12. *J. Bacteriol.* 173:6837–6843.

2515. Lloyd, R. G., and G. J. Sharples. 1993. Processing of recombination intermediates by the RecG and RuvAB proteins of *Escherichia coli*. *Nucleic Acids Res.* 21:1719–1725.

2516. Lloyd, R. G., and G. J. Sharples. 1993. Dissociation of synthetic Holliday junctions by *E. coli* RecG protein. *EMBO J.* 12:17–21.

2517. Lobell, R. B., and R. F. Schleif. 1990. DNA looping and unlooping by AraC protein. *Science* 250:528–532.

2518. Lobner-Olesen, A., T. Atlung, and K. V. Rasmussen. 1987. Stability and replication control of *Escherichia coli* minichromosomes. *J. Bacteriol.* 169:2835–2842.

2519. Lobner-Olesen, A., and E. Boye. 1992. Different effects of *mioC* transcription on initiation of chromosomal and minichromosomal replication in *Escherichia coli*. *Nucleic Acids Res.* 20:3029–3036.

2520. Lobner-Olesen, A., E. Boye, and M. G. Marinus. 1992. Expression of the *Escherichia coli dam* gene. *Mol. Microbiol.* 6:1841–1851.

2521. Lobner-Olesen, A., K. Skarstad, F. G. Hansen, K. von Meyenburg, and E. Boye. 1989. The DnaA protein determines the initiation mass of *Escherichia coli* K-12. *Cell* 57:881–889.

2522. Lobocka, M., J. Hennig, J. Wild, and T. Klopotowski. 1994. Organization and expression of the *Escherichia coli* K-12 *dad* operon encoding the smaller subunit of D-amino acid dehydrogenase and the catabolic alanine racemase. *J. Bacteriol.* 176:1500–1510.

2523. Loenen, W. A., A. S. Daniel, H. D. Braymer, and N. E. Murray. 1987. Organization and sequence of the *hsd* genes of *Escherichia coli* K-12. *J. Mol. Biol.* 198:159–170.

2524. Loewen, P. C. 1984. Isolation of catalase-deficient *Escherichia coli* mutants and genetic mapping of *katE*, a locus that affects catalase activity. *J. Bacteriol.* 157:622–626.

2525. Loewen, P. C., and B. L. Triggs. 1984. Genetic mapping of *katF*, a locus that with *katE* affects the synthesis of a second catalase species in *Escherichia coli*. *J. Bacteriol.* 160:668–675.

2526. Loewen, P. C., B. L. Triggs, C. S. George, and B. E. Hrabarchuk. 1985. Genetic mapping of *katG*, a locus that affects synthesis of the bifunctional catalase-peroxidase hydroperoxidase I in *Escherichia coli*. *J. Bacteriol.* 162:661–667.

2527. Lohman, T. M. 1992. *Escherichia coli* DNA helicases: mechanisms of DNA unwinding. *Mol. Microbiol.* 6:5–14.

2528. Lohmeier, E., D. S. Hagen, P. Dickie, and J. H. Weiner. 1981. Cloning and expression of fumarate reductase gene of *Escherichia coli*. *Can. J. Biochem.* 59:158–164.

2529. Lomax, M. S., and G. R. Greenberg. 1968. Characteristics of the *deo* operon: role in thymine utilization and sensitivity to deoxyribonucleosides. *J. Bacteriol.* 96:501–514.

2530. Lombardo, M. J., D. Bagga, and C. G. Miller. 1991. Mutations in *rpoA* affect expression of anaerobically regulated genes in *Salmonella typhimurium*. *J. Bacteriol.* 173:7511–7518.

2531. Lombardo, M. J., C. G. Miller, and K. E. Rudd. 1993. Physical mapping of the *Escherichia coli pepT* and *potABCD* genes. *J. Bacteriol.* 175:7745–7746.

2532. Lomovskaya, O., and K. Lewis. 1992. *emr*, an *Escherichia coli* locus for multidrug resistance. *Proc. Natl. Acad. Sci. USA* 89:8938–8942.

2533. Lomovskaya, O., K. Lewis, and A. Matin. 1995. EmrR is a negative regulator of the *Escherichia coli* multidrug resistance pump EmrAB. *J. Bacteriol.* 177:2328–2334.

2534. Lomovskaya, O. L., J. P. Kidwell, and A. Matin. 1994. Characterization of the sigma 38-dependent expression of a core *Escherichia coli* starvation gene, *pexB*. *J. Bacteriol.* 176:3928–3935.

2535. Lonetto, M., M. Gribskov, and C. A. Gross. 1992. The sigma 70 family: sequence conservation and evolutionary relationships. *J. Bacteriol.* 174:3843–3849.

2536. Lonetto, M. A., K. L. Brown, K. E. Rudd, and M. J. Buttner. 1994. Analysis of the *Streptomyces coelicolor sigE* gene reveals the existence of a subfamily of eubacterial RNA polymerase sigma factors involved in the regulation of extracytoplasmic functions. *Proc. Natl. Acad. Sci. USA* 91:7573–7577.

2536a. Long, W. S., C. L. Slayman, and K. B. Low. 1978. Production of giant cells of *Escherichia coli*. *J. Bacteriol.* 133:995–1007.

2537. Loomis, C. R., J. P. Walsh, and R. M. Bell. 1985. *sn*-1,2-Diacylglycerol kinase of *Escherichia coli*. Purification, reconstitution, and partial amino- and carboxyl-terminal analysis. *J. Biol. Chem.* 260:4091–4097.

2538. Loomis, W. F., Jr., and B. Magasanik. 1965. Genetic control of catabolite repression of the *lac* operon in *Escherichia coli*. *Biochem. Biophys. Res. Commun.* 20:230–234.

2539. Loomis, W. F., Jr., and B. Magasanik. 1967. The catabolite repression gene of the *lac* operon in *Escherichia coli*. *J. Mol. Biol.* 23:487–494.

2540. Lopes, J. M., and R. P. Lawther. 1986. Analysis and comparison of the internal promoter, pE, of the *ilvGMEDA* operons from *Escherichia coli* K-12 and *Salmonella typhimurium*. *Nucleic Acids Res.* 14:2779–2798.

2541. Lopes, J. M., N. Soliman, P. K. Smith, and R. P. Lawther. 1989. Transcriptional polarity enhances the contribution of the internal promoter, *ilvEp*, in the expression of the *ilvGMEDA* operon in wild-type *Escherichia coli* K12. *Mol. Microbiol.* **3:**1039–1051.

2542. **Lopez, J., and R. E. Webster.** 1985. *fipB* and *fipC:* two bacterial loci required for morphogenesis of the filamentous bacteriophage f1. *J. Bacteriol.* **163:**900–905.

2543. **Lopilato, J., S. Bortner, and J. Beckwith.** 1986. Mutations in a new chromosomal gene of *Escherichia coli* K-12, *pcnB*, reduce plasmid copy number of pBR322 and its derivatives. *Mol. Gen. Genet.* **205:**285–290.

2544. **Lopilato, J. E., J. L. Garwin, S. D. Emr, T. J. Silhavy, and J. R. Beckwith.** 1984. D-Ribose metabolism in *Escherichia coli* K-12: genetics, regulation, and transport. *J. Bacteriol.* **158:**665–673.

2545. **Lorence, M. C., S. D. Maika, and C. S. Rupert.** 1990. Physical analysis of *phr* gene transcription in *Escherichia coli* K-12. *J. Bacteriol.* **172:**6551–6556.

2546. **Lorowitz, W., and D. Clark.** 1982. *Escherichia coli* mutants with a temperature-sensitive alcohol dehydrogenase. *J. Bacteriol.* **152:**935–938.

2547. **Lother, H., and W. Messer.** 1981. Promoters in the *E. coli* replication origin. *Nature* (London) **294:**376–378.

2548. **Louarn, J., F. Cornet, V. Francois, J. Patte, and J. M. Louarn.** 1994. Hyperrecombination in the terminus region of the *Escherichia coli* chromosome: possible relation to nucleoid organization. *J. Bacteriol.* **176:**7524–7531.

2549. **Loubens, I., L. Debarbieux, A. Bohin, J. M. Lacroix, and J. P. Bohin.** 1993. Homology between a genetic locus (*mdoA*) involved in the osmoregulated biosynthesis of periplasmic glucans in *Escherichia coli* and a genetic locus (*hrpM*) controlling pathogenicity of *Pseudomonas syringae. Mol. Microbiol.* **10:**329–340.

2550. **Louie, G. V., P. D. Brownlie, R. Lambert, J. B. Cooper, T. L. Blundell, S. P. Wood, M. J. Warren, S. C. Woodcock, and P. M. Jordan.** 1992. Structure of porphobilinogen deaminase reveals a flexible multidomain polymerase with a single catalytic site. *Nature* (London) **359:**33–39.

2551. **Lovett, S. T., and A. J. Clark.** 1984. Genetic analysis of the *recJ* gene of *Escherichia coli* K-12. *J. Bacteriol.* **157:**190–196.

2552. **Lovett, S. T., and A. J. Clark.** 1985. Cloning of the *Escherichia coli recJ* chromosomal region and identification of its encoded proteins. *J. Bacteriol.* **162:**280–285.

2553. **Lovett, S. T., and R. D. Kolodner.** 1989. Identification and purification of a single-stranded-DNA-specific exonuclease encoded by the *recJ* gene of *Escherichia coli. Proc. Natl. Acad. Sci. USA* **86:**2627–2631.

2554. **Lovett, S. T., and R. D. Kolodner.** 1991. Nucleotide sequence of the *Escherichia coli recJ* chromosomal region and construction of recJ-overexpression plasmids. *J. Bacteriol.* **173:**353–364.

2555. **Lozoya, E., R. Sanchez-Pescador, A. Covarrubias, I. Vichido, and F. Bolivar.** 1980. Tight linkage of genes that encode the two glutamate synthase subunits of *Escherichia coli* K-12. *J. Bacteriol.* **144:**616–621.

2556. **Lu, M., J. L. Campbell, E. Boye, and N. Kleckner.** 1994. SeqA: a negative modulator of replication initiation in *E. coli. Cell* **77:**413–426.

2557. **Lu, M., and N. Kleckner.** 1994. Molecular cloning and characterization of the *pgm* gene encoding phosphoglucomutase of *Escherichia coli. J. Bacteriol.* **176:**5847–5851.

2558. **Lu, M. F., and H. E. Umbarger.** 1987. Effects of deletion and insertion mutations in the *ilvM* gene of *Escherichia coli. J. Bacteriol.* **169:**600–604.

2559. **Lu, Z., and E. C. Lin.** 1989. The nucleotide sequence of *Escherichia coli* genes for L-fucose dissimilation. *Nucleic Acids Res.* **17:**4883–4884.

2560. **Lucht, J. M., W. Boos, and E. Bremer.** 1992. Alignment of genes from the 9-minute region (*araJ* to *tsx*) of the *Escherichia coli* K-12 linkage map to the physical map. *J. Bacteriol.* **174:**1709–1710.

2561. **Lucht, J. M., and E. Bremer.** 1991. Characterization of mutations affecting the osmoregulated *proU* promoter of *Escherichia coli* and identification of 5′ sequences required for high-level expression. *J. Bacteriol.* **173:**801–809.

2562. **Luckey, M., R. Ling, A. Dose, and B. Malloy.** 1991. Role of a disulfide bond in the thermal stability of the LamB protein trimer in *Escherichia coli* outer membrane. *J. Biol. Chem.* **266:**1866–1871.

2563. **Luckey, M., and J. B. Neilands.** 1976. Iron transport in *Salmonella typhimurium* LT-2: prevention, by ferrichrome, of adsorption of bacteriophages ES18 and ES18.h1 to a common cell envelope receptor. *J. Bacteriol.* **127:**1036–1037.

2564. **Luckey, M., J. R. Pollack, R. Wayne, B. N. Ames, and J. B. Neilands.** 1972. Iron uptake in *Salmonella typhimurium:* utilization of exogenous siderochromes as iron carriers. *J. Bacteriol.* **111:**731–738.

2565. **Ludtke, D., J. Bernstein, C. Hamilton, and A. Torriani.** 1984. Identification of the *phoM* gene product and its regulation in *Escherichia coli* K-12. *J. Bacteriol.* **159:**19–25.

2566. **Luecke, H., and F. A. Quiocho.** 1990. High specificity of a phosphate transport protein determined by hydrogen bonds. *Nature* (London) **347:**402–406.

2567. **Lugtenberg, E. J., H. J. Wijsman, and D. Zaane.** 1973. Properties of a D-glutamic acid-requiring mutant of *Escherichia coli. J. Bacteriol.* **114:**499–506.

2568. **Luirink, J., S. High, H. Wood, A. Giner, D. Tollervey, and B. Dobberstein.** 1992. Signal-sequence recognition by an *Escherichia coli* ribonucleoprotein complex. *Nature* (London) **359:**741–743.

2569. **Luirink, J., C. M. ten Hagen-Jongman, C. C. van der Weijden, B. Oudega, S. High, B. Dobberstein, and R. Kusters.** 1994. An alternative protein targeting pathway in *Escherichia coli:* studies on the role of FtsY. *EMBO J.* **13:**2289–2296.

2570. **Luisi, B. F., and P. B. Sigler.** 1990. The stereochemistry and biochemistry of the *trp* repressor-operator complex. *Biochim. Biophys. Acta* **1048:**113–126.

2571. **Luisi-DeLuca, C.** 1995. Homologous pairing of single-stranded DNA and superhelical double-stranded DNA catalyzed by RecO protein from *Escherichia coli. J. Bacteriol.* **177:**566–572.

2572. **Lum, D., C. J. Lee, and B. J. Wallace.** 1993. Location of the *gltP* gene on the physical map of *Escherichia coli* K-12. *J. Bacteriol.* **175:**5735.

2573. **Lundberg, L. G., O. H. Karlstrom, and P. O. Nyman.** 1983. Isolation and characterization of the *dut* gene of *Escherichia coli.* II. Restriction enzyme mapping and analysis of polypeptide products. *Gene* **22:**127–131.

2574. **Lundberg, L. G., H. O. Thoresson, O. H. Karlstrom, and P. O. Nyman.** 1983. Nucleotide sequence of the structural gene for dUTPase of *Escherichia coli* K-12. *EMBO J.* **2:**967–971.

2575. **Lundegaard, C., and K. F. Jensen.** 1994. The DNA damage-inducible *dinD* gene of *Escherichia coli* is equivalent to *orfY* upstream of *pyrE. J. Bacteriol.* **176:**3383–3385.

2576. **Lundrigan, M., and C. F. Earhart.** 1981. Reduction in three iron-regulated outer membrane proteins and protein a by the *Escherichia coli* K-12 *perA* mutation. *J. Bacteriol.* **146:**804–807.

2577. **Lundrigan, M. D., L. C. de Veaux, B. J. Mann, and R. J. Kadner.** 1987. Separate regulatory systems for the repression of *metE* and *btuB* by vitamin B12 in *Escherichia coli. Mol. Gen. Genet.* **206:**401–407.

2578. **Lundrigan, M. D., and C. F. Earhart.** 1984. Gene *envY* of *Escherichia coli* K-12 affects thermoregulation of major porin expression. *J. Bacteriol.* **157:**262–268.

2579. **Lundrigan, M. D., M. J. Friedrich, and R. J. Kadner.** 1989. Nucleotide sequence of the *Escherichia coli* porin thermoregulatory gene *envY. Nucleic Acids Res.* **17:**800.

2580. **Lundrigan, M. D., and R. J. Kadner.** 1986. Nucleotide sequence of the gene for the ferrienterochelin receptor FepA in *Escherichia coli.* Homology among outer membrane receptors that interact with TonB. *J. Biol. Chem.* **261:**10797–10801.

2581. **Lundrigan, M. D., and R. J. Kadner.** 1989. Altered cobalamin metabolism in *Escherichia coli btuR* mutants affects *btuB* gene regulation. *J. Bacteriol.* **171:**154–161.

2582. **Luo, G. X., and J. S. Nishimura.** 1991. Site-directed mutagenesis of *Escherichia coli* succinyl-CoA synthetase. Histidine 142 alpha is a facilitative catalytic residue. *J. Biol. Chem.* **266:**20781–20785.

2583. **Lupo, M., and Y. S. Halpern.** 1970. Gene controlling L-glutamic acid decarboxylase synthesis in *Escherichia coli* K-12. *J. Bacteriol.* **103:**382–386.

2584. **Lupski, J. R., A. A. Ruiz, and G. N. Godson.** 1984. Promotion, termination, and anti-termination in the *rpsU-dnaG-rpoD* macromolecular synthesis operon of *E. coli* K-12. *Mol. Gen. Genet.* **195:**391–401.

2585. **Lupski, J. R., B. L. Smiley, F. R. Blattner, and G. N. Godson.** 1982. Cloning and characterization of the *Escherichia coli* chromosomal region surrounding the *dnaG* gene, with a correlated physical and genetic map of *dnaG* generated via transposon Tn5 mutagenesis. *Mol. Gen. Genet.* **185:**120–128.

2586. **Lupski, J. R., B. L. Smiley, and G. N. Godson.** 1983. Regulation of the *rpsU-dnaG-rpoD* macromolecular synthesis operon and the initiation of DNA replication in *Escherichia coli* K-12. *Mol. Gen. Genet.* **189:**48–57.

2587. **Lutkenhaus, J. F., H. Wolf-Watz, and W. D. Donachie.** 1980. Organization of genes in the *ftsA-envA* region of the *Escherichia coli* genetic map and identification of a new *fts* locus (*ftsZ*). *J. Bacteriol.* **142:**615–620.

2588. **Lutkenhaus, J. F., and H. C. Wu.** 1980. Determination of transcriptional units and gene products from the *ftsA* region of *Escherichia coli. J. Bacteriol.* **143:**1281–1288.

2589. **Lutz, S., R. Bohm, A. Beier, and A. Bock.** 1990. Characterization of divergent NtrA-dependent promoters in the anaerobically expressed gene cluster coding for hydrogenase 3 components of *Escherichia coli. Mol. Microbiol.* **4:**13–20.

2590. **Lutz, S., A. Jacobi, V. Schlensog, R. Bohm, G. Sawers, and A. Bock.** 1991. Molecular characterization of an operon (*hyp*) necessary for the activity of the three hydrogenase isoenzymes in *Escherichia coli. Mol. Microbiol.* **5:**123–135.

2591. **Lynch, A. S., and J. C. Wang.** 1993. Anchoring of DNA to the bacterial cytoplasmic membrane through cotranscriptional synthesis of polypeptides encoding membrane proteins or proteins for export: a mechanism of plasmid hypernegative supercoiling in mutants deficient in DNA topoisomerase I. *J. Bacteriol.* **175:**1645–1655.

2592. **Lynn, S. P., C. E. Bauer, K. Chapman, and J. F. Gardner.** 1985. Identification and characterization of mutants affecting transcription termination at the threonine operon attenuator. *J. Mol. Biol.* **183:**529–541.

2593. **Ma, D., D. N. Cook, M. Alberti, N. G. Pon, H. Nikaido, and J. E. Hearst.** 1993. Molecular cloning and characterization of *acrA* and *acrE* genes of *Escherichia coli. J. Bacteriol.* **175:**6299–6313.

2594. **Ma, J., C. A. J. Newman, and R. S. Hayward.** 1981. Internal promoters of the *rpoBC* operon of *Escherichia coli. Mol. Gen. Genet.* **184:**548–550.

2595. **Maas, W. K.** 1965. Genetic defects affecting an arginine permease and repression of arginine synthesis in *Escherichia coli. Fed. Proc.* **24:**1239–1242.

2596. **Mabuchi, K., H. Kanazawa, T. Kayano, and M. Futai.** 1981. Nucleotide sequence of the gene coding for the delta subunit of proton translocating ATPase of *Escherichia coli. Biochem. Biophys. Res. Commun.* **102:**172–179.

2597. **Macdonald, H., and J. Cole.** 1985. Molecular cloning and functional analysis of the *cysG* and *nirB* genes of *Escherichia coli* K12, two closely-linked genes required for NADH- dependent nitrite reductase activity. *Mol. Gen. Genet.* **200:**328–334.

2598. **Macdonald, H., N. R. Pope, and J. A. Cole.** 1985. Isolation, characterization and complementation analysis of *nirB* mutants of *Escherichia coli* deficient only in NADH-dependent nitrite reductase activity. *J. Gen. Microbiol.* **131:**2771–2782.

2599. **Mackay, W. J., S. Han, and L. D. Samson.** 1994. DNA alkylation repair limits spontaneous base substitution mutations in *Escherichia coli. J. Bacteriol.* **176:**3224–3230.

2600. Mackie, G. A. 1980. Cloning of fragments of lambda *dapB2* DNA and identification of the *dapB* gene product. *J. Biol. Chem.* **255**:8928–8935.

2601. Mackie, G. A. 1981. Nucleotide sequence of the gene for ribosomal protein S20 and its flanking regions. *J. Biol. Chem.* **256**:8177–8182.

2602. Mackie, G. A. 1986. Structure of the DNA distal to the gene for ribosomal protein S20 in *Escherichia coli* K12: presence of a strong terminator and an IS*1* element. *Nucleic Acids Res.* **14**:6965–6981.

2603. Mackie, G. A. 1991. Specific endonucleolytic cleavage of the mRNA for ribosomal protein S20 of *Escherichia coli* requires the product of the *ams* gene in vivo and in vitro. *J. Bacteriol.* **173**:2488–2497.

2604. MacNeil, D. 1981. General method, using Mu-Mud1 dilysogens, to determine the direction of transcription of and generate deletions in the *glnA* region of *Escherichia coli.* *J. Bacteriol.* **146**:260–268.

2604a.Macnab, R. M. 1992. Genetics and biogenesis of bacterial flagella. *Annu. Rev. Genet.* **26**:131–158.

2605. MacNeil, T., D. MacNeil, and B. Tyler. 1982. Fine-structure deletion map and complementation analysis of the *glnA-glnL-glnG* region in *Escherichia coli.* *J. Bacteriol.* **150**:1302–1313.

2606. MacNeil, T., G. P. Roberts, D. MacNeil, and B. Tyler. 1982. The products of *glnL* and *glnG* are bifunctional regulatory proteins. *Mol. Gen. Genet.* **188**:325–333.

2607. MacPherson, A. J., M. C. Jones-Mortimer, and P. J. Henderson. 1981. Identification of the AraE transport protein of *Escherichia coli.* *Biochem. J.* **196**:269–283.

2608. Madern, D., J. Anselme, and M. Hartlein. 1992. Asparaginyl-tRNA synthetase from the *Escherichia coli* temperature-sensitive strain HO202. A proline replacement in motif 2 is responsible for a large increase in Km for asparagine and ATP. *FEBS Lett.* **299**:85–89.

2609. Madiraju, M. V., P. E. Lavery, S. C. Kowalczykowski, and A. J. Clark. 1992. Enzymatic properties of the RecA803 protein, a partial suppressor of *recF* mutations. *Biochemistry* **31**:10529–10535.

2610. Madiraju, M. V., A. Templin, and A. J. Clark. 1988. Properties of a mutant recA-encoded protein reveal a possible role for *Escherichia coli recF*-encoded protein in genetic recombination. *Proc. Natl. Acad. Sci. USA* **85**:6592–6596.

2611. Maeda, S., K. Takayanagi, Y. Nishimura, T. Maruyama, K. Sato, and T. Mizuno. 1991. Activation of the osmoregulated *ompC* gene by the OmpR protein in *Escherichia coli:* a study involving synthetic OmpR-binding sequences. *J. Biochem.* **110**:324–327.

2612. Magasanik, B. 1989. Regulation of transcription of the *glnALG* operon of *Escherichia coli* by protein phosphorylation. *Biochimie* **71**:1005–1012.

2613. Magnuson, K., M. R. Carey, and J. E. Cronan, Jr. 1995. The putative *fabJ* gene of *Escherichia coli* fatty acid synthesis is the *fabF* gene. *J. Bacteriol.* **177**:3593–3595.

2614. Magnuson, K., S. Jackowski, C. O. Rock, and J. E. Cronan, Jr. 1993. Regulation of fatty acid biosynthesis in *Escherichia coli.* *Microbiol. Rev.* **57**:522–542.

2615. Magnuson, K., W. Oh, T. J. Larson, and J. E. Cronan, Jr. 1992. Cloning and nucleotide sequence of the *fabD* gene encoding malonyl coenzyme A-acyl carrier protein transacylase of *Escherichia coli.* *FEBS Lett.* **299**:262–266.

2616. Magota, K., N. Otsuji, T. Miki, T. Horiuchi, S. Tsunasawa, J. Kondo, F. Sakiyama, M. Amemura, T. Morita, H. Shinagawa, and A. Nakata. 1984. Nucleotide sequence of the *phoS* gene, the structural gene for the phosphate-binding protein of *Escherichia coli.* *J. Bacteriol.* **157**:909–917.

2617. Maguin, E., H. Brody, C. W. Hill, and R. D'Ari. 1986. SOS-associated division inhibition gene *sfiC* is part of excisable element e14 in *Escherichia coli.* *J. Bacteriol.* **168**:464–466.

2618. Mahadevan, S., A. E. Reynolds, and A. Wright. 1987. Positive and negative regulation of the *bgl* operon in *Escherichia coli.* *J. Bacteriol.* **169**:2570–2578.

2619. Mahadevan, S., and A. Wright. 1987. A bacterial gene involved in transcription antitermination: regulation at a rho-independent terminator in the *bgl* operon of *E. coli.* *Cell* **50**:485–494.

2620. Mahajan, S. K., C. C. Chu, D. K. Willis, A. Templin, and A. J. Clark. 1990. Physical analysis of spontaneous and mutagen-induced mutants of *Escherichia coli* K-12 expressing DNA exonuclease VIII activity. *Genetics* **125**:261–273.

2621. Mahajan, S. K., N. B. Vartak, and A. R. Datta. 1988. A new pleiotropic mutation causing defective carbohydrate uptake in *Escherichia coli* K-12: isolation, mapping, and preliminary characterization. *J. Bacteriol.* **170**:2568–2574.

2622. Mahan, M. J., and L. N. Csonka. 1983. Genetic analysis of the *proBA* genes of *Salmonella typhimurium:* physical and genetic analyses of the cloned *proB⁺A⁺* genes of *Escherichia coli* and of a mutant allele that confers proline overproduction and enhanced osmotolerance. *J. Bacteriol.* **156**:1249–1262.

2623. Mahdi, A. A., and R. G. Lloyd. 1989. Identification of the *recR* locus of *Escherichia coli* K-12 and analysis of its role in recombination and DNA repair. *Mol. Gen. Genet.* **216**:503–510.

2624. Mahdi, A. A., and R. G. Lloyd. 1989. The *recR* locus of *Escherichia coli* K-12: molecular cloning, DNA sequencing and identification of the gene product. *Nucleic Acids Res.* **17**:6781–6794.

2625. Mahoney, W. C., R. W. Hogg, and M. A. Hermodson. 1981. The amino acid sequence of the D-galactose-binding protein from *Escherichia coli* B/r. *J. Biol. Chem.* **256**:4350–4356.

2626. Maiden, M. C., E. O. Davis, S. A. Baldwin, D. C. Moore, and P. J. Henderson. 1987. Mammalian and bacterial sugar transport proteins are homologous. *Nature* (London) **325**:641–643.

2627. Maiden, M. C., M. C. Jones-Mortimer, and P. J. Henderson. 1988. The cloning, DNA sequence, and overexpression of the gene *araE* coding for arabinose-proton symport in *Escherichia coli* K12. *J. Biol. Chem.* **263**:8003–8010.

2628. Maier, T., A. Jacobi, M. Sauter, and A. Bock. 1993. The product of the *hypB* gene, which is required for nickel incorporation into hydrogenases, is a novel guanine nucleotide-binding protein. *J. Bacteriol.* **175**:630–635.

2629. Maita, T., and G. Matsuda. 1980. The primary structure of L-asparaginase from *Escherichia coli. Hoppe-Seyler's Z. Physiol. Chem.* **361**:105–117.

2630. Maita, T., K. Morokuma, and G. Matsuda. 1979. Amino acid sequences of the tryptic peptides from carboxymethylated L-asparaginase from *Escherichia coli. Hoppe-Seyler's Z. Physiol. Chem.* **360**:1483–1495.

2631. Makaroff, C. A., and H. Zalkin. 1985. Regulation of *Escherichia coli purF.* Analysis of the control region of a *pur* regulon gene. *J. Biol. Chem.* **260**:10378–10387.

2632. Maki, H., T. Horiuchi, and M. Sekiguchi. 1983. Structure and expression of the *dnaQ* mutator and the RNase H genes of *Escherichia coli:* overlap of the promoter regions. *Proc. Natl. Acad. Sci. USA* **80**:7137–7141.

2633. Maki, H., and M. Sekiguchi. 1992. MutT protein specifically hydrolyses a potent mutagenic substrate for DNA synthesis. *Nature* (London) **355**:273–275.

2634. Maki, S., and A. Kornberg. 1988. DNA polymerase III holoenzyme of *Escherichia coli.* I. Purification and distinctive functions of subunits tau and gamma, the *dnaZX* gene products. *J. Biol. Chem.* **263**:6547–6554.

2635. Makino, K., S. K. Kim, H. Shinagawa, M. Amemura, and A. Nakata. 1991. Molecular analysis of the cryptic and functional *phn* operons for phosphonate use in *Escherichia coli* K-12. *J. Bacteriol.* **173**:2665–2612.

2636. Makino, K., H. Shinagawa, M. Amemura, and A. Nakata. 1986. Nucleotide sequence of the *phoB* gene, the positive regulatory gene for the phosphate regulon of *Escherichia coli* K-12. *J. Mol. Biol.* **190**:37–44.

2637. Makino, K., H. Shinagawa, M. Amemura, and A. Nakata. 1986. Nucleotide sequence of the *phoR* gene, a regulatory gene for the phosphate regulon of *Escherichia coli.* *J. Mol. Biol.* **192**:549–556.

2638. Makino, K., H. Shinagawa, and A. Nakata. 1984. Cloning and characterization of the alkaline phosphatase positive regulatory gene (*phoM*) of *Escherichia coli.* *Mol. Gen. Genet.* **195**:381–390.

2639. Makino, K., H. Shinagawa, and A. Nakata. 1985. Regulation of the phosphate regulon of *Escherichia coli* K-12: regulation and role of the regulatory gene *phoR.* *J. Mol. Biol.* **184**:231–240.

2640. Malakooti, J., B. Ely, and P. Matsumura. 1994. Molecular characterization, nucleotide sequence, and expression of the *fliO, fliP, fliQ,* and *fliR* genes of *Escherichia coli.* *J. Bacteriol.* **176**:189–197.

2641. Malakooti, J., Y. Komeda, and P. Matsumura. 1989. DNA sequence analysis, gene product identification, and localization of flagellar motor components of *Escherichia coli.* *J. Bacteriol.* **171**:2728–2734.

2642. Malcolm, B. A., and J. F. Kirsch. 1985. Site-directed mutagenesis of aspartate aminotransferase from *E. coli. Biochem. Biophys. Res. Commun.* **132**:915–921.

2643. Maleszka, R., P. Y. Wang, and H. Schneider. 1982. A Col E1 hybrid plasmid containing *Escherichia coli* genes complementing D-xylose negative mutants of *Escherichia coli* and *Salmonella typhimurium. Can. J. Biochem.* **60**:144–151.

2644. Malhotra, K. T., and R. A. Nicholas. 1992. Substitution of lysine 213 with arginine in penicillin-binding protein 5 of *Escherichia coli* abolishes D-alanine carboxypeptidase activity without affecting penicillin binding. *J. Biol. Chem.* **267**:11386–11391.

2645. Malo, M. S., and R. E. Loughlin. 1990. Promoter elements and regulation of expression of the *cysD* gene of *Escherichia coli* K-12. *Gene* **87**:127–131.

2646. Maloney, P. C. 1990. A consensus structure for membrane transport. *Res. Microbiol.* **141**:374–383.

2647. Maloy, S., and V. Stewart. 1993. Autogenous regulation of gene expression. *J. Bacteriol.* **175**:307–316.

2648. Maloy, S. R., and W. D. Nunn. 1982. Genetic regulation of the glyoxylate shunt in *Escherichia coli* K-12. *J. Bacteriol.* **149**:173–180.

2649. Mandal, T. N., A. A. Mahdi, G. J. Sharples, and R. G. Lloyd. 1993. Resolution of Holliday intermediates in recombination and DNA repair: indirect suppression of *ruvA, ruvB,* and *ruvC* mutations. *J. Bacteriol.* **175**:4325–4334.

2650. Mandrand-Berthelot, M. A., G. Couchoux-Luthaud, C. L. Santini, and G. Giordano. 1988. Mutants of *Escherichia coli* specifically deficient in respiratory formate dehydrogenase activity. *J. Gen. Microbiol.* **134**:3129–3139.

2651. Mandrand-Berthelot, M. A., P. Ritzenthaler, and M. Mata-Gilsinger. 1984. Construction and expression of hybrid plasmids containing the structural gene of the *Escherichia coli* K-12 3-deoxy-2-oxo-D-gluconate transport system. *J. Bacteriol.* **160**:600–606.

2652. Manna, D., and J. Gowrishankar. 1994. Evidence for involvement of proteins HU and RpoS in transcription of the osmoresponsive *proU* operon in *Escherichia coli.* *J. Bacteriol.* **176**:5378–5384.

2653. Mans, R. M., C. W. Pleij, and L. Bosch. 1991. tRNA-like structures. Structure, function and evolutionary significance. *Eur. J. Biochem.* **201**:303–324.

2654. Maples, V. F., and S. R. Kushner. 1982. DNA repair in *Escherichia coli:* identification of the *uvrD* gene product. *Proc. Natl. Acad. Sci. USA* **79**:5616–5620.

2655. Maras, B., G. Sweeney, D. Barra, F. Bossa, and R. A. John. 1992. The amino acid sequence of glutamate decarboxylase from *Escherichia coli.* Evolutionary relationship between mammalian and bacterial enzymes. *Eur. J. Biochem.* **204**:93–98.

2656. Marceau, M., E. McFall, S. D. Lewis, and J. A. Shafer. 1988. D-Serine dehydratase from *Escherichia coli.* DNA sequence and identification of catalytically inactive glycine to aspartic acid variants. *J. Biol. Chem.* **263**:16926–16933.

2657. March, J. B., M. D. Colloms, D. Hart-Davis, I. R. Oliver, and M. Masters. 1989. Cloning and characterization of an *Escherichia coli* gene, *pcnB,* affecting plasmid copy number. *Mol. Microbiol.* **3**:903–910.

2658. **March, P. E.** 1992. Membrane-associated GTPases in bacteria. *Mol. Microbiol.* 6:1253–1257.

2659. **March, P. E., J. Ahnn, and M. Inouye.** 1985. The DNA sequence of the gene (*rnc*) encoding ribonuclease III of *Escherichia coli. Nucleic Acids Res.* 13:4677–4685.

2660. **March, P. E., and M. Inouye.** 1985. GTP-binding membrane protein of *Escherichia coli* with sequence homology to initiation factor 2 and elongation factors Tu and G. *Proc. Natl. Acad. Sci. USA* 82:7500–7504.

2661. **March, P. E., and M. Inouye.** 1985. Characterization of the *lep* operon of *Escherichia coli*. Identification of the promoter and the gene upstream of the signal peptidase I gene. *J. Biol. Chem.* 260:7206–7213.

2662. **Marcus, F., B. Gontero, P. B. Harrsch, and J. Rittenhouse.** 1986. Amino acid sequence homology among fructose-1,6-bisphosphatases. *Biochem. Biophys. Res. Commun.* 135:374–381.

2663. **Marcus, M., and Y. S. Halpern.** 1967. Genetic analysis of glutamate transport and glutamate decarboxylase in *Escherichia coli. J. Bacteriol.* 93:1409–1415.

2664. **Marcus, M., and Y. S. Halpern.** 1969. Genetic analysis of the glutamate permease in *Escherichia coli* K-12. *J. Bacteriol.* 97:1118–1128.

2665. **Marek, L. E., and J. M. Henson.** 1988. Cloning and expression of the *Escherichia coli* K-12 *sad* gene. *J. Bacteriol.* 170:991–994.

2666. **Margison, G. P., D. P. Cooper, and J. Brennand.** 1985. Cloning of the *E. coli* O^6-methylguanine and methylphosphotriester methyltransferase gene using a functional DNA repair assay. *Nucleic Acids Res.* 13:1939–1952.

2667. **Margison, G. P., D. P. Cooper, and P. M. Potter.** 1990. The *E. coli ogt* gene. *Mutat. Res.* 233:15–21.

2668. **Margolin, P., L. Zumstein, R. Sternglanz, and J. C. Wang.** 1985. The *Escherichia coli supX* locus is *topA*, the structural gene for DNA topoisomerase I. *Proc. Natl. Acad. Sci. USA* 82:5437–5441.

2669. **Marinus, M. G.** 1992. Identification of the gene (*aroK*) encoding shikimic acid kinase I of *Escherichia coli. J. Bacteriol.* 174:525–529.

2669a.**Marinus, M. G., N. R. Morris, D. Soll, and T. C. Kwong.** 1975. Isolation and partial characterization of three *Escherichia coli* mutants with altered transfer ribonucleic acid methylases. *J. Bacteriol.* 122:257–265.

2670. **Mark, D. F., J. W. Chase, and C. C. Richardson.** 1977. Genetic mapping of *trxA*, a gene affecting thioredoxin in *Escherichia coli* K12. *Mol. Gen. Genet.* 155:145–152.

2671. **Markham, B. E., J. W. Little, and D. W. Mount.** 1981. Nucleotide sequence of the *lexA* gene of *Escherichia coli* K-12. *Nucleic Acids Res.* 9:4149–4161.

2672. **Markham, G. D., J. DeParasis, and J. Gatmaitan.** 1984. The sequence of *metK*, the structural gene for S-adenosylmethionine synthetase in *Escherichia coli. J. Biol. Chem.* 259:14505–14507.

2673. **Markiewicz, P., L. G. Kleina, C. Cruz, S. Ehret, and J. H. Miller.** 1994. Genetic studies of the *lac* repressor. XIV. Analysis of 4000 altered *Escherichia coli lac* repressors reveals essential and non-essential residues, as well as "spacers" which do not require a specific sequence. *J. Mol. Biol.* 240:421–433.

2674. **Markiewicz, P., C. Malone, J. W. Chase, and L. B. Rothman-Denes.** 1992. *Escherichia coli* single-stranded DNA-binding protein is a supercoiled template-dependent transcriptional activator of N4 virion RNA polymerase. *Genes Dev.* 6:2010–2019.

2675. **Markovitz, A., and N. Rosenbaum.** 1965. A regulator gene that is dominant on an episome and recessive on a chromosome. *Proc. Natl. Acad. Sci. USA* 54:1084–1091.

2676. **Marolewski, A., J. M. Smith, and S. J. Benkovic.** 1994. Cloning and characterization of a new purine biosynthetic enzyme: a non-folate glycinamide ribonucleotide transformylase from *E. coli*. *Biochemistry* 33:2531–2537.

2677. **Marquardt, J. L., D. A. Siegele, R. Kolter, and C. T. Walsh.** 1992. Cloning and sequencing of *Escherichia coli murZ* and purification of its product, a UDP-*N*-acetylglucosamine enolpyruvyl transferase. *J. Bacteriol.* 174:5748–5752.

2678. **Marsh, L., and G. C. Walker.** 1985. Cold sensitivity induced by overproduction of UmuDC in *Escherichia coli. J. Bacteriol.* 162:155–161.

2679. **Martel, A., C. Bouthier de la Tour, and F. Le Goffic.** 1987. Pyridoxal 5′phosphate binding site of *Escherichia coli* beta cystathionase and cystathionine gamma synthase comparison of their sequences. *Biochem. Biophys. Res. Commun.* 147:565–571.

2680. **Martin, E., V. Sagitov, E. Burova, V. Nikiforov, and A. Goldfarb.** 1992. Genetic dissection of the transcription cycle. A mutant RNA polymerase that cannot hold onto a promoter. *J. Biol. Chem.* 267:20175–20180.

2681. **Martin, J. L., J. C. Bardwell, and J. Kuriyan.** 1993. Crystal structure of the DsbA protein required for disulphide bond formation in vivo. *Nature* (London) 365:464–468.

2682. **Martinez, D., and F. Whitehouse, Jr.** 1973. Selective autocytotoxicity in a model system of *Escherichia coli* K-12 recombinants. *J. Bacteriol.* 114:882–884.

2683. **Maruyama, M., T. Horiuchi, H. Maki, and M. Sekiguchi.** 1983. A dominant (*mutD5*) and a recessive (*dnaQ49*) mutator of *Escherichia coli. J. Mol. Biol.* 167:757–771.

2684. **Marvel, C. C., P. J. Arps, B. C. Rubin, H. O. Kammen, E. E. Penhoet, and M. E. Winkler.** 1985. *hisT* is part of a multigene operon in *Escherichia coli* K-12. *J. Bacteriol.* 161:60–71.

2685. **Masai, H., and K. Arai.** 1988. Operon structure of *dnaT* and *dnaC* genes essential for normal and stable DNA replication of *Escherichia coli* chromosome. *J. Biol. Chem.* 263:15083–15093.

2686. **Masai, H., M. W. Bond, and K. Arai.** 1986. Cloning of the *Escherichia coli* gene for primosomal protein i: the relationship to *dnaT*, essential for chromosomal DNA replication. *Proc. Natl. Acad. Sci. USA* 83:1256–1260.

2687. **Maskell, D.** 1992. Protein sequence from downstream of *Escherichia coli galK* is homologous with *galM* from other organisms. *Mol. Microbiol.* 6:2211. (Letter.)

2688. **Mason, S. W., J. Li, and J. Greenblatt.** 1992. Direct interaction between two *Escherichia coli* transcription antitermination factors, NusB and ribosomal protein S10. *J. Mol. Biol.* 223:55–66.

2689. **Masters, M.** 1990. Location of purine genes on the physical map of *Escherichia coli. J. Bacteriol.* 172:1173. (Erratum, 172:2825.)

2690. **Masters, M., M. D. Colloms, I. R. Oliver, L. He, E. J. Macnaughton, and Y. Charters.** 1993. The *pcnB* gene of *Escherichia coli*, which is required for ColE1 copy number maintenance, is dispensable. *J. Bacteriol.* 175:4405–4413.

2691. **Masters, M., J. B. March, I. R. Oliver, and J. F. Collins.** 1990. A possible role for the *pcnB* gene product of *Escherichia coli* in modulating RNA: RNA interactions. *Mol. Gen. Genet.* 220:341–344.

2692. **Masters, M., T. Paterson, A. G. Popplewell, T. Owen-Hughes, J. H. Pringle, and K. J. Begg.** 1989. The effect of DnaA protein levels and the rate of initiation at *oriC* on transcription originating in the *ftsQ* and *ftsA* genes: in vivo experiments. *Mol. Gen. Genet.* 216:475–483.

2693. **Masters, P. S., and J. S. Hong.** 1981. Genetics of the glutamine transport system in *Escherichia coli. J. Bacteriol.* 147:805–819.

2694. **Masterson, C., P. E. Boehmer, F. McDonald, S. Chaudhuri, I. D. Hickson, and P. T. Emmerson.** 1992. Reconstitution of the activities of the RecBCD holoenzyme of *Escherichia coli* from the purified subunits. *J. Biol. Chem.* 267:13564–13572.

2695. **Masuda, Y., K. Miyakawa, Y. Nishimura, and E. Ohtsubo.** 1993. *chpA* and *chpB*, *Escherichia coli* chromosomal homologs of the *pem* locus responsible for stable maintenance of plasmid R100. *J. Bacteriol.* 175:6850–6856.

2696. **Masuda, Y., and E. Ohtsubo.** 1994. Mapping and disruption of the *chpB* locus in *Escherichia coli. J. Bacteriol.* 176:5861–5863.

2697. **Mata-Gilsinger, M., and P. Ritzenthaler.** 1983. Physical mapping of the *exuT* and *uxaC* operators by use of *exu* plasmids and generation of deletion mutants in vitro. *J. Bacteriol.* 155:973–982.

2698. **Mata-Gilsinger, M., P. Ritzenthaler, and C. Blanco.** 1983. Characterization of the operator sites of the *exu* regulon in *Escherichia coli* K-12 by operator-constitutive mutations and repressor titration. *Genetics* 105:829–842.

2699. **Matijasevic, Z., M. Sekiguchi, and D. B. Ludlum.** 1992. Release of *N*-2,3-ethenoguanine from chloroacetaldehyde-treated DNA by *Escherichia coli* 3-methyladenine DNA glycosylase II. *Proc. Natl. Acad. Sci. USA* 89:9331–9334.

2700. **Matin, A.** 1991. The molecular basis of carbon-starvation-induced general resistance in *Escherichia coli. Mol. Microbiol.* 5:3–10.

2701. **Matsuhashi, M., I. N. Maruyama, Y. Takagaki, S. Tamaki, Y. Nishimura, and Y. Hirota.** 1978. Isolation of a mutant of *Escherichia coli* lacking penicillin-sensitive D-alanine carboxypeptidase IA. *Proc. Natl. Acad. Sci. USA* 75:2631–2635.

2702. **Matsuhashi, M., Y. Takagaki, I. N. Maruyama, S. Tamaki, Y. Nishimura, H. Suzuki, U. Ogino, and Y. Hirota.** 1977. Mutants of *Escherichia coli* lacking in highly penicillin-sensitive D-alanine carboxypeptidase activity. *Proc. Natl. Acad. Sci.* 74:2976–2979.

2703. **Matsuhashi, M., S. Tamaki, S. J. Curtis, and J. L. Strominger.** 1979. Mutational evidence for identity of penicillin-binding protein 5 in *Escherichia coli* with the major D-alanine carboxypeptidase IA activity. *J. Bacteriol.* 137:644–647.

2704. **Matsuhisa, A., N. Suzuki, T. Noda, and K. Shiba.** 1995. Inositol monophosphatase activity from the *Escherichia coli suhB* gene product. *J. Bacteriol.* 177:200–205.

2705. **Matsui, M., A. Oka, M. Takanami, S. Yasuda, and Y. Hirota.** 1985. Sites of *dnaA* protein-binding in the replication origin of the *Escherichia coli* K-12 chromosome. *J. Mol. Biol.* 184:529–533.

2706. **Matsumoto, Y., K. Shigesada, M. Hirano, and M. Imai.** 1986. Autogenous regulation of the gene for transcription termination factor *rho* in *Escherichia coli*: localization and function of its attenuators. *J. Bacteriol.* 166:945–958.

2707. **Matsumura, P., J. J. Rydel, R. Linzmeier, and D. Vacante.** 1984. Overexpression and sequence of the *Escherichia coli cheY* gene and biochemical activities of the CheY protein. *J. Bacteriol.* 160:36–41.

2708. **Matsuoka, M., and B. A. McFadden.** 1988. Isolation, hyperexpression, and sequencing of the *aceA* gene encoding isocitrate lyase in *Escherichia coli. J. Bacteriol.* 170:4528–4536.

2709. **Matsuyama, A., H. Yamamoto, and E. Nakano.** 1989. Cloning, expression, and nucleotide sequence of the *Escherichia coli* K-12 *ackA* gene. *J. Bacteriol.* 171:577–580.

2710. **Matsuyama, A., H. Yamamoto-Otake, J. Hewitt, R. T. MacGillivray, and E. Nakano.** 1994. Nucleotide sequence of the phosphotransacetylase gene of *Escherichia coli* strain K12. *Biochim. Biophys. Acta* 1219:559–562.

2711. **Matsuyama, S., J. Akimaru, and S. Mizushima.** 1990. SecE-dependent overproduction of SecY in *Escherichia coli*. Evidence for interaction between two components of the secretory machinery. *FEBS Lett.* 269:96–100.

2712. **Matsuyama, S., Y. Fujita, K. Sagara, and S. Mizushima.** 1992. Overproduction, purification and characterization of SecD and SecF, integral membrane components of the protein translocation machinery of *Escherichia coli. Biochim. Biophys. Acta* 1122:77–84.

2713. **Matsuzawa, H., S. Asoh, K. Kunai, K. Muraiso, A. Takasuga, and T. Ohta.** 1989. Nucleotide sequence of the *rodA* gene, responsible for the rod shape of *Escherichia coli*: *rodA* and the *pbpA* gene, encoding penicillin-binding protein 2, constitute the *rodA* operon. *J. Bacteriol.* 171:558–560.

2714. **Matsuzawa, H., S. Ushiyama, Y. Koyama, and T. Ohta.** 1984. *Escherichia coli* K-12 *tolZ* mutants tolerant to colicins E2, E3, D, Ia, and Ib: defect in generation of the electrochemical proton gradient. *J. Bacteriol.* 160:733–739.

2715. **Mattern, I. E., and J. Pittard.** 1971. Regulation of tyrosine biosynthesis in *Escherichia coli* K-12: isolation and characterization of operator mutants. *J. Bacteriol.* **107**:8–15.

2716. **Matteson, R. J., S. J. Biswas, and D. A. Steege.** 1991. Distinctive patterns of translational reinitiation in the *lac* repressor mRNA: bridging of long distances by out-of-frame translation and RNA secondary structure, effects of primary sequence. *Nucleic Acids Res.* **19**:3499–3506.

2717. **Mattheakis, L. C., and M. Nomura.** 1988. Feedback regulation of the *spc* operon in *Escherichia coli*: translational coupling and mRNA processing. *J. Bacteriol.* **170**:4484–4492.

2718. **Maupin, J. A., and K. T. Shanmugam.** 1990. Genetic regulation of formate hydrogenlyase of *Escherichia coli*: role of the *fhlA* gene product as a transcriptional activator for a new regulatory gene, *fhlB*. *J. Bacteriol.* **172**:4798–4806.

2719. **Maurizi, M. R.** 1991. ATP-promoted interaction between Clp A and Clp P in activation of Clp protease from *Escherichia coli*. *Biochem. Soc. Trans.* **19**:719–723.

2720. **Maurizi, M. R., W. P. Clark, Y. Katayama, S. Rudikoff, J. Pumphrey, B. Bowers, and S. Gottesman.** 1990. Sequence and structure of Clp P, the proteolytic component of the ATP-dependent Clp protease of *Escherichia coli*. *J. Biol. Chem.* **265**:12536–12545.

2721. **Maurizi, M. R., W. P. Clark, S. H. Kim, and S. Gottesman.** 1990. Clp P represents a unique family of serine proteases. *J. Biol. Chem.* **265**:12546–12552.

2722. **Maurizi, M. R., P. Trisler, and S. Gottesman.** 1985. Insertional mutagenesis of the *lon* gene in *Escherichia coli*: *lon* is dispensable. *J. Bacteriol.* **164**:1124–1135.

2723. **Mauzy, C. A., and M. A. Hermodson.** 1992. Structural and functional analyses of the repressor, RbsR, of the ribose operon of *Escherichia coli*. *Protein Sci.* **1**:831–842.

2724. **Mauzy, C. A., and M. A. Hermodson.** 1992. Structural homology between *rbs* repressor and ribose binding protein implies functional similarity. *Protein Sci.* **1**:843–849.

2725. **Maxon, M. E., B. Redfield, X. Y. Cai, R. Shoeman, K. Fujita, W. Fisher, G. Stauffer, H. Weissbach, and N. Brot.** 1989. Regulation of methionine synthesis in *Escherichia coli*: effect of the MetR protein on the expression of the *metE* and *metR* genes. *Proc. Natl. Acad. Sci. USA* **86**:85–89.

2726. **Maxon, M. E., J. Wigboldus, N. Brot, and H. Weissbach.** 1990. Structure-function studies on *Escherichia coli* MetR protein, a putative prokaryotic leucine zipper protein. *Proc. Natl. Acad. Sci. USA* **87**:7076–7079.

2727. **May, G., P. Dersch, M. Haardt, A. Middendorf, and E. Bremer.** 1990. The *osmZ* (*bglY*) gene encodes the DNA-binding protein H-NS (H1a), a component of the *Escherichia coli* K12 nucleoid. *Mol. Gen. Genet.* **224**:81–90.

2728. **May, G., E. Faatz, J. M. Lucht, M. Haardt, M. Bolliger, and E. Bremer.** 1989. Characterization of the osmoregulated *Escherichia coli proU* promoter and identification of ProV as a membrane-associated protein. *Mol. Microbiol.* **3**:1521–1531.

2729. **May, G., E. Faatz, M. Villarejo, and E. Bremer.** 1986. Binding protein dependent transport of glycine betaine and its osmotic regulation in *Escherichia coli* K12. *Mol. Gen. Genet.* **205**:225–233.

2730. **Mayaux, J. F., G. Fayat, M. Fromant, M. Springer, M. Grunberg-Manago, and S. Blanquet.** 1983. Structural and transcriptional evidence for related *thrS* and *infC* expression. *Proc. Natl. Acad. Sci. USA* **80**:6152–6156.

2731. **Mazel, D., S. Pochet, and P. Marliere.** 1994. Genetic characterization of polypeptide deformylase, a distinctive enzyme of eubacterial translation. *EMBO J.* **13**:914–923.

2732. **Mazzarelli, J. M., S. B. Rajur, P. L. Iadarola, and L. W. McLaughlin.** 1992. Interactions between the *trp* repressor and its operator sequence as studied by base analogue substitution. *Biochemistry* **31**:5925–5936.

2733. **McAlister-Henn, L., M. Blaber, R. A. Bradshaw, and S. J. Nisco.** 1987. Complete nucleotide sequence of the *Escherichia coli* gene encoding malate dehydrogenase. *Nucleic Acids Res.* **15**:4993.

2734. **McCalla, D. R., C. Kaiser, and M. H. Green.** 1978. Genetics of nitrofurazone resistance in *Escherichia coli*. *J. Bacteriol.* **133**:10–16.

2735. **McCaman, M. T., and J. D. Gabe.** 1986. Sequence of the promoter and 5′ coding region of *pepN*, and the amino-terminus of peptidase N from *Escherichia coli* K-12. *Mol. Gen. Genet.* **204**:148–152.

2736. **McCaman, M. T., and J. D. Gabe.** 1986. The nucleotide sequence of the *pepN* gene and its over-expression in *Escherichia coli*. *Gene* **48**:145–153.

2737. **McCaman, M. T., A. McPartland, and M. R. Villarejo.** 1982. Genetics and regulation of peptidase N in *Escherichia coli* K-12. *J. Bacteriol.* **152**:848–854.

2738. **McCarthy, J. E., B. Gerstel, B. Surin, U. Wiedemann, and P. Ziemke.** 1991. Differential gene expression from the *Escherichia coli atp* operon mediated by segmental differences in mRNA stability. *Mol. Microbiol.* **5**:2447–2458.

2739. **McCarthy, J. E., H. U. Schairer, and W. Sebald.** 1985. Translational initiation frequency of *atp* genes from *Escherichia coli*: identification of an intercistronic sequence that enhances translation. *EMBO J.* **4**:519–526.

2740. **McCarthy, J. E., W. Sebald, G. Gross, and R. Lammers.** 1986. Enhancement of translational efficiency by the *Escherichia coli atpE* translational initiation region: its fusion with two human genes. *Gene* **41**:201–206.

2741. **McCarty, J. S., and G. C. Walker.** 1991. DnaK as a thermometer: threonine-199 is site of autophosphorylation and is critical for ATPase activity. *Proc. Natl. Acad. Sci. USA* **88**:9513–9517.

2742. **McCarty, J. S., and G. C. Walker.** 1994. DnaK mutants defective in ATPase activity are defective in negative regulation of the heat shock response: expression of mutant DnaK proteins results in filamentation. *J. Bacteriol.* **176**:764–780.

2743. **McClain, M. S., I. C. Blomfield, K. J. Eberhardt, and B. I. Eisenstein.** 1993. Inversion-independent phase variation of type 1 fimbriae in *Escherichia coli*. *J. Bacteriol.* **175**:4335–4344.

2744. **McConville, M. L., and H. P. Charles.** 1979. Mutants of *Escherichia coli* K12 accumulating porphobilinogen: a new locus, *hemC*. *J. Gen. Microbiol.* **111**:193–200.

2745. **McConville, M. L., and H. P. Charles.** 1979. Isolation of haemin-requiring mutants of *Escherichia coli* K12. *J. Gen. Microbiol.* **113**:155–164.

2746. **McCormick, K. A., and B. D. Cain.** 1991. Targeted mutagenesis of the b subunit of F_1F_0 ATP synthase in *Escherichia coli*: Glu-77 through Gln-85. *J. Bacteriol.* **173**:7240–7248.

2747. **McCulloch, R., M. E. Burke, and D. J. Sherratt.** 1994. Peptidase activity of *Escherichia coli* aminopeptidase A is not required for its role in Xer site-specific recombination. *Mol. Microbiol.* **12**:241–251.

2748. **McCulloch, R., L. W. Coggins, S. D. Colloms, and D. J. Sherratt.** 1994. Xer-mediated site-specific recombination at *cer* generates Holliday junctions in vivo. *EMBO J.* **13**:1844–1855.

2749. **McDougall, J., I. Kaasen, and A. R. Strom.** 1993. A yeast gene for trehalose-6-phosphate synthase and its complementation of an *Escherichia coli otsA* mutant. *FEMS Microbiol. Lett.* **107**:25–30.

2750. **McDowall, K. J., R. G. Hernandez, S. Lin-Chao, and S. N. Cohen.** 1993. The *ams-1* and *rne-3071* temperature-sensitive mutations in the *ams* gene are in close proximity to each other and cause substitutions within a domain that resembles a product of the *Escherichia coli mre* locus. *J. Bacteriol.* **175**:4245–4249.

2751. **McEwen, J., and P. Silverman.** 1980. Chromosomal mutations of *Escherichia coli* that alter expression of conjugative plasmid functions. *Proc. Natl. Acad. Sci. USA* **77**:513–517.

2752. **McEwen, J., and P. Silverman.** 1980. Genetic analysis of *Escherichia coli* K-12 chromosomal mutants defective in expression of F-plasmid functions: identification of genes *cpxA* and *cpxB*. *J. Bacteriol.* **144**:60–67.

2753. **McFall, E.** 1975. *Escherichia coli* K-12 mutant forming a temperature-sensitive D-serine deaminase. *J. Bacteriol.* **21**:1074–1077.

2754. **McFall, E., and L. Runkel.** 1983. DNA sequences of the D-serine deaminase control region and N-terminal portion of the structural gene. *J. Bacteriol.* **154**:1508–1512.

2755. **McFarland, N., L. McCarter, S. Artz, and S. Kustu.** 1981. Nitrogen regulatory locus "*glnR*" of enteric bacteria is composed of cistrons *ntrB* and *ntrC*: identification of their protein products. *Proc. Natl. Acad. Sci. USA* **78**:2135–2139.

2756. **McGovern, K., and J. Beckwith.** 1991. Membrane insertion of the *Escherichia coli* MalF protein in cells with impaired secretion machinery. *J. Biol. Chem.* **266**:20870–20876.

2757. **McGovern, K., M. Ehrmann, and J. Beckwith.** 1991. Decoding signals for membrane protein assembly using alkaline phosphatase fusions. *EMBO J.* **10**:2773–2782.

2758. **McGrath, M. E., T. Erpel, C. Bystroff, and R. J. Fletterick.** 1994. Macromolecular chelation as an improved mechanism of protease inhibition: structure of the ecotin-trypsin complex. *EMBO J.* **13**:1502–1507.

2759. **McGrath, M. E., W. M. Hines, J. A. Sakanari, R. J. Fletterick, and C. S. Craik.** 1991. The sequence and reactive site of ecotin. A general inhibitor of pancreatic serine proteases from *Escherichia coli*. *J. Biol. Chem.* **266**:6620–6625.

2760. **McGraw, B. R., and M. G. Marinus.** 1980. Isolation and characterization of Dam$^+$ revertants and suppressor mutations that modify secondary phenotypes of *dam-3* strains of *Escherichia coli* K-12. *Mol. Gen. Genet.* **178**:309–315.

2761. **McKown, R. L., K. A. Orle, T. Chen, and N. L. Craig.** 1988. Sequence requirements of *Escherichia coli attTn7*, a specific site of transposon Tn7 insertion. *J. Bacteriol.* **170**:352–358.

2762. **McLachlan, A. D.** 1978. The double helix coiled coil structure of murein lipoprotein from *Escherichia coli*. *J. Mol. Biol.* **121**:493–506.

2763. **McLaren, R. S., S. F. Newbury, G. S. Dance, H. C. Causton, and C. F. Higgins.** 1991. mRNA degradation by processive 3′-5′ exoribonucleases in vitro and the implications for prokaryotic mRNA decay in vivo. *J. Mol. Biol.* **221**:81–95.

2764. **McMurry, L. M., and S. B. Levy.** 1987. Tn5 insertion in the polynucleotide phosphorylase (*pnp*) gene in *Escherichia coli* increases susceptibility to antibiotics. *J. Bacteriol.* **169**:1321–1324.

2765. **McPherson, M. J., A. J. Baron, D. J. Pappin, and J. C. Wootton.** 1984. Respiratory nitrate reductase of *Escherichia coli*. Sequence identification of the large subunit gene. *FEBS Lett.* **177**:260–264.

2766. **McPherson, M. J., and J. C. Wootton.** 1983. Complete nucleotide sequence of the *Escherichia coli gdhA* gene. *Nucleic Acids Res.* **11**:5257–5266.

2767. **Meador, J., and D. Kennell.** 1990. Cloning and sequencing the gene encoding *Escherichia coli* ribonuclease I: exact physical mapping using the genome library. *Gene* **95**:1–7.

2768. **Meadow, N. D., D. W. Saffen, R. P. Dottin, and S. Roseman.** 1982. Molecular cloning of the *crr* gene and evidence that it is the structural gene for IIIGlc, a phosphocarrier protein of the bacterial phosphotransferase system. *Proc. Natl. Acad. Sci. USA* **79**:2528–2532.

2769. **Mechulam, Y., S. Blanquet, and G. Fayat.** 1987. Dual level control of the *Escherichia coli pheST-himA* operon expression. tRNA^{Phe}-dependent attenuation and transcriptional operator-repressor control by *himA* and the SOS network. *J. Mol. Biol.* **197**:453–470.

2770. **Mechulam, Y., G. Fayat, and S. Blanquet.** 1985. Sequence of the *Escherichia coli pheST* operon and identification of the *himA* gene. *J. Bacteriol.* **163**:787–791.

2771. Mechulam, Y., M. Fromant, P. Mellot, P. Plateau, S. Blanchin-Roland, G. Fayat, and S. Blanquet. 1985. Molecular cloning of the *Escherichia coli* gene for diadenosine 5',5'''-P^1,P^4-tetraphosphate pyrophosphohydrolase. *J. Bacteriol.* **164:**63–69.

2772. Mecsas, J., P. E. Rouviere, J. W. Erickson, T. J. Donohue, and C. A. Gross. 1993. The activity of sigma E, an *Escherichia coli* heat-inducible sigma-factor, is modulated by expression of outer membrane proteins. *Genes Dev.* **7673:**2618–2628.

2773. Mecsas, J., R. Welch, J. W. Erickson, and C. A. Gross. 1995. Identification and characterization of an outer membrane protein, OmpX, in *Escherichia coli* that is homologous to a family of outer membrane proteins including Ail of *Yersinia enterocolitica*. *J. Bacteriol.* **177:**799–804.

2774. Medina, V., R. Pontarollo, D. Glaeske, H. Tabel, and H. Goldie. 1990. Sequence of the *pckA* gene of *Escherichia coli* K-12: relevance to genetic and allosteric regulation and homology of *E. coli* phosphoenolpyruvate carboxykinase with the enzymes from *Trypanosoma brucei* and *Saccharomyces cerevisiae*. *J. Bacteriol.* **172:**7151–7156.

2775. Meek, D. W., and R. S. Hayward. 1984. Nucleotide sequence of the *rpoA-rplQ* DNA of *Escherichia coli*: a second regulatory binding site for protein S4? *Nucleic Acids Res.* **12:**5813–5821.

2776. Meier, U., and H. Mayer. 1985. Genetic location of genes encoding enterobacterial common antigen. *J. Bacteriol.* **163:**756–762.

2777. Meier-Dieter, U., K. Barr, R. Starman, L. Hatch, and P. D. Rick. 1992. Nucleotide sequence of the *Escherichia coli rfe* gene involved in the synthesis of enterobacterial common antigen. Molecular cloning of the *rfe-rff* gene cluster. *J. Biol. Chem.* **267:**746–753.

2778. Meier-Dieter, U., R. Starman, K. Barr, H. Mayer, and P. D. Rick. 1990. Biosynthesis of enterobacterial common antigen in *Escherichia coli*. Biochemical characterization of Tn10 insertion mutants defective in enterobacterial common antigen synthesis. *J. Biol. Chem.* **265:**13490–13497.

2779. Meijer, M., E. Beck, F. G. Hansen, H. E. Bergmans, W. Messer, K. von Meyenburg, and H. Schaller. 1979. Nucleotide sequence of the origin of replication of the *Escherichia coli* K-12 chromosome. *Proc. Natl. Acad. Sci. USA* **76:**580–584.

2780. Meiklejohn, A. L., and J. D. Gralla. 1989. Activation of the *lac* promoter and its variants. Synergistic effects of catabolite activator protein and supercoiling in vitro. *J. Mol. Biol.* **207:**661–673.

2781. Meinnel, T., and S. Blanquet. 1993. Evidence that peptide deformylase and methionyl-tRNA^{fMet} formyltransferase are encoded within the same operon in *Escherichia coli*. *J. Bacteriol.* **175:**7737–7740.

2782. Meinnel, T., and S. Blanquet. 1995. Enzymatic properties of *Escherichia coli* peptide deformylase. *J. Bacteriol.* **177:**1883–1887.

2783. Meinnel, T., J. M. Guillon, Y. Mechulam, and S. Blanquet. 1993. The *Escherichia coli fmt* gene, encoding methionyl-tRNA^{fMet} formyltransferase, escapes metabolic control. *J. Bacteriol.* **175:**993–1000.

2784. Meinnel, T., Y. Mechulam, F. Dardel, J. M. Schmitter, C. Hountondji, S. Brunie, P. Dessen, G. Fayat, and S. Blanquet. 1990. Methionyl-tRNA synthetase from *E. coli*—a review. *Biochimie* **72:**625–632.

2785. Meinnel, T., E. Schmitt, Y. Mechulam, and S. Blanquet. 1992. Structural and biochemical characterization of the *Escherichia coli argE* gene product. *J. Bacteriol.* **174:**2323–2331.

2786. Meins, M., P. Jeno, D. Muller, W. J. Richter, J. P. Rosenbusch, and B. Erni. 1993. Cysteine phosphorylation of the glucose transporter of *Escherichia coli*. *J. Biol. Chem.* **268:**11604–11609.

2787. Mejean, V., C. Iobbi-Nivol, M. Lepelletier, G. Giordano, M. Chippaux, and M. C. Pascal. 1994. TMAO anaerobic respiration in *Escherichia coli*: involvement of the *tor* operon. *Mol. Microbiol.* **11:**1169–1179.

2788. Melefors, O., and A. von Gabain. 1991. Genetic studies of cleavage-initiated mRNA decay and processing of ribosomal 9S RNA show that the *Escherichia coli ams* and *rne* loci are the same. *Mol. Microbiol.* **5:**857–864.

2789. Mellies, J., A. Wise, and M. Villarejo. 1995. Two different *Escherichia coli proP* promoters respond to osmotic and growth phase signals. *J. Bacteriol.* **177:**144–151.

2790. Melville, S. B., and R. P. Gunsalus. 1990. Mutations in *fnr* that alter anaerobic regulation of electron transport-associated genes in *Escherichia coli*. *J. Biol. Chem.* **265:**18733–18736.

2791. Melzer, M., and L. Heide. 1994. Characterization of polyprenyldiphosphate:4-hydroxybenzoate polyprenyltransferase from *Escherichia coli*. *Biochim. Biophys. Acta* **1212:**93–102.

2792. Mende, L. 1978. The primary structure of protein L13 from the large subunit of *Escherichia coli* ribosomes. *FEBS Lett.* **96:**313–316.

2793. Mende, L., B. Timm, and R. Subramanian. 1978. Primary structures of two homologous ribosome-associated DNA-binding proteins of *Escherichia coli*. *FEBS Lett.* **96:**395–398.

2794. Mendelson, I., M. Gottesman, and A. B. Oppenheim. 1991. HU and integration host factor function as auxiliary proteins in cleavage of phage lambda cohesive ends by terminase. *J. Bacteriol.* **173:**1670–1676.

2795. Mendelson, I., H. Haluzi, S. Koby, and A. B. Oppenheim. 1991. Physical map locations of the *himA* and *hip* genes of *Escherichia coli*. *J. Bacteriol.* **173:**5249.

2796. Mendonca, V. M., K. Kaiser-Rogers, and S. W. Matson. 1993. Double helicase II (*uvrD*)-helicase IV (*helD*) deletion mutants are defective in the recombination pathways of *Escherichia coli*. *J. Bacteriol.* **175:**4641–4651.

2797. Mendonca, V. M., H. D. Klepin, and S. W. Matson. 1995. DNA helicases in recombination and repair: construction of a ΔuvrD ΔhelD ΔrecQ mutant deficient in recombination and repair. *J. Bacteriol.* **177:**1326–1335.

2798. Meng, L. M., M. Kilstrup, and P. Nygaard. 1990. Autoregulation of PurR repressor synthesis and involvement of *purR* in the regulation of *purB*, *purC*, *purL*, *purMN* and *guaBA* expression in *Escherichia coli*. *Eur. J. Biochem.* **187:**373–379.

2799. Meng, S. Y., and G. N. Bennett. 1992. Regulation of the *Escherichia coli cad* operon: location of a site required for acid induction. *J. Bacteriol.* **174:**2670–2678.

2800. Meng, S. Y., and G. N. Bennett. 1992. Nucleotide sequence of the *Escherichia coli cad* operon: a system for neutralization of low extracellular pH. *J. Bacteriol.* **174:**2659–2669.

2801. Mengin-Lecreulx, D., C. Parquet, L. R. Desviat, J. Pla, B. Flouret, J. A. Ayala, and J. van Heijenoort. 1989. Organization of the *murE-murG* region of *Escherichia coli*: identification of the *murD* gene encoding the D-glutamic-acid-adding enzyme. *J. Bacteriol.* **171:**6126–6134.

2802. Mengin-Lecreulx, D., L. Texier, M. Rousseau, and J. van Heijenoort. 1991. The *murG* gene of *Escherichia coli* codes for the UDP-N-acetylglucosamine:N-acetylmuramyl-(pentapeptide) pyrophosphoryl-undecaprenol N-acetylglucosamine transferase involved in the membrane steps of peptidoglycan synthesis. *J. Bacteriol.* **173:**4625–4636.

2803. Mengin-Lecreulx, D., L. Texier, and J. van Heijenoort. 1990. Nucleotide sequence of the cell-envelope *murG* gene of *Escherichia coli*. *Nucleic Acids Res.* **18:**2810.

2804. Mengin-Lecreulx, D., and J. van Heijenoort. 1990. Nucleotide sequence of the *murD* gene encoding the UDP-MurNAc-L-Ala-D-Glu synthetase of *Escherichia coli*. *Nucleic Acids Res.* **18:**183.

2805. Mengin-Lecreulx, D., and J. van Heijenoort. 1993. Identification of the *glmU* gene encoding N-acetylglucosamine-1-phosphate uridyltransferase in *Escherichia coli*. *J. Bacteriol.* **175:**6150–6157.

2806. Mengin-Lecreulx, D., and J. van Heijenoort. 1994. Copurification of glucosamine-1-phosphate acetyltransferase and N-acetylglucosamine-1-phosphate uridyltransferase activities of *Escherichia coli*: characterization of the *glmU* gene product as a bifunctional enzyme catalyzing two subsequent steps in the pathway for UDP-N-acetylglucosamine synthesis. *J. Bacteriol.* **176:**5788–5795.

2807. Menon, N. K., C. Y. Chatelus, M. Dervartanian, J. C. Wendt, K. T. Shanmugam, H. D. Peck, Jr., and A. E. Przybyla. 1994. Cloning, sequencing, and mutational analysis of the *hyb* operon encoding *Escherichia coli* hydrogenase 2. *J. Bacteriol.* **176:**4416–4423.

2808. Menon, N. K., J. Robbins, H. D. Peck, Jr., C. Y. Chatelus, E. S. Choi, and A. E. Przybyla. 1990. Cloning and sequencing of a putative *Escherichia coli* [NiFe] hydrogenase-1 operon containing six open reading frames. *J. Bacteriol.* **172:**1969–1977.

2809. Menon, N. K., J. Robbins, J. C. Wendt, K. T. Shanmugam, and A. E. Przybyla. 1991. Mutational analysis and characterization of the *Escherichia coli hya* operon, which encodes [NiFe] hydrogenase 1. *J. Bacteriol.* **173:**4851–4861.

2810. Menzel, R., and M. Gellert. 1983. Regulation of the genes for *E. coli* DNA gyrase: homeostatic control of DNA supercoiling. *Cell* **34:**105–113.

2811. Menzel, R., and M. Gellert. 1987. Fusions of the *Escherichia coli gyrA* and *gyrB* control regions to the galactokinase gene are inducible by coumermycin treatment. *J. Bacteriol.* **169:**1272–1278.

2812. Menzel, R., and M. Gellert. 1987. Modulation of transcription by DNA supercoiling: a deletion analysis of the *Escherichia coli gyrA* and *gyrB* promoters. *Proc. Natl. Acad. Sci. USA* **84:**4185–4189.

2813. Menzel, R., and J. Roth. 1980. Identification and mapping of a second proline permease in *Salmonella typhimurium*. *J. Bacteriol.* **141:**1064–1070.

2814. Merkel, T. J., D. M. Nelson, C. L. Brauer, and R. J. Kadner. 1992. Promoter elements required for positive control of transcription of the *Escherichia coli uhpT* gene. *J. Bacteriol.* **174:**2763–2770.

2815. Merkl, R., M. Kroger, P. Rice, and H. J. Fritz. 1992. Statistical evaluation and biological interpretation of non-random abundance in the *E. coli* K-12 genome of tetra- and pentanucleotide sequences related to VSP DNA mismatch repair. *Nucleic Acids Res.* **20:**1657–1662.

2816. Messer, W., B. Egan, H. Gille, A. Holz, C. Schaefer, and B. Woelker. 1991. The complex of *oriC* DNA with the DnaA initiator protein. *Res. Microbiol.* **142:**119–125.

2817. Messer, W., and W. Vielmetter. 1965. High resolution colony staining for the detection of bacterial growth requirement mutants using naphthol azo-dye techniques. *Biochem. Biophys. Res. Commun.* **21:**182–186.

2818. Metcalf, W. W., P. M. Steed, and B. L. Wanner. 1990. Identification of phosphate starvation-inducible genes in *Escherichia coli* K-12 by DNA sequence analysis of *psi::lacZ*(Mu d1) transcriptional fusions. *J. Bacteriol.* **172:**3191–3200.

2819. Metcalf, W. W., and B. L. Wanner. 1991. Involvement of the *Escherichia coli phn* (*psiD*) gene cluster in assimilation of phosphorus in the form of phosphonates, phosphite, Pi esters, and Pi. *J. Bacteriol.* **173:**587–600.

2820. Metcalf, W. W., and B. L. Wanner. 1993. Mutational analysis of an *Escherichia coli* fourteen-gene operon for phosphonate degradation, using TnphoA' elements. *J. Bacteriol.* **175:**3430–3442.

2821. Metcalf, W. W., and B. L. Wanner. 1993. Evidence for a fourteen-gene, *phnC* to *phnP* locus for phosphonate metabolism in *Escherichia coli*. *Gene* **129:**27–32.

2821a. Metzer, E., and Y. S. Halpern. 1990. In vivo cloning and characterization of the *gabCDTP* gene cluster of *Escherichia coli* K-12. *J. Bacteriol.* **172:**3250–3256.

2822. Metzger, S., I. B. Dror, E. Aizenman, G. Schreiber, M. Toone, J. D. Friesen, M. Cashel, and G. Glaser. 1988. The nucleotide sequence and characterization of the *relA* gene of *Escherichia coli*. *J. Biol. Chem.* **263:**15699–15704.

2823. Metzger, S., E. Sarubbi, G. Glaser, and M. Cashel. 1989. Protein sequences encoded by the relA and the spoT genes of Escherichia coli are interrelated. J. Biol. Chem. 264:9122–9125.

2824. Meyer, C. R., P. Ghosh, S. Nadler, and J. Preiss. 1993. Cloning, expression, and sequence of an allosteric mutant ADPglucose pyrophosphorylase from Escherichia coli B. Arch. Biochem. Biophys. 302:64–71.

2825. Meyer, C. R., P. Ghosh, E. Remy, and J. Preiss. 1992. Cloning, expression, and nucleotide sequence of a mutant glgC gene from Escherichia coli B. J. Bacteriol. 174:4509–4512.

2826. Meyer, E., N. J. Leonard, B. Bhat, J. Stubbe, and J. M. Smith. 1992. Purification and characterization of the purE, purK, and purC gene products: identification of a previously unrecognized energy requirement in the purine biosynthetic pathway. Biochemistry 31:5022–5032.

2827. Meyer, R. R., and P. S. Laine. 1990. The single-stranded DNA-binding protein of Escherichia coli. Microbiol. Rev. 54:342–380.

2828. Meyer, R. R., D. C. Rein, and J. Glassberg. 1982. The product of the lexC gene of Escherichia coli is single-stranded DNA-binding protein. J. Bacteriol. 150:433–435.

2829. Michael, N. P., J. K. Brehm, G. M. Anlezark, and N. P. Minton. 1994. Physical characterisation of the Escherichia coli B gene encoding nitroreductase and its over-expression in Escherichia coli K12. FEMS Microbiol. Lett. 124:195–202.

2830. Michaeli, S., M. Mevarech, and E. Z. Ron. 1984. Regulatory region of the metA gene of Escherichia coli K-12. J. Bacteriol. 160:1158–1162.

2831. Michaeli, S., E. Z. Ron, and G. Cohen. 1981. Construction and physical mapping of plasmids containing the MetA gene of Escherichia coli K-12. Mol. Gen. Genet. 182:349–354.

2832. Michaels, M. L., C. Cruz, and J. H. Miller. 1990. mutA and mutC: two mutator loci in Escherichia coli that stimulate transversions. Proc. Natl. Acad. Sci. USA 87:9211–9215.

2833. Michaels, M. L., C. W. Kim, D. A. Matthews, and J. H. Miller. 1990. Escherichia coli thymidylate synthase: amino acid substitutions by suppression of amber nonsense mutations. Proc. Natl. Acad. Sci. USA 87:3957–3961.

2834. Michaels, M. L., L. Pham, Y. Nghiem, C. Cruz, and J. H. Miller. 1990. MutY, an adenine glycosylase active on G-A mispairs, has homology to endonuclease III. Nucleic Acids Res. 18:3841–3845.

2835. Michaud, C., D. Mengin-Lecreulx, J. van Heijenoort, and D. Blanot. 1990. Overproduction, purification and properties of the uridine-diphosphate-N-acetylmuramoyl-L-alanyl-D-glutamate: meso-2,6-diaminopimelate ligase from Escherichia coli. Eur. J. Biochem. 194:853–861.

2836. Michaud, C., C. Parquet, B. Flouret, D. Blanot, and J. van Heijenoort. 1990. Revised interpretation of the sequence containing the murE gene encoding the UDP-N-acetylmuramyl-tripeptide synthetase of Escherichia coli. Biochem. J. 269:277–278. (Letter.)

2837. Miczak, A., A. K. Chauhan, and D. Apirion. 1991. Two new genes located between 2758 and 2761 kilobase pairs on the Escherichia coli genome. J. Bacteriol. 173:3271–3272.

2838. Miczak, A., R. A. Srivastava, and D. Apirion. 1991. Location of the RNA-processing enzymes RNase III, RNase E and RNase P in the Escherichia coli cell. Mol. Microbiol. 5:1801–1810.

2839. Middendorf, A., H. Schweizer, J. Vreemann, and W. Boos. 1984. Mapping of markers in the gyrA-his region of Escherichia coli. Mol. Gen. Genet. 197:175–181.

2840. Miki, T., Y. Ebina, F. Kishi, and A. Nakazawa. 1981. Organization of the lexA gene of Escherichia coli and nucleotide sequence of the regulatory region. Nucleic Acids Res. 9:529–543.

2841. Miki, T., M. Kimura, S. Hiraga, T. Nagata, and T. Yura. 1979. Cloning and physical mapping of the dnaA region of the Escherichia coli chromosome. J. Bacteriol. 140:817–824.

2842. Miki, T., T. Orita, M. Furuno, and T. Horiuchi. 1988. Control of cell division by sex factor F in Escherichia coli. III. Participation of the groES (mopB) gene of the host bacteria. J. Mol. Biol. 201:327–338.

2843. Miki, T., J. A. Park, K. Nagao, N. Murayama, and T. Horiuchi. 1992. Control of segregation of chromosomal DNA by sex factor F in Escherichia coli. Mutants of DNA gyrase subunit A suppress letD (ccdB) product growth inhibition. J. Mol. Biol. 225:39–52.

2844. Mikulskis, A. V., and G. R. Cornelis. 1994. A new class of proteins regulating gene expression in enterobacteria. Mol. Microbiol. 11:77–86.

2845. Mikuni, O., K. Ito, J. Moffat, K. Matsumura, K. McCaughan, T. Nobukuni, W. Tate, and Y. Nakamura. 1994. Identification of the prfC gene, which encodes peptide-chain-release factor 3 of Escherichia coli. Proc. Natl. Acad. Sci. USA 91:5798–5802.

2846. Mildener, B., T. P. Fondy, R. Engel, and B. E. Tropp. 1981. Effects of halo analogs of glycerol 3-phosphate and dihydroxyacetone phosphate upon Escherichia coli. Antimicrob. Agents Chemother. 19:678–681.

2847. Miles, J. S., and J. R. Guest. 1984. Nucleotide sequence and transcriptional start point of the phosphomannose isomerase gene (manA) of Escherichia coli. Gene 32:41–48.

2848. Miles, J. S., and J. R. Guest. 1984. Complete nucleotide sequence of the fumarase gene fumA, of Escherichia coli. Nucleic Acids Res. 12:3631–3642.

2849. Miles, J. S., and J. R. Guest. 1987. Subgenes expressing single lipoyl domains of the pyruvate dehydrogenase complex of Escherichia coli. Biochem. J. 245:869–874.

2850. Milkman, R. 1994. An Escherichia coli homologue of eukaryotic potassium channel proteins. Proc. Natl. Acad. Sci. USA 91:3510–3514.

2851. Millar, G., and J. R. Coggins. 1986. The complete amino acid sequence of 3-dehydroquinate synthase of Escherichia coli K12. FEBS Lett. 200:11–17.

2852. Millar, G., A. Lewendon, M. G. Hunter, and J. R. Coggins. 1986. The cloning and expression of the aroL gene from Escherichia coli K12. Purification and complete amino acid sequence of shikimate kinase II, the aroL-gene product. Biochem. J. 237:427–437.

2853. Miller, A. D., G. J. Hart, L. C. Packman, and A. R. Battersby. 1988. Evidence that the pyrromethane cofactor of hydroxymethylbilane synthase (porphobilinogen deaminase) is bound to the protein through the sulphur atom of cysteine-242. Biochem. J. 254:915–918.

2854. Miller, A. D., L. C. Packman, G. J. Hart, P. R. Alefounder, C. Abell, and A. R. Battersby. 1989. Evidence that pyridoxal phosphate modification of lysine residues (Lys-55 and Lys-59) causes inactivation of hydroxymethylbilane synthase (porphobilinogen deaminase). Biochem. J. 262:119–124.

2855. Miller, C. G., and G. Schwartz. 1978. Peptidase-deficient mutants of Escherichia coli. J. Bacteriol. 135:603–611.

2856. Miller, H. I. 1984. Primary structure of the himA gene of Escherichia coli: homology with DNA-binding protein HU and association with the phenylalanyl-tRNA synthetase operon. Cold Spring Harbor Symp. Quant. Biol. 49:691–698.

2857. Miller, H. I., and D. I. Friedman. 1980. An E. coli gene product required for lambda site-specific recombination. Cell 20:711–719.

2858. Miller, H. I., M. Kirk, and H. Echols. 1981. SOS induction and autoregulation of the himA gene for site-specific recombination in Escherichia coli. Proc. Natl. Acad. Sci. USA 78:6754–6758.

2859. Miller, H. I., and H. A. Nash. 1981. Direct role of the himA gene product in phage lambda integration. Nature (London) 290:523–526.

2860. Miller, K. W., and H. C. Wu. 1987. Cotranscription of the Escherichia coli isoleucyl-tRNA synthetase (ileS) and prolipoprotein signal peptidase (lsp) genes. Fine-structure mapping of the lsp internal promoter. J. Biol. Chem. 262:389–393.

2861. Miller, W. T., K. A. Hill, and P. Schimmel. 1991. Evidence for a "cysteine-histidine box" metal-binding site in an Escherichia coli aminoacyl-tRNA synthetase. Biochemistry 30:6970–6976.

2862. Miller, W. T., and P. Schimmel. 1992. A metal-binding motif implicated in RNA recognition by an aminoacyl-tRNA synthetase and by a retroviral gene product. Mol. Microbiol. 6:1259–1262.

2863. Miller, W. T., and P. Schimmel. 1992. A retroviral-like metal binding motif in an aminoacyl-tRNA synthetase is important for tRNA recognition. Proc. Natl. Acad. Sci. USA 89:2032–2035.

2863a. Milner, J. L., and J. M. Wood. 1989. Insertion proQ220::Tn5 alters regulation of proline porter II, a transporter of proline and glycine betaine in Escherichia coli. J. Bacteriol. 171:947–951.

2864. Minagawa, J., H. Nakamura, I. Yamato, T. Mogi, and Y. Anraku. 1990. Transcriptional regulation of the cytochrome b562-o complex in Escherichia coli. Gene expression and molecular characterization of the promoter. J. Biol. Chem. 265:11198–11203.

2865. Mineno, J., H. Fukui, Y. Ishino, I. Kato, and H. Shinagawa. 1990. Nucleotide sequence of the araD gene of Escherichia coli K12 encoding the L-ribulose 5-phosphate 4-epimerase. Nucleic Acids Res. 18:6722.

2866. Miner, K. M., and L. Frank. 1974. Sodium-stimulated glutamate transport in osmotically shocked cells and membrane vesicles of Escherichia coli. J. Bacteriol. 117:1093–1098.

2867. Minghetti, K. C., V. C. Goswitz, N. E. Gabriel, J. J. Hill, C. A. Barassi, C. D. Georgiou, S. I. Chan, and R. B. Gennis. 1992. Modified, large-scale purification of the cytochrome o complex (bo-type oxidase) of Escherichia coli yields a two heme/one copper terminal oxidase with high specific activity. Biochemistry 31:6917–6924.

2868. Minoshima, S., and H. Hayashi. 1980. Studies on bacterial chemotaxis. VI. Effect of cheX mutation on the methylation of methyl-accepting chemotaxis protein of Escherichia coli. J. Biochem. 87:1371–1377.

2869. Minton, K. W., H. Tabor, and C. W. Tabor. 1990. Paraquat toxicity is increased in Escherichia coli defective in the synthesis of polyamines. Proc. Natl. Acad. Sci. USA 87:2851–2855.

2870. Miranda-Rios, J., R. Sanchez-Pescador, M. Urdea, and A. A. Covarrubias. 1987. The complete nucleotide sequence of the glnALG operon of Escherichia coli K12. Nucleic Acids Res. 15:2757–2770.

2871. Mironov, A. S., G. D. Nechaeva, and V. V. Sukhodolets. 1989. Interaction of negative (CytT) and positive (cAMP-CRP) regulation in the promoter region of the uridine phosphorylase (udp) gene in Escherichia coli K-12. Genetika 25:438–447.

2872. Misra, R. 1993. A novel ompC mutation of Escherichia coli K-12 that reduces OmpC and OmpF levels in the outer membrane. Mol. Microbiol. 10:1029–1035.

2872a. Misra, R., and S. Benson. 1989. A novel mutation, cog, which results in production of a new porin protein (OmpG) of Escherichia coli. J. Bacteriol. 171:4105–4111.

2873. Misra, R., and P. Reeves. 1985. Molecular characterisation of the Stc⁻ mutation of Escherichia coli K-12. Gene 40:337–342.

2874. Misra, T. K., and D. Apirion. 1980. Gene rne affects the structure of the ribonucleic acid-processing enzyme ribonuclease E of Escherichia coli. J. Bacteriol. 142:359–361.

2875. Missiakas, D., C. Georgopoulos, and S. Raina. 1993. The Escherichia coli heat shock gene htpY: mutational analysis, cloning, sequencing, and transcriptional regulation. J. Bacteriol. 175:2613–2624.

2876. Missiakas, D., C. Georgopoulos, and S. Raina. 1993. Identification and characterization of the *Escherichia coli* gene *dsbB*, whose product is involved in the formation of disulfide bonds in vivo. *Proc. Natl. Acad. Sci. USA* 90:7084–7088.

2877. Missiakas, D., C. Georgopoulos, and S. Raina. 1994. The *Escherichia coli* *dsbC* (*xprA*) gene encodes a periplasmic protein involved in disulfide bond formation. *EMBO J.* 13:2013–2020.

2877a. Mitra, S., B. C. Pal, and R. S. Foote. 1982. O^6-methylguanine-DNA methyltransferase in wild-type and *ada* mutants of *Escherichia coli*. *J. Bacteriol.* 152:534.

2878. Miura, A., J. H. Krueger, S. Itoh, H. A. de Boer, and M. Nomura. 1981. Growth-rate-dependent regulation of ribosome synthesis in *E. coli*: expression of the *lacZ* and *galK* genes fused to ribosomal promoters. *Cell* 25:773–782.

2879. Miura-Masuda, A., and H. Ikeda. 1990. The DNA gyrase of *Escherichia coli* participates in the formation of a spontaneous deletion by *recA*-independent recombination in vivo. *Mol. Gen. Genet.* 220:345–352.

2880. Miyada, C. G., A. H. Horwitz, L. G. Cass, J. Timko, and G. Wilcox. 1980. DNA sequence of the *araC* regulatory gene from *Escherichia coli* B/r. *Nucleic Acids Res.* 8:5267–5274.

2881. Miyada, C. G., X. Soberon, K. Itakura, and G. Wilcox. 1982. The use of synthetic oligodeoxyribonucleotides to produce specific deletions in the *araBAD* promoter of *Escherichia coli* B/r. *Gene* 17:167–177.

2882. Miyajima, A., M. Shibuya, Y. Kuchino, and Y. Kaziro. 1981. Transcription of the *E. coli* *tufB* gene: cotranscription with four tRNA genes and inhibition by guanosine-5'-diphosphate-3'-diphosphate. *Mol. Gen. Genet.* 183:13–19.

2883. Miyamoto, K., K. Nakahigashi, K. Nishimura, and H. Inokuchi. 1991. Isolation and characterization of visible light-sensitive mutants of *Escherichia coli* K12. *J. Mol. Biol.* 219:393–398.

2884. Miyamoto, K., K. Nishimura, T. Masuda, H. Tsuji, and H. Inokuchi. 1992. Accumulation of protoporphyrin IX in light-sensitive mutants of *Escherichia coli*. *FEBS Lett.* 310:246–248.

2885. Miyazaki, T., S. Tanaka, H. Fujita, and H. Itikawa. 1992. DNA sequence analysis of the *dnaK* gene of *Escherichia coli* B and of two *dnaK* genes carrying the temperature-sensitive mutations *dnaK7*(Ts) and *dnaK756*(Ts). *J. Bacteriol.* 174:3715–3722.

2886. Miyoshi, Y., and H. Yamagata. 1976. Sucrose-dependent spectinomycin-resistant mutants of *Escherichia coli*. *J. Bacteriol.* 125:142–148.

2887. Mizote, T., and H. Nakayama. 1989. The *thiM* locus and its relation to phosphorylation of hydroxyethylthiazole in *Escherichia coli*. *J. Bacteriol.* 171:3228–3232.

2888. Mizuno, T., M. Y. Chou, and M. Inouye. 1983. DNA sequence of the promoter region of the *ompC* gene and the amino acid sequence of the signal peptide of pro-OmpC protein of *Escherichia coli*. *FEBS Lett.* 151:159–164.

2889. Mizuno, T., M. Y. Chou, and M. Inouye. 1983. A comparative study on the genes for three porins of the *Escherichia coli* outer membrane. DNA sequence of the osmoregulated *ompC* gene. *J. Biol. Chem.* 258:6932–6940.

2890. Mizuno, T., M. Y. Chou, and M. Inouye. 1984. A unique mechanism regulating gene expression: translational inhibition by a complementary RNA transcript (micRNA). *Proc. Natl. Acad. Sci. USA* 81:1966–1970.

2891. Mizuno, T., and S. Mizushima. 1987. Isolation and characterization of deletion mutants of *ompR* and *envZ*, regulatory genes for expression of the outer membrane proteins OmpC and OmpF in *Escherichia coli*. *J. Biochem.* 101:387–396.

2892. Mizuno, T., and S. Mizushima. 1990. Signal transduction and gene regulation through the phosphorylation of two regulatory components: the molecular basis for the osmotic regulation of the porin genes. *Mol. Microbiol.* 4:1077–1082.

2893. Mizuno, T., E. T. Wurtzel, and M. Inouye. 1982. Osmoregulation of gene expression. II. DNA sequence of the *envZ* gene of the *ompB* operon of *Escherichia coli* and characterization of its gene product. *J. Biol. Chem.* 257:13692–13698.

2894. Mizuno, T., E. T. Wurtzel, and M. Inouye. 1982. Cloning of the regulatory genes (*ompR* and *envZ*) for the matrix proteins of the *Escherichia coli* outer membrane. *J. Bacteriol.* 150:1462–1466.

2895. Mizusawa, S., D. Court, and S. Gottesman. 1983. Transcription of the *sulA* gene and repression by LexA. *J. Mol. Biol.* 171:337–343.

2896. Mizushima, T., S. Natori, and K. Sekimizu. 1992. Inhibition of *Escherichia coli* DNA topoisomerase I activity by phospholipids. *Biochem. J.* 285:503–506.

2897. Mo, J. Y., H. Maki, and M. Sekiguchi. 1991. Mutational specificity of the *dnaE173* mutator associated with a defect in the catalytic subunit of DNA polymerase III of *Escherichia coli*. *J. Mol. Biol.* 222:925–936.

2898. Moffat, K. G., and G. Mackinnon. 1985. Cloning of the *Escherichia coli* K-12 *guaC* gene following its transposition into the RP4::Mu cointegrate. *Gene* 40:141–143.

2899. Mogi, T., H. Yamamoto, T. Nakao, I. Yamato, and Y. Anraku. 1986. Genetic and physical characterization of *putP*, the proline carrier gene of *Escherichia coli* K12. *Mol. Gen. Genet.* 202:35–41. (Erratum, 204:362.)

2899a. Mohan, S., T. M. Kelly, S. S. Eveland, C. R. Raetz, and M. S. Anderson. 1994. An *Escherichia coli* gene (*fabZ*) encoding (3R)-hydroxymyristoyl *acyl* carrier protein dehydrase. Relation to *fabA* and suppression of mutation in lipid A biosynthesis. *J. Biol. Chem.* 269:32896–32903.

2900. Moir, P. D., R. Spiegelberg, I. R. Oliver, J. H. Pringle, and M. Masters. 1992. Proteins encoded by the *Escherichia coli* replication terminus region. *J. Bacteriol.* 174:2102–2110.

2901. Molina, I., M. T. Pellicer, J. Badia, J. Aguilar, and L. Baldoma. 1994. Molecular characterization of *Escherichia coli* malate synthase G. Differentiation with the malate synthase A isoenzyme. *Eur. J. Biochem.* 224:541–548.

2902. Molnar, J., I. B. Holland, and Y. Mandi. 1977. Selection of ion mutants in *Escherichia coli* by treatment with phenothiazines. *Genet. Res.* 30:13–20.

2903. Monticello, R. A., E. Angov, and W. S. Brusilow. 1992. Effects of inducing expression of cloned genes for the F_0 proton channel of the *Escherichia coli* F_1F_0 ATPase. *J. Bacteriol.* 174:3370–3376.

2904. Moolenaar, G. F., C. A. van Sluis, C. Backendorf, and P. van de Putte. 1987. Regulation of the *Escherichia coli* excision repair gene *uvrC*. Overlap between the *uvrC* structural gene and the region coding for a 24 kD protein. *Nucleic Acids Res.* 15:4273–4289.

2905. Moore, R. C., and S. M. Boyle. 1990. Nucleotide sequence and analysis of the *speA* gene encoding biosynthetic arginine decarboxylase in *Escherichia coli*. *J. Bacteriol.* 172:4631–4640.

2906. Moore, S. K., R. T. Garvin, and E. James. 1981. Nucleotide sequence of the *argF* regulatory region of *Escherichia coli* K-12. *Gene* 16:119–132.

2907. Moore, S. K., and E. James. 1979. Mapping of restriction sites in the *argF* gene of *Escherichia coli* by partial endonuclease digestion of end-labeled DNA. *Gene* 5:159–175.

2908. Moralejo, P., S. M. Egan, E. Hidalgo, and J. Aguilar. 1993. Sequencing and characterization of a gene cluster encoding the enzymes for L-rhamnose metabolism in *Escherichia coli*. *J. Bacteriol.* 175:5585–5594.

2909. Moran, M. C., A. J. Mazaitis, R. H. Vogel, and H. J. Vogel. 1979. Clustered *arg* genes on a BamHI segment of the *Escherichia coli* chromosome. *Gene* 8:25–34.

2909a. Morgan, S., A. Korner, K. B. Low, and D. Soll. 1977. Regulation of biosynthesis of aminoacyl-tRNA synthetases and of tRNA in *Escherichia coli*. I. Isolation and characterization of a mutant with elevated levels of tRNA$_1^{Gln}$. *J. Mol. Biol.* 117:1013–1031.

2910. Morimyo, M. 1988. Isolation and characterization of methyl viologen-sensitive mutants of *Escherichia coli* K-12. *J. Bacteriol.* 170:2136–2142.

2911. Morimyo, M., E. Hongo, H. Hama-Inaba, and I. Machida. 1992. Cloning and characterization of the *mvrC* gene of *Escherichia coli* K-12 which confers resistance against methyl viologen toxicity. *Nucleic Acids Res.* 20:3159–3165.

2912. Morinaga, T., G. Funatsu, and M. Funatsu. 1978. Primary structure of protein L14 isolated from *Escherichia coli* ribosomes. *FEBS Lett.* 91:74–77.

2913. Morinaga, T., G. Funatsu, M. Funatsu, and H. G. Wittman. 1976. Primary structure of the 16S rRNA binding protein S15 from *Escherichia coli* ribosomes. *FEBS Lett.* 64:307–309.

2914. Morita, T., M. Amemura, K. Makino, H. Shinagawa, K. Magota, N. Otsuji, and A. Nakata. 1983. Hyperproduction of phosphate-binding protein, PhoS, and pre-PhoS proteins in *Escherichia coli* carrying a cloned *phoS* gene. *Eur. J. Biochem.* 130:427–435.

2915. Moriya, H., H. Kasai, and K. Isono. 1995. Cloning and characterization of the *hrpA* gene in the *terC* region of *Escherichia coli* that is highly similar to the DEAH family RNA helicase genes of *Saccharomyces cerevisiae*. *Nucleic Acids Res.* 23:595–598.

2916. Morona, R., and U. Henning. 1986. New locus (*ttr*) in *Escherichia coli* K-12 affecting sensitivity to bacteriophage T2 and growth on oleate as the sole carbon source. *J. Bacteriol.* 168:534–540.

2917. Morona, R., M. Klose, and U. Henning. 1984. *Escherichia coli* K-12 outer membrane protein (OmpA) as a bacteriophage receptor: analysis of mutant genes expressing altered proteins. *J. Bacteriol.* 159:570–578.

2918. Morona, R., C. Kramer, and U. Henning. 1985. Bacteriophage receptor area of outer membrane protein OmpA of *Escherichia coli* K-12. *J. Bacteriol.* 164:539–543.

2919. Morona, R., and P. Reeves. 1981. Molecular cloning of the *tolC* locus of *Escherichia coli* K-12 with the use of transposon Tn10. *Mol. Gen. Genet.* 184:430–433.

2920. Morona, R., and P. Reeves. 1982. The *tolC* locus of *Escherichia coli* affects the expression of three major outer membrane proteins. *J. Bacteriol.* 150:1016–1023.

2921. Morris, J. F., and E. B. Newman. 1980. Map location of the *ssd* mutation in *Escherichia coli* K-12. *J. Bacteriol.* 143:1504–1505.

2922. Morris, P. W., J. P. Binkley, J. M. Henson, and P. L. Kuempel. 1985. Cloning and location of the *dgsA* gene of *Escherichia coli*. *J. Bacteriol.* 163:785–786.

2923. Morris, T. W., K. E. Reed, and J. E. Cronan, Jr. 1994. Identification of the gene encoding lipoate-protein ligase A of *Escherichia coli*. Molecular cloning and characterization of the *lplA* gene and gene product. *J. Biol. Chem.* 269:16091–16100.

2924. Morris, T. W., K. E. Reed, and J. E. Cronan, Jr. 1995. Lipoic acid metabolism in *Escherichia coli*: the *lplA* and *lipB* genes define redundant pathways for ligation of lipoyl groups to apoprotein. *J. Bacteriol.* 177:1–10.

2925. Morrison, P. T., S. T. Lovett, L. E. Gilson, and R. Kolodner. 1989. Molecular analysis of the *Escherichia coli recO* gene. *J. Bacteriol.* 171:3641–3649.

2926. Mortensen, L., G. Dandanell, and K. Hammer. 1989. Purification and characterization of the *deoR* repressor of *Escherichia coli*. *EMBO J.* 8:325–331.

2927. Motamedi, H., Y. Lee, and F. J. Schmidt. 1984. Tandem promoters preceding the gene for the M1 RNA component of *Escherichia coli* ribonuclease P. *Proc. Natl. Acad. Sci. USA* 81:3959–3963.

2928. Motojima, K., I. Yamato, Y. Anraku, A. Nishimura, and Y. Hirota. 1979. Amplification and characterization of the proline transport carrier of *Escherichia coli* K-12 by using *proT*⁺ hybrid plasmids. *Proc. Natl. Acad. Sci. USA* 76:6255–6259.

2929. Mottl, H., P. Terpstra, and W. Keck. 1991. Penicillin-binding protein 4 of *Escherichia coli* shows a novel type of primary structure among penicillin-interacting proteins. *FEMS Microbiol. Lett.* 62:213–220.

2930. Movva, N. R., K. Nakamura, and M. Inouye. 1980. Regulatory region of the gene for the OmpA protein, a major outer membrane protein of *Escherichia coli*. *Proc. Natl. Acad. Sci. USA* 77:3845–3849.

2931. **Movva, N. R., K. Nakamura, and M. Inouye.** 1980. Gene structure of the OmpA protein, a major surface protein of *Escherichia coli* required for cell-cell interaction. *J. Mol. Biol.* **143:**317–328.

2932. **Movva, R. N., P. Green, K. Nakamura, and M. Inouye.** 1981. Interaction of cAMP receptor protein with the *ompA* gene, a gene for a major outer membrane protein of *Escherichia coli. FEBS Lett.* **128:**186–190.

2933. **Mowbray, S. L., and L. B. Cole.** 1992. 1.7 A X-ray structure of the periplasmic ribose receptor from *Escherichia coli. J. Mol. Biol.* **225:**155–175.

2934. **Moyed, H. S., and K. P. Bertrand.** 1983. *hipA,* a newly recognized gene of *Escherichia coli* K-12 that affects frequency of persistence after inhibition of murein synthesis. *J. Bacteriol.* **155:**768–775.

2935. **Moyed, H. S., and S. H. Broderick.** 1986. Molecular cloning and expression of *hipA,* a gene of *Escherichia coli* K-12 that affects frequency of persistence after inhibition of murein synthesis. *J. Bacteriol.* **166:**399–403.

2936. **Muday, G. K., D. I. Johnson, R. L. Somerville, and K. M. Herrmann.** 1991. The tyrosine repressor negatively regulates *aroH* expression in *Escherichia coli. J. Bacteriol.* **173:**3930–3932.

2937. **Mudd, E. A., H. M. Krisch, and C. F. Higgins.** 1990. RNase E, an endoribonuclease, has a general role in the chemical decay of *Escherichia coli* mRNA: evidence that *rne* and *ams* are the same genetic locus. *Mol. Microbiol.* **4:**2127–2135.

2938. **Mukherjee, J. J., and E. E. Dekker.** 1987. Purification, properties, and N-terminal amino acid sequence of homogeneous *Escherichia coli* 2-amino-3-ketobutyrate CoA ligase, a pyridoxal phosphate-dependent enzyme. *J. Biol. Chem.* **262:**14441–14447.

2939. **Muller, B., I. Burdett, and S. C. West.** 1992. Unusual stability of recombination intermediates made by *Escherichia coli* RecA protein. *EMBO J.* **11:**2685–2693.

2940. **Muller, M. M., A. Vianney, J. C. Lazzaroni, R. E. Webster, and R. Portalier.** 1993. Membrane topology of the *Escherichia coli* TolR protein required for cell envelope integrity. *J. Bacteriol.* **175:**6059–6061.

2941. **Muller, V., C. J. Jones, I. Kawagishi, S. Aizawa, and R. M. Macnab.** 1992. Characterization of the *fliE* genes of *Escherichia coli* and *Salmonella typhimurium* and identification of the FliE protein as a component of the flagellar hook-basal body complex. *J. Bacteriol.* **174:**2298–2304.

2942. **Muller-Hill, B., L. Crapo, and W. Gilbert.** 1968. Mutants that make more *lac* repressor. *Proc. Natl. Acad. Sci. USA* **59:**1259–1264.

2943. **Muller-Hill, B., and P. Kolkhof.** 1994. DNA recognition and the code. *Nature* (London) **369:**614. (Letter.)

2944. **Mulligan, R. C., and P. Berg.** 1981. Factors governing the expression of a bacterial gene in mammalian cells. *Mol. Cell Biol.* **1:**449–459.

2945. **Mullin, D. A., C. L. Woldringh, J. M. Henson, and J. R. Walker.** 1983. Cloning of the *Escherichia coli dnaZX* region and identification of its products. *Mol. Gen. Genet.* **192:**73–79.

2946. **Mulvey, M. R., and P. C. Loewen.** 1989. Nucleotide sequence of *katF* of *Escherichia coli* suggests KatF protein is a novel sigma transcription factor. *Nucleic Acids Res.* **17:**9979–9991.

2947. **Mulvey, M. R., P. A. Sorby, B. L. Triggs-Raine, and P. C. Loewen.** 1988. Cloning and physical characterization of *katE* and *katF* required for catalase HPII expression in *Escherichia coli. Gene* **73:**337–345.

2948. **Munch-Petersen, A., and N. Jensen.** 1990. Analysis of the regulatory region of the *Escherichia coli nupG* gene, encoding a nucleoside-transport protein. *Eur. J. Biochem.* **190:**547–551.

2949. **Munch-Petersen, A., and B. Mygind.** 1976. Nucleoside transport systems in *Escherichia coli* K12: specificity and regulation. *J. Cell Physiol.* **89:**551–559.

2950. **Munch-Petersen, A., B. Mygind, A. Nicolaisen, and N. J. Pihl.** 1979. Nucleoside transport in cells and membrane vesicles from *Escherichia coli* K12. *J. Biol. Chem.* **254:**3730–3737.

2951. **Munch-Petersen, A., P. Nygaard, K. Hammer-Jespersen, and N. Fiil.** 1972. Mutants constitutive for nucleoside-catabolizing enzymes in *Escherichia coli* K12. Isolation, characterization and mapping. *Eur. J. Biochem.* **27:**208–215.

2952. **Munekiyo, R., T. Tsuzuki, and M. Sekiguchi.** 1979. A new locus of *Escherichia coli* that determines sensitivity to bacteriophage ϕX174. *J. Bacteriol.* **138:**1038–1040.

2953. **Munoa, F. J., K. W. Miller, R. Beers, M. Graham, and H. C. Wu.** 1991. Membrane topology of *Escherichia coli* prolipoprotein signal peptidase (signal peptidase II). *J. Biol. Chem.* **266:**17667–17672.

2954. **Munro, A. W., G. Y. Ritchie, A. J. Lamb, R. M. Douglas, and I. R. Booth.** 1991. The cloning and DNA sequence of the gene for the glutathione-regulated potassium-efflux system KefC of *Escherichia coli. Mol. Microbiol.* **5:**607–616.

2955. **Murakami, A., H. Inokuchi, Y. Hirota, H. Ozeki, and H. Yamagishi.** 1980. Characterization of *dnaA* gene carried by lambda transducing phage. *Mol. Gen. Genet.* **180:**235–247.

2956. **Murakami, H., K. Kita, and Y. Anraku.** 1984. Cloning of *cybB,* the gene for cytochrome b561 of *Escherichia coli* K12. *Mol. Gen. Genet.* **198:**1–6.

2957. **Murakami, H., K. Kita, H. Oya, and Y. Anraku.** 1984. Chromosomal location of the *Escherichia coli* cytochrome b556 gene, *cybA. Mol. Gen. Genet.* **196:**1–5.

2958. **Murakami, Y., T. Nagata, W. Schwarz, C. Wada, and T. Yura.** 1985. Novel *dnaG* mutation in a *dnaP* mutant of *Escherichia coli. J. Bacteriol.* **162:**830–832.

2959. **Murakawa, G. J., C. Kwan, J. Yamashita, and D. P. Nierlich.** 1991. Transcription and decay of the *lac* messenger: role of an intergenic terminator. *J. Bacteriol.* **173:**28–36.

2960. **Murakawa, G. J., and D. P. Nierlich.** 1989. Mapping the *lacZ* ribosome binding site by RNA footprinting. Biochemistry **28:**8067–8072.

2961. **Muramatsu, S., and T. Mizuno.** 1989. Nucleotide sequence of the region encompassing the *glpKF* operon and its upstream region containing a bent DNA sequence of *Escherichia coli. Nucleic Acids Res.* **17:**4378.

2962. **Muramatsu, S., and T. Mizuno.** 1989. Nucleotide sequence of the *fabE* gene and flanking regions containing a bent DNA sequence of *Escherichia coli. Nucleic Acids Res.* **17:**3982.

2963. **Muramatsu, S., and T. Mizuno.** 1990. Nucleotide sequence of the region encompassing the *int* gene of a cryptic prophage and the *dnaY* gene flanked by a curved DNA sequence of *Escherichia coli* K12. *Mol. Gen. Genet.* **220:**325–328.

2964. **Muramatsu, T., K. Nishikawa, F. Nemoto, Y. Kuchino, S. Nishimura, T. Miyazawa, and S. Yokoyama.** 1988. Codon and amino-acid specificities of a transfer RNA are both converted by a single post-transcriptional modification. *Nature* (London) **336:**179–181.

2965. **Muranova, T. A., A. V. Muranov, L. F. Markova, and Y. A. Ovchinnikov.** 1978. The primary structure of ribosomal protein L3 from *Escherichia coli* 70 S ribosomes. *FEBS Lett.* **96:**301–305.

2965a. **Murgola, E. J., and E. A. Adelberg.** 1970. Mutants of *Escherichia coli* K-12 with an altered glutamyl-transfer ribonucleic acid. *J. Bacteriol.* **103:**178–183.

2966. **Murgola, E. J., and G. Guarneros.** 1991. Ribosomal RNA and peptidyl-tRNA hydrolase: a peptide chain termination model for lambda bar RNA inhibition. *Biochimie* **73:**1573–1578.

2967. **Murooka, Y., T. Higashiura, and T. Harada.** 1978. Genetic mapping of tyramine oxidase and arylsulfatase genes and their regulation in intergeneric hybrids of enteric bacteria. *J. Bacteriol.* **136:**714–722.

2968. **Murphy, C. K., and J. Beckwith.** 1994. Residues essential for the function of SecE, a membrane component of the *Escherichia coli* secretion apparatus, are located in a conserved cytoplasmic region. *Proc. Natl. Acad. Sci. USA* **91:**2557–2561.

2969. **Murray, N. E., J. A. Gough, B. Suri, and T. A. Bickle.** 1982. Structural homologies among type I restriction-modification systems. *EMBO J.* **1:**535–539.

2970. **Murray, N. E., and W. S. Kelley.** 1979. Characterization of lambda *polA* transducing phages; effective expression of the *E. coli polA* gene. *Mol. Gen. Genet.* **175:**77–87.

2971. **Mutoh, N., K. Inokuchi, and S. Mizushima.** 1982. Amino acid sequence of the signal peptide of OmpF, a major outer membrane protein of *Escherichia coli. FEBS Lett.* **137:**171–174.

2972. **Mutoh, N., T. Nagasawa, and S. Mizushima.** 1981. Specialized transducing bacteriophage lambda carrying the structural gene for a major outer membrane matrix protein of *Escherichia coli* K-12. *J. Bacteriol.* **145:**1085–1090.

2973. **Mutoh, N., and M. I. Simon.** 1986. Nucleotide sequence corresponding to five chemotaxis genes in *Escherichia coli. J. Bacteriol.* **165:**161–166.

2974. **Nadeau, K., A. Das, and C. T. Walsh.** 1993. Hsp90 chaperonins possess ATPase activity and bind heat shock transcription factors and peptidyl prolyl isomerases. *J. Biol. Chem.* **268:**1479–1487.

2975. **Nagai, H., H. Yuzawa, and T. Yura.** 1991. Interplay of two cis-acting mRNA regions in translational control of sigma 32 synthesis during the heat shock response of *Escherichia coli. Proc. Natl. Acad. Sci. USA* **88:**10515–10519.

2976. **Nagano, Y., R. Matsuno, and Y. Sasaki.** 1991. An essential gene of *Escherichia coli* that has sequence similarity to a chloroplast gene of unknown function. *Mol. Gen. Genet.* **228:**62–64.

2977. **Nagaraja, R., and R. A. Weisberg.** 1990. Specificity determinants in the attachment sites of bacteriophages HK022 and lambda. *J. Bacteriol.* **172:**6540–6550.

2978. **Nagasawa, H., Y. Sakagami, A. Suzuki, H. Suzuki, H. Hara, and Y. Hirota.** 1989. Determination of the cleavage site involved in C-terminal processing of penicillin-binding protein 3 of *Escherichia coli. J. Bacteriol.* **171:**5890–5893.

2979. **Nagasawa, S., K. Ishige, and T. Mizuno.** 1993. Novel members of the two-component signal transduction genes in *Escherichia coli. J. Biochem.* **114:**350–357.

2980. **Nagasawa, S., S. Tokishita, H. Aiba, and T. Mizuno.** 1992. A novel sensor-regulator protein that belongs to the homologous family of signal-transduction proteins involved in adaptive responses in *Escherichia coli. Mol. Microbiol.* **6:**799–807.

2981. **Nagase, T., S. Ishii, and F. Imamoto.** 1988. Differential transcriptional control of the two tRNA^{fMet} genes of *Escherichia coli* K-12. *Gene* **67:**49–57.

2982. **Nagel, R., A. Chan, and E. Rosen.** 1994. *Ruv* and *recG* genes and the induced precise excision of Tn*10* in *Escherichia coli. Mutat. Res.* **311:**103–109.

2983. **Naggert, J., M. L. Narasimhan, L. DeVeaux, H. Cho, Z. I. Randhawa, J. E. Cronan, Jr., B. N. Green, and S. Smith.** 1991. Cloning, sequencing, and characterization of *Escherichia coli* thioesterase II. *J. Biol. Chem.* **266:**11044–11050.

2984. **Nagy, P. L., A. Marolewski, S. J. Benkovic, and H. Zalkin.** 1995. Formyltetrahydrofolate hydrolase, a regulatory enzyme that functions to balance pools of tetrahydrofolate and one-carbon tetrahydrofolate adducts in *Escherichia coli. J. Bacteriol.* **177:**1292–1298.

2985. **Nagy, P. L., G. M. McCorkle, and H. Zalkin.** 1993. *purU,* a source of formate for *purT*-dependent phosphoribosyl-*N*-formylglycinamide synthesis. *J. Bacteriol.* **175:**7066–7073.

2986. **Nahlik, M. S., T. J. Brickman, B. A. Ozenberger, and M. A. McIntosh.** 1989. Nucleotide sequence and transcriptional organization of the *Escherichia coli* enterobactin biosynthesis cistrons *entB* and *entA. J. Bacteriol.* **171:**784–790.

2987. **Nahlik, M. S., T. P. Fleming, and M. A. McIntosh.** 1987. Cluster of genes controlling synthesis and activation of 2,3-dihydroxybenzoic acid in production of enterobactin in *Escherichia coli. J. Bacteriol.* **169:**4163–4170.

2988. **Naik, S. M., and H. M. Hassan.** 1990. Use of site-directed mutagenesis to identify an upstream regulatory sequence of *sodA* gene of *Escherichia coli* K-12. *Proc. Natl. Acad. Sci. USA* **87:**2618–2622.

2989. **Naito, S., T. Kitani, T. Ogawa, T. Okazaki, and H. Uchida.** 1984. *Escherichia coli* mutants suppressing replication-defective mutations of the ColE1 plasmid. *Proc. Natl. Acad. Sci. USA* **81**:550–554.

2990. **Nakabeppu, Y., H. Kondo, S. Kawabata, S. Iwanaga, and M. Sekiguchi.** 1985. Purification and structure of the intact Ada regulatory protein of *Escherichia coli* K12, O^6-methylguanine-DNA methyltransferase. *J. Biol. Chem.* **260**:7281–7288.

2991. **Nakabeppu, Y., H. Kondo, and M. Sekiguchi.** 1984. Cloning and characterization of the *alkA* gene of *Escherichia coli* that encodes 3-methyladenine DNA glycosylase II. *J. Biol. Chem.* **259**:13723–13729.

2992. **Nakabeppu, Y., T. Miyata, H. Kondo, S. Iwanaga, and M. Sekiguchi.** 1984. Structure and expression of the *alkA* gene of *Escherichia coli* involved in adaptive response to alkylating agents. *J. Biol. Chem.* **259**:13730–13736.

2993. **Nakabeppu, Y., and M. Sekiguchi.** 1986. Regulatory mechanisms for induction of synthesis of repair enzymes in response to alkylating agents: *ada* protein acts as a transcriptional regulator. *Proc. Natl. Acad. Sci. USA* **83**:6297–6301.

2994. **Nakagawa, J., and M. Matsuhashi.** 1982. Molecular divergence of a major peptidoglycan synthetase with transglycosylase-transpeptidase activities in *Escherichia coli*—penicillin-binding protein 1Bs. *Biochem. Biophys. Res. Commun.* **105**:1546–1553.

2995. **Nakahigashi, K., and H. Inokuchi.** 1990. Nucleotide sequence of the *fadA* and *fadB* genes from *Escherichia coli*. *Nucleic Acids Res.* **18**:4937.

2996. **Nakahigashi, K., and H. Inokuchi.** 1990. Nucleotide sequence between the *fadB* gene and the *rrnA* operon from *Escherichia coli*. *Nucleic Acids Res.* **18**:6439.

2997. **Nakahigashi, K., K. Miyamoto, K. Nishimura, and H. Inokuchi.** 1992. Isolation and characterization of a light-sensitive mutant of *Escherichia coli* K-12 with a mutation in a gene that is required for the biosynthesis of ubiquinone. *J. Bacteriol.* **174**:7352–7359.

2998. **Nakahigashi, K., K. Nishimura, K. Miyamoto, and H. Inokuchi.** 1991. Photosensitivity of a protoporphyrin-accumulating, light-sensitive mutant (*visA*) of *Escherichia coli* K-12. *Proc. Natl. Acad. Sci. USA* **88**:10520–10524.

2999. **Nakajima, N., H. Ozeki, and Y. Shimura.** 1981. Organization and structure of an *E. coli* tRNA operon containing seven tRNA genes. *Cell* **23**:239–249.

3000. **Nakajima, N., H. Ozeki, and Y. Shimura.** 1982. In vitro transcription of the *supB-E* tRNA operon of *Escherichia coli*. Characterization of transcription products. *J. Biol. Chem.* **257**:11113–11120.

3001. **Nakamura, H.** 1979. Novel acriflavin resistance genes, *acrC* and *acrD*, in *Escherichia coli* K-12. *J. Bacteriol.* **139**:8–12.

3002. **Nakamura, H., H. Murakami, I. Yamato, and Y. Anraku.** 1988. Nucleotide sequence of the *cybB* gene encoding cytochrome b561 in *Escherichia coli* K12. *Mol. Gen. Genet.* **212**:1–5.

3003. **Nakamura, H., I. Yamato, Y. Anraku, L. Lemieux, and R. B. Gennis.** 1990. Expression of *cyoA* and *cyoB* demonstrates that the CO-binding heme component of the *Escherichia coli* cytochrome o complex is in subunit I. *J. Biol. Chem.* **265**:11193–11197.

3004. **Nakamura, K., and M. Inouye.** 1979. DNA sequence of the gene for the outer membrane lipoprotein of *E. coli*: an extremely AT-rich promoter. *Cell* **18**:1109–1117.

3005. **Nakamura, K., R. M. Pirtle, I. L. Pirtle, K. Takeishi, and M. Inouye.** 1980. Messenger ribonucleic acid of the lipoprotein of the *Escherichia coli* outer membrane. II. The complete nucleotide sequence. *J. Biol. Chem.* **255**:210–216.

3006. **Nakamura, M., I. N. Maruyama, M. Soma, J. Kato, H. Suzuki, and Y. Horota.** 1983. On the process of cellular division in *Escherichia coli*: nucleotide sequence of the gene for penicillin-binding protein 3. *Mol. Gen. Genet.* **191**:1–9.

3007. **Nakamura, M., M. Yamada, Y. Hirota, K. Sugimoto, A. Oka, and M. Takanami.** 1981. Nucleotide sequence of the *asnA* gene coding for asparagine synthetase of *E. coli* K-12. *Nucleic Acids Res.* **9**:4669–4676.

3008. **Nakamura, S., M. Nakamura, T. Kojima, and H. Yoshida.** 1989. *gyrA* and *gyrB* mutations in quinolone-resistant strains of *Escherichia coli*. *Antimicrob. Agents Chemother.* **33**:254–255.

3009. **Nakamura, Y.** 1980. Hybrid plasmid carrying *Escherichia coli* genes for the primase (*dnaG*) and RNA polymerase sigma factor (*rpoD*); gene organization and control of their expression. *Mol. Gen. Genet.* **178**:487–497.

3010. **Nakamura, Y., T. Kurihara, H. Saito, and H. Uchida.** 1979. Sigma subunit of *Escherichia coli* RNA polymerase affects the function of lambda *N* gene. *Proc. Natl. Acad. Sci. USA* **76**:4593.

3011. **Nakamura, Y., and S. Mizusawa.** 1985. In vivo evidence that the *nusA* and *infB* genes of *E. coli* are part of the same multi-gene operon which encodes at least four proteins. *EMBO J.* **4**:527–532.

3012. **Nakao, T., I. Yamato, and Y. Anraku.** 1987. Nucleotide sequence of *putP*, the proline carrier gene of *Escherichia coli* K12. *Mol. Gen. Genet.* **208**:70–75.

3013. **Nakao, T., I. Yamato, and Y. Anraku.** 1987. Nucleotide sequence of *putC*, the regulatory region for the *put* regulon of *Escherichia coli* K12. *Mol. Gen. Genet.* **210**:364–368.

3014. **Nakashima, K., A. Sugiura, K. Kanamaru, and T. Mizuno.** 1993. Signal transduction between the two regulatory components involved in the regulation of the *kdpABC* operon in *Escherichia coli*: phosphorylation-dependent functioning of the positive regulator, KdpE. *Mol. Microbiol.* **7**:109–116.

3015. **Nakashima, K., A. Sugiura, and T. Mizuno.** 1993. Functional reconstitution of the putative *Escherichia coli* osmosensor, KdpD, into liposomes. *J. Biochem.* **114**:615–621.

3016. **Nakashima, K., A. Sugiura, H. Momoi, and T. Mizuno.** 1992. Phosphotransfer signal transduction between two regulatory factors involved in the osmoregulated *kdp* operon in *Escherichia coli*. *Mol. Microbiol.* **6**:1777–1784.

3017. **Nakata, A., M. Amemura, and H. Shinagawa.** 1984. Regulation of the phosphate regulon in *Escherichia coli* K-12: regulation of the negative regulatory gene *phoU* and identification of the gene product. *J. Bacteriol.* **159**:979–985.

3018. **Nakata, A., H. Shinagawa, and M. Amemura.** 1982. Cloning of alkaline phosphatase isozyme gene (*iap*) of *Escherichia coli*. *Gene* **19**:313–319.

3019. **Nakayama, H.** 1995. Genetic analysis of thiamin pyrophosphate biosynthesis in *Escherichia coli*. *Vitamins* **64**:619–632.

3020. **Nakayama, H., K. Nakayama, R. Nakayama, N. Irino, Y. Nakayama, and P. C. Hanawalt.** 1984. Isolation and genetic characterization of a thymineless death-resistant mutant of *Escherichia coli* K12: identification of a new mutation (*recQ1*) that blocks the RecF recombination pathway. *Mol. Gen. Genet.* **195**:474–480.

3021. **Nakayama, K., N. Irino, and H. Nakayama.** 1985. The *recQ* gene of *Escherichia coli* K12: molecular cloning and isolation of insertion mutants. *Mol. Gen. Genet.* **200**:266–271.

3022. **Nakayama, N., N. Arai, M. W. Bond, Y. Kaziro, and K. Arai.** 1984. Nucleotide sequence of *dnaB* and the primary structure of the *dnaB* protein from *Escherichia coli*. *J. Biol. Chem.* **259**:97–101.

3023. **Nakayama, N., M. W. Bond, A. Miyajima, J. Kobori, and K. Arai.** 1987. Structure of *Escherichia coli dnaC*. Identification of a cysteine residue possibly involved in association with *dnaB* protein. *J. Biol. Chem.* **262**:10475–10480.

3024. **Nakayashiki, T., K. Nishimura, and H. Inokuchi.** 1995. Cloning and sequencing of a previously unidentified gene that is involved in the biosynthesis of heme in *Escherichia coli*. *Gene* **153**:67–70.

3025. **Nanninga, N.** 1991. Cell division and peptidoglycan assembly in *Escherichia coli*. *Mol. Microbiol.* **5**:791–795.

3026. **Naom, I. S., S. J. Morton, D. R. Leach, and R. G. Lloyd.** 1989. Molecular organization of *sbcC*, a gene that affects genetic recombination and the viability of DNA palindromes in *Escherichia coli* K-12. *Nucleic Acids Res.* **17**:8033–8045.

3027. **Nara, F., S. Matsuyama, T. Mizuno, and S. Mizushima.** 1986. Molecular analysis of mutant *ompR* genes exhibiting different phenotypes as to osmoregulation of the *ompF* and *ompC* genes of *Escherichia coli*. *Mol. Gen. Genet.* **202**:194–199.

3028. **Narasimhan, M. L., J. L. Lampi, and J. E. Cronan, Jr.** 1986. Genetic and biochemical characterization of an *Escherichia coli* K-12 mutant deficient in acyl-coenzyme A thioesterase II. *J. Bacteriol.* **165**:911–917.

3029. **Nargang, F. E., C. S. Subrahmanyam, and H. E. Umbarger.** 1980. Nucleotide sequence of *ilvGEDA* operon attenuator region of *Escherichia coli*. *Proc. Natl. Acad. Sci. USA* **77**:1823–1827.

3030. **Naroditskaya, V., M. J. Schlosser, N. Y. Fang, and K. Lewis.** 1993. An *E. coli* gene *emrD* is involved in adaptation to low energy shock. *Biochem. Biophys. Res. Commun.* **196**:803–809.

3031. **Nash, H. A., and A. E. Granston.** 1991. Similarity between the DNA-binding domains of IHF protein and TFIID protein. *Cell* **67**:1037–1038. (Letter.)

3032. **Nashimoto, H., and H. Uchida.** 1985. DNA sequencing of the *Escherichia coli* ribonuclease III gene and its mutations. *Mol. Gen. Genet.* **201**:25–29.

3033. **Nasoff, M. S., H. V. Baker, and R. E. Wolf, Jr.** 1984. DNA sequence of the *Escherichia coli* gene, *gnd*, for 6-phosphogluconate dehydrogenase. *Gene* **27**:253–264.

3034. **Nasoff, M. S., and R. E. Wolf, Jr.** 1980. Molecular cloning, correlation of genetic and restriction maps, and determination of the direction of transcription of *gnd* of *Escherichia coli*. *J. Bacteriol.* **143**:731–741.

3035. **Nass, G., and J. Thomale.** 1974. Alteration of structure of level of threonyl-tRNA-synthetase in Borrelidin resistant mutants of *E. coli*. *FEBS Lett.* **39**:182–186.

3036. **Natarajan, S., W. L. Kelley, and D. Bastia.** 1991. Replication terminator protein of *Escherichia coli* is a transcriptional repressor of its own synthesis. *Proc. Natl. Acad. Sci. USA* **88**:3867–3871.

3037. **Nau, C. D., and J. Konisky.** 1989. Evolutionary relationship between the TonB-dependent outer membrane transport proteins: nucleotide and amino acid sequences of the *Escherichia coli* colicin I receptor gene. *J. Bacteriol.* **171**:1041–1047. (Erratum, **171**:4530.)

3038. **Navarro, C., L. F. Wu, and M. A. Mandrand-Berthelot.** 1993. The *nik* operon of *Escherichia coli* encodes a periplasmic binding-protein-dependent transport system for nickel. *Mol. Microbiol.* **9**:1181–1191.

3039. **Navre, M., and H. K. Schachman.** 1983. Synthesis of aspartate transcarbamoylase in *Escherichia coli*: transcriptional regulation of the *pyrB-pyrI* operon. *Proc. Natl. Acad. Sci. USA* **80**:1207–1211.

3040. **Nazos, P. M., T. K. Antonucci, R. Landick, and D. L. Oxender.** 1986. Cloning and characterization of *livH*, the structural gene encoding a component of the leucine transport system in *Escherichia coli*. *J. Bacteriol.* **166**:565–573.

3041. **Nazos, P. M., M. M. Mayo, T. Z. Su, J. J. Anderson, and D. L. Oxender.** 1985. Identification of *livG*, a membrane-associated component of the branched-chain amino acid transport in *Escherichia coli*. *J. Bacteriol.* **163**:1196–1202.

3042. **Neely, M. N., C. L. Dell, and E. R. Olson.** 1994. Roles of LysP and CadC in mediating the lysine requirement for acid induction of the *Escherichia coli cad* operon. *J. Bacteriol.* **176**:3278–3285.

3043. **Negre, D., J. C. Cortay, I. G. Old, A. Galinier, C. Richaud, I. Saint Girons, and A. J. Cozzone.** 1991. Overproduction and characterization of the *iclR* gene product of *Escherichia coli* K-12 and comparison with that of *Salmonella typhimurium* LT2. *Gene* **97**:29–37.

3044. **Nehrke, K. W., F. Zalatan, and T. Platt.** 1993. NusG alters rho-dependent termination of transcription in vitro independent of kinetic coupling. *Gene Expression* **3**:119–133.

3045. Neidhardt, F. C., T. A. Phillips, R. A. VanBogelen, M. W. Smith, Y. Georgalis, and A. R. Subramanian. 1981. Identity of the B56.5 protein, the A-protein, and the groE gene product of Escherichia coli. J. Bacteriol. 145:513–520.

3046. Neidhardt, F. C., R. A. VanBogelen, and E. T. Lau. 1983. Molecular cloning and expression of a gene that controls the high-temperature regulon of Escherichia coli. J. Bacteriol. 153:597–603.

3047. Neidhardt, F. C., V. Vaughn, T. A. Phillips, and P. L. Bloch. 1983. Gene-protein index of Escherichia coli K-12. Microbiol. Rev. 47:231–284.

3048. Nellemann, L. J., F. Holm, T. Atlung, and F. G. Hansen. 1989. Cloning and characterization of the Escherichia coli phosphoglycerate kinase (pgk) gene. Gene 77:185–191.

3049. Nelson, K., T. S. Whittam, and R. K. Selander. 1991. Nucleotide polymorphism and evolution in the glyceraldehyde-3-phosphate dehydrogenase gene (gapA) in natural populations of Salmonella and Escherichia coli. Proc. Natl. Acad. Sci. USA 88:6667–6671.

3050. Nelson, S. O., J. Lengeler, and P. W. Postma. 1984. Role of IIIGlc of the phosphoenolpyruvate-glucose phosphotransferase system in inducer exclusion in Escherichia coli. J. Bacteriol. 160:360–364.

3051. Nersisian, A. A., I. A. Fedorova, and E. M. Khurges. 1986. Cloning of genes for proline biosynthesis in Escherichia coli. Genetika 22:2713–2720.

3052. Nersisian, A. A., I. L. Mett, Z. R. Badalian, A. V. Kochikian, and A. L. Mett. 1986. Restriction mapping of recombinant plasmids carrying the genes for arginine biosynthesis in Escherichia coli K-12. Genetika 22:1061–1066.

3053. Nesin, M., J. R. Lupski, P. Svec, and G. N. Godson. 1987. Possible new genes as revealed by molecular analysis of a 5-kb Escherichia coli chromosomal region 5′ to the rpsU-dnaG-rpoD macromolecular-synthesis operon. Gene 51:149–161.

3054. Nettleton, C. J., C. Bull, T. O. Baldwin, and J. A. Fee. 1984. Isolation of the Escherichia coli iron superoxide dismutase gene: evidence that intracellular superoxide concentration does not regulate oxygen-dependent synthesis of the manganese superoxide dismutase. Proc. Natl. Acad. Sci. USA 81:4970–4973.

3055. Neuhard, J., and L. Tarpo. 1993. Location of the udk gene on the physical map of Escherichia coli. J. Bacteriol. 175:5742–5743.

3056. Neuhard, J., and E. Thomassen. 1976. Altered deoxyribonucleotide pools in P2 eductants of Escherichia coli K-12 due to deletion of the dcd gene. J. Bacteriol. 126:999–1001.

3057. Neuwald, A. F., D. E. Berg, and G. V. Stauffer. 1992. Mutational analysis of the Escherichia coli serB promoter region reveals transcriptional linkage to a downstream gene. Gene 120:1–9.

3058. Neuwald, A. F., B. R. Krishnan, I. Brikun, S. Kulakauskas, K. Suziedelis, T. Tomcsanyi, T. S. Leyh, and D. E. Berg. 1992. cysQ, a gene needed for cysteine synthesis in Escherichia coli K-12 only during aerobic growth. J. Bacteriol. 174:415–425.

3059. Neuwald, A. F., and G. V. Stauffer. 1985. DNA sequence and characterization of the Escherichia coli serB gene. Nucleic Acids Res. 13:7025–7039.

3060. Neuwald, A. F., and G. V. Stauffer. 1989. An Escherichia coli membrane protein with a unique signal sequence. Gene 82:219–228.

3061. Neuwald, A. F., and G. V. Stauffer. 1990. IS30 activation of an smp′-lacZ gene fusion in Escherichia coli. FEMS Microbiol. Lett. 56:13–17.

3062. Newcomer, M. E., G. L. Gilliland, and F. A. Quiocho. 1981. L-Arabinose-binding protein-sugar complex at 2.4 A resolution. Stereochemistry and evidence for a structural change. J. Biol. Chem. 256:13213–13217.

3063. Newlands, J. T., C. A. Josaitis, W. Ross, and R. L. Gourse. 1992. Both fis-dependent and factor-independent upstream activation of the rrnB P1 promoter are face of the helix dependent. Nucleic Acids Res. 20:719–726.

3064. Newlands, J. T., W. Ross, K. K. Gosink, and R. L. Gourse. 1991. Factor-independent activation of Escherichia coli rRNA transcription. II. Characterization of complexes of rrnB P1 promoters containing or lacking the upstream activator region with Escherichia coli RNA polymerase. J. Mol. Biol. 220:569–583.

3065. Newman, A., and R. S. Hayward. 1980. Cloning of DNA of the rpoBC operon from the chromosome of Escherichia coli K12. Mol. Gen. Genet. 177:527–533.

3066. Newman, E. B., R. D'Ari, and R. T. Lin. 1992. The leucine-Lrp regulon in E. coli: a global response in search of a raison d'etre. Cell 68:617–619.

3067. Newman, E. B., N. Malik, and C. Walker. 1982. L-Serine degradation in Escherichia coli K-12: directly isolated ssd mutants and their intragenic revertants. J. Bacteriol. 150:710–715.

3068. Newman, T., P. Friden, A. Sutton, and M. Freundlich. 1982. Cloning and expression of the ilvB gene of Escherichia coli K-12. Mol. Gen. Genet. 186:378–384.

3069. Newman, T. C., and M. Levinthal. 1980. A new map location for the ilvB locus of Escherichia coli. Genetics 96:59–77.

3070. Newton, G., C. H. Yun, and R. B. Gennis. 1991. Analysis of the topology of the cytochrome d terminal oxidase complex of Escherichia coli by alkaline phosphatase fusions. Mol. Microbiol. 5:2511–2518.

3071. Nghiem, Y., M. Cabrera, C. G. Cupples, and J. H. Miller. 1988. The mutY gene: a mutator locus in Escherichia coli that generates G·C→T·A transversions. Proc. Natl. Acad. Sci. USA 85:2709–2713.

3072. Nichols, B. P., and J. M. Green. 1992. Cloning and sequencing of Escherichia coli ubiC and purification of chorismate lyase. J. Bacteriol. 174:5309–5316.

3073. Nichols, B. P., and G. G. Guay. 1989. Gene amplification contributes to sulfonamide resistance in Escherichia coli. Antimicrob. Agents Chemother. 33:2042–2048.

3074. Nichols, B. P., G. F. Miozzari, M. van Cleemput, G. N. Bennett, and C. Yanofsky. 1980. Nucleotide sequences of the trpG regions of Escherichia coli, Shigella dysenteriae, Salmonella typhimurium and Serratia marcescens. J. Mol. Biol. 142:503–517.

3075. Nichols, B. P., M. van Cleemput, and C. Yanofsky. 1981. Nucleotide sequence of Escherichia coli trpE. Anthranilate synthetase component I contains no tryptophan residues. J. Mol. Biol. 146:45–54.

3076. Nichols, B. P., and C. Yanofsky. 1979. Nucleotide sequences of trpA of Salmonella typhimurium and Escherichia coli: an evolutionary comparison. Proc. Natl. Acad. Sci. USA 76:5244–5248.

3077. Niegemann, E., A. Schulz, and K. Bartsch. 1993. Molecular organization of the Escherichia coli gab cluster: nucleotide sequence of the structural genes gabD and gabP and expression of the GABA permease gene. Arch. Microbiol. 160:454–460.

3078. Nielsen, J., F. G. Hansen, J. Hoppe, P. Friedl, and K. von Meyenburg. 1981. The nucleotide sequence of the atp genes coding for the F0 subunits a, b, c and the F1 subunit delta of the membrane bound ATP synthase of Escherichia coli. Mol. Gen. Genet. 184:33–39.

3079. Nielsen, J., B. B. Jorgensen, K. V. van Meyenburg, and F. G. Hansen. 1984. The promoters of the atp operon of Escherichia coli K12. Mol. Gen. Genet. 193:64–71.

3080. Niersbach, M., F. Kreuzaler, R. H. Geerse, P. W. Postma, and H. J. Hirsch. 1992. Cloning and nucleotide sequence of the Escherichia coli K-12 ppsA gene, encoding PEP synthase. Mol. Gen. Genet. 231:332–336.

3081. Nieto, J. M., M. Carmona, S. Bolland, Y. Jubete, F. de la Cruz, and A. Juarez. 1991. The hha gene modulates haemolysin expression in Escherichia coli. Mol. Microbiol. 5:1285–1293.

3082. Nikaido, H. 1992. Porins and specific channels of bacterial outer membranes. Mol. Microbiol. 6:435–442.

3083. Niki, H., C. Ichinose, T. Ogura, H. Mori, M. Morita, M. Hasegawa, N. Kusukawa, and S. Hiraga. 1988. Chromosomal genes essential for stable maintenance of the mini-F plasmid in Escherichia coli. J. Bacteriol. 170:5272–5278.

3084. Niki, H., R. Imamura, M. Kitaoka, K. Yamanaka, T. Ogura, and S. Hiraga. 1992. E. coli MukB protein involved in chromosome partition forms a homodimer with a rod-and-hinge structure having DNA binding and ATP/GTP binding activities. EMBO J. 11:5101–5109.

3085. Niki, H., R. Imamura, T. Ogura, and S. Hiraga. 1990. Nucleotide sequence of the tolC gene of Escherichia coli. Nucleic Acids Res. 18:5547.

3086. Niki, H., A. Jaffe, R. Imamura, T. Ogura, and S. Hiraga. 1991. The new gene mukB codes for a 177 kd protein with coiled-coil domains involved in chromosome partitioning of E. coli. EMBO J. 10:183–193.

3087. Niki, H., T. Ogura, and S. Hiraga. 1990. Linear multimer formation of plasmid DNA in Escherichia coli hopE (recD) mutants. Mol. Gen. Genet. 224:1–9.

3088. Nikkila, H., R. B. Gennis, and S. G. Sligar. 1991. Cloning and expression of the gene encoding the soluble cytochrome b562 of Escherichia coli. Eur. J. Biochem. 202:309–313.

3089. Nilsson, L., A. Vanet, E. Vijgenboom, and L. Bosch. 1990. The role of FIS in trans activation of stable RNA operons of E. coli. EMBO J. 9:727–734.

3090. Nilsson, O., A. Aberg, T. Lundqvist, and B. M. Sjoberg. 1988. Nucleotide sequence of the gene coding for the large subunit of ribonucleotide reductase of Escherichia coli. Correction. Nucleic Acids Res. 16:4174.

3091. Ninfa, A. J., and R. L. Bennett. 1991. Identification of the site of autophosphorylation of the bacterial protein kinase/phosphatase NRII. J. Biol. Chem. 266:6888–6893.

3092. Ninnemann, O., C. Koch, and R. Kahmann. 1992. The E. coli fis promoter is subject to stringent control and autoregulation. EMBO J. 11:1075–1083.

3093. Nishi, K., F. Morel-Deville, J. W. Hershey, T. Leighton, and J. Schnier. 1988. An eIF-4A-like protein is a suppressor of an Escherichia coli mutant defective in 50S ribosomal subunit assembly. Nature (London) 336:496–498. (Erratum, 340:246, 1989.)

3094. Nishida, M., K. H. Kong, H. Inoue, and K. Takahashi. 1994. Molecular cloning and site-directed mutagenesis of glutathione S-transferase from Escherichia coli. The conserved tyrosyl residue near the N terminus is not essential for catalysis. J. Biol. Chem. 269:32536–32541.

3095. Nishijima, S., Y. Asami, N. Uetake, S. Yamagoe, A. Ohta, and I. Shibuya. 1988. Disruption of the Escherichia coli cls gene responsible for cardiolipin synthesis. J. Bacteriol. 170:775–780.

3096. Nishimura, A. 1989. A new gene controlling the frequency of cell division per round of DNA replication in Escherichia coli. Mol. Gen. Genet. 215:286–293.

3097. Nishimura, A., K. Akiyama, Y. Kohara, and K. Horiuchi. 1992. Correlation of a subset of the pLC plasmids to the physical map of Escherichia coli K-12. Microbiol. Rev. 56:137–151.

3098. Nishimura, K., K. Nakahigashi, and H. Inokuchi. 1992. Location of the ubiA gene on the physical map of Escherichia coli. J. Bacteriol. 174:5762.

3099. Nishimura, K., T. Nakayashiki, and H. Inokuchi. 1993. Cloning and sequencing of the hemE gene encoding uroporphyrinogen III decarboxylase (UPD) from Escherichia coli K-12. Gene 133:109–113.

3100. Nishimura, K., S. Taketani, and H. Inokuchi. 1995. Cloning of a human cDNA for protoporphyrinogen oxidase by complementation in vivo of a hemG mutant of Escherichia coli. J. Biol. Chem. 270:8076–8080.

3101. Nishimura, Y., H. Suzuki, Y. Hirota, and J. T. Park. 1980. A mutant of Escherichia coli defective in penicillin-binding protein 5 and lacking D-alanine carboxypeptidase IA. J. Bacteriol. 143:531–534.

3102. Nishitani, J., and G. Wilcox. 1991. Cloning and characterization of the L-rhamnose regulon in Salmonella typhimurium LT2. Gene 105:37–42.

3103. Nishiyama, K., M. Hanada, and H. Tokuda. 1994. Disruption of the gene encoding p12 (SecG) reveals the direct involvement and important function of SecG in the protein translocation of Escherichia coli at low temperature. EMBO J. 13:3272–3277.

3104. Nishiyama, K., S. Mizushima, and H. Tokuda. 1992. The carboxyl-terminal region of SecE interacts with SecY and is functional in the reconstitution of protein translocation activity in *Escherichia coli. J. Biol. Chem.* 267:7170–7176.

3105. Nishiyama, K., S. Mizushima, and H. Tokuda. 1993. A novel membrane protein involved in protein translocation across the cytoplasmic membrane of *Escherichia coli. EMBO J.* 12:3409–3415.

3106. Noble, J. A., M. A. Innis, E. V. Koonin, K. E. Rudd, F. Banuett, and I. Herskowitz. 1993. The *Escherichia coli hflA* locus encodes a putative GTP-binding protein and two membrane proteins, one of which contains a protease-like domain. *Proc. Natl. Acad. Sci. USA* 90:10866–10870.

3107. Noble, M. A., and E. E. Ishiguro. 1993. Temperature-sensitive mutation in *lytF*, a new gene involved in autolysis of *Escherichia coli. J. Gen. Microbiol.* 139:3109–3113.

3108. Noda, A., J. B. Courtright, P. F. Denor, G. Webb, Y. Kohara, and A. Ishihama. 1991. Rapid identification of specific genes in *E. coli* by hybridization to membranes containing the ordered set of phage clones. *BioTechniques* 10:474, 476–474, 477.

3109. Nogami, T., T. Mizuno, and S. Mizushima. 1985. Construction of a series of *ompF-ompC* chimeric genes by in vivo homologous recombination in *Escherichia coli* and characterization of the translational products. *J. Bacteriol.* 164:797–801.

3110. Noguchi, S., Y. Nishimura, Y. Hirota, and S. Nishimura. 1982. Isolation and characterization of an *Escherichia coli* mutant lacking tRNA-guanine transglycosylase. Function and biosynthesis of queuosine in tRNA. *J. Biol. Chem.* 257:6544–6550.

3111. Nohno, T., Y. Kasai, and T. Saito. 1988. Cloning and sequencing of the *Escherichia coli chlEN* operon involved in molybdopterin biosynthesis. *J. Bacteriol.* 170:4097–4102.

3112. Nohno, T., S. Noji, S. Taniguchi, and T. Saito. 1989. The *narX* and *narL* genes encoding the nitrate-sensing regulators of *Escherichia coli* are homologous to a family of prokaryotic two-component regulatory genes. *Nucleic Acids Res.* 17:2947–2957.

3113. Nohno, T., and T. Saito. 1987. Two transcriptional start sites found in the promoter region of *Escherichia coli* glutamine permease operon, *glnHPQ. Nucleic Acids Res.* 15:2777.

3114. Nohno, T., T. Saito, and J. S. Hong. 1986. Cloning and complete nucleotide sequence of the *Escherichia coli* glutamine permease operon (*glnHPQ*). *Mol. Gen. Genet.* 205:260–269.

3115. Noji, S., T. Nohno, T. Saito, and S. Taniguchi. 1989. The *narK* gene product participates in nitrate transport induced in *Escherichia coli* nitrate-respiring cells. *FEBS Lett.* 252:139–143.

3116. Nomura, T., H. Aiba, and A. Ishihama. 1985. Transcriptional organization of the convergent overlapping *dnaQ-rnh* genes of *Escherichia coli. J. Biol. Chem.* 260:7122–7125.

3117. Nomura, T., N. Fujita, and A. Ishihama. 1987. Expression of the *leuX* gene in *Escherichia coli*. Regulation at transcription and tRNA processing steps. *J. Mol. Biol.* 197:659–670.

3118. Nonet, M. L., C. C. Marvel, and D. R. Tolan. 1987. The *hisT-purF* region of the *Escherichia coli* K-12 chromosome. Identification of additional genes of the *hisT* and *purF* operons. *J. Biol. Chem.* 262:12209–12217.

3119. Nordlund, P., B. M. Sjoberg, and H. Eklund. 1990. Three-dimensional structure of the free radical protein of ribonucleotide reductase. *Nature* (London) 345:593–598.

3120. Normark, S., T. Edlund, T. Grundstrom, S. Bergstrom, and H. Wolf-Watz. 1977. *Escherichia coli* K-12 mutants hyperproducing chromosomal β-lactamase by gene repetitions. *J. Bacteriol.* 132:912–922.

3120a.Normark, S., L. Norlander, T. Grundstrom, G. D. Bloom, P. Boquel, and G. Frelat. 1976. Septum formation-defective mutants of *Escherichia coli. J. Bacteriol.* 128:401–412.

3120b.Norregaard-Madsen, M., B. Mygind, R. Pederson, P. Valentin-Hansen, and L. Sogaard-Andersen. 1994. The gene encoding the periplasmic cyclophilin homologue, PPIase A in *Escherichia coli*, is expressed from four promoters, three of which are activated by the cAMP-CRP complex and negatively regulated by the CytR repressor. *Mol. Microbiol.* 14:989–997.

3121. Norris, U., P. E. Karp, and A. L. Fimmel. 1992. Mutational analysis of the glycine-rich region of the c subunit of the *Escherichia coli* F_0F_1 ATPase. *J. Bacteriol.* 174:4496–4499.

3122. Norris, V., T. Alliotte, A. Jaffe, and R. D'Ari. 1986. DNA replication termination in *Escherichia coli parB* (a *dnaG* allele), *parA*, and *gyrB* mutants affected in DNA distribution. *J. Bacteriol.* 168:494–504.

3123. Norwood, W. I., and J. R. Sadler. 1977. Pseudoreversion of lactose operator-constitutive mutants. *J. Bacteriol.* 130:100–106.

3124. Novel, G., and M. Novel. 1973. Mutants of *E. coli* K 12 unable to grow on methyl-beta-D-glucuronide: map location of *uid A* locus of the structural gene of beta-D-glucuronidase. *Mol. Gen. Genet.* 120:319–335.

3125. Novel, M., and G. Novel. 1976. Regulation of beta-glucuronidase synthesis in *Escherichia coli* K-12: constitutive mutants specifically derepressed for *uidA* expression. *J. Bacteriol.* 127:406–417.

3126. Novotny, M. J., J. Reizer, F. Esch, and M. H. Saier, Jr. 1984. Purification and properties of D-mannitol-1-phosphate dehydrogenase and D-glucitol-6-phosphate dehydrogenase from *Escherichia coli. J. Bacteriol.* 159:986–990.

3127. Nudler, E., A. Goldfarb, and M. Kashlev. 1994. Discontinuous mechanism of transcription elongation. *Science* 265:793–796.

3128. Nuesch, J., and D. Schumperli. 1984. Structural and functional organization of the *gpt* gene region of *Escherichia coli. Gene* 32:243–249.

3129. Nunoshiba, T., E. Hidalgo, C. F. Amabile Cuevas, and B. Demple. 1992. Two-stage control of an oxidative stress regulon: the *Escherichia coli* SoxR protein triggers redox-inducible expression of the *soxS* regulatory gene. *J. Bacteriol.* 174:6054–6060.

3130. Nunoshiba, T., E. Hidalgo, Z. Li, and B. Demple. 1993. Negative autoregulation by the *Escherichia coli* SoxS protein: a dampening mechanism for the *soxRS* redox stress response. *J. Bacteriol.* 175:7492–7494.

3131. Nurse, P., R. J. DiGate, K. H. Zavitz, and K. J. Marians. 1990. Molecular cloning and DNA sequence analysis of *Escherichia coli priA*, the gene encoding the primosomal protein replication factor Y. *Proc. Natl. Acad. Sci. USA* 87:4615–4619.

3132. Nurse, P., J. Wrzesinski, A. Bakin, B. G. Lane, and J. Ofengand. 1995. Purification, cloning, and properties of the tRNA psi55 synthase from *Escherichia coli. RNA* 1:102–112.

3133. Nurse, P., K. H. Zavitz, and K. J. Marians. 1991. Inactivation of the *Escherichia coli priA* DNA replication protein induces the SOS response. *J. Bacteriol.* 173:6686–6693.

3134. Ny, T., and G. R. Bjork. 1980. Cloning and restriction mapping of the *trmA* gene coding for transfer ribonucleic acid (5-methyluridine)-methyltransferase in *Escherichia coli* K-12. *J. Bacteriol.* 142:371–379.

3135. Ny, T., H. R. Lindstrom, T. G. Hagervall, and G. R. Bjork. 1988. Purification of transfer RNA (m^5U^{54})-methyltransferase from *Escherichia coli*. Association with RNA. *Eur. J. Biochem.* 177:467–475.

3136. Nyengaard, N. R., K. K. Mortensen, S. F. Lassen, J. W. Hershey, and H. U. Sperling-Petersen. 1991. Tandem translation of *E. coli* initiation factor IF2 beta: purification and characterization in vitro of two active forms. *Biochem. Biophys. Res. Commun.* 181:1572–1579.

3137. Nygaard, P., and J. M. Smith. 1993. Evidence for a novel glycinamide ribonucleotide transformylase in *Escherichia coli. J. Bacteriol.* 175:3591–3597.

3138. Nystrom, T., and F. C. Neidhardt. 1992. Cloning, mapping and nucleotide sequencing of a gene encoding a universal stress protein in *Escherichia coli. Mol. Microbiol.* 6:3187–3198.

3139. Nystrom, T., and F. C. Neidhardt. 1993. Isolation and properties of a mutant of *Escherichia coli* with an insertional inactivation of the *uspA* gene, which encodes a universal stress protein. *J. Bacteriol.* 175:3949–3956.

3140. Nyunoya, H., and C. J. Lusty. 1983. The *carB* gene of *Escherichia coli*: a duplicated gene coding for the large subunit of carbamoyl-phosphate synthetase. *Proc. Natl. Acad. Sci. USA* 80:4629–4633.

3141. O'Connor, M., R. F. Gesteland, and J. F. Atkins. 1990. Sequence of *E. coli* K12 *valU* operon which contains 3 genes for $tRNA_1^{Val}$ and 1 gene for $tRNA^{Lys}$. *Nucleic Acids Res.* 18:672.

3142. O'Connor, M., N. M. Willis, L. Bossi, R. F. Gesteland, and J. F. Atkins. 1993. Functional tRNAs with altered 3′ ends. *EMBO J.* 12:2559–2566.

3143. O'Day, K., J. Lopilato, and A. Wright. 1991. Physical locations of *bglA* and *serA* on the *Escherichia coli* K-12 chromosome. *J. Bacteriol.* 173:1571.

3144. O'Donnell, M. 1992. Accessory protein function in the DNA polymerase III holoenzyme from *E. coli. Bioessays* 14:105–111.

3145. O'Donovan, G. A., G. Edlin, J. A. Fuchs, J. Neuhard, and E. Thomassen. 1971. Deoxycytidine triphosphate deaminase: characterization of an *Escherichia coli* mutant deficient in the enzyme. *J. Bacteriol.* 105:666–672.

3146. O'Neill, G. P., S. Thorbjarnardottir, U. Michelsen, D. Soll, and G. Eggertsson. 1991. δ-Aminolevulinic acid dehydratase deficiency can cause δ-aminolevulinate auxotrophy in *Escherichia coli. J. Bacteriol.* 173:94–100.

3147. O'Regan, M., R. Gloeckler, S. Bernard, C. Ledoux, I. Ohsawa, and Y. Lemoine. 1989. Nucleotide sequence of the *bioH* gene of *Escherichia coli. Nucleic Acids Res.* 17:8004.

3148. Oeda, K., T. Horiuchi, and M. Sekiguchi. 1981. Molecular cloning of the *uvrD* gene of *Escherichia coli* that controls ultraviolet sensitivity and spontaneous mutation frequency. *Mol. Gen. Genet.* 184:191–199.

3149. Oeda, K., T. Horiuchi, and M. Sekiguchi. 1982. The *uvrD* gene of *E. coli* encodes a DNA-dependent ATPase. *Nature* (London) 298:98–100.

3150. Ogawa, T., and T. Okazaki. 1994. Cell cycle-dependent transcription from the *gid* and *mioC* promoters of *Escherichia coli. J. Bacteriol.* 176:1609–1615.

3151. Ogawa, T., G. G. Pickett, T. Kogoma, and A. Kornberg. 1984. RNase H confers specificity in the *dnaA*-dependent initiation of replication at the unique origin of the *Escherichia coli* chromosome in vivo and in vitro. *Proc. Natl. Acad. Sci. USA* 81:1040–1044.

3152. Ogden, S., D. Haggerty, C. M. Stoner, D. Kolodrubetz, and R. Schleif. 1980. The *Escherichia coli* L-arabinose operon: binding sites of the regulatory proteins and a mechanism of positive and negative regulation. *Proc. Natl. Acad. Sci. USA* 77:3346–3350.

3153. Ogura, T., T. Miki, and S. Hiraga. 1980. Copy-number mutants of the plasmid carrying the replication origin of the *Escherichia coli* chromosome: evidence for a control region of replication. *Proc. Natl. Acad. Sci. USA* 77:3993–3997.

3154. Ogura, T., H. Niki, Y. Kano, F. Imamoto, and S. Hiraga. 1990. Maintenance of plasmids in HU and IHF mutants of *Escherichia coli. Mol. Gen. Genet.* 220:197–203.

3155. Ogura, T., T. Tomoyasu, T. Yuki, S. Morimura, K. J. Begg, W. D. Donachie, H. Mori, H. Niki, and S. Hiraga. 1991. Structure and function of the *ftsH* gene in *Escherichia coli. Res. Microbiol.* 142:279–282.

3156. Oh, B. K., and D. Apirion. 1991. 10Sa RNA, a small stable RNA of *Escherichia coli*, is functional. *Mol. Gen. Genet.* 229:52–56.

3157. Oh, B. K., A. K. Chauhan, K. Isono, and D. Apirion. 1990. Location of a gene (*ssrA*) for a small, stable RNA (10Sa RNA) in the *Escherichia coli* chromosome. *J. Bacteriol.* 172:4708–4709.

3158. Oh, W., and T. J. Larson. 1992. Physical locations of genes in the *rne* (*ams*)-*rpmF*-*plsX*-*fab* region of the *Escherichia coli* K-12 chromosome. *J. Bacteriol.* 174:7873–7874.

3159. Ohara, O., R. L. Dorit, and W. Gilbert. 1989. Direct genomic sequencing of bacterial DNA: the pyruvate kinase I gene of *Escherichia coli*. *Proc. Natl. Acad. Sci. USA* 86:6883–6887.

3160. Ohki, M., F. Tamura, S. Nishimura, and H. Uchida. 1986. Nucleotide sequence of the *Escherichia coli dnaJ* gene and purification of the gene product. *J. Biol. Chem.* 261:1778–1781.

3161. Ohki, R., T. Kawamata, Y. Katoh, F. Hosoda, and M. Ohki. 1992. *Escherichia coli dnaJ* deletion mutation results in loss of stability of a positive regulator, CRP. *J. Biol. Chem.* 267:13180–13184.

3162. Ohki, R., R. Morita, T. Kawamata, H. Uchida, and M. Ohki. 1989. A complete deletion mutant of the *Escherichia coli dnaKdnaJ* operon. *Biochim. Biophys. Acta* 1009:94–98.

3163. Ohlsen, K. L., and J. D. Gralla. 1992. DNA melting within stable closed complexes at the *Escherichia coli rrnB* P1 promoter. *J. Biol. Chem.* 267:19813–19818.

3164. Ohlsen, K. L., and J. D. Gralla. 1992. Melting during steady-state transcription of the *rrnB* P1 promoter in vivo and in vitro. *J. Bacteriol.* 174:6071–6075.

3165. Ohmori, H. 1994. Structural analysis of the *rhlE* gene of *Escherichia coli*. *Jpn. J. Genet.* 69:1–12. (Erratum, 69:425.)

3166. Ohmori, H., M. Kimura, T. Nagata, and Y. Sakakibara. 1984. Structural analysis of the *dnaA* and *dnaN* genes of *Escherichia coli*. *Gene* 28:159–170.

3167. Ohmori, H., M. Saito, T. Yasuda, T. Nagata, T. Fujii, M. Wachi, and K. Nagai. 1995. The *pcsA* gene is identical to *dinD* in *Escherichia coli*. *J. Bacteriol.* 177:156–165.

3167a. Ohnishi, Y. 1974. Genetic analysis of an *Escherichia coli* mutant with a lesion in stable RNA turnover. *Genetics* 76:185–194.

3168. Ohsawa, M., T. Mogi, H. Yamamoto, I. Yamato, and Y. Anraku. 1988. Proline carrier mutant of *Escherichia coli* K-12 with altered cation sensitivity of substrate-binding activity: cloning, biochemical characterization, and identification of the mutation. *J. Bacteriol.* 170:5185–5191.

3169. Ohsumi, M., T. Sekiya, S. Nishimura, and M. Ohki. 1983. Nucleotide sequence of the regulatory region of *malB* operons in *E. coli*. *J. Biochem.* 94:243–247.

3170. Ohta, A., T. Obara, Y. Asami, and I. Shibuya. 1985. Molecular cloning of the *cls* gene responsible for cardiolipin synthesis in *Escherichia coli* and phenotypic consequences of its amplification. *J. Bacteriol.* 163:506–514.

3171. Ohta, A., K. Waggoner, K. Louie, and W. Dowhan. 1981. Cloning of genes involved in membrane lipid synthesis. Effects of amplification of phosphatidylserine synthase in *Escherichia coli*. *J. Biol. Chem.* 256:2219–2225.

3172. Ohta, A., K. Waggoner, A. Radominska-Pyrek, and W. Dowhan. 1981. Cloning of genes involved in membrane lipid synthesis: effects of amplification of phosphatidylglycerophosphate synthase in *Escherichia coli*. *J. Biol. Chem.* 147:552–562.

3173. Ohta, A., M. K. Ina, K. Kusuzaki, N. Kido, Y. Arakawa, and N. Kato. 1991. Cloning and expression of the *rfe-rff* gene cluster of *Escherichia coli*. *Mol. Microbiol.* 5:1853–1862.

3174. Ohta, Y., K. Watanabe, and A. Kimura. 1985. Complete nucleotide sequence of the *E. coli* N-acetylneuraminate lyase. *Nucleic Acids Res.* 13:8843–8852.

3175. Ohyama, T., K. Igarashi, and H. Kobayashi. 1994. Physiological role of the *chaA* gene in sodium and calcium circulations at a high pH in *Escherichia coli*. *J. Bacteriol.* 176:4311–4315.

3176. Ohyama, T., R. Imaizumi, K. Igarashi, and H. Kobayashi. 1992. *Escherichia coli* is able to grow with negligible sodium ion extrusion activity at alkaline pH. *J. Bacteriol.* 174:7743–7749.

3177. Oka, A., H. Sasaki, K. Sugimoto, and M. Takanami. 1984. Sequence organization of replication origin of the *Escherichia coli* K-12 chromosome. *J. Mol. Biol.* 176:443–458.

3178. Oka, A., K. Sugimoto, H. Sasaki, and M. Takanami. 1982. An in vitro method generating base substitutions in preselected regions of plasmid DNA: application to structural analysis of the replication origin of the *Escherichia coli* K-12 chromosome. *Gene* 19:59–69.

3179. Oka, A., K. Sugimoto, M. Takanami, and Y. Hirota. 1980. Replication origin of the *Escherichia coli* K-12 chromosome: the size and structure of the minimum DNA segment carrying the information for autonomous replication. *Mol. Gen. Genet.* 178:9–20.

3180. Okada, Y., M. Wachi, A. Hirata, K. Suzuki, K. Nagai, and M. Matsuhashi. 1994. Cytoplasmic axial filaments in *Escherichia coli* cells: possible function in the mechanism of chromosome segregation and cell division. *J. Bacteriol.* 176:917–922.

3181. Okamoto, K., and M. Freundlich. 1986. Mechanism for the autogenous control of the *crp* operon: transcriptional inhibition by a divergent RNA transcript. *Proc. Natl. Acad. Sci. USA* 83:5000–5004.

3182. Okamura-Ikeda, K., Y. Ohmura, K. Fujiwara, and Y. Motokawa. 1993. Cloning and nucleotide sequence of the *gcv* operon encoding the *Escherichia coli* glycine-cleavage system. *Eur. J. Biochem.* 216:539–548.

3183. Okita, T. W., R. L. Rodriguez, and J. Preiss. 1981. Biosynthesis of bacterial glycogen. Cloning of the glycogen biosynthetic enzyme structural genes of *Escherichia coli*. *J. Biol. Chem.* 256:6944–6952.

3184. Old, I. G., M. G. Hunter, D. T. Wilson, S. M. Knight, C. A. Weatherston, and R. E. Glass. 1988. Cloning and characterization of the genes for the two homocysteine transmethylases of *Escherichia coli*. *Mol. Gen. Genet.* 211:78–87.

3185. Old, I. G., D. Margarita, R. E. Glass, and I. Saint Girons. 1990. Nucleotide sequence of the *metH* gene of *Escherichia coli* K-12 and comparison with that of *Salmonella typhimurium* LT2. *Gene* 87:15–21.

3186. Old, I. G., S. E. Phillips, P. G. Stockley, and I. Saint Girons. 1991. Regulation of methionine biosynthesis in the *Enterobacteriaceae*. *Prog. Biophys. Mol. Biol.* 56:145–185.

3187. Olins, P. O., and M. Nomura. 1981. Translational regulation by ribosomal protein S8 in *Escherichia coli*: structural homology between rRNA binding site and feedback target on mRNA. *Nucleic Acids Res.* 9:1757–1764.

3188. Olins, P. O., and M. Nomura. 1981. Regulation of the S10 ribosomal protein operon in *E. coli*: nucleotide sequence at the start of the operon. *Cell* 26:205–211.

3189. Oliver, D. B. 1985. Identification of five new essential genes involved in the synthesis of a secreted protein in *Escherichia coli*. *J. Bacteriol.* 161:285–291.

3190. Oliver, D. B., and J. Beckwith. 1981. *E. coli* mutant pleiotropically defective in the export of secreted proteins. *Cell* 25:765–772.

3191. Oliver, D. B., and J. Beckwith. 1982. Identification of a new gene (*secA*) and gene product involved in the secretion of envelope proteins in *Escherichia coli*. *J. Bacteriol.* 150:686–691.

3192. Oliver, G., G. Gosset, R. Sanchez-Pescador, E. Lozoya, L. M. Ku, N. Flores, B. Becerril, F. Valle, and F. Bolivar. 1987. Determination of the nucleotide sequence for the glutamate synthase structural genes of *Escherichia coli* K-12. *Gene* 60:1–11.

3193. Oller, A. R., I. J. Fijalkowska, R. L. Dunn, and R. M. Schaaper. 1992. Transcription-repair coupling determines the strandedness of ultraviolet mutagenesis in *Escherichia coli*. *Proc. Natl. Acad. Sci. USA* 89:11036–11040.

3194. Olsen, A., A. Arnqvist, M. Hammar, and S. Normark. 1993. Environmental regulation of curli production in *Escherichia coli*. *Infect. Agents Dis.* 2:272–274.

3195. Olsen, A., A. Arnqvist, M. Hammar, S. Sukupolvi, and S. Normark. 1993. The RpoS sigma factor relieves H-NS-mediated transcriptional repression of *csgA*, the subunit gene of fibronectin-binding curli in *Escherichia coli*. *Mol. Microbiol.* 7:523–536.

3196. Olsen, A., A. Jonsson, and S. Normark. 1989. Fibronectin binding mediated by a novel class of surface organelles on *Escherichia coli*. *Nature* (London) 338:652–655.

3197. Olsen, M. K., E. L. Rosey, and C. S. Tomich. 1993. Isolation and analysis of novel mutants of *Escherichia coli prlA* (*secY*). *J. Bacteriol.* 175:7092–7096.

3198. Olsen, S. G., K. M. Greene, and R. J. Brooker. 1993. Lactose permease mutants which transport (malto)-oligosaccharides. *J. Bacteriol.* 175:6269–6275.

3199. Olson, E. R., D. S. Dunyak, L. M. Jurss, and R. A. Poorman. 1991. Identification and characterization of *dppA*, an *Escherichia coli* gene encoding a periplasmic dipeptide transport protein. *J. Bacteriol.* 173:234–244.

3199a. Ono, M., and M. Kuwano. 1978. Mutation affecting the thermolability of the 50S ribosomal subunit in *Escherichia coli*. *J. Bacteriol.* 134:677–679.

3200. Ono, M., and M. Kuwano. 1980. Chromosomal location of a gene for chemical longevity of messenger ribonucleic acid in a temperature-sensitive mutant of *Escherichia coli*. *J. Bacteriol.* 142:325–326.

3201. Onrust, R., and M. O'Donnell. 1993. DNA polymerase III accessory proteins. II. Characterization of delta and delta'. *J. Biol. Chem.* 268:11766–11772.

3202. Oosawa, K., J. F. Hess, and M. I. Simon. 1988. Mutants defective in bacterial chemotaxis show modified protein phosphorylation. *Cell* 53:89–96.

3203. Oram, M., and L. M. Fisher. 1992. An *Escherichia coli* DNA topoisomerase I mutant has a compensatory mutation that alters two residues between functional domains of the DNA gyrase A protein. *J. Bacteriol.* 174:4175–4178.

3204. Orchard, L. M., and H. L. Kornberg. 1990. Sequence similarities between the gene specifying 1-phosphofructokinase (*fruK*), genes specifying other kinases in *Escherichia coli* K12, and *lacC* of *Staphylococcus aureus*. *Proc. R. Soc. Lond. [Biol.]* 242:87–90.

3205. Ordal, G. W., and J. Adler. 1974. Isolation and complementation of mutants in galactose taxis and transport. *J. Bacteriol.* 117:509–516.

3206. Orndorff, P. E., and S. Falkow. 1985. Nucleotide sequence of *pilA*, the gene encoding the structural component of type 1 pili in *Escherichia coli*. *J. Bacteriol.* 162:454–457.

3207. Orosz, A., I. Boros, and P. Venetianer. 1991. Analysis of the complex transcription termination region of the *Escherichia coli rrnB* gene. *Eur. J. Biochem.* 201:653–659.

3208. Orr, E., N. F. Fairweather, I. B. Holland, and R. H. Pritchard. 1979. Isolation and characterisation of a strain carrying a conditional lethal mutation in the *cou* gene of *Escherichia coli* K12. *Mol. Gen. Genet.* 177:103–112.

3209. Orth, V., M. Chippaux, and M. C. Pascal. 1980. A mutant defective in electron transfer to nitrate in *Escherichia coli* K12. *J. Gen. Microbiol.* 117:257–262.

3210. Ortuno, M. J., and R. P. Lawther. 1987. Effect of the deletion of upstream DNA sequences on expression from the *ilvGp2* promoter of the *ilvGMEDA* operon of *Escherichia coli* K-12. *Nucleic Acids Res.* 15:1521–1542.

3211. Osborne, C., L. M. Chen, and R. G. Matthews. 1991. Isolation, cloning, mapping, and nucleotide sequencing of the gene encoding flavodoxin in *Escherichia coli*. *J. Bacteriol.* 173:1729–1737.

3212. Osborne, R. S., and T. J. Silhavy. 1993. PrlA suppressor mutations cluster in regions corresponding to three distinct topological domains. *EMBO J.* 12:3391–3398.

3213. Ost, K. A., and M. P. Deutscher. 1991. *Escherichia coli orfE* (upstream of *pyrE*) encodes RNase PH. *J. Bacteriol.* 173:5589–5591.

3214. Ostrowski, J., M. J. Barber, D. C. Rueger, B. E. Miller, L. M. Siegel, and N. M. Kredich. 1989. Characterization of the flavoprotein moieties of NADPH-sulfite reductase from *Salmonella typhimurium* and *Escherichia coli*. Physicochemical and catalytic properties, amino acid sequence deduced from DNA sequence of *cysJ*, and comparison with NADPH-cytochrome P-450 reductase. *J. Biol. Chem.* 264:15796–15808.

3215. Ostrowski, J., G. Jagura-Burdzy, and N. M. Kredich. 1987. DNA sequences of the *cysB* regions of *Salmonella typhimurium* and *Escherichia coli*. *J. Biol. Chem.* 262:5999–6005.

3216. Osuna, R., S. E. Finkel, and R. C. Johnson. 1991. Identification of two functional regions in Fis: the N-terminus is required to promote Hin-mediated DNA inversion but not lambda excision. *EMBO J.* 10:1593–1603.

3217. Otsuji, N., T. Soejima, S. Maki, and H. Shinagawa. 1982. Cloning of colicin E1 tolerant *tolC* (*mtcB*) gene of *Escherichia coli* K12 and identification of its gene product. *Mol. Gen. Genet.* 187:30–36.

3218. Otsuka, A. J., M. R. Buoncristiani, P. K. Howard, J. Flamm, C. Johnson, R. Yamamoto, K. Uchida, C. Cook, J. Ruppert, and J. Matsuzaki. 1988. The *Escherichia coli* biotin biosynthetic enzyme sequences predicted from the nucleotide sequence of the bio operon. *J. Biol. Chem.* 263:19577–19585.

3219. Ou, J. T., and L. M. Kuo. 1979. Suppression of the formation of polygenotypic recombinant colonies by a *maf* mutation in mating with HfrH. *Genetics* 93:345–351.

3220. Ouzounis, C. A., and B. J. Blencowe. 1991. Bacterial DNA replication initiation factor PriA is related to proteins belonging to the 'DEAD-box' family. *Nucleic Acids Res.* 19:6953.

3221. Ovchinnikov, Y. A., Y. B. Alakhov, Y. P. Bundulis, M. A. Bundule, N. V. Dovgas, V. P. Kozlov, L. P. Motuz, and L. M. Vinokurov. 1982. The primary structure of elongation factor G from *Escherichia coli*. A complete amino acid sequence. *FEBS Lett.* 139:130–135.

3222. Ovchinnikov, Y. A., G. S. Monastyrskaya, V. V. Gubanov, S. O. Guryev, I. S. Salomatina, T. M. Shuvaeva, V. M. Lipkin, and E. D. Sverdlov. 1982. The primary structure of *E. coli* RNA polymerase, nucleotide sequence of the *rpoC* gene and amino acid sequence of the beta'-subunit. *Nucleic Acids Res.* 10:4035–4044.

3223. Ovchinnikov, Y. A., N. A. Aldanova, V. A. Grinkevich, N. M. Arzamazova, and I. N. Moroz. 1977. The primary structure of a Leu, Ile and Val (LIV)-binding protein from *Escherichia coli*. *FEBS Lett.* 78:313–316.

3224. Ovchinnikov, Y. A., V. M. Lipkin, N. N. Modyanov, O. Y. Chertov, and Y. V. Smirnov. 1977. Primary structure of alpha-subunit of DNA-dependent RNA polymerase from *Escherichia coli*. *FEBS Lett.* 76:108–111.

3225. Ovchinnikov, Y. A., G. S. Monastyrskaya, V. V. Gubanov, S. O. Guryev, O. Y. Chertov, N. N. Modyanov, V. A. Grinkevich, I. A. Makarova, T. V. Marchenko, I. N. Polovnikova, V. M. Lipkin, and E. D. Sverdlov. 1981. The primary structure of *Escherichia coli* RNA polymerase. Nucleotide sequence of the *rpoB* gene and amino acid sequence of the beta-subunit. *Eur. J. Biochem.* 116:621–629.

3226. Overbeeke, N., H. Bergmans, F. van Mansfeld, and B. Lugtenberg. 1983. Complete nucleotide sequence of *phoE*, the structural gene for the phosphate limitation inducible outer membrane pore protein of *Escherichia coli* K12. *J. Mol. Biol.* 163:513–532.

3227. Overdier, D. G., and L. N. Csonka. 1992. A transcriptional silencer downstream of the promoter in the osmotically controlled *proU* operon of *Salmonella typhimurium*. *Proc. Natl. Acad. Sci. USA* 89:3140–3144.

3228. Overduin, P., W. Boos, and J. Tommassen. 1988. Nucleotide sequence of the *ugp* genes of *Escherichia coli* K-12: homology to the maltose system. *Mol. Microbiol.* 2:767–775.

3229. Oxender, D. L. 1972. Membrane transport. *Annu. Rev. Biochem.* 41:777–814.

3230. Oxender, D. L., J. J. Anderson, C. J. Daniels, R. Landick, R. P. Gunsalus, G. Zurawski, E. Selker, and C. Yanofsky. 1980. Structural and functional analysis of cloned DNA containing genes responsible for branched-chain amino acid transport in *Escherichia coli*. *Proc. Natl. Acad. Sci. USA* 77:1412–1416.

3231. Oxender, D. L., J. J. Anderson, C. J. Daniels, R. Landick, R. P. Gunsalus, G. Zurawski, and C. Yanofsky. 1980. Amino-terminal sequence and processing of the precursor of the leucine-specific binding protein, and evidence for conformational differences between the precursor and the mature form. *Proc. Natl. Acad. Sci. USA* 77:2005–2009.

3232. Oxender, D. L., G. Zurawski, and C. Yanofsky. 1979. Attenuation in the *Escherichia coli* tryptophan operon: role of RNA secondary structure involving the tryptophan codon region. *Proc. Natl. Acad. Sci. USA* 76:5524–5528.

3233. Ozaki, M., N. Fujita, A. Wada, and A. Ishihama. 1992. Promoter selectivity of the stationary-phase forms of *Escherichia coli* RNA polymerase and conversion in vitro of the S1 form enzyme into a log-phase enzyme-like form. *Nucleic Acids Res.* 20:257–261.

3234. Ozaki, M., A. Wada, N. Fujita, and A. Ishihama. 1991. Growth phase-dependent modification of RNA polymerase in *Escherichia coli*. *Mol. Gen. Genet.* 230:17–23.

3235. Ozenberger, B. A., T. J. Brickman, and M. A. McIntosh. 1989. Nucleotide sequence of *Escherichia coli* isochorismate synthetase gene *entC* and evolutionary relationship of isochorismate synthetase and other chorismate-utilizing enzymes. *J. Bacteriol.* 171:775–783.

3236. Ozenberger, B. A., M. S. Nahlik, and M. A. McIntosh. 1987. Genetic organization of multiple *fep* genes encoding ferric enterobactin transport functions in *Escherichia coli*. *J. Bacteriol.* 169:3638–3646.

3237. Pace, N. R., and D. Smith. 1990. Ribonuclease P: function and variation. *J. Biol. Chem.* 265:3587–3590.

3238. Padan, E., N. Maisler, D. Taglicht, R. Karpel, and S. Schuldiner. 1989. Deletion of *ant* in *Escherichia coli* reveals its function in adaptation to high salinity and an alternative Na$^+$/H$^+$ antiporter system(s). *J. Biol. Chem.* 264:20297–20302.

3239. Padan, E., and S. Schuldiner. 1993. Na$^+$/H$^+$ antiporters, molecular devices that couple the Na$^+$ and H$^+$ circulation in cells. *J. Bioenerg. Biomembr.* 25:647–669.

3240. Padmanabha, K. P., and M. P. Deutscher. 1991. RNase T affects *Escherichia coli* growth and recovery from metabolic stress. *J. Bacteriol.* 173:1376–1381.

3241. Paek, K. H., and G. C. Walker. 1987. *Escherichia coli dnaK* null mutants are inviable at high temperature. *J. Bacteriol.* 169:283–290.

3242. Paetz, W., and G. Nass. 1973. Biochemical and immunological characterization of threonyl-tRNA synthetase of two borrelidin-resistant mutants of *Escherichia coli* K12. *Eur. J. Biochem.* 35:331–337.

3243. Page, L., L. Griffiths, and J. A. Cole. 1990. Different physiological roles of two independent pathways for nitrite reduction to ammonia by enteric bacteria. *Arch. Microbiol.* 154:349–354.

3244. Pagel, J. M., and G. W. Hatfield. 1991. Integration host factor-mediated expression of the *ilvGMEDA* operon of *Escherichia coli*. *J. Biol. Chem.* 266:1985–1996.

3245. Pagel, J. M., J. W. Winkelman, C. W. Adams, and G. W. Hatfield. 1992. DNA topology-mediated regulation of transcription initiation from the tandem promoters of the *ilvGMEDA* operon of *Escherichia coli*. *J. Mol. Biol.* 224:919–935.

3246. Pahel, G., D. M. Rothstein, and B. Magasanik. 1982. Complex *glnA-glnL-glnG* operon of *Escherichia coli*. *J. Bacteriol.* 150:202–213.

3247. Pahel, G., and B. Tyler. 1979. A new *glnA*-linked regulatory gene for glutamine synthetase in *Escherichia coli*. *Proc. Natl. Acad. Sci. USA* 76:4544–4548.

3248. Pai, C. H. 1974. Biochemical and genetic characterization of dehydrobiotin resistant mutants of *Escherichia coli*. *Mol. Gen. Genet.* 134:345–357.

3249. Pai, C. H., and H. C. Yau. 1975. Chromosomal location of mutations affecting the regulation of biotin synthesis in *Escherichia coli*. *Can. J. Microbiol.* 21:1116–1120.

3250. Palaniappan, C., V. Sharma, M. E. Hudspeth, and R. Meganathan. 1992. Menaquinone (vitamin K2) biosynthesis: evidence that the *Escherichia coli menD* gene encodes both 2-succinyl-6-hydroxy-2,4-cyclohexadiene-1-carboxylic acid synthase and α-ketoglutarate decarboxylase activities. *J. Bacteriol.* 174:8111–8118.

3251. Palm, D., R. Goerl, and K. J. Burger. 1985. Evolution of catalytic and regulatory sites in phosphorylases. *Nature* (London) 313:500–502.

3252. Palm, D., R. Goerl, G. Weidinger, R. Zeier, B. Fischer, and R. Schinzel. 1987. *E. coli* maltodextrin phosphorylase: primary structure and deletion mapping of the C-terminal site. *Z. Naturforsch. Sect. C* 42:394–400.

3253. Palmer, M. L., M. A. Raker, P. J. Kennedy, J. W. Young, W. M. Barnes, R. L. Rodriguez, and H. F. Noller. 1979. Isolation and restriction mapping of plasmids containing ribosomal DNA sequences from the *rrn B* cistron of *E. coli*. *Mol. Gen. Genet.* 172:171–178.

3254. Palmer, T., A. Vasishta, P. W. Whitty, and D. H. Boxer. 1994. Isolation of protein FA, a product of the *mob* locus required for molybdenum cofactor biosynthesis in *Escherichia coli*. *Eur. J. Biochem.* 222:687–692.

3255. Palva, E. T., P. Saris, and T. J. Silhavy. 1985. Gene fusions to the *ptsM/pel* locus of *Escherichia coli*. *Mol. Gen. Genet.* 199:427–433.

3256. Pannekoek, H., I. Noordermeer, and P. van de Putte. 1979. Expression of the cloned *uvrB* gene of *Escherichia coli*: mode of transcription and orientation. *J. Bacteriol.* 139:54–63.

3256a.Pao, C. C., and J. Gallant. 1978. A gene involved in the metabolic control of ppGpp synthesis. *Mol. Gen. Genet.* 158:271–277.

3257. Pao, G. M., L. F. Wu, K. D. Johnson, H. Hofte, M. J. Chrispeels, G. Sweet, N. N. Sandal, and M. H. Saier, Jr. 1991. Evolution of the MIP family of integral membrane transport proteins. *Mol. Microbiol.* 5:33–37.

3258. Pardee, A. B., E. J. Benz, Jr., D. A. St. Peter, J. N. Krieger, M. Meuth, and H. W. Trieshmann, Jr. 1971. Hyperproduction and purification of nicotinamide deamidase, a microconstitutive enzyme of *Escherichia coli*. *J. Biol. Chem.* 246:6792–6796.

3259. Park, J. T. 1993. Turnover and recycling of the murein sacculus in oligopeptide permease-negative strains of *Escherichia coli*: indirect evidence for an alternative permease system and for a monolayered sacculus. *J. Bacteriol.* 175:7–11.

3260. Park, M. H., B. B. Wong, and J. E. Lusk. 1976. Mutants in three genes affecting transport of magnesium in *Escherichia coli*: genetics and physiology. *J. Bacteriol.* 126:1096–1103.

3261. Park, S. F., D. A. Stirling, C. S. Hulton, I. R. Booth, C. F. Higgins, and G. S. Stewart. 1989. A novel, non-invasive promoter probe vector: cloning of the osmoregulated *proU* promoter of *Escherichia coli* K12. *Mol. Microbiol.* 3:1011–1023.

3262. Park, S. J., J. McCabe, J. Turna, and R. P. Gunsalus. 1994. Regulation of the citrate synthase (*gltA*) gene of *Escherichia coli* in response to anaerobiosis and carbon supply: role of the *arcA* gene product. *J. Bacteriol.* 176:5086–5092.

3263. Park, S. K., K. I. Kim, K. M. Woo, J. H. Seol, K. Tanaka, A. Ichihara, D. B. Ha, and C. H. Chung. 1993. Site-directed mutagenesis of the dual translational initiation sites of the *clpB* gene of *Escherichia coli* and characterization of its gene products. *J. Biol. Chem.* 268:20170–20174.

3264. Parker, C. T., A. W. Kloser, C. A. Schnaitman, M. A. Stein, S. Gottesman, and B. W. Gibson. 1992. Role of the *rfaG* and *rfaP* genes in determining the lipopolysaccharide core structure and cell surface properties of *Escherichia coli* K-12. *J. Bacteriol.* 174:2525–2538.

3265. Parker, C. T., E. Pradel, and C. A. Schnaitman. 1992. Identification and sequences of the lipopolysaccharide core biosynthetic genes *rfaQ*, *rfaP*, and *rfaG* of *Escherichia coli* K-12. *J. Bacteriol.* 174:930–934.

3266. Parker, J. 1982. Specific mistranslation in *hisT* mutants of *Escherichia coli*. *Mol. Gen. Genet.* 187:405–409.

3267. Parker, J. 1984. Identification of the *purC* gene product of *Escherichia coli*. *J. Bacteriol.* 157:712–717.

3268. Parker, L. L., and B. G. Hall. 1988. A fourth *Escherichia coli* gene system with the potential to evolve beta-glucoside utilization. *Genetics* 119:485–490.

3269. Parker, L. L., and B. G. Hall. 1990. Characterization and nucleotide sequence of the cryptic *cel* operon of *Escherichia coli* K12. *Genetics* 124:455–471.

3270. Parkinson, J. S. 1980. Novel mutations affecting a signaling component for chemotaxis of *Escherichia coli*. *J. Bacteriol.* 142:953–961.

3271. Parquet, C., B. Flouret, D. Mengin-Lecreulx, and J. van Heijenoort. 1989. Nucleotide sequence of the *murF* gene encoding the UDP-MurNAc-pentapeptide synthetase of *Escherichia coli*. *Nucleic Acids Res.* 17:5379.

3272. Parra, F., P. Britton, C. Castle, M. C. Jones-Mortimer, and H. L. Kornberg. 1983. Two separate genes involved in sulphate transport in *Escherichia coli* K12. *J. Gen. Microbiol.* 129:357–358.

3273. Parsot, C., A. Boyen, G. N. Cohen, and N. Glansdorff. 1988. Nucleotide sequence of *Escherichia coli argB* and *argC* genes: comparison of N-acetylglutamate kinase and N-acetylglutamate-gamma-semialdehyde dehydrogenase with homologous and analogous enzymes. *Gene* 68:275–283.

3274. Parsot, C., P. Cossart, I. Saint-Girons, and G. N. Cohen. 1983. Nucleotide sequence of *thrC* and of the transcription termination region of the threonine operon in *Escherichia coli* K12. *Nucleic Acids Res.* 11:7331–7345.

3275. Pas, H. H., and G. T. Robillard. 1988. S-phosphocysteine and phosphohistidine are intermediates in the phosphoenolpyruvate-dependent mannitol transport catalyzed by *Escherichia coli* EIIMtl. *Biochemistry* 27:5835–5839.

3276. Pascal, M. C., J. F. Burini, and M. Chippaux. 1984. Regulation of the trimethylamine N-oxide (TMAO) reductase in *Escherichia coli*: analysis of *tor*::Mud1 operon fusion. *Mol. Gen. Genet.* 195:351–355.

3277. Pascal, M. C., F. Casse, M. Chippaux, and M. Lepelletier. 1975. Genetic analysis of mutants of *Escherichia coli* K12 and *Salmonella typhimurium* LT2 deficient in hydrogenase activity. *Mol. Gen. Genet.* 141:173–179.

3278. Pascal, M. C., and M. Chippaux. 1982. Involvement of a gene of the *chl E* locus in the regulation of the nitrate reductase operon. *Mol. Gen. Genet.* 185:334–338.

3279. Pascal, M. C., M. Chippaux, A. Abou-Jaoude, H. P. Blaschkowski, and J. Knappe. 1981. Mutants of *Escherichia coli* K12 with defects in anaerobic pyruvate metabolism. *J. Gen. Microbiol.* 124:35–42.

3280. Patel, A. M., H. G. Dallmann, E. N. Skakoon, T. D. Kapala, and S. D. Dunn. 1990. The *Escherichia coli unc* transcription terminator enhances expression of *uncC*, encoding the epsilon subunit of F1-ATPase, from plasmids by stabilizing the transcript. *Mol. Microbiol.* 4:1941–1946.

3281. Patel, A. M., and S. D. Dunn. 1992. RNase E-dependent cleavages in the 5′ and 3′ regions of the *Escherichia coli unc* mRNA. *J. Bacteriol.* 174:3541–3548.

3282. Patil, R. V., and E. E. Dekker. 1992. Cloning, nucleotide sequence, overexpression, and inactivation of the *Escherichia coli* 2-keto-4-hydroxyglutarate aldolase gene. *J. Bacteriol.* 174:102–107.

3283. Pauli, G., and P. Overath. 1972. *ato* operon: a highly inducible system for acetoacetate and butyrate degradation in *Escherichia coli*. *Eur. J. Biochem.* 29:553–562.

3284. Pauza, C. D., M. J. Karels, M. Navre, and H. K. Schachman. 1982. Genes encoding *Escherichia coli* aspartate transcarbamoylase: the *pyrB-pyrI* operon. *Proc. Natl. Acad. Sci. USA* 79:4020–4024.

3285. Payne, J. W., J. S. Morley, P. Armitage, and G. M. Payne. 1984. Transport and hydrolysis of antibacterial peptide analogues in *Escherichia coli*: backbone-modified aminoxy peptides. *J. Gen. Microbiol.* 130:2253–2265.

3286. Peakman, T., S. Busby, and J. Cole. 1990. Transcriptional control of the *cysG* gene of *Escherichia coli* K-12 during aerobic and anaerobic growth. *Eur. J. Biochem.* 191:325–331.

3287. Peakman, T., J. Crouzet, J. F. Mayaux, S. Busby, S. Mohan, N. Harborne, J. Wootton, R. Nicolson, and J. Cole. 1990. Nucleotide sequence, organisation and structural analysis of the products of genes in the *nirB-cysG* region of the *Escherichia coli* K-12 chromosome. *Eur. J. Biochem.* 191:315–323.

3288. Pecher, A., H. P. Blaschkowski, K. Knappe, and A. Bock. 1982. Expression of pyruvate formate-lyase of *Escherichia coli* from the cloned structural gene. *Arch. Microbiol.* 132:365–371.

3289. Pecher, A., F. Zinoni, and A. Bock. 1985. The seleno-polypeptide of formic dehydrogenase (formate hydrogen-lyase linked) from *Escherichia coli*: genetic analysis. *Arch. Microbiol.* 141:359–363.

3290. Peck-Miller, K. A., and S. Altman. 1991. Kinetics of the processing of the precursor to 4.5 S RNA, a naturally occurring substrate for RNase P from *Escherichia coli*. *J. Mol. Biol.* 221:1–5.

3291. Pedersen, H., L. Sogaard-Andersen, B. Holst, and P. Valentin-Hansen. 1991. Heterologous cooperativity in *Escherichia coli*. The CytR repressor both contacts DNA and the cAMP receptor protein when binding to the *deoP2* promoter. *J. Biol. Chem.* 266:17804–17808.

3292. Pedersen, S., J. Skouv, M. Kajitani, and A. Ishihama. 1984. Transcriptional organization of the *rpsA* operon of *Escherichia coli*. *Mol. Gen. Genet.* 196:135–140.

3293. Pegues, J. C., L. S. Chen, A. W. Gordon, L. Ding, and W. G. Coleman, Jr. 1990. Cloning, expression, and characterization of the *Escherichia coli* K-12 *rfaD* gene. *J. Bacteriol.* 172:4652–4660.

3294. Pelletier, A. J., T. M. Hill, and P. L. Kuempel. 1988. Location of sites that inhibit progression of replication forks in the terminus region of *Escherichia coli*. *J. Bacteriol.* 170:4293–4298.

3295. Peng, H., and K. J. Marians. 1993. *Escherichia coli* topoisomerase IV. Purification, characterization, subunit structure, and subunit interactions. *J. Biol. Chem.* 268:24481–24490.

3296. Peng, H. L., M. J. Hsieh, C. L. Zao, and H. Y. Chang. 1994. Nucleotide sequence and expression in *Escherichia coli* of the *Klebsiella pneumoniae deaD* gene. *J. Biochem.* 115:409–414.

3297. Penninckx, M., and D. Gigot. 1979. Synthesis of a peptide form of N-delta-(phosphonoacetyl)-L-ornithine. Its antibacterial effect through the specific inhibition of *Escherichia coli* L-ornithine carbamoyltransferase. *J. Biol. Chem.* 254:6392–6396.

3298. Pereira, R. F., M. J. Ortuno, and R. P. Lawther. 1988. Binding of integration host factor (IHF) to the *ilvGp1* promoter of the *ilvGMEDA* operon of *Escherichia coli* K12. *Nucleic Acids Res.* 16:5973–5989.

3299. Perez-Roger, I., F. Macian, and M. E. Armengod. 1995. Transcription termination in the *Escherichia coli dnaA* gene is not mediated by the internal DnaA box. *J. Bacteriol.* 177:1896–1899.

3300. Peri, K. G., H. Goldie, and E. B. Waygood. 1990. Cloning and characterization of the N-acetylglucosamine operon of *Escherichia coli*. *Biochem. Cell Biol.* 68:123–137.

3301. Peri, K. G., and E. B. Waygood. 1988. Sequence of cloned enzyme IIN-acetylglucosamine of the phosphoenolpyruvate:N-acetylglucosamine phosphotransferase system of *Escherichia coli*. *Biochemistry* 27:6054–6061.

3302. Perreault, J. P., and S. Altman. 1992. Important 2′-hydroxyl groups in model substrates for M1 RNA, the catalytic RNA subunit of RNase P from *Escherichia coli*. *J. Mol. Biol.* 226:399–409.

3303. Perry, K. L., S. J. Elledge, B. B. Mitchell, L. Marsh, and G. C. Walker. 1985. *umuDC* and *mucAB* operons whose products are required for UV light- and chemical-induced mutagenesis: UmuD, MucA, and LexA proteins share homology. *Proc. Natl. Acad. Sci. USA* 82:4331–4335.

3304. Persson, B. C., C. Gustafsson, D. E. Berg, and G. R. Bjork. 1992. The gene for a tRNA modifying enzyme, m^{5}U5^4-methyltransferase, is essential for viability in *Escherichia coli*. *Proc. Natl. Acad. Sci. USA* 89:3995–3998.

3305. Peter, B., Y. M. Man, C. E. Begg, I. Gall, and D. P. Leader. 1988. Mouse cytoskeletal gamma-actin: analysis and implications of the structure of cloned cDNA and processed pseudogenes. *J. Mol. Biol.* 203:665–675.

3306. Peterson, J. W., A. K. Chopra, and R. Prasad. 1991. Fine mapping of the *rrnE*, *purHD*, and *hydGH* operons on the *Escherichia coli* chromosome. *J. Bacteriol.* 173:3274–3275.

3307. Peterson, R. G., F. F. Richards, and R. E. Handschumacher. 1977. Structure of peptide from active site region of *Escherichia coli* L-asparaginase. *J. Biol. Chem.* 252:2072–2076.

3308. Petit, C., C. Cayrol, C. Lesca, P. Kaiser, C. Thompson, and M. Defais. 1993. Characterization of *dinY*, a new *Escherichia coli* DNA repair gene whose products are damage inducible even in a *lexA* (Def) background. *J. Bacteriol.* 175:642–646.

3309. Pettersson, I., S. J. S. Hardy, and A. Liljas. 1976. The ribosomal protein L8 is a complex L7/L12 and L10. *FEBS Lett.* 64:135–138.

3310. Pettigrew, D. W., D. P. Ma, C. A. Conrad, and J. R. Johnson. 1988. *Escherichia coli* glycerol kinase. Cloning and sequencing of the *glpK* gene and the primary structure of the enzyme. *J. Biol. Chem.* 263:135–139.

3311. Pettis, G. S., T. J. Brickman, and M. A. McIntosh. 1988. Transcriptional mapping and nucleotide sequence of the *Escherichia coli fepA-fes* enterobactin region. Identification of a unique iron-regulated bidirectional promoter. *J. Biol. Chem.* 263:18857–18863.

3312. Pettis, G. S., and M. A. McIntosh. 1987. Molecular characterization of the *Escherichia coli* enterobactin cistron *entF* and coupled expression of *entF* and the *fes* gene. *J. Bacteriol.* 169:4154–4162.

3313. Phadnis, S. H., S. Kulakauskas, B. R. Krishnan, J. Hiemstra, and D. E. Berg. 1991. Transposon Tn5 *supF*-based reverse genetic method for mutational analysis of *Escherichia coli* with DNAs cloned in lambda phage. *J. Bacteriol.* 173:896–899.

3314. Phillips, G. J., and S. R. Kushner. 1987. Determination of the nucleotide sequence for the exonuclease I structural gene (*sbcB*) of *Escherichia coli* K12. *J. Biol. Chem.* 262:455–459.

3315. Phillips, G. J., D. C. Prasher, and S. R. Kushner. 1988. Physical and biochemical characterization of cloned *sbcB* and *xonA* mutations from *Escherichia coli* K-12. *J. Bacteriol.* 170:2089–2094.

3316. Phillips, G. J., and T. J. Silhavy. 1992. The *E. coli ffh* gene is necessary for viability and efficient protein export. *Nature* (London) 359:744–746.

3317. Phillips, R. S., and P. D. Gollnick. 1989. Evidence that cysteine 298 is in the active site of tryptophan indole- lyase. *J. Biol. Chem.* 264:10627–10632.

3318. Phoenix, D. A., S. E. Peters, M. A. Ramzan, and J. M. Pratt. 1994. Analysis of the membrane-anchoring properties of the putative amphiphilic alpha-helical anchor at the C-terminus of *Escherichia coli* PBP 6. *Microbiology* 140:73–77.

3319. Pi, J., P. J. Wookey, and A. J. Pittard. 1991. Cloning and sequencing of the *pheP* gene, which encodes the phenylalanine-specific transport system of *Escherichia coli*. *J. Bacteriol.* 173:3622–3629.

3320. Pi, J., P. J. Wookey, and A. J. Pittard. 1993. Site-directed mutagenesis reveals the importance of conserved charged residues for the transport activity of the PheP permease of *Escherichia coli*. *J. Bacteriol.* 175:7500–7504.

3321. **Pichersky, E., L. D. Gottlieb, and J. F. Hess.** 1984. Nucleotide sequence of the triose phosphate isomerase gene of *Escherichia coli. Mol. Gen. Genet.* 195:314–320.

3322. **Picksley, S. M., S. J. Morton, and R. G. Lloyd.** 1985. The *recN* locus of *Escherichia coli* K12: molecular analysis and identification of the gene product. *Mol. Gen. Genet.* 201:301–307.

3323. **Pierard, A., N. Glansdorff, D. Gigot, M. Crabeel, P. Halleux, and L. Thiry.** 1976. Repression of *Escherichia coli* carbamoylphosphate synthase: relationships with enzyme synthesis in the arginine and pyrimidine pathways. *J. Bacteriol.* 127:291–301.

3324. **Pierard, A., N. Glansdorff, and J. Yashphe.** 1972. Mutations affecting uridine monophosphate pyrophosphorylase or the *argR* gene in *Escherichia coli.* Effects on carbamoyl phosphate and pyrimidine biosynthesis and on uracil uptake. *Mol. Gen. Genet.* 118:235–245.

3325. **Pierce, J. R., and C. F. Earhart.** 1986. *Escherichia coli* K-12 envelope proteins specifically required for ferrienterobactin uptake. *J. Bacteriol.* 166:930–936.

3326. **Pierce, J. R., C. L. Pickett, and C. F. Earhart.** 1983. Two *fep* genes are required for ferrienterochelin uptake in *Escherichia coli* K-12. *J. Bacteriol.* 155:330–336.

3327. **Pierson, D. E., and A. Campbell.** 1990. Cloning and nucleotide sequence of *bisC*, the structural gene for biotin sulfoxide reductase in *Escherichia coli. J. Bacteriol.* 172:2194–2198.

3327a.**Pierson, L. S., III, and M. L. Kahn.** 1987. Integration of satellite bacteriophage P4 in *Escherichia coli.* DNA sequences of the phage and host regions involved in site-specific recombination. *J. Mol. Biol.* 196:487–496.

3328. **Piette, J., R. Cunin, A. Boyen, D. Charlier, M. Crabeel, F. Van Vliet, N. Glansdorff, C. Squires, and C. L. Squires.** 1982. The regulatory region of the divergent *argECBH* operon in *Escherichia coli* K-12. *Nucleic Acids Res.* 10:8031–8048.

3329. **Piette, J., R. Cunin, F. Van Vliet, D. Charlier, M. Crabeel, Y. Ota, and N. Glansdorff.** 1982. Homologous control sites and DNA transcription starts in the related *argF* and *argI* genes of *Escherichia coli* K12. *EMBO J.* 1:853–857.

3330. **Piette, J., H. Nyunoya, C. J. Lusty, R. Cunin, G. Weyens, M. Crabeel, D. Charlier, N. Glansdorff, and A. Pierard.** 1984. DNA sequence of the *carA* gene and the control region of *carAB*: tandem promoters, respectively controlled by arginine and the pyrimidines, regulate the synthesis of carbamoyl-phosphate synthetase in *Escherichia coli* K-12. *Proc. Natl. Acad. Sci. USA* 81:4134–4138.

3331. **Pilipcinec, E., T. T. Huisman, P. T. Willemsen, B. J. Appelmelk, F. K. De Graaf, and B. Oudega.** 1994. Identification by Tn*10* transposon mutagenesis of host factors involved in the biosynthesis of K99 fimbriae of *Escherichia coli*: effect of LPS core mutations. *FEMS Microbiol. Lett.* 123:201–206.

3332. **Pinkham, J. L., and T. Platt.** 1983. The nucleotide sequence of the *rho* gene of *E. coli* K-12. *Nucleic Acids Res.* 11:3531–3545.

3333. **Pinner, E., Y. Kotler, E. Padan, and S. Schuldiner.** 1993. Physiological role of *nhaB*, a specific Na$^+$/H$^+$ antiporter in *Escherichia coli. J. Biol. Chem.* 268:1729–1734.

3334. **Pinner, E., E. Padan, and S. Schuldiner.** 1992. Cloning, sequencing, and expression of the *nhaB* gene, encoding a Na$^+$/H$^+$ antiporter in *Escherichia coli. J. Biol. Chem.* 267:11064–11068.

3335. **Pistocchi, R., K. Kashiwagi, S. Miyamoto, E. Nukui, Y. Sadakata, H. Kobayashi, and K. Igarashi.** 1993. Characteristics of the operon for a putrescine transport system that maps at 19 minutes on the *Escherichia coli* chromosome. *J. Biol. Chem.* 268:146–152.

3336. **Pittard, A. J., and B. E. Davidson.** 1991. TyrR protein of *Escherichia coli* and its role as repressor and activator. *Mol. Microbiol.* 5:1585–1592.

3337. **Pittard, J., J. Praszkier, A. Certoma, G. Eggertsson, J. Gowrishankar, G. Narasaiah, and M. J. Whipp.** 1990. Evidence that there are only two tRNA^{Phe} genes in *Escherichia coli. J. Bacteriol.* 172:6077–6083.

3338. **Pla, J., M. Sanchez, P. Palacios, M. Vicente, and M. Aldea.** 1991. Preferential cytoplasmic location of FtsZ, a protein essential for *Escherichia coli* septation. *Mol. Microbiol.* 5:1681–1686.

3339. **Plamann, M. D., W. D. Rapp, and G. V. Stauffer.** 1983. *Escherichia coli* K12 mutants defective in the glycine cleavage enzyme system. *Mol. Gen. Genet.* 192:15–20.

3340. **Plamann, M. D., and G. V. Stauffer.** 1983. Characterization of the *Escherichia coli* gene for serine hydroxymethyltransferase. *Gene* 22:9–18.

3341. **Plamann, M. D., L. T. Stauffer, M. L. Urbanowski, and G. V. Stauffer.** 1983. Complete nucleotide sequence of the *E. coli glyA* gene. *Nucleic Acids Res.* 11:2065–2075.

3342. **Plasterk, R. H., A. Brinkman, and P. van de Putte.** 1983. DNA inversions in the chromosome of *Escherichia coli* and in bacteriophage Mu: relationship to other site-specific recombination systems. *Proc. Natl. Acad. Sci. USA* 80:5355–5358.

3343. **Plasterk, R. H., and P. van de Putte.** 1985. The invertible P-DNA segment in the chromosome of *Escherichia coli. EMBO J.* 4:237–242.

3344. **Plate, C. A., S. A. Seely, and T. G. Laffler.** 1986. Evidence for a protonmotive force related regulatory system in *Escherichia coli* and its effects on lactose transport. *Biochemistry* 25:6127–6132.

3345. **Platko, J. V., D. A. Willins, and J. M. Calvo.** 1990. The *ilvIH* operon of *Escherichia coli* is positively regulated. *J. Bacteriol.* 172:4563–4570.

3346. **Platt, T., J. G. Files, and K. Weber.** 1973. Lac repressor. Specific proteolytic destruction of the NH$_2$-terminal region and loss of the deoxyribonucleic acid-binding activity. *J. Biol. Chem.* 248:110–121.

3347. **Platz, A., and B. M. Sjoberg.** 1980. Construction and characterization of hybrid plasmids containing the *Escherichia coli nrd* region. *J. Bacteriol.* 143:561–568.

3348. **Plumbridge, J.** 1987. Organisation of the *Escherichia coli* chromosome between genes *glnS* and *glnU. V. Mol. Genet.* 209:618–620.

3349. **Plumbridge, J., and D. Soll.** 1987. The effect of *dam* methylation on the expression of *glnS* in *E. coli. Biochimie* 69:539–541.

3350. **Plumbridge, J. A.** 1989. Sequence of the *nagBACD* operon in *Escherichia coli* K12 and pattern of transcription within the *nag* regulon. *Mol. Microbiol.* 3:505–515.

3351. **Plumbridge, J. A.** 1991. Repression and induction of the *nag* regulon of *Escherichia coli* K-12: the roles of *nagC* and *nagA* in maintenance of the uninduced state. *Mol. Microbiol.* 5:2053–2062.

3352. **Plumbridge, J. A.** 1992. A dominant mutation in the gene for the Nag repressor of *Escherichia coli* that renders the *nag* regulon uninducible. *J. Gen. Microbiol.* 138:1011–1017.

3353. **Plumbridge, J. A., O. Cochet, J. M. Souza, M. M. Altamirano, M. L. Calcagno, and B. Badet.** 1993. Coordinated regulation of amino sugar-synthesizing and -degrading enzymes in *Escherichia coli* K-12. *J. Bacteriol.* 175:4951–4956.

3354. **Plumbridge, J. A., J. G. Howe, M. Springer, D. Touati-Schwartz, J. W. Hershey, and M. Grunberg-Manago.** 1982. Cloning and mapping of a gene for translational initiation factor IF2 in *Escherichia coli. Proc. Natl. Acad. Sci. USA* 79:5033–5037.

3355. **Plumbridge, J. A., and M. Springer.** 1980. Genes for the two subunits of phenylalanyl-tRNA synthesis of *Escherichia coli* are transcribed from the same promoter. *J. Mol. Biol.* 144:595–600.

3356. **Plumbridge, J. A., and M. Springer.** 1982. *Escherichia coli* phenylalanyl-tRNA synthetase operon: transcription studies of wild-type and mutated operons on multicopy plasmids. *J. Bacteriol.* 152:661–668.

3357. **Plumbridge, J. A., and M. Springer.** 1983. Organization of the *Escherichia coli* chromosome around the genes for translation initiation factor IF2 (*infB*) and a transcription termination factor (*nusA*). *J. Mol. Biol.* 167:227–243.

3358. **Plumbridge, J. A., M. Springer, M. Graffe, R. Goursot, and M. Grunberg-Manago.** 1980. Physical localisation and cloning of the structural gene for *E. coli* initiation factor IF3 from a group of genes concerned with translation. *Gene* 11:33–42.

3359. **Plunkett, G., V. Burland, D. L. Daniels, and F. R. Blattner.** 1993. Analysis of the *Escherichia coli* genome. III. DNA sequence of the region from 87.2 to 89.2 minutes. *Nucleic Acids Res.* 21:3391–3398.

3360. **Pogliano, J. A., and J. Beckwith.** 1994. SecD and SecF facilitate protein export in *Escherichia coli. EMBO J.* 13:554–561.

3361. **Pogliano, K. J., and J. Beckwith.** 1993. The Cs *sec* mutants of *Escherichia coli* reflect the cold sensitivity of protein export itself. *Genetics* 133:763–773.

3362. **Pogliano, K. J., and J. Beckwith.** 1994. Genetic and molecular characterization of the *Escherichia coli secD* operon and its products. *J. Bacteriol.* 176:804–814.

3363. **Polacco, M. L., and J. E. Cronan, Jr.** 1981. A mutant of *Escherichia coli* conditionally defective in the synthesis of holo-[acyl carrier protein]. *J. Biol. Chem.* 256:5750–5754.

3364. **Polarek, J. W., G. Williams, and W. Epstein.** 1992. The products of the *kdpDE* operon are required for expression of the Kdp ATPase of *Escherichia coli. J. Bacteriol.* 174:2145–2151.

3365. **Polayes, D. A., P. W. Rice, and J. E. Dahlberg.** 1988. DNA polymerase I activity in *Escherichia coli* is influenced by spot 42 RNA. *J. Bacteriol.* 170:2083–2088.

3366. **Pon, C. L., R. A. Calogero, and C. O. Gualerzi.** 1988. Identification, cloning, nucleotide sequence and chromosomal map location of *hns*, the structural gene for *Escherichia coli* DNA-binding protein H-NS. *Mol. Gen. Genet.* 212:199–202.

3367. **Pon, C. L., B. Wittmann-Liebold, and C. Gualerzi.** 1979. Structure–function relationships in *Escherichia coli* initiation factors. II. Elucidation of the primary structure of initiation factor IF-1. *FEBS Lett.* 101:157–160.

3368. **Pontis, E., X. Y. Sun, H. Jornvall, M. Krook, and P. Reichard.** 1991. ClpB proteins copurify with the anaerobic *Escherichia coli* reductase. *Biochem. Biophys. Res. Commun.* 180:1222–1226.

3369. **Poole, R. K., F. Gibson, and G. Wu.** 1994. The *cydD* gene product, component of a heterodimeric ABC transporter, is required for assembly of periplasmic cytochrome c and of cytochrome bd in *Escherichia coli. FEMS Microbiol. Lett.* 117:217–223.

3370. **Poole, R. K., L. Hatch, M. W. Cleeter, F. Gibson, G. B. Cox, and G. Wu.** 1993. Cytochrome bd biosynthesis in *Escherichia coli*: the sequences of the *cydC* and *cydD* genes suggest that they encode the components of an ABC membrane transporter. *Mol. Microbiol.* 10:421–430.

3371. **Poole, R. K., H. D. Williams, J. A. Downie, and F. Gibson.** 1989. Mutations affecting the cytochrome D-containing oxidase complex of *Escherichia coli* K12: identification and mapping of a fourth locus, *cydD. J. Gen. Microbiol.* 135:1865–1874.

3372. **Poon, A. P., and T. S. Dhillon.** 1986. Temperate coliphage HK253: attachment site and restricted transduction of *proAB* mutants of *Escherichia coli* K-12. *J. Virol.* 60:317–319.

3373. **Popkin, P. S., and W. K. Maas.** 1980. *Escherichia coli* regulatory mutation affecting lysine transport and lysine decarboxylase. *J. Bacteriol.* 141:485–492.

3374. **Popp, J. L.** 1989. Sequence and overexpression of the *menD* gene from *Escherichia coli. J. Bacteriol.* 171:4349–4354.

3375. **Porco, A., and T. Isturiz.** 1991. Selection of *lacZ* operon fusions in genes of gluconate metabolism in *E. coli*. characterization of a *gntT::lacZ* fusion. *Acta Cient. Venez.* 42:270–275.

3376. **Poritz, M. A., H. D. Bernstein, K. Strub, D. Zopf, H. Wilhelm, and P. Walter.** 1990. An *E. coli* ribonucleoprotein containing 4.5S RNA resembles mammalian signal recognition particle. *Science* 250:1111–1117.

3377. Portalier, R., J. Robert-Baudouy, and F. Stoeber. 1980. Regulation of *Escherichia coli* K-12 hexuronate system genes: *exu* regulon. *J. Bacteriol.* 143:1095–1107.

3378. Porter, A. C., W. S. Brusilow, and R. D. Simoni. 1983. Promoter for the *unc* operon of *Escherichia coli*. *J. Bacteriol.* 155:1271–1278.

3379. Porter, A. C., C. Kumamoto, K. Aldape, and R. D. Simoni. 1985. Role of the b subunit of the *Escherichia coli* proton-translocating ATPase. A mutagenic analysis. *J. Biol. Chem.* 260:8182–8187.

3380. Portier, C. 1982. Physical localisation and direction of transcription of the structural gene for *Escherichia coli* ribosomal protein S15. *Gene* 18:261–266.

3381. Portier, C., C. Migot, and M. Grumberg-Manago. 1981. Cloning of *E. coli pnp* gene from an episome. *Mol. Gen. Genet.* 183:298–305.

3382. Portier, C., and P. Regnier. 1984. Expression of the *rpsO* and *pnp* genes: structural analysis of a DNA fragment carrying their control regions. *Nucleic Acids Res.* 12:6091–6102.

3383. Post, D. A., B. Hove-Jensen, and R. L. Switzer. 1993. Characterization of the *hemA-prs* region of the *Escherichia coli* and *Salmonella typhimurium* chromosomes: identification of two open reading frames and implications for *prs* expression. *J. Gen. Microbiol.* 139:259–266.

3384. Post, L. E., A. E. Arfsten, G. R. Davis, and M. Nomura. 1980. DNA sequence of the promoter region for the alpha ribosomal protein operon in *Escherichia coli*. *J. Biol. Chem.* 255:4653–4659.

3385. Post, L. E., A. E. Arfsten, F. Reusser, and M. Nomura. 1978. DNA sequences of promoter regions for the *str* and *spc* ribosomal protein operons in *E. coli*. *Cell* 15:215–229.

3386. Post, L. E., and M. Nomura. 1979. Nucleotide sequence of the intercistronic region preceding the gene for RNA polymerase subunit alpha in *Escherichia coli*. *J. Biol. Chem.* 254:10604–10606.

3387. Post, L. E., and M. Nomura. 1980. DNA sequences from the *str* operon of *Escherichia coli*. *J. Biol. Chem.* 255:4660–4666.

3388. Post, L. E., G. D. Strycharz, M. Nomura, H. Lewis, and P. P. Dennis. 1979. Nucleotide sequence of the ribosomal protein gene cluster adjacent to the gene for RNA polymerase subunit beta in *Escherichia coli*. *Proc. Natl. Acad. Sci. USA* 76:1697–1701.

3389. Postle, K., and R. F. Good. 1983. DNA sequence of the *Escherichia coli tonB* gene. *Proc. Natl. Acad. Sci. USA* 80:5235–5239.

3390. Postle, K., and R. F. Good. 1985. A bidirectional rho-independent transcription terminator between the *E. coli tonB* gene and an opposing gene. *Cell* 41:577–585.

3391. Postle, K., and W. S. Reznikoff. 1979. Identification of the *Escherichia coli tonB* gene product in minicells containing *tonB* hybrid plasmids. *J. Mol. Biol.* 131:619–636.

3392. Postma, P. W. 1977. Galactose transport in *Salmonella typhimurium*. *J. Bacteriol.* 129:630–639.

3393. Postma, P. W., and J. W. Lengeler. 1985. Phosphoenolpyruvate:carbohydrate phosphotransferase system of bacteria. *Microbiol. Rev.* 49:232–269.

3394. Potter, P. M., L. Harris, and G. P. Margison. 1989. Mapping of OGT in the *E. coli* chromosome. *Nucleic Acids Res.* 17:10505.

3395. Potter, P. M., K. Kleibl, L. Cawkwell, and G. P. Margison. 1989. Expression of the *ogt* gene in wild-type and *ada* mutants of *E. coli*. *Nucleic Acids Res.* 17:8047–8060.

3396. Potter, P. M., M. C. Wilkinson, J. Fitton, F. J. Carr, J. Brennand, D. P. Cooper, and G. P. Margison. 1987. Characterisation and nucleotide sequence of *ogt*, the O^6-alkylguanine-DNA-alkyltransferase gene of *E. coli*. *Nucleic Acids Res.* 15:9177–9193.

3397. Poulsen, L. K., N. W. Larsen, S. Molin, and P. Andersson. 1989. A family of genes encoding a cell-killing function may be conserved in all gram-negative bacteria. *Mol. Microbiol.* 3:1463–1472.

3398. Poulsen, L. K., N. W. Larsen, S. Molin, and P. Andersson. 1992. Analysis of an *Escherichia coli* mutant strain resistant to the cell-killing function encoded by the *gef* gene family. *Mol. Microbiol.* 6:895–905.

3399. Poulsen, L. K., A. Refn, S. Molin, and P. Andersson. 1991. Topographic analysis of the toxic Gef protein from *Escherichia coli*. *Mol. Microbiol.* 5:1627–1637.

3400. Poulsen, L. K., A. Refn, S. Molin, and P. Andersson. 1991. The *gef* gene from *Escherichia coli* is regulated at the level of translation. *Mol. Microbiol.* 5:1639–1648.

3401. Poulsen, P., J. T. Andersen, and K. F. Jensen. 1989. Molecular and mutational analysis of three genes preceding *pyrE* on the *Escherichia coli* chromosome. *Mol. Microbiol.* 3:393–404.

3402. Poulsen, P., F. Bonekamp, and K. F. Jensen. 1984. Structure of the *Escherichia coli pyrE* operon and control of *pyrE* expression by a UTP modulated intercistronic attentuation. *EMBO J.* 3:1783–1790.

3403. Poulsen, P., and K. F. Jensen. 1991. Three genes preceding *pyrE* on the *Escherichia coli* chromosome are essential for survival and normal cell morphology in stationary culture and at high temperature. *Res. Microbiol.* 142:283–288.

3404. Poulsen, P., K. F. Jensen, P. Valentin-Hansen, P. Carlsson, and L. G. Lundberg. 1983. Nucleotide sequence of the *Escherichia coli pyrE* gene and of the DNA in front of the protein-coding region. *Eur. J. Biochem.* 135:223–229.

3405. Pourcher, T., M. Deckert, M. Bassilana, and G. Leblanc. 1991. Melibiose permease of *Escherichia coli*: mutation of aspartic acid 55 in putative helix II abolishes activation of sugar binding by Na^+ ions. *Biochem. Biophys. Res. Commun.* 178:1176–1181.

3406. Pouyssegur, J. M., and F. R. Stoeber. 1972. A common pathway for hexouronate degradation in *Escherichia coli* K 12. Induction mechanism of 2-keto-3-deoxy-gluconate metabolizing enzymes. *Eur. J. Biochem.* 30:479–494. (In Russian.)

3406a.Powell, K. A., R. Cox, M. McConville, and H. P. Charles. 1973. Mutations affecting porphyrin biosynthesis in *Escherichia coli*. *Enzyme* 16:65–73.

3407. Powers, D. A., and S. Roseman. 1984. The primary structure of *Salmonella typhimurium* HPr, a phosphocarrier protein of the phosphoenolpyruvate:glycose phosphotransferase system. A correction. *J. Biol. Chem.* 259:15212–15214.

3408. Powers, E. L., and L. L. Randall. 1995. Export of periplasmic galactose-binding protein in *Escherichia coli* depends on the chaperone SecB. *J. Bacteriol.* 1771:1906–1907.

3409. Pradel, E., and P. L. Boquet. 1988. Acid phosphatases of *Escherichia coli*: molecular cloning and analysis of *agp*, the structural gene for a periplasmic acid glucose phosphatase. *J. Bacteriol.* 170:4916–4923.

3410. Pradel, E., and P. L. Boquet. 1989. Mapping of the *Escherichia coli* acid glucose-1-phosphatase gene *agp* and analysis of its expression in vivo by use of an *agp-phoA* protein fusion. *J. Bacteriol.* 171:3511–3517.

3411. Pradel, E., and P. L. Boquet. 1991. Utilization of exogenous glucose-1-phosphate as a source of carbon or phosphate by *Escherichia coli* K12: respective roles of acid glucose-1-phosphatase, hexose-phosphate permease, phosphoglucomutase and alkaline phosphatase. *Res. Microbiol.* 142:37–45.

3412. Pradel, E., C. Marck, and P. L. Boquet. 1990. Nucleotide sequence and transcriptional analysis of the *Escherichia coli agp* gene encoding periplasmic acid glucose-1-phosphatase. *J. Bacteriol.* 172:802–807.

3413. Pradel, E., C. T. Parker, and C. A. Schnaitman. 1992. Structures of the *rfaB, rfaI, rfaJ*, and *rfaS* genes of *Escherichia coli* K-12 and their roles in assembly of the lipopolysaccharide core. *J. Bacteriol.* 174:4736–4745.

3414. Pradel, E., and C. A. Schnaitman. 1991. Effect of *rfaH* (*sfrB*) and temperature on expression of *rfa* genes of *Escherichia coli* K-12. *J. Bacteriol.* 173:6428–6431.

3415. Pramanik, A., S. J. Wertheimer, J. J. Schwartz, and I. Schwartz. 1986. Expression of *Escherichia coli infC*: identification of a promoter in an upstream *thrS* coding sequence. *J. Bacteriol.* 168:746–751.

3416. Prasad, I., B. Young, and S. Schaefler. 1973. Genetic determination of the constitutive biosynthesis of phospho-glucosidase A in *Escherichia coli* K-12. *J. Bacteriol.* 114:909–915.

3417. Prasher, D., D. A. Kasunic, and S. R. Kushner. 1983. Physical and genetic characterization of the cloned *sbcB* (exonuclease I) region of the *Escherichia coli* genome. *J. Bacteriol.* 153:903–908.

3418. Prather, N. E., B. H. Himes, and E. J. Murgola. 1983. *supG* and *supL* in *Escherichia coli* code for mutant lysine tRNAs[+]. *Nucleic Acids Res.* 11:8283–8286.

3419. Pratt, D., and S. Subramani. 1983. Nucleotide sequence of the *Escherichia coli* xanthine-guanine phosphoribosyl transferase gene. *Nucleic Acids Res.* 11:8817–8823.

3420. Pratviel-Sosa, F., D. Mengin-Lecreulx, and J. van Heijenoort. 1991. Over-production, purification and properties of the uridine diphosphate N-acetylmuramoyl-L-alanine:D-glutamate ligase from *Escherichia coli*. *Eur. J. Biochem.* 202:1169–1176.

3421. Prescott, C. D., and H. C. Kornau. 1992. Mutations in *E. coli* 16s rRNA that enhance and decrease the activity of a suppressor tRNA. *Nucleic Acids Res.* 20:1567–1571.

3422. Pressler, U., H. Staudenmaier, L. Zimmermann, and V. Braun. 1988. Genetics of the iron dicitrate transport system of *Escherichia coli*. *J. Bacteriol.* 170:2716–2724.

3423. Prince, W. S., and M. R. Villarejo. 1990. Osmotic control of *proU* transcription is mediated through direct action of potassium glutamate on the transcription complex. *J. Biol. Chem.* 265:17673–17679.

3424. Prior, T. I., and H. L. Kornberg. 1988. Nucleotide sequence of *fruA*, the gene specifying enzyme IIfru of the phosphoenolpyruvate-dependent sugar phosphotransferase system in *Escherichia coli* K12. *J. Gen. Microbiol.* 134:2757–2768.

3425. Pritchard, R. H., and S. I. Ahmad. 1971. Fluorouracil and the isolation of mutants lacking uridine phosphorylase in *Escherichia coli*: location of the gene. *Mol. Gen. Genet.* 111:84–88.

3426. Privalle, C. T., and I. Fridovich. 1987. Induction of superoxide dismutase in *Escherichia coli* by heat shock. *Proc. Natl. Acad. Sci. USA* 84:2723–2726.

3427. Privalle, C. T., and I. Fridovich. 1992. Transcriptional and maturational effects of manganese and iron on the biosynthesis of manganese-superoxide dismutase in *Escherichia coli*. *J. Biol. Chem.* 267:9140–9145.

3428. Prodromou, C., P. J. Artymiuk, and J. R. Guest. 1992. The aconitase of *Escherichia coli*. Nucleotide sequence of the aconitase gene and amino acid sequence similarity with mitochondrial aconitases, the iron-responsive-element-binding protein and isopropylmalate isomerases. *Eur. J. Biochem.* 204:599–609.

3429. Prodromou, C., M. J. Haynes, and J. R. Guest. 1991. The aconitase of *Escherichia coli*: purification of the enzyme and molecular cloning and map location of the gene (*acn*). *J. Gen. Microbiol.* 137:2505–2515.

3430. Prody, C. A., and J. B. Neilands. 1984. Genetic and biochemical characterization of the *Escherichia coli* K-12 *fhuB* mutation. *J. Bacteriol.* 157:874–880.

3431. Provence, D. L., and R. Curtiss. 1992. Role of *crl* in avian pathogenic *Escherichia coli*: a knockout mutation of *crl* does not affect hemagglutination activity, fibronectin binding, or curli production. *Infect. Immun.* 60:4460–4467.

3432. Pruss, B. M., J. M. Nelms, C. Park, and A. J. Wolfe. 1994. Mutations in NADH:ubiquinone oxidoreductase of *Escherichia coli* affect growth on mixed amino acids. *J. Bacteriol.* 176:2143–2150.

3433. Pucci, M. J., L. F. Discotto, and T. J. Dougherty. 1992. Cloning and identification of the *Escherichia coli murB* DNA sequence, which encodes UDP-N-acetylenolpyruvoylglucosamine reductase. *J. Bacteriol.* 174:1690–1693.

3434. Pugh, B. F., and M. M. Cox. 1988. General mechanism for RecA protein binding to duplex DNA. *J. Mol. Biol.* 203:479–493.

3435. **Pugsley, A. P.** 1985. *Escherichia coli* K12 strains for use in the identification and characterization of colicins. *J. Gen. Microbiol.* 131:369–376.

3436. **Pugsley, A. P.** 1992. Translocation of a folded protein across the outer membrane in *Escherichia coli. Proc. Natl. Acad. Sci. USA* 89:12058–12062.

3437. **Pugsley, A. P.** 1993. A mutation in the *dsbA* gene coding for periplasmic disulfide oxidoreductase reduces transcription of the *Escherichia coli ompF* gene. *Mol. Gen. Genet.* 237:407–411.

3438. **Pugsley, A. P., and C. Dubreuil.** 1988. Molecular characterization of *malQ*, the structural gene for the *Escherichia coli* enzyme amylomaltase. *Mol. Microbiol.* 2:473–479.

3439. **Pugsley, A. P., M. G. Kornacker, and I. Poquet.** 1991. The general protein-export pathway is directly required for extracellular pullulanase secretion in *Escherichia coli* K12. *Mol. Microbiol.* 5:343–352.

3440. **Pugsley, A. P., F. Moreno, and V. De Lorenzo.** 1986. Microcin-E492-insensitive mutants of *Escherichia coli* K12. *J. Gen. Microbiol.* 132:3253–3259.

3441. **Pugsley, A. P., and P. Reeves.** 1976. Iron uptake in colicin B-resistant mutants of *Escherichia coli* K-12. *J. Bacteriol.* 126:1052–1062.

3442. **Puppe, W., A. Siebers, and K. Altendorf.** 1992. The phosphorylation site of the Kdp-ATPase of *Escherichia coli*: site- directed mutagenesis of the aspartic acid residues 300 and 307 of the KdpB subunit. *Mol. Microbiol.* 6:3511–3520.

3443. **Purewal, A. S.** 1991. Nucleotide sequence of the ethidium efflux gene from *Escherichia coli. FEMS Microbiol. Lett.* 66:229–231.

3444. **Putney, S. D., D. L. Melendez, and P. R. Schimmel.** 1981. Cloning, partial sequencing, and in vitro transcription of the gene for alanine tRNA synthetase. *J. Biol. Chem.* 256:205–211.

3445. **Putney, S. D., N. J. Royal, H. Neuman de Vegvar, W. C. Herlihy, K. Biemann, and P. Schimmel.** 1981. Primary structure of a large aminoacyl-tRNA synthetase. *Science* 213:1497–1501.

3446. **Putney, S. D., and P. Schimmel.** 1981. An aminoacyl tRNA synthetase binds to a specific DNA sequence and regulates its gene transcription. *Nature* (London) 291:632–635.

3447. **Puyo, M. F., P. Calsou, and B. Salles.** 1992. UV resistance of *E. coli* K-12 deficient in cAMP/CRP regulation. *Mutat. Res.* 282:247–252.

3448. **Pyne, C., and A. L. Bognar.** 1992. Replacement of the *folC* gene, encoding folylpolyglutamate synthetase-dihydrofolate synthetase in *Escherichia coli,* with genes mutagenized in vitro. *J. Bacteriol.* 174:1750–1759.

3449. **Qi, S. Y., S. Sukupolvi, and C. D. O'Connor.** 1991. Outer membrane permeability of *Escherichia coli* K12: isolation, cloning and mapping of suppressors of a defined antibiotic-hypersensitive mutant. *Mol. Gen. Genet.* 229:421–427.

3449a. **Quail, M. A., C. E. Dempsey, and J. R. Guest.** 1994. Identification of a fatty acyl responsive regulator (FarR) in *Escherichia coli. FEBS Lett.* 356:183–187.

3450. **Quail, M. A., D. J. Haydon, and J. R. Guest.** 1994. The *pdhR-aceEF-lpd* operon of *Escherichia coli* expresses the pyruvate dehydrogenase complex. *Mol. Microbiol.* 12:95–104.

3451. **Quinones, A., W. R. Juterbock, and W. Messer.** 1991. Expression of the *dnaA* gene of *Escherichia coli* is inducible by DNA damage. *Mol. Gen. Genet.* 227:9–16.

3452. **Quinones, A., and W. Messer.** 1988. Discoordinate gene expression in the *dnaA-dnaN* operon of *Escherichia coli. Mol. Gen. Genet.* 213:118–124.

3453. **Quintilla, F. X., L. Baldoma, J. Badia, and J. Aguilar.** 1991. Aldehyde dehydrogenase induction by glutamate in *Escherichia coli*. Role of 2-oxoglutarate. *Eur. J. Biochem.* 202:1321–1325.

3454. **Quiocho, F. A., G. L. Gilliland, and G. N. Phillips, Jr.** 1977. The 2.8-A resolution structure of the L-arabinose-binding protein from *Escherichia coli*. Polypeptide chain folding, domain similarity, and probable location of sugar-binding site. *J. Biol. Chem.* 252:5142–5149.

3455. **Quirk, S., S. K. Bhatnagar, and M. J. Bessman.** 1990. Primary structure of the deoxyguanosine triphosphate triphosphohydrolase-encoding gene (*dgt*) of *Escherichia coli. Gene* 89:13–18.

3456. **Quirk, S., D. Seto, S. K. Bhatnagar, P. Gauss, L. Gold, and M. J. Bessman.** 1989. Location and molecular cloning of the structural gene for the deoxyguanosine triphosphate triphosphohydrolase of *Escherichia coli. Mol. Microbiol.* 3:1391–1395.

3457. **Rabin, R. S., and V. Stewart.** 1992. Either of two functionally redundant sensor proteins, NarX and NarQ, is sufficient for nitrate regulation in *Escherichia coli* K-12. *Proc. Natl. Acad. Sci. USA* 89:8419–8423.

3458. **Rabin, R. S., and V. Stewart.** 1993. Dual response regulators (NarL and NarP) interact with dual sensors (NarX and NarQ) to control nitrate- and nitrite-regulated gene expression in *Escherichia coli* K-12. *J. Bacteriol.* 175:3259–3268.

3459. **Radding, C. M.** 1991. Helical interactions in homologous pairing and strand exchange driven by RecA protein. *J. Biol. Chem.* 266:5355–5358.

3460. **Radicella, J. P., E. A. Clark, S. Chen, and M. S. Fox.** 1993. Patch length of localized repair events: role of DNA polymerase I in *mutY*-dependent mismatch repair. *J. Bacteriol.* 175:7732–7736.

3460a. **Radke, K. L., and E. C. Siegel.** 1971. Mutation preventing capsular polysaccharide synthesis in *Escherichia coli* K-12 and its effect on bacteriophage resistance. *J. Bacteriol.* 106:432–437.

3461. **Raetz, C. R., G. D. Kantor, M. Nishijima, and M. L. Jones.** 1981. Isolation of *Escherichia coli* mutants with elevated levels of membrane enzymes. A trans-acting mutation controlling diglyceride kinase. *J. Biol. Chem.* 256:2109–2112.

3462. **Raftery, L. A., and M. Yarus.** 1985. Site-specific mutagenesis of *Escherichia coli gltT* yields a weak, glutamic acid-inserting ochre suppressor. *J. Mol. Biol.* 184:343–345.

3463. **Raha, M., I. Kawagishi, V. Muller, M. Kihara, and R. M. Macnab.** 1992. *Escherichia coli* produces a cytoplasmic alpha-amylase, AmyA. *J. Bacteriol.* 174:6644–6652.

3464. **Raha, M., M. Kihara, I. Kawagishi, and R. M. Macnab.** 1993. Organization of the *Escherichia coli* and *Salmonella typhimurium* chromosomes between flagellar regions IIIa and IIIb, including a large non-coding region. *J. Gen. Microbiol.* 139:1401–1407.

3465. **Rahav-Manor, O., O. Carmel, R. Karpel, D. Taglicht, G. Glaser, S. Schuldiner, and E. Padan.** 1992. NhaR, a protein homologous to a family of bacterial regulatory proteins (LysR), regulates *nhaA*, the sodium proton antiporter gene in *Escherichia coli. J. Biol. Chem.* 267:10433–10438.

3466. **Rahman, M. A., J. Moskovitz, J. Strassman, H. Weissbach, and N. Brot.** 1994. Physical map location of the peptide methionine sulfoxide reductase gene on the *Escherichia coli* chromosome. *J. Bacteriol.* 176:1548–1549.

3467. **Rahman, M. A., H. Nelson, H. Weissbach, and N. Brot.** 1992. Cloning, sequencing, and expression of the *Escherichia coli* peptide methionine sulfoxide reductase gene. *J. Biol. Chem.* 267:15549–15551.

3468. **Raibaud, O., and M. Schwartz.** 1980. Restriction map of the *Escherichia coli malA* region and identification of the *malT* product. *J. Bacteriol.* 143:761–771.

3469. **Raibaud, O., D. Vidal-Ingigliardi, and A. Kolb.** 1991. Genetic studies on the promoter of *malT*, the gene that encodes the activator of the *Escherichia coli* maltose regulon. *Res. Microbiol.* 142:937–942.

3470. **Raina, S., and C. Georgopoulos.** 1990. A new *Escherichia coli* heat shock gene, *htrC*, whose product is essential for viability only at high temperatures. *J. Bacteriol.* 172:3417–3426.

3471. **Raina, S., and C. Georgopoulos.** 1991. The *htrM* gene, whose product is essential for *Escherichia coli* viability only at elevated temperatures, is identical to the *rfaD* gene. *Nucleic Acids Res.* 19:3811–3819.

3472. **Raina, S., L. Mabey, and C. Georgopoulos.** 1991. The *Escherichia coli* htrP gene product is essential for bacterial growth at high temperatures: mapping, cloning, sequencing, and transcriptional regulation of *htrP. J. Bacteriol.* 173:5999–6008.

3473. **Raina, S., D. Missiakas, L. Baird, S. Kumar, and C. Georgopoulos.** 1993. Identification and transcriptional analysis of the *Escherichia coli htrE* operon which is homologous to *pap* and related pilin operons. *J. Bacteriol.* 175:5009–5021.

3474. **Raina, S., D. Missiakas, and C. Georgopoulos.** 1995. The *rpoE* gene encoding the sigma E (sigma 24) heat shock sigma factor of *Escherichia coli. EMBO J.* 14:1043–1055.

3475. **Rainwater, S., and P. M. Silverman.** 1990. The Cpx proteins of *Escherichia coli* K-12: evidence that *cpxA, ecfB, ssd,* and *eup* mutations all identify the same gene. *J. Bacteriol.* 172:2456–2461.

3476. **Raji, A., D. J. Zabel, C. S. Laufer, and R. E. Depew.** 1985. Genetic analysis of mutations that compensate for loss of *Escherichia coli* DNA topoisomerase I. *J. Bacteriol.* 162:1173–1179.

3477. **Rakonjac, J., M. Milic, D. Ajdic-Predic, D. Santos, R. Ivanisevic, and D. J. Savic.** 1992. *nov*: a new genetic locus that affects the response of *Escherichia coli* K-12 to novobiocin. *Mol. Microbiol.* 6:1547–1553.

3478. **Raleigh, E. A.** 1992. Organization and function of the *mcrBC* genes of *Escherichia coli* K-12. *Mol. Microbiol.* 6:1079–1086.

3479. **Raleigh, E. A., J. Benner, F. Bloom, H. D. Braymer, E. DeCruz, K. Dharmalingam, J. Heitman, M. Noyer-Weidner, A. Piekarowicz, P. L. Kretz, J. M. Short, and D. Woodcock.** 1991. Nomenclature relating to restriction of modified DNA in *Escherichia coli. J. Bacteriol.* 173:2707–2709.

3480. **Raleigh, E. A., R. Trimarchi, and H. Revel.** 1989. Genetic and physical mapping of the *mcrA* (*rglA*) and *mcrB* (*rglB*) loci of *Escherichia coli* K-12. *Genetics* 122:279–296.

3481. **Raleigh, E. A., and G. Wilson.** 1986. *Escherichia coli* K-12 restricts DNA containing 5-methylcytosine. *Proc. Natl. Acad. Sci. USA* 83:9070–9074.

3482. **Ralling, G., and T. Linn.** 1984. Relative activities of the transcriptional regulatory sites in the *rplKAJLrpoBC* gene cluster of *Escherichia coli. J. Bacteriol.* 158:279–285.

3483. **Ramabhadran, T. V.** 1976. Method for the isolation of *Escherichia coli* relaxed mutants, utilizing near-ultraviolet irradiation. *J. Bacteriol.* 127:1587–1589.

3484. **Ramani, N., L. Huang, and M. Freundlich.** 1992. In vitro interactions of integration host factor with the *ompF* promoter-regulatory region of *Escherichia coli. Mol. Gen. Genet.* 231:248–255.

3485. **Ramirez, R. M., and M. Villarejo.** 1991. Osmotic signal transduction to *proU* is independent of DNA supercoiling in *Escherichia coli. J. Bacteriol.* 173:879–885.

3486. **Randall, L. L.** 1992. Peptide binding by chaperone SecB: implications for recognition of nonnative structure. *Science* 257:241–245.

3487. **Rao, N. N., and A. Torriani.** 1990. Molecular aspects of phosphate transport in *Escherichia coli. Mol. Microbiol.* 4:1083–1090.

3487a. **Rasmussen, L. J., M. G. Marinus, and A. Lobner-Olesen.** 1994. Novel growth rate control of *dam* gene expression in *Escherichia coli. Mol. Microbiol.* 12:631–638.

3488. **Ravi, R. S., S. Sozhamannan, and K. Dharmalingam.** 1985. Transposon mutagenesis and genetic mapping of the *rglA* and *rglB* loci of *Escherichia coli. Mol. Gen. Genet.* 198:390–392.

3489. **Ravnikar, P. D., and R. L. Somerville.** 1986. Localization of the structural gene for threonine dehydrogenase in *Escherichia coli. J. Bacteriol.* 168:434–436.

3490. **Ravnikar, P. D., and R. L. Somerville.** 1987. Structural and functional analysis of a cloned segment of *Escherichia coli* DNA that specifies proteins of a C4 pathway of serine biosynthesis. *J. Bacteriol.* 169:4716–4721.

3491. Rawlings, M., and J. E. Cronan, Jr. 1992. The gene encoding *Escherichia coli* acyl carrier protein lies within a cluster of fatty acid biosynthetic genes. *J. Biol. Chem.* **267:**5751–5754.

3492. Ray, A., and D. Apirion. 1982. Characterization of DNA from the *rne* gene of *Escherichia coli*: uniqueness of the *rne* DNA. *Biochem. Biophys. Res. Commun.* **107:**1361–1367.

3493. Ray, J. M., C. Yanofsky, and R. Bauerle. 1988. Mutational analysis of the catalytic and feedback sites of the tryptophan-sensitive 3-deoxy-D-arabino-heptulosonate-7-phosphate synthase of *Escherichia coli*. *J. Bacteriol.* **170:**5500–5506.

3494. Ray, N. B., and C. K. Mathews. 1992. Nucleoside diphosphokinase: a functional link between intermediary metabolism and nucleic acid synthesis. *Curr. Top. Cell Regul.* **33:**343–357.

3495. RayChaudhuri, D., and J. T. Park. 1992. *Escherichia coli* cell-division gene *ftsZ* encodes a novel GTP-binding protein. *Nature* (London) **359:**251–254.

3496. Rayssiguier, C., D. S. Thaler, and M. Radman. 1989. The barrier to recombination between *Escherichia coli* and *Salmonella typhimurium* is disrupted in mismatch-repair mutants. *Nature* (London) **342:**396–401.

3497. Ream, L. W., and A. J. Clark. 1983. Cloning and deletion mapping of the *recF dnaN* region of the *Escherichia coli* chromosome. *Plasmid* **10:**101–110.

3498. Ream, L. W., L. Margossian, A. J. Clark, F. G. Hansen, and K. von Meyenburg. 1980. Genetic and physical mapping of *recF* in *Escherichia coli* K-12. *Mol. Gen. Genet.* **180:**115–121.

3499. Reaney, S. K., C. Begg, S. J. Bungard, and J. R. Guest. 1993. Identification of the L-tartrate dehydratase genes (*ttdA* and *ttdB*) of *Escherichia coli* and evolutionary relationship with the class I fumarase genes. *J. Gen. Microbiol.* **139:**1523–1530.

3500. Rebeck, G. W., and L. Samson. 1991. Increased spontaneous mutation and alkylation sensitivity of *Escherichia coli* strains lacking the *ogt* O^6-methylguanine DNA repair methyltransferase. *J. Bacteriol.* **173:**2068–2076.

3501. Rech, S., U. Deppenmeier, and R. P. Gunsalus. 1995. Regulation of the molybdate transport operon, *modABCD*, of *Escherichia coli* in response to molybdate availability. *J. Bacteriol.* **177:**1023–1029.

3502. Recny, M. A., C. Grabau, J. E. Cronan, Jr., and L. P. Hager. 1985. Characterization of the alpha-peptide released upon protease activation of pyruvate oxidase. *J. Biol. Chem.* **260:**14287–14291.

3503. Reece, R. J., and A. Maxwell. 1991. DNA gyrase: structure and function. *Crit. Rev. Biochem. Mol. Biol.* **26:**335–375.

3504. Reece, R. J., and A. Maxwell. 1991. The C-terminal domain of the *Escherichia coli* DNA gyrase A subunit is a DNA-binding protein. *Nucleic Acids Res.* **19:**1399–1405.

3505. Reed, K. E., and J. E. Cronan, Jr. 1993. Lipoic acid metabolism in *Escherichia coli*: sequencing and functional characterization of the *lipA* and *lipB* genes. *J. Bacteriol.* **175:**1325–1336.

3506. Reed, K. E., T. W. Morris, and J. E. Cronan, Jr. 1994. Mutants of *Escherichia coli* K-12 that are resistant to a selenium analog of lipoic acid identify unknown genes in lipoate metabolism. *Proc. Natl. Acad. Sci. USA* **91:**3720–3724.

3507. Reed, R. E., and S. Altman. 1983. Repeated sequences and open reading frames in the 3′ flanking region of the gene for the RNA subunit of *Escherichia coli* ribonuclease P. *Proc. Natl. Acad. Sci. USA* **80:**5359–5363.

3508. Reed, R. E., M. F. Baer, C. Guerrier-Takada, H. Donis-Keller, and S. Altman. 1982. Nucleotide sequence of the gene encoding the RNA subunit (M1 RNA) of ribonuclease P from *Escherichia coli*. *Cell* **30:**627–636.

3509. Reeder, T., and R. Schleif. 1991. Mapping, sequence, and apparent lack of function of *araJ*, a gene of the *Escherichia coli* arabinose regulon. *J. Bacteriol.* **173:**7765–7771.

3510. Reeve, E. C. 1968. Genetic analysis of some mutations causing resistance to tetracycline in *Escherichia coli* K12. *Genet. Res.* **11:**303–309.

3511. Regnier, P., M. Grunberg-Manago, and C. Portier. 1987. Nucleotide sequence of the *pnp* gene of *Escherichia coli* encoding polynucleotide phosphorylase. Homology of the primary structure of the protein with the RNA-binding domain of ribosomal protein S1. *J. Biol. Chem.* **262:**63–68.

3512. Regnier, P., and C. Portier. 1986. Initiation, attenuation and RNase III processing of transcripts from the *Escherichia coli* operon encoding ribosomal protein S15 and polynucleotide phosphorylase. *J. Mol. Biol.* **187:**23–32.

3513. Regue, M., J. Remenick, M. Tokunaga, G. A. Mackie, and H. C. Wu. 1984. Mapping of the lipoprotein signal peptidase gene (*lsp*). *J. Bacteriol.* **158:**632–635.

3514. Rehemtulla, A., S. K. Kadam, and K. E. Sanderson. 1986. Cloning and analysis of the *sfrB* (sex factor repression) gene of *Escherichia coli* K-12. *J. Bacteriol.* **166:**651–657.

3515. Reidl, J., and W. Boos. 1991. The *malX malY* operon of *Escherichia coli* encodes a novel enzyme II of the phosphotransferase system recognizing glucose and maltose and an enzyme abolishing the endogenous induction of the maltose system. *J. Bacteriol.* **173:**4862–4876.

3516. Reidl, J., K. Romisch, M. Ehrmann, and W. Boos. 1989. MalI, a novel protein involved in regulation of the maltose system of *Escherichia coli*, is highly homologous to the repressor proteins GalR, CytR, and LacI. *J. Bacteriol.* **171:**4888–4899.

3517. Reinbolt, J., and E. Schiltz. 1973. The primary structure of ribosomal protein S4 from *Escherichia coli*. *FEBS Lett.* **36:**250–252.

3518. Reinbolt, J., D. Tritsch, and B. Wittmann-Liebold. 1979. The primary structure of ribosomal protein S7 from *E. coli* strains K and B. *Biochimie* **61:**501–522.

3519. Reiner, A. M. 1977. Xylitol and D-arabitol toxicities due to derepressed fructose, galactitol, and sorbitol phosphotransferases of *Escherichia coli*. *J. Bacteriol.* **132:**166–173.

3520. Reiss, J., A. Kleinhofs, and W. Klingmuller. 1987. Cloning of seven differently complementing DNA fragments with *chl* functions from *Escherichia coli* K12. *Mol. Gen. Genet.* **206:**352–355.

3521. Reitzer, L. J., and B. Magasanik. 1985. Expression of *glnA* in *Escherichia coli* is regulated at tandem promoters. *Proc. Natl. Acad. Sci. USA* **82:**1979–1983.

3522. Reizer, J., A. Reizer, H. L. Kornberg, and M. H. Saier, Jr. 1994. Sequence of the *fruB* gene of *Escherichia coli* encoding the diphosphoryl transfer protein (DTP) of the phosphoenolpyruvate: sugar phosphotransferase system. *FEMS Microbiol. Lett.* **118:**159–162.

3523. Reizer, J., A. Reizer, and M. H. Saier, Jr. 1990. The cellobiose permease of *Escherichia coli* consists of three proteins and is homologous to the lactose permease of *Staphylococcus aureus*. *Res. Microbiol.* **141:**1061–1067.

3524. Reizer, J., A. Reizer, and M. H. Saier, Jr. 1990. The Na$^+$/pantothenate symporter (PanF) of *Escherichia coli* is homologous to the Na$^+$/proline symporter (PutP) of *E. coli* and the Na+/glucose symporters of mammals. *Res. Microbiol.* **141:**1069–1072.

3525. Reizer, J., A. Reizer, M. H. Saier, Jr., P. Bork, and C. Sander. 1993. Exopolyphosphate phosphatase and guanosine pentaphosphate phosphatase belong to the sugar kinase/actin/hsp 70 superfamily. *Trends Biochem. Sci.* **18:**247–248.

3526. Repoila, F., and C. Gutierrez. 1991. Osmotic induction of the periplasmic trehalase in *Escherichia coli* K12: characterization of the *treA* gene promoter. *Mol. Microbiol.* **5:**747–755.

3527. Retallack, D. M., L. L. Johnson, and D. I. Friedman. 1994. Role for 10Sa RNA in the growth of lambda-P22 hybrid phage. *J. Bacteriol.* **176:**2082–2089.

3528. Reuter, K., S. Chong, F. Ullrich, H. Kersten, and G. A. Garcia. 1994. Serine 90 is required for enzymic activity by tRNA-guanine transglycosylase from *Escherichia coli*. *Biochemistry* **33:**7041–7046.

3529. Reuter, K., R. Slany, F. Ullrich, and H. Kersten. 1991. Structure and organization of *Escherichia coli* genes involved in biosynthesis of the deazaguanine derivative queuine, a nutrient factor for eukaryotes. *J. Bacteriol.* **173:**2256–2264.

3530. Reynolds, A. E., J. Felton, and A. Wright. 1981. Insertion of DNA activates the cryptic *bgl* operon in *E. coli* K12. *Nature* (London) **293:**625–629.

3531. Reynolds, A. E., S. Mahadevan, S. F. LeGrice, and A. Wright. 1986. Enhancement of bacterial gene expression by insertion elements or by mutation in a CAP-cAMP binding site. *J. Mol. Biol.* **191:**85–95.

3532. Ribes, V., K. Romisch, A. Giner, B. Dobberstein, and D. Tollervey. 1990. *E. coli* 4.5S RNA is part of a ribonucleoprotein particle that has properties related to signal recognition particle. *Cell* **63:**591–600.

3533. Rice, M. S., and F. W. Dahlquist. 1991. Sites of deamidation and methylation in Tsr, a bacterial chemotaxis sensory transducer. *J. Biol. Chem.* **266:**9746–9753.

3534. Rice, P. W., and J. E. Dahlberg. 1982. A gene between *polA* and *glnA* retards growth of *Escherichia coli* when present in multiple copies: physiological effects of the gene for spot 42 RNA. *J. Bacteriol.* **152:**1196–1210.

3535. Richardson, K. K., J. Fostel, and T. R. Skopek. 1983. Nucleotide sequence of the xanthine guanine phosphoribosyl transferase gene of *E. coli*. *Nucleic Acids Res.* **11:**8809–8816.

3536. Richardson, K. K., F. C. Richardson, R. M. Crosby, J. A. Swenberg, and T. R. Skopek. 1987. DNA base changes and alkylation following in vivo exposure of *Escherichia coli* to N-methyl-N-nitrosourea or N-ethyl-N-nitrosourea. *Proc. Natl. Acad. Sci. USA* **84:**344–348.

3537. Richaud, C., W. Higgins, D. Mengin-Lecreulx, and P. Stragier. 1987. Molecular cloning, characterization, and chromosomal localization of *dapF*, the *Escherichia coli* gene for diaminopimelate epimerase. *J. Bacteriol.* **169:**1454–1459.

3538. Richaud, C., and C. Printz. 1988. Nucleotide sequence of the *dapF* gene and flanking regions from *Escherichia coli* K12. *Nucleic Acids Res.* **16:**10367.

3539. Richaud, C., F. Richaud, C. Martin, C. Haziza, and J. C. Patte. 1984. Regulation of expression and nucleotide sequence of the *Escherichia coli dapD* gene. *J. Biol. Chem.* **259:**14824–14828.

3540. Richaud, F., C. Richaud, P. Ratet, and J. C. Patte. 1986. Chromosomal location and nucleotide sequence of the *Escherichia coli dapA* gene. *J. Bacteriol.* **166:**297–300.

3541. Richet, E., and O. Raibaud. 1989. MalT, the regulatory protein of the *Escherichia coli* maltose system, is an ATP-dependent transcriptional activator. *EMBO J.* **8:**981–987.

3542. Richter, G., H. Ritz, G. Katzenmeier, R. Volk, A. Kohnle, F. Lottspeich, D. Allendorf, and A. Bacher. 1993. Biosynthesis of riboflavin: cloning, sequencing, mapping, and expression of the gene coding for GTP cyclohydrolase II in *Escherichia coli*. *J. Bacteriol.* **175:**4045–4051.

3543. Richter, G., R. Volk, C. Krieger, H. W. Lahm, U. Rothlisberger, and A. Bacher. 1992. Biosynthesis of riboflavin: cloning, sequencing, and expression of the gene coding for 3,4-dihydroxy-2-butanone 4-phosphate synthase of *Escherichia coli*. *J. Bacteriol.* **174:**4050–4056.

3544. Rieul, C., F. Bleicher, B. Duclos, J. C. Cortay, and A. J. Cozzone. 1988. Nucleotide sequence of the *aceA* gene coding for isocitrate lyase in *Escherichia coli*. *Nucleic Acids Res.* **16:**5689.

3545. Riggs, P. D., A. I. Derman, and J. Beckwith. 1988. A mutation affecting the regulation of a *secA-lacZ* fusion defines a new *sec* gene. *Genetics* **118:**571–579.

3546. Riley, M., and N. Glansdorff. 1983. Cloning the *Escherichia coli* K-12 *argD* gene specifying acetylornithine delta-transaminase. *Gene* **24:**335–339.

3547. Rimmele, M., and W. Boos. 1994. Trehalose-6-phosphate hydrolase of *Escherichia coli*. *J. Bacteriol.* **176:**5654–5664.

3548. Ringquist, S., D. Schneider, T. Gibson, C. Baron, A. Bock, and L. Gold. 1994. Recognition of the mRNA selenocysteine insertion sequence by the specialized translational elongation factor SELB. *Genes Dev.* **8:**376–385.

3549. Rinken, R., B. Thomas, and W. Wackernagel. 1992. Evidence that *recBC*-dependent degradation of duplex DNA in *Escherichia coli recD* mutants involves DNA unwinding. *J. Bacteriol.* **174:**5424–5429.

3550. Rioux, C. R., and R. J. Kadner. 1989. Vitamin B12 transport in *Escherichia coli* K12 does not require the *btuE* gene of the *btuCED* operon. *Mol. Gen. Genet.* **217:**301–308.

3551. Ritter, E., and B. Wittmann-Liebold. 1975. The primary structure of protein L30 from *Escherichia coli* ribosomes. *FEBS Lett.* **60:**153–155.

3552. Ritzenthaler, P., and M. Mata-Gilsinger. 1982. Use of in vitro gene fusions to study the *uxuR* regulatory gene in *Escherichia coli* K-12: direction of transcription and regulation of its expression. *J. Bacteriol.* **150:**1040–1047.

3553. Ritzenthaler, P., M. Mata-Gilsinger, and F. Stoeber. 1980. Construction and expression of hybrid plasmids containing *Escherichia coli* K-12 *uxu* genes. *J. Bacteriol.* **143:**1116–1126.

3554. Ritzenthaler, P., M. Mata-Gilsinger, and F. Stoeber. 1981. Molecular cloning of *Escherichia coli* K-12 hexuronate system genes: *exu* region. *J. Bacteriol.* **145:**181–190.

3555. Rivellini, F., P. Alifano, C. Piscitelli, V. Blasi, C. B. Bruni, and M. S. Carlomagno. 1991. A cytosine- over guanosine-rich sequence in RNA activates rho-dependent transcription termination. *Mol. Microbiol.* **5:**3049–3054.

3556. Rivers, S. L., E. McNairn, F. Blasco, G. Giordano, and D. H. Boxer. 1993. Molecular genetic analysis of the *moa* operon of *Escherichia coli* K-12 required for molybdenum cofactor biosynthesis. *Mol. Microbiol.* **8:**1071–1081.

3557. Roa, B. B., D. M. Connolly, and M. E. Winkler. 1989. Overlap between *pdxA* and *ksgA* in the complex *pdxA-ksgA-apaG-apaH* operon of *Escherichia coli* K-12. *J. Bacteriol.* **171:**4767–4777.

3558. Robbins, J. C., and D. L. Oxender. 1973. Transport systems for alanine, serine, and glycine in *Escherichia coli* K-12. *J. Bacteriol.* **116:**12–18.

3559. Robert-Baudouy, J., R. Portalier, and F. Stoeber. 1981. Regulation of hexuronate system genes in *Escherichia coli* K-12: multiple regulation of the *uxu* operon by *exuR* and *uxuR* gene products. *J. Bacteriol.* **145:**211–220.

3560. Robert-Le Meur, M., and C. Portier. 1992. *E. coli* polynucleotide phosphorylase expression is autoregulated through an RNase III-dependent mechanism. *EMBO J.* **11:**2633–2641.

3561. Roberton, A. M., P. A. Sullivan, M. C. Jones-Mortimer, and H. L. Kornberg. 1980. Two genes affecting glucarate utilization in *Escherichia coli* K12. *J. Gen. Microbiol.* **117:**377–382.

3562. Roberts, D. L., D. W. Bennett, and S. A. Forst. 1994. Identification of the site of phosphorylation on the osmosensor, EnvZ, of *Escherichia coli*. *J. Biol. Chem.* **269:**8728–8733.

3563. Roberts, L. M., and E. C. Reeve. 1970. Two mutations giving low-level streptomycin resistance in *Escherichia coli* K 12. *Genet. Res.* **16:**359–365.

3564. Roberts, P. E. 1992. The galactose/H⁺ symport protein of *Escherichia coli*. Ph.D. thesis. University of Cambridge, Cambridge, England.

3565. Roberts, R. E., C. I. Lienhard, C. G. Gaines, J. M. Smith, and J. R. Guest. 1988. Genetic and molecular characterization of the *guaC-nadC-aroP* region of *Escherichia coli* K-12. *J. Bacteriol.* **170:**463–467.

3566. Robien, M. A., G. M. Clore, J. G. Omichinski, R. N. Perham, E. Appella, K. Sakaguchi, and A. M. Gronenborn. 1992. Three-dimensional solution structure of the E3-binding domain of the dihydrolipoamide succinyltransferase core from the 2-oxoglutarate dehydrogenase multienzyme complex of *Escherichia coli*. *Biochemistry* **31:**3463–3471.

3567. Robinson, A. C., K. J. Begg, and E. MacArthur. 1991. Isolation and characterization of intragenic suppressors of an *Escherichia coli ftsA* mutation. *Res. Microbiol.* **142:**623–631.

3568. Robinson, A. C., D. J. Kenan, G. F. Hatfull, N. F. Sullivan, R. Spiegelberg, and W. D. Donachie. 1984. DNA sequence and transcriptional organization of essential cell division genes *ftsQ* and *ftsA* of *Escherichia coli*: evidence for overlapping transcriptional units. *J. Bacteriol.* **160:**546–555.

3569. Robinson, A. C., D. J. Kenan, J. Sweeney, and W. D. Donachie. 1986. Further evidence for overlapping transcriptional units in an *Escherichia coli* cell envelope-cell division gene cluster: DNA sequence and transcriptional organization of the *ddl ftsQ* region. *J. Bacteriol.* **167:**809–817.

3570. Robinson, C. L., and J. H. Jackson. 1982. New acetohydroxy acid synthase activity from mutational activation of a cryptic gene in *Escherichia coli* K-12. *Mol. Gen. Genet.* **186:**240–246.

3571. Rocha, M., M. Vazquez, A. Garciarrubio, and A. A. Covarrubias. 1985. Nucleotide sequence of the *glnA-glnL* intercistronic region of *Escherichia coli*. *Gene* **37:**91–99.

3572. Rockwell, P., and M. E. Gottesman. 1991. An *Escherichia coli rpoB* mutation that inhibits transcription of catabolite-sensitive operons. *J. Mol. Biol.* **222:**189–196.

3573. Rodel, W., W. Plaga, R. Frank, and J. Knappe. 1988. Primary structures of *Escherichia coli* pyruvate formate-lyase and pyruvate-formate-lyase-activating enzyme deduced from the DNA nucleotide sequences. *Eur. J. Biochem.* **177:**153–158.

3574. Rodolakis, A., F. Casse, and J. Starka. 1974. Morphological mutants of *Escherichia coli* K12. Mapping of the *env C* mutation. *Mol. Gen. Genet.* **130:**177–181.

3575. Rodriguez-Sainz, M. C., C. Hernandez-Chico, and F. Moreno. 1990. Molecular characterization of *pmbA*, an *Escherichia coli* chromosomal gene required for the production of the antibiotic peptide MccB17. *Mol. Microbiol.* **4:**1921–1932.

3576. Roecklein, B., A. Pelletier, and P. Kuempel. 1991. The *tus* gene of *Escherichia coli*: autoregulation, analysis of flanking sequences and identification of a complementary system in *Salmonella typhimurium*. *Res. Microbiol.* **142:**169–175.

3576a.Roecklein, B. A., and P. L. Kuempel. 1992. In vivo characterization of *tus* gene expression in *Escherichia coli*. *Mol. Microbiol.* **6:**1655–1661.

3577. Roeder, W., and R. L. Somerville. 1979. Cloning the *trpR* gene. *Mol. Gen. Genet.* **176:**361–368.

3578. Roehl, R. A., and R. T. Vinopal. 1979. New maltose Blu mutations in *Escherichia coli* K-12. *J. Bacteriol.* **139:**683–685.

3579. Roehl, R. A., and R. T. Vinopal. 1980. Genetic locus, distant from *ptsM*, affecting enzyme IIA/IIB function in *Escherichia coli* K-12. *J. Bacteriol.* **142:**120–130.

3580. Roesser, J. R., Y. Nakamura, and C. Yanofsky. 1989. Regulation of basal level expression of the tryptophan operon of *Escherichia coli*. *J. Biol. Chem.* **264:**12284–12288.

3580a.Rogers, K. C., A. T. Crescenzo, and D. Söll. 1995. Aminoacylation of transfer RNAs with 2-thiouridine derivatives in the wobble position of the anticodon. *Biochimie* **77:**66–74.

3581. Rogers, M. J., T. Ohgi, J. Plumbridge, and D. Soll. 1988. Nucleotide sequences of the *Escherichia coli nagE* and *nagB* genes: the structural genes for the N-acetylglucosamine transport protein of the bacterial phosphoenolpyruvate:sugar phosphotransferase system and for glucosamine-6-phosphate deaminase. *Gene* **62:**197–207.

3582. Rogers, S. D., M. R. Bhave, J. F. Mercer, J. Camakaris, and B. T. Lee. 1991. Cloning and characterization of *cutE*, a gene involved in copper transport in *Escherichia coli*. *J. Bacteriol.* **173:**6742–6748.

3583. Rogers, S. G., and B. Weiss. 1980. Cloning of the exonuclease III gene of *Escherichia coli*. *Gene* **11:**187–195.

3584. Rojiani, M. V., H. Jakubowski, and E. Goldman. 1990. Relationship between protein synthesis and concentrations of charged and uncharged tRNA^{Trp} in *Escherichia coli*. *Proc. Natl. Acad. Sci. USA* **87:**1511–1515.

3585. Rojo, F., J. A. Ayala, M. A. de Pedro, and D. Vazquez. 1984. Analysis of the different molecular forms of penicillin-binding protein 1B in *Escherichia coli ponB* mutants lysogenized with specialized transducing lambda (*ponB*⁺) bacteriophages. *Eur. J. Biochem.* **144:**571–576.

3586. Roland, K. L., F. E. Powell, and C. L. Turnbough, Jr. 1985. Role of translation and attenuation in the control of *pyrBI* operon expression in *Escherichia coli* K-12. *J. Bacteriol.* **163:**991–999.

3587. Rolfes, R. J., and H. Zalkin. 1988. *Escherichia coli* gene *purR* encoding a repressor protein for purine nucleotide synthesis. Cloning, nucleotide sequence, and interaction with the *purF* operator. *J. Biol. Chem.* **263:**19653–19661.

3588. Roman, S. J., B. B. Frantz, and P. Matsumura. 1993. Gene sequence, overproduction, purification and determination of the wild-type level of the *Escherichia coli* flagellar switch protein FliG. *Gene* **133:**103–108.

3589. Romeo, T., and M. Gong. 1993. Genetic and physical mapping of the regulatory gene *csrA* on the *Escherichia coli* K-12 chromosome. *J. Bacteriol.* **175:**5740–5741.

3590. Romeo, T., M. Gong, M. Y. Liu, and A. M. Brun-Zinkernagel. 1993. Identification and molecular characterization of *csrA*, a pleiotropic gene from *Escherichia coli* that affects glycogen biosynthesis, gluconeogenesis, cell size, and surface properties. *J. Bacteriol.* **175:**4744–4755.

3591. Romeo, T., A. Kumar, and J. Preiss. 1988. Analysis of the *Escherichia coli* glycogen gene cluster suggests that catabolic enzymes are encoded among the biosynthetic genes. *Gene* **70:**363–376.

3592. Romeo, T., J. Moore, and J. Smith. 1991. A simple method for cloning genes involved in glucan biosynthesis: isolation of structural and regulatory genes for glycogen synthesis in *Escherichia coli*. *Gene* **108:**23–29.

3593. Romisch, K., J. Webb, J. Herz, S. Prehn, R. Frank, M. Vingron, and B. Dobberstein. 1989. Homology of 54K protein of signal-recognition particle, docking protein and two E. coli proteins with putative GTP-binding domains. *Nature* (London) **340:**478–482.

3594. Roncero, C., and M. J. Casadaban. 1992. Genetic analysis of the genes involved in synthesis of the lipopolysaccharide core in *Escherichia coli* K-12: three operons in the *rfa* locus. *J. Bacteriol.* **174:**3250–3260.

3595. Rondeau, S. S., P. Y. Hsu, and J. A. DeMoss. 1984. Construction in vitro of a cloned *nar* operon from *Escherichia coli*. *J. Bacteriol.* **159:**159–166.

3596. Rood, J. I., A. J. Laird, and J. W. Williams. 1980. Cloning of the *Escherichia coli* K-12 dihydrofolate reductase gene following mu-mediated transposition. *Gene* **8:**255–265.

3597. Roof, S. K., J. D. Allard, K. P. Bertrand, and K. Postle. 1991. Analysis of *Escherichia coli* TonB membrane topology by use of PhoA fusions. *J. Bacteriol.* **173:**5554–5557.

3598. Roof, W. D., K. F. Foltermann, and J. R. Wild. 1982. The organization and regulation of the *pyrBI* operon in *E. coli* includes a rho-independent attenuator sequence. *Mol. Gen. Genet.* **187:**391–400.

3599. Roof, W. D., S. M. Horne, K. D. Young, and R. Young. 1994. *slyD*, a host gene required for phi X174 lysis, is related to the FK506-binding protein family of peptidyl-prolyl cis-trans-isomerases. *J. Biol. Chem.* **269:**2902–2910.

3600. Roovers, M., D. Charlier, A. Feller, D. Gigot, F. Holemans, W. Lissens, A. Pierard, and N. Glansdorff. 1988. *carP*, a novel gene regulating the transcription of the carbamoylphosphate synthetase operon of *Escherichia coli*. *J. Mol. Biol.* **204:**857–865.

3601. **Rosen, B. P.** 1973. Basic amino acid transport in *Escherichia coli:* properties of canavanine-resistant mutants. *J. Bacteriol.* 116:627–635.

3602. **Rosenberg, H.** 1978. Involvement of inner and outer membrane components in the transport of iron and in colicin B action in *Escherichia coli. J. Bacteriol.* 133:661–666.

3603. **Rosenfeld, S. A., P. E. Stevis, and N. W. Ho.** 1984. Cloning and characterization of the *xyl* genes from *Escherichia coli. Mol. Gen. Genet.* 194:410–415.

3604. **Rosner, J. L., T. J. Chai, and J. Foulds.** 1991. Regulation of *ompF* porin expression by salicylate in *Escherichia coli. J. Bacteriol.* 173:5631–5638.

3605. **Rosner, J. L., and J. L. Slonczewski.** 1994. Dual regulation of *inaA* by the multiple antibiotic resistance (*mar*) and superoxide (*soxRS*) stress response systems of *Escherichia coli. J. Bacteriol.* 176:6262–6269.

3606. **Ross, T. K., E. C. Achberger, and H. D. Braymer.** 1987. Characterization of the *Escherichia coli* modified cytosine restriction (*mcrB*) gene. *Gene* 61:277–289.

3607. **Ross, T. K., E. C. Achberger, and H. D. Braymer.** 1989. Nucleotide sequence of the McrB region of *Escherichia coli* K-12 and evidence for two independent translational initiation sites at the *mcrB* locus. *J. Bacteriol.* 171:1974–1981.

3608. **Ross, T. K., and H. D. Braymer.** 1987. Localization of a genetic region involved in McrB restriction by *Escherichia coli* K-12. *J. Bacteriol.* 169:1757–1759.

3609. **Ross, W., J. F. Thompson, J. T. Newlands, and R. L. Gourse.** 1990. E. coli Fis protein activates ribosomal RNA transcription in vitro and in vivo. *EMBO J.* 9:3733–3742.

3610. **Rossi, J., J. Egan, L. Hudson, and A. Landy.** 1981. The *tyrT* locus: termination and processing of a complex transcript. *Cell* 26:305–314.

3611. **Rossi, J. J., and A. Landy.** 1979. Structure and organization of the two tRNA^{Tyr} gene clusters on the *E. coli* chromosome. *Cell* 16:523–534.

3612. **Rossmann, R., M. Sauter, F. Lottspeich, and A. Bock.** 1994. Maturation of the large subunit (HYCE) of *Escherichia coli* hydrogenase 3 requires nickel incorporation followed by C-terminal processing at Arg537. *Eur. J. Biochem.* 220:377–384.

3613. **Rossmann, R., G. Sawers, and A. Bock.** 1991. Mechanism of regulation of the formate-hydrogenlyase pathway by oxygen, nitrate, and pH: definition of the formate regulon. *Mol. Microbiol.* 5:2807–2814.

3614. **Rostas, K., S. J. Morton, S. M. Picksley, and R. G. Lloyd.** 1987. Nucleotide sequence and LexA regulation of the *Escherichia coli recN* gene. *Nucleic Acids Res.* 15:5041–5049.

3615. **Rothery, R. A., J. L. Grant, J. L. Johnson, K. V. Rajagopalan, and J. H. Weiner.** 1995. Association of molybdopterin guanine dinucleotide with *Escherichia coli* dimethyl sulfoxide reductase: effect of tungstate and a *mob* mutation. *J. Bacteriol.* 177:2057–2063.

3616. **Rothstein, D. M., G. Pahel, B. Tyler, and B. Magasanik.** 1980. Regulation of expression from the *glnA* promoter of *Escherichia coli* in the absence of glutamine synthetase. *Proc. Natl. Acad. Sci. USA* 77:7372–7376.

3617. **Rotman, B., and R. Guzman.** 1982. Identification of the *mglA* gene product in the beta-methylgalactoside transport system of *Escherichia coli* using plasmid DNA deletions generated in vitro. *J. Biol. Chem.* 257:9030–9034.

3618. **Rouviere, P. E., A. De Las Penas, J. Mecsas, C. Z. Lu, K. E. Rudd, and C. A. Gross.** 1995. *rpoE*, the gene encoding the second heat-shock sigma factor, sigma E, in *Escherichia coli. EMBO J.* 14:1032–1042.

3619. **Roux, B., and C. T. Walsh.** 1992. *p*-Aminobenzoate synthesis in *Escherichia coli:* kinetic and mechanistic characterization of the amidotransferase PabA. *Biochemistry* 31:6904–6910.

3620. **Rowen, L., J. A. Kobori, and S. Scherer.** 1982. Cloning of bacterial DNA replication genes in bacteriophage lambda. *Mol. Gen. Genet.* 187:501–509.

3621. **Rowland, G. C., P. M. Giffard, and I. R. Booth.** 1984. Genetic studies of the *phs* locus of *Escherichia coli,* a mutation causing pleiotropic lesions in metabolism and pH homeostasis. *FEBS Lett.* 173:295–300.

3622. **Rowland, G. C., P. M. Giffard, and I. R. Booth.** 1985. *phs* locus of *Escherichia coli,* a mutation causing pleiotropic lesions in metabolism, is an *rpoA* allele. *J. Bacteriol.* 164:972–975.

3623. **Rowley, D. L., and R. E. Wolf, Jr.** 1991. Molecular characterization of the *Escherichia coli* K-12 *zwf* gene encoding glucose 6-phosphate dehydrogenase. *J. Bacteriol.* 173:968–977.

3624. **Roy, A., and A. Danchin.** 1981. Restriction map of the *cya* region of the *Escherichia coli* K12 chromosome. *Biochimie* 63:719–722.

3625. **Roy, A., and A. Danchin.** 1982. The *cya* locus of *Escherichia coli* K12: organization and gene products. *Mol. Gen. Genet.* 188:465–471.

3626. **Roy, A., C. Haziza, and A. Danchin.** 1983. Regulation of adenylate cyclase synthesis in *Escherichia coli:* nucleotide sequence of the control region. *EMBO J.* 2:791–797.

3627. **Roy, A. M., and J. Coleman.** 1994. Mutations in *firA*, encoding the second acyltransferase in lipopolysaccharide biosynthesis, affect multiple steps in lipopolysaccharide biosynthesis. *J. Bacteriol.* 176:1639–1646.

3628. **Ruberti, I., F. Crescenzi, L. Paolozzi, and P. Ghelardini.** 1991. A class of *gyrB* mutants, substantially unaffected in DNA topology, suppresses the *Escherichia coli* K12 *ftsZ84* mutation. *Mol. Microbiol.* 31:1065–1072.

3629. **Rubin, E. M., G. A. Wilson, and F. E. Young.** 1980. Expression of thymidylate synthetase activity in *Bacillus subtilis* upon integration of a cloned gene from *Escherichia coli. Gene* 10:227–235.

3630. **Rudd, K. E.** 1992. Alignment of *E. coli* DNA sequences to a revised, integrated genomic restriction map, p. 2.3–2.43. *In* J. Miller (ed.), *A Short Course in Bacterial Genetics: a Laboratory Manual and Handbook for Escherichia coli and Related Bacteria.* Cold Spring Harbor Press, Cold Spring Harbor, N.Y.

3631. **Rudd, K. E.** 1993. Maps, genes, sequences, and computers: an *Escherichia coli* case study. *ASM News* 59:335–341.

3632. **Rudd, K. E., G. Bouffard, and W. Miller.** 1992. Computer analysis of E. coli restriction maps, p. 1–38. *In* K. E. Davies and S. M. Tilghman (ed.), *Genome Analysis,* vol. 4. Cold Spring Harbor Laboratory Press, Cold Spring Harbor, N.Y.

3633. **Rudd, K. E., and R. Menzel.** 1987. *his* operons of *Escherichia coli* and *Salmonella typhimurium* are regulated by DNA supercoiling. *Proc. Natl. Acad. Sci. USA* 84:517–521.

3634. **Rudd, K. E., W. Miller, J. Ostell, and D. A. Benson.** 1990. Alignment of *Escherichia coli* K12 DNA sequences to a genomic restriction map. *Nucleic Acids Res.* 18:313–321.

3635. **Rudd, K. E., W. Miller, C. Werner, J. Ostell, C. Tolstoshev, and S. G. Satterfield.** 1991. Mapping sequenced *E. coli* genes by computer: software, strategies and examples. *Nucleic Acids Res.* 19:637–647.

3636. **Rudd, K. E., H. J. Sofia, E. V. Koonin, G. Plunkett, S. Lazar, and P. E. Rouviere.** 1995. A new family of peptidyl-prolyl isomerases. *Trends Biochem. Sci.* 20:12–14.

3637. **Ruijter, G. J., G. van Meurs, M. A. Verwey, P. W. Postma, and K. van Dam.** 1992. Analysis of mutations that uncouple transport from phosphorylation in enzyme IIGlc of the *Escherichia coli* phosphoenolpyruvate-dependent phosphotransferase system. *J. Bacteriol.* 174:2843–2850.

3638. **Ruiz, S. M., S. Letourneau, and C. G. Cupples.** 1993. Isolation and characterization of an *Escherichia coli* strain with a high frequency of C-to-T mutations at 5-methylcytosines. *J. Bacteriol.* 175:4985–4989.

3639. **Ruiz-Vazquez, R., and E. Cerda-Olmedo.** 1980. An *Escherichia coli* mutant refractory to nitrosoguanidine mutagenesis. *Mol. Gen. Genet.* 178:525–531.

3640. **Rule, G. S., E. A. Pratt, C. C. Chin, F. Wold, and C. Ho.** 1985. Overproduction and nucleotide sequence of the respiratory D-lactate dehydrogenase of *Escherichia coli. J. Bacteriol.* 161:1059–1068.

3641. **Rupprecht, K. R., G. Gordon, M. Lundrigan, R. C. Gayda, A. Markovitz, and C. Earhart.** 1983. *ompT: Escherichia coli* K-12 structural gene for protein a (3b). *J. Bacteriol.* 153:1104–1106.

3642. **Ruscitti, T., and S. Linn.** 1992. DNA polymerase I modulates inducible stable DNA replication in *Escherichia coli. J. Bacteriol.* 174:6311–6313.

3643. **Rusnak, F., M. Sakaitani, D. Drueckhammer, J. Reichert, and C. T. Walsh.** 1991. Biosynthesis of the *Escherichia coli* siderophore enterobactin: sequence of the *entF* gene, expression and purification of EntF, and analysis of covalent phosphopantetheine. *Biochemistry* 30:2916–2927.

3644. **Russel, M., and A. Holmgren.** 1988. Construction and characterization of glutaredoxin-negative mutants of *Escherichia coli. Proc. Natl. Acad. Sci. USA* 85:990–994.

3645. **Russel, M., and P. Model.** 1983. A bacterial gene, *fip,* required for filamentous bacteriophage f1 assembly. *J. Bacteriol.* 154:1064–1076.

3646. **Russel, M., and P. Model.** 1984. Characterization of the cloned *fip* gene and its product. *J. Bacteriol.* 157:526–532.

3647. **Russel, M., and P. Model.** 1985. Thioredoxin is required for filamentous phage assembly. *Proc. Natl. Acad. Sci. USA* 82:29–33.

3648. **Russel, M., and P. Model.** 1985. Direct cloning of the *trxB* gene that encodes thioredoxin reductase. *J. Bacteriol.* 163:238–242.

3649. **Russel, M., and P. Model.** 1988. Sequence of thioredoxin reductase from *Escherichia coli.* Relationship to other flavoprotein disulfide oxidoreductases. *J. Biol. Chem.* 263:9015–9019.

3650. **Russel, M., P. Model, and A. Holmgren.** 1990. Thioredoxin or glutaredoxin in *Escherichia coli* is essential for sulfate reduction but not for deoxyribonucleotide synthesis. *J. Bacteriol.* 172:1923–1929.

3651. **Russell, G. C., and J. R. Guest.** 1990. Overexpression of restructured pyruvate dehydrogenase complexes and site-directed mutagenesis of a potential active-site histidine residue. Biochem. J. 269:443–450.

3652. **Russell, P. W., and P. E. Orndorff.** 1992. Lesions in two *Escherichia coli* type 1 pilus genes alter pilus number and length without affecting receptor binding. *J. Bacteriol.* 174:5923–5935.

3652a.**Russell, R. R. B., and A. J. Pittard.** 1971. New suppressor in *Escherichia coli. J. Bacteriol.* 107:736–740.

3653. **Ryals, J., R. Y. Hsu, M. N. Lipsett, and H. Bremer.** 1982. Isolation of single-site *Escherichia coli* mutants deficient in thiamine and 4-thiouridine syntheses: identification of a *nuvC* mutant. *J. Bacteriol.* 151:899–904.

3654. **Ryden, M., J. Murphy, R. Martin, L. Isaksson, and J. Gallant.** 1986. Mapping and complementation studies of the gene for release factor 1. *J. Bacteriol.* 168:1066–1069.

3655. **Ryden, S. M., and L. A. Isaksson.** 1984. A temperature-sensitive mutant of *Escherichia coli* that shows enhanced misreading of UAG/A and increased efficiency for some tRNA nonsense suppressors. *Mol. Gen. Genet.* 193:38–45.

3656. **Saarma, U., and J. Remme.** 1992. Novel mutants of 23S RNA: characterization of functional properties. *Nucleic Acids Res.* 20:3147–3152.

3657. **Sabe, H., T. Miwa, T. Kodaki, K. Izui, S. Hiraga, and H. Katsuki.** 1984. Molecular cloning of the phosphoenolpyruvate carboxylase gene, *ppc,* of *Escherichia coli. Gene* 31:279–283.

3658. **Sabo, D. L., and E. H. Fischer.** 1974. Chemical properties of *Escherichia coli* lysine decarboxylase including a segment of its pyridoxal 5'-phosphate binding site. *Biochemistry* 13:670–676.

3659. **Sacerdot, C., P. Dessen, J. W. Hershey, J. A. Plumbridge, and M. Grunberg-Manago.** 1984. Sequence of the initiation factor IF2 gene: unusual protein features and homologies with elongation factors. *Proc. Natl. Acad. Sci. USA* 81:7787–7791.

3660. Sacerdot, C., G. Fayat, P. Dessen, M. Springer, J. A. Plumbridge, M. Grunberg-Manago, and S. Blanquet. 1982. Sequence of a 1.26-kb DNA fragment containing the structural gene for *E. coli* initiation factor IF3: presence of an AUU initiator codon. *EMBO J.* 1:311–315.

3661. Sacerdot, C., G. Vachon, S. Laalami, F. Morel-Deville, Y. Cenatiempo, and M. Grunberg-Manago. 1992. Both forms of translational initiation factor IF2 (alpha and beta) are required for maximal growth of *Escherichia coli*. Evidence for two translational initiation codons for IF2 beta. *J. Mol. Biol.* 225:67–80.

3662. Sadosky, A. B., A. Davidson, R. J. Lin, and C. W. Hill. 1989. *rhs* gene family of *Escherichia coli* K-12. *J. Bacteriol.* 171:636–642.

3663. Sadosky, A. B., J. A. Gray, and C. W. Hill. 1991. The RhsD-E subfamily of *Escherichia coli* K-12. *Nucleic Acids Res.* 19:7177–7183.

3664. Saedler, H., A. Gullon, L. Fiethen, and P. Stablinger. 1968. Negative control of the galactose operon in *E. coli. Mol. Gen. Genet.* 102:79–88.

3665. Saffen, D. W., K. A. Presper, T. L. Doering, and S. Roseman. 1987. Sugar transport by the bacterial phosphotransferase system. Molecular cloning and structural analysis of the *Escherichia coli ptsH, ptsI,* and *crr* genes. *J. Biol. Chem.* 262:16241–16253.

3666. Sagara, K., S. Matsuyama, and S. Mizushima. 1994. SecF stabilizes SecD and SecY, components of the protein translocation machinery of the *Escherichia coli* cytoplasmic membrane. *J. Bacteriol.* 176:4111–4116.

3667. Sage, E., and W. A. Haseltine. 1984. High ratio of alkali-sensitive lesions to total DNA modification induced by benzo(a)pyrene diol epoxide. *J. Biol. Chem.* 259:11098–11102.

3668. Saget, B. M., D. E. Shevell, and G. C. Walker. 1995. Alteration of lysine 178 in the hinge region of the *Escherichia coli ada* protein interferes with activation of *ada,* but not *alkA,* transcription. *J. Bacteriol.* 177:1268–1274.

3669. Sahin-Toth, M., R. L. Dunten, A. Gonzalez, and H. R. Kaback. 1992. Functional interactions between putative intramembrane charged residues in the lactose permease of *Escherichia coli. Proc. Natl. Acad. Sci. USA* 89:10547–10551.

3670. Said, B., C. R. Ghosn, L. Vu, and W. D. Nunn. 1988. Nucleotide sequencing and expression of the *fadL* gene involved in long-chain fatty acid transport in *Escherichia coli. Mol. Microbiol.* 2:363–370.

3671. Saier, M. H., Jr., H. Straud, L. S. Massman, J. J. Judice, M. J. Newman, and B. U. Feucht. 1978. Permease-specific mutations in *Salmonella typhimurium* and *Escherichia coli* that release the glycerol, maltose, melibiose, and lactose transport systems from regulation by the phosphoenolpyruvate:sugar phosphotransferase system. *J. Bacteriol.* 133:1358–1367.

3672. Saiki, K., T. Mogi, and Y. Anraku. 1992. Heme O biosynthesis in *Escherichia coli:* the *cyoE* gene in the cytochrome bo operon encodes a protoheme IX farnesyltransferase. *Biochem. Biophys. Res. Commun.* 189:1491–1497.

3673. Sain, B., and N. E. Murray. 1980. The *hsd* (host specificity) genes of *E. coli* K 12. *Mol. Gen. Genet.* 180:35–46.

3674. Saing, K. M., H. Orii, Y. Tanaka, K. Yanagisawa, A. Miura, and H. Ikeda. 1988. Formation of deletion in *Escherichia coli* between direct repeats located in the long inverted repeats of a cellular slime mold plasmid: participation of DNA gyrase. *Mol. Gen. Genet.* 214:1–5.

3675. Saint Girons, I., and D. Margarita. 1985. Evidence for an internal promoter in the *Escherichia coli* threonine operon. *J. Bacteriol.* 161:461–462.

3676. Saint-Girons, I., N. Duchange, G. N. Cohen, and M. M. Zakin. 1984. Structure and autoregulation of the *metJ* regulatory gene in *Escherichia coli. J. Biol. Chem.* 259:14282–14285.

3677. Saint-Girons, I., N. Duchange, M. M. Zakin, I. Park, D. Margarita, P. Ferrara, and G. N. Cohen. 1983. Nucleotide sequence of *metF,* the *E. coli* structural gene for 5–10 methylene tetrahydrofolate reductase and of its control region. *Nucleic Acids Res.* 11:6723–6732.

3678. Saito, H., and H. Uchida. 1977. Initiation of the DNA replication of bacteriophage lambda in *Escherichia coli* K12. *J. Mol. Biol.* 113:1–25.

3679. Saito, M., A. Tsugawa, K. Egawa, and Y. Nakamura. 1986. Revised sequence of the *nusA* gene of *Escherichia coli* and identification of *nusA11* (ts) and *nusA1* mutations which cause changes in a hydrophobic amino acid cluster. *Mol. Gen. Genet.* 205:380–382.

3680. Sak, B. D., A. Eisenstark, and D. Touati. 1989. Exonuclease III and the catalase hydroperoxidase II in *Escherichia coli* are both regulated by the *katF* gene product. *Proc. Natl. Acad. Sci. USA* 86:3271–3275.

3681. Sakakibara, Y. 1988. The *dnaK* gene of *Escherichia coli* functions in initiation of chromosome replication. *J. Bacteriol.* 170:972–979.

3682. Sakakibara, Y. 1995. Suppression of thermosensitive initiation of DNA replication in a *dnaR* mutant of *Escherichia coli* by a rifampin resistance mutation in the *rpoB* gene. *J. Bacteriol.* 177:733–737.

3683. Sakakibara, Y., and T. Mizukami. 1980. A temperature-sensitive *Escherichia coli* mutant defective in DNA replication: *dnaN,* a new gene adjacent to the *dnaA* gene. *Mol. Gen. Genet.* 178:541–553.

3684. Sakakibara, Y., H. Tsukano, and T. Sako. 1981. Organization and transcription of the *dnaA* and *dnaN* genes of *Escherichia coli. Gene* 13:47–55.

3685. Sakamoto, H., N. Kimura, and Y. Shimura. 1983. Processing of transcription products of the gene encoding the RNA component of RNase P. *Proc. Natl. Acad. Sci. USA* 80:6187–6191.

3686. Sako, T. 1991. Novel *prlA* alleles defective in supporting staphylokinase processing in *Escherichia coli. J. Bacteriol.* 173:2289–2296.

3687. Sako, T., and Y. Sakakibara. 1980. Coordinate expression of *Escherichia coli dnaA* and *dnaN* genes. *Mol. Gen. Genet.* 179:521–526.

3688. Sakumi, K., K. Igarashi, M. Sekiguchi, and A. Ishihama. 1993. The Ada protein is a class I transcription factor of *Escherichia coli. J. Bacteriol.* 175:2455–2457.

3689. Sakumi, K., Y. Nakabeppu, Y. Yamamoto, S. Kawabata, S. Iwanaga, and M. Sekiguchi. 1986. Purification and structure of 3-methyladenine-DNA glycosylase I of *Escherichia coli. J. Biol. Chem.* 261:15761–15766.

3690. Salmond, G. P., J. F. Lutkenhaus, and W. D. Donachie. 1980. Identification of new genes in a cell envelope-cell division gene cluster of *Escherichia coli:* cell envelope gene *murG. J. Bacteriol.* 144:438–440.

3691. Salmond, G. P., and S. Plakidou. 1984. Genetic analysis of essential genes in the *ftsE* region of the *Escherichia coli* genetic map and identification of a new cell division gene, *ftsS. Mol. Gen. Genet.* 197:304–308.

3692. Salowe, S. P., and J. Stubbe. 1986. Cloning, overproduction, and purification of the B2 subunit of ribonucleoside-diphosphate reductase. *J. Bacteriol.* 165:363–366.

3693. Sambongi, Y., H. Crooke, J. A. Cole, and S. J. Ferguson. 1994. A mutation blocking the formation of membrane or periplasmic endogenous and exogenous c-type cytochromes in *Escherichia coli* permits the cytoplasmic formation of *Hydrogenobacter thermophilus* holo cytochrome c552. *FEBS Lett.* 344:207–210.

3694. Sambongi, Y., and S. J. Ferguson. 1994. Specific thiol compounds complement deficiency in c-type cytochrome biogenesis in *Escherichia coli* carrying a mutation in a membrane-bound disulphide isomerase-like protein. *FEBS Lett.* 3534:235–238.

3695. Sameshima, J. H., R. C. Wek, and G. W. Hatfield. 1989. Overlapping transcription and termination of the convergent *ilvA* and *ilvY* genes of *Escherichia coli. J. Biol. Chem.* 264:1224–1231.

3696. Sammartano, L. J., R. W. Tuveson, and R. Davenport. 1986. Control of sensitivity to inactivation by H_2O_2 and broad-spectrum near-UV radiation by the *Escherichia coli katF* locus. *J. Bacteriol.* 168:13–21.

3697. Sampei, G., and K. Mizobuchi. 1988. Nucleotide sequence of the *Escherichia coli purF* gene encoding amidophosphoribosyltransferase for de novo purine nucleotide synthesis. *Nucleic Acids Res.* 16:8717.

3698. Sampei, G., and K. Mizobuchi. 1989. The organization of the *purL* gene encoding 5'-phosphoribosylformylglycinamide amidotransferase of *Escherichia coli. J. Biol. Chem.* 264:21230–21238.

3699. Sampson, B. A., R. Misra, and S. A. Benson. 1989. Identification and characterization of a new gene of *Escherichia coli* K-12 involved in outer membrane permeability. *Genetics* 122:491–501.

3700. Samsonov, V. V., E. R. Odoevskaia, and S. P. Sineokii. 1992. Cloning and complementation analysis of the *Escherichia coli gpr* locus, influencing DNA replication of certain lambdoid phages. *Genetika* 28:39–45.

3701. Sanatinia, H., E. C. Kofoid, T. B. Morrison, and J. S. Parkinson. 1995. The smaller of two overlapping *cheA* gene products is not essential for chemotaxis in *Escherichia coli. J. Bacteriol.* 177:2713–2720.

3702. Sancar, A., N. D. Clarke, J. Griswold, W. J. Kennedy, and W. D. Rupp. 1981. Identification of the *uvrB* gene product. *J. Mol. Biol.* 148:63–76.

3703. Sancar, A., B. M. Kacinski, D. L. Mott, and W. D. Rupp. 1981. Identification of the *uvrC* gene product. *Proc. Natl. Acad. Sci. USA* 78:5450–5454.

3704. Sancar, A., and W. D. Rupp. 1983. A novel repair enzyme: UVRABC excision nuclease of *Escherichia coli* cuts a DNA strand on both sides of the damaged region. *Cell* 33:249–260.

3705. Sancar, A., G. B. Sancar, W. D. Rupp, J. W. Little, and D. W. Mount. 1982. LexA protein inhibits transcription of the *E. coli uvrA* gene in vitro. *Nature* (London) 298:96–98.

3706. Sancar, A., C. Stachelek, W. Konigsberg, and W. D. Rupp. 1980. Sequences of the *recA* gene and protein. *Proc. Natl. Acad. Sci. USA* 77:2611–2615.

3707. Sancar, A., R. P. Wharton, S. Seltzer, B. M. Kacinski, N. D. Clarke, and W. D. Rupp. 1981. Identification of the *uvrA* gene product. *J. Mol. Biol.* 148:45–62.

3708. Sancar, A., K. R. Williams, J. W. Chase, and W. D. Rupp. 1981. Sequences of the *ssb* gene and protein. *Proc. Natl. Acad. Sci. USA* 78:4274–4278.

3709. Sancar, G. B., A. Sancar, J. W. Little, and W. D. Rupp. 1982. The *uvrB* gene of *Escherichia coli* has both *lexA*-repressed and *lexA*-independent promoters. *Cell* 28:523–530.

3710. Sancar, G. B., A. Sancar, and W. D. Rupp. 1984. Sequences of the *E. coli uvrC* gene and protein. *Nucleic Acids Res.* 12:4593–4608.

3711. Sancar, G. B., F. W. Smith, M. C. Lorence, C. S. Rupert, and A. Sancar. 1984. Sequences of the *Escherichia coli* photolyase gene and protein. *J. Biol. Chem.* 259:6033–6038.

3712. Sancar, G. B., F. W. Smith, and A. Sancar. 1983. Identification and amplification of the *E. coli phr* gene product. *Nucleic Acids Res.* 11:6667–6678.

3712a. Sanchez, J. C., R. Gimenez, A. Schneider, W. D. Fessner, L. Baldoma, and J. Aguilar. 1994. Activation of a cryptic gene encoding a kinase for l-xylulose opens a new pathway for the utilization of l-lyxose by *Escherichia coli. J. Biol. Chem.* 269:29665–29669.

3712b. Sanchez-Anzaldo, F. J., and F. Bastarrachea. 1974. Genetic characterization of streptomycin-resistant and -dependent mutants of *Escherichia coli* K-12. *Mol. Gen. Genet.* 130:47–64.

3713. Sanchez-Pescador, R., E. Sanvicente, F. Valle, and F. Bolivar. 1982. Recombinant plasmids carrying the glutamate dehydrogenase structural gene from *Escherichia coli* K-12. *Gene* 17:1–8.

3714. Sanders, D. A., B. L. Gillece-Castro, A. M. Stock, A. L. Burlingame, and D. E. Koshland, Jr. 1989. Identification of the site of phosphorylation of the chemotaxis response regulator protein, CheY. *J. Biol. Chem.* 264:21770–21778.

3715. Sandigursky, M., and W. A. Franklin. 1992. DNA deoxyribophosphodiesterase of *Escherichia coli* is associated with exonuclease I. *Nucleic Acids Res.* 20:4699–4703.

3716. Sandler, S. J., B. Chackerian, J. T. Li, and A. J. Clark. 1992. Sequence and complementation analysis of *recF* genes from *Escherichia coli, Salmonella typhimurium, Pseudomonas putida* and *Bacillus subtilis*: evidence for an essential phosphate binding loop. *Nucleic Acids Res.* 20:839–845.

3717. Sandler, S. J., and A. J. Clark. 1994. Mutational analysis of sequences in the *recF* gene of *Escherichia coli* K-12 that affect expression. *J. Bacteriol.* 176:4011–4016.

3718. Sands, J. F., H. S. Cummings, C. Sacerdot, L. Dondon, M. Grunberg-Manago, and J. W. Hershey. 1987. Cloning and mapping of *infA*, the gene for protein synthesis initiation factor IF1. *Nucleic Acids Res.* 15:5157–5168.

3719. Sands, J. F., P. Regnier, H. S. Cummings, M. Grunberg-Manago, and J. W. Hershey. 1988. The existence of two genes between *infB* and *rpsO* in the *Escherichia coli* genome: DNA sequencing and S1 nuclease mapping. *Nucleic Acids Res.* 16:10803–10816.

3720. Sanfacon, H., S. Levasseur, P. H. Roy, and J. Lapointe. 1983. Cloning of the gene for *Escherichia coli* glutamyl-tRNA synthetase. *Gene* 22:175–180.

3721. Sankar, P., J. H. Lee, and K. T. Shanmugam. 1985. Cloning of hydrogenase genes and fine structure analysis of an operon essential for H_2 metabolism in *Escherichia coli*. *J. Bacteriol.* 162:353–360.

3722. Sankar, P., J. H. Lee, and K. T. Shanmugam. 1988. Gene-product relationships of *fhlA* and *fdv* genes of *Escherichia coli*. *J. Bacteriol.* 170:5440–5445.

3723. Sankar, P., and K. T. Shanmugam. 1988. Biochemical and genetic analysis of hydrogen metabolism in *Escherichia coli*: the *hydB* gene. *J. Bacteriol.* 170:5433–5439.

3724. Sankaran, K., and H. C. Wu. 1994. Lipid modification of bacterial prolipoprotein. Transfer of diacylglyceryl moiety from phosphatidylglycerol. *J. Biol. Chem.* 269:19701–19706.

3725. Saporito, S. M., and R. P. Cunningham. 1988. Nucleotide sequence of the *nfo* gene of *Escherichia coli* K-12. *J. Bacteriol.* 170:5141–5145.

3726. Saporito, S. M., B. J. Smith-White, and R. P. Cunningham. 1988. Nucleotide sequence of the *xth* gene of *Escherichia coli* K-12. *J. Bacteriol.* 170:4542–4547.

3727. Saraste, M., N. J. Gay, A. Eberle, M. J. Runswick, and J. E. Walker. 1981. The *atp* operon: nucleotide sequence of the genes for the gamma, beta, and epsilon subunits of *Escherichia coli* ATP synthase. *Nucleic Acids Res.* 9:5287–5296.

3728. Saravani, G. A., and D. R. Martin. 1990. Opacity factor from group A streptococci is an apoproteinase. *FEMS Microbiol. Lett.* 56:35–39.

3729. Sargentini, N. J., and K. C. Smith. 1988. Genetic and phenotypic analyses indicating occurrence of the *recN262* and *radB101* mutations at the same locus in *Escherichia coli*. *J. Bacteriol.* 170:2392–2394.

3730. Sarsero, J. P., P. J. Wookey, P. Gollnick, C. Yanofsky, and A. J. Pittard. 1991. A new family of integral membrane proteins involved in transport of aromatic amino acids in *Escherichia coli*. *J. Bacteriol.* 173:3231–3234.

3731. Sarsero, J. P., P. J. Wookey, and A. J. Pittard. 1991. Regulation of expression of the *Escherichia coli* K-12 *mtr* gene by TyrR protein and Trp repressor. *J. Bacteriol.* 173:4133–4143.

3732. Sarthy, A., S. Michaelis, and J. Beckwith. 1981. Deletion map of the *Escherichia coli* structural gene for alkaline phosphatase, *phoA*. *J. Bacteriol.* 145:288–292.

3733. Sarthy, A., S. Michaelis, and J. Beckwith. 1981. Use of gene fusions to determine the orientation of gene *phoA* on the *Escherichia coli* chromosome. *J. Bacteriol.* 145:293–298.

3734. Sarubbi, E., K. E. Rudd, and M. Cashel. 1988. Basal ppGpp level adjustment shown by new *spoT* mutants affect steady state growth rates and *rrnA* ribosomal promoter regulation in *Escherichia coli*. *Mol. Gen. Genet.* 213:214–222.

3735. Sarubbi, E., K. E. Rudd, H. Xiao, K. Ikehara, M. Kalman, and M. Cashel. 1989. Characterization of the *spoT* gene of *Escherichia coli*. *J. Biol. Chem.* 264:15074–15082.

3736. Sasaki, M., T. Fujiyoshi, K. Shimada, and Y. Takagi. 1982. Fine structure of the *recB* and *recC* gene region of *Escherichia coli*. *Biochem. Biophys. Res. Commun.* 109:414–422.

3737. Sasarman, A., P. Chartrand, M. Lavoie, D. Tardif, R. Proschek, and C. Lapointe. 1979. Mapping of a new *hem* gene in *Escherichia coli* K12. *J. Gen. Microbiol.* 113:297–303.

3738. Sasarman, A., P. Chartrand, R. Proschek, M. Desrochers, D. Tardif, and C. Lapointe. 1975. Uroporphyrin-accumulating mutant of *Escherichia coli* K-12. *J. Bacteriol.* 124:1205–1212.

3739. Sasarman, A., Y. Echelard, J. Letowski, D. Tardif, and M. Drolet. 1988. Nucleotide sequence of the *hemX* gene, the third member of the Uro operon of *Escherichia coli* K12. *Nucleic Acids Res.* 16:11835.

3740. Sasarman, A., J. Letowski, G. Czaika, V. Ramirez, M. A. Nead, J. M. Jacobs, and R. Morais. 1993. Nucleotide sequence of the *hemG* gene involved in the protoporphyrinogen oxidase activity of *Escherichia coli* K12. *Can. J. Microbiol.* 39:1155–1161.

3741. Sasarman, A., A. Nepveu, Y. Echelard, J. Dymetryszyn, M. Drolet, and C. Goyer. 1987. Molecular cloning and sequencing of the *hemD* gene of *Escherichia coli* K-12 and preliminary data on the Uro operon. *J. Bacteriol.* 169:4257–4262.

3742. Sasarman, A., K. E. Sanderson, M. Surdeanu, and S. Sonea. 1970. Hemin-deficient mutants of *Salmonella typhimurium*. *J. Bacteriol.* 102:531–536.

3743. Sasse-Dwight, S., and J. D. Gralla. 1990. Role of eukaryotic-type functional domains found in the prokaryotic enhancer receptor factor sigma 54. *Cell* 62:945–954.

3744. Sastry, S. S., and R. Jayaraman. 1984. Nitrofurantoin-resistant mutants of *Escherichia coli*: isolation and mapping. *Mol. Gen. Genet.* 196:379–380.

3745. Satishchandran, C., Y. N. Hickman, and G. D. Markham. 1992. Characterization of the phosphorylated enzyme intermediate formed in the adenosine 5'-phosphosulfate kinase reaction. *Biochemistry* 31:11684–11688.

3746. Satishchandran, C., G. D. Markham, R. C. Moore, and S. M. Boyle. 1990. Locations of the *speA, speB, speC*, and *metK* genes on the physical map of *Escherichia coli*. *J. Bacteriol.* 172:4748.

3747. Satishchandran, C., J. C. Taylor, and G. D. Markham. 1990. Novel *Escherichia coli* K-12 mutants impaired in S-adenosylmethionine synthesis. *J. Bacteriol.* 172:4489–4496.

3748. Sato, T., and T. Yura. 1981. Regulatory mutations conferring constitutive synthesis of major outer membrane proteins (OmpC and OmpF) in *Escherichia coli*. *J. Bacteriol.* 145:88–96.

3749. Sauer, M., K. Hantke, and V. Braun. 1987. Ferric-coprogen receptor FhuE of *Escherichia coli*: processing and sequence common to all TonB-dependent outer membrane receptor proteins. *J. Bacteriol.* 169:2044–2049.

3750. Sauer, M., K. Hantke, and V. Braun. 1990. Sequence of the *fhuE* outer-membrane receptor gene of *Escherichia coli* K12 and properties of mutants. *Mol. Microbiol.* 4:427–437.

3751. Sauter, M., R. Bohm, and A. Bock. 1992. Mutational analysis of the operon (*hyc*) determining hydrogenase 3 formation in *Escherichia coli*. *Mol. Microbiol.* 6:1523–1532.

3752. Sauter, M., and R. G. Sawers. 1990. Transcriptional analysis of the gene encoding pyruvate formate-lyase-activating enzyme of *Escherichia coli*. *Mol. Microbiol.* 4:355–363.

3753. Sawers, G., and A. Bock. 1988. Anaerobic regulation of pyruvate formate-lyase from *Escherichia coli* K-12. *J. Bacteriol.* 170:5330–5336.

3754. Sawers, G., and A. Bock. 1989. Novel transcriptional control of the pyruvate formate-lyase gene: upstream regulatory sequences and multiple promoters regulate anaerobic expression. *J. Bacteriol.* 171:2485–2498.

3755. Sawers, G., J. Heider, E. Zehelein, and A. Bock. 1991. Expression and operon structure of the *sel* genes of *Escherichia coli* and identification of a third selenium-containing formate dehydrogenase isoenzyme. *J. Bacteriol.* 173:4983–4993.

3756. Sawers, G., and B. Suppmann. 1992. Anaerobic induction of pyruvate formate-lyase gene expression is mediated by the ArcA and FNR proteins. *J. Bacteriol.* 174:3474–3478.

3757. Saxena, P., and J. R. Walker. 1992. Expression of *argU*, the *Escherichia coli* gene coding for a rare arginine tRNA. *J. Bacteriol.* 174:1956–1964.

3758. Schaechter, M., P. Polaczek, and R. Gallegos. 1991. Membrane attachment and DNA bending at the origin of the *Escherichia coli* chromosome. *Res. Microbiol.* 142:151–154.

3759. Schaefler, S. 1967. Inducible system for the utilization of beta-glucosides in *Escherichia coli*. I. Active transport and utilization of beta-glucosides. *J. Bacteriol.* 93:254–263.

3760. Schaffer, S., K. Hantke, and V. Braun. 1985. Nucleotide sequence of the iron regulatory gene *fur*. *Mol. Gen. Genet.* 200:110–113.

3761. Schaller, K., A. Krauel, and V. Braun. 1981. Temperature-sensitive, colicin M-tolerant mutant of *Escherichia coli*. *J. Bacteriol.* 147:135–139.

3762. Schatz, P. J., K. L. Bieker, K. M. Ottemann, T. J. Silhavy, and J. Beckwith. 1991. One of three transmembrane stretches is sufficient for the functioning of the SecE protein, a membrane component of the *E. coli* secretion machinery. *EMBO J.* 10:1749–1757.

3763. Schatz, P. J., P. D. Riggs, A. Jacq, M. J. Fath, and J. Beckwith. 1989. The *secE* gene encodes an integral membrane protein required for protein export in *Escherichia coli*. *Genes Dev.* 3:1035–1044.

3764. Schauder, B., and J. E. McCarthy. 1989. The role of bases upstream of the Shine-Dalgarno region and in the coding sequence in the control of gene expression in *Escherichia coli*: translation and stability of mRNAs in vivo. *Gene* 78:59–72.

3765. Scheideler, M. A., and R. M. Bell. 1991. Characterization of active and latent forms of the membrane-associated sn-glycerol-3-phosphate acyltransferase of *Escherichia coli*. *J. Biol. Chem.* 266:14321–14327.

3766. Schellenberg, G. D., and C. E. Furlong. 1977. Resolution of the multiplicity of the glutamate and aspartate transport systems of *Escherichia coli*. *J. Biol. Chem.* 252:9055–9064.

3767. Schellenberg, G. D., A. Sarthy, A. E. Larson, M. P. Backer, J. W. Crabb, M. Lidstrom, B. D. Hall, and C. E. Furlong. 1984. Xylose isomerase from *Escherichia coli*. Characterization of the protein and the structural gene. *J. Biol. Chem.* 259:6826–6832.

3768. Schellhorn, H. E., and V. L. Stones. 1992. Regulation of *katF* and *katE* in *Escherichia coli* K-12 by weak acids. *J. Bacteriol.* 174:4769–4776.

3769. Schendel, F. J., E. Mueller, J. Stubbe, A. Shiau, and J. M. Smith. 1989. Formylglycinamide ribonucleotide synthetase from *Escherichia coli*: cloning, sequencing, overproduction, isolation, and characterization. *Biochemistry* 28:2459–2471.

3770. Scherrer, R., and H. S. Moyed. 1988. Conditional impairment of cell division and altered lethality in *hipA* mutants of *Escherichia coli* K-12. *J. Bacteriol.* 170:3321–3326.

3771. Scheuermann, R., S. Tam, P. M. Burgers, C. Lu, and H. Echols. 1983. Identification of the epsilon-subunit of *Escherichia coli* DNA polymerase III holoenzyme as the *dnaQ* gene product: a fidelity subunit for DNA replication. *Proc. Natl. Acad. Sci. USA* 80:7085–7089.

3772. Schiltz, E., and J. Reinbolt. 1975. Determination of the complete amino-acid sequence of protein S4 from *Escherichia coli* ribosomes. *Eur. J. Biochem.* 56:467–481.

3773. Schinina, M. E., L. Maffey, D. Barra, F. Bossa, K. Puget, and A. M. Michelson. 1987. The primary structure of iron superoxide dismutase from *Escherichia coli*. *FEBS Lett.* 221:87–90.

3774. Schlagenhauf, E., B. Fol, and J. N. Jansonius. 1990. Crystallization and structure solution at 4 A resolution of the recombinant synthase domain of N-(5′-phosphoribosyl)anthranilate isomerase:indole-3-glycerol-phosphate synthase from *Escherichia coli* complexed to a substrate analogue. *Protein Eng.* 3:173–180.

3775. Schlaman, H. R., E. Risseeuw, M. E. Franke-van Dijk, and P. J. Hooykaas. 1994. Nucleotide sequence corrections of the *uidA* open reading frame encoding beta-glucuronidase. *Gene* 138:259–260.

3776. Schleif, R. 1969. Isolation and characterization of streptolydigin resistant RNA polymerase. *Nature* (London) 223:1068–1069.

3777. Schlensog, V., and A. Bock. 1990. Identification and sequence analysis of the gene encoding the transcriptional activator of the formate hydrogenlyase system of *Escherichia coli*. *Mol. Microbiol.* 4:1319–1327.

3778. Schlensog, V., and A. Bock. 1991. The *Escherichia coli fdv* gene probably encodes *mutS* and is located at minute 58.8 adjacent to the *hyc-hyp* gene cluster. *J. Bacteriol.* 173:7414–7415.

3779. Schlensog, V., S. Lutz, and A. Bock. 1994. Purification and DNA-binding properties of FHLA, the transcriptional activator of the formate hydrogenlyase system from *Escherichia coli*. *J. Biol. Chem.* 269:19590–19596.

3780. Schlesinger, D. H., M. A. Schell, and D. B. Wilson. 1977. The NH$_2$-terminal sequences of galactokinase from *Escherichia coli* and *Saccharomyces cerevisiae*. *FEBS Lett.* 83:45–47.

3781. Schleyer, M., and E. P. Bakker. 1993. Nucleotide sequence and 3′-end deletion studies indicate that the K$^+$-uptake protein *kup* from *Escherichia coli* is composed of a hydrophobic core linked to a large and partially essential hydrophilic C terminus. *J. Bacteriol.* 175:6925–6931.

3782. Schlictman, D., S. Shankar, and A. M. Chakrabarty. 1995. The *Escherichia coli* genes *sspA* and *rnk* can functionally replace the *Pseudomonas aeruginosa* alginate regulatory gene *algR2*. *Mol. Microbiol.* 16:309–320.

3783. Schlindwein, C., G. Giordano, C. L. Santini, and M. A. Mandrand. 1990. Identification and expression of the *Escherichia coli fdhD* and *fdhE* genes, which are involved in the formation of respiratory formate dehydrogenase. *J. Bacteriol.* 172:6112–6121.

3784. Schlindwein, C., and M. A. Mandrand. 1991. Nucleotide sequence of the *fdhE* gene involved in respiratory formate dehydrogenase formation in *Escherichia coli* K-12. *Gene* 97:147–148.

3785. Schlosser, A., A. Hamann, D. Bossemeyer, E. Schneider, and E. P. Bakker. 1993. NAD$^+$ binding to the *Escherichia coli* K$^+$-uptake protein TrkA and sequence similarity between TrkA and domains of a family of dehydrogenases suggest a role for NAD$^+$ in bacterial transport. *Mol. Microbiol.* 9:533–543.

3786. Schlosser, A., S. Kluttig, A. Hamann, and E. P. Bakker. 1991. Subcloning, nucleotide sequence, and expression of *trkG*, a gene that encodes an integral membrane protein involved in potassium uptake via the Trk system of *Escherichia coli*. *J. Bacteriol.* 173:3170–3176.

3787. Schlosser, A., M. Meldorf, S. Stumpe, E. P. Bakker, and W. Epstein. 1995. TrkH and its homolog, TrkG, determine the specificity and kinetics of cation transport by the Trk system of *Escherichia coli*. *J. Bacteriol.* 177:1908–1910.

3788. Schmellik-Sandage, C. S., and E. S. Tessman. 1990. Signal strains that can detect certain DNA replication and membrane mutants of *Escherichia coli*: isolation of a new *ssb* allele, *ssb-3*. *J. Bacteriol.* 172:4378–4385.

3789. Schmid, C., W. Meining, S. Weinkauf, L. Bachmann, H. Ritz, S. Eberhardt, W. Gimbel, T. Werner, H. W. Lahm, H. Nar, et al. 1993. Studies on GTP cyclohydrolase I of *Escherichia coli*. *Adv. Exp. Med. Biol.* 338:157–162.

3790. Schmid, M. B. 1990. More than just "histone-like" proteins. *Cell* 63:451–453.

3791. Schmid, S. R., and P. Linder. 1992. D-E-A-D protein family of putative RNA helicases. *Mol. Microbiol.* 6:283–291.

3792. Schmidt, M. G., E. E. Rollo, J. Grodberg, and D. B. Oliver. 1988. Nucleotide sequence of the *secA* gene and *secA*(Ts) mutations preventing protein export in *Escherichia coli*. *J. Bacteriol.* 170:3404–3414.

3793. Schnaitman, C. A., and J. D. Klena. 1993. Genetics of lipopolysaccharide biosynthesis in enteric bacteria. *Microbiol. Rev.* 57:655–682.

3794. Schnaitman, C. A., C. T. Parker, J. D. Klena, E. L. Pradel, N. B. Pearson, K. E. Sanderson, and P. R. MacClachlan. 1991. Physical maps of the *rfa* loci of *Escherichia coli* K-12 and *Salmonella typhimurium*. *J. Bacteriol.* 173:7410–7411.

3795. Schneider, E., S. Freundlieb, S. Tapio, and W. Boos. 1992. Molecular characterization of the MalT-dependent periplasmic alpha-amylase of *Escherichia coli* encoded by *malS*. *J. Biol. Chem.* 267:5148–5154.

3796. Schneppe, B., G. Deckers-Hebestreit, and K. Altendorf. 1990. Overproduction and purification of the *uncI* gene product of the ATP synthase of *Escherichia coli*. *J. Biol. Chem.* 265:389–395.

3797. Schneppe, B., G. Deckers-Hebestreit, and K. Altendorf. 1991. Detection and localization of the i protein in *Escherichia coli* cells using antibodies. *FEBS Lett.* 292:145–147.

3798. Schneppe, B., G. Deckers-Hebestreit, J. E. McCarthy, and K. Altendorf. 1991. Translation of the first gene of the *Escherichia coli unc* operon. Selection of the start codon and control of initiation efficiency. *J. Biol. Chem.* 266:21090–21098.

3799. Schnetz, K., and B. Rak. 1988. Regulation of the *bgl* operon of *Escherichia coli* by transcriptional antitermination. *EMBO J.* 7:3271–3277.

3800. Schnetz, K., and B. Rak. 1990. Beta-glucoside permease represses the *bgl* operon of *Escherichia coli* by phosphorylation of the antiterminator protein and also interacts with glucose-specific enzyme III, the key element in catabolite control. *Proc. Natl. Acad. Sci. USA* 87:5074–5078.

3801. Schnetz, K., C. Toloczyki, and B. Rak. 1987. Beta-glucoside (*bgl*) operon of *Escherichia coli* K-12: nucleotide sequence, genetic organization, and possible evolutionary relationship to regulatory components of two *Bacillus subtilis* genes. *J. Bacteriol.* 169:2579–2590.

3802. Schnier, J., and K. Isono. 1982. The DNA sequence of the gene *rpsA* of *Escherichia coli* coding for ribosomal protein S1. *Nucleic Acids Res.* 10:1857–1865.

3803. Schnier, J., M. Kimura, K. Foulaki, A. R. Subramanian, K. Isono, and B. Wittmann-Liebold. 1982. Primary structure of *Escherichia coli* ribosomal protein S1 and of its gene *rpsA*. *Proc. Natl. Acad. Sci. USA* 79:1008–1011.

3804. Schnier, J., M. Kitakawa, and K. Isono. 1986. The nucleotide sequence of an *Escherichia coli* chromosomal region containing the genes for ribosomal proteins S6, S18, L9 and an open reading frame. *Mol. Gen. Genet.* 204:126–132.

3805. Schnierow, B. J., M. Yamada, and M. H. Saier, Jr. 1989. Partial nucleotide sequence of the *pts* operon in *Salmonella typhimurium*: comparative analyses in five bacterial genera. *Mol. Microbiol.* 3:113–118.

3806. Schoedon, G., U. Redweik, G. Frank, R. G. Cotton, and N. Blau. 1992. Allosteric characteristics of GTP cyclohydrolase I from *Escherichia coli*. *Eur. J. Biochem.* 210:561–568.

3807. Schoemaker, J. M., and A. Markovitz. 1981. Identification of the gene *lon* (*capR*) product as a 94-kilodalton polypeptide by cloning and deletion analysis. *J. Bacteriol.* 147:46–56.

3808. Schoenlein, P. V., B. B. Roa, and M. E. Winkler. 1989. Divergent transcription of *pdxB* and homology between the *pdxB* and *serA* gene products in *Escherichia coli* K-12. *J. Bacteriol.* 171:6084–6092.

3809. Schofield, M. A., R. Agbunag, M. L. Michaels, and J. H. Miller. 1992. Cloning and sequencing of *Escherichia coli mutR* shows its identity to *topB*, encoding topoisomerase III. *J. Bacteriol.* 174:5168–5170.

3810. Scholle, A., J. Vreemann, V. Blank, A. Nold, W. Boos, and M. D. Manson. 1987. Sequence of the *mglB* gene from *Escherichia coli* K12: comparison of wild-type and mutant galactose chemoreceptors. *Mol. Gen. Genet.* 208:247–253.

3811. Schonbrunner, E. R., and F. X. Schmid. 1992. Peptidyl-prolyl cis-trans isomerase improves the efficiency of protein disulfide isomerase as a catalyst of protein folding. *Proc. Natl. Acad. Sci. USA* 89:4510–4513.

3812. Schreiber, G., S. Metzger, E. Aizenman, S. Roza, M. Cashel, and G. Glaser. 1991. Overexpression of the *relA* gene in *Escherichia coli*. *J. Biol. Chem.* 266:3760–3767.

3813. Schroder, I., R. P. Gunsalus, B. A. Ackrell, B. Cochran, and G. Cecchini. 1991. Identification of active site residues of *Escherichia coli* fumarate reductase by site-directed mutagenesis. *J. Biol. Chem.* 266:13572–13579.

3814. Schroder, I., C. D. Wolin, R. Cavicchioli, and R. P. Gunsalus. 1994. Phosphorylation and dephosphorylation of the NarQ, NarX, and NarL proteins of the nitrate-dependent two-component regulatory system of *Escherichia coli*. *J. Bacteriol.* 176:4985–4992.

3815. Schryvers, A., and J. H. Weiner. 1982. The anaerobic sn-glycerol-3-phosphate dehydrogenase: cloning and expression of the *glpA* gene of *Escherichia coli* and identification of the *glpA* products. *Can. J. Biochem.* 60:224–231.

3816. Schultz, J. E., and A. Matin. 1991. Molecular and functional characterization of a carbon starvation gene of *Escherichia coli*. *J. Mol. Biol.* 218:129–140.

3817. Schultz-Hauser, G., W. Koster, H. Schwarz, and V. Braun. 1992. Iron(III) hydroxamate transport in *Escherichia coli* K-12: FhuB-mediated membrane association of the FhuC protein and negative complementation of *fhuC* mutants. *J. Bacteriol.* 174:2305–2311.

3818. Schulz, A., P. Taggeselle, D. Tripier, and K. Bartsch. 1990. Stereospecific production of the herbicide phosphinothricin (glufosinate) by transamination: isolation and characterization of a phosphinothricin-specific transaminase from *Escherichia coli*. *Appl. Environ. Microbiol.* 56:1–6.

3819. Schwan, W. R., H. S. Seifert, and J. L. Duncan. 1992. Growth conditions mediate differential transcription of *fim* genes involved in phase variation of type 1 pili. *J. Bacteriol.* 174:2367–2375.

3820. Schwartz, I., R. A. Klotsky, D. Elseviers, P. J. Gallagher, M. Krauskopf, M. A. Siddiqui, J. F. Wong, and B. A. Roe. 1983. Molecular cloning and sequencing of *pheU*, a gene for *Escherichia coli* tRNA^{Phe}. *Nucleic Acids Res.* 11:4379–4389.

3821. Schwartz, M. 1966. Location of the maltose A and B loci on the genetic map of *Escherichia coli*. *J. Bacteriol.* 92:1083–1089.

3822. Schwartz, M., M. Roa, and M. Debarbouille. 1981. Mutations that affect *lamB* gene expression at a posttranscriptional level. *Proc. Natl. Acad. Sci. USA* 78:2937–2941.

3823. Schweizer, H., and W. Boos. 1983. Cloning of the *ugp* region containing the structural genes for the *pho* regulon-dependent sn-glycerol-3-phosphate transport system of *Escherichia coli*. *Mol. Gen. Genet.* 192:177–186.

3824. Schweizer, H., and W. Boos. 1984. Characterization of the *ugp* region containing the genes for the *phoB* dependent sn-glycerol-3-phosphate transport system of *Escherichia coli*. *Mol. Gen. Genet.* 197:161–168.

3825. Schweizer, H., W. Boos, and T. J. Larson. 1985. Repressor for the sn-glycerol-3-phosphate regulon of *Escherichia coli* K-12: cloning of the *glpR* gene and identification of its product. *J. Bacteriol.* 161:563–566.

3826. Schweizer, H., T. Grussenmeyer, and W. Boos. 1982. Mapping of two *ugp* genes coding for the *pho* regulon-dependent sn-glycerol-3-phosphate transport system of *Escherichia coli*. *J. Bacteriol.* 150:1164–1171.

3827. Schweizer, H., and T. J. Larson. 1987. Cloning and characterization of the aerobic sn-glycerol-3-phosphate dehydrogenase structural gene glpD of *Escherichia coli* K-12. *J. Bacteriol.* **169:**507–513.

3828. Schweizer, H., G. Sweet, and T. J. Larson. 1986. Physical and genetic structure of the glpD-malT interval of the *Escherichia coli* K-12 chromosome. Identification of two new structural genes of the glp-regulon. *Mol. Gen. Genet.* **202:**488–492.

3829. Schweizer, H. P., and P. Datta. 1989. The complete nucleotide sequence of the tdc region of *Escherichia coli. Nucleic Acids Res.* **17:**3994.

3830. Schweizer, H. P., and P. Datta. 1989. Identification and DNA sequence of tdcR, a positive regulatory gene of the tdc operon of *Escherichia coli. Mol. Gen. Genet.* **218:**516–522.

3831. Schweizer, H. P., and P. Datta. 1990. Physical map location of the tdc operon of *Escherichia coli. J. Bacteriol.* **172:**2825.

3832. Schweizer, H. P., and P. Datta. 1991. Physical linkage and transcriptional orientation of the tdc operon on the *Escherichia coli* chromosome. *Mol. Gen. Genet.* **228:**125–128.

3833. Sclafani, R. A., and J. A. Wechsler. 1981. High frequency of genetic duplications in the dnaB region of the *Escherichia coli* K12 chromosome. *Genetics* **98:**677–689.

3834a. Scofield, M. A., W. S. Lewis, and S. M. Schuster. 1990. Nucleotide sequence of *Escherichia coli* asnB and deduced amino acid sequence of asparagine synthetase B. *J. Biol. Chem.* **265:**12895–12902.

3835. Scripture, J. B., and R. W. Hogg. 1983. The nucleotide sequences defining the signal peptides of the galactose-binding protein and the arabinose-binding protein. *J. Biol. Chem.* **258:**10853–10855.

3836. Scripture, J. B., C. Voelker, S. Miller, R. T. O'Donnell, L. Polgar, J. Rade, B. F. Horazdovsky, and R. W. Hogg. 1987. High-affinity L-arabinose transport operon. Nucleotide sequence and analysis of gene products. *J. Mol. Biol.* **197:**37–46.

3837. Searles, L. L., J. W. Jones, M. J. Fournier, N. Grambow, B. Tyler, and J. M. Calvo. 1986. *Escherichia coli* B/r leuK mutant lacking pseudouridine synthase I activity. *J. Bacteriol.* **166:**341–345.

3838. Seaton, B. L., and L. E. Vickery. 1994. A gene encoding a DnaK/hsp70 homolog in *Escherichia coli. Proc. Natl. Acad. Sci. USA* **91:**2066–2070.

3839. Sedgwick, B. 1982. Genetic mapping of ada and adc mutations affecting the adaptive response of *Escherichia coli* to alkylating agents. *J. Bacteriol.* **150:**984–988.

3840. Sedgwick, B. 1983. Molecular cloning of a gene which regulates the adaptive response to alkylating agents in *Escherichia coli. Mol. Gen. Genet.* **191:**466–472.

3841. Sedgwick, B., and P. Robins. 1980. Isolation of mutants of *Escherichia coli* with increased resistance to alkylating agents: mutants deficient in thiols and mutants constitutive for the adaptive response. *Mol. Gen. Genet.* **180:**85–94.

3842. Sedgwick, S. G., C. Ho, and R. Woodgate. 1991. Mutagenic DNA repair in enterobacteria. *J. Bacteriol.* **173:**5604–5611.

3843. Sedgwick, S. G., D. Lodwick, N. Doyle, H. Crowne, and P. Strike. 1991. Functional complementation between chromosomal and plasmid mutagenic DNA repair genes in bacteria. *Mol. Gen. Genet.* **229:**428–436.

3844. Sedivy, J. M., F. Daldal, and D. G. Fraenkel. 1984. Fructose bisphosphatase of *Escherichia coli:* cloning of the structural gene (fbp) and preparation of a chromosomal deletion. *J. Bacteriol.* **158:**1048–1053.

3845. Seifert, J., N. Kunz, R. Flachmann, A. Laufer, K. D. Jany, and H. G. Gassen. 1990. Expression of the *E. coli* nadB gene and characterization of the gene product L-aspartate oxidase. *Biol. Chem. Hoppe-Seyler* **371:**239–248.

3846. Seifried, S. E., J. B. Easton, and P. H. von Hippel. 1992. ATPase activity of transcription-termination factor rho: functional dimer model. *Proc. Natl. Acad. Sci. USA* **89:**10454–10458.

3847. Selby, C. P., and A. Sancar. 1993. Transcription-repair coupling and mutation frequency decline. *J. Bacteriol.* **175:**7509–7514.

3847a. Rasmussen, L. J., M. G. Marinus, and A. Lobner-Olesen. 1994. Novel growth rate control of dam gene expression in *Escherichia coli. Mol. Microbiol.* **12:**631–638.

3848. Selby, C. P., and A. Sancar. 1993. Molecular mechanism of transcription-repair coupling. *Science* **260:**53–58.

3849. Selby, C. P., and A. Sancar. 1994. Mechanisms of transcription-repair coupling and mutation frequency decline. *Microbiol. Rev.* **58:**317–329.

3850. Selby, C. P., and A. Sancar. 1995. Structure and function of transcription-repair coupling factor. II. Catalytic properties. *J. Biol. Chem.* **270:**4890–4895.

3851. Selby, C. P., and A. Sancar. 1995. Structure and function of transcription-repair coupling factor. I. Structural domains and binding properties. *J. Biol. Chem.* **270:**4882–4889.

3852. Selby, C. P., E. M. Witkin, and A. Sancar. 1991. *Escherichia coli* mfd mutant deficient in "mutation frequency decline" lacks strand-specific repair: in vitro complementation with purified coupling factor. *Proc. Natl. Acad. Sci. USA* **88:**11574–11578.

3853. Seoane, A., A. Sabbaj, L. M. McMurry, and S. B. Levy. 1992. Multiple antibiotic susceptibility associated with inactivation of the prc gene. *J. Bacteriol.* **174:**7844–7847.

3854. Seoane, A. S., and S. B. Levy. 1995. Identification of new genes regulated by the marRAB operon in *Escherichia coli. J. Bacteriol.* **177:**530–535.

3855. Seol, J. H., S. K. Woo, E. M. Jung, S. J. Yoo, C. S. Lee, K. J. Kim, K. Tanaka, A. Ichihara, D. B. Ha, and C. H. Chung. 1991. Protease Do is essential for survival of *Escherichia coli* at high temperatures: its identity with the htrA gene product. *Biochem. Biophys. Res. Commun.* **176:**730–736.

3856. Seol, W., and A. J. Shatkin. 1990. A new gene located between pss and rrnG on the *Escherichia coli* chromosome. *J. Bacteriol.* **172:**4745.

3857. Seol, W., and A. J. Shatkin. 1990. Sequence of the distal end of *E. coli* ribosomal RNA rrnG operon. *Nucleic Acids Res.* **18:**3056.

3858. Seol, W., and A. J. Shatkin. 1991. *Escherichia coli* kgtP encodes an alpha-ketoglutarate transporter. *Proc. Natl. Acad. Sci. USA* **88:**3802–3806.

3859. Seol, W., and A. J. Shatkin. 1992. Site-directed mutants of *Escherichia coli* alpha-ketoglutarate permease (KgtP). *Biochemistry* **31:**3550–3554.

3860. Seol, W., and A. J. Shatkin. 1992. *Escherichia coli* alpha-ketoglutarate permease is a constitutively expressed proton symporter. *J. Biol. Chem.* **267:**6409–6413.

3861. Seol, W., and A. J. Shatkin. 1993. Membrane topology model of *Escherichia coli* alpha-ketoglutarate permease by phoA fusion analysis. *J. Bacteriol.* **175:**565–567.

3862. Serizawa, H., and R. Fukuda. 1987. Structure of the gene for the stringent starvation protein of *Escherichia coli. Nucleic Acids Res.* **15:**1153–1163.

3863. Seto, D., S. K. Bhatnagar, and M. J. Bessman. 1988. The purification and properties of deoxyguanosine triphosphate triphosphohydrolase from *Escherichia coli. J. Biol. Chem.* **263:**1494–1499.

3864. Seufert, W., and W. Messer. 1987. Start sites for bidirectional in vitro DNA replication inside the replication origin, oriC, of *Escherichia coli. EMBO J.* **6:**2469–2472.

3864a. Sevastopoulos, C. G., C. T. Wehr, and D. A. Glaser. 1977. Large-scale automated isolation of *Escherichia coli* mutants with thermosensitive DNA replication. *Proc. Natl. Acad. Sci. USA* **74:**3485–3489.

3864b. Sever, I. S., E. S. Kalyaeva, and O. N. Danilevskaya. 1982. *Sov. Genet.* **18:**965–971.

3865. Shaibe, E., E. Metzer, and Y. S. Halpern. 1985. Metabolic pathway for the utilization of L-arginine, L-ornithine, agmatine, and putrescine as nitrogen sources in *Escherichia coli* K-12. *J. Bacteriol.* **163:**933–937.

3866. Shanmugam, K. T., V. Stewart, R. P. Gunsalus, D. H. Boxer, J. A. Cole, M. Chippaux, J. A. DeMoss, G. Giordano, E. C. Lin, and K. V. Rajagopalan. 1992. Proposed nomenclature for the genes involved in molybdenum metabolism in *Escherichia coli* and *Salmonella typhimurium. Mol. Microbiol.* **6:**3452–3454. (Letter.)

3867. Shao, Z., R. T. Lin, and E. B. Newman. 1994. Sequencing and characterization of the sdaC gene and identification of the sdaCB operon in *Escherichia coli* K12. *Eur. J. Biochem.* **222:**901–907.

3868. Shao, Z., and E. B. Newman. 1993. Sequencing and characterization of the sdaB gene from *Escherichia coli* K-12. *Eur. J. Biochem.* **212:**777–784.

3869. Shapiro, J. A. 1966. Chromosomal location of the gene determining uridine diphoglucose formation in *Escherichia coli* K-12. *J. Bacteriol.* **92:**518–520.

3870. Shapiro, J. A. 1993. A role for the Clp protease in activating Mu-mediated DNA rearrangements. *J. Bacteriol.* **175:**2625–2631.

3871. Sharma, B., and T. M. Hill. 1992. TerF, the sixth identified replication arrest site in *Escherichia coli,* is located within the rcsC gene. *J. Bacteriol.* **174:**7854–7858.

3872. Sharma, R. C., N. J. Sargentini, and K. C. Smith. 1983. New mutation (mmrA1) in *Escherichia coli* K-12 that affects minimal medium recovery and postreplication repair after UV irradiation. *J. Bacteriol.* **154:**743–747.

3873. Sharma, S., W. Dowhan, and R. E. Moses. 1982. Molecular structure of uvrC gene of *Escherichia coli:* identification of DNA sequences required for transcription of the uvrC gene. *Nucleic Acids Res.* **10:**5209–5221.

3874. Sharma, S., A. Ohta, W. Dowhan, and R. E. Moses. 1981. Cloning of the uvrC gene of *Escherichia coli:* expression of a DNA repair gene. *Proc. Natl. Acad. Sci. USA* **78:**6033–6037.

3875. Sharma, S., T. F. Stark, W. G. Beattie, and R. E. Moses. 1986. Multiple control elements for the uvrC gene unit of *Escherichia coli. Nucleic Acids Res.* **14:**2301–2318.

3876. Sharma, V., R. Meganathan, and M. E. Hudspeth. 1993. Menaquinone (vitamin K2) biosynthesis: cloning, nucleotide sequence, and expression of the menC gene from *Escherichia coli. J. Bacteriol.* **175:**4917–4921.

3877. Sharma, V., K. Suvarna, R. Meganathan, and M. E. Hudspeth. 1992. Menaquinone (vitamin K2) biosynthesis: nucleotide sequence and expression of the menB gene from *Escherichia coli. J. Bacteriol.* **174:**5057–5062.

3878. Sharples, G. J., F. E. Benson, G. T. Illing, and R. G. Lloyd. 1990. Molecular and functional analysis of the ruv region of *Escherichia coli* K-12 reveals three genes involved in DNA repair and recombination. *Mol. Gen. Genet.* **221:**219–226.

3879. Sharples, G. J., S. N. Chan, A. A. Mahdi, M. C. Whitby, and R. G. Lloyd. 1994. Processing of intermediates in recombination and DNA repair: identification of a new endonuclease that specifically cleaves Holliday junctions. *EMBO J.* **13:**6133–6142.

3880. Sharples, G. J., and R. G. Lloyd. 1991. Resolution of Holliday junctions in *Escherichia coli:* identification of the ruvC gene product as a 19-kilodalton protein. *J. Bacteriol.* **173:**7711–7715.

3881. Sharples, G. J., and R. G. Lloyd. 1991. Location of a mutation in the aspartyl-tRNA synthetase gene of *Escherichia coli* K12. *Mutat. Res.* **264:**93–96.

3882. Sharples, G. J., and R. G. Lloyd. 1993. An *E. coli* RuvC mutant defective in cleavage of synthetic Holliday junctions. *Nucleic Acids Res.* **21:**3359–3364.

3883. Sharrock, R. A., R. L. Gourse, and M. Nomura. 1985. Defective antitermination of rRNA transcription and derepression of rRNA and tRNA synthesis in the nusB5 mutant of *Escherichia coli. Proc. Natl. Acad. Sci. USA* **82:**5275–5279.

3884. Sharrock, R. A., R. L. Gourse, and M. Nomura. 1985. Inhibitory effect of high-level transcription of the bacteriophage lambda nutL region on transcription of rRNA in *Escherichia coli. J. Bacteriol.* **163:**704–708.

3885. Sharrocks, A. D., and D. P. Hornby. 1991. Transcriptional analysis of the restriction and modification genes of bacteriophage P1. *Mol. Microbiol.* **5:**685–694.

3886. Shattuck-Eidens, D. M., and R. J. Kadner. 1981. Exogenous induction of the *Escherichia coli* hexose phosphate transport system defined by *uhp-lac* operon fusions. *J. Bacteriol.* 148:203–209.

3887. Shattuck-Eidens, D. M., and R. J. Kadner. 1983. Molecular cloning of the *uhp* region and evidence for a positive activator for expression of the hexose phosphate transport system of *Escherichia coli*. *J. Bacteriol.* 155:1062–1070.

3888. Shaw, D. J., and J. R. Guest. 1982. Nucleotide sequence of the *fnr* gene and primary structure of the Enr protein of *Escherichia coli*. *Nucleic Acids Res.* 10:6119–6130.

3889. Shaw, D. J., and J. R. Guest. 1982. Amplification and product identification of the *fnr* gene of *Escherichia coli*. *J. Gen. Microbiol.* 128:2221–2228.

3890. Shaw, D. J., J. R. Guest, R. Meganathan, and R. Bentley. 1982. Characterization of *Escherichia coli men* mutants defective in conversion of *o*-succinylbenzoate to 1,4-dihydroxy-2-naphthoate. *J. Bacteriol.* 152:1132–1137.

3891. Shaw, L., F. Grau, H. R. Kaback, J. S. Hong, and C. Walsh. 1975. Vinylglycolate resistance in *Escherichia coli*. *J. Bacteriol.* 121:1047–1055.

3892. Shea, C. M., and M. A. McIntosh. 1991. Nucleotide sequence and genetic organization of the ferric enterobactin transport system: homology to other periplasmic binding protein-dependent systems in *Escherichia coli*. *Mol. Microbiol.* 5:1415–1428.

3893. Sheldon, R., and S. Brenner. 1976. Regulatory mutants of dihydrofolate reductase in *Escherichia coli* K12. *Mol. Gen. Genet.* 147:91–97.

3894. Shen, W. F., C. Squires, and C. L. Squires. 1982. Nucleotide sequence of the *rrnG* ribosomal RNA promoter region of *Escherichia coli*. *Nucleic Acids Res.* 10:3303–3313.

3895. Shepard, D., R. W. Oberfelder, M. M. Welch, and C. S. McHenry. 1984. Determination of the precise location and orientation of the *Escherichia coli dnaE* gene. *J. Bacteriol.* 158:455–459.

3896. Sheppard, D. E. 1986. Dominance relationships among mutant alleles of regulatory gene *araC* in the *Escherichia coli* B/R L-arabinose operon. *J. Bacteriol.* 168:999–1001.

3897. Sherman, J. M., M. J. Rogers, and D. Soll. 1992. Competition of aminoacyl-tRNA synthetases for tRNA ensures the accuracy of aminoacylation. *Nucleic Acids Res.* 20:2847–2852. (Erratum, 20:1547–1552.)

3898. Sherman, M. Y., and A. L. Goldberg. 1992. Heat shock in *Escherichia coli* alters the protein-binding properties of the chaperonin *groEL* by inducing its phosphorylation. *Nature* (London) 357:167–169.

3899. Shevchik, V. E., G. Condemine, and J. Robert-Baudouy. 1994. Characterization of DsbC, a periplasmic protein of *Erwinia chrysanthemi* and *Escherichia coli* with disulfide isomerase activity. *EMBO J.* 13:2007–2012.

3900. Shevell, D. E., and G. C. Walker. 1991. A region of the Ada DNA-repair protein required for the activation of *ada* transcription is not necessary for activation of *alkA*. *Proc. Natl. Acad. Sci. USA* 88:9001–9005.

3901. Shi, W., M. Bogdanov, W. Dowhan, and D. R. Zusman. 1993. The *pss* and *psd* genes are required for motility and chemotaxis in *Escherichia coli*. *J. Bacteriol.* 175:7711–7714.

3902. Shi, W., Y. Zhou, J. Wild, J. Adler, and C. A. Gross. 1992. DnaK, DnaJ, and GrpE are required for flagellum synthesis in *Escherichia coli*. *J. Bacteriol.* 174:6256–6263.

3903. Shi, X., and G. N. Bennett. 1994. Plasmids bearing *hfq* and the *hns*-like gene *stpA* complement *hns* mutants in modulating arginine decarboxylase gene expression in *Escherichia coli*. *J. Bacteriol.* 176:6769–6775.

3904. Shi, X., and G. N. Bennett. 1994. Effects of *rpoA* and *cysB* mutations on acid induction of biodegradative arginine decarboxylase in *Escherichia coli*. *J. Bacteriol.* 176:7017–7023.

3905. Shi, X., and G. N. Bennett. 1995. Effects of multicopy LeuO on the expression of the acid-inducible lysine decarboxylase gene in *Escherichia coli*. *J. Bacteriol.* 177:810–814.

3906. Shi, X., B. C. Waasdorp, and G. N. Bennett. 1993. Modulation of acid-induced amino acid decarboxylase gene expression by *hns* in *Escherichia coli*. *J. Bacteriol.* 175:1182–1186.

3907. Shiba, K., K. Ito, Y. Nakamura, J. Dondon, and M. Grunberg-Manago. 1986. Altered translation initiation factor 2 in the cold-sensitive *ssyG* mutant affects protein export in *Escherichia coli*. *EMBO J.* 5:3001–3006.

3908. Shiba, K., K. Ito, and T. Yura. 1984. Mutation that suppresses the protein export defect of the *secY* mutation and causes cold-sensitive growth of *Escherichia coli*. *J. Bacteriol.* 160:696–701.

3909. Shiba, K., K. Ito, and T. Yura. 1986. Suppressors of the *secY24* mutation: identification and characterization of additional *ssy* genes in *Escherichia coli*. *J. Bacteriol.* 166:849–856.

3910. Shiba, K., K. Ito, T. Yura, and D. P. Cerretti. 1984. A defined mutation in the protein export gene within the *spc* ribosomal protein operon of *Escherichia coli*: isolation and characterization of a new temperature-sensitive *secY* mutant. *EMBO J.* 3:631–635.

3911. Shimamoto, T., H. Yazyu, M. Futai, and T. Tsuchiya. 1984. Nucleotide sequence of the promoter region of the melibiose operon of *Escherichia coli*. *Biochem. Biophys. Res. Commun.* 121:41–46.

3912. Shimizu, I., and A. Kaji. 1991. Identification of the promoter region of the ribosome-releasing factor cistron (*frr*). *J. Bacteriol.* 173:5181–5187.

3913. Shimmin, L. C., D. Vanderwel, R. E. Harkness, B. R. Currie, C. A. Galloway, and E. E. Ishiguro. 1984. Temperature-sensitive beta-lactam-tolerant mutants of *Escherichia coli*. *J. Gen. Microbiol.* 130:1315–1323.

3914. Shimoike, T., Y. Akiyama, T. Baba, T. Taura, and K. Ito. 1992. SecY variants that interfere with *Escherichia coli* protein export in the presence of normal *secY*. *Mol. Microbiol.* 6:1205–1210.

3915. Shimoike, T., T. Taura, A. Kihara, T. Yoshihisa, Y. Akiyama, K. Cannon, and K. Ito. 1995. Product of a new gene, *syd*, functionally interacts with SecY when overproduced in *Escherichia coli*. *J. Biol. Chem.* 270:5519–5526.

3916. Shinagawa, H., H. Iwasaki, Y. Ishino, and A. Nakata. 1991. SOS-inducible DNA polymerase II of *E. coli* is homologous to replicative DNA polymerase of eukaryotes. *Biochimie* 73:433–435.

3917. Shinagawa, H., T. Kato, T. Ise, K. Makino, and A. Nakata. 1983. Cloning and characterization of the *umu* operon responsible for inducible mutagenesis in *Escherichia coli*. *Gene* 23:167–174.

3918. Shinagawa, H., K. Makino, M. Amemura, S. Kimura, H. Iwasaki, and A. Nakata. 1988. Structure and regulation of the *Escherichia coli ruv* operon involved in DNA repair and recombination. *J. Bacteriol.* 170:4322–4329.

3919. Shinagawa, H., K. Makino, and A. Nakata. 1983. Regulation of the *pho* regulon in *Escherichia coli* K-12. Genetic and physiological regulation of the positive regulatory gene *phoB*. *J. Mol. Biol.* 168:477–488.

3920. Shindo, H., A. Furubayashi, M. Shimizu, M. Miyake, and F. Imamoto. 1992. Preferential binding of *E. coli* histone-like protein HU alpha to negatively supercoiled DNA. *Nucleic Acids Res.* 20:1553–1558.

3920a. Shinozawa, T. 1973. A mutant of *Escherichia coli* K-12 unable to support the multiplication of bacteriophage BF23. *Virology* 54:427–440.

3921. Shiuan, D., and A. Campbell. 1988. Transcriptional regulation and gene arrangement of *Escherichia coli*, *Citrobacter freundii* and *Salmonella typhimurium* biotin operons. *Gene* 67:203–211.

3922. Short, S. A., and J. T. Singer. 1984. Studies on *deo* operon regulation in *Escherichia coli*: cloning and expression of the *deoR* structural gene. *Gene* 31:205–211.

3923. Shrader, T. E., J. W. Tobias, and A. Varshavsky. 1993. The N-end rule in *Escherichia coli*: cloning and analysis of the leucyl, phenylalanyl-tRNA-protein transferase gene *aat*. *J. Bacteriol.* 175:4364–4374.

3924. Shultz, J., M. A. Hermodson, C. C. Garner, and K. M. Herrmann. 1984. The nucleotide sequence of the *aroF* gene of *Escherichia coli* and the amino acid sequence of the encoded protein, the tyrosine-sensitive 3- deoxy-D-arabino-heptulosonate 7-phosphate synthase. *J. Biol. Chem.* 259:9655–9661.

3925. Shultz, J., T. J. Silhavy, M. L. Berman, N. Fiil, and S. D. Emr. 1982. A previously unidentified gene in the *spc* operon of *Escherichia coli* K12 specifies a component of the protein export machinery. *Cell* 31:227–235.

3926. Shurvinton, C. E., R. G. Lloyd, F. E. Benson, and P. V. Attfield. 1984. Genetic analysis and molecular cloning of the *Escherichia coli ruv* gene. *Mol. Gen. Genet.* 194:322–329.

3927. Shuttleworth, H., J. Taylor, and N. Minton. 1986. Sequence of the gene for alkaline phosphatase from *Escherichia coli* JM83. *Nucleic Acids Res.* 14:8689.

3928. Sibilli, L., G. Le Bras, P. Cossart, M. A. Chalvignac, P. A. Briley, and N. Cohen. 1979. The primary structure of *Escherichia coli* K 12 aspartokinase I-homoserine dehydrogenase I: sequence of cyanogen bromide peptide CB 3. *Biochimie* 61:733–739.

3929. Siebert, M., A. Bechthold, M. Melzer, U. May, U. Berger, G. Schroder, J. Schroder, K. Severin, and L. Heide. 1992. Ubiquinone biosynthesis. Cloning of the genes coding for chorismate pyruvate-lyase and 4-hydroxybenzoate octaprenyl transferase from *Escherichia coli*. *FEBS Lett.* 307:347–350.

3930. Siegel, E. C. 1981. Complementation studies with the repair-deficient *uvrD3*, *uvrE156*, and *recL152* mutations in *Escherichia coli*. *Mol. Gen. Genet.* 184:526–530.

3931. Siegel, E. C., and V. Bryson. 1967. Mutator gene of *Escherichia coli* B. *J. Bacteriol.* 94:38–47.

3932. Siegele, D. A., J. C. Hu, W. A. Walter, and C. A. Gross. 1989. Altered promoter recognition by mutant forms of the sigma 70 subunit of *Escherichia coli* RNA polymerase. *J. Mol. Biol.* 206:591–603.

3933. Siegele, D. A., and R. Kolter. 1993. Isolation and characterization of an *Escherichia coli* mutant defective in resuming growth after starvation. *Genes Dev.* 7:2629–2640.

3934. Siggaard-Andersen, M. 1988. Role of *Escherichia coli* beta-ketoacyl-ACP synthase I in unsaturated fatty acid synthesis. *Carlsberg Res. Commun.* 53:371–379.

3935. Siggaard-Andersen, M., M. Wissenbach, J. A. Chuck, I. Svendsen, J. G. Olsen, and P. von Wettstein-Knowles. 1994. The *fabJ*-encoded beta-ketoacyl-[acyl carrier protein] synthase IV from *Escherichia coli* is sensitive to cerulenin and specific for short-chain substrates. *Proc. Natl. Acad. Sci. USA* 91:11027–11031.

3936. Signer, E. R. 1966. Interaction of prophages at the *att*80 site with the chromosome of *Escherichia coli*. *J. Mol. Biol.* 15:243–255.

3937. Silber, K. R., K. C. Keiler, and R. T. Sauer. 1992. Tsp: a tail-specific protease that selectively degrades proteins with nonpolar C termini. *Proc. Natl. Acad. Sci. USA* 89:295–299.

3938. Silber, K. R., and R. T. Sauer. 1994. Deletion of the *prc* (*tsp*) gene provides evidence for additional tail-specific proteolytic activity in *Escherichia coli* K-12. *Mol. Gen. Genet.* 242:237–240.

3939. Silver, S., P. Johnseine, E. Whitney, and D. Clark. 1972. Manganese-resistant mutants of *Escherichia coli*: physiological and genetic studies. *J. Bacteriol.* 110:186–195.

3940. Silverman, P., K. Nat, J. McEwen, and R. Birchman. 1980. Selection of *Escherichia coli* K-12 chromosomal mutants that prevent expression of F-plasmid functions. *J. Bacteriol.* 143:1519–1523.

3941. Silverman, P. M. 1982. Gene *cpxA* is a new addition to the linkage map of *Escherichia coli* K-12. *J. Bacteriol.* 150:425–428.

3942. Silverman, P. M., S. Rother, and H. Gaudin. 1991. Arc and Sfr functions of the *Escherichia coli* K-12 *arcA* gene product are genetically and physiologically separable. *J. Bacteriol.* 173:5648–5652.

3943. Silverman, P. M., L. Tran, R. Harris, and H. M. Gaudin. 1993. Accumulation of the F plasmid TraJ protein in *cpx* mutants of *Escherichia coli*. *J. Bacteriol.* 175:921–925.

3944. Silverstone, A. E., M. Goman, and J. G. Scaife. 1972. ALT: a new factor involved in the synthesis of RNA by *Escherichia coli*. *Mol. Gen. Genet.* 118:223–234.

3945. Simon, G., V. Mejean, C. Jourlin, M. Chippaux, and M. C. Pascal. 1994. The *torR* gene of *Escherichia coli* encodes a response regulator protein involved in the expression of the trimethylamine *N*-oxide reductase genes. *J. Bacteriol.* 176:5601–5606.

3946. Simons, R. W., P. A. Egan, H. T. Chute, and W. D. Nunn. 1980. Regulation of fatty acid degradation in *Escherichia coli*: isolation and characterization of strains bearing insertion and temperature-sensitive mutations in gene *fadR*. *J. Bacteriol.* 142:621–632.

3947. Simons, R. W., K. T. Hughes, and W. D. Nunn. 1980. Regulation of fatty acid degradation in *Escherichia coli*: dominance studies with strains merodiploid in gene *fadR*. *J. Bacteriol.* 143:726–730.

3948. Simpson, E. B., T. W. Hancock, and C. E. Buchanan. 1994. Transcriptional control of *dacB*, which encodes a major sporulation-specific penicillin-binding protein. *J. Bacteriol.* 176:7767–7769.

3949. Singer, M., T. A. Baker, G. Schnitzler, S. M. Deischel, M. Goel, W. Dove, Jaacks, KJ, A. D. Grossman, J. W. Erickson, and C. A. Gross. 1989. A collection of strains containing genetically linked alternating antibiotic resistance elements for genetic mapping of *Escherichia coli*. *Microbiol. Rev.* 53:1–24.

3950. Singer, M., W. A. Walter, B. M. Cali, P. Rouviere, H. H. Liebke, R. L. Gourse, and C. A. Gross. 1991. Physiological effects of the fructose-1,6-diphosphate aldolase *ts8* mutation on stable RNA synthesis in *Escherichia coli*. *J. Bacteriol.* 173:6249–6257.

3951. Singleton, C. K., W. D. Roeder, G. Bogosian, R. L. Somerville, and H. L. Weith. 1980. DNA sequence of the *E. coli trpR* gene and prediction of the amino acid sequence of Trp repressor. *Nucleic Acids Res.* 8:1551–1560.

3952. Sirisena, D. M., P. R. MacLachlan, S. L. Liu, A. Hessel, and K. E. Sanderson. 1994. Molecular analysis of the *rfaD* gene, for heptose synthesis, and the *rfaF* gene, for heptose transfer, in lipopolysaccharide synthesis in *Salmonella typhimurium*. *J. Bacteriol.* 176:2379–2385.

3953. Sirko, A., M. Hryniewicz, D. Hulanicka, and A. Bock. 1990. Sulfate and thiosulfate transport in *Escherichia coli* K-12: nucleotide sequence and expression of the *cysTWAM* gene cluster. *J. Bacteriol.* 172:3351–3357.

3954. Sirko, A. E., M. Zatyka, and M. D. Hulanicka. 1987. Identification of the *Escherichia coli cysM* gene encoding O-acetylserine sulphydrylase B by cloning with mini-Mu-*lac* containing a plasmid replicon. *J. Gen. Microbiol.* 133:2719–2725.

3955. Sivasubramanian, N., and R. Jayaraman. 1980. Mapping of two transcription mutations (*tlnI* and *tlnII*) conferring thiolutin resistance, adjacent to *dnaZ* and *rho* in *Escherichia coli*. *Mol. Gen. Genet.* 180:609–615.

3956. Six, S., S. C. Andrews, G. Unden, and J. R. Guest. 1994. *Escherichia coli* possesses two homologous anaerobic C₄-dicarboxylate membrane transporters (DcuA and DcuB) distinct from the aerobic dicarboxylate transport system (Dct). *J. Bacteriol.* 176:6470–6478.

3957. Sjoberg, B. M., S. Eriksson, H. Jornvall, M. Carlquist, and H. Eklund. 1985. Protein B1 of ribonucleotide reductase. Direct analytical data and comparisons with data indirectly deduced from the nucleotide sequence of the *Escherichia coli* nrdA gene. *Eur. J. Biochem.* 150:423–427.

3958. Skare, J. T., and K. Postle. 1991. Evidence for a TonB-dependent energy transduction complex in *Escherichia coli*. *Mol. Microbiol.* 5:2883–2890.

3959. Skarstad, K., B. Thony, D. S. Hwang, and A. Kornberg. 1993. A novel binding protein of the origin of the *Escherichia coli* chromosome. *J. Biol. Chem.* 268:5365–5370.

3960. Skarstad, K., K. von Meyenburg, F. G. Hansen, and E. Boye. 1988. Coordination of chromosome replication initiation in *Escherichia coli*: effects of different *dnaA* alleles. *J. Bacteriol.* 170:852–858.

3961. Skinner, M. A., and R. A. Cooper. 1982. An *Escherichia coli* mutant defective in the NAD-dependent succinate semialdehyde dehydrogenase. *Arch. Microbiol.* 132:270–275.

3962. Skjold, A. C., and D. H. Ezekiel. 1982. Regulation of D-arabinose utilization in *Escherichia coli* K-12. *J. Bacteriol.* 152:521–523.

3963. Skjold, A. C., and D. H. Ezekiel. 1982. Analysis of lambda insertions in the fucose utilization region of *Escherichia coli* K-12: use of lambda *fuc* and lambda *argA* transducing bacteriophages to partially order the fucose utilization genes. *J. Bacteriol.* 152:120–125.

3964. Skogman, S. G., and J. Nilsson. 1984. Molecular cloning and characterization of the gene for *Escherichia coli* valyl-tRNA synthetase. *Gene* 30:219–226.

3965. Slany, R. K., M. Bosl, P. F. Crain, and H. Kersten. 1993. A new function of S-adenosylmethionine: the ribosyl moiety of AdoMet is the precursor of the cyclopentenediol moiety of the tRNA wobble base queuine. *Biochemistry* 32:7811–7817.

3966. Slater, A. C., M. C. Jones-Mortimer, and H. L. Kornberg. 1981. L-Sorbose phosphorylation in *Escherichia coli* K-12. *Biochim. Biophys. Acta* 646:365–367.

3967. Slater, S. C., M. R. Lifsics, M. O'Donnell, and R. Maurer. 1994. *holE*, the gene coding for the theta subunit of DNA polymerase III of *Escherichia coli*: charac-

terization of a *holE* mutant and comparison with a *dnaQ* (epsilon-subunit) mutant. *J. Bacteriol.* 176:815–821.

3968. Slocum, M. K., and J. S. Parkinson. 1983. Genetics of methyl-accepting chemotaxis proteins in *Escherichia coli*: organization of the *tar* region. *J. Bacteriol.* 155:565–577.

3969. Slocum, M. K., and J. S. Parkinson. 1985. Genetics of methyl-accepting chemotaxis proteins in *Escherichia coli*: null phenotypes of the *tar* and *tap* genes. *J. Bacteriol.* 163:586–594.

3969a.Smallshaw, J. E., and R. A. Kelln. 1992. Cloning, nucleotide sequence and expression of the *Escherichia coli* K-12 *pyrH* gene encoding UMP kinase. *Genetics (Life Sci. Adv.)* 11:59–65.

3970. Smiley, B. L., J. R. Lupski, P. S. Svec, R. McMacken, and G. N. Godson. 1982. Sequences of the *Escherichia coli dnaG* primase gene and regulation of its expression. *Proc. Natl. Acad. Sci. USA* 79:4550–4554.

3971. Smillie, D. A., R. S. Hayward, T. Suzuki, N. Fujita, and A. Ishihama. 1992. Locations of genes encoding alkyl hydroperoxide reductase on the physical map of the *Escherichia coli* K-12 genome. *J. Bacteriol.* 174:3826–3827.

3972. Smith, A. A., and R. C. Greene. 1984. Cloning of the methionine regulatory gene, *metJ*, of *Escherichia coli* K12 and identification of its product. *J. Biol. Chem.* 259:14279–14281.

3973. Smith, B. R., and R. Schleif. 1978. Nucleotide sequence of the L-arabinose regulatory region of *Escherichia coli* K12. *J. Biol. Chem.* 253:6931–6933.

3974. Smith, D. K., T. Kassam, B. Singh, and J. F. Elliott. 1992. *Escherichia coli* has two homologous glutamate decarboxylase genes that map to distinct loci. *J. Bacteriol.* 174:5820–5826.

3975. Smith, D. R., and J. M. Calvo. 1979. Regulation of dihydrofolate reductase synthesis in *Escherichia coli*. *Mol. Gen. Genet.* 175:31–38.

3976. Smith, D. R., and J. M. Calvo. 1980. Nucleotide sequence of the *E. coli* gene coding for dihydrofolate reductase. *Nucleic Acids Res.* 8:2255–2274.

3977. Smith, D. R., and J. M. Calvo. 1982. Nucleotide sequence of dihydrofolate reductase genes from trimethoprim-resistant mutants of *Escherichia coli*. Evidence that dihydrofolate reductase interacts with another essential gene product. *Mol. Gen. Genet.* 187:72–78.

3978. Smith, D. W., W. B. Stine, A. L. Svitil, A. Bakker, and J. W. Zyskind. 1992. *Escherichia coli* cells lacking methylation-blocking factor (leucine-responsive regulatory protein) have precise timing of initiation of DNA replication in the cell cycle. *J. Bacteriol.* 174:3078–3082.

3979. Smith, J. M., and H. A. Daum. 1986. Nucleotide sequence of the *purM* gene encoding 5′-phosphoribosyl-5-aminoimidazole synthetase of *Escherichia coli* K12. *J. Biol. Chem.* 261:10632–10636.

3980. Smith, J. M., and H. A. Daum. 1987. Identification and nucleotide sequence of a gene encoding 5′-phosphoribosylglycinamide transformylase in *Escherichia coli* K12. *J. Biol. Chem.* 262:10565–10569.

3981. Smith, J. M., and J. S. Gots. 1980. *purF-lac* fusion and direction of *purF* transcription in *Escherichia coli*. *J. Bacteriol.* 143:1156–1164.

3982. Smith, M. W., and R. F. Doolittle. 1992. Anomalous phylogeny involving the enzyme glucose-6-phosphate isomerase. *J. Mol. Evol.* 34:544–545. (Letter.)

3983. Smith, M. W., and R. F. Doolittle. 1992. A comparison of evolutionary rates of the two major kinds of superoxide dismutase. *J. Mol. Evol.* 34:175–184.

3984. Smith, M. W., and J. W. Payne. 1992. Expression of periplasmic binding proteins for peptide transport is subject to negative regulation by phosphate limitation in *Escherichia coli*. *FEMS Microbiol. Lett.* 79:183–190.

3985. Smith, R. A., and J. S. Parkinson. 1980. Overlapping genes at the *cheA* locus of *Escherichia coli*. *Proc. Natl. Acad. Sci. USA* 77:5370–5374.

3986. Smith, R. L., J. L. Banks, M. D. Snavely, and M. E. Maguire. 1993. Sequence and topology of the CorA magnesium transport systems of *Salmonella typhimurium* and *Escherichia coli*. Identification of a new class of transport protein. *J. Biol. Chem.* 268:14071–14080.

3987. Smith, T. F., and J. R. Sadler. 1971. The nature of lactose operator constitutive mutations. *J. Mol. Biol.* 59:273–305.

3987a.Snyder, W. B., L. J. B. Davis, P. N. Danese, C. L. Cosma, and T. J. Silhavy. 1995. Overproduction of NlpE, a new outer membrane lipoprotein, suppresses the toxicity of periplasmic LacZ by activation of the Cpx signal transduction pathway. *J. Bacteriol.* 177:4216–4223.

3988. Snyder, W. B., and T. J. Silhavy. 1992. Enhanced export of beta-galactosidase fusion proteins in *prlF* mutants is Lon dependent. *J. Bacteriol.* 174:5661–5668.

3989. Sodano, P., K. V. Chary, O. Bjornberg, A. Holmgren, B. Kren, J. A. Fuchs, and K. Wuthrich. 1991. Nuclear magnetic resonance studies of recombinant *Escherichia coli* glutaredoxin. Sequence-specific assignments and secondary structure determination of the oxidized form. *Eur. J. Biochem.* 200:369–377.

3990. Sodano, P., T. H. Xia, J. H. Bushweller, O. Bjornberg, A. Holmgren, M. Billeter, and K. Wuthrich. 1991. Sequence-specific ¹H n.m.r. assignments and determination of the three-dimensional structure of reduced *Escherichia coli* glutaredoxin. *J. Mol. Biol.* 221:1311–1324.

3991. Soderberg, B. O., A. Holmgren, and C. I. Branden. 1974. Structure of oxidized thioredoxin to 4 with 5 A resolution. *J. Mol. Biol.* 90:143–152.

3992. Sodergren, E. J., and J. A. DeMoss. 1988. *narI* region of the *Escherichia coli* nitrate reductase (*nar*) operon contains two genes. *J. Bacteriol.* 170:1721–1729.

3993. Sodergren, E. J., P. Y. Hsu, and J. A. DeMoss. 1988. Roles of the *narJ* and *narI* gene products in the expression of nitrate reductase in *Escherichia coli*. *J. Biol. Chem.* 263:16156–16162.

3994. Soffer, R. L., and M. Savage. 1974. A mutant of *Escherichia coli* defective in leucyl, phenylalanyl-tRNA-protein transferase. *Proc. Natl. Acad. Sci. USA* 71:1004–1007.

3995. Sofia, H. J., V. Burland, D. L. Daniels, G. Plunkett, and F. R. Blattner. 1994. Analysis of the *Escherichia coli* genome. V. DNA sequence of the region from 76.0 to 81.5 minutes. *Nucleic Acids Res.* 22:2576–2586.

3996. Sogaard-Andersen, L., A. S. Mironov, H. Pedersen, V. V. Sukhodelets, and P. Valentin-Hansen. 1991. Single amino acid substitutions in the cAMP receptor protein specifically abolish regulation by the CytR repressor in *Escherichia coli*. *Proc. Natl. Acad. Sci. USA* 88:4921–4925.

3997. Sogaard-Andersen, L., H. Pedersen, B. Holst, and P. Valentin-Hansen. 1991. A novel function of the cAMP-CRP complex in *Escherichia coli*: cAMP-CRP functions as an adaptor for the CytR repressor in the *deo* operon. *Mol. Microbiol.* 5:969–975.

3998. Sohail, A., M. Lieb, M. Dar, and A. S. Bhagwat. 1990. A gene required for very short patch repair in *Escherichia coli* is adjacent to the DNA cytosine methylase gene. *J. Bacteriol.* 172:4214–4221.

3999. Soll, L. 1980. Pseudovirulent mutants of lambda b221*poriCasnA* resulting from mutations in or near *oriC*, the *E. coli* origin of DNA replication. *Mol. Gen. Genet.* 178:391–396.

4000. Solomon, E., and E. C. Lin. 1972. Mutations affecting the dissimilation of mannitol by *Escherichia coli* K-12. *J. Bacteriol.* 111:566–574.

4001. Son, H. S., and S. G. Rhee. 1987. Cascade control of *Escherichia coli* glutamine synthetase. Purification and properties of PII protein and nucleotide sequence of its structural gene. *J. Biol. Chem.* 262:8690–8695.

4002. Song, W. J., and S. Jackowski. 1992. Cloning, sequencing, and expression of the pantothenate kinase (*coaA*) gene of *Escherichia coli*. *J. Bacteriol.* 174:6411–6417. (Erratum, 175:2792, 1993.)

4003. Song, W. J., and S. Jackowski. 1992. *coaA* and *rts* are allelic and located at kilobase 3532 on the *Escherichia coli* physical map. *J. Bacteriol.* 174:1705–1706.

4004. Sood, P., C. G. Lerner, T. Shimamoto, Q. Lu, and M. Inouye. 1994. Characterization of the autophosphorylation of Era, an essential *Escherichia coli* GTPase. *Mol. Microbiol.* 12:201–208.

4005. Sorensen, K. I., and J. Neuhard. 1991. Dual transcriptional initiation sites from the *pyrC* promoter control expression of the gene in *Salmonella typhimurium*. *Mol. Gen. Genet.* 225:249–256.

4006. Spanjaard, R. A., K. Chen, J. R. Walker, and J. van Duin. 1990. Frameshift suppression at tandem AGA and AGG codons by cloned tRNA genes: assigning a codon to *argU* tRNA and T4 tRNA(Arg). *Nucleic Acids Res.* 18:5031–5036.

4007. Sparkowski, J., and A. Das. 1990. The nucleotide sequence of *greA*, a suppressor gene that restores growth of an *Escherichia coli* RNA polymerase mutant at high temperature. *Nucleic Acids Res.* 18:6443.

4008. Sparkowski, J., and A. Das. 1991. Location of a new gene, *greA*, on the *Escherichia coli* chromosome. *J. Bacteriol.* 173:5256–5257.

4009. Sparling, P. F., and E. Blackman. 1973. Mutation to erythromycin dependence in *Escherichia coli* K-12. *J. Bacteriol.* 116:74–83.

4010. Sparling, P. F., Y. Ikeya, and D. Elliot. 1973. Two genetic loci for resistance to kasugamycin in *Escherichia coli*. *J. Bacteriol.* 113:704–710.

4011. Sparrow, C. P., and C. R. Raetz. 1983. A trans-acting regulatory mutation that causes overproduction of phosphatidylserine synthase in *Escherichia coli*. *J. Biol. Chem.* 258:9963–9967.

4012. Spears, P. A., D. Schauer, and P. E. Orndorff. 1986. Metastable regulation of type 1 piliation in *Escherichia coli* and isolation and characterization of a phenotypically stable mutant. *J. Bacteriol.* 168:179–185.

4013. Spencer, J. B., N. J. Stolowich, C. A. Roessner, and A. I. Scott. 1993. The *Escherichia coli cysG* gene encodes the multifunctional protein, siroheme synthase. *FEBS Lett.* 335:57–60.

4014. Spencer, M. E., M. G. Darlison, P. E. Stephens, I. K. Duckenfield, and J. R. Guest. 1984. Nucleotide sequence of the *sucB* gene encoding the dihydrolipoamide succinyltransferase of *Escherichia coli* K12 and homology with the corresponding acetyltransferase. *Eur. J. Biochem.* 141:361–374.

4015. Spencer, M. E., and J. R. Guest. 1982. Molecular cloning of four tricarboxylic acid cyclic genes of *Escherichia coli*. *J. Bacteriol.* 151:542–552.

4016. Spencer, M. E., and J. R. Guest. 1985. Transcription analysis of the *sucAB*, *aceEF* and *lpd* genes of *Escherichia coli*. *Mol. Gen. Genet.* 200:145–154.

4017. Spencer, P., and P. M. Jordan. 1993. Purification and characterization of 5-aminolaevulinic acid dehydratase from *Escherichia coli* and a study of the reactive thiols at the metal-binding domain. *Biochem. J.* 290:279–287.

4018. Speranza, M. L., G. Valentini, P. Iadarola, M. Stoppini, M. Malcovati, and G. Ferri. 1989. Primary structure of three peptides at the catalytic and allosteric sites of the fructose-1,6-bisphosphate-activated pyruvate kinase from *Escherichia coli*. *Biol. Chem. Hoppe-Seyler* 370:211–216.

4019. Spielmann-Ryser, J., M. Moser, P. Kast, and H. Weber. 1991. Factors determining the frequency of plasmid cointegrate formation mediated by insertion sequence IS3 from *Escherichia coli*. *Mol. Gen. Genet.* 226:441–448.

4020. Spiro, S., K. L. Gaston, A. I. Bell, R. E. Roberts, S. J. Busby, and J. R. Guest. 1990. Interconversion of the DNA-binding specificities of two related transcription regulators, CRP and FNR. *Mol. Microbiol.* 4:1831–1838.

4021. Spiro, S., and J. R. Guest. 1990. FNR and its role in oxygen-regulated gene expression in *Escherichia coli*. *FEMS Microbiol. Rev.* 6:399–428.

4022. Spitzer, E. D., H. E. Jimenez-Billini, and B. Weiss. 1988. β-Alanine auxotrophy associated with *dfp*, a locus affecting DNA synthesis in *Escherichia coli*. *J. Bacteriol.* 170:872–876.

4023. Spitzer, E. D., and B. Weiss. 1985. *dfp* gene of *Escherichia coli* K-12, a locus affecting DNA synthesis, codes for a flavoprotein. *J. Bacteriol.* 164:994–1003.

4024. Sprague, G. F., Jr., R. M. Bell, and J. E. Cronan, Jr. 1975. A mutant of *Escherichia coli* auxotrophic for organic phosphates: evidence for two defects in inorganic phosphate transport. *Mol. Gen. Genet.* 143:71–77.

4025. Spratt, B. G., A. Boyd, and N. Stoker. 1980. Defective and plaque-forming lambda transducing bacteriophage carrying penicillin-binding protein-cell shape genes: genetic and physical mapping and identification of gene products from the *lip-dacA-rodA-pbpA-leuS* region of the *Escherichia coli* chromosome. *J. Bacteriol.* 143:569–581.

4026. Spratt, S. K., P. N. Black, M. M. Ragozzino, and W. D. Nunn. 1984. Cloning, mapping, and expression of genes involved in the fatty acid-degradative multienzyme complex of *Escherichia coli*. *J. Bacteriol.* 158:535–542.

4027. Spratt, S. K., C. L. Ginsburgh, and W. D. Nunn. 1981. Isolation and genetic characterization of *Escherichia coli* mutants defective in propionate metabolism. *J. Bacteriol.* 146:1166–1169.

4028. Sprenger, G. A. 1992. Location of the transketolase (*tkt*) gene on the *Escherichia coli* physical map. *J. Bacteriol.* 174:1707–1708.

4029. Sprenger, G. A. 1993. Nucleotide sequence of the *Escherichia coli* K-12 transketolase (*tkt*) gene. *Biochim. Biophys. Acta* 1216:307–310.

4030. Spring, K. J., P. G. Jerlstrom, D. M. Burns, and I. R. Beacham. 1986. L-Asparaginase genes in *Escherichia coli*: isolation of mutants and characterization of the *ansA* gene and its product. *J. Bacteriol.* 166:135–142.

4031. Springer, M., M. Graffe, J. S. Butler, and M. Grunberg-Manago. 1986. Genetic definition of the translational operator of the threonine-tRNA ligase gene in *Escherichia coli*. *Proc. Natl. Acad. Sci. USA* 83:4384–4388.

4032. Springer, M., J. F. Mayaux, G. Fayat, J. A. Plumbridge, M. Graffe, S. Blanquet, and M. Grunberg-Manago. 1985. Attenuation control of the *Escherichia coli* phenylalanyl-tRNA synthetase operon. *J. Mol. Biol.* 181:467–478.

4033. Springer, M., J. A. Plumbridge, M. Trudel, M. Graffe, and M. Grunberg-Manago. 1982. Transcription units around the gene for *E. coli* translation initiation factor IF3 (*infC*). *Mol. Gen. Genet.* 186:247–252.

4034. Springer, M., M. Trudel, M. Graffe, J. Plumbridge, G. Fayat, J. F. Mayaux, C. Sacerdot, S. Blanquet, and M. Grunberg-Manago. 1983. *Escherichia coli* phenylalanyl-tRNA synthetase operon is controlled by attenuation in vivo. *J. Mol. Biol.* 171:263–279.

4035. Spurio, R., M. Durrenberger, M. Falconi, A. La Teana, C. L. Pon, and C. O. Gualerzi. 1992. Lethal overproduction of the *Escherichia coli* nucleoid protein H-NS: ultramicroscopic and molecular autopsy. *Mol. Gen. Genet.* 231:201–211.

4036. Spyrou, G., E. Haggard-Ljungquist, M. Krook, H. Jornvall, E. Nilsson, and P. Reichard. 1991. Characterization of the flavin reductase gene (*fre*) of *Escherichia coli* and construction of a plasmid for overproduction of the enzyme. *J. Bacteriol.* 173:3673–3679.

4037. Squires, C., A. Krainer, G. Barry, W. F. Shen, and C. L. Squires. 1981. Nucleotide sequence at the end of the gene for the RNA polymerase beta' subunit (*rpoC*). *Nucleic Acids Res.* 9:6827–6840.

4038. Squires, C. H., M. DeFelice, J. Devereux, and J. M. Calvo. 1983. Molecular structure of *ilvIH* and its evolutionary relationship to *ilvG* in *Escherichia coli* K12. *Nucleic Acids Res.* 11:5299–5313.

4039. Squires, C. H., M. DeFelice, S. R. Wessler, and J. M. Calvo. 1981. Physical characterization of the *ilvHI* operon of *Escherichia coli* K-12. *J. Bacteriol.* 147:797–804.

4040. Squires, C. L., S. Pedersen, B. M. Ross, and C. Squires. 1991. ClpB is the *Escherichia coli* heat shock protein F84.1. *J. Bacteriol.* 173:4254–4262.

4041. Srivastava, B. S. 1976. Radiation sensitivity of a mutant of *Escherichia coli* K-12 associated with DNA replication: evidence for a new repair function. *Mol. Gen. Genet.* 143:327–332.

4042. Srivastava, R. A., N. Srivastava, and D. Apirion. 1992. Characterization of the RNA processing enzyme RNase III from wild type and overexpressing *Escherichia coli* cells in processing natural RNA substrates. *Int. J. Biochem.* 24:737–749.

4043. Staab, J. F., M. F. Elkins, and C. F. Earhart. 1989. Nucleotide sequence of the *Escherichia coli entE* gene. *FEMS Microbiol. Lett.* 50:15–19.

4044. Stacey, K. A., and P. Oliver. 1977. A novel pleiotropic mutation in *Escherichia coli* K12 which affects transduction, transformation and rates of mutation. *J. Gen. Microbiol.* 98:569–578.

4045. Stader, J., P. Matsumura, D. Vacante, G. E. Dean, and R. M. Macnab. 1986. Nucleotide sequence of the *Escherichia coli motB* gene and site-limited incorporation of its product into the cytoplasmic membrane. *J. Bacteriol.* 166:244–252.

4046. Stadtman, T. C. 1991. Biosynthesis and function of selenocysteine-containing enzymes. *J. Biol. Chem.* 266:16257–16260.

4047. Stadtman, T. C., J. N. Davis, E. Zehelein, and A. Bock. 1989. Biochemical and genetic analysis of *Salmonella typhimurium* and *Escherichia coli* mutants defective in specific incorporation of selenium into formate dehydrogenase and tRNAs. *Biofactors* 2:35–44.

4048. Stallings, W. C., T. B. Powers, K. A. Pattridge, J. A. Fee, and M. L. Ludwig. 1983. Iron superoxide dismutase from *Escherichia coli* at 3.1-A resolution: a structure unlike that of copper/zinc protein at both monomer and dimer levels. *Proc. Natl. Acad. Sci. USA* 80:3884–3888.

4049. Stalmach, M. E., S. Grothe, and J. M. Wood. 1983. Two proline porters in *Escherichia coli* K-12. *J. Bacteriol.* 156:481–486.

4050. **Stamford, N. P., P. E. Lilley, and N. E. Dixon.** 1992. Enriched sources of *Escherichia coli* replication proteins. The *dnaG* primase is a zinc metalloprotein. *Biochim. Biophys. Acta* **1132:**17–25.

4051. **Stan-Lotter, H., D. M. Clarke, and P. D. Bragg.** 1986. Isolation of a fourth cysteinyl-containing peptide of the alpha-subunit of the F_1 ATPase from *Escherichia coli* necessitates revision of the DNA sequence. *FEBS Lett.* **197:**121–124.

4052. **Staudenmaier, H., B. Van Hove, Z. Yaraghi, and V. Braun.** 1989. Nucleotide sequences of the *fecBCDE* genes and locations of the proteins suggest a periplasmic-binding-protein-dependent transport mechanism for iron(III) dicitrate in *Escherichia coli*. *J. Bacteriol.* **171:**2626–2633.

4053. **Stauffer, G. V., M. D. Plamann, and L. T. Stauffer.** 1981. Construction and expression of hybrid plasmids containing the *Escherichia coli glyA* genes. *Gene* **14:**63–72.

4054. **Stauffer, G. V., and L. T. Stauffer.** 1988. Cloning and nucleotide sequence of the *Salmonella typhimurium* LT2 *metF* gene and its homology with the corresponding sequence of *Escherichia coli*. *Mol. Gen. Genet.* **212:**246–251.

4055. **Stauffer, L. T., A. Ghrist, and G. V. Stauffer.** 1993. The *Escherichia coli gcvT* gene encoding the T-protein of the glycine cleavage enzyme system. *DNA Seq.* **3:**339–346.

4056. **Stauffer, L. T., M. D. Plamann, and G. V. Stauffer.** 1986. Cloning and characterization of the glycine-cleavage enzyme system of *Escherichia coli*. *Gene* **44:**219–226.

4057. **Stauffer, L. T., and G. V. Stauffer.** 1994. Characterization of the *gcv* control region from *Escherichia coli*. *J. Bacteriol.* **176:**6159–6164.

4058. **Stauffer, L. T., P. S. Steiert, J. G. Steiert, and G. V. Stauffer.** 1991. An *Escherichia coli* protein with homology to the H-protein of the glycine cleavage enzyme complex from pea and chicken liver. *DNA Seq.* **2:**13–17.

4059. **Stebbins, C. E., S. Borukhov, M. Orlova, A. Polyakov, A. Goldfarb, and S. A. Darst.** 1995. Crystal structure of the GreA transcript cleavage factor from *Escherichia coli*. *Nature* (London) **373:**636–640.

4060. **Steed, P. M., and B. L. Wanner.** 1993. Use of the *rep* technique for allele replacement to construct mutants with deletions of the *pstSCAB-phoU* operon: evidence of a new role for the PhoU protein in the phosphate regulon. *J. Bacteriol.* **175:**6797–6809.

4061. **Steege, D. A.** 1983. A nucleotide change in the anticodon of an *Escherichia coli* serine transfer RNA results in *supD*-amber suppression. *Nucleic Acids Res.* **11:**3823–3832.

4062. **Steffes, C., J. Ellis, J. Wu, and B. P. Rosen.** 1992. The *lysP* gene encodes the lysine-specific permease. *J. Bacteriol.* **174:**3242–3249.

4063. **Steiert, P. S., L. T. Stauffer, and G. V. Stauffer.** 1990. The *lpd* gene product functions as the L protein in the *Escherichia coli* glycine cleavage enzyme system. *J. Bacteriol.* **172:**6142–6144.

4064. **Steinman, H. M.** 1978. The amino acid sequence of mangano superoxide dismutase from *Escherichia coli* B. *J. Biol. Chem.* **253:**8708–8720.

4065. **Steinman, H. M., and R. L. Hill.** 1973. Sequence homologies among bacterial and mitochondrial superoxide dismutases. *Proc. Natl. Acad. Sci. USA* **70:**3725–3729.

4066. **Steinum, A. L., and E. Seeberg.** 1986. Nucleotide sequence of the *tag* gene from *Escherichia coli*. *Nucleic Acids Res.* **14:**3763–3772.

4067. **Stephens, C. M., and R. Bauerle.** 1991. Analysis of the metal requirement of 3-deoxy-D-arabino-heptulosonate-7-phosphate synthase from *Escherichia coli*. *J. Biol. Chem.* **266:**20810–20817.

4068. **Stephens, P. E., M. G. Darlison, H. M. Lewis, and J. R. Guest.** 1983. The pyruvate dehydrogenase complex of *Escherichia coli* K12. Nucleotide sequence encoding the dihydrolipoamide acetyltransferase component. *Eur. J. Biochem.* **133:**481–489.

4069. **Stephens, P. E., M. G. Darlison, H. M. Lewis, and J. R. Guest.** 1983. The pyruvate dehydrogenase complex of *Escherichia coli* K12. Nucleotide sequence encoding the pyruvate dehydrogenase component. *Eur. J. Biochem.* **133:**155–162.

4070. **Stephens, P. E., H. M. Lewis, M. G. Darlison, and J. R. Guest.** 1983. Nucleotide sequence of the lipoamide dehydrogenase gene of *Escherichia coli* K12. *Eur. J. Biochem.* **135:**519–527.

4071. **Sternglanz, R., S. DiNardo, K. A. Voelkel, Y. Nishimura, Y. Hirota, K. Becherer, L. Zumstein, and J. C. Wang.** 1981. Mutations in the gene coding for *Escherichia coli* DNA topoisomerase I affect transcription and transposition. *Proc. Natl. Acad. Sci. USA* **78:**2747–2751.

4072. **Stevens, F. J., and T. T. Wu.** 1976. Growth on D-lyxose of a mutant strain of *Escherichia coli* K12 using a novel isomerase and enzymes related to D-xylose metabolism. *J. Gen. Microbiol.* **97:**257–265.

4073. **Stevenson, G., B. Neal, D. Liu, M. Hobbs, N. H. Packer, M. Batley, J. W. Redmond, L. Lindquist, and P. Reeves.** 1994. Structure of the O antigen of *Escherichia coli* K-12 and the sequence of its *rfb* gene cluster. *J. Bacteriol.* **176:**4144–4156.

4074. **Steward, K. L., and T. Linn.** 1991. In vivo analysis of overlapping transcription units in the *rplKAJLrpoBC* ribosomal protein-RNA polymerase gene cluster of *Escherichia coli*. *J. Mol. Biol.* **218:**23–31.

4075. **Steward, K. L., and T. Linn.** 1992. Transcription frequency modulates the efficiency of an attenuator preceding the *rpoBC* RNA polymerase genes of *Escherichia coli*: possible autogenous control. *Nucleic Acids Res.* **20:**4773–4779.

4075a.**Stewart, A.** 1995. *Genetic Nomenclature Guide Including Information on Genomic Databases.* Elsevier, New York.

4076. **Stewart, P. S., and R. D'Ari.** 1992. Genetic and morphological characterization of an *Escherichia coli* chromosome segregation mutant. *J. Bacteriol.* **174:**4513–4516.

4077. **Stewart, R. C.** 1993. Activating and inhibitory mutations in the regulatory domain of CheB, the methylesterase in bacterial chemotaxis. *J. Biol. Chem.* **268:**1921–1930.

4078. **Stewart, V.** 1992. Localization of upstream sequence elements required for nitrate and anaerobic induction of *fdn* (formate dehydrogenase-N) operon expression in *Escherichia coli* K-12. *J. Bacteriol.* **174:**4935–4942.

4079. **Stewart, V.** 1993. Nitrate regulation of anaerobic respiratory gene expression in *Escherichia coli*. *Mol. Microbiol.* **9:**425–434.

4080. **Stewart, V., R. Landick, and C. Yanofsky.** 1986. Rho-dependent transcription termination in the tryptophanase operon leader region of *Escherichia coli* K-12. *J. Bacteriol.* **166:**217–223.

4081. **Stewart, V., J. T. Lin, and B. L. Berg.** 1991. Genetic evidence that genes *fdhD* and *fdhE* do not control synthesis of formate dehydrogenase-N in *Escherichia coli* K-12. *J. Bacteriol.* **173:**4417–4423.

4082. **Stewart, V., and C. H. MacGregor.** 1982. Nitrate reductase in *Escherichia coli* K-12: involvement of *chlC*, *chlE*, and *chlG* loci. *J. Bacteriol.* **151:**788–799.

4083. **Stewart, V., and J. Parales, Jr.** 1988. Identification and expression of genes *narL* and *narX* of the *nar* (nitrate reductase) locus in *Escherichia coli* K-12. *J. Bacteriol.* **170:**1589–1597.

4084. **Stewart, V., J. Parales, Jr., and S. M. Merkel.** 1989. Structure of genes *narL* and *narX* of the *nar* (nitrate reductase) locus in *Escherichia coli* K-12. *J. Bacteriol.* **171:**2229–2234.

4085. **Stewart, V., and C. Yanofsky.** 1985. Evidence for transcription antitermination control of tryptophanase operon expression in *Escherichia coli* K-12. *J. Bacteriol.* **164:**731–740.

4086. **Stim, K. P., and G. N. Bennett.** 1993. Nucleotide sequence of the *adi* gene, which encodes the biodegradative acid-induced arginine decarboxylase of *Escherichia coli*. *J. Bacteriol.* **175:**1221–1234.

4087. **Stirling, C. J., S. D. Colloms, J. F. Collins, G. Szatmari, and D. J. Sherratt.** 1989. *xerB*, an *Escherichia coli* gene required for plasmid ColE1 site-specific recombination, is identical to *pepA*, encoding aminopeptidase A, a protein with substantial similarity to bovine lens leucine aminopeptidase. *EMBO J.* **8:**1623–1627.

4088. **Stirling, C. J., G. Stewart, and D. J. Sherratt.** 1988. Multicopy plasmid stability in *Escherichia coli* requires host-encoded functions that lead to plasmid site-specific recombination. *Mol. Gen. Genet.* **214:**80–84.

4089. **Stirling, C. J., G. Szatmari, G. Stewart, M. C. Smith, and D. J. Sherratt.** 1988. The arginine repressor is essential for plasmid-stabilizing site-specific recombination at the ColE1 cer locus. *EMBO J.* **7:**4389–4395.

4090. **Stirling, D. A., C. S. Hulton, L. Waddell, S. F. Park, G. S. Stewart, I. R. Booth, and C. F. Higgins.** 1989. Molecular characterization of the *proU* loci of *Salmonella typhimurium* and *Escherichia coli* encoding osmoregulated glycine betaine transport systems. *Mol. Microbiol.* **3:**1025–1038.

4091. **Stoker, K., L. F. Oltmann, and A. H. Stouthamer.** 1988. Partial characterization of an electrophoretically labile hydrogenase activity of *Escherichia coli* K-12. *J. Bacteriol.* **170:**1220–1226.

4092. **Stoker, K., L. F. Oltmann, and A. H. Stouthamer.** 1989. Randomly induced *Escherichia coli* K-12 Tn5 insertion mutants defective in hydrogenase activity. *J. Bacteriol.* **171:**831–836.

4093. **Stoker, K., W. N. Reijnders, L. F. Oltmann, and A. H. Stouthamer.** 1989. Initial cloning and sequencing of *hydHG*, an operon homologous to *ntrBC* and regulating the labile hydrogenase activity in *Escherichia coli* K-12. *J. Bacteriol.* **171:**4448–4456.

4094. **Stoker, N. G., J. K. Broome-Smith, A. Edelman, and B. G. Spratt.** 1983. Organization and subcloning of the *dacA-rodA-pbpA* cluster of cell shape genes in *Escherichia coli*. *J. Bacteriol.* **155:**847–853.

4095. **Stoker, N. G., J. M. Pratt, and B. G. Spratt.** 1983. Identification of the *rodA* gene product of *Escherichia coli*. *J. Bacteriol.* **155:**854–859.

4096. **Stokes, H. W., P. W. Betts, and B. G. Hall.** 1985. Sequence of the *ebgA* gene of *Escherichia coli*: comparison with the *lacZ* gene. *Mol. Biol. Evol.* **2:**469–477.

4097. **Stokes, H. W., and B. G. Hall.** 1985. Sequence of the *ebgR* gene of *Escherichia coli*: evidence that the EBG and LAC operons are descended from a common ancestor. *Mol. Biol. Evol.* **2:**478–483.

4098. **Stone, D. A. W. Phillips, and J. J. Burchall.** 1977. The amino-acid sequence of the dihydrofolate reductase of a trimethoprim-resistant strain of *Escherichia coli*. *Eur. J. Biochem.* **72:**613–624.

4099. **Stoner, C., and R. Schleif.** 1983. The *araE* low affinity L-arabinose transport promoter. Cloning, sequence, transcription start site and DNA binding sites of regulatory proteins. *J. Mol. Biol.* **171:**369–381.

4100. **Storts, D. R., and A. Markovitz.** 1988. Construction and characterization of mutations in *hupB*, the gene encoding HU-β (HU-1) in *Escherichia coli* K-12. *J. Bacteriol.* **170:**1541–1547.

4101. **Storts, D. R., and A. Markovitz.** 1991. A novel *rho* promoter::Tn10 mutation suppresses an *ftsQ1*(Ts) missense mutation in an essential *Escherichia coli* cell division gene by a mechanism not involving polarity suppression. *J. Bacteriol.* **173:**655–663.

4102. **Storz, G., F. S. Jacobson, L. A. Tartaglia, R. W. Morgan, L. A. Silveira, and B. N. Ames.** 1989. An alkyl hydroperoxide reductase induced by oxidative stress in *Salmonella typhimurium* and *Escherichia coli*: genetic characterization and cloning of *ahp*. *J. Bacteriol.* **171:**2049–2055.

4103. **Storz, G., L. A. Tartaglia, and B. N. Ames.** 1990. Transcriptional regulator of oxidative stress-inducible genes: direct activation by oxidation. *Science* **248:**189–194.

4104. Storz, G., L. A. Tartaglia, S. B. Farr, and B. N. Ames. 1990. Bacterial defenses against oxidative stress. *Trends Genet.* 6:363–368.

4105. Stout, V., and S. Gottesman. 1990. RcsB and RcsC: a two-component regulator of capsule synthesis in *Escherichia coli*. *J. Bacteriol.* 172:659–669.

4106. Stout, V., A. Torres-Cabassa, M. R. Maurizi, D. Gutnick, and S. Gottesman. 1991. RcsA, an unstable positive regulator of capsular polysaccharide synthesis. *J. Bacteriol.* 173:1738–1747.

4107. Stragier, P., O. Danos, and J. C. Patte. 1983. Regulation of diaminopimelate decarboxylase synthesis in *Escherichia coli*. II. Nucleotide sequence of the *lysA* gene and its regulatory region. *J. Mol. Biol.* 168:321–331.

4108. Stragier, P., and J. C. Patte. 1983. Regulation of diaminopimelate decarboxylase synthesis in *Escherichia coli*. III. Nucleotide sequence and regulation of the *lysR* gene. *J. Mol. Biol.* 168:333–350.

4109. Stragier, P., F. Richaud, F. Borne, and J. C. Patte. 1983. Regulation of diaminopimelate decarboxylase synthesis in *Escherichia coli*. I. Identification of a *lysR* gene encoding an activator of the *lysA* gene. *J. Mol. Biol.* 168:307–320.

4110. Straney, R., R. Krah, and R. Menzel. 1994. Mutations in the −10 TATAAT sequence of the *gyrA* promoter affect both promoter strength and sensitivity to DNA supercoiling. *J. Bacteriol.* 176:5999–6006.

4111. Strauch, K. L., K. Johnson, and J. Beckwith. 1989. Characterization of *degP*, a gene required for proteolysis in the cell envelope and essential for growth of *Escherichia coli* at high temperature. *J. Bacteriol.* 171:2689–2696.

4112. Straus, D., W. Walter, and C. A. Gross. 1990. DnaK, DnaJ, and GrpE heat shock proteins negatively regulate heat shock gene expression by controlling the synthesis and stability of sigma 32. *Genes Dev.* 4:2202–2209.

4113. Strom, A. R., and I. Kaasen. 1993. Trehalose metabolism in *Escherichia coli*: stress protection and stress regulation of gene expression. *Mol. Microbiol.* 8:205–210.

4114. Struyve, M., M. Moons, and J. Tommassen. 1991. Carboxy-terminal phenylalanine is essential for the correct assembly of a bacterial outer membrane protein. *J. Mol. Biol.* 218:141–148.

4115. Struyve, M., J. Visser, H. Adriaanse, R. Benz, and J. Tommassen. 1993. Topology of PhoE porin: the 'eyelet' region. *Mol. Microbiol.* 7:131–140.

4116. Studwell-Vaughan, P. S., and M. O'Donnell. 1993. DNA polymerase III accessory proteins. V. Theta encoded by *holE*. *J. Biol. Chem.* 268:11785–11791.

4117. Stuitje, A. R., N. de Wind, J. C. van der Spek, T. H. Pors, and M. Meijer. 1986. Dissection of promoter sequences involved in transcriptional activation of the *Escherichia coli* replication origin. *Nucleic Acids Res.* 14:2333–2344.

4118. Styrvold, O. B., P. Falkenberg, B. Landfald, M. W. Eshoo, T. Bjornsen, and A. R. Strom. 1986. Selection, mapping, and characterization of osmoregulatory mutants of *Escherichia coli* blocked in the choline-glycine betaine pathway. *J. Bacteriol.* 165:856–863.

4119. Su, H., J. Moniakis, and E. B. Newman. 1993. Use of gene fusions of the structural gene *sdaA* to purify L-serine deaminase 1 from *Escherichia coli* K-12. *Eur. J. Biochem.* 211:521–527.

4120. Su, H., and E. B. Newman. 1991. A novel L-serine deaminase activity in *Escherichia coli* K-12. *J. Bacteriol.* 173:2473–2480.

4121. Su, H. S., B. F. Lang, and E. B. Newman. 1989. L-Serine degradation in *Escherichia coli* K-12: cloning and sequencing of the *sdaA* gene. *J. Bacteriol.* 171:5095–5102.

4122. Su, T. Z., H. P. Schweizer, and D. L. Oxender. 1991. Carbon-starvation induction of the *ugp* operon, encoding the binding protein-dependent sn-glycerol-3-phosphate transport system in *Escherichia coli*. *Mol. Gen. Genet.* 230:28–32.

4123. Subrahmanyam, C. S., G. M. McCorkle, and H. E. Umbarger. 1980. Physical location of the *ilvO* determinant in *Escherichia coli* K-12 deoxyribonucleic acid. *J. Bacteriol.* 142:547–555.

4124. Subrahmanyam, C. S., J. D. Noti, and H. E. Umbarger. 1980. Regulation of *ilvEDA* expression occurs upstream of *ilvG* in *Escherichia coli*: additional evidence for an *ilvGEDA* operon. *J. Bacteriol.* 144:279–290.

4125. Sugai, M., and H. C. Wu. 1992. Export of the outer membrane lipoprotein is defective in *secD*, *secE*, and *secF* mutants of *Escherichia coli*. *J. Bacteriol.* 174:2511–2516.

4126. Sugawara, E., and H. Nikaido. 1992. Pore-forming activity of OmpA protein of *Escherichia coli*. *J. Biol. Chem.* 267:2507–2511.

4127. Sugimoto, K., A. Oka, H. Sugisaki, M. Takanami, A. Nishimura, Y. Yasuda, and Y. Hirota. 1979. Nucleotide sequence of *Escherichia coli* K-12 replication origin. *Proc. Natl. Acad. Sci. USA* 76:575–579.

4128. Sugimura, K., and T. Nishihara. 1988. Purification, characterization, and primary structure of *Escherichia coli* protease VII with specificity for paired basic residues: identity of protease VII and OmpT. *J. Bacteriol.* 170:5625–5632.

4129. Sugiura, A., K. Nakashima, K. Tanaka, and T. Mizuno. 1992. Clarification of the structural and functional features of the osmoregulated *kdp* operon of *Escherichia coli*. *Mol. Microbiol.* 6:1769–1776.

4130. Sugiyama, J. E., S. Mahmoodian, and G. R. Jacobson. 1991. Membrane topology analysis of *Escherichia coli* mannitol permease by using a nested-deletion method to create *mtlA-phoA* fusions. *Proc. Natl. Acad. Sci. USA* 88:9603–9607.

4131. Sukharev, S. I., P. Blount, B. Martinac, F. R. Blattner, and C. Kung. 1994. A large-conductance mechanosensitive channel in *E. coli* encoded by *mscL* alone. *Nature* (London) 368:265–268.

4131a. Sullivan, M. A., and R. M. Bock. 1985. Isolation and characterization of antisuppressor mutations in *Escherichia coli*. *J. Bacteriol.* 161:377–384.

4131b. Sullivan, M. A., J. F. Cannon, F. H. Webb, and R. M. Bock. 1985. Antisuppressor mutation in *Escherichia coli* defective in biosynthesis of 5-methylaminomethyl-2-thiouridine. *J. Bacteriol.* 161:368–376.

4132. Sullivan, N. F., and W. D. Donachie. 1984. Overlapping functional units in a cell division gene cluster in *Escherichia coli*. *J. Bacteriol.* 158:1198–1201.

4133. Sullivan, N. F., and W. D. Donachie. 1984. Transcriptional organization within an *Escherichia coli* cell division gene cluster: direction of transcription of the cell separation gene *envA*. *J. Bacteriol.* 160:724–732.

4134. Sullivan, S. L., and M. E. Gottesman. 1992. Requirement for *E. coli* NusG protein in factor-dependent transcription termination. *Cell* 68:989–994.

4135. Sun, L., and J. A. Fuchs. 1994. Regulation of the *Escherichia coli* *nrd* operon: role of DNA supercoiling. *J. Bacteriol.* 176:4617–4626.

4136. Sun, T. P., and R. E. Webster. 1986. *fii*, a bacterial locus required for filamentous phage infection and its relation to colicin-tolerant *tolA* and *tolB*. *J. Bacteriol.* 165:107–115.

4137. Sun, T. P., and R. E. Webster. 1987. Nucleotide sequence of a gene cluster involved in entry of E colicins and single-stranded DNA of infecting filamentous bacteriophages into *Escherichia coli*. *J. Bacteriol.* 169:2667–2674.

4138. Sun, X., J. Harder, M. Krook, H. Jornvall, B. M. Sjoberg, and P. Reichard. 1993. A possible glycine radical in anaerobic ribonucleotide reductase from *Escherichia coli*: nucleotide sequence of the cloned *nrdD* gene. *Proc. Natl. Acad. Sci. USA* 90:577–581.

4139. Sung, C. Y., J. M. Gennity, N. S. Pollitt, and M. Inouye. 1992. A positive residue in the hydrophobic core of the *Escherichia coli* lipoprotein signal peptide suppresses the secretion defect caused by an acidic amino terminus. *J. Biol. Chem.* 267:997–1000.

4140. Sung, M., and R. E. Dalbey. 1992. Identification of potential active-site residues in the *Escherichia coli* leader peptidase. *J. Biol. Chem.* 267:13154–13159.

4141. Sung, Y. C., P. M. Anderson, and J. A. Fuchs. 1987. Characterization of high-level expression and sequencing of the *Escherichia coli* K-12 *cynS* gene encoding cyanase. *J. Bacteriol.* 169:5224–5230.

4142. Sung, Y. C., and J. A. Fuchs. 1988. Characterization of the *cyn* operon in *Escherichia coli* K12. *J. Biol. Chem.* 263:14769–14775.

4143. Sung, Y. C., and J. A. Fuchs. 1992. The *Escherichia coli* K-12 *cyn* operon is positively regulated by a member of the *lysR* family. *J. Bacteriol.* 174:3645–3650.

4144. Sung, Y. C., D. Parsell, P. M. Anderson, and J. A. Fuchs. 1987. Identification, mapping, and cloning of the gene encoding cyanase in *Escherichia coli* K-12. *J. Bacteriol.* 169:2639–2642.

4145. Sunnarborg, A., D. Klumpp, T. Chung, and D. C. LaPorte. 1990. Regulation of the glyoxylate bypass operon: cloning and characterization of *iclR*. *J. Bacteriol.* 172:2642–2649.

4146. Sunshine, M. G., and B. Kelly. 1971. Extent of host deletions associated with bacteriophage P2-mediated eduction. *J. Bacteriol.* 108:695–704.

4147. Sunshine, M. G., and B. Sauer. 1975. A bacterial mutation blocking P2 phage late gene expression. *Proc. Natl. Acad. Sci. USA* 72:2770–2774.

4148. Suppmann, B., and G. Sawers. 1994. Isolation and characterization of hypophosphite–resistant mutants of *Escherichia coli*: identification of the FocA protein, encoded by the *pfl* operon, as a putative formate transporter. *Mol. Microbiol.* 11:965–982.

4149. Surin, B. P., D. A. Jans, A. L. Fimmel, D. C. Shaw, G. B. Cox, and H. Rosenberg. 1984. Structural gene for the phosphate-repressible phosphate-binding protein of *Escherichia coli* has its own promoter: complete nucleotide sequence of the *phoS* gene. *J. Bacteriol.* 157:772–778.

4150. Surin, B. P., H. Rosenberg, and G. B. Cox. 1985. Phosphate-specific transport system of *Escherichia coli*: nucleotide sequence and gene-polypeptide relationships. *J. Bacteriol.* 161:189–198.

4151. Sutherland, L., J. Cairney, M. J. Elmore, I. R. Booth, and C. F. Higgins. 1986. Osmotic regulation of transcription: induction of the *proU* betaine transport gene is dependent on accumulation of intracellular potassium. *J. Bacteriol.* 168:805–814.

4152. Sutherland, P., and L. McAlister-Henn. 1985. Isolation and expression of the *Escherichia coli* gene encoding malate dehydrogenase. *J. Bacteriol.* 163:1074–1079.

4153. Sutton, A., T. Newman, M. Francis, and M. Freundlich. 1981. Valine-resistant *Escherichia coli* K-12 strains with mutations in the *ilvB* operon. *J. Bacteriol.* 148:998–1001.

4154. Sutton, M. R., R. R. Fall, A. M. Nervi, A. W. Alberts, P. R. Vagelos, and R. A. Bradshaw. 1977. Amino acid sequence of *Escherichia coli* biotin carboxyl carrier protein (9100). *J. Biol. Chem.* 252:3934–3940.

4155. Suzuki, H., H. Kumagai, T. Echigo, and T. Tochikura. 1988. Molecular cloning of *Escherichia coli* K-12 *ggt* and rapid isolation of gamma-glutamyltranspeptidase. *Biochem. Biophys. Res. Commun.* 150:33–38.

4156. Suzuki, H., H. Kumagai, T. Echigo, and T. Tochikura. 1989. DNA sequence of the *Escherichia coli* K-12 gamma-glutamyltranspeptidase gene, *ggt*. *J. Bacteriol.* 171:5169–5172.

4157. Suzuki, H., H. Kumagai, and T. Tochikura. 1987. Isolation, genetic mapping, and characterization of *Escherichia coli* K-12 mutants lacking gamma-glutamyltranspeptidase. *J. Bacteriol.* 169:3926–3931.

4158. Suzuki, H., Y. van Heijenoort, T. Tamura, J. Mizoguchi, Y. Hirota, and J. van Heijenoort. 1980. In vitro peptidoglycan polymerization catalysed by penicillin binding protein 1b of *Escherichia coli* K-12. *FEBS Lett.* 110:245–249.

4159. Suzuki, T., A. Itoh, S. Ichihara, and S. Mizushima. 1987. Characterization of the *sppA* gene coding for protease IV, a signal peptide peptidase of *Escherichia coli*. *J. Bacteriol.* 169:2523–2528.

4160. **Svitil, A. L., M. Cashel, and J. W. Zyskind.** 1993. Guanosine tetraphosphate inhibits protein synthesis in vivo. A possible protective mechanism for starvation stress in *Escherichia coli. J. Biol. Chem.* **268:**2307–2311.

4161. **Swanberg, S. L., and J. C. Wang.** 1987. Cloning and sequencing of the *Escherichia coli gyrA* gene coding for the A subunit of DNA gyrase. *J. Mol. Biol.* **197:**729–736.

4162. **Sweet, G., C. Gandor, R. Voegele, N. Wittekindt, J. Beuerle, V. Truniger, E. C. Lin, and W. Boos.** 1990. Glycerol facilitator of *Escherichia coli:* cloning of *glpF* and identification of the *glpF* product. *J. Bacteriol.* **172:**424–430.

4163. **Swidersky, U. E., A. Rienhofer-Schweer, P. K. Werner, F. Ernst, S. A. Benson, H. K. Hoffschulte, and M. Muller.** 1992. Biochemical analysis of the biogenesis and function of the *Escherichia coli* export factor SecY. *Eur. J. Biochem.* **207:**803–811.

4164. **Swift, S., J. Kuhn, and G. S. Stewart.** 1992. Selection and analysis of non-interactive mutants in the *Escherichia coli* tryptophan synthase alpha subunit. *Mol. Gen. Genet.* **233:**129–135.

4165. **Swindle, J., J. Ajioka, D. Dawson, R. Myers, D. Carroll, and C. Georgopoulos.** 1984. The nucleotide sequence of the *Escherichia coli* K12 *nusB* (*groNB*) gene. *Nucleic Acids Res.* **12:**4977–4985.

4166. **Szafranski, P.** 1992. On the evolution of the bacterial major sigma factors. *J. Mol. Evol.* **34:**465–467. (Letter.)

4167. **Szalewska, A., G. Wegrzyn, and K. Taylor.** 1994. Neither absence nor excess of lambda O initiator-digesting ClpXP protease affects lambda plasmid or phage replication in *Escherichia coli. Mol. Microbiol.* **13:**469–474.

4168. **Szekely, E., and M. Simon.** 1983. DNA sequence adjacent to flagellar genes and evolution of flagellar-phase variation. *J. Bacteriol.* **155:**74–81.

4169. **Szentirmai, A., M. Szentirmai, and H. E. Umbarger.** 1968. Isoleucine and valine metabolism of *Escherichia coli.* XV. Biochemical properties of mutants resistant to thiaisoleucine. *J. Bacteriol.* **95:**1672–1679.

4170. **Szumanski, M. B., and S. M. Boyle.** 1990. Analysis and sequence of the *speB* gene encoding agmatine ureohydrolase, a putrescine biosynthetic enzyme in *Escherichia coli. J. Bacteriol.* **172:**538–547.

4171. **Szybalski, E. H., and W. Szybalski.** 1982. A physical map of the *Escherichia coli bio* operon. *Gene* **19:**93–103.

4172. **Ta, D. T., B. L. Seaton, and L. E. Vickery.** 1992. Localization of the ferredoxin (*fdx*) gene on the physical map of the *Escherichia coli* chromosome. *J. Bacteriol.* **174:**5760–5761.

4173. **Ta, D. T., and L. E. Vickery.** 1992. Cloning, sequencing, and overexpression of a [2Fe-2S] ferredoxin gene from *Escherichia coli. J. Biol. Chem.* **267:**11120–11125.

4174. **Tabata, S., A. Higashitani, M. Takanami, K. Akiyama, Y. Kohara, Y. Nishimura, and A. Nishimura.** 1989. Construction of an ordered cosmid collection of the *Escherichia coli* K-12 W3110 chromosome. *J. Bacteriol.* **171:**1214–1218.

4175. **Tabata, S., A. Oka, K. Sugimoto, M. Takanami, S. Yasuda, and Y. Hirota.** 1983. The 245 base-pair *oriC* sequence of the *E. coli* chromosome directs bidirectional replication at an adjacent region. *Nucleic Acids Res.* **11:**2617–2626.

4176. **Tabor, C. W., and H. Tabor.** 1987. The *speEspeD* operon of *Escherichia coli.* Formation and processing of a proenzyme form of S-adenosylmethionine decarboxylase. *J. Biol. Chem.* **262:**16037–16040.

4177. **Tabor, C. W., H. Tabor, and Q. W. Xie.** 1986. Spermidine synthase of *Escherichia coli:* localization of the *speE* gene. *Proc. Natl. Acad. Sci. USA* **83:**6040–6044.

4178. **Tabor, H., E. W. Hafner, and C. W. Tabor.** 1980. Construction of an *Escherichia coli* strain unable to synthesize putrescine, spermidine, or cadaverine: characterization of two genes controlling lysine decarboxylase. *J. Bacteriol.* **144:**952–956.

4179. **Tadmor, Y., R. Ascarelli-Goell, R. Skaliter, and Z. Livneh.** 1992. Overproduction of the beta subunit of DNA polymerase III holoenzyme reduces UV mutagenesis in *Escherichia coli. J. Bacteriol.* **174:**2517–2524.

4180. **Taglicht, D., E. Padan, A. B. Oppenheim, and S. Schuldiner.** 1987. An alkaline shift induces the heat shock response in *Escherichia coli. J. Bacteriol.* **169:**885–887.

4181. **Taglicht, D., E. Padan, and S. Schuldiner.** 1991. Overproduction and purification of a functional Na^+/H^+ antiporter coded by *nhaA* (*ant*) from *Escherichia coli. J. Biol. Chem.* **266:**11289–11294.

4182. **Takagi, J. S., N. Ida, M. Tokushige, H. Sakamoto, and Y. Shimura.** 1985. Cloning and nucleotide sequence of the aspartase gene of *Escherichia coli* W. *Nucleic Acids Res.* **13:**2063–2074.

4183. **Takahagi, M., H. Iwasaki, A. Nakata, and H. Shinagawa.** 1991. Molecular analysis of the *Escherichia coli ruvC* gene, which encodes a Holliday junction-specific endonuclease. *J. Bacteriol.* **173:**5747–5753.

4184. **Takahashi, H.** 1978. Genetic and physiological characterization of *Escherichia coli* K12 mutants (*tabC*) which induce the abortive infection of bacteriophage T4. *Virology* **87:**256–265.

4185. **Takano, K., Y. Nakabeppu, H. Maki, T. Horiuchi, and M. Sekiguchi.** 1986. Structure and function of *dnaQ* and *mutD* mutators of *Escherichia coli. Mol. Gen. Genet.* **205:**9–13.

4186. **Takano, K., T. Nakamura, and M. Sekiguchi.** 1991. Roles of two types of O^6-methylguanine-DNA methyltransferases in DNA repair. *Mutat. Res.* **254:**37–44.

4187. **Takase, I., F. Ishino, M. Wachi, H. Kamata, M. Doi, S. Asoh, H. Matsuzawa, T. Ohta, and M. Matsuhashi.** 1987. Genes encoding two lipoproteins in the *leuS-dacA* region of the *Escherichia coli* chromosome. *J. Bacteriol.* **169:**5692–5699.

4188. **Takata, R., M. Izuhara, and K. Akiyama.** 1992. Processing in the 5′ region of the *pnp* transcript facilitates the site-specific endonucleolytic cleavages of mRNA. *Nucleic Acids Res.* **20:**847–850.

4189. **Takata, R., T. Mukai, M. Aoyagi, and K. Hori.** 1984. Nucleotide sequence of the gene for *Escherichia coli* ribosomal protein S15 (*rpsO*). *Mol. Gen. Genet.* **197:**225–229.

4190. **Takata, R., T. Mukai, and K. Hori.** 1985. Attenuation and processing of RNA from the *rpsO-pnp* transcription unit of *Escherichia coli. Nucleic Acids Res.* **13:**7289–7297.

4191. **Takayanagi, K., S. Maeda, and T. Mizuno.** 1991. Expression of *micF* involved in porin synthesis in *Escherichia coli:* two distinct cis-acting elements respectively regulate *micF* expression positively and negatively. *FEMS Microbiol. Lett.* **67:**39–44.

4192. **Takebe, Y., and Y. Kaziro.** 1982. In vitro construction of the *tufB-lacZ* fusion: analysis of the regulatory mechanism of *tufB* promoter. *Mol. Gen. Genet.* **187:**355–363.

4193. **Takeda, Y., and H. Avila.** 1986. Structure and gene expression of the *E. coli* Mn-superoxide dismutase gene. *Nucleic Acids Res.* **14:**4577–4589.

4194. **Takeda, Y., A. Nishimura, Y. Nishimura, M. Yamada, S. Yasuda, H. Suzuki, and Y. Hirota.** 1981. Synthetic ColE1 plasmids carrying genes for penicillin-binding proteins in *Escherichia coli. Plasmid* **6:**86–98.

4195. **Takiff, H. E., T. Baker, T. Copeland, S. M. Chen, and D. L. Court.** 1992. Locating essential *Escherichia coli* genes by using mini-Tn*10* transposons: the *pdxJ* operon. *J. Bacteriol.* **174:**1544–1553.

4196. **Takiff, H. E., S. M. Chen, and D. L. Court.** 1989. Genetic analysis of the *rnc* operon of *Escherichia coli. J. Bacteriol.* **171:**2581–2590.

4197. **Talarico, T. L., I. K. Dev, W. S. Dallas, R. Ferone, and P. H. Ray.** 1991. Purification and partial characterization of 7,8-dihydro-6-hydroxymethylpterin-pyrophosphokinase and 7,8-dihydropteroate synthase from *Escherichia coli* MC4100. *J. Bacteriol.* **173:**7029–7032.

4198. **Talarico, T. L., P. H. Ray, I. K. Dev, B. M. Merrill, and W. S. Dallas.** 1992. Cloning, sequence analysis, and overexpression of *Escherichia coli folK,* the gene coding for 7,8-dihydro-6-hydroxymethylpterin-pyrophosphokinase. *J. Bacteriol.* **174:**5971–5977.

4199. **Tamaki, S., H. Matsuzawa, and M. Matsuhashi.** 1980. Cluster of *mrdA* and *mrdB* genes responsible for the rod shape and mecillinam sensitivity of *Escherichia coli. J. Bacteriol.* **141:**52–57.

4200. **Tamaki, S., T. Sato, and M. Matsuhashi.** 1971. Role of lipopolysaccharides in antibiotic resistance and bacteriophage adsorption of *Escherichia coli* K-12. *J. Bacteriol.* **105:**968–975.

4201. **Tamura, F., S. Nishimura, and M. Ohki.** 1984. The *E. coli divE* mutation, which differentially inhibits synthesis of certain proteins, is in $tRNA_1^{Ser}$. *EMBO J.* **3:**1103–1107.

4202. **Tamura, J. K., A. D. Bates, and M. Gellert.** 1992. Slow interaction of 5′-adenylyl-βγ-imidodiphosphate with *Escherichia coli* DNA gyrase. Evidence for cooperativity in nucleotide binding. *J. Biol. Chem.* **267:**9214–9222.

4203. **Tamura, T., H. Suzuki, Y. Nishimura, J. Mizoguchi, and Y. Hirota.** 1980. On the process of cellular division in *Escherichia coli:* isolation and characterization of penicillin-binding proteins 1a, 1b, and 3. *Proc. Natl. Acad. Sci. USA* **77:**4499–4503.

4204. **Tanabe, H., J. Goldstein, M. Yang, and M. Inouye.** 1992. Identification of the promoter region of the *Escherichia coli* major cold shock gene, *cspA. J. Bacteriol.* **174:**3867–3873.

4205. **Tanaka, K., S. Muramatsu, H. Yamada, and T. Mizuno.** 1991. Systematic characterization of curved DNA segments randomly cloned from *Escherichia coli* and their functional significance. *Mol. Gen. Genet.* **226:**367–376.

4206. **Tanaka, K., Y. Takayanagi, N. Fujita, A. Ishihama, and H. Takahashi.** 1993. Heterogeneity of the principal sigma factor in *Escherichia coli:* the *rpoS* gene product, sigma 38, is a second principal sigma factor of RNA polymerase in stationary-phase *Escherichia coli. Proc. Natl. Acad. Sci. USA* **90:**3511–3515.

4207. **Tanaka, S., Y. Matsushita, A. Yoshikawa, and K. Isono.** 1989. Cloning and molecular characterization of the gene *rimL* which encodes an enzyme acetylating ribosomal protein L12 of *Escherichia coli* K12. *Mol. Gen. Genet.* **217:**289–293.

4208. **Tanaka, Y., A. Tsujimura, N. Fujita, S. Isono, and K. Isono.** 1989. Cloning and analysis of an *Escherichia coli* operon containing the *rpmF* gene for ribosomal protein L32 and the gene for a 30-kilodalton protein. *J. Bacteriol.* **171:**5707–5712.

4209. **Tang, C. T., R. Engel, and B. E. Tropp.** 1977. L-Glyceraldehyde 3-phosphate, a bactericidal agent. *Antimicrob. Agents Chemother.* **11:**147–153.

4210. **Tao, J. S., and E. E. Ishiguro.** 1989. Nucleotide sequence of the *murE* gene of *Escherichia coli. Can. J. Microbiol.* **35:**1051–1054.

4211. **Tao, K., K. Makino, S. Yonei, A. Nakata, and H. Shinagawa.** 1989. Molecular cloning and nucleotide sequencing of *oxyR,* the positive regulatory gene of a regulon for an adaptive response to oxidative stress in *Escherichia coli:* homologies between OxyR protein and a family of bacterial activator proteins. *Mol. Gen. Genet.* **218:**371–376.

4212. **Tao, K., K. Makino, S. Yonei, A. Nakata, and H. Shinagawa.** 1991. Purification and characterization of the *Escherichia coli* OxyR protein, the positive regulator for a hydrogen peroxide-inducible regulon. *J. Biochem.* **109:**262–266.

4213. **Tapio, S., F. Yeh, H. A. Shuman, and W. Boos.** 1991. The *malZ* gene of *Escherichia coli,* a member of the maltose regulon, encodes a maltodextrin glucosidase. *J. Biol. Chem.* **266:**19450–19458.

4214. **Taraseviciene, L., A. Miczak, and D. Apirion.** 1991. The gene specifying RNase E (*rne*) and a gene affecting mRNA stability (*ams*) are the same gene. *Mol. Microbiol.* **5:**851–855.

4215. **Tardat, B., and D. Touati.** 1991. Two global regulators repress the anaerobic expression of MnSOD in *Escherichia coli*::Fur (ferric uptake regulation) and Arc (aerobic respiration control). *Mol. Microbiol.* **5:**455–465.

4216. Tartaglia, L. A., G. Storz, and B. N. Ames. 1989. Identification and molecular analysis of *oxyR*-regulated promoters important for the bacterial adaptation to oxidative stress. *J. Mol. Biol.* **210:**709–719.

4217. Taschner, P. E., J. G. Verest, and C. L. Woldringh. 1987. Genetic and morphological characterization of *ftsB* and *nrdB* mutants of *Escherichia coli*. *J. Bacteriol.* **169:**19–25.

4218. Taschner, P. E., N. Ypenburg, B. G. Spratt, and C. L. Woldringh. 1988. An amino acid substitution in penicillin-binding protein 3 creates pointed polar caps in *Escherichia coli*. *J. Bacteriol.* **170:**4828–4837.

4219. Tate, C. G., and P. J. Henderson. 1993. Membrane topology of the L-rhamnose-H⁺ transport protein (RhaT) from enterobacteria. *J. Biol. Chem.* **268:**26850–26857.

4220. Tate, C. G., J. A. Muiry, and P. J. Henderson. 1992. Mapping, cloning, expression, and sequencing of the *rhaT* gene, which encodes a novel L-rhamnose-H⁺ transport protein in *Salmonella typhimurium* and *Escherichia coli*. *J. Biol. Chem.* **267:**6923–6932.

4221. Taucher-Scholz, G., and H. Hoffmann-Berling. 1983. Identification of the gene for DNA helicase II of *Escherichia coli*. *Eur. J. Biochem.* **137:**573–580.

4222. Taura, T., T. Baba, Y. Akiyama, and K. Ito. 1993. Determinants of the quantity of the stable SecY complex in the *Escherichia coli* cell. *J. Bacteriol.* **175:**7771–7775.

4223. Taura, T., C. Ueguchi, K. Shiba, and K. Ito. 1992. Insertional disruption of the *nusB* (*ssyB*) gene leads to cold-sensitive growth of *Escherichia coli* and suppression of the *secY24* mutation. *Mol. Gen. Genet.* **234:**429–432.

4224. Tawa, P., and R. C. Stewart. 1994. Mutational activation of CheA, the protein kinase in the chemotaxis system of *Escherichia coli*. *J. Bacteriol.* **176:**4210–4218.

4225. Taylor, A. F. 1992. Movement and resolution of Holliday junctions by enzymes from *E. coli*. *Cell* **69:**1063–1065.

4226. Taylor, A. F., P. G. Siliciano, and B. Weiss. 1980. Cloning of the *dut* (deoxyuridine triphosphatase) gene of *Escherichia coli*. *Gene* **9:**321–336.

4227. Taylor, D. E., Y. Hou, R. J. Turner, and J. H. Weiner. 1994. Location of a potassium tellurite resistance operon (*tehA tehB*) within the terminus of *Escherichia coli* K-12. *J. Bacteriol.* **176:**2740–2742.

4228. Taylor, J. L., J. R. Bedbrook, F. J. Grant, and A. Kleinhofs. 1983. Reconstitution of plant nitrate reductase by *Escherichia coli* extracts and the molecular cloning of the *chlA* gene of *Escherichia coli* K12. *J. Mol. Appl. Genet.* **2:**261–271.

4229. Taylor, R. K., S. Garrett, E. Sodergren, and T. J. Silhavy. 1985. Mutations that define the promoter of *ompF*, a gene specifying a major outer membrane porin protein. *J. Bacteriol.* **162:**1054–1060.

4230. Taylor, R. K., M. N. Hall, L. Enquist, and T. J. Silhavy. 1981. Identification of OmpR: a positive regulatory protein controlling expression of the major outer membrane matrix porin proteins of *Escherichia coli* K-12. *J. Bacteriol.* **147:**255–258.

4231. Taylor, W. E., and R. R. Burgess. 1979. *Escherichia coli* RNA polymerase binding and initiation of transcription on fragments of lambda *rifd 18* DNA containing promoters for lambda genes and for *rrnB*, *tufB*, *rplC,A*, *rplJ,L*, and *rpoB,C* genes. *Gene* **6:**331–365.

4232. Taylor, W. E., D. B. Straus, A. D. Grossman, Z. F. Burton, C. A. Gross, and R. R. Burgess. 1984. Transcription from a heat-inducible promoter causes heat shock regulation of the sigma subunit of *E. coli* RNA polymerase. *Cell* **38:**371–381.

4233. Tecklenburg, M., A. Naumer, O. Nagappan, and P. Kuempel. 1995. The *dif* resolvase locus of the *Escherichia coli* chromosome can be replaced by a 33-bp sequence, but function depends on location. *Proc. Natl. Acad. Sci. USA* **92:**1352–1356.

4234. Tedin, K., and H. Bremer. 1992. Toxic effects of high levels of ppGpp in *Escherichia coli* are relieved by *rpoB* mutations. *J. Biol. Chem.* **267:**2337–2344.

4235. Tei, H., K. Murata, and A. Kimura. 1990. Structure and expression of *cysX*, the second gene in the *Escherichia coli* K-12 *cysE* locus. *Biochem. Biophys. Res. Commun.* **167:**948–955.

4236. Tei, H., K. Murata, and A. Kimura. 1990. Molecular cloning of the *cys* genes (*cysC*, *cysD*, *cysH*, *cysI*, *cysJ*, and *cysG*) responsible for cysteine biosynthesis in *Escherichia coli* K-12. *Biotechnol. Appl. Biochem.* **12:**212–216.

4237. Tei, H., K. Watanabe, K. Murata, and A. Kimura. 1990. Analysis of the *Escherichia coli* K-12 *cysB* gene and its product using the method of gene fusion. *Biochem. Biophys. Res. Commun.* **167:**962–969.

4238. Teo, I., B. Sedgwick, M. W. Kilpatrick, T. V. McCarthy, and T. Lindahl. 1986. The intracellular signal for induction of resistance to alkylating agents in *E. coli*. *Cell* **45:**315–324.

4239. Terada, K., and K. Izui. 1991. Site-directed mutagenesis of the conserved histidine residue of phosphoenolpyruvate carboxylase. His138 is essential for the second partial reaction. *Eur. J. Biochem.* **202:**797–803.

4240. Terada, K., T. Murata, and K. Izui. 1991. Site-directed mutagenesis of phosphoenolpyruvate carboxylase from *E. coli*: the role of His579 in the catalytic and regulatory functions. *J. Biochem.* **109:**49–54.

4241. Terhorst, C., W. Moller, R. Laursen, and B. Wittmann-Liebold. 1973. The primary structure of an acidic protein from 50-S ribosomes of *Escherichia coli* which is involved in GTP hydrolysis dependent on elongation factors G and T. *Eur. J. Biochem.* **34:**138–152.

4242. Tesfa-Selase, F., and W. T. Drabble. 1992. Regulation of the *gua* operon of *Escherichia coli* by the DnaA protein. *Mol. Gen. Genet.* **231:**256–264.

4243. Tessman, I., J. S. Fassler, and D. C. Bennett. 1982. Relative map location of the *rep* and *rho* genes of *Escherichia coli*. *J. Bacteriol.* **151:**1637–1640.

4244. Tetart, F., R. Albigot, A. Conter, E. Mulder, and J. P. Bouche. 1992. Involvement of FtsZ in coupling of nucleoid separation with septation. *Mol. Microbiol.* **6:**621–627.

4245. Tetart, F., and J. P. Bouche. 1992. Regulation of the expression of the cell-cycle gene *ftsZ* by DicF antisense RNA. Division does not require a fixed number of FtsZ molecules. *Mol. Microbiol.* **6:**615–620.

4245a. Theall, G., K. B. Low, and D. Soll. 1979. Regulation of the biosynthesis of aminoacyl-tRNA synthetases and of tRNA in *Escherichia coli*. IV. Mutants with increased levels of leucyl- or seryl-tRNA synthetase. *Mol. Gen. Genet.* **169:**205–211.

4246. Theissen, G., S. E. Behrens, and R. Wagner. 1990. Functional importance of the *Escherichia coli* ribosomal RNA leader box A sequence for post-transcriptional events. *Mol. Microbiol.* **4:**1667–1678.

4247. Thelen, P., T. Tsuchiya, and E. B. Goldberg. 1991. Characterization and mapping of a major Na⁺/H⁺ antiporter gene of *Escherichia coli*. *J. Bacteriol.* **173:**6553–6557.

4248. Thirion, J. P., and M. Hofnung. 1972. On some genetic aspects of phage lambda resistance in *E. coli* K12. *Genetics* **71:**207–216.

4249. Thliveris, A. T., and D. W. Mount. 1992. Genetic identification of the DNA binding domain of *Escherichia coli* LexA protein. *Proc. Natl. Acad. Sci. USA* **89:**4500–4504.

4250. Thomas, A., and R. G. Lloyd. 1980. Altered regulation of the *recA* gene in *Escherichia coli* strains carrying a *recA*-linked suppressor of *lexA*. *Mol. Gen. Genet.* **179:**355–358.

4251. Thomas, C. D., J. Modha, T. M. Razzaq, P. M. Cullis, and A. J. Rivett. 1993. Controlled high-level expression of the *lon* gene of *Escherichia coli* allows overproduction of Lon protease. *Gene* **136:**237–242.

4252. Thomas, J. A., and M. A. Valvano. 1993. Role of *tol* genes in cloacin DF13 susceptibility of *Escherichia coli* K-12 strains expressing the cloacin DF13-aerobactin receptor IutA. *J. Bacteriol.* **175:**548–552.

4253. Thomas, M. S., D. M. Bedwell, and M. Nomura. 1987. Regulation of alpha operon gene expression in *Escherichia coli*. A novel form of translational coupling. *J. Mol. Biol.* **196:**333–345.

4254. Thomas, M. S., and W. T. Drabble. 1984. Molecular cloning and characterisation of the *gua* regulatory region of *Escherichia coli* K12. *Mol. Gen. Genet.* **195:**238–245.

4255. Thomas, M. S., and W. T. Drabble. 1985. Nucleotide sequence and organisation of the *gua* promoter region of *Escherichia coli*. *Gene* **36:**45–53.

4256. Thomas, M. S., and R. E. Glass. 1991. *Escherichia coli rpoA* mutation which impairs transcription of positively regulated systems. *Mol. Microbiol.* **5:**2719–2725.

4257. Thomas, M. S., and M. Nomura. 1987. Translational regulation of the L11 ribosomal protein operon of *Escherichia coli*: mutations that define the target site for repression by L1. *Nucleic Acids Res.* **15:**3085–3096.

4258. Thomas, S. D., and P. M. Jordan. 1986. Nucleotide sequence of the *hemC* locus encoding porphobilinogen deaminase of *Escherichia coli* K12. *Nucleic Acids Res.* **14:**6215–6226.

4259. Thome, B. M., and M. Muller. 1991. Skp is a periplasmic *Escherichia coli* protein requiring SecA and SecY for export. *Mol. Microbiol.* **5:**2815–2821. (Erratum, **6:**1077, 1992.)

4260. Thony, B., D. S. Hwang, L. Fradkin, and A. Kornberg. 1991. *iciA*, an *Escherichia coli* gene encoding a specific inhibitor of chromosomal initiation of replication in vitro. *Proc. Natl. Acad. Sci. USA* **88:**4066–4070.

4261. Thorbjarnardottir, S., H. Uemura, T. Dingermann, T. Rafnar, S. Thorsteinsdottir, D. Soll, and G. Eggertsson. 1985. *Escherichia coli supH* suppressor: temperature-sensitive missense suppression caused by an anticodon change in tRNA₂^Ser. *J. Bacteriol.* **161:**207–211.

4262. Thorbjarnardottir, S. H., R. A. Magnusdottir, and G. Eggertsson. 1978. Mutations determining generalized resistance to aminoglycoside antibiotics in *Escherichia coli*. *Mol. Gen. Genet.* **161:**89–98.

4263. Thorne, G. M., and L. M. Corwin. 1975. Mutations affecting aromatic amino acid transport in *Escherichia coli* and *Salmonella typhimurium*. *J. Gen. Microbiol.* **90:**203–216.

4264. Thorsness, P. E., and D. E. Koshland, Jr. 1987. Inactivation of isocitrate dehydrogenase by phosphorylation is mediated by the negative charge of the phosphate. *J. Biol. Chem.* **262:**10422–10425.

4265. Tian, G., D. Lim, J. Carey, and W. K. Maas. 1992. Binding of the arginine repressor of *Escherichia coli* K12 to its operator sites. *J. Mol. Biol.* **226:**387–397.

4266. Tiedeman, A. A., D. J. DeMarini, J. Parker, and J. M. Smith. 1990. DNA sequence of the *purC* gene encoding 5′-phosphoribosyl-5-aminoimidazole-4-N-succinocarboxamide synthetase and organization of the *dapA*-*purC* region of *Escherichia coli* K-12. *J. Bacteriol.* **172:**6035–6041.

4267. Tiedeman, A. A., J. Keyhani, J. Kamholz, H. A. Daum, J. S. Gots, and J. M. Smith. 1989. Nucleotide sequence analysis of the *purEK* operon encoding 5′-phosphoribosyl-5-aminoimidazole carboxylase of *Escherichia coli* K-12. *J. Bacteriol.* **171:**205–212.

4268. Tiedeman, A. A., and J. M. Smith. 1985. Nucleotide sequence of the *guaB* locus encoding IMP dehydrogenase of *Escherichia coli* K12. *Nucleic Acids Res.* **13:**1303–1316.

4269. Tiedeman, A. A., J. M. Smith, and H. Zalkin. 1985. Nucleotide sequence of the *guaA* gene encoding GMP synthetase of *Escherichia coli* K12. *J. Biol. Chem.* **260:**8676–8679.

4270. Tilly, K., H. Murialdo, and C. Georgopoulos. 1981. Identification of a second *Escherichia coli groE* gene whose product is necessary for bacteriophage morphogenesis. *Proc. Natl. Acad. Sci. USA* 78:1629–1633.

4271. Timms, A. R., H. Steingrimsdottir, A. R. Lehmann, and B. A. Bridges. 1992. Mutant sequences in the *rpsL* gene of *Escherichia coli* B/r: mechanistic implications for spontaneous and ultraviolet light mutagenesis. *Mol. Gen. Genet.* 232:89–96.

4272. Tobe, T., K. Ito, and T. Yura. 1984. Isolation and physical mapping of temperature-sensitive mutants defective in heat-shock induction of proteins in *Escherichia coli. Mol. Gen. Genet.* 195:10–16.

4273. Tobe, T., N. Kusukawa, and T. Yura. 1987. Suppression of *rpoH* (*htpR*) mutations of *Escherichia coli*: heat shock response in *suhA* revertants. *J. Bacteriol.* 169:4128–4134.

4274. Tobey, K. L., and G. A. Grant. 1986. The nucleotide sequence of the *serA* gene of *Escherichia coli* and the amino acid sequence of the encoded protein, D-3-phosphoglycerate dehydrogenase. *J. Biol. Chem.* 261:12179–12183.

4275. Tobin, J. F., and R. F. Schleif. 1987. Positive regulation of the *Escherichia coli* L-rhamnose operon is mediated by the products of tandemly repeated regulatory genes. *J. Mol. Biol.* 196:789–799.

4276. Tokishita, S., A. Kojima, and T. Mizuno. 1992. Transmembrane signal transduction and osmoregulation in *Escherichia coli*: functional importance of the transmembrane regions of membrane- located protein kinase, EnvZ. *J. Biochem.* 111:707–713.

4277. Tokunaga, M., J. M. Loranger, S. Y. Chang, M. Regue, S. Chang, and H. C. Wu. 1985. Identification of prolipoprotein signal peptidase and genomic organization of the *lsp* gene in *Escherichia coli. J. Biol. Chem.* 260:5610–5615.

4278. Tokushige, M., N. Tsujimoto, T. Oda, T. Honda, N. Yumoto, S. Ito, M. Yamamoto, E. H. Kim, and Y. Hiragi. 1989. Role of cysteine residues in tryptophanase for monovalent cation-induced activation. *Biochimie* 71:711–720.

4279. Tolner, B., B. Poolman, B. Wallace, and W. N. Konings. 1992. Revised nucleotide sequence of the *gltP* gene, which encodes the proton-glutamate-aspartate transport protein of *Escherichia coli* K-12. *J. Bacteriol.* 174:2391–2393.

4280. Tomasiewicz, H. G., and C. S. McHenry. 1987. Sequence analysis of the *Escherichia coli dnaE* gene. *J. Bacteriol.* 169:5735–5744.

4281. Tomioka, S., T. Nikaido, T. Miyakawa, and M. Matsuhashi. 1983. Mutation of the *N*-acetylmuramyl-L-alanine amidase gene of *Escherichia coli* K-12. *J. Bacteriol.* 156:463–465.

4282. Tommassen, J., P. de Geus, B. Lugtenberg, J. Hackett, and P. Reeves. 1982. Regulation of the *pho* regulon of *Escherichia coli* K-12. Cloning of the regulatory genes *phoB* and *phoR* and identification of their gene products. *J. Mol. Biol.* 157:265–274.

4283. Tommassen, J., K. Eiglmeier, S. T. Cole, P. Overduin, T. J. Larson, and W. Boos. 1991. Characterization of two genes, *glpQ* and *ugpQ*, encoding glycerophosphoryl diester phosphodiesterases of *Escherichia coli. Mol. Gen. Genet.* 226:321–327.

4284. Tommassen, J., P. Heimstra, P. Overduin, and B. Lugtenberg. 1984. Cloning of *phoM*, a gene involved in regulation of the synthesis of phosphate limitation inducible proteins in *Escherichia coli* K12. *Mol. Gen. Genet.* 195:190–194.

4285. Tommassen, J., M. Koster, and P. Overduin. 1987. Molecular analysis of the promoter region of the *Escherichia coli* K-12 *phoE* gene. Identification of an element, upstream from the promoter, required for efficient expression of *phoE* protein. *J. Mol. Biol.* 198:633–641.

4286. Tommassen, J., and B. Lugtenberg. 1980. Outer membrane protein e of *Escherichia coli* K-12 is co-regulated with alkaline phosphatase. *J. Bacteriol.* 143:151–157.

4287. Tommassen, J., and B. Lugtenberg. 1981. Localization of *phoE*, the structural gene for outer membrane protein e in *Escherichia coli* K-12. *J. Bacteriol.* 147:118–123.

4288. Tommassen, J., P. Overduin, B. Lugtenberg, and H. Bergmans. 1982. Cloning of *phoE*, the structural gene for the *Escherichia coli* phosphate limitation-inducible outer membrane pore protein. *J. Bacteriol.* 149:668–672.

4289. Tomoyasu, T., K. Yamanaka, K. Murata, T. Suzaki, P. Bouloc, A. Kato, H. Niki, S. Hiraga, and T. Ogura. 1993. Topology and subcellular localization of FtsH protein in *Escherichia coli. J. Bacteriol.* 175:1352–1357.

4290. Tomoyasu, T., T. Yuki, S. Morimura, H. Mori, K. Yamanaka, H. Niki, S. Hiraga, and T. Ogura. 1993. The *Escherichia coli* FtsH protein is a prokaryotic member of a protein family of putative ATPases involved in membrane functions, cell cycle control, and gene expression. *J. Bacteriol.* 175:1344–1351.

4291. Toone, W. M., K. E. Rudd, and J. D. Friesen. 1991. *deaD*, a new *Escherichia coli* gene encoding a presumed ATP-dependent RNA helicase, can suppress a mutation in *rpsB*, the gene encoding ribosomal protein S2. *J. Bacteriol.* 173:3291–3302.

4292. Toone, W. M., K. E. Rudd, and J. D. Friesen. 1992. Mutations causing aminotriazole resistance and temperature sensitivity reside in *gyrB*, which encodes the B subunit of DNA gyrase. *J. Bacteriol.* 174:5479–5481.

4292a. Toothman, P. 1981. Restriction alleviation by bacteriophages lambda and lambda reverse. *J. Virol.* 38:621–631.

4293. Torensma, R., M. J. Visser, C. J. Aarsman, M. J. Poppelier, A. C. Fluit, and J. Verhoef. 1993. Monoclonal antibodies that react with live Listeria spp. *Appl. Environ. Microbiol.* 59:2713–2716.

4294. Tormo, A., M. Almiron, and R. Kolter. 1990. *surA*, an *Escherichia coli* gene essential for survival in stationary phase. *J. Bacteriol.* 172:4339–4347.

4295. Torres-Cabassa, A. S., and S. Gottesman. 1987. Capsule synthesis in *Escherichia coli* K-12 is regulated by proteolysis. *J. Bacteriol.* 169:981–989.

4296. Torrey, T. A., T. Atlung, and T. Kogoma. 1984. *dnaA* suppressor (*dasF*) mutants of *Escherichia coli* are stable DNA replication (*sdrA/rnh*) mutants. *Mol. Gen. Genet.* 196:350–355.

4297. Tosa, T., and L. I. Pizer. 1971. Biochemical bases for the antimetabolite action of L-serine hydroxamate. *J. Bacteriol.* 106:972–982.

4298. Touati, D. 1983. Cloning and mapping of the manganese superoxide dismutase gene (*sodA*) of *Escherichia coli* K-12. *J. Bacteriol.* 155:1078–1087.

4299. Touati, D. 1988. Transcriptional and posttranscriptional regulation of manganese superoxide dismutase biosynthesis in *Escherichia coli,* studied with operon and protein fusions. *J. Bacteriol.* 170:2511–2520.

4300. Touati, E., and A. Danchin. 1987. The structure of the promoter and amino terminal region of the pH 2.5 acid phosphatase structural gene (*appA*) of *E. coli*: a negative control of transcription mediated by cyclic AMP. *Biochimie* 69:215–221.

4301. Touati, E., E. Dassa, and P. L. Boquet. 1986. Pleiotropic mutations in *appR* reduce pH 2.5 acid phosphatase expression and restore succinate utilisation in CRP-deficient strains of *Escherichia coli. Mol. Gen. Genet.* 202:257–264.

4302. Touati, E., E. Dassa, J. Dassa, P. L. Boquet, and D. Touati. 1991. Are *appR* and *katF* the same *Escherichia coli* gene encoding a new sigma transcription initiation factor? *Res. Microbiol.* 142:29–36. (Erratum, 142:482.)

4303. Trageser, M., S. Spiro, A. Duchene, E. Kojro, F. Fahrenholz, J. R. Guest, and G. Unden. 1990. Isolation of intact FNR protein (Mr 30,000) of *Escherichia coli. Mol. Microbiol.* 4:21–27.

4304. Tran, P. V., T. A. Bannor, S. Z. Doktor, and B. P. Nichols. 1990. Chromosomal organization and expression of *Escherichia coli pabA. J. Bacteriol.* 172:397–410.

4305. Traub, I., and V. Braun. 1994. Energy-coupled colicin transport through the outer membrane of *Escherichia coli* K-12: mutated TonB proteins alter receptor activities and colicin uptake. *FEMS Microbiol. Lett.* 119:65–70.

4306. Traub, I., S. Gaisser, and V. Braun. 1993. Activity domains of the TonB protein. *Mol. Microbiol.* 8:409–423.

4307. Trchunian, A. A., and A. V. Vasilian. 1993. ATPase activity and K^+ transport in membranes of anaerobically grown *trk*-mutants of *Escherichia coli. Biokhimiia* 58:1062–1070.

4308. Treat, M. L., M. L. Weaver, M. R. Emmett, and J. R. Johnson. 1984. Mutagenesis of the *metJBLF* gene cluster with transposon Tn5: localization of the *metF* transcription unit. *Mol. Gen. Genet.* 193:370–375.

4309. Trempy, J. E., J. E. Kirby, and S. Gottesman. 1994. Alp suppression of Lon: dependence on the *slpA* gene. *J. Bacteriol.* 176:2061–2067.

4310. Trieber, C. A., R. A. Rothery, and J. H. Weiner. 1994. Multiple pathways of electron transfer in dimethyl sulfoxide reductase of *Escherichia coli. J. Biol. Chem.* 269:7103–7109.

4311. Triggs-Raine, B. L., B. W. Doble, M. R. Mulvey, P. A. Sorby, and P. C. Loewen. 1988. Nucleotide sequence of *katG*, encoding catalase HPI of *Escherichia coli. J. Bacteriol.* 170:4415–4419.

4312. Triggs-Raine, B. L., and P. C. Loewen. 1987. Physical characterization of *katG*, encoding catalase HPI of *Escherichia coli. Gene* 52:121–128.

4313. Trimbur, D. E., and R. P. Mortlock. 1991. Isolation and characterization of *Escherichia coli* mutants able to utilize the novel pentose L-ribose. *J. Bacteriol.* 173:2459–2464.

4314. Trisler, P., and S. Gottesman. 1984. *lon* transcriptional regulation of genes necessary for capsular polysaccharide synthesis in *Escherichia coli* K-12. *J. Bacteriol.* 160:184–191.

4315. Tristram, H., and S. Neale. 1968. The activity and specificity of the proline permease in wild-type and analogue-resistant strains of *Escherichia coli. J. Gen. Microbiol.* 50:121–137.

4316. Troup, B., M. Jahn, C. Hungerer, and D. Jahn. 1994. Isolation of the *hemF* operon containing the gene for the *Escherichia coli* aerobic coproporphyrinogen III oxidase by in vivo complementation of a yeast *HEM13* mutant. *J. Bacteriol.* 176:673–680.

4317. Trower, M. K. 1993. PCR cloning, sequence analysis and expression of the *cybC* genes encoding soluble cytochrome b-562 from *Escherichia coli* B strain OP7 and K strain NM522. *Biochim. Biophys. Acta* 1143:109–111.

4318. Trucksis, M., and R. E. Depew. 1981. Identification and localization of a gene that specifies production of *Escherichia coli* DNA topoisomerase I. *Proc. Natl. Acad. Sci. USA* 78:2164–2168.

4319. Trucksis, M., E. I. Golub, D. J. Zabel, and R. E. Depew. 1981. *Escherichia coli* and *Salmonella typhimurium supX* genes specify deoxyribonucleic acid topoisomerase I. *J. Bacteriol.* 147:679–681.

4320. Trun, N. J., and S. Gottesman. 1990. On the bacterial cell cycle: *Escherichia coli* mutants with altered ploidy. *Genes Dev.* 4:2036–2047.

4321. Trun, N. J., and S. Gottesman. 1991. Characterization of *Escherichia coli* mutants with altered ploidy. *Res. Microbiol.* 142:195–200.

4322. Trun, N. J., S. Gottesman, and A. Lobner-Olesen. 1991. Analysis of *Escherichia coli* mutants with altered DNA content. *Cold Spring Harbor Symp. Quant. Biol.* 56:353–358.

4323. Trun, N. J., and T. J. Silhavy. 1987. Characterization and in vivo cloning of *prlC*, a suppressor of signal sequence mutations in *Escherichia coli* K12. *Genetics* 116:513–521.

4324. Truniger, V., and W. Boos. 1994. Mapping and cloning of *gldA*, the structural gene of the *Escherichia coli* glycerol dehydrogenase. *J. Bacteriol.* 176:1796–1800.

4325. Truniger, V., W. Boos, and G. Sweet. 1992. Molecular analysis of the *glpFKX* regions of *Escherichia coli* and *Shigella flexneri. J. Bacteriol.* 174:6981–6991.

4326. Truong, H. T., E. A. Pratt, G. S. Rule, P. Y. Hsue, and C. Ho. 1991. Inactive and temperature-sensitive folding mutants generated by tryptophan substitutions in the membrane-bound d-lactate dehydrogenase of *Escherichia coli*. *Biochemistry* 30:10722–10729.

4327. Tsai-Wu, J. J., H. F. Liu, and A. L. Lu. 1992. *Escherichia coli* MutY protein has both N-glycosylase and apurinic/apyrimidinic endonuclease activities on A·C and A·G mispairs. *Proc. Natl. Acad. Sci. USA* 89:8779–8783.

4328. Tsai-Wu, J. J., J. P. Radicella, and A. L. Lu. 1991. Nucleotide sequence of the *Escherichia coli micA* gene required for A/G-specific mismatch repair: identity of *micA* and *mutY*. *J. Bacteriol.* 173:1902–1910.

4329. Tsaneva, I. R., B. Muller, and S. C. West. 1992. ATP-dependent branch migration of Holliday junctions promoted by the RuvA and RuvB proteins of *E. coli*. *Cell* 69:1171–1180.

4330. Tsaneva, I. R., B. Muller, and S. C. West. 1993. RuvA and RuvB proteins of *Escherichia coli* exhibit DNA helicase activity in vitro. *Proc. Natl. Acad. Sci. USA* 90:1315–1319.

4331. Tsaneva, I. R., and B. Weiss. 1990. *soxR*, a locus governing a superoxide response regulon in *Escherichia coli* K-12. *J. Bacteriol.* 172:4197–4205.

4332. Tsay, J. T., W. Oh, T. J. Larson, S. Jackowski, and C. O. Rock. 1992. Isolation and characterization of the beta-ketoacyl-acyl carrier protein synthase III gene (*fabH*) from *Escherichia coli* K-12. *J. Biol. Chem.* 267:68.

4333. Tse-Dinh, Y. C. 1991. Zinc(II) coordination in *Escherichia coli* DNA topoisomerase I is required for cleavable complex formation with DNA. *J. Biol. Chem.* 266:14317–14320.

4334. Tse-Dinh, Y. C., and R. K. Beran-Steed. 1988. *Escherichia coli* DNA topoisomerase I is a zinc metalloprotein with three repetitive zinc-binding domains. *J. Biol. Chem.* 263:15857–15859.

4335. Tse-Dinh, Y. C., and J. C. Wang. 1986. Complete nucleotide sequence of the *topA* gene encoding *Escherichia coli* DNA topoisomerase I. *J. Mol. Biol.* 191:321–331.

4336. Tso, J. Y., M. A. Hermodson, and H. Zalkin. 1982. Glutamine phosphoribosylpyrophosphate amidotransferase from cloned *Escherichia coli purF*. NH2-terminal amino acid sequence, identification of the glutamine site, and trace metal analysis. *J. Biol. Chem.* 257:3532–3536.

4337. Tso, J. Y., H. Zalkin, M. van Cleemput, C. Yanofsky, and J. M. Smith. 1982. Nucleotide sequence of *Escherichia coli purF* and deduced amino acid sequence of glutamine phosphoribosylpyrophosphate amidotransferase. *J. Biol. Chem.* 257:3525–3531.

4338. Tsuchido, T., R. A. VanBogelen, and F. C. Neidhardt. 1986. Heat shock response in *Escherichia coli* influences cell division. *Proc. Natl. Acad. Sci. USA* 83:6959–6963.

4339. Tsuchihashi, Z., and P. O. Brown. 1992. Sequence requirements for efficient translational frameshifting in the *Escherichia coli dnaX* gene and the role of an unstable interaction between tRNA^Lys and an AAG lysine codon. *Genes Dev.* 6:511–519.

4340. Tsuchihashi, Z., and A. Kornberg. 1989. ATP interactions of the tau and gamma subunits of DNA polymerase III holoenzyme of *Escherichia coli*. *J. Biol. Chem.* 264:17790–17795.

4341. Tsuchihashi, Z., and A. Kornberg. 1990. Translational frameshifting generates the gamma subunit of DNA polymerase III holoenzyme. *Proc. Natl. Acad. Sci. USA* 87:2516–2520.

4342. Tsui, H. C., H. C. Leung, and M. E. Winkler. 1994. Characterization of broadly pleiotropic phenotypes caused by an *hfq* insertion mutation in *Escherichia coli* K-12. *Mol. Microbiol.* 13:35–49.

4343. Tsui, H. C., G. Zhao, G. Feng, H. C. Leung, and M. E. Winkler. 1994. The *mutL* repair gene of *Escherichia coli* K-12 forms a superoperon with a gene encoding a new cell-wall amidase. *Mol. Microbiol.* 11:189–202.

4344. Tsui, H. T., B. S. Mandavilli, and M. E. Winkler. 1992. Nonconserved segment of the MutL protein from *Escherichia coli* K-12 and *Salmonella typhimurium*. *Nucleic Acids Res.* 20:2379.

4345. Tsui, P., V. Helu, and M. Freundlich. 1988. Altered osmoregulation of *ompF* in integration host factor mutants of *Escherichia coli*. *J. Bacteriol.* 170:4950–4953.

4346. Tsui, P., L. Huang, and M. Freundlich. 1991. Integration host factor binds specifically to multiple sites in the *ompB* promoter of *Escherichia coli* and inhibits transcription. *J. Bacteriol.* 173:5800–5807.

4347. Tuan, L. R., R. D'Ari, and E. B. Newman. 1990. The leucine regulon of *Escherichia coli* K-12: a mutation in *rblA* alters expression of L-leucine-dependent metabolic operons. *J. Bacteriol.* 172:4529–4535.

4348. Tucker, S. D., A. S. Gopalakrishnan, R. Bollinger, W. Dowhan, and E. J. Murgola. 1982. Molecular mapping of *glyW*, a duplicate gene for tRNA3^Gly of *Escherichia coli*. *J. Bacteriol.* 152:773–779.

4349. Tucker, S. D., and E. J. Murgola. 1985. Sequence analysis of the *glyW* region in *Escherichia coli*. *Biochimie* 67:1053–1057.

4350. Tuggle, C. K., and J. A. Fuchs. 1986. Regulation of the operon encoding ribonucleotide reductase in *Escherichia coli*: evidence for both positive and negative control. *EMBO J.* 5:1077–1085.

4351. Turlin, E., F. Biville, and F. Gasser. 1991. Complementation of *Methylobacterium organophilum* mutants affected in pyrroloquinoline quinone biosynthesis genes *pqqE* and *pqqF* by cloned *Escherichia coli* chromosomal DNA. *FEMS Microbiol. Lett.* 67:59–63.

4352. Turnbough, C. L., Jr., K. L. Hicks, and J. P. Donahue. 1983. Attenuation control of *pyrBI* operon expression in *Escherichia coli* K-12. *Proc. Natl. Acad. Sci. USA* 80:368–372.

4353. Turnbough, C. L., Jr., K. H. Kerr, W. R. Funderburg, J. P. Donahue, and F. E. Powell. 1987. Nucleotide sequence and characterization of the *pyrF* operon of *Escherichia coli* K12. *J. Biol. Chem.* 262:10239–10245.

4354. Ueguchi, C., and K. Ito. 1992. Multicopy suppression: an approach to understanding intracellular functioning of the protein export system. *J. Bacteriol.* 174:1454–1461.

4355. Ueguchi, C., M. Kakeda, H. Yamada, and T. Mizuno. 1994. An analogue of the DnaJ molecular chaperone in *Escherichia coli*. *Proc. Natl. Acad. Sci. USA* 91:1054–1058.

4356. Ueki, M., M. Wachi, H. K. Jung, F. Ishino, and M. Matsuhashi. 1992. *Escherichia coli mraR* gene involved in cell growth and division. *J. Bacteriol.* 174:7841–7843.

4357. Uemura, H., J. Conley, F. Yamao, J. Rogers, and D. Soll. 1988. *Escherichia coli* glutaminyl-tRNA synthetase: a single amino acid replacement relaxes rRNA specificity. *Protein Seq. Data Anal.* 1:479–485.

4358. Uemura, H., S. Thorbjarnardottir, V. Gamulin, J. Yano, O. S. Andresson, D. Soll, and G. Eggertsson. 1985. *supN* ochre suppressor gene in *Escherichia coli* codes for tRNA^Lys. *J. Bacteriol.* 163:1288–1289.

4359. Uemura, Y., S. Isono, and K. Isono. 1991. Cloning, characterization, and physical location of the *rplY* gene which encodes ribosomal protein L25 in *Escherichia coli* K12. *Mol. Gen. Genet.* 226:341–344.

4360. Ueno-Nishio, S., S. Mango, L. J. Reitzer, and B. Magasanik. 1984. Identification and regulation of the *glnL* operator-promoter of the complex *glnALG* operon of *Escherichia coli*. *J. Bacteriol.* 160:379–384.

4361. Ueshima, R., N. Fujita, and A. Ishihama. 1992. Identification of *Escherichia coli* proteins cross-reacting with antibodies against region 2.2 peptide of RNA polymerase sigma subunit. *Biochem. Biophys. Res. Commun.* 184:634–639.

4362. Umeda, M., and E. Ohtsubo. 1991. Four types of IS*1* with differences in nucleotide sequence reside in the *Escherichia coli* K-12 chromosome. *Gene* 98:1–5.

4363. Umezu, K., K. Nakayama, and H. Nakayama. 1990. *Escherichia coli* RecQ protein is a DNA helicase. *Proc. Natl. Acad. Sci. USA* 87:5363–5367.

4364. Urbanowski, M. L., L. T. Stauffer, L. S. Plamann, and G. V. Stauffer. 1987. A new methionine locus, *metR*, that encodes a *trans*-acting protein required for activation of *metE* and *metH* in *Escherichia coli* and *Salmonella typhimurium*. *J. Bacteriol.* 169:1391–1397.

4365. Ursini, M. V., P. Arcari, and M. De Felice. 1981. Acetohydroxy acid synthase isoenzymes of *Escherichia coli* K-12: a trans-acting regulatory locus of *ilvHI* gene expression. *Mol. Gen. Genet.* 181:491–496.

4366. Usui, M., H. Sembongi, H. Matsuzaki, K. Matsumoto, and I. Shibuya. 1994. Primary structures of the wild-type and mutant alleles encoding the phosphatidylglycerophosphate synthase of *Escherichia coli*. *J. Bacteriol.* 176:3389–3392.

4367. Utsumi, R., S. Katayama, M. Taniguchi, T. Horie, M. Ikeda, S. Igaki, H. Nakagawa, A. Miwa, H. Tanabe, and M. Noda. 1994. Newly identified genes involved in the signal transduction of *Escherichia coli* K-12. *Gene* 140:73–77.

4368. Utsumi, R., Y. Nakamoto, M. Kawamukai, M. Himeno, and T. Komano. 1982. Involvement of cyclic AMP and its receptor protein in filamentation of an *Escherichia coli fic* mutant. *J. Bacteriol.* 151:807–812.

4369. Uzan, M., and A. Danchin. 1976. A rapid test for the *rel A* mutation in *E. coli*. *Biochem. Biophys. Res. Commun.* 69:751–758.

4370. Uzan, M., and A. Danchin. 1978. Correlation between the serine sensitivity and the derepressibility of the *ilv* genes in *Escherichia coli relA⁻* mutants. *Mol. Gen. Genet.* 165:21–30.

4371. Uzan, M., R. Favre, E. Gallay, and L. Caro. 1981. Genetical and structural analysis of a group of lambda *ilv* and lambda *rho* transducing phages. *Mol. Gen. Genet.* 182:462–470.

4372. Valencia, A., P. Chardin, A. Wittinghofer, and C. Sander. 1991. The *ras* protein family: evolutionary tree and role of conserved amino acids. *Biochemistry* 30:4637–4648.

4373. Valentin-Hansen, P. 1982. Tandem CRP binding sites in the *deo* operon of *Escherichia coli* K-12. *EMBO J.* 1:1049–1054.

4374. Valentin-Hansen, P., F. Boetius, K. Hammer-Jespersen, and I. Svendsen. 1982. The primary structure of *Escherichia coli* K12 2-deoxyribose 5-phosphate aldolase. Nucleotide sequence of the *deoC* gene and the amino acid sequence of the enzyme. *Eur. J. Biochem.* 125:561–566.

4375. Valentin-Hansen, P., K. Hammer, J. E. Love Larsen, and I. Svendsen. 1984. The internal regulated promoter of the *deo* operon of *Escherichia coli* K-12. *Nucleic Acids Res.* 12:5211–5224.

4376. Valentin-Hansen, P., K. Hammer-Jespersen, F. Boetius, and I. Svendsen. 1984. Structure and function of the intercistronic regulatory *deoC-deoA* element of *Escherichia coli* K-12. *EMBO J.* 3:179–183.

4377. Valentin-Hansen, P., K. Hammer-Jespersen, and R. S. Buxton. 1979. Evidence for the existence of three promoters for the *deo* operon of *Escherichia coli* K12 in vitro. *J. Mol. Biol.* 133:1–17.

4378. Valentin-Hansen, P., P. Hojrup, and S. Short. 1985. The primary structure of the DeoR repressor from *Escherichia coli* K-12. *Nucleic Acids Res.* 13:5927–5936.

4379. Valentin-Hansen, P., B. Holst, J. Josephsen, K. Hammer, and B. Albrechtsen. 1989. CRP/cAMP- and CytR-regulated promoters in *Escherichia coli* K12: the *cdd* promoter. *Mol. Microbiol.* 3:1385–1390.

4380. Valentin-Hansen, P., B. Holst, L. Sogaard-Andersen, J. Martinussen, J. Nesvera, and S. R. Douthwaite. 1991. Design of cAMP-CRP-activated promoters in *Escherichia coli*. *Mol. Microbiol.* 5:433–437.

4381. Valentin-Hansen, P., J. E. Larsen, P. Hojrup, S. A. Short, and C. S. Barbier. 1986. Nucleotide sequence of the CytR regulatory gene of *E. coli* K-12. *Nucleic Acids Res.* 14:2215–2228.

4382. Valentini, G., M. Stoppini, P. Iadarola, M. Malcovati, G. Ferri, and M. L. Speranza. 1993. Divergent binding sites in pyruvate kinases I and II from *Escherichia coli. Biol. Chem. Hoppe-Seyler* 374:69–74.

4383. Valentini, G., M. Stoppini, M. L. Speranza, M. Malcovati, and G. Ferri. 1991. Bacterial pyruvate kinases have a shorter N-terminal domain. *Biol. Chem. Hoppe-Seyler* 372:91–93.

4384. Vales, L. D., J. W. Chase, and J. B. Murphy. 1979. Orientation of the guanine operon of *Escherichia coli* K-12 by utilizing strains containing *guaB-xse* and *guaB-upp* deletions. *J. Bacteriol.* 139:320–322.

4385. Vales, L. D., B. A. Rabin, and J. W. Chase. 1982. Subunit structure of *Escherichia coli* exonuclease VII. *J. Biol. Chem.* 257:8799–8805.

4386. Vales, L. D., B. A. Rabin, and J. W. Chase. 1983. Isolation and preliminary characterization of *Escherichia coli* mutants deficient in exonuclease VII. *J. Bacteriol.* 155:1116–1122.

4387. Vallari, D. S., and C. O. Rock. 1985. Isolation and characterization of *Escherichia coli* pantothenate permease (*panF*) mutants. *J. Bacteriol.* 164:136–142.

4388. Vallari, D. S., and C. O. Rock. 1987. Isolation and characterization of temperature-sensitive pantothenate kinase (*coaA*) mutants of *Escherichia coli. J. Bacteriol.* 169:5795–5800.

4389. Valle, F., B. Becerril, E. Chen, P. Seeburg, H. Heyneker, and F. Bolivar. 1984. Complete nucleotide sequence of the glutamate dehydrogenase gene from *Escherichia coli* K-12. *Gene* 27:193–199.

4390. Valle, F., E. Sanvicente, P. Seeburg, A. Covarrubias, R. L. Rodriguez, and F. Bolivar. 1983. Nucleotide sequence of the promoter and amino-terminal coding region of the glutamate dehydrogenase structural gene of *Escherichia coli. Gene* 23:199–209.

4391. van Alphen, W., B. Lugtenberg, and W. Berendsen. 1976. Heptose-deficient mutants of *Escherichia coli* K12 deficient in up to three major outer membrane proteins. *Mol. Gen. Genet.* 147:263–269.

4392. van Buul, C. P., and P. H. van Knippenberg. 1985. Nucleotide sequence of the *ksgA* gene of *Escherichia coli:* comparison of methyltransferases effecting dimethylation of adenosine in ribosomal RNA. *Gene* 38:65–72.

4393. van de Klundert, J. A., P. H. van der Meide, P. van de Putte, and L. Bosch. 1978. Mutants of *Escherichia coli* altered in both genes coding for the elongation factor Tu. *Proc. Natl. Acad. Sci. USA* 75:4470–4473.

4394. van de Putte, P., R. Plasterk, and A. Kuijpers. 1984. A Mu gin complementing function and an invertible DNA region in *Escherichia coli* K-12 are situated on the genetic element e14. *J. Bacteriol.* 158:517–522.

4395. van Delft, J. H., B. Marinon, D. S. Schmidt, and L. Bosch. 1987. Transcription of the tRNA-tufB operon of *Escherichia coli:* activation, termination and antitermination. *Nucleic Acids Res.* 15:9515–9530.

4396. van den Berg, E., J. Zwetsloot, I. Noordermeer, H. Pannekoek, B. Dekker, R. Dijkema, and H. van Ormondt. 1981. The structure and function of the regulatory elements of the *Escherichia coli uvrB* gene. *Nucleic Acids Res.* 9:5623–5643.

4397. van der Linden, M. P., H. Mottl, and W. Keck. 1992. Cytoplasmic high-level expression of a soluble, enzymatically active form of the *Escherichia coli* penicillin-binding protein 5 and purification by dye chromatography. *Eur. J. Biochem.* 204:197–202.

4398. van der Meide, P. H., E. Vijgenboom, M. Dicke, and L. Bosch. 1982. Regulation of the expression of *tufA* and *tufB*, the two genes coding for the elongation factor EF-Tu in *Escherichia coli. FEBS Lett.* 139:325–330.

4399. van der Zel, A., H. M. Lam, and M. E. Winkler. 1989. Extensive homology between the *Escherichia coli* K-12 SerC(PdxF) aminotransferase and a protein encoded by a progesterone-induced mRNA in rabbit and human endometria. *Nucleic Acids Res.* 17:8379.

4400. Van Dyk, T. K., A. A. Gatenby, and R. A. LaRossa. 1989. Demonstration by genetic suppression of interaction of GroE products with many proteins. *Nature* (London) 342:451–453.

4401. van Heeswijk, W., O. Kuppinger, M. Merrick, and D. Kahn. 1992. Localization of the *glnD* gene on a revised map of the 200-kilobase region of the *Escherichia coli* chromosome. *J. Bacteriol.* 174:1702–1703.

4402. van Heeswijk, W. C., M. Rabenberg, H. V. Westerhoff, and D. Kahn. 1993. The genes of the glutamine synthetase adenylylation cascade are not regulated by nitrogen in *Escherichia coli. Mol. Microbiol.* 9:443–457.

4403. Van Hove, B., H. Staudenmaier, and V. Braun. 1990. Novel two-component transmembrane transcription control: regulation of iron dicitrate transport in *Escherichia coli. J. Bacteriol.* 172:6749–6758.

4404. van Sluis, C. A., G. F. Moolenaar, and C. Backendorf. 1983. Regulation of the *uvrC* gene of *Escherichia coli* K12: localization and characterization of a damage-inducible promoter. *EMBO J.* 2:2313–2318.

4405. Van Vliet, F., A. Boyen, and N. Glansdorff. 1988. On interspecies gene transfer: the case of the *argF* gene of *Escherichia coli. Ann. Inst. Pasteur Microbiol.* 139:493–496. (Letter.)

4406. Van Vliet, F., M. Crabeel, A. Boyen, C. Tricot, V. Stalon, P. Falmagne, Y. Nakamura, S. Baumberg, and N. Glansdorff. 1990. Sequences of the genes encoding argininosuccinate synthetase in *Escherichia coli* and *Saccharomyces cerevisiae:* comparison with methanogenic archaebacteria and mammals. *Gene* 95:99–104.

4407. Van Vliet, F., R. Cunin, A. Jacobs, J. Piette, D. Gigot, M. Lauwereys, A. Pierard, and N. Glansdorff. 1984. Evolutionary divergence of genes for ornithine and aspartate carbamoyl- transferases—complete sequence and mode of regulation of the *Escherichia coli argF* gene; comparison of *argF* with *argI* and *pyrB. Nucleic Acids Res.* 12:6277–6289.

4408. Vanaman, T. C., S. J. Wakil, and R. L. Hill. 1968. The preparation of tryptic, peptic, thermolysin, and cyanogen bromide peptides from the acyl carrier protein of *Escherichia coli. J. Biol. Chem.* 243:6409–6419.

4409. Vanaman, T. C., S. J. Wakil, and R. L. Hill. 1968. The complete amino acid sequence of the acyl carrier protein of *Escherichia coli. J. Biol. Chem.* 243:6420–6431.

4410. VanBogelen, R. A., P. M. Kelley, and F. C. Neidhardt. 1987. Differential induction of heat shock, SOS, and oxidation stress regulons and accumulation of nucleotides in *Escherichia coli. J. Bacteriol.* 169:26–32.

4411. VanBogelen, R. A., and F. C. Neidhardt. 1990. Ribosomes as sensors of heat and cold shock in *Escherichia coli. Proc. Natl. Acad. Sci. USA* 87:5589–5593.

4412. VanBogelen, R. A., V. Vaughn, and F. C. Neidhardt. 1983. Gene for heat-inducible lysyl-tRNA synthetase (*lysU*) maps near *cadA* in *Escherichia coli. J. Bacteriol.* 153:1066–1068.

4413. Vandekerckhove, J., W. Rombauts, B. Peeters, and B. Wittmann-Liebold. 1975. Determination of the complete amino acid sequence of protein S21 from *Escherichia coli* ribosomes. *Hoppe-Seyler's Z. Physiol. Chem.* 356:1955–1976.

4414. Vandekerckhove, J., W. Rombauts, and B. Wittmann-Liebold. 1977. The complete amino acid sequence of protein S16 from *Escherichia coli. Hoppe-Seyler's Z. Physiol. Chem.* 358:989–1002.

4415. Vanden Boom, T. J., K. E. Reed, and J. E. Cronan, Jr. 1991. Lipoic acid metabolism in *Escherichia coli:* isolation of null mutants defective in lipoic acid biosynthesis, molecular cloning and characterization of the *E. coli lip* locus, and identification of the lipoylated protein of the glycine cleavage system. *J. Bacteriol.* 173:6411–6420.

4416. Vander Horn, P. B., A. D. Backstrom, V. Stewart, and T. P. Begley. 1993. Structural genes for thiamine biosynthetic enzymes (*thiCEFGH*) in *Escherichia coli* K-12. *J. Bacteriol.* 175:982–992.

4416a. Vanderwinkel, E., and M. De Vlieghere. 1968. Physiologie et génétique de l'isocitratase et des malate synthèses chez *Escherichia coli. Eur. J. Biochem.* 5:81–90.

4417. Vanet, A., J. A. Plumbridge, and J. H. Alix. 1993. Cotranscription of two genes necessary for ribosomal protein L11 methylation (*prmA*) and pantothenate transport (*panF*) in *Escherichia coli* K-12. *J. Bacteriol.* 175:7178–7188.

4418. Varshney, U., T. Hutcheon, and J. H. van de Sande. 1988. Sequence analysis, expression, and conservation of *Escherichia coli* uracil DNA glycosylase and its gene (*ung*). *J. Biol. Chem.* 263:7776–7784.

4419. Vasudevan, S. G., W. L. Armarego, D. C. Shaw, P. E. Lilley, N. E. Dixon, and R. K. Poole. 1991. Isolation and nucleotide sequence of the *hmp* gene that encodes a haemoglobin-like protein in *Escherichia coli* K-12. *Mol. Gen. Genet.* 226:49–58.

4420. Veitinger, S., and V. Braun. 1992. Localization of the entire *fec* region at 97.3 minutes on the *Escherichia coli* chromosome. *J. Bacteriol.* 174:3838–3839.

4421. Venkateswaran, P. S., and H. C. Wu. 1972. Isolation and characterization of a phosphonomycin-resistant mutant of *Escherichia coli* K-12. *J. Bacteriol.* 110:935–944.

4422. Verbeek, H., L. Nilsson, G. Baliko, and L. Bosch. 1990. Potential binding sites of the trans-activator FIS are present upstream of all rRNA operons and of many but not all tRNA operons. *Biochim. Biophys. Acta* 1050:302–306.

4423. Verde, P., R. Frunzio, P. P. di Nocera, F. Blasi, and C. B. Bruni. 1981. Identification, nucleotide sequence and expression of the regulatory region of the histidine operon of *Escherichia coli* K-12. *Nucleic Acids Res.* 9:2075–2086.

4424. Verhoef, C., B. Lugtenberg, R. van Boxtel, P. de Graaff, and H. Verheij. 1979. Genetics and biochemistry of the peptidoglycan-associated proteins b and c of *Escherichia coli* K12. *Mol. Gen. Genet.* 169:137–146.

4425. Verkamp, E., V. M. Backman, J. M. Bjornsson, D. Soll, and G. Eggertsson. 1993. The periplasmic dipeptide permease system transports 5-aminolevulinic acid in *Escherichia coli. J. Bacteriol.* 175:1452–1456.

4426. Verkamp, E., and B. K. Chelm. 1989. Isolation, nucleotide sequence, and preliminary characterization of the *Escherichia coli* K-12 *hemA* gene. *J. Bacteriol.* 171:4728–4735.

4427. Verkamp, E., M. Jahn, D. Jahn, A. M. Kumar, and D. Soll. 1992. Glutamyl-tRNA reductase from *Escherichia coli* and *Synechocystis* 6803. Gene structure and expression. *J. Biol. Chem.* 267:8275–8280.

4428. Versalovic, J., T. Koeuth, E. R. McCabe, and J. R. Lupski. 1991. Use of the polymerase chain reaction for physical mapping of *Escherichia coli* genes. *J. Bacteriol.* 173:5253–5255.

4429. Verwoert, I. I., E. C. Verbree, K. H. van der Linden, H. J. Nijkamp, and A. R. Stuitje. 1992. Cloning, nucleotide sequence, and expression of the *Escherichia coli fabD* gene, encoding malonyl coenzyme A-acyl carrier protein transacylase. *J. Bacteriol.* 174:2851–2857.

4430. Vianney, A., T. M. Lewin, W. F. Beyer, Jr., J. C. Lazzaroni, R. Portalier, and R. E. Webster. 1994. Membrane topology and mutational analysis of the TolQ protein of *Escherichia coli* required for the uptake of macromolecules and cell envelope integrity. *J. Bacteriol.* 176:822–829.

4431. Vicente, M., S. R. Kushner, T. Garrido, and M. Aldea. 1991. The role of the 'gearbox' in the transcription of essential genes. *Mol. Microbiol.* 5:2085–2091.

4432. Vik, S. B., D. Lee, and P. A. Marshall. 1991. Temperature-sensitive mutations at the carboxy terminus of the alpha subunit of the *Escherichia coli* F_1F_0 ATP synthase. *J. Bacteriol.* 173:4544–4548.

4433. Vimr, E. R., and F. A. Troy. 1985. Identification of an inducible catabolic system for sialic acids (*nan*) in *Escherichia coli. J. Bacteriol.* 164:845–853.

4434. Vinella, D., and R. D'Ari. 1994. Thermoinducible filamentation in *Escherichia coli* due to an altered RNA polymerase beta subunit is suppressed by high levels of ppGpp. *J. Bacteriol.* 176:966–972.

4435. **Vinella, D., R. D'Ari, and P. Bouloc.** 1992. Penicillin binding protein 2 is dispensable in *Escherichia coli* when ppGpp synthesis is induced. *EMBO J.* 11:1493–1501.

4436. **Vinella, D., A. Jaffe, R. D'Ari, M. Kohiyama, and P. Hughes.** 1992. Chromosome partitioning in *Escherichia coli* in the absence of *dam*-directed methylation. *J. Bacteriol.* 174:2388–2390.

4437. **Vinella, D., D. Joseleau-Petit, D. Thevenet, P. Bouloc, and R. D'Ari.** 1993. Penicillin-binding protein 2 inactivation in *Escherichia coli* results in cell division inhibition, which is relieved by FtsZ overexpression. *J. Bacteriol.* 175:6704–6710.

4438. **Vizan, J. L., C. Hernandez-Chico, I. del Castillo, and F. Moreno.** 1991. The peptide antibiotic microcin B17 induces double-strand cleavage of DNA mediated by *E. coli* DNA gyrase. *EMBO J.* 10:467–476.

4439. **Vlahos, C. J., and E. E. Dekker.** 1988. The complete amino acid sequence and identification of the active-site arginine peptide of *Escherichia coli* 2-keto-4-hydroxyglutarate aldolase. *J. Biol. Chem.* 263:11683–11691.

4440. **Vlahos, C. J., and E. E. Dekker.** 1990. Active-site residues of 2-keto-4-hydroxyglutarate aldolase from *Escherichia coli*. Bromopyruvate inactivation and labeling of glutamate 45. *J. Biol. Chem.* 265:20384–20389.

4441. **Vlazny, D. A., and C. W. Hill.** 1995. A stationary-phase-dependent viability block governed by two different polypeptides from the RhsA genetic element of *Escherichia coli* K-12. *J. Bacteriol.* 177:2209–2213.

4442. **Voelkner, P., W. Puppe, and K. Altendorf.** 1993. Characterization of the KdpD protein, the sensor kinase of the K^+-translocating Kdp system of *Escherichia coli*. *Eur. J. Biochem.* 217:1019–1026.

4443. **Vogel, R. F., K. D. Entian, and D. Mecke.** 1987. Cloning and sequence of the *mdh* structural gene of *Escherichia coli* coding for malate dehydrogenase. *Arch. Microbiol.* 149:36–42.

4444. **Vogler, A. P., and J. W. Lengeler.** 1989. Analysis of the *nag* regulon from *Escherichia coli* K12 and *Klebsiella pneumoniae* and of its regulation. *Mol. Gen. Genet.* 219:97–105.

4445. **Vogler, A. P., and J. W. Lengeler.** 1991. Comparison of the sequences of the *nagE* operons from *Klebsiella pneumoniae* and *Escherichia coli* K12: enhanced variability of the enzyme IIN-acetylglucosamine in regions connecting functional domains. *Mol. Gen. Genet.* 230:270–276.

4446. **Vogler, A. P., S. Trentmann, and J. W. Lengeler.** 1989. Alternative route for biosynthesis of amino sugars in *Escherichia coli* K-12 mutants by means of a catabolic isomerase. *J. Bacteriol.* 171:6586–6592. (Erratum, 172:5521, 1990.)

4447. **Voigt, J., and K. Nagel.** 1990. Isolation and characterization of an inhibitor of ribosome-dependent GTP hydrolysis by elongation factor G. *Eur. J. Biochem.* 194:579–585.

4448. **Volkert, M. R., and L. I. Hajec.** 1991. Molecular analysis of the *aidD6*::Mu d1 (*bla lac*) fusion mutation of *Escherichia coli* K12. *Mol. Gen. Genet.* 229:319–323.

4449. **Volkert, M. R., L. I. Hajec, Z. Matijasevic, F. C. Fang, and R. Prince.** 1994. Induction of the *Escherichia coli aidB* gene under oxygen-limiting conditions requires a functional *rpoS* (*katF*) gene. *J. Bacteriol.* 176:7638–7645.

4450. **Volkert, M. R., L. J. Margossian, and A. J. Clark.** 1981. Evidence that *rnmB* is the operator of the *Escherichia coli recA* gene. *Proc. Natl. Acad. Sci. USA* 78:1786–1790.

4451. **Volkert, M. R., D. C. Nguyen, and K. C. Beard.** 1986. *Escherichia coli* gene induction by alkylation treatment. *Genetics* 112:11–26.

4452. **Voll, M. J., and L. Leive.** 1970. Release of lipopolysaccharide in *Escherichia coli* resistant to the permeability increase induced by ethylenediaminetetraacetate. *J. Biol. Chem.* 245:1108–1114.

4453. **von Freiesleben, U., and K. V. Rasmussen.** 1991. DNA replication in *Escherichia coli gyrB*(Ts) mutants analysed by flow cytometry. *Res. Microbiol.* 142:223–227.

4454. **von Freiesleben, U., K. V. Rasmussen, and M. Schaechter.** 1994. SeqA limits DnaA activity in replication from *oriC* in *Escherichia coli*. *Mol. Microbiol.* 14:763–772.

4455. **von Meyenburg, K., B. B. Jorgensen, J. Nielsen, and F. G. Hansen.** 1982. Promoters of the *atp* operon coding for the membrane-bound ATP synthase of *Escherichia coli* mapped by Tn*10* insertion mutations. *Mol. Gen. Genet.* 188:240–248.

4456. **von Ossowski, I., M. R. Mulvey, P. A. Leco, A. Borys, and P. C. Loewen.** 1991. Nucleotide sequence of *Escherichia coli katE*, which encodes catalase HPII. *J. Bacteriol.* 173:514–520.

4457. **von Wilcken-Bergmann, B., and B. Muller-Hill.** 1982. Sequence of *galR* gene indicates a common evolutionary origin of *lac* and *gal* repressor in *Escherichia coli*. *Proc. Natl. Acad. Sci. USA* 79:2427–2431.

4458. **Vonder Haar, R. A., and H. E. Umbarger.** 1972. Isoleucine and valine metabolism in *Escherichia coli*. XIX. Inhibition of isoleucine biosynthesis by glycyl-leucine. *J. Bacteriol.* 112:142–147.

4459. **Vuorio, R., and M. Vaara.** 1992. Mutants carrying conditionally lethal mutations in outer membrane genes *omsA* and *firA* (*ssc*) are phenotypically similar, and *omsA* is allelic to *firA*. *J. Bacteriol.* 174:7090–7097.

4460. **Vyas, N. K., M. N. Vyas, and F. A. Quiocho.** 1983. The 3 Å resolution structure of a D-galactose-binding protein for transport and chemotaxis in *Escherichia coli*. *Proc. Natl. Acad. Sci. USA* 80:1792–1796.

4461. **Vyas, N. K., M. N. Vyas, and F. A. Quiocho.** 1987. A novel calcium binding site in the galactose-binding protein of bacterial transport and chemotaxis. *Nature* (London) 327:635–638.

4462. **Vyas, N. K., M. N. Vyas, and F. A. Quiocho.** 1988. Sugar and signal-transducer binding sites of the *Escherichia coli* galactose chemoreceptor protein. *Science* 242:1290–1295.

4463. **Wachi, M., M. Doi, Y. Okada, and M. Matsuhashi.** 1989. New *mre* genes *mreC* and *mreD*, responsible for formation of the rod shape of *Escherichia coli* cells. *J. Bacteriol.* 171:6511–6516.

4464. **Wachi, M., M. Doi, S. Tamaki, W. Park, S. Nakajima-Iijima, and M. Matsuhashi.** 1987. Mutant isolation and molecular cloning of *mre* genes, which determine cell shape, sensitivity to mecillinam, and amount of penicillin-binding proteins in *Escherichia coli*. *J. Bacteriol.* 169:4935–4940.

4465. **Wachi, M., M. Doi, T. Ueda, M. Ueki, K. Tsuritani, K. Nagai, and M. Matsuhashi.** 1991. Sequence of the downstream flanking region of the shape-determining genes *mreBCD* of *Escherichia coli*. *Gene* 106:135–136.

4466. **Wachi, M., and M. Matsuhashi.** 1989. Negative control of cell division by *mreB*, a gene that functions in determining the rod shape of *Escherichia coli* cells. *J. Bacteriol.* 171:3123–3127.

4467. **Wachter, E., R. Schmid, G. Deckers, and K. Altendorf.** 1980. Amino acid replacement in dicyclohexylcarbodiimide-reactive proteins from mutant strains of *Escherichia coli* defective in the energy-transducing ATPase complex. *FEBS Lett.* 113:265–270.

4468. **Wackernagel, W., and U. Winkler.** 1972. A mutation in *Escherichia coli* enhancing the UV-mutability of phage lambda but not of its infectious DNA in a spheroplast assay. *Mol. Gen. Genet.* 114:68–79.

4469. **Wackett, L. P., B. L. Wanner, C. P. Venditti, and C. T. Walsh.** 1987. Involvement of the phosphate regulon and the *psiD* locus in carbon-phosphorus lyase activity of *Escherichia coli* K-12. *J. Bacteriol.* 169:1753–1756.

4470. **Wada, A., and T. Sako.** 1987. Primary structures of and genes for new ribosomal proteins A and B in *Escherichia coli*. *J. Biochem.* 101:817–820.

4471. **Wada, A., Y. Yamazaki, N. Fujita, and A. Ishihama.** 1990. Structure and probable genetic location of a `ribosome modulation factor' associated with 100S ribosomes in stationary-phase *Escherichia coli* cells. *Proc. Natl. Acad. Sci. USA* 87:2657–2661.

4472. **Wada, C., and T. Yura.** 1979. *Escherichia coli* mutants incapable of supporting replication of F-like plasmids at high temperature: isolation and characterization of *mafA* and *mafB* mutants. *J. Bacteriol.* 140:864–873.

4473. **Wada, C., and T. Yura.** 1982. Inhibition of initiation of mini-F plasmid replication in temperature-sensitive *mafA* mutants of *Escherichia coli* K-12. *Plasmid* 8:287–298.

4474. **Wada, C., and T. Yura.** 1984. Control of F plasmid replication by a host gene: evidence for interaction of the *mafA* gene product of *Escherichia coli* with the mini-F *incC* region. *J. Bacteriol.* 160:1130–1136.

4475. **Wada, C., T. Yura, and S. Hiraga.** 1977. Replication of Fpoh$^+$ plasmid in *mafA* mutants of *Escherichia coli* defective in plasmid maintenance. *Mol. Gen. Genet.* 152:211–217.

4476. **Wada, M., K. Sekine, and H. Itikawa.** 1986. Participation of the *dnaK* and *dnaJ* gene products in phosphorylation of glutaminyl-tRNA synthetase and threonyl-tRNA synthetase of *Escherichia coli* K-12. *J. Bacteriol.* 168:213–220.

4477. **Wagegg, W., and V. Braun.** 1981. Ferric citrate transport in *Escherichia coli* requires outer membrane receptor protein FecA. *J. Bacteriol.* 145:156–163.

4478. **Wagner, A. F., M. Frey, F. A. Neugebauer, W. Schafer, and J. Knappe.** 1992. The free radical in pyruvate formate-lyase is located on glycine-734. *Proc. Natl. Acad. Sci. USA* 89:996–1000.

4479. **Wagner, L. A., R. F. Gesteland, T. J. Dayhuff, and R. B. Weiss.** 1994. An efficient Shine-Dalgarno sequence but not translation is necessary for *lacZ* mRNA stability in *Escherichia coli*. *J. Bacteriol.* 176:1683–1688.

4480. **Wahab, S. Z., R. Elford, and W. M. Holmes.** 1989. Nucleotide sequence of the *Escherichia coli* tRNA$_3^{Leu}$ gene. *Gene* 81:193–194.

4481. **Waite-Rees, P. A., C. J. Keating, L. S. Moran, B. E. Slatko, L. J. Hornstra, and J. S. Benner.** 1991. Characterization and expression of the *Escherichia coli* Mrr restriction system. *J. Bacteriol.* 173:5207–5219.

4482. **Wakayama, Y., M. Takagi, and K. Yano.** 1984. Gene responsible for protecting *Escherichia coli* from sodium dodecyl sulfate and toluidine blue plus light. *J. Bacteriol.* 159:527–532.

4483. **Walderhaug, M. O., J. W. Polarek, P. Voelkner, J. M. Daniel, J. E. Hesse, K. Altendorf, and W. Epstein.** 1992. KdpD and KdpE, proteins that control expression of the *kdpABC* operon, are members of the two-component sensor-effector class of regulators. *J. Bacteriol.* 174:2152–2159.

4484. **Walker, J. E., N. J. Gay, M. Saraste, and A. N. Eberle.** 1984. DNA sequence around the *Escherichia coli unc* operon. Completion of the sequence of a 17 kilobase segment containing *asnA*, *oriC*, *unc*, *glmS* and *phoS*. *Biochem. J.* 224:799–815.

4485. **Walker, J. R.** 1969. *Escherichia coli ras* locus: its involvement in radiation repair. *J. Bacteriol.* 99:713–719.

4486. **Walker, J. R.** 1970. Defective excision repair of pyrimidine dimers in the ultraviolet-sensitive *Escherichia coli ras*⁻ mutant. *J. Bacteriol.* 103:552–559.

4487. **Wallace, B., Y. J. Yang, J. S. Hong, and D. Lum.** 1990. Cloning and sequencing of a gene encoding a glutamate and aspartate carrier of *Escherichia coli* K-12. *J. Bacteriol.* 172:3214–3220.

4488. **Wallace, B. J., and S. R. Kushner.** 1984. Genetic and physical analysis of the thioredoxin (*trxA*) gene of *Escherichia coli* K-12. *Gene* 32:399–408.

4489. **Wallace, B. J., and J. Pittard.** 1969. Regulator gene controlling enzymes concerned in tyrosine biosynthesis in *Escherichia coli*. *J. Bacteriol.* 97:1234–1241.

4490. **Wallace, R. G., N. Lee, and A. V. Fowler.** 1980. The *araC* gene of *Escherichia coli*: transcriptional and translational start-points and complete nucleotide sequence. *Gene* 12:179–190.

4491. Walter, C., S. Wilken, and E. Schneider. 1992. Characterization of site-directed mutations in conserved domains of MalK, a bacterial member of the ATP-binding cassette (ABC) family. FEBS Lett. 303:41–44. (Erratum, 305:76.)

4492. Walter, E. G., J. H. Weiner, and D. E. Taylor. 1991. Nucleotide sequence and overexpression of the tellurite-resistance determinant from the IncHII plasmid pHH1508a. Gene 101:1–7.

4493. Walter, M. R., W. J. Cook, L. B. Cole, S. A. Short, G. W. Koszalka, T. A. Krenitsky, and S. E. Ealick. 1990. Three-dimensional structure of thymidine phosphorylase from Escherichia coli at 2.8 A resolution. J. Biol. Chem. 265:14016–14022.

4494. Walton, L., C. A. Richards, and L. P. Elwell. 1989. Nucleotide sequence of the Escherichia coli uridine phosphorylase (udp) gene. Nucleic Acids Res. 17:6741.

4495. Wandersman, C., and P. Delepelaire. 1990. TolC, an Escherichia coli outer membrane protein required for hemolysin secretion. Proc. Natl. Acad. Sci. USA 87:4776–4780.

4496. Wandersman, C., F. Moreno, and M. Schwartz. 1980. Pleiotropic mutations rendering Escherichia coli K-12 resistant to bacteriophage TP1. J. Bacteriol. 143:1374–1383.

4497. Wang, A. Y., D. W. Grogan, and J. E. Cronan, Jr. 1992. Cyclopropane fatty acid synthase of Escherichia coli: deduced amino acid sequence, purification, and studies of the enzyme active site. Biochemistry 31:11020–11028.

4498. Wang, E. A., and D. E. Koshland, Jr. 1980. Receptor structure in the bacterial sensing system. Proc. Natl. Acad. Sci. USA 77:7157–7161.

4499. Wang, E. A., K. L. Mowry, D. O. Clegg, and D. E. Koshland, Jr. 1982. Tandem duplication and multiple functions of a receptor gene in bacterial chemotaxis. J. Biol. Chem. 257:4673–4676.

4500. Wang, H. C., and R. C. Gayda. 1990. High-level expression of the FtsA protein inhibits cell septation in Escherichia coli K-12. J. Bacteriol. 172:4736–4740.

4501. Wang, J. C., and K. Becherer. 1983. Cloning of the gene topA encoding for DNA topoisomerase I and the physical mapping of the cysB-topA-trp region of Escherichia coli. Nucleic Acids Res. 11:1773–1790.

4502. Wang, J. C., P. R. Caron, and R. A. Kim. 1990. The role of DNA topoisomerases in recombination and genome stability: a double-edged sword? Cell 62:403–406.

4503. Wang, J. Y., and M. Syvanen. 1992. DNA twist as a transcriptional sensor for environmental changes. Mol. Microbiol. 6:1861–1866.

4504. Wang, L., and B. Weiss. 1992. dcd (dCTP deaminase) gene of Escherichia coli: mapping, cloning, sequencing, and identification as a locus of suppressors of lethal dut (dUTPase) mutations. J. Bacteriol. 174:5647–5653.

4505. Wang, M. D., L. Liu, B. M. Wang, and C. M. Berg. 1987. Cloning and characterization of the Escherichia coli K-12 alanine-valine transaminase (avtA) gene. J. Bacteriol. 169:4228–4234.

4506. Wang, Q. P., and J. M. Kaguni. 1987. Transcriptional repression of the dnaA gene of Escherichia coli by dnaA protein. Mol. Gen. Genet. 209:518–525.

4507. Wang, X. D., P. A. de Boer, and L. I. Rothfield. 1991. A factor that positively regulates cell division by activating transcription of the major cluster of essential cell division genes of Escherichia coli. EMBO J. 10:3363–3372.

4508. Wanner, B. L. 1987. Control of phoR-dependent bacterial alkaline phosphatase clonal variation by the phoM region. J. Bacteriol. 169:900–903. (Erratum, 169:3866.)

4509. Wanner, B. L. 1992. Is cross regulation by phosphorylation of two-component response regulator proteins important in bacteria? J. Bacteriol. 174:2053–2058.

4510. Wanner, B. L. 1993. Gene regulation by phosphate in enteric bacteria. J. Cell Biochem. 51:47–54.

4511. Wanner, B. L., and J. Bernstein. 1982. Determining the phoM map location in Escherichia coli K-12 by using a nearby transposon Tn10 insertion. J. Bacteriol. 150:429–432.

4512. Wanner, B. L., and J. A. Boline. 1990. Mapping and molecular cloning of the phn (psiD) locus for phosphonate utilization in Escherichia coli. J. Bacteriol. 172:1186–1196.

4513. Wanner, B. L., and B. D. Chang. 1987. The phoBR operon in Escherichia coli K-12. J. Bacteriol. 169:5569–5574.

4514. Wanner, B. L., and P. Latterell. 1980. Mutants affected in alkaline phosphatase, expression: evidence for multiple positive regulators of the phosphate regulon in Escherichia coli. Genetics 96:353–366.

4515. Wanner, B. L., and W. W. Metcalf. 1992. Molecular genetic studies of a 10.9-kb operon in Escherichia coli for phosphonate uptake and biodegradation. FEMS Microbiol. Lett. 79:133–139.

4516. Wanner, B. L., A. Sarthy, and J. Beckwith. 1979. Escherichia coli pleiotropic mutant that reduces amounts of several periplasmic and outer membrane proteins. J. Bacteriol. 140:229–239.

4517. Wanner, B. L., S. Wieder, and R. McSharry. 1981. Use of bacteriophage transposon Mu d1 to determine the orientation for three proC-linked phosphate-starvation-inducible (psi) genes in Escherichia coli K-12. J. Bacteriol. 146:93–101.

4518. Wanner, B. L., M. R. Wilmes, and E. Hunter. 1988. Molecular cloning of the wild-type phoM operon in Escherichia coli K-12. J. Bacteriol. 170:279–288.

4519. Ward, D. F., and M. E. Gottesman. 1981. The nus mutations affect transcription termination in Escherichia coli. Nature (London) 292:212–215.

4520. Wargel, R. J., C. A. Hadur, and F. C. Neuhaus. 1971. Mechanism of D-cycloserine action: transport mutants for D-alanine, D-cycloserine, and glycine. J. Bacteriol. 105:1028–1035.

4521. Warne, S. R., J. M. Varley, G. J. Boulnois, and M. G. Norton. 1990. Identification and characterization of a gene that controls colony morphology and auto-aggregation in K12. J. Gen. Microbiol. 136:455–462.

4522. Warren, M. J., C. A. Roessner, P. J. Santander, and A. I. Scott. 1990. The Escherichia coli cysG gene encodes S-adenosylmethionine-dependent uroporphyrinogen III methylase. Biochem. J. 265:725–729.

4523. Washburn, B. K., and S. R. Kushner. 1993. Characterization of DNA helicase II from a uvrD252 mutant of Escherichia coli. J. Bacteriol. 175:341–350.

4524. Watanabe, K., Y. Yamano, K. Murata, and A. Kimura. 1986. The nucleotide sequence of the gene for gamma-glutamylcysteine synthetase of Escherichia coli. Nucleic Acids Res. 14:4393–4400.

4525. Watanabe, T., and E. E. Snell. 1972. Reversibility of the tryptophanase reaction: synthesis of tryptophan from indole, pyruvate, and ammonia. Proc. Natl. Acad. Sci. USA 69:1086–1090.

4526. Watanabe, W., G. Sampei, A. Aiba, and K. Mizobuchi. 1989. Identification and sequence analysis of Escherichia coli purE and purK genes encoding 5'-phosphoribosyl-5-amino-4-imidazole carboxylase for de novo purine biosynthesis. J. Bacteriol. 171:198–204.

4527. Watson, N., and D. Apirion. 1985. Molecular cloning of the gene for the RNA-processing enzyme RNase III of Escherichia coli. Proc. Natl. Acad. Sci. USA 82:849–853.

4528. Watson, N., D. S. Dunyak, E. L. Rosey, J. L. Slonczewski, and E. R. Olson. 1992. Identification of elements involved in transcriptional regulation of the Escherichia coli cad operon by external pH. J. Bacteriol. 174:530–540.

4529. Waugh, D. S., and N. R. Pace. 1990. Complementation of an RNase P RNA (rnpB) gene deletion in Escherichia coli by homologous genes from distantly related eubacteria. J. Bacteriol. 172:6316–6322.

4530. Waugh, R., and D. H. Boxer. 1986. Pleiotropic hydrogenase mutants of Escherichia coli K12: growth in the presence of nickel can restore hydrogenase activity. Biochimie 68:157–166.

4531. Waukau, J., and S. Forst. 1992. Molecular analysis of the signaling pathway between EnvZ and OmpR in Escherichia coli. J. Bacteriol. 174:1522–1527.

4532. Webb, D. C., H. Rosenberg, and G. B. Cox. 1992. Mutational analysis of the Escherichia coli phosphate-specific transport system, a member of the traffic ATPase (or ABC) family of membrane transporters. A role for proline residues in transmembrane helices. J. Biol. Chem. 267:24661–24668.

4533. Webb, G., K. Rohatgi, and J. B. Courtright. 1990. Location of gyrA on the physical map of the Escherichia coli chromosome. J. Bacteriol. 172:6617.

4534. Webb, M. 1970. The mechanism of acquired resistance to Co^{2+} and Ni^{2+} in Gram-positive and Gram-negative bacteria. Biochim. Biophys. Acta 222:440–446.

4535. Weber, K. 1968. New structural model of E. coli aspartate transcarbamylase and the amino-acid sequence of the regulatory polypeptide chain. Nature (London) 218:1116–1119.

4536. Weber, R. F., and P. M. Silverman. 1988. The cpx proteins of Escherichia coli K12. Structure of the cpxA polypeptide as an inner membrane component. J. Mol. Biol. 203:467–478.

4537. Webster, C., L. Gardner, and S. Busby. 1989. The Escherichia coli melR gene encodes a DNA-binding protein with affinity for specific sequences located in the melibiose-operon regulatory region. Gene 83:207–213.

4538. Webster, C., K. Kempsell, I. Booth, and S. Busby. 1987. Organisation of the regulatory region of the Escherichia coli melibiose operon. Gene 59:253–263.

4539. Webster, R. E. 1991. The tol gene products and the import of macromolecules into Escherichia coli. Mol. Microbiol. 5:1005–1011.

4540. Webster, T., H. Tsai, M. Kula, G. A. Mackie, and P. Schimmel. 1984. Specific sequence homology and three-dimensional structure of an aminoacyl transfer RNA synthetase. Science 226:1315–1317.

4541. Webster, T. A., B. W. Gibson, T. Keng, K. Biemann, and P. Schimmel. 1983. Primary structures of both subunits of Escherichia coli glycyl-tRNA synthetase. J. Biol. Chem. 258:10637–10641.

4542. Wei, S. Q., and J. Stader. 1994. A new suppressor of a lamB signal sequence mutation, prlZ1, maps to 69 minutes on the Escherichia coli chromosome. J. Bacteriol. 176:5704–5710.

4543. Wei, T. F., W. Bujalowski, and T. M. Lohman. 1992. Cooperative binding of polyamines induces the Escherichia coli single-strand binding protein-DNA binding mode transitions. Biochemistry 31:6166–6174.

4544. Weichart, D., R. Lange, N. Henneberg, and R. Hengge-Aronis. 1993. Identification and characterization of stationary phase-inducible genes in Escherichia coli. Mol. Microbiol. 10:407–420.

4545. Weickert, M. J., and S. Adhya. 1992. A family of bacterial regulators homologous to Gal and Lac repressors. J. Biol. Chem. 267:15869–15874.

4546. Weickert, M. J., and S. Adhya. 1992. Isorepressor of the gal regulon in Escherichia coli. J. Mol. Biol. 226:69–83.

4547. Weickert, M. J., and S. Adhya. 1993. Control of transcription of gal repressor and isorepressor genes in Escherichia coli. J. Bacteriol. 175:251–258.

4548. Weickert, M. J., R. W. Hogg, and S. Adhya. 1991. Locations and orientations on the Escherichia coli physical map of the mgl operon and galS, a new locus for galactose ultrainduction. J. Bacteriol. 173:7412–7413.

4549. Weidner, U., S. Geier, A. Ptock, T. Friedrich, H. Leif, and H. Weiss. 1993. The gene locus of the proton-translocating NADH: ubiquinone oxidoreductase in Escherichia coli. Organization of the 14 genes and relationship between the derived proteins and subunits of mitochondrial complex I. J. Mol. Biol. 233:109–122.

4550. Weigel, N., D. A. Powers, and S. Roseman. 1982. Sugar transport by the bacterial phosphotransferase system. Primary structure and active site of a general phosphocarrier protein (HPr) from Salmonella typhimurium. J. Biol. Chem. 257:14499–14509.

4551. Weijland, A., K. Harmark, R. H. Cool, P. H. Anborgh, and A. Parmeggiani. 1992. Elongation factor Tu: a molecular switch in protein biosynthesis. *Mol. Microbiol.* **6:**683–688.

4552. Weiner, J. H., G. Shaw, R. J. Turner, and C. A. Trieber. 1993. The topology of the anchor subunit of dimethyl sulfoxide reductase of *Escherichia coli*. *J. Biol. Chem.* **268:**3238–3244.

4553. Weiner, L., J. L. Brissette, and P. Model. 1991. Stress-induced expression of the *Escherichia coli* phage shock protein operon is dependent on sigma 54 and modulated by positive and negative feedback mechanisms. *Genes Dev.* **5:**1912–1923.

4554. Weinreich, M. D., and W. S. Reznikoff. 1992. Fis plays a role in Tn*5* and IS*50* transposition. *J. Bacteriol.* **174:**4530–4537.

4555. Weinreich, M. D., H. Yigit, and W. S. Reznikoff. 1994. Overexpression of the Tn*5* transposase in *Escherichia coli* results in filamentation, aberrant nucleoid segregation, and cell death: analysis of *E. coli* and transposase suppressor mutations. *J. Bacteriol.* **176:**5494–5504.

4556. Weinstock, O., C. Sella, D. M. Chipman, and Z. Barak. 1992. Properties of subcloned subunits of bacterial acetohydroxy acid synthases. *J. Bacteriol.* **174:**5560–5566.

4557. Weiss, B., and R. P. Cunningham. 1985. Genetic mapping of *nth*, a gene affecting endonuclease III (thymine glycol-DNA glycosylase) in *Escherichia coli* K-12. *J. Bacteriol.* **162:**607–610.

4558. Weiss, D. L., D. I. Johnson, H. L. Weith, and R. L. Somerville. 1986. Structural analysis of the *ileR* locus of *Escherichia coli* K12. *J. Biol. Chem.* **261:**9966–9971.

4559. Weissbach, H., and N. Brot. 1991. Regulation of methionine synthesis in *Escherichia coli*. *Mol. Microbiol.* **5:**1593–1597.

4560. Weissborn, A. C., Q. Liu, M. K. Rumley, and E. P. Kennedy. 1994. UTP: alpha-D-glucose-1-phosphate uridylyltransferase of *Escherichia coli*: isolation and DNA sequence of the *galU* gene and purification of the enzyme. *J. Bacteriol.* **176:**2611–2618.

4561. Weissenborn, D. L., N. Wittekindt, and T. J. Larson. 1992. Structure and regulation of the *glpFK* operon encoding glycerol diffusion facilitator and glycerol kinase of *Escherichia coli* K-12. *J. Biol. Chem.* **267:**6122–6131.

4562. Wek, R. C., and G. W. Hatfield. 1986. Nucleotide sequence and in vivo expression of the *ilvY* and *ilvC* genes in *Escherichia coli* K12. Transcription from divergent overlapping promoters. *J. Biol. Chem.* **261:**2441–2450.

4563. Wek, R. C., and G. W. Hatfield. 1986. Examination of the internal promoter, PE, in the *ilvGMEDA* operon of *E. coli* K-12. *Nucleic Acids Res.* **14:**2763–2777.

4564. Wek, R. C., C. A. Hauser, and G. W. Hatfield. 1985. The nucleotide sequence of the *ilvBN* operon of *Escherichia coli*: sequence homologies of the acetohydroxy acid synthase isozymes. *Nucleic Acids Res.* **13:**3995–4010.

4565. Welch, M. M., and C. S. McHenry. 1982. Cloning and identification of the product of the *dnaE* gene of *Escherichia coli*. *J. Bacteriol.* **152:**351–356.

4566. Wende, M., A. Quinones, L. Diederich, W. R. Jueterbock, and W. Messer. 1991. Transcription termination in the *dnaA* gene. *Mol. Gen. Genet.* **230:**486–490.

4567. Weng, M., C. A. Makaroff, and H. Zalkin. 1986. Nucleotide sequence of *Escherichia coli pyrG* encoding CTP synthetase. *J. Biol. Chem.* **261:**5568–5574.

4568. Weng, M. L., and H. Zalkin. 1987. Structural role for a conserved region in the CTP synthetase glutamine amide transfer domain. *J. Bacteriol.* **169:**3023–3028.

4569. Werth, M. T., G. Cecchini, A. Manodori, B. A. Ackrell, I. Schroder, R. P. Gunsalus, and M. K. Johnson. 1990. Site-directed mutagenesis of conserved cysteine residues in *Escherichia coli* fumarate reductase: modification of the spectroscopic and electrochemical properties of the [2Fe-2S] cluster. *Proc. Natl. Acad. Sci. USA* **87:**8965–8969.

4570. Wertheimer, S. J., R. A. Klotsky, and I. Schwartz. 1988. Transcriptional patterns for the *thrS-infC-rplT* operon of *Escherichia coli*. *Gene* **63:**309–320.

4571. Wessler, S. R., and J. M. Calvo. 1981. Control of *leu* operon expression in *Escherichia coli* by a transcription attenuation mechanism. *J. Mol. Biol.* **149:**579–597.

4572. West, S. C., and B. Connolly. 1992. Biological roles of the *Escherichia coli* RuvA, RuvB and RuvC proteins revealed. *Mol. Microbiol.* **6:**2755–2759.

4573. Westh Hansen, S. E., N. Jensen, and A. Munch-Petersen. 1987. Studies on the sequence and structure of the *Escherichia coli* K-12 *nupG* gene, encoding a nucleoside-transport system. *Eur. J. Biochem.* **168:**385–391.

4573a.Westling-Häggström, B., and S. Normark. 1975. Genetic and physiological analysis of an *envB* spherelike mutant of *Escherichia coli* K-12 and characterization of its transductants. *J. Bacteriol.* **123:**75–82.

4574. Weston, L. A., and R. J. Kadner. 1987. Identification of *uhp* polypeptides and evidence for their role in exogenous induction of the sugar phosphate transport system of *Escherichia coli* K-12. *J. Bacteriol.* **169:**3546–3555.

4575. Wetzstein, M., and W. Schumann. 1990. Promoters of major *Escherichia coli* heat shock genes seem non-functional in *Bacillus subtilis*. *FEMS Microbiol. Lett.* **60:**55–58.

4576. Whalen, W. A., and C. M. Berg. 1982. Analysis of an *avtA*::Mu d1(Ap *lac*) mutant: metabolic role of transaminase C. *J. Bacteriol.* **150:**739–746.

4577. Whipp, M. J., D. M. Halsall, and A. J. Pittard. 1980. Isolation and characterization of an *Escherichia coli* K-12 mutant defective in tyrosine- and phenylalanine-specific transport systems. *J. Bacteriol.* **143:**1–7.

4578. Whipp, M. J., and A. J. Pittard. 1995. A reassessment of the relationship between *aroK*- and *aroL*-encoded shikimate kinase enzymes of *Escherichia coli*. *J. Bacteriol.* **177:**1627–1629.

4579. Whitby, M. C., S. D. Vincent, and R. G. Lloyd. 1994. Branch migration of Holliday junctions: identification of RecG protein as a junction specific DNA helicase. *EMBO J.* **13:**5220–5228.

4580. White, M. K., and M. D. Yudkin. 1979. Complementation analysis of eleven tryptophanase mutations in *Escherichia coli*. *J. Gen. Microbiol.* **114:**471–475.

4581. White, P. J., G. Millar, and J. R. Coggins. 1988. The overexpression, purification and complete amino acid sequence of chorismate synthase from *Escherichia coli* K12 and its comparison with the enzyme from *Neurospora crassa*. *Biochem. J.* **251:**313–322.

4582. White, R. J., and P. W. Kent. 1970. An examination of the inhibitory effects of N-iodoacetylglucosamine on *Escherichia coli* and isolation of resistant mutants. *Biochem. J.* **118:**81–87.

4583. White, S., F. E. Tuttle, D. Blankenhorn, D. C. Dosch, and J. L. Slonczewski. 1992. pH dependence and gene structure of *inaA* in *Escherichia coli*. *J. Bacteriol.* **174:**1537–1543.

4584. White-Ziegler, C. A., and D. A. Low. 1992. Thermoregulation of the *pap* operon: evidence for the involvement of RimJ, the N-terminal acetylase of ribosomal protein S5. *J. Bacteriol.* **174:**7003–7012.

4585. Whoriskey, S. K., M. A. Schofield, and J. H. Miller. 1991. Isolation and characterization of *Escherichia coli* mutants with altered rates of deletion formation. *Genetics* **127:**21–30.

4586. Wickner, S., J. Hoskins, and K. McKenney. 1991. Function of DnaJ and DnaK as chaperones in origin-specific DNA binding by RepA. *Nature* (London) **350:**165–167.

4587. Wickner, S., D. Skowyra, J. Hoskins, and K. McKenney. 1992. DnaJ, DnaK, and GrpE heat shock proteins are required in oriP1 DNA replication solely at the RepA monomerization step. *Proc. Natl. Acad. Sci. USA* **89:**10345–10349.

4588. Wijsman, H. J., and H. C. Pafort. 1974. Pleiotropic mutations in *Escherichia coli* conferring tolerance to glycine and sensitivity to penicillin. *Mol. Gen. Genet.* **128:**349–357.

4589. Wild, J., E. Altman, T. Yura, and C. A. Gross. 1992. DnaK and DnaJ heat shock proteins participate in protein export in *Escherichia coli*. *Genes Dev.* **6:**1165–1172.

4590. Wild, J., J. Hennig, M. Lobocka, W. Walczak, and T. Klopotowski. 1985. Identification of the *dadX* gene coding for the predominant isozyme of alanine racemase in *Escherichia coli* K12. *Mol. Gen. Genet.* **198:**315–322.

4591. Wild, J., and T. Klopotowski. 1981. D-Amino acid dehydrogenase of *Escherichia coli* K12: positive selection of mutants defective in enzyme activity and localization of the structural gene. *Mol. Gen. Genet.* **181:**373–378.

4592. Wild, J., and B. Obrepalska. 1982. Regulation of expression of the *dadA* gene encoding D-amino acid dehydrogenase in *Escherichia coli*: analysis of *dadA-lac* fusions and direction of *dadA* transcription. *Mol. Gen. Genet.* **186:**405–410.

4593. Wild, J., B. Zakrzewska, W. Walczak, and T. Klopotowski. 1987. Two distinct types of mutations conferring to *Escherichia coli* K12 capability of D-tryptophan utilization. *Acta Microbiol. Pol.* **36:**17–28.

4594. Wild, J. R., K. F. Foltermann, W. D. Roof, and G. A. O'Donovan. 1981. A mutation in the catalytic cistron of aspartate carbamoyltransferase affecting catalysis, regulatory response and holoenzyme assembly. *Nature* (London) **292:**373–375.

4595. Wilde, R. J., and J. R. Guest. 1986. Transcript analysis of the citrate synthase and succinate dehydrogenase genes of *Escherichia coli* K12. *J. Gen. Microbiol.* **132:**3239–3251.

4596. Williams, K. R., J. B. Murphy, and J. W. Chase. 1984. Characterization of the structural and functional defect in the *Escherichia coli* single-stranded DNA binding protein encoded by the *ssb-1* mutant gene. Expression of the *ssb-1* gene under lambda pL regulation. *J. Biol. Chem.* **259:**11804–11811.

4597. Williams, M. D., J. A. Fuchs, and M. C. Flickinger. 1991. Null mutation in the stringent starvation protein of *Escherichia coli* disrupts lytic development of bacteriophage P1. *Gene* **109:**21–30.

4598. Williams, M. D., T. X. Ouyang, and M. C. Flickinger. 1994. Starvation-induced expression of SspA and SspB: the effects of a null mutation in *sspA* on *Escherichia coli* protein synthesis and survival during growth and prolonged starvation. *Mol. Microbiol.* **11:**1029–1043.

4599. Williams, M. G., M. Fortson, C. C. Dykstra, P. Jensen, and S. R. Kushner. 1989. Identification and genetic mapping of the structural gene for an essential *Escherichia coli* membrane protein. *J. Bacteriol.* **171:**565–568.

4600. Williams, M. V., T. J. Kerr, R. D. Lemmon, and G. J. Tritz. 1980. Azaserine resistance in *Escherichia coli*: chromosomal location of multiple genes. *J. Bacteriol.* **143:**383–388.

4601. Williams, N., D. K. Fox, C. Shea, and S. Roseman. 1986. Pel, the protein that permits lambda DNA penetration of *Escherichia coli*, is encoded by a gene in *ptsM* and is required for mannose utilization by the phosphotransferase system. *Proc. Natl. Acad. Sci. USA* **83:**8934–8938.

4602. Williamson, J. M., M. S. Anderson, and C. R. Raetz. 1991. Acyl-acyl carrier protein specificity of UDP-GlcNAc acyltransferases from gram-negative bacteria: relationship to lipid A structure. *J. Bacteriol.* **173:**3591–3596.

4603. Williamson, R. M., and D. L. Oxender. 1990. Sequence and structural similarities between the leucine-specific binding protein and leucyl-tRNA synthetase of *Escherichia coli*. *Proc. Natl. Acad. Sci. USA* **87:**4561–4565.

4604. Williamson, R. M., and D. L. Oxender. 1992. Premature termination of in vivo transcription of a gene encoding a branched-chain amino acid transport protein in *Escherichia coli*. *J. Bacteriol.* **174:**1777–1782.

4605. **Willins, D. A., C. W. Ryan, J. V. Platko, and J. M. Calvo.** 1991. Characterization of Lrp, an *Escherichia coli* regulatory protein that mediates a global response to leucine. *J. Biol. Chem.* **266:**10768–10774.

4606. **Willis, D. K., K. E. Fouts, S. D. Barbour, and A. J. Clark.** 1983. Restriction nuclease and enzymatic analysis of transposon-induced mutations of the Rac prophage which affect expression and function of *recE* in *Escherichia coli* K-12. *J. Bacteriol.* **156:**727–736.

4607. **Willis, D. K., L. H. Satin, and A. J. Clark.** 1985. Mutation-dependent suppression of *recB21 recC22* by a region cloned from the Rac prophage of *Escherichia coli* K-12. *J. Bacteriol.* **162:**1166–1172.

4608. **Willis, D. K., B. E. Uhlin, K. S. Amini, and A. J. Clark.** 1981. Physical mapping of the *srl recA* region of *Escherichia coli*: analysis of Tn*10* generated insertions and deletions. *Mol. Gen. Genet.* **183:**497–504.

4609. **Willison, J. C.** 1992. An essential gene (*efg*) located at 38.1 minutes on the *Escherichia coli* chromosome. *J. Bacteriol.* **174:**5765–5766.

4610. **Willison, J. C., and G. Tissot.** 1994. The *Escherichia coli efg* gene and the *Rhodobacter capsulatus adgA* gene code for NH₃-dependent NAD synthetase. *J. Bacteriol.* **176:**3400–3402.

4611. **Willsky, G. R., R. L. Bennett, and M. H. Malamy.** 1973. Inorganic phosphate transport in *Escherichia coli*: involvement of two genes which play a role in alkaline phosphatase regulation. *J. Bacteriol.* **113:**529–539.

4612. **Wilmanns, M., J. P. Priestle, T. Niermann, and J. N. Jansonius.** 1992. Three-dimensional structure of the bifunctional enzyme phosphoribosylanthranilate isomerase: indoleglycerolphosphate synthase from *Escherichia coli* refined at 2.0 A resolution. *J. Mol. Biol.* **223:**477–507.

4613. **Wilson, D. M., and T. H. Wilson.** 1992. Asp-51 and Asp-120 are important for the transport function of the *Escherichia coli* melibiose carrier. *J. Bacteriol.* **174:**3083–3086.

4614. **Wilson, H. R., P. T. Chan, and C. L. Turnbough, Jr.** 1987. Nucleotide sequence and expression of the *pyrC* gene of *Escherichia coli* K-12. *J. Bacteriol.* **169:**3051–3058.

4615. **Wilson, R. K., T. Brown, and B. A. Roe.** 1986. Nucleotide sequence of *pheW*; a third gene for *E. coli* tRNA^Phe. *Nucleic Acids Res.* **14:**5937.

4616. **Wilson, R. L., and G. V. Stauffer.** 1994. DNA sequence and characterization of GcvA, a LysR family regulatory protein for the *Escherichia coli* glycine cleavage enzyme system. *J. Bacteriol.* **176:**2862–2868.

4617. **Wilson, R. L., L. T. Stauffer, and G. V. Stauffer.** 1993. Roles of the GcvA and PurR proteins in negative regulation of the *Escherichia coli* glycine cleavage enzyme system. *J. Bacteriol.* **175:**5129–5134.

4618. **Winter, G. P., B. S. Hartley, A. D. McLachlan, M. Lee, and K. H. Muench.** 1977. Sequence homologies between the tryptophanyl tRNA synthetases of *Bacillus stearothermophilus* and *Escherichia coli*. *FEBS Lett.* **82:**348–350.

4619. **Wise, J. G., B. J. Hicke, and P. D. Boyer.** 1987. Catalytic and noncatalytic nucleotide binding sites of the *Escherichia coli* F₁ ATPase. Amino acid sequences of beta-subunit tryptic peptides labeled with 2-azido-ATP. *FEBS Lett.* **223:**395–401.

4620. **Wissenbach, U., B. Keck, and G. Unden.** 1993. Physical map location of the new *artPIQMJ* genes of *Escherichia coli*, encoding a periplasmic arginine transport system. *J. Bacteriol.* **175:**3687–3688.

4621. **Witkin, E. M., and V. Roegner-Maniscalco.** 1992. Overproduction of DnaE protein (alpha subunit of DNA polymerase III) restores viability in a conditionally inviable *Escherichia coli* strain deficient in DNA polymerase I. *J. Bacteriol.* **174:**4166–4168.

4622. **Wittmann, H. G., G. Stoffler, D. Apirion, L. Rosen, K. Tanaka, M. Tamaki, R. Takata, S. Dekio, and E. Otaka.** 1973. Biochemical and genetic studies on two different types of erythromycin resistant mutants of *Escherichia coli* with altered ribosomal proteins. *Mol. Gen. Genet.* **127:**175–189.

4623. **Wittmann-Liebold, B.** 1979. Primary structure of protein L24 from the *Escherichia coli* ribosome. *FEBS Lett.* **108:**75–80.

4624. **Wittmann-Liebold, B., and A. Bosserhoff.** 1981. Primary structure of protein S2 from the *Escherichia coli* ribosome. *FEBS Lett.* **129:**10–16.

4625. **Wittmann-Liebold, B., and B. Greuer.** 1978. The primary structure of protein S5 from the small subunit of the *Escherichia coli* ribosome. *FEBS Lett.* **95:**91–98.

4626. **Wittmann-Liebold, B., and B. Greuer.** 1979. Primary structure of protein L23 from the large subunit of the *Escherichia coli* ribosome. *FEBS Lett.* **108:**69–74.

4627. **Wittmann-Liebold, B., and B. Greuer.** 1980. Amino acid sequence of protein L22 from the large subunit of the *Escherichia coli* ribosome. *FEBS Lett.* **121:**105–112.

4628. **Wittmann-Liebold, B., B. Greuer, and R. Pannenbecker.** 1975. The primary structure of protein L32 from the 50S subunit of *Escherichia coli* ribosomes. *Hoppe-Seyler's Z. Physiol. Chem.* **356:**1977–1979.

4629. **Wittmann-Liebold, B., and E. Marzinzig.** 1977. Primary structure of protein L28 from the large subunit of *Escherichia coli* ribosomes. *FEBS Lett.* **81:**214–217.

4630. **Wittmann-Liebold, B., E. Marzinzig, and A. Lehmann.** 1976. Primary structure of protein S20 from the small ribosomal subunit of *Escherichia coli*. *FEBS Lett.* **68:**110–114.

4631. **Wittmann-Liebold, B., and R. Pannenbecker.** 1976. Primary structure of protein L33 from the large subunit of the *Escherichia coli* ribosome. *FEBS Lett.* **68:**115–118.

4632. **Wittmann-Liebold, B., and C. Seib.** 1979. The primary structure of protein L20 from the large subunit of the *Escherichia coli* ribosome. *FEBS Lett.* **103:**61–65.

4633. **Woisetschlager, M., A. Hodl-Neuhofer, and G. Hogenauer.** 1988. Localization of the *kdsA* gene with the aid of the physical map of the *Escherichia coli* chromosome. *J. Bacteriol.* **170:**5382–5384.

4634. **Woisetschlager, M., and G. Hogenauer.** 1987. The *kdsA* gene coding for 3-deoxy-D-manno-octulosonic acid 8-phosphate synthetase is part of an operon in *Escherichia coli*. *Mol. Gen. Genet.* **207:**369–373.

4635. **Wojtkowiak, D., C. Georgopoulos, and M. Zylicz.** 1993. Isolation and characterization of ClpX, a new ATP-dependent specificity component of the Clp protease of *Escherichia coli*. *J. Biol. Chem.* **268:**22609–22617.

4636. **Wold, M. S., and R. McMacken.** 1982. Regulation of expression of the *Escherichia coli dnaG* gene and amplification of the *dnaG* primase. *Proc. Natl. Acad. Sci. USA* **79:**4907–4911.

4637. **Wolf, R. E., Jr., and J. A. Cool.** 1980. Mapping of insertion mutations in *gnd* of *Escherichia coli* with deletions defining the ends of the gene. *J. Bacteriol.* **141:**1222–1229.

4638. **Wolf-Watz, H., and M. Masters.** 1979. Deoxyribonucleic acid and outer membrane: strains diploid for the *oriC* region show elevated levels of deoxyribonucleic acid-binding protein and evidence for specific binding of the *oriC* region to outer membrane. *J. Bacteriol.* **140:**50–58.

4639. **Wolfe, A. J., B. P. McNamara, and R. C. Stewart.** 1994. The short form of CheA couples chemoreception to CheA phosphorylation. *J. Bacteriol.* **176:**4483–4491.

4640. **Wolfe, P. B., W. Wickner, and J. M. Goodman.** 1983. Sequence of the leader peptidase gene of *Escherichia coli* and the orientation of leader peptidase in the bacterial envelope. *J. Biol. Chem.* **258:**12073–12080.

4641. **Wolfe, S. A., and J. M. Smith.** 1988. Nucleotide sequence and analysis of the *purA* gene encoding adenylosuccinate synthetase of *Escherichia coli* K12. *J. Biol. Chem.* **263:**19147–19153.

4642. **Woo, K. M., K. I. Kim, A. L. Goldberg, D. B. Ha, and C. H. Chung.** 1992. The heat-shock protein ClpB in *Escherichia coli* is a protein-activated ATPase. *J. Biol. Chem.* **267:**20429–20434.

4643. **Wood, D., M. G. Darlison, R. J. Wilde, and J. R. Guest.** 1984. Nucleotide sequence encoding the flavoprotein and hydrophobic subunits of the succinate dehydrogenase of *Escherichia coli*. *Biochem. J.* **222:**519–534.

4644. **Wood, E. R., and S. W. Matson.** 1989. The molecular cloning of the gene encoding the *Escherichia coli* 75-kDa helicase and the determination of its nucleotide sequence and genetic map position. *J. Biol. Chem.* **264:**8297–8303.

4645. **Wood, J. M.** 1981. Genetics of L-proline utilization in *Escherichia coli*. *J. Bacteriol.* **146:**895–901.

4646. **Wood, J. M., and D. Zadworny.** 1979. Characterization of an inducible porter required for L-proline catabolism by *Escherichia coli* K12. *Can. J. Biochem.* **57:**1191–1199.

4647. **Wood, J. M., and D. Zadworny.** 1980. Amplification of the *put* genes and identification of the *put* gene products in *Escherichia coli* K12. *Can. J. Biochem.* **58:**787–796.

4648. **Woodgate, R.** 1992. Construction of a *umuDC* operon substitution mutation in *Escherichia coli*. *Mutat. Res.* **281:**221–225.

4649. **Woodgate, R., and D. G. Ennis.** 1991. Levels of chromosomally encoded Umu proteins and requirements for in vivo UmuD cleavage. *Mol. Gen. Genet.* **229:**10–16.

4650. **Woodgate, R., M. Rajagopalan, C. Lu, and H. Echols.** 1989. UmuC mutagenesis protein of *Escherichia coli*: purification and interaction with UmuD and UmuD′. *Proc. Natl. Acad. Sci. USA* **86:**7301–7305.

4651. **Woods, S. A., J. S. Miles, R. E. Roberts, and J. R. Guest.** 1986. Structural and functional relationships between fumarase and aspartase. Nucleotide sequences of the fumarase (*fumC*) and aspartase (*aspA*) genes of *Escherichia coli* K12. *Biochem. J.* **237:**547–557.

4652. **Woods, S. A., S. D. Schwartzbach, and J. R. Guest.** 1988. Two biochemically distinct classes of fumarase in *Escherichia coli*. *Biochim. Biophys. Acta* **954:**14–26.

4653. **Wookey, P. J., and A. J. Pittard.** 1988. DNA sequence of the gene (*tyrP*) encoding the tyrosine-specific transport system of *Escherichia coli*. *J. Bacteriol.* **170:**4946–4949.

4654. **Wookey, P. J., J. Pittard, S. M. Forrest, and B. E. Davidson.** 1984. Cloning of the *tyrP* gene and further characterization of the tyrosine-specific transport system in *Escherichia coli* K-12. *J. Bacteriol.* **160:**169–174.

4655. **Worsham, P. L., and J. Konisky.** 1981. Use of *cir-lac* operon fusions to study transcriptional regulation of the colicin Ia receptor in *Escherichia coli* K-12. *J. Bacteriol.* **145:**647–650.

4656. **Wower, I., M. P. Kowaleski, L. E. Sears, and R. A. Zimmermann.** 1992. Mutagenesis of ribosomal protein S8 from *Escherichia coli*: defects in regulation of the *spc* operon. *J. Bacteriol.* **174:**1213–1221.

4657. **Wright, M.** 1971. Mutants of *Escherichia coli* lacking endonuclease I, ribonuclease I, or ribonuclease II. *J. Bacteriol.* **107:**87–94.

4658. **Wrzesinski, J., A. Bakin, K. Nurse, B. G. Lane, and J. Ofengand.** 1995. Purification, cloning, and properties of the 16S RNA pseudouridine 516 synthase from *Escherichia coli*. *Biochemistry* **34:**8904–8913.

4659. **Wrzesinski, J., K. Nurse, A. Bakin, B. G. Lane, and J. Ofengand.** A dual-specificity pseudouridine synthase: an *Escherichia coli* synthase purified and cloned on the basis of its specificity for Ψ746 in 23S is also specific for Ψ32 in tRNA^Phe. *RNA* **1:**437–449.

4660. **Wu, A. M., A. B. Chapman, T. Platt, L. P. Guarente, and J. Beckwith.** 1980. Deletions of distal sequence after termination of transcription at the end of the tryptophan operon in *E. coli*. *Cell* **19:**829–836.

4661. **Wu, A. M., G. E. Christie, and T. Platt.** 1981. Tandem termination sites in the tryptophan operon of *Escherichia coli*. *Proc. Natl. Acad. Sci. USA* **78:**2913–2917.

4662. **Wu, B., D. Ang, M. Snavely, and C. Georgopoulos.** 1994. Isolation and characterization of point mutations in the *Escherichia coli* grpE heat shock gene. *J. Bacteriol.* **176:**6965–6973.

4663. **Wu, B., C. Georgopoulos, and D. Ang.** 1992. The essential *Escherichia coli* msgB gene, a multicopy suppressor of a temperature-sensitive allele of the heat shock gene grpE, is identical to dapE. *J. Bacteriol.* **174:**5258–5264.

4664. **Wu, G., H. D. Williams, F. Gibson, and R. K. Poole.** 1993. Mutants of *Escherichia coli* affected in respiration: the cloning and nucleotide sequence of ubiA, encoding the membrane-bound p-hydroxybenzoate:octaprenyltransferase. *J. Gen. Microbiol.* **139:**1795–1805.

4665. **Wu, G., H. D. Williams, M. Zamanian, F. Gibson, and R. K. Poole.** 1992. Isolation and characterization of *Escherichia coli* mutants affected in aerobic respiration: the cloning and nucleotide sequence of ubiG. Identification of an S-adenosyl-methionine-binding motif in protein, RNA, and small-molecule methyltransferases (methoxy-1,4-benzoquinone-O-methyltransferase). *J. Gen. Microbiol.* **138:**2101–2112.

4666. **Wu, J., and B. Weiss.** 1991. Two divergently transcribed genes, soxR and soxS, control a superoxide response regulon of *Escherichia coli*. *J. Bacteriol.* **173:**2864–2871.

4667. **Wu, J., and B. Weiss.** 1992. Two-stage induction of the soxRS (superoxide response) regulon of *Escherichia coli*. *J. Bacteriol.* **174:**3915–3920.

4668. **Wu, J. Y., L. M. Siegel, and N. M. Kredich.** 1991. High-level expression of *Escherichia coli* NADPH-sulfite reductase: requirement for a cloned cysG plasmid to overcome limiting siroheme cofactor. *J. Bacteriol.* **173:**325–333.

4669. **Wu, L. F., and M. A. Mandrand-Berthelot.** 1986. Genetic and physiological characterization of new *Escherichia coli* mutants impaired in hydrogenase activity. *Biochimie* **68:**167–179.

4670. **Wu, L. F., and M. A. Mandrand-Berthelot.** 1987. Regulation of the fdhF gene encoding the selenopolypeptide for benzyl viologen-linked formate dehydrogenase in *Escherichia coli*. *Mol. Gen. Genet.* **209:**129–134.

4671. **Wu, L. F., and M. A. Mandrand-Berthelot.** 1987. Characterization of the product of the cloned fdhF gene of *Escherichia coli*. *J. Gen. Microbiol.* **133:**2421–2426.

4672. **Wu, L. F., M. A. Mandrand-Berthelot, R. Waugh, C. J. Edmonds, S. E. Holt, and D. H. Boxer.** 1989. Nickel deficiency gives rise to the defective hydrogenase phenotype of hydC and fnr mutants in *Escherichia coli*. *Mol. Microbiol.* **3:**1709–1718.

4673. **Wu, L. F., C. Navarro, and M. A. Mandrand-Berthelot.** 1991. The hydC region contains a multi-cistronic operon (nik) involved in nickel transport in *Escherichia coli*. *Gene* **107:**37–42.

4674. **Wu, T. H., E. Grelland, E. Boye, and M. G. Marinus.** 1992. Identification of a weak promoter for the dam gene of *Escherichia coli*. *Biochim. Biophys. Acta* **1131:**47–52.

4675. **Wu, T. H., and M. G. Marinus.** 1994. Dominant negative mutator mutations in the mutS gene of *Escherichia coli*. *J. Bacteriol.* **176:**5393–5400.

4676. **Wu, T. H., D. L. Wood, P. L. Stein, and M. M. Comer.** 1984. Transcription of a gene cluster coding for two aminoacyl-tRNA synthetases and an initiation factor in *Escherichia coli*. *J. Mol. Biol.* **173:**177–209.

4677. **Wu, T. T.** 1976. Growth on D-arabitol of a mutant strain of *Escherichia coli* K12 using a novel dehydrogenase and enzymes related to L-1,2-propanediol and D-xylose metabolism. *J. Gen. Microbiol.* **94:**246–256.

4678. **Wu, Y. F., and P. Datta.** 1992. Integration host factor is required for positive regulation of the tdc operon of *Escherichia coli*. *J. Bacteriol.* **174:**233–240.

4679. **Wubbolts, M. G., P. Terpstra, J. B. van Beilen, J. Kingma, H. A. Meesters, and B. Witholt.** 1990. Variation of cofactor levels in *Escherichia coli*. Sequence analysis and expression of the pncB gene encoding nicotinic acid phosphoribosyltransferase. *J. Biol. Chem.* **265:**17665–17672.

4680. **Wulfing, C., J. Lombardero, and A. Pluckthun.** 1994. An *Escherichia coli* protein consisting of a domain homologous to FK506-binding proteins (FKBP) and a new metal binding motif. *J. Biol. Chem.* **269:**2895–2901.

4681. **Wunderlich, M., and R. Glockshuber.** 1993. In vivo control of redox potential during protein folding catalyzed by bacterial protein disulfide-isomerase (DsbA). *J. Biol. Chem.* **268:**24547–24550.

4682. **Wurgler, S. M., and C. C. Richardson.** 1990. Structure and regulation of the gene for dGTP triphosphohydrolase from *Escherichia coli*. *Proc. Natl. Acad. Sci. USA* **87:**2740–2744. (Erratum, **87:**4022.)

4683. **Wurtzel, E. T., M. Y. Chou, and M. Inouye.** 1982. Osmoregulation of gene expression. I. DNA sequence of the ompR gene of the ompB operon of *Escherichia coli* and characterization of its gene product. *J. Biol. Chem.* **257:**13685–13691.

4684. **Xia, M., Y. Zhu, X. Cao, L. You, and Z. Chen.** 1995. Cloning, sequencing and analysis of a gene encoding *Escherichia coli* proline dehydrogenase. *FEMS Microbiol. Lett.* **1275:**235–242.

4685. **Xiao, H., R. Crombie, Z. Dong, R. Onrust, and M. O'Donnell.** 1993. DNA polymerase III accessory proteins. III. holC and holD encoding chi and psi. *J. Biol. Chem.* **268:**11773–11778.

4686. **Xiao, H., Z. Dong, and M. O'Donnell.** 1993. DNA polymerase III accessory proteins. IV. Characterization of chi and psi. *J. Biol. Chem.* **268:**11779–11784.

4687. **Xiao, H., M. Kalman, K. Ikehara, S. Zemel, G. Glaser, and M. Cashel.** 1991. Residual guanosine 3′,5′-bispyrophosphate synthetic activity of relA null mutants can be eliminated by spoT null mutations. *J. Biol. Chem.* **266:**5980–5990.

4688. **Xie, Q. W., C. W. Tabor, and H. Tabor.** 1989. Spermidine biosynthesis in *Escherichia coli*: promoter and termination regions of the speED operon. *J. Bacteriol.* **171:**4457–4465.

4689. **Xiong, H., and S. B. Vik.** 1995. Construction and plasmid-borne complementation of strains lacking the epsilon subunit of the *Escherichia coli* F_1F_0 ATP synthase. *J. Bacteriol.* **177:**851–853.

4689a.**Xu, F., S. Lin-Chao, and S. N. Cohen.** 1993. The *Escherichia coli* pcnB gene promotes adenylylation of antisense RNAI of ColE1-type plasmids in vivo and degradation of RNAI decay intermediates. *Proc. Natl. Acad. Sci. USA* **90:**6756–6760.

4690. **Xu, J., and R. C. Johnson.** 1995. Identification of genes negatively regulated by Fis: Fis and RpoS comodulate growth-phase-dependent gene expression in *Escherichia coli*. *J. Bacteriol.* **177:**938–947.

4691. **Xu, J., and R. C. Johnson.** 1995. aldB, an RpoS-dependent gene in *Escherichia coli* encoding an aldehyde dehydrogenase that is repressed by Fis and activated by Crp. *J. Bacteriol.* **177:**3166–3175.

4692. **Yagil, E., S. Dolev, J. Oberto, N. Kislev, N. Ramaiah, and R. A. Weisberg.** 1989. Determinants of site-specific recombination in the lambdoid coliphage HK022. An evolutionary change in specificity. *J. Mol. Biol.* **207:**695–717.

4693. **Yaguchi, M.** 1975. Primary structure of protein S18 from the small *Escherichia coli* ribosomal subunit. *FEBS Lett.* **59:**217–220.

4694. **Yaguchi, M., C. Roy, and H. G. Wittmann.** 1980. The primary structure of protein S10 from the small ribosomal subunit of *Escherichia coli*. *FEBS Lett.* **121:**113–116.

4695. **Yaguchi, M., and H. G. Wittmann.** 1978. The primary structure of protein S17 from the small ribosomal subunit of *Escherichia coli*. *FEBS Lett.* **87:**37–40.

4696. **Yaguchi, M., and H. G. Wittmann.** 1978. Primary structure of protein S19 from the small ribosomal subunit of *Escherichia coli*. *FEBS Lett.* **88:**227–230.

4697. **Yamada, H., M. Kitagawa, M. Kawakami, and S. Mizushima.** 1984. The gene coding for lipoprotein signal peptidase (lspA) and that for isoleucyl-tRNA synthetase (ileS) constitute a cotranscriptional unit in *Escherichia coli*. *FEBS Lett.* **171:**245–248.

4698. **Yamada, H., S. Muramatsu, and T. Mizuno.** 1990. An *Escherichia coli* protein that preferentially binds to sharply curved DNA. *J. Biochem.* **108:**420–425.

4699. **Yamada, H., T. Yoshida, K. Tanaka, C. Sasakawa, and T. Mizuno.** 1991. Molecular analysis of the *Escherichia coli* hns gene encoding a DNA-binding protein, which preferentially recognizes curved DNA sequences. *Mol. Gen. Genet.* **230:**332–336.

4700. **Yamada, M., S. Asaoka, M. H. Saier, Jr., and Y. Yamada.** 1993. Characterization of the gcd gene from *Escherichia coli* K-12 W3110 and regulation of its expression. *J. Bacteriol.* **175:**568–571.

4701. **Yamada, M., K. Makino, H. Shinagawa, and A. Nakata.** 1990. Regulation of the phosphate regulon of *Escherichia coli*: properties of phoR deletion mutants and subcellular localization of PhoR protein. *Mol. Gen. Genet.* **220:**366–372.

4702. **Yamada, M., and M. H. Saier, Jr.** 1987. Physical and genetic characterization of the glucitol operon in *Escherichia coli*. *J. Bacteriol.* **169:**2990–2994.

4703. **Yamada, M., and M. H. Saier, Jr.** 1987. Glucitol-specific enzymes of the phosphotransferase system in *Escherichia coli*. Nucleotide sequence of the gut operon. *J. Biol. Chem.* **262:**5455–5463.

4704. **Yamada, M., and M. H. Saier, Jr.** 1988. Positive and negative regulators for glucitol (gut) operon expression in *Escherichia coli*. *J. Mol. Biol.* **203:**569–583.

4705. **Yamada, M., K. Sumi, K. Matsushita, O. Adachi, and Y. Yamada.** 1993. Topological analysis of quinoprotein glucose dehydrogenase in *Escherichia coli* and its ubiquinone-binding site. *J. Biol. Chem.* **268:**12812–12817.

4706. **Yamada, M., Y. Takeda, K. Okamoto, and Y. Hirota.** 1982. Physical map of the nrdA-nrdB-ftsB-glpT region of the chromosomal DNA of *Escherichia coli*. *Gene* **18:**309–318.

4707. **Yamada, M., Y. Yamada, and M. H. Saier, Jr.** 1990. Nucleotide sequence and expression of the gutQ gene within the glucitol operon of *Escherichia coli*. *DNA Seq.* **1:**141–145.

4708. **Yamada, Y., Y. Y. Chang, G. A. Daniels, L. F. Wu, J. M. Tomich, M. Yamada, and M. H. Saier, Jr.** 1991. Insertion of the mannitol permease into the membrane of *Escherichia coli*. Possible involvement of an N-terminal amphiphilic sequence. *J. Biol. Chem.* **266:**17863–17871.

4709. **Yamagata, H., K. Daishima, and S. Mizushima.** 1983. Cloning and expression of a gene coding for the prolipoprotein signal peptidase of *Escherichia coli*. *FEBS Lett.* **158:**301–304.

4710. **Yamagata, H., N. Taguchi, K. Daishima, and S. Mizushima.** 1983. Genetic characterization of a gene for prolipoprotein signal peptidase in *Escherichia coli*. *Mol. Gen. Genet.* **192:**10–14.

4711. **Yamagishi, J., H. Yoshida, M. Yamayoshi, and S. Nakamura.** 1986. Nalidixic acid-resistant mutations of the gyrB gene of *Escherichia coli*. *Mol. Gen. Genet.* **204:**367–373.

4712. **Yamagishi, M., H. Matsushima, A. Wada, M. Sakagami, N. Fujita, and A. Ishihama.** 1993. Regulation of the *Escherichia coli* rmf gene encoding the ribosome modulation factor: growth phase- and growth rate-dependent control. *EMBO J.* **12:**625–630.

4713. **Yamaguchi, K., and M. Inouye.** 1988. Lipoprotein 28, an inner membrane protein of *Escherichia coli* encoded by nlpA, is not essential for growth. *J. Bacteriol.* **170:**3747–3749.

4714. **Yamamoto, K.** 1992. Dissection of functional domains in *Escherichia coli* DNA photolyase by linker-insertion mutagenesis. *Mol. Gen. Genet.* **232:**1–6.

4715. **Yamamoto, K., K. Kusano, N. K. Takahashi, H. Yoshikura, and I. Kobayashi.** 1992. Gene conversion in the *Escherichia coli* RecF pathway: a successive half crossing-over model. *Mol. Gen. Genet.* **234:**1–13.

4716. Yamamoto, T., M. Tomiyama, H. Mita, K. Sode, and I. Karube. 1990. Identification of proteins encoded in *Escherichia coli hydA, hydB* and analysis of the *hydA* locus. *FEMS Microbiol. Lett.* **54**:187–192.

4717. Yamamoto, Y., T. Ogawa, H. Shinagawa, T. Nakayama, H. Matsuo, and H. Ogawa. 1986. Determination of the initiation sites of transcription and translation of the *uvrD* gene of *Escherichia coli. J. Biochem.* **99**:1579–1590.

4718. Yamanaka, H., M. Kameyama, T. Baba, Y. Fujii, and K. Okamoto. 1994. Maturation pathway of *Escherichia coli* heat-stable enterotoxin I: requirement of DsbA for disulfide bond formation. *J. Bacteriol.* **176**:2906–2913.

4719. Yamanaka, K., T. Mitani, T. Ogura, H. Niki, and S. Hiraga. 1994. Cloning, sequencing, and characterization of multicopy suppressors of a *mukB* mutation in *Escherichia coli. Mol. Microbiol.* **13**:301–312.

4720. Yamanaka, K., T. Ogura, E. V. Koonin, H. Niki, and S. Hiraga. 1994. Multicopy suppressors, *mssA* and *mssB*, of an *smbA* mutation of *Escherichia coli. Mol. Gen. Genet.* **243**:9–16.

4721. Yamanaka, K., T. Ogura, K. Murata, T. Suzaki, H. Niki, and S. Hiraga. 1994. Characterization of translucent segments observed in an *smbA* mutant of *Escherichia coli. FEMS Microbiol. Lett.* **116**:61–66.

4722. Yamanaka, K., T. Ogura, H. Niki, and S. Hiraga. 1992. Identification and characterization of the *smbA* gene, a suppressor of the *mukB* null mutant of *Escherichia coli. J. Bacteriol.* **174**:7517–7526.

4723. Yamao, F., H. Inokuchi, A. Cheung, H. Ozeki, and D. Soll. 1982. *Escherichia coli* glutaminyl-tRNA synthetase. I. Isolation and DNA sequence of the *glnS* gene. *J. Biol. Chem.* **257**:11639–11643.

4724. Yamato, I. 1990. Study of the major proline transport system in *Escherichia coli. Seikagaku* **62**:1020–1030.

4725. Yamato, I., and Y. Anraku. 1980. Genetic and biochemical studies of transport systems for branched-chain amino acids in *Escherichia coli* K-12: isolation and properties of mutants defective in leucine-repressible transport activities. *J. Bacteriol.* **144**:36–44.

4726. Yamato, I., and Y. Anraku. 1988. Site-specific alteration of cysteine 281, cysteine 344, and cysteine 349 in the proline carrier of *Escherichia coli. J. Biol. Chem.* **263**:16055–16057.

4727. Yamato, I., M. Kotani, Y. Oka, and Y. Anraku. 1994. Site-specific alteration of arginine 376, the unique positively charged amino acid residue in the mid-membrane-spanning regions of the proline carrier of *Escherichia coli. J. Biol. Chem.* **269**:5720–5724.

4728. Yang, C., D. Carlow, R. Wolfenden, and S. A. Short. 1992. Cloning and nucleotide sequence of the *Escherichia coli* cytidine deaminase (*ccd*) gene. *Biochemistry* **31**:4168–4174.

4729. Yang, J., H. Camakaris, and A. J. Pittard. 1993. Mutations in the *tyrR* gene of *Escherichia coli* which affect TyrR-mediated activation but not TyrR-mediated repression. *J. Bacteriol.* **175**:6372–6375.

4730. Yang, J., S. Ganesan, J. Sarsero, and A. J. Pittard. 1993. A genetic analysis of various functions of the TyrR protein of *Escherichia coli. J. Bacteriol.* **175**:1767–1776.

4731. Yang, J., and J. Pittard. 1987. Molecular analysis of the regulatory region of the *Escherichia coli* K-12 *tyrB* gene. *J. Bacteriol.* **169**:4710–4715.

4732. Yang, S. Y. 1991. Location of the *fadBA* operon on the physical map of *Escherichia coli. J. Bacteriol.* **173**:7405–7406.

4733. Yang, S. Y., X. Y. He Yang, G. Healy-Louie, H. Schulz, and M. Elzinga. 1991. Nucleotide sequence of the *fadA* gene. Primary structure of 3-ketoacyl-coenzyme A thiolase from *Escherichia coli* and the structural organization of the *fadAB* operon. *J. Biol. Chem.* **266**:16255.

4734. Yang, S. Y., J. M. Li, X. Y. He, S. D. Cosloy, and H. Schulz. 1988. Evidence that the *fadB* gene of the *fadAB* operon of *Escherichia coli* encodes 3-hydroxyacyl-coenzyme A (CoA) epimerase, Δ3-cis-Δ2- trans-enoyl-CoA isomerase, and enoyl-CoA hydratase in addition to 3-hydroxyacyl-CoA dehydrogenase. *J. Bacteriol.* **170**:2543–2548.

4735. Yang, S. Y., and H. Schulz. 1983. The large subunit of the fatty acid oxidation complex from *Escherichia coli* is a multifunctional polypeptide. Evidence for the existence of a fatty acid oxidation operon (*fad AB*) in *Escherichia coli. J. Biol. Chem.* **258**:9780–9785.

4736. Yang, S. Y., X. Y. Yang, G. Healy-Louie, H. Schulz, and M. Elzinga. 1990. Nucleotide sequence of the *fadA* gene. Primary structure of 3-ketoacyl-coenzyme A thiolase from *Escherichia coli* and the structural organization of the *fadAB* operon. *J. Biol. Chem.* **265**:10424–10429.

4737. Yang, T. P., and R. E. Depew. 1992. Physical map of the *tolC-htrP* region of the *Escherichia coli* chromosome. *J. Bacteriol.* **174**:1700–1701.

4738. Yang, T. P., and R. E. Depew. 1992. Nucleotide sequence of a region duplicated in *Escherichia coli toc* mutants. *Biochim. Biophys. Acta* **1130**:227–228.

4739. Yang, W., L. Ni, and R. L. Somerville. 1993. A stationary-phase protein of *Escherichia coli* that affects the mode of association between the *trp* repressor protein and operator-bearing DNA. *Proc. Natl. Acad. Sci. USA* **90**:5796–5800.

4740. Yang, X. Y., H. Schulz, M. Elzinga, and S. Y. Yang. 1991. Nucleotide sequence of the promoter and *fadB* gene of the *fadBA* operon and primary structure of the multifunctional fatty acid oxidation protein from *Escherichia coli. Biochemistry* **30**:6788–6795.

4741. Yang, Y., and M. Inouye. 1991. Intermolecular complementation between two defective mutant signal-transducing receptors of *Escherichia coli. Proc. Natl. Acad. Sci. USA* **88**:11057–11061.

4742. Yang, Y. L., and B. Polisky. 1993. Suppression of ColE1 high-copy-number mutants by mutations in the *polA* gene of *Escherichia coli. J. Bacteriol.* **175**:428–437.

4743. Yano, R., H. Nagai, K. Shiba, and T. Yura. 1990. A mutation that enhances synthesis of sigma 32 and suppresses temperature-sensitive growth of the *rpoH15* mutant of *Escherichia coli. J. Bacteriol.* **172**:2124–2130.

4744. Yano, T., S. Kuramitsu, S. Tanase, Y. Morino, K. Hiromi, and H. Kagamiyama. 1991. The role of His143 in the catalytic mechanism of *Escherichia coli* aspartate aminotransferase. *J. Biol. Chem.* **266**:6079–6085.

4745. Yanofsky, C., and V. Horn. 1994. Role of regulatory features of the *trp* operon of *Escherichia coli* in mediating a response to a nutritional shift. *J. Bacteriol.* **176**:6245–6254.

4746. Yanofsky, C., T. Platt, I. P. Crawford, B. P. Nichols, G. E. Christie, H. Horowitz, M. VanCleemput, and A. M. Wu. 1981. The complete nucleotide sequence of the tryptophan operon of *Escherichia coli. Nucleic Acids Res.* **9**:6647–6668.

4747. Yao, Z., and M. A. Valvano. 1994. Genetic analysis of the O-specific lipopolysaccharide biosynthesis region (*rfb*) of *Escherichia coli* K-12 W3110: identification of genes that confer group 6 specificity to *Shigella flexneri* serotypes Y and 4a. *J. Bacteriol.* **176**:4133–4143.

4748. Yazyu, H., S. Shiota-Niiya, T. Shimamoto, H. Kanazawa, M. Futai, and T. Tsuchiya. 1984. Nucleotide sequence of the *melB* gene and characteristics of deduced amino acid sequence of the melibiose carrier in *Escherichia coli. J. Biol. Chem.* **259**:4320–4326.

4749. Ye, Q. Z., J. Liu, and C. T. Walsh. 1990. *p*-Aminobenzoate synthesis in *Escherichia coli*: purification and characterization of PabB as aminodeoxychorismate synthase and enzyme X as aminodeoxychorismate lyase. *Proc. Natl. Acad. Sci. USA* **87**:9391–9395.

4750. Ye, S. Z., and T. J. Larson. 1988. Structures of the promoter and operator of the *glpD* gene encoding aerobic sn-glycerol-3-phosphate dehydrogenase of *Escherichia coli* K-12. *J. Bacteriol.* **170**:4209–4215.

4751. Yerkes, J. H., L. P. Casson, A. K. Honkanen, and G. C. Walker. 1984. Anaerobiosis induces expression of *ant*, a new *Escherichia coli* locus with a role in anaerobic electron transport. *J. Bacteriol.* **158**:180–186.

4752. Yerushalmi, H., M. Lebendiker, and S. Schuldiner. 1995. EmrE, an *Escherichia coli* 12-kDa multidrug transporter, exchanges toxic cations and H⁺ and is soluble in organic solvents. *J. Biol. Chem.* **270**:6856–6863.

4753. Yeung, T., D. A. Mullin, K. S. Chen, E. A. Craig, J. C. Bardwell, and J. R. Walker. 1990. Sequence and expression of the *Escherichia coli recR* locus. *J. Bacteriol.* **172**:6042–6047.

4754. Yi, Q. M., and J. Lutkenhaus. 1985. The nucleotide sequence of the essential cell-division gene *ftsZ* of *Escherichia coli. Gene* **36**:241–247.

4755. Yi, Q. M., S. Rockenbach, J. E. Ward, Jr., and J. Lutkenhaus. 1985. Structure and expression of the cell division genes *ftsQ, ftsA* and *ftsZ. J. Mol. Biol.* **184**:399–412.

4756. Yim, H. H., R. L. Brems, and M. Villarejo. 1994. Molecular characterization of the promoter of *osmY*, an *rpoS*-dependent gene. *J. Bacteriol.* **176**:100–107.

4757. Yim, H. H., and M. Villarejo. 1992. *osmY*, a new hyperosmotically inducible gene, encodes a periplasmic protein in *Escherichia coli. J. Bacteriol.* **174**:3637–3644.

4758. Yin, K. C., A. Blinkowa, and J. R. Walker. 1986. Nucleotide sequence of the *Escherichia coli* replication gene *dnaZX. Nucleic Acids Res.* **14**:6541–6549.

4759. Yoakum, G. H., and L. Grossman. 1981. Identification of *E. coli uvrC* protein. *Nature* (London) **292**:171–173.

4760. Yokota, T., H. Sugisaki, M. Takanami, and Y. Kaziro. 1980. The nucleotide sequence of the cloned *tufA* gene of *Escherichia coli. Gene* **12**:25–31.

4761. Yoo, S. J., J. H. Seol, M. S. Kang, D. B. Ha, and C. H. Chung. 1994. *clpX* encoding an alternative ATP-binding subunit of protease Ti (Clp) can be expressed independently from *clpP* in *Escherichia coli. Biochem. Biophys. Res. Commun.* **203**:798–804.

4762. Yorgey, P., and R. Kolter. 1993. A widely conserved developmental sensor in bacteria? *Trends Genet.* **9**:374–375.

4763. York, M. K., and M. Stodolsky. 1981. Characterization of P1argF derivatives from *Escherichia coli* K12 transduction. I. IS1 elements flank the *argF* gene segment. *Mol. Gen. Genet.* **181**:230–240.

4764. Yoshida, H., T. Kojima, J. Yamagishi, and S. Nakamura. 1988. Quinolone-resistant mutations of the *gyrA* gene of *Escherichia coli. Mol. Gen. Genet.* **211**:1–7.

4765. Yoshida, T., C. Ueguchi, and T. Mizuno. 1993. Physical map location of a set of *Escherichia coli* genes (*hde*) whose expression is affected by the nucleoid protein H-NS. *J. Bacteriol.* **175**:7747–7748.

4766. Yoshida, T., C. Ueguchi, H. Yamada, and T. Mizuno. 1993. Function of the *Escherichia coli* nucleoid protein, H-NS: molecular analysis of a subset of proteins whose expression is enhanced in a *hns* deletion mutant. *Mol. Gen. Genet.* **237**:113–122.

4767. Yoshikawa, A., S. Isono, A. Sheback, and K. Isono. 1987. Cloning and nucleotide sequencing of the genes *rimI* and *rimJ* which encode enzymes acetylating ribosomal proteins S18 and S5 of *Escherichia coli* K12. *Mol. Gen. Genet.* **209**:481–488.

4768. Yoshikawa, H., and N. Ogasawara. 1991. Structure and function of DnaA and the DnaA-box in eubacteria: evolutionary relationships of bacterial replication origins. *Mol. Microbiol.* **5**:2589–2597.

4769. Yoshikawa, M., A. Okuyama, and N. Tanaka. 1975. A third kasugamycin resistance locus, *ksgC*, affecting ribosomal protein S2 in *Escherichia coli* K-12. *J. Bacteriol.* **122**:796–797.

4770. Yoshimoto, T., H. Higashi, A. Kanatani, X. S. Lin, H. Nagai, H. Oyama, K. Kurazono, and D. Tsuru. 1991. Cloning and sequencing of the 7 alpha-hydroxysteroid dehydrogenase gene from *Escherichia coli* HB101 and characterization of the expressed enzyme. *J. Bacteriol.* **173**:2173–2179.

4771. **Yoshimoto, T., H. Nagai, K. Ito, and D. Tsuru.** 1993. Location of the 7 alpha-hydroxysteroid dehydrogenase gene (*hdhA*) on the physical map of the *Escherichia coli* chromosome. *J. Bacteriol.* **175**:5730.

4772. **Yoshimoto, T., H. Tone, T. Honda, K. Osatomi, R. Kobayashi, and D. Tsuru.** 1989. Sequencing and high expression of aminopeptidase P gene from *Escherichia coli* HB101. *J. Biochem.* **105**:412–416.

4773. **Yoshimura, M., H. Inokuchi, and H. Ozeki.** 1984. Identification of transfer RNA suppressors in *Escherichia coli.* IV. Amber suppressor Su^{+6}, a double mutant of a new species of leucine tRNA. *J. Mol. Biol.* **177**:627–644.

4774. **Yoshimura, M., M. Kimura, M. Ohno, H. Inokuchi, and H. Ozeki.** 1984. Identification of transfer RNA suppressors in *Escherichia coli.* III. Ochre suppressors of lysine tRNA. *J. Mol. Biol.* **177**:609–625.

4775. **Yoshimura, T., M. Ashiuchi, N. Esaki, C. Kobatake, S. Y. Choi, and K. Soda.** 1993. Expression of *glr* (*murI, dga*) gene encoding glutamate racemase in *Escherichia coli. J. Biol. Chem.* **268**:24242–24246.

4776. **Young, I. G., B. L. Rogers, H. D. Campbell, A. Jaworowski, and D. C. Shaw.** 1981. Nucleotide sequence coding for the respiratory NADH dehydrogenase of *Escherichia coli.* UUG initiation codon. *Eur. J. Biochem.* **116**:165–170.

4776a. **Young, I. G., L. M. McCann, P. Stroobant, and F. Gibson.** 1971. Characterization and genetic analysis of mutant strains of *Escherichia coli* K-12 accumulating the ubiquinone precursors 2-octaprenyl-6-methoxy-1,4-benzoquinone and 2-octaprenyl-3-methyl-6-methoxy-1,4-benzoquinone. *J. Bacteriol.* **105**:769–778.

4777. **Young, R. A.** 1979. Transcription termination in the *Escherichia coli* ribosomal RNA operon *rrnC. J. Biol. Chem.* **254**:12725–12731.

4778. **Yu, F., S. Inouye, and M. Inouye.** 1986. Lipoprotein-28, a cytoplasmic membrane lipoprotein from *Escherichia coli.* Cloning, DNA sequence, and expression of its gene. *J. Biol. Chem.* **261**:2284–2288.

4779. **Yu, F., H. Yamada, K. Daishima, and S. Mizushima.** 1984. Nucleotide sequence of the *lspA* gene, the structural gene for lipoprotein signal peptidase of *Escherichia coli. FEBS Lett.* **173**:264–268.

4780. **Yu, H., M. J. Schurr, and V. Deretic.** 1995. Functional equivalence of *Escherichia coli* sigma E and *Pseudomonas aeruginosa* AlgU: *E. coli rpoE* restores mucoidy and reduces sensitivity to reactive oxygen intermediates in *algU* mutants of *P. aeruginosa. J. Bacteriol.* **177**:3259–3268.

4781. **Yuan, H. S., S. E. Finkel, J. A. Feng, M. Kaczor-Grzeskowiak, R. C. Johnson, and R. E. Dickerson.** 1991. The molecular structure of wild-type and a mutant Fis protein: relationship between mutational changes and recombinational enhancer function or DNA binding. *Proc. Natl. Acad. Sci. USA* **88**:9558–9562.

4782. **Yuasa, S., and Y. Sakakibara.** 1980. Identification of the *dnaA* and *dnaN* gene products of *Escherichia coli. Mol. Gen. Genet.* **180**:267–273.

4783. **Yudkin, M. D.** 1977. Unstable mutations that relieve catabolite repression of tryptophanase synthesis by *Escherichia coli. J. Bacteriol.* **130**:57–61.

4784. **Yung, B. Y., and A. Kornberg.** 1988. Membrane attachment activates *dnaA* protein, the initiation protein of chromosome replication in *Escherichia coli. Proc. Natl. Acad. Sci. USA* **85**:7202–7205.

4785. **Yung, B. Y., and A. Kornberg.** 1989. The *dnaA* initiator protein binds separate domains in the replication origin of *Escherichia coli. J. Biol. Chem.* **264**:6146–6150.

4786. **Yura, T., and K. Igarashi.** 1968. RNA polymerase mutants of *Escherichia coli.* I. Mutants resistant to streptovaricin. *Proc. Natl. Acad. Sci. USA* **61**:1313–1319.

4787. **Yura, T., H. Mori, H. Nagai, T. Nagata, A. Ishihama, N. Fujita, K. Isono, K. Mizobuchi, and A. Nakata.** 1992. Systematic sequencing of the *Escherichia coli* genome: analysis of the 0–2.4 min region. *Nucleic Acids Res.* **20**:3305–3308.

4788. **Yura, T., T. Tobe, K. Ito, and T. Osawa.** 1984. Heat shock regulatory gene (*htpR*) of *Escherichia coli* is required for growth at high temperature but is dispensable at low temperature. *Proc. Natl. Acad. Sci. USA* **81**:6803–6807.

4789. **Zacharias, M., H. U. Goringer, and R. Wagner.** 1989. Influence of the GCGC discriminator motif introduced into the ribosomal RNA $P2^{-}$ and *tac* promoter on growth-rate control and stringent sensitivity. *EMBO J.* **8**:3357–3363.

4790. **Zacharias, M., H. U. Goringer, and R. Wagner.** 1990. The signal for growth rate control and stringent sensitivity in *E. coli* is not restricted to a particular sequence motif within the promoter region. *Nucleic Acids Res.* **18**:6271–6275.

4791. **Zacharias, M., H. U. Goringer, and R. Wagner.** 1992. Analysis of the Fis-dependent and Fis-independent transcription activation mechanisms of the *Escherichia coli* ribosomal RNA P1 promoter. *Biochemistry* **31**:2621–2628.

4792. **Zakin, M. M., N. Duchange, P. Ferrara, and G. N. Cohen.** 1983. Nucleotide sequence of the *metL* gene of *Escherichia coli.* Its product, the bifunctional aspartokinase II-homoserine dehydrogenase II, and the bifunctional product of the *thrA* gene, aspartokinase I-homoserine dehydrogenase I, derive from a common ancestor. *J. Biol. Chem.* **258**:3028–3031.

4793. **Zakin, M. M., R. C. Greene, A. Dautry-Varsat, N. Duchange, P. Ferrara, M. C. Py, D. Margarita, and G. N. Cohen.** 1982. Construction and physical mapping of plasmids containing the *metJBLF* gene cluster of *E. coli* K12. *Mol. Gen. Genet.* **187**:101–106.

4794. **Zalkin, H.** 1983. Structure, function, and regulation of amidophosphoribosyltransferase from prokaryotes. *Adv. Enzyme Regul.* **21**:225–237.

4795. **Zalkin, H., P. Argos, S. V. Narayana, A. A. Tiedeman, and J. M. Smith.** 1985. Identification of a *trpG*-related glutamine amide transfer domain in *Escherichia coli* GMP synthetase. *J. Biol. Chem.* **260**:3350–3354.

4796. **Zambrano, M. M., and R. Kolter.** 1993. *Escherichia coli* mutants lacking NADH dehydrogenase I have a competitive disadvantage in stationary phase. *J. Bacteriol.* **175**:5642–5647.

4797. **Zaniewski, R., and M. P. Deutscher.** 1982. Genetic mapping of mutation in *Escherichia coli* leading to a temperature-sensitive RNase D. *Mol. Gen. Genet.* **185**:142–147.

4798. **Zavitz, K. H., R. J. DiGate, and K. J. Marians.** 1991. The *priB* and *priC* replication proteins of *Escherichia coli.* Genes, DNA sequence, overexpression, and purification. *J. Biol. Chem.* **266**:13988–13995.

4799. **Zavitz, K. H., and K. J. Marians.** 1992. ATPase-deficient mutants of the *Escherichia coli* DNA replication protein PriA are capable of catalyzing the assembly of active primosomes. *J. Biol. Chem.* **267**:6933–6940.

4800. **Zawadzke, L. E., T. D. Bugg, and C. T. Walsh.** 1991. Existence of two D-alanine:D-alanine ligases in *Escherichia coli:* cloning and sequencing of the *ddlA* gene and purification and characterization of the DdlA and DdlB enzymes. *Biochemistry* **30**:1673–1682.

4801. **Zehnbauer, B. A., E. C. Foley, G. W. Henderson, and A. Markovitz.** 1981. Identification and purification of the Lon^{+} ($capR^{+}$) gene product, a DNA-binding protein. *Proc. Natl. Acad. Sci. USA* **78**:2043–2047.

4802. **Zehnbauer, B. A., and A. Markovitz.** 1980. Cloning of gene *lon* (*capR*) of *Escherichia coli* K-12 and identification of polypeptides specified by the cloned deoxyribonucleic acid fragment. *J. Bacteriol.* **143**:852–863.

4803. **Zeilstra-Ryalls, J., O. Fayet, and C. Georgopoulos.** 1991. The universally conserved GroE (Hsp60) chaperonins. *Annu. Rev. Microbiol.* **45**:301–325.

4804. **Zelwer, C., J. L. Risler, and S. Brunie.** 1982. Crystal structure of *Escherichia coli* methionyl-tRNA synthetase at 2.5 A resolution. *J. Mol. Biol.* **155**:63–81.

4805. **Zengel, J. M., R. H. Archer, and L. Lindahl.** 1984. The nucleotide sequence of the *Escherichia coli fus* gene, coding for elongation factor G. *Nucleic Acids Res.* **12**:2181–2192.

4806. **Zengel, J. M., and L. Lindahl.** 1982. A secondary promoter for elongation factor Tu synthesis in the *str* ribosomal protein operon of *Escherichia coli. Mol. Gen. Genet.* **185**:487–492.

4807. **Zengel, J. M., and L. Lindahl.** 1990. Mapping of two promoters for elongation factor Tu within the structural gene for elongation factor G. *Biochim. Biophys. Acta* **1050**:317–322.

4808. **Zengel, J. M., D. Mueckl, and L. Lindahl.** 1980. Protein L4 of the *E. coli* ribosome regulates an eleven gene r protein operon. *Cell* **21**:523–535.

4809. **Zhang, A., and M. Belfort.** 1992. Nucleotide sequence of a newly-identified *Escherichia coli* gene, *stpA,* encoding an H-NS-like protein. *Nucleic Acids Res.* **20**:6735.

4810. **Zhang, J., and M. P. Deutscher.** 1992. A uridine-rich sequence required for translation of prokaryotic mRNA. *Proc. Natl. Acad. Sci. USA* **89**:2605–2609.

4811. **Zhang, J. R., and M. P. Deutscher.** 1988. Cloning, characterization, and effects of overexpression of the *Escherichia coli rnd* gene encoding RNase D. *J. Bacteriol.* **170**:522–527.

4812. **Zhang, J. R., and M. P. Deutscher.** 1988. *Escherichia coli* RNase D: sequencing of the *rnd* structural gene and purification of the overexpressed protein. *Nucleic Acids Res.* **16**:6265–6278.

4813. **Zhang, S. P., G. Zubay, and E. Goldman.** 1991. Low-usage codons in *Escherichia coli,* yeast, fruit fly and primates. *Gene* **105**:61–72.

4814. **Zhang, Y., J. E. Craig, and M. P. Gallagher.** 1992. Location of the *nupC* gene on the physical map of *Escherichia coli* K-12. *J. Bacteriol.* **174**:5758–5759.

4815. **Zhang, Y., and M. M. Hanna.** 1994. NusA changes the conformation of *Escherichia coli* RNA polymerase at the binding site for the 3′ end of the nascent RNA. *J. Bacteriol.* **176**:1787–1789.

4816. **Zhao, G., A. J. Pease, N. Bharani, and M. E. Winkler.** 1995. Biochemical characterization of *gapB*-encoded erythrose 4-phosphate dehydrogenase of *Escherichia coli* K-12 and its possible role in pyridoxal 5′-phosphate biosynthesis. *J. Bacteriol.* **177**:2804–2812.

4817. **Zhao, G., and M. E. Winkler.** 1994. An *Escherichia coli* K-12 *tktA tktB* mutant deficient in transketolase activity requires pyridoxine (vitamin B_6) as well as the aromatic amino acids and vitamins for growth. *J. Bacteriol.* **176**:6134–6138.

4818. **Zhao, G. P., and R. L. Somerville.** 1992. Genetic and biochemical characterization of the *trpB8* mutation of *Escherichia coli* tryptophan synthase. An amino acid switch at the sharp turn of the trypsin-sensitive "hinge" region diminishes substrate binding and alters solubility. *J. Biol. Chem.* **267**:526–541.

4819. **Zhao, S., C. H. Sandt, G. Feulner, D. A. Vlazny, J. A. Gray, and C. W. Hill.** 1993. Rhs elements of *Escherichia coli* K-12: complex composites of shared and unique components that have different evolutionary histories. *J. Bacteriol.* **175**:2799–2808.

4820. **Zhao, X. J., and K. McEntee.** 1990. DNA sequence analysis of the *recA* genes from *Proteus vulgaris, Erwinia carotovora, Shigella flexneri* and *Escherichia coli* B/r. *Mol. Gen. Genet.* **222**:369–376.

4821. **Zheng, L., and H. D. Braymer.** 1991. Overproduction and purification of McrC protein from *Escherichia coli* K-12. *J. Bacteriol.* **173**:3918–3920. (Erratum, **173**:5933.)

4822. **Zheng, L., X. Wang, and H. D. Braymer.** 1992. Purification and N-terminal amino acid sequences of two polypeptides encoded by the *mcrB* gene from *Escherichia coli* K-12. *Gene* **112**:97–100.

4823. **Zhou, Y. N., W. A. Walter, and C. A. Gross.** 1992. A mutant sigma 32 with a small deletion in conserved region 3 of sigma has reduced affinity for core RNA polymerase. *J. Bacteriol.* **174**:5005–5012.

4824. **Zhou, Z., and M. Syvanen.** 1990. Identification and sequence of the *drpA* gene from *Escherichia coli. J. Bacteriol.* **172**:281–286.

4825. **Zhu, Y., and E. C. Lin.** 1988. A mutant *crp* allele that differentially activates the operons of the *fuc* regulon in *Escherichia coli. J. Bacteriol.* **170**:2352–2358.

4826. **Ziemke, P., and J. E. McCarthy.** 1992. The control of mRNA stability in *Escherichia coli:* manipulation of the degradation pathway of the polycistronic atp mRNA. *Biochim. Biophys. Acta* **1130**:297–306.

4827. Zijderveld, C. A., M. E. Aarsman, and N. Nanninga. 1995. Differences between inner membrane and peptidoglycan-associated PBP1B dimers of *Escherichia coli*. *J. Bacteriol.* 1771:1860–1863.

4828. Zilhao, R., L. Camelo, and C. M. Arraiano. 1993. DNA sequencing and expression of the gene *rnb* encoding *Escherichia coli* ribonuclease II. *Mol. Microbiol.* 8:43–51.

4828a. Zimmerman, R. A., Y. Ikeya, and P. F. Sparling. 1973. Alteration of ribosomal protein S4 by mutation linked to kasugamycin-resistance in *Escherichia coli*. *Proc. Natl. Acad. Sci. USA* 70:71–75.

4829. Zimmermann, L., K. Hantke, and V. Braun. 1984. Exogenous induction of the iron dicitrate transport system of *Escherichia coli* K-12. *J. Bacteriol.* 159:271–277.

4830. Zinkewich-Peotti, K., and J. M. Fraser. 1988. New locus for exopolysaccharide overproduction in *Escherichia coli* K-12. *J. Bacteriol.* 170:1405–1407.

4831. Zinoni, F., A. Birkmann, W. Leinfelder, and A. Bock. 1987. Cotranslational insertion of selenocysteine into formate dehydrogenase from *Escherichia coli* directed by a UGA codon. *Proc. Natl. Acad. Sci. USA* 84:3156–3160.

4832. Zinoni, F., A. Birkmann, T. C. Stadtman, and A. Bock. 1986. Nucleotide sequence and expression of the selenocysteine-containing polypeptide of formate dehydrogenase (formate-hydrogen-lyase-linked) from *Escherichia coli*. *Proc. Natl. Acad. Sci. USA* 83:4650–4654.

4833. Zurawski, G., R. P. Gunsalus, K. D. Brown, and C. Yanofsky. 1981. Structure and regulation of *aroH*, the structural gene for the tryptophan-repressible 3-deoxy-D-arabino-heptulosonic acid-7-phosphate synthetase of *Escherichia coli*. *J. Mol. Biol.* 145:47–73.

4834. Zurawski, G., and S. M. Zurawski. 1985. Structure of the *Escherichia coli* S10 ribosomal protein operon. *Nucleic Acids Res.* 13:4521–4526.

4835. Zwaig, N., R. Nagel de Zwaig, T. Isturiz, and M. Wecksler. 1973. Regulatory mutations affecting the gluconate system in *Escherichia coli*. *J. Bacteriol.* 114:469–473.

4836. Zwiebel, L. J., M. Inukai, K. Nakamura, and M. Inouye. 1981. Preferential selection of deletion mutations of the outer membrane lipoprotein gene of *Escherichia coli* by globomycin. *J. Bacteriol.* 145:654–656.

4837. Zyskind, J. W., and D. W. Smith. 1980. Nucleotide sequence of the *Salmonella typhimurium* origin of DNA replication. *Proc. Natl. Acad. Sci. USA* 77:2460–2464.

4838. Zyskind, J. W., and D. W. Smith. 1992. DNA replication, the bacterial cell cycle, and cell growth. *Cell* 69:5–8.

The Genetic Map of *Salmonella typhimurium,* Edition VIII

KENNETH E. SANDERSON, ANDREW HESSEL, SHU-LIN LIU, AND
KENNETH E. RUDD

110

INTRODUCTION

We present a somewhat modified version of edition VIII of the genetic map of *Salmonella typhimurium* (official designation, *Salmonella enterica* serovar Typhimurium) strain LT2, which was published in 1995 (1717). We list a total of 1,160 genes; 1,081 of these have been located on the circular chromosome, and 29 of these are on pSLT, the 90-kb plasmid which is almost invariably found in LT2 lines; the remaining 50 genes are not yet mapped. The first edition of the map in 1965 contained 133 genes; later editions were published in 1967, 1970, 1972, 1978, and 1983, and edition VII, in 1988, listed a total of 750 genes (1721). A slightly modified version of edition VI of the map, presented originally in 1983, was published in the first edition of the present volumes (1718).

Strain LT2 is one of 25 different *S. typhimurium* isolates established by Lilleengen (1205) and designated LT1 to LT25; these strains represent 25 different phage sensitivity phenotypes. Many of these LT strains were used by Zinder and Lederberg for a study of genetic exchange, leading to the discovery of phage-mediated transduction (2236). M. Demerec obtained the LT strains from Zinder and began an exhaustive series of studies in the mid-1950s. The most important materials used were mutant strains derived from strain LT2, though a few were derived from strain LT7. Since the 1950s, strain LT2 has been the major focus of genetic and biochemical analysis, and thus the map we present here is derived primarily from work with LT2. However, more recently other strains have been used for specific reasons: e.g., some strains of *S. typhimurium* have been studied because they have higher levels of virulence than strain LT2; strain LT2 does not utilize histidine as sole carbon source, so other wild-type strains which do so have been used to study the *hut* operon.

All genes presently known to us are listed in Table 1 (p. 1907), with formerly used or alternative symbols in Table 2 (p. 1939). The references which describe these genes include a selection of

the references from the first six editions which were presented in the first version of the present volumes (1718), those which were in edition VII published in 1988 (1721), and those which have been published since 1988 and which were included in the edition VIII published earlier (1717).

THE GENETIC MAP

The coordinate system used in the first four editions of the linkage map of *S. typhimurium* was determined by F-mediated conjugation; Hfr strains were used in interrupted conjugation experiments to place individual genes and P22-transduction linkage groups on a 138-min time-of-entry linkage map. In edition V, the map was changed to 100 units to correspond to the 100-min linkage map of *Escherichia coli* K-12. This was done to emphasize the similarity of the two genera and to facilitate comparisons. This 100-unit system was also used in editions VI and VII. In those editions, the 100 units of the map were based on "phage lengths" of the transducing phage P22 (1721). P22 can normally encapsulate about 45 kb of DNA; this is approximately 1% of the *Salmonella* chromosome. The map described here is the first edition in which we present a physical map of the chromosome with the interval between genes based directly on the length of the DNA. We report these gene intervals as centisomes (Cs), each one of which represents 1% of the length of DNA in the chromosome, and we also present the DNA segments in kilobase pairs (kb).

This physical map of the chromosome is based on a low-resolution genomic cleavage map determined by digestion with rare-cutting endonucleases and separation of the resulting fragments with pulsed-field gel electrophoresis (PFGE). Preliminary cleavage maps used the enzymes *Xba*I (1224), *Bln*I (1739, 2163), and I-*Ceu*I (hereafter called *Ceu*I) (1223). The overall structure of the chromosome shown in Fig. 1 (p. 1957) is based on a summary of work with these three enzymes (1222), plus some additional unpublished work.

The genomic cleavage map shows the positions of 24 *Xba*I fragments, 12 *Bln*I fragments, and 7 *Ceu*I fragments on a circular molecule of 4,808 kb of DNA (1222). The positions of genes on this cleavage map were determined by three methods. (i) Strains with insertions of the transposon Tn*10* into known genes were analyzed. The presence of *Xba*I and *Bln*I sites in Tn*10* permitted the location of these insertions, and thus of the gene into which Tn*10* is inserted, through digestion of the DNA of the strain by the enzyme, followed by separation of the fragments by PFGE. A total of 109 independent strains with Tn*10* insertions were reported, but some of these were not in genes of known function (1222); together with unpublished work done recently, a total of 75 genes have been located on the physical map solely through analysis of Tn*10*-containing strains. (ii) A total of 12 genes could be placed on the map because their DNA sequence included a site for the endonucleases *Xba*I, *Bln*I, or *Ceu*I. (iii) In some situations, restriction mapping data or data from nucleotide sequences were used to correct or refine the locations of genes on the map. For example, a high-resolution restriction map for a 240-kb region spanning the 91 to 96 Cs region of the map located the positions of several genes (2166); these data were used to modify the locations of some genes in this interval.

The physically mapped Tn*10* insertions and genes (discussed above) were used to align genes whose positions were known only by linkage data, and also were used as "anchors" for DNA sequence (described in detail below).

The chromosome is a closed circle of DNA, and early editions of the linkage map were shown in this way. In editions from V to VII the linkage map was displayed not as a circle but as 10 linear 10-min intervals. The present edition is similar, but because the number of known genes has increased, it is shown as 20 linear 5-Cs intervals (Fig. 1). This figure integrates gene location information discovered by genetic linkage and by physical analyses. The position of any gene shown was determined using the data available in the following preference order: first, by the order of genes found in sequenced DNA segments; second, by Tn*10* or other physical anchors; third, by genetic linkage data; and, last, by extrapolation based on data from *E. coli* K-12 (see below). For display purposes, when gene orientation conflicts arose between physical and genetic data, preference was given to gene linkage information.

Most of the elements shown on the map are structural genes for proteins or, in a smaller number of cases, for rRNA or tRNA. We have not shown control elements for operons such as promoters, mRNA leaders, and terminators, nor have we shown the position of unnamed or poorly characterized chromosomal genes. We have not shown chromosomal elements such as repetitive extragenic palindromic (REP) sequences (771, 1423), but an exception is that we have indicated the positions of the six IS*200* sequences which are mapped on the chromosome of LT2 (1134, 1136, 1724).

NOMENCLATURE

We use the system of nomenclature for genes which was established by Demerec et al. (437). This system uses a three-letter designation for the gene or operon, e.g., *his* for mutations in the genes for histidine synthesis, followed by a capital letter, also in italics, designating the specific gene in that operon, e.g., *hisD* for a mutation in the gene for histidinol dehydrogenase. This system

has become the de facto standard for bacterial genetics. Authors considering a three-letter designation for a new gene in *S. typhimurium* should check the published maps of *S. typhimurium* and *E. coli* to see whether the proposed name has been used previously; they are also encouraged to contact the *Salmonella* Genetic Stock Center (SGSC) to see whether a proposed name has been used previously but not published. Allele number assignments should also be obtained from the SGSC. It is important to clear such numbers with the SGSC so that each mutation is identified by a unique allele number. A mutation is defined by its three-letter designation plus the allele number, i.e., the mutation *his-1* might be initially thought to be *hisD1*, based on its perceived enzyme defect, but is later changed to *hisE1*. Thus, there should be only one *his-1* mutation. Using assigned numbers is especially important for transposon insertions named by the "z– –" sytem proposed by Hong and Ames (824). In this case, the allele number may be the only identifier for a mutation, since all mutations mapped with "z– –" have a single series of allele numbers. As mapping is refined, a mutation's "z– –" designation may change, but it retains its original allele number. It is therefore important that assigned allele numbers be used to avoid confusion of different laboratories using the same number for different insertions. It is also vital that strains be identified by a unique strain designation, which includes two or three capital letters (assigned for *Salmonella* strains by the SGSC to the laboratory, or by the *E. coli* Genetic Stock Center [CGSC] for *E. coli* strains) plus a number. The expanding use of the computer to keep records demands the use of correct strain designations; suffixes and phenotype designations after the strain designation, to refer to derivatives of a parent strain, are often not accepted by programs used for cataloging strains.

When possible, the same name should be used for homologous genes in related species such as *E. coli* and *Salmonella* spp. Many changes have been made in naming genes in these organisms to bring them into correspondence, and each edition of the maps has seen such changes. For example, a system of naming the genes for flagellar synthesis and function was proposed which recognizes the homology of these genera (891) and this system, now widely accepted, is used here. We encourage the trend to the use of these standardized names for *E. coli* and *S. typhimurium* for genes in other bacterial species as well when functional data and amino acid sequence data indicate that the genes are homologous.

PLASMID pSLT IN *S. TYPHIMURIUM* LT2

Because the plasmid content of strains of a species of bacteria is often extremely variable, the plasmids are not normally considered part of the genome; therefore, neither the F-factor in some lines of *E. coli* K-12 nor several types of plasmids frequently found in some wild-type *Salmonella* strains are treated as components of the normal genome. However, the original line of *S. typhimurium* LT2 contains a specific plasmid which we consider part of the genome. This plasmid has been called the virulence plasmid, the 90-kb plasmid, the 60-MDa plasmid, the cryptic plasmid, and pSLT (standing for *salmonella LT*); we use the last designation.

For two reasons, we consider this plasmid to be part of the normal genome of *S. typhimurium* LT2. The first is that the plasmid is an almost invariable part of the genome of LT2 lines. It is carried by all the lines of LT2 which we have tested, except for those few from which it has been intentionally eliminated. This is true even though LT2 has been in culture for many years and has been subjected to innumerable single-colony isolations. Genes on the plasmid which regulate replication, incompatibility, and partitioning (164, 242, 296) enable the plasmid to be maintained stably at low copy number. In addition, pSLT or a closely related plasmid is commonly found in independent *S. typhimurium* isolates from nature.

The second reason for considering pSLT to be part of the genome is that genes on the plasmid influence the phenotype in several ways, and thus mutations of genes on the plasmid may be mistaken for mutations of chromosomal genes. This influence of pSLT on the phenotype was first noted by Smith et al. (1837), who noted that it encoded Fin+ (fertility inhibition) properties, reducing the fertility of Hfr strains of *S. typhimurium;* special measures have been developed which restore fertility up to the level found in *E. coli* K-12 (1720). Many other genes have been detected due to their effect on phenotype. For example, mutations in the *traT* gene of pSLT cause an increase in outer membrane permeability, leading to an antibiotic-supersensitivity phenotype; this gene is listed in Table 1 (p. 1907). Other genes on pSLT, many of which influence the virulence of the strain, are also listed in Table 1 and are shown on a restriction map of pSLT in Fig. 1 (p. 1957).

SEQUENCED GENES OF *S. TYPHIMURIUM*

Gene identification can be largely based on open reading frames, codon usage distributions, and sequence-based evidence of homology (198, 1661). A significant amount of genomic DNA sequence has become available for *S. typhimurium.* By the end of 1994, we could assemble StySeq1, a nonredundant DNA sequence database modeled after the EcoSeq collection of *E. coli* genomic DNA sequences (1690). In fact, the extensive genomic sequence data available for *E. coli* (see, e.g., references 181 and 1853) cover close to two-thirds of the genome. Therefore, many *S. typhimurium* genes were identified at the sequence level as partial, but convincing, matches to one of the *E. coli* protein sequences in the EcoGene subset of SWISS-PROT (101).

StySeq1 has 197 contiguous sequence blocks (contigs) that do not overlap. Together, the nonredundant chromosomal genomic *S. typhimurium* StySeq1 DNA sequence collection is 548,508 bp in length. This represents 11.4% of the *Salmonella* chromosome, estimated to be 4,808 kb in length (1222). Eight contigs are greater than 10,000 bp in length. The longest contig, hisGstyM (Table 3 [p. 1940]), is 33,958 bp in length.

Of the 197 DNA sequences in this collection (Table 3), 191 are ordered and oriented as they might be on the *S. typhimurium* chromosome. This was accomplished in several ways: physical anchor, genetic pin, and extrapolation.

(i) In the physical anchor method, if a Tn*10* insertion, rare restriction site, or other direct physical link could be associated with one of the genes from StySeq, it was anchored and the positions of the other nearby genes were calculated based on a single base-pair anchor point per contig. Gene positions thus have high local precision (1-bp resolution) and coarse genomic map position accuracy, based on analysis by PFGE (ranging from 1,000 bp in some situations to over 10,000 bp, depending on the locations of restriction sites for the enzymes used). Unsequenced genes could also be anchored, and this information was taken into account in mapping the unsequenced genes depicted

in Fig. 1. Sixty-two StySeq1 contigs were positioned in this way. (ii) In the genetic pin method, gene positions were assigned to a number of sequenced and unsequenced genes on the basis of linkage data. The positions of the anchors were used to realign conjugation and transduction mapping distances, allowing the integration of physical and genetic maps presented in Fig. 1. In turn, those genetic map positions were used to pin more sequences to the integrated genomic map. An additional 93 contigs in Table 3 were positioned by this approach. (iii) The extrapolation method was used to position a final 36 contigs by using map position information from *E. coli-Salmonella* gene pairs to extrapolate a position.

The aligned DNA sequences are ordered to coincide with the orientations of homologous sequences in the EcoSeq7 collection of aligned and oriented *E. coli* genes. Although some local inversions are likely to have occurred, most comparisons indicate a close correlation of the orientations and map positions of genes between *E. coli* and *S. typhimurium,* with one major exception. There is a large inversion of a chromosomal segment relative to the corresponding *E. coli* region with endpoints near 26 and 36 min on the genetic map (1721). We could localize this region on the physical map (Fig. 1), and this allowed us to set the orientation of genes in this region of *S. typhimurium* as being opposite to that of the *E. coli* homologs in this inverted segment.

Alignment of all but six of the StySeq1 contigs, at least as a best approximation subject to constant refinement, allowed the genes encoded in the DNA sequences to also be aligned. In this way, 523 protein-coding genes and 15 structural RNA genes were given genomic map positions based locally on DNA sequence information and globally on a combination of physical anchor points, genetic cross data, and extrapolation from the *E. coli* map. Fifty-one unnamed genes (described as *orf* or similar in GenBank entries) were given provisional names beginning with the letter "y" (1690); these genes were excluded from Table 1 but are listed in Table 3.

Until a larger proportion of the *S. typhimurium* genome is sequenced, the genetic linkage data will remain useful as the basis for the map. *E. coli* has a high-resolution genomic restriction map for the entire chromosome, whereas the high-resolution map of *S. typhimurium* is being assembled as parts (2166). Nonetheless, several genes were physically positioned by DNA sequence and high-resolution restriction map data in the 91 to 96 Cs region of the *S. typhimurium* LT2 chromosome (2166) by using a single anchor.

Having access to both *E. coli* and *S. typhimurium* DNA sequence over the same region provides an opportunity to predict frameshift corrections and helps ensure the highest level of accuracy. Other organisms help identify probable new genes in *S. typhimurium* with homologs as well, but the comparison to *E. coli* involves enough preference for third-position changes to establish the correct frame for most protein-coding genes.

MAP REFINEMENT

If errors are detected in this work, we hope that this information will be transmitted to us to allow corrections. To assist in the preparation of future map editions or updates, we would appreciate receiving any information regarding the position and function of new genes, or refinements to the positions or roles of the genes described here. Electronic mail should be addressed to kesander@acs.ucalgary.ca and rudd@ncbi.nlm.nih.gov. StySeq1 and associated files can be retrieved in electronic form from the anonymous ftp site ncbi.nlm.nih.gov in the /repository/Eco/Sty directory. Diskettes will be mailed on request to those without Internet access.

ACKNOWLEDGMENTS

The assistance of numerous investigators in providing strains used for Tn*10* mapping and in furnishing unpublished information is gratefully acknowledged. We are especially grateful to John R. Roth for providing invaluable information and discussions as well as editorial comments. We also acknowledge the excellent editorial comments provided by Charles G. Miller. We thank K. K. Wong and Michael McClelland for their significant contribution to the genomic mapping of *S. typhimurium*. We also thank Lauryl M. J. Nutter for assistance with the final preparation of this manuscript. K.E.R. gives special thanks to Amos Bairoch for providing SWISS-PROT *Salmonella* sequences and many helpful discussions, to Jinghui Zhang for alignment of sequences to EcoSeq, and to Webb Miller for many valuable conversations and algorithm development. During the preparation of this report, K.E.S. was supported by an operating grant and an infrastructure grant from the Natural Sciences and Engineering Research Council, and by grant RO1 AI34829 from the National Institute of Allergy and Infectious Diseases.

LITERATURE CITED

1. Abd-el-al, A., and J. L. Ingraham. 1969. Cold sensitivity and other phenotypes resulting from mutation in *pyrA* gene. *J. Biol. Chem.* **244:**4039–4045.
2. Abdelal, A. T. H., E. Griego, and J. L. Ingraham. 1976. Arginine-sensitive phenotype of mutations in *pyrA Salmonella typhimurium*: role of ornithine carbamyl transferase in the assembly of mutant carbamyl phosphate synthetase. *J. Bacteriol.* **128:**105–113.
3. Abdelal, A. T. H., E. Griego, and J. L. Ingraham. 1978. Arginine auxotrophic phenotype of mutations in *pyrA* of *Salmonella typhimurium*: role of *N*-acetylornithine in the maturation of mutant carbamylphosphate synthetase. *J. Bacteriol.* **134:**528–536.
4. Abdelal, A. T. H., and J. L. Ingraham. 1975. Carbamylphosphate synthetase from *Salmonella typhimurium*. *J. Biol. Chem.* **250:**4410–4417.
5. Abdelal, A. T. H., E. H. Kennedy, and O. Nainan. 1977. Ornithine transcarbamylase from *Salmonella typhimurium*: purification, kinetic analysis, and immunological cross-reactivity. *J. Bacteriol.* **129:**1387–1396.
6. Abdulkarim, F., T. M. Tuohy, R. H. Buckingham, and D. Hughes. 1991. Missense substitutions lethal to essential functions of EF-Tu. *Biochimie* **73:**1457–1464.
7. Abouhamad, W. N., M. Manson, M. M. Gibson, and C. F. Higgins. 1991. Peptide transport and chemotaxis in *Escherichia coli* and *Salmonella typhimurium*: characterization of the dipeptide permease (Dpp) and the dipeptide-binding protein. *Mol. Microbiol.* **5:**1035–1047.
8. Abshire, K. Z., and F. C. Neidhardt. 1993. Analysis of proteins synthesized by *Salmonella typhimurium* during growth within a host macrophage. *J. Bacteriol.* **175:**3734–3743.
9. Abshire, K. Z., and F. C. Neidhardt. 1993. Growth rate paradox of *Salmonella typhimurium* within host macrophages. *J. Bacteriol.* **175:**3744–3748.
10. Aceves-Pina, E., M. V. Ortega, and M. Artis. 1974. Linkage of the *Salmonella typhimurium* chromosomal loci encoding for the cytochrome-linked L-α-glycerophosphate dehydrogenase and amylomaltase activities. *Arch. Microbiol.* **101:**59–70.
11. Adams, D. E., E. M. Shekhtman, E. L. Zechiedrich, M. B. Schmid, and N. R. Cozzarelli. 1992. The role of topoisomerase IV in partitioning bacterial replicons and the structure of catenated intermediates in DNA replication. *Cell* **71:**277–288.
12. Ahmed, N., and R. J. Rowbury. 1971. A temperature-sensitive cell division component in a mutant of *Salmonella typhimurium*. *J. Gen. Microbiol.* **67:**107–115.
13. Ahmed, N., and R. J. Rowbury. 1974. Studies of septation and separation in a mutant of *Salmonella typhimurium*. *Z. Allg. Mikrobiol.* **14:**455–463.
14. Ahmed, S. A., H. Kawasaki, R. Bauerle, H. Morita, and E. W. Miles. 1988. Site-directed mutagenesis of the α subunit of tryptophan synthase from *Salmonella typhimurium*. *Biochem. Biophys. Res. Commun.* **151:**672–678.
15. Ailion, M., T. A. Bobik, and J. R. Roth. 1993. Two global regulatory systems (Crp and Arc) control the cobalamin/propanediol regulon of *Salmonella typhimurium*. *J. Bacteriol.* **175:**7200–7208.
16. Aizawa, S.-I., G. E. Dean, C. J. Jones, R. M. Macnab, and S. Yamaguchi. 1985. Purification and characterization of the flagellar hook-basal body complex of *Salmonella typhimurium*. *J. Bacteriol.* **161:**836–849.
17. Aizawa, S.-I., S. Kato, S. Asakura, H. Kagawa, and S. Yamaguchi. 1980. In-vitro polymerization of polyhook protein from *Salmonella* SJW 880. *Biochim. Biophys. Acta* **625:**291–303.
18. Aksamit, R. R., and D. E. Koshland, Jr. 1974. Identification of the ribose binding protein as the receptor for ribose chemotaxis in *Salmonella typhimurium*. *Biochemistry* **13:**4473–4478.
19. Alami, N., and P. C. Hallenbeck. 1992. Mutations that affect the regulation of *phs* in *Salmonella typhimurium*. *J. Gen. Microbiol.* **138:**1117–1122.
20. Alexander, R. R., and J. M. Calvo. 1969. A *Salmonella typhimurium* locus involved in the regulation of isoleucine, valine, and leucine biosynthesis. *Genetics* **61:**539–556.
21. Alexander, R. R., J. M. Calvo, and M. Freundlich. 1971. Mutants of *Salmonella typhimurium* with an altered leucyl-transfer ribonucleic acid synthetase. *J. Bacteriol.* **106:**213–220.
22. Aliabadi, Z., Y. K. Park, J. L. Slonczewski, and J. W. Foster. 1988. Novel regulatory loci controlling oxygen- and pH-regulated gene expression in *Salmonella typhimurium*. *J. Bacteriol.* **170:**842–851.
23. Aliabadi, Z., F. Warren, S. Mya, and J. W. Foster. 1986. Oxygen-regulated stimulons of *Salmonella typhimurium* identified by Mu d (AP*lac*) operon fusions. *J. Bacteriol.* **165:**780–786.
24. Alifano, P., M. S. Ciampi, A. G. Nappo, C. B. Bruni, and M. S. Carlomagno. 1988. *In vivo* analysis of the mechanisms responsible for strong transcriptional polarity in a "sense" mutant within an intercistronic region. *Cell* **55:**351–360.
25. Alifano, P., C. Piscitelli, V. Blasi, F. Rivellini, A. G. Nappo, C. B. Bruni, and M. S. Carlomagno. 1992. Processing of a polycistronic mRNA requires a 5′ cis element and active translation. *Mol. Microbiol.* **6:**787–798.
26. Alifano, P., F. Rivellini, D. Limauro, C. B. Bruni, and M. S. Carlomagno. 1991. A consensus motif common to all Rho-dependent prokaryotic transcription terminators. *Cell* **64:**553–563.
27. Allen, J. D., and S. M. Parsons. 1979. Nitrocellulose filter binding: quantitation of the histidyl-tRNA-ATP phosphoribosyltransferase complex. *Anal. Biochem.* **92:**22–30.
28. Allen, S. W., A. Senti-Willis, and S. R. Maloy. 1993. DNA sequence of the *putA* gene from *Salmonella typhimurium*: a bifunctional membrane-associated dehydrogenase that binds DNA. *Nucleic Acids Res.* **21:**1676.
29. Aloj, S., C. B. Bruni, H. Edelhoch, and M. M. Rechler. 1973. Physical studies comparing a genetically fused enzyme of the histidine operon with its component enzymes. *J. Biol. Chem.* **248:**5880–5886.
30. Alper, M. D., and B. N. Ames. 1975. Cyclic 3′,5′-adenosine monophosphate phosphodiesterase mutants of *Salmonella typhimurium*. *J. Bacteriol.* **122:**1081–1090.
31. Alper, M. D., and B. N. Ames. 1975. Positive selection of mutants with deletions of the *gal-chl* region of the *Salmonella* chromosome as a screening procedure of mutagens that cause deletions. *J. Bacteriol.* **121:**259–266.
32. Alper, M. D., and B. N. Ames. 1978. Transport of antibiotics and metabolite analogs by systems under cyclic AMP control: positive selection of *Salmonella typhimurium cya* and *crp* mutants. *J. Bacteriol.* **133:**149–157.
33. Alpuche Aranda, C. M., J. A. Swanson, W. P. Loomis, and S. I. Miller. 1992. *Salmonella typhimurium* activates virulence gene transcription within acidified macrophage phagosomes. *Proc. Natl. Acad. Sci. USA* **89:**10079–10083.
34. Altmeyer, R. M., J. K. McNern, J. C. Bossio, I. Rosenshine, B. B. Finlay, and J. E. Galan. 1993. Cloning and molecular characterization of a gene involved in *Salmonella* adherence and invasion of cultured epithelial cells. *Mol. Microbiol.* **7:**89–98.
35. Alvarez-Jacobs, J., M. de la Garza, and M. V. Ortega. 1986. Biochemical and genetic characterization of L-glutamate transport and utilization in *Salmonella typhimurium* LT-2 mutants. *Biochem. Genet.* **24:**195–205.
36. Al-Zarban, S., L. Heffernan, J. Nishitani, L. Ransome, and G. Wilcox. 1984. Positive control of the L-rhamnose genetic system in *Salmonella typhimurium* LT2. *J. Bacteriol.* **158:**603–608.
37. Ames, B. N., and P. E. Hartman. 1963. The histidine operon. *Cold Spring Harbor Symp. Quant. Biol.* **24:**349–356.
38. Ames, B. N., T. H. Tsang, M. Buck, and M. F. Christman. 1983. The leader mRNA of the histidine attenuator region resembles tRNA[His]: possible general regulatory implications. *Proc. Natl. Acad. Sci. USA* **80:**5240–5242.
39. Ames, G. F. 1973. Resolution of bacterial proteins by polyacrylamide gel electrophoresis on slabs. *J. Biol. Chem.* **249:**634–644.
40. Ames, G. F., and J. Lever. 1970. Components of histidine transport: histidine-binding proteins and hisP protein. *Proc. Natl. Acad. Sci. USA* **66:**1096–1103.
41. Ames, G. F., and J. E. Lever. 1972. The histidine-binding protein J is a component of histidine transport. Identification of its structural gene, *hisJ*. *J. Biol. Chem.* **247:**4309–4316.
42. Ames, G. F., and J. R. Roth. 1968. Histidine and aromatic permeases of *Salmonella typhimurium*. *J. Bacteriol.* **96:**1742–1749.
43. Ames, G. F., and E. N. Spudich. 1976. Protein-protein interaction in transport: periplasmic histidine-binding protein J interacts with P protein. *Proc. Natl. Acad. Sci. USA* **73:**1877–1881.
44. Ames, G. F., E. N. Spudich, and H. Nikaido. 1974. Protein composition of the outer membrane of *Salmonella typhimurium*: effect of lipopolysaccharide mutations. *J. Bacteriol.* **117:**406–416.
45. Ames, G. F.-L., D. P. Biek, and E. N. Spudich. 1978. Duplications of histidine transport genes in *Salmonella typhimurium* and their use for the selection of deletion mutants. *J. Bacteriol.* **136:**1094–1108.
46. Ames, G. F.-L., and K. Nikaido. 1978. Identification of a membrane protein as a histidine transport component in *Salmonella typhimurium*. *Proc. Natl. Acad. Sci. USA* **75:**5447–5451.
47. Ames, G. F.-L., and K. Nikaido. 1985. Nitrogen regulation in *Salmonella typhimurium*. Identification of an ntrC protein-binding site and definition of a consensus binding sequence. *EMBO J.* **4:**539–547.

(Continued on page 1961)

TABLE 1 Genes of *S. typhimurium*

Gene symbol	Mnemonic	Former or alternative symbol; enzyme deficiency or other phenotype[a]	Cs[b]	References[c]
accA		Acetyl-CoA carboxylase	6.0	1139, 1717
aceA	Acetate	Growth on acetate or fatty acids; isocitrate lyase (EC 4.1.3.1)	91.3	1717, 2128
aceB	Acetate	Growth on acetate or fatty acids; malate synthase (EC 4.1.3.2)	91.3	1717, 2128
aceE	Acetate	Acetate requirement; pyruvate dehydrogenase (pyruvate:cytochrome b_1 oxidoreductase) (EC 1.2.2.2)	3.7	1140, 1717
aceF	Acetate	Acetate requirement; pyruvate dehydrogenase (pyruvate lipoate oxidoreductase) (EC 1.2.4.1)	3.7	1140, 1717
aceK	Acetate	Isocitrate dehydrogenase kinase/phosphatase	91.3	1717
aciA	Acid-inducible	pH-regulated gene; acid inducible	99.5	562, 1717
aciB	Acid-inducible	pH-regulated gene; acid inducible	92.4	562, 1717
ack	Acetate kinase	Acetate kinase (ATP:acetate phosphotransferase) (EC 2.7.2.1)	50.9	1130, 1183, 1717, 2073
ada		Inducible DNA repair system protecting against methylating and alkylating agents; O^6-methylguanine DNA methyltransferase	50.1	720, 1717, 2078, 2181
adhE		Reduced survival in macrophages; CoA-linked acetaldehyde dehydrogenase and alcohol dehydrogenase	38.4	129, 1717
adk		Adenylate kinase; sensitivity to glycine betaine in high-osmolality media (EC 2.7.4.3)	11.6	707, 1717
ahpC	Alkyl hydroperoxide	Alkyl hydroperoxide reductase, C22 subunit	14.1	522, 1227, 1717, 1930, 1931
ahpF	Alkyl hydroperoxide	Alkyl hydroperoxide reductase, F52a subunit	14.1	522, 1717, 1930, 1931
alaS	Alanine	Alanine tRNA synthetase	62.0	1717
alkB	Alkylation	DNA repair system specific for alkylated DNA	50.1	720, 1717
alr	Alanine racemase	Biosynthetic alanine racemase (EC 5.1.1.1)	92.2	512, 520, 598, 599, 1717
amiA		*N*-Acetylmuramyl-L-alanine amidase activity; putative	53.2	1717, 2178
amiB		*N*-Acetylmuramyl-L-alanine amidase (EC 3.5.1.28)	95.0	1291, 1717
amk		AMP kinase	9.1	1717, 1721
ampC	Ampicillin	β-Lactamase; penicillin resistance (EC 3.5.2.6)	94.5	35, 1717
ampD	Ampicillin	β-Lactamase regulation; putative signalling protein	3.6	861, 1717
amtA		Resistance to 40 mM 3-amino-1,2,4-triazole in the presence of histidine	37.6	1717, 1718, 1721
amyA	Amylase	Cytoplasmic α-amylase (EC 3.2.1.1)	42.6	1008, 1607, 1717
ana		Anaerobic gas production	38.7	777, 1717
aniB	Anaerobically inducible	Induced by anaerobiosis; does not reduce benzyl viologen	93.9	23, 1717
aniC	Anaerobically inducible	Induced by anaerobiosis; does not reduce benzyl viologen	93.9	23, 1717
aniD	Anaerobically inducible	Induced by anaerobiosis; does not reduce benzyl viologen	92.4	23, 1717
aniF	Anaerobically inducible	Induced by anaerobiosis; does not reduce benzyl viologen	68.0	23, 1717
aniG	Anaerobically inducible	Induced by anaerobiosis; does not reduce benzyl viologen	67.8	23, 1717
aniH	Anaerobically inducible	Induced by anaerobiosis; does not reduce benzyl viologen	82.6	23, 1717
aniI	Anaerobically inducible	Induced by anaerobiosis; does not reduce benzyl viologen	40.1	23, 1717
ansB		L-Asparaginase II; regulated by CRP and anaerobiosis	67.1	942, 1717
apbA		Alternative pyrimidine biosynthetic pathway; synthesis of thiamine in presence of exogenous purines	10.0	465, 1717
apeB	Acyl peptide esterase	Acyl amino acid esterase; hydrolyzes *N*-acetyl-L-phenylalanine-β-naphthyl ester	11.0	749, 1717
apeE	Acyl peptide esterase	Membrane-bound acyl amino acid esterase; hydrolyzes *N*-acetyl-L-phenylalanine-β-naphthyl ester	13.5	353, 1717, 1718, 1721
apeR	Acyl peptide esterase	*apeD*; regulatory gene for *apeE*	85.0	1717, 1718, 1721
aphA		Nonspecific acid phosphatase II	NM	1717, 2043
apt		Adenine phosphoribosyltransferase	11.5	1717, 1718, 1721
araA	Arabinose	L-Arabinose isomerase (EC 5.3.1.4)	2.4	168, 831, 1162, 1163, 1208, 1209, 1596, 1717
araB	Arabinose	Ribulokinase (EC 2.7.1.16)	2.5	167, 311, 831, 1163, 1208, 1596, 1717
araC	Arabinose	Regulatory gene for arabinose catabolic enzymes	2.5	169, 345, 346, 831, 1149, 1161–1163, 1208, 1596, 1717
araD	Arabinose	L-Ribulose-phosphate 4-epimerase (EC 5.1.3.4)	2.4	831, 1163, 1207, 1208, 1595, 1596, 1695, 1717
arcA		Cytoplasmic DNA-binding component system for repression of genes during anaerobiosis	100.0	15, 56, 1717
arcB		Membrane-bound sensor of two-component system for repression of genes during anaerobiosis	72.4	15, 56, 1717
argA	Arginine	*argB*; amino acid acetyltransferase (EC 2.3.1.1)	65.1	976, 1419, 1717, 1835, 2083
argB	Arginine	*argC*; *N*-acetyl-γ-glutamate kinase (EC 2.7.2.8)	89.6	1717, 2083
argC	Arginine	*argH*; *N*-acetyl-γ-glutamyl phosphate reductase (EC 1.2.1.38)	89.6	1717, 2083

(Table continues)

TABLE 1 *Continued*

Gene symbol	Mnemonic	Former or alternative symbol; enzyme deficiency or other phenotype[a]	Cs[b]	References[c]
argD	Arginine	argG; acetylornithine aminotransferase (EC 2.6.1.11)	75.1	154, 1106, 1717, 2083
argE	Arginine	argA; acetylornithine deacetylase (EC 3.5.1.16)	89.6	134, 187, 1717, 2083
argG	Arginine	argE; argininosuccinate synthetase (EC 6.3.4.5)	72.0	1106, 1717, 2083
argH	Arginine	argF; argininosuccinate lyase (EC 4.3.2.1)	89.6	577, 1717, 2083
argI	Arginine	Ornithine carbamoyltransferase (EC 2.1.3.3)	97.2	5, 264, 1013, 1016, 1717, 1978
argP	Arginine	Arginine transport	NM	1600, 1717
argQ	Arginine	Arginine tRNA$_2$	62.0	173, 1717
argR	Arginine	L-Arginine regulation	73.4	615, 1015, 1019, 1020, 1240, 1242, 1717
argS	Arginine	Arginyl-tRNA synthetase (EC 6.1.1.19)	41.6	1539, 1717
argT	Arginine	Lysine-arginine-ornithine-binding protein	51.3	47, 73, 74, 762, 763, 1113, 1160, 1717, 1755, 1756, 1901
argU	Arginine	fimU; arginine tRNA$_1$ for rare codons (AGA and AGG); essential for fimbria expression	15.2	1717, 1975
argV	Arginine	Arginine tRNA$_2$	62.0	173, 1717
argX	Arginine	argT, argU, use; arginine tRNA$_1$	85.6	201, 208, 264, 1241, 1717
argY	Arginine	Arginine tRNA$_2$	62.0	173, 1717
argZ	Arginine	Arginine tRNA$_2$	62.0	173, 1717
aroA	Aromatic	3-Enolpyruvylshikimate 5-phosphate synthetase	19.9	310, 322, 482, 655, 806–808, 1437, 1442, 1717, 1741, 1826, 1882
aroB	Aromatic	5-Dehydroquinate synthetase	75.7	655, 1437, 1717
aroC	Aromatic	aroD; chorismate synthetase	52.0	655, 800, 1087, 1437, 1438, 1717, 1945, 1949
aroD	Aromatic	aroE; 5′-dehydroquinate dehydratase (EC 4.2.1.10)	30.5	64, 655, 1437, 1717, 1776, 1846
aroE	Aromatic	aroC; 5-dehydroshikimate reductase	74.4	655, 1437, 1717
aroF	Aromatic	Tyrosine-repressible DAHP synthetase	58.6	432, 1384, 1717, 1875
aroG	Aromatic	Phenylalanine-repressible DAHP synthetase	17.3	1717, 1875, 1933
aroH	Aromatic	Tryptophan-repressible DAHP synthetase	30.4	849, 1717
aroP	Aromatic	Aromatic amino acid transport	3.6	42, 868, 1717
aroT	Aromatic	Transport of tryptophan, phenylalanine, and tyrosine	39.3	1717, 2013, 2014
asd		Aspartate semialdehyde dehydrogenase (EC 1.2.1.11)	77.4	364, 605, 1717
asm		Unable to assimilate low levels of ammonia; deficient in glutamate synthase and glutamine synthase	70.5	445, 592, 1717
asn	Asparagine	Asparagine synthesis	82.4	957, 1717, 2240, 2241
asnU	Asparagine	Asparagine tRNA	44.1	1717
aspA	Aspartate	L-Aspartate ammonia-lyase (EC 4.3.1.1)	94.2	35, 1717
aspC	Aspartate	Aspartate aminotransferase (EC 2.6.1.1)	22.7	1717, 2072
asrA		Anaerobic sulfite reductase	57.1	850, 851, 1717
asrB		Anaerobic sulfite reductase	57.1	850, 851, 1717
asrC		Anaerobic sulfite reductase	57.1	850, 851, 1717
ataA	Attachment	attP22 I; attachment site for prophage P22	7.8	100, 828, 948, 949, 1132, 1287, 1612, 1648, 1717, 2194
atbA	Attachment	attP27 I; attachment site for prophage P27	12.7	100, 1212, 1717
atbB	Attachment	attP27 II; second attachment site for prophage P27	8.1	100, 1717
atcA	Attachment	attP221; attachment site for prophage P221	25.9	1717, 2209
atdA	Attachment	attP14; attachment site for prophage P14 in group C Salmonella spp.	61.1	100, 1717
atp		unc, uncA; membrane-bound (Mg^{2+}, Ca^{2+}) ATPase	84.6	552, 553, 613, 957, 1717, 1883, 2240, 2241
atrB	Acid tolerance response	Defective in pre- but not post-acid shock-induced acid tolerance	32.2	548, 550, 552, 1717
atrD	Acid tolerance response	Defective in both pre- and post-acid shock-induced acid tolerance	68.1	548, 550, 1717
atrF	Acid tolerance response	Defective in both pre- and post-acid shock-induced acid tolerance	13.7	548, 550, 1717
atrG	Acid tolerance response	Defective in both pre- and post-acid shock-induced acid tolerance	NM	550, 1717
atrR	Acid tolerance response	atbR; constitutive acid tolerance; trans-acting regulator of atrB	89.5	550, 1717
ats		Arylsulfatase	NM	1396, 1717, 2183
att15	Attachment	Attachment site for phage e15 in group E Salmonella spp.	46.3	1050, 1309, 1717
att34	Attachment	Attachment site for phage e34 in group E Salmonella spp.	5.7	1309, 1717
attN	Attachment	Attachment site for prophage N in S. montevideo	52.0	411, 1717
avtA		Alanine-valine transaminase, transaminase C (EC 2.6.1.66)	80.2	157, 1717, 2130
aziA	Azide	Resistant to 3 mM sodium azide on L-methionine	3.4	341, 342, 536, 1717
bfp	Bundle-forming pili	Interbacterial linkage, adhesion to epithelial cells	NM	1717, 1854
bioA	Biotin	bio; 7,8-diaminopelargonic acid synthetase	18.1	238, 1717, 1797, 1933

TABLE 1 *Continued*

Gene symbol	Mnemonic	Former or alternative symbol; enzyme deficiency or other phenotype[a]	Cs[b]	References[c]
bioB	Biotin	Biotin synthetase	18.1	1717, 1797
bioC	Biotin	Block prior to pimeloyl-CoA	18.2	1717, 1797
bioD	Biotin	Dethiobiotin synthetase	18.2	1717, 1797
bioF	Biotin	7-Keto-8-aminopelargonic acid synthetase	18.1	1717, 1797
birA	Biotin repressor	Biotin-[acetyl-CoA carboxylase] holoenzyme synthetase	90.0	458, 1717
brnQ		*ilvT;* branched-chain amino acid transport	8.9	1045, 1047, 1307, 1460, 1463, 1717
btuB	B₁₂ utilization	*bfe;* transport of cobalamins	89.7	706, 1372, 1373, 1649, 1650, 1717, 2114
btuC	B₁₂ utilization	Transport of cobalamins	30.4	1717, 1846
btuD	B₁₂ utilization	Transport of cobalamins	30.4	1717
btuE	B₁₂ utilization	Transport of cobalamins	30.4	1717
cadA	Cadaverine	Lysine decarboxylase (EC 4.1.1.18)	56.2	562, 614, 1717
cadC	Cadaverine	Regulation of *cadA*	56.2	562, 1717
capS	Capsule	Capsular polysaccharide synthesis	31.1	1717, 1730
carA		*argD, ars, cap, pyrA;* carbamyl phosphate synthase, glutamine (light) subunit (EC 2.7.2.9)	1.6	1–4, 264, 304, 311, 399, 728, 910, 944, 1023, 1035, 1242, 1243, 1419, 1459, 1717, 1838, 1978, 2197
carB		*pyrA;* carbamyl phosphate synthase, ammonia (heavy) subunit (EC 2.7.2.9)	1.7	1–4, 264, 304, 311, 399, 728, 910, 944, 1023, 1035, 1242, 1243, 1419, 1459, 1717, 1838, 1978, 2197
cbiA		*cobI;* synthesis of vitamin B₁₂ adenosyl cobinamide precursor	43.7	1681, 1717
cbiB		*cobI;* synthesis of vitamin B₁₂ adenosyl cobinamide precursor	43.7	1681, 1717
cbiC		*cobI;* synthesis of vitamin B₁₂ adenosyl cobinamide precursor	43.7	1681, 1717
cbiD		*cobI;* synthesis of vitamin B₁₂ adenosyl cobinamide precursor	43.7	1681, 1717
cbiE		*cobI;* synthesis of vitamin B₁₂ adenosyl cobinamide precursor	43.7	1681, 1717
cbiF		*cobI;* synthesis of vitamin B₁₂ adenosyl cobinamide precursor	43.7	1681, 1717
cbiG		*cobI;* synthesis of vitamin B₁₂ adenosyl cobinamide precursor	43.6	1681, 1717
cbiH		*cobI;* synthesis of vitamin B₁₂ adenosyl cobinamide precursor	43.6	1681, 1717
cbiJ		*cobI;* synthesis of vitamin B₁₂ adenosyl cobinamide precursor	43.6	1681, 1717
cbiK		*cobI;* synthesis of vitamin B₁₂ adenosyl cobinamide precursor	43.6	1681, 1717
cbiL		*cobI;* synthesis of vitamin B₁₂ adenosyl cobinamide precursor	43.6	1681, 1717
cbiM		*cobI;* synthesis of vitamin B₁₂ adenosyl cobinamide precursor	43.6	1681, 1717
cbiN		*cobI;* synthesis of vitamin B₁₂ adenosyl cobinamide precursor	43.5	1681, 1717
cbiO		*cobI;* synthesis of vitamin B₁₂ adenosyl cobinamide precursor	43.5	1681, 1717
cbiP		*cobI;* synthesis of vitamin B₁₂ adenosyl cobinamide precursor	43.5	1681, 1717
cbiQ		*cobI;* synthesis of vitamin B₁₂ adenosyl cobinamide precursor	43.5	1681, 1717
cbiT		*cobI;* synthesis of vitamin B₁₂ adenosyl cobinamide precursor	43.7	1681, 1717
cdd		Cytidine deaminase (EC 3.5.4.5)	47.5	133, 930, 1016, 1717
cheA	Chemotaxis	*cheP;* chemotaxis; ATP-dependent kinase; phosphorylates CheB and CheY	42.0	84–86, 354, 403, 429, 471, 1079, 1118, 1141, 1251, 1260–1263, 1265, 1352, 1717, 1876, 1909, 1916, 1919, 2001, 2002, 2069, 2102, 2176
cheB	Chemotaxis	*cheX;* chemotaxis; bifunctional monomeric protein; C-terminal γ-carboxyl methylesterase and N-terminal transferase	41.9	196, 197, 407, 427–430, 608, 1028, 1118, 1148, 1251, 1616, 1717, 1807, 1808, 1839, 1916, 1917, 1919
cheR	Chemotaxis	Chemotaxis; *S*-adenosylmethionine-dependent methyltransferase	41.9	84–86, 196, 197, 347, 407, 427–430, 1079, 1118, 1148, 1251, 1717, 1873, 1876, 1917–1919, 1951, 2001, 2002, 2102
cheS	Chemotaxis	Chemotaxis	NM	85, 1148, 1717, 2102
cheW	Chemotaxis	Chemotaxis	42.0	85, 354, 429, 1118, 1148, 1717, 1804, 1911, 1919, 2102

(Table continues)

TABLE 1 *Continued*

Gene symbol	Mnemonic	Former or alternative symbol; enzyme deficiency or other phenotype[a]	Cs[b]	References[c]
cheY	Chemotaxis	*cheQ;* chemotaxis; cytoplasmic protein; interacts at flagellar motor to reverse flagellar rotation	41.8	80, 84–86, 314, 354, 403, 427–430, 471, 1028, 1079, 1118, 1148, 1249–1251, 1260–1263, 1265, 1271, 1352, 1717, 1839, 1852, 1876, 1910, 1916, 1919, 2001, 2002, 2069, 2102
cheZ	Chemotaxis	*cheT;* chemotaxis; stimulates phospho-CheY dephosphorylation	41.8	85, 427–430, 471, 608, 1148, 1249, 1251, 1616, 1717, 1839, 1852, 1912, 1915, 1918, 1919, 2002, 2069, 2102
chlF	Chlorate	Resistance; may be part of *moe* operon	19.1	505, 1717, 1718, 1721, 1784, 1904, 1934
chlG	Chlorate	Resistance; affects nitrate reductase, tetrathionate reductase, chlorate reductase, and hydrogen lyase	56.8	1717, 1904, 1935
cil	Citrate lyase	Mutants unable to utilize citrate in anaerobic conditions without additional carbon sources	66.9	1105, 1717
citA	Citrate	Citrate carrier, different from TctI, TctII, TctIII transport systems	17.0	1717, 1795
citB	Citrate	Citrate carrier, different from TctI, TctII, TctIII transport systems	17.0	1717, 1795
cld	Chain length determinant	*rol;* regulation of chain length in the O-units of the LPS	44.9	123, 124, 1717
clmA	Conditional-lethal mutation	Heat- or cold-sensitive mutation	86.4	1717, 1750
clmB	Conditional-lethal mutation	Heat- or cold-sensitive mutation	8.8	1717, 1750
clmC	Conditional-lethal mutation	Heat- or cold-sensitive mutation	72.1	1717, 1750
clmD	Conditional-lethal mutation	Heat- or cold-sensitive mutation	85.5	1717, 1750
clmE	Conditional-lethal mutation	Heat- or cold-sensitive mutation	74.5	1717, 1750
clmG	Conditional-lethal mutation	Heat- or cold-sensitive mutation	12.3	1717, 1721, 1750
clmI	Conditional-lethal mutation	Heat- or cold-sensitive mutation	6.5	1717, 1750
clmJ	Conditional-lethal mutation	Heat- or cold-sensitive mutation	32.4	1717, 1750
clmK	Conditional-lethal mutation	Heat- or cold-sensitive mutation	62.4	1717, 1750
clmM	Conditional-lethal mutation	Heat- or cold-sensitive mutation	99.1	1717, 1750
clmN	Conditional-lethal mutation	Heat- or cold-sensitive mutation	46.9	1717, 1750
clmP	Conditional-lethal mutation	Heat- or cold-sensitive mutation	81.1	1717, 1750
clmQ	Conditional-lethal mutation	Heat- or cold-sensitive mutation	58.1	1717, 1750
clmR	Conditional-lethal mutation	Heat- or cold-sensitive mutation	26.3	1717, 1750
clmS	Conditional-lethal mutation	Heat- or cold-sensitive mutation	95.8	1717, 1750
clmT	Conditional-lethal mutation	Heat- or cold-sensitive mutation	51.0	1717, 1750
cmk		Cytidylate kinase (EC 2.7.4.14)	20.0	136, 1717
coaA	Coenzyme A	CoA synthesis; pantothenate kinase	90.3	15, 480, 1717
cobA	Cobalamin	*cobIV;* ATP:corrinoid adenosyltransferase	38.0	515, 965, 1507, 1717, 1953, 1954
cobB	Cobalamin	DMB ribose phosphate phosphatase	20.6	1717, 2030
cobC	Cobalamin	DMB ribose phosphate phosphatase	15.6	669, 1498, 1509, 1717
cobD	Cobalamin	Synthesis of 1-amino-2-propanol moiety of vitamin B_{12}	15.6	669, 1498, 1717
cobF	Cobalamin	Recessive regulator of *cbi* operon	26.9	57, 1717
cobI	Cobalamin	Operon encoding genes for synthesis of cobinamide intermediate of vitamin B_{12}	43.6	514, 575, 951, 952, 1717
cobII	Cobalamin	Operon encoding genes for synthesis of DMB intermediate of vitamin B_{12}	43.3	514, 951, 952, 1717

TABLE 1 *Continued*

Gene symbol	Mnemonic	Former or alternative symbol; enzyme deficiency or other phenotype[a]	Cs[b]	References[c]
cobIII	Cobalamin	Operon encoding genes for joining cobinamide and DMB intermediates of vitamin B$_{12}$	43.4	514, 951, 952, 1717
cobJ	Cobalamin	cobII; synthesis of 5,6-DMB	43.3	513, 1717
cobK	Cobalamin	cobII; synthesis of 5,6-DMB	43.3	513, 1717
cobL	Cobalamin	cobII; synthesis of 5,6-DMB	43.3	513, 1717
cobM	Cobalamin	cobII; synthesis of 5,6-DMB	43.3	513, 1717
cobR	Cobalamin	cis-Acting, dominant regulator for the cbi operon	44.0	57, 1717
cobS	Cobalamin	cobIII; synthesis of the nucleotide loop that joins DMB to the corrin ring	43.5	513, 1508, 1681, 1717
cobT	Cobalamin	cobIII; synthesis of DMB and transfer of ribose phosphate from NMN to DMB	43.4	513, 1508, 1717, 2030
cobU	Cobalamin	cobIII; synthesis of the nucleotide loop that joins DMB to the corrin ring	43.5	513, 1507, 1508, 1681, 1717
cod		Cytosine deaminase (EC 3.5.4.1)	72.8	134, 1015, 1717, 2127
corA	Cobalt resistance	Magnesium transport; cobalt resistance (high level)	85.8	630, 787, 788, 1717, 1843, 1847–1849
corB	Cobalt resistance	Magnesium transport; cobalt resistance (low level)	58.7	630, 1717
corC	Cobalt resistance	Magnesium transport; cobalt resistance (low level)	16.1	630, 703, 1717
corD	Cobalt resistance	apaG; magnesium transport; cobalt resistance (low level)	2.1	630, 1717
cpdB		cpd; 2′, 3′-cyclic phosphodiesterase (EC 3.1.4.16)	69.5	30, 32, 210, 1221, 1704, 1717, 2027
cpsB	Capsule	M antigen capsular polysaccharide synthesis; homologous to rfbM	45.5	1717, 1903
cpsG	Capsule	rfbL; M antigen capsular polysaccharide synthesis; homologous to rfbK	45.4	1717, 1903
crp		cAMP receptor protein	75.0	15, 32, 53, 187, 382, 441, 455, 825, 982, 1415, 1611, 1625, 1717, 1763, 1767, 2123
crr		Factor III for sugar transport by phosphotransferase IIB′ (ptsG) system	52.9	373–376, 441, 531, 1322–1324, 1365, 1415, 1416, 1578, 1580, 1581, 1673, 1707, 1709, 1717, 1765, 1766, 1801, 2027
cspA	Cold shock protein	Transcriptional activation of cold shock promoters	80.0	1717
cspG	Cold shock protein	Transcriptional activation of cold shock promoters; putative	42.9	1661, 1717
cutE		lnt; apolipoprotein N-acyltransferase	16.1	703, 1717
cwd	Cell wall defect	Sensitive to bile salts; mucoid	38.5	777, 1717
cyaA	cAMP	cya; adenylate cyclase (EC 4.6.1.1)	85.7	32, 187, 455, 518, 825, 982, 1611, 1703, 1710, 1711, 1717, 2015, 2027, 2097, 2123, 2207
cysA	Cysteine	Sulfate-thiosulfate transport; chromate resistance	53.0	107, 117, 199, 316, 469, 470, 847, 873, 874, 976, 1005, 1368, 1375, 1416, 1469, 1532–1534, 1561, 1603, 1717, 2136
cysB	Cysteine	Regulation of L-cysteine transport and biosynthesis	37.7	106, 117, 292, 536, 847, 848, 874, 923–925, 1339, 1368, 1376, 1499–1501, 1503, 1504, 1716, 1717, 1821, 2135
cysC	Cysteine	Adenylylsulfate kinase (EC 2.7.1.25)	64.3	117, 438, 919, 1037–1039, 1375, 1603, 1717
cysD	Cysteine	Sulfate adenylyltransferases (EC 2.7.7.4)	64.3	117, 438, 919, 976, 1037–1039, 1368, 1375, 1603, 1717
cysE	Cysteine	Serine acetyltransferase (EC 2.3.1.30)	81.0	117, 137, 340, 446, 871, 875, 1087, 1095, 1097, 1107, 1368, 1500, 1717
cysG	Cysteine	Bifunctional protein: siroheme synthetase; uroporphyrinogen III methylase	75.5	117, 648, 794, 919, 951, 1717, 1867, 2169

(Table continues)

TABLE 1 *Continued*

Gene symbol	Mnemonic	Former or alternative symbol; enzyme deficiency or other phenotype[a]	Cs[b]	References[c]
cysH	Cysteine	Adenylylsulfate reductase (EC 1.8.99.2)	64.2	117, 438, 919, 1037–1039, 1235, 1376, 1499, 1502, 1503, 1505, 1549, 1717
cysI	Cysteine	Heme protein component of sulfite reductase	64.2	117, 438, 723, 794, 919, 1037–1039, 1235, 1376, 1499, 1502, 1503, 1505, 1717
cysJ	Cysteine	Flavoprotein component of sulfite reductase	64.2	117, 438, 723, 919, 1037–1039, 1235, 1236, 1376, 1499, 1502, 1503, 1505, 1549, 1717
cysK	Cysteine	*trz*; resistance to 1,2,4-triazole; O-acetylserine sulfhydrylase A (EC 4.2.99.8)	52.8	266, 340, 368, 376, 536, 537, 871–874, 876, 1096, 1375, 1376, 1408, 1717, 2136
cysL	Cysteine	Resistance to selenate	53.1	872, 1717
cysM	Cysteine	O-Acetylserine sulfhydrylase B (EC 4.2.99.8)	53.0	537, 873, 874, 1375, 1408, 1717, 2135
cysP	Cysteine	Periplasmic thiosulfate-binding protein	53.1	846, 847, 1717
cysQ	Cysteine	Regulation of 3′-phosphoadenoside 5′-phosphosulfate pools; putative	69.5	1221, 1717
cysU	Cysteine	*cysAa, cysT*; sulfate-thiosulfate transport	53.1	846, 847, 1717
cysW	Cysteine	*cysAb*; sulfate-thiosulfate transport	53.1	847, 1717
cysZ	Cysteine	Sulfate transport; putative	52.8	266, 1717
cytR		Regulatory gene for *deo* operon and *udp* and *cdd* genes	88.6	1086, 1717
dadA		*dad*; D-histidine, D-methionine utilization; D-alanine dehydrogenase (EC 1.4.99.1)	41.1	1371, 1717, 2106, 2142, 2143
dadB		Catabolic alanine racemase (EC 5.1.1.1)	41.1	520, 599, 1717, 2105, 2106
dadR		Insensitivity of *dadA* to catabolite repression	41.1	1717, 2141, 2142
dam		DNA adenine methylase	75.6	659, 713, 1525, 1651, 1717
dapA	Diaminopimelate	Dihydropicolinate synthetase (EC 4.2.1.52)	54.0	1717, 1718, 1721
dapB	Diaminopimelate	Dihydropicolinate reductase	0.5	1717, 1718, 1721
dapC	Diaminopimelate	Tetrahydropicolinate succinylase	5.2	1717, 1718, 1721
dapD	Diaminopimelate	Succinyl-diaminopimelate aminotransferase	5.2	1717, 1718, 1721
dapF	Diaminopimelate	Diaminopimelate epimerase (EC 5.1.1.7)	5.4	1717, 1718, 1721
dcd		dCTP deaminase (EC 3.5.4.13)	45.9	132, 135, 1717
dcm		DNA cytosine methylation	43.1	659, 1717
dcp		Dipeptidyl carboxypeptidase	32.5	724, 1717, 2081
dctA		Transport of dicarboxylic acids	79.9	1010, 1011, 1531, 1717, 1943
ddlA	D-Alanine	D-Alanine:D-alanine ligase (EC 3.6.2.4)	8.7	413, 1062, 1717
deoA	Deoxyribose	*tpp*; thymidine phosphorylase (EC 2.4.2.4)	99.0	175, 796, 798, 935, 936, 1643, 1659, 1717
deoB	Deoxyribose	*drm*; phosphopentomutase (EC 2.7.5.6)	99.0	130, 796, 798, 935, 936, 1659, 1717
deoC	Deoxyribose	*dra*; phosphodeoxyriboaldolase (EC 4.1.2.4)	99.0	130, 377, 796–799, 935–938, 1659, 1717
deoD	Deoxyribose	*pnu, pup*; purine nucleoside phosphorylase (EC 2.4.2.1)	99.0	798, 935, 936, 1658, 1659, 1717
deoK	Deoxyribose	Deoxyribokinase	20.2	798, 1717, 1740
deoP	Deoxyribose	Deoxyribose transport	20.3	798, 1717
deoR	Deoxyribose	Constitutive for enzymes of *deoA, deoB, deoC,* and *deoD*	19.4	174, 611, 1717
dgo		D-Galactonate utilization	83.2	1717
dgt		dGTP triphosphohydrolase (EC 3.1.5.1)	5.1	964, 1717
dhb		2,3-Dihydroxybenzoic acid requirement	19.3	47, 118, 326, 1155, 1160, 1717, 1755, 1901
dhuA	D-Histidine	Utilization; increased activity of histidine-binding protein J	51.2	39–41, 45, 46, 49, 74, 763, 1091, 1110, 1154, 1717
divA	Division	*wrkA*; septum initiation defect	88.4	341, 342, 369, 1717, 1736
divC	Division	*smoA*; septum initiation defect	3.1	65, 1717

TABLE 1 *Continued*

Gene symbol	Mnemonic	Former or alternative symbol; enzyme deficiency or other phenotype[a]	Cs[b]	References[c]
divD	Division	Round cell morphology	55.6	65, 68, 383, 1717, 2174
dml	D-Malate	Utilization	82.2	1717, 1902
dnaA	DNA	DNA initiation	83.6	94, 95, 97–99, 501, 731, 845, 957, 1311, 1313, 1717, 1818
dnaB	DNA	DNA synthesis	92.1	1311–1314, 1717, 2161
dnaC	DNA	DNA synthesis initiation and cell division uncoupling	98.8	12, 13, 135, 1063, 1174, 1311–1313, 1663, 1686, 1717, 1785, 1786, 1869–1871
dnaE	DNA	DNA synthesis	5.9	1139, 1311–1313, 1717
dnaG	DNA	DNA biosynthesis; DNA primase	70.0	509, 1313, 1686, 1717, 1738
dnaJ	DNA	DNA biosynthesis	0.3	1313, 1717
dnaK	DNA	DNA biosynthesis	0.3	522, 1313, 1717
dnaL	DNA	DNA biosynthesis	NM	1313, 1717
dnaN	DNA	DNA biosynthesis; DNA polymerase III, β-subunit	83.6	501, 1313, 1717
dnaQ	DNA	DNA biosynthesis	6.9	1138, 1200, 1312, 1313, 1717, 1820
dnaX	DNA	DNA biosynthesis	11.5	1313, 1717
dnaY	DNA	DNA biosynthesis	NM	1313, 1717
dnaZ	DNA	DNA biosynthesis	NM	501, 1313, 1686, 1717
dor		Deletion of r-determinants from plasmids	55.7	660, 853, 1688, 1717, 2107
dpp		Dipeptide permease	75.9	7, 495, 1717
dsd		D-Serine sensitivity; D-serine dehydratase (EC 4.2.1.14)	52.7	1717, 1718, 1721
dum		dUMP synthesis	NM	135, 1717
earA		Regulates expression of *aniG*	88.2	22, 1717
earC		*trans*-Acting regulatory protein for *aciA*	46.8	562, 1717
eca		Enterobacterial common-antigen synthesis	NM	50, 222, 454, 456, 490, 902, 1257, 1266, 1284, 1397, 1717, 1983, 2065, 2150–2152, 2206, 2231
endA		Endonuclease I	67.1	1717, 1769, 2220
eno	Enolase	Enolase (EC 4.2.1.11)	64.3	617, 1717
ent	Enterochelin	*asc, enb;* enterochelin (dihydroxybenzoylserine trimer)	13.6	146, 305, 329, 1177, 1584, 1585, 1717, 2198
envA	Envelope	Cell division defect, chain formation	3.2	341, 595, 1717
envD	Envelope	Autolysis; drug sensitivity; alterations in cell morphology	17.1	65, 69, 1563, 1717
envZ	Envelope	*ompB, tppB, tppA;* positive regulation of tripeptide permease and outer membrane protein	76.0	146, 631, 632, 644, 765, 927, 959, 1201–1204, 1431, 1521, 1717, 1799, 2040
eutA	Ethanolamine utilization	Vitamin B_{12} adenosyltransferase; required for use of ethanolamine as sole carbon or nitrogen source	53.3	1670–1672, 1717, 1794
eutB	Ethanolamine utilization	Ethanolamine ammonia lyase, subunit I	53.3	525, 526, 1670–1672, 1717, 1794
eutC	Ethanolamine utilization	Ethanolamine ammonia lyase, subunit II	53.2	525, 526, 1670–1672, 1717, 1794
eutD	Ethanolamine utilization	CoA-dependent acetaldehyde dehydrogenase	53.3	1670–1672, 1717
eutE	Ethanolamine utilization	Acetaldehyde CoA reductase	53.3	1670–1672, 1717
eutF	Ethanolamine utilization	Ethanolamine permease or transcription factor for expression of *eut* operon	38.2	1506, 1717
eutG	Ethanolamine utilization	Function unknown	53.3	1717
eutH	Ethanolamine utilization	Function unknown	53.2	1717
eutR	Ethanolamine utilization	Positive regulatory gene for *eut* operon	53.2	1670–1672, 1717, 1794
exbB		Biopolymer uptake; outer membrane protein	69.1	1538, 1717
fabB	Fatty acid biosynthesis	β-Ketoacyl acyl carrier protein synthetase I (EC 2.3.1.41)	51.9	385, 824, 1717
fabI	Fatty acid biosynthesis	*envM;* enoyl-(acyl carrier protein) reductase (EC 1.3.1.9)	37.6	1717, 2038
fabZ	Fatty acid biosynthesis	Fatty acid biosynthesis; putative	4.1	783, 1717
fdhB	Formate dehydrogenase-H	Required for formate dehydrogenase-H and -N	18.0	118, 326, 1717, 1904
fdhF	Formate dehydrogenase-H	*fhl;* formate dehydrogenase-H	92.8	119, 1717, 1904
fdhR	Formate dehydrogenase-H	Positive regulator of *fdhF* expression	75.0	523, 1717
fdhS	Formate dehydrogenase-H	Positive regulator of *fdhF* expression	75.0	523, 1717
fdnA	Formate dehydrogenase-N	Mutants lack formate dehydrogenase-N and nitrate reductase	87.6	118, 120, 1553, 1717
fdnB	Formate dehydrogenase-N	Synthesis or activation of the cytochrome associated with formate dehydrogenase-N	87.7	118, 120, 1553, 1717, 1905

(Table continues)

TABLE 1 *Continued*

Gene symbol	Mnemonic	Former or alternative symbol; enzyme deficiency or other phenotype[a]	Cs[b]	References[c]
fdnC	Formate dehydrogenase-N	Synthesis or activation of the cytochrome associated with formate dehydrogenase-N	87.7	1553, 1717, 1905
fdp		Fructose-1,6-diphosphatase (EC 3.1.3.11)	99.2	1717, 1718, 1721
fhlA	Formate hydrogenlyase	Formate hydrogenlyase; putative	62.8	1359, 1717
fhlD	Formate hydrogenlyase	Mutants lack formate dehydrogenase H activity	81.8	119, 326, 1717
fhuA	Ferric hydroxamate uptake	Outer membrane receptor for ferrichrome	4.9	237, 1067, 1717
fhuB	Ferric hydroxamate uptake	Hydroxamate-dependent iron uptake, cytoplasmic membrane component	4.9	237, 1067, 1717
fhuC	Ferric hydroxamate uptake	Hydroxamate-dependent iron uptake, cytoplasmic membrane component	4.9	237, 1067, 1717
fhuD	Ferric hydroxamate uptake	Hydroxamate-dependent iron uptake, cytoplasmic membrane component	5.0	237, 1067, 1717
fic		Filamentation in presence of cAMP in mutant	75.1	1717, 2026
fimA	Fimbriae	pil, fim; major subunit protein of the fimbriae	15.0	349, 408, 532, 1227, 1228, 1473, 1599, 1717, 1926, 1974, 1997
fimC	Fimbriae	Fimbriae	15.0	795, 1717, 1975
fimD	Fimbriae	Fimbriae	15.1	1717, 1974, 1975
fimF	Fimbriae	Fimbriae	15.1	795, 1717, 1975
fimH	Fimbriae	Adhesin component of the fimbriae	15.1	1227, 1717, 1974, 1975
fimI	Fimbriae	Fimbriae	15.0	795, 1717, 1975
fimW	Fimbriae	Fimbriae	15.2	795, 1717, 1975
fimY	Fimbriae	Fimbriae	15.2	795, 1717, 1975
fimZ	Fimbriae	Fimbriae	15.1	795, 1717, 1975
fis	Factor for inversion stimulation	DNA-binding protein involved in site-specific inversion and recombination, gene regulation, and DNA replication	74.0	538, 1717
flgA	Flagella	flaFI; flagellar synthesis; P-ring formation of the flagellar basal body	26.4	633, 887, 891, 1120, 1122–1124, 1226, 1264, 1717, 1969
flgB	Flagella	flaFII; flagellar synthesis; function unknown	26.4	633, 781, 820, 887, 891, 972, 1102, 1120, 1122, 1264, 1717, 1969
flgC	Flagella	flaFIII; flagellar synthesis; basal-body protein	26.4	633, 820, 887, 891, 892, 972, 1102, 1264, 1717, 1969
flgD	Flagella	flaFIV; flagellar synthesis; hook-capping protein to enable assembly of hook protein subunits	26.4	633, 812, 887, 891, 1115, 1120, 1264, 1468, 1717, 1969
flgE	Flagella	flaFV; flagellar synthesis; hook protein	26.4	16, 17, 812, 823, 887, 891, 1120, 1125, 1264, 1377, 1461, 1717, 1969
flgF	Flagella	flaFVI; flagellar synthesis; basal-body rod protein	26.5	16, 820, 823, 887, 891, 971, 972, 1102, 1120, 1264, 1717, 1969
flgG	Flagella	flaFVII; flagellar synthesis; basal-body rod protein	26.5	16, 820, 823, 887, 891, 972, 1102, 1120, 1264, 1717
flgH	Flagella	flaFVIII; flagellar synthesis; basal-body L-ring protein	26.5	887, 891, 971, 1120, 1264, 1717, 1969
flgI	Flagella	flaFIX; flagellar synthesis; basal-body P-ring protein	26.5	16, 819, 887, 891, 971, 1120, 1264, 1717, 1969
flgJ	Flagella	flaFX; flagellar synthesis; function unknown	26.5	887, 891, 971, 1120, 1264, 1717, 1969
flgK	Flagella	flaW; flagellar synthesis; hook-associated protein 1	26.6	812, 813, 815–817, 821, 822, 891, 895, 994, 1264, 1717, 2189
flgL	Flagella	flaU; flagellar synthesis; hook-associated protein 3	26.6	812, 813, 815–817, 821, 822, 891, 895, 994, 1264, 1717, 2189
flgM	Flagella	flgR, mviS, rflB; negative regulator of σ^F	26.4	633–635, 1119, 1465, 1717, 1754
flgN	Flagella	Initiation of flagellar filament assembly	26.4	1124, 1717

TABLE 1 *Continued*

Gene symbol	Mnemonic	Former or alternative symbol; enzyme deficiency or other phenotype[a]	Cs[b]	References[c]
flhA	Flagella	*flaC;* flagellar synthesis; function unknown	41.7	354, 504, 887, 890, 891, 989, 1074, 1120, 1123, 1264, 1317, 1359, 1361, 1717, 2077, 2085, 2192
flhB	Flagella	*flaM;* flagellar synthesis; assists *fliK* in hook length control	41.7	354, 471, 781, 887, 891, 1118, 1120–1122, 1264, 1361, 1717, 1873, 1915, 2192
flhC	Flagella	*flaE;* flagellar synthesis; regulation of gene expression	42.1	354, 504, 635, 887, 890, 891, 989, 1074, 1118, 1120, 1264, 1317, 1717, 2077, 2192
flhD	Flagella	*flaK;* flagellar synthesis; regulation of gene expression; putative flagellum-specific σ factor	42.1	354, 504, 635, 887, 890, 891, 989, 1074, 1118, 1120, 1122, 1264, 1317, 1717, 2077, 2192
flhE	Flagella	Function unknown; not essential for flagellar synthesis or function	41.7	1361, 1717
fliA	Flagella	*flaL;* flagellar synthesis; alternative σ factor, σ^F, for flagellar expression	42.4	354, 504, 635, 887, 890, 891, 977, 989, 994, 1074, 1119, 1122, 1123, 1264, 1317, 1464, 1465, 1717, 1969, 2077, 2192
fliB	Flagella	*nml;* flagellar synthesis; N-methylation of lysine residues in flagellin	42.5	891, 1026, 1076, 1264, 1717, 1924
fliC	Flagella	*H1;* flagellar synthesis; phase 1 flagellin (filament structural protein)	42.5	274, 507, 508, 582, 583, 585, 634, 743, 814, 821, 829, 832, 881, 884–889, 891–894, 906, 913, 985–988, 994–997, 1001, 1116, 1117, 1157, 1264, 1265, 1278, 1305, 1318, 1454, 1465, 1471, 1556, 1717, 1802, 1804, 1806, 1840, 1841, 1928, 1966, 1967, 2025, 2115, 2116, 2191, 2192, 2208
fliD	Flagella	*flaV;* flagellar synthesis; hook-associated protein 2	42.5	129, 813, 815–817, 821, 822, 891, 895, 993, 994, 1122, 1264, 1717, 2189
fliE	Flagella	*flaAI;* flagellar synthesis; function unknown	42.7	354, 504, 829, 887, 890, 891, 989, 1074, 1102, 1120, 1122, 1264, 1317, 1392, 1607, 1717, 1803, 1965, 2077, 2190–2192
fliF	Flagella	*flaAII.I;* flagellar synthesis; basal-body M-ring protein	42.7	16, 424, 567, 891, 972, 1122, 1264, 1472, 1607, 1717, 2041, 2042, 2186
fliG	Flagella	*flaAII.2, motC, cheV;* flagellar synthesis; motor switching and energizing	42.7	354, 424, 471, 487, 504, 508, 567, 887, 890, 891, 907, 989, 1027, 1034, 1074, 1120, 1148, 1264, 1271, 1481, 1717, 2077, 2102, 2185, 2186, 2192

(Table continues)

TABLE 1 *Continued*

Gene symbol	Mnemonic	Former or alternative symbol; enzyme deficiency or other phenotype[a]	Cs[b]	References[c]
fliH	Flagella	*flaAII.3;* flagellar synthesis; function unknown	42.8	424, 818, 891, 1264, 1717, 2085, 2186
fliI	Flagella	*flaAIII;* flagellar assembly; may be ATPase in protein export pathway	42.8	354, 468, 471, 504, 818, 887, 890, 891, 989, 1074, 1120, 1264, 1717, 1877, 2077, 2085, 2186, 2192
fliJ	Flagella	*flaS;* flagellar synthesis; function unknown	42.8	818, 891, 1120, 1264, 1717, 2085
fliK	Flagella	*flaR;* flagellar synthesis; hook length control and replacement of *flgD* by *flgK*	42.8	17, 781, 818, 891, 1074, 1120, 1121, 1264, 1552, 1717, 1968, 2187
fliL	Flagella	*flaQI;* flagellar synthesis; function unknown	42.8	818, 891, 1034, 1122, 1264, 1608, 1717
fliM	Flagella	*flaQII, cheC, cheU;* flagellar synthesis; motor switching and energizing	42.8	429, 487, 504, 818, 887, 890, 891, 989, 1027–1029, 1034, 1120, 1148, 1264, 1271, 1317, 1481, 1717, 1852, 2102, 2192
fliN	Flagella	*flaN;* flagellar synthesis; motor switching and energizing	42.8	354, 471, 504, 887, 890, 891, 907, 989, 1027, 1034, 1264, 1271, 1317, 1481, 1717, 2085, 2185, 2186, 2192
fliO	Flagella	*flaP;* flagellar synthesis; function unknown	42.9	504, 887, 890, 891, 989, 1264, 1317, 1717, 2192
fliP	Flagella	*flaB;* flagellar synthesis; function unknown	42.9	354, 504, 887, 890, 891, 989, 1074, 1264, 1317, 1717, 2077, 2192
fliQ	Flagella	*flaD;* flagellar synthesis; function unknown	42.9	354, 504, 887, 890, 891, 989, 1074, 1264, 1317, 1717, 2077, 2185, 2192
fliR	Flagella	*flaX;* flagellar synthesis; function unknown	42.9	891, 1264, 1717
fliS	Flagella	Member of axial family of structural proteins	42.6	1008, 1717
fliT	Flagella	Member of axial family of structural proteins	42.6	1008, 1717
fliU	Flagella	Flagellar function; part of basal body, or required for flagellin processing or export	42.5	457, 1717
fliV	Flagella	Flagellar function; part of basal body, or required for flagellin processing or export	42.5	457, 1717
fljA	Flagella	*rhl;* flagellar synthesis; repressor of *fliC*	60.0	583, 891, 1123, 1264, 1717, 2188
fljB	Flagella	*H2;* flagellar synthesis; phase 2 flagellin (filament structural protein)	60.1	487, 507, 508, 582, 634, 814, 884–889, 891–894, 896, 897, 985, 1116, 1117, 1157, 1264, 1278, 1318, 1471, 1556, 1717, 1796, 1802–1804, 1806, 1928, 1966, 1967, 1969, 1982, 2188, 2191, 2192, 2232, 2233
flrB	Fluoroleucine resistance	Leucine or isoleucine regulation or both	16.4	20, 276, 574, 577, 589, 1717
fnr		*oxrA;* anaerobic induction of genes for anaerobic metabolism, such as *pepT*	36.6	648, 942, 1230, 1717, 1904, 1940, 2165
fol	Folate	Trimethoprim resistance; tetrahydrofolate dehydrogenase	1.9	1023, 1717
fpk	Fructose	Fructose phosphate kinase	48.3	620, 1717

TABLE 1 *Continued*

Gene symbol	Mnemonic	Former or alternative symbol; enzyme deficiency or other phenotype[a]	Cs[b]	References[c]
frd	Fumarate reductase	Fumarate reductase (EC 1.3.99.1)	NM	158, 1717
fruA	Fructose	Enzyme II^{Fru} of the phosphotransferase system	48.2	528, 619, 1717
fruB	Fructose	*fruF*; enzyme III^{Fru}-modulator FPr tridomain fusion protein of the phosphotransferase system	48.1	528, 619, 620, 1717
fruK	Fructose	Fructose-1-phosphate kinase	48.1	528, 619, 1717
fruR	Fructose	Regulation of the fructose regulon; regulation of gluconeogenesis; may be the same as *ppsB*	3.0	323, 528, 619, 620, 1717, 2076
ftsZ		*sulB*; essential for cell division	3.1	380, 1717
fucA	Fucose	*fuc*; L-fucose utilization	65.1	1717
fucI	Fucose	L-Fucose utilization	65.1	1717
fucK	Fucose	L-Fucose utilization	65.1	1717
fucP	Fucose	L-Fucose utilization	65.1	1717
fucR	Fucose	L-Fucose utilization	65.1	1717
fumA	Fumarate	Regulatory gene; putative	31.3	1652, 1717
furA	Ferrichrome	*fur*; ferrichrome uptake, regulation of iron uptake; constitutive synthesis of iron-enterochelin	16.9	511, 549, 553, 554, 562, 613, 614, 1717
fusA	Fusidic acid	Protein chain elongation factor EF-G	74.7	958, 1717
galE	Galactose	UDP glucose 4-epimerase (EC 5.1.3.2)	17.8	584, 636, 653, 744, 834, 911, 1259, 1287, 1393, 1432, 1435, 1441, 1442, 1485, 1491, 1523, 1589, 1717, 1742, 2168
galF	Galactose	Modifier of UDP-glucose pyrophosphorylase	45.3	954, 1405, 1406, 1432, 1433, 1717
galK	Galactose	Galactokinase (EC 2.7.1.6)	17.8	31, 584, 834, 845, 1287, 1432, 1717, 1798
galP	Galactose	Specific galactose permease	67.0	1717, 1718, 1721
galR	Galactose	Regulation	65.3	1702, 1717
galS	Galactose	*mglD*; *mgl* repressor and galactose ultrainduction factor	48.0	145, 1717
galT	Galactose	Galactose-1-phosphate uridylyltransferase (EC 2.7.7.10)	17.8	834, 1717
galU	Galactose	Glucose-1-phosphate uridylyltransferase (EC 2.7.7.9)	38.4	506, 777, 1406, 1485, 1717
gapA		Glyceraldehyde-3-phosphate dehydrogenase (EC 1.2.1.12)	41.1	1152, 1717
garA	Gamma resistant	Resistant to γ and UV radiation; large cells; high RNA and protein content; may be equivalent to *rodA*	0.4	882, 1717
garB	Gamma resistant	Resistant to γ and UV radiation; large cells; high RNA and protein content	0.4	882, 1717
gcv	Glycine cleavage	*invD*; defective in the glycine cleavage enzyme system	65.7	1717, 1893
gdhA		*gdh*; glutamate dehydrogenase (EC 1.4.1.4)	29.1	105, 444, 445, 712, 778, 1351, 1675, 1717
gleR		Glycyl-leucyl-resistant regulatory gene for transport of branched-chain amino acids	8.1	1462, 1463, 1717
glgA	Glycogen	Glycogen synthase (EC 2.4.1.21)	77.2	1180, 1181, 1717, 1896
glgC	Glycogen	Glucose-1-phosphate adenylyltransferase (EC 2.7.7.27)	77.2	1180, 1181, 1667, 1717, 1896
glnA	Glutamine	Glutamine synthetase (EC 6.3.1.2)	87.5	47, 163, 229, 230, 592, 593, 729, 933, 1066, 1088–1090, 1109, 1113, 1114, 1316, 1319, 1570, 1609, 1717, 2028, 2039, 2117
glnD	Glutamine	PIIA uridyl transferase	4.3	104, 1717
glnE	Glutamine	Covalent modification of glutamine synthetase; glutamine synthetase adenylyltransferase (EC 2.7.2.42)	69.5	104, 1109, 1717
glnH	Glutamine	Periplasmic glutamine-binding protein	19.1	1113, 1717
glnP	Glutamine	High-affinity glutamine transport	19.1	89, 162, 163, 1717
glpA	Glycerol phosphate	*sn*-Glycerol-3-phosphate dehydrogenase (anaerobic) (EC 1.1.99.5)	49.6	1717, 1718, 1721
glpD	Glycerol phosphate	*sn*-Glycerol-3-phosphate dehydrogenase (aerobic) (EC 1.1.1.8)	77.1	10, 1210, 1717
glpK	Glycerol phosphate	Glycerol kinase (EC 2.7.1.30)	88.5	1449, 1615, 1717, 1926
glpQ	Glycerol phosphate	Glycerol-3-phosphate diesterase	49.4	756, 1717
glpR	Glycerol phosphate	Regulatory gene for *glpD*, *glpK*, and *glpT*	77.1	1717, 1718, 1721
glpT	Glycerol phosphate	*sn*-Glycerol-3-phosphate transport	49.5	757, 1643, 1644, 1709, 1713, 1717
gltA	Glutamate	Requirement	17.1	1717, 1718, 1721
gltB	Glutamate	Glutamate synthetase (EC 2.6.1.53)	72.5	400, 581, 1269, 1717

(Table continues)

TABLE 1 *Continued*

Gene symbol	Mnemonic	Former or alternative symbol; enzyme deficiency or other phenotype[a]	Cs[b]	References[c]
gltC	Glutamate	Growth on glutamate as sole source of carbon	81.7	1717, 1718, 1721
gltD	Glutamate	Glutamate synthase, small subunit	72.7	400, 1269, 1717
gltF	Glutamate	Glutamate-specific transport system	0.1	35, 1717
gltH	Glutamate	Requirement	28.0	1717, 1718, 1721
gltS	Glutamate	Glutamate permease	81.7	35, 1717
glyA	Glycine	Serine hydroxymethyltransferase (EC 2.1.2.1)	56.0	1679, 1717, 1887–1890, 1943, 1944, 2048
glyS	Glycine	Glycyl-tRNA synthetase (EC 6.1.1.14)	80.1	1717, 1943, 1944
glyT	Glycine	*sufS*; glycine tRNA2	90.0	1476, 1717
gnd		Phosphogluconate dehydrogenase (EC 1.1.1.43)	44.9	109, 165, 220, 682, 954, 975, 1398, 1475, 1621, 1717, 2154
gpd		Glucosamine-6-phosphate deaminase	NM	1711, 1717
gpsA		*sn*-Glycerol-3-phosphate dehydrogenase [NAD(P)$^+$] (EC 1.1.1.94)	81.0	1210, 1717
gpt		*gxu*; guanine-hypoxanthine phosphoribosyltransferase (EC 2.4.2.8)	7.4	332, 665, 666, 833, 1483, 1484, 1648, 1717
gsk		Guanosine kinase	11.9	1717, 1718, 1721
guaA	Guanine	GMP synthetase (EC 6.3.4.1)	54.4	534, 611, 664, 1430, 1679, 1717, 1740
guaB	Guanine	IMP dehydrogenase (EC 1.1.1.205)	54.4	534, 611, 664, 943, 1430, 1679, 1717, 1740
guaC	Guanine	GMP reductase (EC 1.6.6.8)	3.5	148, 1717
guaP	Guanine	Guanine uptake	3.5	150, 1717
gyrA	Gyrase	*hisW, nalA*; resistance or sensitivity to nalidixic acid; DNA gyrase	49.3	8, 9, 62, 178, 231, 232, 419, 420, 983, 1313, 1457, 1500, 1632, 1679, 1717, 1752, 2195
gyrB	Gyrase	*hisU, parA*; DNA gyrase	83.3	62, 178, 203, 204, 232, 415–418, 983, 1252, 1455, 1457, 1679, 1717, 1752
hemA	Heme	Glutamyl-tRNA dehydrogenase	38.8	143, 493, 494, 497, 934, 1371, 1571, 1717, 1733, 1812, 2177
hemB	Heme	Heme deficient	8.7	1717, 1733, 2177
hemC	Heme	Heme deficient; urogen I synthase	85.7	1717, 1732, 2177
hemD	Heme	Heme deficient; uroporphyrinogen III cosynthase	85.7	1717, 1731, 2177
hemE	Heme	Accumulation of uroporphyrin III	90.6	449, 1717, 2177
hemF	Heme	Coproporphyrinogen III oxidase	53.2	1717, 2177, 2178
hemG	Heme	Defective in heme synthesis	86.5	1717, 2177
hemH	Heme	*visA*; defective in heme synthesis	11.6	1717, 2177
hemK	Heme	Protoporphyrinogen oxidase; putative (EC 1.3.3.–)	38.8	493, 1717
hemL	Heme	Glutamate-1-semialdehyde aminotransferase	5.0	496, 497, 1717, 2177
hemM	Heme	Glutamyl tRNA dehydrogenase or subunit; putative	38.8	493, 1571, 1717
hemN	Heme	Oxygen-independent coproporphyrinogen III oxidase	86.7	1717, 2177, 2179
hil	Hyperinvasion locus	Hyperinvasion; essential for bacterial entry into epithelial cells	62.8	1158, 1359, 1717
himA		Integration host factor (IHF), α-subunit; site-specific recombination	30.3	1196, 1717
himD		Integration host factor (IHF), β-subunit; site-specific recombination	20.1	1455, 1717
hin	H inversion	*vh2*; flagellar synthesis; regulation of flagellin gene expression by site-specific inversion of DNA	60.1	243, 244, 582, 643, 869, 884–887, 891, 998, 1116, 1117, 1157, 1256, 1264, 1556, 1717, 1772, 1802, 1823, 1981, 1982, 2188, 2232, 2234, 2235
hisA	Histidine	*N*-(5′-phospho-L-ribosylformimino)-5-amino-1-(5′-phosphoribosyl)-4-imidazolecarboxamide isomerase (EC 5.3.1.16)	44.8	26, 37, 165, 232, 280, 281, 335, 682, 736–738, 867, 908, 1071, 1073, 1231, 1295, 1296, 1301, 1610, 1717

TABLE 1 *Continued*

Gene symbol	Mnemonic	Former or alternative symbol; enzyme deficiency or other phenotype[a]	Cs[b]	References[c]
hisB	Histidine	Imidazolegylcerol-phosphate dehydratase and histidinol phosphatase (EC 4.2.1.19, EC 3.1.3.15)	44.8	25, 37, 165, 217, 232, 280, 281, 319, 336, 736–738, 837–841, 1071, 1073, 1231, 1301, 1657, 1717, 1885
hisC	Histidine	Histidinol-phosphate aminotransferase (EC 2.6.1.9)	44.8	24–26, 29, 37, 165, 232, 281, 645, 736–738, 754, 755, 916, 1071, 1073, 1231, 1301–1303, 1619, 1645, 1655–1657, 1717, 1777
hisD	Histidine	Histidinol dehydrogenase (EC 1.1.1.23)	44.7	24, 26, 29, 37, 54, 81, 113, 115, 141, 165, 171, 206, 232, 246, 247, 281, 336, 343, 674, 696, 737, 738, 867, 905, 916, 917, 1070, 1071, 1165, 1619, 1645, 1717, 1748, 1751, 1995, 2159, 2199, 2212–2216
hisF	Histidine	Cyclase	44.8	37, 165, 232, 280, 281, 318, 645, 736, 737, 1231, 1301, 1717
hisG	Histidine	ATP phosphoribosyltransferase (EC 2.4.2.17)	44.7	26, 27, 37, 38, 54, 81, 111–116, 165, 176, 177, 179, 188, 202, 225, 281, 300, 312, 337–339, 351, 372, 530, 597, 645–647, 737, 827, 967, 1055, 1057–1059, 1071, 1084, 1099, 1231, 1301, 1333, 1334, 1380, 1381, 1457, 1542–1545, 1590, 1647, 1682, 1692, 1717, 1770, 1771, 1782, 1842, 1900, 2084, 2086, 2092, 2134, 2149
hisH	Histidine	Amidotransferase (EC 2.4.2.–)	44.8	37, 65, 165, 232, 280, 281, 737, 1071, 1717
hisI	Histidine	*hisE, hisIE;* phosphoribosyl-AMP cyclohydrolase, phosphoribosyl-ATP pyrophosphatase (EC 3.5.4.19, EC 3.6.1.31)	44.8	37, 165, 232, 246, 280–282, 318, 645, 736, 737, 1231, 1301, 1717
hisJ	Histidine	Periplasmic histidine-binding protein J for high-affinity histidine transport system	51.2	39–41, 43, 45, 46, 49, 70, 73, 74, 258, 762, 766, 789, 1111, 1154, 1155, 1160, 1182, 1231, 1292, 1301, 1591, 1717, 2237
hisM	Histidine	Histidine transport; inner membrane protein	51.2	48, 762, 766, 1423, 1554, 1717
hisP	Histidine	High-affinity histidine transport; inner membrane protein	51.2	39–43, 45, 46, 48, 49, 70, 73, 74, 232, 637, 754, 766, 1110, 1160, 1591, 1679, 1717, 1760
hisQ	Histidine	Histidine transport; membrane protein	51.2	45, 46, 48, 70, 74, 762, 766, 1160, 1717

(Table continues)

TABLE 1 *Continued*

Gene symbol	Mnemonic	Former or alternative symbol; enzyme deficiency or other phenotype[a]	Cs[b]	References[c]
hisR	Histidine	tRNA structural gene	85.6	178, 201, 208, 232, 233, 282, 535, 1241, 1457, 1458, 1679, 1680, 1717, 1800, 1810
hisS	Histidine	Histidyl-tRNA synthetase (EC 6.1.1.21)	55.4	178, 232–235, 351, 433–436, 452, 453, 1179, 1189, 1320, 1678, 1679, 1717, 1811, 1942, 2173
hisT	Histidine	Pseudouridine modification of tRNA	51.8	178, 205, 206, 209, 232, 236, 241, 303, 376, 378, 379, 426, 966, 975, 1412, 1413, 1519, 1601, 1654, 1674, 1679, 1714, 1717, 2044
hmpA	Hemoprotein	Dihydropteridine reductase activity (EC 1.6.99.7)	56.0	1717, 1894
hns		*osmZ, bglY, pilG;* histone-like protein; DNA-binding nucleoid-associated protein	38.4	517, 770, 780, 877, 1299, 1516, 1717, 2036
hpt		Hypoxanthine phosphoribosyltransferase (not EC 2.4.2.8)(see *gpt*)	3.8	147, 148, 485, 486, 665, 833, 1484, 1637, 1717
hsdL	Host specificity	*hspLT;* restriction-modification system	8.4	131, 153, 254, 350, 355, 359–361, 409, 421–423, 740, 741, 1470, 1717, 2031
hsdM	Host specificity	*hsdSB;* restriction-modification system; modification component	98.5	253, 254, 256, 362, 588, 1100, 1417, 1717, 1787, 2031, 2065
hsdS	Host specificity	*hsdSB;* restriction-modification system; specificity component	98.4	253, 254, 256, 362, 588, 1100, 1417, 1717, 1787, 2031, 2065
hsdSA	Host specificity	Restriction-modification system (operon)	97.6	131, 153, 254, 255, 350, 355, 359–361, 741, 1040, 1100, 1417, 1470, 1717, 1822, 2031, 2075
htrA	High temperature requirement	*degP;* heat shock (stress)-regulated periplasmic protease; essential for survival in macrophages	5.1	129, 310, 964, 1717
hupA		Histonelike protein HU-2	90.7	773, 779, 1717
hupB		Histonelike protein HU-1	11.4	779, 1298, 1717
hutC	Histidine utilization	Repressor	17.9	161, 192, 238, 625, 714, 716, 717, 1272, 1326, 1573, 1717, 1831–1834, 2039
hutG	Histidine utilization	Formiminoglutamase (EC 3.5.3.8)	17.9	192, 238, 1272, 1326, 1717, 1831–1834, 2039
hutH	Histidine utilization	Histidine ammonia-lyase (EC 4.3.1.3)	17.9	192, 193, 238, 715, 1272, 1326, 1717, 1831-1834, 2039
hutI	Histidine utilization	Imidazolonepropionase (EC 3.5.2.7)	17.9	192, 193, 238, 1272, 1326, 1717, 1831–1834, 2039
hutM	Histidine utilization	Promoter for *hutIGC*	17.9	192, 195, 238, 625, 1272, 1326, 1717, 1831-1834, 2039
hutP	Histidine utilization	Promoter for *hutUH*	17.9	192, 195, 238, 372, 625, 1272, 1326, 1717, 1831–1834, 2039
hutQ	Histidine utilization	Promoter for *hutUH*	17.9	195, 238, 372, 625, 1272, 1326, 1717, 1831–1834, 2039

TABLE 1 *Continued*

Gene symbol	Mnemonic	Former or alternative symbol; enzyme deficiency or other phenotype[a]	Cs[b]	References[c]
hutR	Histidine utilization	Catabolite insensitivity of *hutUH*	17.9	195, 238, 1272, 1326, 1717, 1831–1834, 2039
hutU	Histidine utilization	Urocanate hydratase (EC 4.2.1.49)	17.9	192, 193, 238, 1272, 1326, 1717, 1831, 2039
hydA	Hydrogenase	*aniA, fhlB, hyd;* hydrogenase	62.5	119, 325, 1548, 1717
hydG	Hydrogenase	Hydrogenase regulation	90.8	331, 1717
hydH	Hydrogenase	Hydrogenase regulation	90.7	331, 1717
hyp	Hydrophobic peptide auxotrophy	Hydrophobic polypeptide requirement	52.0	224, 1717
icd		Isocitrate dehydrogenase	27.6	553, 554, 1717
iclR		Constitutive expression of *aceBA* operon	91.2	606, 1414, 1717
ilvA	Isoleucine-valine	*ile;* threonine deaminase (EC 4.2.1.16)	85.3	75, 79, 155, 156, 182, 184, 186, 261, 263, 272, 321, 415, 419, 420, 425, 472, 473, 642, 739, 802, 1185, 1233, 1248, 1717, 1737, 1874, 1985, 2103, 2104, 2224
ilvB	Isoleucine-valine	Acetohydroxy acid synthase I, valine sensitive, large subunit (EC 4.1.3.18)	82.3	404, 417, 419, 420, 642, 1143, 1145, 1478–1480, 1717, 1789, 1790, 2123
ilvC	Isoleucine-valine	*ilvA;* ketol-acid reductoisomerase (EC 1.1.1.86)	85.3	76, 79, 156, 182, 185, 415, 642, 803, 1048, 1717, 1792, 1793
ilvD	Isoleucine-valine	*ilvB;* dihydroxyacid dehydratase (EC 4.2.1.19)	85.3	77–79, 155, 156, 182, 184, 186, 415, 419, 420, 472, 492, 642, 735, 1233, 1717
ilvE	Isoleucine-valine	*ilvC;* branched-chain amino acid aminotransferase (EC 2.6.1.42)	85.2	79, 155, 156, 182, 184, 186, 350, 415, 419, 420, 472, 527, 642, 735, 1044, 1046, 1166, 1232, 1233, 1717
ilvG	Isoleucine-valine	Acetohydroxy acid synthase II, valine insensitive, large subunit (EC 4.1.3.18)	85.2	155, 156, 182, 184, 186, 404, 419, 420, 472, 735, 1142, 1145, 1232, 1233, 1478, 1480, 1587, 1717, 1747, 1789, 1790
ilvH	Isoleucine-valine	Acetohydroxy acid synthase III, valine sensitive, small subunit (EC 4.1.3.18)	3.0	419, 420, 1630, 1717, 1878, 2100
ilvI	Isoleucine-valine	Acetohydroxy acid synthase III, valine sensitive, large subunit (EC 4.1.3.18)	3.0	1630, 1717, 1878, 2100
ilvL	Isoleucine-valine	*ilvGMEDA* operon leader peptide	85.2	1153, 1717, 1986
ilvM	Isoleucine-valine	Acetohydroxy acid synthase II, valine insensitive, small subunit (EC 4.1.3.18)	85.2	404, 1232, 1233, 1717, 1747
ilvN	Isoleucine-valine	Acetohydroxy acid synthase II, valine sensitive, small subunit (EC 4.1.3.18)	82.3	404, 417, 1717
ilvS	Isoleucine-valine	Isoleucyl-tRNA synthetase (EC 6.1.1.5)	1.0	179, 180, 321, 1717
ilvY	Isoleucine	Regulation of *ilvC*	85.3	183, 186, 1717
incR	Incompatibility	Required in *trans* with *parS*	pSLT	295, 296, 1717
infC		Protein chain initiation factor 3 (IF3)	30.2	1225, 1717
inlA	Inositol	Fermentation	92.7	185, 962, 1473, 1717, 1926, 1962
inlB	Inositol	Fermentation	56.5	1717, 1935
inm		Sensitivity to mutagenesis by nitrosoguanidine	80.6	367, 1717
invA	Invasion	Affects invasion but not attachment to cultured epithelial cells; sequence similarity to proteins for protein translocation	63.1	566, 600, 602–604, 639, 685, 970, 1359, 1517, 1717
invD	Invasion	Invasion-related function	NM	600, 603, 1717
invE	Invasion	Triggering of the endocytic uptake of *Salmonella* by epithelial cells; homologous to Mxic of *Shigella*	63.1	638, 639, 685, 1359, 1717
invF	Invasion	Defective in invasion but not in attachment to epithelial cells; similar to *araC*	63.2	1000, 1359, 1717

(Table continues)

TABLE 1 *Continued*

Gene symbol	Mnemonic	Former or alternative symbol; enzyme deficiency or other phenotype[a]	Cs[b]	References[c]
invG	Invasion	Defective in invasion of epithelial cells; member of PulA family of proteins for export of proteins lacking typical signal sequences	63.2	639, 1000, 1359, 1717
invH	Invasion	Defective in invasion of and attachment to epithelial cells	63.4	34, 1359, 1717
iroA	Iron-regulated locus	Induced by alkaline pH; undefined role in iron transport	60.6	554, 562, 1717
IS*200*I		Insertion sequence	65.8	170, 627, 1134–1136, 1717, 1724
IS*200*II		Insertion sequence	75.6	170, 627, 1134–1136, 1717, 1724
IS*200*III		Insertion sequence	94.2	170, 627, 1134–1136, 1717, 1724
IS*200*IV		Insertion sequence	42.4	170, 627, 1134–1136, 1717, 1724
IS*200*V		Insertion sequence	53.6	170, 627, 1134–1136, 1717, 1724
IS*200*VI		Insertion sequence	22.3	170, 627, 1134–1136, 1717, 1724
katG	Catalase	*cls*; hydroperoxidase I (HPI) (EC 1.11.1.6)	89.4	522, 539, 1229, 1378, 1529, 1717, 1977, 1998
kbl		2-Amino-3-ketobutyrate CoA ligase (glycine acetyltransferase) (EC 2.3.1.29)	81.2	1717, 1814
kdsA		Ketodeoxyoctonate synthesis	38.7	279, 650, 651, 726, 732, 1167–1169, 1394, 1492, 1605, 1635–1638, 1640, 1641, 1717, 1937, 2093
kdsB		CMP ketodeoxyoctonate synthetase	20.2	279, 649–651, 1717
lamB	Lambda	Encodes a protein resembling the λ receptor	91.7	568, 1520, 1522, 1717
lepA		GTP-binding protein	57.6	1717, 2071
lepB		*lep*; signal peptidase I	57.6	1717, 2070, 2071
leuA	Leucine	α-Isopropylmalate synthase (EC 4.1.3.12)	2.9	260, 278, 286, 287, 589, 621–623, 742, 760, 990, 1023, 1156, 1297, 1629, 1717, 1773, 2126
leuB	Leucine	β-Isopropylmalate dehydrogenase	2.9	58, 260, 262, 623, 760, 1297, 1717, 1773
leuC	Leucine	α-Isopropylmalate isomerase subunit	2.9	260, 590, 623, 689, 760, 1022, 1024, 1025, 1297, 1677, 1717
leuD	Leucine	α-Isopropylmalate isomerase subunit	2.9	260, 578, 590, 689, 1022, 1024, 1025, 1297, 1676, 1717, 1936
leuS	Leucine	Leucyl-tRNA synthetase (EC 6.1.1.4)	15.8	20, 21, 276, 760, 1717, 1906
leuT	Leucine	Leucine transport	39.4	1602, 1717, 2014
leuU	Leucine	*leuT*; leucine tRNA	85.6	201, 208, 1717
lev	Levomycetin	Levomycetin resistance	NM	478, 1717
lexA		Regulatory gene for SOS functions	91.9	271, 618, 1399, 1717
lgt		*umpA*; prolipoprotein diacylglyceryl transferase	64.9	609, 1717
lig	Ligase	DNA ligase	52.7	1313, 1535, 1717
lip	Lipoic acid	Requirement	15.3	1227, 1717, 1844
livF	Leucine, isoleucine, valine	*livG*; high-affinity branched-chain amino acid transport; membrane component	77.5	1308, 1717
livG	Leucine, isoleucine, valine	*livF*; high-affinity branched-chain amino acid transport	77.5	1308, 1717
livH	Leucine, isoleucine, valine	*livA*; high-affinity branched-chain amino acid transport; membrane component	77.5	1306–1308, 1466, 1717
livJ	Leucine, isoleucine, valine	livB; high-affinity branched-chain amino acid transport; membrane component	77.6	1308, 1467, 1717
livK	Leucine, isoleucine, valine	*livC*; high-affinity branched-chain amino acid transport; membrane component	77.5	1308, 1467, 1717
livM	Leucine, isoleucine, valine	*livEF*; high-affinity branched-chain amino acid transport; membrane component	77.5	1308, 1717
livS	Leucine, isoleucine, valine	*liv*; regulatory gene; high-affinity branched-chain amino acid transport	19.2	1395, 1717
lkyA	Leaky	Leakage of periplasmic proteins	59.8	1717, 2118
lkyB	Leaky	Leakage of periplasmic proteins	59.8	1717, 2118

TABLE 1 *Continued*

Gene symbol	Mnemonic	Former or alternative symbol; enzyme deficiency or other phenotype[a]	Cs[b]	References[c]
lkyC	Leaky	Leakage of periplasmic proteins	22.9	1717, 2118
lkyD	Leaky	Leakage of periplasmic proteins; morphology defect	22.9	298, 369, 594, 595, 1254, 1717, 1959, 2118, 2119
lon	Long form	*capR;* ATP-dependent protease	11.1	302, 463, 1557, 1717, 1730, 2111
lpd		Dihydrolipoamide dehydrogenase (EC 1.8.1.4)	3.8	10, 1140, 1210, 1717
lpp	Lipoprotein	Murein lipoprotein structural gene	30.6	1407, 1717, 2184
lpxA		UDP-*N*-acetylglucosamine acyltransferase	4.2	1717, 1757, 2090
lpxD		*ssc, omsA, firA;* UDP-3-*O*-(*R*-3-hydroxymyristoyl)-glucosamine *N*-acyltransferase; antibiotic supersensitivity (EC 2.3.1.–)	4.0	750, 783, 785, 1687, 1717, 1757, 2089–2091
lrp		Leucine-responsive regulatory protein	19.1	1717, 2100
lysA	Lysine	*lys;* requirement	65.4	371, 489, 1717, 1943, 1944
lysS	Lysine	Lysyl-tRNA synthetase (EC 6.1.1.6)	66.2	1009, 1717
malE	Maltose	*malB;* maltose uptake; periplasmic maltose-binding protein	91.7	402, 1522, 1717
malF	Maltose	Maltose uptake; inner membrane protein	91.6	402, 1717, 1759
malG	Maltose	Maltose uptake; inner membrane protein	91.6	402, 569, 1717
malK	Maltose	Maltose uptake; inner membrane protein	91.7	402, 1717, 1760, 2094, 2095
malM	Maltose	Possible periplasmic protein; function unknown	91.7	1717, 1759
malQ	Maltose	Amylomaltase (EC 1.2.1.25)	77.0	10, 1709, 1711, 1717, 1722
malT	Maltose	Regulation of maltose genes	76.8	1521, 1717
manA	Mannose	*pmi;* mannose-6-phosphate isomerase (EC 5.3.1.8)	31.3	356, 357, 565, 692, 1287, 1652, 1697, 1717, 1921, 2145
mdh		Malate dehydrogenase (EC 1.1.1.37)	73.4	1240, 1717
melA	Melibiose	α-Galactosidase (EC 3.2.1.22)	93.8	1709, 1712, 1717, 1851
melB	Melibiose	Melibiose carrier protein	93.8	1108, 1370, 1712, 1717
mem	Membrane	Sugar transport and membrane protein defective	80.8	1577, 1717
menA	Menaquinone	Menaquinone deficient; defective in trimethylamine oxide reduction; grows on vitamin K_1	88.5	1126, 1129, 1717
menB	Menaquinone	Biosynthesis; grows on vitamins K_1 and K_5	50.1	344, 414, 1126, 1127, 1129, 1664, 1717
menC	Menaquinone	Biosynthesis	50.1	414, 1127, 1129, 1717
menD	Menaquinone	Biosynthesis	50.1	414, 1127, 1129, 1717
menE	Menaquinone	*O*-Succinyl benzoic acid-CoA synthase	50.1	414, 1127, 1664, 1717
metA	Methionine	*metI;* homoserine transsuccinylase (EC 2.3.1.46)	91.1	92, 308, 1150, 1294, 1668, 1683, 1684, 1717, 1829, 1830
metB	Methionine	Cystathionine γ-synthase (EC 4.2.99.9)	89.3	92, 699, 1004, 1653, 1683, 1684, 1717, 1829, 1830, 2047, 2050, 2051
metC	Methionine	Cystathionine γ-lyase (EC 4.4.1.1)	69.1	431, 1385, 1537, 1538, 1683, 1684, 1717, 1830
metD	Methionine	*metP;* high-affinity methionine transport	5.5	89–91, 93, 384, 697, 1567, 1717, 1791
metE	Methionine	Tetrahydropteroyltriglutamate methyltransferase (EC 2.1.1.14)	86.1	294, 1564, 1565, 1717, 1732, 1735, 1768, 1829, 1830, 2056–2058, 2131, 2171, 2172
metF	Methionine	5,10-Methylenetetrahydrofolate reductase (EC 1.1.99.15)	89.4	92, 1717, 1829, 1830, 1891, 1892, 2131
metG	Methionine	Methionyl-tRNA synthetase	47.1	72, 92, 307, 690, 691, 1653, 1717, 1829, 1830
metH	Methionine	Vitamin B_{12}-dependent homocysteine-N^5-methylenetetrahydrofolate transmethylase	91.3	92, 265, 320, 1474, 1565, 1717, 1891, 2005, 2053, 2055, 2056, 2058, 2131, 2132, 2171, 2172

(Table continues)

TABLE 1 *Continued*

Gene symbol	Mnemonic	Former or alternative symbol; enzyme deficiency or other phenotype[a]	Cs[b]	References[c]
metJ	Methionine	Methionine analog resistant; transcriptional repressor of *metE* and *metR*	89.3	92, 93, 306, 307, 386, 1151, 1685, 1717, 1735, 1891, 1944, 2050-2052, 2132, 2172
metK	Methionine	Methionine analog resistant; *S*-adenosylmethionine synthetase	67.0	86, 93, 162, 306, 307, 791–793, 1150, 1151, 1362, 1685, 1717, 1735, 1944, 2220
metL	Methionine	Aspartokinase II-homoserine dehydrogenase II	89.4	1717, 2050
metR	Methionine	Transcriptional activator of *metE* and *metH*	86.1	221, 265, 386, 758, 1294, 1564, 1717, 2054, 2056–2058, 2172
mglA	Methyl galactoside	Membrane-bound protein for methygalactoside transport	48.0	145, 1391, 1633, 1717
mglB	Methyl galactoside	Galactose-binding protein	48.0	145, 1327, 1390, 1391, 1575, 1633, 1717, 1938, 2149, 2238, 2239
mglC	Methyl galactoside	Membrane-bound transport protein	48.0	145, 1391, 1633, 1717
mglE	Methyl galactoside	Transport	NM	145, 1391, 1633, 1717
mgtA	Magnesium transport	Magnesium transport	96.7	787, 1717, 1847–1849
mgtB	Magnesium transport	Magnesium transport	82.1	614, 787, 1717, 1847–1850
mgtC	Magnesium transport	Magnesium transport	82.1	1717, 1850
miaA		Deficient in the nucleotide ms^{2}io^{6}AA, a modified base present in some tRNAs	95.0	189, 211, 251, 510, 975, 1717
miaE		Lack of tRNA (ms^{2}io^{6}A37) hydroxylase	97.2	1558, 1717
min	Minicells	Cell division	NM	1717, 1996
moaA	Molybdenum	*chlA;* molybdenum-containing factor; biosynthesis of molybdopterin	19.0	344, 505, 1717, 1784, 1904, 1933
mob	Molybdenum	*chlB;* molybdenum-containing factor; biosynthesis of molybdopterin guanine dinucleotide	86.4	289, 291, 293, 1717, 1784, 1904
modC	Molybdenum	*chlD;* molybdenum uptake	18.3	31, 1717, 1784, 1904, 1933
moeA	Molybdenum	*chlE;* molybdenum-containing factor; biosynthesis of molybdopterin	19.1	505, 1717, 1784, 1904, 1934
motA	Motility	Nonmotile but flagellate	42.1	471, 502–504, 887, 912, 1074, 1118, 1717, 2192
motB	Motility	Nonmotile but flagellate	42.0	471, 502–504, 887, 912, 1074, 1118, 1717, 2192
mre	Mecillinam resistance	*bac, envB;* round cell morphology; mecillinam resistance	73.6	63, 64, 66–68, 383, 1482, 1717
mscL	Mechanosensitive channel	Large-conductance mechanosensitive channel	74.5	1541, 1717
mta	meso-Tartaric acid	Utilization of and resistance to *meso*-tartaric acid	NM	1315, 1717, 2137
mtlA	Mannitol	D-Mannitol phosphotransferase enzyme IIA	80.4	159, 221, 676, 1176, 1580, 1706, 1711, 1717, 1723
mtlD	Mannitol	Mannitol-1-phosphate dehydrogenase (EC 1.1.1.17)	80.4	159, 947, 1717
murB		UDP-*N*-acetylenolpyruvoylglucosamine reductase	90.0	458, 1717
murI		Glutamate synthase	89.7	1717, 2114
mutG	Mutator	Increased frequency of mutation in host chromosome, not in P22	NM	705, 1717
mutH	Mutator	Mutations inactivate methyl-directed mismatch repair	65.5	713, 1234, 1526, 1618, 1717, 1780, 1781
mutL	Mutator	Mutations inactivate methyl-directed mismatch repair	95.0	1043, 1291, 1525, 1526, 1618, 1717, 1780, 1781, 2221
mutS	Mutator	Mutations inactivate methyl-directed mismatch repair	63.9	709, 710, 1525, 1526, 1618, 1717, 1780, 1781, 2221
mutU	Mutator	Increased frequency of mutation	NM	1618, 1717
mutY	Mutator	*mutB;* increased frequency of mutation with alkylating agents	67.2	448, 713, 1239, 1717, 1780
mviA	Mouse virulence	Mutants have increased virulence in Ity^s, not in Ity^r mice	38.9	144, 1717
mviM	Mouse virulence	*mviB;* affects the virulence of cells in mice	26.3	1717

TABLE 1 *Continued*

Gene symbol	Mnemonic	Former or alternative symbol; enzyme deficiency or other phenotype[a]	Cs[b]	References[c]
mviN	Mouse virulence	Affects the virulence of cells in mice	26.3	1717
mviS	Mouse virulence	Affects the virulence of cells in mice	26.3	284, 1717
nadA	Nicotinamide	*nicA*; requirement; quinolinic acid synthetase	17.2	110, 546, 559–561, 563, 811, 1220, 1717, 1862, 2017, 2226–2228
nadB	Nicotinamide	*nic*; L-aspartate oxidase	57.8	110, 370, 546, 561, 563, 810, 811, 1220, 1717, 2226, 2228
nadC	Nicotinamide	Quinolinic acid PRPP phosphoribosyl transferase	3.6	110, 546, 561, 810, 811, 861, 868, 1220, 1717
nadD	Nicotinamide	NAMN:ATP-ADP transferase	15.5	110, 317, 860, 862, 863, 1717
nadE	Nicotinamide	Essential biosynthetic gene, unsupplementable; NAD synthetase	29.5	865, 1717
nadF	Nicotinamide	NAD kinase I	14.7	317, 1717
nadG	Nicotinamide	NAD kinase II; quinolinate sensitive	74.9	317, 1717
nadR	Nicotinamide	*nadI, pnuA*; NMN transport and repression of transcription of *nadA, nadB*; bifunctional enzyme	99.9	370, 558, 563, 811, 1717, 2226, 2228, 2229
nagA	N-Acetylglucosamine	*nag*; nonutilization	16.5	129, 1393, 1717, 1738
nalB	Nalidixic acid	Resistance or sensitivity	61.6	1429, 1717
nanH	Neuraminidase	Sialidase (EC 3.2.2.18)	23.7	843, 844, 1717, 2003, 2101
nap	Nonspecific acid phosphatase	Deficiency for nonspecific acid phosphatase I	NM	640, 641, 1662, 1717, 2043
nar	Nitrate reductase	*chlC*; nitrate reductase (operon) (EC 1.7.99.4)	38.6	121, 289, 290, 1131, 1646, 1717, 1904
narK	Nitrate reductase	Regulatory gene for *nar* operon	38.6	826, 1717
ndk		Nucleoside diphosphate kinase (EC 2.7.4.6)	55.5	1717, 1718, 1721
neaA	Neamine	Neamine resistance	74.1	540, 1717
newD		Substitute gene for *leuD*	8.0	591, 1022, 1024, 1132, 1133, 1648, 1717, 1936
nfnB		*nfsI*; sensitivity to nitrofurantoin	13.6	1717, 2108
nhoA		N-Hydroxyarylamine O-acetyltransferase (EC 2.3.1.118)	104.0	1717, 2109
nirB	Nitrite reductase	NADH-nitrite oxidoreductase apoprotein, subunit I (EC 1.6.6.4)	75.2	648, 1717, 2169
nirC	Nitrite reductase	Nitrite transport; putative (EC 1.6.6.4)	75.5	1717, 2169
nirD	Nitrite reductase	NADH-nitrite oxidoreductase apoprotein, subunit II	75.5	1717, 2169
nirP	Nitrite reductase	Nitrite permease	7.5	648, 1717
nit	Nitrogen	Nitrogen metabolism	30.1	239, 445, 1717
nlpD		Lipoprotein precursor	63.7	1085, 1717
nmpC	New membrane protein	Outer membrane porin protein	38.7	826, 1717
nol	Norleucine	Norleucine resistance; possible defect in valine uptake or regulation	64.5	790, 1717
nrdA		Ribonucleoside diphosphate reductase, α-subunit (R1) (EC 1.17.4.1)	49.3	977, 1313, 1717, 1904
nrdB		Ribonucleoside diphosphate reductase, β-subunit (R2) (EC 1.17.4.1)	49.4	977, 1717, 1904
nrdE		Ribonucleoside-diphosphate reductase subunit (EC 1.17.4.1)	61.3	977, 1717
nrdF		Ribonucleoside-diphosphate reductase subunit (EC 1.17.4.1)	61.4	977, 1717
nsiA	Nicotinamide starvation inducible	NAD metabolism regulation	77.8	556, 1717
ntrB	Nitrogen regulation	*glnR*; regulation of *glnA* expression and other nitrogen-controlled genes	87.3	47, 1088–1090, 1109, 1112, 1113, 1316, 1319, 1440, 1717, 2117
ntrC	Nitrogen regulation	*glnR*; regulation of *glnA* expression and other nitrogen-controlled genes	87.2	47, 1088–1090, 1112, 1316, 1319, 1440, 1515, 1570, 1717, 1756, 2117
nuoD		NADH ubiquinone oxidoreductase	51.0	71, 1717
nuoE		NADH ubiquinone oxidoreductase	51.0	71, 1717
nuoF		NADH ubiquinone oxidoreductase	51.0	71, 1717
nuoG		NADH ubiquinone oxidoreductase	51.0	71, 1717
nusA	N utilization	Transcription termination; does not support transcription antitermination of N protein of λ	72.0	388, 1717
nuvA		Uridine thiolation factor A activity	9.5	1093, 1717
oadA		Oxalacetate decarboxylase, α-subunit (EC 4.1.1.3)	NM	1717, 2158
oadB		Oxalacetate decarboxylase, β-subunit (EC 4.1.1.3)	NM	1717, 2158
oadG		Oxalacetate decarboxylase, γ-subunit (EC 4.1.1.3)	NM	1717, 2158
oafA	O-antigen factor	O-5, *ofi*; LPS O-factor 5, acetyl group	50.7	1052, 1287, 1717, 1921

(Table continues)

TABLE 1 *Continued*

Gene symbol	Mnemonic	Former or alternative symbol; enzyme deficiency or other phenotype[a]	Cs[b]	References[c]
oafC	O-antigen factor	Determines factor 1 in LPS of group E *Salmonella* spp.	15.2	1287, 1309, 1566, 1717, 1921
oafR	O-antigen factor	Synthesis of LPS O-antigen 122	12.7	1280, 1287, 1717, 1921
ogt		*O*-6-Alkylguanine-DNA-alkyltransferase	36.6	1717
ompA	Outer membrane protein	Outer membrane protein 33K	24.3	146, 412, 573, 1237, 1431, 1510, 1717, 1719, 1799, 1925, 2040
ompC	Outer membrane protein	Outer membrane protein 36K	49.1	146, 309, 462, 644, 670, 959, 984, 1237, 1424, 1431, 1450, 1510, 1521, 1717, 1799, 2020, 2021, 2040, 2225
ompD	Outer membrane protein	Outer membrane protein 34K	33.7	146, 462, 644, 959, 1237, 1431, 1450, 1518, 1717, 1799, 2020, 2021, 2040
ompF	Outer membrane protein	Outer membrane protein 35K	22.8	309, 462, 644, 670, 984, 1450, 1597, 1717, 1734, 2040
ompH	Outer membrane protein	Outer membrane protein 16K, cationic	4.0	782, 783, 1080, 1081, 1717
ompR	Outer membrane protein	*ompB, tppA;* positive regulation of tripeptide permease and of outer membrane protein	76.0	146, 309, 462, 602, 631, 632, 644, 765, 927, 959, 1201–1204, 1431, 1521, 1717, 1799, 2040
opdA	Oligopeptidase	*optA;* endoprotease which hydrolyzes *N*-acetyl-L-Ala$_4$; required for normal phage P22 growth	79.7	364–366, 1717, 2080
oppA	Oligopeptide permease	Oligopeptide-binding protein	38.4	662, 765, 767, 768, 775–777, 804, 921, 1175, 1717, 1886, 1990
oppB	Oligopeptide permease	Oligopeptide transport system	38.3	662, 765, 768, 775, 777, 804, 1555, 1717
oppC	Oligopeptide permease	Oligopeptide transport system	38.3	662, 765, 768, 775, 777, 804, 1555, 1717
oppD	Oligopeptide permease	Oligopeptide transport system	38.3	662, 765, 768, 775, 777, 804, 1717
oppF	Oligopeptide permease	Oligopeptide transport system	38.3	607, 775, 777, 1717
orf11		Putative role in fimbria production	pSLT	580, 1717
orf7		Putative role in fimbria production	pSLT	580, 1717
orf9		Putative role in fimbria production	pSLT	580, 1717
orfE		Function unknown; locates downstream of *spv* operon of pSLT	pSLT	702, 1717
orgA	Oxygen-regulated gene	Noninvasive mutant in low oxygen	62.9	969, 1717
oriC	Origin	*poh;* origin of replication of chromosome	84.9	957, 1717, 1788, 2240, 2241
osmB		Osmotically inducible lipoprotein; resistance to osmotic stress	37.6	1717, 2006
oxdA	Oxygen dependent	Gene activity controlled by *fnr*	67.8	1717, 1940
oxdB	Oxygen dependent	Gene activity controlled by *fnr*	93.2	1717, 1940
oxiA	Oxygen inducible	Induced by anaerobiosis	9.2	23, 929, 1717
oxiB	Oxygen inducible	Induced by anaerobiosis	25.8	23, 1717
oxiC	Oxygen inducible	Induced by anaerobiosis	35.3	23, 1717
oxiE	Oxygen inducible	Induced by anaerobiosis	89.4	23, 1717
oxrF	Oxygen regulation	Regulates expression of *aniH*	NM	22, 1717
oxrG	Oxygen regulation	Regulates expression of *aniC, aniI*	89.4	22, 562, 1717
oxyR	Oxidative stress resistant	Transcriptional activator of oxidative stress response genes	89.6	333, 334, 522, 1164, 1378, 1529, 1717, 1929, 1931, 1932, 1998, 2164
pabA	*p*-Aminobenzoate	Requirement; *p*-aminobenzoate synthase	75.1	1003, 1717
pabB	*p*-Aminobenzoate	*p*-Aminobenzoate synthetase, component I (EC 4.1.3.–)	41.6	661, 1717
pagC	*phoP*-activated gene	*ail, lom; phoP*-activated gene for virulence	NM	1355, 1358, 1598, 1717
panB	Pantothenic acid	Ketopantohydroxymethyl transferase (EC 4.1.2.12)	4.6	396, 440, 1568, 1588, 1717
panC	Pantothenic acid	Pantothenate synthetase (EC 6.3.2.1)	4.5	396, 440, 1568, 1588, 1717

TABLE 1 *Continued*

Gene symbol	Mnemonic	Former or alternative symbol; enzyme deficiency or other phenotype[a]	Cs[b]	References[c]
panD	Pantothenic acid	Ketopantoic acid reductase	4.6	1144, 1488, 1588, 1717
panE	Pantothenic acid	Ketopantoic acid reductase	NM	1588, 1717
panR	Pantothenic acid	Pantothenate excretion; suppression of the thiamine requirement of the *purF* mutation	4.6	467, 1717
panT	Pantothenic acid	Pantothenate transport	NM	1717, 1718, 1721
parA	Partition	Partitioning; homologous to *parA* of phage P1	pSLT	295, 296, 1717
parB	Partition	Partitioning; homologous to *parB* in phage P1	pSLT	295, 296, 1717
parC	Partition	*clmF;* defect in nucleoid segregation; topoisomerase IV subunit; similar to *gyrA*	69.3	11, 1252, 1717, 1749, 1750, 1752
parE	Partition	*clmF;* defect in nucleoid segregation; topoisomerase IV subunit; similar to *gyrB*	69.4	1252, 1717, 1749, 1752, 1872
parF	Partition	*clmF;* partitioning of nucleoid; topoisomerase	69.3	1252, 1717, 1749, 1872
parS	Partition	*incL;* partitioning of pSLT	pSLT	295, 296, 1717
pasA		6-Aminonicotinic acid sensitive	92.9	551, 557, 1717
pasB		6-Aminonicotinic acid sensitive	70.1	551, 557, 1717
pasC		6-Aminonicotinic acid sensitive	23.4	551, 557, 1717
pasD		6-Aminonicotinic acid sensitive	17.9	557, 1717
pasE		6-Aminonicotinic acid sensitive	56.8	557, 1717
pbpA	Penicillin-binding protein	Round cell morphology; mecillinam resistance; penicillin-binding protein 2	15.4	383, 1717
pckA		Phosphoenolpyruvate carboxykinase (ATP) activity (EC 4.1.1.49)	76.0	1201, 1717
pckB		*pck;* phosphoenolpyruvate carboxykinase (ATP) activity (EC 4.1.1.49)	14.5	283, 1717
pclA	Permissive for *cly*	Permissive for lytic growth of P22 *cly*	NM	1717, 2156
pclB	Permissive for *cly*	Permissive for lytic growth of P22 *cly*	NM	1717, 2156
pclC	Permissive for *cly*	Permissive for lytic growth of P22 *cly*	NM	1717, 2156
pde	Phosphodiesterase	2′,3′-Cyclic nucleotide 2′-phosphodiesterase	95.5	1717, 2043
pduA	Propanediol utilization	Propanediol utilization	43.8	15, 194, 315, 950, 1717
pduB	Propanediol utilization	Propanediol utilization	43.8	15, 194, 315, 950, 1717
pduC	Propanediol utilization	Propanediol dehydratase	43.9	15, 194, 950, 1717
pduD	Propanediol utilization	Propanediol dehydratase	44.0	15, 194, 950, 1717
pduE	Propanediol utilization	Vitamin B$_{12}$ adenosyltransferase	44.0	15, 194, 950, 1717
pduF	Propanediol utilization	Facilitated diffusion of propanediol	43.8	15, 194, 315, 950, 1717
pduG	Propanediol utilization	Propanediol utilization	44.0	15, 194, 950, 1717
pduH	Propanediol utilization	Propanediol utilization	44.0	15, 194, 950, 1717
pdxB	Pyridoxine	Requirement	51.8	824, 1717
pefA	Plasmid-encoded fimbriae	Fimbrial biosynthesis; fimbrial/pilin shaft subunits	pSLT	580, 1649, 1717
pefB	Plasmid-encoded fimbriae	Fimbrial biosynthesis; fimbrial regulatory protein	pSLT	580, 1649, 1717
pefC	Plasmid-encoded fimbriae	Fimbrial biosynthesis; outer membrane protein	pSLT	580, 1649, 1717
pefD	Plasmid-encoded fimbriae	Fimbrial biosynthesis; sequence related to periplasmic chaperones	pSLT	580, 1649, 1717
pefI	Plasmid-encoded fimbriae	Short polypeptide related in sequence to *papI* and *sfaC* of *E. coli*	pSLT	580, 1649, 1717
pefK	Plasmid-encoded fimbriae	*orf5;* fimbrial biosynthesis	pSLT	580, 1649, 1717
pefL	Plasmid-encoded fimbriae	*orf6;* fimbrial biosynthesis	pSLT	580, 1649, 1717
pefS	Plasmid-encoded fimbriae	*orf8;* homologous to *dsbA* of *E. coli*	pSLT	580, 1649, 1717
pepA	Peptidase	Peptidase A; similar to aminopeptidase A of *E. coli*	97.4	632, 1340, 1341, 1346, 1350, 1620, 1717, 2204, 2205
pepB	Peptidase	Peptidase B; aminopeptidase	55.8	675, 1341, 1350, 1620, 1717, 2204, 2205
pepD	Peptidase	*ptdD;* peptidase D; a dipeptidase, carnosinase	7.3	828, 1049, 1340, 1341, 1346, 1620, 1717, 2204, 2205
pepE	Peptidase	α-Aspartyl dipeptidase; peptidase E; splits Asp-X peptide bonds	91.4	288, 363, 1717
pepM	Peptidase	Peptidase M; aminopeptidase that removes N-terminal methionine from proteins	4.3	1345, 1349, 1382, 1717, 2153
pepN	Peptidase	*ptdN;* peptidase N; an aminopeptidase, naphthylamidase	24.4	226, 1340, 1341, 1343, 1350, 1620, 1717, 2204, 2205
pepP	Peptidase	*ptdP;* peptidase P; splits X-Pro peptide bonds	66.6	921, 1321, 1342, 1350, 1717, 2200
pepQ	Peptidase	Peptidase Q; splits X-Pro dipeptides	86.2	921, 1321, 1342, 1350, 1717, 2200
pepT	Peptidase	Peptidase T; a tripeptidase	27.3	1347, 1717, 1939, 1941
pfkA		6-Phosphofructokinase (EC 2.7.1.11)	88.2	1717, 1718, 1721
pfl		Pyruvate formate lyase	19.5	806, 808, 1546, 1717, 2165
pgi	Phosphoglucose isomerase	*oxrC, pasA;* regulation of fermentative or biosynthetic enzymes; glucosephosphate isomerase (EC 5.3.1.9)	91.4	565, 928, 1424, 1717
pgn		Poor growth on nutrient plates	3.6	868, 1717
pgtA	Phosphoglycerate	Activator of phosphoglycerate transport	52.1	953, 1713, 1717, 2016, 2201, 2217

(Table continues)

TABLE 1 *Continued*

Gene symbol	Mnemonic	Former or alternative symbol; enzyme deficiency or other phenotype[a]	Cs[b]	References[c]
pgtB	Phosphoglycerate	Protein for signal transmission for phosphoglycerate transport	52.1	953, 1713, 1717, 2201, 2217
pgtC	Phosphoglycerate	Protein for signal transmission for phosphoglycerate transport	52.1	953, 1713, 1717, 2201, 2217
pgtE	Phosphoglycerate	Outer membrane protease E (E protein) precursor (EC 3.4.21.–)	52.0	683, 1717
pgtP	Phosphoglycerate	Transporter for phosphoglycerate transport	52.2	654, 953, 1713, 1717, 2201, 2217
pheA	Phenylalanine	Chorismate mutase (EC 5.4.99.5)	58.3	655, 1437, 1717, 1993
pheR	Phenylalanine	Regulation of pheA	67.7	657, 658, 1717, 1875
pheT	Phenylalanine	Phenylalanyl-tRNA synthetase, β-subunit	30.3	1196, 1717
phoE	Phosphate	Phosphate limitation-inducible outer membrane pore protein	7.6	1717, 1868
phoN	Phosphate	Nonspecific acid phosphatase	94.2	1007, 1031–1033, 1355, 1717, 2043, 2125
phoP	Phosphate	Phosphorylated transcriptional activator; regulator of expression of phoN and virulence genes	27.4	33, 140, 533, 552, 601, 686, 1031–1033, 1354–1357, 1717, 2125
phoQ		Membrane sensor kinase; environmental response regulator in conjunction with PhoP	27.4	33, 686, 1354–1357, 1717
phoS	Phosphatase	Periplasmic phosphate-binding protein	NM	125, 1717, 1746, 2180
phrB	Photoreactivation	Deoxyribodipyrimidine photolyase (EC 4.1.99.3)	16.9	1195, 1717
phsA		aniE, phs; hydrogen sulfide production	44.3	19, 344, 542, 1717, 2087, 2088
phsB		Iron sulfur subunit; electron transfer	44.2	1717
phsC		Membrane-anchoring protein	44.2	1717
phsD		Cytochrome c-containing subunit; electron transfer	44.2	1717
phsE		Function unknown	44.2	1717
phsF		Function unknown	NM	1717
pig	Pigment	Brownish colonies	56.5	1717, 1926
pldA		Outer membrane phospholipase A	86.0	240, 1717
ply	Phage lysogeny	pox; control of P22 lysogeny	94.6	1717, 1718, 1721
pmrA		Polymyxin B resistance; regulation of transcription	93.5	1238, 1665, 1717, 2059–2063
pmrB		Polymyxin B resistance; sensor protein (EC 2.7.3.–)	93.5	1665, 1717
pmrD		Polymyxin B resistance	50.1	1664, 1717
pncA	Pyridine nucleotide cycle	Nicotinamide deamidase (EC 3.5.1.19)	28.9	559, 560, 778, 860, 1042, 1220, 1717
pncB	Pyridine nucleotide cycle	Nicotinic acid phosphoribosyltransferase (EC 2.4.2.11)	23.2	555, 559, 560, 810, 1041, 1042, 1220, 1717, 2082
pncC	Pyridine nucleotide cycle	NMN deamidase; mutations fail to use NMN as a pyridine source	88.8	1717
pncH	Pyridine nucleotide cycle	Nicotinamide used as sole nitrogen source	29.0	778, 1717
pncX	Pyridine nucleotide cycle	6-Aminonicotinamide resistant	29.0	778, 860, 1717
pnuA	Pyridine nucleotide uptake	NMN uptake deficient	99.7	558, 1042, 1717, 1718, 1861, 1864, 1865
pnuB	Pyridine nucleotide uptake	Growth on lower than normal levels of NMN	99.7	1717, 1864, 1865
pnuC	Pyridine nucleotide uptake	NMN uptake deficient	17.2	563, 1717, 1864, 1865, 2017, 2227, 2228
pnuD	Pyridine nucleotide uptake	Restores ability to use NMN to a pnuC mutant	64.0	1717, 1865
pnuE	Pyridine nucleotide uptake	Failure to use exogenous NAD; periplasmic NAD pyrophosphatase	88.3	1536, 1717
pnuF	Pyridine nucleotide uptake	Mutations allow use of quinolinate as pyridine source	2.7	1717
pnuG	Pyridine nucleotide uptake	Mutations allow use of quinolinate as pyridine source	38.9	1717
pnuH	Pyridine nucleotide uptake	Mutations allow use of quinolinate as pyridine source	56.1	1717
pocR		Positive regulator for cob and pdu genes	43.8	15, 194, 315, 1669, 1717
polA	Polymerase	atrC; DNA nucleotidyltransferase (EC 2.7.7.7)	86.5	454, 499, 550, 708, 713, 902, 903, 909, 1077, 1234, 1267, 1268, 1717, 1897, 2133, 2150, 2231
potA	Polyamine transport	Spermidine and putrescine transport; membrane-associated protein	27.3	1347, 1717
poxA	Pyruvate oxidase	Hypersensitivity to antimicrobial agents; lower levels of pyruvate oxidase and acetolactate synthase deficiency in α-ketobutyrate metabolism	94.6	1717, 1895, 2022, 2023, 2074
ppc		Phosphoenolpyruvate carboxylase (EC 4.1.1.31)	89.6	92, 768, 804, 1270, 1700, 1717, 1824, 2008
ppiA		Peptidyl-prolyl cis-trans isomerase A precursor (EC 5.2.1.8)	75.1	1717, 2026
ppsA		Phosphoenolpyruvate synthase	30.5	322, 620, 1717, 1846

TABLE 1 *Continued*

Gene symbol	Mnemonic	Former or alternative symbol; enzyme deficiency or other phenotype[a]	Cs[b]	References[c]
ppsB		Deficiency in phosphoenolpyruvate synthase; may be identical to *fruR*	3.3	277, 1717
praA		Phage P221 receptor function	93.8	1451, 1452, 1717
praB		Phage P221 receptor function	64.7	1451, 1452, 1717
prbA		Phage ES18 receptor function	93.8	1717, 1718, 1721
prbB		Phage ES18 receptor function	35.5	1717, 1718, 1721
prc		Reduced survival in macrophages; similar to *prc* protease of *E. coli*	41.6	129, 1717
prdB		Phage PH51 receptor function	35.6	1717, 1718, 1721
prfA		Protein release factor 1 (RF1)	38.8	493, 494, 498, 1717
prfB		*supT, supK;* protein release factor 2 (RF2)	66.2	87, 1009, 1622, 1623, 1717, 2046
prgH	*phoP*-repressed gene	Influences mouse virulence; defective in macrophage invasion	62.9	140, 1717
prh		Phage HK009 receptor function	94.0	1717, 1718, 1721
prk		Phage HK068 receptor function	35.4	1717, 1718, 1721
proA	Proline	Glutamate to glutamic-γ-semialdehyde	7.8	301, 828, 920, 948, 949, 1132, 1133, 1137, 1275, 1367, 1612, 1648, 1717, 1836, 1845
proB	Proline	Glutamate to glutamic-γ-semialdehyde	7.8	301, 828, 920, 948, 949, 1132, 1133, 1137, 1275, 1367, 1612, 1717, 1836, 1845
proC	Proline	Pyrroline-5-carboxylate reductase (EC 1.5.1.2)	8.9	218, 920, 1367, 1717, 1836, 1845
proL	Proline	*sufB, proW;* frameshift suppressor affecting proline tRNA and correcting +1 frameshifts at runs of C in mRNA	49.1	1717, 1879
proM	Proline	*proT;* proline tRNA	85.6	201, 208, 1717
proP	Proline	Proline permease II; glycine betaine and proline; low affinity	93.5	55, 267, 269, 397, 479, 984, 1329, 1717
proV	Proline	*proU;* high-affinity transport system for glycine betaine and proline; binding protein	61.4	70, 268, 397, 400, 475, 479, 764, 772, 984, 1424, 1513, 1514, 1717, 1908, 1964
proW	Proline	*proU;* high-affinity transport system for glycine betaine and proline; hydrophobic membrane component	61.4	397, 400, 984, 1424, 1513, 1514, 1717, 1908
proX	Proline	*proU;* high-affinity transport system for glycine betaine and proline; glycine betaine-binding protein	61.4	397, 400, 984, 1424, 1513, 1514, 1717, 1908
proY	Proline	Proline transport system	9.0	1717
proZ	Proline	Proline transport system	77.7	491, 1717
prpA	Propionate	*prp;* propionate metabolism	95.7	1717, 1721
prpB	Propionate	Propionate metabolism	8.2	1717
prsA		*prsB;* phosphoribosylpyrophosphate synthetase	38.9	213, 493, 733, 734, 955, 956, 1524, 1571, 1572, 1717
psiA		Phosphate starvation inducible	75.8	564, 1717
psiB		Phosphate starvation inducible	88.9	564, 1717
psiC		Phosphate starvation inducible	10.1	564, 1717
psiD		Phosphate starvation inducible	93.8	564, 1717
psiR		Regulates *psiC* activity	84.8	564, 1717
pss		Outer membrane proteins which protect against oxidative intraleukocyte killing	94.3	1717, 1907
psuA		Suppressor of polarity	NM	1717, 1718, 1721
pta	Phosphotransacetylase	Acetyl-CoA:orthophosphate acetyltransferase (EC 2.3.1.8)	50.8	1130, 1183, 1717, 2073, 2080
ptsF	Phosphotransferase system	*fruA;* fructose phosphotransferase enzyme IIa	48.8	620, 1078, 1580, 1701, 1705, 1706, 1711, 1717
ptsG	Phosphotransferase system	*glu, gpt;* glucose phosphotransferase enzyme IIB'-factor III (*crr*) system (methyl-β-D-glucoside)	26.9	212, 676, 1328, 1574, 1576, 1580, 1626, 1693, 1694, 1701, 1706, 1711, 1717, 1764, 1920, 2068
ptsH	Phosphotransferase system	*carB;* phosphohistidine protein-hexose phosphotransferase (EC 2.7.1.69)	52.9	142, 266, 373, 375, 681, 722, 774, 1038, 1078, 1289, 1366, 1375, 1574, 1580–1583, 1701, 1705, 1706, 1708, 1710, 1711, 1717, 1762, 1763, 1801, 2112, 2121

(Table continues)

TABLE 1 *Continued*

Gene symbol	Mnemonic	Former or alternative symbol; enzyme deficiency or other phenotype[a]	Cs[b]	References[c]
ptsI	Phosphotransferase system	carA; enzyme I of the phosphotransferase system	52.9	266, 373, 375, 676, 722, 1038, 1078, 1104, 1199, 1289, 1366, 1375, 1574, 1580–1582, 1701, 1705, 1706, 1708, 1710–1712, 1717, 1762, 1763, 1801, 2120, 2122
ptsJ	Phosphotransferase system	Enzyme I* of the phosphotransferase system, not expressed in wild type	52.9	324, 1717
ptsM	Phosphotransferase system	manA; mannose-glucose phosphotransferase enzyme IIA (2-deoxyglucose)	NM	1580, 1626, 1703, 1706, 1711, 1717, 1920
purA	Purine	Adenylosuccinate synthetase (EC 6.3.4.4)	95.1	149, 664, 668, 1715, 1717
purB	Purine	Adenylosuccinate lyase (EC 4.3.2.2)	27.5	656, 664, 667, 1717
purC	Purine	Phosphoribosylaminoimidazole-succinocarboxamide synthetase (EC 6.3.2.6)	54.0	664, 1679, 1717
purD	Purine	Phosphoribosylglycinamide synthetase (EC 6.3.1.13)	90.8	129, 330, 482, 664, 1717, 2129
purE	Purine	Phosphoribosylaminoimidazole carboxylase (EC 4.1.1.21)	12.6	664, 1065, 1484, 1717, 2012
purF	Purine	Amidophosphoribosyltransferase (EC 2.4.2.14)	51.7	464, 466, 467, 611, 664, 1065, 1717, 2129
purG	Purine	Phosphoribosylglycinamidine synthetase (EC 6.3.5.3)	56.3	664, 1679, 1717
purH	Purine	Phosphoribosylaminoimidazolecarboxamide formyltransferase (EC 2.1.2.3)	90.8	330, 482, 664, 1717
purI	Purine	Phosphoribosylaminoimidazole synthetase (EC 6.3.3.1)	54.5	405, 1679, 1717
purJ	Purine	IMP cyclohydrolase (EC 3.5.4.10)	90.8	668, 1717
purN	Purine	Cryptic purF analog; synthesis of phosphoribosylamine	4.7	1717, 1721
purR	Purine	Constitutive high expression of pur genes	30.7	1717, 1721
putA	Proline utilization	putB; bifunctional enzyme; proline oxidase and pyrroline-5-carboxylate dehydrogenase	25.6	28, 70, 398, 442, 443, 718, 719, 1290, 1330, 1331, 1496, 1497, 1613, 1614, 1717, 1745
putP	Proline utilization	Major L-proline permease	25.7	55, 70, 269, 397, 451, 479, 719, 1290, 1329, 1330, 1353, 1401, 1455, 1497, 1717
pyrB	Pyrimidine	Aspartate carbamoyltransferase (EC 2.1.3.2)	96.9	139, 335, 399, 541, 543, 544, 940, 944–946, 1013, 1016, 1335, 1336, 1459, 1717, 1770, 1838, 1979, 1980, 2147, 2197
pyrC	Pyrimidine	Dihydroorotase (EC 3.5.2.3)	26.2	139, 264, 940, 944, 1016–1018, 1421, 1459, 1717, 1770, 1838, 1858–1860, 2037, 2147, 2197
pyrD	Pyrimidine	Dihydroorotate oxidase (EC 1.3.3.1)	24.1	139, 264, 576, 940, 944, 1016, 1018, 1459, 1717, 1770, 1838, 1858, 1859, 2037, 2147, 2197
pyrE	Pyrimidine	Orotate phosphoribosyltransferase (EC 2.4.2.10)	81.4	139, 166, 944–946, 1016, 1018, 1087, 1107, 1259, 1420, 1422, 1459, 1717, 1723, 1770, 1838, 2147, 2197
pyrF	Pyrimidine	Orotidine-5′-phosphate decarboxylase (EC 4.1.1.23)	37.6	139, 292, 940, 1016, 1087, 1371, 1459, 1716, 1717, 1770, 1838, 2006, 2007, 2147, 2197

TABLE 1 *Continued*

Gene symbol	Mnemonic	Former or alternative symbol; enzyme deficiency or other phenotype[a]	Cs[b]	References[c]
pyrG	Pyrimidine	CTP synthetase	64.4	133, 931, 1418, 1459, 1717
pyrH	Pyrimidine	UMP kinase	4.4	904, 940, 941, 987, 1014, 1419, 1717, 2222
pyrI	Pyrimidine	Regulatory polypeptide for aspartate transcarbamoylase, regulatory subunit (EC 2.1.3.2)	96.9	541, 944, 1336, 1717
pyrL	Pyrimidine	Aspartate transcarbamoylase leader peptide	96.9	1336, 1717
qor		Quinoline reductase	92.1	1409, 1717
rbsB	Ribose	*rbsP;* ribose-binding protein	85.0	18, 1383, 1717, 1938, 2149
rck	Resistance to complement killing	17-kDa outer membrane protein; sequence similarity to *pagC*	pSLT	580, 711, 745, 746, 1717
recA	Recombination	Recombination deficient; degrades DNA	62.2	50, 108, 222, 250, 454, 456, 473, 490, 660, 698, 864, 902, 1186, 1234, 1257, 1266, 1397, 1487, 1511, 1617, 1717, 1753, 1820, 1857, 1983, 2150–2152, 2206, 2231
recB	Recombination	Recombination deficient; exonuclease V (EC 3.1.11.15)	65.2	250, 473, 490, 864, 1274, 1277, 1617, 1717, 1835, 1857, 1983
recC	Recombination	Recombination deficient; exonuclease V (EC 3.1.11.15)	65.3	250, 473, 490, 864, 1274, 1277, 1617, 1717, 1835, 1983
recD	Recombination	Endonuclease component of RecBCD enzyme	65.2	1274, 1337, 1617, 1717
recF	Recombination	UV sensitive; recombination defective in *recB sbcBC* background	83.5	1717, 1726
recJ	Recombination	UV sensitive; recombination defective in *recB sbcBC* background	66.2	1274, 1276, 1717
recN	Recombination	Recombination defective in *recB sbcBC* background	58.2	1717
recQ	Recombination	Recombination deficient; DNA helicase activity	86.0	240, 1717
relA	RNA relaxed	RC; regulation of RNA synthesis	64.6	82, 416, 911, 1093, 1300, 1413, 1586, 1689, 1714, 1717, 1783, 1899, 2154
repB	Replication	Plasmid partitioning	pSLT	580, 1717
repC	Replication	Plasmid partitioning	pSLT	580, 1717
rfaB	Rough	UDP-D-galactose:LPS α-1,6-D-galactosyltransferase (EC 2.1.4.–)	81.4	228, 992, 1061, 1717, 2160
rfaC	Rough	LPS core defect; LPS heptosyltransferase I	81.3	215, 223, 224, 727, 1288, 1486, 1717, 1719, 1758, 1884, 1922, 1959
rfaD	Rough	D-Glycero-D-manno-heptose epimerase (EC 5.1.3.–)	81.2	852, 992, 1170, 1288, 1717, 1758, 1814, 1922
rfaE	Rough	LPS core defect; proximal heptose deficient	78.5	39, 44, 215, 586, 624, 629, 725, 727, 752, 981, 1246, 1247, 1255, 1282, 1288, 1410, 1489–1491, 1493, 1494, 1604, 1717, 1719, 1742, 1743, 1825, 1884, 1921, 1922, 1927, 1959, 1989
rfaF	Rough	LPS core defect; ADP-heptose–LPS heptosyltransferase II	81.2	138, 725, 752, 786, 845, 852, 981, 1107, 1259, 1288, 1489, 1490, 1493, 1494, 1717, 1723, 1744, 1758, 1814, 1922, 1950
rfaG	Rough	LPS core defect; glucose I transferase	81.4	216, 228, 392, 506, 786, 981, 992, 1107, 1246, 1247, 1273, 1282, 1288, 1388, 1485, 1489, 1490, 1493, 1494, 1697, 1717, 1723, 1742, 1921, 1922, 2145

(Table continues)

TABLE 1 *Continued*

Gene symbol	Mnemonic	Former or alternative symbol; enzyme deficiency or other phenotype[a]	Cs[b]	References[c]
rfaH	Rough	Deficient in LPS core synthesis and in F-factor expression; transcription control factor	86.1	228, 391, 393, 521, 1211, 1246, 1247, 1288, 1489, 1490, 1493, 1494, 1717, 1725, 1921, 1923, 2145
rfaI	Rough	LPS core defect; LPS 1,3-galactosyltransferase (EC 2.4.1.44)	81.4	228, 285, 991, 992, 1288, 1717, 1814
rfaJ	Rough	LPS core defect; LPS 1,2-glucosyltransferase	81.4	228, 285, 786, 991, 992, 1246, 1247, 1282, 1288, 1489, 1490, 1493, 1494, 1717, 1742, 1814, 1921, 2145
rfaK	Rough	LPS core defect; LPS 1,2-*N*-acetylglucosaminetransferase (EC 2.4.1.56)	81.3	981, 1060, 1213, 1246, 1247, 1258, 1282, 1288, 1489, 1490, 1493, 1494, 1717, 1758
rfaL	Rough	LPS core defect; O-antigen ligase	81.3	786, 1060, 1246, 1247, 1258, 1282, 1288, 1387, 1489, 1490, 1493, 1494, 1523, 1717, 1742, 1921, 2144
rfaP	Rough	LPS core defect; heptose phosphorylation	81.4	474, 751, 981, 1061, 1385, 1717, 1921
rfaY	Rough	LPS core defect	81.3	1060, 1258, 1717, 1758, 1814
rfaZ	Rough	LPS core defect	81.3	1060, 1258, 1717, 1758
rfbA	Rough	*musA, musB;* LPS side chain defect; glucose-1-phosphate thymidylyltransferase (EC 2.7.7.24)	45.2	219, 220, 516, 954, 960, 974, 1030, 1050–1052, 1187, 1214, 1247, 1253, 1282, 1288, 1389, 1433, 1434, 1453, 1491, 1639, 1717, 1921, 1922, 1950, 2079, 2110
rfbB	Rough	LPS side chain defect; dTDP-D-glucose-4,6-dehydratase (EC 4.2.1.46)	45.3	219, 220, 954, 1184, 1247, 1279, 1288, 1304, 1433, 1434, 1666, 1717, 1921, 2099
rfbC	Rough	LPS side chain defect; dTDP-4-dehydroxyrhamnose 3,5-epimerase (EC 5.1.3.13)	45.2	954, 1218, 1304, 1717
rfbD	Rough	LPS side chain defect; NADPH:dTDP-4-dehydroxyrhamnose reductase (EC 1.1.1.133)	45.3	219, 220, 954, 1053, 1054, 1218, 1247, 1288, 1304, 1433, 1434, 1717, 1921
rfbF	Rough	LPS side chain defect; glucose-1-phosphate cytidylyltransferase (EC 2.7.7.33)	45.2	219, 220, 954, 1215, 1216, 1247, 1288, 1433, 1434, 1717, 1921
rfbG	Rough	LPS side chain defect; CDP-glucose-4,6-dehydratase (EC 4.2.1.45)	45.2	219, 220, 954, 1216, 1218, 1247, 1433, 1434, 1717, 1921
rfbH	Rough	LPS side chain defect; CDP-6-deoxy-D-xylo-4-hexulose-3-dehydrase	45.1	219, 220, 954, 1216, 1218, 1247, 1400, 1433, 1434, 1717, 1921, 2218
rfbI	Rough	LPS side chain defect; CDP-6-deoxy-Δ3,4-glucoseen reductase	45.2	954, 1216, 1717
rfbJ	Rough	LPS side chain defect; CDP-abequose synthase (EC 4.2.1.–)	45.1	220, 954, 1216, 1218, 1219, 1717
rfbK	Rough	LPS side chain defect; phosphomannomutase (EC 5.4.2.8)	45.0	219, 220, 954, 1218, 1247, 1433, 1434, 1717, 1921

TABLE 1 *Continued*

Gene symbol	Mnemonic	Former or alternative symbol; enzyme deficiency or other phenotype[a]	Cs[b]	References[c]
rfbM	Rough	LPS side chain defect; mannose-1-phosphate guanylyltransferase (EC 2.7.7.22)	45.0	219, 220, 954, 1218, 1247, 1433, 1434, 1717, 1921
rfbN	Rough	LPS side chain defect; rhamnosyltransferase	45.0	219, 220, 954, 1218, 1696, 1697, 1717
rfbP	Rough	*rfbT*; bifunctional enzyme; undecaprenyl-phosphate galactosephosphotransferase; may also relocate O-antigen from cytoplasmic to periplasmic face of cytoplasmic membrane	44.9	219, 954, 1218, 1247, 1287, 1433, 1717, 1921, 2098, 2144
rfbU	Rough	LPS side chain defect; mannosyltransferase	45.0	954, 1218, 1717
rfbV	Rough	LPS side chain defect; abequosyltransferase; putative	45.1	954, 1717
rfbX	Rough	LPS side chain defect; rhamnosyltransferase; putative (EC 2.4.1.–)	45.1	954, 1717
rfc	Rough	*rouC*; O-antigen polymerase	35.7	356, 358, 652, 1247, 1287, 1404, 1411, 1433, 1453, 1559, 1717, 1921, 1952
rfe	Rough	Defect in synthesis of enterobacterial common antigen, the T1 antigen, and O-side chains of *Salmonella* groups L and C1	85.4	245, 932, 981, 1187, 1188, 1281, 1283–1285, 1717
rffM	Rough	*rff*; synthesis of enterobacterial common antigen; UDP-*N*-acetyl-D-mannosaminuronic acid transferase (EC 2.4.1.–)	85.6	459, 1187, 1188, 1241, 1284, 1286, 1325, 1717
rffT	Rough	*rff*; synthesis of enterobacterial common antigen; 4-α-L-fucosyl transferase (EC 2.4.1.–)	85.5	459, 1188, 1241, 1284, 1286, 1717
rft	Rough	"Transient" T1 forms	16.2	160, 245, 801, 1287, 1717, 1729
rfu	Rough	"Transient" T1 forms	NM	245, 1717, 1730
rhaA	Rhamnose	L-Rhamnose isomerase (EC 5.3.1.14)	88.1	36, 500, 1439, 1473, 1715, 1717
rhaB	Rhamnose	L-Rhamnulokinase (EC 2.7.1.5)	88.1	36, 500, 1439, 1473, 1715, 1717
rhaD	Rhamnose	L-Rhamnulose-1-phosphate aldolase (EC 4.1.2.19)	88.1	36, 500, 1439, 1473, 1715, 1717
rhaR	Rhamnose	*rhaC1*; regulation	88.1	36, 500, 1439, 1473, 1715, 1717
rhaS	Rhamnose	*rhaC2*; regulation	88.1	36, 500, 1439, 1473, 1715, 1717
rhaT	Rhamnose	L-Rhamnose transport	88.1	36, 500, 1439, 1473, 1715, 1717
rhlB		RNA helicase; putative	85.4	582, 1360, 1717
rho		*psu*; polarity suppressor; transcription terminator factor Rho	85.3	339, 835, 836, 1158, 1717
rna	RNase	*rnsA*; RNase I	15.2	52, 248, 299, 1172, 1495, 1717, 2113
rnc	RNase	RNase III	57.6	52, 1172, 1495, 1717, 1963
rnhA	RNase H	RNase H (EC 3.1.26.4)	6.9	918, 1717
rnhB	RNase H	RNase HII (EC 3.1.26.4)	5.9	1139, 1717
rnpA	RNase P	RNase P, protein component (EC 3.1.26.5)	83.8	1692, 1717
rnpB	RNase P	RNase P, RNA subunit, M1 RNA	71.1	96, 1717
rodA	Rod	Round cell morphology; mecillinam resistance	15.4	65, 68, 383, 1717
rph	RNase PH	tRNA nucleotidyltransferase; RNase PH (EC 2.7.7.56)	81.5	1422, 1717
rplA	Ribosomal protein, large	50S ribosomal subunit protein L1	90.1	1717, 2230
rplE	Ribosomal protein, large	50S ribosomal subunit protein L5	74.6	297, 1717
rplJ	Ribosomal protein, large	50S ribosomal subunit protein L10	90.2	1551, 1717, 1971, 2018, 2230
rplK	Ribosomal protein, large	50S ribosomal subunit protein L11	90.1	1717, 2230
rplL	Ribosomal protein, large	50S ribosomal subunit protein L7/L12	90.2	1717, 1971, 2018, 2230
rplS	Ribosomal protein, large	50S ribosomal subunit protein L19	57.9	1717
rplX	Ribosomal protein, large	50S ribosomal subunit protein L24	74.6	297, 1717
rpoA	RNA polymerase	*oxrB*; RNA polymerase, α-subunit (EC 2.7.7.6)	74.5	523, 1230, 1717
rpoB	RNA polymerase	*rif*; RNA polymerase, β-subunit (EC 2.7.7.6)	90.2	98, 214, 585, 945, 946, 1419, 1717, 1970, 1971, 2018, 2019, 2024, 2210
rpoC	RNA polymerase	RNA polymerase, β'-subunit (EC 2.7.7.6)	90.2	214, 585, 945, 946, 1717, 1971, 2018, 2019, 2210, 2211
rpoD	RNA polymerase	RNA polymerase, σ^{70} subunit	70.0	509, 753, 1717, 1738
rpoE	RNA polymerase	RNA polymerase, σ^{E} subunit	57.7	1717

(Table continues)

TABLE 1 *Continued*

Gene symbol	Mnemonic	Former or alternative symbol; enzyme deficiency or other phenotype[a]	Cs[b]	References[c]
rpoN	RNA polymerase	*glnF, ntrA;* RNA polymerase, σ^{54} subunit (EC 2.7.7.6)	72.2	230, 524, 581, 612, 1090, 1112, 1316, 1319, 1569, 1570, 1717, 2117
rpoS	RNA polymerase	*katF;* RNA polymerase, σ^{S} subunit	63.6	519, 522, 1085, 1448, 1477, 1717, 1932
rpsD	Ribosomal protein, small	30S ribosomal subunit protein S4	74.5	1230, 1717
rpsE	Ribosomal protein, small	*spcA;* 30S ribosomal subunit protein S5	74.6	1717, 2182
rpsG	Ribosomal protein, small	30S ribosomal subunit protein S7	74.7	958, 1717
rpsL	Ribosomal protein, small	*strA;* 30S ribosomal subunit protein S12	74.7	859, 870, 958, 1717, 1984, 2032, 2034, 2182
rpsP	Ribosomal protein small	30S ribosomal subunit protein S16	57.9	1717
rpsU	Ribosomal protein, small	30S ribosomal subunit protein S21	70.0	509, 1717
rrfB		5S rRNA gene of *rrnB* operon	90.0	458, 1717
rrnA	rRNA	rRNA operon	87.0	52, 257, 1171–1173, 1495, 1717, 1819
rrnB	rRNA	rRNA operon	90.0	52, 257, 1171–1173, 1495, 1717, 1819
rrnC	rRNA	rRNA operon	85.0	52, 257, 1171–1173, 1495, 1717, 1819
rrnD	rRNA	rRNA operon	74.3	257, 1171, 1495, 1717, 1819
rrnE	rRNA	rRNA operon	90.9	52, 257, 1171–1173, 1495, 1717, 1819
rrnG	rRNA	rRNA operon	58.1	257, 1171, 1717, 1819
rrnH	rRNA	rRNA operon	6.2	52, 257, 1171, 1495, 1717, 1819
rsk		Binding site for a regulatory element for virulence traits	pSLT	1717, 2066, 2067
samA	*Salmonella* mutagenesis	Mutagenesis by UV and mutagens; related to *umuDC* operon	pSLT	1064, 1443, 1444, 1717
samB	*Salmonella* mutagenesis	Mutagenesis by UV and mutagens; related to *umuDC* operon	pSLT	1064, 1443, 1444, 1717
sapA	Sensitive to antimicrobial peptides	Resistance to antimicrobial peptides melittin and protamine	37.1	1540, 1717
sapB	Sensitive to antimicrobial peptides	Resistance to antimicrobial peptides melittin and protamine	37.2	1540, 1717
sapC	Sensitive to antimicrobial peptides	Resistance to antimicrobial peptides melittin and protamine	37.2	686, 1540, 1717
sapD	Sensitive to antimicrobial peptides	Resistance to antimicrobial peptides melittin and protamine	37.2	686, 1540, 1717
sapE	Sensitive to antimicrobial peptides	Resistance to antimicrobial peptides melittin and protamine	3.4	686, 1717
sapF	Sensitive to antimicrobial peptides	Resistance to antimicrobial peptides melittin and protamine	37.2	1540, 1717
sapI	Sensitive to antimicrobial peptides	Resistance to antimicrobial peptides melittin and protamine	38.7	686, 1717
sapJ	Sensitive to antimicrobial peptides	Resistance to antimicrobial peptides melittin and protamine	87.7	686, 1717
sapK	Sensitive to antimicrobial peptides	Resistance to antimicrobial peptides melittin and protamine	87.7	686, 1717
sapL		Resistance to antimicrobial peptides melittin and protamine	44.1	686, 1717
sapM	Sensitive to antimicrobial peptides	Resistance to antimicrobial peptides melittin and protamine	6.6	686, 1717
sbcB	Suppressor of *recBC*	Suppressor of *recBC* mutations	41.6	151, 1717
sbcC	Suppressor of *recBC*	Suppressor of *recBC* mutations	9.2	151, 1717
sbcD	Suppressor of *recBC*	Suppressor of *recBC* mutations	9.2	151, 1717
sbcE	Suppressor of *recBC*	Unstable suppressor of *recBC* mutations	57.0	1717
sbp	Sulfate-binding protein	Periplasmic sulfate-binding protein	88.3	616, 914, 1560, 1561, 1717
selA	Selenium	*fdhA;* selenium metabolism; biosynthesis of selenoproteins	80.3	118, 325, 326, 1092, 1547, 1717, 1905
selB	Selenium	*fdhA;* selenium metabolism; translation factor necessary for biosynthesis of selenoproteins	80.3	1717, 1905
selC	Selenium	*fdhC;* selenium metabolism; selenocysteine tRNA	81.5	118, 1717, 1905
selD	Selenium	*selA;* selenium metabolism; biosynthesis of selenoproteins and selenocysteine tRNA	29.0	1717, 1905
serA	Serine	Phosphoglycerate dehydrogenase (EC 1.1.1.95)	66.8	1717, 2045, 2220
serB	Serine	Phosphoserine phosphatase (EC 3.1.3.3)	99.5	558, 1717, 2045
serC	Serine	Requirement	19.6	680, 806, 808, 1717
serD	Serine	Requirement for pyridoxine plus L-serine or glycine	47.2	1717, 1718, 1721

TABLE 1 *Continued*

Gene symbol	Mnemonic	Former or alternative symbol; enzyme deficiency or other phenotype[a]	Cs[b]	References[c]
serV	Serine	Serine tRNA₃	62.0	173, 1717
sgdA	Suppressor of gyrase	Restores *his* attenuation in *gyrA* mutants and suppresses other *gyrA* phenotypes	62.0	173, 1717
sidC	Siderochrome	Siderochrome utilization; ferrichrome transport; albomycin resistance	3.8	1244, 1245, 1717
sidF	Siderochrome	Siderochrome utilization; ferrichrome transport; albomycin resistance	3.8	1244, 1245, 1717
sidK	Siderochrome	Siderochrome utilization; albomycin resistance; receptor of phage ES18 in *S. typhimurium* and of T5 in *S. paratyphi* B	5.0	227, 673, 1244, 1245, 1717
sinR	*Salmonella* insert regulator	Protein in the LysR family of transcription regulators; gene is unique to *Salmonella*	104.0	688, 1717
slt		Soluble lytic transglycosylase	99.9	1717
slyA	Salmolysin	*cyx*; hemolysin, required for survival in macrophages and for virulence	30.8	1197, 1198, 1717
smoB	Smooth	Smooth colony morphology in histidine-constitutive mutants	98.6	34, 1717
smpB	Small protein	Reduced survival in macrophages	59.5	129, 1717
sms		Repair of endogenous alkylation damage; putative; similarity with ATP-dependent proteases Lon and RecA	99.9	563, 1717
smvA		Methyl viologen resistant	38.6	826, 1717
sodB		Iron superoxide dismutase; putative (EC 1.15.1.1)	30.7	1717
spaK	Surface presentation of antigens	*invB*; reduced invasion; secretory pathway	63.1	483, 600, 603, 685, 970, 1359, 1717
spaL	Surface presentation of antigens	*invC*; reduced invasion; secretory pathway, homologous to *spa* in *Shigella*	63.1	483, 600, 603, 639, 685, 1359, 1717
spaM	Surface presentation of antigens	*invI*; reduced invasion; secretory pathway	63.1	352, 685, 1310, 1359, 1717
spaN	Surface presentation of antigens	*invJ*; reduced invasion; secretory pathway	63.0	352, 685, 1310, 1359, 1717
spaO	Surface presentation of antigens	Reduced invasion; secretory pathway	63.0	685, 1359, 1717
spaP	Surface presentation of antigens	Reduced invasion; secretory pathway	63.0	685, 1359, 1717
spaQ	Surface presentation of antigens	Reduced invasion; secretory pathway	63.0	685, 1359, 1717
spaR	Surface presentation of antigens	Reduced invasion; secretory pathway	63.0	685, 1359, 1717
spaS	Surface presentation of antigens	Reduced invasion; secretory pathway	63.0	685, 1359, 1717
spaT	Surface presentation of antigens	Reduced invasion; secretory pathway	62.9	685, 1359, 1717
spcB	Spectinomycin	Nonribosomal resistance	74.5	1717, 2182
spoT	Spot	Guanosine 5′-diphosphate, 3′-diphosphate pyrophosphatase	81.6	418, 1689, 1691, 1717
spvA	*Salmonella* plasmid virulence	*mkaB, vsdB*; hydrophilic protein, 28 kDa, outer membrane protein	pSLT	387, 701, 702, 1085, 1717, 1988
spvB	*Salmonella* plasmid virulence	*vsdC, mkaA, mkfB*; hydrophilic protein, 66 kDa, cytoplasmic protein	pSLT	387, 701, 702, 1085, 1447, 1448, 1456, 1717, 1987, 1988
spvC	*Salmonella* plasmid virulence	*virA, mkaD, mkfA, vsdD*; hydrophilic protein, 28 kDa	pSLT	387, 700–702, 1085, 1445, 1446, 1717, 1988
spvD	*Salmonella* plasmid virulence	*virB, vsdE*; hydrophilic protein, 25 kDa	pSLT	701, 702, 1085, 1445, 1717
spvR	*Salmonella* plasmid virulence	*mkaC, vsdA*; regulation of the *spv* operon; member of LysR family of positive regulatory proteins	pSLT	270, 387, 702, 1085, 1094, 1717, 1988
srlA	Sorbitol	D-Glucitol-specific enzyme II of the phosphotransferase system	62.5	1717, 1728
srlB	Sorbitol	D-Glucitol-specific enzyme III of the phosphotransferase system	62.5	1717, 1728
srlC	Sorbitol	*gut*; regulatory gene	62.4	1056, 1717, 1728
srlD	Sorbitol	Sorbitol-6-phosphate dehydrogenase (EC 1.1.1.140)	62.5	1717, 1721
srlM	Sorbitol	DNA-binding protein which activates transcription of the *srl* operon	62.5	1717, 1721
srlR	Sorbitol	Regulatory gene	62.5	1717, 1721
ssb	Single-strand binding	Single-strand DNA-binding protein	92.3	1313, 1717
stiA	Starvation inducible	*sinA*; induced by starvation for carbon source or other requirements	35.8	547, 1717, 1863, 1866
stiB	Starvation inducible	Induced by starvation for carbon source or other requirements	NM	1717, 1863, 1866
stiC	Starvation inducible	Induced by starvation for carbon source or other requirements	78.2	1717, 1863, 1866
stiD	Starvation inducible	Induced by starvation for carbon source or other requirements	35.9	1717, 1863, 1866
stiE	Starvation inducible	Induced by starvation for carbon source or other requirements	44.0	1717, 1863, 1866
stiF	Starvation inducible	Induced by starvation for carbon source or other requirements	NM	1717, 1863, 1866
stiG	Starvation inducible	Induced by starvation for carbon source or other requirements	88.2	1717, 1863, 1866
stiH	Starvation inducible	Induced by starvation for carbon source or other requirements	57.4	1717, 1863, 1866
stn	Salmonella toxin	*stx*; enterotoxin	NM	328, 329, 1584, 1717
strB	Streptomycin	Low-level resistance plus auxotrophy; nonribosomal	55.7	1530, 1679, 1717, 2173
strC	Streptomycin	Streptomycin resistance; nonribosomal	NM	672, 1717

(Table continues)

TABLE 1 *Continued*

Gene symbol	Mnemonic	Former or alternative symbol; enzyme deficiency or other phenotype[a]	Cs[b]	References[c]
sucA	Succinate	lys, suc; succinate requirement; α-ketoglutarate dehydrogenase, decarboxylase component	17.1	1140, 1717
sufA	Suppressor of frameshifts	Suppressor affecting proline tRNA and correcting +1 frameshifts at runs of C in the mRNA	79.9	1069, 1103, 1643, 1644, 1717
sufB	Suppressor of frameshifts	sufC; recessive suppressor of +1 frameshifts at runs of C in the mRNA	16.8	1069, 1103, 1643, 1644, 1717, 1879
sufD	Suppressor of frameshifts	Frameshift suppressor affecting glycine tRNA and correcting +1 frameshifts at runs of G in the mRNA	65.8	1069, 1642–1644, 1717
sufE	Suppressor of frameshifts	Frameshift suppressor correcting +1 frameshifts at runs of G in the mRNA	90.7	1069, 1643, 1644, 1717
sufF	Suppressor of frameshifts	Recessive frameshift suppressor correcting +1 frameshifts at runs of G in the mRNA	12.4	1643, 1644, 1717
sufG	Suppressor of frameshifts	supI; frameshift suppressor correcting +1 frameshifts at runs of A in the mRNA	16.6	1069, 1072, 1717
sufH	Suppressor of frameshifts	Suppressor	53.2	1069, 1717
sufI	Suppressor of frameshifts	Suppressor	12.5	1069, 1717
sulA	Suppressor of lon	Lacks SOS-induced filamentation	24.4	572, 628, 1717
sumA	Suppressor of missense	Suppressor	95.2	1717, 1718, 1721
supC	Suppressor	Ochre suppressor	38.6	1717, 2157
supD	Suppressor	Amber suppressor; serine insertion	42.9	200, 570, 866, 1717, 2157
supE	Suppressor	supY; amber suppressor; glutamine insertion	16.7	152, 200, 206, 866, 1717, 2157
supG	Suppressor	Ochre suppressor; lysine insertion	NM	1717, 2157
supJ	Suppressor	supH; amber suppressor; leucine insertion	85.1	200, 663, 1717, 2157
supQ	Suppressor	Suppressor of nonsense and deletion mutations of leuD	7.9	591, 1022, 1024, 1025, 1132, 1133, 1717, 1936
supR	Suppressor	Amber suppressor; haploid lethal	85.1	1348, 1717
supS	Suppressor	UGA suppressor; haploid lethal	85.1	1348, 1717
supU	Suppressor	Suppressor of UGA mutations; may be due to alteration of ribosome structure	74.8	968, 1717
tar	Taxis-associated receptor	Chemotaxis transduction polypeptide; aspartate receptor	41.9	401, 427, 428, 545, 1338, 1348, 1369, 1698, 1717, 2096
tcp		Transmembrane receptor for citrate (attractant) and phenol (repellent)	NM	1717, 2193
tctA	Tricarboxylate transport	Membrane protein	60.9	883, 898–901, 1717, 1856, 2138–2140
tctB	Tricarboxylate transport	Membrane protein	60.8	1717, 2138–2140
tctC	Tricarboxylate transport	Tricarboxylate-binding protein	60.8	83, 900, 1717, 1855, 1856, 1972, 1973, 2138-2140
tctD	Tricarboxylate transport	Regulatory protein	60.8	1717, 2138–2140
tctII	Tricarboxylate transport	Transport	17.6	1717, 1718, 1721
tctIII	Tricarboxylate transport	triM, triR; transport	1.3	900, 1717
tdcB		Catabolic threonine dehydratase	70.9	1036, 1717
tdk		Thymidine kinase (EC 2.7.1.21)	38.4	134, 824, 1717
tesA		apeA; thioesterase I (EC 3.1.2.–)	12.3	327, 749, 1344, 1350, 1717
thiA	Thiamine	thiG; thiamine or thiazole moiety	90.5	824, 1144, 1717
thiC	Thiamine	thiA; thiamine or pyrimidine moiety	90.5	1717, 1718, 1721
thiD	Thiamine	Thiamine requirement	50.5	1717, 1718, 1721
thiE	Thiamine	Thiazole type	55.2	1717, 1718, 1721
thiF	Thiamine	Thiazole type	55.3	1717, 1718, 1721
thiH	Thiamine	thiB; thiamine requirement	56.6	1717, 1718, 1721
thiI	Thiamine	thiC; thiazole type	10.5	1717, 1718, 1721
thrA	Threonine	thrC, thrD; aspartokinase I and homoserine dehydrogenase I (EC 2.7.2.4, EC 1.1.1.3)	0.0	642, 939, 1717, 1947
thrB	Threonine	thrA; homoserine kinase (EC 2.7.1.39)	0.0	642, 1717, 1947
thrC	Threonine	thrB; threonine synthase (EC 4.2.99.2)	0.1	642, 1717, 1947
thrT	Threonine	sufJ; threonine tRNA	90.0	205, 207, 209, 231, 1068, 1069, 1717
thrW	Threonine	Threonine tRNA2; sequence contains attachment site for prophage P22; see ataA	7.8	1178
thyA	Thymine	Requirement	65.3	489, 976, 1717, 1835
tip	Taxis-involved protein	Methyl-accepting chemotaxis protein, aspartate receptor	NM	1699, 1717
tkt		Transketolase (EC 2.2.1.1)	NM	484, 1717
tlpA		Prokaryotic coiled-coil protein	pSLT	878, 1082, 1717
tlpB		tlp; loss of protease II	40.4	748, 1717
tlr		Thiolutin resistance; P22 development at high temperature	NM	978–980, 1717

TABLE 1 *Continued*

Gene symbol	Mnemonic	Former or alternative symbol; enzyme deficiency or other phenotype[a]	Cs[b]	References[c]
tonB	T-one	*chr;* regulates levels of some outer membrane proteins; resistance to ES18; determines a salmonellocin; affects iron transport	38.2	146, 227, 237, 381, 461, 730, 777, 1717, 1925
topA	Topoisomerase	*supX, top;* DNA topoisomerase I	37.7	313, 476, 477, 579, 587, 602, 671, 764, 983, 1175, 1206, 1386, 1487, 1511, 1512, 1592–1594, 1631, 1632, 1717, 1752, 1976, 2029, 2195
topB	Topoisomerase	DNA topoisomerase III (EC 5.99.1.2)	29.1	105, 1717
tor		Trimethylamine oxide reductase	82.5	1126–1128, 1717
tppB	Tripeptide permease	Resistance to alafosfalin; tripeptide permease	32.1	309, 632, 765, 927, 928, 1424, 1717
tppR	Tripeptide permease	Regulator of tripeptide permease	4.7	928, 1717
traT	Transfer	Membrane protein cross-reacts immunologically with TraT protein of F plasmid; restores permeability mutants to normal	pSLT	1628, 1717, 1955–1958, 1960, 1961, 2064
treA	Trehalose	*tre;* utilization	40.2	1579, 1580, 1717, 1880
treR	Trehalose	Trehalose regulation	96.6	1717
trkA	Transport of potassium	*sapG;* potassium uptake protein	74.4	686, 1541, 1717
trmA		tRNA (m^{5}U54) methyltransferase (EC 2.1.1.35)	89.7	704, 1717
trmD		tRNA (m^{1}G37) methyltransferase (EC 2.1.1.31)	57.9	172, 1717
trpA	Tryptophan	*trpC;* tryptophan synthetase, α-subunit (EC 4.2.1.20)	38.1	14, 51, 102, 127, 190, 292, 389, 394, 439, 447, 450, 721, 880, 1002, 1193, 1194, 1332, 1364, 1374, 1425, 1428, 1717, 1761, 1828, 1881, 1902, 2155, 2202, 2219
trpB	Tryptophan	*trpD;* tryptophan synthetase, β-subunit (EC 4.2.1.20)	38.1	51, 102, 127, 190, 249, 390, 394, 439, 450, 721, 880, 1012, 1364, 1374, 1717, 1761, 1775, 1828, 1992
trpC	Tryptophan	*trpE; N-*(5-phosphoribosyl) anthranilate isomerase and indole-3-glycerol phosphate synthase (EC 5.1.3.24, EC 4.1.1.48)	38.1	102, 103, 127, 190, 394, 439, 678, 830, 1192, 1374, 1627, 1717, 1775
trpD	Tryptophan	*trpB;* anthranilate phosphoribosyltransferase (EC 2.4.2.18)	38.0	103, 126–128, 190, 191, 394, 439, 460, 677, 678, 693, 694, 830, 879, 1146, 1147, 1293, 1374, 1402, 1403, 1426, 1660, 1717, 1948, 1993, 1994
trpE	Tryptophan	*trpA;* anthranilate synthase (EC 4.1.3.27)	38.0	103, 126, 128, 191, 273, 275, 311, 394, 410, 571, 587, 677–679, 693, 694, 1028, 1147, 1159, 1190, 1191, 1293, 1374, 1402, 1403, 1427, 1624, 1660, 1717, 1991, 1992, 1994, 2203
trpR	Tryptophan	Resistance to 5-methyltryptophan; derepression of tryptophan enzymes	99.9	103, 1717, 1946
trxA	Thioredoxin	Thioredoxin	85.3	1083, 1717
tsr		Chemotaxis receptor; serine specificity	98.5	1717, 1721
tsx	T-six	Nucleoside uptake; receptor for phage T6 (in *E. coli*)	9.1	1717
ttr		Tetrathionate reductase	40.0	291, 1717
tufA		Protein chain elongation factor EF-Tu	74.7	6, 88, 596, 854–858, 958, 1717, 2032–2035

(Table continues)

TABLE 1 *Continued*

Gene symbol	Mnemonic	Former or alternative symbol; enzyme deficiency or other phenotype[a]	Cs[b]	References[c]
tufB		Protein chain elongation factor EF-Tu	90.1	88, 854–858, 1717, 2032, 2034, 2035
tyn		Tyramine oxidase	NM	1396, 1717
tyrA	Tyrosine	Requirement	58.3	1717, 1875, 1993
tyrR	Tyrosine	Regulator gene for *aroF* and *tyrA*	37.1	432, 658, 1437, 1717, 1875
tyrS	Tyrosine	Tyrosyl-tRNA synthetase	30.9	1197, 1198, 1717
tyrT	Tyrosine	*supC, supF;* ochre suppressor; tyrosine tRNA$_1$	38.4	200, 211, 510, 866, 1717, 2157
tyrU	Tyrosine	*supM;* ochre suppressor; tyrosine tRNA$_2$	90.0	1717, 2157
ubiF	Ubiquinone	*cad;* deficient in ubiquinone synthesis; accumulates 2-octaprenyl-3-methyl-6-methoxy-1,4-benzoquinone	16.3	1021, 1717, 2223
ubiG	Ubiquinone	Ubiquinone synthesis	49.3	1717
ubiX	Ubiquinone	Growth stimulation by *p*-hydroxybenzoic acid; 3-octaprenyl-4-hydroxybenzoate decarboxylase	51.3	73, 74, 122, 842, 1717
udk		Uridine kinase (EC 2.7.1.48)	45.7	134, 1717, 2146
udp		Uridine phosphorylase (EC 2.4.2.3)	86.3	133, 1418, 1717
uhpA		Utilization of hexose phosphate	82.1	915, 1717
uhpB		Utilization of hexose phosphate	82.0	915, 1717
uhpC		Utilization of hexose phosphate	82.0	915, 1717
uhpT		Hexosephosphate transport	82.0	485, 486, 915, 1637, 1717
umuC		Induction of mutations by UV; error-prone repair	42.9	488, 759, 1064, 1443, 1444, 1717, 1774, 1817, 1820, 1827, 2009–2011
umuD		Induction of mutations by UV; error-prone repair	42.9	1064, 1443, 1444, 1717, 1774, 1820, 1827, 2009–2011, 2167
upp		Uracil phosphoribosyltransferase (EC 2.4.2.9)	54.2	133, 1717, 2146
urs	Uracil	Uracil catabolism defect	35.2	1717, 2128
ushA	UDP sugar hydrolase	UDP-sugar hydrolase (5′-nucleotidase); silent gene in *Salmonella* spp. of subgenus I	12.2	258, 259, 481, 1363, 1717, 2004
ushB	UDP sugar hydrolase	Inner membrane-associated UDP-sugar hydrolase	88.3	258, 259, 481, 616, 1717
usp	Ureidosuccinate	Permeability to ureidosuccinate	NM	1717, 2223
uvrA	UV	Repair of UV damage to DNA; exinuclease ABC, subunit A	92.3	59–61, 406, 1075, 1101, 1717, 1898, 2016
uvrB	UV	Repair of UV damage to DNA; exinuclease ABC, subunit B	18.4	50, 60, 902, 903, 1234, 1397, 1717, 1815, 1933, 1983
uvrC	UV	Repair of UV damage to DNA; exinuclease ABC, subunit C	42.2	1717, 1816
uvrD	UV	Repair of UV damage to DNA; helicase II (EC 3.6.1.–)	85.8	1234, 1526–1528, 1717, 1780, 1781, 2148, 2231
valS	Valine	Valyl-tRNA synthetase (EC 6.1.1.9)	97.3	1539, 1717
viaA		*ViA;* Vi antigen (in *S. typhi*)	50.3	961, 963, 1717, 1851
xylA	D-Xylose	Xylose isomerase (EC 5.3.1.5)	80.0	626, 1717, 1722, 1723, 1778, 1779
xylB	D-Xylose	Xylulokinase (EC 2.7.1.17)	80.0	626, 1717, 1778, 1779
xylR	D-Xylose	Regulation	80.0	626, 1717, 1778, 1779
xylT	D-Xylose	Transport	80.0	626, 1379, 1717, 1778, 1779

[a]Abbreviations: CoA, coenzyme A; cAMP, cyclic AMP; CRP, cAMP receptor protein; DAHP, 3-deoxy-D-arabinoheptulosonic acid 7-phosphate; DMB, dimethylbenzimidazole; HP, hydrogen peroxide; LPS, Lipopolysaccharide; NAMN, nicotinic acid, mononucleotide; NMN, nicotinamide mononucleotide; PRPP, phosphoribosyl pyrophosphate

[b]Map position in centisomes (Cs) indicates the location of the gene on the map in Fig. 1, from 0 to 100 Cs. NM indicates that gene is on the chromosome, but the map position is not known. The symbol pSLT indicates that the gene is on pSLT, the 90-kb plasmid of LT2; a map of pSLT is shown in Fig. 1 (p. 1957).

[c]Reference 1721 refers to edition VII of the linkage map, in which other references to the indicated gene are given. Also, there are many papers cited in reference 1718 which have important information on the genes of *S. typhimurium*.

TABLE 2 Alternative gene symbols[a]

Former or alternative symbol	Current symbol	Former or alternative symbol	Current symbol
ail	pagC	cobI	cbiQ
aniA	hydA	cobI	cbiT
aniE	phsA	cobII	cobJ
apaG	corD	cobII	cobK
apeA	tesA	cobII	cobL
apeD	apeR	cobII	cobM
argA	argE	cobIII	cobS
argB	argA	cobIII	cobT
argC	argB	cobIII	cobU
argD	carA	cobIV	cobA
argE	argG	cpd	cpdB
argF	argH	cya	cyaA
argG	argD	cysAa	cysU
argH	argC	cysAb	cysW
argT	argX	cysT	cysU
argU	argX	cyx	slyA
aroC	aroE	dad	dadA
aroD	aroC	degP	htrA
aroE	aroD	dra	deoC
ars	carA	drm	deoB
asc	ent	enb	ent
atbR	atrR	envB	mre
atrC	polA	envM	fabI
attP14	atdA	fdhA	selA
attP22 I	ataA	fdhA	selB
attP221	atcA	fdhC	selC
attP27 I	atbA	fhl	fdhF
attP27 II	atbB	fhlB	hydA
bac	mre	fim	fimA
bfe	btuB	fimU	argU
bglY	hns	firA	lpxD
bio	bioA	flaAI	fliE
cad	ubiF	flaAII.2	fliG
cap	carA	flaAII.3	fliH
capR	lon	flaAII.1	fliF
carA	ptsI	flaAIII	fliI
carB	ptsH	flaB	fliP
cheC	fliM	flaC	flhA
cheP	cheA	flaD	fliQ
cheQ	cheY	flaE	flhC
cheT	cheZ	flaFI	flgA
cheU	fliM	flaFII	flgB
cheV	fliG	flaFIII	flgC
cheX	cheB	flaFIV	flgD
chlA	moaA	flaFIX	flgI
chlB	mob	flaFV	flgE
chlC	nar	flaFVI	flgF
chlD	modC	flaFVII	flgG
chlE	moeA	flaFVIII	flgH
chr	tonB	flaFX	flgJ
clmF	parC	flaK	flhD
clmF	parE	flaL	fliA
clmF	parF	flaM	flhB
cls	katG	flaN	fliN
cobI	cbiA	flaP	fliO
cobI	cbiB	flaQI	fliL
cobI	cbiC	flaQII	fliM
cobI	cbiD	flaR	fliK
cobI	cbiE	flaS	fliJ
cobI	cbiF	flaU	flgL
cobI	cbiG	flaV	fliD
cobI	cbiH	flaW	flgK
cobI	cbiJ	flaX	fliR
cobI	cbiK	flgR	flgM
cobI	cbiL	fruA	ptsF
cobI	cbiM	fruF	fruB
cobI	cbiN	fuc	fucA
cobI	cbiO	fur	furA
cobI	cbiP	gdh	gdhA

TABLE 2 *Continued*

Former or alternative symbol	Current symbol	Former or alternative symbol	Current symbol
glnF	rpoN	ompB	envZ
glnR	ntrB	ompB	ompR
glnR	ntrC	omsA	lpxD
glu	ptsG	optA	opdA
gpt	ptsG	orf5	pefK
gut	srlC	orf6	pefL
gxu	gpt	orf8	pefS
H1	fliC	osmZ	hns
H2	fljB	oxrA	fnr
hisE	hisI	oxrB	rpoA
hisIE	hisI	oxrC	pgi
hisU	gyrB	parA	gyrB
hisW	gyrA	pasA	pgi
hsdSB	hsdM	pck	pckB
hsdSB	hsdS	phs	phsA
hspLT	hsdL	pil	fimA
hyd	hydA	pilG	hns
ile	ilvA	pmi	manA
ilvA	ilvC	pnu	deoD
ilvB	ilvD	pnuA	nadR
ilvC	ilvE	poh	oriC
ilvT	brnQ	pox	ply
incL	parS	proT	proM
invB	spaK	proU	proV
invC	spaL	proU	proW
invD	gcv	proU	proX
invI	spaM	proW	proL
invJ	spaN	prp	prpA
katF	rpoS	prsB	prsA
lep	lepB	psu	rho
leuT	leuU	ptdD	pepD
liv	livS	ptdN	pepN
livA	livH	ptdP	pepP
livB	livJ	pup	deoD
livC	livK	putB	putA
livEF	livM	pyrA	carA
livF	livG	pyrA	carB
livG	livF	rbsP	rbsB
lnt	cutE	RC	relA
lom	pagC	rfbL	cpsG
lys	lysA	rfbT	rfbP
lys	sucA	rff	rffM
malB	malE	rff	rffT
manA	ptsM	rflB	flgM
metI	metA	rhaC1	rhaR
metP	metD	rhaC2	rhaS
mglD	galS	rhl	fljA
mkaA	spvB	rif	rpoB
mkaB	spvA	rnsA	rna
mkaC	spvR	rol	cld
mkaD	spvC	rouC	rfc
mkfA	spvC	sapG	trkA
mkfB	spvB	selA	selD
motC	fliG	sinA	stiA
musA	rfbA	smoA	divC
musB	rfbA	spcA	rpsE
mutB	mutY	ssc	lpxD
mviB	mviM	strA	rpsL
mviS	flgM	stx	stn
nadI	nadR	suc	sucA
nag	nagA	sufB	proL
nalA	gyrA	sufC	sufB
nfsI	nfnB	sufJ	thrT
nic	nadB	sufS	glyT
nicA	nadA	sulB	ftsZ
nml	fliB	supC	tyrT
ntrA	rpoN	supF	tyrT
O-5	oafA	supH	supJ
ofi	oafA	supI	supG

(Table continues)

TABLE 2 *Continued*

Former or alternative symbol	Current symbol	Former or alternative symbol	Current symbol
supK . . . . *prfB*		*thrC* . . . . *thrA*	
supM . . . . *tyrU*		*thrD* . . . . *thrA*	
supT . . . . *prfB*		*tlp* . . . . *tlpB*	
supX . . . . *topA*		*top* . . . . *topA*	
supY . . . . *supE*		*tpp* . . . . *deoA*	
thiA . . . . *thiC*		*tppA* . . . . *envZ*	
thiB . . . . *thiH*		*tppA* . . . . *ompR*	
thiC . . . . *thiI*		*tppB* . . . . *envZ*	
thiG . . . . *thiA*		*tre* . . . . *treA*	
thrA . . . . *thrB*		*triM* . . . . *tctIII*	
thrB . . . . *thrC*		*triR* . . . . *tctIII*	

TABLE 2 *Continued*

Former or alternative symbol	Current symbol	Former or alternative symbol	Current symbol
trpA . . . . *trpE*		*ViA* . . . . *viaA*	
trpB . . . . *trpD*		*virA* . . . . *spvC*	
trpC . . . . *trpA*		*virB* . . . . *spvD*	
trpD . . . . *trpB*		*visA* . . . . *hemH*	
trpE . . . . *trpC*		*vsdA* . . . . *spvR*	
trz . . . . *cysK*		*vsdB* . . . . *spvA*	
umpA . . . . *lgt*		*vsdC* . . . . *spvB*	
unc . . . . *atp*		*vsdD* . . . . *spvC*	
uncA . . . . *atp*		*vsdE* . . . . *spvD*	
use . . . . *argX*		*wrkA* . . . . *divA*	
vh2 . . . . *hin*			

[a]The alternative symbols have been used in past publications. It is recommended that their use be abandoned and that the current symbols, listed and described in Table 1 and in the associated references, be used in the future.

TABLE 3 Sequenced genes of *S. typhimurium*[a]

Locus	Ori[b]	Left end[c]	Cs	Code[d]	Length[e]	Accession no.[f]	Reference(s)[g]
STCARAB	(+)	78000	1.622	SP	1,886	X13200	
carA	(+)	78696	1.637	W	382	SG10032/P14845	1035
carB	(+)	79863	1.661	F	8	SG10033/P14846	1035
araDstyM	(+)	115391	2.400	MG	5,947	SS1001	
STYARABAD	(−)	115392	2.400	C	4,790	M11047	
araD	(−)	115954	2.412	W	248	SG10015/P06190	1207
araA	(−)	116841	2.430	W	500	SG10012/P06189	1209
araB	(−)	118354	2.462	W	569	SG10013/P06188	1208
STYARALC	(+)	120053	2.497	C	1,286	J01797	
araC	(+)	120404	2.504	W	281	SG10014/P03022	345, 346
leuDstyM	(+)	137999	2.870	MP	6,168	SS1020	
STLEUD	(−)	138000	2.870	C	1,074	X02528	
leuD	(−)	138454	2.880	W	201	SG10198/P04787	578
STLEUC	(−)	139053	2.892	C	1,432	X51476	
leuC	(−)	139070	2.892	W	464	SG10197/P15717	1677
STLEUB	(−)	140445	2.921	C	1,122	X53376	
leuB	(−)	140467	2.922	W	359	SG10196/P37412	58
STYLEUA	(−)	141545	2.944	C	1,569	X51583	
leuA	(−)	141546	2.944	W	522	SG10195/P15875	623, 1629
STYLEUOP	(−)	143024	2.975	C	497	J01807	
STYLEUP	(−)	143313	2.981	C	855	M12892	
STILVIHO	(+)	144299	3.001	SG	348	X68562	
STFRURG	(+)	144721	3.010	SG	2,497	X55456	
ilvI	(+)	144722	3.010	F	52	SG10129/P40811	926
ilvH	(+)	144904	3.014	W	163	SG10182/P21622	926
fruR	(+)	145679	3.030	W	334	SG10133/P21930	926, 2076
STYNADC	(−)	172126	3.580	SG	1,403	L07292	
nadC	(−)	172264	3.583	W	297	SG10242/P30012	861
ampD	(+)	173245	3.603	F	94	SG10010/P30013	861
ompHstyM	(+)	193281	4.020	MG	2,332	SS1025	
STYOMPH	(+)	193282	4.020	C	992	J05101	
ompH	(+)	193593	4.026	W	161	SG10265/P16974	782, 1080, 1081
STYSSCA	(+)	194063	4.036	C	1,551	M35193	
lpxD	(+)	194082	4.037	W	341	SG10209/P18482	783, 784, 1757, 2089
fabZ	(+)	195213	4.060	F	134	SG10096/P21773	783
STLPXGNA	(+)	200974	4.180	SG	812	Z25462	

TABLE 3 *Continued*

Locus	Ori[b]	Left end[c]	Cs	Code[d]	Length[e]	Accession no.[f]	Reference(s)[g]
lpxA	(+)	200975	4.180	W	262	SG10208/P32200	2089
STPEPMG	(−)	205782	4.280	SG	1,811	X55778	
glnD	(−)	206054	4.286	F	201	SG10144/P23679	1382
pepM	(−)	206623	4.297	W	264	SG10283/P10882	1382, 2153
STYHEML	(−)	240400	5.000	SG	1,658	M60064	
hemL	(−)	240548	5.003	W	426	SG10152/P21267	496
STSPHSPG	(+)	245000	5.096	SH	1,980	X54548	
dgt	(+)	245001	5.096	F	87	SG10473/P40733	964
htrA	(+)	245395	5.104	W	475	SG10173/P26982	964
STYPOL3A	(+)	284000	5.907	SP	3,763	M26046	
rnhB	(+)	284001	5.907	F	55	SG10493/P40675	1139
dnaE	(+)	284159	5.910	W	1,160	SG10087/P14567	1139
accA	(+)	287654	5.983	F	36	SG10496/P40674	1139
rnhAstyM	(+)	329999	6.864	MH	927	SS1031	
STDNAQRN	(−)	330000	6.864	C	545	X57159	
rnhA	(−)	330001	6.864	W	155	SG10360/P23329	
STYPOL3E	(+)	330473	6.873	C	454	M26045	
dnaQ	(+)	330532	6.875	F	100	SG10090/P14566	1138
STGHPT	(+)	356273	7.410	SG	845	X63336	
gpt	(+)	356466	7.414	W	152	SG10147/P26972	
yafA	(+)	357013	7.425	F	35	SG10415/P37722	
STPHOE	(−)	365408	7.600	SG	1,232	X68023	
phoE	(−)	365413	7.600	W	350	SG10291/P30705	1868
STYP22ATB	(+)	375000	7.800	SG	320	M10894	
proA	(+)	375001	7.800	F	23	SG10301/P40861	1178
thrW (ataA)	(+)	375188	7.803	W	RNA	SG30013/tRNA	
STYDDLA	(−)	420000	8.735	SH	1,197	M20793	
ddlA	(−)	420004	8.736	W	364	SG10084/P15051	413
brnQstyM	(+)	429999	8.943	MG	2,645	SS1003	
STYBRNQ	(+)	430000	8.943	C	1,603	D00332	
brnQ	(+)	430220	8.948	W	439	SG10029/P14931	1460
STPROY	(+)	431457	8.974	C	1,188	X74420	
proY	(+)	431551	8.976	W	292	SG10314/P37460	
STTSXOMP	(−)	435000	9.047	SH	2,244	Z26657	
yajD	(+)	435015	9.048	W	109	SG10517/P40777	
tsx	(−)	435987	9.068	W	287	SG10516/P40776	
STU09529	(+)	480000	9.983	SH	1,030	U09529	
apbA	(+)	480102	9.985	W	281	SG10001/P37402	
STHUPB	(+)	547831	11.394	SH	478	X07844	
hupB	(+)	547951	11.397	W	90	SG10175/P05515	1298
STYADK	(+)	555324	11.550	SG	1,590	L26246	
adk	(+)	555536	11.554	W	214	SG10003/P37407	
hemH	(+)	556409	11.573	F	168	SG10151/P37408	
STUSHA0	(+)	585614	12.180	SG	2,607	X04651	
ushA	(+)	585780	12.183	W	550	SG10412/P06196	259
ybaK	(−)	587549	12.220	W	159	SG10416/P37174	259
STNITRD	(−)	652330	13.568	SG	1,690	X17250	
nfnB	(−)	653069	13.583	W	217	SG10246/P15888	2108
STYAHPCFA	(+)	678000	14.101	SP	2,610	J05478	
ahpC	(+)	678166	14.105	W	187	SG10004/P19479	922, 1998, 1999
ahpF	(+)	678971	14.122	W	521	SG10005/P19480	922, 1999

(Table continues)

TABLE 3 *Continued*

Locus	Ori[b]	Left end[c]	Cs	Code[d]	Length[e]	Accession no.[f]	Reference(s)[g]
STYFIMCLUS	(+)	721200	15.000	SG	12,168	L19338	
fimA	(+)	722263	15.022	W	185	SG10275/P37921	
fimI	(+)	722934	15.036	W	164	SG10508/P37922	
fimC	(+)	723472	15.047	W	230	SG10504/P37923	
fimD	(+)	724195	15.062	W	870	SG10505/P37924	
fimH	(+)	726822	15.117	W	335	SG10507/P37925	
fimF	(+)	727839	15.138	W	172	SG10506/P37926	
fimZ	(−)	728403	15.150	W	210	SG10099/P26319	1974
fimY	(−)	729639	15.176	W	240	SG10098/P26318	1974
fimW	(−)	730853	15.201	W	198	SG10509/P37928	
argU	(+)	731705	15.218	W	RNA	SG30004/tRNA	
STU12808	(−)	750048	15.600	SG	1,124	U12808	
cobC	(−)	750261	15.604	W	234	SG10479/P39701	
STPHRG	(+)	814000	16.930	SH	1,353	X60662	
phrB	(+)	814001	16.930	W	450	SG10295/P25078	1195
STYCITA	(+)	815437	16.960	SG	2,119	D90203	
citB	(+)	815438	16.960	F	58	SG10059/P37463	1795
citA	(+)	815667	16.965	W	434	SG10058/P24115	1795
STYNADAPCU	(+)	826000	17.180	SP	2,330	M85180	
nadA	(+)	826142	17.183	W	365	SG10241/P24519	563
pnuC	(+)	827209	17.205	W	322	SG10306/P24520	563
STYGALOPA	(−)	856000	17.804	SP	3,219	M33681	
galK	(−)	856001	17.804	W	380	SG10137/P22713	834
galT	(−)	857147	17.828	W	348	SG10138/P22714	834
galE	(−)	858206	17.850	W	337	SG10135/P22715	834
STYBIOAB	(+)	871000	18.116	SP	117	M21923	
bioA	(−)	871001	18.116	F	5	SG10026/P12677	1797
bioB	(+)	871103	18.118	F	5	SG10027/P12678	1797
STU02273	(+)	920000	19.135	SH	495	U02273	
lrp	(+)	920001	19.135	W	164	SG10210/P37403	
STYAROAPM	(+)	956000	19.884	SG	1,332	M10947	
aroA	(+)	956027	19.884	W	427	SG10019/P07637	1882
STOMPFGE	(−)	1094000	22.754	SH	1,306	Z31594	
ompF	(−)	1094013	22.754	W	363	SG10264/P37432	
STYPNCB	(−)	1114000	23.170	SP	1,837	M55986	
pncB	(−)	1114398	23.178	W	400	SG10305/P22253	2082
STYNEUR	(+)	1140000	23.710	SP	1,803	M55342	
nanH	(+)	1140263	23.716	W	382	SG10244/P29768	395, 843, 2003
ycdD	(−)	1141482	23.741	F	106	SG10500/P40784	843
STPYRDDD	(+)	1160000	24.126	SP	1,286	X55636	
pyrD	(+)	1160149	24.130	W	336	SG10326/P25468	576
sulAstyM	(+)	1169999	24.334	MG	2,236	SS1032	
STOMPA	(−)	1170000	24.334	C	1,400	X02006	
ompA	(−)	1170148	24.338	W	350	SG10263/P02936	573
STYSULA	(−)	1171364	24.363	C	874	M16324	
sulA	(−)	1171557	24.367	W	169	SG10386/P08847	572
putAstyM	(+)	1229999	25.582	MP	6,077	SS1029	
STPUTA	(−)	1230000	25.582	C	4,102	X70843	
putA	(−)	1230140	25.585	W	1,320	SG10321/P10503	28, 719
STPUTPA	(−)	1234022	25.666	C	546	X12569	
STPUTP	(+)	1234473	25.675	C	1,604	X52573	
putP	(+)	1234524	25.676	W	502	SG10322/P10502	719, 1353

TABLE 3 *Continued*

Locus	Ori[b]	Left end[c]	Cs	Code[d]	Length[e]	Accession no.[f]	Reference(s)[g]
STPYRC	(−)	1260000	26.206	SP	1,672	X03928	
pyrC	(−)	1260209	26.211	W	348	SG10325/P06204	1421
yceB	(−)	1261363	26.235	F	103	SG10459/P40822	1421
mviMstyM	(+)	1264987	26.310	MP	15,013	SS1008	
STYFLGA	(−)	1264988	26.310	C	4,943	D25292	
STMVIMN	(+)	1264989	26.310	C	3,120	Z26133	
mviM	(+)	1265263	26.316	W	307	SG10239/P37168	
mviN	(+)	1266451	26.340	W	524	SG10240/P37169	
STU03631	(+)	1268096	26.375	C	893	U03631	
flgN	(−)	1268107	26.375	W	140	SG10113/P37406	1754
flgM	(−)	1268534	26.384	W	97	SG10112/P26477	633, 1754
flgA	(−)	1268919	26.392	W	219	SG10503/P40131	1124
STYFLG	(+)	1269578	26.406	C	1,000	D13703	
STFLGBC	(+)	1269700	26.408	C	878	X52093	
flgB	(+)	1269735	26.409	W	138	SG10101/P16437	820, 973, 1122
flgC	(+)	1270155	26.418	W	134	SG10102/P16438	820, 973, 1122
STYFLGD	(+)	1270377	26.422	C	1,147	D25293	
flgD	(+)	1270571	26.426	W	232	SG10103/P16321	812, 1115
STFLGE	(+)	1271235	26.440	C	1,320	X51737	
flgE	(+)	1271296	26.441	W	403	SG10104/P16322	812
STFLGFG	(+)	1272467	26.466	C	1,687	X52094	
flgF	(+)	1272528	26.467	W	251	SG10105/P16323	812, 820, 973
flgG	(+)	1273297	26.483	W	260	SG10106/P16439	820, 973
STYFLGH	(+)	1274076	26.499	C	2,947	M24466	
flgH	(+)	1274137	26.500	W	232	SG10107/P15929	820, 971
flgI	(+)	1274847	26.515	W	365	SG10108/P15930	819, 971
flgJ	(+)	1275944	26.538	W	316	SG10109/P15931	812, 971
STFLGK	(+)	1276878	26.557	C	1,765	X51738	
flgK	(+)	1276959	26.559	W	553	SG10110/P15932	812, 971, 973
STFLGL	(+)	1278574	26.593	C	1,427	X51739	
flgL	(+)	1278635	26.594	W	317	SG10111/P16326	812
STPTSG	(+)	1293000	26.893	SP	1,434	X74629	
ptsG	(+)	1293001	26.893	W	477	SG10316/P37439	
STYPEPT	(+)	1312178	27.292	SG	2,551	M62725	
potA	(−)	1312179	27.292	F	63	SG10489/P40790	1347
pepT	(+)	1312618	27.301	W	409	SG10284/P26311	1347
STYPHOPQ	(−)	1315063	27.352	SG	2,190	M24424	
phoQ	(−)	1315084	27.352	W	451	SG10294/P14147	1355
phoP	(−)	1316439	27.380	W	224	SG10293/P14146	684, 1355
STYGDHA	(−)	1401000	29.139	SP	1,603	M24021	
topB	(+)	1401001	29.139	F	49	SG10495/P40687	105
gdhA	(−)	1401180	29.143	W	447	SG10140/P15111	105
STYINFCA	(+)	1450000	30.158	SH	570	L11254	
infC	(+)	1450019	30.158	W	180	SG10185/P33321	1225
STHIMA	(+)	1458747	30.340	SG	420	X16739	
pheT	(+)	1458748	30.340	F	17	SG10290/P15434	1196
himA	(+)	1458808	30.341	W	99	SG10155/P15430	1196
STU09502	(−)	1475000	30.678	SH	376	U09502	
sodB	(−)	1475001	30.678	F	56	SG10514/P40726	
STU03842	(+)	1482000	30.824	SG	714	U03842	
slyA	(+)	1482167	30.827	W	146	SG10526/P40676	1197
STPMIPHI	(−)	1502500	31.250	SG	1,650	X57117	
manA	(−)	1502544	31.251	W	391	SG10217/P25081	1652
fumA	(+)	1503919	31.280	F	38	SG10462/P40720	1652
STYDCP	(+)	1562000	32.488	SP	2,841	M84575	
ydfG	(−)	1562001	32.488	F	96	SG10499/P40864	724
dcp	(+)	1562427	32.496	W	680	SG10083/P27236	724

(Table continues)

TABLE 3 *Continued*

Locus	Ori[b]	Left end[c]	Cs	Code[d]	Length[e]	Accession no.[f]	Reference(s)[g]
STYRFC	(−)	1714533	35.660	SG	1,750	M60066	
rfc	(−)	1714600	35.661	W	407	SG10352/P26479	358
STFNR	(+)	1761000	36.626	SP	1,360	U05668	
ogt	(+)	1761001	36.626	F	115	SG10262/P37429	
fnr	(+)	1761546	36.638	W	250	SG10131/P37428	
STSAP	(+)	1785400	37.134	SG	5,714	X74212	
sapA	(+)	1785516	37.136	W	549	SG10377/P36634	1540
sapB	(+)	1787162	37.171	W	321	SG10378/P36668	1540
sapC	(+)	1788114	37.190	W	296	SG10379/P36669	1540
sapD	(+)	1789004	37.209	W	330	SG10380/P36636	1540
sapF	(+)	1789998	37.230	W	268	SG10381/P36638	1540
STYENVM	(+)	1805000	37.542	SH	1,656	M31806	
ycjE	(+)	1805259	37.547	W	99	SG10440/P16656	2038
fabI	(+)	1805676	37.556	W	262	SG10095/P16657	2038
STPYRF	(−)	1808000	37.604	SP	1,459	X05382	
osmB	(+)	1808001	37.604	F	20	SG10272/P37723	2006
yciH	(−)	1808185	37.608	W	108	SG10419/P20770	2006
pyrF	(−)	1808966	37.624	W	245	SG10328/P07691	2006
STYCYSB	(−)	1810000	37.646	SP	1,759	M15040	
cysB	(−)	1810342	37.653	W	324	SG10072/P06614	1501
topA	(−)	1811727	37.682	F	10	SG10515/P40686	1501
STYATPCA	(−)	1824636	37.950	SG	993	L08890	
cobA	(−)	1824854	37.955	W	196	SG10031/P31570	515, 1953
trpEstyM	(+)	1827293	38.005	MP	6,783	SS1035	
STYTRPOP	(+)	1827294	38.005	C	105	M10673	
STYTRPEA	(+)	1827370	38.007	C	294	M24960	
STYTRPE	(+)	1827547	38.011	C	1,563	J01812	
trpE	(+)	1827548	38.011	W	520	SG10392/P00898	1159, 1191, 1426, 2203
STYTRPDC1	(+)	1829109	38.043	C	1,596	M30285	
trpD	(+)	1829110	38.043	W	531	SG10399/P00905	830, 1426
STYTRPDC2	(+)	1830709	38.076	C	1,359	M30286	
trpC	(+)	1830710	38.076	W	452	SG10398/P00910	830
STYTRPCBI	(+)	1832031	38.104	C	82	M24299	
STYTRPBA	(+)	1832077	38.105	C	2,000	J01810	
trpB	(+)	1832078	38.105	W	397	SG10397/P00933	390, 880, 1428, 1761
trpA	(+)	1833271	38.130	W	268	SG10396/P00929	
STTONB	(+)	1838098	38.230	SG	950	X56434	
tonB	(+)	1838219	38.233	W	242	SG10391/P25945	730, 1006
yciA	(−)	1839007	38.249	F	13	SG10418/P25944	730
STOPPAF	(−)	1840000	38.270	SP	6,006	X05491	
oppF	(−)	1840244	38.275	W	334	SG10271/P08007	775
oppD	(−)	1841245	38.295	W	335	SG10270/P04285	769, 775
oppC	(−)	1842264	38.317	W	302	SG10269/P08006	775, 1555
oppB	(−)	1843187	38.336	W	306	SG10268/P08005	775, 1555
oppA	(−)	1844229	38.358	W	542	SG10267/P06202	775, 776
hns-styM	(+)	1846388	38.402	MG	926	SS1009	
STYOSMZ	(+)	1846389	38.402	C	909	M37891	
STHNS	(+)	1846489	38.405	C	827	X14375	
hns	(+)	1846767	38.410	W	137	SG10171/P17428	877, 1299
STYNARK	(+)	1856000	38.602	SH	4,183	D26057	
narK	(−)	1856001	38.602	F	130	SG10245/P37593	
smvA	(+)	1857300	38.629	W	496	SG10384/P37594	
nmpC	(−)	1859335	38.672	F	282	SG10249/P37592	
prfAstyM	(+)	1863999	38.769	MP	5,845	SS1028	
STYHEMAPRF	(−)	1864000	38.769	C	3,341	J04243	

TABLE 3 *Continued*

Locus	Ori[b]	Left end[c]	Cs	Code[d]	Length[e]	Accession no.[f]	Reference(s)[g]
hemK	(−)	1864001	39.769	F	77	SG10534/P40816	493
prfA	(−)	1864231	38.773	W	360	SG10309/P13654	493
hemA	(−)	1865354	38.797	W	418	SG10149/P13581	493
STYPRS	(+)	1866603	38.823	C	2,060	M77236	
hemM	(+)	1866928	38.830	W	207	SG10153/P30752	493, 1571
ychB	(+)	1867548	38.842	W	283	SG10417/P30753	213, 1571
STYPRSA	(+)	1868101	38.854	C	1,740	M19488	
prsA	(+)	1868665	38.866	W	315	SG10315/P15849	213
STYDADB	(−)	1975000	41.077	SP	1,140	K02119	
dadB	(−)	1975010	41.078	W	356	SG10082/P06191	2105
STYGAPA	(+)	1977000	41.119	SH	882	M63369	
gapA	(+)	1977001	41.119	F	294	SG10139/P24165	1152
STYPABB	(+)	2000000	41.597	SH	1,815	M22079	
pabB	(+)	2000207	41.602	W	454	SG10274/P12680	661
flhEstyM	(+)	2003810	41.677	MP	5,097	SS1004	
STYFLHB	(−)	2003811	41.677	C	4,023	D32203	
flhE	(−)	2004025	41.677	W	130	SG10528/P40728	1361
flhA	(−)	2004417	41.689	W	692	SG10529/P40729	1361
flhB	(−)	2006488	41.732	W	383	SG10530/P40727	1361
STYCHEZ	(−)	2007000	41.743	C	1,506	M16691	
cheZ	(−)	2007833	41.760	W	214	SG10057/P07800	1915
STYCHEY	(−)	2008456	41.773	C	452	M12131	
cheY	(−)	2008488	41.774	W	129	SG10056/P06657	761, 1910, 1913, 1914
STYCHER	(−)	2012629	41.860	SG	896	J02757	
cheB	(−)	2012630	41.860	W	349	SG10053/P04042	761, 1807, 1808
cheR	(−)	2012635	41.860	W	288	SG10054/P07801	1809
tar-styM	(+)	2015513	41.920	MG	4,475	SS1033	
STYTAR	(−)	2015514	41.920	C	1,922	J01809	
tar	(−)	2015540	41.921	W	552	SG10387/P02941	1698
STYCHEW	(−)	2017195	41.955	C	770	J02656	
cheW	(−)	2017439	41.960	W	167	SG10055/P06110	1912
STYCHEA	(−)	2017939	41.970	C	2,050	J03611	
cheA	(−)	2017963	41.971	W	671	SG10052/P09384	1909
STYFLIA	(−)	2040000	42.429	SG	1,244	D00497	
fliA	(−)	2040321	42.436	W	239	SG10114/P17168	1464
fliCstyM	(+)	2043999	42.512	MP	14,084	SS1010	
STYFLGH1I	(−)	2044000	42.512	C	1,485	M11332	
fliC	(−)	2044001	42.513	W	490	SG10115/P06179	812, 986
STFLID	(+)	2045409	42.542	C	1,766	X51740	
fliD	(+)	2045730	42.548	W	467	SG10116/P16328	812
STYFLGPRO	(+)	2047047	42.576	C	965	M85241	
fliS	(+)	2047148	42.578	W	135	SG10127/P26609	1008
fliT	(+)	2047555	42.586	W	122	SG10128/P26611	1008
STYFTAA	(+)	2047894	42.593	C	1,700	L01643	
amyA	(+)	2047995	42.596	W	494	SG10011/P26613	1008, 1606
STYAMYA	(+)	2049418	42.625	C	2,940	L13280	
yedD	(−)	2049519	42.627	W	141	SG10420/Q06399	1607
yedE	(+)	2050130	42.640	W	401	SG10421/Q06400	1607
yedF	(+)	2051332	42.665	W	77	SG10422/P31065	1607
STYFLIE	(−)	2052299	42.685	C	647	M84993	
fliE	(−)	2052337	42.686	W	104	SG10117/P26462	971, 1392
STYFLIG	(+)	2052807	42.696	C	2,755	M24462	
fliF	(+)	2052868	42.697	W	560	SG10118/P15928	971
fliG	(+)	2054543	42.732	W	331	SG10119/P15933	567, 1034, 2085
STYFLIHIJ	(+)	2055430	42.750	C	2,654	M62408	
fliH	(+)	2055531	42.752	W	235	SG10120/P15934	1034, 2085
fliI	(+)	2056238	42.767	W	456	SG10121/P26465	468, 2085
fliJ	(+)	2057630	42.796	W	147	SG10122/P26463	2085
STYFLIL	(+)	2058550	42.815	SG	2,013	M24463	

(Table continues)

TABLE 3 *Continued*

Locus	Ori[b]	Left end[c]	Cs	Code[d]	Length[e]	Accession no.[f]	Reference(s)[g]
fliK	(+)	2058551	42.815	F	10	SG10123/P26416	1034, 2085
	(+)	2058671	42.818	W	155	SG10124/P26417	1034
fliL							
fliM	(+)	2059143	42.827	W	334	SG10125/P26418	1034
fliN	(+)	2060144	42.848	W	137	SG10126/P26419	1034
STYUMUDC	(−)	2062151	42.890	SG	2,574	M57431	
cspG	(−)	2062152	42.890	W	40	SG10445/P39818	
umuC	(−)	2062825	42.904	W	422	SG10410/P22494	1827, 2010
umuD	(−)	2064096	42.930	W	139	SG10411/P22493	1827, 2010, 2167
cobTstyM	(+)	2086671	43.400	MP	22,490	SS1005	
STYCOB	(−)	2086672	43.400	C	2,760	L35477	
asnT?	(+)	2087023	43.407	W	RNA	SG30008/tRNA	
yeeG	(−)	2087325	43.414	W	309	SG10484/P40680	
cobT	(−)	2088332	43.435	W	366	SG10062/Q05603	1681
STYVB12AA	(−)	2088675	43.442	C	17,442	L12006	
cobS	(−)	2089429	43.457	W	247	SG10061/Q05602	1681
cobU	(−)	2090169	43.473	W	180	SG10063/Q05599	1681
cbiP	(−)	2090711	43.484	W	506	SG10048/Q05597	1681
cbiO	(−)	2092228	43.516	W	271	SG10047/Q05596	1681
cbiQ	(−)	2093052	43.533	W	225	SG10049/Q05598	1681
cbiN	(−)	2093716	43.547	W	93	SG10046/Q05595	1681
cbiM	(−)	2093999	43.552	W	245	SG10045/Q05594	1681
cbiL	(−)	2094733	43.568	W	237	SG10044/Q05593	1681
cbiK	(−)	2095443	43.582	W	264	SG10043/Q05592	1681
cbiJ	(−)	2096240	43.599	W	263	SG10042/Q05591	1681
cbiH	(−)	2097028	43.615	W	241	SG10041/Q05590	1681
cbiG	(−)	2097753	43.630	W	351	SG10040/Q05631	1681
cbiF	(−)	2098789	43.652	W	257	SG10039/Q05630	1681
cbiT	(−)	2099546	43.668	W	192	SG10050/Q05632	1681
cbiE	(−)	2100114	43.680	W	201	SG10038/Q05629	1681
cbiD	(−)	2100713	43.692	W	379	SG10037/Q05628	1681
cbiC	(−)	2101852	43.716	W	210	SG10036/Q05601	1681
cbiB	(−)	2102495	43.729	W	319	SG10035/Q05600	1681
cbiA	(−)	2103451	43.749	W	459	SG10034/P29946	1634, 1681
STYPDUC	(−)	2105401	43.790	C	3,761	L31414	
pocR	(−)	2105428	43.790	W	303	SG10307/Q05587	1681
pduF	(−)	2106556	43.814	W	264	SG10281/P37451	
pduA	(+)	2107876	43.841	W	97	SG10278/P37448	
pduB	(+)	2108266	43.849	W	233	SG10279/P37449	
pduC	(+)	2108986	43.864	F	59	SG10280/P37450	
STYPBCDEF	(−)	2124655	44.190	SG	3,282	L31538	
phsE	(−)	2125087	44.199	W	390	SG10300/P37604	
phsD	(−)	2126288	44.224	W	84	SG10299/P37603	
phsC	(−)	2126542	44.229	W	199	SG10298/P37602	
phsB	(−)	2127138	44.242	W	198	SG10297/P37601	
phsA	(−)	2127731	44.254	F	68	SG10296/P37600	
hisGstyM	(+)	2148461	44.685	MP	33,958	SS1013	
STYHISOGD	(+)	2148462	44.685	C	4,443	J01804	
STHISOP	(+)	2149090	44.698	C	7,439	X13464	
hisG	(+)	2149366	44.704	W	299	SG10162/P00499	281, 337, 1266, 1562
hisD	(+)	2150368	44.725	W	434	SG10160/P10370	281, 695
hisC	(+)	2151669	44.752	W	359	SG10159/P10369	281
hisB	(+)	2152745	44.774	W	354	SG10158/P10368	281
hisH	(+)	2153809	44.796	W	194	SG10163/P10376	281
hisA	(+)	2154393	44.809	W	246	SG10157/P10372	281
hisF	(+)	2155115	44.824	W	258	SG10161/P10374	281
hisI	(+)	2155885	44.840	W	203	SG10164/P10367	281, 318
SECLDB	(−)	2156446	44.851	C	2,948	Z17278	
cld	(−)	2156576	44.854	W	327	SG10060/Q04866	123, 124
yefA	(−)	2157699	44.877	W	388	SG10439/Q04873	123
STYGNDA	(−)	2158970	44.904	C	1,705	M64332	
gnd	(−)	2159104	44.906	W	468	SG10146/P14062	954, 1621
SERFBB	(−)	2160340	44.932	C	22,080	X56793	
rfbP	(−)	2160674	44.939	W	476	SG10351/P26406	954
rfbK	(−)	2162176	44.970	W	477	SG10349/P26405	954

TABLE 3 *Continued*

Locus	Ori[b]	Left end[c]	Cs	Code[d]	Length[e]	Accession no.[f]	Reference(s)[g]
rfbM	(−)	2163596	45.000	W	479	SG10350/P26404	954
rfbN	(−)	2165036	45.030	W	314	SG10455/P26403	954
rfbU	(−)	2165981	45.050	W	353	SG10446/P26402	954
rfbV	(−)	2167361	45.078	W	333	SG10454/P26401	954
rfbX	(−)	2168367	45.099	W	430	SG10453/P26400	954
rfbJ	(−)	2169741	45.128	W	299	SG10348/P22716	954, 2175
rfbH	(−)	2170668	45.147	W	437	SG10452/P26398	954
rfbG	(−)	2172008	45.175	W	359	SG10347/P26397	954
rfbF	(−)	2173092	45.197	W	257	SG10346/P26396	954
rfbI	(−)	2173862	45.213	W	330	SG10451/P26395	954
rfbC	(−)	2174860	45.234	W	183	SG10450/P26394	954
rfbA	(−)	2175412	45.246	W	292	SG10449/P26393	954
rfbD	(−)	2176338	45.265	W	299	SG10344/P26392	954
rfbB	(−)	2177237	45.284	W	361	SG10345/P26391	954
galF	(−)	2178699	45.314	W	297	SG10136/P26390	954
yefK	(−)	2179770	45.336	W	467	SG10448/P26389	954
yefL	(−)	2181184	45.366	W	406	SG10447/P26388	954
STCPSBG	(−)	2185000	45.445	SH	3,069	X59886	
cpsG	(−)	2185011	45.445	W	456	SG10067/P26341	1903
cpsB	(−)	2186492	45.476	W	480	SG10066/P26340	1903
mglBsty	(−)	2307000	47.983	SH	1,810	SS3004	
mglA	(−)	2307001	47.983	F	42	SG10228/P23924	145
mglB	(−)	2307259	47.988	W	332	SG10229/P23905	145
galS	(−)	2308754	48.019	F	18	SG10521/P41030	145
STFRUF	(+)	2310725	48.060	SG	1,610	X14243	
yeiO	(−)	2310726	48.060	F	22	SG10423/P33027	619
fruB	(+)	2311160	48.069	W	376	SG10132/P17127	619
STRNAP	(+)	2360247	49.090	SG	1,041	X63777	
yejM	(+)	2360248	49.090	F	190	SG10463/P40709	1879
proL	(+)	2360895	49.103	W	RNA	SG30015/tRNA	
STGYRA	(−)	2367940	49.250	SG	261	X78977	
gyrA	(−)	2367941	49.250	F	86	SG10148/P37411	
STNRDABA	(+)	2370000	49.293	SH	4,967	X72948	
ubiG	(+)	2370001	49.293	F	209	SG10405/P37431	977
nrdA	(+)	2370987	49.313	W	761	SG10250/P37426	977
nrdB	(+)	2373385	49.363	F	122	SG10251/P37427	977
STU02281	(−)	2408808	50.100	SG	1,329	U02281	
pmrD	(−)	2409096	50.106	W	85	SG10304/P37589	
menE	(−)	2409451	50.113	F	228	SG10221/P37418	
STYADA	(−)	2411000	50.146	SH	1,266	D90221	
alkB	(−)	2411001	50.146	F	24	SG10006/P37462	720
ada	(−)	2411078	50.147	W	352	SG10002/P26189	720
STYNUOD	(−)	2449676	50.950	SG	3,950	L22504	
nuoG	(−)	2449677	50.950	F	612	SG10257/P33900	71
nuoF	(−)	2451586	50.990	W	431	SG10256/P33901	71
nuoE	(−)	2452878	51.017	W	166	SG10255/P33903	71
nuoD	(−)	2453381	51.027	F	81	SG10254/P33902	71
hisPstyM	(+)	2459999	51.165	MP	5,228	SS1014	
STYHIS3P	(−)	2460000	51.165	C	789	M32273	
STYHISTO	(−)	2460578	51.177	C	4,483	J01805	
hisP	(−)	2460899	51.183	W	258	SG10168/P02915	766
hisM	(−)	2461686	51.200	W	235	SG10167/P02912	766
hisQ	(−)	2462390	51.214	W	228	SG10169/P02913	766
hisJ	(−)	2463259	51.233	W	260	SG10165/P02910	762, 766, 805
argT	(−)	2464279	51.254	W	260	SG10018/P02911	762, 999
STYARGTR	(−)	2465061	51.270	C	467	J01798	
ubiX	(−)	2465352	51.276	F	58	SG10481/P40787	763

(Table continues)

TABLE 3 *Continued*

Locus	Ori[b]	Left end[c]	Cs	Code[d]	Length[e]	Accession no.[f]	Reference(s)[g]
pgtAstyM	(+)	2501999	52.038	MP	7,612	SS1027	
STYPGTA	(−)	2502000	52.038	C	2,680	M13923	
pgtE	(−)	2502268	52.044	W	296	SG10288/P06185	683, 2217
pgtA	(−)	2503374	52.067	W	415	SG10285/P06184	2201, 2217
STYPGTBC	(−)	2504534	52.091	C	3,388	M21279	
pgtB	(−)	2504611	52.093	W	593	SG10286/P37433	953, 2201
pgtC	(−)	2506615	52.134	W	397	SG10287/P37591	953, 2201
STYPGTP	(+)	2507601	52.155	C	2,011	M21278	
pgtP	(+)	2508243	52.168	W	406	SG10289/P12681	654
cysZstyM	(+)	2538995	52.808	MP	11,006	SS1006	
STYCYSPTS	(+)	2538995	52.808	C	3,812	M21450	
cysZ	(+)	2539214	52.812	W	290	SG10081/P12673	266
cysK	(+)	2540250	52.834	W	323	SG10078/P12674	266
ptsH	(+)	2541605	52.862	W	85	SG10317/P07006	266, 1583, 1762, 2121
STYENZI	(+)	2541910	52.868	C	1,728	M76176	
ptsI	(+)	2541911	52.868	W	575	SG10318/P12654	266, 1199, 1762
STCRR	(+)	2543473	52.901	C	733	X05210	
crr	(+)	2543679	52.905	W	169	SG10069/P02908	1416
STU11243	(+)	2544185	52.916	C	4,471	U11243	
yfeI	(−)	2544573	52.924	W	287	SG10475/P40192	
ptsJ	(+)	2545519	52.943	W	430	SG10474/P40193	
yfeJ	(+)	2546826	52.971	W	170	SG10476/P40194	
yfeK	(+)	2547623	52.987	W	120	SG10477/P40195	
yfeL	(+)	2547998	52.995	W	179	SG10478/P40196	
STCYSM	(−)	2548650	53.009	C	1,351	X59595	
cysM	(−)	2548668	53.009	W	303	SG10079/P29848	
cysA	(−)	2549647	53.029	F	117	SG10518/P40860	
cysPsty	(−)	2553600	53.112	SH	1,401	SS3006	
cysU	(−)	2553601	53.112	F	15	SG10520/P41032	846, 847
cysP	(−)	2553645	53.112	W	338	SG10519/P40131	846, 847
yfeF	(−)	2554819	53.137	F	60	SG10424/P37441	846, 847
STYHEMF	(+)	2556414	53.170	SG	2,398	L19503	
amiA	(+)	2556641	53.175	W	289	SG10008/P33772	2178
hemF	(+)	2557513	53.193	W	299	SG10150/P33771	
STYEUTBC	(−)	2560000	53.245	SP	2,526	J05518	
eutC	(−)	2560149	53.248	W	286	SG10093/P19265	526
eutB	(−)	2561028	53.266	W	452	SG10092/P19264	526
STGLYAG	(−)	2691000	55.969	SP	1,903	X15816	
glyA	(−)	2691238	55.974	W	417	SG10145/P06192	1894, 2049
hmpA	(+)	2692817	56.007	F	29	SG10170/P26353	1894
STYASRABC	(+)	2745368	57.100	SG	3,360	M57706	
asrA	(+)	2745678	57.106	W	347	SG10021/P26474	851
asrB	(+)	2746725	57.128	W	272	SG10022/P26475	851
asrC	(+)	2747554	57.145	W	337	SG10023/P26476	851
STLEPG	(−)	2770000	57.612	SH	1,294	X54933	
lepB	(−)	2770196	57.616	W	324	SG10194/P23697	2071
lepA	(−)	2771187	57.637	F	35	SG10193/P23698	2071
STU05669	(−)	2776000	57.737	SH	744	U05669	
rpoE	(−)	2776024	57.738	W	191	SG10370/P37401	
STGENES	(−)	2782000	57.862	SG	2,169	X74933	
rplS	(−)	2782029	57.862	W	115	SG10365/P36240	
trmD	(−)	2782417	57.871	W	255	SG10395/P36245	
yfjA	(−)	2783229	57.887	W	183	SG10425/P36246	
rpsP	(−)	2783799	57.899	W	82	SG10375/P36242	
STYAROF	(−)	2817488	58.600	SG	1,235	M31302	
aroF	(−)	2817489	58.600	W	356	SG10020/P21307	1384
STHINZ	(+)	2889127	60.090	SG	1,149	V01370	

TABLE 3 *Continued*

Locus	Ori[b]	Left end[c]	Cs	Code[d]	Length[e]	Accession no.[f]	Reference(s)[g]
hin	(+)	2889227	60.092	W	190	SG10156/P03013	529, 747, 1805, 2235
STYTCTD	(+)	2924000	60.815	SP	829	M28368	
tctD	(+)	2924155	60.819	W	224	SG10388/P22104	2140
nrdEstyM	(+)	2947999	61.314	MP	5,654	SS1024	
STNRD	(+)	2948000	61.314	C	4,894	X73226	
nrdE	(+)	2948863	61.332	W	714	SG10252/Q08698	977
nrdF	(+)	2950991	61.377	W	319	SG10253/P17424	977, 1908
STPROVW	(+)	2951644	61.390	C	2,010	X52693	
proV	(+)	2952305	61.404	W	400	SG10312/P17328	977, 1514, 1908
proW	(+)	2953500	61.429	F	51	SG10313/P17327	1908
STTRNADNA	(−)	2980960	62.000	SG	650	X64175	
argQ	(−)	2980961	62.000	W	RNA	SG30002/tRNA	
argZ	(−)	2981210	62.005	W	RNA	SG30015/tRNA	
argY	(−)	2981349	62.008	W	RNA	SG30006/tRNA	
argV	(−)	2981488	62.011	W	RNA	SG30005/tRNA	
serV	(−)	2981568	62.013	W	RNA	SG30012/tRNA	
STU16278	(+)	3018943	62.790	SG	1,085	U16278	
fhlA	(+)	3018944	62.790	W	253	SG10480/P40734	
STYORG	(+)	3022790	62.870	SG	1,898	L33855	
orgA	(+)	3022998	62.874	W	412	SG10502/P40823	969
spaTstyM	(+)	3026612	62.950	MG	12,917	SS1018	
STSPA	(−)	3026613	62.950	C	6,387	X73525	
spaT	(−)	3026614	62.950	F	32	SG10472/P40703	685
spaS	(−)	3026848	62.954	W	356	SG10471/P40702	685
spaR	(−)	3027905	62.976	W	263	SG10470/P40701	685
spaQ	(−)	3028700	62.993	W	82	SG10469/P40704	685
spaP	(−)	3028986	62.999	W	224	SG10468/P40700	685
spaO	(−)	3029650	63.013	W	303	SG10467/P40699	685
spaN	(−)	3030561	63.032	W	336	SG10466/P40613	685
spaM	(−)	3031571	63.053	W	147	SG10465/P40612	685
STU08279	(−)	3031903	63.060	C	1,835	U08279	
spaL	(−)	3031996	63.061	W	432	SG10464/P39444	483, 685
spaK	(−)	3033284	63.088	C	134	SG10532/P39443	483
STYINVA	(−)	3033677	63.096	C	2,176	M90846	
invA	(−)	3033712	63.097	W	665	SG10186/P35657	604
STYEPI	(−)	3035753	63.140	C	1,313	M90714	
invE	(−)	3035794	63.140	W	372	SG10187/P35671	638, 1000
STU08280	(−)	3036700	63.159	C	2,830	U08280	
STINVGE	(−)	3036896	63.163	C	1,736	X75302	
invG	(−)	3036909	63.164	W	562	SG10188/P35672	1000
invF	(−)	3038594	63.199	W	216	SG10130/P39437	1000
STINVH	(−)	3047000	63.374	SG	884	Z17242	
invH	(−)	3047328	63.380	W	147	SG10189/P37423	34
STKATFR	(−)	3060000	63.644	SH	1,574	X77752	
rpoS	(−)	3060278	63.650	W	330	SG10372/P37400	1085
nlpD	(−)	3061333	63.672	F	98	SG10513/P40827	1085
STU16303	(+)	3073265	63.920	SP	3,141	U16303	
mutS	(+)	3073812	63.931	W	854	SG10237/P10339	348, 709, 710
STYCYSJIHA	(+)	3085080	64.166	SG	6,050	M23007	
cysJ	(+)	3085470	64.174	W	599	SG10077/P38039	1498, 1505
cysI	(+)	3087269	64.211	W	570	SG10076/P17845	1098, 1505
cysH	(+)	3089063	64.248	W	244	SG10075/P17853	1505
ygcC	(−)	3089885	64.265	W	317	SG10485/P40722	1505
STYLGTX	(−)	3120392	64.900	SG	1,473	L13259	
lgt	(−)	3120516	64.903	W	291	SG10201/Q07293	609, 1727
ygdF	(−)	3121606	64.925	F	85	SG10426/P37178	609

(Table continues)

TABLE 3 *Continued*

Locus	Ori[b]	Left end[c]	Cs	Code[d]	Length[e]	Accession no.[f]	Reference(s)[g]
STYPRF2	(−)	3180492	66.150	SG	1,351	M38590	
lysS	(−)	3180493	66.150	F	31	SG10211/P28354	1009
prfB	(−)	3180595	66.152	W	365	SG10310/P28353	1009
recJ	(−)	3181797	66.177	F	14	SG10332/P28355	1061
SEANSBSQA	(−)	3227000	67.117	SH	147	X69868	942
STYMUTB	(+)	3230495	67.190	SG	1,170	M86634	
mutY	(+)	3230568	67.192	W	350	SG10238/Q05869	448
metCsty	(+)	3322000	69.093	SP	2,186	SS3002	
exbB	(−)	3322001	69.093	F	110	SG10094/P18950	1538
metC	(+)	3322583	69.105	W	395	SG10223/P18949	1538
yghB	(+)	3323910	69.133	F	92	SG10427/P18951	1538
parFstyM	(+)	3328097	69.220	MG	6,409	SS1026	
STU09309	(−)	3328098	69.220	C	3,184	U09309	
ygiK	(+)	3328407	69.226	W	435	SG10523/P40800	
ygiJ	(−)	3329771	69.255	W	465	SG10522/P40799	
STYPARCF	(−)	3330980	69.280	C	3,527	M68936	
parF	(−)	3331242	69.285	W	245	SG10303/P26974	
parC	(−)	3332236	69.306	W	752	SG10276/P26973	1252
STYPARE	(−)	3337000	69.405	SG	1,981	L05544	
parE	(−)	3337037	69.406	W	630	SG10277/P31598	1872
STCPDB	(−)	3341079	69.490	SG	1,311	X54009	
cpdB	(−)	3341080	69.490	F	250	SG10065/P26265	1221
cysQ	(+)	3342097	69.511	F	98	SG10080/P26264	1221
STYUGDOP	(+)	3365000	69.988	SH	4,631	M14427	
ygjD	(−)	3365001	69.988	F	44	SG10491/P40731	509
rpsU	(+)	3365359	69.995	W	71	SG10376/P02379	509
dnaG	(+)	3365810	70.004	W	581	SG10088/P07362	509
rpoD	(+)	3367705	70.044	W	615	SG10369/P07336	509
tdcBsty	(−)	3410000	70.923	SH	72	SS3008	
tdcB	(−)	3410001	70.923	F	24	SG10390/P11954	1036
STYM1R1	(−)	3420000	71.131	SG	555	M10888	
rnpB	(−)	3420128	71.134	W	RNA	SG30001/M1-RNA	96
STYM1R2	(−)	3420705	71.146	SG	225	M10889	
STYNUSAINF	(−)	3459700	71.963	SH	228	M28868	
STYNUSA	(−)	3460000	71.963	SH	1,503	M61008	
nusA	(−)	3460001	71.963	W	500	SG10258/P37430	388
STYSIG54	(+)	3473299	72.240	SG	1,955	M68571	
rpoN	(+)	3473375	72.242	W	477	SG10371/P26979	1569
yhbH	(+)	3474831	72.272	W	95	SG10428/P26983	1569
mdh-styM	(+)	3528999	73.399	MP	2,075	SS1022	
STYLT2MDH	(−)	3529000	73.399	C	1,122	M95049	
mdh	(−)	3529064	73.400	W	312	SG10218/P25077	1240
STYARGR	(+)	3529934	73.418	C	1,141	M75913	
argR	(+)	3530415	73.428	W	156	SG10017/P37170	1242
STU03101	(+)	3559000	74.022	SH	1,757	U03101	
yhdG	(+)	3559286	74.028	W	321	SG10429/P37405	
fis	(+)	3560260	74.049	W	98	SG10100/P37404	
STSAPG	(+)	3578215	74.422	SG	1,785	X80501	
trkA	(+)	3578464	74.427	W	458	SG10533/P39445	1541
mscL	(+)	3579980	74.459	F	7	SG10531/P39446	1541
STYRPOA	(−)	3582000	74.501	SG	1,065	M77750	

TABLE 3 *Continued*

Locus	Ori[b]	Left end[c]	Cs	Code[d]	Length[e]	Accession no.[f]	Reference(s)[g]
rpoA	(−)	3582041	74.502	W	329	SG10367/P00574	1230
rpsD	(−)	3583056	74.523	F	2	SG10525/NA	1230
STYSPCA	(−)	3587806	74.622	SH	201	M36266	
rplE	(−)	3587807	74.622	F	43	SG10362/P37436	
rplX	(−)	3587951	74.625	F	18	SG10366/P37438	
tufAstyM	(+)	3589171	74.650	MG	3,963	SS1036	
STTUFA	(−)	3589172	74.650	C	1,281	X55116	
tufA	(−)	3589198	74.651	W	394	SG10403/P21694	2035
STRPSG	(−)	3590382	74.675	C	2,753	X64591	
fusA	(−)	3590454	74.677	W	704	SG10134/P26229	958
rpsG	(−)	3592665	74.723	W	156	SG10373/P26230	958
STYRPSL	(−)	3592944	74.728	C	605	M68548	
rpsL	(−)	3593051	74.731	W	124	SG10374/P02367	859
STYCRP	(+)	3604000	74.958	SP	959	M13773	
crp	(+)	3604283	74.964	W	210	SG10068/P06170	1767
STYPABAA	(−)	3610327	75.090	SG	1,669	M32355	
argD	(−)	3610328	75.090	F	17	SG10488/P40732	2026
pabA	(−)	3610464	75.093	W	187	SG10273/P06193	1003, 2026
fic	(−)	3611059	75.105	W	200	SG10097/P20751	2026
yhfG	(−)	3611651	75.118	W	55	SG10430/P37771	2026
ppiA	(−)	3611914	75.123	F	26	SG10308/P20753	2026
STYCYSAA	(+)	3629000	75.478	SP	5,280	M64606	
nirD	(+)	3629001	75.478	F	67	SG10487/P40789	2169
nirC	(+)	3629469	75.488	W	269	SG10248/P25926	2169
cysG	(+)	3630290	75.505	W	457	SG10074/P25924	2169
yhfI	(+)	3631997	75.541	W	416	SG10437/P25927	2169
yhfJ	(+)	3633433	75.571	W	264	SG10438/P25928	2169
STENVZ	(−)	3653599	75.990	SG	3,293	X12374	
pckA	(+)	3653600	75.990	F	287	SG10497/P41033	1201
envZ	(−)	3654543	76.010	W	450	SG10091/P08982	1201
ompR	(−)	3655892	76.038	W	239	SG10266/P08981	1201
glgAstyM	(+)	3712737	77.220	MG	1,838	SS1011	
STYGLGC	(−)	3712738	77.220	C	1,410	M17363	
glgA	(−)	3712739	77.220	F	29	SG10141/P05416	1181
glgC	(−)	3712826	77.222	W	431	SG10142/P05415	1181
STGLGCG	(−)	3714118	77.249	C	458	X59281	
livFstyM	(+)	3723795	77.450	MG	7,630	SS1021	
STYLIV	(−)	3723796	77.450	C	4,072	D12589	
yhhV	(−)	3723797	77.450	F	68	SG10457/P40819	1308
livF	(−)	3724127	77.457	W	237	SG10202/P30294	1308
livG	(−)	3724842	77.472	W	255	SG10203/P30293	1308
livM	(−)	3725606	77.488	W	425	SG10207/P30296	1308
livH	(−)	3726880	77.514	W	308	SG10204/P30295	1308
STYLIVBC	(−)	3727808	77.533	C	3,618	D00478	
livK	(−)	3727866	77.535	W	369	SG10206/P17216	1467
livJ	(−)	3729975	77.579	W	365	SG10205/P17215	1467
STYOPDA	(+)	3832000	79.700	SP	2,842	M84574	
opdA	(+)	3832603	79.713	W	239	SG10266/P08981	364
STYCSPSA	(+)	3848000	80.033	SH	481	L23115	
cspA	(+)	3848268	80.039	W	70	SG10070/P37410	
STCYSE	(−)	3892557	80.960	SG	1,497	X59594	
cysE	(−)	3892627	80.961	W	273	SG10073/P29847	
gpsA	(−)	3893535	80.980	F	171	SG10456/P40716	
kbl-styM	(+)	3903999	81.198	MP	10,168	SS1019	
STRFADF	(+)	3904000	81.198	C	2,700	U06472	
kbl	(−)	3904001	81.198	F	162	SG10191/P37419	1814

(Table continues)

TABLE 3 *Continued*

Locus	Ori[b]	Left end[c]	Cs	Code[d]	Length[e]	Accession no.[f]	Reference(s)[g]
rfaD	(+)	3904690	81.212	W	310	SG10335/P37420	1814
rfaF	(+)	3906016	81.240	W	348	SG10336/P37421	1814
STYRFAC	(+)	3906070	81.241	C	1,574	M95927	
rfaC	(+)	3906672	81.254	W	317	SG10334/P26469	1258, 1813
STYRFALK	(+)	3907310	81.267	C	4,280	M73826	
rfaL	(+)	3907665	81.274	W	404	SG10340/P26471	1258
rfaK	(−)	3908936	81.301	W	381	SG10339/P26470	1258
rfaZ	(−)	3910182	81.327	W	269	SG10343/P26473	1258
rfaY	(−)	3911141	81.347	C	232	SG10342/P26472	1258
rfaYsty	(−)	3911590	81.356	C	2,578	SS3007	
rfaJ	(−)	3911935	81.363	W	336	SG10338/P19817	285
rfaI	(−)	3912891	81.383	W	337	SG10337/P19816	285
rfaB	(−)	3913910	81.404	F	81	SG10333/Q06994	285
S56361	(−)	3914809	81.423	SP	815	S56361	
rfaB	(−)	3914810	81.423	F	70	SG10333/Q06994	1061
yibR	(−)	3915089	81.429	W	77	SG10512/P40824	1061
rfaP	(−)	3915354	81.434	F	89	SG10341/Q06995	1061
pyrEstyM	(+)	3915999	81.448	MG	1,047	SS1030	
STOPPRBST	(−)	3916000	81.448	C	645	Z19547	
pyrE	(−)	3916001	81.448	W	213	SG10327/P08870	1422, 1739
M28333	(+)	3916462	81.457	C	585	M28333	
rph	(+)	3916463	81.457	F	108	SG10361/P26155	1422
STYUHPABCT	(−)	3940637	81.960	SG	5,465	M89480	
uhpT	(−)	3940837	81.964	W	463	SG10409/P27670	915
uhpC	(−)	3942369	81.996	W	442	SG10408/P27669	915
uhpB	(−)	3943707	82.024	W	500	SG10407/P27668	915
uhpA	(−)	3945209	82.055	W	196	SG10406/P27667	915
STYMGTBC	(+)	3946406	82.080	SG	4,586	M57715	
mgtC	(+)	3947009	82.093	W	231	SG10232/P22037	1850
mgtB	(+)	3947924	82.112	W	908	SG10231/P22036	1850
STRECF	(−)	4012276	83.450	SG	1,068	X62505	
recF	(−)	4012277	83.450	W	355	SG10331/P24900	1009
STYDNAA	(−)	4019488	83.600	SG	1,946	M17352	
dnaN	(−)	4019489	83.600	F	64	SG10089/P26464	1818
dnaA	(−)	4019679	83.604	W	466	SG10085/P35891	1818
STYORI	(+)	4080550	84.870	SG	552	J01808	
rbsBsty	(+)	4085357	84.970	SH	891	SS3005	
rbsB	(+)	4085358	84.970	W	297	SG10330/P02926	252
STYILVPA	(+)	4098000	85.233	SP	329	J01806	
ilvL	(+)	4098120	85.235	W	32	SG10183/P03060	1153, 1986
ilvEsty	(+)	4098500	85.243	SP	927	SS3001	
ilvE	(+)	4098501	85.243	W	308	SG10181/P15168	527
ilvAstyM	(+)	4099499	85.264	MP	1,756	SS1017	
STYILVAPRI	(+)	4099500	85.264	C	240	M25498	
ilvD	(+)	4099501	85.264	F	68	SG10458/P40810	1233
STYILVA	(+)	4099711	85.269	C	1,545	M26670	
ilvA	(+)	4099712	85.269	W	514	SG10179/P20506	1985
STYILVYCR	(+)	4100000	85.275	SP	378	K03522	
ilvY	(−)	4100001	85.275	F	42	SG10184/P05988	2124
ilvC	(+)	4100295	85.281	F	28	SG10180/P05989	2124
Z21789	(+)	4102040	85.317	SG	2,558	Z21789	
rhlB	(−)	4102041	85.317	F	56	SG10498/P40863	1360
trxA	(+)	4102391	85.324	W	109	SG10402/P00274	880, 1083, 1194, 1428, 1436, 1761
rho	(+)	4103139	85.340	W	419	SG10359/P26980	1083, 1360

TABLE 3 *Continued*

Locus	Ori[b]	Left end[c]	Cs	Code[d]	Length[e]	Accession no.[f]	Reference(s)[g]
STYCARABA	(+)	4112282	85.530	SG	4,378	M95047	
rffT	(+)	4112283	85.530	F	426	SG10354/P37458	
rffM	(+)	4113560	85.557	W	246	SG10353/P37457	
yifK	(+)	4114507	85.576	W	461	SG10431/P37456	
argX	(+)	4115995	85.607	W	RNA	SG30007/tRNA	
hisR	(+)	4116125	85.610	W	RNA	SG30009/tRNA	
leuU	(+)	4116222	85.612	W	RNA	SG30010/tRNA	
proM	(+)	4116351	85.615	W	RNA	SG30011/tRNA	
cyaAstyM	(+)	4120000	85.691	SG	1,661	SS1002	
STCYAG	(+)	4120001	85.691	C	489	X55783	
hemC	(−)	4120001	85.691	F	16	SG10571/NA	
cyaAsty	(+)	4120404	85.699	C	1,257	SS3009	
cyaA	(+)	4120404	85.699	F	419	SG10071/Q05878	809, 2015
		4120001	85.691	F			
STYCRA	(+)	4126226	85.820	SG	1,921	L11043	
uvrD	(+)	4126227	85.820	F	13	SG10414/Q05311	1843
corA	(+)	4126740	85.831	W	316	SG10064/P31138	1843
yigF	(−)	4127739	85.851	W	126	SG10432/P31139	1843
STPLDA	(+)	4133000	85.961	SH	1,254	X76900	
yigI	(−)	4133001	85.961	F	25	SG10461/P40725	240
pldA	(+)	4133241	85.966	W	289	SG10302/P37442	240
recQ	(+)	4134194	85.986	F	22	SG10460/P40724	240
STYMETR	(−)	4137284	86.050	SG	1,203	M17356	
metR	(−)	4137397	86.052	W	276	SG10227/P05984	1564
metE	(+)	4138479	86.075	F	2	SG10483/NA	1564
STHEMN	(+)	4165170	86.630	SG	3,055	U06779	
yihI	(+)	4165987	86.647	W	171	SG10433/P37130	2179
hemN	(+)	4166691	86.662	W	457	SG10154/P37129	2179
glnAstyM	(+)	4208423	87.530	MP	1,965	SS1012	
STYGLNA2	(−)	4208424	87.530	C	395	J01803	
glnA	(−)	4208754	87.536	W	469	SG10143/P06201	933, 2196
STYGLNA	(−)	4208756	87.537	C	1,407	M14536	
STYGLNA1	(−)	4210121	87.565	C	120	J01802	
STYGLNAA	(−)	4210201	87.567	C	188	M11196	
STRHABC2	(−)	4234406	88.070	SG	2,816	X57299	
rhaA	(−)	4234407	88.070	F	70	SG10355/P27031	1439
rhaB	(−)	4234615	88.074	W	489	SG10356/P27030	1439
rhaS	(+)	4236372	88.111	W	277	SG10357/P27029	1439
STYRHAT	(−)	4237806	88.141	SG	2,388	M85157	
rhaR	(+)	4237807	88.141	F	106	SG10524/P40865	2000
rhaT	(−)	4238124	88.147	W	344	SG10358/P27135	2000
STUSHB	(+)	4247000	88.332	SH	1,453	X13380	
sbp	(+)	4247001	88.332	W	311	SG10382/P02906	616, 914, 1560, 1561
ushB	(+)	4247358	88.339	W	251	SG10051/P26219	616
STMETJ	(−)	4295000	89.330	SP	656	X01961	
metJ	(−)	4295063	89.332	W	104	SG10226/P06203	2051
metB	(+)	4295645	89.344	F	4	SG10482/NA	2051
metFstyM	(+)	4296999	89.372	MP	3,726	SS1023	
STMETF	(+)	4297000	89.372	C	1,735	X07689	
metF	(+)	4297435	89.381	W	296	SG10224/P11003	2178
STKATG	(+)	4298339	89.400	C	2,387	X53001	
katG	(+)	4298490	89.403	W	727	SG10190/P17750	1229
trmAstyM	(+)	4310371	89.650	MG	2,604	SS1034	
STYTRMA	(−)	4310372	89.650	C	639	M57569	
trmA	(−)	4310373	89.650	F	102	SG10394/P22038	704
STYBTUB	(+)	4310480	89.652	C	2,496	M89481	
btuB	(+)	4311049	89.664	W	614	SG10030/P37409	2114
murI	(+)	4312838	89.701	F	46	SG10494/P40723	2114

(Table continues)

TABLE 3 *Continued*

Locus	Ori[b]	Left end[c]	Cs	Code[d]	Length[e]	Accession no.[f]	Reference(s)[g]
STYMURBIRA	(+)	4326000	89.975	SP	1,746	L14816	
rrfB	(+)	4326018	89.975	W	RNA	SG30014/rRNA	
murB	(+)	4326317	89.982	W	342	SG10235/P37417	458
birA	(+)	4327342	90.003	W	135	SG10028/P37416	458
STTUFB	(+)	4329681	90.052	SG	1,308	X55117	
tufB	(+)	4329772	90.053	W	394	SG10404/P21694	2035
STRPLJL	(+)	4334412	90.150	SG	1,131	X53072	
rplJ	(+)	4334602	90.154	W	165	SG10363/P17352	1550, 1551, 2230
rplL	(+)	4335166	90.166	W	121	SG10364/P18081	1550, 2230
STRPOB	(+)	4338258	90.230	SG	4,185	X04642	
rpoB	(+)	4338279	90.230	W	1,342	SG10368/P06173	1217, 1971
STYHUPA	(+)	4360500	90.693	SH	584	M22975	
hupA	(+)	4360711	90.697	W	90	SG10174/P15148	773
hydGstyM	(+)	4362999	90.745	MP	4,784	SS1016	
STYHYDGG	(+)	4363000	90.745	C	2,088	M64988	
hydH	(+)	4363001	90.745	F	208	SG10177/P37461	331
hydG	(+)	4363635	90.758	W	441	SG10176/P25852	331
STYPURHD	(−)	4364597	90.778	C	3,187	M66160	
purD	(−)	4364957	90.785	W	429	SG10319/P26977	330
purH	(−)	4366258	90.812	F	508	SG10320/P26978	330
STYMETA	(+)	4379126	91.080	SG	671	M74188	
yjaB	(−)	4379127	91.080	F	102	SG10486/P40677	1294
metA	(+)	4379591	91.090	F	69	SG10222/P37413	1294
STICLR	(−)	4387011	91.244	SG	950	X52950	
iclR	(−)	4387086	91.246	W	274	SG10178/P17430	606
metHsty	(+)	4388000	91.265	SH	1,373	SS3003	
metH	(+)	4388261	91.270	F	371	SG10225/P37586	2055
U01246	(−)	4392589	91.360	SG	1,088	U01246	
yjbB	(+)	4392590	91.360	F	52	SG10492/P40730	363
pepE	(−)	4392824	91.365	W	229	SG10282/P36936	363
STMAL	(−)	4403243	91.582	SP	8,590	X54292	
malG	(−)	4403507	91.587	W	296	SG10214/P26468	569, 1759
malF	(−)	4404412	91.606	W	514	SG10213/P26467	1759
malE	(−)	4407203	91.664	W	396	SG10212/P19576	402, 1759
malK	(+)	4407639	91.673	W	369	SG10215/P19566	402, 1759
lamB	(+)	4408837	91.698	W	452	SG10192/P26466	568, 1759
malM	(+)	4410368	91.730	W	305	SG10216/P26478	1759
STLEXA	(+)	4420513	91.941	SP	921	X63002	
lexA	(+)	4420626	91.943	W	202	SG10200/P29831	618, 1399
dnaBstyM	(+)	4429257	92.123	MP	3,155	SS1007	
STYDNABA	(+)	4429258	92.123	C	2,013	J03390	
qor	(−)	4429259	92.123	F	168	SG10511/P40783	1726, 2161
dnaB	(+)	4429827	92.135	W	471	SG10086/P10338	2161
STYALR	(+)	4431213	92.163	C	1,200	M12847	
alr	(+)	4431274	92.165	W	359	SG10007/P06655	598
STYUVRA	(−)	4435524	92.253	SP	3,623	M93014	
uvrA	(−)	4435800	92.259	W	941	SG10413/P37434	
ssb	(+)	4438874	92.323	F	91	SG10385/P37435	
STYPMTRAB	(−)	4494834	93.487	SP	3,824	L13395	
proP	(+)	4494835	93.487	F	17	SG10510/P40862	1665
pmrB	(−)	4495055	93.491	W	356	SG10025/P36557	1665
pmrA	(−)	4496135	93.514	W	222	SG10024/P36556	1665
yjdB	(−)	4496800	93.527	W	547	SG10434/P36555	1665

TABLE 3 *Continued*

Locus	Ori[b]	Left end[c]	Cs	Code[d]	Length[e]	Accession no.[f]	Reference(s)[g]
STMELB	(+)	4507574	93.752	SP	1,714	X62101	
melA	(+)	4507575	93.752	F	43	SG10219/P30877	1370
melB	(+)	4507790	93.756	W	476	SG10220/P30878	1370
STPHON	(−)	4529711	94.212	SP	1,408	X63599	
phoN	(−)	4529821	94.214	W	250	SG10292/P26976	687, 1007
STYMUTL	(+)	4566000	94.967	SP	2,446	M29687	
amiB	(+)	4566001	94.967	F	100	SG10009/P26366	1291
mutL	(+)	4566313	94.973	W	618	SG10236/P14161	1291
miaA	(+)	4568160	95.012	F	95	SG10233/P37724	1291
STU07843	(+)	4645490	96.620	SG	4,989	U07843	
treR	(−)	4645965	96.630	W	315	SG10393/P36674	
mgtA	(+)	4647296	96.658	F	42	SG10228/P23924	
STPYRBIG	(−)	4658000	96.880	SP	1,812	X05641	
pyrI	(−)	4658031	96.881	W	153	SG10329/P08421	1336
pyrB	(−)	4658505	96.891	W	311	SG10324/P08420	1336
pyrL	(−)	4659476	96.911	W	33	SG10323/P08522	1336
STMIAE	(+)	4672000	97.171	SP	4,531	X73368	
argI	(−)	4672001	97.171	F	218	SG10016/Q08016	1558
yjgD	(+)	4672820	97.188	W	138	SG10435/Q08019	1558
miaE	(+)	4673248	97.197	W	270	SG10234/Q08015	1558
ytgA	(−)	4674295	97.219	W	162	SG10441/Q08020	1558
yjgM	(−)	4674890	97.231	W	167	SG10442/Q08021	1558
yjgN	(+)	4675726	97.249	F	268	SG10443/Q08022	
hsdSstyM	(+)	4732999	98.440	MP	3,010	SS1015	
STHSDS	(−)	4733000	98.440	C	1,650	Y00524	
hsdS	(−)	4733016	98.440	W	469	SG10172/P06187	588, 610
STYHSDMB	(−)	4734420	98.470	C	1,590	L02506	
hsdM	(−)	4734421	98.470	W	529	SG10501/P40813	
STYNADR	(+)	4800788	99.850	SG	2,496	M85181	
sms	(+)	4800789	99.850	F	298	SG10444/P24517	563
nadR	(+)	4801836	99.872	W	409	SG10243/P24518	563, 2229
STYTRPR	(+)	4804154	99.920	SG	748	L13768	
slt	(+)	4804155	99.920	F	54	SG10490/P39434	
trpR	(+)	4804411	99.925	W	108	SG10401/P37444	
yjjX	(−)	4804799	99.933	F	35	SG10436/P39432	
STY326F	(+)	ND[h]	ND	S	326	D12814	
STYNHOA	(+)	ND	ND	S	2,134	D90301	
nhoA	(+)	ND	ND	W	281	SG10247/Q00267	2109
yzzJ	(−)	ND	ND	F	132	SG10500/P40788	1558
STYOADGABA	(+)	ND	ND	S	4,053	M96434	
oadG	(+)	ND	ND	W	83	SG10261/Q03032	2158
oadA	(+)	ND	ND	W	591	SG10259/Q03030	2158
oadB	(+)	ND	ND	W	433	SG10260/Q03031	2158
STYSINRPHO	(+)	ND	ND	S	4,880	L04307	
sinR	(+)	ND	ND	W	315	SG10383/P37459	688
STYTCPCHEM	(+)	ND	ND	S	1,840	L06029	
tcp	(+)	ND	ND	W	547	SG10389/Q02755	2193
SEANSP	(+)	ND	ND	S	1,887	U04851	
ansP	(+)	ND	ND	W	497	SG10527/P40812	

TABLE 3 *Continued*

[a]GenBank LOCUS field, StySeq contig name, or gene name. The StySeq contigs are generally composites created by merging overlapping GenBank entries and are usually named using the first gene present in the contig. This gene name is followed by the organism code "sty" and an "M" to denoted a merged (melded) contig. Three-letter gene names have a "-" instead of a fourth letter, following the conventions established for the EcoSeq data collection. The LOCUS and contig names are mnemonic, but they are subject to change and therefore one should use the accession numbers for retrieval and identification purposes.

[b]Ori, orientation of genes on the chromosome: (+), clockwise; (–), counterclockwise. Orientation of GenBank entries on the chromosome: (+), 5′ to 3′ is the clockwise direction; (–), 5′ to 3′ is the counterclockwise direction. If the orientation is unknown, the orientation of the corresponding *E. coli* sequence is used as the default orientation (see text). All melded contigs are constructed to be in the (+) orientation.

[c]Left end denotes the genomic coordinate (base pairs) of the left (counterclockwise) end of an aligned DNA sequence entry or gene in base pairs and centisomes. A centisome is a physical map unit equal to 1% of a chromosome's length. We estimate 1 Cs for the chromosome of *S. typhimurium* to be 48,080 bp. For genetically pinned genes, the estimated centisome value is used to calculate a left-end genomic coordinate in base pairs. We do not mean to imply that genomic positions are known to single-base-pair accuracy. In contrast, the relative positions of genes within a contig are known to single-base-pair accuracy. Persons wishing to quote a centisome map position for *S. typhimurium* genes should use the centisome values presented in Table 1.

[d]Codes: P, physically anchored contig; G, genetically pinned; H, roughly located by using mapping information from the homologous chromosome of *E. coli*; M, melded StySeq meld; S, single StySeq (GenBank) entry; C, constituent GenBank sequence used in assembly of a StySeq meld; W, whole gene sequence; F, fragment of a gene, partial sequence. StySeq1 master contigs are preceded by blank lines and have two codes.

[e]Length, number of base pairs for DNA sequences or the number of codons (amino acids) for genes. The number of codons is not given for the structural RNA genes, labeled "RNA."

[f]GenBank, SWISS-PROT, and StyGene (SG) accession numbers; completely overlapping (redundant) GenBank entries omitted here are listed in the SWISS-PROT records. Structural RNA genes do not have corresponding SWISS-PROT records; the type of structural RNA gene is indicated instead.

[g]DNA sequence and characterization citations from SWISS-PROT. Additional sequence-related references, including unpublished sources, can be obtained from GenBank and SWISS-PROT records. Additional references are also present in Table 1

[h]ND, not done.

FIGURE 1 Genetic map of *S. typhimurium* LT2. The gene designations used here are described in Table 1. The map is shown as 20 linear segments, representing the circular chromosome, and one linear segment representing the circular plasmid pSLT. The chromosome is composed of 100 centisomes (Cs); a scale covering 5 Cs is shown to the left of each segment. The scale is also designated in kilobases (kb); the entire chromosome is designated as 4,808 kb (1222). In the middle of each segment are vertical lines which show the restriction maps for the endonucleases *Xba*I, *Ceu*I, and *Bln*I, indicated by X, C, and B, respectively; a horizontal line indicates a restriction site. These sites were determined from PFGE (1222, 1224, 2162, 2163). The positions of many genes around the chromosome were anchored using PFGE, through the *Xba*I and *Bln*I sites in Tn*10* transposons which had transposed into these genes (1222, 1224, 2162). Genes which have been mapped in this way are indicated by a short horizontal bar to the left of the *Xba*I map, and the gene named is flagged with a superscript †; e.g., the *carAB* genes at 1.6 Cs, the *leu* genes at 2.8 Cs, and the *pan* genes at 4.5 Cs have been mapped through analysis of Tn*10* insertions in these genes (1222). Other genes are anchored through a specific restriction site which falls into a known gene; e.g., *Ceu*I sites are found only in the *rrn* genes; thus, at 6.2 Cs the *Ceu*I site indicates the location of the *rrn*H operon, at 58.1 Cs it indicates the location of the *rrn*G operon, and so on. In these cases, too, we show a short horizontal bar to the left of the *Xba*I map. The locations of the genes between these fixed points are based on several types of data (see the text). In many cases, the location is based on phage-mediated transduction; the distance between genes was determined by assuming that the lengths of P22 and P1 transducing fragments are 1 and 2 Cs, respectively, and applying the formula developed by Wu (2170) to convert the percentage of joint transduction to map distance in centisomes, with modifications as described in Fig. 2 of Sanderson and Roth (1721). Parentheses around a gene symbol indicate that the location of the gene is known only approximately, usually from conjugation studies or sequence comparison with *E. coli*. An asterisk indicates that the gene has been mapped more precisely, usually by phage-mediated transduction, but that its position with respect to adjacent markers is not known. Arrows to the extreme right of the genes and operons indicate the direction of mRNA transcription at these loci. The endpoints of the region of the chromosome inverted with respect to *E. coli* are indicated by the word "inversion" and by arrows. The region of the chromosome for which high-resolution restriction mapping data have been published (2166) is delimited by the letters HR.

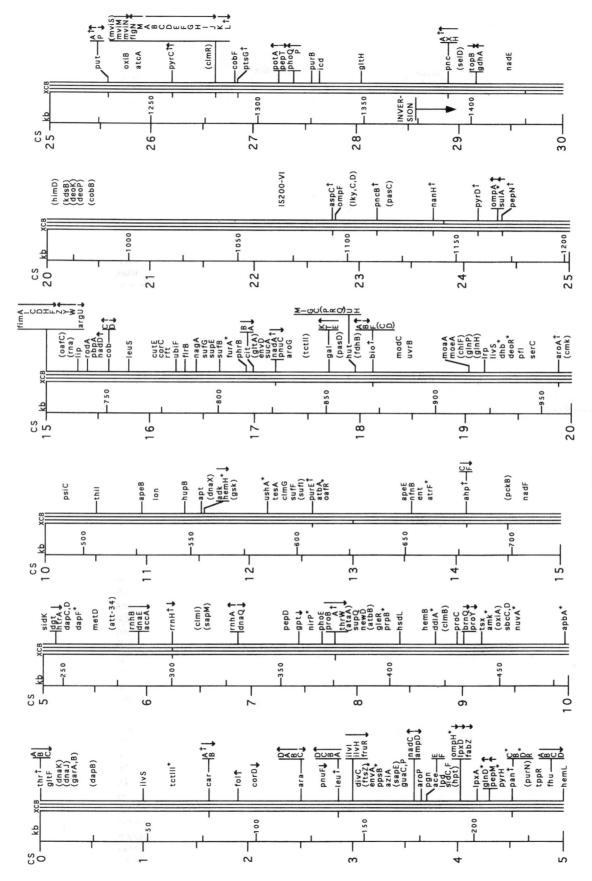

FIGURE 1 Continued

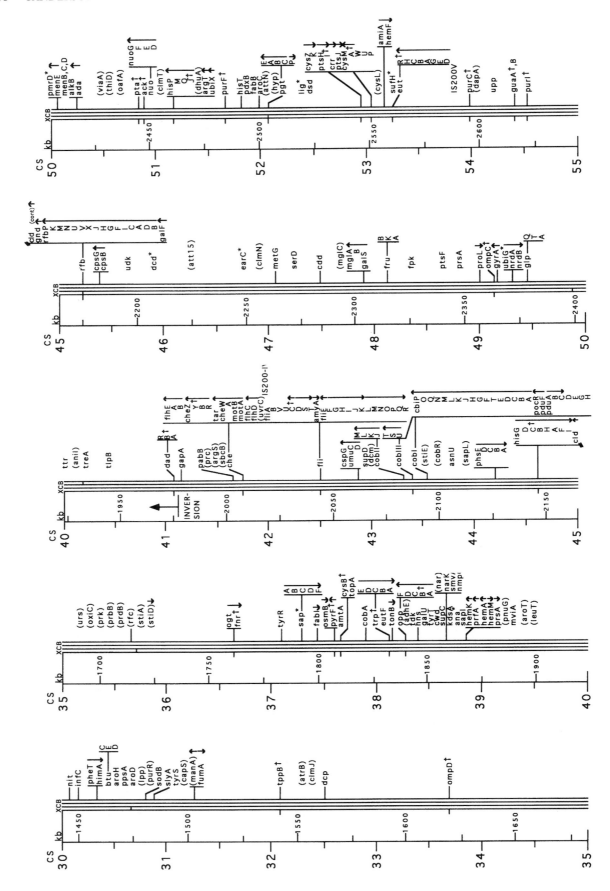

FIGURE 1 Continued

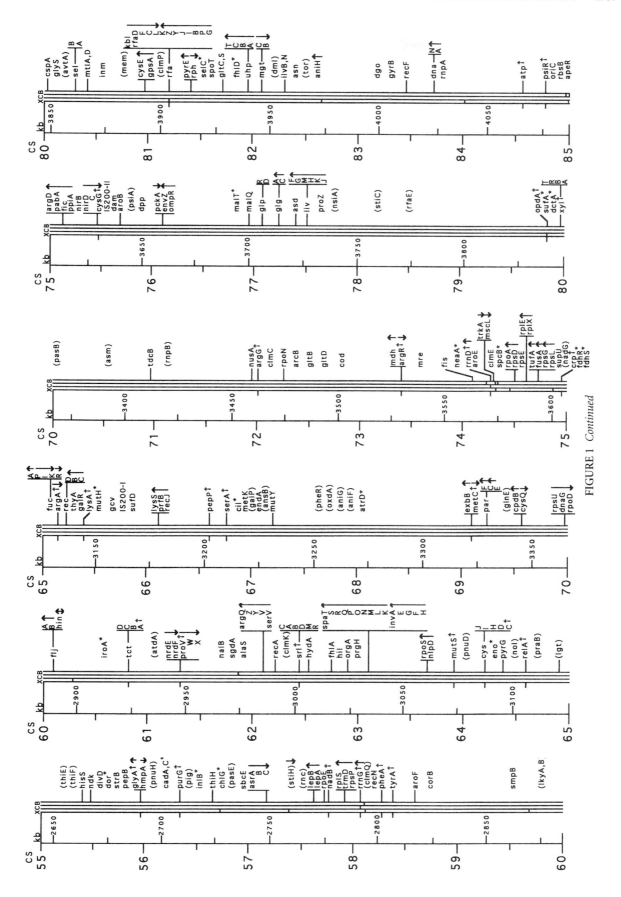

FIGURE 1 *Continued*

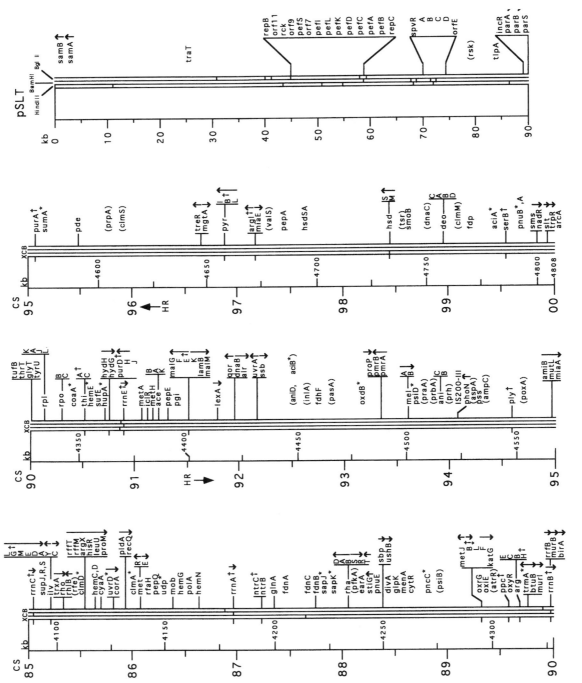

FIGURE 1 Continued

48. **Ames, G. F.-L., K. Nikaido, A. Hobson, and B. Malcolm.** 1985. Overproduction of the membrane-bound components on the histidine permease from *Salmonella typhimurium*: identification of the M protein. *Biochimie* 67:149–154.

49. **Ames, G. F.-L., K. D. Noel, H. Taber, E. N. Spudich, K. Nikaido, J. Afong, and F. Ardeshir.** 1977. Fine-structure map of the histidine transport genes in *Salmonella typhimurium*. *J. Bacteriol.* 129:1289–1297.

50. **Amsden, A. B., D. K. Small, and R. F. Gomez.** 1977. Complex medium toxicity to some DNA repair deficient strains of *Salmonella typhimurium*. *Can. J. Microbiol.* 23:1494–1496.

51. **Anderson, K. S., E. W. Miles, and K. A. Johnson.** 1991. Serine modulates substrate channeling in tryptophan synthase. A novel intersubunit triggering mechanism. *J. Biol. Chem.* 266:8020–8033.

52. **Anderson, P., and J. Roth.** 1981. Spontaneous tandem genetic duplications in *Salmonella typhimurium* arise by unequal recombinations between mRNA (*rrn*) cistrons. *Proc. Natl. Acad. Sci. USA* 78:3113–3117.

53. **Anderson, R. P., and I. Pastan.** 1973. The cyclic AMP receptor of *Escherichia coli*: immunological studies in extracts of *Escherichia coli* and other organisms. *Biochim. Biophys. Acta* 320:577–578.

54. **Anderson, R. P., and J. R. Roth.** 1978. Tandem chromosomal duplications in *Salmonella typhimurium*: fusion of histidine genes to novel promoters. *J. Mol. Biol.* 119:147–166.

55. **Anderson, R. R., R. Menzel, and J. M. Wood.** 1980. Biochemistry and regulation of a second L-proline transport system in *Salmonella typhimurium*. *J. Bacteriol.* 141:1071–1076.

56. **Andersson, D. I.** 1992. Involvement of the Arc system in redox regulation of the Cob operon in *Salmonella typhimurium*. *Mol. Microbiol.* 6:1491–1494.

57. **Andersson, D. I., and J. R. Roth.** 1989. Mutations affecting regulation of cobinamide biosynthesis in *Salmonella typhimurium*. *J. Bacteriol.* 171:6726–6733.

58. **Andreadis, A., and E. R. Rosenthal.** 1992. The nucleotide sequence of *leuB* from *Salmonella typhimurium*. *Biochim. Biophys. Acta* 1129:228–230.

59. **Andreeva, I. V., and A. A. Kiryushkina.** 1994. Contribution of *Salmonella albony uvr*-12 gene in the excision repair pathway. *Genetics* (USSR) 9:177–180.

60. **Andreeva, I. V., A. A. Kiryushkina, and A. M. Amerkhanova.** 1972. Genetic basis of UV-resistance in *Salmonella*. II. *uvrB*-mutants of *Salmonella albony*. *Genetics* (USSR) 8(11):117–121.

61. **Andreeva, I. V., A. A. Kiryushkina, V. N. Pokrovsky, and A. G. Skavronskaya.** 1972. Genetic basis for UV-resistance in *Salmonella*. I. Location of *uvrA* gene on *Salmonella typhimurium* chromosome. *Genetics* (USSR) 8(4):117–121.

62. **Anton, D. N.** 1968. Histidine regulatory mutants in *Salmonella typhimurium*. V. Two new classes of histidine regulatory mutants. *J. Mol. Biol.* 33:533–546.

63. **Anton, D. N.** 1968. Osmotic-sensitive mutant of *Salmonella typhimurium*. *J. Bacteriol.* 109:1273–1283.

64. **Anton, D. N.** 1978. Genetic control of defective cell shape and osmotic sensitivity in a mutant *Salmonella typhimurium*. *Mol. Gen. Genet.* 160:277–286.

65. **Anton, D. N.** 1979. Positive selection of mutants with cell envelope defects of a *Salmonella typhimurium* strain hypersensitive to the products of genes *hisF* and *hisH*. *J. Bacteriol.* 137:1271–1281.

66. **Anton, D. N.** 1981. *envB* mutations confer UV-sensitivity to *Salmonella typhimurium* and UV-resistance to *Escherichia coli*. *Mol. Gen. Genet.* 181:150–152.

67. **Anton, D. N.** 1987. Conditional transduction of *Salmonella typhimurium envB* mutations. *J. Bacteriol.* 169:1767–1771.

68. **Anton, D. N., A. T. de Micheli, and A. M. Palermo.** 1983. Isolation of round-cell mutants of *Salmonella typhimurium*. *Can. J. Microbiol.* 29:170–173.

69. **Anton, D. N., and L. V. Orce.** 1976. Envelope mutation promoting autolysis in *Salmonella typhimurium*. *Mol. Gen. Genet.* 144:97–105.

70. **Antonucci, T. K., and D. L. Oxender.** 1986. The molecular biology of amino-acid transport in bacteria. *Adv. Microb. Physiol.* 28:145–180.

71. **Archer, C. D., X. Wang, and T. Elliott.** 1994. Mutants defective in the energy-conserving NADH dehydrogenase of *Salmonella typhimurium* identified by a decrease in energy-dependent proteolysis after carbon starvation. *Proc. Natl. Acad. Sci. USA* 90:9877–9881.

72. **Archibald, E. R., and L. S. Williams.** 1973. Regulation of methionyl-transfer ribonucleic acid synthetase formation in *Escherichia coli* and *Salmonella typhimurium*. *J. Bacteriol.* 114:1007–1013.

73. **Ardeshir, F., and G. F.-L. Ames.** 1980. Cloning of the histidine transport genes from *Salmonella typhimurium* and characterization of an analogous transport system in *Escherichia coli*. *J. Supramol. Struct.* 13:117–130.

74. **Ardeshir, F., C. F. Higgins, and G. F.-L. Ames.** 1981. Physical map of the *Salmonella typhimurium* histidine transport operon: correlation with the genetic map. *J. Bacteriol.* 147:401–409.

75. **Arfin, S. M., T. Miner, and G. W. Hatfield.** 1974. Synthesis of branched-chain aminoacyl-transfer ribonucleic acid synthetases in a *Salmonella typhimurium* mutant with an altered biosynthetic L-threonine deaminase. *J. Bacteriol.* 120:604–607.

76. **Armstrong, F. B., C. J. R. Hedgecock, J. B. Reary, D. Whitehouse, and C. H. G. Crout.** 1974. Stereochemistry of the reductoisomerase and α,β-dihyroxyacid dehydratase-catalysed steps in valine and isoleucine biosynthesis. *J. Chem. Soc. Chem. Commun.* 9:351–352.

77. **Armstrong, F. B., and H. Ishiwa.** 1971. Isoleucine-valine mutants of *Salmonella typhimurium*. *Genetics* 67:171–182.

78. **Armstrong, F. B., U. S. Muller, J. B. Reary, D. Whitehouse, and D. H. G. Crout.** 1977. Stereoselectivity and stereospecificity of the a,β-dihydroxyacid dehydratase from *Salmonella typhimurium*. *Biochim. Biophys. Acta* 498:282–293.

79. **Armstrong, F. B., and R. P. Wagner.** 1964. Isoleucine-valine requiring mutants of *Salmonella typhimurium*. *Genetics* 50:957–965.

80. **Artymiuk, P. J., D. W. Rice, E. M. Mitchell, and P. Willett.** 1990. Structural resemblance between the families of bacterial signal-transduction proteins and of G proteins revealed by graph theoretical techniques. *Protein Eng.* 4:39–43.

81. **Artz, S., D. Holzschu, P. Blum, and R. Shand.** 1983. Use of M13mp phages to study gene regulation, structure and function: cloning and recombinational analysis of genes of the *Salmonella typhimurium* histidine operon. *Gene* 26:147–158.

82. **Artz, S. W., and J. R. Broach.** 1975. Histidine regulation in *Salmonella typhimurium*. *Proc. Natl. Acad. Sci. USA* 72:3453–3457.

83. **Ashton, D. M., G. D. Sweet, J. M. Somers, and W. W. Kay.** 1980. Citrate transport in *Salmonella typhimurium*: studies with 2-fluoro-L-*erythro*-citrate as a substrate. *Can. J. Biochem.* 58:797–803.

84. **Aswad, D., and D. E. Koshland, Jr.** 1974. Role of methionine in bacterial chemotaxis. *J. Bacteriol.* 118:640–645.

85. **Aswad, D., and D. E. Koshland, Jr.** 1975. Isolation, characterization and complementation of *Salmonella typhimurium* chemotaxis mutants. *J. Mol. Biol.* 97:225–235.

86. **Aswad, D. W., and D. E. Koshland, Jr.** 1975. Evidence for an S-adenosylmethionine requirement in the chemotactic behavior of *Salmonella typhimurium*. *J. Mol. Biol.* 97:207–223.

87. **Atkins, J. F., and S. Ryce.** 1974. UGA and non-triplet suppressor reading of the genetic code. *Nature* (London) 249:527–530.

88. **Aulin, M. R., and D. Hughes.** 1990. Overproduction of release factor reduces spontaneous frameshifting and frameshift suppression by mutant elongation factor Tu. *J. Bacteriol.* 172:6721–6726.

89. **Ayling, P. D.** 1981. Methionine sulfoxide is transported by high-affinity methionine and glutamine transport systems in *Salmonella typhimurium*. *J. Bacteriol.* 148:514–520.

90. **Ayling, P. D., and E. S. Bridgeland.** 1970. Methionine transport systems in *Salmonella typhimurium*. *Heredity* 25:687–688.

91. **Ayling, P. D., and E. S. Bridgeland.** 1972. Methionine transport in wild-type and transport-defective mutants of *Salmonella typhimurium*. *J. Gen. Microbiol.* 73:127–141.

92. **Ayling, P. D., and K. F. Chater.** 1968. The sequence of four structural and two regulatory methionine genes in the *Salmonella typhimurium* map. *Genet. Res.* 12:341–354.

93. **Ayling, P. D., T. Mojica-a, and T. Klopotowski.** 1979. Methionine transport in *Salmonella typhimurium*: evidence for at least one low-affinity transport system. *J. Gen. Microbiol.* 114:227–246.

94. **Backhaus, H., and H. Schmieger.** 1979. Bacterial DNA synthesized under phage control in a DNA defective *Salmonella* mutant and packaged into a special fraction of transducing particles of phage P-22. *Mol. Gen. Genet.* 171:301–306.

95. **Backhaus, H., and H. Schmieger.** 1979. Replication and maturation of phage P-22 in a mutant of *Salmonella typhimurium* temperature sensitive in initiation of DNA replication. *Mol. Gen. Genet.* 171:295–300.

96. **Baer, M., and S. Altman.** 1985. A catalytic RNA and its gene from *Salmonella typhimurium*. *Science* 228:999–1002.

97. **Bagdasarian, M., M. Hryniewicz, M. Zdzienicka, and M. Bagdasarian.** 1975. Integrative suppression of a *dnaA* mutation in *Salmonella typhimurium*. *Mol. Gen. Genet.* 139:213–231.

98. **Bagdasarian, M., M. Izakowska, and M. Bagdasarian.** 1977. Suppression of the DnaA phenotype by mutations in the *rpoB* cistron of ribonucleic acid polymerase in *Salmonella typhimurium* and *Escherichia coli*. *J. Bacteriol.* 130:577–582.

99. **Bagdasarian, M., M. Zdzienicka, and M. Bagdasarian.** 1972. Temperature sensitive initiation of DNA synthesis in a mutant of *Salmonella typhimurium*. *Mol. Gen. Genet.* 117:129–142.

100. **Bagdian, G. B., and P. H. Makela.** 1971. Antigenic conversion by P27. I. Mapping of the prophage attachment site on the *Salmonella* chromosome. *Virology* 43:403–411.

101. **Bairoch, A., and B. Boeckmann.** 1994. The SWISS-PROT protein sequence data bank: current status. *Nucleic Acids Res.* 22:3578–3580.

102. **Balbinder, E., A. J. Blume, A. Weber, and H. Tamaki.** 1968. Polar and antipolar mutants in the tryptophan operon of *Salmonella typhimurium*. *J. Bacteriol.* 95:2217–2229.

103. **Balbinder, E., R. Callahan III, P. P. McCann, C. Cordaro, A. R. Weber, A. M. Smith, and R. Angelosanto.** 1970. Regulatory mutants of the tryptophan operon of *Salmonella typhimurium*. *Genetics* 66:31–53.

104. **Bancroft, S., S. G. Rhee, C. Neumann, and S. Kustu.** 1978. Mutations that alter the covalent modification of glutamine synthetase in *Salmonella typhimurium*. *J. Bacteriol.* 134:1046–1055.

105. **Bansal, A., M. A. Dayton, H. Zalkin, and R. F. Colman.** 1989. Affinity labeling of a glutamyl peptide in the coenzyme binding site of NADP+-specific glutamate dehydrogenase of *Salmonella typhimurium* by 2-[(4-bromo-2,3-dioxobutyl)thio]-1,N6-ethenoadenosine 2′,5′-bisphosphate. *J. Biol. Chem.* 264:9827–9835.

106. **Baptist, E. W., S. G. Hallquist, and N. M. Kredich.** 1982. Identification of the *Salmonella typhimurium cysB* gene product by two-dimensional protein electrophoresis. *J. Bacteriol.* 151:495–499.

107. **Baptist, E. W., and N. M. Kredich.** 1977. Regulation of L-cystine transport in *Salmonella typhimurium*. *J. Bacteriol.* 131:111–118.

108. Barbe, J., A. Villaverde, and R. Guerrero. 1983. Indirect induction of SOS functions in *Salmonella typhimurium*. *Antonie Van Leeuwenhoek J. Microbiol. Serol.* 49:471–484.

109. Barcak, G. J., and R. E. Wolf, Jr. 1988. Comparative nucleotide sequence analysis of growth-rate-regulated *gnd* alleles from natural isolates of *Escherichia coli* and from *Salmonella typhimurium* LT-2. *J. Bacteriol.* 170:372–379.

110. Barker, R. M., and A. A. Yousuf. 1985. Clonal relationships among naturally occurring nicotinamide-requiring *Salmonella typhimurium*. *Genet. Res.* 46:241–250.

111. Barnes, W. M. 1978. DNA sequence from the histidine operon control region: seven histidine codons in a row. *Proc. Natl. Acad. Sci. USA* 75:4281–4285.

112. Barnes, W. M. 1978. DNA sequencing by partial ribo substitution. *J. Mol. Biol.* 119:83–100.

113. Barnes, W. M. 1979. Construction of an M-13 histidine transducing phage: a single stranded cloning vehicle with one *EcoR1* site. *Gene* 5:127–140.

114. Barnes, W. M. 1980. DNA cloning with single-stranded phage vectors, p. 185–200. *In* J. K. Setlow and A. Hollaender (ed.), *Genetic Engineering*, vol. 2. Plenum Publishing Corp., New York.

115. Barnes, W. M. 1981. Cloning and restriction map of the first part of the histidine operon of *Salmonella typhimurium*. *J. Bacteriol.* 147:124–134.

116. Barnes, W. M., and E. Tuley. 1983. DNA sequence changes of mutations in the histidine operon control region that decrease attenuation. *J. Mol. Biol.* 165:443–459.

117. Barrett, E. L., and G. W. Chang. 1979. Cysteine auxotrophs of *Salmonella typhimurium* which grow without cysteine in a hydrogen/carbon dioxide atmosphere. *J. Gen. Microbiol.* 115:513–516.

118. Barrett, E. L., C. E. Jackson, H. T. Fukumoto, and G. W. Chang. 1979. Formate dehydrogenase mutants of *Salmonella typhimurium*: a new medium for their isolation and new mutant classes. *Mol. Gen. Genet.* 177:95–101.

119. Barrett, E. L., H. S. Kwan, and J. Macy. 1984. Anaerobiosis, formate, nitrate, and *pyrA* are involved in the regulation of formate hydrogenlyase in *Salmonella typhimurium*. *J. Bacteriol.* 158:972–977.

120. Barrett, E. L., and D. L. Riggs. 1982. *Salmonella typhimurium* mutants defective in the formate dehydrogenase linked to nitrate reductase. *J. Bacteriol.* 149:554–560.

121. Barrett, E. L., and D. L. Riggs. 1982. Evidence for a second nitrate reductase activity that is distinct from the respiratory enzyme in *Salmonella typhimurium*. *J. Bacteriol.* 150:563–571.

122. Bar-Tana, J., D. J. Howlett, and R. Hertz. 1980. Ubiquinone synthetic pathway in flagellation in *Salmonella typhimurium*. *J. Bacteriol.* 143:637–643.

123. Bastin, D. A., G. Stevenson, P. K. Brown, A. Haase, and P. R. Reeves. 1993. Repeat unit polysaccharides of bacteria: a model for polymerization resembling that of ribosomes and fatty acid synthetase, with a novel mechanism for determining chain length. *Mol. Microbiol.* 7:725–734.

124. Batchelor, R. A., P. Alifano, E. Biffali, S. I. Hull, and R. A. Hull. 1992. Nucleotide sequences of the genes regulating O-polysaccharide antigen chain length (*rol*) from *Escherichia coli* and *Salmonella typhimurium*: protein homology and functional complementation. *J. Bacteriol.* 174:5228–5236.

125. Bauer, K., R. Benz, J. Brass, and W. Boos. 1985. *Salmonella typhimurium* contains an anion-selective outer membrane porin induced by phosphate starvation. *J. Bacteriol.* 161:813–816.

126. Bauerle, R., J. Hess, and S. French. 1987. Anthranilate synthase-anthranilate phosphoribosyltransferase complex and subunits of *Salmonella typhimurium*. *Methods Enzymol.* 142:366–386.

127. Bauerle, R. H., and P. Margolin. 1966. The functional organization of the tryptophan gene cluster in *Salmonella typhimurium*. *Proc. Natl. Acad. Sci. USA* 56:111–118.

128. Bauerle, R. H., and P. Margolin. 1966. A multifunctional enzyme complex in the tryptophan pathway of *Salmonella typhimurium*: comparison of polarity and pseudopolarity mutations. *Cold Spring Harbor Symp. Quant. Biol.* 31:203–214.

129. Baumler, A. J., J. G. Kusters, I. Stojiljkovic, and F. Heffron. 1994. *Salmonella typhimurium* loci involved in survival within macrophages. *Infect. Immun.* 62:1623–1630.

130. Beacham, I. R., A. Eisenstark, P. T. Barth, and R. H. Pritchard. 1968. Deoxynucleoside-sensitive mutants of *Salmonella typhimurium*. *Mol. Gen. Genet.* 102:112–127.

131. Beacham, I. R., and S. Garrett. 1981. Transfer of RP4::Mu to *Salmonella typhimurium*. *J. Gen. Microbiol.* 124:255–228.

132. Beck, C. F., A. R. Eisenshardt, and J. Neuhard. 1975. Deoxycytidine triphosphate deaminase of *Salmonella typhimurium*. *J. Biol. Chem.* 250:609–616.

133. Beck, C. F., and J. L. Ingraham. 1971. Location on the chromosome of *Salmonella typhimurium* of genes governing pyrimidine metabolism. *Mol. Gen. Genet.* 111:303–316.

134. Beck, C. F., J. L. Ingraham, and J. Neuhard. 1972. Location on the chromosome of *Salmonella typhimurium* of genes governing pyrimidine metabolism. II. Uridine kinase, cytosine deaminase, and thymidine kinase. *Mol. Gen. Genet.* 115:208–215.

135. Beck, C. F., J. Neuhard, and E. Thomassen. 1977. Thymidine-requiring mutants of *Salmonella typhimurium* that are defective in deoxyuridine 5′-phosphate synthesis. *J. Bacteriol.* 129:305–316.

136. Beck, C. F., J. Neuhard, E. Thomassen, J. L. Ingraham, and E. Kleker. 1974. *Salmonella typhimurium* mutants defective in cytidine monophosphate kinase (*cmk*). *J. Bacteriol.* 120:1370–1379.

137. Becker, M. A., N. M. Kredich, and G. Tomkins. 1969. The purification and characterization of O-acetylserine sulfhydrylase-A from *Salmonella typhimurium*. *J. Biol. Chem.* 244:2418–2427.

138. Beckmann, I., T. B. Subbaiah, and B. A. D. Stocker. 1964. Rough mutants of *Salmonella typhimurium*. II. Serological and chemical investigations. *Nature* (London) 201:1299–1301.

139. Beckwith, J. R., A. B. Pardee, R. Austrian, and F. Jacob. 1962. Coordination of the synthesis of the enzymes in the pyrimidine pathway of *Escherichia coli*. *J. Mol. Biol.* 5:618–634.

140. Behlau, I., and S. I. Miller. 1993. A PhoP-repressed gene promotes *Salmonella typhimurium* invasion of epithelial cells. *J. Bacteriol.* 175:4475–4484.

141. Benamira, M., U. Singh, and L. J. Marnett. 1992. Site-specific frameshift mutagenesis by a propanodeoxyguanosine adduct positioned in the (CpG)4 hotspot of *Salmonella typhimurium hisD3052* carried on an M13 vector. *J. Biol. Chem.* 267:22392–22400.

142. Beneski, D. A., A. Nakazawa, N. Weigel, P. E. Hartman, and S. Roseman. 1982. Sugar transport by the bacterial phosphotransferase system. Isolation and characterization of a phosphocarrier protein HPr from wild type and mutants of *Salmonella typhimurium*. *J. Biol. Chem.* 257:14492–14498.

143. Benjamin, W. H., Jr., P. Hall, and D. E. Briles. 1991. A *hemA* mutation renders *Salmonella typhimurium* avirulent in mice, yet capable of eliciting protection against intravenous infection with *S. typhimurium*. *Microb. Pathog.* 11:289–295.

144. Benjamin, W. H., Jr., J. Yother, P. Hall, and D. E. Briles. 1991. The *Salmonella typhimurium* locus *mviA* regulates virulence in Itys but not Ityr mice: functional *mviA* results in avirulence; mutant (nonfunctional) *mviA* results in virulence. *J. Exp. Med.* 174:1073–1083.

145. Benner-Luger, D., and W. Boos. 1988. The *mglB* sequence of *Salmonella typhimurium* LT2; promoter analysis by gene fusions and evidence for a divergently oriented gene coding for the *mgl* repressor. *Mol. Gen. Genet.* 214:579–587.

146. Bennett, R. L., and L. I. Rothfield. 1976. Genetic and physiological regulation of intrinsic proteins of the outer membrane of *Salmonella typhimurium*. *J. Bacteriol.* 127:498–504.

147. Benson, C. E., and J. S. Gots. 1975. Genetic modification of substrate specificity of hypoxanthine phosphoribosyltransferase in *Salmonella typhimurium*. *J. Bacteriol.* 121:77–82.

148. Benson, C. E., and J. S. Gots. 1975. Regulation of GMP reductase in *Salmonella typhimurium*. *Biochim. Biophys. Acta* 403:47–57.

149. Benson, C. E., and J. S. Gots. 1976. Occurrence of a regulatory deficiency of purine biosynthesis among *purA* mutants of *Salmonella typhimurium*. *Mol. Gen. Genet.* 145:31–36.

150. Benson, C. E., D. L. Hornick, and J. S. Gots. 1980. Genetic separation of purine transport from phosphoribosyltransferase activity in *Salmonella typhimurium*. *J. Gen. Microbiol.* 121:357–364.

151. Benson, N., and J. R. Roth. 1994. Suppressors of *recB* mutations in *Salmonella typhimurium*. *Genetics* 138:11–28.

152. Benson, N., P. Sugiono, S. Bass, L. V. Mendelman, and P. Youderian. 1986. General selection for specific DNA-binding activities. *Genetics* 114:1–14.

153. Benzinger, R., and I. Kleber. 1971. Transfection of *Escherichia coli* and *Salmonella typhimurium* spheroplasts: host-controlled restriction of infective bacteriophage P22 deoxyribonucleic acid. *J. Virol.* 8:197–202.

154. Berg, C. M., and J. J. Rossi. 1974. Proline excretion and indirect suppression in *Escherichia coli* and *Salmonella typhimurium*. *J. Bacteriol.* 118:928–939.

155. Berg, C. M., and K. J. Shaw. 1981. Organization and regulation of the *ilvGEDA* operon in *Salmonella typhimurium*. *J. Bacteriol.* 145:984–989.

156. Berg, C. M., K. J. Shaw, L. Sarokin, and D. E. Berg. 1981. Probing the organization and regulation of bacterial operons with transposons, p. 121–123. *In* D. Schlessinger (ed.), *Microbiology—1981*. American Society for Microbiology, Washington, D.C.

157. Berg, C. M., W. A. Whalen, and L. B. Archambault. 1983. Role of alanine-valine transaminase in *Salmonella typhimurium* and analysis of an *avtA*::Tn5 mutant. *J. Bacteriol.* 155:1009–1014.

158. Bergstrom, S., F. P. Lindberg, O. Olsson, and S. Normark. 1983. Comparison of the overlapping *frd* and *ampC* operons of *Escherichia coli* with the corresponding DNA sequences in other gram-negative bacteria. *J. Bacteriol.* 155:1297–1305.

159. Berkowitz, D. 1971. D-Mannitol utilization in *Salmonella typhimurium*. *J. Bacteriol.* 105:232–240.

160. Berst, M., C. G. Hellerqvist, B. Lindberg, O. Luderitz, S. Svensson, and O. Westphal. 1969. Structural investigations on T1 lipopolysaccharides. *Eur. J. Biochem.* 11:353–359.

161. Best, E. A., and R. A. Bender. 1990. Cloning of the *Klebsiella aerogenes nac* gene, which encodes a factor required for nitrogen regulation of the histidine utilization (*hut*) operons in *Salmonella typhimurium*. *J. Bacteriol.* 172:7043–7048.

162. Betteridge, P. R., and P. D. Ayling. 1975. The role of methionine transport-defective mutations in resistance to methionine sulfoximine in *Salmonella typhimurium*. *Mol. Gen. Genet.* 138:41–52.

163. Betteridge, P. R., and P. D. Ayling. 1976. The regulation of glutamine transport and glutamine synthetase in *Salmonella typhimurium*. *J. Gen. Microbiol.* 95:324–334.

164. Beyer, W., and L. Geue. 1992. Characterization of the virulence regions in the plasmids of three live *Salmonella* vaccines. *Int. J. Med. Microbiol. Virol. Parasitol. Infect. Dis.* 277:10–21.

165. Bhaduri, S., T. Kasai, D. Schlessinger, and H. J. Raskas. 1980. pMB-9 plasmids bearing the *Salmonella typhimurium his* operon and *gnd* gene. *Gene* 8:239–254.

166. Bhatia, M. B., A. Vinitsky, and C. Grubmeyer. 1990. Kinetic mechanism of orotate phosphoribosyltransferase from *Salmonella typhimurium*. *Biochemistry* 29:10480–10487.

167. Bhattacharya, A. K., and M. Chakravorty. 1971. Induction and repression of L-arabinose isomerase in *Salmonella typhimurium*. *J. Bacteriol.* 106:107–112.

168. Bhattacharya, A. K., and M. Chakravorty. 1974. Effect of antibiotics and antimetabolites on the induction of L-arabinose isomerase in *Salmonella typhimurium*. *Curr. Sci.* 43:499–503.

169. Bhattacharya, A. K., and M. Chakravorty. 1975. Isolation and characterization of an L-arabinose negative mutant of *Salmonella typhimurium*. *Indian J. Exp. Biol.* 13:244–246.

170. Bisercic, M., and H. Ochman. 1993. Natural populations of *Escherichia coli* and *Salmonella typhimurium* harbor the same classes of insertion sequences. *Genetics* 133:449–454.

171. Bitar, K. G., F. R. Firca, and J. C. Loper. 1977. Histidinol dehydrogenase from *Salmonella typhimurium* and *Escherichia coli*. *Biochim. Biophys. Acta* 493:429–440.

172. Bjork, G. R., P. M. Wikstrom, and A. S. Bystrom. 1989. Prevention of translational frameshifting by the modified nucleoside 1-methylguanosine. *Science* 244:986–989.

173. Blanc-Potard, A.-B., and L. Bossi. 1994. Phenotypic suppression of DNA gyrase deficiencies by a deletion lowering the gene dosage of a major tRNA in *Salmonella typhimurium*. *J. Bacteriol.* 176:2216–2226.

174. Blank, J., and P. Hoffee. 1972. Regulatory mutants of the *deo* regulon in *Salmonella typhimurium*. *Mol. Gen. Genet.* 116:291–298.

175. Blank, J. G., and P. A. Hoffee. 1975. Purification and properties of thymidine phosphorylase from *Salmonella typhimurium*. *Arch. Biochem. Biophys.* 168:259–265.

176. Blasi, F., S. M. Aloj, and R. F. Goldberger. 1971. Effect of histidine on the enzyme which catalyzes the first step of histidine biosynthesis in *Salmonella typhimurium*. *Biochemistry* 10:1409–1417.

177. Blasi, F., R. W. Barton, J. S. Kovach, and R. F. Goldberger. 1971. Interaction between the first enzyme for histidine biosynthesis and histidyl transfer ribonucleic acid. *J. Bacteriol.* 106:508–513.

178. Blasi, F., and C. B. Bruni. 1981. Regulation of the histidine operon: translation-controlled transcription termination (a mechanism common to several biosynthetic operons). *Curr. Top. Cell. Regul.* 19:1–45.

179. Blasi, F., C. B. Bruni, A. Avitabile, R. G. Deeley, R. F. Goldberger, and M. M. Meyers. 1973. Inhibition of transcription of the histidine operon *in vitro* by the first enzyme of the histidine pathway. *Proc. Natl. Acad. Sci. USA* 70:2692–2696.

180. Blatt, J. M., and H. E. Umbarger. 1972. On the role of isoleucyl-tRNA synthetase in multivalent repression. *Biochem. Genet.* 6:99–118.

181. Blattner, F. R., V. Burland, G. Plunkett, H. J. Sofia, and D. L. Daniels. 1993. Analysis of the *Escherichia coli* genome. IV. DNA sequence of the region from 89.2 to 92.8 minutes. *Nucleic Acids Res.* 21:5408–5417.

182. Blazey, D. L., and R. O. Burns. 1979. Genetic organization of the *Salmonella typhimurium ilv* gene cluster. *Mol. Gen. Genet.* 177:1–11.

183. Blazey, D. L., and R. O. Burns. 1980. Gene *ilvY* of *Salmonella typhimurium*. *J. Bacteriol.* 142:1015–1018.

184. Blazey, D. L., and R. O. Burns. 1982. Transcriptional activity of the transposable element Tn10 in the *Salmonella typhimurium ilvGEDA* operon. *Proc. Natl. Acad. Sci. USA* 79:5011–5015.

185. Blazey, D. L., and R. O. Burns. 1984. Regulation of *Salmonella typhimurium ilvYC* genes. *J. Bacteriol.* 159:951–957.

186. Blazey, D. L., R. Kim, and R. O. Burns. 1981. Molecular cloning and expression of the *ilvGEDAY* genes from *Salmonella typhimurium*. *J. Bacteriol.* 147:452–462.

187. Blum, P., L. Blaha, and S. Artz. 1986. Reversion and immobilization of phage Mud1 cts (Ap^r *lac*) insertion mutations in *Salmonella typhimurium*. *Mol. Gen. Genet.* 202:327–330.

188. Blum, P., D. Holzschu, H. S. Kwan, D. Riggs, and S. Artz. 1989. Gene replacement and retrieval with recombinant M13mp bacteriophages. *J. Bacteriol.* 171:538–546.

189. Blum, P. H. 1988. Reduced *leu* operon expression in a *miaA* mutant of *Salmonella typhimurium*. *J. Bacteriol.* 170:5125–5133.

190. Blume, A. J., and E. Balbinder. 1966. The tryptophan operon of *Salmonella typhimurium*. Fine structure analysis by deletion mapping and abortive transduction. *Genetics* 53:577–592.

191. Blume, A. J., A. Weber, and E. Balbinder. 1968. Analysis of polar and nonpolar tryptophan mutants by depression kinetics. *J. Bacteriol.* 95:2230–2241.

192. Blumenberg, M., and B. Magasanik. 1979. A study in evolution: the histidine utilization genes of enteric bacteria. *J. Mol. Biol.* 135:23–38.

193. Blumenberg, M., and B. Magasanik. 1981. Physical maps of *Klebsiella aerogenes* and *Salmonella typhimurium hut* genes. *J. Bacteriol.* 145:664–667.

194. Bobik, T. A., M. Ailion, and J. R. Roth. 1992. A single regulatory gene integrates control of vitamin B12 synthesis and propanediol degradation. *J. Bacteriol.* 174:2253–2266.

195. Bochner, B. R., and M. A. Savageau. 1979. Inhibition of growth by imidazol(on)e propionic acid. Evidence *in-vivo* for coordination of histidine catabolism with the catabolism of other amino acids. *Mol. Gen. Genet.* 168:87–96.

196. Borczuk, A., A. Staub, and J. Stock. 1986. Demethylation of bacterial chemoreceptors is inhibited by attractant stimuli in the complete absence of the regulatory domain of the demethylating enzyme. *Biochem. Biophys. Res. Commun.* 141:918–923.

197. Borczuk, A., A. Stock, and J. Stock. 1987. S-Adenosylmethionine may not be essential for signal transduction during bacterial chemotaxis. *J. Bacteriol.* 169:3295–3300.

198. Borodovsky, M., E. V. Koonin, and K. E. Rudd. 1994. New genes in old sequence: a strategy for finding genes in the bacterial genome. *Trends Biochem. Sci.* 19:309–313.

199. Borum, P. R., and K. J. Monty. 1976. Regulatory mutants and control of cysteine biosynthetic enzymes in *Salmonella typhimurium*. *J. Bacteriol.* 125:94–101.

200. Bossi, L. 1983. Context effects: translation of UAG codon by suppressor tRNA is affected by the sequence following UAG in the message. *J. Mol. Biol.* 164:73–87.

201. Bossi, L. 1983. The *hisR* locus of *Salmonella*: nucleotide sequence and expression. *Mol. Gen. Genet.* 192:163–170.

202. Bossi, L., and M. S. Ciampi. 1981. DNA sequences at the sites of three insertions of the transposable element Tn5 in the histidine operon of *Salmonella*. *Mol. Gen. Genet.* 183:406–408.

203. Bossi, L., M. S. Ciampi, and R. Cortese. 1978. Characterization of a *Salmonella typhimurium hisU* mutant defective in tRNA precursor processing. *J. Bacteriol.* 134:612–620.

204. Bossi, L., and R. Cortese. 1977. Biosynthesis of tRNA in histidine regulatory mutants of *Salmonella typhimurium*. *Nucleic Acids Res.* 4:1945–1956.

205. Bossi, L., T. Kohno, and J. R. Roth. 1982. Genetic characterization of the *sufJ* suppressor in *Salmonella typhimurium*. *Genetics* 103:31–42.

206. Bossi, L., and J. R. Roth. 1980. The influence of codon context on genetic code translation. *Nature* (London) 286:123–127.

207. Bossi, L., and D. M. Smith. 1984. Suppressor *sufJ*: a novel type of tRNA mutant that induces translational frameshifting. *Proc. Natl. Acad. Sci. USA* 81:6105–6109.

208. Bossi, L., and D. M. Smith. 1984. Conformational change in the DNA associated with an unusual promotor mutation in a tRNA operon of *Salmonella*. *Cell* 39:643–652.

209. Bossi, M., and J. R. Roth. 1981. Four-base codons ACCA, ACCU, and ACCC are recognized by frameshift suppressor *sufJ*. *Cell* 25:489–496.

210. Botsford, J. L. 1984. Cyclic AMP phosphodiesterase in *Salmonella typhimurium*: characteristics and physiological function. *J. Bacteriol.* 160:826–830.

211. Boudadloun, F., T. Srichaiyo, L. A. Isaksson, and G. R. Bjork. 1986. Influence of modification next to the anticodon in tRNA on codon context sensitivity of translational suppression and accuracy. *J. Bacteriol.* 166:1022–1027.

212. Bouma, C. L., N. D. Meadow, E. W. Stover, and S. Roseman. 1987. II-B^Glc, a glucose receptor of the bacterial phosphotransferase system: molecular cloning of *ptsG* and purification of the receptor from an overproducing strain of *Escherichia coli*. *Proc. Natl. Acad. Sci. USA* 84:930–934.

213. Bower, S. G., B. Hove-Jensen, and R. L. Switzer. 1988. Structure of the gene encoding phosphoribosylpyrophosphate synthetase (*prsA*) in *Salmonella typhimurium*. *J. Bacteriol.* 170:3243–3248.

214. Boyd, D. H., L. M. Porter, B. S. Young, and A. Wright. 1979. The *in-vitro* detection of defects in temperature sensitive RNA polymerases from mutants of *Salmonella typhimurium*. *Mol. Gen. Genet.* 173:279–284.

215. Brade, H., H. Moll, and E. T. Rietschel. 1985. Structural investigations on the inner core region of lipopolysaccharides from *Salmonella minnesota* rough mutants. *Biomed. Mass. Spectrom.* 12:602–609.

216. Brade, L., F. E. Nano, S. Schlecht, S. Schramek, and H. Brade. 1987. Antigenic and immunogenic properties of recombinants from *Salmonella typhimurium* and *Salmonella minnesota* rough mutants expressing in their lipopolysaccharide a genus-specific chlamydial epitope. *Infect. Immun.* 55:482–486.

217. Brady, D. R., and L. L. Houston. 1973. Some properties of the catalytic sites of imidazoleglycerol phosphate dehydratase-histidinol phosphate phosphatase, a bifunctional enzyme from *Salmonella typhimurium*. *J. Biol. Chem.* 248:2588–2592.

218. Brady, R. A., and L. N. Csonka. 1988. Transcriptional regulation of the *proC* gene of *Salmonella typhimurium*. *J. Bacteriol.* 170:2379–2382.

219. Brahmbhatt, H. N., N. B. Quigley, and P. R. Reeves. 1986. Cloning part of the region encoding biosynthetic enzymes for surface antigen (O-antigen) of *Salmonella typhimurium*. *Mol. Gen. Genet.* 203:172–176.

220. Brahmbhatt, H. N., P. Wyk, N. B. Quigley, and P. R. Reeves. 1988. Complete physical map of the *rfb* gene cluster encoding biosynthetic enzymes for the O antigen of *Salmonella typhimurium* LT2. *J. Bacteriol.* 170:98–102.

221. Bramley, H. F., and H. L. Kornberg. 1987. Sequence homologies between proteins of bacterial phosphoenolpyruvate-dependent sugar phosphotransferase systems: identification of possible phosphate-carrying histidine residues. *Proc. Natl. Acad. Sci. USA* 84:4777–4780.

222. Brana, H., and J. Hubacek. 1977. Virulence of *Salmonella typhimurium* infected with the R plasmid and relation with the *rec* mutation. *Folia Microbiol.* 22:451.

223. Branes, L. V., and W. W. Kay. 1983. Lipopolysaccharide core mutants of *Salmonella typhimurium* containing D-*glycero*-D-*manno*-heptose. *J. Bacteriol.* 154:1462–1466.

224. Branes, L. V., J. M. Somers, and W. W. Kay. 1981. Hydrophobic peptide auxotrophy in *Salmonella typhimurium*. *J. Bacteriol.* 147:986–996.

225. Brashear, W. T., and S. M. Parsons. 1975. Evidence against a covalent intermediate in the adenosine triphosphate phosphoribosyltransferase reaction of histidine biosynthesis. *J. Biol. Chem.* 250:6885–6890.

226. Braun, V., K. Gunthner, K. Hantke, and L. Zimmermann. 1983. Intracellular activation of albomycin in *Escherichia coli* and *Salmonella typhimurium*. *J. Bacteriol.* 156:308–315.

227. **Braun, V., K. Hantke, and W. Stauder.** 1977. Identification of the *sid* outer membrane protein in *Salmonella typhimurium* strain SL1027. *Mol. Gen. Genet.* **155:**227–230.

228. **Brazas, R., E. Davie, A. Farewell, and L. I. Rothfield.** 1991. Transcriptional organization of the *rfaGBIJ* locus of *Salmonella typhimurium. J. Bacteriol.* **173:**6168–6173.

229. **Brenchley, J. E., C. A. Baker, and L. G. Patil.** 1975. Regulation of the ammonia assimilatory enzymes in *Salmonella typhimurium. J. Bacteriol.* **124:**182–189.

230. **Brenchley, J. E., D. M. Bedwell, S. M. Dendinger, and J. M. Kuchta.** 1980. Analysis of mutations affecting the regulation of nitrogen utilization in *Salmonella typhimurium*, p. 79–93. *In* J. Mora and R. Palacios (ed.), *Glutamine: Metabolism, Enzymology and Regulation*. International Symposium on Glutamine: Metabolism, Enzymology and Regulation, Queretaro, Mexico. Academic Press, Inc., New York.

231. **Brenchley, J. E., and J. L. Ingraham.** 1973. Characterization of a cold-sensitive *hisW* mutant of *Salmonella typhimurium. J. Bacteriol.* **114:**528–536.

232. **Brenner, M., and B. N. Ames.** 1971. The histidine operon and its regulations, p. 349–387. *In* H. J. Vogel (ed.), *Metabolic Regulation*, vol. 5 *in* D. Greenberg (ed.), *Metabolic Pathways*. Academic Press, Inc., New York.

233. **Brenner, M., and B. N. Ames.** 1972. Histidine regulation in *Salmonella typhimurium*. IX. Histidine transfer ribonucleic acid of the regulatory mutants. *J. Biol. Chem.* **247:**1080–1088.

234. **Brenner, M., F. DeLorenzo, and B. N. Ames.** 1970. Energy charge and protein synthesis. Control of aminoacyl transfer ribonucleic acid synthesis. *J. Biol. Chem.* **245:**450–452.

235. **Brenner, M., J. A. Lewis, D. S. Straus, R. DeLorenzo, and B. N. Ames.** 1972. Histidine regulation in *Salmonella typhimurium*. XIV. Interaction of the histidyl transfer ribonucleic acid synthetase with histidine transfer ribonucleic acid. *J. Biol. Chem.* **247:**4333–4339.

236. **Bresalier, R. S., A. A. Rizzino, and M. Freundlich.** 1975. Reduced maximal levels of derepression of the isoleucine-valine and leucine enzymes in *hisT* mutants of *Salmonella typhimurium. Nature* (London) **253:**279–280.

237. **Brewer, S., M. Tolley, I. P. Trayer, G. C. Barr, C. J. Dorman, C. Hannavy, C. F. Higgins, J. S. Evans, B. A. Levine, and M. R. Wormald.** 1990. Structure and function of X-Pro dipeptide repeats in the TonB proteins of *Salmonella typhimurium* and *Escherichia coli. J. Mol. Biol.* **216:**883–895.

238. **Brill, W. J., and B. Magasanik.** 1969. Genetic and metabolic control of histidase and urocanase in *Salmonella typhimurium*, strain 15–59. *J. Biol. Chem.* **244:**5392–5402.

239. **Broach, J., C. Neumann, and S. Kustu.** 1976. Mutant strains (*nit*) of *Salmonella typhimurium* with a pleiotropic defect in nitrogen metabolism. *J. Bacteriol.* **128:**86–98.

240. **Brok, R. G., E. Brinkman, R. van Boxtel, A. C. Bekkers, H. M. Verheij, and J. Tommassen.** 1994. Molecular characterization of enterobacterial *pldA* genes encoding outer membrane phospholipase A. *J. Bacteriol.* **176:**861–870.

241. **Brown, B. A., S. R. Lax, L. Liang, B. J. Dabney, L. L. Spremulli, and J. M. Ravel.** 1977. Repression of the tyrosine, lysine, and methionine biosynthetic pathways in a *hisT* mutant of *Salmonella typhimurium. J. Bacteriol.* **129:**1168–1170.

242. **Brown, D. J., J. E. Olsen, and M. Bisgaard.** 1992. *Salmonella enterica*: infection, cross infection and persistence within the environment of a broiler parent stock unit in Denmark. *Int. J. Med. Microbiol. Virol. Parasitol. Infect. Dis.* **277:**129–138.

243. **Bruist, M. F., A. C. Glasgow, R. C. Johnson, and M. I. Simon.** 1987. Fis binding to the recombinational enhancer of the Hin DNA inversion system. *Genes Dev.* **1:**762–772.

244. **Bruist, M. F., and M. I. Simon.** 1984. Phase variation and the Hin protein: in vivo activity measurements, protein overproduction, and purification. *J. Bacteriol.* **159:**71–79.

245. **Bruneteau, M., W. A. Volk, P. P. Singh, and O. Luderitz.** 1974. Structural investigations on the *Salmonella* T2 lipopolysaccharide. *Eur. J. Biochem.* **43:**501–508.

246. **Bruni, C. B., M. S. Carlomagno, S. Formisano, and G. Paolella.** 1986. Primary and secondary structural homologies between the HIS4 gene product of *Saccharomyces cerevisiae* and the *hisIE* and *hisD* gene products of *Escherichia coli* and *Salmonella typhimurium. Mol. Gen. Genet.* **203:**389–396.

247. **Bruni, C. B., R. G. Martin, and M. Rechler.** 1972. COOH-terminal amino acid sequence of histidinol dehydrogenase from a *Salmonella typhimurium* mutant. *J. Biol. Chem.* **247:**6671–6678.

248. **Bruni, C. B., M. M. Rechler, and R. G. Martin.** 1973. *Salmonella typhimurium* mutants lacking ribonuclease I: effect on the polarity of histidine mutants. *J. Bacteriol.* **113:**1207–1212.

249. **Brzovic, P. S., A. M. Kayastha, E. W. Miles, and M. F. Dunn.** 1992. Substitution of glutamic acid 109 by aspartic acid alters the substrate specificity and catalytic activity of the beta-subunit in the tryptophan synthase bienzyme complex from *Salmonella typhimurium. Biochemistry* **31:**1180–1190.

250. **Buchmeier, N. A., C. J. Lipps, M. Y. So, and F. Heffron.** 1993. Recombination-deficient mutants of *Salmonella typhimurium* are avirulent and sensitive to the oxidative burst of macrophages. *Mol. Microbiol.* **7:**933–936.

251. **Buck, M., and B. N. Ames.** 1984. A modified nucleotide in tRNA as a possible regulator of aerobiosis: synthesis of *cis*-2-methyl-thioribosylzeatin in the tRNA of *Salmonella. Cell* **36:**523–531.

252. **Buckenmeyer, G. K., and M. A. Hermodson.** 1983. The amino acid sequence of D-ribose-binding protein from *Salmonella typhimurium* ST1. *J. Biol. Chem.* **258:**12957.

253. **Bullas, L. R., and C. Colson.** 1975. DNA restriction and modificiation systems in *Salmonella. Mol. Gen. Genet.* **139:**177–188.

254. **Bullas, L. R., C. Colson, and B. Neufeld.** 1980. DNA restriction and modification systems in *Salmonella*: chromosomally located systems of different serotypes. *J. Bacteriol.* **141:**275–292.

255. **Bullas, L. R., C. Colson, and A. Van Pel.** 1976. DNA restriction and modification systems in *Salmonella*. IV. SQ, a new system derived by recombination between the SB system of *Salmonella typhimurium* and of *Salmonella potsdam. J. Gen. Microbiol.* **95:**166–172.

256. **Bullas, L. R., and J.-I. Ryu.** 1983. *Salmonella typhimurium* LT2 strains which are $r^- m^+$ for all three chromosomally located systems of DNA restriction and modification. *J. Bacteriol.* **156:**471–474.

257. **Burgin, A. B., K. Parodos, D. J. Lane, and N. R. Pace.** 1990. The excision of intervening sequences from *Salmonella* 23S ribosomal RNA. *Cell* **60:**405–414.

258. **Burns, D. M., and I. R. Beacham.** 1985. Rare codons in *E. coli* and *S. typhimurium* signal sequences. *FEBS Lett.* **189:**318–324.

259. **Burns, D. M., and I. R. Beacham.** 1986. Identification and sequence analysis of a silent gene (*ushA*^{0}) in *Salmonella typhimurium. J. Mol. Biol.* **192:**163–175.

260. **Burns, R. O., J. Calvo, P. Margolin, and H. E. Umbarger.** 1966. Expression of the leucine operon. *J. Bacteriol.* **91:**1570–1576.

261. **Burns, R. O., J. G. Hofler, and G. H. Luginbuhl.** 1979. Threonine deaminase from *Salmonella typhimurium* substrate specific patterns in an activator site deficient form of the enzyme. *J. Biol. Chem.* **254:**1074–1079.

262. **Burns, R. O., H. E. Umbarger, and S. R. Gross.** 1963. The biosynthesis of leucine. III. The conversion of α-hydroxy-β-carboxyisocaproate to α-ketoisocaproate. *Biochemistry* **2:**1053–1058.

263. **Burns, R. O., and M. H. Zarlengo.** 1969. Threonine deaminase from *Salmonella typhimurium*. I. Purification and properties. *J. Biol. Chem.* **243:**178–185.

264. **Bussey, L. B., and J. L. Ingraham.** 1982. A regulatory gene (*use*) affecting the expression of *pyrA* and certain other pyrimidine genes. *J. Bacteriol.* **151:**144–152.

265. **Byerly, K. A., M. L. Urbanowski, and G. V. Stauffer.** 1991. The *metR* binding site in the *Salmonella typhimurium metH* gene: DNA sequence constraints on activation. *J. Bacteriol.* **173:**3547–3553.

266. **Byrne, C. R., R. S. Monroe, K. A. Ward, and N. M. Kredich.** 1988. DNA sequences of the *cysK* regions of *Salmonella typhimurium* and *Escherichia coli* and linkage of the *cysK* regions to *ptsH. J. Bacteriol.* **170:**3150–3157.

267. **Cairney, J., I. R. Booth, and C. F. Higgins.** 1985. *Salmonella typhimurium proP* gene encodes a transport system for the osmoprotectant betaine. *J. Bacteriol.* **164:**1218–1223.

268. **Cairney, J., I. R. Booth, and C. F. Higgins.** 1985. Osmoregulation of gene expression in *Salmonella typhimurium*: *proU* encodes an osmotically induced betaine transport system. *J. Bacteriol.* **164:**1224–1232.

269. **Cairney, J., C. F. Higgins, and I. R. Booth.** 1984. Proline uptake through the major transport system of *Salmonella typhimurium* is coupled to sodium ions. *J. Bacteriol.* **160:**22–27.

270. **Caldwell, A. L., and P. A. Gulig.** 1991. The *Salmonella typhimurium* virulence plasmid encodes a positive regulator of a plasmid-encoded virulence gene. *J. Bacteriol.* **173:**7176–7185.

271. **Calero, S., X. Garriga, and J. Barbe.** 1991. One-step cloning system for isolation of bacterial *lexA*-like genes. *J. Bacteriol.* **173:**7345–7350.

272. **Calhoun, D. H., and G. W. Hatfield.** 1973. Autoregulation: a role for a biosynthetic enzyme in the control of gene expression. *Proc. Natl. Acad. Sci. USA* **70:**2757–2761.

273. **Caligiuri, M. G., and R. Bauerle.** 1991. Identification of amino acid residues involved in feedback regulation of the anthranilate synthase complex from *Salmonella typhimurium*. Evidence for an amino-terminal regulatory site. *J. Biol. Chem.* **266:**8328–8335.

274. **Calladine, C. R.** 1975. Construction of bacterial flagella. *Nature* (London) **255:**121–124.

275. **Callahan, R., III, M. M. Dooley, and E. Ballbinder.** 1978. A mutation to 5-methyltryptophan-dependence in the tryptophan (*trp*) operon of *Salmonella typhimurium*. II. Studies of 5-methyltryptophan-dependent mutants and their revertants. *Mol. Gen. Genet.* **165:**129–143.

276. **Calvo, J. M., M. Freundlich, and H. E. Umbarger.** 1969. Regulation of branched-chain amino acid biosynthesis in *Salmonella typhimurium*: isolation of regulatory mutants. *J. Bacteriol.* **97:**1272–1282.

277. **Calvo, J. M., M. Goodman, M. Salgo, and N. Kapes.** 1971. *Salmonella* locus affecting phosphoenolpyruvate synthase activity identified by deletion analysis. *J. Bacteriol.* **106:**286–288.

278. **Calvo, J. M., and H. E. Worden.** 1970. A multisite-mutation map of the leucine operon of *Salmonella typhimurium. Genetics* **64:**199–214.

279. **Capobianco, J. O., R. P. Darveau, R. C. Goldman, P. A. Lartey, and A. G. Pernet.** 1987. Inhibition of exogenous 3-deoxy-D-*manno*-octulosonate incorporation into lipid A precursor of toluene-treated *Salmonella typhimurium. J. Bacteriol.* **169:**4030–4035.

280. **Carlomagno, M. S., F. Blasi, and C. B. Bruni.** 1983. Gene organization in the distal part of the *Salmonella typhimurium* histidine operon and determination and sequence of the operon transcription terminator. *Mol. Gen. Genet.* **191:**413–420.

281. **Carlomagno, M. S., L. Chiariotti, P. Alifano, A. G. Nappo, and C. B. Bruni.** 1988. Structure and function of the *Salmonella typhimurium* and *Escherichia coli* K-12 histidine operons. *J. Mol. Biol.* **203:**585–606.

282. Carlomagno, M. S., A. Riccio, and C. B. Bruni. 1985. Convergently functional, rho-independent terminator in *Salmonella typhimurium*. *J. Bacteriol.* **163**:362–368.

283. Carrillo-Castaneda, G., and M. V. Ortega. 1970. Mutants of *Salmonella typhimurium* lacking phosphoenolpyruvate carboxykinase and α-ketoglutarate dehydrogenase activities. *J. Bacteriol.* **100**:524–530.

284. Carsiotis, M., B. A. Stocker, and I. A. Holder. 1989. *Salmonella typhimurium* virulence in a burned-mouse model. *Infect. Immun.* **57**:2842–2846.

285. Carstenius, P., J. I. Flock, and A. Lindberg. 1990. Nucleotide sequence of *rfaI* and *rfaJ* genes encoding lipopolysaccharide glycosyl transferases from *Salmonella typhimurium*. *Nucleic Acids Res.* **18**:6128.

286. Carter, P. W., J. M. Bartkus, and J. M. Calvo. 1986. Transcription attenuation in *Salmonella typhimurium*: the significance of rare leucine codons in the *leu* leader. *Proc. Natl. Acad. Sci. USA* **83**:8127–8131.

287. Carter, P. W., D. L. Weiss, H. L. Weith, and J. M. Calvo. 1985. Mutations that convert the four leucine codons of the *Salmonella typhimurium* *leu* leader to four threonine codons. *J. Bacteriol.* **162**:943–949.

288. Carter, T. H., and C. G. Miller. 1984. Aspartate-specific peptidases in *Salmonella typhimurium*: mutants deficient in peptidase E. *J. Bacteriol.* **159**:453–459.

289. Casse, F. 1970. Mapping of gene *chl-B* controlling membrane-bound nitrate-reductase and formic hydrogenylase activities in *Escherichia coli* K-12. *Biochem. Biophys. Res. Commun.* **39**:429–436.

290. Casse, F., M. Chippaux, and M.-C. Pascal. 1973. Isolation from *Salmonella typhimurium* LT2 of mutants lacking specifically nitrate reductase activity and mapping of the *chl-C* gene. *Mol. Gen. Genet.* **124**:247–251.

291. Casse, F., M.-C. Pascal, and M. Chippaux. 1972. A mutant of *Salmonella typhimurium* deficient in tetrathionate reductase activity. *Mol. Gen. Genet.* **119**:71–74.

292. Casse, F., M.-C. Pascal, and M. Chippaux. 1973. Comparison between the chromosomal maps of *Escherichia coli* and *Salmonella typhimurium*. Length of the inverted segment in the *trp* region. *Mol. Gen. Genet.* **124**:213–257.

293. Casse, F., M.-C. Pascal, M. Chippaux, and J. Ratouchniak. 1972. Mapping of the *chlB* gene in *Salmonella typhimurium* LT2. *Mol. Gen. Genet.* **119**:67–70.

294. Cauthen, S. E., M. A. Foster, and D. D. Woods. 1966. Methionine synthesis by extracts of *Salmonella typhimurium*. *Biochem. J.* **98**:630–635.

295. Cerin, H., and J. Hackett. 1989. Molecular cloning and analysis of the incompatibility and partition functions of the virulence plasmid of *Salmonella typhimurium*. *Microb. Pathog.* **7**:85–99.

296. Cerin, H., and J. Hackett. 1993. The *parVP* region of the *Salmonella typhimurium* virulence plasmid pSLT contains four loci required for incompatibility and partition. *Plasmid* **30**:30–38.

297. Cerretti, D. P., L. C. Mattheakis, K. R. Kearney, L. Vu, and M. Nomura. 1988. Translational regulation of the *spc* operon in *Escherichia coli*. Identification and structural analysis of the target site for S8 repressor protein. *J. Mol. Biol.* **204**:309–329.

298. Chakraborti, A. S., K. Ishidate, W. R. Cook, J. Zrike, and L. I. Rothfield. 1986. Accumulation of a murein-membrane attachment site fraction when cell division is blocked in *lkyD* and *cha* mutants of *Salmonella typhimurium* and *Escherichia coli*. *J. Bacteriol.* **168**:1422–1429.

299. Chakravorty, M., T. Suryanarayana, and A. K. Datta. 1975. A ribonuclease I deficient mutant (MB24) of *Salmonella typhimurium*. *Indian J. Biochem. Biophys.* **12**:153–157.

300. Chan, C. L., and R. Landick. 1989. The *Salmonella typhimurium* *his* operon leader region contains an RNA hairpin-dependent transcription pause site. Mechanistic implications of the effect on pausing of altered RNA hairpins. *J. Biol. Chem.* **264**:20796–20804.

301. Chan, R. K., and D. Bostein. 1976. Specialized transduction by bacteriophage P-22 in *Salmonella typhimurium*: genetic and physical structure of the transducing genomes and the prophage attachment site. *Genetics* **83**:433–458.

302. Chang, G. W., and K. Fenton. 1974. A simple method for measuring protein degradation in bacteria. *Anal. Biochem.* **59**:185–189.

303. Chang, G. W., J. R. Roth, and B. N. Ames. 1971. Histidine regulation in *Salmonella typhimurium*. VII. Mutations of the *hisT* gene. *J. Bacteriol.* **108**:410–414.

304. Charlier, D., M. Roovers, D. Gigot, N. Huysveld, A. Pierard, and N. Glansdorff. 1993. Integration host factor (IHF) modulates the expression of the pyrimidine-specific promoter of the *carAB* operons of *Escherichia coli* K12 and *Salmonella typhimurium* LT2. *Mol. Gen. Genet.* **237**:273–286.

305. Chary, P., R. Prasad, A. K. Chopra, and J. W. Peterson. 1993. Location of the enterotoxin gene from *Salmonella typhimurium* and characterization of the gene products. *FEMS Microbiol. Lett.* **111**:87–92.

306. Chater, K. F. 1970. Dominance of the wild-type of methionine regulatory genes in *Salmonella typhimurium*. *J. Gen. Microbiol.* **63**:95–109.

307. Chater, K. F., D. A. Lawrence, R. J. Rowbury, and T. S. Gross. 1970. Suppression of methionine transfer RNA synthetase mutants of *Salmonella typhimurium* by methionine regulatory mutations. *J. Gen. Microbiol.* **63**:121–131.

308. Chater, K. F., and R. J. Rowbury. 1970. A genetical study of the feedback-sensitive enzyme of methionine synthesis of *Salmonella typhimurium*. *J. Gen. Microbiol.* **63**:111–120.

309. Chatfield, S. N., C. J. Dorman, C. Hayward, and G. Dougan. 1991. Role of *ompR*-dependent genes in *Salmonella typhimurium* virulence: mutants deficient in both *ompC* and *ompF* are attenuated in vivo. *Infect. Immun.* **59**:449–452.

310. Chatfield, S. N., K. Strahan, D. Pickard, I. G. Charles, C. E. Hormaeche, and G. Dougan. 1992. Evaluation of *Salmonella typhimurium* strains harbouring de-

311. fined mutations in *htrA* and *aroA* in the murine salmonellosis model. *Microb. Pathog.* **12**:145–151.

311. Chelala, C. A., and P. Margolin. 1974. Effects of deletions on cotransduction linkage in *Salmonella typhimurium*: evidence that bacterial chromosome deletions affect the formation of transducing DNA fragments. *Mol. Gen. Genet.* **131**:97–112.

312. Chelsky, D., and S. M. Parsons. 1975. Stereochemical course of the adenosine triphosphate phosphoribosyltransferase reaction in histidine biosynthesis. *J. Biol. Chem.* **250**:5669–5673.

313. Chen, D., R. Bowater, C. J. Dorman, and D. M. Lilley. 1992. Activity of a plasmid-borne *leu-500* promoter depends on the transcription and translation of an adjacent gene. *Proc. Natl. Acad. Sci. USA* **89**:8784–8788.

314. Chen, J. M., G. Lee, R. B. Murphy, P. W. Brandt-Rauf, and M. R. Pincus. 1990. Comparisons between the three-dimensional structures of the chemotactic protein CheY and the normal Gly 12-p21 protein. *Int. J. Pept. Protein Res.* **36**:1–6.

315. Chen, P., D. I. Andersson, and J. R. Roth. 1994. The control region of the *Salmonella pdu/cob* regulon. *J. Bacteriol.* **176**:5474–5482.

316. Cheney, R. W., Jr., and N. M. Kredich. 1975. Fine-structure genetic map of the *cysB* locus in *Salmonella typhimurium*. *J. Bacteriol.* **124**:1273–1281.

317. Cheng, W., and J. R. Roth. 1994. *Salmonella typhimurium* has two NAD kinases. *J. Bacteriol.* **176**:4260–4268.

318. Chiariotti, L., P. Alifano, M. S. Carlomagno, and C. B. Bruni. 1986. Nucleotide sequence of the *Escherichia coli* *hisD* gene of the *Escherichia coli* and *Salmonella typhimurium* *hisIE* region. *Mol. Gen. Genet.* **203**:382–388.

319. Chiariotti, L., A. G. Nappo, M. S. Carlomagno, and C. B. Bruni. 1986. Gene structure in the histidine operon of *Escherichia coli*. Identification and nucleotide sequence of the *hisB* gene. *Mol. Gen. Genet.* **202**:42–47.

320. Child, J. D., and D. A. Smith. 1969. New methionine structural gene in *Salmonella typhimurium*. *J. Bacteriol.* **100**:377–382.

321. Childs, G., F. Sonnenberg, and M. Freundlich. 1977. Detection of messenger RNA from the isoleucine-valine operons of *Salmonella typhimurium* by heterologous DNA-RNA hybridization: involvement of transfer RNA in transcriptional repression. *Mol. Gen. Genet.* **151**:121–126.

322. Chin, A. M., D. A. Feldheim, and M. H. Saier, Jr. 1989. Altered transcriptional patterns affecting several metabolic pathways in strains of *Salmonella typhimurium* which overexpress the fructose regulon. *J. Bacteriol.* **171**:2424–2434.

323. Chin, A. M., B. U. Feucht, and M. H. Saier, Jr. 1987. Evidence for regulation of gluconeogenesis by the fructose phosphotransferase system in *Salmonella typhimurium*. *J. Bacteriol.* **169**:897–899.

324. Chin, A. M., S. Sutrina, D. A. Feldheim, and M. H. Saier, Jr. 1987. Genetic expression of enzyme I* activity of the phophoenolpyruvate: sugar phosphotransferase system in *ptsHI* deletion strains of *Salmonella typhimurium*. *J. Bacteriol.* **169**:894–896.

325. Chippaux, M., F. Casse, and M.-C. Pascal. 1972. Isolation and phenotypes of mutants from *Salmonella typhimurium* defective in formate hydrogenylase activity. *J. Bacteriol.* **110**:766–768.

326. Chippaux, M., M.-C. Pascal, and F. Casse. 1977. Formate hydrogenylase system in *Salmonella typhimurium*. *Eur. J. Biochem.* **72**:149–155.

327. Cho, H., and J. E. Cronan, Jr. 1994. "Protease I" of *Escherichia coli* functions as a thioesterase in vivo. *J. Bacteriol.* **176**:1793–1795.

328. Chopra, A. K., C. W. Houston, J. W. Peterson, R. Prasad, and J. J. Mekalanos. 1987. Cloning and expression of the *Salmonella* enterotoxin gene. *J. Bacteriol.* **169**:5095–5100.

329. Chopra, A. K., J. W. Peterson, C. W. Houston, R. Pericas, and R. Prasad. 1991. Enterotoxin-associated DNA sequence homology between *Salmonella* species and *Escherichia coli*. *FEMS Microbiol. Lett.* **61**:133–138.

330. Chopra, A. K., J. W. Peterson, and R. Prasad. 1991. Nucleotide sequence analysis of *purH* and *purD* genes from *Salmonella typhimurium*. *Biochim. Biophys. Acta* **1090**:351–354.

331. Chopra, A. K., J. W. Peterson, and R. Prasad. 1991. Cloning and sequence analysis of hydrogenase regulatory genes (*hydHG*) from *Salmonella typhimurium*. *Biochim. Biophys. Acta* **1129**:115–118.

332. Chou, J. Y., and R. G. Martin. 1972. Purine phosphoribosyltransferases of *Salmonella typhimurium*. *J. Bacteriol.* **112**:1010–1013.

333. Christman, M. F., R. W. Morgan, F. S. Jacobson, and B. N. Ames. 1985. Positive control of a regulon for defenses against oxidative stress and some heat-shock proteins in *Salmonella typhimurium*. *Cell* **41**:753–762.

334. Christman, M. F., G. Storz, and B. N. Ames. 1989. OxyR, a positive regulator of hydrogen peroxide-inducible genes in *Escherichia coli* and *Salmonella typhimurium*, is homologous to a family of bacterial regulatory proteins. *Proc. Natl. Acad. Sci. USA* **86**:3484–3488.

335. Chumley, F. G., and J. R. Roth. 1980. Rearrangement of the bacterial chromosome using Tn-10 as a region of homology. *Genetics* **94**:1–14.

336. Chumley, F. G., and J. R. Roth. 1981. Genetic fusions that place the lactose genes under histidine operon control. *J. Mol. Biol.* **145**:697–712.

337. Ciampi, M. S., P. Alifano, A. G. Nappo, C. B. Bruni, and M. S. Carlomagno. 1989. Features of the rho-dependent transcription termination polar element within the *hisG* cistron of *Salmonella typhimurium*. *J. Bacteriol.* **171**:4472–4448.

338. Ciampi, M. S., and J. R. Roth. 1988. Polarity effects in the *hisG* gene of *Salmonella* require a site within the coding sequence. *Genetics* **118**:193–202.

339. Ciampi, M. S., M. B. Schmid, and J. R. Roth. 1982. Transposon Tn10 provides a promoter for transcription of adjacent sequences. *Proc. Natl. Acad. Sci. USA* **79**:5016–5020.

340. Ciesla, F. G., M. Filutowicz, and T. Klopotowski. 1980. Involvement of the L-cysteine biosynthetic pathway in azide induced mutagenesis in *Salmonella typhimurium*. *Mutat. Res.* 70:261–268.

341. Ciesla, Z., M. Bagdasarian, W. Szczurkiewicz, M. Przygonska, and T. Klopotowski. 1972. Defective cell division in thermosensitive mutants of *Salmonella typhimurium*. *Mol. Gen. Genet.* 116:107–125.

342. Ciesla, Z., K. Mardarowicz, and T. Klopotowski. 1974. Inhibition of DNA synthesis and cell division in *Salmonella typhimurium* by azide. *Mol. Gen. Genet.* 135:339–348.

343. Ciesla, Z., F. Salvatore, J. R. Broach, S. W. Artz, and B. N. Ames. 1975. Histidine regulation in *Salmonella typhimurium*. XVI. A sensitive radiochemical assay for histidine dehydrogenase. *Anal. Biochem.* 63:44–55.

344. Clark, M. A., and E. L. Barrett. 1987. The *phs* gene and hydrogen sulfide production by *Salmonella typhimurium*. *J. Bacteriol.* 169:2391–2397.

345. Clarke, P., J. H. Lee, K. Burke, and G. Wilcox. 1992. Mutations in the *araC* gene of *Salmonella typhimurium* LT2 which affect both activator and auto-regulatory functions of the AraC protein. *Gene* 117:31–37.

346. Clarke, P., H.-C. Lin, and G. Wilcox. 1982. The nucleotide sequence of the *araC* regulatory gene in *Salmonella typhimurium* LT2. *Gene* 18:157–163.

347. Clarke, S., K. Sparrow, S. Panasenko, and D. E. Koshland, Jr. 1980. In vitro methylation of bacterial chemotaxis proteins: characterization of protein methyltransferase activity in crude extracts of *Salmonella typhimurium*. *J. Supramol. Struct.* 13:315–328.

348. Claverie, J. M. 1993. Detecting frame shifts by amino acid sequence comparison. *J. Mol. Biol.* 234:1140–1157.

349. Clegg, S., B. K. Purcell, and J. Pruckler. 1987. Characterization of genes encoding type 1 fimbriae of *Klebsiella pneumoniae*, *Salmonella typhimurium*, and *Serratia marcescens*. *Infect. Immun.* 55:281–287.

350. Coleman, M. S., W. G. Soucie, and F. B. Armstrong. 1971. Branched chain amino acid aminotransferase of *Salmonella typhimurium*. II. Kinetic comparison with the enzyme from *Salmonella montevideo*. *J. Biol. Chem.* 246:1310–1312.

351. Coleman, W. G., Jr., and L. S. Williams. 1974. First enzyme of histidine biosynthesis and repression control of histidyl-transfer ribonucleic acid synthetase of *Salmonella typhimurium*. *J. Bacteriol.* 120:390–393.

352. Collazo, C. M., M. K. Zierler, and J. E. Galan. 1995. Functional analysis of the *Salmonella typhimurium* invasion genes *invI* and *invJ* and identification of a target of the protein secretion apparatus encoded in the *inv* locus. *Mol. Microbiol.* 15:25–38.

353. Collin-Osdoby, P., and C. G. Miller. 1994. Mutations affecting a regulated, membrane-associated esterase in *Salmonella typhimurium* LT2. *Mol. Gen. Genet.* 243:674–680.

354. Collins, A. L. T., and B. A. D. Stocker. 1976. *Salmonella typhimurium* mutants generally defective in chemotaxis. *J. Bacteriol.* 128:754–765.

355. Collins, A. M., and C. Colson. 1972. Expression of *Escherichia coli* K, B, and phage P1 DNA host specificities in *Salmonella typhimurium*. *J. Gen. Microbiol.* 70:123–128.

356. Collins, L. V., S. Attridge, and J. Hackett. 1991. Mutations at *rfc* or *pmi* attenuate *Salmonella typhimurium* virulence for mice. *Infect. Immun.* 59:1079–1085.

357. Collins, L. V., and J. Hackett. 1991. Sequence of the phosphomannose isomerase-encoding gene of *Salmonella typhimurium*. *Gene* 103:135–136.

358. Collins, L. V., and J. Hackett. 1991. Molecular cloning, characterization, and nucleotide sequence of the *rfc* gene, which encodes an O-antigen polymerase of *Salmonella typhimurium*. *J. Bacteriol.* 173:2521–2529.

359. Colson, A. M., C. Colson, and A. Van Pel. 1969. Host-controlled restriction mutants of *Salmonella typhimurium*. *J. Gen. Microbiol.* 58:57–64.

360. Colson, C., and A. M. Colson. 1971. A new *Salmonella typhimurium* DNA host specificity. *J. Gen. Microbiol.* 69:345–351.

361. Colson, C., A. M. Colson, and A. Van Pel. 1970. Chromosomal location of host specificity in *Salmonella typhimurium*. *J. Gen. Microbiol.* 60:265–271.

362. Colson, C., and A. Van Pel. 1974. DNA restriction and modification systems in *Salmonella*. I. SA and SB, two *Salmonella typhimurium* systems determined by genes with a chromosomal location comparable to that of the *Escherichia coli hsd* genes. *Mol. Gen. Genet.* 129:325–337.

363. Conlin, C. A., K. Hakensson, A. Liljas, and C. G. Miller. 1994. Cloning and nucleotide sequence of the cyclic AMP receptor protein-regulated *Salmonella typhimurium pepE* gene and crystallization of its product, an alpha-aspartyl dipeptidase. *J. Bacteriol.* 176:166–172.

364. Conlin, C. A., and C. G. Miller. 1992. Cloning and nucleotide sequence of *opdA*, the gene encoding oligopeptidase A in *Salmonella typhimurium*. *J. Bacteriol.* 174:1631–1640.

365. Conlin, C. A., N. J. Trun, T. J. Silhavy, and C. G. Miller. 1992. *Escherichia coli prlC* encodes an endopeptidase and is homologous to the *Salmonella typhimurium opdA* gene. *J. Bacteriol.* 174:5881–5887.

366. Conlin, C. A., E. R. Vimr, and C. G. Miller. 1992. Oligopeptidase A is required for normal phage P22 development. *J. Bacteriol.* 174:5869–5880??.

367. Contreras, A., and J. Casadesus. 1983. Transposition of Tn10 to the *xyl-mtl* region (minute 78) of the *Salmonella typhimurium* chromosome. *Microbiol. Espan.* 36:23–33. (In Spanish.)

368. Cook, P. F., and R. T. Wedding. 1976. A reaction mechanism from steady state kinetic studies for O-acetylserine sulfhydrylase from *Salmonella typhimurium*. *J. Biol. Chem.* 251:2023–2029.

369. Cook, W. R., T. J. MacAlister, and L. I. Rothfield. 1986. Compartmentalization of the periplasmic space at division sites in gram-negative bacteria. *J. Bacteriol.* 168:1430–1438.

370. Cookson, B. T., B. M. Olivera, and J. R. Roth. 1987. Genetic characterization and regulation of the *nadB* locus of *Salmonella typhimurium*. *J. Bacteriol.* 169:4285–4293.

371. Cooper, S. 1988. Rate and topography of cell wall synthesis during the division cycle of *Salmonella typhimurium*. *J. Bacteriol.* 170:422–430.

372. Cooper, T. G., and B. Tyler. 1976. Transcription of the *hut* operons of *Salmonella typhimurium*. *J. Bacteriol.* 130:192–199.

373. Cordaro, J. C., R. P. Anderson, E. W. Grogan, Jr., D. J. Wenzel, M. Engler, and S. Roseman. 1974. Promoter-like mutation affecting HPr and enzyme I of the phosphoenolpyruvate:sugar phosphotransferase system in *Salmonella typhimurium*. *J. Bacteriol.* 120:245–252.

374. Cordaro, J. C., and E. Balbinder. 1971. Evidence for the separability of the operator from the first structural gene in the tryptophan operon of *Salmonella typhimurium*. *Genetics* 67:151–169.

375. Cordaro, J. C., T. Melton, J. P. Stratis, M. Atagun, C. Gladding, P. E. Hartman, and S. Roseman. 1976. Fosfomycin resistance: selection method for internal and extended deletions of the phosphoenolpyruvate. *J. Bacteriol.* 128:785–793.

376. Cordaro, J. C., and S. Roseman. 1972. Deletion mapping of the genes coding for HPr and enzyme I of the phosphoenolpyruvate:sugar phosphotransferase system in *Salmonella typhimurium*. *J. Bacteriol.* 112:17–29.

377. Corina, D. L., and D. C. Wilton. 1976. An apparent lack of stereospecificity in the reaction catalyzed by deoxyribose 5-phosphate aldolase due to methyl-group rotation and enolization before product release. *Biochem. J.* 157:573–576.

378. Cortese, R., H. O. Kammen, S. J. Spengler, and B. N. Ames. 1974. Biosynthesis of pseudouridine in transfer ribonucleic acid. *J. Biol. Chem.* 249:1103–1108.

379. Cortese, R., R. Landsberg, R. A. Vonder Haar, H. E. Umberger, and B. N. Ames. 1974. Pleiotrophy of *hisT* mutants blocked in pseudouridine synthesis in tRNA: *leucine* and *isoleucine-valine* operons. *Proc. Natl. Acad. Sci. USA* 71:1857–1861.

380. Corton, J. C., J. E. Ward, Jr., and J. Lutkenhaus. 1987. Analysis of cell division gene *ftsZ* (*sulB*) from gram-negative and gram-positive bacteria. *J. Bacteriol.* 169:1–7.

381. Corwin, L. M., G. R. Fanning, F. Feldman, and P. Margolin. 1966. Mutation leading to increased sensitivity to chromium in *Salmonella typhimurium*. *J. Bacteriol.* 91:1509–1515.

382. Cossart, P., E. A. Groisman, M.-C. Serre, M. J. Casadaban, and B. Gicquel-Sanzey. 1986. *crp* genes of *Shigella flexneri*, *Salmonella typhimurium*, and *Escherichia coli*. *J. Bacteriol.* 167:639–646.

383. Costa, C. S., and D. N. Anton. 1993. Round-cell mutants of *Salmonella typhimurium* produced by transposition mutagenesis: lethality of *rodA* and *mre* mutations. *Mol. Gen. Genet.* 236:387–394.

384. Cottam, A. N., and P. D. Ayling. 1989. Genetic studies of mutants in a high-affinity methionine transport system in *Salmonella typhimurium*. *Mol. Gen. Genet.* 215:358–363.

385. Cottam, P. F., N.-B. He, S. W. Hui, and C. Ho. 1986. Biochemical and morphological properties of membranes of unsaturated fatty acid auxotrophs of *Salmonella typhimurium*: effects of fluorinated myristic acids. *Biochim. Biophys. Acta* 862:413–428.

386. Cowan, J. M., M. L. Urbanowski, M. Talmi, and G. V. Stauffer. 1993. Regulation of the *Salmonella typhimurium metF* gene by the MetR protein. *J. Bacteriol.* 175:5862–5866.

387. Coynault, C., V. Robbe-Saule, M. Y. Popoff, and F. Norel. 1992. Growth phase and SpvR regulation of transcription of *Salmonella typhimurium spvABC* virulence genes. *Microb. Pathog.* 13:133–143.

388. Craven, M. G., A. E. Granston, A. T. Schauer, C. Zheng, T. A. Gray, and D. I. Friedman. 1994. *Escherichia coli*-*Salmonella typhimurium* hybrid *nusA* genes: identification of a short motif required for action of the lambda N transcription antitermination protein. *J. Bacteriol.* 176:1394–1404.

389. Crawford, I. P. 1975. Gene rearrangement in the evolution of the tryptophan synthetic pathway. *Bacteriol. Rev.* 39:87–120.

390. Crawford, I. P., B. P. Nichols, and C. Yanofsky. 1980. Nucleotide sequence of the *trpB* gene in *Escherichia coli* and *Salmonella typhimurium*. *J. Mol. Biol.* 142:489–502.

391. Creeger, E. S., J. F. Chan, and L. I. Rothfield. 1979. Cloning of genes for bacterial glycosyl transferases. II. Selection of a hybrid plasmid carrying the *rfaH* gene. *J. Biol. Chem.* 254:811–815.

392. Creeger, E. S., and L. I. Rothfield. 1979. Cloning of genes for bacterial glycosyl transferases. I. Selection of hybrid plasmids carrying genes for two glucosyl transferases. *J. Biol. Chem.* 254:804–810.

393. Creeger, E. S., T. Schulte, and L. I. Rothfield. 1984. Regulation of membrane glycosyltransferases by the *sfrB* and *rfaH* genes of *Escherichia coli* and *Salmonella typhimurium*. *J. Biol. Chem.* 259:3064–3069.

394. Creighton, T. E. 1974. The functional significance of the evolutionary divergence between the tryptophan operons of *Escherichia coli* and *Salmonella typhimurium*. *J. Mol. Evol.* 4:121–137.

395. Crennell, S. J., E. F. Garman, W. G. Laver, E. R. Vimr, and G. L. Taylor. 1993. Crystal structure of a bacterial sialidase (from *Salmonella typhimurium* LT2) shows the same fold as an influenza virus neuraminidase. *Proc. Natl. Acad. Sci. USA* 90:9852–9856.

396. Cronan, J. E., Jr., K. J. Littel, and S. Jackowski. 1982. Genetic and biochemical analysis of pantothenate biosynthesis in *Escherichia coli* and *Salmonella typhimurium*. *J. Bacteriol.* 149:916–922.

397. Csonka, L. 1982. A third L-proline permease in *Salmonella typhimurium* which functions in media of elevated osmotic strength. *J. Bacteriol.* 151:1433–1443.

398. Csonka, L. N. 1988. Regulation of cytoplasmic proline levels in *Salmonella typhimurium*: effect of osmotic stress on synthesis, degradation, and cellular retention of proline. *J. Bacteriol.* 170:2374–2378.

399. Csonka, L. N., M. M. Howe, J. L. Ingraham, L. S. Pierson III, and C. L. Turnbough, Jr. 1981. Infection of *Salmonella typhimurium* with coliphage Mu d1(Apr lac): construction of *pyr::lac* gene fusions. *J. Bacteriol.* 145:299–305.

400. Csonka, L. N., T. P. Ikeda, S. A. Fletcher, and S. Kustu. 1994. The accumulation of glutamate is necessary for optimal growth of *Salmonella typhimurium* in media of high osmolarity but not induction of the *proU* operon. *J. Bacteriol.* 176:6324–6333.

401. Dahl, M. K., W. Boos, and M. D. Manson. 1989. Evolution of chemotactic-signal transducers in enteric bacteria. *J. Bacteriol.* 171:2361–2371.

402. Dahl, M. K., E. Francoz, W. Saurin, W. Boos, M. D. Manson, and M. Hofnung. 1989. Comparison of sequences from the *malB* regions of *Salmonella typhimurium* and *Enterobacter aerogenes* with *Escherichia coli* K12: a potential new regulatory site in the interoperonic region. *Mol. Gen. Genet.* 218:199–207.

403. Dahlquist, F. W., P. Lovely, and D. E. Koshland, Jr. 1972. Quantitative analysis of bacterial migration in chemotaxis. *Nature* (London) *New Biol.* 236:120–123.

404. Dailey, F. E., J. E. Cronan, Jr., and S. R. Maloy. 1987. Acetohydroxy acid synthase I is required for isoleucine and valine biosynthesis by *Salmonella typhimurium* LT2 during growth on acetate or long-chain fatty acids. *J. Bacteriol.* 169:917–919.

405. Dalal, F. R., R. E. Gots, and J. S. Gots. 1966. Mechanism of adenine inhibition in adenine-sensitive mutants of *Salmonella typhimurium*. *J. Bacteriol.* 91:507–513.

406. Danagulian, K. G., N. N. Sarkisian, D. B. Beglarian, and Z. A. Ktosian. 1979. Some properties of UV-sensitive mutants of *Salmonella derby*. *Biol. Zh. Arm.* 32:1030–1034.

407. Dang, C. V., M. Niwano, J. Ryu, and B. L. Taylor. 1986. Inversion of aerotactic response in *Escherichia coli* deficient in *cheB* protein methylesterase. *J. Bacteriol.* 166:275–280.

408. Darekar, M. R., and H. Eyer. 1973. The role of fimbriae in the processes of infection. Preliminary report. *Zentralbl. Bakteriol. Mikrobiol. Hyg. Abt. 1 Orig. Reihe A* 225:130–134.

409. Dartois, V., O. De Backer, and C. Colson. 1993. Sequence of the *Salmonella typhimurium* StyLT1 restriction-modification genes: homologies with EcoP1 and EcoP15 type-III R-M systems and presence of helicase domains. *Gene* 127:105–110.

410. Das, A., J. Urbanowski, H. Weissbach, J. Nestor, and C. Yanofsky. 1983. In vitro synthesis of the tryptophan operon leader peptides of *Escherichia coli*, *Serratia marcescens*, and *Salmonella typhimurium*. *Proc. Natl. Acad. Sci. USA* 80:2879–2883.

411. Dassa, E. 1974. Localisation chromosomique du prophage N chez *Salmonella montevideo* (sous-group C$_1$). *C.R. Acad. Sci. Ser. D* 278:385–388.

412. Datta, D. B., C. Kramer, and U. Hemming. 1976. Diploidy for a structural gene specifying a major protein of the outer cell envelope membrane from *Escherichia coli* K-12. *J. Bacteriol.* 128:834–841.

413. Daub, E., L. E. Zawadzke, D. Botstein, and C. T. Walsh. 1988. Isolation, cloning, and sequencing of the *Salmonella typhimurium ddlA* gene with purification and characterization of its product, D-alanine:D-alanine ligase (ADP forming). *Biochemistry* 27:3701–3708.

414. Davidson, A. E., H. E. Fukimoto, C. E. Jackson, E. L. Barrett, and G. W. Chang. 1979. Mutants of *Salmonella typhimurium* defective in the reduction of trimethylamine oxide. *FEMS Microbiol. Lett.* 6:417–420.

415. Davidson, J. P., and L. S. Williams. 1979. Regulation of isoleucine and valine biosynthesis in *Salmonella typhimurium*. The effect of *hisU* on repression control. *J. Mol. Biol.* 131:229–236.

416. Davidson, J. P., and L. S. Williams. 1979. Relaxed control of RNA synthesis during nutritional shiftdowns of a *hisU* mutant of *Salmonella typhimurium*. *Biochem. Biophys. Res. Commun.* 88:682–687.

417. Davidson, J. P., and D. J. Wilson. 1991. Evidence for isoleucine as a positive effector of the *ilvBN* operon in *Salmonella typhimurium*. *Biochem. Biophys. Res. Commun.* 178:934–939.

418. Davidson, J. P., D. J. Wilson, and L. S. Williams. 1982. Role of a *hisU* gene in the control of stable RNA synthesis in *Salmonella typhimurium*. *J. Mol. Biol.* 157:237–264.

419. Davis, L., and L. S. Williams. 1982. Altered regulation of isoleucine-valine biosynthesis in a *hisW* mutant of *Salmonella typhimurium*. *J. Bacteriol.* 151:860–866.

420. Davis, L., and L. S. Williams. 1982. Characterization of a cold-sensitive *hisW* mutant of *Salmonella typhimurium*. *J. Bacteriol.* 151:867–887.

421. De Backer, O., and C. Colson. 1991. Transfer of the genes for the StyLTI restriction-modification system of *Salmonella typhimurium* to strains lacking modification ability results in death of the recipient cells and degradation of their DNA. *J. Bacteriol.* 173:1328–1330.

422. De Backer, O., and C. Colson. 1991. Two-step cloning and expression in *Escherichia coli* of the DNA restriction-modification system StyLTI of *Salmonella typhimurium*. *J. Bacteriol.* 173:1321–1327.

423. De Backer, O., and C. Colson. 1991. Identification of the recognition sequence for the M.StyLTI methyltransferase of *Salmonella typhimurium* LT7: an asymmetric site typical of type-III enzymes. *Gene* 97:103–107.

424. Dean, G. E., S.-I. Aizawa, and R. M. Macnab. 1983. *flaAII* (*motC, cheV*) of *Salmonella typhimurium* is a structural gene involved in the energization and switching of the flagellar motor. *J. Bacteriol.* 154:84–91.

425. Decedue, C. J., J. G. Hofler, and R. O. Burns. 1975. Threonine deaminase from *Salmonella typhimurium*. *J. Biol. Chem.* 250:1563–1570.

426. Deeley, R. G., R. F. Goldberger, J. S. Kovach, M. M. Meyers, and K. P. Mullinix. 1975. Interaction between phosphoribosyltransferase and purified histidine tRNA from wild-type *Salmonella typhimurium* and a derepressed *hisT* mutant strain. *Nucleic Acids Res.* 2:545–554.

427. DeFranco, A. L., and D. E. Koshland, Jr. 1980. Multiple methylation in processing of sensory signals during bacterial chemotaxis. *Proc. Natl. Acad. Sci. USA* 77:2429–2433.

428. DeFranco, A. L., and D. E. Koshland, Jr. 1981. Molecular cloning of chemotaxis genes and overproduction of gene products in the bacterial sensing system. *J. Bacteriol.* 147:390–400.

429. DeFranco, A. L., and D. E. Koshland, Jr. 1982. Construction and behavior of strains with mutations in two chemotaxis genes. *J. Bacteriol.* 150:1297–1301.

430. DeFranco, A. L., J. S. Parkinson, and D. E. Koshland, Jr. 1979. Functional homology of chemotaxis genes in *Escherichia coli* and *Salmonella typhimurium*. *J. Bacteriol.* 139:107–114.

431. Delavier-Klutchko, C., and M. Flavin. 1965. Role of a bacterial cystathionine β-cleavage enzyme in disulfide decomposition. *Biochim. Biophys. Acta* 99:375–377.

432. DeLeo, A. B., J. Dayan, and D. B. Sprinson. 1973. Purification and kinetics of tyrosine-sensitive 3-deoxy-D-*arabino*-heptulo-sonic acid 7-phosphate synthetase from *Salmonella*. *J. Biol. Chem.* 248:2344–2353.

433. DeLeo, A. B., and D. B. Sprinson. 1975. 3-Deoxy-D-arabinoheptulosonic acid 7-phosphate synthase mutants of *Salmonella typhimurium*. *J. Bacteriol.* 124:1312–1320.

434. DeLorenzo, F., and B. N. Ames. 1970. Histidine regulation in *Salmonella typhimurium*. VII. Purification and general properties of the histidyl transfer ribonucleic acid synthetase. *J. Biol. Chem.* 245:1710–1716.

435. DeLorenzo, F., P. DiNatale, and A. N. Schechter. 1974. Chemical and physical studies on the structure of the histidyl transfer ribonucleic acid synthetase from *Salmonella typhimurium*. *J. Biol. Chem.* 249:908–913.

436. DeLorenzo, F., S. S. Straus, and B. N. Ames. 1972. Histidine regulation in *Salmonella typhimurium*. X. Kinetic studies of mutant histidyl transfer ribonucleic acid synthetases. *J. Biol. Chem.* 247:2302–2307.

437. Demerec, M., A. E. Adelberg, A. J. Clark, and P. E. Hartman. 1966. A proposal for a uniform nomenclature in bacterial genetics. *Genetics* 54:61–76.

438. Demerec, M., D. H. Gillespie, and K. Mizobuchi. 1963. Genetic structure of the *cysC* region of the *Salmonella* genome. *Genetics* 48:997–1009.

439. Demerec, M., and Z. Hartman. 1956. Tryptophan mutants in *Salmonella typhimurium*. *Carnegie Inst. Wash. Publ.* 612:5–33.

440. Demerec, M., E. L. Lahr, E. Balbinder, T. Miyake, C. Mack, D. MacKay, and J. Ishidsu. 1959. Bacterial genetics. *Carnegie Inst. Wash. Year Book* 58:433–439.

441. den Blaauwen, J. L., and P. W. Postma. 1985. Regulation of cyclic AMP synthese by enzyme III^{Glc} of the phosphoenolpyruvate:sugar phosphotransferase system in *crp* strains of *Salmonella typhimurium*. *J. Bacteriol.* 164:477–478.

442. Dendinger, S., and W. J. Brill. 1970. Regulation of proline degradation of *Salmonella typhimurium*. *J. Bacteriol.* 103:144–152.

443. Dendinger, S., and W. J. Brill. 1972. Effect of the proline analogue baikiain on proline metabolism in *Salmonella typhimurium*. *J. Bacteriol.* 112:1134–1141.

444. Dendinger, S. M., and J. E. Brenchley. 1980. Temperature-sensitive glutamate dehydrogenase mutants of *Salmonella typhimurium*. *J. Bacteriol.* 144:1043–1047.

445. Dendinger, S. M., L. G. Patil, and J. E. Brenchley. 1980. *Salmonella typhimurium* mutants with altered glutamate dehydrogenase and glutamate synthase activities. *J. Bacteriol.* 141:190–198.

446. Denk, D., and A. Bock. 1987. L-Cysteine biosynthesis in *Escherichia coli*: nucleotide sequence and expression of the serine acetyltransferase (*cysE*) gene from the wild-type and a cysteine-excreting mutant. *J. Gen. Microbiol.* 133:515–525.

447. Denney, R. M., and C. Yanofsky. 1974. Isolation and characterization of specialized ϕ80 transducing phages carrying regions of the *Salmonella typhimurium trp* operon. *J. Bacteriol.* 118:505–513.

448. Desiraju, V., W. G. Shanabruch, and A. L. Lu. 1993. Nucleotide sequence of the *Salmonella typhimurium mutB* gene, the homolog of *Escherichia coli mutY*. *J. Bacteriol.* 175:541–543.

449. Desrochers, M., L. Peloquin, and A. Saarman. 1978. Mapping of the *hemE* locus in *Salmonella typhimurium*. *J. Bacteriol.* 135:1151–1153.

450. diCamelli, R. F., and E. Balbinder. 1976. The association of tryptophan synthetase subunits from *Escherichia coli* and *Salmonella typhimurium* in homologous and heterologous combinations. *Genet. Res.* 27:323–333.

451. Dila, D. K., and S. R. Maloy. 1986. Proline transport in *Salmonella typhimurium*: *putP* permease mutants with altered substrate specificity. *J. Bacteriol.* 168:590–594.

452. DiNatale, P., F. Cimino, and F. DeLorenzo. 1974. The pyrophosphate exchange reaction of histidyl-tRNA synthetase from *Salmonella typhimurium*: reaction parameters and inhibition by ribonucleic acid. *FEBS Lett.* 46:175–179.

453. DiNatale, P., A. N. Schechter, G. Castronuovo Lepore, and F. DeLorenzo. 1976. Histidyl transfer ribonucleic acid synthetase from *Salmonella typhimurium*. Interaction with substrates and ATP analogues. *Eur. J. Biochem.* 62:293–298.

454. Diver, W. P., and D. G. MacPhee. 1981. The effects of mutation in the *polA* and *recA* genes on mutagenesis by nitroguanidine in *Salmonella typhimurium*. *Mutat. Res.* **83**:349–359.

455. Dobrogosz, W. J., G. W. Hall, D. K. Sherba, D. O. Silva, J. G. Harman, and T. Melton. 1983. Regulatory interactions among the *cya, crp* and *pts* gene products in *Salmonella typhimurium*. *Mol. Gen. Genet.* **192**:477–486.

456. Dobson, P. P., and G. C. Walker. 1980. Plasmid pKM-101 mediated Weigle reactivation in *Escherichia coli* K-12 and *Salmonella typhimurium* LT-2. Genetic dependence kinetics of induction and effect of chloramphenicol. *Mutat. Res.* **71**:25–42.

457. Doll, L., and G. Frankel. 1993. *fliU* and *fliV*: two flagellar genes essential for biosynthesis of *Salmonella* and *Escherichia coli* flagella. *J. Gen. Microbiol.* **139**:2415–2422.

458. Dombrosky, P. M., M. B. Schmid, and K. D. Young. 1994. Sequence divergence of the *murB* and *rrfB* genes from *Escherichia coli* and *Salmonella typhimurium*. *Arch. Microbiol.* **161**:501–507.

459. Domingue, G., and E. Johnson. 1974. Isolation of subcellular fractions containing immunogenic enterobacterial common antigens. *Z. Immunitaetsforsch. Exp. Klin. Immunol.* **148**:23–38.

460. Dooley, M., R. Torget, and E. Balbinder. 1979. Differences between the anthranilate-5-phosphoribosylpyrophosphate phosphoribosyltransferases of *Salmonella typhimurium* strains LT-2 and LT-7. *J. Gen. Microbiol.* **112**:171–180.

461. Dorman, C. J., G. C. Barr, N. Ni Bhriain, and C. F. Higgins. 1988. DNA supercoiling and the anaerobic and growth phase regulation of *tonB* gene expression. *J. Bacteriol.* **170**:2816–2826.

462. Dorman, C. J., S. Chatfield, C. F. Higgins, C. Hayward, and G. Dougan. 1989. Characterization of porin and *ompR* mutants of a virulent strain of *Salmonella typhimurium*: *ompR* mutants are attenuated in vivo. *Infect. Immun.* **57**:2136–2140.

463. Downs, D., L. Waxman, A. L. Goldberg, and J. Roth. 1986. Isolation and characterization of *lon* mutants in *Salmonella typhimurium*. *J. Bacteriol.* **165**:193–197.

464. Downs, D. M. 1992. Evidence for a new, oxygen-regulated biosynthetic pathway for the pyrimidine moiety of thiamine in *Salmonella typhimurium*. *J. Bacteriol.* **174**:1515–1521.

465. Downs, D. M., and L. Petersen. 1994. *apbA*, a new genetic locus involved in thiamine biosynthesis in *Salmonella typhimurium*. *J. Bacteriol.* **176**:4858–4864.

466. Downs, D. M., and J. R. Roth. 1987. A novel P22 prophage in *Salmonella typhimurium*. *Genetics* **117**:367–380.

467. Downs, D. M., and J. R. Roth. 1991. Synthesis of thiamine in *Salmonella typhimurium* independent of the *purF* function. *J. Bacteriol.* **173**:6597–6604.

468. Dreyfus, G., A. W. Williams, I. Kawagishi, and R. M. Macnab. 1993. Genetic and biochemical analysis of *Salmonella typhimurium* FliI, a flagellar protein related to the catalytic subunit of the F0F1 ATPase and to virulence proteins of mammalian and plant pathogens. *J. Bacteriol.* **175**:3131–3138.

469. Dreyfuss, J. 1974. The characterization of a sulfate- and thiosulfate-transporting system in *Salmonella typhimurium*. *J. Biol. Chem.* **239**:2292–2297.

470. Dreyfuss, J., and A. B. Pardee. 1966. Regulation of sulfate transport in *Salmonella typhimurium*. *J. Bacteriol.* **91**:2275–2280.

471. Driks, A., and D. J. DeRosier. 1990. Additional structures associated with bacterial flagellar basal body. *J. Mol. Biol.* **211**:669–672.

472. Driver, R. P., and R. P. Lawther. 1985. Restriction endonuclease analysis of the *ilvGEDA* operon of members of the family *Enterobacteriaceae*. *J. Bacteriol.* **162**:1317–1319.

473. Droffner, M. L., and N. Yamamoto. 1983. Anaerobic cultures of *Salmonella typhimurium* do not exhibit inducible proteolytic function of the *recA* gene and *recBC* function. *J. Bacteriol.* **156**:962–965.

474. Droge, W., E. Ruschmann, O. Luderitz, and O. Westphal. 1968. Biochemical studies on lipopolysaccharides of Salmonella R mutants. 4. Phosphate groups linked to heptose units and their absence in some lipopolysaccharides. *Eur. J. Biochem.* **4**:134–138.

475. Druger-Liotta, J., V. J. Prange, D. G. Overdier, and L. N. Csonka. 1986. Selection of mutations that alter the osmotic control of transcription of the *Salmonella typhimurium proU* operon. *J. Bacteriol.* **169**:2449–2459.

476. Dubnau, E., A. B. Lenny, and P. Margolin. 1973. Nonsense mutations of the *supX* locus: further characterization of the *supX* mutant phenotype. *Mol. Gen. Genet.* **126**:191–200.

477. Dubnau, E., and P. Margolin. 1972. Suppression of promoter mutations by the pleiotropic *top* mutations. *Mol. Gen. Genet.* **117**:91–112.

478. Dugasheva, L. G., and V. S. Levashev. 1974. Levomycetin resistance marker in *Escherichia coli* K12 and *Salmonella typhimurium*. *Zh. Mikrobiol. Epidemiol. Immunobiol.* **51**:85–90.

479. Dunlap, V. J., and L. N. Csonka. 1985. Osmotic regulation of L-proline transport in *Salmonella typhimurium*. *J. Bacteriol.* **163**:296–304.

480. Dunn, S. D., and E. E. Snell. 1979. Isolation of temperature-sensitive pantothenate kinase mutants of *Salmonella typhimurium* and mapping of the *coaA* gene. *J. Bacteriol.* **140**:805–808.

481. Edwards, C. J., D. J. Innes, D. M. Burns, and I. R. Beacham. 1993. UDP-sugar hydrolase isozymes in *Salmonella enterica* and *Escherichia coli*: silent alleles of *ushA* in related strains of group I *Salmonella* isolates, and of *ushB* in wild-type and K12 strains of *E. coli*, indicate recent and early silencing events, respectively. *FEMS Microbiol. Lett.* **114**:293–298.

482. Edwards, M. F., and B. A. D. Stocker. 1988. Construction of ΔaroA his Δpur strains of *Salmonella typhi*. *J. Bacteriol.* **170**:3991–3995.

483. Eichelberg, K., C. C. Ginocchio, and J. E. Galan. 1994. Molecular and functional characterization of the *Salmonella typhimurium* invasion genes *invB* and *invC*: homology of InvC to the F0F1 ATPase family of proteins. *J. Bacteriol.* **176**:4501–4510.

484. Eidels, L., and M. J. Osborn. 1971. Lipopolysaccharide and aldoheptose biosynthesis in transketolase mutants of *Salmonella typhimurium*. *Proc. Natl. Acad. Sci. USA* **68**:1673–1677.

485. Eidels, L., and M. J. Osborn. 1974. Phosphoheptose isomerase, first enzyme in the biosynthesis of aldoheptose in *Salmonella typhimurium*. *J. Biol. Chem.* **249**:5642–5648.

486. Eidels, L., P. D. Rick, N. P. Stimler, and M. J. Osborn. 1974. Transport of D-arabinose-5-phosphate and D-sedoheptulose-7-phosphate by the hexose phosphate transport system of *Salmonella typhimurium*. *J. Biol. Chem.* **249**:5642–5648.

487. Eisenbach, M., A. Wolf, M. Welch, S. R. Caplan, I. R. Lapidus, R. M. Macnab, H. Aloni, and O. Asher. 1990. Pausing, switching and speed fluctuation of the bacterial flagellar motor and their relation to motility and chemotaxis. *J. Mol. Biol.* **211**:551–563.

488. Eisenstadt, E., M. Wolf, and I. H. Goldberg. 1980. Mutagenesis by neocarzinostatin in *Escherichia coli* and *Salmonella typhimurium*: requirement for *umuC+* or plasmid pKM101. *J. Bacteriol.* **144**:656–660.

489. Eisenstark, A., R. Eisenstark, and S. Cunningham. 1968. Genetic analysis of thymineless (*thy*) mutants in *Salmonella typhimurium*. *Genetics* **58**:493–506.

490. Eisenstark, A., R. Eisenstark, J. Van Dillewijn, and A. Rorsch. 1969. Radiation-sensitive and recombinationless mutants of *Salmonella typhimurium*. *Mutat. Res.* **8**:497–504.

491. Ekena, K., and S. Maloy. 1990. Regulation of proline utilization in *Salmonella typhimurium*: how do cells avoid a futile cycle? *Mol. Gen. Genet.* **220**:492–494.

492. Elliott, C. J., and F. B. Armstrong. 1968. Isoleucine-valine requiring mutants of *Salmonella typhimurium*. II. Strains deficient in dihydroxyacid dehydratase activity. *Genetics* **58**:171–179.

493. Elliott, T. 1989. Cloning, genetic characterization, and nucleotide sequence of the *hemA-prfA* operon of *Salmonella typhimurium*. *J. Bacteriol.* **171**:3948–3960.

494. Elliott, T. 1992. A method for constructing single-copy *lac* fusions in *Salmonella typhimurium* and its application to the *hemA-prfA* operon. *J. Bacteriol.* **174**:245–253.

495. Elliott, T. 1993. Transport of 5-aminolevulinic acid by the dipeptide permease in *Salmonella typhimurium*. *J. Bacteriol.* **175**:325–331.

496. Elliott, T., Y. J. Avissar, G. E. Rhie, and S. I. Beale. 1990. Cloning and sequence of the *Salmonella typhimurium hemL* gene and identification of the missing enzyme in *hemL* mutants as glutamate-1-semialdehyde aminotransferase. *J. Bacteriol.* **172**:7071–7084.

497. Elliott, T., and J. R. Roth. 1989. Heme-deficient mutants of *Salmonella typhimurium*: two genes required for ALA synthesis. *Mol. Gen. Genet.* **216**:303–314.

498. Elliott, T., and X. Wang. 1991. *Salmonella typhimurium prfA* mutants defective in release factor 1. *J. Bacteriol.* **173**:4144–4154.

499. Engler, M., and M. J. Bessman. 1977. The purification of a mutator polymerase from *Salmonella typhimurium*. *Fed. Proc.* **36**:735.

500. Englesberg, E., and L. S. Baron. 1959. Mutation to L-rhamnose resistance and transduction to L-rhamnose utilization in *Salmonella typhosa*. *J. Bacteriol.* **78**:675–686.

501. Engstrom, J., J. Wong, and R. Maurer. 1986. Interaction of DNA polymerase III γ and β subunits *in vivo* in *Salmonella typhimurium*. *Genetics* **113**:499–515.

502. Enomoto, M. 1966. Genetic studies of paralyzed mutants in Salmonella. I. Genetic fine structure of the *mot* loci in *Salmonella typhimurium*. *Genetics* **54**:715–726.

503. Enomoto, M. 1966. Genetic studies of paralyzed mutants in Salmonella. II. Mapping of three *mot* loci by linkage analysis. *Genetics* **54**:1069–1076.

504. Enomoto, M. 1971. Genetic analysis of non-motile mutants in *Salmonella typhimurium*: a new mapping method by abortive transduction. *Genetics* **69**:145–161.

505. Enomoto, M. 1972. Genetic studies of chlorate-resistant mutants of *Salmonella typhimurium*. *Jpn. J. Genet.* **47**:227–234.

506. Enomoto, M., and B. A. D. Stocker. 1974. Transduction by phage P1kc in *Salmonella typhimurium*. *Virology* **60**:503–514.

507. Enomoto, M., and B. A. D. Stocker. 1975. Integration, at *hag* or elsewhere, of H2 (phase 2 flagellin) genes transduced from *Salmonella* to *Escherichia coli*. *Genetics* **81**:595–614.

508. Enomoto, M., and S. Yamaguchi. 1969. Different frequencies of cotransduction of *motC* and H1 in *Salmonella*. *Genet. Res.* **14**:45–52.

509. Erickson, B. D., Z. F. Burton, K. K. Watanabe, and R. R. Burgess. 1985. Nucleotide sequence of the *rpsU-dnaG-rpoD* operon from *Salmonella typhimurium* and a comparison of this sequence with the homologous operon of *Escherichia coli*. *Gene* **40**:67–78.

510. Erickson, J. U., and G. R. Bjork. 1986. Pleiotropic effects induced by modification deficiency next to the anticodon of tRNA from *Salmonella typhimurium* LT2. *J. Bacteriol.* **166**:1013–1021.

511. Ernst, J. F., R. L. Bennett, and L. I. Rothfield. 1978. Constitutive expression of the iron-enterochelin and ferrichrome uptake systems in a mutant strain of *Salmonella typhimurium*. *J. Bacteriol.* **135**:928–934.

512. Esaki, N., and C. T. Walsh. 1986. Biosynthetic alanine racemase of *Salmonella typhimurium*: purification and characterization of the enzyme encoded by the *alr* gene. *Biochemistry* **25**:3261–3267.

513. Escalante-Semerena, J. C., M. G. Johnson, and J. R. Roth. 1992. The CobII and CobIII regions of the cobalamin (vitamin B$_{12}$) biosynthetic operon of *Salmonella typhimurium*. J. Bacteriol. 174:24–29.

514. Escalante-Semerena, J. C., and J. R. Roth. 1987. Regulation of cobalamin biosynthetic operons in *Salmonella typhimurium*. J. Bacteriol. 169:2251–2258.

515. Escalante-Semerena, J. C., S. J. Suh, and J. R. Roth. 1990. cobA function is required for both de novo cobalamin biosynthesis and assimilation of exogenous corrinoids in *Salmonella typhimurium*. J. Bacteriol. 172:273–280.

516. Faelen, M., M. Mergeay, J. Gerits, A. Toussaint, and N. Lefebvre. 1981. Genetic mapping of a mutation conferring sensitivity to bacteriophage Mu in *Salmonella typhimurium* LT2. J. Bacteriol. 146:914–919.

517. Falconi, M., V. McGovern, C. Gualerzi, D. Hillyard, and N. P. Higgins. 1991. Mutations altering chromosomal protein H-NS induce mini-Mu transposition. New Biol. 3:615–625.

518. Fandl, J. P., L. K. Thorner, and S. W. Artz. 1990. Mutations that affect transcription and cyclic AMP-CRP regulation of the adenylate cyclase gene (cya) of *Salmonella typhimurium*. Genetics 125:719–727.

519. Fang, F. C., S. J. Libby, N. A. Buchmeier, P. C. Loewen, J. Switala, J. Harwood, and D. G. Guiney. 1992. The alternative sigma factor katF (rpoS) regulates *Salmonella* virulence. Proc. Natl. Acad. Sci. USA 89:11978–11982.

520. Faraci, W. S., and C. T. Walsh. 1988. Racemization of alanine by the alanine racemases from *Salmonella typhimurium* and *Bacillus stearothermophilus*: energetic reaction profiles. Biochemistry 27:3267–3276.

521. Farewell, A., R. Brazas, E. Davie, J. Mason, and L. I. Rothfield. 1991. Suppression of the abnormal phenotype of *Salmonella typhimurium* rfaH mutants by mutations in the gene for transcription termination factor Rho. J. Bacteriol. 173:5188–5193.

522. Farr, S. B., and T. Kogoma. 1991. Oxidative stress responses in *Escherichia coli* and *Salmonella typhimurium*. Microbiol. Rev. 55:561–585.

523. Fasciano, A., and P. C. Hallenbeck. 1991. Mutations in trans that affect formate dehydrogenase (fdhF) gene expression in *Salmonella typhimurium*. J. Bacteriol. 173:5893–5900.

524. Fasciano, A., and P. C. Hallenbeck. 1992. The role of ntrA in the anaerobic metabolism of *Salmonella typhimurium*. FEMS Microbiol. Lett. 90:101–104.

525. Faust, L. P., and B. M. Babior. 1992. Overexpression, purification, and some properties of the AdoCbl-dependent ethanolamine ammonia-lyase from *Salmonella typhimurium*. Arch. Biochem. Biophys. 294:50–54.

526. Faust, L. R., J. A. Connor, D. M. Roof, J. A. Hoch, and B. M. Babior. 1990. Cloning, sequencing, and expression of the genes encoding the adenosylcobalamin-dependent ethanolamine ammonia-lyase of *Salmonella typhimurium*. J. Biol. Chem. 265:12462–12466.

527. Feild, M. J., D. C. Nguyen, and F. B. Armstrong. 1989. Amino acid sequence of *Salmonella typhimurium* branched-chain amino acid aminotransferase. Biochemistry 28:5306–5310.

528. Feldheim, D. A., A. M. Chin, C. T. Nierva, B. U. Feucht, Y. W. Cao, Y. F. Xu, S. L. Sutrina, and M. H. Saier, Jr. 1990. Physiological consequences of the complete loss of phosphoryl-transfer proteins HPr and FPr of the phosphoenolpyruvate:sugar phosphotransferase system and analysis of fructose (fru) operon expression in *Salmonella typhimurium*. J. Bacteriol. 172:5459–5469.

529. Feng, J. A., R. C. Johnson, and R. E. Dickerson. 1994. Hin recombinase bound to DNA: the origin of specificity in major and minor groove interactions. Science 263:348–355.

530. Fernandez, V. M., R. Martin del Rio, A. R. Tebar, M. M. Guisan, and A. O. Ballesteros. 1975. Derepression and repression of the histidine operon: role of the feedback site of the first enzyme. J. Bacteriol. 124:1366–1373.

531. Feucht, B. U., and M. H. Saier, Jr. 1980. Fine control of adenylate cyclase by the phosphoenolpyruvate and sugar phosphotransferase systems in *Escherichia coli* and *Salmonella typhimurium*. J. Bacteriol. 141:603–610.

532. Feutrier, J., W. W. Kay, and T. J. Trust. 1986. Purification and characterization of fimbriae from *Salmonella enteritidis*. J. Bacteriol. 168:221–227.

533. Fields, P. I., E. A. Groisman, and F. Heffron. 1989. A *Salmonella* locus that controls resistance to microbicidal proteins from phagocytic cells. Science 243:1059–1062.

534. Fields, P. I., R. V. Swanson, C. G. Haidaris, and F. Heffron. 1986. Mutants of *Salmonella typhimurium* that cannot survive within the macrophage are avirulent. Proc. Natl. Acad. Sci. USA 83:5189–5193.

535. Figueroa, N., N. Wills, and L. Bossi. 1991. Common sequence determinants of the response of a prokaryotic promoter to DNA bending and supercoiling. EMBO J. 10:941–949.

536. Filutowicz, M., Z. Ciesla, and T. Klopotowski. 1979. Interference of azide with cysteine biosynthesis in *Salmonella typhimurium*. J. Gen. Microbiol. 113:45–55.

537. Filutowicz, M., A. Wiater, and D. Hulanicka. 1982. Delayed inducibility of sulfite reductase in cysM mutants of *Salmonella typhimurium* under anaerobic conditions. J. Gen. Microbiol. 128:1791–1794.

538. Finkel, S. E., and R. C. Johnson. 1992. The Fis protein: it's not just for DNA inversion anymore. Mol. Microbiol. 6:3257–3265. (Erratum, 7:1023, 1993.)

539. Finn, G. J., and S. Condon. 1975. Regulation of catalase synthesis in *Salmonella typhimurium*. J. Bacteriol. 123:570–579.

540. Fira, D., B. Vasiljevic, and L. Topisirovic. 1990. Altered translational fidelity of a *Salmonella typhimurium* LT2 mutant resistant to the aminoglycoside antibiotic neamine. J. Gen. Microbiol. 136:249–253.

541. Folterman, K. F., D. A. Beck, and J. R. Wild. 1986. In vivo formation of hybrid aspartate transcarbamoylases from native subunits of divergent members of the family Enterobacteriaceae. J. Bacteriol. 167:285–290.

542. Fong, C.-L. W., N. K. Heinzinger, S. Tongklan, and E. L. Barrett. 1993. Cloning of the phs genetic locus from *Salmonella typhimurium* and a role for a phs product in its own induction. J. Bacteriol. 175:6368–6371.

543. Ford, S. R., and R. L. Switzer. 1975. Stimulation of derepressed enzyme synthesis in bacteria by growth on sublethal concentrations of chloramphenicol. Antimicrob. Agents Chemother. 7:555–563.

544. Ford, S. R., and R. L. Switzer. 1975. Stimulation of enzyme synthesis by sublethal concentrations of chloramphenicol is not mediated by ribonucleotide pools. Antimicrob. Agents Chemother. 7:564–570.

545. Foster, D. L., S. L. Mowbray, B. K. Jap, and D. E. Koshland, Jr. 1985. Purification and characterization of the aspartate chemoreceptor. J. Biol. Chem. 260:11706–11710.

546. Foster, J. W. 1981. Pyridine nucleotide cycle of *Salmonella typhimurium*: in vitro demonstration of nicotinamide adenine dinucleotide glycohydrolase, nicotinamide mononucleotide glycohydrolase, and nicotinamide adenine dinucleotide pyrophosphatase activities. J. Bacteriol. 145:1002–1009.

547. Foster, J. W. 1983. Identification and characterization of a relA-dependent starvation-inducible locus (sin) in *Salmonella typhimurium*. J. Bacteriol. 156:424–428.

548. Foster, J. W. 1991. *Salmonella* acid shock proteins are required for the adaptive acid tolerance response. J. Bacteriol. 173:6896–6902.

549. Foster, J. W. 1993. The acid tolerance response of *Salmonella typhimurium* involves transient synthesis of key acid shock proteins. J. Bacteriol. 175:1981–1987.

550. Foster, J. W., and B. Bearson. 1994. Acid-sensitive mutants of *Salmonella typhimurium* identified through a dinitrophenol lethal screening strategy. J. Bacteriol. 176:2596–2602.

551. Foster, J. W., and D. Falconer. 1983. Isolation of pyridine analog supersensitive (pas) mutants of *Salmonella typhimurium*, abstr. K52, p. 51. Abstr. Annu. Meet. Am. Soc. Microbiol. 1983.

552. Foster, J. W., and H. K. Hall. 1990. Adaptive acidification tolerance response of *Salmonella typhimurium*. J. Bacteriol. 172:771–778.

553. Foster, J. W., and H. K. Hall. 1991. Inducible pH homeostasis and the acid tolerance response of *Salmonella typhimurium*. J. Bacteriol. 173:5129–5135.

554. Foster, J. W., and H. K. Hall. 1992. Effect of *Salmonella typhimurium* ferric uptake regulator (fur) mutations on iron- and pH-regulated protein synthesis. J. Bacteriol. 174:4317–4323.

555. Foster, J. W., and E. A. Holley. 1981. Genetic mapping of the *Salmonella typhimurium* pncB locus. J. Bacteriol. 148:394–396.

556. Foster, J. W., and E. A. Holley. 1983. Characterization of a nicotinamide starvation-inducible locus in *Salmonella typhimurium*, abstr. K53, p. 51. Abstr. Annu. Meet. Am. Soc. Microbiol. 1983.

557. Foster, J. W., E. A. Holley, and S. Mya. 1984. NAD metabolism in *Salmonella typhimurium*: isolation of pyridine analogue supersensitive (pas) and pas suppressor mutants. J. Gen. Microbiol. 130:2873–2881.

558. Foster, J. W., E. A. Holley-Guthrie, and F. Warren. 1987. Regulation of NAD metabolism in *Salmonella typhimurium*: genetic analysis and cloning of the nadR repressor locus. Mol. Gen. Genet. 208:279–287.

559. Foster, J. W., D. M. Kinney, and A. G. Moat. 1979. Pyridine nucleotide cycle of *Salmonella typhimurium*: regulation of nicotinic acid phophoribosyl transferase and nicotinamide deamidase. J. Bacteriol. 138:957–961.

560. Foster, J. W., D. M. Kinney, and A. G. Moat. 1979. Pyridine nucleotide cycle of *Salmonella typhimurium*: isolation and characterization of pncA, pncB, and pncC mutants and utilization of exogenous nicotinamide adenine dinucleotide. J. Bacteriol. 137:1165–1175.

561. Foster, J. W., and A. G. Moat. 1978. Mapping and characterization of the nad genes in *Salmonella typhimurium* LT-2. J. Bacteriol. 133:775–779.

562. Foster, J. W., Y. K. Park, I. S. Bang, K. Karem, H. Betts, H. K. Hall, and E. Shaw. 1994. Regulatory circuits involved with pH-regulated gene expression in *Salmonella typhimurium*. Microbiology 140:341–352.

563. Foster, J. W., Y. K. Park, T. Penfound, T. Fenger, and M. P. Spector. 1990. Regulation of NAD metabolism in *Salmonella typhimurium*: molecular sequence analysis of the bifunctional nadR regulator and the nadA-pnuC operon. J. Bacteriol. 172:4187–4196.

564. Foster, J. W., and M. P. Spector. 1986. Phosphate starvation regulon of *Salmonella typhimurium*. J. Bacteriol. 166:666–669.

565. Fraenkel, D., M. J. Osborn, B. K. Horecker, and S. M. Smith. 1963. Metabolism and cell wall structure of a mutant of *Salmonella typhimurium* deficient in phosphoglucose isomerase. Biochem. Biophys. Res. Commun. 11:423–428.

566. Francis, C. L., T. A. Ryan, B. D. Jones, S. J. Smith, and S. Falkow. 1993. Ruffles induced by *Salmonella* and other stimuli direct macropinocytosis of bacteria. Nature (London) 364:639–642.

567. Francis, N. R., V. M. Irikura, S. Yamaguchi, D. J. DeRosier, and R. M. Macnab. 1992. Localization of the *Salmonella typhimurium* flagellar switch protein FliG to the cytoplasmic M-ring face of the basal body. Proc. Natl. Acad. Sci. USA 89:6304–6308.

568. Francoz, E., A. Molla, E. Dassa, W. Saurin, and M. Hofnung. 1990. The maltoporin of *Salmonella typhimurium*: sequence and folding model. Res. Microbiol. 141:1039–1059.

569. Francoz, E., E. Schneider, and E. Dassa. 1990. The sequence of the *malG* gene from *Salmonella typhimurium* and its functional implications. *Res. Microbiol.* 141:633–644.

570. Frankhauser, D. B., and P. E. Hartman. 1971. Direct selection for transduction of suppressor mutations and linkage of *supD* to *fla* genes in *Salmonella. J. Bacteriol.* 108:1427–1430.

571. French, S., K. Martin, T. Patterson, R. Bauerle, and O. L. Miller, Jr. 1985. Electron microscopic visualization of *trp* operon expression in *Salmonella typhimurium. Proc. Natl. Acad. Sci. USA* 82:4638–4642.

572. Freudl, R., G. Braun, N. Honore, and S. T. Cole. 1987. Evolution of the enterobacterial *sulA* gene: a component of the SOS system encoding an inhibitor of cell division. *Gene* 52:31–40.

573. Freudl, R., and S. T. Cole. 1983. Cloning and molecular characterization of the *ompA* gene from *Salmonella typhimurium. Eur. J. Biochem.* 134:497–502.

574. Freundlich, M., and J. M. Trela. 1969. Control of isoleucine, valine, and leucine biosynthesis. VI. Effect of 5′,5′,5′-trifluoroleucine on repression in *Salmonella typhimurium. J. Bacteriol.* 99:101–106.

575. Frey, B., J. McCloskey, W. Kersten, and H. Kersten. 1988. New function of vitamin B$_{12}$: cobamide-dependent reduction of epoxyqueuosine to queuosine in tRNAs of *Escherichia coli* and *Salmonella typhimurium. J. Bacteriol.* 170:2078–2082.

576. Frick, M. M., J. Neuhard, and R. A. Kelln. 1990. Cloning, nucleotide sequence and regulation of the *Salmonella typhimurium pyrD* gene encoding dihydroorotate dehydrogenase. *Eur. J. Biochem.* 194:573–578.

577. Friedberg, D., T. W. Mikulka, J. Jones, and J. M. Calvo. 1974. *flrB*, a regulatory locus controlling branched-chain amino acid biosynthesis in *Salmonella typhimurium. J. Bacteriol.* 118:942–951.

578. Friedberg, D., E. R. Rosenthal, J. W. Jones, and J. M. Calvo. 1985. Characterization of the 3′ end of the leucine operon of *Salmonella typhimurium. Mol. Gen. Genet.* 199:486–494.

579. Friedman, S. B., and P. Margolin. 1968. Evidence for an altered operator specificity: catabolite repression control of the leucine operon in *Salmonella typhimurium. J. Bacteriol.* 95:2263–2269.

580. Friedrich, M. J., N. E. Kinsey, J. Vila, and R. J. Kadner. 1993. Nucleotide sequence of a 13.9 kb segment of the 90 kb virulence plasmid of *Salmonella typhimurium*: the presence of fimbrial biosynthetic genes. *Mol. Microbiol.* 8:543–558.

581. Fuchs, R. L., M. J. Madonna, and J. E. Brenchley. 1982. Identification of the structural genes for glutamine synthase and genetic characterization of this region of the *Salmonella typhimurium* chromosome. *J. Bacteriol.* 149:906–915.

582. Fujita, H., S. Yamaguchi, and T. Iino. 1973. Studies on H-O variants in *Salmonella* in relation to phase variation. *J. Gen. Microbiol.* 76:127–134.

583. Fujita, H., S. Yamaguchi, T. Taira, T. Hirano, and T. Iino. 1987. Isolation and genetic analysis of operator-constitutive mutants of the *H1* operon in *Salmonella typhimurium. J. Gen. Microbiol.* 133:3071–3080.

584. Fukasawa, T., and H. Nikaido. 1961. Galactose mutants of *Salmonella typhimurium. Genetics* 46:1295–1303.

585. Fukuda, R., A. Ishihama, T. Saitoh, and M. Taketo. 1977. Comparative studies of RNA polymerase subunits from various bacteria. *Mol. Gen. Genet.* 154:135–144.

586. Fukushi, K., H. Kudo, H. Asano, and J.-I. Sasaki. 1977. Electron microscopy of endotoxin extracted from rough mutant of *Salmonella. Jpn. J. Med. Sci. Biol.* 30:51–54.

587. Fulcher, C. A., and R. Bauerle. 1978. Re-initiation of tryptophan operon expression in a promoter deletion strain of *Salmonella typhimurium. Mol. Gen. Genet.* 158:239–250.

588. Fuller-Pace, F. V., and N. E. Murray. 1986. Two DNA recognition domains of the specificity polypeptides of a family of type I restriction enzymes. *Proc. Natl. Acad. Sci. USA* 83:9368–9372.

589. Fultz, P. N., K. K. L. Choung, and J. Kemper. 1980. Construction and characterization of *Salmonella typhimurium* strains that accumulate and excrete α and β isopropylmalate. *J. Bacteriol.* 142:513–520.

590. Fultz, P. N., and J. Kemper. 1980. Wild-type isopropylmalate isomerase in *Salmonella typhimurium* is composed of two different subunits. *J. Bacteriol.* 148:210–219.

591. Fultz, P. N., D. Y. Kwoh, and J. Kemper. 1979. *Salmonella typhimurium newD* and *Escherichia coli leuC* genes code for a functional isopropyl malate isomerase in *Salmonella typhimurium-Escherichia coli* hybrids. *J. Bacteriol.* 137:1253–1262.

592. Funanage, V. L., P. D. Ayling, S. M. Dendinger, and J. E. Brenchley. 1978. *Salmonella typhimurium* LT-2 mutants with altered glutamine synthetase levels and amino acid uptake activities. *J. Bacteriol.* 136:588–596.

593. Funanage, V. L., and J. Brenchley. 1977. Characterization of *Salmonella typhimurium* mutants with altered glutamine synthetase activity. *Genetics* 86:513–526.

594. Fung, J., T. J. MacAlister, and L. I. Rothfield. 1978. Role of murein lipoprotein in morphogenesis of the bacterial division septum: phenotypic similarity of *lkyD* and *lpo* mutants. *J. Bacteriol.* 133:1467–1471.

595. Fung, J. C., T. J. MacAlister, R. A. Weigand, and L. I. Rothfield. 1980. Morphogenesis of the bacterial division septum: identification of potential sites of division in *lkyD* mutants of *Salmonella typhimurium. J. Bacteriol.* 143:1019–1024.

596. Furano, A. V. 1978. Direct demonstration of duplicate *tuf* genes in enteric bacteria. *Proc. Natl. Acad. Sci. USA* 75:3104–3108.

597. Gaitanaris, G. A., A. McCormick, B. H. Howard, and M. E. Gottesman. 1986. Reconstitution of an operon from overlapping fragments: use of the λSV2, integrative cloning system. *Gene* 46:1–11.

598. Galakatos, N. G., E. Daub, D. Botstein, and C. T. Walsh. 1986. Biosynthetic *alr* alanine racemase from *Salmonella typhimurium*: DNA and protein sequence determination. *Biochemistry* 25:3255–3260.

599. Galakatos, N. G., and C. T. Walsh. 1987. Specific proteolysis of native alanine racemases from *Salmonella typhimurium*: identification of the cleavage site and characterization of the clipped two-domain proteins. *Biochemistry* 26:8475–8480.

600. Galan, J. E., and R. Curtiss. 1989. Cloning and molecular characterization of genes whose products allow *Salmonella typhimurium* to penetrate tissue culture cells. *Proc. Natl. Acad. Sci. USA* 86:6383–6387.

601. Galan, J. E., and R. Curtiss. 1989. Virulence and vaccine potential of *phoP* mutants of *Salmonella typhimurium. Microb. Pathog.* 6:433–443.

602. Galan, J. E., and R. Curtiss. 1990. Expression of *Salmonella typhimurium* genes required for invasion is regulated by changes in DNA supercoiling. *Infect. Immun.* 58:1879–1885.

603. Galan, J. E., and R. Curtiss. 1991. Distribution of the *invA*, -B, -C, and -D genes of *Salmonella typhimurium* among other *Salmonella* serovars: *invA* mutants of *Salmonella typhi* are deficient for entry into mammalian cells. *Infect. Immun.* 59:2901–2908.

604. Galan, J. E., C. Ginocchio, and P. Costeas. 1992. Molecular and functional characterization of the *Salmonella* invasion gene *invA*: homology of InvA to members of a new protein family. *J. Bacteriol.* 174:4338–4349.

605. Galan, J. E., K. Nakayama, and R. Curtiss. 1990. Cloning and characterization of the *asd* gene of *Salmonella typhimurium*: use in stable maintenance of recombinant plasmids in *Salmonella* vaccine strains. *Gene* 94:29–35.

606. Galinier, A., D. Negre, J. C. Cortay, S. Marcandier, S. R. Maloy, and A. J. Cozzone. 1990. Sequence analysis of the *iclR* gene encoding the repressor of the acetate operon in *Salmonella typhimurium. Nucleic Acids Res.* 18:3656.

607. Gallagher, M. P., S. R. Pearce, and C. F. Higgins. 1989. Identification and localization of the membrane-associated, ATP-binding subunit of the oligopeptide permease of *Salmonella typhimurium. Eur. J. Biochem.* 180:133–141.

608. Galloway, R. J., and B. L. Taylor. 1980. Histidine starvation and adenosine 5′-triphosphate depletion in chemotaxis of *Salmonella typhimurium. J. Bacteriol.* 144:1068–1075.

609. Gan, K., S. D. Gupta, K. Sankaran, M. B. Schmid, and H. C. Wu. 1993. Isolation and characterization of a temperature-sensitive mutant of *Salmonella typhimurium* defective in prolipoprotein modification. *J. Biol. Chem.* 268:16544–16550.

610. Gann, A. A., A. J. Campbell, J. F. Collins, A. F. Coulson, and N. E. Murray. 1987. Reassortment of DNA recognition domains and the evolution of new specificities. *Mol. Microbiol.* 1:13–22.

611. Garber, B. B., and J. S. Gots. 1980. Utilization of 2,6-diaminopurine by *Salmonella typhimurium. J. Bacteriol.* 143:864–871.

612. Garcia, E., S. Bancroft, S. G. Rhee, and S. Kustu. 1977. The product of a newly identified gene, *glnF*, is required for synthesis of glutamine synthetase in *Salmonella. Proc. Natl. Acad. Sci. USA* 74:1662–1666.

613. Garcia-del Portillo, F., J. W. Foster, and B. B. Finlay. 1993. Role of acid tolerance response genes in *Salmonella typhimurium* virulence. *Infect. Immun.* 61:4489–4492.

614. Garcia-del Portillo, F., J. W. Foster, M. E. Maguire, and B. B. Finlay. 1992. Characterization of the micro-environment of *Salmonella typhimurium*-containing vacuoles within MDCK epithelial cells. *Mol. Microbiol.* 6:3289–3297.

615. Gardner, M. M., D. O. Hennig, and R. A. Kelln. 1983. Control of the *arg* gene expression in *Salmonella typhimurium* by the arginine repressor from *Escherichia coli* K-12. *Mol. Gen. Genet.* 189:458–462.

616. Garrett, A. R., L. A. Johnson, and I. R. Beacham. 1989. Isolation, molecular characterization and expression of the *ushB* gene of *Salmonella typhimurium* which encodes a membrane-bound UDP-sugar hydrolase. *Mol. Microbiol.* 3:177–186.

617. Garrido-Pertierra, A. 1980. Isolation and properties of *Salmonella typhimurium* mutants defective in enolase. *Rev. Esp. Fisiol.* 36:33–40.

618. Garriga, X., S. Calero, and J. Barbe. 1992. Nucleotide sequence analysis and comparison of the *lexA* genes from *Salmonella typhimurium, Erwinia carotovora, Pseudomonas aeruginosa* and *Pseudomonas putida. Mol. Gen. Genet.* 236:125–134.

619. Geerse, R. H., F. Izzo, and P. W. Postma. 1989. The PEP:fructose phosphotransferase system in *Salmonella typhimurium*: FPr combines enzyme IIIFru and pseudo-HPr activities. *Mol. Gen. Genet.* 216:517–525.

620. Geerse, R. H., C. R. J. Ruig, A. R. J. Schuitema, and P. W. Postma. 1986. Relationship between pseudo-HPr and the PEP: fructose phosphotransferase system in *Salmonella typhimurium* and *Escherichia coli. Mol. Gen. Genet.* 203:435–444.

621. Gemmill, R. M., J. W. Jones, G. W. Haughn, and J. M. Calvo. 1983. Transcription initiation sites of the leucine operons of *Salmonella typhimurium* and *Escherichia coli. J. Mol. Biol.* 170:39–59.

622. Gemmill, R. M., M. Tripp, S. B. Friedman, and J. M. Calvo. 1984. Promoter mutation causing catabolite repression of the *Salmonella typhimurium* leucine operon. *J. Bacteriol.* 158:948–953.

623. Gemmill, R. M., S. R. Wessler, E. B. Keller, and J. M. Calvo. 1979. *Leu* operon of *Salmonella typhimurium* is controlled by an attenuation mechanism. *Proc. Natl. Acad. Sci. USA* 76:4941–4945.

624. Gemski, P., Jr., and B. A. D. Stocker. 1967. Transduction by bacteriophage P22 in non-smooth mutants of *Salmonella typhimurium. J. Bacteriol.* 93:1588–1597.

625. Gerson, S. L., and B. Magasanik. 1975. Regulation of the *hut* operons of *Salmonella typhimurium* and *Klebsiella aerogenes* by the heterologous *hut* repressors. *J. Bacteriol.* 124:1269–1272.

626. Ghangas, G. S., and D. B. Wilson. 1984. Isolation and characterization of the *Salmonella typhimurium* LT2 xylose regulon. *J. Bacteriol.* 157:158–164.

627. Gibert, I., J. Barbe, and J. Casadesus. 1990. Distribution of insertion sequence IS200 in *Salmonella* and *Shigella*. *J. Gen. Microbiol.* 136:2555–2560.

628. Gibert, I., and J. Casadesus. 1990. *sulA*-independent division inhibition in *his*-constitutive strains of *Salmonella typhimurium*. *FEMS Microbiol. Lett.* 57:205–210.

629. Gibson, B. W., W. Melaugh, N. J. Phillips, M. A. Apicella, A. A. Campagnari, and J. M. Griffiss. 1993. Investigation of the structural heterogeneity of lipooligosaccharides from pathogenic *Haemophilus* and *Neisseria* species and of R-type lipopolysaccharides from *Salmonella typhimurium* by electrospray mass spectrometry. *J. Bacteriol.* 175:2702–2712.

630. Gibson, M. M., D. A. Bagga, C. G. Miller, and M. E. Maguire. 1991. Magnesium transport in *Salmonella typhimurium*: the influence of new mutations conferring Co2+ resistance on the CorA Mg2+ transport system. *Mol. Microbiol.* 5:2753–2762.

631. Gibson, M. M., E. M. Ellis, K. A. Graeme-Cook, and C. F. Higgins. 1987. OmpR and EnvZ are pleiotropic regulatory proteins: positive regulation of the tripeptide permease (*tppB*) of *Salmonella typhimurium*. *Mol. Gen. Genet.* 207:120–129.

632. Gibson, M. M., M. Price, and C. F. Higgins. 1984. Genetic characterization and molecular cloning of the tripeptide permease (*tpp*) genes of *Salmonella typhimurium*. *J. Bacteriol.* 160:122–130.

633. Gillen, K. L., and K. T. Hughes. 1991. Molecular characterization of *flgM*, a gene encoding a negative regulator of flagellin synthesis in *Salmonella typhimurium*. *J. Bacteriol.* 173:6453–6459.

634. Gillen, K. L., and K. T. Hughes. 1991. Negative regulatory loci coupling flagellin synthesis to flagellar assembly in *Salmonella typhimurium*. *J. Bacteriol.* 173:2301–2310.

635. Gillen, K. L., and K. T. Hughes. 1993. Transcription from two promoters and autoregulation contribute to the control of expression of the *Salmonella typhimurium* flagellar regulatory gene *flgM*. *J. Bacteriol.* 175:7006–7015.

636. Gilman, R. H., R. B. Hornick, W. E. Woodward, H. L. Dupont, M. J. Snyder, M. M. Levine, and J. P. Libonti. 1977. Evaluation of a UDP glucose-4-epimeraseless mutant of *Salmonella typhi* as a live oral vaccine. *J. Infect. Dis.* 136:717–723.

637. Gilson, E., C. F. Higgins, M. Hofnung, G. Ferro-Luzzi Ames, and H. Nikaido. 1982. Extensive homology between membrane-associated components of histidine and maltose transport systems of *Salmonella typhimurium* and *Escherichia coli*. *J. Biol. Chem.* 257:9915–9918.

638. Ginocchio, C., J. Pace, and J. E. Galan. 1992. Identification and molecular characterization of a *Salmonella typhimurium* gene involved in triggering the internalization of salmonellae into cultured epithelial cells. *Proc. Natl. Acad. Sci. USA* 89:5976–5980.

639. Ginocchio, C. C., S. B. Olmsted, C. L. Wells, and J. E. Galan. 1994. Contact with epithelial cells induces the formation of surface appendages on *Salmonella typhimurium*. *Cell* 76:717–724.

640. Ginther, C. L., and J. L. Ingraham. 1974. Nucleoside diphosphokinase of *Salmonella typhimurium*. *J. Biol. Chem.* 249:3406–3411.

641. Ginther, C. L., and J. L. Ingraham. 1974. Cold-sensitive mutant of *Salmonella typhimurium* defective in nucleosidediphosphokinase. *J. Bacteriol.* 118:1020–1026.

642. Glanville, E. V., and M. Demerec. 1960. Threonine, isoleucine and isoleucine-valine mutants of *Salmonella typhimurium*. *Genetics* 45:1359–1374.

643. Glasgow, A. C., M. F. Bruist, and M. I. Simon. 1989. DNA-binding properties of the Hin recombinase. *J. Biol. Chem.* 264:10072–10082.

644. Gmeiner, J., and S. Schlecht. 1979. Molecular organization of the outer membrane of *Salmonella typhimurium*. *Eur. J. Biochem.* 93:609–620.

645. Goitein, R. K., and S. M. Parsons. 1980. Possible regulation of the *Salmonella typhimurium* histidine operon by adenosine triphosphate phosphoribosyltransferase: large metabolic effects. *J. Bacteriol.* 144:337–345.

646. Goldberger, R. F. 1974. Autogenous regulation of gene expression. *Science* 183:810–816.

647. Goldberger, R. F., and J. S. Kovach. 1972. Regulation of histidine biosynthesis in *Salmonella typhimurium*. *Curr. Top. Cell. Regul.* 5:285–308.

648. Goldman, B. S., and J. R. Roth. 1993. Genetic structure and regulation of the *cysG* gene in *Salmonella typhimurium*. *J. Bacteriol.* 175:1457–1466.

649. Goldman, R., W. Kohlbrenner, P. Lartey, and A. Pernet. 1987. Antibacterial agents specifically inhibiting lipopolysaccharide synthesis. *Nature* (London) 329:162–164.

650. Goldman, R. C., and E. M. Devine. 1987. Isolation of *Salmonella typhimurium* strains that utilize exogenous 3-deoxy-D-*manno*-octulosonate for synthesis of lipopolysaocharide. *J. Bacteriol.* 169:5060–5065.

651. Goldman, R. C., C. C. Doran, and J. O. Capobianco. 1988. Analysis of lipopolysaccharide biosynthesis in *Salmonella typhimurium* and *Escherichia coli* by using agents which specifically block incorporation of 3-deoxy-D-*manno*-octulosonate. *J. Bacteriol.* 170:2185–2191.

652. Goldman, R. C., and F. Hunt. 1990. Mechanism of O-antigen distribution in lipopolysaccharide. *J. Bacteriol.* 172:5352–5359.

653. Goldman, R. C., and L. Leive. 1980. Heterogeneity of antigenic-side-chain length in lipopolysaccharide from *Escherichia coli* O111 and *Salmonella typhimurium* LT2. *Eur. J. Biochem.* 107:145–153.

654. Goldrick, D., G.-Q. Yu, S.-Q. Jiang, and J.-S. Hong. 1988. Nucleotide sequence and transcription start point of the phosphoglycerate transporter gene of *Salmonella typhimurium*. *J. Bacteriol.* 170:3421–3426.

655. Gollub, E., H. Zalkin, and D. B. Sprinson. 1966. Correlation of genes and enzymes, and studies on regulation of the aromatic pathway in *Salmonella*. *J. Biol. Chem.* 242:5323–5328.

656. Gollub, E. G., and J. S. Gots. 1959. Purine metabolism in bacteria. VI. Accumulations by mutants lacking adenylosuccinase. *J. Bacteriol.* 78:320–325.

657. Gollub, E. G., K. P. Liu, and D. B. Sprinson. 1973. A regulatory gene of phenylalanine biosynthesis (*pheR*) in *Salmonella typhimurium*. *J. Bacteriol.* 115:121–128.

658. Gollub, E. G., K. P. Liu, and D. B. Sprinson. 1973. *tyrR*, a regulatory gene of tyrosine biosynthesis in *Salmonella typhimurium*. *J. Bacteriol.* 115:1094–1102.

659. Gomez-Eichelmann, M. C. 1979. Deoxyribonucleic acid adenine and cytosine methylation in *Salmonella typhimurium* and *Salmonella typhi*. *J. Bacteriol.* 140:574–579.

660. Gomez-Eichelmann, M. C., and H. K. Torres. 1983. Stability of plasmids R1–19 and R100 in hyper-recombinant *Escherichia coli* strains and in *Salmonella typhimurium* strains. *J. Bacteriol.* 154:1493–1497.

661. Goncharoff, P., and B. P. Nichols. 1988. Evolution of aminobenzoate synthases: nucleotide sequences of *Salmonella typhimurium* and *Klebsiella aerogenes pabB*. *Mol. Biol. Evol.* 5:531–548.

662. Goodell, E. W., and C. F. Higgins. 1987. Uptake of cell wall peptides by *Salmonella typhimurium* and *Escherichia coli*. *J. Bacteriol.* 169:3861–3865.

663. Gopinathan, K. P., and A. Garen. 1970. A leucyl-transfer RNA specified by the amber suppressor gene Su6+. *J. Mol. Biol.* 47:393–401.

664. Gots, J. S. 1971. Regulation of purine and pyrimidine metabolism, p. 225–255. *In* H. J. Vogel (ed.), *Metabolic Regulation*, vol. 5 *in* D. Greenberg (ed.), *Metabolic Pathways*. Academic Press, Inc., New York.

665. Gots, J. S., and C. E. Benson. 1974. Genetic control of bacterial purine phosphoribosyltransferases and an aproach to gene enrichment, p. 33–39. *In* O. Sperling, A. de Vries, and J. B. Wyngaarden (ed.), *Purine Metabolism in Man*. Plenum Publishing Corp, New York.

666. Gots, J. S., C. E. Benson, and S. R. Shumas. 1972. Genetic separation of hypoxanthine and guanine-xanthine phosphoribosyltransferase activities by deletion mutations in *Salmonella typhimurium*. *J. Bacteriol.* 112:910–916.

667. Gots, J. S., F. R. Dalal, and S. R. Shumas. 1969. Genetic separation of the inosinic acid cyclohydrolase-transformylase complex of *Salmonella typhimurium*. *J. Bacteriol.* 99:441–449.

668. Gots, J. S., and E. G. Gollub. 1957. Sequential blockade in adenine biosynthesis by genetic loss of an apparent bifunctional deacylase. *Proc. Natl. Acad. Sci. USA* 43:826–834.

669. Grabau, C., and J. R. Roth. 1992. A *Salmonella typhimurium* cobalamin-deficient mutant blocked in 1-amino-2-propanol synthesis. *J. Bacteriol.* 174:2138–2144.

670. Graeme-Cook, K. A., G. May, E. Bremer, and C. F. Higgins. 1989. Osmotic regulation of porin expression: a role for DNA supercoiling. *Mol. Microbiol.* 3:1287–1294.

671. Graf, L. H., Jr., and R. O. Burns. 1973. The *supX/leu-500* mutations and expression of the leucine operon. *Mol. Gen. Genet.* 126:291–301.

672. Graf, L. H., Jr., R. Kim, and R. O. Burns. 1974. New class of streptomycin-resistant mutants incompatible with *supX* suppressor mutations in *Salmonella typhimurium*. *J. Bacteriol.* 120:1315–1321.

673. Graham, A. C., and B. A. D. Stocker. 1977. Genetics of sensitivity of *Salmonella* species to colicin M and bacteriophages T5, T1 and ES18. *J. Bacteriol.* 130:1214–1223.

674. Greeb, J., J. F. Atkins, and J. C. Loper. 1971. Histidinol dehydrogenase (*hisD*) mutants of *Salmonella typhimurium*. *J. Bacteriol.* 106:421–431.

675. Green, L., and C. G. Miller. 1980. Genetic mapping of the *Salmonella typhimurium pepB* locus. *J. Bacteriol.* 143:1524–1526.

676. Grenier, F. C., E. B. Waygood, and M. H. Saier, Jr. 1986. The bacterial phosphotransferase system: kinetic characterization of the glucose, mannitol, glucitol, and N-acetylglucosamine systems. *J. Cell. Biochem.* 31:97–105.

677. Grieshaber, M. 1978. On the evolution of an oligocephalic enzyme. Glutamine chorismate amidotransferase-free anthranilate phosphoribosyl transferases from mutant strains of *Salmonella typhimurium*. *Z. Naturforsch. Teil C Biosci.* 33:235–244.

678. Grieshaber, M., and R. Bauerle. 1972. Structure and evolution of a bifunctional enzyme of the tryptophan operon. *Nature* (London) *New Biol.* 236:232–235.

679. Grieshaber, M., and R. Bauerle. 1974. Monomeric and dimeric forms of component II of the anthranilate synthetase-anthranilate 5-phosphoribosylpyrophosphate phosphoribosyltransferase complex of *Salmonella typhimurium*. Implications concerning the mode of assembly of the complex. *Biochemistry* 13:232–235.

680. Griffin, H. G. 1990. Nucleotide sequence of the *Salmonella serC* gene. *Nucleic Acids Res.* 18:4260.

681. Grill, H., N. Weigel, B. J. Gaffney, and S. Roseman. 1982. Sugar transport by bacterial phosphotransferase system. Radioactive and electron paramagnetic resonance labelling of the *Salmonella typhimurium* phosphocarrier protein (HPr) at the NH2-terminal methionine. *J. Biol. Chem.* 257:24510–24517.

682. Gritzmacher, C. A., and M. Levinthal. 1976. A mutation amplifying the genes carried by the pi histidine plasmid. *ICN-UCLA Symp. Mol. Cell. Biol.* 5:479–485.

683. **Grodberg, J., and J. J. Dunn.** 1989. Comparison of *Escherichia coli* K-12 outer membrane protease OmpT and *Salmonella typhimurium* E protein. *J. Bacteriol.* **171**:2903–2905.

684. **Groisman, E. A., E. Chiao, C. J. Lipps, and F. Heffron.** 1989. *Salmonella typhimurium phoP* virulence gene is a transcriptional regulator. *Proc. Natl. Acad. Sci. USA* **86**:7077–7081.

685. **Groisman, E. A., and H. Ochman.** 1993. Cognate gene clusters govern invasion of host epithelial cells by *Salmonella typhimurium* and *Shigella flexneri. EMBO J.* **12**:3779–3787.

686. **Groisman, E. A., C. Parra-Lopez, M. Salcedo, C. J. Lipps, and F. Heffron.** 1992. Resistance to host antimicrobial peptides is necessary for *Salmonella* virulence. *Proc. Natl. Acad. Sci. USA* **89**:11939–11943.

687. **Groisman, E. A., M. H. Saier, Jr., and H. Ochman.** 1992. Horizontal transfer of a phosphatase gene as evidence for mosaic structure of the *Salmonella* genome. *EMBO J.* **11**:1309–1316.

688. **Groisman, E. A., M. A. Sturmoski, F. R. Solomon, R. Lin, and H. Ochman.** 1993. Molecular, functional, and evolutionary analysis of sequences specific to *Salmonella. Proc. Natl. Acad. Sci. USA* **90**:1033–1037.

689. **Gross, S. R., R. O. Burns, and H. E. Umbarger.** 1963. The biosynthesis of leucine. II. The enzymic isomerization of α-carboxy-β-hydroxyisocaproate and α-hydroxy-β-carboxyisocaproate. *Biochemistry* **2**:1046–1052.

690. **Gross, T. S., and R. J. Rowbury.** 1969. Methionyl transfer RNA synthetase mutants of *Salmonella typhimurium* which have normal control of the methionine biosynthetic enzymes. *Biochim. Biophys. Acta* **184**:233–236.

691. **Gross, T. S., and R. J. Rowbury.** 1971. Biochemical and physiological properties of methionyl-sRNA synthetase mutants of *Salmonella typhimurium. J. Gen. Microbiol.* **65**:5–21.

692. **Grossman, N., M. A. Schmetz, J. Foulds, E. N. Klima, V. Jiminez, L. L. Leive, and K. K. Joiner.** 1987. Lipopolysaccharide size and distribution determine serum resistance in *Salmonella montevideo. J. Bacteriol.* **169**:856–863.

693. **Grove, T. H., and H. R. Levy.** 1975. Anthranilate synthetase-anthranilate 5-phosphoribosylpyrophosphate phosphoribosyltransferase from *Salmonella typhimurium.* Inactivation of glutamine-dependent anthranilate synthetase by agarose-bound anthranilate. *Biochim. Biophys. Acta* **397**:80–93.

694. **Grove, T. H., and H. R. Levy.** 1976. Anthranilate synthase-anthranilate 5-phosphoribosylpyrophosphate phosphoribosyltransferase from *Salmonella typhimurium.* Purification of the enzyme complex and analysis of multiple forms. *Biochim. Biophys. Acta* **445**:464–474.

695. **Grubmeyer, C. T., and W. R. Gray.** 1986. A cysteine residue (cysteine-116) in the histidinol binding site of histidinol dehydrogenase. *Biochemistry* **25**:4778–4784.

696. **Grubmeyer, C. T., S. Insinga, M. Bhatia, and N. Moazami.** 1989. *Salmonella typhimurium* histidinol dehydrogenase: complete reaction stereochemistry and active site mapping. *Biochemistry* **28**:8174–8180.

697. **Grundy, C. E., and P. D. Ayling.** 1992. Fine structure mapping and complementation studies of the *metD* methionine transport system in *Salmonella typhimurium. Genet. Res.* **60**:1–6.

698. **Guerrero, R., and J. Barbe.** 1982. Expression of *recA*-gene dependent SOS functions in *Salmonella typhimurium. Antonie Van Leeuwenhoek J. Microbiol. Serol.* **48**:159–167.

699. **Guggenheim, S., and M. Flavin.** 1969. Cystathionine α-synthase from *Salmonella.* β Elimination and replacement reactions and inhibition by O-succinylserine. *J. Biol. Chem.* **244**:3722–3727.

700. **Gulig, P. A., and V. A. Chiodo.** 1990. Genetic and DNA sequence analysis of the *Salmonella typhimurium* virulence plasmid gene encoding the 28,000-molecular-weight protein. *Infect. Immun.* **58**:2651–2658.

701. **Gulig, P. A., and R. Curtiss.** 1988. Cloning and transposon insertion mutagenesis of virulence genes of the 100-kilobase plasmid of *Salmonella typhimurium. Infect. Immun.* **56**:3262–3271.

702. **Gulig, P. A., H. Danbara, D. G. Guiney, A. J. Lax, F. Norel, and M. Rhen.** 1993. Molecular analysis of *spv* virulence genes of the *Salmonella* virulence plasmids. *Mol. Microbiol.* **7**:825–830.

703. **Gupta, S. D., K. Gan, M. B. Schmid, and H. C. Wu.** 1993. Characterization of a temperature-sensitive mutant of *Salmonella typhimurium* defective in apolipoprotein N-acyltransferase. *J. Biol. Chem.* **268**:16551–16556.

704. **Gustafsson, C., P. H. R. Lindstrom, T. G. Hagervall, K. B. Esberg, and G. R. Bjork.** 1991. The *trmA* promoter has regulatory features and sequence elements in common with the rRNA P1 promoter family of *Escherichia coli. J. Bacteriol.* **173**:1757–1764.

705. **Guterman, S. K., and A. Wright.** 1974. Effect of mutator mutation of *Salmonella typhimurium* on P22 and R factor genes. *J. Bacteriol.* **119**:638–639.

706. **Guterman, S. K., A. Wright, and D. H. Boyd.** 1975. Genes affecting coliphage BF23 and E colicin sensitivity in *Salmonella typhimurium. J. Bacteriol.* **124**:1351–1358.

707. **Gutierrez, J. A., and L. N. Csonka.** 1995. Isolation and characterization of adenylate kinase (*adk*) mutations in *Salmonella typhimurium* which block the ability of glycine betaine to function as an osmoprotectant. *J. Bacteriol.* **177**:390–400.

708. **Gutterson, N. I., and D. E. Koshland, Jr.** 1983. Replacement and amplification of bacterial genes with sequences altered *in vitro. Proc. Natl. Acad. Sci. USA* **80**:4894–4898.

709. **Haber, L. T., P. P. Pang, D. I. Sobell, J. A. Mankovich, and G. C. Walker.** 1988. Nucleotide sequence of the *Salmonella typhimurium mutS* gene required for mismatch repair: homology of MutS and HexA of *Streptococcus pneumoniae. J. Bacteriol.* **170**:197–202.

710. **Haber, L. T., and G. C. Walker.** 1991. Altering the conserved nucleotide binding motif in the *Salmonella typhimurium* MutS mismatch repair protein affects both its ATPase and mismatch binding activities. *EMBO J.* **10**:2707–2715.

711. **Hackett, J., P. Wyk, P. Reeves, and V. Mathan.** 1987. Mediation of serum resistance in *Salmonella typhimurium* by an 11-kilodalton polypeptide encoded by the cryptic plasmid. *J. Infect. Dis.* **155**:540–549.

712. **Haeffner-Gormley, L., Z. D. Chen, H. Zalkin, and R. F. Colman.** 1991. Evaluation of cysteine 283 and glutamic acid 284 in the coenzyme binding site of *Salmonella typhimurium* glutamate dehydrogenase by site-directed mutagenesis and reaction with the nucleotide analogue 2-((4-bromo-2,3-dioxobutyl)thio)-1,N^6-ethenoadenosine 2′,5′-bisphosphate. *J. Biol. Chem.* **266**:5388–5394.

713. **Hafner, L. M., and D. G. MacPhee.** 1991. Precise excision of Tn*10* in *Salmonella typhimurium:* effects of mutations in the *polA, dam, mutH,* and *mutB* genes and of methionine or ethionine in the plating medium. *Mutat. Res.* **263**:179–184.

714. **Hagen, D. C., S. L. Gerson, and B. Magasanik.** 1975. Isolation of super-repressor mutants in the histidine utilization system of *Salmonella typhimurium. J. Bacteriol.* **121**:583–593.

715. **Hagen, D. C., P. J. Lipton, and B. Magasanik.** 1974. Isolation of a *trans*-dominant histidase-negative mutant of *Salmonella typhimurium. J. Bacteriol.* **120**:906–916.

716. **Hagen, D. C., and B. Magasanik.** 1973. Isolation of the self-regulated repressor protein of the *hut* operons of *Salmonella typhimurium. Proc. Natl. Acad. Sci. USA* **70**:808–812.

717. **Hagen, D. C., and B. Magasanik.** 1976. Deoxyribonucleic acid-binding studies on the *hut* repressor and mutant forms of the *hut* repressor on *Salmonella typhimurium. J. Bacteriol.* **127**:837–847.

718. **Hahn, D. R., and S. R. Maloy.** 1986. Regulation of the *put* operon in *Salmonella typhimurium:* characterization of promoter and operator mutations. *Genetics* **114**:687–703.

719. **Hahn, D. R., R. S. Myers, C. R. Kent, and S. R. Maloy.** 1988. Regulation of proline utilization in *Salmonella typhimurium:* molecular characterization of the *put* operon, and DNA sequence of the *put* control region. *Mol. Gen. Genet.* **213**:125–133.

720. **Hakura, A., K. Morimoto, T. Sofuni, and T. Nohmi.** 1991. Cloning and characterization of the *Salmonella typhimurium ada* gene, which encodes O^6-methylguanine-DNA methyltransferase. *J. Bacteriol.* **173**:3663–3672.

721. **Hall, B. G.** 1993. The role of single-mutant intermediates in the generation of *trpAB* double revertants during prolonged selection. *J. Bacteriol.* **175**:6411–6414.

722. **Hall, G. W., D. O. Silva, T. Melton, and W. J. Dobrogosz.** 1980. Catabolite repression and inducer exclusion in *Salmonella typhimurium* mutants possessing deletions in the *pts* locus, abstr. K94, p. 142. *Abstr. Annu. Meet. Am. Soc. Microbiol., 1980.*

723. **Hallenbeck, P. C., M. A. Clark, and E. L. Barrett.** 1989. Characterization of anaerobic sulfite reduction by *Salmonella typhimurium* and purification of the anaerobically induced sulfite reductase. *J. Bacteriol.* **171**:3008–3015.

724. **Hamilton, S., and C. G. Miller.** 1992. Cloning and nucleotide sequence of the *Salmonella typhimurium dcp* gene encoding dipeptidyl carboxypeptidase. *J. Bacteriol.* **174**:1626–1630.

725. **Hammerling, G., V. Lehmann, and O. Luderitz.** 1973. Structural studies on the heptose region of *Salmonella* lipopolysaccharides. *Eur. J. Biochem.* **38**:453–458.

726. **Hammond, S. M., A. Claesson, A. M. Jansson, L.-G. Larsson, B. G. Pring, C. M. Town, and B. Ekstrom.** 1987. A new class of synthetic antibacterials acting on lipopolysaccharide biosynthesis. *Nature* (London) **327**:730–732.

727. **Hampton, M. J., R. A. Floyd, J. B. Clark, and J. H. Lancaster.** 1980. Studies of the fatty acid composition and membrane microviscosity in *Salmonella typhimurium* TA98. *Chem. Phys. Lipids* **27**:177–183.

728. **Han, B. D., W. G. Nolan, H. P. Hopkins, R. T. Jones, J. L. Ingraham, and A. T. Abdelal.** 1990. Effect of growth temperature on folding of carbamoylphosphate synthetases of *Salmonella typhimurium* and a cold-sensitive derivative. *J. Bacteriol.* **172**:5089–5096.

729. **Hanau, R., R. K. Koduri, N. Ho, and J. E. Brenchley.** 1983. Nucleotide sequence of the control regions for the *glnA* and *glnL* genes of *Salmonella typhimurium. J. Bacteriol.* **155**:82–89.

730. **Hannavy, K., G. C. Barr, C. J. Dorman, J. Adamson, L. R. Mazengera, M. P. Gallagher, J. S. Evans, B. A. Levine, I. P. Trayer, and C. F. Higgins.** 1990. TonB protein of *Salmonella typhimurium.* A model for signal transduction between membranes. *J. Mol. Biol.* **216**:897–910.

731. **Hansen, F. G., T. Atlung, R. E. Braun, A. Wright, P. Hughes, and M. Kohiyama.** 1991. Initiator (DnaA) protein concentration as a function of growth rate in *Escherichia coli* and *Salmonella typhimurium. J. Bacteriol.* **173**:5194–5199.

732. **Hansen-Hagge, T., V. Lehmann, U. Seydel, B. Linder, and U. Zahringer.** 1985. Isolation and structural analysis of two lipid A precursors from a KDO deficient mutant of *Salmonella typhimurium* differing in their hexadecanoic acid content. *Arch. Microbiol.* **141**:353–358.

733. **Harlow, K. W., and R. L. Switzer.** 1990. Chemical modification of *Salmonella typhimurium* phosphoribosylpyrophosphate synthetase with 5′-(p-fluorosulfonylbenzoyl)adenosine. Identification of an active site histidine. *J. Biol. Chem.* **265**:5487–5493.

734. **Harlow, K. W., and R. L. Switzer.** 1990. Sulfhydryl chemistry of *Salmonella typhimurium* phosphoribosylpyrophosphate synthetase: identification of two classes of cysteinyl residues. *Arch. Biochem. Biophys.* **276**:466–472.

735. **Harms, E., J.-H. Hsu, C. S. Subrahmanyam, and H. E. Umbarger.** 1985. Comparison of the regulatory regions of *ilvGEDA* operons from several enteric organisms. *J. Bacteriol.* **164**:207–216.

736. Hartman, P. E., Z. Hartman, and D. Serman. 1970. Complementation mapping by abortive transduction of histidine-requiring *Salmonella* mutants. *J. Gen. Microbiol.* 22:354–368.

737. Hartman, P. E., Z. Hartman, R. C. Stahl, and B. N. Ames. 1971. Classification and mapping of spontaneous and induced mutations in the histidine operon of *Salmonella. Adv. Genet.* 17:1–34.

738. Hartman, P. E., J. C. Loper, and D. Serman. 1960. Fine structure mapping by complete transduction between histidine-requiring *Salmonella* mutants. *J. Gen. Microbiol.* 22:323–353.

739. Hatfield, G. W., and R. O. Burns. 1970. Specific binding of leucyl transfer RNA to an immature form of L-threonine deaminase: its implications in repression. *Proc. Natl. Acad. Sci. USA* 66:1027–1035.

740. Hattman, S. 1971. Variation of 6-methylaminopurine content in bacteriophage P22 deoxyribonucleic acid as a function of host specificity. *J. Virol.* 7:690–691.

741. Hattman, S., S. Schlagman, L. Goldstein, and M. Frohlich. 1976. *Salmonella typhimurium* SA host specificity system is based on deoxyribonucleic acid-adenine methylation. *J. Bacteriol.* 127:211–217.

742. Haughn, G. W., S. R. Wessler, R. M. Gemmill, and J. M. Calvo. 1986. High A+T content conserved in DNA sequences upstream of *leuABCD* in *Escherichia coli* and *Salmonella typhimurium. J. Bacteriol.* 166:1113–1117.

743. He, X.-S., M. Rivkina, B. A. D. Stocker, and W. S. Robinson. 1994. Hypervariable region IV of *Salmonella* gene *fliC*[d] encodes a dominant surface epitope and a stabilizing factor for functional flagella. *J. Bacteriol.* 176:2406–2414.

744. Heasley, F. A. 1981. Reducing terminus of O-haptene accumulated in a *Salmonella montevideo galE* mutant. *J. Bacteriol.* 148:624–627.

745. Heffernan, E. J., J. Harwood, J. Fierer, and D. Guiney. 1992. The *Salmonella typhimurium* virulence plasmid complement resistance gene *rck* is homologous to a family of virulence-related outer membrane protein genes, including *pagC* and *ail. J. Bacteriol.* 174:84–91.

746. Heffernan, E. J., S. Reed, J. Hackett, J. Fierer, C. Roudier, and D. Guiney. 1992. Mechanism of resistance to complement-mediated killing of bacteria encoded by the *Salmonella typhimurium* virulence plasmid gene *rck. J. Clin. Invest.* 90:953–964.

747. Heichman, K. A., and R. C. Johnson. 1990. The Hin invertasome: protein-mediated joining of distant recombination sites at the enhancer. *Science* 249:511–517.

748. Heiman, C., and C. G. Miller. 1978. *Salmonella typhimurium* mutants lacking protease II. *J. Bacteriol.* 135:588–594.

749. Heiman, C., and C. G. Miller. 1978. Acylamino acid esterase mutants of *Salmonella typhimurium. Mol. Gen. Genet.* 164:57–62.

750. Helander, I. M., L. Hirvas, J. Tuominen, and M. Vaara. 1992. Preferential synthesis of heptaacyl lipopolysaccharide by the *ssc* permeability mutant of *Salmonella typhimurium. Eur. J. Biochem.* 204:1101–1106.

751. Helander, I. M., M. Vaara, S. Sukupolvi, M. Rhen, S. Saarela, U. Zahringer, and P. H. Makela. 1989. *rfaP* mutants of *Salmonella typhimurium. Eur. J. Biochem.* 185:541–546.

752. Hellerqvist, C. G. 1971. Structural studies of the common-core polysaccharide of the cell-wall lipopolysaccharide from *Salmonella typhimurium. Carbohydr. Res.* 16:39–44.

753. Helmann, J. D., and M. J. Chamberlin. 1987. DNA sequence analysis suggests that expression of flagellar and chemotaxis genes in *Escherichia coli* and *Salmonella typhimurium* is controlled by an alternative σ factor. *Proc. Natl. Acad. Sci. USA* 84:6422–6424.

754. Henderson, G. B., S. Shaltiel, and E. E. Snell. 1974. ω-Aminoalkylagaroses in the purification of the L-histidinolphosphate aminotransferase. *Biochemistry* 13:4335–4338.

755. Henderson, G. B., and E. E. Snell. 1973. Crystalline L-histidinol phosphate aminotransferase from *Salmonella typhimurium. J. Biol. Chem.* 248:1906–1911.

756. Hengge, R., and W. Boos. 1985. Defective secretion of maltose- and ribose-binding proteins caused by a truncated periplasmic protein in *Escherichia coli. J. Bacteriol.* 162:972–978.

757. Hengge, R., T. J. Larson, and W. Boos. 1983. sn-Glycerol-3-phosphate transport in *Salmonella typhimurium. J. Bacteriol.* 155:186–195.

758. Henikoff, S., G. W. Haughn, J. M. Calvo, and J. C. Wallace. 1988. A large family of bacterial activator proteins. *Proc. Natl. Acad. Sci. USA* 85:6602–6606.

759. Herrera, G., A. Urios, V. Aleixandre, and M. Blanco. 1988. UV light-induced mutability in *Salmonella* strains containing the *umuDC* or the *mucAB* operon: evidence for a *umuC* function. *Mutat. Res.* 198:9–13.

760. Hertzgery, K. M., R. Gemmill, J. Jones, and J. M. Calvo. 1979. Cloning of an *Eco*RI generated fragment of the leucine operon of *Salmonella typhimurium. Gene* 8:135–152.

761. Hess, J. F., K. Oosawa, N. Kaplan, and M. I. Simon. 1988. Phosphorylation of three proteins in the signaling pathway of bacterial chemotaxis. *Cell* 53:78–87.

762. Higgins, C. F., and G. F.-L. Ames. 1981. Two periplasmic transport proteins which interact with a common membrane receptor show extensive homology: complete nucleotide sequences. *Proc. Natl. Acad. Sci. USA* 78:6038–6042.

763. Higgins, C. F., and G. F.-L. Ames. 1982. Regulatory regions of two transport operons under nitrogen control: nucleotide sequences. *Proc. Natl. Acad. Sci. USA* 79:1083–1087.

764. Higgins, C. F., C. J. Dorman, D. A. Stirling, L. Waddell, I. R. Booth, G. May, and E. Bremer. 1988. A physiological role for DNA supercoiling in the osmotic regulation of gene expression in *S. typhimurium* and *E. coli. Cell* 52:569–584.

765. Higgins, C. F., and M. M. Gibson. 1984. Peptide transport in bacteria. *Methods Enzymol.* 125:365–377.

766. Higgins, C. F., P. D. Haag, K. Nikaido, F. Ardeshir, G. Garcia, and G. F.-L. Ames. 1982. Complete nucleotide sequence and identification of membrane components of the histidine transport operon of *S. typhimurium. Nature* (London) 298:723–727.

767. Higgins, C. F., and M. M. Hardie. 1983. Periplasmic protein associated with the oligopeptide permeases of *Salmonella typhimurium* and *Escherichia coli. J. Bacteriol.* 155:1434–1438.

768. Higgins, C. F., M. M. Hardie, D. Jamieson, and L. M. Powell. 1983. Genetic map of the *opp* (oligopeptide permease) locus of *Salmonella typhimurium. J. Bacteriol.* 153:830–836.

769. Higgins, C. F., I. D. Hiles, K. Whalley, and D. J. Jamieson. 1985. Nucleotide binding by membrane components of bacterial periplasmic binding protein-dependent transport systems. *EMBO J.* 4:1033–1039.

770. Higgins, C. F., J. C. D. Hinton, C. S. J. Hulton, T. Owen-Hughes, and A. Serirafi. 1990. Protein H1: a role for chromatin structure in the regulation of bacterial gene expression and virulence? *Mol. Microbiol.* 4:2007–2012.

771. Higgins, C. F., R. S. McLaren, and S. F. Newbury. 1988. Repetitive extragenic palindromic sequences, mRNA stability and gene expression: evolution by gene conversion? A review. *Gene* 72:3–14.

772. Higgins, C. F., L. Sutherland, J. Cairney, and I. R. Booth. 1987. The osmotically regulated *proU* locus of *Salmonella typhimurium* encodes a periplasmic betaine-binding protein. *J. Gen. Microbiol.* 133:305–310.

773. Higgins, N. P., and D. Hillyard. 1988. Primary structure and mapping of the *hupA* gene of *Salmonella typhimurium. J. Bacteriol.* 170:5751–5758.

774. Hildenbrand, K., L. Brand, and S. Roseman. 1982. Sugar transport by the bacterial phosphotransferase system. Nanosecond fluorescence studies of the phosphocarrier protein (HPr) labelled at the NH_2-terminal methionine. *J. Biol. Chem.* 257:14518–14525.

775. Hiles, I. D., M. P. Gallagher, D. J. Jamieson, and C. F. Higgins. 1987. Molecular characterization of the oligopeptide permease of *Salmonella typhimurium. J. Mol. Biol.* 195:125–142.

776. Hiles, I. D., and C. F. Higgins. 1986. Peptide uptake by *Salmonella typhimurium.* The periplasmic oligopeptide-binding protein. *Eur. J. Biochem.* 158:561–567.

777. Hiles, I. D., L. M. Powell, and C. F. Higgins. 1987. Peptide transport in *Salmonella typhimurium:* molecular cloning and characterization of the oligopeptide permease genes. *Mol. Gen. Genet.* 206:101–109.

778. Hill-Chappell, J. M., M. P. Spector, and J. W. Foster. 1986. The pyridine nucleotide cycle of *Salmonella typhimurium:* genetic characterization of the *pncXA* operon. *Mol. Gen. Genet.* 205:507–514.

779. Hillyard, D. R., M. Edlund, K. T. Hughes, M. Marsh, and N. P. Higgins. 1990. Subunit-specific phenotypes of *Salmonella typhimurium* HU mutants. *J. Bacteriol.* 172:5402–5407.

780. Hinton, J. C., D. S. Santos, A. Seirafi, C. S. Hulton, G. D. Pavitt, and C. F. Higgins. 1992. Expression and mutational analysis of the nucleoid-associated protein H-NS of *Salmonella typhimurium. Mol. Microbiol.* 6:2327–2337.

781. Hirano, T., S. Yamaguchi, K. Oosawa, and S.-I. Aizawa. 1994. Roles of FliK and FlhB in determination of flagellar hook length in *Salmonella typhimurium. J. Bacteriol.* 176:5439–5449.

782. Hirvas, L., J. Coleman, P. Koski, and M. Vaara. 1990. Bacterial "histone-like protein I" (HLP-I) is an outer membrane constituent? *FEBS Lett.* 262:123–126.

783. Hirvas, L., P. Koski, and M. Vaara. 1990. Primary structure and expression of the Ssc-protein of *Salmonella typhimurium. Biochem. Biophys. Res. Commun.* 173:53–59.

784. Hirvas, L., P. Koski, and M. Vaara. 1991. Identification and sequence analysis of the gene mutated in the conditionally lethal outer membrane permeability mutant SS-C of *Salmonella typhimurium. EMBO J.* 10:1017–1023.

785. Hirvas, L., and M. Vaara. 1992. Effect of Ssc protein mutations on the outer membrane permeability barrier function in *Salmonella typhimurium:* a study using *ssc* mutant alleles made by site-directed mutagenesis. *FEMS Microbiol. Lett.* 69:289–294.

786. Hitchcock, P. J., and T. M. Brown. 1983. Morphological heterogeneity among *Salmonella* lipopolysaccharide chemotypes in silver-stained polyacrylamide gels. *J. Bacteriol.* 154:269–277.

787. Hmiel, S. P., M. D. Snavely, J. B. Florer, M. E. Maguire, and C. G. Miller. 1989. Magnesium transport in *Salmonella typhimurium:* genetic characterization and cloning of three magnesium transport loci. *J. Bacteriol.* 171:4742–4751.

788. Hmiel, S. P., M. D. Snavely, C. G. Miller, and M. E. Maguire. 1986. Magnesium transport in *Salmonella typhimurium:* characterization of magnesium influx and cloning of a transport gene. *J. Bacteriol.* 168:1444–1450.

789. Ho, C., Y.-H. Giza, S. Takahashi, K. E. Ugen, P. F. Cottam, and S. R. Dowd. 1981. A proton nuclear magnetic resonance investigation of histidine-binding protein J of *Salmonella typhimurium:* a model for transport of L-histidine across cytoplasmic membrane. *J. Supramol. Struct.* 13:131–145.

790. Hobson, A. C. 1974. A norleucine-resistant mutant of *Salmonella typhimurium* with a possible defect in valine uptake or regulation. *J. Gen. Microbiol.* 82:425–429.

791. Hobson, A. C. 1974. The regulation of methionine and S-adenosylmethionine biosynthesis and utilization in mutants of *Salmonella typhimurium* with defects in S-adenosylmethionine synthetase. *Mol. Gen. Genet.* 131:263–273.

792. Hobson, A. C. 1976. The synthesis of S-adenosylmethionine by mutants with defects in S-adenosylmethionine synthetase. *Mol. Gen. Genet.* 144:87–95.

793. Hobson, A. C., and D. A. Smith. 1973. S-Adenosylmethionine synthetase in methionine regulatory mutants of *Salmonella typhimurium. Mol. Gen. Genet.* **126:**7–18.

794. Hoeksma, W. D., and D. E. Schoenhard. 1971. Characterization of a thermolabile sulfite reductase from *Salmonella pullorum. J. Bacteriol.* **108:**154–158.

795. Hof, H., J. Stroder, J. P. Buisson, and R. Royer. 1986. Effect of different nitroheterocyclic compounds on aerobic, microaerophilic, and anaerobic bacteria. *Antimicrob. Agents Chemother.* **30:**679–683.

796. Hoffee, P., and B. C. Robertson. 1969. 2-Deoxyribose gene-enzyme complex in *Salmonella typhimurium:* regulation of phosphodeoxyribomutase. *J. Bacteriol.* **97:**1386–1396.

797. Hoffee, P., R. Snyder, C. Sushak, and P. Jargiello. 1974. Deoxyribose-5-P aldolase: subunit structure and composition of active site lysine region. *Arch. Biochem. Biophys.* **164:**736–742.

798. Hoffee, P. A. 1968. 2-Deoxyribose gene-enzyme complex in *Salmonella typhimurium.* I. Isolation and enzymatic characterization of 2-deoxyribose-negative mutants. *J. Bacteriol.* **95:**449–457.

799. Hoffee, P. A. 1968. 2-Deoxyribose-5-phosphate aldolase of *Salmonella typhimurium:* purification and propoerties. *Arch. Biochem. Biophys.* **126:**795–802.

800. Hoffman, G. R., M. J. Walkowicz, J. M. Mason, and J. F. Atkins. 1983. Genetic instability associated with the *aroC321* allele in *Salmonella typhimurium* involves genetic duplication. *Mol. Gen. Genet.* **190:**183–188.

801. Hoffman, J., B. Lindberg, M. Glowacka, M. Derylo, and Z. Lorkiewicz. 1980. Structural studies of the lipopolysaccharide from *Salmonella typhimurium* 902 (ColIb drd2). *Eur. J. Biochem.* **105:**103–107.

802. Hofler, J. G., and R. O. Burns. 1978. Threonine deaminase from *Salmonella typhimurium. J. Biol. Chem.* **253:**1245–1251.

803. Hofler, J. G., D. J. Decedue, G. H. Luginbuhl, J. A. Reynolds, and R. O. Burns. 1975. The subunit structure of α-acetohydroxyacid isomeroreductase from *Salmonella typhimurium. J. Biol. Chem.* **250:**877–882.

804. Hogarth, B. C., and C. F. Higgins. 1983. Genetic organization of the oligopeptide permease (*opp*) locus of *Salmonella typhimurium* and *Escherichia coli. J. Bacteriol.* **153:**1548–1551.

805. Hogg, R. W. 1981. The amino acid sequence of the histidine binding protein of *Salmonella typhimurium. J. Biol. Chem.* **256:**1935–1939.

806. Hoiseth, S. K. 1982. Aromatic-deficient mutants as live *Salmonella* vaccines. Ph.D. thesis. Stanford University, Stanford, Calif.

807. Hoiseth, S. K., and B. A. D. Stocker. 1981. Aromatic-dependent *Salmonella typhimurium* are non-virulent and effective as vaccines. *Nature* (London) **291:**238–239.

808. Hoiseth, S. K., and B. A. D. Stocker. 1985. Genes *aroA* and *serC* of *Salmonella typhimurium* constitute an operon. *J. Bacteriol.* **163:**355–361.

809. Holland, M. M., T. K. Leib, and J. A. Gerlt. 1988. Isolation and characterization of a small catalytic domain released from adenylate cyclase from *Escherichia coli* by digestion with trypsin. *J. Biol. Chem.* **263:**14661–14668.

810. Holley, E. A., and J. W. Foster. 1982. Bacteriophage P22 as a vector for Mu mutagenesis in *Salmonella typhimurium:* isolation of *nad-lac* and *pnc-lac* gene fusions. *J. Bacteriol.* **152:**959–962.

811. Holley, E. A., M. P. Spector, and J. W. Foster. 1985. Regulation of NAD biosynthesis in *Salmonella typhimurium:* expression of *nad-lac* gene fusions and identification of a *nad* regulatory focus. *J. Gen. Microbiol.* **131:**2759–2770.

812. Homma, M., D. J. DeRosier, and R. M. Macnab. 1990. Flagellar hook and hook-associated proteins of *Salmonella typhimurium* and their relationship to other axial components of the flagellum. *J. Mol. Biol.* **213:**819–832.

813. Homma, M., H. Fujita, S. Yamaguchi, and T. Iino. 1984. Excretion of unassembled flagellin by *Salmonella typhimurium* mutants deficient in hook-associated proteins. *J. Bacteriol.* **159:**1056–1059.

814. Homma, M., H. Fujita, S. Yamaguchi, and T. Iino. 1987. Regions of *Salmonella typhimurium* flagellin essential for its polymerization and excretion. *J. Bacteriol.* **169:**291–296.

815. Homma, M., and T. Iino. 1985. Locations of hook-associated proteins in flagellar structures of *Salmonella typhimurium. J. Bacteriol.* **162:**183–189.

816. Homma, M., and T. Iino. 1985. Excretion of unassembled hook-associated proteins by *Salmonella typhimurium. J. Bacteriol.* **164:**1370–1372.

817. Homma, M., T. Iino, K. Kutsukake, and S. Yamaguchi. 1986. *In vitro* reconstitution of flagellar filaments onto hooks of filamentless mutants of *Salmonella typhimurium* by addition of hook-associated proteins. *Proc. Natl. Acad. Sci. USA* **83:**6169–6173.

818. Homma, M., T. Iino, and R. M. Macnab. 1988. Identification and characterization of the products of six region III flagellar genes (*flaAII.3* through *flaQII*) of *Salmonella typhimurium. J. Bacteriol.* **170:**2221–2228.

819. Homma, M., Y. Komeda, T. Iino, and R. M. Macnab. 1987. The *flaFIX* gene product of *Salmonella typhimurium* is a flagellar basal body component with a signal peptide for export. *J. Bacteriol.* **169:**1493–1498.

820. Homma, M., K. Kutsukake, M. Hasebe, T. Iino, and R. M. Macnab. 1990. FlgB, FlgC, FlgF and FlgG. A family of structurally related proteins in the flagellar basal body of *Salmonella typhimurium. J. Mol. Biol.* **211:**465–477.

821. Homma, M., K. Kutsukake, and T. Iino. 1985. Structural genes for flagellar hook-associated proteins in *Salmonella typhimurium. J. Bacteriol.* **163:**464–471.

822. Homma, M., K. Kutsukake, T. Iino, and S. Yamaguchi. 1984. Hook-associated proteins essential for flagellar filament formation in *Salmonella typhimurium. J. Bacteriol.* **157:**100–108.

823. Homma, M., K. Ohnishi, T. Iino, and R. M. Macnab. 1987. Identification of flagellar hook and basal body gene products (FlaFV, FlaFVI, FlaFVII, and FlaFVIII) in *Salmonella typhimurium. J. Bacteriol.* **169:**3617–3624.

824. Hong, J.-S., and B. N. Ames. 1971. Localized mutagenesis of any specific small region of the bacterial chromosome. *Proc. Natl. Acad. Sci. USA* **68:**3158–3162.

825. Hong, J.-S., G. R. Smith, and B. N. Ames. 1971. Adenosine 3'5'-cyclic monophosphate concentration in the bacterial host regulates the viral decision between lysogeny and lysis. *Proc. Natl. Acad. Sci. USA* **68:**2258–2262.

826. Hongo, E., M. Morimyo, K. Mita, I. Machida, H. Hama-Inaba, H. Tsuji, S. Ichimura, and Y. Noda. 1994. The methyl viologen-resistance-encoding gene *smvA* of *Salmonella typhimurium. Gene* **148:**173–174.

827. Hoppe, I., H. M. Johnston, D. Biek, and J. R. Roth. 1979. A refined map of the *hisG* gene of *Salmonella typhimurium. Genetics* **92:**17–26.

828. Hoppe, I., and J. Roth. 1974. Specialized transducing phages derived from *Salmonella* phage P22. *Genetics* **76:**633–654.

829. Horiguchi, T., S. Yamaguchi, K. Yao, T. Tairo, and T. Iino. 1975. Genetic analysis of H1, the structural gene for phase-1 flagellin in *Salmonella. J. Gen. Microbiol.* **91:**139–149.

830. Horowitz, H., J. Van Arsdell, and T. Platt. 1983. Nucleotide sequence of the *trpD* and *trpC* genes of *Salmonella typhimurium. J. Mol. Biol.* **169:**775–797.

831. Horwitz, A. H., L. Heffernan, C. Morandi, J.-H. Lee, J. Timko, and G. Wilcox. 1981. DNA sequence of the *araBAD-araC* controlling region in *Salmonella typhimurium* LT2. *Gene* **14:**309–319.

832. Hotani, H. 1976. Light microscope study of mixed helices in reconstituted *Salmonella* flagella. *J. Mol. Biol.* **106:**151–166.

833. Houlberg, U., and K. J. Jensen. 1983. Role of hypoxanthine and guanine in regulation of *Salmonella typhimurium pur* gene expression. *J. Bacteriol.* **153:**837–845.

834. Houng, H. S., D. J. Kopecko, and L. S. Baron. 1990. Molecular cloning and physical and functional characterization of the *Salmonella typhimurium* and *Salmonella typhi* galactose utilization operons. *J. Bacteriol.* **172:**4392–4398.

835. Housley, P. R., A. D. Leavitt, and H. J. Whitfield. 1981. Genetic analysis of a temperature-sensitive *Salmonella typhimurium rho* mutant with an altered rho-associated polycytidylate-dependent adenosine triphosphatase activity. *J. Bacteriol.* **147:**13–24.

836. Housley, P. R., and H. J. Whitfield. 1982. Transcription termination factor ρ from wildtype and ρ-111 strains of *Salmonella typhimurium. J. Biol. Chem.* **257:**2569–2577.

837. Houston, L. L. 1973. Specialized subregions of the bifunctional *hisB* gene of *Salmonella typhimurium. J. Bacteriol.* **113:**82–87.

838. Houston, L. L. 1973. Purification and properties of a mutant bifunctional enzyme from the *hisB* gene of *Salmonella typhimurium. J. Biol. Chem.* **248:**4144–4149.

839. Houston, L. L. 1973. Evidence for proteolytic degradation of histidinol phosphate phosphatase specified by nonsense mutants of the *hisB* gene of *Salmonella typhimurium. J. Bacteriol.* **116:**88–97.

840. Houston, L. L., and M. E. Graham. 1974. Divalent metal ion effects on a mutant histidinol phosphate phosphatase from *Salmonella typhimurium. Arch. Biochem. Biophys.* **162:**513–522.

841. Houston, L. L., and R. H. Millay, Jr. 1974. Effect of sulfhydryl reagents on the activity of histidinolphosphatase from *Salmonella typhimurium* and baker's yeast. *Biochim. Biophys. Acta* **370:**216–226.

842. Howlett, B. J., and J. Bar-Tana. 1980. Polyprenyl *p*-hydroxybenzoate carboxylase in flagellation of *Salmonella typhimurium. J. Bacteriol.* **143:**644–651.

843. Hoyer, L. L., A. C. Hamilton, S. M. Steenbergen, and E. R. Vimr. 1992. Cloning, sequencing and distribution of the *Salmonella typhimurium* LT2 sialidase gene, *nanH*, provides evidence for interspecies gene transfer. *Mol. Microbiol.* **6:**873–884.

844. Hoyer, L. L., P. Roggentin, R. Schauer, and E. R. Vimr. 1991. Purification and properties of cloned *Salmonella typhimurium* LT2 sialidase with virus-typical kinetic preference for sialyl α2-3 linkages. *J. Biochem.* (Tokyo) **110:**462–467.

845. Hryniewicz, M., M. Bagdasarian, and M. Bagdasarian. 1979. Integration of F factor and cryptic LT-2 plasmid into a specific site of the *Salmonella typhimurium* chromosome. *Acta Biochim. Pol.* **26:**73–82.

846. Hryniewicz, M., A. Sirko, A. Palucha, A. Bock, and D. Hulanicka. 1990. Sulfate and thiosulfate transport in *Escherichia coli* K-12: identification of a gene encoding a novel protein involved in thiosulfate binding. *J. Bacteriol.* **172:**3358–3366.

847. Hryniewicz, M. M., and N. M. Kredich. 1991. The *cysP* promoter of *Salmonella typhimurium:* characterization of two binding sites for CysB protein, studies of in vivo transcription initiation, and demonstration of the anti-inducer effects of thiosulfate. *J. Bacteriol.* **173:**5876–5886.

848. Hryniewicz, M. M., and N. M. Kredich. 1994. Stoichiometry of binding of CysB to the *cysJIH, cysK,* and *cysP* promoter regions of *Salmonella typhimurium. J. Bacteriol.* **176:**3673–3682.

849. Hu, C.-Y., and D. B. Sprinson. 1977. Properties of tyrosine-inhibitable 3-deoxy-D-arabinoheptulosonic acid-7-phosphate synthase from *Salmonella. J. Bacteriol.* **129:**177–183.

850. Huang, C. J., and E. L. Barrett. 1990. Identification and cloning of genes involved in anaerobic sulfite reduction by *Salmonella typhimurium. J. Bacteriol.* **172:**4100–4102.

851. Huang, C. J., and E. L. Barrett. 1991. Sequence analysis and expression of the *Salmonella typhimurium asr* operon encoding production of hydrogen sulfide from sulfite. *J. Bacteriol.* **173:**1544–1553.

852. Hudson, H. P., A. A. Lindberg, and B. A. D. Stocker. 1978. Lipopolysaccharide core defects in *Salmonella typhimurium* mutants which are resistant to Felix O phage but retain smooth character. *J. Gen. Microbiol.* **109**:97–112.

853. Huffman, G. A., and R. H. Rownd. 1984. Transition of deletion mutants of the composite resistance plasmid NR1 in *Escherichia coli* and *Salmonella typhimurium*. *J. Bacteriol.* **159**:488–498.

854. Hughes, D. 1986. The isolation and mapping of EF-Tu mutations in *Salmonella typhimurium*. *Mol. Gen. Genet.* **202**:108–111.

855. Hughes, D. 1987. Mutant forms of *tufA* and *tufB* independently suppress nonsense mutations. *J. Mol. Biol.* **197**:611–615.

856. Hughes, D. 1990. Both genes for EF-Tu in *Salmonella typhimurium* are individually dispensable for growth. *J. Mol. Biol.* **215**:41–51.

857. Hughes, D. 1991. Error-prone EF-Tu reduces *in vivo* enzyme activity and cellular growth rate. *Mol. Microbiol.* **5**:623–630.

858. Hughes, D., J. F. Atkins, and S. Thompson. 1987. Mutants of elongation factor Tu promote ribosomal frameshifting and nonsense readthrough. *EMBO J.* **6**:4235–4239.

859. Hughes, D., and R. H. Buckingham. 1991. The nucleotide sequence of *rpsL* and its flanking regions in *Salmonella typhimurium*. *Gene* **104**:123–124.

860. Hughes, K. T., B. T. Cookson, D. Ladika, B. M. Olivera, and J. R. Roth. 1983. 6-Aminonicotinamide-resistant mutants of *Salmonella typhimurium*. *J. Bacteriol.* **154**:1126–1136.

861. Hughes, K. T., A. Dessen, J. P. Gray, and C. Grubmeyer. 1993. The *Salmonella typhimurium nadC* gene: sequence determination by use of Mud-P22 and purification of quinolinate phosphoribosyltransferase. *J. Bacteriol.* **175**:479–486.

862. Hughes, K. T., D. Ladika, J. R. Roth, and B. M. Olivera. 1983. An indispensable gene for NAD biosynthesis in *Salmonella typhimurium*. *J. Bacteriol.* **155**:213–221.

863. Hughes, K. T., B. M. Olivera, and J. R. Roth. 1983. *nadD*: novel mutants in an essential step in NAD biosynthesis in *Salmonella typhimurium*, abstr. K189, p. 208. *Abstr. Annu. Meet. Am. Soc. Microbiol., 1983.*

864. Hughes, K. T., B. M. Olivera, and J. R. Roth. 1987. Rec dependence of Mu transposition from P22-transduced fragments. *J. Bacteriol.* **169**:403–409.

865. Hughes, K. T., B. M. Olivera, and J. R. Roth. 1988. Structural gene for NAD synthetase in *Salmonella typhimurium*. *J. Bacteriol.* **170**:2113–2120.

866. Hughes, K. T., and J. R. Roth. 1984. Conditionally transposition-defective derivative of Mu d1(Amp Lac). *J. Bacteriol.* **159**:130–137.

867. Hughes, K. T., and J. R. Roth. 1988. Transitory cis complementation: a method for providing transposition functions to defective transposons. *Genetics* **119**:9–12.

868. Hughes, K. T., J. R. Roth, and B. M. Olivera. 1991. A genetic characterization of the *nadC* gene of *Salmonella typhimurium*. *Genetics* **127**:657–670.

869. Hughes, K. T., P. Youderian, and M. I. Simon. 1988. Phase variation in *Salmonella*: analysis of Hin recombinase and *hix* recombination site interaction *in vivo*. *Genes Dev.* **2**:937–948.

870. Hughes, V., and G. G. Meynell. 1977. The contribution of plasmid and host genes to plasmid mediated interference with phage growth. *Genet. Res.* **30**:179–186.

871. Hulanicka, D., and T. Klopotowski. 1972. Mutants of *Salmonella typhimurium* resistant to triazole. *Acta Biochim. Pol.* **19**:251–260.

872. Hulanicka, D., T. Klopotowski, and D. A. Smith. 1972. The effect of triazole on cysteine biosynthesis in *Salmonella typhimurium*. *J. Gen. Microbiol.* **72**:291–301.

873. Hulanicka, M. D., C. Garrett, G. Jagura-Burdzy, and N. M. Kredich. 1986. Cloning and characterization of the *cysAMK* region of *Salmonella typhimurium*. *J. Bacteriol.* **168**:322–327.

874. Hulanicka, M. D., S. G. Hallquist, N. M. Kredich, and T. Mojica-a. 1979. Regulation of O-acetylserine sulfhydrylase B by L-cysteine in *Salmonella typhimurium*. *J. Bacteriol.* **140**:141–146.

875. Hulanicka, M. D., and N. M. Kredich. 1976. A mutation affecting expression of the gene coding for serine transacetylase in *Salmonella typhimurium*. *Mol. Gen. Genet.* **148**:143–148.

876. Hulanicka, M. D., N. M. Kredich, and D. M. Treiman. 1974. The structural gene of O-acetylserine sulfhydrylase A in *Salmonella typhimurium*. Identity with the *trzA* locus. *J. Biol. Chem.* **249**:867–872.

877. Hulton, C. S., A. Seirafi, J. C. Hinton, J. M. Sidebotham, L. Waddell, G. D. Pavitt, T. Owen-Hughes, A. Spassky, H. Buc, and C. F. Higgins. 1990. Histone-like protein H1 (H-NS), DNA supercoiling, and gene expression in bacteria. *Cell* **63**:631–642.

878. Hurme, R., E. Namork, E. L. Nurmiaho-Lassila, and M. Rhen. 1994. Intermediate filament-like network formed *in vitro* by a bacterial coiled coil protein. *J. Biol. Chem.* **269**:10675–10682.

879. Hwang, L. H., and H. Zalkin. 1971. Multiple forms of anthranilate synthetase-anthranilate 5-phosphoribosylpyrophosphate phosphoribosyl-transferase from *Salmonella typhimurium*. *J. Biol. Chem.* **246**:2338–2345.

880. Hyde, C. C., S. A. Ahmed, E. A. Padlan, E. W. Miles, and D. R. Davies. 1988. Three-dimensional structure of the tryptophan synthase alpha 2 beta 2 multienzyme complex from *Salmonella typhimurium*. *J. Biol. Chem.* **263**:17857–17871.

881. Hyman, H. C., and S. Trachtenberg. 1991. Point mutations that lock *Salmonella typhimurium* flagellar filaments in the straight right-handed and left-handed forms and their relation to filament superhelicity. *J. Mol. Biol.* **220**:79–88.

882. Ibe, S. N., A. J. Sinskey, and D. Botstein. 1982. Genetic mapping of mutations in a highly radiation-resistant mutant of *Salmonella typhimurium* LT2. *J. Bacteriol.* **152**:260–268.

883. Iijima, T., and K. Imai. 1975. Genetic locus of *tct* (tricarboxylic acid transport) gene in *Salmonella typhimurium*. *Inst. Ferment. Res. Commun.* **7**:61–64.

884. Iino, T. 1961. A stabilizer of antigenic phases in *Salmonella abortus-equi*. *Genetics* **46**:1465–1469.

885. Iino, T. 1961. Genetic analysis of O-H variation in Salmonella. *Jpn. J. Genet.* **36**:268–275.

886. Iino, T. 1969. Phase specific regulator of flagellar genes (H1 and H2) in Salmonella. *Nat. Inst. Genet.* (Mishima) *Annu. Rep.* **13**:72–73.

887. Iino, T. 1969. Genetics and chemistry of bacterial flagella. *Bacteriol. Rev.* **33**:454–475.

888. Iino, T. 1974. Assembly of *Salmonella* flagellin *in vitro* and *in vivo*. *J. Supramol. Struct.* **2**:372–384.

889. Iino, T. 1977. Genetics of structure and function of bacterial flagella. *Annu. Rev. Genet.* **11**:161–182.

890. Iino, T., and M. Enomoto. 1966. Genetical studies of nonflagellate mutants of *Salmonella*. *J. Gen. Microbiol.* **43**:315–327.

891. Iino, T., Y. Komeda, K. Kutsukake, R. Macnab, P. Matsumura, J. S. Parkinson, M. I. Simon, and S. Yamaguchi. 1988. New unified nomenclature for the flagellar genes of *Escherichia coli* and *Salmonella typhimurium*. *Microbiol. Rev.* **52**:533–535.

892. Iino, T., and T. Oguchi. 1972. A non-chemotactic mutant in *Salmonella*. *Annu. Rep. Natl. Inst. Genet.* **22**:13–14.

893. Iino, T., T. Oguchi, and T. Hirano. 1975. Temporary expression of flagellar phase-1 phase-2 clones of diphasic *Salmonella*. *J. Gen. Microbiol.* **89**:265–276.

894. Iino, T., T. Oguchi, and T. Kuroiwa. 1974. Polymorphism in a flagellar-shape mutant of *Salmonella typhimurium*. *J. Gen. Microbiol.* **81**:37–45.

895. Ikeda, T., M. Homma, T. Iino, S. Asakura, and R. Kamiya. 1987. Localization and stoichiometry of hook-associated proteins within *Salmonella typhimurium* flagella. *J. Bacteriol.* **169**:1168–1173.

896. Ikeda, T., R. Kamiya, and S. Yamaguchi. 1983. Excretion of flagellin by a short-flagella mutant of *Salmonella typhimurium*. *J. Bacteriol.* **153**:506–510.

897. Ikeda, T., R. Kamiya, and S. Yamaguchi. 1984. In vitro polymerization of flagellin excreted by a short-flagellum *Salmonella typhimurium* mutant. *J. Bacteriol.* **159**:787–789.

898. Imai, K. 1975. Isolation of tricarboxylic acid transport-negative mutants of *Salmonella typhimurium*. *J. Gen. Appl. Microbiol.* **21**:127–134.

899. Imai, K. 1975. Transport system for the C_4-dicarboxylic acids in *Salmonella typhimurium*. *Inst. Ferment. Res. Commun.* **7**:53–60.

900. Imai, K., T. Iijima, and I. Banno. 1977. Location of *tct* (tricarboxylic acid transport) genes on the chromosome of *Salmonella typhimurium*. *Inst. Ferment. Res. Commun.* **8**:63–68.

901. Imai, K., T. Iijima, and T. Hasegawa. 1973. Transport of tricarboxylic acids in *Salmonella typhimurium*. *J. Bacteriol.* **114**:961–965.

902. Imray, F. P., and D. G. MacPhee. 1975. Induction of base-pair substitution and frameshift mutations in wild-type and repair-deficient strains of *Salmonella typhimurium* by the photodynamic action of methylene blue. *Mutat. Res.* **27**:299–306.

903. Imray, F. P., and D. G. MacPhee. 1976. Spontaneous and induced mutability of frameshift strains of *Salmonella typhimurium* carrying *uvrB* and *polA* mutations. *Mutat. Res.* **34**:35–42.

904. Ingraham, J. L., and J. Neuhard. 1972. Cold-sensitive mutants of *Salmonella typhimurium* defective in UMP kinase (*pyrH*). *J. Biol. Chem.* **247**:6259–6265.

905. Ino, I., P. E. Hartman, Z. Hartman, and J. Yourno. 1975. Deletions fusing the *hisG* genes in *Salmonella typhimurium*. *J. Bacteriol.* **123**:1254–1264.

906. Inoue, Y. H., K. Kutsukake, T. Iino, and S. Yamaguchi. 1989. Sequence analysis of operator mutants of the phase-1 flagellin-encoding gene, *fliC*, in *Salmonella typhimurium*. *Gene* **85**:221–226.

907. Irikura, V. M., M. Kihara, S. Yamaguchi, H. Sockett, and R. M. Macnab. 1993. *Salmonella typhimurium fliG* and *fliN* mutations causing defects in assembly, rotation, and switching of the flagellar motor. *J. Bacteriol.* **175**:802–810.

908. Isaki, L. S., and M. J. Voll. 1976. Genetic characterization of a φ80 transducing bacteriophage carrying the histidine operon of *Salmonella typhimurium*. *J. Virol.* **19**:313–317.

909. Ishido, M., T. Hase, F. Ito, and Y. Masamune. 1980. Comparative studies of DNA polymerase I of enteric bacteria. *J. Gen. Appl. Microbiol.* **26**:183–202.

910. Ishidsu, J. 1975. Physiological properties of arginine sensitive mutants of *Salmonella typhimurium*. *Jpn. J. Genet.* **50**:99–113.

911. Ishiguro, E. E., D. Vanderwel, and W. Kusser. 1986. Control of lipopolysaccharide biosynthesis and release by *Escherichia coli* and *Salmonella typhimurium*. *J. Bacteriol.* **168**:328–333.

912. Ishihara, A., S. Yamaguchi, and H. Hotani. 1981. Passive rotation of flagella on paralyzed *Salmonella typhimurium* (*mot*) mutants by external rotatory driving force. *J. Bacteriol.* **145**:1082–1084.

913. Ishima, R., K. Akasaka, S. Aizawa, and F. Vonderviszt. 1991. Mobility of the terminal regions of flagellin in solution. *J. Biol. Chem.* **266**:23682–23688.

914. Isihara, H., and R. W. Hogg. 1980. Amino acid sequence of the sulfate-binding protein from *Salmonella typhimurium* LT2. *J. Biol. Chem.* **255**:4614–4618.

915. Island, M. D., B. Y. Wei, and R. J. Kadner. 1992. Structure and function of the *uhp* genes for the sugar phosphate transport system in *Escherichia coli* and *Salmonella typhimurium*. *J. Bacteriol.* **174**:2754–2762.

916. Isono, K., and J. Yourno. 1973. Mutation leading to gene fusion in the histidine operon of *Salmonella typhimurium*. *J. Mol. Biol.* **76**:455–461.

917. Isono, K., and J. Yourno. 1974. Chemical carcinogens as frameshift mutagens: *Salmonella* DNA sequence sensitive to mutagenesis by polycyclic carcinogens. *Proc. Natl. Acad. Sci. USA* **71**:1612–1617.

918. Itaya, M., D. McKelvin, S. K. Chatterjie, and R. J. Crouch. 1991. Selective cloning of genes encoding RNase H from *Salmonella typhimurium, Saccharomyces cerevisiae* and *Escherichia coli rnh* mutant. *Mol. Gen. Genet.* **227:**438–445.

919. Itikawa, H., and M. Demerec. 1967. Ditto deletions in the *cysC* region of the Salmonella chromosome. *Genetics* **55:**63–68.

920. Itikawa, H., and M. Demerec. 1968. *Salmonella typhimurium* proline mutants. *J. Bacteriol.* **95:**1189–1190.

921. Jackson, M. B., J. M. Becker, A. S. Steinfield, and F. Naider. 1976. Oligopeptide transport in proline peptidase mutants of *Salmonella typhimurium. J. Biol. Chem.* **251:**5300–5309.

922. Jacobson, F. S., R. W. Morgan, M. F. Christman, and B. N. Ames. 1989. An alkyl hydroperoxide reductase from Salmonella typhimurium involved in the defense of DNA against oxidative damage. Purification and properties. *J. Biol. Chem.* **264:**1488–1496.

923. Jagura, G., D. Hulanicka, and N. M. Kredich. 1978. Analysis of merodiploids of the *cysB* region in *Salmonella typhimurium. Mol. Gen. Genet.* **165:**31–38.

924. Jagura-Burdzy, G., and D. Hulanicka. 1987. Cloning of *cysB* mutant alleles of *S. typhimurium. Acta Biochim. Pol.* **34:**35–44.

925. Jagura-Burdzy, G., and N. M. Kredich. 1983. Cloning and physical mapping of the *cysB* region of *Salmonella typhimurium. J. Bacteriol.* **155:**578–585.

926. Jahreis, K., P. W. Postma, and J. W. Lengeler. 1991. Nucleotide sequence of the *ilvH-fruR* gene region of *Escherichia coli* K12 and *Salmonella typhimurium* LT2. *Mol. Gen. Genet.* **226:**332–336.

927. Jamieson, D. J., and C. F. Higgins. 1984. Anaerobic and leucine-dependent expression of a peptide transport gene in *Salmonella typhimurium. J. Bacteriol.* **160:**131–136.

928. Jamieson, D. J., and C. F. Higgins. 1986. Two genetically distinct pathways for transcriptional regulation of anaerobic gene expression in *Salmonella typhimurium. J. Bacteriol.* **168:**389–397.

929. Jamieson, D. J., R. G. Sawers, P. A. Rugman, D. H. Boxer, and C. F. Higgins. 1986. Effects of anaerobic regulatory mutations and catabolite repression on regulation of hydrogen metabolism and hydrogenase isoenzyme composition in *Salmonella typhimurium. J. Bacteriol.* **168:**405–411.

930. Janion, C. 1977. On the ability of *Salmonella typhimurium* cells to form deoxycytidine nucleotides. *Mol. Gen. Genet.* **153:**179–183.

931. Janion, C., and E. Popowska. 1975. The reduction of N_4-hydroxycytidine to cytidine by *Salmonella typhimurium* cells. *Nucleic Acids Res.* **2**(Suppl.):s159-s163.

932. Jann, K. S., S. Kanegasaki, G. Goldemann, and P. H. Makela. 1979. On the effect of *rfe* mutation on the biosynthesis of the O-8 and O-9 antigens of *Escherichia coli. Biochem. Biophys. Res. Commun.* **86:**1185–1191.

933. Janson, C. A., P. S. Kayne, R. J. Almassy, M. Grunstein, and D. Eisenberg. 1986. Sequence of glutamine synthetase from *Salmonella typhimurium* and implications for the protein structure. *Gene* **46:**297–300.

934. Janzer, J., H. Stan-Lotter, and K. E. Sanderson. 1981. Isolation and characterization of hemin-permeable, envelope-defective mutants of *Salmonella typhimurium. Can. J. Microbiol.* **27:**226–237.

935. Jargiello, P. 1976. Simultaneous selection of mutants in gluconeogenesis and nucleoside catabolism in *Salmonella typhimurium. Biochim. Biophys. Acta* **444:**321–325.

936. Jargiello, P., and P. Hoffee. 1972. Orientation of the *deo* genes and the *serB* locus in *Salmonella typhimurium. J. Bacteriol.* **111:**296–297.

937. Jargiello, P., M. D. Stern, and P. Hoffee. 1974. 2-Deoxyribose 5-phosphate aldolase: genetic analyses of structure. *J. Mol. Biol.* **88:**671–691.

938. Jargiello, P., C. Sushak, and P. Hoffee. 1976. 2-Deoxyribose 5-phosphate aldolase: isolation and characterization of proteins genetically modified in the active site region. *Arch. Biochem. Biophys.* **177:**630–641.

939. Jegede, V. A., F. Spencer, and J. E. Brenchley. 1976. Thialysine-resistant mutant of *Salmonella typhimurium* with a lesion in the *thrA* gene. *Genetics* **83:**619–632.

940. Jenness, D. D., and H. K. Schachman. 1980. *pyrB* mutations as suppressors of arginine auxotrophy in *Salmonella typhimurium. J. Bacteriol.* **141:**33–40.

941. Jenness, D. D., and H. K. Schachman. 1983. Genetic characterization of the folding domains of the catalytic chains in aspartate transcarbamoylase. *J. Biol. Chem.* **258:**3266–3279.

942. Jennings, M. P., S. P. Scott, and I. R. Beacham. 1993. Regulation of the *ansB* gene of *Salmonella enterica. Mol. Microbiol.* **9:**165–172.

943. Jensen, K. F. 1979. Apparent involvement of purines in the control of expression of *Salmonella typhimurium pyr* genes: analysis of a leaky *guaB* mutant resistant to pyrimidine analogs. *J. Bacteriol.* **138:**731–738.

944. Jensen, K. F. 1989. Regulation of *Salmonella typhimurium pyr* gene expression: effect of changing both purine and pyrimidine nucleotide pools. *J. Gen. Microbiol.* **135:**805–815.

945. Jensen, K. F., R. Fast, O. Karlstrom, and J. N. Larsen. 1986. Association of RNA polymerase having increased K_m for ATP and UTP with hyperexpression of the *pyrB* and *pyrE* genes of *Salmonella typhimurium. J. Bacteriol.* **166:**857–865.

946. Jensen, K. F., J. Neuhard, and L. Shack. 1982. RNA polymerase involvement in the control of *Salmonella typhimurium pyr* gene expression. Isolation and characterization of a fluorouracil resistant mutant with high constitutive expression of *pyrB* and *pyrE* due to a mutation in *rpoBC. EMBO J.* **1:**69–74.

947. Jenson, P., C. Parkes, and D. Berkowitz. 1972. Mannitol sensitivity. *J. Bacteriol.* **111:**351–355.

948. Jessop, A. P. 1972. A specialized transducing phage of P22 for which the ability to form plaques is associated with transduction of the *proAB* region. *Mol. Gen. Genet.* **114:**214–222.

949. Jessop, A. P. 1976. Specialized transducing phages derived from phage P22 that carry the *proAB* region of the host, *Salmonella typhimurium:* genetic evidence for their structure and mode of transduction. *Genetics* **83:**459–475.

950. Jeter, R. M. 1990. Cobalamin-dependent 1,2-propanediol utilization by *Salmonella typhimurium. J. Gen. Microbiol.* **136:**887–896.

951. Jeter, R. M., B. M. Olivera, and J. R. Roth. 1984. *Salmonella typhimurium* synthesizes cobalamin (vitamin B_{12}) de novo under anaerobic growth conditions. *J. Bacteriol.* **159:**206–213.

952. Jeter, R. M., and J. R. Roth. 1987. Cobalamin (vitamin B_{12}) biosynthetic genes of *Salmonella typhimurium. J. Bacteriol.* **169:**3189–3198.

953. Jiang, S.-Q., G.-Q. Yu, Z.-G. Li, and J.-S. Hong. 1988. Genetic evidence for modulation of the activator by two regulatory proteins involved in the exogenous induction of phosphoglycerate transport in *Salmonella typhimurium. J. Bacteriol.* **170:**4304–4308.

954. Jiang, X. M., B. Neal, F. Santiago, S. J. Lee, L. K. Romana, and P. R. Reeves. 1991. Structure and sequence of the *rfb* (O antigen) gene cluster of *Salmonella serovar typhimurium* (strain LT2). *Mol. Microbiol.* **5:**695–713.

955. Jochimsen, B., B. Garber, and J. S. Gots. 1980. Phosphoribosylpyrophosphate (PRPP) synthetase mutant in *Salmonella typhimurium. Adv. Exp. Med. Biol.* **122B:**131–136.

956. Jochimsen, B. U., B. Hove-Jensen, B. B. Garber, and J. S. Gots. 1985. Characterization of a *Salmonella typhimurium* mutant defective in phosphoribosylpyrophosphate synthetase. *J. Gen. Microbiol.* **131:**245–252.

957. Joh, K., and S. Hiraga. 1979. Genetic mapping of the chromosomal replication origin of *Salmonella typhimurium. J. Bacteriol.* **138:**297–304.

958. Johanson, U., and D. Hughes. 1992. Comparison of the complete sequence of the *str* operon in *Salmonella typhimurium* and *Escherichia coli. Gene* **120:**93–98.

959. Johansson, V., A. Aarti, M. Nurminen, and P. H. Makela. 1978. Outer membrane protein-specific bacteriophage of *Salmonella typhimurium. J. Bacteriol.* **107:**183–187.

960. Johnson, B. N., A. Weintraub, A. A. Lindberg, and B. A. Stocker. 1992. Construction of *Salmonella* strains with both antigen O4 (of group B) and antigen O9 (of group D). *J. Bacteriol.* **174:**1911–1915.

961. Johnson, E. M., and L. S. Baron. 1969. Genetic transfer of the Vi antigen from *Salmonella typhosa* to *Escherichia coli. J. Bacteriol.* **99:**358–359.

962. Johnson, E. M., B. Krauskopf, and L. S. Baron. 1965. Genetic mapping of Vi and somatic antigenic determinants in *Salmonella. J. Bacteriol.* **90:**302–308.

963. Johnson, E. M., B. Krauskopf, and L. S. Baron. 1966. Genetic analysis of the *ViA-his* chromosomal region in *Salmonella. J. Bacteriol.* **92:**1457–1463.

964. Johnson, K., I. Charles, G. Dougan, D. Pickard, P. O'Gaora, G. Costa, T. Ali, I. Miller, and C. Hormaeche. 1991. The role of a stress-response protein in *Salmonella typhimurium* virulence. *Mol. Microbiol.* **5:**401–407.

965. Johnson, M. G., and J. C. Escalante-Semerena. 1992. Identification of 5,6-dimethylbenzimidazole as the Co alpha ligand of the cobamide synthesized by *Salmonella typhimurium.* Nutritional characterization of mutants defective in biosynthesis of the imidazole ring. *J. Biol. Chem.* **267:**13302–13305.

966. Johnston, H. M., W. M. Barnes, F. G. Chumley, L. Bossi, and J. R. Roth. 1980. Model for regulation of the histidine operon of *Salmonella. Proc. Natl. Acad. Sci. USA* **77:**508–512.

967. Johnston, H. M., and J. R. Roth. 1979. Histidine mutants requiring adenine: selection of mutants with reduced *hisG* expression in *Salmonella typhimurium. Genetics* **92:**1–16.

968. Johnston, H. M., and J. R. Roth. 1980. UGA suppressor that maps within a cluster of ribosomal protein genes. *J. Bacteriol.* **144:**300–305.

969. Jones, B. D., and S. Falkow. 1994. Identification and characterization of a *Salmonella typhimurium* oxygen-regulated gene required for bacterial internalization. *Infect. Immun.* **62:**3745–3752.

970. Jones, B. D., N. Ghori, and S. Falkow. 1994. *Salmonella typhimurium* initiates murine infection by penetrating and destroying the specialized epithelial M cells of the Peyer's patches. *J. Exp. Med.* **180:**15–23.

971. Jones, C. J., M. Homma, and R. M. Macnab. 1989. L-, P-, and M-ring proteins of the flagellar basal body of *Salmonella typhimurium:* gene sequences and deduced protein sequences. *J. Bacteriol.* **171:**3890–3900.

972. Jones, C. J., and R. M. Macnab. 1990. Flagellar assembly in *Salmonella typhimurium:* analysis with temperature-sensitive mutants. *J. Bacteriol.* **172:**1327–1339.

973. Jones, C. J., R. M. Macnab, H. Okino, and S. Aizawa. 1990. Stoichiometric analysis of the flagellar hook-(basal-body) complex of *Salmonella typhimurium. J. Mol. Biol.* **212:**377–387.

974. Jones, R. T., D. E. Keltzow, and B. A. Stocker. 1972. Genetic transfer of *Salmonella typhimurium* and *Escherichia coli* lipopolysaccharide antigens to *Escherichia coli* K-12. *J. Bacteriol.* **111:**758–770.

975. Jones, W. R., G. J. Barcak, and R. E. Wolf, Jr. 1990. Altered growth-rate-dependent regulation of 6-phosphogluconate dehydrogenase level in *hisT* mutants of *Salmonella typhimurium* and *Escherichia coli. J. Bacteriol.* **172:**1197–1205.

976. Jones-Mortimer, M. C. 1973. Mapping of structural genes for the enzymes of cysteine biosynthesis in *Escherichia coli* K12 and *Salmonella typhimurium* LT2. *Heredity* **31:**213–221.

977. Jordan, A., I. Gibert, and J. Barbe. 1994. Cloning and sequencing of the genes from *Salmonella typhimurium* encoding a new bacterial ribonucleotide reductase. *J. Bacteriol.* **176:**3420–3427.

978. Joshi, A., J. Z. Siddiqui, M. Verma, and M. Chakravorty. 1982. Participation of the host protein(s) in the morphogenesis of bacteriophage P22. *Mol. Gen. Genet.* **186**:44–49.

979. Joshi, A., M. Verma, and M. Chakravorty. 1982. Thiolutin-resistant mutants of *Salmonella typhimurium. Antimicrob. Agents Chemother.* **22**:541–547.

980. Joshi, A. R., and M. Chakravorty. 1979. Bacteriophage P-22 development is temperature sensitive in thiolutin resistant mutants of *Salmonella typhimurium. Biochem. Biophys. Res. Commun.* **89**:1–6.

981. Jousimies, H., and P. H. Makela. 1974. Genetic analysis of *Salmonella minnesota* R mutants with defects in the biosynthesis of the lipopolysaccharide core. *J. Bacteriol.* **119**:753–759.

982. Jovanovich, S. B. 1985. Regulation of a *cya-lac* fusion by cyclic AMP in *Salmonella typhimurium. J. Bacteriol.* **161**:641–649.

983. Jovanovich, S. B., and J. Lebowitz. 1987. Estimation of the effect of coumermycin A₁ on *Salmonella typhimurium* promoters by using random operon fusions. *J. Bacteriol.* **169**:4431–4435.

984. Jovanovich, S. B., M. Martinell, M. T. Record, Jr., and R. R. Burgess. 1988. Rapid response to osmotic upshift by osmoregulated genes in *Escherichia coli* and *Salmonella typhimurium. J. Bacteriol.* **170**:234–539.

985. Joys, T. M. 1976. Identification of the antibody binding site in the phase-1 flagellar protein of *Salmonella typhimurium. Microbios* **15**:221–228.

986. Joys, T. M. 1985. The covalent structure of the phase-1 flagellar filament protein of *Salmonella typhimurium* and its comparison with other flagellins. *J. Biol. Chem.* **260**:15758–15761.

987. Joys, T. M., J. F. Martin, H. L. Wilson, and V. Rankis. 1974. Differences in the primary structure of the phase-1 flagellins of two strains of *Salmonella typhimurium. Biochim. Biophys. Acta* **351**:301–305.

988. Joys, T. M., and V. Rankis. 1972. The primary structure of the phase-1 flagellar protein of *Salmonella typhimurium*. I. The tryptic peptides. *J. Biol. Chem.* **247**:5180–5193.

989. Joys, T. M., and B. A. D. Stocker. 1969. Recombination of *H1*, the gene determining the flagellar antigen-*i* of *Salmonella typhimurium*: mapping of *H1* and *fla* mutations. *J. Gen. Microbiol.* **58**:267–276.

990. Jungwirth, C., S. R. Gross, P. Margolin, and H. E. Umbarger. 1963. The biosynthesis of leucine. I. The accumulation of β-carboxy-β-hydroxyisocaproate by leucine auxotrophs of *Salmonella typhimurium* and *Neurospora crassa. Biochemistry* **2**:1–6.

991. Kadam, S. K., M. S. Peppler, and K. E. Sanderson. 1985. Temperature-sensitive mutants in *rfaI* and *rfaJ*, genes for galactosyltransferase I and glucosyltransferase II, for synthesis of lipopolysaccharide in *Salmonella typhimurium. Can. J. Microbiol.* **31**:861–869.

992. Kadam, S. K., A. Rehemtulla, and K. E. Sanderson. 1985. Cloning of *rfaG, B, I*, and *J* genes for glycosyltransferase enzymes for synthesis of the lipopolysaccharide core of *Salmonella typhimurium. J. Bacteriol.* **161**:277–284.

993. Kagawa, H., T. Nishiyama, and S. Yamaguchi. 1983. Motility development of *Salmonella typhimurium* cells with *flaV* mutations after addition of exogenous flagellin. *J. Bacteriol.* **155**:435–437.

994. Kagawa, H., N. Ono, M. Enomoto, and Y. Komeda. 1984. Bacteriophage Chi sensitivity and motility of *Escherichia coli* K-12 and *Salmonella typhimurium* Fla⁻ mutants possessing the hook structure. *J. Bacteriol.* **157**:649–654.

995. Kagawa, H., K. Owaribe, S. Asakura, and N. Takahashi. 1976. Flagellar hook protein from *Salmonella* SJ25. *J. Bacteriol.* **125**:68–73.

996. Kamiya, R., and S. Asakura. 1974. Formation of a flagella-like but straight polymer of *Salmonella* flagellin. *J. Mol. Biol.* **87**:55–62.

997. Kamiya, R., and S. Asakura. 1976. Helical transformations of *Salmonella* flagella in vitro. *J. Mol. Biol.* **106**:167–186.

998. Kamp, D., and R. Kahmann. 1981. The relationship of two invertible segments in bacteriophage Mu and *Salmonella typhimurium* DNA. *Mol. Gen. Genet.* **184**:564–566.

999. Kang, C. H., W. C. Shin, Y. Yamagata, S. Gokcen, G. F. Ames, and S. H. Kim. 1991. Crystal structure of the lysine-, arginine-, ornithine-binding protein (LAO) from Salmonella typhimurium at 2.7-A resolution. *J. Biol. Chem.* **266**:23893–23899.

1000. Kaniga, K., J. C. Bossio, and J. E. Galan. 1994. The *Salmonella typhimurium* invasion genes *invF* and *invG* encode homologues of the AraC and PulD family of proteins. *Mol. Microbiol.* **13**:555–568.

1001. Kanto, S., H. Okino, S. Aizawa, and S. Yamaguchi. 1991. Amino acids responsible for flagellar shape are distributed in terminal regions of flagellin. *J. Mol. Biol.* **219**:471–480.

1002. Kanzaki, H., P. McPhie, and E. W. Miles. 1991. Effect of single amino acid substitutions at positions 49 and 60 on the thermal unfolding of the tryptophan synthase alpha subunit from *Salmonella typhimurium. Arch. Biochem. Biophys.* **284**:174–180.

1003. Kaplan, J. B., W. K. Merkel, and B. P. Nichols. 1985. Evolution of glutamine amidotransferase genes. Nucleotide sequences of the *pabA* genes from *Salmonella typhimurium, Klebsiella aerogenes* and *Serratia marcescens. J. Mol. Biol.* **183**:327–340.

1004. Kaplan, M. M., and M. Flavin. 1966. Cystathionine γ-synthetase of Salmonella: structural properties of a new enzyme in bacterial methionine biosynthesis. *J. Biol. Chem.* **241**:5781–5789.

1005. Karbonowska, H., A. Wiater, and D. Hulanicka. 1977. Sulfate permease of *Escherichia coli* K12. *Acta Biochim. Pol.* **24**:329–334.

1006. Karlsson, M., K. Hannavy, and C. F. Higgins. 1993. A sequence-specific function for the N-terminal signal-like sequence of the TonB protein. *Mol. Microbiol.* **8**:379–388.

1007. Kasahara, M., A. Nakata, and H. Shinagawa. 1991. Molecular analysis of the *Salmonella typhimurium phoN* gene, which encodes nonspecific acid phosphatase. *J. Bacteriol.* **173**:6760–6765.

1008. Kawagishi, I., V. Muller, A. W. Williams, V. M. Irikura, and R. M. Macnab. 1992. Subdivision of flagellar region III of the *Escherichia coli* and *Salmonella typhimurium* chromosomes and identification of two additional flagellar genes. *J. Gen. Microbiol.* **138**:1051–1065.

1009. Kawakami, K., and Y. Nakamura. 1990. Autogenous suppression of an opal mutation in the gene encoding peptide chain release factor 2. *Proc. Natl. Acad. Sci. USA* **87**:8432–8436.

1010. Kay, W. W., and M. Cameron. 1978. Citrate transport in *Salmonella typhimurium. Arch. Biochem. Biophys.* **190**:270–280.

1011. Kay, W. W., and M. J. Cameron. 1978. Transport of C₄ dicarboxylic acids in *Salmonella typhimurium. Arch. Biochem. Biophys.* **190**:281–289.

1012. Kayastha, A. M., Y. Sawa, S. Nagata, and E. W. Miles. 1991. Site-directed mutagenesis of the beta subunit of tryptophan synthase from *Salmonella typhimurium*. Role of active site glutamic acid 350. *J. Biol. Chem.* **266**:7618–7625.

1013. Kaye, R., J. Barravecchio, and J. Roth. 1974. Isolation of P22 specialized transducing phage following F'-episome fusion. *Genetics* **76**:655–667.

1014. Kelln, R. A. 1984. Evidence for involvement of *pyrH*⁺ of an *Escherichia coli* K-12 F-prime factor in inhibiting construction of hybrid merodiploids with *Salmonella typhimurium. Can. J. Microbiol.* **30**:991–996.

1015. Kelln, R. A., K. F. Foltermann, and G. A. O'Donovan. 1975. Location of the *argR* gene on the chromosome of *Salmonella typhimurium. Mol. Gen. Genet.* **139**:279–284.

1016. Kelln, R. A., J. J. Kinahan, K. F. Foltermann, and G. A. O'Donovan. 1975. Pyrimidine biosynthetic enzymes of *Salmonella typhimurium*, repressed specifically by growth in the presence of cytidine. *J. Bacteriol.* **124**:764–774.

1017. Kelln, R. A., and J. Neuhard. 1988. Regulation of *pyrC* expression in *Salmonella typhimurium*: identification of a regulatory region. *Mol. Gen. Genet.* **212**:287–294.

1018. Kelln, R. A., J. Neuhard, and L. Stauning. 1985. Isolation and characterization of pyrimidine mutants of *Salmonella typhimurium* altered in expression of *pyrC*, *pyrD*, and *pyrE. Can. J. Microbiol.* **31**:981–991.

1019. Kelln, R. A., and G. A. O'Donovan. 1976. Isolation and partial characterization of an *argR* mutant of *Salmonella typhimurium. J. Bacteriol.* **128**:528–535.

1020. Kelln, R. A., and V. L. Zak. 1978. Arginine regulon control in a *Salmonella typhimurium-Escherichia coli* hybrid merodiploid. *Mol. Gen. Genet.* **161**:333–336.

1021. Kelln, R. A., and V. L. Zak. 1980. A mutation in *Salmonella typhimurium* imparting conditional resistance to 5-fluorouracil and a bio-energetic defect: mapping of *cad. Mol. Gen. Genet.* **179**:678–682.

1022. Kemper, J. 1974. Evolution of a new gene substituting for the *leuD* gene of *Salmonella typhimurium*: characterization of *supQ* mutations. *J. Bacteriol.* **119**:937–951.

1023. Kemper, J. 1974. Gene order and co-transduction in the *leu-ara-fol-pyrA* region of the *Salmonella typhimurium* linkage map. *J. Bacteriol.* **117**:94–99.

1024. Kemper, J. 1974. Evolution of a new gene substituting for the *leuD* gene of *Salmonella typhimurium*: origin and nature of *supQ* and *newD* mutations. *J. Bacteriol.* **120**:1176–1185.

1025. Kemper, J., and P. Margolin. 1970. Suppression by gene substitution for the *leuD* gene of *Salmonella typhimurium. Genetics* **63**:263–279.

1026. Kerridge, D. 1966. Flagellar synthesis in *Salmonella typhimurium*: factors affecting the formation of the flagellar epsilon-*N*-methyllysine. *J. Gen. Microbiol.* **42**:71–82.

1027. Khan, I. H., T. S. Reese, and S. Khan. 1992. The cytoplasmic component of the bacterial flagellar motor. *Proc. Natl. Acad. Sci. USA* **89**:5956–5960.

1028. Khan, S., and R. M. Macnab. 1980. The steady state counterclockwise-clockwise ratio of bacterial flagellar motors is regulated by protonmotive force. *J. Mol. Biol.* **138**:563–597.

1029. Khan, S., R. M. Macnab, A. L. DeFranco, and D. E. Koshland, Jr. 1978. Inversion of a behavioral response in bacterial chemotaxis: explanation at the molecular level. *Proc. Natl. Acad. Sci. USA* **75**:4150–4154.

1030. Kiefer, V., G. Schmidt, B. Jann, and K. Jann. 1976. Genetic transfer of *Salmonella* O antigens to *Escherichia coli* O8. *J. Gen. Microbiol.* **92**:311–324.

1031. Kier, L. D., R. Weppelman, and B. N. Ames. 1977. Resolution and purification of three periplasmic phosphatases of *Salmonella typhimurium. J. Bacteriol.* **130**:399–410.

1032. Kier, L. D., R. Weppelman, and B. N. Ames. 1977. Regulation of two phosphatases and a cyclic phosphodiesterase of *Salmonella typhimurium. J. Bacteriol.* **130**:420–428.

1033. Kier, L. D., R. M. Weppelman, and B. N. Ames. 1979. Regulation of nonspecific acid phosphatase in *Salmonella: phoN* and *phoP* genes. *J. Bacteriol.* **138**:155–161.

1034. Kihara, M., M. Homma, K. Kutsukate, and R. M. Macnab. 1989. Flagellar switch of *Salmonella typhimurium*: gene sequences and deduced protein sequences. *J. Bacteriol.* **171**:3247–3257.

1035. Kilstrup, M., C. D. Lu, A. Abdelal, and J. Neuhard. 1988. Nucleotide sequence of the *carA* gene and regulation of the *carAB* operon in *Salmonella typhimurium. Eur. J. Biochem.* **176**:421–429.

1036. Kim, S. S., and P. Datta. 1982. Chemical characterization of biodegradative threonine dehydratases from two enteric bacteria. *Biochim. Biophys. Acta* **706**:27–35.

1037. Kingsman, A. J. 1977. The structure of the *cysCDHIJ* region in unstable cysteine or methionine requiring mutants of *Salmonella typhimurium*. *Mol. Gen. Genet.* **156**:327–332.

1038. Kingsman, A. J., and D. A. Smith. 1978. The nature of genetic instability in auxotrophs of *Salmonella typhimurium* requiring cysteine or methionine and resistant to inhibition by 1,2,4-triazole. *Genetics* **89**:439–452.

1039. Kingsman, A. J., D. A. Smith, and M. D. Hulanicka. 1978. Genetic instability in auxotrophs of *Salmonella typhimurium* requiring cysteine or methionine and resistant to inhibition by 1,2,4-triazole. *Genetics* **89**:419–438.

1040. Kinkeldey, U., D. Von Lieres, and H. Schmieger. 1978. Bacterial donor mutants affecting the efficiency of generalized transduction by *Salmonella typhimurium* phage P-22. *J. Gen. Microbiol.* **108**:227–238.

1041. Kinney, D. M., and J. W. Foster. 1985. Identification of a *cis*-acting regulatory region in the *pncB* locus of *Salmonella typhimurium*. *Mol. Gen. Genet.* **199**:512–517.

1042. Kinney, D. M., J. W. Foster, and A. G. Moat. 1979. Pyridine nucleotide cycle of *Salmonella typhimurium*: in vitro demonstration of nicotinamide mononucleotide deamidase and characterization of *pnuA* mutants defective in nicotinamide mononucleotide transport. *J. Bacteriol.* **140**:607–611.

1043. Kirchner, C. E. J., and M. J. Rudden. 1966. Location of a mutator gene in *Salmonella typhimurium* by cotransduction. *J. Bacteriol.* **92**:1453–1456.

1044. Kiritani, K. 1972. Mutants deficient or altered in branched-chain-amino-acid transferase in *Salmonella typhimurium*. *Jpn. J. Genet.* **47**:91–102.

1045. Kiritani, K. 1974. Mutants of *Salmonella typhimurium* defective in transport of branched-chain amino acids. *J. Bacteriol.* **120**:1093–1101.

1046. Kiritani, K., and N. Inuzuka. 1970. Mutations affecting the branched-chain-amino-acid aminotransferase in *Salmonella typhimurium*. *Jpn. J. Genet.* **45**:293–304.

1047. Kiritani, K., and K. Ohnishi. 1977. Repression and inhibition of transport systems for branched-chain amino acids in *Salmonella typhimurium*. *J. Bacteriol.* **129**:589–598.

1048. Kiritani, K., and K. Ohnishi. 1978. Multiple transport systems for branched-chain amino acids as studied by mutants of *Salmonella typhimurium*. *Jpn. J. Genet.* **53**:265–274.

1049. Kirsh, M., D. R. Dembinski, P. E. Hartman, and C. G. Miller. 1978. *Salmonella typhimurium* peptidase active on carnosine. *J. Bacteriol.* **134**:361–374.

1050. Kishi, K., and S. Iseki. 1973. Genetic analysis of the O antigens in *Salmonella*. III. Inheritance of O antigens 3, 10, and 15 of *Salmonella* group E. *Jpn. J. Genet.* **48**:255–262.

1051. Kishi, K., and S. Iseki. 1973. Genetic analysis of the O antigens in *Salmonella*. I. Heredity of blood group-active O antigens in *Salmonella* groups G, R, and U. *Jpn. J. Genet.* **48**:89–97.

1052. Kishi, K., and S. Iseki. 1973. Genetic analysis of the O antigens in *Salmonella*. II. Heredity of the O antigens 4, 5, and 9 of *Salmonella*. *Jpn. J. Genet.* **48**:133–136.

1053. Kishi, K., and S. Iseki. 1974. Genetic analysis of the O antigens in *Salmonella*. IV. Inheritance of O antigens 1, 3, 19, and 34 of *Salmonella* group E. *Jpn. J. Genet.* **49**:125–129.

1054. Kita, H., and H. Nikaido. 1973. Structure of cell wall lipopolysaccharide of *Salmonella typhimurium*. IV. Anomeric configuration of L-rhamnose residues and its taxonomic implications. *J. Bacteriol.* **113**:672–679.

1055. Kleckner, N., K. Reichardt, and D. Botstein. 1979. Inversions and deletions of the *Salmonella typhimurium* chromosome generated by the translocatable tetracycline resistance element Tn-10. *J. Mol. Biol.* **127**:89–116.

1056. Kleckner, N., J. Roth, and D. Botstein. 1977. Genetic engineering in vivo using translocatable drug-resistance elements: new methods in bacterial genetics. *J. Mol. Biol.* **116**:125–159.

1057. Kleckner, N., D. A. Steele, K. Reichardt, and D. Botstein. 1979. Specificity of insertion by the translocatable tetracycline resistance element Tn-10. *Genetics* **92**:1023–1040.

1058. Kleeman, J. E., and S. M. Parsons. 1975. A sensitive assay for the reverse reaction of the first histidine biosynthetic enzyme. *Anal. Biochem.* **68**:236–241.

1059. Kleeman, J. E., and S. M. Parsons. 1977. Inhibition of histidyl-tRNA-adenosine triphosphate phosphoribosyltransferase complex formation by histidine and by guanosine tetraphosphate. *Proc. Natl. Acad. Sci. USA* **74**:1535–1537.

1060. Klena, J. D., E. Pradel, and C. A. Schnaitman. 1992. Comparison of lipopolysaccharide biosynthesis genes *rfaK*, *rfaL*, *rfaY*, and *rfaZ* of *Escherichia coli* K-12 and *Salmonella typhimurium*. *J. Bacteriol.* **174**:4746–4752.

1061. Klena, J. D., E. Pradel, and C. A. Schnaitman. 1993. The *rfaS* gene, which is involved in production of a rough form of lipopolysaccharide core in *Escherichia coli* K-12, is not present in the *rfa* cluster of *Salmonella typhimurium* LT2. *J. Bacteriol.* **175**:1524–1527.

1062. Knox, J. R., H. S. Liu, C. T. Walsh, and L. E. Zawadzke. 1989. D-Alanine-D-alanine ligase (ADP) from *Salmonella typhimurium*. Overproduction, purification, crystallization and preliminary X-ray analysis. *J. Mol. Biol.* **205**:461–463.

1063. Kobori, J. A., and A. Kornberg. 1982. The *Escherichia coli dnaC* gene product. I. Overproduction of the *dnaC* proteins of *Escherichia coli* and *Salmonella typhimurium* by cloning into a high copy number plasmid. *J. Biol. Chem.* **257**:13757–13762.

1064. Koch, W. H., T. A. Cebula, P. L. Foster, and E. Eisenstadt. 1992. UV mutagenesis in *Salmonella typhimurium* is *umuDC* dependent despite the presence of *samAB*. *J. Bacteriol.* **174**:2809–2815.

1065. Koduri, R., C. E. Benson, and J. S. Gots. 1978. A DNA binding protein with specificity for *pur* genes in enteric bacteria. *Fed. Proc.* **37**:1875.

1066. Koduri, R. K., D. M. Bedwell, and J. E. Brenchley. 1980. Characterization of a *Hind*III-generated DNA fragment carrying the glutamine synthetase gene of *Salmonella typhimurium*. *Gene* **11**:227–237.

1067. Koebnik, R., K. Hantke, and V. Braun. 1993. The TonB-dependent ferrichrome receptor FcuA of *Yersinia enterocolitica*: evidence against a strict co-evolution of receptor structure and substrate specificity. *Mol. Microbiol.* **7**:383–393.

1068. Kohno, T., L. Bossi, and J. R. Roth. 1983. Genetic characterization of the *sufJ* frameshift suppressor in *Salmonella typhimurium*. *Genetics* **103**:31–42.

1069. Kohno, T., L. Bossi, and J. R. Roth. 1983. New suppressors of frameshift mutations in *Salmonella typhimurium*. *Genetics* **103**:23–29.

1070. Kohno, T., and W. R. Gray. 1981. Chemical and genetic studies of L-histidinol dehydrogenase of *Salmonella typhimurium*. Isolation and structure of tryptic peptides. *J. Mol. Biol.* **147**:451–464.

1071. Kohno, T., and J. Roth. 1979. Electrolyte effects on the activity of mutant enzymes in vivo and in vitro. *Biochemistry* **18**:1386–1392.

1072. Kohno, T., and J. R. Roth. 1978. A *Salmonella* frameshift suppressor that acts at runs of adenylic acid residues in the messenger RNA. *J. Mol. Biol.* **126**:37–52.

1073. Kohno, T., M. Schmid, and J. R. Roth. 1980. Effect of electrolytes on growth of mutant bacteria, p. 53–57. *In* D. W. Rains, R. C. Valentine, and A. Hollaender (ed.), *Genetic Engineering of Osmoregulation*. Plenum Publishing Corp., New York.

1074. Komeda, Y., H. Suzuki, J. Ishidsu, and T. Iino. 1975. The role of cAMP in flagellation of *Salmonella typhimurium*. *Mol. Gen. Genet.* **142**:289–298.

1075. Kondratiev, Y. S., G. V. Brukhansky, I. V. Andreeva, and A. G. Skavronskaya. 1977. UV sensitivity and repair of UV damages in *Salmonella* of wild type. *Mol. Gen. Genet.* **158**:211–214.

1076. Konno, R., H. Fujita, T. Horiguchi, and S. Yamaguchi. 1976. Precise position of the *nml* locus on the genetic map of *Salmonella*. *J. Gen. Microbiol.* **93**:182–183.

1077. Korenevskaya, N. F., I. V. Andreeva, A. A. Kiryushkina, L. Y. Lichoded, Y. S. Kondratiev, and A. G. Skavronskaya. 1977. Intergeneric mating transfer of *polA* gene from *Escherichia coli* to *Salmonella typhimurium*. *Mutat. Res.* **45**:351–354.

1078. Kornberg, H. L., and M. C. James-Mortimer. 1977. The phosphotransferase system as a site of cellular control. *Symp. Soc. Gen. Microbiol.* **27**:217–240.

1079. Koshland, D. E., Jr. 1977. Bacterial chemotaxis and some enzymes in energy metabolism. *Symp. Soc. Gen. Microbiol.* **27**:317–331.

1080. Koski, P., L. Hirvas, and M. Vaara. 1990. Complete sequence of the *ompH* gene encoding the 16-kDa cationic outer membrane protein of *Salmonella typhimurium*. *Gene* **88**:117–120.

1081. Koski, P., M. Rhen, J. Kantele, and M. Vaara. 1989. Isolation, cloning, and primary structure of a cationic 16-kDa outer membrane protein of *Salmonella typhimurium*. *J. Biol. Chem.* **264**:18973–18980.

1082. Koski, P., H. Saarilahti, S. Sukupolvi, S. Taira, P. Riikonen, K. Osterlund, R. Hurme, and M. Rhen. 1992. A new alpha-helical coiled coil protein encoded by the *Salmonella typhimurium* virulence plasmid. *J. Biol. Chem.* **267**:12258–12265.

1083. Kotani, H., and K. Nakajima. 1992. Cloning and sequence of thioredoxin gene of *Salmonella typhimurium* LT2. *Nucleic Acids Res.* **20**:1424.

1084. Kovach, J. S., A. O. Ballesteros, M. Meyers, M. Soria, and R. F. Goldberger. 1973. A *cis/trans* test of the effect on the first enzymes for histidine biosynthesis on regulation of the histidine operon. *J. Bacteriol.* **114**:351–356.

1085. Kowarz, L., C. Coynault, V. Robbe-Saule, and F. Norel. 1994. The *Salmonella typhimurium katF* (*rpoS*) gene: cloning, nucleotide sequence, and regulation of *spvR* and *spvABCD* virulence plasmid genes. *J. Bacteriol.* **176**:6852–6860.

1086. Krajewska, E., and D. Shugar. 1975. Pyrimidine nucleoside analogues as inducers of pyrimidine nucleoside catabolizing enzymes in *Salmonella typhimurium*. *Mol. Biol. Rep.* **2**:295–301.

1087. Krajewska-Grynkiewicz, K., and T. Klopotowski. 1979. Altered linkage values in phage P22-mediated transduction caused by distant deletions or insertions in donor chromosomes. *Mol. Gen. Genet.* **176**:87–94.

1088. Krajewska-Grynkiewicz, K., and S. Kustu. 1983. Regulation of transcription of *glnA*, the structural gene encoding glutamine synthetase, in *glnA*::Mud1(Ap^R, *lac*) fusion strains of *Salmonella typhimurium*. *Mol. Gen. Genet.* **192**:187–197.

1089. Krajewska-Grynkiewicz, K., and S. Kustu. 1983. Operon organization of the *glnA*, *ntrB*, and *ntrC* genes of *Salmonella typhimurium*, abstr. H153, p. 131. *Abstr. Annu. Meet. Am. Soc. Microbiol. 1983*.

1090. Krajewska-Grynkiewicz, K., and S. Kustu. 1984. Evidence that nitrogen regulatory gene *ntrC* of *Salmonella typhimurium* is transcribed from the *glnA* promoter as well as from a separate *ntr* promoter. *Mol. Gen. Genet.* **193**:135–142.

1091. Krajewska-Grynkiewicz, K., W. Walczak, and T. Klopotowski. 1971. Mutants of *Salmonella typhimurium* able to utilize D-histidine as a source of L-histidine. *J. Bacteriol.* **105**:28–37.

1092. Kramer, G. F., and B. N. Ames. 1988. Isolation and characterization of a selenium metabolism mutant of *Salmonella typhimurium*. *J. Bacteriol.* **170**:736–743.

1093. Kramer, G. F., J. C. Baker, and B. N. Ames. 1988. Near-UV stress in *Salmonella typhimurium*: 4-thiouridine in tRNA, ppGpp, and ApppGpp as components of an adaptive response. *J. Bacteriol.* **170**:2344–2351.

1094. Krause, M., F. C. Fang, and D. G. Guiney. 1992. Regulation of plasmid virulence gene expression in *Salmonella dublin* involves an unusual operon structure. *J. Bacteriol.* **174**:4482–4489.

1095. Kredich, N. M., M. A. Becker, and G. M. Tomkins. 1969. Purification and characterization of cysteine synthetase, a bifunctional protein complex, from *Salmonella typhimurium. J. Biol. Chem.* **244:**2428–2439.

1096. Kredich, N. M., L. J. Foote, and J. D. Hulanicka. 1975. Studies on the mechanism of inhibition of *Salmonella typhimurium* by 1,2,4-triazole. *J. Biol. Chem.* **250:**7324–7331.

1097. Kredich, N. M., and G. M. Tomkins. 1966. The enzymic synthesis of L-cysteine in *Escherichia coli* and *Salmonella typhimurium. J. Biol. Chem.* **241:**4955–4965.

1098. Krone, F. A., G. Westphal, and J. D. Schwenn. 1991. Characterisation of the gene *cysH* and of its product phospho-adenylylsulphate reductase from *Escherichia coli. Mol. Gen. Genet.* **225:**314–319.

1099. Kronenberg, H. M., T. Vogel, and R. F. Goldberger. 1975. A new and highly sensitive assay for the ATP phosphoribosyltransferase that catalyzes the first step in histidine biosynthesis. *Anal. Biochem.* **65:**380–388.

1100. Kruger, D. H., S. Hansen, and M. Reuter. 1983. The *ocr*+ gene function of bacteriophages T3 and T7 counteracts the *Salmonella typhimurium* DNA restriction systems SA and SB. *J. Virol.* **45:**1147–1149.

1101. Ktsoyan, Z. A., and N. N. Sarkisyan. 1979. Transfection of phage DP-8 DNA in *Salmonella derby* and radio-sensitive mutants. *Biol. Zh. Arm.* **32:**352–356.

1102. Kubori, T., N. Shimamoto, S. Yamaguchi, K. Namba, and S.-I. Aizawa. 1992. Morphological pathway of flagellar assembly in *Salmonella typhimurium. J. Mol. Biol.* **226:**433–446.

1103. Kuchino, Y., Y. Yabusaki, F. Mori, and S. Nishimura. 1984. Nucleotide sequences of three proline tRNAs from *Salmonella typhimurium. Nucleic Acids Res.* **12:**1559–1562.

1104. Kukuruzinska, M. A., W. F. Harrington, and S. Roseman. 1982. Sugar transport by the bacterial phophotransferase system. Studies on the molecular weight and association of enzyme I. *J. Biol. Chem.* **257:**14470–14476.

1105. Kulla, H. G. 1983. Regulatory citrate lyase mutants of *Salmonella typhimurium. J. Bacteriol.* **153:**546–549.

1106. Kuo, T.-T., and B. A. D. Stocker. 1969. Suppressor of proline requirement of *proA* and *proAB* deletion mutants in *Salmonella typhimurium* by mutation to arginine requirement. *J. Bacteriol.* **98:**593–598.

1107. Kuo, T.-T., and B. A. D. Stocker. 1972. Mapping of *rfa* genes in *Salmonella typhimurium* by ES18 and P22 transduction by conjugation. *J. Bacteriol.* **112:**48–63.

1108. Kuroda, M., S. de Waard, K. Mizushima, M. Tsuda, P. Postma, and T. Tsuchiya. 1992. Resistance of the melibiose carrier to inhibition by the phosphotransferase system due to substitutions of amino acid residues in the carrier of *Salmonella typhimurium. J. Biol. Chem.* **267:**18336–18341.

1109. Kustu, S., J. Hirschman, D. Burton, J. Jelesko, and J. C. Meeks. 1984. Covalent modification of bacterial glutamine synthetase: physiological significance. *Mol. Gen. Genet.* **197:**309–317.

1110. Kustu, S. G., and G. F. Ames. 1973. The *hisP* protein, a known histidine transport component in *Salmonella typhimurium*, is also an arginine transport component. *J. Bacteriol.* **116:**107–113.

1111. Kustu, S. G., and G. F. Ames. 1974. The histidine-binding protein J, a histidine transport component, has two different functional sites. *J. Biol. Chem.* **249:**6976–6983.

1112. Kustu, S. G., D. Burton, E. Garcia, L. McCarter, and N. McFarland. 1979. Nitrogen control in *Salmonella:* regulation by the *glnR* and *glnF* gene products. *Proc. Natl. Acad. Sci. USA* **76:**4576–4580.

1113. Kustu, S. G., N. C. McFarland, S. P. Hui, B. Esmon, and G. F.-L. Ames. 1979. Nitrogen control in *Salmonella typhimurium:* co-regulation of synthesis of glutamine synthetase and amino acid transport systems. *J. Bacteriol.* **138:**218–234.

1114. Kustu, S. G., and K. McKereghan. 1975. Mutations affecting glutamine synthetase activity in *Salmonella typhimurium. J. Bacteriol.* **122:**1006–1016.

1115. Kutsukake, K., and H. Doi. 1994. Nucleotide sequence of the *flgD* gene of *Salmonella typhimurium* which is essential for flagellar hook formation. *Biochim. Biophys. Acta* **1218:**443–446.

1116. Kutsukake, K., and T. Iino. 1980. A *trans*-acting factor mediates inversion of a specific DNA segment in flagellar phase variation of a *Salmonella. Nature* (London) **284:**479–481.

1117. Kutsukake, K., and T. Iino. 1980. Inversions of specific DNA segments in flagellar phase variation of *Salmonella* and inversion systems of bacteriophages P1 and Mu. *Proc. Natl. Acad. Sci. USA* **77:**7338–7341.

1118. Kutsukake, K., and T. Iino. 1985. Refined genetic analysis of the region II *che* mutants in *Salmonella typhimurium. Mol. Gen. Genet.* **199:**406–409.

1119. Kutsukake, K., and T. Iino. 1994. Role of the FliA-FlgM regulatory system on the transcriptional control of the flagellar regulon and flagellar formation in *Salmonella typhimurium. J. Bacteriol.* **176:**3598–3605.

1120. Kutsukake, K., T. Iino, Y. Komeda, and S. Yamaguchi. 1980. Functional homology of *fla* genes between *Salmonella typhimurium* and *Escherichia coli. Mol. Gen. Genet.* **178:**59–67.

1121. Kutsukake, K., T. Minamino, and T. Yokoseki. 1994. Isolation and characterization of FliK-dependent flagellation mutants from *Salmonella typhimurium. J. Bacteriol.* **176:**7625–7629.

1122. Kutsukake, K., Y. Ohya, and T. Iino. 1990. Transcriptional analysis of the flagellar regulon of *Salmonella typhimurium. J. Bacteriol.* **172:**741–747.

1123. Kutsukake, K., Y. Ohya, S. Yamaguchi, and T. Iino. 1988. Operon structure of flagellar genes in *Salmonella typhimurium. Mol. Gen. Genet.* **214:**11–15.

1124. Kutsukake, K., T. Okada, T. Yokoseki, and T. Iino. 1994. Sequence analysis of the *flgA* gene and its adjacent region in *Salmonella typhimurium*, and identification of another flagellar gene, *flgN. Gene* **143:**49–54.

1125. Kutsukake, K., T. Suzuki, S. Yamaguchi, and T. Iino. 1979. Role of gene *flaFV* on flagellar hook formation in *Salmonella typhimurium. J. Bacteriol.* **140:**267–275.

1126. Kwan, H. S., and E. L. Barrett. 1983. Roles for menaquinone and the two trimethylamine oxide (TMAO) reductases in TMAO respiration in *Salmonella typhimurium:* Mud(Ap^r *lac*) insertion mutations in *men* and *tor. J. Bacteriol.* **155:**1147–1151.

1127. Kwan, H. S., and E. L. Barrett. 1983. Mutants of *Salmonella typhimurium* defective in trimethylamine oxide reduction, abstr. K46, p. 51. *Abstr. Annu. Meet. Am. Soc. Microbiol., 1983.*

1128. Kwan, H. S., and E. L. Barrett. 1983. Purification and properties of trimethylamine oxide reductase from *Salmonella typhimurium. J. Bacteriol.* **155:**1455–1458.

1129. Kwan, H. S., and E. L. Barrett. 1984. Map locations and functions of *Salmonella typhimurium men* genes. *J. Bacteriol.* **159:**1090–1092.

1130. Kwan, H. S., H. W. Chui, and K. K. Wong. 1988. *ack*::Mu *d*1–8 (Ap^r *lac*) operon fusions of *Salmonella typhimurium* LT2. *Mol. Gen. Genet.* **211:**183–185.

1131. Kwan, H. S., and K. K. Wong. 1986. A general method for isolation of Mu d 1–8(Ap^r *lac*) operon fusions in *Salmonella typhimurium* LT2 from Tn*10* insertion strains: *chlC*::Mud1–8. *Mol. Gen. Genet.* **205:**221–224.

1132. Kwoh, D. Y., and J. Kemper. 1978. Bacteriophage P22-mediated specialized transduction in *Salmonella typhimurium:* high frequency of aberrant prophage excision. *J. Virol.* **27:**519–534.

1133. Kwoh, D. Y., and J. Kemper. 1978. Bacteriophage P22-mediated specialized transduction in *Salmonella typhimurium:* identification of different types of specialized transducing particles. *J. Virol.* **27:**535–550.

1134. Lam, S., and J. R. Roth. 1983. Genetic mapping of IS*200* copies in *Salmonella typhimurium* strain LT2. *Genetics* **105:**801–811.

1135. Lam, S., and J. R. Roth. 1983. IS*200*: a *Salmonella*-specific insertion sequence. *Cell* **34:**951–960.

1136. Lam, S., and J. R. Roth. 1986. Structural and functional studies of insertion element IS*200. J. Mol. Biol.* **187:**157–167.

1137. Lampel, K. A., and M. Riley. 1982. Discontinuity of homology of *Escherichia coli* and *Salmonella typhimurium* DNA in the *lac* region. *Mol. Gen. Genet.* **186:**82–86.

1138. Lancy, E. D., M. R. Lifsics, D. G. Kehres, and R. Maurer. 1989. Isolation and characterization of mutants with deletions in *dnaQ*, the gene for the editing subunit of DNA polymerase III in *Salmonella typhimurium. J. Bacteriol.* **171:**5572–5580.

1139. Lancy, E. D., M. R. Lifsics, P. Munson, and R. Maurer. 1989. Nucleotide sequences of *dnaE*, the gene for the polymerase subunit of DNA polymerase III in *Salmonella typhimurium*, and a variant that facilitates growth in the absence of another polymerase subunit. *J. Bacteriol.* **171:**5581–5586.

1140. Langley, D., and J. R. Guest. 1974. Biochemical and genetic characteristics of deletion and other mutant strains of *Salmonella typhimurium* LT2 lacking α-keto acid dehydrogenase complex activities. *J. Gen. Microbiol.* **82:**319–335.

1141. Lapidus, I. R., M. Welch, and M. Eisenbach. 1988. Pausing of flagellar rotation is a component of bacterial motility and chemotaxis. *J. Bacteriol.* **170:**3627–3632.

1142. LaRossa, R. A., and J. V. Schloss. 1984. The sulfonylurea herbicide sulfometuron methyl is an extremely potent and selective inhibitor of acetolactate synthase in *Salmonella typhimurium. J. Biol. Chem.* **259:**8753–8757.

1143. LaRossa, R. A., and D. R. Smulski. 1984. *ilvB*-encoded acetolactate synthase is resistant to the herbicide sulfometuron methyl. *J. Bacteriol.* **160:**391–394.

1144. LaRossa, R. A., and T. K. Van Dyk. 1989. Leaky pantothenate and thiamin mutations of *Salmonella typhimurium* conferring sulphometuron methyl sensitivity. *J. Gen. Microbiol.* **135:**2209–2222.

1145. LaRossa, R. A., T. K. VanDyk, and D. R. Smulski. 1987. Toxic accumulation of α-ketobutyrate caused by inhibition of the branched-chain amino acid biosynthetic enzyme acetolactate synthase in *Salmonella typhimurium. J. Bacteriol.* **169:**1372–1378.

1146. LaScolea, L. J., Jr., and E. Balbinder. 1972. Restoration of phosphoribosyl transferase activity by partially deleting the *trpB* gene in the tryptophan operon of *Salmonella typhimurium. J. Bacteriol.* **112:**877–885.

1147. LaScolea, L. J., Jr., M. M. Dooley, R. Torget, and E. Balbinder. 1978. A mutation to 5-methyl tryptophan dependence in the *trp* operon of *Salmonella typhimurium.* III. Correlation between phenotype and the properties of the second enzyme for tryptophan biosynthesis in a 5-methyl tryptophan dependent mutant and several 5-methyl tryptophan independent revertants. *Mol. Gen. Genet.* **165:**145–153.

1148. Laszlo, D. J., and B. L. Taylor. 1981. Aerotaxis in *Salmonella typhimurium:* role of electron transport. *J. Bacteriol.* **145:**990–1001.

1149. Lauble, H., Y. Georgalis, and U. Heinemann. 1989. Studies on the domain structure of the *Salmonella typhimurium* AraC protein. *Eur. J. Biochem.* **185:**319–325.

1150. Lawrence, D. A. 1972. Regulation of the methionine feedback sensitive enzyme in mutants of *Salmonella typhimurium. J. Bacteriol.* **109:**8–11.

1151. Lawrence, D. A., D. A. Smith, and R. J. Rowbury. 1968. Regulation of methionine synthesis in *Salmonella typhimurium:* mutants resistant to inhibition by analogues of methionine. *Genetics* **58:**473–492.

1152. Lawrence, J. G., H. Ochman, and D. L. Hartl. 1991. Molecular and evolutionary relationships among enteric bacteria. *J. Gen. Microbiol.* **137**(Pt. 8):1911–1921.

1153. Lawther, R. P., R. C. Wek, J. M. Lopes, R. Pereira, B. E. Taillon, and G. W. Hatfield. 1987. The complete nucleotide sequence of the *ilvGMEDA* operon of *Escherichia coli* K-12. *Nucleic Acids Res.* **15:**2137–2155. (Errata, **15:**9108, 1987, **16:**3602, 1988.)

1154. Lawton, K. G., and H. W. Taber. 1980. Isolation of an F′ carrying a portion of the histidine transport operon of *Salmonella typhimurium*, abstr. H106, p. 126. *Abstr. Annu. Meet. Am. Soc. Microbiol., 1980.*

1155. Lawton, K. G., and H. W. Taber. 1984. Isolation of F′ plasmids carrying a portion of the *Salmonella typhimurium* histidine transport operon. *J. Bacteriol.* **157:**697–702.

1156. Leary, T. R., and G. B. Kohlhaw. 1974. α-Isopropylmalate synthase from *Salmonella typhimurium*. *J. Biol. Chem.* **247:**1089–1095.

1157. Lederberg, J., and T. Iino. 1956. Phase variation in *Salmonella*. *Genetics* **41:**743–757.

1158. Lee, C. A., B. D. Jones, and S. Falkow. 1992. Identification of a *Salmonella typhimurium* invasion locus by selection for hyperinvasive mutants. *Proc. Natl. Acad. Sci. USA* **89:**1847–1851.

1159. Lee, F., K. Bertrand, G. Bennett, and C. Yanofsky. 1978. Comparison of the nucleotide sequences of the initial transcribed regions of the tryptophan operons of *Escherichia coli* and *Salmonella typhimurium*. *J. Mol. Biol.* **121:**193–217.

1160. Lee, G. S., and G. F.-L. Ames. 1984. Analysis of promoter mutations in the histidine transport operon of *Salmonella typhimurium*: use of hybrid M13 bacteriophages for cloning, transformation, and sequencing. *J. Bacteriol.* **159:**1000–1005.

1161. Lee, J.-H., K. Burke, and G. Wilcox. 1986. Mutations resulting in promoter-like sequences which enhance the expression of *araC* in *Salmonella typhimurium*. *Gene* **46:**113–121.

1162. Lee, J.-H., L. Heffernan, and G. Wilcox. 1980. Isolation of *ara-lac* gene fusions in *Salmonella typhimurium* LT-2 by using transducing bacteriophage Mu *d*(Apʳ *lac*). *J. Bacteriol.* **143:**1325–1331.

1163. Lee, J.-H., J. Nishitani, and G. Wilcox. 1984. Genetic characterization of *Salmonella typhimurium* LT2 *ara* mutations. *J. Bacteriol.* **158:**344–346.

1164. Lee, P. C., B. R. Bochner, and B. N. Ames. 1983. AppppA, heat-shock stress, and cell oxidation. *Proc. Natl. Acad. Sci. USA* **80:**7496–7500.

1165. Lee, S. Y., and C. T. Grubmeyer. 1987. Purification and in vitro complementation of mutant histidinol dehydrogenases. *J. Bacteriol.* **169:**3938–3944.

1166. Lee-Peng, F.-C., M. A. Hermodson, and G. B. Kohlhaw. 1979. Transaminase B from *Escherichia coli*: quaternary structure, amino-terminal sequence, substrate specificity, and absence of a separate valine-α-ketoglutarate activity. *J. Bacteriol.* **139:**339–345.

1167. Lehman, V., J. Redmond, A. Egan, and I. Minner. 1978. The acceptor for polar head groups of the lipid A component of *Salmonella* lipopolysaccharides. *Eur. J. Biochem.* **86:**487–496.

1168. Lehman, V., and E. Rupprecht. 1977. Micro-heterogeneity in lipid A demonstrated by a new intermediate in the biosynthesis of 3-deoxy-D-manno-octulosonic acid lipid A. *Eur. J. Biochem.* **81:**443–452.

1169. Lehman, V., E. Rupprecht, and M. J. Osborn. 1977. Isolation of mutants conditionally blocked in the biosynthesis of the 3-deoxy-D-manno-octulosonic acid lipid A part of lipopolysaccharides derived from *Salmonella typhimurium*. *Eur. J. Biochem.* **76:**41–49.

1170. Lehmann, V., G. Hammerling, M. Nurminen, I. Minner, E. Ruschmann, O. Luderitz, T. Kuo, and B. A. D. Stocker. 1973. A new class of heptose-defective mutants of *Salmonella typhimurium*. *Eur. J. Biochem.* **32:**268–275.

1171. Lehner, A. F., S. Harvey, and C. W. Hill. 1984. Mapping and spacer identification of rRNA operons of *Salmonella typhimurium*. *J. Bacteriol.* **160:**682–686.

1172. Lehner, A. F., and C. W. Hill. 1980. Involvement of ribosomal ribonucleic acid operons in *Salmonella typhimurium* chromosomal rearrangements. *J. Bacteriol.* **143:**492–496.

1173. Lehner, A. F., and C. W. Hill. 1985. Merodiploidy in *Escherichia coli-Salmonella typhimurium* crosses: the role of unequal recombination between ribosomal RNA genes. *Genetics* **110:**365–380.

1174. Lemoine, V. R., and R. J. Rowbury. 1977. An effect of F-like plasmids on the maintenance of F*lac* in a *dnaC* mutant of *Salmonella typhimurium*. *Mol. Gen. Genet.* **156:**313–318.

1175. Lenny, A. B., and P. Margolin. 1980. Locations of the *opp* and *supX* genes of *Salmonella typhimurium* and *Escherichia coli*. *J. Bacteriol.* **143:**747–752.

1176. Leonard, J. E., and M. H. Saier, Jr. 1981. Genetic dissection of catalytic activities of the *Salmonella typhimurium* mannitol enzyme II. *J. Bacteriol.* **145:**1106–1109.

1177. Leong, J., and J. B. Neilands. 1976. Mechanisms of siderophore iron transport in enteric bacteria. *J. Bacteriol.* **126:**823–830.

1178. Leong, J. M., S. Nunes-Duenby, C. F. Lesser, P. Youderian, M. M. Susskind, and A. Landy. 1985. The (phi)80 and P22 attachment sites: primary structure and interaction with *Escherichia coli* integration host factor. *J. Biol. Chem.* **260:**4468–4477.

1179. Lepore, G. C., P. DiNatale, L. Guarini, and F. DeLorenzo. 1975. Histidyl-tRNA synthetase from *Salmonella typhimurium*: specificity in the binding of histidine analogues. *Eur. J. Biochem.* **56:**369–374.

1180. Leung, P. S. C., and J. Preiss. 1987. Cloning of the ADPglucose pyrophosphorylase (*glgC*) and glycogen synthase (*glgA*) structural genes from *Salmonella typhimurium* LT2. *J. Bacteriol.* **169:**4349–4354.

1181. Leung, P. S. C., and J. Preiss. 1987. Biosynthesis of bacterial glycogen: primary structure of *Salmonella typhimurium* ADP-glucose synthetase as deduced from the nucleotide sequence of the *glgC* gene. *J. Bacteriol.* **169:**4355–4360.

1182. Lever, J. E. 1972. Purification and properties of a component of histidine transport in *Salmonella typhimurium*. *J. Biol. Chem.* **247:**4317–4326.

1183. Levine, S. M., F. Ardeshir, and G. F.-L. Ames. 1980. Isolation and characterization of acetatekinase and phosphotransacetylase mutants of *Escherichia coli* and *Salmonella typhimurium*. *J. Bacteriol.* **143:**1081–1085.

1184. Levinthal, M., and H. Nikaido. 1969. Consequences of deletion mutations joining two operons of opposite polarity. *J. Mol. Biol.* **42:**511–520.

1185. Levinthal, M., L. S. Williams, and H. E. Umbarger. 1973. Role of threonine deaminase in the regulation of isoleucine and valine biosynthesis. *Nature* (London) *New Biol.* **246:**65–68.

1186. Levy, M. S., P. Pomposiello, and R. Nagel. 1991. *RecA*-dependent increased precise excision of Tn*10* in *Salmonella typhimurium*. *Mutat. Res.* **255:**95–100.

1187. Lew, H. C., P. H. Makela, H.-M. Kuhn, H. Mayer, and H. Nikaido. 1986. Biosynthesis of enterobacterial common antigen requires dTDPglucose pyrophosphorylase determined by a *Salmonella typhimurium rfb* gene and a *Salmonella montevideo rfe* gene. *J. Bacteriol.* **168:**715–721.

1188. Lew, H. C., H. Nikaido, and P. H. Makela. 1978. Biosynthesis of uridine diphosphate *N*-acetylmannosaminuronic acid in *rff* mutants of *Salmonella typhimurium*. *J. Bacteriol.* **136:**227–233.

1189. Lewis, J. A., and B. N. Ames. 1972. Histidine regulation in *Salmonella typhimurium*. XI. The percentage of transfer RNAʰⁱˢ charged *in vivo* and its relation to the repression of the histidine operon. *J. Mol. Biol.* **66:**131–142.

1190. Li, S.-L., J. Hanlon, and C. Yanofsky. 1974. Structural homology of the glutamine amidotransferase subunits of the anthranilate synthetasesof *Escherichia coli*, *Salmonella typhimurium*, and *Serratia marcescens*. *Nature* (London) **248:**48–50.

1191. Li, S.-L., J. Hanlon, and C. Yanofsky. 1974. Separation of anthranilate synthesis components I and II of *Escherichia coli*, *Salmonella typhimurium*, and *Serratia marcescens* and determination of their amino-terminal sequences by Edman degradation. *Biochemistry* **13:**1736–1744.

1192. Li, S.-L., J. Hanlon, and C. Yanofsky. 1975. Amino-terminal sequences of indoleglycerol phosphate synthetase of *Escherichia coli* and *Salmonella typhimurium*. *J. Bacteriol.* **123:**761–764.

1193. Li, S.-L., and C. Yanofsky. 1972. Amino acid sequences of fifty residues from the amino termini of the tryptophan synthetase α chains of several Enterobacteria. *J. Biol. Chem.* **247:**1031–1037.

1194. Li, S.-L., and C. Yanofsky. 1973. Amino acid sequence studies with the tryptophan synthetase α chain of *Salmonella typhimurium*. *J. Biol. Chem.* **248:**1830–1836.

1195. Li, Y. F., and A. Sancar. 1991. Cloning, sequencing, expression and characterization of DNA photolyase from *Salmonella typhimurium*. *Nucleic Acids Res.* **19:**4885–4890.

1196. Li, Z. J., D. Hillyard, and P. Higgins. 1989. Nucleotide sequence of the *Salmonella typhimurium himA* gene. *Nucleic Acids Res.* **17:**8880.

1197. Libby, S. J., W. Goebel, A. Ludwig, N. Buchmeier, F. Bowe, F. C. Fang, D. G. Guiney, J. G. Songer, and F. Heffron. 1994. A cytolysin encoded by *Salmonella* is required for survival within macrophages. *Proc. Natl. Acad. Sci. USA* **91:**489–493.

1198. Libby, S. J., W. Goebel, S. Muir, G. Songer, and F. Heffron. 1990. Cloning and characterization of a cytotoxin gene from *Salmonella typhimurium*. *Res. Microbiol.* **141:**775–783.

1199. LiCalsi, C., T. S. Crocenzi, E. Freire, and S. Roseman. 1991. Sugar transport by the bacterial phosphotransferase system. Structural and thermodynamic domains of enzyme I of *Salmonella typhimurium*. *J. Biol. Chem.* **266:**19519–19527.

1200. Lifsics, M. R., E. D. Lancy, Jr., and R. Maurer. 1992. DNA replication defect in *Salmonella typhimurium* mutants lacking the editing (epsilon) subunit of DNA polymerase III. *J. Bacteriol.* **174:**6965–6973.

1201. Liljestrom, P., I. Laamanen, and E. T. Palva. 1988. Structure and expression of the *ompB* operon, the regulatory locus for the outer membrane porin regulon in *Salmonella typhimurium* LT-2. *J. Mol. Biol.* **201:**663–673.

1202. Liljestrom, P., M. Luokkamaki, and E. T. Palva. 1987. Isolation and characterization of a substitution mutation in the *ompR* gene of *Salmonella typhimurium* LT2. *J. Bacteriol.* **169:**438–441.

1203. Liljestrom, P., P. L. Maattanen, and E. T. Palva. 1982. Cloning of the regulatory locus *ompB* of *Salmonella typhimurium* LT-2. I. Isolation of the *ompR* gene and identification of its gene product. *Mol. Gen. Genet.* **188:**184–189.

1204. Liljestrom, P., P. L. Maattanen, and E. T. Palva. 1982. Cloning of the regulatory locus *ompB* of *Salmonella typhimurium* LT-2. II. Identification of the *envZ* gene product, a protein involved in the expression of the porin proteins. *J. Bacteriol.* **188:**190–194.

1205. Lilleengen, K. 1948. Typing *Salmonella typhimurium* by means of bacteriophage. *Acta Pathol. Microbiol. Scand. Suppl.* **77:**11–125.

1206. Lilley, D. M., and C. F. Higgins. 1991. Local DNA topology and gene expression: the case of the *leu-500* promoter. *Mol. Microbiol.* **5:**779–783.

1207. Lin, H.-C., S.-P. Lei, G. Studnicka, and G. Wilcox. 1985. The *araBAD* operon of *Salmonella typhimurium* LT2. III. Nucleotide sequence of *araD* and its flanking regions, and primary structure of its product, L-ribulose-5-phosphate 4-epimerase. *Gene* **34:**129–134.

1208. Lin, H.-C., S.-P. Lei, and G. Wilcox. 1985. The *araBAD* operon of *Salmonella typhimurium* LT2. I. Nucleotide sequence of *araB* and primary structure of its product, ribulokinase. *Gene* **34:**111–122.

1209. Lin, H.-C., S.-P. Lei, and G. Wilcox. 1985. The *araBAD* operon of *Salmonella typhimurium* LT2. II. Nucleotide sequence of *araA* and primary structure of its product, L-arabinose isomerase. *Gene* **34:**123–128.

1210. Lin, J. J.-C., and P. Wu. 1976. Biosynthesis and assembly of envelope lipoprotein in a glycerol-requiring mutant of *Salmonella typhimurium*. *J. Bacteriol.* **125:**892–904.

1211. Lindberg, A. A., and C.-G. Hellerqvist. 1980. Rough mutants of *Salmonella typhimurium*. Immunochemical and structural analysis of lipopolysaccharides from *rfaH* mutants. *J. Gen. Microbiol.* **116**:25–32.

1212. Lindberg, A. A., C. G. Hellerqvist, G. Bagdian-Motta, and P. H. Makela. 1978. Lipopolysaccharide modification accompanying antigenic conversion by phage P27. *J. Gen. Microbiol.* **107**:279–287.

1213. Lindberg, A. A., and S. Svensson. 1975. *Salmonella typhimurium* mutations conferring resistance to Felix O phage without loss of smooth character: phage attachment and immunochemical and structural analyses of lipopolysaccharides. *J. Gen. Microbiol.* **87**:11–19.

1214. Lindqvist, L., R. Kaiser, P. R. Reeves, and A. A. Lindberg. 1993. Purification, characterization and HPLC assay of *Salmonella* glucose-1-phosphate thymidylyltransferase from the cloned *rfbA* gene. *Eur. J. Biochem.* **211**:763–770.

1215. Lindqvist, L., R. Kaiser, P. R. Reeves, and A. A. Lindberg. 1994. Purification, characterization, and high performance liquid chromatography assay of *Salmonella* glucose-1-phosphate cytidylyltransferase from the cloned *rfbF* gene. *J. Biol. Chem.* **269**:122–126.

1216. Lindqvist, L., E. K. H. Schweda, P. R. Reeves, and A. A. Lindberg. 1994. *In vitro* synthesis of CDP-D-abequose using *Salmonella* enzymes of cloned *rfb* genes: production of CDP-6-deoxy-D-*xylo*-4-hexulose, CDP-3,6-dideoxy-D-*xylo*-hexulose and CDP-3,6-dideoxy-D-galactose, and isolation by HPLC. *Eur. J. Biochem.* **225**:863–872.

1217. Lisitsyn, N. A., G. S. Monastyrskaya, and E. D. Sverdlov. 1988. Genes coding for RNA polymerase beta subunit in bacteria. Structure/function analysis. *Eur. J. Biochem.* **177**:363–369.

1218. Liu, D., A. M. Haase, L. Lindqvist, A. A. Lindberg, and P. R. Reeves. 1994. Glycosyl transferases of O-antigen biosynthesis in *Salmonella enterica*: identification and characterization of transferase genes of groups B, C2, and E1. *J. Bacteriol.* **175**:3408–3413.

1219. Liu, D., N. K. Verma, L. K. Romana, and P. R. Reeves. 1991. Relationships among the *rfb* regions of *Salmonella* serovars A, B, and D. *J. Bacteriol.* **173**:4814–4819.

1220. Liu, G., J. Foster, P. Manlapaz-Ramos, and B. M. Olivera. 1982. Nucleoside salvage pathway for NAD biosynthesis in *Salmonella typhimurium*. *J. Bacteriol.* **152**:1111–1116.

1221. Liu, J., and I. R. Beacham. 1990. Transcription and regulation of the *cpdB* gene in *Escherichia coli* K12 and *Salmonella typhimurium* LT2: evidence for modulation of constitutive promoters by cyclic AMP-CRP complex. *Mol. Gen. Genet.* **222**:161–165.

1222. Liu, S.-L., A. Hessel, and K. E. Sanderson. 1993. The XbaI-BlnI-CeuI genomic cleavage map of *Salmonella typhimurium* LT2 determined by double digestion, end labelling, and pulsed-field gel electrophoresis. *J. Bacteriol.* **175**:4104–4120.

1223. Liu, S.-L., A. Hessel, and K. E. Sanderson. 1994. Genomic mapping with I-CeuI, an intron-encoded endonuclease specific for genes for ribosomal RNA, in *Salmonella* spp., *Escherichia coli*, and other bacteria. *Proc. Natl. Acad. Sci. USA* **90**:6874–6878.

1224. Liu, S.-L., and K. E. Sanderson. 1992. A physical map of the *Salmonella typhimurium* LT2 genome made by using XbaI analysis. *J. Bacteriol.* **174**:1662–1672.

1225. Liveris, D., J. J. Schwartz, R. Geertman, and I. Schwartz. 1993. Molecular cloning and sequencing of *infC*, the gene encoding translation initiation factor IF3, from four enterobacterial species. *FEMS Microbiol. Lett.* **112**:211–216.

1226. Lockman, H. A., and R. Curtiss. 1990. *Salmonella typhimurium* mutants lacking flagella or motility remain virulent in BALB/c mice. *Infect. Immun.* **58**:137–143.

1227. Lockman, H. A., and R. Curtiss. 1992. Isolation and characterization of conditional adherent and non-type 1 fimbriated *Salmonella typhimurium* mutants. *Mol. Microbiol.* **6**:933–945.

1228. Lockman, H. A., and R. Curtiss. 1992. Virulence of non-type 1-fimbriated and nonfimbriated nonflagellated *Salmonella typhimurium* mutants in murine typhoid fever. *Infect. Immun.* **60**:491–496.

1229. Loewen, P. C., and G. V. Stauffer. 1990. Nucleotide sequence of *katG* of *Salmonella typhimurium* LT2 and characterization of its product, hydroperoxidase I. *Mol. Gen. Genet.* **224**:147–151.

1230. Lombardo, M. J., D. Bagga, and C. G. Miller. 1991. Mutations in *rpoA* affect expression of anaerobically regulated genes in *Salmonella typhimurium*. *J. Bacteriol.* **173**:7511–7518.

1231. Loper, J. C., M. Grabnar, R. C. Stahl, Z. Hartman, and P. E. Hartman. 1964. Genes and proteins involved in histidine biosynthesis in *Salmonella*. *Brookhaven Symp. Biol.* **17**:15–52.

1232. Lopes, J. M., and R. P. Lawther. 1986. Analysis and comparison of the internal promoter, pE, of the *ilvGMEDA* operons from *Escherichia coli* K-12 and *Salmonella typhimurium*. *Nucleic Acids Res.* **14**:2779–2798.

1233. Lopes, J. M., and R. P. Lawther. 1989. Physical identification of an internal promoter, *ilvAp*, in the distal portion of the *ilvGMEDA* operon. *Gene* **76**:255–269.

1234. Lorenzo, C., E. Howard, and R. Nagel. 1990. Studies on Tn*10* transposition and excision in DNA-repair mutants of *Salmonella typhimurium*. *Mutat. Res.* **232**:99–104.

1235. Loughlin, R. E. 1975. Polarity of the *cysJIH* operon of *Salmonella typhimurium*. *J. Gen. Microbiol.* **86**:275–282.

1236. Loughlin, R. E. 1976. Identification of nonsense mutations in the *cysJ* gene of *Salmonella typhimurium*. *J. Gen. Microbiol.* **95**:186–187.

1237. Lounatmaa, K. 1979. Ultrastructure of the outer membrane of *Salmonella typhimurium* bacteriocin-resistant mutants deficient in the 33K protein. *J. Bacteriol.* **139**:646–651.

1238. Lounatmaa, K., P. H. Makela, and M. Sarvas. 1976. Effect of polymyxin on the ultrastructure of the outer membrane of wild-type and polymyxin-resistant strains of *Salmonella*. *J. Bacteriol.* **127**:1400–1407.

1239. Lu, A. L., M. J. Cuipa, M. S. Ip, and W. G. Shanabruch. 1990. Specific A/G-to-C.G mismatch repair in *Salmonella typhimurium* LT2 requires the *mutB* gene product. *J. Bacteriol.* **172**:1232–1240.

1240. Lu, C. D., and A. T. Abdelal. 1993. Complete sequence of the *Salmonella typhimurium* gene encoding malate dehydrogenase. *Gene* **123**:143–144.

1241. Lu, C. D., and A. T. Abdelal. 1993. The *Salmonella typhimurium* uracil-sensitive mutation use is in *argU* and encodes a minor arginine tRNA. *J. Bacteriol.* **175**:3897–3899.

1242. Lu, C. D., J. E. Houghton, and A. T. Abdelal. 1992. Characterization of the arginine repressor from *Salmonella typhimurium* and its interactions with the *carAB* operator. *J. Mol. Biol.* **225**:11–24.

1243. Lu, C. D., M. Kilstrup, J. Neuhard, and A. Abdelal. 1989. Pyrimidine regulation of tandem promoters for *carAB* in *Salmonella typhimurium*. *J. Bacteriol.* **171**:5436–5442.

1244. Luckey, M., and J. B. Neilands. 1976. Iron transport in *Salmonella typhimurium* LT-2: prevention, by ferrichrome, of adsorption of bacteriophages ES18 and ES18.h1 to a common cell envelope receptor. *J. Bacteriol.* **127**:1036–1037.

1245. Luckey, M., J. R. Pollack, R. Wayne, B. N. Ames, and J. B. Neilands. 1972. Iron uptake in *Salmonella typhimurium*: utilization of exogenous siderochromes as iron carriers. *J. Bacteriol.* **111**:731–738.

1246. Luderitz, O., A. M. Staub, and O. Westphal. 1966. Immunochemistry of O and R antigens of *Salmonella* and related *Enterobacteriaceae*. *Bacteriol. Rev.* **30**:192–255.

1247. Luderitz, O., O. Westphal, A. M. Staub, and H. Nikaido. 1971. Isolation and chemical and immunological characterization of bacterial lipopolysaccharides, p. 145–233. *In* G. Weinbaum, S. Kadis, and S. J. Ajl (ed.), *Bacterial Endotoxins*, vol. 4 of *Microbial Toxins*. Academic Press, Inc., New York.

1248. Luginbuhl, G. H., J. G. Hofler, C. J. Decedue, and R. O. Burns. 1974. Biodegradative L-threonine deaminase of *Salmonella typhimurium*. *J. Bacteriol.* **120**:559–561.

1249. Lukat, G. S., B. H. Lee, J. M. Mottonen, A. M. Stock, and J. B. Stock. 1991. Roles of the highly conserved aspartate and lysine residues in the response regulator of bacterial chemotaxis. *J. Biol. Chem.* **266**:8348–8354.

1250. Lukat, G. S., A. M. Stock, and J. B. Stock. 1990. Divalent metal ion binding to the CheY protein and its significance to phosphotransfer in bacterial chemotaxis. *Biochemistry* **29**:5436–5442.

1251. Lupas, A., and J. Stock. 1989. Phosphorylation of an N-terminal regulatory domain activates the CheB methylesterase in bacterial chemotaxis. *J. Biol. Chem.* **264**:17337–17342.

1252. Luttinger, A. L., A. L. Springer, and M. B. Schmid. 1991. A cluster of genes that affects nucleoid segregation in *Salmonella typhimurium*. *New Biol.* **3**:687–697.

1253. Lyman, M. B., A. B. D. Stocker, and R. J. Roantree. 1979. Evaluation of the immune response directed against the *Salmonella* antigenic factors O4,5 and O9. *Infect. Immun.* **26**:956–965.

1254. MacAlister, T. J., W. R. Cook, R. Weigand, and L. I. Rothfield. 1987. Membrane-murein attachment at the leading edge of the division septum: a second membrane-murein structure associated with morphogenesis of the gram-negative bacterial division septum. *J. Bacteriol.* **169**:3945–3951.

1255. Macias, E. A., F. Rana, J. Blazyk, and M. C. Modrzakowski. 1990. Bactericidal activity of magainin 2: use of lipopolysaccharide mutants. *Can. J. Microbiol.* **36**:582–584.

1256. Mack, D. P., J. P. Sluka, J. A. Shin, J. H. Griffin, M. I. Simon, and P. B. Dervan. 1990. Orientation of the putative recognition helix in the DNA-binding domain of Hin recombinase complexed with the *hix* site. *Biochemistry* **29**:6561–6567.

1257. MacKay, D. A., A. Eisenstark, R. B. Webb, and M. S. Brown. 1976. Action spectra for lethality in recombinationless strains of *Salmonella typhimurium* and *Escherichia coli*. *Photochem. Photobiol.* **24**:337–343.

1258. MacLachlan, P. R., S. K. Kadam, and K. E. Sanderson. 1991. Cloning, characterization, and DNA sequence of the *rfaLK* region for lipopolysaccharide synthesis in *Salmonella typhimurium* LT2. *J. Bacteriol.* **173**:7151–7163.

1259. MacLachlan, P. R., and K. E. Sanderson. 1985. Transformation of *Salmonella typhimurium* with plasmid DNA: differences between rough and smooth strains. *J. Bacteriol.* **161**:442–445.

1260. Macnab, R., and D. E. Koshland, Jr. 1972. Persistence as a concept in the motility of chemotactic bacteria. *J. Mechanochem. Cell Motil.* **2**:141–148.

1261. Macnab, R., and D. E. Koshland, Jr. 1974. Bacterial motility and chemotaxis: light-induced tumbling response and visualization of individual flagella. *J. Mol. Biol.* **84**:399–406.

1262. Macnab, R., and D. E. Koshland, Jr. 1974. The gradient mechanism in bacterial chemotaxis. *Proc. Natl. Acad. Sci. USA* **69**:2509–2512.

1263. Macnab, R. M. 1977. Bacterial flagella rotating in bundles: a study of helical geometry. *Proc. Natl. Acad. Sci. USA* **74**:221–225.

1264. Macnab, R. M. 1987. Flagella, p. 70–83. *In* F. C. Neidhardt, J. L. Ingraham, K. B. Low, B. Magasanik, M. Schaechter, and H. E. Umbarger (ed.), *Escherichia coli and Salmonella typhimurium: Cellular and Molecular Biology*. American Society for Microbiology, Washington, D.C.

1265. Macnab, R. M., and M. K. Ornston. 1977. Normal-to-curly flagellar transitions and their role in bacterial tumbling. *J. Mol. Biol.* **112**:1–30.

1266. MacPhee, D. G. 1973. Effect of *rec* mutations on the ultraviolet protecting and mutation-enhancing properties of the plasmid R-Utrecht in *Salmonella typhimurium*. *Mutat. Res.* **19**:357–359.

1267. MacPhee, D. G., and M. R. Beazer. 1973. Mutants of *Salmonella typhimurium* deficient in DNA polymerase I: detection by their failure to produce colicin E1. *Mol. Gen. Genet.* **127**:229–240.

1268. MacPhee, D. G., and M. R. Beazer. 1975. Mutants of *Salmonella typhimurium* deficient in DNA polymerase I: further characterization and genetic analysis. *Aust. J. Biol. Sci.* **28**:559–565.

1269. Madonna, J. J., R. L. Fuchs, and J. E. Brenchley. 1985. Fine structure analysis of *Salmonella typhimurium* glutamate synthase genes. *J. Bacteriol.* **161**:353–360.

1270. Maeba, P., and B. D. Sanwal. 1969. Phosphoenolpyruvate carboxylase of *Salmonella*. Some chemical and allosteric properties. *J. Biol. Chem.* **244**:2549–2557.

1271. Magariyama, Y., S. Yamaguchi, and S. Aizawa. 1990. Genetic and behavioral analysis of flagellar switch mutants of *Salmonella typhimurium*. *J. Bacteriol.* **172**:4359–4369.

1272. Magasanik, B. 1978. Regulation of the *hut* system. *Cold Spring Harbor Monogr. Ser.* **1978**:373–388.

1273. Magnussen, K.-E., O. Stendahl, C. Tagesson, L. Edebo, and G. Johansson. 1977. The tendency of smooth and rough *Salmonella typhimurium* bacteria and lipopolysaccharide to hydrophobic and ionic interaction as studied in aqueous polymer 2 phase systems. *Acta Pathol. Microbiol. Scand. Sect. B* **85**:212–218.

1274. Mahan, M. J., J. Casadesus, and J. R. Roth. 1992. The *Salmonella typhimurium* RecJ function permits growth of P22 abc phage on *recBCD+* hosts. *Mol. Gen. Genet.* **232**:470–478.

1275. Mahan, M. J., and L. N. Csonka. 1983. Genetic anlaysis of the *proBA* genes of *Salmonella typhimurium*: physical and genetic analyses of the cloned *proB⁺A⁺* genes of *Escherichia coli* and of a mutant allele that confers proline overproduction and enhanced osmotolerance. *J. Bacteriol.* **156**:1249–1262.

1276. Mahan, M. J., A. Garzon, and J. Casadesus. 1993. Host RecJ is required for growth of P22 *erf* bacteriophage. *J. Bacteriol.* **175**:288–290.

1277. Mahan, M. J., and J. R. Roth. 1989. *recB* and *recC* genes of *Salmonella typhimurium*. *J. Bacteriol.* **171**:612–615.

1278. Mäkelä, P. H. 1964. Genetic homologies between flagellar antigens of *Escherichia coli* and *Salmonella abony*. *J. Gen. Microbiol.* **35**:503–510.

1279. Mäkelä, P. H. 1966. Genetic determination of the O antigens of *Salmonella* groups B (4, 5, 12) and C₁ (6, 7). *J. Bacteriol.* **91**:1115–1125.

1280. Mäkelä, P. H. 1973. Glucosylation of lipopolysaccharide in *Salmonella*: mutants negative for O antigen factor 12². *J. Bacteriol.* **116**:847–856.

1281. Mäkelä, P. H., M. Jahkola, and O. Lüderitz. 1970. A new gene cluster *rfe* concerned with the biosynthesis of *Salmonella* lipopolysaccharide. *J. Gen. Microbiol.* **60**:91–106.

1282. Mäkelä, P. H., and O. Mäkelä. 1966. Salmonella antigen 12₂: genetics of form variation. *Ann. Med. Exp. Biol. Fenn.* **44**:310–317.

1283. Mäkelä, P. H., and H. Mayer. 1974. Participation of lipopolysaccharide genes in the determination of the enterobacterial common antigen: analysis in *Salmonella* groups B and C₁. *J. Bacteriol.* **119**:765–770.

1284. Mäkelä, P. H., and H. Mayer. 1976. Enterobacterial common antigen. *Bacteriol. Rev.* **40**:591–632.

1285. Mäkelä, P. H., H. Mayer, H. Y. Whang, and E. Neter. 1974. Participation of lipopolysaccharide genes in the determination of the enterobacterial common antigen: analysis of R mutants of *Salmonella minnesota*. *J. Bacteriol.* **119**:760–764.

1286. Mäkelä, P. H., G. Schmidt, H. Mayer, H. Nikaido, H. Y. Whang, and E. Neter. 1976. Enterobacterial common antigen in *rfb* deletion mutants of *Salmonella typhimurium*. *J. Bacteriol.* **127**:1141–1149.

1287. Mäkelä, P. H., and B. A. D. Stocker. 1969. Genetics of polysaccharide biosynthesis. *Annu. Rev. Genet.* **3**:291–322.

1288. Mäkelä, P. H., and B. A. D. Stocker. 1981. Genetics of the bacterial cell surface. *Symp. Soc. Gen. Microbiol.* **31**:219–264.

1289. Makover, S., and E. Telep. 1978. The antibacerial potential of a phosphoenolpyruvate sugar phosphotransferase system blocking agent. *J. Antibiot.* **31**:237–238.

1290. Maloy, S. R., and J. R. Roth. 1983. Regulation of proline utilization in *Salmonella typhimurium*: characterization of *put*::Mu d(Ap, *lac*) operon fusions. *J. Bacteriol.* **154**:561–568.

1291. Mankovich, J. A., C. A. McIntyre, and G. C. Walker. 1989. Nucleotide sequence of the *Salmonella typhimurium mutL* gene required for mismatch repair: homology of MutL to HexB of *Streptococcus pneumoniae* and to PMS1 of the yeast *Saccharomyces cerevisiae*. *J. Bacteriol.* **171**:5325–5331.

1292. Manuck, B. A., and C. Ho. 1979. High-resolution proton nuclear magnetic resonance studies of histidine binding proteins J of *Salmonella typhimurium*. An investigation of substrate and membrane interaction sites. *Biochemistry* **18**:566–573.

1293. Marcus, S. L., and E. Balbinder. 1972. Purification of anthranilate-5-phosphoribosyl pyrophosphate phosphoribosyl transferase from *Salmonella typhimurium* using affinity chromatography: resolution of monomeric and dimeric forms. *Biochem. Biophys. Res. Commun.* **47**:438–444.

1294. Mares, R., M. L. Urbanowski, and G. V. Stauffer. 1992. Regulation of the *Salmonella typhimurium metA* gene by the *metR* protein and homocysteine. *J. Bacteriol.* **174**:390–397.

1295. Margolies, M. N., and R. F. Goldberger. 1967. Physical and chemical characterization of the isomerase of histidine biosynthesis in *Salmonella typhimurium*. *J. Biol. Chem.* **242**:256–264.

1296. Margolies, M. N., and R. F. Goldberger. 1968. Correlation between mutation type and the production of cross-reacting material in mutants of the A gene of the histidine operon in *Salmonella typhimurium*. *J. Bacteriol.* **95**:507–519.

1297. Margolin, P. 1963. Genetic fine structure of the leucine operon in *Salmonella*. *Genetics* **48**:441–457.

1298. Marsh, M., and D. R. Hillyard. 1988. Nucleotide sequence of the HU-1 gene of *Salmonella typhimurium*. *Nucleic Acids Res.* **16**:7196.

1299. Marsh, M., and D. R. Hillyard. 1990. Nucleotide sequence of *hns* encoding the DNA-binding protein H-NS of *Salmonella typhimurium*. *Nucleic Acids Res.* **18**:3397.

1300. Martin, R. G. 1968. Polarity in relaxed strains of *Salmonella typhimurium*. *J. Mol. Biol.* **31**:127–134.

1301. Martin, R. G., M. A. Berberich, B. N. Ames, W. W. Davis, R. F. Goldberger, and J. Yourno. 1971. Enzymes and intermediates of histidine biosynthesis in *Salmonella typhimurium*. *Methods Enzymol.* **17**:3–44.

1302. Martin, R. G., and R. F. Goldberger. 1967. Imidazolylacetolphosphate:L-glutamate aminotransferase. Purification and physical properties. *J. Biol. Chem.* **242**:1168–1174.

1303. Martin, R. G., M. J. Voll, and E. Appella. 1967. Imidazolylacetolphosphate:L-glutamate aminotransferase. Composition and substructure. *J. Biol. Chem.* **242**:1175–1181.

1304. Marumo, K., L. Lindqvist, N. Verma, A. Weintraub, P. R. Reeves, and A. A. Lindberg. 1992. Enzymatic synthesis and isolation of thymidine diphosphate-6-deoxy-D-*xylo*-4-hexulose and thymidine diphosphate-L-rhamnose. *Eur. J. Biochem.* **204**:539–545.

1305. Masten, B. J., and T. M. Joys. 1993. Molecular analyses of the *Salmonella g...* flagellar antigen complex. *J. Bacteriol.* **175**:5359–5365.

1306. Matsubara, K., K. Ohnishi, and K. Kiritani. 1987. Location of *livA* gene participating in the high-affinity transport of branched-chain amino acids in *Salmonella typhimurium* LT2. *Jpn. J. Genet.* **62**:189–196.

1307. Matsubara, K., K. Ohnishi, and K. Kiritani. 1988. The third general transport system for branched-chain amino acids in *Salmonella typhimurium*. *J. Gen. Appl. Microbiol.* **34**:183–189.

1308. Matsubara, K., K. Ohnishi, and K. Kiritani. 1992. Nucleotide sequences and characterization of *liv* genes encoding components of the high-affinity branched-chain amino acid transport system in *Salmonella typhimurium*. *J. Biochem. (Tokyo)* **112**:93–101.

1309. Matsuyama, T., and H. Uetake. 1972. Chromosomal locations of *Salmonella* conversion phages: mapping of prophages g₃₄₁, ε¹⁵ and ε²⁴ in *Salmonella anatum*. *Virology* **49**:359–367.

1310. Maurelli, A. T. 1994. Virulence protein export systems in *Salmonella* and *Shigella*: a new family or lost relatives? *Trends Cell. Biol.* **4**:240–242.

1311. Maurer, R., B. C. Osmond, and D. Botstein. 1981. Genetic analysis of DNA replication in *Salmonella typhimurium*. *ICN-UCLA Symp. Mol. Cell. Biol.* **22**:375–386.

1312. Maurer, R., B. C. Osmond, and D. Botstein. 1984. Genetic analysis of DNA replication in bacteria: *dnaB* mutations that suppress *dnaC* mutations and the *dnaQ* mutations that suppress *dnaE* mutations in *Salmonella typhimurium*. *Genetics* **108**:25–38.

1313. Maurer, R., B. C. Osmond, E. Shekhtman, A. Wong, and D. Botstein. 1984. Functional interchangeability of DNA replication genes in *Salmonella typhimurium* and *Escherichia coli* demonstrated by a general complementation procedure. *Genetics* **108**:1–12.

1314. Maurer, R., and A. Wong. 1988. Dominant lethal mutations in the *dnaB* helicase gene of *Salmonella typhimurium*. *J. Bacteriol.* **170**:3682–3688.

1315. May, S. G., and D. C. Old. 1980. meso-Tartrate resistance and phylogenetic relationship of biotypes of *Salmonella typhimurium*. *Genet. Res.* **36**:327.

1316. McCarter, L., K. Krajewska-Grynkiewicz, D. Trinh, G. Wei, and S. Kustu. 1984. Characterization of mutations that lie in the promoter-regulatory region for *glnA*, the structural gene encoding glutamine synthetase. *Mol. Gen. Genet.* **197**:150–160.

1317. McClatchy, J. K., and H. V. Rickenberg. 1967. Heterogeneity of the stability of messenger ribonucleic acid in *Salmonella typhimurium*. *J. Bacteriol.* **93**:115–121.

1318. McDonough, M. W. 1965. Amino acid composition of antigenically distinct Salmonella flagellar proteins. *J. Mol. Biol.* **12**:342–355.

1319. McFarland, N., L. McCarter, S. Artz, and S. Kustu. 1971. Nitrogen regulatory locus "*glnR*" of enteric bacteria is composed of cistrons *ntrB* and *ntrC*: identification of their protein products. *Proc. Natl. Acad. Sci. USA* **78**:2135–2139.

1320. McGinnis, E., and L. S. Williams. 1972. Role of histidine transfer ribonucleic acid in regulation of synthesis of histidyl-transfer ribonucleic acid synthetase of *Salmonella typhimurium*. *J. Bacteriol.* **109**:505–511.

1321. McHugh, G. L., and C. G. Miller. 1974. Isolation and characterization of proline peptidase mutants of *Salmonella typhimurium*. *J. Bacteriol.* **120**:364–371.

1322. Meadow, N. D., and S. Roseman. 1982. Sugar transport by the bacterial phosphotransferase system. Isolation and characterization of a glucose-specific phosphocarrier protein (III^Glc) from *Salmonella typhimurium*. *J. Biol. Chem.* **257**:14526–14537.

1323. Meadow, N. D., J. M. Rosenberg, H. M. Pinkert, and S. Roseman. 1982. Sugar transport by the bacterial phosphotransferase system. Evidence that *crr* is the structural gene for the *Salmonella typhimurium* glucose-specific phosphocarrier protein III^{Glc}. *J. Biol. Chem.* **257**:14538–14542.

1324. **Meadow, N. D., D. W. Saffen, R. P. Dottin, and S. Roseman.** 1982. Molecular cloning of the *crr* gene and evidence that it is the structural gene for III^{Glc}, a phosphocarrier protein of the bacterial phosphotransferase system. *Proc. Natl. Acad. Sci. USA* **79**:2528–2532.

1325. **Meier, U., and H. Mayer.** 1985. Genetic location of genes encoding enterobacterial common antigen. *J. Bacteriol.* **163**:756–762.

1326. **Meiss, H. K., W. J. Brill, and B. Magasanik.** 1969. Genetic control of histidine degradation in *Salmonella typhimurium*, strain LT-2. *J. Biol. Chem.* **244**:5382–5391.

1327. **Melton, T., P. E. Hartman, J. P. Stratis, T. L. Lee, and A. T. Davis.** 1978. Chemotaxis of *Salmonella* to amino acids and some sugars. *J. Bacteriol.* **133**:708–716.

1328. **Melton, T., W. Kundig, P. E. Hartman, and N. Meadow.** 1976. 3-Deoxy-3-fluoro-D-glucose-resistant *Salmonella typhimurium* mutants defective in the phosphoenolpyruvate:glycose phosphotransferase system. *J. Bacteriol.* **128**:794–800.

1329. **Menzel, R., and J. Roth.** 1980. Identification and mapping of a second proline permease in *Salmonella typhimurium*. *J. Bacteriol.* **141**:1064–1070.

1330. **Menzel, R., and J. Roth.** 1981. Regulation of the genes for proline regulation in *Salmonella typhimurium*: autogenous repression by the *putA* gene product. *J. Mol. Biol.* **148**:21–44.

1331. **Menzel, R., and J. Roth.** 1981. Purification of the *putA* gene product. A bifunctional membrane-bound protein from *Salmonella typhimurium* responsible for the two-step oxidation of proline to glutamate. *J. Biol. Chem.* **256**:9755–9761.

1332. **Mergeay, M., and J. Gerits.** 1983. Transduction of *Escherichi coli trp* genes in *Salmonella typhimurium* and effect of *N*-methyl-*N'*-nitro-*N*-nitrosoguanidine on transduction with heterogenotic DNA. *J. Gen. Microbiol.* **129**:321–335.

1333. **Meyers, M., F. Blasi, C. B. Bruni, R. G. Deeley, J. S. Kovach, M. Levinthal, K. P. Mullinix, T. Vogel, and R. G. Goldberger.** 1975. Specific binding of the first enzyme for histidine biosynthesis to the DNA of the histidine operon. *Nucleic Acids Res.* **2**:2021–2036.

1334. **Meyers, M., M. Levinthal, and R. F. Goldberger.** 1975. *trans*-Recessive mutation in the first structural gene of the histidine operon that results in constitutive expression of the operon. *J. Bacteriol.* **124**:1227–1235.

1335. **Michaels, G., and R. A. Kelln.** 1983. Construction and use of *pyr*::*lac* fusion strains to study regulation of pyrimidine biosynthesis in *Salmonella typhimurium*. *Mol. Gen. Genet.* **189**:463–470.

1336. **Michaels, G., R. A. Kelln, and F. E. Nargang.** 1987. Cloning, nucleotide sequence and expression of the *pyrBI* operon of *Salmonella typhimurium* LT2. *Eur. J. Biochem.* **166**:55–61.

1337. **Miesel, L., and J. R. Roth.** 1994. *Salmonella recD* mutations increase recombination in a short sequence transduction assay. *J. Bacteriol.* **176**:4092–4103.

1338. **Milburn, M. V., G. G. Prive, D. L. Milligan, W. G. Scott, J. Yeh, J. Jancarik, D. E. Koshland, Jr., and S. H. Kim.** 1991. Three-dimensional structures of the ligand-binding domain of the bacterial aspartate receptor with and without a ligand. *Science* **254**:1342–1347.

1339. **Miller, B. E., and N. M. Kredich.** 1987. Purification of the *cysB* protein from *Salmonella typhimurium*. *J. Biol. Chem.* **262**:6006–6009.

1340. **Miller, C. G.** 1975. Genetic mapping of *Salmonella typhimurium* peptidase mutations. *J. Bacteriol.* **122**:171–176.

1341. **Miller, C. G., and L. Green.** 1981. Degradation of abnormal proteins in peptidase-deficient mutants of *Salmonella typhimurium*. *J. Bacteriol.* **147**:925–930.

1342. **Miller, C. G., and L. Green.** 1983. Degradation of proline peptidases in peptidase-deficient strains of *Salmonella typhimurium*. *J. Bacteriol.* **153**:350–356.

1343. **Miller, C. G., L. Green, and R. Schultz.** 1982. *Salmonella typhimurium* mutations affecting utilization of L-leucine β-naphthylamide. *Mol. Gen. Genet.* **186**:228–234.

1344. **Miller, C. G., C. Helman, and C. Yen.** 1976. Mutants of *Salmonella typhimurium* deficient in an endoprotease. *J. Bacteriol.* **127**:490–497.

1345. **Miller, C. G., A. M. Kukral, J. L. Miller, and N. R. Movva.** 1989. *pepM* is an essential gene in *Salmonella typhimurium*. *J. Bacteriol.* **171**:5215–5217.

1346. **Miller, C. G., and K. Mackinnon.** 1974. Peptidase mutants of *Salmonella typhimurium*. *J. Bacteriol.* **120**:355–363.

1347. **Miller, C. G., J. L. Miller, and D. A. Bagga.** 1991. Cloning and nucleotide sequence of the anaerobically regulated *pepT* gene of *Salmonella typhimurium*. *J. Bacteriol.* **173**:3554–3558.

1348. **Miller, C. G., and J. R. Roth.** 1971. Recessive-lethal nonsense suppressors in *Salmonella typhimurium*. *J. Mol. Biol.* **59**:63–75.

1349. **Miller, C. G., K. L. Strauch, A. M. Kukral, J. L. Miller, P. T. Wingfield, G. J. Mazzei, R. C. Werlen, P. Graber, and N. R. Movva.** 1987. N-terminal methionine-specific peptidase in *Salmonella typhimurium*. *Proc. Natl. Acad. Sci. USA* **84**:2718–2722.

1350. **Miller, C. G., and D. Zipser.** 1977. Degradation of *Escherichia coli* β-galactosidase fragments in protease-deficient mutants of *Salmonella typhimurium*. *J. Bacteriol.* **130**:347–353.

1351. **Miller, E. S., and J. E. Brenchley.** 1983. Cloning and characterization of *gdhA*, the structural gene for glutamate dehydrogenase of *Salmonella typhimurium*. *J. Bacteriol.* **157**:171–178.

1352. **Miller, J. B., and D. E. Koshland, Jr.** 1977. Membrane fluidity and chemotaxis: effects of temperature and membrane lipid composition on the swimming behavior of *Salmonella typhimurium* and *Escherichia coli*. *J. Mol. Biol.* **111**:183–201.

1353. **Miller, K., and S. Maloy.** 1990. DNA sequence of the *putP* gene from *Salmonella typhimurium* and predicted structure of proline permease. *Nucleic Acids Res.* **18**:3057.

1354. **Miller, S. I.** 1991. PhoP/PhoQ: macrophage-specific modulators of *Salmonella* virulence? *Mol. Microbiol.* **5**:2073–2078.

1355. **Miller, S. I., A. M. Kukral, and J. J. Mekalanos.** 1989. A two-component regulatory system (*phoP phoQ*) controls *Salmonella typhimurium* virulence. *Proc. Natl. Acad. Sci. USA* **86**:5054–5058.

1356. **Miller, S. I., and J. J. Mekalanos.** 1990. Constitutive expression of the *phoP* regulon attenuates *Salmonella* virulence and survival within macrophages. *J. Bacteriol.* **172**:2485–2490.

1357. **Miller, S. I., W. S. Pulkkinen, M. E. Selsted, and J. J. Mekalanos.** 1990. Characterization of defensin resistance phenotypes associated with mutations in the *phoP* virulence regulon of *Salmonella typhimurium*. *Infect. Immun.* **58**:3706–3710.

1358. **Miller, V. L., K. B. Beer, W. P. Loomis, J. A. Olson, and S. I. Miller.** 1992. An unusual *pagC*::Tn*phoA* mutation leads to an invasion- and virulence-defective phenotype in salmonellae. *Infect. Immun.* **60**:3763–3770.

1359. **Mills, D. M., V. Bajaj, and C. A. Lee.** 1995. A 40 kilobase chromosomal fragment encoding *Salmonella typhimurium* invasion genes is absent from the corresponding region of the *Escherichia coli* K-12 chromosome. *Mol. Microbiol.* **15**:749–759.

1360. **Miloso, M., D. Limauro, P. Alifano, F. Rivellini, A. Lavitola, E. Gulletta, and C. B. Bruni.** 1993. Characterization of the *rho* genes of *Neisseria gonorrhoeae* and *Salmonella typhimurium*. *J. Bacteriol.* **175**:8030–8037.

1361. **Minamino, T., T. Iino, and K. Kutsukake.** 1994. Molecular characterization of the *Salmonella typhimurium flhB* operon and its protein products. *J. Bacteriol.* **176**:7630–7637.

1362. **Minson, A. C., and D. A. Smith.** 1972. Methionine regulatory defects in *Salmonella typhimurium* arising from amber-suppressible mutations. *J. Gen. Microbiol.* **70**:471–476.

1363. **Minton, N., J. Gunn, and I. R. Beacham.** 1979. Nucleoside diphosphate sugar hydrolase gene of *Salmonella typhimurium*: chromosomal location determined by intergeneric crosses. *J. Bacteriol.* **137**:1428–1429.

1364. **Miozarri, G. F., and C. Yanofsky.** 1979. Gene fusion during the evolution of the tryptophan operon in Enterobacteriaceae. *Nature* (London) **277**:486–489.

1365. **Misko, T. P., W. J. Mitchell, N. D. Meadow, and S. Roseman.** 1987. Sugar transport by the bacterial phosphotransferase system. *J. Biol. Chem.* **262**:16261–16266.

1366. **Mitchell, W. J., T. P. Misko, and S. Roseman.** 1982. Sugar transport by the bacterial phosphotransferase system. Regulation of other transport systems (lactose and melibiose). *J. Biol. Chem.* **257**:14553–14564.

1367. **Miyake, T., and M. Demerec.** 1960. Proline mutants of *Salmonella typhimurium*. *Genetics* **45**:755–762.

1368. **Mizobuchi, K., M. Demerec, and D. H. Gillespie.** 1962. Cysteine mutants of *Salmonella typhimurium*. *Genetics* **47**:1617–1627.

1369. **Mizuno, T., N. Mutoh, S. M. Panasenko, and Y. Imae.** 1986. Acquisition of maltose chemotaxis in *Salmonella typhimurium* by the introduction of the *Escherichia coli* chemosensory transducer gene. *J. Bacteriol.* **165**:890–895.

1370. **Mizushima, K., S. Awakihara, M. Kuroda, T. Ishikawa, M. Tsuda, and T. Tsuchiya.** 1992. Cloning and sequencing of the *melB* gene encoding the melibiose permease of *Salmonella typhimurium* LT2. *Mol. Gen. Genet.* **234**:74–80.

1371. **Mojica-a, T.** 1975. Transduction by phage PICM *clr*-100 in *Salmonella typhimurium*. *Mol. Gen. Genet.* **138**:113–126.

1372. **Mojica-a, T., and E. Garcia.** 1976. Growth of coliphage BF23 on rough strains of *Salmonella typhimurium*. The *bfe* locus. *Mol. Gen. Genet.* **147**:195–202.

1373. **Mojica-a, T., E. Garcia, and C. Ascaso.** 1976. Mutants of coliphage BF-23 able to propagate on smooth strains of *Salmonella typhimurium*. *Arch. Int. Physiol. Biochim.* **84**:402–403.

1374. **Mojica-a, T., and R. B. Middleton.** 1972. *Salmonella typhimurium-Escherichia coli* hybrids for the tryptophan region. *Genetics* **71**:491–505.

1375. **Monroe, R. S., and N. M. Kredich.** 1988. Isolation of *Salmonella typhimurium cys* genes by transduction with a library of recombinant plasmids packaged in bacteriophage P22HT capsids. *J. Bacteriol.* **170**:42–47.

1376. **Monroe, R. S., J. Ostrowski, M. M. Hryniewicz, and N. M. Kredich.** 1990. In vitro interactions of CysB protein with the *cysK* and *cysJIH* promoter regions of *Salmonella typhimurium*. *J. Bacteriol.* **172**:6919–6929.

1377. **Morgan, D. G., R. M. Macnab, N. R. Francis, and D. J. DeRosier.** 1993. Domain organization of the subunit of the *Salmonella typhimurium* flagellar hook. *J. Mol. Biol.* **229**:79–84.

1378. **Morgan, R. W., M. F. Christman, F. S. Jacobson, G. Storz, and B. N. Ames.** 1986. Hydrogen peroxide-inducible proteins in *Salmonella typhimurium* overlap with heat shock and other stress proteins. *Proc. Natl. Acad. Sci. USA* **83**:8059–8063.

1379. **Mortlock, R. P., and D. C. Old.** 1979. Utilization of D-xylose by wild-type strains of *Salmonella typhimurium*. *J. Bacteriol.* **137**:173–178.

1380. **Morton, D. P., and S. M. Parsons.** 1977. Synergistic inhibition of ATP phosphoribosyltransferase by guanosine tetraphosphate and histidine. *Biochem. Biophys. Res. Commun.* **74**:172–177.

1381. **Morton, D. P., and S. M. Parsons.** 1977. Inhibition of ATP phosphoribosyltransferase by AMP and ADP in the absence and presence of histidine. *Arch. Biochem. Biophys.* **181**:643–648.

1382. **Movva, N. R., D. Semon, C. Meyer, E. Kawashima, P. Wingfield, J. L. Miller, and C. G. Miller.** 1990. Cloning and nucleotide sequence of the *Salmonella typhimurium pepM* gene. *Mol. Gen. Genet.* **223**:345–348.

1383. **Mowbray, S. L., and G. A. Petsko.** 1982. Preliminary X-ray data for the ribose binding protein from *Salmonella typhimurium*. *J. Mol. Biol.* **160**:545–547.

1384. **Muday, G. K., and K. M. Herrmann.** 1990. Regulation of the *Salmonella typhimurium aroF* gene in *Escherichia coli. J. Bacteriol.* 172:2259–2266.

1385. **Muhlradt, P.** 1969. Biosynthesis of *Salmonella* lipopolysaccharide. The in vitro transfer of phosphate to the heptose moiety of the core. *Eur. J. Biochem.* 11:241–248.

1386. **Mukai, F. H., and P. Margolin.** 1963. Analysis of unlinked suppressors of an O° mutation in *Salmonella. Proc. Natl. Acad. Sci. USA* 50:140–148.

1387. **Mulford, C. A., and M. J. Osborn.** 1983. An intermediate step in translocation of lipopolysaccharide to the outer membrane of *Salmonella typhimurium. Proc. Natl. Acad. Sci. USA* 80:1159–1163.

1388. **Muller, E., A. Hinckley, and L. Rothfield.** 1972. Studies of phospholipid-requiring bacterial enzymes. III. Purification and properties of uridine diphosphate glucose: lipopolysaccharide glucosyltransferase I. *J. Biol. Chem.* 247:2614–2622.

1389. **Muller, K.-H., T. J. Trust, and W. W. Kay.** 1988. Unmasking of bacteriophage Mu lipopolysaccharide receptors in *Salmonella enteritidis* confers sensitivity to Mu and permits Mu mutagenesis. *J. Bacteriol.* 170:1076–1081.

1390. **Muller, N., H. G. Heine, and W. Boos.** 1982. Cloning of *mglB*, the structural gene for the galactose-binding protein of *Salmonella typhimurium* and *Escherichia coli. Mol. Gen. Genet.* 185:473–480.

1391. **Muller, N., H. G. Heine, and W. Boos.** 1985. Characterization of the *Salmonella typhimurium mgl* operon and its gene products. *J. Bacteriol.* 163:37–45.

1392. **Muller, V., C. J. Jones, I. Kawagishi, S. Aizawa, and R. M. Macnab.** 1992. Characterization of the *fliE* genes of *Escherichia coli* and *Salmonella typhimurium* and identification of the FliE protein as a component of the flagellar hook-basal body complex. *J. Bacteriol.* 174:2298–2304.

1393. **Munford, R. S., C. L. Hall, and P. D. Rick.** 1980. Size heterogeneity of *Salmonella typhimurium* lipopolysaccharides in outer membranes and culture supernatant membrane fragments. *J. Bacteriol.* 144:630–640.

1394. **Munson, R. S., Jr., N. S. Rasmussen, and M. J. Osborn.** 1978. Biosynthesis of lipid A. Enzymatic incorporation of 3-deoxy-D-mannoctulosonate into a precursor of lipid A in *Salmonella typhimurium. J. Biol. Chem.* 253:1503–1511.

1395. **Murata-Matsubara, K., K. Ohnishi, and K. Kiritani.** 1985. Genetic and biochemical studies of *livS* mutation affecting the regulation of branched-chain amino acid transport in *Salmonella typhimurium. Jpn. J. Genet.* 60:11–25.

1396. **Murooka, Y., and T. Harada.** 1981. Regulation of derepressed synthesis of arylsulfatase by tyramine oxidase in *Salmonella typhimurium. J. Bacteriol.* 145:796–802.

1397. **Murphy-Corb, M., H.-L. Kong, and M. L. Murray.** 1980. Interaction of mutagenic spermidine-nitrous acid reaction products with *uvr*- and *recA*-dependent repair systems in *Salmonella typhimurium. J. Bacteriol.* 142:191–195.

1398. **Murray, M. L., and T. Klopotowski.** 1968. Genetic map position of the gluconate-6-phosphate dehydrogenase gene in *Salmonella typhimurium. J. Bacteriol.* 95:1279–1282.

1399. **Mustard, J. A., A. T. Thliveris, and D. W. Mount.** 1992. Sequence of the *Salmonella typhimurium* LT2 *lexA* gene and its regulatory region. *Nucleic Acids Res.* 20:1813.

1400. **Myers, D. E., B. A. D. Stocker, and R. J. Roantree.** 1980. Mapping of genes determining penicillin-resistance and serum-sensitivity in *Salmonella enteritidis. J. Gen. Microbiol.* 118:367–376.

1401. **Myers, R. S., D. Townsend, and S. Maloy.** 1991. Dissecting the molecular mechanism of ion-solute cotransport: substrate specificity mutations in the *putP* gene affect the kinetics of proline transport. *J. Membr. Biol.* 121:201–214.

1402. **Nagano, H., and H. Zalkin.** 1970. Some physicochemical properties of anthranilate synthase component I from *Salmonella typhimurium. J. Biol. Chem.* 245:3097–3103.

1403. **Nagano, H., H. Zalkin, and E. J. Henderson.** 1970. The anthranilate synthetase-anthranilate-5-phosphoribosyl pyrophosphate phosphoribosyltransferase aggregate. *J. Biol. Chem.* 245:3810–3820.

1404. **Naide, Y., H. Nikaido, P. H. Makela, R. G. Wilkinson, and B. A. D. Stocker.** 1965. Semirough strains of *Salmonella. Proc. Natl. Acad. Sci. USA* 53:147–153.

1405. **Nakake, T.** 1971. Multiple molecular forms of uridine diphosphate glucose pyrophosphorylase from *Salmonella typhimurium*. III. Interconversion between various forms. *J. Biol. Chem.* 246:4404–4411.

1406. **Nakake, T., and H. Nikaido.** 1971. Multiple molecular forms of uridine diphosphate glucose pyrophosphorylase from *Salmonella typhimurium*. II. Genetic determination of multiple forms. *J. Biol. Chem.* 246:4397–4403.

1407. **Nakamura, K., R. M. Pirtle, and M. Inouye.** 1979. Homology of the gene coding for outer membrane lipoprotein within various gram-negative bacteria. *J. Bacteriol.* 137:595–604.

1408. **Nakamura, T., Y. Kon, H. Iwahashi, and Y. Eguchi.** 1983. Evidence that thiosulfate assimilation by *Salmonella typhimurium* is catalyzed by cysteine synthase B. *J. Bacteriol.* 156:656–662.

1409. **Nakayama, N., N. Arai, M. W. Bond, Y. Kaziro, and K. Arai.** 1984. Nucleotide sequence of *dnaB* and the primary structure of the DnaB protein from *Escherichia coli. J. Biol. Chem.* 259:97–101.

1410. **Naumann, D., C. Schultz, J. Born, H. Labischinski, K. Brandenburg, G. von Busse, H. Brade, and U. Seydel.** 1987. Investigations into the polymorphism of lipid A from lipopolysaccharides of *Escherichia coli* and *Salmonella minnesota* by Fourier-transform infrared spectroscopy. *Eur. J. Biochem.* 164:159–169.

1411. **Neal, B. L., P. K. Brown, and P. R. Reeves.** 1993. Use of *Salmonella* phage P22 for transduction in *Escherichia coli. J. Bacteriol.* 175:7115–7118.

1412. **Negre, D., J. C. Cortay, P. Donini, and A. J. Cozzone.** 1988. Inaccurate protein synthesis in a mutant of *Salmonella typhimurium* defective in transfer RNA pseudouridylation. *FEBS Lett.* 234:165–168.

1413. **Negre, D., J. C. Cortay, P. Donini, and A. J. Cozzone.** 1989. Relationship between guanosine tetraphosphate and accuracy of translation in *Salmonella typhimurium. Biochemistry* 28:1814–1819.

1414. **Negre, D., J. C. Cortay, I. G. Old, A. Galinier, C. Richaud, I. Saint Girons, and A. J. Cozzone.** 1991. Overproduction and characterization of the *iclR* gene product of *Escherichia coli* K-12 and comparison with that of *Salmonella typhimurium* LT2. *Gene* 97:29–37.

1415. **Nelson, S. O., B. J. Scholte, and P. W. Postma.** 1982. Phosphoenolpyruvate:sugar phosphotransferase system-mediated regulation of carbohydrate metabolism in *Salmonella typhimurium. J. Bacteriol.* 150:604–615.

1416. **Nelson, S. O., A. R. J. Schuitema, R. Benne, L. H. T. van der Ploeg, J. S. Plijter, F. Aan, and P. W. Postma.** 1984. Molecular cloning, sequencing, and expression of the *crr* gene: the structural gene for III^{Glc} of the bacterial PEP:glucose phosphotransferase system. *EMBO J.* 3:1587–1593.

1417. **Neufeld, B. R., L. R. Bullas, and M. M. Ball.** 1978. Nonparental recombinants in intra-serotype phage P-1 transductions of the *hsd*-SB region of *Salmonella typhimurium. Genetics* 88:S71-S72.

1418. **Neuhard, J., and J. L. Ingraham.** 1968. Mutants of *Salmonella typhimurium* requiring cytidine for growth. *J. Bacteriol.* 95:2431–2433.

1419. **Neuhard, J., K. F. Jensen, and E. Stauning.** 1982. *Salmonella typhimurium* mutants with altered expression of the *pyrA* gene due to changes in RNA polymerase. *EMBO J.* 1:1141–1145.

1420. **Neuhard, J., and R. A. Kelln.** 1988. A chromosomal mutation mediating increased expression of *pyrE* in *Salmonella typhimurium* is located within the proposed attenuator. *Can. J. Microbiol.* 34:686–687.

1421. **Neuhard, J., R. A. Kelln, and E. Stauning.** 1986. Cloning and structural characterization of the *Salmonella typhimurium pyrC* gene encoding dihydroorotase. *Eur. J. Biochem.* 157:335–342.

1422. **Neuhard, J., E. Stauning, and R. A. Kelln.** 1985. Cloning and characterization of the *pyrE* gene and of *PyrE*::Mud1 (Ap^R *lac*) fusions from *Salmonella typhimurium. Eur. J. Biochem.* 146:597–603.

1423. **Newbury, S. F., N. H. Smith, E. C. Robinson, I. D. Hiles, and C. F. Higgins.** 1987. Stabilization of translationally active mRNA by prokaryotic REP sequences. *Cell* 48:297–310.

1424. **Ni Bhriain, N., C. J. Dorman, and C. F. Higgins.** 1989. An overlap between osmotic and anaerobic stress responses: a potential role for DNA supercoiling in the coordinate regulation of gene expression. *Mol. Microbiol.* 3:933–942.

1425. **Nichols, B. P., M. Blumenberg, and C. Yanofsky.** 1981. Comparison of the nucleotide sequence of *trpA* and sequences immediately beyond the *trp* operon of *Klebsiella aerogenes, Salmonella typhimurium,* and *Escherichia coli. Nucleic Acids Res.* 9:1743–1755.

1426. **Nichols, B. P., G. F. Miozzari, M. Van Cleemput, G. N. Bennett, and C. Yanofsky.** 1980. Nucleotide sequences of the *trpG* regions of *Escherichia coli, Shigella dysenteriae, Salmonella typhimurium* and *Serratia marcescens. J. Mol. Biol.* 142:503–518.

1427. **Nichols, B. P., M. Van Cleemput, and C. Yanofsky.** 1981. Nucleotide sequence of *Escherichia coli trpE*: anthranilate synthetase component I contains no tryptophan residues. *J. Mol. Biol.* 146:45–54.

1428. **Nichols, B. P., and C. Yanofsky.** 1979. Nucleotide sequences of *trpA* of *Salmonella typhimurium* and *Escherichia coli*: an evolutionary comparison. *Proc. Natl. Acad. Sci. USA* 76:5244–5248.

1429. **Nicolaidis, A. A., and W. T. Drabble.** 1979. Plasmid replication in *Salmonella typhimurium* LT-2 and *Escherichia coli* K-12: a differential effect of nalidixic acid. *FEMS Microbiol. Lett.* 6:261–264.

1430. **Nijkamp, H. J. J., and P. G. DeHaan.** 1967. Genetic and biochemical studies of the guanosine 5′-monophosphate pathway in *Escherichia coli. Biochim. Biophys. Acta* 145:31–40.

1431. **Nikaido, H., S. Ah Song, L. Shaltiel, and M. Nurminen.** 1977. Outer membrane of *Salmonella typhimurium*. Part 14. Reduced transmembrane diffusion rates in porin deficient mutants. *Biochem. Biophys. Res. Commun.* 76:324–330.

1432. **Nikaido, H., and T. Fukasawa.** 1961. The effect of mutation in a structural gene on the inducibility of the enzymes controlled by other genes of the same operon. *Biochim. Biophys. Res. Commun.* 4:338–342.

1433. **Nikaido, H., M. Levinthal, K. Nikaido, and K. Nakane.** 1967. Extended deletions in the histidine-rough-B region of the Salmonella chromosome. *Proc. Natl. Acad. Sci. USA* 57:1825–1832.

1434. **Nikaido, H., K. Nikaido, and P. H. Makela.** 1966. Genetic determination of enzymes synthesizing O-specific sugars of Salmonella lipopolysaccharides. *J. Bacteriol.* 91:1126–1135.

1435. **Nikaido, H., Y. Takeuchi, S.-I. Ohnishi, and T. Nakae.** 1977. Outer membrane of *Salmonella typhimurium*. ESR studies. *Biochim. Biophys. Acta* 465:152–164.

1436. **Nikkola, M., F. K. Gleason, J. A. Fuchs, and H. Eklund.** 1993. Crystal structure analysis of a mutant *Escherichia coli* thioredoxin in which lysine 36 is replaced by glutamic acid. *Biochemistry* 32:5093–5098.

1437. **Nishioka, Y., M. Demerec, and A. Eisenstark.** 1967. Genetic analysis of aromatic mutants of *Salmonella typhimurium. Genetics* 56:341–351.

1438. **Nishioka, Y., and A. Eisenstark.** 1970. Sequence of genes replicated in *Salmonella typhimurium* as examined by transduction techniques. *J. Bacteriol.* 102:320–333.

1439. **Nishitani, J., and G. Wilcox.** 1991. Cloning and characterization of the L-rhamnose regulon in *Salmonella typhimurium* LT2. *Gene* 105:37–42.

1440. Nixon, B. T., C. W. Ronson, and F. M. Ausubel. 1986. Two-component regulatory systems responsive to environmental stimuli share strongly conserved domains with the nitrogen assimilation regulatory genes *ntrB* and *ntrC*. Proc. Natl. Acad. Sci. USA 83:7850–7854.

1441. Nnalue, N. A., and B. A. D. Stocker. 1986. Some *galE* mutants of *Salmonella choleraesuis* retain virulence. Infect. Immun. 54:635–640.

1442. Nnalue, N. A., and B. A. D. Stocker. 1987. Test of the virulence and live-vaccine efficacy of auxotrophic and *galE* derivatives of *Salmonella choleraesuis*. Infect. Immun. 55:955–962.

1443. Nohmi, T., A. Hakura, Y. Nakai, M. Watanabe, S. Y. Murayama, and T. Sofuni. 1991. *Salmonella typhimurium* has two homologous but different *umuDC* operons: cloning of a new *umuDC*-like operon (*samAB*) present in a 60-megadalton cryptic plasmid of *S. typhimurium*. J. Bacteriol. 173:1051–1063.

1444. Nohmi, T., M. Yamada, M. Watanabe, S. Y. Murayama, and T. Sofuni. 1992. Roles of *Salmonella typhimurium* *umuDC* and *samAB* in UV mutagenesis and UV sensitivity. J. Bacteriol. 174:6948–6955.

1445. Norel, F., C. Coynault, I. Miras, D. Hermant, and M. Y. Popoff. 1989. Cloning and expression of plasmid DNA sequences involved in *Salmonella* serotype *typhimurium* virulence. Mol. Microbiol. 3:733–743.

1446. Norel, F., M. R. Pisano, J. Nicoli, and M. Y. Popoff. 1989. Nucleotide sequence of the plasmid-borne virulence gene *mkfA* encoding a 28 kDa polypeptide from *Salmonella typhimurium*. Res. Microbiol. 140:263–265.

1447. Norel, F., M. R. Pisano, J. Nicoli, and M. Y. Popoff. 1989. Nucleotide sequence of the plasmid-borne virulence gene *mkfB* from *Salmonella typhimurium*. Res. Microbiol. 140:455–457.

1448. Norel, F., V. Robbe-Saule, M. Y. Popoff, and C. Coynault. 1992. The putative sigma factor KatF (RpoS) is required for the transcription of the *Salmonella typhimurium* virulence gene *spvB* in *Escherichia coli*. FEMS Microbiol. Lett. 78:271–276.

1449. Novotny, M. J., W. L. Frederickson, E. B. Waygood, and M. H. Saier, Jr. 1985. Allosteric regulation of glycerol kinase by enzyme III^{Glc} of the phosphotransferase system in *Escherichia coli* and *Salmonella typhimurium*. J. Bacteriol. 162:810–816.

1450. Nurminen, M. 1978. A mild procedure to isolate the 34K, 35K and 36K porins of the outer membrane of *Salmonella typhimurium*. FEMS Microbiol. Lett. 3:331–334.

1451. Nurminen, M., K. Lounatmaa, M. Sarvas, P. H. Makela, and T. Nakae. 1976. Bacteriophage-resistant mutants of *Salmonella typhimurium* deficient in two major outer membrane proteins. J. Bacteriol. 127:941–955.

1452. Nurminen, M., and R. Oertli. 1975. Bacteriophage P221-receptor in *Salmonella typhimurium*: purification and characterization. Abstr. 10th Congr. Fed. Eur. Biochem. Soc. 1975.

1453. Nyman, K., M. Plosila, L. Howden, and P. H. Makela. 1979. Genetic determination of lipopolysaccharide: locus of O specific unit polymerase in Group E of *Salmonella*. Zentralbl. Bakteriol. Parasitenkd. Infektionskr. Hyg. Abt. 1 Orig. Reihe A 243:355–362.

1454. O'Brien, E. J., and P. M. Bennett. 1972. Structure of straight flagella from a mutant *Salmonella*. J. Mol. Biol. 70:133–152.

1455. O'Brien, K., G. Deno, P. Ostrovsky de Spicer, J. F. Gardner, and S. R. Maloy. 1992. Integration host factor facilitates repression of the *put* operon in *Salmonella typhimurium*. Gene 118:13–19.

1456. O'Byrne, C. P., and C. J. Dorman. 1994. The *spv* virulence operon of *Salmonella typhimurium* LT2 is regulated negatively by the cyclic AMP (cAMP)-cAMP receptor protein system. J. Bacteriol. 176:905–912.

1457. O'Byrne, C. P., N. Ni Bhriain, and C. J. Dorman. 1992. The DNA supercoiling-sensitive expression of the *Salmonella typhimurium* *his* operon requires the *his* attenuator and is modulated by anaerobiosis and by osmolarity. Mol. Microbiol. 6:2467–2476.

1458. O'Connor, M., N. M. Willis, L. Bossi, R. F. Gesteland, and J. F. Atkins. 1993. Functional tRNAs with altered 3′ ends. EMBO J. 12:2559–2566.

1459. O'Donovan, G. A., and J. Neuhard. 1970. Pyrimidine metabolism in microorganisms. Bacteriol. Rev. 34:278–343.

1460. Ohnishi, K., A. Hasegawa, K. Matsubara, T. Date, T. Okada, and K. Kiritani. 1988. Cloning and nucleotide sequence of the *brnQ* gene, the structural gene for a membrane-associated component of the LIV-II transport system for branched-chain amino acids in *Salmonella typhimurium*. Jpn. J. Genet. 63:343–357.

1461. Ohnishi, K., M. Homma, K. Kutsukake, and T. Iino. 1987. Formation of flagella lacking outer rings by *flaM*, *flaU*, and *flaY* mutants of *Escherichia coli*. J. Bacteriol. 169:1485–1488.

1462. Ohnishi, K., and K. Kiritani. 1978. Glycyl-L-leucine resistance mutation affecting transport of branched chain amino acids in *Salmonella typhimurium*. Jpn. J. Genet. 53:275–283.

1463. Ohnishi, K., and K. Kiritani. 1980. Close linkage relationship between *gleR-* and *brnQ* loci in *Salmonella typhimurium*. Jpn. J. Genet. 55:67–70.

1464. Ohnishi, K., K. Kutsukake, H. Suzuki, and T. Iino. 1990. Gene *fliA* encodes an alternative sigma factor specific for flagellar operons in *Salmonella typhimurium*. Mol. Gen. Genet. 221:139–147.

1465. Ohnishi, K., K. Kutsukake, H. Suzuki, and T. Iino. 1992. A novel transcriptional regulation mechanism in the flagellar regulon of *Salmonella typhimurium*: an antisigma factor inhibits the activity of the flagellum-specific sigma factor, sigma F. Mol. Microbiol. 6:3149–3157.

1466. Ohnishi, K., K. Murata, and K. Kiritani. 1980. A regulatory transport mutant for branched-chain amino acids in *Salmonella typhimurium*. Jpn. J. Genet. 55:349–360.

1467. Ohnishi, K., A. Nakazima, K. Matsubara, and K. Kiritani. 1990. Cloning and nucleotide sequences of *livB* and *livC*, the structural genes encoding binding proteins of the high-affinity branched-chain amino acid transport in *Salmonella typhimurium*. J. Biochem. (Tokyo) 107:202–208.

1468. Ohnishi, K., Y. Ohto, S. Aizawa, R. M. Macnab, and T. Iino. 1994. FlgD is a scaffolding protein needed for flagellar hook assembly in *Salmonella typhimurium*. J. Bacteriol. 176:2272–2281.

1469. Ohta, N., P. R. Galsworthy, and A. B. Pardee. 1971. Genetics of sulfate transport by *Salmonella typhimurium*. J. Bacteriol. 105:1053–1062.

1470. Okada, M., T. Watanabe, and T. Miyake. 1968. On the nature of the recipient ability of *Salmonella typhimurium* for foreign deoxyribonucleic acids. J. Gen. Microbiol. 50:241–252.

1471. Okazaki, N., S. Matsuo, K. Saito, A. Tominaga, and M. Enomoto. 1993. Conversion of the *Salmonella* phase 1 flagellin gene *fliC* to the phase 2 gene *fljB* on the *Escherichia coli* K-12 chromosome. J. Bacteriol. 175:758–766.

1472. Okino, H., M. Isomura, S. Yamaguchi, Y. Magariyama, S. Kudo, and S.-I. Aizawa. 1989. Release of flagellar filament-hook-rod complex by a *Salmonella typhimurium* mutant defective in the M ring of the basal body. J. Bacteriol. 171:2075–2082.

1473. Old, D. C., P. F. H. Dawes, and R. M. Barker. 1980. Transduction of inositol-fermenting ability demonstrating phylogenetic relationships among strains of *Salmonella typhimurium*. Genet. Res. 35:215–224.

1474. Old, I. G., D. Margarita, R. E. Glass, and I. Saint Girons. 1990. Nucleotide sequence of the *metH* gene of *Escherichia coli* K-12 and comparison with that of *Salmonella typhimurium* LT2. Gene 87:15–21.

1475. Oliver, D. R., J. J. Manis, and H. J. Whitfield. 1974. Evidence for a composite state of an F′ *his* element and a cryptic plasmid in a derivative of *Salmonella typhimurium* LT2. J. Bacteriol. 119:191–201.

1476. O'Mahony, D. J., D. Hughes, S. Thompson, and J. F. Atkins. 1989. Suppression of a –1 frameshift mutation by a recessive tRNA suppressor which causes doublet decoding. J. Bacteriol. 171:3824–3830.

1477. O'Neal, C. R., W. M. Gabriel, A. K. Turk, S. J. Libby, F. C. Fang, and M. P. Spector. 1994. RpoS is necessary for both the positive and negative regulation of starvation survival genes during phosphate, carbon, and nitrogen starvation in *Salmonella typhimurium*. J. Bacteriol. 176:4610–4616.

1478. O'Neill, J. P., and M. Freundlich. 1972. Effect of cyclopentaneglycine on metabolism in *Salmonella typhimurium*. J. Bacteriol. 111:510–515.

1479. O'Neill, J. P., and M. Freundlich. 1972. Two forms of biosynthetic acetohydroxy acid synthetase in *Salmonella typhimurium*. Biochem. Biophys. Res. Commun. 48:437–443.

1480. O'Neill, J. P., and M. Freundlich. 1973. Temperature-sensitive growth inhibition by valine in *Salmonella typhimurium*: alteration of one form of acetohydroxy acid synthetase. J. Bacteriol. 116:98–106.

1481. Oosawa, K., T. Ueno, and S. Aizawa. 1994. Overproduction of the bacterial flagellar switch proteins and their interactions with the MS ring complex in vitro. J. Bacteriol. 176:3683–3691.

1482. Oppezzo, O. J., B. Avanzati, and D. N. Anton. 1991. Increased susceptibility to β-lactam antibiotics and decreased porin content caused by *envB* mutations of *Salmonella typhimurium*. Antimicrob. Agents Chemother. 35:1203–1207.

1483. O'Reilly, C., G. W. Black, R. Laffey, and D. J. McConnell. 1990. Molecular analysis of an IS200 insertion in the *gpt* gene of *Salmonella typhimurium* LT2. J. Bacteriol. 172:6599–6601.

1484. O'Reilly, C., P. D. Turner, P. F. Smith-Keary, and D. J. McConnell. 1984. Molecular cloning of genes involved in purine biosynthetic and salvage pathways of *Salmonella typhimurium*. Mol. Gen. Genet. 196:152–157.

1485. Ornellas, E. P., and B. A. D. Stocker. 1974. Relation of lipopolysaccharide character to P1 sensitivity in *Salmonella typhimurium*. Virology 60:491–502.

1486. Orr, J. C., D. W. Bryant, D. R. McCalla, and M. A. Quilliam. 1985. Dinitropyrene-resistant *Salmonella typhimurium* are deficient in an acetyl-CoA acetyltransferase. Chem. Biol. Interact. 54:281–288.

1487. Orrego, C., and E. Eisenstadt. 1987. An inducible pathway is required for mutagenesis in *Salmonella typhimurium* LT2. J. Bacteriol. 169:2885–2888.

1488. Ortega, M. V., A. Cardenas, and D. Ubiera. 1975. *panD*, a new chromosomal locus of *Salmonella typhimurium* for the biosynthesis of β-alanine. Mol. Gen. Genet. 140:159–164.

1489. Osborn, M. J. 1968. Biochemical characterization of mutants of *Salmonella typhimurium* lacking glucosyl or galactosyl lipopolysaccharide transferases. Nature (London) 217:957–960.

1490. Osborn, M. J. 1969. Structure and biosynthesis of the bacterial cell wall. Annu. Rev. Biochem. 29:501–538.

1491. Osborn, M. J., J. E. Gander, E. Parisi, and J. Carson. 1972. Mechanism and assembly of the outer membrane of *Salmonella typhimurium*. Isolation and characterization of cytoplasmic and outer membrane. J. Biol. Chem. 247:3962–3972.

1492. Osborn, M. J., P. D. Rick, and N. S. Rasmussen. 1980. Mechanism of assembly of the outer membrane of *Salmonella typhimurium*. Translocation and integration of an incomplete mutant lipid A into the outer membrane. J. Biol. Chem. 255:4246–4251.

1493. Osborn, M. J., S. M. Rosen, L. Rothfield, L. D. Zeleznick, and B. L. Horecker. 1964. Lipopolysaccharide of the gram-negative cell wall. Science 145:783–789.

1494. Osborn, M. J., and L. I. Rothfield. 1971. Biosynthesis of the core region of lipopolysaccharides, p. 331–350. In G. Weinbaum, S. Kadis, and S. J. Ajl (ed.), Bacterial Endotoxins, vol. 4 in Microbial Toxins. Academic Press, Inc., New York.

1495. Ostapchuk, P., A. Anilionis, and M. Riley. 1980. Conserved genes in enteric bacteria are not identical. Mol. Gen. Genet. 180:475–478.

1496. Ostrovsky de Spicer, P., and S. Maloy. 1993. PutA protein, a membrane-associated flavin dehydrogenase, acts as a redox-dependent transcriptional regulator. Proc. Natl. Acad. Sci. USA 90:4295–4298.

1497. Ostrovsky de Spicer, P., K. O'Brien, and S. Maloy. 1991. Regulation of proline utilization in Salmonella typhimurium: a membrane-associated dehydrogenase binds DNA in vitro. J. Bacteriol. 173:211–219.

1498. Ostrowski, J., M. J. Barber, D. C. Rueger, B. E. Miller, L. M. Siegel, and N. M. Kredich. 1989. Characterization of the flavoprotein moieties of NADPH-sulfite reductase from Salmonella typhimurium and Escherichia coli. Physicochemical and catalytic properties, amino acid sequence deduced from DNA sequence of cysJ, and comparison with NADPH-cytochrome P-450 reductase. J. Biol. Chem. 264:15796–15808.

1499. Ostrowski, J., and D. Hulanicka. 1979. Constitutive mutation of cysJIH operon in a cysB deletion strain of Salmonella typhimurium. Mol. Gen. Genet. 175:145–150.

1500. Ostrowski, J., and D. Hulanicka. 1981. Effect of DNA gyrase inhibitors on gene expression of the cysteine regulon. Mol. Gen. Genet. 181:363–366.

1501. Ostrowski, J., G. Jagura-Burdzy, and N. M. Kredich. 1987. DNA sequences of the cysB regions of Salmonella typhimurium and Escherichia coli. J. Biol. Chem. 262:5999–6005.

1502. Ostrowski, J., and N. M. Kredich. 1989. Molecular characterization of the cysJIH promoters of Salmonella typhimurium and Escherichia coli: regulation by cysB protein and N-acetyl-L-serine. J. Bacteriol. 171:130–140.

1503. Ostrowski, J., and N. M. Kredich. 1990. In vitro interactions of CysB protein with the cysJIH promoter of Salmonella typhimurium: inhibitory effects of sulfide. J. Bacteriol. 172:779–785.

1504. Ostrowski, J., and N. M. Kredich. 1991. Negative autoregulation of cysB in Salmonella typhimurium: in vitro interactions of CysB protein with the cysB promoter. J. Bacteriol. 173:2212–2218.

1505. Ostrowski, J., J. Y. Wu, D. C. Rueger, B. E. Miller, L. M. Siegel, and N. M. Kredich. 1989. Characterization of the cysJIH regions of Salmonella typhimurium and Escherichia coli B. DNA sequences of cysI and cysH and a model for the siroheme-Fe4S4 active center of sulfite reductase hemoprotein based on amino acid homology with spinach nitrite reductase. J. Biol. Chem. 264:15726–15737.

1506. O'Toole, G. A., and J. C. Escalante-Semerena. 1991. Identification and initial characterization of the eutF locus of Salmonella typhimurium. J. Bacteriol. 173:5168–5172.

1507. O'Toole, G. A., and J. C. Escalante-Semerena. 1993. cobU-dependent assimilation of nonadenosylated cobinamide in cobA mutants of Salmonella typhimurium. J. Bacteriol. 175:6328–6336.

1508. O'Toole, G. A., M. R. Rondon, and J. C. Escalante-Semerena. 1993. Analysis of mutants of Salmonella typhimurium defective in the synthesis of the nucleotide loop of cobalamin. J. Bacteriol. 175:3317–3326.

1509. O'Toole, G. A., J. R. Trzebiatowski, and J. C. Escalante-Semerena. 1994. The cobC gene of Salmonella typhimurium codes for a novel phosphatase involved in the assembly of the nucleotide loop of cobalamin. J. Biol. Chem. 269:26503–26511.

1510. Overbeeke, N., G. Van Scharrenburg, and B. Lugtenburg. 1980. Antigenic relationships between pore proteins of Escherichia coli K12. Eur. J. Biochem. 110:247–254.

1511. Overbye, K. M., S. K. Basu, and P. Margolin. 1983. Loss of DNA topoisomerase I activity alters many cellular functions in Salmonella typhimurium. Cold Spring Harbor Symp. Quant. Biol. 47:785–791.

1512. Overbye, K. M., and P. Margolin. 1981. Role of the supX gene in ultraviolet light-induced mutagenesis in Salmonella typhimurium. J. Bacteriol. 146:170–178.

1513. Overdier, D. G., and L. N. Csonka. 1992. A transcriptional silencer downstream of the promoter in the osmotically controlled proU operon of Salmonella typhimurium. Proc. Natl. Acad. Sci. USA 89:3140–3144.

1514. Overdier, D. G., E. R. Olson, B. D. Erickson, M. M. Ederer, and L. N. Csonka. 1989. Nucleotide sequence of the transcriptional control region of the osmotically regulated proU operon of Salmonella typhimurium and identification of the 5' endpoint of the proU mRNA. J. Bacteriol. 171:4694–4706.

1515. Ow, D. W., V. Sundaresan, D. M. Rothstein, S. E. Brown, and F. M. Ausubel. 1983. Promoters regulated by the glnG (ntrC) and nifA gene products share a heptameric consensus sequence in the −15 region. Proc. Natl. Acad. Sci. USA 80:2524–2528.

1516. Owen-Hughes, T. A., G. D. Pavitt, D. S. Santos, J. M. Sidebotham, C. S. Hulton, J. C. Hinton, and C. F. Higgins. 1992. The chromatin-associated protein H-NS interacts with curved DNA to influence DNA topology and gene expression. Cell 71:255–265.

1517. Pace, J., M. J. Hayman, and J. E. Galan. 1993. Signal transduction and invasion of epithelial cells by S. typhimurium. Cell 72:505–514.

1518. Pai, S. R., Y. Upshaw, and S. P. Singh. 1992. Characterization of monoclonal antibodies to the outer membrane protein (OmpD) of Salmonella typhimurium. Can. J. Microbiol. 38:1102–1107.

1519. Palmer, D. T., P. H. Blum, and S. W. Artz. 1983. Effects of the hisT mutation of Salmonella typhimurium on translation elongation rate. J. Bacteriol. 153:357–363.

1520. Palva, E. T. 1978. Major outer membrane protein in Salmonella typhimurium induced by maltose. J. Bacteriol. 136:286–294.

1521. Palva, E. T. 1979. Relationship between ompB genes of Escherichia coli and Salmonella typhimurium. FEMS Microbiol. Lett. 5:205–210.

1522. Palva, E. T., P. Liljestrom, and S. Harayama. 1981. Cosmid cloning and transposon mutagenesis in Salmonella typhimurium using phage lambda vehicles. Mol. Gen. Genet. 181:153–157.

1523. Palva, E. T., and P. H. Makela. 1980. Lipopolysaccharide heterogeneity in Salmonella typhimurium analyzed by sodium dodecyl sulfate/polyacrylamide gel electrophoresis. Eur. J. Biochem. 107:137–143.

1524. Pandey, N. K., and R. L. Switzer. 1982. Mutant strains of Salmonella typhimurium with defective phosphoribosylpyrophosphate synthetase activity. J. Gen. Microbiol. 128:1863–1871.

1525. Pang, P. P., A. S. Lundberg, and G. C. Walker. 1985. Identification and characterization of the mutL and mutS gene products of Salmonella typhimurium LT2. J. Bacteriol. 163:1007–1015.

1526. Pang, P. P., S. D. Tsen, A. S. Lundberg, and G. C. Walker. 1984. The mutH, mutL, mutS, and uvrD genes of Salmonella typhimurium LT2. Cold Spring Harbor Symp. Quant. Biol. 49:597–602.

1527. Pang, P. P., and G. C. Walker. 1983. The Salmonella typhimurium LT2 uvrD gene is regulated by the lexA gene product. J. Bacteriol. 154:1502–1504.

1528. Pang, P. P., and G. C. Walker. 1983. Identification of the uvrD gene product of Salmonella typhimurium LT2. J. Bacteriol. 153:1172–1179.

1529. Papp-Szabo, E., M. Firtel, and P. D. Josephy. 1994. Comparison of the sensitivities of the Salmonella typhimurium oxyR and katG mutants to killing by human neutrophils. Infect. Immun. 62:2662–2668.

1530. Parada, J. L., and M. V. Ortega. 1975. Lysis of a temperature conditional thiamineless mutant of Salmonella typhimurium by glucose and hexoses. Rev. Assoc. Argent. Microbiol. 7:91–96.

1531. Parada, J. L., M. V. Ortega, and G. Carrillo-Castaneda. 1973. Biochemical and genetic characteristics of the C4-dicarboxylic acids transport system of Salmonella typhimurium. Arch. Mikrobiol. 94:65–76.

1532. Pardee, A. B. 1966. Purification and properties of a sulfate-binding protein from Salmonella typhimurium. J. Biol. Chem. 241:5886–5892.

1533. Pardee, A. B., L. S. Prestidge, M. B. Whipple, and J. Dreyfuss. 1966. A binding site for sulfate and its relation to sulfate transport into Salmonella typhimurium. J. Biol. Chem. 241:3962–3969.

1534. Pardee, A. B., and K. Watanabe. 1968. Location of sulfate-binding protein in Salmonella typhimurium. J. Bacteriol. 96:1049–1054.

1535. Park, U. E., B. M. Olivera, K. T. Hughes, J. R. Roth, and D. R. Hillyard. 1989. DNA ligase and the pyridine nucleotide cycle in Salmonella typhimurium. J. Bacteriol. 171:2173–2180.

1536. Park, U. E., J. R. Roth, and B. M. Olivera. 1988. Salmonella typhimurium mutants lacking NAD pyrophosphatase. J. Bacteriol. 170:3725–3730.

1537. Park, Y. M., and G. V. Stauffer. 1989. Salmonella typhimurium metC operator-constitutive mutations. FEMS Microbiol. Lett. 51:137–141.

1538. Park, Y. M., and G. V. Stauffer. 1989. DNA sequence of the metC gene and its flanking regions from Salmonella typhimurium LT2 and homology with the corresponding sequence of Escherichia coli. Mol. Gen. Genet. 216:164–169.

1539. Parker, J., M. Flashner, W. G. McKeever, and F. C. Neidhardt. 1974. Metabolic regulation of the arginyl and valyl transfer ribonucleic acid synthetases in bacteria. J. Biol. Chem. 249:1044–1053.

1540. Parra-Lopez, C., M. T. Baer, and E. A. Groisman. 1993. Molecular genetic analysis of a locus required for resistance to antimicrobial peptides in Salmonella typhimurium. EMBO J. 12:4053–4062.

1541. Parra-Lopez, C., R. Lin, A. Aspedon, and E. A. Groisman. 1994. A Salmonella protein that is required for resistance to antimicrobial peptides and transport of potassium. EMBO J. 13:3964–3972.

1542. Parsons, S. M., and D. E. Koshland, Jr. 1974. Multiple aggregation states of phosphoribosyladenosine triphosphate synthetase. J. Biol. Chem. 249:4119–4126.

1543. Parsons, S. M., and D. E. Koshland, Jr. 1974. A rapid isolation of phosphoribosyladenosine triphosphate synthetase and comparison to native enzyme. J. Biol. Chem. 249:4104–4109.

1544. Parsons, S. M., and M. Lipsky. 1975. Composition of the first enzyme of histidine biosynthesis isolated from wild-type and mutant operator strains of Salmonella typhimurium. J. Bacteriol. 121:485–490.

1545. Parsons, S. M., and J. Begley. 1975. A unique reactive residue in adenosine triphosphate phosphoribosyltransferase sensitive to five conformation and dissociation sites. J. Biol. Chem. 250:5660–5668.

1546. Pascal, M.-C., F. Casse, and M. Chippaux. 1977. Localization of pfl gene by transductional study of the gal-aroA segment of the Salmonella typhimurium LT2 chromosome. Mol. Gen. Genet. 150:331–334.

1547. Pascal, M.-C., F. Casse, M. Chippaux, and M. Lepelletier. 1973. Genetic analysis of mutants of Salmonella typhimurium deficient in formate dehydrogenase activity. Mol. Gen. Genet. 150:337–340.

1548. Pascal, M.-C., F. Casse, M. Chippaux, and M. Lepelletier. 1975. Genetic analysis of mutants of Escherichia coli K12 and Salmonella typhimurium LT2 deficient in hydrogenase activity. Mol. Gen. Genet. 141:173–179.

1549. Pasternak, C. A., R. J. Ellis, M. C. Jones-Mortimer, and C. E. Crichton. 1965. The control of sulfate reduction in bacteria. Biochem. J. 96:270–275.

1550. Paton, E. B., M. I. Woodmaska, I. V. Kroupskaya, A. N. Zhyvoloup, and G. K. Matsuka. 1990. Evidence for the ability of L10 ribosomal proteins of Salmonella *typhimurium* and *Klebsiella pneumoniae* to regulate *rplJL* gene expression in *Escherichia coli. FEBS Lett.* **265**:129–132.

1551. Paton, E. B., S. B. Zolotukhin, M. I. Woodmaska, I. V. Kroupskaya, and A. N. Zhyvoloup. 1990. The nucleotide sequence of gene *rplJ* encoding ribosomal protein L10 of *Salmonella typhimurium. Nucleic Acids Res.* **18**:2824.

1552. Patterson-Delafield, J., R. J. Martinex, B. A. D. Stocker, and S. Yamaguchi. 1973. A new *fla* gene in *Salmonella typhimurium*—*flaR*—and its mutant phenotype—superhooks. *Arch. Mikrobiol.* **90**:107–120.

1553. Paveglio, M. T., J. S. Tang, R. E. Unger, and E. L. Barrett. 1988. Formate-nitrate respiration in *Salmonella typhimurium*: studies of two *rha*-linked *fdn* genes. *J. Bacteriol.* **170**:213–217.

1554. Payne, G. M., E. N. Spudich, and G. F.-L. Ames. 1985. A mutational hot-spot in the *hisM* gene of the histidine transport operon in *Salmonella typhimurium* is due to deletion of repeated sequences and results in an altered specificity of transport. *Mol. Gen. Genet.* **200**:493–496.

1555. Pearce, S. R., M. L. Mimmack, M. P. Gallagher, U. Gileadi, S. C. Hyde, and C. F. Higgins. 1992. Membrane topology of the integral membrane components, OppB and OppC, of the oligopeptide permease of *Salmonella typhimurium. Mol. Microbiol.* **6**:47–57.

1556. Pearce, U. B., and B. A. D. Stocker. 1967. Phase variation of flagellar antigens in *Salmonella*: abortive transduction studies. *J. Gen. Microbiol.* **49**:335–349.

1557. Pekkel, V. A., M. A. Abramova, and V. S. Levashev. 1974. Role of polyamines in the process of bacterial cell division. *Zh. Mikrobiol. Epidemiol. Immunobiol.* **51**:8–13.

1558. Persson, B. C., and G. R. Bjork. 1993. Isolation of the gene (*miaE*) encoding the hydroxylase involved in the synthesis of 2-methylthio-*cis*-ribozeatin in tRNA of *Salmonella typhimurium* and characterization of mutants. *J. Bacteriol.* **175**:7776–7785.

1559. Peterson, A. A., and E. J. McGroarty. 1985. High-molecular-weight components in lipopolysaccharides of *Salmonella typhimurium, Salmonella minnesota*, and *Escherichia coli. J. Bacteriol.* **162**:738–745.

1560. Pflugrath, J. W., and F. A. Quiocho. 1985. Sulphate sequestered in the sulphate-binding protein of *Salmonella typhimurium* is bound solely by hydrogen bonds. *Nature* (London) **314**:257–260.

1561. Pflugrath, J. W., and F. A. Quiocho. 1988. The 2 A resolution structure of the sulfate-binding protein involved in active transport in *Salmonella typhimurium. J. Mol. Biol.* **200**:163–180.

1562. Piszkiewicz, D., B. E. Tilley, T. Rand-Meir, and S. M. Parsons. 1979. Amino acid sequence of ATP phosphoribosyltransferase of *Salmonella typhimurium. Proc. Natl. Acad. Sci USA* **76**:1589–1592.

1563. Pizarro, R. A., G. O. Boselli, and L. V. Orce. 1985. Biochemical and phenotypical correction of an envelope mutant of *Salmonella typhimurium. Arch. Int. Physiol. Biochim.* **92**:333–337.

1564. Plamann, L. S., and G. V. Stauffer. 1987. Nucleotide sequence of the *Salmonella typhimurium metR* gene and the *metR-metE* control region. *J. Bacteriol.* **169**:3932–3937.

1565. Plamann, L. S., M. L. Urbanowski, and G. V. Stauffer. 1988. *Salmonella typhimurium metE* operator-constitutive mutations. *Gene* **73**:201–208.

1566. Plosila, M., and P. H. Makela. 1972. Mapping of a gene *oafA* determining antigen 1 in *Salmonella* of group E4. *Scand. J. Clin. Lab. Invest.* **29**(Suppl. 122):55.

1567. Poland, J., and P. D. Ayling. 1994. Methionine and glutamine transport sustems in D-methionine utilising revertants of *Salmonella typhimurium. Mol. Gen. Genet.* **194**:219–226.

1568. Pollack, J. R., B. N. Ames, and J. B. Neilands. 1970. Iron transport in *Salmonella typhimurium*: mutants blocked in the biosynthesis of enterobactin. *J. Bacteriol.* **104**:635–639.

1569. Popham, D., J. Keener, and S. Kustu. 1991. Purification of the alternative sigma factor, sigma 54, from *Salmonella typhimurium* and characterization of sigma 54 holoenzyme. *J. Biol. Chem.* **266**:19510–19518.

1570. Popham, D. L., D. Szeto, J. Keener, and S. Kustu. 1989. Function of a bacterial activator protein that binds to transcriptional enhancers. *Science* **243**:629–635.

1571. Post, D. A., B. Hove-Jensen, and R. L. Switzer. 1993. Characterization of the *hemA-prs* region of the *Escherichia coli* and *Salmonella typhimurium* chromosomes: identification of two open reading frames and implications for *prs* expression. *J. Gen. Microbiol.* **139**:259–266.

1572. Post, D. A., and R. L. Switzer. 1991. *prsB* is an allele of the *Salmonella typhimurium prsA* gene: characterization of a mutant phosphoribosylpyrophosphate synthetase. *J. Bacteriol.* **173**:1978–1986.

1573. Postgate, J. R., and V. Krishnapillai. 1977. Expression of *Klebsiella nif* and *his* genes in *Salmonella typhimurium. J. Gen. Microbiol.* **98**:379–385.

1574. Postma, P. W. 1976. Involvement of the phosphotransferase system in galactose transport in *Salmonella typhimurium. FEBS Lett.* **61**:49–53.

1575. Postma, P. W. 1977. Galactose transport in *Salmonella typhimurium. J. Bacteriol.* **129**:630–639.

1576. Postma, P. W. 1981. Defective enzyme II-B^{Glc} of the phosphoenolpyruvate:sugar phosphotransferase system leading to uncoupling of transport and phosphorylation in *Salmonella typhimurium. J. Bacteriol.* **147**:382–389.

1577. Postma, P. W., J. C. Cordaro, and S. Roseman. 1977. Sugar transport. A pleiotrophic membrane mutant of *Salmonella typhimurium. J. Biol. Chem.* **252**:3862–3876.

1578. Postma, P. W., W. Epstein, A. R. J. Schuitema, and S. O. Nelson. 1984. Interaction between III^{Glc} of the phosphoenolpyruvate:sugar phosphotransferase system and glycerol kinase of *Salmonella typhimurium. J. Bacteriol.* **158**:351–353.

1579. Postma, P. W., H. G. Keizer, and P. Koolwijk. 1986. Transport of trehalose in *Salmonella typhimurium. J. Bacteriol.* **168**:1107–1111.

1580. Postma, P. W., and J. W. Lengeler. 1985. Phosphoenolpyruvate:carbohydrate phosphotransferase system of bacteria. *Microbiol. Rev.* **49**:232–269.

1581. Postma, P. W., A. Schuitema, and C. Kwa. 1981. Regulation of methyl β-galactoside permease activity in *pts* and *crr* mutants of *Salmonella typhimurium. Mol. Gen. Genet.* **181**:448–453.

1582. Postma, P. W., and J. B. Stock. 1980. Enzymes II of the phosphotransferase system do not catalyze sugar transport in the absence of phosphorylation. *J. Bacteriol.* **141**:476–484.

1583. Powers, D. A., and S. Roseman. 1984. The primary structure of *Salmonella typhimurium* HPr, a phosphocarrier protein of the phosphoenolpyruvate:glycose phosphotransferase system. A correction. *J. Biol. Chem.* **259**:15212–15214.

1584. Prasad, R., A. K. Chopra, P. Chary, and J. W. Peterson. 1992. Expression and characterization of the cloned *Salmonella typhimurium* enterotoxin. *Microb. Pathog.* **13**:109–121.

1585. Prasad, R., A. K. Chopra, J. W. Peterson, R. Pericas, and C. W. Houston. 1990. Biological and immunological characterization of a cloned cholera toxin-like enterotoxin from *Salmonella typhimurium. Microb. Pathog.* **9**:315–329.

1586. Primakoff, P., and S. W. Artz. 1979. Positive control of *lac* operon expression in-vitro by guanosine-5′-diphosphate-3′-diphosphate. *Proc. Natl. Acad. Sci. USA* **76**:1726–1730.

1587. Primerano, D. A., and R. O. Burns. 1982. Metabolic basis for the isoleucine, pantothenate, or methionine requirement of *ilvG* strains of *Salmonella typhimurium. J. Bacteriol.* **150**:1202–1211.

1588. Primerano, D. A., and R. O. Burns. 1983. Role of acetohydroxy acid isomeroreductase in biosynthesis of pantothenic acid in *Salmonella typhimurium. J. Bacteriol.* **153**:259–269.

1589. Pritchard, D. G., S. C. Nivas, M. D. York, and B. S. Pomeroy. 1979. Effects of *galE* mutant of *Salmonella typhimurium* on experimental salmonellosis in chickens. *Avian Dis.* **22**:562–575.

1590. Prival, M. J., and T. A. Cebula. 1992. Sequence analysis of mutations arising during prolonged starvation of *Salmonella typhimurium. Genetics* **132**:303–310.

1591. Prossnitz, E. 1991. Determination of a region of the HisJ binding protein involved in the recognition of the membrane complex of the histidine transport system of *Salmonella typhimurium. J. Biol. Chem.* **266**:9673–9677.

1592. Pruss, G. J. 1985. DNA topoisomerase I mutants. Increased heterogeneity in linking number and other replicon-dependent changes in DNA supercoiling. *J. Mol. Biol.* **185**:51–63.

1593. Pruss, G. J., and K. Drlica. 1985. DNA supercoiling and suppression of the *leu-500* promoter mutation. *J. Bacteriol.* **164**:947–949.

1594. Pruss, G. J., and K. Drlica. 1986. Topoisomerase I mutants: the gene on pBR322 that encodes resistance to tetracycline affects plasmid DNA supercoiling. *Proc. Natl. Acad. Sci. USA* **83**:8952–8956.

1595. Pueyo, C. 1978. Forward mutations to arabinose resistance in *Salmonella typhimurium* strains. A sensitive assay for mutagenicity testing. *Mutat. Res.* **54**:311–322.

1596. Pueyo, C., and J. Lopez-Barea. 1979. The L-arabinose resistance test with *Salmonella typhimurium* strain SV-3 selects forward mutations at several *ara* genes. *Mutat. Res.* **64**:249–258.

1597. Pugsley, A. P., D. J. Conrad, C. A. Schnaitman, and T. I. Gregg. 1980. In vivo effects of local anesthetics on the production of major outer membrane proteins by *Escherichia coli. Biochim. Biophys. Acta* **599**:1–12.

1598. Pulkkinen, W. S., and S. I. Miller. 1991. A *Salmonella typhimurium* virulence protein is similar to a *Yersinia enterocolitica* invasion protein and a bacteriophage lambda outer membrane protein. *J. Bacteriol.* **173**:86–93.

1599. Purcell, B. K., J. Pruckler, and S. Clegg. 1987. Nucleotide sequences of the genes encoding type 1 fimbrial subunits of *Klebsiella pneumoniae* and *Salmonella typhimurium. J. Bacteriol.* **169**:5831–5834.

1600. Quay, S., and H. N. Christensen. 1974. Basis of transport discrimination of arginine from other basic amino acids in *Salmonella typhimurium. J. Biol. Chem.* **249**:7011–7017.

1601. Quay, S. C., and D. L. Oxender. 1980. Role of tRNA^{leu} in branched-chain amino acid transport. *Cold Spring Harbor Monogr. Ser.* **9B**:481–491.

1602. Quay, S. C., D. L. Oxender, S. Tsuyumu, and H. E. Umbarger. 1975. Separate regulation of transport and biosynthesis of leucine, isoleucine, and valine in bacteria. *J. Bacteriol.* **122**:994–1000.

1603. Qureshi, M. A., D. A. Smith, and A. J. Kingsman. 1975. Mutants of *Salmonella typhimurium* responding to cysteine or methionine: their nature and possible role in the regulation of cysteine biosynthesis. *J. Gen. Microbiol.* **89**:353–370.

1604. Qureshi, N., K. Takayama, D. Heller, and C. Fenselau. 1983. Position of ester groups in the lipid A backbone of lipopolysaccharides obtained from *Salmonella typhimurium. J. Biol. Chem.* **258**:12947–12951.

1605. Raetz, C. R. H., S. Purcell, M. V. Meyer, N. Qureshi, and K. Takayama. 1985. Isolation and characterization of eight lipid A precursors from a 3-deoxy-D-manno-oculosonic acid-deficient mutant of *Salmonella typhimurium. J. Biol. Chem.* **260**:16080–16088.

1606. Raha, M., I. Kawagishi, V. Muller, M. Kihara, and R. M. Macnab. 1992. *Escherichia coli* produces a cytoplasmic alpha-amylase, AmyA. *J. Bacteriol.* **174**:6644–6652.

1607. Raha, M., M. Kihara, I. Kawagishi, and R. M. Macnab. 1993. Organization of the *Escherichia coli* and *Salmonella typhimurium* chromosomes between flagellar regions IIIa and IIIb, including a large non-coding region. *J. Gen. Microbiol.* **139:**1401–1407.

1608. **Raha, M., H. Sockett, and R. M. Macnab.** 1994. Characterization of the *fliL* gene in the flagellar regulon of *Escherichia coli* and *Salmonella typhimurium. J. Bacteriol.* **176:**2308–2311.

1609. **Raju, K. K., D. M. Bedwell, and J. E. Brenchley.** 1980. Characterization of a HinD-III generated NDA fragment carrying the glutamine synthetase gene in *Salmonella typhimurium. Gene* **11:**227–238.

1610. **Rao, R., and M. G. Pereira.** 1975. Isolation of a hybrid F′ factor-carrying *Escherichia coli* lactose region and *Salmonella typhimurium* histidine region, F42–400 (F′$_{ts114}$ *lac*⁺, *his*⁺): its partial characterization and behavior in *Salmonella typhimurium. J. Bacteriol.* **123:**779–791.

1611. **Rao, R. N., and C. V. S. Rao.** 1973. *Salmonella typhimurium* mutants affecting establishment of lysogeny. *Mol. Gen. Genet.* **125:**119–123.

1612. **Rao, R. N., and H. O. Smith.** 1968. Phage P22 lysogens of a *Salmonella typhimurium* mutant deleted at the normal prophage attachment site. *Virology* **36:**328–330.

1613. **Ratzkin, B., M. Grabnar, and J. R. Roth.** 1978. Regulation of the major proline permease gene of *Salmonella typhimurium. J. Bacteriol.* **133:**737–743.

1614. **Ratzkin, B., and J. R. Roth.** 1978. Cluster of genes controlling proline degradation in *Salmonella typhimurium. J. Bacteriol.* **133:**744–754.

1615. **Ravdonikas, L. E.** 1976. Production and characteristics of *Salmonella typhimurium* glycerin mutants. *Zh. Mikrobiol. Epidemiol. Immunobiol.* **12:**29–32.

1616. **Ravid, S., and M. Eisenbach.** 1984. Direction of flagellar rotation in bacterial cell envelopes. *J. Bacteriol.* **158:**222–230.

1617. **Rayssiguier, C., C. Dohet, and M. Radman.** 1991. Interspecific recombination between *Escherichia coli* and *Salmonella typhimurium* occurs by the RecABCD pathway. *Biochimie* **73:**371–374.

1618. **Rayssiguier, C., D. S. Thaler, and M. Radman.** 1989. The barrier to recombination between *Escherichia coli* and *Salmonella typhimurium* is disrupted in mismatch-repair mutants. *Nature* (London) **342:**396–401.

1619. **Rechler, M. M., and C. B. Bruni.** 1971. Properties of a fused protein formed by genetic manipulation. *J. Biol. Chem.* **246:**1806–1813.

1620. **Reeve, C. A., A. T. Bockman, and A. Matin.** 1984. Role of protein degradation in the survival of carbon-starved *Escherichia coli* and *Salmonella typhimurium. J. Bacteriol.* **157:**758–763.

1621. **Reeves, P., and G. Stevenson.** 1989. Cloning and nucleotide sequence of the *Salmonella typhimurium* LT2 *gnd* gene and its homology with the corresponding sequence of *Escherichia coli* K12. *Mol. Gen. Genet.* **217:**182–184.

1622. **Reeves, R. H., and J. R. Roth.** 1971. A recessive UGA suppressor. *J. Mol. Biol.* **56:**523–533.

1623. **Reeves, R. H., and J. R. Roth.** 1975. Transfer ribonucleic acid methylase deficiency found in UGA suppressor strains. *J. Bacteriol.* **124:**332–340.

1624. **Reiners, J. J., Jr., and H. Zalkin.** 1975. Immunological study of anthranilate synthetase. *J. Bacteriol.* **123:**620–630.

1625. **Rephaeli, A. W., and M. H. Saier, Jr.** 1976. Effects of *crp* mutations on adenosine 3′,5′-monophosphate metabolism in *Salmonella typhimurium. J. Bacteriol.* **127:**120–127.

1626. **Rephaeli, A. W., and M. H. Saier, Jr.** 1980. Regulation of genes coding for enzyme constituents of the bacterial phosphotransferase system. *J. Bacteriol.* **141:**658–663.

1627. **Reyes, G. R., and V. Rocha.** 1977. Immuno-chemical comparison of phosphoribosylanthranilate isomerase-indoleglycerol phosphate synthetase among the Enterobacteriaceae. *J. Bacteriol.* **129:**1448–1456.

1628. **Rhen, M., and S. Sukupolvi.** 1988. The role of the *traT* gene of the *Salmonella typhimurium* virulence plasmid for serum resistance and growth within liver macrophages. *Microb. Pathog.* **5:**275–285.

1629. **Ricca, E., and J. M. Calvo.** 1990. The nucleotide sequence of *leuA* from *Salmonella typhimurium. Nucleic Acids Res.* **18:**1290.

1630. **Ricca, E., C. T. Lago, M. Sacco, and M. De Felice.** 1991. Absence of acetohydroxy acid synthase III in *Salmonella typhimurium* is due to early termination of translation within the *ilvI* gene. *Mol. Microbiol.* **5:**1741–1743.

1631. **Richardson, S. M., C. F. Higgins, and D. M. Lilley.** 1988. DNA supercoiling and the *leu-500* promoter mutation of *Salmonella typhimurium. EMBO J.* **7:**1863–1869.

1632. **Richardson, S. M. H., C. F. Higgins, and D. M. J. Lilley.** 1984. The genetic control of DNA supercoiling in *Salmonella typhimurium. EMBO J.* **3:**1745–1752.

1633. **Richarme, G., A. el Yaagoubi, and M. Kohiyama.** 1993. The MglA component of the binding protein-dependent galactose transport system of *Salmonella typhimurium* is a galactose-stimulated ATPase. *J. Biol. Chem.* **268:**9473–9477.

1634. **Richter-Dahlfors, A. A., and D. I. Andersson.** 1992. Cobalamin (vitamin B12) repression of the Cob operon in *Salmonella typhimurium* requires sequences within the leader and the first translated open reading frame. *Mol. Microbiol.* **6:**743–749.

1635. **Rick, P. D., W.-M. Fung, C. Ho, and M. J. Osborn.** 1977. Lipid A mutants of *Salmonella typhimurium. J. Biol. Chem.* **252:**4902–4912.

1636. **Rick, P. D., B. A. Neumeyer, and D. A. Young.** 1983. Effect of altered lipid A synthesis on the synthesis of the OmpA protein in *Salmonella typhimurium. J. Biol. Chem.* **258:**629–635.

1637. **Rick, P. D., and M. J. Osborn.** 1972. Isolation of a mutant of *Salmonella typhimurium* dependent on D-arabinose-5-phosphate for growth and synthesis of 3-deoxy-D-mannoctulosonate (ketodeoxyoctonate). *Proc. Natl. Acad. Sci. USA* **69:**3756–3760.

1638. **Rick, P. D., and M. J. Osborn.** 1977. Lipid A mutants of *Salmonella typhimurium.* Characterization of a conditional lethal mutant in 3-deoxy-D-mannoctulosonate-8-phosphate synthetase. *J. Biol. Chem.* **252:**4895–4903.

1639. **Rick, P. D., S. Wolski, K. Barr, S. Ward, and L. Ramsey-Sharer.** 1988. Accumulation of a lipid-linked intermediate involved in enterobacterial common antigen synthesis in *Salmonella typhimurium* mutants lacking dTDP-glucose pyrophosphorylase. *J. Bacteriol.* **170:**4008–4014.

1640. **Rick, P. D., and D. A. Young.** 1982. Relationship between cell death and altered lipid A synthesis in a temperature-sensitive lethal mutant of *Salmonella typhimurium* that is conditionally defective in 3-deoxy-D-manno-octulosonate-8-phosphate synthesis. *J. Bacteriol.* **150:**456–464.

1641. **Rick, P. D., and D. A. Young.** 1982. Isolation and characterization of a temperature-sensitive lethal mutant of *Salmonella typhimurium* that is conditionally defective in 3-deoxy-D-manno-octulosonate-8-phosphate synthesis. *J. Bacteriol.* **150:**447–455.

1642. **Riddle, D. L., and J. Carbon.** 1972. Frameshift suppression: a nucleotide addition in the anticodon of a glycine transfer RNA. *Nature* (London) *New Biol.* **242:**230–234.

1643. **Riddle, D. L., and J. R. Roth.** 1972. Frameshift suppressors. III. Effects of suppressor mutations on transfer RNA. *J. Mol. Biol.* **66:**495–506.

1644. **Riddle, D. L., and J. R. Roth.** 1972. Frameshift suppressors. II. Genetic mapping and dominance studies. *J. Mol. Biol.* **66:**483–493.

1645. **Riggs, D., and S. Artz.** 1984. The *hisD-hisC* gene border of the *Salmonella typhimurium* histidine operon. *Mol. Gen. Genet.* **196:**526–529.

1646. **Riggs, D. L., and E. Barrett.** 1983. Role for thiosulfate reductase in the chlorate sensitivity of *Salmonella typhimurium,* abstr. K246, p. 218. *Abstr. Annu. Meet. Am. Soc. Microbiol., 1983.*

1647. **Riggs, D. L., R. D. Meuller, H.-S. Kwan, and S. W. Artz.** 1986. Promoter domain mediates guanosine tetraphosphate activation of the histidine operon. *Proc. Natl. Acad. Sci. USA* **83:**9333–9337.

1648. **Riley, M., C. O'Reilly, and D. McConnell.** 1983. Physical map of *Salmonella typhimurium* LT2 DNA in the vicinity of the *proA* gene. *J. Bacteriol.* **157:**655–657.

1649. **Rioux, C. R., M. J. Friedrich, and R. J. Kadner.** 1990. Genes on the 90-kilobase plasmid of *Salmonella typhimurium* confer low-affinity cobalamin transport: relationship to fimbria biosynthesis genes. *J. Bacteriol.* **172:**6217–6222.

1650. **Rioux, C. R., and R. J. Kadner.** 1989. Two outer membrane transport systems for vitamin B₁₂ in *Salmonella typhimurium. J. Bacteriol.* **171:**2986–2993.

1651. **Ritchie, L. J., R. M. Hall, and D. M. Podger.** 1986. Mutant of *Salmonella typhimurium* LT2 deficient in DNA adenine methylation. *J. Bacteriol.* **167:**420–422.

1652. **Rivera, M., A. Bertasso, C. McCaffrey, and N. H. Georgopapadakou.** 1993. Porins and lipopolysaccharide of *Escherichia coli* ATCC 25922 and isogenic rough mutants. *FEMS Microbiol. Lett.* **108:**183–187.

1653. **Rizzino, A., M. Mastanduno, and M. Freundlich.** 1977. Partial derepression of the isoleucine-valine enzymes during methionine starvation in *Salmonella typhimurium. Biochim. Biophys. Acta* **475:**267–275.

1654. **Rizzino, A. A., R. S. Bresalier, and M. Freundlich.** 1974. Derepressed levels of the isoleucine-valine and leucine enzymes in *hisT1504. J. Bacteriol.* **117:**449–455.

1655. **Roberts, J. H., and A. P. Levin.** 1972. Normal synthesis of aminotransferase protein in a pyridoxineless strain of *S. typhimurium. Biochem. Biophys. Res. Commun.* **48:**802–807.

1656. **Roberts, J. H., and A. P. Levin.** 1973. Imidazolylacetolphosphate aminotransferase. Properties of the apoprotein produced in a pyridoxine auxotroph of *Salmonella typhimurium. J. Biol. Chem.* **248:**7748–7753.

1657. **Roberts, J. H., D. R. McCarroll, and A. P. Levin.** 1975. Properties of the imidazolylacetolphosphate aminotransferase produced in a mutant demonstrating no apparent genetic involvement of the structural gene. *J. Bacteriol.* **123:**233–241.

1658. **Robertson, B. C., and P. A. Hoffee.** 1973. Purification and properties of purine nucleoside phosphorylase from *Salmonella typhimurium. J. Biol. Chem.* **148:**2040–2043.

1659. **Robertson, B. C., P. Jargiello, J. Blank, and P. A. Hoffee.** 1970. Genetic regulation of ribonucleoside and deoxyribonucleoside catabolism in *Salmonella typhimurium. J. Bacteriol.* **102:**628–635.

1660. **Robinson, P. B., and H. R. Levy.** 1976. Metal ion requirement and tryptophan inhibition of normal and variant anthranilate synthase-anthranilate 5-phosphoribosylpyrophosphate. *Biochim. Biophys. Acta* **445:**475–485.

1661. **Robison, K., W. Gilbert, and G. M. Church.** 1994. Large scale bacterial gene discovery by similarity search. *Nat. Genet.* **7:**205–214.

1662. **Rodriguez, S. B., and J. L. Ingraham.** 1983. Location on the *Salmonella typhimurium* chromosome of the gene encoding nucleoside diphosphokinase (*ndk*). *J. Bacteriol.* **153:**1101–1103.

1663. **Rodriguez Lemoine, V., and R. J. Rowbury.** 1975. Instability of the Flac⁺ factor in a *dnaC* mutant of *Salmonella typhimurium. J. Gen. Microbiol.* **90:**360–364.

1664. **Roland, K. L., C. R. Esther, and J. K. Spitznagel.** 1994. Isolation and characterization of a gene, *pmrD,* from *Salmonella typhimurium* that confers resistance to polymyxin when expressed in multiple copies. *J. Bacteriol.* **176:**3589–3597.

1665. **Roland, K. L., L. E. Martin, C. R. Esther, and J. K. Spitznagel.** 1993. Spontaneous *pmrA* mutants of *Salmonella typhimurium* LT2 define a new two-component regulatory system with a possible role in virulence. *J. Bacteriol.* **175:**4154–4164.

1666. **Romana, L. K., F. S. Santiago, and P. R. Reeves.** 1991. High level expression and purification of thymidine diphospho-D-glucose 4,6-dehydratase (*rfbB*) from *Salmonella* serovar *typhimurium* LT2. *Biochem. Biophys. Res. Commun.* **174:**846–852.

1667. Romeo, T., and J. Moore. 1991. Comparison of the 5′ flanking regions of the *Salmonella typhimurium* and *Escherichia coli glgC* genes, encoding ADP glucose pyrophosphorylases. *Nucleic Acids Res.* **19**:3452.

1668. Ron, E. A. 1975. Growth rate of *Enterobacteriaceae* at elevated temperatures: limitation by methionine. *J. Bacteriol.* **124**:243–246.

1669. Rondon, M. R., and J. C. Escalante-Semerena. 1992. The *poc* locus is required for 1,2-propanediol-dependent transcription of the cobalamin biosynthetic (*cob*) and propanediol utilization (*pdu*) genes of *Salmonella typhimurium*. *J. Bacteriol.* **174**:2267–2272.

1670. Roof, D. M., and J. R. Roth. 1988. Ethanolamine utilization in *Salmonella typhimurium*. *J. Bacteriol.* **170**:3855–3863.

1671. Roof, D. M., and J. R. Roth. 1989. Functions required for vitamin B_{12}-dependent ethanolamine utilization in *Salmonella typhimurium*. *J. Bacteriol.* **171**:3316–3323.

1672. Roof, D. M., and J. R. Roth. 1992. Autogenous regulation of ethanolamine utilization by a transcriptional activator of the *eut* operon in *Salmonella typhimurium*. *J. Bacteriol.* **174**:6634–6643.

1673. Roseman, S., and N. D. Meadow. 1990. Signal transduction by the bacterial phosphotransferase system. Diauxie and the *crr* gene (J. Monod revisited). *J. Biol. Chem.* **265**:2993–2996.

1674. Rosenfeld, S. A., and J. E. Brenchley. 1980. Regulation of nitrogen utilization in *hisT* mutants of *Salmonella typhimurium*. *J. Bacteriol.* **143**:801–808.

1675. Rosenfeld, S. A., S. M. Dendinger, C. H. Murphy, and J. E. Brenchley. 1982. Genetic characterization of the glutamate dehydrogenase gene (*gdhA*) of *Salmonella typhimurium*. *J. Bacteriol.* **150**:795–803.

1676. Rosenthal, E. R., and J. M. Calvo. 1987. Transcription termination sites at the distal end of the *leu* operon of *Salmonella typhimurium*. *J. Mol. Biol.* **194**:443–452.

1677. Rosenthal, E. R., and J. M. Calvo. 1990. The nucleotide sequence of *leuC* from *Salmonella typhimurium*. *Nucleic Acids Res.* **18**:3072.

1678. Roth, J. R., and B. N. Ames. 1966. Histidine regulatory mutants in *Salmonella typhimurium*. II. Histidine regulatory mutants having altered histidyl-tRNA synthetase. *J. Mol. Biol.* **22**:325–334.

1679. Roth, J. R., D. N. Anton, and P. E. Hartman. 1966. Histidine regulatory mutants in *Salmonella typhimurium*. I. Isolation and general properties. *J. Mol. Biol.* **22**:305–323.

1680. Roth, J. R., and P. E. Hartman. 1965. Heterogeneity in P22 transducing particles. *Virology* **27**:297–307.

1681. Roth, J. R., J. G. Lawrence, M. Rubenfield, S. Kieffer-Higgins, and G. M. Church. 1993. Characterization of the cobalamin (vitamin B_{12}) biosynthetic genes of *Salmonella typhimurium*. *J. Bacteriol.* **175**:3303–3316.

1682. Rothman-Denes, L., and R. G. Martin. 1971. Two mutations in the first gene of the histidine operon of *Salmonella typhimurium* affecting control. *J. Bacteriol.* **106**:227–237.

1683. Rowbury, R. J. 1964. Synthesis of cystathionine and its control in *Salmonella typhimurium*. *Nature* (London) **203**:977–978.

1684. Rowbury, R. J. 1964. The accumulation of O-succinylhomoserine by *Escherichia coli* and *Salmonella typhimurium*. *J. Gen. Microbiol.* **37**:171–180.

1685. Rowbury, R. J., D. A. Lawrence, and D. A. Smith. 1968. Regulation of the methionine-specific aspartokinase and homoserine dehydrogenase of *Salmonella typhimurium*. *J. Gen. Microbiol.* **54**:337–342.

1686. Rowen, L., J. A. Kobori, and S. Scherer. 1982. Cloning of bacterial DNA replication genes. *Mol. Gen. Genet.* **187**:501–509.

1687. Roy, A. M., and J. Coleman. 1994. Mutations in *firA*, encoding the second acyltransferase in lipopolysaccharide biosynthesis, affect multiple steps in lipopolysaccharide biosynthesis. *J. Bacteriol.* **176**:1639–1646.

1688. Roy, S., and M. Chakravorty. 1986. Spontaneous deletions of drug-resistance determinants from *Salmonella typhimurium* in *Escherichia coli*. *J. Med. Microbiol.* **22**:119–123.

1689. Rudd, K., and J. R. Roth. 1983. Further studies on the role of ppGpp in the control of the *his* operon of *Salmonella*, abstr. H163, p. 112. *Abstr. Annu. Meet. Am. Soc. Microbiol., 1983*.

1690. Rudd, K. E. 1993. Maps, genes, sequences, and computers: an *Escherichia coli* case study. *ASM News* **59**:335–341.

1691. Rudd, K. E., B. R. Bochner, M. Cashel, and J. R. Roth. 1985. Mutations in the *spoT* gene of *Salmonella typhimurium*: effects of *his* operon expression. *J. Bacteriol.* **163**:534–542.

1692. Rudd, K. E., and R. Menzel. 1987. *his* operons of *Escherichia coli* and *Salmonella typhimurium* are regulated by DNA supercoiling. *Proc. Natl. Acad. Sci. USA* **84**:517–521.

1693. Ruijter, G. J., P. W. Postma, and K. van Dam. 1990. Adaptation of *Salmonella typhimurium* mutants containing uncoupled enzyme II^{Glc} to glucose-limited conditions. *J. Bacteriol.* **172**:4783–4789.

1694. Ruijter, G. J., P. W. Postma, and K. van Dam. 1991. Energetics of glucose uptake in a *Salmonella typhimurium* mutant containing uncoupled enzyme IIGlc. *Arch. Microbiol.* **155**:234–237.

1695. Ruiz-Vazquez, R., C. Pueyo, and E. Cerda-Olmedo. 1978. A mutagen assay detecting forward mutations in an arabinose-sensitive strain of *Salmonella typhimurium*. *Mutat. Res.* **54**:121–130.

1696. Rundell, K., and C. W. Shuster. 1973. Membrane-associated nucleotide sugar reactions. I. Properties of the first enzyme of O antigen synthesis. *J. Biol. Chem.* **248**:5436–5442.

1697. Rundell, K., and C. W. Shuster. 1975. Membrane-associated nucleotide sugar reactions: influence of mutations affecting lipopolysaccharide of the first enzyme of O-antigen synthesis. *J. Bacteriol.* **123**:928–936.

1698. Russo, A. F., and D. E. Koshland, Jr. 1983. Separation of signal transduction and adaptation functions of the aspartate receptor in bacterial sensing. *Science* **220**:1016–1020.

1699. Russo, A. F., and D. E. Koshland, Jr. 1986. Identification of the tip-encoded receptor in bacterial sensing. *J. Bacteriol.* **165**:276–282.

1700. Sabe, H., T. Miwa, T. Kodaki, K. Izui, S. Hiraga, and H. Katsuki. 1984. Molecular cloning of the phosphoenolpyruvate carboxylase gene, *ppc*, of *Escherichia coli*. *Gene* **31**:279–283.

1701. Saier, M. H., Jr. 1977. Bacterial phosphoenolpyruvate:sugar phosphotransferase systems: structural, functional, and evolutionary interrelationships. *Bacteriol. Rev.* **41**:856–871.

1702. Saier, M. H., Jr., F. R. Bromberg, and S. Roseman. 1973. Characterization of constitutive galactose permease mutants in *Salmonella typhimurium*. *J. Bacteriol.* **113**:512–514.

1703. Saier, M. H., Jr., and B. U. Feucht. 1976. Coordinate regulation of adenylate cyclase and carbohydrate permeases by the phosphoenolpyruvate:sugar phosphotransferase system in *Salmonella typhimurium*. *J. Biol. Chem.* **250**:7078–7080.

1704. Saier, M. H., Jr., B. U. Feucht, and M. T. McCaman. 1975. Regulation of intracellular adenosine cyclic 3′5′-monophosphate levels in *Escherichia coli* and *Salmonella typhimurium*. *J. Biol. Chem.* **250**:7593–7601.

1705. Saier, M. H., Jr., B. U. Feucht, and W. K. Mora. 1977. Sugar phosphate:sugar transphosphorylation and exchange group translocation catalyzed by the enzyme II complexes of the bacterial phosphoenolpyruvate:sugar phosphotransferase system. *J. Biol. Chem.* **252**:8899–8907.

1706. Saier, M. H., Jr., F. C. Grenier, C. A. Lee, and E. B. Waygood. 1985. Evidence for the evolutionary relatedness of the proteins of the bacterial phosphoenolpyruvate:sugar phosphotransferase system. *J. Cell. Biochem.* **27**:43–56.

1707. Saier, M. H., Jr., M. J. Novotny, D. Comeau-Fuhrman, T. Osumi, and J. D. Desai. 1983. Cooperative binding of the sugar substrates and allosteric regulatory protein (enzyme III^{Glc} of the phosphotransferase system) to the lactose and melibiose permeases in *Escherichia coli* and *Salmonella typhimurium*. *J. Bacteriol.* **155**:1351–1357.

1708. Saier, M. H., Jr., and S. Roseman. 1972. Inducer exclusion and repression of enzyme synthesis in mutants of *Salmonella typhimurium* defective in enzyme I of the phosphoenolpyruvate:sugar phosphotransferase system. *J. Biol. Chem.* **247**:972–975.

1709. Saier, M. H., Jr., and S. Roseman. 1976. Sugar transport. Inducer exclusion and regulation of the melibiose, maltose, glycerol, and lactose transport systems by the phosphoenolpyruvate:sugar phosphotransferase system. *J. Biol. Chem.* **251**:6606–6615.

1710. Saier, M. H., Jr., M. R. Schmidt, and M. Leibowitz. 1978. Cyclic AMP-dependent synthesis of fimbriae in *Salmonella typhimurium*: effects of *cya pts* mutations. *J. Bacteriol.* **134**:356–358.

1711. Saier, M. H., Jr., B. D. Simoni, and S. Roseman. 1976. Sugar transport. Properties of mutant bacteria defective in proteins of the phosphoenolpyruvate:sugar phosphotransferase system. *J. Biol. Chem.* **251**:6584–6597.

1712. Saier, M. H., Jr., H. Straud, L. S. Massman, J. J. Judice, M. H. Newman, and B. U. Feucht. 1978. Permease-specific mutations in *Salmonella typhimurium* and *Escherichia coli* that release the glycerol, maltose, melibiose, and lactose transport systems from regulation by the phosphoenolpyruvate:sugar phosphotransferase system. *J. Bacteriol.* **133**:1358–1367.

1713. Saier, M. H., Jr., D. L. Wentzel, B. U. Feucht, and J. J. Judice. 1975. A transport system for phosphoenolpyruvate, 2-phosphoglycerate, and 3-phosphoglycerate in *Salmonella typhimurium*. *J. Biol. Chem.* **250**:5089–5096.

1714. Sales, M., and J. E. Brenchley. 1982. The regulation of the ammonia assimilatory enzymes in Rel$^+$ and Rel$^-$ strains of *Salmonella typhimurium*. *Mol. Gen. Genet.* **186**:263–268.

1715. Sanderson, K. E., and M. Demerec. 1965. The linkage map of *Salmonella typhimurium*. *Genetics* **51**:897–913.

1716. Sanderson, K. E., and C. A. Hall. 1970. F-prime factors of *Salmonella typhimurium* and an inversion between *Salmonella typhimurium* and *Escherichia coli*. *Genetics* **64**:215–228.

1717. Sanderson, K. E., A. Hessel, and K. E. Rudd. 1995. Genetic map of *Salmonella typhimurium*, edition VIII. *Microbiol. Rev.* **59**:241–303.

1718. Sanderson, K. E., and J. A. Hurley. 1987. Linkage map of *Salmonella typhimurium*, p. 877–918. *In* F. C. Neidhardt, J. L. Ingraham, K. B. Low, B. Magasanik, M. Schaechter, and H. E. Umbarger (ed.), *Escherichia coli and Salmonella typhimurium: Cellular and Molecular Biology*. American Society for Microbiology, Washington, D.C.

1719. Sanderson, K. E., J. Janzer, and J. Head. 1981. Influence of lipopolysaccharide and protein in the cell envelope on recipient capacity in conjunction of *Salmonella typhimurium*. *J. Bacteriol.* **148**:283–293.

1720. Sanderson, K. E., S. K. Kadam, and P. R. MacLachlan. 1983. Derepression of F factor function in *Salmonella typhimurium*. *Can. J. Microbiol.* **29**:1205–1212.

1721. Sanderson, K. E., and J. R. Roth. 1988. Linkage map of *Salmonella typhimurium*, edition VII. *Microbiol. Rev.* **52**:485–532.

1722. Sanderson, K. E., and Y. A. Saeed. 1972. Insertion of the F-factor into a cluster of *rfa* (rough A) genes of *Salmonella typhimurium*. *J. Bacteriol.* **112**:64–73.

1723. **Sanderson, K. E., and Y. A. Saeed.** 1972. P22-mediated transduction analysis of the rough A (*rfa*) region of the chromosome of *Salmonella typhimurium. J. Bacteriol.* 112:58–63.

1724. **Sanderson, K. E., P. Sciore, S.-L. Liu, and A. Hessel.** 1993. Location of IS*200* on the genomic cleavage map of *Salmonella typhimurium* LT2. *J. Bacteriol.* 175:7624–7628.

1725. **Sanderson, K. E., and B. A. D. Stocker.** 1981. Gene *rfaH*, which affects lipopolysaccharide core structure in *Salmonella typhimurium,* is also required for expression of F-factor functions. *J. Bacteriol.* 146:535–541.

1726. **Sandler, S. J., B. Chackerian, J. T. Li, and A. J. Clark.** 1992. Sequence and complementation analysis of *recF* genes from *Escherichia coli, Salmonella typhimurium, Pseudomonas putida* and *Bacillus subtilis:* evidence for an essential phosphate binding loop. *Nucleic Acids Res.* 20:839–845.

1727. **Sankaran, K., and H. C. Wu.** 1994. Lipid modification of bacterial prolipoprotein. Transfer of diacylglyceryl moiety from phosphatidylglycerol. *J. Biol. Chem.* 269:19701–19706.

1728. **Sarno, M. V., L. G. Tenn, A. Desai, A. M. Chin, F. C. Grenier, and M. H. Saier, Jr.** 1984. Genetic evidence for glucitol-specific enzyme III, an essential phosphocarrier protein of the *Salmonella typhimurium* glucitol phosphotransferase system. *J. Bacteriol.* 157:953–955.

1729. **Sarvas, M.** 1967. Inheritance of *Salmonella* T1 antigen. *Ann. Med. Exp. Biol. Fenn.* 45:447–471.

1730. **Sarvas, M., M. Malinen, M. Nurminen, and P. H. Makela.** 1976. T-2 lipopolysaccharide antigen of *Salmonella.* Comparison of the properties of T-2 and mucoid forms. *Infect. Immun.* 14:839–842.

1731. **Sasarman, A., and M. Desrochers.** 1976. Uroporphyrinogen III cosynthase-deficient mutant of *Salmonella typhimurium* LT2. *J. Bacteriol.* 128:717–721.

1732. **Sasarman, A., M. Desrochers, S. Sonea, K. E. Sanderson, and M. Surdenu.** 1976. Porphobilinogen-accumulating mutants of *Salmonella typhimurium. J. Gen. Microbiol.* 94:359–366.

1733. **Sasarman, A., K. E. Sanderson, M. Surdenu, and S. Sonea.** 1970. Hemin-deficient mutants of *Salmonella typhimurium. J. Bacteriol.* 102:531–536.

1734. **Sato, T., and T. Yura.** 1979. Chromosomal location and expression of the structural gene for major outer membrane protein Ia of *Escherichia coli* K-12 and of the homologous gene of *Salmonella typhimurium. J. Bacteriol.* 139:468–477.

1735. **Savin, M. A., M. Flavin, and C. Slaughter.** 1972. Regulations of homocysteine biosynthesis in *Salmonella typhimurium. J. Bacteriol.* 111:547–556.

1736. **Sawicka, T., and M. Bagdasarian.** 1976. Temperature sensitive carbon-14 galactose uptake by a division of *Salmonella typhimurium. Bull. Acad. Pol. Sci. Ser. Sci. Biol.* 24:441–444.

1737. **Sawyer, M. E., and F. B. Armstrong.** 1970. Order of the *ilv* genes of *Salmonella montevideo. Mol. Gen. Genet.* 109:370–372.

1738. **Scaife, J. G., J. S. Heilig, L. Rowen, and R. Calendar.** 1979. Gene for the RNA polymerase σ subunit mapped in *Salmonella typhimurium* and *Escherichia coli* by cloning and deletion. *Proc. Natl. Acad. Sci. USA* 76:6510–6514.

1739. **Scapin, G., J. C. Sacchettini, A. Dessen, M. Bhatia, and C. Grubmeyer.** 1993. Primary structure and crystallization of orotate phosphoribosyltransferase from *Salmonella typhimurium. J. Mol. Biol.* 230:1304–1308.

1740. **Schafer, M. P., W. H. Hannon, and A. P. Levin.** 1974. In vivo and in vitro complementation between *guaB* and in vivo complementation between *guaA* auxotrophs of *Salmonella typhimurium. J. Bacteriol.* 117:1270–1279.

1741. **Schafer, R., and T. K. Eisenstein.** 1992. Natural killer cells mediate protection induced by a *Salmonella aroA* mutant. *Infect. Immun.* 60:791–797.

1742. **Schlecht, S., E. Ferber, and I. Fromme.** 1979. The fatty acid composition of lipids from *Salmonella typhimurium* S and R forms. *Zentralbl. Bakteriol. Parasitenkd. Infektionskr. Hyg. Abt. 1 Orig. Reihe A* 245:476–484.

1743. **Schlecht, S., and I. Fromme.** 1980. Growth of *Salmonella typhimurium* R-mutants in submersed cultures. 2. Influence of growth phases on the lipopolysaccharide content of the bacteria and on the chemical composition and serological behavior of the lipopolysaccharides. *Zentralbl. Bakteriol. Parasitenkd. Infektionskr. Hyg. Abt. 1 Orig. Reihe A* 248:352–367.

1744. **Schlecht, S., I. Fromme, E. Ferber, W. Meuller, and J. Gmeiner.** 1980. Chemische und biologische Eigenschaften von Revertanten aus einer *Salmonella typhimurium* Rd$_1$-Mutante. *Zentralbl. Bakteriol. Parasitenkd. Infektionskr. Hyg. Abt. 1 Orig. Reihe A* 247:50–63.

1745. **Schleibinger, H., C. Leberl, and H. Ruden.** 1989. Nitrated polycyclic aromatic hydrocarbons (nitro-PAH) in the suspended substances of the atmosphere. 2. Comparison of the mutagenicity of nitro-PAH and dust extracts of the air in the Ames, SOS repair induction and SCE test. *Zentralbl. Hyg. Umweltmed.* 188:421–438. (In German.)

1746. **Schlesinger, M. J., and R. Olsen.** 1968. Expression and localization of *Escherichia coli* alkaline phosphatase synthesized in *Salmonella typhimurium* cytoplasm. *J. Bacteriol.* 96:1601–1605.

1747. **Schloss, J. V., D. E. Van Dyk, J. F. Vasta, and R. M. Kutny.** 1985. Purification and properties of *Salmonella typhimurium* acetolactate synthase isozyme II from *Escherichia coli* HB101/pDU9. *Biochemistry* 24:4952–4959.

1748. **Schmid, M., and J. R. Roth.** 1980. Circularization of transduced fragments: a mechanism for adding segments to the bacterial chromosome. *Genetics* 94:15–30.

1749. **Schmid, M. B.** 1990. A locus affecting nucleoid segregation in *Salmonella typhimurium. J. Bacteriol.* 172:5416–5424.

1750. **Schmid, M. B., N. Kapur, D. R. Isaacson, P. Lindroos, and C. Sharpe.** 1989. Genetic analysis of temperature-sensitive lethal mutants of *Salmonella typhimurium. Genetics* 123:625–633.

1751. **Schmid, M. B., and J. R. Roth.** 1987. Gene location affects expression level in *Salmonella typhimurium. J. Bacteriol.* 169:2872–2875.

1752. **Schmid, M. B., and J. A. Sawitzke.** 1993. Multiple bacterial topoisomerases: specialization or redundancy? *Bioessays* 15:445–449.

1753. **Schmidt, C., and H. Schmieger.** 1984. Selective transduction of recombinant plasmids with cloned *pac* sites by *Salmonella* phage P22. *Mol. Gen. Genet.* 196:123–128.

1754. **Schmitt, C. K., S. C. Darnell, V. L. Tesh, B. A. Stocker, and A. D. O'Brien.** 1994. Mutation of *flgM* attenuates virulence of *Salmonella typhimurium,* and mutation of *fliA* represses the attenuated phenotype. *J. Bacteriol.* 176:368–377.

1755. **Schmitz, G., P. Durre, G. Mullenbach, and G. F.-L. Ames.** 1987. Nitrogen regulation of transport operons: analysis of promoters *argTr* and *dhuA. Mol. Gen. Genet.* 209:403–407.

1756. **Schmitz, G., K. Nikaido, and G. F. Ames.** 1988. Regulation of a transport operon promoter in *Salmonella typhimurium:* identification of sites essential for nitrogen regulation. *Mol. Gen. Genet.* 215:107–117.

1757. **Schnaitman, C. A., and J. D. Klena.** 1993. Genetics of lipopolysaccharide biosynthesis in enteric bacteria. *Microbiol. Rev.* 57:655–682.

1758. **Schnaitman, C. A., C. T. Parker, J. D. Klena, E. L. Pradel, N. B. Pearson, K. E. Sanderson, and P. R. MacLachlan.** 1991. Physical maps of the *rfa* loci of *Escherichia coli* K-12 and *Salmonella typhimurium. J. Bacteriol.* 173:7410–7411.

1759. **Schneider, E., E. Francoz, and E. Dassa.** 1992. Completion of the nucleotide sequence of the "maltose B" region in *Salmonella typhimurium:* the high conservation of the *malM* gene suggests a selected physiological role for its product. *Biochim. Biophys. Acta* 1129:223–227.

1760. **Schneider, E., and C. Walter.** 1991. A chimeric nucleotide-binding protein, encoded by a *hisP-malK* hybrid gene, is functional in maltose transport in *Salmonella typhimurium. Mol. Microbiol.* 5:1375–1383.

1761. **Schneider, W. P., B. P. Nichols, and C. Yanofsky.** 1981. Procedure for production of hybrid genes and proteins and its use in assessing significance of amino acid differences in homologous tryptophan synthetase α polypeptides. *Proc. Natl. Acad. Sci. USA* 78:2169–2173.

1762. **Schnierow, B. J., M. Yamada, and M. H. Saier, Jr.** 1989. Partial nucleotide sequence of the *pts* operon in *Salmonella typhimurium:* comparative analyses in five bacterial genera. *Mol. Microbiol.* 3:113–118.

1763. **Scholte, B. J., and P. W. Postma.** 1980. Mutation in the *crp* gene of *Salmonella typhimurium* which interferes with inducer exclusion. *J. Bacteriol.* 141:751–757.

1764. **Scholte, B. J., and P. W. Postma.** 1981. Competition between two pathways for sugar uptake by the phosphoenolpyruvate-dependent sugar phosphotransferase system in *Salmonella typhimurium. Eur. J. Biochem.* 114:51–58.

1765. **Scholte, B. J., A. R. Schuitema, and P. W. Postma.** 1981. Isolation of III^{Glc} of the phosphoenolpyruvate-dependent glucose phosphotransferase system of *Salmonella typhimurium. J. Bacteriol.* 148:257–264.

1766. **Scholte, B. J., A. R. J. Schuitema, and P. W. Postma.** 1982. Characterization of factor III^{Glc} in catabolite repression-resistant *crr* mutants of *Salmonella typhimurium. J. Bacteriol.* 149:576–586.

1767. **Schroeder, C. J., and W. J. Dobrogosz.** 1986. Cloning and DNA sequence analysis of the wild-type and mutant cyclic AMP receptor protein genes from *Salmonella typhimurium. J. Bacteriol.* 167:616–622.

1768. **Schulte, L. L., L. T. Stauffer, and G. V. Stauffer.** 1984. Cloning and characterization of the *Salmonella typhimurium metE* gene. *J. Bacteriol.* 158:928–933.

1769. **Schumann, W., and E. G. Bade.** 1977. A *Salmonella typhimurium* endonuclease that converts native DNA to fragments of about 800,000 daltons. *J. Gen. Microbiol.* 101:319–326.

1770. **Schwartz, M., and J. Neuhard.** 1975. Control of expression of the *pyr* genes in *Salmonella typhimurium:* effects of variations in uridine and cytidine nucleotide pools. *J. Bacteriol.* 121:814–822.

1771. **Scott, J. F., J. R. Roth, and S. W. Artz.** 1975. Regulation of histidine operon does not require *hisG* enzyme. *Proc. Natl. Acad. Sci. USA* 72:5021–5025.

1772. **Scott, T. N., and M. I. Simon.** 1982. Genetic analysis of the mechanism of the *Salmonella* phase variation site specific recombination system. *Mol. Gen. Genet.* 188:313–321.

1773. **Searles, L. L., S. R. Wessler, and J. M. Calvo.** 1983. Transcription attenuation is the major mechanism by which the *leu* operon of *Salmonella typhimurium* is controlled. *J. Mol. Biol.* 163:377–394.

1774. **Sedgwick, S. G., C. Ho, and R. Woodgate.** 1991. Mutagenic DNA repair in enterobacteria. *J. Bacteriol.* 173:5604–5611.

1775. **Selker, E., and C. Yanofsky.** 1979. Nucleotide sequence of the *trpC-trpB* intercistronic region from *Salmonella typhimurium. J. Mol. Biol.* 130:135–143.

1776. **Servos, S., S. Chatfield, D. Hone, M. Levine, G. Dimitriadis, D. Pickard, G. Dougan, N. Fairweather, and I. Charles.** 1991. Molecular cloning and characterization of the *aroD* gene encoding 3-dehydroquinase from *Salmonella typhi. J. Gen. Microbiol.* 137:147–152.

1777. **Shaltiel, S., G. B. Henderson, and E. E. Snell.** 1974. Specific ion effects in affinity chromatography. The case of L-histidinolphosphate aminotransferase. *Biochemistry* 13:4330–4335.

1778. **Shamanna, D. K., and K. E. Sanderson.** 1979. Genetics and regulation of D-xylose utilization in *Salmonella typhimurium* LT2. *J. Bacteriol.* 139:71–79.

1779. **Shamanna, D. K., and K. E. Sanderson.** 1979. Uptake and catabolism of D-xylose in *Salmonella typhimurium* LT2. *J. Bacteriol.* 139:64–70.

1780. Shanabruch, W. G., I. Behlau, and G. C. Walker. 1981. Spontaneous mutators of *Salmonella typhimurium* LT2 generated by insertion of transposable elements. *J. Bacteriol.* 147:827–835.

1781. Shanabruch, W. G., R. P. Rein, I. Beglau, and G. C. Walker. 1983. Mutagenesis, by methylating and ethylating agents, in *mutH, mutS*, and *uvrD* mutants of *Salmonella typhimurium. J. Bacteriol.* 153:33–44.

1782. Shand, R. F., P. H. Blum, D. L. Holzschu, M. S. Urdea, and S. W. Artz. 1989. Mutational analysis of the histidine operon promoter of *Salmonella typhimurium. J. Bacteriol.* 171:6330–6337.

1783. Shand, R. F., P. H. Blum, R. D. Mueller, D. L. Riggs, and S. W. Artz. 1989. Correlation between histidine operon expression and guanosine 5′-diphosphate-3′-diphosphate levels during amino acid downshift in stringent and relaxed strains of *Salmonella typhimurium. J. Bacteriol.* 171:737–742.

1784. Shanmugam, K. T., V. Stewart, R. P. Gunsalus, D. H. Boxer, J. A. Cole, M. Chippaux, J. A. DeMoss, G. Giordano, E. C. Lin, and K. V. Rajagopalan. 1992. Proposed nomenclature for the genes involved in molybdenum metabolism in *Escherichia coli* and *Salmonella typhimurium. Mol. Microbiol.* 6:3452–3454. (Letter.)

1785. Shannon, K. P., and R. J. Rowbury. 1972. Alternation in the rate of cell division independent of the rate of DNA synthesis in a mutant of *Salmonella typhimurium. Mol. Gen. Genet.* 115:122–125.

1786. Shannon, K. P., B. G. Spratt, and R. J. Rowbury. 1972. Cell division and the production of cells lacking nuclear bodies in a mutant of *Salmonella typhimurium. Mol. Gen. Genet.* 118:185–197.

1787. Sharp, P. M., J. E. Kelleher, A. S. Daniel, G. M. Cowan, and N. E. Murray. 1992. Roles of selection and recombination in the evolution of type I restriction-modification systems in enterobacteria. *Proc. Natl. Acad. Sci. USA* 89:9836–9840.

1788. Sharp, P. M., D. C. Shields, K. H. Wolfe, and W. H. Li. 1989. Chromosomal location and evolutionary rate variation in enterobacterial genes. *Science* 246:808–810.

1789. Shaw, K. J., and C. M. Berg. 1980. Substrate channeling: α-ketobutyrate inhibition of acetohydroxy acid synthase in *Salmonella typhimurium. J. Bacteriol.* 143:1509–1512.

1790. Shaw, K. J., C. M. Berg, and T. J. Sobol. 1980. *Salmonella typhimurium* mutants defective in acetohydroxy acid synthases I and II. *J. Bacteriol.* 141:1258–1263.

1791. Shaw, N. A., and P. D. Ayling. 1991. Cloning of high-affinity methionine transport genes from *Salmonella typhimurium. FEMS Microbiol. Lett.* 62:127–131.

1792. Shematek, E. M., S. M. Arfin, and W. F. Diven. 1973. A kinetic study of α-acetohydroxy acid isomerase from *Salmonella typhimurium. Arch. Biochem. Biophys.* 158:132–138.

1793. Shematek, E. M., W. F. Diven, and S. N. Arfin. 1973. Subunit structure of α-acetohydroxy acid isomeroreductase from *Salmonella typhimurium. Arch. Biochem. Biophys.* 158:126–131.

1794. Sheppard, D. E., and J. R. Roth. 1994. A rationale for autoinduction of a transcriptional activator: ethanolamine ammonia-lyase (EutBC) and the operon activator (EutR) compete for adenosyl-cobalamin in *Salmonella typhimurium. J. Bacteriol.* 176:1287–1296.

1795. Shimamoto, T., H. Izawa, H. Daimon, N. Ishiguro, M. Shinagawa, Y. Sakano, M. Tsuda, and T. Tsuchiya. 1991. Cloning and nucleotide sequence of the gene (*citA*) encoding a citrate carrier from *Salmonella typhimurium. J. Biochem. (Tokyo)* 110:22–28.

1796. Shirakihara, Y., and T. Wakabayashi. 1979. Three-dimensional reconstruction of straight flagella from a mutant of *Salmonella typhimurium. J. Mol. Biol.* 131:485–507.

1797. Shiuan, D., and A. Campbell. 1988. Transcriptional regulation and gene arrangement of *Escherichia coli, Citrobacter freundii* and *Salmonella typhimurium* biotin operons. *Gene* 67:203–211.

1798. Shuster, C. W., and K. Rundell. 1969. Resistance of *Salmonella typhimurium* mutants to galactose death. *J. Bacteriol.* 100:103–109.

1799. Siitonen, A., V. Johansson, M. Nurminen, and P. H. Makela. 1977. *Salmonella* bacteriophages that use outer membrane protein receptors. *FEMS Microbiol. Lett.* 1:141–144.

1800. Silbert, D. F., G. R. Fink, and B. N. Ames. 1966. Histidine regulatory mutants in *Salmonella typhimurium.* III. A class of regulatory mutants deficient in tRNA for histidine. *J. Mol. Biol.* 22:335–347.

1801. Silva, D. O., and W. J. Dobrogosz. 1978. Proton efflux associated with melibiose permease activity in *Salmonella typhimurium. Biochem. Biophys. Res. Commun.* 81:755–755.

1802. Silverman, M., and M. Simon. 1980. Phase variation genetic analysis of switching mutants. *Cell* 19:845–854.

1803. Silverman, M., and M. I. Simon. 1977. Bacterial flagella. *Annu. Rev. Microbiol.* 31:397–419.

1804. Silverman, M., J. Zieg, M. Hilmen, and M. Simon. 1979. Phase variation in *Salmonella*: genetic analysis of a recombinational switch. *Proc. Natl. Acad. Sci. USA* 76:391–395.

1805. Silverman, M., J. Zieg, G. Mandel, and M. Simon. 1981. Analysis of the functional components of the phase variation system. *Cold Spring Harbor Symp. Quant. Biol.* 45:17–26.

1806. Silverman, M., J. Zieg, and M. Simon. 1979. Flagellar-phase variation: isolation of the *rh1* gene. *J. Bacteriol.* 137:517–523.

1807. Simms, S. A., E. W. Cornman, J. Mottonen, and J. Stock. 1987. Active site of the enzyme which demethylates receptors during bacterial chemotaxis. *J. Biol. Chem.* 262:29–31.

1808. Simms, S. A., M. G. Keane, and J. Stock. 1985. Multiple forms of the CheB methylesterase in bacterial chemosensing. *J. Biol. Chem.* 260:10161–10168.

1809. Simms, S. A., A. M. Stock, and J. B. Stock. 1987. Purification and characterization of the S-adenosylmethionine:glutamyl methyltransferase that modifies membrane chemoreceptor proteins in bacteria. *J. Biol. Chem.* 262:8537–8543.

1810. Singer, C. E., and G. R. Smith. 1972. Histidine regulation in *Salmonella typhimurium*. XIII. Nucleotide sequence of histidine transfer ribonucleic acid. *J. Biol. Chem.* 247:2989–3000.

1811. Singer, C. E., G. R. Smith, R. Cortese, and B. N. Ames. 1972. Mutant tRNA[his] ineffective in repression and lacking two pseudouridine modifications. *Nature (London) New Biol.* 238:72–74.

1812. Singh, A. P., and P. D. Bragg. 1977. Energetics of galactose transport in a cytochrome deficient mutant of *Salmonella typhimurium. J. Supramol. Struct.* 1(Suppl.):136.

1813. Sirisena, D. M., K. A. Brozek, P. R. MacLachlan, K. E. Sanderson, and C. R. Raetz. 1992. The *rfaC* gene of *Salmonella typhimurium*. Cloning, sequencing, and enzymatic function in heptose transfer to lipopolysaccharide. *J. Biol. Chem.* 267:18874–18884.

1814. Sirisena, D. M., P. R. MacLachlan, S.-L. Liu, A. Hessel, and K. E. Sanderson. 1994. Molecular analysis of the *rfaD* gene, for heptose synthesis, and the *rfaF* gene, for heptose transfer, in lipopolysaccharide synthesis in *Salmonella typhimurium. J. Bacteriol.* 176:2379–2385.

1815. Skavronskaya, A. G., I. V. Andreeva, and A. A. Kiryushkina. 1973. Specifically UV-sensitive photoreactivable mutant of *Salmonella abony. Mutat. Res.* 18:259–266.

1816. Skavronskaya, A. G., I. V. Andreeva, and A. A. Kiryushkina. 1974. The *uvrC* gene in Salmonella: phenotypic properties and a preliminary mapping. *Mutat. Res.* 23:275–277.

1817. Skavronskaya, A. G., N. F. Stepanova, and I. V. Andreeva. 1982. UV-mutable hybrids of *Salmonella* incorporating *Escherichia coli* region adjacent to *tryptophan* operon. *Mol. Gen. Genet.* 185:315–318.

1818. Skovgaard, O., and F. G. Hansen. 1987. Comparison of *dnaA* nucleotide sequences of *Escherichia coli, Salmonella typhimurium*, and *Serratia marcescens. J. Bacteriol.* 169:3976–3981.

1819. Skurnik, M., and P. Toivanen. 1991. Intervening sequences (IVSs) in the 23S ribosomal RNA genes of pathogenic *Yersinia enterocolitica* strains. The IVSs in *Y. enterocolitica* and *Salmonella typhimurium* have a common origin. *Mol. Microbiol.* 5:585–593.

1820. Slater, S. C., and R. Maurer. 1991. Requirements for bypass of UV-induced lesions in single-stranded DNA of bacteriophage phi X174 in *Salmonella typhimurium. Proc. Natl. Acad. Sci. USA* 88:1251–1255.

1821. Sledziewska, E., and D. Hulanicka. 1978. Method of isolation of cysteine constitutive mutants of the cysteine regulon in *Salmonella typhimurium. Mol. Gen. Genet.* 165:289–294.

1822. Slocum, H., and H. W. Boyer. 1973. Host specificity of *Salmonella typhimurium* deoxyribonucleic acid restriction and modification. *J. Bacteriol.* 113:724–726.

1823. Sluka, J. P., S. J. Horvath, A. C. Glasgow, M. I. Simon, and P. B. Dervan. 1990. Importance of minor-groove contacts for recognition of DNA by the binding domain of Hin recombinase. *Biochemistry* 29:6551–6561.

1824. Smando, R., E. B. Waygood, and B. D. Sanwal. 1974. Cooperative interactions in the binding of allosteric effectors to phosphoenolpyruvate carboxylase. *J. Biol. Chem.* 249:182–190.

1825. Smit, J., Y. Kamio, and H. Nikaido. 1975. Outer membrane of *Salmonella typhimurium*: chemical analysis and freeze fracture studies with lipopolysaccharide mutants. *J. Bacteriol.* 124:942–958.

1826. Smith, B. P., M. Reina-Guerra, S. K. Hoiseth, B. A. D. Stocker, F. Habasha, E. Johnson, and F. Merritt. 1984. Aromatic-dependent *Salmonella typhimurium* as modified live vaccines for calves. *Am. J. Vet. Res.* 45:59–66.

1827. Smith, C. M., W. H. Koch, S. B. Franklin, P. L. Foster, T. A. Cebula, and E. Eisenstadt. 1990. Sequence analysis and mapping of the *Salmonella typhimurium* LT2 *umuDC* operon. *J. Bacteriol.* 172:4964–4978.

1828. Smith, D., and R. H. Bauerle. 1969. The anthranilate synthetase-5-phosphoribosyl transferase complex of the tryptophan pathway in *Salmonella typhimurium.* Purification in the in vitro assembly of its subunits. *Biochemistry* 8:1451–1459.

1829. Smith, D. A. 1971. S-Amino acid metabolism and its regulation in *Escherichia coli* and *Salmonella typhimurium. Adv. Genet.* 16:141–165.

1830. Smith, D. A., and J. D. Childs. 1966. Methionine genes and enzymes of *Salmonella typhimurium. Heredity* 21:265–268.

1831. Smith, G. R. 1971. Specialized transduction of the *Salmonella hut* operons by coliphage lambda: deletion analysis of the *hut* operons employing lambda *phut. Virology* 45:208–223.

1832. Smith, G. R., Y. S. Halpern, and B. Magasanik. 1971. Genetic and metabolic control of enzymes responsible for histidine degradation in *Salmonella typhimurium. J. Biol. Chem.* 156:3320–3329.

1833. Smith, G. R., and B. Magasanik. 1971. Nature and self-regulated synthesis of the repressor of the *hut* operons in *Salmonella typhimurium. Proc. Natl. Acad. Sci. USA* 68:1493–1497.

1834. Smith, G. R., and B. Magasanik. 1971. The two operons of the histidine utilization system in *Salmonella typhimurium. J. Biol. Chem.* 246:3330–3341.

1835. Smith, G. R., C. M. Roberts, and D. W. Schultz. 1986. Activity of Chi recombinational hotspots in *Salmonella typhimurium. Genetics* 112:429–439.

1836. Smith, H. O., and M. Levine. 1966. Gene order in prophage P22. *Virology* 27:229–231.

1837. Smith, H. R., G. O. Humphreys, N. D. F. Grindley, J. N. Grindley, and E. S. Anderson. 1973. Molecular studies of an *fi⁺* plasmid from strains of *Salmonella typhimurium*. *Mol. Gen. Genet.* 126:143–151.

1838. Smith, J. M., R. A. Kelin, and G. A. O'Donovan. 1980. Repression and derepression of the enzymes of the pyrimidine biosynthetic pathway in *Salmonella typhimurium*. *J. Gen. Microbiol.* 121:27–38.

1839. Smith, J. M., E. H. Rowsell, J. Shioi, and B. L. Taylor. 1988. Identification of a site of ATP requirement for signal processing in bacterial chemotaxis. *J. Bacteriol.* 170:2698–2704.

1840. Smith, N. H., P. Beltran, and R. K. Selander. 1990. Recombination of *Salmonella* phase 1 flagellin genes generates new serovars. *J. Bacteriol.* 172:2209–2216.

1841. Smith, N. H., and R. K. Selander. 1990. Sequence invariance of the antigen-coding central region of the phase 1 flagellar filament gene (*fliC*) among strains of *Salmonella typhimurium*. *J. Bacteriol.* 172:603–609.

1842. Smith, O., M. M. Meyers, T. Vogel, R. D. Deeley, and R. Goldberger. 1974. Defective in vitro binding of histidyl-transfer ribonucleic acid to feedback resistant phosphoribosyl transferase of *Salmonella typhimurium*. *Nucleic Acids Res.* 1:881–888.

1843. Smith, R. L., J. L. Banks, M. D. Snavely, and M. E. Maguire. 1993. Sequence and topology of the CorA magnesium transport systems of *Salmonella typhimurium* and *Escherichia coli*. Identification of a new class of transport protein. *J. Biol. Chem.* 268:14071–14080.

1844. Smith, R. L., J. W. Pelley, and R. M. Jeter. 1991. Characterization of *lip* expression in *Salmonella typhimurium*: analysis of *lip::lac* operon fusions. *J. Gen. Microbiol.* 137:2307–2312.

1845. Smith-Keary, P. F. 1966. Restricted transduction by bacteriophage P22 in *Salmonella typhimurium*. *Genet. Res.* 8:73–82.

1846. Smyer, J. R., and R. M. Jeter. 1989. Characterization of phosphoenolpyruvate synthase mutants in *Salmonella typhimurium*. *Arch. Microbiol.* 153:26–32.

1847. Snavely, M. D., J. B. Florer, C. G. Miller, and M. E. Maguire. 1989. Magnesium transport in *Salmonella typhimurium*: expression of cloned genes for three distinct Mg²⁺ transport systems. *J. Bacteriol.* 171:4752–4760.

1848. Snavely, M. D., J. B. Florer, C. G. Miller, and M. E. Maguire. 1989. Magnesium transport in *Salmonella typhimurium*: ²⁸Mg²⁺ transport by the CorA, MgtA, and MgtB systems. *J. Bacteriol.* 171:4761–4766.

1849. Snavely, M. D., S. A. Gravina, T. T. Cheung, C. G. Miller, and M. E. Maguire. 1991. Magnesium transport in *Salmonella typhimurium*. Regulation of *mgtA* and *mgtB* expression. *J. Biol. Chem.* 266:824–829.

1850. Snavely, M. D., C. G. Miller, and M. E. Maguire. 1991. The *mgtB* Mg²⁺ transport locus of *Salmonella typhimurium* encodes a P-type ATPase. *J. Biol. Chem.* 266:815–823.

1851. Snellings, N. J., E. M. Johnson, and L. S. Baron. 1977. Genetic basis of Vi antigen expression in *Salmonella paratyphi* C. *J. Bacteriol.* 131:57–62.

1852. Sockett, H., S. Yamaguchi, M. Kihara, V. M. Irikura, and R. M. Macnab. 1992. Molecular analysis of the flagellar switch protein FliM of *Salmonella typhimurium*. *J. Bacteriol.* 174:793–806.

1853. Sofia, H. J., V. Burland, D. L. Daniels, G. Plunkett, and F. R. Blattner. 1994. Analysis of the *Escherichia coli* genome. V. DNA sequence of the region from 76.0 to 81.5 minutes. *Nucleic Acids Res.* 22:2576–2586.

1854. Sohel, I., J. L. Puente, W. J. Murray, J. Vuopio-Varkila, and G. K. Schoolnik. 1993. Cloning and characterization of the bundle-forming pilin gene of enteropathogenic *Escherichia coli* and its distribution in *Salmonella* serotypes. *Mol. Microbiol.* 7:563–575.

1855. Somers, J. M., and W. W. Kay. 1983. Genetic fine structure of the tricarboxylate transport (*tct*) locus of *Salmonella typhimurium*. *Mol. Gen. Genet.* 190:20–26.

1856. Somers, J. M., G. D. Sweet, and W. W. Kay. 1981. Fluorocitrate resistant tricarboxylate transport mutants in *Salmonella typhimurium*. *Mol. Gen. Genet.* 181:338–345.

1857. Sonti, R. V., D. H. Keating, and J. R. Roth. 1993. Lethal transposition of Mu*d* phages in Rec- strains of *Salmonella typhimurium*. *Genetics* 133:17–28.

1858. Sorensen, K. I. 1994. Conformational heterogeneity in the *Salmonella typhimurium pyrC* and *pyrD* leader mRNAs produced *in vivo*. *Nucleic Acids Res.* 22:625–631.

1859. Sorensen, K. I., K. E. Baker, R. A. Kelln, and J. Neuhard. 1993. Nucleotide pool-sensitive selection of the transcriptional start site in vivo at the *Salmonella typhimurium pyrC* and *pyrD* promoters. *J. Bacteriol.* 175:4137–4144.

1860. Sorensen, K. I., and J. Neuhard. 1991. Dual transcriptional initiation sites from the *pyrC* promoter control expression of the gene in *Salmonella typhimurium*. *Mol. Gen. Genet.* 225:249–256.

1861. Spears, K. R., R. E. Wooley, J. Brown, O. J. Fletcher, and J. B. Payeur. 1990. Characteristics of *Salmonella* spp. and *Escherichia coli* isolated from broiler flocks classified as "good" or "poor" producers. *Avian Dis.* 34:855–860.

1862. Spector, M. P., Z. Aliabadi, T. Gonzalez, and J. W. Foster. 1986. Global control in *Salmonella typhimurium*: two-dimensional electrophoretic analysis of starvation-, anaerobiosis-, and heat shock-inducible proteins. *J. Bacteriol.* 168:420–424.

1863. Spector, M. P., and C. L. Cubitt. 1992. Starvation-inducible loci of *Salmonella typhimurium*: regulation and roles in starvation-survival. *Mol. Microbiol.* 6:1467–1476.

1864. Spector, M. P., and J. W. Foster. 1983. Isolation and characterization of pyridine nucleotide uptake (*pnu*) mutants of *Salmonella typhimurium*, abstr. K51, p. 51. *Abstr. Annu. Meet. Am. Soc. Microbiol., 1983*.

1865. Spector, M. P., J. M. Hill, E. A. Holley, and J. W. Foster. 1985. Genetic characterization of pyridine nucleotide uptake mutants of *Salmonella typhimurium*. *J. Gen. Microbiol.* 131:1313–1322.

1866. Spector, M. P., Y. K. Park, S. Tirgari, T. Gonzalez, and J. W. Foster. 1988. Identification and characterization of starvation-regulated genetic loci in *Salmonella typhimurium* by using Mu d-directed *lacZ* operon fusions. *J. Bacteriol.* 170:345–351.

1867. Spencer, J. B., N. J. Stolowich, C. A. Roessner, and I. Scott. 1994. The *Escherichia coli cysG* gene encodes the multifunctional protein, siroheme synthase. *FEBS Lett.* 335:57–60.

1868. Spierings, G., R. Elders, B. van Lith, H. Hofstra, and J. Tommassen. 1992. Characterization of the *Salmonella typhimurium phoE* gene and development of *Salmonella*-specific DNA probes. *Gene* 122:45–52.

1869. Spratt, B. G. 1972. Replication of extrachromosomal elements in a DNA synthesis initiation mutant of *Salmonella typhimurium*. *Biochem. Biophys. Res. Commun.* 48:496–501.

1870. Spratt, B. G., and R. J. Rowbury. 1971. Cell division in a mutant of *Salmonella typhimurium* which is temperature-sensitive for DNA synthesis. *J. Gen. Microbiol.* 65:305–314.

1871. Spratt, B. G., and R. J. Rowbury. 1971. Physiological and genetical studies on a mutant of *Salmonella typhimurium* which is temperature-sensitive for DNA synthesis. *Mol. Gen. Genet.* 114:35–49.

1872. Springer, A. L., and M. B. Schmid. 1993. Molecular characterization of the *Salmonella typhimurium parE* gene. *Nucleic Acids Res.* 21:1805–1809.

1873. Springer, W., H. Grimminger, and F. Lingens. 1972. Isoleucine auxotrophy due to feedback hypersensitivity of biosynthetic threonine deaminase. *J. Bacteriol.* 112:259–263.

1874. Springer, W. R., and D. E. Koshland, Jr. 1977. Identification of a protein methyltransferase as the *cheR* gene product in the bacterial sensing system. *Proc. Natl. Acad. Sci. USA* 74:533–537.

1875. Sprinson, D. B., E. G. Gollub, R. C. Hu, and K.-P. Liu. 1976. Regulation of tyrosine and phenylalanine biosynthesis in *Salmonella typhimurium*. *Acta Microbiol. Acad. Sci. Hung.* 23:167–170.

1876. Spudich, J. L., and D. E. Koshland, Jr. 1975. Quantitation of the sensory response in bacterial chemotaxis. *Proc. Natl. Acad. Sci. USA* 72:710–713.

1877. Spudich, J. L., and D. E. Koshland, Jr. 1979. Specific inactivator of flagellar reversal in *Salmonella typhimurium*. *J. Bacteriol.* 139:442–447.

1878. Squires, C. H., M. DeFelice, C. T. Lago, and J. M. Calvo. 1983. *ilvHI* locus of *Salmonella typhimurium*. *J. Bacteriol.* 154:1054–1063.

1879. Sroga, G. E., F. Nemoto, Y. Kuchino, and G. R. Bjork. 1992. Insertion (*sufB*) in the anticodon loop or base substitution (*sufC*) in the anticodon stem of tRNA(Pro)2 from *Salmonella typhimurium* induces suppression of frameshift mutations. *Nucleic Acids Res.* 20:3463–3469.

1880. St. Pierre, M. L., and M. Demerec. 1968. Isolation and mapping of *Salmonella typhimurium* mutants defective in the utilization of trehalose. *J. Bacteriol.* 95:1185–1186.

1881. Stackhouse, T. M., J. J. Onuffer, C. R. Matthews, S. A. Ahmed, and E. W. Miles. 1988. Folding of homologous proteins: conservation of the folding mechanism of the α subunit of tryptophan synthase from *Escherichia coli*, *Salmonella typhimurium*, and five interspecies hybrids. *Biochemistry* 27:824–832.

1882. Stalker, D. M., W. R. Hiatt, and L. Comai. 1985. A single amino acid substitution in the enzyme 5-enolpyruvylshikimate-3-phosphate synthase confers resistance to the herbicide glyphosate. *J. Biol. Chem.* 260:4724–4728.

1883. Stan-Lotter, H., and P. D. Bragg. 1985. Subunit distribution of the sulfhydryl groups of the F₁ adenosine triphosphatase of *Salmonella typhimurium*. *Arch. Biochem. Biophys.* 239:280–285.

1884. Stan-Lotter, H., M. Gupta, and K. E. Sanderson. 1979. The influence of cations on the permeability of the outer membrane of *Salmonella typhimurium* and other gram negative bacteria. *Can. J. Microbiol.* 25:475–485.

1885. Staples, M. A., and L. L. Houston. 1979. Proteolytic degradation of imidazoleglycerolphosphate dehydratase-histidinolphosphate from *Salmonella typhimurium* and the isolation of a resistant bi-functional core enzyme. *J. Biol. Chem.* 254:1395–1401.

1886. Staskawicz, K. N., and N. J. Panopoulos. 1980. Phaseolotoxin transport in *Escherichia coli* and *Salmonella typhimurium* via the oligopeptide permease. *J. Bacteriol.* 142:474–479.

1887. Stauffer, G. V., C. A. Baker, and J. E. Brenchley. 1974. Regulation of serine transhydroxymethylase activity in *Salmonella typhimurium*. *J. Bacteriol.* 120:1017–1025.

1888. Stauffer, G. V., and J. E. Brenchley. 1974. Evidence for the involvement of serine transhydroxymethylase in serine and glycine interconversions in *Salmonella typhimurium*. *Genetics* 77:185–198.

1889. Stauffer, G. V., and J. E. Brenchley. 1977. Influence of methionine biosynthesis on serine transhydroxymethylase regulation in *Salmonella typhimurium* LT2. *J. Bacteriol.* 129:740–749.

1890. Stauffer, G. V., and J. E. Brenchley. 1978. Selection of *Salmonella typhimurium* mutants with altered serine transhydroxymethylase regulation. *Genetics* 88:221–223.

1891. Stauffer, G. V., and L. T. Stauffer. 1988. *Salmonella typhimurium* LT2 *metF* operator mutations. *Mol. Gen. Genet.* 214:32–36.

1892. Stauffer, G. V., and L. T. Stauffer. 1988. Cloning and nucleotide sequence of the *Salmonella typhimurium* LT2 *metF* gene and its homology with the corresponding sequence of *Escherichia coli*. *Mol. Gen. Genet.* 212:246–251.

1893. Stauffer, G. V., L. T. Stauffer, and M. D. Plamann. 1989. The *Salmonella typhimurium* glycine cleavage enzyme system. *Mol. Gen. Genet.* 220:154–156.

1894. Steiert, J. G., M. L. Urbanowski, L. T. Stauffer, M. D. Plamann, and G. V. Stauffer. 1990. Nucleotide sequence of the *Salmonella typhimurium glyA* gene. *DNA Seq.* 1:107–113.

1895. Steinberg, B., and M. Gough. 1975. Altered DNA synthesis in a mutant of *Salmonella typhimurium* that channels bacteriophage P22 toward lysogeny. *J. Virol.* 16:1154–1160.

1896. Steiner, K. E., and J. Preiss. 1977. Biosynthesis of bacterial glycogen: genetic and allosteric regulation of glycogen biosynthesis in *Salmonella typhimurium* LT-2. *J. Bacteriol.* 129:246–256.

1897. Stepanova, N. F., I. V. Andreeva, A. A. Kiryushkina, Y. S. Kondrat'ev, L. Y. Likhoded, and A. G. Skavronskaya. 1977. Intergeneric mating of *Escherichia coli* and *Salmonella typhimurium*. II. Transfer of *polA*-1 mutation from *Escherichia coli* to *Salmonella typhimurium* and its phenotypic expression in the *Salmonella* genome. *Sov. Genet.* 13:1474–1479.

1898. Stepanova, N. F., I. V. Andreeva, and A. G. Skavronskaya. 1977. Intergenus conjugation crossing of *Escherichia coli* and *Salmonella typhimurium*. Communication I. Production of a hybrid of *Salmonella* possessing increased activity of the recipient, in crosses with *Escherichia coli*. *Sov. Genet.* 13:1358–1362.

1899. Stephens, J. C., S. W. Artz, and B. N. Ames. 1975. Guanosine 5′-diphosphate 3′-diphosphate (ppGpp): positive effector for histidine operon transcription and a general signal for amino acid deficiency. *Proc. Natl. Acad. Sci. USA* 72:4389–4393.

1900. Sterboul, C. C., J. E. Kleeman, and S. M. Parsons. 1977. Purification and characterization of a mutant ATP phosphoribosyltransferase hypersensitive to histidine feedback inhibition. *Arch. Biochem. Biophys.* 181:632–642.

1901. Stern, M. J., C. F. Higgins, and G. F.-L. Ames. 1984. Isolation and characterization of *lac* fusions to two nitrogen-regulated promoters. *Mol. Gen. Genet.* 195:219–227.

1902. Stetter, D. W., and R. B. Middleton. 1979. Tryptophan requiring parental strains yield *Salmonella typhimurium* x *Escherichia coli* hybrid recombinants with functional tryptophan operons. *Can. J. Genet. Cytol.* 21:255–260.

1903. Stevenson, G., S. J. Lee, L. K. Romana, and P. R. Reeves. 1991. The *cps* gene cluster of *Salmonella* strain LT2 includes a second mannose pathway: sequence of two genes and relationship to genes in the *rfb* gene cluster. *Mol. Gen. Genet.* 227:173–180.

1904. Stewart, V. 1988. Nitrate respiration in relation to facultative metabolism in enterobacteria. *Microbiol. Rev.* 52:190–232.

1905. Stewart, V., J. T. Lin, and B. L. Berg. 1991. Genetic evidence that genes *fdhD* and *fdhE* do not control synthesis of formate dehydrogenase-N in *Escherichia coli* K-12. *J. Bacteriol.* 173:4417–4423.

1906. Stieglitz, B. I., and J. M. Calvo. 1971. Effect of 4-azaleucine upon leucine metabolism in *Salmonella typhimurium*. *J. Bacteriol.* 108:95–104.

1907. Stinavage, P. S., L. E. Martin, and J. K. Spitznagel. 1990. A 59 kiloDalton outer membrane protein of *Salmonella typhimurium* protects against oxidative intraleukocytic killing due to human neutrophils. *Mol. Microbiol.* 4:283–293.

1908. Stirling, D. A., C. S. Hulton, L. Waddell, S. F. Park, G. S. Stewart, I. R. Booth, and C. F. Higgins. 1989. Molecular characterization of the *proU* loci of *Salmonella typhimurium* and *Escherichia coli* encoding osmoregulated glycine betaine transport systems. *Mol. Microbiol.* 3:1025–1038.

1909. Stock, A., T. Chen, D. Welsh, and J. Stock. 1988. CheA protein, a central regulator of bacterial chemotaxis, belongs to a family of proteins that control gene expression in response to changing environmental conditions. *Proc. Natl. Acad. Sci. USA* 85:1403–1407.

1910. Stock, A., D. E. Koshland, Jr., and J. Stock. 1985. Homologies between the *Salmonella typhimurium* CheY protein and proteins involved in the regulation of chemotaxis, membrane protein synthesis, and sporulation. *Proc. Natl. Acad. Sci. USA* 82:7989–7993.

1911. Stock, A., J. Mottonen, T. Chen, and J. Stock. 1987. Identification of a possible nucleotide binding site in the CheW, a protein required for sensory transduction in bacterial chemotaxis. *J. Biol. Chem.* 262:535–537.

1912. Stock, A., E. Schaeffer, D. E. Koshland, Jr., and J. Stock. 1987. A second type of protein methylation reaction in bacterial chemotaxis. *J. Biol. Chem.* 262:8011–8014.

1913. Stock, A. M., E. Martinez-Hackert, B. F. Rasmussen, A. H. West, J. B. Stock, D. Ringe, and G. A. Petsko. 1993. Structure of the Mg(2+)-bound form of CheY and mechanism of phosphoryl transfer in bacterial chemotaxis. *Biochemistry* 32:13375–13380.

1914. Stock, A. M., J. M. Mottonen, J. B. Stock, and C. E. Schutt. 1989. Three-dimensional structure of CheY, the response regulator of bacterial chemotaxis. *Nature* (London) 337:745–749.

1915. Stock, A. M., and J. Stock. 1987. Purification and characterization of the CheZ protein of bacterial chemotaxis. *J. Bacteriol.* 169:3301–3311.

1916. Stock, A. M., D. C. Wylie, J. M. Mottonen, A. N. Lupas, E. G. Ninfa, A. J. Ninfa, C. E. Schutt, and J. B. Stock. 1988. Phosphoproteins involved in bacterial signal transduction. *Cold Spring Harbor Symp. Quant. Biol.* 53:49–57.

1917. Stock, J., G. Kersulis, and D. E. Koshland, Jr. 1985. Neither methylation nor demethylating enzymes are required for bacterial chemotaxis. *Cell* 42:683–690.

1918. Stock, J. B., and D. E. Koshland, Jr. 1979. A protein methyl esterase involved in bacterial sensing. *Proc. Natl. Acad. Sci. USA* 75:3659–3663.

1919. Stock, J. B., G. S. Lukat, and A. M. Stock. 1991. Bacterial chemotaxis and the molecular logic of intracellular signal transduction networks. *Annu. Rev. Biophys. Biophys. Chem.* 20:109–136.

1920. Stock, J. B., E. B. Waygood, N. D. Meadow, P. W. Postma, and S. Roseman. 1982. Sugar transport by the bacterial phosphotransferase system. The glucose receptors of the *Salmonella typhimurium* phosphotransferase system. *J. Biol. Chem.* 257:14543–14552.

1921. Stocker, B. A. D., and P. H. Mäkelä. 1971. Genetic aspects of biosynthesis and structure of *Salmonella typhimurium* lipopolysaccharide, p. 369–438. *In* G. Weinbaum, S. Kadis, and S. J. Ajl (ed.), *Bacterial Endotoxins*, vol. 4 in *Microbial Toxins*. Academic Press, Inc., New York.

1922. Stocker, B. A. D., and P. H. Makela. 1978. Genetics of the (gram-negative) bacterial surface. *Proc. R. Soc. London Ser. B Biol. Sci.* 202:5–30.

1923. Stocker, B. A. D., B. M. Males, and W. Takano. 1980. *Salmonella typhimurium* mutants of *rfaH* negative phenotype: genetics and antibiotic sensitivities. *J. Gen. Microbiol.* 116:17–24.

1924. Stocker, B. A. D., M. W. McDonough, and R. P. Ambler. 1961. A gene determining presence or absence of ε-N-methyl-lysine in *Salmonella* flagellar protein. *Nature* (London) 189:556–558.

1925. Stocker, B. A. D., M. Nurminen, and P. H. Makela. 1979. Mutants defective in the 33K outer membrane protein of *Salmonella typhimurium*. *J. Bacteriol.* 139:376–383.

1926. Stocker, B. A. D., S. M. Smith, and T. V. Subbaiah. 1963. Mapping in *Salmonella typhimurium* with colicine factors to obtain fertility. *Microb. Genet. Bull.* 19:22–24.

1927. Stocker, B. A. D., R. G. Wilkinson, and P. H. Makela. 1966. Genetic aspects of biosynthesis and structure of *Salmonella* somatic polysaccharide. *Ann. N.Y. Acad. Sci.* 133:334–348.

1928. Stocker, B. A. D., N. D. Zinder, and J. Lederberg. 1953. Transduction of flagellar characters in *Salmonella*. *J. Gen. Microbiol.* 9:410–433.

1929. Storz, G., M. F. Christman, H. Sies, and B. N. Ames. 1987. Spontaneous mutagenesis and oxidative damage to DNA in *Salmonella typhimurium*. *Proc. Natl. Acad. Sci. USA* 84:8917–8921.

1930. Storz, G., F. S. Jacobson, L. A. Tartaglia, R. W. Morgan, L. A. Silveira, and B. N. Ames. 1989. An alkyl hydroperoxide reductase induced by oxidative stress in *Salmonella typhimurium* and *Escherichia coli*: genetic characterization and cloning of *ahp*. *J. Bacteriol.* 171:2049–2055.

1931. Storz, G., L. A. Tartaglia, and B. N. Ames. 1990. Transcriptional regulator of oxidative stress-inducible genes: direct activation by oxidation. *Science* 248:189–194.

1932. Storz, G., and M. B. Toledano. 1994. Regulation of bacterial gene expression in response to oxidative stress. *Methods Enzymol.* 236:196–207.

1933. Stouthamer, A. H. 1969. A genetical and biochemical study of chlorate-resistant mutants of *Salmonella typhimurium*. *Antonie Van Leeuwenhoek J. Microbiol. Serol.* 35:505–521.

1934. Stouthamer, A. H. 1970. Genetics and biochemistry of reductase formation in Enterobacteriaceae. *Antonie Van Leeuwenhoek J. Microbiol. Serol.* 36:181.

1935. Stouthamer, A. H., and C. W. Bettenhaussen. 1970. Mapping of a gene causing resistance to chlorate in *Salmonella typhimurium*. *Antonie Van Leeuwenhoek J. Microbiol. Serol.* 36:555–565.

1936. Stover, C. K., J. Kemper, and R. C. Marsh. 1988. Molecular cloning and characterization of *supQ/newD*, a gene substitution system for the *leuD* gene of *Salmonella typhimurium*. *J. Bacteriol.* 170:3115–3124.

1937. Strain, S. M., I. M. Armitage, L. Anderson, K. Takayama, N. Qureshi, and C. R. H. Raetz. 1985. Location of polar substituents and fatty acid acyl chains on lipid A precursors from a 3-deoxy-D-manno-octulosonic acid-deficient mutant of *Salmonella typhimurium*. Studies by ¹H, ¹³C, and ³¹P nuclear magnetic resonance. *J. Biol. Chem.* 260:16089–16098.

1938. Strange, P. G., and D. E. Koshland, Jr. 1976. Receptor interactions in a signalling system: competition between ribose receptor and galactose receptor in the chemotaxis response. *Proc. Natl. Acad. Sci. USA* 73:762–766.

1939. Strauch, K. L., T. H. Carter, and C. G. Miller. 1983. Overproduction of *Salmonella typhimurium* peptidase T. *J. Bacteriol.* 156:743–751.

1940. Strauch, K. L., J. B. Lenk, B. L. Gamble, and C. G. Miller. 1985. Oxygen regulation in *Salmonella typhimurium*. *J. Bacteriol.* 161:673–680.

1941. Strauch, K. L., and C. G. Miller. 1983. Isolation and characterization of *Salmonella typhimurium* mutants lacking a tripeptidase (peptidase T). *J. Bacteriol.* 154:763–771.

1942. Straus, D. S., and B. N. Ames. 1973. Histidyl transfer ribonucleic acid synthetase mutants requiring a high internal pool of histidine for growth. *J. Bacteriol.* 115:188–197.

1943. Straus, D. S., and G. R. Hoffman. 1975. Selection for a large genetic duplication in *Salmonella typhimurium*. *Genetics* 80:227–237.

1944. Straus, D. S., and L. D. Straus. 1976. Large overlapping tandem genetic duplications in *Salmonella typhimurium*. *J. Mol. Biol.* 103:143–154.

1945. Strugnell, R. A., D. Maskell, N. Fairweather, D. Pickard, A. Cockayne, C. Penn, and G. Dougan. 1990. Stable expression of foreign antigens from the chromosome of *Salmonella typhimurium* vaccine strains. *Gene* 88:57–63.

1946. Stuttard, C. 1972. Location of *trpR* mutations in the *serB-thr* region of *Salmonella typhimurium*. *J. Bacteriol.* 111:368–374.

1947. Stuttard, C. 1973. Genetic analysis of *thr* mutations in *Salmonella typhimurium*. *J. Bacteriol.* 116:1–11.

1948. **Stuttard, C.** 1975. Tryptophan biosynthesis in *Salmonella typhimurium:* location in *trpB* of a genetic difference between strains LT2 and LT7. *J. Bacteriol.* 123:878–887.

1949. **Su, G. F., H. N. Brahmbhatt, J. Wehland, M. Rohde, and K. N. Timmis.** 1992. Construction of stable LamB-Shiga toxin B subunit hybrids: analysis of expression in *Salmonella typhimurium aroA* strains and stimulation of B subunit-specific mucosal and serum antibody responses. *Infect. Immun.* 60:3345–3359.

1950. **Subbaiah, T. V., and B. A. D. Stocker.** 1964. Rough mutants of *Salmonella typhimurium.* I. Genetics. *Nature* (London) 201:1298–1299.

1951. **Subbaramaiah, K., and S. A. Simms.** 1992. Photolabeling of CheR methyltransferase with S-adenosyl-L-methionine (AdoMet). Studies on the AdoMet binding site. *J. Biol. Chem.* 267:8636–8642.

1952. **Sugiyama, T., N. Kido, T. Komatsu, M. Ohta, and N. Kato.** 1991. Expression of the cloned *Escherichia coli* O9 *rfb* gene in various mutant strains of *Salmonella typhimurium. J. Bacteriol.* 173:55–58.

1953. **Suh, S. J., and J. C. Escalante-Semerena.** 1993. Cloning, sequencing and overexpression of *cobA* which encodes ATP:corrinoid adenosyltransferase in *Salmonella typhimurium. Gene* 129:93–97.

1954. **Suh, S. J., and J. C. Escalante-Semerena.** 1995. Purification and initial characterization of the ATP:corrinoid adenosyltransferase encoded by the *cobA* gene of *Salmonella typhimurium. J. Bacteriol.* 177:921–925.

1955. **Sukupolvi, S., I. M. Helander, R. Hukari, M. Vaara, and P. H. Makela.** 1985. Lipopolysaccharides of three classes of supersensitive mutants of *Salmonella typhimurium. FEMS Microbiol. Lett.* 30:341–345.

1956. **Sukupolvi, S., and D. O'Connor.** 1987. Amino acid alterations in a hydrophobic region of the TraT protein of R6–5 increase the outer membrane permeability of enteric bacteria. *Mol. Gen. Genet.* 210:178–180.

1957. **Sukupolvi, S., D. O'Connor, and M. F. Edwards.** 1986. The *traT* protein is able to normalize the phenotype of a plasmid-carried permeability mutation of *Salmonella typhimurium. J. Gen. Microbiol.* 132:2079–2085.

1958. **Sukupolvi, S., P. Riikonen, S. Taira, H. Saarilahti, and M. Rhen.** 1992. Plasmid-mediated serum resistance in *Salmonella enterica. Microb. Pathog.* 12:219–225.

1959. **Sukupolvi, S., and M. Vaara.** 1989. *Salmonella typhimurium* and *Escherichia coli* mutants with increased outer membrane permeability to hydrophobic compounds. *Biochim. Biophys. Acta* 988:377–387.

1960. **Sukupolvi, S., M. Vaara, I. M. Helander, P. Viljanen, and P. H. Makela.** 1984. New *Salmonella typhimurium* mutants with altered outer membrane permeability. *J. Bacteriol.* 159:704–712.

1961. **Sukupolvi, S., R. Vuorio, S. Y. Qi, D. O'Connor, and M. Rhen.** 1990. Characterization of the *traT* gene and mutants that increase outer membrane permeability from the *Salmonella typhimurium* virulence plasmid. *Mol. Microbiol.* 4:49–57.

1962. **Sundaram, T. K.** 1972. *myo*-Inositol catabolism in *Salmonella typhimurium:* enzyme repression dependent on growth history of organism. *J. Gen. Microbiol.* 73:209–219.

1963. **Suryanarayana, T., and C. P. Burma.** 1975. Substrate specificity of *Salmonella typhimurium* RNase III and the nature of products formed. *Biochim. Biophys. Acta* 407:459–468.

1964. **Sutherland, L., J. Cairney, M. J. Elmore, I. R. Booth, and C. F. Higgins.** 1986. Osmotic regulation of transcription: induction of the *proU* betaine transport gene is dependent on accumulation of intracellular potassium. *J. Bacteriol.* 168:805–814.

1965. **Suzuki, H., and T. Iino.** 1966. An assay for newly synthesized intracellular flagellin. *Biochim. Biophys. Acta* 124:212–215.

1966. **Suzuki, H., and T. Iino.** 1973. *In vitro* synthesis of phase-specific flagellin of *Salmonella. J. Mol. Biol.* 81:57–70.

1967. **Suzuki, H., and T. Iino.** 1975. Absence of messenger ribonucleic acid specific for flagellin in non-flagellate mutants of *Salmonella. J. Mol. Biol.* 95:549–556.

1968. **Suzuki, T., and T. Iino.** 1981. Role of the *flaR* gene in flagellar hook formation in *Salmonella* spp. *J. Bacteriol.* 148:973–979.

1969. **Suzuki, T., T. Iino, T. Horiguchi, and S. Yamaguchi.** 1978. Incomplete flagella structures in nonflagellate mutants of *Salmonella typhimurium. J. Bacteriol.* 133:904–915.

1970. **Sverdlov, E. D., N. A. Lisitsyn, S. O. Guryev, and G. S. Monastyrskaia.** 1986. Nucleotide sequence of the *Salmonella typhimurium rpoB* gene encoding the β-subunit of RNA polymerase. *Dokl. Akad. Nauk. SSSR* 287:232–236. (In Russian.)

1971. **Sverdlov, E. D., N. A. Lisitsyn, S. O. Guryev, Y. V. Smirnov, V. M. Rostapshov, and G. S. Monastyrskaya.** 1986. Genes coding the β-subunit of bacterial RNA polymerases. I. Primary structure of the *rpoB* gene EcoRI-C fragment from *Salmonella typhimurium. Bioorg. Khim.* 12:699–707. (In Russian with English abstract.)

1972. **Sweet, G. D., C. M. Kay, and W. W. Kay.** 1984. Tricarboxylate-binding proteins of *Salmonella typhimurium.* Purification, crystallization, and physical properties. *J. Biol. Chem.* 259:1586–1592.

1973. **Sweet, G. D., J. M. Somers, and W. W. Kay.** 1979. Purification and properties of a citrate-binding transport component, the C protein of *Salmonella typhimurium. Can. J. Biochem.* 57:710–716.

1974. **Swenson, D. L., and S. Clegg.** 1992. Identification of ancillary *fim* genes affecting *fimA* expression in *Salmonella typhimurium. J. Bacteriol.* 174:7697–7704.

1975. **Swenson, D. L., K. J. Kim, E. W. Six, and S. Clegg.** 1994. The gene *fimU* affects expression of *Salmonella typhimurium* type 1 fimbriae and is related to the *Escherichia coli* tRNA gene *argU. Mol. Gen. Genet.* 244:216–218.

1976. **Swenson, P. A., L. Riester, and T. V. Palmer.** 1983. Role of the *supX* gene in sensitizing *Salmonella typhimurium* cells to respiration shutoff induced by far ultraviolet irradiation. *Photochem. Photobiol.* 38:305–310.

1977. **Switala, J., B. L. Triggs-Raine, and P. C. Loewen.** 1990. Homology among bacterial catalase genes. *Can. J. Microbiol.* 36:728–731.

1978. **Syvanen, J. M., and J. R. Roth.** 1972. The structural genes for ornithine transcarbamylase in *Salmonella typhimurium* and *Escherichia coli. J. Bacteriol.* 110:66–70.

1979. **Syvanen, J. M., and J. R. Roth.** 1973. Structural genes for catalytic and regulatory subunits of aspartate transcarbamylase. *J. Mol. Biol.* 76:363–378.

1980. **Syvanen, J. M., Y. R. Yang, and M. W. Kirschner.** 1973. Preparation of ^{125}I-catalytic subunit of aspartate transcarbamylase and its use in studies of the regulatory subunit. *J. Biol. Chem.* 248:3762–3768.

1981. **Szekely, E., and M. Simon.** 1981. Homology between the invertible deoxyribonucleic acid sequence that controls flagellar-phase variation in the *Salmonella* sp. and deoxyribonucleic acid sequences in other organisms. *J. Bacteriol.* 148:829–836.

1982. **Szekely, E., and M. Simon.** 1983. DNA sequence adjacent to flagellar genes and evolution of flagellar-phase variation. *J. Bacteriol.* 155:74–81.

1983. **Szwacka, M., Z. Ciesla, and T. Klopotowski.** 1979. Azide induced mutagenesis in gram negative bacteria is *recA* independent and *lexA* independent. *Mutat. Res.* 62:221–226.

1984. **Tai, P. C., D. P. Kessler, and J. Ingraham.** 1966. Cold-sensitive mutations in *Salmonella typhimurium* which affect ribosome synthesis. *J. Bacteriol.* 97:1298–1304.

1985. **Taillon, B. E., R. Little, and R. P. Lawther.** 1988. Analysis of the functional domains of biosynthetic threonine deaminase by comparison of the amino acid sequences of three wild-type alleles to the amino acid sequence of biodegradative threonine deaminase. *Gene* 63:245–252.

1986. **Taillon, M. P., D. A. Gotto, and R. P. Lawther.** 1981. The DNA sequence of the promoter-attenuator of the *ilvGEDA* operon of *Salmonella typhimurium. Nucleic Acids Res.* 9:3419–3432.

1987. **Taira, S., and M. Rhen.** 1989. Identification and genetic analysis of *mkaA*—a gene of the *Salmonella typhimurium* virulence plasmid necessary for intracellular growth. *Microb. Pathog.* 7:165–173.

1988. **Taira, S., P. Riikonen, H. Saarilahti, S. Sukupolvi, and M. Rhen.** 1991. The *mkaC* virulence gene of the *Salmonella* serovar *typhimurium* 96 kb plasmid encodes a transcriptional activator. *Mol. Gen. Genet.* 228:381–384.

1989. **Takayama, K., N. Qureshi, and P. Mascagni.** 1983. Complete structure of lipid A obtained from the lipopolysaccharides of the heptoseless mutant of *Salmonella typhimurium. J. Biol. Chem.* 258:12801–12803.

1990. **Tame, J. R., G. N. Murshudov, E. J. Dodson, T. K. Neil, G. G. Dodson, C. F. Higgins, and A. J. Wilkinson.** 1994. The structural basis of sequence-independent peptide binding by OppA protein. *Science* 264:1578–1581.

1991. **Tamir, H., and P. R. Srinivasan.** 1972. Studies of the mechanism of anthranilate synthase. Evidence for an acyl-enzyme. *J. Biol. Chem.* 247:1153–1155.

1992. **Tanemura, S., and R. Bauerle.** 1977. Internal reinitiation of translation in polar mutants of the *trpB* gene of *Salmonella typhimurium. Mol. Gen. Genet.* 153:135–143.

1993. **Tanemura, S., and R. Bauerle.** 1979. Suppression of a deletion mutation in the glutamine amidotransferase region of the *Salmonella typhimurium trpD* gene by mutations in *pheA* and *tyrA. J. Bacteriol.* 139:573–582.

1994. **Tanemura, S., and R. Bauerle.** 1980. Conditionally expressed missense mutations: the basis for the unusual phenotype of an apparent *trpD* nonsense mutant of *Salmonella typhimurium. Genetics* 95:545–560.

1995. **Tanemura, S., and J. Yourno.** 1969. Frameshift revertant of *Salmonella typhimurium* producing histidinol dehydrogenase with a sequence of four extra amino acid residues. *J. Mol. Biol.* 40:459–466.

1996. **Tankersley, W. G., and J. M. Woodward.** 1974. Induction and isolation of a minicell-producing strain of *Salmonella typhimurium. Proc. Soc. Exp. Biol. Med.* 145:802–805.

1997. **Tannock, G. W., R. V. H. Blumershine, and D. C. Savage.** 1975. Association of *Salmonella typhimurium* with, and its invasion of, the ileal mucosa in mice. *Infect. Immun.* 11:365–370.

1998. **Tartaglia, L. A., G. Storz, and B. N. Ames.** 1989. Identification and molecular analysis of *oxyR*-regulated promoters important for the bacterial adaptation to oxidative stress. *J. Mol. Biol.* 210:709–719.

1999. **Tartaglia, L. A., G. Storz, M. H. Brodsky, A. Lai, and B. N. Ames.** 1990. Alkyl hydroperoxide reductase from *Salmonella typhimurium.* Sequence and homology to thioredoxin reductase and other flavoprotein disulfide oxidoreductases. *J. Biol. Chem.* 265:10535–10540.

2000. **Tate, C. G., J. A. Muiry, and P. J. Henderson.** 1992. Mapping, cloning, expression, and sequencing of the *rhaT* gene, which encodes a novel L-rhamnose-H+ transport protein in *Salmonella typhimurium* and *Escherichia coli. J. Biol. Chem.* 267:6923–6932.

2001. **Taylor, B. L., and D. E. Koshland, Jr.** 1974. Reversal of flagellar rotation in monotrichous and peritrichous bacteria: generation of changes in direction. *J. Bacteriol.* 119:640–642.

2002. **Taylor, B. L., and D. E. Koshland, Jr.** 1975. Intrinsic and extrinsic light responses of *Salmonella typhimurium* and *Escherichia coli. J. Bacteriol.* 123:557–569.

2003. **Taylor, G., E. Vimr, E. Garman, and G. Laver.** 1992. Purification, crystallization and preliminary crystallographic study of neuraminidase from *Vibrio cholerae* and *Salmonella typhimurium* LT2. *J. Mol. Biol.* 226:1287–1290.

2004. Taylor, N. S., and I. R. Beacham. 1976. Synthesis and localization of *Escherichia coli* UDP-glucose hydrolase (5′-nucleotidase), and demonstration of a cytoplasmic inhibitor of this enzyme in *Salmonella typhimurium*. *Biochim. Biophys. Acta* 411:216–221.

2005. Taylor, R. T., and H. Weissbach. 1967. N^5-methyltetrahydrofolate-homocysteine transmethylase. Partial purification and properties. *J. Biol. Chem.* 242:1502–1508.

2006. Theisen, M., R. A. Kelln, and J. Neuhard. 1987. Cloning and characterization of the *pyrF* operon of *Salmonella typhimurium*. *Eur. J. Biochem.* 164:613–619.

2007. Theisen, M., and J. Neuhard. 1990. Translational coupling in the *pyrF* operon of *Salmonella typhimurium*. *Mol. Gen. Genet.* 222:345–352.

2008. Theodore, T. S., and E. Engelsberg. 1964. Mutant of *Salmonella typhimurium* deficient in the carbon dioxide-fixing enzyme phosphoenolpyruvic carboxylase. *J. Bacteriol.* 88:946–955.

2009. Thomas, S. M. 1993. Extreme cold sensitivity of *Salmonella typhimurium umu* clones, effects of the *umu* region and flanking sequences. *Mutat. Res.* 285:95–99.

2010. Thomas, S. M., H. M. Crowne, S. C. Pidsley, and S. G. Sedgwick. 1990. Structural characterization of the *Salmonella typhimurium* LT2 *umu* operon. *J. Bacteriol.* 172:4979–4987.

2011. Thomas, S. M., and S. G. Sedgwick. 1989. Cloning of *Salmonella typhimurium* DNA encoding mutagenic DNA repair. *J. Bacteriol.* 171:5776–5782.

2012. Thomulka, K. W., and J. S. Gots. 1982. Isolation and characterization of purine regulatory mutants of *Salmonella typhimurium* with an episomal *purE-lac* fusion. *J. Bacteriol.* 151:153–161.

2013. Thorne, G. M., and L. M. Corwin. 1972. Genetic locus of a gene affecting leucine transport in *Salmonella typhimurium*. *J. Bacteriol.* 110:784–785.

2014. Thorne, G. M., and L. M. Corwin. 1975. Mutations affecting aromatic amino acid transport in *Escherichia coli* and *Salmonella typhimurium*. *J. Gen. Microbiol.* 90:203–216.

2015. Thorner, L. K., J. P. Fandl, and S. W. Artz. 1990. Analysis of sequence elements important for expression and regulation of the adenylate cyclase gene (*cya*) of *Salmonella typhimurium*. *Genetics* 125:709–717.

2016. Timme, T. L., C. B. Lawrence, and R. E. Moses. 1989. Two new members of the OmpR superfamily detected by homology to a sensor-binding core domain. *J. Mol. Evol.* 28:545–552.

2017. Tirgari, S., M. P. Spector, and J. W. Foster. 1986. Genetics of NAD metabolism in *Salmonella typhimurium* and cloning of the *nadA* and *pnuC* loci. *J. Bacteriol.* 167:1086–1088.

2018. Tittawella, I. P. B. 1984. Evidence for clustering of RNA polymerase and ribosomal protein genes in six species of Enterobacteria. *Mol. Gen. Genet.* 195:215–218.

2019. Tittawella, I. P. B. 1985. *Salmonella typhimurium rpoB* and *rpoC* genes cloned on the λ phages. *FEBS Lett.* 185:33–36.

2020. Tokunaga, H., M. Tokunaga, and T. Nakae. 1979. Characterization of porins from the outer membrane of *Salmonella typhimurium*. 1. Chemical analysis. *Eur. J. Biochem.* 95:433–440.

2021. Tokunaga, M., H. Tokunaga, Y. Okajima, and T. Nakae. 1979. Characterization of porins from the outer membrane of *Salmonella typhimurium*. 2. Physical properties of the functional oligomeric aggregates. *Eur. J. Biochem.* 95:441–448.

2022. Tokuno, S.-I., E. P. Goldschmidt, and M. Gough. 1974. Mutant of *Salmonella typhimurium* that channels infecting bacteriophage P22 toward lysogenization. *J. Bacteriol.* 119:508–513.

2023. Tokuno, S.-I., and M. Gough. 1975. Host influence on the activity of genes *c1* and *c3* in regulating the decision between lysis and lysogeny in bacteriophage P22. *J. Virol.* 16:1184–1190.

2024. Tokuno, S.-I., L. Roth, C. Weinberger, and M. Gough. 1977. Effect of mutant host RNA polymerase on the bi-functional activities of phage P-22 gene C-1. *Mol. Gen. Genet.* 153:205–210.

2025. Trachtenberg, S., and D. J. DeRosier. 1991. A molecular switch: subunit rotations involved in the right-handed to left-handed transitions of *Salmonella typhimurium* flagellar filaments. *J. Mol. Biol.* 220:67–77.

2026. Tran, P. V., T. A. Bannor, S. Z. Doktor, and B. P. Nichols. 1990. Chromosomal organization and expression of *Escherichia coli pabA*. *J. Bacteriol.* 172:397–410.

2027. Tribhuwan, R. C., M. S. Johnson, and B. L. Taylor. 1986. Evidence against direct involvement of cyclic GMP or cyclic AMP in bacterial chemotactic signaling. *J. Bacteriol.* 168:624–630.

2028. Tronick, S. R., J. E. Ciardi, and E. R. Stadtman. 1973. Comparative biochemical and immunological studies of bacterial glutamine synthetases. *J. Bacteriol.* 115:858–868.

2029. Trucksis, M., E. I. Golub, D. J. Zabel, and R. F. Depew. 1981. *Escherichia coli* and *Salmonella typhimurium supX* genes specify deoxyribonucleic acid topoisomerase I. *J. Bacteriol.* 147:679–681.

2030. Trzebiatowski, J. R., G. A. O'Toole, and J. C. Escalante-Semerena. 1994. The *cobT* gene of *Salmonella typhimurium* encodes the NaMN:5,6-dimethylbenzimidazole phosphoribosyltransferase responsible for the synthesis of N^1-(5-phospho-α-D-ribosyl)-5,6-dimethylbenzimidazole, an intermediate in the synthesis of the nucleotide loop of cobalamin. *J. Bacteriol.* 176:3568–3575.

2031. Tsai, S. P., R. J. Hartin, and J. Ryu. 1989. Transformation in restriction-deficient *Salmonella typhimurium* LT2. *J. Gen. Microbiol.* 135:2561–2567.

2032. Tubulekas, I., R. H. Buckingham, and D. Hughes. 1991. Mutant ribosomes can generate dominant kirromycin resistance. *J. Bacteriol.* 173:3635–3643.

2033. Tubulekas, I., and D. Hughes. 1993. Growth and translation elongation rate are sensitive to the concentration of EF-Tu. *Mol. Microbiol.* 8:761–770.

2034. Tubulekas, I., and D. Hughes. 1993. A single amino acid substitution in elongation factor Tu disrupts interaction between the ternary complex and the ribosome. *J. Bacteriol.* 175:240–250.

2035. Tuohy, T. M., S. Thompson, R. F. Gesteland, D. Hughes, and J. F. Atkins. 1990. The role of EF-Tu and other translation components in determining translocation step size. *Biochim. Biophys. Acta* 1050:274–278.

2036. Tupper, A. E., T. A. Owen-Hughes, D. W. Ussery, D. S. Santos, D. J. Ferguson, J. M. Sidebotham, J. C. Hinton, and C. F. Higgins. 1994. The chromatin-associated protein H-NS alters DNA topology *in vitro*. *EMBO J.* 13:258–268.

2037. Turnbough, C. L., Jr., and B. R. Bochner. 1985. Toxicity of the pyrimidine biosynthetic pathway intermediate carbamyl asparate in *Salmonella typhimurium*. *J. Bacteriol.* 163:500–505.

2038. Turnowsky, F., K. Fuchs, C. Jeschek, and G. Hogenauer. 1989. *envM* genes of *Salmonella typhimurium* and *Escherichia coli*. *J. Bacteriol.* 171:6555–6565.

2039. Tyler, B. M., A. B. DeLeo, and B. Magasanik. 1974. Activation of transcription of *hut* DNA by glutamine synthetase. *Proc. Natl. Acad. Sci. USA* 71:225–229.

2040. Ueki, T., T. Mitsui, and H. Nikaido. 1979. X-ray diffraction studies of outer membranes of *Salmonella typhimurium*. *J. Biochem.* 85:173–182.

2041. Ueno, T., K. Oosawa, and S. Aizawa. 1992. M ring, S ring and proximal rod of the flagellar basal body of *Salmonella typhimurium* are composed of subunits of a single protein, FliF. *J. Mol. Biol.* 227:672–677.

2042. Ueno, T., K. Oosawa, and S. Aizawa. 1994. Domain structures of the MS ring component protein (FliF) of the flagellar basal body of *Salmonella typhimurium*. *J. Mol. Biol.* 236:546–555.

2043. Uerkvitz, W., and C. F. Beck. 1981. Periplasmic phosphatases in *Salmonella typhimurium* Lt-2. A biochemical, physiological, and partial genetic analysis of three nucleoside monophosphate dephosphorylating enzymes. *J. Biol. Chem.* 256:382–389.

2044. Umbarger, H. E. 1980. Comments on the role of aminoacyl-tRNA in the regulation of amino acid biosynthesis. *Cold Spring Harbor Monogr. Ser.* 9B:453–467.

2045. Umbarger, H. E., M. A. Umbarger, and P. M. L. Sui. 1963. Biosynthesis of serine in *Escherichia coli* and *Salmonella typhimurium*. *J. Bacteriol.* 85:1431–1439.

2046. Uomini, J. R., and J. R. Roth. 1974. Suppressor-dependent frameshift mutants of bacteriophage P22. *Mol. Gen. Genet.* 134:237–247.

2047. Urbanowski, M. L., L. S. Plamann, and G. V. Stauffer. 1987. Mutations affecting the regulation of the *metB* gene of *Salmonella typhimurium* LT2. *J. Bacteriol.* 169:126–130.

2048. Urbanowski, M. L., M. D. Plamann, and G. V. Stauffer. 1983. Comparison of the regulatory regions of the *glyA* genes of *Escherichia coli* and *Salmonella typhimurium*, abstr. H159, p. 112. *Abstr. Annu. Meet. Am. Soc. Microbiol. 1983*.

2049. Urbanowski, M. L., M. D. Plamann, L. T. Stauffer, and G. V. Stauffer. 1984. Cloning and characterization of the gene for *Salmonella typhimurium* serine hydroxymethyltransferase. *Gene* 27:47–54.

2050. Urbanowski, M. L., and G. V. Stauffer. 1985. Cloning and initial characterization of the *metJ* and *metB* genes from *Salmonella typhimurium* LT2. *Gene* 35:187–197.

2051. Urbanowski, M. L., and G. V. Stauffer. 1985. Nucleotide sequence and biochemical characterization of the *metJ* gene from *Salmonella typhimurium* LT2. *Nucleic Acids Res.* 13:673–685.

2052. Urbanowski, M. L., and G. V. Stauffer. 1986. Autoregulation by tandem promoters of the *Salmonella typhimurium* LT2 *metJ* gene. *J. Bacteriol.* 165:740–745.

2053. Urbanowski, M. L., and G. V. Stauffer. 1986. The *metH* gene from *Salmonella typhimurium* LT2: cloning and initial characterization. *Gene* 44:211–217.

2054. Urbanowski, M. L., and G. V. Stauffer. 1987. Regulation of the *metR* gene of *Salmonella typhimurium*. *J. Bacteriol.* 169:5841–5844.

2055. Urbanowski, M. L., and G. V. Stauffer. 1988. The control region of the *metH* gene of *Salmonella typhimurium* LT2: an atypical *met* promoter. *Gene* 73:193–200.

2056. Urbanowski, M. L., and G. V. Stauffer. 1989. Role of homocysteine in *metR*-mediated activation of the *metE* and *metH* genes in *Salmonella typhimurium* and *Escherichia coli*. *J. Bacteriol.* 171:3277–3281.

2057. Urbanowski, M. L., and G. V. Stauffer. 1989. Genetic and biochemical analysis of the MetR activator-binding site in the *metE metR* control region of *Salmonella typhimurium*. *J. Bacteriol.* 171:5620–5629.

2058. Urbanowski, M. L., L. T. Stauffer, L. S. Plamann, and G. V. Stauffer. 1987. A new methionine locus, *metR*, that encodes a *trans*-acting protein required for activation of *metE* and *metH* in *Escherichia coli* and *Salmonella typhimurium*. *J. Bacteriol.* 169:1391–1397.

2059. Vaara, M. 1980. Increased outer membrane resistance to ethyleneaminetetraacetate and cations in novel lipid A mutants. *J. Bacteriol.* 148:426–434.

2060. Vaara, M. 1981. Effect of ionic strength on polymyxin resistance of *pmrA* mutants of *Salmonella typhimurium*. *FEMS Microbiol. Lett.* 11:321–326.

2061. Vaara, M., and T. Vaara. 1981. Outer membrane permeability barrier disruption by polymyxin in polymyxin-susceptible and -resistant *Salmonella typhimurium*. *Antimicrob. Agents Chemother.* 19:578–583.

2062. Vaara, M., T. Vaara, M. Jensen, I. Helander, M. Nurminen, E. T. Rietschel, and P. H. Makela. 1981. Characterization of the lipopolysaccharide from the polymyxin-resistant *pmrA* mutants of *Salmonella typhimurium*. *FEBS Lett.* 129:145–149.

2063. Vaara, M., T. Vaara, and M. Sarvas. 1979. Decreased binding of polymyxin by polymyxin-resistant mutants of *Salmonella typhimurium*. *J. Bacteriol.* 139:664–667.

2064. Vaara, M., P. Viljanen, S. Sukupolvi, and T. Vaara. 1985. Does polymyxin B nonapeptide increase outer membrane permeability in antibiotic supersensitive enterobacterial mutants? *FEMS Microbiol. Lett.* 26:289–294.

2065. Valtonen, M. V., U. M. Larinkari, M. Posila, V. V. Valtonen, and P. H. Makela. 1976. Effect of enterobacterial common antigen on mouse virulence of Salmonella typhimurium. Infect. Immun. 13:1601–1605.

2066. Van den Bosch, J. L., D. R. Kurlandsky, R. Urdangaray, and G. W. Jones. 1989. Evidence of coordinate regulation of virulence in Salmonella typhimurium involving the rsk element of the 95-kilobase plasmid. Infect. Immun. 57:2566–2568.

2067. Van den Bosch, J. L., D. K. Rabert, D. R. Kurlandsky, and G. W. Jones. 1989. Sequence analysis of rsk, a portion of the 95-kilobase plasmid of Salmonella typhimurium associated with resistance to the bactericidal activity of serum. Infect. Immun. 57:850–857.

2068. van der Vlag, J., K. van Dam, and P. W. Postma. 1994. Quantification of the regulation of glycerol and maltose metabolism by IIA^{Glc} of the phosphoenolpyruvate-dependent glucose phosphotransferase system in Salmonella typhimurium. J. Bacteriol. 176:3518–3526.

2069. Van der Werf, R., and D. E. Koshland, Jr. 1977. Identification of a γ-glutamyl methyl ester in bacterial membrane protein involved in chemotaxis. J. Biol. Chem. 252:2793–2795.

2070. van Dijl, J. M., A. de Jong, H. Smith, S. Bron, and G. Venema. 1991. Lack of specific hybridization between the lep genes of Salmonella typhimurium and Bacillus licheniformis. FEMS Microbiol. Lett. 65:345–351.

2071. van Dijl, J. M., R. van den Bergh, T. Reversma, H. Smith, S. Bron, and G. Venema. 1990. Molecular cloning of the Salmonella typhimurium lep gene in Escherichia coli. Mol. Gen. Genet. 223:233–240.

2072. Van Dyk, T. K., and R. A. LaRossa. 1986. Sensitivity of a Salmonella typhimurium aspC mutant to sulfometuron methyl, a potent inhibitor of acetolactate synthase II. J. Bacteriol. 165:386–392.

2073. Van Dyk, T. K., and R. A. LaRossa. 1987. Involvement of ack-pta operon products in alpha-ketobutyrate metabolism by Salmonella typhimurium. Mol. Gen. Genet. 207:435–440.

2074. Van Dyk, T. K., D. R. Smulski, and Y.-Y. Chang. 1987. Pleiotropic effects of poxA regulatory mutations of Escherichia coli and Salmonella typhimurium, mutations conferring sulfometuron methyl and α-ketobutyrate hypersensitivity. J. Bacteriol. 169:4540–4546.

2075. Van Pel, A., and C. Colson. 1974. DNA restriction and modification systems in Salmonella. II. Genetic complementation between the K and B systems of Escherichia coli and the Salmonella typhimurium system, SB. Mol. Gen. Genet. 135:51–60.

2076. Vartak, N. B., J. Reizer, A. Reizer, J. T. Gripp, E. A. Groisman, L. F. Wu, J. M. Tomich, and M. H. Saier, Jr. 1991. Sequence and evolution of the FruR protein of Salmonella typhimurium: a pleiotropic transcriptional regulatory protein possessing both activator and repressor functions which is homologous to the periplasmic ribose-binding protein. Res. Microbiol. 142:951–963.

2077. Vary, P. S., and B. A. D. Stocker. 1973. Nonsense motility mutants of Salmonella typhimurium. Genetics 73:229–245.

2078. Vaughan, P., and B. Sedgwick. 1991. A weak adaptive response to alkylation damage in Salmonella typhimurium. J. Bacteriol. 173:3656–3662.

2079. Verma, N. K., N. B. Quigley, and P. R. Reeves. 1988. O-antigen variation in Salmonella spp.: rfb gene clusters of three strains. J. Bacteriol. 170:130–107.

2080. Vimr, E. R., L. Green, and C. G. Miller. 1983. Oligopeptidase-deficient mutants of Salmonella typhimurium. J. Bacteriol. 153:1259–1265.

2081. Vimr, E. R., and C. G. Miller. 1983. Dipeptidyl carboxypeptidase-deficient mutants of Salmonella typhimurium. J. Bacteriol. 153:1252–1258.

2082. Vinitsky, A., H. Teng, and C. T. Grubmeyer. 1991. Cloning and nucleic acid sequence of the Salmonella typhimurium pncB gene and structure of nicotinate phosphoribosyltransferase. J. Bacteriol. 173:536–540.

2083. Vogel, H. J., D. F. Baron, and A. Baich. 1963. Induction of acetyl ornithine δ-transaminase during pathway-wide repression, p. 293–300. In H. J. Vogel, V. Bryson, and J. O. Lampen (ed.), Informational Macromolecules. Academic Press, Inc., New York.

2084. Vogel, T., M. Meyers, J. S. Kovach, and R. F. Goldberger. 1972. Specificity of interaction between the first enzyme for histidine biosynthesis and aminoacylated histidine transfer ribonucleic acid. J. Bacteriol. 112:126–130.

2085. Vogler, A. P., M. Homma, V. M. Irikura, and R. M. Macnab. 1991. Salmonella typhimurium mutants defective in flagellar filament regrowth and sequence similarity of FliI to F_0F_1, vacuolar, and archaebacterial ATPase subunits. J. Bacteriol. 173:3564–3572.

2086. Voll, M. J., E. Appella, and R. G. Martin. 1967. Purification and composition studies of phosphoribosyl-adenosine triphosphate:pyrophosphate phosphoribosyl transferase, the first enzyme of histidine biosynthesis. J. Biol. Chem. 242:1760–1767.

2087. Voll, M. J., L. A. Cohen, and J. J. Germida. 1979. his-linked hydrogen sulfide locus of Salmonella typhimurium and its expression in Escherichia coli. J. Bacteriol. 139:1082–1084.

2088. Voll, M. J., L. M. Shiller, and J. Castrilli. 1974. his-linked hydrogen sulfide locus in Salmonella typhimurium. J. Bacteriol. 120:902–905.

2089. Vuorio, R., T. Harkonen, M. Tolvanen, and M. Vaara. 1994. The novel hexapeptide motif found in the acyltransferases LpxA and LpxD of lipid A biosynthesis is conserved in various bacteria. FEBS Lett. 337:289–292.

2090. Vuorio, R., L. Hirvas, and M. Vaara. 1991. The Ssc protein of enteric bacteria has significant homology to the acyltransferase Lpxa of lipid A biosynthesis, and to three acetyltransferases. FEBS Lett. 292:90–94.

2091. Vuorio, R., and M. Vaara. 1992. Mutants carrying conditionally lethal mutations in outer membrane genes omsA and firA (ssc) are phenotypically similar, and omsA is allelic to firA. J. Bacteriol. 174:7090–7097.

2092. Wainscott, V. J., and J. J. Ferretti. 1978. Biochemical-genetic study of the first enzyme of histidine biosynthesis in Salmonella typhimurium: substrate and feedback binding regions. J. Bacteriol. 133:114–121.

2093. Walenga, R. W., and M. J. Osborn. 1980. Biosynthesis of lipid A. In-vivo formation of an intermediate containing 3-deoxy-D-manno-octulosonate in a mutant of Salmonella typhimurium. J. Biol. Chem. 255:4252–4256.

2094. Walter, C., K. Honer zu Bentrup, and E. Schneider. 1992. Large scale purification, nucleotide binding properties, and ATPase activity of the MalK subunit of Salmonella typhimurium maltose transport complex. J. Biol. Chem. 267:8863–8869.

2095. Walter, C., S. Wilken, and E. Schneider. 1992. Characterization of side-directed mutations in conserved domains of MalK, a bacterial member of the ATP-binding cassette (ABC) family. FEBS Lett. 303:41–44.

2096. Wang, E., and D. E. Koshland, Jr. 1980. Receptor structure in the bacterial sensing system. Proc. Natl. Acad. Sci. USA 77:7157–7161.

2097. Wang, J. Y. J., D. O. Clegg, and D. E. Koshland, Jr. 1981. Molecular cloning and amplification of the adenylate cyclase gene. Proc. Natl. Acad. Sci. USA 78:4684–4688.

2098. Wang, L., and P. R. Reeves. 1994. Involvement of the galactosyl-1-phosphate transferase encoded by the Salmonella enterica rfbP gene in O-antigen subunit processing. J. Bacteriol. 176:4348–4356.

2099. Wang, L., L. K. Romana, and P. R. Reeves. 1992. Molecular analysis of a Salmonella enterica group E1 rfb gene cluster: O antigen and the genetic basis of the major polymorphism. Genetics 130:429–443.

2100. Wang, Q., M. Sacco, E. Ricca, C. T. Lago, M. DeFelice, and J. M. Calvo. 1993. Organization of Lrp-binding sites upstream of ilvIH in Salmonella typhimurium. Mol. Microbiol. 7:883–891.

2101. Warner, T. G., R. Harris, R. McDowell, and E. R. Vimr. 1992. Photolabelling of Salmonella typhimurium LT2 sialidase. Identification of a peptide with a predicted structural similarity to the active sites of influenza-virus sialidases. Biochem. J. 285:957–964.

2102. Warrick, H. M., B. L. Taylor, and D. E. Koshland, Jr. 1977. Chemotactic mechanism of Salmonella typhimurium: preliminary mapping and characterization of mutants. J. Bacteriol. 130:223–231.

2103. Wasmuth, J. J., and H. E. Umbarger. 1973. Effect of isoleucine, valine, or leucine starvation on the potential for formation of the branched-chain amino acid biosynthetic enzymes. J. Bacteriol. 116:548–561.

2104. Wasmuth, J. J., and H. E. Umbarger. 1973. Participation of branched-chain amino acid analogues in multivalent repression. J. Bacteriol. 116:562–570.

2105. Wasserman, S. A., E. Daub, P. Grisafi, D. Botstein, and C. T. Walsh. 1984. Catabolic alanine racemase from Salmonella typhimurium: DNA sequence, enzyme purification, and characterization. Biochemistry 23:5182–5187.

2106. Wasserman, S. A., C. T. Walsh, and D. Botstein. 1983. Two alanine racemase genes in Salmonella typhimurium that differ in structure and function. J. Bacteriol. 153:1439–1450.

2107. Watanabe, H., H. Hashimoto, and S. Misuhashi. 1980. Salmonella typhimurium LT-2 mutation affecting the deletion of resistance determinants on R plasmids. J. Bacteriol. 142:145–152.

2108. Watanabe, M., M. Ishidate, Jr., and T. Nohmi. 1990. Nucleotide sequence of Salmonella typhimurium nitroreductase gene. Nucleic Acids Res. 18:1059.

2109. Watanabe, M., T. Sofuni, and T. Nohmi. 1992. Involvement of Cys69 residue in the catalytic mechanism of N-hydroxyarylamine O-acetyltransferase of Salmonella typhimurium. Sequence similarity at the amino acid level suggests a common catalytic mechanism of acetyltransferase for S. typhimurium and higher organisms. J. Biol. Chem. 267:8429–8436.

2110. Watanbe, T., T. Arai, and T. Hattori. 1970. Effects of cell wall polysaccharide on the mating ability of Salmonella typhimurium. Nature (London) 225:70–71.

2111. Waxman, L., and A. L. Goldberg. 1985. Protease La, the lon gene product, cleaves specific fluorogenic peptides in an ATP-dependent reaction. J. Biol. Chem. 260:12022–12028.

2112. Way, J. C., M. A. Davis, D. Morisato, D. E. Roberts, and N. Kleckner. 1984. New Tn10 derivatives for transposon mutagenesis and for construction of lacZ operon fusions by transposition. Gene 32:369–379.

2113. Wehr, C. T. 1973. Isolation and properties of a ribonuclease-deficient mutant of Salmonella typhimurium. J. Bacteriol. 114:96–102.

2114. Wei, B. Y., C. Bradbeer, and R. J. Kadner. 1992. Conserved structural and regulatory regions in the Salmonella typhimurium btuB gene for the outer membrane vitamin B12 transport protein. Res. Microbiol. 143:459–466.

2115. Wei, L.-N., and T. M. Joys. 1985. Covalent structure of three phase-1 flagellar filament proteins of Salmonella. J. Mol. Biol. 186:791–803.

2116. Wei, L.-N., and T. M. Joys. 1986. The nucleotide sequence of the H-1r gene of Salmonella rubislaw. Nucleic Acids Res. 14:8227.

2117. Wei, R. W., and S. Kustu. 1981. Glutamine auxotrophs with mutations in a nitrogen regulatory gene, ntrC, that is near glnA. Mol. Gen. Genet. 183:392–397.

2118. Weigand, R. A., and L. I. Rothfield. 1976. Genetic and physiological classification of periplasmic-leaky mutants of Salmonella typhimurium. J. Bacteriol. 125:340–345.

2119. Weigand, R. A., K. D. Vinci, and L. T. Rothfield. 1976. Morphogenesis of the bacterial division septum: a new class of septation defective mutants. Proc. Natl. Acad. Sci. USA 73:1882–1886.

2120. Weigel, N., A. Kukuruzinska, A. Nakazawa, E. B. Waygood, and S. Roseman. 1982. Sugar transport by the bacterial phosphotransferase system. Phosphoryl transfer reactions catalyzed by enzyme I of *Salmonella typhimurium. J. Biol. Chem.* 257:14477–14491.

2121. Weigel, N., D. A. Powers, and S. Roseman. 1982. Sugar transport by the bacterial phosphotransferase system. Primary structure and active site of a general phosphocarrier protein (HPr) from *Salmonella typhimurium. J. Biol. Chem.* 257:14499–14509.

2122. Weigel, N., E. B. Waygood, M. A. Kukuruzinska, A. Nakazawa, and S. Roseman. 1982. Sugar transport by the bacterial phosphotransferase system. Isolation and characterization of enzyme I from *Salmonella typhimurium. J. Biol. Chem.* 257:14461–14469.

2123. Weinberg, R. A., and R. O. Burns. 1984. Regulation of expression of the *ilvB* operon in *Salmonella typhimurium. J. Bacteriol.* 160:833–841.

2124. Wek, R. C., and G. W. Hatfield. 1986. Nucleotide sequence and in vivo expression of the *ilvY* and *ilvC* genes in *Escherichia coli* K12. Transcription from divergent overlapping promoters. *J. Biol. Chem.* 261:2441–2450.

2125. Weppelman, R., L. D. Kier, and B. N. Ames. 1977. Properties of two phosphatases and a cyclic phosphodiesterase of *Salmonella typhimurium. J. Bacteriol.* 130:411–419.

2126. Wessler, S. R., and J. M. Calvo. 1981. Control of *leu* operon expression in *Escherichia coli* by a transcription attenuation mechanism. *J. Mol. Biol.* 149:579–597.

2127. West, T. P., and G. A. O'Donovan. 1982. Repression of cytosine deaminase by pyrimidines in *Salmonella typhimurium. J. Bacteriol.* 149:1171–1174.

2128. West, T. P., T. W. Traut, M. S. Shanley, and G. A. O'Donovan. 1985. A *Salmonella typhimurium* strain defective in uracil catabolism and beta-alanine synthesis. *J. Gen. Microbiol.* 131:1083–1090.

2129. Westby, C. A., and J. S. Gots. 1969. Genetic blocks and unique features in the biosynthesis of 5′-phosphoribosyl-*N*-formyl-glycinamide in *Salmonella typhimurium. J. Biol. Chem.* 244:2095–2102.

2130. Whalen, W. A., and C. M. Berg. 1984. Gratuitous repression of *avtA* in *Escherichia coli* and *Salmonella typhimurium. J. Bacteriol.* 158:571–574.

2131. Whitehouse, J. M., and D. A. Smith. 1974. Methionine and vitamin B$_{12}$ repression and precursor induction in the regulation of homocysteine methylation in *Salmonella typhimurium. Mol. Gen. Genet.* 120:341–353.

2132. Whitehouse, J. M., and D. A. Smith. 1974. The involvement of methionine regulatory mutants in the suppression of B$_{12}$-dependent homocysteine transmethylase (*metH*) mutants of *Salmonella typhimurium. Mol. Gen. Genet.* 129:259–267.

2133. Whitfield, H. J., and G. Levine. 1973. Isolation and characterization of a mutant of *Salmonella typhimurium* deficient in a major deoxyribonucleic acid polymerase activity. *J. Bacteriol.* 116:54–58.

2134. Whitfield, H. J., Jr. 1971. Purification and properties of the wild type and a feedback resistant phosphoribosyladenosine triphosphate:pyrophosphate phosphoribosyltransferase. *J. Biol. Chem.* 246:899–909.

2135. Wiater, A., M. Filutowicz, and D. Hulanicka. 1982. A new class of mutants of the *cysB* regulatory gene for cysteine biosynthesis in *Salmonella typhimurium. J. Gen. Microbiol.* 128:1785–1790.

2136. Wiater, A., and D. Hulanicka. 1978. The regulatory *cysK* mutant of *Salmonella typhimurium. Acta Biochim. Pol.* 25:281–288.

2137. Wiater, A., and T. Klopotowski. 1972. Mutations rendering *Salmonella typhimurium* resistant to 3-amino triazole in the presence of histidine. *Acta Biochim. Pol.* 19:191–199.

2138. Widenhorn, K. A., W. Boos, J. M. Somers, and W. W. Kay. 1988. Cloning and properties of the *Salmonella typhimurium* tricarboxylate transport operon in *Escherichia coli. J. Bacteriol.* 170:883–888.

2139. Widenhorn, K. A., J. M. Somers, and W. W. Kay. 1988. Expression of the divergent tricarboxylate transport operon (*tctI*) of *Salmonella typhimurium. J. Bacteriol.* 170:3223–3227.

2140. Widenhorn, K. A., J. M. Somers, and W. W. Kay. 1989. Genetic regulation of the tricarboxylate transport operon (*tctI*) of *Salmonella typhimurium. J. Bacteriol.* 171:4436–4441.

2141. Wild, J., M. Filutowicz, and T. Klopotowski. 1978. Utilization of D-amino acids by *dadR* mutants of *Salmonella typhimurium. Arch. Microbiol.* 118:71–77.

2142. Wild, J., and T. Klopotowski. 1975. Insensitivity of D-amino acid dehydrogenase synthesis to catabolite repression in *dadR* mutants of *Salmonella typhimurium. Mol. Gen. Genet.* 136:63–73.

2143. Wild, J., W. Walczak, K. Krajewska-Grynkiewicz, and T. Klopotowski. 1974. D-Amino acid dehydrogenase. The enzyme of the first step of D-histidine and D-methionine racemization in *Salmonella typhimurium. Mol. Gen. Genet.* 128:131–146.

2144. Wilkinson, R. G., P. Gemski, Jr., and B. A. D. Stocker. 1972. Non-smooth mutants of *Salmonella typhimurium*: differentiation by phage sensitivity and genetic mapping. *J. Gen. Microbiol.* 70:527–554.

2145. Wilkinson, R. G., and B. A. D. Stocker. 1968. Genetics and cultural properties of mutants of *Salmonella typhimurium* lacking glucosyl or galactosyl lipopolysaccharide transferases. *Nature* (London) 217:955–957.

2146. Williams, J. C., C. E. Lee, and J. R. Wild. 1980. Genetics and biochemical characterization of distinct transport systems for uracil, uridine and cytidine in *Salmonella typhimurium. Mol. Gen. Genet.* 178:121–130.

2147. Williams, J. C., and G. A. O'Donovan. 1973. Repression of enzyme synthesis of the pyrimidine pathway in *Salmonella typhimurium. J. Bacteriol.* 115:1071–1076.

2148. Williams, R. C., and C. H. Clarke. 1974. The isolation and characterization of a mutant of *Salmonella typhimurium* defective in mutation frequency decline. *Mutat. Res.* 22:243–253.

2149. Willis, R. C., R. G. Morris, C. Cirakoglu, G. D. Schellenberg, N. H. Gerber, and C. E. Furlong. 1974. Preparation of the periplasmic binding proteins from *Salmonella typhimurium* and *Escherichia coli. Arch. Biochem. Biophys.* 161:64–75.

2150. Winfield, S. L., and J. O. Falkinham, III. 1981. Effect of *recA* and *polA* mutations on gene duplications in *Salmonella typhimurium. Mutat. Res.* 91:15–20.

2151. Wing, J. P. 1968. Transduction by phage P22 in a recombination-deficient mutant of *Salmonella typhimurium. Virology* 36:271–276.

2152. Wing, J. P., M. Levine, and H. O. Smith. 1968. Recombination-deficient mutant of *Salmonella typhimurium. J. Bacteriol.* 95:1828–1834.

2153. Wingfield, P., P. Graber, G. Turcatti, N. R. Movva, M. Pelletier, S. Craig, K. Rose, and C. G. Miller. 1989. Purification and characterization of a methionine-specific aminopeptidase from *Salmonella typhimurium. Eur. J. Biochem.* 180:23–32.

2154. Winkler, M. E., D. J. Roth, and P. E. Hartman. 1978. Promoter- and attenuator-related metabolic regulation of the *Salmonella typhimurium* histidine operon. *J. Bacteriol.* 133:830–843.

2155. Winkler, M. E., and C. Yanofsky. 1981. Pausing of RNA polymerase during *in vitro* transcription of the tryptophan operon leader region. *Biochemistry* 20:3738–3744.

2156. Winston, F., and D. Botstein. 1982. Control of lysogenization by phage P22. I. The P22 *cro* gene. *J. Mol. Biol.* 152:209–232.

2157. Winston, F., D. Botstein, and J. H. Miller. 1979. Characterization of amber and ochre suppressors in *Salmonella typhimurium. J. Bacteriol.* 137:433–439.

2158. Woehlke, G., K. Wifling, and P. Dimroth. 1992. Sequence of the sodium ion pump oxaloacetate decarboxylase from *Salmonella typhimurium. J. Biol. Chem.* 267:22798–22803.

2159. Wolf, R. E., Jr., and J. C. Loper. 1970. The differential inactivation of histidinol dehydrogenase from *Salmonella typhimurium* by sulfhydryl reagents. *J. Biol. Chem.* 244:6297–6303.

2160. Wollin, R., E. S. Creeger, L. I. Rothfield, B. A. D. Stocker, and A. A. Lindberg. 1983. *Salmonella typhimurium* mutants defective in UDP-D-galactose:lipopolysaccharide α 1,6-D-galactosyltransferase. *J. Biol. Chem.* 258:3769–3774.

2161. Wong, A., L. Kean, and R. Maurer. 1988. Sequence of the *dnaB* gene of *Salmonella typhimurium. J. Bacteriol.* 170:2668–2675.

2162. Wong, K. K., and M. McClelland. 1992. A *Bln*I restriction map of the *Salmonella typhimurium* LT2 genome. *J. Bacteriol.* 174:1656–1661.

2163. Wong, K. K., and M. McClelland. 1992. Dissection of the *Salmonella typhimurium* genome by use of a Tn*5* derivative carrying rare restriction sites. *J. Bacteriol.* 174:3807–3811.

2164. Wong, K. K., and M. McClelland. 1994. Stress-inducible gene of *Salmonella typhimurium* identified by arbitrarily primed PCR of RNA. *Proc. Natl. Acad. Sci. USA* 91:639–643.

2165. Wong, K. K., K. L. Suen, and H. S. Kwan. 1989. Transcription of *pfl* is regulated by anaerobiosis, catabolite repression, pyruvate, and *oxrA*: *pfl*::Mu dA operon fusions of *Salmonella typhimurium. J. Bacteriol.* 171:4900–4905.

2166. Wong, K. K., R. M. Wong, K. E. Rudd, and M. McClelland. 1994. High-resolution restriction map of a 240-kilobase region spanning 91 to 96 minutes on the *Salmonella typhimurium* LT2 chromosome. *J. Bacteriol.* 176:5729–5734.

2167. Woodgate, R., A. S. Levine, W. H. Koch, T. A. Cebula, and E. Eisenstadt. 1991. Induction and cleavage of *Salmonella typhimurium* UmuD protein. *Mol. Gen. Genet.* 229:81–85.

2168. Wray, C., W. J. Sojka, J. A. Morris, and W. J. B. Morgan. 1977. The immunization of mice and calves with *galE* mutants of *Salmonella typhimurium. J. Hyg.* 79:17–24.

2169. Wu, J. Y., L. M. Siegel, and N. M. Kredich. 1991. High-level expression of *Escherichia coli* NADPH-sulfite reductase: requirement for a cloned *cysG* plasmid to overcome limiting siroheme cofactor. *J. Bacteriol.* 173:325–333.

2170. Wu, T. T. 1966. A model for three-point analysis of random general transduction. *Genetics* 54:405–410.

2171. Wu, W. F., M. L. Urbanowski, and G. V. Stauffer. 1992. Role of the MetR regulatory system in vitamin B$_{12}$-mediated repression of the *Salmonella typhimurium* *metE* gene. *J. Bacteriol.* 174:4833–4837.

2172. Wu, W. F., M. L. Urbanowski, and G. V. Stauffer. 1993. MetJ-mediated regulation of the *Salmonella typhimurium* *metE* and *metR* genes occurs through a common operator region. *FEMS Microbiol. Lett.* 108:145–150.

2173. Wyche, J. H., B. Ely, T. A. Cebula, M. C. Snead, and P. E. Hartman. 1974. Histidyl-transfer ribonucleic acid synthetase in positive control of the histidine operon in *Salmonella typhimurium. J. Bacteriol.* 117:708–716.

2174. Wyche, J. H., J. Kennedy, Z. Hartman, P. E. Hartman, and J. Diven. 1974. Round-cell mutant of *Salmonella typhimurium. J. Bacteriol.* 120:965–969.

2175. Wyk, P., and P. Reeves. 1989. Identification and sequence of the gene for abequose synthase, which confers antigenic specificity on group B salmonellae: homology with galactose epimerase. *J. Bacteriol.* 171:5687–5693.

2176. Wylie, D., A. Stock, C.-Y. Wong, and J. Stock. 1988. Sensory transduction in bacterial chemotaxis involves phosphotransfer between *che* proteins. *Biochem. Biophys. Res. Commun.* 151:891–896.

2177. Xu, K., J. Delling, and T. Elliott. 1992. The genes required for heme synthesis in *Salmonella typhimurium* include those encoding alternative functions for aerobic and anaerobic coproporphyrinogen oxidation. *J. Bacteriol.* 174:3953–3963.

2178. **Xu, K., and T. Elliott.** 1993. An oxygen-dependent coproporphyrinogen oxidase encoded by the *hemF* gene of *Salmonella typhimurium. J. Bacteriol.* 175:4990–4999.

2179. **Xu, K., and T. Elliott.** 1994. Cloning, DNA sequence, and complementation analysis of the *Salmonella typhimurium hemN* gene encoding a putative oxygen-independent coproporphyrinogen III oxidase. *J. Bacteriol.* 176:3196–3203.

2180. **Yagil, E., and E. Hermoni.** 1976. Repression of alkaline phosphatase in *Salmonella typhimurium* carrying a *phoA+ phoR−* episome from *Escherichia coli. J. Bacteriol.* 128:661–664.

2181. **Yamada, M., A. Hakura, T. Sofuni, and T. Nohmi.** 1993. New method for gene disruption in *Salmonella typhimurium*: construction and characterization of an *ada*-deletion derivative of *Salmonella typhimurium* TA1535. *J. Bacteriol.* 175:5539–5547.

2182. **Yamada, T., and J. Davies.** 1971. A genetic and biochemical study of streptomycin and spectinomycin resistance in *Salmonella typhimurium. Mol. Gen. Genet.* 110:197–210.

2183. **Yamada, T., Y. Murooka, and T. Harada.** 1978. Comparative immunological studies of arylsulfatase in bacteria of the family *Enterobacteriaceae*: occurrence of arylsulfatase protein regulated by sulfur compounds and tyramine. *J. Bacteriol.* 133:536–541.

2184. **Yamagata, H., K. Nakamura, and M. Inouye.** 1980. Comparison of the lipoprotein gene among the *Enterobacteriaceae.* DNA sequence of *Erwinia amylovora* lipoprotein gene. *J. Biol. Chem.* 256:2194–2198.

2185. **Yamaguchi, S., S.-I. Aizawa, M. Kihara, M. Isomura, C. J. Jones, and R. M. Macnab.** 1986. Genetic evidence for a switching and energy-transducing complex in the flagellar motor of *Salmonella typhimurium. J. Bacteriol.* 168:1172–1179.

2186. **Yamaguchi, S., H. Fujita, A. Ishihara, S.-I. Aizawa, and R. M. Macnab.** 1986. Subdivision of flagellar genes of *Salmonella typhimurium* into regions responsible for assembly, rotation, and switching. *J. Bacteriol.* 166:187–193.

2187. **Yamaguchi, S., H. Fujita, T. Kuroiwa, and T. Iino.** 1977. Sensitivity of non-flagellate *Salmonella* mutants to the flagellotropic bacteriophage chi. *J. Gen. Microbiol.* 99:209–214.

2188. **Yamaguchi, S., H. Fujita, K. Sugata, T. Taira, and T. Iino.** 1984. Genetic analysis of H2, the structural gene for phase-2 flagellin in *Salmonella. J. Gen. Microbiol.* 130:255–265.

2189. **Yamaguchi, S., H. Fujita, T. Taira, K. Kutsukake, M. Homma, and T. Iino.** 1984. Genetic analysis of three additional *fla* genes in *Salmonella typhimurium. J. Gen. Microbiol.* 130:3339–3342.

2190. **Yamaguchi, S., and T. Iino.** 1969. Genetic determination of the antigenic specificity of flagellar protein in *Salmonella. J. Gen. Microbiol.* 55:59–74.

2191. **Yamaguchi, S., and T. Iino.** 1970. Serological and fingerprinting analyses of mutant flagella antigens in *Salmonella. J. Gen. Microbiol.* 64:311–318.

2192. **Yamaguchi, S., T. Iino, T. Horiguchi, and K. Ohta.** 1972. Genetic analysis of *fla* and *mot* cistrons closely linked to H1 in *Salmonella abortus-equi* and its derivatives. *J. Gen. Microbiol.* 70:59–75.

2193. **Yamamoto, K., and Y. Imae.** 1993. Cloning and characterization of the *Salmonella typhimurium*-specific chemoreceptor Tcp for taxis to citrate and from phenol. *Proc. Natl. Acad. Sci.* 90:217–221.

2194. **Yamamoto, N.** 1978. Somatic O1 antigen conversion of *Salmonella typhimurium* by a type B phage P221*dis* hybrid between P22 and FELS1 phages. *J. Gen. Virol.* 41:367–376.

2195. **Yamamoto, N., and M. L. Droffner.** 1985. Mechanisms determining aerobic or anaerobic growth in the facultative anaerobe *Salmonella typhimurium. Proc. Natl. Acad. Sci. USA* 82:2077–2081.

2196. **Yamashita, M. M., R. J. Almassy, C. A. Janson, D. Cascio, and D. Eisenberg.** 1989. Refined atomic model of glutamine synthetase at 3.5 Å resolution. *J. Biol. Chem.* 264:17681–17690.

2197. **Yan, Y., and M. Demerec.** 1965. Genetic analysis of pyrimidine mutants of *Salmonella typhimurium. Genetics* 52:643–651.

2198. **Yancey, R. J., S. A. L. Breeding, and C. E. Lankford.** 1979. Enterochelin (enterobactin): virulence factor for *Salmonella typhimurium. Infect. Immun.* 24:174–180.

2199. **Yang, H. J., B. Lee, and J. L. Haslam.** 1973. Studies on histidinol dehydrogenase. Preliminary crystallographic data. *J. Mol. Biol.* 81:517–519.

2200. **Yang, S.-L., J. M. Becker, and F. Naider.** 1977. Transport of [14C]Gly-Pro in a proline peptidase mutant of *Salmonella typhimurium. Biochim. Biophys. Acta* 471:135–144.

2201. **Yang, Y.-L., D. Goldrick, and J.-S. Hong.** 1988. Identification of the products and nucleotide sequences of two regulatory genes involved in the exogenous induction of phosphoglycerate transport in *Salmonella typhimurium. J. Bacteriol.* 170:4299–4303.

2202. **Yanofsky, C., S. S.-L. Li, V. Horn, and J. Rowe.** 1977. Structure and properties of a hybrid tryptophan synthetase α chain produced by genetic exchange between *Escherichia coli* and *Salmonella typhimurium. Proc. Natl. Acad. Sci. USA* 74:286–290.

2203. **Yanofsky, C., and M. Van Cleemput.** 1982. Nucleotide sequence of *trpE* of *Salmonella typhimurium* and its homology with the corresponding sequence of *Escherichia coli. J. Mol. Biol.* 154:235–246.

2204. **Yen, C., L. Green, and C. G. Miller.** 1980. Degradation of intracellular protein in *Salmonella typhimurium* peptidase mutants. *J. Mol. Biol.* 143:21–33.

2205. **Yen, C., L. Green, and C. G. Miller.** 1980. Peptide accumulation during growth of peptidase deficient mutants. *J. Mol. Biol.* 143:35–48.

2206. **Yoakum, G., and A. Eisenstark.** 1972. Toxicity of L-tryptophan photoproduct on recombinationless (*rec*) mutants of *Salmonella typhimurium. J. Bacteriol.* 112:653–655.

2207. **Yokoto, T., and J. S. Gots.** 1970. Requirement of adenosine-3′,5′-cyclic phosphate for flagella formation in *Escherichia coli* and *Salmonella typhimurium. J. Bacteriol.* 103:513–516.

2208. **Yoshioka, K., S.-I. Aizawa, and S. Yamaguchi.** 1995. Flagellar filament structure and cell motility of *Salmonella typhimurium* mutants lacking part of the outer domain of flagellin. *J. Bacteriol.* 177:1090–1093.

2209. **Young, B. G., and P. E. Hartman.** 1966. Sites of P22 and P221 prophage integration in *Salmonella typhimurium. Virology* 28:265–270.

2210. **Young, B. S., S. K. Guterman, and A. Wright.** 1976. Temperature-sensitive ribonucleic acid polymerase mutant of *Salmonella typhimurium* with a defect in the β′ subunit. *J. Bacteriol.* 127:1292–1297.

2211. **Young, B. S., and A. Wright.** 1977. Multiple effects of an RNA polymerase β-prime mutation on *in-vitro* transcription. *Mol. Gen. Genet.* 155:191–196.

2212. **Yourno, J.** 1968. Composition and subunit structure of histidinol dehydrogenase from *Salmonella typhimurium. J. Biol. Chem.* 243:3277–3288.

2213. **Yourno, J.** 1970. Nature of the compensating frameshift in the double frameshift mutant hisD3018 R5 of *Salmonella typhimurium. J. Mol. Biol.* 48:437–442.

2214. **Yourno, J.** 1972. Externally suppressible +1 "glycine" frameshift: possible quadruplet isomers for glycine and proline. *Nature* (London) *New Biol.* 239:219–221.

2215. **Yourno, J., and I. Ino.** 1968. Purification and crystallization for histidinol dehydrogenase from *Salmonella typhimurium. J. Biol. Chem.* 243:3273–3276.

2216. **Yourno, J., T. Kohno, and J. R. Roth.** 1970. Enzyme evolution: generation of a bifunctional enzyme by fusion of adjacent genes. *Nature* (London) 228:820–824.

2217. **Yu, C.-Q., and J.-S. Hong.** 1986. Identification and nucleotide sequence of the activator gene of the externally induced phosphoglycerate transport system of *Salmonella typhimurium. Gene* 45:51–57.

2218. **Yuasa, R., M. Levinthal, and H. Nikaido.** 1969. Biosynthesis of cell wall lipopolysaccharide in mutants of *Salmonella.* V. A mutant of *Salmonella typhimurium* defective in the synthesis of cytidine diphosphoabequose. *J. Bacteriol.* 100:433–444.

2219. **Yutani, K., T. Sato, K. Ogasahara, and E. W. Miles.** 1984. Comparison of denaturation of tryptophan synthase alpha-subunits from *Escherichia coli, Salmonella typhimurium,* and an interspecies hybrid. *Arch. Biochem. Biophys.* 229:448–454.

2220. **Zabel, D. J., M. Trucksis, and R. E. Depew.** 1980. *Salmonella typhimurium* mutants with reduced levels of transfer ribonucleic acid-inhibitable endodeoxyribonucleolytic activity. *J. Bacteriol.* 144:173–178.

2221. **Zahrt, T. C., G. C. Mora, and S. Maloy.** 1994. Inactivation of mismatch repair overcomes the barrier to transduction between *Salmonella typhimurium* and *Salmonella typhi. J. Bacteriol.* 176:1527–1529.

2222. **Zak, V. L., and R. A. Kelln.** 1978. 5-Fluoroorotate-resistant mutants of *Salmonella typhimurium. Can. J. Microbiol.* 24:1339–1345.

2223. **Zak, V. L., and R. A. Kelln.** 1981. *Salmonella typhimurium* mutant dependent upon carbamyl aspartate for resistance to 5-fluorouracil is specifically affected in ubiquinone biosynthesis. *J. Bacteriol.* 145:1095–1098.

2224. **Zarlengo, M. H., G. W. Robinson, and R. O. Burns.** 1968. Threonine deaminase from *Salmonella typhimurium.* II. The subunit structure. *J. Biol. Chem.* 243:186–191.

2225. **Zaror, I., I. Gomez, G. Castillo, A. Yudelevich, and A. Venegas.** 1988. Molecular cloning and expression in *E. coli* of a *Salmonella typhi* porin gene. *FEBS Lett.* 229:77–81.

2226. **Zhu, N., B. M. Olivera, and J. R. Roth.** 1988. Identification of a repressor gene involved in the regulation of NAD de novo biosynthesis in *Salmonella typhimurium. J. Bacteriol.* 170:117–125.

2227. **Zhu, N., B. M. Olivera, and J. R. Roth.** 1989. Genetic characterization of the *pnuC* gene, which encodes a component of the nicotinamide mononucleotide transport system in *Salmonella typhimurium. J. Bacteriol.* 171:4402–4409.

2228. **Zhu, N., B. M. Olivera, and J. R. Roth.** 1991. Activity of the nicotinamide mononucleotide transport system is regulated in *Salmonella typhimurium. J. Bacteriol.* 173:1311–1320.

2229. **Zhu, N., and J. R. Roth.** 1991. The *nadI* region of *Salmonella typhimurium* encodes a bifunctional regulatory protein. *J. Bacteriol.* 173:1302–1310.

2230. **Zhyvoloup, A. N., M. I. Woodmaska, I. V. Kroupskaya, and E. B. Paton.** 1990. Nucleotide sequence of the *rplJL* operon and the deduced primary structure of the encoded L10 and L7/L12 proteins of *Salmonella typhimurium* compared to that of *Escherichia coli. Nucleic Acids Res.* 18:4620.

2231. **Ziebell, R., F. P. Imray, and D. G. MacPhee.** 1977. DNA degradation in wild-type and repair-deficient strains of *Salmonella typhimurium* exposed to ultraviolet light or photodynamic treatment. *J. Gen. Microbiol.* 101:143–149.

2232. **Zieg, J., M. Hilmen, and M. Simon.** 1978. Regulation of gene expression by site specific inversion. *Cell* 15:237–244.

2233. **Zieg, J., M. Silverman, M. Hilmen, and M. Simon.** 1977. Recombinational switch for gene expression. *Science* 196:170–172.

2234. **Zieg, J., M. Silverman, M. Hilmen, and M. I. Simon.** 1978. The mechanisms of phase variation. *Cold Spring Harbor Monogr. Ser.* 1978:411–424.

2235. **Zieg, J., and M. Simon.** 1980. Analysis of the nucleotide sequence of an invertible controlling element. *Proc. Natl. Acad. Sci. USA* 77:4196–4200.

2236. **Zinder, N. D., and J. Lederberg.** 1952. Genetic exchange in *Salmonella. J. Bacteriol.* **64:**679–699.

2237. **Zukin, R. S., M. F. Klos, and R. E. Hirsch.** 1986. Conformational dynamics of two histidine-binding proteins of *Salmonella typhimurium. Biophys. J.* **49:**1229–1235.

2238. **Zukin, R. S., and D. E. Koshland, Jr.** 1976. Mg^{2+}, Ca^{2+}-dependent adenosine triphosphate as receptor for divalent cations in bacterial sensory receptor. *Proc. Natl. Acad. Sci. USA* **74:**1932–1936.

2239. **Zukin, R. S., P. G. Strange, L. R. Heavey, and D. E. Koshland, Jr.** 1977. Properties of galactose binding protein of *Salmonella typhimurium* and *Escherichia coli. Biochemistry* **16:**381–386.

2240. **Zyskind, J. W., L. T. Deen, and D. W. Smith.** 1979. Isolation and mapping of plasmids containing the *Salmonella typhimurium* origin of DNA replication. *Proc. Natl. Acad. Sci. USA* **76:**3097–3101.

2241. **Zyskind, J. W., and D. W. Smith.** 1980. Nucleotide sequence of the *Salmonella typhimurium* origin of replication. *Proc. Natl. Acad. Sci. USA* **77:**2460–2464.

Native Insertion Sequence Elements: Locations, Distributions, and Sequence Relationships

RICHARD C. DEONIER

111

INTRODUCTION

Insertion sequences (ISs) are transposable elements whose only genes are related to promotion and regulation of transposition. IS elements typically fall within the 700- to 2,000-bp size range, and they are normal constituents of many bacterial chromosomes and plasmids. IS elements were originally recognized by the phenotypes resulting from their insertion into other genes. An excellent overview of the earlier studies was presented by Starlinger and Saedler (55). The biology, structure, regulation, transposition mechanisms, and other attributes of many IS elements have been comprehensively reviewed by Galas and Chandler (18).

IS elements participate in a range of events. Transpositional processes include insertion, adjacent deletion (a consequence of intramolecular transposition in many cases), and transpositional inversion of adjacent genes. RecA-mediated processes such as inversion, cointegration, or deletion involving pairs of identical IS elements are a natural consequence of the presence of multiple copies of particular IS elements. Precise excision (removal of the element and one of the flanking direct repetitions) is unrelated to transposition. The multiplicity of transpositional and recombinational events associated with IS elements allows them to contribute to plasticity of bacterial chromosomes and plasmids in which they are found.

The magnitude and nature of influences by IS elements on a bacterial chromosome will depend upon the exact types of IS elements that are present. For example, *Salmonella typhimurium* (official designation, *Salmonella enterica* serovar Typhimurium) LT2 appears to contain only IS*200*, and this particular element transposes infrequently. In contrast, *Escherichia coli* K-12 contains 8 to 12 different types of IS element, and they contribute significantly to the collection of mutations that have been obtained for various genes. For example, in the *lacI* gene (excluding a hot spot for point mutation), two IS*1* insertions were detected in a mutant pool containing 24 point mutations (and other deletion or frameshift mutations) (16), and in a collection of 25 *c*I mutations in λ lysogens, 15 were caused by IS insertions (32). Because of their ability to induce mutations and genome rearrangements, the identification of IS elements and determination of their locations on the genome map are necessary for any global view of the genetics of a given bacterial type.

This chapter focuses on the distribution and sequence relationships of chromosomal IS elements in *E. coli* and *Salmonella* spp., particularly in the common laboratory bacteria *E. coli* K-12 and *S. typhimurium* LT2. Nine of the most significant of these elements are listed in Table 1, and representative structures are illustrated in Fig. 1. Similar elements are found in a broad spectrum of other bacteria, including archaebacteria. A detailed discussion of IS*1*, IS*10*, and IS*911* structures and transposition mechanisms is provided in chapter 124 of this volume.

MAPPING IS ELEMENTS

The IS elements that have been mapped in the *E. coli* K-12 and *S. typhimurium* LT2 chromosomes are listed in Table 1. Presence and absence of cleavage sites for several restriction endonucleases

TABLE 1 Insertion sequences from *E. coli* K-12 and *S. typhimurium* LT2 and their sensitivities to cleavage by selected diagnostic restriction endonucleases[a]

Element	Length (bp)	GenBank accession no.[b]	Cleavage by:						
			*Ava*I	*Bam*HI	*Bgl*II	*Bst*EII	*Eco*RI	*Hind*III	*Pst*I
IS*1*	768	V00609	−	−	−	+	−	−	+
IS*2*	1,327	V00610	+	−	−	−	−	+	−
IS*3*	1,258	X02311	−	−	−	+	−	+	+
IS*4*	1,426	J01733	+	−	−	−	−	−	+
IS*5*	1,195	J01734	−	−	+	+	+	−	−
IS*30*	1,221	X00792	−	−	+	−	−	+	−
IS*150*	1,443	X07037	−	−	+	+	−	+	−
IS*186*	1,338	X03123	+	+	−	−	−	−	+
IS*200*	707	X56834	−	−	−	−	+	+	−

[a]Note that not all IS elements of a given type in these bacteria need have identical sequences. For example, there are four different sequence types for IS*1* in the *E. coli* K-12 chromosome (61). Consequently, some variants of these elements may have an altered spectrum of sensitivity to the indicated restriction endonucleases.

[b]Typical type sequence. Additional accession numbers for individual chromosomal elements of each type may also be available.

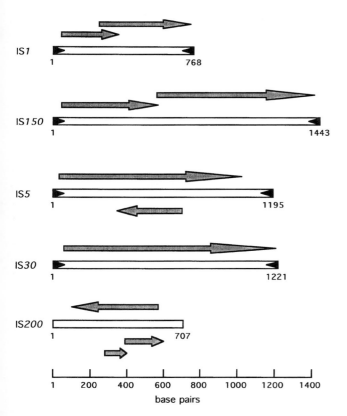

FIGURE 1 Structures of some typical IS elements from *E. coli* and *S. typhimurium*. Shaded arrows designate the most important ORFs, and filled triangles represent the terminal inverted repetitions. Terminal inverted repetitions may be imperfect, and other shorter ORFs that might be present have been omitted (see reference 18 for complete descriptions). IS*150* is a member of the IS*3* family, whose members include IS*3*, IS*2*, IS*150*, IS*911*, and other elements (51). IS*5* is a member of the IS*4* family, whose members include IS*4*, IS*5*, IS*186*, and other elements (45). Data for IS*200*, which lacks terminal inverted repetitions, were supplied by J. Casadesús (personal communication).

are indicated to serve as a guide for preliminary identification of these elements when new transpositions in these particular bacterial strains are identified. Isoforms for some of these elements are known, and these sequence variants may differ in restriction patterns from those given.

IS Elements in the *E. coli* K-12 Chromosome

Because IS elements can transpose, their status as genetic loci is less secure than that of conventional genes. This is emphasized by the absence of some types of elements from particular *E. coli* strains but not from others (49). Nevertheless, the positions of some IS elements of *E. coli* K-12 are sufficiently conserved that they appear in most commonly used contemporary isolates.

Earlier IS mapping experiments were directed at limited regions of the chromosome using heteroduplex or restriction mapping of F′ plasmids (20) or specialized restriction mapping approaches (see, for example, reference 29). More recent studies have mapped most of the chromosomal IS elements in the entire chromosome by screening lambda or cosmid libraries representing the entire genome (3, 4, 58–60). Unfortunately, the *E. coli* K-12 strain W3110 used to make the λ libraries (27) contained chromosomal rearrangements, and strain BHB2600 used to make the cosmid library had a pedigree that involved extended laboratory manipulation. Some of the mapped IS elements may be peculiar to one or the other of these strains, given the propensity of IS elements to transpose and to induce genome rearrangements (e.g., adjacent deletion). Strain MG1655 is considered to be a better representative of wild-type *E. coli* K-12 (43).

The set of IS elements that were present in the original *E. coli* K-12 isolate represents the starting point from which the complement of insertion sequences in any contemporary laboratory strain has evolved, and members of this original, "standard" set are likely to be present in them. During the past 50 years of experimentation with *E. coli* K-12, transpositions have occurred in many lineages, particularly during storage on nutrient agar stabs. Since chromosomes lacking transpositional alterations with respect to at least one IS type may not be available, the likely original sets of IS*1*, IS*2*, IS*3*, IS*5*, and IS*30* were estimated from

TABLE 2 Presence or absence of *Eco*RI fragments containing various IS elements in selected *E. coli* K-12 strains

| Element | Fragment size (kb) | Presence or absence in strain[a]: | | | | | | | | | Basic element[b] |
| | | K-12 (CGSC 5073) | W1485F⁻ | W3110 | | W6F– | χ59 | BHB2600 | χ101F⁻ | HB101 | |
				Reference 58	Reference 3						
IS1	31	1	0	0	0	0	0	0	0	0	
	26	0	1	1	1	0	1	1	1	1	IS1F
	22	1	0	0	0	0	1	1	0	0	
	20	1	1	1	1	1	1	1	1	1	IS1A
	18	0	0	0	0	0	0	1	0	0	
	15	0	0	0	0	0	0	1	0	0	
	15	?	?	1	1	?	?	1	?	1	IS1B
	14	0	0	0	0	0	0	1	0	0	
	14	0	0	0	0	0	0	1	1	0	
	14	1	0	1	1	0	0	1	0	1	IS1E[c]
	13.1	0	0	0	0	0	1	0	1	0	
	12.7	0	0	0	0	0	1	0	0	0	
	12.0	0	1	0	0	0	0	1	0	0	
	10.5	1	0	0	0	0	0	1	0	0	
	8.0	0	0	0	0	0	0	0	0	1	
	4.8	0	0	?	1	0	0	0	0	0	
	3.8	0	1	1	1	1	1	1	?	1	IS1C
	3.7	1	1	1	1	1	1	1	?	1	IS1D
IS2	39	0	0	1	0	0	0	0	0	0	
	25–30	0	0	0	1	1	0	0	0	0	
	20	1	1	1	1	1	1	1	1	1	IS2A
	17	0	0	0	0	1	0	0	1	0	
	12.5	0	1	1	1	1	1	1	1	1	IS2K
	12.0	0	?	1	1	1	0	0	0	0	
	11.0	1	1	1	1	1	1	1	1	1	IS2I
	9.8	0	0	1	1	0	0	0	0	0	
	9.1	0	0	0	1	0	0	0	0	0	
	8.7	1	1	1	1	1	1	1	1	1	IS2F
	8.2	0	0	0	0	0	0	1	1	1	
	7.9	0	0	1	1	0	0	0	0	0	
	7.0	0	0	0	0	1	0	0	0	0	
	6.3	1	?	1	1	1	?	0	?	0	IS2J
	5.5	1	0	0	1	0	0	1	0	1	
	4.9	0	0	1	0	0	0	0	0	0	
	4.0	0	1	1	1	1	1	1	1	1	IS2E
	3.0	1	1	?	1	1	1	1	1	1	IS2D
	2.9	0	0	0	1	0	0	0	0	0	
IS3	18	—	1	1	1	1	0	0	1	1	IS3A
	17	—	0	0	1	0	0	0	0	0	
	13.5	—	1	1	1	1	1	1	1	0	IS3C
	12.5	—	0	0	0	0	0	1	0	0	
	12.3	—	1	1	1	1	1	1	1	0	IS3D
	12.0	—	0	0	0	0	0	1	0	0	
	11.8	—	1	1	1	1	1	1	1	1	IS3B
	10.5	—	0	0	0	0	0	0	0	1	
	8.7	—	0	0	0	0	0	0	0	1	
	8.0	—	0	0	0	1	0	0	0	0	
	5.9	—	1	1	1	0	1	1	1	1	IS3E
	1.9	—	0	0	0	0	1	1	0	0	
IS5	24	—	—	1	1	—	—	1	—	1	IS5T[d]
	20	—	—	1	0	—	—	0	—	1	
	13.5	—	—	1	1	—	—	1	—	0	IS5B
	12.9	—	—	0	1	—	—	1	—	1	"IS5Y"
	11.4	—	—	1	1	—	—	1	—	1	IS5H
	11.0	—	—	1	0	—	—	0	—	1	
	10.5	—	—	1	1	—	—	1	—	1	IS5LO
	8.9	—	—	1	1	—	—	1	—	1	IS5D
	8.0	—	—	1	0	—	—	0	—	0	
	7.7	—	—	0	1	—	—	0	—	0	
	7.6	—	—	0	1	—	—	0	—	0	
	7.4	—	—	1	1	—	—	1	—	1	IS5K
	5.9	—	—	1	1	—	—	1	—	1	IS5R
	5.9	—	—	1	0	—	—	0	—	0	
	5.7	—	—	1	1	—	—	1	—	1	IS5A
	5.5	—	—	0	1	—	—	0	—	0	
	5.1	—	—	1	1	—	—	0	—	0	

TABLE 2 *Continued*

Element	Fragment size (kb)	K-12 (CGSC 5073)	W1485F⁻	W3110 Reference 58	W3110 Reference 3	W6F–	χ59	BHB2600	χ101F⁻	HB101	Basic element[b]
	5.0	—	—	0	0	—	—	1	—	0	
	4.9	—	—	1	0	—	—	0	—	1	
	4.6	—	—	0	0	—	—	0	—	0	
	4.0	—	—	0	1	—	—	1	—	1	"IS5X"
	3.8	—	—	0	1	—	—	0	—	0	
	3.8	—	—	1	1	—	—	0	—	0	
	3.2	—	—	0	0	—	—	0	—	1	
	2.9	—	—	1	1	—	—	1	—	1	IS5E
	2.7	—	—	1	0	—	—	0	—	1	
	1.8	—	—	0	1	—	—	0	—	1	
	1.6	—	—	0	0	—	—	0	—	1	
	1.3	—	—	0	1	—	—	0	—	0	
	1.2	—	—	1	1	—	—	0	—	0	
	1.1	—	—	1	1	—	—	0	—	1	

[a]1, presence; 0, absence; ?, presence uncertain because of hybridization from a vector containing an IS probe; —, data not available. Data for *Eco*RI fragment sizes derived from each strain are taken directly from references 3, 4, 15, 23, and 57 or from maps of cloned fragments (27) for data from references 58–60.
[b]Basic elements are those inferred by parsimony to probably have been present in the original isolate of *E. coli* K-12. Names enclosed in quotation marks are provisional.
[c]May or may not have been ancestral; parsimony unable to discriminate between these alternatives.
[d]The 31-kb fragment from W3110 (58) was considered to be equivalent to the 24-kb fragments from the other strains, based on map location and larger experimental error in this size range.

the contemporary sets of nine laboratory strains by using parsimony arguments and the known pedigrees.

The following procedure was used. Sizes of *Eco*RI fragments containing these elements have been reported (or can be inferred) for a number of *E. coli* K-12 strains (3, 4, 23, 57–60). (Other published data for *E. coli* K-12 restriction fragments containing IS elements were not used because accurate sizes had not been presented.) The data for IS*1*, IS*2*, IS*3*, and IS*5* are summarized in Table 2. If fragments hybridizing to an IS had identical sizes (within estimated experimental error) for two or more different strains, the fragments were assumed to be identical. This assumption was not made if fragments of apparently identical size appeared only in strains widely separated by pedigree. For example, if all strains contained a 20-kb *Eco*RI fragment hybridizing to IS*2*, and if a 20-kb fragment mapping at 8.3 min had been shown to contain IS*2A*, then all strains were assumed to contain IS*2A*. Undoubtedly there are particular strains for which this assumption is wrong, but this would require (i) that IS*2A* be lost and (ii) that another IS*2* transpose to generate a fragment of identical size by insertion or adjacent deletion.

Given the assumptions stated above, the presence or absence of IS elements in the collection of nine strains was scored for IS*1* to IS*3*, and four strains were scored for the presence of IS*5* (Table 2; data for IS*30* not shown). The phylogenetic relationships of these strains are known (Fig. 2), thanks to committed and meticulous detective work by Barbara J. Bachmann (chapter 133; see the legend to Fig. 2 for a discussion regarding nodes of the tree). The phylogenetic tree was used together with the complement of IS elements in the individual strains to infer the most parsimonious set of IS elements present in the original *E. coli* K-12. The scoring at each node in Fig. 2 is shown for a 5.5-kb

*Eco*RI fragment containing IS*2* (Table 2) and found in all three strains studied by Birkenbihl and Vielmetter (3) and in *E. coli* K-12 (*E. coli* Genetic Stock Center strain CGSC 5073 [23]). The assigned score is 1 if the element is inferred to be present in a strain or at a node, 0 if the element is inferred to be absent, or 1,0 if a decision cannot be made. This 5.5-kb fragment was detected in four of the nine strains, but by parsimony arguments, it was considered not to have been present in the original *E. coli* K-12. Assuming that the excision frequency is not greater than the insertion frequency (see below), the most parsimonious explanation for the observed pattern of strains having this fragment invokes four transposition events: insertions in one of the W3110 strains, in BHB2600, in HB101, and in CGSC 5073. If there had been an IS*2* in an identical 5.5-kb fragment of the original *E. coli* K-12, then five excisions would have been required to generate the observed pattern. In some cases, such as IS*2B*, one cannot determine whether the element was ancestral from the set of strains in Fig. 2 and the data in Table 2.

Figure 3 shows the map of the elements inferred to have been present in the original *E. coli* K-12; these elements are listed by name and map location in Table 3. In most cases, map locations were taken from EcoMap6 (K. E. Rudd, personal communication). The possible significance of the apparently uneven distribution is discussed in a later section. Some IS elements mapped in previous studies (3, 58, 59) were not placed on this map because they were judged not to be ancestral. Designation of the basic set for IS*5* was the most problematic because data for fewer strains were directly accessible. In addition, the apparent number of IS*5* transposition events in three of the four strains analyzed was larger than for any of the other elements. Also, W3110 had been the victim of amplification probably caused by IS*5* × IS*5* reciprocal recombination (59). Because the starting

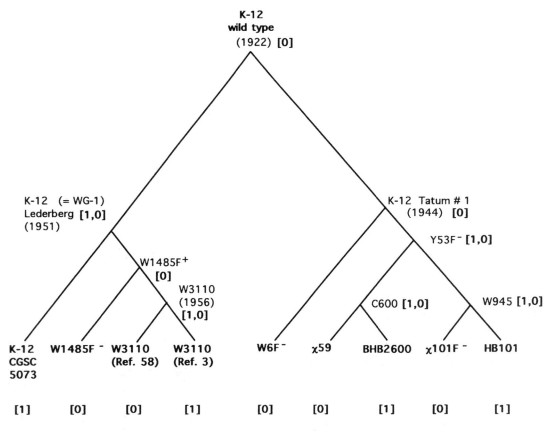

FIGURE 2 Abbreviated phylogenetic tree of some laboratory strains of *E. coli* K-12. Many intermediate steps have been omitted for clarity. Some steps (e.g., K-12 wild type [1922] → WG-1 [1951]) involved no intentional mutagenic steps: the strains either were propagated in the laboratory or were stored. Some nodes on this tree (e.g., for W1485) correspond to times when samples of the same strain were taken from storage for further study or dissemination. The complement of IS elements in W3110 was independently studied by two groups (3, 58). The separate W3110 isolates have diverged in their IS complements since W3110 was first isolated in 1956. Strain χ59 is the laboratory stock of C600 from the laboratory of Roy Curtiss III. HB101 has undergone a complicated construction history (B. Bachmann, personal communication). Its placement on this tree is conjectural, based on recipient markers retained after several conjugational steps. The maximum parsimony argument is illustrated for an IS2 found on a 5.5-kb *Eco*RI fragment in four of the strains by scoring each strain at the bottom (0 if the element is absent and 1 if the element is present). Moving up to higher and higher nodes, scores are 0 if the element is absent in paired branches, 1 if the element is present in paired branches, and 1,0 if its presence is neither supported nor rejected.

point was a "restored" version of the W3110 chromosome implicit in EcoMap6, some IS5 elements mapped in W3110 are excluded. Most of the IS5 elements represented as indeterminate in Table 3 are unlikely to be members of the standard set, because other analyses using strains from different branches of the tree are more consistent with 10 to 11 copies per chromosome (50, 56). This number is consistent with data for strains BHB2600 (3), JE5519 (59), and JE5527 (40).

Additional IS Elements in the *E. coli* K-12 Chromosome

E. coli K-12 carries at least four other types of IS elements, some of whose locations have not yet been determined: IS421 (47), IS600 (38), IS629 (36), and IS911 (44). The latter three elements were first isolated from *Shigella* species. Note that by electrophoretic enzyme analysis, DNA sequence similarity, and other criteria, *Shigella* species are closely related to *E. coli* (see reference 52 and chapter 148).

IS421 is a member of the IS4 family (see below), and four copies are present on chromosomes of some *E. coli* K-12 strains. IS600, IS629, and IS911 are all members of the IS3 family. IS600 and IS629 were first detected in *Shigella sonnei*, and they each hybridize weakly to one chromosomal fragment from *E. coli* K-12 strain JM109, suggesting the presence of fragments or isoforms of these elements in *E. coli* K-12 (37). The sequence of IS629 is closely related to that of IS3411 (25), whose presence was not detected in another *E. coli* K-12 strain. This apparent disagreement may reflect different hybridization stringencies.

IS911 was isolated and identified after it had transposed into bacteriophage λ which was integrated in the *Shigella dysenteriae* chromosome (44). Prère et al. found in chromosomal DNA from an *E. coli* K-12 strain C600 derivative four *Bgl*II-*Pst*I restriction fragments that hybridized to an IS911 probe (44). Subsequent mapping has shown that these fragments are derived from two deleted IS911 elements. One contains IS30A inserted

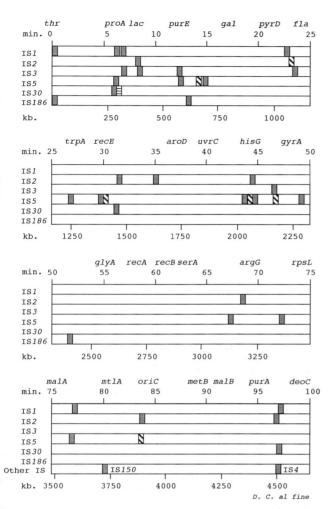

FIGURE 3 Locations of standard IS elements in the chromosome of *E. coli* K-12. Coordinates are physical map minutes (above the position fields) or kilobases (below the position fields). Precise designations, locations, and orientations are presented in Table 3. Shaded boxes represent the standard sets, diagonally hatched boxes represent elements whose status was not resolved by parsimony arguments, and the horizontally hatched box represents a fragment of IS*30*. IS elements appearing as single copies appeared only in the last quadrant of the map (Other IS).

at IS*911* bp 334 (with IS*911* sequences beyond bp 762 deleted); the other contains IS*30*D and an adjacent 327 bp of IS*600* substituting for IS*911* sequences between bp 335 and 454 (M. F. Prère, M. Chandler, and O. Fayet, personal communication). In other words, the IS*911* sequences in the *E. coli* K-12 chromosome map coincidentally with IS*30*A and IS*30*D (the four hybridizing chromosomal fragments arose because the IS*30* elements contain a *Bgl*II site). The fragmentary nature of the IS*911* sequences explains why no IS*911* insertions were detected during the initial use of λ as an IS trap in *E. coli* K-12 (32). The transposition properties of IS*911* are discussed in chapter 124.

IS*200* in the Chromosome of *S. typhimurium* LT2

The IS map of *S. typhimurium* LT2 is simpler because this bacterium appears to contain only IS*200* from among the elements described in this chapter, and because IS*200* does not transpose as frequently as the *E. coli* K-12 elements (7). The six IS*200* elements in the *S. typhimurium* LT2 chromosome were initially mapped by identifying chromosomal Tn*10* insertions that affected mobilities of restriction fragments that contain them (29). Subsequent refinement of this mapping has been achieved by using hybridization of IS*200* to chromosomal *Bln*I and *Xba*I fragments separated by pulsed-field gel electrophoresis and P22-mediated cotransduction frequencies (46).

The locations of the IS*200* elements in *S. typhimurium* LT2 are as follows: IS*200*(I), 65.2 min, clockwise of *cysC*; IS*200*(II), 74.5 min, clockwise of *crp*; IS*200*(III), 94.4 min, counterclockwise of *purA*; IS*200*(IV), 42.5 min, clockwise of *cysB*; IS*200*(V), 53.6 min, clockwise of *ptsG*; and IS*200*(VI), 22.2 min, clockwise of *galT*. IS*200* elements obviously are relatively widely spaced around the *S. typhimurium* LT2 chromosome.

Variations in IS Copy Numbers for *E. coli* K-12 Strains

Patterns of restriction fragments containing IS elements can be affected by genome rearrangements, by mutations that remove or create restriction sites, and by transposition. *E. coli* K-12 laboratory strains exhibit examples of all of these types of alterations.

Perkins et al. (43) have documented and correlated changes in the *Xba*I, *Bln*I, *Not*I, and *Sfi*I restriction maps of *E. coli* K-12 strains AB1157, EMG2 (one representative of supposedly wild-type *E. coli* K-12), MG1655, W1485, and W3110. Even though MG1655 was derived directly from W1485, the MG1655 chromosome contained deletions of 8, 7, 6, 1, and 1 kb and insertions of 12 and 7 kb compared with W1485. One of the deletions removed IS*5*G, which was not included as a standard element in the present compilation (Table 3). Exact correlations of differences in restriction maps of MG1655 and W1485 with IS elements would require mapping with additional enzymes and Southern blotting. It can be seen, however, that the insertion/deletion regions would potentially affect only the following IS elements (standard or nonstandard as defined above, locations given in kilobase coordinates and minutes): IS*2*C (not standard), 1,305 kb, 27.4 min; IS*5*K (standard), 2,299 kb, 46.6 min; IS*5*G (not standard), 2,331 kb, 47.5 min; IS*2*H (not standard), 3,011 kb, 61.5 min; IS*5*LO (standard), 3,147 kb, 64.4 min; and IS*2*I (standard), 3,203 kb, 66.5 min.

Most of the variation in IS element distribution appears to arise from transposition, as witnessed by increases in IS copy numbers. One hundred eighteen clones isolated from W3110 after storage for 30 years in a nutrient agar stab revealed extensive polymorphism in sizes and numbers of restriction fragments containing IS*2*, IS*3*, IS*5*, and IS*30* but little variation in fragments containing IS*1*, IS*4*, IS*150*, and IS*186* (39). A phylogenetic tree for the 118 clones was constructed, and it was estimated that at least 174 IS-associated mutations had occurred during storage. The restriction pattern for IS*1* was strictly conserved in all strains, but 31 new fragments containing IS*5* were detected, indicating that IS*5* was more transpositionally active during storage than were other elements.

The results for the laboratory strains described in Table 2 are consistent with these observations. The numbers of ancestral IS copies estimated by Naas et al. (39) are in reasonable agreement with the numbers of standard elements reported in Table 3, differing by at most one for any element type. The total changes

TABLE 3 Inferred standard sets of selected IS elements[a] in the *E. coli* K-12 chromosome

Element	Standard element[b]							
IS*1*	**A**	**B**	**C**	D	E	F		
	0.4	6.2	6.42	2.6	77.1	97.3		
	←	←	←	→	→	←		
IS*2*	**A**	**D**	E	**F**	**I**	**J**	K	(B)
	8.3	31.6	35.5	44.5	68.6	83.8	96.9	23.8
	←	→	?	→	←	←	→	→
IS*3*	**A**	**B**	**C**	**D**	E			
	6.9	8.5	12.3	23.6	46.6			
	→	←	→	←	→			
IS*5*[c]	**A**	**B**	**D**	E	Y	X	H	K	LO
	6.1	12.4	14.9	26.8	30.0	43.8	44.5	49.2	67.4
	←	←	←	←	←	←	←	←	→
	R	**T**		(C)	(F)	(I)	(J)	(U)	
	72.4	78.6		14.2	30.1	45.2	46.7	83.7	
	←	←		→	→	←	→	←	
IS*30*	**A**	**B**[d]	**C**	**D**				
	6.0	6.2	31.6	97.1				
	→	→	→	←				
IS*186*	A	B	C					
	0.3	13.2	51.8					
	→	→	→					

[a]IS*4* (one copy, 97.0, →) and IS*150* (one copy, 80.3, → [54]) are also present. Unmapped elements are IS*911*, IS*600*, and IS*629* elements or isoforms.

[b]Map positions are in physical map minutes and were taken from EcoMap6. Symbols → and ← indicate clockwise and counterclockwise orientations of the elements relative to the genetic map, respectively. The direction of the arrow on each element corresponds to the transcriptional direction of ORFs for the known or putative transposases. Elements listed in boldface were mapped prior to 1987 and have been confirmed by subsequent studies; see reference 13 for primary references, procedures, and rationales. Data for IS*5*T are taken from reference 54. Parentheses indicate that the status of the element after parsimony analysis is indeterminate.

[c]Copy number in C600 was 10 to 11 (50), copy number in JE5527 was approximately 10 (40), and copy number in W1485 was estimated at 11 (56). These results suggest that the upper limit for the number of ancestral (standard) IS*5* elements is 11.

[d]Only 181 bp of IS*30* are present at this site (60).

in copy numbers for IS*1*, IS*2*, IS*3*, and IS*5* from the presumed original copy numbers in *E. coli* K-12 are presented in Table 4 for strains at the bottom of the pedigree chart in Fig. 2. Although the number of generations along each branch differs, the number of events for each element type can be compared along any chosen pathway. Note that the number of generations of active growth and the length of time in storage will differ for each branch of the tree.

Along the pathway from C600 to BHB2600, there were five IS*1* transpositions and one IS*2* transposition, corresponding to

0.8 IS*1* transposition per ancestral IS*1* and to 0.14 IS*2* transposition per ancestral IS*2*. In contrast, the lineage from the early W1485 isolate to W3110 (3) experienced one IS*1* transposition (0.17 transposition per ancestral IS*1*) and seven IS*2* transpositions (1.0 transposition per ancestral IS*2*). In each lineage, the number of generations during which IS*1* and IS*2* could have transposed was the same, yet in the C600 → BHB2600 lineage, IS*1* appeared to transpose five times more frequently than IS*2*, while in the early W1485 → W3110 (3) lineage (again having the

TABLE 4 Inferred accumulated transposition events for different IS elements in various *E. coli* K-12 sublines

Element	No. of transposition events in strain:							
	W1485F⁻	W3110 Reference 58	W3110 Reference 3	W6F⁻	χ59	BHB2600	χ101	HB101
IS*1*	1	1	1	0	3	7	1	2
IS*2*	1	5	7	5	0	2	2	2
IS*3*	0	1	1	1	1	3	0	2
IS*5*	—[a]	9	10	—	—	1	—	8
IS*30*	0	1 or 2[b]	2 + 1[c]	—	—	1[c]	—	1[c]

[a] —, not tested.

[b] Two elements identified but not mapped (60).

[c] Hybridization intensity lower than that expected for a single element.

TABLE 5 Copy numbers of selected insertion sequences among *E. coli* and *S. typhimurium* isolates

Element	Standard copy no.[a]		ECOR set			SARA set, fraction of strains positive
	E. coli K-12	*S. typhimurium* LT2	Fraction of strains positive	Copy no. range	Avg copy no. (of positives)[b]	
IS1	6	0	60/71	0–27	7.5	9/66[c]
IS2	7	0	43/71	0–17	4.5	0/10[d]
IS3	5	0[d]	48/71	0–6	2.5	31/40
IS4	1	0	28/71	0–14	5.1	0[d]
IS5	11	0[d]	25/71	0–21	3.4	~18/40[d]
IS30	3	0[d]	35/71	0–5	1.9	0/40[d]
IS150	1	?	30/71	0–6	1.8	?
IS186	3	?	?	?	?	?
IS200	0	6	14/72	0–11	3.3	47/66[c]

[a] *E. coli* K-12 copy numbers for IS1, IS2, IS3, and IS5 determined by parsimony; *S. typhimurium* LT2 numbers from reference 28 except as otherwise noted.
[b] Excluding strains with copy number of 0.
[c] For IS1, the number range is 0 to 5, with an average number of 3.1 among those strains having IS1. For IS200, the number range is 0 to 19, with an average of 6.5 for those containing IS200.
[d] M. Biserčić and H. Ochman, personal communication.

same number of generations allowed for each element), IS2 appeared to transpose six times more frequently than IS1.

The difference in apparent IS activity may partly be attributed to periods of active growth for these laboratory strains, during which IS1 can contribute significantly to the IS mutational spectrum (32). This is in sharp contrast to the behavior of IS1 during long-term storage of W3110 in stabs (39). Interestingly, the W3110 lineages in Table 4 show IS2 (5 and 7 transposition events) and IS5 (9 and 10 transposition events) to have been among the most transpositionally active IS elements, as was observed by Naas et al. (39). Moreover, IS30 did not appear to be transpositionally active in these strains, as was the case for a large subset of the W3110 clones isolated after long-term storage. Strains BHB2600 and HB101 (Table 5) do not display the same spectrum of IS transpositions as W3110, perhaps reflecting their different histories.

DISTRIBUTIONS OF IS ELEMENTS IN *E. COLI* AND *SALMONELLA* STRAINS

Earlier studies had focused primarily on IS elements in *E. coli* K-12 and derivatives or in *S. typhimurium* LT2 (see reference 13 for review). There were some published reports that had included a broader range of bacteria (41). A more comprehensive view has emerged with studies of the *E. coli* ECOR and the *Salmonella* SARA collections. Chosen to be more representative of the natural diversity of *E. coli*, members of the ECOR set have been characterized by electrophoretic typing at a number of chromosomal loci, and this allowed construction of phylogenies for members of the set (52). The SARA strains (1) are more closely related to each other than are members of the ECOR set and do not represent the diversity of *Salmonella* spp. in the wild.

IS Elements in the ECOR Strains

Among the *E. coli* ECOR strains, presence or absence of IS1 to IS5, IS30, and IS150 has been determined (19, 21, 49). Results are summarized in Table 3. IS2 to IS5 or IS30 is absent in 40 to 60% of the strains. In contrast, nearly 90% of the strains contain IS1. Three of the ECOR strains (strains 47, 48, and 68) contained no

chromosomal copies of any of these elements. Fourteen of the ECOR strains contained IS200, originally identified in *S. typhimurium* (6, 28).

By using the observed distributions of copy numbers among the ECOR strains, Sawyer et al. (49) tested various mathematical models which differed in the functional dependence of transposition and cell death on IS element copy number n. More than one model could account for the observed distributions for IS2, IS4, IS5, and IS30. In contrast, IS3 and IS150 distributions were most consistent with models implying strong regulation of transposition (e.g., transposition rates proportional to $1/n$). IS1 distributions could be described by a model in which IS1 transposition was unregulated. The results for the laboratory strains presented in Table 2, though representing a much smaller sample, are consistent with this result. Over that portion of the tree not containing *E. coli* K-12 (CGSC 5073), IS3 has transposed 6 times (1.1 per element), whereas IS1, for example, has transposed 13 times (2.1 per element).

The IS distributions also provide information on how IS elements are acquired. The appearance of unrelated IS elements together on chromosomes more frequently than would have been expected for random distributions suggested that IS elements are transmitted horizontally by bacteriophages or plasmids (22). This is also suggested by the distribution of individual elements. For example, IS200 is distributed sporadically among many branches of the ECOR phylogenetic tree (6), as would be expected if it were disseminated horizontally. In contrast to IS200, most (21 of 25) of one major phylogenetic division (the A group) contain chromosomal IS150 elements, suggesting that most of the members of this group inherited IS150 vertically from a common ancestor. However, sparse and sporadic appearances of IS150 in sister groups to group A suggest that IS150 was not present at the root of the *E. coli* phylogenetic tree.

IS elements are found on plasmids as well as chromosomes in the ECOR collection. Approximately half of the 72 ECOR strains possessed one or more IS1 to IS5 or IS30 elements on plasmids, whereas over 90% carried one or more of these elements on their chromosomes. IS150 distributions differed, showing more frequent association with plasmids (21). Clearly, the presence of these elements on plasmids, which may be conjugative or mobilizable, is consistent with horizontal transmission of IS elements to chromosomes.

IS Elements in the SARA Strains

Because of the nature of the SARA collection (see above), the proportions of strains containing various elements are unlikely to apply to other populations. Nevertheless, they give examples of the magnitudes of proportions that are likely to be encountered.

IS200, which was first identified in *S. typhimurium* LT2 (28), appears in the majority of the SARA strains (47 of 66 tested strains [5]). IS1 and IS3 were also detected in a 40-member subset of the SARA collection (6), contrary to previous observations for a more restricted collection of strains (28). IS3 was found in 31 of 40 strains, and in 4 of 40 strains, it appeared together with both IS200 and IS1. IS200 was more frequently encountered (24 of 40 strains) than was IS1 (13 of 40 strains); moreover, the IS200 average copy number among strains possessing the element (6.5) was larger than that for IS1 (3.1). Given the infrequent transposition of IS200, this finding suggests a longer, more stable association for IS200 with *Salmonella* chromosomes than for IS1. This point has been addressed more rigorously from studies of the sequences of these elements, as will be described below. The sporadic distribution of IS1 among branches of the SARA phylogenetic tree suggests that these elements have been transmitted horizontally, an observation consistent with the relatively frequent association of IS1 with plasmids. In contrast, IS200 is rarely found on plasmids.

In addition to IS1, IS3, and IS200, some members of the SARA collection also carry IS5 and IS30 (H. Ochman, personal communication). However, IS2 and IS4 have not been found in this group of bacteria.

IS ELEMENT SEQUENCE DIVERGENCE

The various IS elements of *E. coli* and other bacteria show sequence similarities indicating that different families of IS elements exist. For example, the IS3 family includes the *E. coli* elements IS3, IS2, IS150, and IS3411, as well as *Shigella* elements IS600, IS629, and IS911 (8, 44, 51). The latter three elements (or fragments or isoforms of them) are present in the *E. coli* K-12 chromosome (see above). The IS3 family, which includes elements from organisms not included among the *Enterobacteriaceae*, is characterized by sequence similarities among the putative transposases and by the presence of two tandem open reading frames (ORF A and ORF B), with the downstream ORF B in the −1 frame relative to ORF A. Translational frameshifting has been demonstrated for some of these elements, and other members of this family contain sequences capable of promoting frameshifting (see chapter 124 for more details). A phylogenetic tree relating to some of these elements is presented in Fig. 4.

Another grouping of IS elements is the IS4 family, which includes *E. coli* elements IS4, IS5, IS186, and IS421, in addition to members from other genera (45). The ORF B regions of transposases from both the IS3 and IS4 families show sequence similarities to a portion of the integrase genes of retroviruses (17). The significance of this observation for possible origins of prokaryotic IS elements has not yet been fully explored.

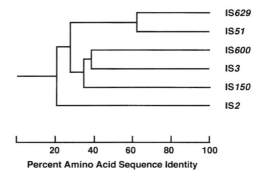

FIGURE 4 Phylogenetic relationships of members of the IS3 family. The organization of principal genetic features of these elements (e.g., IS150) is illustrated in Fig. 1. The relationships were inferred from predicted amino acid sequences of the second (ORF B) reading frames (redrawn from reference 51). IS911 is also a member of this group, and its ORF B shows 34% amino acid identity to IS600 ORF B and 28% identity to IS150 ORF B (44).

Sequence variations among individual members of particular IS types also provide clues about the time and mode of their acquisition by bacteria. For example, predicted protein sequences from IS3 elements of *E. coli*, *Shigella dysenteriae*, *Escherichia fergusonii*, and *Shigella odorifera* differed from each other to approximately the same extent that chromosomal genes from these bacteria differed, and these differences correlated with the phylogenies inferred from chromosomal gene sequences (30). This indicates that IS3 may have been associated with the common ancestor of these bacteria. Since only 23 members of the 72-member ECOR collection do not carry IS3, and since these strains are distributed among separated branches of the phylogenetic tree, mechanisms for removal or "curing" of IS3 appear to have operated in some of the strains since they diverged from the common ancestor.

In contrast to IS3, divergence of IS1 sequences does not parallel the divergence of the chromosomal genes in organisms from which they are isolated (30). This indicates that some of the IS1 elements were evolving independently of the chromosomes in which they are found today, which implies that they have been acquired by horizontal transfer. In the ECOR collection, bacteria have acquired either IS1F or IS1R versions of IS1, but not both, again emphasizing the relative recency and independence of IS1 (30). *E. coli* K-12 is an exception: it contains both IS1F and IS1R versions (61). DNA sequences of IS1 and IS3 elements from various isolates of the ECOR collection usually show little sequence variation, ranging from identity to the *E. coli* K-12 prototype sequences up to one to two substitutions in most cases.

IS200, like IS3, appears to have been present in *S. typhimurium* and *E. coli* since before these species diverged (5). *E. coli* IS200 copies differ from those of *S. typhimurium* by approximately 7%, which is similar to the differences between chromosomal genes from these two organisms. Among ECOR strains tested, sequence divergence among IS200 examples is approximately 3% (5), which is similar to the extent of divergence of chromosomal genes within the ECOR collection.

General recombination also appears to play a role in the evolution of IS elements. A particular IS3 from ECOR strain 63 contained a 107-bp sequence block in ORF A that was only 63% identical to the corresponding IS3 region from *E. coli* K-12, and comparisons of IS3 sequences from *E. coli* K-12, ECOR strain 63, and *Shigella dysenteriae* suggested that the IS3 from ECOR strain 63 was a recombinational composite (30). The same study showed that the middle one-third of IS1 elements isolated from *E. coli*, *E. fergusonii*, *E. hermanii*, and *E. vulneris* shows a lesser percentage of nucleotide sequence identity than do the flanking regions, again indicating that different portions of IS elements may have experienced different evolutionary histories.

INFLUENCES OF IS ELEMENTS

Map Locations

Do the locations and distributions of IS elements on the chromosomes reflect intrinsic properties of various chromosomal regions? Are the apparent clusters and gaps historical accidents, as might occur if an initially random IS insertion were to make more likely interactions of other IS elements with the same regions (e.g., by cointegration with plasmids containing a variety of IS elements)? Might a tendency toward *cis*-acting transposition lead to accumulation within topologically restricted chromosomal domains? A preliminary approach to these issues is to ask whether the relative abundance or paucity of IS elements in particular chromosomal regions in *E. coli* K-12 might have occurred by chance.

The distribution of IS elements (Fig. 2) shows regions with relatively high densities of elements (the interval from 5 to 15 min) and regions devoid of IS elements (the interval from 53 to 67 min). For a random distribution, the expected pattern of elements depends upon the distribution function relating frequency of occurrence to the element density (number of elements per defined interval). Jurka and Savageau (26) showed that regions of high gene density in the *E. coli* K-12 map were a natural consequence of the log normal gene density distribution function. (The normal, or gaussian, distribution represents frequency by using the gene density as the independent variable; however, with the log normal distribution, the frequency is a function of the logarithm of the gene density.) If IS elements were to obey a similar log normal distribution function, then clustering would be expected to occur for them as well.

One approach is to make the simplifying assumption that IS elements are distributed in a Poisson manner and to estimate the expected maximal and minimal spacings between elements. If the expected value for the smallest spacing, $X_{(1)}$, is defined as $E(X_{(1)}) = \mu/n$ (equation 6 in reference 9), where μ is the average spacing (2.86 min) and n is the number of elements (35 elements), then for the present case, $E(X_{(1)}) = 0.08$ min = 3.8 kb. The expected value for the largest interval, $X_{(n)}$, is $E(X_{(n)}) = \mu[(0.5772 + \lambda v(n)] = 11.8$ min = 552 kb (equation 8 in reference 9). The predicted maximum and minimum spacings are not very different from the observed values (0.03 min minimum interval, 14 min maximum interval). Equation 9 from Churchill et al. (9) indicates that the probability of observing by chance a maximum spacing greater than 14 min would be 0.23. Thus, the observed maximum spacing is reasonably probable for a random distribution obtained by using Poisson statistics.

An alternative approach to analyzing the distribution of IS elements is to ask whether the observed distribution differs significantly from a uniform distribution, in which the number of IS elements encountered during progression along the genome increases linearly with distance. The Kolmogorov-Smirnov test associates the maximum deviation of a computed test statistic (equation 2 in reference 9) with the probability that the observed data follow the hypothetical (in this case, uniform) distribution. By using the locations of the 35 unambiguously identified elements in Table 3, one can calculate the probability P that the observed distribution is identical to a uniform distribution. This calculated P value falls between 0.02 and 0.05, indicating that the nonuniformity of the IS distribution along the chromosome is significant but not highly significant.

Clustering of IS elements might result from characteristics of a particular chromosomal region (such as transcriptional activity, superhelix density, or DNA sequence composition), or it might merely reflect historical contingency. For example, if an IS3 were to transpose by chance into a region of the chromosome lacking IS elements, then insertion at that locus of F or other plasmids containing IS3 would be facilitated. These plasmids might carry other IS elements, which then might preferentially transpose into the immediate neighborhood. Alternatively, IS locations might be correlated as a result of acquisition of composite transposons like the one proposed to be responsible for the duplicate gene *argF* in *E. coli* K-12 (24). A third possibility is that IS elements are preferred targets for other types of IS elements (e.g., possible transposition of IS30 into IS911 [Prère et al., personal communication]).

Genome Rearrangements

The homology associated with multiple IS copies provides a pathway for several (presumably *recA*-mediated) processes that lead to chromosome reorganization. IS3A can recombine with IS3B to invert the *lac* region of *E. coli* K-12 (48). Inversions of this type had been noted in Hfr strains (2), and they can be explained by the same type of recombination event. IS5 elements also can cause extensive chromosomal inversions. The *oxa1* mutations (12) and other, similar *oxa* mutations (33) are a result of recombination between IS5Y and a presumably inverted IS5 between *argG* and *xylA*. The correspondence between inversion endpoints with the locations of IS elements in *E. coli* K-12 indicates that recombination between inverted IS elements of the same type is a major inversion pathway in this strain.

The integration of the F plasmid at chromosomal IS2, IS3, or γδ (Tn1000) elements to form Hfr strains (11) has been experimentally important in the development of *E. coli* K-12 genetics (34). At chromosomal sites that have been examined in Hfr strains, there is excellent correlation between Hfr points of origin and chromosomal IS2A and IS3A to C elements (14). Although F can presumably integrate by transpositional cointegration, the *recA* dependence of F integration (10) suggests that most F integration occurs by reciprocal homologous exchange. Similarly, the directly repeated chromosomal IS5A and IS5B elements can function in the excision of F' plasmids from the bacterial chromosome (56). The roles of these and other IS elements in F integration and F' excision have been summarized by Umeda and Ohtsubo (58), and the effects of IS elements and other repeated sequences on genome rearrangements are discussed in chapter 112.

Coordinated Transposition

Transposition of IS elements is particularly evident for *E. coli* strains that have been stored in stab cultures (19, 39). This phenomenon also occurs for *S. typhimurium,* in which copy numbers of IS*200* may increase from 6 to 11 after serial propagation in stabs (C. R. Beuzón and J. Casadesús, personal communication). For IS*200,* the increase appears to occur abruptly. This result differs from observations with IS elements in *E. coli* K-12, which reportedly show linear increases in transposition events with time (32).

In the case of IS*30,* bursts of transposition may arise after rearrangements that form the tandem structure (IS*30*)$_2$ (42). It would be interesting to know how these rearrangements contribute to accumulation of IS-associated events in stab cultures and whether they appear gradually over time or primarily after a threshold period of storage.

Evolutionary Questions

Because IS elements can promote genome rearrangements and create mutations, they potentially could contribute to the evolution of bacterial chromosomes. These effects will be more readily documented as DNA sequences of complete bacterial chromosomes become available. It already appears that *E. coli* K-12 is missing DNA sequences that are presumably present in other *E. coli* strains, since IS*1*A, IS*1*B, IS*1*C, and IS*1*F are not flanked by direct repetitions (61). The loss of direct repetitions might indicate that these elements promoted adjacent deletions of chromosomal DNA. It would be interesting to check members of the ECOR collection for their chromosomal DNA sequences adjacent to the locations of the IS elements in *E. coli* K-12 to determine what may have been lost from *E. coli* K-12.

The horizontal dissemination of IS elements makes it likely that sequence variants of other classes of IS elements will be found in the *E. coli* K-12 and *S. typhimurium* chromosomes. It may prove fruitful to search the *E. coli* K-12 chromosomal DNA sequence (soon to be available) for limited similarities to sequences of other IS elements and for imperfect inverted repetitions of appropriate-size flanking 700- to 2,000-bp regions that just fail to contain appropriate-size ORFs or that include appropriate-size regions containing only a few stop codons. This approach might reveal derelict IS elements that are no longer functional in *E. coli* K-12. The H-rpt sequence found in some Rhs elements (62) might be an example of an IS that is rarely or not at all functional in *E. coli* K-12 (see chapter 112).

The dynamics of inheritance of IS elements over long periods of time is another issue to be addressed. Does the persistence in the population of elements that transpose by "cut and paste" mechanisms differ from that for elements that transpose replicatively? Why are some elements found preferentially on chromosomes, whereas others are more frequently found on plasmids? Is this bias related to the transposition mechanisms?

Similarly, it will be interesting to know what determines the spectrum of IS elements that transpose into particular chromosomal genes. In some cases, this will be attributable to biochemical idiosyncrasies of different elements and the particular target sequences. For example, over half of the IS*2* insertions in bacteriophage P1 fall in a fragment representing <2% of the P1 genome but containing sequence similarities to IS*2* (53). A recent study showed that all IS insertions in the *hemB* gene of *E. coli* K-12 were IS*2* (31). Is this only a reflection of the sequence properties of *hemB,* or is it partly a consequence of the close proximity (8-kb separation along the contour length) of IS*2* as the nearest neighboring IS element?

Some elements like IS*3* appear to have been present in *E. coli* for a very long time, yet approximately one-third of the ECOR strains lack IS*3* on their chromosomes. Do these elements shuttle back and forth between chromosomes on the one hand and plasmids and phages on the other? How are IS elements removed from chromosomes? Is precise excision sufficient, or are there other mechanisms for rectification of chromosomal DNA (e.g., replacement of an IS-containing segment by the corresponding uninterrupted DNA introduced during conjugation with an Hfr)? Precise excision frequencies range from 10^{-7} to 10^{-9}, depending on sequence context (31, 55). Excision frequencies also are influenced by external factors, such as the *ref* gene of bacteriophage P1. When *ref* is derepressed, it stimulates the efficiency of excision of IS*1* from *galT* by a factor of 10^5 (35). Thus, rectification of IS mutations in the bacterial chromosome may depend on particular historical contingencies (e.g., whether they were infected by a particular type of phage or plasmid).

IS Element Chimeras

Some IS elements may be recombinationally composite (30), presumably as a result of general recombination. Other composites formed from pieces of different IS elements are known. The 181-bp IS*30* fragment IS*30*B abuts an end of IS*1*B, suggesting that it may be a remnant of an adjacent deletion event (60). The origin of the IS*911* deletant containing IS*30*D and an IS*600* fragment (see above) is not so readily explained. Are IS elements themselves particularly susceptible to attack by other elements, or are IS elements accumulated on plasmids and phages and then, after transpositional deletion and other events, mobilized to the chromosome?

ACKNOWLEDGMENTS

I thank Barbara J. Bachmann of the *E. coli* Genetic Stock Center for assembling and transmitting additional information on *E. coli* pedigrees, and I thank Kenneth E. Rudd for unpublished mapping information and extensive documentation and discussion relating to EcoMap6. Michael Waterman contributed most helpful information on statistical analysis. Howard Ochman provided unpublished information on IS elements in the SARA collection and helpful insights. J. Casadesús, M.-F. Prère, M. Chandler, and O. Fayet generously provided unpublished data. I thank Elaine Freund for her thoughtful comments on the manuscript.

LITERATURE CITED

1. Beltran, P., S. A. Plock, N. H. Smith, T. S. Whittam, D. C. Old, and R. H. Selander. 1991. Reference collection of strains of the *Salmonella typhimurium* complex from natural populations. *J. Gen. Microbiol.* **137:**601–606.
2. Berg, C. M., and R. C. Curtiss III. 1967. Transposition derivative of an Hfr strain of *Escherichia coli* K-12. *Genetics* **56:**503–525.
3. Birkenbihl, R. P., and W. Vielmetter. 1989. Complete maps of IS1, IS2, IS3, IS4, IS5, IS30, and IS150 locations in *E. coli* K12. *Mol. Gen. Genet.* **220:**147–153.
4. Birkenbihl, R. P., and W. Vielmetter. 1991. Completion of the IS map in *E. coli:* IS*186* positions on the *E. coli* K12 chromosome. *Mol. Gen. Genet.* **226:**318–320.
5. Bisercić, M., and H. Ochman. 1993. The ancestry of insertion sequences common to *Escherichia coli* and *Salmonella typhimurium. J. Bacteriol.* **175:**7863–7868.
6. Bisercić, M., and H. Ochman. 1993. Natural populations of *Escherichia coli* and *Salmonella typhimurium* harbor the same classes of insertion sequences. *Genetics* **133:**449–454.
7. Casadesús, J., C. R. Beuzon, and I. Gilbert. 1992. IS200, basic and applied. *Genetics (Life Sci. Adv.)* **11:**179–186.
8. Chandler, M., and O. Fayet. 1993. Translational frameshifting in the control of transposition in bacteria. *Mol. Microbiol.* **7:**497–503.
9. Churchill, G. A., D. L. Daniels, and M. S. Waterman. 1990. The distribution of restriction enzyme sites in *Escherichia coli. Nucleic Acids Res.* **18:**589–597.
10. Cullum, J., and P. Broda. 1979. Chromosome transfer and Hfr formation by F in *rec*$^+$ and *recA* strains of *Escherichia coli* K-12. *Plasmid* **2:**358–365.
11. Davidson, N., R. C. Deonier, S. Hu, and E. Ohtsubo. 1975. Electron microscope heteroduplex studies of sequence relations among plasmids of *Escherichia coli.* X. Deoxyribonucleic acid sequence organization of F and F-primes, and the sequences involved in Hfr formation, p. 56–65. *In* D. Schlessinger (ed.), *Microbiology—1974.* American Society for Microbiology, Washington, D.C.

12. de Massey, B., J. Patte, J.-M. Louarn, and J.-P. Bouché. 1984. *OriX*: a new origin of replication in *E. coli*. Correction. *Cell* **38**:333.

13. Deonier, R. C. 1987. Locations of native insertion sequence elements, p. 982–989. *In* F. C. Neidhardt, J. L. Ingraham, K. B. Low, B. Magasanik, M. Schaechter, and H. E. Umbarger (ed.), *Escherichia coli and Salmonella typhimurium: Cellular and Molecular Biology*. American Society for Microbiology, Washington, D.C.

14. Deonier, R. C., and R. G. Hadley. 1980. IS2-IS2 and IS3-IS3 relative recombination frequencies in F integration. *Plasmid* **3**:48–64.

15. Deonier, R. C., R. G. Hadley, and M. Hu. 1979. Enumeration and identification of IS3 elements in *Escherichia coli* strains. *J. Bacteriol.* **137**:1421–1424.

16. Farabaugh, P. J., U. Schmeissner, M. Hofer, and J. H. Miller. 1978. Genetic studies of the *lac* repressor. VII. On the molecular nature of spontaneous hotspots in the *lacI* gene of *Escherichia coli*. *J. Mol. Biol.* **126**:847–863.

17. Fayet, O., P. Ramond, P. Polard, M.-F. Prère, and M. Chandler. 1990. Functional similarities between retroviruses and the IS3 family of bacterial insertion sequences. *Mol. Microbiol.* **4**:1771–1777.

18. Galas, D. J., and M. Chandler. 1989. Bacterial insertion sequences, p. 109–162. *In* D. E. Berg and M. M. Howe (ed.), *Mobile DNA*. American Society for Microbiology, Washington, D.C.

19. Green, L., R. D. Miller, D. E. Dykhuizen, and D. L. Hartl. 1984. Distribution of DNA insertion element IS5 in natural isolates of *Escherichia coli*. *Proc. Natl. Acad. Sci. USA* **81**:4500–4504.

20. Hadley, R. G., M. Hu, M. Timmons, K. Yun, and R. C. Deonier. 1983. A partial restriction map of the *proA-purE* region of the *E. coli* K-12 chromosome. *Gene* **22**:281–287.

21. Hall, B. G., L. L. Parker, P. W. Betts, R. F. DuBose, S. A. Sawyer, and D. L. Hartl. 1989. IS103, a new insertion element in *Escherichia coli*: characterization and distribution in natural populations. *Genetics* **121**:423–431.

22. Hartl, D. L., and S. A. Sawyer. 1988. Why do unrelated insertion sequences occur together in the genome of *Escherichia coli*? *Genetics* **118**:537–541.

23. Hu, M., and R. C. Deonier. 1981. Comparison of IS1, IS2 and IS3 copy number in *Escherichia coli* strains K-12, B, and C. *Gene* **16**:161–170.

24. Hu, M., and R. C. Deonier. 1981. Mapping of IS1 elements flanking the *argF* gene region on the *Escherichia coli* K-12 chromosome. *Mol. Gen. Genet.* **181**:222–229.

25. Ishiguro, N., and G. Sato. 1988. Nucleotide sequence of insertion sequence IS3411, which flanks the citrate utilization determinant of transposon Tn3411. *J. Bacteriol.* **170**:1902–1906.

26. Jurka, J., and M. A. Savageau. 1985. Gene density over the chromosome of *Escherichia coli*: frequency, distribution, spatial clustering, and symmetry. *J. Bacteriol.* **163**:806–811.

27. Kohara, Y., K. Akiyama, and K. Isono. 1987. The physical map of the whole *E. coli* chromosome: application of a new strategy for rapid analysis and sorting of a large genomic library. *Cell* **50**:495–508.

28. Lam, S., and J. R. Roth. 1983. IS200: a *Salmonella*-specific insertion sequence. *Cell* **34**:951–960.

29. Lam, S., and J. R. Roth. 1983. Genetic mapping of IS200 copies in *Salmonella typhimurium* strain LT2. *Genetics* **105**:801–811.

30. Lawrence, J. G., H. Ochman, and D. L. Hartl. 1992. The evolution of insertion sequences within enteric bacteria. *Genetics* **131**:9–20.

31. Lewis, L. A., D. Lewis, V. Persaud, S. Gopaul, and B. Turner. 1994. Transposition of IS2 into the *hemB* gene of *Escherichia coli* K-12. *J. Bacteriol.* **176**:2114–2120.

32. Lieb, M. 1981. A fine structure map of spontaneous and induced mutations in the lambda repressor gene, including insertions of IS elements. *Mol. Gen. Genet.* **184**:364–371.

33. Louarn, J. M., J.-P. Bouché, R. Legendre, J. Louarn, and J. Patte. 1985. Characterization and properties of very large inversions of the *Escherichia coli* chromosome along the origin-to-terminus axis. *Mol. Gen. Genet.* **201**:467–476.

34. Low, K. B. 1972. *Escherichia coli* K-12 F-prime factors, old and new. *Bacteriol. Rev.* **36**:587–607.

35. Lu, S. D., D. Lu, and M. Gottesman. 1989. Stimulation of IS1 excision by bacteriophage P1 *ref* function. *J. Bacteriol.* **171**:3427–3432.

36. Matsutani, S., and E. Ohtsubo. 1990. Complete sequence of IS629. *Nucleic Acids Res.* **18**:1899.

37. Matsutani, S., and E. Ohtsubo. 1993. Distribution of the *Shigella sonnei* insertion elements in Enterobacteriaceae. *Gene* **127**:111–115.

38. Matsutani, S., H. Ohtsubo, Y. Maeda, and E. Ohtsubo. 1987. Isolation and characterization of IS elements repeated in the bacterial chromosome. *J. Mol. Biol.* **196**:445–455.

39. Naas, T., M. Blot, W. M. Fitch, and W. Arber. 1994. Insertion sequence-related genetic variation in resting *Escherichia coli* K-12. *Genetics* **136**:721–730.

40. Nakamura, K., and M. Inouye. 1981. Inactivation of the *Serratia marcescens* gene for the lipoprotein in *Escherichia coli* by insertions sequences, IS1 and IS5; sequence analysis of junction points. *Mol. Gen. Genet.* **183**:107–114.

41. Nyman, K., K. Nakamura, H. Ohtsubo, and E. Ohtsubo. 1981. Distribution of the insertion sequence IS1 in Gram-negative bacteria. *Nature* (London) **289**:609–612.

42. Olasz, F., R. Stalder, and W. Arber. 1993. Formation of the tandem repeat (IS30)₂ and its role in IS30-mediated transpositional DNA rearrangements. *Mol. Gen. Genet.* **239**:177–187.

43. Perkins, J. D., J. D. Heath, B. R. Sharma, and G. M. Weinstock. 1993. *Xba*I and *Bln*I genomic cleavage maps of *Escherichia coli* K-12 strain MG1655 and comparative analysis of other strains. *J. Mol. Biol.* **232**:419–445.

44. Prère, M.-F., M. Chandler, and O. Fayet. 1990. Transposition in *Shigella dysenteriae*: isolation and analysis of IS911, a new member of the IS3 group of insertion sequences. *J. Bacteriol.* **172**:4090–4099.

45. Rezsöhazy, R., B. Hallet, J. Delcour, and J. Mahillon. 1993. The IS4 family of insertion sequences: evidence for a conserved transposase motif. *Mol. Microbiol.* **9**:1283–1295.

46. Sanderson, K. E., P. Sciore, S.-L. Liu, and A. Hessel. 1993. Location of IS200 on the genomic cleavage map of *Salmonella typhimurium* LT2. *J. Bacteriol.* **175**:7624–7628.

47. Sato, S., Y. Nakada, and A. Shiratsuchi. 1989. IS421, a new insertions sequence in *Escherichia coli*. *FEBS Lett.* **249**:21–24.

48. Savić, D. J., S. P. Romac, and S. D. Ehrlich. 1983. Inversion in the lactose region of *Escherichia coli* K-12: inversion termini map within IS3 elements α3β3 and β5α5. *J. Bacteriol.* **155**:943–946.

49. Sawyer, S. A., D. E. Dykhuizen, R. F. DuBose, L. Green, T. Mutangadura-Mhlanga, D. F. Wolczyk, and D. L. Hartl. 1987. Distribution and abundance of insertion sequences among natural isolates of *Escherichia coli*. *Genetics* **115**:51–63.

50. Schoner, B., and R. G. Schoner. 1981. Distribution of IS5 in bacteria. *Gene* **16**:347–352.

51. Schwartz, E., M. Droger, and B. Rak. 1988. IS150: distribution, nucleotide sequence, and phylogenetic relationships of a new *E. coli* insertion element. *Nucleic Acids Res.* **16**:6789–6802.

52. Selander, R. K., D. A. Caugent, and T. S. Whittam. 1987. Genetic structure and variation in natural populations of *Escherichia coli*, p. 1625–1648. *In* F. C. Neidhardt, J. L. Ingraham, K. B. Low, B. Magasanik, M. Schaechter, and H. E. Umbarger (ed.), *Escherichia coli and Salmonella typhimurium: Cellular and Molecular Biology*. American Society for Microbiology, Washington, D.C.

53. Sengstag, C., J. C. W. Shepherd, and W. Arber. 1983. The sequence of the bacteriophage P1 genome region serving as a hot target for IS2 insertion. *EMBO J.* **2**:1777–1781.

54. Sofia, H. J., V. Burland, D. L. Daniels, G. Plunkett III, and F. R. Blattner. 1994. Analysis of the *Escherichia coli* genome. V. DNA sequence of the region from 76.0 to 81.5 minutes. *Nucleic Acids Res.* **22**:2576–2586.

55. Starlinger, P., and H. Saedler. 1976. IS-elements in microorganisms. *Curr. Top. Microbiol. Immunol.* **75**:111–152.

56. Timmons, M. S., A. M. Bogardus, and R. C. Deonier. 1983. Mapping of chromosomal IS5 elements that mediate type II F-prime plasmid excision in *Escherichia coli* K-12. *J. Bacteriol.* **153**:395–407.

57. Timmons, M. S., K. Spear, and R. C. Deonier. 1984. IS121 is near *proA* in the chromosomes of *Escherichia coli* K-12 strains. *J. Bacteriol.* **160**:1175–1117.

58. Umeda, M., and E. Ohtsubo. 1989. Mapping of insertion elements IS1, IS2 and IS3 on the *Escherichia coli* K-12 chromosome. Role of the insertion elements in formation of Hfrs and F' factors and in rearrangement of bacterial chromosomes. *J. Mol. Biol.* **208**:601–604.

59. Umeda, M., and E. Ohtsubo. 1990. Mapping of insertion element IS5 in the *Escherichia coli* K12 chromosome. Chromosomal rearrangements mediated by IS5. *J. Mol. Biol.* **213**:229–237.

60. Umeda, M., and E. Ohtsubo. 1990. Mapping of insertion element IS30 in the *Escherichia coli* K12 chromosome. *Mol. Gen. Genet.* **222**:317–322.

61. Umeda, M., and E. Ohtsubo. 1991. Four types of IS1 with differences in nucleotide sequence reside in the *Escherichia coli* K12 chromosome. *Gene* **98**:1–5.

62. Zhao, S., C. H. Sandt, G. Feulner, D. A. Vlazny, J. A. Gray, and C. W. Hill. 1993. *Rhs* elements of *Escherichia coli* K-12: complex composites of shared and unique components that have different evolutionary histories. *J. Bacteriol.* **175**:2799–2808.

Repeated Sequences

SOPHIE BACHELLIER, ERIC GILSON, MAURICE HOFNUNG, AND
CHARLES W. HILL

112

INTRODUCTION

Sequence repetition in *Escherichia coli* and *Salmonella typhimurium* (official designation, *Salmonella enterica* serovar Typhimurium) is encountered in many contexts. Sources of repeated sequences include structural genes, accessory genetic elements, sequence motifs, and genetic duplications. On a random basis, there is a high probability that any sequence of 12 or fewer bases will be repeated at least once in a chromosome this size. Therefore, for a sequence repetition to be considered significant, it must be either substantially larger than 12 bases or present in many copies.

Certain genetic issues relate especially to repeated sequences. One issue has to do with advantages conferred by sequence multiplicity. In the case of structural genes, benefits might include greater expression or differential control. Benefits may also be derived from functional diversity if repeated genes are similar but not identical. In the case of sequence motifs, the same structural domain can be incorporated into diverse gene products or the motif can invoke the same process at many chromosomal locations.

A second issue has to do with the source of the DNA comprising the repeated sequences and also the mechanism for generation of the repetition. As to source, we generally recognize two alternatives. The repeated sequences may be DNA common to the species, or they may derive from DNA introduced through horizontal transfer (e.g., accessory genetic elements [23]). The repeated sequences may have been established early with respect to species divergence, in which case the arrangement would tend to be conserved among individuals throughout the species (e.g.,

rRNA genes). Alternatively, the sequences may have become amplified by recent rearrangement (e.g., tandem duplications), and such arrangements can be highly specific for individual strains or clones. Repeated sequences derived from accessory elements (e.g., insertion sequences [ISs]) can vary considerably as to numbers and chromosomal positions in individuals of the species. As to possible mechanisms for establishment of sequence repetition, illegitimate, site-specific, and homologous recombination all might have roles in different circumstances, and at least in the case of very small repeated sequences, convergent evolution also should be considered.

A third issue concerns genetic interaction between repeated sequences. Repeated sequences provide homology for ectopic recombination, i.e., recombination between dispersed homologous sequences. This will lead to chromosomal rearrangement such as duplication, deletion, inversion, and transposition (Fig. 1). The nature of the rearrangement will depend on the positions and orientations of the repeated sequences and also on whether the repeated sequences are on the same or different chromatids. The schemes drawn in Fig. 1 imply reciprocal crossover events, but nonreciprocal events may happen as well. Specifically in the case of direct repeats, nonreciprocal crossover may leave one product rejoined but the other a nonviable, incomplete fragment. Recombination between inverted repeats, however, must be reciprocal if a circular chromosome is to be recovered. Each type of rearrangement has distinct consequences, with duplication and deletion obviously affecting dosage. Inversion changes gross positional relationships between genes. This may be especially important when the position of a gene relative to the replication origin and terminus is altered.

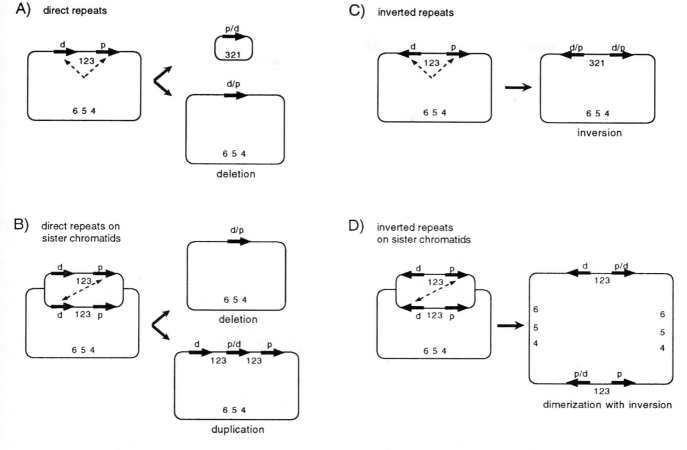

FIGURE 1 Crossover between sequence repeats. These schemes represent fundamental interactions between repeated sequences, and no mechanism is implied. The repeats are shown by the arrows: d (distal) and p (proximal). They may occur as direct repeats (A and B) or as inverted repeats (C and D). The crossover may be between repeats present on the same chromosome (A and C) or on sister chromatids (B and D). For schemes B and D, the structures of the ultimate recombinant products assume completion of the replication cycle. The unique sequences separating the repeats are designated 1 2 3 and 4 5 6; 1 2 3 may be deleted, duplicated, or inverted with respect to 4 5 6, depending on the configuration of the event. The repeated sequences, p and d, need not be identical, in which case the crossovers will produce hybrid sequences p/d and d/p, distinct from either of the parental repeats.

The schemes depicted in Fig. 1 do not exhaust the configurations of possible interactions. For example, the small circle excised in Fig. 1A could reinsert, either into the homologous portion of a different chromosome (yielding a tandem duplication) or into a third copy of the repeated sequence (yielding a transposition). Figure 1D depicts the consequences of a crossover between inverted repeats on sister chromatids. The product is a complex chromosomal dimer that cannot be resolved to monomers by a single additional crossover between the large homologs (i.e., region 4 5 6 with 4 5 6 or region 1 2 3 with 1 2 3) since the large homologs are themselves inverted. It could be resolved by the less frequent crossover between appropriate copies of the repeated sequence (i.e., p [proximal] with d [distal] or p/d with p/d). This dimer would have two replication origins, and a cell containing it would stably inherit two copies of every gene. The genetic properties of such a cell would be striking, and the fact that such a mutant has not been reported likely indicates that a cell harboring such a chromosome is not viable.

While rearrangements will often be deleterious, this is not true in all circumstances. Duplications will be advantageous under special conditions if they amplify a gene whose product is limiting (88, 186, 190). For example, growth on a limiting carbon source such as arabinose selects for cells with a large duplication that enhances transport of not only the compound used for selection, but several other compounds as well (186). An inherent property of a large tandem duplication is that it will be very unstable because of the extensive homology available for looping out. Therefore, when conditions change so that the large duplication becomes a burden, the duplication can be eliminated with high frequency (5, 186). Duplication can serve as the first essential step toward acquisition of new capabilities when it is followed by divergence of the duplicated copies (77, 93, 120).

If the repeated sequences are not identical, recombination between them can either eliminate the differences or produce new combinations. Elimination of diversity can occur by gene conversion (in its strict sense) or by double crossovers between divergent repeated sequences on sister chromatids. Novel joints characteristic of rearrangements will generally be a hybrid gene (Fig. 1).

Clearly, genetic interaction between repeated sequences is a major contributor to genome plasticity. The topic of repeated sequences in *E. coli* and *S. typhimurium* has been reviewed extensively (3, 83, 109, 149, 159–161).

FACTORS AFFECTING ECTOPIC RECOMBINATION

Several interesting mechanistic questions relate to ectopic recombination, and much recent experimentation has focused on this deceptively complex problem. Important issues include which recombination pathways are involved, whether the crossover is reciprocal, and how the crossover frequency depends on physical parameters such as length, sequence, sequence divergence (for nonidentical repeated sequences), and distance between repeats.

recA Dependence

Repetitious structures were recognized early in the study of *E. coli* genetics. Early examples included lysogens with tandem prophages (138) and mutants with large chromosomal duplications (21, 22, 29, 88). A common property of these early mutants was their high degree of genetic instability. It was observed that the duplications were rapidly lost, presumably through crossover between the tandem repeats (Fig. 1A and B). Homologous recombination was directly implicated by the observation that a *recA* mutation stabilized duplications (4, 77, 174). In a specific example, *recA* reduced the 6% segregation frequency of a 164-kb tandem duplication at least 250-fold (77). It was quite surprising, therefore, when later studies using plasmid model systems found that deletion between repeats sometimes occurs efficiently in *recA* strains (33, 127, 130, 132). One way of resolving this apparent contradiction was to hypothesize a fundamental difference in recombination pathways operating in plasmids compared with those operating in the host chromosome. However, when deletion between 787-bp repeats was tested on both plasmid and chromosome, the process was *recA* independent in both settings (127). Interestingly, the deletion frequency on the chromosome was 2 orders of magnitude lower than on the plasmid even when the arrangement of the repeats was otherwise identical. The distance between the points of crossover strongly influences the degree of *recA* dependence (13). If the repeated sequences are small and in tandem so that the crossover positions are spaced at only 300 bp, deletion occurs in a *recA*-independent manner. As the distance is increased, *recA* dependence increases, and at a spacing of 4 kb, deletion is highly *recA* dependent. Apparently, deletion between direct repeats can occur by both *recA*-independent and *recA*-dependent pathways, but the effectiveness of the *recA*-independent pathway diminishes as the separation between the crossover sites increases. One model proposed for nonreciprocal, *recA*-independent deletion hypothesizes that the exchange is initiated between repeats on sister chromatids during passage of the replication fork (127). The authors speculate that the asymmetric association of the leading and lagging strands with respect to the polymerase dimer complex brings a proximal repeat located on one chromatid close to a distal repeat located on the sister chromatid. Strand switching ultimately results in one chromatid with a deletion, while the other retains the parental duplication. Such a mechanism would inherently operate most efficiently if the repeated sequences are small and, more

importantly, if they are spaced with a short periodicity. Deletion between very large repeats or widely spaced repeats could require a different, *recA*-dependent mechanism. The *recA*-independent deletion between 18-bp direct repeats in *Bacillus subtilis* is also strongly dependent on the distance between the repeats (26). It increased 1,000-fold as the distance was increased from 33 to 2,313 bp, and the effect was observed in both plasmid and chromosomal systems.

Size

A priori there is a strong presumption that ectopic recombination frequency will increase with size of the repeated sequence, and this has been the general observation. Below a threshold of 20 to 40 bp, homologous recombination is very inefficient (101, 132, 180, 208). Above this threshold, its frequency increases with length. The segregation frequency of very large chromosomal duplications also increases with size. Segregation occurred at frequencies of 1.4 and 5.9% for 40- and 164-kb duplications, respectively, when measured under otherwise identical conditions (77, 81). At least in this case, the increase in frequency with size was remarkably proportional. Although recombination frequency falls sharply when the repeats are shorter than a threshold of about 20 bp (208), sequence repetitions as small as 8 to 17 bp are observed to recombine specifically (1, 41, 100, 194, 210). In some studies, recombination between very short sequences was observed to be *recA* independent (97, 129, 132).

The major pathway for reciprocal recombination is the RecBCD pathway (101). Consistent with the assumption that chromosomal inversion between inverted repeats must be fully reciprocal in order to produce a viable chromosome (Fig. 1C), inversion is reduced by a *recB* mutation (177). Duplications, on the other hand, which can be formed by nonreciprocal recombination, occur at comparable frequencies in *recB* and *recB*+ backgrounds.

Sequence

Repeat size is actually a deceptive parameter to assess since increase in the size of the repeated sequence will generally require the introduction of completely new sequence, and it may also increase the spacing between the points of exchange. As described above, the spacing factor can be a very significant one (13). Sequence is also clearly important (184). The best understood example of sequence influencing crossover frequency involves the Chi sequence. This sequence motif strongly stimulates the RecBCD recombination pathway (see chapter 119). In the context of ectopic recombination, a Chi sequence located 2 kb away from one of two recombining homologies stimulated crossover 20-fold (45).

Sequence Divergence

Homologous recombination can occur between sequences which are similar but not identical. However, the frequency has been shown to be sensitive to small amounts of sequence divergence, and the effect can be quite strong (68, 180, 181). In one study, 10% sequence divergence reduced recombination 40-fold (180). The mismatch repair system eliminates recombination intermediates that contain mismatched heteroduplex (157, 181). Inactivation of mismatch repair enhances recombination between divergent sequences (131). Reduction of recombination by sequence divergence has important implications for the evolution of microbial genomes (154, 155). It provides a means of suppressing chromosomal rearrangement due to crossover between naturally repeated sequences, since repeats that are somewhat divergent will cause less rearrangement than equivalent perfect repeats (150). By the same token, sequence divergence, once established, would be less prone to elimination by gene conversion. This would be especially important if the divergence has functional significance. There are many examples of protein-encoding genes that show significant sequence similarity indicating a common origin, but the similarity is usually less than 80% (159). This degree of divergence would reduce crossover between these gene pairs to very low levels.

Mutagens

Environmental factors strongly affect ectopic recombination. Mutagens such as UV irradiation, nitrous acid, ethyl methanesulfonate, and niridazole greatly stimulate duplication frequency by recombination between large repeated sequences such the rRNA genes (76) and the *Rhs* elements (122, 189) as well as other duplications whose precise origins are not established (72, 86, 87). The frequency of large duplications induced by moderate doses of UV light is remarkable, with estimates ranging from 5 to 12% of the survivors (72, 76). The SOS response, even in the absence of mutagen damage to DNA, stimulates duplication formation (34).

Nonpermissive Inversions

Chromosomal inversion between inversely oriented homologies has been documented extensively (79, 81, 172). However, not all chromosomal intervals are permissive for inversion (107, 158, 175, 176, 218). In at least some cases, the nonpermissiveness is not due to lethality of the final product since the equivalent structure can be created by other means (175, 176). Furthermore, the same homologies, located at the same chromosomal positions but with direct orientation, can recombine. It appears that repeats placed at certain positions in the chromosome cannot participate in the reciprocal, intrachromosomal crossover necessary for inversion.

LARGE REPEATED SEQUENCES

In *E. coli* and *S. typhimurium,* instances of large repeats that have similarities greater than 95% are rather limited (16). Essential genes that occur as highly conserved repeated sequences are the rRNA genes, some tRNA genes, genes for the small stable RNA component of RNase P (104, 105), and the genes for a protein elongation factor (*tufA* and *tufB*) (2, 215). A number of other protein-encoding genes show clear homology, but their sequence divergence is substantial (159). Accessory elements are a significant source of repetitious DNA. Accessory elements that have been found in multiple copies include ISs and the composite *Rhs* elements.

rRNA Genes

The most significant large repeated sequence in *E. coli* is the array of rRNA genes. Each of the seven *rrn* operons encodes 16S, 23S, and 5S rRNAs, with the 16S and 23S sequences separated by a tRNA spacer (43). Five operons are located clockwise from the origin of replication, and two are counterclockwise (Fig. 2). Each is oriented so that it is transcribed in the same direction as it is replicated. All have tandem promoters, and some operons encode additional tRNA species at their distal end (see chapter 13). Under conditions of rapid growth, *rrn* expression accounts for over half of the cell's transcrip-

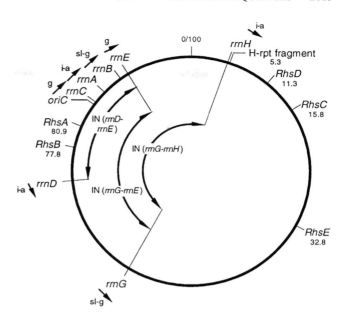

FIGURE 2 Map positions of large repetitions found in the *E. coli* chromosome. The spacer types and orientations of the *rrn* operons are shown. The tRNA spacer types are abbreviated as follows: i-a, $tRNA^{Ile} - tRNA^{Ala}_{1B}$; g, $tRNA^{Glu}_2$; sl-g, $rsl - tRNA^{Glu}_2$. Three inversion mutations between *rrn* operons are indicated. The positions of the five K-12 *Rhs* elements and an H-rpt fragment are shown.

tional activity. The significance of there being seven of these operons is a long-standing question. The fact that both *E. coli* and *S. typhimurium* have seven *rrn* operons (117) indicates that the arrangement is both ancient and selectively maintained. The advantage of seven could simply be one of gene dosage. Alternatively, the operons might encode rRNA molecules with (slightly) different functions, or they might be affected differently by changes in growth conditions. The existence of rRNA sequence divergence has been long recognized (47), but its extent has not been clear. The recent determination of the complete sequences of five of the seven (*rrnA, rrnB, rrnC, rrnE,* and *rrnH*) (17) may bring us closer to an understanding. Divergence is limited, and most is unique to an individual operon, but there is a prominent exception. The five available 23S sequences are divided into two classes (*rrnB-E* and *rrnA-C-H*) by a 13-bp segment containing eight differences (17). This segment comprises a stem-loop (bases 1722 to 1738), and six of the substitutions are actually three pairs of compensating stem changes (Fig. 3). The accumulation of so many compensating changes implies that the divergence of these two versions of this stem-loop is relatively ancient. If both versions have always been

FIGURE 3 Alternative stem-loops of 23S RNA species. Nucleotides 1721 to 1740 of the 23S RNA of *rrnB* and *rrnE* are shown in panel A. The corresponding nucleotides of *rrnA, rrnC,* and *rrnH* are shown in panel B. Positions that differ between the two are enclosed in boxes.

carried within the same chromosome, one or the other should have been eliminated through conversion unless there is a selective advantage to having both (183). An alternative possibility is that the two versions originated in different lineages and came together relatively recently in an ancestor of the present day *E. coli* K-12 chromosome.

Differential Expression. The question of differential expression of individual *rrn* operons has been directly addressed (28). Significant differences exist with regard to promoter activity in minimal medium, response to amino acid starvation, and response to depletion of a transcription factor. Compared with its expression at a standardized location, *rrnH* transcription was specifically reduced when the gene was placed at its normal chromosomal location. Although it is not clear how these observations relate to the fitness of the cell in natural settings, taken collectively they convincingly establish differential expression. An early attempt at assessing the necessity of all seven operons showed that deletion of *rrnE* had no observable detrimental effect even under conditions of maximum growth (42). More recently, individual *rrn* operons were sequentially disrupted within the chromosome and the mutants were tested for whether the expression of other *rrn* operons would increase to compensate for the loss (27). Disruption of one or two operons had minimal effect, but disruption of a third and fourth operon progressively reduced both growth and ribosome content. These experiments support a model for feedback control of rRNA synthesis (96) that compensates for changes in gene copy. The mechanism of compensation involved increases in transcription initiation of the remaining operons and, curiously, in the rate of transcription elongation (27). The problem of *rrn* operon redundancy is complicated by the diversity of the tRNA spacers. Consequently, a separate question concerns the importance of spacer multiplicity. To assess this problem, an *E. coli* mutant was constructed in which three of the $tRNA_2^{Glu}$ spacers were replaced with $tRNA^{Ile}$-$tRNA_{1B}^{Ala}$ spacers, leaving only the *rrnG* operon with a $tRNA_2^{Glu}$ spacer (70). The growth rate of this mutant was approximately half that of the wild type in synthetic medium, and cultures were rapidly overgrown by variants that had converted one of their $tRNA^{Ile}$-$tRNA_{1B}^{Ala}$ spacers to a second $tRNA_2^{Glu}$ spacer. Taken all together, these observations suggest that gene dosage and differential expression are both factors in the natural selection that has established seven *rrn* operons in *E. coli* and *S. typhimurium*.

The *E. coli* $tRNA_2^{Glu}$ spacer exists in two forms, which differ in that a 106-bp sequence (called *rsl*) of one is replaced by a 20-bp sequence in the other. The *rrnG* and *rrnB* operons of strain K-12 contain the *rsl* version, while *rrnC* and *rrnE* contain the alternate version. However, some K-12 derivatives have had the *rsl* form replaced by the alternate, presumably through recombination (71). No function has been assigned to the *rsl* sequence. It is clearly not essential, because simultaneous elimination of *rrnB* and *rrnG*, which eliminates both copies of *rsl*, does not cause loss of viability (27).

Genetic Exchange between rRNA Genes. Since the rRNA genes are such large sequence repeats, the extent of genetic exchange between them is an important issue. The rRNA genes are hot spots for chromosomal rearrangement in both *E. coli* (32, 76, 78) and *S. typhimurium* (5, 118). Rearrangements associated with *rrn* recombination include duplication (5, 76), deletion (32, 82), inversion (81), and transposition (82). Recombination between the *rrnB* and *rrnE* operons has been the focus of several studies. They have the same orientation, and with a spacing of only 39.5 kb, they are the most closely linked *rrn* operons. Recombination between them occurs spontaneously at a frequency of 0.1×10^{-3} to 0.2×10^{-3} (82). If all possible pairwise combinations of the seven *rrn* operons were to recombine at this frequency, 2 to 4% of the cells in an *E. coli* population would bear some sort of chromosomal rearrangement. However, a number of factors, such as orientation, distance, and possibly sequence divergence, probably affect the frequency of specific pairwise interactions. For example, recombination between *rrnD* and either *rrnB* or *rrnE* occurs at a frequency of only 10^{-5} (79). *rrnD* is oriented in opposition to *rrnB* and *rrnE* and is separated from them by about 18 min. Nevertheless, the frequency of 10^{-5} is still large for a spontaneous genotypic change, and it is many orders of magnitude larger than the frequency of base substitution. It should be noted that DNA-damaging agents such as UV irradiation can increase the frequency of *rrn* recombination by at least 2 orders of magnitude (76, 79).

Given the frequency of crossover between *rrn* operons, the apparent stability of the genetic map of the enteric bacteria is remarkable (160, 168). Despite the divergence of the *E. coli* and *S. typhimurium* genomes by roughly 1 million base substitutions, the gross linkage maps, particularly the positions of the *rrn* operons, are highly similar. At least three exchanges between *rrn* operons have, in fact, occurred either in the natural population or without immediate recognition in early laboratory stocks. *E. coli* K-12 and *S. typhimurium* LT2 differ by a reciprocal exchange of spacers between *rrnD* and *rrnB* (117), and there has been a conversion of the *rrnB* spacer carried by the subline of *E. coli* K-12 that contains Cavalli Hfr (see above). In addition, an inversion, IN(*rrnD-rrnE*)1, occurred in another *E. coli* K-12 subline that contains such commonly used strains as W3110 and W3102 (81). IN(*rrnD-rrnE*)1 involves 18 min of the chromosome, and the replication origin is 6 min from the *rrnE* end (Fig. 2). The IN(*rrnD-rrnE*)1 mutation causes a 2.7% growth disadvantage, probably reflecting displacement of genes outside the inversion toward or away from the replication origin (79). A larger inversion, IN(*rrnG-rrnE*), which is much more asymmetric with respect to the origin, has severely detrimental effects (79). Another large and extremely asymmetric inversion, IN(29–78), also has severely detrimental effects (126). The IN(29–78) inversion was shown experimentally to affect relative gene dosages along the rapidly replicating chromosome (31). Gene expression can be affected by the distance of the gene from the origin in a manner that correlates with the predicted gene dosage gradient in a population of partially replicated chromosomes (173). Another possible effect of inversions is to alter the relative positions of the origin and the terminus of replication if they span the origin asymmetrically. This symmetry is important for cell growth (125). Consistent with these ideas is the observation that the IN(*rrnG-rrnH*) inversion is relatively harmless despite its huge size (70). While IN(*rrnG-rrnH*) includes half of the chromosome, it is virtually symmetrical with respect to the origin (Fig. 2). The implication is that although rearrangements mediated by *rrn* operon recombination, as well as those derived by other means, may be present at high levels in all populations, the combination of growth disadvantage and frequent reversion to wild type would prevent the establishment of mutant clones in nature (79). In contrast to the conservation of chromosomal organization seen for *E. coli* and *S. typhimurium,* recent results with *Salmonella typhi* show that its chromosomal organization differs considerably and that a series of inversions and transpositions between rrn operons can account for much of the rearrangement (S.-L. Liu and K. E. Sanderson, personal communication).

The efficiency of mating between *S. typhimurium* and *E. coli* is strongly reduced by the general level of sequence divergence (131, 157). A high proportion of the infrequent recombinants are merodiploid (9). If markers in the intervals between the *rrn* operons are selected in interspecies conjugation, merodiploid recombinants are particularly prevalent (119). The *rrn* operons evidently have two significant roles. First, the *rrn* operons of *E. coli* and *S. typhimurium* are much more conserved than most genes and are more efficient sites for crossover. Second, their repetition provides opportunity for unequal recombination leading to duplication. Consequently, the interspecies merodiploids can retain a full set of recipient genes and thus avoid possible incompatible combinations. ISs, which can be highly conserved between species (14, 15), might serve a similar role in facilitating interspecies transfer by providing long stretches of sequence identity.

tRNA Genes

Many tRNA genes are present in multiple copies in *E. coli,* and they are organized in a variety of ways (51, 103). The example of the tRNA genes in *rrn* operon spacers has already been mentioned. In addition, both *rrnC* and *rrnH* encode tRNA$_1^{Asp}$ at their distal ends (43). Beyond these cases, identical tRNA genes sometimes occur as tandem repeats in the same transcription unit. Examples include tRNA$_{f1}^{Met}$ (99), tRNA$_3^{Gly}$ (92, 103), and tRNA$_1^{Leu}$ (39), all of which occur as tandem triplications, tRNA$_2^{Arg}$, which occurs as a tandem quadruplication (103), and tRNA$_1^{Tyr}$, which occurs as a tandem duplication. Interestingly, genes for tRNA$_3^{Gly}$ and tRNA$_1^{Leu}$ occur in additional single copies, unlinked to the triplicate versions (51). The *metT* operon presents a particularly complex arrangement. In this operon, duplicate copies of tRNA$_1^{Gln}$ and tRNA$_2^{Gln}$ occur in tandem, while duplicates of tRNA$_m^{Met}$ are separated by other tRNA genes. In those cases where the duplicate copies are in the same transcription unit, the advantage of the arrangement would seem to be simply the capacity for more product. When the genes are unlinked, the possibility of differential control should be considered. A correlation seems to exist between tRNA gene multiplicity and codon usage. For example, tRNA$_3^{Gly}$, which translates the preferred glycine codons GGU and GGC (178), is present in four copies, while the species responsible for the rarer codons, GGA and GGG, are present as single copies (85). Similarly tRNA$_1^{Leu}$, present in four copies, translates the preferred leucine codon CAG. The tRNA sequences of *S. typhimurium* and *E. coli* tend to be highly conserved and often identical. Recombination between tandem copies of tRNA genes has been observed (99, 164).

IS Elements

Several IS elements are present in multiple copies in both *E. coli* and *S. typhimurium* (see chapter 111). Independent natural isolates tend to have distinctive IS profiles, although closely related strains show statistically significant conservation of IS position (67, 114, 115, 170). The IS patterns of strain subclones can vary considerably (140, 199). Aside from potential to cause rearrange-

ment through site-specific recombination, IS elements provide sites for homology-dependent chromosomal rearrangement. There are numerous reports of repeated ISs serving as endpoints for deletions, duplications, and large inversions (106, 126, 169, 192, 198, 199). Their role in F-factor integration is particularly important (30). At least for IS*5,* this recombination is homology dependent and *recA* dependent, and it does not depend on element-encoded functions (193). IS elements have been implicated in the creation of an interesting genetic redundancy found in *E. coli* K-12. Ornithine carbamoyltransferase of *E. coli* K-12 is unusual in that this enzyme is encoded by two unlinked loci, *argI* and *argF* (201). While clearly homologous, these genes share only 78% nucleotide identity. The *argF* locus is absent from *E. coli* B and W (116). In K-12, the *argF* locus is flanked by IS*1* elements (90, 216). This duplicate gene likely evolved in a related species and entered a progenitor of strain K-12 facilitated by the IS*1* elements.

Prophage Insertion Sites

Some classes of temperate phage insert into the host chromosome by site-specific mechanisms that recognize both phage and host attachment sites. These sites share a homologous core sequence (24). In the case of lambda, the identity segment is 15 bp. Generally speaking, the crossover occurs within the identity segment, and once integrated, the prophage is flanked by duplicate copies of the homology. A number of phage or phagelike elements integrate within host structural genes. The attachment site of these phages contains the 3′ end of the target gene so that it replaces the portion of the gene displaced by the integration. The consequence of this mechanism is that the target gene remains intact, and the prophage is flanked by duplications of the 3′ end of the gene. Both protein and tRNA genes are used for attachment. For example, e14 (80), phage 21 (24), and Atlas (24, 134) all integrate into protein genes. The gene used by both e14 and 21 is the *icd* (isocitrate dehydrogenase) locus. P22 inserts into a threonine tRNA gene (24), DLP12 inserts into an arginine tRNA gene (123), and P4 inserts into a leucine tRNA gene (151). In each case, the displaced 3′ end is replaced by an equivalent sequence from the phage.

Rhs Elements

A major source of repeated sequence in the *E. coli* K-12 genome is the *Rhs* element family (122). The five *Rhs* elements of strain K-12 have been mapped (Fig. 2) and sequenced (49, 166, 217). These unusual elements are complex composites of distinct components. Some components are conserved, while others are divergent or even unique. The largest *Rhs* element is 9.6 kb long, and collectively the elements comprise 0.8% of K-12 DNA. Recombination between conserved portions of *RhsA* and *RhsB* produces a characteristic duplication that includes 3% of the chromosome (hence the name "rearrangement hot spot") (25, 122). Generation of this specific duplication is *recA* dependent, and there is no indication that it requires a specific *Rhs* function (122).

Rhs **Organization.** The most prominent *Rhs* component is a 3.7-kb *Rhs* core (Fig. 4A), and core homology is present in each of the five K-12 elements (Fig. 4B). This core maintains a single open reading frame (ORF) throughout its length, with the start codon coinciding with the first base of the homology. Remarkably, the respective core ORFs extend up to 177 codons beyond the homology. Thus, the *Rhs* elements are predicted to produce a set of roughly 160-kDa proteins with long, conserved N termini and shorter, dissimilar C termini. The core ORF is immediately followed by another ORF, termed the downstream ORF. Typically, the downstream ORFs contain from 100 to 200 codons, and each is predicted to have a signal sequence for export across the inner membrane (217). In two cases, this capability was proven through protein fusions with alkaline phosphatase (84). Like the adjacent core extensions, most of the downstream ORFs are unique, showing no homology with other downstream ORFs. An additional *Rhs* component is an insertion sequence, the H-rpt. Classification of the H-rpt as an IS is based in part on its homology to IS elements in other organisms. In both *Vibrio cholerae* (191) and *Salmonella enterica* serovar Strasbourg (212), homologous sequences are found at the *rfb* locus, linked to determinants of O-antigen variation. An H-rpt homolog in *Aeromonas salmonicida* has been shown to have transposition activity (69). This element, IS*AS2*, is 57% similar to the H-rpt over 335 amino acids.

No individual *Rhs* element has precisely the structure depicted in Fig. 4A. The left half of the core of *RhsE* is deleted. The H-rpt can be absent, defective, or present in multiple copies. In some elements, the distal portion of the core is repeated one or more times, and each core repetition is accompanied by an additional core extension and downstream ORF (165). Not all

natural *E. coli* strains have *Rhs* elements, and comparison of *E. coli* strains with and without *Rhs* elements has been used to define the boundaries (49). Different *Rhs* elements replace from 10 to 807 bp of the reference chromosome (49, 217), and *RhsD* replaces a 224-bp bacterial interspersed mosaic element (BIME) (166). A general observation of accessory elements is that their ends are related by some kind of sequence repetition, but no sequence similarities are observed when the termini of the *Rhs* elements are compared. This holds both when the left end of an element is compared with its right end and when the ends of different elements are compared.

The *Rhs* Core. The most striking feature of the *Rhs* core ORF is a peptide motif that is repeated 28 times (49). The motif can be written xxGxxRYxYDxxGRL(I or T)xxxx, and in one cluster, it repeats with an average periodicity of 21 amino acids. This large protein is predicted to be strongly hydrophilic, but it has a hydrophobic region near the N terminus that could serve as a membrane anchor. It has been proposed that the core proteins are ligand-binding proteins of the cell surface (217). This idea received strong support from the report of a wall-associated protein (WAPA) of *B. subtilis* (50). WAPA is an abundant, nonessential protein. It is derived from a giant precursor encoded by a 7,002-bp ORF, and its C-terminal domain contains 31 copies

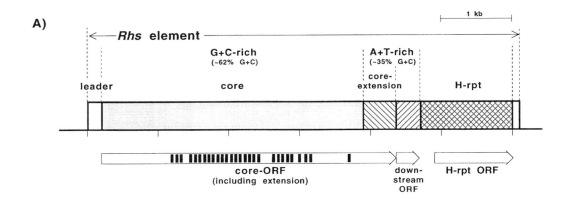

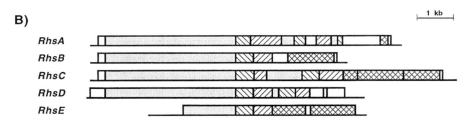

FIGURE 4 Structure of the *Rhs* elements. (A) The various *Rhs* components are shown in their usual positions in an idealized composite element. The boxes denoting different components are identified in the diagram, and the thin line denotes flanking genomic DNA. ORFs are indicated by the arrows, and the repetitions of the peptide motif are marked by bars within the core ORF. (B) Schematic representation of the organization of the five *Rhs* elements of *E. coli* K-12. The symbols are as in panel A. Note that no specific element has exactly the idealized structure depicted in panel A. Partial core repetitions found in *RhsA*, *RhsC*, and *RhsD* generally contain the 3′ end of the core and are followed by unique A+T-rich core extensions. The exceptional core fragment is a small piece from the 5′ end found in *RhsD*.

of a motif that closely resembles the *Rhs* motif. The *Rhs* motif also resembles motifs associated with bacterial cell surface proteins involved with carbohydrate binding (207, 211). If the core protein is a cell surface component, the mechanism of its export is problematic because it does not have a good signal sequence. A possible function of the downstream ORFs, which do have signal peptides, is to assist export of the core protein (84).

Rhs Origins. The *Rhs* cores fall into three distinct subfamilies based upon sequence divergence. The *RhsA-B-C* and the *RhsD-E* subfamilies are about 22% divergent at the nucleotide level, while divergence within the subfamilies is limited to between 1 and 4% (166, 217). The prototype of a third subfamily, *RhsG*, has been detected in other strains of *E. coli*, and partial sequence analysis indicates that *RhsG* is about 22% divergent from each of the other two (84). This degree of mutual divergence is greater than that observed for homologous genes in *E. coli* and *S. typhimurium* (143). The cores of all three subfamilies have G+C contents of about 62%, while the core extensions and downstream ORFs are only 35% G+C. Both of these values are significantly different from the 51% value found for the average *E. coli* gene. Our picture of *Rhs* evolution is a complex one. Apparently, the cores evolved into three subfamilies in a high-GC-content species. Separately, the core extension/downstream ORF combinations evolved in a high-AT-content species, diverging to a much greater degree. Much more recently, the components joined and entered the *E. coli* species. It seems especially significant that despite the relatively ancient divergence of the *RhsA* and *RhsD* cores, both have the same 28 repetitions of the *Rhs* peptide motif (166). This finding suggests that the motif repetitions were in place at the time of divergence, and that strong selective pressures exist for maintaining this aspect of *Rhs* structure.

SMALL EXTRAGENIC HIGHLY REPEATED SEQUENCES

In noncoding regions, the *E. coli* and *S. typhimurium* genomes contain a number of highly repetitive sequences (usually more than 30 to 50 occurrences per genome). Despite the fact that they are highly repeated, they do not constitute more than 2% of the total bacterial DNA. This explains why they were not detected by the classic C_0t analysis, which revealed eukaryotic highly recurrent sequences (19). They were essentially discovered in the last 12 years or so, usually by inspection or computer analysis of sequence data or by hybridization experiments. As the sequencing of the *E. coli* genome progresses to completion, computer analysis should lead to an exhaustive listing of such sequences and may reveal yet unsuspected relationships between them.

This section deals with the six classes of highly repetitive sequences which have been identified so far (Table 1). These sequences are rather short, in comparison with IS sequences, for example, do not usually encode proteins, and are in all cases except one (*iap* sequences) dispersed throughout the chromosome. They will be presented in the following order: BIMEs, with a review of their possible functions, intergenic repeat units (IRUs), box C sequences, RSA sequences, *iap* sequences, and Ter sequences (a class of *rho*-independent terminators).

BIMEs

Structure and Evolution. BIMEs are highly repetitive sequences found initially in the genomes of *E. coli* and *S. typhimurium* (64, 65). About 500 BIMEs are scattered over the whole bacterial

chromosome, where they appear to be homogeneously distributed. BIMEs are found in extragenic locations at the 3′ ends of operons or between two genes of an operon, but rarely upstream of the first gene of an operon, and as far as we know, they are transcribed.

BIMEs are a mosaic combination of several short sequence motifs (64). One of these motifs, called PU (for palindromic unit) or REP (for repetitive extragenic palindromic) (Fig. 5), is present in all BIMEs and was the first to be described as a palindromic repetitive sequence (75). PUs occur usually in clusters in which they are associated with several sequences (called extra-PU sequences), belonging to a total of seven possible motifs.

BIMEs are composed of 10 short DNA motifs. The description of BIMEs in terms of their component motifs requires that each motif be clearly defined and named. This is done in the rest of this section, which primarily establishes the nomenclature.

Two types of PUs, Y and Z, have been distinguished according to the seventh position of the consensus, which is a G and a T, respectively. There are two PU motifs of the Y type, called Y and Y*, and two of the Z type, called Z^1 and Z^2 (8, 63). These motifs differ in size and in sequence (Fig. 5). Y* motifs are smaller than typical PUs (15 to 22 bp), are located between convergent operons, and are used as bidirectional *rho*-independent transcription terminators (see below).

When a BIME contains several PUs, successive occurrences of PUs are separated by short sequences (of up to 40 bp), which we called extra-PU sequences (64) (Table 2). Because of the strict alternation of successive PU orientations within clusters, extra-PU sequences are located either between the head ends of the two flanking PUs (and are called head internal sequences) or between the tail ends of the two flanking PUs (and are called tail internal sequences). Head internal sequences can be separated into two motifs, S (12- to 14-bp-long sequences) and L (32- to 34-bp-long sequences). There is a consensus for the L motif, which contains the consensus of an integration host factor

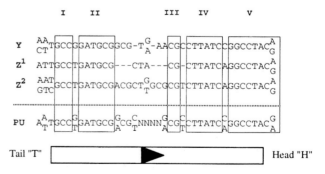

FIGURE 5 Three *E. coli* PU motifs. The consensus sequences of the three PU motifs are aligned. Y-type PUs are 38 bp long and possess a G at the seventh position. Z-type PUs possess a T at position 7 and are subdivided into Z^1 (33 bp long) and Z^2 (40 bp long). Below the three motifs is given the consensus previously found for PUs (57). Sequences similar among the three consensus sequences are boxed and numbered I to V. The lower part of the figure shows a schematic drawing of a PU; the triangle indicates its orientation, from the tail end to the head end. In previous publications, "tail" was named "left" and "head" was named "right."

TABLE 1 Main properties of extragenic highly repetitive sequences

Sequence	Size (bp)	Bacterial hosts	Tropism for location	Exchange
BIME	40–500	Escherichia coli/Shigella spp.; Salmonella typhimurium; Levinea, Citrobacter, Klebsiella, and Enterobacter spp.	IRU (S. typhimurium) Box C (E. coli)	IRU (malE-F in K. oxytoca; glnA-L in S. typhimurium) Box C (araA-D in K. pneumoniae)
Box C	56	Escherichia coli/Shigella spp.; Klebsiella pneumoniae; Rhizobium spp.	BIME (E. coli)	RSA (envM in S. typhimurium)
IRU	127	Escherichia coli/Shigella spp.; Salmonella typhimurium; Citrobacter, Levinea, Erwinia, Klebsiella, Vibrio, Serratia, Yersinia, and Xenorhabdus spp.	BIME (S. typhimurium) IRU (X. luminescens)	BIME (glnA-L in S. typhimurium; malE-F in K. oxytoca)
RSA	152	Escherichia coli, Salmonella typhimurium, Erwinia carotovora		Box C (envM in E. coli)
iap	29	Escherichia coli, Shigella dysenteriae, Salmonella typhimurium		
Ter	30	Escherichia coli		

(IHF)-binding site in its central part (18, 147). The S sequences are less conserved (Table 2). Two short motifs called s and l belong to the tail internal sequences. A third group of tail internal sequences called r is composed of a few sequences ranging from 18 to 31 bp which do not exhibit any sequence similarity. Two external motifs, flanking the tail end of the last PU, are present in a subset of BIMEs (reviewed in reference 65). They are juxtaposed either to a Z-type PU (called the A motif) or to a Y-type PU (called the B motif).

Two major BIME families. BIMEs with more than two PUs can be described as direct repetitions of a given association of PUs and extra-PU motifs. It is noteworthy that even for extra-PU motifs with poor sequence conservation (the S motif or the r sequences), the sequence similarity of two motifs within the same BIME is higher than for sequences originating from two different BIMEs (64).

BIME organization is variable, since these elements may contain different numbers of PUs (from 1 to 12) and since PUs can be associated with several motifs among a total of seven. However, E. coli BIMEs containing at least two PUs belong to two major BIME families, called BIME-1 and BIME-2 (Fig. 6) (8). Members of the BIME-1 family are composed of only two PUs, one Y and one Z^1, associated with an L motif. The external motifs A and B are frequently found in BIME-1. Members of the

BIME-2 family are BIMEs with 2 to 12 PUs, which are Y and Z^2, and are associated with the motifs S and s or l. External motifs are rarely present in BIME-2. Another difference between the two families is that BIME-1 members are mostly located after the last gene of an operon (18, 147), while BIME-2 members are located either between genes or after the last gene of an operon. However, the two families seem evenly distributed on the E. coli chromosome. The presence of two major BIME families on the E. coli chromosome could reflect a functional specialization of these sequences (see below).

BIMEs in other bacteria. BIME motifs were first identified in E. coli because of the high number of chromosomal sequences known in this species. However, BIME-like sequences were identified recently in other bacteria. For example, PUs are known in S. typhimurium and relatives and also in several Klebsiella species and relatives (6, 56, 62). In these bacteria, some structural features of the PU are identical to those of E. coli, but their sequences are species specific. The more divergent PU sequences are found in bacterial species that belong to phylogenetically more distant groups. For example, the major difference between S. typhimurium and E. coli PUs is the presence of an additional base pair in S. typhimurium sequences (underlined in Fig. 7), while Klebsiella PUs exhibit sequence singularities other than this additional base pair (not shown) (6). As in E. coli, there are several

TABLE 2 Extra-PU motifs in BIMEs

Extra-PU motifs[a]	Schematic representation	Motif name and size (bp)	Consensus[b]
Head internal sequences		L, 32–34 S, 12–14	AAAGCATGCAAATTCAATATATATTGCAGRRATCR RWTYSGCAMCRACT
Tail internal sequences		l, 8–9 s, 1 r, 18–30	CGSWGCAC C
Tail external sequences		A, 20 B, 20	TTTGCGTTTGTCATCAGTCT AACAAAGCGCACTTTGTCAR

[a]Internal sequences are between two successive PUs, while external sequences form the border of the BIME, and flank only one PU. Head internal sequences (HIS) and tail internal sequences (TIS) are located respectively between the head ends and the tail ends of flanking PUs. Tail external sequences (TES) flank the tail end of the last PU of the BIME. In previous publications, tail was named "left" and head was named "right."
[b]R = A or G; Y = C or T; M = A or C; W = A or T.

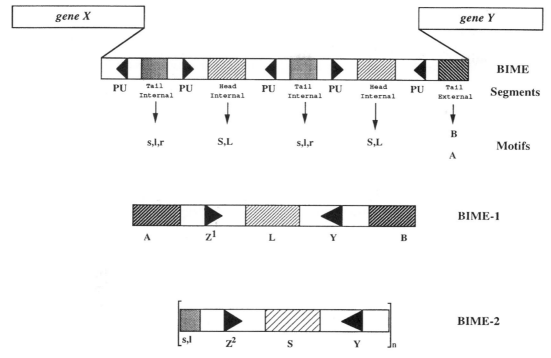

FIGURE 6 Two major BIME families. The upper part represents the schematic organization of BIMEs. Successive occurrences of PUs have alternate orientations and are separated by short extra-PU motifs. According to the ends of the PUs flanking these motifs, they are called head internal or tail internal sequences. In some BIMEs, the tail end of the last PU is flanked by a motif and hence called tail external. The lower part shows the structural organizations of the two major BIME families. BIME-1 is composed of two PUs (Z^1 and Y), separated by the L motif and often associated with the two external motifs. BIME-2 contains a variable number of motifs; the PUs are Z^2 and Y and are associated with the S motif and to either s or l.

PU motifs in *S. typhimurium*; we were able to identify four different PU motifs (S. Bachellier, E. Gilson, and M. Hofnung, unpublished data), three of them being homologous to the *E. coli* Y, Z^1, and Z^2 motifs (Fig. 7). *S. typhimurium* PUs are associated with short sequences in a BIME-like structure. However, the sizes of extra-PU sequences are more variable than in *E. coli*; hence no consensus was determined, and no hybridization was obtained on the *S. typhimurium* DNA with the *E. coli* L motif as a probe (64). In conclusion, while BIMEs are present in *S. typhimurium*, their components, other than the PUs, are different in the two bacteria. This is not the case in *Klebsiella* species, in which the sizes of the extra-PU motifs are homogeneous and in some cases identical to the sizes of *E. coli* motifs (not shown) (6).

BIME intraspecific variations. To gain insight on the DNA rearrangements associated with BIME regions in *E. coli*, we undertook a systematic comparison, in 42 of the 72 strains forming the ECOR collection (144), of a subset of intergenic regions containing BIMEs in *E. coli* K-12. The observed BIME local variations between strains are described below.

BIMEs belonging to the BIME-2 family exhibit a polymorphism of repetition: the motifs present in the BIMEs are the same in all strains, but the number of repetitions of the BIME-2 basic motif combination is different than in *E. coli* K-12. As deduced from the phylogenetic relationships of ECOR strains

(74), there can be either an increase or a decrease of this number of repetitions (Bachellier et al., unpublished data). Different mechanisms could explain such a result. It has been shown that DNA polymerase I (Pol I) can generate a polymorphism of repetition because of DNA strand slippage during a process called reiterative replication (108). It is worth noting that BIME DNA has affinity for Pol I (see below) (61). This affinity could cause pausing of the polymerase at BIME DNA, favoring the slippage reaction. It is also possible that homologous recombination events occur within a single BIME, leading to the deletion of a part of the element.

Members of the BIME-1 family do not vary locally. The BIME structure is invariant, but in two cases, we observed no BIME in the intergenic regions of several ECOR strains, while the *E. coli* K-12 region had one (Bachellier et al., unpublished data). Such a result could be explained either by a deletion of the BIME in some ECOR strains or by an insertion of BIME-1 in K-12 and the other ECOR strains. As in the two intergenic regions the BIMEs are flanked by direct repetitions of non-BIME sequences, the deletions can easily be explained by homologous recombination events between the repeats. Conversely, it can be hypothesized that the direct repeats originate from the duplication of the BIME insertion site, as has been described for transposable elements. However, the sizes of the repeats in the two regions (30 and 34 bp) are larger than the average size of IS target site duplication, for example (3 to 13 bp) (54).

BIME spreading and sequence homogenization.

Spreading. The interspersed distribution of BIMEs on the bacterial chromosome suggests that these sequences have been dispersed. However, their transposition has never been reported, which could indicate that the mechanism(s) used for their spreading has been lost or occurs at a very low frequency. Many hypotheses for BIME spreading can be imagined, according to known mechanisms, such as (i) formation of a duplex DNA molecule (slDNA) in the presence of a stable stem-loop structure during replication (146) or (ii) reverse transcription of BIME-containing mRNAs by Pol I or specialized reverse transcriptases originating from retrons (113, 121) or group II introns (48), which are found in several *E. coli* strains. In addition, since BIMEs do not possess an ORF which could lead to the synthesis of proteins used for their own transposition, it can be hypothesized that there are on the chromosome a few copies of active or complete elements leading to the synthesis of proteins involved in their spreading. Such a phenomenon has already been described for several eukaryotic repeated sequences (for example, L1 [46]). It can also be hypothesized that BIME spreading relies on proteins of other transposable elements, like ISs, and/or on host proteins. It has indeed been shown that IS transposition requires some bacterial proteins, namely, DNA Pol I, DNA gyrase, and IHF (for a review, see reference 54; see also chapter 124), which are known to interact specifically with BIME DNA (see below).

Homogenization. The sequences of the BIME motifs exhibit a high level of species specificity (see previous section). This had already been reported for eukaryotic repeated sequences and attributed to "concerted" evolution (for reviews, see references 36 and 37). Two major models have been established to explain the mechanisms of sequence homogenization. The first is a succession of nucleotide sequence variations of the repeats, but new sequences appear in the genome, from intact copies, through a mechanism of duplicative transposition. The second model, called gene conversion, involves nonreciprocal information transfer between two repeats (reviewed in reference 36). In the case of BIMEs, the presence of different extra-PU motifs could avoid frequent exchanges between nonidentical BIMEs, leading to the fixation of the two major families (BIME-1 and BIME-2).

BIMEs as Multifunctional Genetic Elements. BIME sequences appear to participate in seemingly disparate functions: transcription, translation, chromosome organization, and stability. However, it is still unknown whether BIMEs are essential for bacterial viability. If such a role exists, it does not require all of the BIMEs present on the chromosome since the removal of one has no effect on cell growth (187).

Here, we summarize work on the multiple processes in which BIMEs have been shown to be involved. Finally, we present a model that attempts to explain this functional diversity in terms of different combinations of BIME motifs and of sequence context. This leads to the idea that noncoding repeated DNA can be a source of different functions through a number of sequence variations within or around BIMEs.

BIMEs and gene expression. BIMEs were first described as potential regulatory sequences because of their palindromic nature and ability to form stable stem-loop structures in transcribed RNA (75). In fact, among the BIMEs examined, many but not all participate in the stabilization of the 3′ end of mRNA and subsequently in the expression of the upstream gene; a small subset is implicated in a *rho*-independent transcription termination event, and only one seems to be involved in the translational control at the ribosome binding site.

A subset of BIMEs acts as bidirectional transcription terminators. None of the examined BIME-2 motifs located between two cotranscribed genes *(hisJ-Q, lamB-malM)* act as a transcription terminator (63, 187). The major messenger endpoint of several E. coli operons (for example, *glnA* [128] and *rhaA-D* [136]) was mapped at a typical factor-independent transcription terminator located next to, but clearly distinct from, either BIME-1 or BIME-2. However, members of a subclass of PU, called Y*, act as bidirectional transcription terminators (see above; reviewed in reference 58). Interestingly, all of the known BIMEs containing one Y* are located between two convergent ORFs and account for most of the DNA in these regions.

mRNA turnover and retroregulation. In numerous operons, the gene located upstream of BIMEs has a much higher expression level than the downstream gene. For example, in the *S. typhimurium hisJQMP* and *E. coli malEFG* and *deoCABD* operons, the first gene, followed by one BIME-2 sequence, is expressed at up to 40 times the level of the distal genes (57, 142, 187, 200). The mRNA corresponding to the proximal gene is over-represented compared with the full-size transcript of the operon, with a 3′ end located precisely in the BIME-2 region. Since deletions within the BIME-2 region decrease both the expression of the proximal gene and the amount of the transcript ending at BIME-2, it was proposed that BIMEs act as retroregulators by stabilizing the mRNA of the upstream gene. This stabilization, probably due to a protection of the RNA against a 3′-5′ exonuclease activity, can be explained by the ability of BIME-2 RNA to form complex secondary structures. However, the extent of this BIME-2 effect cannot account for all of the differential expression observed in these operons. Indeed, a total deletion of the BIME-2 sequence between *hisJ* and *hisQ* leads to only a twofold decrease in the expression of *hisJ* (188). This effect on mRNA stability is not confined to BIMEs located between two genes of the same

FIGURE 7 Four PU motifs in *S. typhimurium*. PU sequences of *S. typhimurium* were subdivided, according to size and sequence, into four groups, for which a consensus was found. Groups A, B, and C present sequence similarities with the *E. coli* Y, Z^1, and Z^2 motifs, respectively. Sequences of group D are not homologous to any *E. coli* motif and could constitute a new PU motif specific to the *S. typhimurium* genome. Positions of the sequences which differ from the consensus are in lowercase letters; dashes represent gaps that have been introduced to maximize the alignments; dots indicate when a sequence is not known; the positions underlined in the four consensus sequences are not present in the *E. coli* PUs. Each sequence is identified by the names of the flanking genes (as in Table 7).

A

```
... ....... ... ..CGc GA AgctC TT ATCCGGC CTACg        - adk
gGT GCCGGAT GGC GGCcT GA ACGCC TT ATCCGGC CTACt    adk - visA
tGT GCCGGAT GGC GaCaT AA AtGCC TT ATtCGGC CTACA    araA - araD
caT GCCGGAT GGC GagcT Gc gCGtC TT ATCCGGC CTACc    cchA - cchB
ATT GCCGGAT GGC GGtGc GA gCaCC TT ATCCGGC CTACA    eutG - eutH
ATT GCCGGAT GGC GaCGT A- ACGCC TT ATCCGGC CTACA    lexA -
ATT GCCGGAT GGC aGCGT AA ACGCC TT ATCCaGC CTACA         - livB
tGT GCCGGAT GGC -GCaT AA AtGCC TT ATCCGGC CTACA    malE - malF
Aca GCCGGAT acg GcaGa Ag ttGCg TT CgAtg            mviM - mviN
ATT GCCGGAT GGC GaCGT t- gCGCC TT -TCCGGC CTgaA    mviN - flgN
AGc GCCGGAT GGC GGCGc AA gCGCC TT ATCCGGC CTACA    nrdA - nrdB
tTT GCCGGAT GGC GGCGT GA ACGCC cg ATC.... .....        - nrdE
AGc GCCGGAT GGC GGCtT Ag AtGCC TT gTCCGGC CTACg    parC - parF
Aaa GCCGGAT aat GGCGT GA ACGCC TT ATCCGGC CTACg    pepM - ORF1
ATT GCCGGAT aGC GGCGc AA ACGCC TT ATCCGGC CTACA    ushA -
-------------------------------------------------
        G        A
A T GCCGGAT GGC GGCGT  A ACGCC TT ATCCGGC CTACA
    T                G
```

B

```
ATT GCCTGAT GGC Gt T-AC GC TT ATCAGGC tTgCG        adk - visA
ATT GCCcGgT GGC aC T-GC Gt TT AcCgGGC CTACG        araA - araD
ccT GCCgGAT GGC GC T-tc GC TT ATCcGGC CTACA        araA - araD
ggg GCCTGAT GGC GC T-GC GC TT ATCAGGC CTgCG        argA - recD
ATT GCCTGAT GGt GC g-ca Ga TT -caGGC CTACG         cchA - cchB
tTg aCCTGAT GGC GC T-GC GC TT ATCgGGt tcgaA        cheZ - flhB
AcT GCCTGAT GGC GC T-AC GC TT ATCAGGC CTACA        crr - orf287
cTT GCCTGAT GGC GC T-Gt GC TT ATCAGGC CTACA        cysJ - cysI
ATT GCCTGAT GGC Gt T-AC GC TT ATCAGGC CTgaA        hemA -
gTT GCCcGgT GGC GC T-GC GC TT AcCAGGC CTACA        hemN -
gac GCCTGAT GGC GC T-GC GC TT ATCAGGC CTACG        hisG - hisD
ATT GCCTGAT GGC GC T-Gt GC gT gTCAGGC CTACG        hisJ - hisQ
ccT GCCTGAT GGC GC T-AC GC TT ATCgGac CTggG        ilvH -
gga GCttGAT GGC GC T-AC GC TT ATCAGGC CTACA        leuD -
gTa GCCTGAT GGC GC T-GC GC TT ATCAGGC CTACG        lexA -
tTT GCCcGAT GGC GC T-tc GC TT ATCgGGC CTACG            - livB
Aaa GCCgGAT GGC GC TgAC GC TT ATCcGGC CTACA        livB - livC
tgT GCCTGAT GGC GC T-GC GC TT ATCAGGC CTACG        malE - malF
ATT GCCgGAT Gcg GC T-AC GC TT ATCcGGC tTAtc        melB -
AgT GCCTGAT aGC GC T-Gt GC T- ATCAGGC CTACA        mviN - flgN
AcT GCCTGAT GGC GC T-tc GC TT ATCAGGC CTACA        nadR -
AgT GCCgGgT aGC GC T-At GC TT AcCcGGC CTACG        nrdA - nrdB
tcT GCCTGAT GGC GC --GC aa cc tTaAGGC CTACG        ushA -
-------------------------------------------------
                    G              A
ATT GCCTGAT GGC GC T  C GC TT ATCAGGC CTAC
                    A              G
```

C

```
AAC GCCGGAT GAC GcCGt TcAAA GCcgC c- ATCCGGC CT-gtt     envZ -
AAT GCCGGA- GGC GACG- T-AAA -CGTC TT ATCCGGt tT-ACA     eutG - eutH
TTC GCCGGAT GAC GACGC g-cgt GCGTt TT ATCCGGt CT-ACt     ilvH -
cgC GCCTGAc GGC GACGC T-AAA aCGTC TT ATCAGGC CTtAcA     mviM - mviN
TTT GCCGGAT GGC GAtGC a-AT- GCGTC TT ATCCGGC CT-ACg     nadR -
AAT GCCTGAT G-C GACGC TtAcC GCGTC TT ATCAGGC CT-ACA         - nadR
gTT GCCGGAT GAC GACGC a-gTA GCGTC TT ATCCGGC CT-ACg     parC - parF
TTT GCCTGAT GGC GAtGC c-tTC GCaTC TT ATCAGGC CT-ACA     pepM -
AgC GCCTGAT G-C GACGC T-gcC GCGTC TT ATCAGGC CT-ACA     trpR -
-------------------------------------------------
AAT    G    G         TA            C
    GCC GAT G C GACGC T A    GCGTC TT ATC GGC CT ACA
TTC    T    A         AC            A
```

D

```
AaT GCCGGAT tGC --C TGg CGC TT ATCCGGC CTACA            - adk
cgg GCCGGgT GGC G-C TTg CGC TT ATCCGGC tTgtA       alr -
ATT GCCGGAT GGC Gcg TTg CGC TT AcCgGGC aTAat       cpsB - cpsG
ATC GCCGGAT GGC GGC -GC -aa gT A-CaGGC aaAaA       crr - orf287
Aaa aCCGGgT GGC GGC TTC GCC TT AcCCGGC CTACg       hemL -
tgT GCCGGAT GGC G-C Taa tGC TT ATCCGGC CTACA       hemN -
tTT aCtccAT caCaGGg TTC CGCcTT ATCCGGC CTACA       hisG - hisD
caC GCCGGAT GGC TGt GCC TT gcCCGGC CTACg           hisJ - hisQ
ATC GCCGAAT GGC Gct gGC GCC TT ATtCGGC tTACA       lamB - malM
tcT GCCGGAT GGC GGC TcC GCC TT ATCCGGC CTACA           - murB
ATC GCCGGAT GGC GGC TTC GCC TT ATCCGGC CTACA       uvrD - corA
-------------------------------------------------
        C               T GC
AT GCCGGAT GGC GGC T C   C TT ATCCGGC CTACA
    T               G CG
```

operon but holds for BIMEs located at the distal end of an operon (for example, *glyA* [152] and *gdhA* [11]).

In summary, both BIME-1 and BIME-2 sequences can participate in the differential expression within polycistronic operons by protecting the proximal part of the mRNA against exonucleolytic degradation. This effect depends on the number of PU motifs present within the BIME, since one PU does not stabilize mRNA (142), and on the structure of the transcript outside BIME, since the same BIME has different mRNA stabilization effects according to the operon into which it is inserted (219). Although the increase in the upstream gene expression is modest, it can be of biological importance in some operons. Indeed, the removal of part of the BIME-2 sequence in the *malEFG* operon leads to a decrease in *malE* expression and to a partial defect in maltose utilization (142).

We believe that this stabilization effect cannot account for the high level of BIME sequence conservation for the following reasons. (i) The presence or absence of BIME-2 in the *glpK-X* intergenic region has no effect on transport or on growth on glycerol (196). (ii) Insertion of the *hisJQ* BIME-2 downstream of the *atpH* gene does not affect the half-life of the corresponding mRNA (219). (iii) Since any stem-loop structure is sufficient to stabilize upstream RNA, for example as observed in the regulation of λ*int* expression by *sib* (171) or in the maturation of the 3′ end of the *trp* operon (139), it is difficult to involve RNA stabilization as an explanation for the sequence conservation of BIMEs, in particular for the nonpalindromic extra-PU motifs.

The BIME-2 of the *rplL-rpoB* region includes an RNase III processing site in one PU motif (57). The sequence of this PU is atypical: the upper part of the stem-loop is missing. Interestingly, some loose homology exists between the lower part and a known RNase III site in phage T7 (57). No other evidence exists for an association of a PU with RNase III processing. In particular, the BIME-2 present in the *hisJ-Q* region is not processed by RNase III in vitro (187).

BIMEs and translation initiation. The removal of the *hisJ-Q* BIME affects both the expression of the upstream *hisJ* gene (see above) and the translation of the downstream *hisQ* gene (188). The secondary structures that can be adopted in the RNA of the intergenic region could modify the accessibility of the ribosome binding site.

Functional organization of the bacterial chromatin. The observation that BIMEs specifically bind nucleoid-associated proteins (60) could provide a plausible cause for BIME sequence homogeneity. It has been shown that DNA gyrase (213), DNA Pol I (61), and IHF (18, 147) are able to specifically recognize BIME DNA. Gyrase and Pol I interact with the PU motif, while IHF binds in the center of the L motif (Fig. 8).

BIME-DNA Pol I complexes. Starting from a crude *E. coli* extract, two moieties which specifically protect a BIME-2 DNA against digestion with exonuclease III were purified. One of these involved Pol I. This interaction requires the presence of the PU motif. The other activity is less characterized but has been shown to be devoid of DNA gyrase (61). This finding was the first evidence that Pol I is able to bind intact duplex DNA. Whether BIME-1 DNA is also able to bind Pol I is not known.

The functional significance of this interaction is still a matter of speculation. BIME DNA could serve as a preferred entry site for Pol I, providing a specific pausing site for the polymerization reaction or playing a role in replication fidelity. A possible effect

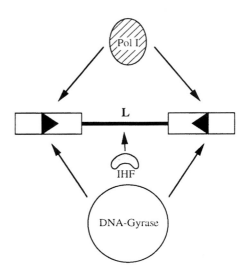

FIGURE 8 Summary of the known DNA-protein interactions occurring at BIME DNA. DNA Pol I and DNA gyrase both interact with the PU motifs, while IHF binds at the center of the L motif.

of the BIME-Pol I interaction is the amplification of BIME regions (see above).

BIME-gyrase complexes. Purified gyrase binds specifically to the PU motif of BIMEs (213). Differences in binding affinity of up to threefold have been observed between the different PU motifs and between BIME-1 and BIME-2, with the order $Y > Z^2 > Z^1$ and BIME-2 > BIME-1 (7, 8). From these findings, the following determinants for an efficient BIME-gyrase interaction are revealed. (i) PU motif appears to be the basic sequence recognized by gyrase. Interestingly, a critical size of 7 to 9 nucleotides (nt) in the central part of the PU seems important for an efficient binding, and sequence variability in this part of the PU can account for the differences in affinity that we observed between the three PU motifs (8). This finding strongly suggests that the two external parts of the PU, which are highly conserved between the different motifs, are directly recognized by gyrase only if a proper spacing between them is respected (7 to 9 nt).

(ii) In BIME-1 DNA, it appears that a particular arrangement of PU sequences impairs an efficient PU DNA-gyrase interaction (7). The L sequence is slightly bent, as revealed by a circular permutation assay (18). This DNA curvature could be unfavorable for a proper wrapping of DNA around gyrase (10). In BIME-2 DNA, the spacing between two PUs seems optimum to allow a proper wrapping around gyrase. The high gyrase affinity for BIME-2 DNA could be due to a binding of one gyrase dimer on a Y and a Z^2 sequence. Gyrase DNase I footprinting on a BIME-2 sequence has revealed a protected region covering both PUs, substantiating this hypothesis (213).

The fact that PU and BIME DNAs are specific binding sites for gyrase does not imply that they represent sites of catalysis. No strong cleavage site induced by oxolinic acid has been mapped within a BIME-2 DNA (213). A binding site in pBR322 that is not a cleavage site has also been mapped (102). This shows that a binding site does not necessarily define a site of cleavage and catalysis. For example, BIMEs could be preferred sites of gyrase binding without catalysis: gyrase could enter into

BIME DNA and then move along DNA or, alternatively, BIME DNA could be a stop sequence for gyrase movement. Evidence for such a linear diffusion of eukaryotic topoisomerase II has been reported (89, 148).

BIME-IHF complexes. Purified IHF binds specifically to all L DNA sequences examined (18, 147). Indeed, the central part of the L motif includes an *ihf* consensus sequence (YAANNN-NTTGATW) (55). For this reason, the L-containing BIMEs were called RIP, for repetitive IHF-binding palindromic elements (147), or RIB, for reiterative IHF-BIME (18). Since all of the L motifs are present in a BIME-1 structure and since the *ihf* consensus is highly conserved within the L sequences (18, 64), most, if not all, of the RIP/RIB elements belong to the BIME-1 family.

IHF is known to play a role in various processes, including site-specific recombination, transcriptional regulation, and replication (reviewed in reference 52). Since IHF binding induces a strong bend of 140° and since an IHF-binding site can be functionally replaced by an intrinsic curved sequence, it is believed that the main function of IHF is to facilitate the formation of higher-order protein-DNA complexes.

Higher-order nucleoprotein complexes at BIME DNA. At least three proteins have been shown to specifically bind BIME DNA in vitro: DNA gyrase, DNA Pol I, and IHF (Fig. 8). Evidence for the occurrence of these interactions in living cells is still lacking, except for IHF–BIME-1 complexes that have been shown to be formed in vivo (43a). A synergistic fixation of gyrase and IHF on BIME-1 has been observed in vitro (7). This cooperativity suggests that the inefficient binding between BIME-1 and gyrase is relieved by IHF. Changing the bend angle in the L motif can lead to a spatial closeness of both PUs, now in a favorable arrangement to wrap around gyrase. In this hypothesis, it is worth noting that the length of a BIME-1 DNA (the consensus has 144 nt) is within the range of the size of DNA wrapped around gyrase (120 to 140 nt) (156). This model is also in good agreement with a previously proposed role for IHF in the formation of higher-order specialized nucleoprotein structures (40) by the proper alignment of distant protein-binding sites. This model does not exclude a direct interaction between both proteins, but evidence for such an interaction is lacking (53). An increase in gyrase binding on BIME-2 DNA has been reported in the presence of HU, another histone-like protein that binds DNA with low specificity and that introduces a DNA curvature (214) (for a review on HU, see reference 38).

On the basis of the specific binding for gyrase, BIMEs have been proposed to be the counterpart of the eukaryotic scaffold-associated regions (58, 213), which are in vivo binding sites for topoisomerase II (98); scaffold-associated-region DNA is believed to be involved in the formation of chromatin loops and independent topological domains (for a recent review, see reference 162). Since BIMEs are almost exclusively located within transcriptional units, BIME-gyrase interactions could also be involved in the removal of positive supercoils generated ahead of RNA polymerase during transcription (18, 124). The presence of an IHF-binding site in some BIMEs suggests the formation of multiprotein-DNA complexes (reviewed in reference 52). The formation of such higher-order structures reinforces the idea that BIMEs play an important role in the architecture of the bacterial chromosome (18).

Redundancy and functional specialization. BIMEs are functionally diverse (see above), and a clear relationship can be drawn between the different BIME functions and variations in their own organization and in their flanking sequences. For example, (i) a subclass of PU (Y* [63]), and not the other PU sequences, is implicated in bidirectional transcription termination; (ii) only BIME-1 DNA binds specifically IHF (18, 147); and (iii) modulations in gyrase binding can be achieved by different combination of BIME motifs and by the binding of histone-like proteins, such as IHF and HU (7).

BIMEs are not the only case of noncoding repeated sequences that can be recruited to achieve different functions according to local sequence variation or genetic context. One can cite the following examples. (i) The repeated uptake sequences of *Haemophilus influenzae* and *Neisseria gonorrhoeae* are involved both in species-specific DNA transformation and in transcription termination when they are part of an inverted structure (66, 110, 195). (ii) Short eukaryotic interspersed repetitive elements, like the Ocr elements of *Xenopus laevis* and the rat ID, mouse B1 and B2, and human *Alu* sequences, appear to constitute modules in complex regulatory elements ensuring the coordinated expression of various genes (137, 167, 205). (iii) A class of repeated simple sequences acts as a functional telomere when located at the very ends of eukaryotic chromosomes or as a transcriptional regulatory element when located at internal sites (reviewed in reference 59).

Short interspersed repetitive sequences appear, thus, as building blocks of a variety of genetic elements. This extends the concept of gene duplication involved in genetic diversity (145) to noncoding repeated elements.

Other Short Repeats

IRUs. The IRUs (179) were initially identified in several members of the family *Enterobacteriaceae* during the analysis of the region adjacent to the 3′ end of the *tls* gene from *E. coli.* This region is homologous to 17 other bacterial regions. Another laboratory identified the same sequences and renamed them ERIC, for enterobacterial repetitive intergenic consensus (91).

General characteristics. IRUs are imperfect palindromic sequences which are about 125 nt long; they may form a stem-loop structure and can be oriented. They have been detected simultaneously in several bacterial species, and their sequences are very homogeneous, which allows establishment of a consensus (91, 179). IRUs were found in a number of *Enterobacteriaceae* (Table 3) as well as in *V. cholerae,* a species not very distant from members of the *Enterobacteriaceae.* These sequences are always extragenic. The majority of these sequences are transcribed. In contrast to BIMEs, IRUs are sometimes associated with promoter regions (in at least six cases). They are dispersed on the bacterial chromosome and have not been detected in bacteriophages or in plasmids. Their total number in bacterial genomes has been estimated to be between 30 and 150 (91, 179), and their number is not necessarily identical in all species. For example, IRUs seem more abundant in *S. typhimurium* than in *E. coli* (Table 4). By screening sequence banks, we found several supplementary IRUs, sometimes in other bacterial species (Table 3). This led us to establish a new consensus of 127 nt for IRUs (Fig. 9). As in the case of a number of BIMEs, some IRUs are located in corresponding intergenic regions of bacteria belonging to different species: *metE-R* (91, 179) and *ahpC-F* in *S. typhimurium* and *E. coli;* and *rpsU-dnaG* in *S. typhimurium, Levinea malonatica, Citrobacter freundii,* and *C. amalonaticus* (203).

TABLE 3 Distribution of IRUs (ERICs) in various bacterial species

Bacterium	No. of sequences	Comments
Citrobacter amalonaticus	1	See reference 203
C. diversus	1	See reference 203
C. freundii	1	See reference 203
Erwinia carotovora	1	1[a]
E. chrysanthemi	2	2[a]
Escherichia coli	14	4[a]
Klebsiella oxytoca	1	S. Bachellier, unpublished sequence[b]
K. pneumoniae	2	1[a]
Salmonella typhimurium	9	1,[b] 2[c]
Serratia marcescens	1	1[a]
Vibrio cholerae	4	3[a,d]
Xenorhabdus luminescens	6	See reference 133[d]
Yersinia pseudotuberculosis	2	

[a]Number of sequences which were not used to build up the first consensus.
[b]The IRU sequence lies at the same position as a BIME in *E. coli*.
[c]In the same region, there is one IRU and one BIME.
[d]A single region contains several IRUs, the numbers and the positions of which vary between two strains of the same species.

Variability of IRU sequences from different bacterial species. The sequences of IRUs do not vary much from one species to another, at least in comparison with BIMEs; for example, IRUs from *E. coli*, *Klebsiella* species, and even *Vibrio* species have homologous sequences, which can be detected by hybridization or by computer search in sequence banks (91, 179). Because of the small number of IRUs known for each bacterial genus, it is not possible to decide whether the variations with respect to the consensus from one species to another reflect a species specificity. It does seem clear, however, that if there is a species specificity for IRUs, it is much less apparent than in the case of BIMEs. Their sequences are indeed mostly conserved in distant species, which can be explained in several ways. They may have appeared recently and been dispersed, for example, via an association with a transposable element or by some other type of horizontal transfer. On the contrary, it is possible to assume that IRUs are ancient elements whose sequences have been conserved either because of gene conversion or because these sequences have a function which is selected for. The fact that IRUs are dispersed on the chromosome suggests a transposition step, but since there is no ORF, they do not carry information for a protein involved in transposition. There are no known functions for IRUs. All of the

TABLE 4 Compilation of IRU (ERIC) sequences

Sequence	Flanking gene(s)		GenBank accession no.	Observations
1[a]	*aceF*	– *lpd*	V01498	
2	*ahpC*	– *ahpF*	D13187	
3		– *aldH*	M38433	
4	*aspS*	–	X53863	
5	f83	– f143	U00039	
6	*fdx*	–	U01827	
7	*hpaC*	– *cts1*	Z37980	
8	*hsdR*	– *hsdM*	X06545	
9	*metE*	– *metR*	J04155	
10	*narK*	– *narG*	X15996	
11	*pgk*	– *fda*	X14436	
12	*pmbA*	– *cybC*	U14003	
13	*rhlB*	– *gppA*	X56310	
14	*rplA*	– *rplJ*	V00339	
15	*rpsB*	– *tsf*	V00343	
16	*yafP*	–	D38582	
17	*ahpC*	– *ahpF*	J05478	
18		– *cysJ*	M23007	
19	*glnA*	–	J01803	
20	*livB*	– *livC*	D00478	A BIME is distant from 30 bp.
21	*metE*	– *metR*	M17356	
22		– *nirC*	M64606	
23	*pepM*	–	X55778	The IRU is inserted within a BIME.
24	*rpsU*	– *dnaG*	M14427	
25	*tdk*	– *osmZ*	–	IRU sequence is given in reference 91.
26	*topA*	– *cysB*	M15040	

[a]The numbers identifying each sequence are the same as in Fig. 9. Sequences 1 through 6 are from *E. coli*; sequences 17 through 26 are from *S. typhimurium*.

data can be interpreted by saying that the IRUs are (or were) very mobile sequences, susceptible to local rearrangements (203).

The Box C Sequences.

Box C sequences as highly repetitive sequences on the *E. coli* genome. The box C sequences were initially described as sequences of 43 nt, located in extragenic positions, transcribed, and composed mainly of G and C; a consensus was deduced from the alignment of the first eight box C sequences identified (12). The 5′ end of the sequence is composed mainly of pyrimidines (mainly C), while the 3′ end presents a large proportion of purines (mainly G). The box C sequences are imperfect palindromes which can therefore be oriented.

Box C sequences were identified in a region which is partially homologous between *E. coli* and *S. typhimurium* (12). The *envM* gene is present in both bacterial species, but the adjacent nucleotide sequences are completely different. Upstream of the *envM* gene in *S. typhimurium*, there is a small ORF which is not present in *E. coli*; instead, there is a short DNA sequence, the box C sequence (12). The center of box C had been previously identified as a repetitive sequence present in five extragenic regions of *E. coli* K-12 (111). The existence of this sequence as a repetitive element was also reported following computer analysis of inverted repetitions on the *E. coli* genome (group IV in reference 16). Box C was used as a probe in hybridization experiments on the genomic DNAs from *S. typhimurium* and *E. coli* (12). The probe hybridizes with several genomic DNA fragments from *E. coli* but not with the genomic DNA from *S. typhimurium*.

Box C sequences exist in other bacterial species. Upon re-examination of the eight regions previously identified by Bergler and colleagues (12), we found four other box C sequences. The regions ahead of *fepB* and *envM* contain two box C sequences as well as the *mtlA-D* and *phnP-Q* intergenic regions. When two box C sequences are located in the same region, their orientation can be direct (*mtlA-D*, *phnP-Q*) or inverted (*fepB*, *envM*). The sequences located between the two box C sequences may vary substantially in length (from 8 bp in the *phnP-Q* region up to 131 bp ahead of *envM*). In a few cases, flanking regions exhibit sequence similarities (S. Bachellier, unpublished data). Searches in sequence banks led us to identify 28 regions containing box C sequences, including four in *Rhizobium* sp., a gram-negative bacterium quite distant from members of the *Enterobacteriaceae* (examples 24–27 in Fig. 10). By aligning all of the sequences (Fig. 10), we defined a new consensus which comprises 56 nt. In addition, two intergenic regions, one from a natural isolate of *E. coli* (ECOR 8 [144]) and the other from *Klebsiella pneumoniae*, which we sequenced in our laboratory, also contained a box C (Bachellier et al., unpublished data) (Table 5).

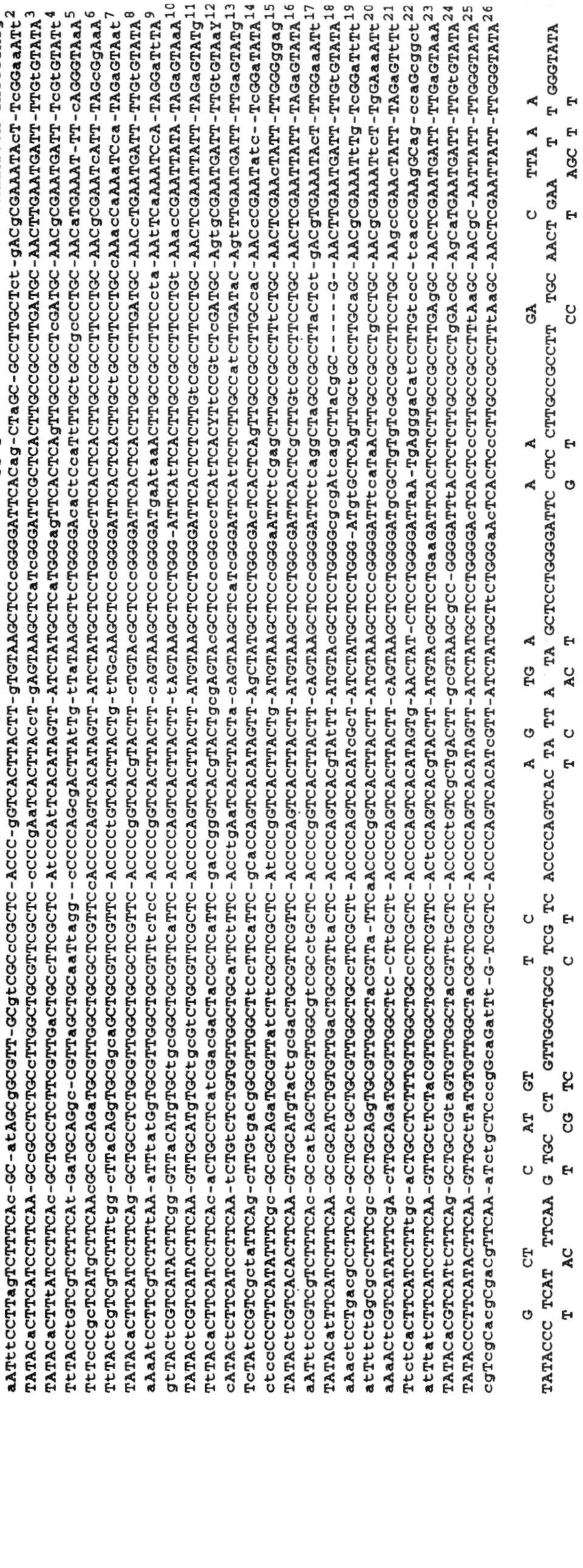

FIGURE 9 Alignment and consensus of IRU sequences. IRU sequences from *E. coli* (sequences 1 to 13) and *S. typhimurium* (sequences 14 to 23) are aligned. Below the sequences is given the consensus, which was deduced from the alignment of all available IRUs, including those from bacteria other than *E. coli* and *S. typhimurium*. The positions which are different from the general consensus are in lowercase letters; points indicate that the sequence is not known; dashes represent deletions introduced to optimize the alignment. The numbers following the sequences refer to the numbering in Table 4.

```
GGcCTGTTCCCCTCACCCTAACCCTCTCCCgA----Gg--GGG-CGAGGGGGAC-tG-TCCG-TGC 1
GG-CTGTTCCCCTCACCCTAACCCTCTCCCCA----AA--GGGGCGAGGGGAC-gG-AttG-TGC 2
GaAcaTTaaCCCTCACCCcAgCCCTCaCCCTG----GA--aGGGGAGAGGGGgC-aGaACgG-CGC 2'
GcGCTTTTCCCCTCACCCTAACCCTCTCCCCA----GA--GGGGCGAGGGGAC-CGtAttG-TGC 3
ctGCaGgaaCtgaacCggTcACCCTCTCCCtG----AA--aGaGCGAGGGGgC-aG-ACCG-aGC 3'
cGGgTTTTCCCCTCACCCTAACCCTCTCCCCA----AA--GGGGCGAGGGGAC-CG-TCCGgTGC 4
GGtcGTGTYCCCCTCACCCTAACCCTCTCCCCA---gGA--GGGGCGAGGGGAC-tG-TCCG-gGC 5
GGcCTGTTCCCCTCACCCTAACCCTCTCCCCA----AA--GGGGCGAGGGGAC-tG-TCtG-aGC 6
GaAtaTTTaCCCTCACCCcggCCCTCTCCCtG----AA--aGGGCGAGGGGga-aa-AgCG-TGC 6'
GGttTGcgCCCCTCACCCTAAtCCTCTCCCCA----tA--GGGGAGAGGGaAC-tG-cCaG-TGC 7
GtGaacgggCCCTCACCCTAACCCTCTCCCCA-----A--GGGGCGAGGGGAC-CG-TCCa-ctC 8
GtttgcgcCCCCTCACCCTAACCCTCTCCC-t----cA--GAG-AGAGGGGAC-CG-TtCG--GC 9
GGcCTTTTCCCCTCACCCcggCCCTCTCCCCG----GA--GGGGCGAGGG-Ag-aa-Aaaa-aGa 9'
cGAgTGggagCacggtttTcACCCTCTtCCCA----GA--GGGGcGAGGGGAC-tc-TCCG-aGt 10
GaGgcTTTtCCCTCACCCaAACCCTCTCC-gt----GA--GGaGAG-GGGcAg-Cgga-TGC 11
GcAtcaggCaatTaACCCc-ACCTTCTCC-at----Gt--GGaGAG-GGtGgg-at-T-gG-att 11'
caGggTTcCCtCTCACCCTAACCCTCTCCC-G----Gt--GGGGCGAGGGGAC-tG-ACCG-aGC 12
tattcGTaaggtTggtttTctCCCTCTCCCCG----t----GGGAGAGGG--C-CG--ggG-TGa 13
GcACgTTcaCCCTCACCCTAACCCTCTCCCtc----AA--GGG-AGAGGGGAC-CG-AtCG-aGC 13'
cGtacGTTCCCCTCACCCTAACCCTCTCCCCA----AA--GGGGCGAGGGGAC-CGtTCaG-TaC 14
acAaTTTTCCCCTCACCCTAACCCTtTtCCCG----AA--GGGGCGAGGGGAC-tG-TCCG-gGC 15
aacggGTTCCCCTCACCCTAACCCTcaCCCCA----AA--GGG-CGAGGGGAC-CG-TtCGtTGC 16
GaAaaaTTaCCCTCACCCCgACCCTCTCCCtG----GA--aGGa-GAGGaGAa-aa--CgG-TGC 17
GGcCTGTTCCCCTCACCCTAACCCTCTCCCCA----AAc-GGGGCGAGGGGAC-tG-ACCG-aGt 17'
cGcCTTTTCCCCTCACCCTAACCCTCTCCCCA----GA--GGGGCGAGGGGAC-CG-AtCG-cGC 18
aatCaGccCCtaTCAaCCgcctttaCgaatCA----AA---taaCGAtaaGgC-aG-TCCa-TcC 18'
GcGCTGgcCCCCTCACCCTAACCCTCTCCCCA----GA--GGGGcGAGGGGAC-CG-AttG-TGC 19
GaA-TaTTgCgCTCgtttTctCCCTCTCCCCA----tt--GGGGtGAGGGG-C-ga-TgCc-TGC 19'
acGgTTTTCCCCTCACCCTAACCCTCTCCCCA----GA--GGGGCGAGGGGAC-CG-ACCG-aGC 20
Gt-CgTTTCCCCTCACCCcggCCCTCTtCCCA----AA--GGGGCGAGGGGga-aa-ACaa-TGC 21
GGcCTGTTCCCCTCACCCTAACCCTCTCCCCA----AA--GGG-CGAGGGGAC-CG-TCCG-TGC 22
GaGgccgTCCCCTCACCCTtcCCCTCTCCCCG----tt---GGGGAGGGGGAC-CG-ACCG-gGC 23
cGACTTTgCCCCTCACCCTAACCCTCTCCCCAc-agGc-GGGGAGGGGGAC-tt-T--G-gGC 24
GGAgaGccCCCCTCACCCc-ACCCTCTCCCCGc-aaGc--GGGGCGAGGG-Ag-aG-cagGcaGC 25
acctaGcgCataTgACCCTAACCCTCTCCCCGtgaaGAacGGGGGAGAGGGGAC--GtTCCa-TaC 26
GGG-TTaaaCCCgaggaCTAgCCCTCTCCCCGc-aaGc--GGGGAGgGGGGACgtG-cCtG-atC 27
tGAaTGTcgatC-CgCCCTAgCCCTCTCCCCtc-tcGA--GGGGAGAGGGttg-gG-AgaG-gGg 28
-------------------------------------------------------------------
GGRCTKTTCCCCTCACCCTAACCCTCTCCCCR    RA   GGGGMGAGGGGAC CG WCCG TGC
```

FIGURE 10 Box C sequences: alignment and consensus. Shown is an alignment of all known box C sequences as well as the consensus which we deduced from it according to the method described in reference 64. This consensus is slightly longer than that in reference 12 (56 bp instead of 43 bp). Sequences 1 to 20 are from the *E. coli* K-12 genome, 21 is from *E. coli* ECOR 8, 22 is from *Shigella flexneri,* 23 originates from the *K. pneumoniae araA-D* intergenic region (Bachellier, unpublished data), 24 to 27 are from *Rhizobium* sp., and sequence 28 is from *Pseudomonas paucimobilis.* The 5' end of the box C sequence 24 is included in the end of the *nolC* gene of *R. fredii*; the stop codon TAA is in italics. The *E. coli* sequences are numbered according to Table 5. Underlined sequences belong to PUs flanking the box C (see text). Positions differing from the consensus (deletions, additions, or different nucleotides) are in lowercase letters.

TABLE 5 Compilation of box C sequences

Sequence[a]	Flanking gene(s)			GenBank accession no.	Observations
1	*adk*	–	*visA*	D90259	Presence of a BIME in the same intergenic region of *S. typhimurium*
2	*entC*	–	*fepB*	M29730	2 box C sequences in inverted orientation, separated by 11 bp, within putative ORF1
3		–	*envM*	M97219	2 box C sequences in direct orientation separated by 131 bp
4	f308	–	f418	L19201	
5	*fimH*	–	f477	U14003	Presence of a BIME in the same intergenic region at a distance of 18 bp
6	*gvcP*	–		L20872	2 box C sequences in inverted orientation separated by 11 bp
7	*glnH*	–	*glnP*	X14180	
8		–	*hemB*	M24488	
9	*lctD*	–	ORF	L13970	2 box C sequences in inverted orientation separated by 38 bp
10		–	*melA*	X04894	
11	*mtlA*	–	*mtlD*	X51359	2 box C sequences in direct repeat, bracketing a BIME
12	ORF2	–	ORF3	X52904	Presence of a BIME in the same intergenic region at a distance of 14 bp ORF2 is downstream from *cstA*
13	ORF	–	*ponB*	D26562	2 box C sequences in direct orientation separated by 12 bp; the first one is poorly conserved
14	o265	–	f361	U18997	Presence of a BIME in the same intergenic region, at a distance of 24 bp
15	o290	–	f90	U00006	f90 is downstream from *lysC*
16	o309	–	f481	U18997	
17	o321	–	o414	U18997	2 box C sequences in inverted orientation separated by 11 bp
18	o732	–	f81	U00039	
19	*phnP*	–	*phnQ*	J05260	2 box C sequences in direct orientation separated by 8 bp within putative *phnQ* gene
20	*pstA*	–	*pstB*	K01992	Presence of a BIME in the same intergenic region at a distance of 29 bp
21	*araA*	–	*araD*		From *E. coli* ECOR 8; the box C sequence flanks a BIME; unpublished
22		–	*recA*	X55561	From *Shigella flexneri*

[a]As in Fig. 10. Sequences 1 to 20 are from *E. coli* K-12.

Like IRUs, the box C sequences seem to be restricted to genomes of bacteria which are very closely related to *E. coli,* but in contrast to IRUs, they are not present in genome of *S. typhimurium* (12). Their total number on the *E. coli* chromosome can be estimated to be between 40 and 45. Their sequences present several interesting features: a quasipalindromic structure and a separation into two regions, one purine rich and the other pyrimidine rich, including a mirror symmetry of 10 bp (see consensus in Fig. 10). It has been shown in vitro that DNA with similar characteristics may form a triple helical structure, also called form H (206). Like BIMEs and IRUs, the box C sequences have been found so far only in chromosomal DNA.

As was noticed for BIMEs, box C sequences may be differently located in various strains of the same species (the region between *araA* and *araD* from *E. coli* K-12 and ECOR 8, the 3′ region of the *recA* gene from *Shigella flexneri* and *E. coli*). This finding suggests strongly that these sequences are (were) mobile.

Functions hypothesized for the box C sequences include an effect on the level of transcription on the gene located downstream and/or a role in the stabilization of the upstream mRNA. Because three of the five sequences identified by Kunisawa and Nakamura (111) are linked to genes involved in the transport of different substrates, these authors suggested that these sequences could play a role in regulating the expression of transport systems. This seems rather unlikely since most of the box C sequences identified later are located near genes whose products are not implicated in transport.

RSA Sequences. RSA sequences were initially found in four extragenic regions of *E. coli* as well as in one region of *S. typhimurium* and one region of *Erwinia carotovora* (K. Mizobuchi, personal communication). In *E. coli,* the region upstream of gene *envM* contains two box C sequences (12) (see also above), whereas in *S. typhimurium,* there is an ORF which could encode 99 amino acid residues (197); the 3′ region of this ORF contains an RSA sequence.

An RSA sequence is also found upstream of the gene *araC* from *Erwinia carotovora,* and in *E. coli,* RSAs are located upstream of *rpsP,* in the intergenic region between *rplT* and *pheS,* and upstream of *cysP* (Table 6). The RSAs are thus in most instances located outside structural genes. We established for them a consensus of 152 bp (Fig. 11). The RSAs could form a large stem-loop structure (Fig. 11) which would be quite stable since their folding energy calculated with the program foldRNA (220) would vary from −39.6 kcal (1 kcal = 4.184 kJ)/mol for the sequence in *S. typhimurium* to −54.7 kcal/mol for the sequence of *Erwinia carotovora.* Since these elements are not perfect palindromes, they can be oriented. The proportion of G and C is about 50%, and the ratio of purines to pyrimidines is 40%. Two new RSA sequences were recently found. The first is located in an intergenic region from *E. coli* located between genes *smbA* and *frr;* the second is upstream of a short ORF located on a recombinant plasmid (pNM506). In both cases, the RSA sequence is extragenic.

RSAs constitute a distinct family of repeated sequence in the genomes of members of the *Enterobacteriaceae.* The sequences of RSAs from distant phylogenetic species (*E. coli* and *Erwinia carotovora,* for example) are homologous. RSA sequences are dispersed on the *E. coli* chromosome, and their sequences do not contain an ORF which could play a role in their dispersion (data not shown). If one assumes a uniform distribution of RSAs on the *E. coli* chromosome, one may estimate that there are about 10 of them. There are no data pertaining to a possible function of RSAs in the bacterial genome.

***iap* Sequences.** Between min 59 and 60 on the *E. coli* genetic map, there is a highly conserved sequence of 29 bp, containing an inverted repeat of 7 bp that appears 14 times, 32 or 33 bp apart, downstream of the *iap* gene coding region (Fig. 12). About 24 kb downstream of the 14 repeats, a second intergenic region containing similar 29-bp sequences occurring seven times with a spacing of 32 bp was first described (95, 141). The same intergenic region appears to contain two additional 29-bp repeats, at a distance of 500 bp from the other repeats (Bachellier, unpublished data). These sequences, called *iap* sequences, correspond to imperfect palindromes with the first half pyrimidine rich and the second half purine rich. This is reminiscent of the structure of box C sequences. The spacing sequences have a constant size but a variable nucleotide composition.

Nucleotide sequences hybridizing with the 29-bp fragment were not detected in other regions of the *E. coli* chromosome. Thus, the 23 repeats of these sequences are clustered in a 1-min region of the *E. coli* genetic map. In that sense, they are not dispersed throughout the genome, but they can be considered as highly repetitive sequences. Hybridizing sequences were also detected in *Shigella dysenteriae* and *S. typhimurium* but not in *K. pneumoniae* or *Pseudomonas aeruginosa.*

The sizes of these sequences are of the order of that of PUs. No function has been proposed for the *iap* repeats.

Ter Sequences. In a search for dispersed recurrent DNA sequences in the *E. coli* genome, Blaisdell and colleagues (16) found eight groups of structural repeat identities on a contig sequence which was 1.6×10^6 bp long. While other groups corresponded to known

---

TABLE 6 Compilation of RSA sequences

Flanking gene(s)[a]			GenBank accession no.	Observations
	-	*cysP*	M32101	
	-	ORF1	S67014	ORF1 is located upstream of *glnB.* The sequence in the database starts at position 38 of the RSA consensus.
ORF	-	*rpsP*	X01818	The first 18 nt of the RSA are included within the ORF.
rplT	-	*pheS*	K02844	
smbA	-	*frr*	D13334	
ORF	-	*envM*	M31806	The 5′ end of the RSA sequence is included within the putative ORF. A box C is present at the same place in *E. coli.*

[a]The first five sequences are from *E. coli;* the last one is from *S. typhimurium.*

GGAGTTCTACCGCAGAGGCGGGGGAACTC < 32 bp >
CGGTTTATCCCCGCTGATGCGGGGAACAC < 32 bp >
CGGTTTATCCCTGCTGGCGCGGGGAACTC < 32 bp >
CGGTTTATCCCCGCTAACGCGGGGAACTC < 32 bp >
CGGTTTATCCCCGCTGGCGCGGGGAACTC < 32 bp >
CGGTTTATCCCCGCTGGCGCGGGGAACTC < 32 bp >
CGGTTTATCCCCGCTGGCGCGGGGAACTC < 32 bp >
CGGTTTATCCCCGCTGGCGCGGGGAACTC < 32 bp >
CGGTTTATCCCCGCTGGCGCGGGGAACTC < 33 bp >
CGGTTTATCCCCGCTGGCGCGGGGAACTC < 33 bp >
CGGTTTATCCCCGCTGGCGCGGGGAACTC < 32 bp >
CGGTTTATCCCCGCTGGCGCGGGGAACTC < 32 bp >
CGGTTTATCCCCGCTGGCGCGGGGAACTC

ATGGTTATCCCCGCTGACGCGGGGAACAT < 32 bp >
CGGTTTATCCCCGCTGATGCAGGGAACAT <500 bp >

TGGTTTATCCCCGCTGGCGCGGGGAACTC < 32 bp >
CGGTTTATCCCCGCTGGCGCGGGGAACAC < 32 bp >
CGGTTTATCCCCGCTGGCGCGGGGAACAC < 32 bp >
CGGTTTATCCCCGCTGGCGCGGGGAACAC < 32 bp >
CGGTTTATCCCCGCTGGCGCGGGGAACAC < 32 bp >
AGGTTTATCCCCGCTGGCGCGGGGAACAC < 32 bp >
CGGTTTATCCCCGCTGGCGCGGGGAACAC

FIGURE 12 Structure of *iap* repeats. The 29-bp repeats located after the *iap* gene as well as in the second region (see text) have been aligned in two separate clusters. The sequences of the two first repeats of the second region appeared in GenBank U29580. Sizes of spacer sequences are given in base pairs (bp) on the right.

extragenic repeats (see above) or to coding sequences, group III corresponded to *rho*-independent terminators. The target sequence was a palindromic sequence of 30 bp in length. This group consisted of 22 members, 3 of which occurred in tandem on the same noncoding region following gene *rnpB* at successive displacement of 83 bp each. This means that the exact palindromic members occur in 20 compact regions. All of the members lie in noncoding regions that are highly variable in length. These sequences are examples of the *rho*-independent terminators reviewed by Rosenberg and Court (163): a C+G-rich region capable of forming a stable stem-loop structure followed closely by a T-rich region that does not bind strongly to the DNA template strand. In the case presented (16), the form has been specialized by the incorporation of an A-rich region at the beginning that is capable, with the following T-rich region, of extending the stem-loop structure. Such structures are capable of acting as bidirectional terminators (153) and bear also similarities to Y* (see above).

BIME, IRU, Box C, RSA, *iap*, and Ter Sequences

Similarities and Differences. The six small elements which we have described have a number of common properties: they are found only on the bacterial chromosome, dispersed and in

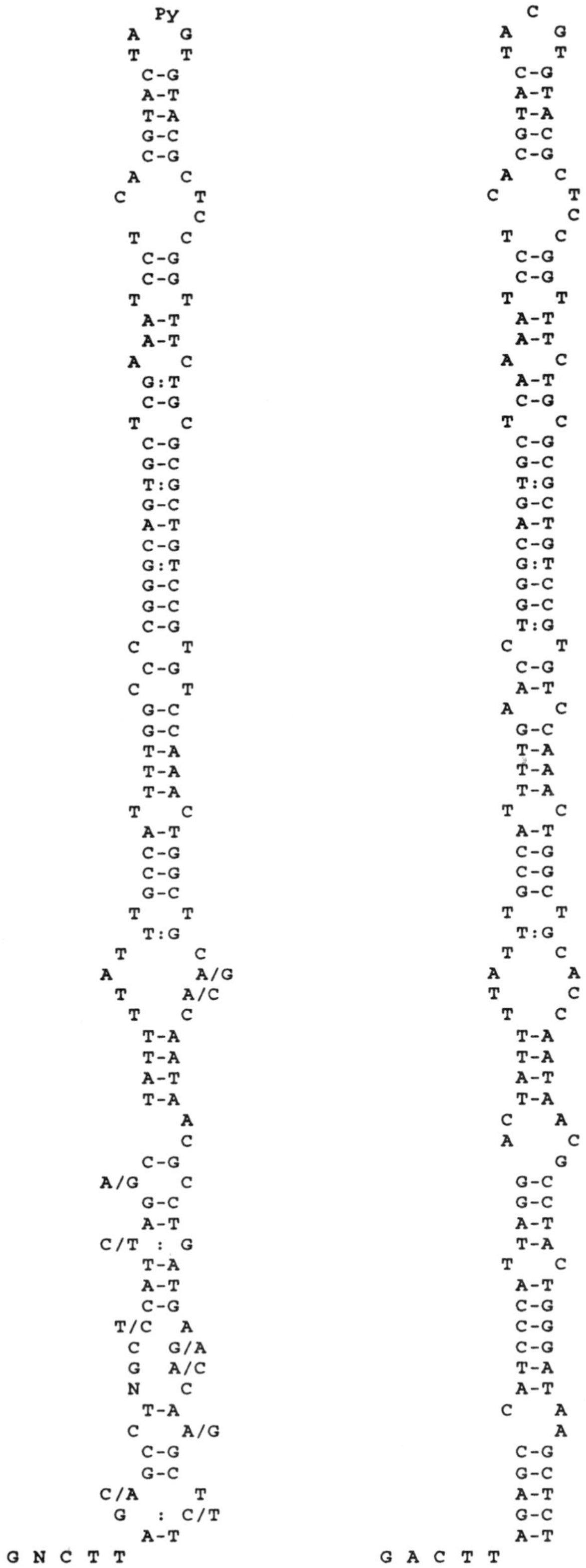

FIGURE 11 RSA sequences: consensus and palindromic sequences. The left side represents the palindromic structure that the consensus of RSA sequences could adopt. N = A, C, G, or T; Py = C or T. At the other variable positions of the consensus, the two possible nucleotides are shown separated by a slant line. On the right side, we show the palindrome which could be formed by the RSA sequence located upstream of the *araC* gene from *Erwinia carotovora* ($\Delta G = -54.7$ kcal/mol).

```
              Py                              C
          A       G                      A       G
          T         T                    T         T
            C-G                            C-G
            A-T                            A-T
            T-A                            T-A
            G-C                            G-C
            C-G                            C-G
          A       C                      A       C
          C         T                    C         T
                      C                              C
          T       C                      T       C
            C-G                            C-G
            C-G                            C-G
          T         T                    T         T
            A-T                            A-T
            A-T                            A-T
          A       C                      A       C
            G:T                            A-T
            C-G                            C-G
          T       C                      T       C
            C-G                            C-G
            G-C                            G-C
            T:G                            T:G
            G-C                            G-C
            A-T                            A-T
            C-G                            C-G
            G:T                            G:T
            G-C                            G-C
            G-C                            G-C
            C-G                            T:G
          C       T                      C       T
            C-G                            C-G
          C       T                      A-T
            G-C                          A       C
            G-C                            G-C
            T-A                            T-A
            T-A                            T-A
            T-A                            T-A
          T       C                      T       C
            A-T                            A-T
            C-G                            C-G
            C-G                            C-G
            G-C                            G-C
          T         T                    T         T
            T:G                            T:G
          T       C                      T       C
        A       A/G                    A       A
        T       A/C                    A       C
        T         C                    T         C
            T-A                            T-A
            T-A                            T-A
            A-T                            A-T
            T-A                            T-A
              A                          C       A
              C                          A       C
            C-G                                    G
        A/G     C                            G-C
            G-C                            G-C
            A-T                            A-T
        C/T : G                            T-A
            T-A                          T       C
            A-T                            A-T
            C-G                            C-G
        T/C     A                          C-G
          C   G/A                          C-G
          G   A/C                          T-A
          N       C                        A-T
            T-A                          C       A
          C   A/G                                A
            C-G                            C-G
            G-C                            G-C
        C/A     T                          A-T
          G  :  C/T                        G-C
            A-T                            A-T
    G N C T T                          G A C T T
```

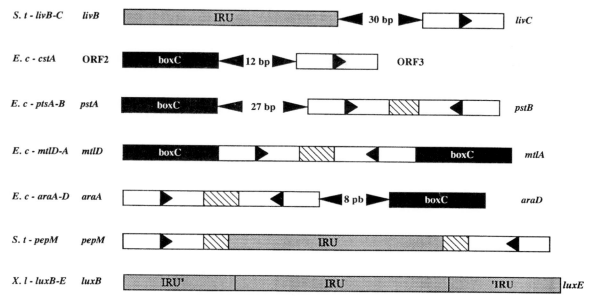

FIGURE 13 Location tropism of extragenic highly repetitive elements. The structures of extragenic regions including several repetitive elements are shown in a schematic form. On the left, the bacteria which contain these elements (*E. c, E. coli; S. t, S. typhimurium; X. l, Xenorhabdus luminescens*) and the names of the adjacent genes are indicated. Distances separating the two elements are indicated between two arrows; the names of the elements are indicated except for BIMEs, which are shown in the same form as in other figures.

extragenic locations, form imperfect palindromes, and are usually transcribed. In addition, four of them show some kind of tropism for location (see below) which may be interpreted as if they had a common mechanism for dispersion.

One of the main differences between these sequences is the number of repeats found for each family. It is quite variable from one species to another, and the four types of elements are not necessarily found in the genomes of all *Enterobacteriaceae*; in *E. coli*, BIMEs are present in hundreds of copies and Ter sequences are present in 60 to 100 copies, while there are about 50 IRUs and box C sequences and about 10 RSAs. In *S. typhimurium*, there is no box C, while the IRUs are more represented than in *E. coli* and the BIMEs are present in a lower number of copies. The differences in repartition of these elements could be due to differences in the mechanisms of dispersion between these elements from one bacterial species to another.

For these six families of sequences, no transposition mechanism has been demonstrated, and none of these elements encodes an ORF likely to direct the synthesis of proteins playing a role in dispersion as is found for ISs. One hypothesis could be that the sequences of these families are the "ghosts" of transposable elements or that they depend on *trans*-acting factors for their transposition. Since these six elements are generally transcribed, they are located on a number of mRNA molecules which could be used as intermediates in transposition (see above). Another large difference between these elements is the species specificity of the sequences between different phylogenetic groups of bacteria. This specificity is high for BIMEs and weak or absent for box C sequences, RSAs, and IRUs. The specificity has not been examined for *iap* and Ter sequences.

Tropism for Location. In a large number of cases, it is interesting that repetitive sequences belonging to two different families are located at the same site. In five intergenic regions of *E. coli* and two from *S. typhimurium*, one finds two repetitive elements, which are adjacent and separated by a few nucleotides (30 at most) or are inserted one into the other (Fig. 13). In *E. coli*, all of these regions carry box C and BIME sequences. The two sequences located between *pstA* and *pstB* are separated by 29 nt, those in the 3′ regions of *cstA* and *fimH* are separated by 14 and 18 nt, respectively, and the BIME and the box C located between *araA* and *araD* from ECOR 8 overlap by 5 nt. In the region between *mtlA* and *mtlD*, a BIME is flanked by two box C sequences in direct repetition (Fig. 13). The two box C sequences of this region are incomplete in that one of their extremities is quite different from the consensus: the box C located 5′ to the BIME is deleted in its 3′ part, while the box C located 3′ to the BIME is deleted in its 5′ region (underlined in Fig. 10). One interpretation of the structure of this region is that the insertion of a BIME led to a duplication of part of the box C sequence. In *S. typhimurium*, one finds IRUs and BIMEs in the same regions. One IRU and one BIME are separated by 30 nt in the region located between *livB* and *livC*, and in the region located immediately after gene *pepM*, an IRU is inserted within a BIME. The frequency with which one finds an association between two of these elements is striking, especially for the box C sequences: in 5 of 18 regions known to carry box C sequences in *E. coli*, they are associated with BIMEs.

How can we explain this location tropism? One may suppose that the mechanisms which are responsible for their dispersion are similar or even identical and/or that the insertion sites have

TABLE 7 Compilation of BIME sequences

Flanking gene(s)[a]		GenBank accession no.	BIME structure[b]	Observations
aceA	– *aceK*	U00006	$Y 1 Z^2 S Y$	
acs	– f104	U00006	$Z^2 s Y$	
aidB	–	L20915	$Z S Y$	
aldB	– f107	U00039	$Y S Z^2$	
alkA	–	K02498	$Y S Z^2$	
alr	– *tyrB*	U00006	$Y L$	
apaH	– *folA*	X04711	Z^2	
appA	–	M58708	Y	
	– *appA*	M58708	Z^2	
araA	– *araD*	M62646	$Z^2 S Y s Z^2 S Y s Z^2 S Y$	
araJ	– ORF	M64787	$Y S Z^2 1 Y S Z^2$	
argA	– *recD*	U29581	PU	
argD	– o696	U18997	Y	
argE	– *ppc*	X55417	$Y< 24 >Z^2> 21 <Y< 24 >Z^2> 21 <Y< 24 >Z^2$	
aroG	–	J01591	Y	
aroG	–	J01591	$Y S Z^2 A$	
arsG	– o260	U00039	$Z^2 L Z^1$	
artM	– *artJ*	X67753	$Z^2 1 Y$	
artJ	–	X67753	$A Z^1 L Y B$	
aslB	– *aslA*	M90498	$Y L Z^2$	
aspC	–	X03629	$A Z^1 L Y'$	The sequence in the database contains only A Z^1; for the end of the sequence see reference 112.
	– *avtA*	Y00490	$B Y L Z^1 A$	
barA	–	D10888	$B Y L Z^1 A$	
bioF	–	J04423	Z^1	Included within the *bioF* gene
caiA	– *caiB*	X73904	$Z^2 S Y$	
	– *caiE*	X73904	$Y S Z^2$	
cbpA	–	D16500	$Z^2 S Y 1 Z^2$	
cca	–	M12788	$Y S Z^2 s Y$	
cdd	–	M60916	Y	
cheZ	–	M13463	Z^2	
chlN	–	M21151	Z^2	
chm	–	X53666	$Z^1 S Y$	From *E. coli* C
	– *codB*	X63656	$Y 1 Z^2 S Y 1 Z^2$	
codA	– *cynR*	X63656	$Y S Z^1$	
	– *cynR*	M93053	$Y S Z^1$	This BIME is present in a duplicated segment between *codA* and *cynR*. This segment constitutes one end of a discontinuity between *E. coli* and *S. typhimurium* chromosomes (the *lac* loop) (20).
cpdB	– o163	U14003	$PU S Z^2 s Y$	
cpxA	– ORF	M36795	$Y 1 Z^2$	
cutE	–	X58070	Y	
	– *cyd*	J03939	$B Y L Z^1 A$	
cysA	– *cysM*	M32101	Z^2	
cysE	–	M34333	Y	
cysI	– *cysH*	Y07525	Z^2	
cysM	–	M32101	$A Z^1 L Y B$	
dacB	–	X59460	$B Y L Z^1 A$	
dacC	–	X06480	Z^1	
dapX	– *purC*	M33928	Y	
dbpA	–	X52647	Z^2	
dedD	– *dedE*	M68934	$Z^2 S Y$	
deoC	– *deoA*	X03224	$Z^1 S Y$	
dppA	– *dppB*	L08399	$A Z^1 L Y B$	
ecoA	–	U00008	$Y< 14 >Z^2> 21 <Y< 14 >Z^2> 21 <Y< 14 >Z^2> 21$ $<Y< 14 >Z^2> 21 <Y< 14 >Z^2> 21 <Y< 14 >Z^2$	
exbD	–	M28819	$Z^1 L Y 1 Z^1 L Y$	
f104	– f338	U18997	$Z^2 S Y$	
f159	– f451	L10328	$Y S Z'$	
f170	– *rrfA*	L19201	Z^2	
f199	– *spf*	L19201	Z^1	
f230	– *araE*	U29581	Y^*	
f239	– f118	U28377	$B Y S Z^1 A$	
f254	– o207	U28379	Y	
f268	– f432	U29581	$A Z^1 L Y B$	
f289	– o128	U14003	$B Y L Z^1 A$	
f310	– f848	U28377	$Y L Z^1 r Y$	
f311	– f510	U14003	$Z^1 S PU$	
f342	– f289	U14003	$A Z^1 L Y B$	
f481	– f506	U18997	$Y S Z^2 s Y S Z^2 s Y S Z^2 s Y$	
f500	– o361	U14003	Z^2	

TABLE 7 *Continued*

Flanking gene(s)[a]		GenBank accession no.	BIME structure[b]	Observations
f577	– f283	U00006	$Y > 9 < Z^2$	
f589	– dppA	U00039	$Y \, l \, Z^2$	
f675	– f560	U28377	Y^*	
fabB	–	M24427	Z^2	
fadA	– ubiB	M59368	$Y \, S \, Z^2 \, l \, Y \, S \, Z^2 \, l \, Y \, S \, Z^2$	
fda	– ORF4	X14436	$A \, Z^1 \, L \, Y \, B$	
fdhE	– o80	L19201	$B \, Y \, L \, Z^1 \, A$	
fepA	– entD	X17426	$B \, Y \, L \, Z^1 \, A$	
fhuB	– gsa	D26562	$Z^2 \, S \, Y$	
fimH	– f477	U14003	Z^2	This intergenic region also contains a box C, at a distance of 18 bp from the PU.
fliC	– fliA	Z36877	$A \, Z^1 \, L \, Y'$	From *E. coli* B38. BIMEs with the same structure ($A \, Z^1 \, L \, Y \, B$) are found in *Shigella dysenteriae* and *Shigella boydii*.
fpg	– kdtB	X06036	Y^*	
fucO	–	X15025	Y	
gabT	– gabP	M88334	Y	
	– galE	X51449	Y	
	– gatY	X79837	$Y \, B$	
gcd	– ORF	D26562	$A \, Z^1 \, L \, Y \, B$	
gdhA	–	X00988	$Y \, L \, Z^1$	
gldA	– f205	U00006	$Y \, L \, Z^1$	
glnA	– glnL	K02176	$Z^2 \, S \, Y$	
glnG	– hemN	L19201	$Z^2 \, l \, Y$	
glnQ	–	X14180	$Z^1 \, L \, Y \, B$	
glnS	–	M10187	$Y \, r \, Z^1 \, L \, Y$	
glu-tRNA–		J01713	Z^2	
glpD	– ORFX	D00425	$A \, Z^1 \, L \, Y \, B$	
glpK	– glpX	Z11767	Y	A BIME with the structure $Z^2 \, S \, Y$ is present in the same intergenic region of *Shigella flexneri* (Z11766).
gltP	– f229	U00006	$Z^2 \, L \, Y \, s \, Z^2 \, L \, Y \, s \, Z^2 \, L \, Y \, s \, Z^2 \, L \, Y \, s \, Z^2 \, L \, Y \, s \, Z^2$	
gltS	– recG	L10328	$Z^1 \, L \, Y$	
glyA	–	J01620	$B \, Y \, L \, Z^1 \, A$	
glyK	–	M18393	Y	
gsk	–	D00798	$Y \, l \, Z^2$	
gusC	– ORF	M14641	$B \, Y \, L \, Z^1 \, A$	
gutD	– gutM	X13463	Z^1	
gyrB	– f135	L10328	$Z^1 \, L \, Y$	
hag	–	M14358	$A \, Z^1 \, L'$	
hpaF	– hpaH	Z37980	$Z^1 \, S \, Y$	
hpaI	– hpaX	Z37980	$Z^1 \, S \, Y \, ; \, Y$	The two Y are in the same orientation, separated by 14 bp.
htrB	–	X61000	Y^*	
hybG	–	U09177	$Z^2 \, l \, Y'$	
hycH	–	X17506	$A \, Z^1 \, S \, Y$	
hydA	–	D14422	$Y \, S \, Z^2$	
hydN	– hydA	D14422	Z^2	
iciA	–	M62865	Y^*	
ilvA	– ilvY	M11689	Z^1	
ilvBN	– o396	L10328	Z^2	
inaA	– nrd	K02672	Z^2	See reference 209.
infB	– P15B	X00153	$Y \, s \, Z^2$	
katE	–	M55161	$B \, Y \, L \, Z^1 \, A$	
lacA	–	J01636	Z^1	
lacY	– lacA	J01636	Y	
lamB	– malM	X04477	$Y \, s \, Z^2 \, S \, Y$	
	– lctP	L13970	Y	
leuQ	– f343b	U14003	$Z^2 \, S \, Y$	
livF	– ugpB	J05516	$A \, Z^1 \, L \, Y \, B$	
livJ	– livK	J05516	$B \, Y \, L \, Z^1 \, A$	
lpd	–	V01498	Y^*	
lspA	– ORF1	X54945	$Z^2 \, S \, Y$	
lysV	–	X17321	Z^1	
malE	– malF	J01648	$Y \, S \, Z^2 \, s \, Y$	
malQ	–	M32793	$Z^2 \, S \, Y$	
malT	–	M13585	Y	
mdoH	–	X64197	$Z^1 \, L \, Y$	
melB	–	U14003	$A \, Z^1 \, L \, Y \, B$	
metJ	– f202	L19201	$Z^1 \, L \, Y \, B$	

(Table continues)

TABLE 7 *Continued*

Flanking gene(s)[a]		GenBank accession no.	BIME structure[b]	Observations
metK	– *galP*	U28377	Y	
metL	– *metF*	K01547	Y	
mglC	–	M59444	Y 1 Z^2 L Y 1 Z^2	
mgtA	– f141	U14003	B Y L Z^1 A	
moaE	–	X07420	Y	
motB	–	J01652	Z^2	Included within *motB*
mrcA	– f186	U18997	B Y L Z^1	
mrr	– ORF	X54198	Y	
msbB	–	M77039	Y	
msyB	–	X59939	Y L Z^1	
mtlA	– *mtlD*	V01503	Z^1 L Y	The BIME is flanked by two box C sequences in direct orientation.
mtrA	– o335	M58338	Z^1	
mutT	–	D26562	Y 1 Z^2 L Y 1 Z^2	
nadR	– f579	U14003	Y 1 Z^2 L Y	
nagD	–	X14135	A Z^1 L Y B	
narP	– *yejP*	U00008	A Z^1 L Y B	
	– *ndh*	V00306	Y 1 Z^1	
ndh	–	V00306	Y	
nlpE	– *yeaF*	U18345	Y 1 Z^2	
nrdA	– *nrdB*	K02672	Z^2 s Y S Z^2	
nrdD	– *ndrG*	L06097	Y S Z^2	
ntr	–	M15328	B Y L Z^1 A	
o56	– f188	U18997	Y L Z^2	
o128	– f138	M87049	Y	
o138	– f97	U14003	Y B	
o147	– f268	U29581	Y,	
o163	– o384	U18997	Z^2 S Y	
o180	– f341	U28377	Y S Z^2 ; Z^1	The Z^2 and the Z^1 are separated by 9 bp and are in direct orientation.
o186	– f470	U14003	A Z^1 L Y B	
o191	– f346	U18997	Z^2 S Y	
o191	– o437	U18997	Z^2	
o191	– f297	U29579	Y s Z^2	
o207	– f506	U28379	Y S Z^2	
o211	– f287	U14003	Y 1 Z^2 S Y 1 Z^2 S Y 1 Z^2 S Y 1 Z^2	
o265	– f361	U18997	Z^2	Presence of a box C in the same intergenic region at a distance of 24 bp
o310	– f541	U28375	B Y L Z^1 A	
o318	– o417a	U14003	Y 1 Z^2	
o320	– o361	L02123	Y	o320 is downstream from *dinG*.
o326	– f219	U28375	Z^2 L Y 1 Z^2	
o328	– o130	U18997	B Y L Z^1 A	
o337	– o440	U00039	Y S Z^2	
o351	– f582	L19201	Y*	
o361	– f125	L02123	Z^2 S Y	o361 is downstream from *dinG*.
o378	– f335	U28377	Y S Z^2 s Y	
o386	– f271	U28377	Y*	
o417	– o103	U00039	A Z^1 L Y B	
o421	– f723	L19201	Z^1 S Y	
o440	– f472	U00039	Z^1 L Y B-15- Y*	
o499	– f111	U00039	B Y L Z^1 A	
o548	– *thdF*	L10328	Z^1	
o569	– f772	L10328	Y*	
o622	– *feoA*	U18997	Y	
o698	– *yjbA*	U00006	Y B	*yjbA* is upstream from *xylE*.
o783	– o672	U18997	Y S Z^2< 11 >Y S Z^2< 11 >Y S Z^2	
ORF	– ORF	D10483	Z^2 S Y	The first ORF is downstream from *araC*.
ORF1	–	M26934	S Z^2	ORF1 is located downstream from *ansA*.
ORF2	– ORF3	X52904	Y	ORF2 is located downstream from *cstA*.
ORFX	– ORFY	M96791	Y 1 Z	ORFY is located upstream from *g30k*.
ORF	–	K01197	Z^1 S Y	ORF is located downstream from *tyrT*.
ORF122–	ORF153	L20897	Z^2	ORF122 is located downstream from *purT*.
pepD	–	M34034	Y*	
pepE	– o543	U00006	Z^2< 11 >Y*	
pfkA	– *sbp*	X02519	A Z^1 L Y B	
pgi	– o80	U00006	Y s Z^2	
pgm	–	U08369	A Z^1 L Y B	
pheU	–	X00912	Z^1	
phnA	– *phnB*	J05260	Y 1 Z^2 S Y 1 Z^2 S Y 1 Z^2 S Y 1 Z^2 S Y 1 Z^2	This sequence originates from *E. coli* B.
phnD	– *phnE*	D90227	Z^2 1 Y	

TABLE 7 *Continued*

Flanking gene(s)[a]		GenBank accession no.	BIME structure[b]	Observations
phnK	– phnL	D90227	Y S Z²	
phoE	–	V00316	Y*	
phoH	– ORF	D10391	Z² S Y	
pho	– phoW	L10328	Y B	
phr	–	K01299	Y L'	
plsB	– ubiA	K00127	Y L Z¹ 1 Y*	
pmsR	–	M89992	B Y L Z¹ A	
polA	– ORF	J01663	Z¹ L Y	
priC	–	D13958	A Z¹ L Y'	
proP	– basS	U14003	Z²	
proX	–	M24856	Y 1 Z²	
pstA	– pstB	L10328	Y Z¹	The BIME is flanked by a box C.
purK	–	M19657	B Y S Z² 1 Y/2 S Z²	
pyrE	– ORF	L10328	Y*	
rcsC	– rcsB	M28242	Y 1 Z² L Y 1 Z²	
recD	–	X04582	Y	
rep	–	X04794	Z²	
	– rf2	M11520	Y	
rhaA	– rhaD	X60472	Y S Z² s Y S Z² s Y S Z² s Y S Z² s Y	
rhlE	– f160	L02123	A Z¹ L Y B	
rnpB	– ORFX	D90212	B Y L Z¹ A	
rol	–	M89934	Z² S Y	
rplY	–	D13326	Y	
rpoD	–	J01687	Y	
ruvB	–	M21298	Y 1 Z² S Y	
sdhB	– sucA	J01619	Z² s Y S Z¹	
sodB	–	J03511	A Z¹ S Y B	
speB	–	M32363	r Y S Z² r Y	
srmB	–	X14152	A Z¹ L Y B	
sseB	–	D10496	Z¹ L Y	
sucB	– sucD	J01619	Y s Z² S Y s Z²	
thrC	– ORF	J01706	Z² L Y	
tktB	–	S65463	Z' Z² 'Z	The Z² PU is included within another PU.
torA	–	X73888	Z² 1 Y S Z²	
trmA	– o119	U00006	Y*	
trpR	– ORF	J01715	Z¹ 1 Y L Z²	
trpS	– f228	U18997	A Z¹ > 25 < Y B	
tsx	– ORF6	M57685	B Y L Z¹ A	ORF6 is downstream from *secF*.
tyrA	– pheA	X02137	Y*	
ubiG	– ORFX	Y00544	Z² S Y	This region (around 48 min) has been studied in several strains, where the sequences downstream of *ubiG* differ: in M87509, the BIME contains 3 PUs, and ORFX is *acs,* located around 90 min in *E. coli* MG1655.
ugpB	– ugpA	X13141	Z¹ S Y	
ung	–	J03725	Y	
ushA	–	X03895	Y*	
uvrB	–	X03722	A Z¹ L Y B	
uvrD	– ORF	L02122	Z² L Y	
uxuB	– uxuR	U14003	Y	
vacB	– o243	U14003	B Y L Z¹	
valS	– o204b	U14003	Y*	
x96	–	M95096	Z² 1 Y	
xerB	– ORF13	X15130	Y S Z²	
xylR	– f274b	U00039	Z² 1 Y, Z²	The last Z² is separated from the Y PU by 49 bp and does not seem to belong to the same BIME.
yehZ	– yohA	U00007	Y	
yohD	– yohE	U00007	Y	
yohK	–	U00007	Y	
yeiI	– yeiJ	U00007	Y > 24 < Z¹	
yeiL	– yeiM	U00007	Z²	
rhsD*		X60997	Z¹ L Y 1 Z² S Y	Segment replacing *rhsD* element in *E. coli* ECOR 39 (166)
pSCIS1IN*		M25018	'Y S Z² s Y	Target region of IS*1*–mediated type I deletions (185)
		I01818	A Z¹ L Y B' ⎫	From "unclassified" organisms. The sequences of the BIME motifs are typical of E. coli or close relatives.
		I03295	A Z¹ L ⎭	
	– adk	L26246	'Y > 12 < D	First PU incompletely sequenced

(Table continues)

TABLE 7 *Continued*

Flanking gene(s)[a]		GenBank accession no.	BIME structure[b]	Observations
adk	– *visA*	L26246	Z^1 1 Y	In *E. coli*, box C at the same location
alr	–	M12847	D	In *E. coli*, Y L
araA	– *araD*	M11047	Z^1> 4 <Y< 8 <Z^1	In *E. coli*, BIME-2 with 6 PUs
argA	– *recD*	–	Z^1> 14 <Y/2	See reference 182.
cchA	– *cchB*	U18560	Y< 11 >Z^2	
cheW	–	J02656	PU	
cheZ	– *flhB*	M16691	Z^1> 7 <PU	In *E. coli*, Z^2
cpsB	– *cpsG*	X59886	D	
crr	– ORF287	U11243	D< 10 >Z^1	
cysI	– *cysH*	M23007	Z^1	In *E. coli*, Z^2
envZ	–	X12374	Z^2	
eutG	– *eutH*	U18560	Y< 10 >Z^1	
	– *hemA*	J04243	Z^1	
hemL	–	M60064	D	
hemN	–	U06779	Z^1> 13 <D	In *E. coli*, BIME-2 with 2 PUs
hisG	– *hisD*	J01804	Z^1> 23 <D	
hisJ	– *hisQ*	V01373	Z^1> 8 <D	
ilvH	– *fruR*	X55456	Z^2> 13 <Z^1	
lamB	– *malM*	X54292	D	In *E. coli*, BIME-2 with 3 PUs
leuD	–	X02528	Z^1	
lexA	–	X63002	Y/2<3>Z^1> 27 <Y	
	– *livB*	D00478	Y<3>Z^1	
livB	– *livC*	D00478	Z^1	An IRU is distant from 30 bp
malE	– *malF*	X54292	Y> 6 <Z^1	In *E. coli*, BIME-2 with 3 PUs
melB	–	S41630	Z^1	In *E. coli*, BIME-1
	– *murB*	J14816	D	
mviM	– *mviN*	Z26133	Z^2<5>Y	
mviN	– *flgN*	Z26133	Z^1> 5 <Y	
	– *nadR*	M85181	Z^2> 19 <Z^1	
nadR	–	M85181	Z^2> 29 <PU	In *E. coli*, BIME with 3 PUs
nrdA	– *nrdB*	X72948	Z^1> 11 <Y	In *E. coli*, BIME-2 with 3 PUs
	– *nrdE*	X73226	'Y	
	– ORF1	L13395	PU	ORF1 is in the 5′ region of *pmrA*.
parC	– *parF*	M68936	Y<5>Z^2	
pepM	– ORF1	X55778	Z^2> 137 <Y	An IRU is inserted between the 2 PUs.
trpR	–	L13768	Z^2	In *E. coli*, BIME with 3 PUs
ushA[0]	–	X04651	Z^1> 12 <Y	In *E. coli*, Y*
uvrD	– *corA*	L11043	PU> 20 <D	In *E. coli*, BIME with 2 PUs

[a] The first 256 BIMEs are from *E. coli* or close relatives, while the last 39 BIMEs are from *S. typhimurium*. Stars indicate two regions where BIMEs are not defined according to flanking genes.
[b] > indicates PU orientation, from tail to head (and inversely for <). *S. typhimurium* PUs are named according to Fig. 7: group A, Y; group B, Z^1; group C, Z^2; group D, D. When the sequence of the PU is not related to any of these groups, it is named PU.

common characteristics. Examination of the nucleotide sequence of the intergenic region containing two repeated sequences did not show any obvious homology; this may indicate that it is the structure of the insertion site rather than its sequence which is important. Another possibility would be that the target of the transposition system is the repeated sequence itself. This would account also for the cases where repetitive sequences insert one into the other.

At this stage, it is difficult to exclude that this tropism might just reflect the limited number of extragenic regions.

Exchange of Repetitive Sequences. Another intriguing type of relations between four of the families of repetitive sequences has been found. Corresponding intergenic regions of two different bacteria are sometimes occupied by two different sequences. Sharples and Lloyd (179) noticed that one IRU of *S. typhimurium* is found in the intergenic region *glnA-L* instead of the BIME which is found in the same region in *E. coli*. Between the genes *malE* and *malF* from *E. coli*, *Enterobacter aerogenes*, *K. pneumoniae*, and *S. typhimurium*, there is a BIME sequence. However, the corresponding region of *Klebsiella oxytoca* contains an IRU

(Bachellier et al., unpublished data). The region ahead of *envM* carries an RSA in *S. typhimurium* and two box C sequences in *E. coli*. The *araA-D* region of *K. pneumoniae* contains a box C (see above and Fig. 10), while *S. typhimurium* and *E. coli* carry a BIME at the same place; it is remarkable that in *E. coli* ECOR 8 there is at the same site one BIME and one box C. Finding one or the other type of these sequences at the same site could suggest that they have similar functions (interaction with proteins or need for a secondary structure at a precise site) or, as supposed previously, that they insert at the same sites. In the case of the region located between *malE* and *malF*, one would have to think that there was independent insertion of the repetitive element in each bacterium or that there was a replacement of the BIME by the IRU in *K. oxytoca*. Two other examples of exchanges have been reported. (i) In *E. coli* K-12, a BIME (Y*) is located in the *pyrE-ttk* intergenic region (Table 7), while a retron is found at the same location in the *E. coli* ECOR 70, 71, and 72 (73). Retrons are not repetitive sequences but are chromosomal elements encoding a reverse transcriptase (reviewed in references 94 and 202). (ii) In *E. coli* ECOR 39, a BIME is located at the same place as the *RhsD* element of *E. coli* K-12 (166) (see above).

A Superfamily? BIMEs, box C sequences, RSAs, IRUs, *iap* sequences, and Ter sequences constitute six families of repetitive sequences which are found in the genomes of a number of gram-negative bacteria. It has been shown that at least some of these repetitive sequences can be useful for the typing of bacteria (35, 204). They share a number of characteristics: they are scattered over the bacterial chromosome (except *iap* sequences), all are transcribed, all are imperfectly palindromic, and four of them present some tropism in their location. Their sizes are less than the size of ISs, and no mechanism for their transposition has yet been demonstrated. It appears that they can be exchanged locally, which may reveal a function or a dispersion mechanism which would be common. All these similarities lead us to propose that these families define a superfamily of extragenic elements which may play a critical role in genome function and evolution.

PERSPECTIVES

Although our knowledge is yet severely limited, the distribution and interaction of both large and small repeated sequences in *E. coli* and *S. typhimurium* populations is an important source of genetic plasticity. More studies of repeated sequences may provide greater insights into the degree of natural diversity and into the selective pressures that maintain the profiles. Such studies might also give clues as to the natural mechanisms that rearrange and/or redistribute the diverse examples of repeated sequences that have been discovered.

ACKNOWLEDGMENTS

C.W.H. acknowledges support by Public Health Service grant GM16329 from the National Institutes of Health. S.B., E.G., and M.H. thank David Perrin and William Saurin for their important contributions to the study of PUs and BIMEs and Ana Cova-Rodrigues for helpful technical assistance in the elaboration of the compilation tables.

LITERATURE CITED

1. Albertini, A. M., M. Hofer, M. P. Calos, and J. H. Miller. 1982. On the formation of spontaneous deletions: the importance of short sequence homologies in the generation of large deletions. *Cell* 29:319–328.
2. An, G., and J. D. Frisen. 1980. The nucleotide sequence of *tufB* and four nearby tRNA structural genes of *Escherichia coli*. *Gene* 12:33–39.
3. Anderson, R. P., and J. R. Roth. 1977. Tandem genetic duplications in phage and bacteria. *Annu. Rev. Microbiol.* 31:473–505.
4. Anderson, R. P., and J. R. Roth. 1978. Tandem chromosomal duplications in *Salmonella typhimurium*: fusion of histidine genes to novel promoters. *J. Mol. Biol.* 119:147–166.
5. Anderson, R. P., and J. R. Roth. 1978. Gene duplication in bacteria: alteration of gene dosage by sister-chromosome exchanges. *Cold Spring Harbor Symp. Quant. Biol.* 119:147–166.
6. Bachellier, S., D. Perrin, M. Hofnung, and E. Gilson. 1993. Bacterial interspersed mosaic elements (BIMEs) are present in the genome of *Klebsiella*. *Mol. Microbiol.* 7:537–544.
7. Bachellier, S., D. Perrin, M. Hofnung, and E. Gilson. Two major bacterial interspersed mosaic element (BIMEs) families with different gyrase and integration host factor binding abilities. Submitted for publication.
8. Bachellier, S., W. Saurin, D. Perrin, M. Hofnung, and E. Gilson. 1994. Structural and functional diversity among bacterial interspersed mosaic elements (BIMEs). *Mol. Microbiol.* 12:61–70.
9. Baron, L. S., Jr., P. Gemski, E. M. Johnson, and J. A. Wohlhieter. 1968. Intergeneric bacterial matings. *Bacteriol. Rev.* 32:362–369.
10. Bates, A. D., and A. Maxwell. 1989. DNA gyrase can supercoil DNA circles as small as 174 base pairs. *EMBO J.* 8:1861–1866.
11. Becerril, B., F. Valle, E. Merino, L. Riba, and F. Bolivar. 1985. Repetitive extragenic palindromic (REP) sequences in the *Escherichia coli gdhA* gene. *Gene* 37:53–62.
12. Bergler, H., G. Högenauer, and F. Turnowski. 1992. Sequences of the *envM* gene and of two mutated alleles in *Escherichia coli*. *J. Gen. Microbiol.* 138:2093–2100.
13. Bi, X., and L. F. Liu. 1994. *recA*-independent and *recA*-dependent intramolecular plasmid recombination. Differential homology requirement and distance effect. *J. Mol. Biol.* 235:414–423.
14. Bisercic, M., and H. Ochman. 1993. The ancestry of insertion sequences common to *Escherichia coli* and *Salmonella typhimurium*. *J. Bacteriol.* 175:7863–7868.
15. Bisercic, M., and H. Ochman. 1993. Natural populations of *Escherichia coli* and *Salmonella typhimurium* harbor the same classes of insertion sequences. *Genetics* 133:449–454.
16. Blaisdell, B. E., K. E. Rudd, A. Matin, and S. Karlin. 1993. Significant dispersed recurrent DNA sequences in the *Escherichia coli* genome: several new groups. *J. Mol. Biol.* 229:833–848.
17. Blattner, F. R., V. Burland, G. Plunkett, III, H. J. Sofia, and D. L. Daniels. 1993. Analysis of the *Escherichia coli* genome. IV. DNA sequence of the region from 89.2 to 92.8 minutes. *Nucleic Acids Res.* 21:5408–5417.
18. Boccard, F., and P. Prentki. 1993. Specific interaction of IHF with RIBs, a class of bacterial repetitive DNA elements located at the 3′ end of transcription units. *EMBO J.* 12:5019–5027.
19. Britten, R. J., and D. E. Kohne. 1968. Repeated sequences in DNA. *Science* 161:529–540.
20. Buvinger, W. E., K. A. Lampel, R. J. Bojanowski, and M. Riley. 1984. Location and analysis of nucleotide sequences at one end of a putative *lac* transposon in the *Escherichia coli* chromosome. *J. Bacteriol.* 159:618–623.
21. Campbell, A. 1963. Segregants from lysogenic heterogenotes carrying recombinant lambda prophages. *Virology* 20:344–356.
22. Campbell, A. 1965. The steric effect in lysogenization by bacteriophage lambda I. Lysogenization of a partially diploid stain of *Escherichia coli* K12. *Virology* 27:329–339.
23. Campbell, A. 1981. Evolutionary significance of accessory DNA elements in bacteria. *Annu. Rev. Microbiol.* 35:55–83.
24. Campbell, A. M. 1992. Chromosomal insertion sites for phages and plasmids. *J. Bacteriol.* 174:7495–7499.
25. Capage, M., and C. W. Hill. 1979. Preferential unequal recombination in the *glyS* region of the *Escherichia coli* chromosome. *J. Mol. Biol.* 127:73–87.
26. Chédin, F., E. Dervyn, R. Dervyn, S. D. Ehrlich, and P. Noirot. 1994. Frequency of deletion formation decreases exponentially with distance between short direct repeats. *Mol. Microbiol.* 12:561–569.
27. Condon, C., S. French, C. Squires, and C. L. Squires. 1993. Depletion of functional ribosomal RNA operons in *Escherichia coli* causes increased expression of the remaining intact copies. *EMBO J.* 12:4305–4315.
28. Condon, C., J. Philips, Z.-Y. Fu, C. Squires, and C. L. Squires. 1992. Comparison of the expression of the seven ribosomal RNA operons in *Escherichia coli*. *EMBO J.* 11:4175–4185.
29. Curtiss, R., III. 1964. A stable partial diploid strain of *Escherichia coli*. *Genetics* 50:679–694.
30. Davidson, N., R. C. Deonier, S. Hu, and E. Ohtsubo. 1975. Electron microscope heteroduplex studies of sequence relations among plasmids of *Escherichia coli*. X. Deoxyribonucleic acid sequence organization of F and of F-primes, and the sequences involved in Hfr formation, p. 56–65. *In* D. Schlessinger (ed.), *Microbiology—1974*. American Society for Microbiology, Washington, D.C.
31. de Massy, B., S. Bejar, J. Louarn, J.-M. Louarn, and J.-P. Bouche. 1987. Inhibition of replication forks exiting the terminus region of the *Escherichia coli* chromosome occurs at two loci separated by 5 min. *Proc. Natl. Acad. Sci. USA* 84:1759–1763.
32. Deonier, R. C., E. Ohtsubo, H. J. Lee, and N. Davidson. 1974. Electron microscope heteroduplex studies of sequence relations among plasmids of *Escherichia coli*. VII. Mapping the ribosomal RNA genes of plasmid F14. *J. Mol. Biol.* 89:619–629.
33. Dianov, G. L., A. V. Kuzminov, A. V. Mazin, and R. I. Salganik. 1991. Molecular mechanisms of deletion formation in *Escherichia coli* plasmids I. Deletion formation mediated by long direct repeats. *Mol. Gen. Genet.* 228:153–159.
34. Dimpfl, J., and H. Echols. 1989. Duplication mutation as an SOS response in *Escherichia coli*: enhanced duplication formation by a constitutively activated RecA. *Genetics* 123:255–260.
35. Dimri, G. P., K. E. Rudd, M. K. Morgan, H. Bayat, and G. Ferro-Luzzi Ames. 1992. Physical mapping of repetitive extragenic palindromic sequences in *Escherichia coli* and phylogenetic distribution among *Escherichia coli* strains and other enteric bacteria. *J. Bacteriol.* 174:4583–4593.
36. Doolittle, W. F. 1994. RNA-mediated gene conversion? *Trends Genet.* 1:64–65.
37. Dover, G., S. Brown, E. Coen, J. Dallas, T. Strachan, and M. Trick. 1982. The dynamics of genome evolution and species differentiation, p. 343–372. *In* G. A. Dover and R. B. Flavell (ed.), *Genome Evolution*. Academic Press, London.
38. Drlica, K., and J. Rouvière-Yaniv. 1987. Histonelike proteins of bacteria. *Microbiol. Rev.* 51:301–319.
39. Duester, G., R. K. Campen, and W. M. Holmes. 1981. Nucleotide sequence of an *Escherichia coli* tRNA$^{\text{Leu1}}$ operon and identification of the transcription promoter signal. *Nucleic Acids Res.* 9:2121–2139.
40. Echols, H. 1986. Multiple DNA-protein interactions governing high-precision DNA transactions. *Science* 233:1050–1056.
41. Edlund, T., and S. Normark. 1981. Recombination between short DNA homologies causes tandem duplication. *Nature* (London) 292:269–271.
42. Ellwood, M., and M. Nomura. 1980. Deletion of a ribosomal ribonucleic acid operon in *Escherichia coli*. *J. Bacteriol.* 143:1077–1080.
43. Ellwood, M., and M. Nomura. 1982. Chromosomal locations of the genes for rRNA in *Escherichia coli* K-12. *J. Bacteriol.* 149:458–468.
44. Engelhorn, M., F. Boccard, C. Murtin, P. Prentki, and J. Geiselmann. 1995. In vivo interaction of the *Escherichia coli* integration host factor with the specific binding sites. *Nucleic Acids Res.* 23:2959–2965.

45. Ennis, D. G., S. K. Amundsen, and G. R. Smith. 1987. Genetic functions promoting homologous recombination in *Escherichia coli*: a study of inversions in phage λ. *Genetics* 115:11–24.

46. Fanning, T. G., and M. F. Singer. 1987. LINE-1: a mammalian transposable element. *Biochim. Biophys. Acta* 910:203–210.

47. Fellner, P., C. Ehresmann, and J. P. Ebel. 1970. Nucleotide sequences present within the 16S ribosomal RNA of *Escherichia coli. Nature* (London) 225:26–29.

48. Ferat, J.-L., M. Le Gouar, and F. Michel. 1994. Multiple group II self-splicing introns in mobile DNA from *Escherichia coli. C. R. Acad. Sci.* 317:141–148.

49. Feulner, G., J. A. Gray, A. Kirschman, A. F. Lehner, A. B. Sadosky, D. A. Vlazny, J. Zhang, S. Zhao, and C. W. Hill. 1990. Structure of the *rhsA* locus from *Escherichia coli* K-12 and comparison of *rhsA* with other members of the *rhs* multigene family. *J. Bacteriol.* 172:446–456.

50. Foster, S. J. 1993. Molecular analysis of three major wall-associated proteins of *Bacillus subtilis* 168: evidence for processing of the product of a gene encoding a 258 kDa precursor two-domain ligand-binding protein. *Mol. Microbiol.* 8:299–310.

51. Fournier, M. J., and H. Ozeki. 1985. Structure and organization of the transfer ribonucleic acid genes of *Escherichia coli* K-12. *Microbiol. Rev.* 49:379–397.

52. Friedman, D. I. 1988. Integration host factor: a protein for all reasons. *Cell* 55:545–554.

53. Friedman, D. I., E. J. Olson, D. Carver, and M. Gellert. 1984. Synergistic effect of *himA* and *gyrB* mutations: evidence that Him functions control expression of *ilv* and *xyl* genes. *J. Bacteriol.* 157:484–489.

54. Galas, D. J., and M. Chandler. 1989. Bacterial insertion sequences, p. 109–162. *In* D. E. Berg and M. M. Howe (ed.), *Mobile DNA.* American Society for Microbiology, Washington, D.C.

55. Gamas, P., M. G. Chandler, P. Prentki, and D. J. Galas. 1987. *Escherichia coli* integration host factor binds specifically to the ends of the insertion sequence IS*1* and to its major insertion hot-spot in pBR322. *J. Mol. Biol.* 195:261–272.

56. Gilson, E., S. Bachellier, S. Perrin, D. Perrin, P. A. D. Grimont, F. Grimont, and M. Hofnung. 1990. Palindromic units highly repetitive DNA sequences exhibit species specificity within *Enterobacteriaceae. Res. Microbiol.* 141:1103–1116.

57. Gilson, E., J.-M. Clément, D. Brutlag, and M. Hofnung. 1984. A family of dispersed repetitive extragenic palindromic DNA sequences in *E. coli. EMBO J.* 3:1417–1421.

58. Gilson, E., J.-M. Clément, D. Perrin, and M. Hofnung. 1987. Palindromic units: a case of highly repetitive DNA sequences in bacteria. *Trends Genet.* 3:226–230.

59. Gilson, E., T. Laroche, and S. Gasser. 1993. Telomeres and the functional architecture of the nucleus. *Trends Cell. Biol.* 3:128–134.

60. Gilson, E., D. Perrin, J.-M. Clément, S. Szmelcman, E. Dassa, and M. Hofnung. 1986. Palindromic units from *E. coli* as binding sites for a chromoid-associated protein. *FEBS Lett.* 206:323–328.

61. Gilson, E., D. Perrin, and M. Hofnung. 1990. DNA polymerase I and a protein complex bind specifically to *E. coli* palindromic units highly repetitive DNA: implications for bacterial chromosome organization. *Nucleic Acids Res.* 18:3941–3952.

62. Gilson, E., D. Perrin, W. Saurin, and M. Hofnung. 1987. Species specificity of bacterial palindromic units. *J. Mol. Evol.* 25:371–373.

63. Gilson, E., J.-P. Rousset, J.-M. Clément, and M. Hofnung. 1986. A subfamily of *E. coli* palindromic units implicated in transcription termination? *Ann. Inst. Pasteur Microbiol.* 137 B:259–270.

64. Gilson, E., W. Saurin, D. Perrin, S. Bachellier, and M. Hofnung. 1991. Palindromic units are part of a new bacterial interspersed mosaic element (BIME). *Nucleic Acids Res.* 19:1375–1383.

65. Gilson, E., W. Saurin, D. Perrin, S. Bachellier, and M. Hofnung. 1991. The BIME family of bacterial highly repetitive sequences. *Res. Microbiol.* 142:217–222.

66. Goodman, S. D., and J. J. Scocca. 1988. Identification and arrangement of the DNA sequence recognized in specific transformation of *Neisseria gonorrhoeae. Proc. Natl. Acad. Sci. USA* 85:6982–6986.

67. Green, L., R. D. Miller, D. E. Dykhuizen, and D. L. Hartl. 1984. Distribution of DNA insertion element IS5 in natural isolates of *Escherichia coli. Proc. Natl. Acad. Sci. USA* 81:4500–4504.

68. Guest, J. R. 1981. Hybrid plasmids containing the citrate synthase gene (*gltA*) of *Escherichia coli* K-12. *J. Gen. Microbiol.* 124:17–23.

69. Gustafson, C. E., S. Chu, and T. J. Trust. 1994. Mutagenesis of the paracrystalline surface protein array of *Aeromonas salmonicida* by endogenous insertion elements. *J. Mol. Biol.* 237:452–463.

70. Harvey, S., and C. W. Hill. 1990. Exchange of spacer regions between rRNA operons in *Escherichia coli. Genetics* 125:683–690.

71. Harvey, S., C. W. Hill, C. Squires, and C. L. Squires. 1988. Loss of the spacer loop sequence from the *rrnB* operon in the *Escherichia coli* K-12 subline that bears the *relA1* mutation. *J. Bacteriol.* 170:1235–1238.

72. Heath, J. D., and G. M. Weinstock. 1991. Tandem duplications of the *lac* region of the *Escherichia coli* chromosome. *Biochimie* 73:343–352.

73. Herzer, P. J., M. Inouye, and S. Inouye. 1992. Retron-EC107 is inserted into the *Escherichia coli* genome by replacing a palindromic 34bp intergenic sequence. *Mol. Microbiol.* 6:355–361.

74. Herzer, P. J., S. Inouye, M. Inouye, and T. S. Whittam. 1990. Phylogenetic distribution of branched RNA-linked multicopy single-stranded DNA among natural isolates of *Escherichia coli. J. Bacteriol.* 172:6175–6181.

75. Higgins, C. F., G. Ferro-Luzzi Ames, W. M. Barnes, J.-M. Clément, and M. Hofnung. 1982. A novel intercistronic regulatory element of prokaryotic operons. *Nature* (London) 298:760–762.

76. Hill, C. W., and G. Combriato. 1973. Genetic duplications induced at very high frequency by ultraviolet irradiation in *Escherichia coli. Mol. Gen. Genet.* 127:197–214.

77. Hill, C. W., J. Foulds, L. Soll, and P. Berg. 1969. Instability of a missense suppressor resulting from a duplication of genetic material. *J. Mol. Biol.* 39:563–581.

78. Hill, C. W., R. H. Grafstrom, B. W. Harnish, and B. S. Hillman. 1977. Tandem duplications resulting from recombination between ribosomal RNA genes in *Escherichia coli. J. Mol. Biol.* 116:407–428.

79. Hill, C. W., and J. A. Gray. 1988. Effects of chromosomal inversion on cell fitness in *Escherichia coli* K-12. *Genetics* 119:771–778.

80. Hill, C. W., J. A. Gray, and H. Brody. 1989. Use of the isocitrate dehydrogenase structural gene for attachment of e14 in *Escherichia coli* K-12. *J. Bacteriol.* 171:4083–4084.

81. Hill, C. W., and B. W. Harnish. 1981. Inversions between ribosomal RNA genes of *Escherichia coli. Proc. Natl. Acad. Sci. USA* 78:7069–7072.

82. Hill, C. W., and B. W. Harnish. 1982. Transposition of a chromosomal segment bounded by redundant rRNA genes into other rRNA genes in *Escherichia coli. J. Bacteriol.* 149:449–457.

83. Hill, C. W., S. Harvey, and J. A. Gray. 1990. Recombination between rRNA genes in *Escherichia coli* and *Salmonella typhimurium*, p. 335–340. *In* K. Drlica and M. Riley (ed.), *The Bacterial Chromosome.* American Society for Microbiology, Washington, D.C.

84. Hill, C. W., C. H. Sandt, and D. A. Vlazny. 1994. *Rhs* elements of *Escherichia coli*: a family of genetic composites each encoding a large mosaic protein. *Mol. Microbiol.* 12:865–871.

85. Hill, C. W., C. Squires, and J. Carbon. 1970. Glycine transfer RNA of *Escherichia coli*. I. Structural genes for two glycine tRNA species. *J. Mol. Biol.* 52:557–569.

86. Hoffmann, G. R., and R. W. Morgan. 1976. The effect of ultraviolet light on the frequency of a genetic duplication in *Salmonella typhimurium. Radiat. Res.* 67:114–119.

87. Hoffmann, G. R., R. W. Morgan, and R. C. Harvey. 1978. Effect of chemical and physical mutagens on the frequency of large genetic duplications in *Salmonella typhimurium*: induction of duplications. *Mutat. Res.* 52:73–80.

88. Horiuchi, T., S. Horiuchi, and A. Novick. 1963. The genetic basis of hypersynthesis of β-galactosidase. *Genetics* 48:157–169.

89. Howard, M. T., M. P. Lee, T.-S. Hsieh, and J. D. Griffith. 1991. *Drosophila* topoisomerase II-DNA interactions are affected by DNA structure. *J. Mol. Biol.* 217:53–62.

90. Hu, M., and R. C. Deonier. 1981. Mapping of IS*1* elements flanking the *argF* gene region of the *Escherichia coli* K-12 chromosome. *Mol. Gen. Genet.* 181:222–229.

91. Hulton, C. S. J., C. F. Higgins, and P. M. Sharp. 1991. ERIC sequences: a novel family of repetitive elements in the genomes of *Escherichia coli, Salmonella typhimurium* and other enterobacteria. *Mol. Microbiol.* 5:825–834.

92. Ilgen, C., L. L. Kirk, and J. Carbon. 1976. Isolation and characterization of large transfer ribonucleic acid precursors from *Escherichia coli. J. Biol. Chem.* 251:922–929.

93. Ingram, V. M. 1963. *The Hemoglobins in Genetics and Evolution.* Columbia University Press, New York.

94. Inouye, M., and S. Inouye. 1992. Retrons and multicopy single-stranded DNA. *J. Bacteriol.* 174:2419–2424.

95. Ishino, Y., W. Shinagawa, K. Makino, M. Amemura, and A. Nakata. 1987. Nucleotide sequence of the *iap* gene, responsible for alkaline phosphatase isozyme conversion in *Escherichia coli*, and identification of the gene product. *J. Bacteriol.* 169:5429–5433.

96. Jinks-Robertson, S., R. L. Gourse, and M. Nomura. 1983. Expression of rRNA and tRNA genes in *Escherichia coli*: evidence for feedback regulation by products of rRNA operons. *Cell* 33:865–876.

97. Jones, I. M., S. B. Primose, and S. D. Ehrlich. 1982. Recombination between short direct repeats in a RecA host. *Mol. Gen. Genet.* 188:486–489.

98. Käs, E., and U. K. Laemmli. 1992. *In vivo* topoisomerase II cleavage of the *Drosophila* histone and satellite III repeats: DNA sequence and structural characteristics. *EMBO J.* 11:705–716.

99. Kenri, T., F. Imamoto, and Y. Kano. 1994. Three tandemly repeated structural genes encoding tRNA$_{f1}$^Met in the *metZ* operon of *Escherichia coli* K-12. *Gene* 138:261–262.

100. King, S. R., M. A. Krolewski, S. L. Marvo, P. J. Lipson, K. L. Pogue-Geile, J. H. Chung, and S. R. Jaskunas. 1982. Nucleotide sequence analysis of in vivo recombinants between bacteriophage λ DNA and pBR322. *Mol. Gen. Genet.* 186:548–557.

101. King, S. R., and J. P. Richardson. 1986. Role of homology and pathway specificity for recombination between plasmids and bacteriophage λ. *Mol. Gen. Genet.* 204:141–147.

102. Kirkegaard, K., and J. C. Wang. 1981. Mapping the topography of DNA wrapped around gyrase by nucleolytic and chemical probing of complexes of unique DNA sequences. *Cell* 23:721–729.

103. Komine, Y., T. Adachi, H. Inokuchi, and H. Ozeki. 1990. Genomic organization and physical mapping of the transfer RNA genes in *Escherichia coli* K12. *J. Mol. Biol.* 212:579–598.

104. Komine, Y., and H. Inokuchi. 1991. Physical map locations of the genes that encode small stable RNAs in *Escherichia coli. J. Bacteriol.* 173:5252.

105. Komine, Y., and H. Inokuchi. 1991. Precise mapping of the *rnpB* gene encoding the RNA component of RNase P in *Escherichia coli* K-12. *J. Bacteriol.* 173:1813–1816.

106. Komoda, Y., M. Enomoto, and A. Tominaga. 1991. Large inversion in *Escherichia coli* K-12 1485IN between inversely oriented IS3 elements near *lac* and *cdd*. *Genetics* 129:639–645.

107. Konrad, E. B. 1977. Method for the isolation of *Escherichia coli* mutants with enhanced recombination between chromosomal duplications. *J. Bacteriol.* 130:167–172.

108. Kornberg, A., L. L. Bertsch, J. F. Jackson, and H. G. Khorana. 1964. Enzymatic synthesis of deoxyribonucleic acid. XVI. Oligonucleotides as templates and the mechanism of their replication. *Proc. Natl. Acad. Sci. USA* 51:315–323.

109. Krawiec, S., and M. Riley. 1990. Organization of the bacterial chromosome. *Microbiol. Rev.* 54:502–539.

110. Kroll, J. S., B. M. Loynds, and P. Langford. 1992. Palindromic *Haemophilus* DNA uptake sequences in presumed transcriptional terminators from *H. influenzae* and *H. parainfluenzae*. *Gene* 114:151–152.

111. Kunisawa, T., and M. Nakamura. 1991. Identification of regulatory building blocks in the *Escherichia coli* genome. *Protein Sequence Data Anal.* 4:43–47.

112. Kuramitsu, S., S. Okuno, T. Ogawa, H. Ogawa, and H. Kagamiyama. 1985. Aspartate aminotransferase of *Escherichia coli*: nucleotide sequence of the *aspC* gene. *J. Bacteriol.* 97:1259–1262.

113. Lampson, B. C., J. Sun, M.-Y. Hsu, J. Vallejo-Ramirez, S. Inouye, and M. Inouye. 1989. Reverse transcriptase in a clinical strain of *Escherichia coli*: production of branched RNA-linked ms DNA. *Science* 243:1033–1038.

114. Lawrence, J. G., D. E. Dykhuizen, R. F. Dubose, and D. L. Hartl. 1989. Phylogenetic analysis using insertion sequence fingerprinting in *Escherichia coli*. *Mol. Biol. Evol.* 6:1–14.

115. Lawrence, J. G., H. Ochman, and D. L. Hartl. 1992. The evolution of insertion sequences within enteric bacteria. *Genetics* 131:9–20.

116. Legrain, C., V. Stalon, and N. Glansdorff. 1976. *Escherichia coli* ornithine carbamoyltransferase isoenzymes: evolutionary significance and the isolation of λargF and λargI transducing bacteriophages. *J. Bacteriol.* 128:35–38.

117. Lehner, A. F., S. Harvey, and C. W. Hill. 1984. Mapping and spacer identification of rRNA operons of *Salmonella typhimurium*. *J. Bacteriol.* 160:682–686.

118. Lehner, A. F., and C. W. Hill. 1980. Involvement of ribosomal ribonucleic acid operons in *Salmonella typhimurium* chromosomal rearrangements. *J. Bacteriol.* 143:492–498.

119. Lehner, A. F., and C. W. Hill. 1985. Merodiploidy in *Escherichia coli-Salmonella typhimurium* crosses: the role of unequal recombination between ribosomal RNA genes. *Genetics* 110:365–380.

120. Lewis, E. B. 1951. Pseudoparallelism and gene evolution. *Cold Spring Harbor Symp. Quant. Biol.* 16:159–174.

121. Lim, D., and W. K. Maas. 1989. Reverse transcriptase-dependent synthesis of a covalently linked, branched DNA-RNA compound in *E. coli* B. *Cell* 56:891–904.

122. Lin, R.-J., M. Capage, and C. W. Hill. 1984. A repetitive DNA sequence, *rhs*, responsible for duplications within the *Escherichia coli* K-12 chromosome. *J. Mol. Biol.* 177:1–18.

123. Lindsey, D. F., D. A. Mullin, and J. R. Walker. 1989. Characterization of the cryptic lambdoid prophage DLP12 of *Escherichia coli* and overlap of the DLP12 integrase gene with the tRNA gene *argU*. *J. Bacteriol.* 171:6197–6205.

124. Liu, L. F., and J. C. Wang. 1987. Supercoiling of the DNA template during transcription. *Proc. Natl. Acad. Sci. USA* 84:7024–7027.

125. Louarn, J., J. Patte, and J.-M. Louarn. 1982. Suppression of *Escherichia coli* *dnaA46* mutations by integration of plasmid R100.1 derivatives: constraints imposed by the replication terminus. *J. Bacteriol.* 151:657–667.

126. Louarn, J. M., J. P. Bouche, F. Lengendre, J. Louarn, and J. Patte. 1985. Characterization and properties of very large inversions of the *E. coli* chromosome along the origin-to-terminus axis. *Mol. Gen. Genet.* 201:467–476.

127. Lovett, S. T., P. T. Drapkin, V. A. Sutera, Jr., and T. J. Gluckman-Peskind. 1993. A sister-strand exchange mechanism for recA-independent deletion of repeated DNA sequences in *Escherichia coli*. *Genetics* 135:631–642.

128. MacFarlane, S. A., and M. Merrick. 1985. The nucleotide sequence of the nitrogen regulation *ntrB* and the *glnA-ntrBC* intergenic region of *Klebsiella pneumoniae*. *Nucleic Acids Res.* 13:7591–7606.

129. Marvo, S. L., S. R. King, and S. R. Jaskunas. 1983. Role of short regions of homology in intermolecular illegitimate recombination events. *Proc. Natl. Acad. Sci. USA* 80:2452–2456.

130. Matfield, M., R. Badawi, and W. J. Brammar. 1985. Rec-dependent and Rec-independent recombination of plasmid-borne duplications in *Escherichia coli* K12. *Mol. Gen. Genet.* 199:518–523.

131. Matic, I., M. Radman, and C. Rayssiguier. 1994. Structure of recombinants from conjugational crosses between *Escherichia coli* donor and mismatch-repair deficient *Salmonella typhimurium* recipients. *Genetics* 136:17–26.

132. Mazin, A. V., A. V. Kuzminov, G. L. Dianov, and R. I. Salganik. 1991. Molecular mechanisms of deletion formation in *Escherichia coli* plasmids. II. Deletion formation mediated by short direct repeats. *Mol. Gen. Genet.* 228:209–214.

133. Meighen, E. A., and R. B. Szittner. 1992. Multiple repetitive elements and organization of the *lux* operons of luminescent terrestrial bacteria. *J. Bacteriol.* 174:5371–5381s.

134. Milkman, R., and A. Stoltzfus. 1988. Molecular evolution of the *Escherichia coli* chromosome. II. Clonal segments. *Genetics* 120:359–366.

135. Mizobuchi, K. 1993. Personal communication. First International Symposium Mapping and Sequencing of Small Genomes, Institut Pasteur, Paris.

136. Moralejo, P., S. M. Egan, E. Hidalgo, and J. Aguilar. 1993. Sequencing and characterization of a gene cluster encoding the enzymes for L-rhamnose metabolism in *Escherichia coli*. *J. Bacteriol.* 175:5585–5594.

137. Morgan, G. T., and K. M. Middleton. 1990. Short interspersed repeats from *Xenopus* that contain multiple octamer motifs are related to known transposable elements. *Nucleic Acids Res.* 18:5781–5785.

138. Morse, M. L., E. Lederberg, and J. Lederberg. 1956. Transduction in *Escherichia coli* K-12. *Genetics* 41:121.

139. Mott, J. E., J. C. Galloway, and T. Platt. 1985. Maturation of *Escherichia coli* tryptophan operon mRNA: evidence for 3′ exonucleolytic processing after rho-independent termination. *EMBO J.* 4:1887–1891.

140. Naas, T., M. Blot, W. M. Fitch, and W. Arber. 1994. Insertion sequence-related genetic variation in resting *Escherichia coli* K-12. *Genetics* 136:721–730.

141. Nakata, A., M. Amemura, and K. Makino. 1989. Unusual nucleotide arrangement with repeated sequences in the *Escherichia coli* K-12 chromosome. *J. Bacteriol.* 171:3553–3556.

142. Newbury, S. F., N. H. Smith, and C. F. Higgins. 1987. Differential mRNA stability controls relative gene expression within a polycistronic operon. *Cell* 51:1131–1143.

143. Nichols, B. P., and C. Yanofsky. 1979. Nucleotide sequences of *trpA* of *Salmonella typhimurium* and *Escherichia coli*: an evolutionary comparison. *Proc. Natl. Acad. Sci. USA* 76:5244–5248.

144. Ochman, H., and R. K. Selander. 1984. Standard reference strains of *Escherichia coli* from natural populations. *J. Bacteriol.* 157:690–693.

145. Ohno, S. 1970. *Evolution by Gene Duplication*. Springer-Verlag, Berlin.

146. Ohshima, A., S. Inouye, and M. Inouye. 1992. *In vivo* duplication of genetic elements by the formation of stem-loop DNA without an RNA intermediate. *Proc. Natl. Acad. Sci. USA* 89:1016–1020.

147. Oppenheim, A. B., K. E. Rudd, I. Mendelson, and D. Teff. 1993. Integration host factor binds to a unique class of complex repetitive extragenic DNA sequences in *Escherichia coli*. *Mol. Microbiol.* 10:113–122.

148. Osheroff, N. 1986. Eukaryotic topoisomerase II. *J. Biol. Chem.* 261:9944–9950.

149. Petes, T. D., and C. W. Hill. 1988. Recombination between repeated sequences in microorganisms. *Annu. Rev. Genet.* 22:147–168.

150. Petit, M.-A., J. Dimpfl, M. Radman, and H. Echols. 1991. Control of large chromosomal duplications in *Escherichia coli* by the mismatch repair system. *Genetics* 129:327–332.

151. Pierson, L. S., III, and M. L. Kahn. 1987. Integration of satellite bacteriophage P4 in *Escherichia coli*: DNA sequences of the phage and host regions. *J. Mol. Biol.* 196:487–496.

152. Plamann, M. D., and G. V. Stauffer. 1985. Characterization of a cis-acting regulatory mutation that maps at the distal end of the *Escherichia coli glyA* gene. *J. Bacteriol.* 161:650–654.

153. Platt, T. 1986. Transcription termination and regulation of gene expression. *Annu. Rev. Biochem.* 55:339–372.

154. Radman, M. 1989. Mismatch repair and the fidelity of genetic recombination. *Genome* 31:68–73.

155. Radman, M. 1991. Avoidance of inter-repeat recombination by sequence divergence and a mechanism of neutral evolution. *Biochimie* 73:357–361.

156. Rau, D. C., M. Gellert, F. Thoma, and A. Maxwell. 1987. Structure of the DNA gyrase-DNA complex as revealed by transient electric dichroism. *J. Mol. Biol.* 193:555–569.

157. Rayssiguier, C., D. S. Thaler, and M. Radman. 1989. The barrier to recombination between *Escherichia coli* and *Salmonella typhimurium* is disrupted in mismatch-repair mutants. *Nature* (London) 342:396–401.

158. Rebollo, J.-E., V. Francois, and J.-M. Louarn. 1988. Detection and possible role of two large nondivisible zones on the *Escherichia coli* chromosome. *Proc. Natl. Acad. Sci. USA* 85:9391–9395.

159. Riley, M. 1984. Arrangement and rearrangement of bacterial genomes, p. 285–315. *In* R. P. Morlock (ed.), *Microorganisms as Model Systems for Studying Evolution*. Plenum, New York.

160. Riley, M., and A. Anilionis. 1978. Evolution of the bacterial genome. *Annu. Rev. Microbiol.* 32:519–560.

161. Riley, M., and S. Krawiec. 1987. Genome organization, p. 967–981. *In* F. C. Neidhardt, J. L. Ingraham, K. B. Low, B. Magasanik, M. Schaechter, and H. E. Umbarger (ed.), *Escherichia coli and Salmonella typhimurium: Cellular and Molecular Biology*, vol. 2. American Society for Microbiology, Washington, D.C.

162. Roberge, M., and S. M. Gasser. 1992. DNA loops: structural and functional properties of scaffold-attached regions. *Mol. Microbiol.* 6:419–423.

163. Rosenberg, M., and D. Court. 1979. Regulatory sequences involved in the promotion and termination of RNA transcription. *Annu. Rev. Genet.* 13:319–353.

164. Russell, R. L., J. N. Abelson, A. Landy, M. L. Gefter, S. Brenner, and J. D. Smith. 1970. Duplicate genes for tyrosine transfer RNA in *Escherichia coli*. *J. Mol. Biol.* 47:1–13.

165. Sadosky, A. B., A. Davidson, R.-J. Lin, and C. W. Hill. 1989. *rhs* gene family of *Escherichia coli* K-12. *J. Bacteriol.* 171:636–642.

166. Sadosky, A. B., J. A. Gray, and C. W. Hill. 1991. The *RhsD-E* subfamily of *Escherichia coli* K-12. *Nucleic Acids Res.* 19:7177–7183.

167. Saksela, K., and D. Baltimore. 1993. Negative regulation of immunoglobulin kappa light-chain gene transcription by a short sequence homologous to the murine B1 repetitive element. *Mol. Cell. Biol.* 13:3698–3705.

168. Sanderson, K. E. 1976. Genetic relatedness in the family Enterobacteriaceae. *Annu. Rev. Microbiol.* **30:**327–349.

169. Savic, D. J., S. P. Romac, and S. D. Ehrlich. 1983. Inversion in the lactose region of *Escherichia coli* K-12: inversion termini map within IS3 elements α3β3 and β5α5. *J. Bacteriol.* **155:**943–946.

170. Sawyer, S. A., D. E. Dykhuizen, R. F. Dubose, L. Green, T. Mutangadura-Mhlanga, D. F. Wolczyk, and D. L. Hartl. 1987. Distribution and abundance of insertion sequences among natural isolates of *Escherichia coli. Genetics* **115:**51–63.

171. Schmeissner, U., K. MacKenney, M. Rosenberg, and D. Court. 1984. Removal of a terminator structure by mRNA processing regulates *int* gene expression. *J. Mol. Biol.* **176:**39–53.

172. Schmid, M. B., and J. R. Roth. 1983. Selection and endpoint distribution of bacterial inversion mutations. *Genetics* **105:**539–557.

173. Schmid, M. B., and J. R. Roth. 1987. Gene location affects expression level in *Salmonella typhimurium. J. Bacteriol.* **169:**2872–2875.

174. Sclafani, R. A., and J. A. Wechsler. 1981. High frequency of genetic duplications in the *dnaB* region of the *Escherichia coli* K12 chromosome. *Genetics* **98:**677–689.

175. Segall, A. M., M. J. Mahan, and J. R. Roth. 1988. Rearrangement of the bacterial chromosome: forbidden inversions. *Science* **241:**1314–1318.

176. Segall, A. M., and J. R. Roth. 1989. Recombination between homologies in direct and inverse orientation in the chromosome of Salmonella: intervals which are nonpermissive for inversion formation. *Genetics* **122:**737–747.

177. Segall, A. M., and J. R. Roth. 1994. Approaches to half-tetrad analysis in bacteria: recombination between repeated, inverse-order chromosomal sequences. *Genetics* **136:**27–39.

178. Sharp, P. M., and W.-H. Li. 1987. The codon adaptation index—a measure of directional synonymous codon usage bias, and its potential applications. *Nucleic Acids Res.* **15:**1281–1295.

179. Sharples, G. J., and R. G. Lloyd. 1990. A novel repeated DNA sequence located in the intergenic regions of bacterial chromosome. *Nucleic Acids Res.* **18:**6503–6508.

180. Shen, P., and H. V. Huang. 1986. Homologous recombination in *Escherichia coli:* dependence on substrate length and homology. *Genetics* **112:**441–457.

181. Shen, P., and H. V. Huang. 1989. Effect of base pair mismatches on recombination via the RecBCD pathway. *Mol. Gen. Genet.* **218:**358–360.

182. Shyamala, V., E. Schneider, and G. Ferro-Luzzi Ames. 1990. Tandem chromosomal duplications: role of REP sequences in the recombination event at the joinpoint. *EMBO J.* **9:**939–946.

183. Smith, G. P. 1973. Unequal crossover and the evolution of multigene families. *Cold Spring Harbor Symp. Quant. Biol.* **38:**507–513.

184. Smith, G. R. 1988. Homologous recombination sites and their recognition, p. 115–154. *In* K. B. Low (ed.), *The Recombination of Genetic Material.* Academic Press, San Diego, Calif.

185. Sommer, H., B. Schumacher, and H. Saedler. 1981. A new type of IS*1*-mediated deletion. *Mol. Gen. Genet.* **184:**300–307.

186. Sonti, R. V., and J. R. Roth. 1989. Role of gene duplications in the adaptation of *Salmonella typhimurium* to growth on limiting carbon sources. *Genetics* **123:**19–28.

187. Stern, M. J., G. Ferro-Luzzi Ames, N. H. Smith, E. C. Robinson, and C. F. Higgins. 1984. Repetitive extragenic palindromic sequences: a major component of the bacterial genome. *Cell* **37:**1015–1026.

188. Stern, M. J., E. Prossnitz, and G. Ferro-Luzzi Ames. 1988. Role of the intercistronic region in post-transcriptional control of gene expression in the histidine transport operon of *Salmonella typhimurium:* involvement of REP sequences. *Mol. Microbiol.* **2:**141–152.

189. Straus, D. S. 1974. Induction by mutagens of tandem gene duplications in the *glyS* region of the *Escherichia coli* chromosome. *Genetics* **78:**823–830.

190. Straus, D. S., and G. R. Hoffmann. 1975. Selection for a large genetic duplication in *Salmonella typhimurium. Genetics* **80:**227–237.

191. Stroeher, U. H., L. E. Karageorgos, R. Morona, and P. A. Manning. 1992. Serotype conversion in *Vibrio cholerae* O1. *Proc. Natl. Acad. Sci. USA* **89:**2566–2570.

192. Timmons, M. S., A. M. Bogardus, and R. C. Deonier. 1983. Mapping of chromosomal IS*5* elements that mediate type II F-prime plasmid excision in *Escherichia coli* K-12. *J. Bacteriol.* **153:**395–407.

193. Timmons, M. S., M. Lieb, and R. C. Deonier. 1986. Recombination between IS*5* elements: requirement for homology and recombination functions. *Genetics* **113:**797–810.

194. Tlsty, T. D., A. M. Albertini, and J. H. Miller. 1984. Gene amplification in the *lac* region of *E. coli. Cell* **37:**217–224.

195. Tomb, J.-F., H. El-Hajj, and H. O. Smith. 1991. Nucleotide sequence of a cluster of genes involved in the transformation of *Haemophilus influenzae* Rd. *Gene* **104:**1–10.

196. Truniger, V., W. Boos, and G. Sweet. 1992. Molecular analysis of the *glpFKX* regions of *Escherichia coli* and *Shigella flexneri. J. Bacteriol.* **174:**6981–6991.

197. Turnowsky, F., C. Fuchs, C. Jeschek, and G. Högenauer. 1989. *envM* genes of *Salmonella typhimurium* and *Escherichia coli. J. Bacteriol.* **171:**6555–6565.

198. Umeda, M., and E. Ohtsubo. 1989. Mapping of insertion elements IS*1*, IS*2* and IS*3* on the *Escherichia coli* K-12 chromosome. *J. Mol. Biol.* **208:**601–614.

199. Umeda, M., and E. Ohtsubo. 1990. Mapping of insertion element IS*5* in the *Escherichia coli* K-12 chromosome. Chromosomal rearrangements mediated by IS*5. J. Mol. Biol.* **213:**229–237.

200. Valentin-Hansen, P., K. Hammer-Jespersen, F. Boetius, and I. Svendsen. 1984. Structure and function of the intercistronic regulatory *deoC-deoA* element of *Escherichia coli* K-12. *EMBO J.* **3:**179–183.

201. Van Vliet, F., R. Cunin, A. Jacobs, J. Piette, D. Gigot, M. Lauwereys, A. Pierard, and N. Glansdorff. 1984. Evolutionary divergence of genes for ornithine and aspartate carbamoyl-transferases—complete sequence and mode of regulation of the *Escherichia coli argF* gene; comparison of *argF* with *argI* and *pyrB. Nucleic Acids Res.* **12:**6277–6289.

202. Varmus, H. E. 1989. Reverse transcription in bacteria. *Cell* **56:**721–724.

203. Versalovic, J., T. Koeuth, R. Britton, K. Geszvain, and J. R. Lupski. 1993. Conservation and evolution of the *rpsU-dnaG-rpoD* macromolecular synthesis operon in bacteria. *Mol. Microbiol.* **8:**343–355.

204. Versalovic, J., T. Koeuth, and J. R. Lupski. 1991. Distribution of repetitive DNA sequences in eubacteria and application to fingerprinting of bacterial genomes. *Nucleic Acids Res.* **19:**6423–6831.

205. Vidal, F., E. Mougneau, N. Glaichenhaus, P. Vaigot, M. Darmon, and F. Cuzin. 1993. Coordinated posttranscriptional control of gene expression by modular elements including *Alu*-like repetitive sequences. *Proc. Natl. Acad. Sci. USA* **90:**208–212.

206. Voloshin, O. N., S. M. Mirkin, V. I. Lyamichev, B. P. Belotserkovskii, and M. D. Frank-Kamenetski. 1988. Chemical probing of homopurine-homopyrimidine mirror repeats in supercoiled DNA. *Nature* (London) **333:**475–476.

207. von Eichel-Streiber, C., M. Sauerborn, and H. K. Kuramitsu. 1992. Evidence for a modular structure of the homologous repetitive C-terminal carbohydrate-binding sites of *Clostridium difficile* toxins and *Streptococcus mutans* glucosyltransferases. *J. Bacteriol.* **174:**6707–6710.

208. Watt, V. M., C. J. Ingles, M. S. Urdea, and W. J. Rutter. 1985. Homology requirements for recombination in *Escherichia coli. Proc. Natl. Acad. Sci. USA* **82:**4768–4772.

209. White, S., F. E. Tuttle, D. Blankenhorn, D. C. Dosch, and J. L. Slonczewski. 1992. pH dependence and gene structure of *inaA* in *Escherichia coli. J. Bacteriol.* **174:**1537–1543.

210. Whoriskey, S. K., V.-H. Nghiem, P.-M. Leong, J.-M. Masson, and J. H. Miller. 1987. Genetic rearrangements and gene amplification in *Escherichia coli:* DNA sequences at the junctures of amplified gene fusions. *Genes Dev.* **1:**227–237.

211. Wren, B. W. 1991. A family of clostridial and streptococcal ligand-binding proteins with conserved C-terminal repeat sequences. *Mol. Microbiol.* **5:**797–803.

212. Xiang, S.-H., M. Hobbs, and P. R. Reeves. 1994. Molecular analysis of the *rfb* gene cluster of a group D2 *Salmonella enterica* strain: evidence for intraspecific gene transfer in O antigen variation. *J. Bacteriol.* **176:**4357–4365.

213. Yang, Y., and G. Ferro-Luzzi Ames. 1988. DNA gyrase binds to the family of prokaryotic repetitive extragenic palindromic sequences. *Proc. Natl. Acad. Sci. USA* **85:**8850–8854.

214. Yang, Y., and G. Ferro-Luzzi Ames. 1990. The family of repetitive extragenic palindromic sequences: interaction with DNA gyrase and histonelike protein HU, p. 211–226. *In* K. Drlica and M. Riley (ed.), *The Bacterial Chromosome.* American Society for Microbiology, Washington, D.C.

215. Yokota, T., H. Sugisaki, M. Takanami, and Y. Kaziro. 1980. The nucleotide sequence of the cloned *tufA* gene of *Escherichia coli. Gene* **12:**25–31.

216. York, M. K., and M. Stodolsky. 1981. Characterization of P1*argF* derivatives from *Escherichia coli* K12 transduction I. IS*1* elements flank the *argF* gene segment. *Mol. Gen. Genet.* **181:**230–240.

217. Zhao, S., C. H. Sandt, G. Feulner, D. A. Vlazny, J. A. Gray, and C. W. Hill. 1993. *Rhs* elements of *Escherichia coli* K-12: complex composites of shared and unique components that have different evolutionary histories. *J. Bacteriol.* **175:**2799–2808.

218. Zieg, J., and S. R. Kushner. 1977. Analysis of genetic recombination between two partially deleted lactose operons of *Escherichia coli* K-12. *J. Bacteriol.* **131:**123–132.

219. Ziemke, P., and J. E. G. McCarthy. 1992. The control of mRNA stability in *Escherichia coli:* manipulation of the degradation pathway of the polycistronic *atp* mRNA. *Biochim. Biophys. Acta* **1130:**297–306.

220. Zuker, M., and P. Stiegler. 1981. Optimal computer folding of large RNA sequences using thermodynamics and auxiliary information. *Nucleic Acids Res.* **9:**133–148.

Cryptic Prophages

ALLAN M. CAMPBELL

113

DEFINITION AND EXAMPLES

The defining characteristic of temperate phages is their ability to enter a prophage state, where most of the phage genes are unexpressed and the prophage replicates as part of the host genome. Classically, the presence of the prophage was manifested by the occasional breakdown (either spontaneous or induced) of latency, so that every culture of a prophage-bearing (lysogenic) strain contains some free infectious phage liberated by lysis of a few cells in the culture.

Prophages occasionally mutate to lose some of the functions essential for lytic growth, in which case the strain no longer liberates infectious particles. Such strains have been called *defective lysogens,* and their prophages are termed *defective prophages* (13). Because a lysogenic strain can also lose its entire prophage, the concept of defective lysogeny became needed only because some defective lysogens retain prophage-determined traits, such as the superinfection immunity that results from expression of repressor genes. The term "cryptic prophage" was introduced to denote a particular type of defective prophage that does not confer immunity to superinfection but nonetheless harbors a partial prophage, demonstrable by recombination with a genetically marked superinfecting phage. This distinction between defective prophages and cryptic prophages has had limited utility, and here we will use the terms as synonyms.

Defective lysogens frequently arise spontaneously during laboratory propagation of lysogenic strains. Lambda prophages disabled by mutation or deletion have been deliberately sought for various purposes. It was early shown that various defective lysogens, apparently with point mutations in their prophages, were blocked at different stages of the phage life cycle (3, 22, 37, 43, 55). Complementation analysis of λ phage functions was performed by synthesizing double lysogens with one defective prophage on the chromosome and a second on an F′ element (27). These inherently awkward approaches were largely superseded with the advent of conditionally lethal mutations (11, 23). A conditionally lethal mutant prophage under nonpermissive conditions (e.g., an amber mutant in a nonsuppressing host) is effectively a defective prophage, although unconditionally defective prophages include some types, such as promoter mutations, that are not accessible as conditional lethals. Numerous X-ray-induced mutants of prophage λ were isolated in the unfulfilled hope that X-ray might preferentially induce deletions, allowing deletion mapping of the prophage (A. del Campillo-Campbell, quoted in reference 12). Successful deletion mapping was accomplished by isolation of nested sets of deletions selected for loss of function of specific genes in or near the prophage (1, 26, 30, 46, 52) and by the complementary approach of studying specialized transducing phages (10).

Specialized transducing phages, which arise by abnormal excision from the lysogenic chromosome of a partial prophage connected to flanking DNA, are frequently defective. Phages that have picked up antibiotic resistance transposons can also be defective, sometimes with deletions of large blocks of phage DNA at the site of transposon insertion. Phage P1, which ordinarily lysogenizes by plasmid formation, spawned such deletion derivatives bearing chloramphenicol resistance derived from Tn9 (P1 *cry*) or tetracycline resistance from Tn10 (P1 *tet*), which survive integrated into the chromosome (50, 51).

Besides these laboratory creations, natural *Escherichia coli* strains frequently contain λ-related sequences presumed to be cryptic prophages (2). The most intensely studied are those present in the K-12 strain, whose chromosome contains, beside λ itself, four elements (DLP12, Rac, Qin, and e14) that are related in sequence either to λ or (in the case of e14) to a λ relative, phage 21. All these elements are shorter than λ (the shortest, e14, being 14 kb compared to the 49 kb of λ). From comparison of the available sequence, plus recombinational studies, their genetic structures can be inferred. Assuming that each element entered the K-12 genome as an active λ-related phage, each element must have lost substantial segments of the functional phage genome through deletion. Insertions of IS elements are also frequent. Additional deletions within some of these prophages have been observed during laboratory propagation of K-12 (53). Figure 1 shows genetic maps of some of these elements.

MODES OF DETECTION

Defective λ lysogens bearing single point mutations remain immune to superinfection by λ. They are distinguishable from λ-resistant host mutants by their sensitivity to virulent mutants

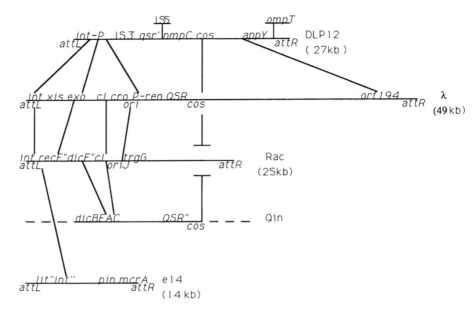

FIGURE 1 Lambdoid prophages of *E. coli* K-12, located at 12 min (DLP12), 17 min (λ), 25 min (e14), 30 min (Rac), and 35 min (Qin). All but λ are defective. Homologous genes or segments of the different phages are connected. (For *cI*Rac, homology and location are assumed on indirect evidence.) Genes are shown above the lines and recognition sites below the lines. Genes in quotation marks are identified only by sequence. Genes affecting host phenotype are *nmpC* (new outer membrane protein), *trgG* (K$^+$ transport), *dicB* (division control), *lit* (phage T4 development), *pin* (DNA inversion), and *mcrA* (restriction). Major sources: DLP12 (41, 44); λ (20); Rac (18, 21; C. C. Chu, L. Satin, A. J. Clark, M. Faubladier, and J.-P. Bouché, personal communication); Qin (8, 24); e14 (35, 39, 47, 50).

of λ (λ *vir*) or heteroimmune lambdoid phages (such as λ *imm*[434]). Many such isolates lyse on induction, in some cases liberating phage-related particles visible in the electron microscope.

Some natural strains likewise lyse on induction, liberating phagelike particles. This is true of *E. coli* 15, whose lysates contain two types of phagelike particles, each packed with a unique DNA species carried by the bacterium in plasmid form (29, 36). Neither type has been shown to form plaques. The phagelike particles and free tails can kill susceptible cells and thus qualify as colicins. This strain also harbors a large plasmid related in DNA sequence to phage P1 and displaceable by P1 prophage, but not packaged into virions.

Most natural prophages are not recognizable by such criteria. Three of the four cryptic prophages of K-12 (e14, Qin, and Rac) have repressor genes, presumably present in ancestral lambdoid phages but with a specificity that has not been encountered among the known lambdoid phages. Nor do they make the strain sensitive to DNA-damaging agents (like UV light) that induce derepression, at least not to the same extent that λ prophage does. However, derepression is induced in these elements, some of which were discovered through observation of its sequelae. In the case of e14, 14-kb circles of the excised element were found in induced cells, and later, chromosomes cured of e14 were detected (9, 31). With Rac, a phage-coded recombinase that can act on bacterial chromosomes is derepressed by conjugation (42). And mutations in the repressor gene (*dicA*) of Qin derepress another gene (*dicB*) whose product inhibits cell division (6).

When a cryptic prophage is bracketed by *attL* and *attR* sites and functional integrase and excisionase are available (either from the cryptic prophage itself or by complementation), the prophage boundaries are precisely defined. This is true of e14 (which excises) and of Rac (which can recombine with active λ, giving a recombinant phage, λ reverse, with the Rac *attP* site and flanking DNA derived from the two ends of the inserted prophage). Rac can be excised by superinfection with λ reverse (7). DLP12 has a full-length *int* gene and a truncated *xis*. Specific excision has not been shown, but a sequence from the *argU* tRNA gene at its left end is repeated on the right and could represent *attR*. No such information is available for Qin, which is delineated only approximately by knowledge of functions and sequences related to lambdoid phages on the one hand and standard bacterial genes on the other. As laboratory deletions frequently remove *att* sites, there is no expectation that all cryptic prophages should retain them.

Another way to detect cryptic prophages is through their interactions with superinfecting phages. Both DLP12 and Qin were found by the recombinational rescue of QSR function by λ phages mutant in one or more of these genes. The first such recombinant for DLP12 (called p4) was unwittingly selected because the recombinant fitted better with the λ packaging constraints when recombination was selected between other markers that changed the genome size. Later, selection for pseudorevertants of double amber mutants or deletions in gene Q gave the same type of recombinant, in which λ genes Q, S, R are replaced by analogs from DLP12 or Qin (33, 38). In both

DLP12 and Qin, the *QSR* analogs have no significant homology to λ, but there is homology in the flanking DNA. In the case of phage P2, infection of *E. coli* B produced recombinants (called P2HyDis) with a different repressor specificity from P2; as *E. coli* B does not spontaneously produce phage, the prophage bearing these determinants may be defective (19). *E. coli* B can be cured of the prophage by superinfection with P2. Furthermore, it has long been known that P2 does not readily insert at the chromosomal site of K-12 that is preferred in *E. coli* C. It turns out that a similar site in K-12 is already occupied by a defective remnant, 639 bp in length, of P2 DNA (5).

Another type of interaction takes place during prophage insertion. The cryptic prophages of K-12 can serve as secondary insertion sites for λ (through homologous recombination), and a Qin prophage in *E. coli* C has been detected by this property (40).

However, surveys at this level could mislead. Without information on the extent of the homology or the part of the genome involved, no certain conclusion can be drawn as to whether some of the hybridizations might register host sequences that have been incorporated by the phage rather than the other way around. As a specific possibility, many phages (including some λ derivatives) have picked up bacterial insertion sequences that are also found in the host chromosome; it would be unwarranted to conclude that the latter are of phage origin.

If the genomes of large double-stranded DNA phages originated as chimeras of various functional modules derived either from the host or from accessory DNA elements (14, 16), phage homology with genome DNA need not indicate the presence of prophage residua. Of the K-12 elements DLP12, Qin, Rac, and e14, the only one that might credibly have a nonphage origin is e14. The others all contain phage genes (frequently incomplete) in the same order found in active phages, and it is hard to imagine a reason for that arrangement other than derivation from an active prophage. With e14, the only known homology with another lambdoid phage is in its attachment sites and putative integrase gene.

Another possible manner of surveying natural strains is to focus on a specific chromosomal attachment site and to look for phage-related DNA in that position. This was effectively done for the Atlas prophages, which were discovered when a DNA segment near *trp* was examined for variation among a collection of natural *E. coli* strains (ECOR collection [45]). Atlas prophages are present in 23 of 72 strains examined (but not in K-12). At least two of these are active (able to form plaques), but most of the others are shorter and probably defective (A. B. Stolzfus, Ph.D. thesis, University of Iowa, Iowa City, 1991).

EFFECTS ON HOST

Cryptic prophages frequently have shown up as loci of mutations affecting host phenotype. This is true of the four cryptic prophages of *E. coli* K-12, although all except Rac were initially detected by other means. The mutations may either inactivate expressed prophage genes or activate unexpressed genes.

An unexpressed gene of DLP12 (*nmpC*), present also in some lambdoid phages, is homologous to (and probably derived from) the gene for the major host outer membrane protein, *ompC* (34). Host *ompC* mutants grow slowly, and some faster growing derivatives that come up have mutations in DLP12 that activate *nmpC*, whose product then can functionally replace OmpC. The *nmpC* gene is not essential for plaque formation by those phages that have it; its product may render the cell surface unable to adsorb phage through replacing OmpC, a common receptor for lambdoid phages.

Two other genes, *appY* and *ompT*, lie at the right end of the DLP12 prophage (41). Overexpression of *appY* cloned on a multicopy plasmid enhances expression of various *E. coli* genes including the alkaline phosphatase gene *appA* (4). The *ompT* gene is expressed from the DLP12 prophage and encodes a protease located in the outer membrane (32). It is not clear that either *appY* or *ompT* is useful to *E. coli*.

The Rac prophage came to light because of its gene for general recombination. In λ, the analogous genes (*exo*, β) promote RecA-independent general recombination and increase the rate of replication. As in λ, the Rac recombination gene, *recE*, is regulated by repression, which can be eliminated by zygotic induction during bacterial conjugation, by mutational inactivation of the repressor gene, or by long deletions that place *recE* under control of nearby bacterial promoters. Zygotic induction, causing a temporary restoration of recombination proficiency to a *recB recC rac⁻* recipient mated to an Hfr *rac⁺* donor, was the characteristic by which Rac was first identified (42). Mutations (*sbcA*) that permanently activate *recE* expression were selected on *recB recC* strains; *recB recC* mutants are hypersensitive to DNA-damaging agents such as mitomycin C, and the *recE* enzyme Exo VIII can substitute for the *recB recC* protein ExoV in increasing resistance.

Another gene in the Rac prophage (*trgG*) encodes a protein that cooperates with the TrgA protein of the host to form one of two systems catalyzing a high rate of K⁺ transport (21). The system is not essential for *E. coli* growth but does have a higher affinity for K⁺ than the other available system (TrgH), which could be advantageous under some circumstances. The conditions used to select for activation of *nmpC* and *recE* suggest possible evolutionary pathways whereby a phage-borne gene might be coopted by the bacterial host. Obviously, the same selective conditions set up in the laboratory must sometimes arise in nature, and the result should be the same in the short run. There is no good evidence that cooption of prophage genes has ever been a step in an effective long-range evolutionary strategy.

The *dic* genes of the Qin prophage have been discussed in terms of cooption because they can affect cell growth, but the observed effects are negative (8). The *dicB* gene interferes with cell division and seems analogous and similarly located to the λ *kil* gene, whose expression is lethal to the cell. The *dicA* and *dicC* genes regulate *dicB* and have some homology to λ *cI* and λ *cro*.

The e14 element has no genes with defined functions analogous to those of other lambdoid phages. Even the integrase and excisionase functions have not been directly shown to reside in e14 (although e14 is at least necessary for their expression [9]), and the identified segment with *int* homology could be a pseudogene (S. J. Schneider, Ph.D. thesis, Stanford University, Stanford, Calif., 1992). Some genes that have been identified are *lit* (which regulates translation of T4 late genes), *mcrA* (a restriction gene), and an invertible segment P with its cognate site-specific recombinase encoded by *pin*. The P-*pin* system is homologous to counterparts in phages P1 and Mu-1, as well as in strains of *Salmonella* that can switch surface proteins (phase variation).

FIGURE 2 Structure of C32, the most common type of λ *cry* prophage. Homologous recombination has replaced prophage DNA from an IS2 near *xis* through *cos* with bacterial DNA from the IS2 at 9 min through the *cos* site of DLP12. From reference 49 with permission.

INTERACTIONS WITH ACTIVE PHAGES

Prophages can interact with superinfecting phages on both the physiological and the genetic level. Some of these affect heterologous phages (such as the exclusion of T4 *rII* mutants by the λ prophage *rex* gene), whereas others act on related phages (such as superinfection immunity or alterations of cell surface affecting phage attachment). The *nmpC* gene in DLP12 is not the ideal example here, because it is unexpressed unless activated by mutation. In lambdoid phage PA-2, *nmpC* is expressed from the prophage and interferes with attachment of phage of the carried type (48). The ε phages of *Salmonella* likewise alter the cell surface by introducing biosynthetic enzymes for surface polysaccharide phage receptors, so that ε15, for example, renders the cell resistant to superinfection by another ε15 (54). If the superinfecting phage gained access to the cell, it would be controlled by the prophage repressor, but in fact it does not get that far. Superinfecting λ mutants defective in certain genes can induce expression of their homologs in repressed or defective prophages at a level which is sometimes adequate to complement the defect (as in the case of gene *R*).

With respect to defective phages, the recombinational interactions have received most attention. The simplest is marker rescue. In general, rescue into an active phage requires homology on both sides of the rescued gene, not necessarily in the gene itself. The cryptic phages of *E. coli* K-12 (like the natural lambdoid phages) include segments homologous to various lambdoid phages. Thus rescue of the *q'* gene from DLP12 is accompanied by rescue of *s'* and *r'*, because there is no λ homology within the *qsr'* segment. Rescue of *cos* into λ is sometimes accompanied by rescue of *qsr'* and sometimes not, because *cos* lies within a λ-homologous segment. Whereas homologous exchange is the rule, there are clear examples where homologous exchange on one side of the rescued marker is accompanied by an apparent heterologous exchange on the other (discussed in reference 15). These are rare events, and little is known about mechanism.

The *recE* gene of Rac can be rescued into a λ deleted for the *rec* genes (28). Here the prophage includes no λ homology to the left of *recE*, but an excised prophage would have λ homology further to the left. The recombinant phage obtained has the structure expected if the prophage had excised and then recombined. Technically, all that can be said is that the event requires site-specific recombination at the *att* site; the actual intermediate might be a tandem double lysogen rather than an excised prophage.

A more complex example comes from homologous recombination between a λ prophage and the cryptic element DLP12. It was first noted that the survivors of a λ lysogen subjected to heavy doses of UV irradiation included a few individuals that

had lost part, but not all, of the prophage. Such individuals (whose incomplete prophages were dubbed "cryptic") were much more frequently obtained from lysogens of a particular λ strain (called λ crypticogen) than from the reference λ used in most laboratories (25). Further study showed that λ crypticogen differs from reference λ by an IS2 insertion upstream of the *xis* gene (56).

The basis of this phenomenology is now understood (49). The IS2, behaving as a portable stretch of homology, recombines with an IS2 in the bacterial chromosome (most frequently the one located at about 9 min), and the λ prophage recombines near *cos* with DLP12 (Fig. 2). Substitution of the bacterial DNA between the IS2 and DLP12*cos* for the λ DNA between IS2 and *cos* produces a strain that still has the λ DNA from *cos* through *attR* and from *attL* through *xis* to the IS2 (Fig. 1) and bears a duplication of the bacterial DNA from 9 to 12 min. This is the most common type of cryptic formed by this mechanism. In principle, homology could be provided by Rac or Qin, but use of those elements as recombination points has not been observed.

The orientation of IS2-*cos* in the λ prophage is the same as that in the 9- to 12-min segment, and one class of possible mechanisms would require that the orientation be the same. However, cryptics with the same duplication are readily generated when λ *crg* is inserted in the opposite orientation, as though the cryptics are formed through double crossover between two copies of the bacterial chromosome (17).

The duplication shown in Fig. 2 is the type most commonly found. Rarer types include crossovers at another IS2 element (producing a longer duplication) and individuals where one crossover is in *cos* and the other appears to take place between heterologous DNA sequences. The rare cryptics coming from wild-type λ are in the latter class. The exact position of the crossover near *cos* varies. It can be either to the right or to the left of *cos* and occasionally takes place to the left of *qsr'*, giving strains that include λ genes *Q* through *att*.

CRYPTIC PROPHAGES AND MOLECULAR PHYLOGENY OF NATURAL STRAINS

Mobile elements such as transposons and prophages can be used as molecular markers to trace the history of the strains that harbor them. Their utility in this respect is restricted to a limited evolutionary timespan and is complicated (especially for phages with unique insertion sites) by reinfection and reinsertion into strains that have lost all or part of an element. Surveys of natural strains have some potential value in showing how frequent reinsertion may have been.

The junction between host DLP12 and *argU* has been found in many natural isolates of *E. coli* and also in bacteria as distant from it as *Salmonella typhimurium* and *Shigella boydii* (41). This could have come about if this part of the cryptic prophage has been preserved for some advantage it imparts to its bearer or if lysogenization at this site has been frequent.

A survey of Atlas prophages (defective and active) is in some ways more informative, because the presence of Atlas could be compared with molecular phylogenies constructed on other bases. The results show that Atlas prophages are distributed among lines not closely related to one another. Detailed analysis of the relationships between prophages and their hosts suggests multiple independent lysogenizations at this site (Stoltzfus, Ph.D. thesis).

SUMMARY AND CONCLUSIONS

Cryptic prophages have been aptly described as genetic debris that clutters the bacterial genome (53). From the cell's perspective, all prophages might be considered junk DNA; cryptic prophages appear to be junk from the phage's perspective as well. Although the available information is somewhat anecdotal, perhaps the most noteworthy feature of cryptic prophages is their prevalence. They are present in most bacterial strains examined and compose several percent of the E. coli K-12 genome. The clear implication is that in nature either the rate of lysogenization is high or the decay rate of cryptic prophages is slow. Available evidence (e.g., from Atlas prophages) indicates a high rate of lysogenization.

In a laboratory strain in pure culture, a cryptic prophage is simply an element doomed to extinction. However, in natural populations cryptic prophages are not genetically dead but can recombine with active phages and contribute to the phage gene pool. Indeed, it is plausible that the most common confrontation allowing generation of new recombinant types is infection by an active phage of a strain harboring a related cryptic prophage (11).

LITERATURE CITED

1. Adhya, S., P. Cleary, and A. Campbell. 1968. A deletion analysis of prophage λ and adjacent genetic regions. Proc. Natl. Acad. Sci. USA 61:956–962.
2. Anilionis, A., and M. Riley. 1980. Conservation and variation of nucleotide sequences within related bacterial genomes: Escherichia coli strains. J. Bacteriol. 143:355–365.
3. Arber, W., and G. Kellenberger. 1958. Study of the properties of seven defective lysogenic strains derived from Escherichia coli K-12 (λ). Virology 5:458–475.
4. Atling, T., A. Nielsen, and F. G. Hansen. 1989. Isolation, characterization, and nucleotide sequence of appY, a regulatory gene for growth phase-dependent gene expression in Escherichia coli. J. Bacteriol. 171:1683–1691.
5. Barreiro, V., and E. Haggård-Ljungquist. 1992. Attachment sites for bacteriophage P2 on the Escherichia coli chromosome: DNA sequences, localization on the physical map, and detection of a P2-like remnant in E. coli K-12 derivatives. J. Bacteriol. 174:4006–4093.
6. Béjar, S., F. Bouché, and J. P. Bouché. 1988. Cell division inhibition gene dicB is regulated by a locus similar to lambdoid bacteriophage immunity loci. Mol. Gen. Genet. 201:146–150.
7. Binding, G., G. Romansky, R. Bitner, and P. Kuempel. 1981. Isolation and properties of Tn10 insertions in the rac locus of Escherichia coli. Mol. Gen. Genet. 183:333–340.
8. Bouché, J.-P., S. Béjar, and K. Cam. 1990. Cooption of prophage genes: new data on the Kim prophage region of the Escherichia coli chromosome, p. 373–380. In K. Drlica and M. Riley (ed.), The Bacterial Chromosome. American Society for Microbiology, Washington, D.C.
9. Brody, H., A. Greener, and C. W. Hill. 1985. Excision and reintegration of the Escherichia coli K-12 chromosomal element e14. J. Bacteriol. 161:1112–1117.
10. Campbell, A. 1959. Ordering of genetic sites in bacteriophage λ by the use of galactose-transducing defective phages. Virology 9:293–305.
11. Campbell, A. 1961. Sensitive mutants of bacteriophage λ. Virology 14:22–32.
12. Campbell, A. 1968. Techniques for studying defective bacteriophages, p. 279–300. In K. Maramorosch and H. Koprowski (ed.), Methods in Virology II. Academic Press, New York.
13. Campbell, A. 1977. Defective bacteriophages and incomplete prophages. Comp. Virol. 8:259–328.
14. Campbell, A. 1988. Phage evolution and speciation, p. 1–14. In R. Calendar (ed.), The Bacteriophages. Plenum Publishing Corp., New York.
15. Campbell, A. 1994. Comparative molecular biology of lambdoid phages. Annu. Rev. Microbiol. 48:193–222.
16. Campbell, A., and D. Botstein. 1983. Evolution of the lambdoid phages, p. 365–380. In R. W. Hendrix, J. W. Roberts, F. W. Stahl, and R. A. Weisberg (ed.), Lambda II. Cold Spring Harbor Laboratory, Cold Spring Harbor, N.Y.
17. Campbell, A., J. Kim-Ha, R. J. Limberger, and S. J. Schneider. 1990. Bacteriophage evolution and population structure, p. 191–199. In M. T. Clegg and S. J. O'Brien (ed.), Molecular Evolution. Wiley-Liss, New York.
18. Chu, C. C., A. Templin, and A. J. Clark. 1989. Suppression of a frameshift mutation in the recE gene of Escherichia coli K-12 occurs by gene fusion. J. Bacteriol. 17:2101–2109.
19. Cohen, D. 1959. A variant of phage P2 originating in Escherichia coli, strain B. Virology 7:112–126.
20. Daniels, D., J. Schroeder, W. Szybalski, F. Sanger, A. Coulson, G. Hong, D. Hill, G. Peterson, and F. Blattner. 1983. Complete annotated lambda sequence, p. 516–576. In R. W. Hendrix, J. W. Roberts, F. W. Stahl, and R. A. Weisberg (ed.), Lambda II. Cold Spring Harbor Laboratory, Cold Spring Harbor, N.Y.
21. Dosch, D. C., G. L. Helmer, S. H. Sutton, F. F. Salvacion, and W. Epstein. 1991. Genetic analysis of potassium transport loci in Escherichia coli: evidence for three constitutive systems moderating uptake of potassium. J. Bacteriol. 173:687–696.
22. Eisen, H. A., C. R. Fuerst, L. Siminovitch, R. Thomas, L. Lambert, L. Pereira da Silva, and F. Jacob. 1966. Genetics and physiology of defective lysogeny in K12 (λ): studies of early mutants. Virology 30:224–241.
23. Epstein, R. H., A. Bolle, C. M. Steinberg, E. Kellenberger, E. Boy de la Tour, R. Chevalley, R. S. Edgar, M. Sussman, G. H. Denhardt, and A. Lielausis. 1963. Physiological studies of conditional lethal mutants of bacteriophage T4D. Cold Spring Harbor Symp. Quant. Biol. 28:375–394.
24. Espion, D., K. Kaiser, and C. Dambly-Chaudière. 1983. A third defective prophage of Escherichia coli K-12 defined by the λ derivative λ qin 111. J. Mol. Biol. 170:611–633.
25. Fischer-Fantuzzi, L., and E. Calef. 1964. A type of λ prophage unable to confer immunity. Virology 23:209–216.
26. Franklin, N. C., W. F. Dove, and C. Yanofsky. 1965. A linear insertion of a prophage into the chromosome of E. coli as shown by deletion mapping. Biochem. Biophys. Res. Commun. 18:910–923.
27. Goodgal, S., and F. Jacob. 1961. Complementation between lysogenic defectives of E. coli K-12 (T). Bacteriol. Proc. (Soc. Am. Bacteriologists) 1970, p. 28.
28. Gottesman, M. M., M. E. Gottesman, S. Gottesman, and M. Gellert. 1974. Characterization of bacteriophage lambda reverse as an Escherichia coli phage carrying a unique set of host-derived recombination functions. J. Mol. Biol. 88:471–487.
29. Grady, L. J., D. B. Cowie, and W. P. Campbell. 1971. Deoxyribonucleic acid hybridization analysis of the defective bacteriophage carried by strain 15 of Escherichia coli. J. Virol. 8:850–855.
30. Gratia, J. P. 1966. Studies on defective lysogeny due to chromosomal deletion in Escherichia coli. I. Single lysogens. Biken J. 9:77–87.
31. Greener, A., and C. W. Hill. 1980. Identification of a novel genetic element in Escherichia coli K-12. J. Bacteriol. 144:312–321.
32. Grodberg, J., and J. J. Dunn. 1988. OmpT encodes the Escherichia coli outer membrane protease that cleaves T7 polymerase during purification. J. Bacteriol. 170:1245–1253.
33. Herskowitz, I., and E. R. Signer. 1974. Substitution mutations in bacteriophage λ with new specificity for late gene expression. Virology 61:112–119.
34. Highton, P. J., Y. Chang, W. R. Marcotte, Jr., and C. A. Schnaitman. 1985. Evidence that the outer membrane porin protein gene nmpC of Escherichia coli K-12 lies within the defective qsr' prophage. J. Bacteriol. 162:256–262.
35. Hiom, K., and S. G. Sedgwick. 1991. Cloning and structural characterization of the mcrA locus of Escherichia coli. J. Bacteriol. 173:7368–7373.
36. Ikeda, H., M. Inuzaki, and J. Tomizawa. 1970. P1-like plasmid in Escherichia coli 15. J. Mol. Biol. 50:457–470.
37. Jacob, F., C. R. Fuerst, and E. L. Wollman. 1957. Recherches sur les bactéries lysogènes défectives. II. Les types physiologique liés aux mutations du prophage. Ann. Inst. Pasteur 93:724–753.
38. Kaiser, K. 1980. The origin of Q-independent derivatives of phage λ. Mol. Gen. Genet. 179:547–554.
39. Kao, C., and L. Snyder. 1988. The lit gene product which blocks bacteriophage T4 late gene expression is a membrane protein encoded by a cryptic DNA element, e14. J. Bacteriol. 170:2056–2062.
40. Lichens-Park, A., C. L. Smith, and M. Syvanen. 1990. Integration of bacteriophage λ into the cryptic lambdoid prophages of Escherichia coli. J. Bacteriol. 172:2201–2208.
41. Lindsey, D. R., D. A. Mullin, and J. P. Walker. 198. Characterization of the cryptic lambdoid prophage DLP12 of Escherichia coli and overlap of the DLP12 integrase gene with the tRNA gene argU. J. Bacteriol. 171:6197–6205.
42. Low, B. 1973. Restoration by the rac locus of recombinant forming ability in recB⁻ and recC⁻ merozygotes of Escherichia coli K-12. Mol. Gen. Genet. 122:119–130.
43. Mount, D., A. W. Harris, C. R. Fuerst, and L. Siminovitch. 1988. Mutations in bacteriophage λ affecting particle morphogenesis. Virology 35:134–149.
44. Muramatsu, S., and T. Mizuno. 1990. Nucleotide sequence of the region encompassing the int gene of a cryptic prophage and the dnaY gene flanked by a curved DNA sequence of Escherichia coli K12. Mol. Gen. Genet. 220:325–328.
45. Ochman, H., and R. K. Selander. 1984. Standard reference strains of Escherichia coli from natural populations. J. Bacteriol. 157:690–693.
46. Pero, J. 1971. Deletion mapping of the site of action of the tof gene product, p. 599–608. In A. D. Hershey (ed.), The Bacteriophage Lambda. Cold Spring Harbor Laboratory, Cold Spring Harbor, N.Y.

47. **Plasterk, R. H. A., and P. van de Putte.** 1985. The invertible P-DNA segment in the chromosome of *Escherichia coli. EMBO J.* **4:**237–242.

48. **Pugsley, A. P., and C. A. Schnaitman.** 1978. Identification of three genes controlling production of new outer membrane pore proteins in *Escherichia coli* K-12. *J. Bacteriol.* **135:**1118–1129.

49. **Redfield, R. J., and A. Campbell.** 1987. Structures of cryptic prophages. *J. Mol. Biol.* **198:**393–404.

50. **Scott, J. R.** 1970. A defective P1 prophage with a chromosomal location. *Virology* **40:**144–151.

51. **Scott, J. R.** 1975. Superinfection immunity and prophage repression in phage P1. *Virology* **65:**173–178.

52. **Shapiro, J. A., and S. L. Adhya.** 1969. The galactose operon of *E. coli* K-12. II. A deletion analysis of operon structure and polarity. *Genetics* **62:**249–264.

53. **Strathern, A., and I. Herskowitz.** 1975. Defective prophage in *Escherichia coli* K12 strains. *Virology* **67:**136–143.

54. **Uetake, H., S. E. Luria, and J. W. Burrows.** 1958. Conversion of somatic antigens in *Salmonella* by phage infection leading to lysis or lysogeny. *Virology* **5:**68–91.

55. **Whitfield, J. F., and R. K. Appleyard.** 1957. Formation of the vegetative pool by induced defective and healthy lysogenic strains of *Escherichia coli. J. Gen. Microbiol.* **17:**453–466.

56. **Zissler, J., E. Mosharrafa, W. Pilacinski, M. Fiandt, and W. Szybalski.** 1977. Position effects of insertion sequence IS2 near the genes for prophage λ insertion and excision, p. 381–387. *In* A. J. Bukhari, J. A. Shapiro, and S. L. Adhya (ed.), *DNA Insertion Elements, Plasmids and Episomes.* Cold Spring Harbor Laboratory, Cold Spring Harbor, N.Y.

Analysis and Predictions from *Escherichia coli* Sequences, or *E. coli* In Silico

A. HÉNAUT AND A. DANCHIN

114

INTRODUCTION

Bacteria have been studied as living entities (in vivo) for 150 years and in cell-free systems (in vitro) for almost half that time. With the availability of DNA sequence information, it has become evident that genes can also be studied as lines of text. A new domain of knowledge and research concerned with this means of studying living systems has arisen; it has been referred to as informatics but also involves standard aspects of mathematics and statistics as well. Beside using computer programs that mechanically perform standard but tedious analysis of the information content of DNA, investigators are now using programs that help generate new knowledge about this information. Thus, in addition to the study of bacteria in vivo and in vitro, there is now an active endeavor studying them "in silico." *Escherichia coli* has been a paradigm for such studies. It is anticipated that the *E. coli* genome sequence will be known by the end of 1997. In parallel, a vast amount of data on a variety of organisms has been collected, and it has become an important task not only to handle this huge quantity of information but also to extract from it the features that pertain to the concrete expression of life in general and to *E. coli* in particular.

For the *E. coli* geneticist, no literature reviewing this new aspect of research exists; information is scattered through a vast number of journals and papers, often presenting independent but redundant approaches. Here we have summarized the less self-evident aspects of the data presented in the literature. Readers interested in features relevant specifically to informatics can find in the databank SEQANALREF an updated bibliography on software dealing with sequence analysis (SEQANALREF, present in the EMBL data library package, contained 3,076 references in release 64 [October 1995]). Major DNA and protein data banks are accessible on the Internet. The appropriate addresses can be found in references 23, 42, and 71.

We shall follow the path that molecular geneticists pursue when they use formal techniques for investigating the significance of genes at the DNA level or of proteins at the polypeptide chain level: acquisition of sequences, analysis of these data, and management of DNA, RNA, and protein sequence data.

METHODS FOR SEQUENCE ACQUISITION

Generation of Long DNA Fragments without Gaps

Sequencing short pieces of DNA (i.e., less than 5 kb) does not require the use of specialized programs for aligning sequence subfragments obtained experimentally. One can start from both ends of the original segment and extend the sequence by using oligonucleotides or by using sequencing of nested deletions. This is not an easy or cost-effective effort when one is sequencing large segments. In this case, it is more efficient to use as the starting material a DNA library generated by shotgun cloning of ultrasonicated DNA or DNA fragmented with frequently cutting restriction enzymes. In this regard, it is important to make use of programs permitting automated alignment of fragments, generating contiguous overlaps (contigs). This task is not trivial, however, because the sequenced DNA segment often contains repeated regions displaying little or no variation. In addition, sequenced fragments contain errors, especially at their extremities, which are the very regions relied on for generating contigs.

Roger Staden has developed a program which has improved over the years in efficiency, speed, and user friendliness. It is now incorporated in many sequence analysis softwares and generates a contiguous sequence from a few hundred fragments (33, 34, 59, 175, 177, 178). Recent availability of sequencing machines has renewed interest in constructing new programs aiming at generating reliable contigs, and this is good news for investigators sequencing genomes of bacteria related to *E. coli* such as *Salmonella typhimurium* (official designation, *Salmonella enterica* serovar Typhimurium). New programs are needed because of limitations in current programs, such as the following: (i) one must often order batches of more than 1,000 sequences; and (ii) one can seldom make use of the original raw data, even though this would be extremely helpful, because most fragments contain a high level of ambiguous assignments, as well as erroneous insertions or deletions (of the order of 1% or more), and also because some fragments corresponded to duplicated regions very similar in sequence. The new programs based on completely new and original principles permit construction of reliable contigs generated from more than 1,000 fragments within a few hours, using a workstation (80) or simply a personal computer (60), even when the sequenced segment contains repetitions. New improvements deal mainly with contig length.

All programs for contig construction proceed in two steps. They start by determining a similarity index between fragments, in order to identify overlaps, and subsequently concatenate those which display regions of strong similarity, using a procedure designed for automatic classification.

(i) Programs differ chiefly in the methods used for detecting similarities. Those described in references 80 and 135 compute a similarity index by dynamic programming (the latter takes into account a global similarity index between sequences, whereas the former uses the highest score from the best homology fragment present in similar sequences). The program described in reference 177 identifies the longest common similar motif, and that described in reference 60 combines an information index computed from the number of words in common between fragments with a score obtained by dynamic programming.

(ii) The programs described in references 80, 135, and 177 use the same method for contig construction: they order the fragments as matching pairs according to their similarity. Two fragments are the most proximate when their similarity (as discussed above) is higher. The method is a sequential one, each fragment being placed with respect to the fragments which have already been positioned with respect to each other. This kind of approach is questionable when sequences are chimeric or repeated, or when they display numerous errors at their extremities (backtracking inappropriate ordering is extremely difficult when the starting clustering process is erroneous). In contrast, the program written by Gleizes and Hénaut (60) proposes a global approach: the fragment collection is taken as a single entity, and the fragments are ordered into each contig *before* alignment, thus allowing early detection of unlikely events.

An important aspect of the procedure used for generating contigs will be the management of the many sequences which are present in the literature and which can be combined to generate a detailed patchwork chromosomal sequence for organisms such as *E. coli*.

At this stage of data acquisition, elimination of parasite data (vector sequences) is an important issue (cf. the many vector sequences present in data libraries [96, 108, 141]). On the whole, sequence information is generally obtained from analysis of short DNA fragments cloned into a variety of vectors, and one must discard the sequences which contain vector sequences.

This is not particularly difficult. In a first round, sequences containing only vector sequences are eliminated, and chimeric sequences containing fragments of the vectors are subsequently treated, with knowledge of characteristic features of the vectors; "clean" sequences, with no remaining vector sequences, are then used for contig generation.

Localization on the Restriction Map

E. coli was the first living organism for which we possessed a detailed physical map. In 1987, Kohara and coworkers, in a seminal paper, described the restriction map of the entire *E. coli* K-12 chromosome, using eight restriction enzymes, *Eco*RI, *Eco*RV, *Hin*dIII, *Bam*HI, *Bgl*I, *Kpn*I, *Pst*I, and *Pvu*II (92). The corresponding map ordered almost 8,000 restriction sites. It was therefore generally difficult, knowing only a list of fragment lengths from a cloned region, to map it by visual comparison with Kohara's map. Many programs permitting direct comparisons between restriction maps have been written, and some have been devoted specifically to *E. coli* (81, 113, 120, 154). The programs of Médigue et al. (113) and Rudd et al. (154) are specifically designed for the *E. coli* genome. They take into account the fact that Kohara's data are not always accurate (i.e., some sites are missing, and sites that are next to each other are often inverted when they originate from different enzymes or are taken as a single site when they come from a single enzyme). Such programs have been used for some time to permit localization of cloned genes, in order to avoid using the tedious and sometimes difficult in vivo genetic mapping strategy.

It is clear, however, that as more and more sequences of genes have become known, these programs have become obsolete. It is now much easier to sequence a short fragment of the cloned gene and compare it with the known sequence (e.g., the BigSeq file proposed by Rudd and coworkers [153]). In principle, however, the programs could be used to map genes on other chromosomes, at the start of genome projects, when a restriction map is known.

Data Libraries and Specialized Databases

Data Libraries. At present, four main libraries constitute a repository of an almost complete set of the DNA sequences generated worldwide (Table 1): the EMBL-EBI data library (146), GenBank (genetic sequences data bank [11], now administered by the National Center for Biotechnology Information), DDBJ (DNA data bank of Japan [122]), and GSDB (genome sequence data bank [1]). These libraries collect all existing information on DNA sequences published or submitted for publication. They are not specialized for individual organisms; in fact, they contain all data sent to them, from sequences shorter than 10 nucleotides (!) to complete individual genomes. As a result, the corresponding information is highly redundant.

Data can be organized into phylogenetically consistent patterns, which permits one to compare a particular organism, in our case *E. coli* and other, related enterobacteria, with near and distant relatives. As a case in point, the Ribosomal Database Project compiles ribosomal sequences and related data from all possible organisms and redistributes them in aligned and phylogenetically ordered form. It also offers various software packages for handling, analyzing, and displaying sequences. In addition, this project offers certain analytic services dealing with phylogenies of rRNAs. The project is still in an intermediate stage of

TABLE 1 Main databases germane to *E. coli-* and *S. typhimurium*-related sequences or literature[a]

Database	No. of entries	Date	Address
SwissProt	40,292	24 Nov. 1994	http://expasy.hcuge.ch
SwissNew	4,144	13 Dec. 1994	http://expasy.hcuge.ch
PIR	71,995	29 Nov. 1994	http://www.gdb.org
EMBL	234,501	11 Dec. 1994	http://www.ebi.ac.uk
EMNEW	27,280	10 Jan. 1995	http://www.ebi.ac.uk
GenBank	238,000	15 Dec. 1994	http://ncbi.nlm.nih.gov
DDBJ	239,689	15 Jan. 1995	gopher.nig.ac.jp
GSDB	102,748	12 Jan. 1995	http://www.ncgr.org/gsdb
NRL3D	4,153	31 Dec. 1994	http://www.gdb.org
PDB	3,091	6 Jan. 1995	http://bach.pdb.bnl.gov
PROSITE	1,029	24 Nov. 1994	http://expasy.hcuge.ch
BLOCKS	770	24 Nov. 1994	http://www.blocks.fhcrc.org
ENZYME	3,546	24 Nov. 1994	http://expasy.hcuge.ch
PRODOM	23,105	25 Nov. 1994	http://www.sanger.ac.uk
EcoSeq	374 contigs	31 Jan. 1995	ftp.ncbi.nlm.nih.gov/repository/Eco
EcoCyc	See text	3 Jan. 1995	ftp.ai.sri.com/hidden/pkarp/ecocyc
ECDC	1,920	26 Jan. 1995	http://susi.bio.uni-giessen.de/usr/local/www/html/ecdc.html
Colibri	2,150	25 Jan. 1995	ftp.pasteur.fr/pub/GenomeDB/colibri
Metalgen	See text	25 Jan. 1995	ftp.pasteur.fr/pub/GenomeDB/metalgen
SEQANALREF	2,579	24 Nov. 1994	http://expasy.hcuge.ch

[a]The number of entries is indicated with the corresponding date. Note that there is an exponential increase in the number of entries as a function of time. Addresses for accessing databases are indicated. Access to the World Wide Web, where possible, is preferred because it represents a lower-level access.

development but may be expected to bring important information on the structures and functions of ribosomes in general and of model organisms in particular (98, 127). R. Christen (christen@ccrv.obs-vlfr.fr, unpublished data) has also developed an rRNA database for bacteria which comprises more than 1,800 individual species.

For protein sequences, data have been collected since the early 1950s. A paradigm is the *Atlas of Protein Sequence and Structure,* developed by Dayhoff and coworkers and published yearly (32). At present, there are two main protein data banks: PIR/NBRF (9) and SwissProt (8). The latter is much less redundant than the former, and Bairoch is very careful in annotating sequences as precisely and completely as possible. This is, understandably, at the expense of exhaustivity.

There also exist libraries which are derivatives of these primary sources of information and dedicated to a specific topic. They contain an "added value" due to the fact that their authors have done considerable work—sometimes measured in years of effort—to organize or annotate the data. The most frequently used are NRL_3D, PROSITE, and BLOCKS. NRL_3D (133) is a PIR/NBRF subproduct; it contains all protein sequences whose structures have been determined at a significant resolution level, generally by X-ray diffraction or high-resolution nuclear magnetic resonance spectroscopy. PROSITE (7) is distributed by the EMBL-EBI. It is a bank of consensus motifs organizing the protein sequences present in SwissProt into classes. For each class, the corresponding annotations are the result of the work of specialists. As a result, they are very rich and precise. The way in which they are indexed permits an easy analysis of protein classes present in the genomes of prokaryotes, such as *E. coli* and *S. typhimurium.* BLOCKS (76) has the same goal as PROSITE, but it provides multiple alignments corresponding to the consensus motifs.

PRODOM is a protein data library, generated from Swiss-Prot, which aims at identifying individual domains present in proteins and assembled in a combinatorial way in different types of proteins (173). PRODOM identifies domains only when they are part of proteins differing in structure or function. Hence, domains can be identified only when they have been shuffled during evolution between different protein types. A multi-domain protein having only homologous counterparts with the same overall structure will appear as a single domain protein for PRODOM. As a consequence, as time elapses, many proteins present in PRODOM as now displaying one domain will be split into several domains.

Searching Data Banks. The total amount of data in sequence libraries is so large that it is impossible to extract appropriate information without the help of a specialized query interface. Many programs aiming at fast interrogation have been written, but none can execute all search criteria that any scientist would like to follow. However, ACNUC, ATLAS, SRS, and Entrez, taken together, cover most of the needs. ACNUC (62) is used mostly for nucleic acid sequences. It has the unique feature of verifying the consistency of the information present in the EMBL data library, GenBank, or PIR before it creates its own index tables and thus is the cleanest and surest search program. In contrast, ATLAS (9) is used mostly for proteins. It allows one to answer immediately queries of the type "which are all the proteins, present in the banks, having the sequence GDSGGP?" or "which are the proteins having less than 15% acidic residues?" SRS (48) permits one to work with almost any type of data library by automatic generation of links between them; from a sequence identified in the EMBL-EBI data library, it is possible to find the records for the protein in PIR or SwissProt to see whether it contains a motif identified in PROSITE or BLOCKS and extract all proteins belonging to the same family. Entrez constitutes the paradigm for the new generation of search softwares. It links sequences present in GenBank or PIR to references present in MEDLINE. In addition, this software links together the sequences which have been considered similar by the program BLAST (5).

BLAST is a program meant to detect similarities between DNA or protein sequences by identifying segments which have some degree of identity, according to a preset correspondence matrix between the letters of the alphabet describing the sequences (such as PAM matrices [160], in the case of proteins). This program does not try to concatenate segments, and it displays those which have an identity score higher than a threshold value. Gaps are not taken into account, and so a list of fragments is produced, some of which can be widely distant in the sequence. Because gaps are not considered, it is possible to calculate a probability a priori for obtaining a given sequence, and to keep it in the output, as significant if this probability is low enough. This criterion has been used in the construction of the PRODOM data library. Because of its structure, BLAST generates both false positives and false negatives. The former correspond to segments which are rich in a single amino acid type, for instance, arginine or proline. The latter correspond to pairs of proteins which are globally similar over a very long segment but dissimilar locally.

Gaps can be taken into account by a second program which is widely used for comparing all data present in libraries, FASTA (and its derivatives) (134). To gain time, this program identifies, in the sequences which are compared with each other, all regions of identity. It considers subsequently an alphabet of equivalence between residues, nucleotides, or amino acids, which can be degenerate (a matrix for accepted point mutations [PAM matrix] or an alphabet of 19 residues wherein lysine and arginine would be equivalent, for instance), and computes a score for each similarity segment. It can easily be seen that speed increases at the expense of sensitivity: two "homologous" segments displaying a large number of conservative replacements will escape identification. In a second step, FASTA tries to chain homologous segments and aligns the regions located in between with a classical algorithm. Thus, FASTA attempts to create an alignment for the largest possible composite segment, creating gaps at appropriate locations. However, because there is a penalty score for gaps, if the gap is too large, only the best aligned segment will be conserved, leaving parts of the sequence aside even though they had been identified as significant in the first round.

BLAST and FASTA are complementary programs: a similarity detected by one program can escape the other, and vice versa. It is therefore important to have an idea of the significance of alignments, in terms of both statistics and biology, and, in many cases, to use both programs. The algorithm permitting detection of local homology written by Smith and Waterman (171) (program BESTFIT of the Genetics Computer Group package [40]) is more sensitive, but it is also much slower and cannot be used for scanning the totality of banks unless run on machines having a dedicated architecture (program BLITZ on a machine using 4096 parallel processors, accessible through an electronic mail server at the EMBL-EBI). The very high number of sequence duplicates in data libraries makes the identification of similarities difficult. Altschul et al. (4) discuss some of the means which can be used to get round this difficulty.

It has not been possible to give a straightforward interpretation of the significance of a sequence alignment. As usual, E. coli provided paradigm studies. Landès et al. (97), analyzing E. coli tRNA synthetases, have shown that one of the most reliable methods, but certainly one not without flaws, is based on a statistical evaluation, using the simulation power of computers, of the score value Q of program BESTFIT (106). The principle of the method is to compare score Q_{ij} obtained by using BESTFIT between sequences i and j with the score obtained with sequence i and a random permutation of the symbols describing sequence j. This operation is repeated a hundred times, permitting calculation of a mean score Q_m and a standard deviation σ. The Z score Z_{ij} for sequences i and j is given by $Z_{ij} = (Q_{ij} - Q_m)/\sigma$. One can extract from the Z score histogram a value Z_0 that permits separation between those sequences which are strongly related to each other and the others (Z scores higher than Z_0 for the former and lower than Z_0 for the latter). To calculate the distance d_{ij} between sequences i and j, one first calculates the probability p_{ij} that a score Z will be higher than score Z_{ij}, assuming that Z scores follow a normal distribution (p_0 is the value of this probability for score Z_0): $d_{ij} = \log(p_{ij}) - \log(p_0)$ if $Z_{ij} < Z_0$ and $d_{ij} = 0$ otherwise.

Databases Dedicated to E. coli. Several data libraries dedicated to E. coli have been maintained, some of them for a long time. Since 1989, Wahl et al. have collected the E. coli sequences present in the EMBL data library and in GenBank (187, 188). In the same way, Rudd et al. have developed a software which contains both data files and programs for collecting, aligning, representing graphically, and analyzing sequences (151–153). This permits them to represent the gene and restriction maps together with sequences obtained from various laboratories after elimination of redundancies. These programs and data sets are available from the National Center for Biotechnology Information at the U.S. National Library of Medicine (Table 1). EcoSeq is a nonredundant collection of genomic DNA sequences; EcoMap is an alignment and integration of EcoSeq with the genomic restriction map of Kohara et al. (92); EcoGene is a database of information about genes and gene products; MapSearch is a restriction map alignment program; PrintMap is a Postscript-based publication-quality map drawing utility; GeneScape is a genomic restriction map editor/browser for Macintosh computers; ChromoScope is a platform-independent network-based application for sequence/map display, alignment, retrieval, and maintenance. Also, StySeq1 (11% of the chromosome of S. typhimurium) has just been completed and is presently available on the same site. It is described in this book (chapter 110). At much the same time, Médigue and coworkers developed a database managed by the relational database management system (DBMS) 4th Dimension on Macintosh (116). The corresponding data structure, which allows multicriterion searches, was exported to develop specialized databases for other genomes such as that of Saccharomyces cerevisiae (170) or Bacillus subtilis (123). Finally a database for managing E. coli sequences, Escherichia coli Genome Database (95), has been developed in Japan. More recently Mori and coworkers have organized E. coli sequence data into a series of flat files, GenoBase, which can be accessed through a graphic interface: clicking with the mouse on a given region of the map (displayed graphically) allows one to display the records containing the related information (DNA and protein sequences, aligned result of a FASTA search) (GenoBase 1.1; K. Takemoto, M. Yano, Y. Akiyama, and H. Mori, unpublished data). This structure does not provide, however, the relational interface permitting multicriterion searches.

A recent trend in the development of databases dedicated to the management of compete genomes is the use of sophisticated informatics techniques allowing scientists to handle not only data but also the knowledge associated with them. Development of software able to guide the user in the choice of appropriate methods for a given purpose would constitute a significant advance. Environments which allow one to model and to manipulate descriptive knowledge generated by a genome sequencing program, to help the user in solving sequence analysis problems through task decomposition and method selection, and finally to display and to manage the set of newly created objects are being constructed. These systems are therefore meant to integrate descriptive knowledge on the entities involved (genes, promoters, maps, etc.) together with methodological knowledge on a large and extendable set of analysis methods. Such object-oriented knowledge bases have been developed by Shin et al. (166) and by Perrière and Gautier (ColiGene [138]).

Finally, there are databases dedicated to the management of bacterial strains. The paragon is the base developed by Berlyn and Letovsky for the management of B. Bachmann's *E. coli* Genetic Stock Center (13, 102; chapter 134 in this volume). A description of the *Salmonella* Genetic Stock Center and related information appear in chapters 135 and 136. There is also a reference collection, EcoR, for natural isolates of *E. coli* isolated from all parts of the world (126).

Linking Gene Databases to Data on Intermediary Metabolism. A few databases are oriented to the intermediary metabolism of *E. coli*. DBEMP (database for enzymes and metabolic pathways) is being developed by E. Selkov et al. (Institute of Biophysics, Puschino, Russia). DBEMP contains information on enzymes and metabolic pathways derived from published articles. Its major strength comes from its comprehensiveness: 10,000 records, based on 6,000 articles and including more than 2,000 enzyme activities from all animal and plant sources. The DBEMP format contains about 300 individual fields, covering all features of enzymology. The software management system has been built specifically for this application; it is developed under MS-DOS and runs on PCs.

METALGEN (150) is a relational database constructed with the relational database management system 4th Dimension (Macintosh). It presents metabolic charts inspired by Boehringer's metabolic chart (118). Metabolic pathways (the actual schemes are predefined) are freely explored by the user while keeping links to the corresponding genes and their regulation. In addition to its dedicated graphics, METALGEN has a built-in possibility to edit pathways and data and to do multicriterion searches. Attached procedures are regularly created and updated for specialized queries. In particular, it permits calculation of optimal pathways, for a given constraint, as a function of the culture medium and permits investigation of the possible phenotypes derived from the presence of one or several mutations, as proposed by the user. METALGEN currently contains information about 17 general metabolic pathways and 111 specialized pathways. They are represented by 300 pictures, 50 of which can be starting points to specialized pathways, using mouse clicks. A total of 550 enzymes or protein complexes that catalyze 602 reactions and transport are described. They involve 587 metabolites and 450 *E. coli* genes. In parallel, it is being linked to information about intermediary metabolism in other model organisms. The stoichiometry of 5,000 biosynthetic pathways

has also been calculated by an embedded program and is included in the database.

Finally, EcoCyc, a sibling of the knowledge base CompoundKB dedicated to compounds of intermediary metabolism (88), is a knowledge base of *E. coli* genes and metabolism that runs on the Macintosh on Unix workstations. Its graphical user interface creates drawings of metabolic pathways, of individual reactions, and of the *E. coli* genetic map (89, 90). Users can call up objects through a variety of queries (such as retrieving an enzyme by a substring search) and then navigate to related entities shown in the resulting display window. For example, a user could zoom in on a region of the genetic map, click on a gene to obtain detailed information about it, and then navigate to the enzyme product of the gene and then to the metabolic pathway containing the enzyme. Metabolic pathway drawings are produced automatically and can be executed in several styles, such as with compound structures present or absent. The EcoCyc knowledge base currently contains information about 38 pathways, 200 enzymes that catalyze 160 reactions, 1,100 metabolic compounds, and 2,030 *E. coli* genes. EcoCyc contains extensive information about each enzyme, including its cofactors, activators and inhibitors (qualified by type), subunit composition, substrate specificity, and molecular weight. Individual values in the knowledge base are extensively annotated with citations to the literature, as are comment fields.

Two-Dimensional Gels. Finally, there exists a large collection of data describing two-dimensional gel electrophoresis of proteins, organized in a specialized database (see chapter 115). It will be important, in the future, to connect the corresponding data to databases describing genes, sequences, and gene products and their functions.

IDENTIFICATION OF SIGNALS IN NUCLEIC ACID SEQUENCES

Identification of Genes

Predicting Genes Coding for Proteins. The first problem addressed by experts when they are facing newly sequenced DNA segments is localization of the regions which code for proteins. Despite the fact that many methods for solving this problem have been published, none are totally successful. In the case of *E. coli*, the problem has been initially simplified, because the codon usage, as derived from the first set of sequenced genes, is strongly biased. Knowing that an open reading frame (i.e., a region of $3n$ nucleotides limited by a translation termination codon, UAA, UAG, or UGA) contains a high frequency of such biased codons strongly suggests that a protein-coding sequence is contained within the open reading frame (with the corresponding reading frame) (176).

This method is suitable only for genes which have a codon bias defined by the first set of genes which have been sequenced and is therefore dangerous to use as an only source of information for defining protein-coding sequences (we reflect here the experimentalist's bias). Indeed, as found by Médigue et al. (114), there is in *E. coli* at least one class of genes which escapes identification: the class of genes exchanged by horizontal transfer. Other methods should therefore be used in this case. Examples are the method of Fichant and Gautier (52) (which is based on the identification of a heterogeneity in the frequency of bases present in each of the three reading frames) and the GenMark (renamed GeneMark, for copyright reasons) software (20), derived from work on periodical Markov chains (91). Both methods are based on the identification of the triplet base composition bias between the three frames inside coding regions.

Predicting the Physiological Activity of Genes Coding for Proteins. Because the genetic code is redundant, the codons specifying a given amino acid may be used in genes at highly variable frequencies. It was established long ago that at least under certain circumstances, there is a relationship between codon preference and expression level of genes (17, 61, 65). It has been therefore interesting to order genes having similar codon usage by using the computation of a χ^2 distance between genes according to their codon usage. This distance has been subsequently used for clustering genes by using factorial correspondence analysis (64, 77, 99). In particular, resorting to this method (35, 73) permitted

Médigue et al. to unambiguously demonstrate that *E. coli* genes fall into three main classes based on codon usage (114).

In this method, classes are characterized by the importance of the bias between synonymous codons. For example, as seen in Table 2, it was observed that in class II genes, leucine is coded by CUA in less than 1% of the cases, whereas CUG accounts for 77% of the codons. The bias is lower in class I genes (3% CUA and 53% CUG) and becomes weak in class III genes (7% CUA and 30% CUG, which nevertheless remains the major leucine codon). The number of avoided codons (frequency of less than 6%) is very high in class II genes (61, 66, 84, 165). In class I genes, there are only five codons of frequency lower than 6%, and there are none in class III genes. Médigue et al. have summarized the clustering of more than 780 genes of *E. coli* into the three most obvious classes (116). The classes are unequal; class I comprises 64% of the genes, class II comprises 25%, and class III comprises 11%. Generally speaking, there is a functional kinship

TABLE 2 Codon usage in the three major classes of *E. coli* genes[a]

Amino Acid	Codon	Class I	Class II	Class III	Amino acid	Codon	Class I	Class II	Class III
Phe	ttt	55.09	29.08	67.14	Leu	ctt	9.70	5.56	19.00
	ttc	44.91	70.92	32.86		ctc	10.40	8.03	9.04
Leu	tta	10.99	3.44	20.09		cta	3.09	0.83	6.81
	ttg	13.02	5.47	15.05		ctg	52.79	76.67	29.99
Ser	tct	13.26	32.41	19.63	Pro	cct	13.71	11.23	28.30
	tcc	15.02	26.56	11.34		ccc	11.19	1.63	16.26
	tca	10.83	4.79	22.09		cca	18.63	15.25	31.50
	tcg	16.88	7.39	10.60		ccg	56.47	71.89	23.94
Tyr	tat	54.42	35.23	69.60	His	cat	56.80	29.77	61.69
	tac	45.58	64.77	30.40		cac	43.20	70.23	38.31
TER	taa				Gln	caa	33.40	18.65	37.06
	tag					cag	66.60	81.35	62.94
Cys	tgt	40.90	38.85	55.71	Arg	cgt	38.99	64.25	26.05
	tgc	59.10	61.15	44.29		cgc	43.23	32.97	21.94
TER	tga					cga	5.52	1.07	12.80
Trp	tgg	100.00	100.00	100.00		cgg	8.97	0.80	13.62
Ile	att	51.20	33.49	47.57	Val	gtt	23.74	39.77	34.33
	atc	44.37	65.94	26.65		gtc	22.48	13.45	18.95
	ata	4.43	0.57	25.78		gta	14.86	19.97	21.78
Met	atg	100.00	100.00	100.00		gtg	38.92	26.81	24.94
Thr	act	14.85	29.08	26.83	Ala	gct	14.52	27.54	22.86
	acc	46.83	53.60	24.45		gcc	27.62	16.14	23.67
	aca	10.52	4.67	27.93		gca	19.63	24.01	31.27
	acg	27.81	12.65	20.80		gcg	38.23	32.30	22.19
Asn	aat	40.87	17.25	64.06	Asp	gat	62.83	46.05	70.47
	aac	59.13	82.75	35.94		gac	37.17	53.95	29.53
Lys	aaa	75.44	78.55	72.21	Glu	gaa	68.33	75.35	66.25
	aag	24.56	21.45	27.79		gag	31.67	24.65	33.75
Ser	agt	13.96	4.52	18.73	Gly	ggt	32.91	50.84	31.79
	agc	30.04	24.33	17.61		ggc	43.17	42.83	24.51
Arg	aga	1.75	0.62	15.63		gga	9.19	1.97	24.75
	agg	1.54	0.29	9.96		ggg	14.74	4.36	18.95

[a]Genes are clustered by using factorial correspondence analysis into three classes. Class I contains genes involved in most metabolic processes. Class II genes correspond to genes highly and continuously expressed during exponential growth. Class III genes are implicated in horizontal transfer of DNA. One can see that the distribution of codons in class III genes is more or less even, whereas it is extremely biased in class II genes (in particular, codons terminated in A are selected against). See reference 114.

between the members of a given class. Genes found in class III are involved in genetic exchange between organisms. A separate analysis has shown that genes present in plasmids belong to this latter class.

It had long been assumed that codon usage in a gene was the result of a selection pressure adjusting the pool of available tRNA molecules to the codon frequency, as a function of the drift toward a mean base composition mediated by mutations. Class II genes were thought to best reflect this adaptation (17, 65, 66, 83, 84, 164). Kurland et al., however, have shown that isoacceptor tRNA concentration and relative abundance vary as a function of the growth rate, and that the tRNA pool is constantly adapted to the mean protein composition of the cell (45–47). Correspondence between tRNA relative abundance and codons used in class II genes, as noticed by Ikemura (83), is true only during exponential growth on a rich medium. The diversity in tRNA composition is much more important when cells are not growing exponentially, and it matches much better the codon frequency of class I or class III genes.

Bacteria are rarely living under exponential growth conditions, and the other states are certainly as important for population survival. Accordingly, the results obtained by Kurland and coworkers suggest a new interpretation for the meaning of codon usage: the preferential usage of a codon in a gene reflects the composition of the tRNA pool under conditions in which the gene is normally expressed, and this composition can be very different from that of class II genes. This hypothesis predicts that all genes specific to a given physiological state tend to use synonymous codons in the same way, the degree of expression having only a secondary effect on modulating the adjustment between codons and tRNAs. Whether this is true will probably be seen in the fine analysis of codon usage in gene families expressed under similar conditions. In conclusion, analysis of codon usage is interesting because it can provide information on the integration of a gene function in *E. coli* physiology.

Predicting Genes for tRNAs and rRNAs. With regard to RNA genes, the situation for *E. coli* is unique, because the sequences as well as the locations in the genome of presumably all rRNA and tRNA genes are known. The programs which have been developed for proposing identification of such sequences therefore are not of much use. However, in the case of tRNAs in particular, there might be genes that are cryptic, or expressed in very special conditions, and thus not yet identified. It would therefore be of interest to use programming language such as the one devised by Searls, which uses a linguistic approach to identify tRNA molecules (161, 162). The program of Fichant and Burks (51), which integrates the knowledge of many tRNA genes into a software constructed ad hoc, permits scanning of the *E. coli* genome for putative sequences. This program is supposed to yield less than 1 (presumably false-positive) positive outcome for 3×10^5 bp. This low value should prompt further investigation by the biologist, when a positive score corresponding to no known gene is encountered, to determine whether this result indicates the presence of a cryptic gene.

Identification of Sequence Signals

Predicting Promoters. Three definitions of a promoter coexist: (i) the region permitting the control of transcription of an operon, as defined by Monod and coworkers; (ii) the binding site of an RNA polymerase, active for transcription initiation; and (iii) a transcription start site, as seen by actual identification of the 5′ terminus of mRNA molecules. The first definition encompasses large regions of DNA and involves in particular those regions where transcription factors interact directly or indirectly with RNA polymerase. Generally speaking, promoter sequences are identified as corresponding to the second definition, and this permitted early investigators to propose the existence of so-called consensus regions believed to identify the binding sites for RNA polymerase, after isolation of a collection of promoter region sequences. According to this definition, *E. coli* promoters encompass a ≈75-nucleotide region upstream of the RNA start site. Counting from the start site, Pribnow initially proposed that a consensus sequence TAT(A/G)AT, comprising six bases, was situated around −10 (144). Subsequently, Maniatis et al. (110) discovered that a second conserved region, TGTTG, was located around −35. Siebenlist et al. (169) further refined the consensus by changing it to TTGACA after having analyzed a collection of promoters. Automatic identification of promoters started with this concept of consensus. However, as more promoters were collected, the very notion of consensus became more and more fuzzy, because for each position of a promoter, an exception could be found. The degeneracy is such that a combination of −10 or −35 motifs can be found every 200 nucleotides in a random sequence (44). This obviously precludes the use of consensus sequences in the actual prediction of promoters. It is therefore clear that one has to find other means to identify promoters (and indeed, when facing the problem, those involved in the sequencing of the *E. coli* genome have in fact combined a consensus approach with individual scanning of the sequence by eye).

Some progress was made when it was recognized that the spacing between the consensus regions was preserved (presumably indicating some sort of physical constraint between RNA polymerase subunits). This demonstrated that the meaning of a consensus motif depended on its surrounding regions. One therefore had to incorporate biological knowledge of the actual process of recognition if one wished to construct predictive methods (6, 72). Hawley and McClure (72) introduced the idea that the significance of a consensus motif in the overall composition of the region was to be found in the global nucleotide composition of the region. Harley and Reynolds (70) demonstrated that the region between −35 and −10 generally spanned 17 ± 1 nucleotides, the maximum variation being between 15 and 21 nucleotides. On the basis of this description, most promoters can be separated into three groups as a function of the spacing between the consensus regions (16, 17, or 18 nucleotides), maximum efficiency being obtained when the spacing is 17 nucleotides (6).

Schneider et al. (159) proposed to give a measure for consensus sequence "plasticity" at a given position (its degree of uncertainty), thus giving a quantitative background to the concept of signal. For this, they used a fundamental concept in information theory, Shannon's entropy (163).

H, Shannon's entropy of system *X*, is described as

$$H(X) = -\sum_{i=1}^{n} p_i \log_2 p_i$$

where p_i represents the probability for the different states of the system (i.e., probability of finding nucleotide A, C, G, or T at a given position *X*). Three properties justify the use of Shannon's entropy $H(X)$ as a measure of the degree of uncertainty: (i) it becomes zero when a given state is known for certain (a single nucleotide type can be present at position X); (ii) for a given number of states, it is maximum when the states have equal probability, and it increases with the number of states (the entropy value is 2 if the four bases have equal probability); (iii) it is additive (i.e., when several independent systems are taken as a whole, their entropies are added [the entropy of a sequence is equal to the sum of the entropies of each position]). It must be emphasized that this measure of information refers not to a single sequence but only to families of sequences. The bit is the entropy unit.

Despite its very primitive character (it is simply the computation of an average value), Shannon's entropy can measure some degree of knowledge about a given sequence. Let us consider a sequence segment X; we shall evaluate the information which is added when the state of the system is becoming known. Before knowing the function of the system (initial state), the probability that a given position is occupied by a given nucleotide is equal to the probability p_i of presence of this nucleotide in the segment. The corresponding Shannon's entropy $H(X)$ is maximum. Once the function Y of the segment is deciphered (for example, it is a promoter), the uncertainty as to the presence of a nucleotide at a given position is reduced, and Shannon's entropy $H(X|Y)$ is lower than the entropy a priori (it becomes zero if there is only one nucleotide possible at each position). If we term $I_{Y \to X}$ the information brought about by the determination of state Y of system X, it can be measured by the diminution of Shannon's entropy when compared with the situation a priori: $I_{Y \to X} = H(X) - H(X|Y)$.

Stormo proposes two additional approaches to measure the information content of a sequence (180, 181). The first one is based on an analogy with a thermodynamic equilibrium, and the second is based on a probabilistic view of the problem (likelihood statistics). The domain of validity of the thermodynamic analogy has been experimentally studied (181). Each of these different viewpoints has its own value. All lead to the same mathematical equations. In what follows, we shall use only the vocabulary of information theory, but one should bear in mind that the underlying justification could rest on assumptions other than those explicitly made in this theory.

In the practical use of the concept of Shannon's entropy, one must take into account the fact that it is relative and refers to families of sequences having the same general properties. To speak of the Shannon information content of a given genome, as is sometimes done, therefore makes no sense.

When considering a set of sequences for which a consensus is sought, one must be sure of the conformity of each sequence with respect to all sequences present in the consensus set. Thus, O'Neill (128) proposed a semiempirical formula derived from the work of Berg and von Hippel (12), who have detected a composition bias in the 58 nucleotides covering the promoter region. This semiempirical formula results in a description of an *E. coli* σ^{70} promoter which is more complete than the description which uses a consensus matrix (176). It takes into account the probability of dissociation of RNA polymerase as a factor permitting discrimination of promoter sequences.

All of these methods, which helped refine the concept of signal and progressively led to abandonment of the idea of consensus (70, 112, 176), permitted O'Neill and Chiafari (132) to construct a prediction method for identification of σ^{70} promoters. Despite the fact that one has to look for a single biological entity, these authors have been led to construct four different programs; they take into account a region spanning 58 nucleotides and split the search between four promoter types, in which 16, 17, and 18 nucleotides, as noticed by Hawley and McClure (72), separate the −35 region from the −10 region. The fourth type corresponds to viral promoters, wherein a 17-nucleotide span is fixed, and it differs from chromosomal counterparts.

We wish to emphasize at this point that the very nature of a promoter remains a mystery; the programs of O'Neill and Chiafari (132) cluster together heuristics (an heuristic is a general exploration procedure, an educated guess based on intuition and expertise, which models reality but explores only the "best" pathways and therefore cannot necessarily achieve its goal or provide an explanation for its possible success), which are applicable only to promoters for which the separation between the −35 and −10 regions is 16 to 18 nucleotides. Not only do they leave open the question of the other types of promoters, but they also assume the existence of four types, while at the same time it is supposed that it is the same σ^{70} holoenzyme which recognizes all types. The true reason underlying the promoter recognition process by RNA polymerase is therefore still unknown (see, however, reference 41).

O'Neill (129) summarized efforts made during 15 years to precisely identify promoters. He studied the logic underlying the organization of the various types of promoters. It seems that in all cases, one could identify a 10-nucleotide-long degenerate palindromic sequence, repeated at least five times and constituting the core sequence of all promoters. However, there is no experimental verification of the importance of degenerate palindromes, and the statistical analyses that led to these conclusions were flawed by the small sample size and too many related promoters, as pointed out by Cardon (24), who used a larger, more complete promoter collection.

The analysis of crystals of a repressor-operator complex permitted identification of some of the constraints operating in the protein-DNA interaction, which could be important in the case of RNA polymerase-promoter interactions (93). This would explain some of the variety of conservation observed along the promoter sequences: some nucleotides interact directly with the protein, whereas others are responsible for minute alterations in the local DNA structure, permitting optimum accommodation of the protein. This change in standpoint, which has witnessed a change in the study of promoter sequences from a simple linear consensus sequence to a three-dimensional structure wherein nucleotides do not play a role in themselves but act through their influence on the spatial configuration, is typical of a major trend in the evolution of ideas in the future analysis of genomes in silico. A comparable evolution is witnessed at present in the direct experimental approach, whereby, for example, supercoiling of DNA is now taken into account as a major determinant of promoter function (53). This change is in fact reflecting a modification in our understanding of the biological meaning of the sequence: in terms of Shannon's entropy, this means that we are changing our baseline. The actual information (in the sense of Shannon) is in fact measured by the deviation from an a priori that we do not know; a major task for biologists will be to propose new ideas about the nature of the sequence a priori (what is a random sequence?) in order to pinpoint the deviation, biologically significant, from this zero level. This places information analysis of genomes in a position where the discussion of Bayesian statistics has been the most passionate. And one must note that this zero level can differ according to the biological process which is considered. As a first approximation, for instance, the zero level for replication machinery is a random sequence having the same base composition. In contrast, for translation it is likely that one should use as the random baseline a sequence, taking into account the triplet structure of the coding sequences.

Predicting Transcription Terminators. Two processes control transcription termination in *E. coli*. Among other features, they differ by the proteins which are involved, in particular by the presence of termination factor Rho. Most operons terminate transcription in a Rho-independent manner at sites which have been defined by the presence of stem-loop structures followed by a run of T's. Identification of terminators is a good example of the variety of viewpoints which must be correlated when one wishes to assign a function to a nucleotide sequence with some efficiency. Despite their distinctive features, true Rho-independent terminators cannot be predicted easily.

At first sight, the efficiency of a termination signal should be governed by the stability of its stem-loop structure and by the length of the run of T's. But many counterexamples exist, and one can describe stem-loop structures with the same thermodynamic stability placed in terminators having very different efficiencies. There also exist strong terminators with no or only a few T's. In the same way, the two sequences proposed to be important by Brendel et al. (21), CGGG(C/G) and TCTG, are absent from many terminators. As in the case of promoters, consensus sequences cannot be used to predict that a sequence acts as a terminator. It seems, however, reasonable to assume that the stem-loop structure, ending in T's, can constitute the core sequence of terminators. D'Aubenton-Carafa et al. (31) have analyzed the structure of terminators, and in particular the role of the run of T's, and have proposed that what is important is the relative position of each T residue with respect to the length of the T run. They have defined an empirical parameter nucleotide which penalizes an interruption of the poly(T) sequence the most when it is located near the end of the stem structure.

The second requisite for the sequence to be a terminator is that it form a hairpin structure in the 50 nucleotides upstream of the poly(T). The efficiency of the hairpin as a terminator is measured by the parameter Y, which is the ratio of its energy stability ΔG (calculated according to the model of Yager and von Hippel [193]) to the total length of the hairpin, including the loop. An empirical linear combination of both parameters permits one to decide whether a sequence is likely to be a Rho-independent terminator. According to the authors, using this criterion gives 5% false positives and 5% false negatives. A direct experimental approach has substantiated some of this description, but it could not be further confirmed because the efficiency of terminators may be strongly dependent on the environmental growth conditions (143, 179).

Rho-independent transcription termination signals illustrate identification of sequence signals without going through a step whereby consensus sequences are looked for. The efficiency of the method of d'Aubenton-Carafa et al. (31) is probably due to the fact that their description takes into account actual physico-chemical processes involved in termination. Another example of such identification is the search for self-splicing intron sequences.

Predicting Introns. Self-splicing introns have been discovered in eukaryotic cells and their viruses. However, it was found early on that bacteriophage T4 harbored several group I introns, and such intervening sequences were also found in archaebacteria and most organisms (25, 94, 145, 168, 192). More recently, mobile elements encoding reverse transcriptase were found in myxobacteria and in *E. coli* B (105). It was further discovered that *E. coli* strains in the EcoR collection often contained group II introns (50). It was therefore of interest to devise programs able to predict the locations of introns in new DNA sequences. Group I introns are defined in biochemical terms by their ability to make a covalent link between a guanosine and their 5′ end during the first phase of excision. Group I introns are so diverse that only seven nucleotides are strictly conserved, and among these only two are located at a fixed distance from each other in the sequence. In contrast, the secondary structure of the intron's core is very well preserved, having six helices always present. There are, however, a large variety of forms, including bulging nucleotides, unpaired bases, variation in length, and, from time to time, large insertions, so that the lengths of the terminal loops as well as the distance between the core and the excision site are impossible to predict (for a review, see reference 119). In spite of these problems, Lisacek et al. (107) showed that it was possible to predict the presence of a group I intron with a probability of 92%, with less than one false positive in 10^6 bases. The recognition process consists of generating and evaluating a large number of possible local solutions, which are then progressively combined into more and more complex structures, up to a stage where a complete core is formed (six to seven putative helices are formed of paired segments, six segments form connecting regions, and three loops form the ends). Identification of a group I intron core rests not on identification of a special motif but, on the contrary, on a global approach whereby all data on primary or secondary sequences simultaneously interplay. This success has been possible because the program of Lisacek et al. (107) is not simply heuristic but reflects in a formal way the very nature, in physicochemical terms, of these objects and permits the direct and unambiguous search of the corresponding pattern in the sequences under study. At present, *E. coli* K-12 seems to be an exception in that it does not appear to harbor any intron.

Initiation of Translation. In 1974, Shine and Dalgarno (167), studying the translation start sites of RNA virus genes, discovered that these sites were complementary to the 3′-OH end of 16S rRNA. They immediately proposed that this corresponded to a specific site selected by the ribosome 30S subunit for identification of the start codon, AUG, GUG, or UUG. This was the first example of a consensus sequence, and it can be surmised that Shine and Dalgarno's publication started the general trend for looking for consensus sequences as determining recognition processes for regulatory sites. When mRNAs were identified, it became more and more evident that something like the Shine-Dalgarno sequence was in general present upstream of the start codon, but that the spacing between these ribosome binding sites and the start codon could vary widely (from 5 to 13 nucleotides). In fact, it has been difficult to establish the definition of a domain that is necessary for initiation of translation (182). Several motifs have been identified either upstream (184) or downstream (174) of the initiation codon in strongly expressed genes. Thanaraj and Pandit (184) have combined the identification of a consensus sequence with information on putative secondary structures; Sprengart et al. performed experiments directly on the translation initiation region of gene 0.3 of bacteriophage T7 (the T7 gene displaying the highest expression level) (174). Finally, Dreyfus (43) randomly cloned translation initiation regions from *E. coli* and compared these regions with in-phase initiation codons of eukaryotic origin. This led him to identify some of the constraints operating in the −20,+15 region of the initiation region. He could demonstrate that this region, although still not well characterized, is sufficient to entirely determine the presence of a start site, despite the fact that the region studied corresponded to genes translated with an extremely variable efficiency. This result was further extended and substantiated by analysis of the expression of more translation initiation regions, including regions with artificially constructed Shine-Dalgarno sequences (10, 147).

Perrière explored the idea that there is a strong correlation between translation initiation signals and the degree of gene expression (137). With this aim in mind, he constructed three sets of genes thought to differ in expression level plus a random sequence batch displaying the same average base composition. He then compared the information content $I_{Y \to X}$, the similarity with the 3′ end of 16S rRNA, the pairing energy between mRNA and rRNA, and the putative secondary structures in the region spanning positions −55 to −1 of the mRNA. This study indicated that the additional sites proposed as favoring translation initiation (139, 174, 184) were not preferentially found in strongly expressed genes. Furthermore, the actual distribution of the analyzed criteria does not differ from what would be found in random sequences. There is a clear contribution of structure to initiation efficiency, however. For example, de Smit and van Duin investigated the influence of secondary structures on the ribosome binding site and found that competition between intramolecular and intermolecular binding forces was a determinant for initiation efficiency (38, 39).

Thus, notwithstanding a significant role of secondary structures, the Shine-Dalgarno sequence is the only signal which has been clearly demonstrated as pertinent in translation initiation, and it is generally closer to the consensus in strongly expressed genes than in the others (the motif described by Sprengart et al. [174] is a specific case because it seems characteristic of phages T4 and T7). In conclusion, although a probably pertinent description of translation initiation regions involving binding of the mRNA to the 3′ end of 16S rRNA has been proposed (82), the level of gene expression in this region cannot be evaluated with only knowledge of the sequences because we do not know yet how to predict reliably the existence of secondary structures in individual mRNAs.

Theoretical Considerations for the Identification of Signals

Signal Description Using Consensus Sequences. Finding signals has been the first application of informatics to sequence analysis in *E. coli*. The vast majority of studies aiming at identifying signals dealt with the implementation of heuristics and not with searches of solutions modeled from a realistic representation of phenomena. They are much faster than exhaustive searches but may miss their goal in a generally unpredictable way. This heu-

ristic approach was usually summarized as consensus sequence identification. The corresponding studies can be ordered as a function of the estimated complexity of the consensus sequence.

(i) The consensus sequence is composed of letters (representing nucleotides or amino acids) present, at a given position, in more than half of the sequences. Sometimes it is even less well characterized, displaying a frequency in the examples as low as 35%. In this situation, to use a consensus motif means to look for an identical or nearly identical motif in a new genomic sequence or in a new protein. The underlying logic of this approach is very similar to that used in syntactic analysis. The best results are obtained with proteins (which justifies the use of the PROSITE database of Bairoch) rather than with nucleic acids.

(ii) The composition diversity at a given position in a consensus sequence is conserved in a multiple alignment (this corresponds to another database, BLOCKS [76]), in a consensus matrix (176), or in the Shannon's entropy of a motif (128). In all such cases, it is possible to use an algorithm of dynamic programming in order to evaluate the similarity between the consensus and the analyzed sequence, thus permitting introduction of gaps (insertions or deletions) when necessary.

Description Using Attributes. The methods described above aim solely at preserving a more or less exact motif in a sequence, implicitly supposed to be the archetype or ancestor from which all experimental sequences would have been derived by mutation (or toward which the sequence could converge). Each position is considered an isolated entity, and no correlation between letters or positions is considered. Several methods can take into account the existing correlations between letters present at differing positions in a given sequence. They have several features in common and can be thought of as methods involving a process of learning from examples meant to solve classification problems (either through discrimination between two sets of objects or through assimilation of objects to a set of objects having some feature in common). They are used when one does not know, a priori, a decision procedure but when one has a sufficient number of positive and negative examples. Following a training step using a set of examples, one builds up a procedure meant to solve the problem. The most significant methods of this type are as follows.

(i) The Perceptron, a classical learning method in pattern recognition research (149), is an input/output automaton whereby the input and output levels are linked in such a way that each input point is linked to all output points. Learning consists in making the efficiency (weight) of each connection (synapse) evolve by "training" in such a way that at the end of the training period, the behavior of the overall system reflects as much as possible the behavior expected from the set of given examples. This allows one to compute, using the training set, a weight for each synapse. The system is subsequently used in a fixed way, with the synapses having the weight that they obtained after the training period. Stormo and coworkers have pioneered the field by using a Perceptron approach for creating matrices permitting identification of the ribosome binding sites in *E. coli* mRNAs (183), and they have been followed by many other authors both for the study of DNA and protein sequences and for generating pertinent patterns for promoter recognition (2, 124). The Perceptron cannot manage the logical "or" rule, however; it can recognize only classes well enough separated (it cannot separate a class for which two properties are simultaneously true or simultaneously false from a class for which one of the properties is true when the second is false).

(ii) The algorithm of back-propagation gradient (generalized delta rule) alleviates some of the limitations of the Perceptron by allowing implementation of the logical "or" (155). However, the convergence of the method toward an optimal solution is not certain. Neural networks represent an easy way to implement this class of algorithms, but they do not permit generation of classifications more efficient than those obtained using nonneural algorithms of equivalent complexity, as illustrated empirically by Hirst and Sternberg (78). However, they have the advantage that because they can be parallelized, they can sometimes be very fast (37, 49, 79, 109, 130, 131).

(iii) An alternative to these approaches consists of using artificial intelligence techniques. Statisticosyntactic learning techniques have been used for identification of translation initiation regions (148) as well as of promoters (158). With such methods, each object's description is summarized as a list of attributes. The knowledge generated through learning is visualized as a set of rules (e.g., there is a purine at position 3 and a C at position 7), taken as arguments in favor of a decision process. In order to take a decision, one follows quantitative logic rules, using appropriate threshold values (for instance, true in more than x% of the examples and less than y% of the counterexamples) (172).

(iv) Classical techniques in data analysis (multidimensionnal scaling) are scarcely used in this domain despite the fact that they provide solutions which seem to be well adapted to the analysis of relationships existing between the positions correlated in short sequences (137).

Such methods represent the most evolved form of investigation in terms of consensus sequences. All have some limitation. In particular, they operate poorly on motifs in which one allows for random insertions or deletions. A significant improvement can be made when the conceptual analysis of the biological problem can be implemented in the structure of the network. But this does not correspond to a true help in discovery, since it is precisely biological intuition which is implemented at the start during the building up of the network. In this respect, one should remember that most neural networks are sensitive to the "ugly duckling" effect described by Watanabe (189), which indicates that a learning image can be blurred when the number of learning examples in the training sets increases, unless the examples have strictly homogeneous properties. But in order to fit this requirement, one must either be extremely lucky or have already understood most of the problem. In addition, there is another difficulty inherent to methods which are not explicitly based on statistical analysis: one cannot check that there are enough degrees of freedom in the system; i.e., the sample may contain many more pieces of information than the number of parameters which are computed. The problem is particularly difficult to tackle in the case of neural networks which are overdimensioned compared with the volume of data. One obtains a system which behaves very well on the training data, because it can effectively "memorize," but does not do well at generalizing and therefore does poorly on test data not included in the training set.

The Perceptron, as well as most neural network learning techniques (and this has often escaped attention), because they involve reversible formal synapses, have a weight which evolves in a quantitative fashion as a function of their actual use. This means that the synapse weight can fluctuate as a function of the training set. As a consequence, when the set size increases, the weight of each synapse may go through an optimum value and then slowly go back toward a more or less average value, thus losing discrimination power. Therefore, as the training set of examples increases in size, the actual efficiency of the consensus matrix created by the learning procedure loses accuracy. In fact, it would be useful to test ancestors of neural networks which evolve in such a way that the effective transmitting capacity of a synapse goes irreversibly to zero when its value falls below a certain threshold value, freezing the learning state (29). This is reminiscent of the importance of the terminating step in recursive algorithms. A relevant example is the prediction of promoter recognition by Horton and Kanehisa (79). These authors used "trimmed" neural networks, in which they reduced the number of weights in the input window during training by deleting weights when they fell below a threshold (low) value. It would be important to compare and develop new approaches with emphasis on the stability of their learning capacity as a function of the training sets.

Structural Descriptions. Such descriptive methods provide us only with heuristics, which can be completely disconnected from the biological reality. A series of studies performed 10 years ago has identified the following bottlenecks (55, 56). (i) Biochemical objects are well accounted for if one uses structural descriptions, whereas such objects can be described only partially in terms of attributes. (ii) There are specific learning methods adapted to the structural mode of description. The knowledge that they generate significantly differs from the knowledge derived from more classical methods and cannot be reduced to the latter (because the descriptions that they use are more complete). (iii) Such modes of description, together with the corresponding learning methods, are complex ("learning" is used here with the meaning found in computer sciences), and this requires large amounts of central processing unit and memory allocation.

Gascuel and Danchin (57) demonstrated the validity of this structural approach when they characterized the structural differences between signal peptides necessary for protein secretion in prokaryotes and eukaryotes. The studies described above further demonstrate that a pertinent description of a signal is possible only if one considers the physicochemical structure and dynamics of the underlying biological process. In these cases, the solution was obtained not after identification of one or several consensus sequences but from a description combining information which does not correspond to the same level (primary sequence, secondary structures), in which the three-dimensional organization plays a major role.

Linguistic Approaches. The case of consensus sequences reflects the inadequacy of context-free grammars in the description of biological signals. One needs a descriptive process that takes into account the actual biological recognition process in order to fit the experimental data. As stated by Collado-Vides (27, 28), a model that is descriptively adequate can be thought of as a classification scheme which takes into account all well-formed signals and discards those which do not fit. A systematic and integrated description has to be consistent with the actual mechanistic explanations of the process. The main illustrations of this approach are the Prolog-derived description of Searls (161, 162) and the linguistic description of units of genetic information (promoters and the like) devised by Collado-Vides (27, 28). Created for the specific situation of σ^{70}-dependent *E. coli* promoters, it is an attempt which is still in its infancy and as yet unable to make accurate predictions. One therefore must still consider other methods permitting investigation of biological sequences.

Chaitin-Kolmogorov Algorithmic Complexity. Biology is more a science of relationships between objects than a science of objects (this is indeed the raison d'être of analyses based on genetics). In particular, the very nature of the information carried, from generation to generation, in the DNA molecules that constitute genes cannot be summarized by Shannon's entropy. For example, "context," "meaning," or "semantics" are, by construction, absent from a figure such as a simple statistical mean (which actually represents Shannon's entropy). When we speak of the "genetic program," biology provides us with a metaphor that displaces the idea of information toward a new field, that of programming and informatics. Is there more insight in these new fields than in the "natural" way of considering information? At least since 1965, Kolmogorov and Chaitin, following Solomonoff and the Russian school of electronicians, have formulated the problem in detail in the following way (for a general description, see references 104 and 195). Let us consider the simple case of a chain of characters such as those found in computer sciences. What can be said about its information content? A way to consider the chain is to try to reduce its length so that it can be accommodated in a memory, for example, using the minimum space, without altering its performance when used in a program (at least in terms of accuracy, if not in terms of time)—in short, without losing its information content. This is called compressing the data. This problem is of very broad and general interest. It is possible, given a chain, to define the shortest formal program (in terms of Turing's universal computation algorithms, i.e., algorithms that can be implemented on any machine operating on integers) that can compress an original chain or restore it given its compression state. The algorithmic complexity, or information value of a chain S, is therefore defined in this model as the minimal length of the program that can represent S in a compressed form.

The complexity of the sequence made of 1 million A's is very low, for there are very short programs (e.g., "for $i = 1$ to 1,000,000; print A; end") which print this sequence out. The (short) programs which permit printing out long sequences can be considered compressed versions of the sequences. In contrast, a sequence of length 1,000,000 having a Chaitin-Kolmogorov complexity equal to 1,000,000 is not compressible: no means allows it to be described in a shorter form. With this definition, it appears that a completely random chain S cannot be compressed, implying that the minimal program required to compress S is identical to S. In this context, the information of a sequence is defined as the measure of its compressibility (one often uses the concept of complexity rather than information in this context). A sequence of symbols is random if and only if it is not compressible.

Represented as sequences of letters, genes and chromosomes have an intermediate information (complexity) content: one finds local repetitions or, in contrast, sequences which are impossible to predict locally. Their complexity is intermediary between randomness and repetition. The overall complexity of sequences originating from higher eukaryotes or prokaryotes is very different, and it links genomic sequences to both sides of

the "uninteresting" fraction of information (repetition or randomness). The complexity of the former is more repetitive and is usually much lower than that of the latter, which looks more random (63). This is quite understandable if one remembers that bacterial or viral genomes are submitted to stringent economy constraints, implying that they must remain very compact. In contrast, the lengths of genomes of higher eukaryotes are much less constrained, and they contain, for instance, many repetitive sequences.

Up to this point, we have considered the Chaitin-Kolmogorov complexity of a sequence, but this can also apply to programs meant to identify meaningful sequences. In addition to the knowledge, present but hidden, in a given sequence, it is possible to propose an evaluation of the quality of a description by measuring its complexity. (i) The description operating at the lowest level consists of a complete list of all known sequences. The corresponding information is not at all compressed, which is equivalent to saying that the sequences are random (the complexity is maximum). (ii) In contrast, structural descriptions such as those used for describing Rho-independent transcription termination, or used for describing the class I intron core, provide a measure of the Chaitin-Kolmogorov complexity of these biological objects, because one now knows the length of a program able to build them up.

(iii) Descriptions using attributes allow the writing of programs much shorter than those which have just been described. These too simple programs are likely to have a complexity lower than that of the signals they attempt to describe. In this case accordingly, they correspond to attempts which must fail.

FURTHER ANALYSIS OF NUCLEOTIDE SEQUENCE INFORMATION

From an abstract point of view, a DNA molecule can be seen as a long sequence possessing meaning, but becoming more and more random as the result of mutagenic processes. In contrast, from the biologist's point of view this sequence is submitted to a variety of constraints and certainly carries many signals, even if it looks random. The self-consistency of a genome is maintained as the result of natural selection, the individual's survival being the result of these constraints and signals.

The paragraphs above have shown that even when it corresponds to well-identified processes, the analysis of known signals is not an easy task. A fortiori, the analysis of other (yet unknown) signals in a DNA molecule will be very difficult, precisely because one does not know their very existence. Furthermore, one has reason to suspect, especially in prokaryotes, that the information carried by different signals can correspond to superimposed signals at a given position of the sequence. The genetic code redundancy for example allows superimposition of independent information at a given position in the sequence. In addition, there exists a further degree of freedom in protein sequences because amino acid residues are functionally similar. Nobody knows, at present, how to measure the contribution of this latter level of degeneracy in the genome.

Lacking a direct experimental approach for evaluating information in sequences, one must rest on methods which have been validated in various fields other than biology, such as signal analysis or linguistics: identification of heterogeneity in nucleotide composition, detection of motifs which are under- or over-represented, as well as study of their distribution, analysis of regularities, etc.

Identification of Significant Motifs

As in the preceding examples, *E. coli* has been a paradigm for the identification of motifs. The basic idea for identifying significant motifs is to design, a priori, a probabilistic model permitting generation of a theoretical genetic sequence (for example, actualizing the probability of chaining appropriate nucleotides) and then to compute the mean frequency (expected value) of a given motif in this model-derived sequence. This latter theoretical motif frequency is subsequently compared with the frequency observed in the real sequence. If the difference between the two frequencies is important, one can surmise that the motif reflects a process of biological significance. This approach asks mathematical questions (what is the statistical significance of the deviation?) as well as biological questions (how can one interpret the observations?).

The Statistical Meaning of Motifs. Knowing the absolute value of the distance between a mean frequency and the observed frequency is not enough to permit one to define the statistical meaning of a motif's frequency. One must also know the standard deviation in the frequency distribution of the motif. A first, classical, approximation is to assume that the standard deviation is equal to the observed mean (Poisson's law).

The simplest method, and the only possible one in the case of complex models, is to generate a set of sequences in which the chaining of letters follows a given probability law and then to determine empirically the frequency distribution of the motifs in these sequences (69, 74). The method is particularly well suited to the identification of motifs located in genes coding for proteins. In this case, the most appropriate theoretical model generates a nucleotide sequence while preserving the amino acid content, the specific codon usage of the gene, and the dinucleotide frequency in the codon chaining. This theoretical model permits elimination of all constraints linked to the nature and the function of the protein as well as of the general features of the considered nucleic acid segment.

Unfortunately, construction of such a simulation is not adapted when the frequency of the motifs is small (i.e., for long motifs). A precise estimate of motif distribution is impossible because this would require generation of extremely long theoretical sequences, an exercise that poses insuperable practical difficulties.

In fact, a rigorous analysis of the problem is even more complicated, because the characteristic features of the frequency distribution of motifs depend not only on the length of the motif but also on possible overlapping of the motif with itself (for example, two TATA motifs can overlap with a two-letter shift). Gardner (54) has nicely illustrated some unexpected consequences of such overlaps and shown that intuition is of no help if one wishes to make a statistical study of motif distribution.

Pevzner et al. (140) have proposed a measure for the evaluation of the statistical meaning of any motif. The standard deviation of the frequency is increased according to a specific law when the motif allows for overlaps. These authors have also calculated the distribution of motifs comprising several words, not necessarily contiguous. Taking into account in this way complex motifs as well as motifs with allowed overlaps is a technical trick which permits one to easily consider the behavior of Markov chains of high order. Such chains correspond to models in which the probability for finding a letter at a given position depends on the letters present upstream. The simplest model is the Markov chain of order 1: it generates a sequence of letters which keeps the composition bias of dinucleotides (or dipeptides in protein sequences) observed in experimental sequences. Periodic Markov chain models (19) were introduced for more accurate description of DNA regions which code for proteins in *E. coli*. Periodic Markov chains of order 1 keep the three periodic variations in dinucleotide frequencies observed in a protein-coding region, as a function of their situation in the reading frame. Kleffe and Borodovsky (91) have established the exact value for the mean and standard deviation of the frequency of a motif in the case of periodical Markov chains. In this model, one identifies 12 symbols: the four letters (A, T, G, C) in the first position, the four letters (A', T', G', C') in the second position, and the four same letters (A", T", G", C") in the third position of a reading frame of period 3. Using this alphabet sequence, ACTGTT . . . is rewritten AC'T"GT'T". . ., the transition probability for a letter doublet being different according to the position in the hypothetical codon. Possessing such a model permits one to calculate directly the exact statistical parameters (means, standard deviation, etc.) of the motif's frequency in sequences coding for proteins, whereas simulations give only approximations of such parameters, even after a long computation time (to generate long random sequences of digits is very costly in terms of computation time).

Burge et al. (22) have modified Pevzner formulas which compute the mean value of a motif's frequency in order to take into account the fact that a DNA molecule is double stranded, imposing the condition that the constraints operating on one strand must also be consistently operating on the complementary strand.

Several authors have been interested in motifs which are prominent not because of their frequency in the sequence but because they display special features. Guibas and Odlyzko (67, 68) have studied repeated motifs, such as ATAT . . . AT, or the periodicity of motifs in sequences. Their work does not, unfortunately, provide an analytic description for the standard deviation of the distribution of such repeated motifs. Karlin and coworkers (86) have solved the problem by using classical tools in nonparametric statistics, which permit one to compare the values of variables for which the theoretical frequency distribution is completely unknown, including their means and standard deviations, but at the cost of efficiency.

From Statistical to Biological Meaning. There is no method which permits one to go from a statistically significant observation to a biologically significant interpretation. In fact, the biological interpretation always faces difficult problems. One has to propose various biological representations of a given situation in order to analyze whether a statistical model may be significant. In the end, the appropriate way to be sure that what has been proposed is meaningful is to build up mutants which should behave in the way predicted from the interpretation of the models. Bacteria will be, for this reason, extremely well-adapted tools for the investigation of the meaning of genomic sequences.

The difficulty can be seen from the very start of sequence analysis: different authors identify different motifs as "significant," even though they have analyzed the same set of sequences.

This is because the statistically significant outcome of a given analysis can vary according to the model chosen for calculating the statistical mean or for generating the statistical baseline (theoretical sequence). Nussinov (125), for instance, noticed a long time ago that there are correlations between a given nucleotide and its neighbors. This can be interpreted in statistical terms by saying that a sequence is, to make a minimum hypothesis, a Markov chain of order 1. But Blaisdell (15) and Phillips et al. (142) have shown that the Markov chain must be at least of order 3 if one wants a theoretical chain to simulate correctly an experimental one. Hénaut and Vigier (75) and Hanai and Wada (69) have obtained results for motifs present inside coding sequences, but their approach uses the fact that the region considered uses the genetic code in order to generate the theoretical reference sequence, therefore implying that the method cannot be generalized to noncoding regions (or to regions coding for RNA molecules). From these studies, it appears that the probabilistic model used for generating the reference sequence must take into account the local composition and significance (such as "this is a protein-coding region") of the DNA molecule if one wants the comparison between the theoretical model and the experimental sequence to be meaningful. The same statement comes from investigation of protein coding regions (19), whereby it was shown that ordinary Markov chain analysis cannot simulate properly any coding region.

Furthermore, the very fact that a given motif is favored or avoided does not mean that the motif itself is the target sequence for a biological process. Selection can operate on a longer or shorter motif comprising the prominent motif. Finding the motif which is actually significant is not trivial. We shall illustrate

this point in the case of the CTAG motif, which has been noticed by several authors to be strongly counterselected in DNA sequences (Table 3). Several authors have focused on the very low frequency of TAG (in and out of coding frame) and CTAG (22, 111, 115) in most organisms. One can therefore ask whether counterselection on TAG is not enough to explain why CTAG is rare. Table 3 indeed shows that all motifs containing TAG are rarer than the theory would predict, the ratio between the expected and observed frequencies varying between 0.6 and 0.8. However, in the case of CTAG, the bias is extremely strong, as low as 0.2. This indicates that selection does not act only on TAG but specifically acts against the presence of CTAG in sequences. At this point, one must verify that CTAG, and not some longer sequence containing this motif, is the actual target. All motifs of five letters containing the CTAG sequence display a ratio of frequency observed over theoretical frequency, situated in the 0.2 range. These results indicate that counterselection does not increase when the sequence length increases. Thus, CTAG is, in itself, the subject of some selective pressure. But it must be emphasized that it is not possible, from these results alone, to determine whether the cause of this effect stems from a special structural property of the DNA molecule having this sequence, as proposed by Médigue et al. (115) or Burge et al. (22).

The latter example demonstrates the importance of having independent statistical results for each studied motif. Studies in which both strands of the DNA molecule are considered equivalent cannot distinguish between selection on a motif on one strand and selection on its complement. In studies like those of Burge et al. (22), one cannot, for instance, distinguish CTA, which does not present a significant bias in E. coli, when one

TABLE 3 Frequency of motifs overlapping two codons comprised in or comprising CTAG[a]

Motif	No. observed	No. expected	Observed/ expected	Motif	No. observed	No. expected	Observed/ expected
CTA	4,982	5,251.8	0.9	TAG	2,675	4,145.7	0.6
CTAT	2,187	2,242.9	1.0	TTAG	844	1,403.1	0.6
CTAC	2,283	2,081.8	1.1	CTAG	80	469.1	0.2
CTAA	1,116	1,142.8	1.0	ATAG	451	649.4	0.7
CTAG	80	469.1	0.2	GTAG	1,300	1,624.1	0.8
CTAGT	16	99.9	0.2	TCTAG	8	108.3	0.1
CTAGC	30	236.0	0.1	CCTAG	20	92.8	0.2
CTAGA	16	82.3	0.2	ACTAG	20	96.0	0.2
CTAGG	18	50.1	0.4	GCTAG	32	172.0	0.2
TAGT	529	938.2	0.6	TAGC	1,169	1,838.1	0.6
TAGA	556	854.9	0.7	TAGG	421	514.5	0.8

[a]Data are derived from analysis of 761 identified genes of *E. coli* (115) comprising 286,895 codons. Genes involved in horizontal transfer are excluded from the analysis. The observed frequency of an oligonucleotide sequence is compared with the frequency it would have if the codon sequence in the gene was fixed only by the sequence of amino acids in the protein, knowing the relative frequency of corresponding synonymous codons in the gene (75). As a first step, the frequency of hexanucleotides overlapping two codons (when in phase) or three codons (when out of phase) is calculated. Each hexanucleotide is subsequently split into shorter sequences of two, three, or four nucleotides, overlapping two codons. As an example, a hexanucleotide 5′ 123-123 3′ can be decomposed into one dinucleotide 3-1, two trinucleotides 23-1 and 3-12, and three tetranucleotides 123-1, 23-12, and 3-123. Thus, each of the 256 tetranucleotides can be described as observed and theoretical (expected) sets. All observed and calculated sets are computed for each of the 761 coding sequences in the sample and added up. In the present situation, 469.1 CTAG motifs were expected, while only 80 (17%) were observed. The difference between observed and calculated values is much larger than for each of the subsequences CTA (95%) and TAG (65%). In contrast, counterselection does not significantly increase when the length of the sequence increases (the ratio observed/expected varies from 7.4% for TCTAG to 36% for CTAGG). One thus concludes that there exists a selection pressure specific to CTAG, which cannot be accounted for by pressure on TAG, and that this selection pressure is enough to explain counterselection of five-letter motifs containing CTAG. The distance between the observed and the theoretical frequency is negligible for dinucleotides CT, TA, and AG.

considers the complementary of the transcribed strand, from its complement TAG, which is strongly counterselected in this case (115). A fine analysis of significant motifs requires construction of homogeneous sets of sequences in which all DNA fragments possess similar biological characters, in particular in terms of coding capacity. This can be done by restricting the analysis to special regions, such as coding sequences or strands having the same orientation with respect to the orientation of the replicating fork, as a function of the likely biological significance of the motifs that one is looking for.

As we have just seen, a statistically significant difference between the observed and expected frequencies was not sufficient to prove that selection was operating on a motif. This is not a necessary condition either, as indicated by the study of nonsense triplets. If TAG is counterselected, this is compensated for by an overrepresentation of TAA and TGA triplets, resulting in the expected frequency for nonsense triplets (91). These triplets do not have a special frequency or distribution along the chromosome, but if one analyzes the interval between two successive such triplets, imposing the condition that this interval be a multiple of 3, a very prominent and well-known figure emerges, corresponding to open reading frames comprising protein-coding sequences. This demonstrates the importance of the reading frame in the analysis of DNA sequences. Curiously, however, no analysis method takes this fact into consideration.

DNA Stretches with Special Structural Properties: Defined Ordered Sequences

Complexity of Substrings in dosDNA. Numerous regions of defined ordered sequence DNA (dosDNA) exist in the genomes of prokaryotes and eukaryotes (190). The analysis of the complexity of such sequences is somewhat similar to that of significant motifs, as already discussed. The minimum analytic criterion is to identify the variety in composition of the sequence. Intuitively, once again, one can consider that a sequence made of identical letters has a minimum complexity and, at the other extreme, that the most complex sequences are those for which the probability for a letter at any given position is equal to that at any other position, all letters being randomly intertwined. We find here the same analysis as that given above for measuring the Chaitin-Kolmogorov complexity. But this analysis is not well fitted for analyzing short sequences: the shortest program permitting one to print out the sequence of the first eight digits of π is still "print 3.14159265," which means that the most condensed form to describe the sequence is the sequence itself (as in the case of random sequences), whereas this would no longer be true for the first 500 digits of π.

To take this situation into account, several authors (156, 157, 191) have proposed a measure, K, for the complexity of sequences of length L when L is small. K takes into account the degree of inhomogeneity of the substring. In this model, a substring in which all four nucleotides are equally represented is maximum, while it is zero if the substring contains only one type of nucleotide. The complexity K is easier to compute than the complexity of Chaitin and Kolmogorov, but this is because it corresponds to a much coarser view of reality. In particular, K does not contain any information on the actual chaining of letters in the substring. A repeated string such as ATGCATGC . . . ATGC, which has a low Chaitin-Kolmogorov complexity

("for i=1 to n; print ATGC; end"), has the same K value as a random sequence having the same composition.

Lebbe and Vignes (100) give a measure of the complexity of substrings more refined than K. For this, they use the analysis of local prediction in a sequence. A substring of length L is of low complexity if the knowledge of a few letters is enough to reconstitute the string. It has a maximum complexity if the knowledge of $L - 1$ letters is not enough to predict the Lth letter. The algorithm predicts the letter located in the middle of the string L, the neighborhood of this position being constituted by the $L - 1$ other letters. A training procedure is performed on the totality of the sequence (of length $N >> L$). The prediction takes into account not only the neighborhoods of the letter which is studied, when they are identical to each other, but also the k neighborhoods which are the most similar to them. The fact that these k neighborhoods are taken into consideration improves the prediction by smoothing out the sampling-induced fluctuations. The local previsibility is measured by the mean of the differences observed between the predicted letter and the letter observed in the real sequence.

Repeated Sequences. Clift et al. (26) have introduced the usage, in molecular biology, of an algorithm able to identify all repeated strings, of any given length, in a sequence. This algorithm has the enormous advantage of asking for a computation time and a memory allocation, which is a linear function of the length of the sequence (18). They have proposed a graphical representation of the results, a "landscape," whereby each repetition is localized along the sequence. Unusually long repetitions appear as peaks and valleys containing sequences which are rare or unique.

Such programs are limited to the identification of exact repeats. But biologists often believe that looking for approximate repetitions might be more rewarding. Leung et al. (103) have solved the problem when the studied motif is made of a succession of exact repeats, chained through short segments where the sequences are permitted to differ. It is indeed necessary to limit the error amount which one admits in approximate repeats if one does not want to find that any one sequence matches the investigated pattern. The constraint imposed in the method of Leung et al. (103) is to fix the minimum length of the exacts repeats (b) and the maximum length of the segments which are allowed to vary (ε). Although the complexity of the algorithm is not a linear function of the length of the sequence, the program can still be run on standard computers if b is large and ε is small compared with the analyzed sequence. This program allowed Blaisdell et al. (16) to discover new classes of repetitive extragenic palindromes (58) in *E. coli*.

Several authors use the concept of information compression (121, 194). It is a measure of Chaitin-Kolmogorov algorithmic complexity as presented above. The compression range of a sequence is very sensitive to the presence of symmetry elements (palindromes, tandem duplications, etc.) or to the type of mono- or oligonucleotide repetitions found in dosDNA. The programs of Milosavljevic and Jurka (121) and of Yee and Allison (194) take only direct repetitions into account, but this reflects simply a technical choice made by the authors, not any specific limit of the method. Restricting investigation to repetitions in a sequence, it can be seen that the more repetitions, the stronger the compression. In addition, the outcome of the compression permits one to take chance into account, i.e., the fact that if repetition of a substring does not allow any compression (because it is too short), this repetition does not have any other algorithmic explanation, chance aside.

Distribution of Motifs along the Sequence. We have seen above that it is necessary to take into account the distribution of nonsense triplets in each reading frame if one wants to discover that they have a special function. Classical methods in statistical analysis designed for the study of the distribution of motifs along a sequence do not perform the investigation at such a deep level. They do go, however, beyond the simple model of an exponential distribution of the intervals between to successive motifs. An important improvement in sensitivity is to study the distance between the extremities of a fragment limited by two motifs and containing k motifs. This statistical analysis allows one to recognize regions exhibiting unusual regular features or unusually dense clustering of motifs. It has been used in molecular biology to demonstrate that the early steps of mutation fixation in a divergent evolution scheme are linked to constraints operating on the DNA itself and not on the proteins (36). Karlin and Macken (87) have generalized this method under the name r-fragment analysis. They have thus shown that in *E. coli*, restriction sites made of 6 bp are more regularly spaced than a random sequence would predict (86). Since then, Karlin and coworkers have extensively used r-fragment analysis, for example to study *Saccharomyces cerevisiae* chromosome III (85), but at least one new algorithm has discovered internal repeats which had escaped the attention of Karlin et al. in this case (101).

Looking for long-distance correlations (several hundreds of nucleotides) requires new methods, such as those derived from the study of fractal signals. A preliminary observation has been made by Peng et al. (136). They have studied the distribution of purines and pyrimidines along sequences by giving the value -1 to a purine and the value $+1$ to a pyrimidine. The analysis of sequences devoid of introns, in particular *E. coli* sequences, does not reveal any long-range correlation. For this reason, a periodical Markov chain model, such as the one devised by Borodovsky and coworkers, reveals most important features in *E. coli* sequences. In contrast, in sequences containing many introns, long-distance correlation are prominent, with specific scale-invariant properties typical of fractal structures. In the case of the purine/pyrimidine analysis, this means that one finds similar organizations whether one looks at segments of 1,000 nucleotides or 10,000 nucleotides, provided that the analysis sliding window, used for smoothing out noise, is changed in the same proportions. The work by Peng et al. (136) has been criticized for technical reasons (see the discussion in reference 85). It seems, however, that despite some inaccuracy at the quantitative level, its conclusions are valid (6a).

Some Biological Results

Trifonov and Brendel (185) have created GNOMIC, a dictionary of the genetic codes wherein motif sequences are associated with their biological meanings. Unfortunately, the explosion of results derived from the many sequences accumulated in data libraries has made their initial attempt impossible to continue. In addition, it was soon observed that a meaningful motif in a given organism could have no significance in another organism, and vice versa. Even in a given organism, there are important differences between genes belonging to a given category and the others (e.g., translation and core intermediary metabolism versus genes

involved in horizontal transfer [74, 114, 115, 142]). It is easy in some cases to relate an observation to a biological feature: GGAGG is counterselected in genes (the ratio between the theoretical and observed frequencies is 4.2), which is certainly linked to its presence in the Shine-Dalgarno consensus sequence. As mentioned above, the selection against CTAG may be related to the fact that the double helix is kinked at the center of this motif. But there remains only little hope that all "significant" motifs correspond to actual biological signals or are linked to the stability of the DNA molecule. Indeed, there is some contradiction between the elusiveness of the concept of consensus sequences, as discussed above, and the idea that signals corresponding to precise motifs would be the target of selection.

A completely different approach consists of considering the motifs as indicative of the presence of very general constraints, having no direct relationship with the meaning of the actual motifs' sequences. There is a strong correlation between the frequency of a codon and that of its complementary counterpart in *E. coli* (3). This correlation is even stronger in the case of oligonucleotides overlapping codons, which is permitted because synonymous codons can be chosen, preserving the nature of the amino acid residue in the coded protein. The correlation between complementary oligonucleotides is also present in phage T7 and lambda genomes, which have different strategies for codon usage. This correlation is true for all three classes of genes in *E. coli* (114, 142, 186).

Bhagwat and McClelland (14) and Merkl et al. (117) have proposed that in *E. coli*, the frequency anomalies for some tetra- and pentanucleotides are the result of a specific process of mismatch repair, the so-called very short patch repair. This process repairs T·G pairs as C·G pairs when the mispair occurs in a motif $5'T(A/T)GG3'/3'G(T/A)CC5'$ or in a motif $5'CT(A/T)G3'/3'GG(T/A)C5'$ at the positions of the boldfaced letters. These tetranucleotides are part of the CC(A/T)GG pentanucleotide sequence that is methylated (on the second C) by the *dcm* gene product, Dcm cytosine methylase. Thus, this system could have evolved to prevent C→T mutagenesis caused by spontaneous or enzyme-mediated deamination of 5-methylcytosine to T. However, if this repair system were very efficient, then each replication error G opposite template T at the above position would be fixed by this specialized repair system to a TA→CG mutation instead of being corrected by the long patch strand-directed mismatch repair system (*mutHLSU*), hence a T→C mutagenesis! With time, this process should lead to a decrease in the presence of tetranucleotides TAGG, CCTA, TTGG, CCAA, CTAG, CTTG, and CAAG and to an enrichment in tetranucleotides CAGG, CTGG, CCAG, and CCTG.

It seems possible to extend this kind of analysis to other repair processes inducing spontaneous mutations. Many over- or underexpressed motifs are part of what Wells and Sinden (190) name dosDNA (see above). Such sequences generally contain elements having some kind of symmetry (palindromes, tandem duplications, and the like) or multiple repetitions of mono- or oligonucleotides. Regions organized as dosDNA can have spatial configurations which significantly differ from the standard B-DNA structure. This implies that dosDNA is associated with specific mutation processes and with instabilities during DNA replication. In humans, this seems to be at the origin of many genetic diseases. Studies performed in the next several years should lead us to understand whether preferred or excluded motifs are indeed the scar of such genetic instabilities. Identifying specific substrings by statistical means will then necessarily be linked to a direct experimental approach in order to understand genome stability.

CONCLUSION

After much debate, and promises which have not been fulfilled, it appears that molecular genome analysis has at last come of age. One can expect that most if not all of the *E. coli* genome will be known by the end of 1997. A measure of the corresponding success is that progress in sequence acquisition no longer seems remarkable, stretches of the order of 100 kb or more being now commonplace. However, although some information can be derived from the knowledge of sequences only, it seems that much work must still be performed if one is to correlate a given sequence and its biological function.

In addition to the information levels discussed in this chapter, several other levels of information should be considered if one wishes to understand the true meaning of biological sequences. It is, for instance, of particular importance to take into account the time-dependent part of biological processes. We have mentioned this fact when discussing the efficiency of transcription terminators. The underlying dynamic processes give their biological meaning to graphical representations wherein a sliding window is used in order to smooth out the fluctuations of a given variable. Indeed, when using a sliding window, one considers implicitly that the phenomenon which is described is homogeneous at the corresponding scale. The interpretation is self-evident when the window comprises a few tens of nucleotides, because this can correspond to the region of interaction of a protein with its particular target. This is no longer so when one uses windows longer than 1,000 nucleotides. In the presence of an unexpected regularity, one must then assume that this is the speed (or, more generally, the inertia) of a sliding machinery, or of a global deformation wave which has to be taken into account, because this would explain why the phenomenon remains unchanged over such a large length. As an illustration, one could visualize a window of a few thousand nucleotides as showing the genome as it is seen by a DNA polymerase complex, while a window of a few hundred nucleotides would correspond to transcription. All of this would require general considerations which will not be discussed here (see reference 30 for a general discussion). In any case, it seems very important to make hypotheses and to test them experimentally by constructing appropriate mutant strains. We think that this will be a major future trend of molecular genetics once the sequences of the complete genomes are known.

We have summarized in this chapter the many informatics approaches which enable geneticists to explore the meaning of the sequences they have been able to generate. This is but the first step in a much deeper analysis for which research is presently starting. One can see from the work which has already been done that the light which can be shed by informatics techniques on sequences can certainly help geneticists in building up rich and promising biological experiments: *E. coli* can be studied in silico before one returns to it in vivo.

ACKNOWLEDGMENTS

We thank Mark Borodovsky, Marie-Odile Delorme, F. Neidhardt, M. Radman, and M. Riley as well as anonymous referees for improvement of the manuscript, Joël Potier for providing us with a large reference data library, Max Dauchet, Simon Diner, and Jean-Paul Delahaye for stressing the importance of Chaitin-Kolmogorov information, and Jean-Loup Risler for improving our knowledge of databases and linked software. This work was supported by grants from the Groupement de Recherche et d'Etudes des Génomes (91CO82) and Groupement de Recherches (GDR 1029), from the Centre National de la Recherche Scientifique.

LITERATURE CITED

1. **Adamson, A., and D. Casey.** 1994. Managing genome sequence data. *Hum. Genome News* **8:**2–7.
2. **Alexandrov, N. N., and A. A. Mironov.** 1987. Recognition of *Escherichia coli* promoters given the DNA primary structure. *Mol. Biol.* (Moscow) **21:**242–249.
3. **Alff Steinberger, C.** 1984. Evidence for a coding pattern on the non coding strand of the *E. coli* genome. *Nucleic Acids Res.* **12:**2235–2239.
4. **Altschul, S. F., M. S. Boguski, W. Gish, and J. C. Wooton.** 1994. Issues in searching molecular sequence databases. *Nat. Genet.* **6:**119–129.
5. **Altschul, S. F., W. Gish, W. Miller, E. W. Myers, and D. J. Lipman.** 1990. Basic local alignment search tool. *J. Mol. Biol.* **215:**403–410.
6. **Aoyama, T., M. Takanami, E. Ohtsuka, Y. Taniyama, R. Marumoto, H. Sato, and M. Ikehara.** 1983. Essential structure of *E. coli* promoter: effect of spacer length between the two consensus on promoter function. *Nucleic Acids Res.* **11:**5855–5864.
6a. **Arnéodo, A., E. Bacny, P. V. Graves, and J. F. Muzy.** 1995. Characterizing long-range correlations in DNA sequences from wavelet analysis. *Phys. Rev. Lett.* **74:**3293–3296.
7. **Bairoch, A.** 1992. PROSITE: a dictionary of sites and patterns in proteins. *Nucleic Acids Res.* **20:**2013–2018.
8. **Bairoch, A., and B. Boeckmann.** 1993. The SWISS-PROT protein sequence data bank. *Nucleic Acids Res.* **19:**2247–2249.
9. **Barker, W. C., D. G. George, H.-W. Mewes, F. Pfeiffer, and A. Tsugita.** 1993. The PIR-International databases. *Nucleic Acids Res.* **21:**3089–3092.
10. **Barrick, D., K. Villanueba, J. Childs, R. Kalil, T. D. Schneider, C. E. Lawrence, L. Gold, and G. D. Stormo.** 1994. Quantitative analysis of ribosome binding sites in *E. coli. Nucleic Acids Res.* **22:**1287–1295.
11. **Benson, D., D. J. Lipman, and J. Ostell.** 1993. GenBank. *Nucleic Acids Res.* **21:**2963–2965.
12. **Berg, O. G., and P. H. von Hippel.** 1987. Selection of DNA binding sites by regulatory proteins. Statistical-mechanical theory and application to operators and promoters. *J. Mol. Biol.* **193:**723–750.
13. **Berlyn, M. B., and S. Letovsky.** 1992. Genome-related datasets within the *E. coli* genetic stock center database. *Nucleic Acids Res.* **20:**6143–6151.
14. **Bhagwat, A. S., and M. McClelland.** 1992. DNA mismatch correction by very short patch repair may have altered the abundance of oligonucleotides in the *E. coli* genome. *Nucleic Acids Res.* **20:**1663–1668.
15. **Blaisdell, B. E.** 1985. Markov chain analysis finds a significant influence of neighboring bases on the occurrence of a base in eucaryotic nuclear DNA sequences both protein-coding and noncoding. *J. Mol. Evol.* **21:**278–288.
16. **Blaisdell, B. E., K. E. Rudd, A. Matin, and S. Karlin.** 1993. Significant dispersed recurrent DNA sequences in the *Escherichia coli* genomes. Several new groups. *J. Mol. Biol.* **229:**833–848.
17. **Blake, R. D., and P. W. Hinds.** 1984. Analysis of the codon bias in *E. coli* sequences. *J. Biomol. Struct. Dyn.* **2:**593–606.
18. **Blumer, A., J. Blumer, A. Ehrenfeucht, D. Haussler, M. T. Chen, and J. Seiferas.** 1985. The smallest automaton recognizing the subwords of a word. *Theor. Comput. Sci.* **40:**31–56.
19. **Borodovskii, M. Y., Y. A. Sprizhitskii, E. I. Golovanov, and A. A. Aleksandrov.** 1986. Statistical patterns in primary structures of the functional regions of the genome in *Escherichia coli-. Mol. Biol.* **20:**826–833.
20. **Borodovsky, M., and J. McIninch.** 1993. GENMARK: parallel gene recognition for both DNA strands. *Comput. Chem.* **17:**123–133.
21. **Brendel, V., G. H. Hamm, and E. N. Trifonov.** 1986. Terminators of transcription with RNA polymerase from *Escherichia coli*: what do they look like and how to find them. *J. Biomol. Struct. Dyn.* **3:**705–723.
22. **Burge, C., A. M. Campbell, and S. Karlin.** 1992. Over- and under-representation of short olignonucleotides in DNA sequences. *Proc. Natl. Acad. Sci. USA* **89:**1358–1362.
23. **Butler, B.** 1994. Nucleic acid sequence analysis software packages. *Curr. Opin. Biotechnol.* **5:**19–23.
24. **Cardon, L. R.** 1992. Expectation maximization algorithm for identifying protein-binding sites with variable lengths from unaligned DNA fragments. *J. Mol. Biol.* **223:**159–170.
25. **Cech, T. R.** 1988. Conserved sequences and structures of group I introns: building an active site for RNA catalysis—a review. *Gene* **73:**259–271.
26. **Clift, B., D. Haussler, R. McConnell, T. D. Schneider, and G. D. Stormo.** 1986. Sequence landscapes. *Nucleic Acids Res.* **14:**141–158.
27. **Collado-Vides, J.** 1992. Grammatical model of the regulation of gene expression. *Proc. Natl. Acad. Sci. USA* **89:**9405–9409.
28. **Collado-Vides, J.** 1993. The elements for a classification of units of genetic information with a combinatorial component. *J. Theor. Biol.* **163:**527–548.
29. **Danchin, A.** 1979. The generation of immune specificity: a general selective model. *Mol. Immunol.* **16:**515–526.

30. **Danchin, A.** 1996. On genomes and cosmologies. *In* J. Collado-Vides, B. Magasanik, and T. Smith (ed.), *Integrative Methods in Molecular Biology.* MIT Press, Cambridge, Mass.

31. **d'Aubenton-Carafa, Y., E. Brody, and C. Thermes.** 1990. Prediction of rho-independent *Escherichia coli* transcription terminators. A statistical analysis of their RNA stem-loop structures. *J. Mol. Biol.* **216:**835–858.

32. **Dayhoff, M. O., R. M. Schwartz, and B. C. Orcutt.** 1978. *Atlas of Protein Sequence and Structure*, p. 345–352. National Biomedical Research Foundation, Washington, D.C.

33. **Dear, S., and R. Staden.** 1991. A sequence assembly and editing program for efficient management of large projects. *Nucleic Acids Res.* **19:**3907–3911.

34. **Dear, S., and R. Staden.** 1992. A standard file format for data from DNA sequencing instruments. *DNA Sequence* **3:**107–110.

35. **Delorme, M.-O., and A. Hénaut.** 1988. Merging of distance matrices and classification by dynamic clustering. *CABIOS* **4:**453–458.

36. **Delorme, M. O., A. Hénaut, and P. Vigier.** 1988. Mutations in the *NAM2* genes of *Saccharomyces cerevisiae* and *Saccharomyces douglasii* are clustered non-randomly as a result of constraints on the nucleic acid sequence and not on the protein. *Mol. Gen. Genet.* **213:**310–314.

37. **Demeler, B., and G. W. Zhou.** 1991. Neural network optimization for *E. coli* promoter prediction. *Nucleic Acids Res.* **19:**1593–1599.

38. **de Smit, M. H., and J. van Duin.** 1990. Secondary structure of the ribosome binding site determines translational efficiency: a quantative analysis. *Proc. Natl. Acad. Sci. USA* **87:**7668–7672.

39. **de Smit, M. H., and J. van Duin.** 1994. Translation initiation on structured messengers. Another role for the Shine-Dalgarno interaction. *J. Mol. Biol.* **235:**173–184.

40. **Devereux, J., P. Haeberli, and O. Smithies.** 1984. A comprehensive set of sequence analysis programs for the VAX. *Nucleic Acids Res.* **12:**387–395.

41. **Dombroski, A. J., W. A. Walter, M. T. Record, D. A. Siegele, and C. A. Gross.** 1992. Polypeptides containing highly conserved regions of transcription initiation factor σ70 exhibit specificity of binding to promoter DNA. *Cell* **70:**501–512.

42. **Doolittle, R. F.** 1994. Protein sequence comparisons: searching databases and aligning sequences. *Curr. Opin. Biotechnol.* **5:**24–28.

43. **Dreyfus, M.** 1988. What constitutes the signal for the initiation of protein synthesis on *Escherichia coli* mRNAs. *J. Mol. Biol.* **204:**79–94.

44. **Dykes, G., R. Bambara, K. Marians, and R. Wu.** 1975. On the statistical significance of primary structural features found in DNA-protein interaction sites. *Nucleic Acids Res.* **2:**327–345.

45. **Ehrenberg, M., and C. G. Kurland.** 1984. Cost of accuracy determined by a maximal growth constraint. *Q. Rev. Biophys.* **17:**45–82.

46. **Emilsson, V., and C. G. Kurland.** 1990. Growth rate dependence of transfer RNA abundance in *Escherichia coli*. *EMBO J.* **9:**4359–4366.

47. **Emilsson, V., A. K. Naslund, and C. G. Kurland.** 1993. Growth-rate dependent accumulation of twelve tRNA species in *Escherichia coli*. *J. Mol. Biol.* **230:**483–491.

48. **Etzold, T., and P. Argos.** 1993. SRS—an indexing and retrieval tool for flat file data libraries. *CABIOS* **9:**49–57.

49. **Ezhov, A. A., Y. A. Kalambet, and D. Y. Cherny.** 1989. Neuron network for the recognition of *E. coli* promoters. *Stud. Biophys.* **129:**183–192.

50. **Férat, J. L., M. Le Gouar, and F. Michel.** 1994. Multiple group II self-splicing introns in mobile DNA from *Escherichia coli*. *C. R. Acad. Sci.* (Paris) **317:**141–148.

51. **Fichant, G., and C. Burks.** 1991. Identifying potential tRNA genes in genomic DNA sequences. *J. Mol. Biol.* **220:**659–671.

52. **Fichant, G., and C. Gautier.** 1987. Statistical methods for predicting protein coding regions in nucleic acids sequences. *CABIOS* **3:**287–295.

53. **Figueroa, N., N. Wills, and L. Bossi.** 1991. Common sequence determinants of the response of a prokaryotic promoter to DNA bending and supercoiling. *EMBO J.* **10:**941–949.

54. **Gardner, M.** 1974. On the paradoxical situations that arise from nontransitive relations. *Sci. Am.* **231:**120–125.

55. **Gascuel, O.** 1985. Structural description, learning and discrimination of these descriptions. *Biochimie* **67:**499–507.

56. **Gascuel, O.** 1993. Inductive learning and biological sequence analysis. The PLAGE program. *Biochimie* **75:**363–370.

57. **Gascuel, O., and A. Danchin.** 1986. Protein export in prokaryotes and eukaryotes: indications for a difference in the mechanism of exportation. *J. Mol. Evol.* **24:**130–142.

58. **Gilson, E., W. Saurin, D. Perrin, S. Bachellier, and M. Hofnung.** 1991. Palindromic units are part of a new bacterial interspersed mosaic element (BIME). *Nucleic Acids Res.* **19:**1375–1383.

59. **Gleeson, T. J., and R. Staden.** 1991. An X windows and UNIX implementation of our sequence analysis package. *CABIOS* **7:**398.

60. **Gleizes, A., and A. Hénaut.** 1994. A global approach for contig construction. *CABIOS* **10:**401–405.

61. **Gouy, M., and C. Gautier.** 1982. Codon usage in bacteria: correlation with gene expressivity. *Nucleic Acids Res.* **10:**7055–7074.

62. **Gouy, M., C. Gautier, M. Attimonelli, C. Lanave, and G. DiPaola.** 1985. ACNUC—a portable retrieval system for nucleic acid sequence databases: logical and physical designs and usage. *CABIOS* **1:**167–172.

63. **Graniero-Porati, M. I., and A. Porati.** 1988. Informational parameters and randomness of mitochondrial DNA. *J. Mol. Evol.* **27:**109–113.

64. **Grantham, R., and C. Gautier.** 1980. Genetic distances from mRNA sequences. *Naturwissenschaften* **67:**93–94.

65. **Grantham, R., C. Gautier, M. Gouy, M. Jacobzone, and R. Mercier.** 1981. Codon catalog usage is a genome strategy modulated for gene expressivity. *Nucleic Acids Res.* **9:**r43-r74.

66. **Gribskov, M., J. Devereux, and R. R. Burgess.** 1984. The codon preference plot: graphic analysis of protein coding sequences and prediction of gene expression. *Nucleic Acids Res.* **12:**539–549.

67. **Guibas, L. J., and A. M. Odlyzko.** 1980. Long repetitive patterns in random sequences. *Wahrcheinlichkeitstheorie* **53:**241–262.

68. **Guibas, L. J., and A. M. Odlyzko.** 1981. Periods in strings. *J. Combinatorial Theory Ser. A* **30:**19–42.

69. **Hanai, R., and A. Wada.** 1989. Novel third-letter bias in *Escherichia coli* codons revealed by rigorous treatment of coding constraints. *J. Mol. Biol.* **207:**655–660.

70. **Harley, C. B., and R. P. Reynolds.** 1987. Analysis of *E. coli* promoter sequences. *Nucleic Acids Res.* **15:**2343–2361.

71. **Harper, R.** 1994. Access to DNA and protein databases on the Internet. *Curr. Opin. Biotechnol.* **5:**4–18.

72. **Hawley, D. K., and W. R. McClure.** 1983. Compilation and analysis of *Escherichia coli* promoter DNA sequences. *Nucleic Acids Res.* **11:**2237–2255.

73. **Hénaut, A., and M. O. Delorme.** 1988. Distance matrix comparison and tree construction. *Pattern Recognition Lett.* **7:**207–213.

74. **Hénaut, A., J. Limaiem, and P. Vigier.** 1985. The origins of the strategy of codon use. *Biochimie* **67:**475–483.

75. **Hénaut, A., and P. Vigier.** 1985. Etude des contraintes qui s'exercent sur la succession des bases dans un polynucléotide: I. La signification de la dégénérescence du code. *C. R. Acad. Sci.* (Paris) **301:**277–282.

76. **Henikoff, S., and J. G. Henikoff.** 1991. Automated assembly of protein blocks for database searching. *Nucleic Acids Res.* **19:**6565–6572.

77. **Hill, M. O.** 1974. Correspondence analysis: a neglected multivariate method. *Appl. Stat.* **23:**340–353.

78. **Hirst, J. D., and M. J. Sternberg.** 1991. Prediction of ATP-binding motifs: a comparison of a perceptron-type neural network and a consensus method. *Protein Eng.* **4:**615–623.

79. **Horton, P. B., and M. Kanehisa.** 1992. An assessment of neural network and statistical approaches for prediction of *E. coli* promoters sites. *Nucleic Acids Res.* **16:**4331–4338.

80. **Huang, X.** 1992. A contig assembly program based on sensitive detection of fragment overlap. *Genomics* **14:**18–25.

81. **Huang, X., and M. S. Waterman.** 1992. Dynamic programing algorithms for restriction map comparison. *CABIOS* **8:**511–520.

82. **Hüttenhofer, A., and H. F. Noller.** 1994. Footprinting mRNA-ribosome complexes with chemical probes. *EMBO J.* **13:**3892–3901.

83. **Ikemura, T.** 1981. Correlation between the abundance of *Escherichia coli* transfer RNAs and occurrence of the respective codons in its protein genes. *J. Mol. Biol.* **146:**1–21.

84. **Ikemura, T.** 1981. Correlation between the abundance of *Escherichia coli* transfer RNA and the occurrence of the respective codons in its protein gene: a proposal for a synonymous codon choice that is optimal for *E. coli* translation system. *J. Mol. Biol.* **151:**389–409.

85. **Karlin, S., B. E. Blaisdell, R. J. Sapolsky, L. Cardon, and C. Burge.** 1993. Assessments of DNA inhomogeneities in yeast chromosome III. *Nucleic Acids Res.* **21:**703–711.

86. **Karlin, S., C. Burge, and A. M. Campbell.** 1992. Statistical analyses of counts and distributions of restriction sites in DNA sequences. *Nucleic Acids Res.* **20:**1363–1370.

87. **Karlin, S., and C. Macken.** 1991. Assessment of inhomogeneities in an *E. coli* physical map. *Nucleic Acids Res.* **19:**4241–4246.

88. **Karp, P.** 1992. A knowledge base of the chemical compounds of intermediary metabolism. *CABIOS* **8:**347–357.

89. **Karp, P., and S. Paley.** 1994. Representation of metabolic knowledge: pathways, p. 203–211. *In* R. Altman, D. Brutlag, P. Karp, R. Lathrop, and D. Searls (ed.), *Second International Conference on Intelligent Systems and Molecular Biology.* AAAI Press, Washington, D.C.

90. **Karp, P., and M. Riley.** 1993. Representation of metabolic knowledge, p. 207–215. *In* L. Hunter, D. Searls, and J. Sharlik (ed.), *First International Conference on Intelligent Systems and Molecular Biology.* AAAI Press, Washington, D.C.

91. **Kleffe, J., and M. Borodovsky.** 1992. First and second moment of counts of words in random texts generated by Markov chains. CABIOS **8:**433–441.

92. **Kohara, Y., K. Akiyama, and K. Isono.** 1987. The physical map of the whole *E. coli* chromosome: application of a new strategy for rapid analysis and sorting of a large genomic library. *Cell* **50:**495–508.

93. **Koudelka, G. B., S. C. Harrison, and M. Ptashne.** 1987. Effect of non-contacted bases on the affinity of 434 operator for 434 repressor and Cro. *Nature* (London) **329:**886–888.

94. **Kuehsel, M. G., R. Strickland, and J. D. Palmer.** 1990. An ancient group I intron shared by eubacteria and chloroplasts. *Science* **250:**1570–1573.

95. **Kunisawa, T., M. Nakamura, H. Watanabe, J. Otsuka, A. Tsugita, L. S. Yeh, D. G. George, and W. C. Barker.** 1990. *Escherichia coli* K12 genomic database. *Protein Sequences Data Anal.* **3:**157–162.

96. **Lamperti, E. D., J. M. Kittelberger, T. F. Smith, and L. Villa-Komaroff.** 1992. Corruption of genomic databases with anomalous sequence. *Nucleic Acids Res.* **20:**2741–2747.

97. Landès, C., A. Hénaut, and J. L. Risler. 1992. A comparison of several similarity indices used in the classification of protein sequences: a multivariate analysis. *Nucleic Acids Res.* 20:3631–3637.

98. Larsen, N., G. J. Olsen, B. L. Maidak, M. J. McCaughey, R. Overbeek, T. J. Macke, T. L. Marsh, and C. R. Woese. 1993. The Ribosomal Database Project. *Nucleic Acids Res.* 21:3021–3023.

99. Lebart, L., A. Morineau, and K. A. Warwick. 1984. *Multivariate Descriptive Statistical Analysis.* John Wiley & Sons, New York.

100. Lebbe, J., and R. Vignes. 1993. Local predictability in biological sequences, algorithm and applications. *Biochimie* 75:371–378.

101. Lefèvre, C., and J.-E. Ikeda. 1994. A fast word search algorithm for the representation of sequence similarity in genomic DNA. *Nucleic Acids Res.* 22:404–411.

102. Letovsky, S., and M. B. Berlyn. 1994. Issues in the development of complex scientific databases. *Biotechnol. Comput.* 5:5–14.

103. Leung, M.-Y., B. E. Blaisdell, C. Burge, and S. Karlin. 1991. An efficient algorithm for identifying matches with errors in multiple long molecular sequences. *J. Mol. Biol.* 221:1367–1378.

104. Li, M., and P. M. B. Vitanyi. 1993. *An Introduction to Kolmogorov Complexity and Its Applications.* Springer-Verlag, New York.

105. Lim, D. 1991. Structure of two retrons of *Escherichia coli* and their common chromosomal insertion site. *Mol. Microbiol.* 5:1863–1872.

106. Lipman, D. J., W. J. Wilbur, T. F. Smith, and S. Waterman. 1984. On the statistical significance of nucleic acid similarities. *Nucleic Acids Res.* 12:215–226.

107. Lisacek, F., Y. Diaz, and F. Michel. 1994. Automatic identification of group I intron cores in genomic DNA sequences. *J. Mol. Biol.* 235:1206–1217.

108. Lopez, R. S., T. Kristensen, and H. Prydz. 1992. Vector sequences contaminating the sequence data bases. *Nature* (London) 355:211–211.

109. Lukashin, A. V., V. V. Anshelevich, B. R. Amirikyan, A. I. Gragerov, and M. D. Frank-Kamenetskii. 1989. Neural network models for promoter recognition. *J. Biomol. Struct. Dyn.* 6:1123–1133.

110. Maniatis, T. M., K. Ptashne, D. Backman, S. Kleid, A. Flashman, A. Jeffrey, and R. Maurer. 1975. Recognition sequences of repressor and polymerase in the operators of bacteriophage lambda. *Cell* 5:109–113.

111. McClelland, M., R. Jones, Y. Patel, and M. Nelson. 1987. Restriction endonucleases for pulsed field mapping of bacterial genomes. *Nucleic Acids Res.* 15:5985–6005.

112. McClure, W. R. 1985. Mechanism and control of transcription initiation in prokaryotes. *Annu. Rev. Biochem.* 54:171–204.

113. Médigue, C., J. P. Bouché, A. Hénaut, and A. Danchin. 1990. Mapping of sequenced genes (700 kbp) in the restriction map of the *Escherichia coli* chromosome. *Mol. Microbiol.* 4:169–187.

114. Médigue, C., T. Rouxel, P. Vigier, A. Hénaut, and A. Danchin. 1991. Evidence for horizontal gene transfer in *Escherichia coli* speciation. *J. Mol. Biol.* 222:851–856.

115. Médigue, C., A. Viari, A. Hénaut, and A. Danchin. 1991. *Escherichia coli* molecular genetic map (1500 kbp): update II. *Mol. Microbiol.* 5:2629–2640.

116. Médigue, C., A. Viari, A. Hénaut, and A. Danchin. 1993. Colibri: a functional data base for the *Escherichia coli* genome. *Microbiol. Rev.* 57:623–654.

117. Merkl, R., M. Kröger, P. Rice, and H. J. Fritz. 1992. Statistical evaluation and biological interpretation of non-random abundance in the *E. coli* K-12 genome of tetra and pentanucleotide sequences related to VSP DNA mismatch repair. *Nucleic Acids Res.* 20:1657–1662.

118. Michal, G. 1982. Biochemical pathways wall chart. Boehringer-Mannheim GmbH. Universitätsdruckerei H. Sturtz AG, Würzburg.

119. Michel, F., and E. Westhof. 1990. Modeling of the three-dimensional architecture of group I catalytic introns based on comparative sequence analysis. *J. Mol. Biol.* 216:581–606.

120. Miller, W., J. Barr, and K. E. Rudd. 1991. Improved algorithms for searching restriction maps. *CABIOS* 7:447–456.

121. Milosavljevic, A., and J. Jurka. 1993. Discovering simple DNA sequences by the algorithmic significance method. *CABIOS* 9:407–411.

122. Miyazawa, S. 1989. DNA Databank of Japan. Present status and future plans, p. 47–62. *In Computers and DNA, SFI Studies in the Sciences of Complexity.* Addison-Wesley, Reading, Mass.

123. Moszer, I., P. Glaser, and A. Danchin. 1995. SubtiList: a relational data base for the Bacillus subtilis genome. *Microbiology* 141:261–268.

124. Nakata, K., M. Kanehisa, and J. J. Maizel. 1988. Discriminant analysis of promoters regions in *Escherichia coli* sequences. *CABIOS* 4:367–371.

125. Nussinov, R. 1981. The universal dinucleotide assymmetry rules in DNA and amino acid codon choice. *Nucleic Acids Res.* 17:237–244.

126. Ochman, H., and R. K. Selander. 1984. Standard reference strains of *Escherichia coli* from natural populations. *J. Bacteriol.* 157:690–693.

127. Olsen, G. J., R. Overbeek, N. Larsen, T. L. Marsh, M. J. McCaughey, M. A. Maciukenas, W.-M. Kuan, T. J. Macke, Y. Xing, and C. R. Woese. 1992. The Ribosomal Database Project. *Nucleic Acids Res.* 20:2199–2200.

128. O'Neill, M. C. 1989. Consensus methods for finding and ranking DNA binding sites. Application to *Escherichia coli* promoters. *J. Mol. Biol.* 207:301–310.

129. O'Neill, M. C. 1989. *Escherichia coli* promoters. I. Consensus as it relates to spacing class, specificity, repeat substructure, and three-dimensional organization. *J. Biol. Chem.* 264:5522–5530.

130. O'Neill, M. C. 1991. Training back-propagation neural networks to define and detect DNA- binding sites. *Nucleic Acids Res.* 19:313–318.

131. O'Neill, M. C. 1992. *Escherichia coli* promoters: neural networks develop distinct descriptions in learning to search for promoters of different spacing classes. *Nucleic Acids Res.* 20:3471–3477.

132. O'Neill, M. C., and F. Chiafari. 1989. *Escherichia coli* promoters. II. A spacing class-dependent promoter search protocol. *J. Biol. Chem.* 264:5531–5534.

133. Pattabiraman, N., K. Namboodiri, A. Lowrey, and B. P. Gaber. 1990. NRL-3D: a sequence structure database derived from the protein data bank (PDB) and searchable within the PIR environment. *Protein Sequences Data Anal.* 3:387–405.

134. Pearson, W. R., and D. J. Lipman. 1988. Improved tools for biological sequence comparison. *Proc. Natl. Acad. Sci. USA* 85:2444–2448.

135. Peltola, H., H. Soderlund, and E. Ukkonen. 1984. SEQAID: a DNA sequence assembling program based on a mathematical model. *Nucleic Acids Res.* 12:307–321.

136. Peng, C. K., S. V. Buldyrev, A. L. Goldberger, S. Havlin, F. Sciortino, M. Simons, and H. E. Stanley. 1992. Long-range correlations in nucleotide sequences. *Nature* (London) 356:170.

137. Perrière, G. 1992. Application of a knowledge representation based upon objects to the modelling of some aspects of *Escherichia coli* gene expression. Doctoral thesis. Université Claude Bernard, Lyon 1, Lyon, France.

138. Perrière, G., and C. Gautier. 1993. ColiGene: object-centered representation for the study of E. coli gene expressivity by sequence analysis. *Biochimie* 75:415–422.

139. Petersen, G. B., P. A. Stockwell, and D. F. Hill. 1988. Messenger RNA recognition in *Escherichia coli*: a possible second site of interaction with 16S ribosomal RNA. *EMBO J.* 7:3957–3962.

140. Pevzner, P. A., M. Y. Borodovsky, and A. A. Mironov. 1989. Linguistics of nucleotides sequences. I. The significance of deviations from mean statistical characteristics and prediction of the frequencies of occurrence of words. *J. Biomol. Struct. Dyn.* 6:1013–1026.

141. Pfeiffer, F., and W. A. Gilbert. 1988. VecBase: a cloning vector sequence data base. *Protein Sequences Data Anal.* 1:269–280.

142. Phillips, G. J., J. Arnold, and R. Ivarie. 1987. The effect of codon usage on the oligonucleotide composition of the E. coli genome and identification of over- and underrepresented sequences by Markov chain analysis. *Nucleic Acids Res.* 15:2627–2638.

143. Platt, T. 1986. Transcription termination and regulation of gene expression. *Annu. Rev. Biochem.* 55::339–372.

144. Pribnow, D. 1975. Nucleotide sequence of an RNA polymerase binding site at an early T7 promoter. *Proc. Natl. Acad. Sci. USA* 72:784–788.

145. Reinhold-Hurek, B., and D. A. Shub. 1992. Self-splicing introns in tRNA genes of widely divergent bacteria. *Nature* (London) 357:173–176.

146. Rice, C. M., R. Fuchs, D. G. Higgins, P. J. Stoehr, and G. N. Cameron. 1993. The EMBL data library. *Nucleic Acids Res.* 21:2967–2971.

147. Ringquist, S., S. Shinedling, D. Barrick, L. Green, J. Binkley, G. D. Stormo, and L. Gold. 1992. Translation initiation in *Escherichia coli*: sequences within the ribosome-binding site. *Mol. Microbiol.* 6:1219–1229.

148. Rodier, F., and J. Sallantin. 1985. Localization of the initiation of translation in messenger RNAs of prokaryotes by learning techniques. *Biochimie* 67:533–539.

149. Rosenblatt, F. 1959. *Principles of Neurodynamics.* Spartan Books, New York.

150. Rouxel, T., A. Danchin, and A. Hénaut. 1993. METALGEN.DB: metabolism linked to the genome of *Escherichia coli*, a graphics-oriented database. *CABIOS* 9:315–324.

151. Rudd, K. E. 1992. Alignment of E. coli DNA sequences to a revised, integrated genomic restriction map, p. 2.3–2.43. *In J. Miller (ed.), A Short Course in Bacterial Genetics: a Laboratory Manual and Handbook for Escherichia coli and Related Bacteria.* Cold Spring Harbor Laboratory Press, Cold Spring Harbor, N.Y.

152. Rudd, K. E. 1993. Maps, genes, sequences, and computers: an *Escherichia coli* case study. *ASM News* 59:335–341.

153. Rudd, K. E., W. Miller, C. Werner, J. Ostell, C. Tolstoshev, and S. G. Satterfield. 1991. Mapping sequenced E. coli genes by computer: software, strategies and examples. *Nucleic Acids Res.* 19:637–647.

154. Rudd, N. G. E., W. Miller, J. Ostell, and D. A. Benson. 1990. Alignment of Escherichia coli K12 DNA sequences to a genomic restriction map. *Nucleic Acids Res.* 18:313–321.

155. Rumelhart, R., and M. McClelland. 1986. *Parallel Distributed Processing Exploration in the Micro-Structure of Cognition.* MIT Press, Cambridge, Mass.

156. Salamon, P., and A. K. Konopka. 1992. A maximum entropy principle for distribution of local complexity in naturally occurring nucleotide sequences. *Comput. Chem.* 16:117–124.

157. Salamon, P., J. C. Wooton, A. K. Konopka, and L. Hansen. 1993. On the robustness of maximum entropy relationships for complexity distributions of nucleotide sequences. *Comput. Chem.* 17:135–148.

158. Sallantin, J., J. Haiech, and F. Rodier. 1985. Search for promoter sites of prokaryotic DNA using learning techniques. *Biochimie* 67:549–553.

159. Schneider, T. D., G. D. Stormo, L. Gold, and A. Ehrenfeucht. 1986. Information content of binding sites in nucleotide sequences. *J. Mol. Biol.* 188:415–431.

160. Schwartz, R. M., and M. O. Dayhoff. 1978. *Atlas of Protein Sequence and Structure.* National Biomedical Research Foundation, Washington, D.C.

161. Searls, D. B. 1988. Representing genetic information with formal grammars, p. 386–391. *In 7th National Conference on Artificial Intelligence,* vol. 1. Morgan Kauffman Publishers, San Mateo, Calif.

162. Searls, D. B. 1989. Investigating the linguistics of DNA with definite clause grammars, p. 189–208. *In E. Lusk and R. Overbeek (ed.), Logic Programming: Proceedings of the North American Conference,* vol. 1. MIT Press, Cambridge, Mass.

163. Shannon, C. E., and W. Weaver. 1949. *The Mathematical Theory of Communication.* University of Illinois Press, Urbana.

164. Sharp, P. M., and W. H. Li. 1986. Codon usage in regulatory genes in *Escherichia coli* does not reflect selection for `rare' codons. *Nucleic Acids Res.* 14:7737–7749.

165. Sharp, P. M., and W. H. Li. 1987. The codon adaptation index—a measure of directional synonymous codon usage bias, and its potential applications. *Nucleic Acids Res.* **15**:1281–1925.

166. Shin, D.-G., C. Lee, J. Zhang, K. E. Rudd, and C. M. Berg. 1992. Redesigning, implementing and integrating *Escherichia coli* genome software tools with an object-oriented database system. *CABIOS* **8**:227–238.

167. Shine, J., and L. Dalgarno. 1974. The 3′-terminal sequence of *Escherichia coli* 16 S ribosomal RNA: complementarity to nonsense triplets and ribosome binding sites. *Proc. Natl. Acad. Sci. USA* **71**:1342–1346.

168. Shub, D. A., J. M. Gott, M. Q. Xu, B. F. Lang, F. Michel, J. Tomaschevski, J. Pedersen-Lane, and M. Belfort. 1988. Structural conservation among three homologous introns of phage T4 and the group I introns of eukaryotes. *Proc. Natl. Acad. Sci. USA* **85**:1151–1155.

169. Siebenlist, U., R. B. Simpson, and W. Gilbert. 1980. *E. coli* RNA polymerase interacts homologously with two different promoters. *Cell* **20**:269–281.

170. Slonimski, P. P., and S. Brouillet. 1993. A data-base of chromosome III of *Saccharomyces cerevisiae. Yeast* **9**:941–1029.

171. Smith, T. F., and M. S. Waterman. 1981. Comparison of bio-sequences. *Adv. Appl. Math.* **2**:482–489.

172. Soldano, H., and J. L. Moisy. 1985. Statistico-syntactic learning techniques. *Biochimie* **67**:493–498.

173. Sonnhammer, E. L. L., and D. Kahn. 1994. Modular arrangement of proteins as inferred from analysis of homology. *Protein Sci.* **3**:482–492.

174. Sprengart, M. L., H. P. Fatscher, and E. Fuchs. 1990. The initiation of translation in *E. coli:* apparent base pairing between the 16S rRNA and downstream sequences of the mRNA. *Nucleic Acids Res.* **18**:1719–1723.

175. Staden, R. 1982. Automation of the computer handling of gel reading data produced by the shotgun method of DNA sequencing. *Nucleic Acids Res.* **10**:4731–4751.

176. Staden, R. 1984. Computer methods to locate signals in nucleic acid sequences. *Nucleic Acids Res.* **12**:505–519.

177. Staden, R. 1984. A computer program to enter DNA gel reading data into a computer. *Nucleic Acids Res.* **12**:499–503.

178. Staden, R. 1987. Computer handling of DNA sequencing projects, p. 173–217. *In* M. J. Bishop and C. J. Rawling (ed.), *Nucleic Acid and Protein Sequence Analysis: a Practical Approach.* IRL Press, Oxford.

179. Stanssens, P., E. Remaut, and W. Fiers. 1986. Inefficient translation initiation cause premature transcription termination in the *lacZ* gene. *Cell* **44**:711–718.

180. Stormo, G. D. 1990. Consensus patterns in DNA. *Methods Enzymol.* **183**:211–221.

181. Stormo, G. D. 1991. Probing information content of DNA-binding sites. *Methods Enzymol.* **208**:458–468.

182. Stormo, G. D., T. D. Schneider, and L. Gold. 1982. Characterization of translational initiation sites in *E. coli. Nucleic Acids Res.* **10**:2971–2996.

183. Stormo, G. D., T. D. Schneider, L. Gold, and A. Ehrenfeucht. 1982. Use of the `Perceptron' algorithm to distinguish translational initiation sites in *E. coli. Nucleic Acids Res.* **10**:2997–3011.

184. Thanaraj, T. A., and M. W. Pandit. 1989. An additional ribosome-binding site on mRNA of highly expressed genes and a bifunctionnal site on the colicin fragment of 16S rRNA from *Escherichia coli:* important determinants of the efficiency of translation initiation. *Nucleic Acids Res.* **17**:2973–2985.

185. Trifonov, E. F., and V. Brendel. 1985. *GENOMIC, a Dictionary of Genetic Codes.* Balaban Publisher, Rehovot, Irsael.

186. Vigier, P., and A. Hénaut. 1986. Etude des contraintes qui s'exercent sur la succession des bases dans un polynucléotide. II. La distribution des tétranucléotides complémentaires dans les gènes d'*Escherichia coli* et des bactériophages lambda et T4. *C. R. Acad. Sci.* (Paris) **302**:1–6.

187. Wahl, R., and M. Kröger. 1995. ECDC—a totally integrated and interactively usable genetic map of *Escherichia coli* K12. *Microbiol. Res.* **150**:7–61.

188. Wahl, R., P. Rice, C. M. Rice, and M. Kröger. 1994. ECD—a totally integrated database of *Escherichia coli. Nucleic Acids Res.* **22**:3450–3455.

189. Watanabe, S. 1968. *Knowing and Guessing.* John Wiley & Sons, New York.

190. Wells, R. D., and R. R. Sinden. 1993. Defined ordered sequence DNA, DNA structure, and DNA-directed mutation, p. 107–138. *In* K. E. Davies and S. T. Warren (ed.), *Genome Analysis,* vol. 7. Cold Spring Harbor Press, Cold Spring Harbor, N.Y.

191. Wooton, J. C., and S. Federhen. 1993. Statistics of local complexity in amino acid sequences and sequence database. *Comput. Chem.* **17**:149–163.

192. Xu, M. Q., S. D. Kathe, H. Goodrich-Blair, S. A. Nierzwicki-Bauer, and D. A. Shub. 1990. Bacterial origin of a chloroplast intron: conserved self-splicing group I intron in cyanobacteria. *Science* **250**:1566–1570.

193. Yager, T. D., and P. H. von Hippel. 1987. Transcript elongation and termination in *Escherichia coli,* p. 1241–1275. *In* F. C. Neidhardt, J. L. Ingraham, K. B. Low, B. Magasanik, M. Schaechter, and H. E. Umbarger (ed.), *Escherichia coli and Salmonella typhimurium: Cellular and Molecular Biology.* American Society for Microbiology, Washington, D.C.

194. Yee, C. N., and L. Allison. 1993. Reconstruction of strings past. *CABIOS* **9**:1–7.

195. Yockey, H. P. 1992. *Information Theory and Molecular Biology.* Cambridge University Press, Cambridge.

Gene-Protein Database of
Escherichia coli K-12, Edition 6

RUTH A. VanBOGELEN, KELLY Z. ABSHIRE, ALEXANDER PERTSEMLIDIS, ROBERT L. CLARK, AND FREDERICK C. NEIDHARDT

115

INTRODUCTION

The gene-protein database is unique among other databases being constructed for *Escherichia coli* because it is configured on a global approach that allows the cell's total complement of polypeptides to be examined at one time (27). Two-dimensional (2-D) polyacrylamide gel electrophoresis (PAGE) permits this global approach by separating complex mixtures of polypeptides into individual polypeptide species (called spots on the 2-D gel) by two independent separation steps, isoelectric focusing and sodium dodecyl sulfate (SDS)-PAGE (38). However, the reason for creating this database is not to build the "master" 2-D gel for *E. coli*, because the small number of investigators routinely using 2-D gels would not justify this enormous venture. The purpose is to provide other investigators with physiological and regulatory data on the entire set of *E. coli* proteins.

The ultimate goal of this database is to catalog when, why, and to what level each protein-encoding gene is expressed. Two projects that tackle this problem are under way. The first project, called the Genome Expression Map, is designed to link each of the protein-encoding genes to a spot on the 2-D gel. The second project, called the Response/Regulation Map, is focused on cataloging the conditions under which each of these genes is expressed and on determining what molecules regulate their expression. This database includes two types of information. First, for identified proteins, we provide the following information: gene name, protein name, EC number, SWISS-PROT accession number, GenBank code, metabolic class, position and orientation of the gene on the chromosome, molecular weight (MW), and pI (calculated from DNA sequence information). This provides sufficient information to allow a user to do a literature search or to access more information in other databases. Second, for identified as well as unidentified proteins,

information obtained from 2-D gels is included: MW and pI of the protein (estimated from its migration on the gels), abundance of individual proteins grown under different conditions, and memberships of proteins in particular regulons and/or stimulons. Some of the other databases (e.g., SWISS-PROT [5]) also provide linkage to this database by including the 2-D spot name (called an alpha-numeric, or A-N, name) in their information list. The entire gene-protein database (including the 2-D gel images) is available electronically and can be obtained through anonymous ftp at the database repository at the National Center for Biotechnology Information (see the section on Information Exchange).

HISTORY OF THE DATABASE

The gene-protein database was begun immediately after the introduction of the 2-D gel method (38). The first set of data (which is still included in the database) was a catalog of 140 individual proteins (21 were identified) that reported variations in the levels of each protein in cultures grown under different growth conditions (43). The first important step in establishing the structure of the database, the alpha-numeric naming system used to uniquely identify each 2-D gel spot, was described in that catalog.

In 1980, the first set of reference 2-D gels was published along with the identities of 81 more proteins (6). Five years into the development of the database, it became apparent that in order to track each 2-D gel spot through numerous gels, a standard cumulative map of each type of 2-D gel had to be established. Each reference 2-D gel was overlaid with a grid to give each spot a unique x and y coordinate. The alpha-numeric naming system was maintained to match proteins among the reference gels.

In 1983, the information in the database was linked to the chromosome in the first gene-protein index (33). In that update and review, the identities of 157 proteins were listed, and in addition, many unidentified proteins were mapped to a small region on the chromosome (33). Throughout the 1980s, many published reports on the responses of proteins observed on 2-D gels used (and added to) the protein identifications found in the index. Most of these reports gave physiological and regulatory information about protein spots on 2-D gels (both those identified and those not identified).

In 1990, all of this information from the previous gene-protein indexes and from many separate reports was gathered together, put into an electronic database, and published as the gene-protein database (58). A year later, edition 4 was published. That edition introduced a new standard 2-D gel that was generated by using a standardized 2-D gel method (61). The switch to this new standardized method was important: it allowed other investigators to reproduce the protein pattern so that they could access and contribute to the information in the database independently of the database laboratory. Edition 5 (63) included the first set of identifications made using the T7 expression system on Kohara clones and also announced that an electronic version of the database had been released to the database repository at the National Center for Biotechnology Information so that it would be available in a more usable form and could be updated more frequently.

This sixth edition of the database introduces several changes necessary to accommodate the input of data from many sources. A new naming system was started, not to replace the alphanumeric naming system but to prevent redundancy in this system. The alpha-numeric names will now be reserved for proteins that have been identified as the product of a particular gene. The new naming system is being used for the Response/Regulation Map and will be used for the Genome Expression Map to name proteins that have been observed but await identification. In the Response/Regulation Map project, as many proteins as possible are matched to proteins already in the database and are assigned that alpha-numeric name, but others are matched only within the Response/Regulation Map project. These will be given a Response/Regulation Map name, which is an R followed by a four-digit number (e.g., R1698). In the Genome Expression Map project, the proteins matched to a single open reading frame (ORF) will be assigned alpha-numeric names and will be added to Table 2 (see p. 2094) under the appropriate gene name and also in the SWISS-PROT (5) and E. coli (24) databases (as a reference between these databases). The proteins that cannot be matched to a single ORF will be given only a Genome Expression Map name, which is an X followed by a four-digit number (e.g., X2404). These proteins appear in Table 2 with reference only to their chromosomal map positions until further analysis allows a match to an ORF. Table 1 (see p. 2076) will continue to serve as a list of all proteins found on 2-D gels, which are listed in order of alphanumeric name, Genome Expression Map name, and Response/Regulation Map name.

DEVELOPMENT OF THE DATABASE

It is predicted that within the next few years, the entire DNA sequence of E. coli will be determined. The next steps in the analysis of E. coli will be (i) to confirm that the proposed ORFs encode proteins, (ii) to determine how these genes are regulated, and (iii) to elucidate the function of these proteins. The plans for this database are designed to assist in this analysis. Two complementary projects to develop this database are under way. Each of these projects will provide a separate data set, and a third data set will be provided by the DNA sequencing projects. Eventually, the three data sets will converge, because each contains information on the same set of 3,500 to 4,000 E. coli proteins.

The initial concept of the Genome Expression Map was published in 1980 (34). All of those early protein identifications were made one at time, primarily using purified proteins as markers to identify the spots. The supply of purified E. coli proteins was quickly exhausted, and so other methods to identify proteins were tried (44). The Genome Expression Map was intended to provide a method for identifying all of the proteins on the E. coli chromosome without relying on biochemists to purify the proteins or geneticists to construct mutants in each protein-encoding gene. Expressing genes carried on plasmids seemed like the ideal approach. At that time, a recombinant plasmid library, constructed by Clarke and Carbon (11), was available. One method for expressing proteins from recombinant plasmids had been described (49), and two more expression methods were developed (34, 47). Although many identifications have been made by using these three expression methods, all of the methods failed to consistently express all of the proteins encoded by the plasmids. The primary reason for the failure was that each of these methods relied on the E. coli transcription system, and gene expression was thus controlled by the cell's own regulatory mechanisms, which do not allow equal transcription of all genes.

The new approach currently used for the Genome Expression Map project focuses on simultaneously identifying many gene products. This will be accomplished by using the sets of ordered clones produced and sequenced by other laboratories (13, 23), by expressing the genes on these clones with a selective expression system, and by matching the proteins produced by each clone to ORFs found on the clone. By using clones that have been mapped to a position on the chromosome and completely sequenced, the cloning is easier (all restriction sites are known), and a list of potential protein products is already generated. The expression system uses phage transcription systems (56), which offer two advantages over the E. coli transcription system. First, because phage RNA polymerases appear to ignore the transcription signals (encoded within the DNA sequence) used by the E. coli RNA polymerase to start and stop transcription, every ORF on a plasmid should be transcribed within a single transcription unit. Second, by taking advantage of the sensitivity of the E. coli RNA polymerase and the resistance of the phage RNA polymerases to the antibiotic rifampin, the plasmid-encoded genes can be expressed exclusively. The sizes of the ordered chromosomal fragments allow 10 to 20 proteins to be identified simultaneously (based on the assumption that the average gene is 1 kb long), and yet this number of proteins is still small enough to allow unambiguous matching of most proteins to ORFs (in most cases) because of the variation in charges and masses of proteins (migration on the 2-D gel) and also because all 20 genes will rarely be expressed from a single strand. The experimental methods used for this project have been described in detail elsewhere (50) and are presented here only briefly.

To express the genes from ordered sets of clones, the E. coli DNA from these clones is moved into a special plasmid vector, and then the recombinant plasmid is transformed into a special E. coli strain. The special vector possesses several important features, including (i) a low-copy replicon to minimize the effects of certain genes that

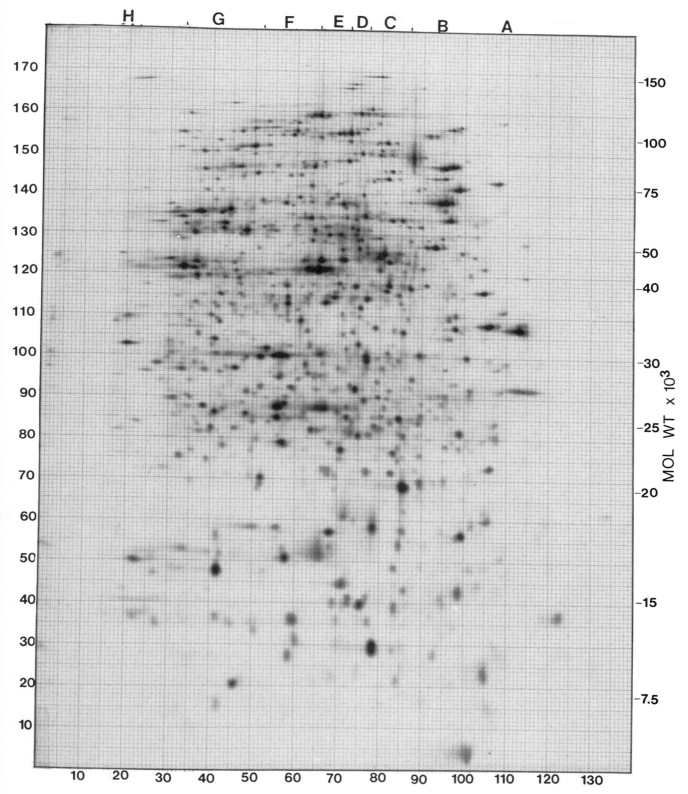

FIGURE 1   2-D polyacrylamide gel of extracts of *E. coli* K-12 W3110 grown aerobically in glucose minimal MOPS (29) medium plus thiamine at 37°C and labeled with [^{35}S]methionine (63). The 2-D gels were run on the Investigator 2-D gel system (Millipore), using 4-8 ampholines in the first dimension (as described in the Investigator manual) and 11.5% acrylamide (as described in the Investigator manual except that pH 8.8 Trizma from Sigma Chemical Co. was used as the 1.5 M Tris for making up the slab gel). A grid overlay provides coordinates for individual spots. Letters from A to H across the top follow the alphanumeric nomenclature system described in reference 43. MW estimates given on the right were made by using protein spots with known MWs (deduced from the sequence).

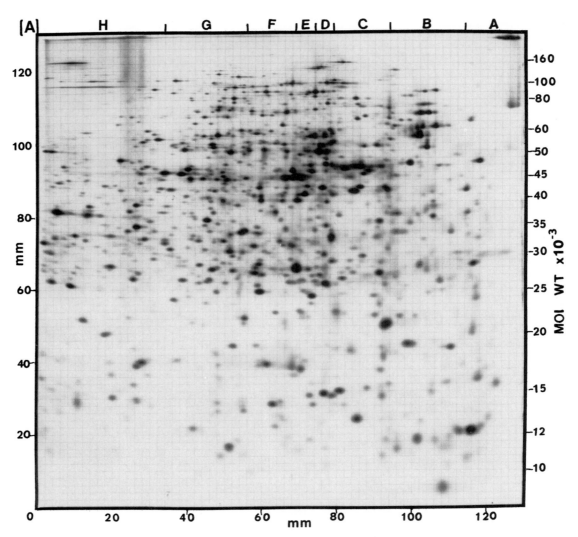

FIGURE 2 2-D polyacrylamide gels of extracts of *E. coli* K-12 W3110 grown aerobically in glucose minimal MOPS (29) medium at 37°C and labeled with $^{35}SO_4$ (63). A grid overlay provides coordinates for individual spots. Letters from A to H define zones of identified proteins versus their migration distances. (A) First dimension, isoelectric focusing to equilibrium (8,200 V·h), 1.6% pH 5 to 7 and 0.4% pH 3.5 to 10 carrier ampholyte mixture; second dimension, 11.5% acrylamide. (B) First dimension, nonequilibrium (NE) pH gradient electrophoresis (39) (1,250 V·h), 2% pH 3.5 to 10 carrier ampholyte mixture; second dimension, 11.5% acrylamide.

are lethal to *E. coli* when present in high copy, (ii) the *lacZ* gene within the multiple cloning site to allow simple screening for plasmids containing inserts, and (iii) two different phage promoters flanking the multiple cloning site (oriented opposite to each other) to provide a means of independently expressing the protein-encoding genes on each DNA strand. Each of the special strains used to express the genes on these plasmids carries one of the phage RNA polymerase genes under the control of an inducible promoter to prevent the expression of the plasmid-encoded genes until the inducer is added, again minimizing the effects of lethal genes. The strains are also *recA* mutants; thus, recombination between the *E. coli* DNA on the plasmid and the chromosome is prevented.

To tag the proteins produced from the plasmid-encoded genes, a mixture of ^{3}H-amino acids is added to a culture in which the phage RNA polymerase has been induced and the *E. coli* RNA

polymerase has been inhibited by rifampin. These ^{3}H-labeled proteins are separated on 2-D gels. Because there is virtually no contamination from chromosomally encoded proteins to serve as landmark spots on the 2-D gels, the ^{3}H-labeled extracts are also comigrated on a 2-D gel with a whole-cell extract made from a culture (strain W3110) labeled with [^{14}C]glucose in order to map each plasmid-encoded protein to a precise location on the reference 2-D images (50).

To match the ORFs found in the DNA sequence to spots on the 2-D gels, standard curves (shown in Fig. 5) were prepared by using the large set of proteins that have been identified on 2-D gels and whose genes have been sequenced. From the sequence of the gene, the amino acid composition is deduced, and from the amino acid composition, the pI and MW of the protein are calculated. Plots of pI versus migration in the first dimension and MW versus migra-

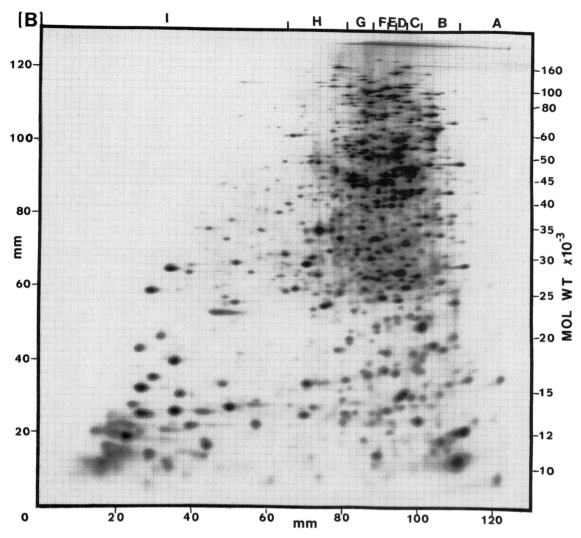

FIGURE 2   *(Continued)*

tion yield the equations that give an estimate of where the products of other genes should migrate. By themselves, these estimates are not sufficient to make a spot identification. However, when the number of candidate spots is reduced to 10 or so through the use of the selective expression system described, matches between ORFs and their protein products can be found. In some cases, no unambiguous assignment of a protein to an ORF can be made. In these cases, the protein will be assigned a Genome Expression Map name until further analysis clarifies which ORF matches the protein. This system of many-at-a-time protein identifications should rapidly increase the information compiled in the Genome Expression Map section of the database.

The Genome Expression Map project specifically addresses the question of whether each ORF identified within the DNA sequence actually expresses a protein. By themselves, the results of such work would be a significant contribution to the study of *E. coli*. However, because the expression of the ORFs will be determined through analysis on 2-D gels, this project also provides the Response/Regulation Map project with the necessary linkage to the chromosome.

The Response/Regulation Map really began with the first publication describing the 2-D gel method (38). O'Farrell used protein extracts from *E. coli* and revealed the proteins synthesized under given growth conditions. Many other global studies that used 2-D gels have since been published. Two factors have limited the growth of this part of the database. First, lack of a standardized 2-D gel method (prior to 1991) hindered other investigators from contributing global studies to the database. Only one independent investigator has ever contributed to this part of the database (15). Second, although 2-D gels can resolve about 1,200 protein spots, the methods used to quantify the spots restrict global quantitation to the most abundant 200 to 600 proteins. Manual methods of quantitation are restricted by the low specific activities of radiolabeled amino acids and the time required to punch out individual spots for counting in scintillation counters. Computer-aided image analysis systems were introduced in the 1980s, but quantitation by this method is limited by the slow processing time of the computers, the immaturity of the software, and the narrow linear optical density range of the X-ray film used to capture the gel image.

With the development of faster computers, better image analysis software, and a new method to measure radioactivity in the spots on the gel (phosphorimagers [41]), it is now possible to quickly quantify 1,000 to 1,200 spots per gel and to match the spots among

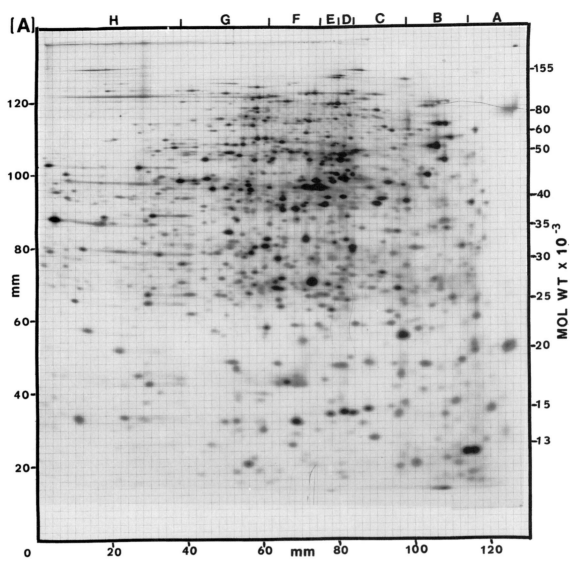

FIGURE 3 2-D polyacrylamide gels of extracts of *E. coli* B/r NC3 grown aerobically in glucose minimal MOPS (29) medium at 37°C and labeled with $^{35}SO_4$ (63). A grid overlay provides coordinates for individual spots. Letters from A to H define zones of identified proteins versus their migration distances. (A) Gel conditions are the same as in Fig. 2A; (B) gel conditions are the same as in Fig. 2B.

multiple images. When these methods of detection and analysis are used, proteins with steady-state levels of more than 50 molecules per cell or those with synthesis rates accounting for 0.04% of the total protein synthesized during a pulse-label can be detected and included in the analysis of each 2-D gel. The results from the two different analysis methods are expressed differently, as discussed in the footnotes to Table 3 (Table 3 on p. 2101; footnotes on p. 2111). The results of four comprehensive analyses were added to this edition of the database, and eventually, all of the partial analyses will be redone and entered into the database along with data from additional experiments. For each of these, only the data are given; the interpretations and conclusions made from these experiments are given elsewhere (see the footnotes to Tables 1 through 4).

The Response/Regulation Map project is cataloging, through 2-D gel analysis, when (the response) and how (the regulation) each individual protein is expressed. Although this catalog will seldom define the exact function for any individual protein, it is expected to provide many of the clues that will direct the study of each protein's function and to provide the physiological data needed to help define regulatory elements contained in the DNA sequence by revealing which proteins belong to a particular regulon. Because the information in the Response/Regulation Map has accumulated over many years and from quantitative and qualitative analyses of the 2D gels, the data are presented in two tables (Tables 3 [p. 2101] and 4 [p. 2112]). Quantitative data for all proteins measured in each experiment are listed in Table 3 and are expressed as a ratio for each test condition. The level of proteins (during the labeling period) is

TABLE 1 Continued

Alpha-numeric (A-N) (a)	R R M Name (b)	Data in other Tables			W3110 coordinates on reference gels								NC3 coordinates				MW calc. from seq. (c)	pI calc. from seq. (d)	MW est. from gels (e)	pI est. from gels (f)	Fig. used for est.
					Large gels				pH 5-7		NE		pH 5-7		NE						
		2	3	4	Fig. 1		Fig. 4		Fig. 2A		Fig. 2B		Fig. 3A		Fig. 3B						
					X	Y	X	Y	X	Y	X	Y	X	Y	X	Y					
B043.0		2							111	91							42,947	4.91	44,256	4.90	2A
B043.1		2							113	91							42,947	4.91	44,256	4.86	2A
B044.8				4					99	91									44,256	5.13	2A
B046.5	R2831	2	3	4	95.50	125.50	600	326	102	93									46,798	5.09	1
B046.7	R1861	2	3	4	93.00	127.00	594	315	99	94	103	94	103	99	110	91	50,303	5.10	47,861	5.14	1
B047.0		2			91.50	128.50			97	95			102	101			52,374	5.16	48,961	5.17	1
B050.3	R1859	2	3	4	97.00	133.50	607	289	104	98	104	98	108	104	110	95	47,997	4.88	52,976	5.06	1
B052.0		2							104	101									55,226	5.04	2A
B056.0				4					104	102									56,705	5.04	2A
B056.5	R0565	2	3	4	95.50	138.00	602	244	102	102	103	101	106	107	112	99	57,117	5.04	57,243	5.09	1
B058.3	R1790	2	3	4	98.50	141.50	613	234	106	104	105	104	110	109			63,532	4.97	61,193	5.03	1
B060.1		2							98	102							59,428	5.07	56,705	5.15	2A
B060.2		2							98	103							59,428	5.07	58,292	5.15	2A
B060.3		2							100	103							59,428	5.07	58,292	5.12	2A
B060.4		2							100	104							59,428	5.07	60,003	5.12	2A
B060.5		2							101	104							59,428	5.07	60,003	5.10	2A
B060.6		2							101	105							59,428	5.07	61,851	5.10	2A
B060.7		2							103	104							59,428	5.07	60,003	5.06	2A
B060.8		2							103	105							59,428	5.07	61,851	5.06	2A
B061.0	R0610	2	3	4	108.00	143.00	647	234	116	106	109	105	116	111			54,833	4.73	63,104	4.84	1
B065.0	R0650	2	3	4	95.50	146.50	601	210	102	108	103	107	106	113	111	104	61,137	5.06	68,203	5.09	1
B066.0	R0660	2	3	4	97.00	146.50	610	205	104	108	104	107	108	113	114	104	68,963	5.07	68,203	5.06	1
B068.0		2			93.00	147.00											73,864	5.37 *	69,016	5.14	1
B068.1		2			93.00	147.50													69,852	5.14	1
B076.0				4					103	110									73,767	5.06	2A
B082.0		2							106	113							70,220	4.90	83,831	5.00	2A
B082.5			3	4	91.50	154.50			98	115			103	118					84,534	5.17	1
B083.0		2	3	4	92.50	155.00			100	113			104	118			87,368	5.17	85,837	5.15	1
B084.0				4					109	113									83,831	4.94	2A
B096.0		2			92.50	160.00													101,300	5.15	1
B105.0				4					105	117									102,176	5.02	2A
B140.0				4					102	121									128,231	5.08	2A
C011.5				4					93	18			96	22					10,065	5.25	2A
C012.8				4					84	24									12,548	5.43	2A
C013.2				4					93	25									13,079	5.25	2A
C013.4	R2296	2	3	4	83.00	28.50	556	748	89	24							15,842	5.21	12,744	5.35	1
C013.5	R0135	2	3	4	78.50	29.50	538	712	85	24	100	24	89	28			15,842	5.21	13,048	5.44	1
C013.6		2							93	27							17,043	5.29	14,163	5.25	2A
C014.3	R2263	2	3		83.50	39.00	557	708	92	30			95	34			18,481	4.90	15,784	5.34	1
C014.7	R2265	2	3	4	77.00	35.50	535	723	80	27			86	33			16,091	5.37	14,832	5.47	1
C014.8	R2307	2	3	4	72.50	41.00	520	703	77	31			81	34	98	32	15,695	5.08	16,286	5.56	1
C015.3	R2241	2	3	4	77.00	41.50	535	700	81	32			87	35	99	32	15,695	5.08	16,407	5.47	1
C015.4		2	3	4	75.50	40.00	530	712	79	30			83	34			10,385	5.33	16,039	5.50	1
C017.2		2			83.50	48.50			92	38			95	41			16,153	5.33	17,886	5.34	1
C017.7	R2794		3	4	84.50	54.00	562	654	93	41									18,828	5.32	1
C018.0	R2206		3	4	78.50	58.00	538	633	84	43			88	48					19,446	5.44	1
C018.1				4					84	46									21,654	5.43	2A
C019.5		2							80	57							19,466	5.40	24,334	5.51	2A
C022.3				4					93	54									23,536	5.25	2A
C022.7	R2203	2	3	4	82.50	71.50	552	577	90	53	100	52	94	59	102	54			21,732	5.36	1
C023.0	R2202		3		77.00	72.00	533	577	80	54			84	58					21,836	5.47	1
C023.4		2			77.00	77.50													23,129	5.47	1
C024.0		2			85.00	80.00													23,816	5.31	1
C024.5		2			86.00	82.00													24,414	5.29	1
C025.2	R2166		3	4	77.50	81.00	537	539	80	61									24,109	5.46	1
C025.3		2							84	65							25,527	5.35	27,142	5.43	2A
C025.4		2			82.50	84.00			88	63							22,999	5.09	25,057	5.36	1
C025.5				4					93	61									25,594	5.25	2A
C025.8		2			67.00	86.00													25,745	5.67	1
C026.5		2							80	63			86	67					26,329	5.51	2A
C026.8	R2165	2	3	4	80.00	82.50	547	533	82	63			87	70					24,571	5.41	1
C027.1	R2160		3	4	80.00	85.50	544	521	83	65									25,569	5.41	1
C028.5		2	3	4	82.00	92.00			87	69			91	73					28,079	5.37	1
C029.1	R2090		3		79.50	90.00	544	502	83	68									27,257	5.42	1
C029.8		2	3	4	81.00	97.50			84	71			86	76			35,710	6.04 *	30,551	5.39	1
C030.7	R2481	2	3	4	77.00	99.00	536	452	78	73	97	72	83	79	101	73	30,417	5.38	31,276	5.47	1
C031.2				4					82	72									30,619	5.47	2A

(Table continues)

TABLE 1 *Continued*

Alpha-numeric (A-N) (a)	R R M Name (b)	Data in other Tables		W3110 coordinates on reference gels								NC3 coordinates				MW calc. from seq. (c)	pI calc. from seq. (d)	MW est. from gels (e)	pI est. from gels (f)	Fig. used for est.
				Large gels				pH 5-7		NE		pH 5-7		NE						
				Fig. 1		Fig. 4		Fig. 2A		Fig. 2B		Fig. 3A		Fig. 3B						
				X	Y	X	Y	X	Y	X	Y	X	Y	X	Y					
C031.3		2						82	86							31,267	5.36	40,118	5.47	2A
C031.5		2		85.00	102.50													33,041	5.31	1
C031.6	R2480	2	3	76.50	100.00	537	447	78	74	97	73	83	80	101	74	30,417	5.38	31,770	5.48	1
C031.9		2		84.00	104.00											33,701	5.18	33,827	5.33	1
C032.4		2		84.50	102.00			86	74			91	81					32,783	5.32	1
C032.7		2		79.50	99.50			82	74			87	80			27,474	5.35	31,522	5.42	1
C033.0		2						89	75									32,397	5.33	2A
C033.1		2						92	75									32,397	5.27	2A
C034.3		3	4	82.00	101.00			87	75			92	81					32,272	5.37	1
C035.6	R2006	3	4	85.50	106.00	566	423	91	78			95	84					34,901	5.30	1
C036.3	R2005	2	3 4	79.00	107.00	543	422	81	80									35,448	5.43	1
C036.7	R2767	3	4	77.00	108.00	536	419											36,002	5.47	1
C037.0			4					84	64									26,726	5.43	2A
C037.5	R2004	3		83.50	106.50	556	419	88	78									35,173	5.34	1
C037.6		2		79.00	116.50													40,961	5.43	1
C038.3			4					81	83									37,841	5.49	2A
C039.0		2						80	83							40,825	5.66 *	37,841	5.51	2A
C039.3	R2457	2	3 4	81.50	116.50	552	377	84	87	98	85	90	91	103	84	37,830	5.28	40,961	5.38	1
C040.3		3	4	82.00	117.50			86	88	98	87	91	93	103	85			41,574	5.37	1
C040.5			4					81	91									44,256	5.49	2A
C040.6			4					85	91									44,256	5.41	2A
C041.0	R1923	2	3 4	80.00	121.50	547	352											44,107	5.41	2A
C042.0			4					92	88									41,715	5.27	2A
C042.5			4									83	97					43690	5.26	3A
C042.6	R1922	2	3 4			558	352	88	91			91	97			51,512	5.30	43,690	5.33	4
C042.9			4					95	88									41,715	5.21	2A
C043.1		3	4	80.00	117.50			82	88			88	94					41,574	5.41	1
C043.5	R1933	3	4			568	347	89	93									44,124	5.28	4
C043.8	R1916	2	3 4	84.50	125.50	561	323	86	93			85	98			45,735	5.35	46,798	5.32	2A
C044.0		3						83	91			88	96					44,256	5.45	2A
C044.2		3						93	92			97	97	107	89			45,152	5.25	2A
C044.6		2	3 4	84.00	124.00			88	92			93	97	105	89	42,068	6.05	45,767	5.33	1
C047.0		2		87.00	128.50													48,961	5.27	1
C048.7		3	4					86	95			92	101					48,037	5.39	2A
C050.5			4					80	98									51,329	5.51	2A
C051.0			4	85.00	133.50			91	99									52,976	5.31	1
C053.0			4					91	100									53,845	5.29	2A
C054.0			4					91	102									56,705	5.29	2A
C054.1			4					87	102									56,705	5.37	2A
C055.0		3	4									95	107					50,241	5.34	3A
C056.0		3	4									98	108					51,099	5.28	3A
C058.3		2	4	87.50	136.00			94	101	100	100					52,541	5.22	55,251	5.26	1
C058.4		2		78.00	139.50			80	103			86	109					58,856	5.45	1
C058.5		3										98	109					51,972	5.28	3A
C059.3		2						83	104							59,286	5.30	60,003	5.45	2A
C059.4		2						88	104							59,286	5.30	60,003	5.35	2A
C060.5		2		78.50	141.50			80	105			86	112					61,193	5.44	1
C060.7		3	4									94	112	106	102			54,682	5.36	3A
C062.5	R2285	2	3 4	82.50	145.50	557	230	87	107	98	106	92	113	105	103	71,399	5.31	66,645	5.36	1
C062.7	R2290	2	3 4	87.50	147.00	577	223	94	107	100	106	99	112	108	103	66,088	5.27	69,016	5.26	1
C065.0		2		87.00	145.00													65,898	5.27	1
C067.0			4					85	107									44256	5.2	2A
C070.0	R2291	2	3 4	87.50	149.00	576	209	94	109	100	109	98	114	108	106	66,088	5.27	72,508	5.26	1
C074.0	R2671	3	4			560	207	88	110									80,547	5.32	4
C075.0	R2672	3	4	79.50	149.50	552	206	82	110			87	115					73,446	5.42	1
C078.0	R1770	2	3 4	81.50	152.50	556	195	85	112	97	111	90	117			77,158	5.34	79,697	5.38	1
C080.0		2	4					80	112			86	117			76,198	5.33	80,175	5.51	2A
C137.0	R1733	2	3 4	90.00	166.00	585	174	95	120	103	119	97	125			135,899	5.18	127,208	5.20	1
C142.0		2						92	120							141,940	5.30	120,850	5.27	2A
C158.0			4					89	122									136,269	5.33	2A
D012.4		2						76	28							12,222	5.57	14,705	5.59	2A
D014.3		2						78	28	99	25	82	30	105	30			14,705	5.55	2A
D014.7	R2308	2	3 4	69.00	40.00	505	707	73	31	95	30	78	34	97	31	15,695	5.08	16,039	5.63	1
D023.3		2						75	58							23,427	5.54	24,625	5.61	2A
D025.5	R2357	3	4	75.00	81.00	527	541	77	61			82	65					24,109	5.51	1
D026.3		2	4					76	63							25,579	5.72	26,329	5.59	2A
D027.1	R2163	2	3	72.00	85.00	517	525	75	64							24,296	5.41	25,396	5.57	1

TABLE 1 *Continued*

Alpha-numeric (A-N) (a)	R R M Name (b)	Data in other Tables	W3110 Fig.1 X	Fig.1 Y	Fig.4 X	Fig.4 Y	pH 5-7 Fig.2A X	Fig.2A Y	NE Fig.2B X	Fig.2B Y	NC3 pH 5-7 Fig.3A X	Fig.3A Y	NE Fig.3B X	Fig.3B Y	MW calc. from seq. (c)	pI calc. from seq. (d)	MW est. from gels (e)	pI est. from gels (f)	Fig. used for est.
D028.0	R2092	3 4	72.00	88.00	518	510	75	66									26,479	5.57	1
D028.3	R2091	3 4	76.00	89.50	532	502	79	67									27,058	5.49	1
D029.8	R2089	3 4			525	494	77	67									25,712	5.49	4
D030.2		4					76	70									29,526	5.59	2A
D031.5	R2010	2 3 4	72.00	102.00	517	446	74	76	94	74	79	81			34,481	5.55	32,783	5.57	1
D032.5	R2011	3 4	75.00	103.00	528	442	77	77	96	76	82	82	99	74			33,301	5.51	1
D033.4	R2008	2 3 4	71.00	105.50	515	429	73	79			78	84					34,630	5.59	1
D035.5		4					79	79									35,001	5.53	2A
D035.7	R2007	3 4	73.50	106.00	525	428	75	79									34,901	5.54	1
D035.8		4	77.00	114.50			79	86									39,753	5.47	1
D036.0		2					77	87			82	92			43,376	5.77 *	40,908	5.57	2A
D036.1		4					79	71									30,063	5.53	2A
D037.2		2	76.00	116.00													40,656	5.49	1
D038.0		4					79	88									41,715	5.53	2A
D038.3	R1999	3 4	73.50	110.50	522	400	75	83									37,414	5.54	1
D040.7	R1929	2 3 4	74.00	117.00	527	373	76	88			81	93			39,036	5.84 *	41,267	5.53	1
D040.9	R1998	3	75.50	112.00	536	389	77	84									38,280	5.50	1
D042.0		4					79	92									45,152	5.53	2A
D044.5		3 4									81	99					43,873	5.61	3A
D044.8		4	75.00	126.50			77	94									47,503	5.51	1
D046.0		3 4									83	101					45,385	5.57	3A
D046.4		2					78	94									47,037	5.55	2A
D046.5		2					76	95							52,661	5.49	48,037	5.59	2A
D047.0		3 4					79	93			83	100					46,077	5.53	2A
D047.5		3									75	103					46,949	5.73	3A
D048.4		4					77	95									48,037	5.57	2A
D048.5		2 4	70.50	128.50			77	95			84	101			49,586	5.43	48,961	5.60	1
D049.2		3	74.50	131.00			75	97			80	103	100	95			50,892	5.52	1
D049.9	R1845	2 3 4	74.00	133.50	527	284	76	99			81	105			51,754	5.52	52,976	5.53	1
D050.0		2 3 4	76.50	134.50			78	100			83	105			51,754	5.52	53,860	5.48	1
D055.0		2	49.00	137.50			74	101			79	107			61,141	5.72	56,729	6.04	1
D057.0		2 4					74	102			79	108			60,426	5.56	56,705	5.63	2A
D058.0		4					75	103									58,292	5.61	2A
D058.1		2					77	101							60,023	5.73	55,226	5.57	2A
D058.2		2					77	102							60,023	5.73	56,705	5.57	2A
D058.3		2					79	102							60,023	5.73	56,705	5.53	2A
D058.5		2 3 4	76.00	139.50			78	104			83	109			57,454	5.33	58,856	5.49	1
D060.5		2 3 4	76.50	142.00			78	105			83	112			57,676	5.32	61,813	5.48	1
D062.5		2	74.00	145.00													65,898	5.53	1
D074.0		3 4	75.50	149.50			77	111	95	109	83	116					73,446	5.50	1
D078.0		4					80	111									76,830	5.51	2A
D078.1	R1773	3 4	77.50	150.00	537	207	82	111									74,412	5.46	1
D079.0		2	76.00	152.00											36,820	5.97	78,576	5.49	1
D084.0	DO840	2 3 4	72.50	154.50	526	190	74	114	94	112	80	119	99	109	77,429	5.45	84,534	5.56	1
D087.5	R1731	3 4	74.50	155.00	533	188	76	114	95	112	82	119					85,837	5.52	1
D088.0	R2424	3 4			540	186	79	114									105,987	5.42	4
D094.0		2 3 4	79.50	157.50			81	115	97	114	87	120			87,330	5.42	92,982	5.42	1
D099.0		3 4	75.00	159.50			77	116	95	115	82	122					99,534	5.51	1
D100.0		2 3	80.00	159.00			82	116	97	115	90	121			97,266	5.41	97,821	5.41	1
D101.5		2	76.50	159.50			79	116			83	121			96,929	5.29	99,534	5.48	1
D157.0	R1698	2 3 4	80.00	169.00	553	172	82	122	97	120	85	128	112	118	150,561	5.36	144,055	5.41	1
E014.0	RO140	2 4	70.00	36.00	510	722	74	28							15,404	5.56	14,973	5.61	1
E014.1		2	73.00	32.00											13,158	5.52	13,806	5.55	1
E017.6		4					70	42									20,604	5.71	2A
E018.0	R2200	3 4			501	637	71	42									21,425	5.61	4
E021.1	R2195	2 3	66.50	72.50	493	577	71	53							23,099	6.31 *	21,942	5.68	1
E021.2	R2938	2	67.50	72.00	496	582	70	54							23,099	6.31 *	21,836	5.66	1
E022.8	R2170	3 4	70.50	77.50	512	555	73	58	95	56	78	62					23,129	5.60	1
E023.0		3 4									78	63					23,842	5.67	3A
E024.8		3									80	69					26,393	5.63	3A
E025.5		2 3 4	70.00	86.50			72	65			77	69			28,716	5.54	25,925	5.61	1
E026.0		2 4					70	62							25,579	5.72	25,952	5.71	2A
E029.8		4					70	70									29,526	5.71	2A
E031.0		2	68.50	101.50											31,615	5.83	32,526	5.64	1
E031.8		4	70.00	97.00			72	73									30,314	5.61	1
E032.1	R2478	4	71.50	96.50	514	473	74	73									30,080	5.58	1
E033.8		4					72	76									33,024	5.67	2A

(*Table continues*)

TABLE 1 Continued

Alpha-numeric (A-N) (a)	RRM Name (b)	Data in other Tables			W3110 Large gels Fig.1 X	Fig.1 Y	Fig.4 X	Fig.4 Y	pH 5-7 Fig.2A X	Fig.2A Y	NE Fig.2B X	Fig.2B Y	NC3 pH 5-7 Fig.3A X	Fig.3A Y	NE Fig.3B X	Fig.3B Y	MW calc. from seq. (c)	pI calc. from seq. (d)	MW est. from gels (e)	pI est. from gels (f)	Fig. used for est.
E036.0		2							72	87	94	84					43,520	5.54	40,908	5.67	2A
E036.6	R2002	2	3	4	71.00	110.50	514	403	73	82			77	88			32,384	6.05	37,414	5.59	1
E038.5	R2450		3	4	69.50	114.50	507	385	71	86	93	85	76	91					39,753	5.62	1
E038.6		2							73	83							38,645	5.55	37,841	5.65	2A
E039.2	R1987		3	4	67.00	113.50	497	392	69	84									39,159	5.67	1
E039.8	R1925	2	3	4	70.50	118.00	512	365	72	88	93	87	77	93			41,383	5.59	41,884	5.60	1
E042.0		2			65.50	121.50	487	348	68	90	91	89	74	96	96	88	43,171	5.52	44,107	5.70	1
E042.5				4					72	92									45,152	5.67	2A
E043.0				4					74	94									47,037	5.63	2A
E046.5		2							72	95							52,661	5.49	48,037	5.67	2A
E048.0				4					73	96									49,080	5.65	2A
E048.7		2	3	4	70.50	130.00			71	97			77	102			56,080	5.58	50,103	5.60	1
E048.8		2	3		68.00	130.00			70	97			75	102			48,399	5.57	50,103	5.65	1
E049.2				4					72	98									51,329	5.67	2A
E049.3	R1847		3	4			516	295	74	98									47,945	5.53	4
E050.0		2	3	4					72	99			77	105			57,502	5.57	52,549	5.67	2A
E053.0		2			70.00	135.00													54,315	5.61	1
E054.0				4					72	102									56,705	5.67	2A
E054.1		2							73	101							56,422	5.66	55,226	5.65	2A
E056.0		2	3	4					74	103			78	108			51,312	8.35	58,292	5.63	2A
E058.0	R1838	2	3	4	70.00	140.00	506	259	72	104	93	102	77	110			64,659	5.56	59,420	5.61	1
E058.1		2							71	102							60,023	5.73	56,705	5.69	2A
E058.2		2							73	101							60,023	5.73	55,226	5.65	2A
E058.3		2							73	102							60,023	5.73	56,705	5.65	2A
E061.0	R2630		3	4			499	258	70	104									53,408	5.62	4
E072.0		2		4	66.50	147.50			70	109							95,526	5.55	69,852	5.68	1
E072.1				4	70.00	148.00			72	109									70,712	5.61	1
E077.5		2	3	4	72.50	148.00			74	109	95	108	79	115	101	106	76,600	5.44	70,712	5.56	1
E079.0	R1782	2	3	4	72.50	152.50	526	195	74	113	94	111	79	118					79,697	5.56	1
E088.0	R2421		3	4			515	185	72	114									107,511	5.54	4
E093.0		2			64.50	159.00			71	116							95,220	5.62	97,821	5.73	1
E106.0		2	3		77.50	160.50			79	117	97	116	84	123			108,172	5.42	103,120	5.46	1
E123.0		2			73.50	163.00			77	119			82	124			116,313	5.55	113,093	5.54	1
E133.0	R2807	2	3	4	73.00	166.00	525	176	75	120	94	118	80	126	104	116	117,666	5.43	127,208	5.55	1
E140.0	R2431		3	4	74.00	167.00	529	173	76	121	95	119	81	127	108	117			132,499	5.53	1
F007.0	R2275	2	3	4	45.50	20.50	400	792	51	17	89	14	56	21	91	18	9,121	5.84	10,584	6.11	1
F010.1	R2365	2	3	4	52.00	11.00	432	825	58	11			61	14					10,290	5.98	1
F010.6	R2913	2		4	57.00	18.00	439	796	60	14			63	17			7,272	5.95	10,146	5.88	1
F011.7		2			52.50	25.00													11,713	5.97	1
F012.2		2							64	22	92	20	67	25	94	24	12,422	5.43	11,549	5.83	2A
F012.3	R2262		3	4			451	755	65	21									10,745	5.86	4
F013.0	R2261		3	4			459	744	64	25									11,923	5.82	4
F014.4				4					59	30									15,765	5.93	2A
F014.7	R1703	2	3	4	59.50	36.00	454	722	63	28	92	26	69	32			15,404	5.56	14,973	5.83	1
F014.9				4					58	32									16,771	5.95	2A
F017.7	R2192		3	4			442	636	59	41									21,446	5.90	4
F018.8	R2193	2	3	4	50.00	58.50	419	637	58	44			63	48	93	44	18,841	5.66	19,522	6.02	1
F019.0				4					60	45									21,408	5.91	2A
F020.9	R2191		3	4	52.00	69.00	425	597	55	51									21,240	5.98	1
F021.3		2	3	4					63	53			68	58			21,128	5.91	23,290	5.85	2A
F021.4		2			63.00	69.00													21,240	5.76	1
F021.5		2		4					58	51			62	57			23,181	5.81	22,817	5.95	2A
F022.5	R2147	2	3	4	57.00	79.00	447	552	59	59	91	57	64	63			22,279	5.74	23,533	5.88	1
F023.3				4					59	56									24,056	5.93	2A
F023.7				4					60	57									24,334	5.91	2A
F023.9				4					66	64									26,726	5.79	2A
F024.4		2			56.50	81.50													24,260	5.89	1
F024.5		2	3	4	66.50	87.50			69	66	95	64	73	70			37,189	6.37 *	26,291	5.68	1
F024.6	R2157		3	4	58.50	85.00	452	528	60	64			64	68	93	62			25,396	5.85	1
F025.1				4					59	61									25,594	5.93	2A
F025.8	R2154		3		63.50	82.00	478	537	68	62			73	66					24,414	5.75	1
F026.0	R2081	2	3	4	57.50	88.50	449	502	59	67			65	71	93	64	23,580	5.81	26,669	5.87	1
F027.0	R2146		3	4	56.00	85.00	441	541	58	64									25,396	5.90	1
F027.1		2			59.00	92.00													28,079	5.84	1
F028.0	R2080	2	3	4	56.00	87.50	448	510	58	66			63	70			37,189	6.37 *	26,291	5.90	1
F028.5		2			54.50	95.00											26,166	5.69	29,390	5.93	1
F028.6				4					67	68									28,512	5.77	2A

TABLE 1 *Continued*

Alpha-numeric (A-N) (a)	R R M Name (b)	Data in other Tables			Fig. 1 X	Fig. 1 Y	Fig. 4 X	Fig. 4 Y	Fig. 2A X	Fig. 2A Y	Fig. 2B X	Fig. 2B Y	Fig. 3A X	Fig. 3A Y	Fig. 3B X	Fig. 3B Y	MW calc. from seq. (c)	pI calc. from seq. (d)	MW est. from gels (e)	pI est. from gels (f)	Fig. used for est.
F028.8			3	4	56.50	95.00			58	72	89	70	64	77					29,390	5.89	1
F028.9		2			57.00	119.00													42,509	5.88	1
F029.0	R2940			4	59.50	92.00	463	486	61	70			67	74					28,079	5.83	1
F029.1		2			61.50	119.00													42,509	5.79	1
F029.4		2							59	70							30,142	5.88	29,526	5.93	2A
F029.7		2	3	4	47.50	99.00			52	75			58	80			29,321	5.94	31,276	6.07	1
F030.2		2	3	4	52.00	100.00			54	75	89	73	60	80			29,321	5.94	31,770	5.98	1
F030.3	R2073		3	4	59.00	92.50	460	483	61	70									28,291	5.84	
F032.0		2							59	74			63	78			35,555	6.27 *	31,787	5.93	2A
F032.3			3	4									62	86	89	79			35,201	5.98	3A
F032.4		2							65	73	90	70					38,705	6.94	31,194	5.81	2A
F032.5	R1991	2	3	4	63.50	102.50	472	438	66	76	91	74	71	82	95	75	33,953	5.83	33,041	5.75	1
F032.7				4					58	79									35,001	5.95	2A
F032.9				4					64	75									32,397	5.83	2A
F033.0		2			56.00	100.50			58	75			63	80			37,189	6.37 *	32,020	5.90	1
F033.1	R2070	2	3	4	66.50	98.50	492	462	69	75			72	80			37,189	6.37 *	31,032	5.68	1
F033.2				4	57.00	100.00			58	75									31,770	5.88	1
F033.4		2			53.00	103.00			55	77			61	82			32,466	5.94	33,301	5.96	1
F033.6		2	3		57.00	102.50			59	77			64	83			34,034	5.81	33,041	5.88	1
F035.0	R1990	2	3	4	62.50	106.00	469	426	65	79							32,813	5.97	34,901	5.77	1
F035.6				4					58	79									35,001	5.95	2A
F035.8	R1708	2	3	4	61.50	108.50	462	412	63	81			69	87			34,740	5.43	36,281	5.79	1
F037.2	R1993		3		65.50	110.00	485	411	68	82			73	87					37,128	5.70	
F037.4		2							68	86							37,521	5.81 *	40,118	5.75	2A
F037.5	R1981		3	4			443	404	59	81			65	87					37,657	5.90	4
F037.8	R1912	2	3	4	58.00	114.50	454	383	60	85			65	91			20,676	6.22	39,753	5.86	1
F037.9		2	3	4					55	87			61	92			42,036	5.99	40,908	6.01	2A
F038.0	R1986		3	4	58.00	113.00	453	390	60	84			65	90					38,864	5.86	1
F038.1		2	3	4					65	85			70	91			38,796	5.71	39,345	5.81	2A
F038.5	R2685	2	3	4	63.00	114.50	471	388	66	86			71	92			20,676	6.22	39,753	5.76	1
F039.0	R2446	2	3		63.00	113.00	472	395	65	84			70	89			36,766	5.71	38,864	5.76	1
F039.6	R1911	2	3	4	63.50	115.50	473	381	66	87			72	93			43,562	5.80	40,354	5.75	1
F039.7	R1910	2	3	4	58.00	117.50	452	373	59	88			65	93			43,562	5.80	41,574	5.86	1
F040.8	R1906		3		58.00	118.50	452	362	59	89			65	95					42,195	5.86	1
F042.2	R1914	2	3	4	63.00	121.50	474	348	66	91	90	89	71	97			42,601	5.59	44,107	5.76	1
F043.5				4					66	90									43,386	5.79	2A
F043.8		2	3	4	66.00	123.50	492	335	68	92	91	91	74	98	96	89	12,520	8.54	45,429	5.69	1
F043.9	R1920		3	4	71.00	124.00	517	334	73	92	94	90	78	98					45,767	5.59	1
F044.0		2			58.00	125.00													46,451	5.86	1
F044.2	R1927		3		71.00	117.00	515	373	73	87									41,267	5.59	1
F044.6		2	3						61	94			67	100					47,037	5.89	2A
F045.0				4	51.50	126.50			57	94									47,503	5.99	1
F045.4		2							60	93							49,403	5.94	46,077	5.91	2A
F045.5		2	3	4									71	98			49,403	5.94	43,136	5.81	3A
F045.6		2	3	4	51.00	126.50			54	95			60	100			48,759	5.97	47,503	6.00	1
F046.6	RO466	2	3	4	62.00	123.50	446	318	65	93			70	99			49,432	6.09 *	45,429	5.78	1
F047.8		2							56	98			62	104			53,799	5.92	51,329	5.99	2A
F048.1		2	3		53.00	129.50			55	97			61	103			46,884	5.93	49,717	5.96	1
F048.8		2		4	55.50	130.50			57	98			62	104			55,629	5.89	50,495	5.91	1
F049.0	R2388		3	4	64.00	128.50	480	315	66	96									48,961	5.74	1
F050.1	R1751	2	3	4			389	252	56	100			63	106			56,182	5.88	54,999	6.16	4
F050.2			3	4					61	100			67	106					53,845	5.89	2A
F050.3	R2576	2	3	4	64.50	131.00	482	302	67	97			72	104			44,009	5.85	50,892	5.73	1
F050.4		2		4	60.00	131.00			60	98			67	104	94	95	52,135	7.40	50,892	5.82	1
F050.6	R1831		3	4			469	271	64	100									50,800	5.77	4
F051.0		2			64.00	134.00											53,200	5.74	53,414	5.74	1
F054.0			3	4	48.00	137.00			51	103			57	108					56,226	6.06	1
F054.1	R1828		3	4	59.00	137.00	461	261	61	102									56,226	5.84	1
F054.4	R1767		3	4	57.50	137.50	454	254	59	102			65	109					56,729	5.87	1
F056.0			3	4	61.00	139.00			62	103	90	102	69	109					58,306	5.80	1
F056.1				4					66	100									53,845	5.79	2A
F056.2			3	4					55	103			61	110					58,292	6.01	2A
F056.5			3	4									58	110					52,860	6.06	3A
F057.0				4	63.50	139.00			65	103									58,306	5.75	1
F058.5		2	3	4	63.00	142.00			65	105	91	103	71	111			65,900	5.74	61,813	5.76	1
F060.3		2	3	4	47.50	144.00			51	107	86	105	57	113	92	103	64,393	6.21	64,464	6.07	1
F063.4			3		62.00	146.50			64	108			69	115					68,203	5.78	1

(Table continues)

TABLE 1 *Continued*

Alpha-numeric (A-N) (a)	R R M Name (b)	Data in other Tables (2)	(3)	(4)	W3110 Large gels Fig. 1 X	Y	Fig. 4 X	Y	pH 5-7 Fig. 2A X	Y	NE Fig. 2B X	Y	NC3 pH 5-7 Fig. 3A X	Y	NE Fig. 3B X	Y	MW calc. from seq. (c)	pI calc. from seq. (d)	MW est. from gels (e)	pI est. from gels (f)	Fig. used for est.
F063.5			3	4									67	115					57,532	5.89	3A
F063.8			3										61	115					57,532	6.00	3A
F064.5			3		64.50	146.50			66	108	91	107	72	114					68,203	5.73	1
F066.0			3	4									71	115					57,532	5.81	3A
F071.5		2			57.00	149.00													73,446	5.87	1
F072.0	R1746	2	3		55.50	146.50	447	213	57	109	88	107					76,101	5.87	68,203	5.91	1
F076.0		2			59.00	151.00													76,432	5.84	1
F080.0		2							56	111			61	117			81,533	6.07	76,830	5.99	2A
F082.5			3	4	65.50	154.00			68	113	91	112	74	119					83,269	5.70	1
F084.0		2	3	4					61	114	90	112	66	120			99,040	5.74	87,826	5.89	2A
F084.1		2	3	4	70.00	154.50			72	113			77	119			95,526	5.55	84,534	5.61	1
F088.0		2	3	4	50.00	155.50			53	115	87	113	59	121			85,000	6.69	87,181	6.02	1
F093.0		2	3	4	60.00	157.00			61	115			61	121	97	111	96,113	5.97	91,465	5.82	1
F099.0		2	3	4	65.50	159.00			68	116	92	115	74	122			99,577	5.77	97,821	5.70	1
F106.0			3	4									74	122					64,776	5.75	3A
F107.0		2	3		52.00	161.00			54	118	87	116	58	124			104,374	6.04	104,997	5.98	1
F113.0		2	3		66.00	160.00			68	117	91	115	74	123			103,091	5.63	101,300	5.69	1
F119.0		2		4					60	118	89	118	67	124			97,852	6.07	107,867	5.91	2A
F178.0			3	4	62.00	171.50			63	124	89	122			97	121			160,510	5.78	1
G010.6		2		4					55	14			58	18			11,082	5.45	9,969	6.01	2A
G010.7	R2366		3	4	42.00	15.00	382	804	49	14									9,906	6.18	1
G011.3				4					45	17									9,872	6.20	2A
G011.4		2			31.00	16.50			35	11			40	18			8,326	9.45 *	9,981	6.41	1
G011.9				4					45	20									10,691	6.20	2A
G012.3				4					40	22									11,549	6.30	2A
G012.4	R2256		3	4			347	763	42	22							10,092	6.37			4
G013.2				4					49	26									13,619	6.12	2A
G013.5	R2898	2	3	4	49.00	35.00	410	725	52	27			58	32			15769	5.84	14,689	6.04	1
G014.1	R2257		3	4			395	724	51	29									14,449	6.13	4
G015.1	R2338		3	4			422	694	53	33									18,011	6.00	4
G015.4				4					48	35									18,137	6.14	2A
G015.8	R2229	2	3	4	41.50	48.00	379	675	49	35							16,792	6.05	17,792	6.19	1
G016.0	R2337		3	4			429	677	54	36									19,535	5.96	4
G016.1	R2897	2			47.00	40.50	400	708									17,620	6.13	16,164	6.08	1
G016.2				4					49	37									18,943	6.12	2A
G016.3		2							37	37			38	40			18,185	7.18 *	18,943	6.36	2A
G016.4	R2227		3		33.50	52.50	346	658	37	40									18,585	6.36	1
G016.5				4					41	36									18,551	6.28	2A
G018.0		2			45.00	51.50													18,535	6.11	1
G018.3	R2911		3	4	41.50	58.50	382	632	55	43									19,522	6.19	1
G019.0				4					40	45									21,408	6.30	2A
G020.9		2							52	56							20,888	6.07	24,056	6.07	2A
G021.0		2		4	35.00	69.50			42	50			47	57			19,092	6.33	21,335	6.33	1
G023.4	R2141		3	4	39.00	75.00	368	570	45	56									22,506	6.25	1
G024.2		2			51.00	81.00			55	61			61	64			24,109	6.00			1
G024.7	R2137		3	4	43.00	78.00	383	555	48	59									23,261	6.16	1
G025.0				4					46	60									25,255	6.18	2A
G025.1				4					52	60									25,255	6.07	2A
G025.2		2							43	64							25,363	6.38	26,726	6.24	2A
G025.3			3		38.00	87.50			43	66			49	71					26,291	6.27	1
G025.7				4					39	62			40	67					25,952	6.32	2A
G025.8	R2144		3		45.50	82.50	397	534	50	62									24,571	6.11	1
G025.9				4	40.00	82.50			46	62									24,571	6.22	1
G026.0				4	37.00	83.00			43	63									24,730	6.29	1
G026.3				4	36.00	85.00													25,396	6.31	1
G026.6				4	50.00	84.00			53	63									25,057	6.02	1
G027.0	R2143		3		48.00	86.00	410	521	52	65			57	69					25,745	6.06	1
G027.1	R2132		3		41.00	86.00	376	518	47	65									25,745	6.20	1
G027.2			3	4					45	71			50	75					30,063	6.20	2A
G027.3		2			50.00	85.00			54	64	88	61	59	69			27,093	5.83	25,396	6.02	1
G028.0	R2150	2	3	4	43.00	87.00	385	516	48	66							27,991	6.07	26,107	6.16	1
G028.1	R2972	2	3	4	48.00	94.50	408	478	50	72			53	77			25,514	6.10	29,165	6.06	1
G028.2		2	3	4	42.00	96.00			47	73			53	78			30,935	5.99	29,847	6.18	1
G029.2	R2513		3	4	50.50	90.50	419	497	54	68									27,458	6.01	1
G029.3	R2973	2	3	4	51.00	97.00	417	471									30,471	5.99	30,314	6.00	1
G029.4				4					40	68									28,512	6.30	2A
G029.5	R2066		3	4	48.50	91.00	409	497	52	69									27,662	6.05	1

TABLE 1 *Continued*

Alpha-numeric (A-N) (a)	R R M Name (b)	Data in other Tables			W3110 coordinates on reference gels — Large gels Fig. 1 X	Y	Fig. 4 X	Y	pH 5-7 Fig. 2A X	Y	NE Fig. 2B X	Y	NC3 pH 5-7 Fig. 3A X	Y	NE Fig. 3B X	Y	MW calc. from seq. (c)	pI calc. from seq. (d)	MW est. from gels (e)	pI est. from gels (f)	Fig. used for est.
G029.6			3	4	44.00	100.50			49	75			55	80					32,020	6.14	1
G029.7		2											44	74					28,726	6.33	3A
G030.0		2											60	76			30,142	5.88	29,716	6.02	3A
G030.1	R2065		3		51.50	92.50	426	491	54	70									28,291	5.99	1
G030.5			3		41.50	100.50			46	76									32,020	6.19	1
G030.9				4					44	71									30,063	6.22	2A
G031.0		2							48	70	87	64					33,813	7.20	29,526	6.14	2A
G031.3		2							41	73							31,369	6.22	31,194	6.28	2A
G031.4		2			41.00	103.00													33,301	6.20	1
G031.5				4					40	71									30,063	6.30	2A
G032.0	R2064		3	4	46.00	97.00	395	466	50	73									30,314	6.10	1
G032.2		2			40.00	104.50													34,093	6.22	1
G032.6		2							55	74							32,464	5.91	31,787	6.01	2A
G032.8			3	4	41.00	104.00			45	79	85	76	51	84					33,827	6.20	1
G032.9		2			40.00	106.50													35,173	6.22	1
G033.2		2			46.00	107.00													35,448	6.10	1
G033.5		2							54	76							44,156	5.63	33,024	6.03	2A
G033.6		2							50	75							33,517	6.01	32,397	6.10	2A
G035.0				4					52	78									34,327	6.07	2A
G036.0		2	3		42.00	108.50			47	81	85	78	53	87			36,831	6.15	36,281	6.18	1
G036.1				4					46	87									40,908	6.18	2A
G036.2	R2503		3	4	52.00	92.50	434	495	55	70									28,291	5.98	1
G036.5		2			41.50	112.50													38,571	6.19	1
G036.7				4					42	79									33,827	6.29	2A
G037.5		2			43.50	117.00											41,351	6.17	41,267	6.15	1
G038.1				4	50.5	110.5			53	83									57769	6.01	1
G038.2	R1965		3	4	40.00	110.00	374	404	45	83									37,128	6.22	1
G038.3				4	45.50	111.00													37,701	6.11	1
G038.4				4					50	67									28,035	6.10	2A
G038.6		2			46.00	96.50			51	83			57	90			36,820	5.97	30,080	6.10	1
G038.8		2		4					39	84							37,236	6.32	38,586	6.32	2A
G039.1	R1974		3	4			363	394	42	84									39,009	6.29	4
G039.5				4					50	90									43,386	6.10	2A
G039.6		2	3						52	88			58	93			41,956	6.04	41,715	6.07	2A
G040.0		2	3	4	48.50	120.00			51	90			56	95			42,843	6.04	43,142	6.05	1
G040.1		2							42	86							41,546	6.40	40,118	6.26	2A
G040.8				4					54	86									40,118	6.03	2A
G041.0			3	4									52	91					38,313	6.18	3A
G041.2	R2434	2	3	4	36.50	119.50	362	355	41	90			46	96					42,824	6.30	1
G041.3	R2436	2	3	4	43.50	119.50	393	354	48	90			54	96			43,282	6.21	42,824	6.15	1
G041.4	R1882	2	3	4	39.50	119.50	376	352	44	90			50	96			41,430	6.30	42,824	6.24	1
G041.9		2			50.0	122.5											52,618	5.80	52680	5.98	1
G042.1	R1747		3	4			426	219	54	108							70,654	5.98			4
G042.2		2		4					50	88					723		43,197	5.99	41,715	6.10	2A
G043.0		2							35	88	82	86					43,209	6.40	41,715	6.40	2A
G043.2		2	3	4	36.50	123.00			40	93			45	99					45,095	6.30	1
G043.6	R1881	2	3	4	36.00	121.50	362	341	38	92							48,578	6.34	44,107	6.31	1
G043.7		2							40	92							48,578	6.34	45,152	6.30	2A
G043.8		2	3	4	33.50	122.00			34	92			38	98			45,306	6.41	44,434	6.36	1
G043.9		2	3	4	46.00	121.00			50	91			56	97			43,783	6.10	43,142	6.10	1
G044.0	R1894	2	3	4	54.00	120.00	438	355	56	90			61	96			47,383	5.85	43,142	5.94	1
G044.1		2							41	87							44,173	6.29	40,908	6.28	2A
G045.5	R2945		3	4	38.00	122.00	373	338	42	92									44,434	6.27	1
G045.6	R2544		3	4	46.50	119.00	404	358	49	89									42,509	6.09	1
G046.0				4					45	95									48,037	6.20	2A
G046.5		2							54	94			50	101			52,750	6.20	47,037	6.03	2A
G047.8	R2947			4	41.00	121.00	380	345	45	91									43,783	6.20	1
G048.0	R1888		3	4	42.00	128.00	426	320	46	96									48,590	6.18	1
G048.1		2			41.00	130.00													50,103	6.20	1
G048.6	R1891		3	4	41.00	123.00	384	336	46	93									45,095	6.20	1
G049.0		2			51.00	131.50													51,296	6.00	1
G049.1				4					40	97									50,176	6.30	2A
G049.2	R1818		3	4	48.50	130.50	415	284	52	98			58	104					50,495	6.05	1
G050.4				4					38	98									51,329	6.34	2A
G050.5	R2975	2	3	4	43.00	132.50	389	277	47	99			53	105			50,554	6.12	52,122	6.16	1
G050.6		2			41.00	133.50													52,976	6.20	1
G050.7		2							36	97							50,664	6.42	50,176	6.38	2A

(*Table continues*)

TABLE 1 *Continued*

Alpha-numeric (A-N) (a)	R R M Name (b)	Data in other Tables			W3110 coordinates on reference gels — Large gels — Fig. 1 X	Fig. 1 Y	Fig. 4 X	Fig. 4 Y	pH 5-7 Fig. 2A X	Fig. 2A Y	NE Fig. 2B X	Fig. 2B Y	NC3 — pH 5-7 Fig. 3A X	Fig. 3A Y	NE Fig. 3B X	Fig. 3B Y	MW calc. from seq. (c)	pI calc. from seq. (d)	MW est. from gels (e)	pI est. from gels (f)	Fig. used for est.
G051.0		2	3						47	100	86	98	51	106	93	96	55,210	6.11	53,845	6.16	2A
G051.1			3	4									45	104					47,751	6.31	3A
G051.8	R1806		3				355	283							89	99			49,142	6.33	4
G052.0	R1809	2	3	4			345	298	40	99	83	97	45	105	87	95	53,897	6.56	47,693	6.38	4
G054.0			3	4					44	102			50	108					56,705	6.22	2A
G054.1		2			27.00	135.00			40	102			42	108			56,041	6.27	54,315	6.49	1
G054.6	R2560		3	4	43.50	135.50	391	269	47	102			52	108					54,779	6.15	1
G054.7	R2978	2	3	4	44.50	135.00	395	271	48	102	85	100	54	108			50,301	6.08	54,315	6.13	1
G055.0		2							38	102									56,705	6.34	2A
G055.3		2	3	4					55	102			61	108					56,705	6.01	2A
G057.0			3	4	39.00	140.50			42	105	83	103	47	112					59,996	6.25	1
G057.1		2		4					39	102			44	108					56,705	6.32	2A
G057.6				4					45	102									56,705	6.20	2A
G057.9				4					42	102									56,705	6.26	2A
G058.0		2			41.00	138.00			44	103			50	110			62,001	6.22	57,243	6.20	1
G058.1				4					53	103									58,292	6.05	2A
G058.5	R2561		3	4	42.50	135.00	388	272	47	101									54,315	6.17	1
G059.4	R2559		3		44.50	136.00	396	266	49	102									55,251	6.13	1
G059.9		2			46.0	128.0			49.0	96.0							60,293	6.59	48961	6.12	1
G060.0		2			50.50	138.50			52	104							65,551	5.99	57,769	6.01	1
G060.1	R1740		3	4			387	199	45	104									88,874	6.17	4
G060.3		2							56	59							60,429	6.02	24,932	5.99	2A
G061.0		2	3		41.00	143.00			45	107	84	105	51	113			63,327	6.23	63,104	6.23	1
G062.8			3										41	115					57,532	6.39	3A
G063.2		2									87	107			92	104	65,829	6.23	74,742	6.65	2B
G065.0		2	3	4					49	110			55	116			74,011	6.16	73,767	6.12	2A
G067.0				4					41	107									63,104	6.02	2A
G070.0		2	3	4	50.00	149.50			53	111	87	109	58	117	96	107	85,199	6.01	73,446	6.02	1
G070.1				4					53	108									68,387	6.05	2A
G071.5			3										57	117					59,515	6.08	3A
G072.0		2	3										60	114			76,101	5.87	56,566	6.02	3A
G072.1				4					43	109									70,960	6.24	2A
G073.4		2							40	109			45	115	92	105	79,583	6.16	70,960	6.30	2A
G073.5		2							45	109			51	115			79,583	6.16	70,960	6.20	2A
G074.0	R2639	2	3	4	50.00	151.50	424	188	53	112	87	110	58	118	96	108	85,199	6.01	77,488	6.02	1
G075.0		2							47	111			50	116			81,238	6.28	76,830	6.16	2A
G076.0	R1737		3	4	41.50	152.50	383	242	45	113			51	119					79,697	6.19	1
G078.0			3						36	114			38	120					87,826	6.38	2A
G080.0				4	41.00	150.00			46	112									74,412	6.20	1
G080.1				4	34.00	150.00			41	112									74,412	6.35	1
G080.5				4					45	112									74,412	6.18	2A
G088.0				4	42.00	152.50			45	114									79,697	6.18	1
G093.0			3						47	117			50	122					102,176	6.16	2A
G094.0		2			47.00	155.00			50	115			57	120			89,802	6.04	85,837	6.08	1
G095.0		2		4	53.00	156.00			55	115							107,683	6.09 *	88,566	5.96	1
G097.0		2	3	4	35.00	159.00			37	117			41	123			105,035	6.42	97,821	6.33	1
G099.0		2			50.00	156.00			53	115	87	113	59	121			52,419	5.39	88,566	6.02	1
G117.0		2	3	4	45.50	162.00			48	119	85	117	52	125			97,852	6.07	108,923	6.11	1
G127.0			3						49	120			51	126					120,850	6.12	2A
H010.5		2			12.5	21.5													10,805	6.79	1
H011.7		2			16.5	25.0													11,713	6.70	1
H013.2	R2585		3	4			314	717	29	26									15,346	6.53	4
H013.8		2	3								77	25			81	29	15,684	7.20	12,961	7.25	2B
H014.0	R2253	2	3	4	41.50	36.50	379	723	27	29	80	25	29	33	84	30	15,771	6.53	15,113	6.19	1
H015.0				4					26	25									11,923	6.58	
H015.6				4					24	29									14,449	6.62	
H015.65				4					30	30									15,765	6.50	
H015.9				4					23	32									16,771	6.64	
H016.0				4					24	33									18,011	6.62	
H017.2		2							28	40							18,921	6.53	19,996	6.54	2A
H021.1		2							26	63							21,092	6.61	26,329	6.58	2A
H021.6		2	3						12	52	82	45	13	57	88	51	40,448	5.40	23,051	6.86	2A
H021.7		2									82	46			88	52	40,448	5.40	20,301	6.95	2B
H023.2		2			32.50	75.50			36	57									22,625	6.38	1
H024.7		2	3								80	60			83	61	26,976	6.47	27,380	7.07	2B
H025.0		2							18	60	79	56	21	64	83	57	27,188	8.64 *	25,255	6.74	2A
H025.1		2									87	56			92	57	27,188	8.64 *	25,137	6.65	2B

TABLE 1 *Continued*

Alpha-numeric (A-N) (a)	RRM Name (b)	Tables 2	Tables 3	Tables 4	Fig.1 X	Fig.1 Y	Fig.4 X	Fig.4 Y	Fig.2A X	Fig.2A Y	Fig.2B X	Fig.2B Y	Fig.3A X	Fig.3A Y	Fig.3B X	Fig.3B Y	MW calc from seq. (c)	pI calc from seq. (d)	MW est. from gels (e)	pI est. from gels (f)	Fig. used for est.
H025.2		2									68	64					30,422	8.21	29,823	7.78	2B
H025.7		2							25	63	87	59	29	67	89	59	27,188	8.64 *	26,329	6.60	2A
H026.6		2							26	63							26,735	6.47	26,329	6.58	2A
H027.0		2									70	66			76	67	26,618	7.08	31,125	7.66	2B
H027.4	R2134		3	4			321	526	27	64									23,219	6.50	4
H027.5		2			28.00	92.50													28,291	6.47	1
H027.9		2							33	66			37	72			30,955	7.46 *	27,578	6.44	2A
H028.0		2	3	4					23	74			25	78			29,775	6.73	31,787	6.64	2A
H029.3		2							31	83							29,315	6.45	37,841	6.48	2A
H029.5				4					31	68									28,512	6.48	2A
H030.1		2									68	69			76	71			33,185	7.78	2B
H030.2		2			22.00	96.00					74	69			80	69	33,382	6.88	29,847	6.59	1
H030.5		2							23	70	79	67							29,526	6.64	2A
H030.8				4					25	70									29,526	6.60	2A
H031.3	R1704	2	3		20.0	102.5	232	427	26	77	79	72			84	76	34,326	6.53	33,041	6.63	1
H031.5				4					34	71									30,063	6.36	2A
H032.0		2									71	72					33,813	7.20	35,382	7.61	2B
H033.0				4					26	76									33,024	6.58	2A
H034.3		2									77	76			83	78	35,507	7.07	38,539	7.25	2B
H036.0		2									66	70					35,205	8.17	33,902	7.90	2B
H036.5		2									67	87					41,098	8.15	48,750	7.84	2B
H037.0		2							25	81	77	78	28	87	83	79	37,505	7.58	36,393	6.60	2A
H037.4		2							24	81							37,437	6.67	36,393	6.62	2A
H038.6		2									76	85			82	84	40,894	6.72	46,711	7.31	2B
H038.7		2											28	93	84	84	40,894	6.72	39,633	6.65	3A
H038.8		2									75	85	16	94	81	84			46,711	7.37	2B
H038.9		2									77	85			83	84			46,711	7.25	2B
H039.0		2									78	86			84	86	38,589	6.82	47,719	7.19	2B
H040.0				4					22	84									38,586	6.66	2A
H040.5		2									82	85			86	84	44,446	8.49 *	46,711	6.95	2B
H040.6		2			20.00	121.00													43,783	6.63	1
H041.0	R1883	2	3	4			350	363	27	89			32	95					42,643	6.36	4
H042.6		2							8	90	79	85					50,680	8.86	43,386	6.94	2A
H042.7		2							13	90	80	87					50,680	8.86	43,386	6.84	2A
H047.3		2	3		67.00	128.50			69	96					76	96	56,747	7.42	48,961	5.67	1
H047.4	R1864	2	3	4			251	328	22	95			27	102			48,079	6.62	45,571	6.85	4
H047.5		2							27	95			32	102			48,079	6.62	48,037	6.56	2A
H049.2	R1868		3	4			316	320	30	97									46,118	6.52	4
H049.3				4					33	97									50,176	6.44	2A
H050.4		2		4					28	94							50,492	6.51	47,037	6.54	2A
H052.0		2							26	99	78	97	30	105			71,143	6.78	52,549	6.58	2A
H052.7		2	3	4					32	101			36	107			49,192	6.47	55,226	6.46	2A
H054.5	R2733	2	3	4	32.00	136.00	348	265	31	102			36	108			56,749	6.47	55,251	6.39	1
H054.6			3	4					29	103			34	110					58,292	6.52	2A
H054.7				4					28	100									53,845	6.54	2A
H055.0				4					31	101									55,226	6.48	2A
H062.0		2							21	105	77	103	27	111			64,474	6.59	61,851	6.68	2A
H063.0		2									70	103			83	101	73,355	8.93 *	68,619	7.66	2B
H068.9		2							20	107							68,875	6.50	66,024	6.70	2A
H080.0		2							10	112	77	109			84	106	71,143	6.78	80,175	6.90	2A
H083.0		2							25	112									80,175	6.60	2A
H083.1		2							29	112	82	110							80,175	6.52	2A
H094.0		2	3	4					31	115	81	112	35	121			86,979	6.35	92,193	6.48	2A
H094.1		2							38	115	82	112					86,979	6.35	92,193	6.34	2A
H096.8		2			5.0	157.0			15	117	77	115	18	123	86	112	81,853	9.25	91,465	6.94	1
H097.3		2	3	4							76	115					95,984	6.72	88,675	7.31	2B
H124.0		2		4					18	119	77	117	22	125	87	114	103,856	6.57	114,077	6.74	2A
I010.3		2									27	22							12,156	10.22	2B
I010.5		2		4							43	13			50	19	11,357	9.53	10,030	9.27	2B
I011.1		2	3								28	6			38	17	7,279	10.18	8,636	10.16	2B
I011.3		2	3								34	10			44	18	9,541	9.77	9,407	9.81	2B
I011.4		2	3								15	9			23	18	6,419	11.06	9,208	10.94	2B
I011.5		2	3								33	11			42	19	9,231	9.89	9,610	9.87	2B
I011.6		2													47	19			9,725	6.16	3B
I011.7		2	3	4							2	12			12	20	6,327	11.11	9,818	11.71	2B
I011.8		2	3								28	14			38	21	10,700	9.75	10,246	10.16	2B
I011.9		2	3	4							21	15			30	20	9,199	10.67	10,468	10.58	2B

Column group notes: Tables 2/3/4 = "Data in other Tables"; Fig.1 and Fig.4 = "W3110 coordinates on reference gels — Large gels"; Fig.2A = "W3110 pH 5-7"; Fig.2B = "W3110 NE"; Fig.3A = "NC3 pH 5-7"; Fig.3B = "NC3 NE".

TABLE 1 *Continued*

Alpha-numeric (A-N) (a)	R R M Name (b)	Data in other Tables	W3110 coordinates on reference gels — Large gels Fig. 1 X	Fig. 1 Y	Fig. 4 X	Fig. 4 Y	pH 5-7 Fig. 2A X	Fig. 2A Y	NE Fig. 2B X	Fig. 2B Y	NC3 coordinates pH 5-7 Fig. 3A X	Fig. 3A Y	NE Fig. 3B X	Fig. 3B Y	MW calc. from seq. (c)	pI calc. from seq. (d)	MW est. from gels (e)	pI est. from gels (f)	Fig. used for est.
I012.0		2 3							20	14			30	21	10,149	10.53	10,246	10.64	2B
I012.1		2 3 4							44	16			52	21	11,742	9.84	10,694	9.21	2B
I012.2		2 3 4							5	16			14	22	8,889	11.45	10,694	11.53	2B
I012.3		2							17	16			27	22			10,694	10.82	2B
I012.6		2 3							22	18			32	24	9,580	9.80	11,161	10.52	2B
I012.8		2 3 4							6	20			15	25	9,007	10.75	11,648	11.47	2B
I012.9		2							6	18			13	25	9,572	11.28	11,161	11.47	2B
I013.0		2 3 4							19	19			30	25	11,210	10.12	11,402	10.70	2B
I013.2		2 3							32	20			40	26	11,574	10.01	11,648	9.93	2B
I013.3		2											26	27			11,757	6.58	3B
I013.5		2 3							39	22			48	27	13,998	9.74	12,156	9.51	2B
I013.7		2							21	22			30	29			12,156	10.58	2B
I013.8		2							10	25			22	32	13,731	11.36	12,961	11.24	2B
I014.1		2 3							26	25			37	30	14,750	9.81	12,961	10.28	2B
I014.3		2							35	26			45	30			13,241	9.75	2B
I014.4		2 3 4							26	32			35	37	17,486	10.27	15,052	10.28	2B
I014.5		2							24	27			34	32			13,527	10.40	2B
I014.6		2 3 4							43	25			52	32	14,739	11.01	12,961	9.27	2B
I014.7		2 3 4											32	40			16,003	6.46	3B
I014.9		2 3							14	28			24	33	14,379	11.11	13,819	11.00	2B
I015.1		2							48	33			58	39	18,887	9.01	15,377	8.97	2B
I016.0		2							41	35					20,625	9.76	16,049	9.39	2B
I016.4		2 3 4							35	39			46	46	20,174	9.63	17,481	9.75	2B
I017.2		2 3 4							26	43			37	50	18,838	9.76	19,040	10.28	2B
I017.5		2 3							70	33			67	37	17,120	7.44	15,377	7.66	2B
I019.5		2 3 4							32	46			42	53	20,572	9.71	20,301	9.93	2B
I020.1		2 3							37	43			47	50	20,572	9.71	19,040	9.63	2B
I021.1		2 3 4							46	53			55	57	22,098	9.90	23,576	9.90	2B
I021.3		2 3 4							75	54			79	56	22,968	6.88	24,085	7.37	2B
I021.4		2 3							51	55			65	57	23,642	8.53	24,605	8.80	2B
I023.0		2 3 4							29	59			40	62	22,259	10.09	26,801	10.10	2B
I026.0		2 3 4							34	64			46	66	24,610	9.82	29,823	9.81	2B
I026.2		2 3							60	63			69	64	28,677	8.31	29,192	8.26	2B
I030.2		2 3							36	70			64	72	29,767	11.03	33,902	9.69	2B
I030.3		2							46	69			60	70	27,896	9.35	33,185	9.09	2B
I033.5		2 3 4							74	76			80	78	35,507	7.07	38,539	7.43	2B
I034.0		2							55	75			68	77			37,724	8.56	2B
I035.7		2							62	77			70	81	41,553	8.91 *	39,371	8.14	2B
I039.5		2							42	83					39,328	9.07	44,756	9.33	2B
I047.2		3											81	92			54,942	5.49	3B
I048.8		2 3							71	94			79	94	47,009	7.21	56,615	7.61	2B
I049.0		2							64	94					52,556	8.88	56,615	8.02	2B
I058.4		3 4							67	101			77	100			65,749	7.84	2B
I063.5		3 4											75	103			71,322	5.61	3B
I065.0		2 4							46	104					65,724	9.28	70,101	9.09	2B
I115.0		2							53	116					97,404	9.28	90,590	8.68	2B
I160.0		2							79	120			95	118	155,148	7.05	98,673	7.13	2B
	R1705	3			428	586											21,739	5.97	4
	R1706	3 4			374	208											79,610	6.24	4
	R1707	3 4			233	301											47,455	6.93	4
	R1709	3 4			775	522											23,457	4.25	4
	R1710	3 4			512	234											61,811	5.55	4
	R1711	3 4			138	529											23,054	7.40	4
	R1712	3 4			304	170											134,514	6.58	4
	R1714	3 4			381	181											113,932	6.20	4
	R1715	3 4			363	182											112,277	6.29	4
	R1716	3			365	191											98,832	6.28	4
	R1717	3			395	183											110,655	6.13	4
	R1719	3			440	182											112,277	5.91	4
	R1720	3 4			480	181											113,932	5.71	4
	R1721	3			453	181											113,932	5.85	4
	R1722	3 4			492	181											113,932	5.65	4
	R1723	3			464	185											107,511	5.79	4
	R1724	3 4			450	189											101,603	5.86	4
	R1725	3 4			493	190											100,203	5.65	4

TABLE 1 *Continued*

RRM Name (b)	Data in other Tables		W3110 coordinates on reference gels — Large gels — Fig. 4 X	Y	MW est. from gels (e)	pI est. from gels (f)	Fig. used for est.
R1726	3	4	551	186	107,511	5.36	4
R1727	3	4	543	180	115,621	5.40	4
R1732	3		638	172	130,431	4.93	4
R1734	3		331	222	68,602	6.45	4
R1735	3		339	225	66,701	6.41	4
R1736	3		375	231	63,314	6.23	4
R1738	3	4	373	194	94,892	6.24	4
R1739	3		405	197	91,202	6.08	4
R1741	3		437	202	85,572	5.93	4
R1742	3		423	203	84,520	5.99	4
R1743	3		432	206	81,506	5.95	4
R1744	3		410	207	80,547	6.06	4
R1745	3	4	393	208	79,610	6.14	4
R1748	3	4	398	224	67,319	6.12	4
R1749	3	4	384	224	67,319	6.19	4
R1750	3	4	411	224	67,319	6.05	4
R1752	3	4	402	254	54,437	6.10	4
R1753	3	4	432	252	54,999	5.95	4
R1754	3	4	439	194	94,892	5.92	4
R1755	3	4	461	192	97,490	5.81	4
R1756	3		483	202	85,572	5.70	4
R1757	3	4	499	202	85,572	5.62	4
R1758	3	4	470	212	76,075	5.76	4
R1759	3	4	482	211	76,928	5.70	4
R1760	3	4	498	211	76,928	5.62	4
R1761	3		459	222	68,602	5.82	4
R1762	3		469	221	69,269	5.77	4
R1763	3		478	228	64,941	5.72	4
R1764	3	4	465	239	59,563	5.79	4
R1765	3		479	240	59,149	5.72	4
R1766	3		507	248	56,227	5.58	4
R1769	3		453	202	85,572	5.85	4
R1771	3	4	525	200	87,748	5.49	4
R1774	3	4	560	216	72,864	5.32	4
R1775	3	4	546	220	69,953	5.39	4
R1776	3		518	227	65,512	5.52	4
R1778	3	4	538	236	60,875	5.43	4
R1779	3		548	243	57,975	5.38	4
R1780	3		552	201	86,648	5.36	4
R1781	3	4	566	228	64,941	5.29	4
R1785	3	4	592	192	97,490	5.16	4
R1786	3	4	612	192	97,490	5.06	4
R1789	3		618	221	69,269	5.03	4
R1793	3	4	651	197	91,202	4.87	4
R1794	3		170	295	47,945	7.25	4
R1795	3	4	149	314	46,524	7.35	4
R1796	3		223	298	47,693	6.98	4
R1797	3		305	261	52,713	6.58	4
R1798	3	4	297	276	50,039	6.62	4
R1799	3		277	301	47,455	6.72	4
R1800	3	4	291	307	47,010	6.65	4
R1801	3	4	345	257	53,654	6.38	4
R1802	3	4	379	257	53,654	6.21	4
R1803	3	4	353	268	51,315	6.34	4
R1804	3		368	270	50,967	6.27	4
R1805	3		341	284	49,028	6.40	4
R1807	3		365	283	49,142	6.28	4
R1808	3	4	329	292	48,213	6.46	4
R1810	3		379	300	47,533	6.21	4
R1811	3		360	301	47,455	6.31	4
R1812	3	4	334	319	46,186	6.43	4
R1813	3		360	294	48,032	6.31	4
R1814	3		447	265	51,878	5.88	4
R1815	3	4	422	266	51,685	6.00	4
R1816	3		406	281	49,380	6.08	4
R1817	3	4	390	282	49,260	6.16	4

RRM Name (b)	Data in other Tables		W3110 coordinates on reference gels — Large gels — Fig. 4 X	Y	MW est. from gels (e)	pI est. from gels (f)	Fig. used for est.
R1819	3	4	444	284	49,028	5.89	4
R1820	3		433	289	48,499	5.94	4
R1821	3	4	435	299	47,613	5.94	4
R1822	3	4	443	300	47,533	5.90	4
R1823	3	4	408	303	47,303	6.07	4
R1824	3	4	384	311	46,730	6.19	4
R1825	3	4	408	314	46,524	6.07	4
R1829	3		470	261	52,713	5.76	4
R1830	3		479	260	52,938	5.72	4
R1832	3	4	456	272	50,639	5.83	4
R1833	3		477	274	50,330	5.73	4
R1835	3	4	455	299	47,613	5.84	4
R1836	3	4	502	298	47,693	5.60	4
R1839	3		481	292	48,213	5.71	4
R1840	3	4	519	259	53,169	5.52	4
R1841	3	4	558	267	51,497	5.33	4
R1842	3	4	548	271	50,800	5.38	4
R1843	3	4	538	274	50,330	5.43	4
R1844	3	4	566	280	49,505	5.29	4
R1846	3		541	294	48,032	5.41	4
R1848	3		530	296	47,859	5.47	4
R1849	3		573	300	47,533	5.25	4
R1850	3	4	551	306	47,082	5.36	4
R1851	3	4	568	307	47,010	5.28	4
R1852	3	4	517	314	46,524	5.53	4
R1853	3	4	574	258	53,408	5.25	4
R1854	3	4	534	310	46,799	5.45	4
R1855	3	4	532	316	46,388	5.46	4
R1856	3	4	541	266	51,685	5.41	4
R1857	3		521	272	50,639	5.51	4
R1858	3	4	577	267	51,497	5.23	4
R1860	3		600	305	47,155	5.12	4
R1862	3	4	155	356	43,325	7.32	4
R1863	3	4	145	377	41,121	7.37	4
R1865	3	4	209	352	43,690	7.05	4
R1866	3	4	222	379	40,887	6.99	4
R1867	3	4	252	350	43,866	6.84	4
R1869	3		292	348	44,039	6.64	4
R1870	3		280	356	43,325	6.70	4
R1871	3	4	283	370	41,908	6.69	4
R1872	3	4	285	377	41,121	6.68	4
R1873	3	4	266	377	41,121	6.77	4
R1875	3		324	324	45,847	6.48	4
R1876	3	4	336	326	45,710	6.42	4
R1877	3	4	370	332	45,288	6.26	4
R1878	3	4	342	338	44,844	6.40	4
R1879	3	4	352	336	44,995	6.35	4
R1880	3		320	340	44,690	6.50	4
R1884	3	4	334	367	42,230	6.43	4
R1885	3		370	376	41,237	6.26	4
R1886	3	4	320	379	40,887	6.50	4
R1889	3	4	402	323	45,915	6.10	4
R1890	3		447	334	45,143	5.88	4
R1892	3		403	343	44,453	6.09	4
R1893	3	4	445	347	44,124	5.89	4
R1896	3	4	431	368	42,124	5.95	4
R1897	3		419	375	41,351	6.01	4
R1898	3		402	378	41,004	6.10	4
R1902	3	4	385	372	41,689	6.18	4
R1903	3	4	469	336	44995	5.77	4
R1905	3	4	503	346	44,207	5.60	4
R1908	3		465	370	41,908	5.79	4
R1909	3		478	373	41,577	5.72	4
R1915	3	4	464	347	44,124	5.79	4
R1917	3	4	573	324	45,847	5.25	4
R1918	3	4	564	332	45,288	5.30	4

(*Table continues*)

TABLE 1 *Continued*

RRM Name (b)	Data in other Tables		W3110 coordinates on reference gels Large gels Fig. 4		MW est. from gels (e)	pI est. from gels (f)	Fig. used for est.	RRM Name (b)	Data in other Tables		W3110 coordinates on reference gels Large gels Fig. 4		MW est. from gels (e)	pI est. from gels (f)	Fig. used for est.
			X	Y							X	Y			
R1924		3	539	365	42,439	5.42	4	R2021		3	650	400	38,206	4.87	4
R1926		3	553	366	42,335	5.35	4	R2022		3	4	135	472	28,162	7.42	4
R1930		3	4	530	333	45,215	5.47	4	R2023		3	165	475	27,797	7.27	4
R1931		3	4	513	348	44,039	5.55	4	R2024		3	4	184	487	26,432	7.18	4
R1932		3	4	544	323	45,915	5.40	4	R2025		3	152	511	24,220	7.34	4
R1934		3	4	569	342	44,533	5.27	4	R2026		3	4	171	493	25,811	7.24	4
R1935		3	614	328	45,571	5.05	4	R2027		3	244	472	28,162	6.88	4
R1936		3	4	624	335	45,069	5.00	4	R2028		3	4	219	508	24,456	7.00	4
R1937		3	583	343	44,453	5.20	4	R2029		3	4	203	510	24,297	7.08	4
R1938		3	4	632	352	43,690	4.96	4	R2030		3	4	194	468	28,661	7.13	4
R1939		3	4	619	352	43,690	5.02	4	R2031		3	4	313	451	30,929	6.54	4
R1940		3	4	581	355	43,417	5.21	4	R2032		3	4	268	454	30,514	6.76	4
R1941		3	4	604	358	43,135	5.10	4	R2033		3	4	311	464	29,175	6.55	4
R1942		3	4	598	372	41,689	5.13	4	R2034		3	4	295	469	28,535	6.63	4
R1943		3	4	585	373	41,577	5.19	4	R2035		3	313	471	28,285	6.54	4
R1946		3	4	672	367	42,230	4.76	4	R2036		3	4	268	472	28,162	6.76	4
R1947		3	4	126	388	39,784	7.46	4	R2037		3	319	481	27,095	6.51	4
R1948		3	4	123	397	38,610	7.48	4	R2038		3	306	485	26,648	6.57	4
R1949		3	4	184	417	35,818	7.18	4	R2039		3	4	290	487	26,432	6.65	4
R1950		3	4	232	387	39,910	6.94	4	R2040		3	4	264	489	26,220	6.78	4
R1951		3	4	199	444	31,914	7.10	4	R2041		3	4	306	494	25,712	6.57	4
R1952		3	4	310	386	40,035	6.55	4	R2042		3	336	448	31,348	6.42	4
R1953		3	4	287	394	39,009	6.67	4	R2043		3	377	461	29,569	6.22	4
R1954		3	314	397	38,610	6.53	4	R2044		3	357	462	29,437	6.32	4
R1955		3	4	295	401	38,069	6.63	4	R2045		3	329	465	29,045	6.46	4
R1956		3	4	289	412	36,533	6.66	4	R2046		3	4	366	470	28,410	6.28	4
R1957		3	286	424	34,805	6.67	4	R2047		3	4	346	472	28,162	6.38	4
R1958		3	304	441	32,343	6.58	4	R2048		3	4	325	474	27,917	6.48	4
R1959		3	292	441	32,343	6.64	4	R2049		3	4	380	474	27,917	6.21	4
R1960		3	4	267	430	33,933	6.77	4	R2050		3	4	353	483	26,869	6.34	4
R1961		3	4	317	408	37,099	6.52	4	R2051		3	330	485	26,648	6.45	4
R1962		3	381	392	39,270	6.20	4	R2052		3	4	334	495	25,614	6.43	4
R1963		3	4	331	399	38,341	6.45	4	R2053		3	4	350	500	25,143	6.36	4
R1964		3	4	360	404	37,657	6.31	4	R2054		3	338	504	24,789	6.42	4
R1966		3	4	348	404	37,657	6.37	4	R2055		3	4	352	508	24,456	6.35	4
R1967		3	4	325	422	35,095	6.48	4	R2056		3	4	364	510	24,297	6.29	4
R1968		3	4	361	428	34,224	6.30	4	R2057		3	392	451	30,929	6.15	4
R1969		3	377	434	33,353	6.22	4	R2058		3	4	431	451	30,929	5.95	4
R1970		3	4	321	440	32,486	6.50	4	R2059		3	446	451	30,929	5.88	4
R1971		3	378	444	31,914	6.22	4	R2060		3	4	409	454	30,514	6.06	4
R1973		3	4	357	396	38,744	6.32	4	R2061		3	4	440	461	29,569	5.91	4
R1975		3	323	406	37,379	6.49	4	R2062		3	424	461	29,569	5.99	4
R1976		3	4	407	385	40,160	6.07	4	R2063		3	407	464	29,175	6.07	4
R1978		3	435	390	39,529	5.94	4	R2067		3	407	510	24,297	6.07	4
R1979		3	4	420	392	39,270	6.01	4	R2068		3	4	384	497	25,422	6.19	4
R1982		3	4	407	405	37,518	6.07	4	R2069		3	4	492	451	30,929	5.65	4
R1983		3	416	434	33,353	6.03	4	R2071		3	4	481	473	28,039	5.71	4
R1984		3	447	435	33,208	5.88	4	R2072		3	470	477	27,559	5.76	4
R1985		3	4	433	438	32,774	5.94	4	R2074		3	493	486	26,539	5.65	4
R1988		3	460	399	38,341	5.81	4	R2075		3	4	479	489	26,220	5.72	4
R1989		3	4	452	410	36,817	5.85	4	R2076		3	4	505	493	25,811	5.59	4
R1994		3	526	385	40,160	5.48	4	R2077		3	484	508	24,456	5.69	4
R1995		3	4	562	387	39,910	5.31	4	R2078		3	4	453	493	25,811	5.85	4
R1996		3	4	575	389	39,657	5.24	4	R2079		3	454	469	28,535	5.84	4
R1997		3	4	522	391	39,400	5.50	4	R2082		3	4	451	462	29,437	5.86	4
R2000		3	4	531	401	38,069	5.46	4	R2083		3	553	450	31,068	5.35	4
R2001		3	4	546	402	37,932	5.39	4	R2084		3	560	459	29,836	5.32	4
R2003		3	568	406	37,379	5.28	4	R2085		3	4	563	471	28,285	5.30	4
R2009		3	551	442	32,200	5.36	4	R2086		3	545	471	28,285	5.39	4
R2012		3	604	388	39,784	5.10	4	R2087		3	521	486	26,539	5.51	4
R2014		3	4	607	421	35,240	5.08	4	R2088		3	571	490	26,116	5.26	4
R2016		3	4	605	432	33,643	5.09	4	R2093		3	547	511	24,220	5.38	4
R2017		3	630	438	32,774	4.97	4	R2094		3	4	535	467	28,789	5.44	4
R2018		3	595	397	38,610	5.14	4	R2096		3	543	462	29,437	5.40	4
R2019		3	4	675	414	36,248	4.75	4	R2097		3	4	528	450	31,068	5.47	4
R2020		3	4	685	434	33,353	4.70	4	R2098		3	532	475	27,797	5.46	4

TABLE 1 *Continued*

RRM Name (b)	Data in other Tables		W3110 coordinates on reference gels Large gels Fig. 4		MW est. from gels (e)	pI est. from gels (f)	Fig. used for est.
			X	Y			
R2099	3		592	450	31,068	5.16	4
R2103	3	4	599	470	28,410	5.12	4
R2104	3	4	602	482	26,981	5.11	4
R2105	3	4	605	491	26,013	5.09	4
R2106	3		618	492	25,912	5.03	4
R2107	3	4	596	497	25,422	5.14	4
R2108	3		597	505	24,704	5.13	4
R2109	3	4	652	459	29,836	4.86	4
R2110	3	4	644	470	28,410	4.90	4
R2112	3	4	691	463	29,306	4.67	4
R2113	3	4	151	522	23,457	7.34	4
R2114	3	4	158	530	23,002	7.31	4
R2115	3	4	158	537	22,670	7.31	4
R2116	3		150	560	21,978	7.35	4
R2117	3	4	175	559	21,997	7.22	4
R2118	3		230	524	23,335	6.95	4
R2119	3	4	241	533	22,852	6.90	4
R2120	3	4	230	540	22,547	6.95	4
R2121	3		216	557	22,038	7.02	4
R2122	3	4	293	521	23,520	6.64	4
R2123	3		267	524	23,335	6.77	4
R2124	3		311	525	23,277	6.55	4
R2125	3		261	535	22,759	6.80	4
R2126	3		314	538	22,628	6.53	4
R2127	3	4	291	546	22,331	6.65	4
R2128	3		257	560	21,978	6.82	4
R2129	3		300	561	21,960	6.60	4
R2130	3	4	286	574	21,798	6.67	4
R2131	3	4	342	513	24,069	6.40	4
R2133	3		339	524	23,335	6.41	4
R2135	3		372	536	22,714	6.25	4
R2136	3		362	541	22,508	6.30	4
R2138	3		352	560	21,978	6.35	4
R2139	3	4	343	568	21,857	6.39	4
R2140	3		379	569	21,846	6.21	4
R2142	3		359	530	23,002	6.31	4
R2145	3		415	534	22,805	6.03	4
R2148	3		417	553	22,130	6.02	4
R2149	3		443	519	23,649	5.90	4
R2151	3		509	513	24,069	5.57	4
R2152	3	4	493	517	23,784	5.65	4
R2153	3	4	496	531	22,951	5.63	4
R2155	3		499	552	22,156	5.62	4
R2162	3	4	560	521	23,520	5.32	4
R2164	3		541	528	23,108	5.41	4
R2167	3		563	540	22,547	5.30	4
R2169	3		575	553	22,130	5.24	4
R2171	3		556	557	22,038	5.34	4
R2172	3		561	567	21,870	5.31	4
R2173	3		512	569	21,846	5.55	4
R2174	3	4	522	530	23,002	5.50	4
R2180	3	4	577	567	21,870	5.23	4
R2181	3		641	530	23,002	4.92	4
R2182	3		641	546	22,331	4.92	4
R2183	3	4	697	553	22,130	4.64	4
R2184	3	4	165	629	21,566	7.27	4
R2185	3		251	593	21,728	6.85	4
R2186	3		286	622	21,645	6.67	4
R2187	3	4	310	635	21,466	6.55	4
R2188	3	4	289	616	21,686	6.66	4
R2189	3	4	352	593	21,728	6.35	4
R2190	3	4	414	582	21,752	6.04	4
R2194	3		465	579	21,765	5.79	4
R2196	3		506	584	21,745	5.58	4
R2197	3		508	595	21,726	5.57	4
R2198	3		465	595	21,726	5.79	4
R2199	3		476	602	21,722	5.73	4
R2201	3	4	567	578	21,771	5.28	4
R2205	3		523	624	21,626	5.50	4
R2207	3		562	635	21,466	5.31	4
R2208	3		634	580	21,760	4.95	4
R2209	3		617	579	21,765	5.03	4
R2210	3		597	586	21,739	5.13	4
R2211	3	4	579	589	21,733	5.22	4
R2212	3		634	623	21,636	4.95	4
R2213	3	4	610	638	21,403	5.07	4
R2214	3		593	639	21,380	5.15	4
R2217	3	4	186	659	20,664	7.17	4
R2218	3	4	163	666	20,284	7.28	4
R2220	3	4	319	658	20,713	6.51	4
R2221	3	4	315	667	20,223	6.53	4
R2222	3	4	311	697	17,699	6.55	4
R2225	3	4	378	640	21,356	6.22	4
R2226	3	4	382	658	20,713	6.20	4
R2228	3	4	368	661	20,563	6.27	4
R2230	3	4	327	693	18,113	6.47	4
R2231	3	4	438	647	21,155	5.92	4
R2232	3		448	662	20,510	5.87	4
R2234	3	4	473	661	20,563	5.75	4
R2235	3	4	516	654	20,893	5.53	4
R2236	3	4	550	655	20,850	5.37	4
R2237	3		520	662	20,510	5.51	4
R2238	3		556	675	19,685	5.34	4
R2239	3	4	516	687	18,691	5.53	4
R2240	3	4	565	696	17,804	5.29	4
R2243	3		621	673	19,829	5.01	4
R2247	3		648	674	19,758	4.88	4
R2248	3	4	641	695	17,908	4.92	4
R2249	3	4	291	714	15,723	6.65	4
R2250	3	4	293	732	13,414	6.64	4
R2251	3	4	302	764	10,027	6.59	4
R2252	3	4	263	734	13,158	6.79	4
R2254	3	4	381	735	13,030	6.20	4
R2255	3		372	742	12,161	6.25	4
R2258	3		423	735	13,030	5.99	4
R2267	3		595	711	16,093	5.14	4
R2269	3	4	205	781	9,679	7.07	4
R2270	3	4	359	768	9,808	6.31	4
R2271	3	4	352	823	20,329	6.35	4
R2272	3	4	342	812	15,254	6.40	4
R2273	3	4	381	771	9,693	6.20	4
R2274	3	4	410	778	9,617	6.06	4
R2276	3		475	791	10,385	5.74	4
R2277	3	4	527	768	9,808	5.48	4
R2280	3		619	800	11,822	5.02	4
R2281	3	4	633	813	15,632	4.96	4
R2282	3	4	661	777	9,609	4.82	4
R2283	3		591	219	70,654	5.16	4
R2284	3		604	220	69,953	5.10	4
R2288	3	4	524	233	62,299	5.49	4
R2289	3	4	582	225	66,701	5.21	4
R2299	3	4	630	752	11,040	4.97	4
R2300	3	4	641	747	11,578	4.92	4
R2301	3	4	652	754	10,841	4.86	4
R2302	3	4	600	792	10,504	5.12	4
R2303	3	4	628	847	40,171	4.98	4
R2304	3	4	649	738	12,652	4.88	4
R2305	3	4	567	722	14,707	5.28	4
R2306	3		566	715	15,598	5.29	4
R2311	3	4	596	727	14,060	5.14	4
R2314	3		506	744	11,923	5.58	4
R2317	3	4	555	687	18,691	5.34	4

(Table continues)

TABLE 1 *Continued*

RRM Name (b)	Data in other Tables		W3110 coordinates on reference gels Large gels Fig. 4		MW est. from gels (e)	pI est. from gels (f)	Fig. used for est.
			X	Y			
R2318	3	4	578	669	20,098	5.23	4
R2319	3	4	623	688	18,599	5.00	4
R2320	3	4	622	711	16,093	5.01	4
R2321	3	4	644	718	15,219	4.90	4
R2322	3	4	668	758	10,475	4.78	4
R2323	3	4	645	785	9,863	4.90	4
R2327	3	4	569	793	10,632	5.27	4
R2328	3	4	586	768	9,808	5.19	4
R2329	3	4	597	763	10,092	5.13	4
R2330	3	4	589	754	10,841	5.17	4
R2331	3		574	757	10,562	5.25	4
R2332	3	4	556	801	12,036	5.34	4
R2334	3	4	557	778	9,617	5.33	4
R2336	3	4	494	656	20,806	5.64	4
R2339	3	4	418	676	19,611	6.02	4
R2340	3		445	626	21,604	5.89	4
R2342	3	4	455	739	12,528	5.84	4
R2343	3	4	504	656	20,806	5.59	4
R2344	3	4	517	636	21,446	5.53	4
R2346	3	4	577	606	21,717	5.23	4
R2347	3	4	549	607	21,715	5.37	4
R2348	3	4	521	576	21,783	5.51	4
R2349	3	4	482	609	21,711	5.70	4
R2350	3	4	536	666	20,284	5.44	4
R2351	3	4	619	628	21,580	5.02	4
R2352	3	4	630	614	21,695	4.97	4
R2354	3	4	526	536	22,714	5.48	4
R2355	3	4	542	543	22,434	5.41	4
R2356	3	4	515	539	22,587	5.54	4
R2358	3		553	527	23,163	5.35	4
R2359	3	4	564	527	23,163	5.30	4
R2360	3	4	517	789	10,176	5.53	4
R2361	3	4	503	775	9,614	5.60	4
R2362	3	4	491	790	10,276	5.66	4
R2363	3	4	485	781	9,679	5.69	4
R2364	3		453	819	18,241	5.85	4
R2369	3		545	825	21,487	5.39	4
R2371	3	4	517	819	18,241	5.53	4
R2372	3	4	527	800	11,822	5.48	4
R2374	3	4	583	728	13,931	5.20	4
R2375	3	4	571	773	9,642	5.26	4
R2384	3	4	475	325	45,779	5.74	4
R2385	3	4	472	321	46,051	5.75	4
R2386	3	4	463	315	46,456	5.80	4
R2387	3	4	468	312	46,661	5.77	4
R2389	3		496	315	46,456	5.63	4
R2390	3	4	501	314	46,524	5.61	4
R2394	3	4	538	331	45,359	5.43	4
R2395	3	4	559	342	44,533	5.32	4
R2396	3		551	333	45,215	5.36	4
R2397	3		550	327	45,641	5.37	4
R2398	3	4	530	324	45,847	5.47	4
R2399	3	4	538	311	46,730	5.43	4
R2400	3	4	541	315	46,456	5.41	4
R2401	3	4	577	313	46,592	5.23	4
R2402	3		590	298	47,693	5.17	4
R2404	3	4	616	318	46,253	5.04	4
R2405	3	4	610	337	44,920	5.07	4
R2406	3	4	626	291	48,306	4.99	4
R2407	3		626	301	47,455	4.99	4
R2408	3		600	386	40,035	5.12	4
R2409	3	4	612	382	40,528	5.06	4
R2410	3	4	623	345	44,290	5.00	4
R2411	3		583	280	49,505	5.20	4
R2412	3	4	553	295	47,945	5.35	4
R2414	3	4	537	282	49,260	5.43	4

RRM Name (b)	Data in other Tables		W3110 coordinates on reference gels Large gels Fig. 4		MW est. from gels (e)	pI est. from gels (f)	Fig. used for est.
			X	Y			
R2415	3	4	602	255	54,168	5.11	4
R2416	3	4	652	274	50,330	4.86	4
R2419	3	4	523	185	107,511	5.50	4
R2420	3	4	519	188	103,034	5.52	4
R2422	3	4	524	208	79,610	5.49	4
R2423	3	4	512	203	84,520	5.55	4
R2425	3	4	538	181	113,932	5.43	4
R2428	3	4	548	197	91,202	5.38	4
R2432	3	4	534	182	112,277	5.45	4
R2433	3	4	388	327	45,641	6.17	4
R2435	3	4	384	351	43,779	6.19	4
R2437	3	4	390	351	43,779	6.16	4
R2442	3		428	348	44,039	5.97	4
R2444	3	4	450	401	38,069	5.86	4
R2445	3	4	462	385	40,160	5.83	4
R2447	3		498	373	41,577	5.62	4
R2449	3	4	507	380	40,768	5.58	4
R2452	3		528	364	42,542	5.47	4
R2453	3	4	535	358	43,135	5.44	4
R2454	3	4	510	354	43,509	5.56	4
R2455	3	4	521	354	43,509	5.51	4
R2456	3	4	548	373	41,577	5.38	4
R2458	3	4	537	381	40,648	5.43	4
R2459	3	4	565	415	36,105	5.29	4
R2460	3	4	580	414	36,248	5.22	4
R2461	3	4	608	407	37,239	5.08	4
R2462	3		601	412	36,533	5.11	4
R2463	3	4	612	430	33,933	5.06	4
R2464	3	4	621	426	34,515	5.01	4
R2465	3	4	672	421	35,240	4.76	4
R2466	3	4	575	481	27,095	5.24	4
R2467	3		542	492	25,912	5.41	4
R2468	3	4	527	514	23,996	5.48	4
R2469	3	4	528	521	23,520	5.47	4
R2470	3	4	510	524	23,335	5.56	4
R2471	3		552	497	25,422	5.36	4
R2472	3	4	559	508	24,456	5.32	4
R2473	3		577	539	22,587	5.23	4
R2474	3		580	521	23,520	5.22	4
R2475	3	4	571	521	23,520	5.26	4
R2476	3	4	595	516	23,853	5.14	4
R2479	3	4	508	470	28,410	5.57	4
R2482	3	4	564	446	31,630	5.30	4
R2483	3	4	525	420	35,385	5.49	4
R2484	3	4	515	413	36,391	5.54	4
R2488	3		470	495	25,614	5.76	4
R2489	3	4	482	450	31,068	5.70	4
R2490	3	4	494	471	28,285	5.64	4
R2491	3		491	475	27,797	5.66	4
R2492	3		467	460	29,702	5.78	4
R2494	3	4	461	453	30,652	5.81	4
R2498	3		444	474	27,917	5.89	4
R2499	3	4	434	478	27,441	5.94	4
R2500	3	4	451	539	22,587	5.86	4
R2501	3	4	435	527	23,163	5.94	4
R2502	3		430	510	24,297	5.96	4
R2504	3	4	444	487	26,432	5.89	4
R2506	3	4	351	550	22,210	6.35	4
R2507	3	4	353	537	22,670	6.34	4
R2508	3	4	359	521	23,520	6.31	4
R2509	3	4	405	560	21,978	6.08	4
R2510	3	4	410	572	21,815	6.06	4
R2511	3		381	502	24,963	6.20	4
R2512	3		394	509	24,376	6.14	4
R2514	3	4	468	524	23,335	5.77	4
R2516	3	4	466	557	22,038	5.78	4

TABLE 1 *Continued*

RRM Name (b)	Data in other Tables		W3110 coordinates on reference gels Large gels Fig. 4 X	Y	MW est. from gels (e)	pI est. from gels (f)	Fig. used for est.
R2517	3	4	343	486	26,539	6.39	4
R2519	3	4	337	464	29,175	6.42	4
R2520	3		335	453	30,652	6.43	4
R2521	3	4	366	460	29,702	6.28	4
R2522	3	4	388	470	28,410	6.17	4
R2523	3	4	373	440	32,486	6.24	4
R2524	3		383	410	36,817	6.19	4
R2525	3	4	379	414	36,248	6.21	4
R2526	3	4	405	421	35,240	6.08	4
R2527	3	4	389	441	32,343	6.16	4
R2528	3	4	421	401	38,069	6.00	4
R2529	3	4	440	384	40,283	5.91	4
R2530	3	4	430	380	40,768	5.96	4
R2531	3		429	355	43,417	5.96	4
R2532	3		425	359	43,039	5.98	4
R2534	3		463	358	43,135	5.80	4
R2537	3	4	394	388	39,784	6.14	4
R2538	3	4	398	395	38,877	6.12	4
R2539	3		386	376	41,237	6.18	4
R2540	3		400	371	41,799	6.11	4
R2543	3		404	366	42,335	6.09	4
R2545	3	4	352	380	40,768	6.35	4
R2546	3	4	351	375	41,351	6.35	4
R2547	3		346	376	41,237	6.38	4
R2548	3		345	381	40,648	6.38	4
R2549	3		345	355	43,417	6.38	4
R2550	3	4	324	367	42,230	6.48	4
R2551	3		355	414	36,248	6.33	4
R2552	3	4	352	422	35,095	6.35	4
R2553	3	4	364	410	36,817	6.29	4
R2554	3		362	389	39,657	6.30	4
R2555	3	4	304	372	41,689	6.58	4
R2556	3	4	298	376	41,237	6.61	4
R2557	3	4	331	337	44,920	6.45	4
R2558	3	4	337	344	44,372	6.42	4
R2562	3	4	428	282	49,260	5.97	4
R2563	3	4	444	293	48,121	5.89	4
R2565	3	4	396	242	58,356	6.13	4
R2566	3	4	412	257	53,654	6.05	4
R2567	3		437	256	53,907	5.93	4
R2568	3	4	450	249	55,907	5.86	4
R2570	3	4	390	231	63,314	6.16	4
R2571	3	4	367	225	66,701	6.27	4
R2572	3		461	283	49,142	5.81	4
R2573	3		478	283	49,142	5.72	4
R2574	3	4	469	253	54,714	5.77	4
R2575	3	4	493	299	47,613	5.65	4
R2577	3		461	302	47,378	5.81	4
R2578	3	4	456	313	46,592	5.83	4
R2580	3	4	528	261	52,713	5.47	4
R2581	3	4	520	264	52,078	5.51	4
R2582	3	4	531	257	53,654	5.46	4
R2583	3	4	178	789	10,176	7.21	4
R2584	3	4	321	678	19,458	6.50	4
R2586	3	4	322	726	14,190	6.49	4
R2587	3		316	707	16,574	6.52	4
R2588	3	4	247	400	38,206	6.87	4
R2589	3		237	394	39,009	6.91	4
R2590	3	4	337	405	37,518	6.42	4
R2591	3	4	279	446	31,630	6.71	4
R2592	3	4	229	483	26,869	6.95	4
R2593	3		191	534	22,805	7.14	4
R2594	3	4	187	560	21,978	7.16	4
R2595	3	4	257	529	23,054	6.82	4
R2596	3	4	265	477	27,559	6.78	4
R2597	3	4	284	494	25,712	6.68	4
R2598	3	4	293	496	25,517	6.64	4
R2599	3	4	357	445	31,772	6.32	4
R2600	3	4	346	441	32,343	6.38	4
R2601	3	4	356	469	28,535	6.33	4
R2603	3	4	373	757	10,562	6.24	4
R2607	3	4	199	751	11,143	7.10	4
R2608	3	4	296	632	21,520	6.62	4
R2609	3	4	276	631	21,537	6.72	4
R2610	3	4	344	578	21,771	6.39	4
R2611	3	4	337	589	21,733	6.42	4
R2612	3	4	277	671	19,967	6.72	4
R2613	3	4	274	681	19,216	6.73	4
R2614	3	4	287	754	10,841	6.67	4
R2615	3	4	246	722	14,707	6.87	4
R2616	3		181	702	17,151	7.19	4
R2617	3		229	702	17,151	6.95	4
R2619	3		247	667	20,223	6.87	4
R2620	3	4	229	669	20,098	6.95	4
R2621	3	4	712	710	16,214	4.56	4
R2622	3	4	694	709	16,335	4.65	4
R2624	3		707	625	21,615	4.59	4
R2625	3	4	681	607	21,715	4.72	4
R2626	3		758	649	21,087	4.34	4
R2627	3	4	772	631	21,537	4.27	4
R2628	3	4	755	620	21,661	4.35	4
R2631	3	4	488	261	52,713	5.67	4
R2632	3		493	273	50,482	5.65	4
R2634	3	4	525	251	55,293	5.49	4
R2636	3		473	174	126,505	5.75	4
R2637	3	4	469	232	62,799	5.77	4
R2638	3	4	447	238	59,988	5.88	4
R2640	3	4	424	192	97,490	5.99	4
R2642	3	4	409	193	96,177	6.06	4
R2643	3	4	411	189	101,603	6.05	4
R2644	3	4	434	189	101,603	5.94	4
R2646	3		432	179	117,345	5.95	4
R2647	3		383	206	81,506	6.19	4
R2648	3	4	413	218	71,372	6.04	4
R2649	3	4	440	217	72,109	5.91	4
R2650	3		459	212	76,075	5.82	4
R2651	3	4	454	216	72,864	5.84	4
R2652	3		468	206	81,506	5.77	4
R2653	3	4	397	199	88,874	6.12	4
R2654	3	4	307	187	104,495	6.57	4
R2655	3	4	305	181	113,932	6.58	4
R2660	3	4	411	324	45,847	6.05	4
R2661	3	4	406	334	45,143	6.08	4
R2663	3		257	426	34,515	6.82	4
R2664	3	4	296	424	34,805	6.62	4
R2665	3	4	445	421	35,240	5.89	4
R2666	3		464	520	23,584	5.79	4
R2667	3		444	480	27,209	5.89	4
R2668	3	4	165	463	29,306	7.27	4
R2669	3	4	515	283	49,142	5.54	4
R2670	3	4	531	273	50,482	5.46	4
R2673	3		305	181	113,932	6.58	4
R2674	3	4	434	189	101,603	5.94	4
R2675	3		551	185	107,511	5.36	4
R2676	3		552	206	81,506	5.36	4
R2677	3	4	353	268	51,315	6.34	4
R2678	3		511	303	47,303	5.56	4
R2679	3	4	577	313	46,592	5.23	4
R2680	3	4	404	366	42,335	6.09	4
R2681	3	4	386	376	41,237	6.18	4
R2682	3	4	362	389	39,657	6.30	4
R2683	3	4	362	389	39,657	6.30	4

(Table continues)

TABLE 1 *Continued*

RRM Name (b)	Data in other Tables		W3110 coordinates on reference gels — Large gels Fig. 4 X	Y	MW est. from gels (e)	pI est. from gels (f)	Fig. used for est.
R2684	3	4	440	384	40,283	5.91	4
R2710	3	4	515	711	16,093	5.54	4
R2711	3	4	612	753	10,939	5.06	4
R2712	3	4	582	782	9,714	5.21	4
R2713	3	4	339	677	19,535	6.41	4
R2716	3	4	110	628	21,580	7.54	4
R2717	3		139	634	21,485	7.40	4
R2718	3	4	170	555	22,082	7.25	4
R2720	3		485	517	23,784	5.69	4
R2721	3	4	564	365	42,439	5.30	4
R2722	3	4	494	545	22,364	5.64	4
R2723	3		560	554	22,105	5.32	4
R2725	3	4	610	299	47,613	5.07	4
R2726	3		575	349	43,953	5.24	4
R2727	3	4	568	269	51,138	5.28	4
R2729	3	4	421	316	46,388	6.00	4
R2730	3	4	373	263	52,283	6.24	4
R2731	3		406	351	43,779	6.08	4
R2734	3	4	469	319	46,186	5.77	4
R2736	3		530	311	46,730	5.47	4
R2737	3	4	547	279	49,632	5.38	4
R2738	3		305	200	87,748	6.58	4
R2740	3	4	547	790	10,276	5.38	4
R2742	3	4	648	652	20,974	4.88	4
R2744	3	4	698	732	13,414	4.63	4
R2746	3	4	759	610	21,709	4.33	4
R2747	3		718	533	22,852	4.53	4
R2748	3	4	615	542	22,470	5.04	4
R2749	3	4	621	557	22,038	5.01	4
R2753	3	4	417	724	14,449	6.02	4
R2754	3	4	332	726	14,190	6.44	4
R2755	3		311	802	12,263	6.55	4
R2756	3		167	763	10,092	7.26	4
R2757	3	4	291	683	19,047	6.65	4
R2758	3	4	272	514	23,996	6.74	4
R2760	3		374	491	26,013	6.24	4
R2761	3	4	239	376	41,237	6.90	4
R2764	3	4	462	392	39,270	5.80	4
R2765	3	4	412	397	38,610	6.05	4
R2766	3	4	383	367	42,230	6.19	4
R2768	3		549	389	39,657	5.37	4
R2769	3	4	641	370	41,908	4.92	4
R2771	3		457	349	43,953	5.83	4
R2772	3	4	471	300	47,533	5.76	4
R2775	3	4	373	249	55,907	6.24	4
R2776	3		368	241	58,747	6.27	4
R2777	3	4	552	233	62,299	5.36	4
R2778	3	4	500	276	50,039	5.61	4
R2779	3		460	800	11,822	5.81	4
R2780	3	4	230	725	14,319	6.95	4
R2781	3		277	532	22,901	6.72	4
R2784	3	4	613	524	23,335	5.05	4
R2786	3	4	531	461	29,569	5.46	4
R2787	3	4	362	257	53,654	6.30	4
R2788	3	4	343	278	49,764	6.39	4
R2789	3	4	474	287	48,703	5.74	4
R2793	3	4	563	765	9,965	5.30	4
R2795	3	4	154	700	17,374	7.33	4
R2796	3	4	565	549	22,239	5.29	4
R2797	3	4	290	358	43,135	6.65	4
R2798	3	4	301	356	43,325	6.60	4
R2799	3	4	364	596	21,726	6.29	4
R2800	3	4	501	457	30,105	5.61	4
R2802	3	4	421	328	45,571	6.00	4
R2803	3	4	556	279	49,632	5.34	4
R2804	3	4	439	281	49,380	5.92	4

RRM Name (b)	Data in other Tables		W3110 coordinates on reference gels — Large gels Fig. 4 X	Y	MW est. from gels (e)	pI est. from gels (f)	Fig. used for est.
R2805	3	4	416	268	51,315	6.03	4
R2808	3	4	741	766	9,908	4.42	4
R2809	3	4	739	753	10,939	4.43	4
R2810	3	4	269	662	20,510	6.76	4
R2811	3	4	211	505	24,704	7.04	4
R2812	3	4	279	502	24,963	6.71	4
R2813	3	4	479	191	98,832	5.72	4
R2816	3		329	356	43,325	6.46	4
R2818	3		602	511	24,220	5.11	4
R2819	3	4	629	488	26,325	4.98	4
R2820		4	525	247	56,557	5.49	4
R2821	3		606	305	47,155	5.09	4
R2822	3	4	380	286	48,808	6.21	4
R2824	3	4	507	270	50,967	5.58	4
R2828	3		495	222	68,602	5.64	4
R2830	3	4	577	421	35,240	5.23	4
R2899	3		646	425	34,660	4.89	4
R2909	3		564	743	12,042	5.30	4
R2914	3		318	797	11,247	6.51	4
R2916	3	4	195	684	18,961	7.12	4
R2919	3	4	199	716	15,472	7.10	4
R2920	3		168	715	15598	7.26	4
R2925	3		598	419	35,529	5.13	4
R2939	3		587	690	18,409	5.18	4
R2941	3		417	525	23,277	6.02	4
R2948	3		623	204	83,492	5.00	4
R2949	3	4	605	314	46,524	5.09	4
R2951	3	4	564	195	93,635	5.30	4
R2958	3		592	489	26,220	5.16	4
R2960	3	4	538	521	23,520	5.43	4
R2964	3		570	496	25,517	5.27	4
R2974	3		444	394	39,009	5.89	4
R2976		4	397	285	48,917	6.12	4
R2979	3	4	419	277	49,899	6.01	4
R2980	3		413	279	49,632	6.04	4
R2983	3		560	335	45,069	5.32	4
R2988	3		575	334	45,143	5.24	4
R2990	3		591	232	62,799	5.16	4
R2991	3	4	586	235	61,337	5.19	4
R2992	3	4	648	811	14,891	4.88	4
R2993	3	4	604	829	24,042	5.10	4
R2995	3	4	606	666	20,284	5.09	4
R2997	3	4	545	591	21,730	5.39	4
R3003	3	4	441	824	20,898	5.91	4
R3004	3	4	301	750	11,248	6.60	4
R3007	3	4	195	478	27,441	7.12	4
R3008	3	4	414	564	21,911	6.04	4
R3010	3	4	420	542	22,470	6.01	4
R3019	3	4	633	420	35,385	4.96	4
R3023	3	4	547	414	36,248	5.38	4
R3029	3	4	584	365	42,439	5.20	4
R3035	3	4	300	520	23,584	6.60	4
R3040	3	4	356	348	44,039	6.33	4
R3041	3	4	345	349	43,953	6.38	4
R3042	3	4	370	347	44,124	6.26	4
R3097	3	4	390	563	21,927	6.16	4
R3101	3	4	471	383	40,406	5.76	4
R3102	3	4	283	473	28,039	6.69	4
R3103	3	4	222	416	35,962	6.99	4
R3109	3	4	410	200	87,748	6.06	4
R3111	3	4	614	837	30,208	5.05	4
R3112	3	4	589	855	50,234	5.17	4
R3114	3	4	650	848	41,321	4.87	4
R3115	3	4	693	835	28,527	4.66	4
R3116	3	4	555	848	41,321	5.34	4
R3117	3	4	598	810	14,543	5.13	4

TABLE 1 *Continued*

RRM Name (b)	Data in other Tables		W3110 coordinates on reference gels Large gels Fig. 4		MW est. from gels (e)	pI est. from gels (f)	Fig. used for est.	RRM Name (b)	Data in other Tables	W3110 coordinates on reference gels Large gels Fig. 4		MW est. from gels (e)	pI est. from gels (f)	Fig. used for est.
			X	Y						X	Y			
R3118	3	4	577	811	14,891	5.23	4							
R3119	3	4	567	812	15,254	5.28	4							
R3120	3	4	457	654	20,893	5.83	4							
R3121	3	4	452	643	21,277	5.85	4							
R3123	3	4	217	519	23,649	7.01	4							
R3124	3	4	500	525	23,277	5.61	4							
R3128	3		493	434	33,353	5.65	4							
R3129	3		511	288	48,600	5.56	4							
R3136	3		182	453	30,652	7.19	4							
R3142	3		535	345	44,290	5.44	4							
R3145	3		465	502	24,963	5.79	4							

[a] Assignment of alphanumeric designations is described in reference 43. The letters reflect the pI range of the gel in which the spot is located, and the numeric value is its MW $\times 10^{-3}$ as estimated from the gel location.

[b] RRM name, spot name used for proteins included in the analysis in the new Response/Regulation Map. The location of each spot can be found in Fig. 4.

[c] MWs of proteins whose DNA sequences are found in SWISS-PROT (Release 28) are determined as described in Fig. 5.

[d] pIs of proteins whose DNA sequences are found in SWISS-PROT (Release 28) are determined as described in Fig. 5.

[e] Estimates of MWs of proteins are based on their migrations in gels and the equation shown in Fig. 5. The reference gel used for each estimate is listed in the last column of this table.

[f] Estimates of pIs of proteins are based on their migrations in gels and the equation shown in Fig. 5. The reference gel used for each estimate is listed in the last column of this table.

cells and are also examining the effects of drug therapies. Many of the human databases are also linked to the human genome project (8).

The database primarily serves two types of applications: (i) for individual proteins, the database lists how the level and/or synthesis rate varies under different conditions and in different mutant strains; and (ii) for diagnosing the physiological state of a culture, the database identifies sets of proteins that are known by 2-D gel analysis to respond to a particular condition. Table 3 lists the responses of individual proteins to several conditions, making it relatively easy to identify the groups of proteins that respond similarly. With the recent developments in image analysis, more proteins can be analyzed per experiment. In all cases, when the protein spot has been identified as the product of a gene (through the Expression Map), subsequent analysis can go much further.

Perhaps one of the best examples of the use of the database to study an individual gene is the universal stress protein. This protein, C013.5, is a fairly abundant protein under the standard growth conditions used for the database (aerobic growth in glucose minimal medium at 37°C) but was observed to be induced by almost all of the stress conditions tested (Table 4). By reverse genetics (using protein purified on 2-D gels), the gene was identified and cloned, the DNA sequence was determined, and mutants were made (36). None of the known regulatory proteins for the stress responses appear to control this gene. Several phenotypes for the null mutant have been observed, which suggests that the protein is involved in regulating the utilization of glucose and the intermediates of glucose metabolism and also in regulating the steps involved in the differentiation of cells into an easily recoverable postexponential state (37). No other studies of either of these processes had ever identified this protein.

The physiological state of a culture is very difficult to diagnose. Many techniques for measuring or examining a single molecule or enzymatic activity have been developed. A more global look at the physiological states of cells can be taken by means of 2-D gel analysis. This approach allows investigators to alternate between 2-D gel analysis of physiological states of cells and genetic and biochemical analyses of individual genes. The best example of this application is the study of the heat shock response. One of the first global studies done by 2-D gels was of the response to a temperature shift (26). Early studies of the responses to changes in temperature had indicated that protein synthesis was unaffected (for shifts from 37 to 42°C in which the growth rate is unchanged). However, examination of pulse-labeled proteins on 2-D gels revealed that the synthesis rates of almost all proteins change transiently (26). The rate of synthesis of a small set of proteins was found to increase dramatically after a temperature shift-up. Later, 2-D gel analysis of a temperature-sensitive mutant revealed that this set of proteins was part of a regulon (30). Many genetic and biochemical studies that characterized the regulatory gene and its protein followed (31). Many of the members of this regulon had previously been characterized through genetic and biochemical analyses and were subsequently identified as heat shock proteins by means of 2-D gel analysis (e.g., see references 55 and 57). Even the signal transduction pathway for this regulon has been partially studied by 2-D gel analysis (60). Many of the stress conditions listed in Table 4 were used as part of the study of inducers of heat shock proteins. This type of global analysis is beginning to play an important part in expanding our information on other regulons as well (e.g., the LRP regulon [15]), which had previously been studied extensively through the biochemical and genetic analyses of one (or a small set) of the regulon members.

The database contains information on the levels of certain proteins at various growth rates that can prove useful in yet another

(Continued on page 2100)

TABLE 2 *E. coli* proteins identified on 2-D gels

Gene name(a)		Protein name (b)	A-N	E.C. Number(c)	Category of Function(d)	Accession number(e)	Sequence code(f)	Kohara phage number(g)			DIR (h)	Map (i)	LG (j)	ID(k)	Donor
aceE		Pyruvate dehydrogenase-lipoamide	F099.0	1.2.4.1	I.A	P06958	Ecoace	112			R	2.80	T	CGM	R. Perham
aceF		Dihydrolipoamide acetyltransferase	C062.7	2.3.1.12	I.A	P06959	Ecoace	112	113		R	2.80	T	CGMP	R. Perham
aceF		Dihydrolipoamide acetyltransferase	C070.0	2.3.1.12	I.A	P06959	Ecoace	112	113		R	2.80	T	CGMP	R. Perham
ackA		Acetate kinase	G041.3	2.7.2.1	I.A	P15046	Ecoacka					49.50	NG		T. Nystrom
ada		O6-methylguanine-DNA methyltransferase	I039.5	2.1.1.63	I.F	P06134	Ecoada	373			F	47.60		CP	T. Lindahl
adhE		Alcohol dehydrogenase	H097.3	1.1.1.1	I.D	P17547	Ecoadhex	251			L	27.75			[18]
adk		Adenylate kinase	F026.0	2.7.4.3	II.B.1	P05082	Ecoadk	153			R	11.30	C	C	G. Cohen
ahpC		Alkyl hydroperoxide reductase	B020.9		I.B		EcoD13187								C. Pasquali, et al. (l)
alaS		Alanine-tRNA ligase	F093.0	6.1.1.7	III.A.4	P00957	Ecoalas	446			L	58.20	C	C	H. Weissbach
amn		AMP nucleosidase	G052.0	3.2.2.4	II.B.4	P15272	Ecoamn	346	347		R	43.00		CGM	V. Schramm
ampC		Lactamase	I035.7	3.5.2.6	VI.E	P00811	Ecoampcfr	650	651		L	94.30		CGP	B. Jaurin
apaH		Diadenosine tetraphosphatase	C031.3		I.A	P05637	Ecoapah	104	105		L	1.00		DG	pLC4-46 [45]
araA	(m)	L-Arabinose isomerase	G054.1	5.3.1.4	I.A	P08202	Ecoaraabd	107			L	1.50		GP	R. Schleif
araB		Ribulokinase	D055.0	2.7.1.16	I.A	P08204	Ecoaraabd	107	108		L	1.50	DG	CP	N. Lee
araC		AraC	H030.2		I.A	P03021	Ecoarabop	107	108		R	1.40	DG	CGP	N. Lee
araD	(m)	Ribulose-phosphate 4-epimerase	G028.1	5.1.3.4	I.A	P08203	Ecoaraabd	107			L	1.50	PTDG		N. Lee
araF	(n)	Arabinose binding protein	F032.0		V.A	P02924	Ecoarafgh	339	340	341	L	44.80		CP	K. Matthews, [12]
argA		N-acetylglucosamine synthase	H052.7	2.3.1.1	II.A.1	P08205	Ecoarga	459			R	29.25		DG	pLC32-1 [45]
argI		Ornithine carbamoyltransferase	F039.0	2.1.3.3	II.A.1	P04391	EcoargI	658	659		L	96.60	TDG	C	W. Lipscomb
argR	(m)	ArgR	D014.3		II.A.1	P15282	Ecoargr	519	520		R	71.00		C	W. Maas
argS		Arginyl-tRNA ligase	E058.0	6.1.1.19	III.A.4	P11875	Ecoargs	337	338		R	40.00	T	CP	[28]
aroF		Phospho-2-keto-3-deoxyheptonate aldolase	F038.1	4.1.2.15	II.A.4	P00888	Ecoarof	438			L	57.60		C	A. DeLucia
asnS	(o)	Asparagine-tRNA ligase	G099.0	6.1.1.22	III.A.4	P17242	Ecotgasns	219			L	20.90	T	C	H. Weissbach
aspC		Aspartate aminotransferase	F039.6	2.6.1.1	II.A.2	P00509	Ecoaspc	218	219		L	20.70	T	CM	T. Yagi
aspC		Aspartate aminotransferase	F039.7	2.6.1.1	II.A.2	P00509	Ecoaspc	218	219		L	20.70	T	CM	T. Yagi
aspS		Asparate-tRNA ligase	F058.5	6.1.1.12	III.A.4	P21889	Ecoasps	337			L	41.00	TC		H. Weissbach
atoA	(m)	Acetate CoA transferase	C026.8	2.8.3.8	I.A							47.80	T	CP	F. Frerman
atoA	(m)	Acetate CoA transferase	G024.2	2.8.3.8	I.A							47.80	T	CP	F. Frerman
atoB	(m)	Acetyl-CoA acetyltransferase	H038.8	2.3.1.9	I.A							47.80		CP	H. Schulz
atoB	(m)	Acetyl-CoA acetyltransferase	H038.9	2.3.1.9	I.A							47.80		CP	H. Schulz
atpA		ATP synthase- F1 sector, alpha subunit	G051.0	3.6.1.3	I.E	P00822	Ecouncc	560	561		L	83.90		C	P. Bragg
atpD		ATP synthase- F1 sector, beta subunit	B046.7	3.6.1.3	I.E	P00824	Ecouncc	560			L	83.90	T	C	P. Bragg
bfr		Bacterioferritin	C014.3		V.A	P11056	Ecobfr	627			L	73.00	C	CM	J. Yariv
bioB		Biotin synthetase	E038.6		II.D.1	P12996	Ecobio	201	202		R	17.55			pLC25-23 [45]
cadA	(m)	Lysine decarboxylase	G075.0	4.1.1.18	I.A	P23892	Ecocadabc	647	648		L	93.70		CP	E. Boeker
carA		Carbamoyl phosphate synthase, alpha subunit	G041.4	6.3.5.5	II.B.2	P00907	Ecocarab	103			R	0.70	C	CG	J. Villafranca
carB		Carbamoyl phosphate synthase, beta subunit	E133.0	6.3.5.5	II.B.2	P00968	Ecocarab	103			R	0.70	T	CG	J. Villafranca
cheY		Chemotactic response	B014.1		V.C	P06143	Ecoche3	338			L	41.30		DG	pLC24-15 [45]
cheZ		Chemotactic response	A028.0		V.C	P07366	Ecoche3	338			L	41.35	T	DG	pLC24-15, pLC1-29 [45]
cirA		Colicin I receptor	B068.1		VI.C	P17315	Ecocir	365	366	367	L	46.42	N		T. Nystrom
clpB		ClpB	F084.1		III.B.3	P03815	Ecoprot	437			L	56.40	PT	P	[55]
clpB	(p)	ClpB	E072.0		III.B.3	P03815	Ecoprot	437			L	56.40	P		[55]
clpP		ClpP	F021.5		III.B.3	P19245	Ecoclppa	148			R	10.00		P-ref	[25]
crp		Cyclic AMP receptor protein	I021.4		I.F	P03020	Ecocrp	625			R	73.50		CG	J. Krakow
crr		Glucose phosphotransferase system enzyme III GLC	B018.7		V.A	P08837	Ecophosys	418			R	52.20	T	C	S. Roseman
cspA		CspA-cold shock protein	F010.6		VI.H	P15277	Ecocspa	602			R	79.00	P	P	[17]
cynS	(m)	Cyanate hydrolase	C013.6	3.5.5.3	I.A	P00816	Ecocynsa	138			R	8.00		C	J. Fuchs
cysD		Sulfate adenylyltransferase	H036.0	2.7.7.4	II.A.3	P21156	Ecocysdnc	451	452		L	59.30		CP	T. Leyh
cysE		CysE	H029.3		II.A.3	P05796	Ecocysxe	675			R	80.9		DG	pLC17-22 [45]
cysN		ATP sulfurylase	C058.3	2.7.7.4	II.A.3	P23845	Ecocysdnc	451	452		L	59.00	C	C	T. Leyh
dacA		Muramoyl-pentapeptide carboxypeptidase	H040.5	3.4.17.8	IV.B	P04287	Ecodaca	168			L	14.80		CZ	U. Schwartz
dapA		Dihydrodipicolinate synthase	G031.3	4.2.1.52	II.A.2	P05640	Ecodapa	423	424		R	53.25		DG	pLC17-31 [45]
dld		Lactate dehydrogenase	H062.0	1.1.1.28	I.D	P06149	Ecodld	363	364		R	46.70		C	C. Ho
dnaA		DnaA	I049.0		III.A.7	P03004	Ecodnaaop	565			L	83.00		C	J. Kaguni
dnaB		DnaB	B047.0		III.A.7	P03005	Ecodnab	636			R	91.80	C	C	R. McMacken
dnaC		DnaC	I030.3		III.A.7	P07905	Ecodnatc	672			L	98.90		C	R. McMacken
dnaG		DnaG--primase	G060.0		III.A.7	P02923	Ecorpsrpo	509			R	67.00	C	C	R. McMacken
dnaJ		DnaJ	H036.5		III.A.7	P08622	Ecodnajk	101	102		R	0.30		CGP	C. Georgopoulos
dnaK		DnaK (HSP-70)	B066.0		III.A.7	P04475	Ecodnak	101	102		R	0.50	T	G	C. Georgopoulos
dnaN		DNA-directed DNA polymerase III-beta	A036.1	2.7.7.7	III.A.7	P00583	Ecodnaaop	565			L	83.00	T		
dnaN		DNA-directed DNA polymerase III-beta	B036.1	2.7.7.7	III.A.7	P00583	Ecodnaaop	565			L	83.00			
dnaQ		DNA-directed DNA polymerase III-epsilon	G027.3	2.7.7.7	III.A.7	P03007	Ecomutd	125	126		R	5.30	C	C	A. Aoyama
dnaT		DnaT	C019.5		III.A.7	P07904	Ecodnatc	671	672		R	98.9		DG	pLC31-39 [45]
dnaX		DNA-directed DNA polymerase III-gamma	H052.0	2.7.7.7	III.A.7	P06710	Ecodnazx	152			R	11.00		C	A. Aoyama
dnaX		DNA-directed DNA polymerase III-tau	H080.0	2.7.7.7	III.A.7	P06710	Ecodnazx	152			R	11.00		C	A. Aoyama
dppA	(q)	DppA	G059.9		V.A	P23847	Ecodppa	602			R	79.95	T	GP	T. Nystrom, E. Olson
dsbA		DsbA	E021.1		III.A.8	P24991	Ecodsf	548				87.00	C	C	J.C.A. Bardwell
dsbA		DsbA	E021.2		III.A.8	P24991	Ecodsf	548				87.00	C	C	J.C.A. Bardwell
dut		Deoxyuridinetriphosphate	C017.2	2.4.2.10	II.B.3	P06968	Ecodutpyr	572	573		R	81.9	T	DG	pLC17-24 [45]
eco		Ecotin	G016.3		II.B.3	P23827	Ecoecoa	372	373		?			C	A. Goldberg
eda	(r)	Hydroxy-2-oxoglutarate aldolase & Phospho-2- keto-3-deoxygluconate aldolase	F022.5	2.14 & 4.1.3	I.A	P10177	Ecoeddeda	336			L	40.70		G & CM	E. Dekker & D. Fraenkel
eno		Enolase	F043.8	4.2.1.11	I.A	P08324	Ecopyrg	455			L	59.60	T	C	F. Wold
entF		Enterobactin	C142.0		V.A	P11454	Ecoentf	164			L	13.4		DG	C.F. Earhart (pCP111)

TABLE 2 *Continued*

Gene name(a)	Protein name (b)	A-N	E.C. Number(c)	Category of Function(d)	Accession number(e)	Sequence code(f)	Kohara phage number(g)		DIR (h)	Map (i)	Basis ID LG (j)	ID(k)	Donor
era	ERA protein	G031.0		I.F	P06616	Ecoera	434		L	55.30		CG	D. Court, C. Lerner
era (s)	ERA protein	H032.0		I.F	P06616	Ecoera	434		L	55.30		CG	D. Court, C. Lerner
fabA	Hydroxydecanoyl-(acyl carrier protein)-hydrolase	H017.2	4.2.1.60	II.E	P18391	Ecofabaa	225		L	21.70		C	J. Cronan, Jr.
fabB	Ketoacyl-[acyl carrier protein] synthase I, beta	F042.2	2.3.1.41	II.E	P14926	Ecofabb	407	408	L	50.30	C	C	J. Cronan, Jr.
fadA (m)	Acetyl-CoA acyltransferase	H038.7	2.3.1.16	I.A	P21151	Ecofada	549	550	L	86.20		CP	H. Schulz
fadA (m)	Acetyl-CoA acyltransferase	H038.6	2.3.1.16	I.A	P21151	Ecofada	549	550	L	86.20		CP	H. Schulz
fadB	Fatty acid oxidation complex	G073.4	1.1.1.35	I.A	P21177	Ecofadab	172	173	L	86.20		CP	H. Schulz
fadB	Fatty acid oxidation complex	G073.5	1.1.1.35	I.A	P21177	Ecofadab	549	550	L	86.20		CP	H. Schulz
fbp	Fructose 1,6-bisphosphatase	G038.6	3.1.3.11	I.B	P09200	Ecofbpase	657		L	96.00		C	F. Marcus, I. Edelstein, J. Rittenhouse
fepA	FepA	D079.0		V.A	P05825	Ecofepaa	163	164	L	13.43	N		T. Nystrom
fldA	F-protein (Flavodoxin)	A019.0		I.C	P23243	Ecoflda	172	173	L	15.70	C	C	R. Matthews
fmt	Methionyl-tRNA formyltransferase	F033.6	2.1.2.9	III.A.4	P23882	Ecofmt			F		T?C	C	H. Weissbach
folA	Dihydrofolate reductase	B020.0	1.5.1.3	II.D.2	P00379	Ecofola	104	105	R	0.90	T	C	D. Smith
frdA (m)	Succinate dehydrogenase(fumarate reductase)-flavoprotein subunit	G063.2	1.3.99.1	I.D	P00363	Ecoampcfr	650	651	L	94.40		CP	J. Weiner
frdB (m)	Succinate dehydrogenase(fumarate reductase)-iron-sulfur protein subunit	H024.7	1.3.99.1	I.D	P00364	Ecoampcfr	650	651	L	94.40		CP	J. Weiner
fumC	Fumarase	H050.4		I.B	P05042	Ecofumc	313		L	35.70		DG	pLC10-19 [45]
fur	Ferric uptake regulation repressor	G015.8		I.F	P06975	Ecofur	172		L	15.70	T	C	J. Neilands
fusA	Protein chain elongation factor, G	D084.0		III.A.8	P02996	Ecostr2	626		L	73.30	T	CG	[43]
gadA (t)	Glutamate decarboxylase	D046.5	4.1.1.15	I.B	P80063	Ecogada			F	81.30		CP	E. Boeker
gadA (t)	Glutamate decarboxylase	E046.5	4.1.1.15	I.B	P80063	Ecogada			F	81.30		CP	E. Boeker
gapA	Glyceraldehyde-3-phosphate dehydrogenase	H034.3	1.2.1.12	I.A	P06977	Ecogap	330	331	R	39.30		C	D. Fraenkel
gapA	Glyceraldehyde-3-phosphate dehydrogenase	I033.5	1.2.1.12	I.A	P06977	Ecogap	330	331	R	39.30		C	D. Fraenkel
gdhA	Glutamate dehydrogenase (NADP+)	G043.6	1.4.1.4	II.A.1	P00370	Ecogdhak	328	329	R	75.50	T		D. Fraenkel
gdhA	Glutamate dehydrogenase (NADP+)	G043.7	1.4.1.4	II.A.1	P00370	Ecogdhak	328	329	R	75.50			D. Fraenkel
glnA	Glutamate-ammonia ligase unadenylylated	D049.9	6.3.1.2	II.A.1	P06711	Ecoglna	546	547	L	86.70	CPT	C	S. Rhee
glnA	Glutamate-ammonia ligase adenylylated	D050.0	6.3.1.2	II.A.1	P06711	Ecoglna	546	547	L	86.70	CPT	C	R. Bender, S. Rhee
glnB	P II (uridylyated)	F012.2		II.A.1	P05826	Ecoglnb	432	433	L	55.00		CP	S. Rhee
glnH	Glutamine-binding protein	H025.0		V.A	P10344	Ecoglnhpq	205	206	L	18.00		C	[12]
glnH	Glutamine-binding protein	H025.1		V.A	P10344	Ecoglnhpq	205	206	L	18.00		C	[12]
glnH	Glutamine-binding protein	H025.7		V.A	P10344	Ecoglnhpq	205	206	L	18.00		C	C. Ho
glnQ	Glutamate permease	H026.6		V.A	P10346	Ecoglnhpq	205		L	18.00		DG	pLC2-18 [45]
glnS	Glutamine-tRNA ligase	G061.0	6.1.1.18	III.A.4	P00962	Ecoglns	172		R	15.00	T	CP	[28]
glpD (m)	Glycerol-3-P dehydrogenase-(NAD+) aerobic	H047.3	1.1.99.5	I.C	P13035	Ecosnglpd	616	617 618	R	75.30	T	C	J. Weiner
glpK (m)	Glycerol kinase	E048.7	2.7.1.30	II.E	P08859	Ecoglyk	539		L	88.40	PT	CP	H. Paulus
glpQ	Glycerol-3-phosphate diesterase	C039.0		I.A	P09394	Ecoglpq	377	378	R	48.60		C	T. Larson
glpR	GlpR	G028.0		I.C	P09392	Ecoglpreg	617	618	L	75.40	T	C	T. Larson
gltA	Citrate synthase	H047.4	4.1.3.7	I.B	P00891	Ecoglta	176		L	16.50		GMP	H. Duckworth
gltA	Citrate synthase	H047.5	4.1.3.7	I.B	P00891	Ecoglta	176		L	16.50		CGM	H. Duckworth
gltD	Glutamate synthase (NADPH) small subunit	F050.4	1.4.1.13	II.A.1	P09832	Ecogltb	523	524	R	69.40	C	CM	W. Orme-Johnson
gltX (t)	Glutamate-tRNA ligase	F047.8	6.1.1.17	III.A.4	P04805	Ecogltx	416	417	L	43.00		C	H. Weissbach
glyA	Serine hydroxymethyltransferase	G043.8	2.1.2.1	II.A.3	P00477	EcoglyA	432	433	L	55.00	C	CG	V. Schirch
glyS	Glycine-tRNA ligase-beta subunit	E077.5	6.1.1.14	III.A.4	P00961	Ecoglys	601		L	79.50	T	CGP	[28]
gnd	Phosphogluconate dehydrogenase	C042.6	1.1.1.44	I.B	P00350	Ecognd	351		L	44.40		CP	D. Fraenkel
gor	Glutathione reductase-NAD-P-H	F045.6	1.6.4.2	II.D.10	P06715	Ecogor	607	608	R	77.50	CT	C	C. Williams
greA	GreA	A017.6		III.A.8	P21346	Ecogreag	520	521	L	69.1		DG	pLC24-10 [45]
grpE	GrpE	B025.3		VI.B	P09372	Ecogrpe	438	439	L	57.00	P	GPM	[3]
grx	Glutaredoxin	B011.0		I.B	P00277	Ecogrx	208		R	18.50		C	A. Holmgren
guaC	GMP reductase	H037.4	1.6.6.8	II.B.1	P15344	Ecoguac	111		R	2.60		DG	pLC37-40 [45]
gyrA	DNA-topoisomerase II subunit A	D101.5	5.99.1.3	II.A.7	P09097	Ecogyra	376		L	48.30	T	C	N. Cozzarelli, R. McMacken A. Morrison, R. Otter
gyrB	DNA-topoisomerase II subunit B	G094.0	5.99.1.3	II.A.7	P06982	Ecorecfa	565	566	L	82.90	TC	C	N. Cozzarelli, R. McMacken R. Otter
hemX	HemX	B043.0		II.D.12	P09127	Ecohemcd	553	554	L	85.75		DG	pLC41-4 [45]
hemX	HemX	B043.1		II.D.12	P09127	Ecohemcd	553	554	L	85.75		DG	pLC41-4 [45]
himA (u)	IHF-integration host factor	I010.5		III.A.7	P06984	Ecohima	322	323	L	37.40		C	H. Nash
hisJ	Histidine binding protein	C028.5		V.A					F	49.90	T	G	G. Ames
hisP	Histidine permease	I026.2		V.A	P07109	Ecohismp	406		L	49.90		G	G. Ames
hisS	Histidine-tRNA ligase	F048.1	6.1.1.21	III.A.4	P04804	Ecohiss	429		L	54.10	C	CP	H. Weissbach, [28]
hns	H-NS	E014.0		III.A.6	P08936	Ecohns	250	251	L	27.10	C	C	Gualerzi
hns	H-NS	F014.7		III.A.6	P08936	Ecohns	128	129	F	6.10	TC	P,C	Gualerzi
holA	DNA-directed DNA polymerase III-delta	F032.4	2.7.7.7	III.A.7	P28630	Ecodelta			L	11.00		C	A. Aoyama
hsdM	DNA methylase M	C059.3		III.A.7	P08957	Ecohsdrm	669	670	L	98.5		DG	pLC31-39 [45]
hsdM	DNA methylase M	C059.4		III.A.7	P08957	Ecohsdrm	669	670	L	98.5		DG	pLC31-39 [45]
hslU (v)	HslU	D048.5			P32168				F	99.20	P	P	[10],[32]
hslV (v)	HslV	G021.0			P31059				F		P	P	[10],[32]
htpG	HtpG	C062.5		VI.H	P10413	Ecohsp	152		R	11.00	T	P	[32]
htpH	HtpH	D033.4							F		T	P	[32]
htpK	HtpK	F010.1							F		P	P	[32]
htpT	HtpT	A029.5							F		T	P	This laboratory
hupA	DNA-binding protein HU-alpha (HU-2)	I011.3		III.A.6	P02342	Ecohu2	532B	533	R			C	A. Subramanian
hupB	Histone-like protein HU-1 (HU-beta, NS1)	I011.5		III.A.6	P02341	Ecohupb	148		R			C	A. Subramanian

(Table continues)

TABLE 2 *Continued*

Gene name(a)	Protein name (b)	A-N	E.C. Number(c)	Category of Function(d)	Accession number(e)	Sequence code(f)	Kohara phage number(g)			DIR (h)	Map (i)	Basis ID LG (j)	ID(k)	Donor
hycA	HycA	G016.1		I.D	P16427	Ecohevop	449			L	58.60	DG		
hypA	Hydrogenase isomerase A	E014.1		I.D	P24189	Ecohyp	449			R	58.70	DG		
hypB	Hydrogenase isomerase B	E031.0		I.D	P24190	Ecohyp	449			R	58.70	DG		
hypC	Hydrogenase isomerase C	A008.0		I.D	P24191	Ecohyp	449			R	58.70	DG		
hypD	Hydrogenase isomerase D	G037.5		I.D	P24192	Ecohyp	449			R	58.70	DG		
hypE	Hydrogenase isomerase E	C031.9			P24193	Ecohyp	449	450			58.70	G		
ibpA	IbpA	G013.5			P29209	Ecoprots	566			R	82.50	P	P	[2A], [32]
ibpB	IbpB	C014.7			P29210	Ecoprots	566			R	82.50	T	P	[2A], [32]
icdE	Isocitrate dehydrogenase-(NADP+)	C043.8	1.1.1.42	I.B	P08200	Ecoicd	239	240		L	25.30	C	CP	H. Reeves
ileS	Isoleucine-tRNA ligase	F107.0	6.1.1.5	III.A.4	P00956	ileSeco	102	103		R	0.40	T	CG	[28]
ilvA	Threonine dehydratase-biosynthetic	F050.1	4.2.1.16	II.A.2	P04968	Ecoilva+	556	557		R	84.70		CM	E. Goldman, G. Hatfield
ilvB	Acetolactate synthase I	D057.0	4.1.3.18	II.A.2	P08142	Ecoilvbpr	567	568	569	L	82.20		C	P.Silverman
ilvE	Branched-chain amino acid aminotransferase	F032.5	2.6.1.42	II.A.2	P00510	Ecoilve	556	557	558	R	84.70	TC	C	E. Goldman, G. Hatfield
ilvN	Acetolactate synthase I	G010.6	4.1.3.18	II.A.2	P08143	Ecoilvbpr	568	569		L	82.20		C	P.Silverman
infB	Protein chain initiation factor, 2B	F119.0		III.A.8	P02995	Econusa	519	520		L	68.90		C	H. Weissbach
infB	Protein chain initiation factor, 2A	G117.0		III.A.8	P02995	Econusa	519	520		L	68.90	T	C	H. Weissbach
infC	Protein chain initiation factor, 3L	I019.5		III.A.8	P02999	Ecohima	322	323		L	37.50		CG	D. Elhardt
infC	Protein chain initiation factor, 3S	I020.1		III.A.8	P02999	Ecothrinf	322	323		L	37.50		CG	D. Elhardt
kbl	2-Amino-3-ketobutyrate CoA ligase	G042.2	2.3.1.29	I.B	P07912	Ecokbltdh	574	575		L	81.10			B. Ernsting, R. Matthews
kdpE	KdpE	G025.2		V.A	P21866	Ecokdpde	173			L	15.8		DG	W. Epstein (pDE1451)
kdsB	Deoxy-manno-octulosonate cytidylyltransferase	C032.7	2.7.7.38	IV.C	P04951	Ecokdsb	218			R	21.70	T	C	W. Kohlbrenner
ksgA	rRNA(m6/2A)methyltransferase	H025.2		VI.E	P06992	Ecoksga+	105	106		R	1.00		C	P. VanKnippenberg
lacI	Lac operon repressor protein	H039.0		I.A	P03023	Ecolac	138	139		R	8.00		C	K. Matthews
lacZ (m)	beta-Galactosidase	E123.0	3.2.1.23	I.A	P00722	Ecolac	138	139		R	8.00	P	CGP	A. Fowler, [43]
leuO	LeuO protein	F035.0		II.A.2	P10151	Ecoleuo	109			R	85.50	TCDG		
leuS	Leucine-tRNA ligase	D100.0	6.1.1.4	III.A.4	P07813	Ecoleus	169			L	15.10	T	CG	[28]
livF	LivF	F028.5		V.A	P22731	Ecolivhmgf	613	614		L	76.20	DG		
livJ	Binding Protein for leucine, isoleucine, and valine	D040.7		V.A	P02917	Ecolivhmgf	612	613		L	76.20	TDG	CGP	J. Anderson
livK	Leucine binding protein	B040.8		V.A	P04816	Ecolivhmgf	612	613		L	76.20	TDG	CGP	D. Oxender
lon	Lon (protease La)	H094.0		I.F	P08177	Ecolon	148			R	9.10		CG	A. Goldberg, A. Markovitz
lon	Lon (protease La)	H094.1		I.F	P08177	Ecolon	148			R	9.10		C	A. Goldberg, A. Markovitz
lpdA (t)	Dihydrolipoamide dehydrogenase	G050.5	1.8.1.4	I.B	P00391	Ecoace	112			R	2.90	TC	C	R. Perham, C. Williams
lpp	Lipoprotein	G011.4		IV.B	P02937	Ecolpp	320			R	36.30	C	CF	M. Inouye
lrp	Leucine regulatory repressor protein	I015.1		II.A.2	P19494	Ecolrrpa	213	214	215	R	20.00		C	J. Calvo
lysA	Diaminopimelate decarboxylase	G043.9	4.1.1.20	II.A.2	P00861	Ecogallys	462			L	53.87		DG	pLC32-1 [45]
lysS	Lysine-tRNA ligase form I	D058.5	6.1.1.6	III.A.4	P13030	Ecoherc	667	668		L	62.00	P	CG	I. Hirschfield, [28]
lysU	Lysine-tRNA ligase form II	D060.5	6.1.1.6	III.A.4	P14825	Ecolysug	646	647		L	93.50	P	CG	I. Hirschfield
malE	Maltose-binding protein	D036.0		V.A	P02928	Ecomalb	634			L	91.50		P	[12]
map	Methionine aminopeptidase	F029.7		III.A.8	P07906	Ecomap	119	120		L	0.40	C	C	A. Ben-Bassat
mdh	Malate dehydrogenase	F030.2	1.1.1.37	I.B	P06994	Ecomdh1	526	527		L	70.71	N		T. Nystrom
metB	O-Succinylhomoserine(thiol)-lyase	G040.1	4.2.99.9	II.A.2	P00935	Ecometlb1	537	538		R	89.00			A. Martel
metC	Cystathionase, beta	G043.0	4.4.1.8	II.A.2	P06721	Ecometc	504	505		R	65.00		CG	A. Martel
metE	Tetrahydropteroyltriglutamate methylase	F088.0	2.1.1.14	II.A.2	P25665	Ecometer	551			R	85.50	PT	GP	C. Berg
metG	Methionine-tRNA ligase	F072.0	6.1.1.10	III.A.4	P00959	Ecometg	360	361		R	46.00	C	C	H. Weissbach
metG	Methionine-tRNA ligase	G072.0	6.1.1.10	III.A.4	P00959	Ecometg	360	361		R	46.00		C	H. Weissbach
metH	Homocysteine N5-methyltetrahydrofolate transmethylase	C137.0	2.1.1.13	II.A.2	P13009	Ecometh	631	632		R	91.00	T	C	R. Matthews
metK	Methionine adenosyltransferase	C044.6	2.5.1.6	I.B	P04384	Ecometk	473	474		R	63.70	C	CG	E. Hafner
mglB (w)	Galactose binding protein	C029.8		V.C	P02927	Ecomglb1	365	366		L	44.90	TC	C	R. Hogg, [12]
mopA	GroEL	B056.5		VI.B	P06139	Ecogroels	649			R	94.00	T	CG	R. Hendrix
mopB	GroES	C015.4		VI.B	P05380	Ecogroels	649			R	94.20	T	G	K. Tilly, [57]
mrr	Mrr	G033.6		III.A.7	P24202	Ecomrr	670			R	98.6		DG	pLC31-34 [45]
murE	meso-Diaminopimelate-adding enzyme	F051.0		IV.B	P22188	Eco2min	109			R	1.20			
mutS	MutS	E093.0		III.A.7	P23909	Ecomuts	449	450		R	58.80	DG	C	P. Modrich
nanA	N-Acetylneuraminate lyase	G032.6	4.1.3.3	I.B	P06995	EconanA	525	526		L	71.5		DG	pLC32-38 [45]
nusA	NusA	B061.0		III.A.5	P03003	Econusa	519	520		L	68.90	T	C	J.Greenblatt via D.Friedman
nusB	NusB	H013.8		III.A.5	P04381	Econusb	149			F	10.40		G	D. Friedman
ompA (x)	OmpA	F033.1		IV.A	P02934	Ecoompa	222	223		L	21.80	C	F	S. Mizushima
ompA (x)	OmpA	F024.0		IV.A	P02934	Ecoompa	222	223		L	21.80	C	F	S. Mizushima
ompA (x)	OmpA	F028.0		IV.A	P02934	Ecoompa	222	223		L	21.80	C	F	S. Mizushima
ompA (x)	OmpA	F033.0		IV.A	P02934	Ecoompa	222	223		L	21.80	C	F	S. Mizushima
ompC	OmpC	A035.5		V.E	P06996	Ecoompc	373			L	47.70	T	CG	L.van Alphen,S.Mizushima
ompF	OmpF	B036.0		V.E	P02931	Ecoompf	218	219		L	20.90	T	CG	L. van Alphen, M. Inouye, S. Mizushima
pckA (m)	Phosphoenolpyruvate carboxykinase (GTP)	E056.0	4.1.1.47	I.B	P22259	Ecopcka	620	621		R	75.10		CP	W. Bridger
pfkA	Phosphofructokinase I	F035.8	2.7.1.11	I.A	P06998	Ecocdha	540	541		R	88.20	T	C	D. Fraenkel
pfkB	Phosphofructokinase II	E036.6	2.7.1.11	I.A	P06999	Ecopfkbk	323	324		R	37.70	T	C	D. Fraenkel
pfl	Formate acetyltransferase	G070.0	2.3.1.54	I.D	P09373	Ecopfl	216			L	20.00	TC	CP	J. Knappe
pfl	Formate acetyltransferase	G074.0	2.3.1.54	I.D	P09373	Ecopfl	216			L	20.00	TC	CP	J. Knappe
pfs	p46	C025.4			P24247	Ecopfs						C	C	S. Quirk
pheS	Phenylalanine-tRNA ligase-alpha subunit	G036.0	6.1.1.20	III.A.4	P08312	Ecothrinf	322	323		L	37.50	T	CP	[28]
pheT	Phenylalanine-tRNA ligase, beta subunit	D094.0	6.1.1.20	III.A.4	P07395	Ecothrinf	322	323		L	37.50	T	CP	[28]
phoA	Alkaline phosphatase	F046.6	3.1.3.1	I.A	P00634	Ecophoaa	140	141		R	8.90	T	CP	M. Schlesinger

TABLE 2 *Continued*

Gene name(a)		Protein name (b)	A-N	E.C. Number(c)	Category of Function(d)	Accession number(e)	Sequence code(f)	Kohara phage number(g)		DIR (h)	Map (i)	Basis ID		Donor
												LG (j)	ID(k)	
phoE		PhoE	B037.0		IV.A	P02932	Ecophoe	127	128	R	5.80	PT	CP	J. Tommassen
phoP		PhoP	C025.3		I.A	P23836	Ecophopq	239		L	25.62		DG	pLC3-2 [45]
pnp		Polyribonucleotide nucleotidyltransferase	C078.0	2.7.7.8	III.A.5	P05055	Ecorpsop	518	519	L	68.80	TC	CG	C. Portier
polA		DNA-directed DNA polymerase I	F113.0	2.7.7.7	III.A.7	P00582	Ecopola	547	548	R	86.60	T	C	W. Kelley
poxB	*(y)*	Pyruvate oxidase	G058.0	1.2.3.6	I.A	P07003	Ecopoxb	210			19.00	C	C	J. Cronan, Jr.
ppc		Phosphoenolpyruvate carboxylase	F084.0	4..1.1.3	I.B	P00864	Ecoppcg	535	536	L	89.40		C	H. Katsuki
ppsA	*(t)*	Pyruvate water dikinase	B083.0	2.7.9.2	I.B	P23538	Ecopepsyn	322		L	37.20	C	CP	W. Bridger
prfA		Protein release factor 1	H021.6		III.A.8	P07011	Ecorrfx	247	248	R	26.70		C	P. Tai
prfA		Protein release factor 1	H021.7		III.A.8	P07011	Ecorfix	247	248	R	26.70		C	P. Tai
priA		protein n' Replication factor Y	H096.8		III.A.7	P17888	Ecopriafy	538		L	89.00	C	C	S. Wickner via K. Zahn
proV		Glycine betaine binding protein	G033.5		V.A	P14175	Ecoprou	444	445	R	58.00		C	M. Villarejo
pspA		PS (phage shock) protein	D026.3										CP	J. Brisette, P. Model
pspA		PS (phage shock) protein	E026.0		VI.H	P23853	Ecopsp						CP	J. Brisette, P. Model
pta		Phosphotransacetylase	E079.0	2.3.1.8	I.A						49.60	NG		T. Nystrom
ptr		Protease III	G095.0		III.B.3	P05458	Ecoptr	459	460	L	60.70	TDG	C	D. Zipser
ptsG		Glucosephosphotransferase enzyme II	H042.6		V.A	P05053	Ecoptsg	236		R	24.40		C	C. Bouma
ptsG		Glucosephosphotransferase enzyme II	H042.7		V.A	P05053	Ecoptsg	236		R	24.40		C	C. Bouma
ptsH		Phosphohistidinoprotein-hexose phosphotransferase	F007.0	2.7.1.69	V.A	P07006	Ecoptsh	418		L	52.20	C	C	S. Roseman, E.B. Waygood
ptsI		Phosphotransferase system enzyme I	B058.3	2.7.3.9	V.A	P08839	Ecophosys	418		R	52.20	TC	C	E.B. Waygood
purC	*(z)*	Phosphoribosylaminoimidazde-succinocarboxamide synthetase	B026.3	6.3.2.6	II.B.1	P21155	Ecopurca	423	424	L	53.00			[40]
purF		Amidophosphoribosyltransferase	E054.1	2.4.2.14	II.B.1	P00496	Ecopurf	406		L	50.00		C	H. Zalkin
pyrB		Aspartate carbamyltransferase-beta	H031.3	2.1.3.2	II.B.2	P00479	Ecopyrbi	658		R	96.50	C	C	W. Lipscomb
pyrE		Orotate phosphoribosyltransferase	D023.3	2.4.2.10	II.B.2	P00495	Ecodutpyr	572	573	R	81.9		DG	pLC17-24 [45]
pyrF		Pyruvate kinase I	G054.7	2.7.1.40	II.B.2	P08244	Ecopk1	255		R	37.80	T	C	M. Malcovati
pyrG		CTP synthetase	G060.3	6.3.4.2	II.B.2	P08398	Ecopyrg	455		L	59.70		DG	pLC11-8, pLC10-47 [45]
pyrI		Asparate carbamoyltransferase-alpha	I017.5	2.1.3.2	II.B.2	P00478	Ecopyrbi	658		L	96.50		C	W. Lipscomb
rbsB		Ribose-binding protein	H027.9		V.A	P02925	Ecorbs	558	559	R	84.30		C	[12]
recA		RecA	C039.3		III.A.7	P03017	Ecoreca	446		L	58.20	T	CGP	J. Little, [19]
rhaD	*(m)*	Rhamnulose-1-phosphate aldolase	G030.0	4.1.2.19	I.A	P32169	Ecorhabad			F	88.00		CMP	D. Feingold
rhaD	*(m)*	Rhamnulose-1-phosphate aldolase	F029.4	4.1.2.19	I.A	P32169	Ecorhabad			F	88.00		CMP	D. Feingold
rho		Transcription termination factor rho	I048.8		III.A.5	P03002	Ecorho	555	556	R	84.80		CG	[7]
rimL		Ribosomal protein L7/L12 acetyltransferase	F037.8		III.A.2	P13857	Ecoriml	270	271	R	31.40	T	C	H. Weissbach
rimL		Ribosomal protein L7/L12 acetyltransferase	F038.5		III.A.2	P13857	Ecoriml	270	271	R	31.40	T	C	H. Weissbach
rlpA		36kd lipoprotein	F037.4		IV.A	P10100	Ecorlpa	168	169	L	15.95		DG	pLC30-18 [45]
rplA		Ribosomal subunit protein L1	I026.0		III.A.2	P02384	Ecorplrpo	gap		R	89.90		CF	M. Nomura
rplB	*(aa)*	Ribosomal subunit protein L2	I030.2		III.A.2	P02387	Ecorpos10	628		L	73.10		CM	M. Nomura
rplC		Ribosomal subunit protein L3	I023.0		III.A.2	P02386	Ecorpos10	628		L	73.30		CF	M. Nomura
rplD		Ribosomal subunit protein L4	I021.1		III.A.2	P02388	Ecorpos10	628		L	73.20		CF	M. Nomura
rplE		Ribosomal subunit protein L5	I016.4		III.A.2	P02389	Ecorpln	628		L	72.80		CF	M. Nomura
rplF		Ribosomal subunit protein L6	I017.2		III.A.2	P02390	Ecorpln	628		L	72.70		CF	M. Nomura
rplI		Ribosomal subunit protein L9	H014.0		III.A.2	P02418	Ecorpsfri	655		R	95.60	C	CF	M. Nomura
rplK		Ribosomal subunit protein L11	I014.1		III.A.2	P02409	Ecorplrpo	gap		R	89.90		CF	M. Nomura
rplL		Ribosomal subunit protein L7	A013.0		III.A.2	P02392	Ecorplrpo	gap		R	90.00	T	CF	M. Nomura, [2]
rplL		Ribosomal Subunit protein L12	B013.0		III.A.2	P02392	Ecorplrpo	gap		R	90.00	T	CF	M. Nomura, [2]
rplQ	*(aa)*	Ribosomal subunit protein L17	I014.9		III.A.2	P02416	Ecorpa	gap		L	72.40		C	M. Nomura
rplU		Ribosomal subunit protein L21	I013.2		III.A.2	P02422	Ecorplrpm			F	69.10		CF	[43]
rplW		Ribosomal subunit protein L23	I013.0		III.A.2	P02424	Ecorpos10	628		L	73.10		C	M. Nomura
rplY		Ribosomal subunit protein L25	I011.8		III.A.2	P02426	rplYeco	371		R	47.10		CF	M. Nomura
rpmA		Ribosomal subunit protein L27	I012.8		III.A.2	P02427				F	69.20		C	M. Nomura
rpmB	*(aa)*	Ribosomal subunit protein L28	I012.2		III.A.2	P02428	Ecorpmbq	572	573	L	81.80		C	M. Nomura
rpmC		Ribosomal subunit protein L29	I011.1		III.A.2	P02429	Ecorpos10	628		L	73.00		C	M. Nomura
rpmD		Ribosomal subunit protein L30	I011.4		III.A.2	P02430	Ecorpln	628		L	72.60		CF	M. Nomura
rpmF	*(aa)*	Ribosomal subunit protein L32	I011.7		III.A.2	P02435	Ecorpmfq	234	235	L	23.40		C	M. Nomura
rpoA		DNA-directed RNA polymerase-alpha	B040.7	2.7.7.6	III.A.5	P00574	Ecorpa	628		L	72.40	T	CG	H. Weissbach, [43]
rpoB		DNA-directed RNA polymerase-beta	D157.0	2.7.7.6	III.A.5	P00575	Ecorplrpo	533	534	R	90.00		CG	H. Weissbach, [43]
rpoC		DNA-directed RNA polymerase-beta prime	I160.0	2.7.7.6	III.A.5	P00577	Ecorplrpo	533	534	R	90.00		CG	H. Weissbach, [43]
rpoD		DNA-directed RNA polymerase-sigma 70	B082.0	2.7.7.6	I.F	P00579	Ecorpsrpo	509	510	R	67.00		C	H. Weissbach
rpoH		DNA-directed RNA polymerase-sigma 32	F033.4	2.7.7.6	I.F	P00580	Ecohtprr	612	613	R	76.10		G	[30]
rpsA		Ribosomal subunit protein S1	B065.0		III.A.2	P02349	Ecorpsa	217		R	20.50	T	CF	M. Nomura, [43]
rpsB		Ribosome subunit protein S2	H027.0		III.A.2	P02351	Ecorpsbts	120		R	4.00		F	[7]
rpsE		Ribosomal subunit protein S5	I014.4		III.A.2	P02356	Ecorpln	628		L	72.70		CF	M. Nomura
rpsF		Ribosomal subunit protein S6B	C014.8		III.A.2	P02358	Ecorpsfri	655		R	95.40	T	CF	M. Nomura, [46]
rpsF		Ribosomal subunit protein S6C	C015.3		III.A.2	P02358	Ecorpsfri	655		R	95.40	T	F	[46]
rpsF		Ribosomal subunit protein S6A	D014.7		III.A.2	P02358	Ecorpsfri	655		R	95.40	T	CF	M. Nomura, [46]
rpsH		Ribosomal subunit protein S8	I013.5		III.A.2	P02361	Ecorpln	628		L	72.80		CF	M. Nomura
rpsI		Ribosomal subunit protein S9	I014.6		III.A.2	P02363	Ecorpsi	525	526	F	70.00		CM	M. Nomura
rpsJ		Ribosomal subunit protein S10	I012.1		III.A.2	P02364	Ecorpos10	628		L	73.30		CF	M. Nomura
rpsK	*(aa)*	Ribosomal subunit protein S11	I013.8		III.A.2	P02366	Ecorpa	628		L	72.50		C	M. Nomura
rpsO		Ribosomal subunit protein S15	I012.0		III.A.2	P02371	Ecorpsop	518	519	L	68.80		CF	M. Nomura
rpsP		Ribosomal subunit protein S16	I011.9		III.A.2	P02372	Ecotrmd	438	439	L	56.80		C	M. Nomura
rpsQ		Ribosomal subunit protein S17	I012.6		III.A.2	P02373	Ecorpos10	628		L	72.90		C	M. Nomura
rpsT	*(aa)*	Ribosomal subunit protein S20	I012.9		III.A.2	P02378	Ecorpsta	102	103	L	0.50		C	M. Nomura
sdhA		Succinate dehydrogenase	F060.3	1.3.99.1	I.C	P10444	Ecoglta	176		R	16.30	PT	C	J. Weiner

(Table continues)

TABLE 2 *Continued*

Gene name(a)	Protein name (b)	A-N	E.C. Number(c)	Category of Function(d)	Accession number(e)	Sequence code(f)	Kohara phage number(g)			DIR (h)	Map (i)	Basis ID LG (j)	ID(k)	Donor
secB	SecB	A015.6		V.D	P15040	Ecosecb	575			L	81.1		DG	pLC19-11 [45]
selA	Selenocysteine synthase	G050.7		III.A.4	P23328	Ecosela	576			L	80.45		DG	pLC11-17 [45]
selB	SelB	H068.9		III.A.8	P14081	Ecoselb	577			L	80.45		DG	pLC11-17 [45]
serA	D-3-phosphoglycerate dehydrogenase	G044.1	1.1.1.95	I.A	P08328	Ecosera	470	471		L	38.29		DG	pLC15-34 [45]
serS	Serine-tRNA ligase	E048.8	6.1.1.11	III.A.4	P09156	Ecosers	215			R	19.20	TC	C	H. Weissbach
slt	Murein transglycosylase-soluble form	H063.0		IV.B	P03810	Ecoslty	675	676		R			C	U. Schwarz
sodA	Superoxide dismutase-manganese protein	I021.3	1.15.1.1	I.C	P00448	Ecosoda	541			R	88.00		CP	J. Fee, P. Reichard
sodB	Superoxide dismutase-iron protein	F021.3	1.15.1.1	I.C	P09157	Ecosodb	317	318	319	R	36.30		C	J. Fee
speC	Ornithine decarboxylase	F080.0	4.1.1.7	II.F	P21169	Ecospec	476			L	64.00		G	S. Boyle
ssb	Single-stranded DNA binding protein ssb	F018.8		III.A.7	P02339	Ecossb	637	638		R	92.10	C	C	R. McMacken, K. Williams
sspA	Stringent starvation protein	D027.1		I.F	P05838	Ecosspg	525	526		L	70.40	G	G	M. D. Williams
sspB	SspB	A025.8			P25663	Ecosspb	525	526		L	70.40	G	G	M. D. Williams
sucA	Oxoglutarate dehydrogenase-lipoamide	G097.0	1.2.4.2	I.B	P07015	Ecoglta	176			R	16.70	CPT	CGM	R. Perham
sucB	Dihydrolipoamide succinyltransferase	F050.3	2.3.1.61	I.B	P07016	Ecoglta	176	177		R	16.70	C	CMP	R. Perham
sucC	Succinate-CoA ligase-(ADP-forming) beta	E039.8	6.2.1.5	I.B	P07460	Ecoglta	176	177		R	16.30	PT	CP	W. Bridger
sucD	Succinate-CoA ligase-(ADP-forming) alpha	H028.0	6.2.1.5	I.B	P07459	Ecoglta	176	177		R	16.30		CP	W. Bridger
tag	1,3-Methyladenine DNA glycosylase	H021.1		III.A.7	P05100	Ecotag	602			R	79.85		DG	pLC1-3 [45]
tar	Methyl-accepting chemotaxis protein II	D058.1		V.C	P07017	Ecotartap	338	339		L	41.60		L	[14]
tar	Methyl-accepting chemotaxis protein II	D058.2		V.C	P07017	Ecotartap	338	339		L	41.60		L	[14]
tar	Methyl-accepting chemotaxis protein II	D058.3		V.C	P07017	Ecotartap	338	339		L	41.60		L	[14]
tar	Methyl-accepting chemotaxis protein II	E058.1		V.C	P07017	Ecotartap	338	339		L	41.60		L	[14]
tar	Methyl-accepting chemotaxis protein II	E058.2		V.C	P07017	Ecotartap	338	339		L	41.60		L	[14]
tar	Methyl-accepting chemotaxis protein II	E058.3		V.C	P07017	Ecotartap	338	339		L	41.60		L	[14]
tdh	Threonine 3-dehydrogenase	G038.8	1.1.1.103	I.A	P07913	Ecokbltdh	574	575		81	81.20		C	B. Ernsting, R. Matthews
thrS	Threonine-tRNA ligase	G065.0	6.1.1.3	III.A.4	P00955	Ecothrinf	322	323		L	37.60		C	[28]
thyA	Thymidine synthase	G029.3	2.1.1.45	II.B.3	P00470	EcothyA	460	461		L	60.90	DG		
tig	Trigger factor	B050.3		V.B	P22257	Ecotig	148			R	9.00	T	C	W. Wickner
tnaA (m)	Tryptophanase	G046.5	4.1.99.1	I.A	P00913	Ecotnaa	563			R	83.30		CP	M. Goldberg
tnpR (ab)	TnpR (resolvase from Tn3)	I016.0			P03011	Ecotn3x				F			C	N. Cozzarelli, J. Dungan
topA	DNA-topoisomerase I	I115.0	5.99.1.2	III.A.7	P06612	Ecotopa	254			R	27.90		C	N. Cozzarelli, F. Dean
trmA	tRNA uracil-5-methyltransferase	G039.6	2.1.1.35	III.A.4	P23003	Ecotrma	534	535		L	89.50		C	T. Ny
trpA	Tryptophan synthase-alpha subunit	E025.5	4.2.1.20	II.A.4	P00928	Ecotrpx	251	252		L	27.70	CT	CGP	E. Wilson Miles, C. Yanofsky
trpB	Tryptophan synthase-beta subunit	G040.0	4.2.1.20	II.A.4	P00932	Ecotrpx	251	252		L	27.70	T	CGP	E. Wilson Miles, C. Yanofsky
trpC	Indole-3-glycerol-phosphate synthase	F045.4	4.1.1.48	II.A.4	P00909	Ecotrpx	252			L	27.70		CGMP	C. Yanofsky
trpC	Indole-3-glycerol-phosphate synthase	F045.5	4.1.1.48	II.A.4	P00909	Ecotrpx	252			L	27.70		CGMP	C. Yanofsky
trpD	Anthranilate phosphoribosyltransferase	H054.5	2.4.2.18	II.A.4	P00904	Ecotrpx	252			L	27.70	C	CGP	C. Yanofsky
trpE	Anthranilate synthase	E050.0	4.1.3.27	II.A.4	P00895	Ecotrpx	252			L	27.70		CGP	C. Yanofsky
trpR	TrpR	D012.4		II.A.4	P03032	Ecotrpr	675	676		R	99.65		DG	
trpS	Tryptophan-tRNA ligase	H037.0	6.1.1.2	III.A.4	P00954	Ecotrps	622	623		L	74.30		CG	K. Muench, C. Yanofsky
trxA	Thioredoxin	B012.0		II.D.10	P00274	Ecotrxa	555	556		L	85.20	C	CG	C. Lunn, M. Russel, C. Williams
trxB	Thioredoxin reductase-NADPH	D031.5	1.6.4.5	I.B	P09625	Ecotrxb	213	214	215	L	20.00	TC	CG	M. Russel, C. Williams
tsf	Protein chain elongation factor, Ts	C030.7		III.A.8	P02997	Ecorpsbts	120			R	4.00	T	CG	[43]
tsf	Protein chain elongation factor, Ts	C031.6		III.A.8	P02997	Ecorpsbts	120			R	4.00	T	CG	[43]
tsr	Methyl-accepting chemotaxis protein I	B060.1		V.C	P02942	Ecotsr	671			R	99.20		L	[14]
tsr	Methyl-accepting chemotaxis protein I	B060.2		V.C	P02942	Ecotsr	671			R	99.20		L	[14]
tsr	Methyl-accepting chemotaxis protein I	B060.3		V.C	P02942	Ecotsr	671			R	99.20		L	[14]
tsr	Methyl-accepting chemotaxis protein I	B060.4		V.C	P02942	Ecotsr	671			R	99.20		L	[14]
tsr	Methyl-accepting chemotaxis protein I	B060.5		V.C	P02942	Ecotsr	671			R	99.20		L	[14]
tsr	Methyl-accepting chemotaxis protein I	B060.6		V.C	P02942	Ecotsr	671			R	99.20		L	[14]
tsr	Methyl-accepting chemotaxis protein I	B060.7		V.C	P02942	Ecotsr	671			R	99.20		L	[14]
tsr	Methyl-accepting chemotaxis protein I	B060.8		V.C	P02942	Ecotsr	671			R	99.20		L	[14]
tufA	Protein chain elongation factor, Tu	E042.0		III.A.8		Ecostr3	626			L	73.30	T	CG	[43]
tyrA	Chorismate mutase T-prephenate dehydrogenase	F037.9	5.4.99.5	II.A.4	P07023	Ecopheab	438			L	57.60		C	R. Duggleby
tyrB	Aromatic amino acid amino transferase	E036.0	2.6.1.57	II.A.4	P04693	Ecotyrb	636	637		R	92.00		CG	S. Kiramitsu
tyrS	Tyrosine-tRNA ligase	G044.0	6.1.1.1	III.A.4	P00951	Ecotyrs	316			L	36.00	C	C	H. Weissbach
uhpA	UhpA	G020.9		V.A	P10940	Ecouhp	568	569		L	82.5		DG	pLC34-43 [45]
umuD	UmuD	A015.1		III.A.7	P04153	Ecoumucd	243	244		R	26.25		DG	G.C Walker (pSE115)
uspA	UspA isoform II	C013.4		VI.H	P28242	Ecouspa	608	609		R	77.00	?		T. Nystrom
uspA	UspA	C013.5		VI.H	P28242	Ecouspa	608	609		R	77.00	T	PDG	T. Nystrom
uvrA	UvrA	H124.0		III.A.7	P07671	Ecouvra	636			L	92.00		C	L. Grossman, A. Sancar
uvrB	UvrB	C080.0		III.A.7	P07025	Ecouvrb	201	202		R	17.60		C	L. Grossman, A. Sancar
uvrC	UvrC	I065.0		III.A.7	P07028	Ecouvrca	340	341		L	42.10		C	L. Grossman, A. Sancar
valS	Valine-tRNA ligase	E106.0	6.1.1.9	II.A.4	P07118	Ecovals	658	659		L	96.80	TDG	CG	[28]
xthA	Exodeoxyribonuclease III-endonuclease II	G028.2	3.1.11.2	III.B.2	P09030	Ecoxtha	328			R	38.30	T	CG	B. Weiss
xylB	Xylulokinase	G041.9	2.7.1.17	I.A	P09099	Ecoxylaba	580	601		L	80.10		C	D. Fraenkel
xylF	Xylose-binding protein	C032.4		?	P37387	?				?	?	PT	P	[28]
yhhF	YhhS	H023.2			P10120	Ecoftsyex	613	612		L	76.50	TDG	DG	
yicC	YicC-orfX of Ecorfpyre	C033.0			P23839	Ecorfpyre	572			R	81.9		DG	pLC17-24 [45]
yicC	YicC-orfX of Ecorfpyre	C033.1			P23839	Ecorfpyre	572			R	81.9		DG	pLC17-24 [45]
yjjB	p-14	B014.8			P18389	Ecodnatc	671	672		L	98.9		DG	pLC28-5, pLC24-13 [45]
zwf	Glucose-6-phosphate dehydrogenase	F048.8	1.1.1.49	I.B	P22992	Ecozwf	gap			L	40.80	C	CM	D. Fraenkel

TABLE 2 *Continued*

Gene name(a)	Protein name (b)	A-N	E.C. Number(c)	Category of Function(d)	Accession number(e)	Sequence code(f)	Kohara phage number(g)	DIR (h)	Map (i)	LG (j)	ID(k)	Donor
	62Kd protein-J Nielson	G057.1									C	J. O. Lampen, J. Nielson
	Antigen 43	B052.0									C	P. Owen
	Antigen 47	B027.1								T	C	P. Owen
	Cold shock protein	B046.5								P	P	[21]
	Cold shock protein	G041.2								T	P	[21]
	Cold shock protein	G055.0									P	[21]
	Cysteine-binding protein	C026.5									C	[12]
	DNA-topoisomerase II prime sub	F044.6	5.99.1.3								C	A. Morrison
	Glutamate-asparate-binding protein	H030.1									CG	[12]
	GP2-(a glycerol-3-phosphate binding protein)	H041.0									CP	W. Boos
	HP33A	I034.0									C	T. Formosa, B. Alberts
	KS protease	H083.0									C	Karen Silber
	KS protease	H083.1									C	Karen Silber
	Lysine-tRNA ligase form III	C058.4	6.1.1.6							P	CG	I. Hirschfield
	Lysine-tRNA ligase form IV	C060.5	6.1.1.6							P	CG	I. Hirschfield
	NADH oxidase	B035.1								T		
	NADPH: flavodoxin/ferridoxin oxidoreduct	H030.5									C	J. Knappe
	Origin binding protein a	G055.3									C	M. Schaechter
	Polyhook	A039.8									C	P. Matsamura
	Protein chain elongation factor, P	C022.7								C	C	C. Ganoza
	R-protein	G029.7									C	R. Matthews
	Rescue	D046.4									C	C. Ganoza
	Ribosomal subunit protein	I010.3										[22]
	Ribosomal subunit protein	I011.6								F		[22]
	Ribosomal subunit protein	I012.3								F		[22]
	Ribosomal subunit protein	I013.3								F		[22]
	Ribosomal subunit protein	I013.7								F		[22]
	Ribosomal subunit protein	I014.3								F		[22]
	Ribosomal subunit protein	I014.5								F		[22]
	Ribosomal subunit protein	I014.7								F		[22]
		A005.1					544	L		G		
		A013.5					109	R		G		
		A018.5					109	R		G		
		A034.0					580	R		G		
		A034.5					580	R		G		
		A090.0					613	L		G		
		B014.3					544	L		G		
		B068.0					613	R		G		
		B096.0					657	R		G		
		C023.4					545	L		G		
		C024.0					657	L		G		
		C024.5					580	R		G		
		C025.8					545	L		G		
		C031.5					580	R		G		
		C036.3					580	R		G		
		C037.6					545	L		G		
		C041.0					109	R		G		
		C047.0					109	L		G		
		C065.0					544	R		G		
		D037.2					545	L		G		
		D062.5					580	R		G		
		E053.0					109	L		G		
		F011.7					657	R		G		
		F021.4					657	L		G		
		F024.4					613	L		G		
		F027.1					544	L		G		
		F028.9					613	L		G		
		F029.1					613	L		G		
		F044.0					580	L		G		
		F071.5					460	L		G		
		F076.0					449	R		G		
		G018.0					449	L		G		
		G031.4					544	L		G		
		G032.2					109	L		G		
		G032.9					544	L		G		
		G033.2					109	L		G		
		G036.5					109	L		G		
		G048.1					109	R		G		
		G049.0					460	L		G		
		G050.6					580	R		G		
		H010.5					107	R		G		
		H011.7					657	R		G		
		H027.5					580	L		G		
		H040.6					580	R		G		

TABLE 2 *Footnotes next page*

TABLE 2 *Continued*

[a]Gene names are from Bachmann (4).

[b]Protein names are from *Enzyme Nomenclature* (35) and Bachmann (4). CoA, coenzyme A.

[c]EC numbers are from *Enzyme Nomenclature* (35).

[d]Category of function is from Riley (48).

[e]SWISS-PROT accession numbers are from Release 28.

[f]Sequence code for the DNA sequence as found in GenBank or EMBL database.

[g]Kohara miniset number as described in reference 23. The placement of genes was found in the Genescape software program.

[h]DIR, direction of transcription of a gene. The direction was found in the Genescape software program. R, clockwise; L, counterclockwise.

[i]Map approximate position of the gene on the *E. coli* genetic map (4).

[j]LG, basis of identification of a protein spot on Fig. 1. Letter designations are as follows: T, visual transfer from Fig. 2A or from 2-D gels of other investigators; N, N-terminal sequence of the protein spot; C, comigration with purified protein; D, deduced from DNA sequence information as described in Fig. 5; G, genetic criterion, e.g., mutants (deletion, insertion, frameshift, nonsense, missense, regulatory), plasmid-bearing strains, and in vitro synthesis of protein; P, physiological criterion, e.g., induction and repression.

[k]ID, basis of identification in Fig. 2 and 3. Letter designations are as follows: C, comigration with purified protein; D, deduced from DNA sequence information as described in Fig. 5; F, migration with polypeptide or purified cellular fraction; G, genetic criterion, e.g., mutants (deletion, insertion, frameshift, nonsense, missense, regulatory), plasmid-bearing strains, and in vitro synthesis of protein; L, selective labeling, e.g., methylation and phosphorylation; M, peptide map similarity; N, N-terminal sequencing of protein; P, physiological criterion, e.g., induction and repression; Z, selective derivatization, e.g., covalent binding of penicillin.

[l]ahpC was identified by C. Pasquali, D. Schaller, R. D. Appel, J.-C. Sanchez, O. Golaz, I. Humphery-Smith, M. Wilkins, G. J. Hughes, S. Frutiger, N. Paquet, and D. F. Hochstrasser.

[m]Proteins repressed by growth in glucose-containing medium or induced by specific growth conditions do not appear on the autoradiograms in Fig. 1 and 2.

[n]Copeland et al. (12) reported arabinose-binding protein to migrate at ca. 86 by 73 (Fig. 1). The purified proteins used by our two groups were supplied by different donors; we have not yet reconciled this discrepancy.

[o]The large dark spot at these coordinates is F088.0, which obscures G099.0. When cells are grown in methionine-containing medium, synthesis of F088.0 is repressed, and G099.0 can be clearly seen. Similarly, gels of extract of a mutant, CBK040 (*metE*::Tn5), which does not produce F088.0, allow the clear identification of G099.0 as asparagine-tRNA ligase.

[p]Squires et al. (55) reported this protein as F68.5 but stated that it was the same as protein E72.0 in reference 58.

[q]DppA comigrates with the 56-kDa protein listed in previous editions of the database (63).

[r]The unexpected finding that this spot corresponds to two different enzymes has not yet been resolved.

[s]H032.0 is the phosphorylated form of Era (52a).

[t]gadA was gadS, gltX was gltM, lpdA was lpd, and ppsA was pps in edition 5 (63). Gene names are updated from Riley (48).

[u]Integration host factor (IHF) subunits alpha and beta have previously been resolved only on urea-SDS gels (H. Nash, personal communication).

[v]hslU was htpI, and hslV was htpO in edition 5 (63). Gene names are updated from Chuang and Blattner (10).

[w]Copeland et al. (12) reported galactose-binding protein to migrate at ca. 77 by 73 (Fig. 1). The small difference in spot patterns is probably due to variations in electrophoresis methods that resulted in slightly different spot patterns.

[x]These proteins are solubilized when the cell extract is prepared by boiling in the presence of SDS and therefore appear faintly, if at all, on the autoradiograms of Fig. 1 and 2.

[y]The faintness of the spot at these coordinates is misleading. Pyruvate oxidase migrates with a much more abundant polypeptide spot (intensity approximately the same as that at 45 by 107 located at these coordinates on other gels.

[z]Parker (40) reported this protein as C027.0, although he reports its migration at 91 by 64, as our group has. The alphanumeric designation was originally B026.3 migrating at 91 by 64 (43).

[aa]These proteins do not appear in the autoradiogram in Fig. 2B. Some nonequilibrium gels have more proteins on the basic side of those shown in Fig. 2B, with which these ribosomal proteins comigrate. Coordinates given are approximate locations of the proteins if they were to appear on the reference gel.

[ab]This spot would appear only when extracts were prepared from cells carrying transposon Tn5.

way. For example, this information was used as the basis for experiments that used a novel approach to estimating the growth rate of *Salmonella typhimurium* (official designation, *Salmonella enterica* serovar Typhimurium) while these bacteria resided within macrophage host cells (1). Within a certain range, the levels of various translation factors and ribosomal proteins vary directly with growth rate (43). The level of ribosomal protein L7/L12 seen on 2-D gels produced from intracellular *S. typhimurium* suggested that the intracellular bacteria were growing rapidly. Prior to these experiments, the growth rates of intracellular bacteria had been estimated solely by counting viable bacteria following lysis of the host cells. The viable-count approach had indicated that intracellular *S. typhimurium* cells were growing quite slowly. These contrasting results led to further experiments in which it was determined that the intracellular bacteria consisted of at least two populations, one not dividing but viable and the other rapidly dividing (1).

A third type of query of the database is used to identify cellular trends for proteins. For example, Savageau used the database to look at the distribution of MW of proteins (52). Similar types of

distributions for pI, amino acid usage, abundances of different classes of proteins, and even consensus sequences within the promoter regions for sets of coregulated genes could be determined by using the information in the database, especially as the number of identified proteins (in the Genome Expression Map project) and the number of conditions (in the Response/Regulation Map project) increases to represent a larger fraction of the total number of *E. coli* proteins. Once 2-D gel databases for other bacterial species are initiated, interesting comparative studies will be possible.

REFERENCE 2-D GELS

Five figures are included in the database: the three reference 2-D gels published in a previous edition of the database (63) (Fig. 1–3), one new reference 2-D gel that represents the Response/Regulation Map (Fig. 4), and a figure that gives the distributions of MW and pI for the proteins identified on these reference gels (Fig. 5). The reference gels are overlaid with grids, and the exact coordinates for each protein in the database are listed in Table 1 under the

(Continued on page 2115)

TABLE 3   Amount of various proteins in different steady-state and transient growth conditions

A-N(a)	R R M (b)	STEADY STATE GROWTH CONDITIONS (c)												GROWTH TRANSITION CONDITIONS (d)					GENE
		AB a'	Ace RL	Gly RL	Ric RL	13.5 RL	15 RL	23 RL	30 RL	42 RL	46 RL	Phn PPM	Phn Ratio	PS PPM	PS-E Ratio	PS-L Ratio	NS PPM	NS-L Ratio	
A013.0	R 1701	2.93	0.8	0.9		2.2			1.0	0.8	0.5	7926	* 1.06	8095	1.54	2.67	14244	?	rplL
A017.6	R 2244											108	* 1.19	136	1.37	1.43	239	1.26	greA
A019.0	R 2623											101	* 2.72	140	1.89	*1.45	228	0.00	fldA
A025.8	R 2931											124	* 0.00	0	-	-	97	?	sspB
A028.0	R 2111											122	* 2.39	206	0.76	0.81	239	0.00	cheZ
A033.7	R 2102											149	* 1.47	203	0.64	*0.52	229	1.17	
A035.5	R 1700											7945	* 1.06	12574	0.72	0.92	11894	1.78	ompC
A036.1	R 1944											2481	* 1.44	4103	0.92	1.32	4591	1.32	dnaN
A041.3	R 2178											190	* 1.91	247	0.35	*0.35	311	0.61	
A048.0	R 1945											90	* 1.45	69	2.45	1.30	148	* ?	
A053.0			1.0	1.3	0.9														
A165.0			0.9	1.0	0.2	0.9	0.2	0.7	0.8	0.7	25.								
B011.0			1.1	1.0	1.0														grx
B012.0		0.6	1.0	0.9	0.9														trxA
B013.0	R 2292	4.5	0.2	0.7		0.2			0.7	0.6	0.3	866	0.57	1080	0.99	?	977	* 0.00	rplL
B014.5	R 2246											250	20.85	1390	6.59	*6.13	504	?	
B015.0	R 2245											800	* 2.47	1872	1.28	*1.21	1200	?	
B018.4			1.2	1.2	0.6	3.4	1.9	1.6	1.3	1.0	1.4								
B018.7			0.7	0.7	0.9	1.2	0.6	0.6	1.1	1.1	1.3								crr
B020.0	R 2345	2.61	0.8	1.1	1.4							267	?	158	2.42	2.13	280	1.04	folA
B020.9	R 2204		1.2	1.1	1.1	0.9	0.6	0.6	0.8	1.3	2.1	4429	* 2.24	7750	1.75	1.92	10745	0.21	ahpC
B024.1	R 2750											200	* 1.71	260	?	*0.82	304	* 0.00	
B024.2	R 2177											302	* 1.20	388	0.95	0.97	366	0.43	
B025.3	R 2176	1.17	0.7	0.8	1.2	1.8			1.4	1.8	3.4	983	* 1.18	1710	1.48	1.67	2205	* 0.40	grpE
B026.3		4.83	0.8	1.1	0.3														purC
B027.0	R 2724											-	>273	0	>463	>492	0	>644	
B028.3	R 2477											367	* 0.65	0	>408	>440	0	>484	
B033.0	R 2100											1673	* 1.22	2219	1.05	1.30	2714	0.79	
B035.1	R 2101	0.55	1.0	1.0	0.7	0.6	0.6	0.5	0.8	1.2	0.9	301	2.64	805	0.63	0.53	681	?	
B036.0	R 2015		0.9	1.3	0.4							5310	* 0.49	9771	0.11	0.08	12470	* 0.07	ompF
B037.0	R 2801											331	* 0.00	281	?	0.00	373	0.00	phoE
B040.5	R 2728											-	* -	0	>650	>848	85	?	
B040.7	R 1928	3.66	0.7	0.9	1.4	0.8	0.8	0.9	1.1	1.0	1.0	2853	* 0.93	5231	0.47	0.37	4499	0.21	rpoA
B046.5	R 2831											-	?	0	-	-	67	?	
B046.7	R 1861	5.6	1.1	1.1	0.9	1.1	1.2	1.2	1.1	0.9	1.6	4515	0.74	6391	0.48	0.31	8926	* 0.03	atpD
B050.0	R 1859	6.1	0.5	0.9	1.5	1.1	1.0	1.3	1.2	0.8	0.6	4191	* 0.94	4308	0.45	0.26	5793	* 0.12	tig
B056.5	R O565	13.50	0.9	1.0	1.2	0.9	0.8	0.6	0.7	1.6	7.2	6255	1.72	8809	1.50	1.42	15115	* 0.08	mopA
B058.3	R 1790	3.11	0.3	1.3								1108	* 4.40	3631	0.92	1.06	5718	* 0.05	ptsI
B061.0	R O610	1.25	0.5	0.6		1.1	1.1	1.1	1.0	0.9	0.5	672	1.86	944	0.66	0.71	1173	* 0.19	nusA
B065.0	R O650	15.10	0.5	0.8	1.7	0.7	0.7	0.9	1.0	0.9	0.4	3878	1.09	4047	0.78	0.80	8415	* ?	rpsA
B066.0	R O660	9.33	0.6	0.8	1.5	0.6	0.6	0.6	0.7	1.4	3.0	5026	* 0.70	4031	1.25	1.26	2500	3.14	dnaK
B082.5		0.2	2.8	1.9															
B083.0		0.95	2.4	2.2															ppsA
C013.4	R 2296											121	* 1.80	178	0.00	0.00	208	3.12	uspA
C013.5	R O135		1.1	1.1	0.4							1668	3.10	5079	1.64	1.60	6793	0.60	uspA
C014.3	R 2263											683	* 1.55	903	1.30	1.39	1225	0.63	bfr
C014.7	R 2265	0.81	1.8	1.1	1.0	1.2			0.7	2.5	8.0	286	* 1.15	308	0.37	?	371	0.00	ibpB
C014.8	2307	0.9	0.5	0.8								2588	* 0.89	4904	0.08	0.18	6986	?	rpsF
C015.3	R 2241	1.2	0.5	0.9								611	* 1.15	1339	0.26	*0.20	2348	* 0.33	rpsF
C015.4	R O154	2.44	0.9	0.9	1.0	0.6			0.5	1.3	5.6	531	2.52	1357	2.06	2.05	1803	1.20	mopB
C017.7	R 2794											373	0.84	801	?	?	0	-	
C018.0	R 2206											1744	* 2.40	2719	3.48	3.35	3487	1.72	
C022.7	R 2203		0.6	0.9	1.4	0.5	0.6	0.8	0.9	1.0	1.1	575	1.09	1023	0.72	0.70	903	0.57	
C023.0	R 2202											501	* 1.49	586	1.05	0.64	756	0.88	
C025.2	R 2166											480	* 2.95	537	3.08	1.69	629	1.54	
C026.8	R 2165											154	38.20	437	14.93	17.75	270	* 0.00	atoA
C027.1	R 2160											700	?	687	0.56	*0.51	1111	17.32	
C028.5		2.5	0.6	0.9	0.3														hisJ
C029.1	R 2090											253	1.63	407	*1.90	1.69	408	1.88	
C029.8			27.	22.															mglB
C030.7	R 2481	1.3	0.6	0.8	1.6	0.8	0.5	0.8	1.0	1.0	0.5	1739	?	1149	*0.50	*0.44	423	?	tsf
C031.6	R 2480	3.4				0.8	0.7	0.7	1.0	1.0	0.6	6068	0.81	7444	0.39	0.37	6903	0.69	tsf
C034.3		2.6	1.0	1.0	0.9	0.4	1.0	1.0	1.0	1.1	0.8								
C035.6	R 2006											1732	* 1.36	2659	0.91	1.13	2808	* 0.37	
C036.3	R 2005											502	* 1.36	728	1.84	1.62	842	1.34	
C036.7	R 2767											143	* 2.01	144	?	?	296	0.55	
C037.5	R 2004											389	1.07	532	0.82	0.90	661	0.92	
C039.3	R 2457	2.1	1.4	1.1		5.6	4.0	1.5	1.0	1.3	4.3	645	* 1.92	1317	1.21	1.02	1168	* 0.32	recA
C040.3		5.5	0.9	1.1	0.9	1.4	0.7	0.6	0.8	1.1	1.9								
C041.0	R 1923											375	* 3.83	1368	0.51	0.51	1247	0.00	
C042.6	R 1922			0.5	1.1							134	?	306	4.98	6.11	435	2.79	gnd
C043.1			1.0	0.7															
C043.5	R 1933											282	* 1.88	358	3.53	*3.00	380	1.60	
C043.8	R 1916											1503	0.19	752	0.37	0.23	1904	* 0.10	icdE
C044.0		2.8	0.6	0.7	0.8														
C044.2			0.6	0.7	0.9	0.8	0.8	1.0	1.1	0.7	0.6								
C044.6		4.4	0.6	0.8	0.9	0.4	0.4	0.7	0.8	1.2	0.6								metK
C048.7		1.4	39.	0.8															
C055.0		1.3	0.4	0.8															
C056.0		0.6	0.5	0.8	0.1	2.0	1.2	1.1	1.1	1.0	0.2								
C058.5		0.4	0.6	0.9		1.5	1.3	1.1	1.0	1.1	1.1								
C060.7		1.3	0.4	0.8		0.5	0.6	0.7	0.8	0.8	0.5								
C062.5	R 2285	1.8	0.6	0.9	1.4	0.4	0.5	0.4	0.5	1.9	2.7	1008	* 0.90	1197	0.64	0.45	1655	0.18	htpG
C062.7	R 2290	2.5	0.6	0.7	2.7	0.7	0.6	0.5	0.8	1.0	0.6	380	0.00	1071	0.76	0.48	870	* ?	aceF
C070.0	R 2291	0.9	0.6	0.7	2.5	0.7	0.5	0.6	0.8	1.1	0.6	457	2.66	615	*0.45	*0.66	512	* 0.36	aceF

(Table continues)

TABLE 3 *Continued*

A-N(a)	R R M (b)	AB a'	Ace RL	Gly RL	Ric RL	13.5 RL	15 RL	23 RL	30 RL	42 RL	46 RL	Phn PPM	Phn Ratio	PS PPM	PS-E Ratio	PS-L Ratio	NS PPM	NS-L Ratio	GENE
						STEADY STATE GROWTH CONDITIONS (c)								GROWTH TRANSITION CONDITIONS (d)					
C074.0	R 2671											218	* ?	205	2.27	4.75	691	0.14	
C075.0	R 2672											216	* 3.58	312	1.17	1.55	688	0.25	
C078.0	R 1770	3.0	0.7	1.0		3.2	3.2	2.0	1.3	0.9	0.5	1387	1.08	1297	0.70	0.59	2002	0.36	pnp
C137.0	R 1733	2.7	0.5	0.9	0.3							214	1.07	317	0.26	0.30	207	* 0.38	metH
D014.7	R 2308	1.3	0.5	0.8	2.5							1804	* 0.82	3188	0.00	0.00	3230	?	rpsF
D025.5	R 2357											1041	* 2.22	1253	2.54	4.34	3123	0.67	
D027.1	R 2163											826	* 1.17	1150	0.87	0.85	1213	0.89	sspA
D028.0	R 2092											493	* 10.38	375	2.95	2.55	456	1.06	
D028.3	R 2091											1064	* 0.98	1255	1.40	1.48	1754	0.32	
D029.8	R 2089											1095	* 2.19	1524	3.71	3.16	2107	3.29	
D031.5	R 2010	0.9	1.1	1.0	1.4	0.6	0.6	0.7	0.9	1.2	1.9	337	* 1.12	647	0.63	*0.50	508	0.53	trxB
D032.5	R 2011	1.3	0.7	0.9	0.8	0.5	0.5	0.6	0.8	1.1	1.0	1708	* 1.23	1709	0.39	0.36	2778	0.17	
D033.4	R 2008											128	* 2.47	317	6.17	4.79	849	?	htpH
D035.7	R 2007											246	* 1.97	327	8.21	8.95	282	* 2.14	
D038.3	R 1999											449	0.82	572	1.19	0.72	623	1.21	
D040.7	R 1929	4.5	5.9	1.9	0.2							3576	* 0.98	4811	0.73	0.50	5802	0.61	livJ
D040.9	R 1998											1638	* 1.00	1651	2.21	2.21	3264	1.39	
D044.5		6.9	0.5	1.3		2.3	1.4	1.3	1.2	1.4	2.0								
D046.0			0.7	1.0	1.3	0.8	0.7	0.6	1.0	1.2	3.0								
D047.0						1.4	1.0	0.7	0.9	1.1	1.0								
D047.5						1.6	0.9	0.9	1.0	1.0	1.5								
D049.2						1.1	1.0	0.8	0.9	1.0	0.9								
D049.9	R 1845	1.6	0.3	0.6	0.2							6385	* 0.76	7105	0.15	0.09	10345	1.25	glnA
D050.0		4.2	0.8	0.7	1.2														glnA
D058.5		1.0	0.5	0.7	1.3	0.8	0.8	0.9	1.0	0.9	0.4								lysS
D060.5		0.2	1.6	1.1	1.3						4.6								lysU
D074.0		2.5	0.5	0.6	1.5	2.1	2.1	1.8	1.4	0.5	0.2								
D078.1	R 1773											904	1.46	841	0.56	0.53	1593	0.29	
D084.0	R O840	16.	0.6	0.8	1.7	1.0	1.0	1.1	1.1	0.9	0.4	3327	* 1.12	6254	0.45	0.66	5501	1.10	fusA
D087.5	R 1731	0.1	0.9	1.0		0.7	1.0	1.0	1.0	0.7	0.6	1597	* 0.92	1475	*0.67	0.46	2279	0.13	
D088.0	R 2424											430	* 1.68	431	0.91	0.79	820	* 0.17	
D094.0		2.0	0.5	0.9	1.1	0.7	0.8	0.8	1.0	1.0	0.5								pheT
D099.0		2.8	6.4	1.8	0.3	1.8	2.0	1.7	1.4	0.8	1.3								
D100.0		1.1	0.6	0.9		0.7	1.0	1.2	0.9	0.9	0.6								leuS
D157.0	R 1698	5.0	0.7	0.9	1.4	1.0	1.0	1.2	1.0	0.9	0.6	1764	* *1.65	2004	*0.34	*0.44	2919	0.26	rpoB
E018.0	R 2200											899	* 1.35	1103	2.96	2.44	2526	3.42	
E021.1	R 2195											562	1.63	987	1.22	1.17	1087	0.52	dsbA
E022.8	R 2170		0.8	1.2	0.9	0.5	0.5	0.7	0.8	0.8	0.5	3532	* 0.99	4833	0.34	*0.26	6245	0.25	
E023.0			0.8	1.0	1.0	0.5	0.6	0.8	0.8	0.8	0.5								
E024.8			0.9	0.9	1.1	1.0	0.8	0.9	0.9	1.2	1.8								
E025.5		1.4	0.8	0.7	0.3														trpA
E032.1	R 2478											469	* 1.12	969	0.21	?	734	0.24	
E036.6	R 2002		0.8	0.8	1.3							207	* 1.37	256	3.89	3.04	354	1.29	pfkB
E038.5	R 2450	7.9	0.8	0.7	0.0	1.4	0.7	0.9	1.1	0.9	1.0	3039	* 1.50	1687	*1.42	*0.88	4887	* 0.29	
E039.2	R 1987											2832	* 1.13	3609	0.69	0.47	4743	0.18	
E039.8	R 1925	1.6	8.7	2.9								984	* 1.28	2881	0.30	*0.33	2202	* 0.10	sucC
E042.0	R O420	55.	0.7	0.9	1.5	0.8	0.7	0.8	1.0	1.0	0.9	8570	1.36	19230	*0.76	*0.94	16267	* 0.38	tufA
E048.7		1.8	1.3	27.	1.3	1.1	1.1	1.4	1.3	1.4	2.6								glpK
E048.8			0.8	1.0	1.0														serS
E049.3	R 1847											1440	* 2.34	2029	7.31	7.00	3082	* 0.23	
E050.0		0.6	0.7	0.8	0.5														trpE
E056.0		0.8	1.6	1.6	0.4														pckA
E058.0	R 1838	0.8	0.5	0.8	1.6	0.7	0.8	0.9	0.9	0.9	0.5	329	0.00		?	?	0	* >276	argS
E061.0	R 2630											221	?	221	*1.22	*0.76	395	0.00	
E077.5		1.6	0.8	1.0	1.6	0.4	0.5	0.7	0.7	1.1	0.8								glyS
E079.0	R 1782	1.5	1.2	1.2	2.9	0.3	0.4	0.5	0.6	1.1	0.5	592	* 3.61	1890	*0.36	*0.32	477	?	pta
E088.0	R 2421											510	?	768	2.51	3.76	1287	0.75	
E106.0		1.3	0.5	0.8	1.5	0.7	0.7	1.1	1.0	1.2	0.8								valS
E133.0	R 2807	4.5	0.7	0.8	0.1	0.6	0.9	1.6	1.5	0.8	2.4	1959	* 0.76	3051	0.00	0.00	3881	* 0.03	carB
E140.0	R 2431	2.7	0.8	0.8	0.4	0.5	0.7	0.9	0.9	1.0	0.6	1765	0.75	3478	?	*0.10	2679	* 0.07	
F007.0	R 2275											1348	2.23	2678	0.79	0.83	2617	0.33	ptsH
F010.1	R 2365											66	3.52	267	*3.38	*1.47	227	* 5.97	htpK
F012.3	R 2262											701	1.06	1024	0.27	0.25	982	20.62	
F013.0	R 2261											1060	1.22	979	2.42	1.78	1780	0.16	
F014.7	R 1703	3.1	1.2	1.2		1.2			1.0	0.7	0.5	1930	* 2.65	3616	3.43	3.89	4208	4.81	hns
F017.7	R 2192											318	2.27	468	1.98	1.94	766	1.36	
F018.8	R 2193	0.5	1.0	1.0	1.3							676	* 0.99	1409	0.20	*0.12	1175	0.00	ssb
F020.9	R 2191											95	* 2.86	168	2.52	2.65	259	* 1.25	
F021.3			1.4	1.0	0.8														sodB
F022.5	R 2147	0.6	1.6	0.8	0.9							1557	* 3.02	2676	3.16	3.14	2958	* 2.92	eda
F024.5		20.	1.0	0.9	0.7	2.4	1.2	0.9	1.1	1.3	2.4								ompA
F024.6	R 2157					1.4	1.1	1.0	1.0	1.1	2.4	709	1.14	1064	0.82	0.74	1000	0.71	
F025.8	R 2154											886	* 1.70	1271	1.99	1.67	1346	1.26	
F026.0	R 2081					1.4	1.0	0.9	1.1	1.0	1.0	353	* 1.52	490	2.53	2.48	549	2.46	adk
F027.0	R 2146											314	* 1.11	502	0.62	*0.85	428	0.70	
F028.0	R 2080											1428	1.08	3125	0.38	?	3448	0.00	ompA
F028.8		3.5	0.5	0.7	0.3														
F029.7		1.1	0.6	0.9															map
F030.2		2.6	0.7	0.9	1.5	0.7	0.5	0.6	0.8	0.9	0.4								mdh
F030.3	R 2073											159	* 0.00	311	2.47	2.19	461	* 0.00	
F032.3		2.5	3.9	2.0	0.1	2.0	1.6	0.6	0.8	2.0	9.0								
F032.5	1991	5.5	0.4	0.9	0.3	1.2	1.1	1.4	1.3	1.2	2.7	976	* 0.59	1246	0.44	0.33	961	0.42	ilvE
F033.1	R 2070											235	1.19	257	0.77	?	329	0.00	ompA
F033.6		0.2	1.3	1.0	1.5														fmt
F035.0	R 1990											344	* 1.15	461	1.94	2.03	461	1.00	leuO

TABLE 3 *Continued*

A-N(a)	R R M (b)	AB a'	Ace RL	Gly RL	Ric RL	13.5 RL	15 RL	23 RL	30 RL	42 RL	46 RL	Phn PPM	Phn Ratio	PS PPM	PS-E Ratio	PS-L Ratio	NS PPM	NS-L Ratio	GENE
						STEADY STATE GROWTH CONDITIONS (c)								GROWTH TRANSITION CONDITIONS (d)					
F035.8	R 1708	0.5	0.9	0.9	0.8							1659	* *1.68	2847	1.82	1.67	3204	1.97	pfkA
F037.2	R 1993											140	?	173	*1.14	*1.02	205	2.01	
F037.5	R 1981											139	* 2.89	174	*8.14	*5.43	212	1.59	
F037.8	R 1912	0.6	0.8	0.9	0.8							208	* 2.00	568	2.08	*3.05	738	* 0.64	rimL
F037.9		0.3	1.0	0.6	0.3														tyrA
F038.0	R 1986	9.4	0.5	0.7	0.3	1.8	1.5	1.2	1.3	1.0	1.1	3211	* 1.32	4367	1.30	1.39	5602	0.79	
F038.1		1.1	1.3	0.6	0.2														aroF
F038.5	R 2685											1187	0.97	1365	1.07	0.77	1448	* 0.62	rimL
F039.0	R 2446		0.9	1.1	0.9	2.8	1.5	1.2	1.2	1.1	1.4	861	?	1379	0.00	0.00	773	?	argI
F039.6	R 1911	0.3	2.0	1.3	0.4							353	* 2.09	713	0.55	0.59	484	0.00	aspC
F039.7	R 1910	3.2	1.2	1.4	0.3							1809	* 1.28	2482	1.22	0.81	2913	0.42	aspC
F040.8	R 1906											243	1.21	373	1.08	0.61	410	0.38	
F042.2	R 1914	19.	0.7	0.9	1.3	0.7	0.7	0.8	0.9	1.0	0.9	10010	?	10589	*0.70	*0.67	11210	* 0.60	fabB
F043.8	R O438	10.	0.5	0.8	1.1	1.4	1.5	1.4	1.0	1.1	1.2	3667	* 1.71	7376	1.74	1.69	10528	* ?	eno
F043.9	R 1920					1.7	1.3	1.2	1.4	0.7	0.5	5294	* 0.81	7702	0.80	0.94	9352	0.61	
F044.2	R 1927											468	* 1.00	451	*1.19	*1.20	570	?	
F044.6			0.7	0.9	1.3														
F045.5			0.5	0.7	0.4														trpC
F045.6		0.5	1.3	0.9	1.0														gor
F046.6	R O466		0.7	1.0	0.9							142	42.86	290	*20.95	*18.74	3304	-	phoA
F047.8		1.1	0.6	0.9	1.3														gltM
F048.1		0.9	0.8	1.0	1.5	0.9	0.7	0.8	0.7	1.1	0.7								hisS
F049.0	R 2388											230	* 7.62	263	*8.04	*5.99	426	0.00	
F050.1	R 1751	0.7	0.4	0.8	0.4							560	* 2.70	934	2.00	2.33	1093	0.24	ilvA
F050.2			0.6	0.5	0.3														
F050.3	R 2576	1.7	4.3	2.0	0.1	2.6	1.9	1.1	0.8	1.2	2.8	1290	0.49	2071	0.41	0.57	1987	0.00	sucB
F050.6	R 1831											261	* 2.55	397	1.77	2.44	635	0.28	
F054.0		0.6	0.8	0.9															
F054.1	R 1828											303	* 2.00	515	1.16	1.25	665	1.65	
F054.4	R 1767	3.5	0.7	0.8	0.1							1539	1.12	2042	?	0.03	2776	* 0.07	
F056.0		1.2	5.5	3.3		0.6	1.1	0.9	0.9	1.2	1,8								
F056.2		2.3	0.5	0.7	2.1	0.6	0.6	0.8	0.9	1.0	0.6								
F056.5		22.	0.9	0.6		0.8	0.6	0.9	1.2	1.3	1.8								
F058.5		1.3	0.7	0.9		1.4	1.1	1.2	1.2	0.8	0.4								aspS
F060.3		1.1	5.8	1.9	0.2	2.1	1.7	1.0	0.9	1.4	2.7								sdhA
F063.4		0.4	1.0	1.0	1.4	0.6	0.9	0.7	1.2	0.9	0.9								
F063.5			0.9	1.0	1.1	0.5	1.0	0.8	0.7	0.8	0.9								
F063.8		1.8	1.2	1.0	0.9	1.2	0.7	0.8	0.8	1.1	0.7								
F064.5		3.1	0.8	1.0	0.9	0.8	0.9	0.8	0.9	0.8	0.7								
F066.0		1.7	0.4	0.8	1.1														
F072.0	R 1746		0.8	0.8	0.6							363	* 1.59	562	0.49	0.42	445	?	metG
F082.5		0.4	3.6	2.4	0.3	5.3	4.0	2.1	1.1	1.1	1.5								
F084.0		4.3	0.4	0.7	0.1	2.2	1.5	1.2	1.2	1.0	1.8							?	ppc
F084.1		0.8	1.2	1.0	0.8	0.2	0.3	0.4	0.5	0.9	4.6								clpB
F088.0		56.	0.3	0.7	0.0	2.7	5.7	4.1	1.0	0.8	0.5							0.62	metE
F093.0			1.3	0.8	0.4														alaS
F099.0		8.5	0.6	0.8	2.8	0.6	0.5	0.6	0.7	0.9	0.4								aceE
F106.0			0.8	1.0	2.2														
F107.0		2.4	0.7	0.8	1.1	1.0	1.2	1.2	1.0	0.7	0.8								ileS
F113.0		0.6	0.7	0.8	1.6													?	polA
F178.0		0.9	1.0	0.9		1.9	2.4	1.5	1.2		0.5								
G010.7	R 2366	*										424	* 2.26	594	2.75	3.00	1103	-	
G012.4	R 2256											627	* 1.38	689	1.55	1.62	1059	1.09	
G013.5	R 2898											-	-	0	-	-	0	2.51	ibpA
G014.1	R 2257											253	* 2.34	260	8.22	10.75	585	?	
G015.1	R 2338											-	-	63	*2.71	*2.49	150	?	
G015.8	R 2229											1430	7.39	2082	7.96	9.69	1604	-	fur
G016.0	R 2337											66	* 8.44	97	14.58	16.48	360	?	
G016.4	R 2227											425	* 1.35	627	1.48	1.55	772	1.93	
G018.3	R 2911											162	?	0	-	-	0	0.23	
G023.4	R 2141											172	* 1.96	252	5.02	6.68	252	0.55	
G024.7	R 2137											330	* 1.51	413	2.80	3.40	328	* 0.83	
G025.3		1.7	1.0	1.0	1.1	1.1	0.8	0.7	1.0	1.2	1.5								
G025.8	R 2144											286	* 1.58	393	1.95	1.89	521	* 0.42	
G027.0	R 2143											851	1.54	1241	1.76	1.79	1602	0.66	
G027.1	R 2132											1785	* 1.47	2190	1.22	0.81	2615	-	
G027.2		0.8	12.	7.5		3.1	3.3	2.3	1.5	0.9	0.3								
G028.0	R 2150											375	0.80	575	1.00	?	649	* 0.00	glpR
G028.1	R 2972		0.5	0.7	0.3							57	?	0	-	-	208	0.92	araD
G028.2		0.4	0.8	0.9	0.9														xthA
G029.2	R 2513											83	2.75	199	0.85	?	283	-	
G029.3	R 2973											163	* 0.00	0	-	-	0	-	thyA
G029.5	R 2066											519	* 2.24	760	1.74	1.45	1036	* 1.07	
G029.6						23.	1.1	1.2	0.9	1.4	0.8								
G030.1	R 2065											342	* 1.28	515	1.40	1.82	571		
G030.5		0.9	1.1	1.1	1.1														
G032.0	R 2064											1711	* 1.80	2125	1.65	1.83	3302	*	
G032.8		5.3	1.1	1.0	0.1	0.6	1.4	0.9	0.6	0.3	0.5								
G036.0		1.0	0.5	0.8	1.0	1.1	0.7	0.7	0.8	1.0	0.6								pheS
G036.2	R 2503											-	* >297	0	>304	*>239	150	* 0.42	
G038.2	R 1965											286	* 3.13	365	5.87	3.33	379	0.22	
G039.1	R 1974											152	* 3.90	173	4.58	6.31	456	* 0.37	
G039.6		0.3	0.8	0.8	0.8														trmA
G040.0		2.8	0.5	0.9	0.3													?	trpB
G041.0			2.7	1.0	1.8	1.7	0.4	0.7	1.0	0.9	2.9								

(Table continues)

TABLE 3 *Continued*

A-N(a)	R R M (b)	AB a'	Ace RL	Gly RL	Ric RL	13.5 RL	15 RL	23 RL	30 RL	42 RL	46 RL	Phn PPM	Phn Ratio	PS PPM	PS-E Ratio	PS-L Ratio	NS PPM	NS-L Ratio	GENE
G041.2	R 2434	1.9	0.8	1.1	1.0	5.9	1.4	1.1	1.0	1.0	0.8	1212	* 2.50	1404	*2.16	*4.43	930	* -	
G041.3	R 2436	1.0	0.7	1.0	5.0	0.2	0.3	0.4	0.7	0.9	0.2	1368	?	0	*>1631	-	0	0.20	ackA
G041.4	R 1882	2.0	0.5	0.7	0.2	2.3	0.9	1.5	1.3	0.8	2.8	1083	1.21	2975	*0.40	0.61	2355	*	carA
G042.1	R 1747											192	* 3.93	343	5.98	5.87	318	* >622	
G043.2		6.7	0.4	0.5	0.1	0.9	1.8	1.7	1.4	1.5	1.0								
G043.6	R 1881											3901	0.79	6782	0.35	0.15	5682	* 1.14	gdhA
G043.8		7.9	0.8	0.6	0.2	1.4	0.9	1.2	1.1	1.1	2.1								glyA
G043.9			1.5	0.8	0.5	0.5	0.4	0.4	0.7	1.0	· 0.9								
G044.0	R 1894	1.8	0.7	1.0	0.9	0.5	0.5	0.7	0.8	1.1	1.1	372	* 1.43	1008	0.73	0.49	902	-	tyrS
G045.5	R 2945											1547	-	0	-	-	0	?	
G045.6	R 2544											1217	* 1.00	1286	0.00	0.00	1838	* 0.00	
G048.0	R 1888											281	1.37	420	1.48	1.47	447	0.87	
G048.6	R 1891											1252	0.39	1142	0.46	0.28	1541	0.33	
G049.2	R 1818	4.4	0.7	1.2	0.0							3105	* 0.94	3568	0.30	0.23	4317		
G050.5	R 2975	4.6	1.9	1.2	1.4	0.9	0.7	0.7	1.0	1.1	1.2		?	0	-	-	0	-	lpd
G051.0		6.9	1.2	1.2	0.8	0.9	1.1	1.2	1.2	0.9	1.7								atpA
G051.1		0.6	1.0	1.2	0.3														
G051.8	R 1806											572	* 0.93	779	0.44	0.45	803	0.98	
G052.0	R 1809											-	>518	327	2.46	2.36	272		amn
G054.0		3.2	2.0	1.4	0.2														
G054.6	R 2560	0.6	3.7	2.1	0.3							703	?	807	0.31	*0.21	1087	* -	
G054.7	R 2978	2.2	0.3	0.7	1.5	1.3	0.8	0.8	0.9	1.0	0.9	52	?	0	-	-	0	0.29	pykF
G055.3			1.0	0.8	0.7														
G057.0		0.5	1.4	1.7	0.9	0.5	0.8	1.1	1.0	0.9	0.7							0.00	
G058.5	R 2561											211	* 1.30	226	0.42	0.29	204	0.96	
G059.4	R 2559											3693	* 1.05	3454	1.53	1.65	5522		
G060.1	R 1740											299	* ?	262	2.47	1.80	304	* 1.35	
G061.0		1.1	0.7	0.9	1.3	0.9	0.8	0.8	0.8	0.9	0.8								glnS
G062.8		0.5	1.2	0.9	0.7	1.6	0.8	0.7	0.7	0.6	1.7								
G065.0		0.8	0.7	0.8	1.6														thrS
G070.0		0.7	0.5	1.3	1.2													0.00	pfl
G071.5		0.5	0.7	1.0	0.9														
G072.0			0.7	0.8	1.0														metG
G074.0	R 2639	3.2	2.5	1.7	0.7	1.6	3.2	3.4	0.9	0.7	1.5	3100	0.42	3355	0.00	0.00	3981	* ?	pfl
G076.0	R 1737	0.1	9.0	1.3	0.5	1.8	1.7	1.1	1.2	1.0	1.2	303	* 2.89	346	4.75	4.82	353	?	
G078.0		0.8	0.5	0.9	1.0	0.6	0.7	1.2	1.1	1.0	0.9								
G093.0		2.1	1.5	1.3	0.7	0.7	0.8	0.9	1.0	1.1	1.1								
G097.0		1.9	4.4	1.0	0.3	2.4	2.6	1.4	0.9	1.1	1.4								sucA
G117.0		1.7	0.7	0.9		1.7	1.6	1.3	1.0	1.0	1.0							?	infB
G127.0		0.2	0.8	0.8		1.0	1.0	1.0	0.8	0.9	0.9								
H013.2	R 2585											622	1.77	940	0.91	0.51	929	0.00	
H013.8		0.2	0.7	0.8	1.1														nusB
H014.0	R 2253	0.2	0.5	0.8	1.6							475	* 1.22	401	*0.86	*0.51	884	>613	rplI
H021.6		0.9	0.9	1.1	1.5														prfA
H024.7			0.6	0.6															frdB
H027.4	R 2134											327	* 1.17	452	0.51	0.45	386	* 0.79	
H028.0			4.4	1.1															sucD
H031.3	R 1704		0.9	0.7	0.8							5970	* *1.03	9918	0.91	0.99	9555	*	pyrB
H041.0	R 1883	1.0	0.4	0.9	0.3							468	?	721	7.47	9.67	3144	?	
H047.3			0.5	0.8	1.0														glpD
H047.4	R 1864	1.1	4.3	1.6	0.0							260	* 0.56	149	?	*0.59	381	*	gltA
H049.2	R 1868											909	0.38	1188	*0.62	0.57	1471	* ?	
H052.7		1.6	0.3	0.6	2.4	0.3	0.4	0.6	0.9	1.3	0.3								
H054.5	R 2733	0.8	0.4	0.6	0.2							394	0.00	215	?	?	0	-	trpD
H054.6		0.6	4.0	2.3	0.8	1.0	2.1	1.8	1.3	1.0	1.5								
H094.0		2.1	1.5	0.7	0.7	0.8	0.9	1.0	1.1	1.1									lon
H097.3			4.6	1.6	0.3														adhE
I011.1			0.5	0.9	1.5														rpmC
I011.3			1.1	1.1	0.9														hupA
I011.4			0.5	0.9	1.7														rpmD
I011.5			1.3	1.2	1.0														hupB
I011.7			0.5	0.8	1.6														rpmF
I011.8			0.6	0.8	1.5														rplY
I011.9			0.4	0.8	1.7														rpsP
I012.0			0.5	0.8	1.8														rpsO
I012.1			0.4	0.7	1.5														rpsJ
I012.2			0.4	0.9	1.2														rpmB
I012.6			0.5	0.7	1.4														rpsQ
I012.8			0.4	0.7	1.7														rpmA
I013.0			0.3	0.8	1.6														rplW
I013.2			0.5	0.7	1.7														rplU
I013.5			0.6	1.0	1.3														rpsH
I014.1			0.5	0.8	1.8														rplK
I014.4			0.4	0.7	1.5	0.7		1.0	0.9	0.5									rpsE
I014.6			0.4	0.9	1.4														rpsI
I014.7			0.5	0.8	1.8	0.6		1.0	0.9	0.4									
I014.9			0.5	0.6	1.4														rplQ
I016.4			0.6	0.8	2.0	0.6		0.9	1.0	0.5									rplE
I017.2			0.6	0.9	1.9	0.7		0.9	1.0	0.4									rplF
I017.5			0.9	0.8	0.9														pyrI
I019.5			0.6	0.8	1.1	2.0		1.3	1.2	1.0									infC
I020.1			0.5	0.8	1.3														infC
I021.1			0.5	0.8	1.7	0.5		0.8	0.9	0.5									rplD
I021.3			1.4	1.0	1.2	1.4		1.0	0.6	0.6									sodA
I021.4			1.2	1.1	0.9	0.9		0.9	0.7	1.3									crp

TABLE 3 *Continued*

A-N(a)	R R M (b)	STEADY STATE GROWTH CONDITIONS (c)												GROWTH TRANSITION CONDITIONS (d)					GENE
		AB a'	Ace RL	Gly RL	Ric RL	13.5 RL	15 RL	23 RL	30 RL	42 RL	46 RL	Phn PPM	Phn Ratio	PS PPM	PS-E Ratio	PS-L Ratio	NS PPM	NS-L Ratio	
I023.0			0.5	0.8	1.8	0.6			0.9	1.0	0.4								rplC
I026.0			0.5	0.8	1.5	0.6			0.9	0.9	0.4								rplA
I026.2		0.1	1.3	1.0	0.8														hisP
I030.2				1.0	1.4														rplB
I033.5			0.6	0.7	0.7	1.5			1.0	1.0	1.2								gapA
I047.2			0.8	0.9	1.3	0.9			0.9	1.0	0.8								
I048.8			0.6	0.9	1.4														rho
I058.4			0.6	0.9	0.0	0.1			0.9	0.0	0.0								
I063.5			0.3	0.5	2.5	5.8			1.6	0.7	0.4								
	R 1705											1034	* *1.57	1518	1.54	1.37	2094	2.16	
	R 1706											235	* *2.16	423	6.32	6.80	1219	?	
	R 1707											765	?	1266	0.69	0.67	1543	0.00	
	R 1709											-	-	3848	0.22	0.15	3313	0.42	
	R 1710											329	* *2.07	462	0.50	0.46	420	* 0.16	
	R 1711											917	* 0.61	1388	0.34	0.24	1503	0.66	
	R 1712											917	* 1.61	1408	0.82	0.56	1097	0.00	
	R 1714											500	* 0.77	554	*0.26	*0.38	695	* ?	
	R 1715											834	0.51	1085	0.37	*0.29	255		
	R 1716											329	* 0.60	255	0.61	0.75	516	?	
	R 1717											364	* 0.78	397	0.59	*1.08	462		
	R 1719											454	0.70	844	1.22	1.67	970	?	
	R 1720											1592	* 0.96	3551	0.12	0.22	434		
	R 1721											267	1.52	460	*0.93	*1.11	491		
	R 1722											2349	* 1.07	1812	*0.86	*0.74	3128	0.30	
	R 1723											1462	1.52	2625	0.38	0.39	1851		
	R 1724											717	?	508	*1.24	0.00	395	*	
	R 1725											544	* 0.74	693	*0.53	0.22	686	0.07	
	R 1726											737	0.96	1531	0.24	0.26	1942	0.00	
	R 1727											426	1.37	1540	*0.18	0.25	675	0.26	
	R 1732											32	?	61	*1.58	*1.04	117	* 1.04	
	R 1734											140	* 1.42	140	0.43	0.47	117	*	
	R 1735											92	1.92	101	0.76	0.77	129	?	
	R 1736											178	* 0.64	188	0.34	0.37	139	1.18	
	R 1738											394	* 0.60	200	1.64	3.62	407	?	
	R 1739											317	* 1.87	375	0.65	0.60	217		
	R 1741											284	* 1.14	384	1.77	2.31	341	0.90	
	R 1742											443	?	676	2.09	2.73	508	0.51	
	R 1743											302	* 0.87	308	0.63	0.58	367	0.53	
	R 1744											709	* 0.99	916	0.70	0.87	1212	0.53	
	R 1745											267	* 0.00	226	*1.76	*2.05	517	0.00	
	R 1748											494	* 0.54	341	0.46	0.36	469	0.13	
	R 1749											406	* 2.19	571	0.58	0.63	634		
	R 1750											686	* 0.95	938	0.31	*0.53	1433		
	R 1752											311	* 1.24	285	1.75	1.92	481	0.16	
	R 1753											705	0.97	684	0.61	0.55	1600	0.24	
	R 1754											567	?	372	3.17	3.37	992	* 0.00	
	R 1755											1806	?	535	4.15	8.27	729	* 0.97	
	R 1756											605	* 1.13	602	0.35	0.37	655		
	R 1757											335	* 1.00	305	2.59	3.28	401	0.13	
	R 1758											326	* 0.69	261	*0.52	*0.28	246	?	
	R 1759											1264	0.59	1567	*0.19	0.25	1484	0.00	
	R 1760											325	* 0.68	285	*0.38	0.21	227	* 0.43	
	R 1761											301	1.56	381	0.69	0.66	356	?	
	R 1762											261	1.23	271	0.94	0.57	379	1.87	
	R 1763											429	* 1.72	911	0.28	0.38	562	0.18	
	R 1764											312	* 0.96	546	0.88	1.11	610	0.21	
	R 1765											185	* 0.87	193	0.55	0.57	203	?	
	R 1766											163	1.95	272	1.46	1.16	347	* 0.59	
	R 1769											295	0.70	158	1.92	1.67	302	0.39	
	R 1771											614	2.01	1148	*0.53	0.80	1451	0.15	
	R 1774											284	* 2.51	732	0.87	1.44	367	* 0.50	
	R 1775											114	* 3.86	178	1.06	1.22	439	* ?	
	R 1776											79	1.21	100	0.77	0.63	130	0.77	
	R 1778											433	* 0.95	717	0.37	0.27	504	* ?	
	R 1779											161	* 1.46	157	*0.74	*0.55	189	?	
	R 1780											89	1.67	199	0.76	0.57	138	?	
	R 1781											389	0.65	241	0.34	0.35	413	* 0.23	
	R 1785											1455	* 2.41	1940	0.84	0.89	3263	* 0.10	
	R 1786											1161	* 1.85	1957	0.50	0.45	0	* >800	
	R 1789											173	?	178	0.96	0.95	318	* ?	
	R 1793											55	* 4.05	176	*0.54	*0.51	154	* 0.00	
	R 1794											99	* 1.78	86	*2.85	2.33	236	* -	
	R 1795											140	?	60	*3.1	?	0	?	
	R 1796											522	?	761	0.44	0.43	556	0.54	
	R 1797											274	* 1.04	453	0.76	0.80	354	?	
	R 1798											75	0.00	147	*1.28	*0.83	152	0.37	
	R 1799											178	* 0.89	243	*0.64	0.66	280	0.63	
	R 1800											86	?		?	?	130	0.00	
	R 1801											268	1.31	339	0.34	0.27	378	0.00	
	R 1802											543	1.23	721	0.41	0.64	906	0.00	
	R 1803											2647	* 1.26	2471	0.25	0.32	3441	-	
	R 1804											2508	* 0.95	2611	0.45	0.47	3713	-	
	R 1805											47	?		?	?	72	?	
	R 1807											645	1.21	648	1.41	1.51	950	-	
	R 1808											84	* 1.90	105	1.02	1.02	106	0.11	

(Table continues)

TABLE 3 *Continued*

A-N(a)	R R M (b)	AB a'	Ace RL	Gly RL	Ric RL	13.5 RL	15 RL	23 RL	30 RL	42 RL	46 RL	Phn PPM	Phn Ratio	PS PPM	PS-E Ratio	PS-L Ratio	NS PPM	NS-L Ratio	GENE
						STEADY STATE GROWTH CONDITIONS (c)									GROWTH TRANSITION CONDITIONS (d)				
	R 1810											239	* 1.27	209	1.53	*0.71	327	1.15	
	R 1811											1939	1.68	1889	0.72	0.52	2828	0.76	
	R 1812											203	* 0.49	199	*2.56	*1.16	546	*	
	R 1813											167	1.82	194	1.90	1.46	173	0.54	
	R 1814											391	* 0.94	218	1.90	1.76	383	0.87	
	R 1815											1216	?	776	1.22	1.00	781	5.81	
	R 1816											616	* 0.47	536	?	0.23	684	-	
	R 1817											4782	* 0.79	5097	0.29	0.25	6506	-	
	R 1819											241	?	595	1.35	1.41	758	* 0.00	
	R 1820											246	1.36	461	0.87	0.76	465	?	
	R 1821											237	1.76	355	0.78	1.16	339	0.00	
	R 1822											485	* 1.66	752	0.83	0.74	1218	0.00	
	R 1823											321	0.71	652	0.14	0.13	639	0.00	
	R 1824											132	0.00	128	?	?	194	0.90	
	R 1825											257	* 1.00	450	0.81	0.61	408	0.00	
	R 1829											393	* ?	214	1.41	1.33	351	1.21	
	R 1830											222	?	218	1.84	0.95	288	* 1.75	
	R 1832											97	* 2.66	147	1.09	0.88	162	?	
	R 1833											475	* 1.38	631	1.01	0.74	759	?	
	R 1835											168	* 0.00	177	?	0.00	252	6.34	
	R 1836											655	1.26	757	2.75	*3.77	1251	0.81	
	R 1839											206	* 1.34	234	2.02	2.01	374	?	
	R 1840											568	* 2.27	499	*2.24	2.21	828	? 0.22	
	R 1841											1711	?	905	*1.65	*1.60	1803	0.00	
	R 1842											2058	* 0.24	652	0.56	0.00	3427	0.00	
	R 1843											2914	1.02	4779	0.06	?	4966	0.05	
	R 1844											1624	* 0.49	1114	*0.45	*0.28	2482	?	
	R 1846											427	* 1.18	401	*16.63	*13.15	446	0.92	
	R 1848											1581	* 3.32	1653	8.14	8.03	3190	0.81	
	R 1849											315	1.18	442	0.50	0.43	456	1.22	
	R 1850											326	* 2.12	557	0.81	0.77	811	0.89	
	R 1851											200	* 1.03	243	0.91	0.42	302	0.00	
	R 1852											2295	* 1.48	2940	0.47	0.34	4317	0.07	
	R 1853											240	* 0.81	231	0.61	0.33	249	* 0.37	
	R 1854											442	* 2.88	1055	1.14	1.14	2506	* 0.00	
	R 1855											3411	* 0.94	4077	0.32	0.25	5868	0.44	
	R 1856											970	* 1.13	1282	*0.19	*0.64	943	* ?	
	R 1857											384	* 0.84	379	1.08	1.02	482	* 0.62	
	R 1858											399	* 1.27	371	*0.56	*0.29	986	* 0.00	
	R 1860											180	* 1.57	214	*1.97	*2.03	261	2.02	
	R 1862											121	?	104	1.63	*1.62	125	* 4.43	
	R 1863											303	* 3.18	326	3.68	1.99	391	0.00	
	R 1865											168	?	203	0.52	0.20	186	1.37	
	R 1866											430	* 1.47	768	0.52	0.36	615	0.00	
	R 1867											122	?	56	?	*1.05	135	0.22	
	R 1869											298	0.56	91	?	*1.66	126	0.35	
	R 1870											165	* 0.82	61	1.80	*2.46	94	?	
	R 1871											123	* 2.70	220	*1.51	1.83	475	* 0.00	
	R 1872											188	?	130	0.00	0.00	220	0.41	
	R 1873											125	1.83	201	0.60	0.47	191	>2385	
	R 1875											265	0.89	292	0.70	0.49	257		
	R 1876											325	0.83	350	*0.83	0.59	377	* 0.10	
	R 1877											1829	0.36	962	0.38	0.28	1747	0.62	
	R 1878											776	* 2.21	1184	0.67	0.78	1301	?	
	R 1879											7772	* 0.69	8500	0.22	0.29	10358	0.84	
	R 1880											252	1.30	538	0.77	0.59	552	1.73	
	R 1884											164	0.00	146	?	?	169	0.00	
	R 1885											271	1.08	280	0.98	0.79	301		
	R 1886											73	* 2.72	87	*2.80	1.63	0	0.00	
	R 1889											410	1.67	793	0.20	0.17	740	* 0.89	
	R 1890											243	?	168	*1.62	*1.00	139	?	
	R 1892											754	* 1.30	723	0.80	0.68	1060		
	R 1893											211	* 2.16	301	*1.84	*1.22	287	0.93	
	R 1896											91	* 2.03	150	0.96	*0.71	160	0.61	
	R 1897											235	* 1.26	354	0.83	0.89	337	* 1.92	
	R 1898											543	* 1.55	626	1.33	1.40	480	0.40	
	R 1902											283	2.20	238	*3.63	*2.51	401	* ?	
	R 1903											519	1.62	1086	2.14	3.87	801	1.17	
	R 1905											3236	* 0.90	5295	*0.77	*0.55	7451	* 0.09	
	R 1908											106	* 1.49	232	0.85	0.67	184	1.25	
	R 1909											404	* 0.87	489	0.84	*0.70	565		
	R 1915											1106	* ?	2661	*0.55	*0.24	1428	0.00	
	R 1917											1648	?	574	*0.30	*0.27	1332	* 0.32	
	R 1918											4003	* 0.68	4812	0.54	0.32	5847	* 0.08	
	R 1924											-	-		?	?	0	-	
	R 1926											3213	* 1.19	4777	1.43	1.37	6444	0.76	
	R 1930											3203	?	3436	1.10	*1.27	5323	0.21	
	R 1931											597	0.00	1108	0.67	*0.59	963	0.57	
	R 1932											306	1.36	454	*2.23	*3.02	530	?	
	R 1934											575	* 0.75	554	0.85	0.73	638	0.00	
	R 1935											445	* 1.23	543	*0.68	0.48	680	0.34	
	R 1936											104	0.00	0	>213	-	0	-	
	R 1937											559	* 0.90	720	0.57	0.44	820	0.37	
	R 1938											190	* 1.27	254	0.74	0.54	319	* 0.22	
	R 1939											158	* 2.88	256	2.73	3.04	329	?	

TABLE 3 *Continued*

Left panel:

R	R M (b)	Phn PPM		Phn Ratio	PS PPM	PS-E Ratio	PS-L Ratio	NS PPM		NS-L Ratio
R	1940	155		0.00	124	1.90	1.56	216		0.69
R	1941	548	*	2.04	819	1.15	0.77	998		?
R	1942	204	*	0.00	230	*0.64	?	247		?
R	1943	369		?	643	0.72	*0.97	467		0.33
R	1946	178		3.45	176	1.26	1.11	221		?
R	1947	86		?	45	8.04	*3.90	133		?
R	1948	138		0.00	100	?	?	143		0.00
R	1949	127	*	0.00	146	*0.81	0.48	170		0.00
R	1950	134	*	?	262	*0.34	0.25	199		?
R	1951	118		0.00	0	-	-	117		?
R	1952	140		2.03	214	*1.17	0.67	235		?
R	1953	100	*	2.75	166	1.67	2.45	125	*	0.34
R	1954	363	*	1.96	529	1.00	0.85	533	*	0.81
R	1955	677	*	3.50	1145	3.19	2.88	1011		0.28
R	1956	830	*	1.96	1728	0.53	0.45	1299		0.15
R	1957	2380		0.83	3151	0.65	0.73	3272	*	0.99
R	1958	311		1.20	322	0.99	0.69	358		0.41
R	1959	275	*	1.18	375	?	*0.30	187		
R	1960	-	*	>156	115	1.10	1.33	317	*	0.65
R	1961	981		2.12	1746	0.54	0.52	2033		?
R	1962	729	*	1.76	1148	1.17	1.03	1135	*	0.71
R	1963	336	*	2.43	383	1.99	1.85	419		0.00
R	1964	186		?	111	*1.41	?	182		0.31
R	1966	418		1.01	866	*0.37	0.34	780		0.00
R	1967	1099	*	0.41	993	0.92	*0.67	1107	*	
R	1968	687		?	418	2.90	2.99	630	*	0.00
R	1969	1154	*	1.14	1732	0.59	0.81	1923	*	?
R	1970	5213	*	0.30	7122	0.03	0.02	6648		0.79
R	1971	579	*	1.18	628	1.25	1.01	618		
R	1973	141	*	1.24	161	1.12	?	182	*	0.00
R	1975	1015		1.70	1897	0.41	0.40	1627		?
R	1976	106	*	3.49	305	1.15	0.46	263		1.37
R	1978	132		?	468	*0.73	?	336		0.00
R	1979	417		?	325	*1.05	0.00	514		0.00
R	1982	331	*	1.67	563	1.08	0.78	515		0.00
R	1983	558	*	0.62	554	0.52	0.37	798		
R	1984	281		?	170	1.03	0.65	179		1.83
R	1985	1623		1.31	3163	0.74	0.45	3248		0.25
R	1988	254		0.92	323	0.58	*0.57	520	*	1.62
R	1989	197		2.37	395	1.45	1.13	498		10.27
R	1994	1200		?	1293	0.96	0.60	2266	*	0.34
R	1995	475	*	1.85	672	4.32	4.15	762		1.28
R	1996	84		2.25	137	1.99	1.13	292		0.86
R	1997	526		0.79	602	0.71	0.67	799	*	0.00
R	2000	1091	*	1.05	1432	1.25	0.83	2240	*	0.18
R	2001	-		>318	183	?	*1.35	102		?
R	2003	450	*	1.64	913	0.80	0.77	833		0.51
R	2009	409	*	1.61	596	0.97	1.07	662		1.11
R	2012	122	*	1.48	159	1.28	0.62	215		1.34
R	2014	921		0.95	1948	0.58	0.49	1733		0.28
R	2016	319	*	3.50	593	0.92	1.24	504		1.15
R	2017	125		?	200	1.02	0.40	307		?
R	2018	68	*	1.92	114	1.24	1.02	123		0.94
R	2019	106	*	3.26	163	1.64	0.72	273		0.00
R	2020	706	*	0.48	560	0.74	0.70	599		?
R	2021	279		1.73	538	0.77	0.70	438		0.66
R	2022	294		0.00	134	?	0.00	128		1.78
R	2023	695	*	1.07	647	0.96	0.49	983		?
R	2024	90	*	0.00	138	*0.59	*1.65	37		0.00
R	2025	1005		?	1632	0.48	0.49	1307		?
R	2026	172		?	240	0.95	0.56	110		0.00
R	2027	165	*	1.18	186	1.03	0.77	277		0.95
R	2028	387	*	1.59	431	0.75	1.23	404		0.00
R	2029	249		0.00	135	*0.74	?	174		2.48
R	2030	143		0.00	126	*1.41	*0.89	65		
R	2031	206		1.64	213	0.80	*0.79	195	*	4.79
R	2032	81	*	3.09	192	3.03	3.36	66	*	?
R	2033	109		?	210	4.37	5.44	291		0.17
R	2034	1400	*	0.82	2177	0.32	0.30	1985		1.13
R	2035	343	*	0.75	398	0.93	?	364		?
R	2036	1465	*	2.14	1901	1.75	2.24	2123		
R	2037	-		?		?	?	108		1.70
R	2038	313	*	1.13	380	0.72	0.74	394		?

Right panel:

R	R M (b)	Phn PPM		Phn Ratio	PS PPM	PS-E Ratio	PS-L Ratio	NS PPM		NS-L Ratio
R	2039	184		1.18	252	0.73	1.22	97		0.00
R	2040	178	*	1.89	234	3.65	2.91	200		?
R	2041	101		?	717	*0.37	*0.25	374		?
R	2042	614	*	1.58	881	1.37	1.05	775		1.59
R	2043	206	*	1.26	284	0.98	0.51	423		
R	2044	316	*	1.64	342	1.27	1.42	345		?
R	2045	949	*	1.70	1393	0.68	0.71	1438		1.06
R	2046	315		2.21	637	1.35	1.43	657	*	0.00
R	2047	529	*	0.68	818	?	0.18	566	*	0.00
R	2048	293	*	12.65	293	*7.66	11.39	273		0.85
R	2049	240	*	1.43	331	1.06	1.19	374		0.00
R	2050	402		1.36	656	0.24	0.19	318		?
R	2051	261	*	1.23	419	0.47	?	335		1.23
R	2052	191	*	3.04	424	0.95	*1.25	433	*	6.46
R	2053	264	*	2.45	324	4.41	3.97	341		5.78
R	2054	490	*	1.45	779	1.73	1.86	661		0.83
R	2055	258	*	1.20	274	*1.09	0.00	312		0.94
R	2056	803	*	1.67	1217	3.43	3.78	1386		3.11
R	2057	195	*	1.36	208	*1.67	*1.46	231	*	0.00
R	2058	276		2.09	921	1.64	0.93	368	*	0.31
R	2059	6643	*	1.28	7433	1.08	1.20	8769		2.16
R	2060	397	*	1.04	696	0.71	*0.81	604		3.42
R	2061	409		13.20	670	10.90	12.63	525		17.37
R	2062	1963	*	1.43	3425	1.17	1.10	4035		?
R	2063	402	*	1.22	485	1.58	1.12	652		0.63
R	2067	116	*	1.82	152	1.85	*1.97	206		0.95
R	2068	106		?	340	0.53	0.44	340		0.00
R	2069	3518	*	0.47	1653	1.52	1.37	2939		4.02
R	2071	838	*	0.85	1187	0.34	0.21	1700	*	0.21
R	2072	381		1.06	693	0.69	0.35	526		0.68
R	2074	170	*	1.55	250	1.31	0.88	353		0.88
R	2075	148		?	175	*4.38	*3.27	173		?
R	2076	195	*	3.66	218	4.23	4.33	314		1.58
R	2077	745		?	617	0.79	0.59	477		0.58
R	2078	343	*	13.14	664	10.63	10.92	525		?
R	2079	937		?		?	?	280		0.00
R	2082	218		1.94	330	3.85	3.49	355		3.11
R	2083	3458		?	1050	0.67	0.95	1832	*	1.70
R	2084	391		1.14	612	0.83	0.56	523		0.68
R	2085	167		?	170	2.25	1.89	319		0.00
R	2086	328	*	1.28	538	1.45	1.49	540		0.70
R	2087	349		1.25	556	1.17	1.19	598		0.67
R	2088	557		1.22	926	0.39	0.43	766		0.59
R	2093	731	*	1.51	563	0.84	0.64	529		1.98
R	2094	298		0.00	147	?	0.00	0		-
R	2096	167	*	1.04	225	0.93	*1.16	266		0.72
R	2097	1813		?	714	0.75	0.74	1224		0.00
R	2098	287		?	229	0.94	0.64	228		1.05
R	2099	466	*	0.55	356	0.82	*0.87	390		0.96
R	2103	173		2.92	281	0.95	*0.74	291		0.13
R	2104	173		?	224	1.43	?	346		0.00
R	2105	150		0.00	111	3.56	3.87	120		?
R	2106	386	*	0.79	537	0.58	0.66	547		0.87
R	2107	174	*	3.43	282	5.66	5.05	263		3.13
R	2108	388	*	1.70	309	1.86	1.46	496		1.10
R	2109	166		2.75	305	1.47	1.39	371		?
R	2110	117	*	2.49	174	2.05	2.15	239		1.50
R	2112	44	*	4.03	103	1.23	1.01	141		?
R	2113	106	*	2.71	120	1.50	*1.42	98		0.00
R	2114	-		?	272	2.31	*1.85	612		15.72
R	2115	582		?	578	0.57	0.26	477		0.00
R	2116	205	*	1.89	241	0.81	0.86	251		?
R	2117	621	*	0.00	789	0.80	0.55	289		0.00
R	2118	181	*	1.17	218	1.09	*0.61	259		0.41
R	2119	158		0.00	99	?	?	176	*	?
R	2120	121	*	2.06	182	0.66	0.36	198		0.00
R	2121	519		0.83	862	0.67	*0.46	918		?
R	2122	672	*	4.17	1474	0.38	0.22	895		
R	2123	166		?	216	?	?	114	*	0.78
R	2124	116	*	1.94	209	1.25	1.33	220		0.62
R	2125	147	*	1.48	165	*1.15	*0.97	184		1.60
R	2126	562		0.97	856	0.58	*0.38	675	*	1.35
R	2127	179	*	2.07	185	1.09	0.92	275		0.20

(Table continues)

TABLE 3 *Continued*

R R M (b)	Phn PPM	Phn Ratio	PS PPM	PS-E Ratio	PS-L Ratio	NS PPM	NS-L Ratio
R 2128	3040	*0.51	3606	0.91	*0.97	4343	?
R 2129	584	1.46	485	0.73	0.71	498	
R 2130	137	*5.15	135	9.88	8.40	313	0.66
R 2131	184	*3.68	201	2.95	*2.23	166	*0.00
R 2133	137	*1.47	200	0.64	*0.73	190	0.45
R 2135	1083	*1.18	1544	0.65	0.65	1404	1.01
R 2136	300	*1.64	435	1.21	1.14	510	0.65
R 2138	223	*1.74	230	1.89	2.11	346	-
R 2139	426	1.83	701	1.49	1.57	732	0.14
R 2140	153	?	152	*0.73	?	232	1.52
R 2142	248	*0.95	195	1.33	*1.27	234	0.84
R 2145	379	*0.99	232	1.12	0.90	364	0.70
R 2148	160	*1.24	490	*0.58	0.68	310	*?
R 2149	7374	1.05	10771	1.29	1.21	14354	1.77
R 2151	565	*1.99	751	1.03	1.03	874	1.51
R 2152	1700	?	2315	*1.63	?	3030	*3.88
R 2153	202	1.50	297	1.65	2.16	362	0.00
R 2155	261	*1.25	365	0.63	*0.44	416	0.91
R 2162	3549	*1.12	5356	?	*0.09	6749	0.00
R 2164	478	*1.24	767	1.28	1.62	801	1.40
R 2167	515	1.65	762	2.60	2.35	859	1.89
R 2169	291	*1.20	413	0.67	*0.53	498	0.53
R 2171	617	1.87	1259	0.96	0.85	1190	1.27
R 2172	348	*1.43	606	0.97	1.33	568	0.58
R 2173	256	*1.66	302	1.48	1.28	458	0.76
R 2174	146	*3.91	385	1.50	1.64	461	0.99
R 2180	184	*2.18	329	1.28	1.26	337	0.75
R 2181	261	*1.40	481	0.89	0.66	519	1.05
R 2182	560	1.20	812	1.38	1.53	924	0.92
R 2183	178	*2.30	292	1.29	1.06	283	?
R 2184	92	0.00	180	*0	*0	112	0.92
R 2185	1136	1.13	1602	0.87	0.73	1497	?
R 2186	936	*1.04	1351	0.93	*0.78	1125	1.56
R 2187	169	*2.93	253	2.90	3.35	582	?
R 2188	663	0.00	606	0.61	0.58	277	0.45
R 2189	55	*3.85	162	1.89	1.44	193	1.42
R 2190	76	2.92	134	2.12	1.91	244	
R 2194	102	?	118	1.96	1.58	191	*0.78
R 2196	141	*1.67	182	0.91	0.80	242	0.75
R 2197	170	*1.69	169	1.98	2.00	271	1.36
R 2198	270	*1.50	332	0.85	0.65	420	0.34
R 2199	168	*?	326	0.71	0.57	313	*?
R 2201	171	*2.81	283	2.85	2.97	283	?
R 2205	111	*1.63	192	1.31	1.01	243	0.64
R 2207	486	*1.62	525	1.93	1.79	755	1.14
R 2208	288	*1.02	607	0.59	*0.44	522	*0.53
R 2209	197	*1.58	212	1.55	1.57	334	0.55
R 2210	493	*1.28	536	1.91	1.75	804	0.72
R 2211	1001	*1.15	1725	0.81	0.97	1744	0.18
R 2212	782	0.85	790	0.80	0.78	1138	0.46
R 2213	1482	*1.50	1873	1.54	1.62	2285	3.10
R 2214	158	*1.05	204	1.03	0.62	292	?
R 2217	130	3.10	271	1.77	1.26	335	-
R 2218	1381	*0.85	2047	*0.03	0.00	2752	-
R 2220	429	0.00	720	0.71	*0.47	531	0.84
R 2221	934	*1.57	1376	1.74	1.62	1738	3.21
R 2222	226	*2.11	436	0.96	0.93	540	?
R 2225	626	0.91	912	0.25	*0.08	901	-
R 2226	135	*3.28	217	4.00	*3.26	309	2.41
R 2228	89	?	227	5.59	4.92	221	-
R 2230	-	?	142	3.18	3.59	224	?
R 2231	65	*5.16	85	18.00	20.56	384	4.40
R 2232	1329	1.48	2231	1.23	1.26	2774	0.75
R 2234	80	*0.00	135	1.09	0.61	150	*0.80
R 2235	937	*1.12	566	2.75	*2.50	1259	4.14
R 2236	1304	2.15	2614	2.25	2.90	2896	?
R 2237	288	0.73	252	0.87	0.93	85	1.25
R 2238	501	*1.66	612	1.06	1.12	730	0.79
R 2239	536	*1.63	750	*1.96	*1.73	407	*4.56
R 2240	269	*1.30	382	2.73	3.43	443	1.21
R 2243	257	*1.29	284	0.98	*0.58	330	1.66
R 2247	173	*1.68	205	0.87	0.84	260	?
R 2248	44	?	68	2.89	4.27	120	*4.17

R R M (b)	Phn PPM	Phn Ratio	PS PPM	PS-E Ratio	PS-L Ratio	NS PPM	NS-L Ratio
R 2252	929	*5.26	1467	0.22	0.26	1397	0.00
R 2254	171	?	314	0.98	0.44	183	0.00
R 2255	323	*0.72	266	1.32	*1.18	450	?
R 2258	461	*1.13	509	0.57	?	687	*1.15
R 2267	734	*0.98	631	2.03	2.13	1058	1.67
R 2269	221	*3.31	293	3.68	*4.59	512	*19.54
R 2270	156	*2.25	334	0.90	0.00	146	0.00
R 2271	91	*0.62	134	*0	*0	92	0.00
R 2272	57	4.99	79	*1.70	*1.14	199	?
R 2273	83	?	50	5.55	*2.77	129	2.64
R 2274	134	*3.42	383	2.27	*1.65	220	*-
R 2276	222	?	265	*2.31	?	0	?
R 2277	57	?	245	0.97	?	77	*0.00
R 2280	1320	1.61	2290	0.55	0.61	1719	?
R 2281	289	?	353	*1.29	0.00	952	?
R 2282	234	*2.64	229	5.99	8.75	88	0.00
R 2283	134	*1.91	262	*0.91	1.40	381	*0.48
R 2284	419	*1.21	424	0.39	0.41	478	0.35
R 2288	107	?	98	0.00	0.00	111	0.24
R 2289	575	0.96	364	1.36	0.81	835	>224
R 2299	1104	0.00	0	>1584	>1645	0	-
R 2300	11	*17.18	0	-	-	0	-
R 2301	-		0	>879	>851	0	-
R 2302	360	*2.65	471	0.00	0.00	239	*2.29
R 2303	1737	1.10	1565	0.00	0.00	2286	0.71
R 2304	-	-	0	>100	>100	0	?
R 2305	-	*>245	0	>1493	>1900	392	?
R 2306	382	?	197	1.60	?	412	?
R 2311	-	?	0	?	?	0	>236
R 2314	64	?	158	*1.23	?	3	?
R 2317	-	-	0	>496	>594	0	>241
R 2318	78	*1.76	79	2.74	3.01	168	3.24
R 2319	66	*2.79	112	4.75	4.69	203	?
R 2320			0	>162	>137	0	*?
R 2321	-	?	0	>221	>181	0	?
R 2322	-	?	0	>810	>881	0	-
R 2323	-	?	0	>1024	*>827	0	-
R 2327	783	*2.24	1696	0.93	0.55	1285	0.32
R 2328			0	*>2430	*>1425	0	-
R 2329	-		0	>1845	>2325	0	-
R 2330	126	6.27	320	*1.23	*1.22	309	*0.00
R 2331	-	-	0	?	>466	0	-
R 2332	-	-	0	>3232	>3421	0	?
R 2334	320	*1.27	466	?	0.00	534	0.00
R 2336	2220	*1.09	3919	0.68	0.30	4355	0.48
R 2339	19	0.00	0	>323	>292	90	?
R 2340	-		-	?	?	245	-
R 2342	673	?	317	0.73	0.00	0	?
R 2343	480	*0.89	416	*0	*0.36	0	?
R 2344	115	?	141	0.88	0.00	146	?
R 2346	85	*4.41	189	1.51	1.54	287	0.87
R 2347	44	*3.62	97	*1.89	*1.80	126	?
R 2348	-	?	0	>102	>96	56	?
R 2349	-	-	0	>224	>197	0	?
R 2350	342	*1.27	141	2.08	2.17	0	0.00
R 2351	505	1.01	629	0.00	0.00	750	?
R 2352	94	?	165	0.37	0.00	0	-
R 2354	567	?	944	0.35	0.00	704	?
R 2355	114	0.00	0	>248	>230	97	?
R 2356	409	?	415	9.41	0.00	708	0.00
R 2358	454	?	350	1.44	1.52	450	0.61
R 2359	159	0.00	108	*6.43	?	0	-
R 2360	154	*1.51	209	1.84	1.35	346	0.00
R 2361	473	*0.72	615	0.32	0.18	487	0.00
R 2362	526	*0.62	258	1.63	0.00	606	?
R 2363	1327	0.00	0	>63	>111	340	?
R 2364	-	*-	121	*2.48	*0.93	110	?
R 2369	249	*3.27	304	0.00	0.00	526	?
R 2371	176	*3.56	337	0.57	0.95	80	-
R 2372	-	-	0	>289	>165	0	-
R 2374	-	?	0	>354	>305	99	*0.00
R 2375	266	*1.47	366	0.00	0.00	350	1.25
R 2384	373	6.51	870	0.92	1.00	1088	?

TABLE 3 Continued

R R M (b)	Phn PPM	Phn Ratio	PS PPM	PS-E Ratio	PS-L Ratio	NS PPM	NS-L Ratio
R 2249	683	1.36	1096	0.24	?	1128	17.21
R 2250	279	?	531	1.63	1.43	389	10.96
R 2251	156	?	245	*5.17	*5.18	567	* ?
R 2389	261	* ?	484	1.79	2.39	399	* -
R 2390	-	>755	0	>901	>640	0	-
R 2394	3075	0.94	3710	0.55	*0.43	6678	* 0.11
R 2395	1301	* 1.45	2336	0.47	0.32	3159	0.06
R 2396	8096	1.11	11507	*0.82	*0.67	6925	1.08
R 2397	-	-	4114	?	?	8446	?
R 2398	249	0.00	350	1.85	1.39	0	0.00
R 2399	475	* 1.63	659	?	*0	1532	* 0.12
R 2400	1790	* 0.44	937	?	?	2967	0.00
R 2401	574	1.23	663	0.00	-	1392	?
R 2402	122	1.47	133	1.43	0.90	185	0.49
R 2404	33	5.11	0	?	*>151	337	* 0.00
R 2405	-	>170	0	-	-	235	0.00
R 2406	220	?	151	*1.73	*1.51	241	0.00
R 2407	110	?	0	-	?	78	?
R 2408	94	* 1.98	188	*2.96	1.55	155	1.06
R 2409	45	?	92	0.00	?	119	0.00
R 2410	207	* 0.40	300	0.00	0.00	139	?
R 2411	129	* 1.82	240	*0.80	*0.56	399	* 0.35
R 2412	124	* 3.99		?	?	563	?
R 2414	3372	* 0.16	1621	?	?	5007	* 1.56
R 2415	162	?	401	0.36	?	316	* 0.00
R 2416	31	* 14.90	63	7.64	6.93	317	* ?
R 2419	326	?	0	>942	?	409	1.06
R 2420	891	* 1.04	1487	0.38	?	1275	0.23
R 2422	191	?	113	1.05	1.15	177	0.00
R 2423	119	* 0.00	0	>100	>144	224	?
R 2425	1326	* 0.64		?	?	1081	* 0.16
R 2428	-	-	0	>116	>245	0	-
R 2432	3727	* 1.07	7428	0.29	0.16	4448	0.15
R 2433	342	* 0.59	375	0.00	0.00	221	0.45
R 2435	844	* 0.00	773	?	*1.42	0	0.18
R 2437	1653	?	2119	3.48	2.23	2986	* 0.57
R 2442	504	0.57	226	*0.95	?	393	* ?
R 2444	222	0.00	334	0.00	0.00	210	0.00
R 2445	218	?	0	>379	*>396	304	* ?
R 2447	95	1.92	208	*0.64	?	197	?
R 2449	433	?	1113	*1.20	*0.48	917	* 0.00
R 2452	216	?	263	?	?	344	1.08
R 2453	228	?	260	?	0.00	254	0.73
R 2454	256	?	336	0.00	?	464	1.09
R 2455	-	-	207	*0	*0	0	?
R 2456	1200	* 0.40	1012	0.57	?	1665	0.18
R 2458	258	* 0.00	352	0.00	?	355	?
R 2459	-	>4825	0	>508	*>356	0	-
R 2460	79	* 9.85	111	*1.80	*1.46	129	?
R 2461	668	1.33	1206	?	0.00	1413	0.11
R 2462	373	* 1.51	944	0.86	0.82	922	1.46
R 2463	165	67.99	127	38.67	32.85	161	1.57
R 2464	133	* 24.86	0	>826	>632	0	?
R 2465	574	* 1.16	671	0.82	0.59	611	>743
R 2466	-	-	0	>461	>458	0	0.98
R 2467	206	?	201	*2.56	*2.08	459	0.88
R 2468	1525	?	1228	0.32	?	1079	* 1.71
R 2469	357	* 2.82	660	*1.26	*1.16	593	* 1.71
R 2470	282	* 1.32	382	0.79	?	435	0.00
R 2471	662	* 0.63	1003	?	?	962	1.76
R 2472	-	?	0	>565	>570	0	>285
R 2473	768	* 1.20	1258	0.74	*0.66	999	0.34
R 2474	395	1.50	685	0.56	0.40	647	0.80
R 2475	520	* 0.58	598	0.00	?	381	?
R 2476	874	* 0.88	1047	0.38	0.24	759	?
R 2479	192	0.00	240	3.08	2.86	408	?
R 2482	238	* 1.64	344	?	?	295	0.00
R 2483	155	?	155	?	*1.59	206	0.00
R 2484	234	?	403	0.00	0.00	320	* 1.42
R 2488	-	-	0	?	-	101	?
R 2489	-	-	0	>450	-	0	-
R 2490	376	0.00	323	*1.25	*0.65	238	?
R 2491	588	1.27	541	0.58	?	1006	?

R R M (b)	Phn PPM	Phn Ratio	PS PPM	PS-E Ratio	PS-L Ratio	NS PPM	NS-L Ratio
R 2385	-	>5848		?	?	899	?
R 2386	362	?	444	?	3.21	930	0.32
R 2387	245	0.00	0	?	>877	335	?
R 2502	-	-	0	*>236	?	0	-
R 2504	-	-	0	>369	*>348	158	
R 2506	113	* 0.00	159	0.62	?	161	0.00
R 2507	464	1.01	730	0.00	0.00	689	
R 2508	349	* 0.00	310	0.00	0.00	362	0.67
R 2509	-	-	168	0.00	0.00	0	?
R 2510	69	?	89	3.62	3.98	192	* 0.00
R 2511	126	* 1.46	0	*>188	>131	179	* 0.43
R 2512	206	0.00	0	>221	>244	205	1.58
R 2514	278	?	540	0.00	0.00	201	?
R 2516	119	* 0.00	435	*0	*0	131	* ?
R 2517	276	* 1.50	545	0.00	0.00	419	?
R 2519	-	-	0	>202	-	0	0.87
R 2520	411	1.07	495	1.20	1.36	583	-
R 2521	-	-	0	>163	?	0	0.00
R 2522	314	* 1.56	221	1.27	0.00	382	?
R 2523	-	>471	0	>848	>917	529	* 0.00
R 2524	705	1.34	632	0.63	*0.39	643	* 2.19
R 2525	373	0.00	446	*0.67	*0.63	335	0.41
R 2526	123	?	118	?	0.00	72	0.00
R 2527	203	* 0.90	175	*0		213	0.00
R 2528	668	* 0.67	1447	*0	*0	1826	*
R 2529	327	* 3.58	915	3.35	*1.48	897	* 0.56
R 2530	249	?	293	0.00	0.00	121	?
R 2531	-	-		?	?	328	* 1.04
R 2532	452	* 0.78	450	*2.08	0.46	337	* 0.09
R 2534	437	1.13	573	0.00	0.00	627	0.00
R 2537	185	?	0	>464	-	0	?
R 2538	1770	* 1.11	1836	0.26	0.54	2942	1.25
R 2539	496	* 1.18	785	1.96	1.24	823	?
R 2540	195	?	238	*0.97	?	278	?
R 2543	421	* 0.90	748	0.96	0.76	757	-
R 2545	113	* 0.00	147	0.00	0.00	0	0.00
R 2546	312	* 0.55	157	*1.13	?	200	?
R 2547	256	* 1.05	247	*1.08	0.84	149	?
R 2548	149	?	198	*0.83	0.51	107	?
R 2549	514	?	338	?	?	213	-
R 2550	148	0.00	153	0.00	0.00	0	0.40
R 2551	·163	* 1.84	266	2.31	2.22	683	-
R 2552	1100	0.00	0	>218	>134	0	0.00
R 2553	-	-	0	>131	?	0	0.00
R 2554	326	?	519	1.18	*0.80	483	*
R 2555	165	1.12	229	0.00	?	182	?
R 2556	635	* 1.42	401	0.88	0.71	472	* 0.00
R 2557	635	?	856	0.00	?	611	-
R 2558	-	-	150	?	?	0	>151
R 2562	513	* 0.58	611	*0.49	0.35	464	0.00
R 2563	237	* 0.00	100	2.80	4.04	123	0.00
R 2565	213	* 1.18	107	3.06	3.48	203	?
R 2566	335	* 2.06	369	*0.46	0.99	634	0.00
R 2567	2518	?	342	0.78	0.55	293	?
R 2568	188	* 0.00	344	?	?	308	0.00
R 2570	70	* 1.12	107	0.84	?	115	0.23
R 2571	233	?	282	0.36	0.30	202	* 0.58
R 2572	1874	1.49	1995	0.78	*0.55	3143	-
R 2573	1144	* 0.88	613	1.33	1.04	1491	0.69
R 2574	245	1.04	366	0.00	0.00	845	0.00
R 2575	197	* ?	0	>5163	>4674	0	-
R 2577	-	-	113	?	?	121	?
R 2578	157	0.00	212	0.00	0.00	197	?
R 2580	551	?	0	*>582	-	0	0.54
R 2581	237	3.30	363	*1.23	*1.88	695	?
R 2582	319	* 2.84	441	*1.00	*0.72	893	0.34
R 2583	63	0.00	0	?	-	123	0.00
R 2584	256	* 1.59	310	0.00	0.00	426	10.49
R 2586	97	* 4.17	247	4.45	4.08	245	?
R 2587	-	-	225	*1.91	*2.02	126	-
R 2588	34	?	0	>150	>97	0	0.00
R 2589	-	-	17	?	?	0	0.69
R 2590	176	?	331	0.00	0.00	227	?

(Table continues)

TABLE 3 Continued

R R M (b)	Phn PPM	Phn Ratio	PS PPM	PS-E Ratio	PS-L Ratio	NS PPM	NS-L Ratio
R 2492	790	1.28	1362	*1.22	*1.47	1152	* 1.36
R 2494	275	0.00	0	?	-	0	>982
R 2498	1344	?	1041	1.27	0.90	1773	1.02
R 2499	-	-	0	>311	?	121	-
R 2500	-	-	0	>456	*>585	0	-
R 2501	73	0.00	151	4.40	3.85	442	* 2.41
R 2597	248	?	866	0.00	0.00	149	* ?
R 2598	145	* 1.45	0	>181	*>202	526	?
R 2599	120	2.02	193	?	0.67	118	0.00
R 2600	1123	* 0.29	783	0.00	0.00	1125	?
R 2601	188	* 0.00	313	0.27	0.00	114	-
R 2603	-	-	0	?	-	0	>1381
R 2607	224	* 2.77	221	5.12	*6.18	305	0.58
R 2608	89	?	597	*0	*0	487	* ?
R 2609	67	0.00	433	?	0.00	88	?
R 2610	32	* 5.09	0	>233	*>243	73	-
R 2611	-	-	0	*>81	?	0	1.50
R 2612	773	* 3.42	1038	3.75	3.67	1167	0.00
R 2613	268	0.00	240	0.45	?	0	>1866
R 2614	-	?	0	*>213	*>121	0	?
R 2615	-	?	291	*2.63	*3.42	318	?
R 2616	684	* 1.69	846	0.44	0.34	1021	-
R 2617	359	?	217	*1.36	*1.07	183	* ?
R 2619	450	?	367	0.48	*0.42	184	0.93
R 2620	181	* 2.37	227	2.18	1.17	340	?
R 2621	1106	* 0.86	976	0.70	0.28	1019	?
R 2622	133	?	262	0.00	0.00	69	0.00
R 2624	62	* 1.39	121	0.90	*0.61	57	* 0.69
R 2625	70	* 2.43	66	*2.31	*2.87	140	* 0.37
R 2626	732	?	373	?	*0.81	375	?
R 2627	681	* 0.27	134	*1.69	*0.77	90	* ?
R 2628	45	* 3.56	114	*3.28	?	310	-
R 2631	295	0.00	97	?	?	0	0.00
R 2632	364	* 1.76	263	1.36	0.67	454	
R 2634	511	1.25	464	0.30	0.16	583	0.39
R 2636	281	0.62	104	*1.93	*1.63	105	2.10
R 2637	-	-	0	>76	>39	0	-
R 2638	209	* 1.23	227	0.79	0.79	453	0.33
R 2640	2229	* 0.56	1278	2.93	3.58	2971	?
R 2642	366	1.29	267	*2.88	*3.01	362	
R 2643	1119	* 0.48	4016	0.06	0.09	1709	* ?
R 2644	1251	* 0.85	1664	0.00	0.00	774	?
R 2646	551	* 1.02	219	?	?	152	
R 2647	453	* 1.11	161	*5.95	*3.00	487	
R 2648	-	-	0	>125	>146	0	* 0.13
R 2649	50	?	91	5.08	3.59	194	0.57
R 2650	174	0.90	205	0.53	0.44	209	* ?
R 2651	-	?	0	>87	>79	0	-
R 2652	326	0.54	143	?	?	319	* ?
R 2653	367	0.00	0	>253	>138	161	2.58
R 2654	712	?	365	*0.79	*1.09	550	0.00
R 2655	977	* 0.69	1051	*3.74	*2.89	745	?
R 2660	337	0.00	145	?	0.76	224	0.00
R 2661	570	* 0.00	315	*1.30	*0.70	610	
R 2663	-	?	231	*1.10	*0.95	298	* ?
R 2664	196	?	349	2.40	*1.31	149	0.00
R 2665	36	0.00	110	*1.61	*1.54	138	?
R 2666	1290	* 1.19	1424	1.46	1.18	1444	0.87
R 2667	-	-	888	?	1.04	332	-
R 2668	-	* >273	0	>987	>395	0	0.00
R 2669	292	0.00	763	0.00	0.00	516	?
R 2670	528	0.00	541	?	0.41	1511	* -
R 2673	-	-	1872	0.00	0.00	0	-
R 2674	-	-	2543	0.00	0.00	0	-
R 2675	-	-	2021	0.00	0.00	0	-
R 2676	-	-	475	0.00	0.00	0	-
R 2677	-	-	2364	0.00	0.00	0	0.00
R 2678	383	?	288	?	*1.49	287	-
R 2679	-	-	1167	0.00	0.00	0	-
R 2680	-	-	775	*0	*0	0	-
R 2681	-	-	1170	0.00	0.00	0	-
R 2682	-	-	779	0.00	0.00	0	-
R 2683	-	-	779	0.00	0.00	0	-
R 2591	147	* 2.08	65	?	?	111	* ?
R 2592	185	* 1.25	487	0.47	0.24	361	?
R 2593	634	?	698	*0.84	0.45	894	1.32
R 2594	-	-	567	*0	*0	258	-
R 2595	869	0.00	0	?	-	0	-
R 2596	301	?	301	0.00	0.00	0	0.00
R 2721	665	0.80	552	1.94	1.17	557	* 0.00
R 2722	-	-	0	>263	*>193	0	0.97
R 2723	-	-	257	?	?	0	-
R 2725	90	* 0.00	0	>212	>182	0	?
R 2726	232	* 1.13	253	1.31	1.24	265	* ?
R 2727	333	* 1.36	220	*3.56	*2.07	801	-
R 2729	104	4.09	0	>411	>558	0	-
R 2730	-	-	0	>438	>461	0	-
R 2731	706	* 1.19	803	1.69	0.80	1270	0.34
R 2734	207	* 12.26	0	?	?	0	-
R 2736	369	* 0.82	723	*1.18	*1.11	707	?
R 2737	1297	?	560	*0	*0	918	* -
R 2738	-	-	111	?	?	0	-
R 2740	568	?	230	0.00	0.00	583	>686
R 2742	72	* 2.76	94	?	*1.84	157	1.04
R 2744	38	0.00	0	?	*>122	0	?
R 2746	-	-	0	?	>400	0	>219
R 2747	-	-	0	?	>65	0	?
R 2748	-	-	220	0.00	*1.03	0	?
R 2749	50	* 3.46	63	?	?	0	>380
R 2753	150	?	166	?	*2.62	0	0.00
R 2754	-	>1092	0	?	>85	0	-
R 2755	-	-	0	-	>171	0	?
R 2756	77	* 1.53	176	?	*0.93	120	-
R 2757	227	0.00	415	?	*0.54	0	-
R 2758	198	* 0.00	235	0.00	?	0	?
R 2760	112	?	106	?	?	59	?
R 2761	121	0.00	0	-	*>129	31	-
R 2764	381	0.00	0	?	*>299	232	?
R 2765	180	0.00	0	?	>401	0	3.99
R 2766	239	0.00	0	?	>484	0	
R 2768	315	* 0.81	227	?	*1.31	0	?
R 2769	137	* 0.00	0	-	*>251	0	-
R 2771	474	0.00	1181	?	1.29	0	* >514
R 2772	-	-	0	?	*>850	0	-
R 2775	-	-	0	?	?	0	>2406
R 2776	-	?	0	?	>207	0	
R 2777	57	0.00	0	?	>298	0	-
R 2778	208	0.00	0	?	>108	123	?
R 2779	-	?	0	?	>1598	0	?
R 2780	-	* >304	0	?	>235	0	* -
R 2781	244	* 0.65	379	?	?	38	
R 2784	316	?	0	?	*>199	0	>565
R 2786	224	0.00	0	?	*>273	0	0.00
R 2787	234	* 0.66	147	0.00	0.00	217	?
R 2788	342	0.56	230	0.00	0.00	143	-
R 2789	-	-	0	?	*>461	0	-
R 2793	-	-	65	?	*0	0	-
R 2795	106	0.00	185	*0	*0	0	?
R 2796	160	* 1.57	228	0.00	0.00	121	0.00
R 2797	-	* -	84	0.00	0.00	107	* ?
R 2798	138	* 2.10	89	0.00	0.00	63	?
R 2799	36	0.00	80	0.00	0.00	31	?
R 2800	267	0.00	0	-	-	291	2.78
R 2802	228	* ?	216	*0	*0	274	?
R 2803	816	0.00	360	?	?	0	0.00
R 2804	624	* 0.76	567	0.00	0.00	340	?
R 2805	-	-	182	*0	*0	424	?
R 2808	57	?	80	*0	*0	113	2.24
R 2809	83	* 3.39	104	?	?	170	* ?
R 2810	31	0.00	714	0.00	0.00	182	?
R 2811	529	0.00	238	?	*0	132	-
R 2812	157	0.00	765	*0	*0	0	?
R 2813	313	?	456	*0	?	523	0.49
R 2816	-	-	0	?	-	45	
R 2818	500	?	0	-	-	458	-
R 2819	-	-	0	-	-	0	>293

TABLE 3 Continued

R	RRM (b)	Phn PPM	Phn Ratio	PS PPM	PS-E Ratio	PS-L Ratio	NS PPM	NS-L Ratio
R	2684	-	-	1068	0.00	0.00	0	-
R	2710	-	-	0	>873	-	0	-
R	2711	-	-	0	*>967	*>900	0	?
R	2712	710	*0.71	249	*2.32	*4.74	58	?
R	2713	-	?	0	>509	>491	0	-
R	2716	46	6.61	206	1.28	1.30	131	-
R	2717	74	?	171	?	*1.27	0	
R	2718	-	-	192	0.50	0.00	0	-
R	2720	2734	?	1260	*0.99	?	1621	-
R	2919	1173	*0.47	-	-	-	0	?
R	2920	-	-	-	-	-	0	-
R	2925	131	-	-	-	-	0	-
R	2939	189	-	0	-	-	0	-
R	2941	165	-	0	-	-	0	
R	2948	353	-	0	-	-	0	-
R	2949	2034	*0.00	-	-	-	0	-
R	2951	703	*0.00	-	-	-	686	*?
R	2958	200	*0.69	0	-	-	294	?
R	2960	166	*3.28	-	-	-	0	-
R	2964	241	-	0	-	-	0	-
R	2974	683	?	0	-	-	0	-
R	2979	204	?	0	-	-	0	0.00
R	2980	134	?	0	-	-	0	0.57
R	2983	2127	?	0	-	-	0	-
R	2988	767	?	0	-	-	0	-
R	2990	292	?	0	-	-	0	-
R	2991	-	?	0	-	-	0	>84
R	2992	57	*12.78	0	-	-	61	*>2944
R	2993	87	*2.72	0	-	-	0	*-
R	2995	79	*2.48	0	-	-	163	*-
R	2997	31	7.87	-	-	-	0	-
R	3003	74	*3.07	0	-	-	198	?
R	3004	408	0.00	0	-	-	448	-
R	3007	-	?	0	-	-	0	>187
R	3008	75	*2.99	0	-	-	0	*?
R	3010	-	>277	0	-	-	120	?
R	3019	152	*3.02	0	-	-	0	?

R	RRM (b)	Phn PPM	Phn Ratio	PS PPM	PS-E Ratio	PS-L Ratio	NS PPM	NS-L Ratio
R	2821	173	?	0	?	?	0	?
R	2822	361	0.00	0	?	-	0	
R	2824	174	0.00	0	-	-	207	?
R	2828	93	?	0	-	?	0	-
R	2830	256	0.00	0	-	-	0	-
R	2899	5431	?	-	-	-	1855	?
R	2909	82	-	0	-	-	56	?
R	2914	17	-	0	-	-	0	-
R	2916	-	-	0	-	-	0	7.54
R	3023	-	>300	0	-	-	0	8.56
R	3029	192	22.00	0	-	-	0	-
R	3035	-	*>718	0	-	-	0	?
R	3040	-	>2896	0	-	-	0	-
R	3041	-	>264	0	-	-	0	-
R	3042	-	*>301	0	-	-	0	-
R	3097	-	-	0	-	-	0	>255
R	3101	-	-	0	-	-	0	>488
R	3102	-	-	0	-	-	0	>275
R	3103	-	-	0	-	-	0	>309
R	3109	-	-	0	-	-	0	>2120
R	3111	-	-	0	-	-	0	*>1306
R	3112	-	-	0	-	-	0	>217
R	3114	-	-	0	-	-	0	>2063
R	3115	-	-	0	-	-	0	>4082
R	3116	-	-	0	-	-	0	>1508
R	3117	-	-	0	-	-	0	*>1419
R	3118	-	-	0	-	-	0	>842
R	3119	628	-	0	-	-	0	>124
R	3120	-	-	0	-	-	0	*>167
R	3121	-	-	0	-	-	0	>1075
R	3123	-	-	0	-	-	0	>554
R	3124	-	-	0	-	-	0	>399
R	3128	353	-	0	-	-	99	>392
R	3129	-	-	0	-	-	0	*>474
R	3136	-	-	-	-	-	0	>2207
R	3142	500	-	0	-	-	0	-
R	3145	275	-	0	-	-	90	-

[a]A-N, alpha-numeric spot naming system.

[b]RRM is a spot naming system for proteins in the Response/Regulation Map project.

[c]Steady-state growth conditions are defined as growth conditions that permit a constant growth rate over at least six generations. The abbreviations used in the table are defined as follows, with reference to the original work. AB a′, abundance of proteins from cultures grown in glucose-minimal MOPS medium (29) at 37°C in a′ units × 10^3 (43). These units multiplied by 0.1 give the percentage of total protein. Ace RL, relative level of proteins from cultures grown in acetate-minimal MOPS medium compared to glucose-minimal MOPS medium at 37°C (43). Gly RL, relative level of proteins from cultures grown in glycerol-minimal MOPS medium compared to glucose-minimal MOPS medium at 37°C (43). Ric RL, relative level of proteins from cultures grown in glucose-rich (amino acid, bases, and vitamins) MOPS medium compared to glucose-minimal MOPS medium at 37°C (43). 13.5 RL, 15 RL, 23 RL, 30 RL, 42 RL, 46 RL, relative levels of proteins from cultures grown in glucose-rich (amino acid, bases, and vitamins) MOPS medium at 13.5, 15, 23, 30, 42, or 46°C compared to 37°C (20). Phn PPM, amount of protein (expressed as parts per million of total protein) in cultures grown in glucose-minimal MOPS medium at 37°C and used to compare the levels of proteins grown with P_i and phosphonate as the phosphate source (VanBogelen, unpublished data). Although the cultures were grown at steady state, the radiolabeling used to quantitate the levels of proteins was a 5-min pulse-chase. The PPM value is the average value found for two gels from each of two separate experiments. Phn Ratio, ratio of PPM of proteins from cultures grown in glucose-minimal (phosphonate) MOPS medium compared to those grown in glucose-minimal MOPS medium with P_i. *, PPM average value (two experiments, two gels each) used to find the ratio had a standard error of >20% but <50%; ?, standard error was >50%, and additional editing of the images is in progress; >, protein was not found in cultures grown with P_i, and the number that follows the > is the PPM value for the phosphonate cultures.

[d]Growth transition conditions are defined as growth conditions that have changed because of depletion of a nutrient, addition of a chemical, or alteration in environmental growth conditions. The abbreviations used in the table are defined as follows, with reference to the original work. PS PPM, amount of the protein (expressed as parts per million of total protein) in cells grown in glucose-minimal MOPS medium at 37°C and used to compare the levels of proteins in cultures grown in media with limited P_i (VanBogelen, unpublished data). Although the cultures were grown at steady state, the radiolabeling used to quantitate the levels of proteins was a 5-min pulse-chase. The PPM value is the average value found for two gels from each of two separate experiments. PS-E Ratio, ratio of PPM of proteins from cultures grown in glucose-minimal medium and radiolabeled early after growth indicated that phosphate was depleted (10 to 30 min after an inflection in growth) compared to those grown in steady-state conditions in glucose-minimal medium. *, PPM average value (two experiments, two gels each) used to find the ratio had a standard error of >20% but <50%; ?, standard error was >50%, and additional editing of the images is in progress; >, protein was not found in cultures grown with P_i. The number is the PPM value for the phosphonate cultures. PS-L Ratio, as described for PS-E except that radiolabeling was done 30 to 60 min after a growth inflection (due to depletion of phosphate) was observed. NS PPM, as described for PS PPM except that this was the control culture for a culture with limited ammonia (the only nitrogen source) (VanBogelen, unpublished data). NS-L Ratio, as described for PS-L except that the cultures were depleted of a nitrogen source.

TABLE 4 Proteins belonging to stimulons and regulons[a]

A-N (b)	R R M (c)	REG (d)	NSI	PSI	CSI	NA	42C	50C	Cd	QN	HP	Dnp	ILE	10C	O2 -	O2
A010.8		LRP														
A013.0	R 1701															
A015.6					Y											
A019.0	R 2623															
A021.5									Y							
A025.8	R 2931															
A028.0	R 2111			Y												
A029.5		HTP				Y										
A032.5				Y												
A035.5	R 1700	LRP														
A041.3	R 2178		Y													
A048.0	R 1945														Y10	
A052.5					Y											
A165.0																
B010.1				Y												
B011.0				Y												
B011.8															Y6	
B013.0	R 2292							Y								
B013.1										Y10	Y					
B013.4										Y						
B014.5	R 2246			Y					Y5	Y10						
B015.0	R 2245			Y	Y										Y3	
B015.5		OXY							Y	Y	Y					
B015.8									Y							
B018.4																Y4
B020.9	R 2204	OXY	Y			Y4	Y5	Y10	Y9	Y10						
B021.7										Y						
B023.3										Y						
B024.1	R 2750															
B025.3	R 2176	HTP	Y	Y		Y2	Y10	Y10	Y10		Y		Y2			
B026.3																
B027.0	R 2724		Y													
B027.1					Y											
B028.3	R 2477		Y													
B030.2					Y											
B030.8									Y							
B032.2					Y											
B033.0	R 2100		Y													
B035.1	R 2101															
B036.0	R 2015	LRP														
B037.0	R 2801		Y													
B040.5	R 2728		Y													
B040.7	R 1928															
B040.8											Y					
B042.0			Y													
B042.2			Y													
B044.8		OXY							Y10	Y10	Y4					
B046.5	R 2831												Y			
B046.7	R 1861															
B050.3	R 1859															
B056.0		LRP														
B056.5	R O565	HTP	Y	Y		Y4	Y10	Y10			Y		Y2			
B058.3	R 1790															
B061.0	R O610													Y5		
B065.0	R O650															
B066.0	R O660	HTP	Y	Y		Y5	Y10	Y10	Y5	Y2	Y5	Y				
B076.0														Y		
B082.5																
B083.0																
B084.0				Y												
B105.0				Y												
B140.0				Y												
C011.5		LRP														
C012.8					Y											
C013.2										Y						
C013.4	R 2296									Y						
C013.5	R O135		Y	Y	Y		Y		Y4	Y2	Y		Y3			
C014.7	R 2265	HTP					Y10	Y10	Y10	Y4						
C014.8	R 2307															
C015.3	R 2241															
C015.4	R O154	HTP	Y	Y		Y5	Y10	Y10	Y3		Y3	Y				
C017.7	R 2794		Y	Y												
C018.0	R 2206		Y	Y												
C018.1										Y						

A-N (b)	R R M (c)	REG (d)	NSI	PSI	CSI	NA	42C	50C	Cd	QN	HP	Dnp	ILE	10C	O2 -	O2
C022.3										Y						
C022.7	R 2203															
C025.2	R 2166															
C025.5				Y												
C026.8	R 2165							Y3								
C027.1	R 2160															
C028.5																Y3
C029.8																
C030.7	R 2481															
C031.2					Y											
C034.3																
C035.6	R 2006			Y												
C036.3	R 2005			Y								Y				
C036.7	R 2767															
C037.0					Y									Y6	Y		
C038.3										Y						
C039.3	R 2457	SOS						Y10			Y3	Y	Y6	Y		
C040.3															Y2	
C040.5				Y												
C040.6											Y					
C041.0	R 1923															
C042.0		LRP														
C042.5																Y
C042.6	R 1922			Y							Y					
C042.9											Y					
C043.1															Y2	
C043.5	R 1933	LRP		Y												
C043.8	R 1916															
C044.6																
C048.7																
C050.5					Y											
C051.0					Y											
C053.0						Y										
C054.0						Y										
C054.1						Y										
C055.0																Y3
C056.0												Y				
C058.3						Y										
C060.7																
C062.5	R 2285	HTP						Y2	Y10	Y10	Y3		Y	Y2		
C062.7	R 2290															Y5
C067.0											Y					
C070.0	R 2291												Y			Y6
C074.0	R 2671			Y												
C075.0	R 2672															
C078.0	R 1770												Y			
C080.0		SOS				Y										
C137.0	R 1733															Y2
C158.0				Y												
D014.7	R 2308															
D025.5	R 2357			Y												
D026.3							Y									
D028.0	R 2092			Y												
D028.3	R 2091			Y												
D029.8	R 2089		Y	Y												
D030.2				Y												
D031.5	R 2010			Y												
D032.5	R 2011															
D033.4	R 2008	HTP		Y	Y	Y3	Y10	Y10				Y				
D035.5				Y												
D035.7	R 2007		Y				Y	Y				Y				
D035.8			Y													
D036.1														Y10		
D038.0				Y									Y			
D038.3	R 1999			Y												
D040.7	R 1929	LRP									Y					
D042.0											Y					
D044.5																
D044.8											Y					
D046.0																
D047.0											Y					
D048.4		LRP														
D048.5		HTP									Y	Y	Y			
D049.9	R 1845	LRP	Y													

TABLE 4 *Continued*

A-N (b)	R R M (c)	REG (d)	NSI	PSI	CSI	NA	42C	50C	Cd	QN	HP	Dnp	ILE	10C	O2 -	O2
D050.0		LRP	Y													
D057.0									Y							
D058.0		LRP														
D058.5																
D060.5		HTP,LRP					Y6	Y5								
D074.0																
D078.0		OXY							Y	Y	Y5					
D078.1	R 1773	OXY				Y	Y3		Y	Y10	Y2					
D084.0																
D087.5	R 1731															
D088.0	R 2424			Y												
D094.0																
D099.0																Y8
D157.0	R 1698															
E014.0	R O140															
E017.6		LRP														
E018.0	R 2200		Y													
E022.8	R 2170															
E023.0																
E025.5																
E026.0							Y									
E029.8					Y											
E031.8			Y													
E032.1	R 2478															
E033.8									Y	Y						
E036.6	R 2002															
E038.5	R 2450															
E039.2	R 1987															
E039.8	R 1925															Y9
E042.5		LRP														
E043.0				Y												
E048.0		LRP														
E048.7											Y					
E049.2				Y												
E049.3	R 1847			Y												
E050.0																
E054.0				Y												
E056.0																
E058.0	R 1838															
E061.0	R 2630			Y												
E072.0		HTP		Y	Y	Y	Y	Y4	Y	Y	Y					
E072.1								Y9								
E077.5																
E079.0	R 1782			Y									Y3				
E088.0	R 2421			Y	Y											
E133.0	R 2807															
E140.0	R 2431															
F007.0	R 2275			Y												
F010.1	R 2365	HTP				Y	Y	Y								
F010.6	R 2913												Y			
F012.3	R 2262		Y													
F013.0	R 2261			Y												
F014.4								Y								
F014.7	R 1703		Y	Y									Y	Y		
F014.9								Y								
F017.7	R 2192		Y													
F018.8	R 2193					Y										
F019.0								Y								
F020.9	R 2191															
F021.3			Y													
F021.5		HTP				Y					Y					
F022.5	R 2147		Y	Y												
F023.3									Y							
F023.7										Y						
F023.9		LRP														
F024.5									Y3	Y2		Y2				
F024.6	R 2157															
F025.1											Y					
F026.0	R 2081				Y											
F027.0	R 2146		Y													
F028.0	R 2080															
F028.6		LRP														
F028.8																
F029.0	R 2940		Y	Y												

A-N (b)	R R M (c)	REG (d)	NSI	PSI	CSI	NA	42C	50C	Cd	QN	HP	Dnp	ILE	10C	O2 -	O2
F030.2															Y4	
F030.3	R 2073															
F032.3																
F032.5	R 1991															
F032.7		LRP														
F032.9			Y													
F033.1	R 2070															
F033.2				Y												
F035.0	R 1990			Y												
F035.6		OXY							Y10	Y	Y6					
F035.8	R 1708			Y	Y											Y6
F037.5	R 1981			Y												
F037.8	R 1912															
F037.9				Y						Y2				Y		
F038.0	R 1986				Y											Y2
F038.1																
F039.0	R 2446								Y							
F039.6	R 1911															
F039.7	R 1910			Y												
F042.2	R 1914			Y												
F043.5				Y												
F043.8					Y											Y2
F043.9	R 1920															
F045.0										Y						
F045.5																
F045.6				Y	Y											
F046.6	R O466															
F048.8												Y				
F049.0	R 2388												Y			
F050.1	R 1751			Y												
F050.2																
F050.3	R 2576												Y			Y6
F050.4		LRP														
F050.6	R 1831	OXY	Y	Y					Y7	Y10	Y10					
F054.1	R 1828			Y												
F054.4	R 1767															
F056.0																
F056.1		LRP														
F056.2																
F056.5																
F057.0																
F058.5					Y											
F060.3																Y10
F063.5																
F066.0																
F082.5																
F084.0												Y	Y			
F084.1		HTP		Y		Y3	Y10	Y10	Y10	Y2						
F088.0					Y											Y3
F093.0																
F099.0													Y			Y4
F106.0																
F119.0				Y									Y			
F178.0																
G010.6									Y	Y						
G010.7	R 2366			Y							Y					
G011.3																
G011.9				Y	Y											
G012.3																
G012.4	R 2256			Y	Y											
G013.2				Y												
G013.5	R 2898	HTP							Y	Y	Y10		Y10			
G014.1	R 2257			Y												
G015.1	R 2338			Y												
G015.4		OXY	Y	Y			Y3	Y3	Y	Y10	Y10					
G015.8	R 2229															
G016.0	R 2337			Y												
G016.2												Y				
G016.5				Y												
G018.3	R 2911											Y				
G019.0									Y							
G021.0		HTP		Y		Y	Y						Y			
G023.4	R 2141			Y									Y			
G024.7	R 2137			Y	Y											

(Table continues)

TABLE 4 *Continued*

A-N (b)	R R M (c)	REG (d)	NSI	PSI	CSI	NA	42C	50C	Cd	QN	HP	Dnp	ILE	10C	O2-	O2
G025.0																
G025.1				Y	Y											
G025.3									Y							
G025.7										Y10						
G025.9																
G026.0																
G026.3												Y				
G026.6				Y												
G027.2									50C Y							
G028.0	R 2150															
G028.1	R 2972															
G028.2									Y							
G029.2	R 2513															
G029.3	R 2973															
G029.4												Y				
G029.5	R 2066															
G029.6																
G030.9			Y													
G031.5															Y7	
G032.0	R 2064									Y						
G032.8																
G035.0		LRP							Y		Y7					
G036.1				Y												
G036.2	R 2503								Y							
G036.7												Y				
G038.1		LRP														
G038.2	R 1965		Y	Y								Y				
G038.3																
G038.4										Y						
G038.8		LRP														
G039.1	R 1974			Y												
G039.5									Y							
G040.0																
G040.8									Y							
G041.0																
G041.2	R 2434							Y								
G041.3	R 2436				Y									Y3		
G041.4	R 1882															
G042.1	R 1747		Y	Y								Y				
G042.2		LRP														
G043.2		LRP														
G043.6	R 1881															
G043.8																
G043.9																
G044.0	R 1894															
G045.5	R 2945								Y							
G045.6	R 2544															
G046.0																
G047.8	R 2947				Y											
G048.0	R 1888		Y	Y								Y				
G048.6	R 1891															
G049.1										Y						
G049.2	R 1818															
G050.4																
G050.5																Y4
G051.1					Y											
G052.0	R 1809		Y	Y	Y											
G054.0																
G054.6	R 2560															
G054.7	R 2978			Y	Y										Y2	
G055.0													Y			
G055.3			Y													
G057.0																
G057.1			Y	Y							Y					
G057.6					Y							Y				
G057.9		LRP														
G058.1			Y													
G058.5	R 2561											Y				
G060.1	R 1740		Y	Y								Y				
G065.0					Y											
G067.0																
G070.0															Y10	
G070.1		LRP			Y											
G072.1		LRP		Y	Y							Y				

A-N (b)	R R M (c)	REG (d)	NSI	PSI	CSI	NA	42C	50C	Cd	QN	HP	Dnp	ILE	10C	O2-	O2
G074.0	R 2639			Y	Y								Y			
G076.0	R 1737		Y	Y												
G080.0							Y2	Y2	Y10	Y10	Y10			Y2		
G080.1							Y2	Y2	Y5	Y10	Y3	Y		Y3		
G080.5												Y				
G088.0												Y				
G095.0					Y							Y				
G097.0																Y6
G117.0														Y		
H013.2	R 2585		Y													
H014.0	R 2253															
H015.0												Y				
H015.6												Y				
H015.65												Y				
H015.9												Y				
H016.0												Y				
H027.4	R 2134		Y													
H028.0																
H029.5												Y				
H030.8												Y				
H031.5												Y				
H033.0			Y													
H040.0		LRP														
H041.0	R 1883		Y	Y												
H047.4	R 1864		Y													Y2
H049.2	R 1868		Y													
H049.3			Y													
H050.4												Y				
H052.7																
H054.5	R 2733															
H054.6																
H054.7												Y				
H055.0		LRP														
H094.0		HTP		Y			Y4	Y10	Y3			Y				
H097.3														Y9		
H124.0										Y						
I010.5		SOS								Y						
I011.7																
I011.9																
I012.1																
I012.2																
I012.8																
I013.0																
I014.4																
I014.6																
I014.7																
I016.4																
I017.2																
I019.5																
I021.1																
I021.3		OXY								Y	Y	Y				Y10
I023.0																
I026.0																
I033.5														Y2		
I058.4																
I063.5																
I065.0		SOS				Y										

TABLE 4 *Continued*

[a]Regulon is a term used to indicate that the genes for the proteins are coregulated (see chapter 84).
[b]A-N, alphanumeric spot naming system.
[c]RRM, spot naming system used for the new Response/Regulation Map project.
[d]REG, regulon. Only four regulons are included so far in the database: HTP, heat shock proteins regulated by σ^{32} (31); OXY, proteins regulated by the OxyR proteins (9); SOS, proteins regulated by the LexA protein (64); and LRP, proteins regulated by the leucine response regulator (Lrp) (115).
[e]Y, protein was visually observed to be induced; Y with a number, induction was measured and found to be that recorded as the number (data are not found in Table 3). Abbreviations in the table are defined as follows, with a reference that describes the original work. NSI, nitrogen starvation conditions. However, changes in proteins were not measured but were only visually observed (VanBogelen and Neidhardt, unpublished observations). PSI, starvation for phosphate. However, changes in proteins were not measured but were only visually observed (VanBogelen, Peruski, and Neidhardt, unpublished observations). CSI, starvation for glucose and changes in proteins were visually recorded (VanBogelen and Neidhardt, unpublished observations); NA, treatment with nalidixic acid (59); 42C, temperature shift from 28 to 42°C (59); 50C, temperature shift from 28 to 50°C (59); Cd, cadmium chloride treatment (59); QN, treatment with the quinone ACDQ (59); HP, treatment with hydrogen peroxide (59); Dnp, treatment with 2,4-dinitrophenol (16); ILE, starvation for a single amino acid, isoleucine (59); 10C, temperature shift from 37 to 10°C (21); O2–, shift from aerobic to anaerobic conditions (53); O2, shift from anaerobic to aerobic conditions (54).

spot name. The coordinates for Fig. 4 are assigned by the computer program. A coarse grid was placed on the figure to locate the spots. The equations listed in Fig. 5 were used to estimate the MWs and pIs of the proteins listed in Table 1.

DATABASE TABLES

The volume of data found in this database is difficult to present as tables, especially considering the numerous starting points for posing questions of the database. Users are encouraged to obtain the electronic version of the database (see section on Information Exchange).

Table 1

Table 1 (p. 2076) gives the positions of protein spots on 2-D gels and the MW and pI for each protein. This table is sorted in order by the spot name, first by alphanumeric names and then by the Response/Regulation Map names. All of the protein spots listed in other tables of the database are listed in Table 1. All of the spots observed in Fig. 1 and 4 have been assigned names, but some have no data entered and thus have not been included in this tabular version of the database. They are listed in the electronic version, ECO2DBASE (see the section on Information Exchange). Table 1 lists the coordinate positions (on Fig. 1 through 4) for the spots, the calculated MW and pI of each identified protein, and an estimated MW and pI for every protein in the table.

Table 2

Table 2 (p. 2094) lists all of the proteins that have been identified as products of particular genes (or ORFs found within the DNA sequence) or are known proteins. The table is sorted by gene name, and it references all of the information in the Expression Map. The following types of information for each protein spot are included: gene name, protein name (if one has been assigned), alphanumeric name, category of function (48), EC number, SWISS-PROT number, GenBank codes, direction of the gene on the chromosome, genetic map location, physical map location (using the Kohara miniset to approximate the location), basis of the identification, and donor of the material used in the identification. Table 2 lists some proteins expressed from a specific Kohara clone but not linked to a gene contained on that clone.

Table 3

Table 3 (p. 2101) lists all proteins included in a global study in which the level or synthesis rates of proteins were measured. Columns 3 to 14 represent steady-state growth conditions; the next 5 columns list growth transition conditions. The table presents the data, and the footnotes give a brief description of the experiment and/or the paper that originally presented the data. Included in this table are the gene names associated with identified proteins.

Table 4

Table 4 (p. 2112) lists the protein spots induced by one or more of the conditions not listed in Table 3. Y indicates that the proteins appeared to be induced, according to visual analysis of the 2-D gels, and Y followed by a number indicates the induction ratio of that protein. This table also lists proteins belonging to one or more regulons (only the HTP, SOS, OXY, and LRP regulons have been included so far).

INFORMATION EXCHANGE

Information exchange is a priority issue for the database. By 1990, information from numerous publications, laboratory notebooks, and the gene-protein index had all been entered into an electronic version of the database. In 1992, the electronic version was deposited at the database repository at the National Center for Biotechnology Information, and updates were submitted to make all of the information accessible to investigators. Large-volume information databases are best used in electronic form, and users are encouraged to obtain the database through anonymous ftp from the repository. The Internet address is ncbi.nlm.gov or 130.14.20.1 in the directory /ncbi/repository/ECO2DBASE. The reference 2-D gels are in the GELS directory, and the database and information files are in the edition6 directory. For those users who do not have access to Internet, a copy of the database can be obtained from the authors (please specify a disk format).

The alphanumeric names of proteins that have been identified have been incorporated into the other databases, including the SWISS-PROT protein database (5) and the ECD database (24), so that users can easily and accurately move among the different databases. A new database for *E. coli* (based on the *Caenorhabditis elegans* database) is being developed. It will serve as an encyclopedia of all the information known about *E. coli*

(Staffan Bergh, personal communication). All of the independent databases are being included in this encyclopedia. The gene-protein database, including the 2-D reference gels, has already been entered.

Other investigators can contribute information to the database. For the Genome Expression Map project, samples of purified proteins can be sent to assist in the identification project. For the Response/Regulation Map, investigators are encouraged to submit physiological and regulatory information from their own 2-D gel analyses (as was done by B. Ernsting and R. Matthews [15]), although this requires that the 2-D gel pattern closely match that of the reference gels.

ACKNOWLEDGMENTS

The Genome Expression Map project is supported by grant DMB-8903787 from the National Science Foundation and grant GM17892 from the National Institutes of Health (NIH). Current work on the Response/Regulation Map is supported through Parke-Davis Pharmaceutical Research. A. Pertsemlidis was supported by NIH grant GM08352–784525–31002.

We thank the many investigators (listed in Table 2) who have contributed biological material for protein identifications. We thank Amos Bairoch for assistance with the gene names and SWISS-PROT accession numbers and Manfried Kroger and Kenn Rudd for their assistance with map positions of genes. We also acknowledge all of the scientists who have worked on the database in the past: David Appleby, Philip L. Bloch, Jacqueline A. Bogan, Madhumita Ghosh, Sherrie Herendeen, M. Elizabeth Hutton, Douglas Irvine, Peggy LeMaux, Steen Pedersen, Teresa A. Phillips, Sankar P. Reddy, Solvejg Reeh, and Vicki Vaughn.

LITERATURE CITED

1. **Abshire, K. Z., and F. C. Neidhardt.** 1993. Growth rate paradox of *Salmonella typhimurium* within host macrophages. *J. Bacteriol.* 175:3744–3748.
2. **Ames, G. F.-L., and K. Nikaido.** 1976. Two-dimensional gel electrophoresis of membrane proteins. *Biochemistry* 15:616–622.
2a. **Allen, S. P., J. O. Polazzi, J. K. Gierse, and A. M. Easton.** 1992. Two novel heat shock genes encoding proteins produced in response to heterologous protein expression in *Escherichia coli. J. Bacteriol.* 174:6938–6947.
3. **Ang, D., G. N. Chandrasekhar, M. Zylicz, and C. Georgopoulos.** 1986. *Escherichia coli grpE* gene codes for heat shock protein B25.3, essential for both lambda DNA replication at all temperatures and host growth at high temperature. *J. Bacteriol.* 167:25–29.
4. **Bachmann, B. J.** 1990. Linkage map of *Escherichia coli* K-12, edition 8. *Microbiol. Rev.* 54:130–197.
5. **Bairoch, A., and B. Boeckmann.** 1993. The SWISS-PROT protein sequence data bank recent developments. *Nucleic Acids Res.* 21:3093–3096.
6. **Bloch, P. L., T. A. Phillips, F. C. Neidhardt.** 1980. Protein identifications of O'Farrell two-dimensional gels: locations of 81 *Escherichia coli* proteins. *J. Bacteriol.* 141:1409–1420.
7. **Blumenthal, R. M., P. G. Lemaux, F C. Neidhardt, and P. P. Dennis.** 1976. The effects of the *relA* gene on the synthesis of aminoacyl-tRNA synthetases and other transcription and translation proteins in *Escherichia coli* A. *Mol. Gen. Genet.* 149:291–296.
8. **Celis, J. E., H. H. Rasmussen, E. Olsen, P. Madsen, H. Leffers, B. Honore, K. Dejgaard, P. Gromov, H. J. Hoffmann, and M. Nielsen.** 1993. The human keratinocyte two-dimensional gel protein database: update 1993. *Electrophoresis* 14:1091–1198.
9. **Christman, M. F., R. W. Morgan, F. S. Jacobson, and B. N. Ames.** 1985. Positive control of a regulon for defenses against oxidative stress and some heat-shock proteins in *Salmonella typhimurium. Cell* 41:753–762.
10. **Chuang, S.-E., and F. R. Blattner.** 1993. Characterization of twenty-six new heat shock genes of *Escherichia coli. J. Bacteriol.* 175:5242–5252.
11. **Clarke, L., and J. Carbon.** 1976. A colony bank containing synthetic ColE1 hybrid plasmids representative of the entire *E. coli* genome. *Cell* 9:91–99.
12. **Copeland, B. R., R. J. Richter, and C E. Furlong.** 1982. Renaturation and identification of periplasmic proteins in two-dimensional gels of *Escherichia coli. J. Biol. Chem.* 257:15065–15071.
13. **Daniels, D. L., G. Plunkett, V. Burland, and F. Blattner.** 1992. DNA sequence of *E. coli.* I. The region from 84.5 to 86.5 minutes. *Science* 257:771–778.
14. **Engstrom, P., and G. L. Hazelbauer.** 1980. Multiple methylation of methyl-accepting chemotaxis proteins during adaptation of *E. coli* to chemical stimuli. *Cell* 20:165–171.
15. **Ernsting, B. R., M. R. Atkinson, A. J. Ninfa, and R. G. Matthews.** 1992. Characterization of the regulon controlled by the leucine-responsive regulatory protein in *Escherichia coli. J. Bacteriol.* 174:1109–1118.
16. **Gage, D. J., and F. C. Neidhardt.** 1993. Adaptation of *Escherichia coli* to the uncoupler of oxidative phosphorylation 2,4-dinitrophenol. *J. Bacteriol.* 175:7105–7108.

17. **Goldstein, J., N. S. Pollitt, and M. Inouye.** 1990. Major cold shock protein of *Escherichia coli. Proc. Natl. Acad. Sci. USA* 87:283–287.
18. **Goodlove, P. E., P. R. Cunningham, J. Parker, and D. P. Clark.** 1989. Cloning and sequence analysis of the fermentative alcohol-dehydrogenase-encoding gene of *Escherichia coli. Gene* 85:209–214.
19. **Gudas, L. J., and D. W. Mount.** 1977. Identification of the *recF* (*tif*) gene product of *Escherichia coli. Proc. Natl. Acad. Sci. USA* 74:5280–5284.
20. **Herendeen, S. H., R. A. VanBogelen, F. C. Neidhardt.** 1979. Levels of major proteins of *Escherichia coli* during growth at different temperatures. *J. Bacteriol.* 139:185–194.
21. **Jones, P. G., R. A. VanBogelen, and F. C. Neidhardt.** 1987. Induction of proteins in response to low temperature in *Escherichia coli. J. Bacteriol.* 169:2092–2095.
22. **Kaltschmidt, E., and H. G. Wittmann.** 1970. Ribosomal proteins. VII. Two-dimensional polyacrylamide gel electrophoresis for fingerprinting of ribosomal proteins. *Anal. Biochem.* 36:401–412.
23. **Kohara, Y., K. Akiyama, and K. Isono.** 1987. The physical map of the whole *E. coli* chromosome: application of a new strategy for rapid analysis and sorting of a large genomic library. *Cell* 50:495–508.
24. **Kroger, M., R. Wahl, and P. Rice.** 1993. Compilation of DNA sequences of *Escherichia coli* (update 1993). *Nucleic Acids Res.* 21:2973–3000.
25. **Kroh, H. E., and L. D. Simon.** 1990. The C1pP component of C1p protease is the sigma-32 dependent heat shock protein F21.5. *J. Bacteriol.* 172:6026–6034.
26. **Lemaux, P. G., S. L. Herendeen, P. L. Bloch, and F. C. Neidhardt.** 1978. Transient rates of synthesis of individual polypeptides in *E. coli* following temperature shifts. *Cell* 13:427–434.
27. **Neidhardt, F. C.** 1987. Multigene systems and regulons, p. 1313–1317. *In* F. C. Neidhardt, J. L. Ingraham, K. B. Low, B. Magasanik, M. Schaecter, and H. E. Umbarger (ed.), *Escherichia coli and Salmonella typhimurium: Cellular and Molecular Biology*, vol. 2. American Society for Microbiology, Washington, D.C.
28. **Neidhardt, F. C., P. L. Bloch, S. Pedersen, and S. Reeh.** 1977. Chemical measurement of steady-state levels of ten aminoacyl-transfer ribonucleic acid synthetases in *Escherichia coli. J. Bacteriol.* 129:378–387.
29. **Neidhardt, F. C., P. L. Bloch, and D. F. Smith.** 1974. Culture media for enterobacteria. *J. Bacteriol.* 119:736–747.
30. **Neidhardt, F. C., and R. A. VanBogelen.** 1981. Positive regulatory gene for temperature-controlled proteins in *Escherichia coli. Biochem. Biophys. Res. Commun.* 100:894–900.
31. **Neidhardt, F. C., and R. A. VanBogelen.** 1987. Heat shock response, p. 1334–1345. *In* F. C. Neidhardt, J. L. Ingraham, K. B. Low, B. Magasanik, M. Schaecter, and H. E. Umbarger (ed.), *Escherichia coli and Salmonella typhimurium: Cellular and Molecular Biology*, vol. 2. American Society for Microbiology, Washington, D.C.
32. **Neidhardt, F. C., R. A. VanBogelen, and V. Vaughn.** 1984. The genetics and regulation of heat-shock proteins. *Annu. Rev. Genet.* 18:295–329.
33. **Neidhardt, F. C., V. Vaughn, T. A. Phillips, and P. L. Bloch.** 1983. Gene-protein index of *Escherichia coli* K-12. *Microbiol. Rev.* 47:231–284.
34. **Neidhardt, F. C., R. Wirth, M. W. Smith, and R. VanBogelen.** 1980. Selective synthesis of plasmid-coded proteins by *Escherichia coli* during recovery from chloramphenicol treatment. *J. Bacteriol.* 143:535–537.
35. **Nomenclature Committee of the International Union of Biochemistry.** 1984. *Enzyme Nomenclature.* Academic Press, Inc., New York.
36. **Nystrom, T., and F. C. Neidhardt.** 1992. Cloning, mapping, and nucleotide sequence of a gene encoding a universal stress protein in *Escherichia coli. Mol. Microbiol.* 6:3187–3198.
37. **Nystrom, T., and F. C. Neidhardt.** 1993. Isolation and properties of a mutant of *Escherichia coli* with an insertional inactivation of the *uspA* gene, which encodes a universal stress protein. *J. Bacteriol.* 175:3949–3956.
38. **O'Farrell, P. H.** 1975. High resolution two-dimensional electrophoresis of proteins. *J. Biol. Chem.* 250:4007–4021.
39. **O'Farrell, P. Z., H. M. Goodman, and P. H. O'Farrell.** 1977. High resolution two-dimensional electrophoresis of basic as well as acidic proteins. *Cell* 12:1133–1142.
40. **Parker, J.** 1984. Identification of the *purC* gene product of *Escherichia coli. J. Bacteriol.* 157:712–717.
41. **Patterson, S. D., and G. I. Latter.** 1993. Evaluation of storage phospho imaging for quantitative analysis of 2-D gels using the Quest II system. *BioComputing* 15:1076–1083.
42. **Patton, W. F., M. F. Lopez, P. Barry, and W. M. Skea.** 1992. A mechanically strong matrix for protein electrophoresis with enhanced silver staining properties. *BioTechniques* 12:580–585.
43. **Pedersen, S., P. L. Bloch, S. Reeh, and F. C. Neidhardt.** 1978. Patterns of protein synthesis in *E. coli*: a catalog of the amount of 140 individual proteins at different growth rates. *Cell* 14:179–190.
44. **Phillips, T. A., P. L. Bloch, and F. C. Neidhardt.** 1980. Protein identifications on O'Farrell two-dimensional gels: locations of 55 additional *Escherichia coli* proteins. *J. Bacteriol.* 144:1024–1033.
45. **Phillips, T. A., V. Vaughn, P. L. Bloch, and F. C. Neidhardt.** 1987. Gene-protein index of *Escherichia coli* K-12, edition 2, p. 919–966. *In* F. C. Neidhardt, J. L. Ingraham, K. B. Low, B. Magasanik, M. Schaecter, and H. E. Umbarger (ed.), *Escherichia coli and Salmonella typhimurium: Cellular and Molecular Biology*, vol. 2. American Society for Microbiology, Washington, D.C.
46. **Reeh, S., and S. Pedersen.** 1979. Post-translational modification of *Escherichia coli* ribosomal protein S6. *Mol. Gen. Genet.* 173:183–187.
47. **Reeve, J.** 1979. The use of minicells for bacteriophage directed polypeptide synthesis. *Methods Enzymol.* 68:493–503.
48. **Riley, M.** 1993. Functions of the gene products of *Escherichia coli. Microbiol. Rev.* 57:362–952.
49. **Sancar, A., A. M. Hack, and W. D. Rupp.** 1979. Simple method for identification of plasmid-coded proteins. *J. Bacteriol.* 137:692–693.

50. **Sankar, P., M. E. Hutton, R. A. VanBogelen, R. L. Clark, and F. C. Neidhardt.** 1993. Expression analysis of cloned chromosomal segments of *Escherichia coli*. *J. Bacteriol.* **175**:5145–5152.

51. **Santaren, J. F.** 1990. Towards establishing a protein database of Drosophila. *Electrophoresis* **11**:254–267.

52. **Savageau, M. A.** 1986. Proteins of *Escherichia coli* come in sizes that are multiples of 14kDa: domain concepts and evolutionary implications. *Proc. Natl. Acad. Sci. USA* **83**:1198–1202.

52a. **Sood, P., C. G. Lerner, T. Shimamoto, Q. Lu, and M. Inouye.** 1994. Characterization of Era, essential *Escherichia coli* GTPase. *Mol. Microbiol.* **12**:201–208.

53. **Smith, M. W., and F. C. Neidhardt.** 1983. Proteins induced by anaerobiosis in *Escherichia coli*. *J. Bacteriol.* **154**:336–343.

54. **Smith, M. W., and F. C. Neidhardt.** 1983. Proteins induced by aerobiosis in *Escherichia coli*. *J. Bacteriol.* **154**:344–350.

55. **Squires, C. L., S. Petersen, B. M. Ross, and C. Squires.** 1991. ClpB is the *Escherichia coli* heat shock protein F84.1. *J. Bacteriol.* **173**:4254–4262.

56. **Studier, F. W., and B. A. Moffatt.** 1986. Use of bacteriophage T7 RNA polymerase of direct selective high-level expression of cloned genes. *J. Mol. Biol.* **189**:113–130.

57. **Tilly, K., R. A. VanBogelen, C. Georgopoulis, and F. C. Neidhardt.** 1983. Identification of the heat-inducible protein C15.4 as the *groES* gene product in *Escherichia coli*. *J. Bacteriol.* **154**:1505–1507.

58. **VanBogelen, R. A., M. E. Hutton, and F. C. Neidhardt.** 1990. Gene protein database of *Escherichia coli* K-12: edition 3. *Electrophoresis* **11**:1131–1166.

59. **VanBogelen, R. A., P. M. Kelley, and F. C. Neidhardt.** 1987. Differential induction of heat shock, SOS, and oxidation stress regulons and accumulation of nucleotides in *Escherichia coli*. *J. Bacteriol.* **169**:26–32.

60. **VanBogelen, R. A., and F. C. Neidhardt.** 1990. Ribosomes as sensors of heat and cold shock in *Escherichia coli*. *Proc. Natl. Acad. Sci. USA* **87**:5589–5593.

61. **VanBogelen, R. A., and F. C. Neidhardt.** 1991. The gene-protein database of *Escherichia coli* K-12: edition 4. *Electrophoresis* **12**:955–994.

62. **VanBogelen, R. A., and E. R. Olson.** Application of 2-D protein gels in biotechnology. *Biotech. Annu. Rev.*, in press.

63. **VanBogelen, R. A., P. Sankar, R. L. Clark, J. A. Bogan, and F. C. Neidhardt.** 1992. The gene-protein database of *Escherichia coli* K-12: edition 5. *Electrophoresis* **13**:1014–1054.

64. **Walker, G. C.** 1987. The SOS response of *Escherichia coli*, p. 1346–1357. *In* F. C. Neidhardt, J. L. Ingraham, K. B. Low, B. Magasanik, M. Schaecter, and H. E. Umbarger (ed.), *Escherichia coli and Salmonella typhimurium: Cellular and Molecular Biology*, vol. 2. American Society for Microbiology, Washington, D.C.

65. **Wanner, B. L.** 1992. Is cross regulation by phosphorylation of two-component response regulator proteins important in bacteria? *J. Bacteriol.* **174**:2053–2058.

Escherichia coli Gene Products: Physiological Functions and Common Ancestries

MONICA RILEY AND BERNARD LABEDAN

116

INTRODUCTION

Knowledge about *Escherichia coli* genes, their gene products, and their roles in cell physiology has reached a point that invites a look at the horizon. We can look at what we know today and see it in a context of some kind of totality of knowledge about *E. coli* genes and their gene products that will be reached in time. Assembled here is a table of *E. coli* K-12 genes whose gene products are known, organized by principal physiological function, with literature references through most of 1994. We have summarized the distribution of gene products among physiological and functional categories. Also assembled is information on the amino acid sequence relatedness of *E. coli* proteins and classification of the functional relatedness of sequence-related pairs. These results are interpreted in the context of evolution, with the objective of understanding more about how genes evolve and ultimately of identifying the relatively small numbers of unique ancestral sequences that probably have generated the families of sequences we observe in *E. coli* today. We also present a compilation of multiple enzymes present in *E. coli* that carry out the same biochemical reaction. Some multiple enzymes are related by sequence and thus are likely to have descended from a common ancestor, but others have no apparent sequence relationship and may have derived from different ancestral sequences independently by convergent evolution, or they may have been acquired from another source by lateral transmission. We discuss the capacity for these multiple enzymes to serve as a rich resource that contributes significantly to the adaptive capability of *E. coli*.

FUNCTIONAL CATEGORIES

Table 1 presents a scheme of categories of cellular functions. Any such classification is arbitrary in that there are many ways to organize the complex function of a cell, and it is artificial in that it seems to create hard boundaries between functional categories when of course cellular functions and processes are complexly intertwined. Also, some categories refer to metabolic pathways, others such as "transport" refer to processes, and still others such as "membranes" refer to cellular structure, categories that are not comparable in kind or mutually exclusive. Thus, one gene product can belong to more than one category. Nevertheless, this classification scheme allows a gross assignment of a major cellular role to each gene and gene product. This allows us to view as a whole the activities and division of labor of the genes and gene products of *E. coli* that have been characterized to date.

Table 2 lists 1,827 characterized gene products of *E. coli* K-12. Since not all known *E. coli* genes have characterized gene products, the list of genes that have been sequenced and mapped (see chapter 109) is longer than the list of genes for which the function of the gene product is known. We followed certain rules in choosing genes and gene products for listing. Most open reading frames with hypothetically translated gene products were excluded, except for a few with strong sequence similarities to other well-characterized gene products whose functions are known. These are noted as "putative" or "possible" functions in the table. Genes were excluded whose presence was simply deduced from the phenotype of a mutant, when the effect of the mutation was not described in enough detail to allow assignment of the gene product to a physiological category. However, genes were included whose effect has been characterized as to the physiological role in the cell even though the nature of the actual gene product (enzyme, regulator, permease, etc.) is not yet clear. Such genes were included in the list, assigned a physiological category, but classified as having only a phenotype known rather than having the type of gene product specified. All of the genes in the compilation by Barbara Bachmann in the first edition of this work (88) have been retained, even though some genes are are known only by phenotype (228 of them). One hesitates to delete information of any kind, but it is possible that "old" genes that were known only by mutant phenotype and have not been subjects of study for, say, 20 years should be removed from any subsequent listing of gene products of *E. coli*.

TABLE 1 Categories of cellular functions

Physiological function	Numbers of genes
Small-molecule metabolism	
Degradation	
Carbon compounds	96
Amino acids and amines	24
Fatty acids	12
Phosphorus compounds	20
Energy metabolism	
Glycolysis	17
Pyruvate dehydrogenase	4
Tricarboxylic acid cycle	16
Pentose phosphate pathway	
Oxidative branch	3
Nonoxidative branch	5
Entner-Doudoroff pathway	3
Respiration	
Aerobic	30
Anaerobic	61
Electron transport	13
Fermentation	25
ATP-proton motive force interconversion	9
Central intermediary metabolism	
General	46
Gluconeogenesis	5
Sugar-nucleotide biosynthesis, conversions	7
Amino sugars	5
Sulfur metabolism	13
Amino acid biosynthesis	
Glutamate family	
Glutamate	2
Glutamine	5
Arginine	10
Proline	3
Aspartate family	
Aspartate	2
Asparagine	3
Lysine	10
Threonine	5
Methionine	8
Serine family	
Glycine	2
Serine	3
Cysteine	4
Aromatic amino acid family	
Common stem	9
Phenylalanine	3
Tyrosine	3
Tryptophan	9
Histidine	9
Pyruvate family	
Alanine	2
Branched-chain family	
Valine and isoleucine	17
Leucine	7
Polyamine biosynthesis	7
Purines, pyrimidines, nucleosides, nucleotides	
Purine ribonucleotide biosynthesis	21
Pyrimidine ribonucleotide biosynthesis	9
2'-Deoxyribonucleotide metabolism	10
Salvage of nucleosides and nucleotides	19
Miscellaneous	8
Biosynthesis of cofactors, prosthetic groups, and carriers	
Biotin	8
Folic acid	9
Lipoate	2
Molybdopterin	9
Pantothenate	4
Pyridoxine	4
Pyridine nucleotide	7
Thiamine	10
Riboflavin	5
Thioredoxin, glutaredoxin, and glutathione	5
Menaquinone and ubiquinone	16

TABLE 1 *Continued*

Physiological function	Numbers of genes
Heme and porphyrin	13
Enterochelin	6
Fatty acid biosynthesis	26
Broad regulatory functions	51
Macromolecules	
Synthesis and modification of macromolecules	
Ribosomal and "stable" RNAs	25
Ribosomal protein synthesis and modification	57
Ribosome maturation and modification	15
tRNAs	80
Aminoacyl tRNA synthetases and their modification	49
Nucleoproteins	7
DNA replication, modification, recombination	91
Protein translation and modification	27
RNA synthesis, modification, transcription	28
Polysaccharides (cytoplasmic)	6
Phospholipids	12
Degradation of macromolecules	
RNA	13
DNA	22
Proteins and peptides	30
Polysaccharides	3
Cell envelope	
Membranes, lipoproteins, porins	31
Surface polysaccharides, lipopolysaccharides	44
Surface structures	55
Murein sacculus and peptidoglycan	37
Processes	
Transport/binding proteins	
Amino acids and amines	57
Cations	62
Anions	15
Carbohydrates, organic alcohols, and acids	92
Nucleosides, purines, and pyrimidines	7
Other	17
Chaperones	7
Cell division	37
Chemotaxis and mobility	12
Protein and peptide secretion	30
Osmotic adaptation	16
Detoxification	10
Cell killing	6
Other	
Phage-related functions and prophages	25
Colicin-related functions	12
Plasmid-related functions	7
Drug/analog sensitivity	46
Radiation sensitivity	5
Adaptations and atypical conditions	11

References of genes that were listed in an earlier compilation of gene products (1652) are updated in Table 2 when possible. The choice of references in Table 2 has some arbitrary features. In the previous compilation of gene products of *E. coli* (1652), early papers on a gene product were cited as well as more recent references. In Table 2 in this chapter, the accent is on more recent work on each gene product. Citations were limited arbitrarily to three per entry. It was not possible in any reasonable time frame to become well enough informed to cite the most meritorious work for each entry; instead, citation to some of the more recent papers on each entry is used. Earlier literature should be accessible by tracing back citations. Genes and gene products that have not received attention in recent years still

(Continued on page 2150)

TABLE 2 *E. coli* genes grouped by function of gene product

Category and gene	Gene product description[a]	Reference(s)
I. SMALL-MOLECULE METABOLISM		
A. Degradation		
1. Carbon compounds		
araA	L-Arabinose isomerase	1097, 1177
araB	L-Ribulokinase	1097, 1177
araC	Activator and repressor protein for *ara* genes	1096, 1177, 1297
araD	L-Ribulosephosphate 4-epimerase	1097
ascB	6-Phospho-β-glucosidase; cryptic	696, 1507
ascG	*ascBF* operon repressor	696, 1507
bglA	Phospho-β-glucosidase A; cryptic	696, 1507, 1769
bglB	Phospho-β-glucosidase B; cryptic	696, 1507, 1769
bglG	Positive regulation of *bgl* operon	696, 1767
bglT	Regulator for phospho-β-glucosidase A biosynthesis	1581, 1582
celD	Negative regulator of *cel* operon	1508
celF	Phospho-β-glucosidase; cryptic	1508
cynR	*cyn* operon positive regulator	52, 1071, 1935
dgd	D-Galactose dehydrogenase	1171
dgoA	2-Oxo-3-deoxygalactonate 6-phosphate aldolase	356
dgoD	Galactonate dehydratase	356
dgoK	2-Oxo-3-deoxygalactonate kinase	356
dgoR	Regulator of *dgo* operon	356
ebgA	Evolved β-D-galactosidase, alpha subunit; cryptic gene	499, 694, 1132
ebgB	Possible homolog of *lacY*	694
ebgC	Evolved β-D-galactosidase, beta subunit; cryptic gene	694, 1132
ebgR	Regulator of *ebg* operon	1196
exuR	Negative regulator of *exu* regulon, *exuT*, *uxaAC*, and *uxuB*	173
fucA	L-Fuculose-1-phosphate aldolase	298, 299, 2273
fucI	L-Fucose isomerase	298, 299
fucK	L-Fuculokinase	298, 299
fucO	L-1,2-Propanediol oxidoreductase	299, 353
fucR	Positive regulator of the *fuc* operon	299, 1192
galE	UDP-galactose 4-epimerase	111, 1019, 2085
galK	Galactokinase	2085
galR	Repressor of *galETK* operon	1219, 1997, 2129
galS	*mgl* repressor, a lactose operon inducer	1219, 1997, 2129
galT	Galactose-1-phosphate uridylyltransferase	535, 2085
galU	Glucose-1-phosphate uridylyltransferase	2135
garA	Glucarate utilization	1663
garB	Glucarate utilization	1663
gatD	Galactitol-1-phosphate dehydrogenase	435, 1108
gatR	Regulator, galactitol metabolism	1108
gcd	Glucose dehydrogenase	333, 2212, 2216
glk	Glucokinase	381
gntV	Gluconokinase, thermosensitive	336, 857
gurB	Utilization of methyl-β-D-glucuronide; *crp?*	1910
gurC	Utilization of methyl-β-D-glucuronide	1910
gurD	Utilization of methyl-β-D-glucuronide	1910
gutD	Glucitol (sorbitol)-6-phosphate dehydrogenase	2215
gutM	Glucitol operon activator	2215
gutR	Regulator for *gut* (*srl*), glucitol operon	2215
kdgK	Ketodeoxygluconokinase	1576
kdgR	Regulator of *kdgK*, *kdgT*, and *eda*	1576
lacA	Thiogalactoside acetyltransferase	56
lacI	Repressor of the *lac* operon	977, 1012, 1718
lacZ	β-D-galactosidase	607
mac	Maltose acetyltransferase, broad specificity	210
malI	Repressor of *malX* and *malY* genes	1629, 1630
malM	Periplasmic protein of *mal* regulon	636, 1692
malP	Maltodextrin phosphorylase	826
malT	Positive regulator of *mal* regulon	1765
malY	Enzyme that may degrade or block biosynthesis of endogenous *mal* inducer	1629
malZ	Maltodextrin glucosidase	1765
manA	Mannose-6-phosphate isomerase	1901
manC	D-Mannose isomerase regulation; utilization of D-lyxose	1901
melA	α-Galactosidase	1147, 1379, 1574
melR	Regulator of melibiose operon	2124
mtlD	Mannitol-1-phosphate dehydrogenase	1980
mtlR	Repressor for *mtl*	536
nlp	Regulatory factor of maltose metabolism; similar to Ner repressor protein of phage Mu	313, 408

TABLE 2 *Continued*

Category and gene	Gene product description[a]	Reference(s)
pac	Penicillin acylase, detaches phenylacetate residue	119, 1301
pga	Penicillin G acylase; precursor polypeptide processed to two nonidentical subunits	310, 1448
pgm	Phosphoglucomutase	914, 1191, 1577
poxA	Regulator for *poxB*	2057
poxB	Pyruvate oxidase	654, 2104
prp	Propionate metabolism	1876
pta	Phosphotransacetylase activity	681, 1587, 2113
rbsK	Ribokinase	49, 785
rbsR	Regulator for *rbs* operon	1187
rhaA	L-Rhamnose isomerase	90, 1341, 1902
rhaB	Rhamnulokinase	90, 1341
rhaD	Rhamnulose-phosphate aldolase	90, 1341
rhaR	Positive regulator for *rhaRS* operon	488, 1995, 1996
rhaS	Positive regulator for *rhaBAD* operon	488, 1995
sfsA	Probable regulator for maltose metabolism	964
treC	Amylotrehalase	189, 1919
treE	Trehalose-6-P phosphatase, catabolic	997, 1919
treF	Cytoplasmic trehalase	94
uidA	β-D-Glucuronidase	171, 889
uidR	Regulator for *uid*	171
uxaA	Altronate hydrolase	1564
uxaB	Altronate oxidoreductase	172
uxaC	Uronate isomerase	1431
uxuA	Mannonate hydrolase	173
uxuB	Mannonate oxidoreductase	173
uxuR	Regulator of *uxuBA* operon	173
xylA	D-Xylose isomerase	110, 1688
xylB	Xylulokinase	191, 1688
xylR	Regulator for *xyl*	1688, 1861
xylR	Putative regulator of *xyl* operon	1554, 1688, 1861

2. Amino acids and amines

adi	Biodegradative arginine decarboxylase	1905
ansA	Cytoplasmic L-asparaginase I, isozyme	814, 898, 1879
ansB	Periplasmic L-asparaginase II; secreted isozyme	188, 894, 2128
asu	Asparagine utilization, as sole nitrogen source	304
cadA	Lysine decarboxylase	1293, 2119
cadC	Transcriptional activator of *cad* operon	1400, 2119
dadA	D-Amino acid dehydrogenase subunit	1180, 1462, 2158
dadB	D-Amino acid dehydrogenase subunit	563, 585, 1462
dadQ	Regulator of *dad* regulon	563
dadX	Alanine racemase; isozyme	670, 1180, 2157
dsdA	D-Serine deaminase	1233–1235
dsdC	Activator for *dsdA*	1493
maoA	Tyramine oxidase	1376, 2220
poaR	Regulation of proline oxidase production	347
putA	Proline dehydrogenase	232, 2183
sdaA	L-Serine deaminase	1233, 1923, 1925
sdaB	L-Serine deaminase, L-SD2	1811, 1924
tdcA	Transcriptional activator of *tdc* operon	589, 1787
tdcB	Threonine dehydratase, catabolic	648, 772, 2199
tdcR	Threonine dehydratase operon activator protein	1786
tdh	Threonine dehydrogenase	71, 366, 509
tnaA	Tryptophanase	637, 1998
tnaL	Tryptophanase leader peptide	642

3. Fatty acids

atoA	Acetyl-CoA:acetoacetyl-CoA transferase beta-subunit	892, 1516
atoB	Acetyl-CoA acetyltransferase	892, 1516
atoC	Positive regulator of *ato*	891, 892, 1516
atoD	Acetyl-CoA:acetoacetyl-CoA transferase alpha-subunit?	892, 1516
fadA	Thiolase I	2227
fadB	3-Hydroxyacyl-CoA dehydrogenase; 3-hydroxyacyl-epimerase; *delta*(3)-*cis-delta*(2)-*trans*-enoyl-CoA isomerase; enoyl-CoA-hydratase	2226, 2229
fadD	Acyl-CoA synthetase	
fadE	Electron transport flavoprotein (ETF) of beta-oxidation	323, 1478
fadH	2,4-Dienoyl-CoA reductase	2244
fadR	Negative regulator for *fad* regulon, and positive activator of *fabA*	164, 451, 743
fatA	Utilization of *trans*-unsaturated fatty acids	445

(Table continues)

TABLE 2 *Continued*

Category and gene	Gene product description[a]	Reference(s)
sbm	Methylmalonyl-CoA mutase (MCM)	1696
4. Phosphorus compounds		
phnF	Putative regulator, *phn* operon	1305, 1306, 2111
phnG	Probably carbon-P lyase subunit	1305, 1306, 2112
phnH	Probably carbon-P lyase subunit	1305, 1306, 2112
phnI	Probably carbon-P lyase subunit	1305, 1306, 2112
phnJ	Probably carbon-P lyase subunit	1305, 1306, 2112
phnK	Probably carbon-P lyase subunit	1305, 1306, 2112
phnL	Probably carbon-P lyase subunit	1305, 1306, 2112
phnM	Probably carbon-P lyase subunit	1304–1306
phnN	Probably accessory to carbon-P lyase	1304–1306
phnO	Putative regulator, *phn* operon	1304–1306
phnP	Probably accessory to carbon-P lyase	1305, 1306, 2112
phoA	Alkaline phosphatase	440, 853, 1577
phoH	PhoB-dependent, ATP-binding *pho* regulon component; induced by P starvation	987, 1303
pldA	Outer membrane phospholipase A	226, 1386
pldB	Lysophospholipase L(2)	946, 1007
pldC	Lysophospholipase L(1)	946
ppk	Polyphosphate kinase	15, 16, 371
ppx	Exopolyphosphatase	16, 371, 1637
psiE	*phoB*-dependent *pho* regulon component; induced by P starvation	1303, 2111
psiF	Induced by phosphate starvation	1303, 2111
B. Energy metabolism		
1. Glycolysis		
eno	Enolase	849, 850, 1880
fba	Fructose-bisphosphate aldolase, class II	23, 97, 1526
fruK	Fructose-1-phosphate kinase	1467
fruL	*fruR* leader peptide	884
fruR	Repressor of *fru* operon and others	609, 883, 1616
fruS	Regulator of *fruA* and *fruF*	187
gapA	Glyceraldehyde-3-phosphate dehydrogenase A	289, 759, 849
gapB	Glyceraldehyde 3-phosphate dehydrogenase B	24, 464
gpmA	Phosphoglyceromutase	389
gpmB	Phosphoglyceromutase	389
pfkA	6-Phosphofructokinase I	446, 1047, 1803
pfkB	6-Phosphofructokinase II; suppressor of *pfkA*	85, 680
pgi	Glucosephosphate isomerase	573
pgk	Phosphoglycerate kinase	1402
pykA	Pyruvate kinase II, glucose stimulated	1280, 2047
pykF	Pyruvate kinase I (formerly F), fructose stimulated	1449, 1873
tpiA	Triosephosphate isomerase	165, 1559
2. Pyruvate dehydrogenase		
aceE	Pyruvate dehydrogenase (decarboxylase component)	656, 657, 1527
aceF	Pyruvate dehydrogenase (dihydrolipoyltransacetylase component)	397, 1780, 1781
lpdA	Lipoamide dehydrogenase (NADH); component of 2-oxodehydrogenase and pyruvate complex, L-protein of glycine cleavage complex	35, 1898
pdhR	Transcriptional regulator for pyruvate dehydrogenase complex	722, 1598
3. Tricarboxylic acid cycle		
acnA	Aconitate hydrase A	674, 1587, 1588
acnB	Aconitate hydrase B	674
fumA	Fumarase A = fumarate hydratase class I; aerobic isozyme	2030, 2186, 2247
fumB	Fumarase B = fumarate hydratase class I; anaerobic isozyme	124, 676, 2186
fumC	Fumarase C = fumarate hydratase class II; isozyme	1159, 2121, 2185
gltA	Citrate synthase	50, 457, 1225
icdC	Isocitrate dehydrogenase, specific for $NADP^+$, chromosomal e14 hybrid	756, 816, 817
icdE	Isocitrate dehydrogenase, specific for $NADP^+$, chromosomal fragment	263, 816, 817
mdh	Malate dehydrogenase	2078
sdhA	Succinate dehydrogenase, flavoprotein subunit	211, 2181
sdhB	Succinate dehydrogenase, iron sulfur protein	348, 401, 1319
sdhC	Succinate dehydrogenase, cytochrome b_{556}	993, 1319, 1370
sdhD	Succinate dehydrogenase, hydrophobic subunit	1319, 2181
sucA	2-Oxoglutarate dehydrogenase (decarboxylase component)	744
sucB	2-Oxoglutarate dehydrogenase (dihydrolipoyltranssuccinase component)	744, 1483
sucC	Succinyl-CoA synthetase, beta subunit	1220, 2177
sucD	Succinyl-CoA synthetase, alpha subunit	1220, 2177

TABLE 2 *Continued*

Category and gene	Gene product description[a]	Reference(s)
5. Pentose phosphate pathway		
a. Oxidative branch		
gnd	Gluconate-6-phosphate dehydrogenase, decarboxylating	271, 1694, 2070
pgl	6-Phosphogluconolactonase	1051
zwf	Glucose-6-phosphate dehydrogenase	1694, 1695
b. Nonoxidative branch		
rpiA	Ribosephosphate isomerase, constitutive	794, 1851
talA	Transaldolase A	94
talB	Transaldolase B	94, 2248
tktA	Transketolase	913, 1877
tktB	Transketolase isozyme	833
6. Entner-Doudoroff pathway		
eda	2-Keto-3-deoxygluconate 6-phosphate aldolase; 2-keto-4-hydroxyglutarate aldolase	487, 1514, 2076
edd	6-Phosphogluconate dehydratase	487, 496
gntR	Regulator of *edd;* transport and phosphorylation of gluconate	87, 336
7. Respiration		
a. Aerobic		
cyoA	Cytochrome *o* ubiquinol oxidase subunit II	1325, 1390, 2054
cyoB	Cytochrome *o* ubiquinol oxidase subunit I	302, 1205, 1326
cyoC	Cytochrome *o* ubiquinol oxidase subunit III	302, 1205, 1326
cyoD	Cytochrome *o* ubiquinol oxidase subunit IV	302, 303, 1326
cyoE	Heme O biosynthesis, protoheme IX farnesyltransferase	302, 303, 1326
dld	D-Lactate dehydrogenase, FAD protein, NADH independent	923, 1702, 2017
glpD	*sn*-Glycerol-3-phosphate dehydrogenase (aerobic)	84, 864, 1784
hyaA	Hydrogenase-1 small subunit	1299, 1591, 1754
hyaB	Hydrogenase-1 large subunit	1299, 1591, 1754
hyaC	Possible membrane-spanning protein of *hya* operon	1299, 1300, 1591
hyaD	Processing of HyaA and HyaB proteins	1299, 1300, 1591
hyaE	Processing of HyaA and HyaB proteins	1299, 1300, 1591
hyaF	Nickel incorporation into hydrogenase-1 proteins	1299, 1300, 1591
lctD	L-Lactate dehydrogenase, FMN protein	459, 862
lctR	Putative *lct* repressor	459
ndh	Respiratory NADH dehydrogenase	571, 843, 888
nuoA	NADH dehydrogenase I chain A	260
nuoB	NADH dehydrogenase I chain B	260
nuoC	NADH dehydrogenase I chain C	260
nuoD	NADH dehydrogenase I chain D	260
nuoE	NADH dehydrogenase I chain E	260
nuoF	NADH dehydrogenase I chain F	260, 1590, 2130
nuoG	NADH dehydrogenase I chain G	260, 1590, 2130
nuoH	NADH dehydrogenase I chain H	260, 1590, 2130
nuoI	NADH dehydrogenase I chain I	260, 1590, 2130
nuoJ	NADH dehydrogenase I chain J	260, 1590, 2130
nuoK	NADH dehydrogenase I chain K	260, 1590, 2130
nuoL	NADH dehydrogenase I chain L	260, 1590, 2130
nuoM	NADH dehydrogenase I chain M	260, 1590, 2130
nuoN	NADH dehydrogenase I chain N	260, 1590, 2130
dmsA	Anaerobic dimethyl sulfoxide reductase subunit A	1727, 1728, 2009
b. Anaerobic		
dmsC	Anaerobic dimethyl sulfoxide reductase subunit C	1727, 1728
dniR	Regulator for nitrite reductase (cytochrome c_{552}) expression	928
fdhD	Affects formate dehydrogenase-N	1904
fdhE	Affects formate dehydrogenase-N	1762, 1904
fdhF	Selenopolypeptide subunit of formate dehydrogenase H (part of formate hydrogen-lyase complex [FHL complex])	292, 1690, 2277
fdnG	Formate dehydrogenase-N, nitrate inducible, major subunit	136, 1133, 1653
fdnH	Formate dehydrogenase-N, nitrate inducible, iron-sulfur subunit	136, 507, 1133
fdnI	Formate dehydrogenase-N, nitrate inducible, cytochrome B_{556}(Fdn) subunit	136, 507, 1133
fdoG	Formate dehydrogenase-O, major subunit	1554, 1752
fdoH	Formate dehydrogenase-O, iron-sulfur subunit	1554, 1752
fdoI	Formate dehydrogenase, cytochrome b_{556} (Fdo) subunit	1554, 1752
frdA	Fumarate reductase, anaerobic, flavoprotein subunit	211, 349, 1289
frdB	Fumarate reductase, anaerobic, iron-sulfur protein subunit	349, 1229, 2145
frdC	Fumarate reductase, anaerobic, membrane anchor polypeptide	349, 1289
frdD	Fumarate reductase, anaerobic, membrane anchor polypeptide	349, 1289
glpA	*sn*-Glycerol-3-phosphate dehydrogenase (anaerobic), large subunit	338, 1080, 1775

(Table continues)

TABLE 2 *Continued*

Category and gene	Gene product description[a]	Reference(s)
glpB	sn-Glycerol-3-phosphate dehydrogenase (anaerobic), membrane anchor subunit	338, 1080, 1775
glpC	sn-Glycerol-3-phosphate dehydrogenase (anaerobic), K-small subunit	338, 1080, 1775
glpE	Protein of glp regulon	864
glpG	Protein of glp regulon	1785
glpR	Repressor of the glp operon	864, 1080, 2268
hybA	Putative small subunit of hydrogenase-2, probable iron-sulfur protein	183, 1298, 1591
hybB	Putative cytochrome component of hydrogenase-2	1298, 1591
hybC	Probable large subunit, hydrogenase-2	1298, 1591
hybD	Probable processing element for hydrogenase-2	1298, 1591
hybE	Member of hyb operon	1298
hybF	May modulate levels of hydrogenase-2	1298
hybG	May effect maturation of large subunit of hydrogenase-2	1298, 1591
hydL	Probable member of hyb operon; pleiotropic effects	1911
hydN	Probably involved in electron transport from formate to hydrogen	
hypA	Pleiotropic effects on three hydrogenase isozymes	1148, 1203, 1591
hypC	Pleiotropic effects on three hydrogenase isozymes	1148, 1203
hypD	Pleiotropic effects on three hydrogenase isozymes	1148, 1203
hypE	Plays structural role in synthesis of hydrogenase	1203
narG	Nitrate reductase, alpha subunit	175, 478, 552
narH	Nitrate reductase, beta subunit	175, 478, 552
narI	Cytochrome b(NR), nitrate reductase, gamma subunit	175, 478, 1206
narJ	Nitrate reductase, delta subunit, assembly function	175, 478
narL	Pleiotropic regulation of anaerobic respiration	489, 1425, 1487
narP	Nitrate/nitrite response regulator	1602
narQ	Sensor for nitrate reductase system, protein histidine kinase	305
narV	Cryptic nitrate reductase II, gamma subunit	176
narW	Cryptic nitrate reductase II, delta subunit, assembly function	176, 177
narX	Nitrate sensor, histidine protein kinase acts on narL	305, 342, 1953
narY	Cryptic nitrate reductase II, beta subunit	176
narZ	Cryptic nitrate reductase II, alpha subunit	176
nirB	Nitrite reductase [NAD(P)H] subunit	706
nirC	Nitrite reductase activity	706
nirD	Nitrite reductase [NAD(P)H] subunit	706
nrfA	Formate-dependent nitrite reductase; tetra-heme cytochrome c_{552}	179, 818
nrfB	Formate-dependent nitrite reductase; a penta-heme cytochrome c	818
nrfC	Formate-dependent nitrite reductase; Fe-S centers	818
nrfD	Formate-dependent nitrate reductase complex; transmembrane protein	818, 1487
nrfE	Formate-dependent nitrite reductase; assembly function?	818, 1487
nrfF	Part of formate-dependent nitrite reductase complex	818, 1487
nrfG	Part of formate-dependent nitrite reductase complex	818, 1487
torA	Trimethylamine N-oxide reductase subunit	1288, 1840, 1841
torC	Trimethylamine N-oxide reductase, cytochrome subunit	848, 1288
torR	Regulator for torA	1513
c. Electron transport		
ackA	Acetate kinase	1256, 1849, 2113
appB	Probable third cytochrome oxidase, subunit II	404
appC	Probable third cytochrome oxidase, subunit I	404
cybB	Cytochrome b_{561}	476, 1325, 1413
cybC	Cytochrome b_{562}	1135, 1413, 2013
cydA	Cytochrome d terminal oxidase, polypeptide subunit I	362, 524, 717
cydB	Cytochrome d terminal oxidase, polypeptide subunit II	362, 524, 717
fdx	[2FE-2S] ferredoxin, electron carrier protein	1790, 1948
fldA	Flavodoxin	1022, 1471
fpr	Ferredoxin-NADP reductase	147, 148
fre	Ferrisiderophore reductase; flavin reductase (NADPH:flavin oxidoreductase)	363, 600, 1882
hmpA	Hemoprotein; ferrisiderophore reductase activity	58, 540, 2064
qor	Quinone oxidoreductase	179
8. Fermentation		
acd	Acetaldehyde-CoA dehydrogenase	326
act	Pyruvate formate-lyase-activating enzyme	1751
adhC	Alcohol dehydrogenase class III; formaldehyde dehydrogenase, glutathione dependent	378
adhE	Alcohol dehydrogenase; CoA-linked acetaldehyde dehydrogenase; pyruvate formate-lyase deactivase	681, 975, 1699
adhR	Regulator for acd and adhE	328
ald	Aldehyde dehydrogenase, NAD linked	753
aldH	Aldehyde dehydrogenase, prefers NADP over NAD	729
fhlA	Formate hydrogen-lyase transcriptional activator for fdhF, hyc, and hyp operons	867, 1261, 1690
fhlB	Regulator for formate hydrogen lyase (FHL complex)	1261, 1690

TABLE 2 *Continued*

Category and gene	Gene product description[a]	Reference(s)
hycA	Transcriptional repression of *hyc* and *hyp* operons	1591, 1750
hycB	Probable small subunit of hydrogenase-3, iron-sulfur protein (part of formate hydrogen lyase (FHL complex)	183, 1591, 1750
hycC	Membrane-spanning protein of hydrogenase-3 (part of FHL complex)	183, 1591, 1750
hycD	Membrane-spanning protein of hydrogenase-3 (part of FHL complex)	183, 1591, 1750
hycE	Large subunit of hydrogenase-3 (part of FHL complex)	1591, 1689, 1750
hycF	Probable iron-sulfur protein of hydrogenase-3 (part of FHL complex)	183, 1591, 1750
hycG	Hydrogenase activity	183, 1591, 1750
hycH	Processing of large subunit (HycE) of hydrogenase-3 (part of FHL complex)	1591, 1750
hydG	Two-component regulation of hydrogenase-3 activity (with HydH)	179, 1148, 1911
hydH	Two-component regulation of hydrogenase-3 activity (sensor kinase of HydG)	1148, 1911
hypB	Guanine-nucleotide-binding protein, functions as nickel donor for large subunit of hydrogenase-3	1148, 1203, 1218
ldhA	D-Lactate dehydrogenase, NAD dependent	325, 1255, 1974
pfl	Pyruvate formate-lyase	1005, 1753
ppc	Phosphoenolpyruvate carboxylase	285, 1978, 1979
ttdA	L-Tartrate dehydratase	1623
ttdB	L-Tartrate dehydratase	1623

9. ATP-proton motive force

atpA	Membrane-bound ATP synthase, F_1 sector, alpha-subunit	895, 915, 1128
atpB	Membrane-bound ATP synthase, F_0 sector, subunit a	239, 895, 2075
atpC	Membrane-bound ATP synthase, F_1 sector, epsilon-subunit	239, 1078, 1291
atpD	Membrane-bound ATP synthase, F_1 sector, beta-subunit	239, 895, 1098
atpE	Membrane-bound ATP synthase, F_0 sector, subunit c	239, 2075, 2267
atpF	Membrane-bound ATP synthase, F_0 sector, subunit b	239, 1273, 2075
atpG	Membrane-bound ATP synthase, F_1 sector, gamma-subunit	239, 813, 895
atpH	Membrane-bound ATP synthase, F_1 sector, delta-subunit	239, 504, 895
atpI	Membrane-bound ATP synthase, dispensable protein	239, 895, 1766

C. Central intermediary metabolism
1. General

aceA	Isocitrate lyase	4, 1006, 1697
aceB	Malate synthase A	256a, 2055
aceK	Isocitrate dehydrogenase kinase/phosphatase	319, 358, 359
agp	Periplasmic glucose-1-phosphatase	405, 1577
appA	Phosphoanhydride phosphorylase; pH 2.5 acid phosphatase	404, 1473
appY	Regulatory protein affecting *appA* and other genes	81
aspA	Aspartate ammonia-lyase (aspartase)	522, 1373, 2185
cpsG	Phosphomannomutase	1242
cxm	Methylglyoxal biosynthesis	982
cynS	Cyanate aminohydrolase, cyanase	52, 679, 1162
cynT	Carbonic anhydrase	52, 679, 1071
dprA	Dihydropteridine reductase	668, 2065
gabC	Regulator for *gabPDT*	109, 1307, 1410
gabD	Succinate-semialdehyde dehydrogenase, NADP-dependent activity	109, 1410
gabT	4-Aminobutyrate aminotransferase activity	109, 1307, 1410
gadA	Glutamate decarboxylase isozyme	1232, 1856
gadB	Glutamate decarboxylase isozyme	1232, 1856
gadR	Regulator for *gadS*	1201, 1856
galM	Galactose-1-epimerase (mutarotase)	198, 1251
gcl	Glyoxylate carboligase	283
gcvA	Positive regulator of *gcv*	2169, 2170
gcvH	H protein of glycine cleavage complex, carrier of aminomethyl moiety	1458, 1894, 1895
gcvP	Glycine decarboxylase, P protein of glycine cleavage system	1458, 1894
gcvT	T protein (tetrahydrofolate dependent) of glycine cleavage system	1458, 1894
glc	Malate synthase G	2055
gldA	Glycerol dehydrogenase, NAD dependent	899, 2016
glpK	Glycerol kinase	140, 439, 2281
glpQ	Glycerophosphodiester phosphodiesterase, periplasmic	1080, 2002
gltB	Glutamate synthase, large subunit	605, 733, 1231
gltD	Glutamate synthase, small subunit	274, 605, 733
gltF	Regulator of *gltBDF* operon, induction of Ntr enzymes	275
hdhA	NAD-dependent 7α-hydroxysteroid dehydrogenase, dehydroxylation of bile acids	2241
iclR	Repressor of *aceBA* operon	359, 1401, 1936
kba	Ketose-bisphosphate aldolase, active on D-tagatose-1,6-diphosphate	1526
kbl	2-Amino-3-ketobutyrate CoA ligase (glycine acetyltransferase)	1357, 1358
lrp	Regulator for leucine (or *lrp*) regulon and high-affinity branched-chain amino acid transport system	516, 1152, 2109
metF	5,10-Methylenetetrahydrofolate reductase	1721
metK	Methionine adenosyltransferase (AdoMet synthetase); methyl and propylamine donor, corepressor of *met* genes	174, 1746, 1747
metX	Methionine adenosyltransferase 2 (AdoMet synthase)	1459, 1747

(Table continues)

TABLE 2 *Continued*

Category and gene	Gene product description[a]	Reference(s)
pntA	Pyridine nucleotide transhydrogenase, alpha subunit	9, 34, 2004
pntB	Pyridine nucleotide transhydrogenase, beta subunit	9, 10, 2004
ppa	Inorganic pyrophosphatase	1064, 1065
pqq	Redox cofactor, functions as cofactor of apoglucose dehydrogenase; cryptic in K-12	155
prrA	γ-Aminobutyraldehyde (pyrroline) dehydrogenase activity	1340, 1584
sad	Succinate-semialdehyde dehydrogenase, NAD dependent	1240
ugpQ	Glycerophosphodiester phosphodiesterase, cytosolic	2002
2. Gluconeogenesis		
fbp	Fructose-bisphosphatase	86, 699
maeA	NAD-linked malate dehydrogenase	354, 1216
maeB	NADP-linked malate dehydrogenase	230, 871
pckA	Phosphoenolpyruvate carboxykinase	285, 286, 1036
ppsA	Phosphoenolpyruvate synthase	286, 1515
3. Sugar-nucleotides		
cpsB	Mannose-1-phosphate guanyltransferase	199, 652, 2010
glmU	N-Acetylglucosamine-1-phosphate uridyltransferase	1296, 2098
rfbA	TDP-glucose pyrophosphorylase	1399, 1902, 2233
rfbB	DTDP-glucose-4,6 dehydratase	1399, 1902, 2233
rfbC	DTDP-4-dehydrorhamnose-3,5 epimerase	1902, 2233
rfbD	DTDP-4-dehydrorhamnose reductase	1003, 1399, 1902
ushA	UDP-sugar hydrolase (5′-nucleotidase)	250
4. Amino sugars		
glmS	L-Glutamine:D-fructose-6-phosphate aminotransferase	89, 543, 2029
nagA	N-Acetylglucosamine-6-phosphate deacetylase	1551, 1552, 2079
nagB	Glucosamine-6-phosphate deaminase	38, 1551, 1552
nagC	Repressor of *nag* operon	1550, 1552, 1553
nagD	N-Acetylglucosamine metabolism	1551
5. Sulfur metabolism		
aslA	Arylsulfatase	395
aslB	Putative arylsulfatase regulatory protein	395
atsA	Putative arylsulfatase	395
cysC	Adenosine 5′-phosphosulfate kinase	1130, 1131
cysD	ATP:sulfurylase (ATP:sulfate adenylyltransferase), subunit 2	1129–1131
cysH	Phosphoadenylylsulfate reductase	1040, 1475
cysI	Sulfite reductase, alpha subunit	314, 1475, 1836
cysJ	Sulfite reductase flavoprotein, beta subunit	314, 1474, 1836
cysN	ATP-sulfurylase (ATP:sulfate adenylyltransferase) subunit 1, probably a GTPase	1129–1131
cysQ	Affects pool of 3′-phosphoadenosine-5′-phosphosulfate in pathway of sulfite synthesis	1405, 2116
sseA	Putative thiosulfate sulfur transferase	94
thdA	Sulfone and sulfoxide oxidase activity	20, 917
D. Amino acid biosynthesis		
1. Glutamate family		
gdhA	NADP-specific glutamate dehydrogenase	733, 1725, 2069
gltH	Glutamate requirement	1238
glnA	Glutamine synthetase	36, 1633, 1833
glnB	Regulatory protein P-II for glutamine synthetase	79, 725
glnD	Uridylyltransferase acts on regulator of *glnA*	79
glnG	Response regulator for *gln* (sensor *glnL*) (nitrogen regulator I [NRI])	1415, 1822, 2127
glnL	Histidine protein kinase sensor for *glnG* regulator (nitrogen regulator II [NRII])	79, 1415, 1633
argA	Amino acid acetyltransferase; N-acetylglutamate synthase	233, 1105, 1249
argB	Acetylglutamate kinase	208, 1510, 2077
argC	N-Acetyl-γ-glutamylphosphate reductase	208, 1510, 2077
argD	Acetylornithine δ-aminotransferase	152, 153, 730
argE	Acetylornithine deacetylase	207, 1287
argF	Ornithine carbamyltransferase 2	331, 792, 1102
argG	Argininosuccinate synthetase	646, 2061
argH	Argininosuccinate lyase	208
argI	Ornithine carbamoyltransferase 1	1050, 1322, 2251
argR	Repressor of *arg* regulon; *cer*-mediated site specific recombination	1274, 1907, 1989
proA	γ-Glutamylphosphate reductase	724, 1791
proB	γ-Glutamate kinase	724, 1791
proC	Pyrroline-5-carboxylate reductase	442, 724
2. Aspartate family		
aspC	Aspartate aminotransferase	1052, 2230, 2231

TABLE 2 *Continued*

Category and gene	Gene product description[a]	Reference(s)
asnA	Asparagine synthetase A	764
asnB	Asparagine synthetase B	814, 1788
asnC	Regulator for *asnA, asnC,* and *gidA*	1013
asd	Aspartate-semialdehyde dehydrogenase	950
dapA	Dihydrodipicolinate synthase	1062, 1817
dapB	Dihydrodipicolinate reductase	205, 1962
dapC	Tetrahydrodipicolinate succinylase	248
dapD	Tetrahydrapicolinate *N*-succinyltransferase	248, 1179, 1645
dapE	*N*-Succinyl-diaminopimelate deacylase	204, 2190
dapF	Diaminopimelate epimerase	755, 1069, 2172
lysA	Diaminopimelate decarboxylase	300, 1918
lysC	Aspartokinase III, lysine sensitive	273
lysR	Positive regulator for *lys*	739
ilvR	Positive regulator for *thr* and *ilv* operons	902
thrA	Aspartokinase I-homoserine dehydrogenase I	918, 1808, 2074
thrB	Homoserine kinase	251, 361, 1509
thrC	Threonine synthase	526, 1512
thrL	*thr* operon leader peptide	394, 595
metA	Homoserine transsuccinylase	479, 1312, 1683
metB	Cystathionine gamma-synthase	781, 1246, 1693
metC	Cystathionine beta-lyase	781, 1246, 1693
metE	Tetrahydropteroyltriglutamate methyltransferase	316
metH	B_{12}-dependent homocysteine-N^5-methyltetrahydrofolate transmethylase, repressor of *metE* and *metF*	99, 179
metJ	Repressor of all *met* genes but *metF*	1538
metL	Aspartokinase II and homoserine dehydrogenase II	1511
metR	Regulator for *metE* and *metH*	1264

3. Serine family
glyA	Serine hydroxymethyltransferase	61, 1831, 1917
sbaA	Regulation of serine and branched-chain amino acid metabolism	392
serA	D-3-Phosphoglycerate dehydrogenase	1771, 1776
serB	3-Phosphoserine phosphatase	2036
serC	3-Phosphoserine aminotransferase	481, 1067
cysB	Positive regulator for cysteine regulon	1224, 1338
cysE	Serine acetyltransferase	1037, 1224, 2152
cysK	Cysteine synthase A, *O*-acetylserine sulfhydrolase A	257, 1480
cysM	Cysteine synthase B, *O*-acetylserine sulfhydrolase B	1480, 1842, 1843

4. Aromatic amino acid family
aroA	5-Enolpyruvylshikimate-3-phosphate synthetase	51, 1485, 1834
aroB	Dehydroquinate synthase	995
aroC	Chorismate synthase	287, 2148
aroD	5-Dehydroquinate dehydratase	291, 995, 996
aroE	Dehydroshikimate reductase	62
aroF	Phospho-2-dehydro-3-deoxyheptonate aldolase (DAHP synthetase, tyrosine repressible)	1622
aroG	Phospho-2-dehydro-3-deoxyheptonate aldolase (DAHP synthetase, phenylalanine repressible)	412, 1899
aroH	Phospho-2-dehydro-3-deoxyheptonate aldolase (DAHP synthetase, tryptophan repressible)	809, 1621, 1622
aroI	Member of *aro* operon	630
aroK	Shikimate kinase I	1179
aroL	Shikimate kinase II	425, 728
aroM	Regulated by *aroR*	425
pheA	Chorismate mutase-P-prephenate dehydratase	603, 604, 1403
pheL	Leader peptide of chorismate mutase-P-prephenate dehydratase	601, 604, 1488
tyrA	Chorismate mutase T and prephenate dehydrogenase	1248, 2027, 2028
tyrB	Tyrosine aminotransferase, tyrosine repressible	614, 1806, 2224
trpA	Tryptophan synthase, A protein	242, 1446, 1710
trpB	Tryptophan synthase, B protein	242, 483, 1710
trpC	*N*-(5-Phosphoribosyl)anthranilate isomerase and indole-3-glycerolphosphate synthetase	2166
trpD	Glutamine amidotransferase and phosphoribosylanthranilate transferase	790, 1826
trpE	Anthranilate synthase	749, 1826
trpL	*trp* operon leader peptide	1073, 1668
trpR	Regulator for *trp* operon and *aroH; trp* aporepressor	72, 1044, 2228
wrbA	*trp* repressor-binding protein; affects association of *trp* repressor and operator	2228

5. Histidine
hisA	*N*-(5′-phospho-L-ribosyl-formimino)-5-amino- 1-(5′-phosphoribosyl)-4-imidazolecarboxamide isomerase	266
hisB	Imidazole glycerol phosphate dehydratase and histidinol phosphate phosphatase	266, 307
hisC	Histidinol phosphate aminotransferase	266
hisD	L-Histidinal:NAD$^+$ oxidoreductase; L-histidinol:NAD$^+$ oxidoreductase	266, 306, 916

(Table continues)

TABLE 2 *Continued*

Category and gene	Gene product description[a]	Reference(s)
hisF	Imidazole glycerol phosphate synthase subunit in heterodimer with HisH = imidazole glycerol phosphate synthase holoenzyme	596, 998, 1650
hisG	ATP phosphoribosyltransferase	266
hisH	Glutamine amidotransferase subunit of heterodimer with HisF = imidazole glycerol phosphate synthase holoenzyme	266, 998, 1650
hisIE	Phosphoribosyl-AMP cyclohydrolase; phosphoribosyl-ATP pyrophosphatase	266, 306
hisL	*his* operon leader peptide	282

6. Pyruvate family
alr	Alanine racemase; isozyme	2106, 2153
avtA	Alanine-α-ketoisovalerate transaminase, transaminase C	2108

7. Branched-chain family
azl	Regulation of *ilv* and *leu* genes; azaleucine resistance	1549
ileR	Negative regulator for *thr* and *ilv* operons	2134
ilvA	Threonine deaminase	497, 542, 1951
ilvB	Acetolactate synthase I, valine sensitive, large subunit	568, 2132, 2141
ilvC	Ketol-acid reductoisomerase	2139
ilvD	Dihydroxyacid dehydratase	548, 804
ilvE	Branched-chain amino acid aminotransferase	846, 934
ilvF	Acetolactate synthase (valine insensitive) activity, probably fifth isozyme, silent in K-12	26
ilvG	Acetolactate synthase II, valine insensitive, large subunit, silent in K-12	804
ilvH	Acetolactate synthase III, valine sensitive, small subunit	1640, 1641, 2132
ilvI	Acetolactate synthase III, valine sensitive, large subunit	1640, 1883, 2132
ilvJ	Acetolactate synthase IV, valine insensitive	878, 1664
ilvL	IlvGEDA operon leader peptide	1085
ilvM	Acetolactate synthase II, valine insensitive, small subunit	804, 2132
ilvN	Acetolactate synthase I, valine sensitive, small subunit	568, 2132, 2141
ilvY	Positive regulator for *ilvC*	2139, 2140
ivbL	*ilvB* operon leader peptide	569, 2141
leuA	2-Isopropylmalate synthase	615, 1864
leuB	3-Isopropylmalate dehydrogenase	1049
leuC	Isopropylmalate isomerase subunit	579
leuD	Isopropylmalate isomerase subunit	579
leuJ	Regulator for *leu* and *ilv* operons	1432
leuL	*leu* operon leader peptide	108, 969
leuO	Probable activator protein for *leuABCD* operon	739

E. Polyamine biosynthesis
pat	Putrescine aminotransferase activity	1584
speA	Biosynthetic arginine decarboxylase	725, 1339
speB	Agmatinase	1946, 1947, 2206
speC	Ornithine decarboxylase isozyme	1496, 2189
speD	S-Adenosylmethionine decarboxylase	1949, 2206
speE	Spermidine synthase = putrescine aminopropyltransferase	1949, 2206
speF	Ornithine decarboxylase isozyme, inducible	954

F. Purines, pyrimidines, nucleosides, and nucleotides
1. Purine ribonucleotide biosynthesis
adk	Adenylate kinase activity; pleiotropic effects on glycerol-3-phosphate acyltransferase activity	623, 1631, 1686
gmk	Guanylate kinase	616
guaA	GMP synthetase	1992
guaB	IMP dehydrogenase	57
guaC	GMP reductase	57
ndk	Nucleoside diphosphate kinase	698
prs	Phosphoribosylpyrophosphate synthetase	206, 793
purA	Adenylosuccinate synthetase	460, 1163, 1860
purB	Adenylosuccinate lyase	726
purC	Phosphoribosylaminoimidazole-succinocarboxamide synthetase = SAICAR synthetase	1990
purD	Phosphoribosylglycinamide synthetase = GAR synthetase	11, 1818
purE	Phosphoribosylaminoimidazole carboxylase = AIR carboxylase, catalytic subunit	1311, 1991, 2118
purF	Amidophosphoribosyltransferase = PRPP amidotransferase	1282, 1729
purH	Phosphoribosylaminoimidazolecarboxamide formyltransferase = AICAR formyltransferase; IMP cyclohydrolase	11, 547
purK	Phosphoribosylaminoimidazole carboxylase = AIR carboxylase, CO_2-fixing subunit	1311, 1991, 2118
purL	Phosphoribosylformyl glycinamide synthetase = FGAM synthetase	1758
purM	Phosphoribosylaminoimidazole synthetase = AIR synthetase	1857
purN	5′-Phosphoribosyl glycinamide (GAR) transformylase 1	844, 845
purR	Repressor for *pur* regulon, *glyA*, *glnB*, *prsA*, *speA*	312, 1782, 1897
purT	5′-Phosphoribosylglycinamide (GAR) transferase 2	1243
purU	Provides formate for *purT*-dependent FGAR synthesis	1384

TABLE 2 *Continued*

Category and gene	Gene product description[a]	Reference(s)
2. Pyrimidine ribonucleotide biosynthesis		
carA	Carbamoyl-phosphate synthetase, glutamine (small) subunit	216, 1365, 1698
carB	Carbamoyl-phosphate synthase large subunit	216, 1365, 1698
pyrB	Aspartate carbamoyltransferase, catalytic subunit	1533, 2204, 2208
pyrC	Dihydro-orotase	231, 311, 2167
pyrD	Dihydro-orotate oxidase	2167
pyrE	Orotate phosphoribosyltransferase	47, 1573
pyrF	Orotidine-5'-phosphate decarboxylase	2026
pyrI	Aspartate carbamoyltransferase, regulatory subunit	2265, 2266
pyrL	*pyrBI* operon leader peptide	1119
3. 2'-Deoxyribonucleotide metabolism		
dcd	2'-Deoxycytidine 5'-triphosphate deaminase	2107, 2133
dut	Deoxyuridine triphosphatase	277, 776, 2107
grx	Glutaredoxin, redox coenzyme for glutathione-dependent ribonucleotide reductase	782, 1706, 1732
mutT	(Deoxy)nucleoside triphosphatase, prefers dGTP, causes AT-GC transversions	17, 145, 1756
nrdA	Ribonucleoside diphosphate reductase, subunit B1	334, 1428, 1933
nrdB	Ribonucleoside-diphosphate reductase subunit B2	1428, 1920, 1933
nrdD	Anaerobic ribonucleoside-triphosphate reductase	147, 707, 1934
thyA	Thymidylate synthetase	1314
tmk	Thymidylate kinase	160
trxB	Thioredoxin reductase	1053, 1328, 1589
4. Salvage of nucleosides and nucleotides		
add	Adenosine deaminase	284
amn	AMP nucleosidase	1114
apaH	Diadenosine tetraphosphatase	525, 904, 1659
apt	Adenine phosphoribosyltransferase	1121–1123, 748a
cdd	Cytidine/deoxycytidine deaminase	2223
codA	Cytosine deaminase	396
cpdB	2':3'-cyclic nucleotide 2'-phosphodiesterase	1164
deoA	Thymidine phosphorylase	541, 2102, 2133
deoB	Phosphopentomutase	255, 541
deoC	2-Deoxyribose-5-phosphate aldolase	2046
deoD	Purine-nucleoside phosphorylase	541
deoR	Regulator for *deo* operon, *tsx, nupG*	44, 393, 1348
gpt	Guanine-hypoxanthine phosphoribosyltransferase	775, 1122, 1123
gsk	Inosine-guanosine kinase	795
hpt	Hypoxanthine phosphoribosyltransferase	775, 1122, 1123
optA	Regulator for *dgt*	1601
tdk	Thymidine kinase	160
udp	Uridine phosphorylase	221, 1316
upp	Uracil phosphoribosyltransferase	48, 1619
5. Miscellaneous nucleoside/nucleotide reactions		
dgt	Deoxyguanosine triphosphate triphosphohydrolase	1600, 1601, 2203
lepA	GTP-binding membrane protein	1237
mrp	Putative ATPase	399
pyrG	CTP synthetase	2144
pyrH	UMP kinase	288, 919, 2219
udk	Uridine/cytidine kinase	2045
xapA	Xanthosine phosphorylase	143, 144
xapR	Regulator for *xapA*	256
G. Biosynthesis of cofactors, prosthetic groups, and carriers		
1. Biotin		
bioA	7,8-Diaminopelargonic acid synthetase	829, 1477
bioB	Biotin synthetase	829, 1477, 1737
bioC	Biotin biosynthesis; reaction prior to pimeloyl-CoA	829, 1477
bioD	Dethiobiotin synthetase	27, 829, 1477
bioF	7-Keto-8-aminopelargonic acid synthetase	829, 1477
bioH	Biotin biosynthesis; reaction prior to pimeloyl CoA	829, 1468
birA	Biotin-(acetyl-CoA carboxylase) holoenzyme synthetase; biotin operon repressor	2, 368, 1331
bisC	Biotin sulfoxide reductase	1542
2. Folic acid		
folA	Dihydrofolate reductase; trimethoprim resistance	523, 570, 805
folC	Dihydrofolate:folylpolyglutamate synthetase; dihydrofolate synthetase	182, 974, 989
folD	5,10-Methylene-tetrahydrofolate dehydrogenase; 5,10-methylene-tetrahydrofolate cyclohydrolase	400

(Table continues)

TABLE 2 *Continued*

Category and gene	Gene product description[a]	Reference(s)
folE	GTP cyclohydrolase I	1599, 1764, 1770
folK	7,8-Dihydro-6-hydroxymethylpterin-pyrophosphokinase	1959, 1960
folP	7,8-Dihydropteroate synthase	390, 1959
pabA	p-Aminobenzoate synthetase, component II	661
pabB	p-Aminobenzoate synthetase, component I	661, 2100, 2235
pabC	Aminodeoxychorismate lyase	660, 661, 2235
3. Lipoate		
lipA	Protein of lipoate biosynthesis	721, 1626
lipB	Protein of lipoate biosynthesis	1626
4. Molybdopterin		
moaA	Molybdopterin biosynthesis, protein A	1612, 1657, 1809
moaB	Molybdopterin biosynthesis, protein B	1657
moaC	Molybdopterin biosynthesis, protein C	1657
moaD	Molybdopterin biosynthesis	765, 1612, 1657
moaE	Molybdopterin-converting factor, subunit 2	1657
mob	Molybdopterin → molybdopterin-guanine dinucleotide	903, 1612, 1657
moeA	Molybdopterin biosynthesis	765, 1424, 1657
moeB	Molybdopterin biosynthesis	765, 1424, 1657
mog	Required for the efficient incorporation of molybdate in molybdoproteins	765, 1657
5. Pantothenate		
coaA	Pantothenate kinase	1866, 1867, 2050
panB	Ketopentoate hydroxymethyltransferase	370, 906
panC	Pantothenate synthetase	370
panD	Aspartate 1-decarboxylase	370
6. Pyridoxine		
pdxA	Pyridoxine biosynthesis	1659
pdxB	Erythronate-4-phosphate dehydrogenase	1067, 1771
pdxH	Pyridoxinephosphate oxidase	1068
pdxJ	Pyridoxine biosynthesis	1066, 1958
7. Pyridine nucleotide		
nadA	Quinolinate synthetase, A protein	546, 1794
nadB	Quinolinate synthetase, B protein	546, 1794
nadC	Quinolinate phosphoribosyltransferase	2011
nadE	NAD synthetase, prefers NH_3 over glutamine	2165
nadR	Probable *nadAB* transcriptional regulator	554
pncA	Nicotinamide deamidase	1499
pncB	Nicotinate phosphoribosyltransferase	2200
8. Thiamine		
thiA	Thiamine thiazole requirement	965, 2051
thiB	Thiamine phosphate pyrophosphorylase	965, 2051
thiC	Thiamine biosynthesis, pyrimidine moiety	2051
thiD	Phosphomethylpyrimidine kinase activity	841, 2051
thiE	Thiamine biosynthesis, thiazole moiety	2051
thiF	Thiamine biosynthesis, thiazole moiety	179, 2051
thiG	Thiamine biosynthesis, thiazole moiety	2051
thiH	Thiamine biosynthesis, thiazole moiety	2051
thiK	Thiamine kinase	842
thiL	Thiamine monophosphate kinase	842
9. Riboflavin		
ribA	GTP cyclohydrolase II	1092, 1599, 1646
ribB	3,4-Dihydroxy-2-butanone-4-phosphate synthase	1647
ribC	Riboflavin synthase, alpha chain	2088
ribD	Deaminase in pathway of riboflavin synthesis	253
ribE	Riboflavin synthase, beta chain	2088
10. Thioredoxin, glutaredoxin, and glutathione		
ggt	γ-Glutamyltranspeptidase	1939
gor	Glutathione oxidoreductase	439, 513, 1048
gshA	γ-Glutamate-cysteine ligase	202
gshB	Glutathione synthetase	415, 957, 1968
trxA	Thioredoxin	1035, 1198, 1706
11. Menaquinone and ubiquinone		
ispA	Geranyltranstransferase (farnesyldiphosphate synthase)	132, 576, 1819

TABLE 2 *Continued*

Category and gene	Gene product description[a]	Reference(s)
ispB	Octaprenyl diphosphate synthase	76
menA	1,4-Dihydroxy-2-naphthoate → dimethylmenaquinone	132, 1815, 1828
menB	Dihydroxynaphtoic acid synthetase	132
menC	*o*-Succinylbenzoyl-CoA synthase	132, 1281, 1814
menD	Menaquinone biosynthesis	1281, 1491, 1562
menE	*o*-Succinylbenzoate-CoA synthase	132
ubiA	*p*-Hydroxybenzoate:octaprenyltransferase	1290, 1835, 2193
ubiB	2-Octaprenylphenol → 2-octaprenyl-6-methoxyphenol	132
ubiC	Chorismate lyase	132, 1409, 1835
ubiD	3-Octaprenyl-4-hydroxy-benzoate → 2-octaprenylphenol	132, 1110
ubiE	2-Octaprenyl-6-methoxy-1,4-benzoquinone → 2-octaprenyl-3-methyl-6-methoxy-1,4-benzoquinone	132
ubiF	2-Octaprenyl-3-methyl-6-methoxy-1,4-benzoquinone → 2-octaprenyl-3-methyl-5-hydroxy-6-methoxy-1,4-benzoquinone	132, 343
ubiG	2-Octaprenyl-3-methyl-5-hydroxy-6-methoxy-1,4-benzoquinone → ubiquinone 8	132, 629, 2194
ubiH	2-Octaprenyl-6-methoxyphenol → 2-octaprenyl-6-methoxy-1,4-benzoquinone	132, 1387
ubiX	Putative polyprenyl *p*-hydroxybenzoate carboxylase	1427

12. Heme and porphyrin

cysG	Uroporphyrinogen III methylase; siroheme biosynthesis	1517, 2115, 2116
hemA	Enzyme in alternate path of synthesis of 5-aminolevulinate	837, 1134, 2067
hemB	5-Aminolevulinate dehydratase = porphobilinogen synthase	1331, 1464, 1872
hemC	Porphobilinogen deaminase = hydroxymethylbilane synthase	691, 1072, 1321
hemD	Uroporphyrinogen III synthase	40
hemE	Uroporphyrinogen decarboxylase	1743
hemF	Coproporphyrinogen III oxidase	365, 2012
hemG	Protoporphyrinogen oxidase activity	1742
hemL	Glutamate-1-semialdehyde aminotransferase	667, 838, 839
hemM	An enzyme in the main pathway of synthesis of 5-aminolevulinate, possibly glutamyl-tRNA dehydrogenase	837
hemX	Putative uroporphyrinogen III methylase	395, 1744
popD	5-Aminolevulinate dehydratase activity	1872
visA	Ferrochetalase: final enzyme of heme biosynthesis	574, 1334

13. Cobalamin (incomplete in K-12)

btuR	Cob(I)alamin adenosyltransferase	517

14. Enterochelin

entA	2,3-Dihydro-2,3-dihydroxybenzoate dehydrogenase, enterochelin biosynthesis	1165, 1481, 2025
entB	2,3-Dihydro-2,3-dihydroxybenzoate synthetase, enterochelin biosynthesis	1703, 1888
entC	Isochorismate synthetase, enterochelin biosynthesis	927, 1166, 1481
entD	Enterochelin synthetase, component D	68, 335, 1888
entE	ATP-dependent activation of 2,3-dihydroxybenzoate	1888, 1889
entF	ATP-dependent serine-activating enzyme	1535, 1704, 1888

H. Fatty acid biosynthesis

aas	2-Acylglycerophospho-ethanolamine acyl transferase; acyl-acyl carrier protein synthetase	800, 876
accA	Acetyl-CoA carboxylase, carboxytransferase component, alpha subunit	1140, 1141
accB	Acetyl-CoA carboxylase, BCCP subunit; carrier of biotin	1139, 1938
accC	Acetyl-CoA carboxylase, biotin carboxylase subunit	1018, 1212
accD	Acetyl-CoA carboxylase, carboxytransferase component, beta subunit	1140–1142
acpP	Acyl carrier protein	779, 780, 1620
acpS	CoA:apo-[acyl-carrier-protein] pantetheinephosphotransferase = holo-[acyl-carrier-protein] synthase	1556, 1583
acs	Acetyl-CoA synthetase	179, 2194
cdh	CDP-diglyceride hydrolase	827
cdsA	CDP-diglyceride synthetase (CTP:phosphatidate cytidylyltransferase)	828
cdsS	Stability of CDP-diglyceride synthetase activity	591
cfa	Cyclopropane fatty acyl synthase	2105
dgkA	Diglyceride kinase	1705
dgkR	Level of diglyceride kinase	1604, 2101
eutB	Ethanolamine-ammonia lyase heavy subunit	911, 912, 1441
eutC	Ethanolamine-ammonia lyase light subunit	911, 912, 1441
fabA	β-Hydroxydecanoyl thioester dehydrase	327, 369, 1812
fabB	3-Oxoacyl-[acyl-carrier-protein] synthase I	960, 1837, 2021
fabD	Malonyl-CoA-[acyl-carrier-protein] transacylase	1213, 2071
fabF	3-Oxoacyl-[acyl-carrier-protein] synthase II	877, 1837
fabG	3-Oxoacyl-[acyl-carrier-protein] reductase	1018, 1620
fabH	3-Oxoacyl-[acyl-carrier-protein] synthase III; acetyl-CoA [acyl-carrier protein] transacylase	2020
fabI	Enoyl-ACP reductase (NADH)	137, 138
hlyC	Acyl carrier protein for processing prohemolysin	856, 2103
tesA	Acyl-CoA thioesterase I; also functions as protease I	308, 309, 821

(Table continues)

TABLE 2 *Continued*

Category and gene	Gene product description[a]	Reference(s)
tesB	Acyl-CoA thioesterase II	1383

II. BROAD REGULATORY FUNCTIONS

Category and gene	Gene product description[a]	Reference(s)
anr	Activator of *ntrL* gene	986
arcA	Negative response regulator of genes in aerobic pathways (sensors for *arcB* and *cpxA*)	863, 865, 1839
arcB	Aerobic respiration sensor response protein; histidine protein kinase/phosphatase (sensor for *arcA*)	861, 866, 869
barA	Sensor regulator, probably activates OmpR by phophorylation	1381
cpxA	Probable inner membrane sensor protein (histidine protein kinase), acting on *arcA*, energy coupling factor, F-pilin formation	2123
creB	Catabolic regulation response regulator	1039, 1170, 1500
creC	Catabolite repression sensor autophosphorylates and phosphorylates PhoB; alternative sensor for *pho* regulon	1305, 2111, 2113
crp	Cyclic AMP receptor protein	103, 750, 1714
csrA	Carbon storage regulator; affects glycogen synthesis, gluconeogenesis, cell size and surface properties	1679, 1680
cstA	Carbon starvation protein	1778
cyaA	Adenylate cyclase	1530
cytR	Regulator for *deo* operon, *udp*, *cdd*, *tsx*, *nupC*, and *nupG*	104, 1520, 1862
ecfA	Energy-coupling factor; pleiotropic effects on active transport coupling to metabolic energy	783
envZ	Protein histidine kinase/phosphatase sensor for *ompR*, modulates expression of *ompF* and *ompC*	553, 2120
era	GTP-binding protein	297, 1111, 1154
fexB	*fexA* (*arcA*) phenotype affected	1112
fnr	Regulatory gene for oxidoreductases and others	867, 1289, 1874
frnH	Regulation of superoxide response regulon	530, 662
fur	Ferric iron uptake; negative regulator	715, 2126
gppA	Guanosine pentaphosphatase; exopolyphosphatase	121, 967, 1637
kdpD	Regulator (sensor) for high-affinity potassium transport system	1394, 1395, 2097
kdpE	Regulator of *kdp* operon (transcriptional effector)	1395, 1557, 2097
lctZ	Pleiotropic effects on components of respiratory chain	364
lexA	Regulator for SOS(*lexA*) regulon	1151
lon	DNA-binding, ATP-dependent protease La; heat shock K protein	441
lytB	Control of stringent response; involved in penicillin tolerance	682
ntrL	Nitrogen-regulatory protein	33
ompR	Response regulator (sensor, *envZ*) affecting *ompC* and *ompF*; outer membrane protein synthesis	832, 1709, 2120
oxyR	Activator, hydrogen peroxide-inducible genes	186, 1971, 1972
phoB	Positive response regulator for *pho* regulon, autophosphorylates and phosporylates, sensor for *phoR*	1221, 1304
phoP	Sensor for *phoQ*	671, 951
phoQ	Response regulator (sensor for *phoP*)	951
phoR	Positive and negative sensor protein for *pho* regulon	1221, 2111, 2213
phoU	Negative regulator for *pho* regulon and putative enzyme in phosphate metabolism	1304, 1618
pus	Effect of suppressors on *relB* mutations	448
relA	Regulation of RNA synthesis; stringent factor; (p)ppGpp synthetase I	1308, 1773, 2205
relB	Negative regulator of translation	1349
relX	Control of synthesis of ppGpp	1497
rpoD	RNA polymerase, σ^{70} subunit; regulation of proteins induced at high temperatures	593, 1046, 2096
rpoE	RNA polymerase, σ^{E} factor; heat shock and oxidative stress	1184
rpoH	RNA polymerase, σ^{32} subunit; regulation of proteins induced at high temperatures	890, 1378, 2035
rpoN	RNA polymerase, σ^{54} or σ^{60} subunit; nitrogen and fermentation regulation	909, 1632
rpoS	RNA polymerase, sigma S (σ^{38}) subunit; synthesis of many growth phase-related proteins	1075, 1408, 1855
soxR	Redox-sensing activator of *soxS*	41, 754, 2196
soxS	Regulation of superoxide response regulon	530, 2196
spf	Spot 42 RNA, inhibition of DNA synthesis	747, 1558, 1642
spoT	(p)ppGpp synthetase II; also guanosine-3′,5′-bispyrophosphate 3′-pyrophosphohydrolase	748, 1309, 1807
sspA	Stringent starvation protein A	577, 2161, 2162
sspB	Stringent starvation protein B	2162
suhA	Induction of heat shock genes	1994
uspT	Histidine protein kinase (sensor) for universal stress protein	1438

III. MACROMOLECULE METABOLISM
A. Synthesis and modification of macromolecules
1. rRNA and "stable" RNAs

Category and gene	Gene product description[a]	Reference(s)
ffs	4.5S RNA	234, 1519, 1563
rrfA	5S rRNA	490, 1472, 2149
rrfB	5S rRNA	490, 1472, 2149
rrfC	5S rRNA	490, 1472, 2149
rrfD	5S rRNA	490, 1472, 2149
rrfE	5S rRNA	490, 1472, 2149
rrfG	5S rRNA	490, 1472, 2149
rrfH	5S rRNA	490, 1472, 2149
rrlA	23S rRNA	1, 95, 1332
rrlB	23S rRNA	1, 95, 1332
rrlC	23S rRNA	1, 95, 1332

TABLE 2 *Continued*

Category and gene	Gene product description[a]	Reference(s)
rrlD	23S rRNA	1, 95, 1332
rrlE	23S rRNA	1, 95, 1332
rrlG	23S rRNA	1, 95, 1332
rrlH	23S rRNA	1, 95, 1332
rrsA	16S rRNA	379, 1083, 1736
rrsB	16S rRNA	379, 1083, 1736
rrsC	16S rRNA	379, 1083, 1736
rrsD	16S rRNA	379, 1083, 1736
rrsE	16S rRNA	379, 1083, 1736
rrsG	16S rRNA	379, 1083, 1736
rrsH	16S rRNA	379, 1083, 1736
rrvD	5S rRNA	480
ssr	6S RNA	801, 1090
ssrA	10Sa RNA, nonribosomal	991, 1447

2. Ribosomal protein synthesis and modification

Category and gene	Gene product description[a]	Reference(s)
prmA	Methylation of 50S ribosomal subunit protein L11	345
prmB	Methylation of 50S ribosomal subunit protein L3	345
rplA	50S ribosomal subunit protein L1, regulates synthesis of L1 and L11	491, 1719
rplB	50S ribosomal subunit protein L2	491, 1682
rplC	50S ribosomal subunit protein L3	1932
rplD	50S ribosomal subunit protein L4, regulates expression of S10 operon	2258–2260
rplE	50S ribosomal subunit protein L5	1188
rplF	50S ribosomal subunit protein L6	1472
rplI	50S ribosomal subunit protein L9	658
rplJ	50S ribosomal subunit protein L10	1687, 2257
rplK	50S ribosomal subunit protein L11	492, 945, 1687
rplL	50S ribosomal subunit protein L7/L12	1531, 2257
rplM	50S ribosomal subunit protein L13	1472
rplN	50S ribosomal subunit protein L14	1472, 1565
rplO	50S ribosomal subunit protein L15	560
rplP	50S ribosomal subunit protein L16	560, 1682
rplQ	50S ribosomal subunit protein L17	1472
rplR	50S ribosomal subunit protein L18	490
rplS	50S ribosomal subunit protein L19	2155
rplT	50S ribosomal subunit protein L20, and regulator	1113, 1932
rplU	50S ribosomal subunit protein L21	897, 1472
rplV	50S ribosomal subunit protein L22	65
rplW	50S ribosomal subunit protein L23	491
rplX	50S ribosomal subunit protein L24	1472
rplY	50S ribosomal subunit protein L25	557
rpmA	50S ribosomal subunit protein L27	897, 2188
rpmB	50S ribosomal subunit protein L28	1472
rpmC	50S ribosomal subunit protein L29	658, 1472
rpmD	50S ribosomal subunit protein L30	1472
rpmE	50S ribosomal subunit protein L31	658
rpmF	50S ribosomal subunit protein L32	1969
rpmG	50S ribosomal subunit protein L33	1472
rpmH	50S ribosomal subunit protein L34	807
rpmI	50S ribosomal subunit protein A	1113
rpmJ	50S ribosomal subunit protein X	2034
rpsA	30S ribosomal subunit protein S1	212, 1082, 1568
rpsB	30S ribosomal subunit protein S2	45
rpsC	30S ribosomal subunit protein S3	212, 235, 463
rpsD	30S ribosomal subunit protein S4	30, 1227, 2138
rpsE	30S ribosomal subunit protein S5	463, 658
rpsF	30S ribosomal subunit protein S6	941, 1900
rpsG	30S ribosomal subunit protein S7, initiates assembly	463, 901, 1227
rpsH	30S ribosomal subunit protein S8, and regulator	1227, 1260, 2195
rpsI	30S ribosomal subunit protein S9	2151
rpsJ	30S ribosomal subunit protein S10	1253
rpsK	30S ribosomal subunit protein S11	1900, 2151
rpsL	30S ribosomal subunit protein S12	30, 150, 1993
rpsM	30S ribosomal subunit protein S13	1945
rpsN	30S ribosomal subunit protein S14	658
rpsO	30S ribosomal subunit protein S15	1227, 1565
rpsP	30S ribosomal subunit protein S16	2138, 2155
rpsQ	30S ribosomal subunit protein S17	658, 2138

(Table continues)

TABLE 2 *Continued*

Category and gene	Gene product description[a]	Reference(s)
rpsR	30S ribosomal subunit protein S18	1336, 1900
rpsS	30S ribosomal subunit protein S19	2137
rpsT	30S ribosomal subunit protein S20	357, 1712, 2138
rpsU	30S ribosomal subunit protein S21	212
rpsV	30S ribosomal subunit protein S22	94, 1216

3. Ribosome maturation and modification

fusB	Pleiotropic effects on RNA synthesis, ribosomes, and ribosomal protein S6	1956
rimB	Maturation of 50S ribosomal subunit	240
rimC	Maturation of 50S ribosomal subunit	240
rimD	Maturation of 50S ribosomal subunit	240
rimE	Modification of ribosomal proteins	1055
rimF	Modification of ribosome	598
rimG	Modification of 30S ribosomal subunit protein S4	2275
rimH	Modification of ribosome	598
rimI	Modification of 30S ribosomal subunit protein S18; acetylation of N-terminal alanine	2239
rimJ	Modification of 30S ribosomal subunit protein S5; acetylation of N-terminal alanine	990, 2239
rimK	Ribosomal protein S6 modification protein	941
rimL	Modification of 30S ribosomal subunit protein L7; acetylation of N-terminal serine	1967
rit	Affects thermolability of 50S ribosomal subunit	1465
rmf	Ribosome modulation factor	2090
strM	Modifies ribosome structure	1731

4. tRNAs

alaT	Alanine tRNA 1B (duplicate of alaUV)	1503, 1964, 1966
alaU	Alanine tRNA 1B (duplicate of alaTV)	1503, 1964, 1966
alaV	Alanine tRNA 1B (duplicate of alaTU)	1503, 1964, 1966
alaW	Alanine tRNA 2 (duplicate of alaX)	1503, 1964, 1966
alaX	Alanine tRNA 2 (duplicate of alaW)	1503, 1964, 1966
argQ	Arginine tRNA 2 (duplicate of argVYZ)	1271
argU	Arginine tRNA 4	294, 1724, 1755
argV	Arginine tRNA 2 (duplicate of argQYZ)	1271
argW	Arginine tRNA 5	1271
argX	Arginine tRNA 3	1271
argY	Arginine tRNA 2 (duplicate of argVQZ)	1271
argZ	Arginine tRNA 2 (duplicate of argVYQ)	1271
asnT	Asparagine tRNA	1016, 1138
asnU	Asparagine tRNA	1016, 1138
asnV	Asparagine tRNA	1016, 1138
aspT	Aspartate tRNA 1 (duplicate of aspVU)	1016
aspU	Aspartate tRNA 1 (duplicate of aspTV)	1016
aspV	Aspartate tRNA 1 (duplicate of aspTU)	1016
divE	tRNASer1, affects cell division	1748, 1963
glnU	Glutamine tRNA 1 (duplicate of glnW)	505, 882, 1670
glnV	Glutamine tRNA 2 (duplicate of glnX)	505, 882, 1670
glnW	Glutamine tRNA 1 (duplicate of glnU)	505, 882, 1670
glnX	Glutamine tRNA 2 (duplicate of glnV)	505, 882, 1670
gltT	Glutamate tRNA 2 (duplicate of gltUVW)	1016
gltU	Glutamate tRNA 2 (duplicate of gltTVW)	1016
gltV	Glutamate tRNA 2 (duplicate of gltTUW)	1016
gltW	Glutamate tRNA 2 (duplicate of gltTUV)	1016
glyT	Glycine tRNA 2	1272
glyU	Glycine tRNA 1	1272
glyV	Glycine tRNA 3 (duplicate of glyXYW)	1272
glyW	Glycine tRNA 3 (duplicate of glyVXY)	1272
glyX	Glycine tRNA 3 (duplicate of glyVWY)	1272
glyY	Glycine tRNA 3 (duplicate of glyXYW)	1272
hisR	Histidine tRNA	761
ileT	Isoleucine tRNA 1 (duplicate of ileUV)	1016
ileU	Isoleucine tRNA 1 (duplicate of ileTV)	1016
ileV	Isoleucine tRNA 1 (duplicate of ileTU)	1016
ileX	Isoleucine tRNA 2	1016
leuP	Leucine tRNA 1 (duplicate of leuQTV)	73, 1016
leuQ	Leucine tRNA 1 (duplicate of leuPTV)	73, 1016
leuT	Leucine tRNA 1 (duplicate of leuQPV)	73, 1016
leuU	Leucine tRNA 2	73, 1016
leuV	Leucine tRNA 1 (duplicate of leuQPT)	73, 1016
leuW	Leucine tRNA 3	73, 1016
leuX	Leucine tRNA 5	73, 1429
leuZ	Leucine tRNA 4	73, 1016
lrs	Level of leucine tRNA	2040

TABLE 2 *Continued*

Category and gene	Gene product description[a]	Reference(s)
lysT	Lysine tRNA (duplicate of *lysWV*)	236, 1271, 1966
lysV	Lysine tRNA (duplicate of *lysTW*)	1016, 1271, 1966
lysW	Lysine tRNA (duplicate of *lysTV*)	1016, 1271, 1966
metT	Methionine tRNA (duplicate of *metU*)	1777
metU	Methionine tRNA (duplicate of *metT*)	1777
metV	Methionine tRNA-fMet2	416, 2063
metW	Methionine tRNA fMet1 (duplicate of *metZ*)	678, 1489, 2063
metY	Methionine tRNA(fMet2)	973
metZ	Methionine tRNA-fMet1 (duplicate of *metW*)	678, 1489, 2063
pheR	Phenylalanine tRNA (replicate of *pheUVW*)	3, 602, 603
pheU	Phenylalanine tRNA (replicate of *pheVWR*)	3, 603, 1534
pheV	Phenylalanine tRNA (replicate of *pheURW*)	3, 1488, 1534
pheW	Phenylalanine tRNA (replicate of *pheUVR*)	3, 602, 2168
proK	Proline tRNA 1	1016
proL	Proline tRNA 2	1016
proM	Proline tRNA 3	1016
selC	Selenocystyl tRNA inserts at UGA	106, 1143, 1752
serT	Serine tRNA 1	762, 1429, 1672
serU	Serine tRNA 2	190, 762, 1089
serV	Serine tRNA 3	762, 1429, 1672
serW	Serine tRNA 5 (duplicate of *serX*)	762, 1429, 1672
serX	Serine tRNA 5 (duplicate of *serW*)	762, 1429, 1672
thrT	Threonine tRNA 3	238, 714, 1985
thrU	Threonine tRNA 4	760, 1490, 1674
tyrT	Tyrosine tRNA 1 (duplicate of *tyrV*)	194, 762, 1315
tyrU	Tyrosine tRNA 2	762
tyrV	Tyrosine tRNA 1 (duplicate of *tyrT*)	762
valT	Valine tRNA 1 (duplicate of *valUXY*)	1965
valU	Valine tRNA 1 (duplicate of *valTXY*)	236, 1965
valV	Valine tRNA 2B	1965
valW	Valine tRNA 2A	1965
valX	Valine tRNA 1 (duplicate of *valTUY*)	1965
valY	Valine tRNA 1 (duplicate of *valTUX*)	1965
5. Aminoacyl tRNA synthetases and their modification		
aat	Leucyl, phenylalanyl-tRNA-protein transferase	1832
alaS	Alanyl-tRNA synthetase	245, 538, 1323
argS	Arginine tRNA synthetase	139, 510, 1153
asnS	Asparagine tRNA synthetase	763, 1211
aspS	Aspartate tRNA synthetase	511, 575, 763
cca	tRNA nucleotidyltransferase	708, 2271
cysS	Cysteine tRNA synthetase	885, 1015, 1494
fmt	10-Formyltetrahydrofolate:L-methionyl-tRNA(fMet) *N*-formyltransferase	677, 1091, 1286
glnS	Glutamine tRNA synthetase	720, 882, 1670
gltE	Glutamate tRNA synthetase; possible regulatory subunit	1077, 1374, 1944
gltM	Level of glutamate tRNA synthetase activity	1374
gltX	Glutamate tRNA synthetase, catalytic subunit	223, 1433, 1944
glyQ	Glycine tRNA synthetase, alpha subunit	2006, 2007
glyS	Glycine tRNA synthetase, beta subunit	383, 971, 2007
hisS	Histidine tRNA synthetase	761
hisT	Pseudouridine synthase I	413, 935, 2023
ileS	Isoleucine tRNA synthetase	1433, 2160
ilvU	Regulator for *ileS* and modifier of isoleucine tRNA 2 and valine tRNA 2	529
leuR	Level of leucine tRNA synthetase	1983
leuS	Leucine tRNA synthetase	2150, 2164
leuY	Level of leucine tRNA synthetase	1983
lysS	Lysine tRNA synthetase, constitutive; suppressor of ColE1 mutation in primer RNA	962, 1118, 1393
lysU	Lysine tRNA synthetase, inducible; heat shock protein	329, 716, 1392
metG	Methionine tRNA synthetase	626, 1279, 2063
miaA	Δ^2-isopentenylpyrophosphate tRNA-adenosine transferase	352
nuvA	Uridine thiolation factor A activity	1161, 1986
nuvC	4-Thiouridine modification of tRNA; near-UV sensitivity and resistance	1161, 1986
pheM	Phenylalanine tRNA synthetase leader peptide	1488, 1881
pheS	Phenylalanine tRNA synthetase, alpha subunit	603, 956, 1534
pheT	Phenylalanine tRNA synthetase, beta subunit	603, 956, 1534
proS	Proline tRNA synthetase	383
prrC	tRNA(Lys)-specific anticodon nuclease	1124, 1340
pth	Peptidyl-tRNA hydrolase	592, 1375
queA	Synthesis of queuine in tRNA; probably *S*-adenosylmethionine:tRNA ribosyltransferase-isomerase	1638, 1852

(Table continues)

TABLE 2 *Continued*

Category and gene	Gene product description[a]	Reference(s)
selA	Selenocysteine synthase: L-seryl-tRNA dehydrated	550, 1752
selD	Selenophosphate synthase, H_2Se added to acrylyl-tRNA	985, 1104, 1752
serR	Level of seryl-tRNA synthetase	383
serS	Serine tRNA synthetase; also charges selenocysteine tRNA with serine	74, 382, 1757
tgt	tRNA-guanine transglycosylase	567, 2031
thrS	Threonine tRNA synthetase	238, 1337, 1677
trmA	tRNA methyltransferase; tRNA (uracil-5-)methyltransferase	683, 1436, 1529
trmB	tRNA methyltransferase; tRNA (guanine-7-)methyltransferase	1241
trmC	tRNA methyltransferase; 5-methylaminoethyl-2-thiouridine biosynthesis	157, 692
trmD	tRNA methyltransferase; tRNA (guanine-7-)methyltransferase	258, 774, 2155
trmE	tRNA methyltransferase; 5-methylaminoethyl-2-thiouridine biosynthesis	157, 692
trmF	tRNA methyltransferase; 5-methylaminoethyl-2-thiouridine biosynthesis	157, 692
trpS	Tryptophan tRNA synthetase	1302, 1490, 1669
tyrS	Tyrosine tRNA synthetase	116, 117, 539
valS	Valine tRNA synthetase	317, 506

6. Nucleoproteins

hns	Histone-like protein HLP-II (HU, BH2, HD, NS); pleiotropic regulator	1265, 2033, 2211
hnsA	DNA-binding protein H-NS	520, 1560
hnsB	DNA-binding protein H-NS	520, 1560, 2238
hupA	DNA-binding protein HU-alpha (HU-2)	473, 647, 1011
hupB	DNA-binding protein HU-beta, NS1 (HU-1)	473, 474, 647
stpA	H-NS-like protein	2261
tpr	A protamine-like protein	194

7. DNA replication, restriction/modification, and recombination

ada	O^6-Methylguanine-DNA methyltransferase; transcription activator/repressor	14, 1207, 1726
aidB	Induced by alkylating agents	2081, 2084
alkA	3-Methyladenine DNA glycosylase II, inducible	920, 1385, 2081
alkB	DNA repair system specific for alkylated DNA	1017, 2081, 2082
cer	Site-specific recombinase	344
dam	DNA adenine methylase	990, 1435, 1654
dcm	DNA cytosine methylase	643, 1146, 2225
dfp	Flavoprotein affecting synthesis of DNA and pantothenate metabolism	1875
dinG	Probably ATP-dependent helicase	1021, 1127
dksA	*dnaK* suppressor protein	940
dnaA	DNA biosynthesis; initiation of chromosome replication; can be transcription regulator	80, 559, 2174
dnaB	Chromosome replication; chain elongation; part of primosome	29, 2092, 2093
dnaC	Chromosome replication; initiation and chain elongation	29, 1245, 2092
dnaE	DNA polymerase III, alpha subunit	664, 1922, 2191
dnaG	DNA biosynthesis; DNA primase	672, 1891
dnaI	DNA biosynthesis	1010, 1970
dnaN	DNA polymerase III, beta subunit	664, 1922, 2191
dnaQ	DNA polymerase III, epsilon subunit	1921, 1922, 2191
dnaR	Thermosensitive initiation of chromosome replication	1723
dnaT	DNA biosynthesis; primosomal protein i	29, 1250
dnaX	DNA polymerase III, tau and gamma subunits; DNA elongation factor III	1921, 2022, 2191
fis	Site-specific DNA inversion stimulation factor; DNA-binding protein; a *trans* activator for transcription	1414, 1476, 2249
fpg	Formamidopyrimidine DNA glycosylase	184, 185, 659
gidA	Glucose-inhibited division; chromosome replication?	77, 1445
gidB	Glucose-inhibited division; chromosome replication?	77, 1445
gyrA	DNA gyrase, subunit A, type II topoisomerase, DNA cleavage with transient covalent bonding	475, 1466, 1625
gyrB	DNA gyrase subunit B, type II topoisomerase, DNA cleavage with transient covalent bonding, ATPase activity	25, 432, 466
helD	DNA helicase IV	1033, 1292, 2182
het	Binding of DNA sequences in *oriC* region to outer membrane; DNA-binding protein?	2176
himA	Integration host factor (IHF), alpha subunit; site-specific recombination	197, 943, 1093
himD	Integration host factor (IHF), beta subunit; site-specific recombination	197, 564, 1093
holA	DNA polymerase III, delta subunit	268, 461, 2191
holB	DNA polymerase III, delta prime subunit	267, 270, 461
holC	DNA polymerase III, chi subunit	2191
holD	DNA polymerase III, psi subunit	269, 2191
holE	DNA polymerase III, theta subunit	664, 1853, 2191
hsdM	Host modification; DNA methylase M	1181, 1580
hsdS	Specificity determinant for *hsdM* and *hsdR*	1181, 1580
iciA	Replication initiation inhibitor, binds to 13-mers at *oriC*	819, 1987
lig	DNA ligase	1103
mfd	Mutation frequency decline; transcription-repair coupling factor	597, 1797
mioC	Initiation of chromosome replication	1178
mmrA	Postreplication repair	1208, 1813
mrr	Restriction of methylated adenine	1615, 2094
mutA	Mutator, transversion specific	1313

TABLE 2 *Continued*

Category and gene	Gene product description[a]	Reference(s)
mutC	Mutator, transversion specific	1313
mutH	Methyl-directed mismatch repair	83, 1756, 2143
mutL	Enzyme in methyl-directed mismatch repair	127, 666, 2198
mutS	Methyl-directed mismatch repair	83, 330, 1145
mutY	Adenine glycosylase; G·C → T·A transversions	82, 2018
ogt	O^6-Alkylguanine-DNA/cysteine-protein methyltransferase	1207, 1569, 1624
parC	DNA topoisomerase IV subunit A	958, 1523, 1524
parE	DNA topoisomerase IV subunit B	868, 1523, 1524
phrA	Photoreactivation	468, 988, 2218
phrB	Deoxyribodipyrimidine photolyase (photoreactivation)	468, 2131
pinO	Calcium-binding protein required for initiation of chromosome replication	687, 688
polA	DNA polymerase I, 3′ → 5′ polymerase, 5′ → 3′ and 3′ → 5′ exonuclease	264, 1024, 1498
polB	DNA polymerase II	293, 811
priA	Primosomal protein N′ (= factor Y)(putative helicase)	29, 1434, 2255
priB	Primosomal replication protein N	28, 29, 2254
priC	Primosomal replication protein N″	29, 2254
recA	DNA strand exchange and renaturation, DNA-dependent ATPase, DNA- and ATP-dependent coprotease	1033, 1617, 1850
recF	ssDNA and dsDNA binding, ATP binding	321, 1033, 2037
recG	DNA helicase, resolution of Holliday junctions, branch migration	1033, 1172, 2147
recN	Protein in recombination and DNA repair	1033, 1173, 1691
recO	Protein interaction with RecR and possibly RecF proteins	1033, 1713, 1734
recQ	ATP-dependent DNA helicase	1033, 2038, 2039
recT	Recombinase, DNA renaturation	322, 697, 1033
rep	*rep* helicase, ssDNA-dependent ATPase	42, 2178, 2179
rob	Right origin-binding protein	1848
rra	Reverses *recBC, sbcA* alleviation of Mcr (formerly Rgl) restriction of glucosyl-free DNA containing hydroxymethyl- and methylcytosine	942
rus	Suppressor of *ruv* mutants; with *recG* processes Holliday junctions	1226
ruvA	Branch migration of Holliday structures; repair	872, 1058, 2019
ruvB	Branch migration of Holliday structures; repair helicase	1362, 1823, 2019
ruvC	Holliday junction nuclease; resolution of structures; repair	129, 873, 1952
seqA	Negative modulator of initiation of replication	1190
ssb	ssDNA-binding protein	247, 1033
tag	3-Methyladenine DNA glycosylase I, constitutive	156, 920
tdi	Transduction, transformation, and rates of mutation	1890
toc	Suppressor of *topA*	466
topA	DNA topoisomerase type I, omega protein	475, 751, 2280
topB	DNA topoisomerase III	449, 751
tus	DNA-binding protein; inhibition of replication at Ter sites	653, 752, 994
umuC	SOS mutagenesis; error-prone repair; forms complex with UmuD and UmuD′	1008, 1865, 2184
umuD	SOS mutagenesis; error-prone repair; processed to UmuD′; forms complex with UmuC	92, 562, 1008
ung	Uracil-DNA glycosylase	545, 2062
uup	Precise excision of insertion element	787
uvrA	Repair of UV damage to DNA; excision nuclease subunit A	320, 1020, 1267
uvrD	DNA-dependent ATPase I and helicase II	1292, 1342, 2117
xerC	Site-specific recombinase, acts on *cer* sequence of ColE1, effects chromosome segregation at cell division	169, 769, 1274
xerD	Site-specific recombinase	170, 769, 1274
8. Protein translation and modification		
def	Peptide deformylase	1266, 1285
dsbA	Protein disulfide isomerase I	905, 2202, 2253
frr	Ribosome-releasing factor	824, 1825
fusA	Protein chain elongation factor EF-G	791, 901, 1335
glnE	Adenylylating enzyme for glutamine synthetase	36
greA	Transcription elongation factor: cleaves 3′ nucleotide of paused mRNA	192, 193, 1868
hha	Hemolysin expression-modulating protein	1411
iap	Alkaline phosphatase isozyme conversion, aminopeptidase	853, 1396
infA	Protein chain initiation factor IF-1	376, 377
infB	Protein chain initiation factor IF-2	1059, 1347, 1715
infC	Protein chain initiation factor IF-3	1371
map	Methionine aminopeptidase	126
pcm	L-Isoaspartate protein carboxylmethyltransferase type II	640
pmbA	Maturation of antibiotic MccB17	1665
pms	Sulfoxide reductase for peptide methionine	1609
ppiA	Peptidyl-prolyl *cis-trans* isomerase A (a rotamase)	346, 719, 1168
ppiB	Peptidyl-prolyl *cis-trans* isomerase B (a rotamase)	346, 719, 1168
prfA	Peptide chain release factor RF-1	367, 604
prfB	Peptide chain release factor RF-2	367, 1318
prfC	Peptide chain release factor RF-3	663, 1317

(Table continues)

TABLE 2 *Continued*

Category and gene	Gene product description[a]	Reference(s)
prfH	Probable peptide chain release factor	1522
selB	Selenocysteinyl-tRNA-specific translation factor	551, 1653, 1752
slyD	Probable rotamase, peptidyl-prolyl *cis-trans* isomerase	1685, 2201
tsf	Protein chain elongation factor EF-Ts	45, 820
tufA	Protein chain elongation factor EF-Tu (duplicate of *tufB*)	994, 1335, 1541
tufB	Protein chain elongation factor EF-Tu (duplicate of *tufA*)	820, 994, 1973
ups	Efficiency of nonsense suppressors	409

9. RNA synthesis, RNA modification, and DNA transcription

baeR	Transcriptional regulatory protein	1380
baeS	Sensor protein	1380
basR	Transcriptional regulatory protein	1380
basS	Sensor protein for *basR*	1380
dbpA	ATP-dependent RNA helicase	578, 831
deaD	Putative ATP-dependent RNA helicase	2005, 2219
evgA	Putative positive transcription regulator	2043
evgS	Putative sensor for EvgA	2043
greB	Transcription elongation factor and transcript cleavage	193
hepA	Probable RNA helicase	1127, 2248
msrA	Methionine sulfoxide reductase	1608
nusA	Transcription termination; L factor	635, 1158
nusB	Transcription termination; L factor	1252, 1423, 1976
nusG	Component in transcription antitermination	1158, 1930
opr	Rate of degradation of aberrant subunit proteins of RNA polymerase	1804
pcnB	Poly(A) polymerase I	931, 2207
pnp	Polynucleotide phosphorylase	1628, 2222
ranA	RNA metabolism	63
rhlB	Putative ATP-dependent RNA helicase	933, 1450
rhlE	Putative ATP-dependent RNA helicase	1450
rho	Transcription termination factor Rho; polarity suppressor	612, 613, 1795
rpoA	RNA polymerase, alpha subunit	723, 851, 1651
rpoB	RNA polymerase, beta subunit	1074, 1805, 2192
rpoC	RNA polymerase, beta prime subunit	860, 1199, 1532
rpoZ	RNA polymerase, omega subunit	617, 618, 830
spoU	Putative rRNA methylase	1022
srmB	ATP-dependent RNA helicase	1416, 1450
tabC	Possible rho factor	272

10. Polysaccharides (cytoplasmic)

glgA	Glycogen synthase	556, 583, 584
glgB	1,4-α-Glucan branching enzyme	1681
glgC	Glucose-1-phosphate adenylyltransferase	627, 758, 1310
glgP	Glycogen phosphorylase	1759, 1760, 2246
glgS	Glycogen biosynthesis, *rpoS* dependent	737
glgX	Probably part of glycogen operon	1681

11. Phospholipids

cls	Cardiolipin synthase, a major membrane phospholipid	771, 1418, 1824
gpsA	sn-Glycerol-3-phosphate dehydrogenase [NAD(P)$^+$]	484
lgt	Phosphatidylglycerol-prolipoprotein diacylglyceryltransferase; a major membrane phospholipid	1735
pgpA	Nonessential phosphatidylglycerophosphate phosphatase, membrane bound	580, 826
pgpB	Nonessential phosphatidylglycerophosphate phosphatase, membrane bound	580, 825
pgsA	Phosphatidylglycerophosphate synthetase = CDP-1,2-diacyl-sn-glycero-3-phosphate phosphatidyltransferase	1417, 2042
plsB	Glycerolphosphate acyltransferase activity	355, 2159
plsC	1-Acyl-sn-glycerol-3-phosphate acyltransferase	339
plsX	Glycerolphosphate auxotrophy in *plsB* background	1081
psd	Phosphatidylserine decarboxylase; phospholipid synthesis	1136, 1137, 1821
pssA	Phosphatidylserine synthase; phospholipid synthesis	423, 1821
pssR	Regulator of *pssA*	1870

B. Degradation of macromolecules
1. RNA

rna	RNase I, cleaves phosphodiester bond between any two nucleotides	1278, 2272
rnb	RNase II, mRNA degradation	462, 970, 2274
rnc	RNase III, dsRNA	297, 1627, 1887
rnd	RNase D, processes tRNA precursor	970, 2263, 2264
rne	RNase E, enzyme complex for RNA processing, mRNA turnover, maturation of 5S RNA	493, 1886
rnhA	RNase HI, degrades RNA of DNA-RNA hybrids, participates in DNA replication	372, 859, 1444
rnhB	RNAse HII, degrades RNA of DNA-RNA hybrids	858
rnpA	RNase P, protein component; processes tRNA, 4.5S RNA	708, 992, 1937
rnpB	RNase P, RNA component; processes tRNA, 4.5S RNA	444, 1886, 1937

TABLE 2 *Continued*

Category and gene	Gene product description[a]	Reference(s)
rnt	RNase T, degrades tRNA	806, 970, 1486
rph	RNase PH	444, 970
srnA	Degradation of stable RNA	1453
stsA	RNase activity	1109

2. DNA

endA	DNA-specific endonuclease I	1149, 1880
hsdR	Host restriction; endonuclease R	1181, 1580
mcrA	Restriction of DNA at 5-methylcytosine residues; at locus of e14 element	766, 767, 1615
mcrB	Component of McrBC 5-methylcytosine restriction system	1041, 1614, 2270
mcrC	Component of McrBC 5-methylcytosine restriction system	450, 1041, 2269
nfo	Endonuclease IV	709, 1120, 1957
nth	Endonuclease III; specific for apurinic and/or apyrimidinic sites	75
recB	DNA helicase, ATP-dependent dsDNA/ssDNA exonuclease V subunit, ssDNA endonuclease, chi sequence recognition	321, 1033, 1492
recC	DNA helicase, ATP-dependent dsDNA/ssDNA exonuclease V subunit, ssDNA endonuclease, chi sequence recognition	321, 1033, 1492
recD	DNA helicase, ATP-dependent dsDNA/ssDNA exonuclease V subunit, ssDNA endonuclease, chi sequence recognition	321, 1033, 1492
recE	Exonuclease VIII, dsDNA exonuclease, $5' \rightarrow 3'$ specific	315, 321, 1033
recJ	ssDNA exonuclease, $5' \rightarrow 3'$ specific	321, 1033, 1189
sbcB	Exonuclease I, $3' \rightarrow 5'$ specific; deoxyribophosphodiesterase	32, 225, 1536
sbcC	ATP-dependent dsDNA exonuclease	453, 1088, 1397
sbcD	ATP-dependent dsDNA exonuclease	453, 631, 1088
uvrB	DNA repair; excision nuclease subunit B	1470, 1792, 1793
uvrC	Repair of UV damage to DNA; excision nuclease subunit C	1150, 1796, 2059
vsr	DNA mismatch endonuclease, patch repair protein	398, 740, 1863
xseA	Exonuclease VII, large subunit	290
xseB	Exonuclease VII, small subunit	2048, 2049
xthA	Exonuclease III	1722, 1738, 1957

3. Proteins, peptides, and glycopeptides

alpA	Transcriptional regulator of *slpA* gene; a prophage P4-like protein	991, 2008
clpA	ATP-binding subunit of serine protease, alternate subunit determines specificity	650, 651
clpB	Probable alternate ATP-binding subunit of serine protease, determines specificity; heat shock proteins F84.1 and F68.5	1504, 1884, 2180
clpP	ATP-dependent proteolytic subunit of *clpA-clpP* serine protease, heat shock protein F21.5	1262, 1263, 2180
clpX	ATP-binding subunit of *clpP* serine protease, alternate subunit determines specificity	649
dcp	Dipeptidyl carboxypeptidase	443
eco	Ecotin, a serine protease inhibitor	1275, 1276
hflC	Protease specific for phage lambda *c*II repressor	102, 301
hflK	Protease specific for phage lambda *c*II repressor	102, 301
hflX	GTP-binding subunit of protease specific for phage lambda *c*II repressor	1422
hlyA	Hemolysin	972, 1194, 1892
htrA	Periplasmic serine protease Do and heat shock protein	1160, 1801
pepA	Aminopeptidase A/I	1906
pepD	Peptidase D, a dipeptidase where amino-terminal residue is histidine	549, 741, 742
pepE	Peptidase E, a dipeptidase where amino-terminal residue is aspartate	179, 350
pepN	Aminopeptidase N	98, 1269
pepP	Aminopeptidase P II	2242
pepQ	Proline dipeptidase	395, 939
pepT	Putative peptidase T	1183
prc	Carboxy-terminal protease for penicillin-binding protein 3	705
prlC	Oligopeptidase A	351
ptrA	Protease III	101
ptrB	Protease II	939
slpA	Regulator of expression of Alp protease; integrase of P4-like prophage	991
sms	Probable ATP-dependent protease	1404
snoB	Affects degradability of *Rhizobium* NifA in *E. coli*	802
snoC	Increases rate of degradation by *lon* pathway of *Rhizobium* NifA in *E. coli*	802
sohA	Putative protease	93
sohB	Putative protease	93
sppA	Protease IV, a signal peptide peptidase	822, 1940

4. Polysaccharides

amyA	Cytoplasmic alpha-amylase	1605, 1606
malQ	4-α-glucanotransferase (amylomaltase)	1594
malS	α-Amylase	565, 1765

(Table continues)

TABLE 2 *Continued*

Category and gene	Gene product description[a]	Reference(s)
C. Cell envelope		
1. Membranes, lipoproteins, and porins		
acrA	Lipoprotein mutants sensitive to drugs	340, 1204, 1389
acrE	Transmembrane protein; mutants sensitive to drugs	1204
cog	Regulator of *ompG*	1327
envN	Envelope protein; osmotically remedial envelope defect	486
envP	Envelope protein; osmotically remedial envelope defect	486
envQ	Envelope protein; osmotically remedial envelope defect	486
envT	Envelope protein; osmotically remedial envelope defect	486
envY	Envelope protein; thermoregulation of porin biosynthesis	1195
fhuA	Outer membrane protein receptor for ferrichrome, colicin M, and phages T1, T5, and φ80	983, 984, 1009
hlpA	Outer membrane protein	773
micF	Regulatory antisense RNA affecting *ompF* expression	46
nlpA	Lipoprotein-28	2217
nlpB	Lipoprotein-34	203
nlpD	Lipoprotein	823
ompA	Outer membrane protein 3a (II*; G; d)	467, 1004, 1658
ompC	Outer membrane protein 1b (Ib; c)	433, 1649, 2120
ompF	Outer membrane protein 1a (Ia; b; F)	655, 1649, 1798
ompG	Outer membrane porin protein	1327
ompP	Outer membrane protease; receptor for phage OX2	959
ompT	Outer membrane protein 3b (a), a protease	100, 703, 959
phoE	Outer membrane pore protein E (E; Ic; NmpAB), structural gene	112, 424, 886
qmeA	Unspecified membrane defect	2154
qmeC	Unspecified membrane defect; tolerance to glycine; penicillin sensitivity	2154
qmeD	Unspecified membrane defect; tolerance to glycine; penicillin sensitivity	2154
qmeE	Unspecified membrane defect	2154
rlpA	A minor lipoprotein	1954
rlpB	A minor lipoprotein	1954
sipB	Suppressor of outer membrane mutant	1597
sipC	Suppressor of outer membrane mutant	1597
sipD	Suppressor of outer membrane mutant	1597
smpA	Membrane protein	1406
2. Surface polysaccharides, lipopolysaccharides, and antigens		
cpsA	Capsular polysaccharide biosynthesis, colanic acid	2010
cpsC	Capsular polysaccharide biosynthesis, colanic acid	2010
cpsD	Capsular polysaccharide biosynthesis, colanic acid	2010
cpsE	Capsular polysaccharide biosynthesis, colanic acid	2010
cpsF	Capsular polysaccharide biosynthesis, colanic acid	2010
envA	UDP-3-O-acyl N-acetylglucosamine deacetylase; lipid A biosynthesis; splits layers of septum during cell division	456, 1678
firA	UDP-3-O-(R-3-hydroxymyristoyl)-glucosamine N-acyltransferase; third step of endotoxin synthesis	2175
kdsB	CTP:CMP-3-deoxy-D-manno-octulosonate transferase	534
kdtA	3-Deoxy-D-manno-octulosonic acid transferase (KDO transferase)	125, 332, 1684
kdtB	Putative enzyme of lipopolysaccharide synthesis	1684
lpcA	Lipopolysaccharide core biosynthesis; resistance to phages T4, T7, and P1; deficiency in conjugation	718, 1961
lpcB	Lipopolysaccharide core biosynthesis	718, 1961
lpxA	UDP-N-acetylglucosamine acetyltransferase; lipid A biosynthesis	373, 586, 1603
lpxB	Lipid A disaccharide synthetase; lipid A biosynthesis	373, 1603
nanA	N-Acetylneuraminate lyase (aldolase)	12, 13
ops	Level of exopolysaccharide production	2276
rcsA	Positive regulator for *ctr* capsule biosynthesis	652, 1915, 1916
rcsB	Positive response regulator for *ctr* capsule biosynthesis, (sensor, *rcsC*)	625, 652, 1915
rcsC	Negative regulator for *ctr* capsule biosynthesis, probable histidine kinase sensor acting on *rcsB*	222, 652, 1915
rcsF	Regulator in solanic acid synthesis; interacts with RcsB	624
rfaB	UDP-D-galactose:(glucosyl)lipopolysaccharide-1,6-D-galactosyltransferase	1505, 1578, 1684
rfaC	Lipopolysaccharide core biosynthesis; heptosyltransferase I	295, 1684
rfaD	ADP-L-glycero-D-mannoheptose-6-epimerase; permits growth at high temperature	949, 1521, 1610
rfaF	Lipopolysaccharide core biosynthesis	1684
rfaG	Lipopolysaccharide core biosynthesis; glucosyltransferase I	1505, 1506, 1684
rfaH	Transcriptional activator affecting biosynthesis of lipopolysaccharide core, F pilin, and hemolysin	91, 1579, 1684
rfaI	UDP-D-galactose:(glucosyl)lipopolysaccharide-α-1,3-D-galactosyltransferase	1505, 1578, 1684
rfaJ	UDP-D-glucose:(galactosyl)lipopolysaccharide glucosyltransferase	1505, 1578, 1684
rfaK	Lipopolysaccharide core biosynthesis; probably hexose transferase	1001, 1002, 1506
rfaL	Lipopolysaccharide core biosynthesis; O-antigen ligase	1001, 1002, 1684
rfaM	Lipopolysaccharide core biosynthesis; glucosyltransferase II	1684
rfaP	Lipopolysaccharide core biosynthesis; phosphorylation of core heptose; attaches phosphate-containing substrate to lipopolysaccharide core	1505, 1506
rfaQ	Lipopolysaccharide core biosynthesis	1001, 1506, 1684
rfaS	Lipopolysaccharide core biosynthesis	1001, 1003, 1578
rfaY	Lipopolysaccharide core biosynthesis	1002, 1684

TABLE 2 *Continued*

Category and gene	Gene product description[a]	Reference(s)
rfaZ	Lipopolysaccharide core biosynthesis	1001, 1002, 1684
rfbX	Hydrophobic protein involved in assembly of O antigen	1902, 2233
rfe	Synthesis of enterobacterial common antigen (ECA): UDP-GlcNAc:undecaprenylphosphate GlcNAc-1-phosphate transferase	1455, 1648, 1902
rffA	Synthesis of enterobacterial common antigen (ECA): TDP-4-keto-6-deoxy-D-glucose:TDP-D-glucosamine transaminase	1043, 1283, 1284
rffC	Synthesis of enterobacterial common antigen (ECA): ECA chain elongation	1043, 1283, 1284
rffD	Synthesis of enterobacterial common antigen (ECA): UDP-ManNAc dehydrogenase (UDP-*N*-acetyl-D-mannosaminuronic acid dehydrogenase)	1043, 1283, 1284
rffE	Synthesis of enterobacterial common antigen (ECA): UDP-GlcNAc-2-epimerase	1043, 1283, 1284
rffM	Synthesis of enterobacterial common antigen (ECA): UDP-ManNAcA:lipid I transferase	107, 1283, 1284
rffT	Synthesis of enterobacterial common antigen (ECA): TDP-Fuc4NAc:lipid II transferase	1043, 1283, 1284
3. Surface structures		
crl	DNA-binding protein affecting expression of cryptic *csgA* gene for surface fibers	70, 1461
csgA	Curlin subunit, coiled surface structures; cryptic	70, 1461
dsbB	Reoxidizes DsbA protein following formation of disulfide bond in P-ring of flagella	105, 1328
dsbC	Protein disulfide isomerase II	1330, 1820
ecpD	Probable pilin chaperone similar to PapD	1611
fimA	Major type 1 subunit fimbrin (pilin)	180, 999, 1469
fimB	Regulator for *fimA*	1270, 1495, 1783
fimC	Periplasmic chaperone, required for type 1 fimbriae	907, 1982
fimD	Outer membrane protein; export and assembly of type 1 fimbriae	1000
fimE	Regulator for *fimA*	181, 1270, 1495
fimF	Fimbrial morphology	1038, 1708
fimG	Fimbrial morphology	1038, 1708
fimH	Minor fimbrial subunit, D-mannose-specific adhesin	907, 1038, 1982
fimZ	Fimbrial Z protein; probable signal transducer	1372
flgA	Flagellar biosynthesis; assembly of basal-body periplasmic P-ring	834, 1209, 1453
flgB	Flagellar biosynthesis, cell-proximal portion of basal-body rod	834, 1209
flgC	Flagellar biosynthesis, cell-proximal portion of basal-body rod	834, 1209
flgD	Flagellar biosynthesis, initiation of hook assembly	834, 1209
flgE	Flagellar biosynthesis, hook protein	834, 1209
flgF	Flagellar biosynthesis, cell-proximal portion of basal-body rod	834, 1209
flgG	Flagellar biosynthesis, cell-distal portion of basal-body rod	834, 1209
flgH	Flagellar biosynthesis, basal-body outer membrane L (lipopolysaccharide layer)-ring protein	834, 908, 1209
flgJ	Flagellar biosynthesis	834, 1209
flgK	Flagellar biosynthesis, hook-filament junction protein	834, 1209
flgL	Flagellar biosynthesis; hook-filament junction protein	834, 1209
flgM	Anti-FliA (anti-sigma) factor; also known as RflB protein; active only when hook assembly not completed	834, 1209
flhA	Flagellar biosynthesis; export of flagellar proteins?	834, 1057, 1209
flhB	Flagellar biosynthesis	834, 1209
flhC	Regulator of flagellar biosynthesis acting on class 2 operons; transcription initiation factor?	834, 1209
flhD	Regulator of flagellar biosynthesis, acting on class 2 operons; transcriptional initiation factor?	834, 1209
flhE	Flagellar biosynthesis	834, 1209
fliA	Flagellar biosynthesis; regulation of late gene expression (class 3a and 3b operons); sigma factor	834, 961, 1209
fliB	Flagellar biosynthesis; in *Salmonella* spp., methylation of lysine residues on the filament protein, flagellin	834, 961, 1209
fliC	Flagellar biosynthesis; flagellin, filament structural protein	69, 834, 1209
fliD	Flagellar biosynthesis; filament capping protein; enables filament assembly	69, 834, 1209
fliE	Flagellar biosynthesis; basal-body component, possibly at (MS-ring)-rod junction	834, 1209, 1364
fliF	Flagellar biosynthesis; basal-body MS (membrane and supramembrane)-ring and collar protein	834, 961, 1209
fliG	Flagellar biosynthesis, component of motor switching and energizing, enabling rotation and determining its direction	976, 1209, 1675
fliH	Flagellar biosynthesis; export of flagellar proteins?	834, 961, 1209
fliI	Flagellar biosynthesis; export of flagellar proteins?	834, 961, 1209
fliJ	Flagellar biosynthesis	834, 961, 1209
fliK	Flagellar biosynthesis, hook length control	834, 961, 1209
fliL	Flagellar biosynthesis	834, 1209, 1223
fliM	Flagellar biosynthesis, component of motor switch and energizing, enabling rotation and determining its direction	976, 1209, 1676
fliN	Flagellar biosynthesis, component of motor switch and energizing, enabling rotation and determining its direction	976, 1209, 1676
fliO	Flagellar biosynthesis	961, 1209, 1222
fliP	Flagellar biosynthesis	961, 1209, 1222
fliQ	Flagellar biosynthesis	961, 1209, 1222
fliR	Flagellar biosynthesis	961, 1209, 1222
fliS	Flagellar biosynthesis; repressor of class 3a and 3b operons (RflA activity)	834, 961, 1209
fliT	Flagellar biosynthesis; repressor of class 3a and 3b operons (RflA activity)	834, 961, 1209
fliU	Involved in secretion of flagellin and motility	455
fliV	Involved in secretion of flagellin and motility	455
flu	Metastable gene affecting surface properties, piliation, and colonial morphology	447

(Table continues)

TABLE 2 *Continued*

Category and gene	Gene product description[a]	Reference(s)
mor	Regulator of switching between two sets of surface properties	2114

4. Murein sacculus and peptidoglycan

amiA	N-Acetylmuramoyl-1-alanine amidase I, septum separation	2000, 2056
amiB	N-Acetylmuramoyl-1-alanine amidase II; a murein hydrolase	2024
bolA	Possible regulator of murein genes	22
dacA	D-Alanyl-D-alanine carboxypeptidase, fraction A; penicillin-binding protein 5	118, 879, 2053
dacB	D-Alanyl-D-alanine carboxypeptidase, fraction B; penicillin-binding protein 4	1023, 1352, 1353
dacC	D-Alanyl-D-alanine carboxypeptidase; penicillin-binding protein 6	229
ddlA	D-Alanine-D-alanine ligase A	21, 2256
ddlB	D-Alanine-D-alanine ligase B	2256
hipA	Frequency of persistence to inhibition of murein or DNA biosynthesis, DNA-binding regulator	158, 159
hipB	Frequency of persistence to inhibition of murein or DNA biosynthesis; regulatory protein interacts with HipA	158
lpp	Murein lipoprotein	847, 1355
mepA	Murein DD-endopeptidase	968
mlt	Membrane-bound lytic murein transglycosylase	503, 2041
mraA	D-Alanine carboxypeptidase	1333
mraB	D-Alanine requirement; cell wall peptidoglycan biosynthesis	1333
mraY	Phospho-N-acetylmuramoyl-pentapeptide transferase?	836
mrbA	UDP-N-acetylglucosaminyl-3-enolpyruvate reductase activity	1333
mrbB	Cell wall peptidylglycan biosynthesis; mutation causes D-alanine auxotrophy	1333
mrbC	Cell wall peptidylglycan biosynthesis	1333
mrcA	Peptidoglycan synthetase; cell wall biosynthesis; penicillin-binding protein 1A	228, 2245
mrcB	Peptidoglycan synthetase; cell wall biosynthesis; penicillin-binding protein 1B	228, 2245
mrdA	Cell shape; peptidoglycan synthetase; penicillin-binding protein 2	78, 852, 1955
mrdB	Rod shape-determining protein; sensitivity to radiation and drugs	118, 835, 1258
mreB	Rod shape-determining protein	454
mreC	Rod shape-determining protein	2087
mreD	Rod shape-determining protein	2087
murB	UDP-N-acetylenolpyruvoylglucosamine reductase	130, 1592
murC	L-Alanine-adding enzyme, UDP-N-acetyl-muramate-alanine ligase	638, 1295
murD	UDP-N-acetylmuramoylalanine-D-glutamate ligase	638, 1294, 1295
murE	*meso*-Diaminopimelate-adding enzyme	638, 1294, 1295
murF	D-Alanine:D-alanine-adding enzyme	482, 638, 1295
murG	Transferase in peptidoglycan synthesis	638, 1295
murH	Peptidoglycan biosynthesis, late stage	384, 638
murI	Glutamate racemase, required for biosynthesis of D-glutamate and peptidoglycan	470, 1593, 2243
murZ	First step in murein biosynthesis; UDP-N-glucosamine 1-carboxyvinyltransferase	638, 1244
pal	Peptidoglycan-associated lipoprotein	296, 1087
slt	Soluble lytic murein transglycosylase	502, 503, 2041

IV. CELL PROCESSES
A. Transport/binding proteins
1. Amino acids and amines

abpS	Low-affinity transport, arginine and ornithine; periplasmic binding protein	278, 279
argP	Transport of arginine, ornithine, and lysine	281
argT	Lysine-, arginine-, ornithine-binding protein	1427
aroP	General aromatic amino acid transport	784
aroT	Transport of aromatic amino acids, alanine and glycine	1988
artI	Arginine periplasmic transport system protein	2173
artJ	Arginine-binding protein	2173
artM	Arginine periplasmic transport system protein	2173
artP	Arginine periplasmic transport system protein	2173
artQ	Arginine periplasmic transport system protein	2173
brnQ	Transport system 1 for isoleucine, leucine, and valine	2221
brnR	Component of transport systems 1 and 2 for isoleucine, leucine, and valine	2221
brnS	Transport system for isoleucine, leucine, and valine	675
brnT	Low-affinity transport system for isoleucine	675
cadB	Transport of lysine/cadaverine	1293, 2119
cycA	Transport of D-alanine, D-serine, and glycine	360, 362, 1661
glnH	Periplasmic glutamine-binding protein	1426
glnP	Glutamine high-affinity transport system; membrane component	1426
glnQ	Glutamine high-affinity transport system	1426
gltP	Glutamate-aspartate symport protein	427, 1999
gltR	Regulator for *gltS*	1239
gltS	Glutamate transport	471, 932
hisJ	Histidine-binding protein of high-affinity histidine transport system	64
hisM	Histidine transport, membrane protein M	1034
hisP	Histidine transport, inner membrane receptor protein P	1034, 1324
livF	Leucine transport protein	7
livG	High-affinity branched-chain amino acid transport system	7

TABLE 2 *Continued*

Category and gene	Gene product description[a]	Reference(s)
livH	High-affinity branched-chain amino acid transport system; membrane component	7
livJ	High-affinity amino acid transport system; periplasmic binding protein	7
livK	High-affinity leucine-specific transport system; periplasmic binding protein	7
livL	High-affinity branched-chain amino acid transport	7
livM	High-affinity branched-chain amino acid transport	7
lysP	Lysine-specific permease; pleiotropic increase in lysine decarboxylase	1400, 1896
lysX	Lysine excretion	893
metD	High-affinity uptake of D- and L-methionine	924, 925
mtr	Tryptophan-specific transport protein	727, 1739, 1741
nagE	Phosphotransferase system enzyme II, specific for *N*-acetylglucosamine	1528, 1671, 2080
pheP	Phenylalanine-specific transport system	1539, 1540
potA	Spermidine/putrescine transport protein	581, 953
potB	Spermidine/putrescine transport protein	581
potC	Spermidine/putrescine transport protein	581
potD	Spermidine/putrescine transport protein	581, 953
potE	Putrescine transport protein	954
potF	Periplasmic putrescine-binding protein	1546
potG	Putrescine transport protein	1546
potH	Putrescine transport protein	1546
potI	Putrescine transport protein	1546
proP	Low-affinity transport system; proline permease II	276, 375
proT	Proline transport	1350, 1351
proV	High-affinity transport system; glycine betaine-binding protein	276, 407, 1908
proW	High-affinity transport system for glycine betaine and proline	276, 407, 1908
sdaC	Probable serine transporter	1810
tdcC	Anaerobically inducible L-threonine, L-serine permease	648, 1787, 1931
tnaB	Low-affinity tryptophan permease	1740, 2232
tyrP	Tyrosine-specific transport system	2187
tyrR	Regulation of *aroF, aroG,* and *tyrA* and aromatic amino acid transport systems	66, 374, 1547
bfr	Bacterioferritin	59, 60
2. Cations		
calA	Calcium transport	219
calC	Calcium transport	219
calD	Calcium transport	219
chaA	Sodium-calcium/proton antiporter	870, 1456, 1484
cirA	Iron-regulated colicin I receptor; porin; requires *tonB* gene product	123, 665
corA	Mg^{2+} transport, system I	632, 1502, 1858
corB	Mg^{2+} transport, system I	632, 1502, 1858
cutE	Copper homeostasis protein	1673
exbB	Uptake of enterochelin; *tonB*-dependent uptake of B colicins	213, 215, 937
exbC	Uptake of enterochelin; sensitivity or resistance to colicins	1596
exbD	Uptake of enterochelin; *tonB*-dependent uptake of B colicins	215, 494, 1567
fecA	Citrate-dependent iron transport, outer membrane receptor	1029, 1893, 2060
fecB	Citrate-dependent iron transport, periplasmic protein	1029, 1893, 2060
fecC	Citrate-dependent iron(III) transport protein, cytosolic	1029, 1893, 2060
fecD	Citrate-dependent iron transport, membrane-bound protein	1029, 1893, 2060
fecE	Citrate-dependent iron(III) transport protein, membrane bound	1029, 1893, 2060
fecI	Regulator for *fec* operon, membrane location	2060
fecR	Regulator for *fec* operon, periplasmic	2060
feoA	Ferrous iron transport protein A	936
feoB	Ferrous iron transport protein B	936
fepA	Receptor for ferric enterobactin (enterochelin) and colicins B and D	1167, 1377, 2171
fepB	Ferric enterobactin (enterochelin) uptake; periplasmic component	498, 1482
fepC	Ferric enterobactin (enterochelin) uptake; cytoplasmic membrane component	1482, 1816
fepD	Ferric enterobactin (enterochelin) uptake	1482, 1816
fepE	Ferric enterobactin (enterochelin) uptake	1482
fepG	Ferric enterobactin transport protein	1816
fes	Enterochelin esterase	220, 2171
fhuB	Hydroxamate-dependent iron uptake, cytoplasmic membrane component	1028, 1030, 1779
fhuC	Hydroxymate-dependent iron uptake, cytoplasmic membrane component	1028, 1030, 1779
fhuD	Hydroxamate-dependent iron uptake, cytoplasmic membrane component	1028, 1030, 1031
fhuE	Outer membrane receptor for ferric iron uptake	1749
fhuF	Ferric hydroxymate transport	704
fiu	Ferric iron uptake, outer membrane protein	380
ftn	Ferritin	808
kch	Putative potassium channel protein	1320
kdpA	High-affinity potassium transport system; probable K^+-stimulated ATPase	39, 1557
kdpB	High-affinity potassium transport system	39, 1557

(Table continues)

TABLE 2 *Continued*

Category and gene	Gene product description[a]	Reference(s)
kdpC	High-affinity potassium transport system	39, 1557
kdpF	Inner membrane protein in potassium transport	39
kefB	K⁺ efflux; NEM-activable K⁺/H⁺ antiporter	500
kefC	K⁺ efflux; NEM-activable K⁺/H⁺ antiporter	500, 1369
kup	Low-affinity potassium transport system	196, 1761
mgt	Mg^{2+} transport, system II	590
molR	Regulation of *mod* locus governing molybdate transport	765, 1094
nhaA	Na⁺/H antiporter, pH dependent	1544, 1545, 1950
nhaB	Na⁺/H⁺ antiporter, pH independent	1544, 1545, 1984
nhaR	Activator of *nhaA*	1484, 1607
nikA	Periplasmic binding protein for nickel	1398, 2197
nikB	Transport of nickel, membrane protein	1398, 2197
nikC	Transport of nickel, membrane protein	1398
nikD	Transport of nickel, ATP-binding protein	1398
nikE	Transport of nickel, ATP-binding protein	1398
panF	Sodium/pantothenate symporter	875, 1635
putP	Major sodium/proline symporter	375, 701, 1635
rsgA	Ferritin-like protein	808, 874
tonB	Energy transducer; uptake of iron and cyanocobalamin; sensitivity to phages, colicins	214, 887, 1847
trkA	Transport of potassium	195, 469, 1763
trkD	Kup protein, transport of potassium, cesium	196, 1761
trkE	Transport of potassium	469
trkG	Potassium uptake	469
trkH	Potassium uptake	469
trpP	Low-affinity tryptophan-specific permease	485
3. Carbohydrates, organic acids, and alcohols		
alu	5-Aminolevulinate uptake	2066
araE	Low-affinity L-arabinose transport system; L-arabinose proton symport	227, 736, 1210
araF	L-arabinose-binding protein	789, 2068
araG	High-affinity L-arabinose transport system	789
araH	High-affinity L-arabinose transport system; membrane protein	789
arbT	Phosphorylation and transport of arbutin; cryptic	1507
ascF	Phosphotransferase enzyme II (*asc*), cryptic, transports specific β-glucosides	696
bglF	β-Glucoside transport, PEP-dependent enzyme II, part of cryptic operon	209, 1768, 1769
bymA	Bypass of maltose permease at *malB*	777
cbt	Dicarboxylate-binding protein	1176
celA	PEP-dependent phosphotransferase enzyme IV for cellobiose, arbutin, and salicin	1508, 1636
celB	PEP-dependent phosphotransferase enzyme II for cellobiose, arbutin, and salicin	1508, 1636
celC	PEP-dependent phosphotransferase enzyme III for cellobiose, arbutin, and salicin	695, 1636
citA	Cryptic gene of citrate transport system	693
citB	Cryptic gene of citrate transport system	693
cmtA	PEP-dependent phosphotransferase enzyme IIBC for mannitol	1878
cmtB	PEP-dependent phosphotransferase enzyme IIA for mannitol	1878
crr	Glucose phosphotransferase enzyme III^Glc	438, 695, 1716
cup	Uptake of carbohydrates	1217
dctA	Uptake of C₄-dicarboxylic acids	1175
dctB	Uptake of C₄-dicarboxylic acids	1175
dcuA	Anaerobic dicarboxylic acid uptake	1845, 1846
dcuB	Anaerobic dicarboxylate transport	1845, 1846
dgoT	Galactonate transport	356
dgsA	Enzyme IIA/IIB of phosphotransferase system	1666
exuT	Transport of hexuronates	1407
fadL	Transport of long-chain fatty acids; sensitivity to phage T2	162, 163, 1720
fruA	Fructose phosphotransferase enzyme II	1467, 1585
fruB	Fructose phosphotransferase enzyme III	608
fruF	Phosphohistidinoprotein-hexose phosphotransferase, fructose specific	608, 673, 1467
frvA	Fructose-like phosphotransferase enzyme IIA; cryptic	1634
frvB	Fructose-like phosphotransferase enzyme IIBC; cryptic	1634
frvR	Putative *frv* operon regulatory protein	1634
fucP	Fucose permease	298, 299
gabP	Transport of γ-aminobutyrate	1410
galP	Galactose permease	1655
gatA	Galactitol-specific enzyme IIA of phosphotransferase system	1106–1108
gatB	Galactitol-specific enzyme IIB of phosphotransferase system	435, 1108
gatC	Galactitol-specific enzyme IIC of phosphotransferase system	1106, 1108
glpF	Facilitated diffusion of glycerol	1942, 2136
glpT	*sn*-Glycerol-3-phosphate permease	495, 1080
glvB	Arbutin-like phosphotransferase enzyme type IIB	1634
glvC	Arbutin-like phosphotransferase enzyme type IIC	1634
glvG	Probable 6-phospho-β-glucosidase	1634

TABLE 2 *Continued*

Category and gene	Gene product description[a]	Reference(s)
gntS	Second system for transport and possible phosphorylation of gluconate	87, 336
gntT	High-affinity transport of gluconate	519, 1382
gutA	D-Glucitol (sorbitol)-specific enzyme II of phosphotransferase system	2214
gutB	D-Glucitol (sorbitol)-specific enzyme II of phosphotransferase system	695, 2214
kdgT	2-Keto-3-deoxy-D-gluconate transport system	1228
kgtP	α-Ketoglutarate permease	1802
lacY	Galactoside permease (M protein)	149, 227, 922
lamB	Phage lambda receptor protein; maltose high-affinity uptake system	566, 1193
lctP	L-Lactate permease	458, 459
malE	Periplasmic maltose-binding protein; substrate recognition for transport and chemotaxis	410, 417, 788
malF	Maltose transport; cytoplasmic membrane protein	403, 410, 417
malG	Active transport of maltose and maltodextrins	403, 410, 417
malK	Maltose permeation	418, 1630, 1639
malX	Phosphotransferase enzyme II, maltose and glucose specific	1629
manX	Mannose phosphotransferase system, protein II-A (III)	514, 515, 1913
manY	Mannose phosphotransferase system: Pel protein II-P; penetration of phage lambda	514, 515, 1913
manZ	Mannose phosphotransferase system, enzyme IIB (IIM)	514, 515, 1913
melB	Melibiose permease II	1575, 1667, 2252
mglA	Methylgalactoside transport and galactose taxis, cytoplasmic membrane protein	778
mglB	Galactose-binding protein; receptor for galactose taxis	778, 1772
mglC	Methylgalactoside transport and galactose taxis	778
mglD	Regulator for methylgalactoside transport	1660
mglR	*mgl* regulator	590
mtlA	Mannitol-specific enzyme II of phosphotransferase system	1566, 1878, 1928
ptsG	Glucosephosphotransferase enzyme II	246, 1700, 1711
ptsH	Phosphohistidinoprotein-hexose phosphotransferase, HPr	246, 438, 1700
ptsI	PEP-protein phosphotransferase system enzyme I	438, 700, 1700
rbsA	D-Ribose high-affinity transport system; membrane-associated protein	244
rbsB	D-Ribose periplasmic binding protein	154, 669
rbsC	D-Ribose high-affinity transport system; membrane-associated protein	122
rbsD	D-Ribose high-affinity transport system; membrane-associated protein	122
rhaT	Rhamnose transport	96, 1975
shiA	Shikimate and dehydroshikimate permease	1548
treB	PEP:CHO phosphotransferase system enzyme II, trehalose specific	997, 1919
ugpA	*sn*-Glycerol 3-phosphate transport system, integral membrane protein	241, 1479, 1926
ugpB	*sn*-Glycerol 3-phosphate transport system; periplasmic binding protein	241, 1479
ugpC	*sn*-Glycerol 3-phosphate transport system, permease	241, 1479
ugpE	*sn*-Glycerol 3-phosphate transport system, integral membrane protein	241, 1479
uhpA	Response regulator, positive activator of *uhpT* transcription (sensor, *uhpB*)	855
uhpB	Regulator of *uhpT*, sensor for histidine protein kinase	854, 855
uhpC	Regulator of *uhpT*	854, 855
uhpR	Regulation of hexose phosphate transport; receptor for glucose 6-phosphate	462, 926
uhpT	Hexose phosphate transport protein	43
xylE	Xylose-proton symport	227, 414
xylF	Xylose binding protein transport system	8, 736, 1861
xylG	Putative xylose transport, ATP-binding protein	1554, 1861
xylH	Putative xylose transport, membrane component	1861
xylU	D-Xylose uptake protein	736, 1054
4. Nucleosides, purines, and pyrimidines		
codB	Cytosine transport	396
nupC	Transport of nucleosides, except guanosine	1367
nupG	Transport of nucleosides	1366, 2146
pnuC	Membrane protein required for NMN transport	554
purP	High-affinity adenine transport	254
tsx	Nucleoside channel; receptor of phage T6 and colicin K	217, 575, 622
uraA	Uracil transport, ABC transporter	94
5. Anions		
cysA	Sulfate permease A protein; chromate resistance	1706, 1842
cysP	Thiosulfate-binding protein	799
cysT	Sulfate, thiosulfate transport system	1842
cysW	Sulfate permease W protein	1842
cysZ	Required for sulfate transport	224
modA	Molybdate uptake	735, 765, 1657
modB	Molybdate uptake	765, 1657, 1789
modC	Molybdate uptake	900, 1657, 1789
narK	Transport of nitrate	436
pit	Low-affinity phosphate transport	501

(*Table continues*)

TABLE 2 *Continued*

Category and gene	Gene product description[a]	Reference(s)
pstA	High-affinity phosphate-specific transport system	1618
pstB	High-affinity phosphate-specific transport system, cytoplasmic membrane protein?	1618
pstC	High-affinity phosphate-specific transport system, cytoplasmic membrane component	1618
pstS	High-affinity phosphate-specific transport system; periplasmic phosphate-binding protein	1221
sbp	Periplasmic sulfate-binding protein	880

6. Other
abc	ABC transporter	34
abs	Sensitivity and permeability to antibiotics and dyes	324
betT	High-affinity choline transport	53, 1070
bioP	Biotin transport	262, 1543
btuB	Receptor for transport of vitamin B_{12}, E colicins, and bacteriophage BF23	123, 1032, 1197
btuC	Vitamin B_{12} transport	1656
btuD	Vitamin B_{12} transport, membrane-associated protein	1656
btuE	Vitamin B_{12} transport	1656
cydC	Putative transport component of cytochrome *d* terminal oxidase	115, 757, 1561
cydD	Putative transport component of cytochrome *d* terminal oxidase, Zn sensitive	115, 757, 1561
htrE	Probable porin protein similar to PapC	1611
mdl	ATP-binding transport protein	34
modD	Molybdate uptake	735, 765, 1789
msbA	Probable ATP-binding transport protein; multicopy suppressor of *htrB*	34, 948
phnC	Binding protein-dependent alkylphosphonate transporter, permease component	1305, 1306, 2112
phnD	Binding protein-dependent alkylphosphonate transporter, periplasmic component	1305, 1306, 2112
phnE	Binding protein-dependent alkylphosphonate transporter, integral membrane component, cryptic in K-12	1305, 1306, 2112

B. Chaperones
cbpA	Curved DNA-binding protein; functions closely related to DnaJ	2032
dnaJ	Chaperone with DnaK; heat shock protein	588, 1076, 1774
dnaK	Chaperone Hsp70; DNA biosynthesis; autoregulated heat shock proteins	319, 1076, 1774
hscA	Member of Hsp70 protein family	966
htpG	Chaperone Hsp90, heat shock protein C62.5	732, 1871
mopA	GroEL, chaperone Hsp60, peptide-dependent ATPase, heat shock protein	243, 620, 1644
mopB	Chaperone affecting head assembly of phages T4 and lambda	1079

C. Cell division
cafA	Bundles of cytoplasmic filaments	1457
cfcA	Frequency of cell division	1419
dicA	Regulator of *dicB*	120
dicB	Inhibition of cell division	120, 261, 1361
dicC	Regulator of *dicB*	120
fcsA	Cell division; septation	1042
fic	Induced in stationary phase, recognized by *rpoS,* affects cell division	1014, 2044
ftsA	Cell division protein, complexes with FtsZ	385, 456
ftsE	Cell division membrane protein	628, 633
ftsH	Inner membrane protein essential for cell division, putative ATPase, chaperone	18, 746, 2003
ftsI	Septum formation; penicillin-binding protein 3; peptidoglycan synthetase	118, 119, 639
ftsJ	Cell division protein	1448, 2003
ftsL	Cell division protein; ingrowth of wall at septum	456, 689
ftsN	Essential cell division protein	388
ftsQ	Cell division protein; ingrowth of wall at septum	385, 2073
ftsW	Cell division; membrane protein involved in shape determination	835
ftsX	Cell division membrane protein	628, 633, 634
ftsY	Cell division membrane protein	628, 633, 634
ftsZ	Cell division; forms circumferential ring; GTP-binding protein and GTPase	1202, 1356, 2086
mbrA	Coupling of cell division and DNA replication	2014, 2015
mbrB	Link between growth rate and partitioning chromosomes	2014, 2015
mbrC	Partitioning chromosomes	2014, 2015
minB	Formation of minute cells containing no DNA; complex locus, position of division septum	1360
minC	Cell division inhibitor, inhibits *ftsZ* ring formation	146, 421, 422
minD	Cell division inhibitor, a membrane ATPase, activates *minC*	420–422
minE	Cell division topological specificity factor, reverses *min* inhibition	421, 422, 456
mukB	Cell division protein involved in chromosome partitioning	768, 770, 1412
mukC	Cell division and chromosome partitioning	770
mukD	Cell division and chromosome partitioning	770
pcsA	Cell division; chromosome segregation	1042
sdiA	Regulator of transcription of *ftsQAZ* gene cluster	2110
sefA	Septum formation	1430
sfiC	Cell division inhibition; at locus of element e14	881, 1214, 1215
sulA	Suppressor of *lon;* inhibits cell division and *ftsZ* ring formation	146, 441
tig	Trigger factor; a molecular chaperone involved in cell division	684
tolC	Outer membrane channel; specific tolerance to colicin E1; segregation of daughter chromosomes	133, 456, 770

TABLE 2 *Continued*

Category and gene	Gene product description[a]	Reference(s)
weeA	Cell elongation	428, 429
D. Chemotaxis and mobility		
cheA	Chemotaxis protein	610, 1346, 1941
cheB	Response regulator for chemotaxis (*cheA* sensor); protein methylesterase	200, 1200, 1903
cheR	Response regulator for chemotaxis; protein glutamate methyltransferase	1707
cheW	Positive regulator of CheA protein activity	610, 1169, 1277
cheY	Chemotaxis protein transmits chemoreceptor signals to flagellar motors	201, 537, 1346
cheZ	Chemotactic response; CheY protein phophatase; antagonist of CheY as switch regulator	178, 803, 1909
motA	Proton conductor component of motor; no effect on switching	166, 167, 1912
motB	Enables flagellar motor rotation, linking torque machinery to cell wall; no effect on switching	166, 168, 1912
tap	Methyl-accepting chemotaxis protein IV, peptide receptor	1230
tar	Methyl-accepting chemotaxis protein II, chemoreceptor for aspartate	594, 1095
trg	Methyl-accepting chemotaxis protein III, ribose receptor	252, 2209, 2210
tsr	Methyl-accepting chemotaxis protein I, serine receptor	606, 1169
E. Protein and peptide secretion		
dppA	Dipeptide transport protein	5, 1463
excD	Export of periplasmic proteins	1086
expA	Expression of a group of export proteins	402
ffh	Protein transport	1537
hlyB	ABC protein translocator, exports hemolysin	619, 1025, 1027
hlyD	With HlyB, protein translocator for hemolysin	1026, 1892, 2262
lepB	Leader peptidase (signal peptidase I)	151, 161, 1236
lspA	Prolipoprotein signal peptidase (SPaseII)	1368
msyB	Acidic protein suppresses mutants lacking function of protein export	2031
oppA	Oligopeptide transport; periplasmic binding protein	54, 952, 955
oppB	Oligopeptide transport	55, 1501
oppC	Oligopeptide transport	55
oppD	Oligopeptide transport	55
oppE	Oligopeptide transport	55
oppF	Oligopeptide transport, ATP hydrolysis	55
prlF	Protein export	979, 1859
sapF	Peptide transport, ABC family of transporters	94
secA	Protein secretion, ATP hydrolysis	218, 318, 1829
secB	Protein export; molecular chaperone	341, 713, 1045
secD	Protein secretion; membrane protein	1257, 1555, 1927
secE	Inner membrane protein, protein secretion (with *secY*)	237, 1420, 1927
secF	Membrane protein, protein secretion function	1257, 1717, 1927
secG	Protein export; membrane protein	1421
secY	Membrane protein, protein secretion (with *secE*)	237, 1420, 1943
ssaE	Suppression of *secA* mutation	1460
ssaG	Suppression of *secA* mutation	1460
ssaH	Suppression of *secA* mutation	1460
ssyA	Suppression of *secY* mutation	1460
ssyB	Suppression of *secY* mutation	1460
ssyD	Suppression of *secY* mutation	1460
F. Osmotic adaptation		
betA	Choline dehydrogenase, a flavoprotein	53, 1070
betB	NAD^+-dependent betaine aldehyde dehydrogenase	53, 521, 1070
betI	Probably repressor of *bet* genes	53, 1070
mdoA	Membrane-derived oligosaccharides; component of glucosyltransferase system	611, 1063
mdoB	Membrane-derived oligosaccharides; phosphoglycerol transferase I activity	611, 1063
mdoG	Periplasmic membrane-derived oligosaccharide synthesis	1063
mdoH	Membrane glycosyltransferase, membrane-derived oligosaccharide synthesis	1063
osmB	Osmotically inducible lipoprotein	738
osmC	Osmotically inducible protein	686
osmY	Hyperosmotically inducible periplasmic protein	2236
otsA	Trehalose-6-phosphate synthase	738, 921, 1919
otsB	Trehalose-6-phosphate phosphatase, biosynthetic	738, 921, 1919
otsR	Regulation of *ots*	997
proX	High-affinity transport system for glycine betaine and proline	276, 407, 1908
treA	Trehalase, periplasmic	685, 738, 1919
treR	Repressor of *treABC*	94
G. Detoxification		
ahpC	Alkyl hydroperoxide reductase, C22 subunit; detoxification of hydroperoxides	1914
ahpF	Alkyl hydroperoxide reductase, F52a subunit; detoxification of hydroperoxides	1914
katC	Regulation of catalase activity	2083

(Table continues)

TABLE 2 *Continued*

Category and gene	Gene product description[a]	Reference(s)
katE	Catalase hydroperoxidase HPII(III)	6, 416, 2083
katG	Catalase-peroxidase hydroperoxidase HPI(I)	6, 1182, 1359
sodA	Superoxide dismutase, manganese	141, 786, 1586
sodB	Superoxide dismutase, iron	142, 530, 1586
thdC	Detoxification of furans and thiophenes	20, 917
thdD	Detoxification of furans and thiophenes	20, 917
thdF	GTP-binding protein in thiophene and furan oxidation	19, 249
H. Cell killing		
cma	Colicin M	711
gef	Polypeptide destructive to membrane potential	1570–1572
gefL	Leader peptide of Gef	1572
kicA	Killing protein	533
kicB	Suppressor of killing protein	533
relF	Polypeptide destructive to membrane potential	430, 621, 1570
V. OTHER		
A. Phage-related functions and prophage		
bfm	Phage BF23 multiplication	1830
dicF	RNA of 65 nucleotides, cell division inhibitor of Kim prophage	528, 1981
esp	Site for efficient packaging of phage T1	472
fipB	Morphogenesis of phage F1	1186
fipC	Morphogenesis of phage F1	1186
gprA	Replication of certain lambdoid phages	1443, 1730
gprB	Replication of certain lambdoid phages	1443, 1730
grpE	Phage lambda replication; host DNA synthesis; heat shock protein; protein repair	588, 1076, 1774
hfq	Host factor I for bacteriophage Qβ replication, a growth-related protein	929, 930
lit	Phage T4 late gene expression; at locus of e14 element	756, 944
msp	Sensitivity or resistance of male strains to male-specific phages R17 and f2	255
mul	Mutability of UV-irradiated phage lambda	2089
nfrA	Bacteriophage N4 receptor, outer membrane protein	978, 980, 981
nfrB	Bacteriophage N4 receptor, outer membrane protein	978, 980, 981
nfrC	Bacteriophage N4 adsorption protein, cytoplasmic	978
nfrD	Bacteriophage N4 adsorption protein	978, 980
nmpC	Outer membrane porin protein; locus of qsr prophage	174
ogr	Regulator of late transcription in phage P2; part of cryptic P2 prophage	1100, 1854
phxB	Adsorption of φX174	1733
pin	Inversion of adjacent DNA; at locus of e14 element	1056
qin	Cryptic lambdoid phage	1144
qsr	Defective prophage qsr′	1157
racC	Defective prophage rac; contains recE and oriJ	1144
rap	Growth of phage lambda	1525
tnm	Transposition of Tn9 and other transposons; development of phage Mu	840, 1174
B. Colicin-related functions		
cet	Tolerance to colicin E2	477
cvpA	Required for colicin V production	527
tolA	Membrane-spanning protein, required for outer membrane integrity	128, 1115, 1117
tolB	Tolerance to colicins E2, E3, A, and K; leakage of periplasmic proteins	1116, 2125
tolD	Tolerance to colicins E2 and E3; ampicillin resistance	512
tolE	Tolerance to colicins E2 and E3; ampicillin resistance	512
tolI	Tolerance to colicins Ia and Ib	265
tolJ	Resistance to colicins L, A, and S4; partial resistance to colicins E and K	411
tolM	Mutant phenotype: high-level tolerance to colicin M	710
tolQ	Inner membrane protein, membrane spanning, maintains integrity of cell envelope; tolerance to group A colicins	213, 938, 2072
tolR	Inner membrane protein, maintains integrity of cell envelope; tolerance to group A colicins	494, 938, 1363
tolZ	Tolerance to colicins E2, E3, D, 1a, and 1b; generation of chemical proton gradient	1259, 2125
C. Plasmid-related functions		
chpAI	Suppressor of growth inhibitor ChpAK	1254
chpAK	Growth inhibitor	1254
chpBI	Suppressor of growth inhibitor ChpBK	1254
chpBK	Growth inhibitor	1254
mafA	Maintenance of F-like plasmids	2091
mafB	Maintenance of F-like plasmids	2091
mprA	Regulator of plasmid mcrB operon (microcin B17 synthesis)	430, 431
D. Drug/analog sensitivity		
acrB	Sensitivity to acriflavine	1204
acrC	Sensitivity to acriflavine	1388
ampC	β-lactamase; penicillin resistance	561

TABLE 2 *Continued*

Category and gene	Gene product description[a]	Reference(s)
ampD	Regulates *ampC*	1155
ampE	Regulates *ampC*	1155
ampG	Regulates β-lactamase synthesis	1156
azaA	Resistance or sensitivity to azaserine	2163
azaB	Resistance or sensitivity to azaserine	2163
bacA	Bacitracin resistance; possibly phosphorylates undecaprenol	259
bcr	Bicyclomycin resistance protein; transmembrane protein	131
can	Canavanine resistance	280
cmlA	Resistance or sensitivity to chloramphenicol	113
dvl	Sensitivity to sodium dodecyl sulfate and toluidine blue plus light	2095
emrA	Multidrug resistance efflux pump	582, 1126
emrB	Multidrug resistance efflux pump	582, 1126
emrE	Membrane protein, methyl viologen resistance	1343
eryD	Erythromycin growth dependence	2156
inm	Susceptibility to mutagenesis by nitrosoguanidine	1701
ksgA	S-Adenosylmethionine-6-N',N'-adenosyl (rRNA) dimethyltransferase; kasugamycin resistance	2058
ksgB	Second-step (high-level) resistance to kasugamycin	555
ksgC	Kasugamycin resistance; affects ribosomal protein S2	2240
ksgD	Kasugamycin resistance	555
lev	Resistance to levallorphan	391
linB	High-level resistance to lincomycin	815
lytA	Tolerance to β-lactams; autolysis defective?	712, 1827
marA	Multiple antibiotic resistance; transcriptional activator of defense systems	67, 690
marB	Multiple antibiotic resistance protein	67, 337, 587
marR	Multiple antibiotic resistance protein; repressor of *mar* operon	67, 337
mng	Resistance or sensitivity to manganese	1838
nalB	Resistance or sensitivity to nalidixic acid	702, 796, 797
nalD	Penetration of nalidixic acid through outer membrane	798
neaB	Resistance to neamine	434
nek	Resistance to neomycin, kanamycin, and other aminoglycoside antibiotics	812
nfnA	Sensitivity to nitrofurantoin	1745, 1799
nfnB	Sensitivity to nitrofurantoin	1745, 1799
nfsA	Nitrofuran reductase I activity	1268, 1345
nfsB	Nitrofuran reductase I activity	1268, 1345
nov	Sensitivity to novobiocin	1613
psu	Pleiotropic suppressor; resistance to oxolinic acid	508
sbmA	Sensitivity to microcin B17, possibly envelope protein	1084, 2237
semA	Sensitivity to microcin E492	1595
sloB	Low growth rate; tolerance to amidinopenicillin and nalidixic acid	1185
strC	Low-level streptomycin resistance	1662
tehA	Tellurite resistance	1977
tehB	Tellurite resistance	1977
tlnA	Resistance or sensitivity to thiolutin	1844
E. Radiation sensitivity		
ior	Radiation sensitivity, particularly gamma rays; recombination ability decreased	518
radA	Sensitivity to gamma and UV radiation and methyl methanesulfonate	452
radC	Sensitivity to radiation	531, 532
ras	Sensitivity to UV and X rays	2099
rer	Resistance to UV and gamma radiation	1885
F. Adaptations and atypical conditions		
crg	Cold-resistant growth	963
cspA	Cold shock protein 7.4, transcriptional activator of *hns*	641, 910, 1099
cspB	Cold shock protein; may affect transcription	910, 1099
dps	Global regulator, starvation conditions	37
htgA	Positive regulator for σ^{32} heat shock promoters, permitting growth at high temperature	419, 1329
ibpA	16-kDa heat shock protein A; belongs to the small heat shock (HSP20) family	31, 249
ibpB	16-kDa heat shock protein B; belongs to the small heat shock (HSP20) family	31, 249
mscL	Mechanosensitive channel	1929
rdgA	Dependence of growth upon *recA* gene product	572
rdgB	Dependence of growth and viability upon *recA* function	572
uspA	Universal stress protein; broad regulatory function?	1438–1440

[a]Abbreviations: ABC, ATP-binding cassette; BCCP, biotin carboxyl carrier protein; CoA, coenyme A; dsDNA, ssDNA, double-stranded DNA and single-stranded DNA; FAD, flavin adenine dinucleotide; FGAR, N-formylglycinamide ribonucleotide-5′-phosphate; FMN, flavin mononucleotide; NEM, N-ethylmaleimide; NMN, nicotinamide mononucleotide; PEP, phosphoenolpyruvate.

carry their original citations. The citations in Table 2 are intended to help the reader enter the literature, not to make any judgment on the priority or scientific value of any paper cited or omitted.

Table 2 shows only one assignment to a category of function for each gene product, even though some gene products play multiple roles in the cell. For instance, in metabolism, the acetyl kinase enzyme functions in aerobic catabolism of acetate as a carbon source, but it also is an important enzyme of anaerobic fermentation. A protein kinase or an adenylylation enzyme can be classified either as an enzyme that modifies proteins or as a regulator. Likewise, a porin can be classified either as a transport entity or as a part of a membrane component of the cell structure. A phosphotransferase enzyme can be classified either as a transport entity or as an enzyme of phosphorus metabolism. In each such case, multiple assignments of functional categories have been made and all are provided in the electronic version of the table. However, in the printed version of Table 2 here, one physiological category has been chosen for each gene product and the entries are ordered by that one category of cellular function. The electronic version of the data will be useful to reveal other functions and also will permit views of many other aspects of the data. To this end, the tabulation of the data in electronic format is sortable for instance alphabetically by gene, alphabetically by gene type, numerically by EC number, and so forth. The data is in a database for PC computers called GenProtEc, available by anonymous ftp from mbl.edu as /pub/ecoli.zip or by mail on an MS-DOS disk from M. Riley on request.

Besides providing a list of the presently known gene products of *E. coli* and entries to the literature for each gene product, Table 2 contains information on how many currently known gene products carry out each kind of functions. Some of this information is shown in Table 1 and is summarized in Tables 3 and 4. The number of gene products assigned to each physiological category is shown in Table 1. These are grouped and summarized in Table 3. Small-molecule

TABLE 3 Distribution of *E. coli* gene products among physiological categories

Category	No. of genes
Small molecules	
Degradation and energy metabolism	316
Central intermediary metabolism	78
Broad regulatory functions	51
Biosynthesis	
Amino acids, polyamines	122
Purines, pyrimidines, nucleosides, and nucleotides	60
Cofactors and prosthetic groups	98
Fatty acids	26
Macromolecules	
Synthesis and modification	406
Degradation	69
Cell envelope	168
Cell processes	
Transport	253
Other, e.g., cell division, chemotaxis, mobility, osmotic adaptation, detoxification, and cell killing	118
Miscellaneous	107
Total	1,894

TABLE 4 Distribution types of gene products among classifiable *E. coli* genes[a]

Type of gene product	No. of genes
Enzymes, leader sequences	844
Transport	256
Regulators	217
RNA	107
Structural components	122
Factors	62
Carriers	8
Total	1,616

[a]Not including 279 genes known only by mutant phenotype.

metabolism involves 435 genes or 22.9% of the presently known whole. Large-molecule metabolism involves an even larger fraction: 643 genes or 33.9% of the whole. Of the 370 genes assigned to cell processes, transport involves 253 genes, by far the largest component. Transport functions alone make up 13.3% of the whole. Some of the miscellaneous group are genes concerned with phages and plasmids; others, such as the many heat shock proteins or drug resistance factors, are poorly characterized gene products that will be better delineated with further study.

The number of *E. coli* gene products that fall in each major category of type of gene product is shown in Table 4. In this view, gene products are classified as either an enzyme, a regulator, a transporter, a protein factor, a membrane component, or an RNA molecule. Assignment of the type of gene product could be made for only 1,616 of the 1,897 genes. The rest are too vaguely defined at present to know what the gene product is. Of the 1,616 gene products, enzymes of metabolism constitute the major fraction. The enzymes and proteins of transport functions and regulatory function constitute two other large categories. The number of gene products in the category of structural elements of the cell is relatively small compared with metabolic functions. This is caused partly by the fact that many membrane components have roles in cellular processes such as transport or cell division and thus were not listed primarily as structural elements and partly by the fact that the enzymes of synthesis of macromolecular components of the cell structure such as peptidoglycans or phospholipids are not classified as structural elements but rather as metabolic functions. Also, we recognize that we still have much to learn about the genetic basis of the structure of the cell and the process of assembly of structural components.

As the remainder of the *E. coli* genes are sequenced and function is assigned to the gene products, the proportions of cellular roles will probably change. More functions of cell structure and its assembly probably will be added. Possibly, more global regulatory mechanisms will surface. There is room for many more aspects of the life process in the unsequenced portion of the *E. coli* genome. Even in the parts already sequenced, the Blattner group has found many non-open reading frames, that is, sequences that do not constitute transcribable genes as we presently understand them. These sequences have no known shape or function and have been dubbed grey holes (395). After we have unlocked all their secrets, the distribution of functions of all *E. coli* gene products may change from what we are seeing today.

RELATIONSHIPS AMONG GENES, ENZYMES, AND REACTIONS

In its time, the historic one gene-one enzyme hypothesis illuminated the relationship of genes to cellular function (114). Later, the word "cistron" was introduced to define the genetic element coding for a gene product that is not subdividable by the *trans* complementation test (134). The cistron emphasized the basic genetic element as the coding entity for a polypeptide chain rather than the genetic unit underlying a functional entity such as an enzyme. Today, with many genes, enzymes, and reactions characterized in *E. coli*, we appreciate the many types of relationships that exist in reality between genes and enzymes and the reactions they catalyze.

In many cases, one gene encodes one polypeptide, which catalyzes one biochemical reaction. However, these relationships are not always one to one to one. Figure 1 diagrams some of the other types of relationships found for reactions, enzymes, and genes in *E. coli*. In the case of isozymes such as fumarase, more than one gene and polypeptide are capable of carrying out one reaction. In a different kind of case, a single polypeptide carries out more than one reaction. Illustrating this is the FadB polypeptide, which catalyzes four separate reactions. Another kind of case is TrpD. The N-terminal part of the TrpD polypeptide associates with the TrpE polypeptide to catalyze one reaction, and the C-terminal part of TrpD catalyzes another reaction. One gene can make two polypeptides when, as in the case of the *speD* gene, the initial gene product is further processed into two nonidentical subunits. Sometimes there is confusion about the relationships of enzymes, reactions, and EC numbers. (An EC number, designated by the Enzyme Commission of the Internatiuonal Union of Biochemistry and Molecular Biology, represents a biochemical reaction and thus is associated with each component of a multimeric enzyme [1518].) Therefore, in the case of a multisubunit enzyme, like succinate dehydrogenase, more than one gene and one polypeptide are required to carry out the one reaction described by one EC number, in this case 1.3.99.1. Finally, levels of organization can be more complex than multimeric enzymes. Multienzyme complexes like pyruvate dehydrogenase contain more than one multimeric enzyme that work together in catalyzing a concerted set of reactions.

Because of the variable relationships between polypeptides and reactions, there are many possible relationships of genes to metabolic reactions. Analysis of mutant phenotypes and genetic complementation tests can be complicated by the variety of possible gene-enzyme-reaction relationships.

MULTIPLE ENZYMES

As is the case for isozymes, some metabolic reactions are carried out in *E. coli* by more than one enzyme. Table 5 lists examples of "multiple" proteins that catalyze the same reaction or very similar reactions. Only enzymes of metabolism of small molecules are shown. Multiple genes and enzymes also exist for metabolism of large molecules such as DNA polymerases, sigma factors, and nucleases, but these are not included in Table 5.

Why does *E. coli* contain more than one enzyme for so many reactions? Are they redundant? Do the multiple enzymes serve in the cell as backup systems in case of loss of one enzyme for a vital function? In fact, one may ask how we were able to isolate mutants lacking one enzyme if another enzyme for that reaction existed in the cell. One answer is that quite a few mutants are

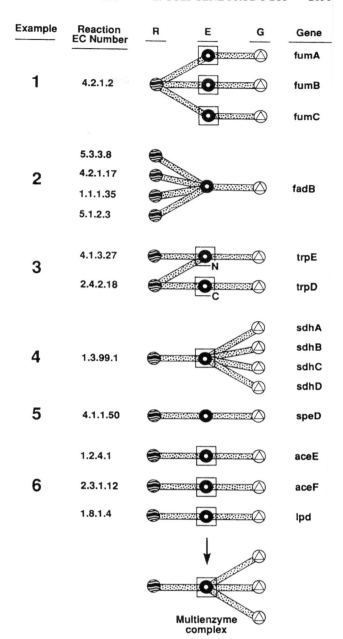

FIGURE 1 Types of gene-enzyme-reaction relationships in *E. coli*. Nodes labelled R, E, and G represent the number of unique reactions, enzymes, and genes, respectively, present in *E. coli* K-12 for each example. A boxed node for an enzyme signifies a dimer.

leaky and that leakiness in many cases is known now to be a consequence of multiple enzymes. Another answer is that the conditions under which many of the pairs of genes are expressed can differ and also the conditions under which the two enzymes are catalytically active can differ. A list of ways that multiple enzymes for the same reaction differ from each other is given in Table 6.

This phenomenon of enzyme repetition and specialization must have the effect of extending the metabolic capabilities of

TABLE 5   Multiple enzyme ion small-molecule metabolism in *E. coli*

Gene product	EC no.	Enzyme name	Reference(s)
AdhC	1.1.1.1	Alcohol dehydrogenase	325, 378
AdhE	1.1.1.1	Alcohol dehydrogenase	
MetL	1.1.1.3	Homoserine dehydrogenase II	2250
ThrA	1.1.1.3	Homoserine dehydrogenase I	
GlpA	1.1.99.5	Glycerol-3-phosphate dehydrogenase	864
GlpD	1.1.99.5	Glycerol-3-phosphate dehydrogenase	
FdhF	1.2.1.2	Formate dehydrogenase H	135, 1752
FdnG	1.2.1.2	Formate dehydrogenase N	
FdoG	1.2.1.2	Formate dehydrogenase O	
GapA	1.2.1.12	Glyceraldehyde 3-phosphate dehydrogenase	464, 1247
GapB	1.2.1.12	Glyceraldehyde 3-phosphate dehydrogenase	
GabD	1.2.1.16	Succinate semialdehyde dehydrogenase (NADP)	1240, 1410
Sad	1.2.1.24	Succinate semialdehyde dehydrogenase (NAD)	
Ald	1.2.1.22	Aldehyde dehydrogenase	729, 753
AldH	1.2.1.22	Aldehyde dehydrogenase	
SdhA	1.3.99.1	Succinate dehydrogenase, flavoprotein subunit	1452
FrdA	1.3.99.1	Fumarate reductase, flavoprotein subunit	
SdhB	1.3.99.1	Succinate dehydrogenase, iron-sulfur protein	1452
FrdB	1.3.99.1	Fumarate reductase, iron-sulfur protein	
SdhC	1.3.99.1	Succinate dehydrogenase, cytochrome b_{556}	1452
FrdC	1.3.99.1	Fumarate reductase, cytochrome b_{556}	
SdhD	1.3.99.1	Succinate dehydrogenase, cytochrome b_{556}	1452
FrdD	1.3.99.1	Fumarate reductase, cytochrome b_{556}	
TorA	1.6.6.9	Trimethylamine-*N*-oxide reductase, inducible	1841
—	1.6.6.9	Trimethylamine-*N*-oxide reductase, constitutive	
Ndh	1.6.99.3	NADH dehydrogenase	260
Nuo	1.6.99.3	NADH dehydrogenase	
DprA	1.6.99.7	Dihydropteridine reductase/ferrisiderophore reductase	540
HmpA	1.6.99.7	Dihydropteridine reductase/ferrisiderophore reductase	
NarG	1.7.99.4	Nitrate reductase, alpha subunit	176
NarZ	1.7.99.4	Nitrate reductase, alpha subunit	
NarH	1.7.99.4	Nitrate reductase, beta subunit	176
NarY	1.7.99.4	Nitrate reductase, beta subunit	
NarI	1.7.99.4	Nitrate reductase, gamma subunit	176
NarV	1.7.99.4	Nitrate reductase, gamma subunit	
NarJ	1.7.99.4	Nitrate reductase, delta subunit	478
NarW	1.7.99.4	Nitrate reductase, delta subunit	
LpdA	1.8.1.4	Lipoamide dehydrogenase	1643
—	1.8.1.4	Lipoamide dehydrogenase	
AppB	1.10.3.-	Cytochrome oxidase subunit II	404, 757
CydB	1.10.3.-	Cytochrome oxidase subunit II	
CyoA	1.10.3.-	Cytochrome oxidase subunit II	
AppC	1.10.3.-	Cytochrome oxidase subunit I	404, 757
CydA	1.10.3.-	Cytochrome oxidase subunit I	
CyoB	1.10.3.-	Cytochrome oxidase subunit I	
KatE	1.11.1.6	Catalase hydroperoxidase	416, 731
KatG	1.11.1.6	Catalase hydroperoxidase	
SodA	1.15.1.1	Superoxide dismutase	786
SodB	1.15.1.1	Superoxide dismutase	
HyaA	1.18.99.1	Hydrogen lyase, alpha subunit (large)	183, 1300, 1591
HybA	1.18.99.1	Hydrogen lyase, alpha subunit (large)	
HycB	1.18.99.1	Hhydrogen lyase, alpha subunit (large)	
HyaB	1.18.99.1	Hydrogen lyase, beta subunit (small)	183, 1300, 1591

TABLE 5 *Continued*

Gene product	EC no.	Enzyme name	Reference(s)
HybC	1.18.99.1	Hydrogen lyase, beta subunit (small)	
HycE	1.18.99.1	Hydrogen lyase, beta subunit (small)	
MetE	2.1.1.14	Homocysteine transmethylase, B$_{12}$ independent	645
MetH	2.1.1.13	Homocysteine transmethylase, B$_{12}$ dependent	
PurN	2.1.2.2	5′-Phosphoribosylglycinamide transformylase	1437
PurT	2.1.2.2	5′-Phosphoribosylglycinamide transformylase	
ArgF	2.1.3.3	Ornithine carbamoyltransferase	331, 1101, 1800
ArgI	2.1.3.3	Ornithine carbamoyltransferase	
TktA	2.2.1.1	Transketolase A	833
TktB	2.2.1.1	Transketolase B	
TalA	2.2.1.2	Transaldolase A	
TalB	2.2.1.2	Transaldolase B	
FabB	2.3.1.41	3-Oxoacyl-ACP synthase I	599, 1212, 2020
FabF	2.3.1.41	3-Oxoacyl-ACP synthase II	
FabH	2.3.1.41	3-Oxoacyl-ACP synthase III	
DeoD	2.4.2.1	Purine-nucleoside phosphorylase I	143
DapA	2.4.2.1	Purine-nucleoside phosphorylase II	
MetK	2.5.1.6	*S*-Adenosylmethionine synthetase	1747
MetX	2.5.1.6	*S*-Adenosylmethionine synthetase	
PfkA	2.7.1.11	6-Phosphofructokinase	388, 558, 734
PfkB	2.7.1.11	6-Phosphofructokinase	
PykA	2.7.1.40	Pyruvate kinase	2047
PykF	2.7.1.40	Pyruvate kinase	
AroK	2.7.1.71	Shikimate kinase I	426, 1179
AroL	2.7.1.71	Shikimate kinase II	
LysC	2.7.2.4	Aspartate kinase III	273
MetL	2.7.2.4	Aspartate kinase II	
ThrA	2.7.2.4	Aspartate kinase I	
RelA	2.7.6.5	(p)ppGpp synthetase I	1807, 2205
SpoT	2.7.6.5	(p)ppGpp synthetase II	
PldB	3.1.1.5	Lysophospholipase L1	946, 947
PldC	3.1.1.5	Lysophospholipase L2	
TreE	3.1.3.12	Trehalose-6-phosphate phosphatase	997, 1919
OtsP	3.1.3.12	Trehalose-6-phosphate phosphatase	
PgpA	3.1.3.27	Phosphatidylglycerophosphate phosphatase	580, 826
PgpB	3.1.3.27	Phosphatidylglycerophosphate phosphatase	
GlpQ	3.1.4.46	Glycerophosphodiester phosphodiesterase, membrane	2001
UgpQ	3.1.4.46	Glycerophosphodiester phosphodiesterase, cytoplasmic	
EbgA	3.2.1.23	β-Galactosidase	694
LacZ	3.2.1.23	β-Galactosidase	
AscB	3.2.1.86	6-Phospho-β-glucosidase	696, 1508, 1769
CelF	3.2.1.86	6-Phospho-β-glucosidase	
BglB	3.2.1.86	6-Phospho-β-glucosidase	
DacA	3.4.16.4	D-Alanyl-D-alanine carboxypeptidase	1354, 2051, 2052
DacB	3.4.16.4	D-Alanyl-D-alanine carboxypeptidase	
DacC	3.4.16.4	D-Alanyl-D-alanine carboxypeptidase	
FolE	3.5.4.16	GTP cyclohydrolase I	1646, 1770
RibA	3.5.4.25	GTP cyclohydrolase II	
AnsA	3.5.1.1	Asparaginase I	894, 1879
AnsB	3.5.1.1	Asparaginase II	

(Table continues)

TABLE 5   *Continued*

Gene product	EC no.	Enzyme name	Reference(s)
AmiA	3.5.1.28	*N*-Acetylmuramoyl-*l*-alanine amidase	2024
AmiB	3.5.1.28	*N*-Acetylmuramoyl-*l*-alanine amidase	
Ppx	3.6.1.11	Exopolyphosphatase	16, 967
GppA	3.6.1.11	Exopolyphosphatase	
GadA	4.1.1.15	Glutamate decarboxylase	1856
GadB	4.1.1.15	Glutamate decarboxylase	
SpeC	4.1.1.17	Ornithine decarboxylase	1496, 2189
SpeF	4.1.1.17	Ornithine decarboxylase	
Adi	4.1.1.19	Arginine decarboxylase	1339, 1905
SpeA	4.1.1.19	Arginine decarboxylase	
Fba	4.1.2.13	Fructose-bisphosphate aldolase class II	1526
Kba	4.1.2.13	Fructose-bisphosphate aldolase class I	
AroF	4.1.2.15	DAHPase	1622
AroG	4.1.2.15	DAHPase	
AroH	4.1.2.15	DAHPase	
IlvB	4.1.3.18	Acetolacetate synthase I, large subunit	387, 1883, 2142
IlvG	4.1.3.18	Acetolacetate synthase II, large subunit	
IlvI	4.1.3.18	Acetolacetate synthase III, large subunit	
IlvN	4.1.3.18	Acetolacetate synthase I, small subunit	26, 387, 1883
IlvM	4.1.3.18	Acetolacetate synthase II, small subunit	
IlvH	4.1.3.18	Acetolacetate synthase III, small subunit	
IlvJ	4.1.3.18	Acetolacetate synthase IV	26, 1664
IlvF	4.1.3.18	Acetolacetate synthase V, possible	
FumA	4.2.1.2	Fumarate hydratase	2186
FumB	4.2.1.2	Fumarate hydratase	
FumC	4.2.1.2	Fumarate hydratase	
AcnA	4.2.1.3	Aconitate hydrase	674
AcnB	4.2.1.3	Aconitate hydrase	
SdaA	4.2.1.13	Serine dehydratase	1811
SdaB	4.2.1.13	Serine dehydratase	
IlvA	4.2.1.16	Threonine dehydratase	406, 497
TdcB	4.2.1.16	Threonine dehydratase	
CysK	4.2.99.8	Cysteine synthase	1842
CysM	4.2.99.8	Cysteine synthase	
Alr	5.1.1.1	Alanine racemase	2157
DadX	5.1.1.1	Alanine racemase	
PheA	5.4.99.5	Chorismate mutase	810, 2028
TyrA	5.4.99.5	Chorismate mutase	
AsnA	6.3.1.1	Asparagine synthetase A	1391, 1788
AsnB	6.3.1.1	Asparagine synthetase B	
DdlA	6.3.2.4	D-Alanine:D-alanine ligase	2256
DdlB	6.3.2.4	D-Alanine:D-alanine ligase	

E. coli. With genes producing a given enzyme under more than one set of conditions and with multiple enzymes being active under different conditions, the bacteria are able to address successfully a wide range of environmental conditions that require enzymes with appropriate properties to become available. For instance, the *speA* gene produces the biosynthetic arginine decarboxylase; gene expression is induced by growth in minimal media and is repressed by putrescine and spermidine. The en-

zyme is located in the periplasm and is inhibited by cyclic AMP (1339). The *adi* gene, on the other hand, produces the degradative arginine decarboxylase, and gene expression is induced under acid conditions and anaerobiosis. The degradative enzyme is located in the cytoplasm (1905). Another example involves the *sodA* and *sodB* genes. The *sodA* gene is induced by oxidative stress under aerobic conditions to produce a manganese-activated superoxide dismutase, whereas the *sodB* gene is expressed

TABLE 6 Ways in which multiple enzymes differ from each other

Conditions of enzyme synthesis
  Constitutive/inducible
  Conditions of induction
    Identity of inducer
    Physical conditions: anaerobic/aerobic, pH

Physical properties of enzyme
  Substrate specificity
  Sensitivity of enzyme to inhibitors/activators
  Heat stability
  Subunit organization

Chemistry of enzyme
  Substrate specificity
  Mechanism of reaction
  Cofactor requirement

Cell location
  Cytoplasmic
  Periplasmic
  Membrane

under both aerobic and anaerobic conditions, constitutively producing an iron-activated superoxide dismutase (786).

In terms of evolution, one can ask if multiple enzymes descended from common ancestors. If so, one might expect to see residual similarities in amino acid sequence. Comparison of sequences of the 75 pairs of isozymes for which the sequence is known for both proteins showed that 44 of 75 pairs are related by sequence, some very closely related, some less so. The other 31 pairs were not demonstrably related by sequence (Table 7). Therefore, somewhat over half of the pairs of the currently known multiple enzymes involved in small-molecule metabolism seem to be related by a common ancestor. The other half either do not share ancestry or have diverged to a point that the relationship is no longer detectable. It is possible that the pairs that are not related by sequence are examples of convergent evolution, that is, descendants of separate ancestral sequences evolving to the same function; alternatively, the gene for one of a pair of isozymes might have been acquired in the past by lateral transfer from another organism.

PROTEIN SEQUENCE RELATIONSHIPS AS A TOOL TO STUDY THE ORIGIN OF *E. COLI* GENES

In the context of evolution, we believe that many of the genes of present-day *E. coli* originated by a process of duplication of ancestral genes followed by divergence, then by further duplication of these genes, followed by more divergence, and so on (1125, 1454). The very early genes and their proteins are visual-

TABLE 7 Sequence relationships between pairs of multiple enzymes

Relationship	No. of pairs
Sequences unrelated (>250 PAM)	31
Sequences related (<250 PAM)	44
Sequence not known for both	14
Total pairs	89

ized as having broad specificity of action, which then narrowed successively in the descendants formed by duplication and divergence (896, 2234). If all descendants of all ancestral sequences still retain detectable vestiges of sequence similarity, we could expect to be able to identify all ancestral relationships and build a set of trees of evolutionary descent of all organisms that extend back to a set of unique ancestral sequences that were parents of all the genes and gene products extant today.

By examining the relationships among protein sequences within one organism, one is identifying pairs of proteins which display significant level of similarity. This is generally interpreted as a demonstration that these proteins are products of paralogous genes, i.e., homologous genes that descended from a common ancestor by duplication and divergence, according to the definitions proposed by Fitch (544), who opposed paralogy (homology in which divergence occurs after gene duplication in the same species) and orthology (divergence of homologous genes through speciation). Moreover, groups of proteins whose sequences are related could also be detected, meaning it is possible to identify present-day genes descending from shared ancestral genes (familial relationships within a genome).

We have engaged in such a study, identifying pairs and groups of *E. coli* proteins whose sequence similarities could indicate shared ancestry. Indeed, a higher percentage of all its protein sequences are available for *E. coli* than for any other organism, providing the opportunity to test for compatibility of the sequence relationships among *E. coli* proteins with accepted ideas of mechanisms of molecular evolution. *E. coli* is also the organism for which the greatest proportion of chromosomal genes sequenced to date correspond to genes previously well characterized in terms of function of gene product and regulation. No other organism provides a better opportunity to determine how many ancestral relationships can be detected and how important the mechanism of duplication and divergence has been during the evolution of a genome and its encoded proteins than the massively sequenced *E. coli* genome does.

Our main goal being to identify descendants of past whole-gene duplications, we undertook to detect any significant similarity extending along either the whole sequence or at least long stretches of each amino acid sequence. We call this kind of similarity "extended sequence similarity," as opposed to "local similarity," i.e., similarity localized to domains or motifs. To do that, we analyzed all the *E. coli* K-12 chromosomal sequences longer than 100 residues present in the SwissProt database by using two different algorithms designed for extended sequence homology searches. In one study, we first used the well-known FASTA program (1518) and retained any pair displaying alignment of segments at least 100 amino acids long and with at least 20% identity. Then, from the obtained FASTA alignments, we excluded those of questionable biological significance by imposing a high threshold on the number of gaps (corresponding to a NAS [Normalized Alignment Score] of at least 180, calculated according to the method of Doolittle et al. [465]). In a subsequent study, the ALLALLDB program (Darwin package available at the CRBG server at the ETH, Zurich, Switzerland) was used to detect any match corresponding to an alignment of at least 100 amino acid residues and separated no more than 250 PAM (percent accepted mutations) units (644). These two approaches gave us very similar results (1060, 1061), which can be summarized as follows.

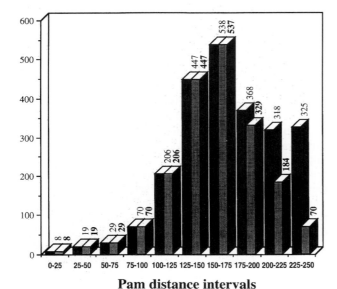

FIGURE 2 Distribution of PAM values among 2,329 pairs of protein sequences. In black is shown the number of pairs for all 2,329 matches. In grey is shown the number of pairs that have amino acid identity values greater than 20%.

The 1,862 protein sequences derived from the sequences of *E. coli* chromosomal genes present in SwissProt database (version 28) (1,339 known genes and 523 open reading frames) were compared in all pairwise combinations. This operation gave us 2,329 matches separated by no more than 250 PAM units. The distribution of the PAM values was found to be Gaussian only when confined to the matches displaying identity values greater than 20% (Fig. 2). The large majority of the excess pairs with identity values less than 20% but PAM values less than 250 corresponded to the more distantly related members of large families (as defined below).

The 2,329 pairs corresponded to a large set of 971 sequences (52.15% of the total) displaying similarities to at least one other sequence of this set. An alphabetical list of these 971 proteins (using the SwissProt mnemonics) that have at least one paralogous partner is given in Table 8; 786 of these sequences code for proteins the function of which is known, amounting to 58.7% of all known and sequenced genes. The rest correspond to 185 open reading frames (34.58% of all open reading frames). Thus, a significant number of the genes already sequenced—more than half—appear to be coding for paralogous proteins, and this proportion is even higher when considering only the genes known to have a functional gene product. Interestingly, when we used only the latter, i.e., genes known to be functional, for the ALLALLDB search, we lost only 44 of them, corresponding to those found to match exclusively with an open reading frame. The high proportion, over half of sequences in alignments, is without doubt a minimum figure, since our arbitrary cutoff criteria exclude some well-known examples of proteins believed to share evolutionary ancestry but corresponding to an alignment of less than 100 amino acids. Indeed, there seem to be biologically significant relationships at even lower levels of similarity, comparable to those we detected between far remote

members of the same large family. However, such putative supplementary paralogous genes will be added only when we obtain better phylogenetic arguments (see below).

This set of 971 paralogous proteins was further analyzed in two complementary ways. (i) We looked at the functional relatedness of the paired proteins, designating each pair as being related, different, or unknown (as in the case of open reading frames or proteins whose cellular function has not been characterized well enough to judge the level of similarity of function). Of the paired 971 proteins, 587 could be related by function to its (best) partner, only 12 were paired with another protein displaying an apparently totally different function, and 336 could not be assessed for lack of information on at least one partner of the pair. The extremely high percentage of similarity of functional relationship among paralogous gene pairs (60.45% of 971, and 98% of the 599 assessable proteins) shows that the sequence relationships between pairs are not accidental but have biological significance. We also used other ways of assessing functional relationships. These are summarized in Table 9. The assignment of the members of the 2,329 pairs to the functional categories reported here in Tables 2 and 3 was examined. Not all members of all pairs had assignments that characterized function, but for those pairs in which both members had been characterized, a large fraction registered as having similar functions (Table 9). Thus, these results strongly suggest that the genes coding for these proteins are paralogous, descendants of duplicate copies of ancestral genes residing in the same genome.

(ii) Many of the proteins were found to be related to more than one other *E. coli* protein and thus are members of groups of interrelated proteins. Besides the 112 pairs (224 sequences), we could distinguish 38 triplets (113 sequences), 41 small groups (281 sequences), and 13 large families (353 sequences). These combinations are listed in Table 10. If each cluster and family were descended from one ancestral gene by duplication and divergence, one could begin to count the numbers of ancestral genes necessary to generate the *E. coli* genome. This leads to a dramatically small number of putative ancestral sequences. Indeed, the 747 sequences belonging to groups larger than pairs could originate from as few as 92 putative ancestral sequences. This number will undoubtedly fall further as additional genes are sequenced, providing partners for some of the single sequences and amalgamating some of the pairs into families. When the full sequence of the *E. coli* chromosome is known, we will be able to count the number of unique ancestral sequences required to generate *E. coli*.

To go a step farther in this analysis, we are reconstructing phylogenetic trees for each of the sequence-related groups and then using the putative ancestral sequences to extract other related sequences from databases of sequences. As long as the sequence relationhips among distantly related proteins can be detected, one can continue to move earlier in the tree of descent, relating ancestral sequences for a given species to even earlier ancestral genes that fed many species, thereby progressively reducing the total number of ancestral sequences as one moves in the direction of the begining of the tree. Ultimately, we will be able to approach the identification of a relatively small number of unique primitive ancestral sequences that gave rise to all contemporary genes. *E. coli* gene sequences will be very useful in this evolutionary context.

TABLE 8 List of the 971 sequences displaying similarities to at least one other *E. coli* K-12 chromosomal sequence

3MG2	ASG2	CPXA	DHSA	FECD	GABP	HISP	LIPA
6PGD	ASLA	CREB	DHSB	FECE	GABT	HISX	LIVF
A	ASLB	CREC	DLDH	FECI	GALE	HMPA	LIVG
AAS	ASNB	CRED	DMSA	FENR	GALR	HNS	LIVH
AAT	ASNC	CRP	DMSB	FEPA	GALS	HOLB	LIVJ
ABC	ASPA	CSTA	DNAA	FEPB	GALU	HSLU	LIVK
ACCC	ATP6	CUTE	DNAB	FEPC	GCL	HTRA	LIVM
ACKA	ATKA	CVPA	DNAC	FEPD	GCSP	HTRB	LON
ACON	ATKB	CYBH	DNAJ	FEPE	GCST	HTRE	LPXA
ACRA	ATPA	CYDA	DNAK	FEPG	GCVA	HTRH	LRP
ACRB	ATPB	CYDB	DNLJ	FES	GENA	HYCB	LVSP
ACUA	ATPF	CYDD	DP3X	FHLA	GENF	HYCC	LYSR
ADA	BAER	CYNR	DPO1	FHUA	GLDA	HYDA	MALE
ADHE	BAES	CYOE	DPO2	FHUB	GLGB	HYDG	MALF
ADI	BARA	CYPB	DPPA	FHUC	GLGC	HYDH	MALG
AGAL	BASR	CYPH	EBGR	FHUD	GLGX	HYDN	MALI
AGP	BASS	CYSA	ECPD	FHUE	GLMS	HYPD	MALK
AK1H	BCCP	CYSB	EDD	FIMB	GLNE	HYPE	MALT
AK2H	BCR	CYSD	EFG	FIMC	GLNH	IBPA	MALY
AK3	BETT	CYSE	EFTA	FIMD	GLNP	IBPB	MALZ
ALD	BGA2	CYSH	EFTB	FIME	GLNQ	ICIA	MARA
ALKH	BGAL	CYSI	EFTS	FIMF	GLPA	IDH	MARR
ALR1	BGLB	CYSJ	EMRA	FIMG	GLPD	IF2	MBHL
ALR2	BGLR	CYSK	EMRB	FIMZ	GLPF	ILVB	MCP1
AMPM	BIOA	CYSM	END3	FIRA	GLPG	ILVC	MCP2
AMPP	BIOB	CYSN	ENTA	FKBX	GLPK	ILVD	MCP3
AMY1	BIOC	CYSP	ENTC	FLGL	GLPR	ILVE	MCP4
AMY2	BIOF	CYSQ	ENTE	FLIA	GLPT	ILVG	MDL
APPB	BISC	CYST	ENTF	FLIC	GLPX	ILVI	MDOH
APPC	BTUB	CYSW	ENVC	FLID	GLTB	ILVY	MDRA
APPY	BTUC	CYTR	ENVD	FLIG	GLTP	IMP	MELB
ARAC	BTUD	DACA	ENVM	FLIR	GLTS	INAA	MELR
ARAD	BTUR	DACC	ENVY	FM1A	GLYA	IPYR	MENB
ARAE	CADB	DAMX	ENVZ	FMT	GPPA	ISPA	MEND
ARAF	CADC	DAPA	ERA	FNR	GREA	K1PF	METB
ARAG	CARA	DAPB	EVGA	FOLC	GREB	K6P2	METC
ARAH	CARB	DAPD	EVGS	FRDA	GRPE	KBL	METE
ARAJ	CCA	DAPE	EX5A	FRDB	GSA	KDPD	METH
ARCA	CDSA	DBPA	EX5B	FRE	GSH2	KDPE	METK
ARCB	CELD	DCDA	EX5C	FRUR	GSHR	KDSB	METR
ARGA	CELF	DCEA	EX7L	FTSA	GUAA	KDTA	METX
ARGB	CFA	DCEB	EXBB	FTSE	GUAC	KEFC	MFD
ARGD	CH60	DCLY	EXBD	FTSN	GUTD	KGTP	MGLA
ARGE	CHAA	DCOR	FABA	FTSW	GUTR	KGUA	MGLC
ARGT	CHEA	DCOS	FABG	FTSY	GYRA	KIRI	MIOC
ARLY	CHEB	DCP	FABH	FUCA	GYRB	KPRS	MIND
AROA	CHEY	DDLA	FADB	FUCK	HDHA	KPY1	MOAA
AROF	CHEZ	DDLB	FADR	FUCO	HELD	KPY2	MOAB
AROG	CIRA	DEAD	FDHD	FUCP	HEM1	KUP	MODB
AROH	CISY	DEDA	FDHF	FUCR	HEMX	LACI	MODC
AROP	CLPA	DEDD	FDNG	FUMA	HEPA	LACY	MOEB
ARTI	CLPB	DEOC	FDNH	FUMB	HFLC	LAMB	MPRA
ARTJ	CLPX	DEOD	FDNI	FUMC	HFLK	LCFA	MRAY
ARTM	CMTA	DEOR	FDOG	FUR	HFLX	LEPA	MREB
ARTP	CN16	DGAL	FDOH	G3P1	HIS4	LEU1	MRED
ARTQ	COAA	DHAB	FDOI	G3P2	HIS6	LEU2	MRP
ASCB	CODB	DHAL	FECA	G6PD	HIS7	LEU3	MSBA
ASCG	CPSB	DHAS	FECB	G6PI	HIS8	LEUO	MSBB
ASG1	CPSG	DHNA	FECC	GABD	HISM	LEXA	MTLD

(Table continues)

TABLE 8 *Continued*

MTR	PABC	PT2B	ROB	TDH	YAAG	YFIB	YIFG
MUKB	PAL	PT2D	RODA	TESB	YAAJ	YFJA	YIFI
MURC	PANF	PT2F	RP32	THD1	YAAL	YGGA	YIFK
MURD	PARC	PT2G	RP70	THD2	YAAM	YGGB	YIGK
MURE	PARE	PT2M	RPOS	THDF	YAAP	YGGC	YIGM
MURF	PBP2	PT2N	RS1	THGA	YAAS	YGIE	YIGN
MURZ	PBP3	PT2P	RS4	THIF	YAAU	YGIF	YIGO
MUTL	PBPA	PT2X	RUVA	THIH	YABE	YHAA	YIGT
MUTS	PBPB	PT3G	RUVB	TKT	YABG	YHAE	YIHG
MUTY	PCNB	PT3M	SBCC	TNAA	YABI	YHBD	YIHI
NADB	PDXB	PTR	SBM	TNAB	YABJ	YHBE	YIHK
NAGA	PEPQ	PUR1	SDHD	TOLA	YABK	YHBF	YIHL
NAGB	PFL	PUR2	SDHL	TOLR	YABM	YHBG	YIHN
NAGR	PFLA	PUR3	SDHM	TONB	YADB	YHBI	YIHO
NARG	PFS	PUR5	SECA	TOP1	YADD	YHDE	YIHP
NARH	PHEA	PURA	SECD	TOP3	YAEA	YHDF	YIHQ
NARI	PHEP	PURR	SECF	TRKA	YAEC	YHDG	YIHU
NARJ	PHNC	PUTP	SECY	TRKD	YAED	YHDH	YIHV
NARK	PHNE	PYRB	SELA	TRKG	YAEE	YHDK	YIHW
NARL	PHNF	PYRE	SELB	TRKH	YAFB	YHFD	YIHX
NARP	PHNJ	RARD	SERA	TRPB	YAFC	YHGA	YIHY
NARQ	PHNK	RBSA	SLT	TRPE	YAFG	YHHI	YIIA
NARV	PHNL	RBSB	SLYD	TRPG	YAJF	YHIA	YIIG
NARW	PHNN	RBSC	SMBA	TRXB	YBBA	YHID	YIII
NARX	PHOB	RBSK	SMS	TTDA	YBDA	YIBC	YIIJ
NARY	PHOE	TBSR	SODF	TTDB	YBEF	YIBF	YIIK
NARZ	PHOP	RCSB	SODM	TTK	YBFB	YIBH	YIIT
NFRB	PHOQ	RCSC	SOHB	TYRA	YBFD	YICC	YIIX
NHAA	PHOR	RECA	SOXS	TYRB	YBIB	YICE	YIJE
NHAB	PHOU	RECE	SPEA	TYRP	YCAC	YICF	YIJG
NHAR	PHRA	RECG	SPOT	TYRR	YCAD	YICI	YIJH
NIRB	PHSG	RECN	SPPA	UBIA	YCAE	YICJ	YIJJ
NIRC	PHSM	RECQ	SRMB	UBIG	YCEE	YICK	YIJL
NLPA	PLSB	RELA	SRP5	UBIH	YCEF	YICL	YIJM
NMPC	PLSC	REP	SSPA	UDHA	YCHE	YICM	YIJO
NOHA	PNP	RF1	STPA	UDP	YCIB	YICO	YIJP
NOHB	PNTA	RF2	SUBI	UGPA	YCIE	YICP	YJBB
NPL	PNTB	RFAD	SUCD	UGPB	YCIF	YICQ	YJBI
NRFC	POTA	RFAH	SUHB	UGPC	YDBC	YIDA	YJBK
NRFE	POTB	RFAI	SURA	UGPE	YDCC	YIDE	YJBN
NTRB	POTC	RFAJ	SYA	UHPA	YDDA	YIDF	YJCC
NTRC	POTD	RFAL	SYC	UHPB	YDDC	YIDJ	YJCD
NUOM	POTE	RFAP	SYD	UHPC	YDDD	YIDK	YJCE
NUPC	POTF	RFAQ	SYE	UHPT	YDEA	YIDL	YJCF
NUPG	POTG	RFE	SYFA	UIDP	YDEB	YIDM	YJCG
NUSA	POTH	RFFE	SYGA	UMUC	YDED	YIDN	YJCP
NUSG	POTI	RFFM	SYGB	UMUD	YDEE	YIDO	YJCQ
ODO2	POXB	RFFT	SYI	UPP	YDEK	YIDP	YJCR
ODP1	PPA	RHAA	SYK1	URK	YEBB	YIDT	YJCT
ODP2	PPB	RHAB	SYK2	USG	YECB	YIDU	YJCU
OGT	PPSA	RHAR	SYK3	USHA	YEDA	YIDW	YJCV
OMPC	PPX	RHAS	SYL	USPA	YEFA	YIDY	YJCW
OMPF	PRC	RHAT	SYM	USPT	YEFB	YIDZ	YJDB
OMPR	PRIA	RHLB	SYN	UVRA	YEIC	YIEA	YJGA
OPDA	PROA	RHLE	SYP	UVRB	YEIE	YIEC	YJGB
OPPA	PROP	RHO	SYQ	UVRC	YFCA	YIEG	YPRA
OTC1	PROV	RIBG	SYS	UVRD	YFEB	YIEH	YPTF
OTC2	PROW	RIMJ	SYT	VISC	YFEC	YIEK	YZFB
OTSB	PSTA	RIMK	SYV	XERC	YFED	YIEL	
OXYR	PSTB	RIML	T1R	XPRB	YFFE	YIEO	
P30	PSTC	RLPA	TALB	XYLE	YFGA	YIEP	
PABA	PSPA	RNE	TDCA	XYLK	YFHA	YIFB	
PABB	PT1	RNPH	TDCC	YAAA	YFHC	YIFC	

TABLE 9 Functional relationships among sequence-related proteins

Functional category	No. of pairs with identical assignment	Total no. of pairs[a]	% of pairs with related function
Physiological categories as listed in Table 1	450	1,636	27.5
Type of gene product as listed in Table 4	1,375	1,523	90.3
Enzymes that share the first two numbers of EC classification[b]	287	408	70.3

[a] The total number of sequence-related pairs for which both members bear assignments in Table 1 or 4.
[b] EC numbers represent categorization of reactions over four levels of specification, assigned by the Enzyme Commission of the International Union of Biochemistry and Molecular Biology (2122).

TABLE 10 Families of paralogous genes

224 sequences having only partner, with PAM[a] values

aat/tyrB	48.32	cysP/sbp	83.71	glne/yjcC	222.42	moaB/yaaG	142.87	slt/yafG	154.82
aceE/tktA	200.97	cysQ/suhB	162.26	glpX/sucD	198.81	mtlD/rpsD	247.56	sodF/sodM	90.78
ackA/yhaA	104.54	dacA/dacC	46.04	gppA/ppx	100.59	nagB/yieK	179.56	sspA/yibF	167.57
acn/leuC	172.72	dapA/nanA	176.95	greA/greB	124.38	narJ/narW	58.25	pheS/yihI	214.90
alr/dadX	102.66	lysA/speA	166.07	guaC/guaB	116.02	nlpA/yaeC	46.03	talB/yijG	131.06
araD/fucA	141.41	dedA/yabI	171.49	hisA/hisF	179.55	nohA/nohB	1.99	topA/topB	169.99
malS/amyA	215.75	sdhB/frdB	123.35	hns/stpA	58.13	nusG/rfah	197.41	trkG/trkH	103.26
appC/cydA	50.93	dnaJ/hydA	245.97	htrB/msbB	179.47	ogt/purA	240.43	ttk/hdK	223.94
argA/argB	192.23	dnaA/dnaC	159.80	hypE/purM	213.07	pal/yfiB	178.26	uspA/yiiT	118.75
argE/dapE	155.27	lig/yicF	190.08	ibpA/ibpB	74.69	pdxB/serA	151.41	yadD/yhgA	67.22
ascB/bglB	66.20	dppA/oppA	187.50	icd/leuB	129.09	act/yijM	152.86	yafB/ydbC	158.25
aslB/yidF	143.37	ecpD/fimC	132.15	ilvC/yptF	228.87	pheA/tyrA	187.74	yceF/yhdE	119.78
asnC/lrp	172.85	edd/ilvD	142.49	ilvE/pabC	164.33	potD/potF	118.46	ychE/ydeB	126.41
bioB/lipA	196.37	nth/mutY	192.02	inaA/rfaP	157.51	relA/spoT	126.97	yciE/yciF	170.33
bioF/kbl	123.00	exbD/tolR	134.40	gmk/phnN	176.91	prfA/prfB	103.29	yebB/yiiX	157.38
cca/pcnB	171.44	fabA/yaeA	174.51	prs/pyrE	218.26	rfaI/rfaJ	120.95	yfeC/yfeD	120.06
cdsA/hyaB	239.19	fepE/yifC	197.10	lexA/umuD	125.85	rfaQ/yibC	221.06	yhdG/yjbN	193.77
gltA/yieH	223.03	fes/yieL	226.06	livJ/livK	25.43	rffM/ycaC	203.83	yhfD/yjcU	109.27
clpX/hslU	90.45	yaaD/slyD	151.37	malE/ugpB	190.62	ribG/yfhC	125.39	yicI/yihQ	177.20
cpsB/yjcP	210.22	fmt/purN	211.85	marR/mprA	167.01	rimJ/rimL	167.28	yijP/yjdB	178.47
crp/fnr	183.24	fur/yjbK	174.87	metK/metX	3.26	rpsA/yaaS	192.62		
ppiB/ppiA	55.50	gapA/gapB	94.82	minD/mrp	221.76	ruvA/uvrC	198.85		
cysD/cysH	162.21	pgi/yicP	225.43	moaA/thiH	194.88	sdhL/sdhM	28.39		

113 sequences which are part of a triplet

gnd/yihU/yhaE	aroF/aroG/aroH	coaA/udk/yggC	sdhA/frdA/nadB	fecB/fepB/fhuD	gyrB/parE/fliD	lon/recA/sms	proS/serS/thrS
melA/celF/yidM	ansA/ansB/polB	cutE/cyoE/ubiA	dnaB/leuA/nfrB	fimD/htrE/yabG	hisB/yaeD/yihX	metB/metC/selA	tdh/yhdH/yjgB
eda/fdhD/yidU	atpA/atpB/rho	cysJ/mioC/recG	enC/pabB/trpE	galU/glgC/yifG	ispA/phoA/yhbD	argI/argF/pyrB	ybfD/ydcC/yhhI
map/pepP/pepQ	ebgA/lacZ/uidA	gadA/gadB/gcvP	fadB/menB/yaaL	genA/genF/cstA	pykF/pykA/upp	sohB/sppA/fabH	
argH/aspA/fumC	cpdB/ushA/yaaA	dodD/pfs/udp	fdnI/fdoI/hyaC	glgB/glgX/malZ	lamB/yieC/yifI	gltX/glns/yadB	

281 sequences belonging to small groups

yacB/fadR/ybgB/phnF/yidP/yidW/yieP/yihL
aas/acs/entE/entF/fadD/plsC/plsB/ubiH/visC/yaaM
acrA/envC/emrA/dcp/prlC/yibH/yjcR
acrB/envD/glyA/nupC/secD/secF/rhaT/yiiI
adhE/ald/betB/aldH/fucD/gabD/gldA/proA
adi/cadA/speC/speF/recE
thrA/metL/lysC/smbA
argD/bioA/gabT/hemL
argT/artI/artJ/glnH/ygiF
asnB/glms/dapB/zwf/purF
bioC/cfa/prc/ubiG/yigO
bisC/dmsA/fdhF/fdnG/fdoG/narG/narZ
cysE/dapD/aroA/murZ/lacA/firA/lpxA/gcvT/kdsB/yieA
fumA/fumB/ttdA/ttdB
galE/rfaD/rffE/yefA/yefB
gcl/ilvB/ilvG/ilvI/poxB/menD
glpA/glpD/murD/murE/murF/murC/folC
grpE/hepA/recN/hsdR/yicC/yiiG
guaA/pabA/trpG/carA/ybiB
gyrA/parC/eftS/hisC/malY
helD/rep/urvD/recD/recB/recC
hflC/hflK/metE
fruK/pfkB/rbsK/yeiC/yihV
tsr/tar/trg/tap/metH/mopA

cysK/cysM/ilvA/tdcB/aslA/sdhD/sbm/ptr/phoU/trpB/yddC/yidJ
ddlA/ddlB/accC/carB/purD/gshB/rimK
deoR/glpR/gutR/fucR/yihW
ndh/gor/nirB/udhA/lpdA/trxB/cysI/pntA
entA/fabG/gutD/hdhA/envM
era/hflX/thdF
fdoH/fdnH/narH/narY/hycB/hydN/dmsB/nrfC/yffE
fpr/fre/hmpA
fimB/fime/xerC/xprB
fimF/fimG/fimA/htrA/htrH/phnJ
fliA/rpoH/rpoD/rpoS/cheZ/otsB
fliC/flgL/alaS/ydeK
fucK/glpK/araB/xylB
mukB/mutS/secA/hemX/sbcC/yidA
nagR/yajF/yfjA/yjcT
nmpC/ompC/ompF/phoE
sucB/aceF/accB
mrdA/ftsI/mrcA/mrcB/alkA
glgP/malP/moeB/thiF
pnp/rph/cpsG/yhbF
lysS/lysU/aspS/asnS/glyQ/yjeA/umuC
yigK/chaA/nrfE
hycC/nuoM/kup/mdoH/rarD/rodA/ftsW/yijE/yedA/yicL/yigM/yggA/ydeD/
  yhbE/ppa/ileS/leuS/metG/valS/cysS

(Table continues)

TABLE 10 *Continued*

353 sequences belonging to large families

abc/araG/artPbtuD/cydD/cysA/fecE/fepC/fhuC/ftsE/glnQ/hisP/livF/livG/malK/mdl/cydC/mglA/modC/msbA/phnC/phnK/phnL/phrA/potA/potG/proV/
  pstB/rbsA/ugpC/uvrA/yjgA/yabJ/ybbA/yddA/yhbG/yjcW/yjgA/xseA

ada/appY/araC/celD/envY/marA/melR/rhaR/rhaS/rob/soxS/yidL/yijO

artM/artQ/cysT/cysW/hisM/glnP/malF/malG/modB/potB/potC/potH/potI/proW/pstA/pstC/ugpA/ugpE/phnE/yabK/yaeE/ybfB/yjbI/yjcF/kup

ascG/cytR/ebgR/fruR/galR/galS/lacI/malI/purR/rbsB/rbsR/araF/mglB

btuB/cirA/fecA/fepA/fhuA/fhuE

btuC/fecC/fecD/fepD/fepG/fhuB/rfe/kefC/trkA/yjcE

cynR/cysB/gcvA/iciA/ilvY/leuO/lysR/metR/nhaR/oxyR/tdcA/yafC/ybeF/yeiE/yfeB/yidZ/fliG/nagA/rhaA

cysN/fusA/tufA/tufB/infB/lepA/selB/yihK/yigN/tnaA

btuR/dpbA/deaD/mfd/priA/recG/recQ/rhlB/rhlE/srmB/uvrB

arcA/baeR/basR/cheB/cheY/creB/evgA/fimZ/hydG/kdpE/narL/narP/ntrC/ompR/phoB/phoP/rcsB/tyrR/uhpA/rhlA/yfhA/yecB/yiiA/hemA/clpA/clpB/
  yifB/ruvB/cadC/kdtA/surA/glyS

arcB/baeS/barA/basS/cpxA/creC/envZ/evgS/hydH/kdpD/ntrB/phoQ/phoR/uspT/rcsC/cheA/polA/nusA/malT/narQ/narX/uhpB/creD/yddD

appB/araE/araH/araJ/aroP/kdpA/atpB/bcr/betT/cadB/codB/cvpA/cydB/e mrB/exbB/fliR/fucP/gabP/glpT/gltS/hypD/kgtP/lacY/livH/livM/lysP/
  melB/mglC/mreD/mraY/mtr/narI/narK/narV/nirC/nhaA/nhaB/nupG/panF/pheP/pntB/potE/proP/nagE/putP/rbsC/rfaL/rffT/secY/tdcC/tnaB/
  tyrP/uhpC/uhpT/uidR/xylE/yaaJ/yaaP/yaaU/yabE/yabM/ybdA/ycaD/ycaE/yceE/yciB/ydeA/ydeE/yfcA/yggB/yhiD/yicE/yicK/yicM/yicO/yicQ/yidE/
  yidK/yidT/yidY/yieG/yieO/yifK/yihG/yihN/yihO/yihP/yihY/yjbB/yjcD/yjcG/yjcV/yprA

agp/kdpB/atpF/damX/dedD/deoC/dnaK/dnaX/fecI/ftsA/ftsN/ftsY/hisD/holB/mreB/mutL/pfl/ppa/pspA/rlpA/rne/ffh/tolA/tonB/yfgA/yhdF/yigT/yijL/yzfB

cmtA/asd/glpF/glpG/gltB/gltP/ppsA/ptsI/bglF/ascF/fruA/ptsG/mtlA/nagE/frdB/crr/cmtB/tesB/usg/ygiE/yhbI/yhiA/yidN/yidO/yiiJ/yiiK/yijH/yijJ/yjcQ

---

[a]PAM value, accepted point mutations (644).

MAP RELATIONSHIPS AMONG FUNCTIONALLY RELATED *E. COLI* GENES

Some 20 years ago, a proposal was put forward that evolution of the *E. coli* genome might have occurred by successive duplications of the entire genome and that as a consequence, functionally and ancestrally related genes might be located either 90° or 180° from each other on the genetic map (2278, 2279). With many more genes now mapped than were at the time, one can test whether there is a tendency for genes related either by cellular function or by type of protein to cluster at 90° and 180° positions.

When the map positions of genes underlying each functional category as defined in Table 1 were examined, they did not lie at regular positions on the circular map. When map positions of genes for enzymes that catalyze similar reactions were examined, again no pattern of gene location was seen. For instance, phosphotransferase enzymes with an alcohol group as acceptor are enzymes with EC numbers beginning with 2.7.1. They were not clustered at 90° or 180° positions, nor were oxidoreductases acting on the CH-OH group of donors with NAD^+ or $NADP^+$ as the acceptor (EC numbers beginning 1.1.1.). Therefore, the idea of whole genome doubling (745, 1344, 1869, 2278, 2279) does not find support in current *E. coli* genetic data.

LITERATURE CITED

1. Aagaard, C., and S. Douthwaite. 1994. Requirement for a conserved, tertiary interaction in the core of 23S ribosomal RNA. *Proc. Natl. Acad. Sci. USA* 91:2989–2993.
2. Abbott, J., and D. Beckett. 1993. Cooperative binding of the *Escherichia coli* repressor of biotin biosynthesis to the biotin operator sequence. *Biochemistry* 32:9649–9656.
3. Abdurashidova, G. G., E. A. Tsvetkova, and E. I. Budowsky. 1991. Direct tRNA-protein interactions in ribosomal complexes. *Nucleic Acids Res.* 19:1909–1915.
4. Abeysinghe, S. I., P. J. Baker, D. W. Rice, H. F. Rodgers, T. J. Stillman, Y. H. Ko, B. A. McFadden, and H. G. Nimmo. 1991. Use of chemical modification in the crystallization of isocitrate lyase from *Escherichia coli. J. Mol. Biol.* 220:13–16.
5. Abouhamad, W. N., M. Manson, M. M. Gibson, and C. F. Higgins. 1991. Peptide transport and chemotaxis in *Escherichia coli* and *Salmonella typhimurium*: characterization of the dipeptide permease (Dpp) and the dipeptide-binding protein. *Mol. Microbiol.* 5:1035–1047.
6. Abril, N., and C. Pueyo. 1990. Mutagenesis in *Escherichia coli* lacking catalase. *Environ. Mol. Mutagen.* 15:184–189.
7. Adams, M. D., L. M. Wagner, T. J. Graddis, R. Landick, T. K. Antonucci, A. L. Gibson, and D. L. Oxender. 1990. Nucleotide sequence and genetic characterization reveal six essential genes for the LIV-I and LS transport systems of *Escherichia coli. J. Biol. Chem.* 265:11436–11443.
8. Ahlem, C., W. Huisman, G. Neslund, and A. S. Dahms. 1982. Purification and properties of a periplasmic D-xylose-binding protein from *Escherichia coli* K-12. *J. Biol. Chem.* 257:2926–2931.
9. Ahmad, S., N. A. Glavas, and P. D. Bragg. 1992. Subunit interactions involved in the assembly of pyridine nucleotide transhydrogenase in the membranes of *Escherichia coli. J. Biol. Chem.* 267:7007–7012.
10. Ahmad, S., N. A. Glavas, and P. D. Bragg. 1992. A mutation at Gly314 of the beta subunit of the *Escherichia coli* pyridine nucleotide transhydrogenase abolishes activity and affects the NADP(H)-induced conformational change. *Eur. J. Biochem.* 207:733–739.
11. Aiba, A., and K. Mizobuchi. 1989. Nucleotide sequence analysis of genes purH and purD involved in the de novo purine nucleotide biosynthesis of *Escherichia coli. J. Biol. Chem.* 264:21239–21246.
12. Aisaka, K., A. Igarashi, K. Yamaguchi, and T. Uwajima. 1991. Purification, crystallization and characterization of N-acetylneuraminate lyase from *Escherichia coli. Biochem. J.* 276:541–546.
13. Aisaka, K., and T. Uwajima. 1986. Cloning and constitutive expression of the N-acetylneuraminate lyase gene of *Escherichia coli. Appl. Environ. Microbiol.* 51:562–565.
14. Akimaru, H., K. Sakumi, T. Yoshikai, M. Anai, and M. Sekiguchi. 1990. Positive and negative regulation of transcription by a cleavage product of Ada protein. *J. Mol. Biol.* 216:261–273.
15. Akiyama, M., E. Crooke, and A. Kornberg. 1992. The polyphosphate kinase gene of *Escherichia coli*. Isolation and sequence of the ppk gene and membrane location of the protein. *J. Biol. Chem.* 267:22556–22561.
16. Akiyama, M., E. Crooke, and A. Kornberg. 1993. An exopolyphosphatase of *Escherichia coli*. The enzyme and its ppx gene in a polyphosphate operon. *J. Biol. Chem.* 268:633–639.

17. **Akiyama, M., H. Maki, M. Sekiguchi, and T. Horiuchi.** 1989. A specific role of MutT protein: to prevent dG.dA mispairing in DNA replication. *Proc. Natl. Acad. Sci. USA* 86:3949–3952.

18. **Akiyama, Y., Y. Shirai, and K. Ito.** 1994. Involvement of FtsH in protein assembly into and through the membrane. II. Dominant mutations affecting FtsH functions. *J. Biol. Chem.* 269:5225–5229.

19. **Alam, K. Y., and D. P. Clark.** 1991. Molecular cloning and sequence of the *thdF* gene, which is involved in thiophene and furan oxidation by *Escherichia coli. J. Bacteriol.* 173:6018–6024.

20. **Alam, K. Y., M. J. Worland, and D. P. Clark.** 1990. Analysis and molecular cloning of genes involved in thiophene and furan oxidation by *E. coli. Appl. Biochem. Biotechnol.* 24–25:843–855.

21. **al-Bar, O. A., C. D. O'Connor, I. G. Giles, and M. Akhtar.** 1992. D-Alanine:D-alanine ligase of *Escherichia coli*. Expression, purification and inhibitory studies on the cloned enzyme. *Biochem. J.* 282:747–752.

22. **Aldea, M., T. Garrido, C. Hernandez-Chico, M. Vicente, and S. R. Kushner.** 1989. Induction of a growth-phase-dependent promoter triggers transcription of bolA, an *Escherichia coli* morphogene. *EMBO J.* 8:3923–3931.

23. **Alefounder, P. R., S. A. Baldwin, R. N. Perham, and N. J. Short.** 1989. Cloning, sequence analysis and over-expression of the gene for the class II fructose 1,6-bisphosphate aldolase of *Escherichia coli. Biochem. J.* 257:529–534.

24. **Alefounder, P. R., and R. N. Perham.** 1989. Identification, molecular cloning and sequence analysis of a gene cluster encoding the class II fructose 1,6-bisphosphate aldolase, 3-phosphoglycerate kinase and a putative second glyceraldehyde 3-phosphate dehydrogenase of *Escherichia coli. Mol. Microbiol.* 3:723–732.

25. **Aleixandre, V., A. Urios, G. Herrera, and M. Blanco.** 1989. New *Escherichia coli* gyrA and gyrB mutations which have a graded effect on DNA supercoiling. *Mol. Gen. Genet.* 219:306–312.

26. **Alexander-Caudle, C., L. M. Latinwo, and J. H. Jackson.** 1990. Acetohydroxy acid synthase activity from a mutation at *ilvF* in *Escherichia coli* K-12. *J. Bacteriol.* 172:3060–3065.

27. **Alexeev, D., S. M. Bury, C. W. Boys, M. A. Turner, L. Sawyer, A. J. Ramsey, H. C. Baxter, and R. L. Baxter.** 1994. Sequence and crystallization of *Escherichia coli* dethiobiotin synthetase, the penultimate enzyme of biotin biosynthesis. *J. Mol. Biol.* 235:774–776.

28. **Allen, G. C., Jr., and A. Kornberg.** 1991. The priB gene encoding the primosomal replication n protein of *Escherichia coli. J. Biol. Chem.* 266:11610–11613.

29. **Allen, G. C., Jr., and A. Kornberg.** 1993. Assembly of the primosome of DNA replication in *Escherichia coli. J. Biol. Chem.* 268:19204–19209.

30. **Allen, P. N., and H. F. Noller.** 1989. Mutations in ribosomal proteins S4 and S12 influence the higher order structure of 16 S ribosomal RNA. *J. Mol. Biol.* 208:457–468.

31. **Allen, S. P., J. O. Polazzi, J. K. Gierse, and A. M. Easton.** 1992. Two novel heat shock genes encoding proteins produced in response to heterologous protein expression in *Escherichia coli. J. Bacteriol.* 174:6938–6947.

32. **Allgood, N. D., and T. J. Silhavy.** 1991. *Escherichia coli* xonA (sbcB) mutants enhance illegitimate recombination. *Genetics* 127:671–680.

33. **Allibert, P., J. C. Willison, and P. M. Vignais.** 1987. Complementation of nitrogen-regulatory (ntr-like) mutations in *Rhodobacter capsulatus* by an *Escherichia coli* gene: cloning and sequencing of the gene and characterization of the gene product. *J. Bacteriol.* 169:260–271.

34. **Allikmets, R., B. Gerrard, D. Court, and M. Dean.** 1993. Cloning and organization of the abc and mdl genes of *Escherichia coli*: relationship to eukaryotic multidrug resistance. *Gene* 136:231–236.

35. **Allison, N., C. H. Williams, Jr., and J. R. Guest.** 1988. Overexpression and mutagenesis of the lipoamide dehydrogenase of *Escherichia coli. Biochem. J.* 256:741–749.

36. **Almassy, R. J., C. A. Janson, R. Hamlin, N. H. Xuong, and D. Eisenburg.** 1986. Novel subunit-subunit interactions in the structure of glutamine synthetase. *Nature* (London) 323:304–309.

37. **Almiron, M., A. J. Link, D. Furlong, and R. Kolter.** 1992. A novel DNA-binding protein with regulatory and protective roles in starved *Escherichia coli. Genes Dev.* 6:2646–2654.

38. **Altamirano, M. M., J. A. Plumbridge, and M. L. Calcagno.** 1992. Identification of two cysteine residues forming a pair of vicinal thiols in glucosamine-6-phosphate deaminase from *Escherichia coli* and a study of their functional role by site-directed mutagenesis. *Biochemistry* 31:1153–1158.

39. **Altendorf, K., A. Siebers, and W. Epstein.** 1992. The KDP ATPase of *Escherichia coli. Ann. N. Y. Acad. Sci.* 671:228–243.

40. **Alwan, A. F., B. I. Mgbeje, and P. M. Jordan.** 1989. Purification and properties of uroporphyrinogen III synthase (co-synthase) from an overproducing recombinant strain of *Escherichia coli* K-12. *Biochem. J.* 264:397–402.

41. **Amabile-Cuevas, C. F., and B. Demple.** 1991. Molecular characterization of the soxRS genes of *Escherichia coli*: two genes control a superoxide stress regulon. *Nucleic Acids Res.* 19:4479–4484.

42. **Amaratunga, M., and T. M. Lohman.** 1993. *Escherichia coli* rep helicase unwinds DNA by an active mechanism. *Biochemistry* 32:6815–6820.

43. **Ambudkar, S. V., V. Anantharam, and P. C. Maloney.** 1990. UhpT, the sugar phosphate antiporter of *Escherichia coli*, functions as a monomer. *J. Biol. Chem.* 265:12287–12292.

44. **Amouyal, M., L. Mortensen, H. Buc, and K. Hammer.** 1989. Single and double loop formation when deoR repressor binds to its natural operator sites. *Cell* 58:545–551.

45. **An, G., D. S. Bendiak, L. A. Mamelak, and J. D. Friesen.** 1981. Organization and nucleotide sequence of a new ribosomal operon in *Escherichia coli* containing the genes for ribosomal protein S2 and elongation factor Ts. *Nucleic Acids Res.* 9:4163–4172.

46. **Andersen, J., and N. Delihas.** 1990. micF RNA binds to the 5′ end of ompF mRNA and to a protein from *Escherichia coli. Biochemistry* 29:9249–9256.

47. **Andersen, J. T., K. F. Jensen, and P. Poulsen.** 1991. Role of transcription pausing in the control of the pyrE attenuator in *Escherichia coli. Mol. Microbiol.* 5:327–333.

48. **Andersen, P. S., J. M. Smith, and B. Mygind.** 1992. Characterization of the upp gene encoding uracil phosphoribosyltransferase of *Escherichia coli* K12. *Eur. J. Biochem.* 204:51–56.

49. **Anderson, A., and R. A. Cooper.** 1970. Biochemical and genetical studies on ribose catabolism in *Escherichia coli* K12. *J. Gen. Microbiol.* 62:335–339.

50. **Anderson, D. H., L. J. Donald, M. V. Jacob, and H. W. Duckworth.** 1991. A mutant of *Escherichia coli* citrate synthase that affects the allosteric equilibrium. *Biochem. Cell Biol.* 69:232–238.

51. **Anderson, K. S., and K. A. Johnson.** 1990. "Kinetic competence" of the 5-enolpyruvoylshikimate-3-phosphate synthase tetrahedral intermediate. *J. Biol. Chem.* 265:5567–5572.

52. **Anderson, P. M., Y. C. Sung, and J. A. Fuchs.** 1990. The cyanase operon and cyanate metabolism. *FEMS Microbiol. Rev.* 7:247–252.

53. **Andresen, P. A., I. Kaasen, O. B. Styrvold, G. Boulnois, and A. R. Strom.** 1988. Molecular cloning, physical mapping and expression of the bet genes governing the osmoregulatory choline-glycine betaine pathway of *Escherichia coli. J. Gen. Microbiol.* 134:1737–1746.

54. **Andrews, J. C., T. C. Blevins, and S. A. Short.** 1986. Regulation of peptide transport in *Escherichia coli*: induction of the *trp*-linked operon encoding the oligopeptide permease. *J. Bacteriol.* 165:428–433.

55. **Andrews, J. C., and S. A. Short.** 1986. opp-lac operon fusions and transcriptional regulation of the *Escherichia coli trp*-linked oligopeptide permease. *J. Bacteriol.* 165:434–442.

56. **Andrews, K. J., and E. C. C. Lin.** 1976. Thiogalactoside transacetylase of the lactose operon as an enzyme for detoxification. *J. Bacteriol.* 128:510–513.

57. **Andrews, S. C., and J. R. Guest.** 1988. Nucleotide sequence of the gene encoding the GMP reductase of *Escherichia coli* K12. *Biochem. J.* 255:35–43.

58. **Andrews, S. C., D. Shipley, J. N. Keen, J. B. Findlay, P. M. Harrison, and J. R. Guest.** 1992. The haemoglobin-like protein (HMP) of *Escherichia coli* has ferrisiderophore reductase activity and its C-terminal domain shares homology with ferredoxin NADP+ reductases. *FEBS Lett.* 302:247–252.

59. **Andrews, S. C., J. M. Smith, J. R. Guest, and P. M. Harrison.** 1989. Amino acid sequence of the bacterioferritin (cytochrome b1) of *Escherichia coli*-K12. *Biochem. Biophys. Res. Commun.* 158:489–496.

60. **Andrews, S. C., J. M. Smith, C. Hawkins, J. M. Williams, P. M. Harrison, and J. R. Guest.** 1993. Overproduction, purification and characterization of the bacterioferritin of *Escherichia coli* and a C-terminally extended variant. *Eur. J. Biochem.* 213:329–338.

61. **Angelaccio, S., S. Pascarella, E. Fattori, F. Bossa, W. Strong, and V. Schirch.** 1992. Serine hydroxymethyltransferase: origin of substrate specificity. *Biochemistry* 31:155–162.

62. **Anton, I. A., and J. R. Coggins.** 1988. Sequencing and overexpression of the *Escherichia coli* aroE gene encoding shikimate dehydrogenase. *Biochem. J.* 249:319–326.

63. **Apirion, D., and N. Watson.** 1975. Mapping and characterization of a mutation in *Escherichia coli* that reduces the level of ribonuclease III specific for double-stranded ribonucleic acid. *J. Bacteriol.* 124:317–324.

64. **Ardeshir, F., and G. F. L. Ames.** 1980. Cloning of the histidine transport genes from *Salmonella typhimurium* and characterization of an analogous transport system in *Escherichia coli. J. Supramol. Struct.* 13:117–130.

65. **Arevalo, M. A., F. Tejedor, F. Polo, and J. P. Ballesta.** 1988. Protein components of the erythromycin binding site in bacterial ribosomes. *J. Biol. Chem.* 263:58–63.

66. **Argaet, V. P., T. J. Wilson, and B. E. Davidson.** 1994. Purification of the *Escherichia coli* regulatory protein TyrR and analysis of its interactions with ATP, tyrosine, phenylalanine, and tryptophan. *J. Biol. Chem.* 269:5171–5178.

67. **Ariza, R. R., S. P. Cohen, N. Bachhawat, S. B. Levy, and B. Demple.** 1994. Repressor mutations in the marRAB operon that activate oxidative stress genes and multiple antibiotic resistance in *Escherichia coli. J. Bacteriol.* 176:143–148.

68. **Armstrong, S. K., M. H. Pettis, L. J. Forrester, and M. A. McIntosh.** 1989. The *Escherichia coli* enterobactin biosynthesis gene, entD: nucleotide sequence and membrane localization of its protein product. *Mol. Microbiol.* 3:757–766.

69. **Arnosti, D. N.** 1990. Regulation of *Escherichia coli* sigma F RNA polymerase by flhD and flhC flagellar regulatory genes. *J. Bacteriol.* 172:4106–4108.

70. **Arnqvist, A., A. Olsen, J. Pfeifer, D. G. Russell, and S. Normark.** 1992. The Crl protein activates cryptic genes for curli formation and fibronectin binding in *Escherichia coli* HB101. *Mol. Microbiol.* 6:2443–2452.

71. **Aronson, B. D., M. Levinthal, and R. L. Somerville.** 1989. Activation of a cryptic pathway for threonine metabolism via specific IS3-mediated alteration of promoter structure in *Escherichia coli. J. Bacteriol.* 171:5503–5511.

72. **Arvidson, D. N., M. Shapiro, and P. Youderian.** 1991. Mutant tryptophan aporepressors with altered specificities of corepressor recognition. *Genetics* 128:29–35.

73. **Asahara, H., H. Himeno, K. Tamura, T. Hasegawa, K. Watanabe, and M. Shimizu.** 1993. Recognition nucleotides of *Escherichia coli* tRNA(Leu) and its

elements facilitating discrimination from tRNA(Ser) and tRNA(Tyr). *J. Mol. Biol.* 231:219–229.

74. **Asahara, H., H. Himeno, K. Tamura, N. Nameki, T. Hasegawa, and M. Shimizu.** 1994. *Escherichia coli* seryl-tRNA synthetase recognizes tRNA(Ser) by its characteristic tertiary structure. *J. Mol. Biol.* 236:738–748.

75. **Asahara, H., P. M. Wistort, J. F. Bank, R. H. Bakerian, and R. P. Cunningham.** 1989. Purification and characterization of *Escherichia coli* endonuclease III from the cloned nth gene. *Biochemistry* 28:4444–4449.

76. **Asai, K., S. Fujisaki, Y. Nishimura, T. Nishino, K. Okada, T. Nakagawa, M. Kawamukai, and H. Matsuda.** 1994. The identification of *Escherichia coli* ispB (cel) gene encoding the octaprenyl diphosphate synthase. *Biochem. Biophys. Res. Commun.* 202:340–345.

77. **Asai, T., M. Takanami, and M. Imai.** 1990. The AT richness and gid transcription determine the left border of the replication origin of the *E. coli* chromosome. *EMBO J.* 9:4065–4072.

78. **Asoh, S., H. Matsuzawa, F. Ishino, J. L. Strominger, M. Matsuhashi, and T. Ohta.** 1986. Nucleotide sequence of the pbpA gene and characteristics of the deduced amino acid sequence of penicillin-binding protein 2 of *Escherichia coli* K12. *Eur. J. Biochem.* 160:231–238.

79. **Atkinson, M. R., and A. J. Ninfa.** 1992. Characterization of *Escherichia coli* glnL mutations affecting nitrogen regulation. *J. Bacteriol.* 174:4538–4548.

80. **Atlung, T., and F. G. Hansen.** 1993. Three distinct chromosome replication states are induced by increasing concentrations of DnaA protein in *Escherichia coli*. *J. Bacteriol.* 175:6537–6545.

81. **Atlung, T., A. Nielsen, and F. G. Hansen.** 1989. Isolation, characterization, and nucleotide sequence of appY, a regulatory gene for growth-phase-dependent gene expression in *Escherichia coli*. *J. Bacteriol.* 171:1683–1691.

82. **Au, K. G., S. Clark, J. H. Miller, and P. Modrich.** 1989. *Escherichia coli* mutY gene encodes an adenine glycosylase active on G-A mispairs. *Proc. Natl. Acad. Sci. USA* 86:8877–8881.

83. **Au, K. G., K. Welsh, and P. Modrich.** 1992. Initiation of methyl-directed mismatch repair. *J. Biol. Chem.* 267:12142–12148.

84. **Austin, D., and T. J. Larson.** 1991. Nucleotide sequence of the glpD gene encoding aerobic *sn*-glycerol 3-phosphate dehydrogenase of *Escherichia coli* K-12. *J. Bacteriol.* 173:101–107.

85. **Babul, J.** 1978. Phosphofructokinases from *Escherichia coli*. Purification and characterization of the nonallosteric enzyme. *J. Biol. Chem.* 253:4350–4355.

86. **Babul, J., and V. Guixe.** 1983. Fructose bisphosphatase from *Escherichia coli*. Purification and characterization. *Arch. Biochem. Biophys.* 225:944–949.

87. **Bachi, B., and H. L. Kornberg.** 1975. Genes involved in the uptake and catabolism of gluconate by *Escherichia coli*. *J. Gen. Microbiol.* 90:321–335.

88. **Bachmann, B. J.** 1987. Linkage map of *Escherichia coli* K-12, edition 7, p. 807–876. *In* F. C. Neidhardt, J. L. Ingraham, K. B. Low, B. Magasanik, M. Schaechter, and H. E. Umbarger (ed.), *Escherichia coli and Salmonella typhimurium: Cellular and Molecular Biology*, vol. 2. American Society for Microbiology, Washington, D.C.

89. **Badet-Denisot, M. A., and B. Badet.** 1992. Chemical modification of glucosamine-6-phosphate synthase by diethyl pyrocarbonate: evidence of histidine requirement for enzymatic activity. *Arch. Biochem. Biophys.* 292:475–478.

90. **Badia, J., L. Baldoma, J. Aguilar, and A. Boronat.** 1989. Identification of the rhaA, rhaB and rhaD gene products from *Escherichia coli* K-12. *FEMS Microbiol. Lett.* 53:253–257.

91. **Bailey, M. J., V. Koronakis, T. Schmoll, and C. Hughes.** 1992. *Escherichia coli* HlyT protein, a transcriptional activator of haemolysin synthesis and secretion, is encoded by the rfaH (sfrB) locus required for expression of sex factor and lipopolysaccharide genes. *Mol. Microbiol.* 6:1003–1012.

92. **Bailone, A., S. Sommer, J. Knezevic, and R. Devoret.** 1991. Substitution of UmuD' for UmuD does not affect SOS mutagenesis. *Biochimie* 73:471–478.

93. **Baird, L., B. Lipinska, S. Raina, and C. Georgopoulos.** 1991. Identification of the *Escherichia coli* sohB gene, a multicopy suppressor of the HtrA (DegP) null phenotype. *J. Bacteriol.* 173:5763–5770.

94. **Bairoch, A., and B. Boeckman.** 1993. The SWISS-PROT protein sequence data bank, recent developments. *Nucleic Acids Res.* 21:3093–3096.

95. **Bakin, A., and J. Ofengand.** 1993. Four newly located pseudouridylate residues in *Escherichia coli* 23S ribosomal RNA are all at the peptidyltransferase center: analysis by the application of a new sequencing technique. *Biochemistry* 32:9754–9762.

96. **Baldoma, L., J. Badia, G. Sweet, and J. Aguilar.** 1990. Cloning, mapping and gene product identification of rhaT from *Escherichia coli* K12. *FEMS Microbiol. Lett.* 60:103–107.

97. **Baldwin, S. A., R. N. Perham, and D. Stribling.** 1978. Purification and characterization of the class-II D-fructose 1,6-bisphosphate aldolase from *Escherichia coli* (Crookes' strain). *Biochem. J.* 169:633–641.

98. **Bally, M., M. Foglino, M. Bruschi, M. Murgier, and A. Lazdunski.** 1986. Nucleotide sequence of the promoter and amino-terminal encoding region of the *Escherichia coli* pepN gene. *Eur. J. Biochem.* 155:565–569.

99. **Banerjee, R. V., N. L. Johnston, J. K. Sobeski, P. Datta, and R. G. Matthews.** 1989. Cloning and sequence analysis of the *Escherichia coli* metH gene encoding cobalamin-dependent methionine synthase and isolation of a tryptic fragment containing the cobalamin-binding domain. *J. Biol. Chem.* 264:13888–13895.

100. **Baneyx, F., and G. Georgiou.** 1990. In vivo degradation of secreted fusion proteins by the *Escherichia coli* outer membrane protease OmpT. *J. Bacteriol.* 172:491–494.

101. **Baneyx, F., and G. Georgiou.** 1991. Construction and characterization of *Escherichia coli* strains deficient in multiple secreted proteases: protease III degrades high-molecular-weight substrates in vivo. *J. Bacteriol.* 173:2696–2703.

102. **Banuett, F., and I. Herskowitz.** 1987. Identification of polypeptides encoded by an *Escherichia coli* locus (hflA) that governs the lysis-lysogeny decision of bacteriophage lambda. *J. Bacteriol.* 169:4076–4085.

103. **Barber, A. M., V. B. Zhurkin, and S. Adhya.** 1993. CRP-binding sites: evidence for two structural classes with 6-bp and 8-bp spacers. *Gene* 130:1–8.

104. **Barbier, C. S., and S. A. Short.** 1992. Amino acid substitutions in the CytR repressor which alter its capacity to regulate gene expression. *J. Bacteriol.* 174:2881–2890.

105. **Bardwell, J. C., J. O. Lee, G. Jander, N. Martin, D. Belin, and J. Beckwith.** 1993. A pathway for disulfide bond formation in vivo. *Proc. Natl. Acad. Sci. USA* 90:1038–1042.

106. **Baron, C., and A. Bock.** 1991. The length of the aminoacyl-acceptor stem of the selenocysteine-specific tRNA(Sec) of *Escherichia coli* is the determinant for binding to elongation factors SELB or Tu. *J. Biol. Chem.* 266:20375–20379.

107. **Barr, K., S. Ward, U. Meier-Dieter, H. Mayer, and P. D. Rick.** 1988. Characterization of an *Escherichia coli* rff mutant defective in transfer of N-acetylmannosaminuronic acid (ManNAcA) from UDP-ManNAcA to a lipid-linked intermediate involved in enterobacterial common antigen synthesis. *J. Bacteriol.* 170:228–233.

108. **Bartkus, J. M., B. Tyler, and J. M. Calvo.** 1991. Transcription attenuation-mediated control of *leu* operon expression: influence of the number of Leu control codons. *J. Bacteriol.* 173:1634–1641.

109. **Bartsch, K., A. von Johnn-Marteville, and A. Schulz.** 1990. Molecular analysis of two genes of the *Escherichia coli* gab cluster: nucleotide sequence of the glutamate:succinic semialdehyde transaminase gene (gabT) and characterization of the succinic semialdehyde dehydrogenase gene (gabD). *J. Bacteriol.* 172:7035–7042.

110. **Batt, C. A., A. C. Jamieson, and M. A. Vandeyar.** 1990. Identification of essential histidine residues in the active site of *Escherichia coli* xylose (glucose) isomerase. *Proc. Natl. Acad. Sci. USA* 87:618–622.

111. **Bauer, A. J., I. Rayment, P. A. Frey, and H. M. Holden.** 1992. The molecular structure of UDP-galactose 4-epimerase from *Escherichia coli* determined at 2.5 Å resolution. *Proteins* 12:372–381.

112. **Bauer, K., M. Struyve, D. Bosch, R. Benz, and J. Tommassen.** 1989. One single lysine residue is responsible for the special interaction between polyphosphate and the outer membrane porin PhoE of *Escherichia coli*. *J. Biol. Chem.* 264:16393–16398.

113. **Baughman, G. A., and S. R. Fahnestock.** 1979. Chloramphenicol resistance mutation in *Escherichia coli* which maps in the major ribosomal protein gene cluster. *J. Bacteriol.* 137:1315–1323.

114. **Beadle, G. W., and E. L. Tatum.** 1941. Genetic control of biochemical reactions in *Neurospora*. *Proc. Natl. Acad. Sci. USA* 27:499–506.

115. **Bebbington, K. J., and H. D. Williams.** 1993. Investigation of the role of the cydD gene product in production of a functional cytochrome d oxidase in *Escherichia coli*. *FEMS Microbiol. Lett.* 112:19–24.

116. **Bedouelle, H.** 1990. Recognition of tRNA(Tyr) by tyrosyl-tRNA synthetase. *Biochimie* 72:589–598.

117. **Bedouelle, H., V. Guez, A. Vidal-Cros, and M. Hermann.** 1990. Overproduction of tyrosyl-tRNA synthetase is toxic to *Escherichia coli*: a genetic analysis. *J. Bacteriol.* 172:3940–3945.

118. **Begg, K. J., A. Takasuga, D. H. Edwards, S. J. Dewar, B. G. Spratt, H. Adachi, T. Ohta, H. Matsuzawa, and W. D. Donachie.** 1990. The balance between different peptidoglycan precursors determines whether *Escherichia coli* cells will elongate or divide. *J. Bacteriol.* 172:6697–6703.

119. **Begg, K. J., T. Tomoyasu, W. D. Donachie, M. Khattar, H. Niki, K. Yamanaka, S. Hiraga, and T. Ogura.** 1992. *Escherichia coli* mutant Y16 is a double mutant carrying thermosensitive ftsH and ftsI mutations. *J. Bacteriol.* 174:2416–2417.

120. **Bejar, S., F. Bouche, and J. P. Bouche.** 1988. Cell division inhibition gene dicB is regulated by a locus similar to lambdoid bacteriophage immunity loci. *Mol. Gen. Genet.* 212:11–19.

121. **Belitskii, B. R., and R. S. Shakulov.** 1988. Cloning of the gpp gene of *Escherichia coli* and the use of recBC, sbcB cells for inserting its mutant allele into the chromosomal structure. *Genetika* 24:1333–1342. (In Russian.)

122. **Bell, A. W., S. D. Buckel, J. M. Groarke, J. N. Hope, D. H. Kingsley, and M. A. Hermodson.** 1986. The nucleotide sequences of the rbsD, rbsA, and rbsC genes of *Escherichia coli* K12. *J. Biol. Chem.* 261:7652–7658.

123. **Bell, P. E., C. D. Nau, J. T. Brown, J. Konisky, and R. J. Kadner.** 1990. Genetic suppression demonstrates interaction of TonB protein with outer membrane transport proteins in *Escherichia coli*. *J. Bacteriol.* 172:3826–3829.

124. **Bell, P. J., S. C. Andrews, M. N. Sivak, and J. R. Guest.** 1989. Nucleotide sequence of the FNR-regulated fumarase gene (fumB) of *Escherichia coli* K-12. *J. Bacteriol.* 171:3494–3503.

125. **Belunis, C. J., and C. R. Raetz.** 1992. Biosynthesis of endotoxins. Purification and catalytic properties of 3-deoxy-D-manno-octulosonic acid transferase from *Escherichia coli*. *J. Biol. Chem.* 267:9988–9997.

126. **Ben-Bassat, A., K. Bauer, S. Y. Chang, K. Myambo, A. Boosman, and S. Chang.** 1987. Processing of the initiation methionine from proteins: properties of the *Escherichia coli* methionine aminopeptidase and its gene structure. *J. Bacteriol.* 169:751–757.

127. Bende, S. M., and R. H. Grafstrom. 1991. The DNA binding properties of the MutL protein isolated from *Escherichia coli*. *Nucleic Acids Res.* 19:1549–1555.

128. Benedetti, H., C. Lazdunski, and R. Lloubes. 1991. Protein import into *Escherichia coli*: colicins A and E1 interact with a component of their translocation system. *EMBO J.* 10:1989–1995.

129. Bennett, R. J., H. J. Dunderdale, and S. C. West. 1993. Resolution of Holliday junctions by RuvC resolvase: cleavage specificity and DNA distortion. *Cell* 74:1021–1031.

130. Benson, T. E., J. L. Marquardt, A. C. Marquardt, F. A. Ekzhorn, and C. T. Walsh. 1993. Overexpression, purification, and mechanistic study of UDP-N-acetyl-enolpyruvylglucosamine reductase. *Biochemistry* 32:2024–2030.

131. Bentley, J., L. S. Hyatt, K. Ainley, J. H. Parish, R. B. Herbert, and G. R. White. 1993. Cloning and sequence analysis of an *Escherichia coli* gene conferring bicyclomycin resistance. *Gene* 127:117–120.

132. Bentley, R., and R. Meganathan. 1987. Biosynthesis of the isoprenoid quinones ubiquinone and menaquinone, p. 512–520. *In* F. C. Neidhardt, J. L. Ingraham, K. B. Low, B. Magasanik, M. Schaechter, and H. E. Umbarger (ed.), *Escherichia coli and Salmonella typhimurium: Cellular and Molecular Biology.* American Society for Microbiology, Washington, D.C.

133. Benz, R., E. Maier, and I. Gentschev. 1993. TolC of *Escherichia coli* functions as an outer membrane channel. *Int. J. Med. Microbiol. Virol. Parasitol. Infect. Dis.* 278:187–196.

134. Benzer, S. 1955. Fine structure of a genetic region in bacteriophage. *Proc. Natl. Acad. Sci. USA* 41:344–354.

135. Berg, B. L., J. Li, J. Heider, and V. Stewart. 1991. Nitrate-inducible formate dehydrogenase in *Escherichia coli* K-12. I. Nucleotide sequence of the *fdnGHI* operon and evidence that opal (UGA) encodes selenocysteine. *J. Biol. Chem.* 266:22380–22385.

136. Berg, B. L., and V. Stewart. 1990. Structural genes for nitrate-inducible formate dehydrogenase in *Escherichia coli* K-12. *Genetics* 125:691–702.

137. Bergler, H., G. Hogenauer, and F. Turnowsky. 1992. Sequences of the envM gene and of two mutated alleles in *Escherichia coli*. *J. Gen. Microbiol.* 138:2093–2100.

138. Bergler, H., P. Wallner, A. Ebeling, B. Leitinger, S. Fuchsbichler, H. Aschauer, G. Kollenz, G. Hogenauer, and F. Turnowsky. 1994. Protein EnvM is the NADH-dependent enoyl-ACP reductase (FabI) of *Escherichia coli*. *J. Biol. Chem.* 269:5493–5496.

139. Berleth, E. S., J. Li, J. A. Braunscheidel, and C. M. Pickart. 1992. A reactive nucleophile proximal to vicinal thiols is an evolutionarily conserved feature in the mechanism of Arg aminoacyl-tRNA protein transferase. *Arch. Biochem. Biophys.* 298:498–504.

140. Bethell, R. C., and G. Lowe. 1988. The stereochemical course of D-glyceraldehyde-induced ATPase activity of glycerokinase from *Escherichia coli*. *Eur. J. Biochem.* 174:387–389.

141. Beyer, W. F., Jr., and I. Fridovich. 1991. In vivo competition between iron and manganese for occupancy of the active site region of the manganese-superoxide dismutase of *Escherichia coli*. *J. Biol. Chem.* 266:303–308.

142. Beyer, W. F., Jr., J. A. Reynolds, and I. Fridovich. 1989. Differences between the manganese- and the iron-containing superoxide dismutases of *Escherichia coli* detected through sedimentation equilibrium, hydrodynamic, and spectroscopic studies. *Biochemistry* 28:4403–4409.

143. Bezirdzhian, K. O., S. M. Kocharian, and Z. I. Akopian. 1986. Isolation of the hexameric form of purine nucleoside phosphorylase from *E. coli*. Comparative study of trimeric and hexameric forms of the enzyme. *Biokhimiia* 51:1085–1092. (In Russian.)

144. Bezirdzhian, K. O., S. M. Kocharian, and Z. I. Akopian. 1987. Hexameric purine nucleoside phosphorylase II from *Escherichia coli* K-12. Physico-chemical and catalytic properties and stabilization with substrates. *Biokhimiia* 52:1624–1631. (In Russian.)

145. Bhatnagar, S. K., L. C. Bullions, and M. J. Bessman. 1991. Characterization of the mutT nucleoside triphosphatase of *Escherichia coli*. *J. Biol. Chem.* 266:9050–9054.

146. Bi, E., and J. Lutkenhaus. 1993. Cell division inhibitors SulA and MinCD prevent formation of the FtsZ ring. *J. Bacteriol.* 175:1118–1125.

147. Bianchi, V., R. Eliasson, M. Fontecave, E. Mulliez, D. M. Hoover, R. G. Matthews, and P. Reichard. 1993. Flavodoxin is required for the activation of the anaerobic ribonucleotide reductase. *Biochem. Biophys. Res. Commun.* 197:792–797.

148. Bianchi, V., P. Reichard, R. Eliasson, E. Pontis, M. Krook, H. Jornvall, and E. Haggard-Ljungquist. 1993. *Escherichia coli* ferredoxin NADP$^+$ reductase: activation of *E. coli* anaerobic ribonucleotide reduction, cloning of the gene (*fpr*), and overexpression of the protein. *J. Bacteriol.* 175:1590–1595.

149. Bibi, E., and H. R. Kaback. 1990. In vivo expression of the *lacY* gene in two segments leads to functional lac permease. *Proc. Natl. Acad. Sci. USA* 87:4325–4329.

150. Bilgin, N., F. Claesens, H. Pahverk, and M. Ehrenberg. 1992. Kinetic properties of *Escherichia coli* ribosomes with altered forms of S12. *J. Mol. Biol.* 224:1011–1027.

151. Bilgin, N., J. I. Lee, H. Y. Zhu, R. Dalbey, and G. von Heijne. 1990. Mapping of catalytically important domains in *Escherichia coli* leader peptidase. *EMBO J.* 9:2717–2722.

152. Billheimer, J. T., H. N. Carnevale, T. Leisinger, T. Eckhardt, and E. E. Jones. 1976. Ornithine delta-transaminase activity in *Escherichia coli*: its identity with acetylornithine delta-transaminase. *J. Bacteriol.* 127:1315–1323.

153. Billheimer, J. T., M. Y. Shen, H. N. Carnevale, H. R. Horton, and E. E. Jones. 1979. Isolation and characterization of acetylornithine delta-transaminase of wild-type *Escherichia coli* W. Comparison with arginine-inducible acetylornithine delta-transaminase. *Arch. Biochem. Biophys.* 195:401–413.

154. Binnie, R. A., H. Zhang, S. Mowbray, and M. A. Hermodson. 1992. Functional mapping of the surface of *Escherichia coli* ribose-binding protein: mutations that affect chemotaxis and transport. *Protein Sci.* 1:1642–1651.

155. Biville, F., E. Turlin, and F. Gasser. 1991. Mutants of *Escherichia coli* producing pyrroloquinoline quinone. *J. Gen. Microbiol.* 137:1775–1782.

156. Bjelland, S., and E. Seeberg. 1987. Purification and characterization of 3-methyladenine DNA glycosylase I from *Escherichia coli*. *Nucleic Acids Res.* 15:2787–2801.

157. Bjork, G. R., and K. Kjellin-Straby. 1978. *Escherichia coli* mutants with defects in the biosynthesis of 5-methylaminomethyl-2-thio-uridine or 1-methylguanosine in their tRNA. *J. Bacteriol.* 133:508–517.

158. Black, D. S., B. Irwin, and H. S. Moyed. 1994. Autoregulation of *hip*, an operon that affects lethality due to inhibition of peptidoglycan or DNA synthesis. *J. Bacteriol.* 176:4081–4091.

159. Black, D. S., A. J. Kelly, M. J. Mardis, and H. S. Moyed. 1991. Structure and organization of *hip*, an operon that affects lethality due to inhibition of peptidoglycan or DNA synthesis. *J. Bacteriol.* 173:5732–5739.

160. Black, M. E., and D. E. Hruby. 1991. Nucleotide sequence of the *Escherichia coli* thymidine kinase gene provides evidence for conservation of functional domains and quaternary structure. *Mol. Microbiol.* 5:373–379.

161. Black, M. T., J. G. Munn, and A. E. Allsop. 1992. On the catalytic mechanism of prokaryotic leader peptidase 1. *Biochem. J.* 282:539–543.

162. Black, P. N. 1988. The *fadL* gene product of *Escherichia coli* is an outer membrane protein required for uptake of long-chain fatty acids and involved in sensitivity to bacteriophage T2. *J. Bacteriol.* 170:2850–2854.

163. Black, P. N. 1991. Primary sequence of the *Escherichia coli fadL* gene encoding an outer membrane protein required for long-chain fatty acid transport. *J. Bacteriol.* 173:435–442.

164. Black, P. N., C. C. DiRusso, A. K. Metzger, and T. L. Heimert. 1992. Cloning, sequencing, and expression of the fadD gene of *Escherichia coli* encoding acyl coenzyme A synthetase. *J. Biol. Chem.* 267:25513–25520.

165. Blacklow, S. C., K. D. Liu, and J. R. Knowles. 1991. Stepwise improvements in catalytic effectiveness: independence and interdependence in combinations of point mutations of a sluggish triosephosphate isomerase. *Biochemistry* 30:8470–8476.

166. Blair, D. F., and H. C. Berg. 1990. The MotA protein of *E. coli* is a proton-conducting component of the flagellar motor. *Cell* 60:439–449.

167. Blair, D. F., and H. C. Berg. 1991. Mutations in the MotA protein of *Escherichia coli* reveal domains critical for proton conduction. *J. Mol. Biol.* 221:1433–1442.

168. Blair, D. F., D. Y. Kim, and H. C. Berg. 1991. Mutant MotB proteins in *Escherichia coli*. *J. Bacteriol.* 173:4049–4055.

169. Blakely, G., S. Colloms, G. May, M. Burke, and D. Sherratt. 1991. *Escherichia coli* XerC recombinase is required for chromosomal segregation at cell division. *New Biol.* 3:789–798.

170. Blakely, G., G. May, R. McCulloch, L. K. Arciszewska, M. Burke, S. T. Lovett, and D. J. Sherratt. 1993. Two related recombinases are required for site-specific recombination at dif and cer in *E. coli* K12. *Cell* 75:351–361.

171. Blanco, C. 1987. Transcriptional and translational signals of the uidA gene in *Escherichia coli* K12. *Mol. Gen. Genet.* 208:490–498.

172. Blanco, C., and M. Mata-Gilsinger. 1986. A DNA sequence containing the control sites for the uxaB gene of *Escherichia coli*. *J. Gen. Microbiol.* 132:697–705.

173. Blanco, C., P. Ritzenthaler, and A. Kolb. 1986. The regulatory region of the uxuAB operon in *Escherichia coli* K12. *Mol. Gen. Genet.* 202:112–119.

174. Blasband, A. J., W. R. Marcotte, Jr., and C. A. Schnaitman. 1986. Structure of the lc and nmpC outer membrane porin protein genes of lambdoid bacteriophage. *J. Biol. Chem.* 261:12723–12732.

175. Blasco, F., C. Iobbi, G. Giordano, M. Chippaux, and V. Bonnefoy. 1989. Nitrate reductase of *Escherichia coli*: completion of the nucleotide sequence of the nar operon and reassessment of the role of the alpha and beta subunits in iron binding and electron transfer. *Mol. Gen. Genet.* 218:249–256.

176. Blasco, F., C. Iobbi, J. Ratouchniak, V. Bonnefoy, and M. Chippaux. 1990. Nitrate reductases of *Escherichia coli*: sequence of the second nitrate reductase and comparison with that encoded by the narGHJI operon. *Mol. Gen. Genet.* 222:104–111.

177. Blasco, F., J. Pommier, V. Augier, M. Chippaux, and G. Giordano. 1992. Involvement of the narJ or narW gene product in the formation of active nitrate reductase in *Escherichia coli*. *Mol. Microbiol.* 6:221–230.

178. Blat, Y., and M. Eisenbach. 1994. Phosphorylation-dependent binding of the chemotaxis signal molecule CheY to its phosphatase, CheZ. *Biochemistry* 33:902–906.

179. Blattner, F. R., V. Burland, G. Plunkett, H. J. Sofia, and D. L. Daniels. 1993. Analysis of the *Escherichia coli* genome. IV. DNA sequence of the region from 89.2 to 92.8 minutes. *Nucleic Acids Res.* 21:5408–5417.

180. Blomfield, I. C., P. J. Calie, K. J. Eberhardt, M. S. McClain, and B. I. Eisenstein. 1993. Lrp stimulates phase variation of type 1 fimbriation in *Escherichia coli* K-12. *J. Bacteriol.* 175:27–36.

181. Blomfield, I. C., M. S. McClain, J. A. Princ, P. J. Calie, and B. I. Eisenstein. 1991. Type 1 fimbriation and *fimE* mutants of *Escherichia coli* K-12. *J. Bacteriol.* 173:5298–5307.

182. **Bognar, A. L., C. Osborne, and B. Shane.** 1987. Primary structure of the *Escherichia coli* folC gene and its folylpolyglutamate synthetase-dihydrofolate synthetase product and regulation of expression by an upstream gene. *J. Biol. Chem.* **262:**12337–12343.

183. **Bohm, R., M. Sauter, and A. Bock.** 1990. Nucleotide sequence and expression of an operon in *Escherichia coli* coding for formate hydrogenlyase components. *Mol. Microbiol.* **4:**231–243.

184. **Boiteux, S., E. Gajewski, J. Laval, and M. Dizdaroglu.** 1992. Substrate specificity of the *Escherichia coli* Fpg protein (formamidopyrimidine-DNA glycosylase): excision of purine lesions in DNA produced by ionizing radiation or photosensitization. *Biochemistry* **31:**106–110.

185. **Boiteux, S., T. R. O'Connor, F. Lederer, A. Gouyette, and J. Laval.** 1990. Homogeneous *Escherichia coli* FPG protein. A DNA glycosylase which excises imidazole ring-opened purines and nicks DNA at apurinic/apyrimidinic sites. *J. Biol. Chem.* **265:**3916–3922.

186. **Bolker, M., and R. Kahmann.** 1989. The *Escherichia coli* regulatory protein OxyR discriminates between methylated and unmethylated states of the phage Mu mom promoter. *EMBO J.* **8:**2403–2410.

187. **Bolshakova, T. N., M. L. Molchanova, R. S. Erlagaeva, Y. A. Grigorenko, and V. N. Gershanovitch.** 1992. A novel mutation FruS, altering synthesis of components of the phosphoenolpyruvate:fructose phosphotransferase system in *Escherichia coli* K12. *Mol. Gen. Genet.* **232:**394–398.

188. **Bonthron, D. T.** 1990. L-Asparaginase II of *Escherichia coli* K-12: cloning, mapping and sequencing of the ansB gene. *Gene* **91:**101–105.

189. **Boos, W., U. Ehmann, H. Forkl, W. Klein, M. Rimmele, and P. Postma.** 1990. Trehalose transport and metabolism in *Escherichia coli. J. Bacteriol.* **172:**3450–3461.

190. **Borel, F., M. Hartlein, and R. Leberman.** 1993. In vivo overexpression and purification of *Escherichia coli* tRNA(ser). *FEBS Lett.* **324:**162–166.

191. **Bork, P., C. Sander, and A. Valencia.** 1992. An ATPase domain common to prokaryotic cell cycle proteins, sugar kinases, actin, and hsp70 heat shock proteins. *Proc. Natl. Acad. Sci. USA* **89:**7290–7294.

192. **Borukhov, S., A. Polyakov, V. Nikiforov, and A. Goldfarb.** 1992. GreA protein: a transcription elongation factor from *Escherichia coli. Proc. Natl. Acad. Sci. USA* **89:**8899–8902.

193. **Borukhov, S., V. Sagitov, and A. Goldfarb.** 1993. Transcript cleavage factors from *E. coli. Cell* **72:**459–466.

194. **Bosl, M., and H. Kersten.** 1991. A novel RNA product of the tyrT operon of *Escherichia coli. Nucleic Acids Res.* **19:**5863–5870.

195. **Bossemeyer, D., A. Borchard, D. C. Dosch, G. C. Helmer, W. Epstein, I. R. Booth, and E. P. Bakker.** 1989. K+-transport protein TrkA of *Escherichia coli* is a peripheral membrane protein that requires other trk gene products for attachment to the cytoplasmic membrane. *J. Biol. Chem.* **264:**16403–16410.

196. **Bossemeyer, D., A. Schlosser, and E. P. Bakker.** 1989. Specific cesium transport via the *Escherichia coli* Kup (TrkD) K$^+$ uptake system. *J. Bacteriol.* **171:**2219–2221.

197. **Boubrik, F., E. Bonnefoy, and J. Rouviere-Yaniv.** 1991. HU and IHF: similarities and differences. In *Escherichia coli,* the lack of HU is not compensated for by IHF. *Res. Microbiol.* **142:**239–247.

198. **Bouffard, G. G., K. E. Rudd, and S. L. Adhya.** 1994. Dependence of lactose metabolism upon mutarotase encoded in the gal operon in *Escherichia coli. J. Mol. Biol.* **244:**269–278.

199. **Boulnois, G., R. Drake, R. Pearce, and I. Roberts.** 1992. Genome diversity at the serA-linked capsule locus in *Escherichia coli. FEMS Microbiol. Lett.* **79:**121–124.

200. **Bourret, R. B., J. Davagnino, and M. I. Simon.** 1993. The carboxy-terminal portion of the CheA kinase mediates regulation of autophosphorylation by transducer and CheW. *J. Bacteriol.* **175:**2097–2101.

201. **Bourret, R. B., S. K. Drake, S. A. Chervitz, M. I. Simon, and J. J. Falke.** 1993. Activation of the phosphosignaling protein CheY. II. Analysis of activated mutants by 19F NMR and protein engineering. *J. Biol. Chem.* **268:**13089–13096.

202. **Bouter, S., P. R. Kerklaan, C. E. Zoetemelk, and G. R. Mohn.** 1988. Biochemical characterization of glutathione-deficient mutants of *Escherichia coli* K12 and *Salmonella* strains TA1535 and TA100. *Biochem. Pharmacol.* **37:**577–581.

203. **Bouvier, J., A. P. Pugsley, and P. Stragier.** 1991. A gene for a new lipoprotein in the dapA-purC interval of *Escherichia coli* chromosome. *J. Bacteriol.* **173:**5523–5531.

204. **Bouvier, J., C. Richaud, W. Higgins, O. Bogler, and P. Stragier.** 1992. Cloning, characterization, and expression of the dapE gene of *Escherichia coli. J. Bacteriol.* **174:**5265–5271.

205. **Bouvier, J., C. Richaud, F. Richaud, J. C. Patte, and P. Stragier.** 1984. Nucleotide sequence and expression of the *Escherichia coli* dapB gene. *J. Biol. Chem.* **259:**14829–14834.

206. **Bower, S. G., K. W. Harlow, R. L. Switzer, and B. Hove-Jensen.** 1989. Characterization of the *Escherichia coli* prsA1-encoded mutant phosphoribosylpyrophosphate synthetase identifies a divalent cation-nucleotide binding site. *J. Biol. Chem.* **264:**10287–10291.

207. **Boyen, A., D. Charlier, J. Charlier, V. Sakanyan, I. Mett, and N. Glansdorff.** 1992. Acetylornithine deacetylase, succinyldiaminopimelate desuccinylase and carboxypeptidase G2 are evolutionarily related. *Gene* **116:**1–6.

208. **Boyen, A., D. Charlier, M. Crabreel, R. Cunin, S. Palchaudhuri, and N. Glansdorff.** 1978. Studies on the control region of the bipolar argECBH operon. I. Effect of regulatory mutations and IS2 insertions. *Mol. Gen. Genet.* **161:**185–196.

209. **Bramley, H. F., and H. L. Kornberg.** 1987. Nucleotide sequences of bglC, the gene specifying enzyme II(bgl) of the PEP:sugar phophotransferase system in *Escherichia coli* K12 and overexpression of the gene product. *J. Gen. Microbiol.* **133:**563–573.

210. **Brand, B., and W. Boos.** 1991. Maltose transacetylase of *Escherichia coli.* Mapping and cloning of its structural, gene, mac, and characterization of the enzyme as a dimer of identical polypeptides with a molecular weight of 20,000. *J. Biol. Chem.* **266:**14113–14118.

211. **Brandsch, R., and V. Bichler.** 1989. Covalent cofactor binding to flavoenzymes requires specific effectors. *Eur. J. Biochem.* **182:**125–128.

212. **Brandt, R., and C. O. Gualerzi.** 1992. Ribosomal localization of the mRNA in the 30S initiation complex as revealed by UV crosslinking. *FEBS Lett.* **311:**199–202.

213. **Braun, V.** 1989. The structurally related exbB and tolQ genes are interchangeable in conferring tonB-dependent colicin, bacteriophage, and albomycin sensitivity. *J. Bacteriol.* **171:**6387–6390.

214. **Braun, V., K. Gunter, and K. Hantke.** 1991. Transport of iron across the outer membrane. *Biol. Methods* **4:**14–22.

215. **Braun, V., and C. Herrmann.** 1993. Evolutionary relationship of uptake systems for biopolymers in *Escherichia coli:* cross-complementation between the TonB-ExbB-ExbD and the TolA-TolQ-TolR proteins. *Mol. Microbiol.* **8:**261–268.

216. **Braxton, B. L., L. S. Mullins, F. M. Raushel, and G. D. Reinhart.** 1992. Quantifying the allosteric properties of *Escherichia coli* carbamyl phosphate synthetase: determination of thermodynamic linked-function parameters in an ordered kinetic mechanism. *Biochemistry* **31:**2309–2316.

217. **Bremer, E., A. Middendorf, J. Martinussen, and P. Valentin-Hansen.** 1990. Analysis of the tsx gene, which encodes a nucleoside-specific channel-forming protein (Tsx) in the outer membrane of *Escherichia coli. Gene* **96:**59–65.

218. **Breukink, E., R. A. Demel, G. de Korte-Kool, and B. de Kruijff.** 1992. SecA insertion into phospholipids is stimulated by negatively charged lipids and inhibited by ATP: a monolayer study. *Biochemistry* **31:**1119–1124.

219. **Brey, R. N., and B. P. Rosen.** 1979. Properties of *Escherichia coli* mutants altered in calcium/proton antiport activity. *J. Bacteriol.* **139:**824–834.

220. **Brickman, T. J., and M. A. McIntosh.** 1992. Overexpression and purification of ferric enterobactin esterase from *Escherichia coli.* Demonstration of enzymatic hydrolysis of enterobactin and its iron complex. *J. Biol. Chem.* **267:**12350–12355.

221. **Brikun, I. A., A. S. Mironov, R. V. Masiliunaite, and V. V. Sukhodolets.** 1990. Cloning of *Escherichia coli* uridine phosphorylase gene: localization of structural and regulatory regions in the cloned fragment and identification of the protein product. *Mol. Gen. Mikrobiol. Virusol.* **27:**7–11. (In Russian.)

222. **Brill, J. A., C. Quinlan-Walshe, and S. Gottesman.** 1988. Fine-structure mapping and identification of two regulators of capsule synthesis in *Escherichia coli* K-12. *J. Bacteriol.* **170:**2599–2611.

223. **Brisson, A., Y. V. Brun, A. W. Bell, P. H. Roy, and J. Lapointe.** 1989. Overproduction and domain structure of the glutamyl-tRNA synthetase of *Escherichia coli. Biochem. Cell Biol.* **67:**404–410.

224. **Britton, P., A. Boronat, D. A. Hartley, M. C. Jones-Mortimer, H. L. Kornberg, and F. Parra.** 1983. Phosphotransferase-mediated regulation of carbohydrate utilization in *Escherichia coli* K12: location of the gsr (tgs) and iex (crr) genes by specialized transduction. *J. Gen. Microbiol.* **129:**349–356.

225. **Brody, R. S.** 1991. Nucleotide positions responsible for the processivity of the reaction of exonuclease I with oligodeoxyribonucleotides. *Biochemistry* **30:**7072–7080.

226. **Brok, R. G., E. Brinkman, R. van Boxtel, A. C. Bekkers, H. M. Verheij, and J. Tommassen.** 1994. Molecular characterization of enterobacterial pldA genes encoding outer membrane phospholipase A. *J. Bacteriol.* **176:**861–870.

227. **Brooker, R. J.** 1991. An analysis of lactose permease "sugar specificity" mutations which also affect the coupling between proton and lactose transport. I. Val177 and Val177/Asn319 permeases facilitate proton uniport and sugar uniport. *J. Biol. Chem.* **266:**4131–4138.

228. **Broome-Smith, J. K., A. Edelman, S. Yousif, and B. G. Spratt.** 1985. The nucleotide sequences of the ponA and ponB genes encoding penicillin-binding protein 1A and 1B of *Escherichia coli* K12. *Eur. J. Biochem.* **147:**437–446.

229. **Broome-Smith, J. K., I. Ionnidas, A. Edelman, S. Yousif, and B. G. Spratt.** 1988. Nucleotide sequences of the penicillin-binding protein 5 and 6 genes of *Escherichia coli. Nucleic Acids Res.* **16:**1617

230. **Brown, D. A., and R. A. Cook.** 1981. Role of metal cofactors in enzyme regulation. Differences in the regulatory properties of the *Escherichia coli* nicotinamide adenine dinucleotide phosphate specific malic enzyme, depending on whether magnesium ion or manganese ion serves as divalent cation. *Biochemistry* **20:**2503–2512.

231. **Brown, D. C., and K. D. Collins.** 1991. Dihydroorotase from *Escherichia coli.* Substitution of Co(II) for the active site Zn(II). *J. Biol. Chem.* **266:**1597–1604.

232. **Brown, E. D., and J. M. Wood.** 1993. Conformational change and membrane association of the PutA protein are coincident with reduction of its FAD cofactor by proline. *J. Biol. Chem.* **268:**8972–8979.

233. **Brown, K. P., W. Finch, I. D. Hickson, and P. T. Emmerson.** 1987. Complete nucleotide sequence of the *Escherichia coli* argA gene. *Nucleic Acids Res.* **15:**10586

234. **Brown, S.** 1989. Time of action of 4.5 S RNA in *Escherichia coli* translation. *J. Mol. Biol.* **209:**79–90.

235. **Bruckner, R. C., and M. M. Cox.** 1989. The histone-like H protein of *Escherichia coli* is ribosomal protein S3. *Nucleic Acids Res.* **17:**3145–3161.

236. **Brun, Y. V., R. Breton, P. Lanouette, and J. Lapointe.** 1990. Precise mapping and comparison of two evolutionarily related regions of the *Escherichia coli* K-12

chromosome. Evolution of valU and lysT from an ancestral tRNA operon. *J. Mol. Biol.* 214:825–843.

237. Brundage, L., C. J. Fimmel, S. Mizushima, and W. Wickner. 1992. SecY, SecE, and band 1 form the membrane-embedded domain of *Escherichia coli* preprotein translocase. *J. Biol. Chem.* 267:4166–4170.

238. Brunel, C., J. Caillet, P. Lesage, M. Graffe, J. Dondon, H. Moine, P. Romby, C. Ehresmann, B. Ehresmann, M. Grunberg-Manago, and M. Springer. 1992. Domains of the *Escherichia coli* threonyl-tRNA synthetase translational operator and their relation to threonine tRNA isoacceptors. *J. Mol. Biol.* 227:621–634.

239. Brusilow, W. S. 1993. Assembly of the *Escherichia coli* F1F0 ATPase, a large multimeric membrane-bound enzyme. *Mol. Microbiol.* 9:419–424.

240. Bryant, R. E., and P. S. Sypherd. 1974. Genetic analysis of cold-sensitive ribosome maturation mutants of *Escherichia coli*. *J. Bacteriol.* 117:1082–1092.

241. Brzoska, P., M. Rimmele, K. Brzostek, and W. Boos. 1994. The *pho* regulon-dependent Ugp uptake system for glycerol-3-phosphate in *Escherichia coli* is *trans* inhibited by P_i. *J. Bacteriol.* 176:15–20.

242. Brzovic, P. S., K. Ngo, and M. F. Dunn. 1992. Allosteric interactions coordinate catalytic activity between successive metabolic enzymes in the tryptophan synthase bienzyme complex. *Biochemistry* 31:3831–3839.

243. Buchner, J., M. Schmidt, M. Fuchs, R. Jaenicke, R. Rudolph, F. X. Schmid, and T. Kiefhaber. 1991. GroE facilitates refolding of citrate synthase by suppressing aggregation. *Biochemistry* 30:1586–1591.

244. Buckel, S. D., A. W. Bell, J. K. Rao, and M. A. Hermodson. 1986. An analysis of the structure of the product of the rbsA gene of *Escherichia coli* K12. *J. Biol. Chem.* 261:7659–7662.

245. Buechter, D. D., and P. Schimmel. 1993. Dissection of a class II tRNA synthetase: determinants for minihelix recognition are tightly associated with domain for amino acid activation. *Biochemistry* 32:5267–5272.

246. Buhr, A., G. A. Daniels, and B. Erni. 1992. The glucose transporter of *Escherichia coli*. Mutants with impaired translocation activity that retain phosphorylation activity. *J. Biol. Chem.* 267:3847–3851.

247. Bujalowski, W., and T. M. Lohman. 1991. Monomers of the *Escherichia coli* SSB-1 mutant protein bind single-stranded DNA. *J. Mol. Biol.* 217:63–74.

248. Bukhari, A. I., and A. L. Taylor. 1971. Genetic analysis of diaminopimelic acid-and lysine-requiring mutants of *Escherichia coli*. *J. Bacteriol.* 105:844–854.

249. Burland, V., G. Plunkett, D. L. Daniels, and F. R. Blattner. 1993. DNA sequence and analysis of 136 kilobases of the *Escherichia coli* genome: organizational symmetry around the origin of replication. *Genomics* 16:551–561.

250. Burns, D. M., and I. R. Beacham. 1987. Altered localisation of the precursor of a secreted protein in *E. coli* by a carboxyl-deletion. *Biochem. Int.* 14:1073–1077.

251. Burr, B., J. Walker, P. Truffa-Bachi, and G. N. Cohen. 1976. Homoserine kinase from *Escherichia coli* K12. *Eur. J. Biochem.* 62:519–526.

252. Burrows, G. G., M. E. Newcomer, and G. L. Hazelbauer. 1989. Purification of receptor protein Trg by exploiting a property common to chemotactic transducers of *Escherichia coli*. *J. Biol. Chem.* 264:17309–17315.

253. Burrows, R., and G. Brown. 1978. Presence in *Escherichia coli* of a deaminase and a reductase invoived in biosynthesis of riboflavin. *J. Bacteriol.* 136:657–667.

254. Burton, K. 1994. Adenine transport in *Escherichia coli*. *Proc. R. Soc. London Ser. B* 255:153–157.

255. Buxton, R. S., K. Hammer-Jespersen, and T. D. Hansen. 1978. Insertion of bacteriophage lambda into the *deo* operon of *Escherichia coli* K-12 and isolation of plaque-forming λdeo^+ transducing bacteriophages. *J. Bacteriol.* 136:668–681.

256. Buxton, R. S., K. Hammer-Jespersen, and P. Valentin-Hansen. 1980. A second purine nucleoside phosphorylase in *Escherichia coli* K12. I. Xanthosine phosphorylase regulatory mutants isolated as secondary-site revertants of a deoD mutant. *Mol. Gen. Genet.* 179:331–340.

256a.Byrne, C., H. W. Stokes, and K. A. Ward. 1988. Nucleotide sequence of the aceB gene encoding malate synthase A in Escherichia coli. *Nucleic Acids Res.* 16:9342.

257. Byrne, C. R., R. S. Monroe, K. A. Ward, and N. M. Kredich. 1988. DNA sequences of the cysK regions of *Salmonella typhimurium* and *Escherichia coli* and linkage of the cysK regions to ptsH. *J. Bacteriol.* 170:3150–3157.

258. Bystrom, A. S., and G. R. Bjork. 1982. The structural gene (trmD) for the tRNA(m1G)methyltransferase is part of a four polypeptide operon in *Escherichia coli* K-12. *Mol. Gen. Genet.* 188:447–454.

259. Cain, B. D., P. J. Norton, W. Eubanks, H. S. Nick, and C. M. Allen. 1993. Amplification of the bacA gene confers bacitracin resistance to *Escherichia coli*. *J. Bacteriol.* 175:3784–3789.

260. Calhoun, M. W., and R. B. Gennis. 1993. Demonstration of separate genetic loci encoding distinct membrane-bound respiratory NADH dehydrogenases in *Escherichia coli*. *J. Bacteriol.* 175:3013–3019.

261. Cam, K., S. Bejar, D. Gil, and J. P. Bouche. 1988. Identification and sequence of gene dicB: translation of the division inhibitor from an in-phase internal start. *Nucleic Acids Res.* 16:6327–6338.

262. Campbell, A., A. del Campillo-Campbell, and R. Chang. 1972. A mutant of *Escherichia coli* that requires high concentrations of biotin. *Proc. Natl. Acad. Sci. USA* 69:676–680.

263. Campbell, A., S. J. Schneider, and B. Song. 1992. Lambdoid phages as elements of bacterial genomes (integrase/phage21/*Escherichia coli* K-12/icd gene). *Genetica* 86:259–267.

264. Cao, Y., R. R. Rowland, and T. Kogoma. 1993. DNA polymerase I and the bypassing of RecA dependence of constitutive stable DNA replication in *Escherichia coli* rnhA mutants. *J. Bacteriol.* 175:7247–7253.

265. Cardelli, J., and J. Konisky. 1974. Isolation and characterization of an *Escherichia coli* mutant tolerant to colicins Ia and Ib. *J. Bacteriol.* 119:379–385.

266. Carlomagno, M. S., L. Chiariotti, P. Alifano, A. G. Nappo, and C. B. Bruni. 1988. Structure and function of the *Salmonella typhimurium* and *Escherichia coli* K-12 histidine operons. *J. Mol. Biol.* 203:586–606.

267. Carter, J. R., M. A. Franden, R. Aebersold, D. R. Kim, and C. S. McHenry. 1993. Isolation, sequencing and overexpression of the gene encoding the theta subunit of DNA polymerase III holoenzyme. *Nucleic Acids Res.* 21:3281–3286.

268. Carter, J. R., M. A. Franden, R. Aebersold, and C. S. McHenry. 1992. Molecular cloning, sequencing, and overexpression of the structural gene encoding the delta subunit of *Escherichia coli* DNA polymerase III holoenzyme. *J. Bacteriol.* 174:7013–7025.

269. Carter, J. R., M. A. Franden, R. Aebersold, and C. S. McHenry. 1993. Identification, isolation, and overexpression of the gene encoding the psi subunit of DNA polymerase III holoenzyme. *J. Bacteriol.* 175:5604–5610.

270. Carter, J. R., M. A. Franden, R. Aebersold, and C. S. McHenry. 1993. Identification, isolation, and characterization of the structural gene encoding the delta' subunit of *Escherichia coli* DNA polymerase III holoenzyme. *J. Bacteriol.* 175:3812–3822.

271. Carter-Muenchau, P., and R. E. Wolf, Jr. 1989. Growth-rate-dependent regulation of 6-phosphogluconate dehydrogenase level mediated by an anti-Shine-Dalgarno sequence located within the *Escherichia coli* gnd structural gene. *Proc. Natl. Acad. Sci. USA* 86:1138–1142.

272. Caruso, M., A. Coppo, A. Manzi, and J. F. Pulitzer. 1979. Host—virus interactions in the control of T4 prereplicative transcription. I. tabC (rho) mutants. *J. Mol. Biol.* 135:959–977.

273. Cassan, M., C. Parsot, G. N. Cohen, and J. C. Patte. 1986. Nucleotide sequence of lysC gene encoding the lysine-sensitive aspartokinase III of *Escherichia coli* K12. Evolutionary pathway leading to three isofunctional enzymes. *J. Biol. Chem.* 261:1052–1057.

274. Castano, I., F. Bastarrachea, and A. A. Covarrubias. 1988. *gltBDF* operon of *Escherichia coli*. *J. Bacteriol.* 170:821–827.

275. Castano, I., N. Flores, F. Valle, A. A. Covarrubias, and F. Bolivar. 1992. gltF, a member of the gltBDF operon of *Escherichia coli*, is involved in nitrogen-regulated gene expression. *Mol. Microbiol.* 6:2733–2741.

276. Cayley, S., B. A. Lewis, and M. T. Record, Jr. 1992. Origins of the osmoprotective properties of betaine and proline in *Escherichia coli* K-12. *J. Bacteriol.* 174:1586–1595.

277. Cedergren-Zeppezauer, E. S., G. Larsson, I. Hoffmann, K. W. Tornroos, S. Al-Karadaghi, and P. O. Nyman. 1988. Crystallization and preliminary investigation of single crystals of deoxyuridine triphosphate nucleotidohydrolase from *Escherichia coli*. *Proteins* 4:71–75.

278. Celis, R. T. 1981. Chain-terminating mutants affecting a periplasmic binding protein involved in the active transport of arginine and ornithine in *Escherichia coli*. *J. Biol. Chem.* 256:773–779.

279. Celis, R. T. 1982. Mapping of two loci affecting the synthesis and structure of a periplasmic protein involved in arginine and ornithine transport in *Escherichia coli* K-12. *J. Bacteriol.* 151:1314–1319.

280. Celis, T. F. 1977. Properties of an *Escherichia coli* K-12 mutant defective in the transport of arginine and ornithine. *J. Bacteriol.* 130:1234–1243.

281. Celis, T. F. R., H. J. Rosenfeld, and W. K. Maas. 1973. Mutant of *Escherichia coli* K-12 defective in the transport of basic amino acids. *J. Bacteriol.* 116:619–626.

282. Chan, C. L., and R. Landick. 1989. The *Salmonella typhimurium* his operon leader region contains an RNA hairpin-dependent transcription pause site. Mechanistic implications of the effect on pausing of altered RNA hairpins. *J. Biol. Chem.* 264:20796–20804.

283. Chang, Y. Y., A. Y. Wang, and J. E. Cronan, Jr. 1993. Molecular cloning, DNA sequencing, and biochemical analyses of *Escherichia coli* glyoxylate carboligase. An enzyme of the acetohydroxy acid synthase-pyruvate oxidase family. *J. Biol. Chem.* 268:3911–3919.

284. Chang, Z. Y., P. Nygaard, A. C. Chinault, and R. E. Kellems. 1991. Deduced amino acid sequence of *Escherichia coli* adenosine deaminase reveals evolutionarily conserved amino acid residues: implications for catalytic function. *Biochemistry* 30:2273–2280.

285. Chao, Y. P., and J. C. Liao. 1993. Alteration of growth yield by overexpression of phosphoenolpyruvate carboxylase and phosphoenolpyruvate carboxykinase in *Escherichia coli*. *Appl. Environ. Microbiol.* 59:4261–4265.

286. Chao, Y. P., R. Patnaik, W. D. Roof, R. F. Young, and J. C. Liao. 1993. Control of gluconeogenic growth by pps and pck in *Escherichia coli*. *J. Bacteriol.* 175:6939–6944.

287. Charles, I. G., H. K. Lamb, D. Pickard, G. Dougan, and A. R. Hawkins. 1990. Isolation, characterization and nucleotide sequences of the aroC genes encoding chorismate synthase from *Salmonella typhi* and *Escherichia coli*. *J. Gen. Microbiol.* 136:353–358.

288. Charlier, D., G. Weyens, M. Roovers, J. Piette, C. Bocquet, A. Pierard, and N. Glansdorff. 1988. Molecular interactions in the control region of the carAB operon encoding *Escherichia coli* carbamoylphosphate synthetase. *J. Mol. Biol.* 204:867–877.

289. Charpentier, B., and C. Branlant. 1994. The *Escherichia coli* gapA gene is transcribed by the vegetative RNA polymerase holoenzyme E sigma 70 and by the heat shock RNA polymerase E sigma 32. *J. Bacteriol.* 176:830–839.

290. Chase, J. W., B. A. Rabin, J. B. Murphy, K. L. Stone, and K. R. Williams. 1986. *Escherichia coli* exonuclease VII. Cloning and sequencing of the gene encoding the large subunit (xseA). *J. Biol. Chem.* 261:14929–14935.

291. Chaudhuri, S., K. Duncan, L. D. Graham, and J. R. Coggins. 1991. Identification of the active-site lysine residues of two biosynthetic 3-dehydroquinases. *Biochem. J.* 275:1–6.

292. Chen, G. T., M. J. Axley, J. Hacia, and M. Inouye. 1992. Overproduction of a selenocysteine-containing polypeptide in *Escherichia coli:* the fdhF gene product. *Mol. Microbiol.* 6:781–785.

293. Chen, H., Y. Sun, T. Stark, W. Beattie, and R. E. Moses. 1990. Nucleotide sequence and deletion analysis of the polB gene of *Escherichia coli.* *DNA Cell Biol.* 9:631–635.

294. Chen, K. S., T. C. Peters, and J. R. Walker. 1990. A minor arginine tRNA mutant limits translation preferentially of a protein dependent on the cognate codon. *J. Bacteriol.* 172:2504–2510.

295. Chen, L., and W. G. Coleman, Jr. 1993. Cloning and characterization of the *Escherichia coli* K-12 rfa-2 (rfaC) gene, a gene required for lipopolysaccharide inner core synthesis. *J. Bacteriol.* 175:2534–2540.

296. Chen, R., and U. Henning. 1987. Nucleotide sequence of the gene for the peptidoglycan-associated lipoprotein of *Escherichia coli* K12. *Eur. J. Biochem.* 163:73–77.

297. Chen, S. M., H. E. Takiff, A. M. Barber, G. C. Dubois, J. C. Bardwell, and D. L. Court. 1990. Expression and characterization of RNase III and Era proteins. Products of the rnc operon of *Escherichia coli.* *J. Biol. Chem.* 265:2888–2895.

298. Chen, Y. M., Z. Lu, and E. C. Lin. 1989. Constitutive activation of the fucAO operon and silencing of the divergently transcribed fucPIK operon by an IS5 element in *Escherichia coli* mutants selected for growth on L-1,2-propanediol. *J. Bacteriol.* 171:6097–6105.

299. Chen, Y. M., Y. Zhu, and E. C. Lin. 1987. The organization of the fuc regulon specifying L-fucose dissimilation in *Escherichia coli* K12 as determined by gene cloning. *Mol. Gen. Genet.* 210:331–337.

300. Chenais, J., C. Richaud, J. Ronceray, H. Cherest, Y. Surdin-Kerjan, and J. C. Patte. 1981. Construction of hybrid plasmids containing the lysA gene of *Escherichia coli* and *Saccharomyces cerevisiae.* *Mol. Gen. Genet.* 182:456–461.

301. Cheng, H. H., P. J. Muhlrad, M. A. Hoyt, and H. Echols. 1988. Cleavage of the cII protein of phage lambda by purified HflA protease: control of the switch between lysis and lysogeny. *Proc. Natl. Acad. Sci. USA* 85:7882–7886.

302. Chepuri, V., and R. B. Gennis. 1990. The use of gene fusions to determine the topology of all of the subunits of the cytochrome o terminal oxidase complex of *Escherichia coli.* *J. Biol. Chem.* 265:12978–12986.

303. Chepuri, V., L. Lemieux, J. Hill, J. O. Alben, and R. B. Gennis. 1990. Recent studies of the cytochrome o terminal oxidase complex of *Escherichia coli.* *Biochim. Biophys. Acta* 1018:124–127.

304. Chesney, R. H., P. Sollitti, and D. R. Vickery. 1985. Identification of a new locus in the *Escherichia coli* cotransduction gap that represents a new genetic component of the L-asparagine utilization system. *J. Gen. Microbiol.* 131:2079–2085.

305. Chiang, R. C., R. Cavicchioli, and R. P. Gunsalus. 1992. Identification and characterization of narQ, a second nitrate sensor for nitrate-dependent gene regulation in *Escherichia coli.* *Mol. Microbiol.* 6:1913–1923.

306. Chiariotti, L., P. Alifano, M. S. Carlomagno, and C. B. Bruni. 1986. Nucleotide sequence of the *Escherichia coli* hisD gene and of the *Escherichia coli* and *Salmonella typhimurium* hisIE region. *Mol. Gen. Genet.* 203:382–388.

307. Chiariotti, L., A. G. Nappo, M. S. Carlomagno, and C. B. Bruni. 1986. Gene structure in the histidine operon of *Escherichia coli.* Identification and nucleotide sequence of the hisB gene. *Mol. Gen. Genet.* 202:42–47.

308. Cho, H., and J. E. Cronan, Jr. 1993. *Escherichia coli* thioesterase I, molecular cloning and sequencing of the structural gene and identification as a periplasmic enzyme. *J. Biol. Chem.* 268:9238–9245.

309. Cho, H., and J. E. Cronan, Jr. 1994. "Protease I" of *Escherichia coli* functions as a thioesterase in vivo. *J. Bacteriol.* 176:1793–1795.

310. Choi, K. S., J. A. Kim, and H. S. Kang. 1992. Effects of site-directed mutations on processing and activities of penicillin G acylase from *Escherichia coli* ATCC 11105. *J. Bacteriol.* 174:6270–6276.

311. Choi, K. Y., and H. Zalkin. 1990. Regulation of *Escherichia coli* pyrC by the purine regulon repressor protein. *J. Bacteriol.* 172:3201–3207.

312. Choi, K. Y., and H. Zalkin. 1994. Role of the purine repressor hinge sequence in repressor function. *J. Bacteriol.* 176:1767–1772.

313. Choi, Y. L., T. Nishida, M. Kawamukai, R. Utsumi, H. Sakai, and T. Komano. 1989. Cloning and sequencing of an *Escherichia coli* gene, nlp, highly homologous to the ner genes of bacteriophages Mu and D108. *J. Bacteriol.* 171:5222–5225.

314. Christner, J. A., E. Munck, P. A. Janick, and L. M. Siegel. 1981. Mossbauer spectroscopic studies of *Escherichia coli* sulfite reductase. Evidence for coupling between the siroheme and Fe4S4 cluster prosthetic groups. *J. Biol. Chem.* 256:2098–2101.

315. Chu, C. C., A. Templin, and A. J. Clark. 1989. Suppression of a frameshift mutation in the recE gene of *Escherichia coli* K-12 occurs by gene fusion. *J. Bacteriol.* 171:2101–2109.

316. Chu, J., R. Shoeman, J. Hart, T. Coleman, A. Mazaitis, N. Kelker, N. Brot, and H. Weissbach. 1985. Cloning and expression of the metE gene in *Escherichia coli.* *Arch. Biochem. Biophys.* 239:467–474.

317. Chu, W. C., and J. Horowitz. 1991. Recognition of *Escherichia coli* valine transfer RNA by its cognate synthetase: a fluorine-19 NMR study. *Biochemistry* 30:1655–1663.

318. Chun, S. Y., and L. L. Randall. 1994. In vivo studies of the role of SecA during protein export in *Escherichia coli.* *J. Bacteriol.* 176:4197–4203.

319. Chung, T., E. Resnik, C. Stueland, and D. C. LaPorte. 1993. Relative expression of the products of glyoxylate bypass operon: contributions of transcription and translation. *J. Bacteriol.* 175:4572–4575.

320. Claassen, L. A., and L. Grossman. 1991. Deletion mutagenesis of the *Escherichia coli* UvrA protein localizes domains for DNA binding, damage recognition, and protein-protein interactions. *J. Biol. Chem.* 266:11388–11394.

321. Clark, A. J. 1991. rec genes and homologous recombination proteins in *Escherichia coli.* *Biochimie* 73:523–532.

322. Clark, A. J., V. Sharma, S. Brenowitz, C. C. Chu, S. Sandler, L. Satin, A. Templin, I. Berger, and A. Cohen. 1993. Genetic and molecular analyses of the C-terminal region of the recE gene from the Rac prophage of *Escherichia coli* K-12 reveal the recT gene. *J. Bacteriol.* 175:7673–7682.

323. Clark, D. 1981. Regulation of fatty acid degradation in *Escherichia coli:* analysis by operon fusion. *J. Bacteriol.* 148:521–526.

324. Clark, D. 1984. Novel antibiotic hypersensitive mutants of *Escherichia coli.* Genetic mapping and chemical characterization. *FEMS Microbiol. Lett.* 21:189–195.

325. Clark, D. P. 1989. The fermentation pathways of *Escherichia coli.* *FEMS Microbiol. Rev.* 5:223–234.

326. Clark, D. P., and J. E. Cronan, Jr. 1980. Acetaldehyde coenzyme A dehydrogenase of *Escherichia coli.* *J. Bacteriol.* 144:179–184.

327. Clark, D. P., D. DeMendoza, M. L. Polacco, and J. E. Cronan, Jr. 1983. Beta-hydroxydecanoyl thio ester dehydrase does not catalyze a rate-limiting step in *Escherichia coli* unsaturated fatty acid synthesis. *Biochemistry* 22:5897–5902.

328. Clark, D. P., and M. L. Rod. 1987. Regulatory mutations that allow the growth of *Escherichia coli* on butanol as carbon source. *J. Mol. Evol.* 25:151–158.

329. Clark, R. L., and F. C. Neidhardt. 1990. Roles of the two lysyl-tRNA synthetases of *Escherichia coli:* analysis of nucleotide sequences and mutant behavior. *J. Bacteriol.* 172:3237–3243.

330. Claverys, J. P., and V. Mejean. 1988. Strand targeting signal(s) for in vivo mutation avoidance by post-replication mismatch repair in *Escherichia coli.* *Mol. Gen. Genet.* 214:574–578.

331. Cleary, M. L., R. T. Garvin, and E. James. 1977. Synthesis of the *Escherichia coli* K12 isoenzymes or ornithine transcarbamylase, performed in vitro. *Mol. Gen. Genet.* 157:155–165.

332. Clementz, T., and C. R. Raetz. 1991. A gene coding for 3-deoxy-D-manno-octulosonic-acid transferase in *Escherichia coli.* Identification, mapping, cloning, and sequencing. *J. Biol. Chem.* 266:9687–9696.

333. Cleton-Jansen, A. M., N. Goosen, O. Fayet, and P. van de Putte. 1990. Cloning, mapping, and sequencing of the gene encoding *Escherichia coli* quinoprotein glucose dehydrogenase. *J. Bacteriol.* 172:6308–6315.

334. Climent, I., B. M. Sjoberg, and C. Y. Huang. 1991. Carboxyl-terminal peptides as probes for *Escherichia coli* ribonucleotide reductase subunit interaction: kinetic analysis of inhibition studies. *Biochemistry* 30:5164–5171.

335. Coderre, P. E., and C. F. Earhart. 1989. The entD gene of the *Escherichia coli* K12 enterobactin gene cluster. *J. Gen. Microbiol.* 135:3043–3055.

336. Coello, N. B., and T. Isturiz. 1992. The metabolism of gluconate in *Escherichia coli:* a study in continuous culture. *J. Basic Microbiol.* 32:309–315.

337. Cohen, S. P., H. Hachler, and S. B. Levy. 1993. Genetic and functional analysis of the multiple antibiotic resistance (mar) locus in *Escherichia coli.* *J. Bacteriol.* 175:1484–1492.

338. Cole, S. T., K. Eiglmeier, S. Ahmed, N. Honore, L. Elmes, W. F. Anderson, and J. H. Weiner. 1988. Nucleotide sequence and gene-polypeptide relationships of the glpABC operon encoding the anaerobic sn-glycerol-3-phosphate dehydrogenase of *Escherichia coli* K-12. *J. Bacteriol.* 170:2448–2456.

339. Coleman, J. 1992. Characterization of the *Escherichia coli* gene for 1-acyl-sn-glycerol-3-phosphate acyltransferase (plsC). *Mol. Gen. Genet.* 232:295–303.

340. Coleman, W. G., Jr., and L. Leive. 1979. Two mutations which affect the barrier function of the *Escherichia coli* K-12 outer membrane. *J. Bacteriol.* 139:899–910.

341. Collier, D. N. 1993. SecB: a molecular chaperone of *Escherichia coli* protein secretion pathway. *Adv. Protein Chem.* 44:151–193.

342. Collins, L. A., S. M. Egan, and V. Stewart. 1992. Mutational analysis reveals functional similarity between NARX, a nitrate sensor in *Escherichia coli* K-12, and the methyl-accepting chemotaxis proteins. *J. Bacteriol.* 174:3667–3675.

343. Collis, C. M., and G. W. Grigg. 1989. An *Escherichia coli* mutant resistant to phleomycin, bleomycin, and heat inactivation is defective in ubiquinone synthesis. *J. Bacteriol.* 171:4792–4798.

344. Colloms, S. D., P. Sykora, G. Szatmari, and D. J. Sherratt. 1990. Recombination at ColE1 cer requires the *Escherichia coli* xerC gene product, a member of the lambda integrase family of site-specific recombinases. *J. Bacteriol.* 172:6973–6980.

345. Colson, C., J. Lhoest, and C. Urlings. 1979. Genetics of ribosomal protein methylation in *Escherichia coli.* III. Map position of two genes, prmA and prmB, governing methylation of proteins L11 and L3. *Mol. Gen. Genet.* 169:245–250.

346. Compton, L. A., J. M. Davis, J. R. Macdonald, and H. P. Bachinger. 1992. Structural and functional characterization of *Escherichia coli* peptidyl-prolyl cis-trans isomerases. *Eur. J. Biochem.* 206:927–934.

347. Condamine, H. 1971. Sur la regulation de la production de proline chez *E. coli* K-12. *Ann. Inst. Pasteur.* (Paris) 120:126–143.

348. Condon, C., R. Cammack, D. S. Patil, and P. Owen. 1985. The succinate dehydrogenase of *Escherichia coli*. Immunochemical resolution and biophysical characterization of a 4-subunit enzyme complex. *J. Biol. Chem.* **260**:9427–9434.

349. Condon, C., and J. H. Weiner. 1988. Fumarate reductase of *Escherichia coli*: an investigation of function and assembly using in vivo complementation. *Mol. Microbiol.* **2**:43–52.

350. Conlin, C. A., T. M. Knox, and C. G. Miller. 1994. Cloning and physical map position of an alpha-aspartyl dipeptidase gene, *pepE*, from *Escherichia coli*. *J. Bacteriol.* **176**:1552–1553.

351. Conlin, C. A., N. J. Trun, T. J. Silhavy, and C. G. Miller. 1992. *Escherichia coli prlC* encodes an endopeptidase and is homologous to the *Salmonella typhimurium opdA* gene. *J. Bacteriol.* **174**:5881–5887.

352. Connolly, D. M., and M. E. Winkler. 1991. Structure of *Escherichia coli* K-12 *miaA* and characterization of the mutator phenotype caused by *miaA* insertion mutations. *J. Bacteriol.* **173**:1711–1721.

353. Conway, T., and L. O. Ingram. 1989. Similarity of *Escherichia coli* propanediol oxidoreductase (*fucO* product) and an unusual alcohol dehydrogenase from *Zymomonas mobilis* and *Saccharomyces cerevisiae*. *J. Bacteriol.* **171**:3754–3759.

354. Cook, R. A. 1983. Distinct metal cofactor-induced conformational states in the NAD-specific malic enzyme of *Escherichia coli* as revealed by proteolysis studies. *Biochim. Biophys. Acta* **749**:198–203.

355. Cooper, C. L., S. Jackowski, and C. O. Rock. 1987. Fatty acid metabolism in sn-glycerol-3-phosphate acyltransferase (*plsB*) mutants. *J. Bacteriol.* **169**:605–611.

356. Cooper, R. A. 1978. The utilization of D-galactonate and D-2-oxo-3-deoxygalactonate by *Escherichia coli* K-12. *Arch. Microbiol.* **118**:119–206.

357. Cormack, R. S., and G. A. Mackie. 1991. Mapping ribosomal protein S20–16 S rRNA interactions by mutagenesis. *J. Biol. Chem.* **266**:18525–18529.

358. Cortay, J. C., F. Bleicher, B. Duclos, Y. Cenatiempo, C. Gautier, J. L. Prato, and A. J. Cozzone. 1989. Utilization of acetate in *Escherichia coli*: structural organization and differential expression of the ace operon. *J. Bacteriol.* **71**:1043–1049.

359. Cortay, J. C., D. Negre, A. Galinier, B. Duclos, G. Perriere, and A. J. Cozzone. 1991. Regulation of the acetate operon in *Escherichia coli*: purification and functional characterization of the IclR repressor. *EMBO J.* **10**:675–679.

360. Cossay, S. D. 1973. D-Serine transport system in *Escherichia coli* K-12. *J. Bacteriol.* **114**:679–684.

361. Cossart, P., M. Katinka, and M. Yaniv. 1981. Nucleotide sequence of the thrB gene of *E. coli*, and its two adjacent regions; the thrAB and thrBC junctions. *Nucleic Acids Res.* **9**:339–347.

362. Cotter, P. A., V. Chepuri, R. B. Gennis, and R. P. Gunsalus. 1990. Cytochrome o (*cyoABCDE*) and d (*cydAB*) oxidase gene expression in *Escherichia coli* is regulated by oxygen, pH, and the *fnr* gene product. *J. Bacteriol.* **172**:6333–6338.

363. Coves, J., V. Niviere, M. Eschenbrenner, and M. Fontecave. 1993. NADPH-sulfite reductase from *Escherichia coli*. A flavin reductase participating in the generation of the free radical of ribonucleotide reductase. *J. Biol. Chem.* **268**:18604–18609.

364. Cox, J. C., and P. Jurtshuk, Jr. 1990. An *Escherichia coli* mutant conditionally altered in respiratory chain components. *Membr. Biochem.* **9**:47–60.

365. Cox, R., and H. P. Charles. 1973. Porphyrin-accumulating mutants of *Escherichia coli*. *J. Bacteriol.* **113**:122–132.

366. Craig, P. A., and E. E. Dekker. 1990. The sulfhydryl content of L-threonine dehydrogenase from *Escherichia coli* K-12: relation to catalytic activity and Mn^{2+} activation. *Biochim. Biophys. Acta* **1037**:30–38.

367. Craigen, W. J., and C. T. Caskey. 1987. The function, structure and regulation of *E. coli* peptide chain release factors. *Biochimie* **69**:1031–1041.

368. Cronan, J. E., Jr. 1989. The *E. coli* bio operon: transcriptional repression by an essential protein modification enzyme. *Cell* **58**:427–429.

369. Cronan, J. E., Jr., W. B. Li, R. Coleman, M. Narasimhan, D. de Mendoza, and J. M. Schwab. 1988. Derived amino acid sequence and identification of active site residues of *Escherichia coli* beta-hydroxydecanoyl thioester dehydrase. *J. Biol. Chem.* **263**:4641–4646.

370. Cronan, J. E., Jr., K. J. Littel, and S. Jackowski. 1982. Genetic and biochemical analyses of pantothenate biosynthesis in *Escherichia coli* and *Salmonella typhimurium*. *J. Bacteriol.* **149**:916–922.

371. Crooke, E., M. Akiyama, N. N. Rao, and A. Kornberg. 1994. Genetically altered levels of inorganic polyphosphate in *Escherichia coli*. *J. Biol. Chem.* **269**:6290–6295.

372. Crouch, R. J. 1990. Ribonuclease H: from discovery to 3D structure. *New Biol.* **2**:771–777.

373. Crowell, D. N., W. S. Reznikoff, and C. R. Raetz. 1987. Nucleotide sequence of the *Escherichia coli* gene for lipid A disaccharide synthase. *J. Bacteriol.* **169**:5727–5734.

374. Cui, J., and R. L. Somerville. 1993. Mutational uncoupling of the transcriptional activation function of the TyrR protein of *Escherichia coli* K-12 from the repression function. *J. Bacteriol.* **175**:303–306.

375. Culham, D. E., B. Lasby, A. G. Marangoni, J. L. Milner, B. A. Steer, R. W. van Nues, and J. M. Wood. 1993. Isolation and sequencing of *Escherichia coli* gene proP reveals unusual structural features of the osmoregulatory proline/betaine transporter, ProP. *J. Mol. Biol.* **229**:268–276.

376. Cummings, H. S., and J. W. Hershey. 1994. Translation initiation factor IF1 is essential for cell viability in *Escherichia coli*. *J. Bacteriol.* **176**:198–205.

377. Cummings, H. S., J. F. Sands, P. C. Foreman, J. Fraser, and J. W. Hershey. 1991. Structure and expression of the infA operon encoding translational initiation factor IF1. Transcriptional control by growth rate. *J. Biol. Chem.* **266**:16491–16498.

378. Cunningham, P. R., and D. P. Clark. 1986. The use of suicide substrates to select mutants of *Escherichia coli* lacking enzymes of alcohol fermentation. *Mol. Gen. Genet.* **205**:487–493.

379. Cunningham, P. R., K. Nurse, C. J. Weitzmann, D. Negre, and J. Ofengand. 1992. G1401: a keystone nucleotide at the decoding site of *Escherichia coli* 30S ribosomes. *Biochemistry* **31**:7629–7637.

380. Curtis, N. A., R. L. Eisenstadt, S. J. East, R. J. Cornford, L. A. Walker, and A. J. White. 1988. Iron-regulated outer membrane proteins of *Escherichia coli* K-12 and mechanism of action of catechol-substituted cephalosporins. *Antimicrob. Agents Chemother.* **32**:1879–1886.

381. Curtis, S. J., and W. Epstein. 1975. Phosphorylation of D-glucose in *Escherichia coli* mutants defective in glucosephosphotransferase, and glucokinase. *J. Bacteriol.* **114**:1189–1199.

382. Cusack, S., C. Berthet-Colominas, M. Hartlein, N. Nassar, and R. Leberman. 1990. A second class of synthetase structure revealed by X-ray analysis of *Escherichia coli* seryl-tRNA synthetase at 2.5 Å. *Nature* (London) **347**:249–255.

383. Cusack, S., M. Hartlein, and R. Leberman. 1991. Sequence, structural and evolutionary relationships between class 2 aminoacyl-tRNA synthetases. *Nucleic Acids Res.* **19**:3489–3498.

384. Dai, D., and E. E. Ishiguro. 1988. *murH*, a new genetic locus in *Escherichia coli* involved in cell wall peptidoglycan biosynthesis. *J. Bacteriol.* **170**:2197–2201.

385. Dai, K., and J. Lutkenhaus. 1992. The proper ratio of FtsZ to FtsA is required for cell division to occur in *Escherichia coli*. *J. Bacteriol.* **174**:6145–6151.

386. Dai, K., Y. Xu, and J. Lutkenhaus. 1993. Cloning and characterization of *ftsN*, an essential cell division gene in *Escherichia coli* isolated as a multicopy suppressor of *ftsA12*(Ts). *J. Bacteriol.* **175**:3790–3797.

387. Dailey, F. E., and J. E. Cronan, Jr. 1986. Acetohydroxy acid synthase I, a required enzyme for isoleucine and valine biosynthesis in *Escherichia coli* K-12 during growth on acetate as the sole carbon source. *J. Bacteriol.* **165**:453–460.

388. Daldal, F. 1984. Nucleotide sequence of gene pfkB encoding the minor phosphofructokinase of *Escherichia coli* K-12. *Gene* **28**:337–342.

389. D'Alessio, G., and J. Josse. 1971. Glyceraldehyde phosphate dehydrogenase, phosphoglycerate kinase, and phosphoglyceromutase of *Escherichia coli*. Simultaneous purification and physical properties. *J. Biol. Chem.* **246**:4319–4325.

390. Dallas, W. S., J. E. Gowen, P. H. Ray, M. J. Cox, and I. K. Dev. 1992. Cloning, sequencing, and enhanced expression of the dihydropteroate synthase gene of *Escherichia coli* MC4100. *J. Bacteriol.* **174**:5961–5970.

391. Dame, J. B., and B. M. Shapiro. 1976. Use of polymyxin B, levallorphan, and tetracaine to isolate novel envelope mutants of *Escherichia coli*. *J. Bacteriol.* **127**:961–972.

392. Danchin, A., and L. Dondon. 1980. Serine sensitivity of *Escherichia coli* K 12: partial characterization of a serine resistant mutant that is extremely sensitive to 2-ketobutyrate. *Mol. Gen. Genet.* **178**:155–164.

393. Dandanell, G., K. Norris, and K. Hammer. 1991. Long-distance deoR regulation of gene expression in *Escherichia coli*. *Ann. N. Y. Acad. Sci.* **646**:19–30.

394. Daniel, J., and I. Saint-Girons. 1982. Attenuation in the threonine operon: effects of amino acids present in the presumed leader peptide in addition to threonine and isoleucine. *Mol. Gen. Genet.* **188**:225–227.

395. Daniels, D. L., G. Plunkett, V. Burland, and F. R. Blattner. 1992. Analysis of the *Escherichia coli* genome: DNA sequence of the region from 84.5 to 86.5 minutes. *Science* **257**:771–778.

396. Danielsen, S., M. Kilstrup, K. Barilla, B. Jochimsen, and J. Neuhard. 1992. Characterization of the *Escherichia coli* codBA operon encoding cytosine permease and cytosine deaminase. *Mol. Microbiol.* **6**:1335–1344.

397. Danson, M. J., G. Hale, P. Johnson, R. N. Perham, J. Smith, and P. Spragg. 1979. Molecular weight and symmetry of the pyruvate dehydrogenase multienzyme complex of *Escherichia coli*. *J. Mol. Biol.* **129**:603–617.

398. Dar, M. E., and A. S. Bhagwat. 1993. Mechanism of expression of DNA repair gene vsr, an *Escherichia coli* gene that overlaps the DNA cytosine methylase gene, dcm. *Mol. Microbiol.* **9**:823–833.

399. Dardel, F., M. Panvert, and G. Fayat. 1990. Transcription and regulation of expression of the *Escherichia coli* methionyl-tRNA synthetase gene. *Mol. Gen. Genet.* **223**:121–133.

400. D'Ari, L., and J. C. Rabinowitz. 1991. Purification, characterization, cloning, and amino acid sequence of the bifunctional enzyme 5,10-methylenetetrahydrofolate dehydrogenase/5,10-methenyltetrahydrofolate cyclohydrolase from *Escherichia coli*. *J. Biol. Chem.* **266**:23953–23958.

401. Darlison, M. G., and J. R. Guest. 1984. Nucleotide sequence encoding the iron-sulphur protein subunit of the succinate dehydrogenase of *Escherichia coli*. *Biochem. J.* **223**:507–517.

402. Dassa, E., and P. L. Boquet. 1981. ExpA: a conditional mutation affecting the expression of a group of exported proteins in *Escherichia coli* K-12. *Mol. Gen. Genet.* **181**:192–200.

403. Dassa, E., and S. Muir. 1993. Membrane topology of MalG, an inner membrane protein from the maltose transport system of *Escherichia coli*. *Mol. Microbiol.* **7**:29–38.

404. Dassa, J., H. Fsihi, C. Marck, M. Dion, M. Kieffer-Bontemps, and P. L. Boquet. 1991. A new oxygen-regulated operon in *Escherichia coli* comprises the genes for a putative third cytochrome oxidase and for pH 2.5 acid phosphatase (appA). *Mol. Gen. Genet.* **229**:341–352.

405. **Dassa, J., C. Marck, and P. L. Boquet.** 1990. The complete nucleotide sequence of the *Escherichia coli* gene *appA* reveals significant homology between pH 2.5 acid phosphatase and glucose-1-phosphatase. *J. Bacteriol.* 172:5497–5500.

406. **Datta, P., T. J. Goss, J. R. Omnaas, and R. V. Patil.** 1987. Covalent structure of biodegradative threonine dehydratase of *Escherichia coli*: homology with other dehydratases. *Proc. Natl. Acad. Sci. USA* 84:393–397.

407. **Dattananda, C. S., and J. Gowrishankar.** 1989. Osmoregulation in *Escherichia coli*: complementation analysis and gene-protein relationships in the *proU* locus. *J. Bacteriol.* 171:1915–1922.

408. **d'Aubenton Carafa, Y., E. Brody, and C. Thermes.** 1990. Prediction of rho-independent *Escherichia coli* transcription terminators. A statistical analysis of their RNA stem-loop structures. *J. Mol. Biol.* 216:835–858.

409. **Davidoff-Abelson, R., and L. Mindich.** 1978. A mutation that increases the activity of nonsense suppressors in *Escherichia coli*. *Mol. Gen. Genet.* 159:161–169.

410. **Davidson, A. L., H. A. Shuman, and H. Nikaido.** 1992. Mechanism of maltose transport in *Escherichia coli*: transmembrane signaling by periplasmic binding proteins. *Proc. Natl. Acad. Sci. USA* 89:2360–2364.

411. **Davies, J. K., and P. Reeves.** 1975. Genetics of resistance to colicins in *Escherichia coli* K-12: cross-resistance among colicins of group A. *J. Bacteriol.* 123:102–117.

412. **Davies, W. D., J. Pittard, and B. E. Davidson.** 1985. Cloning of aroG, the gene coding for phospho-2-keto-3-deoxy-heptonate aldolase (phe) in *Escherichia coli* K12 and subcloning of the aroG promoter in a promoter-detecting plasmid. *Gene* 33:323–331.

413. **Davis, D. R., and C. D. Poulter.** 1991. 1H-15N NMR studies of *Escherichia coli* tRNA(Phe) from hisT mutants: a structural role for pseudouridine. *Biochemistry* 30:4223–4231.

414. **Davis, E. O., and P. J. Henderson.** 1987. The cloning and DNA sequence of the gene xylE for xylose-proton symport in *Escherichia coli* K12. *J. Biol. Chem.* 262:13928–13932.

415. **Daws, T., C. J. Lim, and J. A. Fuchs.** 1989. In vitro construction of gshB::kan in *Escherichia coli* and use of gshB::kan in mapping the gshB locus. *J. Bacteriol.* 171:5218–5221.

416. **Dawson, J. H., A. M. Bracete, A. M. Huff, S. Kadkhodayan, C. M. Zeitler, M. Sono, C. K. Chang, and P. C. Loewen.** 1991. The active site structure of *E. coli* HPII catalase. Evidence favoring coordination of a tyrosinate proximal ligand to the chlorin iron. *FEBS Lett.* 295:123–126.

417. **Dean, D. A., L. I. Hor, H. A. Shuman, and H. Nikaido.** 1992. Interaction between maltose-binding protein and the membrane-associated maltose transporter complex in *Escherichia coli*. *Mol. Microbiol.* 6:2033–2040.

418. **Dean, D. A., J. Reizer, H. Nikaido, and M. H. Saier, Jr.** 1990. Regulation of the maltose transport system of *Escherichia coli* by the glucose-specific enzyme III of the phosphoenolpyruvate-sugar phosphotransferase system. Characterization of inducer exclusion-resistant mutants and reconstitution of inducer exclusion in proteoliposomes. *J. Biol. Chem.* 265:21005–21010.

419. **Dean, D. O., and R. James.** 1991. Identification of a gene, closely linked to dnaK, which is required for high-temperature growth of *Escherichia coli*. *J. Gen. Microbiol.* 137:1271–1277.

420. **de Boer, P. A., R. E. Crossley, A. R. Hand, and L. I. Rothfield.** 1991. The MinD protein is a membrane ATPase required for the correct placement of the *Escherichia coli* division site. *EMBO J.* 10:4371–4380.

421. **de Boer, P. A., R. E. Crossley, and L. I. Rothfield.** 1990. Central role for the *Escherichia coli* minC gene product in two different cell division-inhibition systems. *Proc. Natl. Acad. Sci. USA* 87:1129–1133.

422. **de Boer, P. A., R. E. Crossley, and L. I. Rothfield.** 1992. Roles of MinC and MinD in the site-specific septation block mediated by the MinCDE system of *Escherichia coli*. *J. Bacteriol.* 174:63–70.

423. **DeChavigny, A., P. N. Heacock, and W. Dowhan.** 1991. Sequence and inactivation of the pss gene of *Escherichia coli*. Phosphatidylethanolamine may not be essential for cell viability. *J. Biol. Chem.* 266:5323–5332.

424. **de Cock, H., W. Overeem, and J. Tommassen.** 1992. Biogenesis of outer membrane protein PhoE of *Escherichia coli*. Evidence for multiple SecB-binding sites in the mature portion of the PhoE protein. *J. Mol. Biol.* 224:369–379.

425. **DeFeyter, R. C., and J. Pittard.** 1986. Genetic and molecular analysis of aroL, the gene for shikimate kinase II in *Escherichia coli* K-12. *J. Bacteriol.* 165:226–232.

426. **DeFeyter, R. C., and J. Pittard.** 1986. Purification and properties of shikimate kinase II from *Escherichia coli* K-12. *J. Bacteriol.* 165:331–333.

427. **Deguchi, Y., I. Yamato, and Y. Anraku.** 1989. Molecular cloning of gltS and gltP, which encode glutamate carriers of *Escherichia coli* B. *J. Bacteriol.* 171:1314–1319.

428. **de la Campa, A. G., C. Aldea, C. Hernandez-Chico, A. Tormo, and M. Vicente.** 1988. Segregation of elongation potential in *Escherichia coli* mediated by the wee genetic system. *Curr. Microbiol.* 17:315–319.

429. **de la Campa, A. G., E. Martinez-Salas, A. Tormo, and M. Vicente.** 1984. Co-ordination between elongation and division in *Escherichia coli* mediated by the wee gene product. *J. Gen. Microbiol.* 130:2671–2679.

430. **del Castillo, I., J. M. Gomez, and F. Moreno.** 1990. mprA, an *Escherichia coli* gene that reduces growth-phase-dependent synthesis of microcins B17 and C7 and blocks osmoinduction of *proU* when cloned on a high-copy-number plasmid. *J. Bacteriol.* 172:437–445.

431. **del Castillo, I., J. E. Gonzalez-Pastor, J. L. San Millan, and F. Moreno.** 1991. Nucleotide sequence of the *Escherichia coli* regulatory gene mprA and construction and characterization of mprA-deficient mutants. *J. Bacteriol.* 173:3924–3929.

432. **del Castillo, I., J. L. Vizan, M. C. Rodriguez-Sainz, and F. Moreno.** 1991. An unusual mechanism for resistance to the antibiotic coumermycin A1. *Proc. Natl. Acad. Sci. USA* 88:8860–8864.

433. **Delcour, A. H., J. Adler, and C. Kung.** 1991. A single amino acid substitution alters conductance and gating of OmpC porin of *Escherichia coli*. *J. Membr. Biol.* 119:267–275.

434. **Delcuve, G., T. Cabezon, A. Herzog, M. Cannon, and A. Bollen.** 1978. Resistance to the aminoglycoside antibiotic neamine in *Escherichia coli*. A new mutant whose Nea(R) phenotype results from the cumulative effects of two distinct mutations. *Biochem. J.* 174:1–7.

435. **Delidakis, C. E., M. C. Jones-Mortimer, and H. L. Kornberg.** 1982. A mutant inducible for galactitol utilization in *Escherichia coli* K12. *J. Gen. Microbiol.* 128:601–604.

436. **DeMoss, J. A., and P. Y. Hsu.** 1991. NarK enhances nitrate uptake and nitrite excretion in *Escherichia coli*. *J. Bacteriol.* 173:3303–3310.

437. **Deonarain, M. P., N. S. Scrutton, and R. N. Perham.** 1992. Engineering surface charge. 2. A method for purifying heterodimers of *Escherichia coli* glutathione reductase. *Biochemistry* 31:1498–1504.

438. **De Reuse, H., and A. Danchin.** 1991. Positive regulation of the pts operon of *Escherichia coli*: genetic evidence for a signal transduction mechanism. *J. Bacteriol.* 173:727–733.

439. **de Riel, J. K., and H. Paulus.** 1978. Subunit dissociation in the allosteric regulation of glycerol kinase from *Escherichia coli*. 2. Physical evidence. *Biochemistry* 17:5141–5146.

440. **Derman, A. I., and J. Beckwith.** 1991. *Escherichia coli* alkaline phosphatase fails to acquire disulfide bonds when retained in the cytoplasm. *J. Bacteriol.* 173:7719–7722.

441. **Dervyn, E., D. Canceill, and O. Huisman.** 1990. Saturation and specificity of the Lon protease of *Escherichia coli*. *J. Bacteriol.* 172:7098–7103.

442. **Deutch, A. H., C. J. Smith, K. E. Rushlow, and P. J. Kretschmer.** 1982. *Escherichia coli* delta 1-pyrroline-5-carboxylate reductase: gene sequence, protein overproduction and purification. *Nucleic Acids Res.* 10:7701–7714.

443. **Deutch, C. E., and R. L. Soffer.** 1978. *Escherichia coli* mutants defective in dipeptidyl carboxypeptidase. *Proc. Natl. Acad. Sci. USA* 75:5998–6001.

444. **Deutscher, M. P., G. T. Marshall, and H. Cudny.** 1988. RNase PH: an *Escherichia coli* phosphate-dependent nuclease distinct from polynucleotide phosphorylase. *Proc. Natl. Acad. Sci. USA* 85:4710–4714.

445. **DeVeaux, L. C., J. E. Cronan, Jr., and T. L. Smith.** 1989. Genetic and biochemical characterization of a mutation (fatA) that allows *trans* unsaturated fatty acids to replace the essential *cis* unsaturated fatty acids of *Escherichia coli*. *J. Bacteriol.* 171:1562–1568.

446. **Deville-Bonne, D., F. Bourgain, and J. R. Garel.** 1991. pH dependence of the kinetic properties of allosteric phosphofructokinase from *Escherichia coli*. *Biochemistry* 30:5750–5754.

447. **Diderichsen, B.** 1980. flu, a metastable gene controlling surface properties of *Escherichia coli*. *J. Bacteriol.* 141:858–867.

448. **Diderichsen, B., and L. Desmarez.** 1980. Variations in phenotype of relB mutants of *Escherichia coli* and the effect of pus and sup mutations. *Mol. Gen. Genet.* 180:429–437.

449. **DiGate, R. J., and K. J. Marians.** 1989. Molecular cloning and DNA sequence analysis of *Escherichia coli* topB, the gene encoding topoisomerase III. *J. Biol. Chem.* 264:17924–17930.

450. **Dila, D., E. Sutherland, L. Moran, B. Slatko, and E. A. Raleigh.** 1990. Genetic and sequence organization of the mcrBC locus of *Escherichia coli* K-12. *J. Bacteriol.* 172:4888–4900.

451. **DiRusso, C. C., T. L. Heimert, and A. K. Metzger.** 1992. Characterization of FadR, a global transcriptional regulator of fatty acid metabolism in *Escherichia coli*. Interaction with the fadB promoter is prevented by long chain fatty acyl coenzyme A. *J. Biol. Chem.* 267:8685–8691.

452. **Diver, W. P., N. J. Sargentini, and K. C. Smith.** 1982. A mutation (radA100) in *Escherichia coli* that selectively sensitizes cells grown in rich medium to X- or U.V.-radiation, or methyl methanesulphonate. *Int. J. Radiat. Biol.* 42:339–346.

453. **Dixon, D. A., and S. C. Kowalczykowski.** 1993. The recombination hotspot chi is a regulatory sequence that acts by attenuating the nuclease activity of the *E. coli* RecBCD enzyme. *Cell* 73:87–96.

454. **Doi, M., M. Wachi, F. Ishino, S. Tomioka, M. Ito, Y. Sakagami, A. Suzuki, and M. Matsuhashi.** 1988. Determinations of the DNA sequence of the mreB gene and of the gene products of the mre region that function in formation of the rod shape of *Escherichia coli* cells. *J. Bacteriol.* 170:4619–4624.

455. **Doll, L., and G. Frankel.** 1993. fliU and fliV: two flagellar genes essential for biosynthesis of *Salmonella* and *Escherichia coli* flagella. *J. Gen. Microbiol.* 139:2415–2422.

456. **Donachie, W. D.** 1993. The cell cycle of *Escherichia coli*. *Annu. Rev. Microbiol.* 47:199–230.

457. **Donald, L. J., B. R. Crane, D. H. Anderson, and H. W. Duckworth.** 1991. The role of cysteine 206 in allosteric inhibition of *Escherichia coli* citrate synthase. Studies by chemical modification, site-directed mutagenesis, and ^{19}F NMR. *J. Biol. Chem.* 266:20709–20713.

458. **Dong, J. M., J. H. Scott, A. Aristarkhov, S. Iuchi, and E. C. C. Lin.** 1992. The lct operon encoding L-lactate permease and L-lactate dehydrogenase in *Escherichia coli*, abstr. H-156, p. 209. *In* Abstracts of the 92nd General Meeting of the American Society for Microbiology 1992. American Society for Microbiology, Washington, D.C.

459. Dong, J. M., J. S. Taylor, D. J. Latour, S. Iuchi, and E. C. Lin. 1993. Three overlapping *lct* genes involved in L-lactate utilization by *Escherichia coli*. *J. Bacteriol.* 175:6671–6678.

460. Dong, Q., F. Liu, A. M. Myers, and H. J. Fromm. 1991. Evidence for an arginine residue at the substrate binding site of *Escherichia coli* adenylosuccinate synthetase as studied by chemical modification and site-directed mutagenesis. *J. Biol. Chem.* 266:12228–12233.

461. Dong, Z., R. Onrust, M. Skangalis, and M. O'Donell. 1993. DNA polymerase III accessory proteins. I. holA and holB encoding δ and δ'. *J. Biol. Chem.* 268:11758–11765.

462. Donovan, W. P., and S. R. Kushner. 1986. Polynucleotide phosphorylase and ribonuclease II are required for cell viability and mRNA turnover in *Escherichia coli* K-12. *Proc. Natl. Acad. Sci. USA* 83:120–124.

463. Dontsova, O. A., K. V. Rosen, S. L. Bogdanova, E. A. Skripkin, A. M. Kopylov, and A. A. Bogdanov. 1992. Identification of the *Escherichia coli* 30S ribosomal subunit protein neighboring mRNA during initiation of translation. *Biochimie* 74:363–371.

464. Doolittle, R. F., D. F. Feng, K. L. Anderson, and M. R. Alberro. 1990. A naturally occurring horizontal gene transfer from a eukaryote to a prokaryote. *J. Mol. Evol.* 31:383–388.

465. Doolittle, R. F., D. F. Feng, M. S. Johnson, and M. A. McClure. 1986. Relationships of human protein sequences to those of other organisms. *Cold Spring Harbor Symp. Quant. Biol.* 51:447–455.

466. Dorman, C. J., A. S. Lynch, N. N. Bhriain, and C. F. Higgins. 1989. DNA supercoiling in *Escherichia coli*: topA mutations can be suppressed by DNA amplifications involving the tolC locus. *Mol. Microbiol.* 3:531–540.

467. Dornmair, K., H. Kiefer, and F. Jahnig. 1990. Refolding of an integral membrane protein. OmpA of *Escherichia coli*. *J. Biol. Chem.* 265:18907–18911.

468. Dorrell, N., A. H. Ahmed, and S. H. Moss. 1993. Photoreactivation in a phrB mutant of *Escherichia coli* K-12: evidence for the role of a second protein in photorepair. *Photochem. Photobiol.* 58:831–835.

469. Dosch, D. C., G. L. Helmer, S. H. Sutton, F. F. Salvacion, and W. Epstein. 1991. Genetic analysis of potassium transport loci in *Escherichia coli*: evidence for three constitutive systems mediating uptake potassium. *J. Bacteriol.* 173:687–696.

470. Doublet, P., J. van Heijenoort, and D. Mengin-Lecreulx. 1992. Identification of the *Escherichia coli* murI gene, which is required for the biosynthesis of D-glutamic acid, a specific component of bacterial peptidoglycan. *J. Bacteriol.* 174:5772–5779.

471. Dougherty, T. J., J. A. Thanassi, and M. J. Pucci. 1993. The *Escherichia coli* mutant requiring D-glutamic acid is the result of mutations in two distinct genetic loci. *J. Bacteriol.* 175:111–116.

472. Drexler, H. 1977. Specialized transduction of the biotin region of *Escherichia coli* by phage T1. *Mol. Gen. Genet.* 152:59–63.

473. Dri, A. M., P. L. Moreau, and J. Rouviere-Yaniv. 1992. Role of the histone-like proteins OsmZ and HU in homologous recombination. *Gene* 120:11–16.

474. Dri, A. M., J. Rouviere-Yaniv, and P. L. Moreau. 1991. Inhibition of cell division in hupA hupB mutant bacteria lacking HU protein. *J. Bacteriol.* 173:2852–2863.

475. Drlica, K. 1990. Bacterial topoisomerases and the control of DNA supercoiling. *Trends Genet.* 6:433–437.

476. Drolet, M., L. Peloquin, Y. Echelard, L. Cousineau, and A. Sasarman. 1989. Isolation and nucleotide sequence of the hemA gene of *Escherichia coli* K12. *Mol. Gen. Genet.* 216:347–352.

477. Drury, L. S., and R. S. Buxton. 1988. Identification and sequencing of the *Escherichia coli* cet gene which codes for an inner membrane protein, mutation of which causes tolerance to colicin E2. *Mol. Microbiol.* 2:109–119.

478. Dubourdieu, M., and J. A. DeMoss. 1992. The narJ gene product is required for biogenesis of respiratory nitrate reductase in *Escherichia coli*. *J. Bacteriol.* 174:867–872.

479. Duclos, B., J. C. Cortay, F. Bleicher, E. Z. Ron, C. Richaud, I. Saint Girons, and A. J. Cozzone. 1989. Nucleotide sequence of the metA gene encoding homoserine trans-succinylase in *Escherichia coli*. *Nucleic Acids Res.* 17:2856

480. Duester, G. L., and W. M. Holmes. 1980. The distal end of the ribosomal RNA operon rrnD of *Escherichia coli* contains a tRNA1thr gene, two 5s rRNA genes and a transcription terminator. *Nucleic Acids Res.* 8:3793–3807.

481. Duncan, K., and J. R. Coggins. 1986. The serC-aroA operon of *Escherichia coli*. A mixed function operon encoding enzymes from two different amino acid biosynthetic pathways. *Biochem. J.* 234:49–57.

482. Duncan, K., J. van Heijenoort, and C. T. Walsh. 1990. Purification and characterization of the D-alanyl-D-alanine-adding enzyme from *Escherichia coli*. *Biochemistry* 29:2379–2386.

483. Dunn, M. F., V. Aguilar, P. Brzovic, W. F. Drewe, Jr., K. F. Houben, C. A. Leja, and M. Roy. 1990. The tryptophan synthase bienzyme complex transfers indole between the alpha- and beta-sites via a 25–30 Å long tunnel. *Biochemistry* 29:8598–8607.

484. Edgar, J. R., and R. M. Bell. 1979. Biosynthesis in *Escherichia coli* of sn-glycerol 3-phosphate, a precursor of phospholipid. Palmitoyl-CoA inhibition of the biosynthetic sn-glycerol-3-phosphate dehydrogenase. *J. Biol. Chem.* 254:1016–1021.

485. Edwards, R. M., and M. D. Yudkin. 1982. Location of the gene for the low-affinity tryptophan-specific permease of *Escherichia coli*. *Biochem. J.* 204:617–619.

486. Egan, A. F., and R. R. B. Russell. 1973. Conditional mutations affecting the cell envelope of *Escherichia coli* K-12. *Genet. Res.* 21:139–152.

487. Egan, S. E., R. Fliege, S. Tong, A. Shibata, R. E. Wolf, Jr., and T. Conway. 1992. Molecular characterization of the Entner-Doudoroff pathway in *Escherichia coli*: sequence analysis and localization of promoters for the *edd-eda* operon. *J. Bacteriol.* 174:4638–4646.

488. Egan, S. M., and R. F. Schleif. 1993. A regulatory cascade in the induction of rhaBAD. *J. Mol. Biol.* 234:87–98.

489. Egan, S. M., and V. Stewart. 1991. Mutational analysis of nitrate regulatory gene narL in *Escherichia coli* K-12. *J. Bacteriol.* 173:4424–4432.

490. Egebjerg, J., J. Christiansen, R. S. Brown, N. Larsen, and R. A. Garrett. 1989. Protein L18 binds primarily at the junctions of helix II and internal loops A and B in *Escherichia coli* 5 S RNA. Implications for 5 S RNA structure. *J. Mol. Biol.* 206:651–668.

491. Egebjerg, J., J. Christiansen, and R. A. Garrett. 1991. Attachment sites of primary binding proteins L1, L2 and L23 on 23 S ribosomal RNA of *Escherichia coli*. *J. Mol. Biol.* 222:251–264.

492. Egebjerg, J., S. R. Douthwaite, A. Liljas, and R. A. Garrett. 1990. Characterization of the binding sites of protein L11 and the L10.(L12)4 pentameric complex in the GTPase domain of 23 S ribosomal RNA from *Escherichia coli*. *J. Mol. Biol.* 213:275–288.

493. Ehretsmann, C. P., A. J. Carpousis, and H. M. Krisch. 1992. Specificity of *Escherichia coli* endoribonuclease RNase E: in vivo and in vitro analysis of mutants in a bacteriophage T4 mRNA processing site. *Genes Dev.* 6:149–159.

494. Eick-Helmerich, K., and V. Braun. 1989. Import of biopolymers into *Escherichia coli*: nucleotide sequences of the exbB and exbD genes are homologous to those of the tolQ and tolR genes, respectively. *J. Bacteriol.* 171:5117–5126.

495. Eiglmeier, K., W. Boos, and S. T. Cole. 1987. Nucleotide sequence and transcriptional startpoint of the glpT gene of *Escherichia coli*: extensive sequence homology of the glycerol-3-phosphate transport protein with components of the hexose-6-phosphate transport system. *Mol. Microbiol.* 1:251–258.

496. Eisenberg, R. C., and W. J. Dobrogosz. 1967. Gluconate metabolism in *Escherichia coli*. *J. Bacteriol.* 93:941–949.

497. Eisenstein, E. 1991. Cloning, expression, purification, and characterization of biosynthetic threonine deaminase from *Escherichia coli*. *J. Biol. Chem.* 266:5801–5807.

498. Elkins, M. F., and C. F. Earhart. 1989. Nucleotide sequence and regulation of the *Escherichia coli* gene for ferrienterobactin transport protein FepB. *J. Bacteriol.* 171:5443–5451.

499. Elliott, A. C., S. K, M. L. Sinnott, P. J. Smith, J. Bommuswamy, Z. Guo, B. G. Hall, and Y. Zhang. 1992. The catalytic consequences of experimental evolution. Studies on the subunit structure of the second (ebg) beta-galactosidase of *Escherichia coli*, and on catalysis by ebgab, an experimental evolvant containing two amino acid substitutions. *Biochem. J.* 282:155–164.

500. Elmore, M. J., A. J. Lamb, G. Y. Ritchie, R. M. Douglas, A. Munro, A. Gajewska, and I. R. Booth. 1990. Activation of potassium efflux from *Escherichia coli* by glutathione metabolites. *Mol. Microbiol.* 4:405–412.

501. Elvin, C. M., N. E. Dixon, and H. Rosenberg. 1986. Molecular cloning of the phosphate (inorganic) transport (pit) gene of *Escherichia coli* K12. Identification of the pit+ gene product and physical mapping of the pit-gor region of the chromosome. *Mol. Gen. Genet.* 204:477–484.

502. Engel, H., B. Kazemier, and W. Keck. 1991. Murein-metabolizing enzymes from *Escherichia coli*: sequence analysis and controlled overexpression of the slt gene, which encodes the soluble lytic transglycosylase. *J. Bacteriol.* 173:6773–6782.

503. Engel, H., A. J. Smink, L. van Wijngaarden, and W. Keck. 1992. Murein-metabolizing enzymes from *Escherichia coli*: existence of a second lytic transglycosylase. *J. Bacteriol.* 174:6394–6403.

504. Engelbrecht, S., and W. Junge. 1990. Subunit delta of H(+)-ATPases: at the interface between proton flow and ATP synthesis. *Biochim. Biophys. Acta* 1015:379–390.

505. Englisch-Peters, S., J. Conley, J. Plumbridge, C. Leptak, D. Soll, and M. J. Rogers. 1991. Mutant enzymes and tRNAs as probes of the glutaminyl-tRNA synthetase:tRNA(Gln) interaction. *Biochimie* 73:1501–1508.

506. Englisch-Peters, S., F. von der Haar, and F. Cramer. 1990. Fidelity in the aminoacylation of tRNA(Val) with hydroxy analogues of valine, leucine, and isoleucine by valyl-tRNA synthetases from *Saccharomyces cerevisiae* and *Escherichia coli*. *Biochemistry* 29:7953–7958.

507. Enoch, H. G., and R. L. Lester. 1975. The purification and properties of formate dehydrogenase and nitrate reductase from *Escherichia coli*. *J. Biol. Chem.* 250:6693–6705.

508. Ephrati-Elizur, E. 1993. A mutation in a new gene of *Escherichia coli*, psu, requires secondary mutations for survival: psu mutants express a pleiotropic suppressor phenotype. *J. Bacteriol.* 175:207–213.

509. Epperly, B. R., and E. E. Dekker. 1991. L-Threonine dehydrogenase from *Escherichia coli*. Identification of an active site cysteine residue and metal ion studies. *J. Biol. Chem.* 266:6086–6092.

510. Eriani, G., G. Dirheimer, and J. Gangloff. 1990. Structure-function relationship of arginyl-tRNA synthetase from *Escherichia coli*: isolation and characterization of the argS mutation MA5002. *Nucleic Acids Res.* 18:1475–1479.

511. Eriani, G., G. Dirheimer, and J. Gangloff. 1990. Aspartyl-tRNA synthetase from *Escherichia coli*: cloning and characterisation of the gene, homologies of its translated amino acid sequence with asparaginyl- and lysyl-tRNA synthetases. *Nucleic Acids Res.* 18:7109–7118.

512. Eriksson-Greenberg, K. G., and K. Nordstrom. 1973. Genetics and physiology of a tolE mutant of *Escherichia coli* K-12 and phenotypic suppression of its phenotype by galactose. *J. Bacteriol.* 115:1219–1222.

513. Ermler, U., and G. E. Schulz. 1991. The three-dimensional structure of glutathione reductase from *Escherichia coli* at 3.0 A resolution. *Proteins* **9**:174–179.

514. Erni, B., B. Zanolari, P. Graff, and H. P. Kocher. 1989. Mannose permease of *Escherichia coli*. Domain structure and function of the phosphorylating subunit. *J. Biol. Chem.* **264**:18733–18741.

515. Erni, B., B. Zanolari, and H. P. Kocher. 1987. The mannose permease of *Escherichia coli* consists of three different proteins. Amino acid sequence and function in sugar transport, sugar phosphorylation, and penetration of phage lambda DNA. *J. Biol. Chem.* **262**:5238–5247.

516. Ernsting, B. R., M. R. Atkinson, A. J. Ninfa, and R. G. Matthews. 1992. Characterization of the regulon controlled by the leucine-responsive regulatory protein in *Escherichia coli*. *J. Bacteriol.* **174**:1109–1118.

517. Escalante-Semerena, J. C., S. J. Suh, and J. R. Roth. 1990. *cobA* function is required for both de novo cobalamin biosynthesis and assimilation of exogenous corrinoids in *Salmonella typhimurium*. *J. Bacteriol.* **172**:273–280.

518. Estevenon, A. M., B. Martin, and N. Sicard. 1985. Characterization of a mutation conferring radiation sensitivity, ior, located close to the gene coding for deoxycytidine deaminase in *Escherichia coli*. *Mol. Gen. Genet.* **200**:132–137.

519. Faik, P., and H. L. Kornberg. 1973. Isolation and properties of E. *coli* mutants affected in gluconate uptake. *FEBS Lett.* **32**:260–264.

520. Falconi, M., M. T. Gualtieri, A. La Teana, M. A. Losso, and C. L. Pon. 1988. Proteins from the prokaryotic nucleoid: primary and quaternary structure of the 15-kD *Escherichia coli* DNA binding protein H-NS. *Mol. Microbiol.* **2**:323–329.

521. Falkenberg, P., and A. R. Strom. 1990. Purification and characterization of osmoregulatory betaine aldehyde dehydrogenase of *Escherichia coli*. *Biochim. Biophys. Acta* **1034**:253–259.

522. Falzone, C. J., W. E. Karsten, J. D. Conley, and R. E. Viola. 1988. L-Aspartase from *Escherichia coli*: substrate specificity and role of divalent metal ions. *Biochemistry* **27**:9089–9093.

523. Falzone, C. J., P. E. Wright, and S. J. Benkovic. 1991. Evidence for two interconverting protein isomers in the methotrexate complex of dihydrofolate reductase from *Escherichia coli*. *Biochemistry* **30**:2184–2191.

524. Fang, H., R. J. Lin, and R. B. Gennis. 1989. Location of heme axial ligands in the cytochrome d terminal oxidase complex of *Escherichia coli* determined by site-directed mutagenesis. *J. Biol. Chem.* **264**:8026–8032.

525. Farr, S. B., D. N. Arnosti, M. J. Chamberlin, and B. N. Ames. 1989. An apaH mutation causes AppppA to accumulate and affects motility and catabolite repression in *Escherichia coli*. *Proc. Natl. Acad. Sci. USA* **86**:5010–5014.

526. Farrington, G. K., A. Kumar, S. L. Shames, J. I. Ewaskiewicz, D. E. Ash, and F. C. Wedler. 1993. Threonine synthase of *Escherichia coli*: inhibition by classical and slow-binding analogues of homoserine phosphate. *Arch. Biochem. Biophys.* **307**:165–174.

527. Fath, M. J., H. K. Mahanty, and R. Kolter. 1989. Characterization of a *purF* operon mutation which affects colicin V production. *J. Bacteriol.* **171**:3158–3161.

528. Faubladier, M., K. Cam, and J. P. Bouche. 1990. *Escherichia coli* cell division inhibitor DicF-RNA of the dicB operon. Evidence for its generation in vivo by transcription termination and by RNase III and RNase E-dependent processing. *J. Mol. Biol.* **212**:461–471.

529. Fayerman, J. T., M. C. Vann, L. S. Williams, and H. E. Umbarger. 1979. ilvU, a locus in *Escherichia coli* affecting the derepression of isoleucyl-tRNA synthetase and the RPC-5 chromatographic profiles of tRNAIle and tRNAVal. *J. Biol. Chem.* **254**:9429–9440.

530. Fee, J. A. 1991. Regulation of sod genes in *Escherichia coli*: relevance to superoxide dismutase function. *Mol. Microbiol.* **5**:2599–2610.

531. Felzenszwalb, I., S. Boiteux, and J. Laval. 1992. Molecular cloning and DNA sequencing of the radC gene of *Escherichia coli* K-12. *Mutat. Res.* **273**:263–269.

532. Felzenszwalb, I., N. J. Sargentini, and K. C. Smith. 1986. *Escherichia coli* radC is deficient in the recA-dependent repair of X-ray-induced DNA strand breaks. *Radiat. Res.* **106**:166–170.

533. Feng, J., K. Yamanaka, H. Niki, T. Ogura, and S. Hiraga. 1994. New killing system controlled by two genes located immediately upstream of the mukB gene in *Escherichia coli*. *Mol. Gen. Genet.* **243**:136–147.

534. Fesik, S. W., R. T. Gampe, Jr., E. R. Zuiderweg, W. E. Kohlbrenner, and D. Weigl. 1989. Heteronuclear three-dimensional NMR spectroscopy applied to CMP-KDO synthetase (27.5 kD). *Biochem. Biophys. Res. Commun.* **159**:842–847.

535. Field, T. L., W. S. Reznikoff, and P. A. Frey. 1989. Galactose-1-phosphate uridylyltransferase: identification of histidine-164 and histidine-166 as critical residues by site-directed mutagenesis. *Biochemistry* **28**:2094–2099.

536. Figge, R. M., T. M. Ramseier, and M. H. Saier, Jr. 1994. The mannitol repressor (MtlR) of *Escherichia coli*. *J. Bacteriol.* **176**:840–847.

537. Filimonov, V. V., J. Prieto, J. C. Martinez, M. Bruix, P. L. Mateo, and L. Serrano. 1993. Thermodynamic analysis of the chemotactic protein from *Escherichia coli*, CheY. *Biochemistry* **32**:12906–12921.

538. Filley, S. J., and K. A. Hill. 1993. Amino acid substitutions at position 73 in motif 2 of *Escherichia coli* alanyl-tRNA synthetase. *Arch. Biochem. Biophys.* **307**:46–51.

539. First, E. A., and A. R. Fersht. 1993. Mutational and kinetic analysis of a mobile loop in tyrosyl-tRNA synthetase. *Biochemistry* **32**:13658–13663.

540. Fischer, E., B. Strehlow, D. Hartz, and V. Braun. 1990. Soluble and membrane-bound ferrisiderophore reductases of *Escherichia coli* K-12. *Arch. Microbiol.* **153**:329–336.

541. Fischer, M., and S. A. Short. 1982. The cloning of the *Escherichia coli* K-12 deoxyribonucleoside operon. *Gene* **17**:291–298.

542. Fisher, K. E., and E. Eisenstein. 1993. An efficient approach to identify ilvA mutations reveals an amino-terminal catalytic domain in biosynthetic threonine deaminase from *Escherichia coli*. *J. Bacteriol.* **175**:6605–6613.

543. Fisher, M. T. 1992. Promotion of the in vitro renaturation of dodecameric glutamine synthetase from *Escherichia coli* in the presence of GroEL (chaperonin-60) and ATP. *Biochemistry* **31**:3955–3963.

544. Fitch, W. M. 1970. Distinguishing homologous from analogous proteins. *Syst. Zool.* **19**:99–113.

545. Fix, D. F., D. R. Koehler, and B. W. Glickman. 1990. Uracil-DNA glycosylase activity affects the mutagenicity of ethyl methanesulfonate: evidence for an alternative pathway of alkylation mutagenesis. *Mutat. Res.* **244**:115–121.

546. Flachmann, R., N. Kunz, J. Seifert, M. Gutlich, F. J. Wientjes, A. Laufer, and H. G. Gassen. 1988. Molecular biology of pyridine nucleotide biosynthesis in *Escherichia coli*. Cloning and characterization of quinolinate synthesis genes nadA and nadB. *Eur. J. Biochem.* **175**:221–228.

547. Flannigan, K. A., S. H. Hennigan, H. H. Vogelbacker, J. S. Gots, and J. M. Smith. 1990. Purine biosynthesis in *Escherichia coli* K12: structure and DNA sequence studies of the purHD locus. *Mol. Microbiol.* **4**:381–392.

548. Flint, D. H., M. H. Emptage, M. G. Finnegan, W. Fu, and M. K. Johnson. 1993. The role and properties of the iron-sulfur cluster in *Escherichia coli* dihydroxy-acid dehydratase. *J. Biol. Chem.* **268**:14732–14742.

549. Foglino, M., and A. Lazdunski. 1987. Deletion analysis of the promoter region of the *Escherichia coli* pepN gene, a gene subject in vivo to multiple global controls. *Mol. Gen. Genet.* **210**:523–527.

550. Forchhammer, K., W. Leinfelder, K. Boesmiller, B. Veprek, and A. Bock. 1991. Selenocysteine synthase from *Escherichia coli*. Nucleotide sequence of the gene (selA) and purification of the protein. *J. Biol. Chem.* **266**:6318–6323.

551. Forchhammer, K., K. P. Rucknagel, and A. Bock. 1990. Purification and biochemical characterization of SELB, a translation factor involved in selenoprotein synthesis. *J. Biol. Chem.* **265**:9346–9350.

552. Forget, P. 1974. The bacterial nitrate reductases. Solubilization, purification and properties of the enzyme A of *Escherichia coli* K 12. *Eur. J. Biochem.* **42**:325–332.

553. Forst, S., J. Delgado, and M. Inouye. 1989. Phosphorylation of OmpR by the osmosensor EnvZ modulates expression of the ompF and ompC genes in *Escherichia coli*. *Proc. Natl. Acad. Sci. USA* **86**:6052–6056.

554. Foster, J. W., Y. K. Park, T. Penfound, T. Fenger, and M. P. Spector. 1990. Regulation of NAD metabolism in *Salmonella typhimurium*: molecular sequence analysis of the bifunctional nadR regulator and the nadA-pnuC operon. *J. Bacteriol.* **172**:4187–4196.

555. Fouts, K., and S. D. Barbour. 1981. Transductional mapping of ksgB and a new Tn5-induced kasugamycin resistance gene, ksgD, in *Escherichia coli* K-12. *J. Bacteriol.* **145**:914–919.

556. Fox, J., K. Kawaguchi, E. Greenberg, and J. Preiss. 1976. Biosynthesis of bacterial glycogen. Purification and properties of the *Escherichia coli* B ADPglucose:1,4-alpha-D-glucan 4-alpha-glucosyltransferase. *Biochemistry* **15**:849–857.

557. Fox, J. W., D. P. Owens, and K. P. Wong. 1988. Purification and conformation of ribosomal protein L25 from E. *coli* ribosome. *Int. J. Pept. Protein Res.* **31**:255–264.

558. Fraenkel, D. G., D. Kotlarz, and H. Buc. 1973. Two fructose 6-phosphate kinase activities in *Escherichia coli*. *J. Biol. Chem.* **248**:4865–4866.

559. Fralick, J. A. 1991. Studies on the alteration of chromosome copy number and cell division potential in a dnaA mutant of *Escherichia coli*. *Mol. Gen. Genet.* **229**:175–180.

560. Franceschi, F. J., and K. H. Nierhaus. 1990. Ribosomal proteins L15 and L16 are mere late assembly proteins of the large ribosomal subunit. Analysis of an *Escherichia coli* mutant lacking L15. *J. Biol. Chem.* **265**:16676–16682.

561. Francisco, J. A., C. F. Earhart, and G. Georgiou. 1992. Transport and anchoring of beta-lactamase to the external surface of *Escherichia coli*. *Proc. Natl. Acad. Sci. USA* **89**:2713–2717.

562. Frank, E. G., J. Hauser, A. S. Levine, and R. Woodgate. 1993. Targeting of the UmuD, UmuD′, and MucA′ mutagenesis proteins to DNA by RecA protein. *Proc. Natl. Acad. Sci. USA* **90**:8169–8173.

563. Franklin, F. C. H., W. A. Venables, and H. J. W. Wijsman. 1981. Genetic studies of D-alanine-dehydrogenase-less mutants of *Escherichia coli* K12. *Genet. Res.* **38**:197–208.

564. Freundlich, M., N. Ramani, E. Mathew, A. Sirko, and P. Tsui. 1992. The role of integration host factor in gene expression in *Escherichia coli*. *Mol. Microbiol.* **6**:2557–2563.

565. Freundlieb, S., and W. Boos. 1986. Alpha-amylase of *Escherichia coli*, mapping and cloning of the structural gene, malS, and identification of its product as a periplasmic protein. *J. Biol. Chem.* **261**:2946–2953.

566. Freundlieb, S., U. Ehmann, and W. Boos. 1988. Facilitated diffusion of p-nitrophenyl-alpha-D-maltohexaoside through the outer membrane of *Escherichia coli*. Characterization of LamB as a specific and saturable channel for maltooligosaccharides. *J. Biol. Chem.* **263**:314–320.

567. Frey, B., G. Janel, U. Michelsen, and H. Kersten. 1989. Mutations in the *Escherichia coli* fnr and tgt genes: control of molybdate reductase activity and the cytochrome d complex by fnr. *J. Bacteriol.* **171**:1524–1530.

568. Friden, P., J. Donegan, J. Mullen, P. Tsui, M. Freundlich, R. Weber, P. M. Silverman, and L. Eoyang. 1985. The ilvB locus of *Escherichia coli* K-12 is an operon encoding both subunits of acetohydroxyacid synthase I. *Nucleic Acids Res.* **13**:3979–3993.

569. Friden, P., T. Newman, and M. Freundlich. 1982. Nucleotide sequence of the ilvB promoter-regulatory region: a biosynthetic operon controlled by attenuation and cyclic AMP. *Proc. Natl. Acad. Sci. USA* **79**:6156–6160.

570. Frieden, C. 1990. Refolding of *Escherichia coli* dihydrofolate reductase: sequential formation of substrate binding sites. *Proc. Natl. Acad. Sci. USA* **87**:4413–4416.

571. Friedrich, T., U. Weidner, U. Nehls, W. Fecke, R. Schneider, and H. Weiss. 1993. Attempts to define distinct parts of NADH:ubiquinone oxidoreductase (complex I). *J. Bioenerg. Biomembr.* **25**:331–337.

572. Froelich, B., and W. Epstein. 1981. *Escherichia coli* mutants in which transcription is dependent on *recA* function. *J. Bacteriol.* **147**:1117–1120.

573. Froman, B. E., R. C. Tait, and L. D. Gottlieb. 1989. Isolation and characterization of the phosphoglucose isomerase gene from *Escherichia coli. Mol. Gen. Genet.* **217**:126–131.

574. Frustaci, J. M., and M. R. O'Brian. 1993. The *Escherichia coli visA* gene encodes ferrochelatase, the final enzyme of the heme biosynthetic pathway. *J. Bacteriol.* **175**:2154–2156.

575. Fsihi, H., B. Kottwitz, and E. Bremer. 1993. Single amino acid substitutions affecting the substrate specificity of the *Escherichia coli* K-12 nucleoside-specific Tsx channel. *J. Biol. Chem.* **268**:17495–17503.

576. Fujisaki, S., H. Hara, Y. Nishimura, K. Horiuchi, and T. Nishino. 1990. Cloning and nucleotide sequence of the ispA gene responsible for farnesyl diphosphate synthase activity in *Escherichia coli. J. Biochem.* (Tokyo) **108**:995–1000.

577. Fukuda, R., A. Nishimura, and H. Serizawa. 1988. Genetic mapping of the *Escherichia coli* gene for the stringent starvation protein and its dispensability for normal cell growth. *Mol. Gen. Genet.* **211**:515–519.

578. Fuller-Pace, F. V., S. M. Nicol, A. D. Reid, and D. P. Lane. 1993. DbpA: a DEAD box protein specifically activated by 23s rRNA. *EMBO J.* **12**:3619–3626.

579. Fultz, P. N., D. Y. Kwoh, and J. Kemper. 1979. *Salmonella typhimurium newD* and *Escherichia coli leuC* genes code for a functional isopropylmalate isomerase in *Salmonella typhimurium-Escherichia coli* hybrids. *J. Bacteriol.* **137**:1253–1262.

580. Funk, C. R., L. Zimniak, and W. Dowhan. 1992. The *pgpA* and *pgpB* genes of *Escherichia coli* are not essential: evidence for a third phosphatidylglycerophosphate phosphatase. *J. Bacteriol.* **174**:205–213.

581. Furuchi, T., K. Kashiwagi, H. Kobayashi, and K. Igarashi. 1991. Characteristics of the gene for a spermidine and putrescine transport system that maps at 15 min on the *Escherichia coli* chromosome. *J. Biol. Chem.* **266**:20928–20933.

582. Furukawa, H., J. T. Tsay, S. Jackowski, Y. Takamura, and C. O. Rock. 1993. Thiolactomycin resistance in *Escherichia coli* is associated with the multidrug resistance efflux pump encoded by emrAB. *J. Bacteriol.* **175**:3723–3729.

583. Furukawa, K., M. Tagaya, K. Tanizawa, and T. Fukui. 1993. Role of the conserved Lys-X-Gly-Gly sequence at the ADP-glucose-binding site in *Escherichia coli* glycogen synthase. *J. Biol. Chem.* **268**:23837–23842.

584. Furukawa, K., M. Tagaya, K. Tanizawa, and T. Fukui. 1994. Identification of Lys277 at the active site of *Escherichia coli* glycogen synthase. Application of affinity labeling combined with site-directed mutagenesis. *J. Biol. Chem.* **269**:868–871.

585. Galakatos, N. G., and C. T. Walsh. 1989. Mutations at the interdomain hinge region of the DadB alanine racemase: effects of length and conformational constraint of the linker sequence on catalytic efficiency. *Biochemistry* **28**:8167–8174.

586. Galloway, S. M., and C. R. Raetz. 1990. A mutant of *Escherichia coli* defective in the first step of endotoxin biosynthesis. *J. Biol. Chem.* **265**:6394–6402.

587. Gambino, L., S. J. Gracheck, and P. F. Miller. 1993. Overexpression of the MarA positive regulator is sufficient to confer multiple antibiotic resistance in *Escherichia coli. J. Bacteriol.* **175**:2888–2894.

588. Gamer, J., H. Bujard, and B. Bukau. 1992. Physical interaction between heat shock proteins DnaK, DnaJ, and GrpE and the bacterial heat shock transcription factor sigma 32. *Cell* **69**:833–842.

589. Ganduri, Y. L., S. R. Sadda, M. W. Datta, R. K. Jambukeswaran, and P. Datta. 1993. TdcA, a transcriptional activator of the tdcABC operon of *Escherichia coli*, is a member of the LysR family of proteins. *Mol. Gen. Genet.* **240**:395–402.

590. Ganesan, A. K., and B. Rotman. 1966. Transport systems for galactose and galactosides in *Escherichia coli*. I. Genetic determination and regulation of the methylgalactoside permease. *J. Mol. Biol.* **16**:42–50.

591. Ganong, B. R., and C. R. Raetz. 1983. pH-sensitive CDP-diglyceride synthetase mutants of *Escherichia coli*: phenotypic suppression by mutations at a second site. *J. Bacteriol.* **153**:731–738.

592. Garcia-Villegas, M. R., F. M. De La Vega, J. M. Galindo, M. Segura, R. H. Buckingham, and G. Guarneros. 1991. Peptidyl-tRNA hydrolase is involved in lambda inhibition of host protein synthesis. *EMBO J.* **10**:3549–3555.

593. Gardella, T., H. Moyle, and M. M. Susskind. 1989. A mutant *Escherichia coli* sigma 70 subunit of RNA polymerase with altered promoter specificity. *J. Mol. Biol.* **206**:579–590.

594. Gardina, P., C. Conway, M. Kossman, and M. Manson. 1992. Aspartate and maltose-binding protein interact with adjacent sites in the Tar chemotactic signal transducer of *Escherichia coli. J. Bacteriol.* **174**:1528–1536.

595. Gardner, J. F. 1982. Initiation, pausing, and termination of transcription in the threonine operon regulatory region of *Escherichia coli. J. Biol. Chem.* **257**:3896–3904.

596. Garrick-Silversmith, L., and P. E. Hartman. 1970. Histidine-requiring mutants of *Escherichia coli* K12. *Genetics* **66**:231–244.

597. Garvey, N., E. M. Witkin, and D. E. Brash. 1989. Ultraviolet photoproducts at the ochre suppressor mutation site in the glnU gene of *Escherichia coli*: relevance to "mutation frequency decline." *Mol. Gen. Genet.* **219**:359–364.

598. Garvin, R. T., and L. Gorini. 1975. A new gene for ribosomal restriction in *Escherichia coli. Mol. Gen. Genet.* **137**:73–78.

599. Garwin, J. L., A. L. Klages, and J. E. Cronan, Jr. 1980. Structural, enzymatic, and genetic studies of beta-ketoacyl-acyl carrier protein synthases I and II of *Escherichia coli. J. Biol. Chem.* **255**:11949–11956.

600. Gaudu, P., D. Touati, V. Niviere, and M. Fontecave. 1994. The NAD(P)H:flavin oxidoreductase from *Escherichia coli* as a source of superoxide radicals. *J. Biol. Chem.* **269**:8182–8188.

601. Gavini, N., and B. E. Davidson. 1990. pheAo mutants of *Escherichia coli* have a defective pheA attenuator. *J. Biol. Chem.* **265**:21532–21535.

602. Gavini, N., and B. E. Davidson. 1990. The pheR gene of *Escherichia coli* encodes tRNA(Phe), not a repressor protein. *J. Biol. Chem.* **265**:21527–21531.

603. Gavini, N., and B. E. Davidson. 1991. Regulation of pheA expression by the pheR product in *Escherichia coli* is mediated through attenuation of transcription. *J. Biol. Chem.* **266**:7750–7753.

604. Gavini, N., and L. Pulakat. 1991. Role of translation of the *pheA* leader peptide coding region in attenuation regulation of the *Escherichia coli pheA* gene. *J. Bacteriol.* **173**:4904–4907.

605. Geary, L. E., and A. Meister. 1977. On the mechanism of glutamine-dependent reductive amination of alpha-ketoglutarate catalyzed by glutamate synthase. *J. Biol. Chem.* **252**:3501–3508.

606. Gebert, J. F., B. Overhoff, M. D. Manson, and W. Boos. 1988. The Tsr chemosensory transducer of *Escherichia coli* assembles into the cytoplasmic membrane via a SecA-dependent process. *J. Biol. Chem.* **263**:16652–16660.

607. Gebler, J. C., R. Aebersold, and S. G. Withers. 1992. Glu-537, not Glu-461, is the nucleophile in the active site of (lac Z) beta-galactosidase from *Escherichia coli. J. Biol. Chem.* **267**:11126–11130.

608. Geerse, R. H., C. R. Ruig, A. R. Schuitema, and P. W. Postma. 1986. Relationship between pseudo-HPr and the PEP: fructose phosphotransferase system in *Salmonella typhimurium* and *Escherichia coli. Mol. Gen. Genet.* **203**:435–444.

609. Geerse, R. H., J. van der Pluijm, and P. W. Postma. 1989. The repressor of the PEP:fructose phosphotransferase system is required for the transcription of the pps gene of *Escherichia coli. Mol. Gen. Genet.* **218**:348–352.

610. Gegner, J. A., and F. W. Dahlquist. 1991. Signal transduction in bacteria: CheW forms a reversible complex with the protein kinase CheA. *Proc. Natl. Acad. Sci. USA* **88**:750–754.

611. Geiger, O., F. D. Russo, T. J. Silhavy, and E. P. Kennedy. 1992. Membrane-derived oligosaccharides affect porin osmoregulation only in media of low ionic strength. *J. Bacteriol.* **174**:1410–1413.

612. Geiselmann, J., Y. Wang, S. E. Seifried, and P. H. von Hippel. 1993. A physical model for the translocation and helicase activities of *Escherichia coli* transcription termination protein Rho. *Proc. Natl. Acad. Sci. USA* **90**:7754–7758.

613. Geiselmann, J., T. D. Yager, S. C. Gill, P. Calmettes, and P. H. von Hippel. 1992. Physical properties of the *Escherichia coli* transcription termination factor rho. 1. Association states and geometry of the rho hexamer. *Biochemistry* **31**:111–121.

614. Gelfand, D. H., and R. A. Steinberg. 1977. *Escherichia coli* mutants deficient in the aspartate and aromatic amino acid aminotransferases. *J. Bacteriol.* **130**:429–440.

615. Gemmill, R. M., J. W. Jones, G. W. Haughn, and J. M. Calvo. 1983. Transcription initiation sites of the leucine operons of *Salmonella typhimurium* and *Escherichia coli. J. Mol. Biol.* **170**:39–59.

616. Gentry, D., C. Bengra, K. Ikehara, and M. Cashel. 1993. Guanylate kinase of *Escherichia coli* K-12. *J. Biol. Chem.* **268**:14316–14321.

617. Gentry, D. R., and R. R. Burgess. 1989. rpoZ, encoding the omega subunit of *Escherichia coli* RNA polymerase, is in the same operon as spoT. *J. Bacteriol.* **171**:1271–1277.

618. Gentry, D. R., and R. R. Burgess. 1990. Overproduction and purification of the omega subunit of *Escherichia coli* RNA polymerase. *Protein Expression Purif.* **1**:81–86.

619. Gentschev, I., and W. Goebel. 1992. Topological and functional studies on HlyB of *Escherichia coli. Mol. Gen. Genet.* **232**:40–48.

620. Georgopoulos, C., and D. Ang. 1990. The *Escherichia coli* groE chaperonins. *Semin. Cell Biol.* **1**:19–25.

621. Gerdes, K., L. K. Poulsen, T. Thisted, A. K. Nielsen, J. Martinussen, and P. H. Andreasen. 1990. The hok killer gene family in gram-negative bacteria. *New Biol.* **2**:946–956.

622. Gerlach, P., L. Sogaard-Andersen, H. Pedersen, J. Martinussen, P. Valentin-Hansen, and E. Bremer. 1991. The cyclic AMP (cAMP)-cAMP receptor protein complex functions both as an activator and as a corepressor at the *tsx*-p2 promoter of *Escherichia coli* K-12. *J. Bacteriol.* **173**:5419–5430.

623. Gerstein, M., G. Schulz, and C. Chothia. 1993. Domain closure in adenylate kinase. Joints on either side of two helices close like neighboring fingers. *J. Mol. Biol.* **229**:494–501.

624. Gervais, F. G., and G. R. Drapeau. 1992. Identification, cloning, and characterization of rcsF, a new regulator gene for exopolysaccharide synthesis that suppresses the division mutation ftsZ84 in *Escherichia coli* K-12. *J. Bacteriol.* **174**:8016–8022.

625. Gervais, F. G., P. Phoenix, and G. R. Drapeau. 1992. The rcsB gene, a positive regulator of colanic acid biosynthesis in *Escherichia coli*, is also an activator of ftsZ expression. *J. Bacteriol.* **174**:3964–3971.

626. **Ghosh, G., H. Pelka, L. H. Schulman, and S. Brunie.** 1991. Activation of methionine by *Escherichia coli* methionyl-tRNA synthetase. *Biochemistry* 30:9569–9575.

627. **Ghosh, P., C. Meyer, E. Remy, D. Peterson, and J. Preiss.** 1992. Cloning, expression, and nucleotide sequence of glgC gene from an allosteric mutant of *Escherichia coli* B. *Arch. Biochem. Biophys.* 296:122–128.

628. **Gibbs, T. W., D. R. Gill, and G. P. Salmond.** 1992. Localised mutagenesis of the fts YEX operon: conditionally lethal missense substitutions in the FtsE cell division protein of *Escherichia coli* are similar to those found in the cystic fibrosis transmembrane conductance regulator protein (CFTR) of human patients. *Mol. Gen. Genet.* 234:121–128.

629. **Gibert, I., M. Llagostera, and J. Barbe.** 1988. Regulation of ubiG gene expression in *Escherichia coli*. *J. Bacteriol.* 170:1346–1349.

630. **Gibson, F., and J. Pittard.** 1968. Pathways of biosynthesis of aromatic amino acids and vitamins and their control in microorganisms. *Bacteriol. Rev.* 32:465–492.

631. **Gibson, F. P., D. R. Leach, and R. G. Lloyd.** 1992. Identification of sbcD mutations as cosuppressors of recBC that allow propagation of DNA palindromes in *Escherichia coli* K-12. *J. Bacteriol.* 174:1222–1228.

632. **Gibson, M. M., D. A. Bagga, C. G. Miller, and M. E. Maguire.** 1991. Magnesium transport in *Salmonella typhimurium*: the influence of new mutations conferring Co^{2+} resistance on the CorA Mg^{2+} transport system. *Mol. Microbiol.* 5:2753–2762.

633. **Gill, D. R., and G. P. Salmond.** 1987. The *Escherichia coli* cell division proteins FtsY, FtsE and FtsX are inner membrane-associated. *Mol. Gen. Genet.* 210:504–508.

634. **Gill, D. R., and G. P. Salmond.** 1990. The identification of the *Escherichia coli* ftsY gene product: an unusual protein. *Mol. Microbiol.* 4:575–583.

635. **Gill, S. C., S. E. Weitzel, and P. H. von Hippel.** 1991. *Escherichia coli* sigma 70 and NusA proteins. I. Binding interactions with core RNA polymerase in solution and within the transcription complex. *J. Mol. Biol.* 220:307–324.

636. **Gilson, E., J. P. Rousset, A. Charbit, D. Perrin, and M. Hofnung.** 1986. malM, a new gene of the maltose regulon in *Escherichia coli* K12. I. malM is the last gene of the malK-lamB operon and encodes a periplasmic protein. *J. Mol. Biol.* 191:303–311.

637. **Gish, K., and C. Yanofsky.** 1993. Inhibition of expression of the tryptophanase operon in *Escherichia coli* by extrachromosomal copies of the tna leader region. *J. Bacteriol.* 175:3380–3387.

638. **Glauner, B., J. V. Holtje, and U. Schwarz.** 1988. The composition of the murein of *Escherichia coli*. *J. Biol. Chem.* 263:10088–10095.

639. **Goffin, C., J. A. Ayala, M. Nguyen-Disteche, and J. M. Ghuysen.** 1993. Site-directed mutagenesis of dicarboxylic acid residues at the penicillin-binding module of the *Escherichia coli* penicillin-binding protein 3. *FEMS Microbiol. Lett.* 113:247–251.

640. **Golden, J. W., L. L. Whorff, and D. R. Wiest.** 1991. Independent regulation of nifHDK operon transcription and DNA rearrangement during heterocyst differentiation in the cyanobacterium *Anabaena* sp. strain PCC 7120. *J. Bacteriol.* 173:7098–7105.

641. **Goldstein, J., N. S. Pollitt, and M. Inouye.** 1990. Major cold shock protein of *Escherichia coli*. *Proc. Natl. Acad. Sci. USA* 87:283–287.

642. **Gollnick, P., and C. Yanofsky.** 1990. tRNA(Trp) translation of leader peptide codon 12 and other factors that regulate expression of the tryptophanase operon. *J. Bacteriol.* 172:3100–3107.

643. **Gomez-Eichelmann, M. C., and J. Ramirez-Santos.** 1993. Methylated cytosine at Dcm (CCATGG) sites in *Escherichia coli*: possible function and evolutionary implications. *J. Mol. Evol.* 37:11–24.

644. **Gonnet, G. H., M. A. Cohen, and S. A. Benner.** 1992. Exhaustive matching of the entire protein sequence database. *Science* 256:1443–1445.

645. **Gonzalez, J. C., R. V. Banerjee, S. Huang, J. S. Sumner, and R. G. Matthews.** 1992. Comparison of cobalamin-independent and cobalamin-dependent methionine synthases from *Escherichia coli*: two solutions to the same chemical problem. *Biochemistry* 31:6045–6056.

646. **Gorini, L., W. Gunderson, and M. Burger.** 1961. Genetics of regulation of enzyme synthesis in the arginine biosynthetic pathway of *Escherichia coli*. *Cold Spring Harbor Symp. Quant. Biol.* 26:173–182.

647. **Goshima, N., Y. Inagaki, H. Otaki, H. Tanaka, N. Hayashi, F. Imamoto, and Y. Kano.** 1992. Chimeric HU-IHF proteins that alter DNA-binding ability. *Gene* 118:97–102.

648. **Goss, T. J., H. P. Schweizer, and P. Datta.** 1988. Molecular characterization of the tdc operon of *Escherichia coli* K-12. *J. Bacteriol.* 170:5352–5359.

649. **Gottesman, S., W. P. Clark, V. de Crecy-Lagard, and M. R. Maurizi.** 1993. ClpX, an alternative subunit for the ATP-dependent Clp protease of *Escherichia coli*. Sequence and in vivo activities. *J. Biol. Chem.* 268:22618–22626.

650. **Gottesman, S., W. P. Clark, and M. R. Maurizi.** 1990. The ATP-dependent Clp protease of *Escherichia coli*. Sequence of clpA and identification of a Clp-specific substrate. *J. Biol. Chem.* 265:7886–7893.

651. **Gottesman, S., C. Squires, E. Pichersky, M. Carrington, M. Hobbs, J. S. Mattick, B. Dalrymple, H. Kuramitsu, T. Shiroza, T. Foster, W. P. Clark, B. Ross, C. L. Squires, and M. R. Maurizi.** 1990. Conservation of the regulatory subunit for the Clp ATP-dependent protease in prokaryotes and eukaryotes. *Proc. Natl. Acad. Sci. USA* 87:3513–3517.

652. **Gottesman, S., and V. Stout.** 1991. Regulation of capsular polysaccharide synthesis in *Escherichia coli* K12. *Mol. Microbiol.* 5:1599–1606.

653. **Gottlieb, P. A., S. Wu, X. Zhang, M. Tecklenburg, P. Kuempel, and T. M. Hill.** 1992. Equilibrium, kinetic, and footprinting studies of the Tus-Ter protein-DNA interaction. *J. Biol. Chem.* 267:7434–7443.

654. **Grabau, C., Y. Y. Chang, and J. E. Cronan, Jr.** 1989. Lipid binding by *Escherichia coli* pyruvate oxidase is disrupted by small alterations of the carboxyl-terminal region. *J. Biol. Chem.* 264:12510–12519.

655. **Graeme-Cook, K. A.** 1991. The regulation of porin expression in *Escherichia coli*: effect of turgor stress. *FEMS Microbiol. Lett.* 63:219–223.

656. **Graham, L. D., L. C. Packman, and R. N. Perham.** 1989. Kinetics and specificity of reductive acylation of lipoyl domains from 2-oxo acid dehydrogenase multienzyme complexes. *Biochemistry* 28:1574–1581.

657. **Graham, L. D., and R. N. Perham.** 1990. Interactions of lipoyl domains with the E1p subunits of the pyruvate dehydrogenase multienzyme complex from *Escherichia coli*. *FEBS Lett.* 262:241–244.

658. **Graifer, D. M., G. T. Babkina, N. B. Matasova, S. N. Vladimirov, G. G. Karpova, and V. V. Vlassov.** 1989. Structural arrangement of tRNA binding sites on *Escherichia coli* ribosomes, as revealed from data on affinity labelling with photoactivatable tRNA derivatives. *Biochim. Biophys. Acta* 1008:146–156.

659. **Graves, R. J., I. Felzenszwalb, J. Laval, and T. R. O'Connor.** 1992. Excision of 5′-terminal deoxyribose phosphate from damaged DNA is catalyzed by the Fpg protein of *Escherichia coli*. *J. Biol. Chem.* 267:14429–14435.

660. **Green, J. M., W. K. Merkel, and B. P. Nichols.** 1992. Characterization and sequence of *Escherichia coli* pabC, the gene encoding aminodeoxychorismate lyase, a pyridoxal phosphate-containing enzyme. *J. Bacteriol.* 174:5317–5323.

661. **Green, J. M., and B. P. Nichols.** 1991. p-Aminobenzoate biosynthesis in *Escherichia coli*. Purification of aminodeoxychorismate lyase and cloning of pabC. *J. Biol. Chem.* 266:12971–12975.

662. **Greenberg, J. T., P. Monach, J. H. Chou, P. D. Josephy, and B. Demple.** 1990. Positive control of a global antioxidant defense regulon activated by superoxide-generating agents in *Escherichia coli*. *Proc. Natl. Acad. Sci. USA* 87:6181–6185.

663. **Grentzmann, G., D. Brechemier-Baey, V. Heurgue, L. Mora, and R. H. Buckingham.** 1994. Localization and characterization of the gene encoding release factor RF3 in *Escherichia coli*. *Proc. Natl. Acad. Sci. USA* 91:5848–5852.

664. **Griep, M. A., and C. S. McHenry.** 1992. Fluorescence energy transfer between the primer and the beta subunit of the DNA polymerase III holoenzyme. *J. Biol. Chem.* 267:3052–3059.

665. **Griggs, D. W., K. Kafka, C. D. Nau, and J. Konisky.** 1990. Activation of expression of the *Escherichia coli* cir gene by an iron-independent regulatory mechanism involving cyclic AMP-cyclic AMP receptor protein complex. *J. Bacteriol.* 172:3529–3533.

666. **Grilley, M., K. M. Welsh, S. S. Su, and P. Modrich.** 1989. Isolation and characterization of the *Escherichia coli* mutL gene product. *J. Biol. Chem.* 264:1000–1004.

667. **Grimm, B., A. Bull, and V. Breu.** 1991. Structural genes of glutamate 1-semialdehyde aminotransferase for porphyrin synthesis in a cyanobacterium and *Escherichia coli*. *Mol. Gen. Genet.* 225:1–10.

668. **Grimshaw, C. E., D. A. Matthews, K. I. Varughese, M. Skinner, N. H. Xuong, T. Bray, J. Hoch, and J. M. Whiteley.** 1992. Characterization and nucleotide binding properties of a mutant dihydropteridine reductase containing an aspartate 37-isoleucine replacement. *J. Biol. Chem.* 267:15334–15339.

669. **Groarke, J. M., W. C. Mahoney, J. N. Hope, C. E. Furlong, F. T. Robb, H. Zalkin, and M. A. Hermodson.** 1983. The amino acid sequence of D-ribose-binding protein from *Escherichia coli* K12. *J. Biol. Chem.* 258:12952–12956.

670. **Grogan, D. W.** 1988. Temperature-sensitive murein synthesis in an *Escherichia coli* pdx mutant and the role of alanine racemase. *Arch. Microbiol.* 150:363–367.

671. **Groisman, E. A., F. Heffron, and F. Solomon.** 1992. Molecular genetic analysis of the *Escherichia coli* phoP locus. *J. Bacteriol.* 174:486–491.

672. **Grompe, M., J. Versalovic, T. Koeuth, and J. R. Lupski.** 1991. Mutations in the *Escherichia coli* dnaG gene suggest coupling between DNA replication and chromosome partitioning. *J. Bacteriol.* 173:1268–1278.

673. **Grubl, G., A. P. Vogler, and J. W. Lengeler.** 1990. Involvement of the histidine protein (HPr) of the phosphotransferase system in chemotactic signaling of *Escherichia coli* K-12. *J. Bacteriol.* 172:5871–5876.

674. **Gruer, M. J., and J. R. Guest.** 1994. Two genetically-distinct and differentially-regulated aconitases (AcnA and AcnB) in *Escherichia coli*. *Microbiology* 140:2531–2541.

675. **Guardiola, J., M. DeFelice, R. Klopotowski, and M. Iaccarino.** 1974. Mutations affecting the different transport systems for isoleucine, leucine, and valine in *Escherichia coli* K-12. *J. Bacteriol.* 117:393–405.

676. **Guest, J. R., J. S. Miles, R. E. Roberts, and S. A. Woods.** 1985. The fumarase genes of *Escherichia coli*: location of the fumB gene and discovery of a new gene (fumC). *J. Gen. Microbiol.* 131:2971–2984.

677. **Guillon, J. M., Y. Mechulam, J. M. Schmitter, S. Blanquet, and G. Fayat.** 1992. Disruption of the gene for Met-tRNA(fMet) formyltransferase severely impairs growth of *Escherichia coli*. *J. Bacteriol.* 174:4294–4301.

678. **Guillon, J. M., T. Meinnel, Y. Mechulam, C. Lazennec, S. Blanquet, and G. Fayat.** 1992. Nucleotides of tRNA governing the specificity of *Escherichia coli* methionyl-tRNA(fMet) formyltransferase. *J. Mol. Biol.* 224:359–367.

679. **Guilloton, M. B., J. J. Korte, A. F. Lamblin, J. A. Fuchs, and P. M. Anderson.** 1992. Carbonic anhydrase in *Escherichia coli*. A product of the cyn operon. *J. Biol. Chem.* 267:3731–3734.

680. Guixe, V., and J. Babul. 1988. Influence of ligands on the aggregation of the normal and mutant forms of phosphofructokinase 2 of Escherichia coli. Arch. Biochem. Biophys. 264:519–524.

681. Gupta, S., and D. P. Clark. 1989. Escherichia coli derivatives lacking both alcohol dehydrogenase and phosphotransacetylase grow anaerobically by lactate fermentation. J. Bacteriol. 171:3650–3655.

682. Gustafson, C. E., S. Kaul, and E. E. Ishiguro. 1993. Identification of the Escherichia coli lytB gene, which is involved in penicillin tolerance and control of the stringent response. J. Bacteriol. 175:1203–1205.

683. Gustafsson, C., and G. R. Bjork. 1993. The tRNA-(m5U54)-methyltransferase of Escherichia coli is present in two forms in vivo, one of which is present as bound to tRNA and to a 3'-end fragment of 16 S rRNA. J. Biol. Chem. 268:1326–1331.

684. Guthrie, B., and W. Wickner. 1990. Trigger factor depletion or overproduction causes defective cell division but does not block protein export. J. Bacteriol. 172:5555–5562.

685. Gutierrez, C., M. Ardourel, E. Bremer, A. Middendorf, W. Boos, and U. Ehmann. 1989. Analysis and DNA sequence of the osmoregulated treA gene encoding the periplasmic trehalase of Escherichia coli K12. Mol. Gen. Genet. 217:347–354.

686. Gutierrez, C., and J. C. Devedjian. 1991. Osmotic induction of gene osmC expression in Escherichia coli K12. J. Mol. Biol. 220:959–973.

687. Guzman, E. C., and A. Jimenez-Sanchez. 1991. Location of pinO, a new gene located between tufA and rpsJ, on the physical map of the Escherichia coli chromosome. J. Bacteriol. 173:7409

688. Guzman, E. C., R. H. Pritchard, and A. Jimenez-Sanchez. 1991. A calcium-binding protein that may be required for the initiation of chromosome replication in Escherichia coli. Res. Microbiol. 142:137–140.

689. Guzman, L. M., J. J. Barondess, and J. Beckwith. 1992. FtsL, an essential cytoplasmic membrane protein involved in cell division in Escherichia coli. J. Bacteriol. 174:7716–7728.

690. Hachler, H., S. P. Cohen, and S. B. Levy. 1991. marA, a regulated locus which controls expression of chromosomal multiple antibiotic resistance in Escherichia coli. J. Bacteriol. 173:5532–5538.

691. Hadener, A., P. K. Matzinger, V. N. Malashkevich, G. V. Louie, S. P. Wood, P. Oliver, P. R. Alefounder, A. R. Pitt, C. Abell, and A. R. Battersby. 1993. Purification, characterization, crystallisation and X-ray analysis of selenomethionine-labelled hydroxymethylbilane synthase from Escherichia coli. Eur. J. Biochem. 211:615–624.

692. Hagervall, T. G., and G. R. Bjork. 1984. Genetic mapping and cloning of the gene (trmC) responsible for the synthesis of tRNA (mnm5s2U)methyltransferase in Escherichia coli K12. Mol. Gen. Genet. 196:201–207.

693. Hall, B. G. 1982. Chromosomal mutation for citrate utilization by Escherichia coli K-12. J. Bacteriol. 151:269–273.

694. Hall, B. G., P. W. Betts, and J. C. Wootton. 1989. DNA sequence analysis of artificially evolved ebg enzyme and ebg repressor genes. Genetics 123:635–648.

695. Hall, B. G., and P. M. Sharp. 1992. Molecular population genetics of Escherichia coli: DNA sequence diversity at the celC, crr, and gutB loci of natural isolates. Mol. Biol. Evol. 9:654–665.

696. Hall, B. G., and L. Xu. 1992. Nucleotide sequence, function, activation, and evolution of the cryptic asc operon of Escherichia coli K12. Mol. Biol. Evol. 9:688–706.

697. Hall, S. D., M. F. Kane, and R. D. Kolodner. 1993. Identification and characterization of the Escherichia coli RecT protein, a protein encoded by the recE region that promotes renaturation of homologous single-stranded DNA. J. Bacteriol. 175:277–287.

698. Hama, H., N. Almaula, C. G. Lerner, S. Inouye, and M. Inouye. 1991. Nucleoside diphosphate kinase from Escherichia coli: its overproduction and sequence comparison with eukaryotic enzymes. Gene 105:31–36.

699. Hamilton, W. D., D. A. Harrison, and T. A. Dyer. 1988. Sequence of the Escherichia coli fructose-1,6-bisphosphatase gene. Nucleic Acids Res. 16:8707

700. Han, M. K., S. Roseman, and L. Brand. 1990. Sugar transport by the bacterial phosphotransferase system. Characterization of the sulfhydryl groups and site-specific labeling of enzyme I. J. Biol. Chem. 265:1985–1995.

701. Hanada, K., I. Yamato, and Y. Anraku. 1988. Purification and reconstitution of Escherichia coli proline carrier using a site specifically cleavable fusion protein. J. Biol. Chem. 263:7181–7185.

702. Hane, M. W., and T. H. Wood. 1969. Escherichia coli K-12 mutants resistant to nalidixic acid: genetic mapping and dominance studies. J. Bacteriol. 99:239–241.

703. Hanke, C., J. Hess, G. Schumacher, and W. Goebel. 1992. Processing by OmpT of fusion proteins carrying the HlyA transport signal during secretion by the Escherichia coli hemolysin transport system. Mol. Gen. Genet. 233:42–48.

704. Hantke, K. 1987. Selection procedure for deregulated iron transport mutants (fur) in Escherichia coli K12: fur not only affects iron transport. Mol. Gen. Genet. 210:135–139.

705. Hara, H., Y. Yamamoto, A. Higashitani, H. Suzuki, and Y. Nishimura. 1991. Cloning, mapping, and characterization of the Escherichia coli prc gene, which is involved in C-terminal processing of penicillin-binding protein 3. J. Bacteriol. 173:4799–4813.

706. Harborne, N. R., L. Griffiths, S. J. Busby, and J. A. Cole. 1992. Transcriptional control, translation and function of the products of the five open reading frames of the Escherichia coli nir operon. Mol. Microbiol. 6:2805–2813.

707. Harder, J., H. Follmann, and K. Hantke. 1989. Deoxyribonucleotide synthesis in an Escherichia coli mutant (H 1491) which lacks ribonucleotide reductase subunit B2. Z. Naturforsch. Sect. C 44:715–718.

708. Hardt, W. D., J. Schlegl, V. A. Erdmann, and R. K. Hartmann. 1993. Role of the D arm and the anticodon arm in tRNA recognition by eubacterial and eukaryotic RNase P enzymes. Biochemistry 32:13046–13053.

709. Haring, M., H. Rudiger, B. Demple, S. Boiteux, and B. Epe. 1994. Recognition of oxidized abasic sites by repair endonucleases. Nucleic Acids Res. 22:2010–2015.

710. Harkness, R. E., and V. Braun. 1990. In vitro peptidoglycan synthesis by envelopes from Escherichia coli tolM mutants is inhibited by colicin M. J. Bacteriol. 172:498–500.

711. Harkness, R. E., and V. Braun. 1990. Colicin M is only bactericidal when provided from outside the cell. Mol. Gen. Genet. 222:37–40.

712. Harkness, R. E., and E. E. Ishiguro. 1983. Temperature-sensitive autolysis-defective mutants of Escherichia coli. J. Bacteriol. 155:15–21.

713. Hartl, F. U., S. Lecker, E. Schiebel, J. P. Hendrick, and W. Wickner. 1990. The binding cascade of SecB to SecA to SecY/E mediates preprotein targeting to the E. coli plasma membrane. Cell 63:269–279.

714. Hasegawa, T., M. Miyano, H. Himeno, Y. Sano, K. Kimura, and M. Shimizu. 1992. Identity determinants of E. coli threonine tRNA. Biochem. Biophys. Res. Commun. 184:478–484.

715. Hassan, H. M., and H. C. Sun. 1992. Regulatory roles of Fnr, Fur, and Arc in expression of manganese-containing superoxide dismutase in Escherichia coli. Proc. Natl. Acad. Sci. USA 89:3217–3221.

716. Hassani, M., M. V. Saluta, G. N. Bennett, and I. N. Hirshfield. 1991. Partial characterization of a lysU mutant of Escherichia coli K-12. J. Bacteriol. 173:1965–1970.

717. Hata-Tanaka, A., K. Matsuura, S. Itoh, and Y. Anraku. 1987. Electron flow and heme-heme interaction between cytochromes b-558, b-595 and d in a terminal oxidase of Escherichia coli. Biochim. Biophys. Acta 893:289–295.

718. Havekes, L. M., B. J. J. Lugtenberg, and W. P. M. Hoekstra. 1976. Conjugation deficient E. coli K-12 F− mutants with heptose-less lipopolysaccharide. Mol. Gen. Genet. 146:43–50.

719. Hayano, T., N. Takahashi, S. Kato, N. Maki, and M. Suzuki. 1991. Two distinct forms of peptidylprolyl-cis-trans-isomerase are expressed separately in periplasmic and cytoplasmic compartments of Escherichia coli cells. Biochemistry 30:3041–3048.

720. Hayase, Y., M. Jahn, M. J. Rogers, L. A. Sylvers, M. Koizumi, H. Inoue, E. Ohtsuka, and D. Soll. 1992. Recognition of bases in Escherichia coli tRNA(Gln) by glutaminyl-tRNA synthetase: a complete identity set. EMBO J. 11:4159–4165.

721. Hayden, M. A., I. Huang, D. E. Bussiere, and G. W. Ashley. 1992. The biosynthesis of lipoic acid. Cloning of lip, a lipoate biosynthetic locus of Escherichia coli. J. Biol. Chem. 267:9512–9515.

722. Haydon, D. J., M. A. Quail, and J. R. Guest. 1993. A mutation causing constitutive synthesis of the pyruvate dehydrogenase complex in Escherichia coli is located within the pdhR gene. FEBS Lett. 336:43–47.

723. Hayward, R. S., K. Igarashi, and A. Ishihama. 1991. Functional specialization within the alpha-subunit of Escherichia coli RNA polymerase. J. Mol. Biol. 221:23–29.

724. Hayzer, D. J., and T. Leisinger. 1980. The gene-enzyme relationships of proline biosynthesis in Escherichia coli. J. Gen. Microbiol. 118:287–293.

725. He, B., K. Y. Choi, and H. Zalkin. 1993. Regulation of Escherichia coli glnB, prsA, and speA by the purine repressor. J. Bacteriol. 175:3598–3606.

726. He, B., J. M. Smith, and H. Zalkin. 1992. Escherichia coli purB gene: cloning, nucleotide sequence, and regulation by purR. J. Bacteriol. 174:130–136.

727. Heatwole, V. M., and R. L. Somerville. 1991. The tryptophan-specific permease gene, mtr, is differentially regulated by the tryptophan and tyrosine repressors in Escherichia coli K-12. J. Bacteriol. 173:3601–3604.

728. Heatwole, V. M., and R. L. Somerville. 1992. Synergism between the Trp repressor and Tyr repressor in repression of the aroL promoter of Escherichia coli K-12. J. Bacteriol. 174:331–335.

729. Heim, R., and E. E. Strehler. 1991. Cloning an Escherichia coli gene encoding a protein remarkably similar to mammalian aldehyde dehydrogenases. Gene 99:15–23.

730. Heimberg, H., A. Boyen, M. Crabeel, and N. Glansdorff. 1990. Escherichia coli and Saccharomyces cerevisiae acetylornithine aminotransferase: evolutionary relationship with ornithine aminotransferase. Gene 90:69–78.

731. Heimberger, A., and A. Eisenstark. 1988. Compartmentalization of catalases in Escherichia coli. Biochem. Biophys. Res. Commun. 154:392–397.

732. Heitzer, A., C. A. Mason, M. Snozzi, and G. Hamer. 1990. Some effects of growth conditions on steady state and heat shock induced htpG gene expression in continuous cultures of Escherichia coli. Arch. Microbiol. 155:7–12.

733. Helling, R. B. 1994. Why does Escherichia coli have two primary pathways for synthesis of glutamate? J. Bacteriol. 176:4664–4668.

734. Hellinga, H. W., and P. R. Evans. 1985. Nucleotide sequence and high-level expression of the major Escherichia coli phosphofructokinase. Eur. J. Biochem. 149:363–373.

735. Hemschemeier, S., M. Grund, B. Keuntje, and R. Eichenlaub. 1991. Isolation of Escherichia coli mutants defective in uptake of molybdate. J. Bacteriol. 173:6499–6506.

736. Henderson, P. J., and M. C. Maiden. 1990. Homologous sugar transport proteins in Escherichia coli and their relatives in both prokaryotes and eukaryotes. Philos. Trans. R. Soc. London Ser. B 326:391–410.

737. Hengge-Aronis, R., and D. Fischer. 1992. Identification and molecular analysis of glgS, a novel growth-phase-regulated and rpoS-dependent gene involved in glycogen synthesis in Escherichia coli. Mol. Microbiol. 6:1877–1886.

738. **Hengge-Aronis, R., W. Klein, R. Lange, M. Rimmele, and W. Boos.** 1991. Trehalose synthesis genes are controlled by the putative sigma factor encoded by rpoS and are involved in stationary-phase thermotolerance in *Escherichia coli. J. Bacteriol.* **173**:7918–7924.

739. **Henikoff, S., G. W. Haughn, J. M. Calvo, and J. C. Wallace.** 1988. A large family of bacterial activator proteins. *Proc. Natl. Acad. Sci. USA* **85**:6602–6606.

740. **Hennecke, F., H. Kolmar, K. Brundl, and H. J. Fritz.** 1991. The vsr gene product of *E. coli* K-12 is a strand- and sequence-specific DNA mismatch endonuclease. *Nature* (London) **353**:776–778.

741. **Henrich, B., H. Backes, J. R. Klein, and R. Plapp.** 1992. The promoter region of the *Escherichia coli* pepD gene: deletion analysis and control by phosphate concentration. *Mol. Gen. Genet.* **232**:117–125.

742. **Henrich, B., U. Monnerjahn, and R. Plapp.** 1990. Peptidase D gene (*pepD*) of *Escherichia coli* K-12: nucleotide sequence, transcript mapping, and comparison with other peptide genes. *J. Bacteriol.* **172**:4641–4651.

743. **Henry, M. F., and J. E. Cronan, Jr.** 1992. A new mechanism of transcriptional regulation: release of an activator triggered by small molecule binding. *Cell* **70**:671–679.

744. **Herbert, A. A., and J. R. Guest.** 1969. Studies with ketoglutarate dehydrogenase mutants of *Escherichia coli. Mol. Gen. Genet.* **105**:182–190.

745. **Herdman, M.** 1985. The evolution of bacterial genomes, p. 37–68. *In* T. Cavalier (ed.), *The Evolution of Genome Size.* John Wiley & Sons, Inc., New York.

746. **Herman, C., T. Ogura, T. Tomoyasu, S. Hiraga, Y. Akiyama, K. Ito, R. Thomas, R. D'Ari, and P. Bouloc.** 1993. Cell growth and lambda phage development controlled by the same essential *Escherichia coli* gene, ftsH/hflB. *Proc. Natl. Acad. Sci. USA* **90**:10861–10865.

747. **Hernandez, V. J., and H. Bremer.** 1991. *Escherichia coli* ppGpp synthetase II activity requires spoT. *J. Biol. Chem.* **266**:5991–5999.

748. **Hernandez, V. J., and H. Bremer.** 1993. Characterization of RNA and DNA synthesis in *Escherichia coli* strains devoid of ppGpp. *J. Biol. Chem.* **268**:10851–10862.

748a.**Hershey, H. V., R. Gutstein, and M. W. Taylor.** 1982. Cloning and restriction map of the *E. coli* apt gene. *Gene* **19**:89–92.

749. **Hessing, H. G., C. van Rotterdam, and P. H. Pouwels.** 1987. Expression of the *Escherichia coli* trpE gene in *E. coli* K12 bacteria: maximum level, rate and time of initiation of anthranilate synthetase production. *Mol. Gen. Genet.* **210**:256–261.

750. **Heyduk, E., T. Heyduk, and J. C. Lee.** 1992. Intersubunit communications in *Escherichia coli* cyclic AMP receptor protein: studies of the ligand binding domain. *Biochemistry* **31**:3682–3688.

751. **Hiasa, H., R. J. DiGate, and K. J. Marians.** 1994. Decatenating activity of *Escherichia coli* DNA gyrase and topoisomerases I and III during oriC and pBR322 DNA replication in vitro. *J. Biol. Chem.* **269**:2093–2099.

752. **Hiasa, H., and K. J. Marians.** 1992. Differential inhibition of the DNA translocation and DNA unwinding activities of DNA helicases by the *Escherichia coli* Tus protein. *J. Biol. Chem.* **267**:11379–11385.

753. **Hidalgo, E., Y. M. Chen, E. C. Lin, and J. Aguilar.** 1991. Molecular cloning and DNA sequencing of the *Escherichia coli* K-12 ald gene encoding aldehyde dehydrogenase. *J. Bacteriol.* **173**:6118–6123.

754. **Hidalgo, E., and B. Demple.** 1994. An iron-sulfur center essential for transcriptional activation by the redox-sensing SoxR protein. *EMBO J.* **13**:138–146.

755. **Higgins, W., C. Tardif, C. Richaud, M. A. Krivanek, and A. Cardin.** 1989. Expression of recombinant diaminopimelate epimerase in *Escherichia coli*. Isolation and inhibition with an irreversible inhibitor. *Eur. J. Biochem.* **186**:137–143.

756. **Hill, C. W., J. A. Gray, and H. Brody.** 1989. Use of the isocitrate dehydrogenase structural gene for attachment of e14 in *Escherichia coli* K-12. *J. Bacteriol.* **171**:4083–4084.

757. **Hill, J. J., J. O. Alben, and R. B. Gennis.** 1993. Spectroscopic evidence for a heme-heme binuclear center in the cytochrome bd ubiquinol oxidase from *Escherichia coli. Proc. Natl. Acad. Sci. USA* **90**:5863–5867.

758. **Hill, M. A., K. Kaufmann, J. Otero, and J. Preiss.** 1991. Biosynthesis of bacterial glycogen. Mutagenesis of a catalytic site residue of ADP-glucose pyrophosphorylase from *Escherichia coli. J. Biol. Chem.* **266**:12455–12460.

759. **Hillman, J. D., and D. G. Fraenkel.** 1975. Glyceraldehyde 3-phosphate dehydrogenase mutants of *Escherichia coli. J. Bacteriol.* **122**:1175–1179.

760. **Himeno, H., T. Hasegawa, H. Asahara, K. Tamura, and M. Shimizu.** 1991. Identity determinants of *E. coli* tryptophan tRNA. *Nucleic Acids Res.* **19**:6379–6382.

761. **Himeno, H., T. Hasegawa, T. Ueda, K. Watanabe, K. Miura, and M. Shimizu.** 1989. Role of the extra G-C pair at the end of the acceptor stem of tRNA(His) in aminoacylation. *Nucleic Acids Res.* **17**:7855–7863.

762. **Himeno, H., T. Hasegawa, T. Ueda, K. Watanabe, and M. Shimizu.** 1990. Conversion of aminoacylation specificity from tRNA(Tyr) to tRNA(Ser) in vitro. *Nucleic Acids Res.* **18**:6815–6819.

763. **Hinchman, S. K., S. Henikoff, and S. M. Schuster.** 1992. A relationship between asparagine synthetase A and aspartyl tRNA synthetase. *J. Biol. Chem.* **267**:144–149.

764. **Hinchman, S. K., and S. M. Schuster.** 1992. Overproduction, preparation of monoclonal antibodies and purification of *E. coli* asparagine synthetase A. *Protein Eng.* **5**:279–283.

765. **Hinton, S. M., and D. Dean.** 1990. Biogenesis of molybdenum cofactors. *Crit. Rev. Microbiol.* **17**:169–188.

766. **Hiom, K., and S. G. Sedgwick.** 1991. Cloning and structural characterization of the *mcrA* locus of *Escherichia coli. J. Bacteriol.* **173**:7368–7373.

767. **Hiom, K., S. M. Thomas, and S. G. Sedgwick.** 1991. Different mechanisms for SOS induced alleviation of DNA restriction in *Escherichia coli. Biochimie* **73**:399–405.

768. **Hiraga, S.** 1992. Chromosome and plasmid partition in *Escherichia coli. Annu. Rev. Biochem.* **61**:283–306.

769. **Hiraga, S.** 1993. Chromosome partition in *Escherichia coli. Curr. Opin. Genet. Dev.* **3**:789–801.

770. **Hiraga, S., H. Niki, R. Imamura, T. Ogura, K. Yamanaka, J. Feng, B. Ezaki, and A. Jaffe.** 1991. Mutants defective in chromosome partitioning in *E. coli. Res. Microbiol.* **142**:189–194.

771. **Hiraoka, S., K. Nukui, N. Uetake, A. Ohta, and I. Shibuya.** 1991. Amplification and substantial purification of cardiolipin synthase of *Escherichia coli. J. Biochem.* (Tokyo) **110**:443–449.

772. **Hirose, K., M. Fujita, M. Takeuchi, N. Yumoto, M. Tokushige, and Y. Kawata.** 1992. Cloning and overproduction of biodegradative threonine deaminase from *Escherichia coli* W strain. *Biotechnol. Appl. Biochem.* **15**:134–141.

773. **Hirvas, L., J. Coleman, P. Koski, and M. Vaara.** 1990. Bacterial `histone-like protein I′ (HLP-I) is an outer membrane constituent? *FEBS Lett.* **262**:123–126.

774. **Hjalmarsson, K. J., A. S. Bystrom, and G. R. Bjork.** 1983. Purification and characterization of transfer RNA (guanine-1)methyltransferase from *Escherichia coli. J. Biol. Chem.* **258**:1343–1351.

775. **Hoekstra, W. P. M., and H. G. Vis.** 1977. Characterization of the *E. coli* K12 strain AB1157 as impaired in guanine/xanthine metabolism. *Antonie van Leeuwenhoek* **43**:199–204.

776. **Hoffmann, I., J. Widstrom, M. Zeppezauer, and P. O. Nyman.** 1987. Overproduction and large-scale preparation of deoxyuridine triphosphate nucleotidohydrolase from *Escherichia coli. Eur. J. Biochem.* **164**:45–51.

777. **Hofnung, M., and M. Schwartz.** 1971. Mutations allowing growth on maltose of *Escherichia coli* K12 strains with a deleted malT gene. *Mol. Gen. Genet.* **112**:117–132.

778. **Hogg, R. W., C. Voelker, and I. Von Carlowitz.** 1991. Nucleotide sequence and analysis of the mgl operon of *Escherichia coli* K12. *Mol. Gen. Genet.* **229**:453–459.

779. **Holak, T. A., S. K. Kearsley, Y. Kim, and J. H. Prestegard.** 1988. Three-dimensional structure of acyl carrier protein determined by NMR pseudoenergy and distance geometry calculations. *Biochemistry* **27**:6135–6142.

780. **Holak, T. A., M. Nilges, J. H. Prestegard, A. M. Gronenborn, and G. M. Clore.** 1988. Three-dimensional structure of acyl carrier protein in solution determined by nuclear magnetic resonance and the combined use of dynamical simulated annealing and distance geometry. *Eur. J. Biochem.* **175**:9–15.

781. **Holbrook, E. L., R. C. Greene, and J. H. Krueger.** 1990. Purification and properties of cystathionine gamma-synthase from overproducing strains of *Escherichia coli. Biochemistry* **29**:435–442.

782. **Holmgren, A.** 1979. Glutathione-dependent synthesis of deoxyribonucleotides. Characterization of the enzymatic mechanism of *Escherichia coli* glutaredoxin. *J. Biol. Chem.* **254**:3672–3678.

783. **Hong, J. S.** 1986. ECF locus in *Escherichia coli*: defect in energization for ATP synthesis and active transport. *Methods Enzymol.* **125**:180–186.

784. **Honore, N., and S. T. Cole.** 1990. Nucleotide sequence of the aroP gene encoding the general aromatic amino acid transport protein of *Escherichia coli* K-12: homology with yeast transport proteins. *Nucleic Acids Res.* **18**:653.

785. **Hope, J. N., A. W. Bell, M. A. Hermodson, and J. M. Groarke.** 1986. Ribokinase from *Escherichia coli* K12. Nucleotide sequence and overexpression of the rbsK gene and purification of ribokinase. *J. Biol. Chem.* **261**:7663–7668.

786. **Hopkin, K. A., M. A. Papazian, and H. M. Steinman.** 1992. Functional differences between manganese and iron superoxide dismutases in *Escherichia coli* K-12. *J. Biol. Chem.* **267**:24253–24258.

787. **Hopkins, J. D., M. Clements, and M. Syvanen.** 1983. New class of mutations in *Escherichia coli* (*uup*) that affect precise excision of insertion elements and bacteriophage Mu growth. *J. Bacteriol.* **153**:384–389.

788. **Hor, L. I., and H. A. Shuman.** 1993. Genetic analysis of periplasmic binding protein dependent transport in *Escherichia coli*. Each lobe of maltose-binding protein interacts with a different subunit of the MalFGK2 membrane transport complex. *J. Mol. Biol.* **233**:659–670.

789. **Horazdovsky, B. F., and R. W. Hogg.** 1989. Genetic reconstitution of the high-affinity L-arabinose transport system. *J. Bacteriol.* **171**:3053–3059.

790. **Horowitz, H., G. E. Christie, and T. Platt.** 1982. Nucleotide sequence of the trpD gene, encoding anthranilate synthetase component II of *Escherichia coli. J. Mol. Biol.* **156**:245–256.

791. **Hou, Y., Y. P. Lin, J. D. Sharer, and P. E. March.** 1994. In vivo selection of conditional-lethal mutations in the gene encoding elongation factor G of *Escherichia coli. J. Bacteriol.* **176**:123–129.

792. **Houghton, J. E., G. A. O'Donovan, and J. R. Wild.** 1989. Reconstruction of an enzyme by domain substitution effectively switches substrate specificity. *Nature* (London) **338**:172–174.

793. **Hove-Jensen, B.** 1988. Mutation in the phosphoribosylpyrophosphate synthetase gene (*prs*) that results in simultaneous requirements for purine and pyrimidine nucleosides, nicotinamide nucleotide, histidine, and tryptophan in *Escherichia coli. J. Bacteriol.* **170**:1148–1152.

794. **Hove-Jensen, B., and M. Maigaard.** 1993. *Escherichia coli* rpiA gene encoding ribose phosphate isomerase A. *J. Bacteriol.* **175**:5628–5635.

795. **Hove-Jensen, B., and P. Nygaard.** 1989. Role of guanosine kinase in the utilization of guanosine for nucleotide synthesis in *Escherichia coli. J. Gen. Microbiol.* **135**:1263–1273.

796. Howard, B. M., R. J. Pinney, and J. T. Smith. 1993. Studies on mutational cross-resistance between ciprofloxacin, novobiocin and coumermycin in *Escherichia coli* and *Staphylococcus warneri*. *Microbios* 75:185–195.

797. Howard, B. M., R. J. Pinney, and J. T. Smith. 1994. Antagonism between bactericidal activities of 4-quinolones and coumarins gives insight into 4-quinolone killing mechanisms. *Microbios* 77:121–131.

798. Hrebenda, J., H. Heleszko, K. Brzostek, and J. Bielecki. 1985. Mutation affecting resistance of *Escherichia coli* K12 to nalidixic acid. *J. Gen. Microbiol.* 131:2285–2292.

799. Hryniewicz, M., A. Sirko, A. Palucha, A. Bock, and D. Hulanicka. 1990. Sulfate and thiosulfate transport in *Escherichia coli* K-12: identification of a gene encoding a novel protein involved in thiosulfate binding. *J. Bacteriol.* 172:3358–3366.

800. Hsu, L., S. Jackowski, and C. O. Rock. 1991. Isolation and characterization of *Escherichia coli* K-12 mutants lacking both 2-acyl-glycerophosphoethanolamine acyltransferase and acyl-acyl carrier protein synthetase activity. *J. Biol. Chem.* 266:13783–13788.

801. Hsu, L. M., J. Zagorski, Z. Wang, and M. J. Fournier. 1985. *Escherichia coli* 6S RNA gene is part of a dual-function transcription unit. *J. Bacteriol.* 161:1162–1170.

802. Huala, E., A. L. Moon, and F. M. Ausubel. 1991. Aerobic inactivation of *Rhizobium meliloti* NifA in *Escherichia coli* is mediated by *lon* and two newly identified genes, *snoB* and *snoC*. *J. Bacteriol.* 173:382–390.

803. Huang, C., and R. C. Stewart. 1993. CheZ mutants with enhanced ability to dephosphorylate CheY, the response regulator in bacterial chemotaxis. *Biochim. Biophys. Acta* 1202:297–304.

804. Huang, F., G. Coppola, and D. H. Calhoun. 1992. Multiple transcripts encoded by the *ilvGMEDA* gene cluster of *Escherichia coli* K-12. *J. Bacteriol.* 174:4871–4877.

805. Huang, F. Y., Q. X. Yang, and T. H. Huang. 1991. ^{15}N NMR studies of the conformation of *E. coli* dihydrofolate reductase in complex with folate or methotrexate. *FEBS Lett.* 289:231–234.

806. Huang, S., and M. P. Deutscher. 1992. Sequence and transcriptional analysis of the *Escherichia coli* rnt gene encoding RNase T. *J. Biol. Chem.* 267:25609–25613.

807. Huang, S. C., C. A. Panagiotidis, and E. S. Canellakis. 1990. Transcriptional effects of polyamines on ribosomal proteins and on polyamine-synthesizing enzymes in *Escherichia coli*. *Proc. Natl. Acad. Sci. USA* 87:3464–3468.

808. Hudson, A. J., S. C. Andrews, C. Hawkins, J. M. Williams, M. Izuhara, F. C. Meldrum, S. Mann, P. M. Harrison, and J. R. Guest. 1993. Overproduction, purification and characterization of the *Escherichia coli* ferritin. *Eur. J. Biochem.* 218:985–995.

809. Hudson, G. S., P. Rellos, and B. E. Davidson. 1991. Two promoters control the aroH gene of *Escherichia coli*. *Gene* 102:87–91.

810. Hudson, G. S., V. Wong, and B. E. Davidson. 1984. Chorismate mutase/prephenate dehydrogenase from *Escherichia coli* K12: purification, characterization, and identification of a reactive cysteine. *Biochemistry* 23:6240–6249.

811. Hughes, A. J., Jr., S. K. Bryan, H. Chen, R. E. Moses, and C. S. McHenry. 1991. *Escherichia coli* DNA polymerase II is stimulated by DNA polymerase III holoenzyme auxiliary subunits. *J. Biol. Chem.* 266:4568–4573.

812. Hull, R., J. D. Klinger, and E. E. M. Moody. 1976. Isolation and characterization of mutants of *Escherichia coli* K12 resistant to the new aminoglycoside antibiotic, amikacin. *J. Gen. Microbiol.* 94:389–394.

813. Humbert, R., and K. Altendorf. 1989. Defective gamma subunit of ATP synthase (F1F0) from *Escherichia coli* leads to resistance to aminoglycoside antibiotics. *J. Bacteriol.* 171:1435–1444.

814. Humbert, R., and R. D. Simoni. 1980. Genetic and biomedical studies demonstrating a second gene coding for asparagine synthetase in *Escherichia coli*. *J. Bacteriol.* 142:212–220.

815. Hummel, H., W. Piepersberg, and A. Bock. 1979. Analysis of lincomycin resistance mutations in *Escherichia coli*. *Mol. Gen. Genet.* 169:345–347.

816. Hurley, J. H., A. M. Dean, D. E. Koshland, Jr., and R. M. Stroud. 1991. Catalytic mechanism of NADP(+)-dependent isocitrate dehydrogenase: implications from the structures of magnesium-isocitrate and NADP+ complexes. *Biochemistry* 30:8671–8678.

817. Hurley, J. H., P. E. Thorsness, V. Ramalingam, N. H. Helmers, D. E. Koshland, Jr., and R. M. Stroud. 1989. Structure of a bacterial enzyme regulated by phosphorylation, isocitrate dehydrogenase. *Proc. Natl. Acad. Sci. USA* 86:8635–8639.

818. Hussain, H., J. Grove, L. Griffiths, S. Busby, and J. Cole. 1994. A seven-gene operon essential for formate-dependent nitrite reduction to ammonia by enteric bacteria. *Mol. Microbiol.* 12:153–163.

819. Hwang, D. S., B. Thony, and A. Kornberg. 1992. IciA protein, a specific inhibitor of initiation of *Escherichia coli* chromosomal replication. *J. Biol. Chem.* 267:2209–2213.

820. Hwang, Y. W., M. Carter, and D. L. Miller. 1992. The identification of a domain in *Escherichia coli* elongation factor Tu that interacts with elongation factor Ts. *J. Biol. Chem.* 267:22198–22205.

821. Ichihara, S., Y. Matsubara, C. Kato, K. Akasaka, and S. Mizushima. 1993. Molecular cloning, sequencing, and mapping of the gene encoding protease I and characterization of proteinase and proteinase-defective *Escherichia coli* mutants. *J. Bacteriol.* 175:1032–1037.

822. Ichihara, S., T. Suzuki, M. Suzuki, and S. Mizushima. 1986. Molecular cloning and sequencing of the sppA gene and characterization of the encoded protease IV, a signal peptide peptidase, of *Escherichia coli*. *J. Biol. Chem.* 261:9405–9411.

823. Ichikawa, J. K., C. Li, J. Fu, and S. Clarke. 1994. A gene at 59 minutes on the *Escherichia coli* chromosome encodes a lipoprotein with unusual amino acid repeat sequences. *J. Bacteriol.* 176:1630–1638.

824. Ichikawa, S., and A. Kaji. 1989. Molecular cloning and expression of ribosome releasing factor. *J. Biol. Chem.* 264:20054–20059.

825. Icho, T. 1988. Membrane-bound phosphatases in *Escherichia coli*: sequence of the pgpB gene and dual subcellular localization of the pgpB product. *J. Bacteriol.* 170:5117–5124.

826. Icho, T. 1988. Membrane-bound phosphatases in *Escherichia coli*: sequence of the pgpA gene. *J. Bacteriol.* 170:5110–5116.

827. Icho, T., C. E. Bulawa, and C. R. Raetz. 1985. Molecular cloning and sequencing of the gene for CDP-diglyceride hydrolase of *Escherichia coli*. *J. Biol. Chem.* 260:12092–12098.

828. Icho, T., C. P. Sparrow, and C. R. Raetz. 1985. Molecular cloning and sequencing of the gene for CDP-diglyceride synthetase of *Escherichia coli*. *J. Biol. Chem.* 260:12078–12083.

829. Ifuku, O., H. Miyaoka, N. Koga, J. Kishimoto, S. Haze, Y. Wachi, and M. Kajiwara. 1994. Origin of carbon atoms of biotin. ^{13}C-NMR studies on biotin biosynthesis in *Escherichia coli*. *Eur. J. Biochem.* 220:585–591.

830. Igarashi, K., N. Fujita, and A. Ishihama. 1989. Promoter selectivity of *Escherichia coli* RNA polymerase: omega factor is responsible for the ppGpp sensitivity. *Nucleic Acids Res.* 17:8755–8765.

831. Iggo, R., S. Picksley, J. Southgate, J. McPheat, and D. P. Lane. 1990. Identification of a putative RNA helicase in *E. coli*. *Nucleic Acids Res.* 18:5413–5417.

832. Igo, M. M., J. M. Slauch, and T. J. Silhavy. 1990. Signal transduction in bacteria: kinases that control gene expression. *New Biol.* 2:5–9.

833. Iida, A., S. Teshiba, and K. Mizobuchi. 1993. Identification and characterization of the tktB gene encoding a second transketolase in *Escherichia coli* K-12. *J. Bacteriol.* 175:5375–5383.

834. Iino, T., Y. Komeda, K. Kutsukake, R. M. Macnab, P. Matsumura, J. S. Parkinson, M. I. Simon, and S. Yamaguchi. 1988. New unified nomenclature for the flagellar genes of *Escherichia coli* and *Salmonella typhimurium*. *Microbiol. Rev.* 52:533–535.

835. Ikeda, M., T. Sato, M. Wachi, H. K. Jung, F. Ishino, Y. Kobayashi, and M. Matsuhashi. 1989. Structural similarity among *Escherichia coli* FtsW and RodA proteins and *Bacillus subtilis* SpoVE protein, which function in cell division, cell elongation, and spore formation, respectively. *J. Bacteriol.* 171:6375–6378.

836. Ikeda, M., M. Wachi, H. K. Jung, F. Ishino, and M. Matsuhashi. 1991. The *Escherichia coli* mraY gene encoding UDP-N-acetylmuramoyl-pentapeptide:undecaprenyl-phosphate phospho-N-acetylmuramoyl-pentapeptide transferase. *J. Bacteriol.* 173:1021–1026.

837. Ikemi, M., K. Murakami, M. Hashimoto, and Y. Murooka. 1992. Cloning and characterization of genes involved in the biosynthesis of delta-aminolevulinic acid in *Escherichia coli*. *Gene* 121:127–132.

838. Ilag, L. L., and D. Jahn. 1992. Activity and spectroscopic properties of the *Escherichia coli* glutamate 1-semialdehyde aminotransferase and the putative active site mutant K265R. *Biochemistry* 31:7143–7151.

839. Ilag, L. L., D. Jahn, G. Eggertsson, and D. Soll. 1991. The *Escherichia coli* hemL gene encodes glutamate 1-semialdehyde aminotransferase. *J. Bacteriol.* 173:3408–3413.

840. Ilyina, T. S., E. V. Nechaeva, Y. M. Romanova, and G. B. Smirnov. 1981. Isolation and mapping of *Escherichia coli* K12 mutants defective in Tn9 transposition. *Mol. Gen. Genet.* 181:384–389.

841. Imamura, N., and H. Nakayama. 1981. thiD locus of *Escherichia coli*. *Experientia* 37:1265–1266.

842. Imamura, N., and H. Nakayama. 1982. thiK and thiL loci of *Escherichia coli*. *J. Bacteriol.* 151:708–717.

843. Imlay, J., and I. Fridovich. 1992. Exogenous quinones directly inhibit the respiratory NADH dehydrogenase in *Escherichia coli*. *Arch. Biochem. Biophys.* 296:337–346.

844. Inglese, J., D. L. Johnson, A. Shiau, J. M. Smith, and S. J. Benkovic. 1990. Subcloning, characterization, and affinity labeling of *Escherichia coli* glycinamide ribonucleotide transformylase. *Biochemistry* 29:1436–1443.

845. Inglese, J., J. M. Smith, and S. J. Benkovic. 1990. Active-site mapping and site-specific mutagenesis of glycinamide ribonucleotide transformylase from *Escherichia coli*. *Biochemistry* 29:6678–6687.

846. Inoue, K., S. Kuramitsu, K. Aki, Y. Watanabe, T. Takagi, M. Nishigai, A. Ikai, and H. Kagamiyama. 1988. Branched-chain amino acid aminotransferase of *Escherichia coli*: overproduction and properties. *J. Biochem.* (Tokyo) 104:777–784.

847. Inouye, S., N. Lee, M. Inouye, H. C. Wu, H. Suzuki, Y. Nishimura, H. Iketani, and Y. Hirota. 1977. Amino acid replacement in a mutant lipoprotein of the *Escherichia coli* outer membrane. *J. Bacteriol.* 132:308–313.

848. Iobbi-Nivol, C., H. Crooke, L. Griffiths, J. Grove, H. Hussain, J. Pommier, V. Mejean, and J. A. Cole. 1994. A reassessment of the range of c-type cytochromes synthesized by *Escherichia coli* K-12. *FEMS Microbiol. Lett.* 119:89–94.

849. Irani, M., and P. K. Maitra. 1976. Glyceraldehyde 3-P dehydrogenase, glycerate 3-P kinase and enolase mutants of *Escherichia coli*: genetic studies. *Mol. Gen. Genet.* 145:65–71.

850. Irani, M. H., and P. K. Maitra. 1977. Properties of *Escherichia coli* mutants deficient in enzymes of glycolysis. *J. Bacteriol.* 132:398–410.

851. Ishihama, A. 1992. Role of the RNA polymerase alpha subunit in transcription activation. *Mol. Microbiol.* 6:3283–3288.

852. Ishino, F., W. Park, S. Tomioka, S. Tamaki, I. Takase, K. Kunugita, H. Matsuzawa, S. Asoh, T. Ohta, and B. G. Spratt. 1986. Peptidoglycan synthetic activities in membranes of *Escherichia coli* caused by overproduction of penicillin-binding protein 2 and rodA protein. *J. Biol. Chem.* 261:7024–7031.

853. Ishino, Y. H., H. Shinagawa, K. Makino, M. Amemura, and A. Nakata. 1987. Nucleotide sequence of the *iap* gene, responsible for alkaline phosphatase isozyme conversion in *Escherichia coli*, and identification of the gene product. *J. Bacteriol.* 169:5429–5433.

854. Island, M. D., and R. J. Kadner. 1993. Interplay between the membrane-associated UhpB and UhpC regulatory proteins. *J. Bacteriol.* 175:5028–5034.

855. Island, M. D., B. Y. Wei, and R. J. Kadner. 1992. Structure and function of the *uhp* genes for the sugar phosphate transport system in *Escherichia coli* and *Salmonella typhimurium*. *J. Bacteriol.* 174:2754–2762.

856. Issartel, J. P., V. Koronakis, and C. Hughes. 1991. Activation of *Escherichia coli* prohaemolysin to the mature toxin by acyl carrier protein-dependent fatty acylation. *Nature* (London) 351:759–761.

857. Isturiz, T., E. Palmero, and J. Vitelli-Flores. 1986. Mutations affecting gluconate catabolism in *Escherichia coli*. Genetic mapping of the locus for the thermosensitive gluconokinase. *J. Gen. Microbiol.* 132:3209–3219.

858. Itaya, M. 1990. Isolation and characterization of a second RNase H (RNase HII) of *Escherichia coli* K-12 encoded by the rnhB gene. *Proc. Natl. Acad. Sci. USA* 87:8587–8591.

859. Itaya, M., and R. J. Crouch. 1991. Correlation of activity with phenotypes of *Escherichia coli* partial function mutants of rnh, the gene encoding RNase H. *Mol. Gen. Genet.* 227:433–437.

860. Ito, K., K. Egawa, and Y. Nakamura. 1991. Genetic interaction between the beta′ subunit of RNA polymerase and the arginine-rich domain of *Escherichia coli* nusA protein. *J. Bacteriol.* 173:1492–1501.

861. Iuchi, S. 1993. Phosphorylation/dephosphorylation of the receiver module at the conserved aspartate residue controls transphosphorylation activity of histidine kinase in sensor protein ArcB of *Escherichia coli*. *J. Biol. Chem.* 268:23972–23980.

862. Iuchi, S., A. Aristarkhov, J. M. Dong, J. S. Taylor, and E. C. Lin. 1994. Effects of nitrate respiration on expression of the Arc-controlled operons encoding succinate dehydrogenase and flavin-linked L-lactate dehydrogenase. *J. Bacteriol.* 176:1695–1701.

863. Iuchi, S., V. Chepuri, H. A. Fu, R. B. Gennis, and E. C. Lin. 1990. Requirement for terminal cytochromes in generation of the aerobic signal for the arc regulatory system in *Escherichia coli*: study utilizing deletions and lac fusions of cyo and cyd. *J. Bacteriol.* 172:6020–6025.

864. Iuchi, S., S. T. Cole, and E. C. Lin. 1990. Multiple regulatory elements for the glpA operon encoding anaerobic glycerol-3-phosphate dehydrogenase and the glpD operon encoding aerobic glycerol-3-phosphate dehydrogenase in *Escherichia coli*: further characterization of respiratory control. *J. Bacteriol.* 172:179–184.

865. Iuchi, S., and E. C. Lin. 1992. Purification and phosphorylation of the Arc regulatory components of *Escherichia coli*. *J. Bacteriol.* 174:5617–5623.

866. Iuchi, S., and E. C. Lin. 1992. Mutational analysis of signal transduction by ArcB, a membrane sensor protein responsible for anaerobic repression of operons involved in the central aerobic pathways in *Escherichia coli*. *J. Bacteriol.* 174:3972–3980.

867. Iuchi, S., and E. C. Lin. 1993. Adaptation of *Escherichia coli* to redox environments by gene expression. *Mol. Microbiol.* 9:9–15.

868. Iuchi, S., and E. C. C. Lin. 1987. The narL gene product activates the nitrate reductase operon and represses the fumarate reductase and trimethylamine N-oxide reductase operons in *Escherichia coli*. *Proc. Natl. Acad. Sci. USA* 84:3901–3905.

869. Iuchi, S., Z. Matsuda, T. Fujiwara, and E. C. Lin. 1990. The arcB gene of *Escherichia coli* encodes a sensor-regulator protein for anaerobic repression of the arc modulon. *Mol. Microbiol.* 4:715–727.

870. Ivey, D. M., A. A. Guffanti, J. Zemsky, E. Pinner, R. Karpel, E. Padan, S. Schuldiner, and T. A. Krulwich. 1993. Cloning and characterization of a putative Ca^{2+}/H^{+} antiporter gene from *Escherichia coli* upon functional complementation of Na^{+}/H^{+} antiporter-deficient strains by the overexpressed gene. *J. Biol. Chem.* 268:11296–11303.

871. Iwakura, M., J. Hattori, Y. Arita, M. Tokushige, and H. Katsuki. 1979. Studies on regulatory functions of malic enzymes. VI. Purification and molecular properties of NADP-linked malic enzyme from *Escherichia coli* W. *J. Biochem.* (Tokyo) 85:1355–1365.

872. Iwasaki, H., M. Takahagi, A. Nakata, and H. Shinagawa. 1992. *Escherichia coli* RuvA and RuvB proteins specifically interact with Holliday junctions and promote branch migration. *Genes Dev.* 6:2214–2220.

873. Iwasaki, H., M. Takahagi, T. Shiba, A. Nakata, and H. Shinagawa. 1991. *Escherichia coli* RuvC protein is an endonuclease that resolves the Holliday structure. *EMBO J.* 10:4381–4389.

874. Izuhara, M., K. Takamune, and R. Takata. 1991. Cloning and sequencing of an *Escherichia coli* K12 gene which encodes a polypeptide having similarity to the human ferritin H subunit. *Mol. Gen. Genet.* 225:510–513.

875. Jackowski, S., and J. H. Alix. 1990. Cloning, sequence, and expression of the pantothenate permease (*panF*) gene of *Escherichia coli*. *J. Bacteriol.* 172:3842–3848.

876. Jackowski, S., P. D. Jackson, and C. O. Rock. 1994. Sequence and function of the aas gene in *Escherichia coli*. *J. Biol. Chem.* 269:2921–2928.

877. Jackowski, S., and C. O. Rock. 1987. Altered molecular form of acyl carrier protein associated with beta-ketoacyl-acyl carrier protein synthase II (*fabF*) mutants. *J. Bacteriol.* 169:1469–1473.

878. Jackson, J. H., P. A. Herring, E. B. Patterson, and J. M. Blatt. 1993. A mechanism for valine-resistant growth of *Escherichia coli* K-12 supported by the valine-sensitive acetohydroxy acid synthase IV activity from ilvJ662. *Biochimie* 75:759–765.

879. Jackson, M. E., and J. M. Pratt. 1988. Analysis of the membrane-binding domain of penicillin-binding protein 5 of *Escherichia coli*. *Mol. Microbiol.* 2:563–568.

880. Jacobson, B. L., J. J. He, P. S. Vermersch, D. D. Lemon, and F. A. Quiocho. 1991. Engineered interdomain disulfide in the periplasmic receptor for sulfate transport reduces flexibility. Site-directed mutagenesis and ligand-binding studies. *J. Biol. Chem.* 266:5220–5225.

881. Jaffe, A., R. D'Ari, and V. Norris. 1986. SOS-independent coupling between DNA replication and cell division in *Escherichia coli*. *J. Bacteriol.* 165:66–71.

882. Jahn, M., M. J. Rogers, and D. Soll. 1991. Anticodon and acceptor stem nucleotides in tRNA(Gln) are major recognition elements for E. coli glutaminyl-tRNA synthetase. *Nature* (London) 352:258–260.

883. Jahreis, K., and J. W. Lengeler. 1993. Molecular analysis of two ScrR repressors and of a ScrR-FruR hybrid repressor for sucrose and D-fructose specific regulons from enteric bacteria. *Mol. Microbiol.* 9:195–209.

884. Jahreis, K., P. W. Postma, and J. W. Lengeler. 1991. Nucleotide sequence of the ilvH-fruR gene region of *Escherichia coli* K12 and *Salmonella typhimurium* LT2. *Mol. Gen. Genet.* 226:332–336.

885. Jakubowski, H. 1994. Editing function of *Escherichia coli* cysteinyl-tRNA synthetase: cyclization of cysteine to cysteine thiolactone. *Nucleic Acids Res.* 22:1155–1160.

886. Jap, B. K. 1989. Molecular design of PhoE porin and its functional consequences. *J. Mol. Biol.* 205:407–419.

887. Jaskula, J. C., T. E. Letain, S. K. Roof, J. T. Skare, and K. Postle. 1994. Role of the TonB amino terminus in energy transduction between membranes. *J. Bacteriol.* 176:2326–2338.

888. Jaworowski, A., G. Mayo, D. C. Shaw, H. D. Campbell, and I. G. Young. 1981. Characterization of the respiratory NADH dehydrogenase of *Escherichia coli* and reconstitution of NADH oxidase in ndh mutant membrane vesicles. *Biochemistry* 20:3621–3628.

889. Jefferson, R. A., S. M. Burgess, and D. Hirsh. 1986. Beta-glucuronidase from *Escherichia coli* as a gene-fusion marker. *Proc. Natl. Acad. Sci. USA* 83:8447–8451.

890. Jenkins, D. E., E. A. Auger, and A. Matin. 1991. Role of RpoH, a heat shock regulator protein, in *Escherichia coli* carbon starvation protein synthesis and survival. *J. Bacteriol.* 173:1992–1996.

891. Jenkins, L. S., and W. D. Nunn. 1987. Regulation of the ato operon by the atoC gene in *Escherichia coli*. *J. Bacteriol.* 169:2096–2102.

892. Jenkins, L. S., and W. D. Nunn. 1987. Genetic and molecular characterization of the genes involved in short-chain fatty acid degradation in *Escherichia coli*: the ato system. *J. Bacteriol.* 169:42–52.

893. Jenkins, S. J., C. A. Sparkes, and M. C. Jones-Mortimer. 1974. A gene involved in lysine excretion in *Escherichia coli* K12. *Heredity* 32:409–412.

894. Jennings, M. P., and I. R. Beacham. 1990. Analysis of the *Escherichia coli* gene encoding L-asparaginase II, ansB, and its regulation by cyclic AMP receptor and FNR proteins. *J. Bacteriol.* 172:1491–1498.

895. Jensen, P. R., and O. Michelsen. 1992. Carbon and energy metabolism of atp mutants of *Escherichia coli*. *J. Bacteriol.* 174:7635–7641.

896. Jensen, R. 1976. Enzyme recruitment in evolution of new function. *Annu. Rev. Microbiol.* 30:409–425.

897. Jeong, J. H., M. Kitakawa, S. Isono, and K. Isono. 1993. Cloning and nucleotide sequencing of the genes, rpIU and rpmA, for ribosomal proteins L21 and L27 of *Escherichia coli*. *DNA Seq.* 4:59–67.

898. Jerlstrom, P. G., D. A. Bezjak, M. P. Jennings, and I. R. Beacham. 1989. Structure and expression in *Escherichia coli* K-12 of the L-asparaginase I-encoding ansA gene and its flanking regions. *Gene* 78:37–46.

899. Jin, R. Z., J. C. Tang, and E. C. Lin. 1983. Experimental evolution of a novel pathway for glycerol dissimilation in *Escherichia coli*. *J. Mol. Evol.* 19:429–436.

900. Johann, S., and S. M. Hinton. 1987. Cloning and nucleotide sequence of the chlD locus. *J. Bacteriol.* 169:1911–1916.

901. Johanson, U., and D. Hughes. 1992. Comparison of the complete sequence of the str operon in *Salmonella typhimurium* and *Escherichia coli*. *Gene* 120:93–98.

902. Johnson, D. I., and R. L. Somerville. 1984. New regulatory genes involved in the control of transcription initiation at the thr and ilv promoters of *Escherichia coli* K-12. *Mol. Gen. Genet.* 195:70–76.

903. Johnson, J. L., L. W. Indermaur, and K. V. Rajagopalan. 1991. Molybdenum cofactor biosynthesis in *Escherichia coli*. Requirement of the chlB gene product for the formation of molybdopterin guanine dinucleotide. *J. Biol. Chem.* 266:12140–12145.

904. Johnstone, D. B., and S. B. Farr. 1991. ApppA binds to several proteins in *Escherichia coli*, including the heat shock and oxidative stress proteins DnaK, GroEL, E89, C45 and C40. *EMBO J.* 10:3897–3904.

905. Joly, J. C., and J. R. Swartz. 1994. Protein folding activities of *Escherichia coli* protein disulfide isomerase. *Biochemistry* 33:4231–4236.

906. Jones, C. E., J. M. Brook, D. Buck, C. Abell, and A. G. Smith. 1993. Cloning and sequencing of the *Escherichia coli* panB gene, which encodes ketopantoate hydroxymethyltransferase, and overexpression of the enzyme. *J. Bacteriol.* 175:2125–2130.

907. Jones, C. H., J. S. Pinkner, A. V. Nicholes, L. N. Slonim, S. N. Abraham, and S. J. Hultgren. 1993. FimC is a periplasmic PapD-like chaperone that directs assembly of type 1 pili in bacteria. *Proc. Natl. Acad. Sci. USA* **90**:8397–8401.

908. Jones, C. J., M. Homma, and R. M. Macnab. 1987. Identification of proteins of the outer (L and P) rings of the flagellar basal body of *Escherichia coli*. *J. Bacteriol.* **169**:1489–1492.

909. Jones, D. H., F. C. Franklin, and C. M. Thomas. 1994. Molecular analysis of the operon which encodes the RNA polymerase sigma factor sigma 54 of *Escherichia coli*. *Microbiology* **140**:1035–1043.

910. Jones, P. G., and M. Inouye. 1994. The cold-shock response—a hot topic. *Mol. Microbiol.* **11**:811–818.

911. Jones, P. W., and J. M. Turner. 1984. A model for the common control of enzymes of ethanolamine catabolism in *Escherichia coli*. *J. Gen. Microbiol.* **130**:849–860.

912. Jones, P. W., and J. M. Turner. 1984. Interrelationships between the enzymes of ethanolamine metabolism in *Escherichia coli*. *J. Gen. Microbiol.* **130**:299–308.

913. Josephson, B. L., and D. G. Fraenkel. 1969. Transketolase mutants of *Escherichia coli*. *J. Bacteriol.* **100**:1289–1295.

914. Josse, J. G., and P. Handler. 1964. Phosphoglucomutase. *J. Biol. Chem.* **239**:2741–2751.

915. Jounouchi, M., M. Maeda, and M. Futai. 1993. The alpha subunit of ATP synthase (F0F1): the Lys-175 and Thr-176 residues in the conserved sequence (Gly-X-X-X-X-Gly-Lys-Thr/Ser) are located in the domain required for stable subunit-subunit interaction. *J. Biochem.* (Tokyo) **114**:171–176.

916. Jovanovic, G., T. Kostic, and D. J. Savic. 1990. Nucleotide and amino acid polymorphism in the gene for L-histidinol dehydrogenase of *Escherichia coli* K12. *Nucleic Acids Res.* **18**:3634

917. Juhl, M. J., and D. P. Clark. 1990. Thiophene-degrading *Escherichia coli* mutants possess sulfone oxidase activity and show altered resistance to sulfur-containing antibiotics. *Appl. Environ. Microbiol.* **56**:3179–3185.

918. Jullien, M., S. Baudet, F. Rodier, and G. Le Bras. 1988. Allosteric transition of aspartokinase I-homoserine dehydrogenase I studied by time-resolved fluorescence. *Biochimie* **70**:1807–1814.

919. Justesen, J., and J. Neuhard. 1975. *pyrR* identical to *pyrH* in *Salmonella typhimurium*: control of expression of the *pyr* genes. *J. Bacteriol.* **123**:851–854.

920. Kaasen, I., G. Evensen, and E. Seeberg. 1986. Amplified expression of the *tag*+ and *alkA*+ genes in *Escherichia coli*: identification of gene products and effects on alkylation resistance. *J. Bacteriol.* **168**:642–647.

921. Kaasen, I., P. Falkenberg, O. B. Styrvold, and A. R. Strom. 1992. Molecular cloning and physical mapping of the *otsBA* genes, which encode the osmoregulatory trehalose pathway of *Escherichia coli*: evidence that transcription is activated by *katF* (AppR). *J. Bacteriol.* **174**:889–898.

922. Kaback, H. R. 1990. The lac permease of *Escherichia coli*: a prototypic energy-transducing membrane protein. *Biochim. Biophys. Acta* **1018**:160–162.

923. Kaczorowski, G., L. D. Kohn, and H. R. Kaback. 1978. Purification and properties of D-lactate dehydrogenase from *Escherichia coli* ML 308–225. *Methods Enzymol.* **53**:519–527.

924. Kadner, R. J. 1974. Transport system for L-methionine in *Escherichia coli*. *J. Bacteriol.* **117**:232–241.

925. Kadner, R. J., and W. J. Watson. 1974. Methionine transport in *Escherichia coli*: physiological and genetic evidence for two uptake systems. *J. Bacteriol.* **119**:401–409.

926. Kadner, R. J., and H. Winkler. 1973. Isolation and characterization of mutations affecting the transport of hexose phosphate in *Escherichia coli*. *J. Bacteriol.* **113**:895–900.

927. Kaiser, A., and E. Leistner. 1990. Role of the *entC* gene in enterobactin and menaquinone biosynthesis in *Escherichia coli*. *Arch. Biochem. Biophys.* **276**:331–335.

928. Kajie, S., R. Ideta, I. Yamato, and Y. Anraku. 1991. Molecular cloning and DNA sequence of *dniR*, a gene affecting anaerobic expression of the *Escherichia coli* hexaheme nitrite reductase. *FEMS Microbiol. Lett.* **67**:205–211.

929. Kajitani, M., and A. Ishihama. 1991. Identification and sequence determination of the host factor gene for bacteriophage Q beta. *Nucleic Acids Res.* **19**:1063–1066.

930. Kajitani, M., A. Kato, A. Wada, Y. Inokuchi, and A. Ishihama. 1994. Regulation of the *Escherichia coli* hfq gene encoding the host factor for phage Q beta. *J. Bacteriol.* **176**:531–534.

931. Kalapos, M. P., G. J. Cao, S. R. Kushner, and N. Sarkar. 1994. Identification of a second poly(A) polymerase in *Escherichia coli*. *Biochem. Biophys. Res. Commun.* **198**:459–465.

932. Kalman, M., D. R. Gentry, and M. Cashel. 1991. Characterization of the *Escherichia coli* K12 gltS glutamate permease gene. *Mol. Gen. Genet.* **225**:379–386.

933. Kalman, M., H. Murphy, and M. Cashel. 1991. rhlB, a new *Escherichia coli* K-12 gene with an RNA helicase-like protein sequence motif, one of at least five such possible genes in a prokaryote. *New Biol.* **3**:886–895.

934. Kamitori, S., Y. Odagaki, K. Inoue, S. Kuramitsu, H. Kagamiyama, Y. Matsuura, and T. Higuchi. 1989. Crystallization and preliminary X-ray characterization of branched-chain amino acid aminotransferase from *Escherichia coli*. *J. Biochem.* (Tokyo) **105**:671–672.

935. Kammen, H. O., C. C. Marvel, L. Hardy, and E. E. Penhoet. 1988. Purification, structure, and properties of *Escherichia coli* tRNA pseudouridine synthase I. *J. Biol. Chem.* **263**:2255–2263.

936. Kammler, M., C. Schon, and K. Hantke. 1993. Characterization of the ferrous iron uptake system of *Escherichia coli*. *J. Bacteriol.* **175**:6212–6219.

937. Kampfenkel, K., and V. Braun. 1993. Topology of the ExbB protein in the cytoplasmic membrane of *Escherichia coli*. *J. Biol. Chem.* **268**:6050–6057.

938. Kampfenkel, K., and V. Braun. 1993. Membrane topologies of the TolQ and TolR proteins of *Escherichia coli*: inactivation of TolQ by a missense mutation in the proposed first transmembrane segment. *J. Bacteriol.* **175**:4485–4491.

939. Kanatani, A., T. Masuda, T. Shimoda, F. Misoka, X. S. Lin, T. Yoshimoto, and D. Tsuru. 1991. Protease II from *Escherichia coli*: sequencing and expression of the enzyme gene and characterization of the expressed enzyme. *J. Biochem.* (Tokyo) **110**:315–320.

940. Kang, P. J., and E. A. Craig. 1990. Identification and characterization of a new *Escherichia coli* gene that is a dosage-dependent suppressor of a *dnaK* deletion mutation. *J. Bacteriol.* **172**:2055–2064.

941. Kang, W. K., T. Icho, S. Isono, M. Kitakawa, and K. Isono. 1989. Characterization of the gene rimK responsible for the addition of glutamic acid residues to the C-terminus of ribosomal protein S6 in *Escherichia coli* K12. *Mol. Gen. Genet.* **217**:281–288.

942. Kannan, P. R., and K. Dharmalingam. 1987. Restriction alleviation and enhancement of mutagenesis of the bacteriophage T4 chromosome in recBCsbcA strains of *Escherichia coli*. *Mol. Gen. Genet.* **209**:413–418.

943. Kano, Y., T. Ogawa, T. Ogura, S. Hiraga, T. Okazaki, and F. Imamoto. 1991. Participation of the histone-like protein HU and of IHF in minichromosomal maintenance in *Escherichia coli*. *Gene* **103**:25–30.

944. Kao, C., and L. Snyder. 1988. The *lit* gene product which blocks bacteriophage T4 late gene expression is a membrane protein encoded by a cryptic DNA element, e14. *J. Bacteriol.* **170**:2056–2062.

945. Karaoglu, D., and D. L. Thurlow. 1991. A chemical interference study on the interaction of ribosomal protein L11 from *Escherichia coli* with RNA molecules containing its binding site from 23S rRNA. *Nucleic Acids Res.* **19**:5293–5300.

946. Karasawa, K., I. Kudo, T. Kobayashi, H. Homma, N. Chiba, H. Mizushima, K. Inoue, and S. Nojima. 1991. Lysophospholipase L1 from *Escherichia coli* K-12 overproducer. *J. Biochem.* (Tokyo) **109**:288–293.

947. Karasawa, K., and S. Nojima. 1991. Lysophospholipases from *Escherichia coli*. *Methods Enzymol.* **197**:437–445.

948. Karow, M., and C. Georgopoulos. 1993. The essential *Escherichia coli* msbA gene, a multicopy suppressor of null mutations in the htrB gene, is related to the universally conserved family of ATP-dependent translocators. *Mol. Microbiol.* **7**:69–79.

949. Karow, M., S. Raina, C. Georgopoulos, and O. Fayet. 1991. Complex phenotypes of null mutations in the htr genes, whose products are essential for *Escherichia coli* growth at elevated temperatures. *Res. Microbiol.* **142**:289–294.

950. Karsten, W. E., and R. E. Viola. 1992. Identification of an essential cysteine in the reaction catalyzed by aspartate-beta-semialdehyde dehydrogenase from *Escherichia coli*. *Biochim. Biophys. Acta* **1121**:234–238.

951. Kasahara, M., A. Nakata, and H. Shinagawa. 1992. Molecular analysis of the *Escherichia coli* phoP-phoQ operon. *J. Bacteriol.* **174**:492–498.

952. Kashiwagi, K., A. Miyaji, S. Ikeda, T. Tobe, C. Sasakawa, and K. Igarashi. 1992. Increase of sensitivity to aminoglycoside antibiotics by polyamine-induced protein (oligopeptide-binding protein) in *Escherichia coli*. *J. Bacteriol.* **174**:4331–4337.

953. Kashiwagi, K., S. Miyamoto, E. Nukui, H. Kobayashi, and K. Igarashi. 1993. Functions of potA and potD proteins in spermidine-preferential uptake system in *Escherichia coli*. *J. Biol. Chem.* **268**:19358–19363.

954. Kashiwagi, K., T. Suzuki, F. Suzuki, T. Furuchi, H. Kobayashi, and K. Igarashi. 1991. Coexistence of the genes for putrescine transport protein and ornithine decarboxylase at 16 min on *Escherichia coli* chromosome. *J. Biol. Chem.* **266**:20922–20927.

955. Kashiwagi, K., Y. Yamaguchi, Y. Sakai, H. Kobayashi, and K. Igarashi. 1990. Identification of the polyamine-induced protein as a periplasmic oligopeptide binding protein. *J. Biol. Chem.* **265**:8387–8391.

956. Kast, P., and H. Hennecke. 1991. Amino acid substrate specificity of *Escherichia coli* phenylalanyl-tRNA synthetase altered by distinct mutations. *J. Mol. Biol.* **222**:99–124.

957. Kato, H., T. Tanaka, T. Nishioka, A. Kimura, and J. Oda. 1988. Role of cysteine residues in glutathione synthetase from *Escherichia coli* B. Chemical modification and oligonucleotide site-directed mutagenesis. *J. Biol. Chem.* **263**:11646–11651.

958. Kato, J., Y. Nishimura, R. Imamura, H. Niki, S. Hiraga, and H. Suzuki. 1990. New topoisomerase essential for chromosome segregation in E. coli. *Cell* **63**:393–404.

959. Kaufmann, A., Y. D. Stierhof, and U. Henning. 1994. New outer membrane-associated protease of *Escherichia coli* K-12. *J. Bacteriol.* **176**:359–367.

960. Kauppinen, S., M. Siggaard-Andersen, and P. von Wettstein-Knowles. 1988. Beta-ketoacyl-ACP synthase I of *Escherichia coli*: nucleotide sequence of the fabB gene and identification of the cerulenin binding residue. *Carlsberg Res. Commun.* **53**:357–370.

961. Kawagishi, I., V. Muller, A. W. Williams, V. M. Irikura, and R. M. Macnab. 1992. Subdivision of flagellar region III of the *Escherichia coli* and *Salmonella typhimurium* chromosomes and identification of two additional flagellar genes. *J. Gen. Microbiol.* **138**:1051–1065.

962. Kawakami, K., K. Ito, and Y. Nakamura. 1992. Differential regulation of two genes encoding lysyl-tRNA synthetases in *Escherichia coli*: lysU-constitutive mutations compensate for a lysS null mutation. *Mol. Microbiol.* **6**:1739–1745.

963. Kawamoto, S., S. Tokuyama, K. Aoyama, S. Yashima, and Y. Eguchi. 1984. Genetic mapping of cold resistance gene of *Escherichia coli*. *Agric. Biol. Chem.* **48**:2067–2071.

964. Kawamukai, M., R. Utsumi, K. Takeda, A. Higashi, H. Matsuda, Y. L. Choi, and T. Komano. 1991. Nucleotide sequence and characterization of the *sfs1* gene: *sfs1* is involved in CRP*-dependent mal gene expression in *Escherichia coli*. *J. Bacteriol.* **173**:2644–2648.

965. Kawasaki, T., T. Nakata, and Y. Nose. 1968. Genetic mapping with a thiamine-requiring auxotroph of *Escherichia coli* K-12 defective in thiamine phosphate phosphorylase. *J. Bacteriol.* **95**:1483–1485.

966. Kawula, T. H., and M. J. Lelivelt. 1994. Mutations in a gene encoding a new Hsp70 suppress rapid DNA inversion and *bgl* activation, but not *proU* derepression, in *hns-1* mutant *Escherichia coli*. *J. Bacteriol.* **176**:610–619.

967. Keasling, J. D., L. Bertsch, and A. Kornberg. 1993. Guanosine pentaphosphate phosphohydrolase of *Escherichia coli* is a long-chain exopolyphosphatase. *Proc. Natl. Acad. Sci. USA* **90**:7029–7033.

968. Keck, W., A. M. van Leeuwen, M. Huber, and E. W. Goodell. 1990. Cloning and characterization of *mepA*, the structural gene of the penicillin-insensitive murein endopeptidase from *Escherichia coli*. *Mol. Microbiol.* **4**:209–219.

969. Keller, E. B., and J. M. Calvo. 1979. Alternative secondary structures of leader RNAs and the regulation of the trp, phe, his, thr, and leu operons. *Proc. Natl. Acad. Sci. USA* **76**:6186–6190.

970. Kelly, K. O., and M. P. Deutscher. 1992. The presence of only one of five exoribonucleases is sufficient to support the growth of *Escherichia coli*. *J. Bacteriol.* **174**:6682–6684.

971. Keng, T., and P. Schimmel. 1983. Synthesis of two polypeptide subunits of an aminoacyl tRNA synthetase as a single polypeptide chain. *J. Biomol. Struct. Dyn.* **1**:225–229.

972. Kenny, B., S. Taylor, and I. B. Holland. 1992. Identification of individual amino acids required for secretion within the haemolysin (HlyA) C-terminal targeting region. *Mol. Microbiol.* **6**:1477–1489.

973. Kenri, T., F. Imamoto, and Y. Kano. 1992. Construction and characterization of an *Escherichia coli* mutant deficient in the metY gene encoding tRNA(f2Met): either tRNA(f1Met) or tRNA(f2Met) is required for cell growth. *Gene* **114**:109–114.

974. Keshavjee, K., C. Pyne, and A. L. Bognar. 1991. Characterization of a mutation affecting the function of *Escherichia coli* folylpolyglutamate synthetase-dihydrofolate synthetase and further mutations produced in vitro at the same locus. *J. Biol. Chem.* **266**:19925–19929.

975. Kessler, D., I. Leibrecht, and J. Knappe. 1991. Pyruvate-formate-lyase-deactivase and acetyl-CoA reductase activities of *Escherichia coli* reside on a polymeric protein particle encoded by adhE. *FEBS Lett.* **281**:59–63.

976. Khan, I. H., T. S. Reese, and S. Khan. 1992. The cytoplasmic component of the bacterial flagellar motor. *Proc. Natl. Acad. Sci. USA* **89**:5956–5960.

977. Khoury, A. M., H. S. Nick, and P. Lu. 1991. In vivo interaction of *Escherichia coli* lac repressor N-terminal fragments with the lac operator. *J. Mol. Biol.* **219**:623–634.

978. Kiino, D. R., R. Licudine, K. Wilt, D. H. Yang, and L. B. Rothman-Denes. 1993. A cytoplasmic protein, NfrC, is required for bacteriophage N4 adsorption. *J. Bacteriol.* **175**:7074–7080.

979. Kiino, D. R., G. J. Phillips, and T. J. Silhavy. 1990. Increased expression of the bifunctional protein PrlF suppresses overproduction lethality associated with exported β-galactosidase hybrid proteins in *Escherichia coli*. *J. Bacteriol.* **172**:185–192.

980. Kiino, D. R., and L. B. Rothman-Denes. 1989. Genetic analysis of bacteriophage N4 adsorption. *J. Bacteriol.* **171**:4595–4602.

981. Kiino, D. R., M. S. Singer, and L. B. Rothman-Denes. 1993. Two overlapping genes encoding membrane proteins required for bacteriophage N4 adsorption. *J. Bacteriol.* **175**:7081–7085.

982. Kikuchi, A., and L. Gorini. 1976. Studies of the DNA carrying genes, valS, argI, pyrB, and argF by electron microscopy and by site specific endonucleases. *J. Microsc. Biol. Cell* **27**:1–10.

983. Killmann, H., R. Benz, and V. Braun. 1993. Conversion of the FhuA transport protein into a diffusion channel through the outer membrane of *Escherichia coli*. *EMBO J.* **12**:3007–3016.

984. Killmann, H., and V. Braun. 1992. An aspartate deletion mutation defines a binding site of the multifunctional FhuA outer membrane receptor of *Escherichia coli* K-12. *J. Bacteriol.* **174**:3479–3486.

985. Kim, I. Y., Z. Veres, and T. C. Stadtman. 1992. *Escherichia coli* mutant SELD enzymes. The cysteine 17 residue is essential for selenophosphate formation from ATP and selenide. *J. Biol. Chem.* **267**:19650–19654.

986. Kim, S., and D. L. Wulff. 1990. Location of an ntr-like gene on the physical map of *Escherichia coli*. *J. Bacteriol.* **172**:6619

987. Kim, S. K., K. Makino, M. Amemura, H. Shinagawa, and A. Nakata. 1993. Molecular analysis of the *phoH* gene, belonging to the phosphate regulon in *Escherichia coli*. *J. Bacteriol.* **175**:1316–1324.

988. Kim, S. T., Y. F. Li, and A. Sancar. 1992. The third chromophore of DNA photolyase: Trp-277 of *Escherichia coli* DNA photolyase repairs thymine dimers by direct electron transfer. *Proc. Natl. Acad. Sci. USA* **89**:900–904.

989. Kimlova, L. J., C. Pyne, K. Keshavjee, J. Huy, G. Beebakhee, and A. L. Bognar. 1991. Mutagenesis of the folC gene encoding folylpolyglutamate synthetase-dihydrofolate synthetase in *Escherichia coli*. *Arch. Biochem. Biophys.* **284**:9–16.

990. Kimura, T., T. Asai, M. Imai, and M. Takanami. 1989. Methylation strongly enhances DNA bending in the replication origin region of the *Escherichia coli* chromosome. *Mol. Gen. Genet.* **219**:69–74.

991. Kirby, J. E., J. E. Trempy, and S. Gottesman. 1994. Excision of a P4-like cryptic prophage leads to Alp protease expression in *Escherichia coli*. *J. Bacteriol.* **176**:2068–2081.

992. Kirsebom, L. A., and S. G. Svard. 1993. Identification of a region within M1 RNA of *Escherichia coli* RNase P important for the location of the cleavage site on a wild-type tRNA precursor. *J. Mol. Biol.* **231**:594–604.

993. Kita, K., C. R. Vibat, S. Meinhardt, J. R. Guest, and R. B. Gennis. 1989. One-step purification from *Escherichia coli* of complex II (succinate: ubiquinone oxidoreductase) associated with succinate-reducible cytochrome b556. *J. Biol. Chem.* **264**:2672–2677.

994. Kjeldgaard, M., and J. Nyborg. 1992. Refined structure of elongation factor EF-Tu from *Escherichia coli*. *J. Mol. Biol.* **223**:721–742.

995. Kleanthous, C., D. G. Campbell, and J. R. Coggins. 1990. Active site labeling of the shikimate pathway enzyme, dehydroquinase. Evidence for a common substrate binding site within dehydroquinase and dehydroquinate synthase. *J. Biol. Chem.* **265**:10929–10934.

996. Kleanthous, C., R. Deka, K. Davis, S. M. Kelly, A. Cooper, S. E. Harding, Price, NC, A. R. Hawkins, and J. R. Coggins. 1992. A comparison of the enzymological and biophysical properties of two distinct classes of dehydroquinase enzymes. *Biochem. J.* **282**:687–695.

997. Klein, W., U. Ehmann, and W. Boos. 1991. The repression of trehalose transport and metabolism in *Escherichia coli* by high osmolarity is mediated by trehalose-6-phosphate phosphatase. *Res. Microbiol.* **142**:359–371.

998. Klem, T. J., and V. J. Davisson. 1993. Imidazole glycerol phosphate synthase: the glutamine amidotransferase in histidine biosynthesis. *Biochemistry* **32**:5177–5186.

999. Klemm, P. 1984. The fimA gene encoding the type-1 fimbrial subunit of *Escherichia coli*. Nucleotide sequence and primary structure of the protein. *Eur. J. Biochem.* **143**:395–399.

1000. Klemm, P., and G. Christiansen. 1990. The fimD gene required for cell surface localization of *Escherichia coli* type 1 fimbriae. *Mol. Gen. Genet.* **220**:334–338.

1001. Klena, J. D., R. S. Ashford, and C. A. Schnaitman. 1992. Role of *Escherichia coli* K-12 *rfa* genes and the *rfp* gene of *Shigella dysenteriae* 1 in generation of lipopolysaccharide core heterogeneity and attachment of O antigen. *J. Bacteriol.* **174**:7297–7307.

1002. Klena, J. D., E. Pradel, and C. A. Schnaitman. 1992. Comparison of lipopolysaccharide biosynthesis genes rfaK, rfaL, rfaY, and rfaZ of *Escherichia coli* K-12 and *Salmonella typhimurium*. *J. Bacteriol.* **174**:4746–4752.

1003. Klena, J. D., and C. A. Schnaitman. 1994. Genes for TDP-rhamnose synthesis affect the pattern of lipopolysaccharide heterogeneity in *Escherichia coli* K-12. *J. Bacteriol.* **176**:4003–4010.

1004. Klose, M., S. MacIntyre, H. Schwarz, and U. Henning. 1988. The influence of amino substitutions within the mature part of an *Escherichia coli* outer membrane protein (OmpA) on assembly of the polypeptide into its membrane. *J. Biol. Chem.* **263**:13297–13302.

1005. Knappe, J., and G. Sawers. 1990. A radical-chemical route to acetyl-CoA: the anaerobically induced pyruvate formate-lyase system of *Escherichia coli*. *FEMS Microbiol. Rev.* **6**:383–398.

1006. Ko, Y. H., C. R. Cremo, and B. A. McFadden. 1992. Vanadate-dependent photomodification of serine 319 and 321 in the active site of isocitrate lyase from *Escherichia coli*. *J. Biol. Chem.* **267**:91–95.

1007. Kobayashi, T., I. Kudo, K. Karasawa, H. Mizushima, K. Inoue, and S. Nojima. 1985. Nucleotide sequence of the pldB gene and characteristics of deduced amino acid sequence of lysophospholipase L2 in *Escherichia coli*. *J. Biochem.* (Tokyo) **98**:1017–1025.

1008. Koch, W. H., D. G. Ennis, A. S. Levine, and R. Woodgate. 1992. *Escherichia coli* umuDC mutants: DNA sequence alterations and UmuD cleavage. *Mol. Gen. Genet.* **233**:443–448.

1009. Koebnik, R., and V. Braun. 1993. Insertion derivatives containing segments of up to 16 amino acids identify surface- and periplasm-exposed regions of the FhuA outer membrane receptor of *Escherichia coli* K-12. *J. Bacteriol.* **175**:826–839.

1010. Kohiyama, M., H. Eberle, and D. Sporn. 1983. Gamma-ray induction of deoxyribonucleic acid synthesis in temperature-sensitive DNA initiation mutants of *Escherichia coli*. *Eur. J. Biochem.* **132**:411–415.

1011. Kohno, K., M. Wada, Y. Kano, and F. Imamoto. 1990. Promoters and autogenous control of the *Escherichia coli* hupA and hupB genes. *J. Mol. Biol.* **213**:27–36.

1012. Kolkhof, P. 1992. Specificities of three tight-binding Lac repressors. *Nucleic Acids Res.* **20**:5035–5039.

1013. Kolling, R., A. Gielow, W. Seufert, C. Kucherer, and W. Messer. 1988. AsnC, a multifunctional regulator of genes located around the replication origin of *Escherichia coli*, oriC. *Mol. Gen. Genet.* **212**:99–104.

1014. Komano, T., R. Utsumi, and M. Kawamukai. 1991. Functional analysis of the fic gene involved in regulation of cell division. *Res. Microbiol.* **142**:269–277.

1015. Komatsoulis, G. A., and J. Abelson. 1993. Recognition of tRNA(Cys) by *Escherichia coli* cysteinyl-tRNA synthetase. *Biochemistry* **32**:7435–7444.

1016. Komine, Y., T. Adachi, H. Inokuchi, and H. Ozeki. 1990. Genomic organization and physical mapping of the transfer RNA genes in Escherica coli K12. *J. Mol. Biol.* **212**:579–598.

1017. Kondo, H., Y. Nakabeppu, H. Kataoka, S. Kuhara, S. Kawabata, and M. Sekiguchi. 1986. Structure and expression of the alkB gene of *Escherichia coli* related to the repair of alkylated DNA. *J. Biol. Chem.* 261:15772–15777.

1018. Kondo, H., K. Shiratsuchi, T. Yoshimoto, T. Masuda, A. Kitazono, D. Tsuru, M. Anai, M. Sekiguchi, and T. Tanabe. 1991. Acetyl-CoA carboxylase from *Escherichia coli*: gene organization and nucleotide sequence of the biotin carboxylase subunit. *Proc. Natl. Acad. Sci. USA* 88:9730–9733.

1019. Konopka, J. M., C. J. Halkides, J. L. Vanhooke, D. G. Gorenstein, and P. A. Frey. 1989. UDP-galactose 4-epimerase. Phosphorus-31 nuclear magnetic resonance analysis of NAD^+ and NADH bound at the active site. *Biochemistry* 28:2645–2654.

1020. Koo, H. S., L. Claassen, L. Grossman, and L. F. Liu. 1991. ATP-dependent partitioning of the DNA template into supercoiled domains by *Escherichia coli* UvrAB. *Proc. Natl. Acad. Sci. USA* 88:1212–1216.

1021. Koonin, E. V. 1993. *Escherichia coli* dinG gene encodes a putative DNA helicase related to a group of eukaryotic helicases including Rad3 protein. *Nucleic Acids Res.* 21:1497

1022. Koonin, E. V., and K. E. Rudd. 1993. SpoU protein of *Escherichia coli* belongs to a new family of putative rRNA methylases. *Nucleic Acids Res.* 21:5519

1023. Korat, B., H. Mottl, and W. Keck. 1991. Penicillin-binding protein 4 of *Escherichia coli*: molecular cloning of the dacB gene, controlled overexpression, and alterations in murein composition. *Mol. Microbiol.* 5:675–684.

1024. Kornberg, A., and T. A. Baker. 1992. *DNA Replication*. W. H. Freeman & Co., New York.

1025. Koronakis, V., and C. Hughes. 1993. Bacterial signal peptide-independent protein export: HlyB-directed secretion of hemolysin. *Semin. Cell Biol.* 4:7–15.

1026. Koronakis, V., C. Hughes, and E. Koronakis. 1991. Energetically distinct early and late stages of HlyB/HlyD-dependent secretion across both *Escherichia coli* membranes. *EMBO J.* 10:3263–3272.

1027. Koronakis, V., C. Hughes, and E. Koronakis. 1993. ATPase activity and ATP/ADP-induced conformational change in the soluble domain of the bacterial protein translocator HlyB. *Mol. Microbiol.* 8:1163–1175.

1028. Koster, W. 1991. Iron(III) hydroxamate transport across the cytoplasmic membrane of *Escherichia coli*. *Biol. Methods* 4:23–32.

1029. Koster, W., and B. Bohm. 1992. Point mutations in two conserved glycine residues within the integral membrane protein FhuB affect iron(III) hydroxamate transport. *Mol. Gen. Genet.* 232:399–407.

1030. Koster, W., and V. Braun. 1989. Iron-hydroxamate transport into *Escherichia coli* K12: localization of FhuD in the periplasm and of FhuB in the cytoplasmic membrane. *Mol. Gen. Genet.* 217:233–239.

1031. Koster, W., and V. Braun. 1990. Iron (III) hydroxamate transport into *Escherichia coli*. Substrate binding to the periplasmic FhuD protein. *J. Biol. Chem.* 265:21407–21410.

1032. Koster, W., A. Gudmundsdottir, M. D. Lundrigan, A. Seiffert, and R. J. Kadner. 1991. Deletions or duplications in the BtuB protein affect its level in the outer membrane of *Escherichia coli*. *J. Bacteriol.* 173:5639–5647.

1033. Kowalczykowski, S. C., D. A. Dixon, A. K. Eggleston, S. D. Lauder, and W. M. Rehrauer. 1994. Biochemistry of homologous recombination in *Escherichia coli*. *Microbiol. Rev.* 58:401–465.

1034. Kraft, R., and L. A. Leinwand. 1987. Sequence of the complete P protein gene and part of the M protein gene from the histidine transport operon of *Escherichia coli* compared to that of *Salmonella typhimurium*. *Nucleic Acids Res.* 15:8568

1035. Krause, G., and A. Holmgren. 1991. Substitution of the conserved tryptophan 31 in *Escherichia coli* thioredoxin by site-directed mutagenesis and structure-function analysis. *J. Biol. Chem.* 266:4056–4066.

1036. Krebs, A., and W. A. Bridger. 1980. The kinetic properties of phosphoenolpyruvate carboxykinase of *Escherichia coli*. *Can. J. Biochem.* 58:309–318.

1037. Kredich, N. M., and G. M. Tomkins. 1966. The enzymic synthesis of L-cysteine in *Escherichia coli* and *Salmonella typhimurium*. *J. Biol. Chem.* 241:4955–4965.

1038. Krogfelt, K. A., and P. Klemm. 1988. Investigation of minor components of *Escherichia coli* type 1 fimbriae: protein chemical and immunological aspects. *Microb. Pathog.* 4:231–238.

1039. Kroll, D. J., D. M. Sullivan, A. Gutierrez-Hartmann, and J. P. Hoeffler. 1993. Modification of DNA topoisomerase II activity via direct interactions with the cyclic adenosine-3′,5′-monophosphate response element-binding protein and related transcription factors. *Mol. Endocrinol.* 7:305–318.

1040. Krone, F. A., G. Westphal, and J. D. Schwenn. 1991. Characterisation of the gene cysH and of its product phospho-adenylylsulphate reductase from *Escherichia coli*. *Mol. Gen. Genet.* 225:314–319.

1041. Kruger, T., C. Grund, C. Wild, and M. Noyer-Weidner. 1992. Characterization of the mcrBC region of *Escherichia coli* K-12 wild-type and mutant strains. *Gene* 114:1–12.

1042. Kudo, T., K. Nagai, and G. Tamura. 1977. Characteristics of a cold-sensitive chromosome segregation mutant of *Escherichia coli* K12. *Agric. Biol. Chem.* 41:97–107.

1043. Kuhn, H. M., U. Meier-Dieter, and H. Mayer. 1988. ECA, the enterobacterial common antigen. *FEMS Microbiol. Rev.* 4:195–222.

1044. Kumamoto, A. A., W. G. Miller, and R. P. Gunsalus. 1987. *Escherichia coli* tryptophan repressor binds multiple sites within the aroH and trp operators. *Genes Dev.* 1:556–564.

1045. Kumamoto, C. A. 1990. SecB protein: a cytosolic export factor that associates with nascent exported proteins. *J. Bioenerg. Biomembr.* 22:337–351.

1046. Kumar, A., B. Grimes, N. Fujita, K. Makino, R. A. Malloch, R. S. Hayward, and A. Ishihama. 1994. Role of the sigma 70 subunit of *Escherichia coli* RNA polymerase in transcription activation. *J. Mol. Biol.* 235:405–413.

1047. Kundrot, C. E., and P. R. Evans. 1991. Designing an allosterically locked phosphofructokinase. *Biochemistry* 30:1478–1484.

1048. Kunert, K. J., C. F. Cresswell, A. Schmidt, P. M. Mullineaux, and C. H. Foyer. 1990. Variations in the activity of glutathione reductase and the cellular glutathione content in relation to sensitivity to methylviologen in *Escherichia coli*. *Arch. Biochem. Biophys.* 282:233–238.

1049. Kung, H., and H. Weissbach. 1978. DNA-directed in vitro synthesis of *Escherichia coli* beta-isopropylmalate dehydrogenase. *J. Biol. Chem.* 253:2078–2080.

1050. Kuo, L. C., C. Caron, S. Lee, and W. Herzberg. 1990. Zn^{2+} regulation of ornithine transcarbamoylase. II. Metal binding site. *J. Mol. Biol.* 211:271–280.

1051. Kupor, S. R., and D. G. Fraenkel. 1969. 6-Phosphogluconolactonase mutants of *Escherichia coli* and a maltose blue gene. *J. Bacteriol.* 100:1296–1301.

1052. Kuramitsu, S., K. Hiromi, H. Hayashi, Y. Morino, and H. Kagamiyama. 1990. Pre-steady-state kinetics of *Escherichia coli* aspartate aminotransferase catalyzed reactions and thermodynamic aspects of its substrate specificity. *Biochemistry* 29:5469–5476.

1053. Kuriyan, J., L. Wong, M. Russel, and P. Model. 1989. Crystallization and preliminary x-ray characterization of thioredoxin reductase from *Escherichia coli*. *J. Biol. Chem.* 264:12752–12753.

1054. Kurose, N., K. Watanabe, and A. Kimura. 1986. Nucleotide sequence of the gene responsible for D-xylose uptake in *Escherichia coli*. *Nucleic Acids Res.* 14:7115–7123.

1055. Kushner, S. R., V. F. Maples, and W. S. Champney. 1977. Conditionally lethal ribosomal protein mutants: characterization of a locus required for modification of 50S subunit proteins. *Proc. Natl. Acad. Sci. USA* 74:467–471.

1056. Kutsukake, K., T. Nakao, and T. Iino. 1985. A gene for DNA invertase and an invertible DNA in *Escherichia coli* K-12. *Gene* 34:343–350.

1057. Kutsukake, K., Y. Ohya, and T. Iino. 1990. Transcriptional analysis of the flagellar regulon of *Salmonella typhimurium*. *J. Bacteriol.* 172:741–747.

1058. Kuzminov, A. 1993. RuvA, RuvB and RuvC proteins: cleaning-up after recombinational repairs in *E. coli*. *Bioessays* 15:355–358.

1059. Laalami, S., C. Sacerdot, G. Vachon, K. Mortensen, H. U. Sperling-Petersen, Y. Cenatiempo, and M. Grunberg-Manago. 1991. Structural and functional domains of *E. coli* initiation factor IF2. *Biochimie* 73:1557–1566.

1060. Labedan, B., and M. Riley. 1995. Widespread protein sequence similarities: origins of *Escherichia coli* genes. *J. Bacteriol.* 177:1585–1588.

1061. Labedan, B., and M. Riley. 1995. Gene products of *E. coli*: sequence comparisons and common ancestries. *Mol. Biol. Evol.* 12:980–987.

1062. Laber, B., F. X. Gomis-Ruth, M. J. Romao, and R. Huber. 1992. *Escherichia coli* dihydrodipicolinate synthase. Identification of the active site and crystallization. *Biochem. J.* 288:691–695.

1063. Lacroix, J. M., I. Loubens, M. Tempete, B. Menichi, and J. P. Bohin. 1991. The mdoA locus of *Escherichia coli* consists of an operon under osmotic control. *Mol. Microbiol.* 5:1745–1753.

1064. Lahti, R., M. Perala, P. Heikinheimo, T. Pitkaranta, E. Kukko-Kalske, and J. Heinonen. 1991. Characterization of the 5′ flanking region of the *Escherichia coli* ppa gene encoding inorganic pyrophosphatase: mutations in the ribosome-binding site decrease the level of ppa mRNA. *J. Gen. Microbiol.* 137:2517–2523.

1065. Lahti, R., T. Salminen, S. Latonen, P. Heikinheimo, K. Pohjanoksa, and J. Heinonen. 1991. Genetic engineering of *Escherichia coli* inorganic pyrophosphatase. Tyr55 and Tyr141 are important for the structural integrity. *Eur. J. Biochem.* 198:293–297.

1066. Lam, H. M., E. Tancula, W. B. Dempsey, and M. E. Winkler. 1992. Suppression of insertions in the complex pdxJ operon of *Escherichia coli* K-12 by lon and other mutations. *J. Bacteriol.* 174:1554–1567.

1067. Lam, H. M., and M. E. Winkler. 1990. Metabolic relationships between pyridoxine (vitamin B6) and serine biosynthesis in *Escherichia coli* K-12. *J. Bacteriol.* 172:6518–6528.

1068. Lam, H. M., and M. E. Winkler. 1992. Characterization of the complex pdxH-tyrS operon of *Escherichia coli* K-12 and pleiotropic phenotypes caused by pdxH insertion mutations. *J. Bacteriol.* 174:6033–6045.

1069. Lam, L. K., L. D. Arnold, T. H. Kalantar, J. G. Kelland, P. M. Lane-Bell, M. M. Palcic, M. A. Pickard, and J. C. Vederas. 1988. Analogs of diaminopimelic acid as inhibitors of meso-diaminopimelate dehydrogenase and LL-diaminopimelate epimerase. *J. Biol. Chem.* 263:11814–11819.

1070. Lamark, T., I. Kaasen, M. W. Eshoo, P. Falkenberg, J. McDougall, and A. R. Strom. 1991. DNA sequence and analysis of the bet genes encoding the osmoregulatory choline-glycine betaine pathway of *Escherichia coli*. *Mol. Microbiol.* 5:1049–1064.

1071. Lamblin, A. F., and J. A. Fuchs. 1993. Expression and purification of the cynR regulatory gene product: CynR is a DNA-binding protein. *J. Bacteriol.* 175:7990–7999.

1072. Lander, M., A. R. Pitt, P. R. Alefounder, D. Bardy, C. Abell, and A. R. Battersby. 1991. Studies on the mechanism of hydroxymethylbilane synthase concerning the role of arginine residues in substrate binding. *Biochem. J.* 275:447–452.

1073. Landick, R., J. Carey, and C. Yanofsky. 1985. Translation activates the paused transcription complex and restores transcription of the trp operon leader region. *Proc. Natl. Acad. Sci. USA* 82:4663–4667.

1074. **Landick, R., J. Stewart, and D. N. Lee.** 1990. Amino acid changes in conserved regions of the beta-subunit of *Escherichia coli* RNA polymerase alter transcription pausing and termination. *Genes Dev.* **4:**1623–1636.

1075. **Lange, R., and R. Hengge-Aronis.** 1991. Growth phase-regulated expression of *bolA* and morphology of stationary-phase *Escherichia coli* cells are controlled by the novel sigma factor sigma S. *J. Bacteriol.* **173:**4474–4481.

1076. **Langer, T., C. Lu, H. Echols, J. Flanagan, M. K. Hayer, and F. U. Hartl.** 1992. Successive action of DnaK, DnaJ and GroEL along the pathway of chaperone-mediated protein folding. *Nature* (London) **356:**683–689.

1077. **Lapointe, J., and G. Dulcuve.** 1975. Thermosensitive mutants of *Escherichia coli* K-12 altered in the catalytic subunit and in a regulatory factor of the glutamyl-transfer ribonucleic acid synthetase. *J. Bacteriol.* **122:**352–358.

1078. **LaRoe, D. J., and S. B. Vik.** 1992. Mutations at Glu-32 and His-39 in the epsilon subunit of the *Escherichia coli* F1F0 ATP synthase affect its inhibitory properties. *J. Bacteriol.* **174:**633–637.

1079. **LaRossa, R. A., and T. K. Van Dyk.** 1991. Physiological roles of the DnaK and GroE stress proteins: catalysts of protein folding or macromolecular sponges? *Mol. Microbiol.* **5:**529–534.

1080. **Larson, T. J., J. S. Cantwell, and A. T. van Loo-Bhattacharya.** 1992. Interaction at a distance between multiple operators controls the adjacent, divergently transcribed glpTQ-glpACB operons of *Escherichia coli* K-12. *J. Biol. Chem.* **267:**6114–6121.

1081. **Larson, T. J., D. N. Ludtke, and R. M. Bell.** 1984. sn-Glycerol-3-phosphate auxotrophy of *plsB* strains of *Escherichia coli*: evidence that a second mutation, *plsX*, is required. *J. Bacteriol.* **160:**711–717.

1082. **Laughrea, M., and J. Tam.** 1991. Interaction of ribosomal protein S1 and initiation factor IF3 with the 3′ major domain and the decoding site of the 30S subunit of *Escherichia coli*. *Biochemistry* **30:**11412–11420.

1083. **Laughrea, M., and J. Tam.** 1992. In vivo chemical footprinting of the *Escherichia coli* ribosome. *Biochemistry* **31:**12035–12041.

1084. **Lavina, M., A. P. Pugsley, and F. Moreno.** 1986. Identification, mapping, cloning and characterization of a gene (sbmA) required for microcin B17 action on *Escherichia coli* K12. *J. Gen. Microbiol.* **132:**1685–1693.

1085. **Lawther, R. P., R. C. Wek, J. M. Lopes, R. Pereira, B. E. Taillon, and G. W. Hatfield.** 1987. The complete nucleotide sequence of the ilvGMEDA operon of *Escherichia coli* K-12. *Nucleic Acids Res.* **15:**2137–2155. (Errata, **15:**9108, 1987, and **16:**3602, 1988.)

1086. **Lazzaroni, J. C., N. Fognini-Lefebvre, and R. Portalier.** 1989. Cloning of the excC and excD genes involved in the release of periplasmic proteins by *Escherichia coli* K12. *Mol. Gen. Genet.* **218:**460–464.

1087. **Lazzaroni, J. C., and R. Portalier.** 1992. The excC gene of *Escherichia coli* K-12 required for cell envelope integrity encodes the peptidoglycan-associated lipoprotein (PAL). *Mol. Microbiol.* **6:**735–742.

1088. **Leach, D. R., R. G. Lloyd, and A. F. Coulson.** 1992. The SbcCD protein of *Escherichia coli* is related to two putative nucleases in the UvrA superfamily of nucleotide-binding proteins. *Genetica* **87:**95–100.

1089. **Leclerc, G., C. Sirard, and G. R. Drapeau.** 1989. The *Escherichia coli* cell division mutation ftsM₁ is in serU. *J. Bacteriol.* **171:**2090–2095.

1090. **Lee, C. A., M. J. Fournier, and J. Beckwith.** 1985. *Escherichia coli* 6S RNA is not essential for growth or protein secretion. *J. Bacteriol.* **161:**1156–1161.

1091. **Lee, C. P., B. L. Seong, and U. L. RajBhandary.** 1991. Structural and sequence elements important for recognition of *Escherichia coli* formylmethionine tRNA by methionyl-tRNA transformylase are clustered in the acceptor stem. *J. Biol. Chem.* **266:**18012–18017.

1092. **Lee, C. Y., D. J. O'Kane, and E. A. Meighen.** 1994. Riboflavin synthesis genes are linked with the *lux* operon of *Photobacterium phosphoreum*. *J. Bacteriol.* **176:**2100–2104.

1093. **Lee, E. C., L. M. Hales, R. I. Gumport, and J. F. Gardner.** 1992. The isolation and characterization of mutants of the integration host factor (IHF) of *Escherichia coli* with altered, expanded DNA-binding specificities. *EMBO J.* **11:**305–313.

1094. **Lee, J. H., J. C. Wendt, and K. T. Shanmugam.** 1990. Identification of a new gene, molR, essential for utilization of molybdate by *Escherichia coli*. *J. Bacteriol.* **172:**2079–2087.

1095. **Lee, L., and Y. Imae.** 1990. Role of threonine residue 154 in ligand recognition of the tar chemoreceptor in *Escherichia coli*. *J. Bacteriol.* **172:**377–382.

1096. **Lee, N., C. Francklyn, and E. P. Hamilton.** 1987. Arabinose-induced binding of AraC protein to araI2 activates the araBAD operon promoter. *Proc. Natl. Acad. Sci. USA* **84:**8814–8818.

1097. **Lee, N., W. Gielow, R. Martin, E. Hamilton, and A. Fowler.** 1986. The organization of the araBAD operon of *Escherichia coli*. *Gene* **47:**231–244.

1098. **Lee, R. S., J. Pagan, S. Wilke-Mounts, and A. E. Senior.** 1991. Characterization of *Escherichia coli* ATP synthase beta-subunit mutations using a chromosomal deletion strain. *Biochemistry* **30:**6842–6847.

1099. **Lee, S. J., A. Xie, W. Jiang, J. P. Etchegaray, P. G. Jones, and M. Inouye.** 1994. Family of the major cold-shock protein, CspA (CS7.4), of *Escherichia coli*, whose members show a high sequence similarity with the eukaryotic Y-box binding proteins. *Mol. Microbiol.* **11:**833–839.

1100. **Lee, T. C., and G. E. Christie.** 1990. Purification and properties of the bacteriophage P2 ogr gene product. A prokaryotic zinc-binding transcriptional activator. *J. Biol. Chem.* **265:**7472–7477.

1101. **Legrain, C., P. Halleux, V. Stalon, and N. Glansdorff.** 1972. The dual genetic control of ornithine carbamolytransferase in *Escherichia coli*. A case of bacterial hybrid enzymes. *Eur. J. Biochem.* **27:**93–102.

1102. **Legrain, C., V. Stalon, and N. Glansdorff.** 1976. *Escherichia coli* ornithine carbamoyltransferase isoenzymes: evolutionary significance and the isolation of λargF and λargI transducing bacteriophages. *J. Bacteriol.* **128:**35–38.

1103. **Lehman, I. R.** 1974. DNA ligase: structure, mechanism, and function. *Science* **186:**790–797.

1104. **Leinfelder, W., K. Forchhammer, B. Veprek, E. Zehelein, and A. Bock.** 1990. In vitro synthesis of selenocysteinyl-tRNA(UCA) from seryl-tRNA(UCA): involvement and characterization of the selD gene product. *Proc. Natl. Acad. Sci. USA* **87:**543–547.

1105. **Leisinger, T., and D. Haas.** 1975. N-Acetylglutamate synthase of *Escherichia coli* regulation of synthesis and activity by arginine. *J. Biol. Chem.* **250:**1690–1693.

1106. **Lengeler, J.** 1975. Mutations affecting transport of the hexitols D-mannitol, D-glucitol, and galacitol in *Escherichia coli* K-12: isolation and mapping. *J. Bacteriol.* **124:**26–38.

1107. **Lengeler, J.** 1975. Nature and properties of hexitol transport systems in *Escherichia coli*. *J. Bacteriol.* **124:**39–47.

1108. **Lengeler, J.** 1977. Analysis of mutations affecting the dissmilation of galactitol (dulcitol) in *Escherichia coli* K 12. *Mol. Gen. Genet.* **152:**83–91.

1109. **Lennette, E. T., and D. Apirion.** 1971. Genetic analysis of an *Escherichia coli* syndrome. *J. Bacteriol.* **108:**1322–1328.

1110. **Leppik, R. A., I. G. Young, and F. Gibson.** 1976. Membrane-associated reactions in ubiquinone biosynthesis in *Escherichia coli*. 3-Octaprenyl-4-hydroxybenzoate carboxy-lyase. *Biochim. Biophys. Acta* **436:**800–810.

1111. **Lerner, C. G., and M. Inouye.** 1991. Pleiotropic changes resulting from depletion of Era, an essential GTP-binding protein in *Escherichia coli*. *Mol. Microbiol.* **5:**951–957.

1112. **Lerner, T. J., and N. D. Zinder.** 1982. Another gene affecting sexual expression of *Escherichia coli*. *J. Bacteriol.* **150:**156–160.

1113. **Lesage, P., C. Chiaruttini, M. Graffe, J. Dondon, M. Milet, and M. Springer.** 1992. Messenger RNA secondary structure and translational coupling in the *Escherichia coli* operon encoding translation initiation factor IF3 and the ribosomal proteins, L35 and L20. *J. Mol. Biol.* **228:**366–386.

1114. **Leung, H. B., K. L. Kvalnes-Krick, S. L. Meyer, J. K. deRiel, and V. L. Schramm.** 1989. Structure and regulation of the AMP nucleosidase gene (amn) from *Escherichia coli*. *Biochemistry* **28:**8726–8733.

1115. **Levengood, S. K., W. F. Beyer, Jr., and R. E. Webster.** 1991. TolA: a membrane protein involved in colicin uptake contains an extended helical region. *Proc. Natl. Acad. Sci. USA* **88:**5939–5943.

1116. **Levengood, S. K., and R. E. Webster.** 1989. Nucleotide sequences of the tolA and tolB genes and localization of their products, components of a multistep translocation system in *Escherichia coli*. *J. Bacteriol.* **171:**6600–6609.

1117. **Levengood-Freyermuth, S. K., E. M. Click, and R. E. Webster.** 1993. Role of the carboxyl-terminal domain of TolA in protein import and integrity of the outer membrane. *J. Bacteriol.* **175:**222–228.

1118. **Leveque, F., P. Plateau, P. Dessen, and S. Blanquet.** 1990. Homology of lysS and lysU, the two *Escherichia coli* genes encoding distinct lysyl-tRNA synthetase species. *Nucleic Acids Res.* **18:**305–312.

1119. **Levin, H. L., K. Park, and H. K. Schachman.** 1989. Attenuation in the regulation of the pyrBI operon in *Escherichia coli*. In vivo studies of transcriptional termination. *J. Biol. Chem.* **264:**14638–14645.

1120. **Levin, J. D., A. W. Johnson, and B. Demple.** 1988. Homogeneous *Escherichia coli* endonuclease IV. Characterization of an enzyme that recognizes oxidative damage in DNA. *J. Biol. Chem.* **263:**8066–8071.

1121. **Levine, R. A., and M. W. Taylor.** 1979. Regulation of purine salvage enzymes in *E. coli*. *Adv. Exp. Med. Biol.* **122B:**57–60.

1122. **Levine, R. A., and M. W. Taylor.** 1981. Selection for purine regulatory mutants in an *E. coli* hypoxanthine phosphoribosyl transferase-guanine phosphoribosyl transferase double mutant. *Mol. Gen. Genet.* **181:**313–318.

1123. **Levine, R. A., and M. W. Taylor.** 1982. Mechanism of adenine toxicity in *Escherichia coli*. *J. Bacteriol.* **149:**923–930.

1124. **Levitz, R., D. Chapman, M. Amitsur, R. Green, L. Snyder, and G. Kaufmann.** 1990. The optional *E. coli* prr locus encodes a latent form of phage T4-induced anticodon nuclease. *EMBO J.* **9:**1383–1389.

1125. **Lewis, E. B.** 1951. Pseudoallelism and gene evolution. *Cold Spring Harbor Symp. Quant. Biol.* **16:**159–174.

1126. **Lewis, K.** 1994. Multidrug resistance pumps in bacteria: variations on a theme. *Trends Biochem. Sci.* **19:**119–123.

1127. **Lewis, L. K., M. E. Jenkins, and D. W. Mount.** 1992. Isolation of DNA damage-inducible promoters in *Escherichia coli*: regulation of polB (dinA), dinG, and dinH by LexA repressor. *J. Bacteriol.* **174:**3377–3385.

1128. **Lewis, M. J., J. A. Chang, and R. D. Simoni.** 1990. A topological analysis of subunit alpha from *Escherichia coli* F1F0-ATP synthase predicts eight transmembrane segments. *J. Biol. Chem.* **265:**10541–10550.

1129. **Leyh, T. S., and Y. Suo.** 1992. GTPase-mediated activation of ATP sulfurylase. *J. Biol. Chem.* **267:**542–545.

1130. **Leyh, T. S., J. C. Taylor, and G. D. Markham.** 1988. The sulfate activation locus of *Escherichia coli* K12: cloning, genetic, and enzymatic characterization. *J. Biol. Chem.* **263:**2409–2416.

1131. **Leyh, T. S., T. F. Vogt, and Y. Suo.** 1992. The DNA sequence of the sulfate activation locus from *Escherichia coli* K-12. *J. Biol. Chem.* **267:**10405–10410.

1132. **Li, B. F., D. Holdup, C. A. Morton, and M. L. Sinnott.** 1989. The catalytic consequences of experimental evolution. Transition-state structure during catalysis by the evolved beta-galactosidases of *Escherichia coli* (ebg enzymes) changed by a single mutational event. *Biochem. J.* 260:109–114.

1133. Li, J., and V. Stewart. 1992. Localization of upstream sequence elements required for nitrate and anaerobic induction of *fdn* (formate dehydrogenase-N) operon expression in *Escherichia coli* K-12. *J. Bacteriol.* 174:4935–4942.

1134. Li, J. M., C. S. Russell, and S. D. Cosloy. 1989. Cloning and structure of the hemA gene of *Escherichia coli* K-12. *Gene* 82:209–217.

1135. Li, J. M., C. S. Russell, and S. D. Cosloy. 1989. The structure of the *Escherichia coli* hemB gene. *Gene* 75:177–184.

1136. Li, Q. X., and W. Dowhan. 1988. Structural characterization of *Escherichia coli* phosphatidylserine decarboxylase. *J. Biol. Chem.* 263:11516–11522.

1137. Li, Q. X., and W. Dowhan. 1990. Studies on the mechanism of formation of the pyruvate prosthetic group of phosphatidylserine decarboxylase from *Escherichia coli. J. Biol. Chem.* 265:4111–4115.

1138. Li, S., H. Pelka, and L. H. Schulman. 1993. The anticodon and discriminator base are important for aminoacylation of *Escherichia coli* tRNA(Asn). *J. Biol. Chem.* 268:18335–18339.

1139. Li, S. J., and J. E. Cronan, Jr. 1992. The gene encoding the biotin carboxylase subunit of *Escherichia coli* acetyl-CoA carboxylase. *J. Biol. Chem.* 267:855–863.

1140. Li, S. J., and J. E. Cronan, Jr. 1992. The genes encoding the two carboxyltransferase subunits of *Escherichia coli* acetyl-CoA carboxylase. *J. Biol. Chem.* 267:16841–16847.

1141. Li, S. J., and J. E. Cronan, Jr. 1993. Growth rate regulation of *Escherichia coli* acetyl coenzyme A carboxylase, which catalyzes the first committed step of lipid biosynthesis. *J. Bacteriol.* 175:332–340.

1142. Li, S. J., C. O. Rock, and J. E. Cronan, Jr. 1992. The *dedB* (*usg*) open reading frame of *Escherichia coli* encodes a subunit of acetyl-coenzyme A carboxylase. *J. Bacteriol.* 174:5755–5757.

1143. Li, W. Q., and M. Yarus. 1992. Bar to normal UGA translation by the selenocysteine tRNA. *J. Mol. Biol.* 223:9–15.

1144. Lichens-Park, A., C. L. Smith, and M. Syvanen. 1990. Integration of bacteriophage lambda into the cryptic lambdoid prophages of *Escherichia coli. J. Bacteriol.* 172:2201–2208.

1145. Lieb, M. 1987. Bacterial genes *mutL*, *mutS*, and *dcm* participate in repair of mismatches at 5-methylcytosine sites. *J. Bacteriol.* 169:5241–5246.

1146. Lieb, M. 1991. Spontaneous mutation at a 5-methylcytosine hotspot is prevented by very short patch (VSP) mismatch repair. *Genetics* 128:23–27.

1147. Liljestrom, P. L., and P. Liljestrom. 1987. Nucleotide sequence of the melA gene, coding for alpha-galactosidase in *Escherichia coli* K-12. *Nucleic Acids Res.* 15:2213–2220.

1148. Lin, E. C., and S. Iuchi. 1991. Regulation of gene expression in fermentative and respiratory systems in *Escherichia coli* and related bacteria. *Annu. Rev. Genet.* 25:361–387.

1149. Lin, J. J. 1992. Endonuclease A degrades chromosomal and plasmid DNA of *Escherichia coli* present in most preparations of single stranded DNA from phagemids. *Proc. Natl. Sci. Counc. Repub. China Part B* 16:1–5.

1150. Lin, J. J., and A. Sancar. 1991. The C-terminal half of UvrC protein is sufficient to reconstitute (A)BC excinuclease. *Proc. Natl. Acad. Sci. USA* 88:6824–6828.

1151. Lin, L. L., and J. W. Little. 1989. Autodigestion and RecA-dependent cleavage of Ind⁻ mutant LexA proteins. *J. Mol. Biol.* 210:439–452.

1152. Lin, R., R. D'Ari, and E. B. Newman. 1992. Lambda *placMu* insertions in genes of the leucine regulon: extension of the regulon to genes not regulated by leucine. *J. Bacteriol.* 174:1948–1955.

1153. Lin, S. X., Q. Wang, and Y. L. Wang. 1988. Interactions between *Escherichia coli* arginyl-tRNA synthetase and its substrates. *Biochemistry* 27:6348–6353.

1154. Lin, Y. P., J. D. Sharer, and P. E. March. 1994. GTPase-dependent signaling in bacteria: characterization of a membrane-binding site for era in *Escherichia coli. J. Bacteriol.* 176:44–49.

1155. Lindqvist, S., M. Galleni, F. Lindberg, and S. Normark. 1989. Signalling proteins in enterobacterial AmpC beta-lactamase regulation. *Mol. Microbiol.* 3:1091–1102.

1156. Lindqvist, S., K. Weston-Hafer, H. Schmidt, C. Pul, G. Korfmann, J. Erickson, C. Sanders, H. H. Martin, and S. Normark. 1993. AmpG, a signal transducer in chromosomal beta-lactamase induction. *Mol. Microbiol.* 9:703–715.

1157. Lindsey, D. F., D. A. Mullin, and J. R. Walker. 1989. Characterization of the cryptic lambdoid prophage DLP12 of *Escherichia coli* and overlap of the DLP12 integrase gene with the tRNA gene argU. *J. Bacteriol.* 171:6197–6205.

1158. Linn, T., and J. Greenblatt. 1992. The NusA and NusG proteins of *Escherichia coli* increase the in vitro readthrough frequency of a transcriptional attenuator preceding the gene for the beta subunit of RNA polymerase. *J. Biol. Chem.* 267:1449–1454.

1159. Liochev, S. I., and I. Fridovich. 1992. Fumarase C, the stable fumarase of *Escherichia coli*, is controlled by the soxRS regulon. *Proc. Natl. Acad. Sci. USA* 89:5892–5896.

1160. Lipinska, B., M. Zylicz, and C. Georgopoulos. 1990. The HtrA (DegP) protein, essential for *Escherichia coli* survival at high temperatures, is an endopeptidase. *J. Bacteriol.* 172:1791–1797.

1161. Lipsett, M. N. 1978. Enzymes producing 4-thiouridine in *Escherichia coli* tRNA: approximate chromosomal locations of the genes and enzyme activities in a 4-thiouridine-deficient mutant. *J. Bacteriol.* 135:993–997.

1162. Little, R. M., and P. M. Anderson. 1987. Structural properties of cyanase. Denaturation, renaturation, and role of sulfhydryls and oligomeric structure in catalytic activity. *J. Biol. Chem.* 262:10120–10126.

1163. Liu, F., Q. Dong, and H. J. Fromm. 1992. Site-directed mutagenesis of the phosphate-binding consensus sequence in *Escherichia coli* adenylosuccinate synthetase. *J. Biol. Chem.* 267:2388–2392.

1164. Liu, J., and I. R. Beacham. 1990. Transcription and regulation of the cpdB gene in *Escherichia coli* K12 and *Salmonella typhimurium* LT2: evidence for modulation of constitutive promoters by cyclic AMP-CRP complex. *Mol. Gen. Genet.* 222:161–165.

1165. Liu, J., K. Duncan, and C. T. Walsh. 1989. Nucleotide sequence of a cluster of *Escherichia coli* enterobactin biosynthesis genes: identification of entA and purification of its product 2,3-dihydro-2,3-dihydroxybenzoate dehydrogenase. *J. Bacteriol.* 171:791–798.

1166. Liu, J., N. Quinn, G. A. Berchtold, and C. T. Walsh. 1990. Overexpression, purification, and characterization of isochorismate synthase (EntC), the first enzyme involved in the biosynthesis of enterobactin from chorismate. *Biochemistry* 29:1417–1425.

1167. Liu, J., J. M. Rutz, J. B. Feix, and P. E. Klebba. 1993. Permeability properties of a large gated channel within the ferric enterobactin receptor, FepA. *Proc. Natl. Acad. Sci. USA* 90:10653–10657.

1168. Liu, J., and C. T. Walsh. 1990. Peptidyl-prolyl cis-trans-isomerase from *Escherichia coli*: a periplasmic homolog of cyclophilin that is not inhibited by cyclosporin A. *Proc. Natl. Acad. Sci. USA* 87:4028–4032.

1169. Liu, J. D., and J. S. Parkinson. 1991. Genetic evidence for interaction between the CheW and Tsr proteins during chemoreceptor signaling by *Escherichia coli. J. Bacteriol.* 173:4941–4951.

1170. Liu, J. S., E. A. Park, A. L. Gurney, W. J. Roesler, and R. W. Hanson. 1991. Cyclic AMP induction of phosphoenolpyruvate carboxykinase (GTP) gene transcription is mediated by multiple promoter elements. *J. Biol. Chem.* 266:19095–19102.

1171. Ljungcrantz, P., L. Bulow, and K. Mosbach. 1990. Construction and characterization of a recombinant trienzyme, galactose dehydrogenase/betagalactosidase/galactokinase. *FEBS Lett.* 275:91–94.

1172. Lloyd, R. G. 1991. Conjugational recombination in resolvase-deficient ruvC mutants of *Escherichia coli* K-12 depends on recG. *J. Bacteriol.* 173:5414–5418.

1173. Lloyd, R. G., and C. Buckman. 1991. Overlapping functions of recD, recJ and recN provide evidence of three epistatic groups of genes in *Escherichia coli* recombination and DNA repair. *Biochimie* 73:313–320.

1174. Llyina, T. S., Y. M. Romanova, and G. B. Smirnov. 1981. The effect of tnm mutations of *Escherichia coli* K12 on transposition of various movable genetic elements. *Mol. Gen. Genet.* 183:376–379.

1175. Lo, T. C. Y., M. K. Rayman, and B. D. Sanwal. 1972. Transport of succinate in *Escherichia coli*. I. Biochemical and genetic studies of transport in whole cells. *J. Biol. Chem.* 247:6323–6331.

1176. Lo, T. C. Y., and B. D. Sanwal. 1975. Genetic analysis of mutants of *Escherichia coli* defective in dicarboxylate transport. *Mol. Gen. Genet.* 140:303–307.

1177. Lobell, R. B., and R. F. Schleif. 1991. AraC-DNA looping: orientation and distance-dependent loop breaking by the cyclic AMP receptor protein. *J. Mol. Biol.* 218:45–54.

1178. Lobner-Olesen, A., and E. Boye. 1992. Different effects of mioC transcription on initiation of chromosomal and minichromosomal replication in *Escherichia coli. Nucleic Acids Res.* 20:3029–3036.

1179. Lobner-Olesen, A., and M. G. Marinus. 1992. Identification of the gene (*aroK*) encoding shikimic acid kinase I of *Escherichia coli. J. Bacteriol.* 174:525–529.

1180. Lobocka, M., J. Hennig, J. Wild, and T. Klopotowski. 1994. Organization and expression of the *Escherichia coli* K-12 dad operon encoding the smaller subunit of D-amino acid dehydrogenase and the catabolic alanine racemase. *J. Bacteriol.* 176:1500–1510.

1181. Loenen, W. A., A. S. Daniel, H. D. Braymer, and N. E. Murray. 1987. Organization and sequence of the hsd genes of *Escherichia coli* K-12. *J. Mol. Biol.* 198:159–170.

1182. Loewen, P. C., J. Switala, M. Smolenski, and B. L. Triggs-Raine. 1990. Molecular characterization of three mutations in katG affecting the activity of hydroperoxidase I of *Escherichia coli. Biochem. Cell Biol.* 68:1037–1044.

1183. Lombardo, M. J., C. G. Miller, and K. E. Rudd. 1993. Physical mapping of the *Escherichia coli* pepT and potABCD genes. *J. Bacteriol.* 175:7745–7746.

1184. Lonetto, M. A., K. L. Brown, K. E. Rudd, and M. J. Buttner. 1994. Analysis of the *Streptomyces coelicolor* sigE gene reveals the existence of a subfamily of eubacterial RNA polymerase sigma factors involved in the regulation of extracytoplasmic functions. *Proc. Natl. Acad. Sci. USA* 91:7573–7577.

1185. Long, W. S., C. L. Slayman, and K. B. Low. 1978. Production of giant cells of *Escherichia coli. J. Bacteriol.* 133:995–1007.

1186. Lopez, J., and R. E. Webster. 1985. fipB and fipC: two bacterial loci required for morphogenesis of the filamentous bacteriophage f1. *J. Bacteriol.* 163:900–905.

1187. Lopilato, J. E., J. L. Garwin, S. D. Emr, T. J. Silhavy, and J. R. Beckwith. 1984. D-Ribose metabolism in *Escherichia coli* K-12: genetics, regulation, and transport. *J. Bacteriol.* 158:665–673.

1188. Lotti, M., M. Noah, M. Stoffler-Meilicke, and G. Stoffler. 1989. Localization of proteins L4, L5, L20 and L25 on the ribosomal surface by immuno-electron microscopy. *Mol. Gen. Genet.* 216:245–253.

1189. Lovett, S. T., and R. D. Kolodner. 1989. Identification and purification of a single-stranded-DNA-specific exonuclease encoded by the recJ gene of *Escherichia coli. Proc. Natl. Acad. Sci. USA* 86:2627–2631.

1190. Lu, M., J. L. Campbell, E. Boye, and N. Kleckner. 1994. SeqA: a negative modulator of replication initiation in *E. coli. Cell* 77:413–426.

1191. Lu, M., and N. Kleckner. 1994. Molecular cloning and characterization of the *pgm* gene incoding phosphoglucomutase of *Escherichia coli*. *J. Bacteriol.* **176**:5847–5851.

1192. Lu, Z., and E. C. Lin. 1989. The nucleotide sequence of *Escherichia coli* genes for L-fucose dissimilation. *Nucleic Acids Res.* **17**:4883–4884.

1193. Luckey, M., R. Ling, A. Dose, and B. Malloy. 1991. Role of a disulfide bond in the thermal stability of the LamB protein trimer in *Escherichia coli* outer membrane. *J. Biol. Chem.* **266**:1866–1871.

1194. Ludwig, A., A. Schmid, R. Benz, and W. Goebel. 1991. Mutations affecting pore formation by haemolysin from *Escherichia coli*. *Mol. Gen. Genet.* **226**:198–208.

1195. Lundrigan, M. D., and C. F. Earhart. 1984. Gene *envY* of *Escherichia coli* K-12 affects thermoregulation of major porin expression. *J. Bacteriol.* **157**:262–268.

1196. Lundrigan, M. D., M. J. Friedrich, and R. J. Kadner. 1989. Nucleotide sequence of the *Escherichia coli* porin thermoregulatory gene *envY*. *Nucleic Acids Res.* **17**:800

1197. Lundrigan, M. D., and R. J. Kadner. 1989. Altered cobalamin metabolism in *Escherichia coli* *btuR* mutants affects *btuB* gene regulation. *J. Bacteriol.* **171**:154–161.

1198. Lundstrom, J., G. Krause, and A. Holmgren. 1992. A Pro to His mutation in active site of thioredoxin increases its disulfide-isomerase activity 10-fold. New refolding systems for reduced or randomly oxidized ribonuclease. *J. Biol. Chem.* **267**:9047–9052.

1199. Luo, Y., and J. S. Krakow. 1992. Characterization and epitope mapping of monoclonal antibodies directed against the beta' subunit of the *Escherichia coli* RNA polymerase. *J. Biol. Chem.* **267**:18175–18181.

1200. Lupas, A., and J. Stock. 1989. Phosphorylation of an N-terminal regulatory domain activates the CheB methylesterase in bacterial chemotaxis. *J. Biol. Chem.* **264**:17337–17342.

1201. Lupo, M., and Y. S. Halpern. 1970. Gene controlling L-glutamic acid decarboxylase synthesis in *Escherichia coli* K-12. *J. Bacteriol.* **103**:382–386.

1202. Lutkenhaus, J. 1993. FtsZ ring in bacterial cytokinesis. *Mol. Microbiol.* **9**:403–409.

1203. Lutz, S., A. Jacobi, V. Schlensog, R. Bohm, G. Sawers, and A. Bock. 1991. Molecular characterization of an operon (hyp) necessary for the activity of the three hydrogenase isoenzymes in *Escherichia coli*. *Mol. Microbiol.* **5**:123–135.

1204. Ma, D., D. N. Cook, M. Alberti, N. G. Pon, H. Nikaido, and J. E. Hearst. 1993. Molecular cloning and characterization of *acrA* and *acrE* genes of *Escherichia coli*. *J. Bacteriol.* **175**:6299–6313.

1205. Ma, J., L. Lemieux, and R. B. Gennis. 1993. Genetic fusion of subunits I, II, and III of the cytochrome bo ubiquinol oxidase from *Escherichia coli* results in a fully assembled and active enzyme. *Biochemistry* **32**:7692–7697.

1206. MacGregor, C. H. 1975. Anaerobic cytochrome b_1 in *Escherichia coli*: association with and regulation of nitrate reductase. *J. Bacteriol.* **121**:1111–1116.

1207. Mackay, W. J., S. Han, and L. D. Samson. 1994. DNA alkylation repair limits spontaneous base substitution mutations in *Escherichia coli*. *J. Bacteriol.* **176**:3224–3230.

1208. Mackey, B. M., and C. M. Derrick. 1986. Peroxide sensitivity of cold-shocked *Salmonella typhimurium* and *Escherichia coli* and its relationship to minimal medium recovery. *J. Appl. Bacteriol.* **60**:501–511.

1209. Macnab, R. M. 1992. Genetics and biogenesis of bacterial flagella. *Annu. Rev. Genet.* **26**:131–158.

1210. MacPherson, A. J., M. C. Jones-Mortimer, and P. J. Henderson. 1981. Identification of the AraE transport protein of *Escherichia coli*. *Biochem. J.* **196**:269–283.

1211. Madern, D., J. Anselme, and M. Hartlein. 1992. Asparaginyl-tRNA synthetase from the *Escherichia coli* temperature-sensitive strain HO202. A proline replacement in motif 2 is responsible for a large increase in Km for asparagine and ATP. *FEBS Lett.* **299**:85–89.

1212. Magnuson, K., S. Jackowski, C. O. Rock, and J. E. Cronan, Jr. 1993. Regulation of fatty acid biosynthesis in *Escherichia coli*. *Microbiol. Rev.* **57**:522–542.

1213. Magnuson, K., W. Oh, T. J. Larson, and J. E. Cronan, Jr. 1992. Cloning and nucleotide sequence of the fabD gene encoding malonyl coenzyme A-acyl carrier protein transacylase of *Escherichia coli*. *FEBS Lett.* **299**:262–266.

1214. Maguin, E., H. Brody, C. W. Hill, and R. D'Ari. 1986. SOS-associated division inhibition gene sfiC is part of excisable element e14 in *Escherichia coli*. *J. Bacteriol.* **168**:464–466.

1215. Maguin, E., J. Lutkenhaus, and R. D'Ari. 1986. Reversibility of SOS-associated division inhibition in *Escherichia coli*. *J. Bacteriol.* **166**:733–738.

1216. Mahajan, S. K., C. C. Chu, D. K. Willis, A. Templin, and A. J. Clark. 1990. Physical analysis of spontaneous and mutagen-induced mutants of *Escherichia coli* K-12 expressing DNA exonuclease VIII activity. *Genetics* **125**:261–273.

1217. Mahajan, S. K., N. B. Vartak, and A. R. Datta. 1988. A new pleiotropic mutation causing defective carbohydrate uptake in *Escherichia coli* K-12: isolation, mapping, and preliminary characterization. *J. Bacteriol.* **170**:2568–2574.

1218. Maier, T., A. Jacobi, M. Sauter, and A. Bock. 1993. The product of the *hypB* gene, which is required for nickel incorporation into hydrogenases, is a novel guanine nucleotide-binding protein. *J. Bacteriol.* **175**:630–635.

1219. Majumdar, A., S. Rudikoff, and S. Adhya. 1987. Purification and properties of Gal repressor:pL-galR fusion in pKC31 plasmid vector. *J. Biol. Chem.* **262**:2326–2331.

1220. Majumdar, R., J. R. Guest, and W. A. Bridger. 1991. Functional consequences of substitution of the active site (phospho)histidine residue of *Escherichia coli* succinyl-CoA synthetase. *Biochim. Biophys. Acta* **1076**:86–90.

1221. Makino, K., H. Shinagawa, M. Amemura, T. Kawamoto, M. Yamada, and A. Nakata. 1989. Signal transduction in the phosphate regulon of *Escherichia coli* involves phosphotransfer between PhoR and PhoB proteins. *J. Mol. Biol.* **210**:551–559.

1222. Malakooti, J., B. Ely, and P. Matsumura. 1994. Molecular characterization, nucleotide sequence, and expression of the *fliO*, *fliP*, *fliQ*, and *fliR* genes of *Escherichia coli*. *J. Bacteriol.* **176**:189–197.

1223. Malakooti, J., Y. Komeda, and P. Matsumura. 1989. DNA sequence analysis, gene product identification, and localization of flagellar motor components of *Escherichia coli*. *J. Bacteriol.* **171**:2728–2734.

1224. Malo, M. S., and R. E. Loughlin. 1990. Promoter elements and regulation of expression of the cysD gene of *Escherichia coli* K-12. *Gene* **87**:127–131.

1225. Man, W. J., Y. Li, C. D. O'Connor, and D. C. Wilton. 1991. Conversion of citrate synthase into citryl-CoA lyase as a result of mutation of the active-site aspartic acid residue to glutamic acid. *Biochem. J.* **280**:521–526.

1226. Mandal, T. N., A. A. Mahdi, G. J. Sharples, and R. G. Lloyd. 1993. Resolution of Holliday intermediates in recombination and DNA repair: indirect suppression of *ruvA*, *ruvB*, and *ruvC* mutations. *J. Bacteriol.* **175**:4325–4334.

1227. Mandiyan, V., S. Tumminia, J. S. Wall, J. F. Hainfeld, and M. Boublik. 1989. Protein-induced conformational changes in 16 S ribosomal RNA during the initial assembly steps of the *Escherichia coli* 30 S ribosomal subunit. *J. Mol. Biol.* **210**:323–336.

1228. Mandrand-Berthelot, M. A., P. Ritzenthaler, and M. Mata-Gilsinger. 1984. Construction and expression of hybrid plasmids containing the structural gene of the *Escherichia coli* K-12 3-deoxy-2-oxo-D-gluconate transport system. *J. Bacteriol.* **160**:600–606.

1229. Manodori, A., G. Cecchini, I. Schroder, R. P. Gunsalus, M. T. Werth, and M. K. Johnson. 1992. [3Fe-4S] to [4Fe-4S] cluster conversion in *Escherichia coli* fumarate reductase by site-directed mutagenesis. *Biochemistry* **31**:2703–2712.

1230. Manson, M. D., V. Blank, G. Brade, and C. F. Higgins. 1986. Peptide chemotaxis in *E. coli* involves the Tap signal transducer and the dipeptide permease. *Nature* (London) **321**:253–256.

1231. Mantsala, P., and H. Zalkin. 1976. Active subunits of *Escherichia coli* glutamate synthase. *J. Bacteriol.* **126**:539–541.

1232. Maras, B., G. Sweeney, D. Barra, F. Bossa, and R. A. John. 1992. The amino acid sequence of glutamate decarboxylase from *Escherichia coli*. Evolutionary relationship between mammalian and bacterial enzymes. *Eur. J. Biochem.* **204**:93–98.

1233. Marceau, M., S. D. Lewis, C. L. Kojiro, K. Mountjoy, and J. A. Shafer. 1990. Disruption of active site interactions with pyridoxal 5'-phosphate and substrates by conservative replacements in the glycine-rich loop of *Escherichia coli* D-serine dehydratase. *J. Biol. Chem.* **265**:20421–20429.

1234. Marceau, M., S. D. Lewis, C. L. Kojiro, and J. A. Shafer. 1989. Contribution of a conserved arginine near the active site of *Escherichia coli* D-serine dehydratase to cofactor affinity and catalytic activity. *J. Biol. Chem.* **264**:2753–2757.

1235. Marceau, M., S. D. Lewis, and J. A. Shafer. 1988. The glycine-rich region of *Escherichia coli* D-serine dehydratase. Altered interactions with pyridoxal 5'-phosphate produced by substitution of aspartic acid for glycine. *J. Biol. Chem.* **263**:16934–16941.

1236. March, P. E., and M. Inouye. 1985. Characterization of the lep operon of *Escherichia coli*. Identification of the promoter and the gene upstream of the signal peptidase I gene. *J. Biol. Chem.* **260**:7206–7213.

1237. March, P. E., and M. Inouye. 1985. GTP-binding membrane protein of *Escherichia coli* with sequence homology to initiation factor 2 and elongation factors Tu and G. *Proc. Natl. Acad. Sci. USA* **82**:7500–7504.

1238. Marcus, M., and Y. S. Halpern. 1967. Genetic analysis of glutamate transport and glutamate decarboxylase in *Escherichia coli*. *J. Bacteriol.* **93**:1409–1415.

1239. Marcus, M., and Y. S. Halpern. 1969. Genetic analysis of the glutamate permease in *Escherichia coli* K-12. *J. Bacteriol.* **97**:1118–1128.

1240. Marek, L. E., and J. M. Henson. 1988. Cloning and expression of the *Escherichia coli* K-12 sad gene. *J. Bacteriol.* **170**:991–994.

1241. Marinus, M. G., N. R. Morris, D. Soll, and T. C. Kwong. 1975. Isolation and partial characterization of three *Escherichia coli* mutants with altered transfer ribonucleic acid methylases. *J. Bacteriol.* **122**:257–265.

1242. Marolda, C. L., and M. A. Valvano. 1993. Identification, expression, and DNA sequence of the GDP-mannose biosynthesis genes encoded by the O7 rfb gene cluster of strain VW187 (*Escherichia coli* O7:K1). *J. Bacteriol.* **175**:148–158.

1243. Marolewski, A., J. M. Smith, and S. J. Benkovic. 1994. Cloning and characterization of a new purine biosynthetic enzyme: a non-folate glycinamide ribonucleotide transformylase from *E. coli*. *Biochemistry* **33**:2531–2537.

1244. Marquardt, J. L., D. A. Siegele, R. Kolter, and C. T. Walsh. 1992. Cloning and sequencing of *Escherichia coli* murZ and purification of its product, a UDP-N-acetylglucosamine enolpyruvyl transferase. *J. Bacteriol.* **174**:5748–5752.

1245. Marszalek, J., and J. M. Kaguni. 1992. Defective replication activity of a dominant-lethal dnaB gene product from *Escherichia coli*. *J. Biol. Chem.* **267**:19334–19340.

1246. Martel, A., C. Bouthier de la Tour, and F. Le Goffic. 1987. Pyridoxal 5'phosphate binding site of *Escherichia coli* beta cystathionase and cystathionine gamma synthase comparison of their sequences. *Biochem. Biophys. Res. Commun.* **147**:565–571.

1247. Martin, W., H. Brinkmann, C. Savonna, and R. Cerff. 1993. Evidence for a chimeric nature of nuclear genomes: eubacterial origin of eukaryotic glyceraldehyde-3-phosphate dehydrogenase genes. *Proc. Natl. Acad. Sci. USA* **90**:8692–8696.

1248. Maruya, A., M. J. O'Connor, and K. Backman. 1987. Genetic separability of the chorismate mutase and prephenate dehydrogenase components of the *Escherichia coli tyrA* gene product. *J. Bacteriol.* 169:4852–4853.

1249. Marvil, D. K., and T. Leisinger. 1977. N-acetylglutamate synthase of *Escherichia coli*: purification, characterization, and molecular properties. *J. Biol. Chem.* 252:3295–3303.

1250. Masai, H., and K. Arai. 1988. Initiation of lagging-strand synthesis for pBR322 plasmid DNA replication in vitro is dependent on primosomal protein i encoded by dnaT. *J. Biol. Chem.* 263:15016–15023.

1251. Maskell, D. 1992. Protein sequence from downstream of *Escherichia coli* galK is homologous with galM from other organisms. *Mol. Microbiol.* 6:2211. (Letter.)

1252. Mason, S. W., and J. Greenblatt. 1991. Assembly of transcription elongation complexes containing the N protein of phage lambda and the *Escherichia coli* elongation factors NusA, NusB, NusG, and S10. *Genes Dev.* 5:1504–1512.

1253. Mason, S. W., J. Li, and J. Greenblatt. 1992. Direct interaction between two *Escherichia coli* transcription antitermination factors, NusB and ribosomal protein S10. *J. Mol. Biol.* 223:55–66.

1254. Masuda, Y., K. Miyakawa, Y. Nishimura, and E. Ohtsubo. 1993. *chpA* and *chpB*, *Escherichia coli* chromosomal homologs of the *pem* locus responsible for stable maintenance of plasmid R100. *J. Bacteriol.* 175:6850–6856.

1255. Mat-Jan, F., K. Y. Alam, and D. P. Clark. 1989. Mutants of *Escherichia coli* deficient in the fermentative lactate dehydrogenase. *J. Bacteriol.* 171:342–348.

1256. Matsuyama, A., H. Yamamoto, and E. Nakano. 1989. Cloning, expression, and nucleotide sequence of the *Escherichia coli* K-12 ackA gene. *J. Bacteriol.* 171:577–580.

1257. Matsuyama, S., Y. Fujita, K. Sagara, and S. Mizushima. 1992. Overproduction, purification and characterization of SecD and SecF, integral membrane components of the protein translocation machinery of *Escherichia coli*. *Biochim. Biophys. Acta* 1122:77–84.

1258. Matsuzawa, H., S. Asoh, K. Kunai, K. Muraiso, A. Takasuga, and T. Ohta. 1989. Nucleotide sequence of the *rodA* gene, responsible for the rod shape of *Escherichia coli*: rodA and the pbpA gene, encoding penicillin-binding protein 2, constitute the *rodA* operon. *J. Bacteriol.* 171:558–560.

1259. Matsuzawa, H., S. Ushiyama, Y. Koyama, and T. Ohta. 1984. *Escherichia coli* K-12 tolZ mutants tolerant to colicins E2, E3, D, Ia, and Ib: defect in generation of the electrochemical proton gradient. *J. Bacteriol.* 160:733–739.

1260. Mattheakis, L., L. Vu, F. Sor, and M. Nomura. 1989. Retroregulation of the synthesis of ribosomal proteins L14 and L24 by feedback repressor S8 in *Escherichia coli*. *Proc. Natl. Acad. Sci. USA* 86:448–452.

1261. Maupin, J. A., and K. T. Shanmugam. 1990. Genetic regulation of formate hydrogenlyase of *Escherichia coli*: role of the *fhlA* gene product as a transcriptional activator for a new regulatory gene, *fhlB*. *J. Bacteriol.* 172:4798–4806.

1262. Maurizi, M. R., W. P. Clark, Y. Katayama, S. Rudikoff, J. Pumphrey, B. Bowers, and S. Gottesman. 1990. Sequence and structure of Clp P, the proteolytic component of the ATP-dependent Clp protease of *Escherichia coli*. *J. Biol. Chem.* 265:12536–12545.

1263. Maurizi, M. R., W. P. Clark, S. H. Kim, and S. Gottesman. 1990. Clp P represents a unique family of serine proteases. *J. Biol. Chem.* 265:12546–12552.

1264. Maxon, M. E., J. Wigboldus, N. Brot, and H. Weissbach. 1990. Structure-function studies on *Escherichia coli* MetR protein, a putative prokaryotic leucine zipper protein. *Proc. Natl. Acad. Sci. USA* 87:7076–7079.

1265. May, G., P. Dersch, M. Haardt, A. Middendorf, and E. Bremer. 1990. The osmZ (bglY) gene encodes the DNA-binding protein H-NS (H1a), a component of the *Escherichia coli* K12 nucleoid. *Mol. Gen. Genet.* 224:81–90.

1266. Mazel, D., S. Pochet, and P. Marliere. 1994. Genetic characterization of polypeptide deformylase, a distinctive enzyme of eubacterial translation. *EMBO J.* 13:914–923.

1267. Mazur, S. J., and L. Grossman. 1991. Dimerization of *Escherichia coli* UvrA and its binding to undamaged and ultraviolet light damaged DNA. *Biochemistry* 30:4432–4443.

1268. McCalla, D. R., C. Kaiser, and M. H. L. Green. 1978. Genetics of nitrofurazone resistance in *Escherichia coli*. *J. Bacteriol.* 133:10–16.

1269. McCaman, M. T., and J. D. Gabe. 1986. The nucleotide sequence of the pepN gene and its overexpression in *Escherichia coli*. *Gene* 48:145–153.

1270. McClain, W. H., S. I. C. Blomfield, and B. I. Eisenstein. 1991. Roles of *fimB* and *fimE* in site-specific DNA inversion associated with phase variation of type 1 fimbriae in *Escherichia coli*. *J. Bacteriol.* 173:5308–5314.

1271. McClain, W. H., K. Foss, R. A. Jenkins, and J. Schneider. 1990. Nucleotides that determine *Escherichia coli* tRNA(Arg) and tRNA(Lys) acceptor identities revealed by analyses of mutant opal and amber suppressor tRNAs. *Proc. Natl. Acad. Sci. USA* 87:9260–9264.

1272. McClain, W. H., K. Foss, R. A. Jenkins, and J. Schneider. 1991. Rapid determination of nucleotides that define tRNA(Gly) acceptor identity. *Proc. Natl. Acad. Sci. USA* 88:6147–6151.

1273. McCormick, K. A., and B. D. Cain. 1991. Targeted mutagenesis of the b subunit of F1F0 ATP synthase in *Escherichia coli*: Glu-77 through Gln-85. *J. Bacteriol.* 173:7240–7248.

1274. McCulloch, R., L. W. Coggins, S. D. Colloms, and D. J. Sherratt. 1994. Xer-mediated site-specific recombination at cer generates Holliday junctions in vivo. *EMBO J.* 13:1844–1855.

1275. McGrath, M. E., T. Erpel, M. F. Browner, and R. J. Fletterick. 1991. Expression of the protease inhibitor ecotin and its co-crystallization with trypsin. *J. Mol. Biol.* 222:139–142.

1276. McGrath, M. E., W. M. Hines, J. A. Sakanari, R. J. Fletterick, and C. S. Craik. 1991. The sequence and reactive site of ecotin. A general inhibitor of pancreatic serine proteases from *Escherichia coli*. *J. Biol. Chem.* 266:6620–6625.

1277. McNally, D. F., and P. Matsumura. 1991. Bacterial chemotaxis signaling complexes: formation of a CheA/CheW complex enhances autophosphorylation and affinity for CheY. *Proc. Natl. Acad. Sci. USA* 88:6269–6273.

1278. Meador, J., B. Cannon, V. J. Cannistraro, and D. Kennell. 1990. Purification and characterization of *Escherichia coli* RNase I. Comparisons with RNase M. *Eur. J. Biochem.* 187:549–553.

1279. Mechulam, Y., F. Dardel, D. Le Corre, S. Blanquet, and G. Fayat. 1991. Lysine 335, part of the KMSKS signature sequence, plays a crucial role in the amino acid activation catalysed by the methionyl-tRNA synthetase from *Escherichia coli*. *J. Mol. Biol.* 217:465–475.

1280. Medina, V., R. Pontarollo, D. Glaeske, H. Tabel, and H. Goldie. 1990. Sequence of the *pckA* gene of *Escherichia coli* K-12: relevance to genetic and allosteric regulation and homology of *E. coli* phosphoenolpyruvate carboxykinase with the enzymes from *Trypanosoma brucei* and *Saccharomyces cerevisiae*. *J. Bacteriol.* 172:7151–7156.

1281. Meganathan, R., and R. Bentley. 1983. Thiamine pyrophosphate requirement for o-succinylbenzoic acid synthesis in *Escherichia coli* and evidence for an intermediate. *J. Bacteriol.* 153:739–746.

1282. Mei, B. G., and H. Zalkin. 1990. Amino-terminal deletions define a glutamine amide transfer domain in glutamine phosphoribosylpyrophosphate amidotransferase and other PurF-type amidotransferases. *J. Bacteriol.* 172:3512–3514.

1283. Meier-Dieter, U., K. Barr, R. Starman, L. Hatch, and P. D. Rick. 1992. Nucleotide sequence of the *Escherichia coli* rfe gene involved in the synthesis of enterobacterial common antigen. Molecular cloning of the rfe-rff gene cluster. *J. Biol. Chem.* 267:746–753.

1284. Meier-Dieter, U., R. Starman, K. Barr, H. Mayer, and P. D. Rick. 1990. Biosynthesis of enterobacterial common antigen in *Escherichia coli*. Biochemical characterization of Tn10 insertion mutants defective in enterobacterial common antigen synthesis. *J. Biol. Chem.* 265:13490–13497.

1285. Meinnel, T., and S. Blanquet. 1993. Evidence that peptide deformylase and methionyl-tRNA(fMet) formyltransferase are encoded within the same operon in *Escherichia coli*. *J. Bacteriol.* 175:7737–7740.

1286. Meinnel, T., J. M. Guillon, Y. Mechulam, and S. Blanquet. 1993. The *Escherichia coli fmt* gene, encoding methionyl-tRNA(fMet) formyltransferase, escapes metabolic control. *J. Bacteriol.* 175:993–1000.

1287. Meinnel, T., E. Schmitt, Y. Mechulam, and S. Blanquet. 1992. Structural and biochemical characterization of the *Escherichia coli argE* gene product. *J. Bacteriol.* 174:2323–2331.

1288. Mejean, V., C. Iobbi-Nivol, M. Lepelletier, G. Giordano, M. Chippaux, and M. C. Pascal. 1994. TMAO anaerobic respiration in *Escherichia coli*: involvement of the tor operon. *Mol. Microbiol.* 11:1169–1179.

1289. Melville, S. B., and R. P. Gunsalus. 1990. Mutations in fnr that alter anaerobic regulation of electron transport-associated genes in *Escherichia coli*. *J. Biol. Chem.* 265:18733–18736.

1290. Melzer, M., and L. Heide. 1994. Characterization of polyprenyldiphosphate:4-hydroxybenzoate polyprenyltransferase from *Escherichia coli*. *Biochim. Biophys. Acta* 1212:93–102.

1291. Mendel-Hartvig, J., and R. A. Capaldi. 1991. Catalytic site nucleotide and inorganic phosphate dependence of the conformation of the epsilon subunit in *Escherichia coli* adenosinetriphosphatase. *Biochemistry* 30:1278–1284.

1292. Mendonca, V. M., K. Kaiser-Rogers, and S. W. Matson. 1993. Double helicase II (uvrD)-helicase IV (helD) deletion mutants are defective in the recombination pathways of *Escherichia coli*. *J. Bacteriol.* 175:4641–4651.

1293. Meng, S. Y., and G. N. Bennett. 1992. Nucleotide sequence of the *Escherichia coli* cad operon: a system for neutralization of low extracellular pH. *J. Bacteriol.* 174:2659–2669.

1294. Mengin-Lecreulx, D., C. Parquet, L. R. Desviat, J. Pla, B. Flouret, J. A. Ayala, and J. van Heijenoort. 1989. Organization of the *murE-murG* region of *Escherichia coli*: identification of the *murD* gene encoding the D-glutamic-acid-adding enzyme. *J. Bacteriol.* 171:6126–6134.

1295. Mengin-Lecreulx, D., L. Texier, M. Rousseau, and J. van Heijenoort. 1991. The *murG* gene of *Escherichia coli* codes for the UDP-N-acetylglucosamine: N-acetylmuramyl-(pentapeptide) pyrophosphoryl-undecaprenol N-acetylglucosamine transferase involved in the membrane steps of peptidoglycan synthesis. *J. Bacteriol.* 173:4625–4636.

1296. Mengin-Lecreulx, D., and J. van Heijenoort. 1993. Identification of the *glmU* gene encoding N-acetylglucosamine-1-phosphate uridyltransferase in *Escherichia coli*. *J. Bacteriol.* 175:6150–6157.

1297. Menon, K. P., and N. L. Lee. 1990. Activation of ara operons by a truncated AraC protein does not require inducer. *Proc. Natl. Acad. Sci. USA* 87:3708–3712.

1298. Menon, N. K., C. Y. Chatelus, M. Dervartanian, J. C. Wendt, K. T. Shanmugam, H. D. Peck, Jr., and A. E. Przybyla. 1994. Cloning, sequencing, and mutational analysis of the *hyb* operon encoding *Escherichia coli* hydrogenase 2. *J. Bacteriol.* 176:4416–4423.

1299. Menon, N. K., J. Robbins, H. D. Peck, Jr., C. Y. Chatelus, E. S. Choi, and A. E. Przybyla. 1990. Cloning and sequencing of a putative *Escherichia coli* [NiFe] hydrogenase-1 operon containing six open reading frames. *J. Bacteriol.* 172:1969–1977.

1300. Menon, N. K., J. Robbins, J. C. Wendt, K. T. Shanmugam, and A. E. Przybyla. 1991. Mutational analysis and characterization of the *Escherichia coli hya* operon, which encodes [NiFe] hydrogenase 1. *J. Bacteriol.* 173:4851–4861.

1301. Merino, E., P. Balbas, F. Recillas, B. Becerril, F. Valle, and F. Bolivar. 1992. Carbon regulation and the role in nature of the *Escherichia coli* penicillin acylase (pac) gene. *Mol. Microbiol.* 6:2175–2182.

1302. Merle, M., V. Trezeguet, P. V. Graves, D. Andrews, K. H. Muench, and B. Labouesse. 1986. Tryptophanyl adenylate formation by tryptophanyl-tRNA synthetase from *Escherichia coli. Biochemistry* 25:1115–1123.

1303. Metcalf, W. W., P. M. Steed, and B. L. Wanner. 1990. Identification of phosphate starvation-inducible genes in *Escherichia coli* K-12 by DNA sequence analysis of *psi::lacZ*(Mu d1) transcriptional fusions. *J. Bacteriol.* 172:3191–3200.

1304. Metcalf, W. W., and B. L. Wanner. 1991. Involvement of the *Escherichia coli phn* (*psiD*) gene cluster in assimilation of phosphorus in the form of phosphonates, phosphite, P$_i$ esters, and P$_i$. *J. Bacteriol.* 173:587–600.

1305. Metcalf, W. W., and B. L. Wanner. 1993. Evidence for a fourteen-gene, phnC to phnP locus for phosphonate metabolism in *Escherichia coli. Gene* 129:27–32.

1306. Metcalf, W. W., and B. L. Wanner. 1993. Mutational analysis of an *Escherichia coli* fourteen-gene operon for phosphonate degradation, using Tn*phoA*′ elements. *J. Bacteriol.* 175:3430–3442.

1307. Metzer, E., and Y. S. Halpern. 1990. In vivo cloning and characterization of the *gabCTDP* gene cluster of *Escherichia coli* K-12. *J. Bacteriol.* 172:3250–3256.

1308. Metzger, S., I. B. Dror, E. Aizenman, G. Schreiber, M. Toone, J. D. Friesen, M. Cashel, and G. Glaser. 1988. The nucleotide sequence and characterization of the relA gene of *Escherichia coli. J. Biol. Chem.* 263:15699–15704.

1309. Metzger, S., E. Sarubbi, G. Glaser, and M. Cashel. 1989. Protein sequences encoded by the relA and the spoT genes of *Escherichia coli* are interrelated. *J. Biol. Chem.* 264:9122–9125.

1310. Meyer, C. R., P. Ghosh, E. Remy, and J. Preiss. 1992. Cloning, expression, and nucleotide sequence of a mutant glgC gene from *Escherichia coli* B. *J. Bacteriol.* 174:4509–4512.

1311. Meyer, E., N. J. Leonard, B. Bhat, J. Stubbe, and J. M. Smith. 1992. Purification and characterization of the purE, purK, and purC gene products: identification of a previously unrecognized energy requirement in the purine biosynthetic pathway. *Biochemistry* 31:5022–5032.

1312. Michaeli, S., and E. Z. Ron. 1984. Expression of the metA gene of *Escherichia coli* K-12 in recombinant plasmids. *FEMS Microbiol. Lett.* 23:125–129.

1313. Michaels, M. L., C. Cruz, and J. H. Miller. 1990. mutA and mutC: two mutator loci in *Escherichia coli* that stimulate transversions. *Proc. Natl. Acad. Sci. USA* 87:9211–9215.

1314. Michaels, M. L., C. W. Kim, D. A. Matthews, and J. H. Miller. 1990. *Escherichia coli* thymidylate synthase: amino acid substitutions by suppression of amber nonsense mutations. *Proc. Natl. Acad. Sci. USA* 87:3957–3961.

1315. Michelsen, U., M. Bosl, T. Dingermann, and H. Kersten. 1989. The *tyrT* locus of *Escherichia coli* exhibits a regulatory function for glycine metabolism. *J. Bacteriol.* 171:5987–5994.

1316. Mikhailov, A. M., E. A. Smirnova, V. L. Tsuprun, I. V. Tagunova, B. K. Vainshtein, E. V. Linkova, A. A. Komissarov, Z. Z. Siprashvili, and A. S. Mironov. 1992. Isolation, crystallization in the macrogravitation field, preliminary X-ray investigation of uridine phosphorylase from *Escherichia coli* K-12. *Biochem. Int.* 26:607–615.

1317. Mikuni, O., K. Ito, J. Moffat, K. Matsumura, K. McCaughan, T. Nobukuni, W. Tate, and Y. Nakamura. 1994. Identification of the prfC gene, which encodes peptide-chain-release factor 3 of *Escherichia coli. Proc. Natl. Acad. Sci. USA* 91:5798–5802.

1318. Mikuni, O., K. Kawakami, and Y. Nakamura. 1991. Sequence and functional analysis of mutations in the gene encoding peptide-chain-release factor 2 of *Escherichia coli. Biochimie* 73:1509–1516.

1319. Miles, J. S., and J. R. Guest. 1987. Molecular genetic aspects of the citric acid cycle of *Escherichia coli. Biochem. Soc. Symp.* 54:45–65.

1320. Milkman, R. 1994. An *Escherichia coli* homologue of eukaryotic potassium channel proteins. *Proc. Natl. Acad. Sci. USA* 91:3510–3514.

1321. Miller, A. D., L. C. Packman, G. J. Hart, P. R. Alefounder, C. Abell, and A. R. Battersby. 1989. Evidence that pyridoxal phosphate modification of lysine residues (Lys-55 and Lys-59) causes inactivation of hydroxymethylbilane synthase (porphobilinogen deaminase). *Biochem. J.* 262:119–124.

1322. Miller, A. W., and L. C. Kuo. 1990. Ligand-induced isomerizations of *Escherichia coli* ornithine transcarbamoylase. An ultraviolet difference analysis. *J. Biol. Chem.* 265:15023–15027.

1323. Miller, W. T., Y. M. Hou, and P. Schimmel. 1991. Mutant aminoacyl-tRNA synthetase that compensates for a mutation in the major identity determinant of its tRNA. *Biochemistry* 30:2635–2641.

1324. Mimura, C. S., A. Admon, K. A. Hurt, and G. F. Ames. 1990. The nucleotide-binding site of HisP, a membrane protein of the histidine permease. Identification of amino acid residues photoaffinity labeled by 8-azido-ATP. *J. Biol. Chem.* 265:19535–19542.

1325. Minagawa, J., T. Mogi, R. B. Gennis, and Y. Anraku. 1992. Identification of heme and copper ligands in subunit I of the cytochrome bo complex in *Escherichia coli. J. Biol. Chem.* 267:2096–2104.

1326. Minghetti, K. C., V. C. Goswitz, N. E. Gabriel, J. J. Hill, C. A. Barassi, C. D. Georgiou, S. I. Chan, and R. B. Gennis. 1992. Modified, large-scale purification of the cytochrome o complex (bo-type oxidase) of *Escherichia coli* yields a two

heme/one copper terminal oxidase with high specific activity. *Biochemistry* 31:6917–6924.

1327. Misra, R., and S. A. Benson. 1989. A novel mutation, *cog*, which results in production of a new porin protein (OmpG) of *Escherichia coli* K-12. *J. Bacteriol.* 171:4105–4111.

1328. Missiakas, D., C. Georgopoulos, and S. Raina. 1993. Identification and characterization of the *Escherichia coli* gene dsbB, whose product is involved in the formation of disulfide bonds in vivo. *Proc. Natl. Acad. Sci. USA* 90:7084–7088.

1329. Missiakas, D., C. Georgopoulos, and S. Raina. 1993. The *Escherichia coli* heat shock gene *htpY*: mutational analysis, cloning, sequencing, and transcriptional regulation. *J. Bacteriol.* 175:2613–2624.

1330. Missiakas, D., C. Georgopoulos, and S. Raina. 1994. The *Escherichia coli* dsbC (xprA) gene encodes a periplasmic protein involved in disulfide bond formation. *EMBO J.* 13:2013–2020.

1331. Mitchell, L. W., and E. K. Jaffe. 1993. Porphobilinogen synthase from *Escherichia coli* is a Zn(II) metalloenzyme stimulated by Mg(II). *Arch. Biochem. Biophys.* 300:169–177.

1332. Mitchell, P., M. Osswald, D. Schueler, and R. Brimacombe. 1990. Selective isolation and detailed analysis of intra-RNA cross-links induced in the large ribosomal subunit of E. coli: a model for the tertiary structure of the tRNA binding domain in 23S RNA. *Nucleic Acids Res.* 18:4325–4333.

1333. Miyakawa, T., H. Matsuzawa, M. Matsuhashi, and Y. Sugino. 1972. Cell wall peptidoglycan mutants of *Escherichia coli* K-12: existence of two clusters of genes mra and mrb, for cell wall peptidoglycan biosynthesis. *J. Bacteriol.* 112:950–958.

1334. Miyamoto, K., K. Nishimura, T. Masuda, H. Tsuji, and H. Inokuchi. 1992. Accumulation of protoporphyrin IX in light-sensitive mutants of *Escherichia coli. FEBS Lett.* 310:246–248.

1335. Moazed, D., J. M. Robertson, and H. F. Noller. 1988. Interaction of elongation factors EF-G and EF-Tu with a conserved loop in 23S RNA. *Nature* (London) 334:362–364.

1336. Moine, H., C. Bienaime, M. Mougel, J. Reinbolt, J. P. Ebel, C. Ehresmann, and B. Ehresmann. 1988. Crosslinking of ribosomal protein S18 to 16 S RNA in E. coli ribosomal 30 S subunits by the use of a reversible crosslinking agent: trans-diamminedichloroplatinum(II). *FEBS Lett.* 228:1–6.

1337. Moine, H., P. Romby, M. Springer, M. Grunberg-Manago, J. P. Ebel, B. Ehresmann, and C. Ehresmann. 1990. *Escherichia coli* threonyl-tRNA synthetase and tRNA(Thr) modulate the binding of the ribosome to the translational initiation site of the thrS mRNA. *J. Mol. Biol.* 216:299–310.

1338. Monroe, R. S., J. Ostrowski, M. M. Hryniewicz, and N. M. Kredich. 1990. In vitro interactions of CysB protein with the cysK and cysJIH promoter regions of *Salmonella typhimurium. J. Bacteriol.* 172:6919–6929.

1339. Moore, R. C., and S. M. Boyle. 1990. Nucleotide sequence and analysis of the speA gene encoding biosynthetic arginine decarboxylase in *Escherichia coli. J. Bacteriol.* 172:4631–4640.

1340. Morad, I., D. Chapman-Shimshoni, M. Amitsur, and G. Kaufmann. 1993. Functional expression and properties of the tRNA(Lys)-specific core anticodon nuclease encoded by *Escherichia coli* prrC. *J. Biol. Chem.* 268:26842–26849.

1341. Moralejo, P., S. M. Egan, E. Hidalgo, and J. Aguilar. 1993. Sequencing and characterization of a gene cluster encoding the enzymes for L-rhamnose metabolism in *Escherichia coli. J. Bacteriol.* 175:5585–5594.

1342. Morel, P., J. A. Hejna, S. D. Ehrlich, and E. Cassuto. 1993. Antipairing and strand transferase activities of E. coli helicase II (UvrD). *Nucleic Acids Res.* 21:3205–3209.

1343. Morimyo, M., E. Hongo, H. Hama-Inaba, and I. Machida. 1992. Cloning and characterization of the mvrC gene of *Escherichia coli* K-12 which confers resistance against methyl viologen toxicity. *Nucleic Acids Res.* 20:3159–3165.

1344. Morowitz, H. J., and D. C. Wallace. 1973. Genome size and life cycle of the mycoplasm. *Ann. N. Y. Acad. Sci.* 225:62–73.

1345. Morozov, G. I., L. I. Nosova, S. F. Biketov, A. G. Valiaev, and I. V. Domaradskii. 1994. Biochemical bases for the effect of combining kanamycin and nitrofuran resistance genes in *Escherichia coli* cells. *Mol. Gen. Mikrobiol. Virusol.* 61:11–14. (In Russian.)

1346. Morrison, T. B., and J. S. Parkinson. 1994. Liberation of an interaction domain from the phosphotransfer region of CheA, a signaling kinase of *Escherichia coli. Proc. Natl. Acad. Sci. USA* 91:5485–5489.

1347. Mortensen, K. K., N. R. Nyengaard, J. W. Hershey, S. Laalami, and H. U. Sperling-Petersen. 1991. Superexpression and fast purification of E. coli initiation factor IF2. *Biochimie* 73:983–989.

1348. Mortensen, L., G. Dandanell, and K. Hammer. 1989. Purification and characterization of the deoR repressor of *Escherichia coli. EMBO J.* 8:325–331.

1349. Mosteller, R. D. 1978. Evidence that glucose starvation-sensitive mutants are altered in the relB locus. *J. Bacteriol.* 133:1034–1037.

1350. Motojima, K., I. Yamato, and Y. Anraku. 1978. Proline transport carrier-defective mutants of *Escherichia coli* K-12: properties and mapping. *J. Bacteriol.* 136:5–9.

1351. Motojima, K., I. Yamato, Y. Anraku, A. Nishimura, and Y. Hirota. 1979. Amplification and characterization of the proline transport carrier of *Escherichia coli* K-12 by using proT$^+$ hybrid plasmids. *Proc. Natl. Acad. Sci. USA* 76:6255–6259.

1352. Mottl, H., and W. Keck. 1991. Purification of penicillin-binding protein 4 of *Escherichia coli* as a soluble protein by dye-affinity chromatography. *Eur. J. Biochem.* 200:767–773.

1353. Mottl, H., P. Nieland, G. de Kort, J. J. Wierenga, and W. Keck. 1992. Deletion of an additional domain located between SXXK and SXN active-site fingerprints in penicillin-binding protein 4 from *Escherichia coli. J. Bacteriol.* 174:3261–3269.

1354. Mottl, H., P. Terpstra, and W. Keck. 1991. Penicillin-binding protein 4 of *Escherichia coli* shows a novel type of primary structure among penicillin-interacting proteins. *FEMS Microbiol. Lett.* **62**:213–220.

1355. Movva, N. R., E. Katz, P. L. Asdourian, Y. Hirota, and M. Inouye. 1978. Gene dosage effects of the structural gene for a lipoprotein of the *Escherichia coli* outer membrane. *J. Bacteriol.* **133**:81–84.

1356. Mukherjee, A., and J. Lutkenhaus. 1994. Guanine nucleotide-dependent assembly of FtsZ into filaments. *J. Bacteriol.* **176**:2754–2758.

1357. Mukherjee, J. J., and E. E. Dekker. 1987. Purification, properties, and N-terminal amino acid sequence of homogeneous *Escherichia coli* 2-amino-3-ketobutyrate CoA ligase, a pyridoxal phosphate-dependent enzyme. *J. Biol. Chem.* **262**:14441–14447.

1358. Mukherjee, J. J., and E. E. Dekker. 1990. 2-Amino-3-ketobutyrate CoA ligase of *Escherichia coli*: stoichiometry of pyridoxal phosphate binding and location of the pyridoxyllysine peptide in the primary structure of the enzyme. *Biochim. Biophys. Acta* **1037**:24–29.

1359. Mukhopadhyay, S., and H. E. Schellhorn. 1994. Induction of *Escherichia coli* hydroperoxidase I by acetate and other weak acids. *J. Bacteriol.* **176**:2300–2307.

1360. Mulder, E., M. El'Bouhali, E. Pas, and C. L. Woldringh. 1990. The *Escherichia coli* minB mutation resembles gyrB in defective nucleoid segregation and decreased negative supercoiling of plasmids. *Mol. Gen. Genet.* **221**:87–93.

1361. Mulder, E., C. L. Woldringh, F. Tetart, and J. P. Bouche. 1992. New *minC* mutations suggest different interactions of the same region of division inhibitor MinC with proteins specific for *minD* and *dicB* coinhibition pathways. *J. Bacteriol.* **174**:35–39.

1362. Muller, B., I. R. Tsaneva, and S. C. West. 1993. Branch migration of Holliday junctions promoted by the *Escherichia coli* RuvA and RuvB proteins. I. Comparison of RuvAB-and RuvB-mediated reactions. *J. Biol. Chem.* **268**:17179–17184.

1363. Muller, M. M., A. Vianney, J. C. Lazzaroni, R. E. Webster, and R. Portalier. 1993. Membrane topology of the *Escherichia coli* TolR protein required for cell envelope integrity. *J. Bacteriol.* **175**:6059–6061.

1364. Muller, V., C. J. Jones, I. Kawagishi, S. Aizawa, and R. M. Macnab. 1992. Characterization of the *fliE* genes of *Escherichia coli* and *Salmonella typhimurium* and identification of the FliE protein as a component of the flagellar hook-basal body complex. *J. Bacteriol.* **174**:2298–2304.

1365. Mullins, L. S., C. J. Lusty, and F. M. Raushel. 1991. Alterations in the energetics of the carbamoyl phosphate synthetase reaction by site-directed modification of the essential sulfhydryl group. *J. Biol. Chem.* **266**:8236–8240.

1366. Munch-Petersen, A., and N. Jensen. 1990. Analysis of the regulatory region of the *Escherichia coli* nupG gene, encoding a nucleoside-transport protein. *Eur. J. Biochem.* **190**:547–551.

1367. Munch-Peterson, A., B. Mygind, A. Nicolaisen, and N. J. Pihl. 1979. Nucleoside transport in cells and membrane vesicles from *Escherichia coli* K-12. *J. Biol. Chem.* **254**:3730–3737.

1368. Munoa, F. J., K. W. Miller, R. Beers, M. Graham, and H. C. Wu. 1991. Membrane topology of *Escherichia coli* prolipoprotein signal peptidase (signal peptidase II). *J. Biol. Chem.* **266**:17667–17672.

1369. Munro, A. W., G. Y. Ritchie, A. J. Lamb, R. M. Douglas, and I. R. Booth. 1991. The cloning and DNA sequence of the gene for the glutathione-regulated potassium-efflux system KefC of *Escherichia coli*. *Mol. Microbiol.* **5**:607–616.

1370. Murakami, H., K. Kita, H. Oya, and Y. Anraku. 1984. Chromosomal location of the *Escherichia coli* cytochrome b556 gene, cybA. *Mol. Gen. Genet.* **196**:1–5.

1371. Muralikrishna, P., and E. Wickstrom. 1989. Inducible high expression of the *Escherichia coli* infC gene subcloned behind a bacteriophage T7 promoter. *Gene* **80**:369–374.

1372. Muramatsu, S., and T. Mizuno. 1990. Nucleotide sequence of the region encompassing the int gene of a cryptic prophage and the dnaY gene flanked by a curved DNA sequence of *Escherichia coli* K12. *Mol. Gen. Genet.* **220**:325–328.

1373. Murase, S., J. S. Takagi, Y. Higashi, H. Imaishi, N. Yumoto, and M. Tokushige. 1991. Activation of aspartase by site-directed mutagenesis. *Biochem. Biophys. Res. Commun.* **177**:414–419.

1374. Murgola, E. J., and E. A. Adelberg. 1970. Mutants of *Escherichia coli* K-12 with an altered glutamyl-transfer ribonucleic acid. *J. Bacteriol.* **103**:178–183.

1375. Murgola, E. J., and G. Guarneros. 1991. Ribosomal RNA and peptidyl-tRNA hydrolase: a peptide chain termination model for lambda bar RNA inhibition. *Biochimie* **73**:1573–1578.

1376. Murooka, Y., T. Higashiura, and T. Harada. 1978. Genetic mapping of tyramine oxidase and arylsulfatase genes and their regulation in intergeneric hybrids of enteric bacteria. *J. Bacteriol.* **136**:714–722.

1377. Murphy, C. K., V. I. Kalve, and P. E. Klebba. 1990. Surface topology of the *Escherichia coli* K-12 ferric enterobactin receptor. *J. Bacteriol.* **172**:2736–2746.

1378. Nagai, H., H. Yuzawa, and T. Yura. 1991. Interplay of two cis-acting mRNA regions in translational control of sigma 32 synthesis during the heat shock response of *Escherichia coli*. *Proc. Natl. Acad. Sci. USA* **88**:10515–10519.

1379. Nagao, Y., T. Nakada, M. Imoto, T. Shimamoto, S. Sakai, M. Tsuda, and T. Tsuchiya. 1988. Purification and analysis of the structure of alpha-galactosidase from *Escherichia coli*. *Biochem. Biophys. Res. Commun.* **151**:236–241.

1380. Nagasawa, S., K. Ishige, and T. Mizuno. 1993. Novel members of the two-component signal transduction genes in *Escherichia coli*. *J. Biochem.* (Tokyo) **114**:350–357.

1381. Nagasawa, S., S. Tokishita, H. Aiba, and T. Mizuno. 1992. A novel sensor-regulator protein that belongs to the homologous family of signal-transduction proteins involved in adaptive responses in *Escherichia coli*. *Mol. Microbiol.* **6**:799–807.

1382. Nagel de Zwaig, R., N. Zwaig, T. Isturiz, and R. S. Sanchez. 1973. Mutations affecting gluconate metabolism in *Escherichia coli*. *J. Bacteriol.* **114**:463–468.

1383. Naggert, J., M. L. Narasimhan, L. DeVeaux, H. Cho, Z. I. Randhawa, J. E. Cronan, Jr., B. N. Green, and S. Smith. 1991. Cloning, sequencing, and characterization of *Escherichia coli* thioesterase II. *J. Biol. Chem.* **266**:11044–11050.

1384. Nagy, P. L., G. M. McCorkle, and H. Zalkin. 1993. *purU*, a source of formate for *purT*-dependent phosphoribosyl-N-formylglycinamide synthesis. *J. Bacteriol.* **175**:7066–7073.

1385. Nakabeppu, Y., H. Kondo, and M. Sekiguchi. 1984. Cloning and characterization of the alkA gene of *Escherichia coli* that encodes 3-methyladenine DNA glycosylase II. *J. Biol. Chem.* **259**:13723–13729.

1386. Nakagawa, Y., M. Setaka, and S. Nojima. 1991. Detergent-resistant phospholipase A1 from *Escherichia coli* membranes. *Methods Enzymol.* **197**:309–315.

1387. Nakahigashi, K., K. Miyamoto, K. Nishimura, and H. Inokuchi. 1992. Isolation and characterization of a light-sensitive mutant of *Escherichia coli* K-12 with a mutation in a gene that is required for the biosynthesis of ubiquinone. *J. Bacteriol.* **174**:7352–7359.

1388. Nakamura, H. 1979. Novel acriflavin resistance genes, acrC and acrD, in *Escherichia coli* K-12. *J. Bacteriol.* **139**:8–12.

1389. Nakamura, H. 1979. Specific proline accumulation in an acrA mutant of *Escherichia coli* K12 grown in salt-hypertonic medium. *J. Gen. Microbiol.* **113**:425–427.

1390. Nakamura, H., I. Yamato, Y. Anraku, L. Lemieux, and R. B. Gennis. 1990. Expression of cyoA and cyoB demonstrates that the CO-binding heme component of the *Escherichia coli* cytochrome o complex is in subunit I. *J. Biol. Chem.* **265**:11193–11197.

1391. Nakamura, M., M. Yamada, Y. Hirota, K. Sugimoto, A. Oka, and M. Takanami. 1981. Nucleotide sequence of the asnA gene coding for asparagine synthetase of *E. coli* K-12. *Nucleic Acids Res.* **9**:4669–4676.

1392. Nakamura, Y., and K. Ito. 1993. Control and function of lysyl-tRNA synthetases: diversity and co-ordination. *Mol. Microbiol.* **10**:225–231.

1393. Nakamura, Y., and K. Kawakami. 1992. Overproduction and purification of lysyl-tRNA synthetase encoded by the herC gene of *E. coli*. *Biochimie* **74**:581–584.

1394. Nakashima, K., A. Sugiura, and T. Mizuno. 1993. Functional reconstitution of the putative *Escherichia coli* osmosensor, KdpD, into liposomes. *J. Biochem.* (Tokyo) **114**:615–621.

1395. Nakashima, K., A. Sugiura, H. Momoi, and T. Mizuno. 1992. Phosphotransfer signal transduction between two regulatory factors involved in the osmoregulated kdp operon in *Escherichia coli*. *Mol. Microbiol.* **6**:1777–1784.

1396. Nakata, A., M. Amemura, and K. Makino. 1989. Unusual nucleotide arrangement with repeated sequences in the *Escherichia coli* K-12 chromosome. *J. Bacteriol.* **171**:3553–3556.

1397. Naom, I. S., S. J. Morton, D. R. Leach, and R. G. Lloyd. 1989. Molecular organization of sbcC, a gene that affects genetic recombination and the viability of DNA palindromes in *Escherichia coli* K-12. *Nucleic Acids Res.* **17**:8033–8045.

1398. Navarro, C., L. Wu, and M. Mandrand-Berthelot. 1993. The nik operon of *Escherichia coli* encodes a periplasmic binding-protein-dependent transport system for nickel. *Mol. Microbiol.* **9**:1181–1191.

1399. Neal, B. L., G. C. Tsiolis, M. W. Heuzenroeder, P. A. Manning, and P. R. Reeves. 1991. Molecular cloning and expression in *Escherichia coli* K-12 of chromosomal genes determining the O antigen of an *E. coli* O2:K1 strain. *FEMS Microbiol. Lett.* **82**:345–351.

1400. Neely, M. N., C. L. Dell, and E. R. Olson. 1994. Roles of LysP and CadC in mediating the lysine requirement for acid induction of the *Escherichia coli* cad operon. *J. Bacteriol.* **176**:3278–3285.

1401. Negre, D., J. C. Cortay, I. G. Old, A. Galinier, C. Richaud, I. Saint Girons, and A. J. Cozzone. 1991. Overproduction and characterization of the iclR gene product of *Escherichia coli* K-12 and comparison with that of *Salmonella typhimurium* LT2. *Gene* **97**:29–37.

1402. Nellemann, L. J., F. Holm, T. Atlung, and F. G. Hansen. 1989. Cloning and characterization of the *Escherichia coli* phosphoglycerate kinase (pgk) gene. *Gene* **77**:185–191.

1403. Nelms, J., R. M. Edwards, J. Warwick, and I. Fotheringham. 1992. Novel mutations in the pheA gene of *Escherichia coli* K-12 which result in highly feedback inhibition-resistant variants of chorismate mutase/prephenate dehydratase. *Appl. Environ. Microbiol.* **58**:2592–2598.

1404. Neuwald, A. F., D. E. Berg, and G. V. Stauffer. 1992. Mutational analysis of the *Escherichia coli* serB promoter region reveals transcriptional linkage to a downstream gene. *Gene* **120**:1–9.

1405. Neuwald, A. F., B. R. Krishnan, I. Brikun, S. Kulakauskas, K. Suziedelis, T. Tomcsanyi, T. S. Leyh, and D. E. Berg. 1992. cysQ, a gene needed for cysteine synthesis in *Escherichia coli* K-12 only during aerobic growth. *J. Bacteriol.* **174**:415–425.

1406. Neuwald, A. F., and G. V. Stauffer. 1989. An *Escherichia coli* membrane protein with a unique signal sequence. *Gene* **82**:219–228.

1407. Newman, T., P. Friden, A. Sutton, and M. Freundlich. 1982. Cloning and expression of the ilvB genes of *Escherichia coli* K-12. *Mol. Gen. Genet.* **186**:378–384.

1408. Nguyen, L. H., D. B. Jensen, N. E. Thompson, D. R. Gentry, and R. R. Burgess. 1993. In vitro functional characterization of overproduced *Escherichia coli* katF/rpoS gene product. *Biochemistry* **32**:11112–11117.

1409. **Nichols, B. P., and J. M. Green.** 1992. Cloning and sequencing of *Escherichia coli ubiC* and purification of chorismate lyase. *J. Bacteriol.* 174:5309–5316.

1410. **Niegemann, E., A. Schulz, and K. Bartsch.** 1993. Molecular organization of the *Escherichia coli gab* cluster: nucleotide sequence of the structural genes gabD and gabP and expression of the GABA permease gene. *Arch. Microbiol.* 160:454–460.

1411. **Nieto, J. M., M. Carmona, S. Bolland, Y. Jubete, F. de la Cruz, and A. Juarez.** 1991. The hha gene modulates haemolysin expression in *Escherichia coli. Mol. Microbiol.* 5:1285–1293.

1412. **Niki, H., A. Jaffe, R. Imamura, T. Ogura, and S. Hiraga.** 1991. The new gene mukB codes for a 177 kd protein with coiled-coil domains involved in chromosome partitioning of *E. coli. EMBO J.* 10:183–193.

1413. **Nikkila, H., R. B. Gennis, and S. G. Sligar.** 1991. Cloning and expression of the gene encoding the soluble cytochrome b562 of *Escherichia coli. Eur. J. Biochem.* 202:309–313.

1414. **Nilsson, L., H. Verbeek, E. Vijgenboom, C. van Drunen, A. Vanet, and L. Bosch.** 1992. FIS-dependent *trans* activation of stable RNA operons of *Escherichia coli* under various growth conditions. *J. Bacteriol.* 174:921–929.

1415. **Ninfa, A. J., and B. Magasanik.** 1986. Covalent modification of the glnG product, NRI, by the glnL product, NRII, regulates the transcription of the glnALG operon in *Escherichia coli. Proc. Natl. Acad. Sci. USA* 83:5909–5913.

1416. **Nishi, K., F. Morel-Deville, J. W. Hershey, T. Leighton, and J. Schnier.** 1988. An eIF-4A-like protein is a suppressor of an *Escherichia coli* mutant defective in 50S ribosomal subunit assembly *Nature* (London) 336:496–498. (Erratum, 340:246, 1989.)

1417. **Nishijima, M., C. E. Bulawa, and C. R. Raetz.** 1981. Two interacting mutations causing temperature-sensitive phosphatidylglycerol synthesis in *Escherichia coli* membranes. *J. Bacteriol.* 145:113–121.

1418. **Nishijima, S., Y. Asami, N. Uetake, S. Yamagoe, A. Ohta, and I. Shibuya.** 1988. Disruption of the *Escherichia coli cls* gene responsible for cardiolipin synthesis. *J. Bacteriol.* 170:775–780.

1419. **Nishimura, A.** 1989. A new gene controlling the frequency of cell division per round of DNA replication in *Escherichia coli. Mol. Gen. Genet.* 215:286–293.

1420. **Nishiyama, K., S. Mizushima, and H. Tokuda.** 1992. The carboxyl-terminal region of SecE interacts with SecY and is functional in the reconstitution of protein translocation activity in *Escherichia coli. J. Biol. Chem.* 267:7170–7176.

1421. **Nishiyama, K., S. Mizushima, and H. Tokuda.** 1993. A novel membrane protein involved in protein translocation across the cytoplasmic membrane of *Escherichia coli. EMBO J.* 12:3409–3415.

1422. **Noble, J. A., M. A. Innis, E. V. Koonin, K. E. Rudd, F. Banuett, and I. Herskowitz.** 1993. The *Escherichia coli* hflA locus encodes a putative GTP-binding protein and two membrane proteins, one of which contains a protease-like domain. *Proc. Natl. Acad. Sci. USA* 90:10866–10870.

1423. **Nodwell, J. R., and J. Greenblatt.** 1993. Recognition of boxA antiterminator RNA by the *E. coli* antitermination factors NusB and ribosomal protein S10. *Cell* 72:261–268.

1424. **Nohno, T., Y. Kasai, and T. Saito.** 1988. Cloning and sequencing of the *Escherichia coli* chlEN operon involved in molybdopterin biosynthesis. *J. Bacteriol.* 170:4097–4102.

1425. **Nohno, T., S. Noji, S. Taniguchi, and T. Saito.** 1989. The narX and narL genes encoding the nitrate-sensing regulators of *Escherichia coli* are homologous to a family of prokaryotic two-component regulatory genes. *Nucleic Acids Res.* 17:2947–2957.

1426. **Nohno, T., T. Saito, and J. S. Hong.** 1986. Cloning and complete nucleotide sequence of the *Escherichia coli* glutamine permease operon (glnHPQ). *Mol. Gen. Genet.* 205:260–269.

1427. **Nonet, M. L., C. C. Marvel, and D. R. Tolan.** 1987. The hisT-purF region of the *Escherichia coli* K-12 chromosome. Identification of additional genes of the hisT and purF operons. *J. Biol. Chem.* 262:12209–12217.

1428. **Nordlund, P., and H. Eklund.** 1993. Structure and function of the *Escherichia coli* ribonucleotide reductase protein R2. *J. Mol. Biol.* 232:123–164.

1429. **Normanly, J., T. Ollick, and J. Abelson.** 1992. Eight base changes are sufficient to convert a leucine-inserting tRNA into a serine-inserting tRNA. *Proc. Natl. Acad. Sci. USA* 89:5680–5684.

1430. **Normark, S., L. Norlander, T. Grundstrom, G. D. Bloom, P. Boquel, and G. Frelat.** 1976. Septum formation-defective mutant of *Escherichia coli. J. Bacteriol.* 128:401–412.

1431. **Novel, G., and F. Stoeber.** 1973. Individualité de la D-glucuronate-cetol isomerase d'*Escherichia coli* K12. *Biochimie* 55:1057–1070.

1432. **Nummer, B. A., S. F. Barefoot, and E. L. Kline.** 1992. Effects of the flrA regulatory locus on biosynthesis and excretion of amino acids in *Escherichia coli* B/r. *Biochem. Biophys. Res. Commun.* 183:343–349.

1433. **Nureki, O., M. Tateno, T. Niimi, T. Kohno, T. Muramatsu, H. Kanno, Y. Muto, R. Giege, and S. Yokoyama.** 1991. Mechanisms of molecular recognition of tRNAs by aminoacyl-tRNA synthetases. *Nucleic Acids Symp. Ser.* 20:165–166.

1434. **Nurse, P., K. H. Zavitz, and K. J. Marians.** 1991. Inactivation of the *Escherichia coli* priA DNA replication protein induces the SOS response. *J. Bacteriol.* 173:6686–6693.

1435. **Nwosu, V. U.** 1992. Overexpression of the wild-type gene coding for *Escherichia coli* DNA adenine methylase (dam). *Biochem. J.* 283:745–750.

1436. **Ny, T., H. R. Lindstrom, T. G. Hagervall, and G. R. Bjork.** 1988. Purification of transfer RNA (m5U54)-methyltransferase from *Escherichia coli*. Association with RNA. *Eur. J. Biochem.* 177:467–475.

1437. **Nygaard, P., and J. M. Smith.** 1993. Evidence for a novel glycinamide ribonucleotide transformylase in *Escherichia coli. J. Bacteriol.* 175:3591–3597.

1438. **Nystrom, T., and F. C. Neidhardt.** 1992. Cloning, mapping and nucleotide sequencing of a gene encoding a universal stress protein in *Escherichia coli. Mol. Microbiol.* 6:3187–3198.

1439. **Nystrom, T., and F. C. Neidhardt.** 1993. Isolation and properties of a mutant of *Escherichia coli* with an insertional inactivation of the *uspA* gene, which encodes a universal stress protein. *J. Bacteriol.* 175:3949–3956.

1440. **Nystrom, T., and F. C. Neidhardt.** 1994. Expression and role of the universal stress protein, UspA, of *Escherichia coli* during growth arrest. *Mol. Microbiol.* 11:537–544.

1441. **O'Brien, R. J., J. A. Fox, M. G. Kopczynski, and B. M. Babior.** 1985. The mechanism of action of ethanolamine ammonia-lyase, an adenosylcobalamin-dependent enzyme. Evidence that the hydrogen transfer mechanism involves a second intermediate hydrogen carrier in addition to the cofactor. *J. Biol. Chem.* 260:16131–16136.

1442. **Oda, Y., S. Iwai, E. Ohtsuka, M. Ishikawa, M. Ikehara, and H. Nakamura.** 1993. Binding of nucleic acids to *E. coli* RNase HI observed by NMR and CD spectroscopy. *Nucleic Acids Res.* 21:4690–4695.

1443. **Odoevskaya, E. R., and S. P. Sineokii.** 1987. Isolation and genetic study of the bacterial mutations gpr blocking the replication of certain lambdoid phages. *Sov. Genet.* 23:432–440.

1444. **Ogasahara, K., K. Hiraga, W. Ito, E. W. Miles, and K. Yutani.** 1992. Origin of the mutual activation of the alpha and beta 2 subunits in the alpha 2 beta 2 complex of tryptophan synthase. Effect of alanine or glycine substitutions at proline residues in the alpha subunit. *J. Biol. Chem.* 267:5222–5228.

1445. **Ogasawara, N., and H. Yoshikawa.** 1992. Genes and their organization in the replication origin region of the bacterial chromosome. *Mol. Microbiol.* 6:629–634.

1446. **Ogura, T., T. Tomoyasu, T. Yuki, S. Morimura, K. J. Begg, W. D. Donachie, H. Mori, H. Niki, and S. Hiraga.** 1991. Structure and function of the ftsH gene in *Escherichia coli. Res. Microbiol.* 142:279–282.

1447. **Oh, B. K., and D. Apirion.** 1991. 10Sa RNA, a small stable RNA of *Escherichia coli*, is functional. *Mol. Gen. Genet.* 229:52–56.

1448. **Oh, S. J., Y. C. Kim, Y. W. Park, S. Y. Min, I. S. Kim, and H. S. Kang.** 1987. Complete nucleotide sequence of the penicillin G acylase gene and the flanking regions, and its expression in *Escherichia coli. Gene* 56:87–97.

1449. **Ohara, O., R. L. Dorit, and W. Gilbert.** 1989. Direct genomic sequencing of bacterial DNA: the pyruvate kinase I gene of *Escherichia coli. Proc. Natl. Acad. Sci. USA* 86:6883–6887.

1450. **Ohmori, H.** 1994. Structural analysis of the rhlE gene of *Escherichia coli. Jpn. J. Genet.* 69:1–12.

1451. **Ohnishi, K., M. Homma, K. Kutsukake, and T. Iino.** 1987. Formation of flagella lacking outer rings by *flaM, flaU,* and *flaY* mutants of *Escherichia coli. J. Bacteriol.* 169:1485–1488.

1452. **Ohnishi, T.** 1987. Structure of the succinate-ubiquinone oxidoreductase (complex II). *Curr. Top. Bioenerg.* 15:37–65.

1453. **Ohnishi, Y.** 1974. Genetic analysis of an *Escherichia coli* mutant with a lesion in stable RNA turnover. *Genetics* 76:185–194.

1454. **Ohno, S.** 1970. *Evolution by Gene Duplication.* Springer-Verlag, New York.

1455. **Ohta, M., K. Ina, K. Kusuzaki, N. Kido, Y. Arakawa, and N. Kato.** 1991. Cloning and expression of the rfe-rff gene cluster of *Escherichia coli. Mol. Microbiol.* 5:1853–1862.

1456. **Ohyama, T., K. Igarashi, and H. Kobayashi.** 1994. Physiological role of the *chaA* gene in sodium and calcium circulations at a high pH in *Escherichia coli. J. Bacteriol.* 176:4311–4315.

1457. **Okada, Y., M. Wachi, A. Hirata, K. Suzuki, K. Nagai, and M. Matsuhashi.** 1994. Cytoplasmic axial filaments in *Escherichia coli* cells: possible function in the mechanism of chromosome segregation and cell division. *J. Bacteriol.* 176:917–922.

1458. **Okamura-Ikeda, K., Y. Ohmura, K. Fujiwara, and Y. Motokawa.** 1993. Cloning and nucleotide sequence of the gcv operon encoding the *Escherichia coli* glycine-cleavage system. *Eur. J. Biochem.* 216:539–548.

1459. **Old, I. G., I. Saint Girons, and C. Richaud.** 1993. Physical mapping of the scattered methionine genes on the *Escherichia coli* chromosome. *J. Bacteriol.* 175:3689–3691.

1460. **Oliver, D. B.** 1985. Identification of five new essential genes involved in the synthesis of a secreted protein in *Escherichia coli. J. Bacteriol.* 161:285–291.

1461. **Olsen, A., A. Jonsson, and S. Normark.** 1989. Fibronectin binding mediated by a novel class of surface organelles on *Escherichia coli. Nature* (London) 338:652–655.

1462. **Olsiewski, P. J., G. J. Kaczorowski, and C. Walsh.** 1980. Purification and properties of D-amino acid dehydrogenase, an inducible membrane-bound iron-sulfur flavoenzyme from *Escherichia coli* B. *J. Biol. Chem.* 255:4487–4494.

1463. **Olson, E. R., D. S. Dunyak, L. M. Jurss, and R. A. Poorman.** 1991. Identification and characterization of dppA, an *Escherichia coli* gene encoding a periplasmic dipeptide transport protein. *J. Bacteriol.* 173:234–244.

1464. **O'Neill, G. P., S. Thorbjarnardottir, U. Michelsen, S. Palsson, D. Soll, and G. Eggertsson.** 1991. delta-Aminolevulinic acid dehydratase deficiency can cause delta-aminolevulinate auxotrophy in *Escherichia coli. J. Bacteriol.* 173:94–100.

1465. **Ono, M., and M. Kuwano.** 1978. Mutation affecting the thermolability of the 50S ribosomal subunit in *Escherichia coli. J. Bacteriol.* 134:677–679.

1466. Oram, M., and L. M. Fisher. 1992. An *Escherichia coli* DNA topoisomerase I mutant has a compensatory mutation that alters two residues between functional domains of the DNA gyrase A protein. *J. Bacteriol.* 174:4175–4178.

1467. Orchard, L. M., and H. L. Kornberg. 1990. Sequence similarities between the gene specifying 1-phosphofructokinase (fruK), genes specifying other kinases in *Escherichia coli* K12, and lacC of *Staphylococcus aureus*. *Proc. R. Soc. London Ser. B* 242:87–90.

1468. O'Regan, M., R. Gloeckler, S. Bernard, C. Ledoux, I. Ohsawa, and Y. Lemoine. 1989. Nucleotide sequence of the bioH gene of *Escherichia coli*. *Nucleic Acids Res.* 17:8004

1469. Orndorff, P. E., and S. Falkow. 1984. Organization and expression of genes responsible for type 1 piliation in *Escherichia coli*. *J. Bacteriol.* 159:736–744.

1470. Orren, D. K., and A. Sancar. 1990. Formation and enzymatic properties of the UvrB DNA complex. *J. Biol. Chem.* 265:15796–15803.

1471. Osborne, C., L. M. Chen, and R. G. Matthews. 1991. Isolation, cloning, mapping, and nucleotide sequencing of the gene encoding flavodoxin in *Escherichia coli*. *J. Bacteriol.* 173:1729–1737.

1472. Osswald, M., B. Greuer, and R. Brimacombe. 1990. Localization of a series of RNA-protein cross-link sites in the 23S and 5S ribosomal RNA from *Escherichia coli*, induced by treatment of 50S subunits with three different bifunctional reagents. *Nucleic Acids Res.* 18:6755–6760.

1473. Ostanin, K., E. H. Harms, P. E. Stevis, R. Kuciel, M. M. Zhou, and R. L. Van Etten. 1992. Overexpression, site-directed mutagenesis, and mechanism of *Escherichia coli* acid phosphatase. *J. Biol. Chem.* 267:22830–22836.

1474. Ostrowski, J., M. J. Barber, D. C. Rueger, B. E. Miller, L. M. Siegel, and N. M. Kredich. 1989. Characterization of the flavoprotein moieties of NADPH-sulfite reductase from *Salmonella typhimurium* and *Escherichia coli*. Physicochemical and catalytic properties, amino acid sequence deduced from DNA sequence of cysJ, and comparison with NADPH-cytochrome P-450 reductase. *J. Biol. Chem.* 264:15796–15808.

1475. Ostrowski, J., J. Y. Wu, D. C. Rueger, B. E. Miller, L. M. Siegel, and N. M. Kredich. 1989. Characterization of the cysJIH regions of *Salmonella typhimurium* and *Escherichia coli* B. DNA sequences of cysI and cysH and a model for the siro-heme-Fe4S4 active center of sulfite reductase hemoprotein based on amino acid homology with spinach nitrite reductase. *J. Biol. Chem.* 264:15726–15737.

1476. Osuna, R., S. E. Finkel, and R. C. Johnson. 1991. Identification of two functional regions in Fis: the N-terminus is required to promote Hin-mediated DNA inversion but not lambda excision. *EMBO J.* 10:1593–1603.

1477. Otsuka, A. J., M. R. Buoncristiani, P. K. Howard, J. Flamm, C. Johnson, R. Yamamoto, K. Uchida, C. Cook, J. Ruppert, and J. Matsuzaki. 1988. The *Escherichia coli* biotin biosynthetic enzyme sequences predicted from the nucleotide sequence of the bio operon. *J. Biol. Chem.* 263:19577–19585.

1478. Overath, P., G. Pauli, and H. U. Schairer. 1969. Fatty acid degradation in *Escherichia coli*, an inducible acyl-CoA synthetase, the mapping of old-mutations, and the isolation of regulatory mutants. *Eur. J. Biochem.* 7:559–574.

1479. Overduin, P., W. Boos, and J. Tommassen. 1988. Nucleotide sequence of the ugp genes of *Escherichia coli* K-12: homology to the maltose system. *Mol. Microbiol.* 2:767–775.

1480. Owais, W. M., and R. Gharaibeh. 1990. Cloning of the *E. coli* O-acetylserine sulfhydrylase gene: ability of the clone to produce a mutagenic product from azide and O-acetylserine. *Mutat. Res.* 245:151–155.

1481. Ozenberger, B. A., T. J. Brickman, and M. A. McIntosh. 1989. Nucleotide sequence of *Escherichia coli* isochorismate synthetase gene entC and evolutionary relationship of isochorismate synthetase and other chorismate-utilizing enzymes. *J. Bacteriol.* 171:775–783.

1482. Ozenberger, B. A., M. S. Nahlik, and M. A. McIntosh. 1987. Genetic organization of multiple fep genes encoding ferric enterobactin transport functions in *Escherichia coli*. *J. Bacteriol.* 169:3638–3646.

1483. Packman, L. C., and R. N. Perham. 1986. Chain folding in the dihydrolipoyl acyltransferase components of the 2-oxo-acid dehydrogenase complexes from *Escherichia coli*. Identification of a segment involved in binding the E3 subunit. *FEBS Lett.* 206:193–198.

1484. Padan, E., and S. Schuldiner. 1993. Na$^+$/H$^+$ antiporters, molecular devices that couple the Na$^+$ and H$^+$ circulation in cells. *J. Bioenerg. Biomembr.* 25:647–669.

1485. Padgette, S. R., D. B. Re, C. S. Gasser, D. A. Eichholtz, R. B. Frazier, C. M. Hironaka, E. B. Levine, D. M. Shah, R. T. Fraley, and G. M. Kishore. 1991. Site-directed mutagenesis of a conserved region of the 5-enolpyruvylshikimate-3-phosphate synthase active site. *J. Biol. Chem.* 266:22364–22369.

1486. Padmanabha, K. P., and M. P. Deutscher. 1991. RNase T affects *Escherichia coli* growth and recovery from metabolic stress. *J. Bacteriol.* 173:1376–1381.

1487. Page, L., L. Griffiths, and J. A. Cole. 1990. Different physiological roles of two independent pathways for nitrite reduction to ammonia by enteric bacteria. *Arch. Microbiol.* 154:349–354.

1488. Pages, D., and R. H. Buckingham. 1990. Mutants of pheV in *Escherichia coli* affecting control by attenuation of the pheS, T and pheA operons. Two distinct mechanisms for de-attenuation. *J. Mol. Biol.* 216:17–24.

1489. Pak, M., L. Pallanck, and L. H. Schulman. 1992. Conversion of a methionine initiator tRNA into a tryptophan-inserting elongator tRNA in vivo. *Biochemistry* 31:3303–3309.

1490. Pak, M., J. M. Willis, and L. H. Schulman. 1994. Analysis of acceptor stem base pairing on tRNA(Trp) aminoacylation and function in vivo. *J. Biol. Chem.* 269:2277–2282.

1491. Palaniappan, C., V. Sharma, M. E. Hudspeth, and R. Meganathan. 1992. Menaquinone (vitamin K$_2$) biosynthesis: evidence that the *Escherichia coli* menD gene encodes both 2-succinyl-6-hydroxy-2,4-cyclohexadiene-1-carboxylic acid synthase and α-ketoglutarate decarboxylase activities. *J. Bacteriol.* 174:8111–8118.

1492. Palas, K. M., and S. R. Kushner. 1990. Biochemical and physical characterization of exonuclease V from *Escherichia coli*. Comparison of the catalytic activities of the RecBC and RecBCD enzymes. *J. Biol. Chem.* 265:3447–3454.

1493. Palchaudhuri, S., V. Patel, and E. McFall. 1988. DNA sequence of the D-serine deaminase activator gene dsdC. *J. Bacteriol.* 170:330–334.

1494. Pallanck, L., S. Li, and L. H. Schulman. 1992. The anticodon and discriminator base are major determinants of cysteine tRNA identity in vivo. *J. Biol. Chem.* 267:7221–7223.

1495. Pallesen, L., O. Madsen, and P. Klemm. 1989. Regulation of the phase switch controlling expression of type 1 fimbriae in *Escherichia coli*. *Mol. Microbiol.* 3:925–931.

1496. Panagiotidis, C. A., S. Blackburn, K. B. Low, and E. S. Canellakis. 1987. Biosynthesis of polyamines in ornithine decarboxylase, arginine decarboxylase, and agmatine ureohydrolase deletion mutants of *Escherichia coli* strain K-12. *Proc. Natl. Acad. Sci. USA* 84:4423–4427.

1497. Pao, C. C., and J. Gallant. 1978. A gene involved in the metabolic control of ppGpp synthesis. *Mol. Gen. Genet.* 158:271–277.

1498. Papanicolaou, C., and L. S. Ripley. 1989. Polymerase-specific differences in the DNA intermediates of frameshift mutagenesis. In vitro synthesis errors of *Escherichia coli* DNA polymerase I and its large fragment derivative. *J. Mol. Biol.* 207:335–353.

1499. Pardee, A. B., E. J. Benz, D. A. St. Peter, J. N. Krieger, M. Meuth, and H. W. Trieschmann. 1971. Hyperproduction and purification of nicotinamide deamidase, a microconstitutive enzyme of *Escherichia coli*. *J. Biol. Chem.* 246:6792–6796.

1500. Park, E. A., A. L. Gurney, S. E. Nizielski, P. Hakimi, Z. Cao, A. Moorman, and R. W. Hanson. 1993. Relative roles of CCAAT/enhancer-binding protein beta and cAMP regulatory element-binding protein in controlling transcription of the gene for phosphoenolpyruvate carboxykinase (GTP). *J. Biol. Chem.* 268:613–619.

1501. Park, J. T. 1993. Turnover and recycling of the murein sacculus in oligopeptide permease-negative strains of *Escherichia coli*: indirect evidence for an alternative permease system and for a monolayered sacculus. *J. Bacteriol.* 175:7–11.

1502. Park, M. H., B. R. Wong, and J. E. Lusk. 1976. Mutants in three genes affecting transport of magnesium in *Escherichia coli*: genetics and physiology. *J. Bacteriol.* 126:1096–1103.

1503. Park, S. J., Y. M. Hou, and P. Schimmel. 1989. A single base pair affects binding and catalytic parameters in the molecular recognition of a transfer RNA. *Biochemistry* 28:2740–2746.

1504. Park, S. K., K. I. Kim, K. M. Woo, J. H. Seol, K. Tanaka, A. Ichihara, D. B. Ha, and C. H. Chung. 1993. Site-directed mutagenesis of the dual translational initiation sites of the clpB gene of *Escherichia coli* and characterization of its gene products. *J. Biol. Chem.* 268:20170–20174.

1505. Parker, C. T., A. W. Kloser, C. A. Schnaitman, M. A. Stein, S. Gottesman, and B. W. Gibson. 1992. Role of the rfaG and rfaP genes in determining the lipopolysaccharide core structure and cell surface properties of *Escherichia coli* K-12. *J. Bacteriol.* 174:2525–2538.

1506. Parker, C. T., E. Pradel, and C. A. Schnaitman. 1992. Identification and sequences of the lipopolysaccharide core biosynthetic genes rfaQ, rfaP, and rfaG of *Escherichia coli* K-12. *J. Bacteriol.* 174:930–934.

1507. Parker, L. L., and B. G. Hall. 1988. A fourth *Escherichia coli* gene system with the potential to evolve beta-glucoside utilization. *Genetics* 119:485–490.

1508. Parker, L. L., and B. G. Hall. 1990. Characterization and nucleotide sequence of the cryptic cel operon of *Escherichia coli* K12. *Genetics* 124:455–471.

1509. Parsot, C. 1986. Evolution of biosynthetic pathways: a common ancestor for threonine synthase, threonine dehydratase and D-serine dehydratase. *EMBO J.* 5:3013–3019.

1510. Parsot, C., A. Boyen, G. N. Cohen, and N. Glansdorff. 1988. Nucleotide sequence of *Escherichia coli* argB and argC genes: comparison of N-acetylglutamate kinase and N-acetylglutamate-gamma-semialdehyde dehydrogenase with homologous and analogous enzymes. *Gene* 68:275–283.

1511. Parsot, C., and G. N. Cohen. 1988. Cloning and nucleotide sequence of the *Bacillus subtilis* hom gene coding for homoserine dehydrogenase. Structural and evolutionary relationships with *Escherichia coli* aspartokinases-homoserine dehydrogenases I and II. *J. Biol. Chem.* 263:14654–14660.

1512. Parsot, C., P. Cossart, I. Saint-Girons, and G. N. Cohen. 1983. Nucleotide sequence of thrC and of the transcription termination region of the threonine operon in *Escherichia coli* K12. *Nucleic Acids Res.* 11:7331–7345.

1513. Pascal, M. C., M. Lepelletier, G. Giordano, and M. Chippaux. 1991. A regulatory mutant of the trimethylamine N-oxide reductase of *Escherichia coli* K12. *FEMS Microbiol. Lett.* 62:297–300.

1514. Patil, R. V., and E. E. Dekker. 1992. Cloning, nucleotide sequence, overexpression, and inactivation of the *Escherichia coli* 2-keto-4-hydroxyglutarate aldolase gene. *J. Bacteriol.* 174:102–107.

1515. Patnaik, R., W. D. Roof, R. F. Young, and J. C. Liao. 1992. Stimulation of glucose catabolism in *Escherichia coli* by a potential futile cycle. *J. Bacteriol.* 174:7527–7532.

1516. **Pauli, G., and P. Overath.** 1972. ato operon: a highly inducible system for acetoacetate and butyrate degradation in *Escherichia coli. Eur. J. Biochem.* **29:**553–562.

1517. **Peakman, T., J. Crouzet, J. F. Mayaux, S. Busby, S. Mohan, N. Harborne, J. Wootton, R. Nicolson, and J. Cole.** 1990. Nucleotide sequence, organisation and structural analysis of the products of genes in the nirB-cysG region of the *Escherichia coli* K-12 chromosome. *Eur. J. Biochem.* **191:**315–323.

1518. **Pearson, W. R., and D. J. Lipman.** 1988. Improved tools for biological sequence comparisons. *Proc. Natl. Acad. Sci. USA* **85:**2444–2448.

1519. **Peck-Miller, K. A., and S. Altman.** 1991. Kinetics of the processing of the precursor to 4.5 S RNA, a naturally occurring substrate for RNase P from *Escherichia coli. J. Mol. Biol.* **221:**1–5.

1520. **Pedersen, H., L. Sogaard-Andersen, B. Holst, and P. Valentin-Hansen.** 1991. Heterologous cooperativity in *Escherichia coli.* The CytR repressor both contacts DNA and the cAMP receptor protein when binding to the deoP2 promoter. *J. Biol. Chem.* **266:**17804–17808.

1521. **Pegues, J. C., L. S. Chen, A. W. Gordon, L. Ding, and W. G. Coleman, Jr.** 1990. Cloning, expression, and characterization of the *Escherichia coli* K-12 rfaD gene. *J. Bacteriol.* **172:**4652–4660.

1522. **Pel, H. J., M. Rep, and L. A. Grivell.** 1992. Sequence comparison of new prokaryotic and mitochondrial members of the polypeptide chain release factor family predicts a five-domain model for release factor structure. *Nucleic Acids Res.* **20:**4423–4428.

1523. **Peng, H., and K. J. Marians.** 1993. *Escherichia coli* topoisomerase IV. Purification, characterization, subunit structure, and subunit interactions. *J. Biol. Chem.* **268:**24481–24490.

1524. **Peng, H., and K. J. Marians.** 1993. Decatenation activity of topoisomerase IV during oriC and pBR322 DNA replication in vitro. *Proc. Natl. Acad. Sci. USA* **90:**8571–8575.

1525. **Perez-Morga, D., and G. Guarneros.** 1990. A short DNA sequence from lambda phage inhibits protein synthesis in *Escherichia coli* rap. *J. Mol. Biol.* **216:**243–250.

1526. **Perham, R. N.** 1990. The fructose-1,6-bisphosphate aldolases: same reaction, different enzymes. *Biochem. Soc. Trans.* **18:**185–187.

1527. **Perham, R. N., L. C. Packman, and S. E. Radford.** 1987. 2-Oxo acid dehydrogenase multi-enzyme complexes: in the beginning and halfway there. *Biochem. Soc. Symp.* **54:**67–81.

1528. **Peri, K. G., and E. B. Waygood.** 1988. Sequence of cloned enzyme IIN-acetylglucosamine of the phosphoenolpyruvate:N-acetylglucosamine phosphotransferase system of *Escherichia coli. Biochemistry* **27:**6054–6061.

1529. **Persson, B. C., C. Gustafsson, D. E. Berg, and G. R. Bjork.** 1992. The gene for a tRNA modifying enzyme, m5U54-methyltransferase, is essential for viability in *Escherichia coli. Proc. Natl. Acad. Sci. USA* **89:**3995–3998.

1530. **Peterkofsky, A.** 1988. Redistribution of phosphate pools and the regulation of *Escherichia coli* adenylate cyclase activity. *Arch. Biochem. Biophys.* **265:**227–233.

1531. **Petersen, C.** 1990. *Escherichia coli* ribosomal protein L10 is rapidly degraded when synthesized in excess of ribosomal protein L7/L12. *J. Bacteriol.* **172:**431–436.

1532. **Petersen, S. K., and F. G. Hansen.** 1991. A missense mutation in the rpoC gene affects chromosomal replication control in *Escherichia coli. J. Bacteriol.* **173:**5200–5206.

1533. **Peterson, C. B., and H. K. Schachman.** 1991. Role of a carboxyl-terminal helix in the assembly, interchain interactions, and stability of aspartate transcarbamoylase. *Proc. Natl. Acad. Sci. USA* **88:**458–462.

1534. **Peterson, E. T., and O. C. Uhlenbeck.** 1992. Determination of recognition nucleotides for *Escherichia coli* phenylalanyl-tRNA synthetase. *Biochemistry* **31:**10380–10389.

1535. **Pettis, M. H., and M. A. McIntosh.** 1987. Molecular characterization of the *Escherichia coli* enterobactin cistron entF and coupled expression of entF and the fes gene. *J. Bacteriol.* **169:**4154–4162.

1536. **Phillips, G. J., D. C. Prasher, and S. R. Kushner.** 1988. Physical and biochemical characterization of cloned sbcB and xonA mutations from *Escherichia coli* K-12. *J. Bacteriol.* **170:**2089–2094.

1537. **Phillips, G. J., and T. J. Silhavy.** 1992. The *E. coli* ffh gene is necessary for viability and efficient protein export. *Nature* (London) **359:**744–746.

1538. **Phillips, S. E., I. Manfield, I. Parsons, B. E. Davidson, J. B. Rafferty, W. S. Somers, D. Margarita, G. N. Cohen, I. Saint-Girons, and P. G. Stockley.** 1989. Cooperative tandem binding of met repressor of *Escherichia coli. Nature* (London) **341:**711–715.

1539. **Pi, J., P. J. Wookey, and A. J. Pittard.** 1991. Cloning and sequencing of the pheP gene, which encodes the phenylalanine-specific transport system of *Escherichia coli. J. Bacteriol.* **173:**3622–3629.

1540. **Pi, J., P. J. Wookey, and A. J. Pittard.** 1993. Site-directed mutagenesis reveals the importance of conserved charged residues for the transport activity of the PheP permease of *Escherichia coli. J. Bacteriol.* **175:**7500–7504.

1541. **Pieper, U., H. J. Ehbrecht, A. Fliess, B. Schick, F. Jurnak, and A. Pingoud.** 1990. Genetic engineering, isolation and characterization of a truncated *Escherichia coli* elongation factor Tu comprising domains 2 and 3. *Biochim. Biophys. Acta* **1087:**147–156.

1542. **Pierson, D. E., and A. Campbell.** 1990. Cloning and nucleotide sequence of bisC, the structural gene for biotin sulfoxide reductase in *Escherichia coli. J. Bacteriol.* **172:**2194–2198.

1543. **Piffeteau, A., M. Zamboni, and M. Gaudry.** 1982. Biotin transport by a biotin-deficient strain of *Escherichia coli. Biochim. Biophys. Acta* **688:**29–36.

1544. **Pinner, E., Y. Kotler, E. Padan, and S. Schuldiner.** 1993. Physiological role of nhaB, a specific Na$^+$/H$^+$ antiporter in *Escherichia coli. J. Biol. Chem.* **268:**1729–1734.

1545. **Pinner, E., E. Padan, and S. Schuldiner.** 1992. Cloning, sequencing, and expression of the nhaB gene, encoding a Na$^+$/H$^+$ antiporter in *Escherichia coli. J. Biol. Chem.* **267:**11064–11068.

1546. **Pistocchi, R., K. Kashiwagi, S. Miyamoto, E. Nukui, Y. Sadakata, H. Kobayashi, and K. Igarashi.** 1993. Characteristics of the operon for a putrescine transport system that maps at 19 minutes on the *Escherichia coli* chromosome. *J. Biol. Chem.* **268:**146–152.

1547. **Pittard, A. J., and B. E. Davidson.** 1991. TyrR protein of *Escherichia coli* and its role as repressor and activator. *Mol. Microbiol.* **5:**1585–1592.

1548. **Pittard, J., and B. J. Wallace.** 1966. Gene controlling uptake of shikimic acid by *Escherichia coli. J. Bacteriol.* **92:**1070–1075.

1549. **Pledger, W. J., and H. E. Umbarger.** 1973. Isoleucine and valine metabolism in *Escherichia coli.* XXI. Mutations affecting derepression and valine resistance. *J. Bacteriol.* **114:**183–194.

1550. **Plumbridge, J., and A. Kolb.** 1991. CAP and Nag repressor binding to the regulatory regions of the nagE-B and manX genes of *Escherichia coli. J. Mol. Biol.* **217:**661–679.

1551. **Plumbridge, J. A.** 1989. Sequence of the nagBACD operon in *Escherichia coli* K12 and pattern of transcription within the nag regulon. *Mol. Microbiol.* **3:**505–515.

1552. **Plumbridge, J. A.** 1991. Repression and induction of the nag regulon of *Escherichia coli* K-12: the roles of nagC and nagA in maintenance of the uninduced state. *Mol. Microbiol.* **5:**2053–2062.

1553. **Plumbridge, J. A.** 1992. A dominant mutation in the gene for the Nag repressor of *Escherichia coli* that renders the nag regulon uninducible. *J. Gen. Microbiol.* **138:**1011–1017.

1554. **Plunkett, G., V. Burland, D. L. Daniels, and F. R. Blattner.** 1993. Analysis of the *Escherichia coli* genome. III. DNA sequence of the region from 87.2 to 89.2 minutes. *Nucleic Acids Res.* **21:**3391–3398.

1555. **Pogliano, K. J., and J. Beckwith.** 1994. Genetic and molecular characterization of the *Escherichia coli* secD operon and its products. *J. Bacteriol.* **176:**804–814.

1556. **Polacco, M. L., and J. E. Cronan, Jr.** 1981. A mutant of *Escherichia coli* conditionally defective in the synthesis of holo-[acyl carrier protein]. *J. Biol. Chem.* **256:**5750–5754.

1557. **Polarek, J. W., G. Williams, and W. Epstein.** 1992. The products of the kdpDE operon are required for expression of the Kdp ATPase of *Escherichia coli. J. Bacteriol.* **174:**2145–2151.

1558. **Polayes, D. A., P. W. Rice, and J. E. Dahlberg.** 1988. DNA polymerase I activity in *Escherichia coli* is influenced by spot 42 RNA. *J. Bacteriol.* **170:**2083–2088.

1559. **Pompliano, D. L., A. Peyman, and J. R. Knowles.** 1990. Stabilization of a reaction intermediate as a catalytic device: definition of the functional role of the flexible loop in triosephosphate isomerase. *Biochemistry* **29:**3186–3194.

1560. **Pon, C. L., R. A. Calogero, and C. O. Gualerzi.** 1988. Identification, cloning, nucleotide sequence and chromosomal map location of hns, the structural gene for *Escherichia coli* DNA-binding protein H-NS. *Mol. Gen. Genet.* **212:**199–202.

1561. **Poole, R. K., L. Hatch, M. W. J. Cleeter, F. Gibson, G. B. Cox, and G. Wu.** 1994. Cytochrome *bd* biosynthesis in *Escherichia coli:* the sequences of the cydC and cydD genes suggest that they encode the components of an ABC membrane transporter. *Mol. Microbiol.* **10:**421–430.

1562. **Popp, J. L.** 1989. Sequence and overexpression of the menD gene from *Escherichia coli. J. Bacteriol.* **171:**4349–4354.

1563. **Poritz, M. A., H. D. Bernstein, K. Strub, D. Zopf, H. Wilhelm, and P. Walter.** 1990. An *E. coli* ribonucleoprotein containing 4.5S RNA resembles mammalian signal recognition particle. *Science* **250:**1111–1117.

1564. **Portalier, R. C., J. M. Robert-Baudouy, and F. R. Stoeber.** 1972. Genetic and biochemical characterization of mutations affecting altronic hydrolyase structural gene in *Escherichia coli* K-12. *Mol. Gen. Genet.* **118:**335–350.

1565. **Portier, C., L. Dondon, and M. Grunberg-Manago.** 1990. Translational autocontrol of the *Escherichia coli* ribosomal protein S15. *J. Mol. Biol.* **211:**407–414.

1566. **Portlock, S. H., Y. Lee, J. M. Tomich, and L. K. Tamm.** 1992. Insertion and folding of the amino-terminal amphiphilic signal sequences of the mannitol and glucitol permeases of *Escherichia coli. J. Biol. Chem.* **267:**11017–11022.

1567. **Postle, K., and J. T. Skare.** 1988. *Escherichia coli* TonB protein is exported from the cytoplasm without proteolytic cleavage of its amino terminus. *J. Biol. Chem.* **263:**11000–11007.

1568. **Potapov, A. P., and A. R. Subramanian.** 1992. Effect of *E. coli* ribosomal protein S1 on the fidelity of the translational elongation step: reading and misreading of poly(U) and poly(dT). *Biochem. Int.* **27:**745–753.

1569. **Potter, P. M., M. C. Wilkinson, J. Fitton, F. J. Carr, J. Brennand, D. P. Cooper, and G. P. Margison.** 1987. Characterisation and nucleotide sequence of ogt, the O6-alkylguanine-DNA-alkyltransferase gene of *E. coli. Nucleic Acids Res.* **15:**9177–9193.

1570. **Poulsen, L. K., N. W. Larsen, S. Molin, and P. Andersson.** 1989. A family of genes encoding a cell-killing function may be conserved in all gram-negative bacteria. *Mol. Microbiol.* **3:**1463–1472.

1571. **Poulsen, L. K., A. Refn, S. Molin, and P. Andersson.** 1991. Topographic analysis of the toxic Gef protein from *Escherichia coli. Mol. Microbiol.* **5:**1627–1637.

1572. **Poulsen, L. K., A. Refn, S. Molin, and P. Andersson.** 1991. The gef gene from *Escherichia coli* is regulated at the level of translation. *Mol. Microbiol.* **5:**1639–1648.

1573. Poulsen, P., F. Bonekamp, and K. F. Jensen. 1984. Structure of the *Escherichia coli* pyrE operon and control of pyrE expression by a UTP modulated intercistronic attenuation. *EMBO J.* 3:1783–1790.

1574. Pourcher, T., M. Bassilana, H. K. Sarkar, H. R. Kaback, and G. Leblanc. 1990. Melibiose permease and alpha-galactosidase of *Escherichia coli*: identification by selective labeling using a T7 RNA polymerase/promoter expression system. *Biochemistry* 29:690–696.

1575. Pourcher, T., M. Bassilana, H. K. Sarkar, H. R. Kaback, and G. Leblanc. 1992. Melibiose permease of *Escherichia coli*: mutation of histidine-94 alters expression and stability rather than catalytic activity. *Biochemistry* 31:5225–5231.

1576. Pouyssegur, J., and F. Stoeber. 1974. Genetic control of the 2-keto-3-deoxy-D-gluconate metabolism in *Escherichia coli* K-12: *kdg* regulon. *J. Bacteriol.* 117:641–651.

1577. Pradel, E., and P. L. Boquet. 1991. Utilization of exogenous glucose-1-phosphate as a source of carbon or phosphate by *Escherichia coli* K12: respective roles of acid glucose-1-phosphatase, hexose-phosphate permease, phosphoglucomutase and alkaline phosphatase. *Res. Microbiol.* 142:37–45.

1578. Pradel, E., C. T. Parker, and C. A. Schnaitman. 1992. Structures of the *rfaB, rfaI, rfaJ*, and *rfaS* genes of *Escherichia coli* K-12 and their roles in assembly of the lipopolysaccharide core. *J. Bacteriol.* 174:4736–4745.

1579. Pradel, E., and C. A. Schnaitman. 1991. Effect of *rfaH* (*sfrB*) and temperature on expression of *rfa* genes of *Escherichia coli* K-12. *J. Bacteriol.* 173:6428–6431.

1580. Prakash, A., B. Valinluck, and J. Ryu. 1991. Genomic hsd-Mu(lac) operon fusion mutants of *Escherichia coli* K-12. *Gene* 99:9–14.

1581. Prasad, I., and S. Schaefler. 1974. Regulation of the beta-glucoside system in *Escherichia coli* K-12. *J. Bacteriol.* 120:638–650.

1582. Prasad, I., B. Young, and S. Schaefler. 1973. Genetic determination of the constitutive biosynthesis of phospho-beta-glucosidase A in *Escherichia coli* K-12. *J. Bacteriol.* 114:909–915.

1583. Prescott, D. J., and P. R. Vatelos. 1972. Acyl carrier protein. *Adv. Enzymol. Relat. Areas Mol. Biol.* 36:269–311.

1584. Prieto-Santos, M. I., J. Martin-Checa, R. Balana-Fouce, and A. Garrido-Pertierra. 1986. A pathway for putrescine catabolism in *Escherichia coli*. *Biochim. Biophys. Acta* 880:242–244.

1585. Prior, T. I., and H. L. Kornberg. 1988. Nucleotide sequence of fruA, the gene specifying enzyme IIfru of the phosphoenolpyruvate-dependent sugar phosphotransferase system in *Escherichia coli* K12. *J. Gen. Microbiol.* 134:2757–2768.

1586. Privalle, C. T., and I. Fridovich. 1992. Transcriptional and maturational effects of manganese and iron on the biosynthesis of manganese-superoxide dismutase in *Escherichia coli*. *J. Biol. Chem.* 267:9140–9145.

1587. Prodromou, C., P. J. Artymiuk, and J. R. Guest. 1992. The aconitase of *Escherichia coli*. Nucleotide sequence of the aconitase gene and amino acid sequence similarity with mitochondrial aconitases, the iron-responsive-element-binding protein and isopropylmalate isomerases. *Eur. J. Biochem.* 204:599–609.

1588. Prodromou, C., M. J. Haynes, and J. R. Guest. 1991. The aconitase of *Escherichia coli*: purification of the enzyme and molecular cloning and map location of the gene (acn). *J. Gen. Microbiol.* 137:2505–2515.

1589. Prongay, A. J., D. R. Engelke, and C. H. Williams, Jr. 1989. Characterization of two active site mutations of thioredoxin reductase from *Escherichia coli*. *J. Biol. Chem.* 264:2656–2664. (Erratum, 264:12113.)

1590. Pruss, B. M., J. M. Nelms, C. Park, and A. J. Wolfe. 1994. Mutations in NADH:ubiquinone oxidoreductase of *Escherichia coli* affect growth on mixed amino acids. *J. Bacteriol.* 176:2143–2150.

1591. Przybyla, A. E., J. Robbins, N. Menon, and H. D. Peck, Jr. 1992. Structure-function relationships among the nickel-containing hydrogenases. *FEMS Microbiol. Rev.* 88:109–135.

1592. Pucci, M. J., L. F. Discotto, and T. J. Dougherty. 1992. Cloning and identification of the *Escherichia coli* murB DNA sequence, which encodes UDP-*N*-acetylenolpyruvoylglucosamine reductase. *J. Bacteriol.* 174:1690–1693.

1593. Pucci, M. J., J. Novotny, L. F. Discotto, and T. J. Dougherty. 1994. The *Escherichia coli* Dga (MurI) protein shares biological activity and structural domains with the *Pediococcus pentosaceus* glutamate racemase. *J. Bacteriol.* 176:528–530.

1594. Pugsley, A. P., and C. Dubreuil. 1988. Molecular characterization of malQ, the structural gene for the *Escherichia coli* enzyme amylomaltase. *Mol. Microbiol.* 2:473–479.

1595. Pugsley, A. P., F. Moreno, and V. de Lorenzo. 1986. Microcin-E492-insensitive mutants of *Escherichia coli* K12. *J. Gen. Microbiol.* 132:3253–3259.

1596. Pugsley, A. P., and P. Reeves. 1977. The role of colicin receptors in the uptake of ferrienterochelin by *Escherichia coli* K-12. *Biochem. Biophys. Res. Commun.* 74:903–911.

1597. Qi, S. Y., S. Sukupolvi, and C. D. O'Connor. 1991. Outer membrane permeability of *Escherichia coli* K12: isolation, cloning and mapping of suppressors of a defined antibiotic-hypersensitive mutant. *Mol. Gen. Genet.* 229:421–427.

1598. Quail, M. A., D. J. Haydon, and J. R. Guest. 1994. The pdhR-aceEF-lpd operon of *Escherichia coli* expresses the pyruvate dehydrogenase complex. *Mol. Microbiol.* 12:95–104.

1599. Quirk, S., and M. J. Bessman. 1991. dGTP triphosphohydrolase, a unique enzyme confined to members of the family Enterobacteriaceae. *J. Bacteriol.* 173:6665–6669.

1600. Quirk, S., S. K. Bhatnagar, and M. J. Bessman. 1990. Primary structure of the deoxyguanosine triphosphate triphosphohydrolase-encoding gene (dgt) of *Escherichia coli*. *Gene* 89:13–18.

1601. Quirk, S., D. Seto, S. K. Bhatnagar, P. Gauss, L. Gold, and M. J. Bessman. 1989. Location and molecular cloning of the structural gene for the deoxyguanosine triphosphate triphosphohydrolase of *Escherichia coli*. *Mol. Microbiol.* 3:1391–1395.

1602. Rabin, R. S., and V. Stewart. 1993. Dual response regulators (NarL and NarP) interact with dual sensors (NarX and NarQ) to control nitrate- and nitrite-regulated gene expression in *Escherichia coli* K-12. *J. Bacteriol.* 175:3259–3268.

1603. Radika, K., and C. R. Raetz. 1988. Purification and properties of lipid A disaccharide synthase of *Escherichia coli*. *J. Biol. Chem.* 263:14859–14867.

1604. Raetz, C. R., G. D. Kantor, M. Nishijima, and M. L. Jones. 1981. Isolation of *Escherichia coli* mutants with elevated levels of membrane enzymes. A trans-acting mutation controlling diglyceride kinase. *J. Biol. Chem.* 256:2109–2112.

1605. Raha, M., I. Kawagishi, V. Muller, M. Kihara, and R. M. Macnab. 1992. *Escherichia coli* produces a cytoplasmic alpha-amylase, AmyA. *J. Bacteriol.* 174:6644–6652.

1606. Raha, M., M. Kihara, I. Kawagishi, and R. M. Macnab. 1993. Organization of the *Escherichia coli* and *Salmonella typhimurium* chromosomes between flagellar regions IIIa and IIIb, including a large non-coding region. *J. Gen. Microbiol.* 139:1401–1407.

1607. Rahav-Manor, O., O. Carmel, R. Karpel, D. Taglicht, G. Glaser, S. Schuldiner, and E. Padan. 1992. NhaR, a protein homologous to a family of bacterial regulatory proteins (LysR), regulates nhaA, the sodium proton antiporter gene in *Escherichia coli*. *J. Biol. Chem.* 267:10433–10438.

1608. Rahman, M. A., J. Moskovitz, J. Strassman, H. Weissbach, and N. Brot. 1994. Physical map location of the peptide methionine sulfoxide reductase gene on the *Escherichia coli* chromosome. *J. Bacteriol.* 176:1548–1549.

1609. Rahman, M. A., H. Nelson, H. Weissbach, and N. Brot. 1992. Cloning, sequencing, and expression of the *Escherichia coli* peptide methionine sulfoxide reductase gene. *J. Biol. Chem.* 267:15549–15551.

1610. Raina, S., and C. Georgopoulos. 1991. The htrM gene, whose product is essential for *Escherichia coli* viability only at elevated temperatures, is identical to the rfaD gene. *Nucleic Acids Res.* 19:3811–3819.

1611. Raina, S., D. Missiakas, L. Baird, S. Kumar, and C. Georgopoulos. 1993. Identification and transcriptional analysis of the *Escherichia coli* htrE operon which is homologous to *pap* and related pilin operons. *J. Bacteriol.* 175:5009–5021.

1612. Rajagopalan, K. V., and J. L. Johnson. 1992. The pterin molybdenum cofactors. *J. Biol. Chem.* 267:10199–10202.

1613. Rakonjac, J., M. Milic, D. Ajdic-Predic, D. Santos, R. Ivanisevic, and D. J. Savic. 1992. nov: a new genetic locus that affects the response of *Escherichia coli* K-12 to novobiocin. *Mol. Microbiol.* 6:1547–1553.

1614. Raleigh, E. A. 1992. Organization and function of the mcrBC genes of *Escherichia coli* K-12. *Mol. Microbiol.* 6:1079–1086.

1615. Raleigh, E. A., J. Benner, F. Bloom, H. D. Braymer, E. DeCruz, K. Dharmalingam, J. Heitman, M. Noyer Weidner, A. Piekarowicz, and P. L. Kretz. 1991. Nomenclature relating to restriction of modified DNA in *Escherichia coli*. *J. Bacteriol.* 173:2707–2709.

1616. Ramseier, T. M., D. Negre, J. C. Cortay, M. Scarabel, A. J. Cozzone, and M. H. Saier, Jr. 1993. In vitro binding of the pleiotropic transcriptional regulatory protein, FruR, to the fru, pps, ace, pts and icd operons of *Escherichia coli* and *Salmonella typhimurium*. *J. Mol. Biol.* 234:28–44.

1617. Rao, B. J., and C. M. Radding. 1993. Homologous recognition promoted by RecA protein via non-Watson-Crick bonds between identical DNA strands. *Proc. Natl. Acad. Sci. USA* 90:6646–6650.

1618. Rao, N. N., and A. Torriani. 1990. Molecular aspects of phosphate transport in *Escherichia coli*. *Mol. Microbiol.* 4:1083–1090.

1619. Rasmussen, U. B., B. Mygind, and P. Nygaard. 1986. Purification and some properties of uracil phosphoribosyltransferase from *Escherichia coli* K12. *Biochim. Biophys. Acta* 881:268–275.

1620. Rawlings, M., and J. E. Cronan, Jr. 1992. The gene encoding *Escherichia coli* acyl carrier protein lies within a cluster of fatty acid biosynthetic genes. *J. Biol. Chem.* 267:5751–5754.

1621. Ray, J. M., and R. Bauerle. 1991. Purification and properties of tryptophan-sensitive 3-deoxy-D-arabino-heptulosonate-7-phosphate synthase from *Escherichia coli*. *J. Bacteriol.* 173:1894–1901.

1622. Ray, J. M., C. Yanofsky, and R. Bauerle. 1988. Mutational analysis of the catalytic and feedback sites of the tryptophan-sensitive 3-deoxy-D-arabino-heptulosonate-7-phosphate synthase of *Escherichia coli*. *J. Bacteriol.* 170:5500–5506.

1623. Reaney, S. K., C. Begg, S. J. Bungard, and J. R. Guest. 1993. Identification of the L-tartrate dehydratase genes (ttdA and ttdB) of *Escherichia coli* and evolutionary relationship with the class I fumarase genes. *J. Gen. Microbiol.* 139:1523–1530.

1624. Rebeck, G. W., and L. Samson. 1991. Increased spontaneous mutation and alkylation sensitivity of *Escherichia coli* strains lacking the ogt O^6-methylguanine DNA repair methyltransferase. *J. Bacteriol.* 173:2068–2076.

1625. Reece, R. J., and A. Maxwell. 1991. Probing the limits of the DNA breakage-reunion domain of the *Escherichia coli* DNA gyrase A protein. *J. Biol. Chem.* 266:3540–3546.

1626. Reed, K. E., and J. E. Cronan, Jr. 1993. Lipoic acid metabolism in *Escherichia coli*: sequencing and functional characterization of the lipA and lipB genes. *J. Bacteriol.* 175:1325–1336.

1627. Regnier, P., and M. Grunberg-Manago. 1990. RNase III cleavages in non-coding leaders of *Escherichia coli* transcripts control mRNA stability and genetic expression. *Biochimie* 72:825–834.

1628. **Regnier, P., M. Grunberg-Manago, and C. Portier.** 1987. Nucleotide sequence of the pnp gene of *Escherichia coli* encoding polynucleotide phosphorylase. Homology of the primary structure of the protein with the RNA-binding domain of ribosomal protein S1. *J. Biol. Chem.* **262:**63–68.

1629. **Reidl, J., and W. Boos.** 1991. The *malX malY* operon of *Escherichia coli* encodes a novel enzyme II of the phosphotransferase system recognizing glucose and maltose and an enzyme abolishing the endogenous induction of the maltose system. *J. Bacteriol.* **173:**4862–4876.

1630. **Reidl, J., K. Romisch, M. Ehrmann, and W. Boos.** 1989. MalI, a novel protein involved in regulation of the maltose system of *Escherichia coli*, is highly homologous to the repressor proteins GalR, CytR, and LacI. *J. Bacteriol.* **171:**4888–4899.

1631. **Reinstein, J., M. Brune, and A. Wittinghofer.** 1988. Mutations in the nucleotide binding loop of adenylate kinase of *Escherichia coli*. *Biochemistry* **27:**4712–4720.

1632. **Reitzer, L. J., R. Bueno, W. D. Cheng, S. A. Abrams, D. M. Rothstein, T. P. Hunt, B. Tyler, and B. Magasanik.** 1987. Mutations that create new promoters suppress the sigma 54 dependence of *glnA* transcription in *Escherichia coli*. *J. Bacteriol.* **169:**4279–4284.

1633. **Reitzer, L. J., B. Movsas, and B. Magasanik.** 1989. Activation of *glnA* transcription by nitrogen regulator I (NRI)-phosphate in *Escherichia coli*: evidence for a long-range physical interaction between NRI-phosphate and RNA polymerase. *J. Bacteriol.* **171:**5512–5522.

1634. **Reizer, J., V. Michotey, A. Reizer, and M. H. Saier, Jr.** 1994. Novel phosphotransferase system genes revealed by bacterial genome analysis: unique, putative fructose- and glucoside-specific systems. *Protein Sci.* **3:**440–450.

1635. **Reizer, J., A. Reizer, and M. H. Saier, Jr.** 1990. The Na$^+$/pantothenate symporter (PanF) of *Escherichia coli* is homologous to the Na$^+$/proline symporter (PutP) of *E. coli* and the Na$^+$/glucose symporters of mammals. *Res. Microbiol.* **141:**1069–1072.

1636. **Reizer, J., A. Reizer, and M. H. Saier, Jr.** 1990. The cellobiose permease of *Escherichia coli* consists of three proteins and is homologous to the lactose permease of *Staphylococcus aureus*. *Res. Microbiol.* **141:**1061–1067.

1637. **Reizer, J., A. Reizer, M. H. Saier, Jr., P. Bork, and C. Sander.** 1993. Exopolyphosphate phosphatase and guanosine pentaphosphate phosphatase belong to the sugar kinase/actin/hsp 70 superfamily. *Trends Biochem. Sci.* **18:**247–248.

1638. **Reuter, K., R. Slany, F. Ullrich, and H. Kersten.** 1991. Structure and organization of *Escherichia coli* genes involved in biosynthesis of the deazaguanine derivative queuine, a nutrient factor for eukaryotes. *J. Bacteriol.* **173:**2256–2264.

1639. **Reyes, M., and H. A. Shuman.** 1988. Overproduction of MalK protein prevents expression of the *Escherichia coli mal* regulon. *J. Bacteriol.* **170:**4598–4602.

1640. **Ricca, E., D. A. Aker, and J. M. Calvo.** 1989. A protein that binds to the regulatory region of the *Escherichia coli ilvIH* operon. *J. Bacteriol.* **171:**1658–1664.

1641. **Ricca, E., D. Limauro, C. T. Lago, and M. de Felice.** 1988. Enhanced acetohydroxy acid synthase III activity in an *ilvH* mutant of *Escherichia coli* K-12. *J. Bacteriol.* **170:**5197–5199.

1642. **Rice, P. W., D. A. Polayes, and J. E. Dahlberg.** 1987. Spot 42 RNA of *Escherichia coli* is not an mRNA. *J. Bacteriol.* **169:**3850–3852.

1643. **Richarme, G.** 1989. Purification of a new dihydrolipoamide dehydrogenase from *Escherichia coli*. *J. Bacteriol.* **171:**6580–6585.

1644. **Richarme, G., and M. Kohiyama.** 1994. Amino acid specificity of the *Escherichia coli* chaperone GroEL (heat shock protein 60). *J. Biol. Chem.* **269:**7095–7098.

1645. **Richaud, C., F. Richaud, C. Martin, C. Haziza, and J. C. Patte.** 1984. Regulation of expression and nucleotide sequence of the *Escherichia coli dapD* gene. *J. Biol. Chem.* **259:**14824–14828.

1646. **Richter, G., H. Ritz, G. Katzenmeier, R. Volk, A. Kohnle, F. Lottspeich, D. Allendorf, and A. Bacher.** 1993. Biosynthesis of riboflavin: cloning, sequencing, mapping, and expression of the gene coding for GTP cyclohydrolase II in *Escherichia coli*. *J. Bacteriol.* **175:**4045–4051.

1647. **Richter, G., R. Volk, C. Krieger, H. W. Lahm, U. Rothlisberger, and A. Bacher.** 1992. Biosynthesis of riboflavin: cloning, sequencing, and expression of the gene coding for 3,4-dihydroxy-2-butanone 4-phosphate synthase of *Escherichia coli*. *J. Bacteriol.* **174:**4050–4056.

1648. **Rick, P. D., G. L. Hubbard, and K. Barr.** 1994. Role of the *rfe* gene in the synthesis of the O8 antigen in *Escherichia coli* K-12. *J. Bacteriol.* **176:**2877–2884.

1649. **Ried, G., I. Hindennach, and U. Henning.** 1990. Role of lipopolysaccharide in assembly of *Escherichia coli* outer membrane proteins OmpA, OmpC, and OmpF. *J. Bacteriol.* **172:**6048–6053.

1650. **Rieder, G., M. J. Merrick, H. Castorph, and D. Kleiner.** 1994. Function of hisF and hisH gene products in histidine biosynthesis. *J. Biol. Chem.* **269:**14386–14390.

1651. **Riftina, F., E. DeFalco, and J. S. Krakow.** 1989. Monoclonal antibodies as probes of the topological arrangement of the alpha subunits of *Escherichia coli* RNA polymerase. *Biochemistry* **28:**3299–3305.

1652. **Riley, M.** 1993. Functions of the gene products of *Escherichia coli*. *Microbiol. Rev.* **57:**862–952.

1653. **Ringquist, S., D. Schneider, T. Gibson, C. Baron, A. Bock, and L. Gold.** 1994. Recognition of the mRNA selenocysteine insertion sequence by the specialized translational elongation factor SELB. *Genes Dev.* **8:**376–385.

1654. **Ringquist, S., and C. L. Smith.** 1992. The *Escherichia coli* chromosome contains specific, unmethylated dam and dcm sites. *Proc. Natl. Acad. Sci. USA* **89:**4539–4543.

1655. **Riordan, C., and H. L. Kornberg.** 1977. Location of galP, a gene which specifies galactose permease activity, on the *Escherichia coli* linkage map. *Proc. R. Soc. London Ser. B* **198:**401–410.

1656. **Rioux, C. R., and R. J. Kadner.** 1989. Vitamin B12 transport in *Escherichia coli* K12 does not require the btuE gene of the btuCED operon. *Mol. Gen. Genet.* **217:**301–308.

1657. **Rivers, S. L., E. McNairn, F. Blasco, G. Giordano, and D. H. Boxer.** 1993. Molecular genetic analysis of the moa operon of *Escherichia coli* K-12 required for molybdenum cofactor biosynthesis. *Mol. Microbiol.* **8:**1071–1081.

1658. **Rizo, J., F. J. Blanco, B. Kobe, M. D. Bruch, and L. M. Gierasch.** 1993. Conformational behavior of *Escherichia coli* OmpA signal peptides in membrane mimetic environments. *Biochemistry* **32:**4881–4894.

1659. **Roa, B. B., D. M. Connolly, and M. E. Winkler.** 1989. Overlap between *pdxA* and *ksgA* in the complex *pdxA-ksgA-apaG-apaH* operon of *Escherichia coli* K-12. *J. Bacteriol.* **171:**4767–4777.

1660. **Robbins, A. R.** 1975. Regulation of the *Escherichia coli* methylgalactoside transport system by gene mglD. *J. Bacteriol.* **123:**69–74.

1661. **Robbins, J. C., and D. L. Oxender.** 1973. Transport system for alanine, serine, and glycine in *Escherichia coli* K-12. *J. Bacteriol.* **116:**12–18.

1662. **Roberts, L. M., and E. C. R. Reeve.** 1970. Two mutations giving low-level streptomycin resistance in *Escherichia coli* K12. *Genet. Res.* **16:**359–365.

1663. **Robertson, A. M., P. A. Sullivan, and M. C. Jones-Mortimer.** 1980. Two genes affecting glucarate utilization in *Escherichia coli* K12. *J. Gen. Microbiol.* **117:**377–382.

1664. **Robinson, C. L., and J. H. Jackson.** 1982. New acetohydroxy acid synthase activity from mutational activation of a cryptic gene in *Escherichia coli* K-12. *Mol. Gen. Genet.* **186:**240–246.

1665. **Rodriguez-Sainz, M. C., C. Hernandez-Chico, and F. Moreno.** 1990. Molecular characterization of pmbA, an *Escherichia coli* chromosomal gene required for the production of the antibiotic peptide MccB17. *Mol. Microbiol.* **4:**1921–1932.

1666. **Roehl, R. A., and R. T. Vinopal.** 1980. Genetic locus, distant from *ptsM*, affecting enzyme IIA/IIB function in *Escherichia coli* K-12. *J. Bacteriol.* **142:**120–130.

1667. **Roepe, P. D., and H. R. Kaback.** 1990. Isolation and functional reconstitution of soluble melibiose permease from *Escherichia coli*. *Biochemistry* **29:**2572–2577.

1668. **Roesser, J. R., and C. Yanofsky.** 1991. The effects of leader peptide sequence and length on attenuation control of the trp operon of *E. coli*. *Nucleic Acids Res.* **19:**795–800.

1669. **Rogers, M. J., T. Adachi, H. Inokuchi, and D. Soll.** 1992. Switching tRNA(Gln) identity from glutamine to tryptophan. *Proc. Natl. Acad. Sci. USA* **89:**3463–3467.

1670. **Rogers, M. J., T. Adachi, H. Inokuchi, and D. Soll.** 1994. Functional communication in the recognition of tRNA by *Escherichia coli* glutaminyl-tRNA synthetase. *Proc. Natl. Acad. Sci. USA* **91:**291–295.

1671. **Rogers, M. J., T. Ohgi, J. Plumbridge, and D. Soll.** 1988. Nucleotide sequences of the *Escherichia coli* nagE and nagB genes: the structural genes for the N-acetylglucosamine transport protein of the bacterial phosphoenolpyruvate: sugar phosphotransferase system and for glucosamine-6-phosphate deaminase. *Gene* **62:**197–207.

1672. **Rogers, M. J., and D. Soll.** 1988. Discrimination between glutaminyl-tRNA synthetase and seryl-tRNA synthetase involves nucleotides in the acceptor helix of tRNA. *Proc. Natl. Acad. Sci. USA* **85:**6627–6631.

1673. **Rogers, S. D., M. R. Bhave, J. F. Mercer, J. Camakaris, and B. T. Lee.** 1991. Cloning and characterization of cutE, a gene involved in copper transport in *Escherichia coli*. *J. Bacteriol.* **173:**6742–6748.

1674. **Rojiani, M. V., H. Jakubowski, and E. Goldman.** 1989. Effect of variation of charged and uncharged tRNA(Trp) levels on ppGpp synthesis in *Escherichia coli*. *J. Bacteriol.* **171:**6493–6502.

1675. **Roman, S. J., B. B. Frantz, and P. Matsumura.** 1993. Gene sequence, overproduction, purification and determination of the wild-type level of the *Escherichia coli* flagellar switch protein FliG. *Gene* **133:**103–108.

1676. **Roman, S. J., M. Meyers, K. Volz, and P. Matsumura.** 1992. A chemotactic signaling surface on CheY defined by suppressors of flagellar switch mutations. *J. Bacteriol.* **174:**6247–6255.

1677. **Romby, P., C. Brunel, J. Caillet, M. Springer, M. Grunberg-Manago, E. Westhof, C. Ehresmann, and B. Ehresmann.** 1992. Molecular mimicry in translational control of *E. coli* threonyl-tRNA synthetase gene. Competitive inhibition in tRNA aminoacylation and operator-repressor recognition switch using tRNA identity rules. *Nucleic Acids Res.* **20:**5633–5640.

1678. **Romeis, T., U. Kohlrausch, K. Burgdorf, and J. V. Holtje.** 1991. Murein chemistry of cell division in *Escherichia coli*. *Res. Microbiol.* **142:**325–332.

1679. **Romeo, T., and M. Gong.** 1993. Genetic and physical mapping of the regulatory gene csrA on the *Escherichia coli* K-12 chromosome. *J. Bacteriol.* **175:**5740–5741.

1680. **Romeo, T., M. Gong, M. Y. Liu, and A. M. Brun-Zinkernagel.** 1993. Identification and molecular characterization of csrA, a pleiotropic gene from *Escherichia coli* that affects glycogen biosynthesis, gluconeogenesis, cell size, and surface properties. *J. Bacteriol.* **175:**4744–4755.

1681. **Romeo, T., A. Kumar, and J. Preiss.** 1988. Analysis of the *Escherichia coli* glycogen gene cluster suggests that catabolic enzymes are encoded among the biosynthetic genes. *Gene* **70:**363–376.

1682. **Romero, D. P., J. A. Arredondo, and R. R. Traut.** 1990. Identification of a region of *Escherichia coli* ribosomal protein L2 required for the assembly of L16 into the 50 S ribosomal subunit. *J. Biol. Chem.* **265:**18185–18191.

1683. Ron, E. Z., S. Alajem, D. Biran, and N. Grossman. 1990. Adaptation of *Escherichia coli* to elevated temperatures: the metA gene product is a heat shock protein. *Antonie Leeuwenhoek* 58:169–174.

1684. Roncero, C., and M. J. Casadaban. 1992. Genetic analysis of the genes involved in synthesis of the lipopolysaccharide core in *Escherichia coli* K-12: three operons in the rfa locus. *J. Bacteriol.* 174:3250–3260.

1685. Roof, W. D., S. M. Horne, K. D. Young, and R. Young. 1994. slyD, a host gene required for phi X174 lysis, is related to the FK506-binding protein family of peptidyl-prolyl cis-trans-isomerases. *J. Biol. Chem.* 269:2902–2910.

1686. Rose, T., P. Glaser, W. K. Surewicz, H. H. Mantsch, J. Reinstein, K. Le Blay, A. M. Gilles, and O. Barzu. 1991. Structural and functional consequences of amino acid substitutions in the second conserved loop of *Escherichia coli* adenylate kinase. *J. Biol. Chem.* 266:23654–23659.

1687. Rosendahl, G., and S. Douthwaite. 1993. Ribosomal proteins L11 and L10.(L12)4 and the antibiotic thiostrepton interact with overlapping regions of the 23 S rRNA backbone in the ribosomal GTPase centre. *J. Mol. Biol.* 234:1013–1020.

1688. Rosenfeld, S. A., P. E. Stevis, and N. W. Ho. 1984. Cloning and characterization of the xyl genes from *Escherichia coli*. *Mol. Gen. Genet.* 194:410–415.

1689. Rossmann, R., M. Sauter, F. Lottspeich, and A. Bock. 1994. Maturation of the large subunit (HYCE) of *Escherichia coli* hydrogenase 3 requires nickel incorporation followed by C-terminal processing at Arg537. *Eur. J. Biochem.* 220:377–384.

1690. Rossmann, R., G. Sawers, and A. Bock. 1991. Mechanism of regulation of the formate-hydrogenlyase pathway by oxygen, nitrate, and pH: definition of the formate regulon. *Mol. Microbiol.* 5:2807–2814.

1691. Rostas, K., S. J. Morton, S. M. Picksley, and R. G. Lloyd. 1987. Nucleotide sequence and LexA regulation of the *Escherichia coli* recN gene. *Nucleic Acids Res.* 15:5041–5049.

1692. Rousset, J. P., E. Gilson, and M. Hofnung. 1986. malM, a new gene of the maltose regulon in *Escherichia coli* K12. II. Mutations affecting the signal peptide of the MalM protein. *J. Mol. Biol.* 191:313–320.

1693. Rowbury, R. J., and D. D. Woods. 1964. O-succinyl-homoserine as an intermediate in the synthesis of cystathionine by *Escherichia coli*. *J. Gen. Microbiol.* 36:341–358.

1694. Rowley, D. L., A. J. Pease, and R. E. Wolf, Jr. 1991. Genetic and physical analyses of the growth rate-dependent regulation of *Escherichia coli* zwf expression. *J. Bacteriol.* 173:4660–4667.

1695. Rowley, D. L., and R. E. Wolf, Jr. 1991. Molecular characterization of the *Escherichia coli* K-12 zwf gene encoding glucose 6-phosphate dehydrogenase. *J. Bacteriol.* 173:968–977.

1696. Roy, I., and P. F. Leadlay. 1992. Physical map location of the new *Escherichia coli* gene sbm. *J. Bacteriol.* 174:5763–5764.

1697. Rua, J., A. G. Robertson, and H. G. Nimmo. 1992. Identification of the histidine residue in *Escherichia coli* isocitrate lyase that reacts with diethylpyrocarbonate. *Biochim. Biophys. Acta* 1122:212–218.

1698. Rubio, V., J. Cervera, C. J. Lusty, E. Bendala, and H. G. Britton. 1991. Domain structure of the large subunit of *Escherichia coli* carbamoyl phosphate synthetase. Location of the binding site for the allosteric inhibitor UMP in the COOH-terminal domain. *Biochemistry* 30:1068–1075.

1699. Rudolph, F. B., D. L. Purich, and H. J. Fromm. 1968. Coenzyme A-linked aldehyde dehydrogenase from *Escherichia coli*. I. Partial purification, properties, and kinetic studies of the enzyme. *J. Biol. Chem.* 243:5539–5545.

1700. Ruijter, G. J., G. van Meurs, M. A. Verwey, P. W. Postma, and K. van Dam. 1992. Analysis of mutations that uncouple transport from phosphorylation in enzyme IIGlc of the *Escherichia coli* phosphoenolpyruvate-dependent phosphotransferase system. *J. Bacteriol.* 174:2843–2850.

1701. Ruiz-Vazquez, R., and E. Cerda-Olmedo. 1980. An *Escherichia coli* mutant refractory to nitrosoguanidine mutagenesis. *Mol. Gen. Genet.* 178:625–631.

1702. Rule, G. S., E. A. Pratt, C. C. Chin, F. Wold, and C. Ho. 1985. Overproduction and nucleotide sequence of the respiratory D-lactate dehydrogenase of *Escherichia coli*. *J. Bacteriol.* 161:1059–1068.

1703. Rusnak, F., J. Liu, N. Quinn, G. A. Berchtold, and C. T. Walsh. 1990. Subcloning of the enterobactin biosynthetic gene entB: expression, purification, characterization, and substrate specificity of isochorismatase. *Biochemistry* 29:1425–1435.

1704. Rusnak, F., M. Sakaitani, D. Drueckhammer, J. Reichert, and C. T. Walsh. 1991. Biosynthesis of the *Escherichia coli* siderophore enterobactin: sequence of the entF gene, expression and purification of EntF, and analysis of covalent phosphopantetheine. *Biochemistry* 30:2916–2927.

1705. Russ, E., U. Kaiser, and H. Sandermann, Jr. 1988. Lipid-dependent membrane enzymes. Purification to homogeneity and further characterization of diacylglycerol kinase from *Escherichia coli*. *Eur. J. Biochem.* 171:335–342.

1706. Russel, M., P. Model, and A. Holmgren. 1990. Thioredoxin or glutaredoxin in *Escherichia coli* is essential for sulfate reduction but not for deoxyribonucleotide synthesis. *J. Bacteriol.* 172:1923–1929.

1707. Russell, C. B., R. C. Stewart, and F. W. Dahlquist. 1989. Control of transducer methylation levels in *Escherichia coli*: investigation of components essential for modulation of methylation and demethylation reactions. *J. Bacteriol.* 171:3609–3618.

1708. Russell, P. W., and P. E. Orndorff. 1992. Lesions in two *Escherichia coli* type 1 pilus genes alter pilus number and length without affecting receptor binding. *J. Bacteriol.* 174:5923–5935.

1709. Russo, F. D., J. M. Slauch, and T. J. Silhavy. 1993. Mutations that affect separate functions of OmpR the phosphorylated regulator of porin transcription in *Escherichia coli*. *J. Mol. Biol.* 231:261–273.

1710. Ruvinov, S. B., and E. W. Miles. 1992. Subunit communication in the tryptophan synthase alpha 2 beta 2 complex. Effects of beta subunit ligands on proteolytic cleavage of a flexible loop in the alpha subunit. *FEBS Lett.* 299:197–200.

1711. Ruyter, G. J., P. W. Postma, and K. van Dam. 1991. Control of glucose metabolism by enzyme IIGlc of the phosphoenolpyruvate-dependent phosphotransferase system in *Escherichia coli*. *J. Bacteriol.* 173:6184–6191.

1712. Ryden-Aulin, M., Z. Shaoping, P. Kylsten, and L. A. Isaksson. 1993. Ribosome activity and modification of 16S MHA are influenced by deletion of ribosomal protein S20. *Mol. Microbiol.* 7:983–992.

1713. Ryder, L., M. C. Whitby, and R. G. Lloyd. 1994. Mutation of recF, recJ, recO, recQ, or recR improves Hfr recombination in resolvase-deficient ruv recG strains of *Escherichia coli*. *J. Bacteriol.* 176:1570–1577.

1714. Ryu, S., J. Kim, S. Adhya, and S. Garges. 1993. Pivotal role of amino acid at position 138 in the allosteric hinge reorientation of cAMP receptor protein. *Proc. Natl. Acad. Sci. USA* 90:75–79.

1715. Sacerdot, C., G. Vachon, S. Laalami, F. Morel-Deville, Y. Cenatiempo, and M. Grunberg-Manago. 1992. Both forms of translational initiation factor IF2 (alpha and beta) are required for maximal growth of *Escherichia coli*. Evidence for two translational initiation codons for IF2 beta. *J. Mol. Biol.* 225:67–80.

1716. Saffen, D. W., K. A. Presper, T. L. Doering, and S. Roseman. 1987. Sugar transport by the bacterial phosphotransferase system. Molecular cloning and structural analysis of the *Escherichia coli* ptsH, ptsI, and crr genes. *J. Biol. Chem.* 262:16241–16253.

1717. Sagara, K., S. Matsuyama, and S. Mizushima. 1994. SecF stabilizes SecD and SecY, components of the protein translocation machinery of the *Escherichia coli* cytoplasmic membrane. *J. Bacteriol.* 176:4111–4116.

1718. Sahin-Toth, M., R. L. Dunten, A. Gonzalez, and H. R. Kaback. 1992. Functional interactions between putative intramembrane charged residues in the lactose permease of *Escherichia coli*. *Proc. Natl. Acad. Sci. USA* 89:10547–10551.

1719. Said, B., J. R. Cole, and M. Nomura. 1988. Mutational analysis of the L1 binding site of 23S rRNA in *Escherichia coli*. *Nucleic Acids Res.* 16:10529–10545.

1720. Said, B., C. R. Ghosn, L. Vu, and W. D. Nunn. 1988. Nucleotide sequencing and expression of the fadL gene involved in long-chain fatty acid transport in *Escherichia coli*. *Mol. Microbiol.* 2:363–370.

1721. Saint-Girons, I., N. Duchange, M. M. Zakin, I. Park, D. Margarita, P. Ferrara, and G. N. Cohen. 1983. Nucleotide sequence of metF, the *E. coli* structural gene for 5–10 methylene tetrahydrofolate reductase and of its control region. *Nucleic Acids Res.* 11:6723–6732.

1722. Sak, B. D., A. Eisenstark, and D. Touati. 1989. Exonuclease III and the catalase hydroperoxidase II in *Escherichia coli* are both regulated by the katF gene product. *Proc. Natl. Acad. Sci. USA* 86:3271–3275.

1723. Sakakibara, Y. 1992. Novel *Escherichia coli* mutant, dnaR, thermosensitive in initiation of chromosome replication. *J. Mol. Biol.* 226:979–987.

1724. Sakamoto, K., G. Kawai, T. Niimi, T. Satoh, M. Sekine, Z. Yamaizumi, S. Nishimura, T. Miyazawa, and S. Yokoyama. 1993. A modified uridine in the first position of the anticodon of a minor species of arginine tRNA, the argU gene product, from *Escherichia coli*. *Eur. J. Biochem.* 216:369–375.

1725. Sakamoto, N., A. M. Kotre, and M. A. Savageau. 1975. Glutamate dehydrogenase from *Escherichia coli*: purification and properties. *J. Bacteriol.* 124:775–783.

1726. Sakashita, H., T. Sakuma, T. Ohkubo, M. Kainosho, K. Sakumi, M. Sekiguchi, and K. Morikawa. 1993. Folding topology and DNA binding of the N-terminal fragment of Ada protein. *FEBS Lett.* 323:252–256.

1727. Sambasivarao, D., D. G. Scraba, C. Trieber, and J. H. Weiner. 1990. Organization of dimethyl sulfoxide reductase in the plasma membrane of *Escherichia coli*. *J. Bacteriol.* 172:5938–5948.

1728. Sambasivarao, D., and J. H. Weiner. 1991. Dimethyl sulfoxide reductase of *Escherichia coli*: an investigation of function and assembly by use of in vivo complementation. *J. Bacteriol.* 173:5935–5943.

1729. Sampei, G., and K. Mizobuchi. 1988. Nucleotide sequence of the *Escherichia coli* purF gene encoding amidophosphoribosyltransferase for de novo purine nucleotide synthesis. *Nucleic Acids Res.* 16:8717

1730. Samsonov, V. V., E. R. Odoevskaia, and S. P. Sineokii. 1992. Cloning and complementation analysis of the *Escherichia coli* gpr locus, influencing DNA replication of certain lamdoid phages. *Genetika* 28:39–45. (In Russian.)

1731. Sanchez-Anzaldo, F. J., and F. Bastarrachea. 1974. Genetic characterization of streptomycin-resistant and dependent mutants of *Escherichia coli* K12. *Mol. Gen. Genet.* 130:47–64.

1732. Sandberg, V. A., B. Kren, J. A. Fuchs, and C. Woodward. 1991. *Escherichia coli* glutaredoxin: cloning and overexpression, thermodynamic stability of the oxidized and reduced forms, and report of an N-terminal extended species. *Biochemistry* 30:5475–5484.

1733. Sanders, D. A., B. L. Gillece-Castro, A. M. Stock, A. L. Burlingame, and D. E. Koshland, Jr. 1989. Identification of the site of phosphorylation of the chemotaxis response regulator protein, CheY. *J. Biol. Chem.* 264:21770–21778.

1734. Sandler, S. J., and A. J. Clark. 1994. RecOR suppression of recF mutant phenotypes in *Escherichia coli* K-12. *J. Bacteriol.* 176:3661–3672.

1735. Sankaran, K., and H. C. Wu. 1994. Lipid modification of bacterial prolipoprotein. Transfer of diacylglyceryl moiety from phosphatidylglycerol. *J. Biol. Chem.* 269:19701–19706.

1736. **Santer, M., U. Santer, K. Nurse, A. Bakin, P. Cunningham, M. Zain, D. O'Connell, and J. Ofengand.** 1993. Functional effects of a G to U base change at position 530 in a highly conserved loop of *Escherichia coli* 16S RNA. *Biochemistry* 32:5539–5547.

1737. **Sanyal, I., G. Cohen, and D. H. Flint.** 1994. Biotin synthase: purification, characterization as a [2Fe-2S]cluster protein, and in vitro activity of the *Escherichia coli* bioB gene product. *Biochemistry* 33:3625–3631.

1738. **Saporito, S. M., M. Gedenk, and R. P. Cunningham.** 1989. Role of exonuclease III and endonuclease IV in repair of pyrimidine dimers initiated by bacteriophage T4 pyrimidine dimer-DNA glycosylase. *J. Bacteriol.* 171:2542–2546.

1739. **Sarsero, J. P., and A. J. Pittard.** 1991. Molecular analysis of the TyrR protein-mediated activation of *mtr* gene expression in *Escherichia coli* K-12. *J. Bacteriol.* 173:7701–7704.

1740. **Sarsero, J. P., P. J. Wookey, P. Gollnick, C. Yanofsky, and A. J. Pittard.** 1991. A new family of integral membrane proteins involved in transport of aromatic amino acids in *Escherichia coli. J. Bacteriol.* 173:3231–3234.

1741. **Sarsero, J. P., P. J. Wookey, and A. J. Pittard.** 1991. Regulation of expression of the *Escherichia coli* K-12 *mtr* gene by TyrR protein and Trp repressor. *J. Bacteriol.* 173:4133–4143.

1742. **Sasarman, A., P. Chartrand, M. Lavoie, D. Tardif, R. Proschek, and C. Lapointe.** 1979. Mapping of a new hem gene in *Escherichia coli* K12. *J. Gen. Microbiol.* 113:297–303.

1743. **Sasarman, A., P. Chartrand, R. Proschek, M. Desrochers, D. Tardif, and C. Lapointe.** 1975. Uroporphyrin-accumulating mutant in *Escherichia coli* K-12. *J. Bacteriol.* 124:1205–1212.

1744. **Sasarman, A., Y. Echelard, J. Letowski, D. Tardif, and M. Drolet.** 1988. Nucleotide sequence of the hemX gene, the third member of the Uro operon of *Escherichia coli* K12. *Nucleic Acids Res.* 16:11835

1745. **Sastry, S. S., and R. Jayaraman.** 1985. Inhibitors of nitrofuran reduction in *Escherichia coli*: evidence for their existence, partial purification, binding of nitrofurantoin in vitro, and implications for nitrofuran resistance. *Arch. Biochem. Biophys.* 236:252–259.

1746. **Satishchandran, C., J. C. Taylor, and G. D. Markham.** 1990. Novel *Escherichia coli* K-12 mutants impaired in S-adenosylmethionine synthesis. *J. Bacteriol.* 172:4489–4496.

1747. **Satishchandran, C., J. C. Taylor, and G. D. Markham.** 1993. Isozymes of S-adenosylmethionine synthetase are encoded by tandemly duplicated genes in *Escherichia coli. Mol. Microbiol.* 9:835–846.

1748. **Sato, T., M. Ohki, T. Yura, and K. Ito.** 1979. Genetic studies of an *Escherichia coli* K-12 temperature-sensitive mutant defective in membrane protein synthesis. *J. Bacteriol.* 138:305–313.

1749. **Sauer, M., K. Hantke, and V. Braun.** 1990. Sequence of the fhuE outer-membrane receptor gene of *Escherichia coli* K12 and properties of mutants. *Mol. Microbiol.* 4:427–437.

1750. **Sauter, M., R. Bohm, and A. Bock.** 1992. Mutational analysis of the operon (hyc) determining hydrogenase 3 formation in *Escherichia coli. Mol. Microbiol.* 6:1523–1532.

1751. **Sauter, M., and R. G. Sawers.** 1990. Transcriptional analysis of the gene encoding pyruvate formate-lyase-activating enzyme of *Escherichia coli. Mol. Microbiol.* 4:355–363.

1752. **Sawers, G., J. Heider, E. Zehelein, and A. Bock.** 1991. Expression and operon structure of the *sel* genes of *Escherichia coli* and identification of a third selenium-containing formate dehydrogenase isoenzyme. *J. Bacteriol.* 173:4983–4993.

1753. **Sawers, G., and B. Suppmann.** 1992. Anaerobic induction of pyruvate formate-lyase gene expression is mediated by the ArcA and FNR proteins. *J. Bacteriol.* 174:3474–3478.

1754. **Sawers, R. G., and D. H. Boxer.** 1986. Purification and properties of membrane-bound hydrogenase isoenzyme 1 from anaerobically grown *Escherichia coli* K12. *Eur. J. Biochem.* 156:265–275.

1755. **Saxena, P., and J. R. Walker.** 1992. Expression of *argU*, the *Escherichia coli* gene coding for a rare arginine tRNA. *J. Bacteriol.* 174:1956–1964.

1756. **Schaaper, R. M., B. I. Bond, and R. G. Fowler.** 1989. A.T- - - -C.G transversions and their prevention by the *Escherichia coli* mutT and mutHLS pathways. *Mol. Gen. Genet.* 219:256–262.

1757. **Schatz, D., R. Leberman, and F. Eckstein.** 1991. Interaction of *Escherichia coli* tRNA(Ser) with its cognate aminoacyl-tRNA synthetase as determined by footprinting with phosphorothioate-containing tRNA transcripts. *Proc. Natl. Acad. Sci. USA* 88:6132–6136.

1758. **Schendel, F. J., E. Mueller, J. Stubbe, A. Shiau, and J. M. Smith.** 1989. Formylglycinamide ribonucleotide synthetase from *Escherichia coli*: cloning, sequencing, overproduction, isolation, and characterization. *Biochemistry* 28:2459–2471.

1759. **Schinzel, R.** 1991. Active site lysine promotes catalytic function of pyridoxal 5′-phosphate in alpha-glucan phosphorylases. *J. Biol. Chem.* 266:9428–9431.

1760. **Schinzel, R., and D. Palm.** 1990. *Escherichia coli* maltodextrin phosphorylase: contribution of active site residues glutamate-637 and tyrosine-538 to the phosphorolytic cleavage of alpha-glucans. *Biochemistry* 29:9956–9962.

1761. **Schleyer, M., and E. P. Bakker.** 1993. Nucleotide sequence and 3′-end deletion studies indicate that the K⁺-uptake protein Kup from *Escherichia coli* is composed of a hydrophobic core linked to a large and partially essential hydrophilic C terminus. *J. Bacteriol.* 175:6925–6931.

1762. **Schlindwein, C., and M. A. Mandrand.** 1991. Nucleotide sequence of the fdhE gene involved in respiratory formate dehydrogenase formation in *Escherichia coli* K-12. *Gene* 97:147–148.

1763. **Schlosser, A., A. Hamann, D. Bossemeyer, E. Schneider, and E. P. Bakker.** 1993. NAD⁺ binding to the *Escherichia coli* K(⁺)-uptake protein TrkA and sequence similarity between TrkA and domains of a family of dehydrogenases suggest a role for NAD⁺ in bacterial transport. *Mol. Microbiol.* 9:533–543.

1764. **Schmid, C., W. Meining, S. Weinkauf, L. Bachmann, H. Ritz, S. Eberhardt, W. Gimbel, T. Werner, H. W. Lahm, and H. Nar.** 1993. Studies on GTP cyclohydrolase I of *Escherichia coli. Adv. Exp. Med. Biol.* 338:157–162.

1765. **Schneider, E., S. Freundlieb, S. Tapio, and W. Boos.** 1992. Molecular characterization of the MalT-dependent periplasmic alpha-amylase of *Escherichia coli* encoded by malS. *J. Biol. Chem.* 267:5148–5154.

1766. **Schneppe, B., G. Deckers-Hebestreit, and K. Altendorf.** 1990. Overproduction and purification of the uncI gene product of the ATP synthase of *Escherichia coli. J. Biol. Chem.* 265:389–395.

1767. **Schnetz, K., and B. Rak.** 1990. Beta-glucoside permease represses the bgl operon of *Escherichia coli* by phosphorylation of the antiterminator protein and also interacts with glucose-specific enzyme III, the key element in catabolite control. *Proc. Natl. Acad. Sci. USA* 87:5074–5078.

1768. **Schnetz, K., S. L. Sutrina, M. H. Saier, Jr., and B. Rak.** 1990. Identification of catalytic residues in the beta-glucoside permease of *Escherichia coli* by site-specific mutagenesis and demonstration of interdomain cross-reactivity between the beta-glucoside and glucose systems. *J. Biol. Chem.* 265:13464–13471.

1769. **Schnetz, K., C. Toloczyki, and B. Rak.** 1987. Beta-glucoside (*bgl*) operon of *Escherichia coli* K-12: nucleotide sequence, genetic organization, and possible evolutionary relationship to regulatory components of two *Bacillus subtilis* genes. *J. Bacteriol.* 169:2579–2590.

1770. **Schoedon, G., U. Redweik, G. Frank, R. G. Cotton, and N. Blau.** 1992. Allosteric characteristics of GTP cyclohydrolase I from *Escherichia coli. Eur. J. Biochem.* 210:561–568.

1771. **Schoenlein, P. V., B. B. Roa, and M. E. Winkler.** 1989. Divergent transcription of *pdxB* and homology between the *pdxB* and *serA* gene products in *Escherichia coli* K-12. *J. Bacteriol.* 171:6084–6092.

1772. **Scholle, A., J. Vreeman, V. Blank, A. Nold, W. Boos, and M. D. Manson.** 1987. Sequence of the mglB gene from *Escherichia coli* K12: comparison of wild-type and mutant galactose chemoreceptors. *Mol. Gen. Genet.* 208:247–253.

1773. **Schreiber, G., S. Metzger, E. Aizenman, S. Roza, M. Cashel, and G. Glaser.** 1991. Overexpression of the relA gene in *Escherichia coli. J. Biol. Chem.* 266:3760–3767.

1774. **Schroder, H., T. Langer, F. U. Hartl, and B. Bukau.** 1993. DnaK, DnaJ and GrpE form a cellular chaperone machinery capable of repairing heat-induced protein damage. *EMBO J.* 12:4137–4144.

1775. **Schryvers, A., and J. H. Weiner.** 1981. The anaerobic sn-glycerol-3-phosphate dehydrogenase of *Escherichia coli*. Purification and characterization. *J. Biol. Chem.* 256:9959–9965.

1776. **Schuller, D. J., C. H. Fetter, L. J. Banaszak, and G. A. Grant.** 1989. Enhanced expression of the *Escherichia coli* serA gene in a plasmid vector. Purification, crystallization, and preliminary X-ray data of D-3 phosphoglycerate dehydrogenase. *J. Biol. Chem.* 264:2645–2648.

1777. **Schulman, L. H., and H. Pelka.** 1990. An anticodon change switches the identity of *E. coli* tRNA(mMet) from methionine to threonine. *Nucleic Acids Res.* 18:285–289.

1778. **Schultz, J. E., and A. Matin.** 1991. Molecular and functional characterization of a carbon starvation gene of *Escherichia coli. J. Mol. Biol.* 218:129–140.

1779. **Schultz-Hauser, G., W. Koster, H. Schwarz, and V. Braun.** 1992. Iron(III) hydroxamate transport in *Escherichia coli* K-12: FhuB-mediated membrane association of the FhuC protein and negative complementation of *fhuC* mutants. *J. Bacteriol.* 174:2305–2311.

1780. **Schulze, E., A. H. Westphal, G. Obmolova, A. Mattevi, W. G. Hol, and A. de Kok.** 1991. The catalytic domain of the dihydrolipoyl transacetylase component of the pyruvate dehydrogenase complex from *Azotobacter vinelandii* and *Escherichia coli*. Expression, purification, properties and preliminary X-ray analysis. *Eur. J. Biochem.* 201:561–568.

1781. **Schulze, E., A. H. Westphal, M. Veenhuis, and A. de Kok.** 1992. Purification and cellular localization of wild type and mutated dihydrolipoyltransacetylases from *Azotobacter vinelandii* and *Escherichia coli* expressed in *E. coli. Biochim. Biophys. Acta* 1120:87–96.

1782. **Schumacher, M. A., J. R. Macdonald, J. Bjorkman, S. L. Mowbray, and R. G. Brennan.** 1993. Structural analysis of the purine repressor, an *Escherichia coli* DNA-binding protein. *J. Biol. Chem.* 268:12282–12288.

1783. **Schwan, W. R., H. S. Seifert, and J. L. Duncan.** 1992. Growth conditions mediate differential transcription of *fim* genes involved in phase variation of type 1 pili. *J. Bacteriol.* 174:2367–2375.

1784. **Schweizer, H., and T. J. Larson.** 1987. Cloning and characterization of the aerobic sn-glycerol-3-phosphate dehydrogenase structural gene glpD of *Escherichia coli* K-12. *J. Bacteriol.* 169:507–513.

1785. **Schweizer, H., G. Sweet, and T. J. Larson.** 1986. Physical and genetic structure of the glpD-malT interval of the *Escherichia coli* K-12 chromosome. Identification of two new structural genes of the glp-regulon. *Mol. Gen. Genet.* 202:488–492.

1786. **Schweizer, H. P., and P. Datta.** 1989. Identification and DNA sequence of tdcR, a positive regulatory gene of the tdc operon of *Escherichia coli. Mol. Gen. Genet.* 218:516–522.

1787. Schweizer, H. P., and P. Datta. 1991. Physical linkage and transcriptional orientation of the tdc operon on the *Escherichia coli* chromosome. *Mol. Gen. Genet.* 228:125–128.

1788. Scofield, M. A., W. S. Lewis, and S. M. Schuster. 1990. Nucleotide sequence of *Escherichia coli* asnB and deduced amino acid sequence of asparagine synthetase B. *J. Biol. Chem.* 265:12895–12902.

1789. Scott, D., and N. K. Amy. 1989. Molybdenum accumulation in chlD mutants of *Escherichia coli*. *J. Bacteriol.* 171:1284–1287.

1790. Seaton, B. L., and L. E. Vickery. 1994. A gene encoding a DnaK/hsp70 homolog in *Escherichia coli*. *Proc. Natl. Acad. Sci. USA* 91:2066–2070.

1791. Seddon, A. P., K. Y. Zhao, and A. Meister. 1989. Activation of glutamate by gamma-glutamate kinase: formation of gamma-cis-cycloglutamyl phosphate, an analog of gamma-glutamyl phosphate. *J. Biol. Chem.* 264:11326–11335.

1792. Seeley, T. W., and L. Grossman. 1989. Mutations in the *Escherichia coli* UvrB ATPase motif compromise excision repair capacity. *Proc. Natl. Acad. Sci. USA* 86:6577–6581.

1793. Seeley, T. W., and L. Grossman. 1990. The role of *Escherichia coli* UvrB in nucleotide excision repair. *J. Biol. Chem.* 265:7158–7165.

1794. Seifert, J., N. Kunz, R. Flachmann, A. Laufer, K. D. Jany, and H. G. Gassen. 1990. Expression of the *E. coli* nadB gene and characterization of the gene product L-aspartate oxidase. *Biol. Chem. Hoppe-Seyler* 371:239–248.

1795. Seifried, S. E., K. P. Bjornson, and P. H. von Hippel. 1991. Structure and assembly of the *Escherichia coli* transcription termination factor rho and its interactions with RNA. II. Physical chemical studies. *J. Mol. Biol.* 221:1139–1151.

1796. Selby, C. P., and A. Sancar. 1990. Structure and function of the (A)BC excinuclease of *Escherichia coli*. *Mutat. Res.* 236:203–211.

1797. Selby, C. P., E. M. Witkin, and A. Sancar. 1991. *Escherichia coli* mfd mutant deficient in "mutation frequency decline" lacks strand-specific repair: in vitro complementation with purified coupling factor. *Proc. Natl. Acad. Sci. USA* 88:11574–11578.

1798. Sen, K., and H. Nikaido. 1991. Trimerization of an in vitro synthesized OmpF porin of *Escherichia coli* outer membrane. *J. Biol. Chem.* 266:11295–11300.

1799. Sengupta, S., M. S. Rahman, U. Mukherjee, J. Basak, A. K. Pal, and S. N. Chatterjee. 1990. DNA damage and prophage induction and toxicity of nitrofurantoin in *Escherichia coli* and *Vibrio cholerae* cells. *Mutat. Res.* 244:55–60.

1800. Sens, D., W. Natter, and E. James. 1977. Evolutionary drift of the argF and argI genes. Coding for isoenzyme forms of ornithine transcarbamylase in *E. coli* K12. *Cell* 10:275–285.

1801. Seol, J. H., S. K. Woo, E. M. Jung, S. J. Yoo, C. S. Lee, K. J. Kim, K. Tanaka, A. Ichihara, D. B. Ha, and C. H. Chung. 1991. Protease Do is essential for survival of *Escherichia coli* at high temperatures: its identity with the htrA gene product. *Biochem. Biophys. Res. Commun.* 176:730–736.

1802. Seol, W., and A. J. Shatkin. 1991. *Escherichia coli* kgtP encodes an alpha-ketoglutarate transporter. *Proc. Natl. Acad. Sci. USA* 88:3802–3806.

1803. Serre, M. C., and J. R. Garel. 1990. Role of the C-terminal region in the allosteric properties of *Escherichia coli* phosphofructokinase-1. *Eur. J. Biochem.* 189:487–492.

1804. Sever, I. S., E. S. Kalyaeva, O. N. Danilevskaya, and Z. M. Gorlenko. 1982. Decreased degradation of beta beta' RNA polymerase subunits and abnormal proteins in a mutant *E. coli*. *Mol. Gen. Genet.* 188:494–498.

1805. Severinov, K., A. Mustaev, M. Kashlev, S. Borukhov, V. Nikiforov, and A. Goldfarb. 1992. Dissection of the beta subunit in the *Escherichia coli* RNA polymerase into domains by proteolytic cleavage. *J. Biol. Chem.* 267:12813–12819.

1806. Seville, M., M. G. Vincent, and K. Hahn. 1988. Modeling the three-dimensional structures of bacterial aminotransferases. *Biochemistry* 27:8344–8349.

1807. Seyfzadeh, M., J. Keener, and M. Nomura. 1993. spoT-dependent accumulation of guanosine tetraphosphate in response to fatty acid starvation in *Escherichia coli*. *Proc. Natl. Acad. Sci. USA* 90:11004–11008.

1808. Shames, S. L., D. E. Ash, F. C. Wedler, and J. J. Villafranca. 1984. Interaction of aspartate and aspartate-derived antimetabolites with the enzymes of the threonine biosynthetic pathway of *Escherichia coli*. *J. Biol. Chem.* 259:15331–15339.

1809. Shanmugan, K. T., V. Stewart, R. P. Gunsalus, D. H. Boxer, J. A. Cole, M. Chippaux, J. A. DeMoss, G. Giordano, E. C. C. Lin, and K. V. Rajagopalan. 1992. Proposed nomenclature for the genes involved in molybdenum metabolism in *Escherichia coli* and *Salmonella typhimurium*. *Mol. Microbiol.* 6:3452–3454.

1810. Shao, Z., R. T. Lin, and E. B. Newman. 1994. Sequencing and characterization of the sdaC gene and identification of the sdaCB operon in *Escherichia coli* K12. *Eur. J. Biochem.* 222:901–907.

1811. Shao, Z., and E. B. Newman. 1993. Sequencing and characterization of the sdaB gene from *Escherichia coli* K-12. *Eur. J. Biochem.* 212:777–784.

1812. Sharma, A., B. S. Henderson, J. M. Schwab, and J. L. Smith. 1990. Crystallization and preliminary X-ray analysis of beta-hydroxydecanoyl thiol ester dehydrase from *Escherichia coli*. *J. Biol. Chem.* 265:5110–5112.

1813. Sharma, R. C., and K. C. Smith. 1987. Comparison of the rep-38 and mmrA1 mutations of *Escherichia coli*. *Mutat. Res.* 184:23–28.

1814. Sharma, V., R. Meganathan, and M. E. S. Hudspeth. 1993. Menaquinone (vitamin K₂) biosynthesis: cloning, nucleotide sequence, and expression of the menC gene from *Escherichia coli*. *J. Bacteriol.* 175:4917–4921.

1815. Sharma, V., K. Suvarna, R. Meganathan, and M. E. S. Hudspeth. 1992. Menaquinone (vitamin K₂) biosynthesis: nucleotide sequence and expression of the menB gene from *Escherichia coli*. *J. Bacteriol.* 174:5057–5062.

1816. Shea, C. M., and M. A. McIntosh. 1991. Nucleotide sequence and genetic organization of the ferric enterobactin transport system: homology to other periplasmic binding protein-dependent systems in *Escherichia coli*. *Mol. Microbiol.* 5:1415–1428.

1817. Shedlarski, J. G., and C. Gilvarg. 1970. The pyruvate-aspartic semialdehyde condensing enzyme of *Escherichia coli*. *J. Biol. Chem.* 245:1362–1373.

1818. Shen, Y., J. Rudolph, M. Stern, J. Stubbe, K. A. Flannigan, and J. M. Smith. 1990. Glycinamide ribonucleotide synthetase from *Escherichia coli*: cloning, overproduction, sequencing, isolation, and characterization. *Biochemistry* 29:218–227.

1819. Sherman, M. M., L. A. Petersen, and C. D. Poulter. 1989. Isolation and characterization of isoprene mutants of *Escherichia coli*. *J. Bacteriol.* 171:3619–3628.

1820. Shevchik, V. E., G. Condemine, and J. Robert-Baudouy. 1994. Characterization of DsbC, a periplasmic protein of *Erwinia chrysanthemi* and *Escherichia coli* with disulfide isomerase activity. *EMBO J.* 13:2007–2012.

1821. Shi, W., M. Bogdanov, W. Dowhan, and D. R. Zusman. 1993. The pss and psd genes are required for motility and chemotaxis in *Escherichia coli*. *J. Bacteriol.* 175:7711–7714.

1822. Shiau, S. P., P. Chen, and L. J. Reitzer. 1993. Effects of insertions and deletions in glnG (ntrC) of *Escherichia coli* on nitrogen regulator I-dependent DNA binding and transcriptional activation. *J. Bacteriol.* 175:190–199.

1823. Shiba, T., H. Iwasaki, A. Nakata, and H. Shinagawa. 1993. *Escherichia coli* RuvA and RuvB proteins involved in recombination repair: physical properties and interactions with DNA. *Mol. Gen. Genet.* 237:395–399.

1824. Shibuya, I., and S. Hiraoka. 1992. Cardiolipin synthase from *Escherichia coli*. *Methods Enzymol.* 209:321–330.

1825. Shimizu, I., and A. Kaji. 1991. Identification of the promoter region of the ribosome-releasing factor cistron (frr). *J. Bacteriol.* 173:5181–5187.

1826. Shimizu, R. W., and J. L. Botsford. 1981. Regulation of anthranilate synthase in *Escherichia coli* growing in glucose-limited chemostats. *J. Gen. Microbiol.* 123:351–354.

1827. Shimmin, L. C., D. Vanderwel, R. E. Harkness, B. R. Currie, C. A. Galloway, and E. E. Ishiguro. 1984. Temperature-sensitive beta-lactam-tolerant mutants of *Escherichia coli*. *J. Gen. Microbiol.* 130:1315–1323.

1828. Shineberg, B., and I. G. Young. 1976. Biosynthesis of bacterial menaquinones: the membrane-associated 1,4-dihydroxy-2-naphthoate octaprenyltransferase of *Escherichia coli*. *Biochemistry* 15:2754–2758.

1829. Shinkai, A., L. H. Mei, H. Tokuda, and S. Mizushima. 1991. The conformation of SecA, as revealed by its protease sensitivity, is altered upon interaction with ATP, presecretory proteins, everted membrane vesicles, and phospholipids. *J. Biol. Chem.* 266:5827–5833.

1830. Shinozawa, T. 1973. A mutant of *Escherichia coli* K-12 unable to support the multiplication of bacteriophage BF23. *Virology* 54:427–440.

1831. Shostak, K., and V. Schirch. 1988. Serine hydroxymethyltransferase: mechanism of the racemization and transamination of D- and L-alanine. *Biochemistry* 27:8007–8014.

1832. Shrader, T. E., J. W. Tobias, and A. Varshavsky. 1993. The N-end rule in *Escherichia coli*: cloning and analysis of the leucyl, phenylalanyl-tRNA-protein transferase gene aat. *J. Bacteriol.* 175:4364–4374.

1833. Shrake, A., M. T. Fisher, P. J. McFarland, and A. Ginsburg. 1989. Partial unfolding of dodecameric glutamine synthetase from *Escherichia coli*: temperature-induced, reversible transitions of two domains. *Biochemistry* 28:6281–6294.

1834. Shuttleworth, W. A., C. D. Hough, K. P. Bertrand, and J. N. Evans. 1992. Overproduction of 5-enolpyruvylshikimate-3-phosphate synthase in *Escherichia coli*: use of the T7 promoter. *Protein Eng.* 5:461–466.

1835. Siebert, M., A. Bechthold, M. Melzer, U. May, U. Berger, G. Schroder, J. Schroder, K. Severin, and L. Heide. 1992. Ubiquinone biosynthesis. Cloning of the genes coding for chorismate pyruvate-lyase and 4-hydroxybenzoate octaprenyl transferase from *Escherichia coli*. *FEBS Lett.* 307:347–350.

1836. Siegel, L. M., D. C. Rueger, M. J. Barber, R. J. Krueger, N. R. Orme-Johnson, and W. H. Orme-Johnson. 1982. *Escherichia coli* sulfite reductase hemoprotein subunit. Prosthetic groups, catalytic parameters, and ligand complexes. *J. Biol. Chem.* 257:6343–6350.

1837. Siggaard-Andersen, M. 1988. Role of *Escherichia coli* beta-ketoacyl-ACP synthase I in unsaturated fatty acid synthesis. *Carlsberg Res. Commun.* 53:371–379.

1838. Silver, S., P. Johnseine, E. Whitney, and D. Clark. 1972. Manganese-resistant mutants of *Escherichia coli*: physiological and genetic studies. *J. Bacteriol.* 110:186–195.

1839. Silverman, P. M., S. Rother, and H. Gaudin. 1991. Arc and Sfr functions of the *Escherichia coli* K-12 arcA gene product are genetically and physiologically separable. *J. Bacteriol.* 173:5648–5652.

1840. Silvestro, A., J. Pommier, and G. Giordano. 1988. The inducible trimethylamine-N-oxide reductase of *Escherichia coli* K12: biochemical and immunological studies. *Biochim. Biophys. Acta* 954:1–13.

1841. Silvestro, A., J. Pommier, M. C. Pascal, and G. Giordano. 1989. The inducible trimethylamine N-oxide reductase of *Escherichia coli* K12: its localization and inducers. *Biochim. Biophys. Acta* 999:208–216.

1842. Sirko, A., M. Hryniewicz, D. Hulanicka, and A. Bock. 1990. Sulfate and thiosulfate transport in *Escherichia coli* K-12: nucleotide sequence and expression of the cysTWAM gene cluster. *J. Bacteriol.* 172:3351–3357.

1843. Sirko, A. E., M. Zatyka, and M. D. Hulanicka. 1987. Identification of the *Escherichia coli* cysM gene encoding O-acetylserine sulphhydrylase B by cloning with mini-Mu-lac containing a plasmid replicon. *J. Gen. Microbiol.* 133:2719–2725.

1844. Sivasubramanian, N., and R. Jayaraman. 1980. Mapping of two transcription mutations (tlnI and tlnII) conferring thiolutin resistance, adjacent to dnaZ and rho in *Escherichia coli*. *Mol. Gen. Genet.* 180:609–615.

1845. Six, S., S. C. Andrews, R. E. Roberts, G. Unden, and J. R. Guest. 1993. Construction and properties of *Escherichia coli* mutants defective in two genes encoding homologous membrane proteins with putative roles in anaerobic C4-dicarboxylic acid transport. *Biochem. Soc. Trans.* 21:342S

1846. Six, S., S. C. Andrews, G. Unden, and J. R. Guest. 1994. *Escherichia coli* possesses two homologous anaerobic C4-dicarboxylate membrane transporters (DcuA and DcuB) distinct from the aerobic dicarboxylate transport system (Dct). *J. Bacteriol.* 176:6470–6478.

1847. Skare, J. T., B. M. Ahmer, C. L. Seachord, R. P. Darveau, and K. Postle. 1993. Energy transduction between membranes. TonB, a cytoplasmic membrane protein, can be chemically cross-linked in vivo to the outer membrane receptor FepA. *J. Biol. Chem.* 268:16302–16308.

1848. Skarstad, K., B. Thony, D. S. Hwang, and A. Kornberg. 1993. A novel binding protein of the origin of the *Escherichia coli* chromosome. *J. Biol. Chem.* 268:5365–5370.

1849. Skarstedt, M. T., and E. Silverstein. 1976. *Escherichia coli* acetate kinase mechanism studied by net initial rate, equilibrium, and independent isotopic exchange kinetics. *J. Biol. Chem.* 251:6775–6783.

1850. Skiba, M. C., and K. L. Knight. 1994. Functionally important residues at a subunit interface site in the RecA protein from *Escherichia coli*. *J. Biol. Chem.* 269:3823–3828.

1851. Skinner, A. J., and R. A. Cooper. 1974. Genetic studies on ribose 5-phosphate isomerase mutants of *Escherichia coli* K-12. *J. Bacteriol.* 118:1183–1185.

1852. Slany, R. K., M. Bosl, P. F. Crain, and H. Kersten. 1993. A new function of S-adenosylmethionine: the ribosyl moiety of AdoMet is the precursor of the cyclopentenediol moiety of the tRNA wobble base queuine. *Biochemistry* 32:7811–7817.

1853. Slater, S. C., M. R. Lifsics, M. O'Donnell, and R. Maurer. 1994. *holE*, the gene coding for the theta subunit of DNA polymerase III of *Escherichia coli*: characterization of a *holE* mutant and comparison with a *dnaQ* (epsilon-subunit) mutant. *J. Bacteriol.* 176:815–821.

1854. Slettan, A., K. Gebhardt, E. Kristiansen, N. K. Birkeland, and B. H. Lindqvist. 1992. *Escherichia coli* K-12 and B contain functional bacteriophage P2 *ogr* genes. *J. Bacteriol.* 174:4094–4100.

1855. Small, P., D. Blankenhorn, D. Welty, E. Zinser, and J. L. Slonczewski. 1994. Acid and base resistance in *Escherichia coli* and *Shigella flexneri*: role of *rpoS* and growth pH. *J. Bacteriol.* 176:1729–1737.

1856. Smith, D. K., T. Kassam, B. Singh, and J. F. Elliott. 1992. *Escherichia coli* has two homologous glutamate decarboxylase genes that map to distinct loci. *J. Bacteriol.* 174:5820–5826.

1857. Smith, J. M., and H. A. Daum. 1986. Nucleotide sequence of the purM gene encoding 5′-phosphoribosyl-5-aminoimidazole synthetase of *Escherichia coli* K12. *J. Biol. Chem.* 261:10632–10636.

1858. Smith, R. L., J. L. Banks, M. D. Snavely, and M. E. Maguire. 1993. Sequence and topology of the CorA magnesium transport systems of *Salmonella typhimurium* and *Escherichia coli*. Identification of a new class of transport protein. *J. Biol. Chem.* 268:14071–14080.

1859. Snyder, W. B., and T. J. Silhavy. 1992. Enhanced export of beta-galactosidase fusion proteins in *prlF* mutants is Lon dependent. *J. Bacteriol.* 174:5661–5668.

1860. Soans, C., and H. J. Fromm. 1991. Studies of ligand binding to *Escherichia coli* adenylosuccinate synthetase. *Arch. Biochem. Biophys.* 291:107–112.

1861. Sofia, H. J., V. Burland, D. L. Daniels, G. Plunkett, and F. R. Blattner. 1994. Analysis of the *Escherichia coli* genome. V. DNA sequence of the region from 76.0 to 81.5 minutes. *Nucleic Acids Res.* 22:2576–2586.

1862. Sogaard-Andersen, L., H. Pedersen, B. Holst, and P. Valentin-Hansen. 1991. A novel function of the cAMP-CRP complex in *Escherichia coli*: cAMP-CRP functions as an adaptor for the CytR repressor in the deo operon. *Mol. Microbiol.* 5:969–975.

1863. Sohail, A., M. Lieb, M. Dar, and A. S. Bhagwat. 1990. A gene required for very short patch repair in *Escherichia coli* is adjacent to the DNA cytosine methylase gene. *J. Bacteriol.* 172:4214–4221.

1864. Somers, J. M., A. Amzallag, and R. B. Middleton. 1973. Genetic fine structure of the leucine operon of *Escherichia coli* K-12. *J. Bacteriol.* 113:1268–1272.

1865. Sommer, S., J. Knezevic, A. Bailone, and R. Devoret. 1993. Induction of only one SOS operon, umuDC, is required for SOS mutagenesis in *Escherichia coli*. *Mol. Gen. Genet.* 239:137–144.

1866. Song, W. J., and S. Jackowski. 1992. coaA and rts are allelic and located at kilobase 3532 on the *Escherichia coli* physical map. *J. Bacteriol.* 174:1705–1706.

1867. Song, W. J., and S. Jackowski. 1992. Cloning, sequencing, and expression of the pantothenate kinase (coaA) gene of *Escherichia coli*. *J. Bacteriol.* 174:6411–6417.

1868. Sparkowski, J., and A. Das. 1990. The nucleotide sequence of greA, a suppressor gene that restores growth of an *Escherichia coli* RNA polymerase mutant at high temperature. *Nucleic Acids Res.* 18:6443

1869. Sparrow, A. H., and A. F. Nauman. 1976. Evolution of genome size by DNA doublings. *Science* 192:524–529.

1870. Sparrow, C. P., and C. R. Raetz. 1983. A trans-acting regulatory mutation that causes overproduction of phosphatidylserine synthase in *Escherichia coli*. *J. Biol. Chem.* 258:9963–9967.

1871. Spence, J., and C. Georgopoulos. 1989. Purification and properties of the *Escherichia coli* heat shock protein, HtpG. *J. Biol. Chem.* 264:4398–4403.

1872. Spencer, P., and P. M. Jordan. 1993. Purification and characterization of 5-aminolaevulinic acid dehydratase from *Escherichia coli* and a study of the reactive thiols at the metal-binding domain. *Biochem. J.* 290:279–287.

1873. Speranza, M. L., G. Valentini, P. Iadarola, M. Stoppini, M. Malcovati, and G. Ferri. 1989. Primary structure of three peptides at the catalytic and allosteric sites of the fructose-1,6-bisphosphate-activated pyruvate kinase from *Escherichia coli*. *Biol. Chem. Hoppe-Seyler* 370:211–216.

1874. Spiro, S., and J. R. Guest. 1990. FNR and its role in oxygen-regulated gene expression in *Escherichia coli*. *FEMS Microbiol. Rev.* 6:399–428.

1875. Spitzer, E. D., H. E. Jimenez-Billini, and B. Weiss. 1988. β-Alanine auxotrophy associated with *dfp*, a locus affecting DNA synthesis in *Escherichia coli*. *J. Bacteriol.* 170:872–876.

1876. Spratt, S. K., C. L. Ginsburg, and W. D. Nunn. 1981. Isolation and genetic characterization of *Escherichia coli* mutants defective in propionate metabolism. *J. Bacteriol.* 146:1166–1169.

1877. Sprenger, G. A. 1993. Nucleotide sequence of the *Escherichia coli* K-12 transketolase (tkt) gene. *Biochim. Biophys. Acta* 1216:307–310.

1878. Sprenger, G. A. 1993. Two open reading frames adjacent to the *Escherichia coli* K-12 transketolase (tkt) gene show high similarity to the mannitol phosphotransferase system enzymes from *Escherichia coli* and various gram-positive bacteria. *Biochim. Biophys. Acta* 1158:103–106.

1879. Spring, K. J., P. G. Jerlstrom, D. M. Burns, and I. R. Beacham. 1986. L-Asparaginase genes in *Escherichia coli*: isolation of mutants and characterization of the ansA gene and its protein product. *J. Bacteriol.* 166:135–142.

1880. Spring, T. G., and F. Wold. 1971. The purification and characterization of *Escherichia coli* enolase. *J. Biol. Chem.* 246:6797–6802.

1881. Springer, M., M. Graffe, J. F. Mayaux, F. Dardel, G. Fayat, S. Blanquet, and M. Grunberg-Manago. 1987. Open reading frames in the control regions of the phenylalanyl-tRNA synthetase operon of *E. coli*. *Biochimie* 69:1065–1070.

1882. Spyrou, G., E. Haggard-Ljungquist, M. Krook, H. Jornvall, E. Nilsson, and P. Reichard. 1991. Characterization of the flavin reductase gene (fre) of *Escherichia coli* and construction of a plasmid for overproduction of the enzyme. *J. Bacteriol.* 173:3673–3679.

1883. Squires, C. H., H. M. DeFelice, J. Devereux, and J. M. Calvo. 1983. Molecular structure of ilvIH and its evolutionary relationship to ilvG in *Escherichia coli* K12. *Nucleic Acids Res.* 11:5299–5313.

1884. Squires, C. L., S. Pedersen, B. M. Ross, and C. Squires. 1991. ClpB is the *Escherichia coli* heat shock protein F84.1. *J. Bacteriol.* 173:4254–4262.

1885. Srivastava, B. S. 1976. Radiation sensitivity of a mutant of *Escherichia coli* K-12 associated with DNA replication: evidence for a new repair function. *Mol. Gen. Genet.* 143:327–332.

1886. Srivastava, R. A., N. Srivastava, and D. Apirion. 1991. RNA processing enzymes RNase III, E and P in *Escherichia coli* are not ribosomal enzymes. *Biochem. Int.* 25:57–65.

1887. Srivastava, R. A., N. Srivastava, and D. Apirion. 1992. Characterization of the RNA processing enzyme RNase III from wild type and overexpressing *Escherichia coli* cells in processing natural RNA substrates. *Int. J. Biochem.* 24:737–749.

1888. Staab, J. F., and C. F. Earhart. 1990. EntG activity of *Escherichia coli* enterobactin synthetase. *J. Bacteriol.* 172:6403–6410.

1889. Staab, J. F., M. F. Elkins, and C. F. Earhart. 1989. Nucleotide sequence of the *Escherichia coli* entE gene. *FEMS Microbiol. Lett.* 50:15–19.

1890. Stacey, K. A., and P. Oliver. 1977. Novel pleiotropic mutation in *Escherichia coli* K12 which affects transduction, transformation and rates of mutation. *J. Gen. Microbiol.* 98:569–578.

1891. Stamford, N. P., P. E. Lilley, and N. E. Dixon. 1992. Enriched sources of *Escherichia coli* replication proteins. The dnaG primase is a zinc metalloprotein. *Biochim. Biophys. Acta* 1132:17–25.

1892. Stanley, P., V. Koronakis, and C. Hughes. 1991. Mutational analysis supports a role for multiple structural features in the C-terminal secretion signal of *Escherichia coli* haemolysin. *Mol. Microbiol.* 5:2391–2403.

1893. Staudenmaier, H., B. Van Hove, Z. Yaraghi, and V. Braun. 1989. Nucleotide sequences of the fecBCDE genes and locations of the proteins suggest a periplasmic-binding-protein-dependent transport mechanism for iron(III) dicitrate in *Escherichia coli*. *J. Bacteriol.* 171:2626–2633.

1894. Stauffer, L. T., M. D. Plamann, and G. V. Stauffer. 1986. Cloning and characterization of the glycine-cleavage enzyme system of *Escherichia coli*. *Gene* 44:219–226.

1895. Stauffer, L. T., P. S. Steiert, J. G. Steiert, and G. V. Stauffer. 1991. An *Escherichia coli* protein with homology to the H-protein of the glycine cleavage enzyme complex from pea and chicken liver. *DNA Seq.* 2:13–17.

1896. Steffes, C., J. Ellis, J. Wu, and B. P. Rosen. 1992. The lysP gene encodes the lysine-specific permease. *J. Bacteriol.* 174:3242–3249.

1897. Steiert, J. G., R. J. Rolfes, H. Zalkin, and G. V. Stauffer. 1990. Regulation of the *Escherichia coli* glyA gene by the purR gene product. *J. Bacteriol.* 172:3799–3803.

1898. Steiert, P. S., L. T. Stauffer, and G. V. Stauffer. 1990. The lpd gene product functions as the L protein in the *Escherichia coli* glycine cleavage enzyme system. *J. Bacteriol.* 172:6142–6144.

1899. Stephens, C. M., and R. Bauerle. 1992. Essential cysteines in 3-deoxy-D-arabino-heptulosonate-7-phosphate synthase from *Escherichia coli*. Analysis by chemical modification and site-directed mutagenesis of the phenylalanine-sensitive isozyme. *J. Biol. Chem.* 267:5762–5767.

1900. Stern, S., T. Powers, L. M. Changchien, and H. F. Noller. 1988. Interaction of ribosomal proteins S5, S6, S11, S12, S18 and S21 with 16 S rRNA. *J. Mol. Biol.* **201**:683–695.

1901. Stevens, F. J., P. W. Stevens, J. G. Hovis, and T. T. Wu. 1981. Some properties of D-mannose isomerase from *Escherichia coli* K12. *J. Gen. Microbiol.* **124**:219–223.

1902. Stevenson, G., B. Neal, D. Liu, M. Hobbs, N. H. Packer, M. Batley, J. W. Redmond, L. Lindquist, and P. Reeves. 1994. Structure of the O antigen of *Escherichia coli* K-12 and the sequence of its *rfb* gene cluster. *J. Bacteriol.* **176**:4144–4156.

1903. Stewart, R. C., and F. W. Dahlquist. 1988. N-terminal half of CheB is involved in methylesterase response to negative chemotactic stimuli in *Escherichia coli*. *J. Bacteriol.* **170**:5728–5738.

1904. Stewart, V., J. T. Lin, and B. L. Berg. 1991. Genetic evidence that genes *fdhD* and *fdhE* do not control synthesis of formate dehydrogenase-N in *Escherichia coli* K-12. *J. Bacteriol.* **173**:4417–4423.

1905. Stim, K. P., and G. N. Bennett. 1993. Nucleotide sequence of the *adi* gene, which encodes the biodegradative acid-induced arginine decarboxylase of *Escherichia coli*. *J. Bacteriol.* **175**:1221–1234.

1906. Stirling, C. J., S. D. Colloms, J. F. Collins, G. Szatmari, and D. J. Sherratt. 1989. xerB, an *Escherichia coli* gene required for plasmid ColE1 site-specific recombination, is identical to pepA, encoding aminopeptidase A, a protein with substantial similarity to bovine lens leucine aminopeptidase. *EMBO J.* **8**:1623–1627.

1907. Stirling, C. J., G. Szatmari, G. Stewart, M. C. Smith, and D. J. Sherratt. 1988. The arginine repressor is essential for plasmid-stabilizing site-specific recombination at the ColE1 cer locus. *EMBO J.* **7**:4389–4395.

1908. Stirling, D. A., C. S. Hulton, L. Waddell, S. F. Park, M. H. Stewart, I. R. Booth, and C. F. Higgins. 1989. Molecular characterization of the proU loci of *Salmonella typhimurium* and *Escherichia coli* encoding osmoregulated glycine betaine transport systems. *Mol. Microbiol.* **3**:1025–1038.

1909. Stock, A., E. Schaeffer, D. E. Koshland, Jr., and J. Stock. 1987. A second type of protein methylation reaction in bacterial chemotaxis. *J. Biol. Chem.* **262**:8011–8014.

1910. Stoeber, F., A. Lagarde, G. Nemoz, G. Novel, M. Novel, R. Portalier, J. Pouyssegur, and J. Robert-Baudouy. 1974. Le metabolisme des hexuronides et des hexuronates chez *Escherichia coli* K-12. Aspects physiologiques et genetiques de sa regulation. *Biochimie* **56**:119–213.

1911. Stoker, K., W. N. Reijnders, L. F. Oltmann, and A. H. Stouthamer. 1989. Initial cloning and sequencing of *hydHG*, an operon homologous to *ntrBC* and regulating the labile hydrogenase activity in *Escherichia coli* K-12. *J. Bacteriol.* **171**:4448–4456.

1912. Stolz, B., and H. C. Berg. 1991. Evidence for interactions between MotA and MotB, torque-generating elements of the flagellar motor of *Escherichia coli*. *J. Bacteriol.* **173**:7033–7037.

1913. Stolz, B., M. Huber, Z. Markovic-Housley, and B. Erni. 1993. The mannose transporter of *Escherichia coli*. Structure and function of the IIABMan subunit. *J. Biol. Chem.* **268**:27094–27099.

1914. Storz, G., F. S. Jacobson, L. A. Tartaglia, R. W. Morgan, L. A. Silveira, and B. N. Ames. 1989. An alkyl hydroperoxide reductase induced by oxidative stress in *Salmonella typhimurium* and *Escherichia coli*: genetic characterization and cloning of *ahp*. *J. Bacteriol.* **171**:2049–2055.

1915. Stout, V., and S. Gottesman. 1990. RcsB and RcsC: a two-component regulator of capsule synthesis in *Escherichia coli*. *J. Bacteriol.* **172**:659–669.

1916. Stout, V., A. Torres-Cabassa, M. R. Maurizi, D. Gutnick, and S. Gottesman. 1991. RcsA, an unstable positive regulator of capsular polysaccharide synthesis. *J. Bacteriol.* **173**:1738–1747.

1917. Stover, P., M. Zamora, K. Shostak, M. Gautam-Basak, and V. Schirch. 1992. *Escherichia coli* serine hydroxymethyltransferase. The role of histidine 228 in determining reaction specificity. *J. Biol. Chem.* **267**:17679–17687.

1918. Stragier, P., O. Danos, and J. C. Patte. 1983. Regulation of diaminopimelate decarboxylase synthesis in *Escherichia coli*. II. Nucleotide sequence of the lysA gene and its regulatory region. *J. Mol. Biol.* **168**:321–331.

1919. Strom, A. R., and I. Kaasen. 1993. Trehalose metabolism in *Escherichia coli*: stress protection and stress regulation of gene expression. *Mol. Microbiol.* **8**:205–210.

1920. Stubbe, J. 1990. Ribonucleotide reductases: amazing and confusing. *J. Biol. Chem.* **265**:5329–5332.

1921. Studwell-Vaughan, P. S., and M. O'Donnell. 1991. Constitution of the twin polymerase of DNA polymerase III holoenzyme. *J. Biol. Chem.* **266**:19833–19841.

1922. Stukenberg, P. T., P. S. Studwell-Vaughan, and M. O'Donnell. 1991. Mechanism of the sliding beta-clamp of DNA polymerase III holoenzyme. *J. Biol. Chem.* **266**:11328–11334.

1923. Su, H., J. Moniakis, and E. B. Newman. 1993. Use of gene fusions of the structural gene sdaA to purify L-serine deaminase 1 from *Escherichia coli* K-12. *Eur. J. Biochem.* **211**:521–527.

1924. Su, H., and E. B. Newman. 1991. A novel L-serine deaminase activity in *Escherichia coli* K-12. *J. Bacteriol.* **173**:2473–2480.

1925. Su, H., S. B. F. Lang, and E. B. Newman. 1989. L-Serine degradation in *Escherichia coli* K-12: cloning and sequencing of the sdaA gene. *J. Bacteriol.* **171**:5095–5102.

1926. Su, T. Z., H. P. Schweizer, and D. L. Oxender. 1991. Carbon-starvation induction of the ugp operon, encoding the binding protein-dependent sn-glycerol-3-phosphate transport system in *Escherichia coli*. *Mol. Gen. Genet.* **230**:28–32.

1927. Sugai, M., and H. C. Wu. 1992. Export of the outer membrane lipoprotein is defective in *secD*, *secE*, and *secF* mutants of *Escherichia coli*. *J. Bacteriol.* **174**:2511–2516.

1928. Sugiyama, J. E., S. Mahmoodian, and G. R. Jacobson. 1991. Membrane topology analysis of *Escherichia coli* mannitol permease by using a nested-deletion method to create mtlA-phoA fusions. *Proc. Natl. Acad. Sci. USA* **88**:9603–9607.

1929. Sukharev, S. I., P. Blount, B. Martinac, F. R. Blattner, and C. Kung. 1994. A large-conductance mechanosensitive channel in *E. coli* encoded by mscL alone. *Nature* (London) **368**:265–268.

1930. Sullivan, S. L., and M. E. Gottesman. 1992. Requirement for *E. coli* NusG protein in factor-dependent transcription termination. *Cell* **68**:989–994.

1931. Sumantran, V. N., H. P. Schweizer, and P. Datta. 1990. A novel membrane-associated threonine permease encoded by the *tdcC* gene of *Escherichia coli*. *J. Bacteriol.* **172**:4288–4294.

1932. Sumpter, V. G., W. P. Tate, P. Nowotny, and K. H. Nierhaus. 1991. Modification of histidine residues on proteins from the 50S subunit of the *Escherichia coli* ribosome. Effects on subunit assembly and peptidyl transferase centre activity. *Eur. J. Biochem.* **196**:255–260.

1933. Sun, L., and J. A. Fuchs. 1992. *Escherichia coli* ribonucleotide reductase expression is cell cycle regulated. *Mol. Biol. Cell* **3**:1095–1105.

1934. Sun, X., J. Harder, M. Krook, H. Jornvall, B. M. Sjoberg, and P. Reichard. 1993. A possible glycine radical in anaerobic ribonucleotide reductase from *Escherichia coli*: nucleotide sequence of the cloned nrdD gene. *Proc. Natl. Acad. Sci. USA* **90**:577–581.

1935. Sung, Y. C., and J. A. Fuchs. 1992. The *Escherichia coli* K-12 cyn operon is positively regulated by a member of the lysR family. *J. Bacteriol.* **174**:3645–3650.

1936. Sunnarborg, A., D. Klumpp, T. Chung, and D. C. LaPorte. 1990. Regulation of the glyoxylate bypass operon: cloning and characterization of iclR. *J. Bacteriol.* **172**:2642–2649.

1937. Surratt, C. K., B. J. Carter, R. C. Payne, and S. M. Hecht. 1990. Metal ion and substrate structure dependence of the processing of tRNA precursors by RNase P and M1 RNA. *J. Biol. Chem.* **265**:22513–22519.

1938. Sutton, M. R., R. R. Fall, A. M. Nervi, A. W. Alberts, P. R. Vagelos, and R. A. Bradshaw. 1977. Amino acid sequence of *Escherichia coli* biotin carboxyl carrier protein (9100). *J. Biol. Chem.* **252**:3934–3940.

1939. Suzuki, H., H. Kumagai, T. Echigo, and T. Tochikura. 1988. Molecular cloning of *Escherichia coli* K-12 ggt and rapid isolation of gamma-glutamyltranspeptidase. *Biochem. Biophys. Res. Commun.* **150**:33–38.

1940. Suzuki, T., A. Itoh, S. Ichihara, and S. Mizushima. 1987. Characterization of the sppA gene coding for protease IV, a signal peptide peptidase of *Escherichia coli*. *J. Bacteriol.* **169**:2523–2528.

1941. Swanson, R. V., S. C. Schuster, and M. I. Simon. 1993. Expression of CheA fragments which define domains encoding kinase, phosphotransfer, and CheY binding activities. *Biochemistry* **32**:7623–7629.

1942. Sweet, G., C. Gandor, R. Voegele, N. Wittekindt, J. Beuerle, V. Truniger, E. C. Lin, and W. Boos. 1990. Glycerol facilitator of *Escherichia coli*: cloning of glpF and identification of the glpF product. *J. Bacteriol.* **172**:424–430.

1943. Swidersky, U. E., A. Rienhofer-Schweer, P. K. Werner, F. Ernst, S. A. Benson, H. K. Hoffschulte, and M. Muller. 1992. Biochemical analysis of the biogenesis and function of the *Escherichia coli* export factor SecY. *Eur. J. Biochem.* **207**:803–811.

1944. Sylvers, L. A., K. C. Rogers, M. Shimizu, E. Ohtsuka, and D. Soll. 1993. A 2-thiouridine derivative in tRNAGlu is a positive determinant for aminoacylation by *Escherichia coli* glutamyl-tRNA synthetase. *Biochemistry* **32**:3836–3841.

1945. Syu, W. J., and L. Kahan. 1989. Epitope characterization by modifications of antigens and by mapping on resin-bound peptides. Discriminating epitopes near the C-terminus and N-terminus of *Escherichia coli* ribosomal protein S13. *J. Immunol. Methods* **118**:153–160.

1946. Szumanski, M. B., and S. M. Boyle. 1990. Analysis and sequence of the speB gene encoding agmatine ureohydrolase, a putrescine biosynthetic enzyme in *Escherichia coli*. *J. Bacteriol.* **172**:538–547.

1947. Szumanski, M. B., and S. M. Boyle. 1992. Influence of cyclic AMP, agmatine, and a novel protein encoded by a flanking gene on speB (agmatine ureohydrolase) in *Escherichia coli*. *J. Bacteriol.* **174**:758–764.

1948. Ta, D. T., and L. E. Vickery. 1992. Cloning, sequencing, and overexpression of a [2Fe-2S] ferredoxin gene from *Escherichia coli*. *J. Biol. Chem.* **267**:11120–11125.

1949. Tabor, C. W., and H. Tabor. 1987. The speEspeD operon of *Escherichia coli*. Formation and processing of a proenzyme form of S-adenosylmethionine decarboxylase. *J. Biol. Chem.* **262**:16037–16040.

1950. Taglicht, D., E. Padan, and S. Schuldiner. 1991. Overproduction and purification of a functional Na^+/H^+ antiporter coded by nhaA (ant) from *Escherichia coli*. *J. Biol. Chem.* **266**:11289–11294.

1951. Taillon, B. E., R. Little, and R. P. Lawther. 1988. Analysis of the functional domains of biosynthetic threonine deaminase by comparison of the amino acid sequences of three wild-type alleles to the amino acid sequence of biodegradative threonine deaminase. *Gene* **63**:245–252.

1952. Takahagi, M., H. Iwasaki, A. Nakata, and H. Shinagawa. 1991. Molecular analysis of the *Escherichia coli* ruvC gene, which encodes a Holliday junction-specific endonuclease. *J. Bacteriol.* **173**:5747–5753.

1953. Takahashi, K., T. Hattori, H. Shindo, S. Noji, T. Nohno, and S. Taniguchi. 1993. Studies on phosphorylated transcriptional regulator (NarL) for *E. coli* nar operon by ^{31}P-NMR spectroscopy. *Biochem. Mol. Biol. Int.* **31**:161–168.

1954. Takase, I., F. Ishino, M. Wachi, H. Kamata, M. Doi, S. Asoh, H. Matsuzawa, T. Ohta, and M. Matsuhashi. 1987. Genes encoding two lipoproteins in the leuS-dacA region of the Escherichia coli chromosome. J. Bacteriol. 169:5692–5699.

1955. Takasuga, A., H. Adachi, F. Ishino, M. Matsuhashi, T. Ohta, and H. Matsuzawa. 1988. Identification of the penicillin-binding active site of penicillin-binding protein 2 of Escherichia coli. J. Biochem. (Tokyo) 104:822–826.

1956. Takata, R., and L. A. Isaksson. 1978. The temperature sensitive mutant 72c. II. Accumulation at high temperature of ppGpp and pppGpp in the presence of protein synthesis. Mol. Gen. Genet. 161:15–21.

1957. Takeuchi, M., R. Lillis, B. Demple, and M. Takeshita. 1994. Interactions of Escherichia coli endonuclease IV and exonuclease III with abasic sites in DNA. J. Biol. Chem. 269:21907–21914.

1958. Takiff, H. E., T. Baker, T. Copeland, S. M. Chen, and D. L. Court. 1992. Locating essential Escherichia coli genes by using mini-Tn10 transposons: the pdxJ operon. J. Bacteriol. 174:1544–1553.

1959. Talarico, T. L., I. K. Dev, W. S. Dallas, R. Ferone, and P. H. Ray. 1991. Purification and partial characterization of 7,8-dihydro-6-hydroxymethylpterin-pyrophos-phokinase and 7,8-dihydropteroate synthase from Escherichia coli MC4100. J. Bacteriol. 173:7029–7032.

1960. Talarico, T. L., P. H. Ray, I. K. Dev, B. M. Merrill, and W. S. Dallas. 1992. Cloning, sequence analysis, and overexpression of Escherichia coli folK, the gene coding for 7,8-dihydro-6-hydroxymethylpterin-pyrophosphokinase. J. Bacteriol. 174:5971–5977.

1961. Tamaki, S., T. Sato, and M. Matsuhashi. 1971. Role of lipopolysaccharides in antibiotic resistance and bacteriophage adsorption of Escherichia coli K-12. J. Bacteriol. 105:968–975.

1962. Tamir, H., and C. Gilvarg. 1974. Dihydrodipicolinic acid reductase. J. Biol. Chem. 249:3034–3040.

1963. Tamura, F., S. Nishimura, and M. Ohki. 1984. The E. coli divE mutation, which differentially inhibits synthesis of certain proteins, is in tRNASer1. EMBO J. 3:1103–1107.

1964. Tamura, K., H. Asahara, H. Himeno, T. Hasegawa, and M. Shimizu. 1991. Identity elements of Escherichia coli tRNA(Ala). J. Mol. Recognit. 4:129–132.

1965. Tamura, K., H. Himeno, H. Asahara, T. Hasegawa, and M. Shimizu. 1991. Identity determinants of E. coli tRNA(Val). Biochem. Biophys. Res. Commun. 177:619–623.

1966. Tamura, K., H. Himeno, H. Asahara, T. Hasegawa, and M. Shimizu. 1992. In vitro study of E. coli tRNA(Arg) and tRNA(Lys) identity elements. Nucleic Acids Res. 20:2335–2339.

1967. Tanaka, S., Y. Matsushita, A. Yoshikawa, and K. Isono. 1989. Cloning and molecular characterization of the gene rimL which encodes an enzyme acetylating ribosomal protein L12 of Escherichia coli K12. Mol. Gen. Genet. 217:289–293.

1968. Tanaka, T., H. Kato, T. Nishioka, and J. Oda. 1992. Mutational and proteolytic studies on a flexible loop in glutathione synthetase from Escherichia coli B: the loop and arginine 233 are critical for the catalytic reaction. Biochemistry 31:2259–2265.

1969. Tanaka, Y., A. Tsujimura, N. Fujita, S. Isono, and K. Isono. 1989. Cloning and analysis of an Escherichia coli operon containing the rpmF gene for ribosomal protein L32 and the gene for a 30-kilodalton protein. J. Bacteriol. 171:5707–5712.

1970. Tang, M. S., and C. E. Helmstetter. 1980. Coordination between chromosome replication and cell division in Escherichia coli. J. Bacteriol. 141:1148–1156.

1971. Tao, K., K. Makino, S. Yonei, A. Nakata, and H. Shinagawa. 1989. Molecular cloning and nucleotide sequencing of oxyR, the positive regulatory gene of a regulon for an adaptive response to oxidative stress in Escherichia coli: homologies between OxyR protein and a family of bacterial activator proteins. Mol. Gen. Genet. 218:371–376.

1972. Tao, K., K. Makino, S. Yonei, A. Nakata, and H. Shinagawa. 1991. Purification and characterization of the Escherichia coli OxyR protein, the positive regulator for a hydrogen peroxide-inducible regulon. J. Biochem. (Tokyo) 109:262–266.

1973. Tapio, S., and L. A. Isaksson. 1991. Base 2661 in Escherichia coli 23S rRNA influences the binding of elongation factor Tu during protein synthesis in vivo. Eur. J. Biochem. 202:981–984.

1974. Tarmy, E. M., and N. O. Kaplan. 1968. Kinetics of Escherichia coli B D-lactate dehydrogenase and evidence for pyruvate-controlled change in conformation. J. Biol. Chem. 243:2587–2596.

1975. Tate, C. G., and P. J. Henderson. 1993. Membrane topology of the L-rhamnose-H+ transport protein (RhaT) from enterobacteria. J. Biol. Chem. 268:26850–26857.

1976. Taura, T., C. Ueguchi, K. Shiba, and K. Ito. 1992. Insertional disruption of the nusB (ssyB) gene leads to cold-sensitive growth of Escherichia coli and suppression of the secY24 mutation. Mol. Gen. Genet. 234:429–432.

1977. Taylor, D. E., Y. Hou, R. J. Turner, and J. H. Weiner. 1994. Location of a potassium tellurite resistance operon (tehA tehB) within the terminus of Escherichia coli K-12. J. Bacteriol. 176:2740–2742.

1978. Terada, K., and K. Izui. 1991. Site-directed mutagenesis of the conserved histidine residue of phosphoenolpyruvate carboxylase. His138 is essential for the second partial reaction. Eur. J. Biochem. 202:797–803.

1979. Terada, K., T. Murata, and K. Izui. 1991. Site-directed mutagenesis of phosphoenolpyruvate carboxylase from E. coli: the role of His579 in the catalytic and regulatory functions. J. Biochem. (Tokyo) 109:49–54.

1980. Teschner, W., M. C. Serre, and J. R. Garel. 1990. Enzymatic properties, renaturation and metabolic role of mannitol-1-phosphate dehydrogenase from Escherichia coli. Biochimie 72:33–40.

1981. Tetart, F., and J. P. Bouche. 1992. Regulation of the expression of the cell-cycle gene ftsZ by DicF antisense RNA. Division does not require a fixed number of FtsZ molecules. Mol. Microbiol. 6:615–620.

1982. Tewari, R., J. I. MacGregor, T. Ikeda, J. R. Little, S. J. Hultgren, and S. N. Abraham. 1993. Neutrophil activation by nascent FimH subunits of type 1 fimbriae purified from the periplasm of Escherichia coli. J. Biol. Chem. 268:3009–3015.

1983. Theall, G., K. B. Low, and D. Soll. 1979. Regulation of the biosynthesis of aminoacyl-tRNA synthetases and of tRNA in Escherichia coli. IV. Mutants with increased levels of leucyl- or seryl-tRNA synthetase. Mol. Gen. Genet. 169:205–211.

1984. Thelen, P., T. Tsuchiya, and E. B. Goldberg. 1991. Characterization and mapping of a major Na+/H+ antiporter gene of Escherichia coli. J. Bacteriol. 173:6553–6557.

1985. Theobald, A., M. Springer, M. Grunberg-Manago, J. P. Ebel, and R. Giege. 1988. Tertiary structure of Escherichia coli tRNA(3Thr) in solution and interaction of this tRNA with the cognate threonyl-tRNA synthetase. Eur. J. Biochem. 175:511–524.

1986. Thomas, G., and A. Favre. 1980. 4-Thiouridine triggers both growth delay induced by near-ultraviolet light and photoprotection. Eur. J. Biochem. 113:67–74.

1987. Thony, B., D. S. Hwang, L. Fradkin, and A. Kornberg. 1991. iciA, an Escherichia coli gene encoding a specific inhibitor of chromosomal initiation of replication in vitro. Proc. Natl. Acad. Sci. USA 88:4066–4070.

1988. Thorne, G. M., and L. M. Corwin. 1975. Mutations affecting aromatic amino acid transport in Escherichia coli and Salmonella typhimurium. J. Gen. Microbiol. 90:203–216.

1989. Tian, G., D. Lim, J. Carey, and W. K. Maas. 1992. Binding of the arginine repressor of Escherichia coli K12 to its operator sites. J. Mol. Biol. 226:387–397.

1990. Tiedeman, A. A., D. J. DeMarini, J. Parker, and J. M. Smith. 1990. DNA sequence of the purC gene encoding 5'-phosphoribosyl-5-aminoimidazole-4-N-succino-carboxamide synthetase and organization of the dapA-purC region of Escherichia coli K-12. J. Bacteriol. 172:6035–6041.

1991. Tiedeman, A. A., J. Keyhani, J. Kamholz, H. A. Daum, J. S. Gots, and J. M. Smith. 1989. Nucleotide sequence analysis of the purEK operon encoding 5'-phosphoribosyl-5-aminoimidazole carboxylase of Escherichia coli K-12. J. Bacteriol. 171:205–212.

1992. Tiedeman, A. A., J. M. Smith, and H. Zalkin. 1985. Nucleotide sequence of the guaA gene encoding GMP synthetase of Escherichia coli K12. J. Biol. Chem. 260:8676–8679.

1993. Timms, A. R., and B. A. Bridges. 1993. Double, independent mutational events in the rpsL gene of Escherichia coli: an example of hypermutability? Mol. Microbiol. 9:335–342.

1994. Tobe, T., N. Kusukawa, and T. Yura. 1987. Suppression of rpoH (htpR) mutations of Escherichia coli: heat shock response in suhA revertants. J. Bacteriol. 169:4128–4134.

1995. Tobin, J. F., and R. F. Schleif. 1987. Positive regulation of the Escherichia coli L-rhamnose operon is mediated by the products of tandemly repeated regulatory genes. J. Mol. Biol. 196:789–799.

1996. Tobin, J. F., and R. F. Schleif. 1990. Purification and properties of RhaR, the positive regulator of the L-rhamnose operons of Escherichia coli. J. Mol. Biol. 211:75–89.

1997. Tokeson, J. P., S. Garges, and S. Adhya. 1991. Further inducibility of a constitutive system: ultrainduction of the gal operon. J. Bacteriol. 173:2319–2327.

1998. Tokushige, M., N. Tsujimoto, T. Oda, T. Honda, N. Yumoto, S. Ito, M. Yamamoto, E. H. Kim, and Y. Hiragi. 1989. Role of cysteine residues in tryptophanase for monovalent cation-induced activation. Biochimie 71:711–720.

1999. Tolner, B., B. Poolman, B. Wallace, and W. N. Konings. 1992. Revised nucleotide sequence of the gltP gene, which encodes the proton-glutamate-aspartate transport protein of Escherichia coli K-12. J. Bacteriol. 174:2391–2393.

2000. Tomioka, S., T. Nikaido, T. Miyakawa, and M. Matsuhashi. 1983. Mutation of the N-acetylmuramyl-L-alanine amidase gene of Escherichia coli K-12. J. Bacteriol. 156:463–465.

2001. Tommassen, J., K. Eiglmeier, S. T. Cole, P. Overduin, T. J. Larson, and W. Boos. 1991. Characterization of two genes, glpQ and ugpQ, encoding glycerophosphoryl diester phosphodiesterases of Escherichia coli. Mol. Gen. Genet. 226:321–327.

2002. Tommassen, J., K. Eiglmeier, S. T. Cole, P. Overduin, T. J. Larson, and W. Boos. 1991. Characterization of two genes, glpQ and ugpQ, encoding glycerophosphoryl diester phosphodiesterases of Escherichia coli. Mol. Gen. Genet. 226:321–327.

2003. Tomoyasu, T., T. Yuki, S. Morimura, H. Mori, K. Yamanaka, H. Niki, S. Hiraga, and T. Ogura. 1993. The Escherichia coli FtsH protein is a prokaryotic member of a protein family of putative ATPases involved in membrane functions, cell cycle control, and gene expression. J. Bacteriol. 175:1344–1351.

2004. Tong, R. C., N. A. Glavas, and P. D. Bragg. 1991. Topological analysis of the pyridine nucleotide transhydrogenase of Escherichia coli using proteolytic enzymes. Biochim. Biophys. Acta 1080:19–28.

2005. Toone, W. M., K. E. Rudd, and J. D. Friesen. 1991. deaD, a new Escherichia coli gene encoding a presumed ATP-dependent RNA helicase, can suppress a mutation in rpsB, the gene encoding ribosomal protein S2. J. Bacteriol. 173:3291–3302.

2006. Toth, M. J., and P. Schimmel. 1986. Internal structural features of E. coli glycyl-tRNA synthetase examined by subunit polypeptide chain fusions. J. Biol. Chem. 261:6643–6646.

2007. Toth, M. J., and P. Schimmel. 1990. Deletions in the large (beta) subunit of a hetero-oligomeric aminoacyl-tRNA synthetase. *J. Biol. Chem.* 265:1000–1004.

2008. Trempy, J. E., J. E. Kirby, and S. Gottesman. 1994. Alp suppression of Lon: dependence on the *slpA* gene. *J. Bacteriol.* 176:2061–2067.

2009. Trieber, C. A., R. A. Rothery, and J. H. Weiner. 1994. Multiple pathways of electron transfer in dimethyl sulfoxide reductase of *Escherichia coli*. *J. Biol. Chem.* 269:7103–7109.

2010. Trisler, P., and S. Gottesman. 1984. *lon* transcriptional regulation of genes necessary for capsular polysaccharide synthesis in *Escherichia coli* K-12. *J. Bacteriol.* 160:184–191.

2011. Tritz, G. J., T. S. Matney, J. L. R. Chandler, and R. K. Gholson. 1970. Chromosomal location of the C gene involved in the biosynthesis of nicotinamide adenine dinucleotide in *Escherichia coli* K-12. *J. Bacteriol.* 104:45–49.

2012. Troup, B., M. Jahn, C. Hungerer, and D. Jahn. 1994. Isolation of the *hemF* operon containing the gene for the *Escherichia coli* aerobic coproporphyrinogen III oxidase by in vivo complementation of a yeast HEM13 mutant. *J. Bacteriol.* 176:673–680.

2013. Trower, M. K. 1993. PCR cloning, sequence analysis and expression of the *cybC* genes encoding soluble cytochrome b-562 from *Escherichia coli* B strain OP7 and K strain NM522. *Biochim. Biophys. Acta* 1143:109–111.

2014. Trun, N. J., and S. Gottesman. 1990. On the bacterial cell cycle: *Escherichia coli* mutants with altered ploidy. *Genes Dev.* 4:2036–2047.

2015. Trun, N. J., and S. Gottesman. 1991. Characterization of *Escherichia coli* mutants with altered ploidy. *Res. Microbiol.* 142:195–200.

2016. Truniger, V., and W. Boos. 1994. Mapping and cloning of *gldA*, the structural gene of the *Escherichia coli* glycerol dehydrogenase. *J. Bacteriol.* 176:1796–1800.

2017. Truong, H. T., E. A. Pratt, G. S. Rule, P. Y. Hsue, and C. Ho. 1991. Inactive and temperature-sensitive folding mutants generated by tryptophan substitutions in the membrane-bound d-lactate dehydrogenase of *Escherichia coli*. *Biochemistry* 30:10722–10729.

2018. Tsai-Wu, J. J., H. F. Liu, and A. L. Lu. 1992. *Escherichia coli* MutY protein has both N-glycosylase and apurinic/apyrimidinic endonuclease activities on A.C and A.G mispairs. *Proc. Natl. Acad. Sci.* 89:8779–8783.

2019. Tsaneva, I. R., B. Muller, and S. C. West. 1993. RuvA and RuvB proteins of *Escherichia coli* exhibit DNA helicase activity in vitro. *Proc. Natl. Acad. Sci. USA* 90:1315–1319.

2020. Tsay, J. T., W. Oh, T. J. Larson, S. Jackowski, and C. O. Rock. 1992. Isolation and characterization of the beta-ketoacyl-acyl carrier protein synthase III gene (fabH) from *Escherichia coli* K-12. *J. Biol. Chem.* 267:6807–6814.

2021. Tsay, J. T., C. O. Rock, and S. Jackowski. 1992. Overproduction of beta-ketoacyl-acyl carrier protein synthase I imparts thiolactomycin resistance to *Escherichia coli* K-12. *J. Bacteriol.* 174:508–513.

2022. Tsuchihashi, Z., and A. Kornberg. 1989. ATP interactions of the tau and gamma subunits of DNA polymerase III holoenzyme of *Escherichia coli*. *J. Biol. Chem.* 264:17790–17795.

2023. Tsui, H. C., P. J. Arps, D. M. Connolly, and M. E. Winkler. 1991. Absence of *hisT*-mediated tRNA pseudouridylation results in a uracil requirement that interferes with *Escherichia coli* K-12 cell division. *J. Bacteriol.* 173:7395–7400.

2024. Tsui, H. C., G. Zhao, G. Feng, H. C. Leung, and M. E. Winkler. 1994. The mutL repair gene of *Escherichia coli* K-12 forms a superoperon with a gene encoding a new cell-wall amidase. *Mol. Microbiol.* 11:189–802.

2025. Tummuru, M. K., T. J. Brickman, and M. A. McIntosh. 1989. The in vitro conversion of chorismate to isochorismate catalyzed by the *Escherichia coli* entC gene product. Evidence that EntA does not contribute to isochorismate synthase activity. *J. Biol. Chem.* 264:20547–20551.

2026. Turnbough, C. L., Jr., K. H. Kerr, W. R. Funderburg, J. P. Donahue, and F. E. Powell. 1987. Nucleotide sequence and characterization of the pyrF operon of *Escherichia coli* K12. *J. Biol. Chem.* 262:10239–10245.

2027. Turnbull, J., W. W. Cleland, and J. F. Morrison. 1990. Chorismate mutase-prephenate dehydrogenase from *Escherichia coli*. 1. Kinetic characterization of the dehydrogenase reaction by use of alternative substrates. *Biochemistry* 29:10245–10254.

2028. Turnbull, J., and J. F. Morrison. 1990. Chorismate mutase-prephenate dehydrogenase from *Escherichia coli*. 2. Evidence for two different active sites. *Biochemistry* 29:10255–10261.

2029. Ubom, G. A., S. G. Rhee, J. B. Hunt, and P. B. Chock. 1991. Distance changes at the regulatory and catalytic sites on *Escherichia coli* glutamine synthetase: a spin label study on the effect of substrate(s) binding. *Biochim. Biophys. Acta* 1077:91–98.

2030. Ueda, Y., N. Yumoto, M. Tokushige, K. Fukui, and H. Ohya-Nishiguchi. 1991. Purification and characterization of two types of fumarase from *Escherichia coli*. *J. Biochem.* (Tokyo) 109:728–733.

2031. Ueguchi, C., and K. Ito. 1992. Multicopy suppression: an approach to understanding intracellular functioning of the protein export system. *J. Bacteriol.* 174:1454–1461.

2032. Ueguchi, C., M. Kakeda, H. Yamada, and T. Mizuno. 1994. An analogue of the DnaJ molecular chaperone in *Escherichia coli*. *Proc. Natl. Acad. Sci. USA* 91:1054–1058.

2033. Ueguchi, C., and T. Mizuno. 1993. The *Escherichia coli* nucleoid protein H-NS functions directly as a transcriptional repressor. *EMBO J.* 12:1039–1046.

2034. Ueguchi, C., M. Wittekind, M. Nomura, Y. Akiyama, and K. Ito. 1989. The secY-rpmJ region of the spc ribosomal protein operon in *Escherichia coli*: structural alterations affecting secY expression. *Mol. Gen. Genet.* 217:1–5.

2035. Ueshima, R., N. Fujita, and A. Ishihama. 1989. DNA supercoiling and temperature shift affect the promoter activity of the *Escherichia coli* rpoH gene encoding the heat-shock sigma subunit of RNA polymerase. *Mol. Gen. Genet.* 215:185–189.

2036. Umbarger, H. E., M. A. Umbarger, and P. M. L. Siu. 1963. Biosynthesis of serine in *Escherichia coli* and *Salmonella typhimurium*. *J. Bacteriol.* 85:1431–1439.

2037. Umezu, K., N. W. Chi, and R. D. Kolodner. 1993. Biochemical interaction of the *Escherichia coli* RecF, RecO, and RecR proteins with RecA protein and single-stranded DNA binding protein. *Proc. Natl. Acad. Sci. USA* 90:3875–3879.

2038. Umezu, K., and H. Nakayama. 1993. RecQ DNA helicase of *Escherichia coli*. Characterization of the helix-unwinding activity with emphasis on the effect of single-stranded DNA-binding protein. *J. Mol. Biol.* 230:1145–1150.

2039. Umezu, K., K. Nakayama, and H. Nakayama. 1990. *Escherichia coli* RecQ protein is a DNA helicase. *Proc. Natl. Acad. Sci. USA* 87:5363–5367.

2040. Ursini, M. V., P. Arcari, and M. de Felice. 1981. Acetohydroxy acid synthase isoenzymes of *Escherichia coli* K-12: a trans-acting regulatory locus of ilvHI gene expression. *Mol. Gen. Genet.* 181:491–496.

2041. Ursinus, A., and J. V. Holtje. 1994. Purification and properties of a membrane-bound lytic transglycosylase from *Escherichia coli*. *J. Bacteriol.* 176:338–343.

2042. Usui, M., H. Sembongi, H. Matsuzaki, K. Matsumoto, and I. Shibuya. 1994. Primary structures of the wild-type and mutant alleles encoding the phosphatidylglycerophosphate synthase of *Escherichia coli*. *J. Bacteriol.* 176:3389–3392.

2043. Utsumi, R., S. Katayama, M. Ikeda, S. Igaki, H. Nakagawa, A. Miwa, M. Taniguchi, and M. Noda. 1992. Cloning and sequence analysis of the evgAS genes involved in signal transduction of *Escherichia coli* K-12. *Nucleic Acids Symp. Ser.* 422:149–150.

2044. Utsumi, R., S. Kusafuka, T. Nakayama, K. Tanaka, Y. Takayanagi, H. Takahashi, M. Noda, and M. Kawamukai. 1993. Stationary phase-specific expression of the fic gene in *Escherichia coli* K-12 is controlled by the rpoS gene product (sigma 38). *FEMS Microbiol. Lett.* 113:273–278.

2045. Valentin-Hansen, P. 1978. Uridine-cytidine kinase from *Escherichia coli*. *Methods Enzymol.* 51:308–314.

2046. Valentin-Hansen, P., K. Hammer-Jespersen, and R. S. Buxton. 1979. Evidence for the existence of three promoters for the deo operon of *Escherichia coli* K12 in vitro. *J. Mol. Biol.* 133:1–17.

2047. Valentini, G., M. Stoppini, M. L. Speranza, M. Malcovati, and G. Ferri. 1991. Bacterial pyruvate kinases have a shorter N-terminal domain. *Biol. Chem. Hoppe-Seyler* 372:91–93.

2048. Vales, L. D., B. A. Rabin, and J. W. Chase. 1982. Subunit structure of *Escherichia coli* exonuclease VII. *J. Biol. Chem.* 257:8799–8805.

2049. Vales, L. D., B. A. Rabin, and J. W. Chase. 1983. Isolation and preliminary characterization of *Escherichia coli* mutants deficient in exonuclease VII. *J. Bacteriol.* 155:1116–1122.

2050. Vallari, D. S., and C. O. Rock. 1987. Isolation and characterization of temperature-sensitive pantothenate kinase (coaA) mutants of *Escherichia coli*. *J. Bacteriol.* 169:5795–5800.

2051. Vander Horn, P. B., A. D. Backstrom, V. Stewart, and T. P. Begley. 1993. Structural genes for thiamine biosynthetic enzymes (thiCEFGH) in *Escherichia coli* K-12. *J. Bacteriol.* 175:982–992.

2052. van der Linden, M. P., L. de Haan, M. A. Hoyer, and W. Keck. 1992. Possible role of *Escherichia coli* penicillin-binding protein 6 in stabilization of stationary-phase peptidoglycan. *J. Bacteriol.* 174:7572–7578.

2053. van der Linden, M. P., H. Mottl, and W. Keck. 1992. Cytoplasmic high-level expression of a soluble, enzymatically active form of the *Escherichia coli* penicillin-binding protein 5 and purification by dye chromatography. *Eur. J. Biochem.* 204:197–202.

2054. van der Oost, J., P. Lappalainen, A. Musacchio, A. Warne, L. Lemieux, J. Rumbley, R. B. Gennis, R. Aasa, T. Pascher, B. G. Malmstrom, and M. Saraste. 1992. Restoration of a lost metal-binding site: construction of two different copper sites into a subunit of the E. coli cytochrome o quinol oxidase complex. *EMBO J.* 11:3209–3217.

2055. Vanderwinkel, E., and M. De Vlieghere. 1968. Physiologie et genetique de l'isocitrase et des malate synthases chez *Escherichia coli*. *Eur. J. Biochem.* 5:81–90.

2056. Vanderwinkel, E., M. De Vlieghere, P. Charles, and V. Baptist. 1987. Nature of the interactions involved in the lipid-protein complexes of the *Escherichia coli* N-acetylmuramoyl-L-alanine amidase. *Biochim. Biophys. Acta* 913:238–244.

2057. Van Dyk, T. K., D. R. Smulski, and Y. Y. Chang. 1987. Pleiotropic effects of poxA regulatory mutations of *Escherichia coli* and *Salmonella typhimurium*, mutations conferring sulfometuron methyl and alpha-ketobutyrate hypersensitivity. *J. Bacteriol.* 169:4540–4546.

2058. van Gemen, B., J. Twisk, and P. H. van Knippenberg. 1989. Autogenous regulation of the *Escherichia coli* ksgA gene at the level of translation. *J. Bacteriol.* 171:4002–4008.

2059. Van Houten, B., and A. Snowden. 1993. Mechanism of action of the *Escherichia coli* UvrABC nuclease: clues to the damage recognition problem. *Bioessays* 15:51–59.

2060. Van Hove, B., H. Staudenmaier, and V. Braun. 1990. Novel two-component transmembrane transcription control: regulation of iron dicitrate transport in *Escherichia coli* K-12. *J. Bacteriol.* 172:6749–6758.

2061. van Vliet, F., M. Crabeel, A. Boyen, C. Tricot, V. Stalon, P. Falmagne, Y. Nakamura, S. Baumberg, and N. Glansdorff. 1990. Sequences of the genes encoding argininosuccinate synthetase in *Escherichia coli* and *Saccharomyces cerevisiae*: comparison with methanogenic archaebacteria and mammals. *Gene* 95:99–104.

2062. **Varshney, U., T. Hutcheon, and J. H. van de Sande.** 1988. Sequence analysis, expression, and conservation of *Escherichia coli* uracil DNA glycosylase and its gene (ung). *J. Biol. Chem.* **263:**7776–7784.

2063. **Varshney, U., and U. L. RajBhandary.** 1992. Role of methionine and formylation of initiator tRNA in initiation of protein synthesis in *Escherichia coli. J. Bacteriol.* **174:**7819–7826.

2064. **Vasudevan, S. G., W. L. Armarego, D. C. Shaw, P. E. Lilley, N. E. Dixon, and R. K. Poole.** 1991. Isolation and nucleotide sequence of the hmp gene that encodes a haemoglobin-like protein in *Escherichia coli* K-12. *Mol. Gen. Genet.* **226:**49–58.

2065. **Vasudevan, S. G., D. C. Shaw, and W. L. Armarego.** 1988. Dihydropteridine reductase from *Escherichia coli. Biochem. J.* **255:**581–588.

2066. **Verkamp, E., V. M. Backman, J. M. Bjornsson, D. Soll, and G. Eggertsson.** 1993. The periplasmic dipeptide permease system transports 5-aminolevulinic acid in *Escherichia coli. J. Bacteriol.* **175:**1452–1456.

2067. **Verkamp, E., M. Jahn, D. Jahn, A. M. Kumar, and D. Soll.** 1992. Glutamyl-tRNA reductase from *Escherichia coli* and *Synechocystis* 6803. Gene structure and expression. *J. Biol. Chem.* **267:**8275–8280.

2068. **Vermersch, P. S., J. J. Tesmer, D. D. Lemon, and F. A. Quiocho.** 1990. A Pro to Gly mutation in the hinge of the arabinose-binding protein enhances binding and alters specificity. Sugar-binding and crystallographic studies. *J. Biol. Chem.* **265:**16592–16603.

2069. **Veronese, F. M., E. Boccu, and L. Conventi.** 1975. Glutamate dehydrogenase from *Escherichia coli*: induction, purification and properties of the enzyme. *Biochim. Biophys. Acta* **377:**217–228.

2070. **Veronese, F. M., E. Boccu, and A. Fontana.** 1976. Isolation and properties of 6-phosphogluconate from *Escherichia coli*. Some comparisons with the thermophilic enzyme from *Bacillus stearothermophilus. Biochemistry* **15:**4026–4033.

2071. **Verwoert, I. I., E. C. Verbree, K. H. van der Linden, H. J. Nijkamp, and A. R. Stuitje.** 1992. Cloning, nucleotide sequence, and expression of the *Escherichia coli* fabD gene, encoding malonyl coenzyme A-acyl carrier protein transacylase. *J. Bacteriol.* **174:**2851–2857.

2072. **Vianney, A., T. M. Lewin, W. F. Beyer, Jr., J. C. Lazzaroni, R. Portalier, and R. E. Webster.** 1994. Membrane topology and mutational analysis of the TolQ protein of *Escherichia coli* required for the uptake of macromolecules and cell envelope integrity. *J. Bacteriol.* **176:**822–829.

2073. **Vicente, M., P. Palacios, A. Dopazo, T. Garrido, J. Pla, and M. Aldea.** 1991. On the chronology and topography of bacterial cell division. *Res. Microbiol.* **142:**253–257.

2074. **Vickers, L. P., G. K. Ackers, and J. W. Ogilvie.** 1978. Aspartokinase I-homoserine dehydrogenase I of *Escherichia coli* K12. Concentration-dependent dissociation to dimers in the presence of L-threonine. *J. Biol. Chem.* **253:**2155–2160.

2075. **Vik, S. B., and N. N. Dao.** 1992. Prediction of transmembrane topology of F0 proteins from *Escherichia coli* F1F0 ATP synthase using variational and hydrophobic moment analyses. *Biochim. Biophys. Acta* **1140:**199–207.

2076. **Vlahos, C. J., and E. E. Dekker.** 1988. The complete amino acid sequence and identification of the active-site arginine peptide of *Escherichia coli* 2-keto-4-hydroxyglutarate aldolase. *J. Biol. Chem.* **263:**11683–11691.

2077. **Vogel, H. J., and R. H. Vogel.** 1974. Enzymes of arginine biosynthesis and their repressive control. *Adv. Enzymol. Relat. Areas Mol. Biol.* **40:**65–90.

2078. **Vogel, R. F., K. D. Entian, and D. Mecke.** 1987. Cloning and sequence of the mdh structural gene of *Escherichia coli* coding for malate dehydrogenase. *Arch. Microbiol.* **149:**36–42.

2079. **Vogler, A. P., and J. W. Lengeler.** 1989. Analysis of the nag regulon from *Escherichia coli* K12 and *Klebsiella pneumoniae* and of its regulation. *Mol. Gen. Genet.* **219:**97–105.

2080. **Vogler, A. P., and J. W. Lengeler.** 1991. Comparison of the sequences of the nagE operons from *Klebsiella pneumoniae* and *Escherichia coli* K12: enhanced variability of the enzyme IIN-acetylglucosamine in regions connecting functional domains. *Mol. Gen. Genet.* **230:**270–276.

2081. **Volkert, M. R.** 1988. Adaptive response of *Escherichia coli* to alkylation damage. *Environ. Mol. Mutagen.* **11:**241–255.

2082. **Volkert, M. R., and L. I. Hajec.** 1991. Molecular analysis of the aidD6::Mu d1 (bla lac) fusion mutation of *Escherichia coli* K12. *Mol. Gen. Genet.* **229:**319–323.

2083. **Volkert, M. R., P. C. Loewen, J. Switala, D. Crowley, and M. Conley.** 1994. The delta (*argF-lacZ*)205(U169) deletion greatly enhances resistance to hydrogen peroxide in stationary-phase *Escherichia coli. J. Bacteriol.* **176:**1297–1302.

2084. **Volkert, M. R., D. C. Nguyen, and K. C. Beard.** 1986. *Escherichia coli* gene induction by alkylation treatment. *Genetics* **112:**11–26.

2085. **Vorgias, C. E., H. G. Lemaire, and K. S. Wilson.** 1991. Overexpression and purification of the galactose operon enzymes from *Escherichia coli. Protein Expression Purif.* **2:**330–338.

2086. **Voskuil, J. L., C. A. Westerbeek, C. Wu, A. H. Kolk, and N. Nanninga.** 1994. Epitope mapping of *Escherichia coli* cell division protein FtsZ with monoclonal antibodies. *J. Bacteriol.* **176:**1886–1893.

2087. **Wachi, M., M. Doi, Y. Okada, and M. Matsuhashi.** 1989. New mre genes mreC and mreD, responsible for formation of the rod shape of *Escherichia coli* cells. *J. Bacteriol.* **171:**6511–6516.

2088. **Wacker, H., R. A. Harvey, C. H. Winestock, and G. W. E. Plaut.** 1964. 4-(1′-D-Ribitylamino)-5-amino-2,6-dihydroxypyrimidine, the second product of the riboflavin synthetase reaction. *J. Biol. Chem.* **239:**3493–3497.

2089. **Wackernagel, W., and U. Winkler.** 1972. A mutation of *Escherichia coli* enhancing the UV-mutability of phage lamda but not of its infectious DNA in a spheroplast assay. *Mol. Gen. Genet.* **114:**68–79.

2090. **Wada, A., Y. Yamazaki, N. Fujita, and A. Ishihama.** 1990. Structure and probable genetic location of a "ribosome modulation factor" associated with 100S ribosomes in stationary-phase *Escherichia coli* cells. *Proc. Natl. Acad. Sci. USA* **87:**2657–2661.

2091. **Wada, C., and T. Yura.** 1979. Mutants of *Escherichia coli* incapable of supporting replication of F-like plasmids at high temperature: isolation and characterization of mafA and mafB mutants. *J. Bacteriol.* **140:**864–873.

2092. **Wahle, E., R. S. Lasken, and A. Kornberg.** 1989. The dnaB-dnaC replication protein complex of *Escherichia coli*. II. Role of the complex in mobilizing dnaB functions. *J. Biol. Chem.* **264:**2469–2475.

2093. **Wahle, E., R. S. Lasken, and A. Kornberg.** 1989. The dnaB-dnaC replication protein complex of *Escherichia coli*. I. Formation and properties. *J. Biol. Chem.* **264:**2463–2468.

2094. **Waite-Rees, P. A., C. J. Keating, L. S. Moran, B. E. Slatko, L. J. Hornstra, and J. S. Benner.** 1991. Characterization and expression of the *Escherichia coli* Mrr restriction system. *J. Bacteriol.* **173:**5207–5219.

2095. **Wakayama, Y., M. Takagi, and K. Yano.** 1984. Gene responsible for protecting *Escherichia coli* from sodium dodecyl sulfate and toluidine blue plus light. *J. Bacteriol.* **159:**527–532.

2096. **Waldburger, C., T. Gardella, R. Wong, and M. M. Susskind.** 1990. Changes in conserved region 2 of *Escherichia coli* sigma 70 affecting promoter recognition. *J. Mol. Biol.* **215:**267–276.

2097. **Walderhaug, M. O., J. W. Polarek, P. Voelkner, J. M. Daniel, J. E. Hesse, K. Altendorf, and W. Epstein.** 1992. KdpD and KdpE, proteins that control expression of the *kdpABC* operon, are members of the two-component sensor-effector class of regulators. *J. Bacteriol.* **174:**2152–2159.

2098. **Walker, J. E., N. J. Gay, and A. N. Eberle.** 1984. DNA sequence around the *Escherichia coli* unc operon. *Biochem. J.* **224:**799–815.

2099. **Walker, J. R.** 1970. Defective excision repair of pyrimidine dimers in the ultraviolet sensitive *Escherichia coli* ras⁻ mutant. *J. Bacteriol.* **103:**552–559.

2100. **Walsh, C. T., M. D. Erion, A. E. Walts, J. J. Delany, and G. A. Berchtold.** 1987. Chorismate aminations: partial purification of *Escherichia coli* PABA synthase and mechanistic comparison with anthranilate synthase. *Biochemistry* **26:**4734–4745.

2101. **Walsh, J. P., C. R. Loomis, and R. M. Bell.** 1986. Regulation of diacylglycerol kinase biosynthesis in *Escherichia coli*. A trans-acting dgkR mutation increases transcription of the structural gene. *J. Biol. Chem.* **261:**11021–11027.

2102. **Walter, M. R., W. J. Cook, L. B. Cole, S. A. Short, G. W. Koszalka, T. A. Krenitsky, and S. E. Ealick.** 1990. Three-dimensional structure of thymidine phosphorylase from *Escherichia coli* at 2.8 A resolution. *J. Biol. Chem.* **265:**14016–14022.

2103. **Wandersman, C., and P. Delepelaire.** 1990. TolC, an *Escherichia coli* outer membrane protein required for hemolysin secretion. *Proc. Natl. Acad. Sci. USA* **87:**4776–4780.

2104. **Wang, A. Y., Y. Y. Chang, and J. E. Cronan, Jr.** 1991. Role of the tetrameric structure of *Escherichia coli* pyruvate oxidase in enzyme activation and lipid binding. *J. Biol. Chem.* **266:**10959–10966.

2105. **Wang, A. Y., D. W. Grogan, and J. E. Cronan, Jr.** 1992. Cyclopropane fatty acid synthase of *Escherichia coli*: deduced amino acid sequence, purification, and studies of the enzyme active site. *Biochemistry* **31:**11020–11028.

2106. **Wang, E., and C. Walsh.** 1978. Suicide substrates for the alanine racemase of *Escherichia coli* B. *Biochemistry* **17:**1313–1321.

2107. **Wang, L., and B. Weiss.** 1992. dcd (dCTP deaminase) gene of *Escherichia coli*: mapping, cloning, sequencing, and identification as a locus of suppressors of lethal dut (dUTPase) mutations. *J. Bacteriol.* **174:**5647–5653.

2108. **Wang, M. D., L. Liu, B. M. Wang, and C. M. Berg.** 1987. Cloning and characterization of the *Escherichia coli* K-12 alanine-valine transaminase (*avtA*) gene. *J. Bacteriol.* **169:**4228–4234.

2109. **Wang, Q., and J. M. Calvo.** 1993. Lrp, a major regulatory protein in *Escherichia coli*, bends DNA and can organize the assembly of a higher-order nucleoprotein structure. *EMBO J.* **12:**2495–2501.

2110. **Wang, X. D., P. A. de Boer, and L. I. Rothfield.** 1991. A factor that positively regulates cell division by activating transcription of the major cluster of essential cell division genes of *Escherichia coli. EMBO J.* **10:**3363–3372.

2111. **Wanner, B. L.** 1993. Gene regulation by phosphate in enteric bacteria. *J. Cell Biochem.* **51:**47–54.

2112. **Wanner, B. L., and W. W. Metcalf.** 1992. Molecular genetic studies of a 10.9-kb operon in *Escherichia coli* for phosphonate uptake and biodegradation. *FEMS Microbiol. Lett.* **79:**133–139.

2113. **Wanner, B. L., and M. R. Wilmes-Riesenberg.** 1992. Involvement of phosphotransacetylase, acetate kinase, and acetyl phosphate synthesis in control of the phosphate regulon in *Escherichia coli. J. Bacteriol.* **174:**2124–2130.

2114. **Warne, S. R., J. M. Varley, G. J. Boulnois, and M. G. Norton.** 1990. Identification and characterization of a gene that controls colony morphology and auto-aggregation in *Escherichia coli* K12. *J. Gen. Microbiol.* **136:**455–462.

2115. **Warren, M. J., C. A. Roessner, P. J. Santander, and A. I. Scott.** 1990. The *Escherichia coli* cysG gene encodes S-adenosylmethionine-dependent uroporphyrinogen III methylase. *Biochem. J.* **265:**725–729.

2116. **Warren, M. J., N. J. Stolowich, P. J. Santander, C. A. Roessner, B. A. Sowa, and A. I. Scott.** 1990. Enzymatic synthesis of dihydrosirohydrochlorin (precorrin-2) and of a novel pyrrocorphin by uroporphyrinogen III methylase. *FEBS Lett.* **261:**76–80.

2117. **Washburn, B. K., and S. R. Kushner.** 1991. Construction and analysis of deletions in the structural gene (*uvrD*) for DNA helicase II of *Escherichia coli*. *J. Bacteriol.* 173:2569–2575.

2118. **Watanabe, W., G. Sampei, A. Aiba, and K. Mizobuchi.** 1989. Identification and sequence analysis of *Escherichia coli purE* and *purK* genes encoding 5′-phosphoribosyl-5-amino-4-imidazole carboxylase for de novo purine biosynthesis. *J. Bacteriol.* 171:198–204.

2119. **Watson, N., D. S. Dunyak, E. L. Rosey, J. L. Slonczewski, and E. R. Olson.** 1992. Identification of elements involved in transcriptional regulation of the *Escherichia coli cad* operon by external pH. *J. Bacteriol.* 174:530–540.

2120. **Waukau, J., and S. Forst.** 1992. Molecular analysis of the signaling pathway between EnvZ and OmpR in *Escherichia coli*. *J. Bacteriol.* 174:1522–1527.

2121. **Weaver, T. M., D. G. Levitt, and L. J. Banaszak.** 1993. Purification and crystallization of fumarase C from *Escherichia coli*. *J. Mol. Biol.* 231:141–144.

2122. **Webb, E. C.** 1992. *Enzyme Nomenclature 1992.* Academic Press, Inc., New York.

2123. **Weber, R. F., and P. M. Silverman.** 1988. The cpx proteins of *Escherichia coli* K12. Structure of the cpxA polypeptide as an inner membrane component. *J. Mol. Biol.* 203:467–478.

2124. **Webster, C., L. Gardner, and S. Busby.** 1989. The *Escherichia coli melR* gene encodes a DNA-binding protein with affinity for specific sequences located in the melibiose-operon regulatory region. *Gene* 83:207–213.

2125. **Webster, R. E.** 1991. The tol gene products and the import of macromolecules into *Escherichia coli*. *Mol. Microbiol.* 5:1005–1011.

2126. **Wee, S., J. B. Neilands, M. L. Bittner, B. C. Hemming, B. L. Haymore, and R. Seetharam.** 1988. Expression, isolation and properties of Fur (ferric uptake regulation) protein of *Escherichia coli* K 12. *Biol. Methods* 1:62–68.

2127. **Weglenski, P., A. J. Ninfa, S. Ueno-Nishio, and B. Magasanik.** 1989. Mutations in the *glnG* gene of *Escherichia coli* that result in increased activity of nitrogen regulator I. *J. Bacteriol.* 171:4479–4485.

2128. **Wehner, A., E. Harms, M. P. Jennings, I. R. Beacham, C. Derst, P. Bast, and K. H. Rohm.** 1992. Site-specific mutagenesis of *Escherichia coli* asparaginase II. None of the three histidine residues is required for catalysis. *Eur. J. Biochem.* 208:475–480.

2129. **Weickert, M. J., and S. Adhya.** 1993. Control of transcription of *gal* repressor and isorepressor genes in *Escherichia coli*. *J. Bacteriol.* 175:251–258.

2130. **Weidner, U., S. Geier, A. Ptock, T. Friedrich, H. Leif, and H. Weiss.** 1993. The gene locus of the proton-translocating NADH: ubiquinone oxidoreductase in *Escherichia coli*. Organization of the 14 genes and relationship between the derived proteins and subunits of mitochondrial complex I. *J. Mol. Biol.* 233:109–122.

2131. **Weinfeld, M., and M. C. Paterson.** 1988. DNA cyclobutane pyrimidine dimers with a cleaved internal phosphodiester bond can be photoenzymatically reversed by *Escherichia coli* PhrB photolyase. *Nucleic Acids Res.* 16:5693

2132. **Weinstock, O., C. Sella, D. M. Chipman, and Z. Barak.** 1992. Properties of subcloned subunits of bacterial acetohydroxy acid synthases. *J. Bacteriol.* 174:5560–5566.

2133. **Weiss, B., and L. Wang.** 1994. De novo synthesis of thymidylate via deoxycytidine in *dcd* (dCTP deaminase) mutants of *Escherichia coli*. *J. Bacteriol.* 176:2194–2199.

2134. **Weiss, D. L., D. I. Johnson, H. L. Weith, and R. L. Somerville.** 1986. Structural analysis of the ileR locus of *Escherichia coli* K12. *J. Biol. Chem.* 261:9966–9971.

2135. **Weissborn, A. C., Q. Liu, M. K. Rumley, and E. P. Kennedy.** 1994. UTP:alpha-D-glucose-1-phosphate uridylyltransferase of *Escherichia coli*: isolation and DNA sequence of the *galU* gene and purification of the enzyme. *J. Bacteriol.* 176:2611–2618.

2136. **Weissenborn, D. L., N. Wittekindt, and T. J. Larson.** 1992. Structure and regulation of the glpFK operon encoding glycerol diffusion facilitator and glycerol kinase of *Escherichia coli* K-12. *J. Biol. Chem.* 267:6122–6131.

2137. **Weitzmann, C., S. J. Tumminia, M. Boublik, and J. Ofengand.** 1991. A paradigm for local conformational control of function in the ribosome: binding of ribosomal protein S19 to *Escherichia coli* 16S rRNA in the presence of S7 is required for methylation of m2G966 and blocks methylation of m5C967 by their respective methyltransferases. *Nucleic Acids Res.* 19:7089–7095.

2138. **Weitzmann, C. J., P. R. Cunningham, K. Nurse, and J. Ofengand.** 1993. Chemical evidence for domain assembly of the *Escherichia coli* 30S ribosome. *FASEB J.* 7:177–180.

2139. **Wek, R. C., and G. W. Hatfield.** 1986. Nucleotide sequence and in vivo expression of the ilvY and ilvC genes in *Escherichia coli* K12. Transcription from divergent overlapping promoters. *J. Biol. Chem.* 261:2441–2450.

2140. **Wek, R. C., and G. W. Hatfield.** 1988. Transcriptional activation at adjacent operators in the divergent-overlapping ilvY and ilvC promoters of *Escherichia coli*. *J. Mol. Biol.* 203:643–663.

2141. **Wek, R. C., C. A. Hauser, and G. W. Hatfield.** 1985. The nucleotide sequence of the ilvBN operon of *Escherichia coli*: sequence homologies of the acetohydroxy acid synthase isozymes. *Nucleic Acids Res.* 13:3995–4010.

2142. **Wek, R. C., J. H. Sameshima, and G. W. Hatfield.** 1987. Rho-dependent transcriptional polarity in the ilvGMEDA operon of wild-type *Escherichia coli* K12. *J. Biol. Chem.* 262:15256–15261.

2143. **Welsh, K. M., A. L. Lu, S. Clark, and P. Modrich.** 1987. Isolation and characterization of the *Escherichia coli* mutH gene product. *J. Biol. Chem.* 262:15624–15629.

2144. **Weng, M. L., and H. Zalkin.** 1987. Structural role for a conserved region in the CTP synthetase glutamine amide transfer domain. *J. Bacteriol.* 169:3023–3028.

2145. **Werth, M. T., G. Cecchini, A. Manodori, B. A. Ackrell, I. Schroder, R. P. Gunsalus, and M. K. Johnson.** 1990. Site-directed mutagenesis of conserved cysteine residues in *Escherichia coli* fumarate reductase: modification of the spectroscopic and electrochemical properties of the [2Fe-2S] cluster. *Proc. Natl. Acad. Sci. USA* 87:8965–8969.

2146. **Westh Hansen, S. E., N. Jensen, and A. Munch-Petersen.** 1987. Studies on the sequence and structure of the *Escherichia coli* K-12 nupG gene, encoding a nucleoside-transport system. *Eur. J. Biochem.* 168:385–391.

2147. **Whitby, M. C., L. Ryder, and R. G. Lloyd.** 1993. Reverse branch migration of Holliday junctions by RecG protein: a new mechanism for resolution of intermediates in recombination and DNA repair. *Cell* 75:341–350.

2148. **White, P. J., G. Millar, and J. R. Coggins.** 1988. The overexpression, purification and complete amino acid sequence of chorismate synthase from *Escherichia coli* K12 and its comparison with the enzyme from *Neurospora crassa*. *Biochem. J.* 251:313–322.

2149. **White, S. A., M. Nilges, A. Huang, A. T. Brunger, and P. B. Moore.** 1992. NMR analysis of helix I from the 5S RNA of *Escherichia coli*. *Biochemistry* 31:1610–1621.

2150. **Whittaker, J. J., and J. H. Jackson.** 1992. Maintenance of repression control of the ilvGMEDA operon in a temperature-sensitive leucyl-transfer RNA synthetase mutant of *Escherichia coli* K-12 at a restrictive temperature. *Biochem. Biophys. Res. Commun.* 187:1106–1112.

2151. **Wiener, L., D. Schuler, and R. Brimacombe.** 1988. Protein binding sites on *Escherichia coli* 16S ribosomal RNA: RNA regions that are protected by proteins S7, S9 and S19, and by proteins S8, S15 and S17. *Nucleic Acids Res.* 16:1233–1250.

2152. **Wigley, D. B., J. P. Derrick, and W. V. Shaw.** 1990. The serine acetyltransferase from *Escherichia coli*. Over-expression, purification and preliminary crystallographic analysis. *FEBS Lett.* 277:267–271.

2153. **Wijsman, H. J. W.** 1972. The characterization of an alanine racemase mutant of *Escherichia coli*. *Genet. Res.* 20:269–277.

2154. **Wijsman, H. J. W., and H. C. Pafort.** 1974. Pleiotropic mutations in *Escherichia coli* conferring tolerance to glycine and sensitivity to penicillin. *Mol. Gen. Genet.* 128:349–357.

2155. **Wikstrom, P. M., L. K. Lind, D. E. Berg, and G. R. Bjork.** 1992. Importance of mRNA folding and start codon accessibility in the expression of genes in a ribosomal protein operon of *Escherichia coli*. *J. Mol. Biol.* 224:949–966.

2156. **Wild, D. G.** 1988. Reversion from erythromycin dependence in *Escherichia coli*: strains altered in ribosomal sub-unit association and ribosome assembly. *J. Gen. Microbiol.* 134:1251–1263.

2157. **Wild, J., J. Henning, M. Lobocka, W. Walczak, and T. Klopotowski.** 1985. Identification of the dadX gene coding for the predominant isozyme of alanine racemase in *Escherichia coli* K12. *Mol. Gen. Genet.* 198:315–322.

2158. **Wild, J., and B. Obrepalska.** 1982. Regulation of expression of the dadA gene encoding D-amino acid dehydrogenase in *Escherichia coli*: analysis of dadA-lac fusions and direction of dadA transcription. *Mol. Gen. Genet.* 186:405–410.

2159. **Wilkison, W. O., J. P. Walsh, J. M. Corless, and R. M. Bell.** 1986. Crystalline arrays of the *Escherichia coli* sn-glycerol-3-phosphate acyltransferase, an integral membrane protein. *J. Biol. Chem.* 261:9951–9958.

2160. **Williams, J. S., and P. R. Rosevear.** 1991. Nuclear overhauser effect studies of the conformations of Mg(alpha, beta-methylene)ATP bound to *E. coli* isoleucyl-tRNA synthetase. *Biochem. Biophys. Res. Commun.* 176:682–689.

2161. **Williams, M. D., J. A. Fuchs, and M. C. Flickinger.** 1991. Null mutation in the stringent starvation protein of *Escherichia coli* disrupts lytic development of bacteriophage P1. *Gene* 109:21–30.

2162. **Williams, M. D., T. X. Ouyang, and M. C. Flickinger.** 1994. Starvation-induced expression of SspA and SspB: the effects of a null mutation in sspB on *Escherichia coli* protein synthesis and survival during growth and prolonged starvation. *Mol. Microbiol.* 11:1029–1043.

2163. **Williams, M. V., T. J. Kerr, R. D. Lemmon, and G. J. Tritz.** 1980. Azaserine resistance in *Escherichia coli*: chromosomal location of multiple genes. *J. Bacteriol.* 143:383–388.

2164. **Williamson, R. M., and D. L. Oxender.** 1990. Sequence and structural similarities between the leucine-specific binding protein and leucyl-tRNA synthetase of *Escherichia coli*. *Proc. Natl. Acad. Sci. USA* 87:4561–4565.

2165. **Willison, J. C., and G. Tissot.** 1994. The *Escherichia coli efg* gene and the *Rhodobacter capsulatus adgA* gene code for NH3-dependent NAD synthetase. *J. Bacteriol.* 176:3400–3402.

2166. **Wilmanns, M., J. P. Priestle, T. Niermann, and J. N. Jansonius.** 1992. Three-dimensional structure of the bifunctional enzyme phosphoribosylanthranilate isomerase: indoleglycerolphosphate synthase from *Escherichia coli* refined at 2.0 A resolution. *J. Mol. Biol.* 223:477–507.

2167. **Wilson, H. R., and C. L. Turnbough, Jr.** 1990. Role of the purine repressor in the regulation of pyrimidine gene expression in *Escherichia coli* K-12. *J. Bacteriol.* 172:3208–3213.

2168. **Wilson, R. K., T. Brown, and B. A. Roe.** 1986. Nucleotide sequence of pheW; a third gene for *E. coli* tRNAPhe. *Nucleic Acids Res.* 14:5937

2169. **Wilson, R. L., and G. V. Stauffer.** 1994. DNA sequence and characterization of GcvA, a LysR family regulatory protein for the *Escherichia coli* glycine cleavage enzyme system. *J. Bacteriol.* 176:2862–2868.

2170. **Wilson, R. L., P. S. Steiert, and G. V. Stauffer.** 1993. Positive regulation of the *Escherichia coli* glycine cleavage enzyme system. *J. Bacteriol.* 175:902–904.

2171. **Winkelmann, G., A. Cansier, W. Beck, and G. Jung.** 1994. HPLC separation of enterobactin and linear 2,3-dihydroxybenzoylserine derivatives: a study on mutants of *Escherichia coli* defective in regulation (fur), esterase (fes) and transport (fepA). *Biometals* **7:**149–154.

2172. **Wiseman, J. S., and J. S. Nichols.** 1984. Purification and properties of diaminopimelic acid epimerase from *Escherichia coli*. *J. Biol. Chem.* **259:**8907–8914.

2173. **Wissenbach, U., B. Keck, and G. Unden.** 1993. Physical map location of the new *artPIQMJ* genes of *Escherichia coli*, encoding a periplasmic arginine transport system. *J. Bacteriol.* **175:**3687–3688.

2174. **Woelker, B., and W. Messer.** 1993. The structure of the initiation complex at the replication origin, oriC, of *Escherichia coli*. *Nucleic Acids Res.* **21:**5025–5033.

2175. **Woisetschlager, M., and G. Hogenauer.** 1987. The kdsA gene coding for 3-deoxy-D-manno-octulosonic acid 8-phosphate synthetase is part of an operon in *Escherichia coli*. *Mol. Gen. Genet.* **207:**369–373.

2176. **Wolf-Watz, H., and M. Masters.** 1979. Deoxyribonucleic acid and outer membrane: strains diploid for the oriC region show elevated levels of deoxyribonucleic acid-binding protein and evidence for specific binding of the oriC region to outer membrane. *J. Bacteriol.* **140:**50–58.

2177. **Wolodko, W. T., E. R. Brownie, and W. A. Bridger.** 1980. Subunits of succinyl-coenzyme A synthetase: coordination of production in *Escherichia coli* and discovery of a factor that precludes refolding. *J. Bacteriol.* **143:**231–237.

2178. **Wong, I., M. Amaratunga, and T. M. Lohman.** 1993. Heterodimer formation between *Escherichia coli* Rep and UvrD proteins. *J. Biol. Chem.* **268:**20386–20391.

2179. **Wong, I., and T. M. Lohman.** 1992. Allosteric effects of nucleotide cofactors on *Escherichia coli* Rep helicase-DNA binding. *Science* **256:**350–355.

2180. **Woo, K. M., K. I. Kim, A. L. Goldberg, D. B. Ha, and C. H. Chung.** 1992. The heat-shock protein ClpB in *Escherichia coli* is a protein-activated ATPase. *J. Biol. Chem.* **267:**20429–20434.

2181. **Wood, D., M. G. Darlison, R. J. Wilde, and J. R. Guest.** 1984. Nucleotide sequence encoding the flavoprotein and hydrophobic subunits of the succinate dehydrogenase of *Escherichia coli*. *Biochem. J.* **222:**519–534.

2182. **Wood, E. R., and S. W. Matson.** 1989. The molecular cloning of the gene encoding the *Escherichia coli* 75-kDa helicase and the determination of its nucleotide sequence and gentic map position. *J. Biol. Chem.* **264:**8297–8303.

2183. **Wood, J. M.** 1987. Membrane association of proline dehydrogenase in *Escherichia coli* is redox dependent. *Proc. Natl. Acad. Sci. USA* **84:**373–377.

2184. **Woodgate, R., M. Rajagopalan, C. Lu, and H. Echols.** 1989. UmuC mutagenesis protein of *Escherichia coli*: purification and interaction with UmuD and UmuD′. *Proc. Natl. Acad. Sci. USA* **86:**7301–7305.

2185. **Woods, S. A., J. S. Miles, R. E. Roberts, and J. R. Guest.** 1986. Structural and functional relationships between fumarase and aspartase. Nucleotide sequences of the fumarase (fumC) and aspartase (aspA) genes of *Escherichia coli* K12. *Biochem. J.* **237:**547–557.

2186. **Woods, S. A., S. D. Schwartzbach, and J. R. Guest.** 1988. Two biochemically distinct classes of fumarase in *Escherichia coli*. *Biochim. Biophys. Acta* **954:**14–26.

2187. **Wookey, P. J., and A. J. Pittard.** 1988. DNA sequence of the gene (tyrP) encoding the tyrosine-specific transport system of *Escherichia coli*. *J. Bacteriol.* **170:**4946–4949.

2188. **Wower, J., S. S. Hixson, and R. A. Zimmermann.** 1989. Labeling the peptidyltransferase center of the *Escherichia coli* ribosome with photoreactive tRNA(Phe) derivatives containing azidoadenosine at the 3′ end of the acceptor arm: a model of the tRNA-ribosome complex. *Proc. Natl. Acad. Sci. USA* **86:**5232–5236.

2189. **Wright, J. M., C. Satishchandran, and S. M. Boyle.** 1986. Transcription of the speC (ornithine decarboxylase) gene of *Escherichia coli* is repressed by cyclic AMP and its receptor protein. *Gene* **44:**37–45.

2190. **Wu, B., C. Georgopoulos, and D. Ang.** 1992. The essential *Escherichia coli* msgB gene, a multicopy suppressor of a temperature-sensitive allele of the heat shock gene grpE, is identical to dapE. *J. Bacteriol.* **174:**5258–5264.

2191. **Wu, C. A., E. L. Zechner, A. J. Hughes, Jr., M. A. Franden, C. S. McHenry, and K. J. Marians.** 1992. Coordinated leading- and lagging-strand synthesis at the *Escherichia coli* DNA replication fork. IV. Reconstitution of an asymmetric, dimeric DNA polymerase III holoenzyme. *J. Biol. Chem.* **267:**4064–4073.

2192. **Wu, F. Y., W. J. Huang, R. B. Sinclair, and L. Powers.** 1992. The structure of the zinc sites of *Escherichia coli* DNA-dependent RNA polymerase. *J. Biol. Chem.* **267:**25560–25567.

2193. **Wu, G., H. D. Williams, F. Gibson, and R. K. Poole.** 1993. Mutants of *Escherichia coli* affected in respiration: the cloning and nucleotide sequence of ubiA, encoding the membrane-bound p-hydroxybenzoate:octaprenyltransferase. *J. Gen. Microbiol.* **139:**1795–1805.

2194. **Wu, G., H. D. Williams, M. Zamanian, F. Gibson, and R. K. Poole.** 1992. Isolation and characterization of *Escherichia coli* mutants affected in aerobic respiration: the cloning and nucleotide sequence of ubiG. Identification of an S-adenosylmethionine-binding motif in protein, RNA, and small-molecule methyltransferases. *J. Gen. Microbiol.* **138:**2101–2112.

2195. **Wu, H., I. Wower, and R. A. Zimmermann.** 1993. Mutagenesis of ribosomal protein S8 from *Escherichia coli*: expression, stability, and RNA-binding properties of S8 mutants. *Biochemistry* **32:**4761–4768.

2196. **Wu, J., and B. Weiss.** 1991. Two divergently transcribed genes, soxR and soxS, control a superoxide response regulon of *Escherichia coli*. *J. Bacteriol.* **173:**2864–2871.

2197. **Wu, L. F., C. Navarro, and M. A. Mandrand-Berthelot.** 1991. The hydC region contains a multi-cistronic operon (nik) involved in nickel transport in *Escherichia coli*. *Gene* **107:**37–42.

2198. **Wu, T. H., C. H. Clarke, and M. G. Marinus.** 1990. Specificity of *Escherichia coli* mutD and mutL mutator strains. *Gene* **87:**1–5.

2199. **Wu, Y., R. V. Patil, and P. Datta.** 1992. Catabolite gene activator protein and integration host factor act in concert to regulate tdc operon expression in *Escherichia coli*. *J. Bacteriol.* **174:**6918–6927.

2200. **Wubbolts, M. G., P. Terpstra, J. B. van Beilen, J. Kingma, H. A. Meesters, and B. Witholt.** 1990. Variation of cofactor levels in *Escherichia coli*. Sequence analysis and expression of the pncB gene encoding nicotinic acid phosphoribosyltransferase. *J. Biol. Chem.* **265:**17665–17672.

2201. **Wulfing, C., J. Lombardero, and A. Pluckthun.** 1994. An *Escherichia coli* protein consisting of a domain homologous to FK506-binding proteins (FKBP) and a new metal binding motif. *J. Biol. Chem.* **269:**2895–2901.

2202. **Wunderlich, M., R. Jaenicke, and R. Glockshuber.** 1993. The redox properties of protein disulfide isomerase (DsbA) of *Escherichia coli* result from a tense conformation of its oxidized form. *J. Mol. Biol.* **233:**559–566.

2203. **Wurgler, S. M., and C. C. Richardson.** 1990. Structure and regulation of the gene for dGTP triphosphohydrolase from *Escherichia coli*. *Proc. Natl. Acad. Sci. USA* **87:**2740–2744.

2204. **Xi, X. G., F. van Vliet, M. M. Ladjimi, B. de Wannemaeker, C. de Staercke, N. Glansdorff, A. Pierard, R. Cunin, and G. Herve.** 1991. Heterotropic interactions in *Escherichia coli* aspartate transcarbamylase. Subunit interfaces involved in CTP inhibition and ATP activation. *J. Mol. Biol.* **220:**789–799.

2205. **Xiao, H., M. Kalman, K. Ikehara, S. Zemel, G. Glaser, and M. Cashel.** 1991. Residual guanosine 3′,5′-bispyrophosphate synthetic activity of relA null mutants can be eliminated by spoT null mutations. *J. Biol. Chem.* **266:**5980–5990.

2206. **Xie, Q. W., C. W. Tabor, and H. Tabor.** 1989. Spermidine biosynthesis in *Escherichia coli*: promoter and termination regions of the speED operon. *J. Bacteriol.* **171:**4457–4465.

2207. **Xu, F., S. Lin-Chao, and S. N. Cohen.** 1993. The *Escherichia coli* pcnB gene promotes adenylylation of antisense RNAI of ColE1-type plasmids in vivo and degradation of RNAI decay intermediates. *Proc. Natl. Acad. Sci. USA* **90:**6756–6760.

2208. **Xu, W., and E. R. Kantrowitz.** 1991. Function of serine-52 and serine-80 in the catalytic mechanism of *Escherichia coli* aspartate transcarbamoylase. *Biochemistry* **30:**2535–2542.

2209. **Yaghmai, R., and G. L. Hazelbauer.** 1992. Ligand occupancy mimicked by single residue substitutions in a receptor: transmembrane signaling induced by mutation. *Proc. Natl. Acad. Sci. USA* **89:**7890–7894.

2210. **Yaghmai, R., and G. L. Hazelbauer.** 1993. Strategies for differential sensory responses mediated through the same transmembrane receptor. *EMBO J.* **12:**1897–1905.

2211. **Yamada, H., T. Yoshida, K. Tanaka, C. Sasakawa, and T. Mizuno.** 1991. Molecular analysis of the *Escherichia coli* hns gene encoding a DNA-binding protein, which preferentially recognizes curved DNA sequences. *Mol. Gen. Genet.* **230:**332–336.

2212. **Yamada, M., S. Asaoka, M. H. Saier, Jr., and Y. Yamada.** 1993. Characterization of the gcd gene from *Escherichia coli* K-12 W3110 and regulation of its expression. *J. Bacteriol.* **175:**568–571.

2213. **Yamada, M., K. Makino, H. Shinagawa, and A. Nakata.** 1990. Regulation of the phosphate regulon of *Escherichia coli*: properties of phoR deletion mutants and subcellular localization of PhoR protein. *Mol. Gen. Genet.* **220:**366–372.

2214. **Yamada, M., and M. H. Saier, Jr.** 1987. Glucitol-specific enzymes of the phosphotransferase system in *Escherichia coli*. Nucleotide sequence of the gut operon. *J. Biol. Chem.* **262:**5455–5463.

2215. **Yamada, M., and M. H. Saier, Jr.** 1988. Positive and negative regulators for glucitol (gut) operon expression in *Escherichia coli*. *J. Mol. Biol.* **203:**569–583.

2216. **Yamada, M., K. Sumi, K. Matsushita, O. Adachi, and Y. Yamada.** 1993. Topological analysis of quinoprotein glucose dehydrogenase in *Escherichia coli* and its ubiquinone-binding site. *J. Biol. Chem.* **268:**12812–12817.

2217. **Yamaguchi, K., and M. Inouye.** 1988. Lipoprotein 28, an inner membrane protein of *Escherichia coli* encoded by nlpA, is not essential for growth. *J. Bacteriol.* **170:**3747–3749.

2218. **Yamamoto, K.** 1992. Dissection of functional domains in *Escherichia coli* DNA photolyase by linker-insertion mutagenesis. *Mol. Gen. Genet.* **232:**1–6.

2219. **Yamanaka, K., T. Ogura, E. V. Koonin, H. Niki, and S. Hiraga.** 1994. Multicopy suppressors, mssA and mssB, of an smbA mutation of *Escherichia coli*. *Mol. Gen. Genet.* **243:**9–16.

2220. **Yamashita, M., and Y. Murooka.** 1984. Use of lac gene fusions to study regulation of tyramine oxidase which is involved in derepression of latent arylsulfatase in *Escherichia coli*. *Agric. Biol. Chem.* **48:**1459–1470.

2221. **Yamato, I., M. Ohki, and Y. Anraku.** 1979. Genetic and biochemical studies of transport systems for branched-chain amino acids in *Escherichia coli*. *J. Bacteriol.* **138:**24–32.

2222. **Yancey, S. D., and S. R. Kushner.** 1990. Isolation and characterization of a new temperature-sensitive polynucleotide phosphorylase mutation in *Escherichia coli* K-12. *Biochimie* **72:**835–843.

2223. **Yang, C., D. Carlow, R. Wolfenden, and S. A. Short.** 1992. Cloning and nucleotide sequence of the *Escherichia coli* cytidine deaminase (ccd) gene. *Biochemistry* **31:**4168–4174.

2224. **Yang, J., and J. Pittard.** 1987. Molecular analysis of the regulatory region of the *Escherichia coli* K-12 *tyrB* gene. *J. Bacteriol.* 169:4710–4715.

2225. **Yang, M. K., S. C. Ser, and C. H. Lee.** 1989. Involvement of *E. coli* dcm methylase in Tn3 transposition. *Proc. Natl. Sci. Counc. Repub. China Part B* 13:276–283.

2226. **Yang, S. Y., J. M. Li, X. Y. He, S. D. Cosloy, and H. Schulz.** 1988. Evidence that the *fadB* gene of the *fadAB* operon of *Escherichia coli* encodes 3-hydroxyacyl-coenzyme A (CoA) epimerase, delta 3-*cis*-delta 2-*trans*-enoyl-CoA isomerase, and enoyl-CoA hydratase in addition to 3-hydroxyacyl-CoA dehydrogenase. *J. Bacteriol.* 170:2543–2548.

2227. **Yang, S. Y., X. Y. Yang, G. Healy-Louie, H. Schulz, and M. Elzinga.** 1990. Nucleotide sequence of the fadA gene. Primary structure of 3-ketoacyl-coenzyme A thiolase from *Escherichia coli* and the structural organization of the fadAB operon. *J. Biol. Chem.* 265:10424–10429.

2228. **Yang, W., L. Ni, and R. L. Somerville.** 1993. A stationary-phase protein of *Escherichia coli* that affects the mode of association between the trp repressor protein and operator-bearing DNA. *Proc. Natl. Acad. Sci. USA* 90:5796–5800.

2229. **Yang, X. Y., H. Schulz, M. Elzinga, and S. Y. Yang.** 1991. Nucleotide sequence of the promoter and fadB gene of the fadBA operon and primary structure of the multifunctional fatty acid oxidation protein from *Escherichia coli*. *Biochemistry* 30:6788–6795.

2230. **Yano, T., S. Kuramitsu, S. Tanase, Y. Morino, K. Hiromi, and H. Kagamiyama.** 1991. The role of His143 in the catalytic mechanism of *Escherichia coli* aspartate aminotransferase. *J. Biol. Chem.* 266:6079–6085.

2231. **Yano, T., S. Kuramitsu, S. Tanase, Y. Morino, and H. Kagamiyama.** 1992. Role of Asp222 in the catalytic mechanism of *Escherichia coli* aspartate aminotransferase: the amino acid residue which enhances the function of the enzyme-bound coenzyme pyridoxal 5'-phosphate. *Biochemistry* 31:5878–5887.

2232. **Yanofsky, C., V. Horn, and P. Gollnick.** 1991. Physiological studies of tryptophan transport and tryptophanase operon induction in *Escherichia coli*. *J. Bacteriol.* 173:6009–6017.

2233. **Yao, Z., and M. A. Valvano.** 1994. Genetic analysis of the O-specific lipopolysaccharide biosynthesis region (*rfb*) of *Escherichia coli* K-12 W3110: identification of genes that confer group 6 specificity to *Shigella flexneri* serotypes Y and 4a. *J. Bacteriol.* 176:4133–4143.

2234. **Ycas, M.** 1974. On earlier states of the biochemical system. *J. Theor. Biol.* 44:145–160.

2235. **Ye, Q. Z., J. Liu, and C. T. Walsh.** 1990. *p*-Aminobenzoate synthesis in *Escherichia coli*: purification and characterization of PabB as aminodeoxychorismate synthase and enzyme X as aminodeoxychorismate lyase. *Proc. Natl. Acad. Sci. USA* 87:9391–9395.

2236. **Yim, H. H., and M. Villarejo.** 1992. osmY, a new hyperosmotically inducible gene, encodes a periplasmic protein in *Escherichia coli*. *J. Bacteriol.* 174:3637–3644.

2237. **Yorgey, P., and R. Kolter.** 1993. A widely conserved developmental sensor in bacteria? *Trends Genet.* 9:374–375.

2238. **Yoshida, T., C. Ueguchi, H. Yamada, and T. Mizuno.** 1993. Function of the *Escherichia coli* nucleoid protein, H-NS: molecular analysis of a subset of proteins whose expression is enhanced in a hns deletion mutant. *Mol. Gen. Genet.* 237:113–122.

2239. **Yoshikawa, A., S. Isono, A. Sheback, and K. Isono.** 1987. Cloning and nucleotide sequencing of the genes rimI and rimJ which encode enzymes acetylating ribosomal proteins S18 and S5 of *Escherichia coli* K12. *Mol. Gen. Genet.* 209:481–488.

2240. **Yoshikawa, M., A. Okuyama, and N. Tanaka.** 1975. A third kasugamycin resistance locus, *ksgC*, affecting ribosomal protein S2 in *Escherichia coli* K-12. *J. Bacteriol.* 122:796–797.

2241. **Yoshimoto, T., H. Higashi, A. Kanatani, X. S. Lin, H. Nagai, H. Oyama, K. Kurazono, and D. Tsuru.** 1991. Cloning and sequencing of the 7 alpha-hydroxysteroid dehydrogenase gene from *Escherichia coli* HB101 and characterization of the expressed enzyme. *J. Bacteriol.* 173:2173–2179.

2242. **Yoshimoto, T., H. Tone, T. Honda, K. Osatomi, R. Kobayashi, and D. Tsuru.** 1989. Sequencing and high expression of aminopeptidase P gene from *Escherichia coli* HB101. *J. Biochem.* (Tokyo) 105:412–416.

2243. **Yoshimura, T., M. Ashiuchi, N. Esaki, C. Kobatake, S. Y. Choi, and K. Soda.** 1993. Expression of glr (murI, dga) gene encoding glutamate racemase in *Escherichia coli*. *J. Biol. Chem.* 268:24242–24246.

2244. **You, S. Y., S. Cosloy, and H. Schulz.** 1989. Evidence for the essential function of 2,4-dienoyl-coenzyme A reductase in the beta-oxidation of unsaturated fatty acids in vivo. Isolation and characterization of an *Escherichia coli* mutant with a defective 2,4-dienoyl-coenzyme A reductase. *J. Biol. Chem.* 264:16489–16495.

2245. **Yousif, S. Y., J. K. Broome-Smith, and B. G. Spratt.** 1985. Lysis of *Escherichia coli* by beta-lactam antibiotics: deletion analysis of the role of penicillin-binding proteins 1A and 1B. *J. Gen. Microbiol.* 131:2839–2845.

2246. **Yu, F., Y. Jen, E. Takeuchi, M. Inouye, H. Nakayama, M. Tagaya, and T. Fukui.** 1988. Alpha-glucan phosphorylase from *Escherichia coli*. Cloning of the gene, and purification and characterization of the protein. *J. Biol. Chem.* 263:13706–13711.

2247. **Yumoto, N., and M. Tokushige.** 1988. Characterization of multiple fumarase proteins in *Escherichia coli*. *Biochem. Biophys. Res. Commun.* 153:1236–1243.

2248. **Yura, T., H. Mori, H. Nagai, T. Nagata, A. Ishihama, N. Fujita, K. Isono, K. Mizobuchi, and A. Nakata.** 1992. Systematic sequencing of the *Escherichia coli* genome: analysis of the 0–2.4 min region. *Nucleic Acids Res.* 20:3305–3308.

2249. **Zacharias, M., H. U. Goringer, and R. Wagner.** 1992. Analysis of the Fis-dependent and Fis-independent transcription activation mechanisms of the *Escherichia coli* ribosomal RNA P1 promoter. *Biochemistry* 31:2621–2628.

2250. **Zakin, M. M., N. Duchange, P. Ferrara, and G. N. Cohen.** 1983. Nucleotide sequence of the metL gene of *Escherichia coli*. Its product the bifunctional aspartokinase II-homoserine dehydrogenase II, and the bifunctional product of the thrA gene, aspartokinase I-homoserine dehydrogenase I, derive from a common ancestor. *J. Biol. Chem.* 258:3028–3031.

2251. **Zambidis, I., and L. C. Kuo.** 1990. Substrate specificity and protonation state of *Escherichia coli* ornithine transcarbamoylase as determined by pH studies. Binding of carbamoyl phosphate. *J. Biol. Chem.* 265:2620–2623.

2252. **Zani, M. L., T. Pourcher, and G. Leblanc.** 1993. Mutagenesis of acidic residues in putative membrane-spanning segments of the melibiose permease of *Escherichia coli*. II. Effect on cationic selectivity and coupling properties. *J. Biol. Chem.* 268:3216–3221.

2253. **Zapun, A., J. C. Bardwell, and T. E. Creighton.** 1993. The reactive and destabilizing disulfide bond of DsbA, a protein required for protein disulfide bond formation in vivo. *Biochemistry* 32:5083–5092.

2254. **Zavitz, K. H., R. J. DiGate, and K. J. Marians.** 1991. The priB and priC replication proteins of *Escherichia coli*. Genes, DNA sequence, overexpression, and purification. *J. Biol. Chem.* 266:13988–13995.

2255. **Zavitz, K. H., and K. J. Marians.** 1991. Dissecting the functional role of PriA protein-catalysed primosome assembly in *Escherichia coli* DNA replication. *Mol. Microbiol.* 5:2869–2873.

2256. **Zawadzke, L. E., T. D. Bugg, and C. T. Walsh.** 1991. Existence of two D-alanine:D-alanine ligases in *Escherichia coli*: cloning and sequencing of the ddlA gene and purification and characterization of the DdlA and DdlB enzymes. *Biochemistry* 30:1673–1682.

2257. **Zecherle, G. N., A. Oleinikov, and R. R. Traut.** 1992. The proximity of the C-terminal domain of *Escherichia coli* ribosomal protein L7/L12 to L10 determined by cysteine site-directed mutagenesis and protein-protein cross-linking. *J. Biol. Chem.* 267:5889–5896.

2258. **Zengel, J. M., and L. Lindahl.** 1990. *Escherichia coli* ribosomal protein L4 stimulates transcription termination at a specific site in the leader of the S10 operon independent of L4-mediated inhibition of translation. *J. Mol. Biol.* 213:67–78.

2259. **Zengel, J. M., and L. Lindahl.** 1991. Ribosomal protein L4 of *Escherichia coli*: in vitro analysis of L4-mediated attenuation control. *Biochimie* 73:719–727.

2260. **Zengel, J. M., and L. Lindahl.** 1993. Domain I of 23S rRNA competes with a paused transcription complex for ribosomal protein L4 of *Escherichia coli*. *Nucleic Acids Res.* 21:2429–2435.

2261. **Zhang, A., and M. Belfort.** 1992. Nucleotide sequence of a newly-identified *Escherichia coli* gene, stpA, encoding an H-NS-like protein. *Nucleic Acids Res.* 20:6735

2262. **Zhang, F., J. A. Sheps, and V. Ling.** 1993. Complementation of transport-deficient mutants of *Escherichia coli* alpha-hemolysin by second-site mutations in the transporter hemolysin B. *J. Biol. Chem.* 268:19889–19895.

2263. **Zhang, J. R., and M. P. Deutscher.** 1988. *Escherichia coli* RNase D: sequencing of the rnd structural gene and purification of the overexpressed protein. *Nucleic Acids Res.* 16:6265–6278.

2264. **Zhang, J. R., and M. P. Deutscher.** 1988. Transfer RNA is a substrate for RNase D in vivo. *J. Biol. Chem.* 263:17909–17912.

2265. **Zhang, Y., and E. R. Kantrowitz.** 1991. The synergistic inhibition of *Escherichia coli* aspartate carbamoyltransferase by UTP in the presence of CTP is due to the binding of UTP to the low affinity CTP sites. *J. Biol. Chem.* 266:22154–22158.

2266. **Zhang, Y., and E. R. Kantrowitz.** 1992. Probing the regulatory site of *Escherichia coli* aspartate transcarbamoylase by site-specific mutagenesis. *Biochemistry* 31:792–798.

2267. **Zhang, Y., M. Oldenburg, and R. H. Fillingame.** 1994. Suppressor mutations in F1 subunit epsilon recouple ATP-driven H$^+$ translocation in uncoupled Q42E subunit c mutant of *Escherichia coli* F1F0 ATP synthase. *J. Biol. Chem.* 269:10221–10224.

2268. **Zhao, N., W. Oh, D. Trybul, K. S. Thrasher, T. J. Kingsbury, and T. J. Larson.** 1994. Characterization of the interaction of the *glp* repressor of *Escherichia coli* K-12 with single and tandem *glp* operator variants. *J. Bacteriol.* 176:2393–2397.

2269. **Zheng, L., and H. D. Braymer.** 1991. Overproduction and purification of McrC protein from *Escherichia coli* K-12. *J. Bacteriol.* 173:3918–3920.

2270. **Zheng, L., X. Wang, and H. D. Braymer.** 1992. Purification and N-terminal amino acid sequences of two polypeptides encoded by the mcrB gene from *Escherichia coli* K-12. *Gene* 112:97–100.

2271. **Zhu, L., and M. P. Deutscher.** 1987. tRNA nucleotidyltransferase is not essential for *Escherichia coli* viability. *EMBO J.* 6:2473–2477.

2272. **Zhu, L. Q., T. Gangopadhyay, K. P. Padmanabha, and M. P. Deutscher.** 1990. *Escherichia coli rna* gene encoding RNase I: cloning, overexpression, subcellular distribution of the enzyme, and use of an *rna* deletion to identify additional RNases. *J. Bacteriol.* 172:3146–3151.

2273. **Zhu, Y., and E. C. Lin.** 1989. L-1,2-Propanediol exits more rapidly than L-lactaldehyde from *Escherichia coli*. *J. Bacteriol.* 171:862–867.

2274. **Zilhao, R., L. Camelo, and C. M. Arraiano.** 1993. DNA sequencing and expression of the gene rnb encoding *Escherichia coli* ribonuclease II. *Mol. Microbiol.* **8:**43–51.

2275. **Zimmerman, R. A., Y. Ikeya, and P. F. Sparling.** 1973. Alteration of ribosomal protein S4 by mutation linked to kasugamycin-resistance in *Escherichia coli. Proc. Natl. Acad. Sci. USA* **70:**71–75.

2276. **Zinkewich-Peotti, K., and J. M. Fraser.** 1988. New locus for exopolysaccharide overproduction in *Escherichia coli* K-12. *J. Bacteriol.* **170:**1405–1407.

2277. **Zinoni, F., J. Heider, and A. Bock.** 1990. Features of the formate dehydrogenase mRNA necessary for decoding of the UGA codon as selenocysteine. *Proc. Natl. Acad. Sci. USA* **87:**4660–4664.

2278. **Zipkas, D., and M. Riley.** 1975. Proposal concerning mechanism of evolution of the genome of *Escherichia coli. Proc. Natl. Acad. Sci. USA* **72:**1354–1358.

2279. **Zipkas, D., L. Solomon, and M. Riley.** 1978. Relationship between gene function and gene location. *J. Mol. Evol.* **11:**47–56.

2280. **Zumstein, L., and J. C. Wang.** 1986. Probing the structural domains and function in vivo of *Escherichia coli* DNA topoisomerase I by mutagenesis. *J. Mol. Biol.* **191:**333–340.

2281. **Zwaig, N., W. S. Kistler, and E. C. Lin.** 1970. Glycerol kinase, the pacemaker for the dissimilation of glycerol in *Escherichia coli. J. Bacteriol.* **102:**753–759.

Escherichia coli Protein Sequences: Functional and Evolutionary Implications

EUGENE V. KOONIN, ROMAN L. TATUSOV, AND KENNETH E. RUDD

117

INTRODUCTION

As of August 1994, the sequence of approximately 60% of the *Escherichia coli* chromosome, which consists of about 4,720 kbp, was available; this information is being accumulated in at least three independent databases (28, 46, 56). Translation of the genes contained in the EcoSeq7 database (46) results in 2,328 protein sequences that in turn make up about 60% of all proteins encoded in the *E. coli* genome. Obviously, an enormous amount of valuable information on the physiology of the bacterial cell and its evolution is encrypted in these sequences. Extracting this information to the fullest and using it as a platform for future, rationally designed experiments is one of the principal components of the *E. coli* genome project (13). In fact, to a considerable extent, this is what makes the determination of the complete genome sequence such a worthy task; of course, this is true of any genome project, not only *E. coli* (8). The availability of the complete set of protein sequences from one organism, and especially the possibility to compare such sequence sets from different, evolutionarily distant organisms, e.g., *E. coli* and *Saccharomyces cerevisiae* or *E. coli* and *Bacillus subtilis,* will allow one to address fundamental questions that even very recently seemed to be completely out of reach. Here are a few of these basic questions. What is the extent of functional prediction based on protein sequence, and how precise can this prediction be? What is the minimal set of conserved, housekeeping genes that supports the functioning of a bacterial cell? What specific genes define the physiological differences between bacteria? How conserved or how divergent are different functional groups of proteins between bacteria and eukaryotes and between different bacteria? What is the extent of gene duplication in the bacterial genome evolution, and what is typically duplicated—a single gene, an operon, or an even larger group of genes? Which functions benefit from diversification and are performed by multiple related proteins, and which are secured by unique proteins? To what extent can we realistically hope to reconstruct the genome organization of the common ancestor of all bacteria and that of the hypothetical progenote (58), the ultimate common ancestor of bacteria, eukaryotes, and members of the *Archaea*?

Our ability to adequately address these and other important questions depends both on the completeness of the available sequence data and on the power of computer methods for sequence analysis. Obviously, the final answers still belong to the future, if only because not a single complete bacterial genome sequence is currently available, but it is not a remote future any more. The collection of 60% of the *E. coli* protein sequences seems to be an appropriate starting point for a pilot project on comprehensive analysis of the set of proteins encoded in a bacterial genome (E. V. Koonin, R. L. Tatusov, and K. E. Rudd, *Proc. Natl. Acad. Sci. USA,* in press). In this chapter, we summarize the preliminary results of such a project.

STRATEGY AND METHODOLOGY OF COMPUTER-AIDED ANALYSIS OF THE *E. COLI* PROTEIN SEQUENCE SET

The existing approaches to molecular sequence analysis can be classified into two groups—intrinsic and extrinsic methods (11). Intrinsic analysis explores statistical properties of a sequence without explicitly comparing with other sequences. With regard to protein sequences, these are amino acid composition and charge; clustering of charged, hydrophobic, or other residues; compositional complexity, which is indicative of the globular or nonglobular structure of a protein; and others. The extrinsic analysis deals with various aspects of sequence comparison and sequence conservation between different proteins.

In this chapter, we briefly consider some intrinsic properties of *E. coli* proteins, but, by and large, we concentrate on sequence comparison. We pursue two complementary views of the bacterial genome—"from the inside" and "from the outside."

The "outside view" pertains to sequence conservation between *E. coli* proteins and proteins from other organisms. The availability of about 60% of *E. coli* protein sequences gives one the opportunity to derive reliable estimates of the number of highly conserved proteins that contain ancient regions dating back to the divergence of eubacteria, eukaryotes, and the *Archaea;* those that are conserved between *E. coli* and distantly related bacteria; and variable proteins that are conserved only in closely related bacteria or are found in *E. coli* alone. It is of obvious importance to establish correlations between the level of sequence conservation in evolution and protein functions.

The core of our analysis includes the programs of the BLAST family, which perform database searches for sequence similarity by using individual sequences as queries and producing ungapped pairwise alignments (3, 4), and the MoST program, which screens databases for conserved motifs by using ungapped multiple-alignment blocks as the input (54).

The BLAST algorithm calculates the probability (*P* value) of high-scoring sequence segment pairs to be observed by chance on the basis of the Karlin-Altschul statistics for sequence comparison (see reference 3 and references therein). The lengths of the alignments are determined automatically to obtain the highest statistical significance and therefore may vary from as few as 15 to 20 amino acid residues to as many as several hundred, depending on the level of similarity and distribution of conserved regions in related proteins. The use of the *P* value as the criterion of significance in sequence comparisons may produce false positives when a combination of several segments with low scores results in an artifactually low *P* value. Therefore, in our analysis, we defined all the cutoffs in terms of the similarity score as such rather than the *P* value. The BLASTP program is used to screen amino acid sequence databases for similarity to a protein sequence query, and the TBLASTN program is used to screen nucleotide databases translated in all six reading frames (3).

The MoST program converts an ungapped multiple-alignment block, which may be derived directly from a BLAST output by using the CAP program (54) or generated by another multiple-alignment construction method into a position-dependent weight matrix. The matrix is used for database screening, and the statistical significance of the similarity score with the matrix is determined for each segment in the database by using the recently developed theory (54). Typically, the length of alignment blocks used to generate position-dependent weight matrices varies between 12 and 40 amino acid residues.

Thus, both principal methods of sequence similarity search used in our study are oriented at detecting at least one contiguous conserved region per protein sequence. This approach may result in some marginally significant similarities being missed, but it greatly reduces the likelihood of false positives because of the existence of rigorous statistical theory.

In different types of sequence database searches, spurious "hits" with low-complexity (compositionally biased) regions present in many proteins are frequently observed (59, 60). Therefore we routinely used the SEG program (59), in conjunction with both BLAST and MoST, to filter out such regions. However, since functionally important conserved protein segments may in some cases have a compositional bias, for those proteins that failed to show significant similarity to other sequences in the initial BLAST searches, the analysis was repeated without filtering.

Recent analyses of large sets of yeast and bacterial proteins have shown that careful examination of relatively weak similarities detected in database searches performed with different, complementary computer methods is critical for increasing the rate of functional prediction and revealing evolutionary relationships (7–11, 25, 45). Accordingly, after comparing all the available *E. coli* protein sequences with the nonredundant amino acid sequences database (National Center for Biotechnology Information) by using BLASTP and listing all the alignments with similarity scores above 90 (corresponding to $P \approx 10^{-3}$) as authentic relationships, we assessed the relevance of all the alignments in the "twilight zone" (scores between 60 and 90). To this end, we used analysis of conserved motifs with the CAP and MoST programs and multiple-alignment analysis with the MACAW program (51), as well as considerations of functional relevance. In addition, all the *E. coli* protein sequences were searched for conserved motifs typical of *E. coli* protein clusters by using CAP and MoST (see below).

The "inside" view includes delineation of clusters of paralogous proteins within the *E. coli* protein sequence set. (Paralogs are proteins that share significant sequence similarity and, by inference, common ancestry but have distinct functions, as opposed to orthologs—related proteins that share both common ancestry and common function [14]. Accordingly, all related proteins encoded by different genes in a same organism are paralogs, whereas "the same" proteins in different organisms are orthologs. Paralogs should have evolved by intragenomic gene duplication and orthologs evolve by direct, vertical descent. The distinction between orthologs and paralogs is not universally appreciated, and they are often both simply called homologs; however, we believe that this distinction is very important for all attempts to understand the genome evolution, and we will use it throughout this text.) The existence of related genes that presumably have evolved by duplication in *E. coli* has been recognized for a long time (43, 44), but it is only now, with the sequence of a major part of the genome available, that it is becoming feasible to produce a nearly complete catalog of the clusters of paralogs. The fraction of such clusters that is represented in the current database of *E. coli* proteins probably is considerably higher than 60%, since many proteins among the 40% encoded in the not yet sequenced portions of the genome will join already known clusters. For clustering *E. coli* proteins, we used a single-linkage, "greedy" clustering algorithm. A cluster was defined as a group of protein sequences connected by similarity scores above a chosen cutoff but not requiring that each pair of sequences within a cluster had such a score. We found that a BLAST score of 70 or higher, even though not highly statistically significant, almost always corresponded to relationships between *E. coli* proteins that could be confirmed by subsequent multiple alignment analysis, search for conserved motifs, and functional assessment. Accordingly, this score was used as the cutoff for clustering. The "greedy" algorithm runs into a problem when proteins containing two or more distinct conserved domains bring together otherwise unrelated clusters. This was accounted for by deriving conserved motifs typical of each cluster and including the distinct domains of multidomain proteins in different clusters.

Motif conservation was also used as the criterion for delineating higher-rank groups of paralogous *E. coli* proteins which may include more distantly related sequences; we called such groups superclusters. A supercluster was defined as a group of proteins containing at least one unique, conserved motif. A supercluster may

TABLE 4 Some highly conserved proteins in *E. coli*[a]

Protein	Length (amino acid residues)	Best hit with distant bacteria	Score, best alignment[b]	Best eukaryotic/archaeal hit	Score, best alignment	Function or activity	Paralogs in *E. coli*	Comment
Acs	652	BSSRFAD_2	111; 36%/69	PBLACOASYN_1	1,245; 57%/384	Ac-CoA[c] synthetase	Aas, EntE, EntF, FadD, YdiD', YaaM	No bacterial ortholog
GlgP	809	PHSM_STRPN	110; 44%/47 (partial)	PHS2_HUMAN	1,227; 51%/442	Glycogen phosphorylase	MalP	No bacterial ortholog
FumC	467	FUMH_BACSU	1,000; 57%/339	FUMH_RAT	1,152; 55%/397	Fumarase	ArgH, AspA, PurB	Mitochondrial
Ppc	883	CAPP_CORGL	261; 48%/93	CAPP_MEDSA	1,016; 50%/401	PEP carboxylase	Not known	Cytoplasmic chloroplast origin ?
FdhF	715	S36605 (*Synechococcus*)	374; 30%/312	FDHA_METFO	947; 48%/377	Formate dehydrogenase	Nine dehydrogenases	Archaeal ortholog
GcvP		None	NA	GCSP_PEA	928; 58%/284	Gly dehydrogenase	Not known	mitochondrial
YiaY	382	MEDH_BACMT	887; 46%/375	ADH4_YEAST	919; 61%/287	Alcohol dehydrogenase	Eight dehydrogenases	
TrpB	397	TRPB_LACLA	567; 51%/301	TRP1_ARATH	860; 58%/279	Trp synthase	Not known	Chloroplast
AcnA	891	ACON_BACSU	275; 47%/115 (partial)	IREB_RABIT	841; 42%/449	Aconitase	YacI, LeuC, LeuD	Cytoplasmic
NagB	266	MC068_1	151; 41%/78 (partial)	HUMORFKG1E_1	776; 56%/258	Glucosamine-6-phosphate isomerase	YieK	
Udk	213	None	NA	MUSURKI_1	201; 40%/90	Uridine kinase	Not known	
RpoB	1,342	BACRPLL_3	651; 68%/171	RPC2_YEAST	199; 40%/110	RNA polymerase β	Not known (none ?)	
PrfC	529	EFG_ANANI	226; 29%/154	EFGM_YEAST	179; 30%/126	Peptide release factor	14 GTPases	No ortholog
PolB	783	None	NA	DPOA_HUMAN	141; 38%/71	DNA polymerase II	Not known (none ?)	No bacterial homologs

[a]The table includes the top 10 hits between *E. coli* proteins and their homologs from eukaryotes or *Archaea* and, for comparison, four proteins with moderate similarity to eukaryotic homologs.

[b]For each protein, the highest BLASTP score, the percentage of identical amino acid residues in the best ungapped alignment, and the length of this alignment are indicated. NA, not applicable.

[c]Ac-CoA, acetyl coenzyme A.

like this, the only possibility, besides conservation since the time of bacterial-archaeal radiation, is horizontal gene transfer.

However, a more general and perhaps a more important observation is that the mean similarity score between *E. coli* proteins and their eukaryotic or archaeal homologs is 174 (median of the score distribution, 109), whereas a considerably higher mean score of 230 (median, 146) was observed with homologs from distantly related bacteria. A comparison of the two score distributions suggests that the majority of the "ancient conserved regions" probably date back to the original point of radiation between bacteria, eukaryotes, and the *Archaea*. Indeed, the distribution for eukaryotic (archaeal) homologs of *E. coli* proteins is shifted toward lower scores as compared with the distribution for homologs from distantly related bacteria, just as would be expected if the regions that are conserved in *E. coli* proteins and their homologs from eukaryotes and the *Archaea* have undergone longer evolution (Fig. 6). In accord with this, moderately conserved *E. coli* proteins typically have higher similarity scores with the orthologs from distantly related bacteria than with those from eukaryotes or the *Archaea* (see examples in Table 4).

What can one surmise about the 340 *E. coli* proteins (about 14% of the total) that are not similar to any other protein in the current databases? Notably, the majority of the proteins in this set (about 70%) have not been functionally characterized. Those whose functions are known represent a spectrum of activities (Table 5), some of which simply have not yet been discovered in other organisms, whereas others (e.g., dGTP triphosphatase [39]) may be specific for *E. coli* and closely related enterobacteria.

SEQUENCE CONSERVATION IN *E. COLI* PROTEINS: VIEW FROM WITHIN

Clustering of the *E. coli* proteins sequences by using the "greedy" algorithm described above, together with motif searches designed to resolve the multidomain protein problem and to delineate superclusters, showed that about one-half of the proteins belong to 299 clusters and 70 superclusters of paralogs (Fig. 7A). Most of the clusters are small, with only 2 to 4 members, but there are several clusters with more than 10 members (Fig. 8). The observed distribution of the cluster size appears to be compatible with a stochastic duplication model; a simple simulation with uniform duplication and retention rates produced a distribution

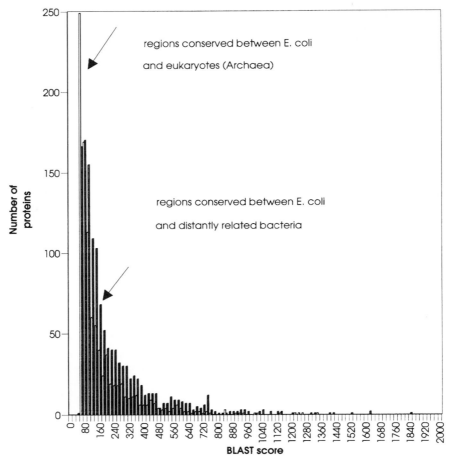

FIGURE 6 Distribution of similarity scores between *E. coli* proteins and their homologs from distantly related organisms. For each *E. coli* protein, the highest score with a protein from eukaryotes/*Archaea* or from distantly related bacteria (see text) produced by BLASTP is included.

similar to the one in Fig. 8 (data not shown) but, obviously, without the larger clusters. To explain the existence of these clusters, one has to postulate that for some functions, duplication with subsequent diversification provides a significant selective advantage. Inspection of the small and large clusters from the functional point of view clearly indicates two types of functions that may be subject to such a selection: metabolite transport and regulation of gene expression. Large clusters are mostly composed of transport proteins and proteins involved in different types of regulation (Table 6), whereas metabolic enzymes typically form small

TABLE 5   Some *E. coli* proteins with known function in search of relatives

Protein	Length	Function	Proteins with analogous function in other organisms
FucI	591	L-Fucose isomerase	Unknown
RecC	1,122	Exonuclease V subunit, DNA repair and recombination	Numerous groups of exonucleases
Dgt	505	dGTPase	Unknown
PepD	485	β-Ala-His dipeptidase	Unknown but other families of dipeptidases exist
SbcB	475	Exonuclease I, DNA repair	Numerous groups of exonucleases
SelA	463	Selenocysteine synthase	Unknown
CreD	450	Inner membrane protein	Unknown
HipA	440	Resistance to inhibitors of peptidoglycan and DNA synthesis	Unknown
ProX	330	Glycine betaine-binding periplasmic protein	Several families of solute-binding proteins
HolA	343	DNA polymerase III δ subunit	Unknown
Gsk	434	IMP-GMP kinase	Numerous nucleotide kinases

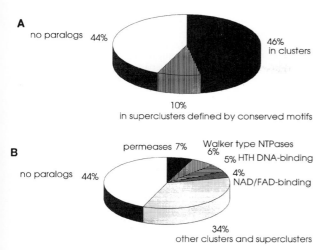

FIGURE 7 Clustering of paralogous *E. coli* proteins. (A) Clusters and superclusters. (B) The four largest superclusters.

clusters (Table 7). Strikingly, there seem to be essential functional classes of proteins in *E. coli* that are not prone to gene duplication, notably catalytic subunits of DNA and RNA polymerases and ribosomal proteins. It is likely that duplication is selected against in these cases, but so far we do not understand the nature of this apparent negative selection.

The extent of apparent duplication in certain functional classes of *E. coli* genes becomes even more dramatic when one considers superclusters defined by the conservation of sequence motifs. The four largest superclusters account for almost one-quarter of all known *E. coli* proteins (Fig. 7B). Figure 9 shows the aligned sequences of the (predicted) ATP/GTP-binding P-loops (23, 49, 57) that form the motif defining the second largest supercluster of paralogous proteins in *E. coli*.

The distribution of the cluster members on the *E. coli* chromosome shows a clear prominence of closely spaced, related genes (Fig. 10). About 50 pairs of such genes appear to be the result of actual tandem duplications, without intervening genes. All but three of these pairs are transcribed in the same direction and belong to the same operon (data not shown). These tandem

duplications may be placed in the same category of evolutionary events with intragenic duplications, of which several cases are apparent in *E. coli*, e.g., in such members of the ATP-binding cassette (ABC) transporter ATPase cluster as AraG and RbsA (16, 21) or in ATPase subunits of ATP-dependent proteases such as ClpA and ClpB (17).

Many of the duplications in the evolution of the *E. coli* chromosome apparently have involved more than one gene. The most striking example of such a cassette duplication is presented by the *dpp, opp,* and *nik* operons, which encode proteins mediating dipeptide, oligopeptide, and nickel transport, respectively, and contain five paralogous genes in a row (1, 30). There are several examples of apparent three- and two-gene cassette duplications in the *E. coli* chromosome (data not shown). Not surprisingly, most of them include genes coding for transport and regulatory proteins (see above).

A weak, large-scale periodicity in the distribution of paralogous genes along the chromosome, with the distance between paralogous genes tending to be a multiple of 6 to 7 min of the *E. coli* chromosome, was observed (Fig. 10). A similar periodicity was noticed previously in a study with a much smaller set of related *E. coli* genes (26). One possible explanation for this apparent periodicity involves duplication of large segments of the chromosome early in evolution, whereas another line of thinking may relate the periodicity to the nucleoid superstructure; additional analysis is needed to critically assess these possibilities.

Our analysis of *E. coli* protein clusters revealed 91 proteins that consist of two or more distinct, conserved domains belonging to different clusters (evidently, there are a lot more multidomain proteins altogether, but in many of them, one or both domains do not have paralogs in *E. coli*). Most of these are accounted for by the two-component, sensor-receiver signal-transducing system (37, 38), the membrane sugar phosphotransferase and transport (PTS) system (41, 48), and various proteins, in which the DNA-binding, helix-turn-helix (HTH) domain is combined with other domains, e.g., repressors containing the HTH domain and a sugar-binding domain (32, 53, 55).

Other examples of proteins with distinct conserved domains are less well known, and here we discuss one which demonstrates not only intricate multidomain organization of some proteins but also the potential of sequence analysis in predicting new protein func-

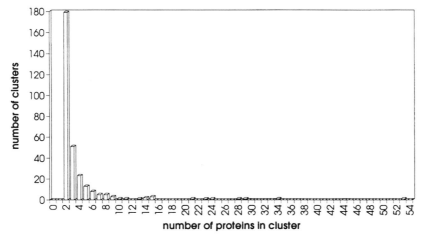

FIGURE 8 Distribution of the number of sequences in *E. coli* protein clusters.

TABLE 6 The 10 largest clusters of paralogous proteins in *E. coli*

Cluster	No. of proteins	Function	Homologs in distantly related bacteria	Homologs in eukaryotes and the *Archaea*
ABC transporter ATPases	54	ATP-dependent membrane transport; DNA repair	Yes	Yes
Permeases (AraE related)	34	Membrane transport	Yes	Yes
HTH proteins (Ada related)	29	Transcription regulation	Yes	No
Receiver domains	28	Membrane signal transduction	Yes	Yes
Sensor domains	24	Membrane signal transduction	Yes	Yes
Permeases (ArtM related)	24	Membrane transport	Yes	Yes
HTH proteins (CynR related)	21	Transcription regulation	Yes	No
Sugar-binding domains	15	Metabolite transport	Yes	
DEAD/H helicases	15	RNA/DNA duplex unwinding	Yes	Yes
GTPases	15	GTP-dependent processes	Yes	Yes

tions. Lon is a protease involved in ATP-dependent proteolysis of abnormal and short-lived proteins in *E. coli* and other organisms (reviewed in reference 17). The ATPase domain containing the typical conserved motifs has been located in the N-terminal portion of Lon (12), whereas the catalytic serine of the protease domain has been identified in the C-terminal portion (5). In addition, a region of significant similarity has been detected between regions from the C-terminal domains of Lon and the *E. coli* protein Sms, which has been implicated in the resistance to methyl methanesulfonate but otherwise remains functionally uncharacterized (31). When the region of conservation between Sms, its *B. subtilis* homolog, and Lon proteins from different species was used to derive a position-dependent weight matrix and search the database (54), a related

motif was identified in several other proteins, including *E. coli* proteins HtrA, HhoA, HhoB, and YifB. Subsequent analysis involving additional database searches and multiple alignment resulted in the delineation of a group of proteins containing ATPase and protease domains (Fig. 11). HtrA, HhoA, and HhoB are serine proteases that are related to the classical, chymotrypsin-like proteases more closely than Lon is and contain the characteristic histidine-aspartate-serine catalytic triad (6, 27; S. Bass, Q. Gu, and A. Goddard, GenBank accession number U1566 1). In Lon, HtrA, HhoA, and HhoB, the highly conserved motif is located directly after the catalytic serine, suggesting that it may be a specific form of a protein-binding site (Fig. 11A). Putative catalytic serines could also be identified in YifB and Sms (Fig. 11A). Like Lon, YifB and Sms contain

TABLE 7 Examples of small clusters of paralogous enzymes in *E. coli*

Cluster	Proteins	Enzymatic activity	Homologs in distantly related bacteria	Homologs in eukaryotes/*Archaea*	Comment
Acetyltransferase	AccB	Biotin carboxyl carrier protein	Yes	Yes	AceF and SucB are subunits of analogous enzymatic complexes and share highly significant similarity, whereas the conservation in AccB is limited to a motif around the biotin-binding site
	AceF	Dihydrolipoamide acetyltransferase	Yes	Yes	
	SucB	Dihydrolipoamide acetyltransferase	Yes	Yes	
Acetate kinase	AckA	Acetate kinase	Yes	Yes	AckA is involved in acetyl coenzyme A formation; the activity of YhaA' can be predicted from sequence similarity
	YhaA'	??	Yes	Yes	
Acid phosphatase	Agp	Glucose-1-phosphatase	No	No	Genes located close to each other, even though not a tandem duplication; periplasmic enzymes
	AppA	pH 2.5 acid phosphatase	No	No	
Alanine racemase	Alr	Alanine racemase	Yes	No	Anabolic enzyme involved in cell wall peptidoglycan synthesis
	DadX	Alanine racemase	Yes	No	Catabolic enzyme
Aminotransferase	AspC	Aspartate aminotransferase	Yes	Yes	Biosynthetic enzymes with different substrate specificity. Both sequences are more similar to eukaryotic homologs than to those from distantly related bacteria
	TyrB	Aromatic amino acid aminotransferase	No	Yes	

```
                           Cluster   Function/activity
Abc        33:  IYGVIGASGAGKSTLIRCVN    yes     transport
UvrA       26:  LIVVTGLSGSGKSSLAFDTL    yes     repair
UvrA      635:  FTCITGVSGSGKSTLINDTL    yes     repair
MutS      609:  MLIITGPNMGGKSTYMRQTA    no      repair
RecN       24:  MTVITGETGAGKSIAIDALG    no      repair/recombination
RecF       25:  FNFLVGANGSGKTSVLEAIY    no      repair/recombination
SbcC       32:  LFAITGPTGAGKTTLLDAIC    no      repair/recombination
RecB       18:  ERLIEASAGTGKTFTIAALY    yes     recombination/helicase
DeaD       62:  DVLGMAQTGSGKTAAFSLPL    yes     helicase
MalT       34:  LALITSPAGYGKTTLISQWA    no      transcription regulation
ClpA      209:  NPLLVGESGVGKTAIAEGLA    yes     ATPase subunit of protease
ClpA      490:  SFLFAGPTGVGKTEVTVQLS    yes     ATPase subunit of protease
Lon       351:  ILCLVGPPGVGKTSLGQSIA    no      ATPase subunit of protease
DnaA      171:  PLFLYGGTGLGKTHLLHAVG    no      replication initiation
DnaC      101:  SFIFSGKPGTGKNHLAAAIC    no      replication initiation
DnaB      226:  LIIVAARPSMGKTTFAMNLV    no      replication/helicase
DnaX       40:  AYLFSGTRGVGKTSIARLLA    yes     replication
MiaA       12:  AIFLMGPTASGKTALAIELR    no      tRNA modification
HydG      164:  TVLIHGDSGTGKELVARAIH    yes     transcription regulation
Sms        97:  AILIGGNPGAGKSTLLLQTL    no      ATP-dependent protease ?
RecA       62:  IVEIYGPESSGKTTLTLQVI    no      repair/recombination
MinD        5:  IIVVTSGKGGGGKTTSSAAIA   yes     chromosome partitioning
HopB      217:  LVLVTGPTGSGKTVTLYSAL    yes     secretion
CysN       29:  RFLTCGSVDDGKSTLIGRLL    yes     GTPase
Ffh       102:  VVLMAGLQGAGKTTSVGKLG    yes     GTPase
HypB      106:  VLNLVSSPGSGKTTLLTETL    no      GTPase ?
YeiR        4   TNLITGFLGSGKTTSILHLL    yes     GTPase ?
AtpA      164:  RELIIGDRQTGKTALAIDAI    no      H⁺ ATPase
Rho       172:  RGLIVAPPKAGKTMLLQNIA    no      transcription/helicase
AroK       34:  NIFLVGPMGAGKSTIGRQLA    yes     shikimate kinase
CoaA       90:  IISIAGSVAVGKSTTARVLQ    yes     panthotenate kinase
Udk        10:  IIGIAGASASGKSLIASTLY    no      uridine kinase
Adk         2:  RIILLGAPGAGKGTQAQFIM    no      adenylate kinase
consensus       .UUU.O....GK$.U...U.
```

FIGURE 9 Alignment of the purine nucleoside triphosphate-binding P-loops from proteins belonging to the "Walker type" ATPase/GTPase supercluster. For each sequence is indicated whether it belongs to a cluster defined on the basis of pairwise similarity or was included in the supercluster based on motif conservation. Question marks indicate proteins for which function was predicted in the course of our analysis. The consensus includes amino acid residues conserved in most of the proteins containing the motif; U indicates a bulky hydrophobic residue; O indicates a small residue (G, A, or S); $ indicates serine or threonine; and a dot indicates any residue. The residues conforming with the consensus are highlighted by boldface type.

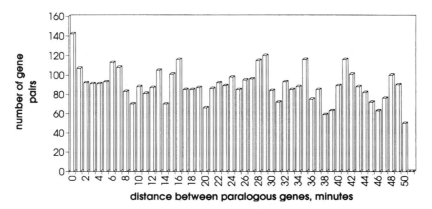

FIGURE 10 Distribution of genes encoding paralogous proteins along the *E. coli* chromosome.

```
                          RecA/Sms-specific region
                          _____

consensus                 g................U.tg...Ud.uUg.GGU
Dmc1  yeast  (79-264)     GFIPATVQ-LDIRQRVYSLSTGSKQLDSILG-GGI
Rad51 yeast  (143-328)    GFVTAADF-HMRRSELICLTTGSKNLDTLLG-GGV
Rad57 yeast  (83-275)     LEICEKNS-ISPDNGPECFTTADVAMDELLG-GGI
UvsX  T4     (21-197)     TASKFFN--EKDVVR-TKIPMMNIALSGEIT-GGM
RecA  B.subt (20-193)     GKGSIMKLGEKTDTRISTVPSGSLALDTALGIGGY
RecA  M.tube (23-197)     GKGSVMRLGDEARQPISVIPTGSIALDVALGIGGL
RecA         (23-196)     GKGSIMRLGEDRSMDVETISTGSLSLDIALGAGGL

Sms          (59-433)     GVAKVQKLSDISLEELPRFSTGFKEFDRVLG-GGV
Sms B.subt   (55-430)     TVQKPSPITSIETSEEPRVKTQLGEFNRVLG-GGV
```

```
                                              ATPase domain
                          _____

                             A                         B                    C
                          _____       _____           _____

consensus                 ..g.uU.UuGp..o$gK$.u...U...    ....UUUUD.u..u       .....uuU...n.u.
Dmc1  yeast               MTMSITEVFGEFRCGKTQMSHTLCVT 67  LSSGDYRLIVVDSIMAN 27  LAEEFNVAVFLTNQVQ
Rad51 yeast               ETGSITELFGEFRTGKSQLCHTLAVT 67  MSESRFSLIVVDSVMAL 27  LADQFGVAVVVTNQVV
Rad57 yeast               FTHGITEIFGESSTGKSQLLMQLALS 72  RSKGSIKLVIIDSISHH 29  LAHDYSLSVVVANQVG
UvsX  T4                  MQSGLLILAGPSKSFKSNFGLTMVSS 55  IERGEKVVVFIDSLGNL 33  YFSTKNIPCIAINHTY
RecA  B.subt              PRGRIIEVYGPESSGKTTVALHAIAE 49  VRSAAVDIVVIDSVAAL 31  AINKSKTIAIFINQIR
RecA  M.tub              PRGRVIEIYGPESSGKTTVALHAVAN 50  IRSGSIDMIVIDSVAAL 31  ALNNSGTTAIFINQLR
RecA                      PMGRIVEIYGPESSGKTTLTLQVIAA 50  ARSGAVDVIVVDSVAAL 31  NLKQSNTLLIFINQIR

Sms                       VPGSAILIGGNPGAGKSTLLLQTLCK 48  AEEEQPKLMVIDSIQVM 24  FAKTRGVAIVMVGHVT
Sms B. subt               VKGSLVLIGGDPGIGKSTLLLQVSAQ 48  IQEMNPSFVVVDSIQTV 24  IAKTKGIPIFIVGHVT

Lon       (347-741)       IKGPILCLVGPPGVGKTSLGQSIAKA 38  KVGVKNPLFLLDEIDKM 32  YDLSDVMFVATSNSMN
Lon human (517-919)       TQGKILCFYGPPGVGKTSIARSIARA 39  KTKTENPLILIDEVDKI 32  VDLSKVLFICTANVTD
Lon1 M.xant (359-752)     LKGPVLCFVGPPGVGKTSLARSIARA 38  KAGSNNPVFLLDEIDKM 32  YDLSKVMFICTANTMH
Lon  B.brev (346-739)     MRGPILCLVGPPGVGKTSLARSVARA 39  QAGTINPVFLLDEIDKL 32  YDLTNVMFITTANSLD

YifB      (222-355)            NLLLIGPPGTGKTMLASRINGL 60  AHNGVLFLDELPEF 26  TYPARFQLVAAMN
Bchi R.caps. (47-195)         GVLVFGDRGTGKSTAVRALAAL 74  ANRGYLYIDECNLL 26  RHPARFVLVGSGN
Ccs E. grac (37-192)          GVMIMGDRGTGKSTIVRALVDL 81  ANRGILYVDEVNLL 26  CHPARFILVGSGN
Mcm3 yeast (404-517)          NILMVGDPSTAKSQLLRFVLNT 39  ADRGVVCIDEFDKM 26  TLNARCSVIAAAN

YgaA      (235-353)           NVLISGETGTGKELVAKAIHEA 46  ADNGTLFLDEIGEL 24  CLRVDVRVLAATN

FtsH      (186-298)           GVLMVGPPGTGKTLLAKAIAGE 34  AAPCIIFIDEIDAV 29  EGNEGIIVIAATN
```

```
                                          protease domain
                          _____

                          catalytic site (?)              protein-binding site ?
                          _____              _____

                                 !
consensus                 u..o....g.soo..U..        .uus.u....u...u ...geUou.g..u.u.......u...jooU...uup.o
Sms B.subt   138          KVAGGVKLDEPAIDLAIVI  0    SIASSFRDTPPNPAD-CFIGEVGLTGEVRRVSRIEQRVKEAAKLGFKRMIIPAAN
Sms          138          NVVGGVKVTETSADLALLL  0    AMVSSLRDRPLPQDL-VVFGEVGLAGEIRPVPSGQERISEAAKHGFRRAIVPAAN

Lon          191          VPEGATPKDGPSAGIAMCT  0    ALVSCLTGNPVRADV-AMTGEITLRGQVLPIGGLKEKLLAAHRGGIKTVLIPFEN
Lon human    200          VPEGATPKDGPSAGCAIVT  0    ALLSLAMGRPVRQNL-AMTGEVSLTGKILPVGGIKEKTIAAKRAGVTCIVLPAEN
Lon1 M.xant  191          LPEGAIPKDGPSAGVTICT  0    ALVSALTRVLIRRDV-AMTGEITLRGRVLPIGGLKEKTLAAHRAGIKTVLIPKAN
Lon B.brev   191          VPEGAIPKDGPSAGITMAT  0    ALVSALTGIPVKKEV-GMTGEITLRGRVLPIGGLKEKCMSAHRAGLTTIILPKDN
Lon C.eleg   (36-108)     LEQIGRTYNGVSMALPFVL  0    LIISAIKKNSLRKDY-VATGDVSLAGAVLTVDYINNKIVGAINAGLKGVVIPAEN

YifB         (55-153)     ARDRVRSAIINSGYEYPAK  26   LAASEQLTANKLDEY-ELVGELALTGALRGVPGAISSATEAIKSGRKIIVAKDNE

HtrA         (225-324)    IQTDAAINRGNSSGALVNL  30   NMVKNLTSQMVEYGQ-VKRGELGIMGTELNSELAKAMKVDAQRGAFVSQVLPNSS
HhoA         (203-302)    IQTDASINRGNSSGALLNL  27   NMARTLAQQLIDFGE-IKRGLLGIKGTEMSADIAKAFNLDVQRGAFVSEVLPGSG
HhoB         (187-288)    LQTDASINHGNSSGALVNS  27   QLATKIMDKLIRdGA-VIRGYIGIGGREIAPLHAQGGGIDQLQGIVVNEVSPDGP
HtrA R.hens  (236-334)    IQIDAAVNRGNSGGPTFDL  27   ATANEVVQQLIEKGL-VQRGWLGVQIQPVTKEISDSIGLKEAKGALITDPLK-GP
HtrA B.abor  (246-345)    IQIDAAVNKGNSGGPAFDL  27   STAKQVVDQLIKKGS-VERGWIGVQIQPVTKDIAASLGLAEEKGAIVASPQDDGP
HtrH B.abor  (209-309)    IQTDAAINPGNSGGALIDM  27   NMVRAVVDAALQGSTRFERPYIGATFQGITPDLAESLGMEKPYGALITAVVKDGP
HtrA C.trac  (92-191)     VTTDAAINPGNSGRSIVKI  27   LMAKRVIDQLISDGQ-VTRGFLGVTLQPIDSELATCYKLEKCTERLVTDVVKGSP
Spro M.para (204-224)     IQADAPIKPGDSGGPMVNS
PV8  S.aure (226-246)     MQYDLSTTGGNSGSPVFNE
EtA  S.aure (222-242)     LRYYGFTVPGNSGSGIFNS
EtB  S.aure (206-226)     SQYFGYTEVGNSGSGIFNL
Gsep B.lich (156-176)     LQYAMDTYGGQSGSPVFEQ
```

clearly defined ATPase domains (Fig. 11A) (24, 31). Thus, it is possible that YifB and Sms are two new *E. coli* ATP-dependent proteases. However, the putative catalytic serine is replaced by alanine in the *B. subtilis* homolog of Sms, raising the alternative possibility that Sms has lost the protease activity. Another notable aspect of this example is the inversion of the ATPase and putative protease

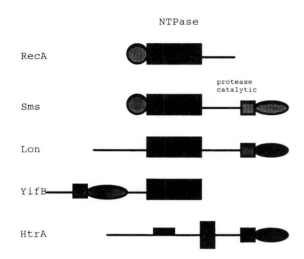

FIGURE 11 A group of multidomain proteins containing ATPase and protease domains. Two putative new ATP-dependent proteases in *E. coli*. (A) (Facing page) Alignment of conserved domain sequences. The C-terminal conserved motif comprising the putative protein-binding site was initially identified by using the MoST program as described in the text. The multiple alignment was constructed by using the MACAW program. For each protein, the positions of the first and last amino acid residues in the sequence are indicated in parentheses. The distances between the conserved blocks are indicated by numbers. The consensus shows amino acid residues that are conserved in all of the aligned sequences (capital letters) or in most of them (lowercase letters); the designations in the consensus are as in Fig. 9. The exclamation mark shows the (predicted) catalytic serine. A group of bacterial serine proteases including the *Staphylococcus* exfoliative toxins (EtA and EtB) share similarity with HtrA in the catalytic domain but not in the putative newly identified protein-binding domain. Organism name abbreviations: B. subt, *Bacillus subtilis;* M. tube, *Mycobacterium tuberculosis;* M. xant, *Myxococcus xanthus;* R. caps., *Rhodobacter capsulata;* E. grac, *Euglena gracilis* (chloroplast); B. brev, *Bacillus brevis;* C. eleg, *Caenorhabditis elegans;* R. hens, *Rochalimea henselae;* B. abor, *Brucella abortus;* C. trac, *Chlamydia trachomatis;* M. pseu, *Mycobacterium pseudotuberculosis;* S. aure, *Staphylococcus aureus;* B. lich, *Bacillus licheniformis.* The sequences of the *E. coli* proteins (except for HhoA and HhoB) were from the EcoSeq7 database; the other sequences were from the SWISS-PROT (Dmc1 yeast, P25453; Rad51 yeast, P25451; UvsX T4, P04529; RecA B. subt, P16971; RAD57 yeast, P25301; Bchi R. caps, P26239; CCS E. grac, P31205; Mcm3 yeast, P24279; Lon human, P36776; Lon1 M. xant, P36773; Lon B. brev, P36772; PV8 [V8 protease] S. aure, P04188; EtA S. aure, P09331; EtB S. aure, P09332; Gsep [glutamyl-specific endopeptidase] B. lich, P80057) or from the GenBank (Lon C. eleg, Z36719 [CEC06C3_9]; Sms B. subt, D26185 [BAC180K_149]; HhoA and HhoB, U15661; HtrA R. hens, L20127; HtrA B. abor, U07352; HtrH [htrA homolog] B. abor, U07351; Spro [serine protease] M. para, Z23092). (B) (Above) Tentative scheme of domain organization. Zn indicates the predicted N-terminal Zn finger domain in Sms (31).

domains in YifB compared with Lon and Sms (Fig. 11B). Domain shuffling is also typical of some other clusters of multidomain proteins in *E. coli,* e.g. the PTS system (41).

THE OUTSIDE AND INSIDE VIEWS OF THE *E. COLI* PROTEIN SEQUENCE SET: A JUXTAPOSITION

E. coli proteins belonging to clusters of paralogs generally show a higher evolutionary conservation than those proteins that do not have paralogs—the fraction containing ancient conserved regions is almost twice as high for the proteins in clusters (Fig. 12A and B). Predictably, the pattern of conservation is even more pronounced if analyzed for whole clusters—in most of them, at least one member is significantly similar to a eukaryotic or archaeal protein (Fig. 12C). Thus, most of the clusters correspond to critical functions that should have already been represented in the common ancestor of bacteria, eukaryotes, and the *Archaea,* even if, in some cases, by only a single member. This being so, it is particularly striking that the universal bacterial transcription regulators (HTH proteins) are not detectably similar to any proteins from eukaryotes or the *Archaea* (with the exception of the very limited similarity to homeo-domains, which in our study could be detected only when one specific family of *E. coli* HTH proteins was used to construct a motif for database search). This lack of ancient conserved regions in one of the largest *E. coli* protein superclusters may reflect the different modes of

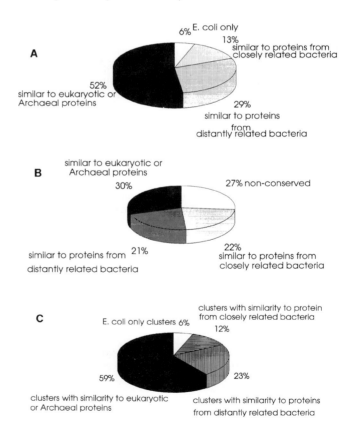

FIGURE 12 Sequence conservation and clustering of *E. coli* proteins. (A) Levels of sequence conservation in proteins belonging to clusters of paralogs. (B) Levels of sequence conservation in proteins not belonging to clusters. (C) Conserved and nonconserved clusters.

transcription regulation in bacteria, eukaryotes, and the *Archaea*. Similarly, the PTS system is composed of several clusters of highly conserved paralogous proteins (domains) without eukaryotic homologs, suggesting that certain important pathways of metabolite transport may be restricted to bacteria (41).

CONCLUSIONS

With increasingly sensitive computer methods becoming available and sequence databases growing rapidly, even a relatively straightforward analysis of the 2,328 proteins forming about 60% of all *E. coli* gene products produced a wealth of information. For more than 90% of these proteins, either functional information or significant sequence similarity or both are available. A surprisingly high fraction—about 86%—are similar to other proteins in current databases; about two-thirds show conservation at least at the level of distantly related bacteria, and about 40% contain regions of conservation with eukaryotic or archaeal proteins. Even though gene transfer between endosymbiotic organellar genomes and the eukaryotic nuclear genome and perhaps also other forms of horizontal gene flow should be taken into account, it may be concluded that *E. coli* proteins generally have been highly conserved through very long periods of evolution. About 47% of the *E. coli* proteins belong to 286 clusters of paralogs defined on the basis of significant pairwise similarity, and an additional 10% could be included in paralogous superclusters, based on motif conservation. The majority of clusters have only two members, but there are several very large superclusters. Combined, the four largest superclusters, which include various permeases, purine NTPases with the conserved "Walker-type" motif, HTH regulatory proteins, and dinucleotide-binding proteins, account for about 25% of all known *E. coli* proteins. Genes encoding paralogous proteins appear to be nonrandomly distributed along the chromosome, with a prevalence of tandem duplications but also an apparent large-scale periodicity. Proteins belonging to paralogous clusters typically show higher conservation in evolution than do proteins that do not have paralogs within *E. coli*. Most of the clusters include proteins with ancient conserved regions and, accordingly, correspond to critical functions that should have already been encoded by the common ancestor of bacteria, eukaryotes, and the *Archaea*. Sequence similarities detected in the course of the analysis of the *E. coli* protein sequence set allow the prediction of possible functions for a number of functionally uncharacterized gene products. The complete *E. coli* chromosome sequence can be readily expected to be available within 2 years. It is our hope that the pilot project outlined in this chapter will set the stage for the complete, systematic information analysis of the whole genome.

ACKNOWLEDGMENTS

We are grateful to Amos Bairoch, Peer Bork, and John Wootton for helpful discussions and critical reading of the manuscript, and to Peer Bork and John Wootton for communicating their results prior to publication.

LITERATURE CITED

1. Abouhamad, W. N., M. Manson, M. M. Gibson, and C. F. Higgins. 1991. Peptide transport and chemotaxis in Escherichia coli and Salmonella typhimurium: characterization of the dipeptide permease (Dpp) and the dipeptide-binding protein. *Mol. Microbiol.* 5:1035–1047.
2. Altendorf, K., A. Siebers, and W. Epstein. 1992. The KDP ATPase of Escherichia coli. *Ann. N. Y. Acad. Sci.* 671:228–243.
3. Altschul, S. F., M. S. Boguski, W. Gish, and J. C. Wootton. 1994. Issues in searching molecular sequence databases. *Nat. Genet.* 6:119–129
4. Altschul, S. F., W. Gish, W. Miller, E. W. Myers, and D. J. Lipman. 1990. Basic local alignment search tool. *J. Mol. Biol.* 215:403–410.
5. Amerik, A. Y., V. K. Antonov, A. E. Gorbalenya, S. A. Kotova, T. V. Rotanova, and E. V. Shimbarevich. 1991. Site-directed mutagenesis of La protease. A catalytically active serine residue. *FEBS Lett.* 287:211–214.
6. Bazan, J. F., and R. J. Fletterick. 1990. Structural and catalytic models of trypsin-like viral proteases. *Semin. Virol.* 1:311–322.
7. Bork, P., C. Ouzounis, G. Casari, R. Schneider, C. Sander, M. Dolan, W. Gilbert, and P. M. Gillevet. 1995. Exploring the Mycoplasma capricolum genome: a small bacterium reveals its physiology. *Mol. Microbiol.* 16:955–967.
8. Bork, P., C. Ouzounis, and C. Sander. 1994. From genome sequences to protein function. *Curr. Opin. Struct. Biol.* 4:393–403.
9. Bork, P., C. Ouzounis, C. Sander, M. Scharf, R. Schneider, and E. Sonnhammer. 1992. Comprehensive sequence analysis of the 182 ORFs of yeast chromosome III. *Protein Sci.* 1:1677–1690.
10. Borodovsky, M., E. V. Koonin, and K. E. Rudd. 1994. New genes in old sequence: a strategy for finding genes in the bacterial genome. *Trends Biochem. Sci.* 19:309–313.
11. Borodovsky, M., K. E. Rudd, and E. V. Koonin. 1994. Intrinsic and extrinsic approaches for detecting genes in a bacterial genome. *Nucleic Acids Res.* 22:4756–4767.
12. Chin, D. T., S. A. Goff, T. Webster, T. Smith, and A. L. Goldberg. 1988. Sequence of the lon gene in Escherichia coli. A heat-shock gene which encodes the ATP-dependent protease La. *J. Biol. Chem.* 263:11718–11728.
13. Daniels, D., G. Plunkett, V. Burland, and F. R. Blattner. 1992. Analysis of the Escherichia coli genome: DNA sequence of the region from 84.5 to 86.5 minutes. Science 257:771–778.
14. Fitch, W. M., and E. Margoliash. 1970. The usefulness of amino acid and nucleotide sequences in evolutionary studies. *Evol. Biol.* 4:67–77.
15. Goffeau, A., K. Nakai, P. Slonimski, and J. L. Risler. 1993. The membrane proteins encoded by yeast chromosome III genes. *FEBS Lett.* 325:112–117.
16. Gorbalenya, A. E., and E. V. Koonin. 1990. Superfamily of UvrA- related NTP-binding proteins. Implications for rational classification of recombination/repair systems. *J. Mol. Biol.* 213:583–591.
17. Gottesman, S., and M. R. Maurizi. 1992. Regulation by proteolysis: energy-dependent proteases and their targets. *Microbiol. Rev.* 56:592–621.
18. Gray, M. W. 1989. The evolutionary origin of organelles. *Trends Genet.* 5:294–299
19. Green, P. 1994. Ancient conserved regions in gene sequences. *Curr. Opin. Struct. Biol.* 4:404–412.
20. Green, P., D. J. Lipman, L. Hillier, R. Waterston, D. J. States, and J. M. Claverie. 1993. Ancient conserved regions in new gene sequences and the protein databases. Science 259:1711–1716.
21. Higgins, C. F., I. D. Hiles, G. P. Salmond, D. R. Gill, J. A. Downie, I. J. Evans, I. B. Holland, L. Gray, S. D. Buckel, A. W. Bell, et al. 1986. A family of related ATP-binding subunits coupled to many distinct biological processes in bacteria. *Nature* (London) 323:448–450.
22. Holm, L., and C. Sander. 1994. Searching protein structure databases has come of age. *Proteins Struct. Funct. Genet.* 19:165–173.
23. Koonin, E. V. 1993. A superfamily of ATPases with diverse functions containing either classical or deviant ATP-binding motif. *J. Mol. Biol.* 229:1165–1174.
24. Koonin, E. V. 1993. A common set of conserved motifs in a vast variety of putative nucleic acid-dependent ATPases including MCM proteins involved in the initiation of eukaryotic DNA replication. *Nucleic Acids Res.* 21:2541–2547.
25. Koonin, E. V., P. Bork, and C. Sander. 1994. Yeast chromosome III: new gene functions. *EMBO J.* 13:493–503.
26. Kunisawa, T., and J. Otsuka. 1988. Periodic distribution of homologous genes or gene segments on the Escherichia coli K12 genome. *Protein Sequences Data Anal.* 1:263–267.
27. Lipinska, B., M. Zylicz, and C. Georgopoulos. 1990. The HtrA (DegP) protein, essential for Escherichia coli survival at high temperatures, is an endopeptidase. *J. Bacteriol.* 172:1791–1797.
28. Medigue, C., A. Viari, A. Henaut, and A. Danchin. 1993. Colibri: a functional data base for the Escherichia coli genome. *Microbiol. Rev.* 57:623–654.
29. Moszer, I., P. Glaser, and A. Danchin. 1991. Multiple IS insertion sequences near the replication terminus in Escherichia coli K-12. *Biochimie* 73:1361–1374.
30. Navarro, C., L. F. Wu, and M. A. Mandrand-Berthelot. 1993. The nik operon of Escherichia coli encodes a periplasmic binding-protein-dependent transport system for nickel. *Mol. Microbiol.* 9:1181–1191.
31. Neuwald, A. F., D. E. Berg, and G. V. Stauffer. 1992. Mutational analysis of the Escherichia coli serB promoter region reveals transcriptional linkage to a downstream gene. *Gene* 120:1–9.
32. Nichols, J. C., N. K. Vyas, F. A. Quiocho, and K. S. Matthews. 1993. Model of lactose repressor core based on alignment with sugar-binding proteins is concordant with genetic and chemical data. *J. Biol. Chem.* 268:17602–17612.
33. Niki, H., A. Jaffe, R. Imamura, T. Ogura, and S. Hiraga. 1991. The new gene mukB codes for a 177 kd protein with coiled-coil domains involved in chromosome partitioning of E. coli. *EMBO J.* 10:183–193.
34. Oliver, G., G. Gosset, R. Sanchez-Pescador, E. Lozoya, L. M. Ku, N. Flores, B. Becerill, F. Valle, and F. Bolivar. 1987. Determination of the nucleotide sequence for the glutamate synthase structural genes of Escherichia coli K-12. *Gene* 60:1–11.
35. Olsen, G. J., C. R. Woese, and R. Overbeek. 1994. The winds of (evolutionary) change: breathing new life into microbiology. *J. Bacteriol.* 176:1–6.
36. Palmer, J. D. 1985. Comparative organization of chloroplast genomes. *Annu. Rev. Genet.* 19:325–354.
37. Pao, G. M., R. Tam, L. S. Lipschitz, and M. H. Saier, Jr. 1994. Response regulators: structure, function and evolution. *Res. Microbiol.* 145:356–362.
38. Parkinson, J. S., and E. C. Kofoid. 1992. Communication modules in bacterial signaling proteins. *Annu. Rev. Genet.* 26:71–112.
39. Quirk, S., and M. J. Bessman. 1991. dGTP triphosphohydrolase, a unique enzyme confined to the members of the family Enterobacteriaceae. *J. Bacteriol.* 173:6665–6669.

40. Rahfeld, J. U., K. P. Rucknagel, B. Schelbert, B. Ludwig, J. Hacker, K. Mann, and G. Fischer. 1994. Confirmation of the existence of a third family among peptidyl-prolyl cis/trans isomerases. Amino acid sequence and recombinant production of parvulin. *FEBS Lett.* **352:**180–184.

41. Reizer, A., G. M. Pao, and M. H. Saier, Jr. 1991. Evolutionary relationships among the permease proteins of the bacterial phosphoenolpyruvate:sugar phosphotransferase system. Construction of phylogenetic trees and possible relatedness to proteins of eukaryotic mitochondria. *J. Mol. Evol.* **33:**179–193.

41a.Reuven, N. B., E. V. Koonin, K. E. Rudd, and M. P. Deutscher. 1995. The gene for the longest known *Escherichia coli* protein is a member of helicase superfamily II. *J. Bacteriol.* **177:**5393–5400.

42. Riley, M. 1993. Functions of the gene products of *Escherichia coli. Microbiol. Rev.* **57:**862–952.

43. Riley, M., and A. Anilionis. 1978. Evolution of the bacterial genome. *Annu. Rev. Microbiol.* **32:**519–560.

44. Riley, M., and S. Krawiec. 1987. Genome organization, p. 967–981. *In* F. C. Neidhardt, J. L. Ingraham, K. B. Low, B. Magasanik, M. Schaechter, and H. E. Umbarger (ed.), *Escherichia coli and Salmonella typhimurium: Cellular and Molecular Biology.* American Society for Microbiology, Washington, D.C.

45. Robison, K., W. Gilbert, and G. M. Church. 1994. Large scale bacterial gene discovery by similarity search. *Nat. Genet.* **7:**205–214.

46. Rudd, K. E. 1993. Maps, genes, sequences, and computers: an *Escherichia coli* case study. *ASM News* **59:**335–341.

47. Rudd, K. E., H. E. Sofia, E. V. Koonin, S. Lazar, G. Plunkett III, and P. E. Rouviere. 1995. A new family of peptidyl-prolyl isomerases. *Trends Biochem. Sci.* **20:**12–14.

48. Saier, M. H., Jr. 1994. Computer-aided analyses of transport protein sequences: gleaning evidence concerning function, structure, biogenesis, and evolution. *Microbiol. Rev.* **58:**71–93.

49. Saraste, M., P. R. Sibbald, and A. Wittinghofer. 1990. The P-loop—a common motif in ATP- and GTP-binding proteins. *Trends Biochem. Sci.* **15:**430–434.

50. Schaffner, A. R., and J. Sheen. 1992. Maize C4 photosynthesis involves differential regulation of phosphoenolpyruvate carboxylase genes. *Plant J.* **2:**221–232.

51. Schuler, G. D., S. F. Altschul, and D. J. Lipman. 1991. A workbench for multiple alignment construction and analysis. *Proteins Struct. Funct. Genet.* **9:**180–190.

52. Smith, M. W., D.-F. Feng, and R. F. Doolittle. 1992. Evolution by acquisition: the case for horizontal gene transfers. *Trends Biochem. Sci.* **17:**489–493.

53. Tam, R., and M. H. Saier, Jr. 1993. Structural, functional and evolutionary relationships among extracellular solute-binding receptors of bacteria. *Microbiol. Rev.* **57:**320–346.

54. Tatusov, R. L., S. F. Altschul, and E. V. Koonin. 1994. Detection of conserved segments in proteins: iterative scanning of sequence databases with alignment blocks. *Proc. Natl. Acad. Sci. USA* **91:**12091–12095.

55. Titgemeyer, F., J. Reizer, A. Reizer, and M. H. Saier, Jr. 1994. Evolutionary relationships between sugar kinases and transcriptional repressors in bacteria. *Microbiology* **140:**2349–2354.

56. Wahl, R., P. Rice, C. M. Rice, and M. Kröger. 1994. ECD—a totally integrated database of *Escherichia coli* K12. *Nucleic Acids Res.* **22:**3450–3455.

57. Walker, J. E., M. Saraste, M. J. Runswick, and N. J. Gay. 1982. Distantly related sequences in the a- and b-subunits of ATP synthase, myosin, kinases and other ATP-requiring enzymes and a common nucleotide binding fold. *EMBO J.* **1:**945–951.

58. Woese, C. R., and G. E. Fox. 1977. Progenotes and the origin of the cytoplasm. *J. Mol. Evol.* **10:**1–6.

59. Wootton, J. C. 1994. Non-globular domains in protein sequences: automated segmentation using complexity measures. *Comput. Chem* **18:**269–285.

60. Wootton, J. C. 1994. Sequences with `unusual' amino acid composition. *Curr. Opin. Struct. Biol.* **4:**413–421.

Mutagenesis
FRANKLIN HUTCHINSON

118

INTRODUCTION

Mutagenesis is better understood in *Escherichia coli* than in any other organism for several reasons: the biology of *E. coli* is known more completely than that of any other organism; mutagenesis is readily observed because the cell is genetically haploid, so a mutation in only one copy of a gene is needed for phenotypic expresssion; the SOS response in *E. coli* (chapter 89, this volume) has greatly aided dissection of mutagenic processes; the use of coliphages has made possible studies of mutagenesis in the simpler viral genomes, with the possibility of treating DNA and cell separately; and it is easy to manipulate the large numbers of *E. coli* cells needed for analyzing the relatively rare process of mutagenesis. The rapid and extensive development of molecular biology has made possible the extrapolation of insights from this prokaryote to other organisms, including humans.

The purpose of this chapter is to provide a coherent and readily understood picture of what is known about mutagenesis of *E. coli.* The literature is enormous, and no attempt at completeness of coverage has been made. Those aspects that are generally accepted are presented as simply as possible, with minimal reference to the original literature. This approach can make some areas appear better understood than is in fact the case, but this disadvantage is outweighed by the ease with which the main outlines of mutational processes can be presented, without requiring the reader to plow through the mass of results that went into determining each part of the picture. Matters that are not well understood, or about which there is controversy, are documented much more extensively, with detailed references to the original literature.

There is an extensive review of mutagenesis by G. Walker in the recent text *DNA Repair and Mutagenesis* (64). Other reviews of *E. coli* mutagenesis include those of Eisenstadt (57), Miller (117, 119), Sedgwick (160), and Bridges (20). More specialized reviews include those on frameshift mutagenesis (144), spontaneous mutation (47), replication fidelity (55), DNA replication and mutagenesis (106), the SOS response (182), T4 bacteriophage mutagenesis (49, 50), and DNA damage in untreated cells (105). Finally, two "golden oldies," Drake's 1970 book (45) and Witkin's 1976 review (188), are still worth perusal for their summing up of particular aspects of mutagenesis and for their insights.

Mutations are usually classified as base substitutions or base changes; deletions, or the loss of any number of base pairs; insertions, or the addition of any number of base pairs; inversions, in which a piece of the linear DNA molecule is cut out and reinserted in reverse order; duplications, a form of insertion in which the added bases are the same as a sequence already in the genome, usually that immediately adjacent to the insert; and complex mutations, which may include any combination of these mutations in a single genome. Base substitutions are sometimes further classified as transitions, a purine changed to a purine or a pyrimidine changed to a pyrimidine, and transversions, a purine changed to a pyrimidine or vice versa. It is convenient (and conventional) to use the term *frameshift* for the loss or addition of one or two base pairs; those situations in which the term frameshift is used to include larger deletions or insertions that change the reading frame during the process of translation are generally self-evident.

HISTORICAL ASPECTS OF *E. COLI* MUTAGENESIS

Until the 1930s, bacteriologists were primarily interested in the properties of a bacterial culture as a whole (176). The extraordinary changes that take place in cultures grown under various conditions, from emergence of colonies utilizing lactose in a culture unable to grow on that sugar (enzyme induction) to smooth colonies appearing in a culture that made only rough colonies (mutation), were ascribed to "bacterial variation." As late as 1945, Dubos could write in a standard text (52): "Except in a few suggestive cases, . . . there is no evidence that bacterial variation behaves according to Mendelian laws."

A shift in focus to properties attributable to the individual cells became significant in the 1930s, in large part because of studies of radiation effects on bacteria by O. Rahn, F. L. Gates, R. W. G. Wyckoff, D. E. Lea, and J. W. Gowen, among others (176). In 1943, the Luria-Delbrück experiment (108) measured the appearance of mutants resistant to coliphage T1 in growing cultures of *E. coli*. The observed distributions were inconsistent with mutations induced by the presence of the phage (that is, unlike enzyme induction) but consistent with random appearance during cell replication, encouraging workers to regard bacterial genetics as similar to the well-established genetics of *Drosophila melanogaster* and maize.

Conclusive evidence that DNA is the carrier of genetic information was provided by the demonstration that a heritable characteristic could be transferred from one type of *Pneumococcus* cell to another by transfer of DNA and only DNA (7). This epochal finding was brought to the attention of many by the Hershey-Chase experiment, showing that propagation of a coliphage requires injection of the ^{32}P-labeled phage DNA, but not the ^{35}S-labeled protein, into the *E. coli* host cell (72). Genetic mapping in coliphage T4 with resolution on the order of a single base by Benzer (14) made possible the demonstration of the triplet nature of the genetic code (37). Certain mutations were dubbed "nonsense" because they made triplet codons that did not code for an amino acid (15), and these codons were then shown to be chain terminators (TAG, TAA, and TGA) whose effects were suppressed by mutant tRNAs that inserted an amino acid at these codons (175).

With these fundamental results, an understanding of mutagenesis on a molecular level became feasible.

DETERMINATION OF MUTAGENESIS IN *E. COLI*

There are two general types of mutation. A forward mutation is a change in the normal or wild-type gene sequence, most commonly detected as a change (usually decrease) in gene product activity. In addition to changes that affect a protein, forward mutations can change the genes for tRNA or rRNA and also the sequences that control the initiation and termination of RNA transcription or DNA replication as well as ribosomal binding. A revertant is a cell with a mutated gene or sequence, as described above, that has regained a phenotype resembling wild type because of a second mutation.

Although *E. coli* is genetically haploid, most cells contain two or more copies of the genome. A forward mutation that inactivates a gene cannot readily be identified until the mutated gene has segregated by cell division and the level of unmutated gene product has decreased sufficiently by inactivation and dilution by cell division to allow phenotypic expression. Offsetting these difficulties, studies of forward mutagenesis have the advantage of including all kinds of mutations.

Reversions are frequently easier to detect and more readily quantitated. On the other hand, only specific mutations that revert a mutated gene back to active form are detected.

Reversion Mutations

Many investigations of mutagenesis in *E. coli*, especially between 1950 and 1980, used strains with a mutation that makes growth dependent on an additive in the medium (auxotrophs) and studied revertants that can grow without the supplement (prototrophs). Such revertants are readily identified, and phenotypic expression is fast, with no need for gene segregation or loss of gene product activity.

In the usual reversion assay, mutagenized cells are spread on plates with agar-containing medium on which prototrophic revertants, but not the auxotroph, will grow, enriched with a small amount of the additive (semienriched medium or SEM plates) to allow sufficient DNA replication to "fix" the mutation. (On occasion, fixation occurs without additions because traces of essential nutrients in the cells, agar, or medium allow some replication.) If quantitative mutation frequencies are wanted, it is essential to determine the number of viable cells under the same conditions used to score mutants, because the probabilities of survival and of mutation are greatly influenced by details of the course of events as the cell recovers from genotoxic insult. Since reversion of a particular mutation is a rather rare event, it is usual to put 10^6 to 10^9 mutagenized cells on one 100-mm-diameter petri plate, and this can give rise to a cell density artifact (188): the fraction of cells that divide often enough to form microcolonies can depend on the number of cells plated, and nutrients leaking from dead cells unable to divide can increase the number of divisions for surviving cells. The difficulty is particularly severe for bacterial cells such as *E. coli* K-12 that, when treated with genotoxic agents, form filaments that are very large compared with the size of a normal cell and rapidly deplete available nutrients.

If quantitative mutation frequencies are wanted, it is essential to determine the number of viable cells under the same conditions used to score mutants, because the probabilities of survival and of mutation are greatly influenced by details of the course of events as the cell recovers from genotoxic insult.

Characterization of Revertant Mutations. A cell that is auxotrophic because of a base change mutation can revert by restoration of the original base sequence or by some other mutation: one that inserts a tolerated amino acid at the mutated site, or a second-site mutation that changes another amino acid in the protein, or even in another protein that interacts with the mutated molecule, to give increased product activity. For an auxotroph with a nonsense mutation, a base change in a tRNA gene can make a suppressor tRNA (see below). Frameshifts may revert to give the original base sequence, or reversion may be effected by nearby frameshifts that restore the original translational reading frame without introducing stop codons in the coding sequence or unacceptable changes in amino acids.

Revertants that have reverted because of different mutations can be distinguished sometimes by phenotype (such as colony morphology) and sometimes by appearance or disappearance of a suppressible stop codon or of a restriction enzyme site in the DNA. Sequencing of the DNA is always possible but a lot of work. In some cases, the most effective method is blotting: a labeled oligonucleotide (typically 12 to 20 bases long), with the exact sequence of an expected mutation and annealed to revertant DNA, is significantly more difficult to elute at high washing temperatures than oligonucleotides differing by only a single base (122).

A strain can be annoying to work with if revertants form colonies of variable size and appearance that slowly arise during several days of incubation (i.e., are "leaky"). Thus, it is common to search for strains with uniform and nonleaky revertants. In the early 1950s, Evelyn Witkin isolated and characterized two *E. coli* B strains, WP2(*trp*) and WU36(*tyr*), that have a low spontaneous reversion rate to prototrophy but are induced to revert by typical mutagens at a high rate, with revertants that are easy to assay and form colonies of uniform size and appearance (186). Revertants of these strains are typically 90% suppressor mutations and 10% base changes in the mutated codon. Various derivatives of these strains have been widely used in many experiments that have contributed much to present knowledge, but the implied assumption that the observed mutagenesis is representative has on occasion turned out to be incorrect.

Nonsense Suppressors. Some base substitutions change the codon for an amino acid to a nonsense codon that is a signal to stop translation, and the mutant polypeptide is truncated. Genes with these extra stop codons—TAG (amber), TAA (ochre), or TGA (UGA or opal)—can revert either by a base change in the stop codon or by the formation of a suppressor mutation in a tRNA gene that alters the tRNA anticodon to the complement of the stop codon and inserts the related amino acid in the gene product. Note that various amino acids are inserted by different suppressors; even with the restriction to single base changes in the tRNA gene, an amber (TAG) suppressor can insert any one of seven amino acids: lysine (AAG), glutamine (CAG), valine (GAG), serine (TCG), tryptophan (TGG), leucine (TTG), or tyrosine (TAC or TAT). The amino acids inserted by specific tRNA suppressors are listed in section 20 of reference 119. In some cases, insertion of any amino acid will give a protein of sufficient activity to score as a revertant, but in other cases, only tRNAs inserting the original amino acid will suppress the mutation.

Amber suppressors are the most common, ochre suppressors are less common, and UGA (or opal) suppressors are rarely found because TGA is the most common terminator codon and TGA suppressors can make a cell nonviable.

Ames Test. In a comprehensive application of reversion to detect mutagenesis, tester strains of *Salmonella typhimurium* (official designation, *Salmonella enterica* serovar Typhimurium) with various mutations in the *his* gene have been selected by Ames and coworkers (4, 114) for the ability to detect different base substitutions and frameshifts. These strains have various DNA repair deficiencies to minimize loss of mutagenic lesions from the DNA and thus increase sensitivity to mutagenic effects; they have genetic defects that make cell membranes more permeable to chemical agents; many of the strains incorporate plasmid pKM101, which greatly increases mutability by agents making DNA lesions which block polymerases. Metabolic processing in mammalian cells that converts many chemicals (such as benzpyrene) to mutagenic form can be mimicked by adding mammalian microsomal fractions to the assay plates (114). Simplicity and the broad understanding of bacterial mutagenesis have made the Ames test the most widely used procedure for detection of mutagens. *Important:* As the Ames test is usually used, results are generally not quantitative except when mutagenic treatment has negligible effects on colony-forming ability.

Six *E. coli* Lac⁻ strains with mutations in the *lacZ* gene have been constructed, each reverting to Lac⁺ only by one of the six possible base changes (39). Since mutagenesis typically depends strongly on adjoining bases, obtaining data for a given base change in only one sequence context is a significant limitation. A second set revert only by specific frameshifts in runs of A's (T's in the complementary strand) or G's (or C's) or in an alternating -G-C- sequence (38).

Forward Mutations

Forward mutations in genes for which there is an assay are readily detected: mutagenize a culture, grow it under nonselective conditions for phenotypic expression, and assay. If there is a much larger fraction of mutants in the mutagenized culture than in an untreated culture processed in a similar way, it is customarily assumed that most of the mutations in the treated culture arose from the mutagen. However, the number of mutants per colony former in the culture after growth does not give a quantitative measure of the number of mutational events because individual treated cells can, and frequently do, grow and divide at greatly different rates.

If treated cells are spread on agar before division, quantitative mutation rates are obtained if all survivors grow and a suitable assay detects mutant cells. A widely used assay utilizes the well-studied *lac* operon (12, 119) and depends on the ability of β-galactosidase, product of the *lacZ* gene, to cleave the indicator 5-bromo-4-chloro-3-indolyl-β-D-galactoside (X-Gal) to give a blue color. Wild-type *E. coli* cells have an active LacI repressor that turns off the *lac* operon, including *lacZ*, and therefore form white colonies. Mutagenized *E. coli* cells on agar with X-Gal form mostly white *lacI*⁺ colonies, with a few all-blue colonies from cells that were *lacI* (defective for the LacI repressor) before treatment, and colonies with white and blue sectors for those in which the *lacI* gene was mutagenized (119), the sectors corresponding to segregation of *lacI*⁺ and *lacI* genes.

A related assay may be used to detect mutations in a 290+-bp fragment of the *lacZ* gene, typically inserted in M13 phage or on single- or double-stranded DNA that will replicate in cells after transfection. The LacZ(α) polypeptide from this fragment complements the product of a suitably modified *lacZ* chromosomal gene to produce active β-galactosidase and form a blue product; mutations in the α-complementing fragment inactivate the enzyme and give a white colony (115, 119).

Coliphages are very useful in mutagenesis studies. The molecular biology of bacteriophage T4 has been extensively explored (85,

111); the *r*II locus is a favorite target for mutagenesis (49, 50) and was used in much of the early work on fine-structure genetic mapping and on the genetic code (see above). Since T4 does not use host cell replication or repair enzymes, much recent work has shifted to the well-known phage λ, which depends on both (71). The clear-plaque phenotype produced by a mutation in the *cI* repressor gene of λ is readily distinguished from the wild-type turbid form, and quantitative determinations are possible for mutagenesis of the gene on phage DNA integrated into the host genome as a prophage, for phage DNA as a plasmid inside the cell, as free phage (thus avoiding action of the mutagenic agent on the host cell), and as free DNA in solution that can be repackaged into phage for assay (78).

Base Sequence Changes in Mutagenesis

Information on base sequence changes was obtained indirectly as early as 1966, by sequencing the mutant protein and deducing the base changes through a knowledge of the genetic code (194).

Most frequently, changes in DNA base sequence are determined by sequencing, often by amplifying the gene DNA by PCR (129, 130) and then sequencing either by chain termination methods (150) or by base-specific cleavage (112). While any assayable gene can be used, most determinations of sequence changes in mutagenesis use certain genes that have been repeatedly sequenced: the procedures are well worked out in the literature, special cell strains and phage and/or plasmids are usually available, and there are comparative data for other mutagens. A computer database of sequence changes in various *E. coli* genes contained over 13,000 mutations (December 1993) and is available on Internet (77).

The *lacI* gene for the Lac repressor is the most widely used for sequencing mutations, partly because it was the first for which extensive data were available (35), partly because the lactose operon is well understood (12, 119), and partly because there is a quick method for determining certain mutational sequence changes (see below). The LacZ(α)-complementing factor is available on single-stranded M13 viruses, and sequencing can be done easily, without PCR, from amplified viral DNA and with widely available primers (115). Phage λ has been extensively studied (71), and its *cI* repressor gene can be readily sequenced after being mutagenized on a prophage in the host *E. coli* genome, on a plasmid in host cells or on free phage, or on free DNA in solution (78). The *gpt* gene for the *E. coli* xanthine guanine phosphoribosyltransferase enzyme is readily expressed and assayed in both bacterial and mammalian cells (128) and is particularly useful for studies comparing bacterial and mammalian mutagenesis.

An ingenious method for rapidly determining sequence changes at nonsense mutations in a *lacI* gene on an F′ episome was developed by Jeffrey Miller and his associates before DNA sequencing became routine (34, 35). Each amber and ochre mutation is precisely located by crosses with appropriate deletion mutants (120); the base change is deduced from the knowledge of the particular amber or ochre mutation and the gene sequence. The advantage is that much larger numbers of mutations can be characterized than is presently feasible with sequencing; the disadvantage is that only certain mutations can be characterized: all single-base changes except A:T to G:C, but only in certain sequence contexts, and a very few tandem base changes.

MUTAGENESIS BY EXOGENOUS AGENTS

Dependence on Mutagenic Exposure

Genotoxic agents kill and mutagenize cells by damaging DNA. In many cases, survival of bacteria is roughly exponential with dose, and mutant fraction increases more or less linearly with dose. That is, an increment of dose dD changes the number n of viable cells by

$$dn = -\sigma\, n\, dD \tag{1}$$

where σ specifies the effect of the agent on cell survival. The same increment of dose changes the number of mutants m

$$dm = \eta\, n\, dD - \sigma\, m\, dD \tag{2}$$

where η specifies the effectiveness of the mutagen in inducing new mutations and the second term is the number of mutants in the culture killed by the dose increment. Equation 1 is readily integrated.

$$n = N \exp(-\sigma D) \tag{3}$$

Substituting equation 3 in equation 2 gives a linear first-order differential equation with a well-known solution that can be written

$$m/n = \eta D + M/N \tag{4}$$

where N and M are the numbers of viable and mutant cells, respectively, in the original culture. In words, assuming one-hit effects, the same lesion can either mutate or kill a cell but not both; equation 2 incorporates the assumption that mutated and unmutated cells have the same probability of surviving.

The numbers of viable bacteria and mutants sometimes show more complicated dependence on dose; for example, mutation sometimes depends on the square of dose (66, 188).

The SOS Response

The role of the SOS response in *E. coli* mutagenesis can be most simply described experimentally in terms of mutagenesis of coliphage λ, whose DNA is repaired and replicated by essentially the same enzymes as the cell chromosome. Phage λ, irradiated with UV light and adsorbed to *E. coli* host cells, produces mutated phage only when the host cells are also irradiated (184). Irradiating the host cell induces the SOS response needed for UV mutagenesis, and the event that triggers the SOS response is inhibition of DNA replication, principally by replication complexes blocked at lesions in the template (chapter 89).

Mutagenesis of a gene on the cell chromosome occurs by the same process: UV both induces the SOS response, as described above, and forms photoproducts in the mutated gene, equivalent to irradiating the phage. No UV-induced mutations are observed in *rec*A cells (124) that lack the SOS response or in *umu*C or *umu*D cells (87, 173) that do not make the SOS-dependent gene products essential for UV mutagenesis.

Wild-type *S. typhimurium* cells are not very mutable by UV but become so with the introduction of plasmid R46 (126). The same effect is produced by plasmid pKM101, derived from R46, which codes for two genes, *mucA* and *mucB*, that are homologous to *umuD* and *umuC*, respectively, and have similar functions (137; chapter 89).

Polymerase at a Lesion in Template DNA

The principal cause of mutation induced by an agent that damages DNA is a polymerase encountering a lesion in a template (Fig. 1). First, the polymerase extends the growing 3' end by addition of a nucleotide opposite the lesion. Next, there is competition between removal of that nucleotide by the 3'→5' proofreading exonuclease of the holoenzyme and extension by addition of still another nucleotide. If addition of either nucleotide is slow, progression of the polymerase is blocked. In most cases, the rate-limiting process is the second step, addition to the nucleotide opposite the lesion (106, 167).

Unassisted Translesion Synthesis. Some agents mutagenize in the absence of an SOS response; the polymerase can synthesize past the lesions, presumably because the lesions make good enough base pairs with normal bases to allow chain extension. The base analog bromouracil is similar stereochemically to T, with a bromine atom replacing the methyl group, and forms a good base pair with A; it also pair with G at low frequency, and when this pair eludes repair, a T:A-to-C:G mutation is formed. O^6-Methylguanine (Fig. 2), formed by the action on G of methylating agents

(a)

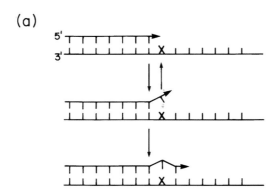

(b)

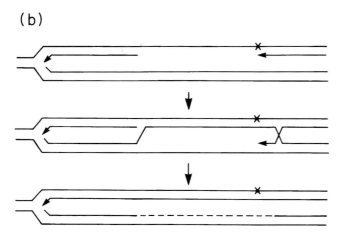

FIGURE 1 Bypass of a lesion (X) in template DNA blocking a replication complex (arrowhead). (a) Bypass by translesion synthesis, with and without SOS assistance. (b) Bypass by postreplication repair. Replication is reinitiated downstream (first figure); the gap is filled in by recombination with the sister strand (second figure), followed by DNA synthesis (dashed line) and resolution of the Holliday-type junction (see chapter 121).

such as N-methyl-N-nitrosourea or N-methyl-N'-nitro-N-nitrosoguanidine, pairs readily with either C or T (102, 191) with efficiencies that depend on sequence; the latter pairing mutates G:C to A:T.

SOS-Assisted Translesion Synthesis. Unassisted translesion synthesis is so slow as to be essentially unobservable, in most cases, for lesions such as the most common UV photoproducts in double-stranded DNA: cis-syn dipyrimidine cyclobutane dimers and pyrimidine (6–4) pyrimidinones (Fig. 2). Polymerase III (Pol III) holoenzyme can replicate past such lesions only with the aid of a complex of SOS proteins UmuC and UmuD' (UmuD' is the processed form of the UmuD protein; see chapter 89) and RecA (65a, 142); no other SOS operon need be induced (172). Even though other polymerases can replicate the genome in the absence of Pol III α polymerase (the product of the dnaE gene), there is evidence that SOS-assisted translesion synthesis probably occurs only with the Pol III holoenzyme with the α component (21, 25, 127). Many DNA lesions, including abasic sites (missing bases in the DNA), thymine glycols (Fig. 2), and bulky groups (such as aflatoxin) attached to a base, also require SOS assistance for translesion synthesis. The process is discussed in detail, with extensive references to the literature, in reference 64.

Translesion synthesis is mutagenic when wrong bases, or wrong numbers of bases, are inserted in the newly synthesized strand, and the SOS-assisted process accounts for nearly all "inducible error-prone repair" cited so frequently in the older literature; there is probably also a small component arising from SOS-related recombination.

Some agents make two kinds of lesions in DNA: those that are readily synthesized past and also those for which translesion synthesis is significant only with SOS assistance. Ionizing radiation, for example, induces mutations in phage λ assayed in uninduced cells or in umuC or recA hosts without SOS functions, showing the presence of readily bypassed lesions; a 7- to 10-fold increase in mutants per survivor for phage assayed in SOS-induced host cells reveals additional lesions that are mutagenic with SOS-assisted bypass (19, 180).

Postreplication Repair. In most cases, a blocked polymerase reinitiates replication at a site thousands of bases downstream (147); the gap between the 3' end of the newly formed strand at the lesion and the reinitiation site is filled by a process involving recombination with the sister duplex (Fig. 1b) (148; chapter 121). The process enables the error-free duplication of the genome but does not eliminate lesions from the DNA. Postreplication repair is essentially normal in umuDC cells with little or no UV-induced mutagenesis (86) and so must be essentially error free.

Competition between Modes of Bypass. Each of the three processes—unassisted translesion synthesis, SOS-assisted translesion synthesis, and postreplication repair—can occur at any particular lesion. The relative contribution of each determines both the frequency and the kind of mutation and depends on the lesion, DNA sequence, and state of the cell. There is competition between SOS-assisted translesion synthesis and postreplication repair at UV photoproducts (normally not bypassed without SOS assistance): in umuDC cells with no translesion synthesis (and without mutation), a small decrease in survival (87) is attributable to loss of bypass.

Competition between unassisted and SOS-assisted translesion synthesis occurs in single-strand DNA at a UV-induced trans-syn T↔T cyclobutane dimer (Fig. 2), a photoproduct found in single-stranded but not double-stranded DNA: without SOS, it causes a

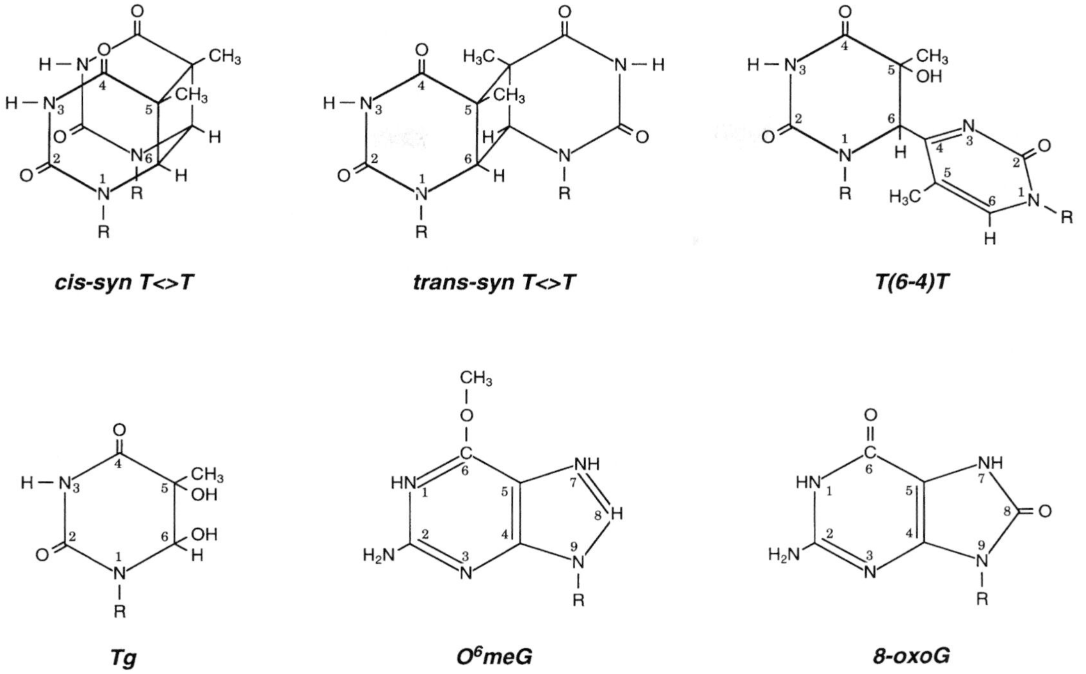

FIGURE 2 Typical mutagenic lesions in DNA. R is a sugar in the phosphate-deoxyribose backbone. T<>T, cyclobutane thymine dimer, *cis-syn* and *trans-syn* forms; T(6–4)T, the T-T pyrimidine-pyrimidinone (6–4) adduct; Tg, thymine glycol (there are four stereoisomers); O^6meG, O^6-methyl-guanine; 8-oxoG, 8-oxo-7,8-dihydroguanine.

TABLE 1 Mutations induced by a lesion at a defined site in circular single-strand DNA transfected into *E. coli*

Lesion	SOS in cell	Translesion frequency	Mutational specificity	Reference(s)
Cyclobutane dimer				
cis-syn T ↔ T	−	0.005		
	+	0.0186	Of 2,608 mutations sequenced: TT → TT, 2,443; TT → TA, 130; TT → TC, 28; TT → CT, 6	10, 98
trans-syn T ↔ T	−	0.139	Of 389 mutations sequenced: TT → TT, 373; −T, 16	9
	+	0.285	Of 605 mutations sequenced: TT → TT, 536; −T, 33; TT → AT, 16; TT → CT, 4; TT → TC, 1; +frameshift, 9; −frameshift, 4; adjacent base changes, 2	9
Pyrimidine-pyrimidinone (6-4)				
T(6-4)T	−	0.019		
	+	0.22	Of 185 mutations sequenced: TT → TT, 16; TT → TC, 158; TT → CT, 2; TT → GT, 2; TT → TG, 1; TT → AC, 3; TT → CC, 2; −T, 1	100
Abasic site (X)				
-G-X-T-G-	−	0.006		
	+	0.069	Of 104 mutations sequenced: X → T, 52; X → A, 24; X → C, 19; X → G, 1; XT → AA, 1; −1 frameshift, 7	99
-G-T-X-G-	−	0.003		
	+	0.05	Of 100 mutations sequenced: X → T, 77; X → A, 4; X → C, 14; X → G, 1; −1 frameshift, 4	99
O^6-Methylguanine	−	High	G → A, 0.4%; In repair-inhibited cells, G → A, ~20%	107
8-OxoG	−	High	G → T, 0.7%	192
Tg[a] sequence 5′-C-Tg-A-3′	−	High[b]	T → C, 0.2–0.4%	11
	+	High[b]	T → C, 0.2–0.4%	11

[a]Tg, thymine glycol.
[b]In uninduced *E. coli*, Tg blocks polymerases very effectively exept at 5′C-Tg-Pur sequences (58), as in this example. In other sequence contexts in SOS-induced *E. coli*, the chance that replication forks bypass Tg rises to 60 to 70%, with the indicated fraction of transitions induced (58).

low frequency of single T deletions; SOS-assisted translesion synthesis causes a much higher frequency of mutations, mostly single base changes with a few –T frameshifts (9). Possibly, *lacI* mutations induced by UV in *umuC* cells that were allowed to multiply many times (28) come from a low frequency of translesion synthesis past photoproducts that have been repeatedly bypassed by postreplication repair but not eliminated from the DNA in the excision-deficient cells.

An attractive hypothesis is that hindering error-free postreplication repair might allow more time for translesion synthesis and therefore mutation; postreplication repair of one lesion might be reduced by a second lesion on the complementary strand and within a distance comparable to the size of the gap formed (Fig. 1b) (18, 159). This might account for the quadratic dependence on dose of mutations in cells in which the SOS response is fully induced—e.g., for UV (188) or gamma rays (66).

It is helpful to realize that for most DNA lesions, only a very small fraction cause either cell killing or mutation. For example, a UV fluence of 1 J/m^2 to *E. coli* unable to excise photoproducts forms about 60 photoproducts in the genome (147) and kills about half the cells, so only 1 photoproduct in 100 is lethal. *E. coli* cells lacking both excision and postreplication repair are killed by an estimated one to two UV photoproducts per genome (74), so the ability of excision-defective cells to survive with tens of photoproducts in the DNA is ascribed to postreplication repair. A UV fluence of 1 J/m^2 induces 0.012 photoproducts in a typical 1,000-bp gene and 1×10^{-5} to 10×10^{-5} mutations per survivor, so only 1 photoproduct in 1,000 causes a mutation.

Mutagenesis by a Defined Lesion at a Specific Site

Useful data on the frequency and kinds of mutations induced by a particular lesion can be obtained by incorporating the lesion at a specific site in a gene on a DNA construction that can replicate after transfection into cells (169). Mutation frequency is the ratio of cells with mutant genes to cells with plasmids (identified by a marker gene, e.g., for drug resistance). The sequence changes in newly replicated DNA can be determined readily (Table 1), in some cases by changes in a suppressible mutation or restriction enzyme site. Modified bases are frequently incorporated in single-stranded circular DNA, which eliminates (or greatly reduces) effects of DNA repair. The use of single-stranded DNA also avoids a serious complication with double-stranded DNA: selective replication of the strand without the lesion. However, extrapolation from data for single-stranded DNA to the effects of a modified base in double-stranded DNA must be done with care. Also, the sequence surrounding the lesion has important effects (see, for example, the two abasic sites listed in Table 1).

Table 1 confirms an earlier deduction (116) that lesions determine the kinds of mutations. The mutations are nearly always at the site of a lesion, only rarely at an adjoining base pair.

REPAIR AND MUTAGENESIS

Removal of Mutagenic Lesions from DNA

Removal of lesions from DNA by repair increases the number of surviving cells n and decreases the number of mutants m induced by a dose D of mutagen, as may be seen by comparing, for example, the response to UV of *E. coli* with and without nucleotide excision repair that removes photoproducts from the cell genome. Different lesions may be removed at different rates, and effects on cell survival and mutagenesis can be estimated, in cases

for which equations 3 and 4 are valid, by considering both σ and η to be the sum of terms

$$\sigma = \sigma_1 + \sigma_2 + \ldots \quad \eta = \eta_1 + \eta_2 + \ldots \quad (5)$$

A decrease in the ith product from its initial value to some smaller value reduces the corresponding σ_i and η_i accordingly.

Also, rate of removal usually differs from one site in the DNA to another. Consider sites a and b, at which there are p_a and p_b lesions per site. If fractions dp_a/p_a and dp_b/p_b are removed during some time interval, and if lesions at site b are removed at a rate k more rapidly than at a,

$$dp_b/p_b = k \, dp_a/p_a \quad (6)$$

and by integration,

$$p_b/P_b = (p_a/P_a)^k \quad (7)$$

where P_a and P_b are the initial numbers of lesions at the sites. Assume, for example, that site a is an average site at which 99% of the lesions are removed (typical for products of alkylating agents) and site b is repaired at half the average rate. After repair, occupancy at site b is $p_b/P_b = (0.01)^{1/2} = 0.1$: a factor of 2 in repair rate leaves 10-fold more lesions than average at the slowly repaired site.

Preferential Repair of the Transcribed Strand. A DNA strand that is being actively transcribed is preferentially repaired by the *uvrABC* excision nuclease (chapter 121). The effect is to reduce mutations with the initiating lesion in the transcribed strand; for example, in the *E. coli lacI* gene mutagenized with UV, mutations were mostly at PyPy sequences (sites of photodimers) in the nontranscribed strand (133). In cells deficient in the factor that couples excision repair to transcription, mutations were mostly at PyPy sequences in the transcribed strand (133), in agreement with the observation that in the absence of the factor, RNA polymerase stalled at a lesion blocks nucleotide excision repair (161).

Mutation Frequency Decline. The high UV-induced reversion rates for WP2(*trp*) and WU36(*tyr*) strains decrease by a factor of 10 in irradiated cells incubated under conditions such that protein synthesis does not occur (186). This effect, named mutation frequency decline (MFD) (44), occurs only for mutations in cells with nucleotide excision repair and (to date) only in tRNA genes (reviewed in references 162 and 189).

For certain C:G sites in tRNA genes at which UV-induced mutations showed MFD, those induced by ethyl methanesulfonate (EMS) did not show MFD; conversely, for a site at which EMS mutations showed MFD, UV mutations did not (17). Bockrath et al. (17) deduced that MFD involves selective nucleotide excision of initiating lesions from the transcribed strand: C-containing dimers at sites showing MFD of UV mutations, and O^6-ethylguanine at the site showing MFD of EMS-induced mutations. This conclusion was supported by the finding that an *mfd* mutant selected as excision competent but without MFD (187) lacks preferential excision from an actively transcribing strand (163) as discussed in the previous section. However, there is still a paradox (162, 189): although selective excision in a gene for a protein requires active transcription, MFD is observed when protein synthesis is blocked, presumably shutting down transcription of tRNA genes! A second complication is that UV-induced mutants showing MFD respond anomalously to photoreactivation (next section).

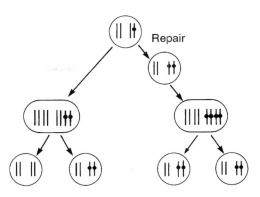

FIGURE 4 Effect of nucleotide excision repair on UV-induced mutagenesis in diploid yeast cells. Excision-proficient cells were irradiated and allowed to divide once; tetrad analysis of the daughter cells showed the genetic composition to be that on the right (mutation shown by a black dot), suggesting the pathway as shown (82). Excision-deficient cells gave daughter cells with the genetic composition on the left, showing that excision repair is required to form mutations in both strands of irradiated DNA before replication (83, 89).

Mutagenesis Caused by Repair

For mutagenesis by physical or chemical agents, it is sometimes asssumed that the actual changes in DNA base sequence take place during semiconservative replication of the genome. There is, however, solid evidence that some of these changes occur during repair and before replication.

Mutagenesis by Nucleotide Excision Repair. Nucleotide excision repair is largely error free, since cells with such repair have both higher survival and lower mutation rates than excision-deficient cells given the same exposure to an agent making bulky DNA lesions. However, nucleotide excision repair can also cause changes in base sequence.

Photoreactivation (chapter 121) can be used as a tool to remove cyclobutane dimers in DNA in UV-irradiated *E. coli* at a known time. At short times after UV exposure, photoreactivation of excision-competent *E. coli* greatly reduces the number of mutants per survivor by removal of these dimers (thought to be the principal mutagenic photoproduct [79]) and then has a decreasing effect at longer times as shown schematically in Fig. 3; photoreactivation becomes ineffective when photoproduct excision is essentially over (as expected) and, in most cases, much before significant DNA synthesis (typically delayed as a result of UV exposure). For cells deficient in nucleotide excision repair, photoreactivation causes large decreases in mutants per survivor at long times after irradiation and loses effectiveness only after much of the DNA is replicated (Fig. 3). This pattern has been shown for forward mutations to streptomycin resistance (22, 131) and for reversion of mutations in the protein structural genes *trp* (132) and *tyr* (90) and is thought to be typical for most UV mutations. The simplest interpretation of these results is that base sequence changes occur during nucleotide excision, and no satisfactory alternative has been proposed. (UV-induced reversions that show MFD can be reversed by photoreactivation any time before DNA replication, even in excision-competent cells [90, 132].)

Additional evidence for mutagenesis by excision repair obtained in studies of yeast cells is presented here for two reasons. It is of great importance that mutations can be fixed before

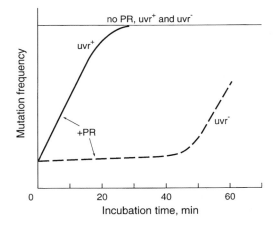

FIGURE 3 Effect on mutation frequency of photoreactivation (PR) after various incubation times following UV irradiation of *E. coli uvr*+ (excision-competent) and *uvr* (excision-defective) cells (schematic). See text for details.

replication of the genome, the process thought by many to actually change the DNA base sequence. Further, the mechanisms by which excision repair causes mutation are not established.

Diploid yeast cells in G_1 growth phase were irradiated with UV and allowed to divide once, and then recessive lethal mutations were determined by tetrad analysis; for cells without excision repair, the segregation pattern was that on the left in Fig. 4, whereas excision-competent cells showed the pattern on the right, implying that because of excision, a mutation exists in both strands of the DNA double helix before replication (82, 83, 89). In another experiment, *arg* or *lys* yeast cells were mutagenized and revertants were determined on plates either with minimal medium or with minimal medium supplemented with enough of the required amino acid to permit a few cell divisions. With UV as the mutagen, excision-competent cells showed comparable numbers of revertants per survivor on either kind of plate, so mutation was equally likely with little DNA replication or with several rounds; excision-defective cells showed significant reversion only with some DNA replication on supplemented plates (88). The conclusion is that changes in base sequence in yeast genes can occur before DNA replication as the result of nucleotide excision repair.

A possible mechanism for mutagenesis by nucleotide excision repair is as follows. The *E. coli* UvrABC nuclease cuts out a single-stranded 12- to 14-mer oligonucleotide that includes the lesion (chapter 121); during replication to close the gap in the DNA, translesion synthesis over a second lesion in the template strand could cause a mutation. Naively, few pairs of lesions so near one another would be expected. However, closely spaced UV photoproducts, one in each strand, occur at a frequency about 1% that of all photoproducts (6, 95), in numbers that vary with the 1.2 to 1.4 power of the fluence (96), and are acted upon by the *uvr* nucleotide excision enzymes (177). Only a very small fraction of lesions cause mutations (see above), so observed mutation rates are not inconsistent with the abundance of closely spaced lesions. Pathways for repair of such closely spaced

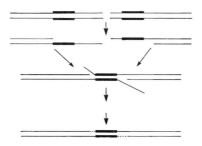

FIGURE 5 Repair of DNA double-strand breaks by single-strand annealing. Single-strand exonucleases digest strands (second line) to expose direct repeats (heavy lines), which anneal together (third line). After unpaired ends are removed, gaps are ligated (dotted lines) to complete the joining. There is strong evidence for this pathway in *E. coli* (32, 157), mammalian (146), nematode (139), and yeast (61) cells, as well as in frog oocytes (84).

lesions have not been identified, but in addition to translesion synthesis, a reasonable possibility is the scheme for repair of psoralen cross-links between strands (chapter 121).

In excision-proficient cells, mutations could occur by excision repair, replication, or both. Note that repair of closely spaced lesions first excises selectively the lesion in the transcribed strand, and so will generate a mutation at the nearby lesion in the nontranscribed strand, in agreement with data on strand-selective repair.

Many more mutations, mostly C:G to T:A, were found after photoreactivation of UV-irradiated plasmids treated in vitro with a protein fraction (fraction II) from *E. coli uvr*⁺ cells than with the fraction from *uvr* mutant cells; this mutagenesis did not depend on *umuC* or *recA* (30, 31). The relation of this process to in vivo nucleotide excision repair is as yet problematical.

Other Error-Prone Repair. Other kinds of repair also cause changes in base sequence. The clearest example is the repair of double-strand breaks in DNA. A number of experimental results (e.g., reference 33) show indiscriminate ligation of free DNA ends with each other, which can cause gross alterations in sequence such as deletions and inversions. Figure 5 illustrates a pathway for repair of double-strand breaks that is characterized by deletion of DNA between two direct sequence repeats, as well as one of the repeats; the mechanism is strongly supported by results of studies using *E. coli* and several other kinds of cells.

PATHWAYS FOR FORMING POINT MUTATIONS

Base Substitutions

The most common type of mutation is that in which one base pair has been replaced by another. Base substitutions are induced by mutagens mainly by translesion synthesis, with or without the assistance of SOS-associated proteins. The kind and site of base substitutions are governed by the lesion (Table 1). Tandem mutations such as CC to TT are induced primarily by mutagens that form lesions involving two adjacent bases on the same strand of the DNA double helix (76), such as UV [pyrimidine dimers or (6–4) products (136)] or *cis*-platinum(II) (cross-links between adjacent purines on the same strand [54]).

Some lesions do not provide the polymerase with any information concerning the original base and are called noninstructive, the prototype being the abasic site. There is evidence that the base just following the lesion influences the nucleotide inserted (92), suggesting that a base added while the lesion (X) is unstacked (N in Fig. 6b, lower diagram) moves opposite the lesion when the DNA changes to the upper configuration (dislocation mutagenesis). Some lesions have a low mutation rate (Table 1), suggesting they may usually be instructive. An interesting case is a common mutation induced by UV, C:G to T:A, by insertion of A opposite a C in the most frequent photoproduct, the cyclobutane pyrimidine dimer; C's in a cyclobutane ring can deaminate to form U's (62), so it is unclear if insertion of A is primarily a polymerase error or a correct pairing of an A with a U.

Most base substitutions change an amino acid in the gene product, so-called missense mutations. Some base substitutions do not change the amino acid and are silent: transitions in the third position of all codons (except TGG), and any base change in the third position of about half of all codons. A third type changes a triplet coding for an amino acid to one coding for a stop codon, a mutation suppressible by a tRNA suppressor (see above).

Frameshifts

Deletion or insertion of one or two base pairs, called a frameshift here, is the second most common type of point mutation. Most such frameshifts occur in monotonic runs of two or more of the same bases; frameshifts in runs of alternating bases, such as -GCGCGC-, are nearly always deletions or duplications of the repeating unit (-GC-). While a replication fork is blocked at a lesion, looping out of a base in a monotonic run could allow slippage of one strand with respect to the other (174), making a newly synthesized strand with one more or one less base (Fig. 6a). If translesion synthesis past the blocking lesion also introduces a base change, a frameshift with an accompanying base substitution, one of the more common complex mutations, is made. Unfortunately, data to support this suggested pathway for making a frameshift are still inconclusive, a generation after it was proposed.

There are several lines of evidence for a mechanism (Fig. 6b) making −1 frameshifts that could explain why mutagens damaging DNA produce more of these than +1 frameshifts. At a stalled replication fork, a base (N) is usually added fairly quickly opposite the lesion. Addition of another nucleotide to complete translesion synthesis is typically much slower, which gives an opportunity for the lesion to loop out (Fig. 6b); if the terminal base on the growing strand forms a Watson-Crick pair with the template base beyond the lesion, rapid extension is likely and the new strand is one base shorter.

Some compounds such as acridines induce in *E. coli* both −1 and +1 frameshifts in runs of identical bases, and essentially only frameshifts, by complexing with DNA and without necessarily causing damage (27, 171). It was originally suggested that acridines generate frameshifts by intercalating between stacked DNA bases (103). However, realistic molecular models for one-base deletions at 5′PyPu sequences induced by the acridine proflavine during in vitro replication by *E. coli* Pol I (Klenow fragment) suggest that these frameshifts arise from stabilization of a looped-out base sandwiched between two proflavines (16). While this model was not intended to account for other mutagenic processes, the findings do support the possibility that looped-out bases stabilized by acridines cause frameshifts in monotonic runs.

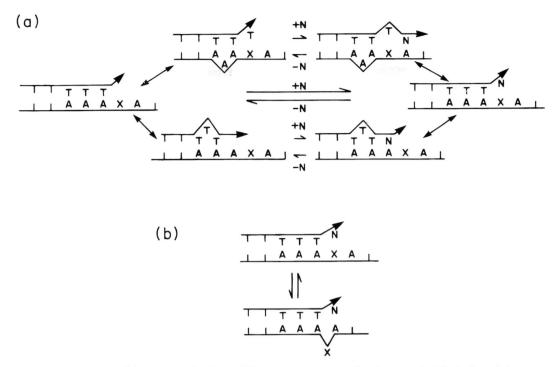

FIGURE 6 Possible pathways for frameshift mutagenesis at a replication complex blocked at a lesion (X). (a) Strand slippage (174). If a base on the strand being newly synthesized loops out, a +1 frameshift will occur; if the looped-out base is on the template, the result is a −1 frameshift. (b) Rearrangement of a newly added nucleotide (N) opposite a template lesion (92, 93, 97, 167, 183). Delay in adding another nucleotide can give time for the lesion (X) to loop out. If the terminal base (N) makes a Watson-Crick pair with the base beyond the lesion, the growing chain will be quickly extended.

In coliphage T4, acridines induce frameshifts both in monotonic runs of bases (134) and at hot spots that are the sites of action of T4 topoisomerase II (145). This enzyme has similarities to mammalian topoisomerase II, and acridines may induce frameshifts in mammalian cells by the same pathway, but there is no evidence for frameshifts at the sites for *E. coli* gyrase (50), which is quite a different enzyme.

Complex Mutations

Two or more mutations in the same gene are observed fairly often, at a frequency far higher than could be expected from two or more independent events. Further, the individual mutations are not randomly distributed throughout the gene but located more closely together, suggesting that most are induced by a single initiating event. Mechanisms have been suggested above for two specific types of such mutants: adjacent base changes or tandem mutation (see the section on base substitutions), and a base change with an accompanying frameshift in a run of the same base (see the section on frameshifts). Excluding these, an additional 10 closely spaced (within a few base pairs) complex mutations in *E. coli* are in the Yale database (77), of which two may arise from a process in which nearly inverted repeats become more exact inverted repeats (Fig. 7c; see the section below on strand switching and copy choice), a process suggested on the basis of similar mutations in phage T4 (40).

In addition, there are 70 other *E. coli* mutants in the Yale database with two mutations (nearly all base changes), and there are 3 with three base changes each, in various genes and induced by

various agents. The separations range randomly from a few base pairs to a few hundred, with a mean of about 75. The mechanism is unknown.

GROSS DNA REARRANGEMENTS

Gross DNA rearrangements include large (hundreds of base pairs and more) deletions and insertions, inversions, translocations, and duplications. Among mutations induced by most mutagens, such rearrangements, deletions excepted, are found only at very low frequencies. Nevertheless, such rearrangements often involve a number of genes and can have a profound effect on the cell. They are covered in more detail in chapter 120. Transposition of mobile elements such as insertion sequence elements has been extensively studied (chapter 124). In one case, UV greatly increases excision of transposon Tn*10* (104).

Deletions

Much the most common large-scale alteration found among mutants in a particular gene is the deletion, ranging in size from 10 or so base pairs to hundreds of kilobases, a limitation set, presumably, by the requirement that the cell be viable.

Large deletions affecting two or more genes can be detected by genetic mapping and have been reported after treatment with ionizing radiation and with nitrous acid, at a frequency about 10% of that of mutations in a single gene: in the *his* operon in *S. typhimurium* (69), the *gal-chl* region of *S. typhimurium* (3), and the *E. coli lac* operon (158) (also in phage T4 treated with

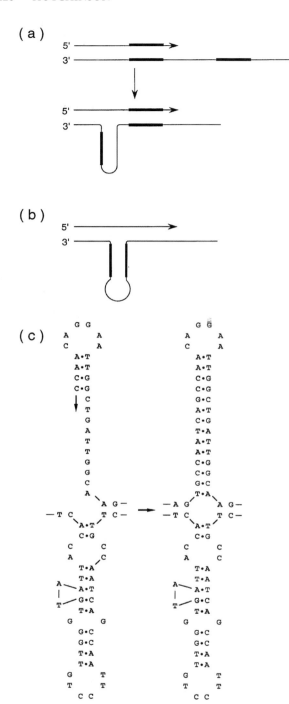

FIGURE 7 Strand switching or copy choice. See text for the reasons that the diagrams are thought to provide only formal representations and should not be taken as depicting actual mechanisms. (a) Deletion of DNA between two direct repeats (heavy lines), together with one of the repeats, by a growing polynucleotide chain mispairing with a downstream repeat as in Streisinger's slipped pairing (2). (b) Single-strand template DNA with two inverted repeats (heavy lines) forms a hairpin that allows a replication complex to jump over to the other arm, deleting the hairpin (67). (c) A template with inverted pseudorepeats goes into a hairpin and allows the replicating strand to fold back and copy itself, forming exact inverted repeats (40).

nitrous acid [179] and gamma rays [31a]). No such deletions were found among more than 100 sequenced mutations induced by gamma rays in the *cI* repressor gene in λ phage and prophage (180); possibly, deletions formed were not detected because they killed the phage. In an *E. coli* strain with over 100 kb of dispensable DNA surrounding a target gene, about 10% of gamma ray-induced mutations were deletions of 0.5 to more than 80 kb, centered around a ColE1 origin of DNA replication (141); the deletion frequency decreased by at least an order of magnitude when this origin was removed. Large deletions are apparently induced by agents that damage both DNA strands at closely apposed sites: ionizing radiation, which makes double-strand breaks and sites which are sensitive to *Aspergillus* S1 endonuclease (65), and nitrous acid, which forms interstrand cross-links (51). The frequency of large deletions in bacteria treated with most other agents is so small that it is unclear whether those observed are induced or of spontaneous origin.

Fragmentary information suggests two mechanisms for forming large deletions. At low doses, most deletions might be associated with specific DNA sequences (such as the ColE1 origin of replication; see above) and induced by a single lesion (75). For higher doses, and for genomic regions where very large deletions still give viable cells, most deletions probably are made by the classical two-hit mechanism: one end of a DNA double-strand break is misrepaired with an end from another break with loss of the intervening segment, a mechanism with strong support from data obtained from studies using *Neurospora crassa* (41).

Some papers report deletions ranging from tens to many hundreds of base pairs induced in genes on various plasmids in *E. coli* cells by agents such as UV (118, 155). Many such deletions eliminate both the DNA between two direct repeats 3 to 9 bp long, as well as one of the repeats (155), which suggests that the deletions arise from repair of double-strand breaks by single-strand annealing (Fig. 5). In the Yale mutational database (77) with more than 13,000 mutations in *E. coli*, all data sets with a significant fraction of these deletions are for genes on plasmids; none are for a gene on the chromosome. The implication is that plasmids may be more susceptible than chromosomes to formation of double-strand breaks and/or less likely to be repaired error free.

Template Switching or Copy Choice

Deletions, as well as other sequence changes, are sometimes interpreted in terms of a mechanism in which a replication complex is assumed to switch from one template to another or to another part of the same template, as diagramed in Fig. 7. In a formal sense, there is probably much truth in such strand switching/copy choice schemes. The mechanism implied by the term "strand switching/copy choice" and by the diagrams, however, has little experimental backing; two recent papers reporting data that supported copy choice mechanisms have been retracted (24, 56). There are, however, some in vitro data on the process shown in Fig. 7c (135). It seems likely that in most cases, results such as those diagramed in Fig. 7 are attained in quite different ways. The overall effect of postreplication repair (Fig. 1b) could readily be described in terms of copy choice, although the actual mechanism is quite different. The process shown in Fig. 7a, deletion of one of two direct repeats and the intervening DNA, can occur by repair of a double-strand break (Fig. 5), for which there is a wealth of data.

ways. The overall effect of postreplication repair (Fig. 1b) could readily be described in terms of copy choice, although the actual mechanism is quite different. The process shown in Fig. 7a, deletion of one of two direct repeats and the intervening DNA, can occur by repair of a double-strand break (Fig. 5), for which there is a wealth of data.

MUTATIONAL SPECTRA

The number of different kinds of mutations and of the locations of these sequence changes is sometimes referred to as a mutational spectrum. Table 2 shows the spectrum of kinds of mutations induced by spontaneous processes in two E. coli genes. Mutational spectra induced by genotoxic agents differ enormously: methylating agents such as N-methyl-N′-nitro-N-nitrosoguanidine induce nearly all G:C-to-A:T transitions in runs of two or more consecutive G's, whereas ionizing radiation induces nearly all kinds of mutations without any one dominant type. The observed spectrum depends on a complex array of factors: on the lesions in the DNA and the changes by repair and by decay of unstable species, on the mutagenic processes involved in replication and repair, and on the assay.

Lesions in the DNA

Lesions in the DNA depend not only on the mutagen but also on accessibility of the DNA (for chemicals) and on base sequence; for example, UV-induced dimers require adjacent pyrimidines in the same strand, and methylation by N-methyl-N-nitrosourea of G to make O^6-methylguanine is strongly dependent on flanking bases (123, 164). Some lesions are unstable and transform with time into other compounds (e.g., deamination of C in cyclobutane pyrimidine dimers) (62). Repair affects not only the total number of lesions but also the distribution in kind and location (see section above on repair and mutagenesis). The extent to which repair is expressed can depend on the treatment level, for example, as in the SOS response (chapter 89) or in the ada genes (chapter 121); specific repair pathways can be blocked by defects in repair genes such as uvr or rec.

Mutagenic Processes

The probability that a lesion at a replication fork induces a mutation, and the kind of mutation, depends on competition between mutagenic translesion synthesis and nonmutagenic postreplication repair. An additional contribution comes from repair processes such as nucleotide excision that change the base sequence in the process of eliminating the lesion.

Assay

Mutation assays usually depend on alterations in gene product activity and detect with high probability those mutations, such as frameshifts, deletions, or stop codons, that make large changes. However, the most common mutation is a base substitution that changes an amino acid, which can affect activity by a little or a lot, and may or may not be detected in a given assay, particularly in a gene with a strong promoter (101).

Valuable information has been obtained from mutational spectra about mechanisms of mutagenesis and identification of mutagenic agents. Several statistical methods for analysis are available (1, 13, 138). Any such analysis, in addition to purely statistical considerations, must take into account the many and sometimes subtle biological effects that influence these spectra.

SPONTANEOUS MUTAGENESIS

Frequency and Spectrum of Spontaneous Mutations

The mutation frequencies and spectra for untreated cells that are given in most papers are most accurately thought of as background: useful, certainly, for comparison with the same parameters after mutagenic treatment, but dependent on the history of the cultures and frequently differing greatly from one experiment to another. Therefore, care must be taken in drawing conclusions regarding spontaneous mutagenesis from such published data.

A meaningful spontaneous mutation frequency and spectrum can be determined by starting with several cultures so small that there is little chance of a preexisting mutant and allowing each to grow to a known size N under defined conditions that do not discriminate between mutated and unmutated cells. The mutation rate per replication ϕ is given by $\phi = 0.4343$ $f/\log(N\phi)$, where f is the median of the measured mutation frequencies (46), assuming that phenotypic lag is small. (Note that the equation must be solved by iteration.) A mutation that occurs early in growth (or is in a culture initially) gives rise to a "jackpot" with a high observed mutation rate, so a grossly outlying value should be excluded. In determining the spectrum of spontaneous mutations, allowance must be made for a mutation that occurs late enough that the overall mutation frequency is not substantially affected but early enough that the mutated cell

TABLE 2 Spontaneous mutation spectra in genes in E. coli[a]

Reference for data set	Gene	Mutations	Transitions		Transversions				Frameshifts		bp 620 ±TGGC hot spot	Deletions	Duplications	Insertion sequences
			C→T	T→C	C→A	C→G	T→A	T→G	−N	+N				
154	lacI[b]	414	137	38	23	12	35	48	18	ND		71	31	1
153	lacI	172	6	3	3	0	3	5	4	1	117	22	3	7
68	lacI	729	45	5	13	3	7	9	28	7	525	72	7	8
59	lacI	140	←			(24 point mutations)				→	94	19	1	2
170	λ cI	52	14	5	5	1	7	2	5	3		1	0	9

[a] C → T indicates a C:G-to-T:A base change, etc.
[b] Only lacI mutations in bp 30 to 242 were detected by the assay used, which excluded the ±TGGC hot spot at about bp 620. In this assay, +1 frameshifts are not detected (ND) (154).

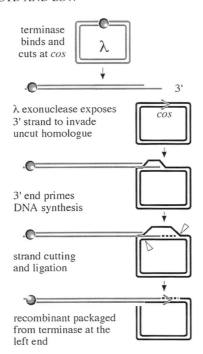

terminase binds and cuts at *cos*

λ exonuclease exposes 3' strand to invade uncut homologue

3' end primes DNA synthesis

strand cutting and ligation

recombinant packaged from terminase at the left end

FIGURE 4 Model of break-join recombination catalyzed by the λ Red system. Terminase bound at *cos* is shown as a shaded circle. Arrowheads indicate sites of strand nicking. New DNA synthesis primed by the invading 3' end is shown as a dashed line.

(217) proposed a nonreciprocal break-join model for the exchange in which a 3' single-strand tail generated at the right end by λ exonuclease invades an intact homolog to form a D-loop (Fig. 4). Appropriate nicks and ligations, coupled with limited DNA synthesis to extend the 3' end, lead to a recombinant molecule which can be packaged from the terminase-bound *cos* at the left end (216, 231). The invaded homolog makes a minor contribution to the recombinant molecule at the right end. If replication is allowed, exchanges are more evenly distributed. This is explained by the fact that the tips of rolling circles are randomly distributed across the chromosome and can therefore initiate exchanges at any point (217).

When λ's own systems for recombination are disabled by *red* and *int* mutations, recombinants are produced by the enzymes of its host, *E. coli.* This normally requires *recA,* and to a lesser extent *recB* and *recC,* and also mutation of λ *gam,* which inhibits ExoV. In *recBC sbcA* strains, λ recombines efficiently without RecA and relies instead on the RecE and RecT proteins, which are functional analogs of λ's Redα and Redβ proteins (27, 87). Indeed, recombination in this background is almost indistinguishable from that catalyzed by λ's Red system (231). Recombination in *recBC sbcBC* strains is efficient and proceeds by a double-chain break and rejoin mechanism. However, in this case it requires RecA and RecJ. The 5'-to-3' exonuclease activity of RecJ, coupled with RecQ or another helicase, is thought to provide the initiating 3' tail. Recent studies have demonstrated an additional need for the RecF, RecO, and RecR proteins and to a lesser extent RecQ, if a small reading frame encoding a protein of 15 kDa is deleted from λ's *ninR* region (190, 191). Overexpression of the 15-kDa protein from plasmid constructs allows

recombination of λΔ*nin5* phages in *recBC sbcBC* strains mutant for *recF, recO,* or *recR* but does nothing for the host (191).

Recombination of λ *red gam* phages in wild-type hosts is thought to be initiated by RecBCD enzyme entering the DNA at the right end. However, exchanges are distributed across the chromosome, even when replication is blocked. Presumably, the potent dsDNA exonuclease activity (ExoV) of RecBCD enzyme causes extensive degradation of the DNA before recombination can initiate. However, the situation changes dramatically when a Chi sequence (5'-GCTGGTGG-3') is present in the λ DNA such that RecBCD entering at the right end encounters Chi from the 3' end. In this case, exchanges are focused near Chi and decrease in a gradient extending leftward for some 10 kb (102).

Chi DNA Sequences and RecBCD Activity

The Chi octamer occurs with a surprisingly high frequency in the *E. coli* chromosome and is found on average every 5 kb or so along the DNA. However, its orientation is nonrandom. The majority are encountered from the 3' end as drawn above when looking in the direction of the replication origin (*oriC*). The disparity, which is evident in all DNA segments examined, varies considerably and can reach as much as 9:1 (19; A. Kerr and R. G. Lloyd, unpublished analysis). Chi is not present in λ but may arise by mutation at several loci. Much of what is known about Chi sequences and their genetic activity has been derived from studies with Chi⁺ phages (for reviews, see references 210 and 213). Chi interacts specifically with RecBCD and is inactive in strains lacking this enzyme (218). RecBCD is a potent DNA helicase and exonuclease that acts on molecules with flush or nearly flush duplex ends (228). An early model of RecBCD-Chi interaction suggested that when RecBCD encounters a Chi from the 3' end, the strand containing Chi is cleaved specifically to the right of the Chi sequence and is displaced as the enzyme continues to unwind the DNA to initiate recombination by invading an intact homolog (210).

A rather different model in which Chi acts to modulate the nuclease activity of RecBCD is supported by more recent studies (40, 41, 182). According to this model, RecBCD unwinds and rapidly degrades the strand ending 3' as it tracks along the molecule, occasionally nicking the strand ending 5'. When it encounters Chi in the correct orientation, it pauses momentarily and the nuclease activity is modulated, possibly through loss of the RecD subunit, such that the strand ending 3' is no longer degraded but is displaced as a single-strand tail as the RecBC(D) enzyme resumes duplex unwinding and nicking of the 5'-ending strand. The 3' tail exposed by RecBCD then acts as a substrate for initiation as before (see reference 92 for a review). The model is consistent with the effect of Chi on λ crosses and with the phenotype of *recD* mutants, which lack ExoV activity but retain RecBC function (2). λ recombination occurs with a high frequency in *recD* strains and is insensitive to Chi (23, 230). The exchanges observed are also focused near the initiating dsDNA end when replication is blocked, as might be expected in the absence of ExoV degradation.

RecBCD enzyme is a potent exonuclease (ExoV) and is responsible for the rapid degradation of the chromosome in *recA* cells following irradiation with UV light (74). Modulation of this activity is likely to be important therefore in times of stress. Recent studies have shown that Chi sequences protect linear DNA from ExoV degradation, both in *cis* and in *trans,* which is